MyMathLab Online Course for Precalculus: A Unit Circle Approach by Ratti, McWaters, Skrzypek (access code required)

Achieve Your Potential

Success in math can make a difference in life. MyMathLab is a learning program with resources to help students achieve their potential in this course and beyond. MyMathLab helps you get up to speed on course material and understand how math will play a role in future careers.

Visualization and Conceptual Understanding

These new MyMathLab resources aid in visual thinking, conceptual understanding and connecting mathematical concepts.

Guided Visualizations

Engaging interactive figures bring mathematical concepts to life, helping you visualize the concepts through directed explorations and purposeful manipulation. Guided Visualizations can be assigned to encourage active learning, critical thinking, and conceptual understanding.

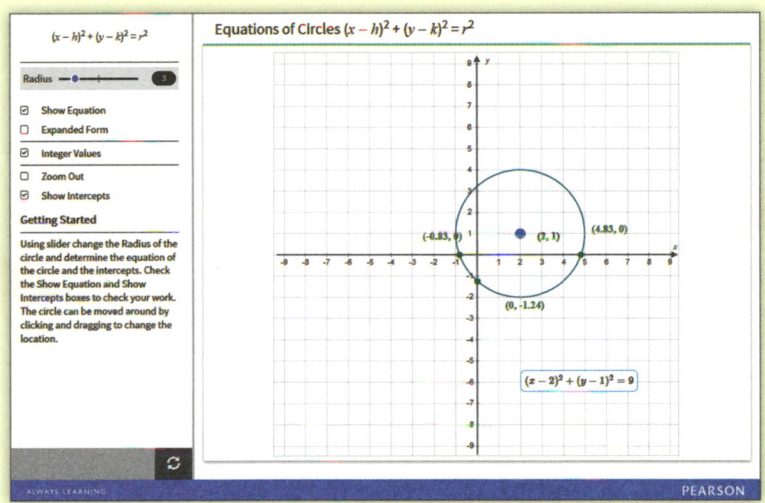

Setup and Solve Exercises

Stepped-out exercises ask students to first describe how they will set up and approach the problem. This reinforces conceptual understanding of the process applied in solving the problem and promotes long term retention of the skill. Access to the eText is available for additional support.

www.mymathlab.com

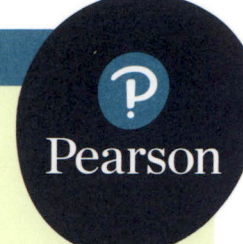

Connect the Concepts and Relate the Math

Preparedness and Maintaining Skills

One of the biggest challenges in Precalculus is being adequately prepared for the course with prerequisite knowledge. MyMathLab's learning resources help refresh knowledge of topics previously learned that are necessary to be successful in Precalculus. Brushing up on these essential algebra skills in each section can dramatically help increase success.

🕐 Due	Assignment
	Q Getting Ready for Chapter 1 Quiz
	H Getting Ready for Chapter 1 Homework
	Q Getting Ready for Chapter 2 Quiz
	H Getting Ready for Chapter 2 Homework
	Q Getting Ready for Chapter 3 Quiz
	H Getting Ready for Chapter 3 Homework
	Q Getting Ready for Chapter 4 Quiz
	H Getting Ready for Chapter 4 Homework

Integrated Review

MyMathLab provides content to refresh skills and understanding of prerequisite topics through skill review quizzes and personalized homework. With additional video and content support in MyMathLab, students receive just the help needed to be prepared to learn the new material.

Ongoing Review

Exercises written to help maintain skills and prepare for the next section encourage review of key mathematical concepts throughout the course. This ongoing review keeps knowledge fresh, prepares for new material learning, and promotes knowledge retention for the current course and for future courses.

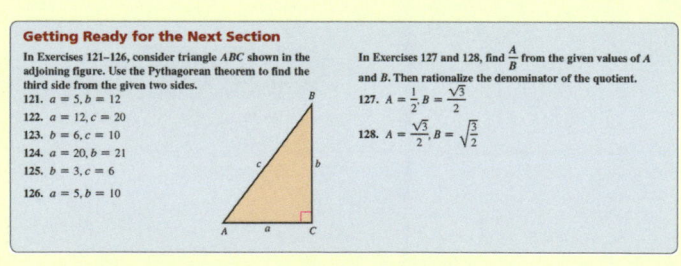

Getting Ready for the Next Section

In Exercises 121–126, consider triangle ABC shown in the adjoining figure. Use the Pythagorean theorem to find the third side from the given two sides.

121. $a = 5, b = 12$
122. $a = 12, c = 20$
123. $b = 6, c = 10$
124. $a = 20, b = 21$
125. $b = 3, c = 6$
126. $a = 5, b = 10$

In Exercises 127 and 128, find $\frac{A}{B}$ from the given values of A and B. Then rationalize the denominator of the quotient.

127. $A = \frac{1}{2}, B = \frac{\sqrt{3}}{2}$

128. $A = \frac{\sqrt{3}}{2}, B = \sqrt{\frac{3}{2}}$

Precalculus

A Unit Circle Approach

Third Edition

J. S. Ratti
University of South Florida

Marcus McWaters
University of South Florida

Lesław A. Skrzypek
University of South Florida

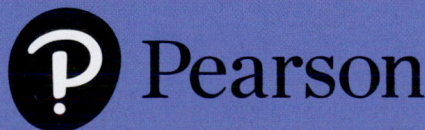

Director, Portfolio Management: Anne Kelly

Portfolio Management Assistant–Mathematics: Ashley Gordon

Content Producer: Rachel S. Reeve

Managing Producer: Scott Disanno

Producer: Jonathan Wooding

Manager, Courseware QA: Mary Durnwald

Senior Content Developer: Eric Gregg

Manager, Content Development: Kristina Evans

Product Marketing Manager: Claire Kozar

Executive Field Marketing Manager: Peggy Sue Lucas

Marketing Assistant: Jennifer Myers

Senior Author Support/Technology Specialist: Joe Vetere

Manager, Rights and Permissions: Gina Cheselka

Manufacturing Buyer: Carol Melville, LSC Communications

Program Design Lead: Barbara T. Atkinson

Associate Director of Design: Blair Brown

Text Design, Production Coordination, Composition, and Illustrations: Cenveo® Publisher Services

Cover Image: FloridaStock/Shutterstock

Library of Congress Cataloging-in-Publication Data

Names: Ratti, J. S. | McWaters, Marcus. | Skrzypek, Lesław.
Title: Precalculus : a unit circle approach.
Description: Third edition / J.S. Ratti, University of South Florida, Marcus McWaters, University of South Florida, Lesław Skrzypek, University of South Florida. | Boston : Pearson Education, [2018] | Includes index.
Identifiers: LCCN 2016035816 | ISBN 9780134433042 (student edition : hardcover : alk. paper)
Subjects: LCSH: Precalculus–Textbooks. | Functions–Textbooks. | Trigonometry–Textbooks. | Algebra–Textbooks.
Classification: LCC QA331 .R358 2018 | DDC 515–dc23 LC record available at https://lccn.loc.gov/2016035816

2   17

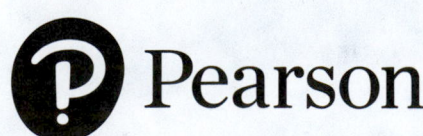

ISBN 13: 978-0-13-443304-2
ISBN 10: 0-13-443304-1

FOREWORD

Our goal in writing this text is to provide material that will actively engage students in the process of learning the mathematical skills necessary for success with calculus. As instructors ourselves, we are familiar with the challenges of making mathematics useful and interesting without sacrificing the solid mathematics essential for conceptual understanding. Our efforts, however, have been aided considerably by the many suggestions we have received from users of the previous editions of the text.

In this book, you will find a strong emphasis on both concept development and real-life applications. Just-in-time review throughout the text ensures that all students are brought to the same level before being introduced to new concepts. We provide numerous applications to motivate students to apply the concepts and skills they learn in precalculus to other courses, including the physical and biological sciences, engineering, economics, and to on-the-job and everyday problem solving. Students are given ample opportunities throughout this book to think about important mathematical ideas and to practice and apply algebraic and trigonometric skills.

We consistently emphasize why the material being covered is important and how it can be applied. By thoroughly developing mathematical concepts with clearly defined terminology, students see the "why" behind those concepts, paving the way for a deeper understanding, better retention, less reliance on rote memorization, and ultimately more success.

Our own experience in using the text at the University of South Florida has been exceptionally rewarding. Success rates have increased significantly since we began using our text. Where our previous pass/fail rates were a frequent source of complaint, they are now cited as evidence of successful change. We have used years of performance data from student homework, quizzes, and exams to identify areas of weakness, and, as a result, we have refined our examples, exercises, and exposition to better help students overcome these problem areas. Because students are accustomed to information being delivered by electronic media, the introduction of MyMathLab® into our courses was completely seamless. We hope that your experience will be as successful as ours.

Marcus McWaters

Lesław Skrzypek

J. S. Ratti

CONTENTS

Foreword iii

Preface xiii

Resources xvi

Acknowledgments xix

Dedication xx

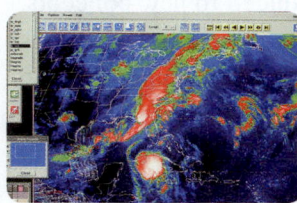

Chapter 1

GRAPHS AND FUNCTIONS 1

1.1 Graphs of Equations 2

The Coordinate Plane ■ Scales on a Graphing Utility ■ The Distance Formula ■ The Midpoint Formula ■ Graph of an Equation ■ Intercepts ■ Symmetry ■ Circles ■ Semicircles

1.2 Lines 22

Slope of a Line ■ Point–Slope Form ■ Slope–Intercept Form ■ Equations of Horizontal and Vertical Lines ■ General Form of the Equation of a Line ■ Parallel and Perpendicular Lines ■ Modeling Data Using Linear Regression

1.3 Functions 39

Functions ■ Function Notation ■ Representations of Functions ■ The Domain of a Function ■ The Range of a Function ■ Graphs of Functions ■ Function Information from Its Graph ■ Average Rate of Change ■ Building Functions

1.4 A Library of Functions 60

Linear Functions ■ Increasing and Decreasing Functions ■ Relative Maximum and Minimum Values ■ Even–Odd Functions and Symmetry ■ Piecewise Functions ■ Graphing Piecewise Functions ■ Basic Functions

1.5 Transformations of Functions 78

Transformations ■ Vertical and Horizontal Shifts ■ Reflections ■ Stretching or Compressing ■ Multiple Transformations in Sequence

1.6 Combining Functions; Composite Functions 99

Combining Functions ■ Composition of Functions ■ Domain of Composite Functions ■ Decomposition of a Function ■ Applications of Composite Functions

iv

1.7 Inverse Functions 115

Inverses ■ Finding the Inverse Function ■ Finding the Range of a One-to-One Function ■ Applications

Summary Definitions, Concepts, and Formulas ■ Review Exercises ■ Practice Test A ■ Practice Test B

Chapter 2

POLYNOMIAL AND RATIONAL FUNCTIONS   137

2.1 Quadratic Functions 138

Quadratic Functions ■ Standard Form of a Quadratic Function ■ Graphing a Quadratic Function $f(x) = ax^2 + bx + c$ ■ Applications

2.2 Polynomial Functions 153

Polynomial Functions ■ Power Functions ■ End Behavior of Polynomial Functions ■ Zeros of a Function ■ Zeros and Turning Points ■ Graphing a Polynomial Function

2.3 Dividing Polynomials and the Rational Zeros Test 173

The Division Algorithm ■ The Remainder and Factor Theorems ■ The Rational Zeros Test

2.4 Rational Functions 187

Rational Functions ■ Vertical and Horizontal Asymptotes ■ Translations of $f(x) = \dfrac{1}{x}$ ■ Graphing Rational Functions ■ Oblique Asymptotes and End Behavior ■ Graph of a Revenue Curve

2.5 Polynomial and Rational Inequalities 210

Polynomial Inequalities ■ Rational Inequalities

2.6 Zeros of a Polynomial Function 225

Descartes's Rule of Signs ■ Bounds on the Real Zeros ■ Complex Zeros of Polynomials ■ Conjugate Pairs Theorem

2.7 Variation 235

Direct Variation ■ Inverse Variation ■ Joint and Combined Variation

Summary Definitions, Concepts, and Formulas ■ Review Exercises ■ Practice Test A ■ Practice Test B ■ Cumulative Review Exercises Chapters 1–2

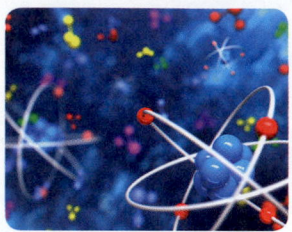

Chapter 3

EXPONENTIAL AND LOGARITHMIC FUNCTIONS   252

3.1   Exponential Functions   253

Exponential Functions ■ Evaluate Exponential Functions ■ Graphing Exponential Functions ■ Transformations on Exponential Functions ■ Simple Interest ■ Compound Interest ■ Continuous Compound Interest Formula ■ The Natural Exponential Function ■ Natural Exponential Growth and Decay

3.2   Logarithmic Functions   271

Logarithmic Functions ■ Evaluating Logarithms ■ Basic Properties of Logarithms ■ Domains of Logarithmic Functions ■ Graphs of Logarithmic Functions ■ Common Logarithm ■ Natural Logarithm ■ Investments ■ Newton's Law of Cooling

3.3   Rules of Logarithms   287

Rules of Logarithms ■ Number of Digits ■ Change of Base ■ Growth and Decay ■ Half-Life ■ Radiocarbon Dating

3.4   Exponential and Logarithmic Equations and Inequalities   301

Solving Exponential Equations ■ Applications of Exponential Equations ■ Solving Logarithmic Equations ■ Logarithmic and Exponential Inequalities

3.5   Logarithmic Scales; Modeling   313

pH Scale ■ Earthquake Intensity ■ Loudness of Sound ■ Star Brightness ■ Modeling

Summary Definitions, Concepts, and Formulas ■ Review Exercises ■ Practice Test A ■ Practice Test B ■ Cumulative Review Exercises Chapters 1–3

Chapter 4

TRIGONOMETRIC FUNCTIONS   336

4.1   Angles and Their Measure   337

Angles ■ Angle Measure ■ Degree Measure ■ Radian Measure ■ Relationship Between Degrees and Radians ■ Complements and Supplements ■ Length of an Arc of a Circle ■ Area of a Sector ■ Linear and Angular Speed

4.2   The Unit Circle; Trigonometric Functions of Real Numbers   351

The Unit Circle ■ Trigonometric Functions of Real Numbers ■ Finding Exact Trigonometric Function Values ■ Trigonometric Functions of an Angle ■ Trigonometric Function Values of Common Angles ■ Evaluating Trigonometric Functions Using a Calculator

4.3   Trigonometric Functions of Angles   367

Trigonometric Function Values of an Angle ■ Signs of the Trigonometric Functions ■ Reference Angle ■ Fundamental Trigonometric Identities ■ Even–Odd Properties of Trigonometric Functions

4.4   Graphs of the Sine and Cosine Functions   381

Properties of Sine and Cosine ■ Domain and Range of Sine and Cosine ■ Zeros of Sine and Cosine Functions ■ Even–Odd Properties of the Sine and Cosine Functions ■ Periodic Functions ■ Graphs of Sine and Cosine Functions ■ Graph of the Sine Function ■ Graph of the Cosine Function ■ Five Key Points ■ Amplitude and Period ■ Phase Shift ■ Vertical Shifts ■ Modeling with Sinusoidal Curves ■ Simple Harmonic Motion

4.5   Graphs of the Other Trigonometric Functions   404

Tangent Function ■ Graph of $y = \tan x$ ■ Graphs of the Reciprocal Functions

4.6   Inverse Trigonometric Functions   415

The Inverse Sine Function ■ The Inverse Cosine Function ■ The Inverse Tangent Function ■ Other Inverse Trigonometric Functions ■ Evaluating Inverse Trigonometric Functions ■ Composition of Trigonometric and Inverse Trigonometric Functions

Summary Definitions, Concepts, and Formulas ■ Review Exercises ■ Practice Test A ■ Practice Test B ■ Cumulative Review Exercises Chapters 1–4

Chapter 5

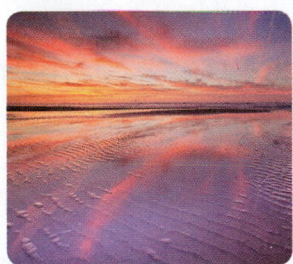

ANALYTIC TRIGONOMETRY   436

5.1   Trigonometric Identities   437

Fundamental Trigonometric Identities ■ Simplifying a Trigonometric Expression ■ Trigonometric Equations and Identities ■ Process of Verifying Trigonometric Identities ■ Methods of Verifying Trigonometric Identities

5.2   Sum and Difference Formulas   448

Sum and Difference Formulas for Cosine ■ Cofunction Identities ■ Sum and Difference Formulas for Sine ■ Sum and Difference Formulas for Tangent ■ Reduction Formula

5.3   Double-Angle and Half-Angle Formulas   461

Double-Angle Formulas ■ Power-Reducing Formulas ■ Alternating Current ■ Half-Angle Formulas

5.4   Product-to-Sum and Sum-to-Product Formulas   473

Product-to-Sum Formulas ■ Sum-to-Product Formula ■ Verify Trigonometric Identities ■ Analyzing Touch-Tone Phones

5.5 Trigonometric Equations 481

Trigonometric Equations ■ Trigonometric Equations of the Form $a\sin(x - c) = k$, $a\cos(x - c) = k$, and $a\tan(x - c) = k$ ■ Equations Involving Multiple Angles ■ Trigonometric Equations and the Zero-Product Property ■ Equations with More Than One Trigonometric Function ■ Extraneous Solutions

Summary Definitions, Concepts, and Formulas ■ Review Exercises ■ Practice Test A ■ Practice Test B ■ Cumulative Review Exercises Chapters 1–5

Chapter 6

APPLICATIONS OF TRIGONOMETRIC FUNCTIONS 499

6.1 Right-Triangle Trigonometry 500

Trigonometric Ratios and Functions ■ Evaluating Trigonometric Functions ■ Complements ■ Solving Right Triangles ■ Applications

6.2 The Law of Sines 513

Solving Oblique Triangles ■ The Law of Sines ■ Solving AAS and ASA Triangles ■ Solving SSA Triangles—the Ambiguous Case ■ Bearings ■ Area of a Triangle

6.3 The Law of Cosines 527

The Law of Cosines ■ Derivation of the Law of Cosines ■ Solving SAS Triangles ■ Solving SSS Triangles ■ Heron's Area Formula

6.4 Vectors 537

Vectors ■ Geometric Vectors ■ Equivalent Vectors ■ Adding Vectors ■ Algebraic Vectors ■ Unit Vectors ■ Vectors in **i, j** Form ■ Vector in Terms of Magnitude and Direction ■ Applications of Vectors

6.5 The Dot Product 549

The Dot Product ■ The Angle Between Two Vectors ■ Orthogonal Vectors ■ Projection of a Vector ■ Decomposition of a Vector ■ Work

6.6 Polar Coordinates 560

Polar Coordinates ■ Multiple Representations ■ Sign of r ■ Converting Between Polar and Rectangular Forms ■ Converting Equations Between Rectangular and Polar Forms ■ The Graph of a Polar Equation

6.7 Polar Form of Complex Numbers; De Moivre's Theorem 577

Geometric Representation of Complex Numbers ■ The Absolute Value of a Complex Number ■ Polar Form of a Complex Number ■ Product and Quotient in Polar Form ■ Powers of Complex Numbers in Polar Form ■ Roots of Complex Numbers

Summary Definitions, Concepts, and Formulas ■ Review Exercises ■ Practice Test A ■ Practice Test B ■ Cumulative Review Exercises Chapters 1–6

Chapter 7

SYSTEMS OF EQUATIONS AND INEQUALITIES   593

7.1   Systems of Equations in Two Variables   594

System of Equations ■ Graphical Method ■ Substitution Method ■ Elimination Method ■ Applications

7.2   Systems of Linear Equations in Three Variables   608

Systems of Linear Equations ■ Number of Solutions of a Linear System ■ Nonsquare Systems ■ Geometric Interpretation ■ An Application to CAT Scans

7.3   Systems of Inequalities   619

Graph of a Linear Inequality in Two Variables ■ Systems of Linear Inequalities in Two Variables ■ Applications: Linear Programming ■ Nonlinear Inequality ■ Nonlinear Systems

7.4   Matrices and Systems of Equations   634

Definition of a Matrix ■ Using Matrices to Solve Linear Systems ■ Gaussian Elimination ■ Gauss–Jordan Elimination

7.5   Determinants and Cramer's Rule   649

The Determinant of a 2×2 Matrix ■ Minors and Cofactors ■ The Determinant of an $n \times n$ Matrix ■ Cramer's Rule

7.6   Partial-Fraction Decomposition   660

Partial Fractions ■ $Q(x)$ Has Only Distinct Linear Factors ■ $Q(x)$ Has Repeated Linear Factors ■ $Q(x)$ Has Distinct Irreducible Quadratic Factors ■ $Q(x)$ Has Repeated Irreducible Quadratic Factors

7.7   Matrix Algebra   672

Equality of Matrices ■ Matrix Addition and Scalar Multiplication ■ Matrix Multiplication ■ Computer Graphics

7.8   The Matrix Inverse   685

The Multiplicative Inverse of a Matrix ■ Finding the Inverse of a Matrix ■ A Rule for Finding the Inverse of a 2×2 Matrix ■ Solving Systems of Linear Equations by Using Matrix Inverses ■ Applications of Matrix Inverses ■ Cryptography

Summary Definitions, Concepts, and Formulas ■ Review Exercises ■ Practice Test A ■ Practice Test B ■ Cumulative Review Exercises Chapters 1–7

Chapter 8

ANALYTIC GEOMETRY   707

8.1   Conic Sections: Overview   708

8.2   The Parabola   710
Geometric Definition of a Parabola ■ Equation of a Parabola ■ Translations of Parabolas ■ Reflecting Property of Parabolas

8.3   The Ellipse   725
Definition of Ellipse ■ Equation of an Ellipse ■ Translations of Ellipses ■ Applications

8.4   The Hyperbola   739
Definition of Hyperbola ■ The Asymptotes of a Hyperbola ■ Graphing a Hyperbola with Center $(0, 0)$ ■ Translations of Hyperbolas ■ Applications

8.5   Rotation of Axes   757
Identifying a Conic ■ Rotation of Axes ■ Eliminating the xy-Term ■ Identifying Conics by Using the Discriminant

8.6   Polar Equations of Conics   768
Definition of a Conic ■ Polar Equation of a Conic ■ Graphing a Conic ■ Applications

8.7   Parametric Equations   779
Parametric Equations ■ Graphing a Plane Curve ■ Eliminating the Parameter ■ Finding Parametric Equations

Summary Definitions, Concepts, and Formulas ■ Review Exercises ■ Practice Test A ■ Practice Test B ■ Cumulative Review Exercises Chapters 1–8

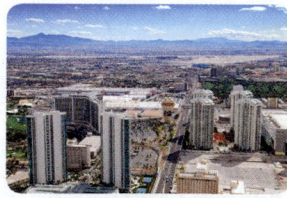

Chapter 9

FURTHER TOPICS IN ALGEBRA   793

9.1   Sequences and Series   794
Sequences ■ Recursive Formulas ■ Factorial Notation ■ Summation Notation ■ Series

9.2   Arithmetic Sequences; Partial Sums   805
Arithmetic Sequence ■ Sum of a Finite Arithmetic Sequence

9.3   Geometric Sequences and Series   813
Geometric Sequence ■ Finding the Sum of a Finite Geometric Sequence ■ Annuities ■ Infinite Geometric Series

9.4   Mathematical Induction

Mathematical Induction ■ Determining the Statement P_{k+1} from the Statement P_k

9.5   The Binomial Theorem   831

Pascal's Triangle ■ The Binomial Theorem ■ Binomial Coefficients

9.6   Counting Principles   839

Fundamental Counting Principle ■ Permutations ■ Combinations ■ Distinguishable Permutations ■ Deciding Whether to Use Permutations, Combinations, or the Fundamental Counting Principle

9.7   Probability   850

The Probability of an Event ■ The Additive Rule ■ Mutually Exclusive Events ■ The Complement of an Event ■ Experimental Probabilities

Summary Definitions, Concepts, and Formulas ■ Review Exercises ■ Practice Test A ■ Practice Test B ■ Cumulative Review Exercises Chapters 1–9

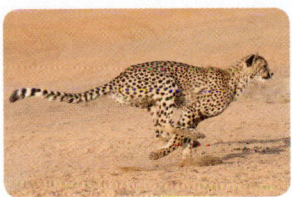

Chapter 10

AN INTRODUCTION TO CALCULUS   865

10.1 Finding Limits Using Tables and Graphs   866

The Limit Concept ■ Limits Using Tables ■ Finding Limits Graphically

10.2 Finding Limits Algebraically   875

Properties of Limits ■ Algebraic Methods ■ Limits of Difference Quotients

10.3 Infinite Limits and Limits at Infinity   885

Infinite Limit ■ Limits at Infinity

10.4 Introduction to Derivatives   893

Instantaneous Rates of Change ■ Tangent Line to the Graph of a Function ■ Derivative of a Function ■ Applications

10.5 Area and the Integral   905

Area ■ Definition of Integral ■ Evaluating Integrals Using a Graphing Utility

Summary Definitions, Concepts, and Formulas ■ Review Exercises ■ Practice Test A ■ Practice Test B

Appendix A

REVIEW 917

A.1 The Real Numbers; Integer Exponents 918

Classifying Sets of Numbers ■ Real Number Line ■ Inequalities ■ Sets ■ Intervals ■ Absolute Value ■ Distance Between Two Points on a Real Number Line ■ Algebraic Expressions ■ Integer Exponents ■ Rules of Exponents

A.2 Polynomials 928

Polynomial Vocabulary ■ Special Products ■ Factoring Polynomials

A.3 Rational Expressions 933

Rational Expressions ■ Lowest Terms for a Rational Expression ■ Multiplication and Division of Rational Expressions ■ Addition and Subtraction of Rational Expressions ■ Complex Fractions

A.4 Radicals and Rational Exponents 942

Square Roots ■ Other Roots ■ Simplifying Radical Expressions ■ Like Radicals ■ Rationalizing ■ Use Conjugates ■ Rational Exponents

A.5 Topics in Geometry 950

Triangles ■ Pythagorean Theorem and Its Converse ■ Geometry Formulas

A.6 Equations 954

Definitions ■ Equivalent Equations ■ Solving Linear Equations in One Variable ■ Solving Quadratic Equations ■ Factoring Method ■ The Quadratic Formula ■ Solving Polynomial Equations by Factoring ■ Rational Equations ■ Equations Involving Radicals ■ Equations That Are Quadratic in Form

A.7 Inequalities 964

Inequalities ■ Linear Inequalities ■ Combining Two Inequalities ■ Using Test Points to Solve Inequalities ■ Inequalities Involving Absolute Value

A.8 Complex Numbers 976

Complex Numbers ■ Addition and Subtraction ■ Multiplying and Dividing Complex Numbers ■ Powers of i

Answers to Selected Exercises A-1

Credits C-1

Index I-1

Students begin precalculus classes with widely varying backgrounds. Some haven't taken a math course in several years and may need to spend time reviewing prerequisite topics, while others are ready to jump right into new and challenging material. We have provided review material in the Appendix and in some of the early sections of other chapters in such a way that it can be used or omitted as is appropriate for your course. In addition, students may follow several different paths after completing a precalculus course. Many will continue their study of mathematics in courses such as finite mathematics, statistics, and calculus. For others, this course may be their last mathematics course. Responding to the current and future needs of all of these students was essential in creating this text. Overall, we present our content in a systematic way that illustrates how to study and what to review. We believe that if students use this textbook well, they will succeed in this course. The changes that occur in this edition are a result of the thoughtful feedback we have received from students and instructors who have used previous editions of the text. This feedback crucially enhances our own experiences, and we are extremely grateful to the many contributors whose insights are reflected in this new edition.

Key Content Changes

Concepts and Vocabulary Exercises. Each exercise section begins with exercises that assess the student's grasp of the definitions and ideas introduced in that section. These true/false and fill-in-the-blank exercises help to rapidly identify gaps in comprehension of the material in that section.

Exercises Preparing Students for Material in the Next Section. Each exercise section ends with a set of exercises that provide a review of the concepts and skills that will be used in the following section.

Additional Review Material. We have provided additional review material in Chapters 1 and 2 that users of the previous editions identified as material they had to supplement or as requiring reference to the Appendix.

Graph and Data-Related Examples and Exercises. We have introduced examples and exercises throughout the text that demonstrate how to extract information about real-world situations from a graphic representation of that situation, as well as how to recover algebraic or trigonometric formulations of a graph by using key characteristics of that graph.

Modeling. A section on modeling data using linear regression was added in Chapter 1, as well as a section in Chapter 3 on building linear, exponential, logarithmic, and power models from data.

Polynomial and Rational Inequalities. A new section on polynomial and rational inequalities was added in Chapter 2.

Trigonometric Equations. The section on trigonometric equations was completely rewritten and is now placed as the last section of Chapter 5.

Systems of Inequalities. The section on systems of inequalities in Chapter 7 was relocated to follow the section on systems of linear equations in three variables.

Identifying Material Particularly Useful for Calculus. We now identify concepts and exercises that are particularly useful for calculus with a new symbol, $\int$.

Basic Calculus Concepts introduced. We added a new chapter, "An Introduction to Calculus," which introduces the basic concepts of limit, derivative, and integral.

CHAPTER 1 In Section **1.2**, we added the two-intercept form of the equation of line and modeling data using linear regression. Section **1.3** added a discussion of the range of a function. In Section **1.4**, we present an example showing how to write a piecewise function from a set of data points. Section **1.5** added a step-by-step process explaining how to graph multiple transformations in sequence. In Section **1.6**, we rewrote the explanation of the domain of composite functions and added a method for computing the average rate of change of a composite function. We also added an example of composing a function with a piecewise function.

CHAPTER 2 In Section **2.1**, we added a schematic showing how to solve quadratic inequalities graphically. We also added exercises using quadratic functions modeling data. In Section **2.2**, we provide a discussion of the behavior of a polynomial function near its zeros to help the student understand why a polynomial crosses or bounces off the x-axis in relation to the multiplicity of its zero. We added an example of graphing a polynomial given in factored form. In Section **2.3**, we expanded the long division and synthetic division material (this material was previously provided in the Appendix). In addition, we added several exercises asking students to sketch a polynomial in nonfactored form. In Section **2.4**, we expanded the discussion of graphing rational functions to include the behavior of a rational function near both its zeros and its vertical asymptotes. Section **2.5** is a new section on polynomial and rational inequalities. The sign of an expression is determined geometrically from its graph (graphic method) or algebraically (test point method).

CHAPTER 3 In Section **3.1**, we added a schematic showing various transformations of the graph of $f(x) = a^x$ and provided exercises on graphing transformations of the graph of $f(x) = a^x$ early in the exercises. In Section **3.2**, we added a schematic showing various transformations of the graph of $f(x) = \log_a x$ and provided exercises on graphing transformations of the graph of $f(x) = \log_a x$ early in the exercises. In Section **3.4**, we added exercises on solving inequalities. In Section **3.5**, we included material on building linear, exponential, logarithmic, and power models from data.

CHAPTER 4 In Section **4.4**, we provided an alternative way of graphing trigonometric functions by using transformations of functions and added a subsection on even–odd properties of the trigonometric functions.

CHAPTER 5 The section on trigonometric equations was completely rewritten and is now placed as the last section of Chapter 5.

CHAPTER 7 The section on systems of inequalities was relocated to follow the section on systems of linear equations in three variables.

CHAPTER 8 Sections **8.2, 8.3,** and **8.4** each have additional examples showing how to obtain the equation of the conic discussed in that section from key characteristics of its graph.

REVIEW APPENDIX The material on synthetic division has been moved to Section 2.3.

Features

CHAPTER OPENER Each chapter opener includes a description of applications (one of them illustrated) relevant to the content of the chapter and the list of topics that will be covered. In one page, students see what they are going to learn and why they are learning it.

SECTION OPENER with APPLICATION Each section opens with a list of prerequisite topics, complete with section and page references, which students can **review** prior to starting the section. The **Objectives** of the section are also clearly stated and numbered, and then referenced again in the margin of the lesson at the point where the objective's topic is taught. An **Application** then follows containing a motivating anecdote or an interesting problem. An example later in the section relating to this application and identified by the same icon (◆) is then solved using the mathematics covered in the section. These applications utilize material from a variety of fields: the physical and biological sciences (including health sciences), economics, art and architecture, history, and more.

EXAMPLES and PRACTICE PROBLEMS Examples include a wide range of computational, conceptual, and modern applied problems carefully selected to build confidence, competency, and understanding. Every example has a title indicating its purpose and presents a detailed solution containing annotated steps. All examples are followed by a **Practice Problem** for students to try so that they can check their understanding of the concept covered. Answers to the Practice Problems are provided just before the section exercises.

ADDITIONAL PEDAGOGICAL FEATURES

Definitions, Theorems, Properties, and *Rules* are all boxed and titled for emphasis and ease of reference.

Warnings appear as appropriate throughout the text to apprise students of common errors and pitfalls that can trip them up in their thinking or calculations.

Summary of Main Facts boxes summarize information related to equations and their graphs, such as those of the conic sections.

MARGIN NOTES

Side Notes provide hints for handling newly introduced concepts.

Recall notes remind students of a key idea learned earlier in the text that will help them work through a current problem.

Technology Connections give students tips on using calculators to solve problems, check answers, and reinforce concepts. Note that the use of graphing calculators is optional in this text.

Do You Know? features provide students with additional interesting information on topics to keep them engaged in the mathematics presented.

Historical Notes give students information on key people or ideas in the history and development of mathematics.

EXERCISES The heart of any textbook is its exercises, so we have tried to ensure that the quantity, quality, and variety of exercises meet the needs of all students. Exercises are carefully graded to strengthen the skills developed in the section and are organized using the following categories. **Concepts and Vocabulary** exercises begin each exercise set with problems that assess the student's grasp of the definitions and ideas introduced in that section. These true/false and fill-in-the-blank exercises help to rapidly identify gaps in comprehension of the material in that section. **Basic Skills** exercises develop fundamental skills—each odd-numbered exercise is closely paired with its consecutive even-numbered exercise. **Applying the Concepts** features use the section's material to solve real-world problems—all are titled and relevant to the topics of the section. **Beyond the Basics** exercises provide more challenging problems that give students an opportunity to reach beyond the material covered in the section—these are generally more theoretical in nature and are suitable for honors students, special assignments, or extra credit. **Critical Thinking/Discussion/Writing** exercises, appearing as appropriate, are designed to develop students' higher-level thinking skills. Calculator problems, identified by the symbol, are included where needed. **Getting Ready for the Next Section** exercises end each exercise set with problems that provide a review of concepts and skills that will be used in the following section.

END-OF-CHAPTER The chapter-ending material begins with a **Summary of Definitions, Concepts, and Formulas** consisting of a brief description of key topics organized by section, encouraging students to reread sections rather than memorize definitions out of context. **Review Exercises** provide students with an opportunity to practice what they have learned in the chapter. Then students are given two chapter test options. They can take **Practice Test A** in the usual open-ended format and/or **Practice Test B**, covering the same topics, in a multiple-choice format. All tests are designed to increase student comprehension and verify that students have mastered the skills and concepts in the chapter. Mastery of these materials should indicate a true comprehension of the chapter and the likelihood of success on the associated in-class examination. **Cumulative Review Exercises** appear at the end of every chapter, starting with Chapter 2, to remind students that mathematics is not modular and that what is learned in the first part of the book will be useful in later parts of the book and on the final examination.

Get the Most Out of
MyMathLab®

MyMathLab is the leading online homework, tutorial, and assessment program for teaching and learning mathematics, built around Pearson's best-selling content. MyMathLab helps students and instructors improve results; it provides engaging experiences and personalized learning for each student so learning can happen in any environment. Plus, it offers flexible and time-saving course management features to allow instructors to easily manage their classes while remaining in complete control, regardless of course format.

Preparedness

One of the biggest challenges in many mathematics courses is making sure students are adequately prepared with the prerequisite skills needed to successfully complete their course work. MyMathLab offers a variety of content and course options to support students with just-in-time remediation and key-concept review.

- **MyMathLab with Integrated Review**—available for Developmental Mathematics through Calculus—can be used for just-in-time prerequisite review or co-requisite courses. These courses provide videos on review topics, along with pre-made, assignable skills-check quizzes and personalized review homework assignments.

- Getting Ready exercises in Precalculus MyMathLab courses cover key pre-requisite concepts that the authors know students will need to be success-ful. When Getting Ready exercises are used in quick quizzes coupled with a personalized homework assignment, students benefit from getting efficient, effective just-in-time review.

Used by more than 37 million students worldwide, MyMathLab delivers consistent, measurable gains in student learning outcomes, retention, and subsequent course success.

www.mymathlab.com

Resources for Success

MyMathLab® Online Course
for *Precalculus: A Unit Circle Approach*
by Ratti, McWaters, and Skrzypek (access code required)

MyMathLab is available to accompany Pearson's market-leading text offerings. To give students a consistent tone, voice, and teaching method each text's flavor and approach is tightly integrated throughout the accompanying MyMathLab course, making learning the material as seamless as possible.

Video Assessment Questions

Video Assessment questions are assignable MyMathLab exercises tied to Example Solution videos. The questions are designed to check students' understanding of the important math concepts covered in the video. Pairing the videos and Video Assessment questions provides an active learning environment where students can work at their own pace.

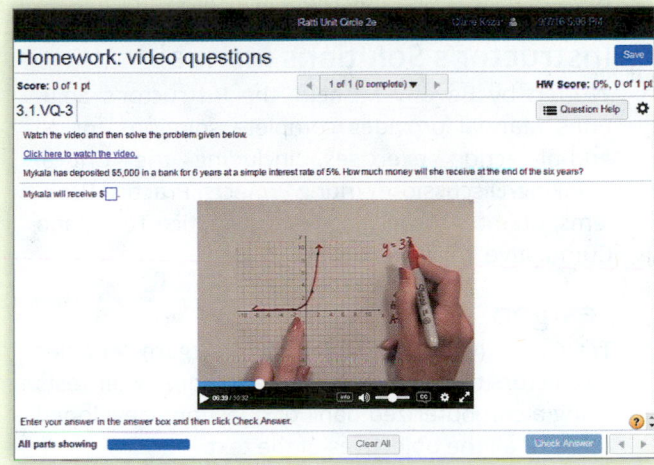

Video Notebook with Worksheets

The Video Notebook with Worksheets gives students a structured place to take notes and work on the example problems as they watch the videos. Definitions and important concepts are highlighted, and worksheets tied to Integrated Review learning objectives are included. The notebook is also available in MyMathLab for download.

Concepts and Vocabulary Exercises

Each exercise section begins with exercises that assess the student's grasp of the definitions and ideas introduced in that section. These true/false and fill-in-the-blank exercises help to rapidly identify gaps in comprehension of the material in that section and are assignable in MyMathLab and Learning Catalytics.

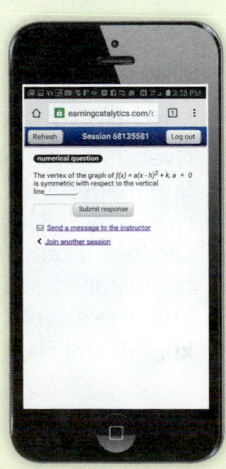

www.mymathlab.com

Resources for Success

Pearson

Instructor Resources

Additional resources can be downloaded from www.pearsonhighered.com, or hardcopy resources can be ordered from your sales representative.

Annotated Instructor's Edition
ISBN: 978-0-13-443835-1 / 0-13-443835-3
Answers are included on the same page beside the text exercises where possible for quick reference.

Instructor's Solutions Manual
Written by Beverly Fusfield, the Instructor's Solutions Manual provides complete solutions for all end-of-section exercises, including the Critical Thinking/Discussion/Writing Projects, Practice Problems, Chapter Review exercises, Practice Tests, and Cumulative Review problems.

Testgen®
TestGen® (www.pearsoned.com/testgen) enables instructors to build, edit, print, and administer tests using a computerized bank of questions developed to cover all the objectives of the text.

Instructor's Testing Manual
The Instructor's Testing Manual includes diagnostic pretests, chapter tests, and additional test items, grouped by section, with answers provided.

PowerPoint® Lecture Slides
The PowerPoint Lecture Slides feature presentations written and designed specifically for this text, including figures and examples from the text.

Video Lectures
Over 25 hours of video instruction feature Section Summaries and Example Solutions. Section Summaries cover key definitions and procedures for most sections. Example Solutions walk students through the detailed solution process for many examples in the textbook. Optional subtitles are available in English and Spanish.

Student Resources

Additional resources to help student success.

Student's Solutions Manual
ISBN: 978-0-13-443846-7 / 0-13-443846-9
Written by Beverly Fusfield, the Student's Solutions Manual provides detailed worked-out solutions to the odd-numbered end-of-section and Chapter Review exercises and solutions to all of the Practice Problems, Practice Tests, and Cumulative Review problems. Also available in MyMathLab.

Video Notebook with Worksheets
ISBN: 978-0-13-472237-5 / 0-13-472237-X
The Video Notebook with Worksheets gives students a structured place to take notes and work out the example problems as they watch the videos. Definitions and important concepts are highlighted and worksheets tied to Integrated Review learning objectives are included. The Video Notebook is available as PDF files and customizable Word files in MyMathLab and can also be packaged with the textbook and MyMathLab access code.

www.mymathlab.com

ACKNOWLEDGMENTS

We would like to express our gratitude to the reviewers of this third edition (marked with an asterisk) and previous editions who provided such invaluable insights and comments. Their contributions helped shape the development of the text and carry out the vision stated in the Preface.

Mario Barrientos, Angelo State University

Kate Bella, Manchester Community College

Sarah Bennet, University of Wisconsin–Barron County

Jeremy Brandl, Fresno City College*

Karen Briggs, North Georgia College and State
University

Katina Davis, Wayne Community College

Stefaan Delcroix, California State University–Fresno*

Nicole Dowd, Gainesville State College

Hussain Elaloui-Talibi, Tuskegee University

Cathy Famiglietti, University of California–Irvine

Sandi Fay, University of California–Riverside

Thomas Fitzkee, Francis Marion University

Diego Grilli, Hillsborough Community College*

Olivier Heubo-Kwegna, Saginaw Valley State University

Leif Jordan, College of the Desert

Diana Klimek, New Jersey Institute of Technology

Julie Kostka, Austin Community College

Kristina Kraakmo, Valencia College*

Dr. Carole King Krueger, University of Texas–Arlington

Lance Lana, University of Colorado–Denver

Janice Lowe, Valdosta State University*

Craig McBride, University of Washington–Tacoma*

Christy Schmidt, Northwest Vista

Comlan de Souza, California State University–Fresno

Linda Snellings Neal, Wright State University

Bob Strozak, Old Dominion University

German Vargas, College of Coastal Georgia

Andrea Wichman, Leeward Community College*

Marti Zimmerman, University of Louisville

Our sincerest thanks also go to the legion of dedicated individuals who worked tirelessly to make this book possible. We would like to express our gratitude to our typist, Beverly DeVine-Hoffmeyer, for her amazing patience and skill. We must also thank Dr. Praveen Rohatgi, Dr. Nalini Rohatgi, and Dr. Bhupinder Bedi for the consulting they provided on all material relating to medicine. We particularly want to thank Professors Mile Krajcevski and Scott Rimbey for many helpful discussions and suggestions, especially for improving the exercise sets. Further gratitude is due to Irena Andreevska, Gokarna Aryal, Ferenc Tookos, and Christine Fitch for their assistance on the answers to the exercises in the text. In addition, we would like to thank Beverly Fusfield, Douglas Ewert, Nathan Kidwell, John Morin, and Perian Herring for their meticulous accuracy in checking the text. Thanks are due as well to Chere Bemelmans and Cenveo for their excellent production work. Finally, our thanks are extended to the professional and remarkable staff at Pearson. In particular, we would like to thank Anne Kelly, Editor in Chief; Ashley Gordon, Portfolio Management Assistant–Mathematics; Rachel Reeve, Content Producer; Scott Disanno, Managing Producer; Peggy Lucas, Marketing Manager; Claire Kozar, Product Marketing Manager; Jennifer Meyers, Marketing Assistant; Barbara Atkinson, Designer; Jonathan Wooding, Producer; Kristina Evans, Manager: Content Development, Math; Eric Gregg, Senior Content Developer, Math; and Joseph Vetere, Senior Author Support/Technology Specialist.

We invite all who use this book to send suggestions for improvements to Marcus McWaters at mmm@usf.edu.

Dedication

To Our Wives,
Lata, Debra, and Leslie

CHAPTER 1

Graphs and Functions

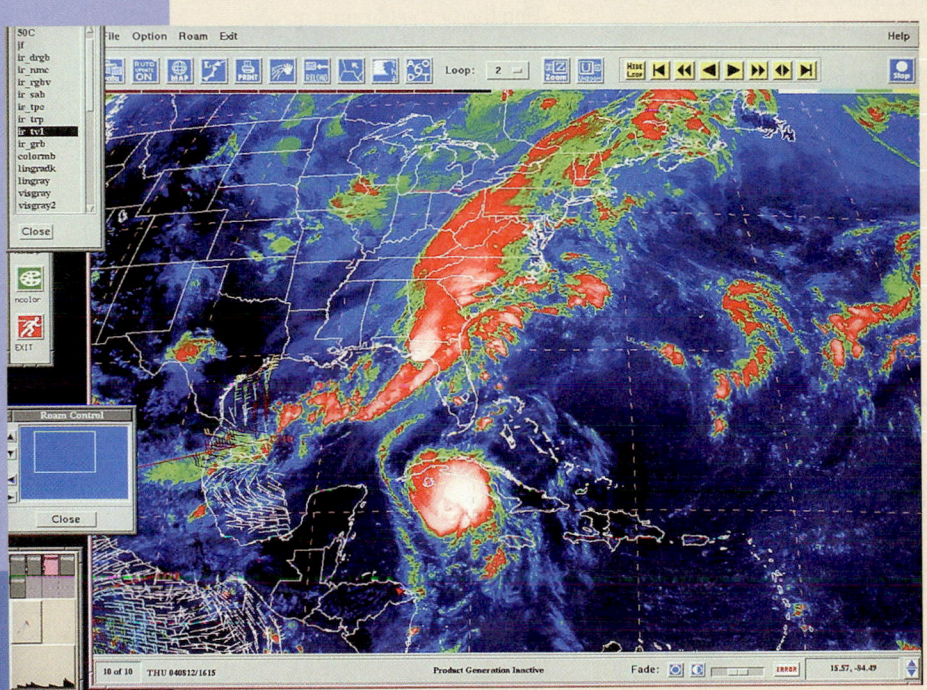

TOPICS

1.1 Graphs of Equations

1.2 Lines

1.3 Functions

1.4 A Library of Functions

1.5 Transformations of Functions

1.6 Combining Functions; Composite Functions

1.7 Inverse Functions

Hurricanes are tracked with the use of coordinate grids and potential paths are shown using spaghetti models, different colors indicating different potential paths. Data from fields as diverse as medicine and sports are related and analyzed by means of functions. The material in this chapter will introduce you to the versatile concepts that are the everyday tools of all dynamic industries.

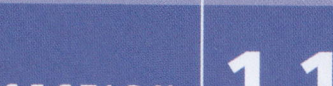

SECTION 1.1

Graphs of Equations

BEFORE STARTING THIS SECTION, REVIEW

1 The number line (Appendix A.1, page 918)

2 The Pythagorean Theorem (Appendix A.5, page 951)

3 Equivalent equations (Appendix A.6, page 954)

4 Completing squares (Appendix A.6, page 957)

5 Interval Notation (Appendix A.1, page 920)

OBJECTIVES

1 Plot points in the Cartesian coordinate plane.

2 Find the distance between two points.

3 Find the midpoint of a line segment.

4 Sketch a graph by plotting points.

5 Find the intercepts of a graph.

6 Find the symmetries in a graph.

7 Find the equation of a circle.

◆ A Fly on the Ceiling

One day the French mathematician René Descartes noticed a fly buzzing around on a ceiling made of square tiles. He watched the fly and wondered how he could mathematically describe its location. Finally, he realized that he could describe the fly's position by its distance from the walls of the room. Descartes had just discovered the coordinate plane! In fact, the coordinate plane is sometimes called the Cartesian plane in his honor. The discovery led to the development of analytic geometry, the first blending of algebra and geometry.

Although the basic idea of graphing with coordinate axes dates all the way back to Apollonius in the second century B.C., Descartes, who lived in the 1600s, gets the credit for coming up with the two-axis system we use today. In Example 2, we will see how the Cartesian plane helps visualize data on credit card interest rates.

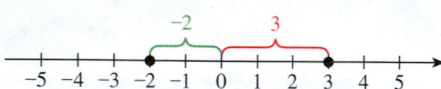

Figure 1.1 Coordinate line.

The Coordinate Plane

We associate the points on a geometric line with the real numbers using a number line or coordinate line as shown in Figure 1.1. Each point on the coordinate line is identified with its signed distance from the zero point. Similarly, every point in a plane will be uniquely identified by an ordered pair of numerical coordinates, which represent the signed distances from the point to two fixed perpendicular coordinate lines that intersect at their zero points.

A pair of real numbers in which the order is specified is called an **ordered pair** of real numbers. The ordered pair (a, b) has **first component** a and **second component** b. Two ordered pairs (x, y) and (a, b) are **equal**, and we write $(x, y) = (a, b)$, if and only if $x = a$ and $y = b$.

The two fixed coordinate lines that form the coordinate plane are called **coordinate axes**. The horizontal line (with positive numbers to the right) is usually called the ***x*-axis**, and the vertical line (with positive numbers up) is usually called the ***y*-axis**. Their point of intersection is called the **origin**, and the plane they form is sometimes called the ***xy*-plane**. See Figure 1.2. The axes divide the plane into four regions called **quadrants**, which are numbered counterclockwise as shown in Figure 1.3. The points on the axes themselves do not belong to any of the quadrants.

Figure 1.2 Coordinate plane.

René Descartes

(1596–1650)
Descartes was born at La Haye, near Tours in southern France. He is often called the father of modern science. Descartes established a new, clear way of thinking about philosophy and science by accepting only those ideas that could be proved by or deduced from first principles. He took as his philosophical starting point the statement **Cogito ergo sum:** "I think; therefore, I am." Descartes made major contributions to modern mathematics, including the Cartesian coordinate system and the theory of equations.

The notation $P(a, b)$, or $P = (a, b)$, designates the point P whose *first component* is a and whose *second component* is b. In an xy-plane, the first component, a, is called the x-coordinate of $P(a, b)$ and the second component, b, is called the y-coordinate of $P(a, b)$. The signs of the x- and y-coordinates for each quadrant are shown in Figure 1.3. The point corresponding to the ordered pair (a, b) is called the **graph of the ordered pair (a, b)**. However, keeping the Cartesian plane in mind, we frequently ignore the distinction between an ordered pair and its graph. Building on this correspondence, we employ the Cartesian plane as a universal translation tool between the languages of geometry and algebra. We will explore the interplay between geometry and algebra that the Cartesian plane offers throughout this book.

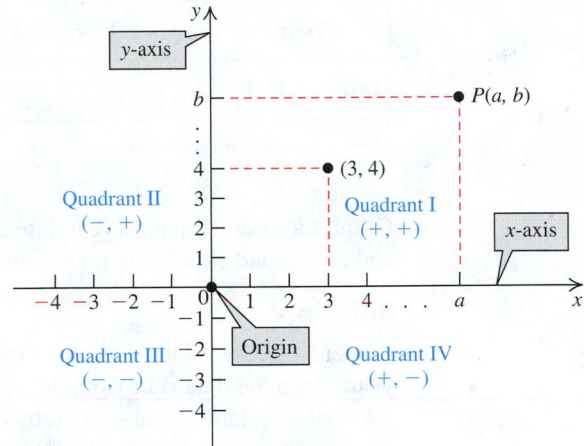

Figure 1.3 Quadrants in a plane.

EXAMPLE 1   **Graphing Points**

Graph the following points in the xy-plane:

$A(3, 1)$, $B(-2, 4)$, $C(-3, -4)$, $D(2, -3)$, and $E(-3, 0)$.

Solution

Figure 1.4 shows a coordinate plane, along with the graph of the given points. These points are located by moving left, right, up, or down starting from the origin $(0, 0)$.

$A(3, 1)$	3 units right, 1 unit up	$D(2, -3)$	2 units right, 3 units down
$B(-2, 4)$	2 units left, 4 units up	$E(-3, 0)$	3 units left
$C(-3, -4)$	3 units left, 4 units down		

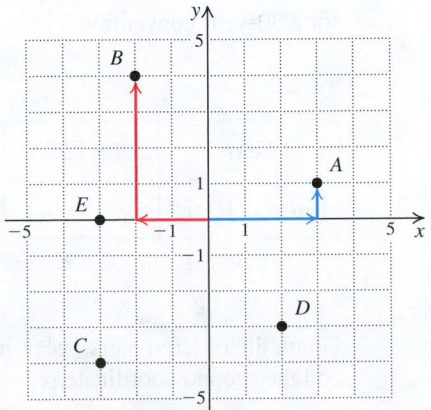

Figure 1.4 Graphing points.

Practice Problem 1 Graph the points in the *xy*-plane:

$$P(-2, 2), Q(4, 0), R(5, -3), S(0, -3), \text{ and } T\left(-2, \frac{1}{2}\right).$$

EXAMPLE 2 **Graphing Data on Credit Card Interest Rates in the US**

The data in Table 1.1 show the average credit card interest rates in the United States over the years 2005–2014.

TABLE 1.1

Year	2005	2006	2007	2008	2009	2010	2011	2012	2013	2014
Interest Rate	14.54	14.73	14.68	13.57	14.31	14.26	13.09	12.96	12.95	13.02

Source: federalreserve.gov

Graph the ordered pairs (year, interest rate), where the first coordinate represents a year and the second coordinate represents the interest rate in that year.

Solution

We let *t* represent the years 2005 through 2014 and % represent the interest rate in each year. Since the data start from the year 2005, we show a break in the *t*-axis. Alternatively, we could declare a year—say, 2004—as 0. Similar comments apply to the %-axis. The graph of the points $(2005, 14.54), (2006, 14.73), \ldots, (2014, 13.02)$ is shown in Figure 1.5. The figure depicts the credit card interest rate for every year since 2005.

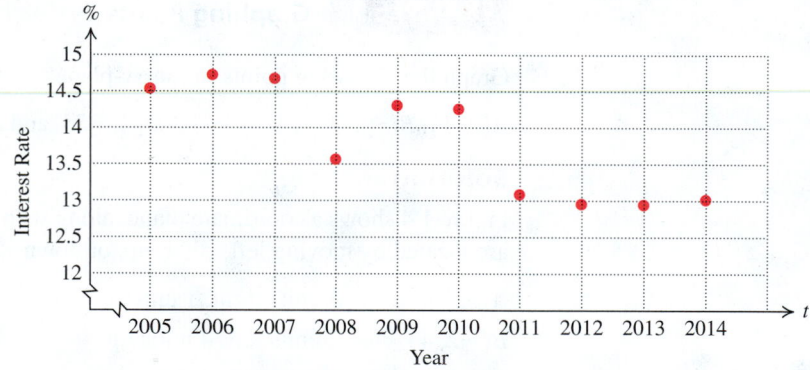

Figure 1.5 Credit card interest rates.

Practice Problem 2 The data in Table 1.2 show the average mortgage interest rates for a 30-year conventional loan in the United States over the years 2005–2014.

TABLE 1.2

Year	2005	2006	2007	2008	2009	2010	2011	2012	2013	2014
Interest Rate	5.87	6.41	6.34	6.03	5.04	4.69	4.45	3.66	3.98	4.22

Source: freddiemac.com

Graph the ordered pairs (year, interest rate), where the first coordinate represents a year and the second coordinate represents the interest rate in that year.

The display in Figure 1.5 is called the scatter diagram of the data. There are numerous other ways of visualizing the data. Two such ways are shown in Figure 1.6(a) and Figure 1.6(b).

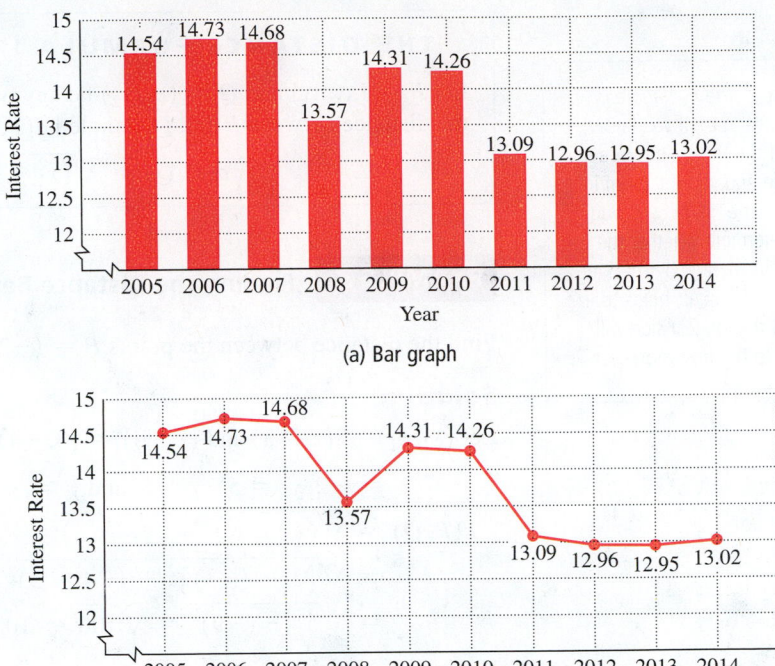

(a) Bar graph

(b) Line graph

Figure 1.6 Two methods of visualizing data.

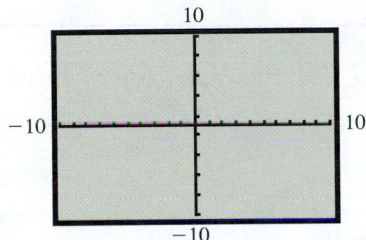

Figure 1.7 Viewing rectangle.

Scales on a Graphing Utility

When drawing a graph, you can use different scales for the x- and y-axes. Similarly, the scale can be set separately for each coordinate axis on a graphing utility. Once scales are set, you get a **viewing rectangle**, where your graphs are displayed. For example, in Figure 1.7, the scale for the x-axis is 1 (the distance between each tick mark), whereas the scale for the y-axis is 2. Read more about viewing rectangles in your graphing calculator manual.

2 Find the distance between two points.

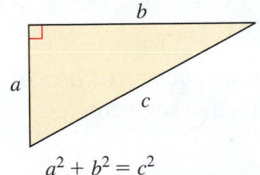

$$a^2 + b^2 = c^2$$

Figure 1.8 Pythagorean Theorem.

The Distance Formula

If a Cartesian coordinate plane has the same unit of measurement, such as inches or centimeters, on both axes, we can calculate the distance between any two points in that plane using this unit of measurement.

Recall that the Pythagorean Theorem states that in a **right triangle** with hypotenuse of length c and the legs of lengths a and b,

$$a^2 + b^2 = c^2, \quad \text{Pythagorean Theorem}$$

as shown in Figure 1.8.

Suppose we want to compute the distance between the two points $P(x_1, y_1)$ and $Q(x_2, y_2)$. We draw a horizontal line through the point Q and a vertical line through the point P to form the right triangle PQS, as shown in Figure 1.9. Also, recall that the distance between two points (two numbers) s and t on a number line is given by $|s - t|$.

The length of the horizontal side of the triangle is $|x_2 - x_1|$, and the length of the vertical side is $|y_2 - y_1|$. The distance between P and Q, denoted $d(P, Q)$, is the length of the hypotenuse.

$$[d(P, Q)]^2 = |x_2 - x_1|^2 + |y_2 - y_1|^2 \quad \text{Pythagorean Theorem}$$
$$d(P, Q) = \sqrt{|x_2 - x_1|^2 + |y_2 - y_1|^2} \quad \text{Take the square root of both sides.}$$
$$d(P, Q) = \sqrt{(x_2 - x_1)^2 + (y_2 - y_1)^2} \quad |a|^2 = a^2$$

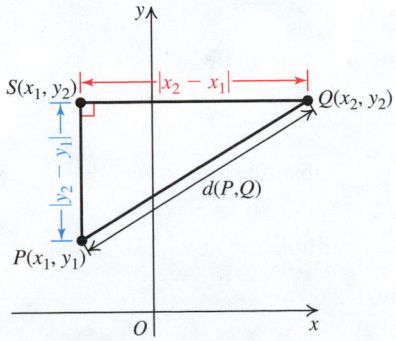

Figure 1.9 Visualizing the distance.

SIDE
NOTE

(The converse of the Pythagorean Theorem is also true). If you know the length of the three sides of a given triangle and you put them in increasing order $a \leq b \leq c$, you can easily determine whether the triangle is a right triangle by checking whether $a^2 + b^2 = c^2$ holds true. Observe that the longest side will always have to be the hypotenuse.

THE DISTANCE FORMULA IN THE COORDINATE PLANE

Let $P = (x_1, y_1)$ and $Q = (x_2, y_2)$ be any two points in the coordinate plane. Then the distance between P and Q, denoted $d(P, Q)$, is given by the **distance formula**:

$$d(P, Q) = \sqrt{(x_2 - x_1)^2 + (y_2 - y_1)^2}$$

EXAMPLE 3   **Finding the Distance Between Two Points**

Find the distance between the points $P = (-2, 5)$ and $Q = (3, -4)$.

Solution

Let $(x_1, y_1) = (-2, 5)$ and $(x_2, y_2) = (3, -4)$. Then

$$x_1 = -2, y_1 = 5, x_2 = 3, \text{ and } y_2 = -4.$$

$$\begin{aligned}
d(P, Q) &= \sqrt{(x_2 - x_1)^2 + (y_2 - y_1)^2} & &\text{Distance formula} \\
&= \sqrt{[3 - (-2)]^2 + (-4 - 5)^2} & &\text{Substitute the values for } x_1, x_2, y_1, y_2. \\
&= \sqrt{5^2 + (-9)^2} = \sqrt{106} \approx 10.3 & &\text{Use a calculator.}
\end{aligned}$$

Practice Problem 3 Find the distance between the points $(-5, 2)$ and $(-4, 1)$.

3 Find the midpoint of a line segment.

The Midpoint Formula

Recall that M is the **midpoint** on a line segment $\overline{PQ}$ if $d(P, M) = d(M, Q)$. The midpoint between two points (two numbers) s and t on a number line is given by $\dfrac{s + t}{2}$.

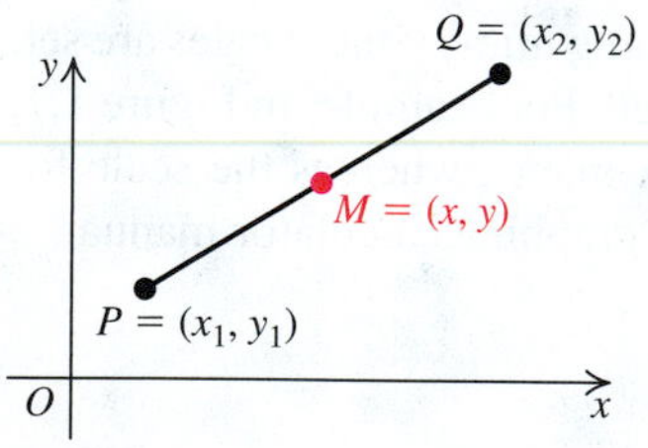

Figure 1.10 Midpoint of a segment.

THE MIDPOINT FORMULA

The coordinates of the midpoint $M = (x, y)$ on the line segment joining $P = (x_1, y_1)$ and $Q = (x_2, y_2)$ are given by (see Figure 1.10):

$$M = (x, y) = \left(\frac{x_1 + x_2}{2}, \frac{y_1 + y_2}{2} \right).$$

In physics, the midpoint of a line segment is interpreted as the barycenter (center of gravity) of a system of two equal weights placed at the endpoints of a line segment.

SIDE
NOTE

The coordinates of the midpoint of a line segment are found by taking the average of the x-coordinates and the average of the y-coordinates of the endpoints.

EXAMPLE 4   **Finding the Midpoint of a Line Segment**

Find the midpoint of the line segment joining the points $P(-3, 6)$ and $Q(1, 4)$.

Solution

Let $(x_1, y_1) = (-3, 6)$ and $(x_2, y_2) = (1, 4)$. Then

$$x_1 = -3, y_1 = 6, x_2 = 1, \text{ and } y_2 = 4.$$

$$\begin{aligned}
\text{Midpoint} &= \left(\frac{x_1 + x_2}{2}, \frac{y_1 + y_2}{2} \right) = \left(\frac{-3 + 1}{2}, \frac{6 + 4}{2} \right) & &\text{Substitute values for } x_1, x_2, y_1, y_2. \\
&= (-1, 5) & &\text{Simplify.}
\end{aligned}$$

The midpoint is $M = (-1, 5)$.

Practice Problem 4 Find the midpoint of the line segment whose endpoints are $(5, -2)$ and $(6, -1)$.

4 Sketch a graph by plotting points.

Graph of an Equation

An equation is an algebraic equality relating one or more quantities. An equation involving two unknown quantities describes a relation between these two quantities and specifies how one quantity changes with respect to the other quantity. The two changing (or varying) quantities are often represented by *variables*. The following equations are examples of relationships between two variables:

$$y = 2x + 1; \quad x^2 + y^2 = 4; \quad y = x^2; \quad x = y^2; \quad F = \frac{9}{5}C + 32;$$
$$\text{and} \quad q = -3p^2 + 30.$$

An ordered pair (a, b) is said to **satisfy** an equation with variables x and y if, when a is substituted for x and b is substituted for y in the equation, the resulting statement is true. For example, the ordered pair $(2, 5)$ satisfies the equation $y = 2x + 1$ because replacing x with 2 and y with 5 yields $5 = 2(2) + 1$, and simplifies to $5 = 5$, which is a true statement. The ordered pair $(5, -2)$ does not satisfy this equation since replacing x with 5 and y with -2 yields $-2 = 2(5) + 1$, and simplifies to $-2 = 11$, which is false. An ordered pair that satisfies an equation is called a **solution** of the equation.

In an equation involving x and y, if the value of y can be found given the value of x, then we say that y is the **dependent variable** and x is the **independent variable**. In the equation $y = 2x + 1$, for any real number x, there is a corresponding value of y. Hence, we have infinitely many solutions of the equation $y = 2x + 1$. When these solutions are graphed or plotted as points in the coordinate plane, they constitute the *graph of the equation*.

The graph of an equation is a geometric visualization of its solution set. In short, the coordinate plane allows us to investigate algebraic equations geometrically and to read many properties directly from their graphs.

Graph of an Equation

> The **graph of an equation** in two variables, such as x and y, is the graph of all ordered pairs (a, b) in the coordinate plane that satisfy the equation.

EXAMPLE 5   **Sketching a Graph by Plotting Points**

Sketch the graph of $y = x^2 - 3$.

Solution

The equation has infinitely many solutions. To find a few, we choose some values of x between -3 and 3. Then we find the corresponding values of y as shown in Table 1.3.

TABLE 1.3

x	$y = x^2 - 3$	(x, y)
-3	$y = (-3)^2 - 3 = 9 - 3 = 6$	$(-3, 6)$
-2	$y = (-2)^2 - 3 = 4 - 3 = 1$	$(-2, 1)$
-1	$y = (-1)^2 - 3 = 1 - 3 = -2$	$(-1, -2)$
0	$y = 0^2 - 3 = 0 - 3 = -3$	$(0, -3)$
1	$y = 1^2 - 3 = 1 - 3 = -2$	$(1, -2)$
2	$y = 2^2 - 3 = 4 - 3 = 1$	$(2, 1)$
3	$y = 3^2 - 3 = 9 - 3 = 6$	$(3, 6)$

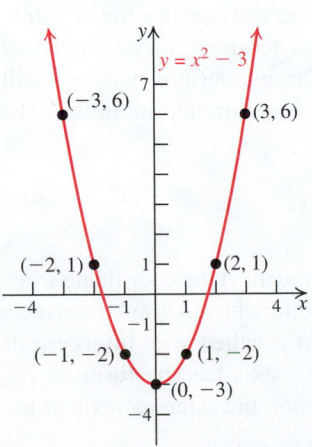

Figure 1.11 Graph of an equation.

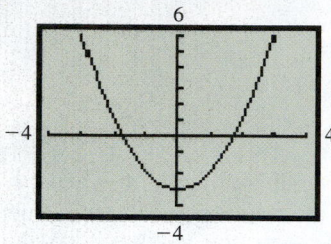

We plot the solutions (x, y) and join them with a smooth curve to sketch the graph of $y = x^2 - 3$ shown in Figure 1.11.

Practice Problem 5 Sketch the graph of $y = -x^2 + 1$.

The bowl-shaped curve sketched in Figure 1.11 is called a *parabola*. You can find parabolic shapes in everyday settings, such as the path of a thrown ball or in the reflector behind a car's headlight.

Example 5 suggests how to sketch the graph of any equation by plotting points. We summarize the steps for this technique.

SKETCHING A GRAPH BY PLOTTING POINTS

Step 1 Make a representative table of solutions of the equation.

Step 2 Plot the solutions as ordered pairs in the Cartesian coordinate plane.

Step 3 Connect the solutions in Step 2 with a smooth curve.

Comment This technique has obvious pitfalls. For instance, many different curves pass through the same four points in Figure 1.12. Assume that these points are solutions of a given equation. We cannot guarantee that any curve through these points is the actual graph of the equation. However, in general, the more solutions plotted, the more accurate the graph.

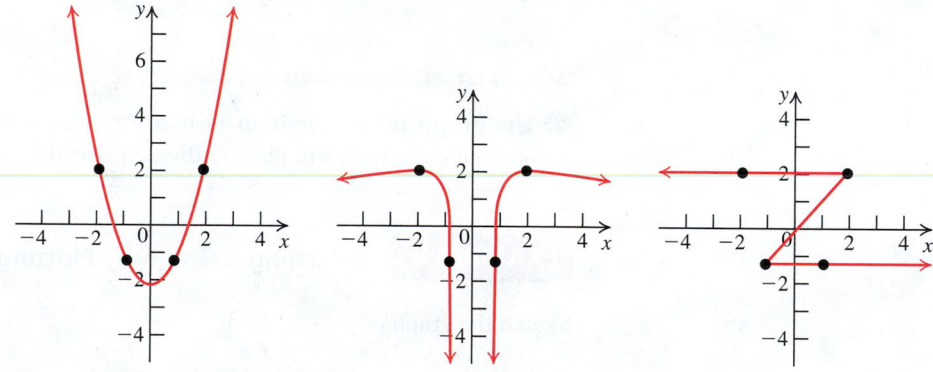

Figure 1.12 Several graphs through the same four points.

One way to graph an equation that reduces plotting pitfalls is to plot enough points so that the graph becomes self-evident. Graphing calculators and computer programs are especially good for this purpose. A better way is to identify the shape of the curve the equation represents based on its algebraic features. For instance, we can identify the equation in Example 5 as a quadratic equation. We will soon learn how to graph the parabola that is the graph of the quadratic equation $y = ax^2 + bx + c$. Throughout this book we will investigate various techniques that will allow us to graph an equation that minimizes the amount of point plotting.

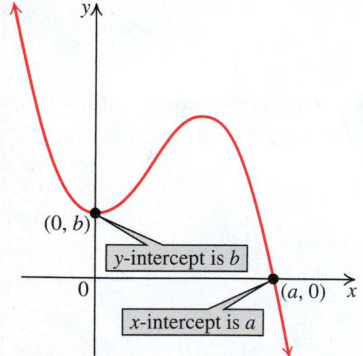

Figure 1.13 Intercepts of a graph.

5 Find the intercepts of a graph

Intercepts

Let's examine the points where a graph intersects (crosses or touches) the coordinate axes. Since all points on the x-axis have a y-coordinate of 0, any point where a graph intersects the x-axis has the form $(a, 0)$. See Figure 1.13. The number a is called an **x-intercept** of the graph. Similarly, any point where a graph intersects the y-axis has the form $(0, b)$, and the number b is called a **y-intercept** of the graph. Together, the x-intercepts and the y-intercepts are referred to as the **intercepts** of the graph.

FINDING THE INTERCEPTS OF THE GRAPH OF AN EQUATION

Step 1   To find the x-intercepts of the graph of an equation, set $y = 0$ in the equation and solve for x.

Step 2   To find the y-intercepts of the graph of an equation, set $x = 0$ in the equation and solve for y.

WARNING

Do not try to calculate the x-intercept by setting $x = 0$. An x-intercept is the x-coordinate of a point where the graph touches or crosses the x-axis, so the y-coordinate must be 0.

EXAMPLE 6   Finding Intercepts

Find the x- and y-intercepts of the graph of the equation $y = x^2 - x - 2$.

Solution

Step 1   Set $y = 0$ in the equation and solve for x.

$$0 = x^2 - x - 2 \qquad \text{Set } y = 0.$$
$$0 = (x + 1)(x - 2) \qquad \text{Factor.}$$
$$x + 1 = 0 \quad \text{or} \quad x - 2 = 0 \qquad \text{Zero-product property}$$
$$x = -1 \quad \text{or} \qquad x = 2 \qquad \text{Solve each equation for } x.$$

The x-intercepts are -1 and 2.

Step 2   Set $x = 0$ in the equation and solve for y.

$$y = 0^2 - 0 - 2 \qquad \text{Set } x = 0.$$
$$y = -2 \qquad \text{Solve for } y.$$

The y-intercept is -2.
The graph of the equation $y = x^2 - x - 2$ is shown in Figure 1.14.

Practice Problem 6   Find the intercepts of the graph of $y = 2x^2 + 3x - 2$.

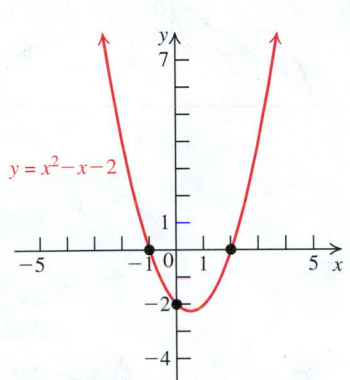

Figure 1.14 Intercepts of a graph.

$y = x^2 - x - 2$

6 Find the symmetries in a graph.

Figure 1.15

Symmetry

The graphs of equations may exhibit various repeating patterns. Extreme examples are fractals, which exhibit prominent self-similar repeating patterns (see Figure 1.15). Recognizing the various repeating patterns helps us to sketch the graphs of equations very efficiently. One such pattern is called **symmetry**. An object has reflectional **symmetry** if it can be divided into two identical pieces that are *mirror images* of each other. As shown in Figure 1.16(a), if a line ℓ is an **axis of symmetry**, or **line of symmetry**, we can construct the mirror image of any point P not on ℓ by first drawing the perpendicular line segment from P to ℓ. Then we extend this segment an equal distance on the other side to a point P' so that the line ℓ perpendicularly bisects the line segment $\overline{PP'}$. See Figure 1.16(a). We say that the point P' is the *symmetric image* of the point P about the line ℓ.

Two points M and M' are **symmetric about a point** Q if Q is the midpoint of the line segment $\overline{MM'}$. Figure 1.16(b) illustrates the symmetry about the origin O. Symmetry lets us use information about part of the graph to draw the remainder of the graph.

The following three types of symmetries occur frequently.

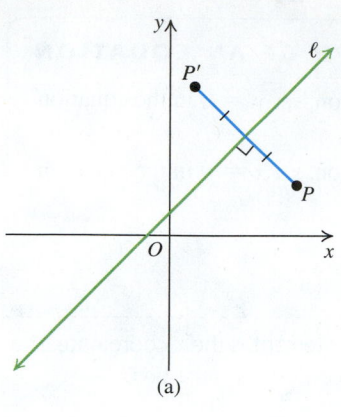

(a)

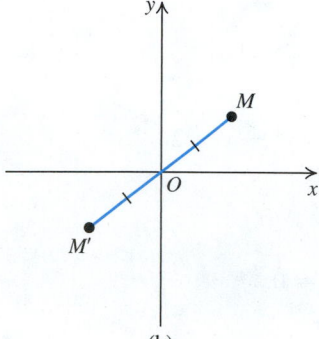

(b)

Figure 1.16 (a) Symmetric points about a line; (b) Symmetric points about the origin O.

SYMMETRIES

1. A graph is **symmetric with respect to (or about) the y-axis** if for every point (x, y) on the graph the point $(-x, y)$ is also on the graph. See Figure 1.17(a) and 1.18(a).

2. A graph is **symmetric with respect to (or about) the x-axis** if for every point (x, y) on the graph the point $(x, -y)$ is also on the graph. See Figure 1.17(b) and 1.18(b).

3. A graph is **symmetric with respect to (or about) the origin** if for every point (x, y) on the graph the point $(-x, -y)$ is also on the graph. See Figure 1.17(c) and 1.18(c).

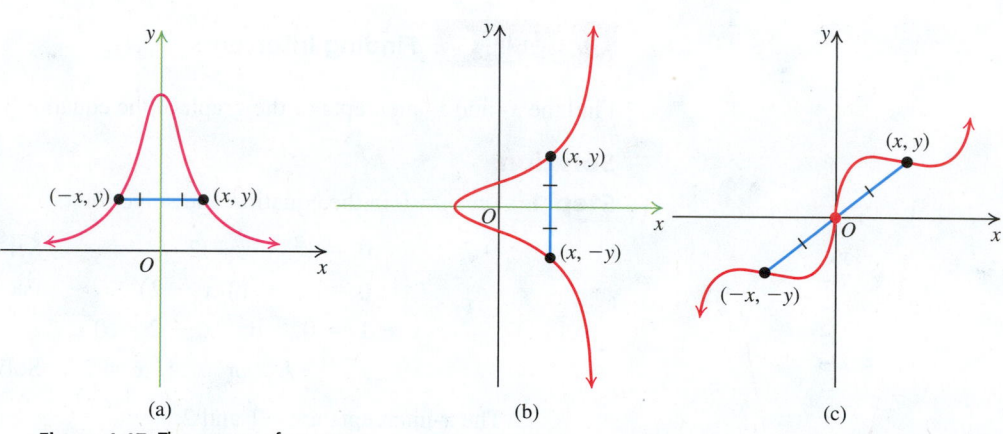

(a)   (b)   (c)

Figure 1.17 Three types of symmetry.

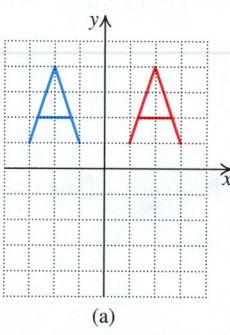

(a)

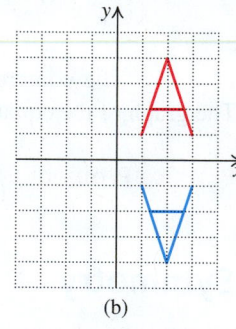

(b)

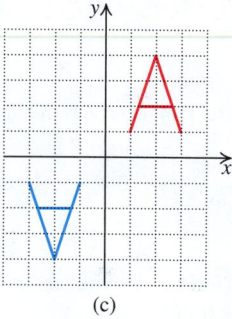

(c)

Figure 1.18 Three types of symmetry.

TESTS FOR SYMMETRY

1. The graph of an equation is symmetric about the y-axis if replacing x with $-x$ results in an equivalent equation.

2. The graph of an equation is symmetric about the x-axis if replacing y with $-y$ results in an equivalent equation.

3. The graph of an equation is symmetric about the origin if replacing x with $-x$ and y with $-y$ results in an equivalent equation.

SIDE NOTE

Note that if *only* even powers of x appear in an equation, then the graph is symmetric with respect to the y-axis since for any integer n, $(-x)^{2n} = x^{2n}$.

EXAMPLE 7   **Checking for Symmetry**

Determine whether the graph of the equation $y = \dfrac{1}{x^2 + 5}$ is symmetric with respect to the y-axis.

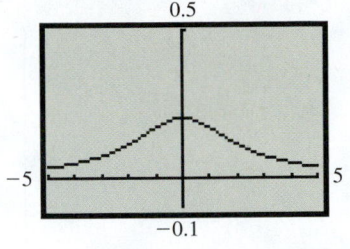

Solution

Replace x with $-x$ to see if $(-x, y)$ also satisfies the equation.

$$y = \frac{1}{x^2 + 5}$$ Original equation

$$y = \frac{1}{(-x)^2 + 5}$$ Replace x with $-x$.

$$y = \frac{1}{x^2 + 5}$$ Simplify: $(-x)^2 = x^2$.

Since replacing x with $-x$ gives us the original equation, the graph of $y = \dfrac{1}{x^2 + 5}$ is symmetric with respect to the y-axis.

Practice Problem 7 Determine whether the graph of $x^2 - y^2 = 1$ is symmetric with respect to the y-axis.

PROCEDURE
IN ACTION

EXAMPLE 8 **Sketching a Graph Using Symmetry**

OBJECTIVE

Use symmetry to sketch the graph of an equation.

Step 1 Test for all three symmetries.
About the x-axis: Replace y with $-y$.
About the y-axis: Replace x with $-x$.
About the origin: Replace x with $-x$ and y with $-y$.
Simplify and compare the result to the original equation.

EXAMPLE

Use symmetry to sketch the graph of $y = 4x - x^3$.

1.

x-axis	y-axis	origin
Replace y with $-y$	Replace x with $-x$	Replace x with $-x$ and y with $-y$
$-y = 4x - x^3$ Simplify and compare	$y = 4(-x) - (-x)^3$ Simplify and compare	$-y = 4(-x) - (-x)^3$ Simplify and compare
$y = -4x + x^3$	$y = -4x + x^3$	$-y = -4x + x^3$ $y = 4x - x^3$
No	**No**	**Yes**

Step 2 Make a table of values using any symmetries found in Step 1.

2. Origin symmetry: If (x, y) is on the graph, so is $(-x, -y)$.
Use only positive x values in the table.

x	0	0.5	1	1.5	2	2.5
$y = 4x - x^3$	0	1.875	3	2.625	0	-5.625

Step 3 Plot the points from the table and draw a smooth curve through them.

3.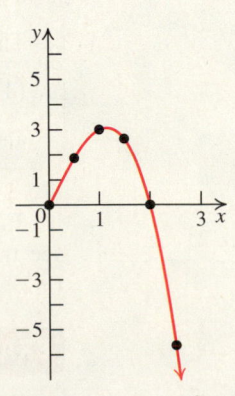

Graph of $y = 4x - x^3, x \geq 0$

continued

Step 4 Extend the portion of the graph found in Step 3 using symmetries.

4.

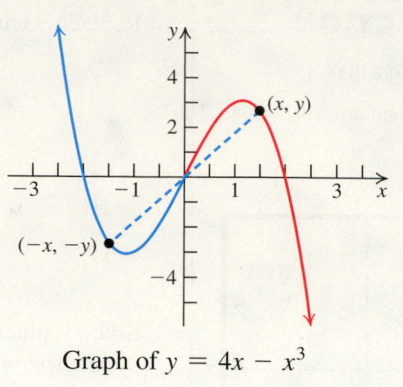

Graph of $y = 4x - x^3$

Practice Problem 8 Use symmetry to sketch the graph of $y = x^4 - 4x^2$.

7 Find the equation of a circle.

Circles

The Cartesian coordinate plane allows us to describe various geometric curves using algebraic equations. We illustrate this situation in the case of a circle. We begin with the geometric definition.

Circle

> A **circle** is a set of points in a Cartesian coordinate plane that are at a fixed distance r from a specified point (h, k). The fixed distance r is called the **radius** of the circle, and the specified point (h, k) is called the **center** of the circle.

Standard Form

A point $P(x, y)$ is on the circle if and only if its distance from the center $C(h, k)$ is r. Using the notation for the distance between the points P and C, we have

$$d(P, C) = r$$
$$\sqrt{(x - h)^2 + (y - k)^2} = r \qquad \text{Distance formula}$$
$$(x - h)^2 + (y - k)^2 = r^2 \qquad \text{Square both sides.}$$

The equation $(x - h)^2 + (y - k)^2 = r^2$ is an equation of a circle with radius r and center (h, k). A point (x, y) is on the circle of radius r and center $C(h, k)$ if and only if it satisfies this equation. Figure 1.19 is the graph of a circle with center $C(h, k)$ and radius r.

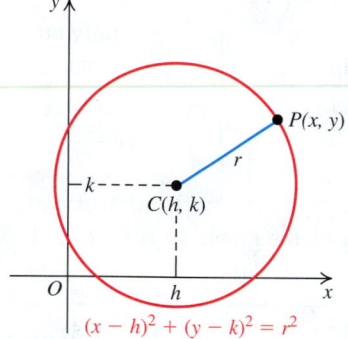

Figure 1.19

The Standard Form for the Equation of a Circle

> The equation of a circle with center (h, k) and radius r is
>
> (1) $$(x - h)^2 + (y - k)^2 = r^2$$
>
> and is called the **standard form** of an equation of a circle.

Employing the Cartesian coordinate plane as a geometric visualization tool, we see that the graph of any equation in x and y that can be written in the standard form $(x - h)^2 + (y - k)^2 = r^2$ is a circle with center (h, k) and radius r.

EXAMPLE 9 **Finding the Equation of a Circle**

Find the standard form of the equation of the circle with center $(7, -3)$ and which passes through the point $P = (5, -2)$.

TECHNOLOGY CONNECTION

To graph the equation $x^2 + y^2 = 1$ on a graphing calculator, we first solve for y.

$y^2 = 1 - x^2$ Subtract x^2 from both sides.

$y = \pm\sqrt{1 - x^2}$ Square root property

We then graph the two equations

$$Y_1 = \sqrt{1 - x^2} \text{ and}$$
$$Y_2 = -\sqrt{1 - x^2}$$

in the same window. The graph of Y_1 is the upper semicircle ($y \geq 0$), and the graph of Y_2 is the lower semicircle ($y \leq 0$). The calculator graph does not quite look like a circle. We use the Square option to make the display look like a circle.

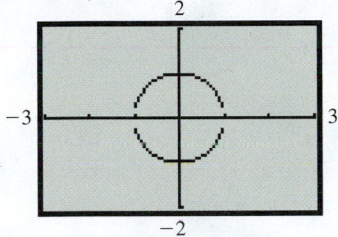

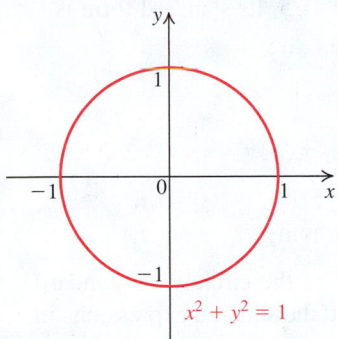

Figure 1.20 The unit circle.

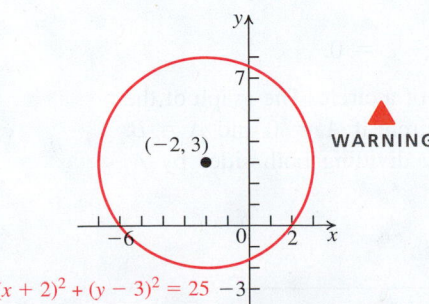

$(x + 2)^2 + (y - 3)^2 = 25$

Figure 1.21 Circle with radius 5 and center at $(-2, 3)$.

Solution

To find the standard form of the equation of a circle, we need to specify its center and radius. The center is given as $(7, -3)$.

$$(x - h)^2 + (y - k)^2 = r^2 \quad \text{Standard form}$$
$$(x - 7)^2 + (y - (-3))^2 = r^2 \quad \text{Replace } h \text{ with 7 and } k \text{ with } -3.$$
$$(2) \qquad (x - 7)^2 + (y + 3)^2 = r^2 \quad -(-3) = 3$$

To find the radius of this circle, we use the fact that the circle passes through the point $P = (5, -2)$.

Since the point $P = (5, -2)$ lies on the circle, its coordinates satisfy equation (2). So,

$$(5 - 7)^2 + (-2 + 3)^2 = r^2 \quad \text{Replace } x \text{ with 5 and } y \text{ with } -2.$$
$$5 = r^2 \quad \text{Simplify.}$$

Replacing r^2 with 5 in equation (2) gives the required standard form

$$(x - 7)^2 + (y + 3)^2 = 5.$$

Practice Problem 9 Find the standard form of the equation of the circle with center $(3, -6)$ and radius 10.

If an equation in two variables can be written in standard form (1), then its graph is a circle with center (h, k) and radius r.

EXAMPLE 10   Graphing a Circle

Specify the center and radius and graph each circle.

a. $x^2 + y^2 = 1$

b. $(x + 2)^2 + (y - 3)^2 = 25$

Solution

a. The equation $x^2 + y^2 = 1$ can be rewritten as

$$(x - 0)^2 + (y - 0)^2 = 1^2.$$

Comparing this equation with equation (1), we conclude that the given equation is an equation of a circle with center $(0, 0)$ and radius 1. The graph is shown in Figure 1.20. This circle is called the **unit circle**.

b. Rewriting the equation $(x + 2)^2 + (y - 3)^2 = 25$ as

$$[x - (-2)]^2 + (y - 3)^2 = 5^2, \quad x + 2 = x - (-2)$$

we see that the graph of this equation is a circle with center $(-2, 3)$ and radius 5. The graph is shown in Figure 1.21.

Practice Problem 10 Graph the equation $(x - 2)^2 + (y + 1)^2 = 36$.

⚠️ **WARNING**

It is easy to make sign errors when identifying the center of the circle. The direct method of finding the coordinates of the center is to set the expression inside each of the squares to zero.

$$(x + 2)^2 + (y - 3)^2 = 25.$$

$x + 2 = 0$ gives $x = -2$ and $y - 3 = 0$ gives $y = 3$. We identify the center of this circle as $(-2, 3)$.

Semicircles

Letting $h = 0$ and $k = 0$ in equation (1), we have

(3)
$$x^2 + y^2 = r^2$$

Equation (3) is the standard form for a circle with center at the origin and radius r.
　　We solve equation (3) for y:

$$y^2 = r^2 - x^2 \qquad \text{Subtract } x^2 \text{ from both sides.}$$
$$y = \pm\sqrt{r^2 - x^2} \qquad \text{Square root property (page 942).}$$
$$\text{or} \qquad y = \sqrt{r^2 - x^2} \quad \text{and} \quad y = -\sqrt{r^2 - x^2}.$$

Similarly, solving equation (3) for x, we have two equations,

$$x = \sqrt{r^2 - y^2} \quad \text{and} \quad x = -\sqrt{r^2 - y^2}.$$

The graphs of these four equations are **semicircles** (half circles), shown in Figure 1.22.

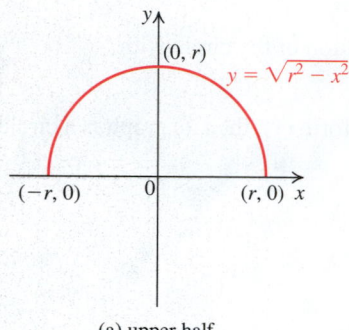

(a) upper half

Figure 1.22 Semicircles.

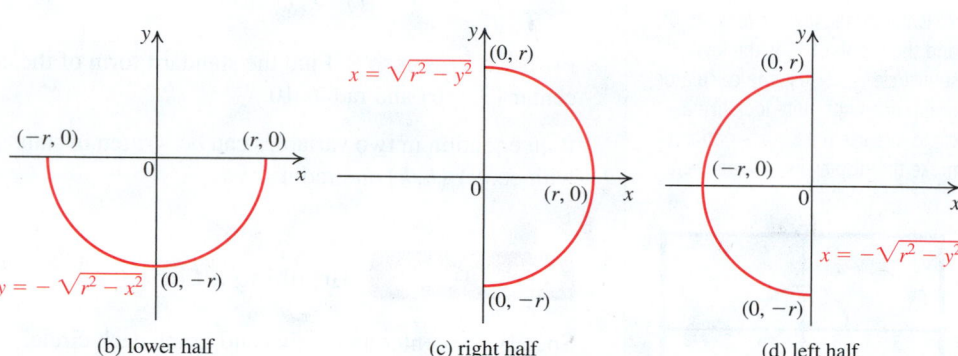

(b) lower half　　　　　　(c) right half　　　　　　(d) left half

General Form

Consider a circle with center $C = (2, -3)$ and radius $r = 5$. Its standard form is

$$(x - 2)^2 + (y + 3)^2 = 5^2.$$

Expanding $(x - 2)^2$ and $(y + 3)^2$, we have

$$(x^2 - 4x + 4) + (y^2 + 6y + 9) = 25 \qquad (a \pm b)^2 = a^2 \pm 2ab + b^2$$
$$x^2 + y^2 - 4x + 6y + 13 = 25 \qquad \text{Rewrite.}$$
$$x^2 + y^2 - 4x + 6y - 12 = 0 \qquad \text{Simplify.}$$

This last equation is the **general form** of the equation of the circle with standard form $(x - 2)^2 + (y + 3)^2 = 5^2$. In general, we expand the squared expressions in the standard equation of a circle,

$$(x - h)^2 + (y - k)^2 = r^2,$$

and then simplify, we obtain an equation of the form

(4)
$$x^2 + y^2 + ax + by + c = 0.$$

Equation (4) is called the *general form* of the equation of a circle. The graph of the equation $Ax^2 + By^2 + Cx + Dy + E = 0$ is also a circle if $A \neq 0$ and $A = B$. You can convert this equation to the general form by dividing both sides by A. See Exercises 97 and 98.

General Form of the Equation of a Circle

The **general form** of the equation of a circle is

$$x^2 + y^2 + ax + by + c = 0.$$

On the other hand, if we are given an equation in general form, we can convert it to standard form by completing the squares on the x- and y-terms. This gives

(5)
$$(x - h)^2 + (y - k)^2 = d.$$

If $d > 0$, the graph of equation (5) is a circle with center (h, k) and radius $\sqrt{d}$. If $d = 0$, the graph of equation (5) is the point (h, k). If $d < 0$, there is no graph.

EXAMPLE 11   **Converting the General Form to Standard Form**

Find the center and radius of the circle with equation

(5)
$$x^2 + y^2 - 6x + 8y + 10 = 0.$$

Solution

Complete the squares on both the x-terms and y-terms to get standard form.

$$x^2 + y^2 - 6x + 8y + 10 = 0 \qquad \text{Original equation}$$
$$(x^2 - 6x) + (y^2 + 8y) = -10 \qquad \text{Group the } x\text{-terms and } y\text{-terms.}$$
$$(x^2 - 6x + 9) + (y^2 + 8y + 16) = -10 + 9 + 16 \qquad \text{Complete the squares by adding 9 and 16 to both sides.}$$
$$(x - 3)^2 + (y + 4)^2 = 15 \qquad \text{Factor and simplify.}$$
$$(x - 3)^2 + [y - (-4)]^2 = (\sqrt{15})^2$$

The last equation tells us that we have $h = 3$, $k = -4$, and $r = \sqrt{15}$. Therefore, the circle has center $(3, -4)$ and radius $\sqrt{15} \approx 3.9$.

Practice Problem 11 Find the center and radius of the circle with equation
$x^2 + y^2 + 4x - 6y - 12 = 0.$

RECALL

To make $x^2 + bx$ a perfect square, add $\left(\frac{1}{2}b\right)^2$ to get
$$x^2 + bx + \left(\frac{1}{2}b\right)^2 = \left(x + \frac{b}{2}\right)^2$$
This completes the square.

Answers to Practice Problems

1.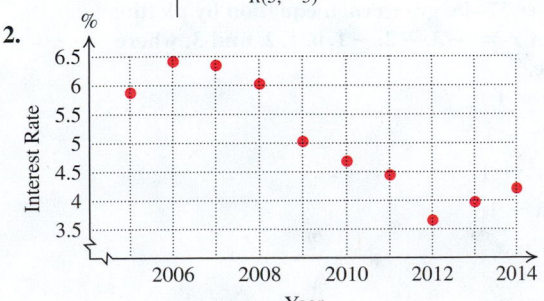

2.

3. $\sqrt{2}$ **4.** $\left(\frac{11}{2}, -\frac{3}{2}\right)$ **5.**

6. x-intercepts: $-2, \frac{1}{2}$, y-intercept: -2 **7.** Symmetric

8. 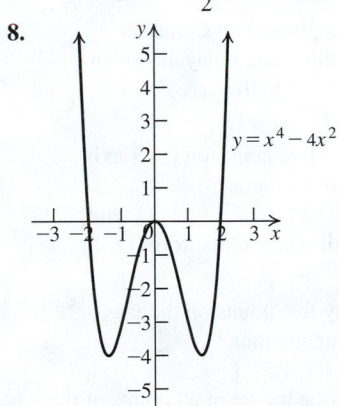 **9.** $(x - 3)^2 + (y + 6)^2 = 100$

10. 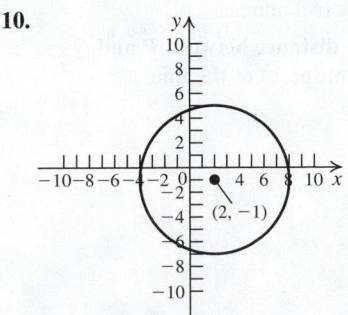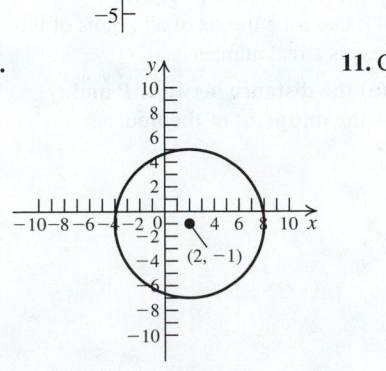 **11.** Center: $(-2, 3)$; radius: 5

Concepts and Vocabulary

1. The graph of an equation in two variables such as x and y is the set of all ordered pairs (a, b) _____.

2. The distance between the points $P = (x_1, y_1)$ and $Q = (x_2, y_2)$ is given by the formula $d(P, Q) =$

_____ .

3. The coordinates of the midpoint $M = (x, y)$ of the line segment joining $P = (x_1, y_1)$ and $Q = (x_2, y_2)$ are given by

$M =$ _____ .

4. The standard form of the equation of a circle with center (h, k) and radius r is _____ .

5. **True or False.** A point with a negative first coordinate and a positive second coordinate lies in the fourth quadrant.

6. **True or False.** For any points (x_1, y_1) and (x_2, y_2):
$\sqrt{(x_1 - x_2)^2 + (y_1 - y_2)^2} = \sqrt{(x_2 - x_1)^2 + (y_2 - y_1)^2}$

7. **True or False.** The center of the circle given by the equation $(x + 3)^2 + (y + 4)^2 = 9$ is the point $(3,4)$.

8. **True or False.** If $(-2, 4)$ is a point on a graph that is symmetric with respect to the y-axis, then the point $(2, 4)$ is also on the graph.

Building Skills

9. Plot and label each of the given points in a Cartesian coordinate plane and state the quadrant, if any, in which each point is located. $(2, 2), (3, -1), (-1, 0), (-2, -5), (0, 0), (-7, 4), (0, 3), (-4, 2)$

10. Plot and label each of the given points in a Cartesian coordinate plane and state the quadrant, if any, in which each point is located. $(1, 4), (3, -2), (1, 0), (-1, -2), (0, 0), (-2, 1), (0, 2), (-3, 1).$

11. **a.** Write the coordinates of any five points on the x-axis. What do these points have in common?
   b. Graph the points $(-2, 1), (0, 1), (0.5, 1), (1, 1),$ and $(2, 1)$. Describe the set of all points of the form $(x, 1)$, where x is a real number.

12. **a.** Write the coordinates of any five points on the y-axis. What do these points have in common?
   b. Graph the points $(-1, 1), (-1, 1.5), (-1, 2), (-1, 3),$ and $(-1, 4)$. Describe the set of all points of the form $(-1, y)$, where y is a real number.

In Exercises 13–20, find (a) the distance between P and Q and (b) the coordinates of the midpoint of the line segment PQ.

13. $P(2, 1), Q(3, 5)$

14. $P(1, 4), Q(3, 2)$

15. $P(4, 5), Q(1, -2)$

16. $P(2, 3), Q(-1, 2)$

17. $P(-1, -5), Q(2, -3)$

18. $P(-4, 1), Q(-7, -9)$

19. $P(x, y), Q(-2, 3)$

20. $P(t, k), Q(k, t)$

In Exercises 21–24, determine whether the given points are collinear. Points are *collinear* if they can be labeled $P, Q,$ and R so that $d(P, Q) + d(Q, R) = d(P, R)$. If the points are noncollinear, find the perimeter of the triangle PQR.

21. $(0, 0), (1, 2), (-1, -2)$

22. $(3, 4), (0, 0), (-3, -4)$

23. $(9, 6), (0, -3), (3, 1)$

24. $(-2, 3), (3, 1), (2, -1)$

In Exercises 25–30, identify the triangle PQR as an *isosceles* (two sides of equal length), *equilateral* (three sides of equal length), or *scalene* (three sides of different lengths) triangle.

25. $P(-5, 5), Q(-1, 4), R(-4, 1)$

26. $P(3, 2), Q(6, 6), R(-1, 5)$

27. $P(-4, 8), Q(0, 7), R(-3, 5)$

28. $P(6, 6), Q(-1, -1), R(-5, 3)$

29. $P(1, -1), Q(-1, 1), R(-\sqrt{3}, -\sqrt{3})$

30. $P(-0.5, -1), Q(-1.5, 1), R(\sqrt{3} - 1, \sqrt{3}/2)$

In Exercises 31–36, determine whether the given points are on the graph of the equation.

Equation	Points
31. $y = x - 1$	$(-3, -4), (1, 0), (4, 3), (2, 3)$
32. $2y = 3x + 5$	$(-1, 1), (0, 2), \left(-\dfrac{5}{3}, 0\right), (1, 4)$
33. $y = \sqrt{x + 1}$	$(3, 2), (0, 1), (8, -3), (8, 3)$
34. $y = \dfrac{1}{x}$	$\left(-3, \dfrac{1}{3}\right), (1, 1), (0, 0), \left(2, \dfrac{1}{2}\right)$
35. $x^2 - y^2 = 1$	$(1, 0), (0, -1), (2, \sqrt{3}), (2, -\sqrt{3})$
36. $y^2 = x$	$(1, -1), (1, 1) (0, 0), (2, -\sqrt{2})$

In Exercises 37–46, graph each equation by plotting points. Let $x = -3, -2, -1, 0, 1, 2,$ and 3, where applicable.

37. $y = x + 1$

38. $y = 2x - 1$

39. $y = |x| + 1$

40. $y = |x + 1|$

41. $y = 4 - x^2$

42. $y = x^2 - 4$

43. $y = \sqrt{9 - x^2}$

44. $y = -\sqrt{9 - x^2}$

45. $y = x^3$

46. $y = -x^3$

In Exercises 47–56, find
a. x and y-intercepts.
b. symmetries (if any) about the x-axis, the y-axis, and the origin.

47.

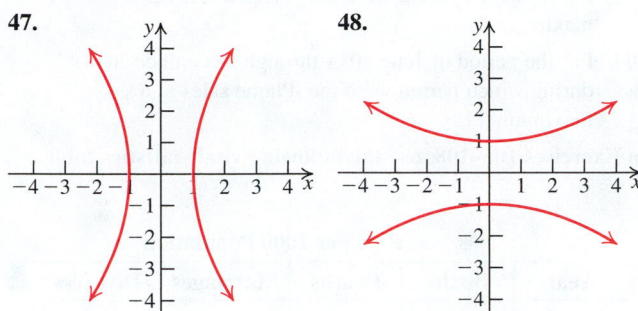

48.

49.

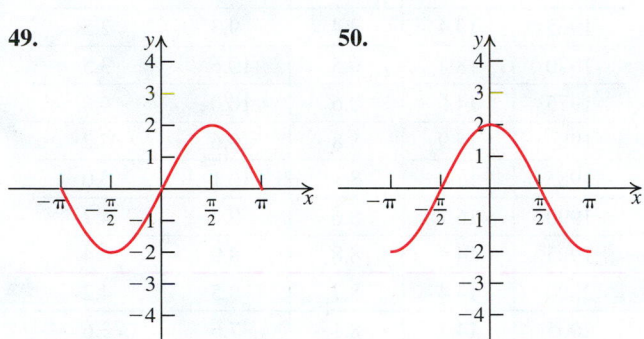

50.

51.

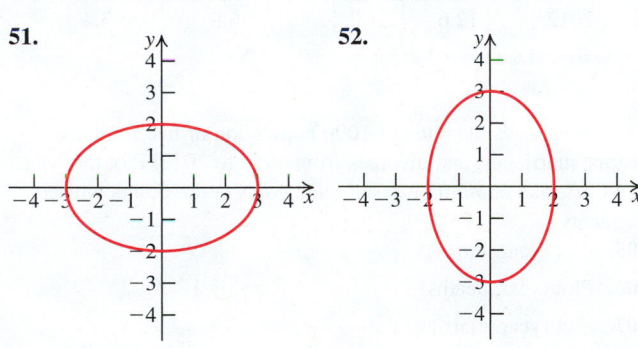

52.

53.

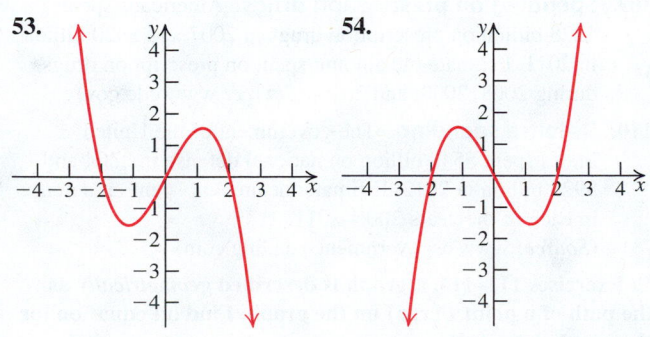

54.

55.

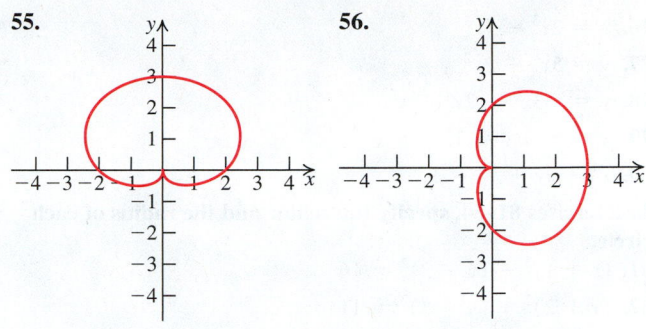

56.

In Exercises 57–60, complete the given graph so that it has the indicated symmetry.

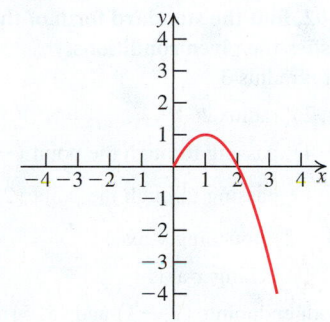

57. symmetry about the x-axis

58. symmetry about the y-axis

59. symmetry about the origin

60. symmetry about the x-axis and symmetry about the y-axis

In Exercises 61–72, find the x- and y-intercepts of the graph of each equation.

61. $3x + 4y = 12$

62. $\dfrac{x}{5} + \dfrac{y}{3} = 1$

63. $2x + 3y = 5$

64. $\dfrac{x}{2} - \dfrac{y}{3} = 1$

65. $y = \dfrac{x + 2}{x - 1}$

66. $x = \dfrac{y - 2}{y + 1}$

67. $y = x^2 - 6x + 8$

68. $x = y^2 - 5y + 6$

69. $x^2 + y^2 = 4$

70. $y = \sqrt{9 - x^2}$

71. $y = \sqrt{x^2 - 1}$

72. $xy = 1$

In Exercises 73–80, test each equation for symmetry with respect to the x-axis, the y-axis, and the origin.

73. $y = x^2 + 1$

74. $x = y^2 + 1$

75. $y = x^3 + x$

76. $y = 2x^3 - x$

77. $y = 5x^4 + 2x^2$

78. $y = -3x^6 + 2x^4 + x^2$

79. $x^2y^2 + 2xy = 1$,

80. $x^4 - xy + y^4 = 1$,

In Exercises 81–84, specify the center and the radius of each circle.

81. $(x + 1)^2 + (y - 3)^2 = 16$

82. $(x + 2)^2 + (y + 3)^2 = 11$

83. $(x - 1)^2 + y^2 = 4$

84. $x^2 + (y + 2)^2 = 13$

In Exercises 85–92, find the standard form of the equation of a circle that satisfies the given conditions.

85. Center $(3, 1)$; radius 3

86. Center $(-1, 2)$; radius 2

87. Center $(3, -4)$; passing through the point $(-1, 5)$

88. Center $(-1, 1)$; passing through the point $(2, 5)$

89. Center $(-3, -2)$; touching y-axis

90. Center $(1, 2)$; touching x-axis

91. Diameter with endpoints $(2, -3)$ and $(8, 5)$

92. Diameter with endpoints $(7, 4)$ and $(-3, 6)$

In Exercises 93–100, determine whether the given equation represents a circle. If it does, a. Find the center and radius of each circle. b. Find the x- and y-intercepts of the graph of each circle.

93. $x^2 + y^2 + 4x - 6y + 9 = 0$,

94. $x^2 + y^2 - 4x + 2y + 4 = 0$,

95. $x^2 + y^2 - 2x - 2y - 4 = 0$

96. $x^2 + y^2 - 4x - 2y - 15 = 0$

97. $2x^2 + 2y^2 + 4y = 0$

98. $3x^2 + 3y^2 + 6x = 0$

99. $x^2 + y^2 - 2x + 4y + 5 = 0$

100. $x^2 + y^2 + 4x - 6y + 17 = 0$

Applying the Concepts

The following graph shows the percentage of smartphone sales in the United States by two leading platforms.

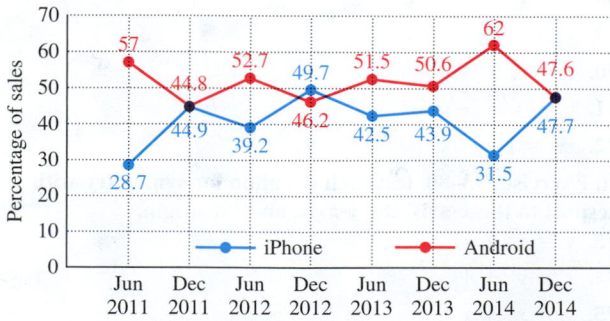

101. Use the appropriate line graph to determine the percentage of Android sales in June 2013.

102. Use the appropriate line graph to determine the percentage of iPhone sales in December 2012.

103. For the period of June 2011 through December 2014, during which period were the Android sales at a maximum?

104. For the period of June 2011 through December 2014, during which period were the iPhone sales at a maximum?

In Exercises 105–108, use the following vital statistics table.

Year	Rate per 1000 Population			
	Births	Deaths	Marriages	Divorces
1955	25.0	9.3	9.3	2.3
1960	23.7	9.5	8.5	2.2
1965	19.4	9.4	9.3	2.5
1970	18.4	9.5	10.6	3.5
1975	14.6	8.6	10.0	4.8
1980	15.9	8.8	10.6	5.2
1985	15.8	8.8	10.1	5.0
1990	16.7	8.6	9.4	4.7
1995	14.8	8.8	8.9	4.4
2000	14.4	8.7	8.5	4.2
2005	14.2	8.1	7.6	3.6
2010	13.0	8.0	6.8	3.6
2012	12.6	7.4	6.8	3.4

Source: U.S. Census Bureau.

The table shows the rate per 1000 population of live births, deaths, marriages, and divorces from 1955 to 2012. Plot the data in a Cartesian coordinate system and connect the points with line segments.

105. Plot (year, births).

106. Plot (year, deaths).

107. Plot (year, marriages).

108. Plot (year, divorces).

109. Spending on prescription drugs. Americans spent $228 billion on prescription drugs in 2007 and $320 billion in 2011. Estimate the amount spent on prescription drugs during 2008, 2009, and 2010. (*Source:* www.cdc.gov)

110. Defense spending. The government of the United States spent $548 billion on national defense in 2004 and $925 billion in 2012. Estimate the amount spent on defense in each of the years 2005–2011.
(*Source:* www.usgovernmentspending.com)

In Exercises 111–114, a graph is described *geometrically* as the path of a point $P(x, y)$ on the graph. Find an equation for the graph described.

111. Geometry. $P(x, y)$ is on the graph if and only if the distance from $P(x, y)$ to the x-axis is equal to its distance from the y-axis.

112. Geometry. $P(x, y)$ is the same distance from the two points $(1, 2)$ and $(3, -4)$.

113. Geometry. $P(x, y)$ is the same distance from the point $(2, 0)$ and the y-axis.

114. Geometry. $P(x, y)$ is the same distance from the point $(0, 4)$ and the x-axis.

115. Distance. A pilot is flying from Dullsville to Middale to Pleasantville. With reference to an origin, Dullsville is located at $(2, 4)$, Middale at $(8, 12)$, and Pleasantville at $(20, 3)$, all numbers being in 100-mile units.
  a. Locate the positions of the three cities on a Cartesian coordinate plane.
  b. Compute the distance traveled by the pilot.
  c. Compute the direct distance between Dullsville and Pleasantville.

116. Docking distance. A rope is attached to the bow of a sailboat that is 24 feet from the dock. The rope is drawn in over a pulley 10 feet higher than the bow at the rate of 3 feet per second. Find the distance from the boat to the dock after t seconds.

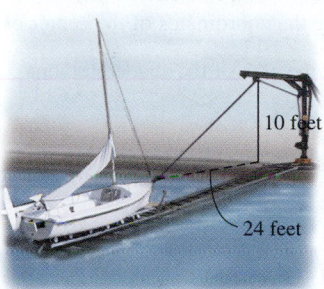

117. Corporate profits. The equation $P = -0.5t^2 - 3t + 8$ describes the monthly profits (in millions of dollars) of ABCD Corp. for the year 2012, with $t = 0$ representing July 2012.
  a. How much profit did the corporation make in March 2012?
  b. How much profit did the corporation make in October 2012?
  c. Sketch the graph of the equation.
  d. Find the t-intercepts. What do they represent?
  e. Find the P-intercept. What does it represent?

118. Female students in colleges. The equation
$$P = -0.002t^2 + 0.51t + 17.5$$
models the approximate number (in millions) of female college students in the United States for the academic years 2005–2009, with $t = 0$ representing 2005.
  a. Sketch the graph of the equation.
  b. Find the P-intercept. What does it represent?
  (*Source: Statistical Abstracts of the United States*)

119. Motion. An object is thrown up from the top of a building that is 320 feet high. The equation $y = -16t^2 + 128t + 320$

gives the object's height (in feet) above the ground at any time t (in seconds) after the object is thrown.
  a. What is the height of the object after 0, 1, 2, 3, 4, 5, and 6 seconds?
  b. Sketch the graph of the equation $y = -16t^2 + 128t + 320$.
  c. What part of the graph represents the physical aspects of the problem?
  d. What are the intercepts of this graph, and what do they mean?

120. Diving for treasure. A treasure-hunting team of divers is placed in a computer-controlled diving cage. The equation $d = \dfrac{40}{3}t - \dfrac{2}{9}t^2$ describes the depth d (in feet) that the cage will descend in t minutes.
  a. Sketch the graph of the equation $d = \dfrac{40}{3}t - \dfrac{2}{9}t^2$.
  b. What part of the graph represents the physical aspects of the problem?
  c. What is the total time of the entire diving experiment?

Beyond the Basics

121. Use coordinates to prove that the diagonals of a parallelogram bisect each other.
  [*Hint:* Choose $(0, 0)$, $(a, 0)$, (b, c), and $(a + b, c)$ as the vertices of a parallelogram.]

122. Let $A(2, 3)$, $B(5, 4)$, and $C(3, 8)$ be three points in a coordinate plane. Find the coordinates of the point D such that the points A, B, C, and D form a parallelogram with
  a. AB as one of the diagonals.
  b. AC as one of the diagonals.
  c. BC as one of the diagonals.
  [*Hint:* Use Exercise 121.]

123. Prove that if the diagonals of a quadrilateral bisect each other, the quadrilateral is a parallelogram. [*Hint:* Choose $(0, 0)$, $(a, 0)$, (b, c), and (x, y) as the vertices of a quadrilateral and show that $x = a + b, y = c$.]

124. Use the notations in the accompanying figures to prove the midpoint formula $M(x, y) = \left(\dfrac{x_1 + x_2}{2} + \dfrac{y_1 + y_2}{2} \right)$.

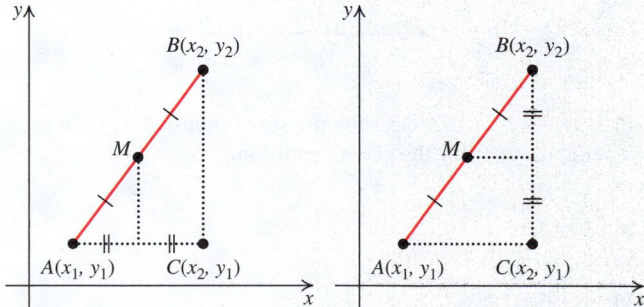

125. **Trisecting a line segment.** Let $A(x_1, y_1)$ and $B(x_2, y_2)$ be two points in a coordinate plane.

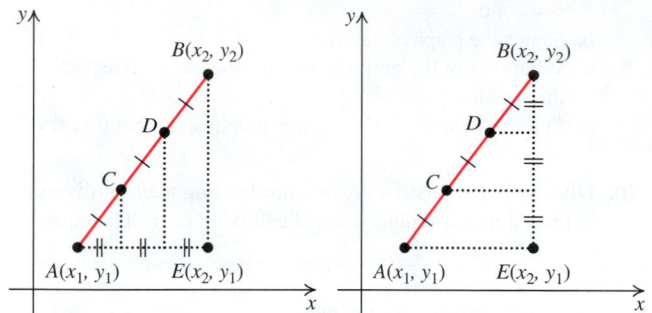

a. Use the accompanying figures to show that the point $C\left(\dfrac{2x_1 + x_2}{3}, \dfrac{2y_1 + y_2}{3}\right)$ is one-third of the way from A to B.

b. Use the accompanying figures to show that the point $D\left(\dfrac{x_1 + 2x_2}{3}, \dfrac{y_1 + 2y_2}{3}\right)$ is two-thirds of the way from A to B.

The points C and D are called the points of trisection of segment AB.

c. Find the points of trisection of the line segment joining $(-1, 2)$ and $(4, 1)$.

126. Let $A = (x_1, y_1)$ and $B(x_2, y_2)$, and $C(x_3, y_3)$ be vertices of a triangle in a coordinate plane. Show that a centroid of this triangle has coordinates

$$M\left(\frac{x_1 + x_2 + x_3}{3}, \frac{y_1 + y_2 + y_3}{3}\right).$$

The centroid of a triangle is the point of intersection of the medians in a triangle. Use the fact that the centroid divides the medians in a 2:1 ratio and apply the trisecting line segment problem to the accompanying figure.

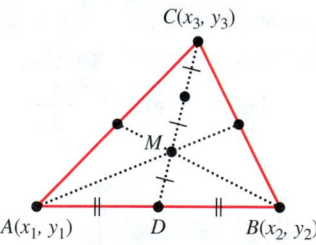

In Exercises 127–132, describe the set of points $P(x, y)$ in the xy-plane that satisfy the given condition.

127. $x = 0$

128. $y = 0$

129. $xy = 0$

130. $xy \neq 0$

131. $xy > 0$

132. $xy < 0$

Critical Thinking / Discussion / Writing

133. Sketch the graph of $y^2 = 2x$ and explain how this graph is related to the graphs of $y = \sqrt{2x}$ and $y = -\sqrt{2x}$.

134. Show that a graph that is symmetric with respect to the x-axis and y-axis must be symmetric with respect to the origin. Give an example to show that the converse is not true.

135. a. Show that a circle with diameter having endpoints $A(0, 1)$ and $B(6, 8)$ intersects the x-axis at the roots of the equation $x^2 - 6x + 8 = 0$.

b. Show that a circle with diameter having endpoints $A(0, 1)$ and $B(a, b)$ intersects the x-axis at the roots of the equation $x^2 - ax + b = 0$.

c. Use graph paper, ruler, and compass to approximate the roots of the equation $x^2 - 3x + 1 = 0$.

136. a. Show that a circle with diameter having endpoints $A(1, 0)$ and $B(10, 7)$ intersects the y-axis at the roots of the equation $y^2 - 7y + 10 = 0$.

b. Show that a circle with diameter having endpoints $A(1, 0)$ and $B(a, b)$ intersects the y-axis at the roots of the equation $y^2 - by + a = 0$.

137. The figure shows two circles, each with radius r.

a. Write the coordinates of the center of each circle.

b. Find the area of the shaded region.

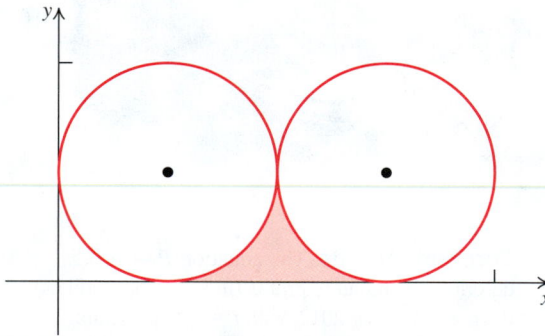

138. Based on the equation $3^2 + 4^2 = 5^2$, explain how a rope with 12 uniformly spaced knots and ends tied together (shown in the picture below) can be used to construct a right angle.

Getting Ready for the Next Section

In Exercises 139–142, perform the indicated operations.

139. $\dfrac{5-3}{6-2}$

140. $\dfrac{1-2}{-2-2}$

141. $\dfrac{2-(-3)}{3-13}$

142. $\dfrac{3-1}{-2-(-6)}$

In Exercises 143–146, solve each equation for the specified variable.

143. $2x + 3y = 6$ for y

144. $\dfrac{x}{2} - \dfrac{y}{5} = 3$ for y

145. $y - 2 - \dfrac{2}{3}(x + 1) = 0$ for y

146. $0.1x + 0.2y - 1 = 0$ for y

In Exercises 147–150, write the negative reciprocal of each number.

147. 2

148. -3

149. $-\dfrac{2}{3}$

150. $\dfrac{4}{3}$

Lines

BEFORE STARTING THIS SECTION, REVIEW

1 Evaluating algebraic expressions (Appendix A.1, page 923)

2 Graphs of equations (Section 1.1, page 2)

OBJECTIVES

1 Find the slope of a line.

2 Write the point–slope form of the equation of a line.

3 Write the slope–intercept form of the equation of a line.

4 Recognize the equations of horizontal and vertical lines.

5 Recognize the general form of the equation of a line.

6 Find equations of parallel and perpendicular lines.

7 Model data using linear regression

◆ A Texan's Tall Tale

Gunslinger Wild Bill Longley's *first* burial took place on October 11, 1878, after he was hanged before a crowd of thousands in Giddings, Texas.

Rumors persisted that Longley's hanging had been a hoax and that he had somehow faked his death and escaped execution. In 2001, Longley's descendants had his grave opened to determine whether the remains matched Wild Bill's description: a tall white male, age 27. Both the skeleton and some personal effects suggested that this was indeed Wild Bill. Modern science lent a hand too: The DNA of Wild Bill's sister's descendant Helen Chapman was a perfect match.

Now the notorious gunman could be buried back in the Giddings cemetery—for the *second* time. How did the scientists conclude from the skeletal remains that Wild Bill was approximately 6 feet tall? (See Example 8.)

---

1 Find the slope of a line.

Slope of a Line

In this section, we study the equation of a straight line, its properties and graph. Geometrically, the line is determined by two points which prescribe its steepness. Using Cartesian coordinates, we translate this notion to algebra when we measure the steepness of a line by a number called its **slope**.

Algebraically, a linear equation describes a relation between two variables, where the change in one variable is directly proportional to the change in the other variable.

Consider two points $P(x_1, y_1)$ and $Q(x_2, y_2)$ on a line, as shown in Figure 1.23. We say that the **rise** is the change in y-coordinates between the points (x_1, y_1) and (x_2, y_2) and that the **run** is the corresponding change in the x-coordinates. A positive run indicates change to the right, as shown in Figure 1.23; a negative run indicates change to the left. Similarly, a positive rise indicates upward change; a negative rise indicates downward change.

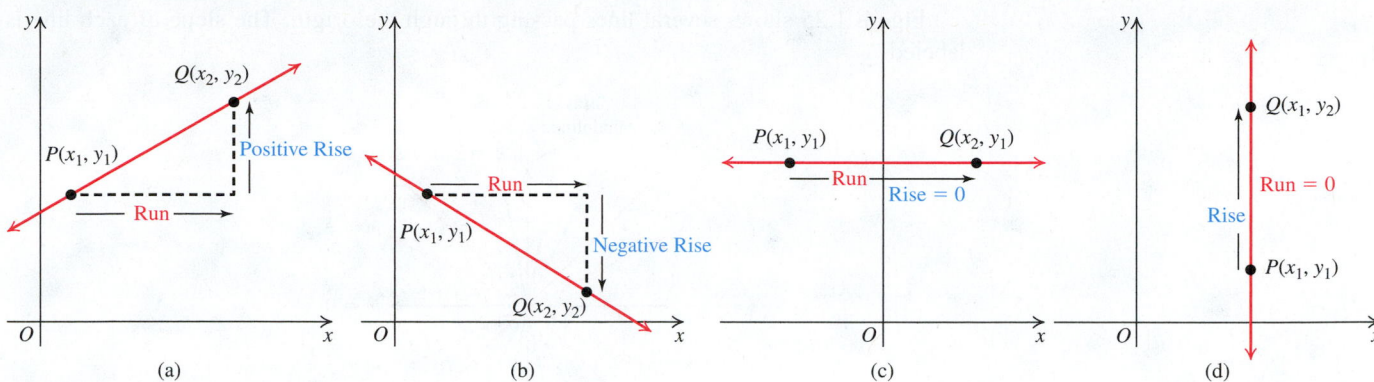

Figure 1.23 Slope of a line.

The symbols Δy (read "delta y") and Δx (read delta x") are used to indicate a "change in y" and a "change in x," respectively. So the slope is sometimes denoted by $m = \dfrac{\Delta y}{\Delta x}$.

Slope of a Line

The **slope** of a nonvertical line that passes through the points $P(x_1, y_1)$ and $Q(x_2, y_2)$ is denoted by m and is defined by

$$m = \frac{\text{rise}}{\text{run}} = \frac{\text{change in } y}{\text{change in } x} = \frac{\Delta y}{\Delta x} = \frac{y_2 - y_1}{x_2 - x_1}, \quad x_1 \neq x_2$$

The slope of a vertical line is undefined.

Since $m = \dfrac{\text{rise}}{\text{run}} = \dfrac{\text{change in } y\text{-coordinates}}{\text{change in } x\text{-coordinates}}$, if the change in x is one unit to the right (x increases by 1 unit), then the change in y equals the slope of the line. The slope of a

line is therefore the **change in y per unit change in x**. In other words, the slope of a line measures the rate of change of y with respect to x. Roofs, staircases, graded landscapes, and mountainous roads all have slopes. For example, if the pitch (slope) of a section of the roof is 0.4, then for every horizontal distance of 10 feet in that section, the roof ascends 4 feet.

EXAMPLE 1   Finding and Interpreting the Slope of a Line

Sketch the graph of the line that passes through $P(1, -1)$ and $Q(3, 3)$. Find and interpret the slope of the line.

Solution

The graph of the line is sketched in Figure 1.24. The slope m of this line is given by

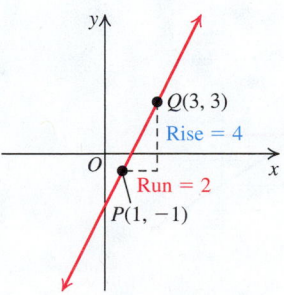

Figure 1.24 Interpreting slope.

$$m = \frac{\text{rise}}{\text{run}} = \frac{\text{change in } y\text{-coordinates}}{\text{change in } x\text{-coordinates}}$$

$$= \frac{(y\text{-coordinate of } Q) - (y\text{-coordinate of } P)}{(x\text{-coordinate of } Q) - (x\text{-coordinate of } P)}$$

$$= \frac{(3) - (-1)}{(3) - (1)} = \frac{3 + 1}{3 - 1} = \frac{4}{2} = 2.$$

Interpretation

A slope of 2 means that the value of y increases two units for every one unit increase in the value of x.

Practice Problem 1 Find and interpret the slope of the line containing the points $(-7, 5)$ and $(6, -3)$.

Figure 1.25 shows several lines passing through the origin. The slope of each line is labeled.

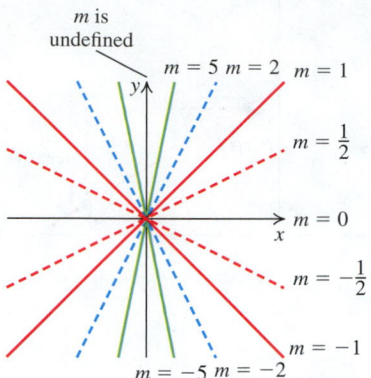

Figure 1.25 Slopes of lines.

MAIN FACTS ABOUT SLOPES OF LINES

1. Scanning graphs from left to right, lines with positive slopes rise and lines with negative slopes fall.
2. The greater the absolute value of the slope, the steeper the line.
3. The slope of a vertical line is undefined.
4. The slope of a horizontal line is zero.

WARNING

Be careful when using the formula for finding the slope of a line joining the points (x_1, y_1) and (x_2, y_2). Be sure to subtract coordinates in the same order.

You can use either

$$m = \frac{y_2 - y_1}{x_2 - x_1} \quad or \quad m = \frac{y_1 - y_2}{x_1 - x_2},$$

but it is incorrect to use

$$m = \frac{y_1 - y_2}{x_2 - x_1} \quad or \quad m = \frac{y_2 - y_1}{x_1 - x_2}$$

Figure 1.26 shows how to construct a line with given slope, $\pm\frac{2}{3}$, with rise $= \pm 2$ and run $= 3$.

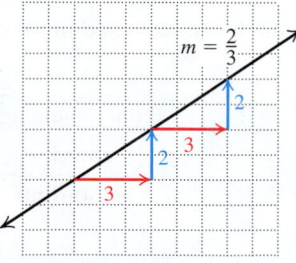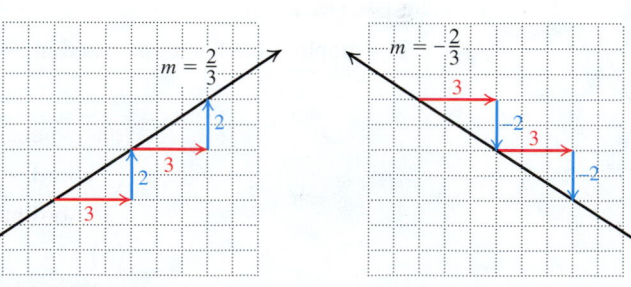

Figure 1.26 Constructing a line.

2 Write the point–slope form of the equation of a line.

Point–Slope Form

We now find the equation of a line ℓ that passes through the point $A(x_1, y_1)$ and has slope m. Let $P(x, y)$, with $x \neq x_1$, be any point in the plane. Then $P(x, y)$ is on the line ℓ if and only if the slope of the line passing through $P(x, y)$ and $A(x_1, y_1)$ is m. This is true if and only if

$$\frac{y - y_1}{x - x_1} = m.$$

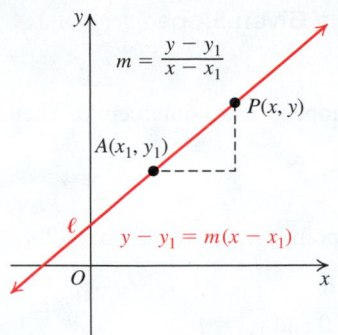

$$m = \frac{y - y_1}{x - x_1}$$

$P(x, y)$

$A(x_1, y_1)$

ℓ

$y - y_1 = m(x - x_1)$

Figure 1.27 Point–slope form.

See Figure 1.27.

Multiplying both sides by $x - x_1$ gives $y - y_1 = m(x - x_1)$, which is also satisfied when $x = x_1$ and $y = y_1$.

The Point–Slope Form of the Equation of a Line

If a line has slope m and passes through the point (x_1, y_1), then the **point–slope form** of an equation of the line is

$$y - y_1 = m(x - x_1).$$

EXAMPLE 2   Finding an Equation of a Line with Given Point and Slope

Find the point–slope form of the equation of the line passing through the point $(1, -2)$ with slope $m = 3$. Then solve for y.

Solution

$$
\begin{aligned}
y - y_1 &= m(x - x_1) && \text{Point–slope form} \\
y - (-2) &= 3(x - 1) && \text{Substitute } x_1 = 1, y_1 = -2, \text{ and } m = 3. \\
y + 2 &= 3x - 3 && \text{Simplify.} \\
y &= 3x - 5 && \text{Solve for } y.
\end{aligned}
$$

Practice Problem 2 Find the point–slope form of the equation of the line passing through the point $(-2, -3)$ and with slope $-\dfrac{2}{3}$. Then solve for y.

EXAMPLE 3   Finding an Equation of a Line Passing Through Two Given Points

Find the point–slope form of the equation of the line ℓ passing through the points $(-2, 1)$ and $(3, 7)$. Then solve for y.

Solution

We first find the slope m of the line ℓ.

$$m = \frac{7 - 1}{3 - (-2)} = \frac{6}{3 + 2} = \frac{6}{5} \qquad m = \frac{y_2 - y_1}{x_2 - x_1}$$

We use $m = \dfrac{6}{5}$ and either of the two given points when substituting into the slope–intercept form:

$$y - y_1 = m(x - x_1)$$

With $(x_1, y_1) = (3, 7)$

$$y - 7 = \frac{6}{5}(x - 3)$$

$$y = \frac{6}{5}x + \frac{17}{5} \qquad \text{Simplify.}$$

With $(x_1, y_1) = (-2, 1)$

$$y - 1 = \frac{6}{5}(x - (-2))$$

$$y = \frac{6}{5}x + \frac{17}{5} \qquad \text{Simplify.}$$

──────Same result──────

Practice Problem 3 Find the point–slope form of the equation of a line passing through the points $(-3, -4)$ and $(-1, 6)$. Then solve for y.

SIDE NOTE

You can avoid operations on fractions

$$y - 7 = \frac{6}{5}(x - 3) \qquad \text{Multiply by 5.}$$

$$5(y - 7) = 6(x - 3)$$
$$5y - 35 = 6x - 18$$
$$5y = 6x + 17$$

and at the end divide by 5

$$y = \frac{6}{5}x + \frac{17}{5}.$$

3 Write the slope–intercept form of the equation of a line.

Slope–Intercept Form

One of most convenient forms of an equation describing a straight line is called the point–slope form, where we know the slope of the line and the given point is the y-intercept.

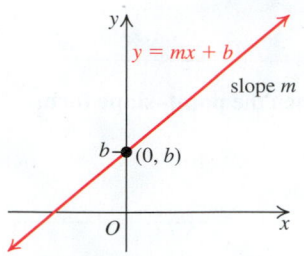

Figure 1.28 Slope–intercept form.

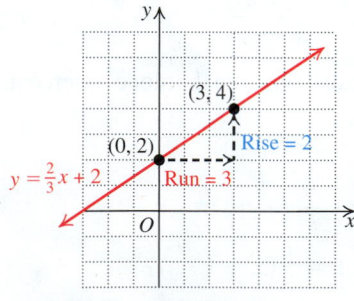

Figure 1.29 Locating a second point on a line.

SIDE
NOTE

To graph a line with integer slope, such as 3, write $3 = \dfrac{3}{1}$ where the rise = 3 and run = 1.

EXAMPLE 4   **Finding an Equation of a Line with a Given Slope and y-intercept**

Find the point–slope form of the equation of the line with slope m and y-intercept b. Then solve for y.

Solution

Since the line has y-intercept b, the line passes through the point $(0, b)$. See Figure 1.28.

$$y - y_1 = m(x - x_1) \qquad \text{Point–slope form}$$
$$y - b = m(x - 0) \qquad \text{Substitute } x_1 = 0 \text{ and } y_1 = b.$$
$$y - b = mx \qquad \text{Simplify.}$$
$$y = mx + b \qquad \text{Solve for } y.$$

Practice Problem 4 Find the point–slope form of the equation of the line with slope 2 and y-intercept -3. Then solve for y.

Slope–Intercept Form of the Equation of a Line

The **slope–intercept form** of the equation of the line with slope m and y-intercept b is

$$y = mx + b.$$

The linear equation in the form $y = mx + b$ displays the slope m (the coefficient of x) and the y-intercept b (the constant term). The number m tells which way and how much the line is tilted; the number b tells where the line intersects the y-axis.

EXAMPLE 5   **Graph Using the Slope and y-intercept**

Graph the line whose equation is $y = \dfrac{2}{3}x + 2$.

Solution

The equation

$$y = \frac{2}{3}x + 2$$

is in the slope-intercept form with slope $\dfrac{2}{3}$ and y-intercept 2. To sketch the graph, find two points on the line and draw a line through the two points. Use the y-intercept as one of the points, and then use the slope to locate a second point. Since $m = \dfrac{2}{3}$, let 2 be the rise and 3 be the run. From the point $(0, 2)$, move three units to the right (run) and two units up (rise). This gives $(0 + \text{run}, 2 + \text{rise}) = (0 + 3, 2 + 2) = (3, 4)$ as the second point.

The line we want joins the points $(0, 2)$ and $(3, 4)$ and is shown in Figure 1.29.

As an *alternative solution*, we can locate the second point on the line by choosing some value of x and finding the corresponding value of y directly from the given equation. Choosing $x = 3$ will give us the corresponding value of $y = \dfrac{2}{3} \cdot 3 + 2 = 4$. This gives $(3, 4)$ as a second point. Note that some choices of x are better than others. Choosing $x = 1$ would give us the point $\left(1, \dfrac{8}{3}\right)$, which would not be easy to plot accurately on a Cartesian grid.

Practice Problem 5 Graph the line with slope $-\dfrac{2}{3}$ that contains the point $(0, 4)$.

Equations of Horizontal and Vertical Lines

4 Recognize the equations of horizontal and vertical lines.

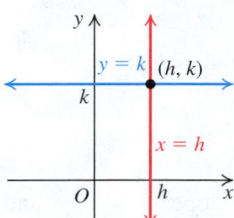

Figure 1.30 Horizontal and vertical lines.

Consider the horizontal line through the point (h, k). The y-coordinate of every point on this line is k. So we can write its equation as $y = k$. This line has slope $m = 0$ and y-intercept k. Similarly, an equation of the vertical line through the point (h, k) is $x = h$. This line has undefined slope and x-intercept h. See Figure 1.30.

> **HORIZONTAL AND VERTICAL LINES**
>
> An equation of a horizontal line through (h, k) is $y = k$.
>
> An equation of a vertical line through (h, k) is $x = h$.

TECHNOLOGY CONNECTION

Calculator graph of $y = 2$

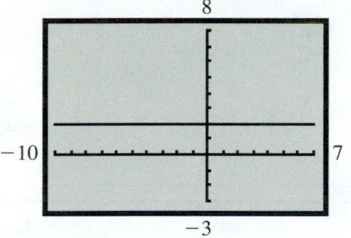

EXAMPLE 6 Recognizing Horizontal and Vertical Lines

Discuss the graph of each equation in the xy-plane.

a. $y = 2$ **b.** $x = 4$

Solution

a. The equation $y = 2$ may be considered as an equation in two variables x and y by writing

$$0 \cdot x + y = 2.$$

Any ordered pair of the form $(x, 2)$ is a solution of this equation. Some solutions are $(-1, 2)$, $(0, 2)$, $(2, 2)$, and $(7, 2)$. It follows that the graph is a line parallel to the x-axis and two units above it, as shown in Figure 1.31. Its slope is 0.

SIDE NOTE

Sometimes it is easier to remember that

(1) given the equation $y = 2$, there is no change in y since y is constant. The line has to be horizontal ($\Delta y = 0$).

(2) given the equation $x = 4$, there is no change in x since x is constant. The line has to be vertical ($\Delta x = 0$).

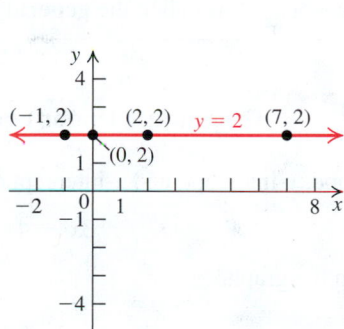

Figure 1.31 Horizontal line.

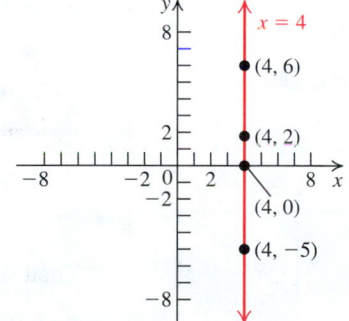

Figure 1.32 Vertical line.

b. The equation $x = 4$ may be written as

$$x + 0 \cdot y = 4.$$

Any ordered pair of the form $(4, y)$ is a solution of this equation. Some solutions are $(4, -5)$, $(4, 0)$, $(4, 2)$, and $(4, 6)$. The graph is a line parallel to the y-axis and four units to the right of it, as shown in Figure 1.32. Its slope is undefined.

TECHNOLOGY CONNECTION

You cannot graph $x = 4$ by using the $Y =$ key on your calculator.

Practice Problem 6 Sketch the graphs of the lines $x = -3$ and $y = 7$.

General Form of the Equation of a Line

5 Recognize the general form of the equation of a line.

An equation of the form

$$ax + by + c = 0,$$

where a, b, and c are constants and a and b are not both zero, is called the *general form* of a linear equation. Consider two possible cases: $b \neq 0$ and $b = 0$.

Suppose $b \neq 0$: we can isolate y on one side and rewrite the equation in slope–intercept form.

$$ax + by + c = 0 \qquad \text{Original equation}$$
$$by = -ax - c \qquad \text{Subtract } ax + c \text{ from both sides.}$$
$$y = -\frac{a}{b}x - \frac{c}{b}. \qquad \text{Divide by } b.$$

The result is an equation of a line in slope–intercept form with slope $m = -\dfrac{a}{b}$ and y-intercept equal $-\dfrac{c}{b}$.

Suppose $b = 0$: we know that $a \neq 0$ since not both a and b are zero, and we can solve for x.

$$ax + 0 \cdot y + c = 0 \qquad \text{Replace } b \text{ with } 0.$$
$$ax + c = 0 \qquad \text{Simplify.}$$
$$x = -\frac{c}{a} \qquad \text{Solve for } x.$$

The graph of the equation $x = -\dfrac{c}{a}$ is a vertical line.

General Form of the Equation of a Line

> The graph of every linear equation
>
> $$ax + by + c = 0,$$
>
> where a, b, and c are constants and a and b are not both zero, is a line. The equation $ax + by + c = 0$ is called the **general form** of the equation of a line.

Calculator graph of $y = \dfrac{3}{4}x + 3$

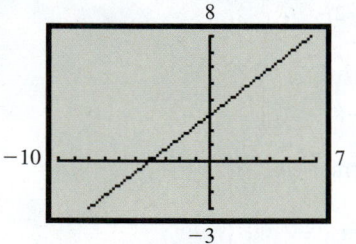

EXAMPLE 7 Graphing a Linear Equation Using Intercepts

Find the slope, y-intercept, and x-intercept of the line with equation

$$3x - 4y + 12 = 0.$$

Then sketch the graph.

Solution

First, solve for y to write the equation in slope–intercept form:

$$3x - 4y + 12 = 0 \qquad \text{Original equation}$$
$$4y = 3x + 12 \qquad \text{Subtract } 3x + 12; \text{ multiply by } -1.$$
$$y = \frac{3}{4}x + 3 \qquad \text{Divide by } 4.$$

This equation tells us that the slope m is $\dfrac{3}{4}$ and the y-intercept is 3. To find the x-intercept, set $y = 0$ in the original equation and obtain $3x + 12 = 0$, or $x = -4$. So the x-intercept is -4.

We can sketch the graph of the equation if we can find two points on the graph. So we use the intercepts and sketch the line joining the points $(-4, 0)$ and $(0, 3)$. The graph is shown in Figure 1.33.

Alternatively, we can verify that the **two-intercept form** of the equation of the line with x-intercept a and y-intercept b is

$$\frac{x}{a} + \frac{y}{b} = 1.$$

Figure 1.33 Graphing a line using intercepts.

We can convert the general form of a linear equation such as $3x - 4y + 12 = 0$ to the two-intercept form by isolating the variables from the constant part and then dividing both sides of the equation by the constant.

$$3x - 4y + 12 = 0 \qquad \text{Original equation}$$
$$3x - 4y = -12 \qquad \text{Separate variables.}$$
$$\frac{3x}{-12} + \frac{-4y}{-12} = 1 \qquad \text{Divide by constant.}$$
$$\frac{x}{-4} + \frac{y}{3} = 1 \qquad \text{Simplify.}$$

The last equation shows that the x-intercept is -4 and the y-intercept is 3.

Practice Problem 7 Sketch the graph of the equation $3x + 4y = 24$.

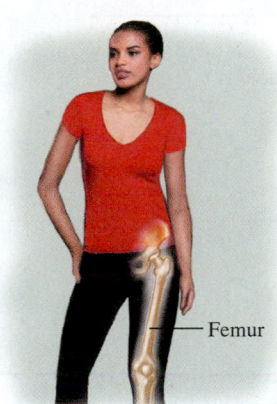
Femur

Forensic scientists use various clues to determine the identity of deceased individuals. Scientists know that certain bones serve as excellent sources of information about a person's height. The length of the *femur,* the long bone that stretches from the hip (pelvis) socket to the kneecap (patella), can be used to estimate a person's height. Knowing both the gender and race of the individual increases the accuracy of this relationship between the length of the bone and the height of the person.

◆ **EXAMPLE 8** **Inferring Height from the Femur**

The height H of a human male is related to the length x of his femur by the formula

$$H = 2.6x + 65,$$

where all measurements are in centimeters. The femur of Wild Bill Longley measured between 45 and 46 centimeters. Estimate the height of Wild Bill (in feet).

Solution

Substituting $x = 45$ and $x = 46$ into the preceding formula, we have possible heights

$$H_1 = (2.6)(45) + 65 = 182$$
$$\text{and} \quad H_2 = (2.6)(46) + 65 = 184.6.$$

Therefore, Wild Bill's height was between 182 and 184.6 centimeters.

Converting to feet, we conclude that his height was between $\dfrac{182}{30.48} \approx 5.97$ ft and $\dfrac{184.6}{30.48} \approx 6.06$ ft. He was approximately 6 feet tall.

RECALL

1 foot ≈ 30.48 centimeters

Practice Problem 8 Suppose the femur of a person measures between 43 and 44 centimeters. Estimate the height of the person in meters.

6 Find equations of parallel and perpendicular lines.

Parallel and Perpendicular Lines

We can use slope to decide whether two lines are parallel, perpendicular, or neither.

Parallel and Perpendicular Lines

Let ℓ_1 *and* ℓ_2 be two distinct lines with slopes m_1 *and* m_2, respectively. Then

ℓ_1 is parallel to ℓ_2 if and only if $m_1 = m_2$.

ℓ_1 is perpendicular to ℓ_2 if and only if $m_1 \cdot m_2 = -1$.

Any two vertical lines are parallel, as are any two horizontal lines. Any horizontal line is perpendicular to any vertical line.

In Exercises 135 and 136, you are asked to prove these assertions. Figure 1.34 shows examples of the slopes of a pair of parallel and a pair of perpendicular lines.

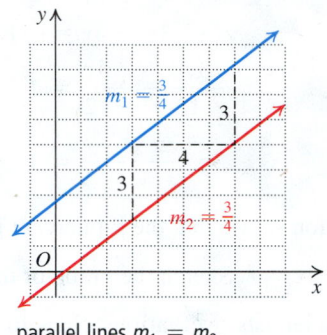

parallel lines $m_1 = m_2$.

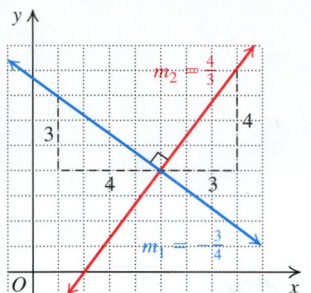

Perpendicular lines $m_1 \cdot m_2 = -1$

Figure 1.34

PROCEDURE
IN ACTION

EXAMPLE 9   **Finding Equations of Parallel and Perpendicular Lines**

OBJECTIVE

Let $\ell : ax + by + c = 0$. Find an equation of each line through the point (x_1, y_1):

(a) ℓ_1 parallel to ℓ

(b) ℓ_2 perpendicular to ℓ

Step 1 Find the slope m of ℓ.

Step 2 Write slope of ℓ_1 and ℓ_2.

The slope m_1 of ℓ_1 is m.

The slope m_2 of ℓ_2 is $-\dfrac{1}{m}$.

Step 3 Write equations of ℓ_1 and ℓ_2.

Use point–slope form to write equations of ℓ_1 and ℓ_2.
Simplify to write equations in the requested form.

EXAMPLE

Let $\ell : 2x - 3y + 6 = 0$. Find a general form of the equation of each line through the point $(2, 8)$:

(a) ℓ_1 parallel to ℓ

(b) ℓ_2 perpendicular to ℓ

1. $\ell : 2x - 3y + 6 = 0$ Original equation

$$-3y = -2x - 6 \quad \text{Subtract } 2x + 6.$$

$$y = \frac{2}{3}x + 2 \quad \text{Divide by } -3.$$

The slope of ℓ is $m = \dfrac{2}{3}$.

2. $m_1 = \dfrac{2}{3}$ ℓ_1 parallel to ℓ

$m_2 = -\dfrac{3}{2}$ ℓ_2 perpendicular to ℓ

3. $\ell_1 : y - 8 = \dfrac{2}{3}(x - 2)$ Point–slope form

$2x - 3y + 20 = 0$ Simplify.

$\ell_2 : y - 8 = -\dfrac{3}{2}(x - 2)$ Point–slope form

$3x + 2y - 22 = 0$ Simplify.

Practice Problem 9 Find the general form of the equation of the line

a. through $(-2, 5)$ and parallel to the line containing $(2, 3)$ and $(5, 7)$.

b. through $(3, -4)$ and perpendicular to the line $4x + 5y + 1 = 0$.

SUMMARY OF **MAIN FACTS**

Form	Equation	Slope	Other Information
General	$ax + by + c = 0$ $a \neq 0$, or $b \neq 0$	$-\dfrac{a}{b}$ if $b \neq 0$ Undefined if $b = 0$	x-intercept is $-\dfrac{c}{a}$, $a \neq 0$. y-intercept is $-\dfrac{c}{b}$, $b \neq 0$.
Point–slope	$y - y_1 = m(x - x_1)$	m	Line passes through (x_1, y_1).
Slope–intercept	$y = mx + b$	m	x-intercept is $-\dfrac{b}{m}$ if $m \neq 0$; y-intercept is b.
Two–intercept	$\dfrac{x}{a} + \dfrac{y}{b} = 1$	$-\dfrac{b}{a}$	x-intercept is a; y-intercept is b.
Vertical line through (h, k)	$x = h$	undefined	When $h \neq 0$, no y-intercept; x-intercept is h. When $h = 0$, the graph is the y-axis.
Horizontal line through (h, k)	$y = k$	0	When $k \neq 0$, no x-intercept; y-intercept is k. When $k = 0$, the graph is the x-axis.

Parallel and perpendicular lines: Two lines with slopes m_1 and m_2 are (a) parallel if $m_1 = m_2$ and (b) perpendicular if $m_1 \cdot m_2 = -1$.

7 Use linear regression.

Modeling Data Using Linear Regression

Many relationships between variables can be modeled using linear equations. The success of modeling many phenomena by linear equations can be easily explained. On a local scale, many nonlinear equations may be reduced to linear equations by assuming that the variables change only to a small extent. We experience this in our daily life when we perceive the surface of the Earth locally as flat. That means that any short-term data can be accurately modeled by linear models. A line that best fits the given data is called a **regression line**. To illustrate the method, we use Table 1.4, which gives the data for the number of registered motorcycles in the United States for the years 2006–2012.

TABLE 1.4

Year	2006	2007	2008	2009	2010	2011	2012
Registered motorcycles (millions)	6.7	7.1	7.8	7.9	8.4	8.8	9.5

The scatter diagram for the motorcycle data is shown in Figure 1.35(a), and the line used to model the data is added to the scatter diagram in Figure 1.35(b). The years 2006–2012 are represented by the numbers 0–6, respectively.

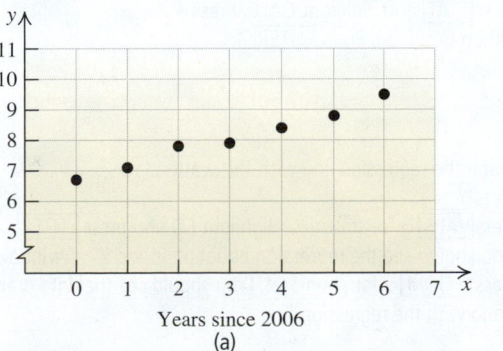
Years since 2006
(a)

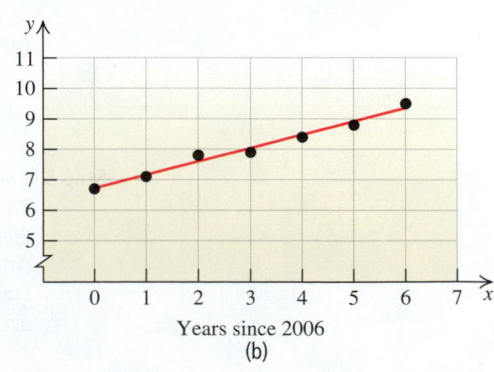
Years since 2006
(b)

Figure 1.35

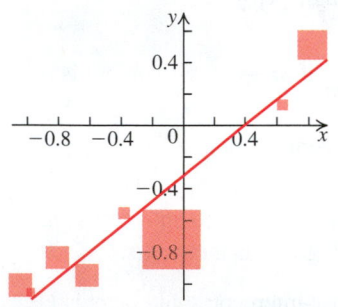

Figure 1.36 Least-squares fit of the curve $y = bx + a$ to 10 given data points.

The line is called the **regression line** or the **least-squares line**. See Figure 1.36. The slope-intercept form of the equation for the line in Figure 1.35(b) is $y = 0.44x + 6.70$, where x is the number of years after 2006, and the slope and intercept values are rounded to the nearest hundredth. The equation is found by using either a graphing calculator or statistical formulas.

EXAMPLE 10 Predictions Using Linear Regression

Use the equation $y = 0.44x + 6.70$ to predict the number of registered motorcycles in the United States for 2015.

Solution

Because 2015 is nine years after 2006, we set $x = 9$; then $y = 0.44(9) + 6.70 \approx 10.66$.

So according to the regression prediction, 10.66 million motorcycles will be registered in the United States for 2015. The quality of a prediction based on linear regression depends on how close the relationship between the data quantities is to being linear.

Practice Problem 10 Repeat Example 10 to predict the number of registered motorcycles in the United States for 2016.

The following Technology Connection shows how a graphing calculator displays the scatter diagram, regression line, and regression line equation.

TECHNOLOGY
CONNECTION

You can use a graphing utility to investigate the linear regression model for data points that lie on or near a straight line. On the TI-84, the data for Example 10 are entered in five steps.

Step 1 Enter the data.
Press STAT 1 for STAT Edit. Enter the values of the independent variable x in L1 and the corresponding values of the dependent variable y in L2.

Step 2 Turn on STAT Plot
Press 2nd Y = 1 for Plot 1. Select *On*; then choose *Type*. [⋰]
Choose *X* List: L1; *Y* list: L2; and mark: [·].

Step 3 Graph the scatterplot. Press Graph .

Step 4 Find the equation of the regression line.
Press STAT and highlight CALC. Press 4 for LinReg ($ax + b$). Press ENTER .

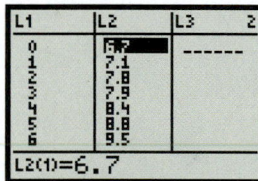

Step 5 Graph the regression line with the scatterplot.
Press Y = .
Press VARS 5 for statistics. Highlight EQ and press ENTER .
(You should see the regression equation in the $Y =$ window.)
Press ZOOM 9 for ZoomStat. (You should see the data graphed along with the regression line.)

Answers to Practice Problems

1. $-\dfrac{8}{13}$, for every 13 units increase in x, the y-value decreases by 8 units.   2. $y + 3 = -\dfrac{2}{3}(x + 2)$; $y = -\dfrac{2}{3}x - \dfrac{13}{3}$

3. $y + 4 = 5(x + 3)$; $y = 5x + 11$ 4. $y + 3 = 2x$; $y = 2x - 3$

5. 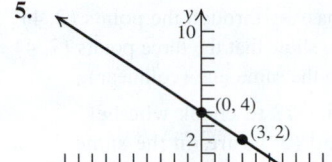   6.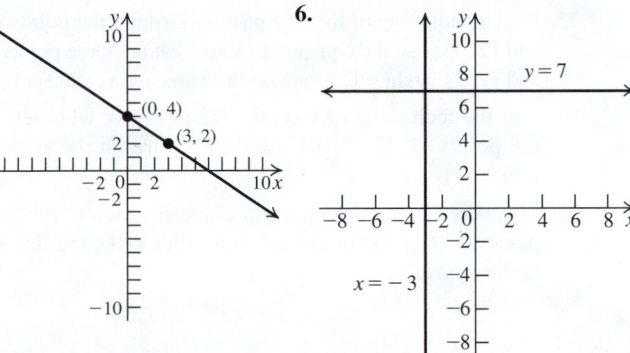

7.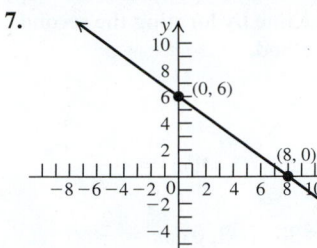

8. Between 1.768 and 1.794 meters
9. a. $4x - 3y + 23 = 0$
   b. $5x - 4y - 31 = 0$
10. 11.1 million

![SECTION 1.2] **Exercises**

Concepts and Vocabulary

1. The slope of a horizontal line is _____; the slope of vertical line is _____ .

2. The slope of the line passing through the points $P = (x_1,y_1)$ and $Q = (x_2,y_2)$ is given by the formula $m =$ _____ .

3. Every line parallel to the line $y = 3x - 2$ has slope, m, equal to _____ .

4. Every line perpendicular to the line $y = 3x - 2$ has slope, m, equal to _____ .

5. **True or False.** The slope of the line $y = -\dfrac{1}{4}x + 5$ is equal to $\dfrac{1}{4}$.

6. **True or False.** The y-intercept of the line $y = 2x - 3$ is equal to 3.

7. **True or False.** The graph of the line $y = 4$ is a horizontal line.

8. **True or False.** The graph of the line $x = -5$ is a vertical line.

Building Skills

In Exercises 9–16, find the slope of the line through the given pair of points. Without plotting any points, state whether the line is rising, falling, horizontal, or vertical.

9. $(1, 3), (4, 7)$
10. $(0, 4), (2, 0)$
11. $(3, -2), (6, -2)$
12. $(-3, 7), (-3, -4)$
13. $(0.5, 2), (3, -3.5)$
14. $(3, -2), (2, -3)$
15. $(\sqrt{2}, 1), (1 + \sqrt{2}, 5)$
16. $(1 - \sqrt{3}, 0), (1 + \sqrt{3}, 3\sqrt{3})$

In the figure, identify the line with the given slope m.
17. $m = 1$
18. $m = -1$
19. $m = 0$
20. m is undefined.

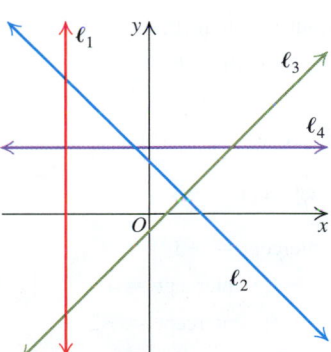

In the figure below, find the slope of each line. (The scale is the same on both axes.)

21. ℓ_1 22. ℓ_2
23. ℓ_3 24. ℓ_4

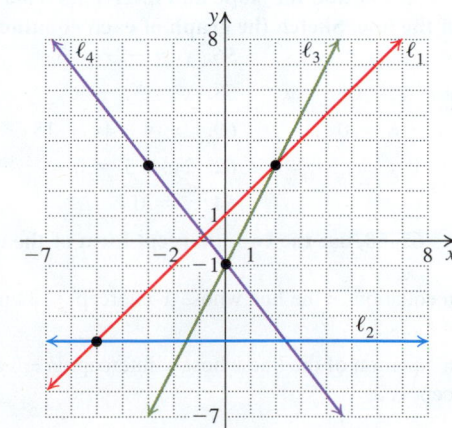

In Exercises 25–32, find an equation in slope–intercept form of the line that passes through the given point and has slope *m*. Also sketch the graph of the line by locating the second point with the rise-and-run method.

25. $(0, 5); m = 3$

26. $(0, 9); m = -2$

27. $(0, 4); m = \dfrac{1}{2}$ **28.** $(0, 4); m = -\dfrac{1}{2}$

29. $(2, 1); m = -\dfrac{3}{2}$ **30.** $(-1, 0); m = \dfrac{2}{5}$

31. $(5, -4); m = 0$ **32.** $(5, -4); m$ is undefined.

In Exercises 33–54, use the given conditions to find an equation in slope–intercept form of each nonvertical line. Write vertical lines in the form $x = h$.

33. Passing through $(0, 1)$ and $(1, 0)$

34. Passing through $(0, 1)$ and $(1, 3)$

35. Passing through $(-1, 3)$ and $(3, 3)$

36. Passing through $(-5, 1)$ and $(2, 7)$

37. Passing through $(-2, -1)$ and $(1, 1)$

38. Passing through $(-1, -3)$ and $(6, -9)$

39. Passing through $\left(\dfrac{1}{2}, \dfrac{1}{4}\right)$ and $(0, 2)$

40. Passing through $(4, -7)$ and $(4, 3)$

41. A vertical line through $(5, 1.7)$

42. A horizontal line through $(1.4, 1.5)$

43. A horizontal line through $(0, 0)$

44. A vertical line through $(0, 0)$

45. $m = 0; y$-intercept $= 14$

46. $m = 2; y$-intercept $= 5$

47. $m = -\dfrac{2}{3}; y$-intercept $= -4$

48. $m = -6; y$-intercept $= -3$

49. x-intercept $= -3; y$-intercept $= 4$

50. x-intercept $= -5; y$-intercept $= -2$

51. Parallel to $y = 5$; passing through $(4, 7)$

52. Parallel to $x = 5$; passing through $(4, 7)$

53. Perpendicular to $x = -4$; passing through $(-3, -5)$

54. Perpendicular to $y = -4$; passing through $(-3, -5)$

In Exercises 55–64, find the slope and intercepts from the equation of the line. Sketch the graph of each equation.

55. $y = 3x - 2$ **56.** $y = -2x + 3$

57. $x + 2y - 4 = 0$ **58.** $x = 3y - 9$

59. $3x - 2y + 6 = 0$ **60.** $2x = -4y + 15$

61. $x - 5 = 0$ **62.** $2y + 5 = 0$

63. $x = 0$ **64.** $y = 0$

In Exercises 65–68, use the two-intercept form of the equation of a line.

65. Find an equation of the line whose x-intercept is 4 and y-intercept is 3.

66. Find an equation of the line whose x-intercept is -3 and y-intercept is 2.

67. Find the x- and y-intercepts of the graph of the equation $2x + 3y = 6$.

68. Repeat Exercise 67 for the equation $3x - 4y + 12 = 0$.

In Exercises 69–72, find the x and y-intercepts and sketch the graph of the equation

69. $2x - 3y = 12$ **70.** $3x - 4y = 12$

71. $-5x + 2y = 10$ **72.** $-4x + 5y = 20$

73. Find an equation of the line passing through the points $(2, 4)$ and $(7, 9)$. Use the equation to show that the three points $(2, 4)$ and $(7, 9)$, and $(-1, 1)$ are on the same line (collinear).

74. Use the technique of Exercise 73 to check whether the points $(7, 2)$, $(2, -3)$, and $(5, 1)$ are on the same line.

75. Write the slope–intercept form of the equation of the line that passes through the origin and is parallel to the red line shown in the figure.

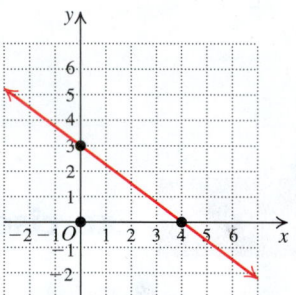

76. Write the slope–intercept form of the equation of the line that passes through the origin and is perpendicular to the red line shown in the figure in Exercise 75.

77. Write the slope–intercept form of the equation of the blue line shown in the figure that is parallel to the red line.

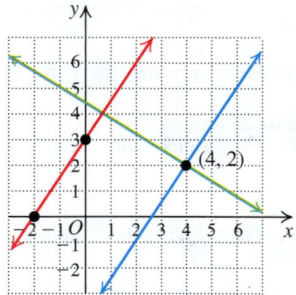

78. Write the slope–intercept form of the equation of the green line shown in the figure that is perpendicular to the red line in Exercise 77.

In Exercises 79–88, determine whether each pair of lines is parallel, perpendicular, or neither.

79. $y = 3x - 1$ and $y = 3x + 2$

80. $y = 2x + 2$ and $y = -2x + 2$

81. $y = 2x - 4$ and $y = -\dfrac{1}{2}x + 4$

82. $y = 3x + 1$ and $y = \dfrac{1}{3}x - 1$

83. $3x + 8y = 7$ and $5x - 7y = 0$

84. $10x + 2y = 3$ and $5x + y = -1$

85. $x = 4y + 8$ and $y = -4x + 1$

86. $y = 3x + 1$ and $6y + 2x = 0$

87. $x = -1$ and $2x + 7 = 9$

88. $2x + 3y = 7$ and $y = 2$

In Exercises 89–104, find the equation of the line in slope–intercept form satisfying the given conditions.

89. Parallel to a line with slope 3; passing through $(2, -3)$.

90. Parallel to a line with slope -2; passing through $(-1, 3)$.

91. Perpendicular to a line with slope $-\dfrac{1}{2}$; passing through $(-1, 2)$.

92. Perpendicular to a line with slope $\dfrac{1}{3}$; passing through $(2, -1)$.

93. Parallel to a line joining points $(1, -2)$ and $(-3, 2)$; passing through $(-2, -5)$.

94. Parallel to a line joining points $(-2, 1)$ and $(3, 5)$; passing through $(1, 2)$.

95. Perpendicular to a line joining points $(-3, 2)$ and $(-4, -1)$; passing through $(1, -2)$.

96. Perpendicular to a line joining points $(2, 1)$ and $(4, -1)$; passing through $(-1, 2)$.

97. Parallel to $y = 6x + 5$; y-intercept 4.

98. Parallel to $y = -\dfrac{1}{2}x + 5$; y-intercept 2.

99. Perpendicular to $y = 6x + 5$; y-intercept 4.

100. Perpendicular to $y = -\dfrac{1}{2}x + 5$; y-intercept -4.

101. Parallel to $x + y = 1$; passing through $(1, 1)$.

102. Parallel to $-2x + 3y = 7$; passing through $(1, 0)$.

103. Perpendicular to $3x - 9y = 18$; passing through $(-2, 4)$.

104. Perpendicular to $-2x + y = 14$; passing through $(0, 2)$.

Applying the Concepts

105. Dance Class. A dance teacher runs a modern dance class for 12 weeks. The class meets each week in a rented dance studio and the teacher has to pay upfront all the costs associated with renting the studio. Each student pays the teacher a single fee for the course. The graph represents the profit the teacher makes depending on the number of students.

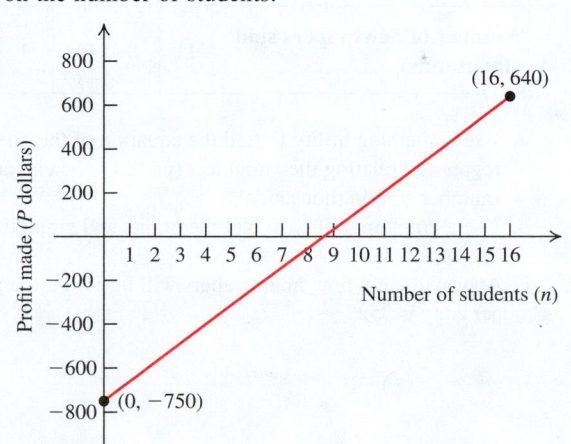

a. What does the y-intercept represent?

b. What does the x-intercept represent?

c. What is the teacher's profit if there are 16 students in the class?

d. Write a linear equation for the profit P the teacher makes depending on the number of students n.

106. Prepaid Cellphone. A prepaid cellphone is loaded with a prepaid SIM card with initial airtime credit for calls. The company charges a flat fee per minute of calling within the continental United States. The graph represents the prepaid amount depending on usage of the cellphone.

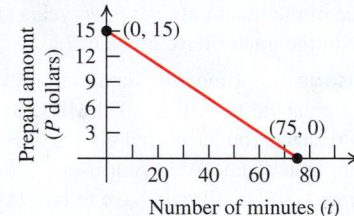

a. What does the y-intercept represent?

b. What does the x-intercept represent?

c. Write a linear equation for the amount left on the SIM card P depending on the time the cellphone was used.

d. What is the cost per minute of calling from this cellphone?

107. Jet takeoff. Upon takeoff, a jet climbs to 4 miles as it passes over 40 miles of land below it. Find the slope of the jet's climb.

108. Road gradient. Driving down the Smoky Mountains in Tennessee, Samantha descends 2000 feet in elevation while moving 4 miles horizontally away from a high point on the straight road. Find the slope (gradient) of the road (1 mile = 5280 feet).

109. Trimming the bush. John noticed that it is springtime and that the hedge by his window has been growing. It grew 8 inches in two weeks. How often does John need to trim his hedge to be sure that it doesn't grow more than 6 inches between trimmings?

110. Taking a bath. Miguel is preparing to take a bath. He sees that it takes 2 minutes to raise the bathtub water level by 5 inches. His bathtub is 36 inches deep, so he has 31 more inches left before water overflows. How much time does he have left before the water overflows?

In Exercises 111–115, write an equation of a line in slope–intercept form. To describe this situation, interpret (a) the variables x and y and (b) the meaning of the slope and the y-intercept.

111. Christmas savings account. You currently have $130 in a Christmas savings account. You deposit $7 per week into this account. (Ignore interest.)

112. Golf club charges. Your golf club charges a yearly membership fee of $1000 and $35 per session of golf.

113. Paying off a refrigerator. You bought a new refrigerator for $700. You made a down payment of $100 and promised to pay $15 per month. (Ignore interest.)

114. Converting currency. In 2012, according to the International Monetary Fund, 1 U.S. dollar equaled 50.5 Indian rupees. Convert currency from rupees to dollars.

115. Life expectancy. In 2010, the life expectancy of a female born in the United States was 80.8 years and was increasing at a rate of 0.17 years per year. (Assume that this rate of increase remains constant.)

116. Depreciating a tractor. The value v of a tractor purchased for $14,000 and depreciated linearly at the rate of $1400 per year is given by $v = -1400t + 14,000$, where t represents the number of years since the purchase. Find the value of the tractor after (a) two years and (b) six years. When will the tractor have no value?

117. Data usage. Your monthly data limit on your cellphone plan is 1 gigabyte (1 GB = 1024 MB). By the end of June 13, you realized you have used up 435 MB of your quota. Assuming you plan to keep your usage patterns, will your limit be enough, or will you have to buy extra data by the end of June?

118. Car trip. You set out on a trip from your hometown to a vacation spot located 520 miles away. You started driving at 9 A.M. and by 12 P.M. you have realized you traveled 195 miles. Assuming your driving patterns continue, what time will you arrive at your destination?

119. Playing a concert. A famous band is considering playing a concert and charging $40,000 plus $5 per person attending the concert. Write an equation relating the income y of the band to the number x of people attending the concert.

120. Car rental. The U Drive car rental agency charges $30.00 per day and $0.25 per mile.
   a. Find an equation relating the cost C of renting a car from U Drive for a one-day trip covering x miles.
   b. Draw the graph of the equation in part (a).
   c. How much does it cost to rent the car for one day covering 60 miles?
   d. How many miles were driven if the one-day rental cost was $47.75?

121. Fahrenheit and Celsius. In the Fahrenheit temperature scale, water boils at 212°F and freezes at 32°F. In the Celsius scale, water boils at 100°C and freezes at 0°C. Assume that the Fahrenheit temperature F and Celsius temperature C are linearly related.
   a. Find the equation in the slope–intercept form relating F and C, with C as the independent variable.
   b. What is the meaning of slope in part (a)?
   c. Find the Fahrenheit temperatures, to the nearest degree, corresponding to 40°C, 25°C, −5°C, and −10°C.
   d. Find the Celsius temperatures, to the nearest degree, corresponding to 100°F, 90°F, 75°F, −10°F, and −20°F.
   e. The normal body temperature in humans ranges from 97.6°F to 99.6°F. Convert this temperature range to degrees Celsius.
   f. When is the Celsius temperature the same numerical value as the Fahrenheit temperature?

122. Cost. The cost of producing four modems is $210.20, and the cost of producing ten modems is $348.80.
   a. Write a linear equation in slope–intercept form relating cost to the number of modems produced. Sketch a graph of the equation.
   b. What is the practical meaning of the slope and the intercepts of the equation in part (a)?
   c. Predict the cost of producing 12 modems.

123. Female prisoners. The number of females in a state's prisons rose from 2425 in 2005 to 4026 in 2011.
   a. Find a linear equation relating the number y of women prisoners to the year t. (Take $t = 0$ to represent 2005.)
   b. Draw the graph of the equation from part (a).
   c. How many women prisoners were there in 2008?
   d. Predict the number of women prisoners in 2017.

124. Viewers of a TV show. After five months, the number of viewers of *The Weekly Show* on TV was 5.73 million. After eight months, the number of viewers of the show rose to 6.27 million. Assume that the linear model applies.
   a. Write an equation expressing the number of viewers V after x months.
   b. Interpret the slope and the intercepts of the line in part (a).
   c. Predict the number of viewers of the show after 11 months.

125. Health insurance coverage. In Michigan, 11.7% of the population was not covered by health insurance in 2000. In 2005, the percentage of uninsured people rose to 12.7%. Use the linear model to predict the percentage of people in Michigan that were not covered in 2010. (*Source:* Michigan Department of Community Health)

126. Oil consumption. The total world consumption of oil increased from 82.7 million barrels per day in 2004 to 84.2 million barrels per day in 2009. Create a linear model involving a relation between the year and the daily oil consumption in that year. Let $t = 0$ represent the year 2004. (*Source:* www.cia.gov)

127. Newspaper sales. The table shows the number of newspapers (in thousands) that sell at a given price (in nickels) per copy for the college newspaper *Midstate Oracle* at Midstate College.

Price (nickels)	1	2	3	4	5
Number of newspapers sold (thousands)	11	8	6	4	3

   a. Use a graphing utility to find the equation of the line of regression relating the variable x (price in nickels) and y (number sold in thousands).
   b. Use a graphing utility to plot the points and graph the regression line.
   c. Approximately how many papers will be sold if the price per copy is 35¢?

128. **Sales and advertising.** Bildwell Computers keeps a careful record of its weekly advertising expenses (in thousands of dollars) and the number of computers (in thousands) it sells the following week. The table shows the data the company collected.

Weekly advertising expenses (in thousands of dollars)	100	200	300	400	500	600
Sales the following week (in thousands)	21	29	36	49	58	67

a. Use a graphing utility to find the equation of the line of regression equation relating the variables x (weekly advertising expenses in thousands of dollars) and y (number of computers sold, in thousands).

b. Use a graphing utility to plot the points and graph the regression line.

c. Approximately how many thousands of computers will be sold if the company spends \$700,000 on advertising?

Beyond the Basics

129. The graph of a line has slope $m = 3$. If $P(-2, 3)$ and $Q(1, c)$ are points on the graph, find c.

130. Find a real number c such that the line $3x - cy - 2 = 0$ has a y-intercept of -4.

In Exercises 131–132, show that the three given points are collinear by using (a) slopes and (b) the distance formula.

131. $(0, 1)$, $(1, 3)$, and $(-1, -1)$

132. $(1, 0.5)$, $(2, 0)$, and $(0.5, 0.75)$

133. Show that the triangle with vertices $A(1, 1)$, $B(-1, 4)$, and $C(5, 8)$ is a right triangle by using (a) slopes and (b) the Pythagorean Theorem.

134. Show that the four points, $A(-4, -1)$, $B(1, 2)$, $C(3, 1)$, and, $D(-2, -2)$ are the vertices of a parallelogram.

[*Hint:* Opposite sides of a parallelogram are parallel.]

135. In the accompanying figure, note that lines ℓ_1 and ℓ_2 have slopes m_1 and m_2. Assume that ℓ_1 and ℓ_2 are parallel. Deduce that $ABCD$ is a parallelogram and that triangles ABD and CDB are congruent, with $AB = CD$. Deduce that $m_1 = m_2$.

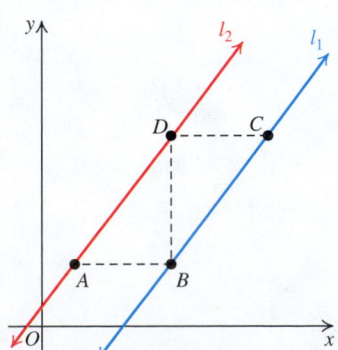

136. In the accompanying figure, note that lines ℓ_1 and ℓ_2 have slopes m_1 and m_2. Assume that ℓ_1 and ℓ_2 are perpendicular.

Show that the right triangles OKA and BLO are similar and from that deduce that $m_1 \cdot m_2 = -1$.

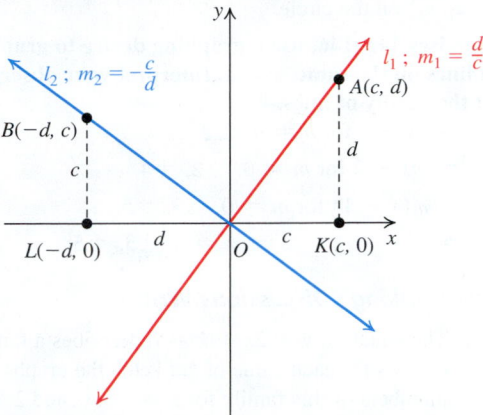

137. **Geometry.** Use a coordinate plane to prove that if two sides of a quadrilateral are congruent (equal in length) and parallel, then so are the other two sides.

138. **Geometry.** Use a coordinate plane to prove that the midpoints of the four sides of a quadrilateral are the vertices of a parallelogram.

139. **Geometry.** Find the coordinates of the point B of a line segment AB, given that $A = (2, 2)$, $d(A, B) = 12.5$, and the slope of the line segment AB is $\frac{4}{3}$.

140. **Geometry.** Show that an equation of a circle with (x_1, y_1) and (x_2, y_2) as endpoints of a diameter can be written in the form

$$(x - x_1)(x - x_2) + (y - y_1)(y - y_2) = 0.$$

[*Hint:* Use the fact that when the two endpoints of a diameter and any other point on the circle are used as vertices for a triangle, the angle opposite the diameter is a right angle.]

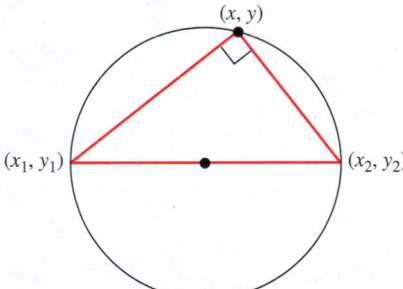

141. **Geometry.** Find an equation of the tangent line to the circle $x^2 + y^2 = 169$ at the point $(-5, 12)$. (See the accompanying figure.)

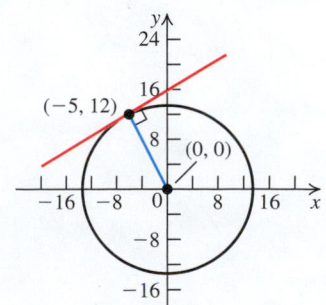

142. Geometry. Show that $xx_1 + yy_1 = a^2$ is an equation of the tangent line to the circle $x^2 + y^2 = a^2$ at the point (x_1, y_1) on the circle.

 In Exercises 143–146, use a graphing device to graph the given lines on the same screen. Interpret your observations about the family of lines.

143. $y = 2x + b$ for $b = 0, \pm 2, \pm 4$

144. $y = mx + 2$ for $m = 0, \pm 2, \pm 4$

145. $y = m(x - 1)$ for $m = 0, \pm 3, \pm 5$

146. $y = m(x + 1) - 2$ for $m = 0, \pm 3, \pm 5$

Critical Thinking / Discussion / Writing

147. a. The equation $y + 2x + k = 0$ describes a family of lines for each value of k. Sketch the graphs of the members of this family for $k = -3, 0,$ and 2. What is the common characteristic of the family?

  b. Repeat part (a) for the family of lines $y + kx + 4 = 0$.

148. Explain how you can determine the sign of the x-coordinate of the point of intersection of the lines $y = m_1 x + b_1$ and $y = m_2 x + b_2$, if

  a. $m_1 > m_2 > 0$ and $b_1 > b_2$.
  b. $m_1 > m_2 > 0$ and $b_1 < b_2$.
  c. $m_1 < m_2 < 0$ and $b_1 > b_2$.
  d. $m_1 < m_2 < 0$ and $b_1 < b_2$.

Getting Ready for the Next Section

In Exercises 149–152, solve each equation for x.

149. $x^2 - 4 = 0$

150. $1 - x^2 = 0$

151. $x^2 - x - 2 = 0$

152. $x^2 + 2x - 3 = 0$

In Exercises 153–156, simplify each expression.

153. $(3(a + h) + 1) - (3a + 1)$

154. $(2(a + h)^2 + 1) - (2a^2 + 1)$

155. $\dfrac{-(a + h)^2 + a^2}{h}$

156. $\dfrac{1}{h}\left(\dfrac{1}{a + h} - \dfrac{1}{a}\right)$

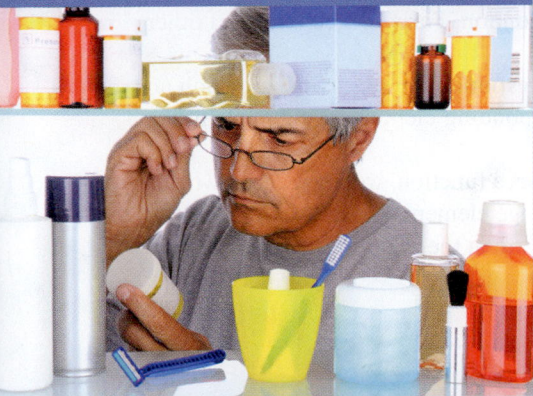

Functions

BEFORE STARTING THIS SECTION, REVIEW

1 Graph of an equation (Section 1.1, page 7)

2 Circles (Section 1.1, page 12)

OBJECTIVES

1 Use functional notation and find function values.

2 Find the domain and range of a function.

3 Identify the graph of a function.

4 Get information about a function from its graph.

5 Find the average rate of change of a function.

6 Solve applied problems by using functions.

◆ The Ups and Downs of Drug Levels

Medicines in the form of a pill have a much harder time getting to work than you do. First, they are swallowed and dumped into a pool of acid in your stomach. Then they dissolve; leave the stomach; and begin to be absorbed, mostly through the lining of the small intestine. Next, the blood around the intestine carries the medicines to the liver, an organ designed to metabolize (break down) and remove foreign material from the blood. Because of this property of the liver, most drugs have a tough time getting past it. The amount of drug that makes it into your bloodstream, compared with the amount that you put in your mouth, is called the drug's *bioavailability*. If a drug has low bioavailability, then much of the drug is destroyed before it reaches the bloodstream. The amount of drug you take in a pill is calculated to correct for this deficiency so that the amount you need actually ends up in your blood.

Once the drugs are past the liver, the bloodstream carries them throughout the rest of the body in about one minute. The study of the ways drugs are absorbed and move through the body is called *pharmacokinetics,* or PK for short. PK measures the ups and downs of drug levels in your body. In Example 13, we consider an application of *functions* to the levels of drugs in the bloodstream.

(Adapted from *The Ups and Downs of Drug Levels* by Bob Munk from Positively Aware, May/June 2001. Copyright © 2001 by Test Positive Aware Network. Reprinted with permission.

1 Use functional notation and find function values.

Functions

There are many situations where one quantity depends directly on another quantity. We want to highlight the predictability of this direct correspondence. The value of the first quantity determines the value of the second quantity. For example, if you are paid $10 per hour, then the relation between the number of hours, x, that you work and the amount of money, y, that you earn may be expressed by the equation $y = 10x$. Replacing x with 40 yields $y = 400$ and indicates that 40 hours of work corresponds to a $400 paycheck. A special relationship such as $y = 10x$ in which to each element x in one set there corresponds a unique element y in another set is called a *function*. We say that your pay is a function of the number of hours you work. Because the value of y depends on the given value of x, y is called the **dependent variable** and x is called the **independent variable**. Any symbol may be used for the dependent or independent variables.

SIDE
NOTE

The range of a function from a set X to a set Y does not necessarily have to be the entire set Y.

Function

A **function** from a set X to a set Y is a rule that assigns to each element of X exactly one element of Y. The set X is the **domain** of the function. The set of those elements of Y that correspond (are assigned) to the elements of X is the **range** of the function.

As an example of an assignment that is not a function, consider the correspondence between U.S. and European airports. A U.S. airport is connected to a European airport if there is a direct flight between them. This is not a function because, for instance, the New York airport is connected by a direct flight to London, Paris, Barcelona, and other airports.

A rule of assignment may be described by a **correspondence diagram**, in which an arrow points from each domain element to the corresponding element or elements in the range.

To decide whether a correspondence diagram defines a function, we must check whether *any* domain element is paired with more than one range element. If, for some element of X, there are two or more corresponding elements of Y, then the relation is not a function. See the correspondence diagrams in Figure 1.37.

SIDE
NOTE

If a correspondence diagram defines a function, then the situation

(a) is allowed

(b) is not allowed

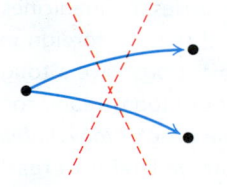

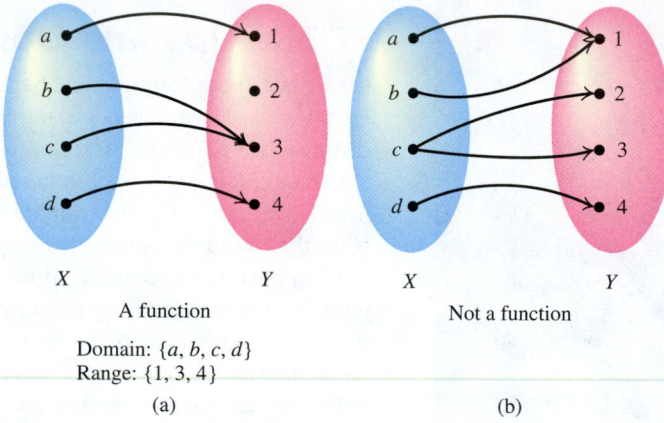

Domain: $\{a, b, c, d\}$
Range: $\{1, 3, 4\}$

(a)   (b)

Figure 1.37 Correspondence diagrams.

MAIN FACTS FROM THE DEFINITION OF A FUNCTION

Let f be a function from a set X to a set Y.

1. The domain of f is the entire set X. That is, *each* element of X must be assigned to some element of Y.

2. The range of f is the set of those elements in Y that are assigned to some element of X. So the range of f may not be the entire set Y.

3. It *is permissible* to assign the same element of Y to two or more elements of X. See Figure 1.37(a).

4. It *is not permissible* to assign an element of X to two or more elements of Y. See Figure 1.37(b).

DO YOU
KNOW?

The functional notation $y = f(x)$ was first used by the great Swiss mathematician Leonhard Euler in the *Commentarii Academia Petropolitanae*, published in 1735.

Function Notation

We usually use single letters such as f, F, g, G, h, and H as the name of a function. Some special function names use more than one letter, like *sin* (the sine function) and *log* (the logarithm function). If we choose f as the name of a function, then for each x in the domain of f, there corresponds a unique y in its range. The number y is denoted by $f(x)$, read as "f of x" or as "f at x." We call $f(x)$ the **value of f at the number x** and say that f *assigns* the value $f(x)$ to x.

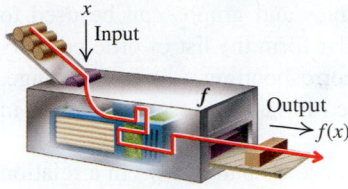

x ↓ Input

f

Output → $f(x)$

Figure 1.38 The function f, as a machine.

If the function shown in Figure 1.37(a) is named f, then the correspondence diagram may be described by:

$$f(a) = 1, f(b) = 3, f(c) = 3, \text{and } f(d) = 4.$$

Another way to illustrate the concept of a function is by visualizing a "machine," as shown in Figure 1.38. If x is in the domain of a function, then the machine accepts x as an "input" and produces the value $f(x)$ as the "output." A hand-held bar code scanner that reads the bar codes and outputs the price of items is an example of a function.

WARNING

In writing $y = f(x)$, do not confuse f, which is the name of the function, with $f(x)$, which is a number in the range of f that the function assigns to x.
The symbol $f(x)$ does not mean "f times x."

Representations of Functions

Functions can be prescribed in many ways: directly by describing the assignment to each element of the domain, by formula, or by algorithm. We will discuss briefly the different ways of describing functions.

Functions Defined by Ordered Pairs If the value of a function f at x is y, we can write this assignment as an ordered pair (x, y). A function can be defined by listing all the ordered pairs that correspond to all assignments made by this function. In general, any set of ordered pairs is called a **relation**. The set of first components is the **domain of the relation**, and the set of second components is the **range of the relation**. Not every relation is a function. Only relations in which each element of the domain corresponds to exactly one element of the range is a function. In other words, a function is a relation in which no two distinct ordered pairs have the same first component.

EXAMPLE 1   **Functions Defined by Ordered Pairs**

Determine whether each relation defines a function.

a. $r = \{(-1, 2), (1, 3), (5, 2), (-1, -3)\}$
b. $s = \{(-2, 1), (0, 2), (2, 1), (-1, -3)\}$

Solution

a. The domain of the relation r is: $\{-1, 1, 5\}$ and its range is: $\{2, 3, -3\}$. It is not a function because the ordered pairs $(-1, 2)$ and $(-1, -3)$ have the same first component. See the correspondence diagram for relation r in Figure 1.39.

b. The domain of the relation s is: $\{-2, -1, 0, 2\}$ and its range is: $\{-3, 1, 2\}$. The relation s is a function because no two ordered pairs of s have the same first component. Figure 1.40 shows the correspondence diagram for the function s.

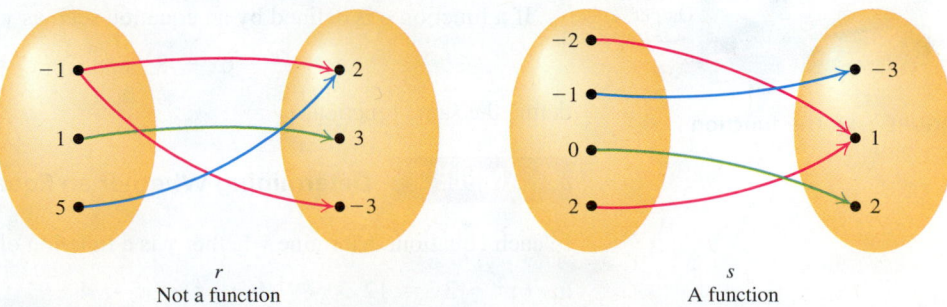

r
Not a function

s
A function

Figure 1.39   **Figure 1.40**

Practice Problem 1 Repeat Example 1 for the ordered pairs:
a. $R = \{(2, 1), (-2, 1), (3, 2)\}$
b. $S = \{(2, 5), (3, -2), (3, 5)\}$

TABLE 1.5

TABLE 1.5

x	$y = r(x)$
-1	2
1	3
5	2
-1	-3

Relation r

TABLE 1.6

x	$y = s(x)$
-2	1
-1	-3
0	2
2	1

Relation s

SIDE NOTE

The definition of a function is independent of the letters used to denote the function and the variables. For example,

$$s(x) = x^2, g(y) = y^2,$$
and $h(t) = t^2$

represent the same function.

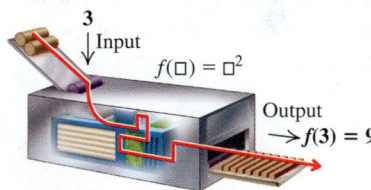

FIGURE 1.42 The function f, as a machine.

Functions Defined by Tables and Graphs Tables and graphs can be used to describe relations and functions. We can display in a tabular form the list of ordered pairs (x, y) of a relation, where x is in the domain and the corresponding y is in the range. For example, the relations r and s of Example 1 described in tabular form are shown in Tables 1.5 and 1.6, respectively.

Alternatively, we can display geometrically the list of ordered pairs (x, y) in a relation by plotting each pair (x, y) as a point in the coordinate plane. The geometric display is the graph of the relation. Graphs of relations r and s are given in Figure 1.41.

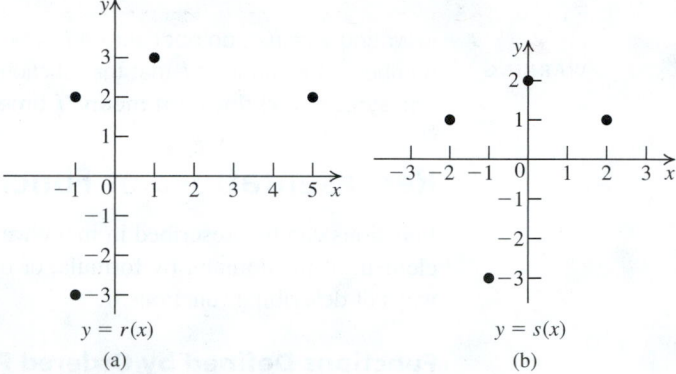

$y = r(x)$
(a)

$y = s(x)$
(b)

Figure 1.41 Graphs of relations.

The graph of the points in the table is called a **scatterplot.**

Functions Defined by Equations When the relation defined by an equation in two variables is a function, we can often solve the equation for the dependent variable in terms of the independent variable. For example, the equation $y - x^2 = 0$ can be solved for y in terms of x:

$$y = x^2.$$

Now we can replace the dependent variable, in this case, y, with functional notation $f(x)$ and express the function as

$$f(x) = x^2 \quad (\text{read "}f \text{ of } x \text{ equals } x^2\text{")}.$$

Here $f(x)$ plays the role of y and is the value of the function f at x. For example, if $x = 3$ is an element (input value) in the domain of f, the corresponding element (output value) in the range is found by replacing x with 3 in the equation

$$f(x) = x^2,$$

so that $\qquad\qquad f(3) = 3^2 = 9.$ Replace x with 3.

We say that the value of the function f at 3 is 9 or that 9 is the **evaluated function value** $f(3)$. See Figure 1.42. In other words, the number 9 in the range corresponds to the number 3 in the domain, and the ordered pair $(3, 9)$ is an ordered pair of the function f.

If a function g is defined by an equation such as $y = x^2 - 6x + 8$, the notations

$$y = x^2 - 6x + 8 \quad \text{and} \quad g(x) = x^2 - 6x + 8$$

define the same function.

EXAMPLE 2   **Determining Whether an Equation Defines a Function**

In each equation, determine whether y is a function of x.

a. $6x^2 - 3y = 12$ **b.** $y^2 - x^2 = 4$

Solution

Solve each equation for y in terms of x. If more than one value of y corresponds to the same value of x, then y is not a function of x.

a.
$$6x^2 - 3y = 12 \qquad \text{Original equation}$$
$$6x^2 - 3y + 3y - 12 = 12 + 3y - 12 \qquad \text{Add } 3y - 12 \text{ to both sides.}$$
$$6x^2 - 12 = 3y \qquad \text{Simplify.}$$
$$2x^2 - 4 = y \qquad \text{Divide by 3.}$$
$$y = 2x^2 - 4 \qquad \text{Interchange sides}$$

The last equation shows that only one value of y corresponds to each value of x. For example, if $x = 0$, then $y = 0 - 4 = -4$. So y is a function of x.

b.
$$y^2 - x^2 = 4 \qquad \text{Original equation}$$
$$y^2 - x^2 + x^2 = 4 + x^2 \qquad \text{Add } x^2 \text{ to both sides.}$$
$$y^2 = x^2 + 4 \qquad \text{Simplify.}$$
$$y = \pm\sqrt{x^2 + 4} \qquad \text{Square root property}$$

The last equation shows that two values of y correspond to each value of x. For example, if $x = 0$, then $y = \pm\sqrt{0^2 + 4} = \pm\sqrt{4} = \pm 2$. Both $y = 2$ and $y = -2$ correspond to $x = 0$. Therefore, y is not a function of x. However, the equation $y^2 - x^2 = 4$ defines two functions:

$f_1(x) = \sqrt{x^2 + 4}$; domain: $(-\infty, \infty)$; range: $[2, \infty)$, and

$f_2(x) = -\sqrt{x^2 + 4}$; domain: $(-\infty, \infty)$; range $(-\infty, -2]$. See Figure 1.43.

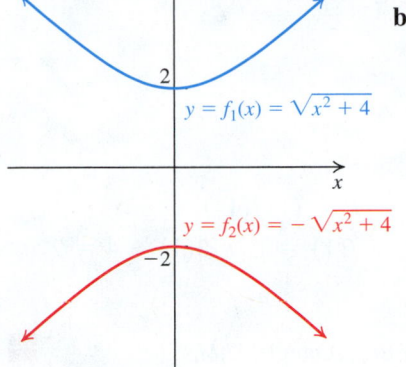

Figure 1.43

Practice Problem 2 In each equation, determine whether y is a function of x.

a. $2x^2 - y^2 = 1$ **b.** $x - 2y = 5$

EXAMPLE 3 Evaluating a Function

Let g be the function defined by the equation

$$y = x^2 - 6x + 8.$$

Find each function value.

a. $g(3)$ **b.** $g(-2)$ **c.** $g\left(\dfrac{1}{2}\right)$ **d.** $g(a + 2)$ **e.** $g(x + h)$

Solution

Since the function is named g, we replace y with $g(x)$ and write

$$g(x) = x^2 - 6x + 8 \qquad \text{Replace } y \text{ with } g(x).$$

In this notation, the independent variable x is a placeholder. We can write $g(x) = x^2 - 6x + 8$ as

$$g(\;) = (\;)^2 - 6(\;) + 8.$$

a. $g(x) = (x)^2 - 6(x) + 8 \qquad$ The given equation
$g(3) = (3)^2 - 6(3) + 8 \qquad$ Replace x with 3 at each occurrence of x.
$\quad\;\; = 9 - 18 + 8 = -1 \qquad$ Simplify.

The statement $g(3) = -1$ means that the value of the function g at 3 is -1.
Just as in part **a**, we can evaluate g at any value x in its domain:

$g(x) = x^2 - 6x + 8 \qquad\qquad\qquad\qquad$ Original equation

b. $g(-2) = (-2)^2 - 6(-2) + 8 = 24 \qquad$ Replace x with -2 and simplify.

c. $g\left(\dfrac{1}{2}\right) = \left(\dfrac{1}{2}\right)^2 - 6\left(\dfrac{1}{2}\right) + 8 = \dfrac{21}{4} \qquad$ Replace x with $\dfrac{1}{2}$ and simplify.

d. $g(a + 2) = (a + 2)^2 - 6(a + 2) + 8 \qquad$ Replace x with $(a + 2)$ in $g(x)$.
$\qquad\qquad = a^2 + 4a + 4 - 6a - 12 + 8 \qquad$ Recall: $(x + y)^2 = x^2 + 2xy + y^2$.
$\qquad\qquad = a^2 - 2a \qquad\qquad\qquad\qquad\quad$ Simplify.

e. $g(x + h) = (x + h)^2 - 6(x + h) + 8$ Replace x with $(x + h)$ in $g(x)$.
$\qquad\qquad = x^2 + 2xh + h^2 - 6x - 6h + 8$ Simplify.

Practice Problem 3 Let g be the function defined by the equation $y = -2x^2 + 5x$. Find each function value.

a. $g(0)$ **b.** $g(-1)$ **c.** $g(x + h)$

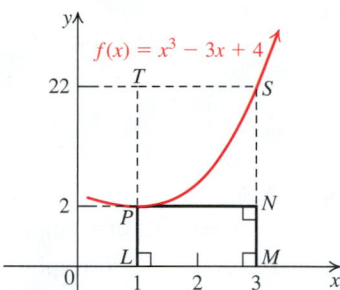
Figure 1.44

EXAMPLE 4   Finding the Area of a Rectangle

In Figure 1.44, find the area of the rectangle $PLMN$.

Solution

Area of the rectangle $PLMN$ = (length)(height)
$\qquad\qquad = (3 - 1)(f(1))$ $P = (1, f(1))$
$\qquad\qquad = (2)(2)$ $f(1) = 1^3 - 3(1) + 4 = 2$
$\qquad\qquad = 4$ square units

Practice Problem 4 In Figure 1.44, find the area of the rectangle $TLMS$.

2 Find the domain and range of a function.

The Domain of a Function

Sometimes a function is given by an equation without a predefined domain. In such cases, we use the following agreement.

> ### AGREEMENT ON DOMAIN
>
> If the domain of a function that is defined by an equation is not specified, then we agree that the domain of the function is the largest set of real numbers that results in real numbers as outputs.

When we use our agreement to find the domain of a function, first we usually find the values of the variable that do not result in real number outputs. Then we exclude those numbers from the domain. Remember that

1. division by zero is undefined.

2. the square root (or any even root) of a negative number is not a real number.

EXAMPLE 5   Finding the Domain of a Function

Find the domain of each function.

a. $f(x) = \dfrac{1}{1 - x^2}$ **b.** $g(x) = \sqrt{x + 1}$

c. $h(x) = \dfrac{1}{\sqrt{x - 1}}$ **d.** $P(x) = \sqrt{x^2 - x - 6}$

Solution

a. The function f is not defined when the denominator $1 - x^2$ is 0. Because $1 - x^2 = 0$ if $x = 1$ or $x = -1$, the domain of f is the set $\{x \mid x \neq -1$ and $x \neq 1\}$, which in interval notation is written as $(-\infty, -1) \cup (-1, 1) \cup (1, \infty)$.

b. Because the square root of a negative number is not a real number, $\sqrt{x + 1}$ is a real number only if $x + 1 \geq 0$, or $x \geq -1$. The domain of g is $\{x \mid x \geq -1\}$, or in interval notation $[-1, \infty)$.

c. The function $h(x) = \dfrac{1}{\sqrt{x-1}}$ has *two* restrictions. The square root of a negative number is not a real number, so $\sqrt{x-1}$ is a real number only if $x - 1 \geq 0$. However, we cannot allow $x - 1 = 0$ because $\sqrt{x-1}$ is in the denominator. Therefore, we must have $x - 1 > 0$, or $x > 1$. The domain of h is $\{x \mid x > 1\}$, or in interval notation $(1, \infty)$.

d. The function $P(x)$ is defined when the expression under the radical sign is nonnegative. You can use the test-point method (See Appendix 1.7) to see that $x^2 - x - 6 = (x + 2)(x - 3) \geq 0$ on the two intervals $(-\infty, -2]$ and $[3, \infty)$. So the domain of the function $P(x)$ in interval notation is $(-\infty, -2] \cup [3, \infty)$.

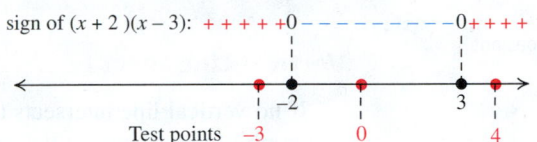

Practice Problem 5 Find the domain of each function.

a. $f(x) = \dfrac{1}{\sqrt{1-x}}.$ **b.** $g(x) = \sqrt{x^2 + 2x - 3}$

The Range of a Function

Suppose a function f is defined by an equation. A number y is in the range of f if the equation $f(x) = y$ has at least one solution.

EXAMPLE 6   **Finding the Range of a Function**

Let $f(x) = x^2$ with domain $X = [3, 5]$.

a. Is 10 in the range of f?

b. Is 4 in the range of f?

c. Find the range of f.

Solution

a. We find possible solutions of the equation

$$f(x) = 10$$
$$x^2 = 10 \qquad \color{blue}{\text{Replace } f(x) \text{ with } x^2.}$$
$$x = \pm\sqrt{10} \qquad \color{blue}{\text{Square root property}}$$

Because $3 < \sqrt{10} < 5$, the number $x = \sqrt{10}$ is in the interval $X = [3, 5]$. So the equation $f(x) = 10$ has at least one solution in the domain of f. Therefore, 10 is in the range of f.

b. The solutions of the equation $x^2 = 4$ are $x = \pm 2$. Neither of these numbers is in the domain $X = [3, 5]$. So 4 is not in the range of f.

c. The range of f is the interval $[9, 25]$ because for each number y in the interval $[9, 25]$, the number $x = \sqrt{y}$ is in the interval $[3, 5]$ so that $f(x) = f(\sqrt{y}) = (\sqrt{y})^2 = y$.

$$9 \leq y \leq 25$$
$$3 = \sqrt{9} \leq \sqrt{y} \leq \sqrt{25} = 5 \qquad \color{red}{\text{Since } y \text{ is positive, we can take}}$$
$$3 \leq x \leq 5 \qquad \qquad \color{red}{\text{square roots of both sides.}}$$

Practice Problem 6 Repeat Example 6 for $f(x) = x^2$ with domain $X = [-3, 3]$. ▪

Finding the range of a function such as $f(x) = \dfrac{x+1}{x-2}$ is more complicated. We discuss this problem in Section 1.7. In calculus, we will learn how to find the range of any function given by an equation.

3 Identify the graph of a function.

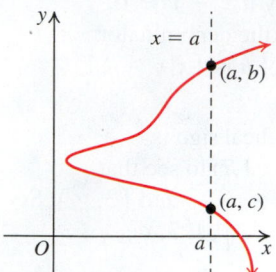

Figure 1.45 Graph does not represent a function.

Graphs of Functions

The *graph of a function f* is the set of ordered pairs $(x, f(x))$ such that x is in the domain of f. That is, the graph of f is the graph of the equation $y = f(x)$. We sketch the graph of $y = f(x)$ by plotting points and joining them with a smooth curve. The graph of a function provides valuable visual information about the function.

Figure 1.45 shows that not every curve in the plane is the graph of a function. In this figure, a vertical line $x = a$ intersects the curve at two distinct points (a, b) and (a, c). This curve cannot be the graph of $y = f(x)$ for any function f because having $f(a) = b$ and $f(a) = c$ means that f assigns two different range values to the same domain element a. We express this statement, called the vertical line test, as follows:

Vertical-Line Test

If no vertical line intersects the graph of a relation at more than one point, then the graph of the relation is the graph of a function.

EXAMPLE 7 **Testing for Functions**

Determine which graphs in Figure 1.46 are graphs of functions.

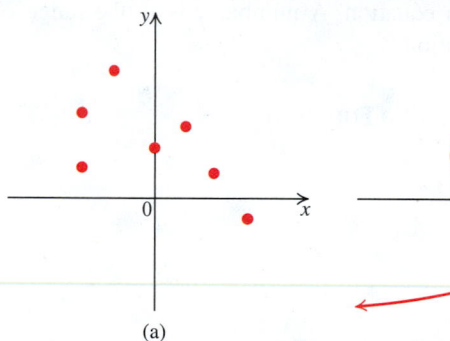
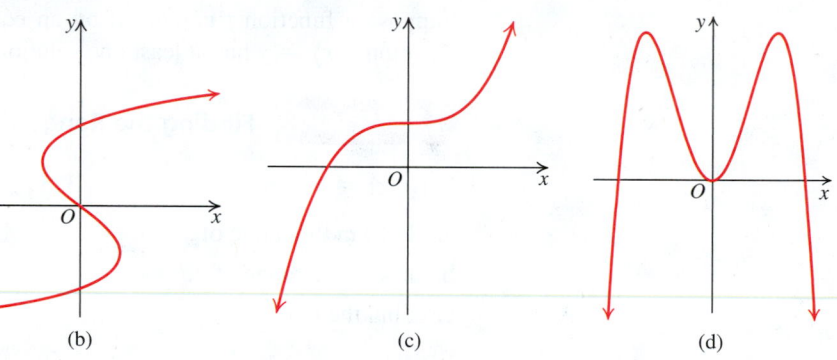

(a) (b) (c) (d)

Figure 1.46 The vertical-line test.

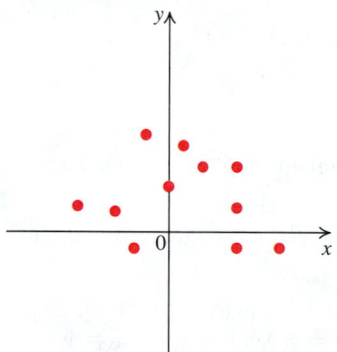

Figure 1.47

4 Get information about a function from its graph.

Solution

The graphs in Figures 1.46(a) and 1.46(b) are not graphs of functions because a vertical line can be drawn through the two points farthest to the left in Figure 1.46(a) and the y-axis is one of many vertical lines that contain more than one point on the graph in Figure 1.46(b). The graphs in Figures 1.46(c) and 1.46(d) are the graphs of functions because no vertical line intersects either graph at more than one point.

Practice Problem 7 Decide whether the graph in Figure 1.47 is the graph of a function.

Function Information from Its Graph

Given a graph in the xy-plane, we use the vertical-line test to determine whether the graph is that of a function. We can also obtain the following additional information from the graph of a function.

1. **Point on a graph**
   A point (a, b) on the graph of f means that a is in the domain of f and the value of f at a is b; that is, $f(a) = b$. We can visually determine whether a given point is on the graph of a function. In Figure 1.48, the point (a, b) is on the graph of f and the point (c, d) is not on the graph of f.

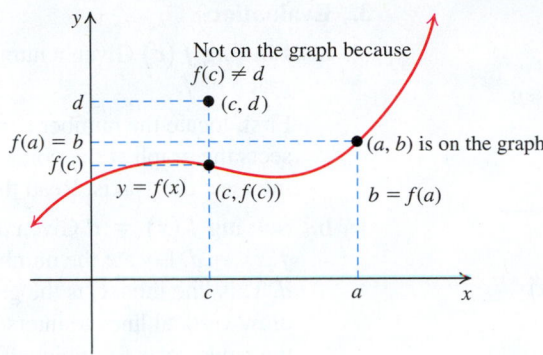

Figure 1.48

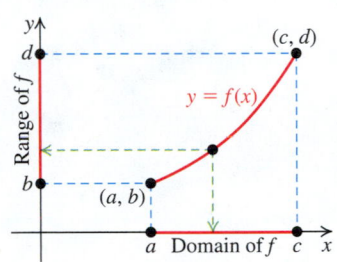

Figure 1.49 Domain and range of f.

2. **Domain and range from a graph**

To determine the *domain* of a function, we look for the portion on the x-axis that is used in graphing f. We can find this portion by projecting (collapsing) the graph on the x-axis. This projection is the domain of f.

The range of f is the projection of its graph on the y-axis. See Figure 1.49.

EXAMPLE 8   **Finding the Domain and Range from a Graph**

Use the graphs in Figure 1.50 to find the domain and range of each function.

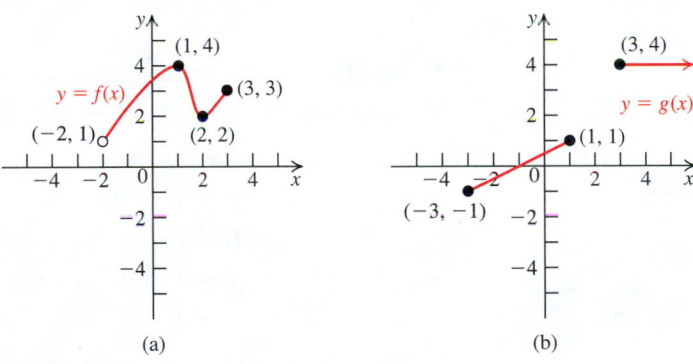

Figure 1.50

Solution

a. In Figure 1.50(a), the open circle at $(-2, 1)$ indicates that the point $(-2, 1)$ does not belong to the graph of f, while the full circle at the point $(3, 3)$ indicates that the point $(3, 3)$ is part of the graph. When we project the graph of $y = f(x)$ onto the x-axis, we obtain the interval $(-2, 3]$. So, the domain of f in interval notation is $(-2, 3]$. Similarly, the projection of the graph of f onto the y-axis gives the interval $(1, 4]$, which is the range of f.

b. The projection of the graph of g in Figure 1.50(b) onto the x-axis is made up of the two intervals $[-3, 1]$ and $[3, \infty)$. So the domain of g is $[-3, 1] \cup [3, \infty)$.

To find the range of g, we project its graph onto the y-axis. The projection of the line segment joining $(-3, -1)$ and $(1, 1)$ onto the y-axis is the interval $[-1, 1]$. The projection of the horizontal ray starting at the point $(3, 4)$ onto the y-axis is just a single point at $y = 4$. Therefore, the range of g is $[-1, 1] \cup \{4\}$.

Practice Problem 8 Find the domain and range of a function whose graph is given in Figure 1.51.

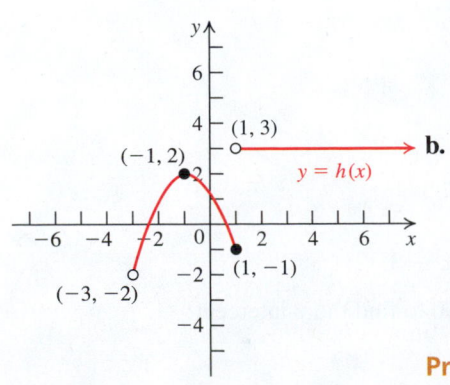

Figure 1.51

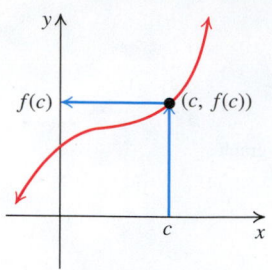

Figure 1.52 Locating $f(c)$ graphically.

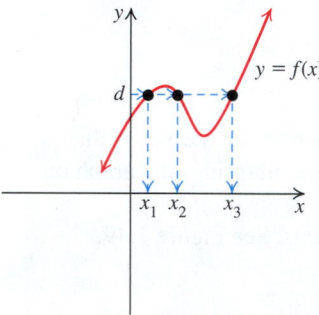

Figure 1.53 Graphically solving $f(x) = d$.

3. **Evaluations**

   a. **Finding $f(c)$** Given a number c in the domain of f, we find $f(c)$ from the graph of f.

      First, locate the number c on the x-axis. Draw a vertical line through c. It intersects the graph at the point $(c, f(c))$. Through this point, draw a horizontal line to intersect the y-axis. Read the value on the y-axis. See Figure 1.52.

   b. **Solving $f(x) = d$** Given d in the range of f, find values of x for which $f(x) = d$. Locate the number d on the y-axis. Draw a horizontal line through d. This line intersects the graph at point(s) (x, d). Through each of these points, draw vertical lines to intersect the x-axis. Read the values on the x-axis. These are the values of x for which $f(x) = d$. Figure 1.53 shows three solutions: x_1, x_2, and x_3 of the equation $f(x) = d$.

EXAMPLE 9 **Examining the Graph of a Function**

Let $f(x) = x^2 - 2x - 3$.

a. Is the point $(1, -3)$ on the graph of f?

b. Find all values of x so that $(x, 5)$ is on the graph of f.

c. Find all y-intercepts of the graph of f.

d. Find all x-intercepts of the graph of f.

Solution

a. We check whether $(1, -3)$ satisfies the equation $y = x^2 - 2x - 3$.

$$-3 \overset{?}{=} (1)^2 - 2(1) - 3 \qquad \text{Replace } x \text{ with 1 and } y \text{ with } -3.$$
$$-3 \overset{?}{=} -4 \quad \text{No!}$$

So $(1, -3)$ is *not* on the graph of f. See Figure 1.54.

b. Substitute 5 for y in $y = x^2 - 2x - 3$ and solve for x.

$$5 = x^2 - 2x - 3$$
$$0 = x^2 - 2x - 8 \qquad \text{Subtract 5 from both sides.}$$
$$0 = (x - 4)(x + 2) \qquad \text{Factor.}$$
$$x - 4 = 0 \quad \text{or} \quad x + 2 = 0 \qquad \text{Zero-product property}$$
$$x = 4 \qquad \text{or} \qquad x = -2. \qquad \text{Solve for } x.$$

The point $(x, 5)$ is on the graph of f only when $x = -2$ or $x = 4$. Both points $(-2, 5)$ and $(4, 5)$ are on the graph of f. See Figure 1.54.

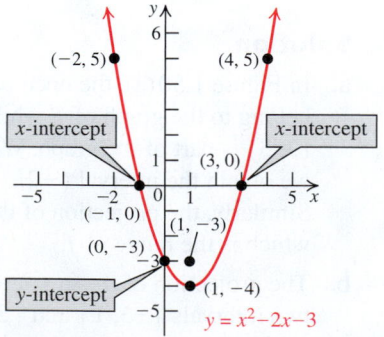

Figure 1.54

c. Find all points (x, y) with $x = 0$ in $y = x^2 - 2x - 3$.

$$y = 0^2 - 2(0) - 3 \qquad \text{Replace } x \text{ with 0 to find the } y\text{-intercept.}$$
$$y = -3 \qquad \text{Simplify.}$$

The only y-intercept is -3. See Figure 1.54.

d. Find all points (x, y) with $y = 0$ in $y = x^2 - 2x - 3$.

$$0 = x^2 - 2x - 3 \qquad \text{Replace } y \text{ with zero to find the } x\text{-intercepts.}$$
$$0 = (x + 1)(x - 3) \qquad \text{Factor.}$$
$$x + 1 = 0 \quad \text{or} \quad x - 3 = 0 \qquad \text{Zero-product property}$$
$$x = -1 \quad \text{or} \quad x = 3. \qquad \text{Solve for } x.$$

The x-intercepts of the graph of f are -1 and 3. See Figure 1.54.

Practice Problem 9 Let $f(x) = x^2 + 4x - 5$.

a. Is the point $(2, 7)$ on the graph of f?

b. Find all values of x so that $(x, -8)$ is on the graph of f.

c. Find the y-intercept of the graph of f.

d. Find the x-intercepts, if any, of the graph of f.

SUMMARY OF **MAIN FACTS**

A function is usually described in one or more of the following six ways:

- A correspondence diagram
- A set of ordered pairs
- A table of values
- An equation or a formula
- A scatterplot or a graph
- In words "y is a function of x."

Input	Output
x	y or $f(x)$
First coordinate	Second coordinate
Independent variable	Dependent variable
Domain is the set of all inputs.	Range is the set of all outputs.

5 Find the average rate of change of a function.

Average Rate of Change

Suppose your salary increases from \$25,000 a year to \$40,000 a year over a five-year period. You then have

$$\text{Change in salary: } \$40{,}000 - \$25{,}000 = \$15{,}000$$

Average rate of change in salary:

$$\frac{\$40{,}000 - \$25{,}000}{5} = \frac{\$15{,}000}{5} = \$3000 \text{ per year.}$$

Regardless of when you actually received the raises during the five-year period, the final salary is the same as if you received your average annual increase, \$3000, *each* year.

Average rate of change (ARC) tells us how fast, on average, the function changes on a given interval. Finding average rates of change is important in many situations. For instance, we may be interested in knowing the average speed (miles per hour) or average gas mileage (miles per gallon) or how fast particular stock values are increasing (dollars per month).

 The Average Rate of Change of a Function

Let $(a, f(a))$ and $(b, f(b))$ be two points on the graph of a function f. Then the **average rate of change** of $f(x)$ as x changes from a to b is defined by

$$\text{ARC of } f \text{ from } a \text{ to } b = \frac{\text{change in } y}{\text{change in } x} = \frac{\Delta y}{\Delta x} = \frac{f(b) - f(a)}{b - a}, \quad b \neq a.$$

The similarity of the average rate of change formula to the slope of the line formula is not a coincidence. Simply stated, a straight line has a constant ARC that is equal to its slope.

Geometrically, the average rate of change of f from $x = a$ to $x = b$ is the slope of the **secant line** joining the points $P(a, f(a))$ and $Q(b, f(b))$ on the graph of f. See Figure 1.55.

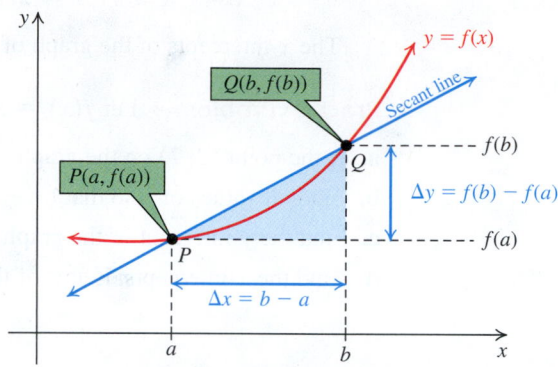

Figure 1.55 Secant line and average rate of change.

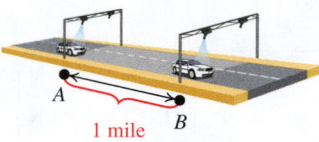

Consider the situation where police are checking the speed of incoming cars at two points A and B that are 1 mile away from each other. On a street with a speed limit of 50 miles per hour, one car was clocked at point A driving 40 miles per hour and the same car was clocked one minute later at point B driving 45 miles per hour. The surprised driver was stopped and issued a traffic citation for speeding. We can solve this mystery by computing the average speed of this driver between point A and B.

$$\text{Average Speed} = \frac{\text{change in distance}}{\text{change in time}} = \frac{1 \text{ mile}}{1 \text{ minute}} = \frac{1 \text{ mile}}{\dfrac{1}{60} \text{ hour}} = 60 \text{ miles per hour.}$$

PROCEDURE
IN ACTION

EXAMPLE 10   Finding the Average Rate of Change

OBJECTIVE

Find the average rate of change of a function f as x changes from a to b.

Step 1 Find $f(a)$ and $f(b)$.

Step 2 Use the values from Step 1 in the definition of average rate of change.

EXAMPLE

Find the average rate of change of $f(x) = 2 - 3x^2$ as x changes from $a = 1$ to $b = 3$.

1. $f(1) = 2 - 3(1)^2 = -1$ $a = 1$
    $f(3) = 2 - 3(3)^2 = -25$ $b = 3$

2. $\dfrac{f(b) - f(a)}{b - a} = \dfrac{-25 - (-1)}{3 - 1} = -\dfrac{24}{2}$
    $= -12$

Practice Problem 10 Find the average rate of change of $f(x) = 1 - x^2$ as x changes from 2 to 4.

EXAMPLE 11   Finding the Average Rate of Change

Find the average rate of change of $f(x) = 2x^2 - 3$ as x changes from $x = c$ to $x = c + h, h \neq 0$.

Solution

$$\text{Average rate of change} = \frac{f(c+h) - f(c)}{(c+h) - c} \qquad \text{Definition}$$

$$= \frac{[2(c+h)^2 - 3] - [2(c)^2 - 3]}{c + h - c} \qquad \text{Find } f(c+h) \text{ and } f(c).$$

$$= \frac{[2(c^2 + 2ch + h^2) - 3] - (2c^2 - 3)}{h} \qquad \text{Expand the binomial square.}$$

$$= \frac{4ch + 2h^2}{h} = \frac{\cancel{h}(4c + 2h)}{\cancel{h}} \qquad \text{Simplify; } h \neq 0.$$

$$= 4c + 2h$$

Practice Problem 11 Find the average rate of change of $f(x) = 1 - x^2$ as x changes from $x = c$ to $x = c + h, h \neq 0$.

The average rate of change calculated in Example 11 is called a *difference quotient*. Average rate of change is an important concept that helps us to study the behavior of functions. This concept will be further improved upon in calculus when the study of the average rate of change of a function on intervals that are arbitrarily small leads to the concept of the instantaneous rate of change. This, in turn, will lead to the notion of the derivative of the function, a fundamental idea in differential calculus.

Difference Quotient

For a function f, the **difference quotient** is

$$\frac{f(x+h) - f(x)}{h}, \quad h \neq 0.$$

EXAMPLE 12 **Evaluating and Simplifying a Difference Quotient**

Let $f(x) = 2x^2 - 3x + 5$. Find and simplify $\dfrac{f(x+h) - f(x)}{h}, h \neq 0$.

Solution

First, we find $f(x+h)$.

$$f(x) = 2x^2 - 3x + 5 \qquad \text{Original equation}$$
$$f(x+h) = 2(x+h)^2 - 3(x+h) + 5 \qquad \text{Replace } x \text{ with } (x+h).$$
$$= 2(x^2 + 2xh + h^2) - 3(x+h) + 5 \qquad (x+h)^2 = x^2 + 2xh + h^2$$
$$= 2x^2 + 4xh + 2h^2 - 3x - 3h + 5 \qquad \text{Use the distributive property.}$$

Then we substitute into the difference quotient.

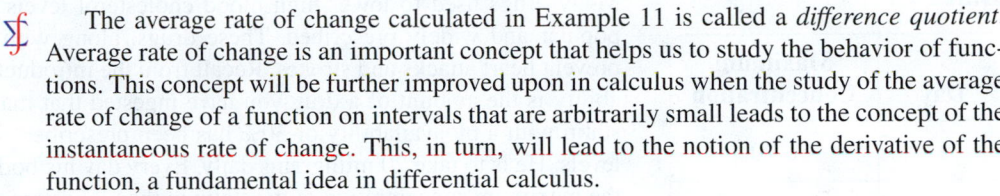

$$\frac{f(x+h) - f(x)}{h} = \frac{\overbrace{(2x^2 + 4xh + 2h^2 - 3x - 3h + 5)}^{f(x+h)} - \overbrace{(2x^2 - 3x + 5)}^{f(x)}}{h}$$

$$= \frac{2x^2 + 4xh + 2h^2 - 3x - 3h + 5 - 2x^2 + 3x - 5}{h} \qquad \text{Subtract.}$$

$$= \frac{4xh - 3h + 2h^2}{h} \qquad \text{Simplify.}$$

$$= \frac{\cancel{h}(4x - 3 + 2h)}{\cancel{h}} \qquad \text{Factor out } h.$$

$$= 4x - 3 + 2h \qquad \text{Remove the common factor } h.$$

Practice Problem 12 Repeat Example 11 for $f(x) = -x^2 + x - 3$.

6 Solve applied problems by using functions.

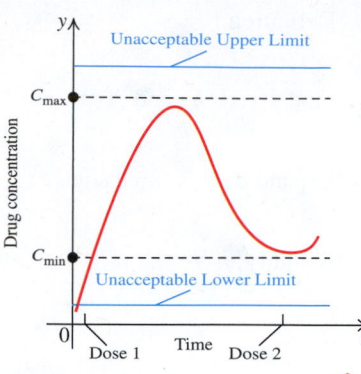

Figure 1.56 Drug levels.

TABLE 1.7

Day	Maximum Concentration
1	6.000
2	9.000
3	10.500
4	11.250
5	11.625
6	11.813
7	11.906
8	11.953
9	11.977
10	11.988

Building Functions

Level of Drugs in Bloodstream When you take a dose of medication, the drug level in the blood goes up quickly and soon reaches its *peak,* called $C_{\max}$ (the maximum concentration). As your liver or kidneys remove the drug, your blood's drug levels drop until the next dose enters your bloodstream. The lowest drug level is called the *trough,* or $C_{\min}$ (the minimum concentration). The ideal dose should be strong enough to be effective without causing too many side effects. We start by setting upper and lower limits on the drug's blood levels, shown by the two horizontal lines on a pharmacokinetics (PK) graph. See Figure 1.56. The upper line represents the drug level at which people start to develop serious side effects. The lower line represents the minimum drug level that provides the desired effect.

◆ **EXAMPLE 13** **Cholesterol-Reducing Drugs**

Many drugs used to lower high blood cholesterol levels are called *statins* and are very popular and widely prescribed. These drugs, along with proper diet and exercise, help prevent heart attacks and strokes. Recall from the introduction to this section that bioavailability is the amount of a drug you have ingested that makes it into your bloodstream. A statin with a bioavailability of 30% has been prescribed for Boris to treat his cholesterol levels. He is to take 20 milligrams daily. Every day his body filters out half the statin. Find the maximum concentration of the statin in the bloodstream on each of the first ten days of using the drug and graph the result.

Solution

Since the statin has 30% bioavailability and Boris takes 20 milligrams per day, the maximum concentration in the bloodstream is 30% of 20 milligrams, or $20(0.3) = 6$ milligrams from each day's prescription. Because half the statin is filtered out of the body each day, the daily maximum concentration is

$$\frac{1}{2}(\text{previous days maximum concentration}) + 6.$$

So, if $C(n)$ denotes the maximum concentration on the nth day then

$$(1) \qquad C(n) = \frac{1}{2}C(n-1) + 6, \ C(0) = 0.$$

Equation (1) is an example of a **recursive definition** of a function.

Table 1.7 shows the maximum concentration of the drug for each of the first ten days. After the first day, each number in the second column is computed (to three decimal places) by adding 6 to one-half the number above it. That is:

$$C(1) = \frac{1}{2}C(0) + 6 = \frac{1}{2}(0) + 6 = 6.000$$

$$C(2) = \frac{1}{2}C(1) + 6 = \frac{1}{2}(6) + 6 = 9.000$$

$$C(3) = \frac{1}{2}C(2) + 6 = \frac{1}{2}(9) + 6 = 10.500$$

$$\vdots$$

$$C(9) = \frac{1}{2}C(8) + 6 = \frac{1}{2}(11.953) + 6 = 11.977$$

$$C(10) = \frac{1}{2}C(9) + 6 = \frac{1}{2}(11.977) + 6 = 11.988$$

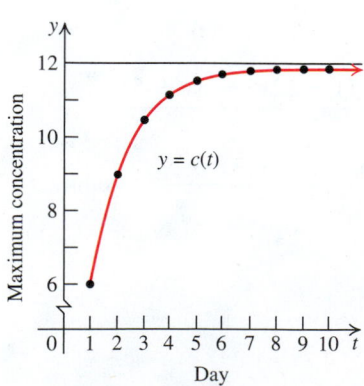

Figure 1.57 Maximum drug concentration.

The graph is shown in Figure 1.57.

The graph shows that the maximum concentration of the statin in the bloodstream approaches 12 milligrams if Boris continues to take one pill every day.

Practice Problem 13 In Example 13, use Table 1.7 to compute the function value $C(11)$.

EXAMPLE 14   Cost of a Fiber Optics Cable

Two points A and B are opposite each other on the banks of a straight river that is 500 feet wide. The point D is on the same side as B but is 1200 feet up the river from B. The local Internet cable company wishes to lay a fiber optics cable from A to D. The cost per foot of cable is $5 per foot under water and $3 per foot on land. To save money, the company lays the cable under water from A to P and then on land from P to D. In Figure 1.58, write the total cost C as a function of x.

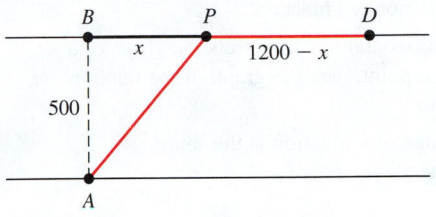

Figure 1.58

Solution

In Figure 1.58, we have

$$d(A, P) = AP = \sqrt{(500)^2 + x^2} \text{ feet.} \qquad \text{Pythagorean Theorem}$$
$$d(P, D) = PD = 1200 - x \text{ feet.}$$

So, the total cost C is given by

$$C = 5(AP) + 3(PD)$$
$$= 5\sqrt{(500)^2 + x^2} + 3(1200 - x).$$

Practice Problem 14 Repeat Example 14 assuming that the cost is 30% more under water than it is on land.

Answers to Practice Problems

1. a. Yes

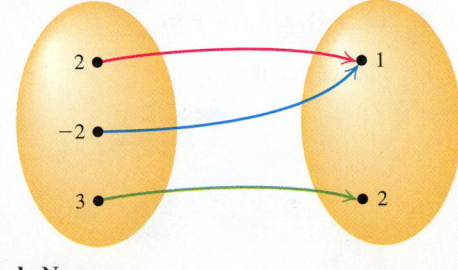

b. No

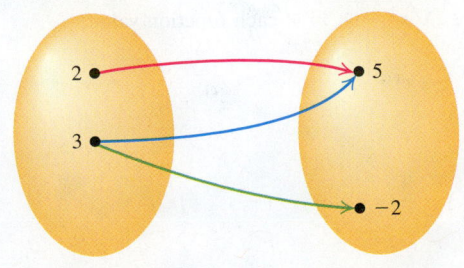

2. a. No   **b.** Yes   **3. a.** 0   **b.** −7
c. $-2x^2 - 4hx + 5x - 2h^2 + 5h$
4. 44 square units   **5. a.** $(-\infty, 1)$ **b.** $(-\infty, -3] \cup [1, +\infty)$
6. a. No   **b.** Yes   **c.** $[0, 9]$ **7.** No   **8.** Domain: $(-3, \infty)$;
range: $(-2, 2] \cup \{3\}$ **9. a.** Yes   **b.** −3, −1   **c.** −5
d. −5, 1   **10.** −6
11. $-2c - h$ **12.** $-2x + 1 - h$
13. $C(11) = 11.994$
14. If $c = $ cost per foot on land, then

$$C = c\left[1.3\sqrt{(500)^2 + x^2} + 1200 - x\right].$$

SECTION 1.3 Exercises

Concepts and Vocabulary

1. In the functional notation $y = f(x)$, x is the _____ variable.

2. If $f(-2) = 7$, then -2 is in the _____ of the function f and 7 is in the _____ of f.

3. If the point $(9, -14)$ is on the graph of a function f, then $f(9) = $ _____.

4. The average rate of change of f as x changes from $x = a$ to $x = b$ is _____, $a \neq b$.

5. **True or False.** Every relation is a function.

6. **True or False.** If no horizontal line intersects the graph of a relation at more than one point, then the graph of the relation is the graph of a function.

7. **True or False.** The range of a function is the set of all values assigned to the elements in the domain of a function.

8. **True or False.** The average rate of change of a linear function is equal to its slope.

Building Skills

In Exercises 9–14, determine the domain and range of each relation. Explain whether each relation is or is not a function.

9.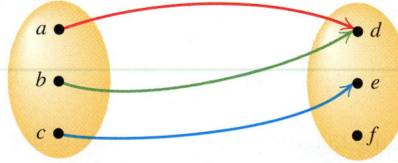

10.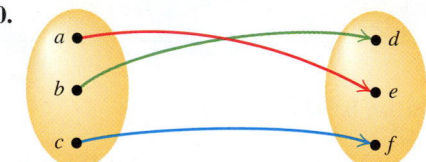

11.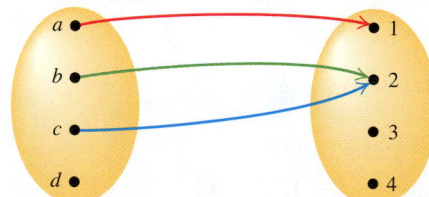

12.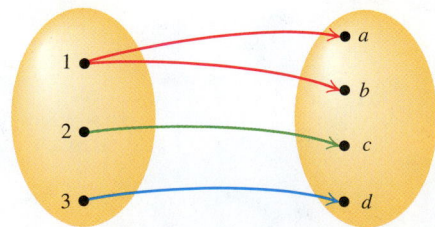

13.

x	-3	-1	0	1	2	3
y	-8	0	1	0	-3	-8

14.

x	0	3	8	0	3	8
y	-1	-2	-3	1	2	2

In Exercises 15–28, determine whether each equation defines y as a function of x.

15. $x + y = 2$

16. $x = y - 1$

17. $y = \dfrac{1}{x}$

18. $xy = -1$

19. $x = |y|$

20. $x = |y - 1|$

21. $y = \dfrac{1}{\sqrt{2x - 5}}$

22. $y = \dfrac{1}{\sqrt{x^2 - 1}}$

23. $2 - y = 3x$

24. $3x - 5y = 15$

25. $x^2 + y^2 = 8$

26. $x = y^2$

27. $x^2 + y^3 = 5$

28. $x + y^3 = 8$

In Exercises 29–32, let $f(x) = x^2 - 3x + 1$, $g(x) = \dfrac{2}{\sqrt{x}}$, and $h(x) = \sqrt{2 - x}$.

29. Find $f(0), g(0), h(0), f(a)$, and $f(-x)$.

30. Find $f(1), g(1), h(1), g(a)$, and $g(x^2)$.

31. Find $f(-1), g(-1), h(-1), h(c)$, and $h(-x)$.

32. Find $f(4), g(4), h(4), g(2 + k)$, and $f(a + k)$.

33. Let $f(x) = \dfrac{2x}{\sqrt{4 - x^2}}$. Find each function value.

   a. $f(0)$
   b. $f(1)$
   c. $f(2)$
   d. $f(-2)$
   e. $f(-x)$

34. Let $g(x) = 2x + \sqrt{x^2 - 4}$. Find each function value.

   a. $g(0)$
   b. $g(1)$
   c. $g(2)$
   d. $g(-3)$
   e. $g(-x)$

35. In the figure below, find the sum of the areas of the shaded rectangles.

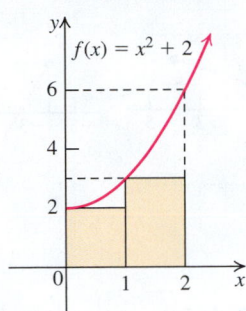

36. Repeat Exercise 35 in the figure below.

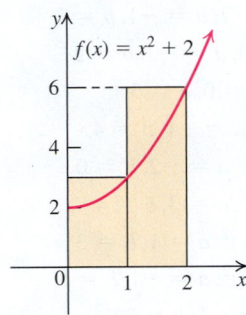

In Exercises 37–52, find the domain of each function.

37. $f(x) = -8x + 7$

38. $f(x) = 2x^2 - 11$

39. $f(x) = \dfrac{1}{x - 9}$

40. $f(x) = \dfrac{1}{x + 9}$

41. $h(x) = \dfrac{2x}{x^2 - 1}$

42. $h(x) = \dfrac{x - 3}{x^2 - 4}$

43. $G(x) = \dfrac{3x}{\sqrt{4 - x}}$

44. $f(x) = \dfrac{-2x}{\sqrt{x + 2}}$

45. $h(x) = \dfrac{\sqrt{x - 1}}{x - 2}$

46. $H(x) = \dfrac{\sqrt{2 - x}}{x - 1}$

47. $F(x) = \dfrac{x + 4}{x^2 + 3x + 2}$

48. $F(x) = \dfrac{1 - x}{x^2 + 5x + 6}$

49. $g(x) = \dfrac{\sqrt{x^2 + 1}}{x}$

50. $g(x) = \dfrac{1}{x^2 + 1}$

51. $s(x) = \sqrt{1 - x^2}$

52. $K(x) = \sqrt{x^2 - 4}$

In Exercises 53–58, use the vertical-line test to determine whether the given graph represents a function.

53.

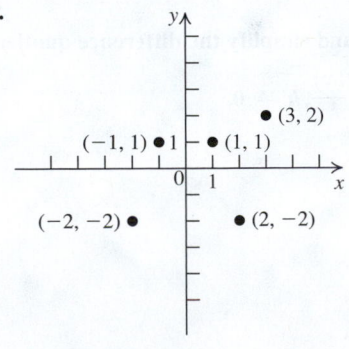

54.

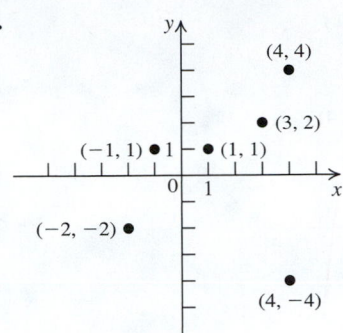

55.

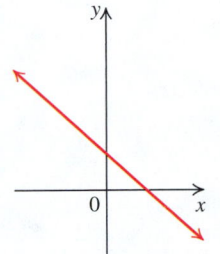

56.

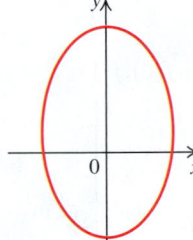

57.

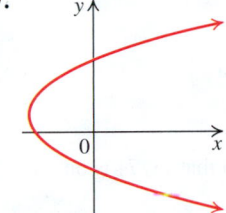

58.

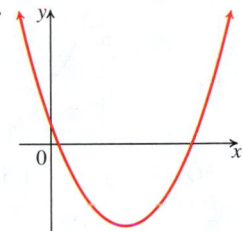

In Exercises 59–62, the graph of a function is given. Find the indicated function values.

59. $f(-4), f(-1), f(3), f(5)$

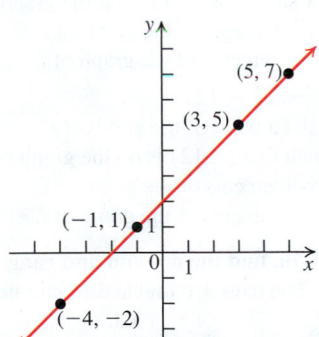

60. $g(-2), g(1), g(3), g(4)$

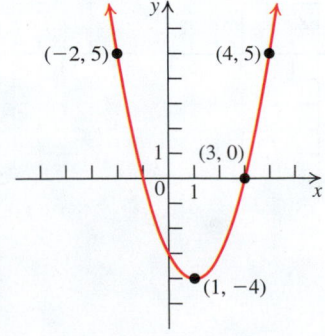

61. $h(-2), h(-1), h(0), h(1)$

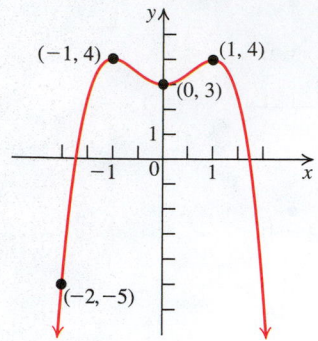

62. $f(-1), f(0), f(1)$

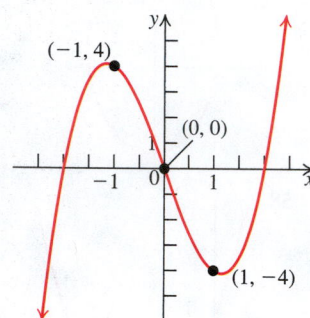

63. Let $h(x) = x^2 - x + 1$. Find x such that $(x, 7)$ is on the graph of h.

64. Let $H(x) = x^2 + x + 8$. Find x such that $(x, 7)$ is on the graph of H.

65. Let $f(x) = -2(x + 1)^2 + 7$.
   a. Is $(1, 1)$ a point of the graph of f?
   b. Find all x such that $(x, 1)$ is on the graph of f.
   c. Find all y-intercepts of the graph of f.
   d. Find all x-intercepts of the graph of f.

66. Let $f(x) = -3x^2 - 12x$.
   a. Is $(-2, 10)$ a point of the graph of f?
   b. Find x such that $(x, 12)$ is on the graph of g.
   c. Find all y-intercepts of the graph of f.
   d. Find all x-intercepts of the graph of f.

In Exercises 67–70, find the domain and range of each function from its graph. The axes are marked in one-unit intervals.

67.

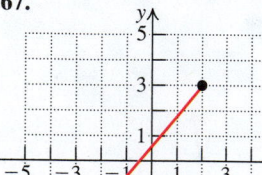

68.

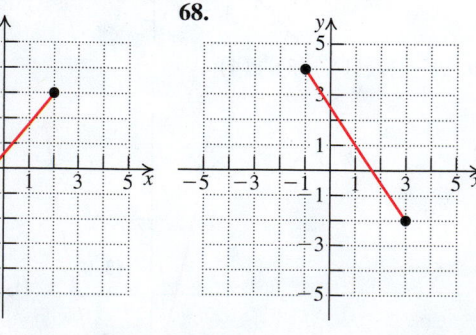

69.

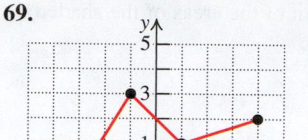

70.

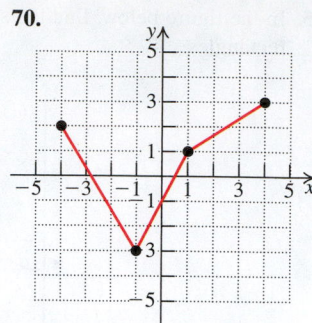

In Exercises 71–82, find the average rate of change of the function as x changes from a to b.

71. $f(x) = -2x + 7; a = -1, b = 3$

72. $f(x) = 4x - 9; a = -2, b = 2$

73. $g(x) = 2x^2; a = 0, b = 5$

74. $g(x) = -4x^2; a = -1, b = 4$

75. $h(x) = x^2 - 1; a = -2, b = 0$

76. $h(x) = 2 - x^2; a = 3, b = 4$

77. $f(x) = (3 - x)^2; a = 1, b = 3$

78. $f(x) = (x - 2)^2; a = -1, b = 5$

79. $g(x) = x^3; a = -1, b = 3$

80. $g(x) = -x^3; a = -1, b = 3$

81. $h(x) = \dfrac{1}{x}; a = 2, b = 6$

82. $h(x) = \dfrac{4}{x + 3}; a = -2, b = 4$

In Exercises 83–90, find and simplify the difference quotient of the form $\dfrac{f(x) - f(a)}{x - a}, x \neq a$.

83. $f(x) = 2x, a = 3$

84. $f(x) = 3x + 2, a = 2$

85. $f(x) = -x^2, a = 1$

86. $f(x) = 2x^2, a = -1$

87. $f(x) = 3x^2 + x, a = 2$

88. $f(x) = -2x^2 + x, a = 3$

89. $f(x) = \dfrac{4}{x}, a = 1$

90. $f(x) = -\dfrac{4}{x}, a = 1$

In Exercises 91–100, find and simplify the difference quotient of the form $\dfrac{f(x + h) - f(x)}{h}, h \neq 0$.

91. $f(x) = -2x + 3$

92. $f(x) = 3x + 2$

93. $f(x) = x^2$

94. $f(x) = x^2 - x$

95. $f(x) = 3x^2 - 2x + 5$

96. $f(x) = 2x^2 + 3x$

97. $f(x) = 4$

98. $f(x) = -3$

99. $f(x) = \dfrac{1}{x}$

100. $f(x) = -\dfrac{1}{x}$

Applying the Concepts

In Exercises 101–104, state whether the given relation is a function and explain why.

101. To each day of the year there corresponds the high temperature in your hometown on that day.

102. To each year since 1950 there corresponds the cost of a first-class postage stamp on January 1 of that year.

103. To each letter of the word *TUNA* there corresponds the states whose names start with that letter.

104. To each day of the week there corresponds the first names of people currently living in the United States born on that day of the week.

105. Square tiles. The area $A(x)$ of a square tile is a function of the length x of the square's side. Write a function rule for the area of a square tile. Find and interpret $A(4)$.

106. Cube. The volume $V(x)$ of a cube is a function of the length x of the cube's edge. Write a function rule for the volume of a cube. Evaluate the function for a cube with sides of length 3 inches.

107. Surface area. Is the total surface area S of a cube a function of the edge length x of the cube? If it is not a function, explain why not. If it is a function, write the function rule $S(x)$ and evaluate $S(3)$.

108. Measurement. One meter equals about 39.37 inches. Write a function rule for converting inches to meters. Evaluate the function for 59 inches.

109. Motion of a projectile. A stone thrown upward with an initial velocity of 128 feet per second will attain a height of h feet in t seconds, where

$$h(t) = 128t - 16t^2, \quad 0 \le t \le 8.$$

a. What is the domain of h?

b. Find $h(2)$, $h(4)$, and $h(6)$.

c. How long will it take the stone to hit the ground?

d. Sketch a graph of $y = h(t)$.

110. Housing affordability in the United States. The following table lists the median price of existing homes during 2000–2014.

Year	Price in thousands of dollars
2000	147.3
2001	156.6
2002	167.6
2003	180.2
2004	195.2
2005	219.0
2006	221.9
2007	219.1
2008	195.2
2009	172.1
2010	173.2
2011	166.2
2012	176.8
2013	197.1
2014	208.5

Find the average rate of change of a median-priced home during each period.

a. 2000–2004. Use your result to estimate the median sale price of a home in 2002. Compare your estimate with the actual data.

b. 2010–2014. Use your result to estimate the median sale price of a home in 2015.

111. Drug prescription. A certain drug has been prescribed to treat an infection. The patient receives an injection of 16 milliliters of the drug every 4 hours. During this same 4-hour period, the kidneys filter out one-fourth of the drug. Find the concentration of the drug after 4 hours, 8 hours, 12 hours, 16 hours, and 20 hours. Sketch the graph of the concentration of the drug in the bloodstream as a function of time.

112. Drug prescription. Every day Jane takes one 500-milligram aspirin tablet that has 80% bioavailability. During the same day, three-fourths of the aspirin is metabolized. Find the maximum concentration of the aspirin in the bloodstream on each of the first ten days of using the drug and graph the result.

113. Numbers. The sum of two numbers x and y is 28. Write the product P of these numbers as a function of x.

114. Area of a rectangle. The dimensions of a rectangle are x and y, and its perimeter is 60 meters. Write the area A of the rectangle as a function of x.

115. Box volume. A closed box with a square base of side x inches is to hold 64 cubic inches. Write the surface area S of the box as a function of x.

116. Inscribed Rectangle. In the figure, a rectangle is inscribed in a semicircle of diameter $2r$.
   a. Write the perimeter P of the rectangle as a function of x.
   b. Write the area A of the rectangle as a function of x.

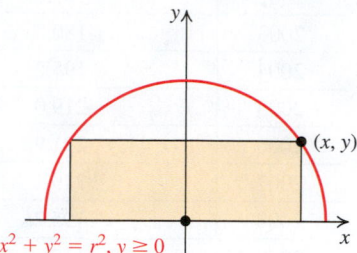

117. Piece of wire. A piece of wire 20 inches long is to be cut into two pieces (see the figure). The piece of length x is to form a circle, and the other is to form a square. Write A, the sum of the areas of the two figures, as a function of x.

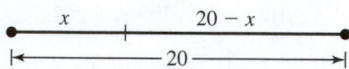

118. Area of metal. An open cylindrical tank with circular base of radius r is to be constructed of metal to contain a volume of 64 cubic inches. Write the surface area A of the metal as a function of r.

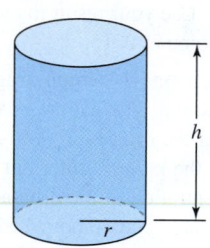

119. Cost of pool. A 288-cubic foot pool is to be built with a square bottom of side length x. The sides are built with tiles and the bottom with cement. The cost per unit of tiles is $6 per square foot while the cost of the bottom is $2 per square foot. Write the total cost C as a function of x.

120. Distance between cars. Two cars are traveling along two roads that cross each other at right angles at point A. Both cars are traveling toward A for t seconds at 30 feet per second. Initially, their distances from A are 1500 feet and 2100 feet, respectively. Write the distance d between the cars as a function of t.

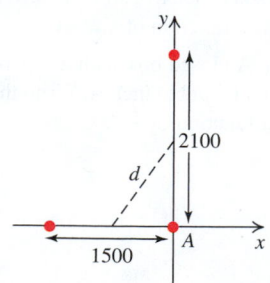

121. Distance. Write the distance d from the point $(2, 1)$ to the point (x, y) on the graph of $y = f(x) = x^3 - 3x + 6$ as a function of x.

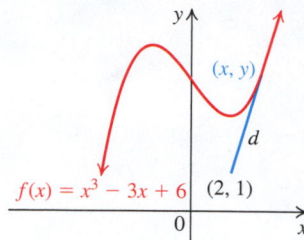

122. Time of travel. An island is at point A, 5 miles from the nearest point B on a straight beach. A store is at point C, 8 miles up the beach from point B. Julio can row at 4 miles per hour and walk 5 miles per hour. He rows to the point P, x miles up the beach from point B, and walks from P to C. Write the total time T of travel from A to C as a function of x.

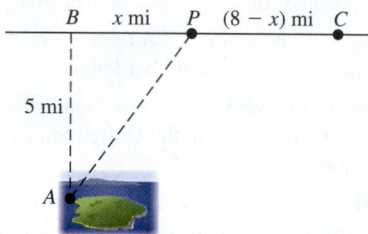

Beyond the Basics

In Exercises 123–128, find an expression for $f(x)$ by solving for y and replacing y with $f(x)$. Then find the domain of f and compute $f(4)$.

123. $x = \dfrac{y}{y - 1}$ **124.** $xy = x - y$

125. $x = \dfrac{2}{y - 4}$ **126.** $xy - 3 = 2y$

127. $(x^2 + 1)y + x = 2$ **128.** $yx^2 - \sqrt{x} = -2y$

In Exercises 129–134, state whether f and g represent the same function. Explain your reasons.

129. $f(x) = 3x - 4, 0 \le x \le 8$; $g(x) = 3x - 4$

130. $f(x) = 3x^2$; $g(x) = 3x^2, -5 \le x \le 5$

131. $f(x) = x - 1$; $g(x) = \dfrac{x^2 - 1}{x + 1}$

132. $f(x) = \dfrac{x + 2}{x^2 - x - 6}$; $g(x) = \dfrac{1}{x - 3}$

133. $f(x) = x^2 + 1$, $g(x) = 8x - 14$, both with domain $\{3, 5\}$.

134. $f(x) = 2x^2 + 8$, $g(x) = 3x + 7$, both with domain $\{1, 2\}$.

135. Let $f(x) = ax^2 + ax - 3$. If $f(2) = 15$, find a.

136. Let $g(x) = x^2 + bx + b^2$. If $g(6) = 28$, find b.

137. Let $h(x) = \dfrac{3x + 2a}{2x - b}$. If $h(6) = 0$ and $h(3)$ is undefined, find a and b.

138. Let $f(x) = 2x - 3$. Find $f(x^2)$ and $[f(x)]^2$.

139. If $g(x) = x^2 - \dfrac{1}{x^2}$, show that $g(x) + g\left(\dfrac{1}{x}\right) = 0$.

140. If $f(x) = \dfrac{x - 1}{x + 1}$, show that $f\left(\dfrac{x - 1}{x + 1}\right) = -\dfrac{1}{x}$.

141. If $f(x) = \dfrac{x + 3}{4x - 5}$ and $t = \dfrac{3 + 5x}{4x - 1}$, show that $f(t) = x$.

Critical Thinking / Discussion / Writing

142. Write an equation of a function with each domain. Answers will vary.

   a. $[2, \infty)$ **b.** $(2, \infty)$

   c. $(-\infty, 2]$ **d.** $(-\infty, 2)$

143. Consider the graph of the function
$$y = f(x) = ax^2 + bx + c, a \neq 0.$$

   a. Write an equation whose solution yields the x-intercepts.

   b. Write an equation whose solution is the y-intercept.

   c. Write (if possible) a condition under which the graph of $y = f(x)$ has no x-intercepts.

   d. Write (if possible) a condition under which the graph of $y = f(x)$ has no y-intercepts.

144. Give an example (if possible) of a function matching each description. Some have many different correct answers.

   a. Its graph is symmetric with respect to the y-axis.

   b. Its graph is symmetric with respect to the x-axis.

   c. Its graph is symmetric with respect to the origin.

   d. Its graph consists of a single point.

   e. Its graph is a horizontal line.

   f. Its graph is a vertical line.

145. Let $X = \{a, b\}$ and $Y = \{1, 2, 3\}$

   a. How many functions are there from X to Y?

   b. How many functions are there from Y to X?

146. If a set X has m elements and a set Y has n elements, how many functions can be defined from X to Y? Justify.

Getting Ready for the Next Section

In Exercises 147–150, find the slope–intercept form of the equation of the line passing through the given points.

147. $(0, 0)$ and $(2, -2)$

148. $(-1, 3)$ and $(4, -2)$

149. $(-3, 2)$ and $(1, 4)$

150. $(3, -5)$ and $(8, -3)$

In Exercises 151–154, find (a) $f(-x)$ and (b) $-f(x)$.

151. $f(x) = 2x^2 - 3x$

152. $f(x) = 3x^4 - 7x^2 + 5$

153. $f(x) = x^3 - 2x$

154. $f(x) = 2x^3 - 5x^2 + x$

A Library of Functions

BEFORE STARTING THIS SECTION, REVIEW

1 Equations of horizontal and vertical lines (Section 1.2, page 27)

2 Absolute value (Appendix A.1, page 921)

3 Graph of an equation (Section 1.1, page 7)

OBJECTIVES

1 Define linear functions.

2 Discuss important properties of functions.

3 Evaluate and graph piecewise functions.

4 Graph basic functions.

◆ The Megatooth Shark

The giant "Megatooth" shark (*Carcharodon megalodon*), once estimated to be 100 to 120 feet in length, is the largest meat-eating fish that ever lived. Its actual length has been the subject of scientific controversy. Sharks first appeared in the oceans more than 400 million years ago, almost 200 million years before the dinosaurs. A shark's skeleton is made of cartilage that decomposes rather quickly, making complete shark fossils rare. Consequently, scientists rely on calcified vertebrae, fossilized teeth, and small skin scales to reconstruct the evolutionary record of sharks.

The great white shark is the closest living relative to the giant Megatooth shark and has been used as a model to reconstruct the Megatooth. John Maisey, curator at the American Museum of Natural History in New York City, used a partial set of Megatooth teeth to make a comparison with the jaws of the great white shark. In Example 1, we will use a formula to calculate the length of the Megatooth on the basis of the height of the tooth of the largest known Megatooth specimen.

1 Define linear functions.

Linear Functions

We know from Section 1.2 that the graph of a linear equation $y = mx + b$ is a straight line with slope m and y-intercept b. A *linear function* has a similar definition.

Linear Functions

Let m and b be real numbers. The function $f(x) = mx + b$ is called a **linear function**. If $m = 0$, the function $f(x) = b$ is called a **constant function**. If $m = 1$ and $b = 0$, the resulting function $f(x) = x$ is called the **identity function**. See Figure 1.59.

The domain of a linear function is the interval $(-\infty, \infty)$ because the expression $mx + b$ is defined for any real number x. Figure 1.59 shows that the graph extends indefinitely to the left and right. The range of a nonconstant linear function also is the interval $(-\infty, \infty)$ because the graph extends indefinitely upward and downward. The range of a constant function $f(x) = b$ is the single real number b. See Figure 1.59(c).

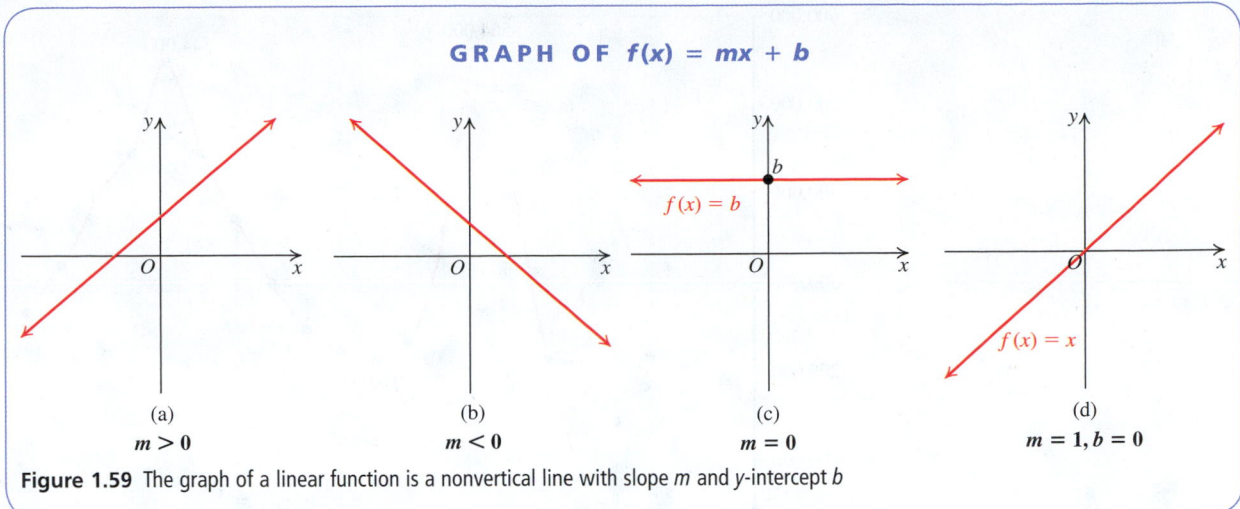

GRAPH OF $f(x) = mx + b$

(a) $m > 0$

(b) $m < 0$

(c) $m = 0$

(d) $m = 1, b = 0$

Figure 1.59 The graph of a linear function is a nonvertical line with slope m and y-intercept b

EXAMPLE 1 Determining the Length of the "Megatooth" Shark

The largest known "Megatooth" specimen is a tooth that has a total height of 15.6 centimeters. Calculate the length of the Megatooth shark from which it came by using the formula

$$\text{Shark length} = (0.96)(\text{height of tooth}) - 0.22,$$

where shark length is measured in meters and tooth height is measured in centimeters.

Solution

Shark length $= (0.96)(\text{height of tooth}) - 0.22$ Formula

Shark length $= (0.96)(15.6) - 0.22$ Replace height of tooth with 15.6.

$= 14.756$ Simplify.

The shark's length is approximately 14.8 meters (48.4 feet).

Practice Problem 1 If the Megatooth specimen was a tooth measuring 16.4 cm, what was the length of the Megatooth shark from which it came?

2 Discuss important properties of functions.

Increasing and Decreasing Functions

In most applications, functions describe the relation between two quantities with variables representing their respective numerical values, with one quantity (the dependent variable) being directly determined by the other (the independent variable). As part of the study of such functions, we explore how the output values of the given function change with respect to the input values. When input values are increasing, the output values of the function may *increase, decrease, or remain constant*. We are also interested in the largest and smallest output values of the function within a given range of input values (relative maximum or minimum). We will formulate these concepts algebraically, but we first look at a graph to help us visualize them. Figure 1.60 represents the function that models regular weekday ridership numbers in the New York City Metro Subway as a function of time.

We can see from the graph that the number of passengers is at an all-time low of 10,000 at about 2 A.M., and that the number of passengers increases during the morning commute period, peaking at 564,000 at about 8 A.M. After the morning rush hour peak, the number of passengers falls, reaching a mid-day low of 210,000 at about 11 A.M., and increasing again to reach the afternoon rush hour period peak of 544,000 at about 5 P.M. Finally, the number of passengers decreases from 5 P.M. to 12 A.M.

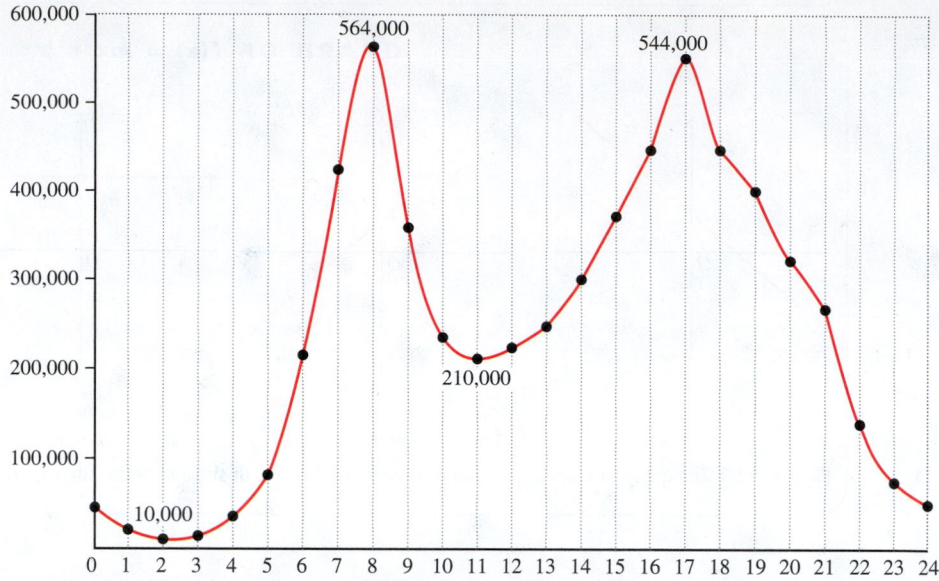

Figure 1.60 Weekday ridership numbers in New York Metro Subway as a function of time.

∫ Increasing, Decreasing, and Constant Functions

Let f be a function and let x_1 and x_2 be any two numbers in an open interval (a, b) contained in the domain of f. See Figure 1.61. The symbols a and b may represent real numbers, $-\infty$, or ∞. Then

  (i) f is called an **increasing function** on (a, b) if $x_1 < x_2$ implies $f(x_1) < f(x_2)$.

  (ii) f is called a **decreasing function** on (a, b) if $x_1 < x_2$ implies $f(x_1) > f(x_2)$.

  (iii) f is **constant** on (a, b) if $x_1 < x_2$ implies $f(x_1) = f(x_2)$.

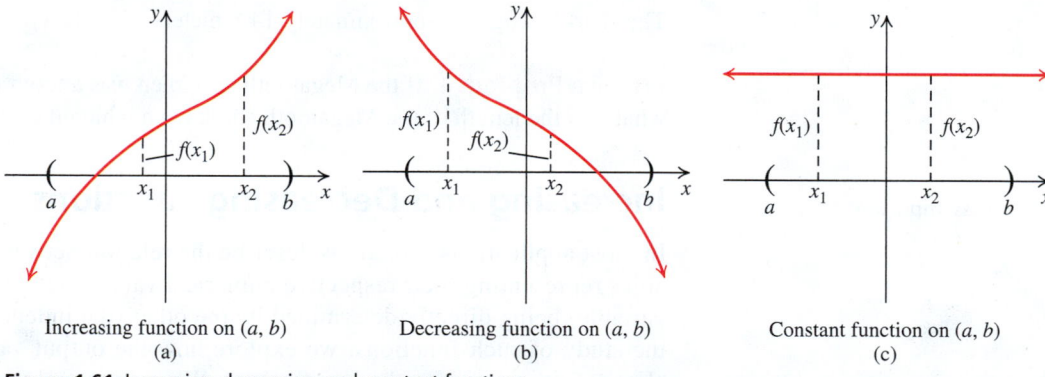

Increasing function on (a, b) Decreasing function on (a, b) Constant function on (a, b)
(a)   (b)   (c)

Figure 1.61 Increasing, decreasing, and constant functions.

SIDE
NOTE

Equivalently, the function f is increasing (decreasing) if for any two numbers in its domain the corresponding Average Rate of Change is positive (negative).

Geometrically, the graph of an increasing function on an open interval (a, b) rises as x-values increase on (a, b). This is because, by definition, as x-values increase from x_1 to x_2, the y-values also increase, from $f(x_1)$ to $f(x_2)$. Similarly, the graph of a decreasing function on (a, b) falls as x-values increase. The graph of a constant function is horizontal, or flat, over the interval (a, b).

Functions can increase, decrease, or remain constant on different intervals within their domains. The function that assigns the height of a ball tossed in the air to the length of time it is in the air is an example of a function that increases, but also decreases, over different intervals of its domain. See Figure 1.62.

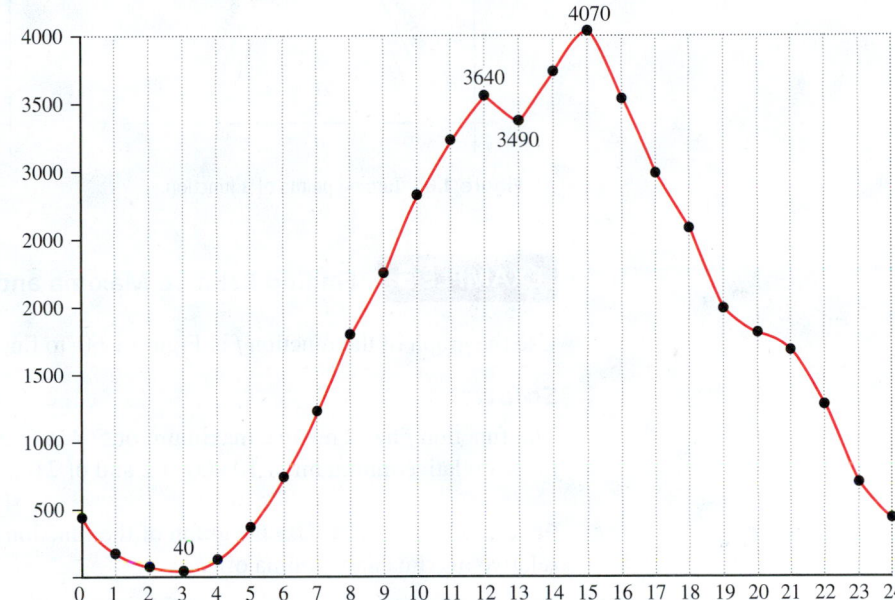

Figure 1.62 The function $h(t)$ increases on (0, 3) and decreases on (3, 6).

SIDE NOTE

When specifying the intervals over which a function f is increasing, decreasing, or constant, you need to use the intervals in the *domain* of f, *not* in the range of f.

EXAMPLE 2 Tracking the Behavior of a Function

From the graph of the function f in Figure 1.60, find the intervals over which f is increasing, is decreasing, or is constant.

Solution

The function f is increasing on the open intervals $(2, 8)$ and $(11, 17)$. The function f is decreasing on the open intervals $(0, 2)$, $(8, 11)$, and $(17, 24)$.

Practice Problem 2 Using the graph in Figure 1.63, find the intervals over which g is increasing, is decreasing, or is constant.

Figure 1.63 Weekend ridership numbers in St. Louis as a function of time.

Relative Maximum and Minimum Values

If a function f has a value at a point a that is greater than or equal (less than or equal) to the values of f at all nearby points on both sides of a, that value is called a *relative maximum (relative minimum)*. To locate the relative maxima and minima of a function on an open interval, we need to determine where the graph attains its highest and lowest points, respectively.

Relative Maximum and Relative Minimum

If a is in the domain of a function f, we say that the value $f(a)$ is a **relative maximum of f** if there is an interval (x_1, x_2) containing a such that

$$f(a) \geq f(x) \text{ for every } x \text{ in the interval } (x_1, x_2).$$

We say that the value $f(a)$ is a **relative minimum of f** if there is an interval (x_1, x_2) containing a such that

$$f(a) \leq f(x) \text{ for every } x \text{ in the interval } (x_1, x_2).$$

Points on a graph where a function changes direction from increasing to decreasing or from decreasing to increasing are called **turning points**. See Figure 1.64. Turning points help us find the relative maxima or minima.

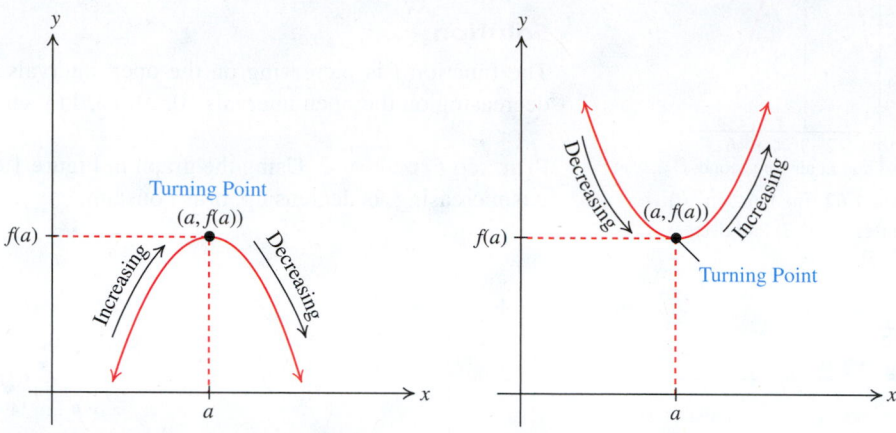

Figure 1.64 Turning points of a function.

EXAMPLE 3 **Finding Relative Maxima and Minima**

Use the graph of the function *f* in Figure 1.60 to find the relative maxima and minima of *f*.

Solution

The function *f* has a relative maximum of 564,000 at 8 and of 544,000 at 17. The function *f* has a relative minimum of 10,000 at 2 and of 210,000 at 11.

Practice Problem 3 Use the graph of the function *g* in Figure 1.63 to find the relative maxima and minima of *g*.

A graph is a powerful visual method of tracking the behavior of a function, including finding its maxima and minima. In Calculus you will learn the general technique for finding the maxima and minima of functions based on just their equation. For the present, however, we will settle for an approximation of these points by using a graphing utility.

EXAMPLE 4 **Approximating Relative Extrema**

Use a graphing utility to approximate the relative maximum and the relative minimum point on the graph of the function $f(x) = x^3 - x^2$.

Solution

The graph of *f* is shown in Figure 1.65. By using the TRACE and ZOOM features of a graphing utility, we see that the function *f* has a

Relative minimum is estimated to be 0.67 at -0.15.

Relative maximum is estimated to be 0 at 0.

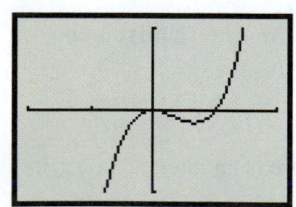

Figure 1.65

Practice Problem 4 Repeat Example 4 for $g(x) = -2x^3 + 3x^2$.

Even–Odd Functions and Symmetry

This topic, *even–odd functions*, uses the ideas of symmetry discussed in Section 1.1.

Even-Odd Functions

A function f is called an **even function** if, for each x in the domain of f, $-x$ is also in the domain of f and

$$f(-x) = f(x).$$

The graph of an even function is symmetric with respect to the y-axis.

A function f is an **odd function** if for each x in the domain of f, $-x$ is also in the domain of f and

$$f(-x) = -f(x).$$

The graph of an odd function is symmetric with respect to the origin.

EXAMPLE 5   **Testing for Evenness and Oddness**

Determine whether each function is even, odd, or neither.

a. $g(x) = x^3 - 4x$ **b.** $f(x) = x^2 + 5$ **c.** $h(x) = 2x^3 + x^2$

Solution

a. $g(x)$ is an odd function because

$$\begin{aligned} g(-x) &= (-x)^3 - 4(-x) &&\text{Replace } x \text{ with } -x. \\ &= -x^3 + 4x &&\text{Simplify.} \\ &= -(x^3 - 4x) &&\text{Distributive property} \\ &= -g(x). \end{aligned}$$

b. $f(x)$ is an even function because

$$\begin{aligned} f(-x) &= (-x)^2 + 5 &&\text{Replace } x \text{ with } -x. \\ &= x^2 + 5 &&\text{Simplify.} \\ &= f(x). \end{aligned}$$

c. $\begin{aligned} h(-x) &= 2(-x)^3 + (-x)^2 &&\text{Replace } x \text{ with } -x. \\ &= -2x^3 + x^2 &&\text{Simplify.} \end{aligned}$

Now comparing $h(x) = 2x^3 + x^2$, $h(-x) = -2x^3 + x^2$, and $-h(x) = -2x^3 - x^2$, we note that $h(-x) \neq h(x)$ and $h(-x) \neq -h(x)$. Hence, $h(x)$ is neither even nor odd.

Practice Problem 5   Repeat Example 5 for the functions:

a. $g(x) = 3^4 - 5x^2$ **b.** $f(x) = 4x^5 + 2x^3$ **c.** $h(x) = 2x + 1$ ■■

3 Evaluate and graph piecewise functions.

Piecewise Functions

In the definition of some functions, different rules for assigning output values are used on different parts of the domain. Such functions are called **piecewise functions**. One example of a piecewise function is a line graph where data points are connected by straight-line segments.

The graph in Figure 1.66 represents the relative number of news headlines for popular Hollywood figures over a period of one year. Line graphs are powerful tools to visualize data and to study trends.

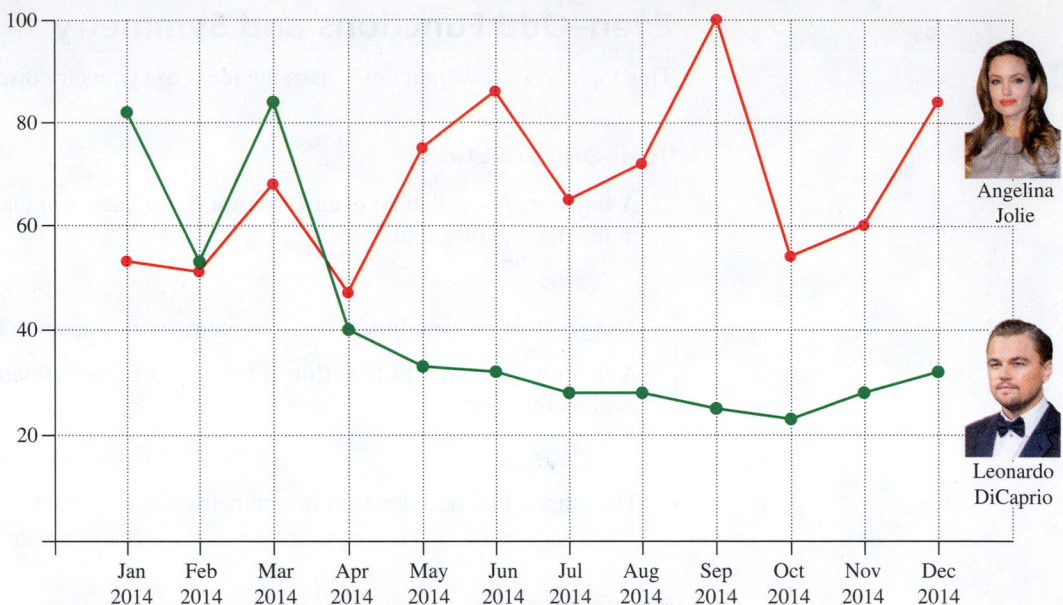

Figure 1.66 *Source:* Google trends.

In Example 8, we will see how such graphs are created using piecewise functions.

PROCEDURE
IN ACTION

EXAMPLE 6 **Evaluating a Piecewise Function**

OBJECTIVE

Evaluate $F(a)$ for a piecewise function F.

EXAMPLE

Let

$$F(x) = \begin{cases} x^2 & \text{if } x < 1 \\ 2x + 1 & \text{if } x \geq 1 \end{cases}$$

Find $F(0)$ and $F(2)$.

Step 1 Determine which line of the function's definition applies to the number a.

1. Let $a = 0$. Because $0 < 1$, use the first line, $F(x) = x^2$.
Let $a = 2$. Because $2 > 1$, use the second line, $F(x) = 2x + 1$.

Step 2 Evaluate $F(a)$ using the line chosen in Step 1.

2. $F(0) = (0)^2 = 0$
$F(2) = 2(2) + 1 = 5$

Practice Problem 6

Let $f(x) = \begin{cases} x^2 & \text{if } x \leq -1 \\ 2x & \text{if } x > -1 \end{cases}$. Find $f(-2)$ and $f(3)$.

EXAMPLE 7 **Finding and Evaluating a Piecewise Function**

In Peach County, Georgia, a section of the interstate highway has a speed limit of 55 miles per hour (mph). If you are caught speeding between 56 and 74 mph, your fine is $50 plus $3 for every mile per hour over 55 mph. For 75 mph and higher, your fine is $150 plus $5 for every mile per hour over 75 mph.

a. Find a piecewise function that gives your fine.

b. What is the fine for driving 60 mph?

c. What is the fine for driving 90 mph?

Solution

a. Let $f(x)$ be the piecewise function that represents your fine for speeding at x miles per hour. We express $f(x)$ as a piecewise function:

$$f(x) = \begin{cases} 50 + 3(x - 55), & 56 \le x < 75 \\ 150 + 5(x - 75), & x \ge 75 \end{cases}$$

b. The first line of the function means that if $56 \le x < 75$, your fine is

$$f(x) = 50 + 3(x - 55).$$

Suppose you are caught driving 60 mph. Then we have

$$
\begin{aligned}
f(x) &= 50 + 3(x - 55) && \text{Expression used for } 56 \le x < 75 \\
f(60) &= 50 + 3(60 - 55) && \text{Substitute 60 for } x. \\
&= 65 && \text{Simplify.}
\end{aligned}
$$

Your fine for speeding at 60 mph is $65.

c. The second line of the function means that if $x \ge 75$, your fine is $f(x) = 150 + 5(x - 75)$. Suppose you are caught driving 90 mph. Then we have

$$
\begin{aligned}
f(x) &= 150 + 5(x - 75) && \text{Expression used for } x \ge 75 \\
f(90) &= 150 + 5(90 - 75) && \text{Substitute 90 for } x. \\
&= 225 && \text{Simplify.}
\end{aligned}
$$

Your fine for speeding at 90 mph is $225.

Practice Problem 7 Repeat Example 7 if the speeding penalties are changed to $50 plus $4 for every mile per hour over 55 mph and $200 plus $5 for every mile per hour over 75 mph.

EXAMPLE 8   Writing a Piecewise Function from the Set of Points

TABLE 1.8

Construct the line graph from the data in Table 1.8 and express the function representing the line graph as a piecewise function.

Solution

Draw three points $(1, 1)$, $(3, 5)$ and $(5, 2)$ and connect them with straight lines (see Figure 1.67). The graph of f is made up of two parts:

Time	Value
1	1
3	5
5	2

a. A line segment passing through $(1, 1)$ and $(3, 5)$ over the interval $[1, 3]$. The slope of this line is

$$m = \frac{\Delta y}{\Delta x} = \frac{5 - 1}{3 - 1} = \frac{4}{2} = 2.$$

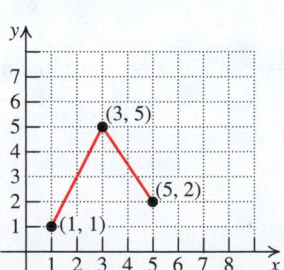

Figure 1.67

Then

$$
\begin{aligned}
y - y_1 &= m(x - x_1) && \text{Point–slope form} \\
y - y_1 &= 2(x - x_1) && \text{Replace } m \text{ with 2.} \\
y - 1 &= 2(x - 1) && \text{Point } (x_1, y_1) = (1, 1) \text{ is on the line;} \\
& && \text{replace both } x_1 \text{ and } y_1 \text{ with 1.} \\
y - 1 &= 2x - 2 && \text{Distribute.} \\
y &= 2x - 1 && \text{Solve for } y.
\end{aligned}
$$

So $f(x) = 2x - 1$ for $1 \le x \le 3$.

b. A line segment passing through $(3, 5)$ and $(5, 2)$ over the interval $[3, 5]$. The slope of this line is

$$m = \frac{\Delta y}{\Delta x} = \frac{2 - 5}{5 - 3} = \frac{-3}{2} = -\frac{3}{2}.$$

Then

$$y - y_1 = m(x - x_1) \qquad \text{Point–slope form}$$

$$y - y_1 = -\frac{3}{2}(x - x_1) \qquad \text{Replace } m \text{ with } -\frac{3}{2}.$$

$$y - 5 = -\frac{3}{2}(x - 3) \qquad \text{Point } (x_1, y_1) = (3, 5) \text{ is on the line;}$$
$$\text{replace } x_1 \text{ with 3 and } y_1 \text{ with 5.}$$

$$y - 5 = -\frac{3}{2}x + \frac{9}{2} \qquad \text{Distribute.}$$

$$y = -\frac{3}{2}x + \frac{19}{2} \qquad \text{Solve for } y.$$

TABLE 1.9

Time	Value
1	5
3	2
5	4

So $f(x) = -\frac{3}{2}x + \frac{19}{2}$ for $3 \le x \le 5$.

Observe that the two line segments that form the graph of this function are joined together without a gap, so we can attach the point $(3, 5)$ to either part. Combining (a) and (b), respectively we have

$$g(x) = \begin{cases} 2x - 1 & \text{if } 1 \le x \le 3 \\ -\frac{3}{2}x + \frac{19}{2} & \text{if } 3 < x \le 5 \end{cases}$$

Practice Problem 8 Repeat Example 8 for Table 1.9.

Graphing Piecewise Functions

Let's consider how to graph a piecewise function. The **absolute value function**, $f(x) = |x|$, can be expressed as a piecewise function by using the definition of absolute value:

$$f(x) = |x| = \begin{cases} -x & \text{if } x < 0 \\ x & \text{if } x \ge 0 \end{cases}$$

The first line in the function means that if $x < 0$, we use the equation $y = f(x) = -x$. So, if $x = -3$, then

$$y = f(-3) = -(-3) \qquad \text{Substitute } -3 \text{ for } x \text{ in } f(x) = -x.$$

$$= 3 \qquad \text{Simplify.}$$

Thus, $(-3, 3)$ is a point on the graph of $y = |x|$. However, if $x \ge 0$, we use the second line in the function, which is $y = f(x) = x$. So, if $x = 2$, then

$$y = f(2) = 2 \qquad \text{Substitute 2 for } x \text{ in } f(x) = x$$

So $(2, 2)$ is a point on the graph of $y = |x|$. The two pieces $y = -x$ and $y = x$ are linear functions. We graph the appropriate parts of these lines ($y = -x$ for $x < 0$ and $y = x$ for $x \ge 0$) to form the graph of $y = |x|$. See Figure 1.68.

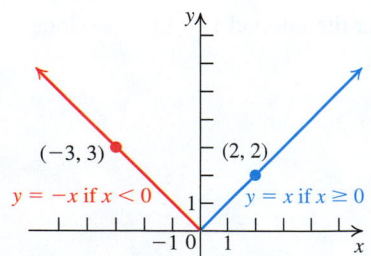

Figure 1.68 Graph of $y = |x|$.

EXAMPLE 9 **Graphing a Piecewise Function**

Let

$$F(x) = \begin{cases} -2x + 1 & \text{if } x < 1 \\ 3x + 1 & \text{if } x \ge 1 \end{cases}$$

Sketch the graph of $y = F(x)$.

Solution

In the definition of F, the formula changes at $x = 1$. We call such numbers the **transition points** of the formula. For the function F, the only transition point is 1. Generally, to graph a piecewise function, we graph the function separately over the open intervals determined by the transition points and then graph the function at the transition points themselves. For the function $y = F(x)$, we graph the equation $y = -2x + 1$ on the interval $(-\infty, 1)$. See Figure 1.69(a). Next, we graph the equation $y = 3x + 1$ on the interval $(1, \infty)$ and at the transition point 1, where $y = F(1) = 3(1) + 1 = 4$. See Figure 1.69(b).

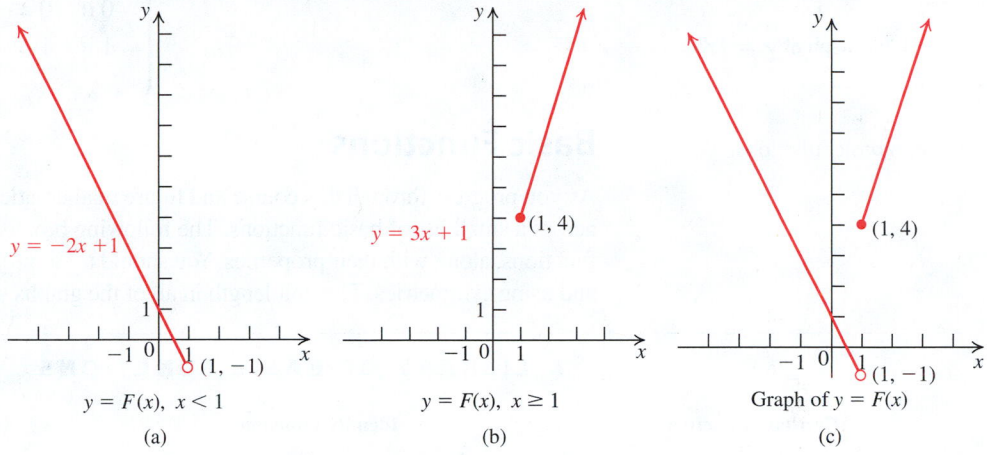

$y = F(x)$, $x < 1$
(a)

$y = F(x)$, $x \geq 1$
(b)

Graph of $y = F(x)$
(c)

Figure 1.69 A piecewise function.

Combining these portions, we obtain the graph of $y = F(x)$, shown in Figure 1.69(c). When we come to the end of the part of the graph we are working with, we draw

(i) a closed circle if that point is included.

(ii) an open circle if the point is excluded.

You may find it helpful to think of this procedure as following the graph of $y = -2x + 1$ when x is less than 1 and then jumping to the graph of $y = 3x + 1$ when x is equal to or greater than 1.

Practice Problem 9 Let $F(x) = \begin{cases} -3x & \text{if } x \leq -1 \\ 2x & \text{if } x > -1 \end{cases}$. Sketch the graph of $y = F(x)$. ■

Some piecewise functions are called **step functions**. Their graphs look like the steps of a staircase.

The **greatest integer function** is denoted by $[\![x]\!]$, or $\text{int}(x)$, where $[\![x]\!] = $ the greatest integer less than or equal to x. For example,

$$[\![2]\!] = 2, \quad [\![2.3]\!] = 2, \quad [\![2.7]\!] = 2, \quad [\![2.99]\!] = 2$$

because 2 is the greatest integer less than or equal to 2, 2.3, 2.7, and 2.99. Similarly,

$$[\![-2]\!] = -2, \quad [\![-1.9]\!] = -2, \quad [\![-1.1]\!] = -2, \quad [\![-1.001]\!] = -2$$

because -2 is the greatest integer less than or equal to -2, -1.9, -1.1, and -1.001. In general, if m is an integer such that $m \leq x < m + 1$, then $[\![x]\!] = m$. In other words, if x is between two consecutive integers m and $m + 1$, then $[\![x]\!]$ is assigned the smaller integer m.

EXAMPLE 10 Graphing a Step Function

Graph the greatest integer function $f(x) = [\![x]\!]$.

Solution

Choose a typical closed interval between two consecutive integers—say, the interval $[2, 3]$. We know that between 2 and 3, the greatest integer function's value is 2. In symbols, if $2 \leq x < 3$, then $[\![x]\!] = 2$. Similarly, if $1 \leq x < 2$, then $[\![x]\!] = 1$. Therefore, the

---

TECHNOLOGY CONNECTION

To graph the greatest integer function, enter $Y_1 = \text{int}(x)$. If you have your calculator set to graph in **connected** mode, you will get the following incorrect graph:

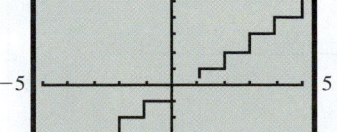

For the graph to appear correctly, change from the **connected** to the **dot** mode:

Note that the step from $0 \leq x \leq 1$ is obscured by the x-axis.

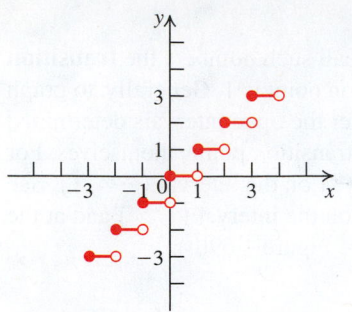

Figure 1.70 Graph of $y = [x]$.

values of $[x]$ are constant between each pair of consecutive integers and jump by one unit at each integer. The graph of $f(x) = [x]$ is shown in Figure 1.70.

Practice Problem 10 Find the values of $f(x) = [x]$ for $x = -3.4$ and $x = 4.7$. ◼

The greatest integer function f can be interpreted as a piecewise function:

$$f(x) = [x] = \begin{cases} \vdots \\ -2 \text{ if } -2 \le x < -1 \\ -1 \text{ if } -1 \le x < 0 \\ 0 \text{ if } 0 \le x < 1 \\ 1 \text{ if } 1 \le x < 2 \\ \vdots \end{cases}$$

4 Graph basic functions.

Basic Functions

As you progress through this course and future mathematics courses, you will repeatedly come across a small list of basic functions. The following box lists some of these common algebraic functions, along with their properties. You should try to produce these graphs by plotting points and using symmetries. The unit length in all of the graphs shown is the same on both axes.

A LIBRARY OF BASIC FUNCTIONS

Constant Function
$f(x) = c$

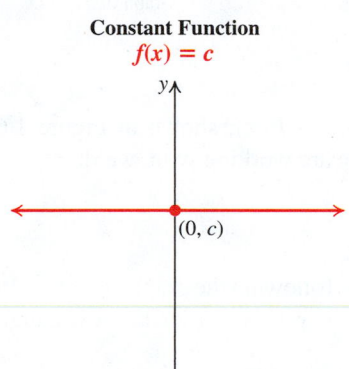

Domain: $(-\infty, \infty)$
Range: $\{c\}$
Constant on $(-\infty, \infty)$
Even function (y-axis symmetry)

Identity Function
$f(x) = x$

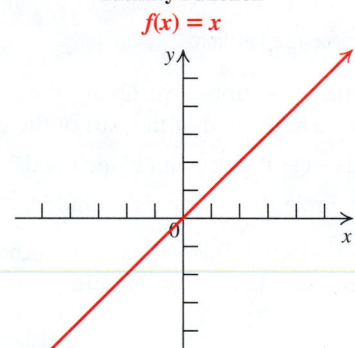

Domain: $(-\infty, \infty)$
Range: $(-\infty, \infty)$
Increasing on $(-\infty, \infty)$
Odd function (origin symmetry)

Squaring Function
$f(x) = x^2$

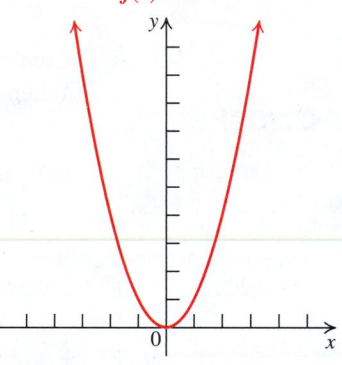

Domain: $(-\infty, \infty)$
Range: $[0, \infty)$
Decreasing on $(-\infty, 0)$
Increasing on $(0, \infty)$
Even function (y-axis symmetry)

Cubing Function
$f(x) = x^3$

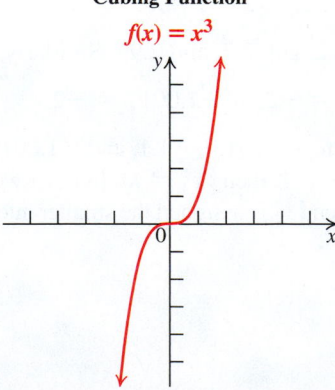

Domain: $(-\infty, \infty)$
Range: $(-\infty, \infty)$
Increasing on $(-\infty, \infty)$
Odd function (origin symmetry)

Absolute Value Function
$f(x) = |x|$

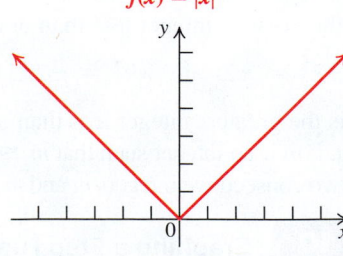

Domain: $(-\infty, \infty)$
Range: $[0, \infty)$
Decreasing on $(-\infty, 0)$
Increasing on $(0, \infty)$
Even function (y-axis symmetry)

Square Root Function
$f(x) = \sqrt{x} = x^{1/2}$

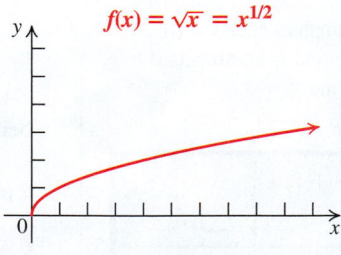

Domain: $[0, \infty)$
Range: $[0, \infty)$
Increasing on $(0, \infty)$
Neither even nor odd (no symmetry)

Cube Root Function

$$f(x) = \sqrt[3]{x} = x^{1/3}$$

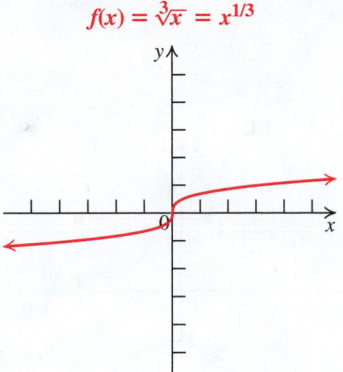

Domain: $(-\infty, \infty)$
Range: $(-\infty, \infty)$
Increasing on $(-\infty, \infty)$
Odd function (origin symmetry)

Reciprocal Function

$$f(x) = \frac{1}{x}$$

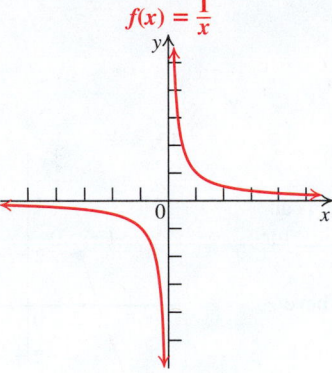

Domain: $(-\infty, 0) \cup (0, \infty)$
Range: $(-\infty, 0) \cup (0, \infty)$
Decreasing on $(-\infty, 0) \cup (0, \infty)$
Odd function (origin symmetry)

Reciprocal Square Function

$$f(x) = \frac{1}{x^2}$$

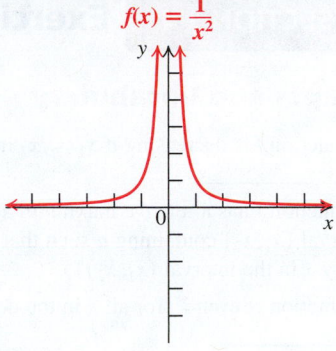

Domain: $(-\infty, 0) \cup (0, \infty)$
Range: $(0, \infty)$
Increasing on $(-\infty, 0)$
Decreasing on $(0, \infty)$
Even function (y-axis symmetry)

Rational Power Functions

$$f(x) = x^{\frac{3}{2}} = \left(x^{\frac{1}{2}}\right)^3$$

Domain: $[0, \infty)$
Range: $[0, \infty)$
Increasing on $(0, \infty)$
Neither even nor odd
(no symmetry)

$$f(x) = x^{\frac{2}{3}} = \left(x^{\frac{1}{3}}\right)^2$$

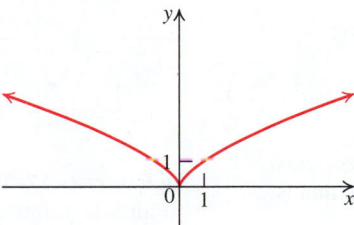

Domain: $(-\infty, \infty)$
Range: $[0, \infty)$
Decreasing on $(-\infty, 0)$
Increasing on $(0, \infty)$
Even function (y-axis symmetry)

Greatest Integer Function

$$f(x) = [\![x]\!]$$

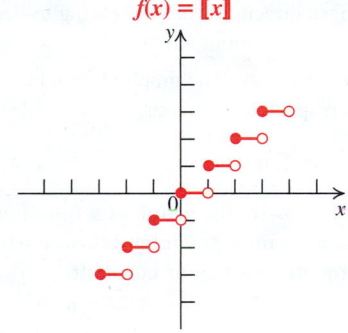

Domain: $(-\infty, \infty)$
Range: $\{\ldots -3, -2, -1, 0, 1\ 2, 3, \ldots\}$
Neither even nor odd (no symmetry)

Answers to Practice Problems

1. 15.524 m

2. Decreasing on $(0, 3)$ and $(12, 13)$ and $(15, 24)$, increasing on $(3, 12)$ and $(13, 15)$.

3. Relative maximum of 3640 at 12 and of 4070 at 15. Relative minimum of 40 at 3 and of 3490 at 13.

4. Relative minimum of 0 at 0. Relative maximum of 1 at 1.

5. a. even   **b.** odd   **c.** neither

6. $f(-2) = 4, f(3) = 6$

7. a $f(x) = \begin{cases} 50 + 4(x - 55), 56 \le x < 75 \\ 200 + 5(x - 75), x \ge 75 \end{cases}$ **b.** \$70   **c.** \$275

8.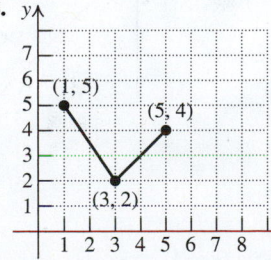

$$g(x) = \begin{cases} -\dfrac{3}{2}x + \dfrac{13}{2} & \text{if } 1 \le x \le 3 \\ x - 1 & \text{if } 3 < x \le 5 \end{cases}$$

9. $f(x) = -3x, x \le -1$
$f(x) = 2x, x > -1$

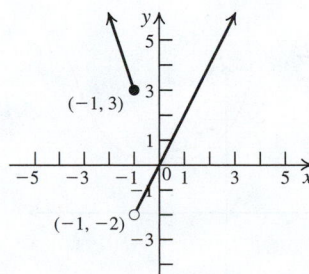

10. $[\![-3.4]\!] = -4; [\![4.7]\!] = 4$

SECTION 1.4 **Exercises**

Concepts and Vocabulary

1. A function f is decreasing if $x_1 < x_2$ implies that
   _____ .

2. A function f has a relative maximum at $x = a$ if there is an interval (x_1, x_2) containing a such that _____ for every x in the interval (x_1, x_2).

3. A function is even if, for all x in the domain of f, we have
   _____ .

4. A function that uses different rules for assigning output values on different parts of the domain is called a
   _____ .

5. **True or False.** Functions can increase, decrease, or remain constant on different intervals within their domains.

6. **True or False.** The average rate of change of an increasing function is positive.

7. **True or False.** At a point on a graph where a function changes direction from increasing to decreasing, the function has a relative minimum.

8. **True or False.** The graph of an odd function is symmetric with respect to the y-axis.

Building Skills

In Exercises 9–16, the graph of a function is given. For each function, determine the intervals over which the function is increasing, decreasing, or constant.

9.

10.

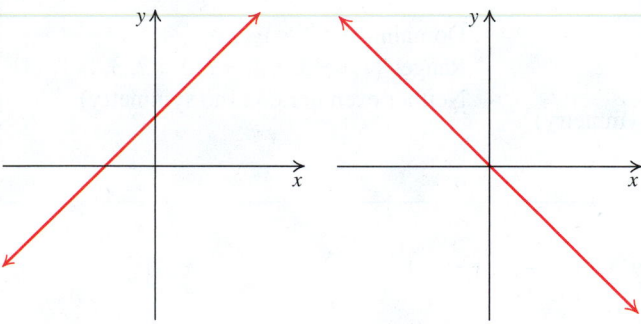

11.

12.

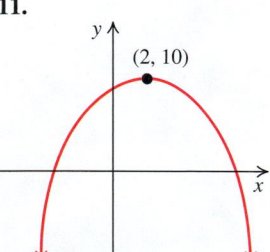

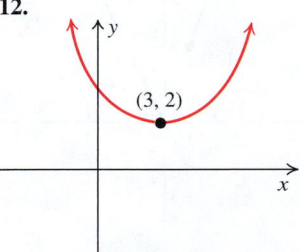

13.

14.

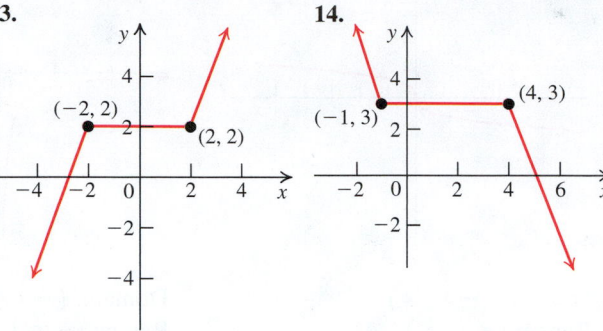

15.

16.

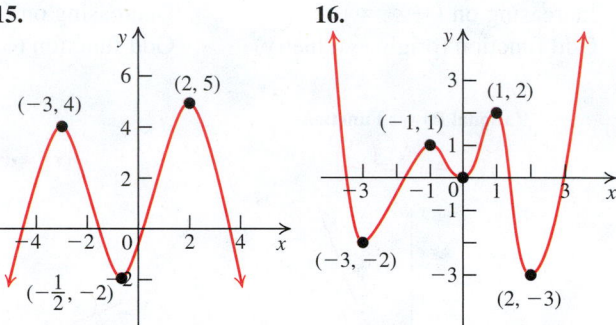

In Exercises 17–20, locate relative maximum and relative minimum points on the graph. State whether each relative extremum point is a turning point.

17. The graph of Exercise 9

18. The graph of Exercise 10

19. The graph of Exercise 11

20. The graph of Exercise 12

21. The graph of Exercise 13

22. The graph of Exercise 14

23. The graph of Exercise 15

24. The graph of Exercise 16

In Exercises 25–34, the graph of a function is given. State whether the function is odd, even, or neither.

25.

26.

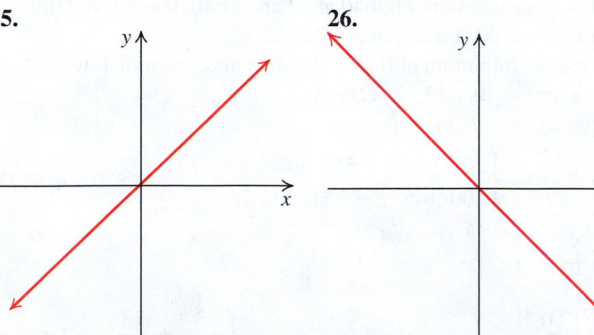

27.

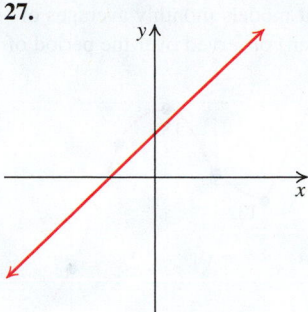

28.

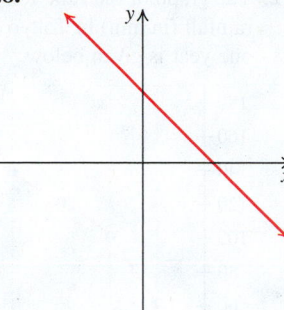

29.

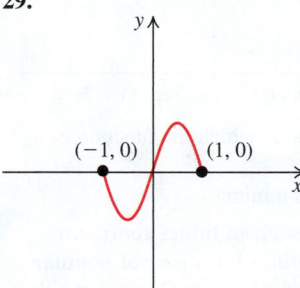

30.

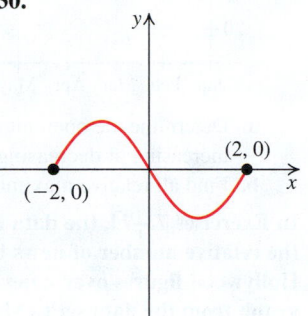

31.

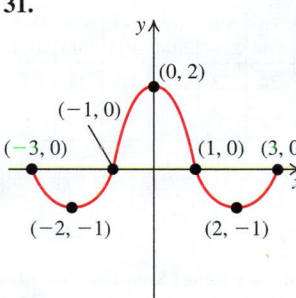

32.

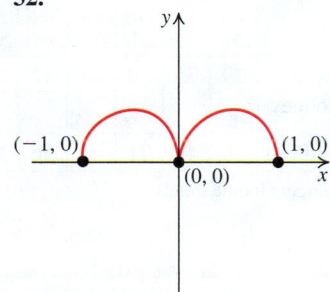

33.

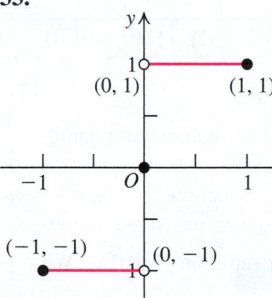

34.

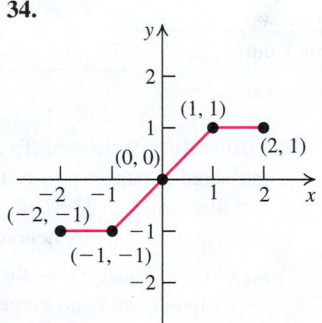

In Exercises 35–48, determine algebraically whether the given function is odd, even, or neither.

35. $f(x) = 2x^4 + 4$

36. $g(x) = 3x^4 - 5$

37. $f(x) = 5x^3 - 3x$

38. $g(x) = 2x^3 + 4x$

39. $f(x) = 2x + 4$

40. $g(x) = 3x + 7$

41. $f(x) = \dfrac{1}{x^2 + 4}$

42. $g(x) = \dfrac{x^2 + 2}{x^4 + 1}$

43. $f(x) = \dfrac{x^3}{x^2 + 1}$

44. $g(x) = \dfrac{x^4 + 3}{2x^3 - 3x}$

45. $f(x) = \dfrac{x}{x^5 - 3x^3}$

46. $g(x) = \dfrac{x^3 + 2x}{2x^5 - 3x}$

47. $f(x) = \dfrac{x^2 - 2x}{5x^4 + 7}$

48. $g(x) = \dfrac{3x^2 + 7}{x - 3}$

49. Let

$$f(x) = \begin{cases} x & \text{if } x \geq 2 \\ 2 & \text{if } x < 2 \end{cases}$$

  a. Find $f(1), f(2),$ and $f(3)$.
  b. Sketch the graph of $y = f(x)$.

50. Let

$$g(x) = \begin{cases} 2x & \text{if } x < 0 \\ x & \text{if } x \geq 0 \end{cases}$$

  a. Find $g(-1), g(0),$ and $g(1)$.
  b. Sketch the graph of $y = g(x)$.

51. Let

$$f(x) = \begin{cases} 1 & \text{if } x > 0 \\ -1 & \text{if } x < 0 \end{cases}$$

  a. Find $f(-15)$ and $f(12)$.
  b. Sketch the graph of $y = f(x)$.
  c. Find the domain and the range of f.

52. Let

$$g(x) = \begin{cases} 2x + 4 & \text{if } x > 1 \\ x + 2 & \text{if } x \leq 1 \end{cases}$$

  a. Find $g(-3), g(1),$ and $g(3)$.
  b. Sketch the graph of $y = g(x)$.
  c. Find the domain and the range of g.

In Exercises 53–62, sketch the graph of each piecewise function. From the graphs find the range of each function.

53. $f(x) = \begin{cases} x^2 & \text{if } x \geq 2 \\ 3x - 2 & \text{if } x < 2 \end{cases}$

54. $f(x) = \begin{cases} -2x & \text{if } x < 0 \\ x^2 & \text{if } x \geq 0 \end{cases}$

55. $g(x) = \begin{cases} |x| & \text{if } x < 1 \\ x^2 & \text{if } x \geq 1 \end{cases}$

56. $f(x) = \begin{cases} |x| & \text{if } -2 \leq x < 1 \\ \sqrt{x} & \text{if } x \geq 1 \end{cases}$

57. $g(x) = \begin{cases} \dfrac{1}{x} & \text{if } x < 0 \\ \sqrt{x} & \text{if } x \geq 0 \end{cases}$

58. $h(x) = \begin{cases} \dfrac{1}{x} & \text{if } x \geq 2 \\ x & \text{if } x < 2 \end{cases}$

59. $f(x) = \begin{cases} [\![x]\!] & \text{if } x < 1 \\ \sqrt[3]{x} & \text{if } x \geq 1 \end{cases}$

60. $f(x) = \begin{cases} \dfrac{1}{x} & \text{if } x < 0 \\ [\![x]\!] & \text{if } x \geq 0 \end{cases}$

61. $f(x) = \begin{cases} 2x + 3 & \text{if } x < -2 \\ x + 1 & \text{if } -2 \leq x < 1 \\ -x + 3 & \text{if } x \geq 1 \end{cases}$

62. $g(x) = \begin{cases} -2x + 1 & \text{if } x \leq -1 \\ 2x + 1 & \text{if } -1 < x < 2 \\ x + 2 & \text{if } x \geq 2 \end{cases}$

In Exercises 63–66, express the function representing the line graph constructed from the following data points as a piecewise function.

63.

Time	Value
1	2
2	5
5	1

64.

Time	Value
1	5
2	2
5	3

65.

Time	Value
1	1
4	3
6	5

66.

Time	Value
1	5
4	4
6	1

In Exercises 67–70, express the function representing the line graph as a piecewise function.

67.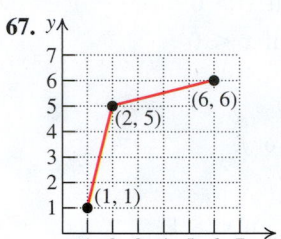

68.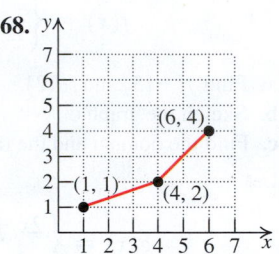

69.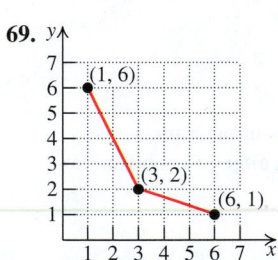

70.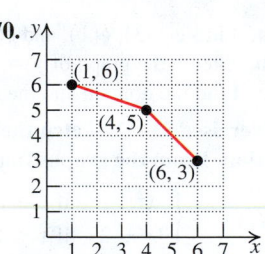

Applying the Concepts

71. The graph of the function f that models the number of flu cases (per 100,000) observed in urban areas in Pennsylvania in flu seasons between 2006 and 2014 is given below.

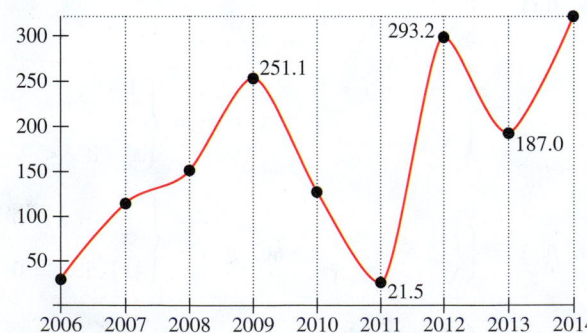

a. Determine the open intervals on which the function is increasing or decreasing.

b. Find all relative maxima and minima.

72. The graph of the function f that models monthly averages of rainfall (in mm) in Tokyo (Japan) observed over the period of one year is given below.

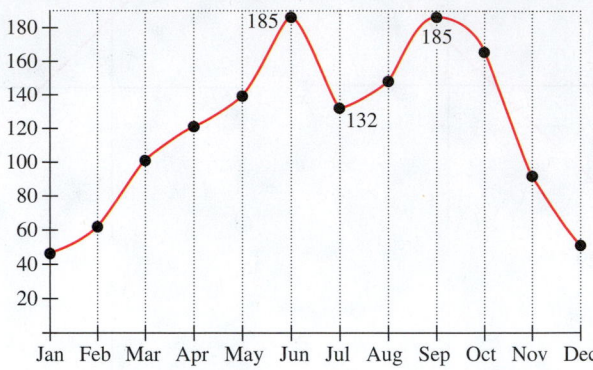

a. Determine the open intervals on which the function is increasing or decreasing.

b. Find all relative maxima and minima.

In Exercises 73–74, the data in the given tables represent the relative number of news headlines for a pair of popular Hollywood figures over a period of one year. Construct a line graph from the data set to visualize headline trends.

73.

	Jan 2014	Feb 2014	Mar 2014	Apr 2014	May 2014	Jun 2014	Jul 2014	Aug 2014	Sep 2014	Oct 2014	Nov 2014	Dec 2014
George Clooney	29	35	30	48	42	22	24	23	100	58	24	31
Brad Pitt	56	49	71	44	56	52	45	62	71	74	55	68

Source: Google trends

74.

	Jan 2014	Feb 2014	Mar 2014	Apr 2014	May 2014	Jun 2014	Jul 2014	Aug 2014	Sep 2014	Oct 2014	Nov 2014	Dec 2014
Scarlett Johansson	56	51	68	95	54	42	58	68	60	44	44	51
Mila Kunis	53	53	70	47	74	86	67	72	100	56	61	88

Source: Google trends

75. Converting volume. To convert the volume of a liquid measured in ounces to a volume measured in liters, we use the fact that 1 liter $\approx$ 33.81 ounces. Let x denote the volume measured in ounces and y denote the volume measured in liters.

a. Write an equation for the linear function $y = f(x)$. What are the domain and range of f?

b. Compute $f(3)$. What does it mean?

c. How many liters of liquid are in a typical soda can containing 12 ounces of liquid?

76. Boiling point and elevation. The boiling point B of water (in degrees Fahrenheit) at elevation h (in thousands of feet) above sea level is given by the linear function $B(h) = -1.8h + 212$.

a. Find and interpret the intercepts of the function $y = B(h)$.

b. Find the domain of this function.

c. Find the elevation at which water boils at 98.6°F. Why is this height dangerous to humans?

77. Pressure at sea depth. The pressure P in atmospheres (atm) at a depth d feet is given by the linear function

$$P(d) = \frac{1}{33}d + 1.$$

a. Find and interpret the intercepts of $y = P(d)$.
b. Find $P(0)$, $P(10)$, $P(33)$, and $P(100)$.
c. Find the depth at which the pressure is 5 atmospheres.

78. Speed of sound. The speed V of sound (in feet per second) in air at temperature T (in degrees Fahrenheit) is given by the linear function $V(T) = 1055 + 1.1T$.
a. Find the speed of sound at 90°F.
b. Find the temperature at which the speed of sound is 1100 feet per second.

79. Manufacturer's cost. A manufacturer of printers has a total cost per day consisting of a fixed overhead of $6000 plus product costs of $50 per printer.
a. Express the total cost C as a function of the number x of printers produced.
b. Draw the graph of $y = C(x)$. Interpret the y-intercept.
c. How many printers were manufactured on a day when the total cost was $11,500?

80. Supply function. When the price p of a commodity is $10 per unit, 750 units are demanded. For every $1 increase in the unit price, the supply q increases by 100 units.
a. Write the linear function $q = f(p)$.
b. Find the supply when the price is $15 per unit.
c. What is the price at which 1750 units can be supplied?

81. Apartment rental. Suppose a two-bedroom apartment in a neighborhood near your school rents for $900 per month. If you move in after the first of the month, the rent is prorated; that is, it is reduced linearly.
a. Suppose you move into the apartment x days after the first of the month. (Assume that the month has 30 days.) Express the rent R as a function of x.
b. Compute the rent if you move in six days after the first of the month.
c. When did you move into the apartment if the landlord charged you a rent of $600?

82. College admissions. The average SAT scores (mathematics and critical reading) for incoming students at Central State College (CSC) have been rising linearly in recent years. In 2010, the average SAT score was 1120; in 2012, it was 1150.
a. Express the average SAT score at CSC as a function of time.
b. If the trend continues, what will be the average SAT score of incoming students at CSC in 2018?
c. If the trend continues, when will the average SAT score at CSC be 1300?

83. Breathing capacity. Suppose the average maximum breathing capacity for humans drops linearly from 100% at age 20 to 40% at age 80. What age corresponds to 50% capacity?

84. Drug dosage for children. If a is the adult dosage of a medicine and t is the child's age, then the *Friend's rule* for the child's dosage y is given by the formula $y = \frac{2}{25}ta$.

a. Suppose the adult dosage is 60 milligrams. Find the dosage for a five-year-old child.
b. How old would a child have to be in order to be prescribed an adult dosage?

85. Air pollution. In Ballerenia, the average number y of deaths per month was observed to be linearly related to the concentration x of sulfur dioxide in the air. Suppose there are 30 deaths when $x = 150$ milligrams per cubic meter and 50 deaths when $x = 420$ milligrams per cubic meter.
a. Write y as a function of x.
b. Find the number of deaths when $x = 350$ milligrams per cubic meter.
c. If the number of deaths per month is 45, what is the concentration of sulfur dioxide in the air?

86. Child shoe sizes. Children's shoe sizes start with size 0, having an insole length L of $3\frac{11}{12}$ inches, and each full size being $\frac{1}{3}$ inch longer.
a. Write the equation $y = L(S)$, where S is the shoe size.
b. Find the length of the insole of a child's shoe of size 4.
c. What size (to the nearest half-size) shoe will fit a child whose insole length is 6.1 inches?

87. State income tax. Suppose a state's income tax code states that the tax liability T on x dollars of taxable income is as follows:

$$T(x) = \begin{cases} 0.04x & \text{if } 0 \le x < 20{,}000 \\ 800 + 0.06(x - 20{,}000) & \text{if } x \ge 20{,}000 \end{cases}$$

a. Graph the function $y = T(x)$.
b. Find the tax liability on each taxable income.
  (i) $12,000
  (ii) $20,000
  (iii) $50,000
c. Find your taxable income if you had each tax liability.
  (i) $600
  (ii) $1400
  (iii) $2300

88. Federal income tax. The tax table for the 2015 U.S. income tax for a single taxpayer is as follows:

If taxable income is over	But not over	The tax is	
$0	$9,225	10%	
9,225	37,450	$922.50 + 15% of the amount over—	9,225
37,450	90,750	5,156.25 + 25% of the amount over—	37,450
90,750	189,300	18,481.25 + 28% of the amount over—	90,750
189,300	411,500	46,075.25 + 33% of the amount over—	189,300
411,500	413,200	119,401.25 + 35% of the amount over—	411,500
413,200	No Limit	119,996.25 + 39.6% of the amount over—	413,200

Source: Internal Revenue Service.

a. Express the information given in the table as a six-part piecewise-defined function f, using x as a variable representing taxable income.

b. Find the tax liability on each taxable income. (Round your answer to the nearest dollar.)

    (i) $35,000

    (ii) $100,000

    (iii) $500,000

c. Find your taxable income if you had each tax liability. (Round your answer to the nearest dollar.)

    (i) $3,500

    (ii) $12,700

    (iii) $35,000

Beyond the Basics

In Exercises 89–90, find the value of a such that the graph of the piecewise defined function f does not have any gaps (the two formulas that define f agree at the transition point).

89. $f(x) = \begin{cases} 2x - 1 & \text{if } 1 \le x \le 3 \\ a - 3x & \text{if } 3 < x \le 5. \end{cases}$

90. $f(x) = \begin{cases} 1 - x & \text{if } 1 \le x \le 3 \\ ax + 3 & \text{if } 3 < x \le 5. \end{cases}$

In Exercises 91 and 92, a function is given. For each function,

a. find the domain and range.

b. find the intervals over which the function is increasing, decreasing, or constant.

c. state whether the function is odd, even, or neither.

91. $f(x) = x - [\![x]\!]$ **92.** $f(x) = \dfrac{1}{[\![x]\!]}$

93. The Windchill Index (WCI). Suppose the outside air temperature is T degrees Fahrenheit and the wind speed is v miles per hour. A formula for the WCI based on observations is as follows:

$$WCI = \begin{cases} T, & 0 \le v \le 4 \\ 91.4 + (91.4 - T)(0.0203v - 0.304\sqrt{v} - 0.474), & 4 < v < 45 \\ 1.6T - 55, & v \ge 45 \end{cases}$$

a. Find the WCI to the nearest degree if the outside air temperature is 40°F and

    (i) $v = 2$ miles per hour.

    (ii) $v = 16$ miles per hour.

    (iii) $v = 50$ miles per hour.

b. Find the air temperature to the nearest degree if

    (i) the WCI is −58°F and $v = 36$ miles per hour.

    (ii) the WCI is −10°F and $v = 49$ miles per hour.

94. Postage-rate function. The U.S. Postal Service uses the function $f(x) = -[\![-x]\!]$, called the **ceiling integer function**, to determine postal charges. For instance, if you have an envelope that weighs 3.01 ounces, you will be charged the rate for $f(3.01) = -[\![-3.01]\!] = -(-4) = 4$ ounces. In 2012, it cost 45¢ for the first ounce (or a fraction of it) and

20¢ for each additional ounce (or a fraction of it) to mail a first-class letter in the United States. (*Source*: www.stamps.com/usps/postage-rate-increase.)

a. Express the cost C (in cents) of mailing a letter weighing x ounces.

b. How much will it cost to mail a 2.3 oz first-class letter within the United States via USPS?

c. Graph the function $y = C(x)$ for $0 \le x \le 6$.

95. Parking cost. The cost C of parking a car at the metropolitan airport is $4 for the first hour and $2 for each additional hour or fraction thereof. Write the cost function $C(x)$, where x is the number of parking hours, in terms of the greatest integer function.

96. Car rentals. The weekly cost of renting a compact car from U-Rent is $150. There is no charge for driving the first 100 miles, but there is a 20¢ charge for each mile (or fraction thereof) driven over 100 miles.

a. Express the cost C of renting a car for a week and driving it for x miles.

b. Sketch the graph of $y = C(x)$.

c. How many miles were driven if the weekly rental cost was $190?

Critical Thinking / Discussion / Writing

In Exercises 97–98, select the graph that best describes the situation.

97. An airplane flying from Tampa, Florida, to Atlanta, Georgia.

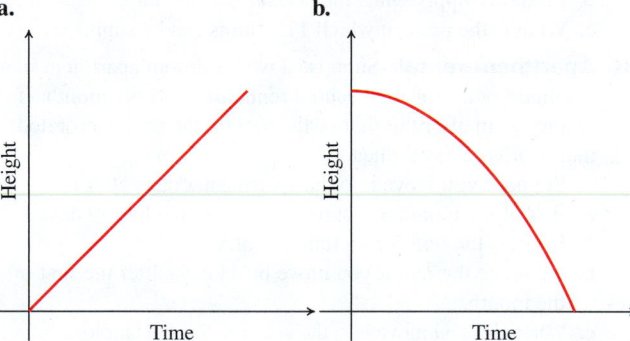

a.   b.

c.   d.

98. Cooling a jar of hot water by dumping a lot of ice.

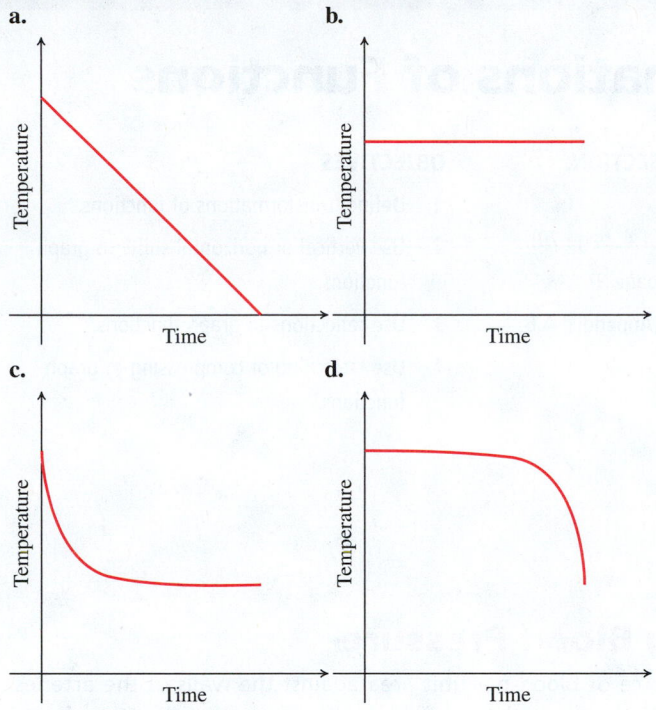

99. Let the domain of a function f be $(-\infty, \infty)$ and let f be increasing on the interval $(0, 1)$ and decreasing on the interval $(1, \infty)$. What information can we deduce about the behavior of the function f on interval $(-\infty, 0)$ if
a. we assume additionally that f is even?
b. we assume additionally that f is odd?

100. Let the domain of a function f be $(-\infty, \infty)$ and let f have a relative maximum at $x = 1$ and relative minimum at $x = 3$. What information can we deduce about the relative maxima and minima of the function f on the interval $(-\infty, 0)$ if
a. we assume additionally that f is even?
b. we assume additionally that f is odd?

101. Using the absolute value function $|x|$, construct the function that would output the same number if the input number is positive and zero if the input number is negative.

102. Using the absolute value function $|x|$, construct the function that would output the same number if the input number is negative and zero if the input number is positive.

103. Using the greatest integer function $[\![x]\!]$, construct the function that would round a positive input number to the nearest integer (the numbers that end with 0.5 are rounded up).

104. Using the greatest integer function $[\![x]\!]$, construct the function that can detect whether a given number is an integer. That is, construct the function that outputs one if the input number is an integer and zero otherwise.

Getting Ready for the Next Section

For Exercises 105–108, recall that the graph of a function f is the graph of the set of ordered pairs $(x, f(x))$.

105. If we add 3 to each y-coordinate of the graph of f, we will obtain the graph of $y =$ _____ .

106. If we subtract 2 from each x-coordinate of the graph of f, we will obtain the graph of $y =$ _____ .

107. If we replace each x-coordinate with its opposite in the graph of f, we will obtain the graph of $y =$ _____ .

108. If we replace each y-coordinate with its opposite in the graph of f, we will obtain the graph of $y =$ _____ .

In Exercises 109–112, sketch the graphs of f and g on the same coordinate axes.

109. $f(x) = x^2$ and $g(x) = x^2 + 1$

110. $f(x) = |x|$ and $g(x) = |x + 2|$

111. $f(x) = \sqrt{x}$ and $g(x) = \sqrt{-x}$

112. $f(x) = x^2$ and $g(x) = -x^2$

Transformations of Functions

BEFORE STARTING THIS SECTION, REVIEW

1 Basic functions (Section 1.4, page 70)

2 Symmetry (Section 1.1, page 9)

3 Completing the square (Appendix A.6, page 957)

OBJECTIVES

1 Define transformations of functions.

2 Use vertical or horizontal shifts to graph functions.

3 Use reflections to graph functions.

4 Use stretching or compressing to graph functions.

◆ Measuring Blood Pressure

Blood pressure is the force of blood per unit area against the walls of the arteries. Blood pressure is recorded as two numbers: the systolic pressure (as the heart beats) and the diastolic pressure (as the heart relaxes between beats). If your blood pressure is "120 over 80," it is rising to a maximum of 120 millimeters of mercury as the heart beats and is falling to a minimum of 80 millimeters of mercury as the heart relaxes.

The most precise way to measure blood pressure is to place a small glass tube in an artery and let the blood flow out and up as high as the heart can pump it. That was the way Stephen Hales (1677–1761), the first investigator of blood pressure, learned that a horse's heart could pump blood 8 feet, 3 inches, up a tall tube. Such a test, however, consumed a great deal of blood. Jean Louis Marie Poiseuille (1797–1869) greatly improved the process by using a mercury-filled manometer, which allowed a smaller, narrower tube.

In Example 10, we investigate Poiseuille's Law for arterial blood flow.

1 Define transformations of functions.

Transformations

We will investigate how the graph of a function $y = f(x)$ changes when we apply basic (linear) algebraic operations to the variables x and y. The graph of the resulting function is called a **transformation** of the graph of f. Using the library of functions we explored in the previous section as a starting point, we will be able to significantly expand the variety of functions that can be accurately graphed without repeatedly plotting points. For example, the graph of a general quadratic equation $y = ax^2 + bx + c$ can always be viewed as a transformation of the graph of the basic function $y = x^2$.

2 Use vertical or horizontal shifts to graph functions.

Vertical and Horizontal Shifts

A transformation that changes only the position of a graph but not its shape is called a **rigid transformation**. We first consider rigid transformations involving **vertical shifts** and **horizontal shifts** of a graph.

EXAMPLE 1 **Graphing Vertical Shifts**

Let $f(x) = |x|$, $g(x) = |x| + 2$, and $h(x) = |x| - 3$. Sketch the graphs of these functions on the same coordinate plane. Describe how the graphs of g and h relate to the graph of f.

Solution

Make a table of values and graph the equations $y = f(x)$, $y = g(x)$, and $y = h(x)$.

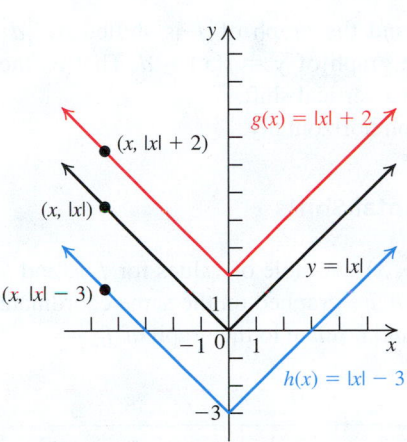

Figure 1.71 Vertical shifts of $y = |x|$.

TABLE 1.10

x	$y = |x|$	$g(x) = |x| + 2$	$h(x) = |x| - 3$
-5	5	7	2
-3	3	5	0
-1	1	3	-2
0	0	2	-3
1	1	3	-2
3	3	5	0
5	5	7	2

Notice that in Table 1.10 and Figure 1.71, each point $(x, |x|)$ on the graph of f has a corresponding point $(x, |x| + 2)$ on the graph of g, and $(x, |x| - 3)$ on the graph of h. So the graph of $g(x) = |x| + 2$ is the graph of $y = |x|$ shifted two units up, and the graph of $y = |x| - 3$ is the graph of $y = |x|$ shifted three units down.

Practice Problem 1 Let

$$f(x) = x^3, g(x) = x^3 + 1, \text{ and } h(x) = x^3 - 2.$$

Sketch the graphs of all three functions on the same coordinate plane. Describe how the graphs of g and h relate to the graph of f.

Example 1 illustrates the concept of the vertical (up or down) shift of a graph.

TECHNOLOGY 📈
CONNECTION

Many graphing calculators use abs(x) for $|x|$. The graphs of $y = |x|$, $y = |x| - 3$, and $y = |x| + 2$ illustrate that these graphs are identical in shape. They differ only in vertical placement.

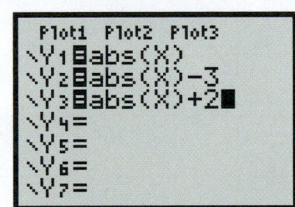

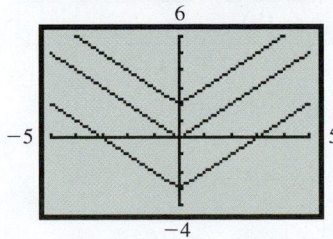

VERTICAL SHIFT

Let $d > 0$. The graph of $g(x) = f(x) + d$ is the graph of $y = f(x)$ shifted d units *up*, and the graph of $h(x) = f(x) - d$ is the graph of $y = f(x)$ shifted d units *down*. See Figure 1.72. If (x, y) is a point on the graph of $y = f(x)$, then the corresponding point $(x, y + d)$ is on the graph of $y = f(x) + d$ and the point $(x, y - d)$ is on the graph of $y = f(x) - d$.

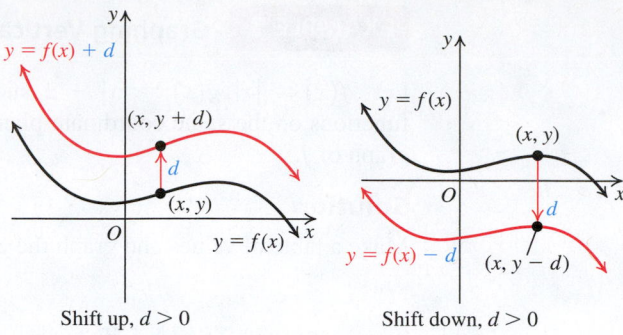

Shift up, $d > 0$ Shift down, $d > 0$

Figure 1.72 Vertical shifts.

So, if (x, y) is a point on the graph of $y = f(x)$ and the graph of f is shifted by $|d|$ units vertically, then the point $(x, y + d)$ is on the graph of $y = f(x) + d$. That is, the x-coordinate of a point is unchanged when we apply a vertical shift.

Next, we consider the operation that shifts a graph horizontally.

EXAMPLE 2 **Writing Functions for Horizontal Shifts**

Let $f(x) = x^2$, $g(x) = (x - 2)^2$, and $h(x) = (x + 3)^2$. A table of values for f, g, and h is given in Table 1.11. The three functions f, g, and h are graphed on the same coordinate plane in Figure 1.73. Describe how the graphs of g and h relate to the graph of f.

TABLE 1.11

x	$y = x^2$	$y = (x - 2)^2$	x	$y = x^2$	$y = (x + 3)^2$
-4	**16**	36	-4	16	1
-3	**9**	25	-3	9	0
-2	**4**	16	-2	4	1
-1	**1**	9	-1	1	4
0	**0**	4	0	**0**	9
1	**1**	1	1	**1**	16
2	4	0	2	**4**	25
3	9	1	3	**9**	36
4	16	4	4	**16**	49

(a)   (b)

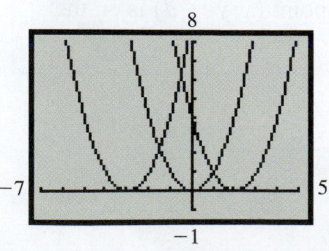
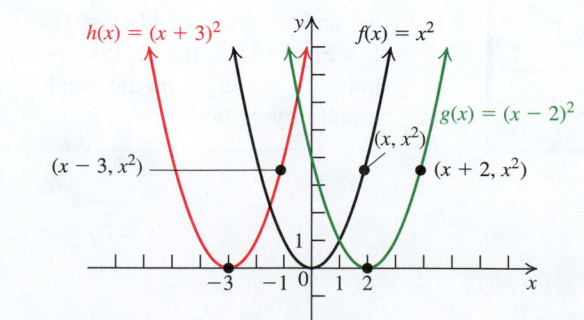

Figure 1.73 Horizontal shifts.

Solution

First, notice that all three functions are squaring functions.

a. Each point (x, x^2) on the graph of f has a corresponding point $(x + 2, x^2)$ on the graph of g because $g(x + 2) = (x + 2 - 2)^2 = x^2$. So the graph of $g(x) = (x - 2)^2$ is just the graph of $f(x) = x^2$ shifted two units to the *right*. Noticing that $f(0) = 0$ and $g(2) = 0$ will help you remember which way to shift the graph. Table 1.11(a) and Figure 1.73 illustrate these ideas.

b. Each point (x, x^2) on the graph of f has a corresponding point $(x - 3, x^2)$ on the graph of h because $h(x - 3) = (x - 3 + 3)^2 = x^2$. So the graph of $h(x) = (x + 3)^2$ is just the graph of $f(x) = x^2$ shifted three units to the *left*. Noticing that $f(0) = 0$ and $h(-3) = 0$ will help you remember which way to shift the graph. Table 1.11(b) and Figure 1.73 confirm these considerations.

Practice Problem 2 Let

$$f(x) = x^3, g(x) = (x - 1)^3, \text{ and } h(x) = (x + 2)^3.$$

Sketch the graphs of all three functions on the same coordinate plane. Describe how the graphs of g and h relate to the graph of f.

HORIZONTAL SHIFTS

Let $c > 0$. The graph of $y = f(x - c)$ is the graph of $y = f(x)$ shifted c units to the right. The graph of $y = f(x + c)$ is the graph of $y = f(x)$ shifted c units to the left. See Figure 1.74. If (x, y) is a point on the graph of f, then the corresponding point $(x + c, y)$ is on the graph of $y = f(x - c)$, and the point $(x - c, y)$ is on the graph of $y = f(x + c)$.

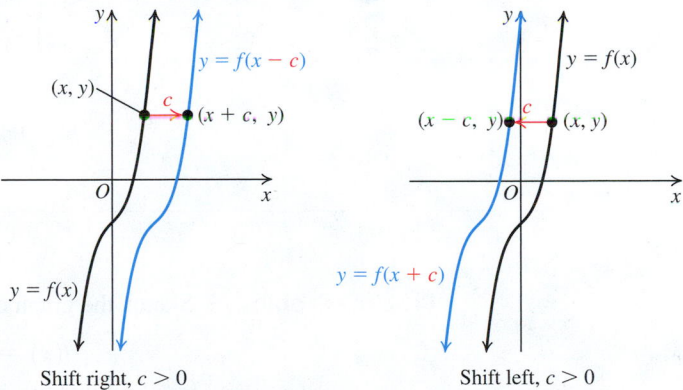

Shift right, $c > 0$ Shift left, $c > 0$

Figure 1.74 Horizontal shifts.

Replacing x with $x - c$ in the equation $y = f(x)$ results in a function whose graph is the graph of f shifted c units to the right ($c > 0$). Similarly, replacing x with $x + c$ in the equation $y = f(x)$ results in a function whose graph is the graph of f shifted c units to the left ($c > 0$).

PROCEDURE
IN ACTION

EXAMPLE 3 **Graphing Combined Vertical and Horizontal Shifts**

OBJECTIVE

Sketch the graph of $g(x) = f(x \pm c) \pm d$, *where f is a function whose graph is known.*

Step 1 Identify and graph the known function f.

Step 2 Identify the constants d and c.

Step 3 Let $c > 0$;
(i) graph $y = f(x - c)$ by shifting the graph of f horizontally c units to the right
(ii) graph $y = f(x + c)$ by shifting the graph of f horizontally c units to the left.

Step 4 Let $d > 0$;
(i) graph $y = f(x \pm c) + d$ by shifting the graph of $y = f(x \pm c)$ vertically up d units.
(ii) graph $y = f(x \pm c) - d$ by shifting the graph of $y = f(x \pm c)$ vertically down d units.

EXAMPLE

Sketch the graph of $g(x) = \sqrt{x + 2} - 3$.

1. Choose $f(x) = \sqrt{x}$. Domain: $[0, \infty)$; range: $[0, \infty)$
The graph of $y = \sqrt{x}$ is shown in Step 4.

2. $g(x) = \sqrt{x + 2} - 3$, so $c = 2$ and $d = 3$.

3. Since $c = 2 > 0$, the graph of $y = \sqrt{x + 2}$ is the graph of f shifted horizontally two units to the left. See the blue graph in Step 4.

4. Graph $y = \sqrt{x + 2} - 3$ by shifting the graph of $y = \sqrt{x + 2}$ three units down. See the red graph. Domain: $[-2, \infty)$; range: $[-3, \infty)$.

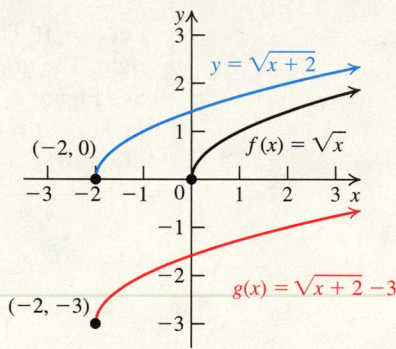

Horizontal and vertical shifts

Practice Problem 3 Sketch the graph of

$$f(x) = \sqrt{x - 2} + 3.$$

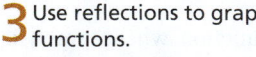

3 Use reflections to graph functions.

Reflections

We now consider rigid transformations that reflect a graph about a coordinate axis.

Comparing the Graphs of $y = f(x)$ and $y = -f(x)$ Consider the graph of $f(x) = x^2$, shown in Figure 1.75. Table 1.12 gives some values for $f(x)$ and $g(x) = -f(x)$. Note that the y-coordinate of each point in the graph of $g(x) = -f(x)$ is the opposite of the y-coordinate of the corresponding point on the graph of $f(x)$. So the graph of $y = -x^2$ is the reflection of the graph of $y = x^2$ about the x-axis. This means that the points (x, x^2) and $(x, -x^2)$ are the same distance from, but on opposite sides of, the x-axis. See Figure 1.75.

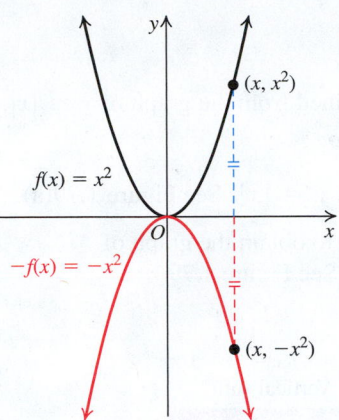

Figure 1.75 Reflection about the x-axis.

TABLE 1.12

x	$f(x) = x^2$	$g(x) = -f(x) = -x^2$
-3	9	-9
-2	4	-4
-1	1	-1
0	0	0
1	1	-1
2	4	-4
3	9	-9

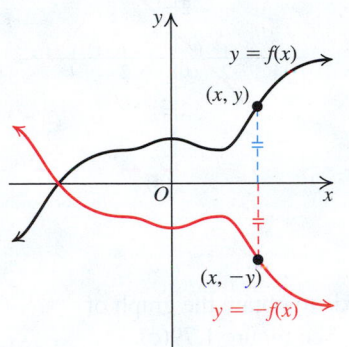

Figure 1.76 Reflection about the x-axis.

> ### REFLECTION ABOUT THE x-AXIS
>
> The graph of $g(x) = -f(x)$ is a reflection of the graph of $y = f(x)$ about the x-axis. If a point (x, y) is on the graph of f, then the point $(x, -y)$ is on the graph of g. See Figure 1.76.

Comparing the Graphs of $y = f(x)$ and $y = f(-x)$ To compare the graphs of $f(x)$ and $g(x) = f(-x)$, consider the graph of $f(x) = \sqrt{x}$. Then $f(-x) = \sqrt{-x}$. See Figure 1.77. Table 1.12 gives some values for $f(x)$ and $g(x) = f(-x)$. The domain of f is $[0, \infty)$, and the domain of g is $(-\infty, 0]$. Each point (x, y) on the graph of f has a corresponding point $(-x, y)$ on the graph of g. So the graph of $y = f(-x)$ is the reflection of the graph of $y = f(x)$ about the y-axis. This means that the points $(x, \sqrt{x})$ and $(-x, \sqrt{x})$ are the same distance from, but on opposite sides of, the y-axis. See Figure 1.77.

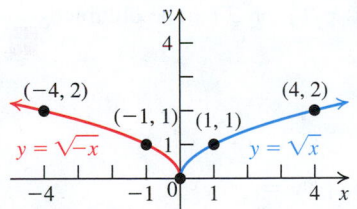

Figure 1.77 Graphing $y = \sqrt{-x}$.

TABLE 1.12

x	$f(x) = \sqrt{x}$	$-x$	$g(x) = \sqrt{-x}$
-4	Undefined	4	2
-1	Undefined	1	1
0	0	0	0
1	1	-1	Undefined
4	2	-4	Undefined

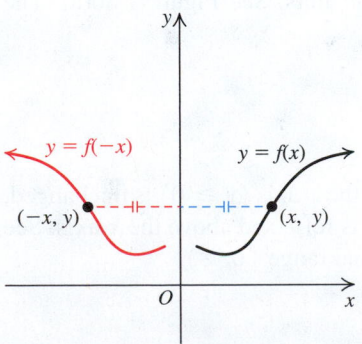

Figure 1.78 Reflection about the y-axis.

> ### REFLECTION ABOUT THE y-AXIS
>
> The graph of $g(x) = f(-x)$ is a reflection of the graph of $y = f(x)$ about the y-axis. If a point (x, y) is on the graph of f, then the point $(-x, y)$ is on the graph of g. See Figure 1.78.

EXAMPLE 4 **Combining Transformations**

Explain how the graph of $y = -|x - 2| + 3$ can be obtained from the graph of $y = |x|$.

Solution

Start with the graph of $y = |x|$. Follow the point $(0, 0)$ on $y = |x|$. See Figure 1.79(a).

Step 1 Shift the graph of $y = |x|$ two units to the right to obtain the graph of $y = |x - 2|$. The point $(0, 0)$ moves to $(2, 0)$. See Figure 1.79(b).

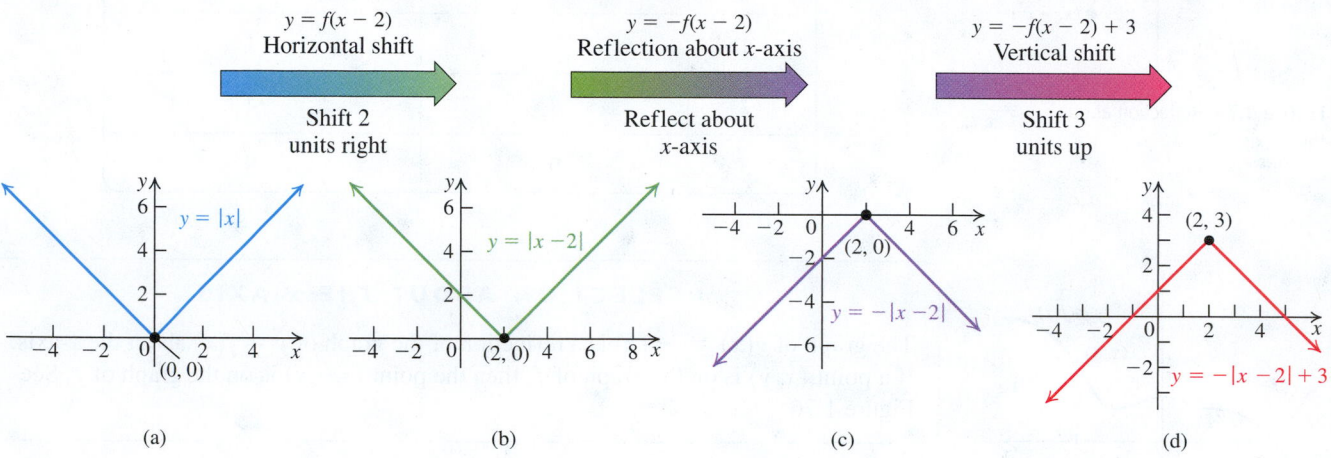

Figure 1.79 Transformations of $y = |x|$.

Step 2 Reflect the graph of $y = |x - 2|$ about the x-axis to obtain the graph of $y = -|x - 2|$. The point $(2, 0)$ remains $(2, 0)$. See Figure 1.79(c).

Step 3 Finally, shift the graph of $y = -|x - 2|$ three units up to obtain the graph of $y = -|x - 2| + 3$. The point $(2, 0)$ moves to $(2, 3)$. See Figure 1.79(d).

Practice Problem 4 Explain how the graph of $y = -(x - 1)^2 + 2$ can be obtained from the graph of $y = x^2$.

EXAMPLE 5 **Graphing $y = |f(x)|$**

Use the graph of $f(x) = (x + 1)^2 - 4$ to sketch the graph of $y = |f(x)|$.

Solution

The graph of $y = (x + 1)^2 - 4$ is obtained by shifting the graph of $y = x^2$ left by one unit and then shifting the resulting graph down by four units. See Figure 1.80(a). The function f has domain $(-\infty, \infty)$, and range $[-4, \infty)$.

We know that

$$|y| = \begin{cases} y & \text{if } y \geq 0 \\ -y & \text{if } y < 0. \end{cases}$$

This means that the portion of the graph on or above the x-axis ($y \geq 0$) is unchanged, while the portion of the graph below the x-axis ($y < 0$) is reflected above the x-axis. See Figure 1.80(b). The function $|f|$ has domain $(-\infty, \infty)$ and range $[0, \infty)$.

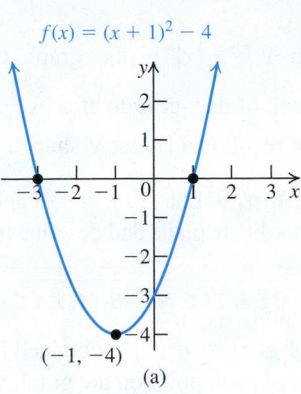

Figure 1.80 Graphing $y = |f(x)|$.

Practice Problem 5 Use the graph of $f(x) = 2x - 4$ to sketch the graph of $y = |f(x)|$.

4 Use stretching or compressing to graph functions.

Stretching or Compressing

We now look at transformations that distort the shape of a graph, these are called **nonrigid transformations**. We consider the relationship of the graphs of $y = af(x)$ and $y = f(bx)$ to the graph of $y = f(x)$.

Comparing the Graphs of $y = f(x)$ and $y = af(x)$

EXAMPLE 6 Stretching or Compressing a Graph Vertically

Let $f(x) = |x|$, $g(x) = 2|x|$, and $h(x) = \dfrac{1}{2}|x|$. Sketch the graphs of f, g, and h on the same coordinate plane and describe how the graphs of g and h are related to the graph of f.

Solution

The graphs of $y = |x|$, $y = 2|x|$, and $y = \dfrac{1}{2}|x|$ are sketched in Figure 1.81. Table 1.13 gives some typical function values.

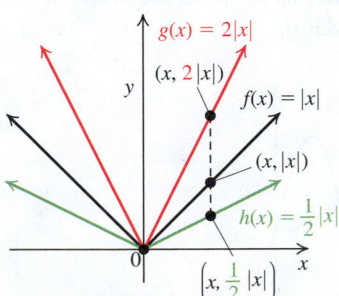

Figure 1.81 Vertical stretch and compression.

TABLE 1.13

x	$f(x) = |x|$	$g(x) = 2|x|$	$h(x) = \dfrac{1}{2}|x|$
-2	2	4	1
-1	1	2	$\dfrac{1}{2}$
0	0	0	0
1	1	2	$\dfrac{1}{2}$
2	2	4	1

The graph of $y = 2|x|$ is the graph of $y = |x|$ vertically stretched (expanded) by multiplying each of its y-coordinates by 2. It is twice as high as the graph of $|x|$ at every real number x. The result is a taller V-shaped curve. See Figure 1.81.

The graph $y = \frac{1}{2}|x|$ is the graph of $y = |x|$ vertically compressed (shrunk) by multiplying each of its y-coordinates by $\frac{1}{2}$. It is half as high as the graph of $|x|$ at every real number x. The result is a flatter V-shaped curve. See Figure 1.81.

Practice Problem 6 Let $f(x) = \sqrt{x}$ and $g(x) = 2\sqrt{x}$. Sketch the graphs of f and g on the same coordinate plane and describe how the graph of g is related to the graph of f. ■

VERTICAL STRETCHING OR COMPRESSING

The graph of $g(x) = af(x)$ is obtained from the graph of $y = f(x)$ by multiplying the y-coordinate of each point on the graph of $y = f(x)$ by a and leaving the x-coordinate unchanged. The result (see Figure 1.82) is as follows:

1. A **vertical stretch** away from the x-axis if $a > 1$
2. A **vertical compression** toward the x-axis if $0 < a < 1$.

If $a < 0$, first graph $y = |a|f(x)$ by stretching or compressing the graph of $y = f(x)$ vertically. Then reflect the resulting graph about the x-axis.

So, if (x, y) is a point on the graph of f, then the point (x, ay) is on the graph of g.

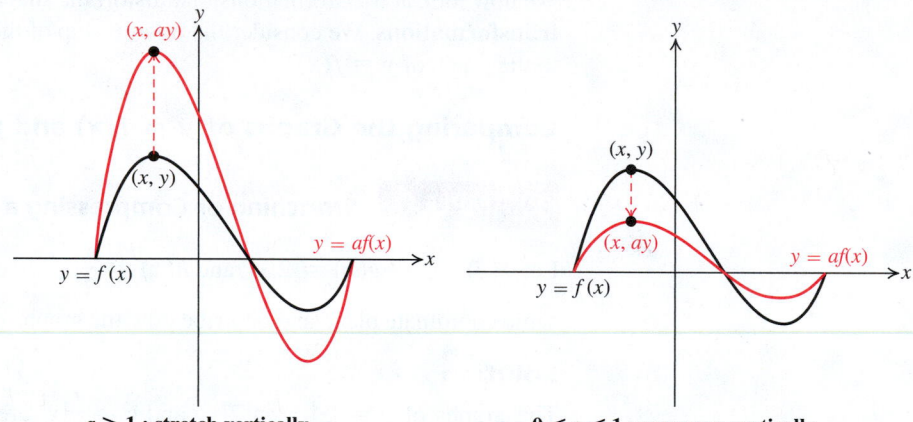

$a > 1$: stretch vertically $0 < a < 1$: compress vertically

Figure 1.82 Vertical stretch or compression.

Comparing the Graphs of $y = f(x)$ and $y = f(bx)$ Given a function $y = f(x)$, let's explore the effect of the constant b in graphing the function $y = f(bx)$. Consider the graphs of $y = f(x)$ and $y = f(2x)$ in Figure 1.83(a). Multiplying the *independent* variable x by 2 compresses the graph $y = f(x)$ horizontally toward the y-axis. Because the value $2x$ is twice the value of x, a point on the x-axis will be only half as far from the origin when $y = f(2x)$ has the same y value as $y = f(x)$. See Figure 1.83(a).

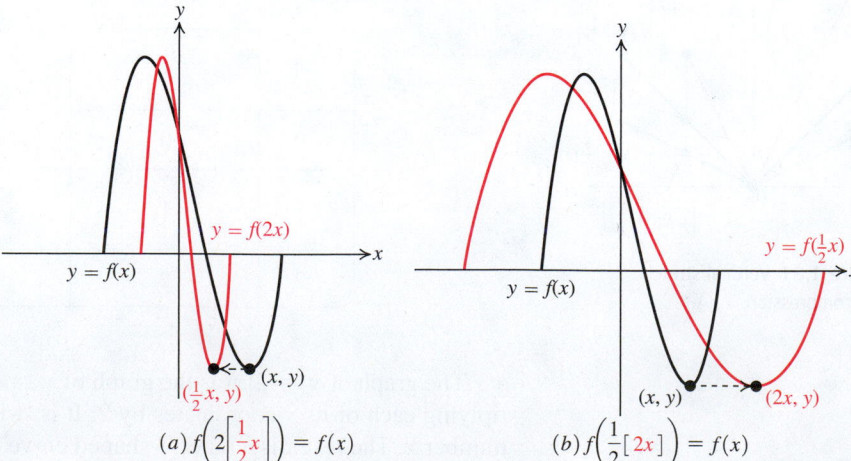

$(a)\, f\left(2\left[\frac{1}{2}x\right]\right) = f(x)$ $(b)\, f\left(\frac{1}{2}[2x]\right) = f(x)$

Figure 1.83 Horizontal stretch or compression.

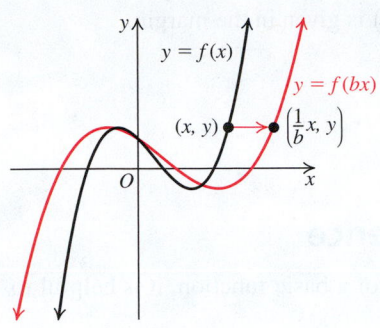

0 < b < 1 : stretch horizontally

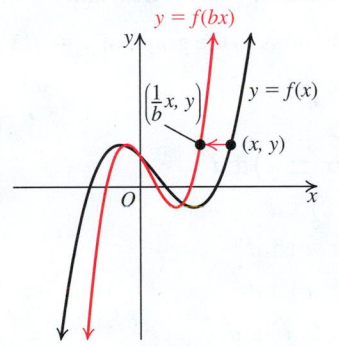

b > 1 : compress horizontally

Figure 1.84 Horizontal stretch and compression.

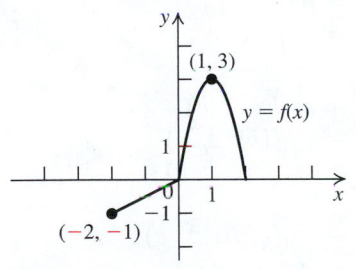

Figure 1.85

Now consider the graphs of $y = f(x)$ and $y = f\left(\frac{1}{2}x\right)$ in Figure 1.83(b). Multiplying the *independent* variable x by $\frac{1}{2}$ stretches the graph of $y = f(x)$ horizontally away from the y-axis. The value $\frac{1}{2}x$ is half the value of x so that a point on the x-axis will be twice as far from the origin for $y = f\left(\frac{1}{2}x\right)$ to have the same y value as $y = f(x)$. See Figure 1.83(b).

> ### HORIZONTAL STRETCHING OR COMPRESSING
>
> Let $b > 0$. The graph of $g(x) = f(bx)$ is obtained from the graph of $y = f(x)$ by multiplying the x-coordinate of each point on the graph of $y = f(x)$ by $\frac{1}{b}$ and leaving the y-coordinate unchanged. The result (see Figure 1.84) is as follows:
>
> 1. A **horizontal stretch** away from the y-axis if $0 < b < 1$.
> 2. A **horizontal compression** toward the y-axis if $b > 1$.
>
> If $b < 0$, first graph $f(|b|x)$ by stretching or compressing the graph of $y = f(x)$ horizontally. Then reflect the graph of $y = f(|b|x)$ about the y-axis.
>
> If (x, y) is a point on the graph of $y = f(x)$, then the point $\left(\frac{1}{b}x, y\right)$ is on the graph of g.

EXAMPLE 7 **Stretching or Compressing a Function Horizontally**

The graph of the function $y = f(x)$ whose equation is not given is shown in Figure 1.85. Sketch the following graphs.

a. $f\left(\frac{1}{2}x\right)$ **b.** $f(2x)$ **c.** $f(-2x)$

Solution

Note that the domain of f is $[-2, 2]$ and its range is $[-1, 3]$

a. To graph $y = f\left(\frac{1}{2}x\right)$, we stretch the graph of $y = f(x)$ horizontally by a factor of 2. In other words, we transform each point (x, y) in Figure 1.85 to the point $(2x, y)$ in Figure 1.86(a).

b. To graph $y = f(2x)$, we compress the graph of $y = f(x)$ horizontally by a factor of $\frac{1}{2}$. Therefore, we transform each point (x, y) in Figure 1.85 to $\left(\frac{1}{2}x, y\right)$ in Figure 1.86(b).

c. To graph $y = f(-2x)$, we reflect the graph of $y = f(2x)$ in Figure 1.86(b) about the y-axis. So we transform each point (x, y) in Figure 1.86(b) to the point $(-x, y)$ in Figure 1.86(c).

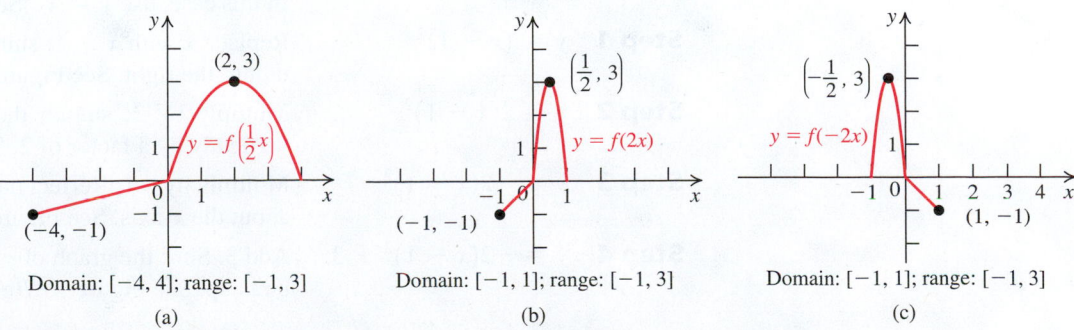

Domain: $[-4, 4]$; range: $[-1, 3]$ Domain: $[-1, 1]$; range: $[-1, 3]$ Domain: $[-1, 1]$; range: $[-1, 3]$

(a)   (b)   (c)

Figure 1.86

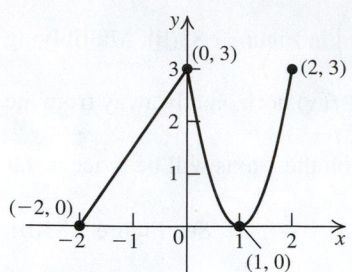

Practice Problem 7 The graph of a function $y = f(x)$ is given in the margin. Sketch the graphs of the following functions.

a. $f\left(\dfrac{1}{2}x\right)$ **b.** $f(2x)$

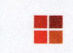

Multiple Transformations in Sequence

When graphing requires more than one transformation of a basic function, it is helpful to perform transformations in the following order:

1. Horizontal shift   2. Stretch or Compression   3. Reflection   4. Vertical shift

Below we present one possible order of applying transformations to the graph of $y = f(x)$. Here we assume the numbers a, b, c, and d are positive.

Graph	$y = \pm af(\pm bx \pm c) \pm d$
Case I	$y = af(bx \pm c) \pm d$
Case II	$y = -af(bx \pm c) \pm d$
Case III	$y = af(-bx \pm c) \pm d$
Case IV	$y = -af(-bx \pm c) \pm d$

Step 0	Identify the basic graph.	$y = f(x)$
Step 1	Horizontal shift	$y = f(x \pm c)$
Step 2	Horizontal stretch or compression	$y = f(bx \pm c)$
	Vertical stretch or compression	$y = a \cdot f(bx \pm c)$
Step 3	Reflections	

	Case I	*none*	$y = af(bx \pm c)$
	Case II	*about the x-axis*	$y = -af(bx \pm c)$
	Case III	*about the y-axis*	$y = af(-bx \pm c)$
	Case IV	*about both the x-axis and the y-axis*	$y = -af(-bx \pm c)$

Step 4	Vertical shift	$y = \pm af(\pm bx \pm c) \pm d$

EXAMPLE 8   **Combining Transformations**

Sketch the graph of the function $f(x) = -2(x - 1)^2 + 3$.

Solution

Begin with the basic function $y = x^2$. Then apply the necessary transformations in a sequence of steps. The result of each step is shown in Figure 1.87. We follow the point $(2, 4)$ from the graph of $y = x^2$.

Step 0 $y = x^2$ Identify a related function whose graph is familiar. In this case, use $y = x^2$. See Figure 1.87(a).

Step 1 $y = (x - 1)^2$ Replace x with $x - 1$; shift the graph of $y = x^2$ one unit to the right. See Figure 1.87(b).

Step 2 $y = 2(x - 1)^2$ Multiply by 2; stretch the graph of $y = (x - 1)^2$ vertically by a factor of 2. See Figure 1.87(c).

Step 3 $y = -2(x - 1)^2$ Multiply by -1. Reflect the graph of $y = 2(x - 1)^2$ about the x-axis. See Figure 1.87(d).

Step 4 $y = -2(x - 1)^2 + 3$ Add 3. Shift the graph of $y = -2(x - 1)^2$ three units up. See Figure 1.87(e).

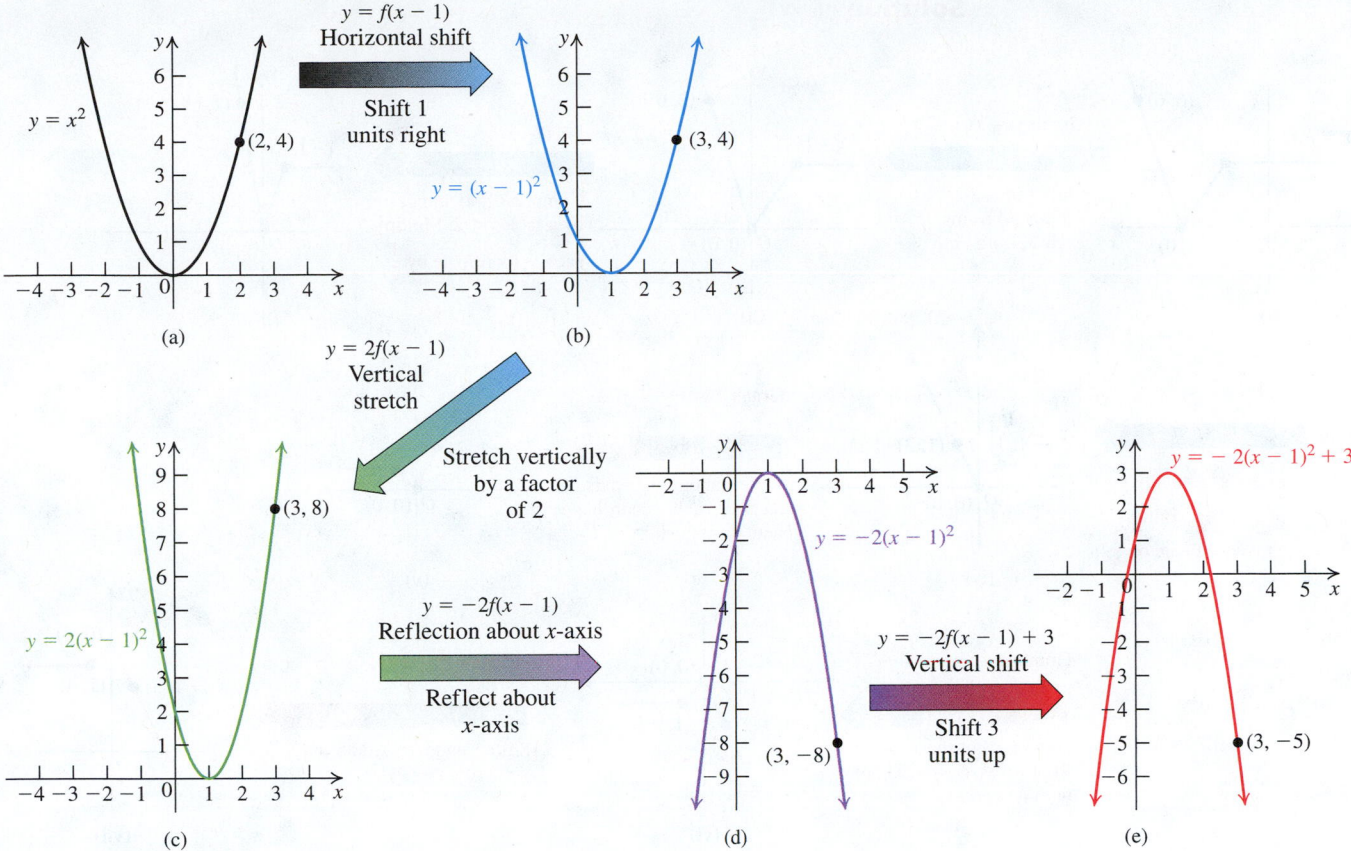

Figure 1.87 Multiple transformations.

Practice Problem 8 Sketch the graph of the function

$$f(x) = 3\sqrt{x + 1} - 2.$$

WARNING

To Graph	Incorrect Sequence	Correct Sequence
$y = f(2x - 1)$	$y = f(x) \rightarrow y = f(2x)$ $\rightarrow y = f(2x - 1)$	$y = f(x) \rightarrow y = f(x - 1)$ $\rightarrow y = f(2x - 1)$
$y = f(2 - x)$	$y = f(x) \rightarrow y = f(-x)$ $\rightarrow y = f(2 - x)$	$y = f(x) \rightarrow y = f(2 + x)$ $\rightarrow y = f(2 - x)$
$y = -f(x - 1) + 3$	$y = f(x) \rightarrow y = f(x - 1)$ $\rightarrow y = f(x - 1) + 3$ $\rightarrow y = -f(x - 1) + 3$	$y = f(x) \rightarrow y = f(x - 1)$ $\rightarrow y = -f(x - 1)$ $\rightarrow y = -f(x - 1) + 3$

EXAMPLE 9   **Using a Sequence of Transformations**

Use the graph of $y = f(x)$ in Figure 1.88(i) to graph $y = -\dfrac{1}{3}f(4 - 2x) + 5$.

Solution

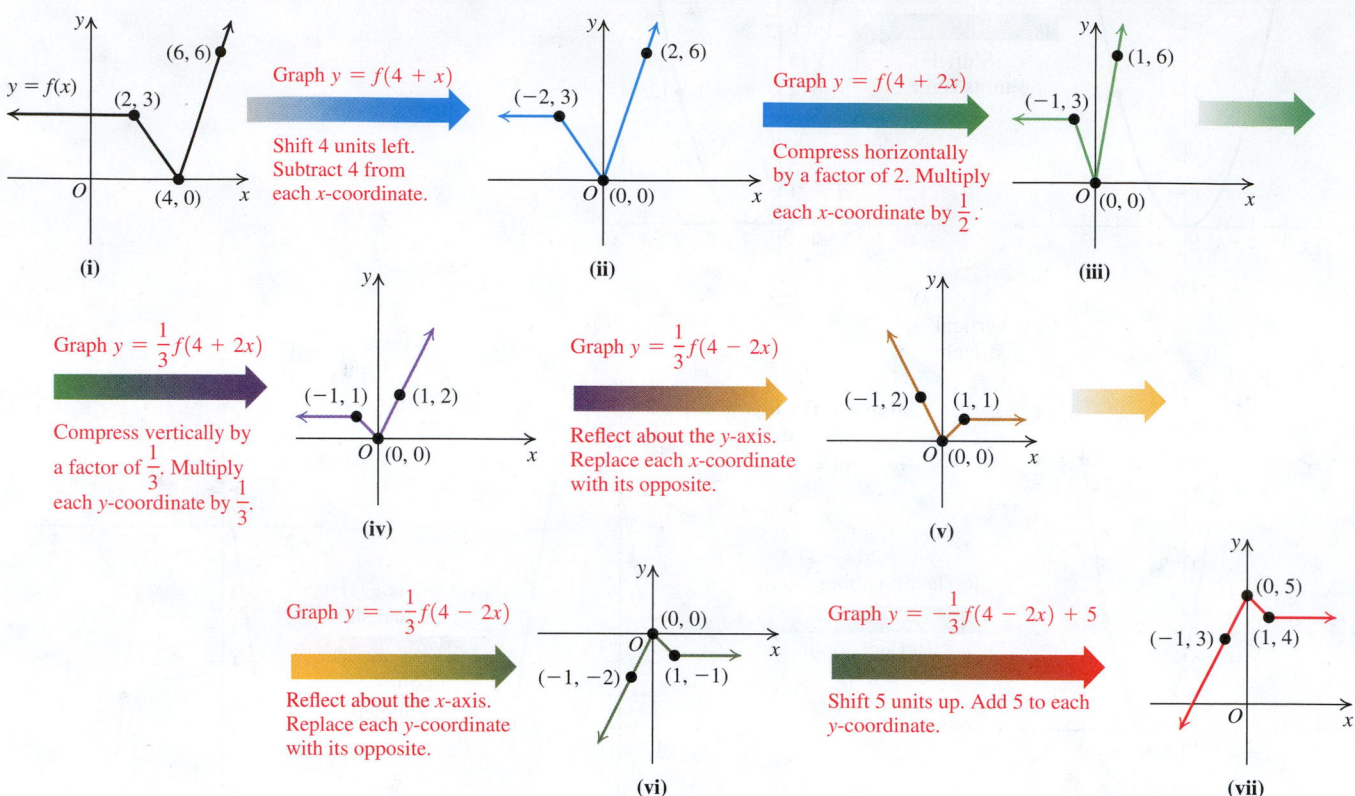

Figure 1.88 Graph of $y = -\dfrac{1}{3}f(4 - 2x) + 5$.

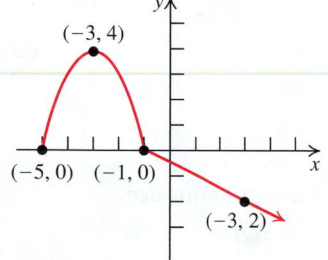

Figure 1.89

Practice Problem 9 Use the graph of $y = f(x)$ in Figure 1.89 to graph

$$y = -\frac{1}{2}f(2x - 1) + 3.$$

SUMMARY OF MAIN FACTS

Summary of Transformations of $y = f(x)$

To graph	Draw the graph of *f* and	Make these changes to the equation $y = f(x)$	Change the graph point (x, y) to
Vertical shift (for $d > 0$)			
$y = f(x) + d$	*Shift the graph of f* up d units.	Add d to $f(x)$.	$(x, y + d)$
$y = f(x) - d$	down d units.	Subtract d from $f(x)$.	$(x, y - d)$

To graph	Draw the graph of f and	Make these changes to the equation $y = f(x)$	Change the graph point (x, y) to
Horizontal shift (for $c > 0$)	Shift the graph of f		
$y = f(x - c)$	to the right c units.	Replace x with $x - c$.	$(x + c, y)$
$y = f(x + c)$	to the left c units.	Replace x with $x + c$.	$(x - c, y)$
Reflection about the x-axis			
$y = -f(x)$	Reflect the graph of f about the x-axis.	Multiply $f(x)$ by -1.	$(x, -y)$
Reflection about the y-axis			
$y = f(-x)$	Reflect the graph of f about the y-axis.	Replace x with $-x$.	$(-x, y)$
Vertical stretching or compressing: $y = af(x)$	Multiply each y-coordinate of $y = f(x)$ by a. The graph of $y = f(x)$ is stretched vertically away from the x-axis if $a > 1$ and is compressed vertically toward the x-axis if $0 < a < 1$. If $a < 0$, sketch the graph of $y = \lvert a \rvert f(x)$, then reflect it about the x-axis.	Multiply $f(x)$ by a.	(x, ay)
Horizontal stretching or compressing: $y = f(bx)$	Multiply each x-coordinate of $y = f(x)$ by $\frac{1}{b}$. The graph of $y = f(x)$ is stretched away from the y-axis if $0 < b < 1$ and is compressed toward the y-axis if $b > 1$. If $b < 0$, sketch the graph of $y = f(\lvert b \rvert x)$, then reflect it about the y-axis.	Replace x with bx.	$\left(\dfrac{x}{b}, y \right)$

Summary for Combining Transformations in Sequence

To graph $y = af(bx + c) + d$, with $b > 0$ and $c > 0$

• Starting with the graph of $y = f(x)$, perform the indicated sequence of transformations. Here is one possible sequence:

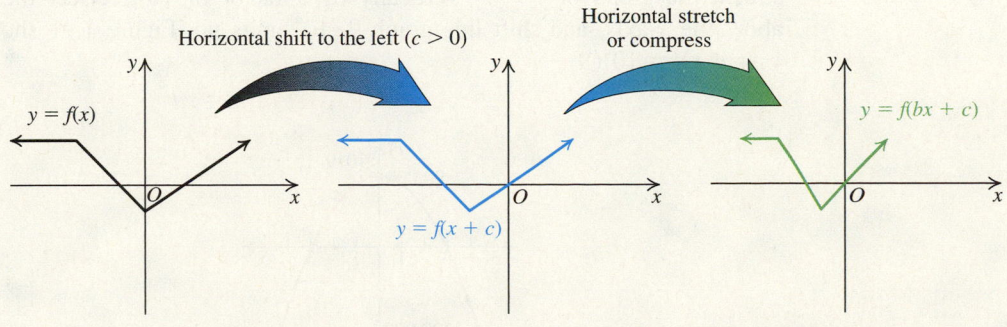

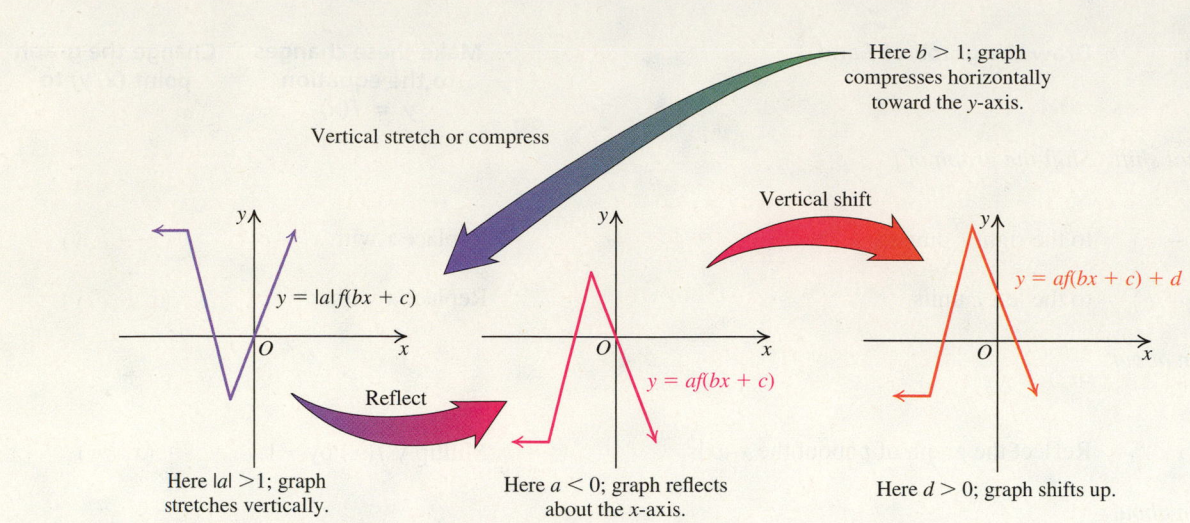

Here $b > 1$; graph compresses horizontally toward the y-axis.

Vertical stretch or compress

Vertical shift

$y = |a|f(bx + c)$

$y = af(bx + c) + d$

$y = af(bx + c)$

Reflect

Here $|a| > 1$; graph stretches vertically.

Here $a < 0$; graph reflects about the x-axis.

Here $d > 0$; graph shifts up.

Note: If $c < 0$, then the horizontal shift should be $|c|$ units to the right; if $b < 0$ graph $y = af(|b|x + c) + d$, but reflect this graph about the y-axis before completing the last step, the vertical shift, d.

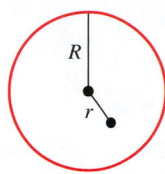

Figure 1.90

◆ **EXAMPLE 10** **Using Poiseuille's Law for Arterial Blood Flow**

For an artery with radius R, the velocity v of the blood flow at a distance r from the center of the artery is given (in appropriate units) by

$$v = c(R^2 - r^2), 0 \leq r \leq R$$

where c is a constant that is determined for a particular artery. See Figure 1.90.

To emphasize that the velocity v depends on the distance r from the center of the artery, we write $v = v(r)$, or $v(r) = c(R^2 - r^2)$. Starting with the graph of $y = r^2$, sketch the graph of $y = v(r)$, where the artery has radius $R = 3$ and $c = 10^4$.

Solution

$v(r) = c(R^2 - r^2)$	Original equation
$v(r) = 10^4(9 - r^2)$	Substitute $c = 10^4$ and $R = 3$.
$v(r) = 9 \cdot 10^4 - 10^4 r^2$	Distributive property

Sketch the graph of $y = v(r)$ in Figure 1.91 through a sequence of transformations: stretch the graph of $y = r^2$ vertically by a factor of 10^4, reflect the resulting graph about the x-axis, and shift this graph $9 \cdot 10^4$ units up. Figure 1.91 shows the graph of $y = v(r) = 10^4(9 - r^2)$.

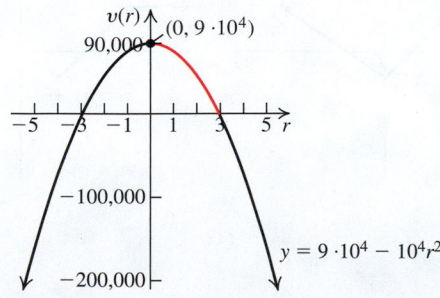

Figure 1.91 Velocity of flow, r units from the artery's center.

The velocity of the blood decreases from the center of the artery ($r = 0$) to the artery wall ($r = 3$), where it ceases to flow. The portions of the graph in Figure 1.91 corresponding to negative values of r, or values of r greater than 3, do not have any physical significance. Only the section of the graph with $0 \leq r \leq 3$ has a physical meaning.

Practice Problem 10 Suppose in Example 9 that the artery has radius $R = 3$ and that $c = 10^3$. Sketch the graph of $y = v(r)$.

Answers to Practice Problems

1. The graph of g is the graph of f shifted one unit up; the graph of h is the graph of f shifted two units down.

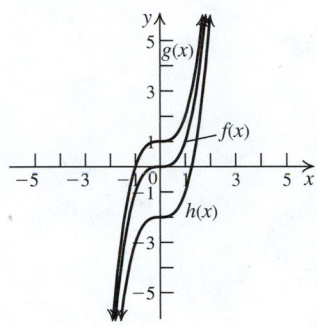

2. The graph of g is the graph of f shifted one unit to the right; the graph of h is the graph of f shifted two units to the left.

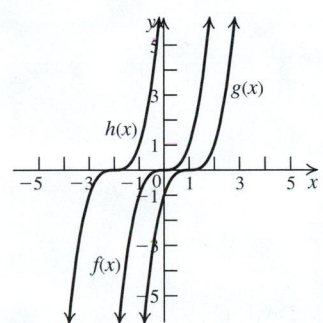

3.

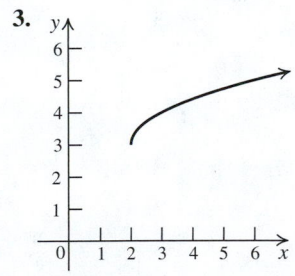

4. Shift the graph of $y = x^2$ one unit to the right, reflect the resulting graph about the x-axis, and shift it two units up.

5.

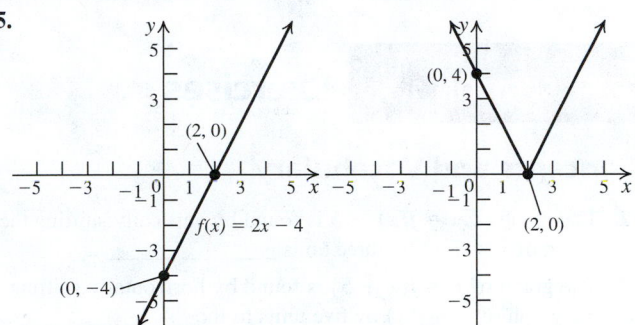

6. The graph of g is the graph of f vertically stretched by multiplying each of its y-coordinates by 2.

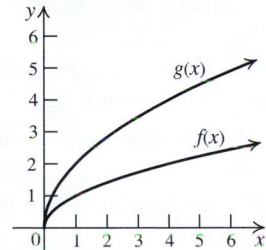

7. a.   **b.**

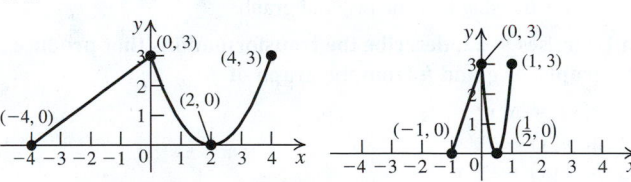

8.

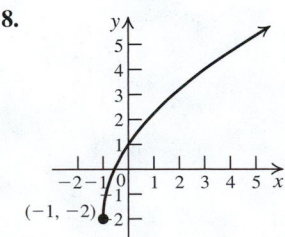

9. $y = f(x) \rightarrow f(x-1) \rightarrow f(2x-1) \rightarrow \frac{1}{2}f(2x-1) \rightarrow$

$\frac{1}{2}f(2x-1) + 3$

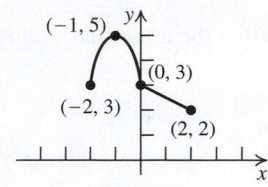

(−1, 5)
(0, 3)
(−2, 3)
(2, 2)

10.

10,000
5000
−5   3   −1 0   1 2 3 4 5 r
−5000
−10,000
−15,000
−20,000

SECTION 1.5 **Exercises**

Concepts and Vocabulary

1. The graph of $y = f(x) - 3$ is found by vertically shifting the graph of $y = f(x)$ by three units _____ .

2. The graph of $y = f(x + 5)$ is found by horizontally shifting the graph of $y = f(x)$ by five units to the _____ .

3. The graph of $y = f(-x)$ is found by reflecting the graph of $y = f(x)$ about the _____ .

4. The graph of $y = -f(x)$ is found by reflecting the graph of $y = f(x)$ about the _____ .

5. True or False. The graph of $y = f(bx)$ is a horizontal compression of the graph $y = f(x)$ if $b < 1$.

6. True or False. The graphs of $y = f(x)$ and $y = f(x) + 1$ cannot be the same.

7. True or False. The graphs of $y = f(x)$ and $y = f(-x)$ cannot be the same.

8. True or False. Combining horizontal and vertical shifts preserves the shape of the original graph.

In Exercises 9–22, describe the transformations that produce the graphs of g and h from the graph of f.

9. $f(x) = \sqrt{x}$
 a. $g(x) = \sqrt{x} + 2$
 b. $h(x) = \sqrt{x} - 1$

10. $f(x) = |x|$
 a. $g(x) = |x| + 1$
 b. $h(x) = |x| - 2$

11. $f(x) = x^2$
 a. $g(x) = (x + 1)^2$
 b. $h(x) = (x - 2)^2$

12. $f(x) = \dfrac{1}{x}$
 a. $g(x) = \dfrac{1}{x + 2}$
 b. $h(x) = \dfrac{1}{x - 3}$

13. $f(x) = \sqrt{x}$
 a. $g(x) = \sqrt{x + 1} - 2$
 b. $h(x) = \sqrt{x - 1} + 3$

14. $f(x) = x^2$
 a. $g(x) = -x^2$
 b. $h(x) = (-x)^2$

15. $f(x) = |x|$
 a. $g(x) = -|x|$
 b. $h(x) = |-x|$

16. $f(x) = \sqrt{x}$
 a. $g(x) = 2\sqrt{x}$
 b. $h(x) = \sqrt{2x}$

17. $f(x) = \dfrac{1}{x}$
 a. $g(x) = \dfrac{2}{x}$
 b. $h(x) = \dfrac{1}{2x}$

18. $f(x) = x^3$
 a. $g(x) = (x - 2)^3 + 1$
 b. $h(x) = -(x + 1)^3 + 2$

19. $f(x) = \sqrt{x}$
 a. $g(x) = -\sqrt{x} + 1$
 b. $h(x) = \sqrt{-x} + 1$

20. $f(x) = [\![x]\!]$
 a. $g(x) = [\![x - 1]\!] + 2$
 b. $h(x) = 3[\![x]\!] - 1$

21. $f(x) = \sqrt[3]{x}$
 a. $g(x) = \sqrt[3]{x} + 1$
 b. $h(x) = \sqrt[3]{x + 1}$

22. $f(x) = \sqrt[3]{x}$
 a. $g(x) = 2\sqrt[3]{1 - x} + 4$
 b. $h(x) = -\sqrt[3]{x - 1} + 3$

In Exercises 23–34, match each function with its graph (a)–(l).

23. $y = -|x| + 1$

24. $y = -\sqrt{-x}$

25. $y = \sqrt{x^2}$

26. $y = \dfrac{1}{2}|x|$

27. $y = \sqrt{x+1}$ **28.** $y = 2|x| - 3$
29. $y = 1 - 2\sqrt{x}$ **30.** $y = -|x-1| + 1$
31. $y = (x-1)^2$ **32.** $y = -x^2 + 3$
33. $y = -2(x-3)^2 - 1$ **34.** $y = 3 - \sqrt{1-x}$

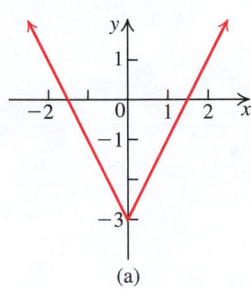
(a)

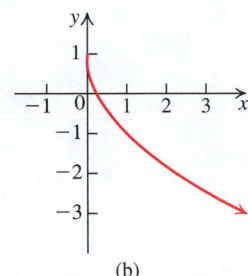
(b)

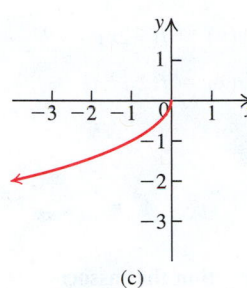
(c)

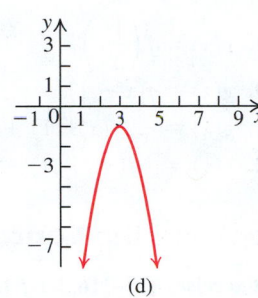
(d)

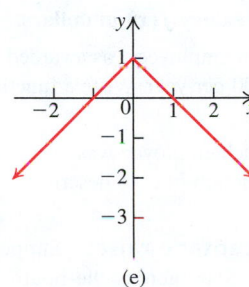
(e)

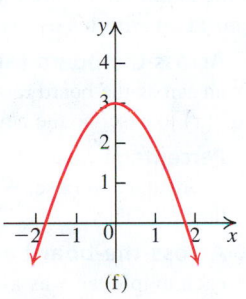
(f)

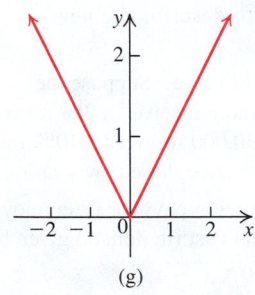
(g)

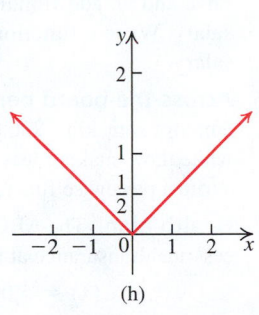
(h)

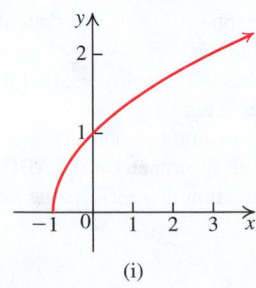
(i)

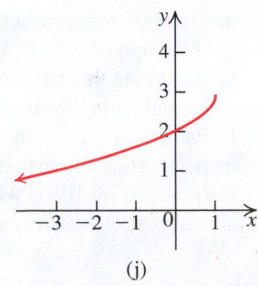
(j)

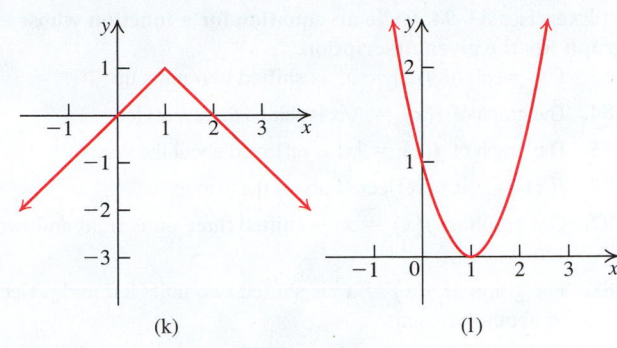
(k)   (l)

In Exercises 35–62, graph each function by starting with a function from the library of functions and then using the techniques of shifting, compressing, stretching, and/or reflecting.

35. $f(x) = x^2 - 2$ **36.** $f(x) = x^2 + 3$
37. $g(x) = \sqrt{x} + 1$ **38.** $g(x) = \sqrt{x} - 4$
39. $f(x) = |x| + 2$ **40.** $f(x) = |x| - 1$
41. $f(x) = x^3 + 2$ **42.** $f(x) = x^3 - 1$
43. $f(x) = \dfrac{1}{x} + 1$ **44.** $f(x) = \dfrac{1}{x} - 2$
45. $f(x) = (x-3)^3$ **46.** $f(x) = (x+2)^3$
47. $f(x) = \sqrt{x-1}$ **48.** $f(x) = \sqrt{x+2}$
49. $h(x) = |x+1|$ **50.** $h(x) = |x-2|$
51. $f(x) = (x+1)^3$ **52.** $f(x) = (x-3)^3$
53. $f(x) = \dfrac{1}{x-3}$ **54.** $f(x) = \dfrac{1}{x+2}$
55. $f(x) = \sqrt{-x}$ **56.** $f(x) = -\sqrt{x}$
57. $f(x) = -x^2$ **58.** $f(x) = -x^3$
59. $f(x) = 2x^2$ **60.** $f(x) = \dfrac{1}{3}x^2$
61. $f(x) = 2|x|$ **62.** $f(x) = \dfrac{1}{3}|x|$

In Exercises 63–74, graph each function by starting with a function from the library of functions and then combining shifting and reflecting techniques.

63. $f(x) = (x-2)^2 + 1$ **64.** $f(x) = (x-3)^2 - 5$
65. $f(x) = 5 - (x-3)^2$ **66.** $f(x) = 2 - (x+1)^2$
67. $f(x) = \sqrt{x+1} - 3$ **68.** $f(x) = \sqrt{x-2} + 1$
69. $f(x) = \sqrt{1-x} + 2$ **70.** $f(x) = -\sqrt{x+2} - 3$
71. $f(x) = |x-1| - 2$ **72.** $f(x) = -|x+3| + 1$
73. $f(x) = \dfrac{1}{x-1} + 3$ **74.** $f(x) = 2 - \dfrac{1}{x+2}$

In Exercises 75–82, graph each function by starting with a function from the library of functions and then combining shifting, compressing, stretching, and/or reflecting techniques.

75. $f(x) = 2(x+1)^2 - 1$ **76.** $f(x) = \dfrac{1}{3}(x+1)^2 + 2$
77. $f(x) = 2 - \dfrac{1}{2}(x-3)^2$ **78.** $f(x) = 1 - 3(x-3)^2$
79. $f(x) = 2\sqrt{x+1} - 3$ **80.** $f(x) = \sqrt{2x-2} + 1$
81. $f(x) = -2|x-1| + 2$ **82.** $f(x) = -\dfrac{1}{2}|3 - x| - 1$

In Exercises 83–94, write an equation for a function whose graph fits the given description.

83. The graph of $f(x) = x^3$ is shifted two units up.

84. The graph of $f(x) = \sqrt{x}$ is shifted three units left.

85. The graph of $f(x) = |x|$ is reflected about the x-axis.

86. $f(x) = \sqrt{x}$ is reflected about the y-axis.

87. The graph of $f(x) = x^2$ is shifted three units right and two units up.

88. The graph of $f(x) = x^2$ is shifted two units left and reflected about the x-axis.

89. The graph of $f(x) = \sqrt{x}$ is shifted three units left, reflected about the x-axis, and shifted two units down.

90. The graph of $f(x) = \sqrt{x}$ is shifted two units down, reflected about the x-axis, and compressed vertically by a factor of $\frac{1}{2}$.

91. The graph of $f(x) = x^3$ is shifted four units left, stretched vertically by a factor of 3, reflected about the y-axis, and shifted two units up.

92. The graph of $f(x) = x^3$ is reflected about the x-axis, shifted one unit up, shifted one unit left, and reflected about the y-axis.

93. The graph of $f(x) = |x|$ is shifted four units right, compressed horizontally by a factor of 2, reflected about the x-axis, and shifted three units down.

94. The graph of $f(x) = |x|$ is shifted two units right, reflected about the y-axis, stretched horizontally by a factor of 2 and shifted three units down.

In Exercises 95–104, graph the function $y = g(x)$ given the following graph of $y = f(x)$.

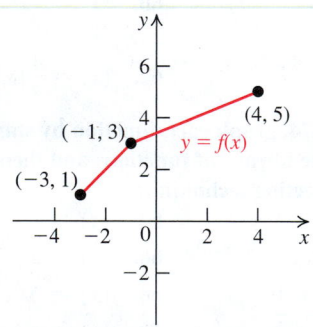

95. $g(x) = f(x) - 1$

96. $g(x) = f(x) + 3$

97. $g(x) = -f(x)$

98. $g(x) = f(-x)$

99. $g(x) = f(2x)$

100. $g(x) = f\left(\frac{1}{2}x\right)$

101. $g(x) = f(x + 1)$

102. $g(x) = f(x - 2)$

103. $g(x) = 2f(x)$

104. $g(x) = \frac{1}{2}f(x)$

In Exercises 105–112, graph (a) $y = g(x)$ and (b) $y = |g(x)|$, given the following graph of $y = f(x)$.

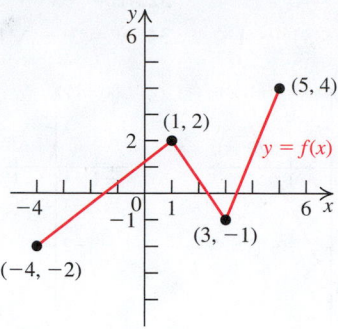

105. $g(x) = f(x) + 1$

106. $g(x) = -2f(x)$

107. $g(x) = f\left(\frac{1}{2}x\right)$

108. $g(x) = f(-2x)$

109. $g(x) = f(x - 1)$

110. $g(x) = f(2 - x)$

111. $g(x) = -2f(x + 1) + 3$

112. $g(x) = -f(-x + 1) - 2$

Applying the Concepts

In Exercises 113–116, let f be the function that associates the employee number x of each employee of the ABC Corporation with his or her annual salary $f(x)$ in dollars.

113. **Across-the-board raise.** Each employee was awarded an across-the-board raise of $800 per year. Write a function $g(x)$ to describe the new salary.

114. **Percentage raise.** Suppose each employee was awarded a 5% raise. Write a function $h(x)$ to describe the new salary.

115. **Across-the-board and percentage raises.** Suppose each employee was awarded a $500 across-the-board raise and an additional 2% of his or her increased salary. Write a function $p(x)$ to describe the new salary.

116. **Across-the-board percentage raise.** Suppose the employees making $30,000 or more received a 2% raise, while those making less than $30,000 received a 10% raise. Write a piecewise function to describe these new salaries.

117. **Health plan.** The ABC Corporation pays for its employees' health insurance at an annual cost (in dollars) given by

$$C(x) = 5{,}000 + 10\sqrt{x - 1},$$

where x is the number of employees covered.

a. Use transformations of the graph of $y = \sqrt{x}$ to sketch the graph of $y = C(x)$.

b. Assuming that the company has 400 employees, find its annual outlay for the health coverage.

118. In Exercise 117, assume that the insurance company increases the annual cost of health insurance for the ABC Corporation by 10%. Write a function that reflects the new annual cost of health coverage.

119. Demand. The weekly demand for paper hats produced by Mythical Manufacturers is given by

$$x(p) = 109{,}561 - (p + 1)^2,$$

where x represents the number of hats that can be sold at a price of p cents each.

a. Use transformations on the graph of $y = p^2$ to sketch the graph of $y = x(p)$.
b. Find the price at which 69,160 hats can be sold.
c. Find the price at which no hats can be sold.

120. Revenue. The weekly demand for cashmere sweaters produced by Wool Shop, Inc., is $x(p) = -3p + 600$. The revenue is given by $R(p) = -3p^2 + 600p$. Describe how to sketch the graph of $y = R(p)$ by applying transformations to the graph of $y = p^2$. [*Hint:* First, write $R(p)$ in the form $-3(p - h)^2 + k$.]

121. Daylight. At 60° north latitude, the graph of $y = f(t)$ gives the number of hours of daylight. (On the t-axis, note that $1 = $ January and $12 = $ December.)

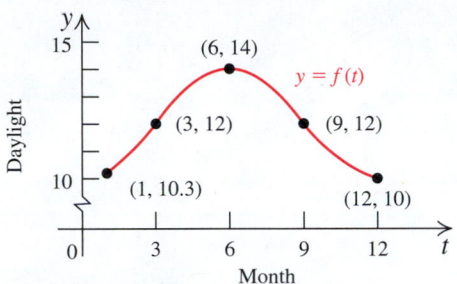

Sketch the graph of $y = f(t) - 12$.

122. Use the graph of $y = f(t)$ of Exercise 121 to sketch the graph of $y = 24 - f(t)$. Interpret the result.

Beyond the Basics

In Exercises 123 and 124, the graphs of $y = f(x)$ and $y = g(x)$ are given. Find an equation for $g(x)$ if the graph of g is obtained from the graph of f by a sequence of transformations.

123.

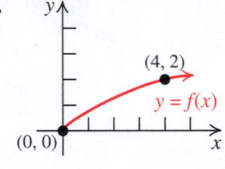
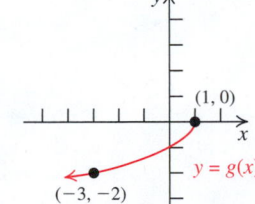

124.

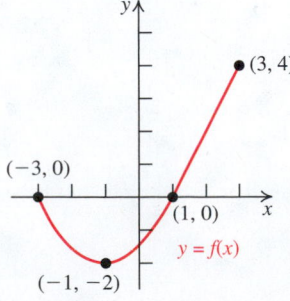
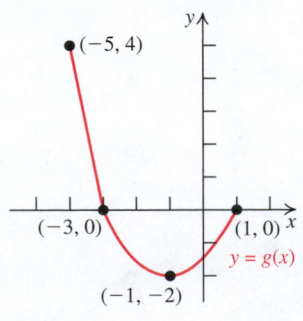

In Exercises 125–132, by completing the square on each quadratic expression, use transformations on $y = x^2$ to sketch the graph of $y = f(x)$.

125. $f(x) = x^2 + 4x$
  [*Hint:* $x^2 + 4x = (x^2 + 4x + 4) - 4 = (x + 2)^2 - 4.$]

126. $f(x) = x^2 - 6x$ **127.** $f(x) = -x^2 + 2x$

128. $f(x) = -x^2 - 2x$ **129.** $f(x) = 2x^2 - 4x$

130. $f(x) = 2x^2 + 6x + 3.5$

131. $f(x) = -2x^2 - 8x + 3$

132. $f(x) = -2x^2 + 2x - 1$

In Exercises 133–138, sketch the graph of each function.

133. $y = |2x + 3|$ **134.** $y = |\,[\![x]\!]\,|$

135. $y = |4 - x^2|$ **136.** $y = \left| \dfrac{1}{x} \right|$

137. $y = \sqrt{|x|}$ **138.** $y = [\![\,|x|\,]\!]$

Critical Thinking / Discussion / Writing

139. The x-intercepts of the function f are $-1, 0, 2$. Find the corresponding x-intercepts for the following functions.
a. $y = f(x + 2)$
b. $y = f(x - 2)$
c. $y = -f(x)$
d. $y = f(-x)$
e. $y = f(2x)$
f. $y = f\left(\dfrac{1}{2}x\right)$

140. The y-intercept of the function f is 2. Find the corresponding y-intercepts for the following functions.
a. $y = f(x) + 2$
b. $y = f(x) - 2$
c. $y = -f(x)$
d. $y = f(-x)$
e. $y = 2f(x)$
f. $y = \dfrac{1}{2}f(x)$

141. Let a function f have domain $[-1, 3]$ and range $[-2, 1]$. Find the corresponding domain and range for the following functions.
a. $y = f(x + 2)$
b. $y = f(x) - 2$
c. $y = -f(x)$
d. $y = f(-x)$
e. $y = 2f(x)$
f. $y = \dfrac{1}{2}f(x)$

142. Let a function f have a relative maximum at $x = 1$ and a relative minimum at $x = 2$. Find the corresponding relative maxima and minima for the following functions.
a. $y = f(x + 2)$
b. $y = f(x) - 2$
c. $y = -f(x)$
d. $y = f(-x)$
e. $y = 2f(x)$
f. $y = \dfrac{1}{2}f(x)$

Getting Ready for the Next Section

In Exercises 143–146, simplify each expression.

143. $(x^2 + 2x) + (6x^3 - 2x + 5)$

144. $(x^3 + 2) - (2x^3 + 5x - 3)$

145. $(x - 2)(x^2 + 2x + 4)$

146. $(x^2 + x + 1)(x^2 - x + 1)$

In Exercises 147–150, find the domain of each function.

147. $f(x) = \dfrac{2x - 3}{x^2 - 5x + 6}$

148. $f(x) = \dfrac{x - 2}{x^2 - 4}$

149. $f(x) = \sqrt{2x - 3}$

150. $f(x) = \dfrac{1}{\sqrt{5 - 2x}}$

In Exercises 151–154, simplify $f(2x)$, $f(x + 1)$, $f(2x - 1)$ for the following functions.

151. $f(x) = x^2 - 3$

152. $f(x) = x^2 - 2x$

153. $f(x) = \sqrt{4 - x}$

154. $f(x) = \dfrac{x - 2}{x + 1}$

SECTION **1.6**

Combining Functions; Composite Functions

BEFORE STARTING THIS SECTION, REVIEW

1   Domain of a function (Section 1.3, page 44)

2   Area of a circle (Appendix A.5, page 952)

3   Rational inequalities (Appendix A.7, page 969)

OBJECTIVES

1   Define basic operations on functions.

2   Form composite functions.

3   Find the domain of a composite function.

4   Decompose a function.

5   Apply composition to practical problems.

◆ BP Oil Spill

The *Deepwater Horizon* was a floating drilling rig that could operate in waters up to 10,000 feet deep. British Petroleum (BP) was the operator and principal developer of the exploratory well situated in the Gulf of Mexico about 41 miles off the Louisiana coast. On April 20, 2010, at approximately 9:45 P.M., high-pressure methane gas from the well ignited and exploded, engulfing the platform of the rig. At the time, 126 crew members were on board. Of these, 11 workers were never found despite extensive search operations and are believed to have died in the explosion.

The total estimated volume of oil leaked was approximately 5 million barrels, making it the largest accidental oil spill in history. The spill directly impacted 68,000 square miles (about the size of Oklahoma). By June 2010, oil had washed up on 1074 miles of the coasts of Louisiana, Mississippi, Alabama, and Florida.

On June 16, 2010, BP announced a $20 billion fund to settle claims arising from the disaster. The fund was to be used only for natural resource restoration and compensation for individual and business claimants. As of July 2013, BP and its partners in the oil well, Transocean and Halliburton, were on trial to determine payouts and fines under the Clean Water Act and the Natural Resources Damages Assessment.

In Exercise 113, we calculate the area covered by an oil spill.

1 Define basic operations on functions.

SIDE
NOTE

A mutual fund is a type of professionally managed investment fund that pools money from many investors to purchase securities.

By participating in a 401k retirement account the investor has a choice of allocating some percentage of the contributions to different mutual funds by purchasing a corresponding number of shares.

Combining Functions

Many complex phenomena (situations) can be described naturally by combining the input from two (or more) independent sources. For example, the total gain of a particular 401k retirement account over a period of time, t, may be described as a sum of gains from two separate mutual funds $A(t)$ and $B(t)$.

$$\text{Gain}(t) = \text{FundA}(t) + \text{FundB}(t).$$

The formula reflects the fact that the total gain depends on the performance of each mutual fund. Similarly, the total value of a retirement account allocated in a particular mutual fund can be described as a product of a unit share price and the number of shares in the account.

$$\text{Value}(t) = \text{Share Price}(t) \cdot \text{Number of Shares}(t)$$

This formula also shows that the unit share price is changing, depending on the market, and that the number of shares changes with every contribution to the retirement account.

It is important to realize how complex functions can be built from the basic functions we described in our library of functions. Two functions, f and g, whose output values $f(x)$ and $g(x)$ are real numbers, can be combined to form new functions $f + g$, $f - g$, $f \cdot g$, and $\frac{f}{g}$ in a manner similar to the way we add, subtract, multiply, and divide real numbers, by performing the corresponding operations directly on their outputs $f(x)$ and $g(x)$.

For example, if $f(x) = x^2 - 2x + 4$ and $g(x) = x + 2$, then

$$f(x) + g(x) = (x^2 - 2x + 4) + (x + 2) = x^2 - x + 6.$$

This new function $y = x^2 - x + 6$ is called the sum function $f + g$.

Sum, Difference, Product, and Quotient of Functions

Let f and g be two functions. The **sum** $f + g$, the **difference** $f - g$, the **product** fg, and the **quotient** $\dfrac{f}{g}$ are functions defined as follows:

- Sum $\quad\quad (f + g)(x) = f(x) + g(x)$
- Difference $(f - g)(x) = f(x) - g(x)$
- Product $\quad fg(x) = f(x) \cdot g(x)$; also written as $(fg)(x)$
- Quotient $\quad \dfrac{f}{g}(x) = \dfrac{f(x)}{g(x)}$, provided that $g(x) \neq 0$; also written as $\left(\dfrac{f}{g}\right)(x)$

The **domain** of each of the new functions consists of those values of x that are common to the domains of f and g, except that for $\dfrac{f}{g}$, all x for which $g(x) = 0$ must also be excluded.

EXAMPLE 1 Combining Functions

Let $f(x) = x^2 - 6x + 8$ and $g(x) = x - 2$. Find each of the following functions or function values.

a. $(f - g)(4)$ **b.** $\left(\dfrac{f}{g}\right)(3)$

c. $(f + g)(x)$ **d.** $(f - g)(x)$

e. $(fg)(x)$ **f.** $\left(\dfrac{f}{g}\right)(x)$

Solution

a. $f(4) = 4^2 - 6(4) + 8 = 0$ Replace x with 4 in $f(x)$.

       $g(4) = 4 - 2 = 2$ Replace x with 4 in $g(x)$.

   $(f - g)(4) = f(4) - g(4)$ Definition of difference

          $= 0 - 2 = -2$ Replace $f(4)$ with 0, $g(4)$ with 2.

b. $f(3) = 3^2 - 6(3) + 8 = -1$ Replace x with 3 in $f(x)$.

       $g(3) = 3 - 2 = 1$ Replace x with 3 in $g(x)$.

   $\left(\dfrac{f}{g}\right)(3) = \dfrac{f(3)}{g(3)} = \dfrac{-1}{1} = -1$ Replace $f(3)$ with -1, $g(3)$ with 1.

c. $(f + g)(x) = f(x) + g(x)$ Definition of sum

          $= (x^2 - 6x + 8) + (x - 2)$ Add $f(x)$ and $g(x)$.

          $= x^2 - 5x + 6$ Combine like terms.

d. $(f - g)(x) = f(x) - g(x)$ Definition of difference

          $= (x^2 - 6x + 8) - (x - 2)$ Subtract $g(x)$ from $f(x)$.

          $= x^2 - 7x + 10$ Combine like terms.

e. $(fg)(x) = f(x) \cdot g(x)$ Definition of product

$\qquad = (x^2 - 6x + 8)(x - 2)$ Multiply $f(x)$ and $g(x)$.

$\qquad = x^2(x - 2) - 6x(x - 2) + 8(x - 2)$ Distributive property

$\qquad = x^3 - 2x^2 - 6x^2 + 12x + 8x - 16$ Distributive property

$\qquad = x^3 - 8x^2 + 20x - 16$ Combine like terms.

f. $\left(\dfrac{f}{g}\right)(x) = \dfrac{f(x)}{g(x)}, \quad g(x) \neq 0$ Definition of quotient

$\qquad = \dfrac{x^2 - 6x + 8}{x - 2}, \quad x - 2 \neq 0$ Replace $f(x)$ with $x^2 - 6x + 8$ and $g(x)$ with $x - 2$.

$\qquad = \dfrac{(x - 2)(x - 4)}{x - 2}, \quad x \neq 2$ Factor the numerator.

$\qquad = x - 4, \quad x \neq 2$

Because f and g are polynomials, the domain of both f and g is the set of all real numbers, or in interval notation, $(-\infty, \infty)$. So the domain for $f + g, f - g,$ and fg is $(-\infty, \infty)$. However, for $\dfrac{f}{g}$, we must exclude 2 because $g(2) = 0$. The domain for $\dfrac{f}{g}$ is the set of all real numbers $x, x \neq 2$ or in interval notation, $(-\infty, 2) \cup (2, \infty)$.

Practice Problem 1 Let $f(x) = 3x - 1$ and $g(x) = x^2 + 2$. Find $(f + g)(x)$, $(f - g)(x), (fg)(x),$ and $\left(\dfrac{f}{g}\right)(x)$.

▲
WARNING

When rewriting a function such as

$$\left(\dfrac{f}{g}\right)(x) = \dfrac{(x - 2)(x - 4)}{x - 2}, \quad x \neq 2,$$

by removing the common factor, remember to specify "$x \neq 2$."
We identify the domain before we simplify and write

$$\left(\dfrac{f}{g}\right)(x) = x - 4, \quad x \neq 2.$$

SIDE
NOTE

An alternative way of finding the domain of combined functions is to directly investigate the output formula they produce. However, we have to be careful to not simplify the resulting expression before finding the domain.

EXAMPLE 2   **Finding the Domain**

Let $f(x) = \sqrt{x + 2}$ and $g(x) = \sqrt{5 - x}$. Find the domain of $f + g, f - g, fg,$ and $\dfrac{f}{g}$.

Solution

Recall (see page 942) that $\sqrt{u}$ is defined only if $u \geq 0$.

So $f(x) = \sqrt{x + 2}$ is defined if $x + 2 \geq 0$ or $x \geq -2$.   Solve for x.

The domain of $f = D_1 = [-2, \infty)$ Interval notation

Similarly, $g(x) = \sqrt{5 - x}$ is defined only if $5 - x \geq 0$

$\qquad\qquad\qquad\qquad$ or $\quad 5 \geq x$ Solve for x.

$\qquad\qquad\qquad\qquad$ or $\quad x \leq 5$ Rewrite.

The domain of $g = D_2 = (-\infty, 5]$ Interval notation

To find the required domains, we first find $D_1 \cap D_2$ from the graphs of D_1 and D_2. See Figure 1.92.

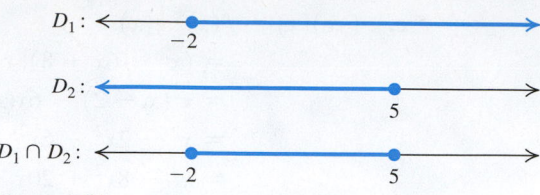

Figure 1.92

The domain of $f + g, f - g$, and fg is $D_1 \cap D_2 = [-2, 5]$. For the domain of $\dfrac{f}{g}$, we must

exclude from $D_1 \cap D_2$ those numbers where $g(x) = \sqrt{5 - x} = 0$. Solving this equation

for x, we have $x = 5$. So the domain of $\dfrac{f}{g}$ is $[-2, 5)$.

Practice Problem 2 Let $f(x) = \sqrt{x - 1}$ and $g(x) = \sqrt{3 - x}$. Find the domain of

$fg, \dfrac{f}{g}$, and $\dfrac{g}{f}$.

2 Form composite functions.

Composition of Functions

Another way of combining two functions is for the output of one function to be fed directly into the second function as input. This is an especially useful way of constructing new functions and further expands the number of functions we can build from the basic functions described in our library of functions. Figure 1.93 shows what happens when you apply one function $g(x) = x^2 - 1$ to an input (domain) value, x, and then apply a second function $f(x) = \sqrt{x}$ to the output value, $g(x)$, from the first function. In this case, the output from the first function becomes the input for the second function.

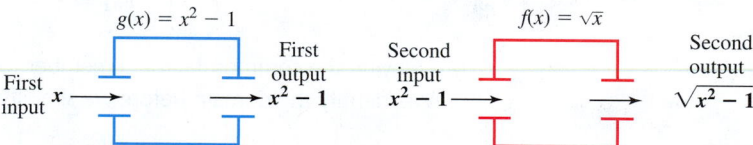

Figure 1.93 Composition of functions.

Composition of Functions

If f and g are two functions, the composition of the function f with the function g is written as $f \circ g$ and is defined by the equation

$$(f \circ g)(x) = f(g(x))$$

We read $(f \circ g)(x)$ as "f composed with g of x" and $f(g(x))$ as "f of g of x". The domain of $f \circ g$ consists of those values x in the domain of g for which $g(x)$ is in the domain of f.

Figure 1.94 may help clarify the definition of $f \circ g$.

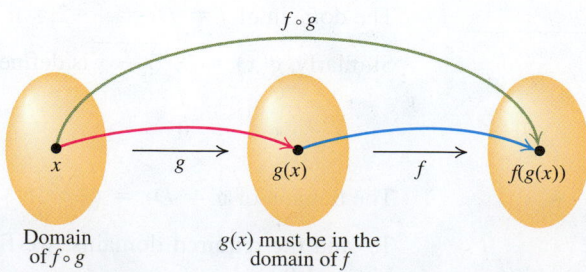

Figure 1.94 Composition of the function f with g.

Note that the range of the inside function g (the first function to be evaluated) needs to be within the domain of the outside function f. Less formally, the composition $f \circ g$ has to make sense in terms of inputs and outputs.

Evaluating $(f \circ g)(x)$ By definition,

```
                 inner function
                      ↓
     (f ∘ g)(x) = f(g(x))
                      ↑
                 outer function
```

Thus, to evaluate $(f \circ g)(x)$, we must

(i) evaluate the inner function g at x.
(ii) use the result $g(x)$ as the input for the outer function f.

EXAMPLE 3 Evaluating a Composite Function

Let $f(x) = x^3$ and $g(x) = x + 1$. Find each composite function value.

a. $(f \circ g)(1)$ **b.** $(g \circ f)(1)$ **c.** $(f \circ f)(-1)$

SIDE NOTE

When computing $(f \circ g)(x) = f(g(x))$, you must replace every x in $f(x)$ with $g(x)$.

Solution

a. $(f \circ g)(1) = f(g(1))$ Definition of $f \circ g$
$\qquad\qquad\quad = f(2)$ $g(1) = 1 + 1 = 2$
$\qquad\qquad\quad = 2^3 = 8$ Evaluate $f(2)$ and simplify.

b. $(g \circ f)(1) = g(f(1))$ Definition of $g \circ f$
$\qquad\qquad\quad = g(1)$ $f(1) = 1^3 = 1$
$\qquad\qquad\quad = 1 + 1 = 2$ Evaluate $g(1)$ and simplify.

c. $(f \circ f)(-1) = f(f(-1))$ Definition of $f \circ f$
$\qquad\qquad\quad = f(-1)$ $f(-1) = (-1)^3 = -1$
$\qquad\qquad\quad = (-1)^3 = -1$ Evaluate $f(-1)$ and simplify.

Practice Problem 3 Let $f(x) = -5x$ and $g(x) = x^2 + 1$. Find each composite function value.

a. $(f \circ g)(0)$ **b.** $(g \circ f)(0)$

EXAMPLE 4 Finding Composite Functions

Let $f(x) = 2x + 1$ and $g(x) = x^2 - 3$. Find each composite function.

a. $(f \circ g)(x)$ **b.** $(g \circ f)(x)$ **c.** $(f \circ f)(x)$

Solution

a. $(f \circ g)(x) = f(g(x))$ Definition of $f \circ g$
$\qquad\qquad\quad = f(x^2 - 3)$ Replace $g(x)$ with $x^2 - 3$.
$\qquad\qquad\quad = 2(x^2 - 3) + 1$ Replace x with $x^2 - 3$ in $f(x) = 2x - 1$.
$\qquad\qquad\quad = 2x^2 - 5$ Simplify.

b. $(g \circ f)(x) = g(f(x))$ Definition of $g \circ f$
$\qquad\qquad\quad = g(2x + 1)$ Replace $f(x)$ with $2x + 1$.
$\qquad\qquad\quad = (2x + 1)^2 - 3$ Replace x with $2x + 1$ in $g(x) = x^2 - 3$.
$\qquad\qquad\quad = 4x^2 + 4x + 1 - 3$ $(A + B)^2 = A^2 + 2AB + B^2$
$\qquad\qquad\quad = 4x^2 + 4x - 2$ Simplify.

c. $(f \circ f)(x) = f(f(x))$ Definition of $f \circ f$

                   $= f(2x + 1)$ Replace $f(x)$ with $2x + 1$.

                   $= 2(2x + 1) + 1$ Replace x with $2x + 1$ in $f(x) = 2x + 1$.

                   $= 4x + 3$ Simplify.

Practice Problem 4 Let $f(x) = 2 - x$ and $g(x) = 2x^2 + 1$. Find each composite function.

a. $(g \circ f)(x)$ **b.** $(f \circ g)(x)$ **c.** $(g \circ g)(x)$ 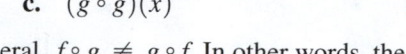

Parts **a.** and **b.** of Example 3 illustrate that in general, $f \circ g \neq g \circ f$. In other words, the composition of functions is not commutative.

⚠ **WARNING**   Note that $f(g(x))$ is not $f(x) \cdot g(x)$. In $f(g(x))$, the output of g is *used as an input* for f; in contrast, in $f(x) \cdot g(x)$, the output of $f(x)$ is *multiplied* by the output of $g(x)$.

 Find the domain of a composite function.

Domain of Composite Functions

Let f and g be two functions. The domain of the composite function $f \circ g$ consists of those values of x in the domain of g for which $g(x)$ is in the domain of f. So to find the domain of $f \circ g$, we first find the domain of g. Then from the domain of g, we exclude those values of x for which $g(x)$ is not in the domain of f.

PROCEDURE
IN ACTION

EXAMPLE 5   Finding the Domain of a Composite Function

OBJECTIVE

Find the domain of $f \circ g$ from the defining equations for f and g.

EXAMPLE

Find the domain of $f \circ g$ if

$$f(x) = \frac{2}{x - 3} \quad \text{and} \quad g(x) = \frac{5}{x + 1}.$$

Step 1 Let A be the domain of the inner function g. We may find A by first finding the values of x for which g is not defined.

1. $g(x) = \dfrac{5}{x + 1}$ is not defined if the denominator $x + 1 = 0$, or $x = -1$. So $A = \{x \mid x \neq -1\}$, or $A = (-\infty, -1) \cup (-1, \infty)$.

Step 2 Let B be the set of values of x for which $g(x)$ is in the domain of f. As before, we may do this by finding the values of x for which $g(x)$ is not in the domain of f.

2. $f(g(x)) = \dfrac{2}{g(x) - 3}$, so $f(g(x))$ is not defined if the denominator

$$g(x) - 3 = 0$$

$$\frac{5}{x + 1} - 3 = 0 \quad \text{Replace } g(x) \text{ with } \frac{5}{x + 1}.$$

$$5 - 3(x + 1) = 0 \quad \text{Multiply by } (x + 1)$$

$$5 - 3x - 3 = 0 \quad \text{Simplify}$$

$$x = \frac{2}{3} \quad \text{Solve for } x.$$

$$B = \left\{ x \,\middle|\, x \neq \frac{2}{3} \right\}, \text{ or } B = \left(-\infty, \frac{2}{3} \right) \cup \left(\frac{2}{3}, \infty \right)$$

Step 3 The domain of $f \circ g$ is $A \cap B$. The set $A \cap B$ is the values of x that are in the domain of g *and* for which $g(x)$ is in the domain of f.

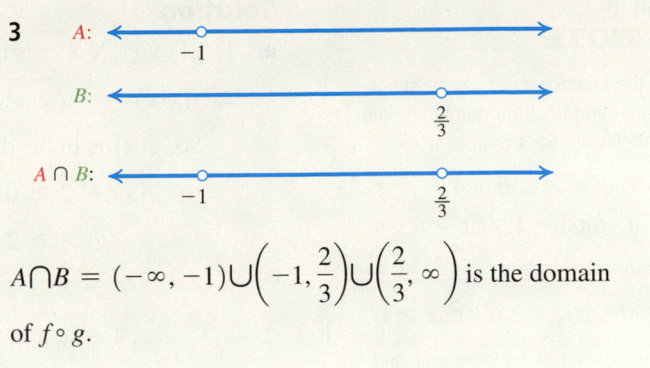

$A \cap B = (-\infty, -1) \cup \left(-1, \dfrac{2}{3}\right) \cup \left(\dfrac{2}{3}, \infty\right)$ is the domain of $f \circ g$.

Practice Problem 5 Let $f(x) = \sqrt{x + 1}$ and $g(x) = \dfrac{2}{x - 3}$. Find the domain of $f \circ g$.

An alternative way to find the domain of the composition of two functions is to directly investigate the output formula they produce. However, we have to be careful to not simplify the resulting expressions before finding the domain.

To find the domain of $f \circ g$ if $f(x) = \dfrac{2}{x - 3}$ and $g(x) = \dfrac{5}{x + 1}$, we first derive the formula for the composition. We start with the outer function

$$f(\square) = \frac{2}{\square - 3}$$

and we insert the expression for the inner function

$$f \circ g(x) = f(g(x)) = f\left(\frac{5}{x + 1}\right) = \frac{2}{\frac{5}{x + 1} - 3}.$$

In the formula, we have two fractions

$$\frac{5}{x + 1} \quad \text{and} \quad \frac{2}{\frac{5}{x + 1} - 3}.$$

which are not defined if either of their denominators is 0. To find the domain of $f \circ g$, we need to exclude every x that is a zero of one of the above denominators.

$$x + 1 = 0 \quad \text{or} \quad \frac{5}{x + 1} - 3 = 0.$$

The first equation gives the solution $x = -1$, and the second gives the solution $x = \dfrac{2}{3}$.

The domain of $f \circ g$ is all real numbers except -1 and $\frac{2}{3}$. In set notation, the domain is $(-\infty, -1) \cup \left(-1, \frac{2}{3}\right) \cup \left(\frac{2}{3}, \infty\right)$.

EXAMPLE 6 **Finding the Domain of Composite Functions**

Let $f(x) = \sqrt{x - 2}$ and $g(x) = \sqrt{4 - x}$. Find the following functions and their domains using the procedure given in Example 5.

a. $f \circ g$

b. $g \circ f$

SIDE
NOTE

If the function $f \circ g$ can be simplified, determine the domain *before* simplifying. For example, if

$f(x) = x^2$ and $g(x) = \sqrt{x}$,

$(f \circ g)(x) = f(g(x)) = f(\sqrt{x})$

Before simplifying:

$(f \circ g)(x) = (\sqrt{x})^2, x \geq 0$

The domain of f is $(-\infty, \infty)$, and $(f \circ g)(x)$ is defined on $[0, \infty)$. After simplifying:

$(f \circ g)(x) = x$

Here it appears that the domain of $f \circ g$ is $(-\infty, \infty)$. However, this is not correct; it is $[0, \infty)$.

Solution

a. **1.** $g(x) = \sqrt{4 - x}$ is defined if $4 - x \geq 0$ or $x \leq 4$. So $A = (-\infty, 4]$.

2. $f(g(x)) = \sqrt{g(x) - 2}$ Replace x with $g(x)$ in $f(x) = \sqrt{x - 2}$.

So, $g(x)$ is in the domain of f if:

$g(x) - 2 \geq 0$

$g(x) \geq 2$ Add 2 to both sides.

$\sqrt{4 - x} \geq 2$ Replace $g(x)$ with $\sqrt{4 - x}$.

$4 - x \geq 4$ Square both sides.

$-x \geq 0$ Subtract 4 from both sides.

$x \leq 0$ Solve for x.

So $B = (-\infty, 0]$.

3. $A \cap B = (-\infty, 4] \cap (-\infty, 0]$

$= (-\infty, 0]$ See figure.

So, the domain of $f \circ g$ is $(-\infty, 0]$

$A \cap B = (-\infty, 0]$

b. To find the domain of $g \circ f$, we interchange the role of g and f.

1. $f(x) = \sqrt{x - 2}$ is defined if $x - 2 \geq 0$ or $x \geq 2$. So $A = [2, \infty)$.

2. $g(f(x)) = \sqrt{4 - f(x)}$ Replace x with $f(x)$ in $g(x) = \sqrt{4 - x}$.

So, $f(x)$ is in the domain of g if:

$4 - f(x) \geq 0$

$4 \geq f(x)$ Add $f(x)$ to both sides.

$f(x) \leq 4$ Rewrite

$\sqrt{x - 2} \leq 4$ Replace $f(x)$ with $\sqrt{x - 2}$.

$x - 2 \leq 16$ Square both sides (if $0 < a \leq b$, then $a^2 \leq b^2$).

$x \leq 18$ Solve for x.

So $B = (-\infty, 18]$.

3. $A \cap B = [2, \infty) \cap (-\infty, 18]$

$= [2, 18]$ See figure.

So, the domain of $g \circ f = [2, 18]$.

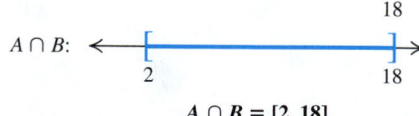

$A \cap B = [2, 18]$

Practice Problem 6 Let $f(x) = \sqrt{x - 1}$ and $g(x) = \sqrt{4 - x^2}$. Find each of the following domains.

a. $f \circ g$ **b.** $g \circ f$

Decomposition of a Function

In the composition of two functions, we combine two functions and create a new function. However, sometimes it is more useful to use the concept of composition to *decompose* a function into simpler functions. For example, consider the function $H(x) = \sqrt{x^2 - 2}$. A way to write $H(x)$ as the composition of two functions is to let $f(x) = \sqrt{x}$ and $g(x) = x^2 - 2$ so that $f(g(x)) = f(x^2 - 2) = \sqrt{x^2 - 2} = H(x)$.

A function may be decomposed into simpler functions in several different ways by choosing various "inner" functions.

PROCEDURE
IN ACTION

| **EXAMPLE 7** | **Decomposing a Function** |

OBJECTIVE

Write a function H as a composite of simpler functions f and g so that H = f ∘ g.

EXAMPLE

Write

$$H(x) = \frac{1}{\sqrt{2x^2 + 1}}$$

as f ∘ g for some functions f and g.

Step 1 Define $g(x)$ as any expression in the defining equation for H.

1. Let $g(x) = 2x^2 + 1$.

Step 2 To get $f(x)$ from the defining equation for H, (1) replace the letter H with f and (2) replace the expression chosen for $g(x)$ with x.

2. $H(x) = \dfrac{1}{\sqrt{2x^2 + 1}}$

becomes

$$f(x) = \frac{1}{\sqrt{x}}$$

Replace the expression $2x^2 + 1$ from Step 1 with x and replace H with f.

Step 3 Now we have

$$H(x) = f(g(x))$$
$$= (f \circ g)(x).$$

3. $f(g(x)) = f(2x^2 + 1)$ $f(x) = \dfrac{1}{\sqrt{x}}, g(x) = 2x^2 + 1$

$$= \frac{1}{\sqrt{2x^2 + 1}}$$
$$= H(x)$$

Note: Other choices include $g(x) = 2x^2$ and $f(x) = \dfrac{1}{\sqrt{x + 1}}$.

Practice Problem 7 Repeat Example 7, letting $g(x) = \sqrt{2x^2 + 1}$. Find $f(x)$ and write $H(x) = (f \circ g)(x)$.

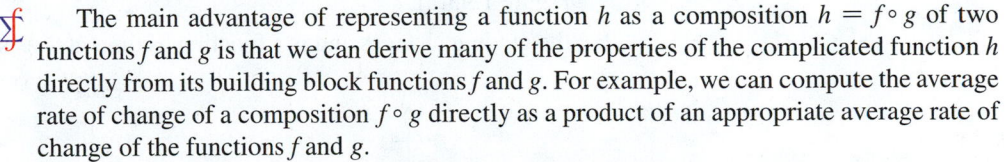

∫   The main advantage of representing a function h as a composition $h = f \circ g$ of two functions f and g is that we can derive many of the properties of the complicated function h directly from its building block functions f and g. For example, we can compute the average rate of change of a composition $f \circ g$ directly as a product of an appropriate average rate of change of the functions f and g.

ARC of $f \circ g$ from a to $b = \dfrac{f(g(b)) - f(g(a))}{b - a}$

$$= \frac{f(g(b)) - f(g(a))}{g(b) - g(a)} \cdot \frac{g(b) - g(a)}{b - a}$$

$$= (\text{ARC of } f \text{ from } g(a) \text{ to } g(b)) \cdot (\text{ARC of } g \text{ from } a \text{ to } b),$$

as long as $a \neq b$ and $g(a) \neq g(b)$.

| **EXAMPLE 8** | **Average Rate of Change of a Composite Function** |

If $f(x) = x^2 + 1$ and $g(x) = 2x - 1$, find the average rate of change of the composite function $f \circ g$ as x changes from $x = -1$ to $x = 1$.

SIDE
NOTE

The average rate of change of $f \circ g$ in the example can also be computed first by finding $f \circ g = 4x^2 - 4x + 2$ and then computing the average rate of change directly.

Solution

First, we compute the component related to the average rate of change of the inner function. $g(1) = 2 \cdot (1) - 1 = 1$ and $g(-1) = 2 \cdot (-1) - 1 = -3$.

$$\text{ARC of } g \text{ from } -1 \text{ to } 1 = \frac{g(1) - g(-1)}{1 - (-1)} = \frac{1 - (-3)}{1 - (-1)} = \frac{4}{2} = 2.$$

To compute the component related to the average rate of change of the outer function, we need to use the range for f as from $g(-1) = -3$ to $g(1) = 1$.

We have $f(1) = (1)^2 + 1 = 2$ and $f(-3) = (-3)^2 + 1 = 10$.

$$\text{ARC of } f \text{ from } g(-1) = -3 \text{ to } g(1) = 1 = \frac{f(1) - f(-3)}{1 - (-3)} = \frac{2 - 10}{1 - (-3)} = \frac{-8}{4} = -2.$$

Finally, we can compute the average rate of change of the composite function as

ARC of $f \circ g$ from -1 to 1

$$= (\text{ARC of } f \text{ from } g(-1) = -3 \text{ to } g(1) = 1) \cdot (\text{ARC of } g \text{ from } -1 \text{ to } 1)$$
$$= (-2) \cdot 2 = -4.$$

Practice Problem 8 Let $f(x) = 2 - x^2$ and $g(x) = 3x - 1$. Find the average rate of change of the composite function $f \circ g$ as x changes from $x = -1$ to $x = 2$.

5 Apply composition to practical problems.

Applications of Composite Functions

In a typical real-life example, we start with a given function $y = f(x)$ representing the known static relation between two variables, and we investigate what happens when the input variable x varies in time with $x = g(t)$.

EXAMPLE 9 **Calorie Burn Rate on a Treadmill**

According to the American College of Sports Medicine, the following equation can be used to estimate the amount of calories burned per minute by a 170-pound man walking on a treadmill at his preferred speed of 3.5 mph in relation to the grade g (expressed as a decimal):

Calories per minute burn rate: $\quad CBR = 0.39 \cdot (12.89 + 168.98 \cdot g)$.

Estimate the number of calories burned during a 1-hour exercise period assuming that:

a. the grade g is a constant function of time $g(t) = 0.03$ (grade of 3%) as shown in Figure 1.95(a).

b. the grade g changes and is a function of time $g(t)$ as shown in Figure 1.95(b).

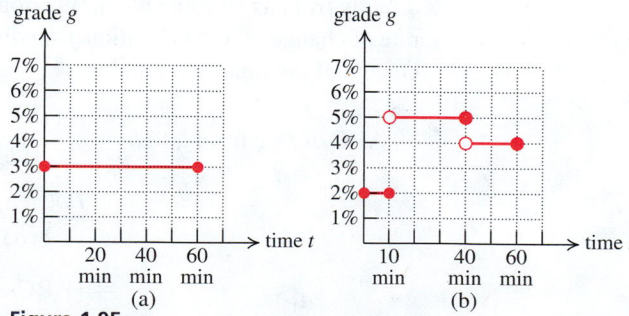

Figure 1.95

Solution

a. Since g is constant over the entire time period, the man will burn calories at the rate given by the equation:

Calories per minute burn rate: $CBR = 0.39 \cdot (12.89 + 168.98 \cdot 0.03) \approx 7.00$.

Since the exercise lasts 1 hour (60 minutes), he will burn $7 \cdot 60 = 420$ calories.

b. We can express g as a piecewise function:

$$g(t) = \begin{cases} 0.02 & \text{if } 0 \le t \le 10 \\ 0.05 & \text{if } 10 < t \le 40 \\ 0.04 & \text{if } 40 < t \le 60. \end{cases}$$

On each of the intervals, the man will burn calories at a different rate. That rate can be computed based on the main formula as:

$$CBR(t) = \begin{cases} 0.39 \cdot (12.89 + 168.98 \cdot 0.02) = 6.35 \text{ calories per minute} & \text{if } 0 \le t \le 10 \\ 0.39 \cdot (12.89 + 168.98 \cdot 0.05) = 8.32 \text{ calories per minute} & \text{if } 10 < t \le 40 \\ 0.39 \cdot (12.89 + 168.98 \cdot 0.04) = 7.66 \text{ calories per minute} & \text{if } 40 < t \le 60. \end{cases}$$

Taking into consideration the length of the intervals, we can estimate that the total number of burned calories is

$$6.35 \cdot 10 + 8.32 \cdot 30 + 7.66 \cdot 20 = 63.2 + 249.6 + 153.2 \approx 466.$$

Practice Problem 9 Repeat Example 9 for a 140-pound woman walking on a treadmill at her prefered speed of 3.0 mph using the following formula for the rate of calories burned per minute:

Calories per minute burn rate: $CBR = 0.32 \cdot (11.55 + 144.84 \cdot g)$.

The general formula appropriate for speeds between 1.9 to 3.7 mph is given by

$$\text{Calories per minute burn rate} = \frac{(3.5 + 0.1 \cdot s + 1.8 \cdot s \cdot g) \cdot w}{200},$$

where s is speed expressed in meters per minute, w is weight expressed in kilograms, and g is a grade expressed in decimal notation.

EXAMPLE 10 **Applying Composition to Sales**

A car dealer offers an 8% discount off the manufacturer's suggested retail price (MSRP) of x dollars for any new car on his lot. At the same time, the manufacturer offers a $4000 rebate for each purchase of a new car.

a. Write a function $r(x)$ that represents the price after only the rebate.

b. Write a function $d(x)$ that represents the price after only the dealer's discount.

c. Write the functions $(r \circ d)(x)$ and $(d \circ r)(x)$. What do they represent?

d. Calculate $(d \circ r)(x) - (r \circ d)(x)$. Interpret this expression.

Solution

a. The MSRP is x dollars, and the manufacturer's rebate is $4000 for each car; so

$$r(x) = x - 4000$$

represents the price of a car after the rebate.

b. The dealer's discount is 8% of x, or $0.08x$; therefore,

$$d(x) = x - 0.08x = 0.92x$$

represents the price of the car after the dealer's discount.

c. **(i)** $(r \circ d)(x) = r(d(x))$ Apply the dealer's discount first.

  $= r(0.92x)$ $d(x) = 0.92x$

  $= 0.92x - 4000$ Evaluate $r(0.92x)$ using $r(x) = x - 4000$.

Then $(r \circ d)(x) = 0.92x - 4000$ represents the price when the dealer's discount is applied first.

(ii) $(d \circ r)(x) = d(r(x))$ Apply the manufacturer's rebate first.

$\qquad\qquad = d(x - 4000)$ $r(x) = x - 4000$

$\qquad\qquad = 0.92(x - 4000)$ Evaluate $d(x - 4000)$ using $d(x) = 0.92x$.

$\qquad\qquad = 0.92x - 3680$

Thus, $(d \circ r)(x) = 0.92x - 3680$ represents the price when the manufacturer's rebate is applied first.

d. $(d \circ r)(x) - (r \circ d)(x) = (0.92x - 3680) - (0.92x - 4000)$ Subtract function values.

$\qquad\qquad\qquad\qquad\qquad = \$320$

This equation shows that any car, regardless of its price, will cost $320 more if you apply the rebate first and then the discount.

Practice Problem 10 Repeat Example 10 if the dealer offers a 6% discount and the manufacturer offers a $4500 rebate.

Answers to Practice Problems

1. $(f + g)(x) = x^2 + 3x + 1$
$(f - g)(x) = -x^2 + 3x - 3$
$(fg)(x) = 3x^3 - x^2 + 6x - 2$
$\left(\dfrac{f}{g}\right)(x) = \dfrac{3x - 1}{x^2 + 2}$

2. Domains $f \cdot g = [1, 3]$,
$\dfrac{f}{g} = [1, 3) \dfrac{g}{f} = (1, 3]$

3. a. $(f \circ g)(0) = -5$
b. $(g \circ f)(0) = 1$

4. a. $(g \circ f)(x) = 2x^2 - 8x + 9$
b. $(f \circ g)(x) = 1 - 2x^2$
c. $(g \circ g)(x) = 8x^4 + 8x^2 + 3$

5. $(-\infty, 1] \cup (3, \infty)$
6. a. $[-\sqrt{3}, \sqrt{3}]$
b. $[1, 5]$
7. $f(x) = \dfrac{1}{x}$
8. $ARC = -3$
9. a. 305 calories   **b.** 338 calories
10. a. $r(x) = x - 4500$
b. $d(x) = 0.94x$
c. (i) $(r \circ d)(x) = 0.94x - 4500$
(ii) $(d \circ r)(x) = 0.94x - 4230$
d. $(d \circ r)(x) - (r \circ d)(x) = 270$

SECTION 1.6   Exercises

Concepts and Vocabulary

1. $(f \cdot g)(x) = $ _____ .

2. The domain of the function $f + g$ consists of those values of x that are _____ to the domains of f and g.

3. The composition of the function f with the function g is written as $f \circ g$ and is defined by $f \circ g(x) = $ _____ .

4. The domain of the composite function $f \circ g$ consists of those values of x in the domain of g for which $g(x)$ _____ .

5. **True or False.** We always have $f \circ g = g \circ f$.

6. **True or False.** If $f(1) = 2$ and $g(2) = 1$, then $(f \circ g)(2) = 2$.

7. **True or False.** The domain of $f \cdot g$ and the domain of $\frac{f}{g}$ are always the same.

8. **True or False.** A function may be decomposed into simpler functions in several different ways.

Building Skills

In Exercises 9–20, use the graphs of f and g shown in the figure to evaluate each expression.

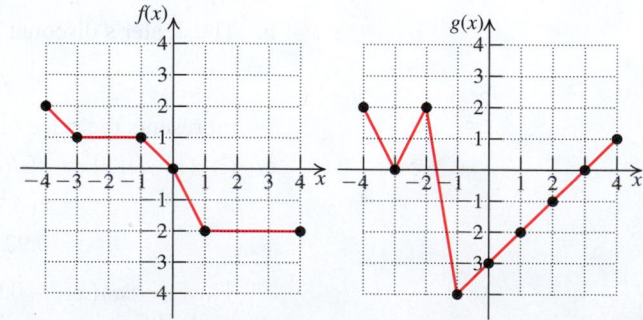

9. $(f + g)(-2)$

10. $(f + g)(2)$

11. $(f - g)(4)$

12. $(f - g)(-1)$

13. $(f \cdot g)(-1)$

14. $(f \cdot g)(2)$

15. $\left(\dfrac{f}{g}\right)(-2)$

16. $\left(\dfrac{f}{g}\right)(2)$

17. $(f \circ g)(1)$

18. $(g \circ f)(1)$

19. $(f \circ g)(-3)$

20. $(g \circ f)(-3)$

In Exercises 21–24, functions f and g are given. Find each of the given values.

a. $(f + g)(-1)$

b. $(f - g)(0)$

c. $(f \cdot g)(2)$

d. $\left(\dfrac{f}{g}\right)(1)$

21. $f(x) = 2x; g(x) = -x$

22. $f(x) = 1 - x^2; g(x) = x + 1$

23. $f(x) = \dfrac{1}{\sqrt{x + 2}}; g(x) = 2x + 1$

24. $f(x) = \dfrac{x}{x^2 - 6x + 8}; g(x) = 3 - x$

In Exercises 25–38, functions f and g are given. Find each of the following functions and state its domain.

a. $f + g$

b. $f - g$

c. $f \cdot g$

d. $\dfrac{f}{g}$

e. $\dfrac{g}{f}$

25. $f(x) = x - 3; g(x) = x^2$

26. $f(x) = 2x - 1; g(x) = x^2$

27. $f(x) = x^3 - 1; g(x) = 2x^2 + 5$

28. $f(x) = x^2 - 4; g(x) = x^2 - 6x + 8$

29. $f(x) = 2x - 1; g(x) = \sqrt{x}$

30. $f(x) = x - 1; g(x) = \sqrt{x}$

31. $f(x) = x - 6; g(x) = \sqrt{x - 3}$

32. $f(x) = x + 2; g(x) = \sqrt{1 - x}$

33. $f(x) = 1 - \dfrac{2}{x + 1}; g(x) = \dfrac{1}{x}$

34. $f(x) = 1 - \dfrac{1}{x}; g(x) = \dfrac{1}{x}$

35. $f(x) = \dfrac{2}{x + 1}; g(x) = \dfrac{x}{x + 1}$

36. $f(x) = \dfrac{5x - 1}{x + 1}; g(x) = \dfrac{4x + 10}{x + 1}$

37. $f(x) = \dfrac{x^2}{x + 1}; g(x) = \dfrac{2x}{x^2 - 1}$

38. $f(x) = \dfrac{x - 3}{x^2 - 25}; g(x) = \dfrac{x - 3}{x^2 + 9x + 20}$

In Exercises 39–42, find the domain of each function.

a. $f \cdot g$

b. $\dfrac{f}{g}$

39. $f(x) = \sqrt{x - 1}; \quad g(x) = \sqrt{5 - x}$

40. $f(x) = \sqrt{x - 2}; \quad g(x) = \sqrt{x + 2}$

41. $f(x) = \sqrt{x + 2}; \quad g(x) = \sqrt{9 - x^2}$

42. $f(x) = \sqrt{x^2 - 4}; \quad g(x) = \sqrt{25 - x^2}$

In Exercises 43 and 44, use each diagram to evaluate $(g \circ f)(x)$. Then evaluate $(g \circ f)(2)$ and $(g \circ f)(-3)$.

43.

$x \rightarrow \boxed{f(x) = x^2 - 1} \rightarrow \boxed{g(x) = 2x + 3} \rightarrow$

44.

$x \rightarrow \boxed{f(x) = |x + 1|} \rightarrow \boxed{g(x) = 3x^2 - 1} \rightarrow$

In Exercises 45–56, let $f(x) = 2x + 1$ and $g(x) = 2x^2 - 3$. Evaluate each expression.

45. $(f \circ g)(2)$

46. $(g \circ f)(2)$

47. $(f \circ g)(-3)$

48. $(g \circ f)(-5)$

49. $(f \circ g)(0)$

50. $(g \circ f)\left(\dfrac{1}{2}\right)$

51. $(f \circ g)(-c)$

52. $(f \circ g)(c)$

53. $(g \circ f)(a)$

54. $(g \circ f)(-a)$

55. $(f \circ f)(1)$

56. $(g \circ g)(-1)$

In Exercises 57–62, the functions f and g are given. Find $f \circ g$ and its domain.

57. $f(x) = \dfrac{2}{x + 1}; g(x) = \dfrac{1}{x}$

58. $f(x) = \dfrac{1}{x - 1}; g(x) = \dfrac{2}{x + 3}$

59. $f(x) = \sqrt{x - 3}; g(x) = 2 - 3x$

60. $f(x) = \dfrac{x}{x - 1}; g(x) = 2 + 5x$

61. $f(x) = |x|; g(x) = x^2 - 1$

62. $f(x) = 3x - 2; g(x) = |x - 1|$

In Exercises 63–78, the functions f and g are given. Find each composite function and describe its domain.

a. $f \circ g$

b. $g \circ f$

c. $f \circ f$

d. $g \circ g$

63. $f(x) = 2x - 3; g(x) = x + 4$

64. $f(x) = x - 3; g(x) = 3x - 5$

65. $f(x) = 1 - 2x; g(x) = 1 + x^2$

66. $f(x) = 2x - 3; g(x) = 2x^2$

67. $f(x) = 2x^2 + 3x; g(x) = 2x - 1$

68. $f(x) = x^2 + 3x; g(x) = 2x$

69. $f(x) = x^2; g(x) = \sqrt{x}$

70. $f(x) = x^2 + 2x; g(x) = \sqrt{x + 2}$

71. $f(x) = \dfrac{1}{2x - 1}; g(x) = \dfrac{1}{x^2}$

72. $f(x) = x - 1; g(x) = \dfrac{x}{x + 1}$

73. $f(x) = \sqrt{x - 1}; g(x) = \sqrt{4 - x}$

74. $f(x) = x^2 - 4; g(x) = \sqrt{4 - x^2}$

75. $f(x) = \dfrac{1 - x}{x + 2}; g(x) = \dfrac{x + 3}{x - 4}$

76. $f(x) = \dfrac{x + 2}{x - 3}; g(x) = \dfrac{x + 1}{x - 1}$

77. $f(x) = 1 + \dfrac{1}{x}$; $g(x) = \dfrac{1+x}{1-x}$

78. $f(x) = \sqrt[3]{x+1}$; $g(x) = x^3 + 1$

In Exercises 79–82, let $g(x)$ be a piecewise function given below. For each f, find the domain of the composite function $f \circ g$.

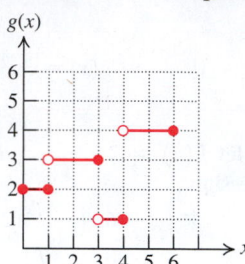

79. $f(x) = \dfrac{3}{x-3}$

80. $f(x) = \dfrac{2}{2-x}$

81. $f(x) = \sqrt{3-x}$

82. $f(x) = \sqrt{x-2}$

In Exercises 83–86, let $g(x)$ be a piecewise function given below. For each f, write the composite function $f \circ g$ as a piecewise function.

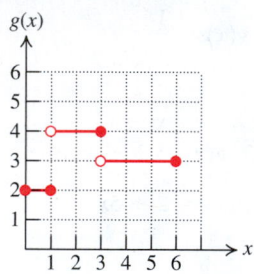

83. $f(x) = x^2 - 2$

84. $f(x) = 1 - x^2$

85. $f(x) = \dfrac{1}{2+x}$

86. $f(x) = \dfrac{1}{1-x}$

In Exercises 87–96, express the given function H as a composition of two functions f and g such that $H(x) = (f \circ g)(x)$.

87. $H(x) = \sqrt{x+2}$

88. $H(x) = |3x+2|$

89. $H(x) = (x^2 - 3)^{10}$

90. $H(x) = \sqrt{3x^2 + 5}$

91. $H(x) = \dfrac{1}{3x-5}$

92. $H(x) = \dfrac{5}{2x+3}$

93. $H(x) = \sqrt[3]{x^2 - 7}$

94. $H(x) = \sqrt[4]{x^2 + x + 1}$

95. $H(x) = \dfrac{1}{|x^3 - 1|}$

96. $H(x) = \sqrt[3]{1 + \sqrt{x}}$

In Exercises 97–102, the functions f and g are given. Find the average rate of change of the composite function $f \circ g$ as x changes from a to b.

97. $f(x) = x^2 + 2$; $g(x) = 1 - 2x$; $a = 1, b = 2$

98. $f(x) = 1 - x^2$; $g(x) = 1 + 3x$; $a = -1, b = 1$

99. $f(x) = x^3 + 2$; $g(x) = 1 - x^2$; $a = 1, b = 2$

100. $f(x) = 1 - x^3$; $g(x) = x^2 + 1$; $a = -1, b = 0$

101. $f(x) = \dfrac{1}{4+x}$; $g(x) = x^2 - 1$; $a = 1, b = 2$

102. $f(x) = \dfrac{1}{2+x}$; $g(x) = x^2 + 1$; $a = 0, b = 2$

Applying the Concepts

103. Cost, revenue, and profit. A retailer purchases x shirts from a wholesaler at a price of $12 per shirt. Her selling price for each shirt is $22. The state has a 7% sales tax. Interpret each of the following functions.
a. $f(x) = 12x$
b. $g(x) = 22x$
c. $h(x) = g(x) + 0.07g(x)$
d. $P(x) = g(x) - f(x)$

104. Cost, revenue, and profit. The demand equation for a product is given by $x = 5000 - 5p$, where x is the number of units produced and sold at price p (in dollars) per unit. The cost (in dollars) of producing x units is given by $C(x) = 4x + 12{,}000$. Express each of the following as a function of price.
a. Cost
b. Revenue
c. Profit

105. Cost and revenue. A manufacturer of radios estimates that his daily cost of producing x radios is given by the equation $C = 350 + 5x$. The equation $R = 25x$ represents the revenue in dollars from selling x radios.
a. Write and simplify the profit function
$$P(x) = R(x) - C(x).$$
b. Find $P(20)$. What does the number $P(20)$ represent?
c. How many radios should the manufacturer produce and sell to have a daily profit of $500?
d. Find the composite function $(R \circ x)(C)$. What does this function represent?
[*Hint:* To find $x(C)$, solve $C = 350 + 5x$ for x.]

106. Mail order. You order merchandise worth x dollars from D-Bay Manufacturers. The company charges you sales tax of 4% of the purchase price plus a shipping-and-handling fee of $3 plus 2% of the after-tax purchase price.
a. Write a function $g(x)$ that represents the sales tax.
b. What does the function $h(x) = x + g(x)$ represent?
c. Write a function $f(x)$ that represents the shipping-and-handling fee.
d. What does the function $T(x) = h(x) + f(x)$ represent?

107. Consumer issues. You work in a department store in which employees are entitled to a 30% discount on their purchases. You also have a coupon worth $5 off any item. Let x represent the retail price of an item.
a. Write a function $f(x)$ that models discounting an item by 30%.
b. Write a function $g(x)$ that models applying the coupon.
c. Use a composition of your two functions from (a) and (b) to model your cost for an item assuming that the clerk applies the discount first and then the coupon.
d. Use a composition of your two functions from (a) and (b) to model your cost for an item assuming

that the clerk applies the coupon first and then the discount.

e. Use the composite functions from (c) and (d) to find how much more an item costs assuming that the clerk applies the coupon first.

108. Consumer issues. You work in a department store in which employees are entitled to a 20% discount on their purchases. In addition, the store is offering a 10% discount on all items. Let x represent the retail price of an item.

a. Write a function $f(x)$ that models discounting an item by 20%.

b. Write a function $g(x)$ that models discounting an item by 10%.

c. Use a composition of your two functions from (a) and (b) to model taking the 20% discount first.

d. Use a composition of your two functions from (a) and (b) to model taking the 10% discount first.

e. Use the composite functions from (c) and (d) to decide which discount you would prefer to take first.

109. Test grades. Professor Harsh gave a test to his college algebra class, and nobody got more than 80 points (out of 100) on the test. One problem worth 8 points had insufficient data, so nobody could solve that problem. The professor adjusted the grades for the class by (1) increasing everyone's score by 10% and (2) giving everyone 8 bonus points. Let x represent the original score of a student.

a. Write statements (1) and (2) as functions $f(x)$ and $g(x)$, respectively.

b. Find $(f \circ g)(x)$ and explain what it means.

c. Find $(g \circ f)(x)$ and explain what it means.

d. Evaluate $(f \circ g)(70)$ and $(g \circ f)(70)$.

e. Does $(f \circ g)(x) = (g \circ f)(x)$?

f. Suppose a score of 90 or better results in an A. What is the lowest original score that will result in an A if the professor uses

   (i) $(f \circ g)(x)$?

   (ii) $(g \circ f)(x)$?

110. Sales commissions. Henrita works as a salesperson in a department store. Her weekly salary is $200 plus a 3% bonus on weekly sales over $8000. Suppose her sales in a week are x dollars.

a. Let $f(x) = 0.03x$. What does this mean?

b. Let $g(x) = x - 8000$. What does this mean?

c. Which composite function, $(f \circ g)(x)$ or $(g \circ f)(x)$, represents Henrita's bonus?

d. What was Henrita's salary for the week in which her sales were $17,500?

e. What were Henrita's sales for the week in which her salary was $521?

111. Enhancing a fountain. A circular fountain has a radius of x feet. A circular fence is installed around the fountain at a distance of 30 feet from its edge.

a. Write the function $f(x)$ that represents the area of the fountain.

b. Write the function $g(x)$ that represents the entire area enclosed by the fence.

c. What area does the function $g(x) - f(x)$ represent?

d. The cost of the fence was $4200, installed at the rate of $10.50 per running foot (per perimeter foot). You are to prepare an estimate for paving the area between the fence and the fountain at $1.75 per square foot. To the nearest dollar, what should your estimate be?

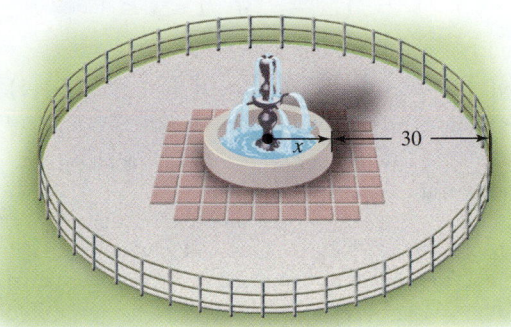

112. Running track design. An outdoor running track with semicircular ends and parallel sides is constructed. The length of each straight portion of the sides is 180 meters. The track has a uniform width of 4 meters throughout. The inner radius of each semicircular end is x meters.

a. Write the function $f(x)$ that represents the area enclosed within the *outer* edge of the running track.

b. Write the function $g(x)$ that represents the area of the field enclosed within the *inner* edge of the running track.

c. What does the function $f(x) - g(x)$ represent?

d. Suppose the inner perimeter of the track is 900 meters.

   (i) Find the area of the track.

   (ii) Find the outer perimeter of the track.

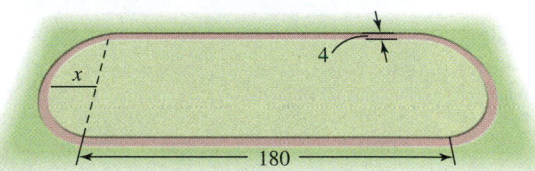

113. Oil Spill Suppose oil spills from a tanker into the Pacific Ocean and the area of the spill is a perfect circle. The radius of this oil slick increases at the rate of 2 miles per hour.

a. Express the area of the oil slick as a function of time.

b. Calculate the area covered by the oil slick in six hours.

114. Volume of a balloon. The volume V of a spherical balloon of radius r inches is given by the formula $V = f(r) = \dfrac{4}{3}\pi r^3$.

Suppose the balloon is being inflated and its radius is increasing at the rate of 2 inches per second. That is, $r = g(t) = 2t$, where t is time in seconds.

a. Find $(f \circ g)(t)$.

b. Determine V as a function of time.

c. Compare parts (a) and (b).

Beyond the Basics

115. Let $f\{(-2, 3), (1, 2), (2, 1), (3, 0)\}$; $g = \{(-2, 0), (0, 2), (1, -2), (3, 2)\}$. Find each function.

a. $f + g$ **b.** fg **c.** $\dfrac{f}{g}$ **d.** $f \circ g$

116. Let

$$f(x) = \begin{cases} 2x & \text{if } -2 \le x \le 1 \\ x + 1 & \text{if } 1 < x \le 3 \end{cases}$$

$$g(x) = \begin{cases} x + 1 & \text{if } -3 \le x < 2 \\ 2x - 1 & \text{if } 2 \le x \le 3 \end{cases}$$

Find $f + g$ and its domain.

117. If $g(x) = 2x + 1$ and $(f \circ g)(x) = x^2 - 2$, find $f(5)$.

118. If $g(x) = 1 - 3x$ and $(f \circ g)(x) = \sqrt{5 + x}$, find $f(4)$.

119. Let $h(x)$ be any function whose domain contains $-x$ whenever it contains x. Show that
  a. $f(x) = h(x) + h(-x)$ is an even function.
  b. $g(x) = h(x) - h(-x)$ is an odd function.
  c. $h(x)$ can always be expressed as the sum of an even function and an odd function.

120. Use Exercise 119 to write each of the following functions as the sum of an even function and an odd function.
  a. $h(x) = x^2 - 2x + 3$
  b. $h(x) = [\![x]\!] + x$

Critical Thinking / Discussion / Writing

121. Let $f(x) = \dfrac{1}{[\![x]\!]}$ and $g(x) = \sqrt{x(2 - x)}$. Find the domain of each of the following functions.
  a. $f(x)$ **b.** $g(x)$
  c. $f(x) + g(x)$ **d.** $\dfrac{f(x)}{g(x)}$

122. For each of the following functions, find the domain of $f \circ f$.
  a. $f(x) = \dfrac{1}{\sqrt{-x}}$
  b. $f(x) = \dfrac{1}{\sqrt{1 - x}}$

123. **Even and odd functions.** State whether each of the following functions is an odd function, an even function, or neither. Justify your statement.
  a. The sum of two even functions
  b. The sum of two odd functions
  c. The sum of an even function and an odd function
  d. The product of two even functions
  e. The product of two odd functions
  f. The product of an even function and an odd function

124. **Even and odd functions.** State whether the composite function $f \circ g$ is an odd function, an even function, or neither in the following situations.
  a. f and g are odd functions.
  b. f and g are even functions.
  c. f is odd and g is even.
  d. f is even and g is odd.

Getting Ready for the Next Section

125. Solve the equation $x = 2y + 3$ for y.

126. Solve the equation $x = y^2 + 1, y \ge 0$ for y.

127. Solve the equation $x^2 + y^2 = 4, x \le 0$ for x.

128. Solve the equation $2x - \dfrac{1}{y} = 3$ for y.

129. Simplify the expression $\dfrac{1 + \frac{1}{1 + x}}{2 - \frac{1}{1 + x}}$.

130. Simplify the expression $\dfrac{2 + \frac{x}{1 - x}}{1 - \frac{x}{1 - x}}$.

Inverse Functions

BEFORE STARTING THIS SECTION, REVIEW

1 Vertical-line test (Section 1.3, page 46)

2 Domain and range of a function (Section 1.3, page 44)

3 Composition of functions (Section 1.6, page 102)

4 Symmetry (Section 1.1, page 9)

5 Solving linear and quadratic equations (Appendix A.6, page 957)

OBJECTIVES

1 Define an inverse function.

2 Find the inverse function.

3 Use inverse functions to find the range of a function.

4 Apply inverse functions in the real world.

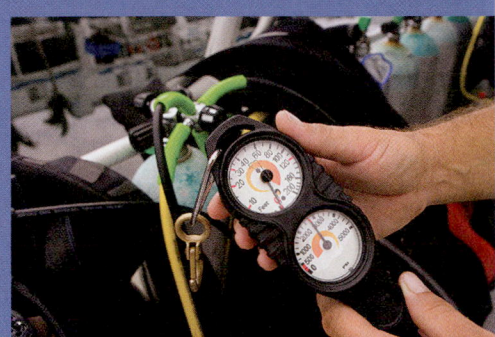

◆ Water Pressure on Underwater Devices

The pressure at any depth in the ocean is the pressure at sea level (1 atm) plus the pressure due to the weight of the overlying water: the deeper you go, the greater the pressure. Pressure is a vital consideration in the design of all tools and vehicles that work beneath the water's surface. Every 11 feet of depth increases the pressure by approximately 5 pounds per square inch (1 atmosphere ≈ 15 psi). Computing the pressure precisely requires a somewhat complicated equation that takes into account the water's salinity, temperature, and slight compressibility. However, a basic formula is given by "pressure equals depth times 5, divided by 11 plus 15." Thus, we have pressure (psi) = depth (ft) × 5(psi per atm)/11(ft per atm) + 15, where *psi* means "pounds per square inch" and *atm* abbreviates "atmosphere."

For example, at 99 feet below the surface, the pressure is 60 psi. At 1 mile (5280 feet) below the surface, the pressure is $\frac{5280 \times 5}{11} + 15 = 2415$ psi. Suppose the pressure gauge on a diving bell breaks and shows a reading of 1800 psi. We can determine the bell's depth when the gauge failed by using an inverse function. We will return to this problem in Example 10.

1 Define an inverse function.

Inverses

Given two variables, say x and y, we want to explore the relations between them. When the variable y can be expressed as a function of x

$$y = f(x),$$

we may ask whether a reverse relation may also be true. That is, can we also express x as a function of y

$$x = g(y)?$$

In doing so, we are trying to reverse the procedure given by the function f that assigned to each x a unique value y. The hand-held air pump may help you visualize this concept. Normally, we operate the air pump to inflate objects, but with a switch of a button you can reverse the operation and use the air pump to deflate objects.

Figure 1.96 Inflate.   **Figure 1.97** Deflate.

Suppose we are considering a cube and use two variables: the volume of the cube, V, and the length of the edge of the cube, E. The formula

$$V = (E)^3,$$

tells us how the volume of the cube depends on the length of its edge and it expresses V as a function of E. What if we want to construct a cube with a given volume? What would be the length of the corresponding edge? Fortunately, we can solve the volume equation for E and obtain

$$E = \sqrt[3]{V},$$

which gives us E as a function of V.

Not all procedures are reversible: remember your favorite jeans that got shrunk in the dryer? Our first task is to learn when we can reverse the procedures given by functions. We need to avoid the following situation where two points in the domain have the same value.

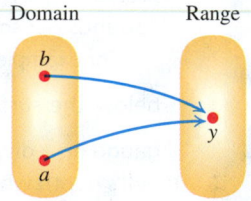

Reversing such functions would require assigning two different values a and b to y, which violates the property required for an assignment to be a function. This leads us to investigate a special type of function called a one-to-one function.

One-to-One Function

- If different elements in the domain of f are assigned to different values, then the function f is **one-to-one**. That is, if $x_1 \neq x_2$, then $f(x_1) \neq f(x_2)$.

- If there are two distinct elements in the domain of f that are assigned the same value, then the function f is **not one-to-one**.

Let f be a one-to-one function. Then the preceding definition says that for any two numbers x_1 and x_2 in the domain of f, if $x_1 \neq x_2$, then $f(x_1) \neq f(x_2)$, or equivalently: if $f(x_1) = f(x_2)$ then $x_1 = x_2$. That is, *f is a one-to-one function if different x-values correspond to different y-values.*

Figure 1.98(a) represents a one-to-one function, and Figure 1.98(b) represents a function that is not one-to-one. The relation shown in Figure 1.98(c) is not a function.

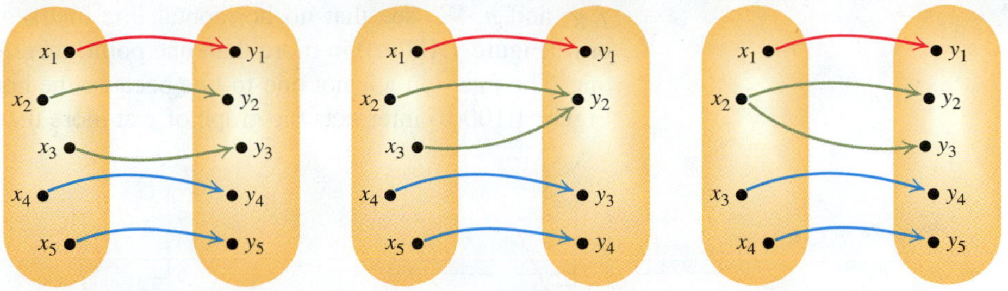

Domain   Range   Domain   Range   Domain   Range

One-to-one function: Each *y*-value corresponds to only one *x*-value.

(a)

Not a one-to-one function: One *y*-value, y_2, corresponds to two different *x*-values.

(b)

Not a function: One *x*-value, x_2, corresponds to two different *y*-values.

(c)

Figure 1.98 Deciding whether a diagram represents a function, a one-to-one function, or neither.

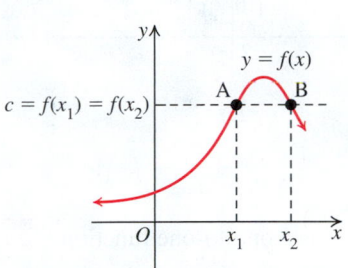

Figure 1.99 The function *f* is *not* one-to-one.

Recall that with a one-to-one function, different *x*-values correspond to different *y*-values. Consequently, if a function *f* is *not* one-to-one, then there are at least two numbers x_1 and x_2 in the domain of *f* such that $x_1 \neq x_2$ and $f(x_1) = f(x_2)$. Geometrically, this means that the horizontal line $y = c$, where $c = f(x_1) = f(x_2)$, intersects the graph of *f* in at least two points $A(x_1, f(x_1))$ and $B(x_2, f(x_2))$ See Figure 1.99.

Horizontal-Line Test

- If no horizontal line intersects the graph of the function more than once, then the function is **one-to-one**.

- If some horizontal line intersects the graph of the function more than once, then the function is **not one-to-one**.

EXAMPLE 1   **Test for the One-to-One Property of a Function**

Determine which of the following functions are one-to-one.

a. $f(x) = 2x + 5$ **b.** $g(x) = x^2 - 1$ **c.** $h(x) = 2\sqrt{x}$

Solution

Algebraic Approach Let x_1 and x_2 be two numbers in the domain of each function.

a. Suppose $f(x_1) = f(x_2)$. Then, $2x_1 + 5 = 2x_2 + 5$

$$2x_1 = 2x_2 \qquad \text{Subtract 5 from each side.}$$
$$x_1 = x_2 \qquad \text{Divide both sides by 2.}$$

So, $f(x_1) = f(x_2)$ implies $x_1 = x_2$, the function *f* is one-to-one.

b. Suppose $g(x_1) = g(x_2)$. Then, $x_1^2 - 1 = x_2^2 - 1$

$$x_1^2 = x_2^2 \qquad \text{Add 1 to both sides.}$$
$$x_1 = \pm x_2 \qquad \text{Square root property}$$

This means that for two distinct numbers x_1 and $x_2 = -x_1$, $g(x_1) = g(x_2)$, so *g* is not a one-to-one function.

c. Suppose $h(x_1) = h(x_2)$. Then $2\sqrt{x_1} = 2\sqrt{x_2}$

$$\sqrt{x_1} = \sqrt{x_2} \qquad \text{Divide both sides by 2.}$$
$$x_1 = x_2 \qquad \text{Square both sides.}$$

So, $h(x_1) = h(x_2)$ implies $x_1 = x_2$, the function *h* is one-to-one.

Geometric Approach (Horizontal-line test) Figure 1.100 shows the graphs of functions f, g, and h. We see that no horizontal line intersects the graphs of f (Figure 1.100(a)) or h (Figure 1.100(c)) in more than one point; therefore, the functions f and h are one-to-one. The function g is not one-to-one because the horizontal line $y = 3$ (among others) in Figure 1.100(b) intersects the graph of g at more than one point.

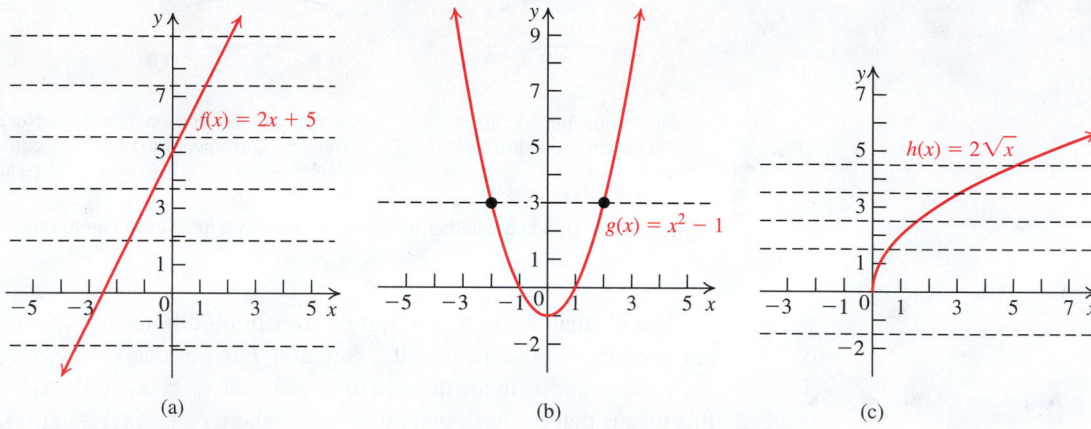

(a) (b) (c)

Figure 1.100 Horizontal-line test.

Practice Problem 1 Determine whether $f(x) = (x - 1)^2$ is a one-to-one function. ▪

SIDE NOTE

The reason that only a one-to-one function can have an inverse is that if $x_1 \neq x_2$ but $f(x_1) = y$ and $f(x_2) = y$, then $f^{-1}(y)$ must be x_1 as well as x_2. This is not possible because $x_1 \neq x_2$.

Inverse Function

Let f represent a one-to-one function. Then if y is in the range of f, there is only one value of x in the domain of f such that $f(x) = y$. We define the inverse of f, called the **inverse function of f**, denoted f^{-1}, by $f^{-1}(y) = x$ if and only if $y = f(x)$.

From this definition we have the following:

Domain of f = Range of f^{-1} and   Range of f = Domain of f^{-1}

WARNING

The notation $f^{-1}(x)$ does not mean $\dfrac{1}{f(x)}$. The expression $\dfrac{1}{f(x)}$ represents the reciprocal of $f(x)$ and is sometimes written as $[f(x)]^{-1}$.

Sometimes f^{-1} has a much more complicated structure than f. As an example, consider the function $f(x) = x + x^7$. This function has an inverse because it is an increasing function, but it is impossible to represent its inverse f^{-1} in a concise algebraic formula.

We can use this to our advantage to further expand our library of functions and be able to include some functions that otherwise would be very difficult to define directly.

A modern application of this phenomenon lies in the area of cryptography where the main idea is to create a one-to-one function (to encode messages) that is hard to inverse (to decode messages). The most widely used algorithm for secure data transmission is called the RSA algorithm. It uses the product of two large prime numbers as a base for the encoding scheme and relies on the difficulty of factoring a given number into the product of two large prime numbers (which would be necessary for a decoding scheme).

EXAMPLE 2   **Relating the Values of a Function and Its Inverse**

Assume that f is a one-to-one function.

a. If $f(3) = 5$, find $f^{-1}(5)$.   **b.** If $f^{-1}(-1) = 7$, find $f(7)$.

Solution

By definition, $f^{-1}(y) = x$ if and only if $y = f(x)$.

a. Let $x = 3$ and $y = 5$. Now reading the definition from right to left, $5 = f(3)$ if and only if $f^{-1}(5) = 3$. So, $f^{-1}(5) = 3$.

b. Let $y = -1$ and $x = 7$. Now $f^{-1}(-1) = 7$ if and only if $f(7) = -1$. So, $f(7) = -1$.

Practice Problem 2 Assume that f is a one-to-one function.

a. If $f(-3) = 12$, find $f^{-1}(12)$.   **b.** If $f^{-1}(4) = 9$, find $f(9)$.

Consider the following input–output diagram and Figure 1.101 for $f^{-1} \circ f$.

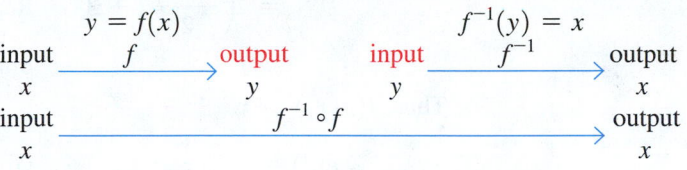

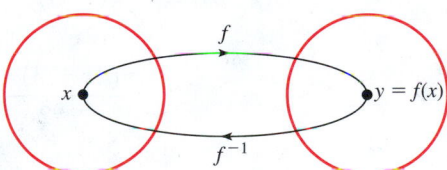

Figure 1.101 $f^{-1}(f(x)) = x$.

Figure 1.101 suggests the following:

INVERSE FUNCTION PROPERTY

Let f denote a one-to-one function. Then

1. $f^{-1}(f(x)) = x$ for every x in the domain of f.
2. $f(f^{-1}(x)) = x$ for every x in the domain of f^{-1}.

Further, if g is any function such that (for all values of x in the domain of the inner function)

$$f(g(x)) = x \quad \text{and} \quad g(f(x)) = x, \quad \text{then} \quad g = f^{-1}.$$

One interpretation of the equation $(f^{-1} \circ f)(x) = x$ is that f^{-1} undoes anything that f does to x. For example, let

$$f(x) = x + 2 \qquad \textcolor{blue}{f \text{ adds } 2 \text{ to any input } x.}$$

To undo what f does to x, we should subtract 2 from x. That is, the inverse of f should be

$$g(x) = x - 2. \qquad \textcolor{blue}{g \text{ subtracts } 2 \text{ from } x.}$$

Let's verify that $g(x) = x - 2$ is indeed the inverse function of x.

$$
\begin{aligned}
(f \circ g)(x) &= f(g(x)) && \text{Definition of } f \circ g \\
&= f(x - 2) && \text{Replace } g(x) \text{ with } x - 2. \\
&= (x - 2) + 2 && \text{Replace } x \text{ with } x - 2 \text{ in } f(x) = x + 2. \\
&= x && \text{Simplify.}
\end{aligned}
$$

We leave it for you to check that $(g \circ f)(x) = x$.

EXAMPLE 3   Verifying Inverse Functions

Verify that the following pairs of functions are inverses of each other:

$$
f(x) = 2x + 3 \quad \text{and} \quad g(x) = \frac{x - 3}{2}
$$

Solution

$$
\begin{aligned}
(f \circ g)(x) &= f(g(x)) && \text{Definition of } f \circ g \\
&= f\left(\frac{x - 3}{2}\right) && \text{Replace } g(x) \text{ with } \frac{x - 3}{2}. \\
&= 2\left(\frac{x - 3}{2}\right) + 3 && \text{Replace } x \text{ with } \frac{x - 3}{2} \text{ in } f(x) = 2x + 3. \\
&= x && \text{Simplify.}
\end{aligned}
$$

Thus, $f(g(x)) = x$.

$$
\begin{aligned}
g(x) &= \frac{x - 3}{2} && \text{Given function } g \\
(g \circ f)(x) &= g(f(x)) && \text{Definition of } g \text{ of } f \\
&= g(2x + 3) && \text{Replace } f(x) \text{ with } 2x + 3. \\
&= \frac{(2x + 3) - 3}{2} && \text{Replace } x \text{ with } 2x + 3 \text{ in } g(x) = \frac{x - 3}{2}. \\
&= x && \text{Simplify.}
\end{aligned}
$$

Thus, $g(f(x)) = x$. Since $f(g(x)) = g(f(x)) = x$, f and g are inverses of each other.

Practice Problem 3 Verify that $f(x) = 3x - 1$ and $g(x) = \dfrac{x + 1}{3}$ are inverses of each other.

In Example 3, notice how the functions f and g neutralize (undo) each other's effect. The function f takes an input x, *multiplies* it by 2, and *adds* 3; g neutralizes (or undoes) this effect by *subtracting* 3 and *dividing* by 2. This process is illustrated in Figure 1.102. Notice that g reverses the operations performed by f *and* the order in which they are done.

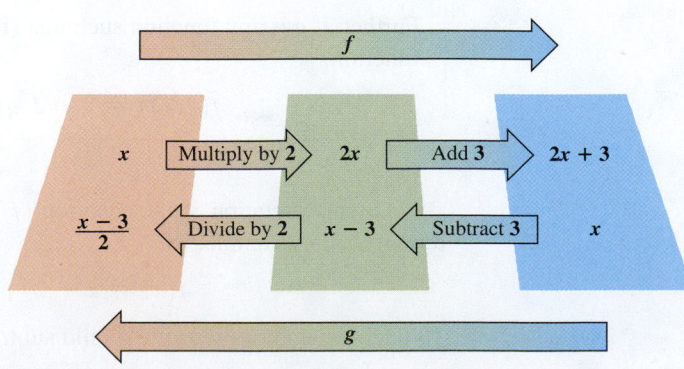

Figure 1.102 Function g undoes f.

2 Find the inverse function.

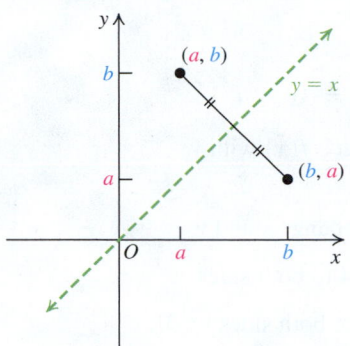

Figure 1.103 Relationship between points (*a*, *b*) and (*b*, *a*).

Finding the Inverse Function

All information about the inverse function f^{-1} can be derived directly from the properties of f itself. As an example, we will discuss the way the graph of f^{-1} can be derived directly from the graph of f.

Let $y = f(x)$ be a one-to-one function; then f has an inverse function. Suppose (a, b) is a point on the graph of f. Then $b = f(a)$. This means that $a = f^{-1}(b)$; so (b, a) is a point on the graph of f^{-1}. The points (a, b) and (b, a) are symmetric with respect to the line $y = x$, as shown in Figure 1.103. (See Exercises 109 and 110.) That is, if the graph paper is folded along the line $y = x$, the points (a, b) and (b, a) will coincide. Therefore, we have the following property.

> ## SYMMETRY PROPERTY OF THE GRAPHS OF f AND f^{-1}
>
> The graph of a one-to-one function f and the graph of f^{-1} are symmetric with respect to the line $y = x$.

EXAMPLE 4 **Finding the Graph of f^{-1} from the Graph of f**

The graph of a function f is shown in Figure 1.104. Sketch the graph of f^{-1}.

Solution

By the horizontal-line test, f is a one-to-one function, so it has an inverse function.

The graph of f consists of two line segments: one joining the points $(-3, -5)$ and $(-1, 2)$ and the other joining the points $(-1, 2)$ and $(4, 3)$.

The graph of f^{-1} is the reflection of the graph of f about the line $y = x$. The reflections of the points $(-3, -5), (-1, 2)$, and $(4, 3)$ about the line $y = x$ are $(-5, -3), (2, -1)$, and $(3, 4)$, respectively.

SIDE NOTE

It is helpful to notice that points on the x-axis, $(a, 0)$, reflect points on the y-axis, $(0, a)$, and conversely. Also notice that points on the line $y = x$ are unaffected by reflection about the line $y = x$.

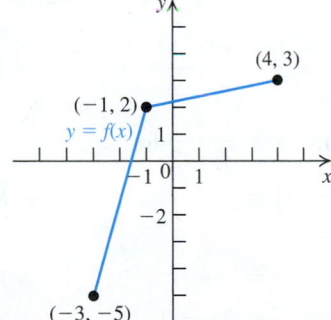

Figure 1.104 Graph of f.

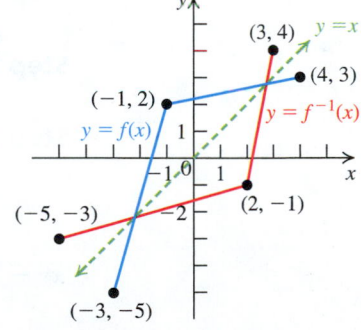

Figure 1.105 Graph of f^{-1}.

The graph of f^{-1} consists of two line segments: one joining the points $(-5, -3)$ and $(2, -1)$ and the other joining the points $(2, -1)$ and $(3, 4)$. See Figure 1.105.

Practice Problem 4 Use the graph of a function f in Figure 1.106 to sketch the graph of f^{-1}.

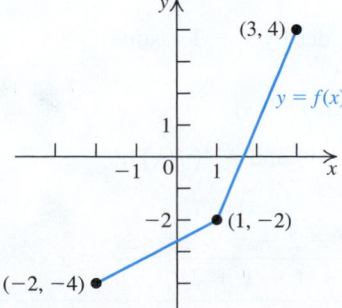

Figure 1.106 Graphing f^{-1} from the graph of f.

The symmetry between the graphs of f and f^{-1} about the line $y = x$ tells us that we can find an equation for the inverse function $y = f^{-1}(x)$ from the equation of a one-to-one function $y = f(x)$ by interchanging the roles of x and y in the equation $y = f(x)$. This results in the equation $x = f(y)$. Then we solve the equation $x = f(y)$ for y in terms of x to get $y = f^{-1}(x)$.

PROCEDURE
IN ACTION

EXAMPLE 5 Finding an Equation for f^{-1}

OBJECTIVE	EXAMPLE
Find the inverse of a one-to-one function f.	*Find the inverse of $f(x) = 3x - 4$.*
Step 1 Replace $f(x)$ with y in the equation defining $f(x)$.	**1.** $y = 3x - 4$ Replace $f(x)$ with y.
Step 2 Interchange x and y.	**2.** $x = 3y - 4$ Interchange x and y.
Step 3 Solve the equation in Step 2 for y.	**3.** $x + 4 = 3y$ Add 4 to both sides.
	$\dfrac{x + 4}{3} = y$ Divide both sides by 3.
Step 4 Write $f^{-1}(x)$ for y.	**4.** $f^{-1}(x) = \dfrac{x + 4}{3}$ We usually end with $f^{-1}(x)$ on the left.

Practice Problem 5 Find the inverse of $f(x) = -2x + 3$.

SIDE
NOTE

Interchanging x and y in Step 2 is done so we can write f^{-1} in the usual format with x as the independent variable and y as the dependent variable.

EXAMPLE 6 Finding the Inverse Function

Find the inverse of the one-to-one function

$$f(x) = \frac{x + 1}{x - 2}, \quad x \neq 2.$$

Solution

Step 1 $y = \dfrac{x + 1}{x - 2}$ Replace $f(x)$ with y.

Step 2 $x = \dfrac{y + 1}{y - 2}$ Interchange x and y.

Step 3   Solve $x = \dfrac{y + 1}{y - 2}$ for y.   This is the most challenging step.

$x(y - 2) = y + 1$ Multiply both sides by $y - 2$.

$xy - 2x = y + 1$ Distributive property

$xy - 2x + 2x - y = y + 1 + 2x - y$ Add $2x - y$ to both sides.

$xy - y = 2x + 1$ Simplify.

$y(x - 1) = 2x + 1$ Factor out y.

$y = \dfrac{2x + 1}{x - 1}$ Divide both sides by $x - 1$ assuming that $x \neq 1$.

Step 4 $f^{-1}(x) = \dfrac{2x + 1}{x - 1}, x \neq 1$ Replace y with $f^{-1}(x)$.

To see if our calculations are accurate, we compute $f(f^{-1}(x))$ and $f^{-1}(f(x))$.

$$f^{-1}(f(x)) = f^{-1}\left(\frac{x + 1}{x - 2}\right) = \frac{2\left(\dfrac{x + 1}{x - 2}\right) + 1}{\dfrac{x + 1}{x - 2} - 1} = \frac{2x + 2 + x - 2}{x + 1 - x + 2} = \frac{3x}{3} = x$$

Multiply the numerator and the denominator by $x - 2$ and simplify.

You should also check that $f(f^{-1}(x)) = x$.

Practice Problem 6 Find the inverse of the one-to-one function $f(x) = \dfrac{x}{x+3}$, $x \neq -3$.

3 Use inverse functions to find the range of a function.

Finding the Range of a One-to-One Function

It is not always easy to determine the range of a function that is defined by an equation. However, for a one-to-one function, we can find the range of f by finding the domain of f^{-1}.

EXAMPLE 7 Finding the Domain and Range of a One-to-One Function

Find the domain and range of the function $f(x) = \dfrac{x+1}{x-2}$ of Example 6.

Solution

The domain of $f(x) = \dfrac{x+1}{x-2}$ is the set of all real numbers x such that $x \neq 2$.

In interval notation, the domain of f is $(-\infty, 2) \cup (2, \infty)$.

From Example 6, $f^{-1}(x) = \dfrac{2x+1}{x-1}$, $x \neq 1$; therefore,

$$\text{Range of } f = \text{Domain of } f^{-1} = \{x \mid x \neq 1\}.$$

In interval notation, the range of f is $(-\infty, 1) \cup (1, \infty)$.

Practice Problem 7 Find the domain and the range of the function

$$f(x) = \dfrac{x}{x+3}.$$

If a function f is not one-to-one, it may still be possible to define an inverse for a function derived from f by restricting the domain of f to an appropriate interval. We saw in Example 1(b) that $g(x) = x^2 - 1$ is not a one-to-one function, so g does not have an inverse function. However, the horizontal-line test shows that the function

$$G(x) = x^2 - 1, \quad x \geq 0$$

with domain $[0, \infty)$ is one-to-one. See Figure 1.107. Therefore, G has an inverse function G^{-1}.

TECHNOLOGY CONNECTION

With a graphing calculator, you can also see that the graphs of $g(x) = x^2 - 1$, $x \geq 0$ and $g^{-1}(x) = \sqrt{x+1}$ are symmetric about the line $y = x$.

Enter Y_1, Y_2, and Y_3 as $g(x)$, $g^{-1}(x)$, and $f(x) = x$, respectively.

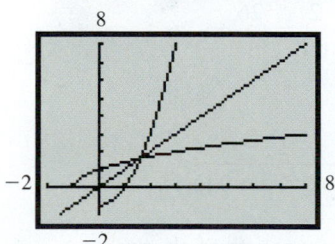

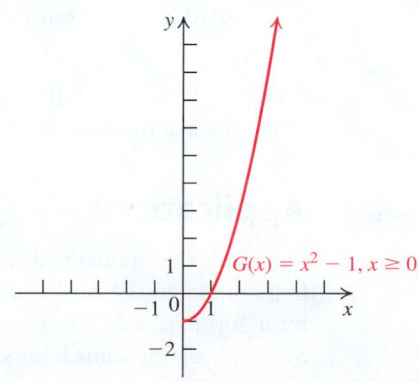

Figure 1.107 The function G has an inverse.

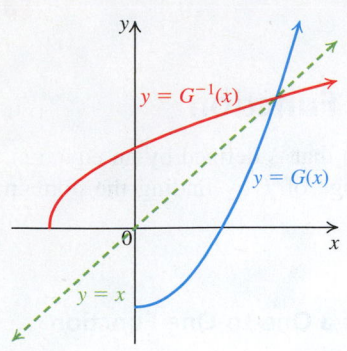

Figure 1.108 Graphs of G and G^{-1}.

Finding an Inverse Function

Find the inverse of $G(x) = x^2 - 1, x \geq 0$.

Solution

Step 1	$y = x^2 - 1, \quad x \geq 0$	Replace $G(x)$ with y.
Step 2	$x = y^2 - 1, \quad y \geq 0$	Interchange x and y.
Step 3	$x + 1 = y^2, \quad y \geq 0$	Add 1 to both sides.
	$y = \sqrt{x + 1}$, since $y \geq 0$	Solve for y.
Step 4	$G^{-1}(x) = \sqrt{x + 1}$	Replace y with $G^{-1}(x)$.

The graphs of G and G^{-1} are shown in Figure 1.108.

Practice Problem 8 Find the inverse of $G(x) = x^2 - 1, x \leq 0$.

The average rate of change of the inverse function f^{-1} can be derived directly as a reciprocal of a corresponding rate of change of f

$$\text{ARC of } f^{-1} \text{ from } f(a) \text{ to } f(b) = \frac{f^{-1}(f(b)) - f^{-1}(f(a))}{f(b) - f(a)}$$

$$= \frac{b - a}{f(b) - f(a)} = \frac{1}{\frac{f(b) - f(a)}{b - a}}$$

$$= \frac{1}{\text{ARC of } f \text{ from } a \text{ to } b},$$

as long as $a \neq b$ and $f(a) \neq f(b)$. The main advantage of this formula is that we do not have to derive a formula for f^{-1} to compute its average rate of change; all we need is a formula for f.

Average Rate of Change of an Inverse Function

If $f(x) = (2x + 1)^3 + 1$ find the average rate of change of the inverse function f^{-1} from $f(0) = 2$ to $f(1) = 28$.

SIDE NOTE

The average rate of change of f^{-1} in Example 9 can also be computed by first finding

$$f^{-1}(x) = \frac{\sqrt[3]{x - 1} - 1}{2}$$

and then computing the average rate of change directly.

Solution

First we compute the corresponding average rate of f.

$$\text{ARC of } f \text{ from } 0 \text{ to } 1 = \frac{f(1) - f(0)}{1 - 0} = \frac{28 - 2}{1 - 0} = \frac{26}{1} = 26.$$

We can compute the average rate of change of f^{-1} as

$$\text{ARC of } f^{-1} \text{ from } f(0) = 2 \text{ to } f(1) = 28 = \frac{1}{\text{ARC of } f \text{ from } 0 \text{ to } 1} = \frac{1}{26}.$$

Practice Problem 9 If $f(x) = 2 - (3x - 1)^3$ find the average rate of change of the inverse function f^{-1} from $f(0) = 3$ to $f(1) = -6$.

4 Apply inverse functions in the real world.

Applications

Functions that model real-life situations are frequently expressed as formulas with letters that remind you of the variable they represent. In finding the inverse of a function expressed by a formula, interchanging the letters could be very confusing. Accordingly, we omit Step 2, keep the same letters, and just solve the formula for the other variable.

◆ EXAMPLE 10   **Water Pressure on Underwater Devices**

At the beginning of this section, we gave the formula for finding the pressure p (in pounds per square inch) at a depth d (in feet) below the surface of water. This formula can be written as $p = \dfrac{5d}{11} + 15$. Suppose the pressure gauge on a diving bell breaks and shows a reading of 1800 psi. How far below the surface was the bell when the gauge failed?

Solution

We want to find the unknown depth in terms of the known pressure. This depth is given by the inverse of the function $p = \dfrac{5d}{11} + 15$. To find the inverse, we solve the given equation for d.

$$p = \frac{5d}{11} + 15 \qquad \text{Original equation}$$

$$11p = 5d + 165 \qquad \text{Multiply both sides by 11.}$$

$$d = \frac{11p}{5} - 33 \qquad \text{Solve for } d.$$

Now we can use this formula to find the depth when the gauge reads 1800 psi. We let $p = 1800$.

$$d = \frac{11p}{5} - 33 \qquad \text{Depth from pressure equation}$$

$$d = \frac{11(1800)}{5} - 33 \qquad \text{Replace } p \text{ with 1800.}$$

$$d = 3927$$

The device was 3927 feet below the surface when the gauge failed.

Practice Problem 10 In Example 10, suppose the pressure gauge showed a reading of 1650 psi. Determine the depth of the bell when the gauge failed.

Answers to Practice Problems

1. Not one-to-one; the horizontal line $y = 1$ intersects the graph at two different points.

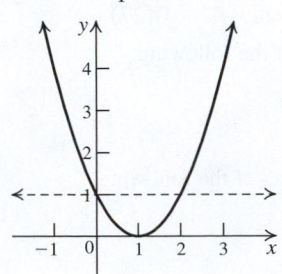

2. a. -3 **b.** 4

3. $(f \circ g)(x) = 3\left(\dfrac{x + 1}{3}\right) - 1 = x + 1 - 1 = x$ and

$(g \circ f)(x) = \dfrac{3x - 1 + 1}{3} = \dfrac{3x}{3} = x$; therefore, f and g are

inverses of each other.

4.

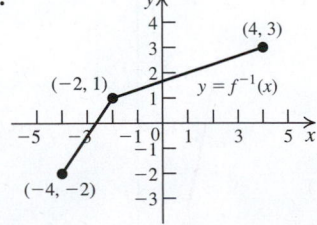

5. $f^{-1}(x) = \dfrac{3 - x}{2}$

6. $f^{-1}(x) = \dfrac{3x}{1 - x}, x \neq 1$

7. Domain: $(-\infty, -3) \cup (-3, \infty)$; range: $(-\infty, 1) \cup (1, \infty)$

8. $G^{-1}(x) = -\sqrt{x + 1}$ **9.** $ARC = -\dfrac{1}{9}$ **10.** 3597 ft

SECTION 1.7 **Exercises**

Concepts and Vocabulary

1. A function f is one-to-one if for any two different numbers x_1 and x_2, $x_1 \neq x_2$, in the domain of f we have _____ .

2. A function f is one-to-one if every horizontal line intersects the graph of f at no more than _____ .

3. $f^{-1} \circ f(x) =$ _____ .

4. The graph of f^{-1} is a reflection of the graph of f about the line _____ .

5. **True or False.** If a function f is one-to-one, then it has inverse f^{-1}.

6. **True or False.** $f^{-1}(x) = \dfrac{1}{f(x)}$.

7. **True or False.** The domain of f^{-1} equals the domain of f.

8. **True or False.** If a point (a, b) is on the graph of a one-to one function f, then the point (b, a) is on the graph of f^{-1}.

Building Skills

In Exercises 9–16, the graph of a function is given. Use the horizontal-line test to determine whether the function is one-to-one.

9.   10.

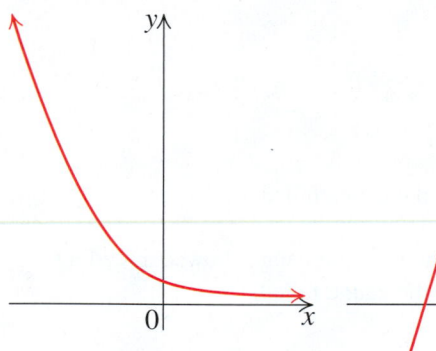

11.   12.

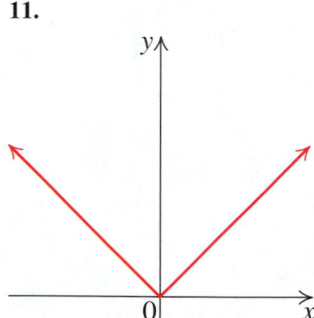

13.   14.

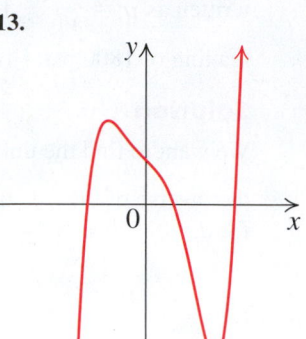

15.   16.

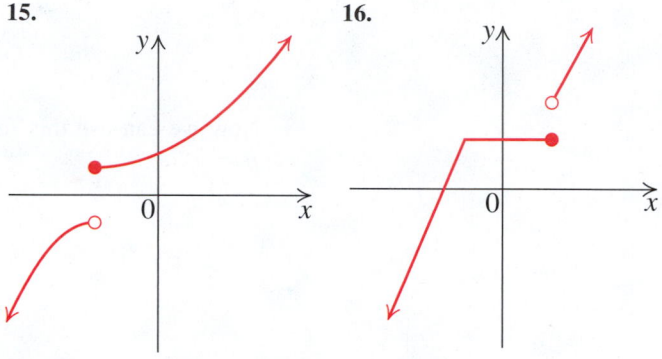

In Exercises 17–24, assume that the function f is one-to-one with domain: $(-\infty, \infty)$.

17. If $f(2) = 7$, find $f^{-1}(7)$.

18. If $f^{-1}(4) = -7$, find $f(-7)$.

19. If $f(-1) = 2$, find $f^{-1}(2)$.

20. If $f^{-1}(-3) = 5$, find $f(5)$.

21. For $f(x) = 2x - 3$, find each of the following.
    a. $f(3)$ **b.** $f^{-1}(3)$
    c. $(f \circ f^{-1})(19)$ **d.** $(f \circ f^{-1})(5)$

22. For $f(x) = x^3$, find each of the following.
    a. $f(2)$ **b.** $f^{-1}(8)$
    c. $(f \circ f^{-1})(15)$ **d.** $(f^{-1} \circ f)(27)$

23. For $f(x) = x^3 + 1$, find each of the following.
    a. $f(1)$
    b. $f^{-1}(2)$
    c. $(f \circ f^{-1})(269)$

24. For $g(x) = \sqrt[3]{2x^3 - 1}$, find each of the following.
    a. $g(1)$
    b. $g^{-1}(1)$
    c. $(g^{-1} \circ g)(135)$

25. Determine which of the functions described below are one-to-one. The function that assigns to each shirt in a store its
    a. color.
    b. price.
    c. bar code.

26. Determine which of the functions described below are one-to-one. The function that assigns to each person in the United States their
    a. Social Security number.
    b. first name.
    c. last name.

In Exercises 27–30, determine the order and write down the steps to undo the following sequence of actions.

27. First: Open the fridge door. Second: Take the milk out of the fridge.

28. First: Put socks on. Second: Put shoes on.

29. First: Wake up. Second: Put makeup on.

30. First: Dig a hole. Second: Plant a tree.

In Exercises 31–34 determine the order and steps to undo the following sequence of actions. Write the original function and the inverse function associated with the final result of these actions.

31. First: Multiply by 2. Second: Add 3.

32. First: Subtract 2. Second: Multiply by 3.

33. First: Cube it. Second: Add 2.

34. First: Subtract 3. Second: Cube it.

In Exercises 35–42, show that f and g are inverses of each other by verifying that $f(g(x)) = x = g(f(x))$.

35. $f(x) = 3x + 1; g(x) = \dfrac{x - 1}{3}$

36. $f(x) = 2 - 3x; g(x) = \dfrac{2 - x}{3}$

37. $f(x) = x^3; g(x) = \sqrt[3]{x}$

38. $f(x) = \dfrac{1}{x}; g(x) = \dfrac{1}{x}$

39. $f(x) = 2x^5 + 1; g(x) = \sqrt[5]{\dfrac{x - 1}{2}}$

40. $f(x) = (1 - 3x)^3; g(x) = \dfrac{1 - \sqrt[3]{x}}{3}$

41. $f(x) = \dfrac{x - 1}{x + 2}; g(x) = \dfrac{1 + 2x}{1 - x}$

42. $f(x) = \dfrac{3x + 2}{x - 1}; g(x) = \dfrac{x + 2}{x - 3}$

In Exercises 43–50, the graph of f is given. Sketch the graph of f^{-1}.

43.

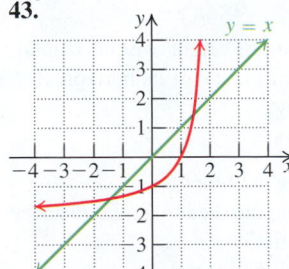

44.

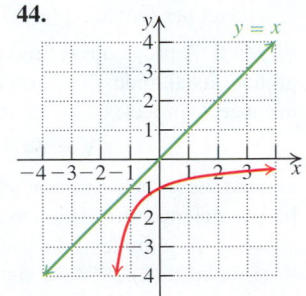

45.

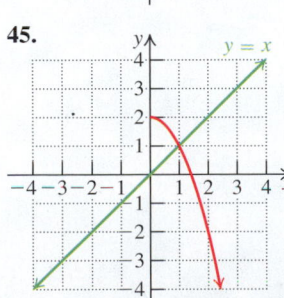

46.

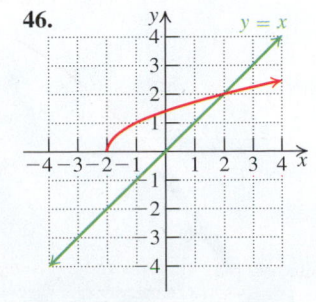

47.

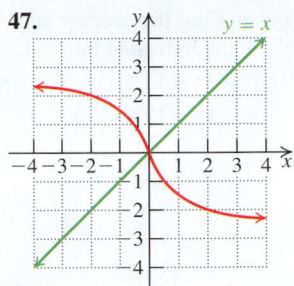

48.

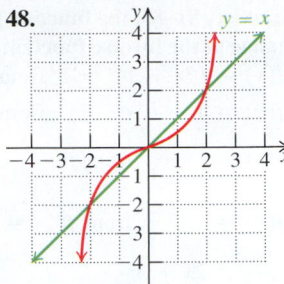

49.

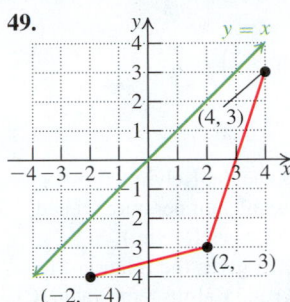

50.

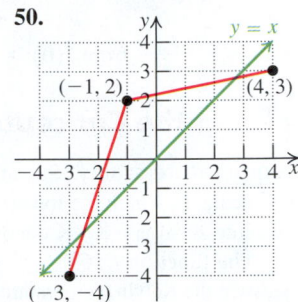

In Exercises 51–60, the function f is one-to-one. Find the f^{-1} and verify your answer. Also find the domain and range of the given function f.

51. $f(x) = 3x - 1$

52. $f(x) = 2x + 3$

53. $f(x) = \sqrt[3]{\dfrac{x + 1}{3}} + 2$

54. $f(x) = \sqrt[3]{\dfrac{x - 2}{3}} - 1$

55. $f(x) = (3x - 1)^3 + 2$

56. $f(x) = (2x + 1)^3 - 3$

57. $f(x) = \dfrac{2}{1 + x}$

58. $f(x) = 1 - \dfrac{1}{x + 1}$

59. $f(x) = \dfrac{x + 1}{x - 2}$

60. $f(x) = \dfrac{1 - 2x}{1 + x}$

In Exercises 61–64, sketch the graph of f using appropriate transformations and confirm that f is one-to-one by using the horizontal line test. Find f^{-1}. Find the domain and range of f and f^{-1}.

61. $f(x) = x^2 + 1, x \geq 0$

62. $f(x) = x^2 - 4, x \leq 0$

63. $f(x) = -x^2 + 2, x \leq 0$

64. $f(x) = -x^2 - 1, x \geq 0$

In Exercises 65–76, sketch the graph of the function and its inverse on the same coordinate axes.

65. $f(x) = 15 - 3x$

66. $g(x) = 2x + 5$

67. $f(x) = \sqrt{4 - x^2}, x \geq 0$

68. $f(x) = -\sqrt{9 - x^2}, x \geq 0$

69. $f(x) = \sqrt{x} + 3$

70. $f(x) = 4 - \sqrt{x}$

71. $g(x) = \sqrt[3]{x} + 1$

72. $h(x) = \sqrt[3]{1 - x}$

73. $f(x) = \dfrac{1}{x - 1}, x \neq 1$

74. $g(x) = 1 - \dfrac{1}{x}, x \neq 0$

75. $f(x) = 2 + \sqrt{x + 1}$

76. $f(x) = -1 + \sqrt{x + 2}$

In Exercises 77–80, assume that the given function is one-to-one. Find the inverse of the function. Also find the domain and the range of the given function.

77. $f(x) = \dfrac{x + 1}{x - 2}, x \neq 2$

78. $g(x) = \dfrac{x + 2}{x + 1}, x \neq -1$

79. $f(x) = \dfrac{1 - 2x}{1 + x}, x \neq -1$

80. $h(x) = \dfrac{x - 1}{x - 3}, x \neq 3$

In Exercises 81–86, the function f is given. Find the average rate of change of the inverse function f^{-1} for the indicated interval.

81. $f(x) = (3x - 1)^3 + 2$; from $f(0) = 1$ to $f(1) = 10$.

82. $f(x) = 1 - (2x - 1)^3$; from $f(0) = 2$ to $f(1) = 0$.

83. $f(x) = \dfrac{3}{x - 1}$; from $f(4) = 1$ to $f(9) = \dfrac{3}{8}$.

84. $f(x) = 1 - \dfrac{2}{x}$; from $f(1) = -1$ to $f(3) = \dfrac{1}{3}$.

85. $f(x) = \dfrac{2x + 3}{x - 2}$; from $f(3) = 9$ to $f(5) = \dfrac{13}{3}$.

86. $f(x) = \dfrac{3x + 1}{x + 1}$; from $f(0) = 1$ to $f(2) = \dfrac{7}{3}$.

Applying the Concepts

87. **Temperature scales.** Scientists use the Kelvin temperature scale, in which the lowest possible temperature (called absolute zero) is 0 K. (K denotes degrees Kelvin.)

    The function $K(C) = C + 273$ gives the relationship between the Kelvin temperature (K) and Celsius temperature(C).

    a. Find the inverse function of $K(C) = C + 273$. What does it represent?

    b. Use the inverse function from (a) to find the Celsius temperature corresponding to 300 K.

    c. A comfortable room temperature is 22°C. What is the corresponding Kelvin temperature?

88. **Temperature scales.** The boiling point of water is 373 K, or 212°F; the freezing point of water is 273 K, or 32°F. The relationship between Kelvin and Fahrenheit temperatures is linear.

    a. Write a linear function expressing $K(F)$ in terms of F.

    b. Find the inverse of the function in part (a). What does it mean?

    c. A normal human body temperature is 98.6°F. What is the corresponding Kelvin temperature?

89. **Composition of functions.** Use Exercises 87 and 88 and the composition of functions to

    a. write a function that expresses F in terms of C.

    b. write a function that expresses C in terms of F.

90. **Celsius and Fahrenheit temperatures.** Show that the functions in (a) and (b) of Exercise 89 are inverse functions.

91. **Currency exchange.** Alisha went to Europe last summer. She discovered that when she exchanged her U.S. dollars for euros, she received 25% fewer euros than the number of dollars she exchanged. (She got 75 euros for every 100 U.S. dollars.) When she returned to the United States, she got 25% more dollars than the number of euros she exchanged.

    a. Write each conversion function.

    b. Show that in part (a), the two functions are not inverse functions.

    c. Does Alisha gain or lose money after converting both ways?

92. **Hourly wages.** Anwar is a short-order cook in a diner. He is paid $4 per hour plus 5% of all food sales per hour. His average hourly wage w in terms of the food sales of x dollars is $w = 4 + 0.05x$.

    a. Write the inverse function. What does it mean?

    b. Use the inverse function to estimate the hourly sales at the diner if Anwar averages $12 per hour.

93. **Hourly wages.** In Exercise 92, suppose in addition that Anwar is guaranteed a minimum wage of $7 per hour.

    a. Write a function expressing his hourly wage w in terms of food sales per hour. [*Hint:* Use a piecewise function.]

    b. Does the function in part (a) have an inverse? Explain.

    c. If the answer in part (b) is yes, find the inverse function. If the answer is no, restrict the domain so that the new function has an inverse.

94. **Simple pendulum.** If a pendulum is released at a certain point, the **period** is the time the pendulum takes to swing along its path and return to the point from which it was released. The period T (in seconds) of a simple pendulum is a function of its length l (in feet) and is given by $T = 1.11\sqrt{l}$.

    a. Find the inverse function. What does it mean?

    b. Use the inverse function to calculate the length of the pendulum assuming that its period is two seconds.

    c. The convention center in Portland, Oregon, has the longest pendulum in the United States. The pendulum's length is 70 feet. Find the period.

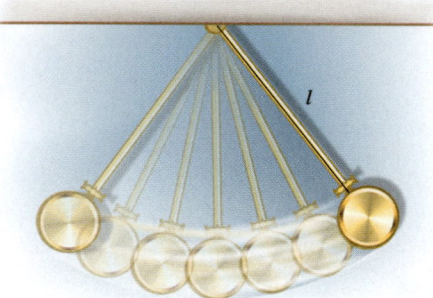

95. **Water supply.** Suppose x is the height of the water above the opening at the base of a water tank. The velocity V of water that flows from the opening at the base is a function of x and is given by $V(x) = 8\sqrt{x}$.

    a. Find the inverse function. What does it mean?

    b. Use the inverse function to calculate the height of the water in the tank when the flow is **(i)** 30 feet per second and **(ii)** 20 feet per second.

96. **Physics.** A projectile is fired from the origin over horizontal ground. Its altitude y (in feet) is a function of its horizontal distance x (in feet) and is given by

$$y = 64x - 2x^2.$$

    a. Find the inverse function where the function is increasing.

    b. Use the inverse function to compute the horizontal distance when the altitude of the projectile is **(i)** 32 feet, **(ii)** 256 feet, and **(iii)** 512 feet.

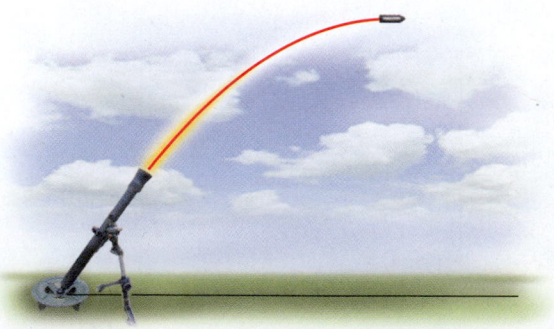

97. Loan repayment. Chris purchased a car at 0% interest for five years. After making a down payment, she agreed to pay the remaining $36,000 in monthly payments of $600 per month for 60 months.
a. What does the function $f(x) = 36,000 - 600x$ represent?
b. Find the inverse of the function in part (a). What does the inverse function represent?
c. Use the inverse function to find the number of months remaining to make payments if the balance due is $22,000.

98. Demand function. A marketing survey finds that the number x (in millions) of computer chips the market will purchase is a function of its price p (in dollars) and is estimated by $x = 8p^2 - 32p + 1200, 0 < p \le 2$.
a. Find the inverse function. What does it mean?
b. Use the inverse function to estimate the price of a chip if the demand is 1180.5 million chips.

99. Physics. According to Newton's Law of Universal Gravitation, the attractive force between two bodies with masses M and m depends on the distance r between them according to the formula

$$F(r) = G\frac{Mm}{r^2},$$

where G is a gravitational constant.
a. Explain why F is a one-to-one function of r.
b. Express r as a function of F.
c. Write the algebraic expression for F^{-1} in terms of x.

100. Geometry. The volume of a sphere depends on its radius according to the Formula

$$V(r) = \frac{4}{3}\pi r^3.$$

a. Explain why V is a one-to-one function of r.
b. Express r as a function of V.
c. Write the algebraic expression for V^{-1} in terms of x.

Beyond the Basics

In Exercises 101 and 102, show that f and g are inverses of each other by verifying that $f(g(x)) = x = g(f(x))$.

101.

x	1	3	4
$f(x)$	3	5	2

x	3	5	2
$g(x)$	1	3	4

102.

x	1	2	3	4
$f(x)$	-2	0	-3	1

x	-2	0	-3	1
$g(x)$	1	2	3	4

103. Let $f(x) = \sqrt{4 - x^2}$.
a. Sketch the graph of $y = f(x)$.
b. Is f one-to-one?
c. Find the domain and the range of f.

104. Let $f(x) = \frac{[\![x]\!] + 2}{[\![x]\!] - 2}$, where $[\![x]\!]$ is the greatest integer function.
a. Find the domain of f.
b. Is f one-to-one?

105. Let $g(x) = \sqrt{1 - x^2}, 0 \le x \le 1$.
a. Show that g is one-to-one.
b. Find the inverse of g.
c. Find the domain and the range of g.

106. The graph of $f(x) = x^3$ is given in Section 1.4.
a. Sketch the graph of $g(x) = (x - 1)^3 + 2$ by using transformations of the graph of f.
b. Find $g^{-1}(x)$.
c. Sketch the graph of $g^{-1}(x)$ by reflecting the graph of g in part (a) about the line $y = x$.

107. Show that the function $f(x) = 1 + 2x^3 + 3x^5 + 4x^7$ is one-to-one by noticing that it is a sum of increasing functions. Observe that it is impossible to find $f^{-1}(x)$ algebraically and explain how you know f^{-1} exists.

108. The function $f(x) = x^3 + x + 1$ is one-to-one.
a. Use the formula on page 124 to find the average rate of change of the inverse function f^{-1} as x changes from $f(1) = 3$ to $f(2) = 11$.
b. Explain why it is difficult to compute the average rate of change of f^{-1} directly.

109. Let $P(3, 7)$ and $Q(7, 3)$ be two points in the plane.
a. Show that the midpoint M of the line segment PQ lies on the line $y = x$.
b. Show that the line $y = x$ is the perpendicular bisector of the line segment $\overline{PQ}$; that is, show that the line $y = x$ contains the midpoint found in (a) and makes a right angle with $\overline{PQ}$. You will have shown that P and Q are symmetric about the line $y = x$.

110. Use the procedure outlined in Exercise 109 to show that the points (a, b) and (b, a) are symmetric about the line $y = x$. (See Figure 1.103 on page 121.)

111. Inverse of composition of functions. We show that if f and g have inverses, then

$$(f \circ g)^{-1} = g^{-1} \circ f^{-1}. \text{ (Note the order.)}$$

Let $f(x) = 2x - 1$ and $g(x) = 3x + 4$.
a. Find the following:
(i) $f^{-1}(x)$ (ii) $g^{-1}(x)$
(iii) $(f \circ g)(x)$ (iv) $(g \circ f)(x)$
(v) $(f \circ g)^{-1}(x)$ (vi) $(g \circ f)^{-1}(x)$
(vii) $(f^{-1} \circ g^{-1})(x)$ (viii) $(g^{-1} \circ f^{-1})(x)$
b. From part (a), conclude that
(i) $(f \circ g)^{-1} = g^{-1} \circ f^{-1}$.
(ii) $(g \circ f)^{-1} = f^{-1} \circ g^{-1}$.

112. Show that $f(x) = x^{2/3}, x \le 0$ is a one-to-one function. Find $f^{-1}(x)$ and verify that $f^{-1}(f(x)) = x$ for every x in the domain of f.

Critical Thinking / Discussion / Writing

113. Does every odd function have an inverse? Explain.

114. Is there an even function that has an inverse? Explain.

115. Does every increasing or decreasing function have an inverse? Explain.

116. A relation R is a set of ordered pairs (x, y). The inverse of R is the set of ordered pairs (y, x).

a. Give an example of a function whose inverse relation is not a function.

b. Give an example of a relation R whose inverse is a function.

Getting Ready for the Next Section

In Exercises 117–120, factor the given expression.

117. $x^2 - x - 12$

118. $x^2 - 5x + 6$

119. $x^2 + 2x - 8$

120. $x^2 + 7x + 10$

In Exercises 121–124, solve each quadratic equation.

121. $x^2 - 7x + 12 = 0$

122. $x^2 - x - 6 = 0$

123. $3x^2 + 7x + 2 = 0$

124. $x^2 - 4x + 1 = 0$

In Exercises 125–128, use transformations on the graph $y = x^2$ to sketch the graph of each function.

125. $y = (x + 2)^2 - 3$

126. $y = (x - 1)^2 + 3$

127. $y = -(x + 1)^2 + 2$

128. $y = -(x - 3)^2 - 1$

SUMMARY Definitions, Concepts, and Formulas

1.1 Graphs of Equations

i. Ordered Pair. A pair of numbers in which the order is specified is called an ordered pair of numbers.

ii. The Distance Formula. The distance between two points $P(x_1, y_1)$ and $Q(x_2, y_2)$, denoted by $d(P, Q)$, is given by

$$d(P, Q) = \sqrt{(x_2 - x_1)^2 + (y_2 - y_1)^2}.$$

iii. The Midpoint Formula. The coordinates of the midpoint $M(x, y)$ of the line segment joining $P(x_1, y_1)$ and $Q(x_2, y_2)$ are given by

$$M = (x, y) = \left(\frac{x_1 + x_2}{2}, \frac{y_1 + y_2}{2} \right).$$

iv. The graph of an equation in two variables, say, x and y, is the set of all ordered pairs (a, b) in the coordinate plane that satisfy the given equation. A graph of an equation, then, is a picture of its solution set.

v. Sketching the graph of an equation by plotting points
Step 1 Make a representative table of solutions of the equation.
Step 2 Plot the solutions in Step 1 as ordered pairs in the coordinate plane.
Step 3 Connect the representative solutions in Step 2 by a smooth curve.

vi. Intercepts
1. The x-intercept is the x-coordinate of a point on the graph where the graph touches or crosses the x-axis.

To find the x-intercepts, set $y = 0$ in the equation and solve for x.
2. The y-intercept is the y-coordinate of a point on the graph where the graph touches or crosses the y-axis. To find the y-intercepts, set $x = 0$ in the equation and solve for y.

vii. Symmetry
A graph is symmetric with respect to
a. the y-axis if for every point (x, y) on the graph the point $(-x, y)$ is also on the graph. That is, replacing x with $-x$ in the equation produces an equivalent equation.
b. the x-axis if for every point (x, y) on the graph the point $(x, -y)$ is also on the graph. That is, replacing y with $-y$ in the equation produces an equivalent equation.
c. the *origin* if for every point (x, y) on the graph the point $(-x, -y)$ is also on the graph. That is, replacing x with $-x$ and y with $-y$ in the equation produces an equivalent equation.

viii. Circle
A circle is a set of points in the coordinate plane that is at a fixed distance r from a fixed point (h, k). The fixed distance r is called the radius of the circle, and the fixed point (h, k) is called the center of the circle. The equation

$$(x - h)^2 + (y - k)^2 = r^2$$

is called the *standard equation* of a circle.

1.2 Lines

i. The slope m of a nonvertical line through the points $P(x_1, y_1)$ and $Q(x_2, y_2)$ is

$$m = \frac{\text{rise}}{\text{run}} = \frac{(y\text{-coordinate of } Q) - (y\text{-coordinate of } P)}{(x\text{-coordinate of } Q) - (x\text{-coordinate of } P)}$$

$$= \frac{y_2 - y_1}{x_2 - x_1}.$$

The slope of a vertical line is undefined. The slope of a horizontal line is 0.

ii. Equation of a line

$$y - y_1 = m(x - x_1) \quad \text{Point–slope form}$$
$$y = mx + b \quad \text{Slope–intercept form}$$
$$\frac{x}{a} + \frac{y}{b} = 1 \quad \text{Two intercept form}$$
$$y = k \quad \text{Horizontal line}$$
$$x = k \quad \text{Vertical line}$$
$$ax + by + c = 0 \quad \text{General form}$$

iii. Parallel and Perpendicular Lines
Two distinct lines with respective slopes m_1 and m_2 are
1. parallel if $m_1 = m_2$.
2. perpendicular if $m_1 m_2 = -1$.

1.3 Functions

i. A set of ordered pairs is called a **relation.**

ii. A **function** from a set X to a set Y is a rule that assigns to each element of X one and only one corresponding element of Y. The set X is the **domain** of the function. The set of those elements of Y that correspond (are assigned) to the elements of X is the **range** of the function.

iii. If a function f is defined by an equation, then its domain is the largest set of real numbers for which $f(x)$ is a real number.

iv. Vertical-line test. If each vertical line intersects a graph at no more than one point, the graph is the graph of a function.

v. The **average rate of change** of $f(x)$ as x changes from a to b is defined by $\dfrac{f(b) - f(a)}{b - a}, b \neq a$.

vi. The quantity $\dfrac{f(x + h) - f(x)}{h}, h \neq 0$, is called the difference quotient.

1.4 A Library of Functions

i. A function f is **increasing**, **decreasing**, or **constant** on an interval depending on whether its graph is respectively rising, falling, or staying the same as you move from left to right on the graph.

ii. A function f is **even** if $f(-x) = f(x)$. The graph of an even function is symmetric with respect to the y-axis. A function f is **odd** if $f(-x) = -f(x)$. The graph of an odd function is symmetric with respect to the origin.

iii. For **piecewise functions**, different rules are used in different parts of the domain.

iv. Basic Functions

Identity function	$f(x) = x$
Constant function	$f(x) = c$
Squaring function	$f(x) = x^2$
Cubing function	$f(x) = x^3$
Absolute value function	$f(x) = |x|$
Square root function	$f(x) = \sqrt{x}$
Cube root function	$f(x) = \sqrt[3]{x}$
Reciprocal function	$f(x) = \dfrac{1}{x}$
Reciprocal square function	$f(x) = \dfrac{1}{x^2}$
Rational power function	$f(x) = x^{m/n}$
Greatest integer function	$f(x) = [\![x]\!]$

1.5 Transformations of Functions

i. Vertical and horizontal shifts: Let f be a function and c be a positive number.

To Graph	Shift the graph of f by c units
$f(x) + c$	up
$f(x) - c$	down
$f(x - c)$	right
$f(x + c)$	left

ii. Reflections:

To Graph	Reflect the graph of f about the
$-f(x)$	x-axis
$f(-x)$	y-axis

iii. Stretching and compressing: The graph of $g(x) = af(x)$ for $a > 0$ has the same shape as the graph of $f(x)$ and is steeper if $a > 1$ and flatter if $0 < a < 1$. If a is negative, the graph of $y = |a| f(x)$ is reflected about the x-axis to obtain the graph of $y = af(x)$.

The graph of $g(x) = f(bx)$ for $b > 0$ is obtained from the graph of f by stretching away from the y-axis if $0 < b < 1$ and compressing horizontally toward the y-axis if $b > 1$. If b is negative, the graph of $y = f(|b|x)$ is reflected about the y-axis to obtain the graph of $y = f(bx)$.

1.6 Combining Functions; Composite Functions

i. Given two functions f and g, for all values of x for which both $f(x)$ and $g(x)$ are defined, the functions can be combined to form **sum**, **difference**, **product**, and **quotient** functions. See page 100 for the domains of these functions.

ii. The **composition** of f and g is defined by $(f \circ g)(x) = f(g(x))$.

The input of f is the output of g. The domain of $f \circ g$ is the set of all x's in the domain of g such that $g(x)$ is in the domain of f.

In general $f \circ g \neq g \circ f$.

1.7 Inverse Functions

i. A function f is **one-to-one** if for any x_1 and x_2 in the domain of f, $f(x_1) = f(x_2)$ implies $x_1 = x_2$.

ii. **Horizontal-line test.** If each horizontal line intersects the graph of a function f in at most one point, then f is a one-to-one function.

iii. **Inverse function.** Let f be a one-to-one function. Then g is the inverse of f, and we write $g = f^{-1}$ if $(f \circ g)(x) = x$ for every x in the domain of g and $(g \circ f)(x) = x$ for every x in the domain of f. The graph of f^{-1} is the reflection of the graph of f about the line $y = x$.

REVIEW EXERCISES

Basic Concepts and Skills

In Exercises 1–8, state whether the given statement is true or false.

1. $(0, 5)$ is the midpoint of the line segment joining $(-3, 1)$ and $(3, 11)$.

2. The equation $(x + 2)^2 + (y + 3)^2 = 5$ is the equation of a circle with center $(2, 3)$ and radius 5.

3. If a graph is symmetric with respect to the x-axis and the y-axis, then it must be symmetric with respect to the origin.

4. If a graph is symmetric with respect to the origin, then it must be symmetric with respect to the x-axis and the y-axis.

5. In the graph of the equation of the line $3y = 4x + 9$, the slope is 4 and the y-intercept is 9.

6. If the slope of a line is 2, then the slope of any line perpendicular to it is $\dfrac{1}{2}$.

7. The slope of a vertical line is undefined.

8. The equation $(x - 2)^2 + (y + 3)^2 = -25$ is the equation of a circle with center $(2, -3)$ and radius 5.

In Exercises 9–14, find

a. the distance between the point P and Q.

b. the coordinates of the midpoint of the line segment $\overline{PQ}$.

c. the slope of the line containing the points P and Q.

9. $P(3, 5), Q(-1, 3)$ 10. $P(-3, 5), Q(3, -1)$

11. $P(4, -3), Q(9, -8)$ 12. $P(2, 3), Q(-7, -8)$

13. $P(2, -7), Q(5, -2)$ 14. $P(-5, 4), Q(10, -3)$

15. Show that the points $A(0, 5)$, $B(-2, -3)$, and $C(3, 0)$ are the vertices of a right triangle.

16. Show that the points $A(1, 2)$, $B(4, 8)$, $C(7, -1)$, and $D(10, 5)$ are the vertices of a rhombus.

17. Which of the points $(-6, 3)$ and $(4, 5)$ is closer to the origin?

18. Which of the points $(-6, 4)$ and $(5, 10)$ is closer to the point $(2, 3)$?

19. Find a point on the x-axis that is equidistant from the points $(-5, 3)$ and $(4, 7)$.

20. Find a point on the y-axis that is equidistant from the points $(-3, -2)$ and $(2, -1)$.

In Exercises 21–24, specify whether the given graph has axis symmetry or origin symmetry (or neither).

21.   22.

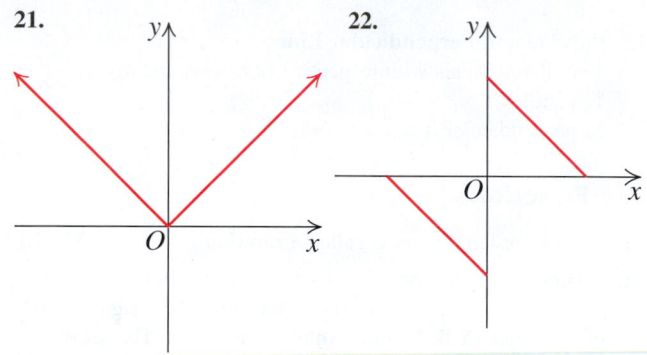

23.   24.

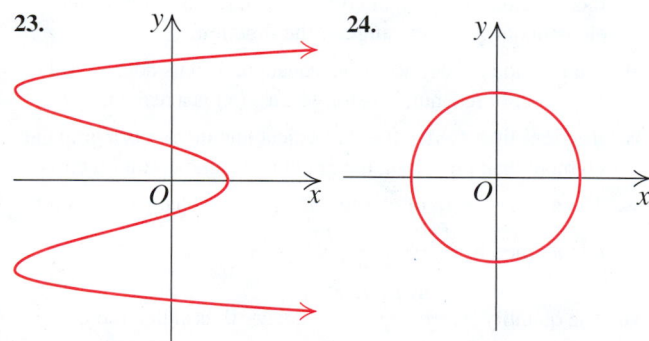

In Exercises 25–34, sketch the graph of the given equation. List all intercepts and describe any symmetry of the graph.

25. $x + 2y = 4$ 26. $3x - 4y = 12$

27. $y = -2x^2$ 28. $x = -y^2$

29. $y = x^3$ 30. $x = -y^3$

31. $y = x^2 + 2$ 32. $y = 1 - x^2$

33. $x^2 + y^2 = 16$ 34. $y = x^4 - 4$

In Exercises 35–37, find the standard form of the equation of the circle that satisfies the given conditions.

35. Center $(2, -3)$, radius 5

36. Diameter with endpoints $(5, 2)$ and $(-5, 4)$

37. Center $(-2, -5)$, touching the y-axis

In Exercises 38–42, describe and sketch the graph of the given equation and give the x- and y- intercepts.

38. $2x - 5y = 10$ **39.** $\dfrac{x}{2} - \dfrac{y}{5} = 1$

40. $(x + 1)^2 + (y - 3)^2 = 16$

41. $x^2 + y^2 - 2x + 4y - 4 = 0$

42. $3x^2 + 3y^2 - 6x - 6 = 0$

In Exercises 43–47, find the slope–intercept form of the equation of the line that satisfies the given conditions.

43. Passing through $(1, 2)$ with slope -2

44. x-intercept 2, y-intercept 5

45. Passing through $(1, 3)$ and $(-1, 7)$

46. Passing through $(1, 2)$, parallel to $2x + 3y = 7$.

47. Passing through $(1, 3)$, perpendicular to $3x - 4y = 12$.

48. Determine whether the lines in each pair are parallel, perpendicular, or neither.
   a. $y = 3x - 2$ and $y = 3x + 2$
   b. $3x - 5y + 7 = 0$ and $5x - 3y + 2 = 0$
   c. $ax + by + c = 0$ and $bx - ay + d = 0$
   d. $y + 2 = \dfrac{1}{3}(x - 3)$ and $y - 5 = 3(x - 3)$

In Exercises 49–58, graph each equation. State which equations determine y as a function of x.

49. $x = y^2$ **50.** $y = x^2$

51. $x - y = 1$ **52.** $y = \sqrt{x - 2}$

53. $x^2 + y^2 = 0.04$ **54.** $x = -\sqrt{y}$

55. $x = 1$ **56.** $y = 2$

57. $y = |x + 1|$ **58.** $x = y^2 + 1$

In Exercises 59–76, let $f(x) = 3x + 1$ and $g(x) = x^2 - 2$. Find each of the following.

59. $f(-2)$ **60.** $g(-2)$

61. x if $f(x) = 4$ **62.** x if $g(x) = 2$

63. $(f + g)(1)$ **64.** $(f - g)(-1)$

65. $(f \cdot g)(-2)$ **66.** $(g \cdot f)(0)$

67. $(f \circ g)(3)$ **68.** $(g \circ f)(-2)$

69. $(f \circ g)(x)$ **70.** $(g \circ g)(x)$

71. $f(a + h)$ **72.** $g(a - h)$

73. $\dfrac{f(x + h) - f(x)}{h}$ **74.** $\dfrac{g(x + h) - g(x)}{h}$

75. Let $f = \{(0, 1), (1, 3), (2, 5)\}$;
   $g = \{(-1, 0), (1, 2), (2, 3)\}$. Find $f \circ g$.

76. Let $f(x) = \sqrt{x + 1}$; $g(x) = \dfrac{1}{x}$. Find $f \circ g$ and its domain.

In Exercises 77–82, graph each function and state its domain and range. Determine the intervals over which the function is increasing, decreasing, or constant.

77. $f(x) = -3$ **78.** $f(x) = x^2 - 2$

79. $g(x) = \sqrt{3x - 2}$ **80.** $h(x) = \sqrt{36 - x^2}$

81. $f(x) = \begin{cases} x + 1 & \text{if } x \geq 0 \\ -x + 1 & \text{if } x < 0 \end{cases}$

82. $g(x) = \begin{cases} x & \text{if } x \geq 0 \\ x^2 & \text{if } x < 0 \end{cases}$

In Exercises 83–86, use transformations to graph each pair of functions on the same coordinate axes.

83. $f(x) = \sqrt{x}, g(x) = \sqrt{x + 1}$

84. $f(x) = |x|, g(x) = 2|x - 1| + 3$

85. $f(x) = \dfrac{1}{x}, g(x) = -\dfrac{1}{x - 2}$

86. $f(x) = x^2, g(x) = (x + 1)^2 - 2$

In Exercises 87–92, state whether each function is odd, even, or neither. Discuss the symmetry of each graph.

87. $f(x) = x^2 - x^4$ **88.** $f(x) = x^3 + x$

89. $f(x) = |x| + 3$ **90.** $f(x) = 3x + 5$

91. $f(x) = \sqrt{x}$ **92.** $f(x) = \dfrac{2}{x}$

In Exercises 93–96, express each function as a composition of two functions.

93. $f(x) = \sqrt{x^2 - 4}$ **94.** $g(x) = (x^2 - x + 2)^{50}$

95. $h(x) = \sqrt{\dfrac{x - 3}{2x + 5}}$ **96.** $H(x) = (2x - 1)^3 + 5$

In Exercises 97–100, determine whether the given function is one-to-one. If the function is one-to-one, find its inverse and sketch the graph of the function and its inverse on the same coordinate axes.

97. $f(x) = x + 2$ **98.** $f(x) = -2x + 3$

99. $f(x) = \sqrt[3]{x} - 2$ **100.** $f(x) = 8x^3 - 1$

In Exercises 101 and 102, assume that f is a one-to-one function. Find f^{-1} and find the domain and range of f.

101. $f(x) = 4 + \sqrt{x - 1}$

102. $f(x) = -3 + \sqrt{x + 2}$

103. For the following graph of a function f
   a. write a formula for f as a piecewise function.
   b. find the domain and range of f.
   c. find the intercepts of the graph of f.
   d. draw the graph of $y = f(-x)$.
   e. draw the graph of $y = -f(x)$.
   f. draw the graph of $y = f(x) + 1$.
   g. draw the graph of $y = f(x + 1)$.
   h. draw the graph of $y = 2f(x)$.
   i. draw the graph of $y = f(2x)$.
   j. draw the graph of $y = f\left(\dfrac{1}{2}x\right)$.
   k. explain why f is one-to-one.
   l. draw the graph of $y = f^{-1}(x)$.

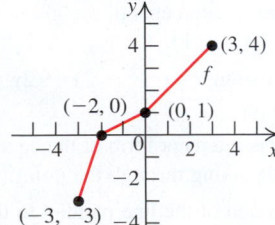

Applying the Concepts

104. Scuba diving. The pressure P on the body of a scuba diver increases linearly as she descends to greater depths d in seawater. The pressure at a depth of 10 feet is

19.2 pounds per square inch, and the pressure at a depth of 25 feet is 25.95 pounds per square inch.

a. Write the equation relating P and d (with d as independent variable) in slope–intercept form.

b. What is the meaning of the slope and the y-intercept in part (a)?

c. Determine the pressure on a scuba diver who is at a depth of 160 feet.

d. Find the depth at which the pressure on the body of the scuba diver is 104.7 pounds per square inch.

105. Waste disposal. The Sioux City council considered the following data regarding the cost of disposing waste material:

Year	2007	2011
Waste (in pounds)	87,000	223,000
Cost	$54,000	$173,000

Assume that the cost C (in dollars) is linearly related to the waste w (in pounds).

a. Write the linear equation relating C and w in slope–intercept form.

b. Explain the meaning of the slope and the intercepts of the equation in part (a).

c. Suppose the city has projected 609,000 pounds of waste for the year 2023. Calculate the projected cost of disposing of the waste for the year 2023.

d. Suppose the city's projected budget for waste collection in 2023 is 1 million dollars. How many pounds of waste can the city handle?

106. Checking your speedometer. Zoe checks the accuracy of her speedometer by driving a few measured (not odometer) miles at a constant 60 miles per hour while her friend checks his watch at the beginning and end of the measured distance.

a. How does a watch tell Zoe's friend whether the speedometer is accurate? What mathematics is involved?

b. Why shouldn't Zoe use her odometer to measure the number of miles?

107. Playing blackjack. Chloe, a bright mathematician, read a book about counting cards in blackjack and went to Las Vegas to try her luck. She started with $100 and discovered that the amount of money she had at time t hours after the start of the game could be expressed by the function $f(t) = 100 + 55t - 3t^2$.

a. What amount of money had Chloe won or lost in the first two hours?

b. What was the average rate at which Chloe was winning or losing money during the first two hours?

c. Did she lose all of her money? If so, when?

d. If she played until she lost all of her money, what was the average rate at which she was losing her money?

108. Volume discounting. Major cola distributors sell a case (containing 24 cans) of cola to the retailer at a price of $4. They offer a discount of 20% for purchases over 100 cases and a discount of 25% for purchases over 500 cases. Write a piecewise function that describes this pricing scheme. Take x as the number of cases purchased and $f(x)$ as the price paid.

109. Air pollution. The daily level L of carbon monoxide in a city is a function of the number of automobiles in the city. Suppose $L(x) = 0.5\sqrt{x^2 + 4}$, where x is the number of automobiles (in hundred thousands). Suppose further that the number of automobiles in a given city is growing according to the formula $x(t) = 1 + 0.002t^2$, where t is time in years measured from now.

a. Form a composite function describing the daily pollution level as a function of time.

b. What pollution level is expected in five years?

110. Toy manufacturing. After being in business for t years, a toy manufacturer is making $x = 5000 + 50t + 10t^2$ GI Jimmy toys per year. The sale price p in dollars per toy has risen according to the formula $p = 10 + 0.5t$. Write a formula for the manufacturer's yearly revenue as

a. a function of time t.

b. a function of price p.

PRACTICE TEST A

1. Determine which symmetries the graph of the equation $3x + 2xy^2 = 1$ has.

2. Find the x- and y-intercepts of the graph of $y = x^2(x - 3)(x + 1)$.

3. Graph the equation $x^2 + y^2 - 2x - 2y = 2$ and give the x- and y-intercepts.

4. Write the slope–intercept form of the equation of the line with slope -1 and passing through the point $(2, 7)$.

5. Write an equation of the line parallel to the line $8x - 2y = 7$ and passing through $(2, -1)$.

6. Use $f(x) = -2x + 1$ and $g(x) = x^2 + 3x + 2$ to find $(fg)(2)$.

7. Use $f(x) = 2x - 3$ and $g(x) = 1 - 2x^2$ to evaluate $g(f(2))$.

8. Use $f(x) = x^2 - 2x$ to find $(f \circ f)(x)$.

9. If $f(x) = \begin{cases} x^3 - 2 & \text{if } x \le 0 \\ 1 - 2x^2 & \text{if} > 0 \end{cases}$, find (a) $f(-1)$, (b) $f(0)$, and (c) $f(1)$.

10. Find the domain of the function $f(x) = \dfrac{\sqrt{x}}{\sqrt{1 - x}}$.

11. Find the domain of the function $f(x) = \sqrt{x^2 + x - 6}$.

12. Determine the average rate of change of the function $f(x) = 2x + 7$ between $x = 1$ and $x = 4$.

13. Determine whether the function $f(x) = 2x^4 - \dfrac{3}{x^2}$ is even, odd, or neither.

14. Find the intervals where the function shown is increasing or decreasing.

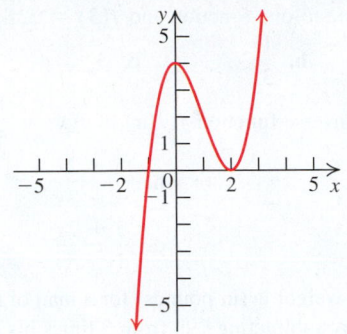

15. Suppose the graph of f is given. Describe how the graph of $y = f(x - 3)$ can be obtained from the graph of f.

16. A ball is thrown upward from the ground. After t seconds, the height h (in feet) above the ground is given by $h = 25 - (2t - 5)^2$. How many seconds does it take for the ball to reach a height of 25 feet?

17. If f is a one-to-one function and $f(2) = 7$, find $f^{-1}(7)$.

18. Find the inverse function $f^{-1}(x)$ of the one-to-one function $f(x) = 1 + \dfrac{1}{x}$.

19. Now that Jo has saved \$1000 for a car, her dad takes over and deposits \$100 into her account each month. Write an equation that relates the total amount of money, A, in Jo's account to the number of months, x, that Jo's dad had been putting money into her account.

20. The cost C in dollars for renting a car for one day is a function of the number of miles traveled, m. For a car renting for \$30.00 per day and \$0.25 per mile, this function is given by

$$C(m) = 0.25m + 30.$$

    a. Find the cost of renting the car for one day and driving 230 miles.

    b. If the charge for renting the car for one day is \$57.50, how many miles were driven?

PRACTICE TEST B

1. The graph of the equation $|x| + 2|y| = 2$ is symmetric with respect to which of the following?
    I. the origin   **II.** the x-axis   **III.** the y-axis
    a. I only   **b.** II only
    c. III only   **d.** I, II, and III

2. Which are the x- and y-intercepts of the graph of $y = x^2 - 9$?
    a. x-intercept 3, y-intercept -9
    b. x-intercepts ± 3, y-intercept -9
    c. x-intercept 3, y-intercept 9
    d. x-intercepts ± 3, y-intercepts ± 9

3. The slope of a line is undefined if the line is parallel to
    a. the x-axis   **b.** the line $x - y = 0$
    c. the line $x + y = 0$ **d.** the y-axis

4. Which is an equation of the circle with center $(0, -5)$ and radius 7?
    a. $x^2 + y^2 - 10y = 0$ **b.** $x^2 + (y - 5)^2 = 7$
    c. $x^2 + (y - 5)^2 = 49$ **d.** $x^2 + y^2 + 10y = 24$

5. Which is an equation of the line passing through $(2, -3)$ and having slope -1?
    a. $y + 3 = -(x + 2)$ **b.** $y - 3 = -(x - 2)$
    c. $y + 3 = -(x - 2)$ **d.** $y + 3 = -(x + 2)$

6. What is a second point on the line through $(3, 2)$ and having slope $-\dfrac{1}{2}$?
    a. $(2, 4)$ **b.** $(5, 1)$ **c.** $(7, 2)$ **d.** $(4, 4)$

7. Which is an equation of the line parallel to the line $6x - 3y = 5$ and passing through $(-1, 2)$?
    a. $y - 2 = -2(x + 1)$ **b.** $y - 2 = 6(x + 1)$
    c. $y - 2 = -6(x + 1)$ **d.** $y - 2 = 2(x + 1)$

8. Use $f(x) = 3x - 5$ and $g(x) = 2 - x^2$ to find $(f \circ g)(x)$.
    a. $-9x^2 + 30x - 23$ **b.** $-3x^2 + 1$
    c. $3x^2 + 1$ **d.** $9x^2 - 30x + 23$

9. Use $f(x) = 2x^2 - x$ to find $(f \circ f)(x)$.
    a. $8x^4 - 8x^3 + x$ **b.** $8x^4 + x$
    c. $4x^4 - 4x^3 + x^2 - x$ **d.** $4x^4 - x^3 + x^2$
    e. $-x^4 + 6x^2 - x$

10. If $g(t) = \dfrac{1 - t}{1 + t}$, find $g(a - 1)$.
    a. $\dfrac{2 - a}{2 + a}$ **b.** $\dfrac{1 - a}{1 + a}$
    c. $\dfrac{2 - a}{a}$ **d.** -1

11. Which of the following is the domain of the function $f(x) = \sqrt{x} + \sqrt{1 - x}$?
    a. $[0, 1]$ **b.** $(-\infty, -1] \cup [0, +\infty)$
    c. $(-\infty, 1]$ **d.** $[0, +\infty)$

12. What is the domain of the function $f(x) = \sqrt{x^2 + 6x - 7}$?
    a. $(-\infty, -7) \cup (1, \infty)$.   **b.** $(-\infty, -7] \cup [1, \infty)$
    c. $[-7, 1]$ **d.** $(1, \infty)$

13. Which description of the behavior of the graph shown here is correct?
    a. increasing on $(0, 3)$; decreasing on $(-3, 0) \cup (3, 4)$
    b. increasing on $(-1, 3)$; decreasing on $(2, -1) \cup (3, 1)$
    c. increasing on $(-3, 3)$; decreasing on $(2, -1) \cup (3, 1)$
    d. increasing on $(1.2, 4)$; decreasing on $(-3, -1.4)$

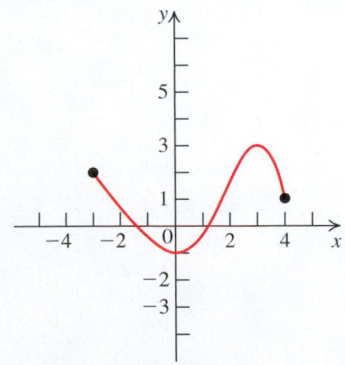

14. Suppose the graph of f is given. Describe how to obtain the graph of $y = f(x + 4)$ from the graph of f.
   a. Shift the graph of f four units to the left.
   b. Shift the graph of f four units down.
   c. Shift the graph of f four units up.
   d. Shift the graph of f four units to the right.
   e. Reflect the graph of f about the x-axis.

15. Suppose the graph of f is given. Describe how to obtain the graph of $y = -2f(x - 4)$ from the graph of f.
   a. Shift the graph of f four units to the left, shrink it vertically by a factor of $\frac{1}{2}$, and reflect it about the x-axis.
   b. Shift the graph of f four units to the right, stretch it vertically by a factor of 2, and reflect it about the x-axis.
   c. Shift the graph of f four units to the left, shrink it vertically by a factor of $\frac{1}{2}$, and reflect it about the x-axis.
   d. Shift the graph of f four units to the right, then stretch it vertically by a factor of 2, and reflect it about the y-axis.

16. Which of the following is a one-to-one function?
   a. $f(x) = |x + 3|$ b. $f(x) = x^2$
   c. $f(x) = \sqrt{x^2 + 9}$ d. $f(x) = x^3 + 2$

17. If f is a one-to-one function and $f(3) = 5$, then $f^{-1}(5) =$
   a. $\frac{1}{5}$ b. $\frac{1}{3}$ c. 3   d. -5

18. Find the inverse function $f^{-1}(x)$ of $f(x) = \dfrac{1 - 3x}{5 + 2x}$.
   a. $\dfrac{1 + 3x}{5 - 2x}$ b. $\dfrac{5 + 2x}{1 - 3x}$
   c. $\dfrac{1 - 5x}{3 + 2x}$ d. $\dfrac{3 + 2x}{1 - 5x}$

19. The ideal weight w (in pounds) for a man of height x inches is found by subtracting 190 from 5 times his height. Write a linear equation that gives the ideal weight of a man of height x inches. Then find the ideal weight of a man whose height is 70 inches.
   a. $w = \dfrac{x + 190}{5}$, 160 lb   b. $w = 5x - 190$, 160 lb.
   c. $w = 190 - \dfrac{x}{5}$, 176 lb   d. $w = \dfrac{x + 190}{5}$, 52 lb

20. Cassie rented a compact sedan at \$25 per day + \$0.20 per mile. How many miles can Cassie travel for \$50?
   a. 125   b. 1025   c. 250   d. 175

CHAPTER 2

Polynomial and Rational Functions

TOPICS

2.1 Quadratic Functions

2.2 Polynomial Functions

2.3 Dividing Polynomials and the Rational Zeros Test

2.4 Rational Functions

2.5 Polynomial and Rational Inequalities

2.6 Zeros of a Polynomial Function

2.7 Variation

Polynomials and rational functions are used in navigation, in computer-aided geometric design, in the study of supply and demand in business, in the design of suspension bridges, in computations determining the force of gravity on planets, and in virtually every area of inquiry requiring numerical approximations. Their applicability is so widespread that it is hard to imagine an area in which they have not made an impact.

Quadratic Functions

BEFORE STARTING THIS SECTION, REVIEW

1 Linear functions (Section 1.4, page 60)

2 Completing the square (Appendix A.6, page 957)

3 Quadratic formula (Appendix A.6, page 957)

4 Transformations (Section 1.5, page 78)

OBJECTIVES

1 Graph a quadratic function in standard form.

2 Graph any quadratic function.

3 Solve problems modeled by quadratic functions.

◆ The Gateway Arch

The Gateway Arch, at 630 feet, is the world's tallest arch. Located on the west bank of the Mississippi River in St. Louis, Missouri, it stands as a monument to the westward expansion of the United States, begun 200 years ago by Lewis and Clark. Built in the form of a flattened catenary arch, it is the tallest human-made monument in the United States and the hallmark of the Jefferson National Expansion Memorial. The arch was designed by Finnish American architect Eero Saarinen and German American structural engineer Hannskarl Bendel in 1947 and has become an internationally famous symbol of St. Louis. In Exercise 102 we investigate the dimensions of this famous structure.

---

1 Graph a quadratic function in standard form.

Quadratic Functions

Quadratic Function

A function of the form

$$f(x) = ax^2 + bx + c,$$

where a, b, and c are real numbers with $a \neq 0$, is called a **quadratic function**.

Note that the squaring function $s(x) = x^2$ (whose graph is a parabola, see page 70) is the quadratic function $f(x) = ax^2 + bx + c$ with $a = 1$, $b = 0$, and $c = 0$. We will show that we can graph any quadratic function by transforming the graph of $s(x) = x^2$. Since the graph of $y = x^2$ is a parabola, the graph of any quadratic function is also a parabola.

First, we compare the graphs of $f(x) = ax^2$ for $a > 0$ and $g(x) = ax^2$ for $a < 0$. These are quadratic functions with $b = 0$ and $c = 0$.

$f(x) = ax^2, a > 0$	$g(x) = ax^2, a < 0$
The graph of $f(x) = ax^2$ is obtained from the graph of the squaring function $s(x) = x^2$ by multiplying the y-values by a, vertically stretching or compressing it. Here are three such graphs on the same axes for $$a = 2, a = 1, \text{ and } a = \frac{1}{2}.$$	The graph of $g(x) = ax^2$ is the reflection of the graph of $f(x) = |a|x^2$ about the x-axis. Here are three graphs, for $$a = -2, a = -1, \text{ and } a = -\frac{1}{2}.$$

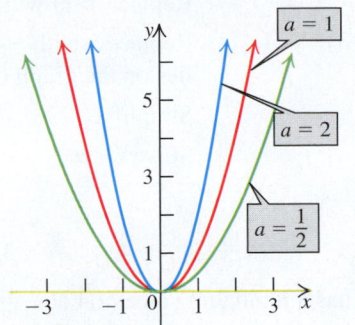

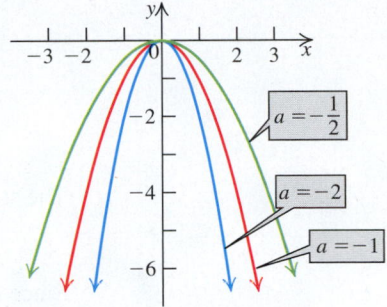

Standard Form of a Quadratic Function

Any quadratic function $f(x) = ax^2 + bx + c$ can be rewritten in the form $f(x) = a(x - h)^2 + k$, called the **standard form** of the quadratic function f. By using transformations (see Section 1.5), we can obtain the graph of $f(x) = a(x - h)^2 + k$ from the graph of the squaring function $s(x) = x^2$ as follows:

$s(x) = x^2$ The squaring function
↓

$g(x) = (x - h)^2$ Horizontal translation to the right if $h > 0$ and to the
↓   left if $h < 0$

$G(x) = a(x - h)^2$ Vertical stretching or compressing. If $a < 0$, the graph
↓   is also reflected about the x-axis.

$f(x) = a(x - h)^2 + k$ Vertical translation, up if $k > 0$ and down if $k < 0$

These transformations show that the graph of a quadratic function $f(x) = a(x - h)^2 + k$ is a parabola that opens up if $a > 0$ and down if $a < 0$. Its domain is $(-\infty, \infty)$. The parabola is symmetric with respect to the vertical line $x = h$. The line of symmetry $x = h$ is called the *axis* (or *axis of symmetry*) of the parabola. The point (h, k) where the axis meets the parabola is called the **vertex** of the parabola. The vertex of any parabola that opens up is the *lowest point* of the graph, and the vertex of any parabola that opens down is the *highest point* of the graph. Therefore, k is the **minimum value** for any parabola that opens up, and k is the **maximum value** for any parabola that opens down. The standard form is particularly useful because it immediately identifies the vertex (h, k).

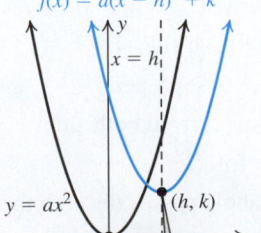

Figure 2.1 Quadratic functions.

The Standard Form of a Quadratic Function

The quadratic function

$$f(x) = a(x - h)^2 + k, \ a \neq 0$$

is in **standard form**. The graph of f is a parabola with **vertex** (h, k). (See Figure 2.1.) The parabola is symmetric with respect to the line $x = h$, called the **axis** of the parabola. If $a > 0$, the parabola opens up, and if $a < 0$, the parabola opens down. If $a > 0$, k is the **minimum value of** f, and if $a < 0$, k is the **maximum value of** f.

EXAMPLE 1 Writing the Equation of a Quadratic Function

Find the standard form of the quadratic function f whose graph has vertex $(-3, 4)$ and passes through the point $(-4, 7)$. Does f have a maximum or a minimum value?

Solution

$f(x) = a(x - h)^2 + k$	Standard form of quadratic function
$f(x) = a[x - (-3)]^2 + 4$	Replace h with -3 and k with 4.
$y = a(x + 3)^2 + 4$	Replace $f(x)$ with y, and simplify.
$7 = a(-4 + 3)^2 + 4$	Replace x with -4 and y with 7 because $(-4, 7)$ lies on the graph of $y = f(x)$.
$7 = a + 4$	Simplify.
$a = 3.$	Solve for a.

The standard form is

$$f(x) = 3(x + 3)^2 + 4. \quad a = 3, h = -3, k = 4$$

Since $a = 3 > 0$, f has a minimum value of 4 at $x = -3$.

Practice Problem 1 Find the standard form of the quadratic function f whose graph has vertex $(1, -5)$ and passes through the point $(3, 7)$. Does f have a maximum or a minimum value?

PROCEDURE
IN ACTION

EXAMPLE 2 Graphing a Quadratic Function in Standard Form

OBJECTIVE

Sketch the graph of $f(x) = a(x - h)^2 + k$.

Step 1 The graph is a parabola because it has the form $f(x) = a(x - h)^2 + k$. Identify a, h, and k.

Step 2 Determine how the parabola opens. If $a > 0$, the parabola opens *up*. If $a < 0$, it opens *down*.

Step 3 Find the vertex (h, k). If $a > 0$ (or $a < 0$), the function f has a minimum (or a maximum) value k at $x = h$.

Step 4 Find the x-intercepts (if any). Set $f(x) = 0$ and solve the equation $a(x - h)^2 + k = 0$ for x. If the solutions are real numbers, they are the x-intercepts. If not, the parabola lies above the x-axis (when $a > 0$) or below the x-axis (when $a < 0$).

EXAMPLE

Sketch the graph of $f(x) = -3(x + 2)^2 + 12$.

1. The graph of $f(x) = -3(x + 2)^2 + 12$
$$= \underset{\uparrow \atop a}{(-3)}[x - \underset{\uparrow \atop h}{(-2)}]^2 + \underset{\uparrow \atop k}{12}$$

is a parabola; $a = -3$, $h = -2$, and $k = 12$.

2. Since $a = -3 < 0$, the parabola opens down.

3. The vertex $(h, k) = (-2, 12)$. Since the parabola opens down, the function f has a maximum value of 12 at $x = -2$.

4.

$0 = -3(x + 2)^2 + 12$	Set $f(x) = 0$.
$3(x + 2)^2 = 12$	Add $3(x + 2)^2$ to both sides.
$(x + 2)^2 = 4$	Divide both sides by 3.
$x + 2 = \pm 2$	Square root property
$x = -2 \pm 2$	Subtract 2 from both sides.
$x = 0 \quad \text{or} \quad x = -4$	Solve for x.

The x-intercepts are 0 and -4. The parabola passes through the points $(0, 0)$ and $(-4, 0)$.

(continued)

Step 5 Find the y-intercept. Replace x with 0. Then $f(0) = ah^2 + k$ is the y-intercept.

5. $f(0) = -3(0 + 2)^2 + 12 = -3(4) + 12 = 0.$ The y-intercept is 0. As already seen, the parabola passes through the origin $(0, 0)$.

Step 6 Sketch the graph. Plot the points found in Steps 3–5 and join them to form a parabola. Show the axis $x = h$ of the parabola by drawing a dashed vertical line.

If there are no x-intercepts, draw the half of the parabola that passes through the vertex and a second point, such as the y-intercept. Then use the axis of symmetry to draw the other half.

6. The axis of the parabola is the vertical line $x = -2$. The graph of the parabola is shown in the figure.

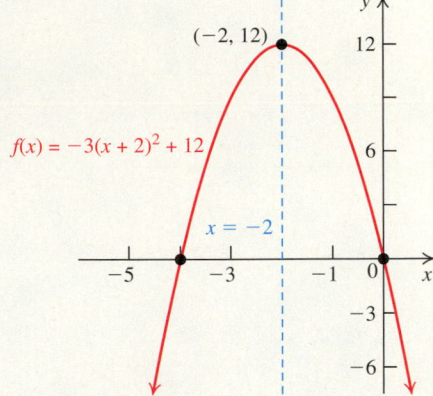

Practice Problem 2 Graph the quadratic function $f(x) = -2(x + 1)^2 + 3$. Find the vertex, maximum or minimum value, and x- and y-intercepts.

![WARNING triangle icon] **WARNING**

It is easy to make sign errors when identifying the vertex of a parabola from its standard form. The direct method is to set the expression being squared (inside the parentheses) to zero.

$$f(x) = -3(x + 2)^2 + 12$$

$x + 2 = 0$ gives $x = -2$. Hence, the first coordinate of the vertex is $h = -2$. The second coordinate of the vertex is $k = f(-2) = 12$. The vertex is

$$(h, k) = (-2, 12).$$

2 Graph any quadratic function.

Graphing a Quadratic Function $f(x) = ax^2 + bx + c$

A quadratic function $f(x) = ax^2 + bx + c$ can be changed to the standard form $f(x) = a(x - h)^2 + k$ by *completing the square*.

To write $x^2 - 4x + 3$ in standard form, we use $\left(\dfrac{-4}{2}\right)^2 = (-2)^2 = 4$, the square of one-half the coefficient of the term $-4x$.

$$x^2 - 4x + 3 = x^2 - 4x + 4 - 4 + 3 \qquad \text{Add and subtract } 4; 4 = \left(\dfrac{-4}{2}\right)^2$$

$$= (x^2 - 4x + 4) - 4 + 3 \qquad \text{Group the first three terms.}$$
$$= (x - 2)^2 - 1 \qquad x^2 - 4x + 4 = (x - 2)^2; -4 + 3 = -1$$

To write $5x^2 - 20x + 8$ in standard form, we first group $5x^2 - 20x$.

$$5x^2 - 20x + 8 = (5x^2 - 20x) + 8 \qquad \text{Group } 5x^2 - 20x.$$

$$= 5(x^2 - 4x) + 8 \qquad \text{Factor out 5 from } 5x^2 - 20x.$$

$$= 5(x^2 - 4x + 4 - 4) + 8 \qquad \text{Add and subtract 4 within the}$$
$$\text{parentheses; } 4 = \left(\dfrac{-4}{2}\right)^2$$

$$= 5(x^2 - 4x + 4) + 5(-4) + 8 \qquad \text{Distributive property}$$
$$= 5(x - 2)^2 - 12 \qquad x^2 - 4x + 4 = (x - 2)^2; \text{simplify.}$$

The general procedure for completing the square is given next.

Converting $f(x) = ax^2 + bx + c$ to Standard Form

$$f(x) = ax^2 + bx + c \qquad \text{Original function}$$

$$= a\left(x^2 + \frac{b}{a}x\right) + c \qquad \text{Factor out } a \text{ from } ax^2 + bx.$$

$$= a\left(x^2 + \frac{b}{a}x + \frac{b^2}{4a^2} - \frac{b^2}{4a^2}\right) + c \qquad \text{To complete the square, add and subtract}$$
$$\left(\frac{1}{2} \cdot \frac{b}{a}\right)^2 = \frac{b^2}{4a^2} \text{ within the parentheses.}$$

$$= a\left(x^2 + \frac{b}{a}x + \frac{b^2}{4a^2}\right) - a \cdot \frac{b^2}{4a^2} + c \qquad \text{Distribute } a \text{ to } -\frac{b^2}{4a^2}; \text{ regroup terms.}$$

$$= a\left(x + \frac{b}{2a}\right)^2 + \left(c - \frac{b^2}{4a}\right) \qquad \left(x + \frac{b}{2a}\right)^2 = x^2 + \frac{b}{a}x + \frac{b^2}{4a^2}$$

Comparing this form with the standard form $f(x) = a(x - h)^2 + k$, we have:

$$h = -\frac{b}{2a} \qquad x\text{-coordinate of the vertex}$$

$$\text{and} \quad k = c - \frac{b^2}{4a} \qquad y\text{-coordinate of the vertex}$$

Notice that $f(h) = a(h - h)^2 + k = k$ Replace x with h.

Finding the Vertex of $f(x) = ax^2 + bx + c, a \neq 0$

To find the vertex (h, k) of $f(x) = ax^2 + bx + c$,

1. Find the x-coordinate $h = -\dfrac{b}{2a}$ of the vertex.

2. Calculate $k = f\left(-\dfrac{b}{2a}\right)$ to find its y-coordinate.

PROCEDURE
IN ACTION

EXAMPLE 3   **Graphing a Quadratic Function**

OBJECTIVE

Graph $f(x) = ax^2 + bx + c, a \neq 0$.

Step 1 Identify a, b, and c.

Step 2 Determine how the parabola opens.
If $a > 0$, the parabola opens *up*;
if $a < 0$, the parabola opens *down*.

Step 3 Find the vertex (h, k). Use the formulas:

$$h = -\frac{b}{2a}$$

$$k = f\left(-\frac{b}{2a}\right)$$

EXAMPLE

Sketch the graph of $f(x) = 2x^2 + 8x - 10$.

1. In the equation $y = f(x) = 2x^2 + 8x - 10$,
$a = 2, b = 8$, and $c = -10$.

2. Since $a = 2 > 0$, the parabola opens up.

3. $h = -\dfrac{b}{2a} = -\dfrac{8}{2(2)} = -2 \qquad a = 2, b = 8$

$k = f(h) = f(-2)$

$\quad = 2(-2)^2 + 8(-2) - 10 \qquad$ Replace h with -2.

$\quad = -18 \qquad$ Simplify.

The vertex is $(-2, -18)$.
The function f has a minimum value of -18 at $x = -2$.

(continued)

Step 4 Find the x-intercepts (if any). Let $f(x) = 0$ and solve $ax^2 + bx + c = 0$. If the solutions are real numbers, they are the x-intercepts. If not, the parabola lies entirely above the x-axis (when $a > 0$) or entirely below the x-axis (when $a < 0$).

4. $2x^2 + 8x - 10 = 0$ Set $f(x) = 0$.

$\quad\quad 2(x^2 + 4x - 5) = 0$ Factor out 2.

$\quad\quad 2(x + 5)(x - 1) = 0$ Factor.

$\quad\quad x + 5 = 0$ or $x - 1 = 0$ Zero-product property

$\quad\quad\quad\quad x = -5$ or $x = 1$ Solve for x.

The x-intercepts are at the points $(-5, 0)$ and $(1, 0)$.

Step 5 Find the y-intercept. Let $x = 0$. The result, $f(0) = c$, is the y-intercept.

5. Set $x = 0$ to obtain $f(0) = 2(0)^2 + 8(0) - 10 = -10$. The y-intercept is -10, and the graph passes through the point $(0, -10)$.

Step 6 The parabola is symmetric with respect to its axis, $x = -\dfrac{b}{2a}$. Use this symmetry to find additional points.

6. The axis of symmetry is $x = -2$. The symmetric image of $(0, -10)$ with respect to the axis $x = -2$ is $(-4, -10)$.

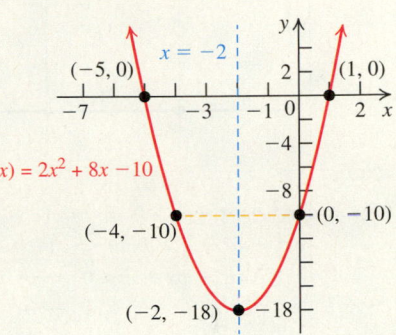

$f(x) = 2x^2 + 8x - 10$

Step 7 Draw a parabola through the points found in Steps 3–6.

If there are no x-intercepts, draw the half of the parabola that passes through the vertex and a second point, such as the y-intercept. Then use the axis of symmetry to draw the other half.

7. The parabola passing through the points in Steps 3–6 is sketched in the figure.

SIDE
NOTE

Standard form of a quadratic equation allows us to quickly find the x-intercepts, if any. As an example, we can solve

$x^2 - 4x - 9 = 0$ Rewrite in

$(x - 2)^2 - 13 = 0$ standard form.

$\quad\ (x - 2)^2 = 13$ Simplify.

$\quad\quad\quad\quad\quad\quad$ Square root

$\quad\ x - 2 = \pm\sqrt{13}$ property

$\quad\quad\ x = 2 \pm \sqrt{13}$ Simplify.

Practice Problem 3 Graph the quadratic function $f(x) = 3x^2 - 3x - 6$. Find the vertex, maximum or minimum value, and x- and y-intercepts.

As the graph suggests, when $a > 0$, every value of y, with $y \geq k$, is the second coordinate for some point on the graph of $f(x) = ax^2 + bx + c = a(x - h)^2 + k$. So when $a > 0$, the interval $[k, \infty)$ is the range of f. Similarly, when $a < 0$, the range of f is the interval $(-\infty, k]$. We can summarize this as:

FINDING THE RANGE OF $f(x) = ax^2 + bx + c$ HAVING VERTEX (h, k)

1. If $a > 0$, the range of $f(x) = ax^2 + bx + c$ is $[k, \infty)$.

2. If $a < 0$, the range of $f(x) = ax^2 + bx + c$ is $(-\infty, k]$.

Below we present a comprehensive table that will facilitate solving quadratic inequalities.

SOLVING QUADRATIC INEQUALITIES USING THE GRAPH OF $f(x) = ax^2 + bx + c$

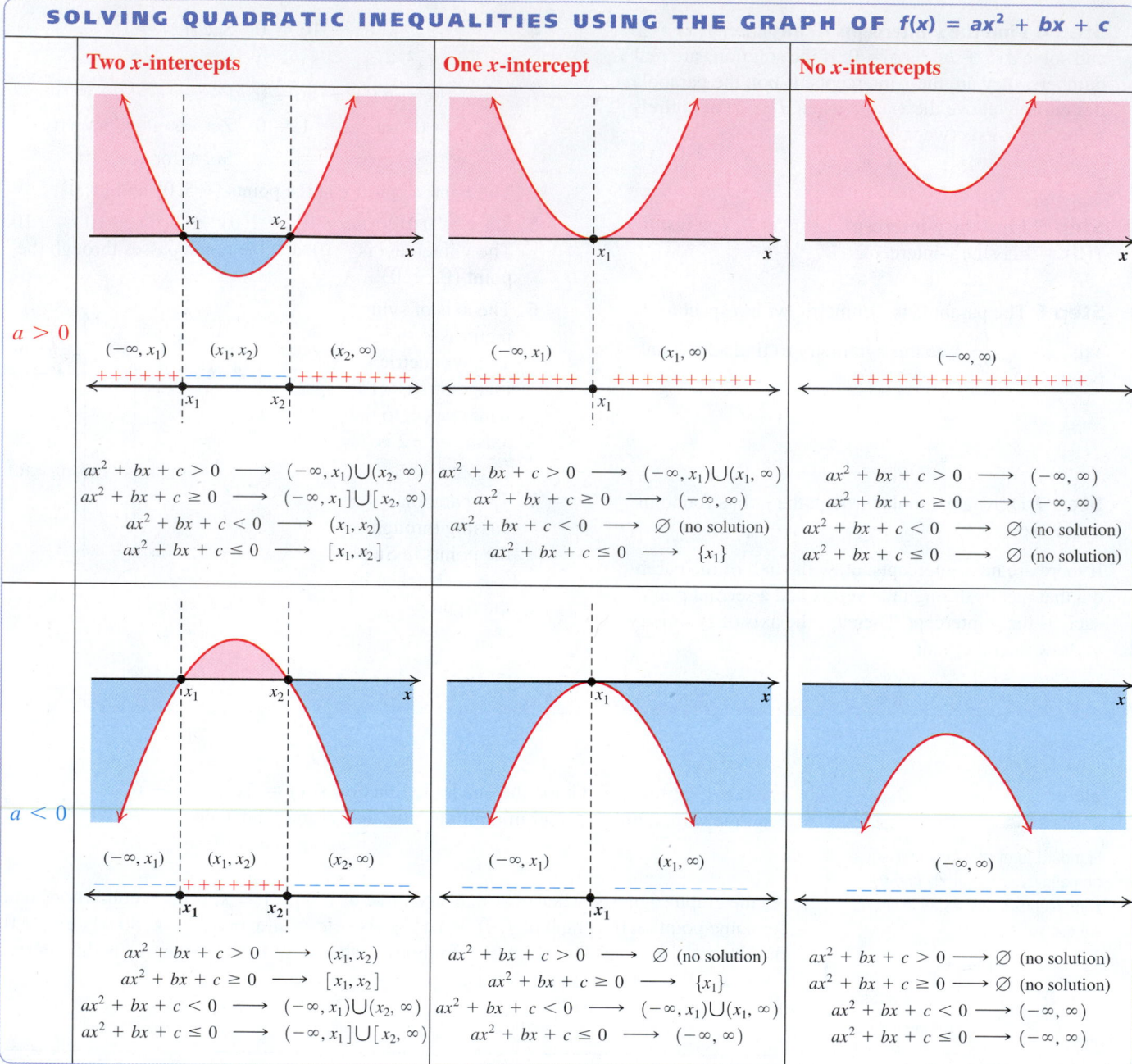

Two x-intercepts	One x-intercept	No x-intercepts

$a > 0$

$(-\infty, x_1)$ (x_1, x_2) (x_2, ∞)

$ax^2 + bx + c > 0 \longrightarrow (-\infty, x_1) \cup (x_2, \infty)$
$ax^2 + bx + c \geq 0 \longrightarrow (-\infty, x_1] \cup [x_2, \infty)$
$ax^2 + bx + c < 0 \longrightarrow (x_1, x_2)$
$ax^2 + bx + c \leq 0 \longrightarrow [x_1, x_2]$

$(-\infty, x_1)$ (x_1, ∞)

$ax^2 + bx + c > 0 \longrightarrow (-\infty, x_1) \cup (x_1, \infty)$
$ax^2 + bx + c \geq 0 \longrightarrow (-\infty, \infty)$
$ax^2 + bx + c < 0 \longrightarrow \varnothing$ (no solution)
$ax^2 + bx + c \leq 0 \longrightarrow \{x_1\}$

$(-\infty, \infty)$

$ax^2 + bx + c > 0 \longrightarrow (-\infty, \infty)$
$ax^2 + bx + c \geq 0 \longrightarrow (-\infty, \infty)$
$ax^2 + bx + c < 0 \longrightarrow \varnothing$ (no solution)
$ax^2 + bx + c \leq 0 \longrightarrow \varnothing$ (no solution)

$a < 0$

$(-\infty, x_1)$ (x_1, x_2) (x_2, ∞)

$ax^2 + bx + c > 0 \longrightarrow (x_1, x_2)$
$ax^2 + bx + c \geq 0 \longrightarrow [x_1, x_2]$
$ax^2 + bx + c < 0 \longrightarrow (-\infty, x_1) \cup (x_2, \infty)$
$ax^2 + bx + c \leq 0 \longrightarrow (-\infty, x_1] \cup [x_2, \infty)$

$(-\infty, x_1)$ (x_1, ∞)

$ax^2 + bx + c > 0 \longrightarrow \varnothing$ (no solution)
$ax^2 + bx + c \geq 0 \longrightarrow \{x_1\}$
$ax^2 + bx + c < 0 \longrightarrow (-\infty, x_1) \cup (x_1, \infty)$
$ax^2 + bx + c \leq 0 \longrightarrow (-\infty, \infty)$

$(-\infty, \infty)$

$ax^2 + bx + c > 0 \longrightarrow \varnothing$ (no solution)
$ax^2 + bx + c \geq 0 \longrightarrow \varnothing$ (no solution)
$ax^2 + bx + c < 0 \longrightarrow (-\infty, \infty)$
$ax^2 + bx + c \leq 0 \longrightarrow (-\infty, \infty)$

When solving quadratic inequalities, the main characteristics of the related quadratic function $f(x) = ax^2 + bx + c$ are the number of x-intercepts and whether the graph of f opens up or down. We illustrate how to effectively use the above table with three examples.

The function $f(x) = x^2 + 2x - 3 = (x + 3)(x - 1)$ has two x-intercepts, $x_1 = -3$ and $x_2 = 1$, and $a = 1 > 0$. We find the solution set of the inequality $f(x) = x^2 + 2x - 3 \geq 0$ from the table as $(-\infty, -3] \cup [1, \infty)$.

The function $g(x) = -2x^2 + 4x - 2 = -2(x - 1)^2$ has only one x-intercept, $x_1 = 1$, and $a = -2 < 0$. We find the solution set of the inequality $g(x) = -2x^2 + 4x - 2 < 0$ from the table as $(-\infty, 1) \cup (1, \infty)$.

The function $h(x) = 3x^2 + 6x + 4 = 3(x + 1)^2 + 1$ has no x-intercepts and $a = 3 > 0$. We find the solution set of the inequality $h(x) = 3x^2 + 6x + 4 > 0$ from the table as $(-\infty, \infty)$.

EXAMPLE 4 **Graph a Quadratic Function and Solve Related Inequality**

Sketch the graph of the function $f(x) = -2x^2 + 8x - 5$. Using the graph of f,

a. Find the domain and range of f.

b. Solve the inequality $-2x^2 + 8x - 5 > 0$.

Solution

In the equation $f(x) = -2x^2 + 8x - 5$, we identify $a = -2, b = 8$, and $c = -5$. We find h as

$$h = \frac{-b}{2 \cdot a} = \frac{-8}{2 \cdot (-2)} = 2,$$

and calculate k as

$$k = f(h) = f(2) = -2(2)^2 + 8(2) - 5 = 3.$$

The vertex of f is $(2, 3)$. Since $a = -2$, the graph of f is a parabola that opens down. Rewriting f in standard form produces

$$-2x^2 + 8x - 5 = a(x - h)^2 + k = -2(x - 2)^2 + 3.$$

To find the x-intercepts, we solve the equation $f(x) = 0$, as follows.

$$-2x^2 + 8x - 5 = 0$$

$-2(x - 2)^2 + 3 = 0$ Rewrite in standard form.

$-2(x - 2)^2 = -3$ Simplify.

$(x - 2)^2 = \dfrac{3}{2}$ Divide both sides by -2 and simplify.

$x - 2 = \pm\sqrt{\dfrac{3}{2}}$ Square root property

$x = 2 \pm \sqrt{\dfrac{3}{2}}$ Add 2 to both sides and simplify.

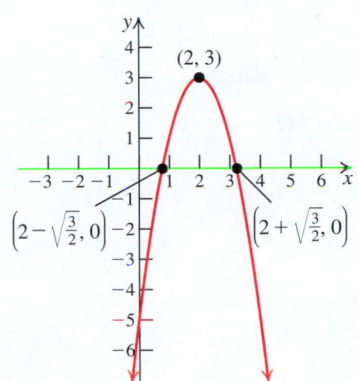

Figure 2.2

The x-intercepts of f are $x_1 = 2 - \sqrt{\dfrac{3}{2}} \approx 0.775$ and $x_2 = 2 + \sqrt{\dfrac{3}{2}} \approx 3.225$. See Figure 2.2.

a. The domain of f is $(-\infty, \infty)$. Because the parabola opens down, the maximum value of f is the y-coordinate, 3, of its vertex $(2, 3)$. So the range of f is $(-\infty, 3]$.

b. To solve $-2x^2 + 8x - 5 > 0$, we need to determine for which values of x the graph of f is above the x-axis. Because $a < 0$ and the parabola opens down, the graph of f is above the x-axis between the two intercepts over the interval (x_1, x_2). See Figure 2.2 and the table on page 144. Because the inequality $>$ is strict, the x-intercepts of $f(x)$ are not included in the solution set. The solution set of the inequality $-2x^2 + 8x - 5 > 0$ is the interval $\left(2 - \sqrt{\dfrac{3}{2}}, 2 + \sqrt{\dfrac{3}{2}}\right)$.

Practice Problem 4 Graph the function $f(x) = 3x^2 - 6x - 1$, find the domain and range of f, and solve the inequality $3x^2 - 6x - 1 \leq 0$.

3 Solve problems modeled by quadratic functions.

Applications

Many applications of quadratic functions involve finding the maximum or minimum value of the function.

◆ **EXAMPLE 5** **Tracking a Moving Object**

The height h, in feet, of a ball thrown upward at an initial speed of v_0 ft/s from a point h_0 ft above the ground is given by the function

$$h(t) = -\frac{1}{2}gt^2 + v_0 t + h_0,$$

where t is the time in seconds after the ball was released, and $g = 32$ ft/s^2 is the acceleration due to gravity on Earth. If you have thrown the ball upward from a cliff 50 feet high and the maximum height that it reached was 99 feet,

a. Find the height of the ball as a function of t.

b. How long did it take for the ball to reach its highest point?

Solution

a. With $g = 32$ and $h_0 = 50$, the height of the ball is governed by the equation

$$(1) \qquad h(t) = -16t^2 + v_0 t + 50. \qquad -\frac{1}{2}g = -\frac{1}{2}\cdot 32 = -16$$

To find v_0, we observe that this function is a quadratic function with $a = -16 < 0$. It attains its maximum value at the point were

$$t = -\frac{v_0}{2(-16)} = \frac{v_0}{32} \qquad \text{Use the formula } t = -\frac{b}{2a} \text{ for the first}$$
coordinate of the vertex of a parabola.

Given that the maximum height of the ball was 99 feet, and $h(t) = -16t^2 + v_0 t + 50$,

$$99 = h\left(\frac{v_0}{32}\right) = -16\left(\frac{v_0}{32}\right)^2 + v_0\left(\frac{v_0}{32}\right) + 50 \qquad \text{Replace } t \text{ with } \frac{v_0}{32} \text{ in } h(t).$$

Solving for v_0 yields

$$\frac{v_0^2}{64} + 50 = 99 \qquad\qquad -16\left(\frac{v_0}{32}\right)^2 + v_0\left(\frac{v_0}{32}\right) = \frac{v_0^2}{64}$$

$$\frac{v_0^2}{64} = 99 - 50 = 49 \qquad\qquad \text{Subtract 50 from both sides.}$$

$$v_0^2 = 64 \cdot 49 \qquad\qquad \text{Multiply both sides by 64.}$$

$$v_0 = \sqrt{64 \cdot 49} = 8 \cdot 7 = 56 \qquad v_0 \text{ is positive since the ball was thrown upward.}$$

$$h(t) = -16t^2 + 56t + 50 \qquad \text{Replace } v_0 \text{ in equation (1) by 56.}$$

b. We again use the formula for the first coordinate of the vertex of a parabola to conclude that the ball reached its highest point at time

$$t = -\frac{56}{2(-16)} = \frac{56}{32} = 1.75 \text{ seconds}$$

after it was released.

Practice Problem 5 Repeat Example 5 assuming that the height of the cliff was 100 feet and the ball went up to a maximum height of 244 feet above the ground.

EXAMPLE 6 **Finding a Maximum Area**

A farmer wants to build a rectangular grazing pen using a straight stone wall for one side. He has 100 feet of fence to use for the other three sides. What is the maximum area that can be enclosed? What are the dimensions of the pen?

Solution

Let x = length of the side perpendicular to the wall (see Figure 2.3) and y = length of the side parallel to the wall. Then,

$$2x + y = 100 \qquad \text{The three sides use 100 feet of fence.}$$
$$y = 100 - 2x \qquad \text{Subtract } 2x \text{ from both sides.}$$
$$A(x) = \text{area of the pen} = xy \qquad \text{Area} = (\text{length})(\text{width})$$
$$A(x) = x(100 - 2x) \qquad \text{Replace } y \text{ with } 100 - 2x.$$

The function $A(x) = x(100 - 2x) = -2x^2 + 100x$ gives the area of the pen, so we want to find the maximum value for $A(x)$. The vertex for the parabola is (h, k), where

$$h = -\frac{b}{2a} = \frac{-100}{2(-2)} = \frac{100}{4} = 25 \qquad A(x) = -2x^2 + 100x, \text{ so } a = -2, b = 100$$
$$k = A(25) = -2(25)^2 + 100(25) = 1250. \qquad k = \text{maximum value}$$

The maximum area is 1250 square feet, which occurs when $x = h = 25$.
The length of the side perpendicular to the wall is 25 feet.
The length of the side parallel to the wall is $y = 100 - 2x = 100 - 2(25) = 50$ feet.

Practice Problem 6 A children's center receives a donation of 1000 feet of fence to enclose a rectangular playground. What is the maximum area that can be enclosed? What are the playground's dimensions?

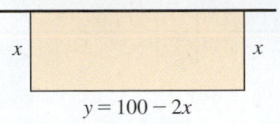

$y = 100 - 2x$

Figure 2.3

Answers to Practice Problems

1. $y = 3(x - 1)^2 - 5$; f has minimum value -5
2. The graph is a parabola. $a = -2, h = -1, k = 3$. It opens downward. Vertex $(-1, 3)$. It has maximum value 3; x-intercepts: $\pm\sqrt{\dfrac{3}{2}} - 1$; y-intercept: 1

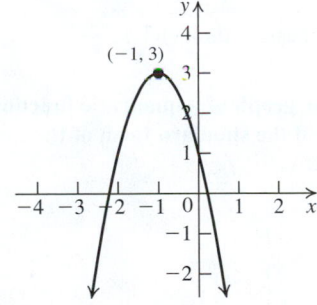

3. The graph is a parabola. It opens upward. Vertex $\left(\dfrac{1}{2}, \dfrac{-27}{4}\right)$; it has a minimum value $-\dfrac{27}{4}$; x-intercepts: $-1, 2$; y-intercept: -6

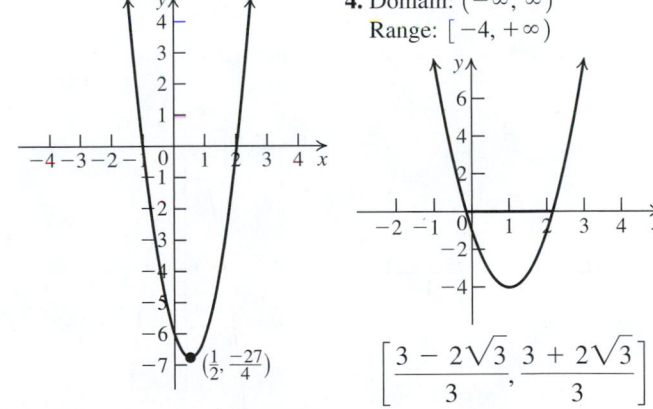

4. Domain: $(-\infty, \infty)$
Range: $[-4, +\infty)$

$$\left[\frac{3 - 2\sqrt{3}}{3}, \frac{3 + 2\sqrt{3}}{3}\right]$$

5. a. $h(t) = -16t^2 + 96t + 100$ (feet) **b.** 3 seconds
6. Maximum area is 62,500 square feet. The playground is a square of side length 250 feet on each side.

SECTION 2.1 **Exercises**

Concepts and Vocabulary

1. The graph of $f(x) = ax^2 + bx + c, a \neq 0$ is a
_____ .

2. The vertex of the graph $f(x) = a(x - h)^2 + k, a \neq 0$ is _____ .

3. The graph of $f(x) = a(x - h)^2 + k, a \neq 0$ is symmetric with respect to the vertical line _____ .

4. The x-coordinate of the vertex of $f(x) = ax^2 + bx + c, a \neq 0$ is given by _____ .

5. **True or False.** The domain of $f(x) = ax^2 + bx + c, a \neq 0$ is $(-\infty, \infty)$.

6. **True or False.** The graph of the $f(x) = 2x^2$ is wider than the graph of $g(x) = x^2$.

7. **True or False.** If $a > 0$, then the graph of $f(x) = ax^2 + bx + c$ opens down.

8. **True or False.** If $a < 0$, then $f(x) = ax^2 + bx + c$ has a maximum.

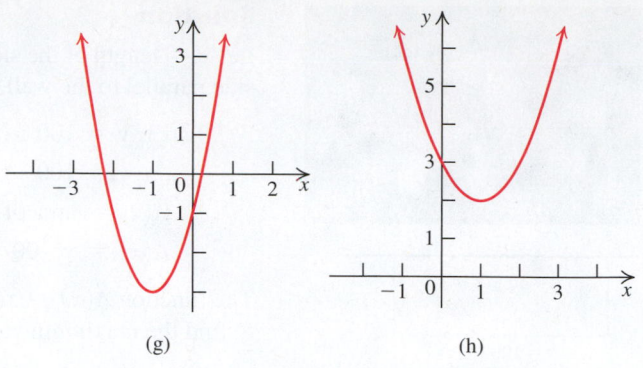

(g)   (h)

Building Skills

In Exercises 9–16, match each quadratic function with its graph.

9. $y = -\dfrac{1}{3}x^2$

10. $y = -3x^2$

11. $y = -3(x + 1)^2$

12. $y = 2(x + 1)^2$

13. $y = (x - 1)^2 + 2$

14. $y = (x - 1)^2 - 3$

15. $y = 2(x + 1)^2 - 3$

16. $y = -3(x + 1)^2 + 2$

In Exercises 17–20, find a quadratic function of the form $y = ax^2$ that passes through the given point.

17. $(2, -8)$

18. $(-3, 3)$

19. $(2, 20)$

20. $(-3, -6)$

In Exercises 21–30, find the quadratic function $y = f(x)$ that has the given vertex and whose graph passes through the given point. Write the function in standard form.

21. Vertex $(0, 0)$; passing through $(-2, 8)$

22. Vertex $(2, 0)$; passing through $(1, 3)$

23. Vertex $(-3, 0)$; passing through $(-5, -4)$

24. Vertex $(0, 1)$; passing through $(-1, 0)$

25. Vertex $(2, 5)$; passing through $(3, 7)$

26. Vertex $(-3, 4)$; passing through $(0, 0)$

27. Vertex $(2, -3)$; passing through $(-5, 8)$

28. Vertex $(-3, -2)$; passing through $(0, -8)$

29. Vertex $\left(\dfrac{1}{2}, \dfrac{1}{2}\right)$; passing through $\left(\dfrac{3}{4}, -\dfrac{1}{4}\right)$

30. Vertex $\left(-\dfrac{3}{2}, -\dfrac{5}{2}\right)$; passing through $\left(1, \dfrac{55}{8}\right)$

In Exercises 31–34, the graph of a quadratic function $y = f(x)$ is given. Find the standard form of the function.

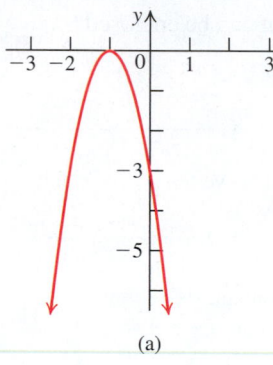

(a)

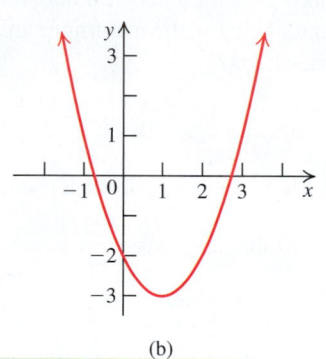

(b)

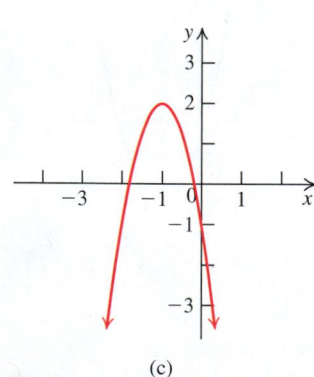

(c)

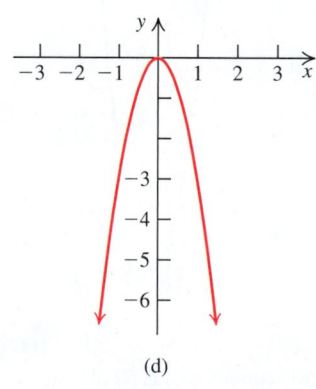

(d)

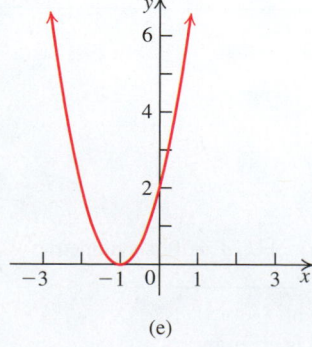

(e)

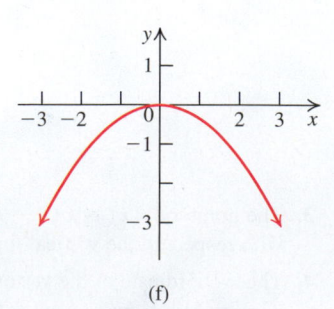

(f)

31.

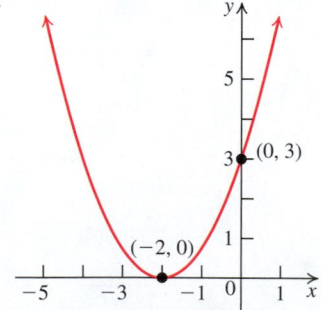

32.

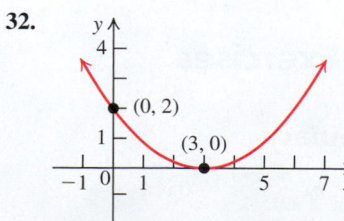

33.

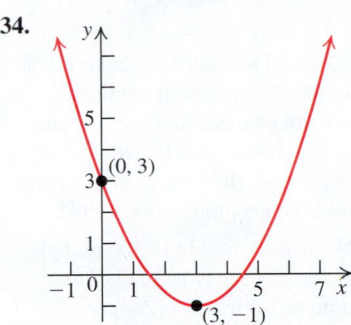

34.

In Exercises 35–44, graph each function by starting with the graph of $y = x^2$ and using transformations.

35. $f(x) = 3x^2$

36. $f(x) = \dfrac{1}{3}x^2$

37. $g(x) = (x - 4)^2$

38. $g(x) = (x + 3)^2$

39. $f(x) = -2x^2 - 4$

40. $f(x) = -x^2 + 3$

41. $g(x) = (x - 3)^2 + 2$

42. $g(x) = (x + 1)^2 - 3$

43. $f(x) = -3(x - 2)^2 + 4$

44. $f(x) = -2(x + 1)^2 + 3$

In Exercises 45–52, graph the given function by writing it in the standard form $y = a(x - h)^2 + k$ and then using transformations on $y = x^2$. Find the vertex, the axis of symmetry, and the x- and y-intercepts.

45. $y = x^2 + 4x$

46. $y = x^2 - 2x + 2$

47. $y = 6x - 10 - x^2$

48. $y = 8 + 3x - x^2$

49. $y = 2x^2 - 8x + 9$

50. $y = 3x^2 + 12x - 7$

51. $y = -3x^2 + 18x - 11$

52. $y = -5x^2 - 20x + 13$

In Exercises 53–60, (a) determine whether the graph of the given quadratic function opens up or down, (b) find the vertex $(h, k) = \left(-\dfrac{b}{2a}, f\left(-\dfrac{b}{2a} \right) \right)$, (c) find the axis of symmetry, (d) find the x- and y-intercepts, and (e) sketch the graph of the function.

53. $y = x^2 - 8x + 15$

54. $y = x^2 + 8x + 13$

55. $y = x^2 - x - 6$

56. $y = x^2 + x - 2$

57. $y = x^2 - 2x + 4$

58. $y = x^2 - 4x + 5$

59. $y = 6 - 2x - x^2$

60. $y = 2 + 5x - 3x^2$

In Exercises 61–68, a quadratic function f is given. (a) Determine whether the given quadratic function has a maximum value or a minimum value. Then find this value. (b) Find the range of f.

61. $f(x) = x^2 - 4x + 3$

62. $f(x) = -x^2 + 6x - 8$

63. $f(x) = -4 + 4x - x^2$

64. $f(x) = x^2 - 6x + 9$

65. $f(x) = 2x^2 - 8x + 3$

66. $f(x) = 3x^2 + 12x - 5$

67. $f(x) = -4x^2 + 12x + 7$

68. $f(x) = 8x - 5 - 2x^2$

In Exercises 69–82, solve the given quadratic inequality by sketching the graph of the corresponding quadratic function.

69. $x^2 - 4 \le 0$

70. $4 - x^2 \ge 0$

71. $x^2 - 4x + 3 < 0$

72. $x^2 + x - 2 > 0$

73. $x^2 - 3x - 7 \ge 0$

74. $x^2 + x - 8 \le 0$

75. $x^2 - 4x + 9 \ge 0$

76. $x^2 - 6x + 9 \le 0$

77. $-x^2 + 2x - 2 > 0$

78. $-x^2 + 6x - 9 < 0$

79. $-x^2 - 3x + 10 \ge 0$

80. $-x^2 - x + 12 \le 0$

81. $-6x^2 + 13x - 7 < 0$

82. $-5x^2 + 9x - 4 > 0$

Applying the Concepts

Note: The domain of a function that models a real-life problem may consist of only integer values. However, in solving these problems we frequently find it convenient to extend the domain to a suitable interval of real numbers containing the relevant integers.

83. Height of a baseball. Ignoring air resistance, the height of a baseball that is hit 3 feet above the ground at an angle of 45° with an initial velocity of 132 feet per second is given by the equation

$$h(x) = -\frac{32}{(132)^2}x^2 + x + 3,$$

where x is the horizontal distance (in feet) the ball traveled after being hit. Baseball fans will notice that air resistance severely distorts the actual distance a baseball travels. Use the graph to answer the following questions.

a. Approximately how many feet did the ball travel horizontally before hitting the ground?

b. Approximately how high did the ball go?

c. Use the function h to find answers to (a) and (b) rounded to the nearest integer.

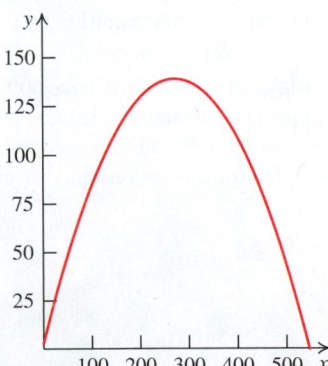

84. Apartment rentals. The number of apartments that can be rented in a 100-unit apartment complex decreases by one for each increase of $25 in rent. The revenue generated from rent after x increases of $25 is given by the equation $R(x) = -25x^2 + 1700x + 80,000$. Use the graph to answer the following questions.

a. Approximately what was the maximum revenue from the apartments?

b. Approximately how many $25 increases generated the maximum revenue?

c. Use the function R to find answers to (a) and (b).

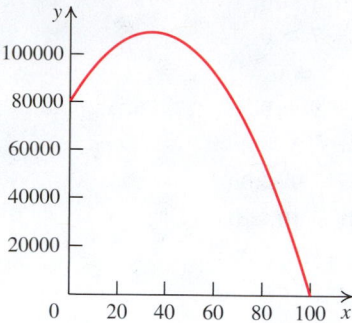

85. Maximizing revenue. A revenue function is given by $R(x) = 15 + 114x - 3x^2$, where x is the number of units produced and sold. Find the value of x for which the revenue is maximum.

86. Maximizing revenue. A demand function is given by $p = 200 - 4x$, where p is the price per unit and x is the number of units produced and sold (the demand). Find x for which the revenue is maximum. [Recall that $R(x) = x \cdot p$.]

87. Minimizing cost. The total cost of a product is $C = 200 - 50x + x^2$, where x is the number of units produced. Find the value of x for which the total cost is minimum.

88. Maximizing profit. A manufacturer produces x items of a product at a total cost of $(50 + 2x)$ dollars. The demand function for the product is $p = 100 - x$. Find the value of x for which the profit is maximum. How much is the maximum profit? [Recall that Profit = Revenue − Cost.]

89. Geometry. Find the dimensions of a rectangle of maximum area if the perimeter of the rectangle is 80 units. What is the maximum area?

90. Enclosing area. A rancher with 120 meters of fence intends to enclose a rectangular region along a river (which serves as a natural boundary requiring no fence). Find the maximum area that can be enclosed.

91. Enclosing playing fields. You have 600 meters of fencing, and you plan to lay out two identical playing fields. The fields can be side-by-side and separated by a fence, as shown in the figure. Find the dimensions and the maximum area of each field.

92. Enclosing area on a budget. You budget $2400 for constructing a rectangular enclosure that consists of a high surrounding fence and a lower inside fence that divides the enclosure in half. The high fence costs $8 per foot, and the low fence costs $4 per foot. Find the dimensions and the maximum area of each half of the enclosure.

93. Maximizing apple yield. From past surveys and records, an orchard owner in northern Michigan has determined that if 26 apple trees per acre are planted, each tree yields 500 apples, on average. The yield decreases by ten apples per tree for each additional tree planted. How many trees should be planted per acre to achieve the maximum total yield?

94. Buying and selling beef. A steer weighing 300 pounds gains 8 pounds per day and costs $1 a day to keep. You bought this steer today at a market price of $1.50 per pound. However, the market price is falling 2¢ per pound per day. When should you sell the steer to maximize your profit?

95. Going on a tour. A group of students plans a tour. The charge per student is $72 if 20 students go on the trip. If more than 20 students participate, the charge per student is reduced by $2 times the number of students over 20. Find the number of students that will furnish the maximum revenue. What is the maximum revenue?

96. Computer disks. A 5-inch-radius computer disk contains information in units called *bytes* that are arranged in concentric tracks on the disk. Assume that m bytes per inch can be put on each track. The tracks are uniformly spread (that is, there is a fixed number, say, p, of tracks per inch, measured radially across the disk). If the number of bytes on each track must be the same, where should the innermost track be located to get the maximum number of bytes on the disk?

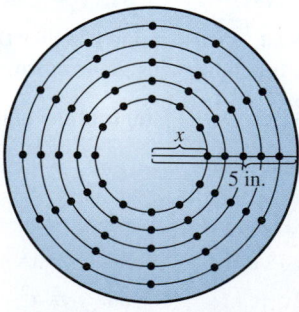

97. Motion on the Moon. On the Moon, the acceleration due to gravity is $g = 1.6 \text{ m/s}^2$. If an astronaut has thrown a stone from the 5-meter-high platform on his spacecraft directly upward, and the stone went up to a maximum height of 25 meters,

a. Find the height of the stone as a function of time t with $t = 0$ corresponding to the moment of releasing the stone.

b. How long did it take for the stone to reach its highest point?

98. Repeat Exercise 97 assuming that the platform was 7 meters high and the stone went up to the maximum height of 52 meters.

99. Motion of a projectile. A projectile is fired straight up with a velocity of 64 feet per second. Its altitude (height) h after t seconds is given by

$$h(t) = -16t^2 + 64t.$$

a. What is the maximum height of the projectile?
b. When does the projectile hit the ground?

100. Motion of a projectile. A projectile is shot upward with a velocity of 64 feet per second. Its altitude (height) h after t seconds is given by

$$h(t) = -16t^2 + 64t + 400.$$

a. What is the maximum height of the projectile?
b. When does the projectile hit the ground?

101. Architecture. A window is to be constructed in the shape of a rectangle surmounted by a semicircle. If the perimeter of the window is 18 feet, find its dimensions for the maximum area.

102. Architecture. The shape of the Gateway Arch in St. Louis, Missouri, is a *catenary* curve, which closely resembles a parabola. The function

$$y = f(x) = -0.00635x^2 + 4x$$

models the shape of the arch, where y is the height in feet and x is the horizontal distance from the base of the left side of the arch in feet.
a. Graph the function $y = f(x)$.
b. What is the width of the arch at the base?
c. What is the maximum height of the arch?

103. Football. The altitude or height h (in feet) of a punted football can be modeled by the quadratic function

$$h = -0.01x^2 + 1.18x + 2,$$

where x (in feet) is the horizontal distance from the point of impact with the punter's foot. (See accompanying figure.)

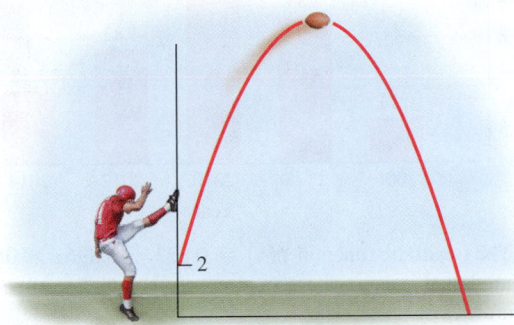

a. Find the vertex of the parabola.
b. If the opposing team fails to catch the ball, how far away from the punter does the ball hit the ground (to the nearest foot)?
c. What is the maximum height of the punted ball (to the nearest foot)?
d. The nearest defensive player is 6 feet from the point of impact. How high must the player reach to block the punt (to the nearest foot)?
e. How far downfield has the ball traveled when it reaches a height of 7 feet for the second time?

104. Field hockey. A *scoop* in field hockey is a pass that propels the ball from the ground into the air. Suppose a player makes a scoop with an upward velocity of 30 feet per second. The function

$$h = -16t^2 + 30t$$

models the altitude or height h in feet at time t in seconds. Will the ball ever reach a height of 16 feet?

105. Car sales The graph shows the number of light vehicle sales (in millions) in the United States in the month of August 2006–2014.

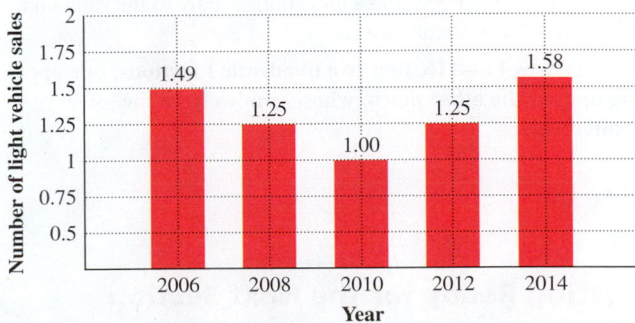

The quadratic function $f(x) = 1.51 - 0.22x + 0.029x^2$ models this data, where x represents the number of years after 2006.
a. Find the standard form of this quadratic function.
b. According to this function, in which year were vehicle sales at a minimum? Round to the nearest year and discuss whether the result fits the original data.
c. Use this function to estimate the number of vehicle sales in 2011.

106. Foreclosed properties The graph shows the number of properties with foreclosure filings (in millions) in the United States in the indicated year, 2006–2014.

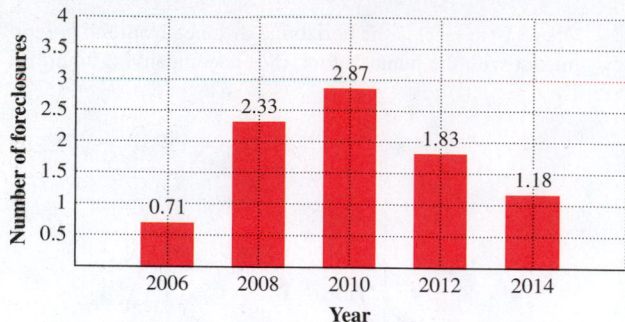

The quadratic function $f(x) = 0.822 + 0.896x - 0.11x^2$ models this data, where x represents the number of years after 2006.

a. Find the standard form of this quadratic function.

b. According to this function, in which year were the foreclosure filings at a maximum? Round to the nearest year and discuss whether the result fits the original data.

c. Use this function to estimate the number of foreclosure filings in 2013.

Beyond the Basics

In Exercises 107–112, find a quadratic function of the form $f(x) = ax^2 + bx + c$ that satisfies the given conditions.

107. The graph of f is obtained by shifting the graph of $y = 3x^2$ two units horizontally to the left and three units vertically up.

108. The graph of f passes through the point $(1, 5)$ and has vertex $(-3, -2)$.

109. The graph of f has vertex $(1, -2)$ and y-intercept 4.

110. The graph of f has x-intercepts 2 and 6 and y-intercept 24.

111. The graph of f has x-intercept 7 and y-intercept 14, and $x = 3$ is the axis of symmetry.

112. The graph of f is obtained by shifting the graph of $y = x^2 + 2x + 2$ three units horizontally to the right and two units vertically down.

In Exercises 113–116, find two quadratic functions, one opening up and the other down, whose graphs have the given x-intercepts.

113. $-2, 6$

114. $-3, 5$

115. $-7, -1$

116. $2, 10$

117. Find the minimum value of the expression

$$2x^2 + y^2 - 4x + 6y + 15.$$

Hint: Consider the expression as a sum of two quadratic functions.

118. Find the maximum value of the expression

$$-3x^2 - y^2 - 12x + 2y - 11.$$

Hint: Consider the expression as a sum of two quadratic functions.

Critical Thinking/Discussion/Writing

119. Symmetry about the line $x = h$. The graph of a polynomial function $y = f(x)$ is symmetric about the line $x = h$ if $f(h + p) = f(h - p)$ for every number p. For $f(x) = ax^2 + bx + c, a \neq 0$, set $f(h + p) = f(h - p)$

and show that $h = -\dfrac{b}{2a}$. This exercise proves that the graph of a quadratic function is symmetric about its axis $x = -\dfrac{b}{2a}$.

120. In the graph of $y = 2x^2 - 8x + 9$, find the coordinates of the point symmetric to the point $(-1, 19)$ about the axis of symmetry.

121. Let $f(x) = a(x - h)^2 + k, a \neq 0$ and $g(x) = mx + b$, $m \neq 0$.

a. Show that the graph of $y = (f \circ g)(x)$ is a parabola and find its vertex.

b. Show that the graph of $y = (g \circ f)(x)$ is a parabola and find its vertex.

122. Let $f(x) = ax^2 + bx + c, a \neq 0$ with discriminant $b^2 - 4ac$ of the equation $f(x) = 0$. Discuss the relationship between the location of the vertex of $y = f(x)$ and the sign of the discriminant.

123. Show that, for any two positive numbers x and a,

$$x^2 + (a - x)^2 \geq \frac{1}{2}a^2$$

by finding the minimum value of the function $f(x) = x^2 + (a - x)^2 = 2x^2 - 2ax + a^2$.

124. Show that, for any two positive numbers x and a,

$$x(a - x) \leq \frac{1}{4}a^2$$

by finding the maximum value of the function $f(x) = x(a - x) = -x^2 + ax$.

Getting Ready for the Next Section

In Exercises 125–130, simplify each expression. Write your answers without negative exponents.

125. -5^0

126. $(-4)^{-2}$

127. $(2x^2)(-3x^4)$

128. $(2x^2)(3x)(-x^3)$

129. $x^2\left(3 - \dfrac{3}{4}\right)$

130. $x^4\left(1 + \dfrac{3}{x^2} - \dfrac{5}{x^3}\right)$

In Exercises 131–136, factor each expression.

131. $4x^2 - 9$

132. $x^2 + 6x + 9$

133. $15x^2 + 11x - 12$

134. $14x^2 - 3x - 2$

135. $x^2(x - 1) - 4(x - 1)$

136. $9x^2(2x + 7) - 25(2x + 7)$

SECTION 2.2

Polynomial Functions

BEFORE STARTING THIS SECTION, REVIEW

1 Graphs of equations (Section 1.1, page 2)

2 Solving linear equations (Appendix A.6, page 955)

3 Solving quadratic equations (Appendix A.6, page 956)

4 Transformations (Section 1.5, page 78)

5 Relative maximum and minimum values (Section 1.4, page 63)

OBJECTIVES

1 Learn properties of the graphs of polynomial functions.

2 Determine the end behavior of polynomial functions.

3 Find the zeros of a polynomial function by factoring.

4 Identify the relationship between degrees, real zeros, and turning points.

5 Graph polynomial functions.

Johannes Kepler (1571–1630)
Kepler was born in Weil in southern Germany and studied at the University of Tübingen. He studied with Michael Mastlin (professor of mathematics), who privately taught him the Copernican theory of the sun-centered universe while hardly daring to recognize it openly in his lectures. Kepler knew that to work out a detailed version of Copernicus's theory, he had to have access to the observations of the Danish astronomer Tycho Brahe (1546–1601). Kepler's correspondence with Brahe resulted in his appointment as Brahe's assistant. Brahe died about 18 months after Kepler's arrival. During those 18 months, Kepler learned enough about Brahe's work to create his own major project. He produced the famous three laws of planetary motion published in 1609 in his book *Nova Astronomia*.

◆ Kepler's Wedding Reception

Kepler is best known as an astronomer who discovered the three laws of planetary motion. However, Kepler's primary field was not astronomy, but mathematics. Legend has it that at his own wedding reception, Kepler observed that the Austrian vintners could quickly and mysteriously compute the volumes of a variety of wine barrels. Each barrel had a hole, called a *bunghole*, in the middle of its side. The vintner would insert a rod, called the *bungrod*, into the hole until it hit the far corner. Then he would announce the volume. During the reception, Kepler occupied his mind with the problem of finding the volume of a wine barrel with a bungrod.

In Example 11, we show how he solved the problem for barrels that are perfect cylinders. But barrels are not perfect cylinders; they are wider in the middle and narrower at the ends, and in between, the sides make a graceful curve that appears to be an arc of a circle. What could be the formula for the volume of such a shape? Kepler tackled this problem in his important book *Nova Stereometria Doliorum Vinariorum* (1615) and developed a related mathematical theory.

1 Learn properties of the graphs of polynomial functions.

Polynomial Functions

A **polynomial function of degree n** is a function of the form

$$f(x) = a_n x^n + a_{n-1} x^{n-1} + \cdots + a_2 x^2 + a_1 x + a_0,$$

where n is a nonnegative integer and the *coefficients* $a_n, a_{n-1}, \ldots, a_2, a_1, a_0$ are real numbers with $a_n \neq 0$. The term $a_n x^n$ is called the **leading term**, the number a_n (the coefficient of x^n) is called the **leading coefficient**, and a_0 is the **constant term**. A constant function $f(x) = a(a \neq 0)$, which may be written as $f(x) = ax^0$, for $x \neq 0$ is a polynomial of degree 0. Because the zero function $f(x) = 0$ can be written in several ways, such as $0 = 0x^6 + 0x = 0x^5 + 0 = 0x^4 + 0x^3 + 0$, no degree is assigned to it.

In Section 1.4, we discussed linear functions, which are polynomial functions of degree 0 and 1. In Section 2.1, we studied quadratic functions, which are polynomial functions of degree 2. Polynomials of degree 3, 4, and 5 are also called **cubic, quartic,** and **quintic polynomials**, respectively. In this section, we concentrate on graphing polynomial functions of degree 3 or more.

Here are some common properties shared by all polynomial functions.

Common Properties of Polynomial Functions

1. The domain of a polynomial function is the set of all real numbers.

2. The graph of a polynomial function is a **continuous** and **smooth** curve. The graph is continuous if it has no holes or gaps and can be drawn on a sheet of paper without lifting the pencil (see Figure 2.4(a)). The graph is smooth if it does not contain any corners or cusps (see Figure 2.4(c)). For the more precise definitions, we have to wait until the calculus course.

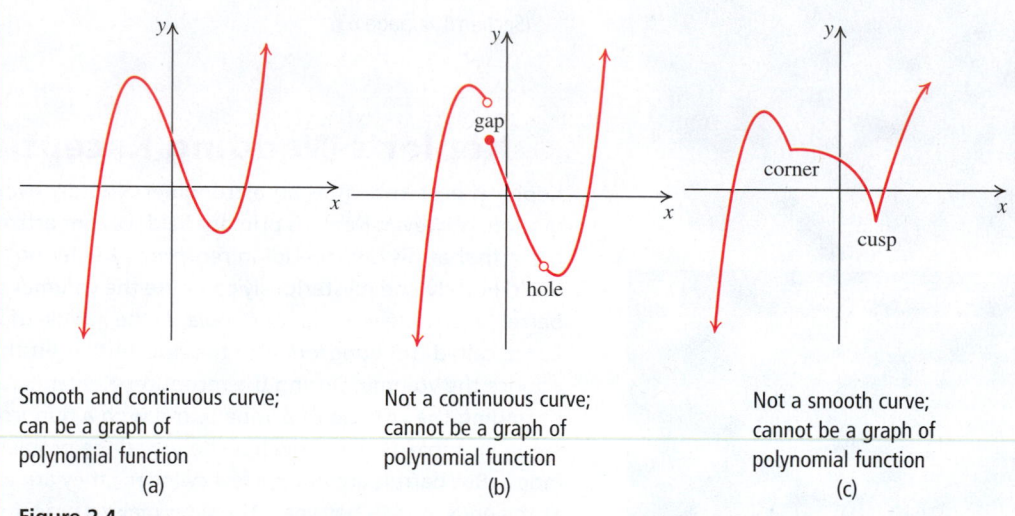

Smooth and continuous curve; can be a graph of polynomial function
(a)

Not a continuous curve; cannot be a graph of polynomial function
(b)

Not a smooth curve; cannot be a graph of polynomial function
(c)

Figure 2.4

EXAMPLE 1 Polynomial Functions

State which functions are polynomial functions. For each polynomial function, find its degree, the leading term, and the leading coefficient.

a. $f(x) = 5x^4 - 2x + 7$

b. $g(x) = 7x^2 - x + 1,\ 1 \le x \le 5$

Solution

a. $f(x) = 5x^4 - 2x + 7$ is a polynomial function. Its degree is 4, the leading term is $5x^4$, and the leading coefficient is 5.

b. $g(x) = 7x^2 - x + 1,\ 1 \le x \le 5$ is not a polynomial function because its domain is not $(-\infty, \infty)$.

Practice Problem 1 Repeat Example 1 for each function.

a. $f(x) = \dfrac{x^2 + 1}{x - 1}$

b. $g(x) = 2x^7 + 5x^2 - 17$

Power Functions

A function of the form $f(x) = ax^n$ is the simplest nth-degree polynomial function and is called a power function.

Power Function

A function of the form

$$f(x) = ax^n$$

is called a **power function of degree n**, where a is a nonzero real number and n is a positive integer.

The graph of a power function can be obtained by stretching or shrinking the graph of $y = x^n$ (and reflecting it about the x-axis if $a < 0$). So the basic shape of the graph of a power function $f(x) = ax^n$ is similar to the shape of the graphs of the equations $y = \pm x^n$.

Power Function of Even Degree The graph of $y = x^n$ (when n is even) has the same basic shape as the graph of $y = x^2$. See Figure 2.5(a). Notice that each of these graphs passes through the points $(-1, 1)$, $(0, 0)$, and $(1, 1)$. As the exponent n increases, the graph of $f(x) = x^n$ becomes flatter around $x = 0$ and becomes almost vertical at $x = -1$ and $x = 1$. The graphs of $y = x^2$, $y = x^4$, and $y = x^6$ are compared in Figure 2.5(a). The graph of $y = x^n$ is always located in the shaded regions shown in Figure 2.5(b), and the larger the exponent n, the more closely the graph will resemble the dark red curve ($\sqcup$) shown in Figure 2.5(b).

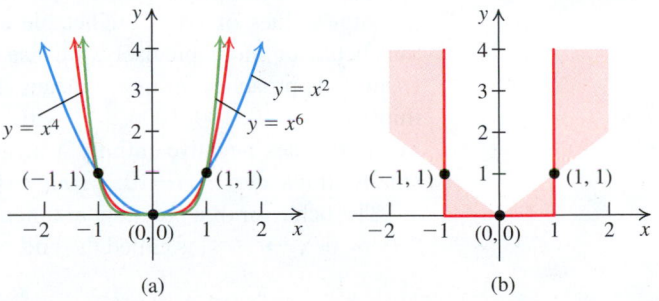

(a)   (b)

Figure 2.5 Power functions of even degree.

Finally, it is important to notice the following hierarchy of the power functions

$$x^2 > x^4 > x^6 > x^8 > \cdots, \quad \text{for } 0 < x < 1$$
$$x^2 < x^4 < x^6 < x^8 < \cdots, \quad \text{for } x > 1.$$

Observe that the hierarchy on $[0, 1]$ is reversed on $[1, \infty)$.

While it may appear that, for large n, the graph of $y = x^n$ coincides with the x-axis, that is not the case and merely reflects the shortcomings of the resolutions of plotting devices. In fact, the graph of $y = x^n$ only touches the x-axis at the origin.

Power Function of Odd Degree The graph of $y = x^n$ (when n is odd) has the same basic shape as the graph of $y = x^3$. See Figure 2.6(a). Notice that each of these graphs passes through the points $(-1, -1)$, $(0, 0)$, and $(1, 1)$. As the exponent n increases, the graph of $f(x) = x^n$ becomes flatter around $x = 0$ and becomes almost vertical at $x = -1$ and $x = 1$. The graphs of $y = x^3$, $y = x^5$, and $y = x^7$ are compared in Figure 2.6(a). The graph of $y = x^n$ is always located in the shaded regions depicted in Figure 2.6(b). The larger the exponent n, the more closely the graph will resemble the dark red curve ($\lrcorner$) shown in Figure 2.6(b).

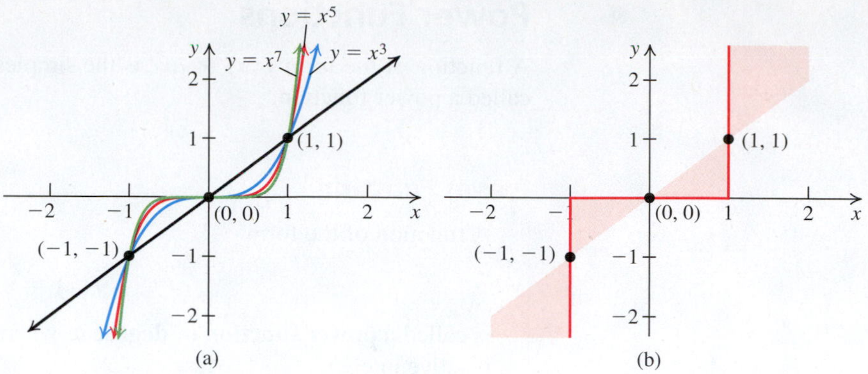

Figure 2.6 Power functions of odd degree.

Finally, it is important to notice the following hierarchy of the power functions

$$x > x^3 > x^5 > x^7 > \ldots, \quad \text{for } 0 < x < 1$$
$$x < x^3 < x^5 < x^7 < \ldots, \quad \text{for } x > 1.$$

Observe that the hierarchy on $[0, 1]$ is reversed on $[1, \infty)$.

While it may appear that, for large n, the graph of $y = x^n$ coincides with the x-axis, that is not the case and merely reflects the shortcomings of the resolutions of plotting devices. In fact, the graph of $y = x^n$ only crosses the x-axis at the origin.

End Behavior of Polynomial Functions

2 Determine the end behavior of polynomial functions.

The polynomial function $y = f(x)$ may oscillate quite significantly. However, its behavior for large values of $|x|$ is predictable and is dominated by its leading term. To describe this behavior more precisely, we use the $-\infty$ and ∞ notations. The notation $x \rightarrow \infty$ (read "x approaches infinity") means that x is increasing (i.e., getting larger and larger-think $x = 1, 10, 100, 1000, \ldots$) without bound. Similarly, the notation $x \rightarrow -\infty$ (read "x approaches negative infinity") means that x is decreasing (i.e., getting smaller and smaller-think $x = -1, -10, -100, -1000, \ldots$) without bound.

The behavior of a polynomial function $y = f(x)$ for very large values of $|x|$, that is, as $x \rightarrow \infty$ or $x \rightarrow -\infty$, is called the **end behavior** of the function.

Symbol	Meaning
$y = f(x) \rightarrow \infty$, as $x \rightarrow \infty$	the values of $f(x)$ increase without bound when x increases without bound
$y = f(x) \rightarrow -\infty$, as $x \rightarrow \infty$	the values of $f(x)$ decrease without bound when x increases without bound
$y = f(x) \rightarrow \infty$, as $x \rightarrow -\infty$	the values of $f(x)$ increase without bound when x decreases without bound
$y = f(x) \rightarrow -\infty$, as $x \rightarrow -\infty$	the values of $f(x)$ decrease without bound when x decreases without bound

SIDE NOTE

Here is a scheme for remembering the axis directions associated with the symbols $-\infty$ and ∞.

$x \rightarrow -\infty$ Left
$x \rightarrow \infty$ Right
$y \rightarrow -\infty$ Down
$y \rightarrow \infty$ Up

In the next box, we describe the behavior of the power function $y = f(x) = ax^n$ as $x \rightarrow \infty$ or $x \rightarrow -\infty$. Every polynomial function exhibits end behavior similar to one of the four functions $y = x^2, y = -x^2, y = x^3,$ or $y = -x^3$.

END BEHAVIOR FOR THE GRAPH OF $y = f(x) = ax^n$

	n Even		*n* Odd	
	Case 1 $a > 0$	**Case 2** $a < 0$	**Case 3** $a > 0$	**Case 4** $a < 0$

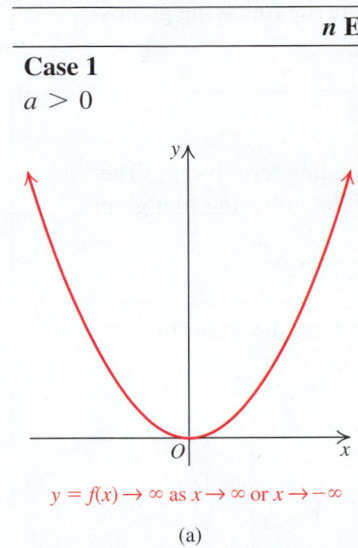
$y = f(x) \to \infty$ as $x \to \infty$ or $x \to -\infty$

(a)

The graph rises to the left and right, similar to $y = x^2$.

$y = f(x) \to -\infty$ as $x \to \infty$ or $x \to -\infty$

(b)

The graph falls to the left and right, similar to $y = -x^2$.

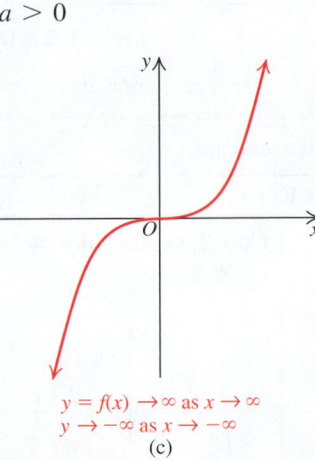
$y = f(x) \to \infty$ as $x \to \infty$
$y \to -\infty$ as $x \to -\infty$

(c)

The graph rises to the right and falls to the left, similar to $y = x^3$.

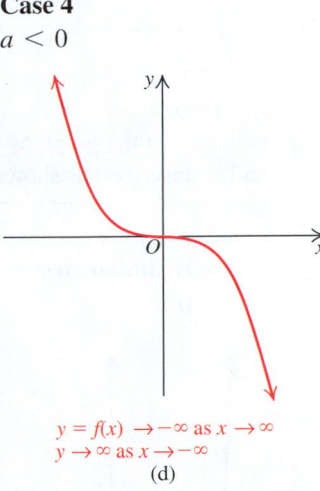
$y = f(x) \to -\infty$ as $x \to \infty$
$y \to \infty$ as $x \to -\infty$

(d)

The graph rises to the left and falls to the right, similar to $y = -x^3$.

TECHNOLOGY CONNECTION

Calculator graph of
$P(x) = 2x^3 + 5x^2 - 7x + 11$

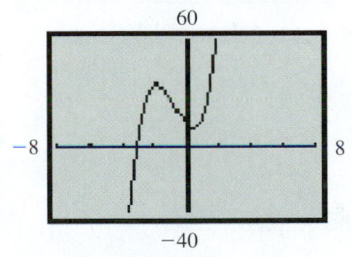

In the next example, we show that the end behavior of the given polynomial function is determined by its leading term. If a polynomial $P(x)$ has approximately the same values as $f(x) = ax^n$ when $|x|$ is very large, we write $P(x) \approx ax^n$ when $|x|$ is very large.

EXAMPLE 2 Understanding the End Behavior of a Polynomial Function

Let $P(x) = 2x^3 + 5x^2 - 7x + 11$ be a polynomial function of degree 3. Show that $P(x) \approx 2x^3$ when $|x|$ is very large.

Solution

$$P(x) = 2x^3 + 5x^2 - 7x + 11 \qquad \text{Given polynomial}$$

$$= x^3\left(2 + \frac{5}{x} - \frac{7}{x^2} + \frac{11}{x^3}\right) \qquad \text{Distributive property}$$

When $|x|$ is very large, the terms $\dfrac{5}{x}$, $-\dfrac{7}{x^2}$, and $\dfrac{11}{x^3}$ are close to 0. Table 2.1 shows some values of $\dfrac{5}{x}$ as x increases. (You should compute the values of $-\dfrac{7}{x^2}$ and $\dfrac{11}{x^3}$ for $x = 10, 10^2,$ $10^3,$ and 10^4.) When $|x|$ is very large, we have

$$P(x) = x^3\left(2 + \frac{5}{x} - \frac{7}{x^2} + \frac{11}{x^3}\right) \approx x^3(2 + 0 - 0 + 0)$$

$$= 2x^3.$$

So the polynomial $P(x) \approx 2x^3$ when $|x|$ is very large.

Practice Problem 2 Let $P(x) = 4x^3 + 2x^2 + 5x - 17$. Show that $P(x) \approx 4x^3$ when $|x|$ is very large.

TABLE 2.1

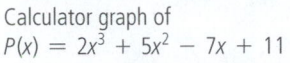

x	$\dfrac{5}{x}$
10	0.5
10^2	0.05
10^3	0.005
10^4	0.0005

It is possible to use the technique of Example 2 to show that the end behavior of any polynomial function is determined by its leading term.

The end behaviors of polynomial functions are shown in the following graphs.

THE LEADING-TERM TEST

Let $f(x) = a_n x^n + a_{n-1} x^{n-1} + \cdots + a_1 x + a_0\, (a_n \neq 0)$ be a polynomial function. Its leading term is $a_n x^n$. The behavior of the graph of $y = f(x)$ as $x \to \infty$ or $x \to -\infty$ has end behavior similar to one of the following four graphs and is described as shown in each case.

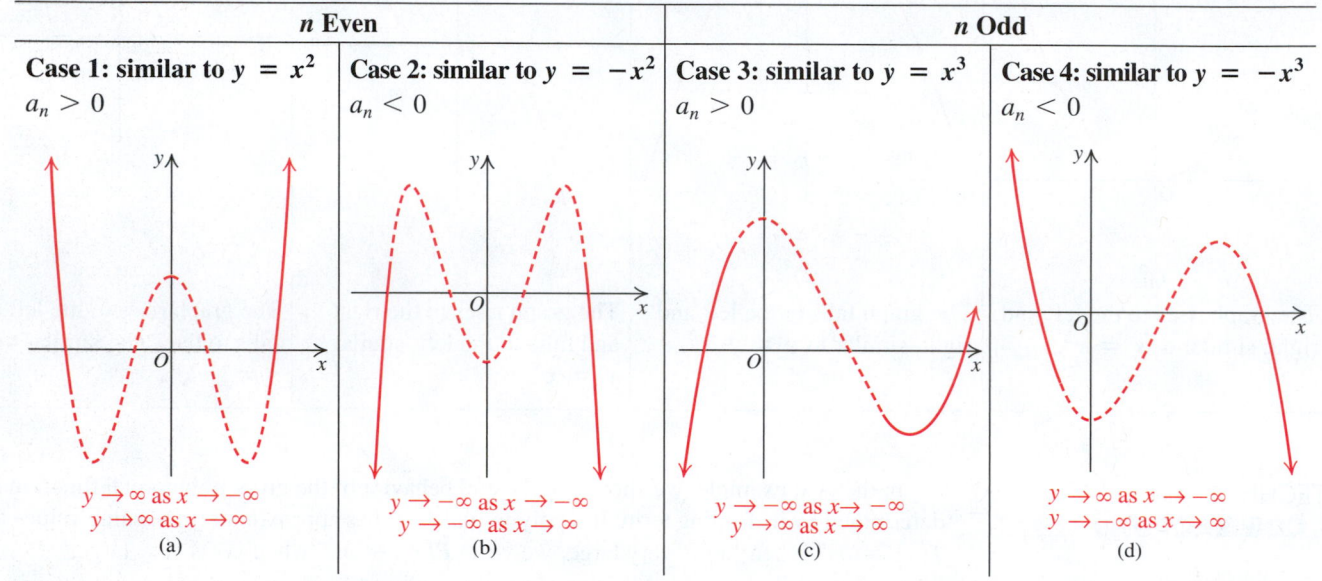

n Even		**n Odd**	
Case 1: similar to $y = x^2$ $a_n > 0$	**Case 2: similar to $y = -x^2$** $a_n < 0$	**Case 3: similar to $y = x^3$** $a_n > 0$	**Case 4: similar to $y = -x^3$** $a_n < 0$
$y \to \infty$ as $x \to -\infty$ $y \to \infty$ as $x \to \infty$	$y \to -\infty$ as $x \to -\infty$ $y \to -\infty$ as $x \to \infty$	$y \to -\infty$ as $x \to -\infty$ $y \to \infty$ as $x \to \infty$	$y \to \infty$ as $x \to -\infty$ $y \to -\infty$ as $x \to \infty$
(a)	(b)	(c)	(d)

This test does not describe the middle portion of each graph, shown by the dashed lines.

Note that: even-degree polynomial functions have the same end behavior on both ends of the x-axis. Odd-degree polynomial functions have the opposite end behavior on opposite ends of the x-axis.

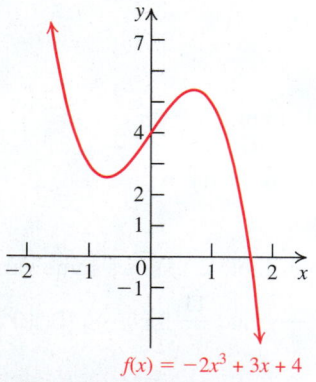

Figure 2.7 End behavior of $f(x) = -2x^3 + 3x + 4$.

$f(x) = -2x^3 + 3x + 4$

3 Find the zeros of a polynomial function by factoring.

EXAMPLE 3   Using the Leading-Term Test

Use the leading-term test to determine the end behavior of the graph of

$$y = f(x) = -2x^3 + 3x + 4.$$

Solution

Here the degree $n = 3$ (odd) and the leading coefficient $a_n = -2 < 0$. Case 4 applies. The graph of f rises to the left and falls to the right. See Figure 2.7. This end behavior is described as $y \to \infty$ as $x \to -\infty$ and $y \to -\infty$ as $x \to \infty$.

Practice Problem 3 What is the end behavior of $f(x) = -2x^4 + 5x^2 + 3$?

Zeros of a Function

Let f be a function. An input c in the domain of f that produces output 0 is called a *zero* of the function f. For example, if $f(x) = (x - 3)^2$, then 3 is a zero of f because $f(3) = (3 - 3)^2 = 0$. In other words, a number c is a **zero** of f if $f(c) = 0$. The zeros of a function f can be obtained by solving the equation $f(x) = 0$. One way to solve a polynomial equation $f(x) = 0$ is to factor $f(x)$ and then use the zero-product property.

If c is a real number and $f(c) = 0$, then c is called a **real zero** of f. Geometrically, this means that the graph of f has an x-intercept at $x = c$.

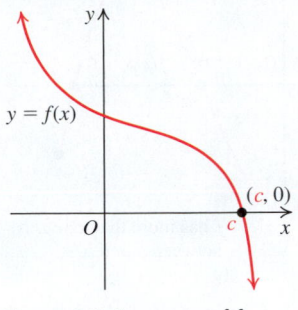

Figure 2.8 A zero at c of f.

> ### REAL ZEROS OF POLYNOMIAL FUNCTIONS
>
> If f is a polynomial function and c is a real number, then the following statements are equivalent:
>
> 1. c is a **zero** of f.
> 2. c is a **solution** (or **root**) of the equation $f(x) = 0$.
> 3. c is an **x-intercept** of the graph of f. The point $(c, 0)$ is on the graph of f. See Figure 2.8.
> 4. $(x - c)$ is a factor of f.

EXAMPLE 4   Finding the Zeros of a Polynomial Function

Find all real zeros of each polynomial function.

a. $f(x) = x^4 - 2x^3 - 3x^2$ **b.** $g(x) = x^3 - 2x^2 + x - 2$

Solution

a. We factor $f(x)$ and then solve the equation $f(x) = 0$.

$$
\begin{aligned}
f(x) &= x^4 - 2x^3 - 3x^2 && \text{Given function}\\
&= x^2(x^2 - 2x - 3) && \text{Factor out the common factor, } x^2.\\
&= x^2(x + 1)(x - 3) && \text{Factor } (x^2 - 2x - 3).\\
x^2(x + 1)(x - 3) &= 0 && \text{Set } f(x) = 0.\\
x^2 = 0, \quad x + 1 &= 0, \quad \text{or} \quad x - 3 = 0 && \text{Zero-product property}\\
x = 0, \quad x &= -1, \quad \text{or} \quad x = 3 && \text{Solve each equation.}
\end{aligned}
$$

The zeros of $f(x)$ are 0, -1, and 3. (See the calculator graph of f in the margin.)

b. By grouping terms, we factor $g(x)$ to obtain

$$
\begin{aligned}
g(x) &= x^3 - 2x^2 + x - 2\\
&= x^2(x - 2) + 1 \cdot (x - 2) && \text{Group terms.}\\
&= (x - 2)(x^2 + 1) && \text{Factor out } x - 2.\\
0 &= (x - 2)(x^2 + 1) && \text{Set } g(x) = 0.\\
x - 2 &= 0, \quad \text{or} \quad x^2 + 1 = 0 && \text{Zero-product property}
\end{aligned}
$$

Solving for x, we obtain 2 as the only real zero of $g(x)$ since $x^2 + 1 > 0$ for all real numbers x. The calculator graph of g in the margin supports our conclusion.

Practice Problem 4 Find all real zeros of $f(x) = 2x^3 - 3x^2 + 4x - 6$.

Finding the zeros of a polynomial of degree 3 or more is one of the most important problems of mathematics. You will learn more about this topic in Sections 2.3 and 2.6. For now, we will approximate the zeros by using an *Intermediate Value Theorem*. Recall that the graph of a polynomial function $f(x)$ is a continuous curve. Suppose you locate two numbers a and b such that the function values $f(a)$ and $f(b)$ have opposite signs (one positive and the other negative). Then the continuous graph of f connecting the points $(a, f(a))$ and $(b, f(b))$ must cross the x-axis (at least once) somewhere between a and b. In other words, the function f has a zero between a and b. See Figure 2.9.

TECHNOLOGY
CONNECTION

$f(x) = x^4 - 2x^3 - 3x^2$

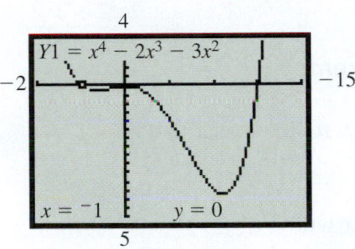

We can verify that the zeros of f are 0, -1, and 3 using the TRACE function.

$g(x) = x^3 - 2x^2 + x - 2$

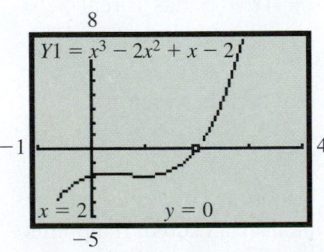

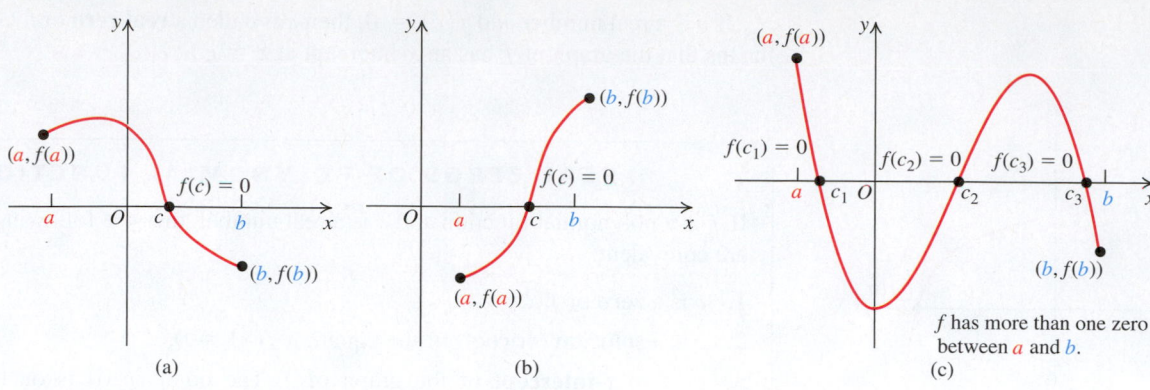

Figure 2.9 Each function has at least one zero between a and b.

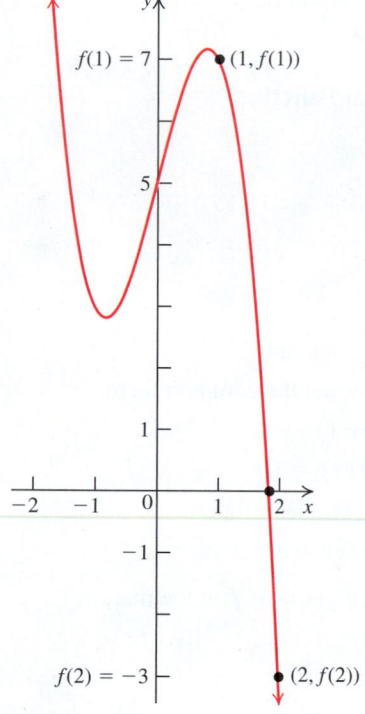

Figure 2.10 A zero lies between 1 and 2.

AN INTERMEDIATE VALUE THEOREM

Let a and b be two numbers with $a < b$. If f is a polynomial function such that $f(a)$ and $f(b)$ have opposite signs, then there is at least one number c, with $a < c < b$, for which $f(c) = 0$.

Notice that Figure 2.9(c) shows more than one zero of f. That is why this Intermediate Value Theorem says that f has *at least one* zero between a and b.

EXAMPLE 5 **Using an Intermediate Value Theorem**

Show that the function $f(x) = -2x^3 + 4x + 5$ has a real zero between 1 and 2.

Solution

$$
\begin{aligned}
f(1) &= -2(1)^3 + 4(1) + 5 \qquad &\text{Replace } x \text{ with 1 in } f(x). \\
&= -2 + 4 + 5 = 7 &\text{Simplify.} \\
f(2) &= -2(2)^3 + 4(2) + 5 &\text{Replace } x \text{ with 2.} \\
&= -2(8) + 8 + 5 = -3 &\text{Simplify.}
\end{aligned}
$$

Since $f(1)$ is positive and $f(2)$ is negative, they have opposite signs. So by this Intermediate Value Theorem, the polynomial function $f(x) = -2x^3 + 4x + 5$ has a real zero between 1 and 2. See Figure 2.10.

Practice Problem 5 Show that $f(x) = 2x^3 - 3x - 6$ has a real zero between 1 and 2.

In Example 5, if you want to find the zero between 1 and 2 more accurately, you can divide the interval $[1, 2]$ into tenths and find the value of the function at each of the points $1, 1.1, 1.2, \ldots, 1.9$, and 2. You will find that

$$f(1.8) = 0.536 \quad \text{and} \quad f(1.9) = -1.118.$$

So by this Intermediate Value Theorem, f has a zero between 1.8 and 1.9. If you divide the interval $[1.8, 1.9]$ into tenths, you will find that $f(1.83) > 0$ and $f(1.84) < 0$, so f has a zero between 1.83 and 1.84. By continuing this process, you can find the zero of f between 1.8 and 1.9 to any desired degree of accuracy.

4 Identify the relationship between degrees, real zeros, and turning points.

Zeros and Turning Points

In Section 2.6, you will learn the Fundamental Theorem of Algebra. One of its consequences is the following fact:

REAL ZEROS OF A POLYNOMIAL FUNCTION

A polynomial function of degree n with real coefficients has, *at most*, n real zeros.

SIDE NOTE

A polynomial of degree n can have no more than n real zeros. In Example 6, parts b and c, we see two polynomials of degree 3 each having *fewer* than 3 zeros.

EXAMPLE 6   **Finding the Number of Real Zeros**

Find the number of distinct real zeros of the following polynomial functions of degree 3.

a. $f(x) = (x - 1)(x + 2)(x - 3)$ **b.** $g(x) = (x + 1)(x^2 + 1)$

c. $h(x) = (x - 3)^2(x + 1)$

Solution

a. $f(x) = (x - 1)(x + 2)(x - 3) = 0$ Set $f(x) = 0$.

$x - 1 = 0$,   or $x + 2 = 0$,   or $x - 3 = 0$ Zero-product property

$x = 1$ or $x = -2$ or $x = 3$ Solve for x.

So $f(x)$ has three real zeros: 1, -2, and 3.

b. $g(x) = (x + 1)(x^2 + 1) = 0$ Set $g(x) = 0$.

$x + 1 = 0$ or $x^2 + 1 = 0$ Zero-product property

$x = -1$ Solve for x.

Because $x^2 + 1 > 0$ for all real x, the equation $x^2 + 1 = 0$ has no real solution. So -1 is the only real zero of $g(x)$.

c. $h(x) = (x - 3)^2(x + 1) = 0$ Set $h(x) = 0$.

$(x - 3)^2 = 0$,   or $x + 1 = 0$ Zero-product property

$x = 3$,   or $x = -1$ Solve for x.

The real zeros of $h(x)$ are 3 and -1. Thus, $h(x)$ has two distinct real zeros.

Practice Problem 6 Find the distinct real zeros of

$$f(x) = (x + 1)^2(x - 3)(x + 5).$$

In Example 6c, the factor $(x - 3)$ appears twice in $h(x)$. We say that 3 is a zero repeated *twice*, or that 3 is a *zero of multiplicity* 2. Notice that the graph of $y = h(x)$ in Figure 2.11 *touches* the x-axis at $x = 3$, where h has a zero of multiplicity 2 (an even number). In contrast, the graph of h *crosses* the x-axis at $x = -1$, where it has a zero of multiplicity 1 (an odd number). We next consider the geometric significance of the multiplicity of a zero.

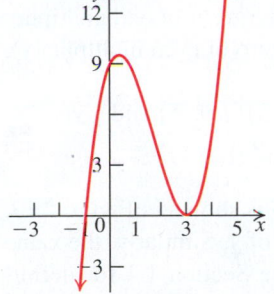

Figure 2.11 Graph of $h(x) = (x - 3)^2(x + 1)$.

Multiplicity of a Zero

If c is a zero of a polynomial function $f(x)$ and the corresponding factor $(x - c)$ occurs exactly m times when $f(x)$ is factored as $f(x) = (x - c)^m q(x)$, with $q(c) \neq 0$, then c is called a **zero of multiplicity m**.

1. If m is odd, the graph of f crosses the x-axis at $x = c$.

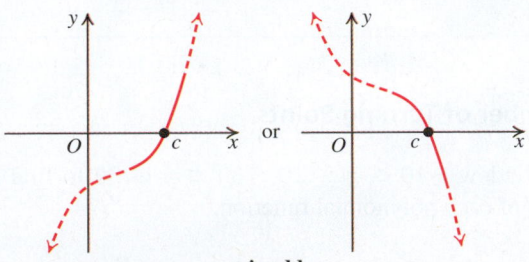

m is odd.
Graph crosses the x-axis.

2. If m is even, the graph of f touches but does not cross the x-axis at $x = c$.

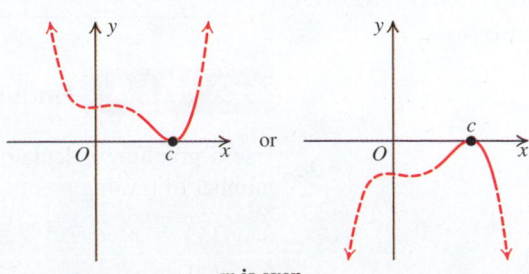

m is even.
Graph touches but does not cross the x-axis.

SIDE
NOTE

As the multiplicity m of a zero c of a polynomial function increases the graph of f becomes more curved around $x = c$ as illustrated below

If $m = 1$

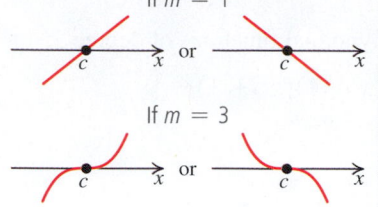

If $m = 3$

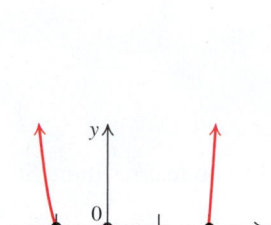

The multiplicity of a zero determines the precise manner in which the graph of a polynomial function crosses or touches the x-axis.

> **BEHAVIOR OF A POLYNOMIAL FUNCTION $f(x)$ NEAR ITS ZERO**
>
> If c is a zero of multiplicity m for a polynomial function $f(x)$ with corresponding factorization $f(x) = (x - c)^m q(x), q(c) \neq 0$ then, near $x = c, f(x)$ looks very much like the graph of $A(x - c)^m$, where A is the constant $A = q(c)$.

The above observation will allow us to draw the graph of a polynomial function that is already fully factored with enhanced precision.

EXAMPLE 7 Finding the Zeros and Their Multiplicities

Find the zeros of the polynomial function $f(x) = x^2(x + 1)(x - 2)$ and give the multiplicity of each zero.

Solution

The polynomial $f(x)$ is already in factored form.

$$f(x) = x^2(x + 1)(x - 2) = 0 \qquad \text{Set } f(x) = 0.$$
$$x^2 = 0, \quad \text{or} \quad x + 1 = 0, \quad \text{or} \quad x - 2 = 0 \qquad \text{Zero-product property}$$
$$x = 0 \quad \text{or} \quad x = -1 \qquad \text{or} \quad x = 2 \qquad \text{Solve for } x.$$

The function has three distinct zeros: $0, -1,$ and 2. The zero $x = 0$ has multiplicity 2, and each of the zeros -1 and 2 has multiplicity 1. The graph of f is given in Figure 2.12. Notice that the graph of f in Figure 2.12 crosses the x-axis at -1 and 2 (the zeros of odd multiplicity), while it touches but does not cross the x-axis at $x = 0$ (the zero of even multiplicity).

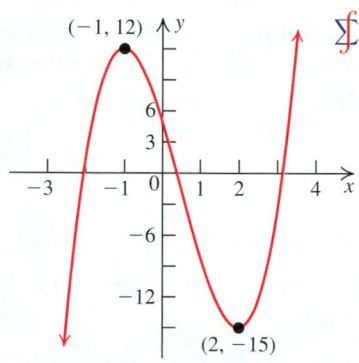

$f(x) = x^2(x + 1)(x - 2)$

Figure 2.12

Practice Problem 7 Write the zeros of $f(x) = (x - 1)^2(x + 3)(x + 5)$ and give the multiplicity of each.

Turning Points The graph of $f(x) = 2x^3 - 3x^2 - 12x + 5$ is shown in Figure 2.13. The value $f(-1) = 12$ is a **local (or relative) maximum** value of f. Similarly, the value $f(2) = -15$ is a **local (or relative) minimum** value of f. (See Section 1.4 for definitions.) The graph points corresponding to the local maximum and local minimum values of a polynomial function are the **turning points**. At each turning point, the graph changes direction from increasing to decreasing or from decreasing to increasing. One of the successes of calculus is finding the maximum and minimum values for many functions.

The graph of $f(x) = 2x^3 - 3x^2 - 12x + 5$ has turning points at $(-1, 12)$ and $(2, -15)$.

Figure 2.13 Turning points.

> **NUMBER OF TURNING POINTS**
>
> If $f(x)$ is a polynomial of degree n, then the graph of f has, *at most*, $(n - 1)$ turning points.

EXAMPLE 8 Finding the Number of Turning Points

Use a graphing calculator and the window $-10 \leq x \leq 10; -30 \leq y \leq 30$ to find the number of turning points of the graph of each polynomial function.

a. $f(x) = x^4 - 7x^2 - 18$

b. $g(x) = x^3 + x^2 - 12x$

c. $h(x) = x^3 - 3x^2 + 3x - 1$

Solution

The graphs of f, g, and h are shown in Figure 2.14.

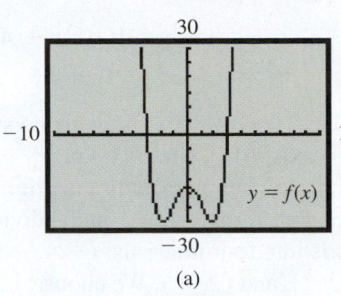

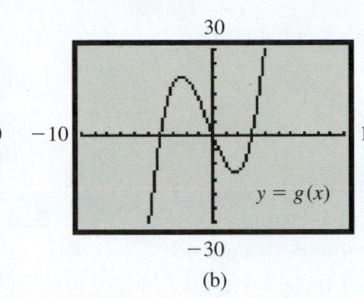

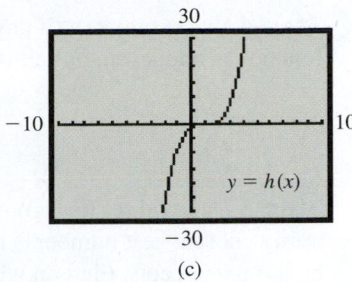

(a)   (b)   (c)

Figure 2.14 Turning points on a calculator graph

a. Function f has three turning points: two local minima and one local maximum.

b. Function g has two turning points: one local maximum and one local minimum.

c. Function h has no turning points: it is increasing on the interval $(-\infty, \infty)$.

Practice Problem 8 Find the number of turning points of the graph of $f(x) = -x^4 + 3x^2 - 2$.

5 Graph polynomial functions.

Graphing a Polynomial Function

You can graph a polynomial function by constructing a table of values for a large number of points, plotting the points, and connecting the points with a smooth curve. (In fact, a graphing calculator uses this method.) Generally, however, this process is a poor way to graph a function by hand because it is tedious and error-prone, and so may give misleading results. Similarly, a graphing calculator may give misleading results if the viewing window is not chosen appropriately. Next, we discuss a better strategy for graphing a polynomial function.

PROCEDURE
IN ACTION

EXAMPLE 9 Graphing a Polynomial Function

OBJECTIVE

Sketch the graph of a polynomial function.

Step 1 Determine the end behavior. Apply the leading-term test.

EXAMPLE

Sketch the graph of $f(x) = -x^3 - 4x^2 + 4x + 16$.

1. Since the degree, 3, of f is odd and the leading coefficient, -1, is negative, the end behavior (Case 4, page 158) is similar to that of $y = -x^3$ as shown in this sketch.

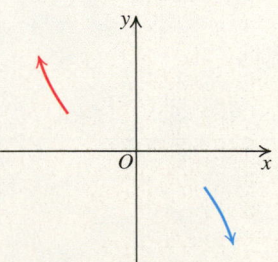

Step 2 Find the real zeros of the polynomial function. Set $f(x) = 0$. Solve the equation $f(x) = 0$ by factoring. The real zeros with their multiplicities will give you the x-intercepts of the graph and determine whether or not the graph crosses the x-axis there.

2. $-x^3 - 4x^2 + 4x + 16 = 0$

$-(x + 4)(x + 2)(x - 2) = 0$

$x = -4, x = -2,$ or $x = 2$

There are three zeros, each of multiplicity 1, so the graph crosses the x-axis at each zero.

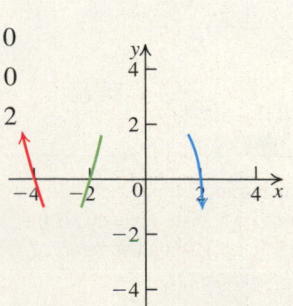

Step 3 Find the y-intercept by computing $f(0)$.

3. The y-intercept is $f(0) = 16$. The graph passes through the point $(0, 16)$.

Step 4 Use symmetry (if any) Check whether the function is odd, even, or neither.

4. $f(-x) = -(-x)^3 - 4(-x)^2 + 4(-x) + 16$
$= x^3 - 4x^2 - 4x + 16$

Since $f(-x) \neq f(x)$, there is no symmetry about the y-axis. Also, since $f(-x) \neq -f(x)$, there is no symmetry with respect to the origin.

Step 5 Determine the sign of $f(x)$ by using arbitrarily chosen "test numbers" in the intervals defined by the x-intercepts. Find on which of these intervals the graph lies above or below the x-axis.

5. The three zeros $-4, -2$, and 2 divide the x-axis into four intervals: $(-\infty, -4), (-4, -2),$ $(-2, 2),$ and $(2, \infty)$. We choose $-5, -3, 0,$ and 3 as test numbers.

		-4		-2		2	
		-5		-3		0	3
Test number	-5		-3		0		3
Value of f	21		-5		16		-35
Sign graph	$+++$		$---$		$+++$		$---$
Graph of f	Above x-axis		Below x-axis		Above x-axis		Below x-axis
Graph point	$(-5, 21)$		$(-3, -5)$		$(0, 16)$		$(3, -35)$

Step 6 Draw the graph using points from Steps 1–5. To check whether the graph is drawn correctly, use the fact that the number of turning points is less than the degree of the polynomial.

6. The graph connects the points we have found with a smooth curve. The number of turning points is 2, which is less than 3, the degree of f.

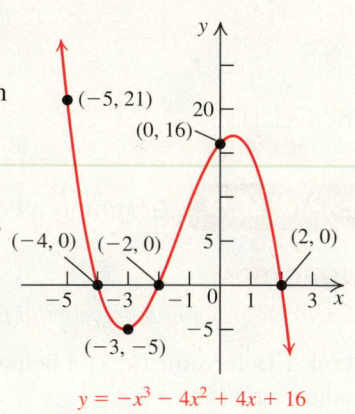

$y = -x^3 - 4x^2 + 4x + 16$

Practice Problem 9 Graph $f(x) = -x^4 + 5x^2 - 4$.

EXAMPLE 10 **Graph a Polynomial Given in Factored Form**

Graph the polynomial function $f(x) = (1 - x)(x + 2)^3(x - 3)^2$.

SIDE NOTE

The leading-term of a product of two polynomials $f(x) \cdot g(x)$ is the product of the leading terms of $f(x)$ and $g(x)$.

Solution

Step 1 We can deduce the end behavior of the polynomial by writing its leading term as the product of the leading terms of each factor

$$f(x) = (1 - x)(x + 2)^3(x - 3)^2 \approx (-x)(x)^3(x)^2 = -x^6.$$

The end behavior of $f(x)$ (Case 2, page 158) is similar to that of $y = -x^2$.

Step 2 The polynomial $f(x)$ is already in factored form

$$f(x) = (1 - x)(x + 2)^3(x - 3)^2 = 0 \qquad \text{Set } f(x) = 0$$

$$1 - x = 0 \quad \text{or} \quad x + 2 = 0 \quad \text{or} \quad x - 3 = 0 \qquad \text{Zero-product property}$$

$$x = 1 \quad \text{or} \quad x = -2 \quad \text{or} \quad x = 3 \qquad \text{Solve for } x.$$

The function has three distinct zeros: $-2, 1,$ and 3. The zero $x = -2$ has multiplicity 3, so it crosses the x-axis. Near this zero, the function looks like

$$f(x) = (1 - x)(x + 2)^3(x - 3)^2 \approx (1-(-2))(x + 2)^3((-2) - 3)^2 = 75(x + 2)^3.$$

The zero $x = 1$ has multiplicity 1, so it crosses the x-axis. Near this zero, the function looks like

$$f(x) = (1 - x)(x + 2)^3(x - 3)^2 \approx (1 - x)(1 + 2)^3(1 - 3)^2 = -108(x - 1).$$

The zero $x = 3$ has multiplicity 2, so it touches the x-axis. Near this zero, the function looks like

$$f(x) = (1 - x)(x + 2)^3(x - 3)^2 \approx (1 - 3)(3 + 2)^3(x - 3)^2 = -250(x - 3)^2.$$

Step 3 We compute the y-intercept as:

$$f(0) = (1 - 0)(0 + 2)^3(0 - 3)^2 = 1 \cdot 2^3 \cdot (-3)^2 = 1 \cdot 8 \cdot 9 = 72.$$

Step 4 $f(-x) = (1 - (-x))((-x) + 2)^3((-x) - 3)^2$

$$= (1 + x)(-x + 2)^3(-x - 3)^2$$

So $f(-x) \neq f(x)$, and $f(-x) \neq -f(x)$. There are no symmetries.

Step 5 Although the information we obtained from the previous steps allows us to sketch the graph of the polynomial function f, we can further check our work using appropriately chosen "test numbers." The three zeros $-2, 1,$ and 3 divide the x-axis into four intervals $(-\infty, -2), (-2, 1), (1, 3),$ and $(3, \infty)$. We choose $-3, 0, 2,$ and 4 as test numbers. We evaluate $f(-3) = -144 < 0$, $f(0) = 72 > 0$, $f(2) = -64 < 0$, and $f(4) = -648 < 0$. We gather all the information about the polynomial function below.

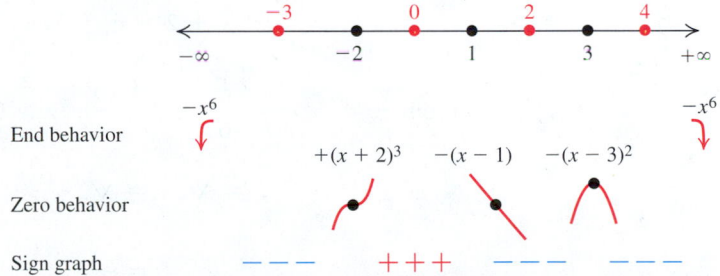

Step 6 We can sketch the graph of the polynomial function f by connecting the pieces of information we gathered above (End behavior and Zero behavior) with a smooth curve (See Figure 2.15). Finally, we can check our graph using the sign graph information.

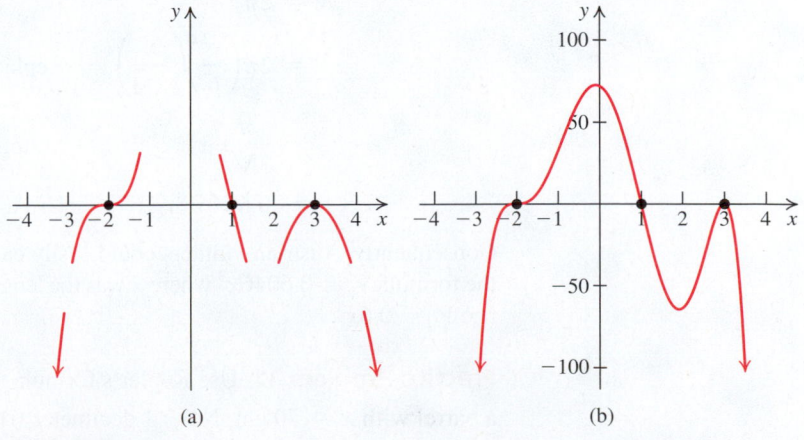

(a)   (b)

Figure 2.15

Practice Problem 10 Graph the polynomial function $f(x) = x(x + 2)^3(x - 1)^2$.

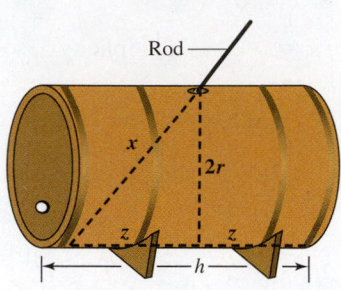

Figure 2.16 A wine barrel.

 Volume of a Wine Barrel Kepler first analyzed the problem of finding the volume of a cylindrical wine barrel. (Review the introduction to this section and see Figure 2.16.) The volume V of a cylinder with radius r and height h is given by the formula $V = \pi r^2 h$. In Figure 2.16, $h = 2z$, so $V = 2\pi r^2 z$. Let x represent the value measured by the rod. Then by the Pythagorean Theorem,

$$x^2 = 4r^2 + z^2. \quad (2r)^2 = 4r^2$$

Kepler's challenge was to calculate V, given only x—in other words, to express V as a function of x. To solve this problem, Kepler observed the key fact that Austrian wine barrels were all made with the same height-to-diameter ratio. Kepler assumed that the winemakers would have made a smart choice for this ratio, one that would maximize the volume for a fixed value of r. He calculated the height-to-diameter ratio to be $\sqrt{2}$.

EXAMPLE 11 **Volume of a Wine Barrel**

Express the volume V of the wine barrel in Figure 2.16 as a function of x. Assume that $\dfrac{\text{height}}{\text{diameter}} = \dfrac{2z}{2r} = \dfrac{z}{r} = \sqrt{2}$.

Solution

We have the following relationships:

$$V = 2\pi r^2 z \qquad \text{Volume with radius } r \text{ and height } 2z$$

$$\frac{z}{r} = \sqrt{2} \qquad \text{Given ratio}$$

$$x^2 = 4r^2 + z^2 \qquad \text{Pythagorean Theorem}$$

$$\frac{x^2}{r^2} = 4 + \frac{z^2}{r^2} \qquad \text{Divide both sides by } r^2.$$

$$\frac{x^2}{r^2} = 4 + 2 = 6 \qquad \text{Since } \frac{z}{r} = \sqrt{2}, \frac{z^2}{r^2} = \left(\frac{z}{r}\right)^2 = (\sqrt{2})^2 = 2.$$

From $\dfrac{x^2}{r^2} = 6$, $r^2 = \dfrac{x^2}{6}$, so $r = \dfrac{x}{\sqrt{6}}$ $(r > 0)$. Using $\dfrac{z}{r} = \sqrt{2}$, we obtain

$$z = \sqrt{2}\, r = \sqrt{2}\,\frac{x}{\sqrt{6}} \qquad \text{Replace } r \text{ with } \frac{x}{\sqrt{6}}.$$

$$= \frac{x}{\sqrt{3}} \qquad \frac{\sqrt{2}}{\sqrt{6}} = \frac{\sqrt{2}}{\sqrt{2}\cdot\sqrt{3}} = \frac{1}{\sqrt{3}}$$

Now

$$V = 2\pi r^2 z$$

$$= 2\pi\left(\frac{x^2}{6}\right)\left(\frac{x}{\sqrt{3}}\right) \qquad \text{Replace } z \text{ with } \frac{x}{\sqrt{3}} \text{ and } r^2 \text{ with } \frac{x^2}{6}.$$

$$= \frac{\pi}{3\sqrt{3}} x^3 \qquad \text{Simplify.}$$

$$\approx 0.6046 x^3 \qquad \text{Use a calculator.}$$

Consequently, Austrian vintners could easily calculate the volume of a wine barrel by using the formula $V = 0.6046 x^3$ where x was the length measured by the rod. For a quick estimate, use $V \approx 0.6 x^3$.

Practice Problem 11 Use Kepler's formula to compute the volume of wine (in liters) in a barrel with $x = 70$ cm. Note: 1 decimeter (dm) $= 10$ cm and 1 liter $= (1 \text{ dm})^3$.

Answers to Practice Problems

1. a. Not a polynomial function
b. A polynomial function; the degree is 7, the leading term is $2x^7$, and the leading coefficient is 2.

2. $P(x) = 4x^3 + 2x^2 + 5x - 17 = x^3\left(4 + \dfrac{2}{x} + \dfrac{5}{x^2} - \dfrac{17}{x^3}\right)$

When $|x|$ is large, the terms $\dfrac{2}{x}, \dfrac{5}{x^2}$, and $-\dfrac{17}{x^3}$ are close to 0. There-fore, $P(x) \approx x^3(4 + 0 + 0 - 0) = 4x^3$.

3. $y \to -\infty$ as $x \to -\infty$ and $y \to -\infty$ as $x \to \infty$ **4.** $\dfrac{3}{2}$
5. $f(1) = -7, f(2) = 4$. Since $f(1)$ and $f(2)$ have opposite signs, by the Intermediate Value Theorem, f has a real zero between 1 and 2.
6. $-5, -1$, and 3 **7.** -5 and -3 with multiplicity 1 and 1 with multiplicity 2 **8.** 3

9.

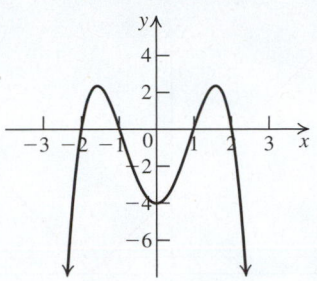

10.

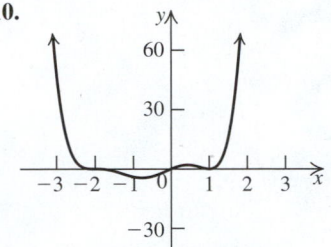

11. 207.378 liters

SECTION 2.2 **Exercises**

Concepts and Vocabulary

1. Consider the polynomial function $f(x) = 2x^5 - 3x^4 + x - 6$. The degree of this polynomial is _____, its leading term is _____, its leading coefficient is _____, and its constant term is _____.

2. A number c for which $f(c) = 0$ is called a(n) _____ of the polynomial function f.

3. If c is a zero of even multiplicity for a polynomial function f, then the graph of f _____ the x-axis at c.

4. If c is a zero of odd multiplicity for a polynomial function f, then the graph of f _____ the x-axis at c.

5. **True or False.** A curve that has a sharp corner can be a graph of a polynomial function.

6. **True or False.** The end behavior of any polynomial function is determined by its leading term.

7. **True or False.** The polynomial function of degree n has, at most, $n - 1$ zeroes.

8. **True or False.** The graph of a polynomial function of degree n has, at most, $n - 1$ turning points.

Building Skills

In Exercises 9–14, for each polynomial function, find the degree, the leading term, and the leading coefficient.

9. $f(x) = 2x^5 - 5x^2$

10. $f(x) = 3 - 5x - 7x^4$

11. $f(x) = \dfrac{2x^3 + 7x}{3}$

12. $f(x) = -7x + 11 + \sqrt{2}x^3$

13. $f(x) = \pi x^4 + 1 - x^2$

14. $f(x) = 5$

In Exercises 15–22, explain why the given function is not a polynomial function.

15. $f(x) = x^2 + 3|x| - 7$

16. $f(x) = 2x^3 - 5x, \quad 0 \le x \le 2$

17. $f(x) = \dfrac{x^2 - 1}{x + 5}$

18. $f(x) = \begin{cases} 2x + 3, & x \ne 0 \\ 1, & x = 0 \end{cases}$

19. $f(x) = 5x^2 - 7\sqrt{x}$

20. $f(x) = x^{4.5} - 3x^2 + 7$

21. $f(x) = \dfrac{x^2 - 1}{x - 1}, x \ne 1$

22. $f(x) = 3x^2 - 4x + 7x^{-2}$

In Exercises 23–28, explain why the given graph cannot be the graph of a polynomial function.

23.   **24.**

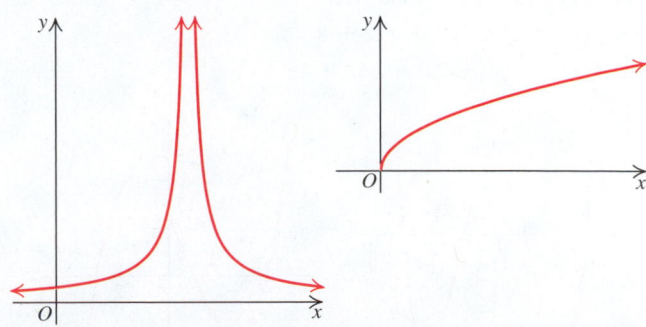

25.

26.

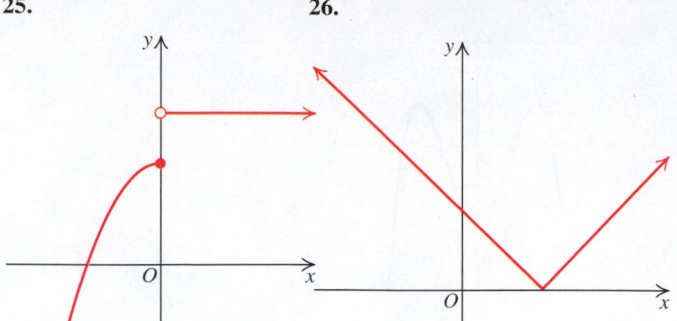

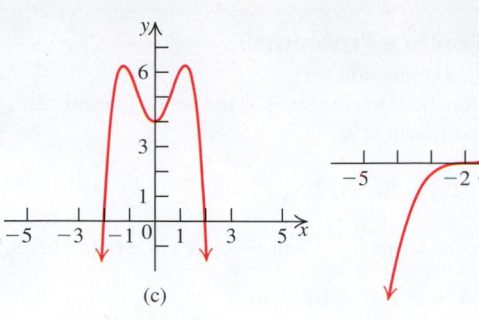

(c)

(d)

27.

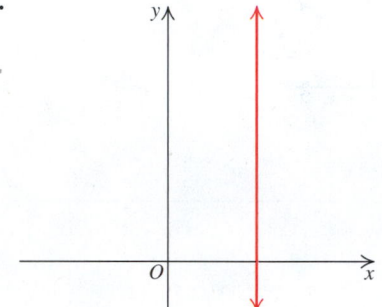

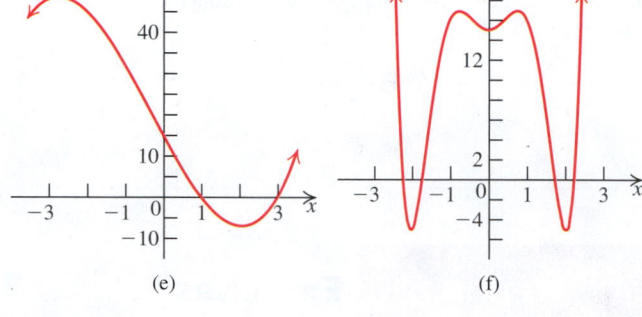

(e)

(f)

In Exercises 35–42, describe the end behavior of the polynomial function f.

35. $f(x) = x - x^3$

36. $f(x) = 2x^3 - 2x^2 + 1$

37. $f(x) = 4x^4 + 2x^3 + 1$

38. $f(x) = -x^4 + 3x^3 + x$

39. $f(x) = (x + 2)^2(2x - 1)$

40. $f(x) = (x - 2)^3(2x + 1)$

41. $f(x) = (x + 2)^2(4 - x)$

42. $f(x) = (x + 3)^3(2 - x)$

28.

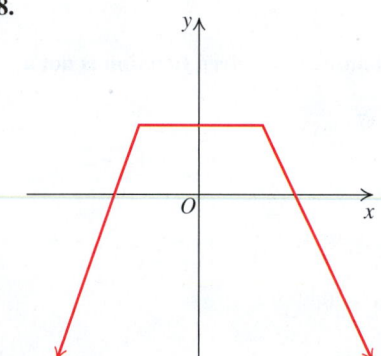

In Exercises 29–34, match the polynomial function with its graph. Use the leading-term test and the y-intercept.

29. $f(x) = -x^4 + 3x^2 + 4$

30. $f(x) = x^6 - 7x^4 + 7x^2 + 15$

31. $f(x) = x^4 - 10x^2 + 9$

32. $f(x) = x^3 + x^2 - 17x + 15$

33. $f(x) = x^3 + 6x^2 + 12x + 8$

34. $f(x) = -x^3 - 2x^2 + 11x + 12$

In Exercises 43–50, Find the real zeros of f and state the multiplicity for each zero. State whether the graph of f crosses or touches the x-axis at each zero.

43. $f(x) = 3(x - 1)(x + 2)(x - 3)$

44. $f(x) = -5(x + 1)(x + 2)(x - 3)$

45. $f(x) = (x + 2)^2(2x - 1)$

46. $f(x) = (x - 2)^3(2x + 1)$

47. $f(x) = x^2(x^2 - 9)(3x + 2)^3$

48. $f(x) = -x^3(x^2 - 4)(3x - 2)^2$

49. $f(x) = (x^2 + 1)(3x - 2)^2$

50. $f(x) = (x^2 + 1)(x + 1)(x - 2)$

In Exercises 51–54, use the Intermediate Value Theorem to show that each polynomial $P(x)$ has a real zero in the specified interval. Approximate this zero to two decimal places.

51. $P(x) = x^4 - x^3 - 10; [2, 3]$

52. $P(x) = x^4 - x^2 - 2x - 5; [1, 2]$

53. $P(x) = x^5 - 9x^2 - 15; [2, 3]$

54. $P(x) = x^5 + 5x^4 + 8x^3 + 4x^2 - x - 5; [0, 1]$

(a)

(b)

In Exercises 55–62, find the polynomial with a leading coeffi-cient of either 1 or −1 and with smallest possible degree that matches the given graph.

55.

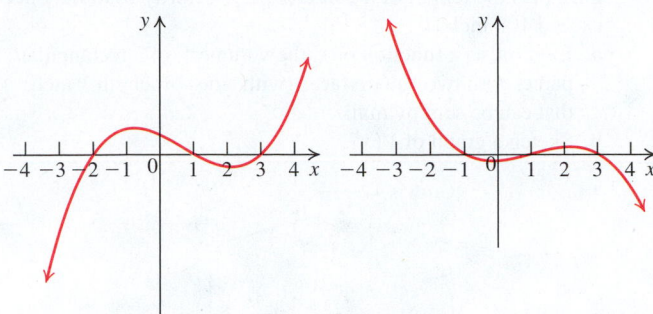

56.

57.

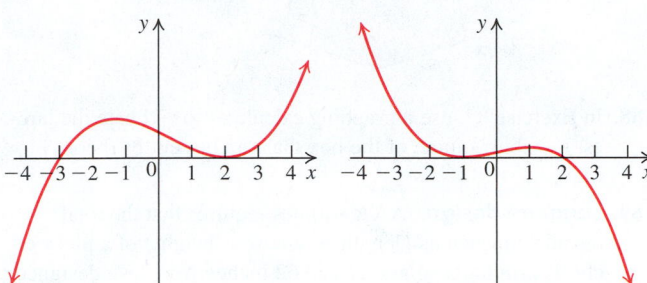

58.

59.

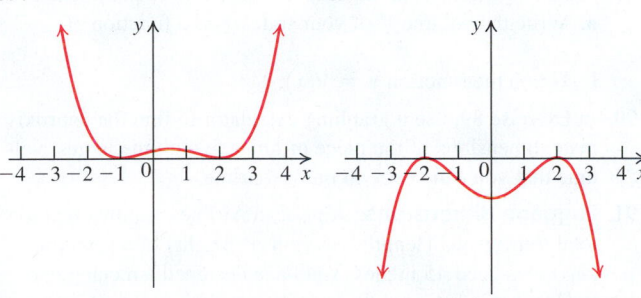

60.

61.

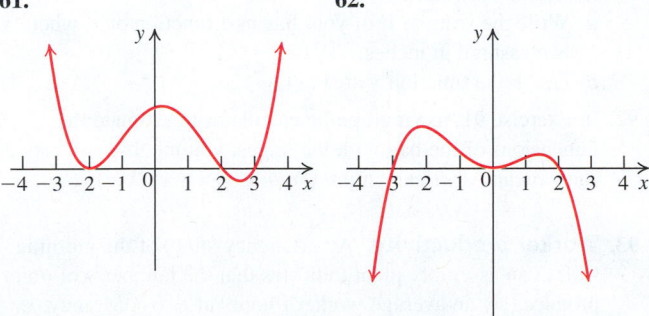

62.

In Exercises 63–74, sketch the graph of the polynomial function f using the techniques described in this section.

63. $f(x) = 2(x + 1)(x - 2)(x + 4)$

64. $f(x) = -(x - 1)(x + 3)(x - 4)$

65. $f(x) = (x - 1)^2(x + 3)(x - 4)$

66. $f(x) = -x^2(x + 1)(x - 2)$

67. $f(x) = -x^2(x - 3)^2$

68. $f(x) = (x - 2)^2(x + 3)^2$

69. $f(x) = (x - 1)^2(x + 2)^3(x - 3)$

70. $f(x) = (x - 1)^3(x + 1)^2(x - 3)$

71. $f(x) = -x^2(x^2 - 1)(x + 1)$

72. $f(x) = x^2(x^2 - 4)(x + 2)$

73. $f(x) = x^2(x^2 + 1)(x - 2)$

74. $f(x) = x(x^2 + 9)(x - 2)^2$

Applying the Concepts

75. One very important feature of any car is the power of the engine. The power P (in horsepower HP) required to over-come its aerodynamic drag by a typical passenger car is a power function of its velocity v (in mph)

$$P = \frac{0.04825}{746}v^3.$$

  a. Compute the horsepower required for a car to cruise on a highway at 40 mph.

  b. Compute the horsepower required for a car to cruise on a highway at 65 mph.

  c. Compute the horsepower required for a car to cruise on a highway at 80 mph.

  d. Show that doubling the speed requires increasing the avail-able power by a factor of 8.

J. B. S. Haldane wrote in his essay *On Being the Right Size*: **"You can drop a small mouse down a thousand-yard mine shaft; and, on arriving at the bottom, it gets a slight shock and walks away, provided that the ground is fairly soft. A rat is killed, a man is broken, a horse splashes."**

76. If an object falls down through air and reaches its terminal velocity, we can estimate its kinetic energy E (in joules, J) at the time of contact with the surface of the earth by the formula

$$E = 12800 \cdot L^4$$

where L represents the length of the object (expressed in meters).

  a. Compute the kinetic energy at the time of contact for a human assuming $L_H = 1.75$ m.

  b. Compute the kinetic energy at the time of contact for a cat assuming $L_C = \dfrac{L_H}{3} = 0.5833$ m (a third the length of a human). By what factor did the energy change?

  c. Compute the kinetic energy at the time of contact for a mouse assuming $L_M = \dfrac{L_H}{25} = 0.07$ m (1/25 the length of a human). By what factor did the energy change?

  d. Show that reducing the size of the animal by a factor of two reduces the impact energy by a factor of 16.

77. Drug reaction. Let $f(x) = 3x^2(4 - x)$.

  a. Find the zeros of f and their multiplicities.

  b. Sketch the graph of $y = f(x)$.

  c. How many turning points are in the graph of f?

  d. A patient's reaction $R(x)(R(x) \geq 0)$ to a certain drug, the size of whose dose is x, is observed to be $R(x) = 3x^2(4 - x)$. What is the domain of the function $R(x)$? What portion of the graph of $f(x)$ constitutes the graph of $R(x)$?

78. In Exercise 77, use a graphing calculator to estimate the value of x for which $R(x)$ is a maximum.

79. Cough and air velocity. The normal radius of the wind-pipe of a human adult at rest is approximately 1 centimeter. When you cough, your windpipe contracts to, say, a radius r. The velocity v at which the air is expelled depends on the radius r and is given by the formula $v = a(1 - r)r^2$, where a is a positive constant.
   a. What is the domain of v?
   b. Sketch the graph of $v(r)$.

80. In Exercise 79, use a graphing calculator to estimate the value of r for which v is a maximum.

81. Maximizing revenue. A manufacturer of slacks has discovered that the number x of khakis sold to retailers per week at a price p dollars per pair is given by the formula

$$p = 27 - \left(\frac{x}{300}\right)^2.$$

   a. Write the weekly revenue $R(x)$ as a function of x.
   b. What is the domain of $R(x)$?
   c. Graph $y = R(x)$.

82. a. In Exercise 81, use a graphing calculator to estimate the number of slacks sold to maximize the revenue.
   b. What is the maximum revenue from part (a)?

83. Maximizing production. An orange grower in California hires migrant workers to pick oranges during the season. He has 12 employees, and each can pick 400 oranges per hour. He has discovered that if he adds more workers, the production per worker decreases due to lack of supervision. When x new workers (above the 12) are hired, each worker picks $400 - 2x^2$ oranges per hour.
   a. Write the number $N(x)$ of oranges picked per hour as a function of x.
   b. Find the domain of $N(x)$.
   c. Graph the function $y = N(x)$.

84. a. In Exercise 83, estimate the maximum number of oranges that can be picked per hour.
   b. Find the number of employees involved in part (a).

85. Making a box. From a rectangular 8×15 piece of card-board, four congruent squares with sides of length x are cut out, one at each corner. (See the accompanying figure.) The resulting crosslike piece is then folded to form a box.
   a. Find the volume V of the box as a function of x.
   b. Sketch the graph of $y = V(x)$.

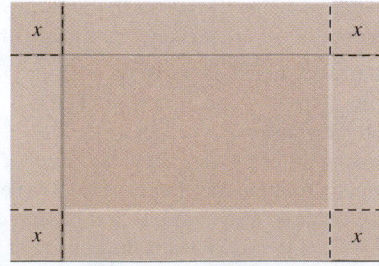

86. In Exercise 85, use a graphing calculator to estimate the largest possible value of the volume of the box.

87. Mailing a box. According to Postal Service regulations, the girth plus the length of a parcel sent by Priority Mail may not exceed 108 inches.
   a. Express, as a function of x, the volume V of a rectangular parcel with two square faces (with sides of length x inches) that can be sent by mail.
   b. Sketch a graph of $V(x)$.

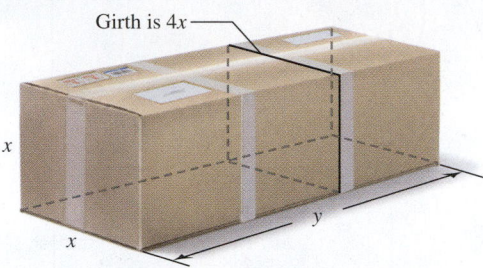

Girth is $4x$

88. In Exercise 87, use a graphing calculator to estimate the largest possible volume of the box that can be sent by Priority Mail.

89. Luggage design. AAA Airlines requires that the total outside dimensions (length + width + height) of a piece of checked-in luggage not exceed 62 inches. You have designed a suitcase in the shape of a box whose height (x inches) equals its width. Your suitcase meets AAA's requirements.
   a. Write the volume V of your suitcase as a function of x.
   b. Graph the function $y = V(x)$.

90. In Exercise 89, use a graphing calculator to find the approximate dimensions of the piece of luggage with the largest volume that you can check in on this airline.

91. Luggage dimensions. American Airlines requires that the total dimensions (length + width + height) of a carry-on bag not exceed 45 inches. You have designed a rectangular boxlike bag whose length is twice its width x. Your bag meets American's requirements.
   a. Write the volume V of your bag as a function of x, where x is measured in inches.
   b. Graph the function $y = V(x)$.

92. In Exercise 91, use a graphing calculator to estimate the dimensions of the bag with the largest volume that you can carry on an American flight. (*Source:* www.aa.com)

93. Worker productivity. An efficiency study of the morning shift at an assembly plant indicates that the number y of units produced by an average worker t hours after 6 A.M. may be modeled by the formula $y = -t^3 + 6t^2 + 16t$. Sketch the graph of $y = f(t)$.

94. In Exercise 93, estimate the time in the morning when a worker is most efficient.

95. Texting The graph shows the annual number of text messages (in trillions) sent in the United States, 2008–2014.

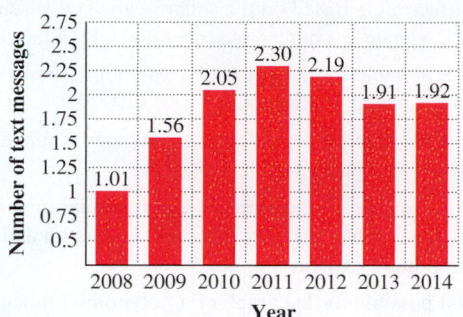

The cubic function $f(x) = 0.963 + 0.88x - 0.192x^2 + 0.0117x^3$ models this data, where x represents the number of years after 2008.

a. Use this function to estimate the number of text messages sent in 2015.

 b. Use a graphing calculator to estimate the number of turning points and the number of zeros.

 c. Use a graphing calculator to estimate in which year the number of text messages was at the maximum. Round to the nearest year and discuss whether it fits the original data.

96. Global warming The graph shows global temperature anomalies (in Celsius) expressed as a deviation from the 1951–1980 global temperature mean.

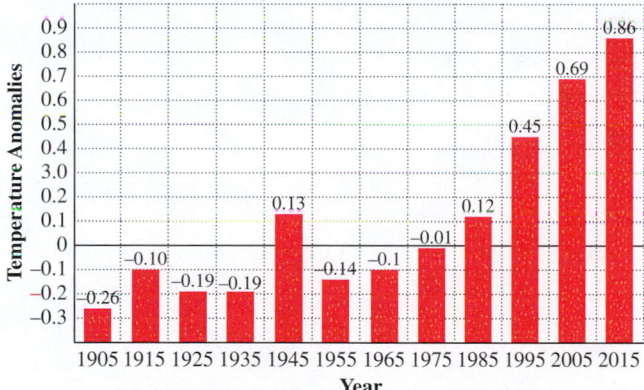

Source: http://data.giss.nasa.gov/gistemp

The cubic function $f(x) = -0.23 + 0.0081x - 0.0002x^2 + 0.000002x^3$ models this data, where x represents the number of years after 1905.

a. Use this function to estimate the temperature anomaly in 2025.

 b. Use a graphing calculator to estimate the number of turning points and the number of zeros.

 c. Use a graphing calculator to estimate in which year the anomaly was 0. Round to the nearest year and discuss whether it fits the original data.

Beyond the Basics

In Exercises 97–106, use transformations of the graph of $y = x^4$ or $y = x^5$ to graph each function $f(x)$. Find the zeros of $f(x)$ and their multiplicity.

97. $f(x) = (x - 1)^4$

98. $f(x) = -(x + 1)^4$

99. $f(x) = x^4 + 2$

100. $f(x) = x^4 + 81$

101. $f(x) = -(x - 1)^4$

102. $f(x) = \frac{1}{2}(x - 1)^4 - 8$

103. $f(x) = x^5 + 1$

104. $f(x) = (x - 1)^5$

105. $f(x) = 8 - \frac{(x + 1)^5}{4}$

106. $f(x) = 81 + \frac{(x + 1)^5}{3}$

107. Prove that if $0 < x < 1$, then $0 < x^4 < x^2$. This shows that the graph of $y = x^4$ is flatter than the graph of $y = x^2$ on the interval $(0, 1)$.

108. Prove that if $x > 1$, then $x^4 > x^2$. This shows that the graph of $y = x^4$ is higher than the graph of $y = x^2$ on the interval $(1, \infty)$.

 109. Sketch the graph of $y = x^2 - x^4$. Use a graphing calculator to estimate the maximum vertical distance between the graphs $y = x^2$ and $y = x^4$ on the interval $(0, 1)$. Also estimate the value of x at which this maximum distance occurs.

 110. Repeat Exercise 109 by sketching the graph of $y = x^3 - x^5$.

In Exercises 111–114, write a polynomial with the given characteristics.

111. A polynomial of degree 3 that has exactly two x-intercepts

112. A polynomial of degree 3 that has three distinct x-intercepts and whose graph rises to the left and falls to the right

113. A polynomial of degree 4 that has exactly two distinct x-intercepts and whose graph falls to the left and right

114. A polynomial of degree 4 that has exactly three distinct x-intercepts and whose graph rises to the left and right

In Exercises 115–118, find the smallest possible degree of a polynomial with the given graph. Explain your reasoning.

115.
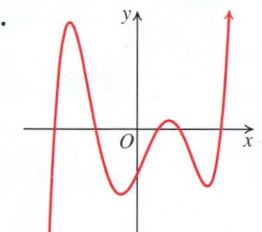

116.
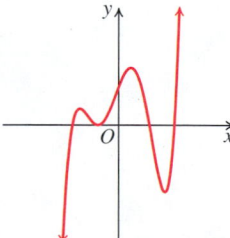

117.
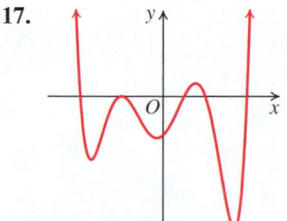

118.
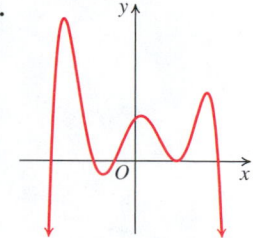

For large values of x the behavior of a polynomial function is determined by its leading term, while for small values of x the behavior of a polynomial function is determined by the constant term, together with the term that has the smallest exponent. Use this argument to determine the behavior of the given polynomial function for large values of x as well as for small values of x and, using this information, sketch its graph.

119. $f(x) = 10x^4 + x + 1$

120. $f(x) = x^3 - x + 1$

121. $f(x) = x^4 - x^2 + 1$

122. $f(x) = x^3 - x^2 + 1$

Critical Thinking/Discussion/Writing

Square-cube law. The principle was first described in 1638 by Galileo Galilei and states that if a solid is proportionally scaled up, then its surface area (or the area of any of its cross sections) increases by the square of the scaling factor, while its volume increases by the cube of the scaling factor.

123. **Giants** The weight of a body is proportional to its volume, while the strength of any bone is proportional to its cross-sectional area. Based on the square-cube law, discuss what will happen if a human is scaled up by a factor of 20.

124. **Lilliputians** Body heat production is proportional to the body's volume while heat loss is proportional to the body's surface area. Based on the square-cube law, discuss what will happen if a human is scaled down by a factor of 70.

125. Is it possible for the graph of a polynomial function to have no y-intercepts? Explain.

126. Is it possible for the graph of a polynomial function to have no x-intercepts? Explain.

127. Is it possible for the graph of a polynomial function of degree 3 to have exactly one local maximum and no local minimum? Explain.

128. Is it possible for the graph of a polynomial function of degree 4 to have exactly one local maximum and exactly one local minimum? Explain.

129. Let P and Q be polynomial functions of degree m and n, respectively. Is the composition $P \circ Q$ a polynomial function? If so, what is the degree of $P \circ Q$? If not, provide an example where the composition fails to be a polynomial.

130. Let P and Q be polynomial functions of degree m and n with leading coefficients a_m and b_n, respectively. Find the leading coefficient of the compositions $P \circ Q$ and $Q \circ P$.

Getting Ready for the Next Section

In Exercises 131–134, evaluate each function at the indicated values of x.

131. $f(x) = x^3 - 3x^2 + 2x - 9; x = -2$

132. $g(x) = x^4 + 2x^3 - 20x - 5; x = 3$

133. $g(x) = (-7x^5 + 2x)(x - 13); x = 13$

134. $f(x) = (x^8 - 15x^7 + 5x^3)(x + 12); x = -12$

In Exercises 135–138, factor each expression.

135. $x^2 - 3x - 10$

136. $-2x^2 + x + 1$

137. $(2x + 3)(x^2 + 6x - 7)$

138. $(x - 9)(6x^2 - 7x - 3)$

Dividing Polynomials and the Rational Zeros Test

BEFORE STARTING THIS SECTION, REVIEW

1 Multiplying polynomials (Appendix A.2, page 929)

2 Evaluating a function (Section 1.3, page 43)

3 Factoring trinomials (Appendix A.2, page 931)

OBJECTIVES

1 Learn the Division Algorithm.

2 Use the Remainder and Factor Theorems.

3 Use the Rational Zeros Test.

◆ Greenhouse Effect and Global Warming

The **greenhouse effect** refers to the fact that Earth is surrounded by an atmosphere that acts like the transparent cover of a greenhouse, allowing sunlight to filter through while trapping heat. It is becoming increasingly clear that we are experiencing *global warming*. This means that Earth's atmosphere is becoming warmer near its surface. Scientists who study climate agree that global warming is due largely to the emission of "greenhouse gases," such as carbon dioxide (CO_2), chlorofluorocarbons (CFCs), methane (CH_4), ozone (O_3), and nitrous oxide (N_2O). The temperature of Earth's surface increased by 1°F over the twentieth century. The late 1990s and early 2000s were some of the hottest years ever recorded. Projections of future warming suggest a global temperature increase of between 2.5°F and 10.4°F by the year 2100.

Global warming is also projected to result in long-term changes in climate, including a rise in average temperatures, unusual rainfalls, and more storms and floods. The sea level may rise so much that people will have to move away from coastal areas. Some regions of the world may become too dry for farming.

The production and consumption of energy is one of the major causes of greenhouse gas emissions. In Example 7, we look at the pattern for the consumption of petroleum in the United States.

1 Learn the Division Algorithm.

RECALL

A polynomial $P(x)$ is an expression of the form

$$a_n x^n + \cdots + a_1 x + a_0.$$

The Division Algorithm

Dividing polynomials is similar to dividing integers. The fact that 5 divides 10 evenly is expressed as $\dfrac{10}{5} = 2$ or as $10 = 5 \cdot 2$. By comparison, in equation

(1) $x^3 - 1 = (x - 1)(x^2 + x + 1).$ Factor, using difference of cubes.

Dividing both sides by $x - 1$, we obtain

$$\frac{x^3 - 1}{x - 1} = x^2 + x + 1, \quad x \neq 1.$$

Polynomial Factor _____

A polynomial $D(x)$ is a **factor** of a polynomial $F(x)$ if there is a polynomial $Q(x)$ such that $F(x) = D(x) \cdot Q(x)$.

Next, consider the equation

(2) $x^3 - 2 = (x - 1)(x^2 + x + 1) - 1$.   Subtract 1 from both sides of equation (1).

Dividing both sides by $(x - 1)$, we can rewrite equation (2) in the form

$$\frac{x^3 - 2}{x - 1} = x^2 + x + 1 - \frac{1}{x - 1}, \quad x \neq 1.$$

Here we say that when the **dividend** $x^3 - 2$ is divided by the **divisor** $x - 1$, the **quotient** is $x^2 + x + 1$ and the **remainder** is -1. This result is an example of the *Division Algorithm*:

THE DIVISION ALGORITHM

If a polynomial $F(x)$ is divided by a polynomial $D(x)$, and $D(x)$ is not the zero polynomial, then there are unique polynomials $Q(x)$ and $R(x)$ such that

$$\frac{F(x)}{D(x)} = Q(x) + \frac{R(x)}{D(x)}$$

or

$$F(x) = D(x) \cdot Q(x) + R(x).$$

         Dividend   Divisor   Quotient   Remainder

Either $R(x)$ is the *zero polynomial* or the degree of $R(x)$ is less than the degree of $D(x)$.

The Division Algorithm says that the dividend is equal to the divisor times the quotient plus the remainder. The expression $\dfrac{F(x)}{D(x)}$ is **improper** if

$$\text{degree of } F(x) \geq \text{degree of } D(x).$$

However, the expression $\dfrac{R(x)}{D(x)}$ is always **proper** because by the Division Algorithm, degree of $R(x) <$ degree of $D(x)$.

For an improper expression $\dfrac{F(x)}{D(x)}$, you may use the **long division** and the **synthetic division** processes reviewed in the next few examples.

PROCEDURE
IN ACTION

EXAMPLE 1 Long Division

OBJECTIVE

Find the quotient and remainder when one polynomial is divided by another.

Step 1 Write the terms in the dividend and the divisor in descending powers of the variable.

EXAMPLE

Find the quotient and remainder when $2x^2 + x^5 + 7 + 4x^3$ is divided by $x^2 + 1 - x$.

1. *Dividend:* $x^5 + 4x^3 + 2x^2 + 7$ *Divisor:* $x^2 - x + 1$

Step 2 Insert terms with zero coefficients in the dividend and the divisor for any missing powers of the variable.

2. $x^5 + 0x^4 + 4x^3 + 2x^2 + 0x + 7$

$$\frac{x^5}{x^2} = x^3 \quad \text{Divide first terms.}$$

Step 3 Divide the first term in the dividend by the first term in the divisor to obtain the first term in the quotient.

3.
$$x^2 - x + 1 \overline{)x^5 + 0x^4 + 4x^3 + 2x^2 + 0x + 7}$$
with x^3 above.

Step 4 Multiply the divisor by the first term in the quotient and subtract the product from the dividend.

4.
$$x^2 - x + 1 \overline{)x^5 + 0x^4 + 4x^3 + 2x^2 + 0x + 7}$$
$$(-)(x^5 - x^4 + x^3) \longleftarrow x^3(x^2 - x + 1)$$
$$x^4 + 3x^3 + 2x^2 + 0x + 7 \quad \text{Remainder}$$

Subtract.

$$\frac{x^4}{x^2} = x^2 \qquad \frac{4x^3}{x^2} = 4x$$

Step 5 Treat the remainder obtained in Step 4 as a new dividend and repeat Steps 3 and 4. Continue this process until a remainder is obtained that is of lower degree than the divisor.

5.
$$x^2 - x + 1 \overline{)x^5 + 0x^4 + 4x^3 + 2x^2 + 0x + 7} \qquad \frac{5x^2}{x^2} = 5$$
quotient $x^3 + x^2 + 4x + 5$
$$(-)(x^5 - x^4 + x^3)$$
$$x^4 + 3x^3 + 2x^2 + 0x + 7 \longleftarrow \text{New dividend}$$
$$(-)(x^4 - x^3 + x^2) \longleftarrow x^2(x^2 - x + 1)$$
$$4x^3 + x^2 + 0x + 7 \qquad \text{Remainder}$$
$$(-)(4x^3 - 4x^2 + 4x) \longleftarrow 4x(x^2 - x + 1)$$
$$5x^2 - 4x + 7 \qquad \text{Remainder}$$
$$(-)(5x^2 - 5x + 5) \longleftarrow 5(x^2 - x + 1)$$
$$x + 2 \qquad \text{Remainder}$$

Step 6 Write the quotient and the remainder.

6. Quotient $= x^3 + x^2 + 4x + 5$ Remainder $= x + 2$

Practice Problem 1 Find the quotient and remainder when $3x^3 + 4x^2 + x + 7$ is divided by $x^2 + 1$.

PROCEDURE
IN ACTION

EXAMPLE 2 **Synthetic Division**

OBJECTIVE

Divide a polynomial $F(x)$ by $x - a$.

Step 1 Arrange the coefficients of $F(x)$ in order of *descending powers of x*, supplying zero as the coefficient of each missing power.

Step 2 Replace the divisor $x - a$ with a. Place a in the position to the left of the coefficients.

Step 3 Bring the first (leftmost) coefficient down below the line. Multiply it by a and write the resulting product one column to the right and above the line.

EXAMPLE

Divide $2x^4 - 3x^2 + 5x - 63$ by $x + 3$.

1.
$$\underline{\quad}\ |\ 2 \quad 0 \quad -3 \quad 5 \quad -63$$

2. Rewrite $x + 3 = x - (-3); a = -3$
$$-3|\ 2 \quad 0 \quad -3 \quad 5 \quad -63$$

3. $-3|\ 2 \quad 0 \quad -3 \quad 5 \quad -63$
$$\qquad -6$$
$$\qquad 2$$

$\otimes$ means "multiply by -3."

(continued)

Step 4 Add the product obtained in the previous step to the coefficient directly above it and write the resulting sum directly below it and below the line. This sum is the "newest" number below the line.

4. $\underline{-3}$ | $\;2\quad\;\;0\quad-3\quad\;\;\;5\quad-63$
$\qquad\quad\;\;-6$
$\qquad\quad\;\;+\downarrow$
$\qquad\quad\;\;2\quad-6$

Step 5 Multiply the newest number below the line by a, write the resulting product one column to the right and above the line, and repeat Step 4.

5. $\underline{-3}$ | $\;2\quad\;\;0\quad-3\quad\;\;\;5\quad-63$
$\qquad\quad\;\;-6\quad\;18$
$\qquad\quad\;\;\otimes\;+$
$\qquad\quad\;\;2\quad-6\quad\;15$

Step 6 Repeat Step 5 until a product is added to the constant term. Separate the last number below the line with a short vertical line.

6. $\underline{-3}$ | $\;2\quad\;\;0\quad-3\quad\;\;\;5\quad-63$
$\qquad\quad\;\;-6\quad\;18\quad-45\quad\;120$
$\qquad\quad\;\;\otimes\;+\;\otimes\;+$
$\qquad\quad\;\;2\quad-6\quad\;15\quad-40\quad\;\;57$

Step 7 The last number below the line is the remainder, and the other numbers, reading from left to right, are the coefficients of the quotient, which has degree one less than that of the dividend $F(x)$.

7. $\underline{-3}$ | $\;2\qquad\;\;0\qquad-3\qquad\;\;\;5\quad-63$ Quotient: degree 4
$\qquad\qquad\quad\;-6\qquad\;18\qquad-45\qquad120$
$\qquad\quad\;\;2\quad-6\qquad\;\;15\qquad-40\;\;|\;57$ Quotient: degree 3
$\qquad\quad\;\uparrow\qquad\uparrow\qquad\;\;\uparrow\qquad\;\;\uparrow\qquad\;\;\uparrow$
$\qquad\quad\;\;2x^3 - 6x^2 + 15x - 40$
$\qquad\qquad\qquad\text{Quotient}$ $\qquad\qquad$ Remainder

Practice Problem 2 Divide $2x^3 - 7x^2 + 5$ by $x - 3$.

EXAMPLE 3 Using Long Division and Synthetic Division

Use both long division and synthetic division to find the quotient and remainder when $2x^4 + x^3 - 16x^2 + 18$ is divided by $x + 2$.

Solution

Long Division

$$
\begin{array}{r}
2x^3 - 3x^2 - 10x + 20 \\
x + 2\overline{)\,2x^4 + \;\;x^3 - 16x^2 + 0x + 18} \\
\underline{2x^4 + 4x^3} \\
-3x^3 - 16x^2 + 0x + 18 \\
\underline{-3x^3 - \;\;6x^2} \\
-10x^2 + 0x + 18 \\
\underline{-10x^2 - 20x} \\
20x + 18 \\
\underline{20x + 40} \\
-22
\end{array}
$$

Synthetic Division

$\underline{-2}$ | $\;2\quad\;\;1\quad-16\quad\;\;\;0\quad\;\;18$
$\qquad\qquad\;\;-4\quad\;\;\;6\quad\;\;20\quad-40$
$\qquad\;\;2\quad-3\quad-10\quad\;20\;\;|{-22}$

The quotient is $2x^3 - 3x^2 - 10x + 20$, and the remainder is -22. So the result is

$$\frac{2x^4 + x^3 - 16x^2 + 18}{x + 2} = 2x^3 - 3x^2 - 10x + 20 + \frac{-22}{x + 2}$$

$$= 2x^3 - 3x^2 - 10x + 20 - \frac{22}{x + 2}.$$

Practice Problem 3 Use synthetic division to find the quotient and remainder when $2x^3 + x^2 - 18x - 7$ is divided by $x - 3$.

2 Use the Remainder and Factor Theorems.

The Remainder and Factor Theorems

In Example 3, we discovered that when the polynomial

$$F(x) = 2x^4 + x^3 - 16x^2 + 18 \text{ is divided by } x + 2,$$

the remainder is -22, a constant.

Now evaluate F at -2. We have:

$$F(-2) = 2(-2)^4 + (-2)^3 - 16(-2)^2 + 18 \qquad \text{Replace } x \text{ with } -2 \text{ in } F(x).$$
$$= 2(16) - 8 - 16(4) + 18 = -22 \qquad \text{Simplify.}$$

We see that $F(-2) = -22$, the remainder we found when we divided $F(x)$ by $x + 2$. The following theorem shows that this result is not a coincidence.

THE REMAINDER THEOREM

If a polynomial $F(x)$ is divided by $x - a$, then the remainder R is given by

$$R = F(a).$$

Note that in the statement of the Remainder Theorem, $F(x)$ plays the dual role of representing a polynomial and the function defined by the same polynomial.

The Remainder Theorem is a consequence of the Division Algorithm. In this case, the divisor $D(x)$ is a linear polynomial $D(x) = x - a$, so the remainder $R(x)$ has to be either the zero polynomial (which has no degree) or a polynomial of degree less than the degree of $D(x) = x - a$. It follows that the remainder has to be a constant polynomial $R(x) = R$, and we write

$$F(x) = (x - a)Q(x) + R.$$

In particular, if we evaluate F at a, we obtain

$$F(x) = (x - a)Q(x) + R$$
$$F(a) = (a - a)Q(a) + R = 0 \cdot Q(a) + R = 0 + R = R. \qquad \text{Substitute } x = a, \text{ and simplify.}$$

This proves the Remainder Theorem.

EXAMPLE 4 **Using the Remainder Theorem**

Find the remainder when the polynomial

$$F(x) = 2x^5 - 4x^3 + 5x^2 - 7x + 2$$

is divided by $x - 1$.

Solution

We could find the remainder by using long division or synthetic division, but a quicker way is to evaluate $F(x)$ when $x = 1$. By the Remainder Theorem, $F(1)$ is the remainder.

$$F(1) = 2(1)^5 - 4(1)^3 + 5(1)^2 - 7(1) + 2 \qquad \text{Replace } x \text{ with } 1 \text{ in } F(x).$$
$$= 2 - 4 + 5 - 7 + 2 = -2$$

The remainder is -2.

Practice Problem 4 Use the Remainder Theorem to find the remainder when

$$F(x) = x^{110} - 2x^{57} + 5 \text{ is divided by } x - 1.$$

Sometimes the Remainder Theorem is used in the other direction: to evaluate a polynomial at a specific value of x. This use is illustrated in the next example.

EXAMPLE 5 **Using the Remainder Theorem**

Let $f(x) = x^4 + 3x^3 - 5x^2 + 8x + 75$. Find $f(-3)$.

Solution

One way of solving this problem is to evaluate $f(x)$ when $x = -3$.

$f(-3) = (-3)^4 + 3(-3)^3 - 5(-3)^2 + 8(-3) + 75$ Replace x with -3.

$= 6$ Simplify.

Another way is to use synthetic division to find the remainder when the polynomial $f(x)$ is divided by $(x + 3)$.

$$
\begin{array}{r|rrrrr}
-3 & 1 & 3 & -5 & 8 & 75 \\
   &   & -3 & 0 & 15 & -69 \\
\hline
   & 1 & 0 & -5 & 23 & \;|\;6
\end{array}
$$

The remainder is 6. By the Remainder Theorem, $f(-3) = 6$.

Practice Problem 5 Use synthetic division to find $f(-2)$, where

$$f(x) = x^4 + 10x^2 + 2x - 20.$$

The Factor Theorem allows us to see the equivalence between finding zeros of a polynomial $F(x)$ and finding its factors. By the Remainder Theorem

$$F(x) = (x - a)Q(x) + F(a).$$

If $F(a) = 0$, then $F(x) = (x - a)Q(x)$ and $(x - a)$ is a factor of $F(x)$. Conversely, if $(x - a)$ is a factor of $F(x)$, then $F(x) = (x - a)Q(x)$, and if we evaluate F at a, we obtain

$$F(x) = (x - a)Q(x)$$
$$F(a) = (a - a)Q(a) = 0 \cdot Q(a) = 0.$$ Substitute $x = a$. Simplify.

This proves the Factor Theorem.

THE FACTOR THEOREM

A polynomial $F(x)$ has $(x - a)$ as a factor if and only if $F(a) = 0$.

The Factor Theorem shows that if $F(x)$ is a polynomial, then the following problems are equivalent:

1. Factoring $F(x)$ and applying the zero-product property
2. Finding the zeros of the function $F(x)$ defined by the polynomial expression
3. Solving (or finding the roots of) the polynomial equation $F(x) = 0$

Note that if a is a zero of the polynomial function $F(x)$, then $F(x) = (x - a)Q(x)$. Any solution of the equation $Q(x) = 0$ is also a zero of $F(x)$. Since the degree of $Q(x)$ is less than the degree of $F(x)$, it is simpler to solve the equation $Q(x) = 0$ to find other possible zeros of $F(x)$. The equation $Q(x) = 0$ is called the **depressed equation** of $F(x)$.

The next example illustrates how to use the Factor Theorem and the depressed equation to solve a polynomial equation.

EXAMPLE 6 **Using the Factor Theorem**

Given that 2 is a zero of the function $f(x) = 3x^3 + 2x^2 - 19x + 6$, solve the polynomial equation $3x^3 + 2x^2 - 19x + 6 = 0$.

Solution

Since 2 is a zero of $f(x)$, we have $f(2) = 0$. The Factor Theorem tells us that $(x - 2)$ is a factor of $f(x)$. Next, we use synthetic division to divide $f(x)$ by $(x - 2)$.

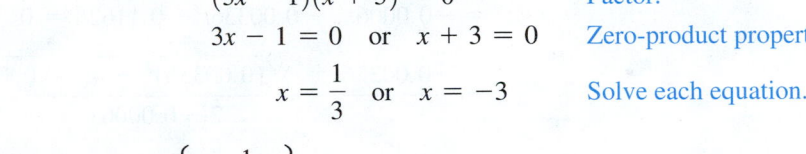

The coefficients of the quotient give the depressed equation $3x^2 + 8x - 3$. Because the remainder is 0, we have:

$$f(x) = 3x^3 + 2x^2 - 19x + 6 = (x - 2)\underbrace{(3x^2 + 8x - 3)}_{\text{quotient}}.$$

Any solution of the depressed equation $3x^2 + 8x - 3 = 0$ is a zero of f. Because this equation is of degree 2, any method of solving a quadratic equation may be used to solve it.

$$3x^2 + 8x - 3 = 0 \qquad \text{Depressed equation}$$
$$(3x - 1)(x + 3) = 0 \qquad \text{Factor.}$$
$$3x - 1 = 0 \quad \text{or} \quad x + 3 = 0 \qquad \text{Zero-product property}$$
$$x = \frac{1}{3} \quad \text{or} \quad x = -3 \qquad \text{Solve each equation.}$$

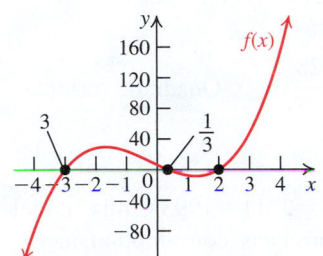

Figure 2.17

The solution set is $\left\{ -3, \dfrac{1}{3}, 2 \right\}$.

In Example 6 we were able to completely factor the polynomial function $f(x) = 3x^3 + 2x^2 - 19x + 6 = (x - 2)(3x - 1)(x + 3)$. Applying the techniques from Section 2.2 allows us to sketch the graph of $y = f(x)$ (see Figure 2.17).

Practice Problem 6 Solve the equation $3x^3 - x^2 - 20x - 12 = 0$, given that one solution is -2.

EXAMPLE 7 **Crude Oil and Petroleum Consumption**

The total crude oil and petroleum products consumption C (in billion barrels) in the United States during 1993–2011 is given in the following table. (Data are rounded to the nearest tenth.)

Year	Consumption	Year	Consumption
1993	6.3	2003	7.3
1995	6.5	2005	7.6
1997	6.8	2007	7.6
1999	7.1	2009	6.9
2001	7.2	2011	6.9

Source: www.eia.gov

These data can be modeled by the function

$$C(t) = -0.0006t^3 + 0.00576t^2 + 0.10284t + 6.33488,$$

where $t = 0$ represents 1993.

The model indicates that $C(4) = 6.8$ billion barrels were consumed in 1997 ($t = 4$). Find another year between 1997 and 2011 when the *model* estimates consumption of 6.8 billion barrels.

Solution

We are given that

$$C(4) = 6.8$$
$$C(t) = (t - 4)Q(t) + 6.8 \qquad \text{Division Algorithm and Remainder Theorem}$$
$$C(t) - 6.8 = (t - 4)Q(t) \qquad \text{Subtract 6.8 from both sides.}$$

Therefore, 4 is a zero of

$$F(t) = C(t) - 6.8 = -0.0006t^3 + 0.00576t^2 + 0.10284t - 0.46512.$$

Finding another year between 1997 and 2011 when the crude oil and petroleum consumption was 6.8 billion barrels requires us to find another zero of $F(t)$ that is between 4 and 18. Because 4 is a zero of $F(t)$, we use synthetic division to find the quotient $Q(t)$. This yields

$$Q(t) = -0.0006t^2 + 0.00336t + 0.11628.$$

We now solve the depressed (quadratic) equation $Q(t) = 0$ using the quadratic formula.

$$Q(t) = -0.0006t^2 + 0.00336t + 0.11628 = 0$$

$$t = \frac{-0.00336 \pm \sqrt{(0.00336)^2 - 4(-0.0006)(0.11628)}}{2(-0.0006)} \qquad \text{Quadratic formula}$$

$$t = -11.4 \quad \text{or} \quad t = 17 \qquad \text{Use a calculator.}$$

Because t needs to be between 4($=1997 - 1993$) and 18($=2011 - 1993$), this model tells us that in 2010 ($t = 17$), crude oil and petroleum products consumption in the United States was also 6.8 billion barrels.

Practice Problem 7 The value of $C(x) = 0.23x^3 - 4.255x^2 + 0.345x + 41.05$ is 10 when $x = 3$. Find another positive integer x for which $C(x) = 10$.

3 Use the Rational Zeros Test.

The Rational Zeros Test

We use the Factor Theorem to determine whether a number a is a zero of a polynomial function $F(x)$. We use the following test (see Exercise 114 for a proof) to find *possible* rational zeros of a polynomial function with integer coefficients.

THE RATIONAL ZEROS TEST

If $F(x) = a_n x^n + a_{n-1}x^{n-1} + \cdots + a_2 x^2 + a_1 x + a_0$ is a polynomial function with integer coefficients ($a_n \neq 0, a_0 \neq 0$) and $\frac{p}{q}$ is a rational number in lowest terms that is a zero of $F(x)$, then

1. p is a factor of the constant term a_0.
2. q is a factor of the leading coefficient a_n.

EXAMPLE 8 **Using the Rational Zeros Test**

Find all rational zeros of $F(x) = 2x^3 + 5x^2 - 4x - 3$.

Solution

First, we list all *possible* rational zeros of $F(x)$.

$$\text{Possible rational zeros} = \frac{p}{q} = \frac{\text{Factors of the constant term, } -3}{\text{Factors of the leading coefficient, } 2}$$

Factors of -3: $\pm 1, \pm 3$

Factors of 2: $\pm 1, \pm 2$

All combinations can be found by using just the *positive* factors of the leading coefficient. The possible rational zeros are

$$\frac{\pm 1}{1}, \frac{\pm 1}{2}, \frac{\pm 3}{1}, \frac{\pm 3}{2}$$

or

$$\pm 1, \pm \frac{1}{2}, \pm 3, \pm \frac{3}{2}.$$

We use synthetic division to determine whether a rational zero exists among these eight candidates. We start by testing 1. If 1 is not a rational zero, we will test other possibilities.

$$
\begin{array}{r|rrrr}
1 & 2 & 5 & -4 & -3 \\
  &   & 2 & 7 & 3 \\
\hline
  & 2 & 7 & 3 & 0
\end{array}
$$

coefficients of the depressed equation

The zero remainder tells us that $x = 1$ is a zero of $F(x)$ and $(x - 1)$ is a factor of $F(x)$, with the other factor being $2x^2 + 7x + 3$. Solving the depressed equation

$$2x^2 + 7x + 3 = 0,$$

we have

$$(2x + 1)(x + 3) = 0 \qquad \text{Factor.}$$

$$2x + 1 = 0 \quad \text{or} \quad x + 3 = 0 \qquad \text{Zero-product property}$$

$$x = -\frac{1}{2} \quad \text{or} \quad x = -3 \qquad \text{Solve each equation.}$$

The solution set is $\left\{ 1, -\frac{1}{2}, -3 \right\}$. The rational zeros of F are -3, $-\frac{1}{2}$, and 1.

The complete factorization of f is $f(x) = (x - 1)(2x + 1)(x + 3)$.

Practice Problem 8 Find all rational zeros of $f(x) = 2x^3 + 3x^2 - 6x - 8$. ◼◼◻

In Example 8 we were able to completely factor the polynomial function $f(x) = 2x^3 + 5x^2 - 4x - 3 = (x - 1)(2x + 1)(x + 3)$. Applying the techniques from Section 2.2 allows us to sketch the graph of $y = f(x)$ (see Figure 2.18).

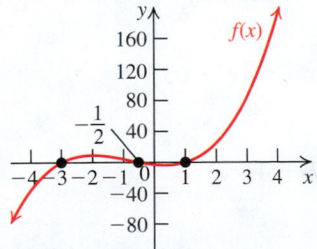

Figure 2.18

EXAMPLE 9 Leaning Ladder

A box with a square base having area 63 square feet and height 3 feet stands in front of a wall. A ladder of length 16 feet leans against the wall and just touches the box at an edge. How high on the wall is the top of the ladder?

Solution

We draw a picture to visualize the situation. See Figure 2.19. Let the variables x and y represent the vertical and horizontal distances from the top and bottom of the ladder to the box, respectively.

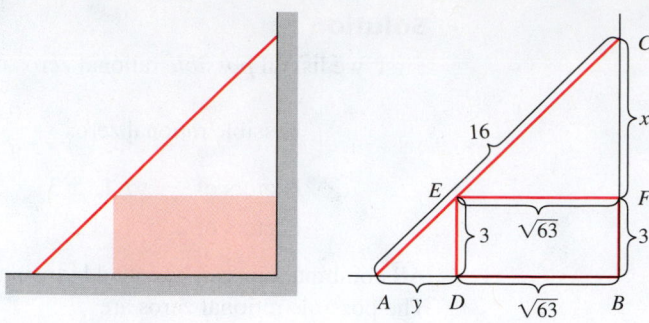

Figure 2.19

From the similar right triangles, $\triangle EFC$ and $\triangle ABC$, we have

$$\frac{\sqrt{63}}{x} = \frac{y + \sqrt{63}}{x + 3}, \quad \text{or} \quad y + \sqrt{63} = \sqrt{63} \cdot \frac{x + 3}{x}.$$

Applying the Pythagorean Theorem to the right triangle $\triangle ABC$ gives

$$(x + 3)^2 + (y + \sqrt{63})^2 = 16^2,$$

$$(x + 3)^2 + \left(\sqrt{63} \cdot \frac{x + 3}{x}\right)^2 = 16^2, \quad \color{blue}{\text{Replacing } y + \sqrt{63} \text{ with } \sqrt{63} \cdot \frac{x + 3}{x}}$$

$$(x + 3)^2 x^2 + 63(x + 3)^2 = 16^2 x^2. \quad \color{blue}{\text{Multiply by } x^2, (x^2 > 0) \text{ and simplify}}$$

Expanding and simplifying the last equation gives

$$P(x) = x^4 + 6x^3 - 184x^2 + 378x + 567 = 0.$$

We can use the factorization of $567 = 3^4 \cdot 7$ to check for all possible rational zeros of $P(x)$. Using synthetic division for numbers

$$\pm 1, \pm 3, \pm(3 \cdot 7), \pm 3^2, \pm(3^2 \cdot 7), \pm 3^3, \pm(3^3 \cdot 7), \pm 3^4, \pm(3^4 \cdot 7)$$

we see that $x = -1$ is a rational zero of $P(x)$ and

$$P(x) = (x + 1)(x^3 + 5x^2 - 189x + 567).$$

We use the same factorization of $567 = 3^4 \cdot 7$ to check for all possible rational zeros of $Q(x) = x^3 + 5x^2 - 189x + 567$. Using synthetic division, we see that $x = 9$ is a zero of $Q(x)$ and

$$Q(x) = x^3 + 5x^2 - 189x + 567 = (x - 9)(x^2 + 14x - 63).$$

Finally, we factor

$$x^2 + 14x - 63 = (x + 7 - 4\sqrt{7})(x + 7 + 4\sqrt{7})$$

and obtain

$$P(x) = x^4 + 6x^3 - 184x^2 + 378x + 567 = (x + 1)(x - 9)(x + 7 - 4\sqrt{7})(x + 7 + 4\sqrt{7}).$$

The positive zeros of $P(x)$ are $x = 9$ and $x = -7 + 4\sqrt{7}$. Since height equals $x + 3$, the top of the ladder touches the wall at either 12 or $-4 + 4\sqrt{7} \approx 6.5830\ldots$ feet above the ground.

Practice Problem 9 Repeat Example 9 for a box with a square base having area 180 square feet, height 2 feet, and a ladder of length 21 feet.

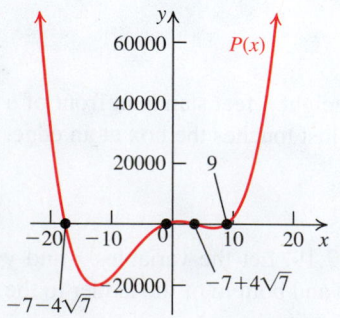

Figure 2.20

In Example 9 we were able to completely factor the polynomial function $P(x) = x^4 + 6x^3 - 184x^2 + 378x + 567$. Applying the techniques from Section 2.2 allows us to sketch the graph of $y = P(x)$ (see Figure 2.20).

SECTION 2.3   Exercises

Concepts and Vocabulary

1. Consider the equation $x^3 - 3x + 5 = (x^2 + 1)x - 4x + 5$. If both sides are divided by $x^2 + 1$, then the dividend is _____, the divisor is _____, the quotient is _____, and the remainder is _____.

2. Consider the division $\dfrac{x^2 + 4x}{x - 2} = x + 6 + \dfrac{12}{x - 2}$. The dividend is _____, the divisor is _____, the quotient is _____, and the remainder is _____.

3. The Remainder Theorem states that if a polynomial $F(x)$ is divided by $(x - a)$ then the remainder $R =$ _____.

4. The Factor Theorem states that $(x - a)$ is a factor of a polynomial $F(x)$ if and only if _____ $= 0$.

5. **True or False.** You can use long division to divide a polynomial by $x^2 + 1$

6. **True or False.** You cannot use synthetic division to divide a polynomial by $x^2 + 1$.

7. **True or False.** Possible rational zeros of $P(x) = 9x^3 - 9x^2 - x + 1$ are $\pm 1, \pm 3, \pm 9$.

8. **True or False.** When a polynomial of degree $n + 1$ is divided by a polynomial of degree n, the remainder is a polynomial of degree 1.

Building Skills

In Exercises 9–16, use long division to find the quotient and the remainder.

9. $\dfrac{6x^2 - x - 2}{2x + 1}$

10. $\dfrac{4x^3 - 2x^2 + x - 3}{2x - 3}$

11. $\dfrac{3x^4 - 6x^2 + 3x - 7}{x + 1}$

12. $\dfrac{x^6 + 5x^3 + 7x + 3}{x^2 + 2}$

13. $\dfrac{4x^3 - 4x^2 - 9x + 5}{2x^2 - x - 5}$

14. $\dfrac{y^5 + 3y^4 - 6y^2 + 2y - 7}{y^2 + 2y - 3}$

15. $\dfrac{z^4 - 2z^2 + 1}{z^2 - 2z + 1}$

16. $\dfrac{6x^4 + 13x - 11x^3 - 10 - x^2}{3x^2 - 5 - x}$

In Exercises 17–26, use synthetic division to find the quotient and the remainder when the first polynomial is divided by the second polynomial.

17. $x^3 - x^2 - 7x + 2$; $x - 1$
18. $2x^3 - 3x^2 - x + 2$; $x + 2$
19. $x^3 + 4x^2 - 7x - 10$; $x + 2$
20. $x^3 + x^2 - 13x + 2$; $x - 3$
21. $x^4 - 3x^3 + 2x^2 + 4x + 5$; $x - 2$
22. $x^4 - 5x^3 - 3x^2 + 10$; $x - 1$
23. $2x^3 + 4x^2 - 3x + 1$; $x - \dfrac{1}{2}$
24. $3x^3 + 8x^2 + x + 1$; $x - \dfrac{1}{3}$
25. $2x^3 - 5x^2 + 3x + 2$; $x + \dfrac{1}{2}$
26. $3x^3 - 2x^2 + 8x + 2$; $x + \dfrac{1}{3}$

In Exercises 27–30, use synthetic division and the Remainder Theorem to find each function value. Check your answer by evaluating the function at the given x-value.

27. $f(x) = x^3 + 3x^2 + 1$
   a. $f(1)$
   b. $f(-1)$
   c. $f\left(\dfrac{1}{2}\right)$
   d. $f(10)$

28. $g(x) = 2x^3 - 3x^2 + 1$
   a. $g(-2)$
   b. $g(-1)$
   c. $g\left(-\dfrac{1}{2}\right)$
   d. $g(7)$

29. $h(x) = x^4 + 5x^3 - 3x^2 - 20$
   a. $h(1)$
   b. $h(-1)$
   c. $h(-2)$
   d. $h(2)$

⚡ **30.** $f(x) = x^4 + 0.5x^3 - 0.3x^2 - 20$
    a. $f(0.1)$
    b. $f(0.5)$
    c. $f(1.7)$
    d. $f(-2.3)$

In Exercises 31–38, use the Factor Theorem to show that the first polynomial is a factor of the second polynomial. Check your answer by synthetic division.

31. $x - 1;\ 2x^3 + 3x^2 - 6x + 1$

32. $x - 3;\ 3x^3 - 9x^2 - 4x + 12$

33. $x + 1;\ 5x^4 + 8x^3 + x^2 + 2x + 4$

34. $x + 3;\ 3x^4 + 9x^3 - 4x^2 - 9x + 9$

35. $x - 2;\ x^4 + x^3 - x^2 - x - 18$

36. $x + 3;\ x^5 + 3x^4 + x^2 + 8x + 15$

37. $x + 2;\ x^6 - x^5 - 7x^4 + x^3 + 8x^2 + 5x + 2$

38. $x - 2;\ 2x^6 - 5x^5 + 4x^4 + x^3 - 7x^2 - 7x + 2$

In Exercises 39–42, find the value of k for which the first polynomial is a factor of the second polynomial.

39. $x + 1;\ x^3 + 3x^2 + x + k$ [*Hint:* Let
    $f(x) = x^3 + 3x^2 + x + k$. Then $f(-1) = 0$.]

40. $x - 1;\ -x^3 + 4x^2 + kx - 2$

41. $x - 2;\ 2x^3 + kx^2 - kx - 2$

42. $x - 1;\ k^2 - 3kx^2 - 2kx + 6$

In Exercises 43–46, use the Factor Theorem to show that the first polynomial *is not* a factor of the second polynomial. [*Hint:* Show that the remainder is not zero.]

43. $x - 2;\ -2x^3 + 4x^2 - 4x + 9$

44. $x + 3;\ -3x^3 - 9x^2 + 5x + 12$

45. $x + 2;\ 4x^4 + 9x^3 + 3x^2 + x + 4$

46. $x - 3;\ 3x^4 - 8x^3 + 5x^2 + 7x - 3$

In Exercises 47–50, find the set of possible rational zeros of the given function.

47. $f(x) = 3x^3 - 4x^2 + 5$

48. $g(x) = 2x^4 - 5x^2 - 2x + 1$

49. $h(x) = 4x^4 - 9x^2 + x + 6$

50. $F(x) = 6x^6 + 5x^5 + x - 35$

In Exercises 51–66, find all rational zeros of the given polynomial function.

51. $f(x) = x^3 - x^2 - 4x + 4$

52. $f(x) = x^3 + x^2 + 2x + 2$

53. $f(x) = x^3 - 4x^2 + x + 6$

54. $f(x) = x^3 + 3x^2 + 2x + 6$

55. $g(x) = 2x^3 + x^2 - 13x + 6$

56. $g(x) = 3x^3 - 2x^2 + 3x - 2$

57. $g(x) = 6x^3 + 13x^2 + x - 2$

58. $g(x) = 2x^3 + 3x^2 + 4x + 6$

59. $h(x) = 3x^3 + 7x^2 + 8x + 2$

60. $h(x) = 2x^3 + x^2 + 8x + 4$

61. $h(x) = x^4 - x^3 - x^2 - x - 2$

62. $h(x) = 2x^4 + 3x^3 + 8x^2 + 3x - 4$

63. $F(x) = x^4 - x^3 - 13x^2 + x + 12$

64. $G(x) = 3x^4 + 5x^3 + x^2 + 5x - 2$

65. $F(x) = x^4 - 2x^3 + 10x^2 - x + 1$

66. $G(x) = x^6 + 2x^4 + x^2 + 2$

In Exercises 67–74, first find all rational zeros of f, then use the depressed equation (see Example 6) to find all roots of the equation $f(x) = 0$.

67. $f(x) = x^3 + 5x^2 - 8x + 2$

68. $f(x) = x^3 - 7x^2 - 5x + 3$

69. $f(x) = x^4 - 3x^3 + 3x - 1$

70. $f(x) = x^4 - 6x^3 - 7x^2 + 54x - 18$

71. $f(x) = 2x^3 - 9x^2 + 6x - 1$

72. $f(x) = 2x^3 - 3x^2 - 4x - 1$

73. $f(x) = x^4 + x^3 - 5x^2 - 3x + 6$

74. $f(x) = x^4 - 2x^3 - 5x^2 + 4x + 6$

In Exercises 75–80, use the Rational Zeros Test and synthetic division to factor completely the given polynomial function.

75. $f(x) = 2x^3 - 5x^2 + x + 2$

76. $f(x) = 2x^3 + x^2 - 4x - 3$

77. $f(x) = 2x^3 + 3x^2 - 17x - 30$

78. $f(x) = 2x^3 - 3x^2 - 23x + 12$

79. $f(x) = x^4 - x^3 - 9x^2 + 11x + 6$

80. $f(x) = x^4 - 6x^3 + x^2 + 30x - 8$

In Exercises 81–88, sketch the graph of each polynomial function.

81. $f(x) = x^3 + 2x^2 - 19x - 20$

82. $f(x) = x^3 + x^2 - 14x - 24$

83. $f(x) = x^3 - x^2 - 16x - 20$

84. $f(x) = x^3 + 2x^2 - 15x - 36$

85. $f(x) = x^4 - 22x^2 - 24x + 45$

86. $f(x) = x^4 + 3x^3 - 13x^2 - 51x - 36$

87. $f(x) = 6x^3 + 13x^2 + x - 2$

88. $f(x) = 6x^3 + 17x^2 + x - 10$

Applying the Concepts

89. Geometry. The area of a rectangle is
$(2x^4 - 2x^3 + 5x^2 - x + 2)$ square centimeters. Its length is $(x^2 - x + 2)$ cm. Find its width.

90. Geometry. The volume of a rectangular solid is $(x^4 + 3x^3 + x + 3)$ cubic inches. Its length and width are $(x + 3)$ and $(x + 1)$ inches, respectively. Find its height.

91. Cell phone sales. The weekly revenue from one type of cell phone sold at a mall outlet is a quadratic function, R, of its price, x, in dollars. The weekly revenue is \$3000 if the phone is priced at \$40 or \$60 (fewer phones are sold at \$60) and is \$2400 if the phone is priced at \$30.
    a. Find an equation for $R(x) - 3000$.
    b. Find an equation for $R(x)$.
    c. Find the maximum weekly revenue and the price that generates this revenue.

92. Ticket sales. The evening revenue from a theater is a quadratic function, R, of its ticket price, x, in dollars. The evening revenue is \$4725 if the tickets are priced at \$8 or \$12 (fewer tickets are sold at \$12) and is \$4125 if the tickets are priced at \$6.
   a. Find an equation for $R(x) - 4725$.
   b. Find an equation for $R(x)$.
   c. Find the maximum weekly revenue and the price that generates this revenue.

93. Total energy consumption. The total energy consumption (in quads) in the United States from 1991 to 2011 can be approximated by the function

$$C(t) = -0.0006t^3 - 0.0613t^2 + 2.0829t + 82.904,$$

where $t = 0$ represents 1991. The model indicates that 97.6 quads of energy were consumed in 2002. Find another year after 2002 when the model estimates 97.6 quads of total energy consumption in the United States. Round your answer to the nearest year. (The model was modified from the data obtained from www.eia.gov.)

94. Lottery sales. In the lottery game called *Florida Lotto*, players should select 6 numbers out of 53. Various prizes are offered for matching three through six numbers drawn by the lottery. Florida Lotto sales (in millions of dollars) from 1998 through 2011 can be approximated by the function

$$s(t) = -0.1779t^3 - 2.8292t^2 + 42.0240t + 698.6831,$$

where $t = 0$ represents 1998. The model indicated that \$737.7 million in Florida Lotto sales occurred in 1999. Find another year after 1999 when the model indicates that \$737.7 million in Florida Lotto sales occurred. (The model was slightly modified from the data obtained from www.floridafiscalportal.state.fl.us.)

95. Marine Corps. The total number of U.S. Marine Corps, M (in thousands), during 1990–2010 can be modeled by the function

$$M(t) = -0.0027t^3 + 0.3681t^2 - 5.8645t + 195.2782,$$

where $t = 0$ represents 1990, $t = 1$ represents 1991, and so on. The model estimates the total number of U.S. Marine Corps in 1992 as 185,000. Find another year between 1992 and 2010 when the model indicates that the number of U.S. Marine Corps was approximately 185,000. Round your answer to the nearest year. (The model was generated using slightly modified data from www.census.gov.)

96. Unemployment. U.S. unemployment, U (in percent), for 2002–2012 can be modeled by the function

$$U(t) = -0.0374t^3 + 0.5934t^2 - 2.0553t + 6.7478,$$

where $t = 0$ represents 2002, $t = 1$ represents 2003, and so on. The model indicates a 7.7% level of unemployment in 2008 ($t = 6$). Find another year after 2008 with about the same unemployment level. Round your answer to the nearest year. (The model was slightly modified from data obtained from www.multipl.com.)

97. Cost function. The total cost of a product (in dollars) is given by $C(x) = 3x^3 - 6x^2 + 108x + 100$, where x is the demand for the product. Find the value of x that gives a total cost of \$628.

98. Profit. For the product in Exercise 97, the demand function is $p(x) = 330 + 10x - x^2$. Find the value of x that gives the profit of \$910.

99. Production cost. The total cost per month, in thousands of dollars, of producing x printers (with x measured in hundreds) is given by $C(x) = x^3 - 15x^2 + 5x + 50$. How many units must be produced so that the total monthly cost is \$125,000?

100. Break even. If the demand function for the printers in Exercise 99 is $p(x) = -3x + 3 + \dfrac{74}{x}$, find the number of printers that must be produced and sold to break even.

101. Shipping boxes. When two shipping boxes in the shape of a cube were put side-by-side, it turned out that one was 3 inches taller than the other. If the combined volume of these two boxes is 1843 cubic inches, what are their respective sizes?

102. Building a Pool. A builder is commissioned to build a diving pool. The pool consists of a deep water section in the shape of a cube joined by a shallow water section of the same width that is 3 meters long and 1 meter deep. If the pool is to hold 536 cubic meters of water, what are the dimensions of the deep water section?

103. Making a box. A square piece of tin 18 inches on each side is to be made into a box without a top by cutting a square from each corner and folding up the flaps to form the sides. What size corners should be cut so that the volume of the box is 432 cubic inches?

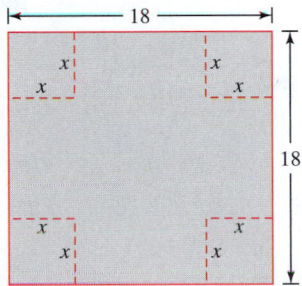

104. Leaning Ladder. A box with a square base having area 27 square feet and height 1 foot stands in front of a wall. A ladder of length 8 feet leans against the wall and just touches the box at an edge. How high on the wall is the top of the ladder?

Beyond the Basics

105. Show that $\dfrac{1}{2}$ is a root of multiplicity 2 of the equation

$$4x^3 + 8x^2 - 11x + 3 = 0.$$

106. Show that $-\dfrac{2}{3}$ is a root of multiplicity 2 of the equation

$$9x^3 + 3x^2 - 8x - 4 = 0.$$

107. Show that

$$b^3 - a^3 = (b - a)(a^2 + ab + b^2)$$

by considering the polynomial $P(x) = x^3 - a^3$ and dividing it by $x - a$.

108. Show that

$$b^4 - a^4 = (b - a)(a^3 + a^2b + ab^2 + b^3)$$

by considering the polynomial $P(x) = x^4 - a^4$ and dividing it by $x - a$.

109. Show that there is no real number a such that $x^2 - 1$ is a factor of $x^{10} + ax + 4$.

110. Show that $b^2 - a^2$ divides $b^{2n} - a^{2n}$ by considering the polynomial $P(x) = x^{2n} - a^{2n}$ and computing remainders when the polynomial $P(x)$ is divided, respectively, by $x - a$ and $x + a$.

111. When the polynomial $f(x)$ is divided by $x - 3$, the remainder is 2, and when the polynomial $g(x)$ is divided by $x - 3$, the remainder is -5. What is the remainder when the polynomial $f(x) \cdot g(x)$ is divided by $x - 3$?

112. When the polynomial $f(x)$ is divided by $x + 2$, the remainder is -2. What is the remainder when the polynomial $[f(x)]^3 + 1$ is divided by $x + 2$?

113. Find all rational roots of the equation

$$2x^4 + 2x^3 + \frac{1}{2}x^2 + 2x - \frac{3}{2} = 0.$$

[*Hint:* Multiply both sides of the equation by the lowest common denominator.]

114. To prove the Rational Zeros Test, we assume the **Fundamental Theorem of Arithmetic**, which states that "Every integer has a unique prime factorization." Supply reasons for each step in the following proof (F is a polynomial function of order n with integer coefficients and nonzero constant):

(i) $F\left(\dfrac{p}{q}\right) = 0$ and $\dfrac{p}{q}$ is in lowest terms.

(ii) $a_n\left(\dfrac{p}{q}\right)^n + a_{n-1}\left(\dfrac{p}{q}\right)^{n-1} + \cdots + a_1\left(\dfrac{p}{q}\right) + a_0 = 0$

(iii) $a_n p^n + a_{n-1} p^{n-1}q + \cdots + a_1 pq^{n-1} + a_0 q^n = 0$

(iv) $a_n p^n + a_{n-1} p^{n-1}q + \cdots + a_1 pq^{n-1} = -a_0 q^n$

(v) p is a factor of the left side of (iv).

(vi) p is a factor of $-a_0 q^n$.

(vii) p is a factor of a_0.

(viii) $a_{n-1}p^{n-1}q + \cdots + a_1 pq^{n-1} + a_0 q^n = -a_n p^n$

(ix) q is a factor of the left side of the equation in (viii).

(x) q is a factor of $-a_n p^n$.

(xi) q is a factor of a_n.

Critical Thinking/Discussion/Writing

In Exercises 115–116, find all integer values of c for which the given polynomial has at least one rational zero.

115. $f(x) = x^3 - cx + 2$

116. $f(x) = x^3 - cx - 6$

In Exercises 117–120, show that the given number is irrational. [*Hint:* To show that $x = 1 + \sqrt{3}$ is irrational, form the equivalent equations $x - 1 = \sqrt{3}, (x - 1)^2 = 3$, and $x^2 - 2x - 2 = 0$ and show that the last equation has no rational roots.]

117. $\sqrt{3}$

118. $\sqrt[3]{4}$

119. $3 - \sqrt{2}$

120. $(9)^{\frac{2}{3}}$

121. Assuming that n is a positive integer, find the values of n for which each of the following is true.
  a. $x + a$ is a factor of $x^n + a^n$.
  b. $x + a$ is a factor of $x^n - a^n$.
  c. $x - a$ is a factor of $x^n + a^n$.
  d. $x - a$ is a factor of $x^n - a^n$.

122. Use Exercise 121 to prove the following:
  a. $7^{11} - 2^{11}$ is divisible by 5.
  b. $(19)^{20} - (10)^{20}$ is divisible by 261.

Getting Ready for the Next Section

In Exercises 123–126, graph the line given by the equation.

123. $x = 3$

124. $y = 2$

125. $y = 2x + 1$

126. $y = x - 1$

In Exercises 127–130, evaluate each expression for the given value of the variable.

127. $\dfrac{-3}{2x + 1}; x = 2$

128. $\dfrac{7 - x}{2x^2 + 3x}; x = -1$

129. $\dfrac{2x + 3}{5 - 2x^2}; x = -3$

130. $\dfrac{x^2 + 4x - 1}{9 - x^3}; x = 2$

Rational Functions

BEFORE STARTING THIS SECTION, REVIEW

1 Domain of a function (Section 1.3, page 40)

2 Zeros of a function (Section 2.2, page 158)

3 Symmetry (Section 1.1, page 9)

4 Quadratic formula (Appendix A.6, page 957)

OBJECTIVES

1 Define a rational function.

2 Define vertical and horizontal asymptotes.

3 Graph translations of $f(x) = \dfrac{1}{x}$.

4 Find vertical and horizontal asymptotes (if any).

5 Graph rational functions.

6 Graph rational functions with oblique asymptotes.

7 Graph a revenue curve.

Figure 2.21 A revenue curve.

◆ **Federal Taxes and Revenues**

Federal governments levy income taxes on their citizens and corporations to pay for defense, infrastructure, and social services. The relationship between tax rates and tax revenues shown in Figure 2.21 is a "revenue curve."

All economists agree that a zero tax rate produces no tax revenues and that no one would bother to work with a 100% tax rate, so tax revenue would be zero. It then follows that in a given economy, there is some optimal tax rate T between zero and 100% at which people are willing to maximize their output and still pay their taxes. This optimal rate brings in the most revenue for the government. However, since no one knows the actual shape of the revenue curve, it is impossible to find the exact value of T.

Notice in Figure 2.21 that between the two extreme rates are two rates (such as a and b in the figure) that will collect the same amount of revenue. In a democracy, politicians can argue that taxes are currently too high (at some point b in Figure 2.21) and should therefore be reduced to encourage incentives and harder work (this is supply-side economics); at the same time, the government will generate more revenue. Others can argue that we are well to the left of T (at some point a in Figure 2.21), so the tax rate should be raised (for the rich) to generate more revenue. In Example 10, we sketch the graph of a revenue curve for a hypothetical economy.

1 Define a rational function.

Rational Functions

Recall that a sum, difference, or product of two polynomial functions is also a polynomial function. However, the quotient of two polynomial functions is generally *not* a polynomial function. We call the quotient of two polynomial functions a *rational function*.

Rational Function

A function f that can be expressed in the form

$$f(x) = \frac{N(x)}{D(x)}$$

where the numerator $N(x)$ and the denominator $D(x)$ are polynomials and $D(x)$ is not the zero polynomial is called a **rational function**. The domain of f consists of all real numbers for which $D(x) \neq 0$.

Examples of rational functions are

$$f(x) = \frac{x-1}{2x+1}, \quad g(y) = \frac{5y}{y^2+1}, \quad h(t) = \frac{3t^2 - 4t + 1}{t - 2}, \quad \text{and} \quad F(z) = 3z^4 - 5z^2 + 1.$$

($F(z)$ has the constant polynomial function $D(z) = 1$ as its denominator.) By contrast,

$$S(x) = \frac{\sqrt{1-x^2}}{x+1} \quad \text{and} \quad G(x) = \frac{|x|}{x}$$ are not rational functions.

EXAMPLE 1 **Finding the Domain of a Rational Function**

Find the domain of each rational function.

a. $f(x) = \dfrac{3x^2 - 12}{x - 1}$ **b.** $g(x) = \dfrac{x}{x^2 - 6x + 8}$ **c.** $h(x) = \dfrac{x^2 - 4}{x - 2}$

Solution

We eliminate all x for which $D(x) = 0$.

a. $D(x) = x - 1 = 0$, if $x = 1$. The domain of $f(x) = \dfrac{3x^2 - 12}{x - 1}$ is the set of all real numbers x except $x = 1$; or in interval notation, $(-\infty, 1) \cup (1, \infty)$.

b. The domain of g is the set of all real numbers x for which the denominator $D(x)$ is nonzero.

$$
\begin{aligned}
D(x) = x^2 - 6x + 8 &= 0 && \text{Set } D(x) = 0. \\
(x - 2)(x - 4) &= 0 && \text{Factor } D(x). \\
x - 2 = 0 \quad \text{or} \quad x - 4 &= 0 && \text{Zero-product property} \\
x = 2 \quad \text{or} \quad x &= 4 && \text{Solve for } x.
\end{aligned}
$$

The domain of g is the set of all real numbers x except 2 and 4, or in interval notation, $(-\infty, 2) \cup (2, 4) \cup (4, \infty)$.

c. The domain of $h(x) = \dfrac{x^2 - 4}{x - 2}$ is the set of all real numbers x except 2, or in interval notation, $(-\infty, 2) \cup (2, \infty)$.

Practice Problem 1 Find the domain of the rational function $f(x) = \dfrac{x - 3}{x^2 - 4x - 5}$. ▪▪

RECALL

The graph of a polynomial function has no holes, gaps, or sharp corners.

In Example 1, part **c**, note that the functions $h(x) = \dfrac{x^2 - 4}{x - 2} = \dfrac{(x - 2)(x + 2)}{x - 2}$ and $H(x) = x + 2$ are not equal: the domain of $H(x)$ is the set of all real numbers, but the domain of $h(x)$ is the set of all real numbers x except 2. The graph of $y = H(x)$ is a line

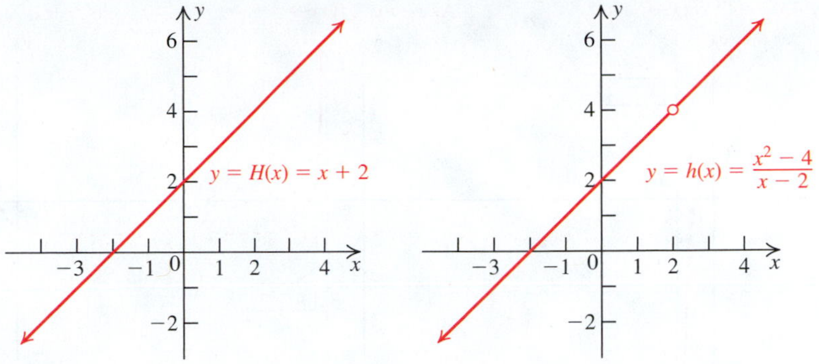

Figure 2.22 Functions $H(x)$ and $h(x)$ have different graphs.

with slope 1 and y-intercept 2. The graph of $y = h(x)$ is the same line as $y = H(x)$, except that the point $(2, 4)$ is missing. See Figure 2.22.

As with the quotient of integers, if the polynomials $N(x)$ and $D(x)$ have no common factors, then the rational function $f(x) = \dfrac{N(x)}{D(x)}$ is said to be in **lowest terms**. The zeros of $N(x)$ and $D(x)$ will play an important role in graphing the rational function f.

2 Define vertical and horizontal asymptotes.

Vertical and Horizontal Asymptotes

The graph of a rational function may consist of several pieces. Each piece, however, is a continuous and smooth curve. By convention the pieces are separated by dashed lines through the points that are not in its domain. It is important, therefore, to study the behavior of a rational function near the points that are excluded from its domain.

Consider the function $f(x) = \dfrac{1}{x}$. The domain of this function is $(-\infty, 0) \cup (0, \infty)$. The function f does not assign any value to $x = 0$ as $f(0)$ is undefined. However, we more closely examine the behavior of $f(x)$ near the excluded value $x = 0$. We start by evaluating $f(x)$ for values of x that are near 0.

	x approaches 0 from the left					x approaches 0 from the right				
x	-1	-0.1	-0.01	-0.001	-0.0001	0	0.0001	0.001	0.01	0.1	1
$f(x) = \frac{1}{x}$	-1	-10	-100	-1000	-10000		10000	1000	100	10	1

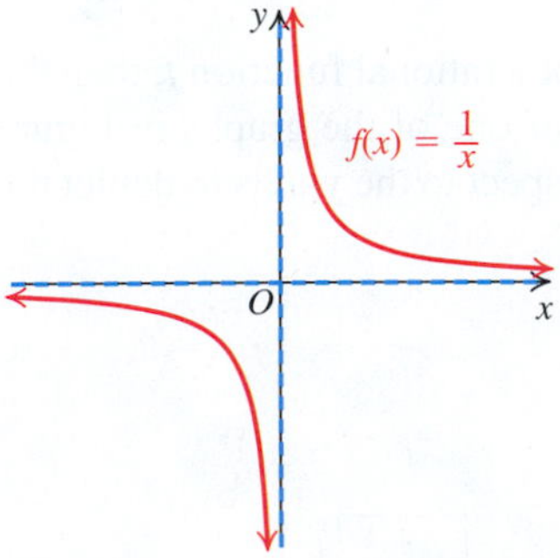

Figure 2.23 Reciprocal function, $f(x) = \dfrac{1}{x}$.

We see that as x is getting closer and closer to 0 from the right (that is, with $x > 0$), the values $f(x)$ increase without bound. In symbols, we write "as $x \to 0^+, f(x) \to \infty$." This statement is read "as x approaches 0 *from the right*, $f(x)$ approaches infinity."

Similarly, from the table, we conclude that "as $x \to 0^-, f(x) \to -\infty$" and say that "as x approaches 0 *from the left*, $f(x)$ approaches negative infinity."

The graph of $f(x) = \dfrac{1}{x}$ is the *reciprocal function* from Chapter 1 (see Figure 2.23). The vertical line $x = 0$ (the y-axis) is a *vertical asymptote*.

Vertical Asymptote

The line $x = a$ is called a **vertical asymptote** of the graph of a function f if $f(x) \to \infty$ as $x \to a^+$ or as $x \to a^-$ or if $f(x) \to -\infty$ as $x \to a^+$ or as $x \to a^-$.

This definition says that if the line $x = a$ is a vertical asymptote of the graph of a function f, then the graph of f *near* $x = a$ behaves like one of the graphs in Figure 2.24.

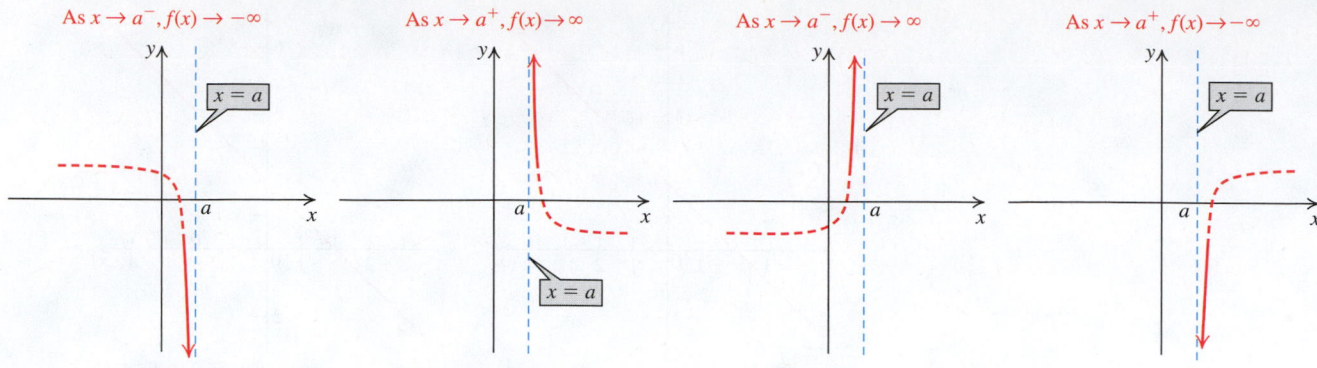

Figure 2.24 Behavior of a function near a vertical asymptote.

Geometrically speaking, a rational function $f(x)$ has a vertical asymptote $x = a$ if its graph approaches the graph of $x = a$ from the left or right side of a.

If $x = a$ is a vertical asymptote for a rational function f, then a is not in its domain; so the graph of f *does not cross* the line $x = a$.

Just as with polynomials, we study the end behavior of rational functions. Consider again the function $f(x) = \dfrac{1}{x}$. We can examine the behavior of f for very large values of $|x|$; that is, as $x \to \infty$ or $x \to -\infty$.

	x approaches $-\infty$						x approaches ∞			
x	-10000	-1000	-100	-10	-1	0	1	10	100	1000	10000
$f(x) = \frac{1}{x}$	-0.0001	-0.001	-0.01	-0.1	-1		1	0.1	0.01	0.001	0.0001

We can see that the larger x is, the closer $f(x) = \dfrac{1}{x}$ is to 0. We conclude that as $x \to \infty, f(x) \to 0$. This statement is read "as x approaches ∞, $f(x)$ approaches 0." Similarly as $x \to -\infty, f(x) \to 0$. That means, geometrically speaking, that for very large values of $|x|$ the graph of $y = \dfrac{1}{x}$ approaches the graph of the horizontal line $y = 0$.

The x-axis, with equation $y = 0$, is a *horizontal asymptote* of the graph of $f(x) = \dfrac{1}{x}$ (see Figure 2.23).

Horizontal Asymptote

The line $y = k$ is a **horizontal asymptote** of the graph of a function f if $f(x) \to k$ as $x \to \infty$ or if $f(x) \to k$ as $x \to -\infty$.

If the line $y = k$ is a horizontal asymptote of the graph of a rational function f, then the end behavior of the graph of f as $x \to \infty$ is similar to that of one of the graphs in Figure 2.25. We can reflect the graphs shown in Figure 2.25 with respect to the y-axis to depict the end behavior of a rational function f as $x \to -\infty$.

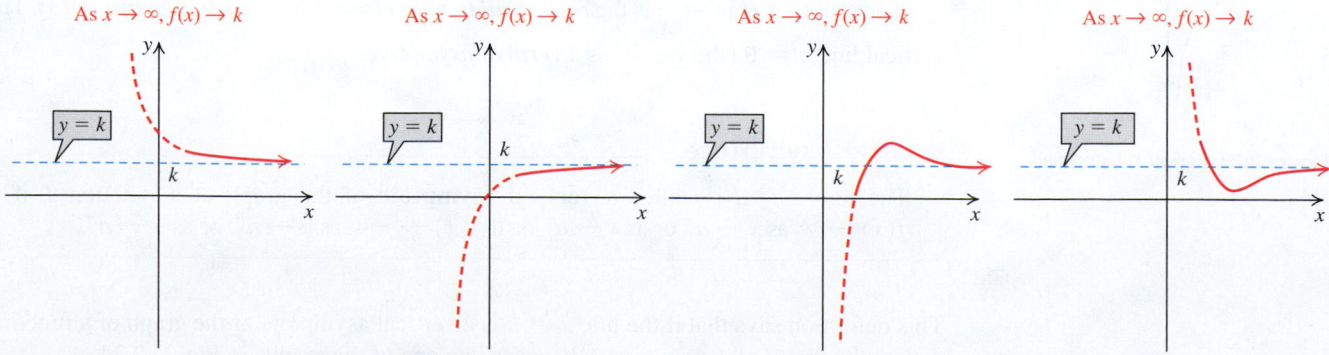

Figure 2.25 End behavior of a function near a horizontal asymptote.

Geometrically speaking, a rational function $f(x)$ has a horizontal asymptote $y = a$ if its graph approaches the graph of $y = a$ as $x \to -\infty$ or as $x \to \infty$.

The graph of a function $y = f(x)$ can cross its horizontal asymptote (in contrast to a vertical asymptote). See Example 7, page 198.

We summarize the symbolic notation that we used in the following table.

Symbol	Read As	Meaning
$x \to a^-$	x approaches a from the left	x is getting arbitrarily close to a with $x < a$
$x \to a^+$	x approaches a from the right	x is getting arbitrarily close to a with $x > a$
$x \to -\infty$	x approaches negative infinity	x decreases without bound
$x \to \infty$	x approaches infinity	x increases without bound

Symbol	Read As	Meaning
$f(x) \to L$	$f(x)$ approaches L	$f(x)$ is getting arbitrarily close to L
$f(x) \to -\infty$	$f(x)$ approaches negative infinity	$f(x)$ decreases without bound
$f(x) \to \infty$	$f(x)$ approaches infinity	$f(x)$ increases without bound

3 Graph translations of $f(x) = \dfrac{1}{x}$.

Translations of $f(x) = \dfrac{1}{x}$

We can graph any function of the form

$$g(x) = \frac{ax + b}{cx + d}$$

starting with the graph of $f(x) = \dfrac{1}{x}$ and using the techniques of vertical stretching and compressing, shifting, and/or reflecting.

> **EXAMPLE 2** **Graphing Rational Functions using Translations**

Graph each rational function, identify the vertical and horizontal asymptotes, and state the domain and range.

a. $g(x) = \dfrac{-2}{x + 1}$ **b.** $h(x) = \dfrac{3x - 2}{x - 1}$

Solution

a. Let $f(x) = \dfrac{1}{x}$. We can write $g(x)$ in terms of $f(x)$:

$$g(x) = \frac{-2}{x + 1}$$

$$= -2\left(\frac{1}{x + 1}\right)$$

$$= -2f(x + 1) \qquad \text{Replace } x \text{ by } x + 1 \text{ in } f(x) = \frac{1}{x}.$$

The graph of $y = f(x + 1)$ is the graph of $y = f(x)$ shifted one unit to the left. This moves the vertical asymptote one unit to the left. The graph of $y = -2f(x + 1)$ is the graph of $y = f(x + 1)$ stretched vertically two units and then reflected about the x-axis. The graph is shown in Figure 2.26. The domain of g is $(-\infty, -1) \cup (-1, \infty)$

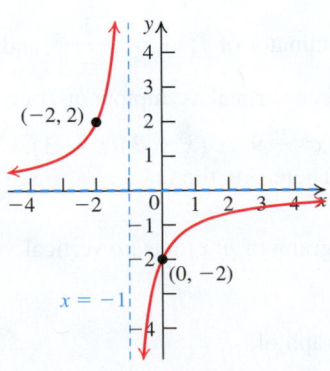

Figure 2.26 Graph of $g(x) = \dfrac{-2}{x + 1}$.

and the range is $(-\infty, 0)\cup(0, \infty)$. The graph has vertical asymptote $x = -1$ and horizontal asymptote $y = 0$ (the x-axis).

b. Using long division, we have

$$h(x) = \frac{1}{x - 1} + 3 \qquad \begin{array}{r} 3 \\ x - 1\overline{)3x - 2} \\ \underline{3x - 3} \\ 1 \end{array}$$

Then $h(x) = f(x - 1) + 3$ Replace x with $x - 1$ in $f(x) = \dfrac{1}{x}$

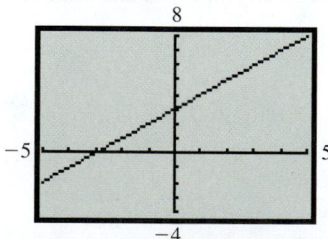

Figure 2.27 Graph of $h(x) = \dfrac{3x - 2}{x - 1}$.

We see that the graph of $y = h(x)$ is the graph of $f(x) = \dfrac{1}{x}$ shifted horizontally one unit to the right and then vertically up three units. The graph is shown in Figure 2.27. The domain of h is $(-\infty, 1)\cup(1, \infty)$ and the range is $(-\infty, 3)\cup(3, \infty)$. The graph of h has vertical asymptote $x = 1$ and horizontal asymptote $y = 3$.

Practice Problem 2 Repeat Example 2 with

a. $g(x) = \dfrac{3}{x - 2}$.   **b.** $h(x) = \dfrac{2x + 5}{x + 1}$.

4 Find vertical and horizontal asymptotes (if any).

> ### LOCATING VERTICAL ASYMPTOTES OF RATIONAL FUNCTIONS
>
> If $f(x) = \dfrac{N(x)}{D(x)}$ is a rational function, where $N(x)$ and $D(x)$ do not have a common factor and a is a real zero of $D(x)$, then the line with equation $x = a$ is a vertical asymptote of the graph of f.

TECHNOLOGY CONNECTION

Calculator graph for

$$h(x) = \frac{x^2 - 9}{x - 3}$$

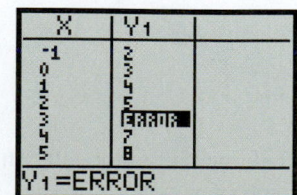

A calculator graph does not show the hole in the graph, but a calculator table explains the situation.

This means that the vertical asymptotes (if any) are found by locating the real zeros of the denominator.

EXAMPLE 3   Finding Vertical Asymptotes

Find all vertical asymptotes of the graph of each rational function.

a. $f(x) = \dfrac{1}{x - 1}$ **b.** $g(x) = \dfrac{1}{x^2 - 9}$ **c.** $h(x) = \dfrac{1}{x^2 + 1}$

Solution

a. There are no common factors in the numerator and denominator of $f(x) = \dfrac{1}{x - 1}$, and the only *zero* of the denominator is 1. Therefore, $x = 1$ is a vertical asymptote of $f(x)$.

b. The rational function $g(x)$ is in lowest terms. Factoring $x^2 - 9 = (x + 3)(x - 3)$, we see that the *zeros* of the denominator are -3 and 3. Therefore, the lines $x = -3$ and $x = 3$ are the two vertical asymptotes of $g(x)$.

c. Because the denominator $x^2 + 1$ has no real zeros, the graph of $h(x)$ has *no* vertical asymptotes.

Practice Problem 3 Find the vertical asymptotes of the graph of:

$$f(x) = \frac{x + 1}{x^2 + 3x - 10}$$

The next example illustrates that the graph of a rational function may have gaps (missing points) with or without vertical asymptotes.

EXAMPLE 4 **Rational Functions Whose Graphs Have a Hole**

Find all vertical asymptotes of the graph of each rational function.

a. $h(x) = \dfrac{x^2 - 9}{x - 3}$ **b.** $g(x) = \dfrac{x + 2}{x^2 - 4}$

Solution

a. $h(x) = \dfrac{x^2 - 9}{x - 3}$ Given function

$= \dfrac{(x + 3)(x - 3)}{x - 3}$ Factor the numerator: $x^2 - 9 = (x + 3)(x - 3)$.

$= x + 3$, if $x \neq 3$ Simplify.

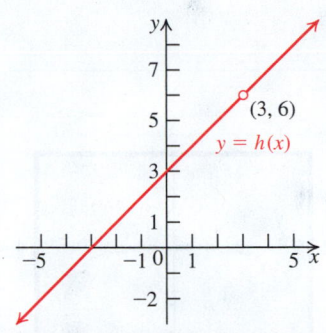

Figure 2.28 Graph with a hole

The graph of $h(x)$ is the line $y = x + 3$ with a gap (or hole) at $x = 3$. See Figure 2.28.

The graph of $h(x)$ has no vertical asymptote at $x = 3$, the zero of the denominator, because the numerator and the denominator have the common factor $(x - 3)$, both with multiplicity 1.

b. $g(x) = \dfrac{x + 2}{x^2 - 4}$ Given function

$= \dfrac{x + 2}{(x + 2)(x - 2)}$ Factor the denominator.

$= \dfrac{1}{x - 2}, x \neq -2$ Simplify.

The graph of $g(x)$ has a hole at $x = -2$ and a vertical asymptote $x = 2$. See Figure 2.29.

Practice Problem 4 Find all vertical asymptotes of the graph of $f(x) = \dfrac{3 - x}{x^2 - 9}$.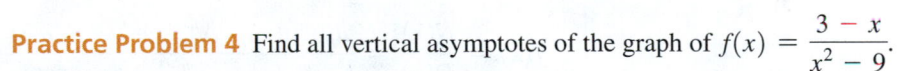

Based on the above discussion, we have

$$g(x) = \frac{x + 2}{x^2 - 4} = \frac{1}{x - 2}, x \neq -2,$$

and we can also describe the behavior of the function $g(x)$ near the hole at $x = -2$. Replacing x by -2 in the last equation $\dfrac{1}{x - 2}$ gives the value $\dfrac{1}{-2 - 2} = -\dfrac{1}{4}$. We conclude that

$$g(x) \rightarrow -\frac{1}{4}, \quad \text{as} \quad x \rightarrow -2.$$

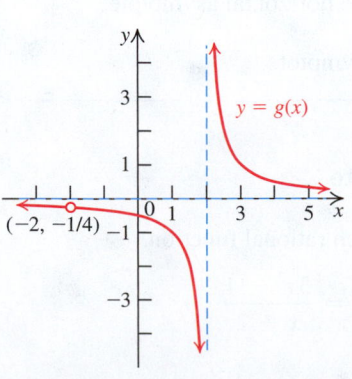

Figure 2.29 Graph with a hole and an asymptote

The graph of $g(x) = \dfrac{x + 2}{x^2 - 4}$, shown in Figure 2.29, has a horizontal asymptote: $y = 0$. This can be verified algebraically. Divide the numerator and the denominator of g by x^2, the highest power of x in the denominator.

$$g(x) = \frac{x + 2}{x^2 - 4} = \frac{\dfrac{x + 2}{x^2}}{\dfrac{x^2 - 4}{x^2}} = \frac{\dfrac{x}{x^2} + \dfrac{2}{x^2}}{\dfrac{x^2}{x^2} - \dfrac{4}{x^2}} = \frac{\dfrac{1}{x} + \dfrac{2}{x^2}}{1 - \dfrac{4}{x^2}}$$

$$\frac{a \pm b}{c} = \frac{a}{c} \pm \frac{b}{c}$$ Simplify each expression.

As $|x| \to \infty$, the expressions $\dfrac{1}{x}, \dfrac{2}{x^2}$, and $\dfrac{4}{x^2}$ all approach 0, so

$$g(x) \to \frac{0 + 0}{1 - 0} = \frac{0}{1} = 0.$$

Since $g(x) \to 0$ as $|x| \to \infty$, the line $y = 0$ (the x-axis) is a horizontal asymptote.

TECHNOLOGY CONNECTION

Graphing calculators give very different graphs for

$$g(x) = \frac{1}{x^2 - 9}$$

depending on whether the mode is set to *connected* or *dot*.

In connected mode, the calculator connects the dots between plotted points, producing vertical lines at $x = -3$ and $x = 3$. Dot mode avoids graphing these vertical lines by plotting the same points, but not connecting them. However, dot mode fails to display continuous sections of the graph.

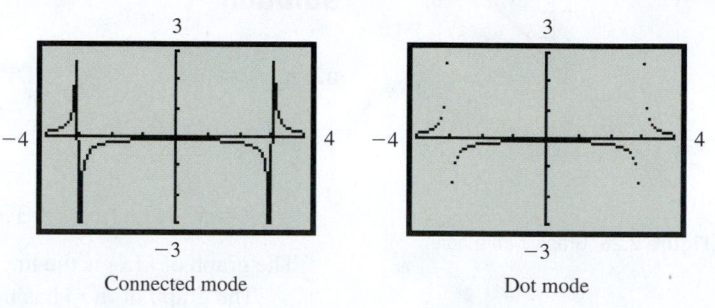

Connected mode   Dot mode

Note that although a rational function can have more than one vertical asymptote (see Example 6), it can have, at most, one horizontal asymptote. **We can find the horizontal asymptote (if any) of a rational function by dividing the numerator and denominator by the highest power of x that appears in the denominator and investigating the resulting expression as $|x| \to \infty$.**

Alternatively, one may use the following rules.

RULES FOR LOCATING HORIZONTAL ASYMPTOTES

Let f be a rational function given by

$$f(x) = \frac{N(x)}{D(x)} = \frac{a_n x^n + a_{n-1} x^{n-1} + \cdots + a_2 x^2 + a_1 x + a_0}{b_m x^m + b_{m-1} x^{m-1} + \cdots + b_2 x^2 + b_1 x + b_0}, \quad a_n \neq 0, b_m \neq 0$$

To find whether the graph of f has one horizontal asymptote or no horizontal asymptote, we compare the degree of the numerator, n, with that of the denominator, m:

1. If $n < m$, then the x-axis ($y = 0$) is the horizontal asymptote.

2. If $n = m$, then the line with equation $y = \dfrac{a_n}{b_m}$ is the horizontal asymptote.

3. If $n > m$, then the graph of f has *no* horizontal asymptote.

EXAMPLE 5 **Finding the Horizontal Asymptote**

Find the horizontal asymptote (if any) of the graph of each rational function.

a. $f(x) = \dfrac{5x + 2}{1 - 3x}$ **b.** $g(x) = \dfrac{2x}{x^2 + 1}$ **c.** $h(x) = \dfrac{3x^2 - 1}{x + 2}$

Solution

a. The numerator and denominator of $f(x) = \dfrac{5x + 2}{1 - 3x}$ are both of degree 1. The leading coefficient of the numerator is 5, and that of the denominator is -3. By Rule 2, the line $y = \dfrac{5}{-3} = -\dfrac{5}{3}$ is the horizontal asymptote of the graph of f.

b. For the function $g(x) = \dfrac{2x}{x^2 + 1}$, the degree of the numerator is 1 and that of the denominator is 2. By Rule 1, the line $y = 0$ (the x-axis) is the horizontal asymptote.

c. The degree of the numerator, 2, of $h(x)$ is greater than the degree of its denominator, 1. By Rule 3, the graph of h has no horizontal asymptote.

Practice Problem 5 Find the horizontal asymptote (if any) of the graph of each function.

a. $f(x) = \dfrac{2x - 5}{3x + 4}$ **b.** $g(x) = \dfrac{x^2 + 3}{x - 1}$

5 Graph rational functions.

SIDE NOTE

If the point is not in its domain, then the rational function has either a vertical asymptote or a hole at this point.

Graphing Rational Functions

If the polynomials $N(x)$ and $D(x)$ have no common factor, then the rational function $f(x) = \dfrac{N(x)}{D(x)}$ is said to be in lowest terms. The zeros of $N(x)$ and $D(x)$ play an important role in graphing a rational function.

Just as the power functions $A(x - a)^m$ help us to describe the behavior of a polynomial or rational function near its zero a, the reciprocal functions $\dfrac{A}{(x - a)^m}$ help us to describe the behavior of rational functions near the vertical asymptote $x = a$. We summarize these two facts next.

BEHAVIOR OF $f(x) = \dfrac{N(x)}{D(x)}$ **NEAR A ZERO**

Assume f is in lowest terms and a is a zero of f, then

1. f can be factored as $f(x) = \dfrac{(x - a)^m S(x)}{D(x)}$, where $S(a) \neq 0$ and $D(a) \neq 0$.

2. Near $x = a, f(x) \approx A(x - a)^m$, where A is the constant with $A = \dfrac{S(a)}{D(a)}$.

We read $f(x) \approx A(x - a)^m$ as "near $x = a, f(x)$ behaves like $A(x - a)^m$." This also implies that, near its zero a, the graph of $f(x)$ looks very much like the graph of the power function $A(x - a)^m$.

BEHAVIOR OF $f(x) = \dfrac{N(x)}{D(x)}$ **NEAR A VERTICAL ASYMPTOTE**

Assume f is in lowest terms and $x = a$ is a vertical asymptote, then

1. f can be factored as $f(x) = \dfrac{N(x)}{(x - a)^m Q(x)}$, where $N(a) \neq 0$ and $Q(a) \neq 0$.

2. Near $x = a, f(x) \approx \dfrac{A}{(x - a)^m}$, where A is the constant with $A = \dfrac{N(a)}{Q(a)}$.

We read $f(x) \approx \dfrac{A}{(x - a)^m}$ as "near $x = a, f(x)$ behaves like $\dfrac{A}{(x - a)^m}$." This also implies that, near the vertical asymptote $x = a$, the graph of $f(x)$ looks very much like the graph of the reciprocal function $\dfrac{A}{(x - a)^m}$.

Depending on the sign of the constant A and the multiplicity m (odd or even) of the zero, a, of $D(x)$ near its asymptote $x = a$, every rational function exhibits behavior similar to one of the four functions $y = \dfrac{1}{x - a}, y = -\dfrac{1}{x - a}, y = \dfrac{1}{(x - a)^2}, y = -\dfrac{1}{(x - a)^2}.$

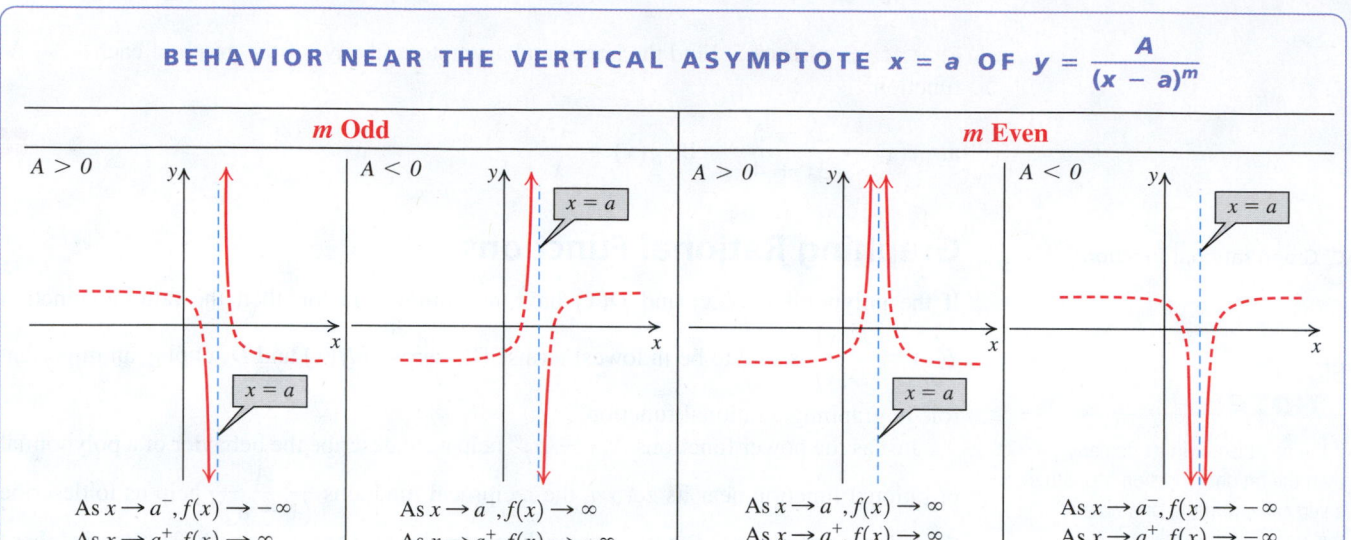

BEHAVIOR NEAR THE VERTICAL ASYMPTOTE $x = a$ OF $y = \dfrac{A}{(x - a)^m}$

m Odd		m Even	
$A > 0$	$A < 0$	$A > 0$	$A < 0$
As $x \to a^-, f(x) \to -\infty$ As $x \to a^+, f(x) \to \infty$	As $x \to a^-, f(x) \to \infty$ As $x \to a^+, f(x) \to -\infty$	As $x \to a^-, f(x) \to \infty$ As $x \to a^+, f(x) \to \infty$	As $x \to a^-, f(x) \to -\infty$ As $x \to a^+, f(x) \to -\infty$

In the next procedure, we assume the rational function is in lowest terms. Otherwise, reduce the rational function to lowest terms by factoring both the numerator $N(x)$ and denominator $D(x)$, cancelling any common factors, and plotting the "hole" (or holes) for the canceled factor (or factors).

PROCEDURE
IN ACTION

EXAMPLE 6 **Graphing a Rational Function**

OBJECTIVE

Graph $f(x) = \dfrac{N(x)}{D(x)}$, *where $f(x)$ is in lowest terms.*

Step 1 Find the intercepts. Since f is in lowest terms, the x-intercepts are found by solving the equation $N(x) = 0$. The y-intercept, if there is one, is $f(0)$.

EXAMPLE

Sketch the graph of: $f(x) = \dfrac{2x^2 - 2}{x^2 - 9}$

1.

$$2x^2 - 2 = 0 \qquad \text{Set } N(x) = 0.$$
$$2(x - 1)(x + 1) = 0 \qquad \text{Factor.}$$
$$x - 1 = 0 \quad \text{or} \quad x + 1 = 0 \qquad \text{Zero-product property}$$
$$x = 1 \quad \text{or} \qquad x = -1 \qquad \text{Solve for } x.$$

The x-intercepts are -1 and 1, so the graph passes through the points $(-1, 0)$ and $(1, 0)$.

$$f(0) = \frac{2(0)^2 - 2}{(0)^2 - 9} \qquad \text{Replace } x \text{ with } 0 \text{ in } f(x).$$

$$= \frac{2}{9} \qquad \text{Simplify.}$$

The y-intercept is $\dfrac{2}{9}$, so the graph of f passes through the point $\left(0, \dfrac{2}{9}\right)$.

Step 2 Find the vertical asymptotes (if any) and determine the behavior near the asymptotes. Solve $D(x) = 0$ to find the vertical asymptotes of the graph. Sketch the vertical asymptotes. Use the multiplicity of the zeros of $D(x)$ to determine the behavior of the rational function near the asymptotes.

2.
$$x^2 - 9 = 0 \qquad \text{Set } D(x) = 0.$$
$$(x - 3)(x + 3) = 0 \qquad \text{Factor.}$$
$$x - 3 = 0 \quad \text{or} \quad x + 3 = 0 \qquad \text{Zero-product property}$$
$$x = 3 \quad \text{or} \qquad x = -3 \qquad \text{Solve for } x.$$

The vertical asymptotes of the graph of f are the lines $x = -3$ and $x = 3$.

$$f(x) = \frac{2x^2 - 2}{x^2 - 9} = \frac{2x^2 - 2}{(x - 3)(x + 3)}.$$

Near the vertical asymptote $x = -3$ (the corresponding factor is $x + 3$), the graph of f looks like

$$f(x) = \frac{2x^2 - 2}{(x - 3)(x + 3)} \approx \frac{2(-3)^2 - 2}{(-3 - 3)(x + 3)} = -\frac{8/3}{(x + 3)}.$$

Near the vertical asymptote $x = 3$ (the corresponding factor is $x - 3$), the graph of f looks like

$$f(x) = \frac{2x^2 - 2}{(x - 3)(x + 3)} \approx \frac{2(3)^2 - 2}{(x - 3)(3 + 3)} = \frac{8/3}{(x + 3)}.$$

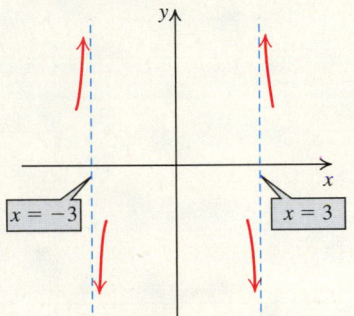

Step 3 Find the horizontal asymptote (if any). Use the rules on page 194 for finding the horizontal asymptote of a rational function.

3. Since $n = m = 2$, by Rule 2 the horizontal asymptote is

$$y = \frac{\text{Leading coefficient of } N(x)}{\text{Leading coefficient of } D(x)} = \frac{2}{1} = 2.$$

Step 4 Test for symmetry. If $f(-x) = f(x)$, then f is symmetric with respect to the y-axis. If $f(-x) = -f(x)$, then f is symmetric with respect to the origin.

4. $f(-x) = \dfrac{2(-x)^2 - 2}{(-x)^2 - 9} = \dfrac{2x^2 - 2}{x^2 - 9} = f(x)$. The graph of f is symmetric about the y-axis. This is the only symmetry.

Step 5 Locate the graph relative to the horizontal asymptote (if any). If $y = k$ is a horizontal asymptote, divide $N(x)$ by $D(x)$ and write $f(x) = \dfrac{N(x)}{D(x)} = k + \dfrac{R(x)}{D(x)}$. Use a sign graph for $f(x) - k = \dfrac{R(x)}{D(x)}$ and "test" numbers associated with the zeros of $R(x)$ and $D(x)$ to determine intervals where the graph of f is above the asymptote $y = k$ ($f(x) - k$ is *positive*) and where it is below $y = k$ ($f(x) - k$ is *negative*). The graph of f intersects the line $y = k$ if $f(x) - k = 0$.

5. $f(x) = \dfrac{2x^2 - 2}{x^2 - 9} = 2 + \dfrac{16}{x^2 - 9}$; $R(x) = 16$ has no zeros and $D(x)$ has zeros -3 and 3. These zeros divide the x-axis into three intervals (see the figure). We choose test points -4, 0, and 4 to test the sign of $\dfrac{16}{x^2 - 9}$.

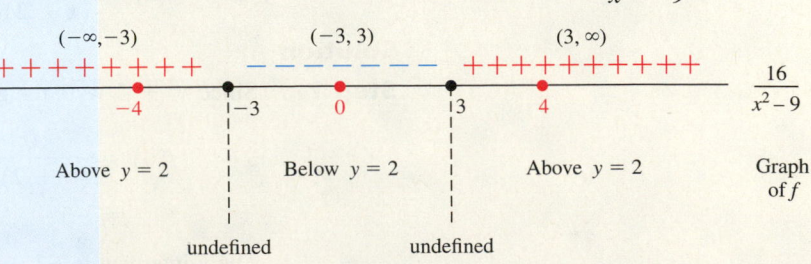

We now have the following information on the end behavior of the rational function f.

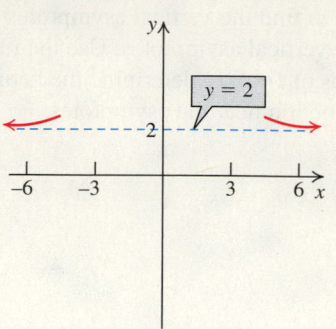

Step 6 Sketch the graph. Plot some points and graph the asymptotes found in Steps 1–5; use symmetry to sketch the graph of f.

6. We can sketch the graph of the rational function f by connecting the pieces of information we gathered above with a smooth curve.

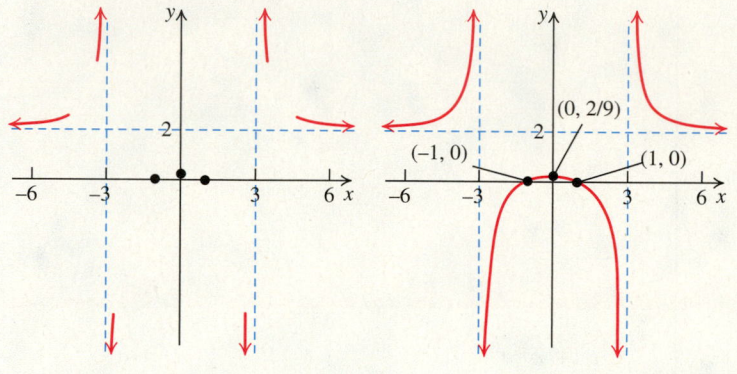

Practice Problem 6 Sketch the graph of $f(x) = \dfrac{2x}{x^2 - 1}$.

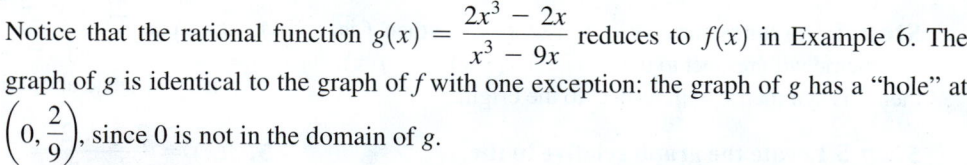

Notice that the rational function $g(x) = \dfrac{2x^3 - 2x}{x^3 - 9x}$ reduces to $f(x)$ in Example 6. The graph of g is identical to the graph of f with one exception: the graph of g has a "hole" at $\left(0, \dfrac{2}{9}\right)$, since 0 is not in the domain of g.

EXAMPLE 7   Graphing a Rational Function

Sketch the graph of $f(x) = \dfrac{x^2 + 2}{(x + 2)(x - 1)}$.

Solution

Step 1   Since $x^2 + 2 > 0$, the graph has no x-intercepts.

$$f(0) = \frac{0^2 + 2}{(0 + 2)(0 - 1)} \qquad \text{Replace } x \text{ with 0 in } f(x).$$

$$= -1 \qquad \text{Simplify.}$$

The y-intercept is -1.

Step 2   Set $(x + 2)(x - 1) = 0$. Solving for x, we have $x = -2$ or $x = 1$. The vertical asymptotes are the lines $x = -2$ and $x = 1$.

Near the vertical asymptote $x = -2$ (the corresponding factor is $x + 2$), the graph of f looks like

$$f(x) = \frac{x^2 + 2}{(x + 2)(x - 1)} \approx \frac{(-2)^2 + 2}{(x + 2)(-2 - 1)} = -\frac{6}{3(x + 2)} = -\frac{2}{(x + 2)}.$$

Near the vertical asymptote $x = 1$ (the corresponding factor is $x - 1$), the graph of f looks like (see Figure 2.30)

$$f(x) = \frac{x^2 + 2}{(x + 2)(x - 1)} \approx \frac{(1)^2 + 2}{(1 + 2)(x - 1)} = \frac{3}{3(x - 1)} = \frac{1}{(x - 1)}.$$

Figure 2.30

Step 3 $f(x) = \dfrac{x^2 + 2}{x^2 + x - 2}$. By Rule 2, page 194, the horizontal asymptote is $y = \dfrac{1}{1} = 1$ because the leading coefficient for both the numerator and the denominator is 1.

Step 4   **Symmetry.** None

Step 5 $f(x) = \dfrac{x^2 + 2}{(x + 2)(x - 1)} = 1 + \dfrac{4 - x}{(x + 2)(x - 1)}$; $R(x) = 4 - x$ has one zero, 4, and $D(x) = (x + 2)(x - 1)$ has two zeros, -2 and 1. These zeros divide the x-axis into four intervals (see the figure). We choose test points $-3, 0, 2,$ and 5 to test the sign of $\dfrac{4 - x}{(x + 2)(x - 1)}$.

$$f(x) - 1 = \frac{4 - x}{(x + 2)(x - 1)}$$

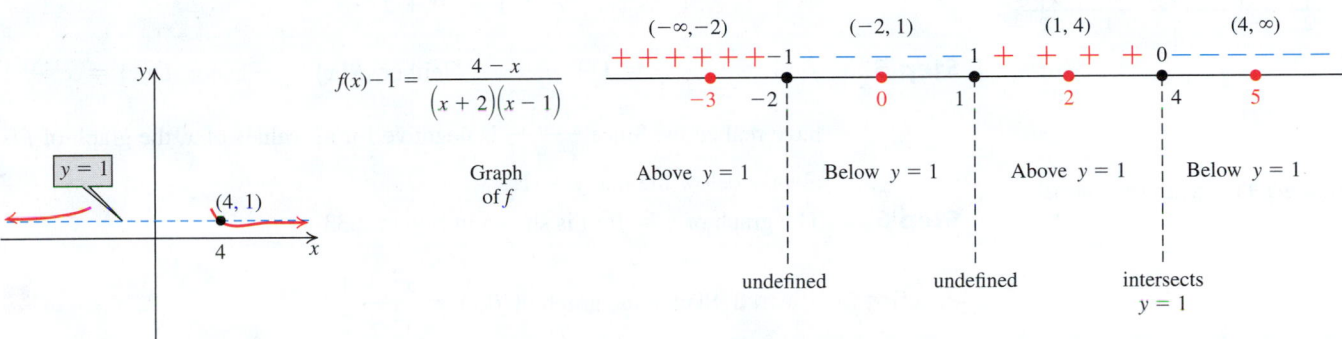

Graph of f

Figure 2.31

We now have the following information on the end behavior of the rational function f. Notice that the graph of f crosses the horizontal asymptote at the point $(4, 1)$. (See Figure 2.31.)

Step 6   We can draw the graph of the rational function f by connecting the pieces of information we gathered above with a smooth curve. (see Figure 2.32.)

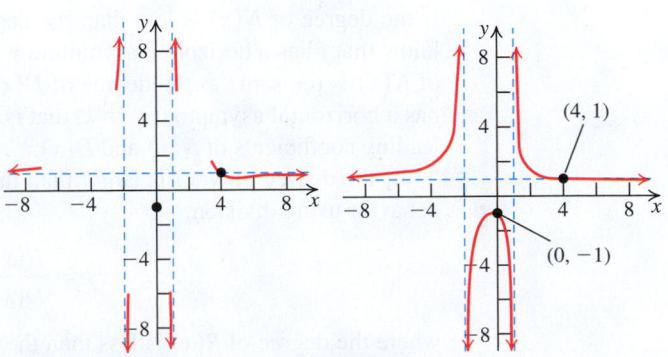

Figure 2.32 Graph crossing horizontal asymptote.

Practice Problem 7 Sketch the graph of $f(x) = \dfrac{2x^2 - 1}{2x^2 + x - 3}$.

We know that the graphs of polynomial functions are continuous (all in one piece). The next example shows that the graph of a rational function (other than a polynomial function) can also be continuous.

EXAMPLE 8   Graphing a Rational Function

Sketch a graph of $f(x) = \dfrac{x^2}{x^2 + 1}$.

Solution

Step 1   Now $f(0) = 0$ and solving $f(x) = 0$ gives $x = 0$. Therefore, 0 is both the x-intercept and the y-intercept for the graph of f.

Step 2   Because $x^2 + 1 > 0$ for all x, there are no real zeros for the denominator, so there are no vertical asymptotes.

Step 3   By Rule 2 for locating horizontal asymptotes, the horizontal asymptote is $y = 1$.

Step 4 $f(-x) = \dfrac{(-x)^2}{(-x)^2 + 1} = \dfrac{x^2}{x^2 + 1} = f(x)$. The graph is symmetric with respect to the y-axis.

         Since the function f is symmetric with respect to the y-axis and $x = 0$ is a zero of f, note the behavior of f near $x = 0$. Using Rule 2 on page 195, we find that near $x = 0$ the graph of f looks like

$$f(x) = \frac{x^2}{x^2 + 1} \approx \frac{x^2}{0 + 1} \approx x^2.$$

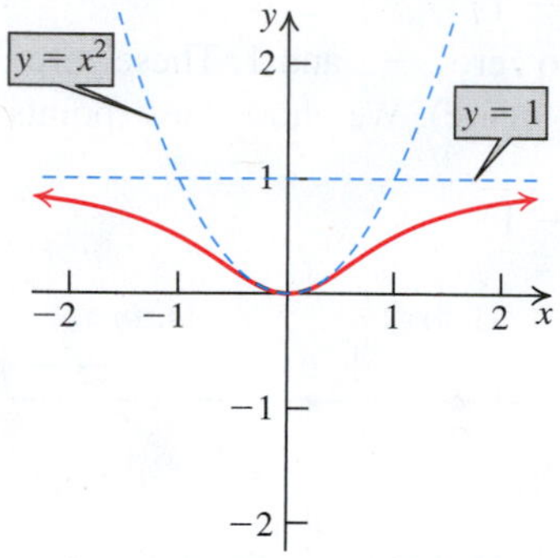

Step 5 $f(x) = \dfrac{x^2}{x^2 + 1} = 1 + \dfrac{-1}{x^2 + 1}$. Neither $R(x) = -1$ nor $D(x) = x^2 + 1$ have real zeros. Since $\dfrac{-1}{x^2 + 1}$ is negative for all values of x, the graph of f is always below the line $y = 1$.

Step 6   The graph of $y = f(x)$ is shown in Figure 2.33.

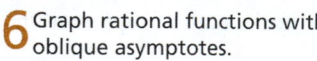

Figure 2.33 A continuous rational function.

Practice Problem 8   Sketch the graph of $f(x) = \dfrac{x^2 + 1}{x^2 + 2}$.

6   Graph rational functions with oblique asymptotes.

Oblique Asymptotes and End Behavior

To describe the end behavior of the rational function f we return to the representation

$$f(x) = \frac{N(x)}{D(x)}.$$

If the degree of $N(x)$ is less than the degree of $D(x)$, then from Rule 1 on page 194 we know that f has a horizontal asymptote $y = 0$, that is, $f(x) \to 0$ as $|x| \to \infty$. If the degree of $N(x)$ is the same as the degree of $D(x)$, then from Rule 2 on page 194 we know that f has a horizontal asymptote $y = k$, that is $f(x) \to k$, as $|x| \to \infty$ where k is the ratio of the leading coefficients of $N(x)$ and $D(x)$.

If the degree of $N(x)$ is larger than the degree of $D(x)$, we can obtain the precise end behavior using division:

$$f(x) = \frac{N(x)}{D(x)} = Q(x) + \frac{R(x)}{D(x)},$$

where the degree of $R(x)$ is less than the degree of $D(x)$. Rule 1 on page 194 tells us that $\dfrac{R(x)}{D(x)}$ has a horizontal asymptote $y = 0$, that is, $\dfrac{R(x)}{D(x)} \to 0$, as $|x| \to \infty$. Therefore, as $x \to \pm\infty$,

$$f(x) \approx Q(x).$$

That means that as $|x|$ gets very large, the graph of f behaves like the graph of the polynomial $Q(x)$.

END BEHAVIOR OF THE RATIONAL FUNCTION $f(x) = \dfrac{N(x)}{D(x)}$

1. If degree of $N(x) <$ degree of $D(x)$, then $f(x) \approx 0$ ($y = 0$ is a horizontal asymptote).

2. If degree of $N(x) =$ degree of $D(x)$, then
$$f(x) = \frac{a_n x^n + \cdots}{b_n x^n + \cdots} \approx \frac{a_n}{b_n} \qquad \left(y = \frac{a_n}{b_n} \text{ is a horizontal asymptote}\right).$$

3. If degree of $N(x) >$ degree of $D(x)$, then $f(x) = Q(x) + \dfrac{R(x)}{D(x)} \approx Q(x)$.

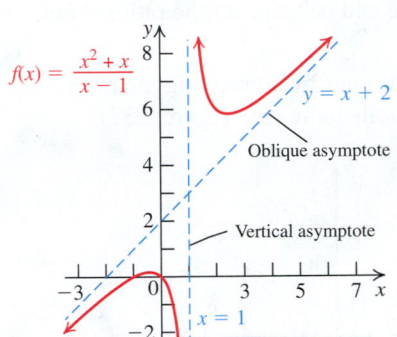

$f(x) = \dfrac{x^2 + x}{x - 1}$

$y = x + 2$

Oblique asymptote

Vertical asymptote

$x = 1$

Figure 2.34 Graph with vertical and oblique asymptotes.

If the degree of $N(x)$ is exactly one more than the degree of $D(x)$, then $Q(x)$ will have the linear form $mx + b$. In this case, the graph of f is said to have an **oblique** (or **slant**) asymptote. Consider, for example, the function

$$f(x) = \frac{x^2 + x}{x - 1} = x + 2 + \frac{2}{x - 1}. \qquad \text{Use either long or synthetic division.}$$

Now as $x \to \pm\infty$, the expression $\dfrac{2}{x - 1} \to 0$ so that the graph of f approaches the graph of the oblique asymptote—the line $y = x + 2$, as shown in Figure 2.34. The graph of f is above the line $y = x + 2$ on $(1, \infty)$ and below it on $(-\infty, 1)$.

EXAMPLE 9 **Graphing a Rational Function with an Oblique Asymptote**

Sketch the graph of $f(x) = \dfrac{x^2 - 4}{x + 1}$.

Solution

Note that $f(x)$ is in lowest terms.

Step 1 **Intercepts:** $f(0) = \dfrac{0 - 4}{0 + 1} = -4$, so the y-intercept is -4. Set $x^2 - 4 = 0$. We have $x = -2$ and $x = 2$, so the x-intercepts are -2 and 2.

Step 2 **Vertical asymptotes:** Set $x + 1 = 0$. We get $x = -1$, so the line $x = -1$ is a vertical asymptote.

Near the vertical asymptote $x = -1$ (the corresponding factor is $x + 1$), the graph of f looks like (see Figure 2.35)
$$f(x) = \frac{x^2 - 4}{x + 1} \approx \frac{(-1)^2 - 4}{x + 1} = -\frac{3}{x + 1}.$$

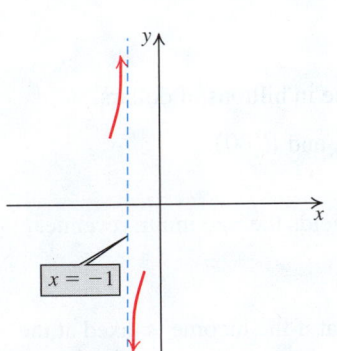

$x = -1$

Figure 2.35

Step 3 **Asymptotes:** Since the degree of the numerator is greater than the degree of the denominator, $f(x)$ has no horizontal asymptote. However, by long division,
$$f(x) = \frac{x^2 - 4}{x + 1} = x - 1 + \frac{-3}{x + 1} = x - 1 - \frac{3}{x + 1}.$$

As $x \to \pm\infty$, the expression $\dfrac{3}{x + 1} \to 0$, so the line $y = x - 1$ is an oblique asymptote: the graph gets close to the line $y = x - 1$ as $x \to \infty$ and as $x \to -\infty$.

Step 4 **Symmetry:** You should check that there is no y-axis or origin symmetry.

Step 5 **Asymptote:** Use the intervals determined by the zeros of the numerator and of the denominator of $f(x) - (x - 1) = \dfrac{x^2 - 4}{x + 1} - (x - 1) = \dfrac{-3}{x + 1}$

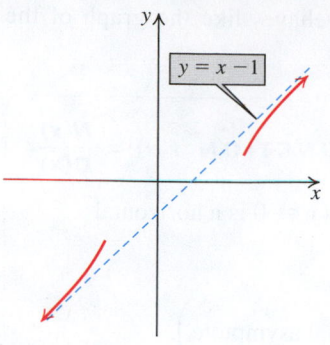

Figure 2.36

to create a sign graph. The numerator -3 has no zero and denominator $x + 1$ has one zero, -1. The zero, -1, divides the x-axis into two intervals (see the figure). We choose test points -3, and 0 to test the sign of

$$f(x) - (x - 1) = \frac{-3}{x + 1}.$$

The graph is above the line $y = x - 1$ on $(-\infty, 1)$ and below it on $(-1, \infty)$.

Together we get the information on the end behavior of the rational function f (see Figure 2.36).

Step 6 We can draw the graph of the rational function f by connecting the pieces of information we gathered above with a smooth curve (see Figure 2.37).

Figure 2.37

Practice Problem 9 Sketch the graph of $f(x) = \dfrac{x^2 + 2}{x - 1}$.

7 Graph a revenue curve.

Graph of a Revenue Curve

EXAMPLE 10 **Graphing a Revenue Curve**

The revenue curve for an economy of a country is given by

$$R(x) = \frac{x(100 - x)}{x + 10},$$

where x is the tax rate in percent and $R(x)$ is the tax revenue in billions of dollars.

a. Find and interpret $R(10)$, $R(20)$, $R(30)$, $R(40)$, $R(50)$, and $R(60)$.

b. Sketch the graph of $y = R(x)$ for $0 \le x \le 100$.

c. Use a graphing calculator to estimate the tax rate that yields the maximum revenue.

Solution

a. $R(10) = \dfrac{10(100 - 10)}{10 + 10} = \$45$ billion. This means that if the income is taxed at the rate of 10%, the total revenue for the government will be \$45 billion.

Similarly, $R(20) \approx \$53.3$ billion,

$$R(30) = \$52.5 \text{ billion,}$$
$$R(40) = \$48 \text{ billion,}$$
$$R(50) \approx \$41.7 \text{ billion, and}$$
$$R(60) \approx \$34.3 \text{ billion.}$$

TECHNOLOGY
CONNECTION

TRACE

$$y = \frac{100x - x^2}{x + 10}$$

to approximate the maximum revenue.

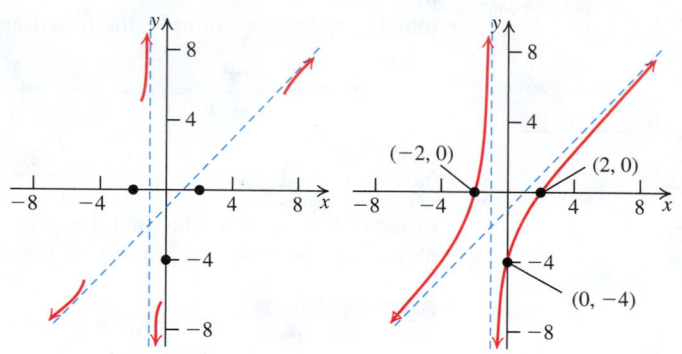

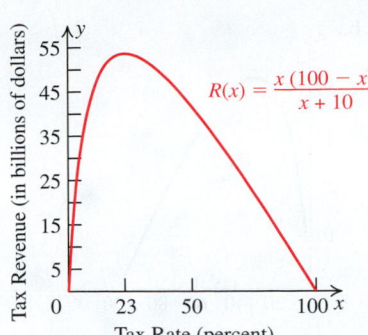

Figure 2.38

b. The graph of the function $y = R(x)$ for $0 \leq x \leq 100$ is shown in Figure 2.38.

c. From the calculator graph of

$$Y = \frac{100x - x^2}{x + 10},$$

by using the TRACE feature, you can see that the tax rate of about 23% produces the maximum tax revenue of about $53.7 billion for the government.

Practice Problem 10 Repeat Example 10 for $R(x) = \dfrac{x(100 - x)}{x + 20}$.

Answers to Practice Problems

1. Domain: $(-\infty, -1) \cup (-1, 5) \cup (5, \infty)$

2. a. Domain: $(-\infty, 2) \cup (2, \infty)$; Range: $(-\infty, 0) \cup (0, \infty)$; Vertical asymptote: $x = 2$; Horizontal asymptote: $y = 0$

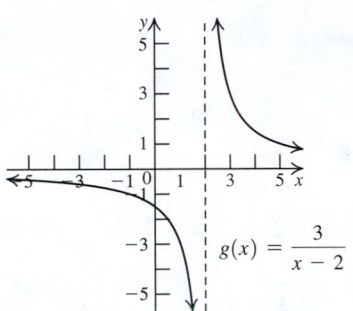

b. Domain: $(-\infty, -1) \cup (-1, \infty)$; Range: $(-\infty, 2) \cup (2, \infty)$; Vertical asymptote: $x = -1$; Horizontal asymptote: $y = 2$

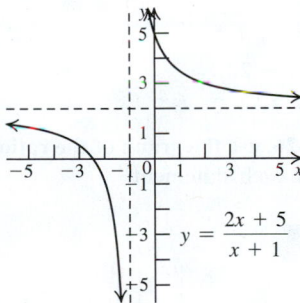

3. $x = -5$ and $x = 2$ **4.** $x = -3$ **5.a.** $y = \dfrac{2}{3}$ **b.** No horizontal asymptote **6.** x-intercept: 0; y-intercept: 0; vertical asymptotes: $x = -1$ and $x = 1$; Horizontal asymptote: x-axis. The graph is symmetric with respect to the origin. The graph is above the x-axis on $(-1, 0) \cup (1, \infty)$ and below the x-axis on $(-\infty, -1) \cup (0, 1)$.

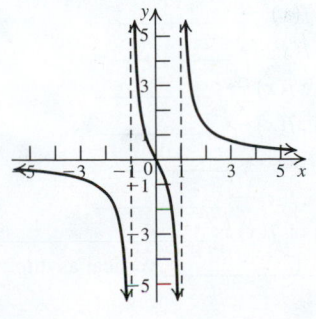

7. x-intercepts: $\pm \dfrac{\sqrt{2}}{2}$; y-intercept: $\dfrac{1}{3}$; vertical asymptotes: $x = -\dfrac{3}{2}$ and $x = 1$; horizontal asymptote: $y = 1$; symmetry: none. The graph is above the line $y = 1$ on $\left(-\infty, -\dfrac{3}{2}\right) \cup (1, 2)$ and below it on $\left(-\dfrac{3}{2}, 1\right) \cup (2, \infty)$. The graph crosses the horizontal asymptote at $(2, 1)$.

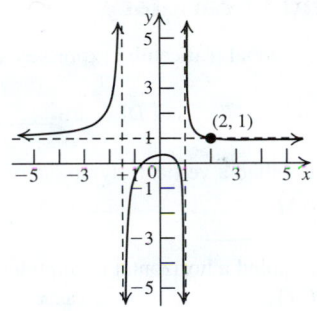

8. No x-intercept; y-intercept: $\dfrac{1}{2}$; no vertical asymptote; horizontal asymptote: $y = 1$. The graph is symmetric with respect to the y-axis. The graph is below the line $y = 1$ on $(-\infty, \infty)$.

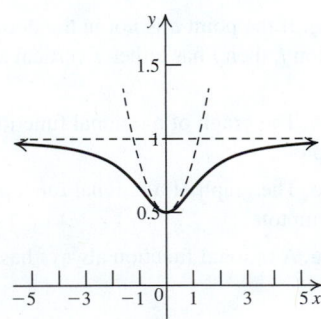

9. No x-intercept; y-intercept: -2; vertical asymptote: $x = 1$; no horizontal asymptote; $y = x + 1$ is an oblique asymptote. Symmetry: none. The graph is above the line $y = x + 1$ on $(1, \infty)$ and below it on $(-\infty, 1)$.

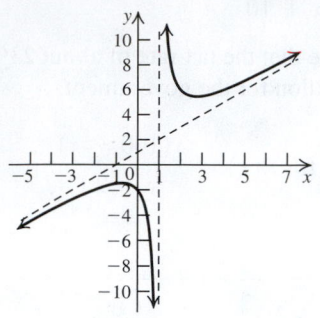

10. a. $R(10) = \$30$ billion
$R(20) = \$40$ billion
$R(30) = \$42$ billion
$R(40) = \$40$ billion
$R(50) = \$35.7$ billion
$R(60) = \$30$ billion

b.

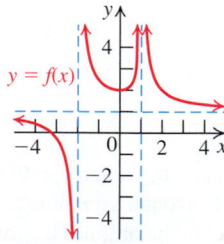

c. 29%

SECTION 2.4 **Exercises**

Concepts and Vocabulary

1. A function f is rational if it can be expressed as a quotient

$f(x) = \dfrac{N(x)}{D(x)}$ where $N(x)$ and $D(x)$ are _____.

2. A line $x = a$ is called a vertical asymptote of the rational function f if $|f(x)| \to$ _____ as $x \to$ _____ or as $x \to$ _____.

3. A line $y = k$ is called a horizontal asymptote of the rational function f if $|f(x)| \to$ _____ as $x \to$ _____ or as $x \to$ _____.

4. If a is a zero of the denominator $D(x)$ of the rational function

$f(x) = \dfrac{N(x)}{D(x)}$ expressed in lowest terms, then _____

is a _____ asymptote of f.

5. True or False. If the point a is not in the domain of the rational function f, then f has either a vertical asymptote or a hole at a.

6. True or False. The graph of a rational function can cross its vertical asymptote.

7. True or False. The graph of a rational function can cross its horizontal asymptote.

8. True or False. A rational function always has a horizontal asymptote.

Building Skills

In Exercises 9–16, find the domain of each rational function.

9. $f(x) = \dfrac{x - 3}{x + 4}$

10. $f(x) = \dfrac{x + 1}{x - 1}$

11. $g(x) = \dfrac{x - 1}{x^2 + 1}$

12. $g(x) = \dfrac{x + 2}{x^2 + 4}$

13. $h(x) = \dfrac{x - 3}{x^2 - x - 6}$

14. $h(x) = \dfrac{x - 7}{x^2 - 6x - 7}$

15. $F(x) = \dfrac{2x + 3}{x^2 - 6x + 8}$

16. $F(x) = \dfrac{3x - 2}{x^2 - 3x + 2}$

In Exercises 17–26, use the graph of the rational function $f(x)$ to complete each statement.

17. As $x \to 1^+$, $f(x) \to$ _____.

18. As $x \to 1^-$, $f(x) \to$ _____.

19. As $x \to -2^+$, $f(x) \to$ _____.

20. As $x \to -2^-$, $f(x) \to$ _____.

21. As $x \to \infty$, $f(x) \to$ _____.

22. As $x \to -\infty$, $f(x) \to$ _____.

23. The domain of $f(x)$ is _____.

24. There are _____ vertical asymptotes.

25. The equations of the vertical asymptotes of the graph are _____ and _____.

26. The equation of the horizontal asymptote of the graph is _____.

In Exercises 27–34, graph each rational function as a translation of $f(x) = \dfrac{1}{x}$. Identify the vertical and horizontal asymptotes, and state the domain and range.

27. $f(x) = \dfrac{3}{x - 4}$

28. $g(x) = \dfrac{-4}{x + 3}$

29. $f(x) = \dfrac{-x}{3x + 1}$

30. $g(x) = \dfrac{2x}{4x - 2}$

31. $g(x) = \dfrac{-3x + 2}{x + 2}$

32. $f(x) = \dfrac{-x + 1}{x - 3}$

33. $g(x) = \dfrac{5x - 3}{x - 4}$

34. $f(x) = \dfrac{2x + 12}{x + 5}$

In Exercises 35–44, find the vertical asymptotes, if any, of the graph of each rational function.

35. $f(x) = \dfrac{x}{x - 1}$

36. $f(x) = \dfrac{x + 3}{x - 2}$

37. $g(x) = \dfrac{(x + 1)(2x - 2)}{(x - 3)(x + 4)}$

38. $g(x) = \dfrac{(2x - 1)(x + 2)}{(2x + 3)(3x - 4)}$

39. $h(x) = \dfrac{x^2 - 1}{x^2 + x - 6}$

40. $h(x) = \dfrac{x^2 - 4}{3x^2 + x - 4}$

41. $f(x) = \dfrac{x^2 - 6x + 8}{x^2 - x - 12}$

42. $f(x) = \dfrac{x^2 - 9}{x^3 - 4x}$

43. $g(x) = \dfrac{2x + 1}{x^2 + x + 1}$

44. $g(x) = \dfrac{x^2 - 36}{x^2 + 5x + 9}$

In Exercises 45–52, find the horizontal asymptote, if any, of the graph of each rational function.

45. $f(x) = \dfrac{x + 1}{x^2 + 5}$

46. $f(x) = \dfrac{2x - 1}{x^2 - 4}$

47. $g(x) = \dfrac{2x - 3}{3x + 5}$

48. $g(x) = \dfrac{3x + 4}{-4x + 5}$

49. $h(x) = \dfrac{x^2 - 49}{x + 7}$

50. $h(x) = \dfrac{x + 3}{x^2 - 9}$

51. $f(x) = \dfrac{2x^2 - 3x + 7}{3x^3 + 5x + 11}$

52. $f(x) = \dfrac{3x^3 + 2}{x^2 + 5x + 11}$

In Exercises 53–58, match the rational function with its graph.

53. $f(x) = \dfrac{2}{x - 3}$

54. $f(x) = \dfrac{x - 2}{x + 3}$

55. $f(x) = \dfrac{1}{x^2 - 2x}$

56. $f(x) = \dfrac{x}{x^2 + 1}$

57. $f(x) = \dfrac{x^2 + 2x}{x - 3}$

58. $f(x) = \dfrac{x^2}{x^2 - 4}$

a.

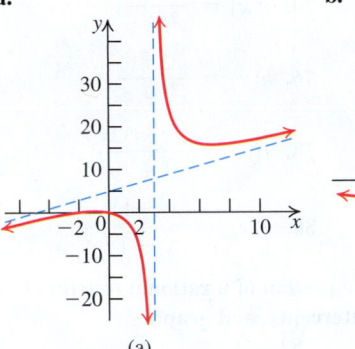

(a)

b.

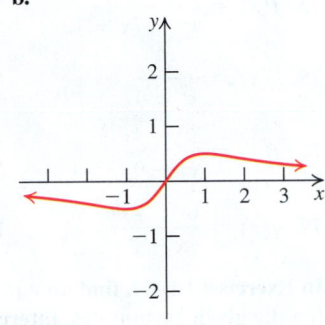

(b)

c.

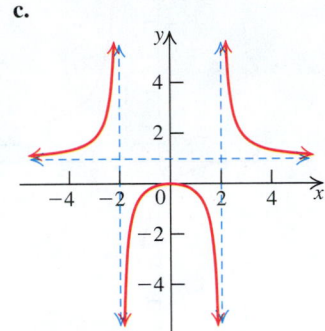

(c)

d.

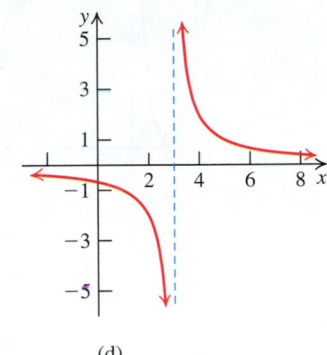

(d)

e.

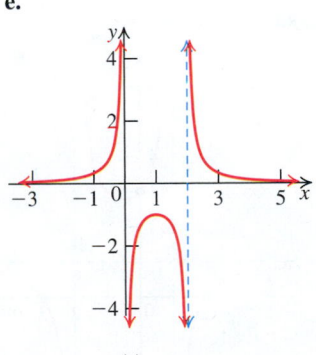

(e)

f.

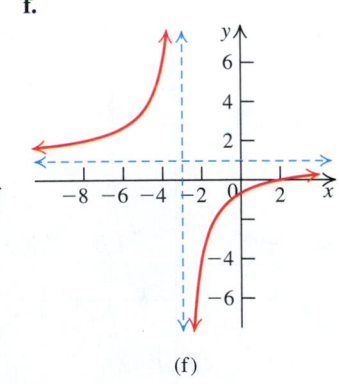

(f)

In Exercises 59–80, use the six-step procedure on pages 196–198 to graph each rational function.

59. $f(x) = \dfrac{2x}{x - 3}$

60. $f(x) = \dfrac{-x}{x - 1}$

61. $f(x) = \dfrac{x}{x^2 - 4}$

62. $f(x) = \dfrac{x}{1 - x^2}$

63. $h(x) = \dfrac{-2x^2}{x^2 - 9}$

64. $h(x) = \dfrac{4 - x^2}{x^2}$

65. $f(x) = \dfrac{2}{x^2 - 2}$

66. $f(x) = \dfrac{-2}{x^2 - 3}$

67. $g(x) = \dfrac{x + 1}{(x - 2)(x + 3)}$

68. $g(x) = \dfrac{x - 1}{(x + 1)(x - 2)}$

69. $f(x) = \dfrac{2x}{x^2 + 3x + 2}$

70. $f(x) = \dfrac{2x}{x^2 - 6x + 8}$

71. $f(x) = \dfrac{x - 1}{x^2(x + 1)}$

72. $f(x) = \dfrac{x - 1}{x(x + 1)^2}$

73. $f(x) = \dfrac{x - 3}{x^2 - 7x + 12}$

74. $f(x) = \dfrac{x + 2}{x^2 - x - 6}$

75. $h(x) = \dfrac{x^2}{x^2 + 1}$

76. $h(x) = \dfrac{2x^2}{x^2 + 4}$

77. $f(x) = \dfrac{x^3 - 4x}{x^3 - 9x}$

78. $f(x) = \dfrac{x^3 + 32x}{x^3 + 8x}$

79. $g(x) = \dfrac{(x - 2)^2}{x - 2}$

80. $g(x) = \dfrac{(x - 1)^2}{x - 1}$

In Exercises 81–84, find an equation of a rational function having the given asymptotes, intercepts, and graph.

81.

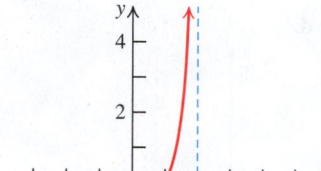

82.

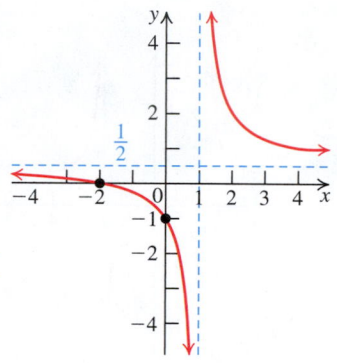

83.

84.

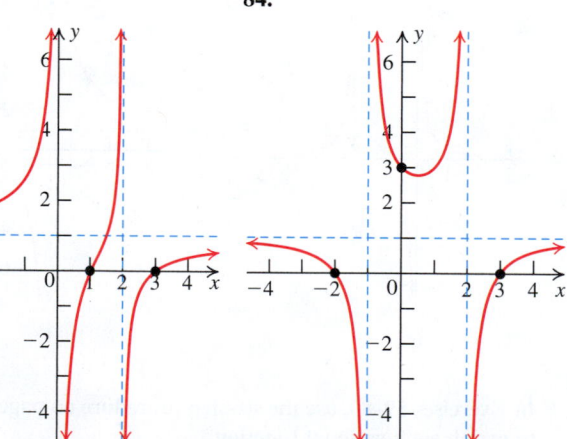

In Exercises 85–92, find the oblique asymptote and sketch the graph of each rational function.

85. $f(x) = \dfrac{2x^2 + 1}{x}$

86. $f(x) = \dfrac{x^2 - 1}{x}$

87. $g(x) = \dfrac{x^3 - 1}{x^2}$

88. $g(x) = \dfrac{2x^3 + x^2 + 1}{x^2}$

89. $h(x) = \dfrac{x^2 - x + 1}{x + 1}$

90. $h(x) = \dfrac{2x^2 - 3x + 2}{x - 1}$

91. $f(x) = \dfrac{x^3 - 2x^2 + 1}{x^2 - 1}$

92. $h(x) = \dfrac{x^3 - 1}{x^2 - 4}$

Applying the Concepts

For Exercises 93 and 94, use the following definition: Given a cost function C, the average cost of producing the first x items is found by the formula:

$$\overline{C}(x) = \dfrac{C(x)}{x}, x > 0.$$

93. Average cost. The Genuine Trinket Co. manufactures "authentic" trinkets for gullible tourists. Fixed daily costs are $2000, and it costs $0.50 to produce each trinket.
  a. Write the cost function C for producing x trinkets.
  b. Write the average cost $\overline{C}$ of producing x trinkets.
  c. Find and interpret $\overline{C}(100)$, $\overline{C}(500)$, and $\overline{C}(1000)$.
  d. Find and interpret the horizontal asymptote of the rational function $\overline{C}(x)$.

94. Average cost. The monthly cost C of producing x portable CD players is given by

$$C(x) = -0.002x^2 + 6x + 7000.$$

  a. Write the average cost function $\overline{C}(x)$.
  b. Find and interpret $\overline{C}(100)$, $\overline{C}(500)$, and $\overline{C}(1000)$.
  c. Find and interpret the oblique asymptote of $\overline{C}(x)$.

95. Biology: birds collecting seeds. A bird is collecting seed from a field that contains 100 grams of seed. The bird collects x grams of seed in t minutes, where

$$t = f(x) = \dfrac{4x + 1}{100 - x}, 0 < x < 100.$$

  a. Sketch the graph of $t = f(x)$.
  b. How long does it take the bird to collect
    (i) 50 grams?
    (ii) 75 grams?
    (iii) 95 grams?
    (iv) 99 grams?
  c. Complete the following statements if applicable:
    (i) As $x \to 100^-$, $f(x) \to$ _____.
    (ii) As $x \to 100^+$, $f(x) \to$ _____.
  d. Does the bird ever collect all of the seed from the field?

96. Criminology. Suppose the city of Las Vegas decides to be crime-free. The estimated cost of catching and convicting $x\%$ of the criminals is given by

$$C(x) = \dfrac{1000}{100 - x} \text{ million dollars.}$$

  a. Find and interpret $C(50)$, $C(75)$, $C(90)$, and $C(99)$.
  b. Sketch a graph of $C(x)$, $0 \le x < 100$.
  c. What happens to $C(x)$ as $x \to 100^-$?
  d. What percentage of the criminals can be caught and convicted for $30 million?

97. Environment. Suppose an environmental agency decides to get rid of the impurities from the water of a polluted river. The estimated cost of removing $x\%$ of the impurities is given by

$$C(x) = \dfrac{3x^2 + 50}{x(100 - x)} \text{ billion dollars.}$$

  a. How much will it cost to remove 50% of the impurities?

b. Sketch the graph of $y = C(x)$.

c. Estimate the percentage of the impurities that can be removed at a cost of $30 billion.

98. Biology. The growth function $g(x)$ describes the growth rate of organisms as a function of some nutrient concentration x. Suppose

$$g(x) = a\frac{x}{k + x}, \quad x \geq 0,$$

where a and k are positive constants.

a. Find the horizontal asymptote of the graph of $g(x)$. Use the asymptote to explain why a is called the *saturation level* of the nutrient.

b. Show that k is the half-saturation constant; that is, show that if $x = k$, then $g(x) = \frac{a}{2}$.

99. Population of bacteria. The population P (in thousands) of a colony of bacteria at time t (in hours) is given by

$$P(t) = \frac{8t + 16}{2t + 1}, \quad t \geq 0.$$

a. Find the initial population of the colony. (Find the population at $t = 0$ hours.)

b. What is the long-term behavior of the population? (What happens when $t \to \infty$?)

100. Drug concentration. The concentration c (in milligrams per liter) of a drug in the bloodstream of a patient at time $t \geq 0$ (in hours since the drug was injected) is given by

$$c(t) = \frac{5t}{t^2 + 1}, \quad t \geq 0.$$

a. By plotting points or by using a graphing calculator, sketch the graph of $y = c(t)$.

b. Find and interpret the horizontal asymptote of the graph of $y = c(t)$.

c. Find the approximate time when the concentration of drug in the bloodstream is maximal.

d. At what time is the concentration level equal to 2 milligrams per liter?

101. Book publishing. The printing and binding cost for a college algebra book is $10. The editorial cost is $200,000. The first 2500 books are samples and are given free to professors. Let x be the number of college algebra books produced.

a. Write a function f describing the average cost of saleable books.

b. Find the average cost of a saleable book if 10,000 books are produced.

c. How many books must be produced to bring the average cost of a saleable book under $20?

d. Find the vertical and horizontal asymptotes of the graph of $y = f(x)$. How are they related to this situation?

102. A "phony" sale. A jewelry merchant at a local mall wants to have a "p percent off" sale. She marks up the stock by q percent to break even with respect to the prices before the sale.

a. Show that $q = \dfrac{p}{1 - p}$, $0 < p < 1$.

b. Sketch the graph of q in part **(a)**.

c. How much should she mark up the stock to break even if she wants to have a "25% off" sale?

Beyond the Basics

In Exercises 103–112, use transformations of the graph of $y = \dfrac{1}{x}$ or $y = \dfrac{1}{x^2}$ to graph each rational function $f(x)$.

103. $f(x) = -\dfrac{2}{x}$

104. $f(x) = \dfrac{1}{2 - x}$

105. $f(x) = \dfrac{1}{(x - 2)^2}$

106. $f(x) = \dfrac{1}{x^2 + 2x + 1}$

107. $f(x) = \dfrac{1}{(x - 1)^2} - 2$

108. $f(x) = \dfrac{1}{(x + 2)^2} + 3$

109. $f(x) = \dfrac{1}{x^2 + 12x + 36}$

110. $f(x) = \dfrac{3x^2 + 18x + 28}{x^2 + 6x + 9}$

111. $f(x) = \dfrac{x^2 - 2x + 2}{x^2 - 2x + 1}$

112. $f(x) = \dfrac{2x^2 + 4x - 3}{x^2 + 2x + 1}$

113. Graphing the reciprocal of a polynomial function. Let $f(x)$ be a polynomial function and $g(x) = \dfrac{1}{f(x)}$. Justify the following statements about the graphs of $f(x)$ and $g(x)$:

a. If c is a zero of $f(x)$, then $x = c$ is a vertical asymptote of the graph of $g(x)$.

b. If $f(x) > 0$, then $g(x) > 0$, and if $f(x) < 0$, then $g(x) < 0$.

c. The graphs of f and g intersect for those values (and only those values) of x for which $f(x) = g(x) = \pm 1$.

d. When f is increasing (decreasing, constant) on an interval of its domain, g is decreasing (increasing, constant) on that interval.

114. Use Exercise 113 to sketch the graphs of $f(x) = x^2 - 4$ and $g(x) = \dfrac{1}{x^2 - 4}$ on the same coordinate axes.

115. Recall that the inverse of a function $f(x)$ is written as $f^{-1}(x)$, while the reciprocal of $f(x)$ can be written as either $\dfrac{1}{f(x)}$ or $[f(x)]^{-1}$. The point of this exercise is to make clear that the reciprocal of a function has nothing to do with the inverse of a function. Let $f(x) = 2x + 3$. Find both $[f(x)]^{-1}$ and $f^{-1}(x)$. Compare the two functions. Graph all three functions on the same coordinate axes.

116. Let $f(x) = \dfrac{x - 1}{x + 2}$. Find $[f(x)]^{-1}$ and $f^{-1}(x)$. Graph all three functions on the same coordinate axes.

117. Sketch the graph of $g(x) = \dfrac{2x^3 + 3x^2 + 2x - 4}{x^2 - 1}$. Discuss the end behavior of $g(x)$.

118. Sketch the graph of $f(x) = \dfrac{x^3 - 2x^2 + 1}{x - 2}$. Discuss the end behavior of $f(x)$.

In Exercises 119 and 120, the graph of the rational function $f(x)$ crosses the horizontal asymptote. Graph $f(x)$ and find the point of intersection of the curve with its horizontal asymptote.

119. $f(x) = \dfrac{x^2 + x - 2}{x^2 - 2x - 3}$ **120.** $f(x) = \dfrac{4 - x^2}{2x^2 - 5x - 3}$

In Exercises 121–124, find an equation of a rational function f satisfying the given conditions.

121. Vertical asymptote: $x = 3$
Horizontal asymptote: $y = -1$
x-intercept: 2

122. Vertical asymptote: $x = -1, x = 1$
Horizontal asymptote: $y = 1$
x-intercept: 0

123. Vertical asymptote: $x = 0$
Oblique asymptote: $y = -x$
x-intercepts: $-1, 1$

124. Vertical asymptote: $x = 3$
Oblique asymptote: $y = x + 4$
y-intercept: 2; $f(4) = 14$

In Exercises 125–130, make up a rational function $f(x)$ that has all of the characteristics given in the exercise.

125. Has a vertical asymptote at $x = 2$, a horizontal asymptote at $y = 1$, and a y-intercept at $(0, -2)$

126. Has vertical asymptotes at $x = 2$ and $x = -1$, a horizontal asymptote at $y = 0$, a y-intercept at $(0, 2)$, and an x-intercept at 4

127. Has $f\left(\dfrac{1}{2}\right) = 0$; $f(x) \to 4$ as $x \to \pm\infty$,

$f(x) \to \infty$ as $x \to 1^-$, and $f(x) \to \infty$ as $x \to 1^+$.

128. Has $f(0) = 0$; $f(x) \to 2$ as $x \to \pm\infty$; has no vertical asymptotes and is symmetric about the y-axis

129. Has $y = 3x + 2$ as an oblique asymptote; has a vertical asymptote at $x = 1$

130. Is it possible for the graph of a rational function to have both a horizontal and an oblique asymptote? Explain.

Critical Thinking/Discussion/Writing

131. The total resistance, R_T, of two resistors connected in parallel depends on their individual resistances R_1 and R_2 according to the formula

$$\frac{1}{R_T} = \frac{1}{R_1} + \frac{1}{R_2}$$

Assume the resistance of the second resistor is $R_2 = 5\Omega$.
a. Express the total resistance R_T as a rational function

of R_1.

b. Graph this function using the techniques from this section.

c. Using the graph, determine the theoretical minimal and maximal total resistance R_T of the system and explain the significance of the horizontal asymptote.

132. Ostwald's dilution law for electrolytes,

$$K = \frac{\alpha^2}{1 - \alpha} \cdot c,$$

is used to compute the dissociation constant K knowing the degree of dissociation α and the total concentration c of the electrolyte. Assume $c = 2$.
a. Graph K as a function of α using the techniques from this section.
b. What happens when $\alpha \to 0^+$? Deduce the behavior K for very weak electrolytes (where α is close to 0).
c. What happens when $\alpha \to 1^-$? Does Ostwald's law hold for strong electrolytes (where a is close to 1)?

133. Give an example of a rational function that intersects its horizontal asymptote at:
a. no points
b. one point
c. two points

134. Give an example of a rational function that intersects its oblique asymptote at:
a. no points
b. one point
c. two points

135. Describe a rational function $R(x) = \dfrac{N(x)}{D(x)}$ that has asymptote $y = ax + b$ and $R(x)$ intersects the asymptote at n points whose x-coordinates are $c_1, c_2, \ldots, c_n$. Discuss the relationship between the degree of $D(x)$ and the number of points of intersection of $R(x)$ and $y = ax + b$.

136. Consider the rational function

$$f(x) = x^2 + \frac{1}{x^2}.$$

a. Describe the end behavior and the behavior near $x = 0$.
b. Use the formula

$$x^2 + \frac{1}{x^2} = \left(x - \frac{1}{x}\right)^2 + 2$$

to show that $f(x) \ge 2$ and find all values of x such that $f(x) = 2$.
c. Sketch the graph of f.

Getting Ready for the Next Section

In Exercises 137–140, simplify each expression.

137. $\dfrac{x-1}{x+3} - 3$

138. $\dfrac{2-x}{x+1} - 4$

139. $\dfrac{5}{x-2} - \dfrac{4}{x+1}$

140. $\dfrac{2}{x+3} - \dfrac{3}{x-5}$

In Exercises 141–144, solve each quadratic inequality.

141. $(x-1)(x+2) > 0$

142. $(x+2)(x+4) < 0$

143. $x^2 + 3x - 10 \geq 0$

144. $x^2 - 6x - 7 \leq 0$

Polynomial and Rational Inequalities

BEFORE STARTING THIS SECTION, REVIEW

1 Simplifying Rational Expressions (Appendix A.3, page 936)

2 Solving Quadratic Inequalities (Section 2.1, page 144)

3 Graphing Polynomial Functions (Section 2.2, page 163)

4 Graphing Rational Functions (Section 2.4, page 196)

OBJECTIVES

1 Solve polynomial inequalities.

2 Solve rational inequalities.

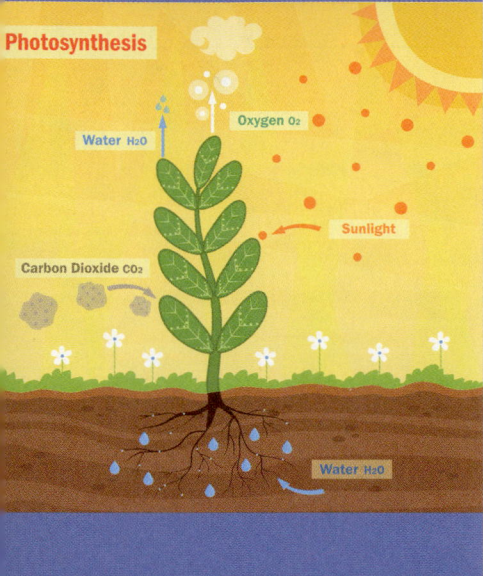

Photosynthesis

Oxygen O₂

Water H₂O

Sunlight

Carbon Dioxide CO₂

Water H₂O

◆ What Is Photosynthesis?

Photosynthesis is a process used by plants, algae, and some bacteria to convert the energy from sunlight into chemical energy. During this process, light energy transfers electrons from water (H_2O) to carbon dioxide (CO_2) to produce carbohydrates (glucose), which is later converted to ATP (adenosine triphosphate), a common form in which energy is stored in living systems. The by-product of this process is oxygen (O_2). Photosynthesis generates all the breathable oxygen in the atmosphere and renders plants rich in nutrients.

The scientist uses photosynthesis-irradiance curves (PI) to study the efficiencies of photosynthesis for different species. In Exercise 85 we ask you to compare the PI curves of two different species.

In previous section, we discussed the basic properties of polynomial and rational functions and their graphs. This section investigates inequalities involving polynomial and rational functions.

The importance of inequalities has it roots in estimations. Few things in the real word can be precisely determined. The most common problems we encounter are related to determining when a given variable falls within the prescribed range.

1 Solve polynomial inequalities.

Polynomial Inequalities

An inequality involving polynomials is called a **polynomial inequality**. For example,

$$x^3 + 2x^2 - 5x - 21 \geq 4x - 2x^2 + 15, \quad 20 - x^2 > x, \quad 3x - x^3 \leq 2, \text{ and } x^4 < 2x^2$$

are polynomial inequalities. Polynomial inequalities can be rearranged into one of the forms

$$P(x) \geq 0, \quad P(x) > 0, \quad P(x) \leq 0, \quad P(x) < 0,$$

where $P(x)$ is a polynomial function. This simple observation reduces solving polynomial inequalities to the study of the sign of the polynomial function. The key property is the continuity of the polynomial function $P(x)$. An immediate consequence of the Intermediate

Value Theorem (page 160) is that a polynomial function can only change sign at its zeros. We summarize this fact with the following rule:

> ### SIGN OF POLYNOMIAL FUNCTION
>
> The polynomial function $P(x)$ will have a constant sign (either always positive or always negative) on each interval formed by the real zeros of $P(x)$ plotted on a number line. If the polynomial function has no real zeros, then it is always positive or always negative.

We present two basic techniques to determine the sign of any polynomial function $P(x)$ and then solve related polynomial inequalities. We can determine the sign of a polynomial (i) geometrically from its graph (Graph Method), or (ii) algebraically by evaluating the polynomial at appropriate test points (Test Points Method).

PROCEDURE
IN ACTION

EXAMPLE 1 **Solve a Polynomial Inequality**

OBJECTIVE

Solve a polynomial inequality
$P(x) > 0$, $P(x) \geq 0$, $P(x) < 0$ or $P(x) \leq 0$.

Step 1 Arrange the inequality so that 0 is on the right-hand side.

Step 2 Factor $P(x)$ and solve the associated equation $P(x) = 0$. The real solutions of this equation are called boundary points.

Step 3 Locate the boundary points on the number line. These boundary points divide the number line into intervals.

EXAMPLE

Solve: $x^3 + 2x + 7 \leq 3x^2 + 6x - 5$

$x^3 + 2x + 7 \leq 3x^2 + 6x - 5$ Original inequality

$P(x) = x^3 - 3x^2 - 4x + 12 \leq 0$ Subtract $3x^2 + 6x - 5$ from both sides and simplify.

Solve: $x^3 - 3x^2 - 4x + 12 = 0$ Possible rational zeros $\pm 1, \pm 2, \pm 3, \pm 4, \pm 6, \pm 12$

$(x - 2)(x^2 - x - 6) = 0$ $x = 2$ is a zero and use synthetic division to factor.

$(x - 2)(x + 2)(x - 3) = 0$ Factor depressed term.

$x - 2 = 0$ or $x + 2 = 0$ or $x - 3 = 0$ Zero-product property

$x = 2$ or $x = -2$ or $x = 3$ Solve for x.

The boundary points are $-2, 2, 3$.

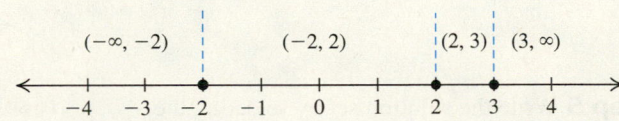

Step 4a (**Graph Method**) Graph $P(x)$ using the end behavior and multiplicity (cross or touch) of the zeros of $P(x)$ and determined the sign of $P(x)$ on the intervals determined by the boundary points.

Since the degree of $P(x)$ is 3 (odd) and the leading coefficient is 1 (positive), the end behavior is given by Case 3 on page 158. All three zeros are of multiplicity 1 (odd), so the polynomial crosses the x-axis at each zero.

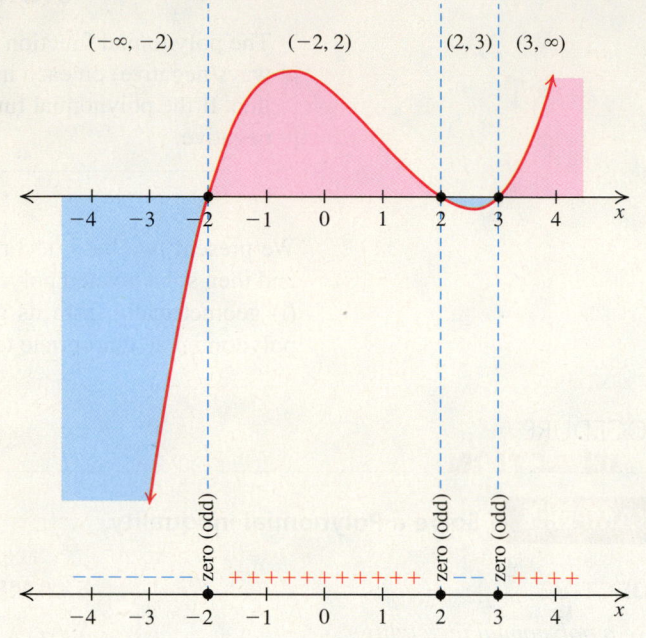

Step 4b (**Test Points Method**) Select a test point in each interval determined by the boundary points and evaluate $P(x)$ at each test point to find the sign of $P(x)$ on that interval.

Interval	Test Point	Value of $P(x)$	Result
$(-\infty, -2)$	-3	$P(-3) = (-3)^3 - 3(-3)^2$ $-4(-3) + 12 = -30$	Negative
$(-2, 2)$	0	$P(0) = (0)^3 - 3(0)^2$ $-4(0) + 12 = 12$	Positive
$(2, 3)$	2.5	$P(2.5) = (2.5)^3 - 3(2.5)^2$ $-4(2.5) + 12 = -1.125$	Negative
$(3, \infty)$	4	$P(4) = (4)^3 - 3(4)^2$ $-4(4) + 12 = 12$	Positive

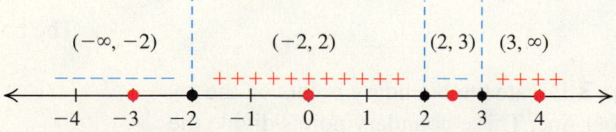

Step 5 Write the solution set by selecting the intervals in Step 4 on which the value of $P(x)$ satisfies the given inequality. If the inequality is $P(x) \geq 0$ or $P(x) \leq 0$, include the corresponding boundary points (otherwise, exclude them). Graph the solution set.

To solve the inequality $P(x) \leq 0$, we must find the intervals on which the function is negative, and since the inequality involves " $\leq$ " we also include the zeros of $P(x)$ (boundary points) in the solution set. The solution set is $(-\infty, -2] \cup [2, 3]$.

Practice Problem 1 Solve $x^3 + 2x^2 - 5x - 21 \geq 4x - 2x^2 + 15$.

EXAMPLE 2 **Using Graphs to Solve a Polynomial Inequality**

Solve $(x + 3)(x + 1)^2(x - 2) < 0$.

Solution

Step 1 The inequality has a 0 on the right-hand side. We proceed to the next step.

Step 2 $P(x) = (x + 3)(x + 1)^2(x - 2)$ is already in factored form. We find the solutions of the associated equation

$$(x + 3)(x + 1)^2(x - 2) = 0 \qquad \text{Replace } < \text{ with } =.$$
$$x + 3 = 0, \quad x + 1 = 0, \quad x - 2 = 0 \qquad \text{Zero-product property}$$
$$x = -3, \quad x = -1, \quad x = 2 \qquad \text{Solve for } x.$$

The zeros (boundary points) are -3, -1, and 2.

Step 3 The zeros of $P(x)$ divide the number line into four intervals: $(-\infty, -3)$, $(-3, -1), (-1, 2)$, and $(2, \infty)$.

Step 4 To determine the end behavior, we find the leading term of $P(x)$.

Leading term of $P(x) = (x + 3)(x + 1)^2(x - 2) \approx (x)(x)^2(x) = x^4$

We see that $P(x) = (x + 3)(x + 1)^2(x - 2) = x^4 + \ldots$ is a polynomial function of degree 4 and the leading coefficient $a_4 = 1 > 0$. Case 1 (page 158) applies. Because -3 and 2 are zeros of odd multiplicity, the graph of $P(x)$ crosses the x-axis at these points. Since -1 is a zero of even multiplicity, the graph of $P(x)$ touches but does not cross the x-axis at $x = -1$. See Figure 2.39.

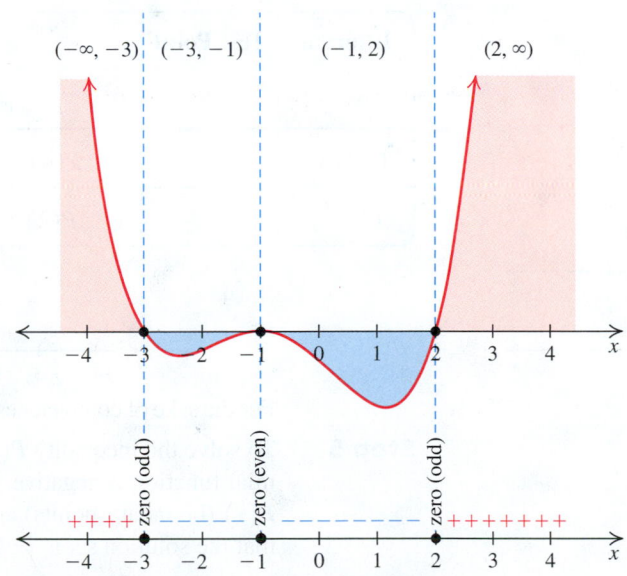

Figure 2.39

Step 5 To solve the inequality $P(x) < 0$, we find the intervals on which the polynomial function is negative. Because the inequality $<$ is strict, the zeros of $P(x)$ (boundary points) are not included in the solution set. We see from Figure 2.39 that the solution set is $(-3, -1) \cup (-1, 2)$.

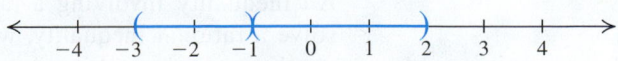

Practice Problem 2 Solve $x(x + 2)^2(x - 1) > 0$.

Once we rearrange the polynomial inequality so that 0 is on the right-hand side, we set the left-hand side equal to 0 and solve the resulting equation by factoring. Some factors may not have real solutions. However, each such factor will always have a constant sign ($+$ or $-$) on intervals determined by its boundary points. The next example illustrates this.

EXAMPLE 3 **Using Test Points to Solve a Polynomial Inequality**

Solve $(x - 1)(x + 1)(x^2 + 2) \leq 0$.

Solution

Step 1 The inequality has a 0 on one side. We proceed to the next step.

Step 2 Note that the quadratic term $x^2 + 2$ has no real zeros and is always positive. Since $P(x) = (x - 1)(x + 1)(x^2 + 2)$ is already in factored form, we find the solutions of the associated equation;

$$(x - 1)(x + 1)(x^2 + 2) = 0 \quad \text{Replace} \leq \text{with} =.$$
$$x - 1 = 0, \quad \text{or } x + 1 = 0, \quad \text{or } x^2 + 2 = 0 \quad \text{Zero-product property}$$
$$x = 1, \quad \text{or } x = -1, \quad \text{or } x^2 = -2 \quad \text{Solve for } x.$$

Since the equation $x^2 = -2$ has no real solutions, the zeros (boundary points) are -1 and 1.

Step 3 The zeros of $P(x)$ divide the number line into three intervals: $(-\infty, -1)$, $(-1, 1)$, and $(1, \infty)$.

Step 4 We select convenient test points in each of these three intervals. We pick -2, 0, and 2. Next, we compute the value of $P(x) = (x - 1)(x + 1)(x^2 + 2)$ at each test point to determine the sign of $P(x)$.

Interval	Test Point	Value of $P(x)$	Result
$(-\infty, -1)$	-2	$P(-2) = (-2 - 1)(-2 + 1)((-2)^2 + 2) = 18$	Positive
$(-1, 1)$	0	$P(0) = (0 - 1)(0 + 1)(0^2 + 2) = -2$	Negative
$(1, \infty)$	2	$P(2) = (2 - 1)(2 + 1)(2^2 + 2) = 18$	Positive

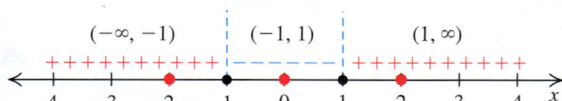

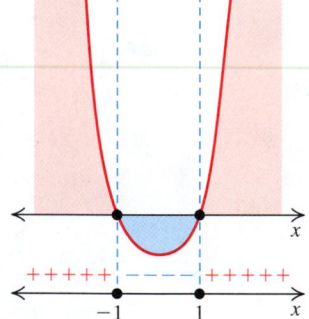

Figure 2.40

For the sake of completeness, we also reproduce the graph of $P(x)$ in Figure 2.40.

Step 5 To solve the inequality $P(x) \leq 0$, we find the intervals on which the polynomial function is negative. Because the inequality $\leq$ is not strict, the zeros of $P(x)$ (boundary points) are included in the solution set. We see from Step 4 that the solution set is $[-1, 1]$.

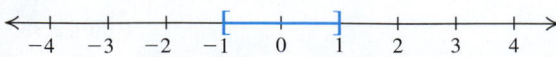

Practice Problem 3 Solve $x(x - 2)(x^2 + 2x + 2) \geq 0$.

2 Solve rational inequalities

Rational Inequalities

An inequality involving a rational expression is called a **rational inequality**. To solve a rational inequality, we follow a procedure similar to that we used to solve a polynomial inequality. An additional consideration in a rational inequality is the polynomial in the denominator. The sign of a rational function can be determined by the following rule:

SIGN OF A QUOTIENT

The sign of the quotient $\dfrac{N(x)}{D(x)}$ is the same as the sign of the product $N(x) \cdot D(x)$, as long as $D(x) \neq 0$.

We see that a rational function can change sign at the zeros of the numerator as well as the zeros of the denominator. We summarize this fact next:

SIGN OF A RATIONAL FUNCTION

A rational function $R(x) = \dfrac{N(x)}{D(x)}$ will have a constant sign (either always positive or always negative) on each interval formed by plotting the real zeros of $N(x)$ and $D(x)$ on a number line. If neither $N(x)$ nor $D(x)$ has real zeros, then the corresponding rational function $R(x)$ is always positive or is always negative.

In essence, the boundary points for the rational functions are determined by their zeros and vertical asymptotes (or holes).

PROCEDURE
IN ACTION

EXAMPLE 4 Solve a Rational Inequality

OBJECTIVE

Solve a rational inequality of the form:
$R(x) > 0,\ R(x) \geq 0,\ R(x) < 0$ or $R(x) \leq 0$.

EXAMPLE

Solve: $\dfrac{2}{x-1} \geq \dfrac{1}{x-2}$

Step 1 Arrange the inequality so that 0 is on the right-hand side.

$\dfrac{2}{x-1} \geq \dfrac{1}{x-2}$ Original inequality

$\dfrac{2}{x-1} - \dfrac{1}{x-2} \geq 0$ Subtract $\dfrac{1}{x-2}$.

$\dfrac{2\cdot(x-2)}{(x-1)\cdot(x-2)} - \dfrac{1\cdot(x-1)}{(x-2)\cdot(x-1)} \geq 0$ Use $(x-1)(x-2)$ as a common denominator.

$\dfrac{2(x-2)-(x-1)}{(x-2)(x-1)} \geq 0$ Combine fractions.

$R(x) = \dfrac{x-3}{(x-2)(x-1)} \geq 0$ Simplify the numerator.

Step 2 Factor $R(x)$ and find the real zeros of the numerator and the denominator. These numbers will form the boundary points.

Set Numerator $= 0$: $x - 3 = 0$

$x = 3$ Solve for x.

Set Denominator $= 0$: $(x-2)(x-1) = 0$

$x - 2 = 0$ or $x - 1 = 0$ Zero-product property

$x = 2$ or $x = 1$ Solve for x.

The boundary points are 1, 2, and 3. The lines $x = 1$ and $x = 2$ are the vertical asymptotes of $R(x)$.

Step 3 Locate the boundary points on the number line. These boundary points divide the number line into intervals.

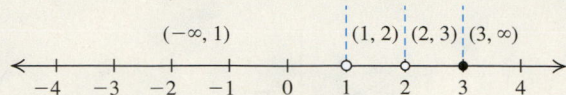

The boundary points 1 and 2 are marked as open circles to indicate that they do not belong to the domain of the rational function $R(x)$.

Step 4a (Graph Method) To determine the sign of the rational function $R(x) = \frac{N(x)}{D(x)}$, form the corresponding polynomial $P(x) = N(x) \cdot D(x)$. Graph $P(x)$ using the end behavior and multiplicity (cross or touch) of the zeros of $P(x)$ and determine the sign of $P(x)$ on the intervals determined by the boundary points. The sign of $P(x)$ is the same as the sign of $R(x)$ on the intervals found in Step 3.

$R(x) = \dfrac{x-3}{(x-2)(x-1)}$. The corresponding polynomial is $P(x) = (x-3)(x-2)(x-1)$. Since the degree of $P(x)$ is 3 (odd) and the leading coefficient is 1 (positive), the end behavior is given by Case 3 (page 158). All three zeros are of multiplicity 1 (odd), so the polynomial crosses the x-axis at each zero.

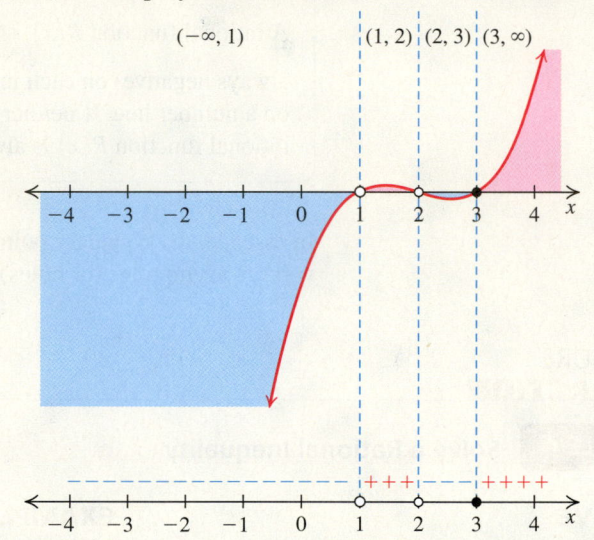

Step 4b (Test Points Method) Select test points in each interval, determined by the boundary points, and evaluate $P(x)$ at each test point to find the sign of $P(x)$ on that interval.

Interval	Test Point	Value of $P(x)$	Result
$(-\infty, 1)$	0	$R(0) = \dfrac{0-3}{(0-2)(0-1)} = \dfrac{-3}{2} = -3/2$	Negative
$(1, 2)$	1.5	$R(1.5) = \dfrac{1.5-3}{(1.5-2)(1.5-1)} = \dfrac{-1.5}{-0.25} = 6$	Positive
$(2, 3)$	2.5	$R(2.5) = \dfrac{2.5-3}{(2.5-2)(2.5-1)}$ $= \dfrac{-0.5}{0.75} = -2/3$	Negative
$(3, \infty)$	4	$R(4) = \dfrac{4-3}{(4-2)(4-1)} = \dfrac{1}{6} = 1/6$	Positive

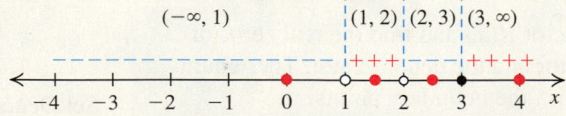

Step 5 Write the solution set by selecting the intervals in Step 4 that satisfy the given inequality. If the inequality is $R(x) \geq 0$ or $R(x) \leq 0$, include only the boundary points corresponding to the zeros of the numerator of the rational function $R(x)$, (otherwise, exclude them). Graph the solution set.

To solve the inequality $R(x) \geq 0$, we must find the intervals on which the function is positive; since the inequality involves " $\geq$," we also include the zeros of the numerator of $R(x)$ (boundary points) in the solution set. The only boundary point that is a zero of the numerator of $R(x)$ is 3. The solution set is $(1, 2) \cup [3, \infty)$.

Practice Problem 4 Solve $\dfrac{3}{x} \le \dfrac{1}{x-4}$.

EXAMPLE 5 Using Graphs to Solve a Rational Inequality

Solve $\dfrac{x+11}{x+2} > 4$.

Solution

Step 1   We first rewrite $\dfrac{x+11}{x+2} > 4$ to get 0 on the right-hand side.

$$\dfrac{x+11}{x+2} > 4 \qquad \text{Original inequality}$$

$$\dfrac{x+11}{x+2} - 4 > 0 \qquad \text{Subtract 4 from both sides.}$$

$$\dfrac{x+11}{x+2} - \dfrac{4\cdot(x+2)}{(x+2)} > 0 \qquad \text{Use } (x+2) \text{ as a common denominator.}$$

$$\dfrac{(x+11)-4(x+2)}{x+2} > 0 \qquad \text{Combine fractions.}$$

$$R(x) = \dfrac{-3(x-1)}{x+2} > 0 \qquad \text{Simplify the numerator.}$$

Step 2 $R(x) = \dfrac{-3(x-1)}{x+2}$ is already in factored form. To find the boundary points we set both the numerator and the denominator of $R(x)$ equal to 0.

numerator = 0	denominator = 0	
$-3(x-1) = 0$	$x + 2 = 0$	Solve for x.
$x = 1$	$x = -2$	

The boundary points are -2 and 1. The line $x = -2$ is a vertical asymptote of $R(x)$.

Step 3   The boundary points divide the number line into three intervals: $(-\infty, -2)$, $(-2, 1)$, and $(1, \infty)$.

Step 4   To determine the sign of $R(x) = \dfrac{-3(x-1)}{x+2}$, we form the corresponding polynomial $P(x) = -3(x-1)(x+2)$. The sign rule on page 215 tells us that $P(x)$ has the same sign as $R(x)$ on the intervals found in Step 3. $P(x)$ is a quadratic polynomial with leading coefficient $a_2 = -3 < 0$. The graph of $P(x)$ is a parabola that opens down. See Figure 2.41.

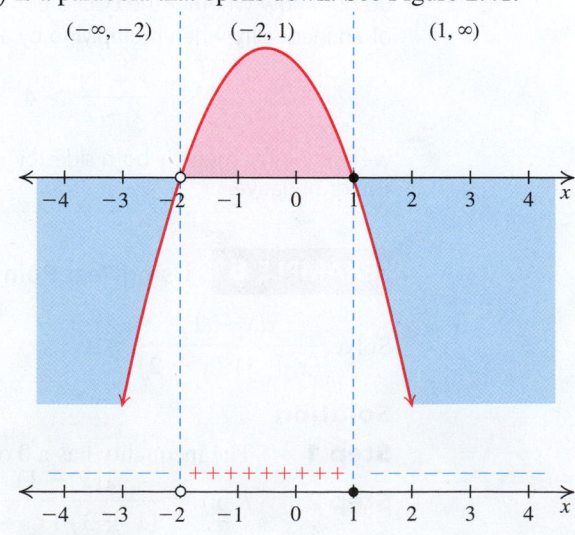

Figure 2.41

Alternatively, we can a sketch the graph of the rational function $R(x) = \dfrac{-3(x - 1)}{x + 2}$ using the procedure from the previous section and draw conclusions from there. We reproduce the graph of $R(x)$ in Figure 2.42.

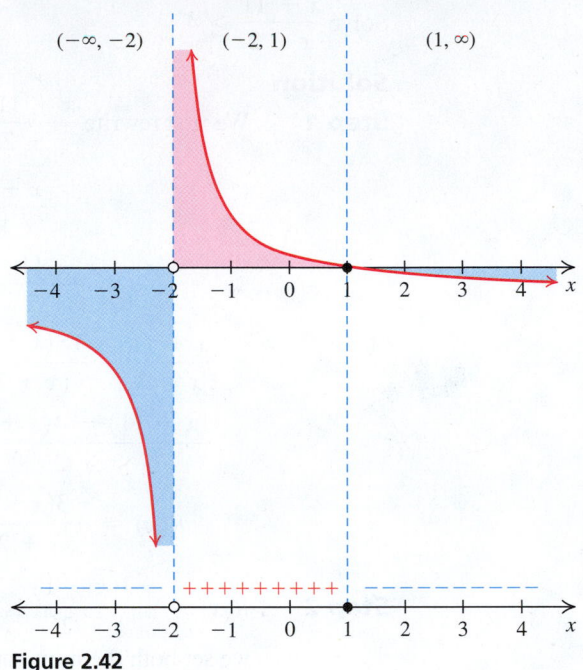

Figure 2.42

Step 5   To solve the inequality $R(x) \geq 0$, we identify the intervals on which the rational function is positive. Because the inequality $>$ is strict, the zeros of $R(x)$ (boundary points) are not included in the solution set. We see from Figure 2.41 or Figure 2.42 that the solution set is $(-2, 1)$.

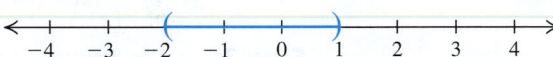

Practice Problem 5   Solve $\dfrac{5(x + 1)}{x + 3} < 3.$

WARNING   You cannot solve a rational inequality by multiplying both sides by the LCD (least common denominator), as you would with a rational equation. When the LCD contains a variable, you do not know whether the LCD is positive or negative. Remember that you reverse the sense of an inequality when multiplying by a negative expression. In Example 5

$$\frac{x + 11}{x + 2} > 4 \qquad \qquad x + 11 > 4(x + 2)$$

we do not just multiply both sides by $x + 2$ because we do not know whether $x + 2$ is positive or negative.

EXAMPLE 6   **Using Test Points to Solve a Rational Inequality**

Solve $\dfrac{x(x - 1)}{(x + 3)^2(x - 2)} \leq 0.$

Solution

Step 1   The inequality has a 0 on the right-hand side. We proceed to the next step.

Step 2 $R(x) = \dfrac{x(x - 1)}{(x + 3)^2(x - 2)}$ is already in factored form. To find the boundary points, we set both the numerator and the denominator of $R(x)$ equal to 0.

numerator = 0   denominator = 0

$x(x - 1) = 0$ $(x + 3)^2(x - 2) = 0$

$x = 0$ or $x - 1 = 0$ $x + 3 = 0$ or $x - 2 = 0$ Zero-product property

$x = 0$ or $x = 1$ $x = -3$ or $x = 2$ Solve for x.

The boundary points are $-3, 0, 1$, and 2. The lines $x = -3$ and $x = 2$ are vertical asymptotes of $R(x)$.

Step 3   The boundary points divide the number line into five intervals $(-\infty, -3)$, $(-3, 0)$, $(0, 1)$, $(1, 2)$ and $(2, \infty)$.

Step 4   We select convenient test points in each of these five intervals. We choose $-4, -1, 0.5, 1.5$, and 3. Next, we compute the value of $R(x) = \dfrac{x(x - 1)}{(x + 3)^2(x - 2)}$ at each test point to determine the sign of $R(x)$.

Interval	Test Point	Value of $P(x)$	Result
$(-\infty, -3)$	-4	$R(-4) = \dfrac{(-4)(-4 - 1)}{(-4 + 3)^2(-4 - 2)} = \dfrac{20}{-6} = -10/3$	Negative
$(-3, 0)$	-1	$R(-1) = \dfrac{(-1)(-1 - 1)}{(-1 + 3)^2(-1 - 2)} = \dfrac{2}{-12} = -1/6$	Negative
$(0, 1)$	0.5	$R(0.5) = \dfrac{(0.5)(0.5 - 1)}{(0.5 + 3)^2(0.5 - 2)} = \dfrac{-1/4}{-147/8} = 2/147$	Positive
$(1, 2)$	1.5	$R(1.5) = \dfrac{(1.5)(1.5 - 1)}{(1.5 + 3)^2(1.5 - 2)} = \dfrac{3/4}{-81/8} = -2/27$	Negative
$(2, \infty)$	3	$R(3) = \dfrac{(3)(3 - 1)}{(3 + 3)^2(3 - 2)} = \dfrac{6}{36} = 1/6$	Positive

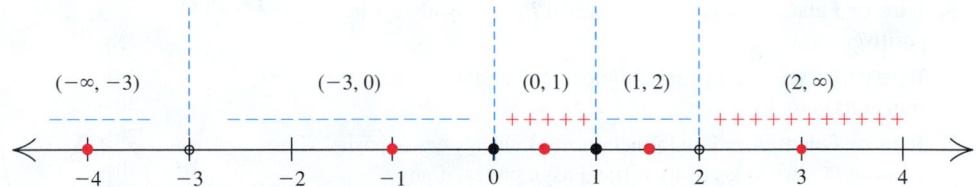

For the sake of completeness, we sketch the graph of $R(x)$ in Figure 2.43.

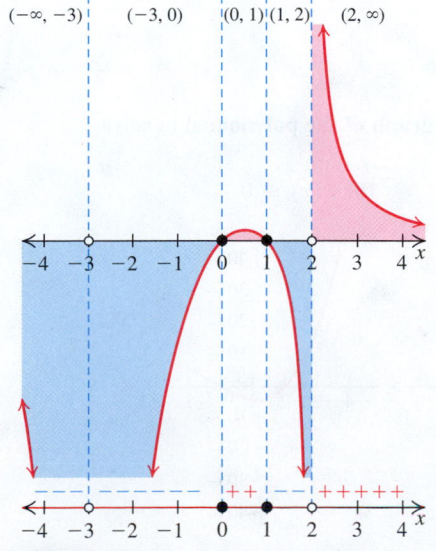

Figure 2.43

HINT:

The zeros of the denominator of a rational function $R(x)$ (vertical asymptotes or holes) are never included in the solution set of a rational inequality because division by 0 is undefined.

Step 5 To solve the inequality $R(x) \leq 0$, we find the intervals on which the rational function is negative. Because the inequality $\leq$ is not strict, the zeros of $R(x)$ (numbers 0 and 1) are included in the solution set. We see from Step 4 that the solution set is $(-\infty, -3) \cup (-3, 0] \cup [1, 2)$.

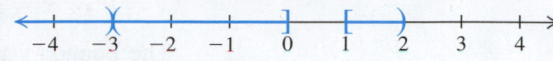

Practice Problem 6 Solve $\dfrac{x(x + 2)^2}{(x + 3)(x - 1)} \geq 0$.

Answers to Practice Problems

1. $[-4, -3] \cup [3, \infty)$ **2.** $(-\infty, -2) \cup (-2, 0) \cup (1, \infty)$ **3.** $(-\infty, 0] \cup [2, \infty)$ **4.** $(-\infty, 0) \cup (4, 6]$ **5.** $(-3, 2)$ **6.** $(-3, 0] \cup (1, \infty)$

SECTION 2.5 **Exercises**

Concepts and Vocabulary

1. For the polynomial inequalities $P(x) \geq 0$ or $P(x) \leq 0$, the zeros of the polynomial function $P(x)$ are always _____ in the solution set.

2. For the polynomial inequalities $P(x) > 0$ or $P(x) < 0$, the zeros of the polynomial function $P(x)$ are always _____ from the solution set.

3. To solve a rational inequality, we begin by locating all _____ and _____ asymptotes (if they exist).

4. For the rational inequalities $R(x) \geq 0, R(x) \leq 0, R(x) > 0$ or $R(x) < 0$, the zeros of the denominator of the rational function $R(x)$ (vertical asymptotes or holes) are always _____ from the solution set.

5. **True or False.** A polynomial function $P(x)$ may always be positive.

6. **True or False.** A polynomial function $P(x)$ always changes sign at its zero.

7. **True or False.** A rational function $R(x)$ always has opposite signs on the two sides of its vertical asymptotes, if any.

8. **True or False.** The sign of the quotient $\dfrac{N(x)}{D(x)}$ is the same as the sign of the product $N(x) \cdot D(x)$, as long as $D(x) \neq 0$.

Building Skills

In Exercises 9–16, use the graph of the polynomial to solve the indicated inequality.

9. $P(x) \geq 0$ **10.** $P(x) \leq 0$

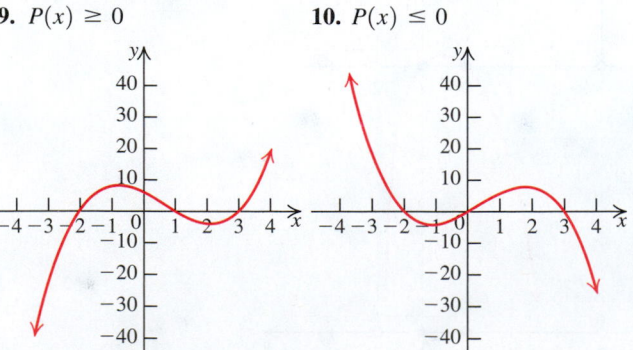

11. $P(x) > 0$ **12.** $P(x) < 0$

13. $P(x) > 0$ **14.** $P(x) < 0$

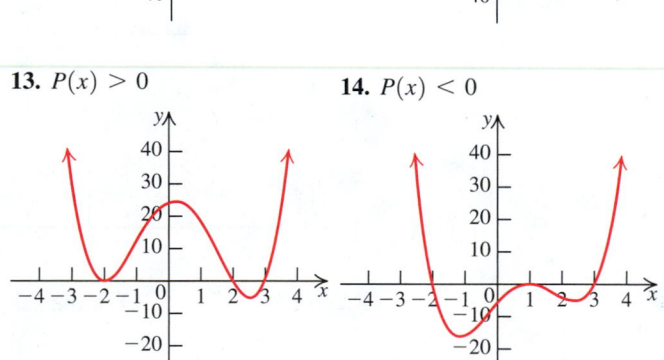

15. $P(x) \geq 0$ **16.** $P(x) \leq 0$

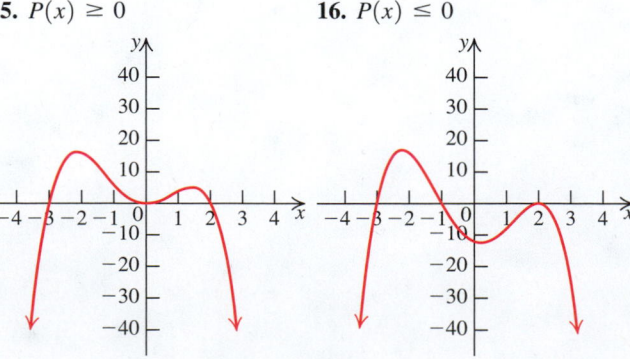

In Exercises 17–30, solve each polynomial inequality.

17. $(x - 3)(x - 4) > 0$

18. $(x + 2)(x - 5) < 0$

19. $(x + 1)(x - 8) \geq 0$

20. $(x + 4)(x + 6) \leq 0$

21. $2(x - 1)(x + 5)(x - 3) < 0$

22. $3(x + 1)(x - 2)(x - 4) > 0$

23. $(1 - x)(x - 4)(x + 7) \leq 0$

24. $(x + 2)(4 - x)(x + 5) \geq 0$

25. $x^2(x + 2)(x - 1) < 0$

26. $(x + 2)(x - 1)^2(x - 6) > 0$

27. $x^2(x - 1)(x - 3) \geq 0$

28. $(x + 4)(x - 3)^2(x - 1) \leq 0$

29. $2x(x - 1)(x^2 + 1) > 0$

30. $-3x(x + 2)(x^2 + 3) > 0$

In Exercises 31–42, solve each polynomial inequality.

31. $x^2 - x > 2$

32. $20 - x^2 > x$

33. $x^2 + 10 \geq 3x$

34. $x^2 + 4 \leq 2x$

35. $x^3 + 4x^2 + x - 6 > 0$

36. $x^3 + x^2 - 10x + 8 < 0$

37. $x^3 - 12x \geq 16$

38. $3x - x^3 \leq 2$

39. $x^3 - 3x^2 + 3x + 5 \leq 2x^2 + x - 3$

40. $x^3 - 4x^2 - 9x + 13 \geq -2x^2 + 2x + 1$

41. $x^4 \geq 3x^2$

42. $x^4 \leq 2x^2$

In Exercises 43–50, use the graph of the rational function to solve the indicated inequality.

43. $R(x) \geq 0$

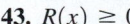
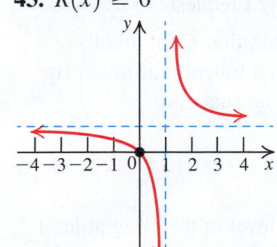

44. $R(x) \leq 0$

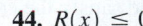
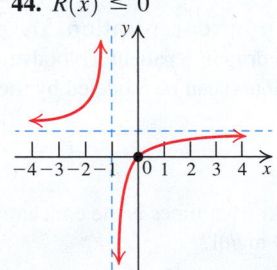

45. $R(x) > 0$

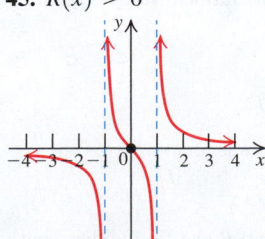

46. $R(x) < 0$

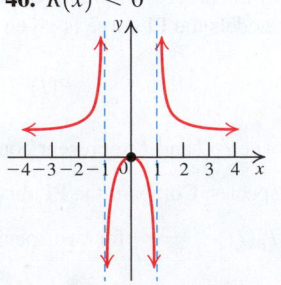

47. $R(x) \geq 0$

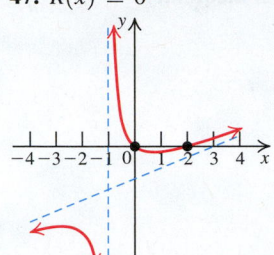

48. $R(x) \leq 0$

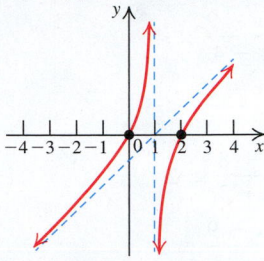

49. $R(x) > 0$

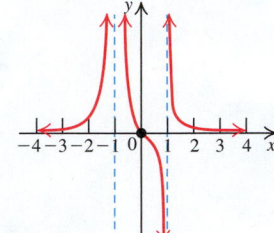

50. $R(x) < 0$

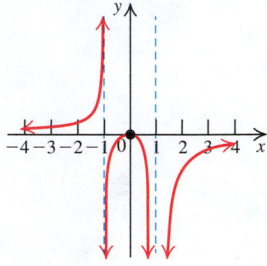

In Exercises 51–64, solve each rational inequality.

51. $\dfrac{x - 1}{x + 2} > 0$

52. $\dfrac{x + 3}{x - 4} < 0$

53. $\dfrac{x + 2}{x - 8} \geq 0$

54. $\dfrac{x + 5}{x - 1} \leq 0$

55. $\dfrac{3x - 2}{4 - 3x} \geq 0$

56. $\dfrac{1 - 2x}{3 + 2x} \geq 0$

57. $\dfrac{(x - 1)(x + 2)}{x + 1} > 0$

58. $\dfrac{x(x - 3)}{x + 2} < 0$

59. $\dfrac{(x + 5)(x - 2)}{x + 3} \geq 0$

60. $\dfrac{x(x - 1)}{x + 4} \leq 0$

61. $\dfrac{(x + 1)^2(x - 1)}{(x + 4)(x - 3)} \geq 0$

62. $\dfrac{(x + 4)(x + 2)}{(x - 2)^2(x - 1)} \geq 0$

63. $\dfrac{(2 - x)^3(x - 3)}{x + 1} \leq 0$

64. $\dfrac{(x - 4)(1 - x)^3}{x - 2} \leq 0$

In Exercises 65–76, solve each rational inequality.

65. $\dfrac{5x - 9}{x - 4} > 4$

66. $\dfrac{2x - 17}{x - 5} < 3$

67. $\dfrac{x + 8}{x + 3} \geq 2$

68. $\dfrac{4x - 9}{x - 3} \leq 3$

69. $\dfrac{2}{x - 3} > \dfrac{1}{x + 2}$

70. $\dfrac{1}{x + 2} < \dfrac{2}{x - 5}$

71. $\dfrac{x}{x + 2} \geq \dfrac{x + 1}{x - 5}$

72. $\dfrac{x - 1}{x + 1} \leq \dfrac{x - 2}{x + 3}$

73. $x + \dfrac{8}{x} \leq 6$

74. $x - \dfrac{12}{x} \geq 1$

75. $\dfrac{3x}{x^2 - 4} \geq 0$

76. $\dfrac{x^2 + 1}{x^2 - 4} \leq 0$

Applying the Concepts

77. **Skid marks and speed.** Under certain conditions, such as a wet blacktop surface, the distance (in feet) it takes a car traveling at speed v (in mph) to stop can be estimated by the equation

$$d = 0.05v^2 + v.$$

The skid marks at one accident where those conditions were met were over 75 feet long. The accident occurred in a 25-mile-per-hour speed zone. Was the driver going over the speed limit?

78. **Physics.** A stone is thrown vertically upward from the edge of a bridge 95 feet high. The height of the stone (in feet) after t seconds is given by

$$h = -16t^2 + 32t + 95.$$

For what times is the height of the stone at least 110 feet?

79. **Lip Gloss sales.** The total profit function TP for a company producing Q units of lip gloss (in thousands) is given by

$$TP = -Q^3 + 18Q^2 - 11Q - 102.$$

Find the value of Q for which the company makes a profit.

80. **Making a box.** An open-top box is made out of a rectangular 9 by 6 piece of cardboard by cutting out four congruent squares with sides of length x, one at each corner, and folding up the remaining sides (see the accompanying figure). For

what range of x is the volume of the resulting open-top box at least 20?

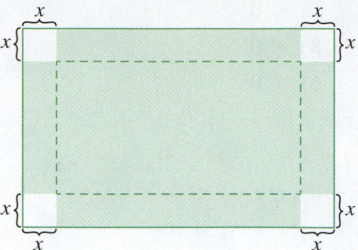

81. **Headphones production.** The total cost of producing Q units of headphones (in thousands) by factory A is given by

$$TC_A = Q^3 - 6Q^2 + 14Q + 10,$$

while the total cost of producing the same number of headphones by factory B is given by

$$TC_B = 5Q^2 - 20Q + 34.$$

For what number of headphones is it more cost effective to produce them at factory A rather than at factory B?

82. **Blackjack.** A gambler playing blackjack in a casino using four decks of cards noticed that 20% of the cards that had been dealt were jacks, queens, kings, or aces. After x cards had been dealt, he knew that the likelihood L that the next card dealt would be a jack, queen, king, or ace was

$$L = \frac{64 - 0.2x}{208 - x}.$$

For what values of x is this likelihood greater than 50%?

83. **Excursion ferry.** A company is looking into establishing an excursion ferry connection between two tourist locations 40 miles apart on a river bank. The speed of the river current is 2 mph, and the company would like to limit the round trip to at most 4.5 hours. Assuming that the power of the ferry is constant and results in the speed in still water being v, what range of v would satisfy the company's request?

84. **Drug concentration.** The concentration C (in mg/dl) of a drug in a patient's bloodstream as a function of time t (in hours) can be modeled by the rational function

$$C = \frac{40t}{t^2 + 16}.$$

At what times is the concentration level of this drug at least 3 mg/dl?

85. **Photosynthesis.** The PI curve is a graphical representation of the effect of light intensity (irradiance), I, on the efficiency of the process of photosynthesis. The rational function that models the PI curve is given by

$$P(I) = \frac{C \cdot I}{K + I},$$

where C and K represent constants dependent on a given species. Compare the PI curves $P_A(I) = \dfrac{3I}{5 + I}$ and $P_B(I) = \dfrac{2I}{2 + I}$ for two species A and B and find under what light conditions species A has more effective photosynthesis than species B.

86. Rocket engine. The propulsive energy E of a rocket moving at speed v with an exhaust velocity v_e can be modeled by the rational function

$$E = \frac{2x}{1 + x^2}, \qquad \text{where } x = \frac{v}{v_e} \text{ is the velocity ratio.}$$

Find the range of the velocity ratio x that guarantees that the efficiency E of this rocket is at least 80%.

87. The graph shows the number of vinyl album sales (in millions of units) over a 20-year period (1993–2013).

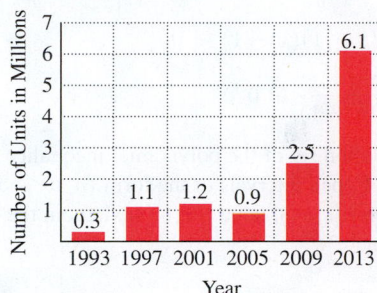

The cubic function
$f(x) = 0.29 + 0.436x - 0.066x^2 + 0.00295x^3$
models the data, where $x = 0$ represents the year 1993. Use a graphing calculator to estimate the years in which the number of vinyl album sales first exceeded 1.4 million.

88. The graph shows the average price (adjusted for inflation) of a music album over a 40-year period (1974–2014).

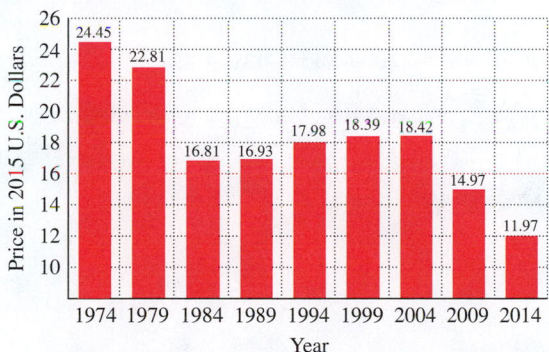

The cubic function
$f(x) = 25.269 - 1.2405x + 0.064x^2 - 0.001036x^3$
models the data, where $x = 0$ represents the year 1974. Use a graphing calculator to estimate the years in which the price of the music album was less than \$17.65.

Beyond the Basics

In Exercises 89–98, find the domain of each function.

89. $f(x) = \sqrt{x^2 - 4x - 12}$

90. $f(x) = \sqrt{15 - 2x - x^2}$

91. $f(x) = \sqrt{4x - x^3}$

92. $f(x) = \sqrt{8x + x^4}$

93. $f(x) = \sqrt{\dfrac{2 - x}{x + 5}}$

94. $f(x) = \sqrt{\dfrac{x + 3}{x - 2}}$

95. $f(x) = \sqrt{1 + \dfrac{1}{x - 3}}$

96. $f(x) = \sqrt{2 - \dfrac{3x + 11}{x + 5}}$

97. $f(x) = \sqrt{1 + \dfrac{3}{x^2 - 4}}$

98. $f(x) = \sqrt{x + \dfrac{1}{x} - \dfrac{5}{2}}$

In Exercises 99–104, Find the values of x for which the range of the given polynomial or rational function is within the prescribed interval.

99. $P(x) = x^3 - 2x^2 - 11x + 12$; range $[-10, 18]$

100. $P(x) = x^3 - 3x^2 - 4x + 20$; range $[8, 14]$

101. $R(x) = \dfrac{x + 2}{x - 3}$; range $[2, 6]$

102. $R(x) = \dfrac{3 - x}{x - 7}$; range $[1, 3]$

103. $R(x) = \dfrac{5 - x}{x^2 - 1}$; range $[0, 1]$

104. $R(x) = \dfrac{2x - 1}{x^2 - 4}$; range $[0, 1]$

105. Show that, for any positive number x,

$$x + \frac{1}{x} \geq 2$$

by showing that the solution set of the above inequality contains the interval $(0, \infty)$.

106. Show that, for any real number x,

$$x^4 + (x - 2)^4 \geq 2$$

by showing that the solution set of the above inequality is the interval $(-\infty, \infty)$. [*Hint:* $(x - 1)^2$ is a factor of $x^4 + (x - 2)^4 - 2$.]

107. Show that, for any two positive numbers x and a, with $x < a$,

$$\frac{1}{x} + \frac{1}{a - x} \geq \frac{4}{a}$$

by showing that the solution set of the above inequality contains the interval $(0, a)$.

108. Show that, for any two positive numbers x and a

$$x^3 + 2a^3 \geq 3xa^2$$

by showing that the solution set of the above inequality contains the interval $[0, \infty)$.

Critical Thinking/Discussion/Writing

In Exercises 109–114, give an example of a polynomial inequality with the indicated solution set (answers may vary).

109. $(-4, 5)$

110. $(-\infty, -2] \cup [4, \infty)$

111. $(-1, 2) \cup (3, \infty)$

112. $[-3, 1] \cup \{4\}$

113. $(-1, 0) \cup (4, 5)$

114. $(-\infty, -2] \cup [0, 1] \cup [4, \infty)$

In Exercises 115–120, give an example of a rational inequality with the indicated solution set (answers may vary).

115. $[-2, 6)$

116. $(-\infty, -1) \cup [3, \infty)$

117. $[-1, 2) \cup [3, \infty)$

118. $(-1, 2] \cup (3, \infty)$

119. $(-1, 2) \cup [3, \infty)$

120. $(-\infty, -4) \cup [0, 1]$

121. Give an example of a polynomial P for which the solution set of $P(x) \geq 0$ is $(-\infty, -2] \cup [1, \infty)$ and the solution set of $P(x) \leq 0$ is $[-2, 1] \cup \{3\}$.

122. Assume that the solution set of the polynomial inequality $P(x) \geq 0$ is $[2, 4] \cup [6, \infty)$. What can you say about the solution set of the indicated inequality?

   a. $P(-x) \geq 0$

   b. $P(2x) \geq 0$

   c. $P\left(\dfrac{1}{2}x\right) \geq 0$

In Exercises 123–126, describe how the solution set of each inequality changes as the number a changes.

123. $(x - a)(x - 1) \geq 0$

124. $\dfrac{x - a}{x - 1} \geq 0$

125. $(x - a)(x + 1)(x - 1) \geq 0$

126. $\dfrac{x - a}{(x - 1)(x + 1)} \geq 0$

127. If the solution set of the polynomial inequality $P(x) > 0$ contains k open intervals of the form (a, b) and one interval of the form (c, ∞), what can be said about the degree of the polynomial P?

128. If the solution set of the polynomial inequality $P(x) < 0$ contains k open intervals of the form (a, b), what can be said about the degree of the polynomial P?

Getting Ready for the Next Section

In Exercises 129–132, simplify each expression.

129. $(x - i)(x + i)$

130. $(5x - 2i)(5x + 2i)$

131. $(x - 1 + 2i)(x - 1 - 2i)$

132. $(x - 2 + 3i)(x - 2 - 3i)$

In Exercises 133–136, use long division to verify that the polynomial $D(x)$ is a factor of a polynomial $P(x)$ and find a polynomial $Q(x)$ such that $P(x) = D(x) \cdot Q(x)$.

133. $P(x) = x^4 + x^3 + 38x - 40;\ D(x) = x^2 - 2x + 10$

134. $P(x) = x^4 - 3x^3 + 10x^2 + 44x - 120;$
$D(x) = x^2 - 4x + 20$

135. $P(x) = x^4 - 7x^3 + 9x^2 + 13x - 4;$
$D(x) = x^2 - 4x + 1$

136. $P(x) = x^4 + 3x^3 - 8x^2 - 32x - 24;$
$D(x) = x^2 - 2x - 4$

Zeros of a Polynomial Function

BEFORE STARTING THIS SECTION, REVIEW

1 Long division (page 174)

2 Synthetic division (page 175)

3 Rational zeros test (Section 2.3, page 180)

4 Complex numbers (Appendix A.8, page 976)

5 Conjugate of a complex number (Appendix A.8, page 978)

OBJECTIVES

1 Find the possible number of positive and negative zeros of polynomials.

2 Find the bounds on the real zeros of polynomials.

3 Learn basic facts about the complex zeros of polynomials.

4 Use the Conjugate Pairs Theorem to find zeros of polynomials.

Gerolamo Cardano (1501–1576)

Cardano was trained as a physician, but his range of interests included philosophy, music, gambling, mathematics, astrology, and the occult sciences. Cardano was a colorful man, full of contradictions. His lifestyle was deplorable, and his personal life was filled with tragedies. His wife died in 1546. He saw one son executed for wife poisoning. He cropped the ears of a second son who attempted the same crime. In 1570, he was imprisoned for six months for heresy when he published the horoscope of Jesus. Cardano spent the last few years in Rome, where he wrote his autobiography, *De Propria Vita*, in which he did not spare the most shameful revelations.

◆ Cardano Accused of Stealing Formula

In sixteenth-century Italy, getting and keeping a teaching job at a university was difficult. University appointments were mostly temporary. One way a professor could convince the administration that he was worthy of continuing in his position was by winning public challenges. Two contenders for a position would present each other with a list of problems, and later in a public forum, each person would present his solutions to the other's problems. As a result, scholars often kept secret their new methods of solving problems.

In 1535, in a public contest primarily about solving the cubic equation, Nicolo Tartaglia (1499–1557) won over his opponent Antonio Maria Fiore. Gerolamo Cardano (1501–1576), in the hope of learning the secret of solving a cubic equation, invited Tartaglia to visit him. After many promises and much flattery, Tartaglia revealed his method on the promise, probably given under oath, that Cardano would keep it confidential. When Cardano's book *Ars Magna* (The Great Art) appeared in 1545, the formula and the method of proof were fully disclosed. Angered at this apparent breach of a solemn oath and feeling cheated out of the rewards of his monumental work, Tartaglia accused Cardano of being a liar and a thief for stealing his formula. Thus began the bitterest feud in the history of mathematics, carried on with name-calling and mudslinging of the lowest order. As Tartaglia feared, the formula (see Exercise 85) has forever since been known as Cardano's formula for the solution of the cubic equation.

Descartes's Rule of Signs

1 Find the possible number of positive and negative zeros of polynomials.

In Section 2.3, we used the Rational Zeros Test to find the possible rational zeros of a polynomial function with integer coefficients. We now investigate briefly some other useful tests for the real zeros of polynomial functions. One such test is called *Descartes's Rule of Signs*.

If a polynomial (with nonzero coefficients) is written in descending powers, we say that a **variation of sign** occurs when the signs of two consecutive terms differ. For example, in the polynomial $2x^5 - 5x^3 - 6x^2 + 7x + 3$, the signs of the terms are $+ \; - \; - \; + \; +$. Therefore, there are two variations of sign, as follows:

There are three variations of sign in

$$x^5 - 3x^3 + 2x^2 + 5x - 7.$$

DESCARTES'S RULE OF SIGNS

Let $F(x)$ be a polynomial function with real coefficients and with nonzero terms written in descending powers.

1. The number of positive zeros of F is equal to the number of variations of sign of $F(x)$ or is less than that number by an even integer.

2. The number of negative zeros of F is equal to the number of variations of sign of $F(-x)$ or is less than that number by an even integer.

When using Descartes's Rule, a zero of multiplicity m should be counted as m zeros.

EXAMPLE 1   **Using Descartes's Rule of Signs**

Find the possible number of positive and negative zeros of

$$f(x) = x^5 - 3x^3 - 5x^2 + 9x - 7.$$

Solution

There are three variations of sign in $f(x)$:

$$f(x) = x^5 - 3x^3 - 5x^2 + 9x - 7$$

The number of positive zeros of $f(x)$ must be either 3 or $3 - 2 = 1$. Next,

$$f(-x) = (-x)^5 - 3(-x)^3 - 5(-x)^2 + 9(-x) - 7$$
$$= -x^5 + 3x^3 - 5x^2 - 9x - 7.$$

The polynomial $f(-x)$ has two variations of sign; therefore, the number of negative zeros of $f(x)$ is either 2 or $2 - 2 = 0$.

Practice Problem 1 Find the possible number of positive and negative zeros of

$$f(x) = 2x^5 + 3x^2 + 5x - 1.$$

---

2 Find the bounds on the real zeros of polynomials.

Bounds on the Real Zeros

If a polynomial function $F(x)$ has no zeros greater than a number k, then k is called an **upper bound** on the zeros of $F(x)$. Similarly, if $F(x)$ has no zeros less than a number k, then k is called a **lower bound** on the zeros of $F(x)$. These bounds have the following rules.

RULES FOR BOUNDS

Let $F(x)$ be a polynomial function with real coefficients and a positive leading coefficient. Suppose $F(x)$ is synthetically divided by $x - k$.

1. If $k > 0$ and each number in the last row is zero or positive, then k is an upper bound on the zeros of $F(x)$.

2. If $k < 0$ and numbers in the last row alternate in sign (zeros in the last row can be regarded as positive or negative), then k is a lower bound on the zeros of $F(x)$.

EXAMPLE 2 **Finding the Bounds on the Zeros**

Find upper and lower bounds on the zeros of

$$F(x) = x^4 - x^3 - 5x^2 - x - 6.$$

Solution

By the Rational Zeros Test, the possible rational zeros of $F(x)$ are $\pm 1, \pm 2, \pm 3,$ and ± 6. There is only one variation of sign in $F(x) = x^4 - x^3 - 5x^2 - x - 6$, so $F(x)$ has only one positive zero. First, we try synthetic division by $x - k$ with $k = 1, 2, 3, \ldots$. The first integer that makes all numbers in the last row 0 or positive is an upper bound on the zeros of $F(x)$.

```
1| 1  -1  -5  -1  -6    With k = 1        2| 1  -1  -5  -1   -6    With k = 2
        1   0  -5  -6                              2   2  -6  -14
   ─────────────────────                      ──────────────────────
   1   0  -5  -6  -12                          1   1  -3  -7  -20
```

```
           3| 1  -1  -5  -1  -6       Synthetic division with k = 3
                   3   6   3   6
              ─────────────────────
              1   2   1   2   0  ←──── All numbers are 0 or positive.
```

By the rule for bounds, 3 is an upper bound on the zeros of $F(x)$.

We now try synthetic division by $x - k$ with $k = -1, -2, -3, \ldots$. The first negative integer for which the numbers in the last row alternate in sign is a lower bound on the zeros of $F(x)$.

```
           -1| 1  -1  -5  -1  -6       With k = -1
                  -1   2   3  -2
              ─────────────────────
              1  -2  -3   2  -8
```

```
           -2| 1  -1  -5  -1  -6       With k = -2
                  -2   6  -2   6
              ─────────────────────                These numbers alternate in
              1  -3   1  -3   0  ←──── sign if we count 0 as positive.
```

By the rule of bounds, -2 is a lower bound on the zeros of $F(x)$.
We conclude that all real zeros of $F(x)$ are located in the interval $[-2, 3]$

Practice Problem 2 Find upper and lower bounds on the zeros of

$$f(x) = x^3 - 3x^2 - 2x + 2.$$

SIDE NOTE

A graphing calculator may be used to obtain the approximate location of the real zeros. For a rational zero, verify by synthetic division.

STEPS FOR GATHERING INFORMATION AND LOCATING THE REAL ZEROS OF A POLYNOMIAL FUNCTION

Step 1 Find the maximum number of real zeros by using the degree of the polynomial function.

Step 2 Find the possible number of positive and negative zeros by using Descartes's Rule of Signs.

Step 3 Write the set of possible rational zeros.

Step 4 Test the smallest positive integer in the set in Step 3, then the next larger, and so on, until an integer zero or an upper bound of the zeros is found.
    a. If a zero is found, use the function represented by the depressed equation in further calculations.
    b. If an upper bound is found, discard all larger numbers in the set of possible rational zeros.

Step 5 Test the positive fractions that remain in the set in Step 3 after considering any bound that has been found.

Step 6 Use modified Steps 4 and 5 for negative numbers in the set in Step 3.

Step 7 If a depressed equation is quadratic, use any method (including the quadratic formula) to solve this equation.

3 Learn basic facts about the complex zeros of polynomials.

Complex Zeros of Polynomials

The Factor Theorem connects the concept of *factors* and *zeros* of a polynomial, and we continue to investigate this connection.

Because the equation $x^2 + 1 = 0$ has no real roots, the polynomial function $P(x) = x^2 + 1$ has no real zeros. However, if we replace x with the complex number i, we have

$$P(i) = i^2 + 1 = (-1) + 1 = 0.$$

Consequently, i is a complex number for which $P(i) = 0$; that is, i is a *complex zero* of $P(x)$. In Section 2.2, we stated that a polynomial of degree n can have, at most, n real zeros. We now extend our number system to allow the coefficients of polynomials and variables to represent complex numbers. When we want to emphasize that complex numbers may be used in this way, we call the polynomial $P(x)$ a **complex polynomial**. If $P(z) = 0$ for a complex number z, we say that z is a **zero** or a **complex zero** of $P(x)$. In the complex number system, every nth-degree polynomial equation has *exactly n* roots and every nth-degree polynomial can be factored into exactly n linear factors. This fact follows from the Fundamental Theorem of Algebra.

> ## RECALL
>
> A number z of the form $z = a + bi$ is a **complex number**, where a and b are real numbers and $i = \sqrt{-1}$.

FUNDAMENTAL THEOREM OF ALGEBRA

Every polynomial

$$P(x) = a_n x^n + a_{n-1} x^{n-1} + \cdots + a_1 x + a_0 \quad (n \geq 1, a_n \neq 0)$$

with complex coefficients $a_n, a_{n-1}, \ldots, a_1, a_0$ has at least one complex zero.

> ## RECALL
>
> If $b = 0$, the complex number $a + bi$ is a real number. So the set of real numbers is a subset of the set of complex numbers.

This theorem was proved by the mathematician Karl Friedrich Gauss at age 20. The proof is beyond the scope of this book, but if we are allowed to use complex numbers, we can use the theorem to prove that every polynomial has a complete factorization.

FACTORIZATION THEOREM FOR POLYNOMIALS

If $P(x)$ is a complex polynomial of degree $n \geq 1$, it can be factored into n (not necessarily distinct) linear factors of the form

$$P(x) = a(x - r_1)(x - r_2) \cdots (x - r_n),$$

where $a, r_1, r_2, \ldots, r_n$ are complex numbers.

To prove the Factorization Theorem for Polynomials, we notice that if $P(x)$ is a polynomial of degree 1 or higher with leading coefficient a, then the Fundamental Theorem of Algebra states that there is a zero r_1 for which $P(r_1) = 0$. By the Factor Theorem, which also holds for complex polynomials, $(x - r_1)$ is a factor of $P(x)$ and

$$P(x) = (x - r_1)q_1(x),$$

where the degree of $q_1(x)$ is $n - 1$ and its leading coefficient is a. If $n - 1$ is positive, then we apply the Fundamental Theorem to $q_1(x)$; this gives us a zero, say, r_2, of $q_1(x)$. By the Factor Theorem, $(x - r_2)$ is a factor of $q_1(x)$ and

$$q_1(x) = (x - r_2)q_2(x),$$

where the degree of $q_2(x)$ is $n - 2$ and its leading coefficient is a. Then

$$P(x) = (x - r_1)(x - r_2)q_2(x).$$

This process can be continued until $P(x)$ is completely factored as

$$P(x) = a(x - r_1)(x - r_2) \cdots (x - r_n),$$

where a is the leading coefficient and $r_1, r_2, \ldots, r_n$ are zeros of $P(x)$. The proof is now complete.

In general, the numbers $r_1, r_2, \ldots, r_n$ may not be distinct. As with the polynomials we encountered previously, if the factor $(x - r)$ appears m times in the complete factorization of $P(x)$, then we say that r is a zero of *multiplicity m*. The polynomial $P(x) = (x - 2)^3(x - i)^2$ has zeros:

$$2 \text{ (of multiplicity 3) and } i \text{ (of multiplicity 2)}.$$

The polynomial $P(x) = (x - 2)^3(x - i)^2$ has only 2 and i as its zeros, although it is of degree 5.

> ### NUMBER OF ZEROS THEOREM
>
> Any polynomial of degree n has exactly n zeros, provided a zero of multiplicity k is counted k times.

EXAMPLE 3 Constructing a Polynomial Whose Zeros Are Given

Find a polynomial $P(x)$ of degree 4 with a leading coefficient of 2 and zeros $-1, 3, i,$ and $-i$. Write $P(x)$

a. in completely factored form.　　b. by expanding the product found in part **a**.

Solution

a. Since $P(x)$ has degree 4, we write:

$$P(x) = a(x - r_1)(x - r_2)(x - r_3)(x - r_4). \quad \text{Completely factored form}$$

Now replace the leading coefficient a with 2 and the zeros $r_1, r_2, r_3,$ and r_4 with $-1, 3, i,$ and $-i$ (in any order). Then

$$P(x) = 2[x - (-1)](x - 3)(x - i)[x - (-i)]$$
$$= 2(x + 1)(x - 3)(x - i)(x + i). \quad \text{Simplify.}$$

b. $P(x) = 2(x + 1)(x - 3)(x - i)(x + i)$
$\quad\quad = 2(x + 1)(x - 3)(x^2 + 1) \quad\quad$ Expand $(x - i)(x + i); i^2 = -1.$
$\quad\quad = 2(x + 1)(x^3 - 3x^2 + x - 3) \quad\quad$ Expand $(x - 3)(x^2 + 1).$
$\quad\quad = 2(x^4 - 2x^3 - 2x^2 - 2x - 3) \quad\quad$ Expand $(x + 1)(x^3 - 3x^2 + x - 3).$
$\quad\quad = 2x^4 - 4x^3 - 4x^2 - 4x - 6 \quad\quad$ Distributive property

Practice Problem 3 Find a polynomial $P(x)$ of degree 4 with a leading coefficient of 3 and zeros $-2, 1, 1 + i, 1 - i$. Write $P(x)$

a. in completely factored form.　　b. by expanding the product found in part **a**.

4 Use the Conjugate Pairs Theorem to find zeros of polynomials.

RECALL

The conjugate of a complex number $z = a + bi$ is $\bar{z} = a - bi$. For two complex numbers z_1 and z_2, we have (see Appendix A.8)
$$\overline{z_1 + z_2} = \bar{z_1} + \bar{z_2}$$
$$\overline{z_1 z_2} = \bar{z_1}\,\bar{z_2}.$$

Conjugate Pairs Theorem

In Example 3, we constructed a polynomial that had both i and its conjugate, $-i$, among its zeros. Our next result says that, for all polynomials with real coefficients, nonreal zeros occur in conjugate pairs.

> ### CONJUGATE PAIRS THEOREM
>
> If $P(x)$ is a polynomial function whose coefficients are real numbers and if $z = a + bi$ is a zero of P, then its conjugate, $\bar{z} = a - bi$, is also a zero of P.

To prove the Conjugate Pairs Theorem, let

$$P(x) = a_n x^n + a_{n-1} x^{n-1} + \cdots + a_1 x + a_0,$$

where $a_n, a_{n-1}, \ldots, a_1, a_0$ are real numbers. Now suppose that $P(z) = 0$. Then we must show that $P(\bar{z}) = 0$ also. We will use the fact that the conjugate of a real number $a = a + 0i$ is identical to the real number a (since $\bar{a} = a - 0i = a$) and then use the conjugate sum and product properties $\overline{z_1 + z_2} = \overline{z_1} + \overline{z_2}$ and $\overline{z_1 z_2} = \overline{z_1}\,\overline{z_2}$.

$$P(z) = a_n z^n + a_{n-1} z^{n-1} + \cdots + a_1 z + a_0 \qquad \text{Replace } x \text{ with } z.$$
$$P(\bar{z}) = a_n (\bar{z})^n + a_{n-1}(\bar{z})^{n-1} + \cdots + a_1 \bar{z} + a_0 \qquad \text{Replace } z \text{ with } \bar{z}.$$
$$= \overline{a_n}\,\overline{z^n} + \overline{a_{n-1}}\,\overline{z^{n-1}} + \cdots + \overline{a_1}\,\overline{z} + \overline{a_0} \qquad (\bar{z})^k = \overline{z^k}, \overline{a_k} = a_k$$
$$= \overline{a_n z^n} + \overline{a_{n-1} z^{n-1}} + \cdots + \overline{a_1 z} + \overline{a_0} \qquad \overline{z_1}\,\overline{z_2} = \overline{z_1 z_2}$$
$$= \overline{a_n z^n + a_{n-1} z^{n-1} + \cdots + a_1 z + a_0} \qquad \overline{z_1} + \overline{z_2} = \overline{z_1 + z_2}$$
$$= \overline{P(z)} = \overline{0} = 0 \qquad P(z) = 0$$

We see that if $P(z) = 0$, then $P(\bar{z}) = 0$, and the theorem is proved.

This theorem has several uses. If, for example, we know that $2 - 5i$ is a zero of a polynomial with real coefficients, then we know that $2 + 5i$ is also a zero. Because nonreal zeros occur in conjugate pairs, we know that there will always be an *even* number of nonreal zeros. Therefore, any polynomial of odd degree with real coefficients has at least one zero that is a real number.

> ### ODD-DEGREE POLYNOMIALS WITH REAL ZEROS
> Any polynomial $P(x)$ of odd degree with real coefficients must have at least one real zero.

For example, the polynomial $P(x) = 5x^7 + 2x^4 + 9x - 12$ has at least one zero that is a real number because $P(x)$ has degree 7 (odd degree) and has real numbers as coefficients.

SIDE NOTE

In problems like Example 4, once you determine all the zeros, you can use the Factorization Theorem to write a polynomial with those zeros.

EXAMPLE 4 Using the Conjugate Pairs Theorem

A polynomial $P(x)$ of degree 9 with real coefficients has the following zeros: 2, of multiplicity 3; $4 + 5i$, of multiplicity 2; and $3 - 7i$. First determine the number of zeros, then write the zeros of $P(x)$.

Solution

Since complex zeros occur in conjugate pairs, the conjugate $4 - 5i$ of $4 + 5i$ is a zero of multiplicity 2 and the conjugate $3 + 7i$ of $3 - 7i$ is a zero of $P(x)$. Counting multiplicities, this gives nine zeros of $P(x)$:

$$2, 2, 2, 4 + 5i, 4 + 5i, 4 - 5i, 4 - 5i, 3 + 7i, \text{ and } 3 - 7i.$$

Practice Problem 4 A polynomial $P(x)$ of degree 8 with real coefficients has the following zeros: -3 and $2 - 3i$, each of multiplicity 2, and i. First determine the number of zeros, then write the zeros of $P(x)$.

Every polynomial of degree n has exactly n zeros and can be factored into a product of n linear factors. If the polynomial has real coefficients, then, by the Conjugate Pairs Theorem, its nonreal zeros occur as conjugate pairs. Consequently, if $r = a + bi$ is a zero, then so is $\bar{r} = a - bi$, and both $x - r$ and $x - \bar{r}$ are linear factors of the polynomial. Multiplying these factors, we have:

$$(x - r)(x - \bar{r}) = x^2 - (r + \bar{r})x + r\bar{r} \qquad \text{Use FOIL.}$$
$$= x^2 - 2ax + (a^2 + b^2) \qquad r + \bar{r} = 2a; r\bar{r} = a^2 + b^2$$

The quadratic polynomial $x^2 - 2ax + (a^2 + b^2)$ has real coefficients and is *irreducible* (cannot be factored any further) *over the real numbers* because its discriminant $(-2a)^2 - 4(a^2 + b^2) = -4b^2 < 0$. This result shows that each pair of nonreal conjugate zeros are the zeros for one *irreducible* quadratic factor.

> ### FACTORIZATION THEOREM FOR A POLYNOMIAL WITH REAL COEFFICIENTS
>
> Every polynomial with real coefficients can be uniquely factored over the real numbers as a product of linear factors and/or irreducible quadratic factors.

EXAMPLE 5 **Finding the Zeros of a Polynomial from a Given Complex Zero**

Given that $2 - i$ is a zero of $P(x) = x^4 - 6x^3 + 14x^2 - 14x + 5$, find the remaining zeros.

Solution

Since $P(x)$ has real coefficients, the conjugate $\overline{2 - i} = 2 + i$ is also a zero. By the Factorization Theorem, the linear factors $[x - (2 - i)]$ and $[x - (2 + i)]$ appear in the factorization of $P(x)$. Consequently, their product

$$
\begin{aligned}
[x - (2 - i)][x - (2 + i)] &= (x - 2 + i)(x - 2 - i) \\
&= [(x - 2) + i][(x - 2) - i] && \text{Regroup.} \\
&= (x - 2)^2 - i^2 && (A + B)(A - B) = A^2 - B^2 \\
&= x^2 - 4x + 4 + 1 && \text{Expand } (x - 2)^2, i^2 = -1. \\
&= x^2 - 4x + 5 && \text{Simplify.}
\end{aligned}
$$

is also a factor of $P(x)$. We divide $P(x)$ by $x^2 - 4x + 5$ to find the other factor.

RECALL

Dividend =
Quotient · Divisor + Remainder

$$
\require{enclose}
\begin{array}{r}
x^2 - 2x + 1 \quad \leftarrow \text{Quotient} \\
x^2 - 4x + 5 \enclose{longdiv}{x^4 - 6x^3 + 14x^2 - 14x + 5} \quad \leftarrow \text{Dividend} \\
\underline{x^4 - 4x^3 + 5x^2} \\
-2x^3 + 9x^2 - 14x \\
\underline{-2x^3 + 8x^2 - 10x} \\
x^2 - 4x + 5 \\
\underline{x^2 - 4x + 5} \\
0 \quad \leftarrow \text{Remainder}
\end{array}
$$

$$
\begin{aligned}
P(x) &= (x^2 - 2x + 1)(x^2 - 4x + 5) && P(x) = \text{Quotient} \cdot \text{Divisor} \\
&= (x - 1)(x - 1)(x^2 - 4x + 5) && \text{Factor } x^2 - 2x + 1. \\
&= (x - 1)(x - 1)[x - (2 - i)][x - (2 + i)] && \text{Factor } x^2 - 4x + 5.
\end{aligned}
$$

The zeros of $P(x)$ are 1 (of multiplicity 2), $2 - i$, and $2 + i$.

Practice Problem 5 Given that $2i$ is a zero of $P(x) = x^4 - 3x^3 + 6x^2 - 12x + 8$, find the remaining zeros. ▪▪

EXAMPLE 6 **Finding the Zeros of a Polynomial**

Find all zeros of the polynomial $P(x) = x^4 - x^3 + 7x^2 - 9x - 18$.

Solution

Since the degree of $P(x)$ is 4, $P(x)$ has four zeros. The Rational Zeros Test tells us that the possible rational zeros are

$$\pm 1, \pm 2, \pm 3, \pm 6, \pm 9, \pm 18.$$

Testing these possible zeros by synthetic division, we find that 2 is a zero.

$$
\begin{array}{r|rrrrr}
2 & 1 & -1 & 7 & -9 & -18 \\
  &   & 2 & 2 & 18 & 18 \\
\hline
  & 1 & 1 & 9 & 9 & 0
\end{array}
$$

$P(2) = 0$, so $(x - 2)$ is a factor of $P(x)$ and the quotient is $x^3 + x^2 + 9x + 9$.
We can solve the depressed equation $x^3 + x^2 + 9x + 9 = 0$ by factoring.

$$x^2(x + 1) + 9(x + 1) = 0 \qquad \text{Group terms, distributive property}$$
$$(x^2 + 9)(x + 1) = 0 \qquad \text{Distributive property}$$
$$x^2 + 9 = 0 \quad \text{or} \quad x + 1 = 0 \qquad \text{Zero-product property}$$
$$x^2 = -9 \quad \text{or} \quad x = -1 \qquad \text{Solve each equation.}$$
$$x = \pm\sqrt{-9} = \pm 3i \qquad \text{Solve } x^2 = -9 \text{ for } x.$$

The four zeros of $P(x)$ are $-1, 2, 3i,$ and $-3i$. The complete factorization of $P(x)$ is

$$P(x) = x^4 - x^3 + 7x^2 - 9x - 18 = (x + 1)(x - 2)(x - 3i)(x + 3i).$$

Practice Problem 6 Find all zeros of the polynomial

$$P(x) = x^4 - 8x^3 + 22x^2 - 28x + 16.$$

Answers to Practice Problems

1. Number of positive zeros: 1; number of negative zeros: 0 or 2
2. Upper bound: 4; lower bound: -1
3. **a.** $P(x) = 3(x + 2)(x - 1)(x - 1 - i)(x - 1 + i)$
   b. $P(x) = 3x^4 - 3x^3 - 6x^2 + 18x - 12$

4. There are eight zeros: $-3, -3, 2 + 3i, 2 + 3i, 2 - 3i, 2 - 3i,$ $i, -i$
5. The zeros are $1, 2, 2i,$ and $-2i$.
6. The zeros are $2, 4, 1 + i,$ and $1 - i$.

SECTION 2.6 **Exercises**

Concepts and Vocabulary

1. If the nonzero terms of a polynomial are written in descending order, then a variation of sign occurs when the signs of two _____ terms differ.

2. The number of positive zeros of a polynomial function $P(x)$ is equal to the number of _____ of sign of $P(x)$ or is less than that number by _____.

3. The Fundamental Theorem of Algebra States that a polynomial function of degree $n \geq 1$ has at least one _____ zero.

4. If P is a polynomial function with real coefficients and if $z = a + bi$ is a zero of P, then _____ is also a zero of P.

5. **True or False.** Every polynomial of odd degree with real coefficients has at least one real zero.

6. **True or False.** There is a polynomial with real coefficients that has $2 + 3i$ as its only nonreal zero.

7. **True or False.** The quadratic polynomial $x^2 + 1$ is irreducible (cannot be factored any farther) over the real numbers.

8. **True or False.** $\sqrt{-16} = \pm 4$.

Basic Skills

In Exercises 9–16, determine the possible number of positive and negative zeros of the given function by using Descartes's Rule of Signs.

9. $f(x) = 5x^3 - 2x^2 - 3x + 4$

10. $g(x) = 3x^3 + x^2 - 9x - 3$

11. $f(x) = 2x^3 + 5x^2 - x + 2$

12. $g(x) = 3x^4 + 8x^3 - 5x^2 + 2x - 3$

13. $h(x) = 2x^5 - 5x^3 + 3x^2 + 2x - 1$

14. $F(x) = 5x^6 - 7x^4 + 2x^3 - 1$

15. $G(x) = -3x^4 - 4x^3 + 5x^2 - 3x + 7$

16. $H(x) = -5x^5 + 3x^3 - 2x^2 - 7x + 4$

17. $f(x) = x^4 + 2x^2 + 4$

18. $f(x) = 3x^4 + 5x^2 + 6$

19. $g(x) = 2x^5 + x^3 + 3x$

20. $g(x) = 2x^5 + 4x^3 + 5x$

21. $h(x) = -x^5 - 2x^3 + 4$

22. $h(x) = 2x^5 + 3x^3 + 1$

In Exercises 23–30, determine upper and lower bounds on the zeros of the given function.

23. $f(x) = 3x^3 - x^2 + 9x - 3$

24. $g(x) = 2x^3 - 3x^2 - 14x + 21$

25. $F(x) = 3x^3 + 2x^2 + 5x + 7$

26. $G(x) = x^3 + 3x^2 + x - 4$

27. $h(x) = x^4 + 3x^3 - 15x^2 - 9x + 31$

28. $H(x) = 3x^4 - 20x^3 + 28x^2 + 19x - 13$

29. $f(x) = 6x^4 + x^3 - 43x^2 - 7x + 7$

30. $g(x) = 6x^4 + 23x^3 + 25x^2 - 9x - 5$

In Exercises 31–40, find all solutions of the equation in the complex number system.

31. $x^2 + 25 = 0$ 32. $9x^2 + 16 = 0$

33. $(x - 2)^2 + 9 = 0$

34. $(x - 1)^2 + 27 = 0$

35. $x^2 + 4x + 4 = -9$

36. $x^2 + 2x + 1 = -16$

37. $x^3 - 8 = 0$

38. $x^4 - 1 = 0$

39. $(x - 2)(x - 3i)(x + 3i) = 0$

40. $(x - 1)(x - 2i)(2x - 6i)(2x + 6i) = 0$

In Exercises 41–46, find the remaining zeros of a polynomial $P(x)$ with real coefficients and with the specified degree and zeros.

41. Degree 3; zeros: $2, 3 + i$

42. Degree 3; zeros: $2, 2 - i$

43. Degree 4; zeros: $0, 1, 2 - i$

44. Degree 4; zeros: $-i, 1 - i$

45. Degree 6; zeros: $0, 5, i, 3i$

46. Degree 6; zeros: $2i, 4 + i, i - 1$

In Exercises 47–50, find the polynomial $P(x)$ with real coefficients having the specified degree, leading coefficient, and zeros.

47. Degree 4; leading coefficient 2; zeros: $5 - i, 3i$

48. Degree 4; leading coefficient -3; zeros: $2 + 3i, 1 - 4i$

49. Degree 5; leading coefficient 7; zeros: 5 (multiplicity 2), 1, $3 - i$

50. Degree 6; leading coefficient 4; zeros: 3, 0 (multiplicity 3), $2 - 3i$

In Exercises 51–54, use the given zero to find all the zeros for each function.

51. $P(x) = x^4 + x^3 + 9x^2 + 9x$, zero: $3i$

52. $P(x) = x^4 - 2x^3 + x^2 + 2x - 2$, zero: $1 - i$

53. $P(x) = x^5 - 5x^4 + 2x^3 + 22x^2 - 20x$, zero: $3 - i$

54. $P(x) = 2x^5 - 11x^4 + 19x^3 - 17x^2 + 17x - 6$, zero: i

In Exercises 55–64, find all zeros of each polynomial function.

55. $P(x) = x^3 - 9x^2 + 25x - 17$

56. $P(x) = x^3 - 5x^2 + 7x + 13$

57. $P(x) = 3x^3 - 2x^2 + 22x + 40$

58. $P(x) = 3x^3 - x^2 + 12x - 4$

59. $P(x) = 2x^4 - 10x^3 + 23x^2 - 24x + 9$

60. $P(x) = 9x^4 + 30x^3 + 14x^2 - 16x + 8$

61. $P(x) = x^4 - 4x^3 - 5x^2 + 38x - 30$

62. $P(x) = x^4 + x^3 + 7x^2 + 9x - 18$

63. $P(x) = 2x^5 - 11x^4 + 19x^3 - 17x^2 + 17x - 6$

64. $P(x) = x^5 - 2x^4 - x^3 + 8x^2 - 10x + 4$

In Exercises 65–68, find an equation of a polynomial function of least degree having the given complex zeros, intercepts, and graph.

65. f has complex zero $3i$

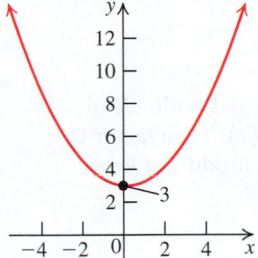

66. f has complex zero $-i$

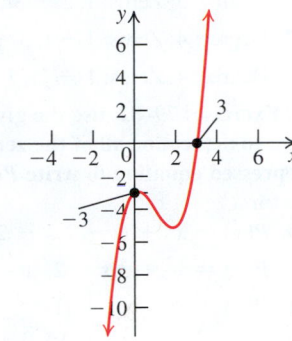

67. f has complex zeros i and $2i$

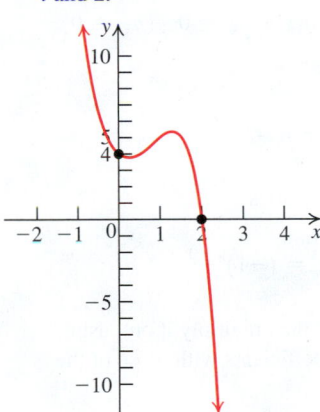

68. f has complex zero $-2i$

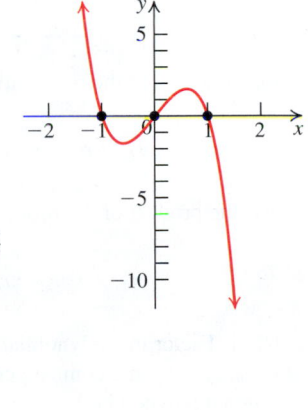

Beyond the Basics

69. The solutions -1 and 1 of the equation $x^2 = 1$ are called the square roots of 1. The solutions of the equation $x^3 = 1$ are called the cube roots of 1. Find the cube roots of 1. How many are there?

70. The solutions of the equation $x^n = 1$, where n is a positive integer, are called the nth roots of 1 or "nth roots of unity." Explain the relationship between the solutions of the equation $x^n = 1$ and the zeros of the complex polynomial $P(x) = x^n - 1$. How many nth roots of 1 are there?

71. Show that the polynomial function
$P(x) = x^6 + 2x^4 + 3x^2 + 4$ has six nonreal zeros.

72. Show that the polynomial function
$F(x) = x^5 + x^3 + 2x + 1$ has four nonreal zeros.

73. Show that the polynomial function $f(x) = x^7 + 5x + 3$ has six nonreal zeros.

74. Show that the polynomial function $g(x) = x^3 - x^2 - 1$ has two nonreal zeros.

In Exercises 75–78, find an equation with real coefficients of a polynomial function f that has the given characteristics. Then write the end behavior of the graph of $y = f(x)$.

75. Degree: 3; Zeros: $2, 1 + 2i$; y-intercept: 40

76. Degree: 3; Zeros: $1, 2 - 3i$; y-intercept: -26

77. Degree: 4; Zeros: $1, -1, 3 + i$; y-intercept: 20

78. Degree: 4; Zeros: $1 - 2i, 3 - 2i$; y-intercept: 130

In Exercises 79–82, use the given zero and synthetic division to determine all of the zeros of $P(x)$. Then factor the depressed equation to write $P(x)$ as a product of linear factors.

79. $P(x) = x^2 + (i - 2)x - 2i$, $x = -i$

80. $P(x) = x^2 + 3ix - 2$, $x = -2i$

81. $P(x) = x^3 - (3 + i)x^2 - (4 - 3i)x + 4i$, $x = i$

82. $P(x) = x^3 - (4 + 2i)x^2 + (7 + 8i)x - 14i$, $x = 2i$

Critical Thinking/Discussion/Writing

83. Show that if $r_1, r_2, \ldots, r_n$ are the roots of the equation
$$a_n x^n + a_{n-1} x^{n-1} + \cdots + a_1 x + a_0 = 0 \quad (a_n \neq 0),$$
then the sum of the roots satisfies
$$r_1 + r_2 + \cdots + r_n = -\frac{a_{n-1}}{a_n}$$
and the product of the roots satisfies
$$r_1 \cdot r_2 \cdot \cdots \cdot r_n = (-1)^n \frac{a_0}{a_n}.$$
[*Hint:* Factor the polynomial; then multiply it out, using $r_1, r_2, \ldots, r_n$, and compare coefficients with those of the original polynomial.]

84. Show that if r_1, r_2 are the roots of the equation
$$ax^2 + bx + c = 0 \quad (a \neq 0)$$
then $r_1^2 + r_2^2 = \dfrac{b^2 - 2ac}{a^2}$
[*Hint:* $x^2 + y^2 = (x + y)^2 - 2xy$].

◆ 85. In his book *Ars Magna,* Cardano explained how to solve cubic equations. He considered the following example:
$$x^3 + 6x = 20.$$

a. Explain why this equation has exactly one real solution.

b. Cardano explained the method as follows: "I take two cubes v^3 and u^3 whose difference is 20 and whose product is 2, that is, a third of the coefficient of x. Then, I say that $x = v - u$ is a solution of the equation." Show that if $v^3 - u^3 = 20$ and $vu = 2$, then $x = v - u$ is indeed the solution of the equation $x^3 + 6x = 20$.

c. Solve the system
$$v^3 - u^3 = 20$$
$$vu = 2$$
to find u and v.

d. Consider the equation $x^3 + px = q$, where p is a positive number. Using your work in parts (a), (b), and (c) as a guide, show that the unique solution of this equation is
$$x = \sqrt[3]{\frac{q}{2} + \sqrt{\left(\frac{q}{2}\right)^2 + \left(\frac{p}{3}\right)^3}} - \sqrt[3]{-\frac{q}{2} + \sqrt{\left(\frac{q}{2}\right)^2 + \left(\frac{p}{3}\right)^3}}.$$

e. Consider an arbitrary cubic equation
$$x^3 + ax^2 + bx + c = 0.$$
Show that the substitution $x = y - \dfrac{a}{3}$ allows you to write the cubic equation as
$$y^3 + py = q.$$

86. Use Cardano's method to solve the equation
$x^3 + 6x^2 + 10x + 8 = 0$.

Getting Ready for the Next Section

87. **a.** Find the equation of the line in slope–intercept form that passes through the point $(-5, 3)$ and has slope $-\dfrac{2}{3}$.

b. From the equation in part (a), find y when $x = 1$.

88. **a.** Find the equation of the line in slope–intercept form that passes through the point $(-2, -3)$ and is perpendicular to the line $4x + 5y = 6$.

b. From the equation in part (a), find y when $x = 10$.

89. Find k if the graph of the equation $y = kx^2 + 1$ passes through the point $(-2, 7)$. Then find y when $x = 2$.

90. Find k if the graph of the equation $y = \dfrac{k}{x^2}$ passes through the point $\left(\dfrac{1}{2}, 12\right)$. Then find y when $x = 2$.

Variation

BEFORE STARTING THIS SECTION, REVIEW

1 Equation of a line (Section 1.2, page 25)

2 Circumference and area of a circle (Appendix A.5, page 952)

OBJECTIVES

1 Solve direct variation problems.

2 Solve inverse variation problems.

3 Solve joint and combined variation problems.

Sir Isaac Newton 1642–1727

Isaac Newton was a mathematician and physicist, one of the foremost scientific intellects of all time. He entered Cambridge University in 1661 and was elected a Fellow of Trinity College in 1667 and Lucasian Professor of Mathematics in 1669. He remained at the university, lecturing most years, until 1696.

As a firm opponent of the attempt by King James II to make the universities into Catholic institutions, Newton was elected Member of Parliament for the University of Cambridge to the Convention Parliament of 1689, and he sat again in 1701–1702.

Meanwhile, in 1696, he moved to London as warden of the Royal Mint. He became master of the Mint in 1699, an office he retained until his death. He was elected a Fellow of the Royal Society of London in 1671, and in 1703, he became president, being annually reelected for the rest of his life. One of his major works, *Opticks*, appeared the next year; he was knighted in Cambridge in 1705.

◆ Newton and the Apple

According to legend, Isaac Newton was sitting under an apple tree, and when an apple fell on his head, he suddenly thought of the Law of Universal Gravitation. As in many such legends, the story may not be true in detail, but it contains elements of the truth.

Before we tell you a more realistic version of this story, let's recall some terminology associated with the motion of objects.

$$\text{Recall that speed} = \frac{\text{distance}}{\text{time}}.$$

(i) The **velocity** of an object is the speed of an object in a specific direction.

(ii) **Acceleration**: If the velocity of an object is changing, we say that the object is accelerating. Acceleration is the rate of change of velocity.

(iii) **Force** = Mass × Acceleration

What really happened with the apple? Perhaps the correct version of the story is that upon observing an apple falling from a tree, Newton concluded that the apple accelerated because the velocity of the apple changed from when it began hanging on the tree (zero velocity) and then moved to the ground. He decided that a force must act on the apple to cause this acceleration. He called this force "gravity" and the associated acceleration (Force = Mass × Acceleration) the "acceleration due to gravity." He conjectured further that if the force of gravity reaches the top of the apple tree, it might reach even farther; might it reach all the way to the moon? In that case, to keep the moon moving in a circular orbit, rather than wandering into outer space, Earth must exert a gravitational force on the moon. Newton realized that the force that brought the apple to the ground and the force that holds the moon in orbit were the same. He concluded that any two objects in the universe exert gravitational attraction on each other, whereupon he gave us the *Law of Universal Gravitation*. (See Example 6.)

Some scientists believe that it was Kepler's Third Law of Planetary Motion (see Exercise 53), not an apple, that led Newton to his Law of Universal Gravitation.

1 Solve direct variation problems.

Direct Variation

Two types of relationships between quantities (variables) occur so frequently in mathematics that they are given special names: *direct variation* and *inverse variation*.

Direct variation describes any relationship between two quantities in which any increase (or decrease) in one causes a proportional increase (or decrease) in the other. For example, the sales tax in Hillsborough County, Florida, in 2015 was 7%, so that the sales tax on a $1000 purchase was $70. If we *double* the purchase to $2000, the sales tax also *doubles*, to $140. If we *reduce* the purchase by *half* to $500, the sales tax is also *reduced* by half, to $35. We say that the sales tax *varies directly* as the purchase price. The equation

$$\text{Sales tax} = (0.07)(\text{Purchase price})$$

expresses the fact that the sales tax was a constant (0.07) multiple of the purchase price.

SIDE
NOTE

The statements
"*y* varies directly as *x*,"
"*y* varies as *x*,"
"*y* is directly proportional to *x*,"
"*y* is proportional to *x*,"
and "$y = kx, k \neq 0$"
have the same meaning.

Direct Variation

A quantity y is said to **vary directly** as the quantity x, or y is **directly proportional** to x, if there is a constant k such that $y = kx$. This constant k is called the **constant of variation** or the **constant of proportionality**.

PROCEDURE
IN ACTION

EXAMPLE 1 Solving the Variation Problems

OBJECTIVE	EXAMPLE
Solve variation problems.	*Suppose y varies as x and $y = 20$ when $x = \dfrac{4}{3}$. Find y when $x = 8$.*
Step 1 Write the equation with the constant of variation, k.	**1.** $y = kx$ *y varies as x.*
Step 2 Substitute the given values of the variables into the equation in Step 1 to find the value of the constant k.	**2.** $20 = k\left(\dfrac{4}{3}\right)$ *Replace y with 20 and x with $\dfrac{4}{3}$.*
	$20\left(\dfrac{3}{4}\right) = k\left(\dfrac{4}{3}\right)\left(\dfrac{3}{4}\right)$ *Multiply both sides by $\dfrac{3}{4}$.*
	$15 = k$ *Simplify to solve for k.*
Step 3 Rewrite the equation in Step 1 with the value of k from Step 2.	**3.** $y = 15x$ *Replace k with 15.*
Step 4 Use the equation from Step 3 to answer the question posed in the problem.	**4.** $y = 15(8)$ *Replace x with 8.*
	$y = 120$ *Simplify.*
	So $y = 120$ when $x = 8$.

Practice Problem 1 Suppose y varies directly as x. If y is 6 when x is 30, find y when $x = 120$.

EXAMPLE 2 Direct Variation in Electrical Circuits

The current in a circuit connected to a 220-volt battery is 50 amperes. If the current is directly proportional to the voltage of the attached battery, what voltage battery is needed to produce a current of 75 amperes?

Solution

Let I = current in amperes and V = voltage in volts of the battery.

Step 1 $\qquad I = kV \qquad\qquad$ I varies directly as V.

Step 2 $\qquad 50 = k(220) \qquad$ Substitute $I = 50$, $V = 220$.

$$\frac{50}{220} = k \qquad\qquad$$ Solve for k.

$$\frac{5}{22} = k \qquad\qquad$$ Simplify.

Step 3 $\qquad I = \frac{5}{22}V \qquad$ Replace k with $\frac{5}{22}$ in $I = kV$.

Step 4 $\qquad 75 = \frac{5}{22}V \qquad$ Substitute $I = 75$ in Step 3.

$$\frac{22}{5} \cdot 75 = V \qquad$$ Solve for V.

$$330 = V \qquad\qquad$$ Simplify.

A battery of 330 volts is needed to produce 75 amperes of current.

Practice Problem 2 In Example 2, if the current in the circuit is 60 amperes, what voltage battery is needed to produce a current of 75 amperes?

We can generalize the concept of direct variation to variation with powers.

Direct Variation with Powers

"A quantity y varies directly as the nth power of x" means

$$y = kx^n,$$

where k is a nonzero constant and $n > 0$.

Note that when $n = 1$, y varies directly as x. In the formula for the area of a circle, $A = \pi r^2$, A varies directly as the square (or second power) of the radius r, with π as the constant of variation.

EXAMPLE 3 Solving a Problem Involving Direct Variation with Powers

Suppose you had forgotten the formula for the volume of a sphere but were told that the volume V of a sphere varies directly as the cube of its radius r. In addition, you are given that $V = 972\pi$ when $r = 9$. Find V when $r = 6$.

Solution

Step 1 $\qquad V = kr^3 \qquad\qquad$ V varies directly as r^3.

Step 2 $\quad 972\pi = k(9)^3 \qquad$ Replace V with 972π and r with 9.

$\qquad\quad 972\pi = k(729) \qquad$ $9^3 = 729$

$$k = \frac{972\,\pi}{729} \qquad\qquad$$ Solve for k.

$$= \frac{4}{3}\,\pi \qquad\qquad$$ Simplify.

Step 3 $\qquad V = \frac{4}{3}\pi r^3 \qquad$ Substitute $k = \frac{4}{3}\pi$ in Step 1.

Step 4 $\qquad V = \dfrac{4}{3}\pi(6)^3 \qquad\qquad$ Replace r with 6.

$\qquad\qquad\qquad = 288\,\pi$ cubic units

Practice Problem 3 If y varies directly as the square of x and $y = 48$ when $x = 2$, find y when $x = 5$.

2 Solve inverse variation problems.

Inverse Variation

Inverse Variation

> A quantity y **varies inversely** as the quantity x, or y is **inversely proportional** to x, if
>
> $$y = \frac{k}{x}$$
>
> where k is a nonzero constant.

Suppose a plane takes four hours to fly from Atlanta to Denver at an average speed of 300 miles per hour. If we *increase* the average speed to 600 miles per hour (twice the previous speed), the flight time *decreases* to two hours (half the previous time). For the Atlanta–Denver trip, the time of flight varies inversely as the speed of the plane:

$$\text{time} = \frac{\text{distance}}{\text{speed}}$$

$$= \frac{k}{\text{speed}}, \text{ where } k = \text{ distance between Atlanta and Denver}$$

We see that the distance is the constant of variation when time and speed are the variables.

EXAMPLE 4 **Solving an Inverse Variation Problem**

Suppose y varies inversely as x and $y = 35$ when $x = 11$. Find y if $x = 55$.

Solution

Step 1 $\qquad y = \dfrac{k}{x} \qquad\qquad$ y varies inversely as x.

Step 2 $\qquad 35 = \dfrac{k}{11} \qquad\qquad$ Substitute $y = 35,\ x = 11$.

$\qquad\qquad 35 \cdot 11 = k \qquad\qquad$ Solve for k.

$\qquad\qquad\quad 385 = k \qquad\qquad$ Simplify.

Step 3 $\qquad y = \dfrac{385}{x} \qquad\qquad$ Replace k with 385 in $y = \dfrac{k}{x}$.

Step 4 $\qquad y = \dfrac{385}{55} \qquad\qquad$ Replace x with 55.

$\qquad\qquad\quad y = 7 \qquad\qquad$ Simplify.

Practice Problem 4 A varies inversely as B, and $A = 12$ when $B = 5$. Find A when $B = 3$.

Inverse Variation with Powers

A quantity y varies inversely as the nth power of x means that $y = \dfrac{k}{x^n}$, where k is a nonzero constant and $n > 0$.

EXAMPLE 5   Solving a Problem Involving Inverse Variation with Powers

The intensity of light varies inversely as the square of the distance from the light source. If Rita doubles her distance from a lamp, what happens to the intensity of light at her new location?

Solution

Let I be the intensity of light at a distance d from the light source. Since I is inversely proportional to the square of d, we have

$$I = \frac{k}{d^2}.$$

If we replace d with $2d$, the intensity I_1 at the new location is given by

$$I_1 = \frac{k}{(2d)^2} \qquad \text{Replace } d \text{ with } 2d \text{ in } I = \frac{k}{d^2}.$$

$$I_1 = \frac{k}{4d^2} \qquad (2d)^2 = 2^2 d^2 = 4d^2$$

Therefore,

$$I_1 = \frac{I}{4} \qquad \text{Replace } \frac{k}{d^2} \text{ with } I.$$

This equation tells us that if Rita doubles her distance from the light source, the intensity of light at the new location will be one-fourth the intensity at the original location.

Practice Problem 5 If y varies inversely as the square root of x and $y = \dfrac{3}{4}$ when $x = 16$, find x when $y = 2$.

3 Solve joint and combined variation problems.

Joint and Combined Variation

Sometimes different types of variations occur in a problem that has more than one independent variable.

Joint Variation

The expression z *varies jointly as* x *and* y means that $z = kxy$ for some nonzero constant k. Similarly, if n and m are positive numbers, then the expression z *varies jointly as the nth power of* x *and mth power of* y means that $z = kx^n y^m$ for some nonzero constant k.

For example, the volume V of a right circular cylinder with radius r and height h is given by

$$V = \pi r^2 h.$$

Using the vocabulary of variation, we say, "The volume V of a right circular cylinder varies jointly as its height h and the square of its radius r." The constant of variation is π.

We can combine joint variation and inverse variation. For example, the relationship

$$w = \frac{kx^2 y^3}{\sqrt{z}}$$

can be described as "w varies jointly as the square of x and cube of y and inversely as the square root of z." The same relationship also can be stated as "w varies directly as $x^2 y^3$ and inversely as $\sqrt{z}$."

EXAMPLE 6 Newton's Law of Universal Gravitation

Newton's Law of Universal Gravitation says that every object in the universe attracts every other object with a force acting along the line of the centers of the two objects. Also, this attracting force is directly proportional to the product of the two masses and inversely proportional to the square of the distance between the two objects.

a. Write the law symbolically.

b. Estimate the value of g (the acceleration due to gravity) near the surface of Earth. Use the following estimates: radius of Earth $R_E = 6.38 \times 10^6$ meters, and mass of Earth $M_E = 5.98 \times 10^{24}$ kilograms.

Solution

a. Let m_1 and m_2 be the masses of the two objects and r be the distance between their centers. See Figure 2.44. Let F denote the gravitational force between the objects.

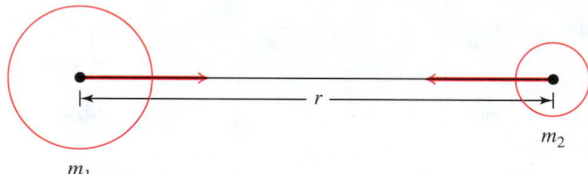

Figure 2.44 Gravitational attraction.

Then according to Newton's Law of Universal Gravitation,

$$F = G \cdot \frac{m_1 m_2}{r^2}.$$

The constant of proportionality G is called the *universal gravitational constant*. It is "universal" because it is thought to be the same at all places and all times. If the masses m_1 and m_2 are measured in kilograms, r is measured in meters, and the force F is measured in newtons, then the value of G is approximately 6.67×10^{-11} meters cubed per kilogram per second squared. This value was experimentally verified by Henry Cavendish in 1795.

b. *Estimating the value of g* (acceleration due to gravity) *near the surface of Earth.*

We use Newton's Laws: Force = Mass × Acceleration = $m \cdot g$

$$\text{Force} = G \cdot \frac{m_1 m_2}{r^2}$$

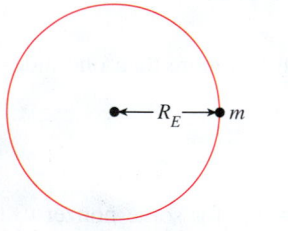

Figure 2.45 Finding g on Earth.

For an object of mass m near the surface of Earth (see Figure 2.45):

$$\text{Force} = G \cdot \frac{m_1 m_2}{r^2} \qquad \text{Law of Universal Gravitation}$$

$$\not{m} \cdot g = G \cdot \frac{\not{m} M_E}{R_E^2} \qquad \text{Force} = m \cdot g; \text{ replace } m_1 \text{ with } m \text{ and } m_2 \text{ with } M_E \text{ and } r \text{ with } R_E.$$

$$g = G \cdot \frac{M_E}{R_E^2} \qquad \text{Divide both sides by } m.$$

$$= \frac{(6.67 \times 10^{-11} \, \text{m}^3/\text{kg}/\text{sec}^2) \cdot (5.98 \times 10^{24} \, \text{kg})}{(6.38 \times 10^6 \, \text{m})^2} \qquad \text{Substitute appropriate values for } G, M_E, \text{ and } R_E.$$

$$\approx 9.8 \, \text{m}/\text{sec}^2 \qquad \text{Use a calculator.}$$

Practice Problem 6 The mass of Mars is about 6.42×10^{23} kilograms, and its radius is about 3397 kilometers. What is the acceleration due to gravity near the surface of Mars?

We summarize the steps involved in solving most variation problems.

> ### SOLVING VARIATION PROBLEMS
>
> **Step 1** Write the equation that models the problem.
>
>   **a.** y varies directly with x. $y = kx$
>
>   **b.** y varies with the nth power of x. $y = kx^n$
>
>   **c.** y varies inversely with x. $y = \dfrac{k}{x}$
>
>   **d.** y varies inversely with the nth power of x. $y = \dfrac{k}{x^n}$
>
>   **e.** z varies jointly with the nth power of x and the mth power of y. $z = kx^n y^m$
>
>   **f.** z varies directly with the nth power of x and inversely with the mth power of y. $z = \dfrac{kx^n}{y^m}$
>
> **Step 2** Substitute the given values in the equation in Step 1 and solve for k.
>
> **Step 3** Rewrite the equation in Step 1 with the value of k from Step 2.
>
> **Step 4** Use the equation from Step 3 to answer the question posed in the problem.

If a variation problem uses a more complicated relationship than those listed in Step 1, simply write the appropriate equation involving k and the variables and then proceed with Steps 2–4.

Answers to Practice Problems

1. $y = 24$ **2.** 275   **3.** $y = 300$ **4.** $A = 20$ **5.** $x = \dfrac{9}{4}$ **6.** $\approx 3.7 \, \text{m/sec}^2$

SECTION 2.7 Exercises

Concepts and Vocabulary

1. y varies directly as x if _____.

2. y varies inversely as x if _____.

3. y varies directly as the nth power of x if _____.

4. z varies jointly as x and y if _____.

5. **True or False.** In the equation $P = 2T$, P varies directly as T.

6. **True or False.** In the equation $A = 3r^2$, A varies directly as the third power of r.

7. **True or False.** In the equation $D = 2t + t^2$, D varies directly as the second power of t.

8. **True or False.** In the equation $Q = \dfrac{\pi \omega^3}{\sqrt{u}}$, Q varies jointly as the third power of ω and inversely as the square root of u.

Building Skills

In Exercises 9–24, use the four-step procedure to solve for the variable requested.

9. x varies directly as y, and $x = 15$ when $y = 30$. Find x if $y = 28$.

10. y varies directly as x, and $y = 3$ when $x = 2$. Find y when $x = 7$.

11. s varies directly as the square of t, and $s = 64$ when $t = 2$. Find s when $t = 5$.

12. y varies directly as the cube of x, and $y = 270$ when $x = 3$. Find x when $y = 80$.

13. r varies inversely as u, and $r = 3$ when $u = 11$. Find r if $u = \dfrac{1}{3}$.

14. y varies inversely as z, and $y = 24$ when $z = \frac{1}{6}$. Find y if $z = 1$.

15. B varies inversely with the cube of A. If $B = 1$ when $A = 2$, find B when $A = 4$.

16. y varies inversely with the cube root of x, and $y = 10$ when $x = 2$. Find x if $y = 40$.

17. z varies jointly as x and y, and $z = 42$ when $x = 2$ and $y = 3$. Find y if $z = 56$ and $x = 2$.

18. m varies directly as q and inversely as p, and $m = \frac{1}{2}$ when $p = 26$ and $q = 13$. Find m when $p = 14$ and $q = 7$.

19. z varies directly as the square of x, and $z = 32$ when $x = 4$. Find z if $x = 5$.

20. u varies inversely as the cube of t, and $u = 9$ when $t = 2$. Find u if $t = 6$.

21. P varies jointly as T and the square of Q, and $P = 36$ when $T = 17$ and $Q = 6$. Find P when $T = 4$ and $Q = 9$.

22. a varies jointly as b and the square root of c, and $a = 9$ when $b = 13$ and $c = 81$. Find a when $b = 5$ and $c = 9$.

23. z varies directly as the square root of x and inversely as the square of y, and $z = 24$ when $x = 16$ and $y = 3$. Find x when $z = 27$ and $y = 2$.

24. z varies jointly as u and the cube of v and inversely as the square of w; $z = 9$ when $u = 4$, $v = 3$, and $w = 2$. Find w when $u = 27$, $v = 2$, and $z = 8$.

In Exercises 25–28, solve for the variable requested without determining the constant of variation, k. Use the fact that if $x_1 = ky_1$ and $x_2 = ky_2$, then $\dfrac{x_1}{y_1} = k = \dfrac{x_2}{y_2}$ so that $\dfrac{x_1}{y_1} = \dfrac{x_2}{y_2}$.

25. If y is proportional to x and if $y = 12$ when $x = 16$, find y when $x = 8$.

26. If z varies directly as w and if $z = 17$ when $w = 22$, find z when $w = 110$.

27. If y is directly proportional to x and if $y = 100$ for the value x_0 of x, find y when x_0 is doubled—that is, when $x = 2x_0$.

28. If x varies directly as the square root of y and if $x = 2$ when $y = 9$, find y when $x = 3$.

Applying the Concepts

29. Hubble constant. The American astronomer Edwin Powell Hubble (1889–1953) is renowned for having determined that there are other galaxies in the universe beyond the Milky Way. In 1929, he stated that the galaxies observed at a particular time recede from each other at a speed that is directly proportional to the distance between them. The constant of proportionality is denoted by the letter H. Write Hubble's statement in the form of an equation.

30. Malthusian doctrine. The British scientist Thomas Robert Malthus (1766–1834) was a pioneer of population science and economics. In 1798, he stated that populations grow faster than the means that can sustain them. One of his arguments was that the rate of change, R, of a given population is directly proportional to the size P of the population. Write the equation that describes Malthus's argument.

31. Converting units of length. Use the fact that 1 foot ≈ 30.5 centimeters.
 a. Write an equation that expresses the fact that a length measured in centimeters is directly proportional to the length measured in feet.
 b. Convert the following measurements into centimeters.
  (i) 8 feet
  (ii) 5 feet 4 inches
 c. Convert the following measurements into feet.
  (i) 57 centimeters
  (ii) 1 meter 24 centimeters

32. Converting weight. Use the fact that 1 kilogram ≈ 2.20 pounds.
 a. Write an equation that expresses the fact that a weight measured in pounds is directly proportional to the weight measured in kilograms.
 b. Convert the following measurements to pounds.
  (i) 125 grams
  (ii) 4 kilograms
  (iii) 2.4 kilograms
 c. Convert the following measurements to kilograms.
  (i) 27 pounds
  (ii) 160 pounds

33. Protein from soybeans. The quantity P of protein obtained from dried soybeans is directly proportional to the quantity Q of dried soybeans used. If 20 grams of dried soybeans produces 7 grams of protein, how much protein can be produced from 100 grams of soybeans?

34. Wages. A person's weekly wages W are directly proportional to the number of hours h worked per week. If Samantha earned $600 for a 40-hour week, how much would she earn if she worked only 25 hours in a particular week? What does the constant of proportionality mean here?

35. Physics. The distance d that an object falls varies directly as the square of the time t during which it is falling. If an object falls 64 feet in 2 seconds, how long will it take the object to fall 9 feet?

36. Physics. Hooke's Law states that the force F required to stretch a spring by x units is directly proportional to x. If a force of 10 pounds stretches a spring by 4 inches, find the force required to stretch a spring by 6 inches.

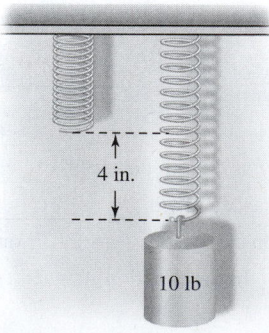

4 in.

10 lb

37. Chemistry. Boyle's Law states that at a constant temperature, the pressure P of a compressed gas is inversely proportional to its volume V. If the pressure is 20 pounds per square inch when the volume of the gas is 300 cubic inches, what is the pressure when the gas is compressed to 100 cubic inches?

38. Chemistry. In the Kelvin temperature scale, the lowest possible temperature (called absolute zero) is 0 K, where K denotes degrees Kelvin. The relationship between Kelvin temperature (T_K) and Celsius temperature (T_C) is given by $T_K = T_C + 273$.

The pressure P exerted by a gas varies directly as its temperature T_K and inversely as its volume V. Assume that at a temperature of 260 K, a gas occupies 13 cubic inches at a pressure of 36 pounds per square inch.
  a. Find the volume of the gas when the temperature is 300 K and the pressure is 40 pounds per square inch.
  b. Find the pressure when the temperature is 280 K and the volume is 39 cubic inches.

39. Weight. The weight of an object varies inversely as the square of the object's distance from the center of Earth. The radius of Earth is 3960 miles.
  a. If an astronaut weighs 120 pounds on the surface of Earth, how much does she weigh 6000 miles above the surface of Earth?
  b. If a miner weighs 200 pounds on the surface of Earth, how much does he weigh 10 miles below the surface of Earth?

40. Making a profit. Suppose you wanted to make a profit by buying gold by weight at one altitude and selling at another altitude for the same price per unit weight. Should you buy or sell at the higher altitude? (Use the information from Exercise 39.)

In Exercises 41 and 42, use Newton's Law of Universal Gravitation. (See Example 6.)

41. Gravity on the moon. The mass of the moon is about 7.4×10^{22} kilograms, and its radius is about 1740 kilometers. How much is the acceleration due to gravity on the surface of the moon?

42. Gravity on the sun. The mass of the sun is about 2×10^{30} kilograms, and its radius is 696,000 kilometers. How much is the acceleration due to gravity on the surface of the sun?

43. Illumination. The intensity I of illumination from a light source is inversely proportional to the square of the distance d from the source. Suppose the intensity is 320 candela at a distance of 10 feet from a light source.
  a. What is the intensity at 5 feet from the source?
  b. How far away from the source will the intensity be 400 candela?

44. Speed and skid marks. Police estimate that the speed s of a car in miles per hour varies directly as the square root of d, where d in feet is the length of the skid marks left by a car traveling on a dry concrete pavement. A car traveling 48 miles per hour leaves skid marks of 96 feet.
  a. Write an equation relating s and d.
  b. Use the equation in part (**a**) to estimate the speed of a car whose skid marks stretched (i) 60 feet, (ii) 150 feet, and (iii) 200 feet.
  c. Suppose you are driving 70 miles per hour and slam on your brakes. How long will your skid marks be?

45. Simple pendulum. *Periodic motion* is motion that repeats itself over successive equal intervals of time. The time required for one complete repetition of the motion is called the *period*. The period of a simple pendulum varies directly as the square root of its length. What is the effect on the length if the period is doubled?

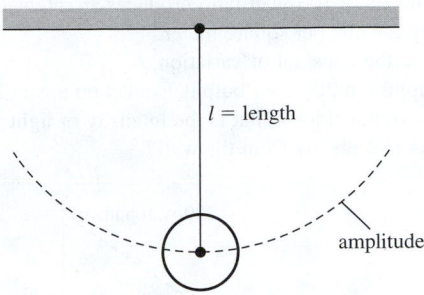

(*Note:* The amplitude of a pendulum has no effect on its period. This is what makes pendulums such good time-keepers. Because they invariably lose energy due to friction, their amplitude decreases but their period remains constant.)

46. Biology. The volume V of a lung is directly proportional to its internal surface area A. A lung with volume 400 cubic centimeters from a certain species has an average internal surface area of 100 square centimeters. Find the volume of a lung of a member of this species if the lung's internal surface area is 120 square centimeters.

47. Horsepower. The horsepower H of an automobile engine varies directly as the square of the piston radius R and the number N of pistons.
  a. Write the given information in equation form.
  b. What is the effect on the horsepower if the piston radius is doubled?
  c. What is the effect on the horsepower if the number of pistons is doubled?
  d. What is the effect on the horsepower if the radius of the pistons is cut in half and the number of pistons is doubled?

48. Safe load. The safe load that a rectangular beam can support varies jointly as the width and square of the depth of the beam and inversely as its length. A beam 4 inches wide, 6 inches deep, and 25 feet long can support a safe load of 576 pounds. Find the safe load for a beam that is of the same material but is 6 inches wide, 10 inches deep, and 20 feet long.

Beyond the Basics

49. Energy from a windmill. The energy E from a windmill varies jointly as the square of the length l of the blades and the cube of the wind velocity v.
  a. Express the given information as an equation with k as the constant of variation.
  b. If blades of length 10 feet and wind velocity 8 miles per hour generate 1920 watts of electric power, find k.
  c. How much electric power would be generated if the blades were 8 feet long and the wind velocity was 25 miles per hour?
  d. If the velocity of the wind doubles, what happens to E?
  e. If the length of the blades doubles, what happens to E?
  f. What happens to E if both the length of the blades and the wind velocity are doubled?

50. Intensity of light. The intensity of light, I_d, at a distance d from the source of light varies directly as the intensity I of the source and inversely as d^2. It is known that at a distance of 2 meters, a 100-watt bulb produces an intensity of approximately 2 watts per square meter.

a. Find the constant of variation, k.

b. Suppose a 200-watt bulb is located on a wall 2 meters above the floor. What is the intensity of light at a point A that is 3 meters from the wall?

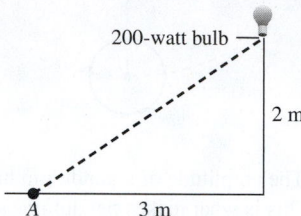

200-watt bulb

2 m

A 3 m

c. What is the intensity of illumination at point A in the figure if the bulb is raised by 1 meter?

51. Metabolic rate. Metabolism is the sum total of all physical and chemical changes that take place within an organism. According to the laws of thermodynamics, all of these changes will ultimately release heat, so the metabolic rate is a measure of heat production by an animal. Biologists have found that the normal resting metabolic rate of a mammal is directly proportional to the $\frac{3}{4}$ power of its body weight.

The resting metabolic rate of a person weighing 75 kilograms is 75 watts.

a. Find the constant of proportionality, k.

b. Estimate the resting metabolic rate of a brown bear weighing 450 kilograms.

c. What is the effect on metabolic rate if the body weight is multiplied by 4?

d. What is the approximate weight of an animal with a metabolic rate of 250 watts?

52. Comparing gravitational forces. The masses of the sun, Earth, and the moon are 2×10^{30} kilograms, 6×10^{24} kilograms, and 7.4×10^{22} kilograms, respectively. The Earth–sun distance is about 400 times the Earth–moon distance. Use Newton's Law of Universal Gravitation to compare the gravitational attraction between the sun and Earth with that between Earth and the moon.

53. Kepler's Third Law. Suppose an object of mass M_1 orbits around an object of mass M_2. Let r be the average distance in meters between the centers of the two objects and let T be the *orbital period* (the time in seconds the object completes one orbit). Kepler's Third Law states that T^2 is directly proportional to r^3 and inversely proportional to $M_1 + M_2$. The constant of proportionality is $\dfrac{4\pi^2}{G}$.

a. Write the equation that expresses Kepler's Third Law.

b. Earth orbits the sun once each year at a distance of about 1.5×10^8 kilometers. Find the mass of the sun. Use these estimates:

Mass of Earth + Mass of sun ≈ Mass of sun,
$G = 6.67 \times 10^{-11}$, 1 year $= 3.15 \times 10^7$ seconds

54. Use Kepler's Third Law. The moon orbits Earth in about 27.3 days at an average distance of about 384,000 kilometers. Find the mass of Earth. [*Hint:* Convert the orbital period, defined in Exercise 53, to seconds; also use Mass of Earth + Mass of moon ≈ Mass of Earth.]

55. Spreading a rumor. Let P be the population of a community. The rate R (per day) at which a rumor spreads in the community is jointly proportional to the number N of people who have already heard the rumor and the number $(P - N)$ of people who have not yet heard the rumor.

A rumor about the president of a college is spread in his college community of 10,000 people. Five days after the rumor started, 1000 people had heard it, and it was spreading at the rate of 45 additional people per day.

a. Write an equation relating R, P, and N.

b. Find the constant k of variation.

c. Find the rate at which the rumor was spreading when one-half the college community had heard it.

d. How many people had heard the rumor when it was spreading at the rate of 100 people per day?

Critical Thinking/Discussion/Writing

56. Electricity. The current I in an electric circuit varies directly as the voltage V and inversely as the resistance R. If the resistance is increased by 20%, what percent increase must occur in the voltage to increase the current by 30%?

57. Precious stones. The value of a precious stone is proportional to the square of its weight.

a. Calculate the loss incurred by cutting a diamond worth $1000 into two pieces whose weights are in the ratio 2: 3.

b. A precious stone worth $25,000 is accidentally dropped and broken into three pieces, the weights of which are in the ratio 5:9:11. Calculate the loss incurred due to breakage.

c. A diamond breaks into five pieces, the weights of which are in the ratio 1:2:3:4:5. If the resulting loss is $85,000, find the value of the original diamond. Also calculate the value of a diamond whose weight is twice that of the original diamond.

58. Bus service. The profit earned in running a bus service is jointly proportional to the distance and the number of passengers over a certain fixed number. The profit is $80 when 30 passengers are carried a distance of 40 km and is $180 when 35 passengers are carried 60 km. What is the minimum number of passengers that results in no loss?

59. Weight of a sphere. The weight of a sphere is directly proportional to the cube of its radius. A metal sphere has a hollow space about its center in the form of a concentric sphere, and its weight is $\frac{7}{8}$ times the weight of a solid sphere of the same radius and material. Find the ratio of the inner to the outer radius of the hollow sphere.

60. Train speed. A locomotive engine can go 24 miles per hour, and its speed is reduced by a quantity that varies directly as the square root of the number of cars it pulls. Pulling four cars, its speed is 20 miles per hour. Find the greatest number of cars the engine can pull.

Getting Ready for the Next Section

In Exercises 61–68, simplify each expression.

61. 2^3 **62.** 3^{-2} **63.** $\left(\dfrac{1}{2}\right)^3$ **64.** $\left(\dfrac{1}{2}\right)^{-4}$

65. $2^{x-1} \cdot 2^{3-x}$ **66.** $5^{2x-3} \cdot 5^{3-x}$ **67.** $\dfrac{2^{3x-2}}{2^{x-5}}$ **68.** $(2^{x-1})^x$

SUMMARY Definitions, Concepts, and Formulas

2.1 Quadratic Functions

i. A **quadratic function** f is a function of the form
$$f(x) = ax^2 + bx + c, a \neq 0.$$

ii. The **standard form** of a quadratic function is
$$f(x) = a(x - h)^2 + k, a \neq 0.$$

iii. The graph of a quadratic function is a transformation of the graph of $y = x^2$.

iv. The graph of a quadratic function is a parabola with vertex
$$(h, k) = \left(-\frac{b}{2a}, f\left(-\frac{b}{2a}\right)\right).$$

v. The maximum (if $a < 0$) or minimum (if $a > 0$) value of a quadratic function $f(x) = ax^2 + bx + c$ occurs at the vertex of the parabola.

2.2 Polynomial Functions

i. A function f of the form
$$f(x) = a_n x^n + a_{n-1} x^{n-1} + \cdots + a_1 x + a_0, a_n \neq 0$$
is a polynomial function of degree n.

ii. The graph of a polynomial function is smooth and continuous.

iii. The end behavior of the graph of a polynomial function depends upon the sign of the leading coefficient and the degree of the polynomial.

iv. A real number c is a **zero** of a function f if $f(c) = 0$. Geometrically, c is an x-intercept of the graph of $y = f(x)$.

v. If in the factorization of a polynomial function $f(x)$ the factor $(x - a)$ occurs exactly m times, then a is a zero of **multiplicity** m. If m is odd, the graph of $y = f(x)$ crosses the x-axis at a; if m is even, the graph touches but does not cross the x-axis at a.

vi. If the degree of a polynomial function $f(x)$ is n, then $f(x)$ has, at most, n real zeros and the graph of $f(x)$ has at most $(n - 1)$ turning points.

vii. **An Intermediate Value Theorem:** Let $f(x)$ be a polynomial function and a and b be two numbers such that $a < b$. If $f(a)$ and $f(b)$ have opposite signs, then there is at least one number c, with $a < c < b$, for which $f(c) = 0$.

viii. See pages 163–164 for graphing a polynomial function.

2.3 Dividing Polynomials and the Rational Zeros Test

i. **Division Algorithm:** If a polynomial $F(x)$ is divided by a polynomial $D(x) \neq 0$, there are unique polynomials $Q(x)$ and $R(x)$ such that $F(x) = D(x)Q(x) + R(x)$, where either $R(x) = 0$ or $\deg R(x) < \deg D(x)$. In words, "The dividend equals the product of the divisor and the quotient plus the remainder."

ii. **Remainder Theorem:** If a polynomial $F(x)$ is divided by $(x - a)$, the remainder is $F(a)$.

iii. **Factor Theorem:** A polynomial function $F(x)$ has $(x - a)$ as a factor if and only if $F(a) = 0$.

iv. **Rational Zeros Test:** If $\frac{p}{q}$ is a rational zero in lowest terms for a polynomial function with integer coefficients, then p is a factor of the constant term and q is a factor of the leading coefficient.

2.4 Rational Functions

i. A function $F(x) = \frac{N(x)}{D(x)}$, where $N(x)$ and $D(x)$ are polynomials and $D(x) \neq 0$, is called a rational function. The domain of $F(x)$ is the set of all real numbers except the real zeros of $D(x)$.

ii. The line $x = a$ is a vertical asymptote of the graph of f if $f(x) \to \infty$ or $f(x) \to -\infty$ as $x \to a^+$ or as $x \to a^-$.

iii. If $\frac{N(x)}{D(x)}$ is in lowest terms, then the graph of $F(x)$ has vertical asymptotes at the real zeros of $D(x)$.

iv. The line $y = k$ is a horizontal asymptote of the graph of f if $f(x) \to k$ as $x \to \infty$ or if $f(x) \to k$ as $x \to -\infty$.

v. A procedure for graphing rational functions is given on page 196.

2.5 Polynomial and Rational Inequalities

i. Graphs or test numbers can be used to solve polynomial or rational inequalities.

ii. **Sign of Polynomial Function:** The polynomial $P(x)$ has a constant sign on each interval formed by the real zeros of $P(x)$ on a number line.

iii. A procedure for solving polynomial inequalities is given on page 211.

iv. **Sign of Rational Function:** The rational function
$$R(x) = \frac{N(x)}{D(x)}$$
has a constant sign on each interval formed by the real zeros of $N(x)$ and $D(x)$ plotted on a number line.

v. A procedure for solving rational inequalities is given on page 215.

2.6 Zeros of a Polynomial Function

i. **Descartes's Rule of Signs.** Let $F(x)$ be a polynomial function with real coefficients.
   a. The number of positive zeros of F is equal to the number of variations of sign of $F(x)$ or is less than that number by an even integer.
   b. The number of negative zeros of F is equal to the number of variations of sign of $F(-x)$ or is less than that number by an even integer.

ii. **Rules for Bounds on the Zeros.** Suppose a polynomial $F(x)$ is synthetically divided by $x - k$.
   a. If $k > 0$ and each number in the last row is zero or positive, then k is an upper bound on the zeros of $F(x)$.

b. If $k < 0$ and the numbers in the last row alternate in sign, then k is a lower bound on the zeros of $F(x)$ where 0 can be regarded as either positive or negative.

iii. The Fundamental Theorem of Algebra. An nth-degree polynomial equation has at least one complex zero.

iv. Factorization Theorem for Polynomials. If $P(x)$ is a polynomial of degree $n \geq 1$, it can be factored into n (not necessarily distinct) linear factors of the form $P(x) = a(x - r_1)(x - r_2) \cdots (x - r_n)$, where a, $r_1, r_2, \ldots, r_n$ are complex numbers.

v. Number of Zeros Theorem. A polynomial of degree n has exactly n complex zeros provided that a zero of multiplicity k is counted k times.

vi. Conjugate Pairs Theorem. If $a + bi$ is a zero of the polynomial function P (with real coefficients), then $a - bi$ is also a zero of P.

2.7 Variation

k is a nonzero constant called the constant of variation.

Variation	Equation
y varies directly with x.	$y = kx$
y varies with the nth power of x.	$y = kx^n$
y varies inversely with x.	$y = \dfrac{k}{x}$
y varies inversely with the nth power of x.	$y = \dfrac{k}{x^n}$
z varies jointly with the nth power of x and the mth power of y.	$z = kx^ny^m$
z varies directly with the nth power of x and inversely with the mth power of y.	$z = \dfrac{kx^n}{y^m}$

REVIEW EXERCISES

Building Skills

In Exercises 1–10, graph each quadratic function by finding (i) whether the parabola opens up or down, (ii) its vertex, (iii) its axis, (iv) its x-intercepts, (v) its y-intercept, and (vi) the intervals over which the function is increasing and decreasing.

1. $y = (x - 1)^2 + 2$ **2.** $y = (x + 2)^2 - 3$

3. $y = -2(x - 3)^2 + 4$ **4.** $y = -\dfrac{1}{2}(x + 1)^2 + 2$

5. $y = -2x^2 + 3$ **6.** $y = 2x^2 + 4x - 1$

7. $y = 2x^2 - 4x + 3$ **8.** $y = -2x^2 - x + 3$

9. $y = 3x^2 - 2x + 1$ **10.** $y = 3x^2 - 5x + 4$

In Exercises 11–14, determine whether the given quadratic function has a maximum or a minimum value and then find that value.

11. $f(x) = 3 - 4x + x^2$ **12.** $f(x) = 8x - 4x^2 - 3$

13. $f(x) = -2x^2 - 3x + 2$ **14.** $f(x) = \dfrac{1}{2}x^2 - \dfrac{3}{4}x + 2$

In Exercises 15–18, graph each polynomial function by using transformations on the appropriate function $y = x^n$.

15. $f(x) = (x + 1)^3 - 2$ **16.** $f(x) = (x + 1)^4 + 2$

17. $f(x) = (1 - x)^3 + 1$ **18.** $f(x) = x^4 + 3$

In Exercises 19–24, for each polynomial function f,

i. determine the end behavior of f.

ii. determine the zeros of f. State the multiplicity of each zero. Determine whether the graph of f crosses or only touches the axis at each x-intercept.

iii. find the x- and y-intercepts of the graph of f.

iv. use test numbers to find the intervals over which the graph of f is above or below the x-axis.

v. find any symmetry.

vi. sketch the graph of $y = f(x)$.

19. $f(x) = x(x - 1)(x + 2)$ **20.** $f(x) = x^3 - x$

21. $f(x) = -x^2(x - 1)^2$ **22.** $f(x) = -x^3(x - 2)^2$

23. $f(x) = -x^2(x^2 - 1)$

24. $f(x) = -(x - 1)^2(x^2 + 1)$

In Exercises 25–28, divide by using long division.

25. $\dfrac{6x^2 + 5x - 13}{3x - 2}$ **26.** $\dfrac{8x^2 - 14x + 15}{2x - 3}$

27. $\dfrac{8x^4 - 4x^3 + 2x^2 - 7x + 165}{x + 1}$

28. $\dfrac{x^3 - 3x^2 + 4x + 7}{x^2 - 2x + 6}$

In Exercises 29–32, divide by using synthetic division.

29. $\dfrac{x^3 - 12x + 3}{x - 3}$ **30.** $\dfrac{-4x^3 + 3x^2 - 5x}{x - 6}$

31. $\dfrac{2x^4 - 3x^3 + 5x^2 - 7x + 165}{x + 1}$

32. $\dfrac{3x^5 - 2x^4 + x^2 - 16x - 132}{x + 2}$

In Exercises 33–36, a polynomial function $f(x)$ and a constant c are given. Find $f(c)$ by (i) evaluating the function and (ii) using synthetic division and the Remainder Theorem.

33. $f(x) = x^3 - 3x^2 + 11x - 29; c = 2$

34. $f(x) = 2x^3 + x^2 - 15x - 2; c = -2$

35. $f(x) = x^4 - 2x^2 - 5x + 10; c = -3$

36. $f(x) = x^5 + 2; c = 1$

In Exercises 37–40, a polynomial function $f(x)$ and a constant c are given. Use synthetic division to show that c is a zero of $f(x)$. Use the result to find all zeros of $f(x)$.

37. $f(x) = x^3 - 7x^2 + 14x - 8; c = 2$

38. $f(x) = 2x^3 - 3x^2 - 12x + 4; c = -2$

39. $f(x) = 3x^3 + 14x^2 + 13x - 6; c = \dfrac{1}{3}$

40. $f(x) = 4x^3 + 19x^2 - 13x + 2; c = \dfrac{1}{4}$

In Exercises 41 and 42, use the Rational Zeros Test to list all possible rational zeros of $f(x)$.

41. $f(x) = x^4 + 3x^3 - x^2 - 9x - 6$

42. $f(x) = 9x^3 - 36x^2 - 4x + 16$

In Exercises 43–46, use Descartes's Rule of Signs and the Rational Zeros Test to find all real zeros of each polynomial function.

43. $f(x) = 5x^3 + 11x^2 + 2x$

44. $f(x) = x^3 + 2x^2 - 5x - 6$

45. $f(x) = x^3 + 3x^2 - 4x - 12$

46. $f(x) = 2x^3 - 9x^2 + 12x - 5$

In Exercises 47–52, find all of the zeros of $f(x)$, real and nonreal, that are not given.

47. $f(x) = x^3 - 7x + 6$; one zero is 2.

48. $f(x) = x^4 + x^3 - 3x^2 - x + 2$; 1 is a zero of multiplicity 2.

49. $f(x) = x^4 - 2x^3 + 6x^2 - 18x - 27$; two zeros are -1 and 3.

50. $f(x) = 4x^3 - 19x^2 + 32x - 15$; one zero is $2 - i$.

51. $f(x) = x^4 + 2x^3 + 9x^2 + 8x + 20$; one zero is $-1 + 2i$.

52. $f(x) = x^5 - 7x^4 + 24x^3 - 32x^2 + 64$; $2 + 2i$ is a zero of multiplicity 2.

In Exercises 53–58, solve each equation in the complex number system.

53. $x^3 - x^2 - 4x + 4 = 0$

54. $2x^3 + x^2 - 12x - 6 = 0$

55. $4x^3 - 7x - 3 = 0$

56. $x^3 + x^2 - 8x - 6 = 0$

57. $x^4 - x^3 - x^2 - x - 2 = 0$

58. $x^4 - x^3 - 13x^2 + x + 12 = 0$

In Exercises 59 and 60, show that the given equation has no rational roots.

59. $x^3 + 13x^2 - 6x - 2 = 0$

60. $3x^4 - 9x^3 - 2x^2 - 15x - 5 = 0$

In Exercises 61 and 62, use the Intermediate Value Theorem to find the value of the real root between 1 and 2 of each equation to two decimal places.

61. $x^3 + 6x^2 - 28 = 0$

62. $x^3 + 3x^2 - 3x - 7 = 0$

In Exercises 63–70, graph each rational function by following the six-step procedure outlined in Section 2.5.

63. $f(x) = 1 + \dfrac{1}{x}$

64. $f(x) = \dfrac{2 - x}{x}$

65. $f(x) = \dfrac{x}{x^2 - 1}$

66. $f(x) = \dfrac{x^2 - 9}{x^2 - 4}$

67. $f(x) = \dfrac{x^3}{x^2 - 9}$

68. $f(x) = \dfrac{x + 1}{x^2 - 2x - 8}$

69. $f(x) = \dfrac{x^4}{x^2 - 4}$

70. $f(x) = \dfrac{x^2 + x - 6}{x^2 - x - 12}$

In Exercises 71–78, solve each inequality.

71. $(x - 2)(x - 3)(x + 2) \le 0$

72. $(x - 1)(x + 1)^2(x + 2) \ge 0$

73. $x^3 - 4x^2 + 7x - 7 > x^2 - 4$

74. $x^3 - 2x^2 + 4x + 8 < 3x^2 + 2x$

75. $\dfrac{(x - 1)(x + 3)}{x + 2} \ge 0$

76. $\dfrac{(x + 1)^2(x - 4)}{x - 2} \le 0$

77. $\dfrac{2x - 3}{x + 4} < 3$

78. $\dfrac{2}{x + 4} > \dfrac{3}{x}$

Applying the Concepts

79. **Variation.** If y varies directly as x and if $y = 12$ when $x = 4$, find y when $x = 5$.

80. **Variation.** If p varies inversely as q and if $p = 4$ when $q = 3$, find p when $q = 4$.

81. **Variation.** If s varies directly as the square of t and if $s = 20$ when $t = 2$, find s when $t = 3$.

82. **Variation.** If y varies inversely as x^2 and if $y = 3$ when $x = 8$, find x when $y = 12$.

83. **Missile path.** A missile fired from the origin of a coordinate system follows a path described by the equation

$$y = -\frac{1}{10}x^2 + 20x,$$ where the x-axis is at ground level.

Sketch the missile's path and determine what its maximum altitude is and where it hits the ground.

84. **Minimizing area.** Suppose a wire 20 centimeters long is to be cut into two pieces, each of which will be formed into a square. Find the size of each piece that minimizes the total area.

85. **Minimizing length.** A farmer wants to fence off three identical adjoining rectangular pens, each 400 square feet in area. (See the figure.) What should be the width and length of each pen so that the least amount of fence is used?

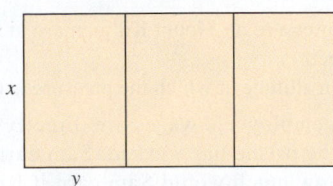

86. Minimizing cost. Suppose the outer boundary of the pens in Exercise 85 requires heavy fence that costs $5 per foot and two internal partitions cost $3 per foot. What dimensions x and y will minimize the cost?

87. Electric circuit. In the circuit shown in the figure, the voltage $V = 100$ volts and the resistance $R = 50$ ohms. We want to determine the size of the remaining resistor (x ohms). The power absorbed by the circuit is given by $p(x) = \dfrac{V^2 x}{(R + x)^2}$.

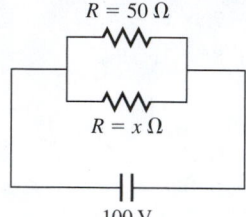

$R = 50\ \Omega$

$R = x\ \Omega$

100 V

   a. Graph the function $y = p(x)$.

   b. For what values of x is the power absorbed by the circuit greater than 42?

   c. Use a graphing calculator to find the value of x that maximizes the power absorbed.

88. Maximizing area. A sheet of paper is 100 square inches in area. The margins at the top and bottom are 1.5 inches, and the margin on each side is 1 inch.

   a. What should the width of the paper be if the printed area is to be at least 56 square inches?

   b. What should the dimensions of the paper be if the printed area is to be a maximum?

89. Maximizing profit. A manufacturer makes and sells printers to retailers at $24 per unit. The total daily cost C in dollars of producing x printers is given by

$$C(x) = 150 + 3.9x + \frac{3}{1000}x^2.$$

   a. Write the profit P as a function of x.

   b. Find the number of printers the manufacturer should produce and sell to achieve maximum profit.

   c. Find the average cost $\overline{C}(x) = \dfrac{C(x)}{x}$. Graph $y = \overline{C}(x)$.

90. Meteorology. The function $p = \dfrac{69.1}{a + 2.3}$ relates the atmospheric pressure p in inches of mercury to the altitude a in miles from the surface of Earth.

   a. Find the pressure on Mount Kilimanjaro at an altitude of 19,340 feet.

   b. Is there an altitude at which the pressure is 0?

91. Wages. An employee's wages are directly proportional to the time he or she has worked. Sam earned $280 for 40 hours. How much would Sam earn if he worked for 35 hours?

92. Car's stopping distance. The distance required for a car to come to a stop after its brakes are applied is directly proportional to the square of its speed. If the stopping distance for a car traveling 30 miles per hour is 25 feet, what is the stopping distance for a car traveling 66 miles per hour?

93. Illumination. The amount of illumination from a source of light varies directly as the intensity of the source and inversely as the square of the distance from the source. At what distance from a light source of intensity 300 candlepower will the illumination be one-half the illumination 6 inches away from the source?

94. Chemistry. *Charles's Law* states that at a constant pressure, the volume V of a gas is directly proportional to its temperature T (in Kelvin degrees). If a bicycle tube is filled with 1.2 cubic feet of air at a temperature of 295 K, what will be the volume of the air in the tube if the temperature rises to 310 K while the pressure stays the same?

95. Electric circuits. The current I (measured in amperes) in an electrical circuit varies inversely as the resistance R (measured in ohms) when the voltage is held constant. The current in a certain circuit is 30 amperes when the resistance is 300 ohms.

   a. Find the current in the circuit if the resistance is decreased to 250 ohms.

   b. What resistance will yield a current of 60 amperes?

96. Safe load. The safe load that a circular column can support varies directly as the square root of its radius and inversely as the square of its length. A pillar with radius 4 inches and length 12 feet can safely support a 20-ton load. Find the load that a pillar of the same material with diameter 6 inches and length 10 feet can safely support.

97. Spread of disease. An infectious cold virus spreads in a community at a rate R (per day) that is jointly proportional to the number of people who are infected with the virus and the number of people in the community who are not infected yet. After the tenth day of the start of a certain infection, 15% of the total population of 20,000 people of Pollutville had been infected and the virus was spreading at the rate of 255 people per day.

   a. Find the constant of proportionality, k.

   b. At what rate is the disease spreading when one-half the population is infected?

   c. Find the number of people infected when the rate of infection reached 95 people per day.

98. Coulomb's law. The electric charge is measured in coulombs. (The charges on an electron and a proton, which are equal and opposite, are approximately 1.602×10^{-19} coulomb.) Coulomb's Law states that the force F between two particles is jointly proportional to their charges q_1 and q_2 and inversely proportional to the square of the distance between the two particles. Two charges are acted upon by a repulsive force of 96 units. What is the force if the distance between the particles is quadrupled?

PRACTICE TEST A

1. Find the x-intercepts of the graph of $f(x) = x^2 - 6x + 2$.

2. Graph $f(x) = 3 - (x + 2)^2$.

3. Find the vertex of the parabola described by $y = -7x^2 + 14x + 3$.

4. Find the domain of the function
$$f(x) = \frac{x^2 - 1}{x^2 + 3x - 4}.$$

5. Find the quotient and remainder of
$$\frac{x^3 - 2x^2 - 5x + 6}{x + 2}.$$

6. Graph the polynomial function $P(x) = x^5 - 4x^3$.

7. Find all of the zeros of $f(x) = 2x^3 - 2x^2 - 8x + 8$, given that 2 is one of the zeros.

8. Find the quotient of $\dfrac{-6x^3 + x^2 + 17x + 3}{2x + 3}$.

9. Use the Remainder Theorem to find the value $P(-2)$ of the polynomial $P(x) = x^4 + 5x^3 - 7x^2 + 9x + 17$.

10. Find all rational roots of the equation $x^3 - 5x^2 - 4x + 20 = 0$ and then find the irrational roots if there are any.

11. Find the zeros of the polynomial function $f(x) = x^4 + x^3 - 15x^2$.

12. For $P(x) = 2x^{18} - 5x^{13} + 6x^3 - 5x + 9$, list all possible rational zeros found by the Rational Zeros Test, but do not check to see which values are actually zeros.

13. Describe the end behavior of $f(x) = (x + 3)^3(x - 5)^2$.

14. Find the zeros and the multiplicity of each zero for $f(x) = (x^2 - 4)(x + 2)^2$.

15. Determine how many positive and how many negative real zeros the polynomial function $P(x) = 3x^6 + 2x^3 - 7x^2 + 8x$ can have.

16. Find the horizontal and the vertical asymptotes of the graph of   ;
$$f(x) = \frac{2x^2 + 3}{x^2 - x - 20}.$$

17. Solve the inequality
$$\frac{1}{x + 4} \le \frac{3}{x - 2}.$$

18. Assume that "y is directly proportional to x and inversely proportional to the square of t." Suppose $y = 6$ when $x = 8$ and $t = 2$. Find y if $x = 12$ and $t = 3$.

19. The cost C of producing x thousand units of a product is given by
$$C = x^2 - 30x + 335 \text{ (dollars)}.$$
Find the value of x for which the cost is minimum.

20. From a rectangular 8×17 piece of cardboard, four congruent squares with sides of length x are cut out, one at each corner. The sides can then be folded to form a box. Find the volume V of the box as a function of x.

PRACTICE TEST B

1. Find the x-intercepts of the graph of $f(x) = x^2 + 5x + 3$.

   a. $\dfrac{-5 \pm i\sqrt{13}}{2}$ **b.** $\dfrac{-5 \pm \sqrt{13}}{2}$

   c. $\dfrac{-5 \pm i\sqrt{37}}{2}$ **d.** $\dfrac{-5 \pm \sqrt{37}}{2}$

2. Which is the graph of $f(x) = 4 - (x - 2)^2$?

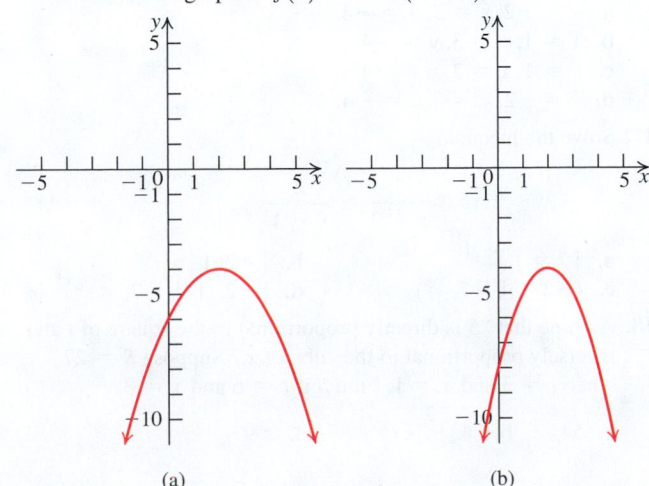

(a)   (b)

(c)   (d)

3. Find the vertex of the parabola described by $y = 6x^2 + 12x - 5$.

   a. $(-1, -11)$ **b.** $(1, -5)$

   c. $(-1, 13)$ **d.** $(1, 13)$

4. Which of the following is *not* in the domain of the function
$$f(x) = \frac{x^2}{x^2 + x - 6}?$$

   I. -3

   II. 0

   III. 2

  a.   I and II   **b.**   I and III

  c.   II and III   **d.**   I only

5. Find the quotient and remainder when $x^3 - 8x + 6$ is divided by $x + 3$.

  a. $x^2 - 8; 2$ **b.** $x^2 - 8; 0$

  c. $x^2 - 3x + 1; x + 3$ **d.** $x^2 - 3x + 1; 3$

6. Which is the graph of the polynomial $P(x) = x^4 + 2x^3$?

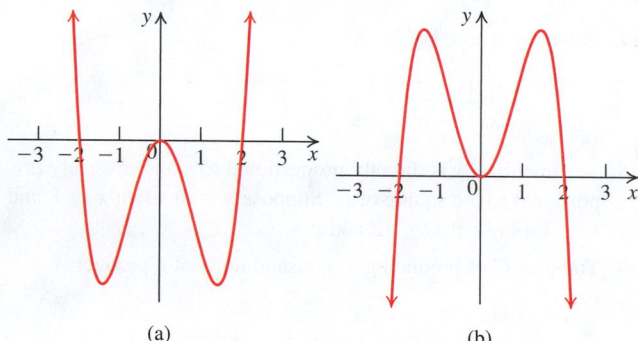

(a)   (b)

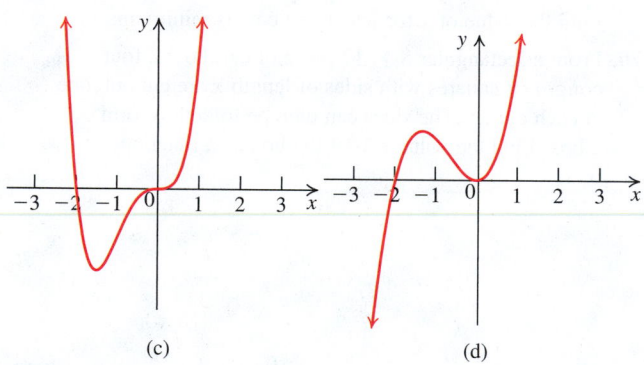

(c)   (d)

7. Find all of the zeros of $f(x) = 3x^3 - 26x^2 + 61x - 30$ given that 3 is one of the zeros (that is, $f(3) = 0$).

  a. $3, -5, -\dfrac{2}{3}$ **b.** $3, 2, \dfrac{5}{3}$

  c. $3, 5, \dfrac{2}{3}$ **d.** $3, -2, \dfrac{5}{3}$

8. Find the quotient $\dfrac{-10x^3 + 21x^2 - 17x + 12}{2x - 3}$.

  a. $x^2 - 3x + 4$ **b.** $-5x^2 + 3x - 4$

  c. $x^2 + 3x - 4$ **d.** $-5x^2 - 4$

9. Use the Remainder Theorem to find the value $P(-3)$ of the polynomial $P(x) = x^4 + 4x^3 + 7x^2 + 10x + 15$.

  a.   13   **b.**   15

  c.   21   **d.**   6

10. Find all rational roots of the equation $-x^3 + x^2 + 8x - 12 = 0$ and then find the irrational roots if there are any.

  a. -3 and 2   **b.** -3 and $\sqrt{2}$

  c. $-1, -2,$ and 3   **d.** -3 and $\sqrt{3}$

11. Find the zeros of the polynomial function $f(x) = x^3 + x^2 - 30x$.

  a. $x = -6, x = 5, x = 0$ **b.** $x = 0, x = -6$

  c. $x = 4, x = 5$ **d.** $x = 0, x = 4, x = 5$

12. For $P(x) = x^{30} - 4x^{25} + 6x^2 + 60$, list all possible rational zeros found by the Rational Zeros Test (but do not check to see which values are actually zeros).

  a. $\pm 2, \pm 3, \pm 4, \pm 5, \pm 6, \pm 8, \pm 10, \pm 12, \pm 15, \pm 20,$ $\pm 30,$ and ± 60

  b. $\pm 1, \pm 2, \pm 3, \pm 4, \pm 5, \pm 6, \pm 10, \pm 12, \pm 15, \pm 18,$ $\pm 20, \pm 24, \pm 30,$ and ± 60

  c. $\pm 1, \pm 3, \pm 4, \pm 5, \pm 6, \pm 12, \pm 15, \pm 20, \pm 30,$ and ± 60

  d. $\pm 1, \pm 2, \pm 3, \pm 4, \pm 5, \pm 6, \pm 10, \pm 12, \pm 15, \pm 20,$ $\pm 30,$ and ± 60

13. Which of the following correctly describes the end behavior of $f(x) = (x + 1)^2 (x - 2)^2$?

  a. $\begin{cases} y \to \infty \text{ as } x \to -\infty \\ y \to -\infty \text{ as } x \to \infty \end{cases}$ **b.** $\begin{cases} y \to -\infty \text{ as } x \to -\infty \\ y \to -\infty \text{ as } x \to \infty \end{cases}$

  c. $\begin{cases} y \to \infty \text{ as } x \to -\infty \\ y \to \infty \text{ as } x \to \infty \end{cases}$ **d.** $\begin{cases} y \to -\infty \text{ as } x \to -\infty \\ y \to \infty \text{ as } x \to \infty \end{cases}$

14. Find the zeros and the multiplicity of each zero for $f(x) = (x^2 - 1)(x + 1)^2$.

  a. $\begin{cases} \text{zero } 1, & \text{multiplicity } 1 \\ \text{zero } -1, & \text{multiplicity } 2 \end{cases}$

  b. $\begin{cases} \text{zero } 1, & \text{multiplicity } 1 \\ \text{zero } -1, & \text{multiplicity } 3 \end{cases}$

  c. $\begin{cases} \text{zero } 1, & \text{multiplicity } 2 \\ \text{zero } -1, & \text{multiplicity } 2 \end{cases}$

  d. $\begin{cases} \text{zero } i, & \text{multiplicity } 2 \\ \text{zero } -1, & \text{multiplicity } 2 \end{cases}$

15. Determine how many positive and how many negative real zeros the polynomial $P(x) = x^5 - 4x^3 - x^2 + 6x - 3$ can have.

  a.   positive 3, negative 2

  b.   positive 2 or 0, negative 2 or 0

  c.   positive 3 or 1, negative 2 or 0

  d.   positive 2 or 0, negative 3 or 1

16. The horizontal and the vertical asymptotes of the graph of
$$f(x) = \frac{x^2 + x - 2}{x^2 + x - 12} \text{ are}$$

  a. $x = -2, y = 3, y = -4$.

  b. $x = 1, y = 3, y = -4$.

  c. $y = 1, x = 3, x = -4$.

  d. $y = -2, x = 3, x = -4$.

17. Solve the inequality
$$\frac{3}{x + 2} \geq \frac{2}{x - 1}.$$

  a. $[7, \infty)$ **b.** $[-2, 1]$

  c. $(-2, 1) \cup [7, \infty)$ **d.** $[-2, 1] \cup [7, \infty)$

18. Assume that "S is directly proportional to the square of t and inversely proportional to the cube of x." Suppose $S = 27$ when $t = 3$ and $x = 1$. Find S if $t = 6$ and $x = 3$.

  a.   54   **b.**   4   **c.** $\dfrac{4}{3}$ **d.**   9

19. The cost C of producing x thousand units of a product is given by

$$C = x^2 - 24x + 319 \text{ (dollars)}.$$

Find the value of x for which the cost is minimum.

  a. 319   **b.** 12   **c.** 175   **d.** 24

20. From a rectangular 10×12 piece of cardboard, four congruent squares with sides of length x are cut out, one at each

corner. The sides can then be folded to form a box. Find the volume V of the box as a function of x.

  a. $V = 2x(10 - x)(12 - x)$
  b. $V = 2x(10 - 2x)(12 - 2x)$
  c. $V = x(10 - x)(12 - x)$
  d. $V = x(10 - 2x)(12 - 2x)$

CUMULATIVE REVIEW EXERCISES CHAPTERS 1–2

1. Find the distance between the points $P(-1, 3)$ and $Q(2, 5)$.

2. Find the midpoint of the line segment joining $P(2, -5)$ and $Q(-8, -3)$.

3. Find the x- and y-intercepts of the graph of the equation $y = x^2 - 2x - 8$ and sketch the graph.

In Exercises 4 and 5, find the slope and intercepts of the line and sketch the graph.

4. $x + 3y - 6 = 0$ **5.** $x = 2y - 6$

6. Find an equation in standard form of the circle with center $(2, -3)$ and radius 4.

7. Find the center and radius of the circle with equation

$$x^2 + y^2 + 2x - 4y - 4 = 0.$$

In Exercises 8 and 9, find the slope–intercept form of the line satisfying the given condition.

8. The line has slope 3 and passes through $(1, -2)$.

9. The line is parallel to $2x + 3y = 5$ and passes through $(1, 3)$.

In Exercises 10 and 11, find the domain of each function.

10. $f(x) = \dfrac{1}{2x + 3}$ **11.** $p(x) = \dfrac{1}{\sqrt{4 - 2x}}$

12. Let $f(x) = x^2 - 2x + 3$. Find $f(-2), f(3), f(x + h)$, and $\dfrac{f(x + h) - f(x)}{h}$.

13. Let $f(x) = \sqrt{x}$ and $g(x) = x^2 + 1$. Find each of the following.

  a. $f(g(x))$ **b.** $g(f(x))$
  c. $f(f(x))$ **d.** $g(g(x))$

14. Let $f(x) = \begin{cases} 3x + 2 & \text{if } x \le 2 \\ 4x - 1 & \text{if } 2 < x \le 3. \\ 6 & \text{if } x > 3 \end{cases}$

  a. Find $f(1), f(3)$, and $f(4)$.
  b. Sketch the graph of $y = f(x)$.

15. Let $f(x) = 2x - 3$. Find $f^{-1}(x)$.

16. Use transformations on $y = \sqrt{x}$ to sketch the graph of each function.

  a. $f(x) = \sqrt{x + 2}$
  b. $g(x) = -2\sqrt{x + 1} + 3$

17. Use the Rational Zeros Test to list all possible rational zeros of $f(x) = 2x^4 - 3x^2 + 5x - 6$.

In Exercises 18–21, (a) Graph each function; (b) solve inequality $f(x) \ge 0$.

18. $f(x) = 2x^2 - 4x + 1$

19. $f(x) = -x^2 + 2x + 3$

20. $f(x) = (x - 1)^2(x + 2)$

21. $f(x) = \dfrac{x^2 - 1}{x^2 - 4}$

22. Let $f(x) = x^4 - 3x^3 + 2x^2 + 2x - 4$. Given that $1 + i$ is a zero of f, find all zeros of f.

23. Suppose that y varies as the square root of x and that $y = 6$ when $x = 4$. Find y if $x = 9$.

24. A drug manufactured by a pharmaceutical company is sold in bulk at a price of $150 per unit. The total production cost (in dollars) for x units in one week is

$$C(x) = 0.02x^2 + 100x + 3000.$$

How many units of the drug must be manufactured and sold in a week to maximize the profit? What is the maximum profit?

25. The profit (in dollars) for a product is given by

$$P(x) = 0.02x^3 + 48.8x^2 - 2990x + 25,000,$$

where x is the number of units produced and sold. One break-even point occurs when $x = 10$. Use synthetic division to find another break-even point for the product.

CHAPTER 3

Exponential and Logarithmic Functions

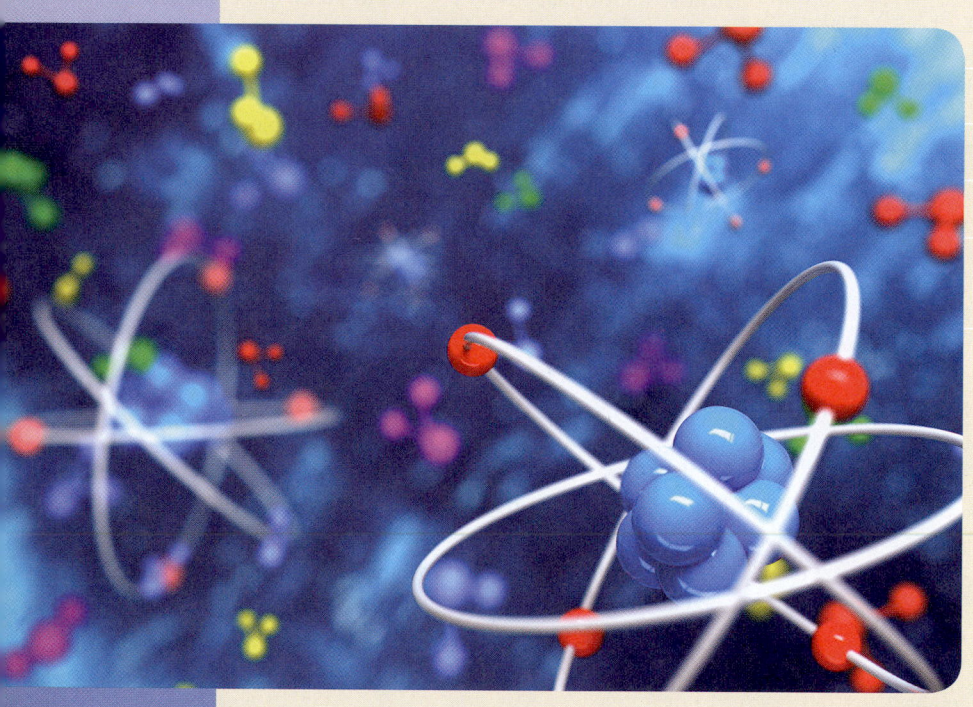

TOPICS

3.1 Exponential Functions

3.2 Logarithmic Functions

3.3 Rules of Logarithms

3.4 Exponential and Logarithmic Equations and Inequalities

3.5 Logarithmic Scales; Modeling

The world of finance, the growth of many populations, musical pitch, the molecular activity that results in an atomic explosion, and countless everyday events are modeled with exponential and logarithmic functions.

Exponential Functions

BEFORE STARTING THIS SECTION, REVIEW

1 Integer exponents (Appendix A.1, page 923)

2 Rational exponents (Appendix A.4, page 946)

3 Graphing and transformations (Section 1.5, page 88)

4 One-to-one functions (Section 1.7, page 116)

5 Increasing and decreasing functions (Section 1.4, page 61)

OBJECTIVES

1 Define an exponential function.

2 Graph exponential functions.

3 Develop formulas for simple and compound interest.

4 Understand the number e.

5 Define the natural exponential function.

6 Model using exponential functions.

TABLE 3.1 Grains of wheat on a chessboard

Square Number	Grains Placed on This Square
1	$1\ (=2^0)$
2	$2\ (=2^1)$
3	$4\ (=2^2)$
4	$8\ (=2^3)$
5	$16\ (=2^4)$
6	$32\ (=2^5)$
.	.
.	.
.	.
63	2^{62}
64	2^{63}

◆ Fooling a King

One night in northwest India, a wise man named Shashi invented a new game called "Shatranj" (chess). The next morning he took it to King Rai Bhalit, who was so impressed that he said to Shashi, "Name your reward." Shashi merely requested that 1 grain of wheat be placed on the first square of the chessboard, 2 grains on the second, 4 on the third, 8 on the fourth, and so on, for all 64 squares. The king agreed to his request, thinking the man was an eccentric fool for asking for only a few grains of wheat when he could have had gold, jewels, or even his daughter's hand in marriage.

You may know the end of this story. Much more wheat was needed to satisfy Shashi's request. The king could never fulfill his promise to Shashi, so instead he had him beheaded.

The details of this rapidly growing wheat phenomenon are given in the margin. Table 3.1 shows the number of grains of wheat that need to be stacked on each square of the chessboard.

These data can be modeled by the function

$$g(n) = 2^{n-1},$$

where $g(n)$ is the number of grains of wheat and n is the number of the square (or the nth square) on the chessboard. Each square has twice the number of grains as the previous square. The function g is an example of an *exponential function* with base 2, with n restricted to 1, 2, 3, . . . , 64—the number of the chessboard squares. The name *exponential function* comes from the fact that the variable n occurs in the exponent. The base 2 is the *growth factor*.

When the growth rate of a quantity is directly proportional to the existing amount, the growth can be modeled by an exponential function. For example, exponential functions are often used to model the growth of investments and populations, the cell division of living organisms, and the decay of radioactive material. Example 11 discusses the growth of the world population.

1 Define an exponential function.

Exponential Functions

Exponential Function

A function f of the form

$$f(x) = a^x, a > 0, \text{ and } a \neq 1,$$

is called an **exponential function with base a and exponent x.** Its domain is $(-\infty, \infty)$.

In the definition of the exponential function, we rule out the base $a = 1$ because in this case, the function is simply the constant function $f(x) = 1^x$, or $f(x) = 1$. We exclude negative bases so that the domain includes all real numbers. For example, a cannot be -2 because $(-2)^{1/2} = \sqrt{-2}$ is not a real number. Other functions that have variables in the exponent, such as $f(x) = 4 \cdot 3^x$, $g(x) = -5 \cdot 4^{3x-2}$, and $h(x) = c \cdot a^x$ ($a > 0$ and $a \neq 1$), are also called exponential functions.

Evaluate Exponential Functions

We can evaluate exponential functions by using the laws of exponents and/or calculators, as in the next example. (See Appendices A.1 and A.4 for the definitions and the laws of exponents involving a^x when x is a rational number.)

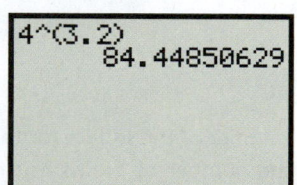
EXAMPLE 1 **Evaluating Exponential Functions**

a. Let $f(x) = 3^{x-2}$. Find $f(4)$.

b. Let $g(x) = -2 \cdot 10^x$. Find $g(-2)$.

c. Let $h(x) = \left(\dfrac{1}{9}\right)^x$. Find $h\left(-\dfrac{3}{2}\right)$.

d. Let $F(x) = 4^x$. Find $F(3.2)$.

Solution

a. $f(4) = 3^{4-2} = 3^2 = 9$

b. $g(-2) = -2 \cdot 10^{-2} = -2 \cdot \dfrac{1}{10^2} = -2 \cdot \dfrac{1}{100} = -0.02$

c. $h\left(-\dfrac{3}{2}\right) = \left(\dfrac{1}{9}\right)^{-\frac{3}{2}} = (9^{-1})^{-\frac{3}{2}} = 9^{\frac{3}{2}} = (\sqrt{9})^3 = 27$

d. $F(3.2) = 4^{3.2} \approx 84.44850629$ Use a calculator.

Practice Problem 1 Let $f(x) = \left(\dfrac{1}{4}\right)^x$. Find $f(2), f(0), f(-1), f\left(\dfrac{5}{2}\right)$, and $f\left(-\dfrac{3}{2}\right)$. ▪

The domain of an exponential function is $(-\infty, \infty)$. In Example 1, we evaluated exponential functions at some rational numbers. But what is the meaning of a^x when x is an irrational number? For example, what do expressions such as $3^{\sqrt{2}}$ and 2^π mean? It turns out that the definition of a^x (with $a > 0$ and x irrational) requires methods discussed in calculus. However, we can understand the basis for the definition from the following discussion. Suppose we want to define the number 2^π. We use several numbers that approximate π. A calculator shows that $\pi \approx 3.14159265 \ldots$. We successively approximate 2^π by using the rational powers shown in Table 3.2.

TABLE 3.2 Values of $f(x) = 2^x$ for rational values of x that approach π

x	3	3.1	3.14	3.141	3.1415
2^x	$2^3 = 8$	$2^{3.1} = 8.5\ldots$	$2^{3.14} = 8.81\ldots$	$2^{3.141} = 8.821\ldots$	$2^{3.1415} = 8.8244\ldots$

It can be shown that the powers $2^3, 2^{3.1}, 2^{3.14}, 2^{3.141}, 2^{3.1415}, \ldots$ approach exactly one number. We define 2^π as that number. Table 3.2 shows that $2^\pi \approx 8.82$ (correct to two decimal places).

For our work with exponential functions, we need the following fundamental facts:

1. Exponential functions $f(x) = a^x$ are defined for all real numbers x.

2. The graph of an exponential function is a continuous (unbroken) curve.

3. The rules of exponents hold for all real number exponents and positive bases.

RULES OF EXPONENTS

Let a, b, x, and y be real numbers with $a > 0$ and $b > 0$.

$$a^x \cdot a^y = a^{x+y} \qquad (a^x)^y = a^{xy}$$

$$\frac{a^x}{a^y} = a^{x-y} \qquad a^0 = 1$$

$$(ab)^x = a^x b^x \qquad a^{-x} = \frac{1}{a^x} = \left(\frac{1}{a}\right)^x$$

---

2 Graph exponential functions.

Graphing Exponential Functions

Let's see how to sketch the graph of an exponential function. Although the domain is the set of all real numbers, we usually evaluate the functions only for the integer values of x (for ease of computation). To evaluate a^x for noninteger values of x, use your calculator. Recall that the graph of a function f is the graph of the equation $y = f(x)$. We can use either $f(x) = a^x$ or $y = a^x$ to represent a given function.

EXAMPLE 2   **Graphing an Exponential Function with Base $a > 1$**

Graph the exponential function $f(x) = 3^x$.

Solution

First, make a table of a few values of x and the corresponding values of y.

x	-3	-2	-1	0	1	2	3
$y = 3^x$	$\dfrac{1}{27}$	$\dfrac{1}{9}$	$\dfrac{1}{3}$	1	3	9	27

Next, plot the points (see Figure 3.1(a)) and draw a smooth curve through them. See Figure 3.1(b).

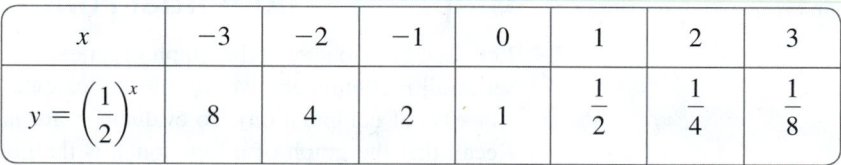

Figure 3.1 The graph of $y = a^x$ with $a > 1$.

The graph in Figure 3.1 (b) is typical of the graphs of exponential functions $f(x) = a^x$ when $a > 1$. Note that the x-axis is the horizontal asymptote of the graph of $y = a^x$.

Practice Problem 2 Sketch the graph of

$$f(x) = 2^x.$$

Let's now sketch the graph of an exponential function $f(x) = a^x$ when $0 < a < 1$.

EXAMPLE 3 **Graphing an Exponential Function $f(x) = a^x$, with $0 < a < 1$**

Sketch the graph of $y = \left(\dfrac{1}{2}\right)^x$ by making a table of values.

Solution

Make a table similar to the one in Example 2.

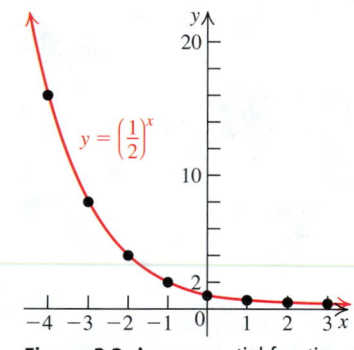

Figure 3.2 An exponential function $y = a^x$ with $0 < a < 1$.

x	-3	-2	-1	0	1	2	3
$y = \left(\dfrac{1}{2}\right)^x$	8	4	2	1	$\dfrac{1}{2}$	$\dfrac{1}{4}$	$\dfrac{1}{8}$

Plotting these points and drawing a smooth curve through them, we get the graph of $y = \left(\dfrac{1}{2}\right)^x$, shown in Figure 3.2. In this graph, as x increases in the positive direction, $y = \left(\dfrac{1}{2}\right)^x$ decreases toward 0. The x-axis is its horizontal asymptote. This is a typical graph for an exponential function $y = a^x$ with $0 < a < 1$.

Practice Problem 3 Sketch the graph of

$$f(x) = \left(\dfrac{1}{3}\right)^x.$$

In general, exponential functions have two basic shapes determined by the base a. If $a > 1$, the graph of $y = a^x$ is rising as in Figure 3.1 (b). If $0 < a < 1$, the graph is falling as in Figure 3.2. To get the correct shape for the graph of $f(x) = a^x$, it is helpful to graph the three points $\left(-1, \dfrac{1}{a}\right)$, $(0, 1)$, and $(1, a)$.

RECALL

The graph of $y = f(-x)$ is the reflection about the y-axis of the graph of $y = f(x)$.

Since $y = \left(\dfrac{1}{2}\right)^x = (2^{-1})^x = 2^{-x}$, the graph of $y = \left(\dfrac{1}{2}\right)^x$ can also be obtained by reflecting the graph of $y = 2^x$ about the y-axis. See Figure 3.3.

In Figure 3.3, we sketch the graphs of four exponential functions on the same set of axes. These graphs illustrate some general properties of exponential functions.

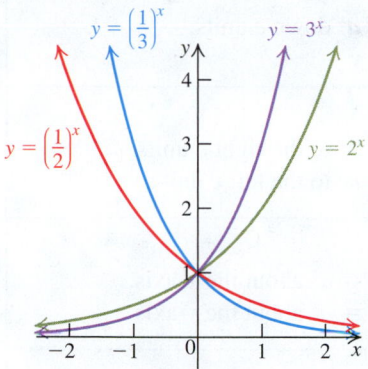

Figure 3.3 Graphs of some exponential functions.

RECALL

A function f is one-to-one if $f(x_1) = f(x_2)$ implies $x_1 = x_2$.

RECALL

A line $y = k$ is a horizontal asymptote for the graph of f if $f(x) \to k$ as $x \to \infty$ or as $x \to -\infty$.

PROPERTIES OF EXPONENTIAL FUNCTIONS

Let $y = f(x) = a^x, a > 0, a \neq 1$.

1. The domain of f is $(-\infty, \infty)$.

2. The range of f is $(0, \infty)$: The entire graph lies above the x-axis.

3. Note that $f(x + 1) = a^{x+1} = a \cdot a^x = af(x)$. This means that the y-values change by a factor of a for each unit increase in x.

4. For $a > 1$, the **growth factor** is a. Because the y-values *increase* by a factor of a for each unit increase in x,
   (i) f is an increasing function, so the graph rises to the right.
   (ii) as $x \to \infty$, $y \to \infty$.
   (iii) as $x \to -\infty$, $y \to 0$.

5. For $0 < a < 1$, the **decay factor** is a. Because the y-values *decrease* by a factor of a for each unit increase in x,
   (i) f is a decreasing function, so the graph falls to the right.
   (ii) as $x \to -\infty$, $y \to \infty$.
   (iii) as $x \to \infty$, $y \to 0$.

6. Each exponential function f is one-to-one. So,
   (i) if $a^m = a^n$, then $m = n$.
   (ii) f has an inverse.

7. The graph of f has no x-intercepts, so it never crosses the x-axis. No value of x will cause $f(x) = a^x$ to equal 0.

8. The graph of f is a smooth and continuous curve, and it passes through the points $\left(-1, \dfrac{1}{a}\right)$, $(0, 1)$, and $(1, a)$.

9. The x-axis $(y = 0)$ is a horizontal asymptote for the graph of every exponential function of the form $f(x) = a^x$.

10. The graph of $y = a^{-x}$ is the reflection about the y-axis of the graph of $y = a^x$.

Transformations on Exponential Functions

In Section 1.5, we discussed transformations on the graph of the function $y = f(x)$. The transformations on the graph of $y = e^x$ are summarized in Table 3.3.

TABLE 3.3 Transformations on the graph of $y = a^x$

Transformation (in each case $c > 0$)	Change the graph point (x, y) on $y = a^x$ to	Description
Vertical shift $y = a^x + c$	$(x, y + c)$	Shift the graph of $y = a^x$ up c units. Horizontal asymptote: $y = c$
$y = a^x - c$	$(x, y - c)$	Shift the graph of $y = a^x$ down c units. Horizontal asymptote: $y = -c$
Horizontal shift $y = a^{x-c}$ $y = a^{x+c}$	$(x + c, y)$ $(x - c, y)$	Shift the graph of $y = a^x$ to the right c units. Shift the graph of $y = a^x$ to the left c units.
Reflection $y = -a^x$ $y = a^{-x}$	$(x, -y)$ $(-x, y)$	Reflect the graph of $y = a^x$ about the x-axis. Reflect the graph of $y = a^x$ about the y-axis.
Vertical stretching or compressing $y = ca^x$	(x, cy)	Vertically stretch the graph of $y = a^x$ if $c > 1$. Vertically compress the graph of $y = a^x$ if $0 < c < 1$.
Horizontal stretching or compressing $y = a^{cx}$	$\left(\dfrac{x}{c}, y\right)$	Horizontally compress the graph of $y = a^x$ if $c > 1$. Horizontally stretch the graph of $y = a^x$ if $0 < c < 1$.

EXAMPLE 4 Using Transformations on Exponential Functions

Sketch the graph of $y = 3^{-x} + 2$. Find the domain, range, and the horizontal asymptote of f.

Solution

Figure 3.4 shows a sequence of transformations on the graph of $y = 3^x$ used to produce the graph of f.

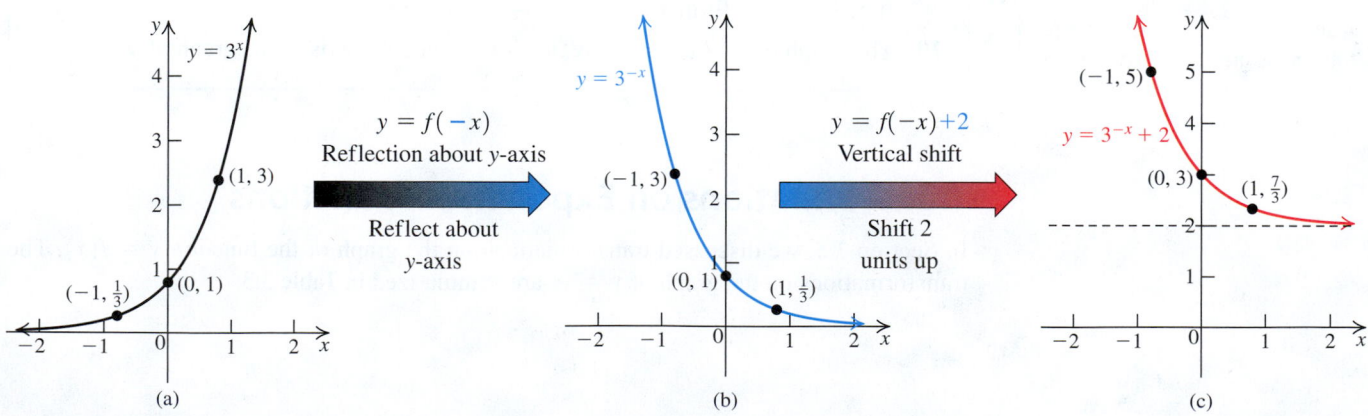

Figure 3.4 Graphing $f(x) = 3^{-x} + 2$.

From the graph of f in Figure 3.4(c), we see that the domain of f is $(-\infty, \infty)$, its range is $(2, \infty)$, and the horizontal asymptote is the line $y = 2$.

Practice Problem 4 Sketch the graph of $f(x) = 2^{x-1} + 3$.

EXAMPLE 5   **Finding Exponential Functions**

Find the exponential function of the form $f(x) = ca^x$ whose graph contains the points $(-1, 18)$ and $\left(4, \dfrac{2}{27}\right)$.

Solution

Write the function in the form $y = f(x) = ca^x$.

$18 = ca^{-1}$ $(-1, 18)$ is on the graph of $y = a^x$.

$18a = c$ Multiply both sides by a; simplify.

$\dfrac{2}{27} = ca^4$ $\left(4, \dfrac{2}{27}\right)$ is on the graph of $y = a^x$.

$\dfrac{2}{27} = 18aa^4$ Replace c with $18a$ in the previous equation,

$\dfrac{2}{27} = 18a^5$ $a \cdot a^4 = a^5$

$a^5 = \dfrac{1}{243}$ Divide both sides by 18.

$a = \sqrt[5]{\dfrac{1}{243}} = \dfrac{1}{3}$ Solve for a.

To find c, we substitute $a = \dfrac{1}{3}$ in the equation $18a = c$. So,

$$18\left(\frac{1}{3}\right) = c \quad \text{or} \quad c = 6.$$

Replacing $c = 6$ and $a = \dfrac{1}{3}$ in $f(x) = ca^x$, we have:

$$f(x) = 6\left(\frac{1}{3}\right)^x.$$

Practice Problem 5 Repeat Example 5 with the graph points $(-2, 16)$ and $\left(3, \dfrac{1}{2}\right)$.

3 Develop formulas for simple and compound interest.

Simple Interest

Let's first review some terminology concerning interest.

Simple Interest

A fee charged for borrowing a lender's money is the **interest**, denoted by I.

The original, or initial, amount of money borrowed is the **principal**, denoted by P.

The period of time during which the borrower pays back the principal plus the interest is the **time**, denoted by t.

The **interest rate** is the percent charged for the use of the principal for the given period. The interest rate, denoted by r, is expressed as a decimal. Unless stated otherwise, the period is assumed to be one year; that is, r is an *annual* rate.

The amount of interest computed only on the principal is called **simple interest**.

RECALL

$8\% = 8$ percent

$= 8$ per hundred

$= \dfrac{8}{100}$

$= 0.08$

When money is deposited with a bank, the bank becomes the borrower. For example, depositing $1000 in an account at 8% interest means that the principal P is $1000 and the interest rate r is 0.08 for the bank as the borrower.

> **SIMPLE INTEREST FORMULA**
>
> The simple interest I on a principal P at a rate r (expressed as a decimal) per year for t years is
>
> $$I = Prt. \qquad (1)$$

EXAMPLE 6 **Calculating Simple Interest**

Juanita has deposited $8000 in a bank for five years at a simple interest rate of 6%.

a. How much interest will she receive?

b. How much money will be in her account at the end of five years?

Solution

a. $P = \$8000$, $r = 0.06$, and $t = 5$.

$$\begin{aligned} I &= Prt & &\text{Equation (1)} \\ &= \$8000\,(0.06)(5) & &\text{Substitute values.} \\ &= \$2400 & &\text{Simplify.} \end{aligned}$$

b. In five years, the amount A she will receive is the principal plus the interest earned:

$$\begin{aligned} A &= P + I \\ &= \$8000 + \$2400 \\ &= \$10{,}400 \end{aligned}$$

Practice Problem 6 Find the amount that will be in a bank account if $10,000 is deposited at a simple interest rate of 7.5% for two years.

Given P and r, the amount $A(t)$ due in t years and calculated at simple interest is found by using the formula

$$A(t) = P + Prt. \qquad (2)$$

Equation (2) is a linear function of t. Simple interest problems are examples of **linear growth**, which take place when the growth of a quantity occurs at a constant rate and so can be modeled by a linear function.

Compound Interest

In the real world, simple interest is rarely used for periods of more than one year. Instead, we use **compound interest**—the interest paid on both the principal and the accrued (previously earned) interest.

To illustrate compound interest, suppose $1000 is deposited in a bank account paying 4% annual interest. At the end of one year, the account will contain the original $1000 plus the 4% interest earned on the $1000:

$$\$1000 + (0.04)(\$1000) = \$1040$$

Similarly, if P represents the initial amount deposited at an interest rate r (expressed as a decimal) per year, then the amount A_1 in the account after one year is

$$\begin{aligned} A_1 &= P + rP & &\text{Principal } P \text{ plus interest earned} \\ &= P(1 + r) & &\text{Factor out } P. \end{aligned}$$

During the second year, the account earns interest on the new principal A_1. The amount A_2 in the account after the second year will be equal to A_1 plus the interest on A_1.

$$\begin{aligned} A_2 &= A_1 + rA_1 &&\quad A_1 \text{ plus interest earned on } A_1\\ &= A_1(1 + r) &&\quad \text{Factor out } A_1.\\ &= P(1 + r)(1 + r) &&\quad A_1 = P(1 + r)\\ &= P(1 + r)^2 \end{aligned}$$

The amount A_3 in the account after the third year is

$$\begin{aligned} A_3 &= A_2 + rA_2 &&\quad A_2 \text{ plus interest earned on } A_2\\ &= A_2(1 + r) &&\quad \text{Factor out } A_2.\\ &= P(1 + r)^2(1 + r) &&\quad A_2 = P(1 + r)^2\\ &= P(1 + r)^3 &&\quad (1 + r)^2(1 + r) = (1 + r)^3 \end{aligned}$$

In general, the amount A in the account after t years is given by

$$A = P(1 + r)^t. \qquad (3)$$

We say that this type of interest is **compounded annually** because it is paid once a year.

EXAMPLE 7 **Calculating Compound Interest**

Juanita deposits $8000 in a bank at the interest rate of 6% compounded annually for five years.

a. How much money will she have in her account after five years?

b. How much interest will she receive?

Solution

a. Here $P = \$8000$, $r = 0.06$, and $t = 5$; so

$$\begin{aligned} A &= P(1 + r)^t\\ &= \$8000(1 + 0.06)^5 &&\quad P = \$8000, r = 0.06, t = 5\\ &= \$8000(1.06)^5\\ &= \$10{,}705.80 &&\quad \text{Use a calculator.} \end{aligned}$$

b. Interest $= A - P = \$10{,}705.80 - \$8000 = \$2705.80$.

Practice Problem 7 Repeat Example 7 assuming that the bank pays 7.5% interest compounded annually.

Comparing Examples 6 and 7, we see that compounding Juanita's interest made her money grow faster. We expect this because the function $A(t) = P(1 + r)^t$ is an exponential function with base $(1 + r)$.

Banks and other financial institutions usually pay savings account interest more than once a year. They pay a smaller amount of interest more frequently. Suppose the quoted annual interest rate r (also called the **nominal rate**) is compounded n times per year (at equal intervals) instead of annually. Then for each period, the interest rate is $\dfrac{r}{n}$, and there are $n \cdot t$ periods in t years. Accordingly, we can restate the formula $A(t) = P(1 + r)^t$ as follows:

COMPOUND INTEREST FORMULA

$$A = P\left(1 + \frac{r}{n}\right)^{nt} \qquad (4)$$

$A =$ amount after t years

$P =$ principal

$r =$ annual interest rate (expressed as a decimal number)

$n =$ number of times interest is compounded each year

$t =$ number of years

The total amount accumulated after t years, denoted by A, is also called the **future value** of the investment.

EXAMPLE 8 **Using Different Compounding Periods to Compare Future Values**

If $100 is deposited in a bank that pays 5% annual interest, find the future value A after one year if the interest is compounded

(i) annually. **(ii)** semiannually. **(iii)** quarterly. **(iv)** monthly. **(v)** daily.

Solution

In the following computations, $P = \$100$, $r = 0.05$, and $t = 1$. Only n, the number of times interest is compounded each year, changes. Since $t = 1$, $nt = n(1) = n$.

(i) Annual Compounding: $\qquad A = P\left(1 + \dfrac{r}{n}\right)^{nt}$ $\qquad n = 1; t = 1$

$$A = \$100(1 + 0.05) = \$105.00$$

(ii) Semiannual Compounding: $\quad A = P\left(1 + \dfrac{r}{2}\right)^{2}$ $\qquad n = 2; t = 1$

$$A = \$100\left(1 + \frac{0.05}{2}\right)^{2}$$

$$\approx \$105.06 \qquad\qquad \text{Use a calculator.}$$

(iii) Quarterly Compounding: $\qquad A = P\left(1 + \dfrac{r}{4}\right)^{4}$ $\qquad n = 4; t = 1$

$$A = \$100\left(1 + \frac{0.05}{4}\right)^{4}$$

$$\approx \$105.09 \qquad\qquad \text{Use a calculator.}$$

(iv) Monthly Compounding: $\qquad A = P\left(1 + \dfrac{r}{12}\right)^{12}$ $\qquad n = 12; t = 1$

$$A = \$100\left(1 + \frac{0.05}{12}\right)^{12}$$

$$\approx \$105.12 \qquad\qquad \text{Use a calculator.}$$

(v) Daily Compounding: $\qquad A = P\left(1 + \dfrac{r}{365}\right)^{365}$ $\qquad n = 365; t = 1$

$$A = \$100\left(1 + \frac{0.05}{365}\right)^{365}$$

$$\approx \$105.13 \qquad\qquad \text{Use a calculator.}$$

DO YOU KNOW?

Compounding that occurs 1, 2, 4, 12, and 365 times a year is known as compounding annually, semiannually, quarterly, monthly, and daily, respectively.

SIDE NOTE

You may make errors such as forgetting parentheses when entering complicated expressions in your calculator. Look at the answer to see whether it is reasonable and makes sense.

Practice Problem 8 Repeat Example 8 assuming that $5000 is deposited at a 6.5% annual rate.

The next example shows that we can solve equation (4) for the interest rate r.

EXAMPLE 9 **Computing Interest Rate**

Carmen has $9000 to invest. She needs $20,000 at the end of 8 years. If the interest is compounded quarterly, find the rate r needed.

Solution

Have $P = \$9000, A = \$20,000, t = 8,$ and $n = 4$. Substituting these values in equation (4), we have

$$\$20,000 = \$9000\left(1 + \frac{r}{4}\right)^{4\cdot8}$$

or

$$\left(1 + \frac{r}{4}\right)^{32} = \frac{20,000}{9000} \qquad \text{Rewrite, with } 4\cdot8 - 32 \text{ and divide both sides by 9000.}$$

$$= \frac{20}{9} \qquad \text{Simplify the right side.}$$

Taking the 32nd root or $\frac{1}{32}$ power of both sides, we have

$$1 + \frac{r}{4} = \left(\frac{20}{9}\right)^{1/32}$$

$$\frac{r}{4} = \left(\frac{20}{9}\right)^{1/32} - 1 \qquad \text{Subtract 1 from both sides.}$$

$$r = 4\left[\left(\frac{20}{9}\right)^{1/32} - 1\right] \qquad \text{Multiply both sides by 4.}$$

$$\approx 0.1010692264 \qquad \text{Use a calculator.}$$

$$\approx 10.107\%$$

Carmen needs an interest rate of about 10.107%.

Check $\qquad 9000\left(1 + \dfrac{0.10107}{4}\right)^{32} = 20{,}000.12073 \approx 20{,}000.$

Practice Problem 9 Repeat Example 9 if the interest is compounded monthly.

4 Understand the number e.

Continuous Compound Interest Formula

Notice that in Example 8 the future value A increases with n, the number of compounding periods. (Of course P, r, and t are fixed.) The question is: If n increases indefinitely (100, 1000, 10,000 times, and so on), does the amount A also increase indefinitely? Let's see why the answer is no. Let $h = \dfrac{n}{r}$. Then we have

$$A = P\left(1 + \frac{r}{n}\right)^{nt} \qquad \text{Equation (4)}$$

$$= P\left[\left(1 + \frac{r}{n}\right)^{n/r}\right]^{rt} \qquad nt = \frac{n}{r}\cdot rt$$

$$= P\left[\left(1 + \frac{1}{h}\right)^{h}\right]^{rt} \quad (5) \qquad h = \frac{n}{r}, \text{ so } \frac{1}{h} = \frac{r}{n}$$

Table 3.4 shows the expression $\left(1 + \dfrac{1}{h}\right)^{h}$ as h takes on increasingly larger values.

Leonhard Euler

(1707–1783)
Leonhard Euler was the son of a Calvinist minister from Switzerland. At 13, Euler entered the University of Basel, pursuing a career in theology. At the university, Euler was tutored by Johann Bernoulli, of the famous Bernoulli family of mathematicians. Euler's interest and skills led him to abandon his theological studies and take up mathematics. In 1741, Euler moved to the Berlin Academy, where he stayed until 1766. He then returned to St. Petersburg, where he remained for the rest of his life.

Euler contributed to many areas of mathematics, including number theory, combinatorics, and analysis, as well as applications to areas such as music and naval architecture. He wrote over 1100 books and papers and left so much unpublished work that it took 47 years after he died for all of his work to be published. During Euler's life, his papers accumulated so quickly that he kept a large pile of articles awaiting publication. The Berlin Academy published the papers on top of this pile, so later results often were published before results they depended on or superseded. The project of publishing his collected works, undertaken by the Swiss Society of Natural Science, is still going on and will require more than 75 volumes.

TABLE 3.4

h	$\left(1 + \dfrac{1}{h}\right)^h$
1	2
2	2.25
10	2.59374
100	2.70481
1000	2.71692
10,000	2.71815
100,000	2.71827
1,000,000	2.71828

Table 3.4 suggests that as h gets larger and larger, $\left(1 + \dfrac{1}{h}\right)^h$ gets closer and closer to a fixed number. This observation can be proven, and the fixed number is denoted by e in honor of the famous mathematician Leonhard Euler (pronounced "oiler"). The number e, an irrational number, is sometimes called the **Euler number**.

> The value of e to 15 decimal places is
> $$e \approx 2.718281828459045.$$

We sometimes write
$$e = \lim_{h \to \infty}\left(1 + \frac{1}{h}\right)^h,$$

which means that when h is very large, $\left(1 + \dfrac{1}{h}\right)^h$ has a value very close to e. Note that as n gets very large, the quantity $h = \dfrac{n}{r}$ also gets very large because r is fixed. Therefore, the compounded amount $A = P\left(1 + \dfrac{r}{n}\right)^{nt} = P\left[\left(1 + \dfrac{r}{n}\right)^{\frac{n}{r}}\right]^{rt}$ approaches Pe^{rt}.

Continuous Compounding When interest is compounded continuously, the amount A after t years is given by the following formula:

> ### CONTINUOUS COMPOUND INTEREST FORMULA
> $$A = Pe^{rt} \qquad (6)$$
> A = amount after t years
> P = principal
> r = annual interest rate (expressed as a decimal number)
> t = number of years

TECHNOLOGY CONNECTION

Graphs of $y_1 = e$ and $y_2 = \left(1 + \dfrac{1}{x}\right)^x$ in the window $0 < x < 20, 0 < y < 3$

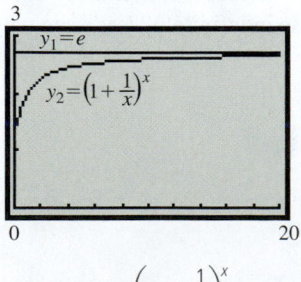

As $x \to \infty$, $\left(1 + \dfrac{1}{x}\right)^x \to e$.

5 Define the natural exponential functions.

6 Model using exponential functions.

EXAMPLE 10 **Calculating Continuous Compound Interest**

Find the amount when a principal of $8300 is invested at a 7.5% annual rate of interest compounded continuously for eight years and three months.

Solution

We use formula (6), with $P = \$8300$ and $r = 0.075$. We convert eight years and three months to 8.25 years.

$$A = \$8300e^{(0.075)(8.25)} \qquad \text{Use the formula } A = Pe^{rt}.$$
$$\approx \$15{,}409.83 \qquad \text{Use a calculator.}$$

Practice Problem 10 Repeat Example 10 assuming that $9000 is invested at a 6% annual rate.

The Natural Exponential Function

The exponential function

$$f(x) = e^x$$

with base e is so prevalent in the sciences that it is often referred to as *the* exponential function or the **natural exponential function**. We use a calculator to find e^x to two decimal places for $x = -2, -1, 0, 1,$ and 2 in Table 3.5.

The graph of $f(x) = e^x$ is sketched in Figure 3.5 by using the ordered pairs in Table 3.5.

TABLE 3.5

x	e^x
-2	0.14
-1	0.37
0	1
1	2.72
2	7.39

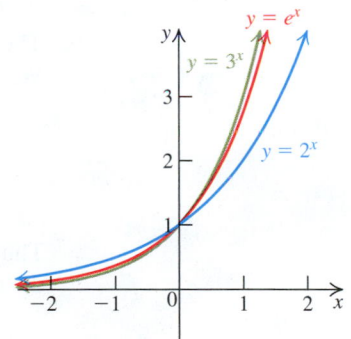

Figure 3.5 The graph of $y = e^x$ is between the graphs $y = 2^x$ and $y = 3^x$.

Since $2 < e < 3$, the graph of $y = e^x$ lies between the graphs $y = 2^x$ and $y = 3^x$. The function $f(x) = e^x$ has all the properties of exponential functions with base $a > 1$ listed on page 257.

We can apply the transformations from page 258 to the natural exponential function.

Natural Exponential Growth and Decay

Numerous applications of the exponential function are based on the fact that many different quantities have a growth (or decay) rate proportional to their size. Using calculus, we can show that in such cases, we obtain the following mathematical models:

MODELS FOR NATURAL EXPONENTIAL GROWTH AND DECAY

For $k > 0$,

Exponential growth: $A(t) = A_0 e^{kt}$, and

Exponential decay: $A(t) = A_0 e^{-kt}$, where

$$A(t) = \text{the amount at time } t$$
$$A_0 = A(0), \text{ the initial amount (the amount at } t = 0)$$
$$k = \text{relative rate of growth or decay}$$
$$t = \text{time}$$

◆ **EXAMPLE 11** **Exponential Growth**

In the year 2000, the human population of the world was approximately 6 billion. Assume the annual rate of growth from 1990 onwards at 2.1%. Using the exponential growth model, estimate the population of the world in the following years.

a. 2030 **b.** 1990

Solution

a. The year 2000 corresponds to $t = 0$. So $A_0 = 6$ billion, $k = 0.021$, and 2030 corresponds to $t = 30$.

$$A(30) = 6e^{(0.021)(30)} \qquad A(t) = A_0 e^{kt}$$
$$A(30) \approx 11.265663 \qquad \text{Use a calculator.}$$

Thus, the model predicts that if the rate of growth is 2.1% per year, over 11.26 billion people will be in the world in 2030.

b. The year 1990 corresponds to $t = -10$ (because 1990 is ten years prior to 2000). We have

$$A(-10) = 6e^{(0.021)(-10)} \qquad A(t) = A_0 e^{kt}$$
$$= 6e^{(-0.21)} \qquad \text{Simplify.}$$
$$\approx 4.8635055 \qquad \text{Use a calculator.}$$

Thus, the model estimates that the world had over 4.86 billion people in 1990, assuming the growth rate had been 2.1% per year. (The actual population in 1990 was 5.28 billion.)

Practice Problem 11 Repeat Example 11 assuming that the annual rate of growth was 2.3%.

EXAMPLE 12 **Exponential Decay**

You buy a fishing boat for $22,000. Your boat depreciates (exponentially) at the annual rate of 15% of its value. Find the depreciated value of your boat at the end of five years.

Solution

We use the exponential decay model: $A(t) = A_0 e^{-kt}$, where $A(t)$ represents the value of the boat after t years.

Here, $A_0 = 22,000$, $k = 0.15$, and $t = 5$. So,

$$A(t) = 22,000\, e^{-kt}$$
$$A(5) = 22,000\, e^{-(0.15)(5)} \qquad \text{Replace } k \text{ with } 0.15 \text{ and } t \text{ with } 5.$$
$$\approx 10,392.06416 \qquad \text{Use a calculator.}$$

So, to the nearest cent, the value of your boat at the end of five years will be $10,392.06.

Practice Problem 12 Repeat Example 12 with $k = 18\%$ and $t = 6$ years.

Answers to Practice Problems

1. $f(2) = \dfrac{1}{16}$

$f(0) = 1$

$f(-1) = 4$

$f\left(\dfrac{5}{2}\right) = \dfrac{1}{32}$

$f\left(-\dfrac{3}{2}\right) = 8$

2.

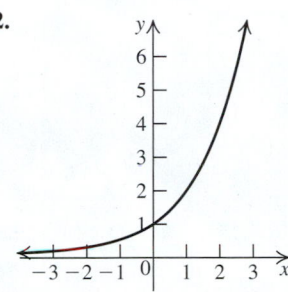

3.

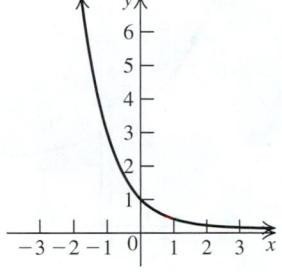

4.

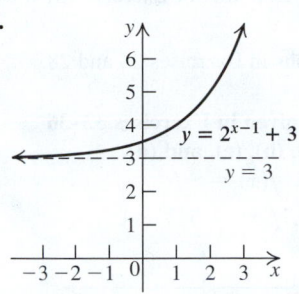

$y = 2^{x-1} + 3$

$y = 3$

Domain: $(-\infty, \infty)$; Range $(3, \infty)$; horizontal asymptote: $y = 3$

5. $4\left(\dfrac{1}{2}\right)^x$ **6.** $\$11,500$

7. a. $A = \$11,485.03$ **b.** $I = \$3,485.03$

8. (i) $\$5325.00$ **(ii)** $\$5330.28$ **(iii)** $\$5333.01$ **(iv)** $\$5334.86$

(v) $\$5335.76$ **9.** 10.023% **10.** $\$14,764.48$

11. a. 11.96229 **b.** 4.76720

12. $\$7471.10$

SECTION 3.1 Exercises

Concepts and Vocabulary

1. For the exponential function $f(x) = ca^x$, $a > 0$, $a \neq 1$, the domain is _____ , and for $c > 0$ the range is _____ .

2. The graph of $f(x) = 3^x$ has y-intercept _____ and has _____ x-intercept.

3. The horizontal asymptote of the graph of $y = \left(\dfrac{1}{3}\right)^x$ is the _____ .

4. The exponential function $f(x) = a^x$ is increasing if _____ and is decreasing if _____

5. **True or False.** The graphs of $y = 2^x$ and $y = \left(\dfrac{1}{2}\right)^x$ are symmetric with respect to the x-axis.

6. **True or False.** For the equation $y = a^x$ $(a > 0, a \neq 1)$, $y \to \infty$ as $x \to \infty$.

7. **True or False.** Let $y = e^{-x}$, then $y \to 0$ as $x \to \infty$.

8. **True or False.** The graph of $y = e^x + 1$ is obtained by shifting the graph of $y = e^x$ horizontally one unit to the left.

Building Skills

In Exercises 9–16, explain whether the given equation defines an exponential function. Give reasons for your answers. Write the base for each exponential function.

9. $y = x^3$

10. $y = 4^x$

11. $y = 2^{-x}$

12. $y = (-5)^x$

13. $y = x^x$

14. $y = 4^2$

15. $y = 0^x$

16. $y = (1.8)^x$

In Exercises 17–22, evaluate each exponential function for the given values. (Use a calculator if necessary.)

17. $f(x) = 5^{x-1}$, $f(0)$

18. $f(x) = -2^{x+1}$, $f(-2)$

19. $g(x) = 3^{1-x}$; $g(3.2)$, $g(-1.2)$

20. $g(x) = \left(\dfrac{1}{2}\right)^{x+1}$; $g(2.8)$, $g(-3.5)$

21. $h(x) = \left(\dfrac{2}{3}\right)^{2x-1}$; $h(1.5)$, $h(-2.5)$

22. $f(x) = 3 - 5^x$; $f(2)$, $f(-1)$

In Exercises 23–30, sketch the graph of the given function by making a table of values. (Use a calculator if necessary.)

23. $f(x) = 4^x$

24. $g(x) = 10^x$

25. $g(x) = \left(\dfrac{3}{2}\right)^{-x}$

26. $h(x) = 7^{-x}$

27. $h(x) = \left(\dfrac{1}{4}\right)^x$ **28.** $f(x) = \left(\dfrac{1}{10}\right)^x$

29. $f(x) = (1.3)^{-x}$ **30.** $g(x) = (0.7)^{-x}$

31. How are the graphs in Exercises 23 and 27 related? Can we obtain the graph of Exercise 27 from that of Exercise 23? If so, how?

32. Repeat Exercise 31 for the graphs in Exercises 24 and 28.

Match each exponential function given in Exercises 33–36 with one of the graphs labeled (a), (b), (c), and (d).

33. $f(x) = 5^x$ **34.** $f(x) = -5^x$

35. $f(x) = 5^{-x}$ **36.** $f(x) = 5^{-x} + 1$

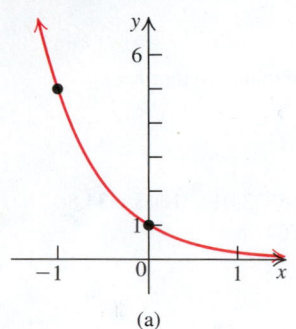

(a)

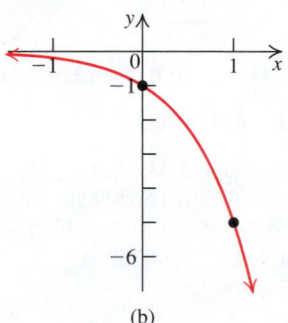

(b)

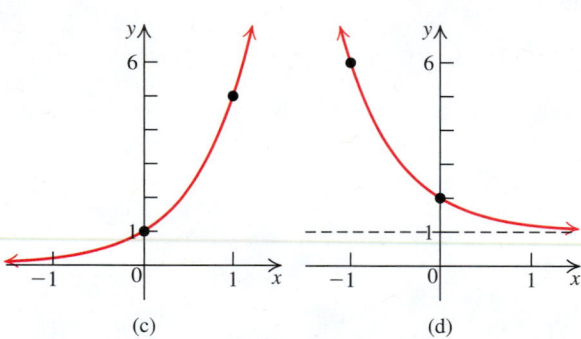

(c)   (d)

In Exercises 37–46, start with the graph of the appropriate basic exponential function f and use transformations to sketch the graph of the function g. State the domain and range of g and the horizontal asymptote of its graph.

37. $g(x) = 3^{x-1}$ **38.** $g(x) = 3^x - 1$

39. $g(x) = 4^{-x}$ **40.** $g(x) = -4^x$

41. $g(x) = -2.5^{x-1} + 4$ **42.** $g(x) = \dfrac{1}{2} \cdot 5^{1-x} - 2$

43. $g(x) = -e^{x-2} + 3$ **44.** $g(x) = 3 + e^{2-x}$

45. $g(x) = e^{|x|}$ **46.** $g(x) = -e^{|x|}$

In Exercises 47–54, find the function of the form $f(x) = ca^x$ that contains the two given graph points.

47. $(0, 1)$ and $(2, 16)$ **48.** $(0, 1)$ and $\left(-2, \dfrac{1}{9}\right)$

49. $(0, 3)$ and $(2, 12)$ **50.** $(0, 5)$ and $(1, 15)$

51. $(1, 1)$ and $(2, 5)$ **52.** $(1, 1)$ and $\left(2, \dfrac{1}{5}\right)$

53. $(1, 5)$ and $(2, 125)$ **54.** $(-1, 4)$ and $(1, 16)$

In Exercises 55–58, write an equation of the form $f(x) = a^x + b$ from the given graph. Then compute $f(2)$.

55.

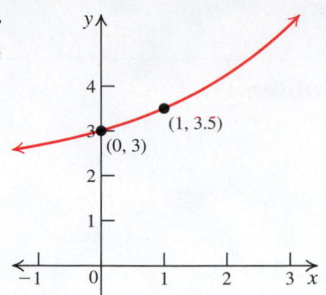

56.

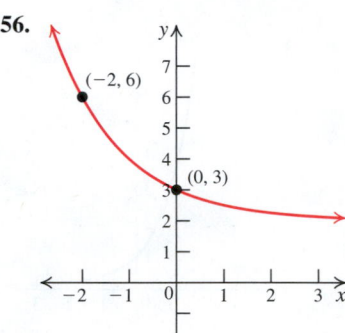

57.

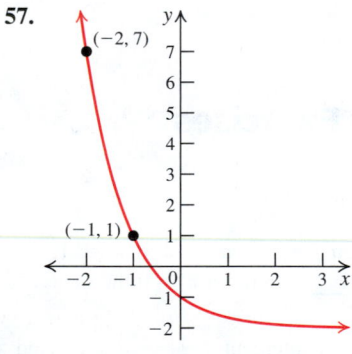

58.

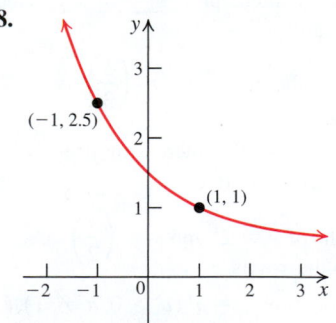

In Exercises 59–62, write an equation of each graph in the final position.

59. The graph of $y = 2^x$ is shifted two units left and then five units up.

60. The graph of $y = 3^x$ is reflected about the y-axis and then shifted three units right.

61. The graph of $y = \left(\dfrac{1}{2}\right)^x$ is stretched vertically by a factor of 2 and then shifted five units down.

62. The graph of $y = 2^{-x}$ is reflected about the x-axis and then shifted three units up.

In Exercises 63–66, find the simple interest for each value of principal P, rate r per year, and time t.

63. $P = \$5000$, $r = 10\%$, $t = 5$ years

64. $P = \$10{,}000$, $r = 5\%$, $t = 10$ years

65. $P = \$7800$, $r = 6\frac{7}{8}\%$, $t = 10$ years and 9 months

66. $P = \$8670$, $r = 4\frac{1}{8}\%$, $t = 6$ years and 8 months

In Exercises 67–70, find (a) the future value of the given principal P and (b) the interest earned in the given period.

67. $P = \$3500$ at 6.5% compounded annually for 13 years

68. $P = \$6240$ at 7.5% compounded monthly for 12 years

69. $P = \$7500$ at 5% compounded continuously for 10 years

70. $P = \$8000$ at 6.5% compounded daily for 15 years

In Exercises 71–74, find the principal P that will generate the given future value A.

71. $A = \$10{,}000$ at 8% compounded annually for 10 years

72. $A = \$10{,}000$ at 8% compounded quarterly for 10 years

73. $A = \$10{,}000$ at 8% compounded daily for 10 years

74. $A = \$10{,}000$ at 8% compounded continuously for 10 years

In Exercises 75–82, starting with the graph of $y = e^x$, use transformations to sketch the graph of each function and state its horizontal asymptote.

75. $f(x) = e^{-x}$ **76.** $f(x) = -e^x$

77. $f(x) = e^{x-2}$ **78.** $f(x) = e^{2-x}$

79. $f(x) = 1 + e^x$ **80.** $f(x) = 2 - e^{-x}$

81. $f(x) = -e^{x-2} + 3$ **82.** $g(x) = 3 + e^{2-x}$

Applying the Concepts

83. Metal cooling. Suppose a metal block is cooling so that its temperature T (in °C) is given by $T = 200 \cdot 4^{-0.1t} + 25$, where t is in hours.
 a. Find the temperature after
  (i) 2 hours.   **(ii)** 3.5 hours.
 b. How long has the cooling been taking place if the block now has a temperature of 125°C?
 c. Find the eventual temperature $(t \rightarrow \infty)$.

84. Ethnic population. The population (in thousands) of people of East Indian origin in the United States is approximated by the function

$$p(t) = 1600(2)^{0.1047t},$$

where t is the number of years since 2010.
 a. Find the population of this group in 2018.
 b. Predict the population in 2025.

85. Price appreciation. In 2016, the median price of a house in Tampa was $190,000. Assuming a rate of increase of 3% per year, what can we expect the price of such a house to be in 2021?

86. Investment. How much should a mother invest at the time her son is born to provide him with $80,000 at age 21? Assume that the interest is 7% compounded quarterly.

87. Depreciation. Trans Trucking Co. purchased a truck for $80,000. The company depreciated the truck at the end of each year at the rate of 15% of its current value. What is the value of the truck at the end of the fifth year?

88. Manhattan Island purchase. In 1626, Peter Du Minuit purchased Manhattan Island from Native Americans for 60 Dutch guilders (about $24). Suppose the $24 was invested in 1626 at a 6% rate. How much money would that investment be worth in 2006 if the interest was
 a. simple interest.
 b. compounded annually.
 c. compounded monthly.
 d. compounded continuously.

89. Investment. Ms. Ann Scheiber retired from government service in 1941 with a monthly pension of $83 and $5000 in savings. At the time of her death in January 1995 at the age of 101, Ms. Scheiber had turned the $5000 into $22 million through shrewd investments in the stock market. She bequeathed all of it to Yeshiva University in New York City. What annual rate of return compounded annually would turn $5000 to a whopping $22 million in 54 years?

90. Population. The population of Sometown, USA, was 12,000 in 2000 and grew to 15,000 in 2010. Assume that the population will continue to grow exponentially at the same constant rate. What will be the population of Sometown in 2020? $\left[\textit{Hint: Show that } e^k = \left(\dfrac{5}{4}\right)^{1/10}. \right]$

91. Medicine. Tests show that a new ointment X helps heal wounds. If A_0 square millimeters is the area of the original wound, then the area A of the wound after n days of application of the ointment X is given by $A = A_0 e^{-0.43n}$. If the area of the original wound was 10 square millimeters, find the area of the wound after ten days of application of the new ointment.

92. Cooling. The temperature T (in °C) of coffee at time t minutes after its removal from the microwave is given by the equation

$$T = 25 + 73e^{-0.28t}.$$

Find the temperature of the coffee at each time listed.
 a. $t = 0$ **b.** $t = 10$
 c. $t = 20$ **d.** after a long time

93. Interest rate. Mary purchases a twelve-year bond for $60,000.00. At the end of 12 years, she will redeem the bond for approximately $95,600.00. If the interest is compounded quarterly, what was the interest rate on the bond?

94. Best deal. Fidelity Federal offers three types of investments: (i) 9.7% compounded annually, (ii) 9.6% compounded monthly, and (iii) 9.5% compounded continuously. Which investment is the best deal?

95. Paper stacking. Suppose we have a large sheet of paper 0.015 centimeter thick and we tear the paper in half and put the pieces on top of each other. We keep tearing and stacking in this manner, always tearing each piece in half. How high will the resulting pile of paper be if we continue the process of tearing and stacking
 a. 30 times?
 b. 40 times?
 c. 50 times? [*Hint:* Use a calculator.]

96. Doubling. A jar with a volume of 1000 cubic centimeters contains bacteria that doubles in number every minute. If the jar is full in 60 minutes, how long did it take for the container to be half full?

Beyond the Basics

97. Let $f(x) = e^x$. Show that

a. $\dfrac{f(x + h) - f(x)}{h} = e^x\left(\dfrac{e^h - 1}{h}\right)$.

b. $f(x + y) = f(x)f(y)$.

c. $f(-x) = \dfrac{1}{f(x)}$.

98. Let $f(x) = 3^x + 3^{-x}$ and $g(x) = 3^x - 3^{-x}$. Find each of the following:

a. $f(x) + g(x)$

b. $f(x) - g(x)$

c. $[f(x)]^2 - [g(x)]^2$

d. $[f(x)]^2 + [g(x)]^2$

99. You need the following definition for this exercise:
Let $0! = 1$; $1! = 1$; $2! = 2 \cdot 1$; and, in general,
$n! = n(n - 1)(n - 2) \cdots 3 \cdot 2 \cdot 1$. The symbol $n!$ is read
"n factorial." Now let

$$S_n = 2 + \frac{1}{2!} + \frac{1}{3!} + \frac{1}{4!} + \cdots + \frac{1}{n!}.$$

Compute S_n for $n = 5, 10$, and 15. Compare the resulting values with the value of e on page 264.

100. The **hyperbolic cosine function** (abbreviated as **cosh x**) and hyperbolic sine function (abbreviated as **sinh x**) are defined by $\cosh x = \dfrac{e^x + e^{-x}}{2}$ and $\sinh x = \dfrac{e^x - e^{-x}}{2}$.

a. Is $\cosh x$ odd, even, or neither?

b. Is $\sinh x$ odd, even, or neither?

c. Show that $\cosh x + \sinh x = e^x$.

d. Show that $(\cosh x)^2 - (\sinh x)^2 = 1$.

101. If an investment of P dollars returns A dollars after one year, the **effective annual interest rate** or **annual yield** y is defined by the equation

$$A = P(1 + y).$$

Show that if the sum of P dollars is invested at a nominal rate r per year, compounded m times per year, the effective annual yield y is given by the equation

$$y = \left(1 + \frac{r}{m}\right)^m - 1.$$

102. The relative rate of growth of $A(t) = Pe^{kt}$ in the interval $[t, t + h]$ is defined by $\dfrac{A(t + h) - A(t)}{A(t)}$.

a. Show that $\dfrac{A(t + 1) - A(t)}{A(t)} = e^k - 1$.

b. Show that $\dfrac{A(t + h) - A(t)}{A(t)} = e^{kh} - 1$.

In Exercises 103–106, describe an order in which a sequence of transformations is applied to the graph of $y = e^x$ to obtain the graph of the given equation.

103. $y = 3e^{2x-1}$ **104.** $y = -2e^{3x+2}$

105. $y = 5e^{2-3x} + 4$ **106.** $y = -2e^{3-4x} - 1$

Critical Thinking / Discussion / Writing

107. Discuss why we do not allow a to be 1, 0, or a negative number in the definition of an exponential function of the form $f(x) = a^x$.

108. Consider the function $g(x) = (-3)^x$, which is not an exponential function. What are the possible rational numbers x for which $g(x)$ is defined?

109. Discuss the end behavior of the graph of $y = \dfrac{b}{1 + ca^x}$, where $a, b, c > 0$ and $a \neq 1$. (Consider the cases $a < 1$ and $a > 1$.)

110. Write a summary of the types of graphs (with sketches) for the exponential functions of the form $y = ca^x, c \neq 0, a > 0$, and $a \neq 1$.

111. Give a convincing argument to show that the equation $2^x = k$ has exactly one solution for every $k > 0$. Support your argument with graphs.

112. Find all solutions of the equation $2^x = 2x$. How do you know you found all solutions?

Getting Ready for the Next Section

In Exercises 113–120, simplify each expression.

113. 10^0 **114.** 10^{-1}

115. $(-8)^{1/3}$ **116.** $(25)^{1/2}$

117. $\left(\dfrac{1}{7}\right)^{-2}$ **118.** $\left(\dfrac{1}{9}\right)^{-1/2}$

119. $(18)^{1/2}$ **120.** $\left(\dfrac{1}{12}\right)^{-1/2}$

In Exercises 121–126, find the inverse of each function. Then find the domain and the range of f^{-1}.

121. $f(x) = 3x + 4$ **122.** $f(x) = \dfrac{1}{2}x - 5$

123. $f(x) = \sqrt{x}, x \geq 0$ **124.** $f(x) = x^2 + 1, x \geq 0$

125. $f(x) = \dfrac{1}{x - 1}$ **126.** $f(x) = \sqrt[3]{x}$

Logarithmic Functions

BEFORE STARTING THIS SECTION, REVIEW

1 Finding the inverse of a one-to-one function (Section 1.7, page 122)

2 Polynomial and rational inequalities (Section 2.5, page 210)

3 Graphing and transformations (Section 1.5, page 88)

OBJECTIVES

1 Define logarithmic functions.

2 Evaluate logarithms.

3 Derive basic properties of logarithmic functions.

4 Find the domains of logarithmic functions.

5 Graph logarithmic functions.

6 Use natural logarithms in applications.

◆ Detecting Possible Fraud

Listed below is the size (in square kilometers) of the first 50 countries of the world in two columns. One of the columns contains the real data, and the other column contains fictitious data. Can you guess which column contains the real data? In Example 8, we show how to detect the real data.

Country	Size I	Size II	Country	Size I	Size II
Afghanistan	608,239	645,807	Bermuda	54	53
Albania	29,562	28,748	Bhutan	47,560	46,650
Algeria	2,481,743	2,381,741	Bolivia	998,582	1,098,581
American Samoa	201	197	Bosnia and Herzegovina	49,857	51,129
Andorra	465	464	Botswana	601,780	581,730
Angola	997,852	1,246,700	Bouvet Island	48	49
Anguilla	98	96	Brazil	8,645,521	8,544,418
Antarctica	15,207,012	13,209,000	British Virgin Islands	201	151
Antigua & Barbuda	445	442	Brunei Darussalam	6,123	5,765
Antilles, Netherlands	801	800	Bulgaria	99,949	110,994
Arabia, Saudi	2,250,790	2,149,690	Burkina Faso	269,013	267,950
Argentina	2,875,510	2,777,409	Burundi	27,735	27,834
Armenia	29,851	29,743	Cambodia	201,137	181,035
Aruba	201	193	Cameroon	485,447	475,442
Australia	7,801,663	7,682,557	Canada	9,976,231	9,976,137
Austria	83,760	83,858	Cape Verde	4,121	4,033
Azerbaijan	86,640	86,530	Cayman Islands	260	259
Bahamas	20,112	13,962	Central African Republic	623,541	622,436
Bahrain	695	694	Chad	999,494	1,284,000
Bangladesh	99,763	142,615	Chile	756,589	755,482
Barbados	423	431	China	9,806,441	9,806,391
Belarus	207,599	207,600	Christmas Island	241	135
Belgium	31,618	30,518	Cocos (Keeling) Islands	21	14
Belize	23,105	22,966	Colombia	986,779	1,141,748
Benin	99,879	112,622	Comoros	2,001	1,862

1 Define logarithmic functions.

Logarithmic Functions

Recall from Section 4.1 that every exponential function

$$f(x) = a^x, a > 0, a \neq 1,$$

is a *one-to-one* function and therefore has an inverse function. Let's find the inverse of the function $f(x) = 3^x$. If we think of an exponential function f as "putting an exponent on" a particular base, then the inverse function f^{-1} must "lift the exponent off" to undo the effect of f, as illustrated here:

$$x \xrightarrow{\ f\ } 3^x \xrightarrow{\ f^{-1}\ } x$$

Let's try to find the inverse of $f(x) = 3^x$ by the procedure outlined in Section 2.9.

Step 1 Replace $f(x)$ in the equation $f(x) = 3^x$ with y: $y = 3^x$.

Step 2 Interchange x and y in the equation in Step 1: $x = 3^y$.

Step 3 Solve the equation for y in Step 2.

In Step 3, we need to solve the equation $x = 3^y$ for y. Finding y is easy when $3 = 3^y$ or $27 = 3^y$, but what if $5 = 3^y$? There actually is a real number y that solves the equation, $5 = 3^y$, but we have not introduced the name for this number yet. (We were in a similar position concerning the solution of $y^3 = 7$ before we introduced the name for the solution, namely, $\sqrt[3]{7}$.) We can say that y is "the exponent on 3 that gives 5." Instead, we use the word *logarithm* (or *log* for short) to describe this exponent. We use the notation $\log_3 5$ (read as "log of 5 with base 3") to express the solution of the equation $5 = 3^y$. This log notation is a new way to write information about exponents: $\mathbf{y = \log_3 x}$ **means** $\mathbf{x = 3^y}$. In other words, $\log_3 x$ *is the exponent to which 3 must be raised to obtain* x. The inverse function of the exponential function $f(x) = 3^x$ is the *logarithmic function* with base 3. So

$$f^{-1}(x) = \log_3 x \quad \text{or} \quad y = \log_3 x, \text{ where } y = f^{-1}(x).$$

The previous discussion could be repeated for any base a instead of 3, provided that $a > 0$ and $a \neq 1$.

Logarithmic Function

For $x > 0$, $a > 0$, and $a \neq 1$,

$$y = \log_a x \quad \text{if and only if} \quad x = a^y.$$

The function $f(x) = \log_a x$ is called the **logarithmic function with base a**.

The definition of the logarithmic function says that the two equations

$$y = \log_a x \quad (\textbf{logarithmic form}) \text{ and}$$
$$x = a^y \quad (\textbf{exponential form})$$

are equivalent. For instance, $\log_3 81 = 4$ because 3 must be raised to the fourth power to obtain 81.

EXAMPLE 1 **Converting from Exponential to Logarithmic Form**

Write each exponential equation in logarithmic form.

a. $4^3 = 64$ **b.** $\left(\dfrac{1}{2}\right)^4 = \dfrac{1}{16}$ **c.** $a^{-2} = 7$

Solution

a. $4^3 = 64$ is equivalent to $\log_4 64 = 3$.

b. $\left(\dfrac{1}{2}\right)^4 = \dfrac{1}{16}$ is equivalent to $\log_{1/2} \dfrac{1}{16} = 4$.

c. $a^{-2} = 7$ is equivalent to $\log_a 7 = -2$.

Practice Problem 1 Write each exponential equation in logarithmic form.

a. $2^{10} = 1024$ **b.** $9^{-1/2} = \dfrac{1}{3}$ **c.** $p = a^q$

SIDE
NOTE

To change from exponential form to logarithmic form, use the pattern

$$a^u = v \text{ means } \log_a v = u.$$

The exponent u is the logarithm.

EXAMPLE 2 **Converting from Logarithmic Form to Exponential Form**

Write each logarithmic equation in exponential form.

a. $\log_3 243 = 5$ **b.** $\log_2 5 = x$ **c.** $\log_a N = x$

Solution

a. $\log_3 243 = 5$ is equivalent to $243 = 3^5$.

b. $\log_2 5 = x$ is equivalent to $5 = 2^x$.

c. $\log_a N = x$ is equivalent to $N = a^x$.

Practice Problem 2 Write each logarithmic equation in exponential form.

a. $\log_2 64 = 6$ **b.** $\log_v u = w$

2 Evaluate logarithms.

Evaluating Logarithms

The technique of converting from logarithmic form to exponential form can be used to evaluate some logarithms by inspection.

EXAMPLE 3 **Evaluating Logarithms**

Find the value of each of the following logarithms.

a. $\log_5 25$ **b.** $\log_2 16$ **c.** $\log_{1/3} 9$

d. $\log_7 7$ **e.** $\log_6 1$ **f.** $\log_4 \dfrac{1}{2}$

Solution

Logarithmic Form	Exponential Form	Value
a. $\log_5 25 = y$	$25 = 5^y$ or $5^2 = 5^y$	$y = 2$
b. $\log_2 16 = y$	$16 = 2^y$ or $2^4 = 2^y$	$y = 4$
c. $\log_{1/3} 9 = y$	$9 = \left(\dfrac{1}{3}\right)^y$ or $3^2 = 3^{-y}$	$y = -2$
d. $\log_7 7 = y$	$7 = 7^y$ or $7^1 = 7^y$	$y = 1$
e. $\log_6 1 = y$	$1 = 6^y$ or $6^0 = 6^y$	$y = 0$
f. $\log_4 \dfrac{1}{2} = y$	$\dfrac{1}{2} = 4^y$ or $2^{-1} = 2^{2y}$	$y = -\dfrac{1}{2}$

Practice Problem 3 Evaluate.

a. $\log_3 9$ **b.** $\log_9 \dfrac{1}{3}$ **c.** $\log_{1/2} 32$

3 Derive basic properties of logarithmic functions.

Basic Properties of Logarithms

We use the definition of the logarithm as an exponent to derive two basic properties for $a > 0$ and $a \neq 1$,

1. Because $a^1 = a$, we have $\log_a a = 1$.

2. Because $a^0 = 1$, we have $\log_a 1 = 0$.

The next two properties follow from the fact that a logarithmic function is the inverse of the appropriate exponential function.

RECALL

If f^{-1} is the inverse of f, then
$$f^{-1}(f(x)) = x$$
and
$$f(f^{-1}(x)) = x.$$

Let $f(x) = a^x$ and $f^{-1}(x) = \log_a x$; then

3. $x = f^{-1}(f(x)) = f^{-1}(a^x) = \log_a a^x$

and

4. $x = f(f^{-1}(x)) = f(\log_a x) = a^{\log_a x}.$

We summarize these basic properties of logarithms.

Basic Properties of Logarithms

For any base $a > 0$, with $a \neq 1$,

1. $\log_a a = 1.$
2. $\log_a 1 = 0.$
3. $\log_a a^x = x$ for any real number x.
4. $a^{\log_a x} = x$ for any $x > 0$.

EXAMPLE 4   Using Basic Properties of Logarithms

Evaluate.

a. $\log_3 3$ **b.** $5^{\log_5 7}$

Solution

a. Because $\log_a a = 1$ (property (1)), we have $\log_3 3 = 1.$
b. Because $a^{\log_a x} = x$ (property (4)), we have $5^{\log_5 7} = 7.$

Practice Problem 4 Evaluate.

a. $\log_5 1$ **b.** $\log_3 3^5$ **c.** $7^{\log_7 5}$

---

4 Find the domains of logarithmic functions.

Domains of Logarithmic Functions

Because the exponential function $f(x) = a^x$ has domain $(-\infty, \infty)$ and range $(0, \infty)$, its inverse function $y = \log_a x$ **has domain $(0, \infty)$ and range $(-\infty, \infty)$**. Therefore, the logarithms of 0 and of negative numbers are not defined; so expressions such as $\log_a(-2)$ and $\log_a(0)$ are meaningless.

EXAMPLE 5   Finding the Domain

Find the domain.

a. $f(x) = \log_3(2 - x)$

b. $g(x) = \log_2\left(\dfrac{x + 1}{x - 3}\right)$

Solution

a. Because the domain of a logarithmic function is $(0, \infty)$, the expression $(2 - x)$ must be positive. The domain of f is the set of all real numbers x, where

$$2 - x > 0$$
$$2 > x \qquad \text{Solve for } x.$$

So, the domain of f is $(-\infty, 2)$.

b. The domain of g consists of all real numbers x for which $\dfrac{x+1}{x-3} > 0$. We can solve this inequality using the test-point method (page 214).

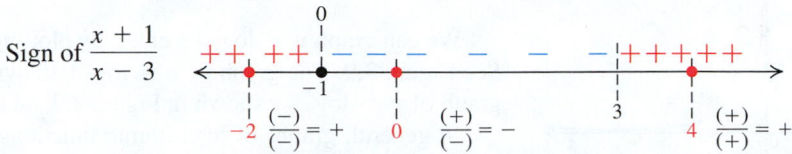

Sign of $\dfrac{x+1}{x-3}$

$-2 \quad \dfrac{(-)}{(-)} = +$ $0 \quad \dfrac{(+)}{(-)} = -$ $4 \quad \dfrac{(+)}{(+)} = +$

Figure 3.6 Solving $\dfrac{x+1}{x-3} > 0$ by the test-point method.

From Figure 3.6, we see that $\dfrac{x+1}{x-3} > 0$ on $(-\infty, -1) \cup (3, \infty)$. So the domain of g is $(-\infty, -1) \cup (3, \infty)$.

Practice Problem 5 Find the domain of $f(x) = \log_{10} \sqrt{1-x}$.

5 Graph logarithmic functions.

Graphs of Logarithmic Functions

We now look at other properties of logarithmic functions and begin with an example showing how to graph a logarithmic function.

EXAMPLE 6   Sketching a Graph

Sketch the graph of $y = \log_3 x$.

RECALL

Remember that "$\log_3 x$" means the exponent on 3 that gives x.

Plotting points (Method 1) To find selected ordered pairs on the graph of $y = \log_3 x$, we choose the x-values to be powers of 3. We can easily compute the logarithms of these values by using Property 3, namely, $\log_3 3^x = x$. We compute the y values in Table 3.6.

TABLE 3.6

x	$y = \log_3 x$	(x, y)	
$\dfrac{1}{27} = 3^{-3}$	$y = \log_3 \dfrac{1}{27} = \log_3 3^{-3} = -3$	$\left(\dfrac{1}{27}, -3\right)$	
$\dfrac{1}{9} = 3^{-2}$	$y = \log_3 \dfrac{1}{9} = \log_3 3^{-2} = -2$	$\left(\dfrac{1}{9}, -2\right)$	
$\dfrac{1}{3} = 3^{-1}$	$y = \log_3 \dfrac{1}{3} = \log_3 3^{-1} = -1$	$\left(\dfrac{1}{3}, -1\right)$	
$1 = 3^0$	$y = \log_3 1 = 0$	$(1, 0)$	Property (2)
$3 = 3^1$	$y = \log_3 3 = 1$	$(3, 1)$	Property (3)
$9 = 3^2$	$y = \log_3 9 = \log_3 3^2 = 2$	$(9, 2)$	

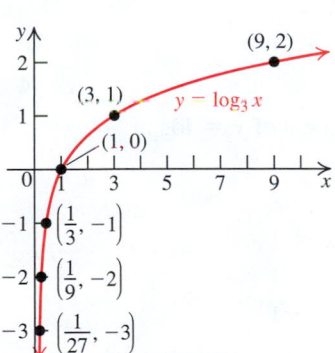

Figure 3.7 Graph $y = \log_3 x$ by plotting points.

Plotting these ordered pairs and connecting them with a smooth curve give us the graph of $y = \log_3 x$, shown in Figure 3.7.

Using the inverse function (Method 2) Because $y = \log_3 x$ is the inverse of the function $y = 3^x$, we first graph the exponential function $y = 3^x$, then reflect the graph of $y = 3^x$ about the line $y = x$. Both graphs are shown in Figure 3.8. Note that the horizontal asymptote $y = 0$ of the graph of $y = 3^x$ is reflected as the vertical asymptote $x = 0$ for the graph of $y = \log_3 x$.

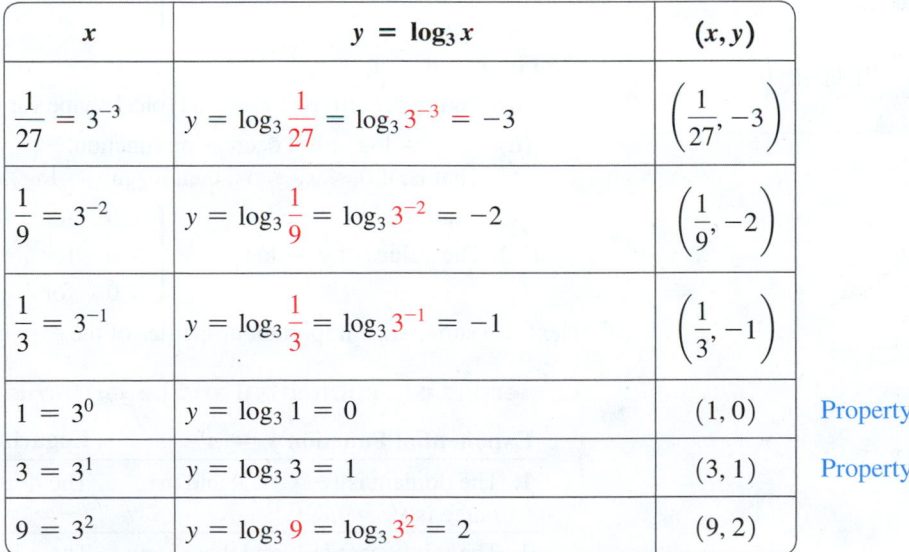

Figure 3.8 Graph $y = \log_3 x$ by using the inverse function $y = 3^x$.

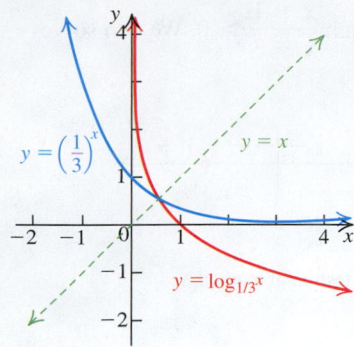

Figure 3.9 Graph of $y = \log_{1/3} x$ by using its inverse $y = \left(\frac{1}{3}\right)^x$.

Practice Problem 6 Sketch the graph of

$$y = \log_2 x.$$

We can graph $y = \log_{1/3} x$ either by plotting points or by using the graph of its inverse. See Figure 3.9. The graph of $y = \log_3 x$ shown in Figure 3.8 is, on one hand, rising. The graph of $y = \log_{1/3} x$ shown in Figure 3.9, on the other hand, is falling.

In general, graphs of logarithmic functions have two basic shapes, determined by the base a. See Figure 3.10.

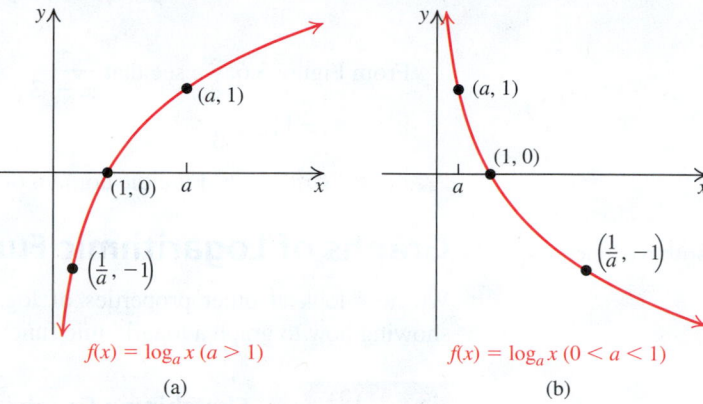

$f(x) = \log_a x \ (a > 1)$ $f(x) = \log_a x \ (0 < a < 1)$

(a)   (b)

Figure 3.10 Basic shapes of $y = \log_a x$.

SIDE NOTE

To get the correct shape for the graph of a logarithmic function $y = \log_a x$, it is helpful to graph the three points

$$(a, 1), (1, 0), \text{ and } \left(\frac{1}{a}, -1\right).$$

1. For $a > 1$:

   (i) Figure 3.10 (a) represents a typical shape for the graph of $y = \log_a x$.

   (ii) $f(x) = \log_a x$ is an increasing function.
   That is, if $0 < x_1 < x_2$, then $\log_a x_1 < \log_a x_2$.

   (iii) The values of $y = \log_a x$ are $\begin{cases} < 0 & \text{for} & 0 < x < 1 \\ = 0 & \text{for} & x = 1 \\ > 0 & \text{for} & x > 1 \end{cases}$

2. For $0 < a < 1$:

   (i) Figure 3.10 (b) represents a typical shape for the graph of $y = \log_a x$.

   (ii) $f(x) = \log_a x$ is a decreasing function.
   That is, if $0 < x_1 < x_2$, then $\log_a x_1 > \log_a x_2$.

   (iii) The values of $y = \log_a x$ are $\begin{cases} > 0 & \text{for} & 0 < x < 1 \\ = 0 & \text{for} & x = 1 \\ < 0 & \text{for} & x > 1 \end{cases}$

Next, we summarize important properties of the exponential and logarithmic functions.

Properties of Exponential and Logarithmic Functions

Exponential Function $y = a^x$	Logarithmic Function $y = \log_a x$
1. The domain is $(-\infty, \infty)$, and the range is $(0, \infty)$.	The domain is $(0, \infty)$, and the range is $(-\infty, \infty)$.
2. The y-intercept is 1, and there is no x-intercept.	The x-intercept is 1, and there is no y-intercept.
3. The x-axis $(y = 0)$ is the horizontal asymptote.	The y-axis $(x = 0)$ is the vertical asymptote.
4. The function is one-to-one; that is, $a^u = a^v$ if and only if $u = v$.	The function is one-to-one; that is, $\log_a u = \log_a v$ if and only if $u = v$.
5. The function is increasing if $a > 1$ and decreasing if $0 < a < 1$.	The function is increasing if $a > 1$ and decreasing if $0 < a < 1$.

Once we know the shape of a logarithmic graph, we can shift it vertically or horizontally, stretch it, compress it, check answers with it, and interpret the graph. The transformations on the graph of $y = \log_a x$ are summarized in Table 3.7.

TABLE 3.7 Transformations on the graph of $y = \log_a x$

Transformation (in each case $c > 0$)	Change the graph point (x, y) on $y = \log_a x$ to	Description
Vertical shift $y = \log_a x + c$ $y = \log_a x - c$	$(x, y + c)$ $(x, y - c)$	Shift the graph of $y = \log_a x$ up c units. Shift the graph of $y = \log_a x$ down c units.
Horizontal shift $y = \log_a(x - c)$ $y = \log_a(x + c)$	$(x + c, y)$ $(x - c, y)$	Shift the graph of $y = \log_a x$ to the right c units. Vertical asymptote: $x = c$. Shift the graph of $y = \log_a x$ to the left c units. Vertical asymptote: $x = -c$.
Reflection $y = -\log_a x$ $y = -\log_a(-x)$	$(x, -y)$ $(-x, y)$	Reflect the graph of $y = \log_a x$ about the x-axis. Reflect the graph of $y = \log_a x$ about the y-axis.
Vertical stretching or compressing $y = c \log_a x$	(x, cy)	Vertically stretch the graph of $y = \log_a x$ if $c > 1$. Vertically compress the graph of $y = \log_a x$ if $0 < c < 1$.
Horizontal stretching or compressing $y = \log_a(cx)$	$\left(\dfrac{x}{c}, y\right)$	Horizontally compress the graph of $y = \log_a x$ if $c > 1$. Horizontally compress the graph of $y = \log_a x$ if $0 < c < 1$.

EXAMPLE 7 **Using Transformations on $f(x) = \log_3 x$**

Start with the graph of $f(x) = \log_3 x$ and use transformations to sketch the graph of each function.

a. $f(x) = \log_3 x + 2$
b. $f(x) = \log_3 (x - 1)$
c. $f(x) = -\log_3 x$
d. $f(x) = \log_3 (-x)$

State the domain and range and the vertical asymptote for the graph of each function.

Solution

We start with the graph of $f(x) = \log_3 x$ and use the transformations shown in Figure 3.11.

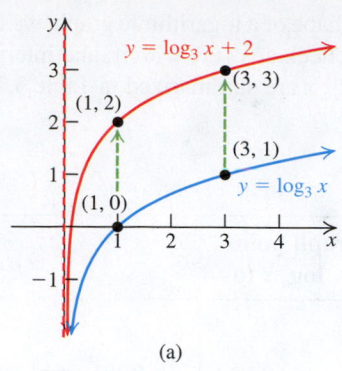

(a)

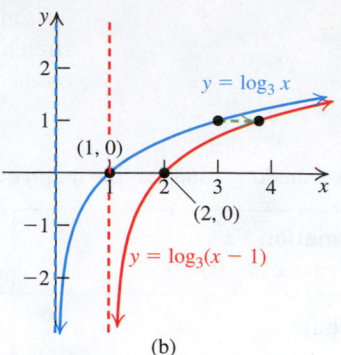

(b)

a. $f(x) = \log_3 x + 2$

Adding 2 to $\log_3 x$ shifts the graph 2 units up. Domain: $(0, \infty)$; range: $(-\infty, \infty)$; vertical asymptote: $x = 0$ (y-axis)

b. $f(x) = \log_3(x - 1)$

Replacing x with $x - 1$ in $\log_3 x$ shifts the graph 1 unit right. Domain: $(1, \infty)$; range $(-\infty, \infty)$; vertical asymptote: $x = 1$

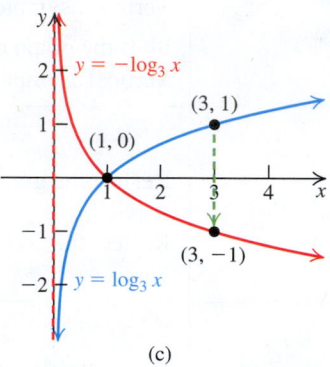

(c)

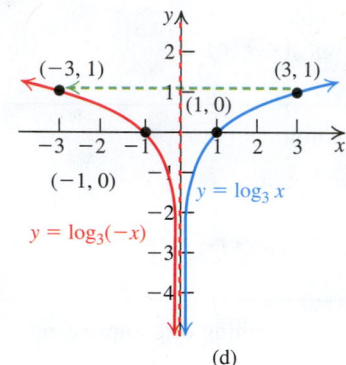

(d)

c. $f(x) = -\log_3(x)$

Multiplying $\log_3 x$ by -1 reflects the graph about the x-axis.
Domain: $(0, \infty)$; range: $(-\infty, \infty)$; vertical asymptote: $x = 0$

d. $f(x) = \log_3(-x)$

Replacing x with $-x$ in $\log_3 x$ reflects the graph about the y-axis.
Domain: $(-\infty, 0)$; range: $(-\infty, \infty)$; vertical asymptote: $x = 0$

Figure 3.11 Transformations on $y = \log_3 x$.

Practice Problem 7 Use transformations of the graph $f(x) = \log_2 x$ to sketch the graph of

$$y = -\log_2(x - 3).$$

TECHNOLOGY
CONNECTION

You can find the common logarithms of any positive number with the $\boxed{\text{LOG}}$ key on your calculator. The exact key sequence will vary with different calculators. On most calculators, if you want to find log 3.7, press

3.7 $\boxed{\text{LOG}}$ $\boxed{\text{ENTER}}$

or

$\boxed{\text{LOG}}$ 3.7 $\boxed{\text{ENTER}}$

and see the display 0.5682017.
   Although the display reads 0.5682017, this number really is an approximate value; you should therefore write $\log_{10} 3.7 \approx 0.5682017$.

Common Logarithm

The logarithm with base 10 is called the **common logarithm** and is denoted by omitting the base, so

$$\log x = \log_{10} x.$$

$$y = \log x \ (x > 0) \qquad \text{if and only if} \qquad x = 10^y.$$

Applying the basic properties of logarithms (see page 274) to common logarithms, we have the following:

1. $\log 10 = 1$

2. $\log 1 = 0$

3. $\log 10^x = x$, x any real number

4. $10^{\log x} = x$, $x > 0$

We use property (3) to evaluate the common logarithms of numbers that are powers of 10. For example,

$$\log 1000 = \log 10^3 = 3$$
$$\text{and} \quad \log 0.01 = \log 10^{-2} = -2.$$

We use a calculator to find the common logarithms of numbers that are not powers of 10 by pressing the LOG key.

The graph of $y = \log x$ is similar to the graph in Figure 3.10(a) because the base for $\log x$ is $10 > 1$.

Frank Benford
(1883–1948)

Frank Benford was an electrical engineer and a physicist who is best known for the probability distribution (Benford's Law) about the occurrence of digits in lists of real datasets. Auditors often use the law to check for fraudulent data. Benford graduated from the University of Michigan in 1910 and worked for General Electric until he retired in July 1948. He died suddenly at his home on December 4, 1948.

EXAMPLE 8 Benford's Law

Benford's Law states that the chances of the digit m occurring as the first decimal digit (from 1 to 9) from many real-life data sources is

$$P_m = \log(m + 1) - \log m.$$

This means that about $100P_m\%$ of the data can be expected to have m as the first digit.

a. Find P_1. Interpret your result.

b. Use part **a** to decide which column (Size I or Size II) represents the actual data for the size of countries listed on page 271.

Solution

a. $P_m = \log(m + 1) - \log m$ Benford's Law
$\quad P_1 = \log 2 - \log 1$ $m = 1$
$\quad\quad = \log 2$ $\log 1 = \log_{10} 1 = 0$
$\quad\quad \approx 0.301$ Use a calculator.

This means that about 30.1% of the data are expected to have 1 as the first digit.

b. We count the number of countries having 1 as the first digit in the columns with headings Size I and Size II. We find the following:

Number of countries with 1 as the first digit:

$$\text{Size I column} = 1$$
$$\text{Size II column} = 16$$

30% of 50 countries $= (0.30)(50) = 15$. So according to Benford's Law, the Size II column represents the actual data.

Practice Problem 8 Find P_2 and interpret your result.

Natural Logarithm

In most applications in calculus and the sciences, the convenient base for logarithms is the number e. The logarithm with base e is called the **natural logarithm** and is denoted by $\ln x$ (read "*ell en x*") so that

$$\ln x = \log_e x.$$

$$y = \ln x \, (x > 0) \quad \text{if and only if} \quad x = e^y.$$

Applying the basic properties of logarithms (see page 274) to natural logarithms, we have the following:

1. $\ln e = 1$
2. $\ln 1 = 0$

3. $\ln e^x = x$, x any real number

4. $e^{\ln x} = x$, $x > 0$

We can use property (3) to evaluate the natural logarithms of powers of e.

EXAMPLE 9 **Evaluating the Natural Logarithm Function**

Evaluate each expression.

a. $\ln e^4$ **b.** $\ln \dfrac{1}{e^{2.5}}$ **c.** $\ln 3$

Solution

a. $\ln e^4 = 4$ Property (3)

b. $\ln \dfrac{1}{e^{2.5}} = \ln e^{-2.5} = -2.5$ Property (3)

c. $\ln 3 \approx 1.0986123$ Use a calculator.

Practice Problem 9 Evaluate each expression.

a. $\ln \dfrac{1}{e}$ **b.** $\ln 2$

Summary of Basic Properties of Logarithms

Base $a > 0, a \neq 1$	Base 10	Base e
1. $\log_a a = 1$	$\log 10 = 1$	$\ln e = 1$
2. $\log_a 1 = 0$	$\log 1 = 0$	$\ln 1 = 0$
3. $\log_a a^x = x$	$\log 10^x = x$	$\ln e^x = x$
4. $a^{\log_a x} = x$	$10^{\log x} = x$	$e^{\ln x} = x$

6 Use natural logarithms in applications.

Investments

We recall (page 264) the continuous compound interest formula:

$$A = Pe^{rt} \qquad (3)$$

We express equation (3) in logarithmic form:

$$\frac{A}{P} = e^{rt} \qquad \text{Divide both sides of (3) by } P.$$

$$\ln \frac{A}{P} = rt \qquad \text{Logarithmic form}$$

CONTINUOUS COMPOUND INTEREST

Exponential Form   **Logarithmic Form**

$$A = Pe^{rt} \qquad\qquad\qquad \ln \frac{A}{P} = rt$$

The exponential form is used when we need to find A or P; logarithmic form is useful in calculating r or t.

EXAMPLE 10   **Doubling Your Money**

a. How long will it take to double your money if it earns 6.5% compounded continuously?

b. At what rate of return, compounded continuously, would your money double in five years?

Solution

If P dollars is invested and you want to double it, then the final amount $A = 2P$.

a. $\ln\left(\dfrac{A}{P}\right) = rt$ Continuous compounding formula, logarithmic form

$\ln\left(\dfrac{2P}{P}\right) = 0.065t$ $A = 2P, r = 0.065$

$\ln 2 = 0.065t$ $\dfrac{2P}{P} = 2$

$\dfrac{\ln 2}{0.065} = t$ Divide both sides by 0.065.

$t \approx 10.66$ Use a calculator.

It will take approximately 11 years to double your money.

b. $\ln\left(\dfrac{A}{P}\right) = rt$ Continuous compounding formula, logarithmic form

$\ln\left(\dfrac{2P}{P}\right) = r(5)$ $A = 2P, t = 5$

$\ln 2 = 5r$ $\dfrac{2P}{P} = 2$

$\dfrac{\ln 2}{5} = r$ Divide both sides by 5.

$r \approx 0.1386$ Use a calculator.

Your investment will double in five years at the approximate rate of 13.86%.

Practice Problem 10 Repeat Example 10 for tripling $(A = 3P)$ your money.   ■

Newton's Law of Cooling

When a cool drink is removed from a refrigerator, the drink warms to the temperature of the room. When removed from the oven, a pizza baked at a high temperature cools to the temperature of the room.

In situations such as these, the rate at which an object's temperature changes at any given time is proportional to the difference between its temperature and the temperature of the surrounding medium. This observation is called *Newton's Law of Cooling*, although, as in the case of a cool drink, it applies to warming as well.

Newton's Law of Cooling states that

$$T(t) = T_s + (T_0 - T_s)e^{-kt} \qquad (6)$$

where $T(t)$ is the temperature of the object at time t, T_s is the surrounding temperature, T_0 is the value of $T(t)$ at $t = 0$, and k is a positive constant that depends on the object.

We can express equation (6) in logarithmic form:

$T(t) - T_s = (T_0 - T_s)e^{-kt}$ Subtract T_s from both sides of equation (6).

$\dfrac{T(t) - T_s}{T_0 - T_s} = e^{-kt}$ Divide both sides by $T_0 - T_s$.

$\ln\left(\dfrac{T(t) - T_s}{T_0 - T_s}\right) = -kt$ Logarithmic form

NEWTON'S LAW OF COOLING

Exponential Form

$$T(t) = T_s + (T_0 - T_s)e^{-kt}$$

Logarithmic Form

$$\ln\left(\frac{T(t) - T_s}{T_0 - T_s}\right) = -kt$$

EXAMPLE 11   **McDonald's Hot Coffee**

The local McDonald's franchise has discovered that when coffee is poured from a coffee-maker whose contents are 180°F into a noninsulated pot, after 1 minute the coffee cools to 165°F if the room temperature is 72°F. How long should the employees wait before pouring the coffee from this noninsulated pot into cups to deliver it to customers at 125°F?

Solution

From the given data, we have $T = 165$ when $t = 1$, $T_0 = 180$, and $T_s = 72$.

$$\ln\left(\frac{165 - 72}{180 - 72}\right) = \ln\left(\frac{93}{108}\right) = -k \cdot 1 \qquad \text{Substitute given data in the logarithmic form.}$$

$$k = -\ln\left(\frac{93}{108}\right) \qquad \text{Solve for } k.$$

With this value of k, we find t when $T = 125$.

$$\ln\left(\frac{125 - 72}{180 - 72}\right) = \ln\left(\frac{53}{108}\right) = -\left(-\ln\left(\frac{93}{108}\right)\right) \cdot t \qquad \text{Substitute data in logarithmic form.}$$

$$\left[\frac{1}{\ln\left(\frac{93}{108}\right)}\right]\ln\left(\frac{53}{108}\right) = t \qquad \text{Solve for } t.$$

$$t \approx 4.76 \qquad \text{Use a calculator.}$$

The employees should wait approximately 4.76 minutes (realistically, about 5 minutes) to deliver the coffee at 125°F to the customers.

Practice Problem 11 Repeat Example 11 assuming that the coffee is to be delivered to the customers at 120°F.

Answers to Practice Problems

1. a. $\log_2 1024 = 10$ **b.** $\log_9\left(\frac{1}{3}\right) = -\frac{1}{2}$ **c.** $\log_a p = q$

2. a. $2^6 = 64$ **b.** $v^w = u$ **3. a.** 2 **b.** $-1/2$ **c.** -5

4. a. 0 **b.** 5 **c.** 5 **5.** $(-\infty, 1)$

6.

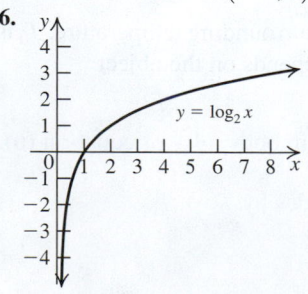

7. Shift the graph of $y = \log_2 x$ three units right and reflect the graph about the x-axis.

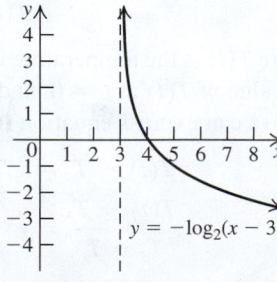

8. $P_2 = \log 3 - \log 2 \approx 0.176$. About 17.6% of the data is expected to have 2 as the first digit.

9. a. -1 **b.** ≈ 0.693

10. a. ≈ 17 years **b.** 21.97%

11. ≈ 5.4 min

SECTION 3.2 Exercises

Concepts and Vocabulary

1. The domain of the function $y = \log_a x$ is _____, and its range is _____.

2. The logarithmic form $y = \log_a x$ is equivalent to the exponential form _____.

3. The logarithm with base 10 is called the _____ logarithm, and the logarithm with base e is called the _____ logarithm.

4. $a^{\log_a x} =$ _____, and $\log_a a^x =$ _____.

5. **True or False.** The graph of $y = \log_a x, a > 0, a \neq 1$, is the graph of an increasing function.

6. **True or False.** The graph of $y = \log_a x, a > 0$, and $a \neq 1$, has no horizontal asymptote.

7. **True or False.** The domain of $f(x) = \log_a x, a > 0$, and $a \neq 1$ is $(-\infty, \infty)$.

8. **True or False.** If $y = \log_a x$, then $x = y$.

Building Skills

In Exercises 9–20, write each exponential equation in logarithmic form.

9. $5^2 = 25$
10. $(49)^{-1/2} = \dfrac{1}{7}$
11. $\left(\dfrac{1}{16}\right)^{-1/2} = 4$
12. $(a^2)^2 = a^4$
13. $10^0 = 1$
14. $10^4 = 10{,}000$
15. $(10)^{-1} = 0.1$
16. $3^x = 5$
17. $a^2 + 2 = 7$
18. $a^e = \pi$
19. $2a^3 - 3 = 10$
20. $5 \cdot 2^{ct} = 11$

In Exercises 21–32, write each logarithmic equation in exponential form.

21. $\log_2 32 = 5$
22. $\log_7 49 = 2$
23. $\log_{10} 100 = 2$
24. $\log_{10} 10 = 1$
25. $\log_{10} 1 = 0$
26. $\log_a 1 = 0$
27. $\log_{10} 0.01 = -2$
28. $\log_{1/5} 5 = -1$
29. $3 \log_8 2 = 1$
30. $1 + \log 1000 = 4$
31. $\ln 2 = x$
32. $\ln \pi = a$

In Exercises 33–42, evaluate each expression without using a calculator.

33. $\log_5 125$
34. $\log_9 81$
35. $\log 10{,}000$
36. $\log_3 \dfrac{1}{3}$
37. $\log_2 \dfrac{1}{8}$
38. $\log_4 \dfrac{1}{64}$
39. $\log_3 \sqrt{27}$
40. $\log_{27} 3$
41. $\log_{16} 2$
42. $\log_5 \sqrt{125}$

In Exercises 43–54, evaluate each expression.

43. $\log_3 1$
44. $\log_{1/2} 1$
45. $\log_7 7$
46. $\log_{1/9} \dfrac{1}{9}$
47. $\log_6 6^7$
48. $\log_{1/2}\left(\dfrac{1}{2}\right)^5$
49. $3^{\log_3 5}$
50. $7^{\log_7 \frac{1}{2}}$
51. $2^{\log_2 7} + \log_5 5^{-3}$
52. $3^{\log_3 5} - \log_2 2^{-3}$
53. $4^{\log_4 6} - \log_4 4^{-2}$
54. $10^{\log x} - e^{\ln y}$

In Exercises 55–70, find the domain of each function.

55. $f(x) = \log_2(x + 1)$
56. $g(x) = \log_3(x - 8)$
57. $f(x) = \log_3 \sqrt{x - 1}$
58. $g(x) = \log_4 \sqrt{3 - x}$
59. $f(x) = \log(x - 2) + \log(2x - 1)$
60. $g(x) = \ln\sqrt{x + 5} - \ln(x + 1)$
61. $h(x) = \ln(x - 1)/\ln(2 - x)$
62. $f(x) = \ln(x - 3) + \ln(2 - x)$
63. $f(x) = \ln|x|$
64. $f(x) = \sqrt{\ln x}$
65. $f(x) = \log_2\left(\dfrac{x}{3} - 4\right)$
66. $f(x) = \log_3\left(2 - \dfrac{x}{5}\right)$
67. $f(x) = \ln\left(\dfrac{x}{x + 1}\right)$
68. $f(x) = \ln\left(\dfrac{x - 2}{x}\right)$
69. $f(x) = \log_3\left(\dfrac{x - 2}{x + 1}\right)$
70. $f(x) = \log_2\left(\dfrac{x + 3}{x - 2}\right)$

71. Match each logarithmic function with one of the graphs labeled **a–f**.
    a. $f(x) = \log x$
    b. $f(x) = -\log|x|$
    c. $f(x) = -\log(-x)$
    d. $f(x) = \log(x - 1)$
    e. $f(x) = (\log x) - 1$
    f. $f(x) = \log(-1 - x)$

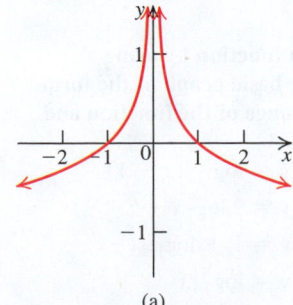

(a)

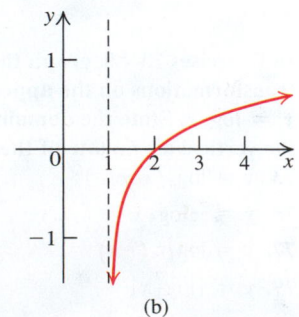

(b)

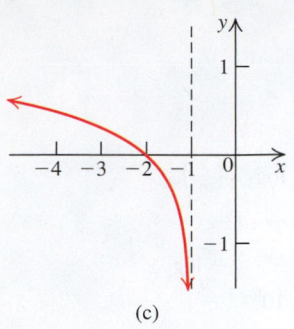

(c)

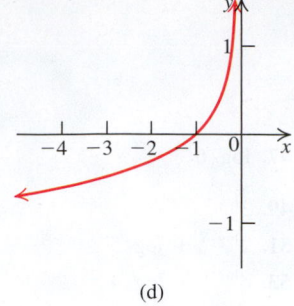

(d)

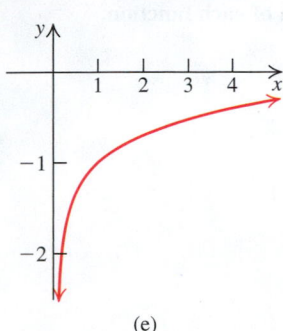

(e)

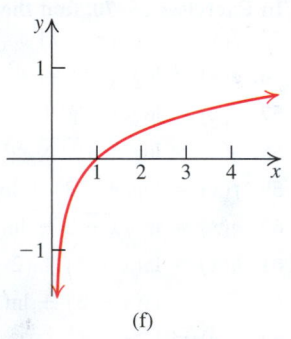

(f)

72. For each given graph, find a function of the form $f(x) = \log_a x$ that represents the graph.

a.
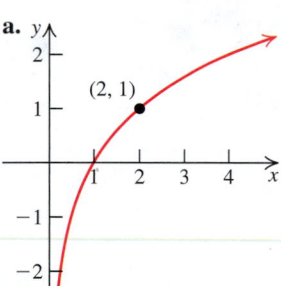

b.
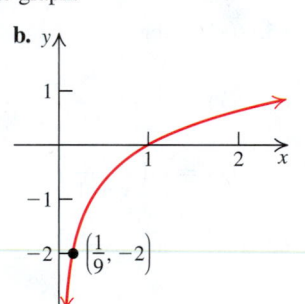

c.
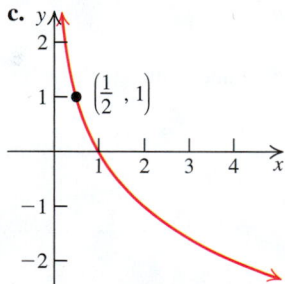

d.
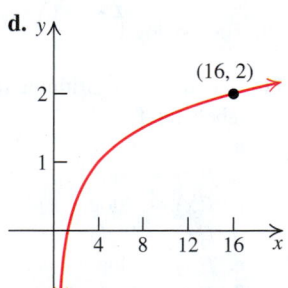

In Exercises 73–80, graph the given function by using transformations on the appropriate basic graph of the form $y = \log_a x$. State the domain and range of the function and the vertical asymptote of the graph.

73. $y = \log_4 (x + 3)$ **74.** $y = \log_{1/2} (x - 1)$

75. $y = -\log_5 x$ **76.** $y = 2 \log_7 x$

77. $y = \log_{1/5} (-x)$ **78.** $y = 1 + \log_{1/5} (-x)$

79. $y = \left| \log_3 x \right|$ **80.** $y = \log_3 \left| x \right|$

In Exercises 81–88, begin with the graph of $f(x) = \log_2 x$ and use transformations to sketch the graph of each function. Find the domain and range of the function and the vertical asymptote of the graph.

81. $y = \log_2(x - 1)$ **82.** $y = \log_2(-x)$

83. $y = \log_2(3 - x)$ **84.** $y = \log_2 x$

85. $y = 2 + \log_2(3 - x)$ **86.** $y = 4 - \log_2(3 - x)$

87. $y = \log_2 \left| x \right|$ **88.** $y = \log_2 x^2$

In Exercises 89–94, evaluate each expression.

89. $\log_4 (\log_3 81)$ **90.** $\log_4 \left[\log_3 (\log_2 8) \right]$

91. $\log_{\sqrt{2}} 2$ **92.** $\log_{2\sqrt{2}} 8$

93. $\log_{\sqrt{2}} 4$ **94.** $\log_{\sqrt{3}} 27$

In Exercises 95–100, begin with the graph of $y = \ln x$ and use transformations to sketch the graph of each of the given functions.

95. $y = \ln (x + 2)$ **96.** $y = \ln (2 - x)$

97. $y = -\ln (2 - x)$ **98.** $y = 2 \ln x$

99. $y = 3 - 2 \ln x$ **100.** $y = 1 - \ln (1 - x)$

Applying the Concepts

In Exercises 101–114, use the model $A = A_0 e^{kt}$ or $\ln \dfrac{A}{A_0} = kt$.

101. **Doubling your money.** How long would it take to double your money if you invested P dollars at the rate of 8% compounded continuously?

102. **Investment goal.** How long would it take to grow your investment from $10,000 to $120,000 at the rate of 10% compounded continuously?

103. **Rate for doubling your money.** At what annual rate of return, compounded continuously, would your investment double in six years?

104. **Rate-of-investment goal.** At what annual rate of return, compounded continuously, would your investment grow from $8000 to $50,000 in 25 years?

105. **Population of the United States.** The U.S. population in 2010 was 308 million. The U.S. Census Bureau reported that the population grew 9.7% since 2000 (the slowest growth rate since the Great Depression)
   a. What was the population of the United States in 2000?
   b. What was the annual rate of growth between 2000 and 2010?
   c. Assuming the same rate of growth, determine the year when the population of the United States will be 400 million.

106. **Population of Canada.** The population of Canada in 2010 was 35 million, and has grown 12.33% since 2000.
   a. What was the population of Canada in 2000?
   b. What was the annual rate of growth between 2000 and 2010?
   c. Assuming the same rate of growth, determine the time from 2010 until Canada's population doubles.
   d. Find the time from 2010 until Canada's population reaches 50 million.

107. Newton's Law of Cooling. A thermometer is taken from a room at 75°F to the outdoors, where the temperature is 20°F. The reading on the thermometer drops to 50°F after one minute.
  a. Find the reading on the thermometer after
    (i) Five minutes.
    (ii) Ten minutes.
    (iii) One hour.
  b. How long will it take for the reading to drop to 22°F?

108. Newton's Law of Cooling. The last bit of ice in a picnic cooler has melted ($T_0 = 32°F$). The temperature in the park is 85°F. After 30 minutes, the temperature in the cooler is 40°F. How long will it take for the temperature inside the cooler to reach 50°F?

109. Cooking salmon. A salmon filet initially at 50°F is cooked in an oven at a constant temperature of 400°F. After ten minutes, the temperature of the filet rises to 160°F. How long does it take until the salmon is medium rare at 220°F?

110. The time of murder. A forensic specialist took the temperature of a victim's body lying in a street at 2:10 A.M. and found it to be 85.7°F. At 2:40 A.M., the temperature of the body was 84.8°F. When was the murder committed if the air temperature during the night was 55°F? [Remember, normal body temperature is 98.6°F.]

111. Water contamination. A chemical is spilled into a reservoir of pure water. The concentration of chemical in the contaminated water is 4%. In one month, 20% of the water in the reservoir is replaced with clean water.
  a. What will be the concentration of the contaminant one year from now?
  b. For water to be safe for drinking, the concentration of this contaminant cannot exceed 0.01%. How long will it be before the water is safe for drinking?

112. Toxic chemicals in a lake. In a lake, one-fourth of the water is replaced by clean water every year. Sixteen thousand cubic meters of soluble toxic chemical spill takes place in the lake. Let $T(n)$ represent the amount of toxin left after n years.
  a. Find a formula for $T(n)$.
  b. How much toxin will be left after 12 years?
  c. When will 80% of the toxin be eliminated?

113. Sheep population. A herd of sheep doubles in size every three years. There are now 1500 sheep in the herd.
  a. Find an exponential function of the form $P = P_0 e^{rt}$ to model the herd's growth.
  b. How many sheep will be in the herd seven years from now (round your answer)?
  c. In how many years will the herd have 15,000 sheep?

114. Shark population. A school of sharks is losing one-ninth of its population every two years. The colony currently has 150 sharks.
  a. Find an exponential function of the form $P = P_0 e^{-rt}$ to model the school's decline.
  b. How many sharks will be in the school four years from now?
  c. In how many years (to the nearest tenth) will the colony have 35 sharks?

115. Benford's Law. The law states that the probability that the first decimal-digit of a raw data sample (from 1 to 9) is given by $P_m = \log(m + 1) - \log m$. That is, about $(100P_m)\%$ of the data can be expected to have m as the first digit.
  a. What percent of the data can be expected to have 3 as the first digit?
  b. Find $P_1 + P_2 + \cdots + P_9$. Interpret your result.

116. Prime Number Theorem. A natural number $p \geq 2$ is *prime* if the only positive divisors of p are 1 and p. An important topic in number theory involves counting prime numbers $\{2, 3, 5, 7, 11, 13, \ldots\}$. For any natural number $n \geq 2$, the quantity of prime numbers less than or equal to n is denoted by $\pi(n)$ (read "pie of n"). For example, $\pi(2) = 1$, $\pi(3) = 2$, $\pi(4) = 2$, $\pi(5) = 3$, and $\pi(6) = 3$. The *Prime Number Theorem* states that for large n, $\pi(n)$ can be approximated by $\dfrac{n}{\ln(n)}$.
  a. Find $\pi(20)$ by listing the primes that are less than or equal to 20.
  b. Find $\pi(50)$ by listing primes and compare your answer with $\dfrac{50}{\ln(50)}$.
  c. Estimate $\pi(1,000,000)$ by using the Prime Number Theorem.

Beyond the Basics

117. Finding the domain. Find the domain of each function.
  a. $f(x) = \log_2(\log_3 x)$
  b. $f(x) = \log(\ln(x - 1))$
  c. $f(x) = \ln(\log(x - 1))$
  d. $f(x) = \log(\log(\log(x - 1)))$

118. Finding the inverse. Find the inverse of each function in Exercise 117.

In Exercises 119 and 120, describe an order for the sequence of transformations on the graph of $y = \log x$ to produce the graph of the given equation.

119. $y = -3\log\left(\dfrac{1}{2}x - 2\right) + 4$

120. $y = -4 \log(2 - 3x) + 1$

121. **Present value of an investment.** Recall that if P dollars is invested in an account at an interest rate r compounded continuously, then the amount A (called the *future value of P*) in the account t years from now will be $A = Pe^{rt}$. Solving the equation for P, we get $P = Ae^{-rt}$. In this formulation, P is called the present value of the investment.

  a. Find the present value of $100,000 at 7% compounded continuously for 20 years.

  b. Find the interest rate r compounded continuously that is needed to have $50,000 be the present value of $75,000 in ten years.

122. Your uncle is 40 years old, and he wants to have an annual pension of $50,000 each year at age 65. What is the present value of his pension if the money can be invested at all times at a continuously compounded interest rate of

  a. 5%?

  b. 8%?

  c. 10%?

Critical Thinking / Discussion / Writing

In Exercises 123 and 124 evaluate each expression without using a calculator.

123. $2^{\log_2 3} - 3^{\log_3 2}$

124. $(\log_3 4 + \log_2 9)^2 - (\log_3 4 - \log_2 9)^2$

125. Solve for x: $\log_3 [\log_4 (\log_2 x)] = 0$

126. Write as a piecewise function.

  a. $f(x) = |\log x|$

  b. $g(x) = |\ln(x - 1)| + |\ln(x - 2)|$

127. **Inequalities involving logarithms.**

  a. If a is positive, is the statement "$a < b$ if and only if $\log a < \log b$" always true?

  b. What property of logarithmic functions is used whenever the statement in part (a) is true?

128. Write a summary of the types of graphs (with sketches) for the logarithmic functions of the form $y = c \log_a x, c \neq 0, a > 0$, and $a \neq 1$.

Getting Ready for the Next Section

In Exercises 129–134, write each expression in the form a^n where n is a positive integer.

129. $a^2 \cdot a^7$

130. $(a^2)^3$

131. $\sqrt{a^8}$

132. $\sqrt[3]{a^6}$

133. $\left(\dfrac{243}{32}\right)^{4/5}$

134. $\sqrt[5]{(32)^{-3}}$

In Exercises 135 and 136, perform the indicated operations and write the answer in scientific notation.

135. $(4.7 \times 10^7)(8.1 \times 10^5)$

136. $\dfrac{7.2 \times 10^6}{2.4 \times 10^{-3}}$

In Exercises 137–140, verify that each equation is true by evaluating each side without using a calculator.

137. $\log_3 81 = \log_3 3 + \log_3 27$

138. $\log_2 8 = \log_2 128 - \log_2 16$

139. $\log_2 16 = 2\log_2 4$

140. $\log_4 64 = \dfrac{\log_2 64}{\log_2 4}$

Rules of Logarithms

BEFORE STARTING THIS SECTION, REVIEW

1 Definition of logarithm (Section 3.2, page 272)

2 Basic properties of logarithms (Section 3.2, page 274)

3 Rules of exponents (Section 3.1, page 255)

OBJECTIVES

1 Learn the rules of logarithms.

2 Estimate a large number.

3 Change the base of a logarithm.

4 Apply logarithms to growth and decay.

King Tut's Golden Mask

◆ The Boy King Tut

Today the most famous pharaoh of Ancient Egypt is King Nebkheperure Tutankhamun, popularly called "King Tut." He was only 9 years old when he became a pharaoh. In 2005, a team of Egyptian scientists headed by Dr. Zahi Hawass determined that the pharaoh was 19 years old when he died (around 1346 B.C.).

Tutankhamun was a short-lived boy king who, unlike the great Egyptian kings Khufu (builder of the Great Pyramid), Amenhotep III (builder of temples throughout Egypt), and Ramesses II (prolific builder and usurper), accomplished nothing significant during his reign. In fact, little was known about him prior to Howard Carter's discovery of his tomb (and the amazing treasures it held) in the Valley of the Kings on November 4, 1922. Carter, with his benefactor Lord Carnarvon at his side, entered the tomb's burial chamber on November 26, 1922. Lord Carnarvon died seven weeks after entering the burial chamber, giving rise to the theory of the "curse" of King Tut.

Work on the tomb continued until 1933. The tomb contained a pristine mummy of an Egyptian king, lying intact in his original burial furniture. He was accompanied by a small slice of the royal world of the pharaohs: golden chariots, statues of gold and ebony, a fleet of miniature ships to accommodate his trip to the hereafter, his throne of gold, bottles of perfume, precious jewelry, and more. The "Treasures of Tutankhamun" exhibition, first shown at the British Museum in London in 1972, traveled to many countries, including the United States.

In Example 8, we discuss the age of a work of art found in King Tut's tomb.

1 Learn the rules of logarithms.

Rules of Logarithms

Recall the basic properties of logarithms from Section 3.2 for $a > 0, a \neq 1$:

$$\log_a a = 1 \qquad (1)$$
$$\log_a 1 = 0 \qquad (2)$$
$$\log_a a^x = x, x \text{ real} \qquad (3)$$
$$a^{\log_a x} = x, x > 0 \qquad (4)$$

In this section, we will discuss some important rules of logarithms that are helpful in calculations, simplifications, and applications.

RULES OF LOGARITHMS

Let M, N, and a be positive real numbers with $a \neq 1$ and let r be any real number.

Rule	Description	Examples
1. *Product Rule:* $\log_a (MN) = \log_a M + \log_a N$	The logarithm of the product of two (or more) numbers is the sum of the logarithms of the numbers.	$\ln (5 \cdot 7) = \ln 5 + \ln 7$ $\log (3x) = \log 3 + \log x$ $\log_2 (5 \cdot 17) = \log_2 5 + \log_2 17$
2. *Quotient Rule:* $\log_a \left(\dfrac{M}{N} \right) = \log_a M - \log_a N$	The logarithm of the quotient of two numbers is the difference of the logarithms of the numbers.	$\ln \dfrac{5}{7} = \ln 5 - \ln 7$ $\log \dfrac{5}{x} = \log 5 - \log x$
3. *Power Rule:* $\log_a M^r = r \log_a M$	The logarithm of a number to the power r is r times the logarithm of the number.	$\ln 5^7 = 7 \ln 5$ $\log 5^{3/2} = \dfrac{3}{2} \log 5$ $\log_2 7^{-3} = -3 \log_2 7$

Rules 1, 2, and 3 follow from the corresponding rules of the exponents:

$$a^u \cdot a^v = a^{u+v} \qquad \text{Product rule}$$

$$\frac{a^u}{a^v} = a^{u-v} \qquad \text{Quotient rule}$$

$$(a^u)^r = a^{u\,r} \qquad \text{Power rule}$$

For example, to prove the product rule for logarithms, we let

$$\log_a M = u \qquad \text{and} \qquad \log_a N = v.$$

The corresponding exponential forms of these equations are

$$M = a^u \qquad \text{and} \qquad N = a^v.$$

Then

$$MN = a^u \cdot a^v$$
$$MN = a^{u+v} \qquad \textcolor{blue}{\text{Product rule of exponents}}$$
$$\log_a MN = u + v \qquad \textcolor{blue}{\text{Logarithmic form}}$$
$$\log_a MN = \log_a M + \log_a N \qquad \textcolor{blue}{\text{Replace } u \text{ with } \log_a M \text{ and } v \text{ with } \log_a N.}$$

This proves the product rule of logarithms. In Exercise 122, you are asked to prove the quotient rule and the power rule similarly by using the corresponding rules for exponents.

EXAMPLE 1   Using Rules of Logarithms to Evaluate Expressions

Given that $\log_5 z = 3$ and $\log_5 y = 2$, evaluate each expression.

a. $\log_5 (yz)$ **b.** $\log_5 (125y^7)$ **c.** $\log_5 \sqrt{\dfrac{z}{y}}$ **d.** $\log_5 (z^{1/30} y^5)$

Solution

a. $\log_5 (yz) = \log_5 y + \log_5 z \qquad \textcolor{blue}{\text{Product rule}}$

$\qquad\qquad\quad = 2 + 3 = 5 \qquad \textcolor{blue}{\text{Use the given values and simplify.}}$

RECALL

Radicals can also be written as exponents. Recall that
$$\sqrt{x} = x^{1/2}$$
and $\sqrt[3]{x} = x^{1/3}$.
In general,
$$\sqrt[m]{x^n} = x^{n/m}.$$

b. $\log_5(125y^7) = \log_5 125 + \log_5 y^7$ Product rule
$$= \log_5 5^3 + \log_5 y^7 \qquad 125 = 5^3$$
$$= 3 + 7\log_5 y \qquad \text{Power rule and } \log_a a = 1$$
$$= 3 + 7(2) = 17 \qquad \text{Use the given values and simplify.}$$

c. $\log_5 \sqrt{\dfrac{z}{y}} = \log_5\left(\dfrac{z}{y}\right)^{1/2}$ Rewrite radical as exponent.

$$= \frac{1}{2}\log_5 \frac{z}{y} \qquad \text{Power rule}$$

$$= \frac{1}{2}\left(\log_5 z - \log_5 y\right) \qquad \text{Quotient rule}$$

$$= \frac{1}{2}(3 - 2) = \frac{1}{2} \qquad \text{Use the given values and simplify.}$$

d. $\log_5\left(z^{1/30}y^5\right) = \log_5 z^{1/30} + \log_5 y^5$ Product rule

$$= \frac{1}{30}\log_5 z + 5\log_5 y \qquad \text{Power rule}$$

$$= \frac{1}{30}(3) + 5(2) \qquad \text{Use the given values.}$$

$$= 0.1 + 10 = 10.1 \qquad \text{Simplify.}$$

Practice Problem 1 Evaluate each expression for the given values in Example 1.

a. $\log_5(y/z)$ **b.** $\log_5(y^2 z^3)$

In many applications in more advanced mathematics courses, the rules of logarithms are used in both directions; that is, the rules are read from left to right and from right to left. For example, by the product rule of logarithms, we have

$$\log_2 3x = \log_2 3 + \log_2 x.$$

The expression $\log_2 3 + \log_2 x$ is the *expanded form* of $\log_2 3x$, while $\log_2 3x$ is the *condensed form,* or the *single logarithmic form,* of $\log_2 3 + \log_2 x$.

EXAMPLE 2   **Writing Expressions in Expanded Form**

Write each expression in expanded form. Assume that all expressions containing variables represent positive numbers.

a. $\log_2 \dfrac{x^2(x-1)^3}{(2x+1)^4}$ **b.** $\log_c \sqrt{x^3 y^2 z^5}$

Solution

a. $\log_2 \dfrac{x^2(x-1)^3}{(2x+1)^4} = \log_2 x^2(x-1)^3 - \log_2(2x+1)^4$ Quotient rule

$$= \log_2 x^2 + \log_2(x-1)^3 - \log_2(2x+1)^4 \qquad \text{Product rule}$$
$$= 2\log_2 x + 3\log_2(x-1) - 4\log_2(2x+1) \qquad \text{Power rule}$$

b. $\log_c \sqrt{x^3 y^2 z^5} = \log_c(x^3 y^2 z^5)^{1/2}$ $\sqrt{a} = a^{1/2}$

$$= \frac{1}{2}\log_c(x^3 y^2 z^5) \qquad \text{Power rule}$$

$$= \frac{1}{2}[\log_c x^3 + \log_c y^2 + \log_c z^5] \qquad \text{Product rule}$$

$$= \frac{1}{2}[3\log_c x + 2\log_c y + 5\log_c z] \qquad \text{Power rule}$$

$$= \frac{3}{2}\log_c x + \log_c y + \frac{5}{2}\log_c z \qquad \text{Distributive property}$$

Practice Problem 2 Write each expression in expanded form. Assume that all expressions containing variables represent positive numbers.

a. $\ln \dfrac{2x - 1}{x + 4}$ b. $\log \sqrt{\dfrac{4xy}{z}}$

EXAMPLE 3   **Writing Expressions in Condensed Form**

Write each expression in condensed form.

a. $\log 3x - \log 4y$ b. $2\ln x + \dfrac{1}{2}\ln(x^2 + 1)$

c. $2\log_2 5 + \log_2 9 - \log_2 75$ d. $\dfrac{1}{3}[\ln x + \ln(x + 1) - \ln(x^2 + 1)]$

Solution

a. $\log 3x - \log 4y = \log\left(\dfrac{3x}{4y}\right) \qquad \text{Quotient rule}$

b. $2\ln x + \dfrac{1}{2}\ln(x^2 + 1) = \ln x^2 + \ln(x^2 + 1)^{1/2} \qquad \text{Power rule}$

$$= \ln(x^2\sqrt{x^2 + 1}) \qquad \text{Product rule; } \sqrt{a} = a^{1/2}$$

c. $2\log_2 5 + \log_2 9 - \log_2 75 = \log_2 5^2 + \log_2 9 - \log_2 75 \qquad \text{Power rule}$

$$= \log_2(25 \cdot 9) - \log_2 75 \qquad 5^2 = 25; \text{ Product rule}$$

$$= \log_2 \frac{25 \cdot 9}{75} \qquad \text{Quotient rule}$$

$$= \log_2 3 \qquad \frac{25 \cdot 9}{75} = \frac{9}{3} = 3$$

d. $\dfrac{1}{3}[\ln x + \ln(x + 1) - \ln(x^2 + 1)] = \dfrac{1}{3}[\ln x(x + 1) - \ln(x^2 + 1)] \qquad \text{Product rule}$

$$= \frac{1}{3}\ln\left[\frac{x(x + 1)}{x^2 + 1}\right] \qquad \text{Quotient rule}$$

$$= \ln \sqrt[3]{\frac{x(x + 1)}{x^2 + 1}} \qquad \text{Power rule; } a^{1/3} = \sqrt[3]{a}$$

Practice Problem 3 Write in condensed form: $\dfrac{1}{2}[\log(x + 1) + \log(x - 1)]$.

Be careful when using the rules of logarithms to simplify expressions. For example, there is no property that allows you to rewrite $\log_a(x + y)$.

In general,

$$\log_a(x + y) \neq \log_a x + \log_a y. \qquad (5)$$

To illustrate this statement, let $x = 100$, $y = 10$, and $a = 10$. Then the value of the left side of (5) is

$$\log(100 + 10) = \log(110) \approx 2.0414. \qquad \text{Use a calculator.}$$

The value of the right side of (5) is

$$
\begin{aligned}
\log 100 + \log 10 &= \log 10^2 + \log 10 \\
&= 2 \log 10 + \log 10 && \text{Power rule} \\
&= 2(1) + 1 = 3. && \log 10 = \log_{10} 10 = 1
\end{aligned}
$$

Therefore, $\log(100 + 10) \neq \log 100 + \log 10$.

WARNING

To avoid common errors, be aware that in general,

$$\log_a(M + N) \neq \log_a M + \log_a N.$$

$$\log_a(M - N) \neq \log_a M - \log_a N.$$

$$(\log_a M)(\log_a N) \neq \log_a MN.$$

$$\log_a\left(\frac{M}{N}\right) \neq \frac{\log_a M}{\log_a N}.$$

$$\frac{\log_a M}{\log_a N} \neq \log_a M - \log_a N.$$

$$(\log_a M)^r \neq r \log_a M.$$

2 Estimate a large number.

Number of Digits

In estimating the magnitude of a positive number K, we look for the number of digits needed to express K. If K has a fractional part, then "number of digits" means the number of digits to the left of the decimal point. For example, 357.29 and 231.4796 are viewed as three-digit numbers in this context.

A number K written in the form $K = s \times 10^n$, where $1 \leq s < 10$ and n is an integer, is said to be in scientific notation. For example, $432.1 = 4.321 \times 10^2$ and $0.56 = 5.6 \times 10^{-1}$ are both in scientific notation.

The exponent n represents the magnitude of K and is closely related to the common logarithm. In fact, we have:

$$
\begin{aligned}
\log K &= \log(s \times 10^n) && \text{Take log of both sides of } K = s \times 10^n. \\
&= \log s + \log 10^n && \text{Product rule} \\
&= \log s + n \log 10 && \text{Power rule} \\
&= \log s + n && \log 10 = \log_{10} 10 = 1
\end{aligned}
$$

Since $1 \leq s < 10$, we have

$$
\begin{aligned}
\log 1 \leq \log s &< \log 10 && \log_{10} \text{ is an increasing function.} \\
0 \leq \log s &< 1 && \log 1 = 0 \text{ and } \log 10 = 1
\end{aligned}
$$

This says that the exponent n is the largest integer less than or equal to $\log K$. We note that a number m is a three-digit number, if

$$10^2 = 100 \leq m < 1000 = 10^3$$

$$\text{or} \quad \log 10^2 \leq \log m < \log 10^3$$

$$2 \leq \log m < 3 \qquad\qquad \log 10^x = \log_{10} 10^x = x$$

Similarly, a number K has $(n + 1)$ digits if $n \leq \log K < n + 1$.

DIGITS AND COMMON LOGARITHMS

A positive number K has $(n + 1)$ digits if and only if $\log K$ is in the interval $[n, n + 1)$.

For example, the number 8765.43, which is a four-digit number, has its common logarithm between 3 and 4. Conversely, if $\log N$ is approximately 57.3, then N is a 58-digit number.

EXAMPLE 4 **Estimating a Large Number**

Write an estimate of the number $K = e^{700}$ in scientific notation.

Solution

$$K = e^{700}$$
$$\log K = \log(e^{700})$$
$$= 700 \log e$$
$$\approx 304.0061373 \qquad \text{Use a calculator.}$$

Since $\log K$ lies between the integers 304 and 305, the number K requires 305 digits to the left of the decimal point. Also, by definition of the common logarithm, we have

$$K \approx 10^{304.0061373}$$
$$= 10^{0.0061373} \times 10^{304}$$
$$\approx 1.014232 \times 10^{304} \qquad \text{Use a calculator.}$$

Practice Problem 4 Repeat Example 4 for $K = (234)^{567}$.

---

3 Change the base of a logarithm.

Change of Base

Calculators usually come with two types of log keys: a key for natural logarithms (base e, $\boxed{\text{LN}}$) and a key for common logarithms (base 10, $\boxed{\text{LOG}}$). These two types of logarithms are frequently used in applications. Sometimes, however, we need to evaluate logarithms for bases other than e and 10. The *change-of-base formula* helps us evaluate these logarithms.

Suppose we are given $\log_b x$ and we want to find an equivalent expression in terms of logarithms with base a. We let

$$u = \log_b x. \qquad (6)$$

In exponential form, we have

$$x = b^u. \qquad (7)$$

Then

$$\log_a x = \log_a b^u \qquad \text{Take } \log_a \text{ of each side of (7).}$$
$$\log_a x = u \log_a b \qquad \text{Power rule}$$
$$u = \frac{\log_a x}{\log_a b} \qquad \text{Solve for } u.$$
$$\log_b x = \frac{\log_a x}{\log_a b} \qquad (8) \qquad \text{Substitute for } u \text{ from (6).}$$

Equation (8) is the change-of-base formula. By choosing $a = 10$ and $a = e$, we can state the change-of-base formula as follows:

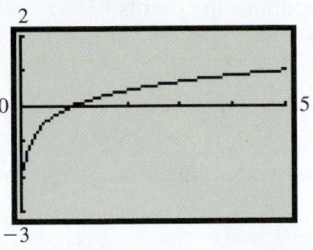

CHANGE-OF-BASE FORMULA

Let a, b, and x be positive real numbers with $a \neq 1$ and $b \neq 1$. Then $\log_b x$ can be converted to a different base as follows:

$$\log_b x = \underbrace{\frac{\log_a x}{\log_a b}}_{(\text{base } a)} = \underbrace{\frac{\log x}{\log b}}_{(\text{base } 10)} = \underbrace{\frac{\ln x}{\ln b}}_{(\text{base } e)}$$

EXAMPLE 5   Using a Change of Base to Compute Logarithms

Compute $\log_5 13$ by changing to **a.** common logarithms and **b.** natural logarithms.

Solution

a. $\log_5 13 = \dfrac{\log 13}{\log 5}$ Change to base 10.

≈ 1.59369 Use a calculator.

b. $\log_5 13 = \dfrac{\ln 13}{\ln 5}$ Change to base e.

≈ 1.59369 Use a calculator.

Practice Problem 5 Find $\log_3 15$.

EXAMPLE 6   Modeling with Logarithmic Functions

Find

a. an equation of the form $y = c + b \log x$ whose graph contains the points $(2, 3)$ and $(4, -5)$.

b. an equation of the form $y = c + b \log_a x$ whose graph contains the points $(2, 3)$ and $(4, -5)$, where a, b, and c are integers.

Solution

Substitute $(2, 3)$ and $(4, -5)$ in the equation $y = c + b \log x$ to obtain:

$$3 = c + b \log 2 \qquad (9)$$
$$-5 = c + b \log 4 \qquad (10)$$

a. Find b: Subtract equation (10) from equation (9) to obtain

$$8 = b \log 2 - b \log 4 = b(\log 2 - \log 4) = b\left(\log \tfrac{2}{4}\right)$$

$$8 = b \log\left(\frac{1}{2}\right) = b \log 2^{-1} = -b \log 2$$

$$b = -\frac{8}{\log 2} \qquad \text{Solve for } b.$$

Find c:

$$c = 3 - b \log 2 \qquad \text{From equation (9)}$$

$$= 3 - \left(-\frac{8}{\log 2}\right)\log 2 \qquad \text{Replace } b \text{ with } -\frac{8}{\log 2}$$

$$= 3 + 8 \qquad \text{Simplify.}$$

$$= 11$$

b. Use the change-of-base formula to find a, b, and c

Substitute b and c from part **a** in the equation $y = c + b \log x$ to obtain:

$$y = 11 - \frac{8}{\log 2}(\log x) = 11 - 8\left(\frac{\log x}{\log 2}\right) \qquad b = -\frac{8}{\log 2}, c = 11$$

$$y = 11 - 8 \log_2 x \qquad \log_2 x = \frac{\log x}{\log 2}$$

Then $a = 2$, $b = -8$, and $c = 11$.

Practice Problem 6 Repeat Example 6 for the graph that contains the points $(3, 3)$ and $(9, 1)$.

4 Apply logarithms to growth and decay.

Growth and Decay

Recall (page 266) that exponential growth (or decay) occurs when a quantity grows (or decreases) at a rate proportional to its size.

$$\text{Growth Formula: } A(t) = A_0 e^{kt} \text{ if } k > 0$$
$$\text{Decay Formula: } A(t) = A_0 e^{kt} \text{ if } k < 0 \qquad (11)$$

where $A(t) = $ the amount of substance (or population) at time t,

$A_0 = A(0)$ is the initial amount, and

$t = $ time.

Half-Life

The **half-life** of any quantity whose value decreases with time is the time required for the quantity to decay to half its initial value.

Radioactive substances undergo exponential decay. Krypton-85, for example, has a half-life of about ten years; this means that any initial quantity of krypton-85 will dissipate or decay to half that amount in about ten years.

If h is the half-life of a substance undergoing exponential decay at the rate k, then any mass A_0 decays to $\frac{1}{2}A_0$ in time h. See Figure 3.12. This fact leads to a half-life formula.

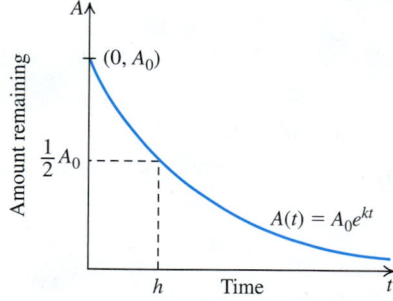

Figure 3.12 Time h is the half-life for the substance with this decay curve.

$$\frac{1}{2}A_0 = A_0 e^{kh} \qquad A(t) = A_0 e^{kt} \text{ with } t = h, \text{ and } A(h) = \frac{1}{2}A_0$$

$$\frac{1}{2} = e^{kh} \qquad \text{Divide both sides by } A_0.$$

$$\ln\left(\frac{1}{2}\right) = \ln\left(e^{kh}\right) \qquad \text{Take the natural log of both sides.}$$

$$-\ln 2 = kh \qquad \ln\left(\frac{1}{2}\right) = \ln\left(2^{-1}\right) = -\ln 2; \ln\left(e^x\right) = x$$

$$h = -\frac{\ln 2}{k} \qquad \text{Solve for } h.$$

HALF-LIFE FORMULA

The half-life h of a substance undergoing exponential decay at a rate k $(k < 0)$ is given by the formula

$$h = -\frac{\ln 2}{k}.$$

EXAMPLE 7 **Finding the Half-Life of a Substance**

In an experiment, 18 grams of the radioactive element sodium-24 decayed to 6 grams in 24 hours. Find its half-life to the nearest hour.

Solution

$$A(t) = A_0 e^{kt} \qquad \text{Exponential growth and decay model}$$

$$6 = 18 e^{k(24)} \qquad A_0 = 18, A(24) = 6$$

$$\frac{1}{3} = e^{k(24)} \qquad \text{Divide both sides by 18.}$$

$$\ln\left(\frac{1}{3}\right) = \ln\left(e^{k(24)}\right) \qquad \text{Take the natural log of both sides.}$$

$$-\ln 3 = 24k \qquad \ln\left(\frac{1}{3}\right) = \ln\left(3^{-1}\right) = -\ln 3 : \ln\left(e^x\right) = x$$

$$k = -\frac{\ln 3}{24} \qquad \text{Solve for } k.$$

So $\qquad A(t) = 18 e^{-\frac{\ln 3}{24} t}$.

To find the half-life of sodium-24, we use the formula

$$h = -\frac{\ln 2}{k} \qquad \text{Half-life formula}$$

$$h = -\frac{\ln 2}{\left(-\dfrac{\ln 3}{24}\right)} \qquad \text{Replace } k \text{ with } -\frac{\ln 3}{24}.$$

$$h = \frac{24 \ln 2}{\ln 3} \approx 15 \text{ hours} \qquad \text{Simplify; use a calculator.}$$

Willard Frank Libby

(1908–1980)
Libby was born in Grand Valley, Colorado. In 1927, he entered the University of California at Berkeley, where he earned his BSc and PhD degrees in 1931 and 1933, respectively. He then became an instructor in the Department of Chemistry at the University of California (Berkeley) in 1933. He was awarded a Guggenheim Memorial Foundation Fellowship in 1941, but the fellowship was interrupted when Libby went to Columbia University on the Manhattan District Project upon America's entry into World War II.

Libby was a physical chemist and a specialist in radiochemistry, particularly in hot-atom chemistry, tracer techniques, and isotope tracer work. He became well known at the University of Chicago for his work on natural carbon-14 (a radiocarbon) and its use in dating archaeological artifacts and on natural tritium and its use in hydrology and geophysics. He was awarded the Nobel Prize in Chemistry in 1960.

Practice Problem 7 In an experiment, 100 grams of the radioactive element strontium-90 decayed to 66 grams in 15 years. Find its half-life to the nearest year.

Radiocarbon Dating

Ordinary carbon, called carbon-12 (^{12}C), is stable and does not decay. However, carbon-14 (^{14}C) is a form of carbon that decays radioactively with a half-life of 5700 years. Sunlight constantly produces carbon-14 in Earth's atmosphere. When a living organism breathes or eats, it absorbs carbon-12 and carbon-14. After the organism dies, no more carbon-14 is absorbed, so the age of its remains can be calculated by determining how much carbon-14 has decayed. The method of radiocarbon dating was developed by the American scientist W. F. Libby.

EXAMPLE 8 **King Tut's Treasure**

In 1960, a group of specialists from the British Museum in London investigated whether a piece of art containing organic material found in Tutankhamun's tomb had been made during his reign or (as some historians claimed) whether it belonged to an earlier period. We know that King Tut died in 1346 B.C. and ruled Egypt for ten years. What percent of the amount of carbon-14 originally contained in the object should be present in 1960 if the object was made during Tutankhamun's reign?

Solution

Because the half-life of carbon-14 is approximately 5700 years, we can rewrite the equation $A(t) = A_0e^{kt}$ as

$$A(5700) = A_0e^{5700k}.$$

Now we can find the value of k.

$$\frac{1}{2}A_0 = A_0e^{5700k} \qquad \text{Replace } A(5700) \text{ with } \frac{1}{2}A_0.$$

$$\frac{1}{2} = e^{5700k} \qquad \text{Divide both sides by } A_0.$$

$$\ln\left(\frac{1}{2}\right) = 5700k \qquad \text{Logarithmic form}$$

$$k = \frac{\ln\left(\frac{1}{2}\right)}{5700} \qquad \text{Solve for } k.$$

$$\approx -0.0001216 \qquad \text{Use a calculator.}$$

Substituting this value of k into equation (11), we obtain

$$A(t) = A_0e^{-0.0001216t} \qquad (12)$$

The time t that elapsed between King Tut's death and 1960 is

$$t = 1960 + 1346 \qquad \text{1346 B.C. to A.D. 1960}$$

$$= 3306.$$

$$A(3306) = A_0e^{-0.0001216(3306)} \qquad \text{Replace } t \text{ with 3306 in (12).}$$

$$\approx 0.66897A_0.$$

Therefore, the percent of the original amount of carbon-14 remaining in the object (after 3306 years) is 66.897%.

Because King Tut began ruling Egypt ten years before he died, the time t_1 that elapsed from the beginning of his reign to 1960 is 3316 years.

$$A(3316) = A_0e^{-0.0001216(3316)} \qquad \text{Replace } t \text{ with 3316 in (12).}$$

$$\approx 0.66816A_0.$$

Therefore, a piece of art made during King Tut's reign would have between 66.816% and 66.897% of carbon-14 remaining in 1960.

Practice Problem 8 Repeat Example 8 assuming that the art object was made 194 years before King Tut's death.

Answers to Practice Problems

1. a. -1 **b.** 13

2. a. $\ln(2x-1) - \ln(x+4)$

b. $\log 2 + \frac{1}{2}\log x + \frac{1}{2}\log y - \frac{1}{2}\log z$

3. $\log\sqrt{x^2-1}$

4. $2.215088 \times 10^{1343}$

5. $\dfrac{\log 15}{\log 3} = \dfrac{\ln 15}{\ln 3} \approx 2.46497$

6. a. $y = 5 - \dfrac{2}{\log 3}\log x$ **b.** $y = 5 - 2\log_3 x$

7. 25 years   **8.** $\approx 65.34\%$

SECTION 3.3   Exercises

Concepts and Vocabulary

1. $\log_a MN =$ _____ $+$ _____ .
2. _____ $= \log_a M - \log_a N$.
3. $\log_a M^r =$ _____ .
4. The change-of-base formula using base e is $\log_a M =$ ____ .
5. **True or False.** $\log_a (u + v) = \log_a u + \log_a v$.
6. **True or False.** $\log u - \log v = \dfrac{\log u}{\log v}$
7. **True or False.** $\ln(2\sqrt{x}) = \dfrac{1}{2}\ln(2x)$.
8. **True or False.** $\log \dfrac{x}{10} = \log x - 1$.

Building Skills

In Exercises 9–20, given that $\log x = 2$, $\log y = 3$, $\log 2 \approx 0.3$, and $\log 3 \approx 0.48$, evaluate each expression without using a calculator.

9. $\log 6$
10. $\log 4$
11. $\log 5 \left[Hint: 5 = \dfrac{10}{2} \right]$
12. $\log (3x)$
13. $\log \left(\dfrac{2}{x} \right)$
14. $\log x^2$
15. $\log (2x^2 y)$
16. $\log xy^3$
17. $\log \sqrt[3]{x^2 y^4}$
18. $\log (\log x^2)$
19. $\log \sqrt[3]{48}$
20. $\log_2 3$

In Exercises 21–38, write each expression in expanded form. Assume that all expressions containing variables represent positive numbers.

21. $\ln [x(x - 1)]$
22. $\ln \dfrac{x - 2}{x + 3}$
23. $\ln \dfrac{x + 1}{(x - 2)^2}$
24. $\ln \dfrac{x(x + 1)}{(x - 1)^2}$
25. $\log \dfrac{\sqrt{x^2 + 1}}{x + 3}$
26. $\log \dfrac{x^2}{\sqrt{x + 1}}$
27. $\log_3 \dfrac{(x - 1)(x + 1)}{x^2 - 4}$
28. $\log_4 \left(\dfrac{x^2 - 9}{x^2 - 6x + 8} \right)^{\frac{2}{3}}$
29. $\log_b x^2 y^3 z$
30. $\log_b \sqrt{xyz}$

31. $\ln \left[\dfrac{x\sqrt{x - 1}}{x^2 + 2} \right]$
32. $\ln \left[\dfrac{\sqrt{x - 2}\,\sqrt[3]{x + 1}}{x^2 + 3} \right]$
33. $\ln \left[\dfrac{(x + 1)^2}{(x - 3)\sqrt{x + 4}} \right]$
34. $\ln \left[\dfrac{2x + 3}{(x + 4)^2 (x - 3)^4} \right]$
35. $\ln \left[(x + 1)\sqrt{\dfrac{x^2 + 2}{x^2 + 5}} \right]$
36. $\ln \left[\dfrac{\sqrt[3]{2x + 1}\,(x + 1)}{(x - 1)^2 (3x + 2)} \right]$
37. $\ln \left[\dfrac{x^3 (3x + 1)^4}{\sqrt{x^2 + 1}(x + 2)^{-5}(x - 3)^2} \right]$
38. $\ln \left[\dfrac{(x + 1)^{\frac{1}{2}}(x^2 - 2)^{\frac{2}{5}}}{(2x - 1)^{\frac{3}{2}}(x^2 + 2)^{-\frac{4}{5}}} \right]$

In Exercises 39–54, write each expression in condensed form.

39. $\log_2 x + \log_2 7$
40. $\log_2 x - \log_2 3$
41. $\log (3x + 2) - \log x$
42. $2 \log x + \log y$
43. $\dfrac{1}{2}\ln x + 2 \ln y$
44. $3 \log x + 4 \log y$
45. $\dfrac{1}{2}\log x - \log y + \log z$
46. $\dfrac{1}{2}(\log x + \log y)$
47. $\dfrac{1}{5}(\log_2 z + 2 \log_2 y)$
48. $\dfrac{1}{3}(\log x - 2 \log y + 3 \log z)$
49. $\ln x + 2 \ln y + 3 \ln z$
50. $2 \ln x - 3 \ln y + 4 \ln z$
51. $3 \log x - \log y + \dfrac{1}{2}\log (x^2 + 4)$
52. $\log x + \log (x^2 + 1) - \log (x - 1) - \log (2x^2 + 3)$
53. $2 \ln x - \dfrac{1}{2}\ln (x^2 + 1)$
54. $2 \ln x + \dfrac{1}{2}\ln (x^2 - 1) - \dfrac{1}{2}\ln (x^2 + 1)$

In Exercises 55–60, write an estimate of each number in scientific notation.

55. e^{500}

56. π^{650}

57. 324^{756}

58. 723^{416}

59. 456^{789}

60. 1234^{5678}

61. Which is larger, 234^{567} or 567^{234}?

62. Which is larger, 4321^{8765} or 8765^{4321}?

63. Find the number of digits in $17^{200} \cdot 53^{67}$.

64. Find the number of digits in $67^{200} \div 23^{150}$.

In Exercises 65–72, use the change-of-base formula and a calculator to evaluate each logarithm.

65. $\log_2 5$

66. $\log_4 11$

67. $\log_{1/2} 3$

68. $\log_{\sqrt{3}} 12.5$

69. $\log_{\sqrt{5}} \sqrt{17}$

70. $\log_{15} 123$

71. $\log_2 7 + \log_4 3$

72. $\log_2 9 - \log_{\sqrt{2}} 5$

In Exercises 73–80, find the value of each expression without using a calculator.

73. $\log_3 \sqrt{3}$

74. $\log_{1/4} 4$

75. $\log_3(\log_2 8)$

76. $2^{\log_2 2}$

77. $5^{2 \log_5 3 + \log_5 2}$

78. $e^{3 \ln 2 - 2 \ln 3}$

79. $\log 4 + 2 \log 5$

80. $\log_2 160 - \log_2 5$

In Exercises 81–88, find an equation of the form $y = c + b \log_a x$ whose graph contains the two given points.

81. $(10, 1)$ and $(1, 2)$

82. $(4, 10)$ and $(2, 12)$

83. $(e, 1)$ and $(1, 2)$

84. $(3, 1)$ and $(9, 2)$

85. $(5, 4)$ and $(25, 7)$

86. $(4, 3)$ and $(8, 5)$

87. $(2, 4)$ and $(4, 9)$

88. $(1, 1)$ and $(5, 7)$

In Exercises 89–92, determine the half-life of each substance that decays from A_0 to A in time t.

89. $A_0 = 50$, $A = 23$, $t = 12$ years.

90. $A_0 = 200$, $A = 65$, $t = 10$ years.

91. $A_0 = 10.3$, $A = 3.8$, $t = 15$ hours.

92. $A_0 = 20.8$, $A = 12.3$, $t = 40$ minutes.

Applying the Concepts

In Exercises 93–96, assume the exponential growth model $A(t) = A_0 e^{kt}$ and a world population of 7 billion in 2011.

93. **Rate of population growth.** If the world adds about 90 million people in 2012, what is the rate of growth?

94. **Population.** Use the rate of growth from Exercise 93 to estimate the year in which the world population will be
a. 12 billion.
b. 20 billion.

95. **Population growth.** If the population must stay below 20 billion during the next 100 years, what is the maximum acceptable annual rate of growth? (Round your answer to six decimal places.)

96. **Population decline.** If the world population must decline below 5 billion during the next 25 years, what must be the annual rate of decline? (Round your answer to six decimal places.)

97. **Continuous compounding.** If $1000 is deposited in a bank at 10% interest compounded continuously, how many years will it take for the money to grow to $3500?

98. **Continuous compounding.** At what rate of interest compounded continuously will an investment double in six years?

99. **Half-life.** Iodine-131, a radioactive substance that is effective in locating brain tumors, has a half-life of only eight days. A hospital purchased 20 grams of the substance but had to wait five days before it could be used. How much of the substance was left after five days?

100. **Half-life.** Plutonium-241 has a half-life of 13 years. A laboratory purchased 10 grams of the substance but did not use it for two years. How much of the substance was left after two years?

101. **Half-life.** Tritium is used in nuclear weapons to increase their power. It decays at the rate of 5.5% per year. Calculate the half-life of tritium.

102. **Half-life.** Sixty grams of magnesium-28 decayed into 7.5 grams after 63 hours. What is the half-life of magnesium-28?

Effective medicine dosage. Use the following model in Exercises 103–106: The concentration $C(t)$ of a drug administered to a patient intravenously jumps to its highest level almost immediately. The concentration subsequently decays exponentially according to the law

$$C(t) = C_0 e^{-kt},$$

where $C(0) = C_0$. The physician administering the drug would like to have

$$m < C(t) < M,$$

where m is the concentration below which the drug is ineffective and M is the concentration above which the drug is dangerous.

103. **Drug concentration.** Suppose that immediately after a certain drug is administered, the concentration is 5 milligrams per milliliter. Ten hours later, the concentration drops to 1.5 milligrams per milliliter. Determine the value of k for this drug.

104. **Drug concentration.** For the drug of Exercise 103, suppose the maximum safe level is $M = 12$ milligrams per milliliter and the minimum effective level is $m = 3$ milligrams per milliliter. If the initial concentration is M, find the maximum possible time between doses of this drug.

105. **Half-life.** The half-life of Valium is 36 hours. Suppose a patient receives 16 milligrams of the drug at 8 A.M. How much Valium is in the patient's blood at 4 P.M. the same day?

106. **Half-life.** The half-life of aspirin in the blood is 12 hours. Estimate the time required for the aspirin to decay to 90% of the original amount ingested.

In Exercises 107–110, use the following amortization formula:

$$P = \frac{r \cdot M}{1 - \left(1 + \dfrac{r}{n}\right)^{-nt}} \div n$$

where P = the payment, r = the annual interest rate, M = the mortgage amount, t = the number of years, and n = the number of payments per year.

107. **Finding P.** What is the monthly payment on a mortgage of $120,000 with a 6% interest rate for 20 years? How much interest will be paid over 20 years?

108. **Finding t.** The Garcia family wanted to take out a mortgage for $150,000 at 8% interest with monthly payments. The family can afford monthly payments of $1200. How long would they have to make payments to pay off their mortgage, and how much interest would they be paying?

109. **Finding M.** The First National Bank offers Andy an 8.5% interest rate on a 30-year mortgage to be paid back in monthly payments. The most Andy can afford to pay in monthly payments is $850.00. What mortgage amount can Andy afford?

110. **Finding r.** Lisa is 40 years old and needs to take out a mortgage of $120,000. She can afford monthly payments up to $850. She wants her mortgage paid off when she retires at age 65. What interest rate on the mortgage will satisfy all her requirements? [*Hint:* Start with $r = 7\%$ and compute the monthly payment, increasing r by 0.001 each time.]

Beyond the Basics

111. Show that
$$\log_b(\sqrt{x^2 + 1} - x) + \log_b(\sqrt{x^2 + 1} + x) = 0.$$

112. Show that
$$\log_b(\sqrt{x + 1} + \sqrt{x}) = -\log_b(\sqrt{x + 1} - \sqrt{x}).$$

113. Show that $(\log_b a)(\log_a b) = 1$.

114. Show that $\log \dfrac{a}{b} + \log \dfrac{b}{a} = 0$.

115. Write $\log_2 x = \dfrac{\ln x}{\ln 2}$ by the change-of-base formula. Then use a graphing calculator to sketch the graph of $y = \log_2 x$.

116. Sketch the graph of $y = \log_5(x + 3)$.

117. Simplify each expression.

a. $\log\left(\dfrac{a}{b}\right) + \log\left(\dfrac{b}{a}\right) + \log\left(\dfrac{c}{a}\right) + \log\left(\dfrac{a}{c}\right)$

b. $\log\left(\dfrac{a^2}{bc}\right) + \log\left(\dfrac{b^2}{ca}\right) + \log\left(\dfrac{c^2}{ab}\right)$

c. $\log_2 3 \cdot \log_3 4$

d. $\log_a b \cdot \log_b c \cdot \log_c a$

118. Find the number of digits in N if $\log_2(\log_2 N) = 4$.

119. Find the domain of
$$f(x) = \log_4(\log_5(\log_3(18x - x^2 - 77))).$$

120. Let $f(x) = \log_a x$. Show that

a. $f\left(\dfrac{1}{x}\right) = -f(x) = \log_{1/a} x$.

b. $\dfrac{f(x + h) - f(x)}{h} = \log_a\left(1 + \dfrac{h}{x}\right)^{1/h}, h \neq 0.$

121. State whether each of the following is true or false for all permissible values of the variable.

a. $\log(x + 2) = \log x + \log 2$
b. $\log 2x = \log x + \log 2$
c. $\log_2 3x = 3 \log_2 x$
d. $(\log_3 2x)^4 = 4 \log_3 2x$
e. $\log_5 x^2 = 2 \log_5 |x|$
f. $\log \dfrac{x}{3} = \log x - \log 3$
g. $\log \dfrac{x}{4} = \dfrac{\log x}{\log 4}$
h. $\ln\left(\dfrac{1}{x}\right) = -\ln x$
i. $\log_2 x^2 = (\log_2 x)(\log_2 x)$
j. $\log|10x| = 1 + \log|x|$

122. a. Prove the quotient rule for logarithms:
$$\log_a \dfrac{M}{N} = \log_a M - \log_a N.$$

b. Prove the power rule for logarithms: $\log_a M^r = r \log_a M$.

123. If $\log\left(\dfrac{a + b}{3}\right) = \dfrac{1}{2}(\log a + \log b)$, show that
$$a^2 + b^2 - 7ab = 0.$$

124. Evaluate $\dfrac{1}{1 + \log_c ab} + \dfrac{1}{1 + \log_a bc} + \dfrac{1}{1 + \log_b ca}$.

[*Hint:* Convert to common logarithms and then simplify.]

Critical Thinking / Discussion / Writing

125. **What went wrong?** Find the error in the following argument:

$$3 < 4$$
$$3 \log \dfrac{1}{2} < 4 \log \dfrac{1}{2}$$
$$\log\left(\dfrac{1}{2}\right)^3 < \log\left(\dfrac{1}{2}\right)^4$$
$$\log \dfrac{1}{8} < \log \dfrac{1}{16}$$
$$\dfrac{1}{8} < \dfrac{1}{16}$$
$$2 < 1$$

126. By the power rule for logarithms, $\log x^2 = 2 \log x$. However, the graphs of $f(x) = \log x^2$ and $g(x) = 2 \log x$ are not identical. Explain why.

127. **Prime Number.** As of 2008, the largest known prime number was $2^{43112609} - 1$. How many digits does this prime number have?

[*Hint:* First, argue that if $p = 2^m - 1$ is a prime number, then p and 2^m have the same number of digits.]

128. Find the domain of $g(x) = \dfrac{1}{\log(1 - x)} + \dfrac{1}{\ln(x + 2)}$.

Getting Ready for the Next Section

In Exercises 129–132, simplify each expression and write your result without exponents.

129. $11 \cdot 3^0$

130. $-4 \cdot 5^x \cdot 5^{-x}$

131. $4^x \cdot 2^{-2x+1}$

132. $(7^x)^2 \cdot (7^2)^{-x}$

In Exercises 133–136, let $t = 5^x$ and write each expression in the form $at^2 + bt + c = 0$, where a, b, and c are real numbers.

133. $5^{2x} - 5^x = -1$

134. $3 \cdot 5^{2x} - 2 \cdot 5^x = 7$

135. $\dfrac{5^x + 3 \cdot 5^{-x}}{5^x} = \dfrac{1}{4}$

136. $\dfrac{5^{-x} + 5^x}{5^{-x} - 5^x} = \dfrac{1}{2}$

In Exercises 137–140, solve for x.

137. $2x - (11 + x) = 8x + (7 + 2x)$

138. $4x - 1 + (6x - 2) = 3 - (5x + 1)$

139. $x^2 + 3x - 1 = 3$

140. $2x^2 - 7x = 3x + 48$

In Exercises 141–144, solve for x.

141. $2x < 7 + x$

142. $5x - 2 \le 19 - (1 - x)$

143. $12x > 30 - 3x$

144. $4x - 17 \ge 6 - (5 + 2x)$

Exponential and Logarithmic Equations and Inequalities

BEFORE STARTING THIS SECTION, REVIEW

1 One-to-one property of exponential functions (Section 3.1, page 257)

2 One-to-one property of logarithmic functions (Section 3.2, page 276)

3 Rules of logarithms (Section 3.3, page 288)

OBJECTIVES

1 Solve exponential equations.

2 Solve applied problems involving exponential equations.

3 Solve logarithmic equations.

4 Use the logistic growth model.

5 Use logarithmic and exponential inequalities.

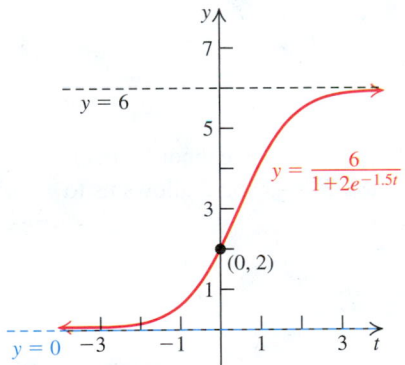

Figure 3.13 A logistic curve.

◆ Logistic Growth Model

In Sections 3.1 and 3.3, we discussed the population growth formula

$$A(t) = A_0 e^{kt}. \quad (1)$$

Equation (1) is a realistic model for a biological population that has plenty of food, enough space to grow, and no threat from predators.

However, most populations are constrained by limitations on resources such as the plants that populations eat. In the 1830s, the Belgian scientist P. F. Verhulst added another constraint to the population growth model: *There is a limited sustainable maximum population M of a species that the habitat can support.* In population biology, *M* is called the *carrying capacity*.

Verhulst replaced equation (1) with

$$P(t) = \frac{M}{1 + ae^{-kt}}, \quad (2)$$

where *k* is the rate of growth and *a* is a positive constant. In 1840, using the data from the first five U.S. censuses, he made a prediction of the U.S. population for the year 1940. As things turned out, it was off by less than 1%.

The mathematical model represented by equation (2) is called the *logistic growth model* or the *Verhulst model*. Verhulst introduced the term *logistique* to refer to the "loglike" qualities of the *S*-shaped graph of equation (2). See Figure 3.13.

To sketch the graph of equation (2), we notice that the *y*-intercept is

$$P(0) = \frac{M}{1 + ae^0} = \frac{M}{1 + a}.$$

As $t \to \infty$, the quantity $ae^{-kt} \to 0$ (because *k* is positive). Consequently, the denominator $1 + ae^{-kt} \to 1$; therefore, $P(t) \to M$ and the line $y = M$ is a horizontal asymptote.

Similarly, as $t \to -\infty$, $e^{-kt} \to \infty$. In this case, the denominator $1 + ae^{-kt} \to \infty$ and $P(t) \to 0$. The *x*-axis, or $y = 0$, is another horizontal asymptote. See the graph of

$$y = \frac{6}{1 + 2e^{-1.5t}}$$ in Figure 3.13.

In Example 9 we use the logistic function defined by equation (2).

1 Solve exponential equations.

Solving Exponential Equations

An equation containing terms of the form a^x ($a > 0, a \neq 1$) is called an **exponential equation**. Here are some examples of simple exponential equations:

$$2^x = 15$$
$$9^x = 3^{x+1}$$
$$7^x = 13^{2x}$$
$$5 \cdot 2^{x-3} = 17$$

If both sides of an equation can be expressed as a power of the same base, then we can use the one-to-one property of exponential functions to solve the equation.

RECALL

The one-to-one property of exponential and logarithmic functions states that
- if $a^u = a^v$, then $u = v$.
- if $\log_a u = \log_a v$, then $u = v$.

EXAMPLE 1   Solving an Exponential Equation

Solve each equation.

a. $25^x = 125$ **b.** $9^x = 3^{x+1}$

Solution

a.
$$(5^2)^x = 5^3 \qquad \text{Rewrite 25 as } 5^2 \text{ and 125 as } 5^3.$$
$$5^{2x} = 5^3 \qquad \text{Power rule for exponents}$$
$$2x = 3 \qquad \text{One-to-one property}$$
$$x = \frac{3}{2} \qquad \text{Solve for } x.$$

b.
$$(3^2)^x = 3^{x+1} \qquad \text{Rewrite 9 as } 3^2.$$
$$3^{2x} = 3^{x+1} \qquad \text{Power rule of exponents}$$
$$2x = x + 1 \qquad \text{One-to-one property}$$
$$x = 1 \qquad \text{Solve for } x.$$

Practice Problem 1 Solve each equation.

a. $3^x = 243$ **b.** $8^x = 4$

We can use logarithms to solve those equations for which both sides cannot be readily expressed with the same base. For example, the fact that $\log 2^x = x \log 2$ allows us to solve the equation $2^x = 9$ by first taking the log of both sides.

$$2^x = 9$$
$$\log 2^x = \log 9$$
$$x \log 2 = \log 9$$
$$x = \frac{\log 9}{\log 2}$$

An approximation $x \approx 3.1699$ is found by using a calculator.

The general procedure for solving exponential equations using logarithms is given next.

PROCEDURE
IN ACTION

EXAMPLE 2 **Solving Exponential Equations Using Logarithms**

OBJECTIVE	EXAMPLE
Solve exponential equations when both sides are not expressed with the same base.	*Solve for x:* $5 \cdot 2^{x-3} = 17$
Step 1 Isolate the exponential expression on one side of the equation.	**1.** $2^{x-3} = \dfrac{17}{5} = 3.4$ Solve for 2^{x-3}.
Step 2 Take the common or natural logarithm of both sides.	**2.** $\ln 2^{x-3} = \ln(3.4)$ Take ln of both sides.
Step 3 Use the power rule, $\log_a M^r = r \log_a M$.	**3.** $(x-3)\ln 2 = \ln(3.4)$ Power rule of logarithms
Step 4 Solve for the variable.	**4.** $x - 3 = \dfrac{\ln(3.4)}{\ln 2}$ Divide both sides by ln 2.
	$x = \dfrac{\ln(3.4)}{\ln 2} + 3 \approx 4.766$ Solve for x; use a calculator.

Practice Problem 2 Solve for x: $7 \cdot 3^{x+1} = 11$.

EXAMPLE 3 **Solving an Exponential Equation with Different Bases**

Solve the equation $5^{2x-3} = 3^{x+1}$ and approximate the answer to three decimal places.

Solution

When different bases are involved, we begin with Step 2 of the procedure above.

$$\ln 5^{2x-3} = \ln 3^{x+1} \qquad \text{Take the natural log of both sides.}$$
$$(2x-3)\ln 5 = (x+1)\ln 3 \qquad \text{Power rule of logarithms}$$
$$2x \ln 5 - 3 \ln 5 = x \ln 3 + \ln 3 \qquad \text{Distributive property}$$
$$2x \ln 5 - x \ln 3 = \ln 3 + 3 \ln 5 \qquad \text{Collect terms containing } x \text{ on one side.}$$
$$x(2 \ln 5 - \ln 3) = \ln 3 + 3 \ln 5 \qquad \text{Factor out } x \text{ on the left side.}$$
$$x = \frac{\ln 3 + 3 \ln 5}{2 \ln 5 - \ln 3} \qquad \text{Solve for } x \text{ to obtain an } exact \text{ solution.}$$
$$\approx 2.795 \qquad \text{Use a calculator.}$$

SIDE
NOTE

When using a calculator to evaluate such expressions, make sure you enclose the numerator and denominator in parentheses.

Practice Problem 3 Solve the equation $3^{x+1} = 2^{2x}$.

EXAMPLE 4 **Solving an Exponential Equation of Quadratic Form**

Solve the equation $3^x - 8 \cdot 3^{-x} = 2$.

Solution

$$3^x(3^x - 8 \cdot 3^{-x}) = 2(3^x) \qquad \text{Multiply both sides by } 3^x.$$
$$3^{2x} - 8 \cdot 3^0 = 2 \cdot 3^x \qquad \text{Distributive property; } 3^x \cdot 3^{-x} = 3^0$$
$$3^{2x} - 8 = 2 \cdot 3^x \qquad 3^0 = 1$$
$$3^{2x} - 2 \cdot 3^x - 8 = 0 \qquad \text{Subtract } 2 \cdot 3^x \text{ from both sides.}$$

The last equation is quadratic in form. To see this, we let $y = 3^x$; then $y^2 = (3^x)^2 = 3^{2x}$. So

$$3^{2x} - 2 \cdot 3^x - 8 = 0$$

$$y^2 - 2y - 8 = 0 \qquad \text{Substitute: } y = 3^x, y^2 = 3^{2x}.$$

$$(y + 2)(y - 4) = 0 \qquad \text{Factor.}$$

$$y + 2 = 0 \quad \text{or} \quad y - 4 = 0 \qquad \text{Zero-product property}$$

$$y = -2 \quad \text{or} \qquad y = 4 \qquad \text{Solve for } y.$$

$$3^x = -2 \quad \text{or} \qquad 3^x = 4 \qquad \text{Replace } y \text{ with } 3^x.$$

But $3^x = -2$ is not possible because $3^x > 0$ for all real numbers x. Solve $3^x = 4$.

$$\ln 3^x = \ln 4 \qquad \text{Take the natural log of both sides.}$$

$$x \ln 3 = \ln 4 \qquad \text{Power rule of logarithms}$$

$$x = \frac{\ln 4}{\ln 3} \qquad \text{An exact solution}$$

$$\approx 1.262 \qquad \text{Use a calculator.}$$

Practice Problem 4 Solve the equation $e^{2x} - 4e^x - 5 = 0$.

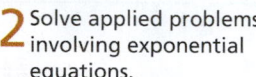

Applications of Exponential Equations

2 Solve applied problems involving exponential equations.

EXAMPLE 5 Solving a Population Growth Problem

The following table shows the approximate population and annual growth rate of the United States and Pakistan in 2010.

Country	Population	Annual Population Growth Rate
United States	308 million	0.9%
Pakistan	185 million	2.1%

Source: www.cia.gov

Use the alternate population model $P(t) = P_0(1 + r)^t$, where P_0 is the initial population and t is the time in years since 2010. Assume that the growth rate for each country stays the same.

a. Estimate the population of each country in 2020.

b. In what year will the population of the United States be 350 million?

c. In what year will the population of Pakistan be the same as the population of the United States?

Solution

The given model is

$$P(t) = P_0(1 + r)^t.$$

a. In 2010, the initial year, the U.S. population was $P_0 = 308$. In ten years, 2020, the population of the United States will be

$$308(1 + 0.009)^{10} \approx 336.87 \text{ million} \qquad P_0 = 308, r = 0.009, t = 10$$

In the same way, the population of Pakistan in 2020 will be

$$185(1 + 0.021)^{10} \approx 227.73 \text{ million} \qquad P_0 = 185, r = 0.021, t = 10$$

b. To find the year when the population of the United States will be 350 million, we solve for t in the equation

$$350 = 308(1 + 0.009)^t \qquad \text{\color{blue}{$P(t) = 350, P_0 = 308, r = 0.009$}}$$

$$\frac{350}{308} = (1.009)^t \qquad \text{\color{blue}{Divide both sides by 308.}}$$

$$\ln\left(\frac{350}{308}\right) = \ln(1.009)^t \qquad \text{\color{blue}{Take natural logs of both sides.}}$$

$$\ln\left(\frac{350}{308}\right) = t\ln(1.009) \qquad \text{\color{blue}{Power rule for logarithms}}$$

$$t = \frac{\ln\left(\dfrac{350}{308}\right)}{\ln(1.009)} \approx 14.27 \qquad \text{\color{blue}{Solve for t and use a calculator.}}$$

The population of the United States will be 350 million approximately 14.27 years after 2010—that is, sometime in 2025.

c. To find when the population of the two countries will be the same, we solve for t in the equation

$$308(1.009)^t = 185(1.021)^t \qquad \text{\color{blue}{Equate populations for both countries.}}$$

$$\frac{308}{185} = \left(\frac{1.021}{1.009}\right)^t \qquad \text{\color{blue}{Divide both sides by $185(1.009)^t$ and rewrite.}}$$

$$\ln\left(\frac{308}{185}\right) = \ln\left(\frac{1.021}{1.009}\right)^t \qquad \text{\color{blue}{Take natural logs of both sides.}}$$

$$\ln\left(\frac{308}{185}\right) = t\ln\left(\frac{1.021}{1.009}\right) \qquad \text{\color{blue}{Power rule for logarithms}}$$

$$t = \frac{\ln\left(\dfrac{308}{185}\right)}{\ln\left(\dfrac{1.021}{1.009}\right)} \qquad \text{\color{blue}{Solve for t.}}$$

$$\approx 43.12 \text{ years} \qquad \text{\color{blue}{Use a calculator.}}$$

Therefore, the two populations will be equal in about 43.12 years, that is, during 2054.

Practice Problem 5 Repeat Example 5 assuming that the rates of growth for the United States and Pakistan are 1.1% and 3.3%, respectively.

3 Solve logarithmic equations.

Solving Logarithmic Equations

Equations that contain terms of the form $\log_a x$ are called **logarithmic equations**. Here are some examples of logarithmic equations.

$$\log_2 x = 4$$
$$\log_3(2x - 1) = \log_3(x + 2)$$
$$\log_2(x - 3) + \log_2(x - 4) = 1$$

To solve an equation such as $\log_2 x = 4$, we rewrite it in the equivalent exponential form:

$$\log_2 x = 4 \text{ means } x = 2^4 = 16$$

Thus, $x = 16$ is a solution of the equation $\log_2 x = 4$. Because the domain of logarithmic functions consists of positive numbers, we must check apparent solutions of logarithmic equations in the original equation.

EXAMPLE 6 **Solving a Logarithmic Equation**

Solve $4 + 3 \log_2 x = 1$.

Solution

$4 + 3 \log_2 x = 1$	Original equation
$3 \log_2 x = 1 - 4$	Subtract 4 from both sides.
$3 \log_2 x = -3$	Simplify.
$\log_2 x = -1$	Divide both sides by 3.
$x = 2^{-1}$	Exponential form
$x = \dfrac{1}{2}$	$a^{-n} = \dfrac{1}{a^n}$

Check Substitute $x = \dfrac{1}{2} = 2^{-1}$ into $4 + 3 \log_2 x = 1$.

$$4 + 3 \log_2 2^{-1} \overset{?}{=} 1$$

$$4 + 3(-1) \log_2 2 \overset{?}{=} 1 \qquad \text{Power rule for logarithms}$$

$$4 - 3 \overset{?}{=} 1 \qquad \log_2 2 = 1$$

$$1 \overset{?}{=} 1 \qquad \text{Yes}$$

This check shows that the solution set is $\left\{\dfrac{1}{2}\right\}$.

Practice Problem 6 Solve $1 + 2 \ln x = 4$.

If each side of an equation can be expressed as a single logarithm with the same base, then we can use the one-to-one property of logarithms to solve the equation.

RECALL

The Product Rule says that
$\log_a M + \log_a N = \log_a MN$.

EXAMPLE 7 **Using the One-to-One Property of Logarithms**

Solve $\log_4 x + \log_4 (x + 1) = \log_4 (x - 1) + \log_4 6$.

Solution

$\log_4 x + \log_4 (x + 1) = \log_4 (x - 1) + \log_4 6$	Original equation
$\log_4 [x(x + 1)] = \log_4 [6(x - 1)]$	Product rule for logarithms
$x(x + 1) = 6(x - 1)$	One-to-one property
$x^2 + x = 6x - 6$	Distributive property
$x^2 - 5x + 6 = 0$	Add $-6x + 6$ to both sides.
$(x - 2)(x - 3) = 0$	Factor.
$x - 2 = 0 \quad \text{or} \quad x - 3 = 0$	Zero-product property
$x = 2 \quad \text{or} \quad x = 3$	Solve for x.

Now we check these possible solutions in the original equation.

Check $x = 2$

$$\log_4 2 + \log_4 (2 + 1) \overset{?}{=} \log_4 (2 - 1) + \log_4 6$$

$$\log_4 2 + \log_4 3 \overset{?}{=} \log_4 1 + \log_4 6$$

$$\log_4 (2 \cdot 3) \overset{?}{=} \log_4 (1 \cdot 6) \qquad \text{Yes}$$

Check $x = 3$

$$\log_4 3 + \log_4 (3 + 1) \overset{?}{=} \log_4 (3 - 1) + \log_4 6$$

$$\log_4 3 + \log_4 4 \overset{?}{=} \log_4 2 + \log_4 6$$

$$\log_4 (3 \cdot 4) \overset{?}{=} \log_4 (2 \cdot 6) \qquad \text{Yes}$$

The solution set is $\{2, 3\}$.

Practice Problem 7 Solve $\ln (x + 5) + \ln (x + 1) = \ln (x - 1)$.

EXAMPLE 8 Using the Product and Quotient Rules

Solve the following equations.

a. $\log_2(x - 3) + \log_2(x - 4) = 1$

b. $\log_2(x + 4) - \log_2(x + 3) = 1$

Solution

a.

$\log_2(x - 3) + \log_2(x - 4) = 1$	Original equation
$\log_2[(x - 3)(x - 4)] = 1$	Product rule for logarithms
$(x - 3)(x - 4) = 2^1$	Exponential form
$x^2 - 7x + 10 = 0$	Write in standard form.
$(x - 2)(x - 5) = 0$	Factor.
$x - 2 = 0$ or $x - 5 = 0$	Zero-product property
$x = 2$ or $x = 5$	Solve for x.

We check the possible solutions in the original equation.

Check $x = 2$	**Check $x = 5$**
$\log_2(2 - 3) + \log_2(2 - 4) \overset{?}{=} 1$	$\log_2(5 - 3) + \log_2(5 - 4) \overset{?}{=} 1$
$\log_2(-1) + \log_2(-2) \overset{?}{=} 1$ No	$\log_2 2 + \log_2 1 \overset{?}{=} 1$
	$1 + 0 \overset{?}{=} 1$
Because logarithms are not defined for negative numbers, $x = 2$ is an extraneous solution of the original equation.	$1 \overset{?}{=} 1$ Yes

The solution set is $\{5\}$.

b.

$\log_2(x + 4) - \log_2(x + 3) = 1$	Original equation
$\log_2 \dfrac{x + 4}{x + 3} = 1$	Quotient rule
$\dfrac{x + 4}{x + 3} = 2^1$	Exponential form
$x + 4 = 2(x + 3)$	Multiply both sides by $x + 3$.
$(x + 2) = 0$	Simplify.
$x = -2$	Solve for x.

RECALL

Remember that $\log_3 x$ means "the exponent on 3 that gives x."

We check the possible solution in the original equation.

Check $x = -2$

$$\log_2(-2 + 4) - \log_2(-2 + 3) \overset{?}{=} 1$$
$$\log_2 2 - \log_2 1 \overset{?}{=} 1$$
$$1 - 0 \overset{?}{=} 1 \qquad \text{Yes}$$

The solution set is $\{-2\}$.

Practice Problem 8 Solve $\log_3(x - 8) + \log_3 x = 2$.

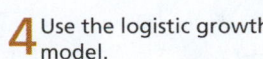

4 Use the logistic growth model.

EXAMPLE 9 Using Logarithms in the Logistic Growth Model

Suppose the carrying capacity M of the human population on Earth is 35 billion. In 1987, the world population was about 5 billion. Use the logistic growth model of P. F. Verhulst to calculate the average rate, k, of growth of the population given that the population was about 6 billion in 2003.

$$P(t) = \frac{M}{1 + ae^{-kt}} \qquad (2) \qquad \text{From the section introduction}$$

Pierre François Verhulst

(1804–1849)
Pierre Verhulst was born and educated in Brussels, Belgium. He received his PhD from the University of Ghent in 1825. Influenced by Lambert Quetelet, he became interested in social statistics. Verhulst's research on the law of population growth is important. Before Quetelet and Verhulst, scientists believed that an increasing population as a function of time was given by

$$P(t) = P_0(1 + k)^t \quad \text{or}$$
$$P(t) = P_0 e^{kt}.$$

Verhulst showed that forces tend to prevent the population growth according to these laws. He discovered the model given by equation (2).

Solution

If 1987 represents $t = 0$, then $P(0) = 5$ and $M = 35$; so equation (2) becomes

$$5 = \frac{35}{1 + ae^{-k(0)}}$$

$$5 = \frac{35}{1 + a}$$

$$5(1 + a) = 35 \qquad \text{Multiply both sides by } 1 + a.$$

$$a = 6 \qquad \text{Solve for } a.$$

Equation (2), with $M = 35$ and $a = 6$, becomes

$$P(t) = \frac{35}{1 + 6e^{-kt}}. \qquad (3)$$

We now solve equation (3) for k given that $t = 16$ (for 2003) and $P(16) = 6$.

$$6 = \frac{35}{1 + 6e^{-16k}}$$

$$6 + 36e^{-16k} = 35 \qquad \text{Multiply both sides by } 1 + 6e^{-16k}$$
$$\text{and distribute.}$$

$$e^{-16k} = \frac{29}{36} \qquad \text{Isolate } e^{-16k}.$$

$$-16k = \ln\left(\frac{29}{36}\right) \qquad \text{Logarithmic form}$$

$$k = -\frac{1}{16}\ln\left(\frac{29}{36}\right) \qquad \text{Solve for } k.$$

$$k \approx 0.0135 = 1.35\% \qquad \text{Use a calculator.}$$

The average growth rate of the world population was approximately 1.35%.

Practice Problem 9 Repeat Example 9 assuming that the estimated world population in 2005 was about 6.5 billion.

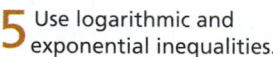
Use logarithmic and exponential inequalities.

Logarithmic and Exponential Inequalities

The properties of logarithms and exponents, together with the same techniques we used to solve polynomials, are used to solve inequalities involving logarithmic and exponential expressions.

The exponential and logarithmic functions with base $a > 1$ are increasing functions. That is, if $x_1 < x_2$, then $a^{x_1} < a^{x_2}$, and if $0 < x_1 < x_2$, then $\log_a x_1 < \log_a x_2$. As their graphs suggest, it is also true for base $a > 1$ that if $a^{x_1} < a^{x_2}$, then $x_1 < x_2$ and if $\log_a x_1 < \log_a x_2$, then $x_1 < x_2$. See Figures 3.14 and 3.15.

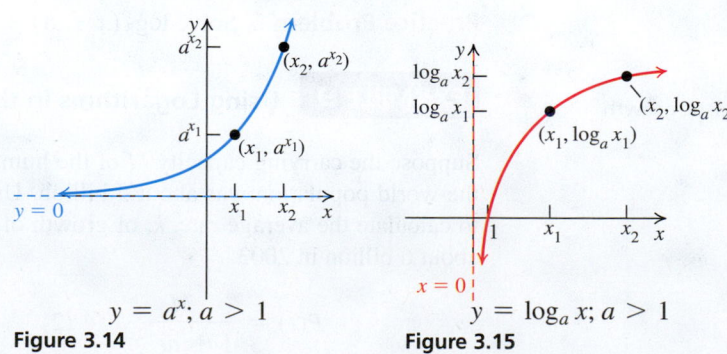

Figure 3.14 $\quad y = a^x; a > 1$

Figure 3.15 $\quad y = \log_a x; a > 1$

When working with inequalities, it is important to remember that $y = \log_a x$, $a > 1$, is negative for $0 < x < 1$.

EXAMPLE 10   Solving an Inequality Involving an Exponential Expression

Solve: $5(0.7)^x + 3 < 18$

Solution

$5(0.7)^x + 3 < 18$	Given inequality
$(0.7)^x < 3$	Isolate $(0.7)^x$ on one side.
$\ln(0.7)^x < \ln 3$	Take the natural log of both sides.
$x \ln(0.7) < \ln 3$	Power rule for logarithms
$x > \dfrac{\ln 3}{\ln(0.7)} \approx -3.080$	Divide both sides by $\ln(0.7)$; reverse the sense of the inequality because $\ln(0.7) < 0$.

RECALL

If $A < B$ and $C < 0$, then $AC > BC$.

Practice Problem 10 Solve: $3(0.5)^x + 7 > 19$

EXAMPLE 11   Solving an Inequality Involving a Logarithmic Expression

Solve: $\log(2x - 5) \le 1$

Solution

We first identify the domain for $\log(2x - 5)$. Because $2x - 5$ must be positive, we solve $2x - 5 > 0$; so $2x > 5$ or $x > \dfrac{5}{2}$.

Then	$\log(2x - 5) \le 1$	Given inequality
	$10^{\log(2x-5)} \le 10^1$	If $u < v$, then $10^u < 10^v$.
	$2x - 5 \le 10$	$a^{\log_a x} = x$
	$x \le \dfrac{15}{2}$	Solve for x.

For $\log(2x - 5)$ to be a real number, we must have $x > \dfrac{5}{2}$. Combining this with the fact that $x \le \dfrac{15}{2}$, we find that the solution set for $\log(2x - 5) \le 1$ is the interval $\left(\dfrac{5}{2}, \dfrac{15}{2}\right]$.

Practice Problem 11 Solve: $\ln(1 - 3x) > 2$

Answers to Practice Problems

1. **a.** $x = 5$ **b.** $x = \dfrac{2}{3}$ 2. $x = \dfrac{\ln\left(\dfrac{11}{7}\right)}{\ln 3} - 1 \approx -0.59$

3. $x = \dfrac{\ln 3}{2\ln 2 - \ln 3} \approx 3.82$ 4. $x = \ln 5 \approx 1.609$

5. **a.** United States: 343.61 million; Pakistan: 255.96 million
b. Sometime in 2022 (after 11.69 yr) **c.** In 23.68 yr (sometime in 2033)   6. $x = e^{3/2}$ 7. $\varnothing$ 8. $x = 9$
9. The average growth rate was approximately 1.74%.

10. $x < -2$ 11. $x < \dfrac{1 - e^2}{3}$

SECTION 3.4 **Exercises**

Concepts and Vocabulary

1. An equation that contains terms of the form a^x is called a(n) _____ equation.

2. An equation that contains terms of the form $\log_a x$ is called a(n) _____ equation.

3. The equation $y = \dfrac{M}{1 + ae^{-bx}}$ represents a(n) _____ model.

4. If $a^u = a^v$, then _____.

5. **True or False.** A logistic curve always has two horizontal asymptotes.

6. **True or False.** Because the domain of $f(x) = \log_a x$ is the interval $(0, \infty)$, a logarithmic equation cannot have a negative solution.

7. **True or False.** The equation $8^{2x} = 4^{3x}$ is true for all real values of x.

8. **True or False.** If $0 < x_1 < x_2$, then $\log_a x_1 < \log_a x_2$.

Building Skills

In Exercises 9–24, solve each equation.

9. $2^x = 16$
10. $3^x = 243$
11. $8^x = 32$
12. $5^{x-1} = 1$
13. $4^{|x|} = 128$
14. $9^{|x|} = 243$
15. $5^{-|x|} = 625$
16. $3^{-|x|} = 81$
17. $\ln x = 0$
18. $\ln (x - 1) = 1$
19. $\log_2 x = -1$
20. $\log_2 (x + 1) = 3$
21. $\log_3 |x| = 2$
22. $\log_2 |x + 1| = 3$
23. $\dfrac{1}{2} \log x - 2 = 0$
24. $\dfrac{1}{3} \log (x + 1) - 1 = 0$

In Exercises 25–60, solve each exponential equation. Write the exact answer with natural logarithms and then approximate the result correct to three decimal places.

25. $2^x = 3$
26. $3^x = 5$
27. $2^{2x+3} = 15$
28. $3^{2x+5} = 17$
29. $e^{x+1} = 3$
30. $e^{2x-1} = 5$
31. $5 \cdot 2^x - 7 = 10$
32. $3 \cdot 5^x + 4 = 11$
33. $3 \cdot 4^{2x-1} + 4 = 14$
34. $2 \cdot 3^{4x-5} - 7 = 10$
35. $2e^{x-2} + 3 = 7$
36. $3e^{3x-5} + 1 = 6$
37. $5^{1-x} = 2^x$
38. $3^{2x-1} = 2^{x+1}$
39. $2^{1-x} = 3^{4x+6}$
40. $5^{2x+1} = 3^{x-1}$
41. $2 \cdot 3^{x-1} = 5^{x+1}$
42. $5 \cdot 2^{2x+1} = 7 \cdot 3^{x-1}$
43. $(1.065)^t = 2$
44. $(1.0725)^t = 2$
45. $2^{2x} - 4 \cdot 2^x = 21$
46. $4^x - 4^{-x} = 2$
47. $9^x - 6 \cdot 3^x + 8 = 0$
48. $\dfrac{3^x + 5 \cdot 3^{-x}}{3} = 2$
49. $3^{3x} - 4 \cdot 3^{2x} + 2 \cdot 3^x = 8$

50. $2^{3x} + 3 \cdot 2^{2x} - 2^x = 3$
51. $e^{2x} - 2e^x - 3 = 0$
52. $e^x + 2e^{-x} = 3$
53. $\dfrac{3^x - 3^{-x}}{3^x + 3^{-x}} = \dfrac{1}{4}$
54. $\dfrac{e^x - e^{-x}}{e^x + e^{-x}} = \dfrac{1}{3}$
55. $\dfrac{4}{2 + 3^x} = 1$
56. $\dfrac{7}{2^x - 1} = 3$
57. $\dfrac{17}{5 - 3^x} = 7$
58. $\dfrac{15}{3 + 2 \cdot 5^x} = 4$
59. $\dfrac{5}{2 + 3^x} = 4$
60. $\dfrac{7}{3 + 5 \cdot 2^x} = 4$

In Exercises 61–78, solve each logarithmic equation.

61. $3 + \log (2x + 5) = 2$
62. $1 + \log (3x - 4) = 0$
63. $\log (x^2 - x - 5) = 0$
64. $\log (x^2 - 6x + 9) = 0$
65. $\log_4 (x^2 - 7x + 14) = 1$
66. $\log_4 (x^2 + 5x + 10) = 1$
67. $\ln (2x - 3) - \ln (x + 5) = 0$
68. $\log (x + 8) + \log (x - 1) = 1$
69. $\log x + \log (x + 9) = 1$
70. $\log_5 (3x - 1) - \log_5 (2x + 7) = 0$
71. $\log_a (5x - 2) - \log_a (3x + 4) = 0$
72. $\log (x - 1) + \log (x + 2) = 1$
73. $\log_6 (x + 2) + \log_6 (x - 3) = 1$
74. $\log_2 (3x - 2) - \log_2 (5x + 1) = 3$
75. $\log_3 (2x - 7) - \log_3 (4x - 1) = 2$
76. $\log_4 \sqrt{x + 3} - \log_4 \sqrt{2x - 1} = \dfrac{1}{4}$
77. $\log_7 3x + \log_7 (2x - 1) = \log_7 (16x - 10)$
78. $\log_3 (x + 1) + \log_3 2x = \log_3 (3x + 1)$

In Exercises 79–86, find a and k and then evaluate the function. Round your answer to three decimal places.

79. Let $f(x) = 20 + a(2^{kx})$ with $f(0) = 50$ and $f(1) = 140$. Find $f(2)$.

80. $f(x) = 40 + a(4^{kx})$ with $f(0) = -216$ and $f(2) = 39$. Find $f(1)$.

81. $f(x) = 16 + a(3^{kx})$ with $f(0) = 21$ and $f(4) = 61$. Find $f(2)$.

82. $f(x) = 50 + a(2^{kx})$ with $f(0) = 34$ and $f(4) = 46$. Find $f(2)$.

83. Let $f(x) = \dfrac{10}{3 + ae^{kx}}$ with $f(0) = 2$ and $f(1) = \dfrac{1}{2}$. Find $f(2)$.

84. Let $f(x) = \dfrac{6}{a + 2e^{kx}}$ with $f(0) = 1$ and $f(1) = 0.8$.
Find $f(2)$.

85. Let $f(x) = \dfrac{4}{a + 4e^{kx}}$ with $f(0) = 2$ and $f(1) = 9$.
Find $f(2)$.

86. Let $f(x) = \dfrac{a}{1 + 3e^{kx}}$ with $f(0) = -1$ and $f(2) = -2$.
Find $f(1)$.

In Exercises 87–100, solve each inequality.

87. $5(0.3)^x + 1 \le 11$

88. $(0.1)^x - 4 > 15$

89. $-3(1.2)^x + 11 \ge 8$

90. $-7(0.4)^x + 19 < 5$

91. $\log(5x + 15) < 2$

92. $\log(2x + 0.9) > -1$

93. $\ln(x - 5) \ge 1$

94. $\ln(4x + 10) \le 2$

95. $\log_2(3x - 7) < 3$

96. $\log_2(5x - 4) > 4$

97. $1 + 4e^{2x} > 9$

98. $23 - 2e^{4x} < 20$

99. $\dfrac{50}{1 + 2e^{-x}} \ge 10$

100. $\dfrac{40}{1 + 3e^{-2x}} \le 25$

Applying the Concepts

101. Investment. Find the time required for an investment of $10,000 to grow to $18,000 at an annual interest rate of 6% if the interest is compounded
a. Yearly.
b. Quarterly.
c. Monthly.
d. Daily.
e. Continuously.

102. Investment. How long will it take for an investment of $100 to double in value if the annual rate of interest is 7.2% compounded
a. Yearly.
b. Quarterly.
c. Monthly.
d. Continuously.

103. Inflation. In the fictional land of Sardonia, the price of a car costing $20,000 in U.S. currency will double in eight years. What will the car cost in five years? What is the annual rate of inflation?

104. Doubling time. An amount P dollars is deposited in a regular bank account. How long does it take to double the initial investment (called the **doubling time**) if
a. The interest rate is r compounded annually.
b. The interest rate is r (per year) compounded continuously.

105. Light absorption in physics. The light intensity I (in lumens) at a depth of x feet in Lake Elizabeth is given by
$$\log\left(\frac{I}{12}\right) = -0.025x.$$
a. Find the light intensity at a depth of 30 feet.
b. At what depth is the light intensity 4 lumens?

106. Rule of 70. Bankers use the **rule of 70** to estimate the doubling time for money invested at different rates. The rule of 70 states that
$$\text{Doubling time} \approx \frac{70}{100r}\text{ years,}$$
where r is the annual interest rate (expressed as a decimal number). Explain why this formula works.

107. Epidemic outbreak. The number of people in a community who became infected during an epidemic t weeks after its outbreak is given by the function
$$f(t) = \frac{20,000}{1 + ae^{-kt}},$$
where 20,000 people of the community are susceptible to the disease. Assuming that 1000 people were infected initially and 8999 had been infected by the end of the fourth week,
a. Find the number of people infected after eight weeks.
b. After how many weeks will 12,400 people be infected?

108. Spreading a rumor. A jealous student started a malicious rumor on an isolated college campus of 5000 students. The number of people who know of the rumor within t days after it was started is given by the function
$$R(t) = \frac{5000}{1 + ae^{-kt}}.$$
Half the students had known of the rumor within ten days.
a. Find a and k and graph the function $y = R(t)$.
b. How many students would have known of the rumor within 15 days after it started?
c. How many students altogether will know the rumor?

109. Biological growth. In his laboratory experiment in 1934, G. F. Gause placed paramecia (unicellular microorganisms) in 5 cubic centimeters of a saline (salt) solution with a constant amount of food and measured their growth on a daily basis. He found that the population $P(t)$ after t days was approximated by
$$P(t) = \frac{4490}{1 + e^{5.4094 - 1.0255t}}, \quad t \ge 0.$$
a. What was the initial population of the paramecia?
b. What was the carrying capacity of the medium?
c. Graph the equation $y = P(t)$.

110. Facebook users. In 2004, there were 1 million Facebook users, and in 2012, there were 845 million users. Use the model $f(t) = \dfrac{20,000}{20 + ae^{-kt}}$, where t is the number of years since 2004 and $f(t)$ is the number of users in millions, to:
a. Find a and k.
b. Estimate the number of users in 2015.
c. Estimate the year in which the number of users will be 2 billion.
(*Source:* www.benposter.com/facebook-user-growth-chart-2004-2010)

111. Earth science. The atmospheric pressure P (pounds per square inch) at height h miles above sea level is given by

$$P = 14.7e^{-0.21h}.$$

Find the height at which P is at most 8 psi.

112. Learning. A toy manufacturer estimates that the number of toys, N, a trainee can assemble after t days of training is given by

$$N = \frac{100}{1 + 19e^{-0.2t}}.$$

How many days of training does it take so that the trainee can assemble at least 40 toys per day?

Beyond the Basics

113. Solve for t: $P = \dfrac{M}{1 + e^{-kt}}.$

114. Find k so that
a. $2^x = e^{kx}.$
b. $e^x = 2^{kx}.$

115. If $\dfrac{\log x}{2} = \dfrac{\log y}{3} = \dfrac{\log z}{5}(=k)$, show that
a. $xy = z.$ **b.** $y^2z^2 = x^8.$
[*Hint:* First, express $x, y,$ and z in terms of k.]

116. The present value P of an annuity with payment size R, periodic rate i, and term of n payments is given by the formula

$$P(1 + i)^n = R\frac{(1 + i)^n - 1}{i}.$$

Solve this equation for n.

In Exercises 117–124, solve each equation for x.
117. $(\log x)^2 = \log x$
118. $(\log_2 x)(\log_2 8x) = 10$
119. $(\log_3 x)(\log_3 3x) = 2$
120. $\log_2 x + \log_4 x = 6$
121. $\log_4 x^2(x - 1)^2 - \log_2(x - 1) = 1$
122. $\dfrac{\log(7x - 12)}{\log x} = 2$
123. $\dfrac{\log(3x - 5)}{2} = \log x$

124. $\log(x - 4) - \log x = \log\left(\dfrac{1}{10 - x}\right)$

In Exercises 125–132, use the four-step procedure (page 122) to find $f^{-1}(x)$ for the given function $f(x)$.
125. $f(x) = 3^x + 5$
126. $f(x) = 2^{-x} + 4$
127. $f(x) = 3 \cdot 4^x + 7$
128. $f(x) = 2 \cdot 3^{x-1} - 5$
129. $f(x) = 1 + \log_2(x - 1)$
130. $f(x) = 7 + 2\log_6(3x - 1)$
131. $f(x) = \dfrac{1}{2}\ln\left(\dfrac{x - 1}{x + 1}\right)$
132. $f(x) = \dfrac{1}{2}\log\left(\dfrac{1 + x}{1 - x}\right)$

In Exercises 133–136, find the smallest integer n for which the given property holds.
133. $7^n > 43^{67}$ **134.** $9^n > e^{321}$
135. $8^{1/n} < 1.01$ **136.** $12^{1/n} < 1.001$
137. Find an integer n so that 31^n has 567 digits.
[*Hint:* $566 \le \log 31^n < 567.$]
138. Find an integer n so that n^{123} has 456 digits.
139. Evaluate $\dfrac{x^{\log y}}{x^{\log z}} \cdot \dfrac{y^{\log z}}{y^{\log x}} \cdot \dfrac{z^{\log x}}{z^{\log y}}.$
140. Evaluate $\dfrac{1}{\log_{xy} xyz} + \dfrac{1}{\log_{yz} xyz} + \dfrac{1}{\log_{zx} xyz}.$

Critical Thinking / Discussion / Writing

141. Logistic function. For $f(t) = \dfrac{P}{1 + ae^{-kt}}$, it can be shown that the maximum rate of growth occurs when $f(t) = \dfrac{P}{2}$. Find the time when the rate of growth is maximum.

142. Solve each of the following equations for x.
a. $\log_4(x - 1)^2 = 3$
b. $2\log_4(x - 1) = 3$
c. $2\log_4|x - 1| = 3$
Explain why these three equations do not have identical solutions.

Getting Ready for the Next Section

In Exercises 143–146, convert each equation from exponential to logarithmic form.
143. $9^2 = 81$ **144.** $3^{-2} = \dfrac{1}{9}$
145. $2 \cdot 10^x + 1 = 7$ **146.** $3e^{2x} + 5 = 17$

In Exercises 147–150, convert each equation from logarithmic to exponential form.
147. $\log_2 64 = 6$ **148.** $\log_{\frac{1}{2}} 16 = -4$
149. $\log\left(\dfrac{A}{2}\right) = 3$ **150.** $\ln\left(\dfrac{A}{P}\right) = kt$

In Exercises 151–158, solve each equation for x.
151. $\log_5 x = -3$ **152.** $\log_3 \dfrac{1}{27} = x - 1$
153. $\log_x 1000 = 3$ **154.** $\log_2(x^2 - 6x + 10) = 1$
155. $2^{x+1} = 32$ **156.** $3^{2x-1} = 7$
157. $3^{x+1} = 5^{2x-3}$
158. $\log(x^2 + x) = \log(3x + 3)$

Logarithmic Scales; Modeling

BEFORE STARTING THIS SECTION, REVIEW

1 Exponential form for logarithms (Section 3.2, page 272)

2 Properties of logarithms (Section 3.2, page 274)

3 Rules for logarithms (Section 3.3, page 288)

OBJECTIVES

1 Define pH.

2 Define the Richter scale for measuring earthquake intensity.

3 Define scales for measuring sound.

4 Define magnitude of star brightness.

5 Build models from data.

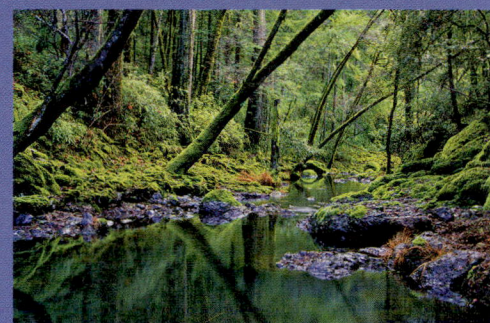

◆ A **logarithmic scale** is a scale in which logarithms are used in the measurement of quantities. Suppose we have a quantity that has a small positive range of variation, for example, from $0.00001 = 10^{-5}$ to $0.0001 = 10^{-4}$, then the common logarithms of such numbers would be between -5 and -4. Similarly, if a quantity has a large positive range of variation from $10,000 = 10^4$ to $100,000 = 10^5$, the common logarithms of these numbers would be between 4 and 5. So, a logarithmic scale serves to make data more manageable by expanding small variations and compressing large ones.

In Example 2 we will see how a small difference on a logarithmic scale results in an impressive increase in the acidity of acid rain compared to ordinary rain.

1 Define pH.

pH Scale

In chemistry, the acidity of a substance is related to the presence of positively charged hydrogen ions. pH represents the effective concentration (activity) of hydrogen ions H^+ in water. Since H^+ concentrations are much smaller than other dissolved species in water, the activity of hydrogen ions is expressed most conveniently in logarithmic units.

---

pH SCALE FORMULA

We define the pH scale of a solution by the formula

$$pH = -\log\left[H^+\right],$$

where $\left[H^+\right]$ is the concentration of H^+ ions in moles per liter. (A mole is a unit of measurement, equal to 6.023×10^{23} atoms.)

The pH value is a measure of the acidity or alkalinity of a solution. The pH value for pure water at 25°C (77°F) is 7.

Søren Sørensen

(1868–1939)
The term *pH* was originally derived from the French term "pouvoir hydrogène." In English, this means "hydrogen power." The term *pH* is always written with a lower-case p and an upper-case H. The pH scale was defined by the Danish biochemist Sören Sörensen in 1909.

Because pH is defined as $-\log[H^+]$, pH decreases as $[H^+]$ increases (which will happen if acid is added to the water). *Acidic* solutions have pH values less than 7, and *alkaline* (also called *basic*) solutions have pH values greater than 7. The smaller the number on the pH scale, the more acidic the substance is.

EXAMPLE 1 Calculating pH and [H⁺]

a. Calculate to the nearest tenth the pH value of grapefruit juice if $[H^+]$ of grapefruit juice is 6.32×10^{-4}.

b. Find the hydrogen concentration $[H^+]$ in beer if its pH value is 4.82.

Solution

a.
$$
\begin{aligned}
\text{pH} &= -\log[H^+] && \text{Definition of pH} \\
&= -\log(6.32 \times 10^{-4}) && \text{Given } [H^+] = 6.32 \times 10^{-4} \\
&= -(\log 6.32 + \log 10^{-4}) && \text{Product rule} \\
&= -\log 6.32 + 4 && -\log 10^{-4} = -(-4)\log 10 = 4 \\
&\approx -0.8007 + 4 && \text{Use a calculator.} \\
&= 3.1993 \approx 3.2 && \text{Simplify.}
\end{aligned}
$$

b.
$$
\begin{aligned}
4.82 &= -\log[H^+] && \text{Replace pH with 4.82.} \\
\log[H^+] &= -4.82 && \text{Rewrite equation.} \\
[H^+] &= 10^{-4.82} && \text{Exponential form} \\
&= 10^{0.18} \times 10^{-5} && -4.82 = -5 + 0.18 \\
&\approx 1.51 \times 10^{-5} && \text{Use a calculator.}
\end{aligned}
$$

Practice Problem 1

a. The $[H^+]$ of a solution is 2.68×10^{-6}. Find its pH value.

b. Find $[H^+]$ of sea water if its pH is 8.47.

EXAMPLE 2 Acid Rain

DO YOU KNOW?

Acid rain is rain, snow, or fog that is polluted by acid in the atmosphere and damages the environment. Two culprits that acidify rain are sulfur dioxide (SO_2) and nitrogen oxide (N_2O). Rain measuring between 0 and 5 on the pH scale is acidic and therefore called "acid rain."

How much more acidic is acid rain with a pH value of 3 than ordinary rain with a pH value of 6?

Solution

We have

$$3 = \text{pH}_{\text{acid rain}} = -\log[H^+]_{\text{acid rain}} \qquad \text{Definition of pH}$$

so that

$$[H^+]_{\text{acid rain}} = 10^{-3} \qquad \text{Exponential form of log } [H^+] = -3$$

Similarly,

$$[H^+]_{\text{ordinary rain}} = 10^{-6}.$$

Therefore,

$$\frac{[H^+]_{\text{acid rain}}}{[H^+]_{\text{ordinary rain}}} = \frac{10^{-3}}{10^{-6}} = 10^{-3+6} = 10^3 = 1000.$$

So, the hydrogen ion concentration in this acid rain is 1000 times greater than that in ordinary rain. That is, this acid rain is 1000 times more acidic than the ordinary rain.

SIDE NOTE

Notice that each one-point increase in pH value corresponds to a factor of 10 increase in H⁺ concentration.

Practice Problem 2 How much more acidic is an acid rain with a pH value of 2.8 than an ordinary rain with a pH value of 6.2?

Table 3.8 lists a few typical pH values.

TABLE 3.8

Solution	pH Value
Battery acid	1
Lemon juice	2
Stomach acid	2–3
Vinegar	3
Milk	6–7
Baking soda, sea water	8–9
Milk of Magnesia	9–10
Ammonia	10–11
Drain opener	10–12
Lye	13

Maximum acidity   Neutral   Maximum alkalinity

0   7   14 x

0 = Maximum acidity
7 = Neutral point
14 = Maximum alkalinity (opposite of acidity)

2 Define the Richter scale for measuring earthquake intensity.

Charles Francis Richter

(1900–1985)
Charles Richter was born in Hamilton, Ohio, and received his PhD from California Institute of Technology in 1928. He developed his scale to measure the strength of earthquakes in 1935. Earlier scales had been developed by Michele Stefano Conte de Rossi in the 1880s and by Giuseppe Mercalli in 1902, but both used a descriptive scale defined in terms of damage to buildings and the behavior and response of the population. Richter's scale is an absolute one, based on the amplitude of the waves produced by the earthquake. He defined the magnitude of an earthquake as the logarithm to the base 10 of the maximum amplitude of the waves, measured in microns. This means that waves whose amplitudes differ by a factor of 100 will differ by 2 points on the Richter scale.

Earthquake Intensity

An earthquake is the vibration, sometimes violent, of Earth's surface that follows a release of energy in Earth's crust. Earth's crust is composed of about 20 huge plates that float on the molten material beneath the crust. These plates slowly move over, under, and past each other. Sometimes the movement is gradual. At other times, the plates are locked together, unable to release the accumulating energy. When this energy grows strong enough, the plates break free and vibrations called "seismic waves" or earthquakes are generated.

The Richter scale was invented in the 1930s by the American scientist Dr. Charles Richter to measure the magnitude of an earthquake. It is based on the idea of comparing the intensity of an earthquake with that of a zero-level earthquake. He defined the zero-level earthquake as an earthquake whose seismographic reading measures one micron (1000 microns = 1 millimeter, or 1,000,000 microns = 1 meter) at a distance of 100 kilometers (about 62 miles) from the epicenter of the earthquake.

Let us denote the intensity of a zero-level earthquake by I_0. A zero-level earthquake is just noticeable and is the *threshold level* below which we would not be aware of a quake.

RICHTER SCALE

The magnitude M of an earthquake is a function of its intensity I and is defined by

$$M = \log\left(\frac{I}{I_0}\right)$$

where I_0 is the intensity of the zero-level earthquake.

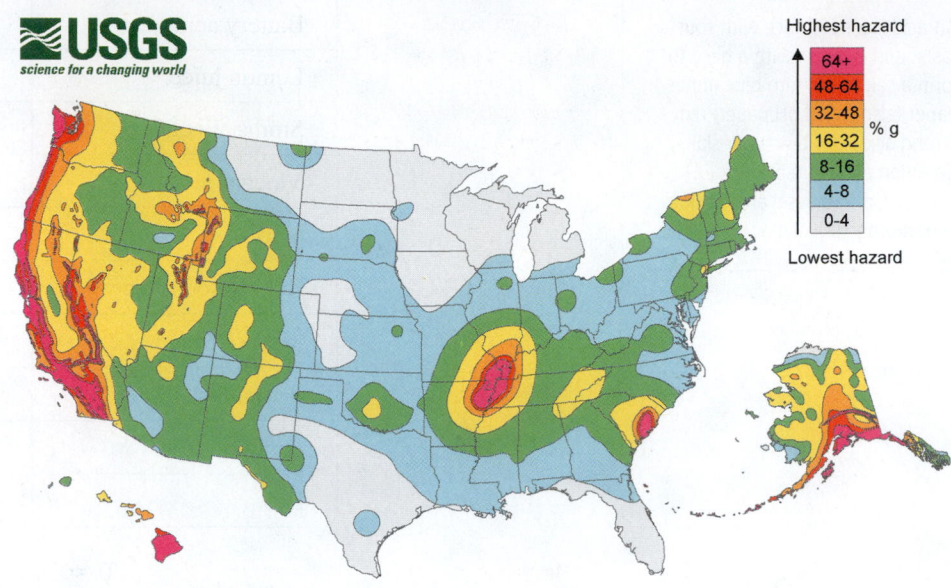

EXAMPLE 3   **Intensity of an Earthquake**

The magnitude of an earthquake is 4.0 on the Richter scale. What is the intensity of this earthquake?

Source: US Geological Survey (USGS).

Solution

$$M = \log\left(\frac{I}{I_0}\right)$$ Definition of magnitude

$$4 = \log\left(\frac{I}{I_0}\right)$$ Replace M with 4.

$$\frac{I}{I_0} = 10^4$$ Exponential form

$$I = 10^4 I_0 = 10{,}000\, I_0$$ Multiply both sides by I_0.

The intensity of this earthquake is 10,000 times the intensity of I_0 (the zero-level earthquake). Notice that a one point increase on the Richter scale corresponds to a factor of 10 increase in intensity.

Practice Problem 3 The magnitude of an earthquake is 6.5 on the Richter scale. What is the intensity of this earthquake?

EXAMPLE 4   **Comparing Two Earthquakes**

Compare the intensity of the Mexico City earthquake of 1985, which registered 8.1 on the Richter scale, to that of the San Francisco area earthquake of 1989, which measured 6.9 on the Richter scale.

Solution

Let I_M and I_S denote the intensities of the Mexico City and the San Francisco earthquakes, respectively. We have

$$8.1 = \log\left(\frac{I_M}{I_0}\right) \qquad 6.9 = \log\left(\frac{I_S}{I_0}\right)$$ Definition of magnitude

$$\frac{I_M}{I_0} = 10^{8.1} \qquad \frac{I_S}{I_0} = 10^{6.9}$$ Exponential form

$$I_M = 10^{8.1} I_0 \qquad I_S = 10^{6.9} I_0.$$ Multiply both sides by I_0.

DO YOU KNOW?

To get some idea of the destructive power of an earthquake, the energy released at the epicenter of a zero-level earthquake is equivalent to more than 10,000 atomic bombs of the kind that leveled Hiroshima, Japan, in 1945 exploding simultaneously.

Divide I_M by I_S and simplify to get

$$\frac{I_M}{I_S} = \frac{10^{8.1}I_0}{10^{6.9}I_0} = \frac{10^{8.1}}{10^{6.9}} = 10^{8.1-6.9} = 10^{1.2}$$

$$I_M = 10^{1.2} I_S \approx 16\, I_S.$$

This equation shows that the intensity of the Mexico City earthquake was about 16 times that of the San Francisco area earthquake.

Practice Problem 4 The magnitudes of the earthquakes of Mozambique (2006) and Southern California (2005) were 7.0 and 5.2, respectively. Compare their intensities. ◼◼

Energy of an Earthquake The magnitude M of an earthquake is related to the released energy E (measured in joules) and is approximated by the equation

$$\log E = 4.4 + 1.5\, M$$

We can rewrite this equation in the exponential form as

$$E = 10^{4.4+1.5M} = 10^{0.4} \times 10^4 \times 10^{1.5M}$$

or $\qquad E \approx (2.5 \times 10^4) \times 10^{1.5M} \qquad 10^{0.4} \approx 2.5.$

ENERGY OF AN EARTHQUAKE

The energy E (in joules) released by an earthquake of magnitude M (Richter scale) is given by

$$\log E = 4.4 + 1.5M$$

or

$$E \approx (2.5 \times 10^4) \times 10^{1.5M}.$$

The energy released by a zero-level earthquake can be calculated by substituting $M = 0$ in the formula. We have

$$E = 2.5 \times 10^4 \text{ joules.}$$

TABLE 3.9 Major Earthquakes ($M \geq 7.5$) of the 20th & 21st Century

Date	Location	Magnitude
April 1906	San Francisco	7.8
December 1908	Messina, Italy	7.5
December 1920	Gansu, China	8.6
September 1923	Sagami Bay, Japan	8.3
February 1931	New Zealand	7.9
August 1950	Assam, India	8.7
July 1976	Tangshan, China	8.0
September 1985	Mexico City	8.1
June 1990	Iran	7.7
May 1997	Iran	7.5
December 2004	Sumatra, Indonesia	9.0
March 2005	Sumatra, Indonesia	8.6
February 2010	Maule, Chile	8.8
March 2011	Tohoku, Japan	9.0

(*Source:* National Earthquake Information Center, U.S. Geological Survey.)

If two earthquakes have magnitude M_1 and M_2, we can compute the ratio of their corresponding energies E_1 and E_2 as follows:

$$\frac{E_1}{E_2} = \frac{(2.5 \times 10^4) \times 10^{1.5M_1}}{(2.5 \times 10^4) \times 10^{1.5M_2}} = \frac{10^{1.5M_1}}{10^{1.5M_2}} = 10^{1.5M_1 - 1.5M_2} = 10^{1.5(M_1 - M_2)}.$$

EXAMPLE 5 **Comparing Two Earthquakes**

Compare the estimated energies released by the San Francisco earthquake of 1906 and the Northridge earthquake of 1994 with magnitude of 6.7.

Solution

Let M_{1906} and E_{1906} represent the magnitude and energy released by the 1906 quake of San Francisco. Attach similar meaning to M_{1994} and E_{1994}.

With $M_{1906} = 7.8$ (Table 3.9) and $M_{1994} = 6.7$, and the equation $\dfrac{E_1}{E_2} = 10^{1.5(M_1 - M_2)}$,

we have

$$\frac{E_{1906}}{E_{1994}} = 10^{1.5(7.8 - 6.7)} = 10^{1.5(1.1)} \approx 45 \qquad \text{Use a calculator}$$

The 1906 earthquake released more than 45 times as much energy as the 1994 earthquake.

Practice Problem 5 Compare the energies released by the Japan earthquake of 2011 and the Iran earthquake of 1997 (see Table 3.9).

3 Define scales for measuring sound.

Loudness of Sound

The *intensity* of a sound wave is defined as the amount of power the wave transmits through a given area. The intensity of a sound is measured in watts per square meter (abbreviated W/m^2). The faintest sound that the human ear can detect has an intensity of $10^{-12} \ \text{W/m}^2$. This faintest sound is known as the *threshold of hearing* (TOH). The most intense sound which the human ear can safely detect without suffering any physical damage is more than one billion times the TOH.

Since the range of intensities that the human ear can detect is so large, a logarithmic scale called *decibels* (abbreviated dB) is used to measure the loudness of a sound. A decibel (as the name suggests), is one-tenth of a *bel,* a unit named after the telephone inventor Alexander Graham Bell.

LOUDNESS OF SOUND AND DECIBELS

The *loudness* (or relative intensity) L of a sound measured in decibels is related to its intensity I by the formula

$$L = 10 \log \left(\frac{I}{I_0} \right),$$

or

$$I = I_0 \times 10^{L/10}$$

where $I_0 = 10^{-12} \ \text{W/m}^2$ is the intensity of TOH.

If we substitute $I = I_0$ in the formula for loudness, we obtain

$$\text{Loudness of TOH} = 10 \log \frac{I_0}{I_0} = 10 \log 1 = 10(0) = 0.$$

So, the threshold of hearing has 0 dB loudness.

Table 3.10 lists approximate loudness of some common sounds.

TABLE 3.10 Approximate Loudness of Common Sounds

Source	Intensity in W/m²	Loudness in dB
Threshold of hearing	10^{-12}	0
Rustling leaves	10^{-11}	10
Whisper	10^{-10}	20
Background noise in average home	10^{-8}	40
Normal conversation	10^{-6}	60
Busy street traffic	10^{-5}	70
Vacuum cleaner	10^{-4}	80
Portable audio player at maximum level	10^{-2}	100
Front rows of rock concert	10^{-1}	110
Threshold of pain	10	130
Jet aircraft 50 m away	10^2	140

EXAMPLE 6 **Computing Loudness**

Find the decibel level of a TV that has an intensity of 250×10^{-7} W/m² at a distance of 12 feet from the TV.

Solution

We are given $I = 250 \times 10^{-7}$ W/m², and we know $I_0 = 10^{-12}$ W/m². We have

$$
\begin{aligned}
L &= 10 \log\left(\frac{I}{I_0}\right) && \text{Formula for loudness} \\
&= 10 \log\left(\frac{250 \times 10^{-7}}{10^{-12}}\right) && I = 250 \cdot 10^{-7} \text{ and } I_0 = 10^{-12} \\
&= 10 \log(250 \times 10^5) && \frac{10^{-7}}{10^{-12}} = 10^{-7+12} = 10^5 \\
&= 10[\log 250 + \log 10^5] && \text{Product rule for logarithms} \\
&= 10[\log 250 + 5 \log 10] && \text{Power rule for logarithms} \\
&= 10[\log 250 + 5] && \log 10 = \log_{10} 10 = 1 \\
&\approx 74. && \text{Calculator and rounding}
\end{aligned}
$$

So, the decibel level of this TV at a distance of 12 feet is approximately 74 dB.

Practice Problem 6 Find the decibel level of a TV that has an intensity of 200×10^{-7} W/m² at a distance of 15 feet from the TV.

EXAMPLE 7 Comparing Intensities

How much more intense is a 65-dB sound than a 42-dB sound?

Solution

$$I = I_0 \times 10^{\frac{L}{10}}$$ Formula for intensity

$$I_{65} = I_0 \times 10^{\frac{65}{10}}$$ I_{65} is the intensity of a 65-dB sound.

$$I_{42} = I_0 \times 10^{\frac{42}{10}}$$ I_{42} is the intensity of a 42-dB sound.

$$\frac{I_{65}}{I_{42}} = \frac{I_0 \times 10^{\frac{65}{10}}}{I_0 \times 10^{\frac{42}{10}}}$$ Divide I_{65} by I_{42}.

$$= \frac{I_0 \times 10^{6.5}}{I_0 \times 10^{4.2}}$$

$$= 10^{6.5-4.2}$$ Remove I_0 and use the Quotient Rule for Exponents.

$$= 10^{2.3}$$ Simplify.

$$\approx 199.53$$ Use a calculator.

$$\approx 200$$

Thus, a sound of 65 dB is about 200 times more intense than a sound of 42 dB.

Practice Problem 7 How much more intense is a 75-dB sound than a 55-dB sound?

EXAMPLE 8 Computing Sound Intensity

Calculate the intensity in watts per square meter of a sound of 73 dB.

Solution

$$I = I_0 \times 10^{\frac{L}{10}}$$ Formula for intensity

$$I = 10^{-12} \times 10^{\frac{73}{10}}$$ Replace I_0 with 10^{-12} and L with 73.

$$= 10^{-12} \times 10^{7.3}$$

$$= 10^{-4.7}$$ Simplify.

$$= 10^{0.3} \times 10^{-5}$$ $-4.7 = 0.3 - 5$

$$\approx 1.995 \times 10^{-5}$$ Use a calculator.

$$\approx 2 \times 10^{-5}\, \text{W/m}^2.$$

Practice Problem 8 What is the intensity of a 48-dB sound?

Musical Pitch The pitch of a sound is determined by its frequency. For example, on a piano the pitch of the note A above middle C may be denoted by A440. This means that A vibrates at 440 Hertz (cycles per second). In music, an interval whose higher note frequency is twice that of its lower note is an **octave**. Two notes that are an octave apart sound essentially "the same," but one has a higher pitch. For this reason, notes an octave apart are given the same name. An octave jump is usually divided into 12 approximately equal *semitones* on a log scale. In an "equal-tempered" scale, a semitone is further divided into 100 *cents*. So, there are 1200 cents in an octave.

CHANGE IN PITCH

If f is a frequency measured in Hertz (Hz) and f_0 is a reference frequency, then the change in pitch $P(f)$ in cents from f_0 to f is given by:

$$P(f) = 1200 \log_2\left(\frac{f}{f_0}\right).$$

We note that the pitch difference of 1 octave means $f = 2f_0$, so

$$P(2f_0) = 1200 \log_2\left(\frac{2f_0}{f_0}\right) = 1200 \log_2 2 = 1200 \text{ cents},$$

which agrees with our previous assertion.

Most people are not sensitive to pitch differences of less than about 2 cents.

EXAMPLE 9 **Equal-Tempered Scale**

In an "equal-tempered" scale for a given reference frequency f_0 of a note, find the frequency of the next higher-pitched semitone.

Solution

For the given reference frequency f_0, let f be the frequency of the next higher-pitched semitone. Since in an equal-tempered scale every semitone is 100 cents, we have $P(f) = 100$.

$$P(f) = 1200 \log_2\left(\frac{f}{f_0}\right) \qquad \text{Change in pitch formula}$$

$$100 = 1200 \log_2\left(\frac{f}{f_0}\right) \qquad \text{Replace } P(f) \text{ with 100.}$$

$$\frac{1}{12} = \log_2\left(\frac{f}{f_0}\right) \qquad \text{Divide by 1200 and simplify.}$$

$$\frac{f}{f_0} = 2^{\frac{1}{12}} = \sqrt[12]{2} \qquad \text{Exponential form}$$

$$f = \sqrt[12]{2}\, f_0 \qquad \text{Multiply by } f_0.$$

Practice Problem 9 In an "equal-tempered" scale, the notes are denoted as:

$$\text{A A\# B C C\# D D\# E F F\# G G\# A}$$

If the frequency of A is 440 Hz, estimate the frequency of B.

The next example shows that an equal-tempered scale does not always produce the best harmonics.

EXAMPLE 10 **Harmony in Music**

A jump of one *perfect fifth* gives a frequency increase of 50%. An equal-tempered *fifth* is 700 cents. Find the difference in cents (rounded to the nearest cent). Is the difference noticeable?

Solution

If the reference frequency is f_0 and it increases by 50%, then $f = f_0 + 0.5f_0 = (1 + 0.5)f_0 = 1.5f_0$. So,

$$P(f) = 1200 \log_2(1.5 f_0/f_0) \qquad \text{Replace } f \text{ with } 1.5\, f_0.$$

$$= 1200 \log_2(1.5) \qquad \text{Simplify.}$$

$$= 1200 \frac{\log(1.5)}{\log 2} \qquad \text{Change to base 10.}$$

$$\approx 701.955 \qquad \text{Use a calculator.}$$

$$\approx 702.$$

Since a fifth on an equal-tempered scale is 700 cents, the difference is a little less than 2 cents. This difference is barely noticeable to most listeners.

Practice Problem 10 Repeat Example 10 if a perfect *major* third gives a frequency increase of 25% and an equal-tempered *major* third is 400 cents.

4 Define magnitude of star brightness.

Star Brightness

Two thousand years ago Greek astronomers Hipparchus and Ptolemy created a system to quantify the *apparent brightness* (wattage received on the surface of the Earth) of stars. They classified star brightness on a scale from magnitude 1 for the brightest visible star to magnitude 6 for the dimmest visible star. Other stars were assigned magnitudes based on their brightness compared with those of magnitude 1 or 6. This system was refined considerably, leading to a much larger range of magnitude values and the following definition.

APPARENT MAGNITUDE

If two stars of magnitudes m_1 and m_2 have apparent brightness b_1 and b_2, respectively, then

$$m_2 - m_1 = 2.5 \log\left(\frac{b_1}{b_2}\right).$$

Although this definition of apparent magnitude was originally intended for comparing the brightness of stars, it can also be used for objects such as planets and their moons. The Hubble telescope can detect stars with apparent magnitude 30.

EXAMPLE 11 Comparing Star Brightness

a. Compare the brightness of a magnitude 1 star with that of a magnitude 6 star.

b. Find the magnitude m of a star that is 650 times as bright as one of magnitude 7.25.

Solution

a.

$m_2 - m_1 = 2.5 \log\left(\dfrac{b_1}{b_2}\right)$	Apparent magnitude formula
$6 - 1 = \dfrac{5}{2} \log\left(\dfrac{b_1}{b_2}\right)$	$m_2 = 6, m_1 = 1,$ and $2.5 = \dfrac{5}{2}$
$2 = \log\left(\dfrac{b_1}{b_2}\right)$	Multiply both sides by $\dfrac{2}{5}$ and simplify.
$\dfrac{b_1}{b_2} = 10^2$	Exponential form
$b_1 = 10^2 b_2 = 100 b_2$	Multiply both sides by b_2.

So a star of magnitude 1 is 100 times brighter than a star of magnitude 6.

b. Letting $m_2 = m$, $m_1 = 7.25$ and $b_2 = 650 b_1$ in the formula, we have

$$m - 7.25 = 2.5 \log\left(\frac{b_1}{650 b_1}\right)$$

$$m - 7.25 = 2.5 \log\left(\frac{1}{650}\right) = 2.5 \log(650)^{-1}$$

$m - 7.25 = -2.5 \log(650)$	Power rule for logs
$m = 7.25 - 2.5 \log(650)$	Add 7.25 to both sides.
$m \approx 0.2177.$	Use a calculator.

Practice Problem 11

a. Compare the brightness of a magnitude 0 star with that of a magnitude 2 star.

b. Find the magnitude *m* of a star that is 50% brighter than a star of magnitude 4.6.

5 Build models from data.

Modeling

In Section 1.2, we discussed modeling the relationship between two variables (*x* and *y*) when the scatterplot of the corresponding data points appears to be linear. The linear least-squares technique ("regression line" $y = ax + b$) is applied to best fit a line for such a set of points. However, if the scatterplot of the corresponding data points does not appear to be linear, then it is common practice to transform the data in such a way that the resulting plot resembles a straight line. Logarithms can be helpful in transforming the data.

We can apply logarithm transformations to either the *y* variable, the *x* variable, or both the *x* and *y* variables at the same time. If we find that one of these transformations results in a linear relationship that produces a regression line, we can deduce that the relationship between the original variables is exponential, logarithmic, or power, respectively. The transformations and resulting models are summarized in the following table:

x variable	*y* variable	Regression line	Model
x	y	$y = ax + b$	**Linear:** $y = ax + b$
x	$\ln y$	$\ln y = ax + b$	**Exponential:** $y = ce^{ax}$, with $c = e^b$
$\ln x$	y	$y = a \ln x + b$	**Logarithmic:** $y = a \ln x + b$
$\ln x$	$\ln y$	$\ln y = a \ln x + b$	**Power:** $y = cx^a$ with $c = e^b$

The graphical representations of scattered data and the corresponding models are shown in Figure 3.16.

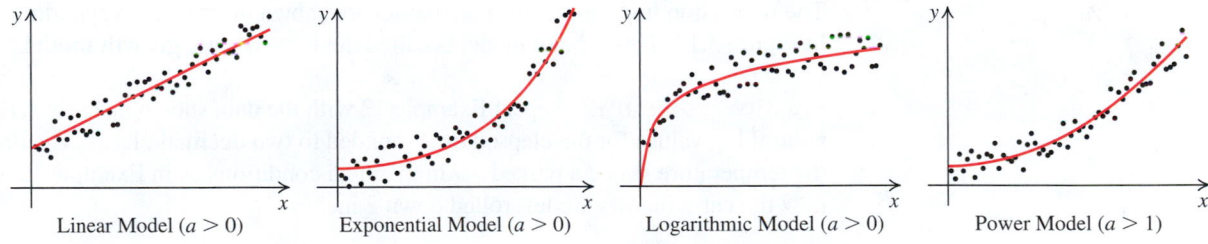

Linear Model ($a > 0$)   Exponential Model ($a > 0$)   Logarithmic Model ($a > 0$)   Power Model ($a > 1$)

Figure 3.16

The resulting models can be used to summarize the data, to predict unobserved values, or to understand the mechanisms that produce the observed values.

EXAMPLE 12   **Temperature Inside a Parked Car**

The data in Table 3.11 show the temperature inside a parked car on a hot day (90°F) with the windows closed and taken in 5-minute intervals. The initial temperature in the cabin was 75°F.

TABLE 3.11

Elapsed Time	5	10	15	20	25	30	35	40	45	50	55	60	65	70	75	80	85	90
Temperature	101	110	118	122	126	130	133	136	136	138	140	141	142	142	144	144	145	145
ln (Elapsed Time)	1.61	2.30	2.71	3.00	3.22	3.40	3.56	3.69	3.81	3.91	4.01	4.09	4.17	4.25	4.32	4.38	4.44	4.50

Source: Based on RACQ study.

Set x as the elapsed time and y as the temperature inside a car. Apply the logarithmic transformation to x. Create a scatterplot of data (y versus x) as well as a scatterplot of transformed data (y vs. $\ln x$) and find the logarithmic model that fits these data.

Solution

The corresponding scatterplots of the data are presented in Figure 3.17.

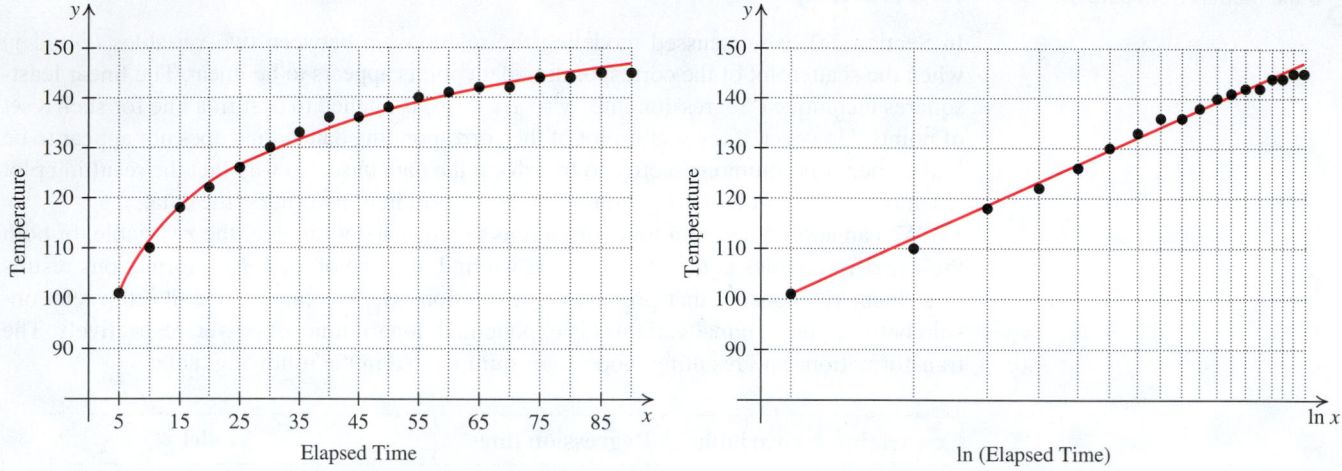

Figure 3.17

From the scatterplot of the transformed data, we see that the logarithmic model was a good choice. Using a graphing calculator (see page 32), we find the regression line (in the variables $Y = y$ and $X = \ln x$) for the transformed data:

$$Y = 74.92 + 15.98\,X.$$

From that we find the coefficients of the logarithmic model that fit the original data:

$$y = 74.92 + 15.98\,\ln x.$$

The regression line and logarithmic model are shown on their corresponding scatterplots in Figure 3.17. This type of model is called the **logarithmic growth model**.

Practice Problem 12 Repeat Example 12 with the data shown in Table 3.12. Use natural log values for the elapsed time rounded to two decimal places. The data show the temperature inside a parked car in identical conditions as in Example 12, except now the car windows are left rolled down 2 inches.

TABLE 3.12

Elapsed Time	5	10	15	20	25	30	35	40	45	50	55	60	65	70	75	80	85	90
Temperature	92	98	103	108	111	114	116	118	116	118	121	121	122	122	123	123	123	125

Source: Based on RACQ study.

In the next example, we investigate Kleiber's Law. In 1932, Max Kleiber observed that, for the vast majority of animals, an animal's metabolic rate q varies directly with the $\frac{3}{4}$ power of the animal's mass M, that is, $q = c \cdot M^{3/4}$.

SIDE NOTE

The basal metabolic rate is the amount of energy expended while at rest to support only the functioning of the vital organs.

EXAMPLE 13 **Body Weight versus Metabolic Rate for Small Animals**

The data in Table 3.13 show the body weight (in grams, g) and the corresponding basal metabolism (in watts, W) for fifteen different species of small animals.

TABLE 3.13

Body Weight (g)	16	67	120	206	336	490	598	828	1120	1551	1782	2250	2750	3257	3600
Metabolism (W)	0.13	0.33	0.58	1.08	1.19	1.24	1.78	2.21	2.84	3.23	3.01	5.07	6.81	6.05	7.74
ln (Body Weight)	2.77	4.20	4.79	5.33	5.82	6.19	6.39	6.72	7.02	7.35	7.48	7.72	7.92	8.09	8.18
ln (Metabolism)	−2.04	−1.11	−0.545	0.077	0.174	0.215	0.577	0.793	1.04	1.17	1.10	1.62	1.92	1.80	2.05

Source: Based on data from "Size and Power ln Mammals" by A. Hausner, *J. exp. Biol.* 160, 25–54 (1991).

Set x as the body weight and y as the metabolism rate. Apply the logarithmic transformation to both x and y. Create a scatterplot of data (y vs. x) as well as a scatterplot of transformed data (ln y vs. ln x) and find the power model that fits these data.

Solution

The corresponding scatterplots of the data are presented in Figure 3.18.

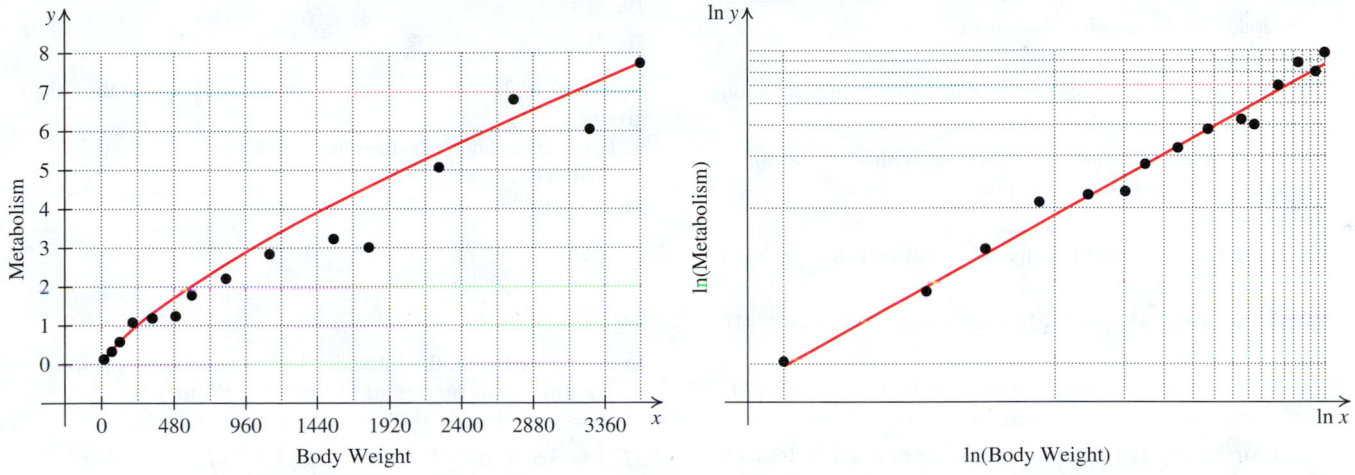

Figure 3.18

From the scatterplot of the transformed data, we see that the power model is a good choice. Using a graphing calculator (see page 32), we find the regression line (in variables $Y = \ln y$ and $X = \ln x$) for the transformed data:

$$Y = -4.09 + 0.73X.$$

From that, we find the coefficients of the power model that fit the original data:

$$y = e^{-4.09}x^{0.73} \approx 0.0167x^{0.73}.$$

The regression line and power model are shown with their corresponding scatterplots in Figure 3.18. Note that the power exponent 0.73 we obtain is very close to the $\frac{3}{4} = 0.75$ exponent in Kleiber's Law. This type of model is called the **power growth model**.

Practice Problem 13 Repeat Example 13 with the data shown in Table 3.14.

TABLE 3.14

Body Weight (kg)	1.1	1.5	2.4	3.6	4.0	4.3	5.7	6.5	6.9	7.7	13.6	15.9	25	30	45.2
Metabolism (W)	2.23	2.79	3.65	7.74	5.63	6.49	5.11	6.52	8.23	11.03	14.98	12.54	14.08	34.46	46.85

Source: Based on data from "Size and Power ln Mammals" by Hausner, *J. exp. Biol.* 160, 25–54 (1991).

SECTION 3.5   Exercises

Concepts and Vocabulary

1. On the Richter scale the magnitude of an earthquake
$M = $ _____ . The energy E released by an earthquake of magnitude M is given by $\log E = $ _____ .

2. The loudness L of sound of intensity I is given by
$L = $ _____ , and $I_0 = 10^{-12}$ W/m^2.

3. If f_0 is a reference frequency and f is the frequency of a note, then the change in pitch $P(f) = $ _____ .

4. If two stars have magnitudes m_1 and m_2 with apparent brightness b_1 and b_2, respectively, then
$m_2 - m_1 = $ _____ .

5. True or False. If the pH value of a solution is bigger than 7, the solution is acidic.

6. True or False. The acidity of a solution increases as its pH value increases.

7. True or False. For a one-unit increase in the magnitude of an earthquake, its intensity increases tenfold.

8. True or False. The apparent brightness of a star increases if its magnitude decreases.

Building Skills

In Exercises 9–12, the concentration [H$^+$] of a substance is given. Calculate the pH value of each substance and classify each substance as an acid or a base.

9. $[\text{H}^+] = 10^{-8}$

10. $[\text{H}^+] = 10^{-4}$

11. $[\text{H}^+] = 2.3 \times 10^{-5}$

12. $[\text{H}^+] = 4.7 \times 10^{-9}$

In Exercises 13–16, use the following information. Negative ions, designated by the notation [OH$^-$], are always present in any acid or base. The concentration, [OH$^-$], of these ions is related to [H$^+$] by the equation $[\text{OH}^-] \cdot [\text{H}^+] = 10^{-14}$ moles per liter. Find the concentrations of [OH$^-$] and [H$^+$] in moles per liter for the substances with the following pH values.

13. pH $= 6$

14. pH $= 8$

15. pH $= 9.5$

16. pH $= 3.7$

In Exercises 17–20, the magnitude M of an earthquake is given.
a. Find the earthquake intensity I in terms of the zero-level earthquake intensity I_0.
b. Find the energy released by the earthquake.

17. $M = 5$

18. $M = 2$

19. $M = 7.8$

20. $M = 3.7$

In Exercises 21–24, the energy E released by an earthquake is given.
a. Find the earthquake's magnitude.
b. Find the earthquake's intensity.

21. $E = 10^{13.4}$ joules

22. $E = 10^{10.4}$ joules

23. $E = 10^{12}$ joules

24. $E = 10^{9}$ joules

In Exercises 25–28, the intensity I of a sound is given. Find the loudness L of the sound ($I_0 = 10^{-12}$ W/m^2)

25. $I = 10^{-8}$ W/m^2 **26.** $I = 10^{-10}$ W/m^2

27. $I = 3.5 \times 10^{-7}$ W/m^2 **28.** $I = 2.37 \times 10^{-5}$ W/m^2

In Exercises 29–32, a loudness L is given. Find the intensity of the sound.

29. $L = 80$ dB   **30.** $L = 90$ dB

31. $L = 64.7$ dB   **32.** $L = 37.4$ dB

In Exercises 33–36, use the "equal temper" scale with the 12 semitones A, A#, B, C, C#, D, D#, E, F, F#, G, and G# and the frequency of A is 440 Hz. On a piano keyboard, the distance between two white keys that are side by side is a whole note if there is a black key between them, and it is a semitone if there is no black key between them. A black key (such as the one between C and D) is called a sharp. The black key between C and D is denoted by C#.

Remember that the change in frequency $P(f)$ between any two consecutive semitones is 100 cents.

33. Find the frequencies of A# and C.

34. Find the frequencies of D and E.

35. A whole tone is a frequency ratio of 10:9. That is, $\dfrac{f}{f_0} = \dfrac{10}{9}$.

An equal-tempered whole tone is 200 cents. Find the difference in cents between a whole tone and "an equal-tempered whole tone." Is the difference noticeable?

36. Repeat Exercise 35 if a whole tone is a frequency ratio of $9:8$.

37. The magnitudes of two stars, A and B, are 4 and 20, respectively. Compare the brightness of these stars.

38. Repeat Exercise 37 for two stars with magnitudes -2 and 5.

39. The magnitude of star A is 2 more than that of star B. How is their corresponding brightness related?

40. The magnitude of star A is 1 more than that of star B. How is their corresponding brightness related?

Applying the Concepts

41. pH of human blood. The hydrogen ion concentration $[H^+]$ of blood is approximately 3.98×10^{-8} moles per liter. Find the pH value of human blood to the nearest tenth. Is human blood acidic or basic?

42. pH of milk. The hydrogen ion concentration $[H^+]$ of milk is approximately 3.16×10^{-7} moles per liter. Find the pH value of milk to the nearest tenth. Is milk acidic or basic?

43. pH of some common substances. The pH value of a sample of each substance is given. Find $[H^+]$ of the substance.
  a. Wine grapes: pH $= 3.15$
  b. Water: pH $= 7.2$
  c. Eggs: pH $= 7.78$
  d. Vinegar: pH $= 3$

44. pH of some common substances. The pH value of a sample of each substance is given. Find $[H^+]$ of the substance.
  a. Battery acid: pH $= 1.0$
  b. Baking soda: pH $= 8.7$
  c. Ammonia: pH $= 10.6$
  d. Stomach acid: pH $= 2.3$

45. Acid rain. In the Netherlands, the average pH value of the rainfall is 3.8. Find the average concentration of hydrogen ions $[H^+]$ in the rainfall. How much more acidic is this rain than ordinary rain with a pH value of 6?

46. pH of a solution. Suppose the pH value of a solution A is 1.0 more than the pH value of a solution B. How are the concentrations of hydrogen ions in the two solutions related?

47. Comparing pH of solutions. Suppose the hydrogen ion concentration in a solution A is 100 times that in solution B. How are the pH values of the two solutions related?

48. pH of a solution. Suppose the pH value of a solution is increased by 1.5. How much change does this represent in the hydrogen ion concentration of this solution? Does the increase in pH make the solution more acidic or more basic?

49. pH of a solution. Suppose the hydrogen ion concentration of a solution is increased 50 times. How much change does this represent in the pH value of this solution? Does this increase in $[H^+]$ make the solution more acidic or more basic?

50. China earthquake. The Great China Earthquake of 1920 registered 8.6 on the Richter scale.

  a. What was the intensity of this earthquake?
  b. How many joules of energy were released?

51. San Francisco earthquake. Repeat Exercise 50 for the Great San Francisco Earthquake of 1906, which registered 7.8 on the Richter scale.

52. Comparing earthquakes. Table 3.9 lists two earthquakes in Japan.
  a. Compare the intensities of these earthquakes.
  b. Compare the energies released by these quakes.

53. Comparing earthquakes. Suppose earthquake A registers one more point on the Richter scale than earthquake B.
  a. How are their corresponding intensities related?
  b. How are their released energies related?

54. Comparing earthquakes. Repeat Exercise 53, if earthquake A registers 1.5 more points on the Richter scale than earthquake B.

55. Comparing magnitudes. If one earthquake is 150 times as intense as another, what is the difference in the Richter scale readings of the two earthquakes?

56. Comparing magnitudes. If the energy released by one earthquake is 150 times that of another, what is the difference in the Richter scale readings of the two earthquakes?

In Exercises 57–64, use Table 3.10 as necessary.

57. dB and intensity. What is the loudness, in decibels, of a radio with intensity 5.2×10^{-5} W/m^2?

58. dB of a jet. What is the loudness of a jet aircraft with intensity 2.5×10^2 W/m^2?

59. Comparing two sounds. How much more intense is a sound of the threshold of pain (130 dB) than a conversation at 65 dB?

60. Comparing two sounds. How much more intense is a 75-dB sound than a 62-dB sound?

61. Comparing two sounds. Suppose a sound is 1000 times as intense as one at the threshold of pain for the human ear. Find the loudness of this sound in decibels.

62. Comparing intensities. Show that if two sounds of intensity I_1 and I_2 register decibel levels of L_1 and L_2 respectively, then $\dfrac{I_2}{I_1} = 10^{0.1(L_2 - L_1)}$.

63. Comparing intensities. Suppose one sound registers one decibel more than another. How are the intensities of the two sounds related?

64. Comparing sounds. The noise level of a street sound outside a new office building in downtown Chicago is measured to be approximately 70 dB. Using special insulation material, the noise level inside the office building is reduced to 29 dB. Compare the intensities of sounds outside and inside the building.

65. Violin tuning. A violinist is tuning her A-string to match concert pitch. The concert's A is a frequency of 440 Hz. Her A-string is improperly tuned to 441 Hz. Find the difference in cents.

66. Violin tuning. Repeat Exercise 65 if her A-string is improperly tuned to 438 Hz.

In Exercises 67–72, use the following table of approximate apparent magnitudes of some celestial objects.

Object	The Sun	Full Moon	Venus	Sirius	Vega	Saturn	Neptune
Magnitude	−27	−13	−4	−1	0.03	1.47	7.8

67. Sun–Moon. How many times brighter is the Sun than the full Moon?

68. Moon–Vega. How many times brighter is the full Moon than Vega?

69. Sun–Venus. How many times brighter is the Sun than Venus?

70. Venus–Neptune. How many times brighter is Venus than Neptune?

71. Saturn–Star. What is the magnitude of a star that is 560 times brighter than Saturn?

72. Neptune–Star. What is the magnitude of a star that is one billion times brighter than Neptune?

 73. Solar panels. The data in the following table show the number of annual solar PV cells installations (in megawatts [MW]) over the period 2004–2014.

Year	2004	2005	2006	2007	2008	2009
PV Installations	58	79	105	160	298	382

Year	2010	2011	2012	2013	2014
PV Installations	852	1922	3369	4776	6201

Source: Solar Energy Industries Association, www.seia.org

Find the exponential model that fits the data, letting the year 2004 correspond to $x = 1$. *This type of model is called the exponential growth model.*

 74. CD sales. The data in this table show annual CD album sales in the United States (in millions of units) over the period, 2004–2014.

Year	2004	2005	2006	2007	2008	2009	2010	2011	2012	2013	2014
CD Sales	651	599	553	449	361	295	237	224	193	165	141

Source: Nielsen Soundscan.

Find the exponential model that fits the data, letting the year 2004 correspond to $x = 1$. *This type of model is called the exponential decay model.*

 75. NFL draft. The data in the following table show the estimated Career Value Rating in relation to the Draft Pick Number for the NFL averaged over the last 25 years.

Pick Number	1	5	10	15	20	25	30	40	50	100	200
Career Value Rating	64.8	51.4	44	38.2	36	32.1	33.1	27.7	27.4	18.2	10.5

Source: http://www.pro-football-reference.com

Find the logarithmic model that fits the data. *This type of model is called the logarithmic decay model.*

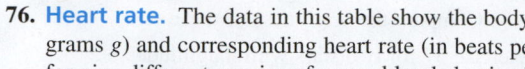 **76. Heart rate.** The data in this table show the body weight (in grams g) and corresponding heart rate (in beats per minute) for nine different species of warm-blooded animals at rest.

Weight	25	60	200	341	1100	2000	5000	90000
Heart Rate	670	450	420	378	190	150	120	60

Find the power model that fits the data. *This type of model is called the power decay model.*

Beyond the Basics

77. Decibel levels. The decibel level of Professor Stout's voice decreases with the distance from the professor according to the formula

$$L = 10 \log\left(\frac{4 \times 10^4}{r^2}\right),$$

where L is the decibel level and r is the distance in feet from the professor to the listener.

a. Find the decibel levels at distances of 10, 25, 50, and 100 feet.

b. How far must a listener be from the professor so that the decibel level drops to 0?

c. Express L in the form $L = a + b \log r$, for suitable constants a and b.

d. Express r as a function of L.

78. Sound intensity. The intensity of sound is inversely proportional to the square of the distance. That is, $I = \dfrac{K}{r^2}$, where I is the intensity at a distance r from the source and K is a constant.

a. Show that if L_1 and L_2 are decibel readings of a sound at distances r_1 and r_2 (in meters), respectively, then

$$L_1 - L_2 = 20 \log\left(\frac{r_2}{r_1}\right).$$

b. A military jet at 30 meters has 140 decibel level of loudness. How loud will you hear this jet sound at a distance of 100 meters? 300 meters?

c. Express $\dfrac{r_2}{r_1}$ in the exponential form.

79. Nuclear bomb. Compare the energy released by a one-megaton nuclear bomb (about 5×10^{15} joules) to the energy released by a great earthquake of magnitude 8.

80. Adriana was calculating the pitch of a sound and mistakenly transposed the frequency f of the sound with the reference frequency f_0. The correct answer is 40 cents. What was her answer?

81. If the A-string of a violin is tuned to 880 Hz with an error of ± 10 cents, what is the frequency range of the A-string?

82. Show that the formula for star brightness can be written as $\dfrac{b_1}{b_2} = 10^{0.4(m_2 - m_1)}$.

83. A star of magnitude m is 176 times brighter than a star of magnitude 3.42. Find m.

84. Star luminosity. The luminosity L of a star is the power output at a star's surface (measured in watts). The star's apparent brightness b is the wattage received on Earth's surface. The apparent magnitude m of a star depends on its absolute magnitude M (related to L) and its distance D in parsecs (1 parsec $\approx$ 3.26 light years) by the formula:

$$m - M = 2.5 \log \frac{L}{b} = 5 \log \frac{D}{10}.$$

a. Given m and M for a star, show that its distance D from Earth is $D = 10^{1+(m-M)/5}$ parsecs.

b. For the star Alpha Centauri, $m = 0$ and $M = 4.39$. How far is it from Earth?

c. Compare the luminosity and brightness of Alpha Centauri.

Getting Ready for the Next Section

In Exercises 85–88, assume that the relationship between two variables d and x is given by the equation

$$\pi d = 180x.$$

85. Find d if $x = \dfrac{\pi}{4}$.

86. Find d if $x = \dfrac{\pi}{3}$.

87. Find x if $d = 30$.

88. Find x if $d = 90$.

In Exercises 89–92, state the quadrant, if any, in which each point is located

89. $\left(-\dfrac{1}{2}, \dfrac{\sqrt{3}}{2}\right)$

90. $\left(\dfrac{1}{2}, -\dfrac{\sqrt{3}}{2}\right)$

91. $\left(\dfrac{1}{2}, \dfrac{\sqrt{3}}{2}\right)$

92. $\left(-\dfrac{1}{2}, -\dfrac{\sqrt{3}}{2}\right)$

SUMMARY Definitions, Concepts, and Formulas

3.1 Exponential Functions

i. A function $f(x) = a^x$, with $a > 0$ and $a \neq 1$, is called an exponential function with base a.

Rules of exponents: $a^x a^y = a^{x+y}$, $\dfrac{a^x}{a^y} = a^{x-y}$, $(a^x)^y = a^{xy}$,

$$a^0 = 1, a^{-x} = \frac{1}{a^x}$$

ii. Exponential functions are one-to-one: If $a^u = a^v$, then $u = v$.

iii. If $a > 1$, then $f(x) = a^x$ is an increasing function; $f(x) \to \infty$ as $x \to \infty$ and $f(x) \to 0$ as $x \to -\infty$.

iv. If $0 < a < 1$, then $f(x) = a^x$ is a decreasing function; $f(x) \to 0$ as $x \to \infty$ and $f(x) \to \infty$ as $x \to -\infty$.

v. The graph of $f(x) = a^x$ has y-intercept 1, and the x-axis is a horizontal asymptote.

vi. Simple interest formula. If P dollars is invested at an interest rate r per year for t years, then the simple interest is given by the formula $I = Prt$. The future value $A(t) = P + Prt$.

vii. Compound interest. P dollars invested at an annual rate r compounded n times per year for t years amounts to

$$A(t) = P\left(1 + \frac{r}{n}\right)^{nt}.$$

viii. The Euler constant $e = \lim\limits_{h \to \infty}\left(1 + \dfrac{1}{h}\right)^h \approx 2.718$.

ix. Continuous compounding. P dollars invested at an annual rate r compounded continuously for t years amounts to $A = Pe^{rt}$.

x. The function $f(x) = e^x$ is the natural exponential function.

3.2 Logarithmic Functions

i. For $a > 0$ and $a \neq 1$, $y = \log_a x$ if and only if $x = a^y$.

ii. Basic properties: $\log_a a = 1$, $\log_a 1 = 0$,

$$\log_a a^x = x, \quad a^{\log_a x} = x \qquad \text{Inverse properties}$$

iii. The domain of $\log_a x$ is $(0, \infty)$, the range is $(-\infty, \infty)$, and the y-axis is a vertical asymptote. The x-intercept is 1.

iv. Logarithmic functions are one-to-one: If $\log_a x = \log_a y$, then $x = y$.

v. If $a > 1$, then $f(x) = \log_a x$ is an increasing function; $f(x) \to \infty$ as $x \to \infty$ and $f(x) \to -\infty$ as $x \to 0^+$.

vi. If $0 < a < 1$, then $f(x) = \log_a x$ is a decreasing function; $f(x) \to -\infty$ as $x \to \infty$ and $f(x) \to \infty$ as $x \to 0^+$.

vii. The common logarithmic function is $y = \log x$ (base 10); the natural logarithmic function is $y = \ln x$ (base e).

3.3 Rules of Logarithms

i. Rules of logarithms:

$$\log_a MN = \log_a M + \log_a N \qquad \text{Product rule}$$

$$\log_a \frac{M}{N} = \log_a M - \log_a N \qquad \text{Quotient rule}$$

$$\log_a M^r = r \log_a M \qquad \text{Power rule}$$

ii. Change-of-base formula:

$$\log_b x = \frac{\log_a x}{\log_a b} = \frac{\log x}{\log b} = \frac{\ln x}{\ln b}$$

$$\text{(base } a\text{)} \quad \text{(base 10)} \quad \text{(base } e\text{)}$$

3.4 Exponential and Logarithmic Equations and Inequalities

An *exponential equation* is an equation in which a variable occurs in one or more exponents.

A *logarithmic equation* is an equation that involves the logarithm of a function of the variable.

Exponential and logarithmic equations are solved by using some or all of the following techniques:

i. Using the one-to-one property of exponential and logarithmic functions

ii. Converting from exponential to logarithmic form or vice versa

iii. Using the Product, Quotient, and Power rules for exponents and logarithms

We use similar techniques to solve logarithmic and exponential inequalities

3.5 Logarithmic Scales; Modeling

A *logarithmic scale* is a scale in which logarithms are used to measure quantities.

i. $pH = -\log[H^+]$, where $[H^+]$ is the concentration of H^+ ions in moles per liter.

ii. The Richter scale is used to measure the *magnitude M* of an earthquake.

iii. $M = \log\left(\dfrac{I}{I_0}\right)$, where I is the intensity of the earthquake and I_0 is the zero-level earthquake.

iv. The *energy E* released by an earthquake of magnitude M is given by $\log E = 4.4 + 1.5M$.

v. The *intensity* of a sound wave is defined as the amount of power the wave transmits through a given area. The *loudness* L of a sound of intensity I is given by: $L = 10\log\left(\dfrac{I}{I_0}\right)$, where $I_0 = 10^{-12}$ W/m^2 is the intensity of the threshold of hearing. Equivalently, $I = 10^{L/10}I_0$.

vi. If f is a frequency and f_0 is a reference frequency, then the change in pitch $P(f)$ in cents is given by
$$P(f) = 1200\log_2\left(\dfrac{f}{f_0}\right).$$

vii. If two stars of magnitudes m_1 and m_2 have apparent brightness b_1 and b_2, respectively, then $m_2 - m_1 = 2.5\log\left(\dfrac{b_1}{b_2}\right)$.

REVIEW EXERCISES

Concepts and Vocabulary

In Exercises 1–10, state whether the given statement is true or false.

1. The function $f(x) = a^x$ is exponential if $a > 0$.

2. The graph of $f(x) = 4^x$ approaches the x-axis as $x \to -\infty$.

3. The domain of $f(x) = \log(2 - x)$ is $(-\infty, 2]$.

4. The equation $u = 10^v$ means that $\log u = v$.

5. The inverse of $f(x) = \ln x$ is $g(x) = e^x$.

6. The graph of $y = a^x$ $(a > 0, a \neq 1)$ always contains the points $(0, 1)$ and $(1, a)$.

7. $\ln(M + N) = \ln M + \ln N$

8. $\log\sqrt{300} = 1 + \dfrac{1}{2}\log 3$

9. The functions $f(x) = 2^{-x}$ and $g(x) = \left(\dfrac{1}{2}\right)^x$ have the same graph.

10. $\ln u = \dfrac{\log u}{\log e}$

Building Skills

In Exercises 11–18, match the function with its graph in (a)–(h).

11. $f_1(x) = \log_2 x$

12. $f_2(x) = 2^x$

13. $f_3(x) = \log_2(3 - x)$

14. $f_4(x) = \log_{1/2}(x - 1)$

15. $f_5(x) = -\log_2 x$

16. $f_6(x) = \left(\dfrac{1}{2}\right)^x$

17. $f_7(x) = 3 - 2^{-x}$

18. $f_8(x) = \dfrac{6}{1 + 2e^{-x}}$

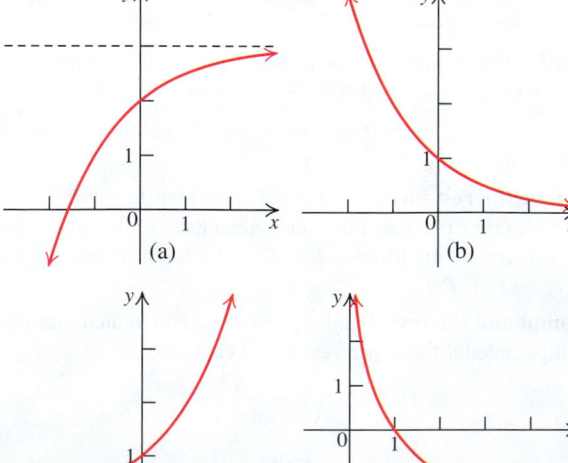

(a)

(b)

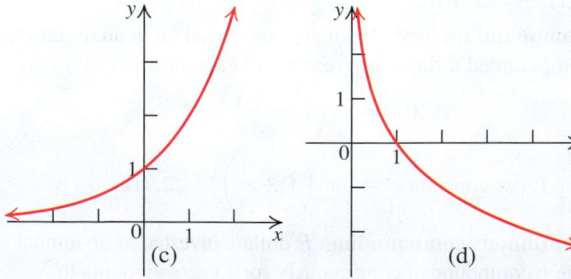

(c)

(d)

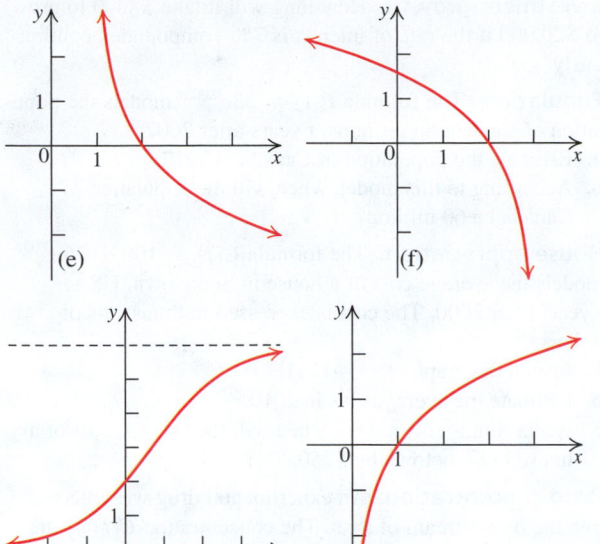

(e) (f)

(g) (h)

In Exercises 19–30, graph each function using transformations on an appropriate graph. Determine the domain, range, and asymptotes (if any).

19. $f(x) = 2^{-x}$

20. $g(x) = 2^{-0.5x}$

21. $h(x) = 3 + 2^{-x}$

22. $f(x) = 5^{|x|}$

23. $g(x) = 5^{-|x|}$

24. $h(x) = e^{-x+1}$

25. $f(x) = \ln(-x)$

26. $g(x) = 2 \ln |x|$

27. $h(x) = 2 \ln(x - 1)$

28. $f(x) = 2 - \left(\dfrac{1}{2}\right)^x$

29. $g(x) = 2 - \ln(-x)$

30. $h(x) = 3 + 2 \ln(5 + x)$

In Exercises 31–34, sketch the graph of the given function using these two steps:

a. Find the intercepts. **b.** Find the end behavior of f.

31. $f(x) = 3 - 2e^{-x}$

32. $f(x) = \dfrac{5}{2 + 3e^{-x}}$

33. $f(x) = e^{-x^2}$

34. $f(x) = 3 - \dfrac{6}{1 + 2e^{-x}}$

In Exercises 35–38, find a and k and then evaluate the function.

35. Let $f(x) = a(2^{kx})$ with $f(0) = 10$ and $f(3) = 640$. Find $f(2)$.

36. Let $f(x) = 50 - a(5^{kx})$ with $f(0) = 10$ and $f(2) = 0$. Find $f(1)$.

37. Let $f(x) = \dfrac{4}{1 + ae^{-kx}}$ with $f(0) = 1$ and $f(3) = \dfrac{1}{2}$. Find $f(4)$.

38. Let $f(x) = 6 - \dfrac{3}{1 + ae^{-kx}}$ with $f(0) = 5$ and $f(4) = 4$. Find $f(10)$.

In Exercises 39 and 40, find an exponential function of the form $f(x) = ca^x$ with the given graph.

39.

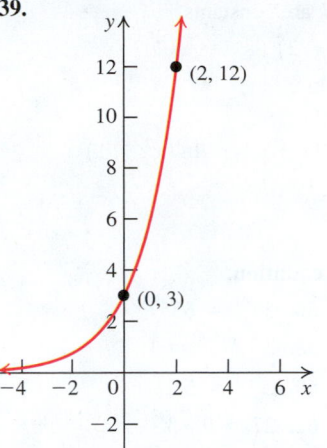

40.

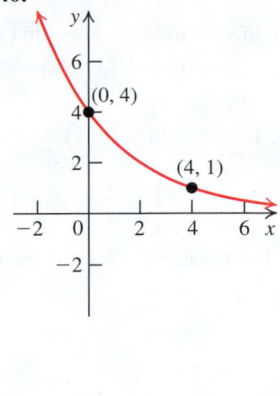

In Exercises 41 and 42, find a logarithmic function of the form $y = \log_a(x - c)$ with the given graph.

41.

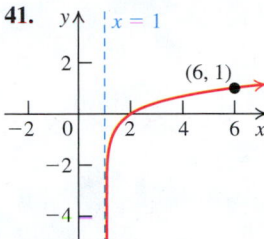

42.

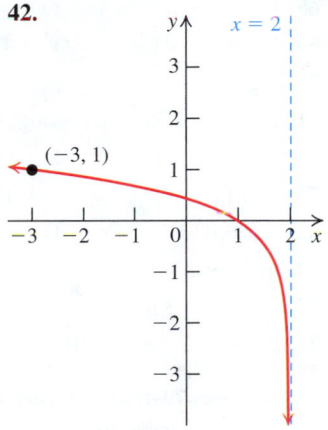

43. Start with the graph of $y = 2^x$. Find an equation of the graph that results when you

a. shift the graph right one unit, shift up three units, and reflect about the x-axis.

b. shift the graph right one unit, reflect about the x-axis, and shift up three units.

44. Start with the graph of $y = \ln x$. Find an equation of the graph that results when you

a. shift the graph left one unit, stretch horizontally by a factor of 2, and compress vertically by a factor of 3.

b. compress horizontally to $\dfrac{1}{3}$, shift left by one unit, and reflect about the y-axis.

In Exercises 45–48, write each logarithm in expanded form.

45. $\ln(xy^2z^3)$

46. $\log(x^3\sqrt{y - 1})$

47. $\ln\left[\dfrac{x\sqrt{x^2 + 1}}{(x^2 + 3)^2}\right]$

48. $\ln\sqrt{\dfrac{x^3 + 5}{x^3 - 7}}$

In Exercises 49–54, write y as a function of x.

49. $\ln y = \ln x + \ln 3$

50. $\ln y = \ln(C) + kx$; C and k are constants.

51. $\ln y = \ln(x - 3) - \ln(y + 2)$

52. $\ln y = \ln x - \ln(x^2 y) - 2\ln y$

53. $\ln y = \dfrac{1}{2}\ln(x - 1) + \dfrac{1}{2}\ln(x + 1) - \ln(x^2 + 1)$

54. $\ln(y - 1) = \dfrac{1}{x} + \ln(y)$

In Exercises 55–78, solve each equation.

55. $3^x = 81$

56. $5^{x-1} = 625$

57. $2^{x^2+2x} = 16$

58. $3^{x^2-6x+8} = 1$

59. $3^x = 23$

60. $2^{x-1} = 5.2$

61. $273^x = 19$

62. $27 = 9^x \cdot 3^{x^2}$

63. $3^{2x} = 7^x$

64. $2^{x+4} = 3^{x+1}$

65. $(1.7)^{3x} = 3^{2x-1}$

66. $3(2^{x+5}) = 5(7^{2x-3})$

67. $\log_3(x + 2) - \log_3(x - 1) = 1$

68. $\log_3(x + 12) - \log_3(x + 4) = 2$

69. $\log_2(x + 2) + \log_2(x + 4) = 3$

70. $\log_5(3x + 7) + \log_5(x - 5) = 2$

71. $\log_5(x^2 - 5x + 6) - \log_5(x - 2) = 1$

72. $\log_3(x^2 - x - 6) - \log_3(x - 3) = 1$

73. $\log_6(x - 2) + \log_6(x + 1) =$
$\log_6(x + 4) + \log_6(x - 3)$

74. $\ln(x - 2) - \ln(x + 2) = \ln(x - 1) - \ln(2x + 1)$

75. $2\ln 3x = 3\ln x$

76. $2\log x = \ln e$

77. $2^x - 8 \cdot 2^{-x} - 7 = 0$

78. $3^x - 24 \cdot 3^{-x} = 10$

In Exercises 79–82, solve each inequality.

79. $3(0.2)^x + 5 \le 20$

80. $-4(0.2)^x + 15 < 6$

81. $\log(2x + 7) < 2$

82. $\ln(3x + 5) \le 1$

Applying the Concepts

83. **Doubling your money.** How much time is required for a $1000 investment to double in value if interest is earned at the rate of 6.25% compounded annually?

84. **Tripling your money.** How much time (to the nearest month) is required to triple your money if the interest rate is 100% compounded continuously?

85. **Half-life.** The half-life of a certain radioactive substance is 20 hours. How long does it take for this substance to fall to 25% of its original value?

86. **Plutonium-210.** Find the half-life of plutonium-210 assuming that its decay equation is $Q = Q_0 e^{-5 \cdot 10^{-3}t}$, where t is in days.

87. **Comparing rates.** You have $7000 to invest for seven years. Which investment will provide the greater return, 5% compounded yearly or 4.75% compounded monthly?

88. **Investment growth.** How long will it take $8000 to grow to $20,000 if the rate of interest is 7% compounded continuously?

89. **Population.** The formula $P(t) = 33e^{0.003t}$ models the population of Canada, in millions, t years after 2007.
    a. Estimate the population of Canada in 2017.
    b. According to this model, when will the population of Canada be 60 million?

90. **House appreciation.** The formula $C(t) = 100 + 25e^{0.03t}$ models the average cost of a house in Sometown, USA, t years after 2000. The cost is expressed in thousands of dollars.
    a. Sketch the graph of $y = C(t)$.
    b. Estimate the average cost in 2010.
    c. According to this model, when will the average cost of a house in Sometown be $250,000?

91. **Drug concentration.** An experimental drug was injected into the bloodstream of a rat. The concentration $C(t)$ of the drug (in micrograms per milliliter of blood) after t hours was modeled by the function

$$C(t) = 0.3e^{-0.47t}, 0.5 \le t \le 10.$$

    a. Graph the function $y = C(t)$ for $0.5 \le t \le 10$.
    b. When will the concentration of the drug be 0.029 micrograms per milliliter? (1 microgram $= 10^{-6}$ gram)

92. **X-ray intensity.** X-ray technicians are shielded by a wall insulated with lead. The equation $x = \dfrac{1}{152}\log\left(\dfrac{I_0}{I}\right)$ measures the thickness x (in centimeters) of the lead insulation required to reduce the initial intensity I_0 of X-rays to the desired intensity I.
    a. What thickness of lead is required to reduce the intensity of X-rays to one-tenth their initial intensity?
    b. What thickness of lead is required to reduce the intensity of X-rays to $\dfrac{1}{40}$ their initial intensity?
    c. How much is the intensity reduced if the lead insulation is 10 centimeters thick?

93. **Cooling tea.** Chai (tea) is made by adding boiling water (212°F) to the chai mix. Suppose you make chai in a room with the air temperature at 75°F. According to Newton's Law of Cooling, the temperature of the chai t minutes after it is boiled is given by a function of the form $f(t) = 75 + ae^{-kt}$. After one minute, the temperature of the chai falls to 200°F. How long will it take for the chai to be drinkable at 150°F?

94. **Spread of influenza.** Approximately $P(t) = \dfrac{6}{1 + 5e^{-0.7t}}$ thousand people caught a new form of influenza within t weeks of its outbreak.
    a. Sketch the graph of $y = P(t)$.
    b. How many people had the disease initially?
    c. How many people contracted the disease within four weeks?
    d. If the trend continues, how many people in all will contract the disease?

95. Population density. The population density x miles from the center of a town called Greenville is approximated by the equation $D(x) = 5e^{0.08x}$, in thousands of people per square mile.
   a. What is the population density at the center of Greenville?
   b. What is the population density 5 miles from the center of Greenville?
   c. Approximately how far from the center of Greenville would the density be 15,000 people per square mile?

96. Population. In 2000, the population of the United States was 280 million and the number of vehicles was 200 million. If the population of the United States is growing at the rate of 1% per year while the number of vehicles is growing at the rate of 3%, in what year will there be an average of one vehicle per person?

97. Light intensity. The *Bouguer–Lambert Law* states that the intensity I of sunlight filtering down through water at a depth x (in meters) decreases according to the exponential decay function $I = I_0 e^{-kx}$, where I_0 is the intensity at the surface and $k > 0$ is an absorption constant that depends on the murkiness of the water. Suppose the absorption constant of Carrollwood Lake was experimentally determined to be $k = 0.73$. How much light has been absorbed by the water at a depth of 2 meters?

98. Signal strength. The strength of a TV signal usually fades due to a damping effect of cable lines. If I_0 is the initial strength of the signal, then its strength I at a distance x miles is measured by the formula $I = I_0 e^{-kx}$, where $k > 0$ is a damping constant that depends on the type of wire used. Suppose the damping constant has been measured experimentally to be $k = 0.003$. What percent of the signal is lost at a distance of 10 miles? 20 miles?

99. Walking speed in a city. In 1976, Marc and Helen Bernstein discovered that in a city with population p, the average speed s (in feet per second) that a person walks on main streets can be approximated by the formula

$$s(p) = 0.04 + 0.86 \ln p.$$

   a. What is the average walking speed of pedestrians in Tampa (population 470,000)?
   b. What is the average walking speed of pedestrians in Bowman, Georgia (population 450)?
   c. What is the estimated population of a town in which the estimated average walking speed is 4.6 feet per second?

100. Drinking and driving. Just after Eric had his last drink, the alcohol level in his bloodstream was 0.26 (milligram of alcohol per milliliter of blood). After one-half hour, his alcohol level was 0.18. The alcohol level $A(t)$ in a person follows the exponential decay law

$$A(t) = A_0 e^{-kt} \ (t \text{ in hours}),$$

where $k > 0$ depends on the individual.

   a. What is the value of A_0 for Eric?
   b. What is the value of k for Eric?
   c. If the legal driving limit for alcohol level is 0.08, how long should Eric wait (after his last drink) before he will be able to drive legally?

101. Bacteria culture. The mass $m(t)$, in grams, of a bacteria culture grows according to the logistic growth model

$$m(t) = \frac{6}{1 + 5e^{-0.7t}},$$

where t is time measured in days.
   a. What is the initial mass of the culture?
   b. What happens to the mass in the long run?
   c. When will the mass reach 5 grams?

102. A learning model. The number of units $n(t)$ produced per day after t days of training is given by

$$n(t) = 60(1 - e^{-kt}),$$

where $k > 0$ is a constant that depends on the individual.
   a. Estimate k for Rita, who produced 20 units after one day of training.
   b. How many units will Rita produce per day after ten days of training?
   c. How many days should Rita be trained if she is expected to produce 40 units per day?

103. Music. A jump of one perfect minor third gives a frequency increase of 20%. A perfect minor third is 300 cents. Find the difference in cents. Is the difference noticeable?

104. Star brightness. The magnitude of star A is 3 more than that of star B. How is their corresponding brightness related?

105. Acid rain. In Scotland, the lowest recorded pH level of rainfall was 2.4. What was the concentration of hydrogen ions $[H^+]$ in that rainfall? How much more acidic was this rain than a rain with a pH value of 5.6?

106. Comparing earthquakes. The 1997 earthquake in Iran registered 7.5 on the Richter scale and the 2010 earthquake in Chile registered 8.8 on the Richter scale.
   a. Compare the intensities of these earthquakes.
   b. Compare the energies released by these quakes.

107. Comparing two sounds. How much more intense is a 115-dB sound than a 95-dB sound?

108. Comparing intensities. If the intensity of one sound is 5000 times another, what is the difference in the decibel levels of the two sounds?

109. Sirius–Saturn. How many times brighter is Sirius than Saturn?

110. Nuclear bomb. What magnitude of earthquake releases energy equivalent to that of a one-megaton nuclear bomb (5×10^{15} joules)?

PRACTICE TEST A

1. Solve the equation $5^{-x} = 125$.

2. Solve the equation $\log_2 x = 5$.

3. State the range of $y = -e^x + 1$ and find the asymptote of its graph.

4. Evaluate $\log_2 \dfrac{1}{8}$.

5. Solve the exponential equation $\left(\dfrac{1}{4}\right)^{2-x} = 4$.

6. Evaluate $\log 0.001$ without using a calculator.

7. Rewrite the expression $\ln 3 + 5 \ln x$ in condensed form.

8. Solve the equation $2^{x+1} = 5$.

9. Solve the equation $e^{2x} + e^x - 6 = 0$.

10. Rewrite the expression $\ln \dfrac{2x^3}{(x+1)^5}$ in expanded logarithmic form.

11. Evaluate $\ln e^{-5}$.

12. Give the equation for the graph obtained by shifting the graph of $y = \ln x$ three units up and one unit right.

13. Sketch the graph of $y = 3^{x-1} + 2$.

14. State the domain of the function $f(x) = \ln(-x) + 4$.

15. Write $3 \ln x + \ln(x^3 + 2) - \dfrac{1}{2} \ln(3x^2 + 2)$ in condensed form.

16. Solve the equation $\log x = \log 6 - \log(x - 1)$.

17. Find x if $\log_x 9 = 2$.

18. Suppose $15,000 is invested in a savings account paying 7% interest per year. Write the formula for the amount in the account after t years if the interest is compounded quarterly.

19. Suppose the number of Hispanic people living in the United States is approximated by $H = 15,000 e^{0.2t}$, where $t = 0$ represents 1960. According to this model, in which year did the Hispanic population in the United States reach 1.5 million?

20. Compare the intensity of the earthquake in 1960 in Morocco of magnitude 5.8 to that of the 6.2 magnitude earthquake in 1972 in Nicaragua.

PRACTICE TEST B

1. Solve the equation $3^{-x} = 9$.

   a. $\{2\}$ **b.** $\{-2\}$ **c.** $\left\{\dfrac{1}{2}\right\}$ **d.** $\left\{-\dfrac{1}{2}\right\}$ **e.** $\{\ln 2\}$

2. Solve the equation $\log_5 x = 2$.

   a. $\{10\}$ **b.** $\{25\}$ **c.** $\left\{\dfrac{5}{2}\right\}$ **d.** $\left\{\dfrac{2}{5}\right\}$ **e.** $\{2\}$

3. State the range and asymptote of $y = e^{-x} - 1$.
   a. $(0, \infty); y = -1$ **b.** $(-1, \infty); y = -1$
   c. $(-\infty, \infty); y = -1$ **d.** $(-1, \infty); y = 1$
   e. $(\infty, 1); y = 1$

4. Evaluate $\log_4 64$.
   a. 16   **b.** 8   **c.** 2   **d.** 3   **e.** 4

5. Find the solution of the exponential equation $\left(\dfrac{1}{3}\right)^{1-x} = 3$.

   a. $\left\{-\dfrac{1}{3}\right\}$ **b.** $\left\{\dfrac{1}{3}\right\}$

   c. $\{1\}$ **d.** $\{2\}$

6. Evaluate $\log 0.01$.
   a. -1.99999999 **b.** -2 **c.** 2   **d.** 100

7. Which of the following expressions is equivalent to $\ln 7 + 2 \ln x$?
   a. $\ln(7 + 2x)$ **b.** $\ln(7x^2)$
   c. $\ln(9x)$ **d.** $\ln(14x)$

8. Solve the equation $2^{x^2} = 3^x$.

   a. $\{0\}$ **b.** $\left\{\dfrac{\ln 3}{\ln 2}\right\}$

   c. $\left\{0, \dfrac{\ln 3}{\ln 2}\right\}$ **d.** $\{0, \ln 3 - \ln 2\}$

9. Solve the equation $e^{2x} - e^x - 6 = 0$.
   a. $\{-\ln 2\}$ **b.** $\{-\ln 3\}$ **c.** $\{\ln 6\}$ **d.** $\{\ln 3\}$

10. Rewrite the expression $\ln \dfrac{3x^2}{(x+1)^{10}}$ in expanded logarithmic form.

    a. $\ln 6x - 10 \ln x + 1$
    b. $2 \ln 3x - 10 \ln(x + 1)$
    c. $2 \ln 3 + 2 \ln x - 10 \ln(x + 1)$
    d. $\ln 3 + 2 \ln x - 10 \ln(x + 1)$

11. Find $\ln e^{3x}$.
    a. 3   **b.** $3x$ **c.** $3 + x$ **d.** x

12. The equation for the graph obtained by shifting the graph of $y = \log_3 x$ two units up and three units left is
    a. $y = \log_3(x - 3) + 2$.   **b.** $y = \log_3(x + 3) - 2$.
    c. $y = \log_3(x - 3) - 2$.   **d.** $y = \log_3(x + 3) + 2$.

13. Which of the following graphs is the graph of $y = 5^{x+1} - 4$?

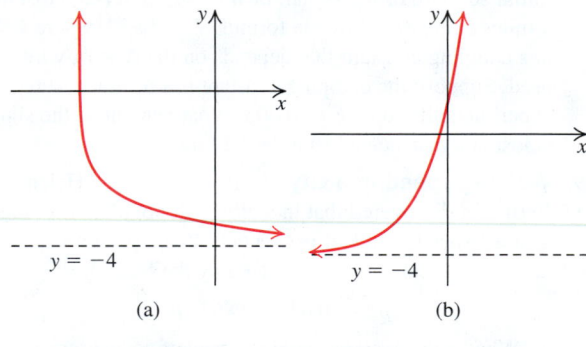

(a)   (b)

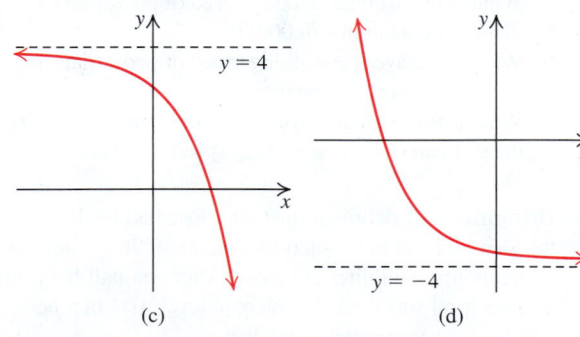

(c)   (d)

14. Find the domain of the function $f(x) = \ln(1 - x) + 3$.
    a. $(-\infty, 1)$ **b.** $(1, \infty)$ **c.** $(-\infty, -1)$
    d. $(-\infty, 3)$ **e.** $(3, \infty)$

15. Write $\ln x - 2 \ln(x^2 + 1) + \dfrac{1}{2} \ln(x^4 + 1)$ in condensed form.

    a. $\ln \dfrac{x\sqrt{x^4 + 1}}{(x^2 + 1)^2}$ **b.** $\ln \dfrac{x}{x^2 + 1}$

    c. $\ln(x - (x^2 + 1)^2 + (x^4 + 1)^{1/2})$

    d. $2 \ln \dfrac{x(1 + x^4)}{x^2 + 1}$

16. Solve the equation $\log x = \log 12 - \log (x + 1)$.

a. $\left\{\dfrac{11}{2}\right\}$ b. $\left\{\dfrac{13}{2}\right\}$ c. $\{3, -4\}$

d. $\{3\}$ e. $\left\{\dfrac{2}{12}\right\}$

17. Find x if $\log_x 16 = 4$.

a. 4 b. 2 c. 64 d. $\dfrac{1}{4}$ e. 16

18. Suppose \$12,000 is invested in a savings account paying 10.5% interest per year. Write the formula for the amount in the account after t years assuming that the interest is compounded monthly.

a. $A = 12,000(1.105)^t$ b. $A = 12,000(1.525)^{2t}$
c. $A = 12,000(1.2625)^{4t}$ d. $A = 12,000(1.00875)^{12t}$

19. The population of a certain city is growing according to the model $P = 10,000 \log_5 (t + 5)$, where t is time in years. If $t = 0$ corresponds to the year 2000, what will the population of the city be in 2020?

a. 30,000 b. 20,000
c. 50,000 d. 10,000

20. Compare the intensity of the earthquake in 1994 in Northridge, California, of magnitude 6.7 to that of the 7.0 earthquake in 1988 in Armenia.

a. The Northridge earthquake was about twice as intense.
b. The Armenia earthquake was about twice as intense.
c. The Northridge earthquake was about a thousand times as intense.
d. The Armenia earthquake was about a thousand times as intense.

CUMULATIVE REVIEW EXERCISES CHAPTERS 1–3

1. Find the x- and y-intercepts of the graph of the equation $x^2 + (y - 1)^2 = 1$. Check for symmetry with respect to both axes and the origin.

2. Graph the equation $y = x^3$.

3. Find the center and radius of the circle $x^2 + 2x + y^2 - 4y - 20 = 0$. Graph the circle.

4. Write the slope–intercept form of the equation of the line that passes through $(-2, 4)$ and is perpendicular to the line $2x + 3y = 17$.

5. Find the domain of $f(x) = \log \dfrac{x - 2}{x + 3}$.

6. Evaluate the function $f(x) = \begin{cases} 2x - 3, & x < 1 \\ 2x^2 + 1, & x \geq 1 \end{cases}$ at each value specified.
a. $f(-2)$ b. $f(0)$ c. $f(3)$.

7. Identify the basic function and use transformations to sketch the graph of the function $f(x) = 3\sqrt{x + 1} - 2$.

8. Let $f(x) = 3\sqrt{x}$ and $g(x) = -\dfrac{1}{x}$. Find and write the domain of $(f \circ g)(x)$ in interval notation.

9. Determine whether the function has an inverse function; if so, find the inverse function.
a. $f(x) = 2x^3 + 1$
b. $f(x) = |x|$
c. $f(x) = \ln x$.

10. Divide
a. $\dfrac{2x^3 + 3x + 1}{x^2 + 1}$ using long division
b. $\dfrac{2x^4 + 7x^3 - 2x + 3}{x + 3}$ using synthetic division

11. Find a polynomial function of least degree with integer coefficients that has the given zeros.
a. $1, 2, -3$
b. $1, -1, i, -i$

12. Use rules of logarithms to rewrite the following expressions in expanded form.
a. $\log_2 5x^3$
b. $\log_a \sqrt[3]{\dfrac{xy^2}{z}}$
c. $\ln \dfrac{3\sqrt{x}}{5y}$

13. Use rules of logarithms to rewrite the following expressions in condensed form.
a. $\dfrac{1}{2}(\log x + \log y)$ b. $3 \ln x - 2 \ln y$

14. Use a calculator to evaluate the expression. Round your answer to three decimal places.
a. $\log_9 17$ b. $\log_3 25$ c. $\log_{1/2} 0.3$

15. Solve each equation.
a. $\log_6 (x + 3) + \log_6 (x - 2) = 1$
b. $5^{x^2 - 4x + 5} = 25$
c. $3.1^{x-1} = 23$

16. Find all possible rational zeros of
$$f(x) = x^4 - x^3 - 2x^2 - 2x + 4.$$
Also find upper and lower bounds for the zeros of f.

17. Let $f(x) = (x - 1)^3(x + 2)^2(x - 3)$.
a. Find the zeros of f along with their multiplicities.
b. What is the behavior of f near each zero?
c. What is the end behavior of f?
d. Sketch the graph of f.

18. Let $f(x) = \dfrac{(x - 1)(x + 2)}{(x - 3)(x + 4)}$.
a. Find the vertical asymptotes of f (if any).
b. Find the horizontal asymptotes of f (if any).
c. Find the intervals over which $f(x) - 1 > 0$ or $f(x) - 1 < 0$.
d. Sketch the graph of f.

CHAPTER | 4

Trigonometric Functions

TOPICS

4.1 Angles and Their Measure

4.2 The Unit Circle; Trigonometric Functions of Real Numbers

4.3 Trigonometric Functions of Angles

4.4 Graphs of the Sine and Cosine Functions

4.5 Graphs of the Other Trigonometric Functions

4.6 Inverse Trigonometric Functions

Angles are vital for activities ranging from taking photographs to designing layouts for city streets. Scientists use angles and basic trigonometry to measure remote distances such as the distance between mountain peaks and the distances between planets. Angles are universally used to specify locations on Earth via longitude and latitude. In this chapter, we begin the study of trigonometry and its many uses.

Angles and Their Measure

BEFORE STARTING THIS SECTION, REVIEW

1 Circumference and area of a circle (Appendix A.5, page 952)

OBJECTIVES

1 Learn vocabulary associated with angles.

2 Use degree and radian measure of an angle.

3 Convert between degree and radian measure.

4 Find complements and supplements.

5 Find the length of an arc of a circle.

6 Find the area of a sector.

7 Compute linear and angular speed.

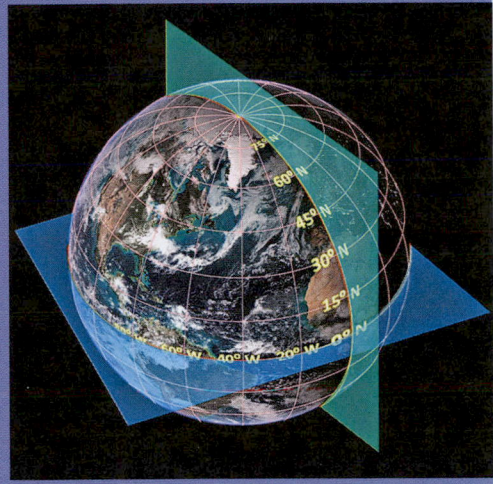

◆ Latitude and Longitude

Any location on Earth can be described by two numbers, its *latitude* and its *longitude*. To understand how these two location numbers are assigned, we think of Earth as being a perfect sphere.

Lines of latitude are parallel circles of different size around the sphere representing Earth. The longest circle is the equator, with latitude 0, and at the poles, the circles shrink to a point. Lines of longitude (also called *meridians*) are circles of identical size that pass through the North Pole and the South Pole as they circle the globe. Each of these circles crosses the equator. The equator is divided into 360 degrees, and the longitude of a location is the number of degrees assigned to the location where the meridian meets the equator. For historical reasons, the meridian near the old Royal Astronomical Observatory in Greenwich, England, is chosen as 0 longitude. Today this prime meridian is marked with a band of brass that stretches across the yard of the Observatory.

Longitude is measured from the prime meridian, with positive values going east (0 to 180) and negative values going west (0 to −180). Both ±180-degree longitudes share the same line, in the middle of the Pacific Ocean. At the equator, the distance on Earth's surface for each one degree of latitude or longitude is just over 69 miles.

Every location on Earth is then identified by the meridian and the latitude lines that pass through it. In Example 8, we explain how latitude values are assigned and how they can be used to compute distances between cities having the same longitude.

1 Learn vocabulary associated with angles.

Angles

A **ray** is a part of a line made up of a point, called the **endpoint**, and all of the points on one side of the endpoint. An **angle** is formed by rotating a ray about its endpoint. The angle's **initial side** is the ray's original position, and the angle's **terminal side** is the ray's position after the rotation. The endpoint is called the **vertex** of the angle. A curved arrow drawn near the angle's vertex indicates both the direction and amount of rotation from the initial side to the terminal side. See Figure 4.1.

The Greek letters α (alpha), β (beta), γ (gamma), and θ (theta) are often used to name angles. If the rotation is counterclockwise, the result is a **positive angle**; if the rotation

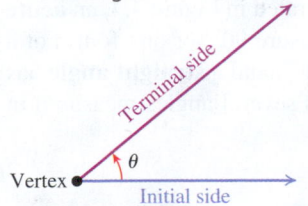

Figure 4.1 An angle.

is clockwise, the result is a **negative angle**. See Figure 4.2 (a) and (b). Rotation in both directions is unrestricted. Angles that have the same initial and terminal sides are called **coterminal** angles. In Figure 4.2 (c), α and β are different angles but are coterminal.

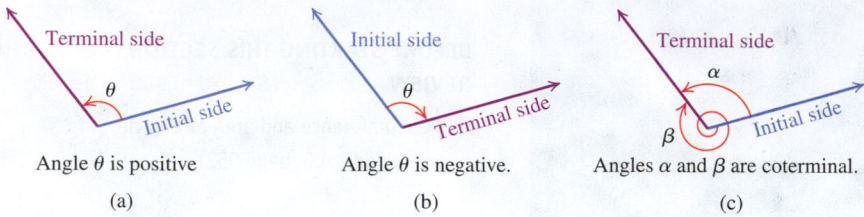

Angle θ is positive

(a)

Angle θ is negative.

(b)

Angles α and β are coterminal.

(c)

Figure 4.2 Positive, negative, and coterminal angles.

An angle in a rectangular coordinate system is in **standard position** if its vertex is at the origin and its initial side lies on the positive x-axis. All of the angles in Figure 4.3 are in standard position. An angle in standard position is **quadrantal** if its terminal side lies on a coordinate axis; it is said to **lie in a quadrant** if its terminal side lies in that quadrant. See Figure 4.3.

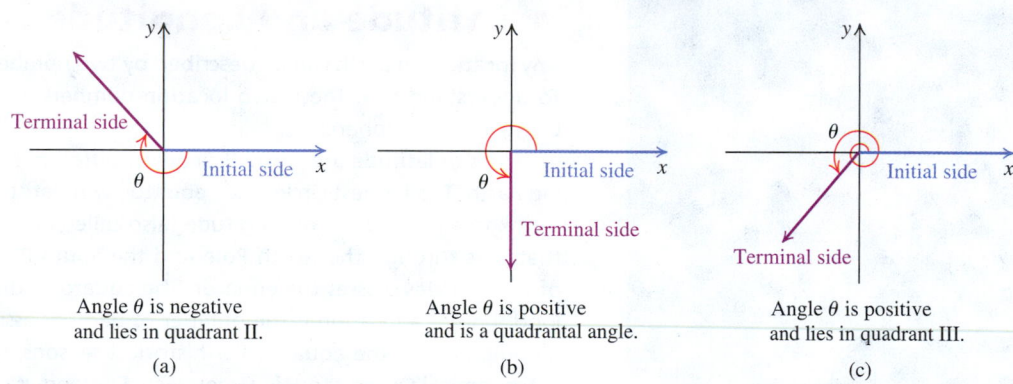

Angle θ is negative
and lies in quadrant II.

(a)

Angle θ is positive
and is a quadrantal angle.

(b)

Angle θ is positive
and lies in quadrant III.

(c)

Figure 4.3 Angles in standard position.

2 Use degree and radian measure of an angle.

Angle Measure

We measure angles by determining the amount of rotation from the initial side to the terminal side, indicated by a curved arrow that also shows direction. Two units of measurement for angles are *degrees* and *radians*.

Degree Measure

A measure of **one degree** (denoted by 1°) is assigned to an angle resulting from a rotation $\frac{1}{360}$ of a complete revolution counterclockwise about the vertex. An angle formed by rotating the initial side counterclockwise one full rotation so that the terminal and initial sides coincide has measure 360 degrees, written as 360°.

Angles are classified according to their measures. As illustrated in Figure 4.4, an **acute angle** has measure between 0° and 90°, a **right angle** has measure 90° (or one-fourth of a revolution), an **obtuse angle** has measure between 90° and 180°, and a **straight angle** has measure 180° (or half of a revolution). Figure 4.4 also shows several angles measured in degrees.

Figure 4.4 Degree measure of various angles.

The measure of an angle has no numerical limit because the terminal side can be rotated without limitation.

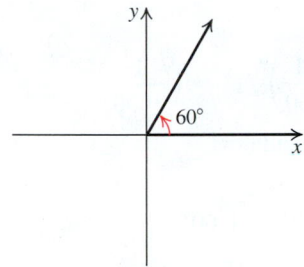

Figure 4.5

EXAMPLE 1 **Drawing an Angle in Standard Position**

Draw each angle in standard position. State the quadrant in which it lies.

a. 60°   **b.** 135°   **c.** −240°   **d.** 405°

Solution

a. Since $60 = \dfrac{2}{3}(90)$, a 60° angle is $\dfrac{2}{3}$ of a 90° angle. It lies in QI. See Figure 4.5.

b. Since $135 = 90 + 45$, a 135° angle is a counterclockwise rotation of 90°, followed by half of a 90° counterclockwise rotation. It lies in QII. See Figure 4.6.

c. Since $-240 = -180 - 60$, a −240° angle is a clockwise rotation of 180°, followed by a clockwise rotation of 60°. It lies in QII. See Figure 4.7.

d. Since $405 = 360 + 45$, a 405° angle is one complete counterclockwise rotation, followed by half of a 90° counterclockwise rotation. It lies in QI. See Figure 4.8.

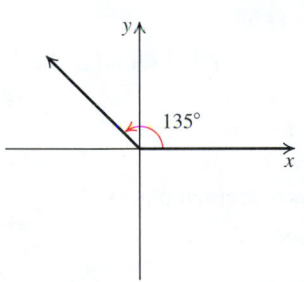

Figure 4.6

Practice Problem 1 Repeat Example 1 for a 225° angle.

Today it is common to divide degrees into fractional parts using *decimal degree notation* such as 30.5°. Traditionally, however, degrees were expressed in terms of *minutes* and *seconds*. One **minute**, denoted $1'$, is defined as $\dfrac{1}{60}(1°)$, and one **second**, denoted $1''$, is defined as $\dfrac{1}{60}(1')$. An angle measuring 27 degrees, 14 minutes, and 39 seconds is written as $27°14'39''$. Since $\dfrac{1}{60} \cdot \dfrac{1}{60} = \dfrac{1}{3600}$, we have $1'' = \dfrac{1}{60}(1') = \dfrac{1}{60}\left[\dfrac{1}{60}(1°)\right] = \dfrac{1}{3600}(1°)$.

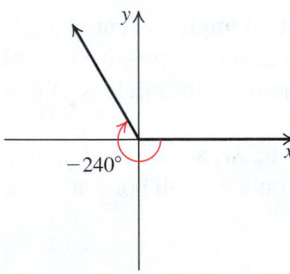

Figure 4.7

RELATIONSHIP BETWEEN DEGREES, MINUTES, AND SECONDS

$$1' = \frac{1}{60}(1°) = \left(\frac{1}{60}\right)° \qquad 1° = 60' \qquad 1° = 3600''$$

$$1'' = \frac{1}{3600}(1°) = \left(\frac{1}{3600}\right)° \qquad 1'' = \frac{1}{60}(1') \qquad 1' = 60''$$

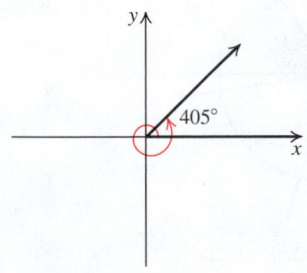

Figure 4.8

We use these relationships to convert between degree, minute, second (DMS) notation and decimal degree (DD) notation. For example,

$$30' = 30 \cdot 1' = 30 \cdot \frac{1}{60}(1°) = 0.5(1°) = 0.5°.$$

So

$$18°30' = 18° + 30' = 18° + 0.5° = 18.5°.$$

Similarly,

$$0.2° = \frac{2}{10}(1°) = \frac{2}{10}(60') = 12',$$

$$39.2° = 39° + 0.2° = 39° + 12' = 39°12'.$$

EXAMPLE 2   Converting Between DMS and DD Notation

a. Convert 24°8′15″ to decimal degree notation, rounded to two decimal places.

b. Convert 67.526° to DMS notation, rounded to the nearest second.

Solution

a. $24°8'15'' = 24° + 8 \cdot 1' + 15 \cdot 1''$ Rewrite.

$= 24° + 8\left(\frac{1}{60}\right)° + 15\left(\frac{1}{3600}\right)°$ $1' = \left(\frac{1}{60}\right)°, 1'' = \left(\frac{1}{3600}\right)°$

$\approx 24.14°$ Use a calculator.

b. $67.526° = 67° + 0.526 \cdot 1°$ Rewrite.

$= 67° + 0.526(60')$ Replace 1° with 60′.

$= 67° + 31.56'$ Use a calculator.

$= 67° + 31' + 0.56 \cdot 1'$ Rewrite.

$= 67° + 31' + 0.56(60'')$ Replace 1′ with 60″.

$= 67° + 31' + 33.6'' \approx 67°31'34''$ Use a calculator and round to the nearest second.

Practice Problem 2

a. Convert 13°9′22″ to decimal degree notation, rounded to two decimal places.

b. Convert 41.275° to DMS notation, rounded to the nearest second.

Radian Measure

An angle whose vertex is at the center of a circle is called a **central angle**. A central angle intercepts the arc of the circle from the initial side to the terminal side. A positive central angle that intercepts an arc of the circle of length equal to the radius of the circle is said to have measure **1 radian**. See Figure 4.9.

Because the circumference of a circle of radius r is $2\pi r \approx 6.28r$, six arcs of length r can be consecutively marked off on a circle of radius r, leaving only a small portion of the circumference uncovered. See Figure 4.10.

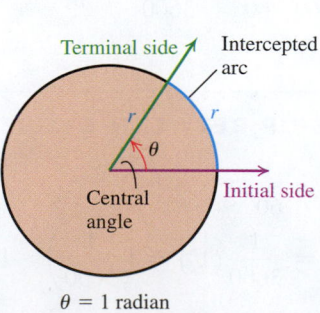

$\theta = 1$ radian

Figure 4.9 One Radian.

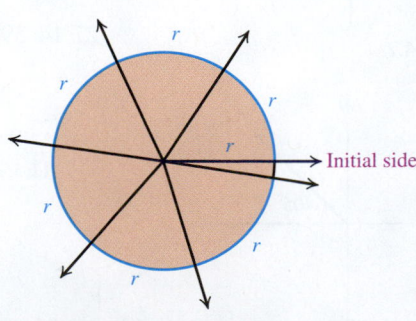

Circumference $= 2\pi r \approx 6.28r$

Figure 4.10

Each of the positive central angles in Figure 4.10 that intercepts an arc of length r has measure 1 radian; so the measure of one complete revolution is a little more than 6 radians. *In this book, we assume (unless otherwise specified) that all central angles are positive.* From geometry, we know that the measure of a central angle θ $(0 \leq \theta \leq 2\pi)$ is proportional to the length, s, of the arc it intercepts. This fact leads to the following relation.

> ### RADIAN MEASURE OF A CENTRAL ANGLE
>
> The radian measure θ of a central angle that intercepts an arc of length s on a circle of radius r is given by
>
> $$\theta = \frac{s}{r} \text{ radians.}$$

Notice that when $s = r$, we have $\theta = \dfrac{r}{r} = 1$ radian.

Relationship Between Degrees and Radians

3 Convert between degree and radian measure.

To see the relationship between degrees and radians, consider the central angle in Figure 4.11 with degree measure $\theta = 180°$ in a circle of radius r. The length of the arc, s, that is intercepted by this angle is half the circumference of the circle. So

$$s = \frac{1}{2}(\text{circumference})$$

$$s = \frac{1}{2}(2\pi r) = \pi r \qquad \text{Circumference} = 2\pi r$$

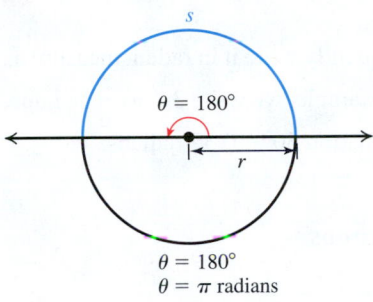

Figure 4.11

We can now find the radian measure for $\theta = 180°$.

$$\theta = \frac{s}{r} \text{ radians} \qquad \text{Radian measure of a central angle}$$

$$180° = \frac{\pi r}{r} \text{ radians} \qquad \text{Replace with } 180° \text{ and } s \text{ with } \pi r.$$

$$180° = \pi \text{ radians} \qquad \text{Simplify.}$$

> ### CONVERTING BETWEEN DEGREES AND RADIANS
>
> The relationship
>
> $$180° = \pi \text{ radians}$$
>
> allows us to convert between the degree and the radian measure of an angle.
>
> **Degrees to Radians:**
>
> $$180° = \pi \text{ radians} \qquad \text{Note that } 180° = (180) \cdot 1°.$$
>
> $$1° = \frac{\pi}{180} \text{ radians} \approx 0.0174533 \qquad \text{Divide both sides by 180.}$$
>
> $$\theta° = \frac{\theta\pi}{180} \text{ radians} \qquad \text{Multiply both sides by } \theta.$$
>
> **Radians to Degrees:**
>
> $$\pi \text{ radians} = 180°$$
>
> $$1 \text{ radian} = \left(\frac{180}{\pi}\right)° \approx 57.29578° \qquad \text{Divide both sides by } \pi.$$
>
> $$\theta \text{ radians} = \left(\frac{180\,\theta}{\pi}\right)° \qquad \text{Multiply both sides by } \theta.$$
>
> These conversions apply to both positive and negative angles.

EXAMPLE 3 **Converting from Degrees to Radians**

Convert each angle from degrees to radians.

a. $30°$ **b.** $90°$ **c.** $-225°$ **d.** $55°$

Solution

To convert degrees to radians, multiply degrees by $\dfrac{\pi}{180°}$.

a. $30° = 30° \cdot \dfrac{\pi}{180°} = \dfrac{30\pi}{180} = \dfrac{\pi}{6}$ radians

b. $90° = 90° \cdot \dfrac{\pi}{180°} = \dfrac{90\pi}{180} = \dfrac{\pi}{2}$ radians

c. $-225° = -225° \cdot \dfrac{\pi}{180°} = \dfrac{-225\pi}{180} = -\dfrac{5\pi}{4}$ radians

d. $55° = 55° \cdot \dfrac{\pi}{180°} = \dfrac{55\pi}{180} \approx 0.96$ radians   Use a calculator.

Practice Problem 3 Convert $-45°$ to radians.

If an angle is a fraction of a complete revolution, we frequently write it in radian measure as a fractional multiple of π rather than as a decimal. For example, we write $30°$ as $\dfrac{\pi}{6}$ radians, which is the exact value, instead of writing the approximation $30° \approx 0.52$ radians.

EXAMPLE 4 **Converting from Radians to Degrees**

Convert each angle in radians to degrees.

a. $\dfrac{\pi}{3}$ radians   **b.** $-\dfrac{3\pi}{4}$ radians   **c.** 1 radian

Solution

To convert radians to degrees, multiply radians by $\dfrac{180°}{\pi}$.

a. $\dfrac{\pi}{3}$ radians $= \dfrac{\pi}{3} \cdot \dfrac{180°}{\pi} = \left(\dfrac{180}{3}\right)° = 60°$

b. $-\dfrac{3\pi}{4}$ radians $= -\dfrac{3\pi}{4} \cdot \dfrac{180°}{\pi} = \left(-\dfrac{3}{4}\right)180° = -135°$

c. 1 radian $= 1 \cdot \dfrac{180°}{\pi} \approx 57.3°$ Use a calculator.

Practice Problem 4 Convert $\dfrac{3\pi}{2}$ radians to degrees.

Figure 4.12 includes the degree and radian measures of some commonly used angles. With practice, you should be able to make these conversions without using the figure.

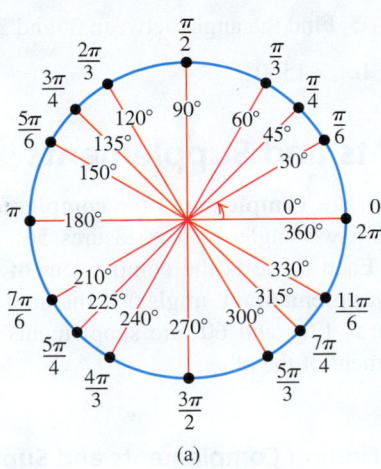

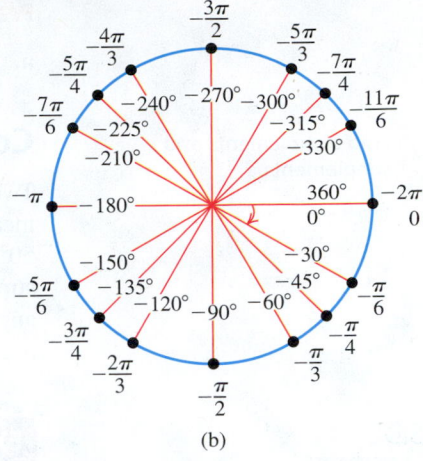

(a)   (b)

Figure 4.12 Frequently used angles.

Recall that two angles are coterminal if they have the same initial and terminal sides. We can find a coterminal angle for a given angle θ in radian measure by adding or subtracting 2π (one revolution) to θ. Similarly, a coterminal angle for an angle θ in degree measure is obtained by adding or subtracting $360°$ to θ.

An angle θ has infinitely many coterminal angles:

$$\theta + 2n\pi, \qquad \theta \text{ measured in radians}$$

or

$$\theta + n \cdot 360°, \qquad \theta \text{ measured in degrees}$$

where n is an integer.

EXAMPLE 5 **Finding Coterminal Angles**

Find the angle between 0 and 2π radians that is coterminal with

a. $\dfrac{19\pi}{4}$ **b.** $-\dfrac{17\pi}{6}$

Solution

a. $\dfrac{19\pi}{4} = 4\pi + \dfrac{3\pi}{4} = 2(2\pi) + \dfrac{3\pi}{4}$, more than two revolutions, so we subtract the equivalent of two revolutions $(2 \cdot 2\pi = 4\pi)$ to obtain:

$$\frac{19\pi}{4} - 4\pi = \frac{3\pi}{4}.$$

The angle of $\dfrac{3\pi}{4}$ (between 0 and 2π) is coterminal with the given angle $\dfrac{19\pi}{4}$.

b. $-\dfrac{17}{6}\pi = -2\pi - \dfrac{5}{6}\pi$, about one and a half negative revolutions, so we add the equivalent of two revolutions to obtain a positive measure:

$$-\frac{17}{6}\pi + 4\pi = \frac{7\pi}{6}; \text{ and } 0 < \frac{7\pi}{6} < 2\pi.$$

The angle $\dfrac{7\pi}{6}$ is coterminal with the given angle $-\dfrac{17}{6}\pi$.

Practice Problem 5 Find the angle between 0° and 360° that is coterminal with

a. 1230° **b.** −1560°

4 Find complements and supplements.

Complements and Supplements

Two positive angles are **complements** (or **complementary angles**) if the sum of their measures is 90°. So two angles with measures 50° and 40° are complements because 50° + 40° = 90°. Each angle is the complement of the other. Two positive angles are **supplements** (or **supplementary angles**) if the sum of their measures is 180°. So two angles with measures 120° and 60° are supplements because 120° + 60° = 180°. Each angle is the supplement of the other.

SIDE NOTE

A positive angle with measure ≥ 90° does not have a complement. A positive angle with measure ≥ 180° does not have a supplement.

EXAMPLE 6 **Finding Complements and Supplements**

Find the complement and the supplement of the given angle or explain why the angle has no complement or supplement.

a. 73° **b.** 110°

Solution

a. If θ represents the complement of 73°, then $\theta + 73° = 90°$; so $\theta = 90° - 73° = 17°$. The complement of 73° is 17°.
If α represents the supplement of 73°, then $\alpha + 73° = 180°$; so $\alpha = 180° - 73° = 107°$. The supplement of 73° is 107°.

b. There is no complement of a 110° angle because 110° > 90°.
If β represents the supplement of 110°, then $\beta = 180° - 110° = 70°$. The supplement of 110° is 70°.

Practice Problem 6 Find the complement and the supplement of 67° or explain why 67° does not have a complement or a supplement.

With radian measure, two positive angles are complements when the sum of their measures is $\dfrac{\pi}{2}$ radians; they are supplements when that sum is π radians.

5 Find the length of an arc of a circle.

Length of an Arc of a Circle

Recall that the formula for the radian measure θ of a central angle that intercepts an arc of length s is $\theta = \dfrac{s}{r}$. See Figure 4.13. Multiplying both sides of this equation by r yields $s = r\theta$.

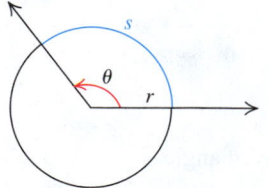

Central angles have positive measure
Figure 4.13

> **ARC LENGTH FORMULA**
>
> The length s of an arc intercepted by a central angle with radian measure θ in a circle of radius r is given by
>
> $$s = r\theta.$$

EXAMPLE 7 **Finding Arc Length of a Circle**

A circle has a radius of 18 inches. Find the length of the arc intercepted by a central angle with measure 210°.

Section 4.1 ■ Angles and Their Measure 345

Solution

We must first convert the central angle measure from degrees to radians.

$$s = r\theta \qquad \text{Arc length formula, } \theta \text{ in radians}$$

$$s = 18\left(\frac{7\pi}{6}\right) \qquad r = 18; \theta = 210° = 210°\left(\frac{\pi}{180°}\right) = \frac{7\pi}{6} \text{ radians}$$

$$= 21\pi \approx 65.97 \text{ inches} \qquad \text{Simplify and use a calculator.}$$

Practice Problem 7 A circle has a radius of 2 meters. Find the length of the arc intercepted by a central angle with measure 225°.

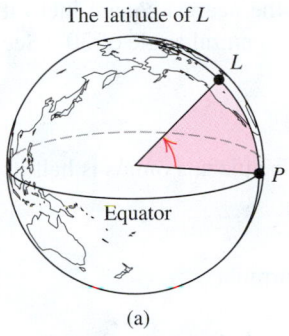

The latitude of L

(a)

◆ **EXAMPLE 8** **Finding the Distance Between Cities**

We discussed longitude and latitude in the introduction to this section. We determine the latitude of a location L by first finding the point of intersection, P, of the meridian through L and the equator. The latitude of L is the angle formed by rays drawn from the center of Earth to points L and P, with the ray through P being the initial ray. See Figure 4.14(a).

Billings, Montana, is due north of Grand Junction, Colorado. Find the distance between Billings (latitude 45°48′ N) and Grand Junction (latitude 39° 7′ N). Use 3960 miles as the radius of Earth. The N in 45°48′ N means that the location is north of the equator.

Solution

Because Billings is due north of Grand Junction, the same meridian passes through both cities. The distance between the cities is the length of the arc, s, on this meridian intercepted by the central angle, θ, the difference in their latitudes. See Figure 4.14(b).

The measure of angle θ is $45°48′ - 39°7′ = 6°41′$. To use the arc length formula, we must convert this angle to radians.

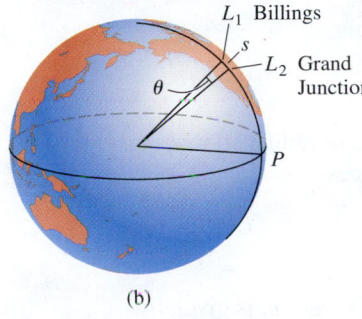

Distance on a meridian

(b)

Figure 4.14 Distance between cities.

$$\theta = 6°41′ \approx 6.6833° = 6.6833\left(\frac{\pi}{180}\right) \text{ radian} \approx 0.117 \text{ radian}$$

We use this value of θ and $r = 3960$ miles in the arc length formula:

$$s \approx (3960)(0.117) \text{ miles} \approx 463 \text{ miles} \qquad s = r\theta$$

The distance between Billings and Grand Junction is about 463 miles.

Practice Problem 8 Chicago, Illinois, is due north of Pensacola, Florida. Find the distance between Chicago (latitude 41°51′ N) and Pensacola (latitude 30°25′ N). Use $r = 3960$ miles.

6 Find the area of a sector.

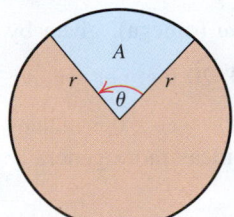

Figure 4.15 A sector of a circle.

Area of a Sector

A **sector** of a disk is a region bounded by the two sides of a central angle and the intercepted arc. Suppose θ is the radian measure of a central angle of a circle of radius r. See Figure 4.15. To find a formula for the area A of the sector formed by θ, we use the proportion

$$\frac{\text{area of a sector}}{\text{area of a circle}} = \frac{\text{length of the intercepted arc}}{\text{circumference of the circle}} \qquad \text{Fact from geometry}$$

$$\frac{A}{\pi r^2} = \frac{r\theta}{2\pi r} \qquad \begin{array}{l}\text{Area of a circle} = \pi r^2, \\ \text{circumference} = 2\pi r \\ \text{Arc length formula } s = r\theta\end{array}$$

$$A = \pi r^2\left(\frac{r\theta}{2\pi r}\right) \qquad \text{Multiply both sides by } \pi r^2.$$

$$= \frac{1}{2}r^2\theta \qquad \text{Simplify.}$$

AREA OF A SECTOR

The area A of a sector of a circle of radius r formed by a central angle with radian measure θ is

$$A = \frac{1}{2}r^2\theta.$$

Figure 4.16 A slice of pizza.

EXAMPLE 9 **Finding the Area of a Sector of a Circle**

How many square inches of pizza have you eaten (rounded to the nearest square inch) if you eat a sector of an 18-inch-diameter pizza whose edges form a central angle of 30°? See Figure 4.16.

Solution

We must convert 30° to radians to use the sector area formula. The pizza's radius is half its diameter, or 9 inches, and $30° = \dfrac{\pi}{6}$ radians.

$$A = \frac{1}{2}r^2\theta \qquad \text{Area of a sector formula}$$

$$= \frac{1}{2}\,9^2\left(\frac{\pi}{6}\right) \qquad \text{Replace } r \text{ with 9 and } \theta \text{ with } -$$

$$\approx 21 \text{ square inches} \qquad \text{Use a calculator.}$$

You have eaten about 21 square inches of pizza.

Practice Problem 9 Find the area of a sector of a circle of radius 10 inches formed by an angle of 60°. Round the answer to two decimal places.

---

7 Compute linear and angular speed.

Linear and Angular Speed

If a skater skates 100 feet in 5 seconds, then her *average speed*, v, is given by

$$v = \frac{\text{distance}}{\text{time}} = \frac{100 \text{ feet}}{5 \text{ seconds}} = 20 \text{ feet per second.}$$

Suppose now that the skater is skating around the circumference of a circular skating rink with radius 40 feet. See Figure 4.17. In five seconds, the skater travels an arc of length 100 feet and moves through an angle θ, where

$$\theta = \frac{s}{r} = \frac{100}{40} = 2.5 \text{ radians.}$$

We say that the skater is traveling at an (average) *angular speed* ω (omega), given by

$$\omega = \frac{\theta}{t} = \frac{2.5}{5} = 0.5 \text{ radians per second.}$$ We call the average speed, 20 feet per second, the (average) *linear speed* to distinguish it from the (average) angular speed, 0.5 radians per second. Notice that $v = 20 = 40(0.5) = r\omega$. We define these ideas more generally as follows.

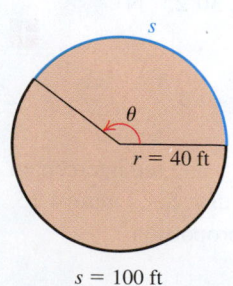

$s = 100 \text{ ft}$
$\theta = 2.5 \text{ radians}$

Figure 4.17

LINEAR AND ANGULAR SPEED

Suppose an object travels around a circle of radius r. If the object travels through a central angle of θ radians and an arc of length s, in time t, then

1. $v = \dfrac{s}{t}$ is the (average) **linear speed** of the object.

2. $\omega = \dfrac{\theta}{t}$ is the (average) **angular speed** of the object.

Further, because $s = r\theta$, by replacing s with $r\theta$ in **1** and then using **2** to replace $\dfrac{\theta}{t}$ with ω, we have

3. $v = r\omega$.

EXAMPLE 10 **Finding Angular and Linear Speed from Revolutions per Minute**

A model plane is attached to a swivel so that it flies in a circular path at the end of a 12-foot wire at the rate of 15 revolutions per minute. Find the angular speed and the linear speed of the plane.

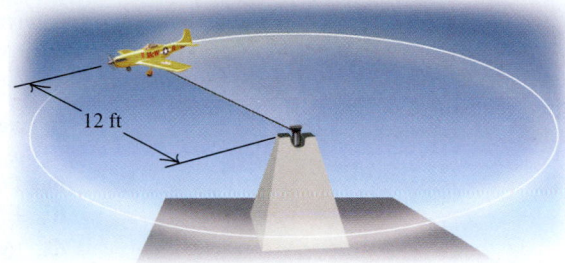

12 ft

Solution

Because angular speed is measured in radians per unit of time, we first convert revolutions per minute into radians per minute. Recall that

$$1 \text{ revolution} = 2\pi \text{ radians.}$$

Then 15 revolutions $= 15 \cdot 2\pi = 30\pi$ radians. So the angular speed ω is 30π radians per minute. Linear speed is given by

$$v = r\omega$$
$$= 12 \cdot 30\pi \text{ feet per minute} \quad \text{Replace } r \text{ with 12 feet and } \omega \text{ with } 30\pi$$
$$\text{radians per minute.}$$
$$\approx 1131 \text{ feet per minute} \quad \text{Use a calculator.}$$

Practice Problem 10 Find the angular speed and the linear speed of the plane in Example 10 assuming that the wire is 10 feet long and the plane flies at a rate of 18 revolutions per minute.

Answers to Practice Problems

1.

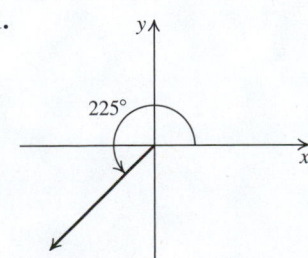

225°

2. **a.** 13.16°
   b. 41°16′30″

3. $-\dfrac{\pi}{4}$ radians   4. 270°

5. **a.** 150°   **b.** 240°

6. Complement $= 23°$; supplement $= 113°$

7. $\dfrac{5\pi}{2} \approx 7.85$ meters

8. ≈ 790 miles

9. $\dfrac{50\pi}{3} \approx 52.36$ square inches

10. $\omega = 36\pi$ radians per minute; $v = 360\pi \approx 1131$ feet per minute

SECTION 4.1 Exercises

Concepts and Vocabulary

1. A negative angle is formed by rotating the initial side in a(n) _____ direction.

2. An angle is in standard position if its vertex is at the origin of a coordinate system and its initial side lies on the _____ .

3. Two angles with the same initial and terminal sides are _____ angles.

4. To convert degrees to radians, multiply degrees by _____ .

5. **True or False.** One radian is smaller than one degree.

6. **True or False.** The circumference of a circle with radius 1 meter is approximately 6.28 meters.

7. **True or False.** Every acute angle has a complement and a supplement.

8. **True or False.** If $r = 5$ ft and $\theta = 30°$, then $s = 5 \cdot 30 = 150$ feet.

Building Skills

In Exercises 9–16, draw each angle in standard position.

9. $30°$

10. $150°$

11. $-120°$

12. $-330°$

13. $\dfrac{5\pi}{3}$

14. $\dfrac{11\pi}{6}$

15. $-\dfrac{4\pi}{3}$

16. $-\dfrac{13\pi}{4}$

In Exercises 17–22, convert each angle to decimal degree notation. Round your answers to two decimal places.

17. $70°45'$

18. $38°38'$

19. $23°42'30''$

20. $45°50'50''$

21. $-15°42'57''$

22. $-70°18'13''$

In Exercises 23–28, convert each angle to DMS notation. Round your answers to the nearest second.

23. $27.32°$

24. $120.64°$

25. $13.347°$

26. $110.433°$

27. $19.0511°$

28. $82.7272°$

In Exercises 29–36, convert each angle from degrees to radians. Express each answer as a multiple of π.

29. $20°$

30. $40°$

31. $-180°$

32. $-210°$

33. $315°$

34. $330°$

35. $-510°$

36. $-420°$

In Exercises 37–44, convert each angle from radians to degrees.

37. $\dfrac{\pi}{12}$

38. $\dfrac{3\pi}{8}$

39. $-\dfrac{5\pi}{9}$

40. $-\dfrac{3\pi}{10}$

41. $\dfrac{5\pi}{3}$

42. $\dfrac{11\pi}{6}$

43. $-\dfrac{11\pi}{4}$

44. $-\dfrac{7\pi}{3}$

In Exercises 45–48, convert each angle from degrees to radians. Round your answers to two decimal places.

45. $12°$

46. $127°$

47. $-84°$

48. $-175°$

In Exercises 49–52, convert each angle from radians to degrees. Round your answers to two decimal places.

49. 0.94

50. 5

51. -8.21

52. -6.28

In Exercises 53–58, find the angle between 0 and 2π radians that is coterminal with the given angle.

53. $-\dfrac{\pi}{4}$

54. $-\dfrac{2\pi}{3}$

55. $\dfrac{7\pi}{4}$

56. $-\dfrac{5\pi}{3}$

57. $\dfrac{16\pi}{3}$

58. $\dfrac{23\pi}{6}$

In Exercises 59–64, find the angle between 0° and 360° that is coterminal with the given angle.

59. $-65°$

60. $-120°$

61. $-200°$

62. $-280°$

63. $700°$

64. $1270°$

In Exercises 65–70, find the complement and the supplement of the given angle or explain why the angle has no complement or supplement.

65. $47°$

66. $75°$

67. $120°$

68. $160°$

69. $210°$

70. $-50°$

In Exercises 71–90, use the following notations: θ = central angle of a circle, r = radius of a circle, s = length of the intercepted arc, v = linear velocity, ω = angular velocity, A = area of the sector of a circle, and t = time.

In each case, find the missing quantity. Find the exact answer.

71. $r = 25$ in, $s = 7$ in, $\theta = ?$

72. $r = 5$ ft, $s = 6$ ft, $\theta = ?$

73. $r = 10.5$ cm, $s = 22$ cm, $\theta = ?$

74. $r = 60$ m, $s = 120$ m, $\theta = ?$

75. $r = 3$ m, $\theta = 25°$, $s = ?$

76. $r = 0.7$ m, $\theta = 357°$, $s = ?$

77. $r = 6.5$ m, $\theta = 12$ radians, $s = ?$

78. $r = 6$ m, $\theta = \dfrac{\pi}{6}$ radians, $s = ?$

79. $\theta = \dfrac{1}{2}$ radians, $s = 3$ centimeters, $r = ?$

80. $\theta = \dfrac{1}{5}$ radians, $s = 2$ feet, $r = ?$

81. $\theta = 30°$, $s = \dfrac{3\pi}{2}$ meters, $r = ?$

82. $\theta = 120°$, $s = 20$ inches, $r = ?$

83. $r = 10$ in, $\theta = 4$ rad, $A = ?$

84. $r = 1.5$ ft, $\theta = 60°$, $A = ?$

85. $A = 20$ ft^2, $\theta = 60°$, $r = ?$

86. $A = 60$ m^2, $\theta = 2$ rad, $r = ?$

87. $r = 6$ m, $\omega = 10$ rad/min, $v = ?$

88. $r = 3.2$ ft, $\omega = 5$ rad/sec, $v = ?$

89. $v = 20$ ft/sec, $r = 10$ ft, $\omega = ?$

90. $v = 10$ m/min, $r = 6$ m, $\omega = ?$

Applying the Concepts

91. Wheel ratios. An automobile is pushed so that its wheels turn three-quarters of a revolution. If the tires have a radius of 15 inches, how many inches does the car move?

92. Nautical miles. A nautical mile is the length of an arc of the equator intercepted by a central angle of $1'$. Using 3960 miles for the value of the radius of Earth, express 1 nautical mile in terms of standard (**statute**) miles.

93. Angles on a clock. What is the radian measure of the smaller central angle made by the hands of a clock at 4:00? Express your answer as a rational multiple of π.

94. Angles on a clock. What is the radian measure of the larger central angle made by the hands of a clock at 7:00? Express your answer as a rational multiple of π.

95. Diameter of a pizza. You are told that a slice of pizza whose edges form a 25° angle with an outer crust edge 4 inches long was found in a gym locker. What was the diameter of the original pizza? Round your answer to the nearest inch.

96. Arc of a pendulum. How far does the tip of a 5-foot pendulum travel as it swings through an angle of 30°? Round your answer to two decimal places.

97. Lifting a shark. A shark is suspended by a wire wound around a pulley of radius 4 inches. The shark must be raised 6 inches more to be completely off the ground. Through how many degrees must the pulley be rotated counterclockwise to accomplish the 6-inch lift? Round your answer to the nearest degree.

98. Security cameras. A security camera rotates through an angle of 120°. To the nearest foot, what is the arc width of the field of view 40 feet from the camera?

99. Pulley wheels. Two pulleys are connected by a belt so that when one pulley rotates, the linear speeds of the belt and both pulleys are the same. See the figure. The radius of the smaller pulley is 2 inches, and the radius of the larger pulley is 5 inches. A point on the belt travels at a rate of 600 inches per minute.
   a. Find the angular speed of the larger pulley.
   b. Find the angular speed of the smaller pulley.

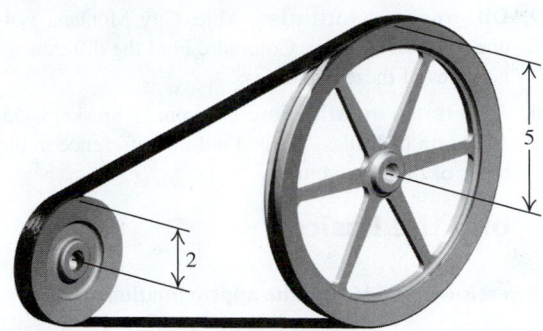

100. Ferris wheel speed. A Ferris wheel in Vienna, Austria, has a diameter of approximately 61 meters. Assume that it takes 90 seconds for the Ferris wheel to make one complete revolution.
   a. Find the angular speed of the Ferris wheel. Round your answer to two decimal places.
   b. Find the linear speed of the Ferris wheel in meters per second. Round your answer to two decimal places.

101. Bicycle speed. A bicycle's wheels are 24 inches in diameter. If the bike is traveling at a rate of 25 miles per hour, find the angular speed of the wheels. Round your answer to two decimal places.

102. Bicycle speed. A bicycle's wheels are 30 inches in diameter. If the angular speed of the wheels is 11 radians per second, find the speed of the bicycle in inches per second.

103. Hard disk speed. Hard disks are circular disks that store data in your computer. Suppose a circular hard disk is 3.75 inches in diameter. If the disk rotates 7200 revolutions per minute, what is the linear speed of a point on the edge of the disk (in inches per minute)?

104. Linear speed at the equator. Earth rotates on an axis that goes through both the North and South Poles. It makes one complete revolution in 24 hours. Find the linear speed (in miles per hour) of a location on the equator. (Use 3960 miles as the radius of Earth.)

In Exercises 105–110, the latitude of any location on Earth is the angle formed by the two rays drawn from the center of Earth to the location and to the point of intersection of the meridian for the location with the equator. The ray through the location is the initial ray. Use 3960 miles as the radius of Earth.

105. Distance between cities. Indianapolis, Indiana, is due north of Montgomery, Alabama. Find the distance between Indianapolis (latitude 39°44′ N) and Montgomery (latitude 32°23′ N).

106. Distance between cities. Pittsburgh, Pennsylvania, is due north of Charleston, South Carolina. Find the distance between Pittsburgh (latitude 40°30′ N) and Charleston (latitude 32°54′ N).

107. **Distance between cities.** Amsterdam, Netherlands, is due north of Lyon, France. Find the distance between Amsterdam (latitude 52°23′ N) and Lyon (latitude 45°42′ N).

108. **Distance between cities.** Adana, Turkey (north latitude 36°59′ N), is due north of Jerusalem, Israel (latitude 31°47′ N). Find the distance between Adana and Jerusalem.

109. **Difference in latitudes.** Miles City, Montana, is 440 miles due north of Boulder, Colorado. Find the difference in the latitudes of these two cities.

110. **Difference in latitudes.** Lincoln, Nebraska, is 554 miles due north of Dallas, Texas. Find the difference in the latitudes of these two cities.

Beyond the Basics

In Exercises 111–116, use the approximation $\pi \approx \dfrac{22}{7}$.

111. A racetrack is in the form of a ring with inner circumference 352 meters and outer circumference 396 meters. Find the width of the track.

112. The diameter of a wheel is 84 centimeters. How many complete revolutions must it take to cover 792 meters?

113. The diameter of a tire is 140 centimeters. How many revolutions per minute will achieve a speed of 66 kilometers per hour?

114. A car has two wipers that do not overlap. Each wiper has a blade length of 14 inches that sweeps through an angle of 110°. Find the total area of the windshield cleaned at each sweep of the blades.

115. An umbrella top is made from a material in the shape of a circle of diameter 3 feet. It has eight equally spaced ribs. Find the area between two consecutive ribs of the umbrella.

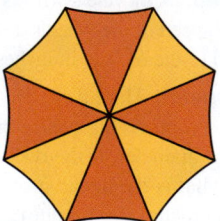

116. A conical drinking cup is made from a circular piece of paper of radius $r = 7$ cm by cutting out a sector of central angle

$\dfrac{\pi}{4}$ and joining the edges OA and OB. Find the volume of the cup. (*Hint:* Volume of a cone $= \dfrac{1}{3}\pi R^2 h$.)

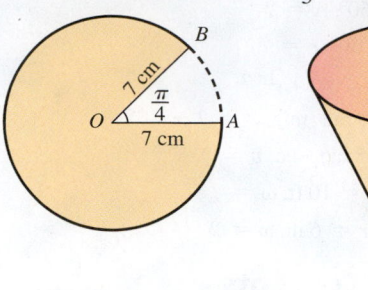

Critical Thinking / Discussion / Writing

117. Arrange the following radian measures from smallest to largest: $\pi, \dfrac{3\pi}{2}, 3, 4$.

118. Which line of latitude has a smaller radius, the line of latitude through a city of latitude 30°40′13″ or the line of latitude through a city of latitude 40°30′13″?

119. Assume that θ is a central angle in a circle of radius r that intercepts an arc of length s. Show that if θ has radian measure less than $\dfrac{\pi}{2}$, then the central angle that intercepts an arc of length $\left(\dfrac{\pi r}{2} - s\right)$ and θ are complementary angles.

120. In about 250 B.C., the philosopher Eratosthenes estimated the radius of Earth based on what might appear to be scant information. He knew that the city of Syene (modern Aswan) is directly south of the city of Alexandria and that Syene and Alexandria are (in modern units) about 500 miles apart. He also knew that in Syene at noon on a midsummer day, an upright rod casts no shadow, but the noon sun makes a 7°12′ angle with a vertical rod at Alexandria and that then the sun's rays at Syene are effectively parallel to the sun's rays at Alexandria. Explain how Eratosthenes could measure the circumference of Earth by measuring the length of the shadow in Alexandria at noon on the summer solstice when there was no shadow in Syene.

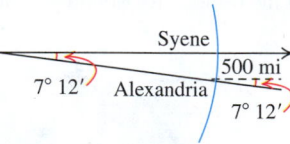

Getting Ready for the Next Section

In Exercises 121–126, consider triangle ABC shown in the adjoining figure. Use the Pythagorean theorem to find the third side from the given two sides.

121. $a = 5, b = 12$

122. $a = 12, c = 20$

123. $b = 6, c = 10$

124. $a = 20, b = 21$

125. $b = 3, c = 6$

126. $a = 5, b = 10$

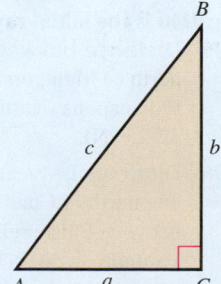

In Exercises 127 and 128, find $\dfrac{A}{B}$ from the given values of A and B. Then rationalize the denominator of the quotient.

127. $A = \dfrac{1}{2}, B = \dfrac{\sqrt{3}}{2}$

128. $A = \dfrac{\sqrt{3}}{2}, B = \sqrt{\dfrac{3}{2}}$

The Unit Circle; Trigonometric Functions of Real Numbers

BEFORE STARTING THIS SECTION, REVIEW

1 Equation of a circle (Section 1.1, page 12)

2 Arc length formula (Section 4.1, page 344)

3 Symmetry (Section 1.1, page 10)

4 Equations of a line (Section 1.2, page 31)

OBJECTIVES

1 Review properties of the unit circle.

2 Define trigonometric functions of real numbers.

3 Find exact trigonometric function values using a terminal point on the unit circle.

4 Define trigonometric functions of angles.

5 Find exact trigonometric function values of common angles.

6 Approximate trigonometric function values using a calculator.

◆ Blood Pressure

In 1628, William Harvey wrote in his book *De Motu Cordis (On the Motion of the Heart)*, "Just as the king is the first and highest authority in the state, so the heart governs the whole body!" He thereby announced to the world that the heart pumped blood through the entire body. There was a time, however, when even the fact that blood circulated at all was not accepted: the arteries were thought to carry air, not blood.

The first measurements of blood pressure were made in the early eighteenth century by Stephen Hales, an English botanist, physiologist, and clergyman. He found that pressure from a horse's heart filled a vertical glass tube with the horse's blood to a height of 8 feet 3 inches. He accomplished this by inserting a narrow brass pipe directly into an artery and fitting a 9-foot-long vertical glass to the pipe.

Fortunately, today a much safer, more convenient, and simplified method for measuring blood pressure is available everywhere. In Exercise 127 we use trigonometric or circular functions to find blood pressure.

1 Review properties of the unit circle.

The Unit Circle

Recall the definition of the unit circle (see Section 1.1).

THE UNIT CIRCLE

The **unit circle** is the circle of radius 1 with its center at the origin. In the *xy*-plane, it is the set of points (x, y) that satisfy the equation

$$x^2 + y^2 = 1.$$

We list some properties of the unit circle that are useful in the development of the trigonometric functions.

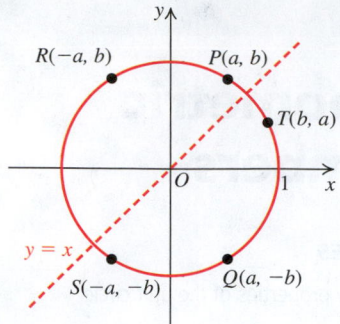

Figure 4.18

1. **Circumference.** The circumference of the unit circle is 2π. This means that the arc length for one revolution around the unit circle is 2π units.

2. **Symmetry.** If $P = (a, b)$ is a point on the unit circle, then the following *symmetric* images of P are also on the unit circle:

   a. $Q = (a, -b)$ about the x-axis   **b.** $R = (-a, b)$ about the y-axis

   c. $S = (-a, -b)$ about the origin   **d.** $T = (b, a)$ about the line $y = x$.

See Figure 4.18.

EXAMPLE 1   Points on the Unit Circle

Is the point $P = \left(\dfrac{2}{3}, -\dfrac{\sqrt{5}}{3}\right)$ on the unit circle? If it is, then use symmetry to find four other points on the circle.

Solution

We show that the coordinates of the point $P = \left(\dfrac{2}{3}, -\dfrac{\sqrt{5}}{3}\right)$ satisfy the equation $x^2 + y^2 = 1$.

$$x^2 + y^2 = \left(\frac{2}{3}\right)^2 + \left(-\frac{\sqrt{5}}{3}\right)^2 \qquad \text{Replace } x \text{ with } \frac{2}{3} \text{ and } y \text{ with } -\frac{\sqrt{5}}{3}.$$

$$= \frac{4}{9} + \frac{5}{9} = \frac{9}{9} = 1.$$

Therefore, the point P is on the unit circle.

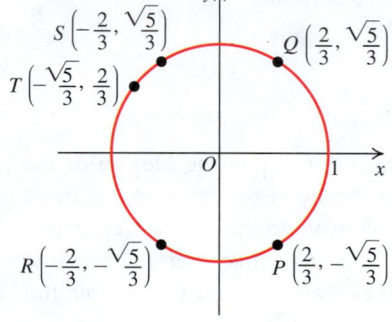

Figure 4.19

Since $P = \left(\dfrac{2}{3}, -\dfrac{\sqrt{5}}{3}\right)$ is on the unit circle, the symmetric points $Q = \left(\dfrac{2}{3}, \dfrac{\sqrt{5}}{3}\right)$, $R = \left(-\dfrac{2}{3}, -\dfrac{\sqrt{5}}{3}\right)$, $S = \left(-\dfrac{2}{3}, \dfrac{\sqrt{5}}{3}\right)$ and $T = \left(-\dfrac{\sqrt{5}}{3}, \dfrac{2}{3}\right)$ are also on the unit circle. See Figure 4.19.

Practice Problem 1 Repeat Example 1 for the point $P = \left(\dfrac{1}{2}, \dfrac{1}{2}\right)$.

EXAMPLE 2   Finding Points on the Unit Circle

Find all points on the unit circle whose y-coordinate is $\dfrac{3}{5}$.

Solution

We need to find the x-coordinates of all points (if any) of the form $\left(x, \dfrac{3}{5}\right)$ that satisfy the equation $x^2 + y^2 = 1$.

We substitute $y = \dfrac{3}{5}$ in the equation $x^2 + y^2 = 1$ and solve for x.

$$x^2 + y^2 = 1 \qquad \text{The unit circle equation}$$

$$x^2 + \left(\frac{3}{5}\right)^2 = 1 \qquad \text{Replace } y \text{ with } \frac{3}{5}.$$

$$x^2 + \frac{9}{25} = 1$$

$$x^2 = 1 - \frac{9}{25} = \frac{16}{25}$$

$$x = \pm\sqrt{\frac{16}{25}} = \pm\frac{4}{5}$$

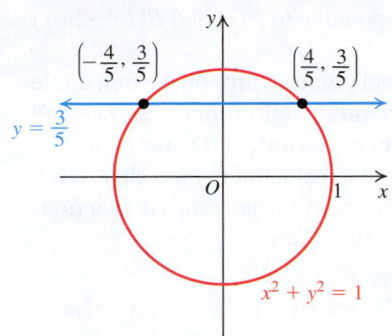

Figure 4.20

The points on the unit circle whose y-coordinate is $\dfrac{3}{5}$ are $\left(\dfrac{4}{5}, \dfrac{3}{5}\right)$ and $\left(-\dfrac{4}{5}, \dfrac{3}{5}\right)$. These are the points of intersection of the unit circle and the line $y = \dfrac{3}{5}$. See Figure 4.20.

Practice Problem 2 Find all points on the unit circle whose x-coordinate is $-\dfrac{2}{3}$.

2 Define trigonometric functions of real numbers.

Trigonometric Functions of Real Numbers

With each real number t, we associate a unique point on the unit circle. We know that the real numbers can be associated with the points on a number line, labeled as the t-axis in Figure 4.21(a). Let $x^2 + y^2 = 1$ be the unit circle in the Cartesian coordinate system, shown in Figure 4.21(b).

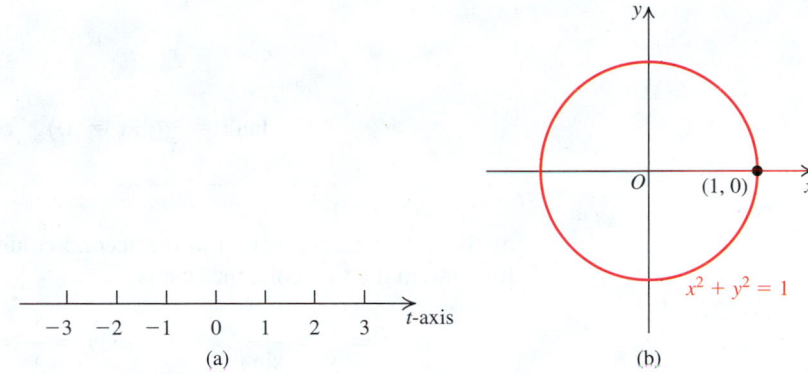

Figure 4.21 Number line and a unit circle.

RECALL

A line is *tangent* to a circle if it intersects the circle in exactly one point.

In Figure 4.22(a), the number line is drawn tangent to the unit circle with its 0 at the point $(1, 0)$. The unit length on the number line is the same as on the coordinate axes, and the positive numbers are above the x-axis. For a real number t on the tangent number line, we "wrap" the segment from 0 to t around the unit circle in a counterclockwise direction if t is positive or clockwise direction if t is negative. See Figure 4.22(b) and (c).

In this way, each real number t corresponds to a unique point $P(t) = (x, y)$ on the unit circle and is often called the **terminal point** associated with the real number t. In Figure 4.22(a), it is the endpoint of the arc from $(1, 0)$ to $P(t) = (x, y)$. Because the circumference of the unit circle is 2π, if $t > 2\pi$, you must go around the unit circle more than once to arrive at the point $P(t)$. Because the arc length between t and $t \pm 2\pi$ is 2π units, the points corresponding to $P(t)$ and $P(t \pm 2\pi)$ coincide on the unit circle. In fact, for

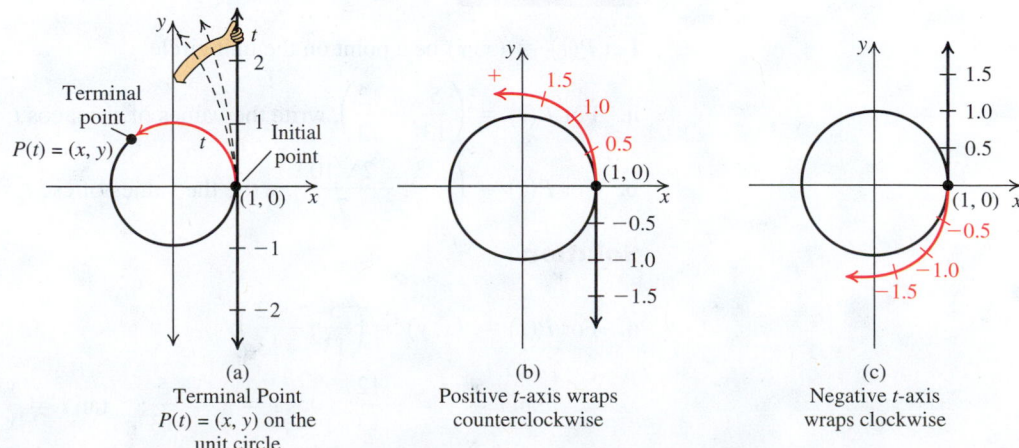

Figure 4.22 Wrapping t-axis on the unit circle.

any integer n and for any real number t, the points corresponding to $P(t)$ and $P(t + 2n\pi)$ coincide on the unit circle.

The correspondence between real numbers and terminal points of arcs on the unit circle is used to define six important functions, called the **trigonometric functions**. The full names of these functions are **sine, cosine, tangent, cosecant, secant,** and **cotangent**. The customary abbreviations for the values of these functions at a real number t are **sin t, cos t, tan t, csc t, sec t,** and **cot t**, respectively. Because the definitions of trigonometric functions are based on the unit circle, they are often called **circular functions**.

Unit Circle Definitions of the Trigonometric Functions of Real Numbers

Let t be any real number and let $P(t) = (x, y)$ be the terminal point on the unit circle associated with t. Then

$$\sin t = y \qquad\qquad \csc t = \frac{1}{y} \quad (y \neq 0)$$

$$\cos t = x \qquad\qquad \sec t = \frac{1}{x} \quad (x \neq 0)$$

$$\tan t = \frac{y}{x} \quad (x \neq 0) \qquad \cot t = \frac{x}{y} \quad (y \neq 0)$$

Note that each function in the second column is the *reciprocal* of the corresponding function in the first column. That is,

$$\csc t = \frac{1}{y} = \frac{1}{\sin t} \qquad \sec t = \frac{1}{x} = \frac{1}{\cos t} \qquad \cot t = \frac{x}{y} = \frac{1}{\tan t}$$

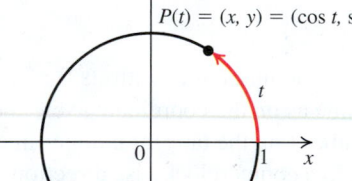

$P(t) = (x, y) = (\cos t, \sin t)$

Note that $\quad \tan t = \dfrac{y}{x} = \dfrac{\sin t}{\cos t}, \quad$ and $\quad \cot t = \dfrac{x}{y} = \dfrac{\cos t}{\sin t}.$

From the definitions of the tangent and secant functions, we see that these functions are not defined when $x = 0$, because denominators can never be 0. Similarly, the cotangent and cosecant functions are not defined when $y = 0$.

Figure 4.23 Unit Circle definitions of cos t and sin t.

COSINE AND SINE

The terminal point $P(t) = (x, y)$ on the unit circle associated with a real number t has coordinates $(\cos t, \sin t)$ because $x = \cos t$ and $y = \sin t$. See Figure 4.23.

EXAMPLE 3 **Using the Definition to Write Trigonometric Function Values**

Let $P(t) = (x, y)$ be a point on the unit circle.

a. For $P(t) = \left(\dfrac{5}{13}, -\dfrac{12}{13}\right)$, write the values of sin t, cos t, and tan t.

b. For $P(t) = \left(-\dfrac{3}{7}, -\dfrac{2\sqrt{10}}{7}\right)$, write the values of csc t, sec t, and cot t.

Solution

a. For $P(t) = (x, y) = \left(\dfrac{5}{13}, -\dfrac{12}{13}\right)$.

$$\sin t = y = -\frac{12}{13}; \quad \cos t = x = \frac{5}{13}; \quad \tan t = \frac{y}{x} = \frac{-12/13}{5/13} = -\frac{12}{5}.$$

b. For $p(t) = (x, y) = \left(-\dfrac{3}{7}, -\dfrac{2\sqrt{10}}{7}\right)$;

$$\csc t = \frac{1}{y} = -\frac{7}{2\sqrt{10}} = -\frac{7\sqrt{10}}{20}; \qquad \sec t = \frac{1}{x} = -\frac{7}{3};$$

$$\cot t = \frac{x}{y} = -\frac{-3/7}{-2\sqrt{10}/7} = \frac{3\sqrt{10}}{20}.$$

Practice Problem 3 Repeat Example 3 for

a. $P(t) = \left(-\dfrac{4}{5}, \dfrac{3}{5}\right).$ **b.** $P(t) = \left(\dfrac{1}{3}, \dfrac{2\sqrt{2}}{3}\right).$

3 Find exact trigonometric function values using a terminal point on the unit circle.

Finding Exact Trigonometric Function Values

To find the trigonometric function values of a real number t, we need to find the coordinates of the terminal point $P(t) = (x, y)$ on the unit circle. The exact values of x and y can only be found for some special values of t. For all other values of t, we use a calculator and settle for approximate values.

EXAMPLE 4 **Evaluating Trigonometric Functions**

Find the values (if any) of the six trigonometric functions at each value of t.

a. $t = 0$ **b.** $t = \dfrac{\pi}{2}$ **c.** $t = \pi$ **d.** $t = \dfrac{3\pi}{2}$ **e.** $t = -3\pi$

Solution

For each value of t, first find the corresponding terminal point $P(t) = (x, y)$ on the unit circle. (See Figure 4.24.) Then use the definitions of the trigonometric functions. Note that the length of the unit circle (one wrap) is 2π.

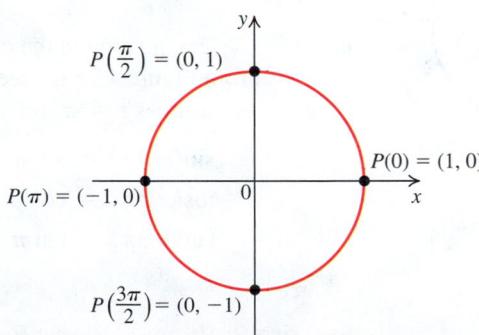

Figure 4.24 Points corresponding to various values of t.

a. $t = 0$ corresponds to the point $P(0) = (x, y) = (1, 0)$. So,

$$\sin 0 = y = 0 \qquad \csc 0 = \frac{1}{y} = \frac{1}{0} \text{ is undefined}$$

$$\cos 0 = x = 1 \qquad \sec 0 = \frac{1}{x} = \frac{1}{1} = 1$$

$$\tan 0 = \frac{y}{x} = \frac{0}{1} = 0 \qquad \cot 0 = \frac{x}{y} = \frac{1}{0} \text{ is undefined}$$

b. Length of $\frac{1}{4}$ (one wrap) $= \frac{1}{4}(2\pi) = \frac{\pi}{2}$, so $t = \frac{\pi}{2}$ corresponds to the terminal point

$P\left(\dfrac{\pi}{2}\right) = (x, y) = (0, 1)$.

$$\sin\frac{\pi}{2} = y = 1 \qquad\qquad \csc\frac{\pi}{2} = \frac{1}{y} = \frac{1}{1} = 1$$

$$\cos\frac{\pi}{2} = x = 0 \qquad\qquad \sec\frac{\pi}{2} = \frac{1}{x} = \frac{1}{0} \text{ is undefined}$$

$$\tan\frac{\pi}{2} = \frac{y}{x} = \frac{1}{0} \text{ is undefined} \qquad \cot\frac{\pi}{2} = \frac{x}{y} = \frac{0}{1} = 0$$

c. Length of $\frac{1}{2}$ (one wrap) $= \frac{1}{2}(2\pi) = \pi$, so $t = \pi$ corresponds to the terminal point
$P(\pi) = (x, y) = (-1, 0)$.

$$\sin\pi = y = 0 \qquad\qquad \csc\pi = \frac{1}{y} = \frac{1}{0} \text{ is undefined}$$

$$\cos\pi = x = -1 \qquad\qquad \sec\pi = \frac{1}{x} = \frac{1}{-1} = -1$$

$$\tan\pi = \frac{y}{x} = \frac{0}{-1} = 0 \qquad \cot\pi = \frac{x}{y} = \frac{-1}{0} \text{ is undefined}$$

d. Length of $\frac{3}{4}$ (one wrap) $= \frac{3}{4}(2\pi) = \frac{3\pi}{2}$, so $t = \frac{3\pi}{2}$ corresponds to the terminal

point $P\left(\dfrac{3\pi}{2}\right) = (x, y) = (0, -1)$.

$$\sin\frac{3\pi}{2} = y = -1 \qquad\qquad \csc\frac{3\pi}{2} = \frac{1}{y} = \frac{1}{-1} = -1$$

$$\cos\frac{3\pi}{2} = x = 0 \qquad\qquad \sec\frac{3\pi}{2} = \frac{1}{x} = \frac{1}{0} \text{ is undefined}$$

$$\tan\frac{3\pi}{2} = \frac{y}{x} = \frac{-1}{0} \text{ is undefined} \qquad \cot\frac{3\pi}{2} = \frac{x}{y} = \frac{0}{-1} = 0$$

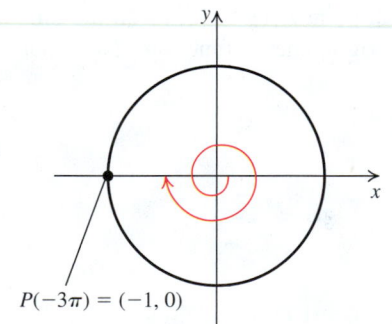

$P(-3\pi) = (-1, 0)$

Figure 4.25

e. If $|t| > 2\pi$, go around the circle more than once to find the point corresponding
to t. From Figure 4.25 we see that $t = -3\pi$ corresponds to exactly the same point
$(-1, 0)$, as does $t = \pi$. So, $P(-3\pi) = P(\pi) = (-1, 0)$. Then,

$$\sin(-3\pi) = \sin\pi = 0 \qquad\qquad \csc(-3\pi) = \csc\pi \text{ is undefined}$$
$$\cos(-3\pi) = \cos\pi = -1 \qquad\qquad \sec(-3\pi) = \sec\pi = -1$$
$$\tan(-3\pi) = \tan\pi = 0 \qquad\qquad \cot(-3\pi) = \cot\pi \text{ is undefined}$$

Practice Problem 4 Repeat Example 4 for $t = \dfrac{5\pi}{2}$.

4 Define trigonometric functions of angles.

Trigonometric Functions of an Angle

We have defined the trigonometric functions of real numbers. That is, the domains of these
functions consist of real numbers. We now define new functions whose domains are sets of
angles. These definitions are based on the following relationship between the real numbers
and angles measured in radians.

Let θ be a positive angle in standard position having radian measure t. Let (x, y) be the
point where the terminal side of θ intersects the unit circle. If θ is positive, then the arc length
formula says that the arc intercepted by θ has length $s = r \cdot (\text{radian measure of } \theta) = r \cdot t$.
Because $r = 1$ for the unit circle, $s = 1 \cdot t = t$. (See Figure 4.26(a).) If θ is negative with

radian measure t then $-\theta$ is positive and has radian measure $-t$. θ and $-\theta$ are congruent. They subtend arcs of the same length s, where $s =$ radian measure of $-\theta = -t$. (See Figure 4.26(b).)

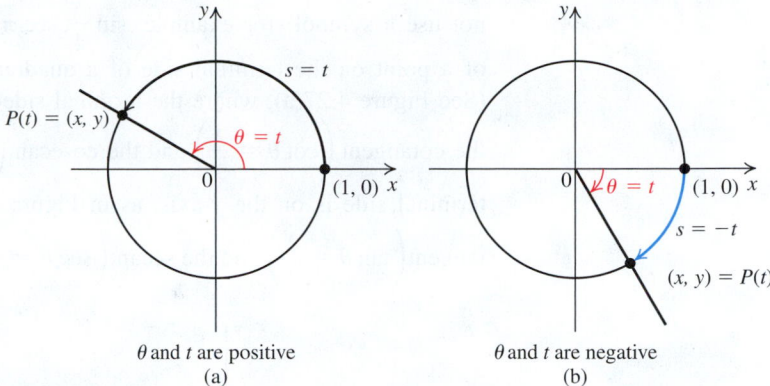

| θ and t are positive | θ and t are negative |
| (a) | (b) |

Figure 4.26

From Figure 4.26 we note that the point $P(t) = (x, y)$ on the unit circle that corresponds to a real number t is also the point on the terminal side of the angle θ with radian measure t.

This leads to the following definition of trigonometric functions of an *angle*.

Trigonometric Function of an Angle

If θ is an angle with radian measure t, then

$$\sin\theta = \sin t \qquad \cos\theta = \cos t \qquad \tan\theta = \tan t$$
$$\csc\theta = \csc t \qquad \sec\theta = \sec t \qquad \cot\theta = \cot t$$

If θ is given in degrees, convert θ to radians before using these equations.

EXAMPLE 5 **Finding the Trigonometric Function Values of a Quadrantal Angle**

Find the trigonometric function values of $90°$.

Solution

Since $90° = \dfrac{\pi}{2}$ radians, we can use the results of Example 4b for $t = \dfrac{\pi}{2}$ and $\theta = 90°$.

$$\sin 90° = \sin\left(\frac{\pi}{2}\text{ radians}\right) = \sin\left(\frac{\pi}{2}\right) = 1$$

$$\cos 90° = \cos\left(\frac{\pi}{2}\text{ radians}\right) = \cos\left(\frac{\pi}{2}\right) = 0$$

$$\tan 90° = \tan\left(\frac{\pi}{2}\text{ radians}\right) = \tan\left(\frac{\pi}{2}\right) \text{ is undefined}$$

$$\csc 90° = \csc\left(\frac{\pi}{2}\text{ radians}\right) = \csc\left(\frac{\pi}{2}\right) = 1$$

$$\sec 90° = \sec\left(\frac{\pi}{2}\text{ radians}\right) = \sec\left(\frac{\pi}{2}\right) \text{ is undefined}$$

$$\cot 90° = \cot\left(\frac{\pi}{2}\text{ radians}\right) = \cot\left(\frac{\pi}{2}\right) = 0$$

Practice Problem 5 Find the trigonometric function values for $-270°$.

If an angle θ is measured in degrees, we use the degree symbol to indicate a trigonometric function value, such as $\cos 60°$ and $\tan 28°$. If θ is measured in radians, we do not use a symbol (for example, $\sin \frac{\pi}{4}$, $\sec \pi$, and $\cos 16$). Notice that the 0 coordinate of a point on the terminal side of a quadrantal angle cannot appear in a denominator. (See Figure 4.27(a), where the terminal side is on the x-axis.) The y-coordinate is 0; so the cotangent $\left(\cot \theta = \frac{x}{y} \right)$ and the cosecant $\left(\csc \theta = \frac{r}{y} \right)$ functions are undefined. If the terminal side is on the y-axis, as in Figure 4.27(b), then the x-coordinate is 0 and the tangent $\left(\tan \theta = \frac{y}{x} \right)$ and the secant $\left(\sec \theta = \frac{r}{x} \right)$ functions are undefined.

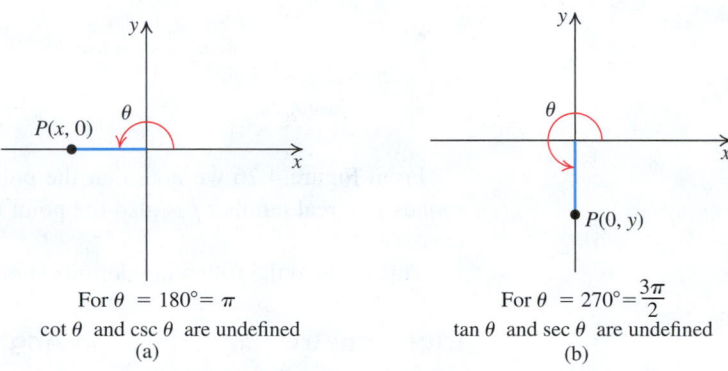

For $\theta = 180° = \pi$
$\cot \theta$ and $\csc \theta$ are undefined
(a)

For $\theta = 270° = \frac{3\pi}{2}$
$\tan \theta$ and $\sec \theta$ are undefined
(b)

Figure 4.27 Undefined trigonometric function values.

5 Find exact trigonometric function values of common angles.

Trigonometric Function Values of Common Angles

We can find the exact values for the trigonometric functions of quadrantal angles whenever they are defined (see Examples 4 and 5). We now find the exact trigonometric function values of **common acute angles:** $\frac{\pi}{6} = 30°$, $\frac{\pi}{4} = 45°$, and $\frac{\pi}{3} = 60°$.

Trigonometric Function Values of $t = \frac{\pi}{4} = 45°$ Let $P(t) = (x, y)$ be the terminal point on the unit circle associated with a real number t. For $t = \frac{\pi}{4}$, the point $P\left(\frac{\pi}{4}\right) = (x, y)$ bisects the arc from $(1, 0)$ to $(0, 1)$ on the unit circle (see Figure 4.28), we conclude that the point $P\left(\frac{\pi}{4}\right) = (x, y)$ lies on the line $y = x$.

That is for $t = \frac{\pi}{4}$, $y = x$. We then have:

$$x^2 + y^2 = 1 \qquad \text{Equation of the unit circle}$$
$$x^2 + x^2 = 1 \qquad y = x \text{ when } t = \frac{\pi}{4}, \text{ replace } y \text{ with } x.$$
$$2x^2 = 1 \qquad \text{Simplify.}$$
$$x^2 = \frac{1}{2} \qquad \text{Divide both sides by 2.}$$
$$x = \sqrt{\frac{1}{2}} = \frac{1}{\sqrt{2}} \cdot \frac{\sqrt{2}}{\sqrt{2}} = \frac{\sqrt{2}}{2} \qquad x \text{ is positive because } P(x, y) \text{ is in quadrant I.}$$

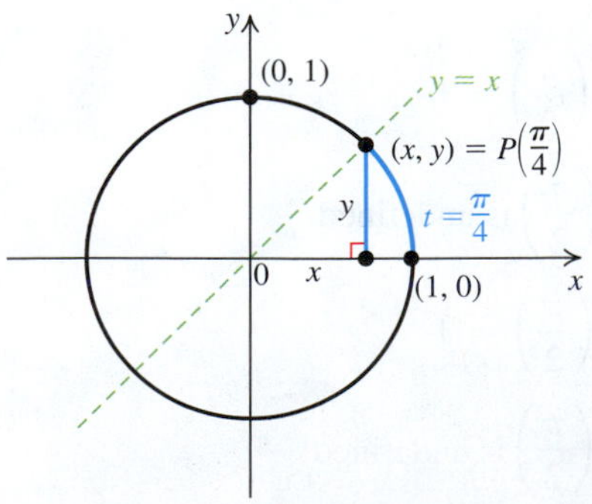

So, $y = x = \frac{\sqrt{2}}{2}$, when $t = \frac{\pi}{4}$.

Therefore $(x, y) = \left(\frac{\sqrt{2}}{2}, \frac{\sqrt{2}}{2} \right)$ is the point on the unit circle associated with $t = \frac{\pi}{4}$.

Figure 4.28 Terminal Point $P\left(\frac{\pi}{4}\right) = \left(\frac{\sqrt{2}}{2}, \frac{\sqrt{2}}{2} \right)$.

So,

$$\sin 45° = \sin \frac{\pi}{4} = y = \frac{\sqrt{2}}{2}; \quad \cos 45° = \cos \frac{\pi}{4} = x = \frac{\sqrt{2}}{2}; \quad \tan 45° = \tan \frac{\pi}{4} = \frac{y}{x} = 1$$

$$\csc 45° = \csc\frac{\pi}{4} = \frac{1}{y} = \sqrt{2}; \ \sec 45° = \sec\frac{\pi}{4} = \frac{1}{x} = \sqrt{2}; \ \cot 45° = \cot\frac{\pi}{4} = \frac{x}{y} = 1$$

RECALL

An *equilateral triangle* is a triangle with all three sides of equal length. The measure of each angle of an equilateral triangle is 60°.

Trigonometric Function Values of $\frac{\pi}{3} = 60°$ and $\frac{\pi}{6} = 30°$ Now consider an equilateral triangle with side length 1, as shown in Figure 4.29(a). The (dashed) altitude in Figure 4.29(b) divides this triangle into two congruent right triangles.

We find the length h of the altitude by applying the Pythagorean Theorem to one of the triangles.

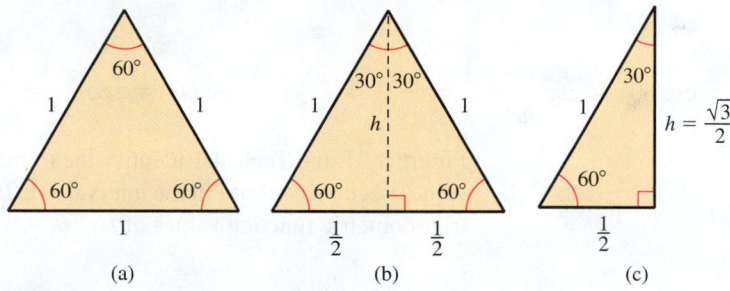

Figure 4.29 A $30° - 60° - 90°$ triangle.

We can use either triangle to find the exact values of all six trigonometric functions for both 30° and 60° angles. We have

$$1^2 = h^2 + \left(\frac{1}{2}\right)^2 \qquad \text{Pythagorean Theorem applied to Figure 4.29(b)}$$

$$h^2 = 1^2 - \left(\frac{1}{2}\right)^2 = 1 - \frac{1}{4} \qquad \text{Subtract } \left(\frac{1}{2}\right)^2 \text{ from both sides and interchange sides.}$$

$$h^2 = \frac{3}{4} \qquad \text{Simplify.}$$

$$h = \frac{\sqrt{3}}{2} \qquad h \text{ is positive.}$$

The resulting right triangle whose acute angles measure 30° and 60° is given in Figure 4.29(c). We call this a **30°– 60°– 90° triangle**.

Place the triangle of Figure 4.29(c) so that the angle of $60° = \frac{\pi}{3}$ is in standard position. See Figure 4.30(a).

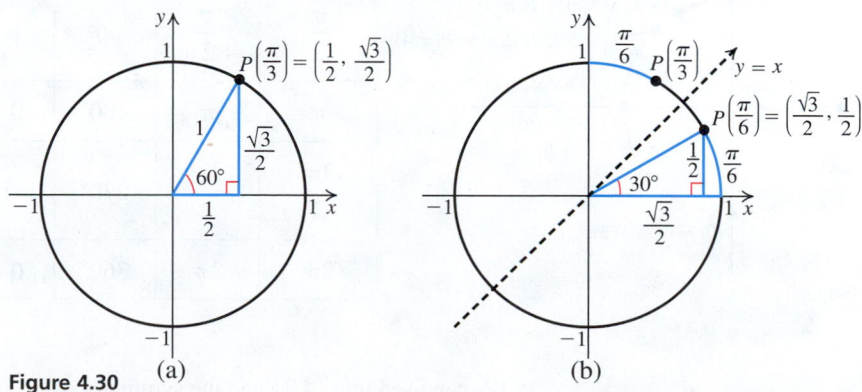

Figure 4.30

SIDE
NOTE

Note that the point $P\left(\frac{\pi}{6}\right)$ is the symmetric image of the point $P\left(\frac{\pi}{3}\right)$ about the line $y = x$.

The point corresponding to $t = \frac{\pi}{3} = 60°$ is $P\left(\frac{\pi}{3}\right) = \left(\frac{1}{2}, \frac{\sqrt{3}}{2}\right)$. Similarly, if the triangle of Figure 4.29(c) is placed so that the $30° = \frac{\pi}{3}$ angle is in standard position the point $P\left(\frac{\pi}{6}\right) = \left(\frac{\sqrt{3}}{2}, \frac{1}{2}\right)$ is the point on the unit circle associated with $t = \frac{\pi}{6} = 30°$. See Figure 4.30(b). We then have:

$$\sin 30° = \sin\frac{\pi}{6} = y = \frac{1}{2}; \quad \cos 30° = \cos\frac{\pi}{6} = x = \frac{\sqrt{3}}{2}; \quad \tan 30° = \tan\frac{\pi}{6} = \frac{y}{x} = \frac{1}{\sqrt{3}} = \frac{\sqrt{3}}{3}$$

$$\csc 30° = \csc\frac{\pi}{6} = \frac{1}{y} = 2; \quad \sec 30° = \sec\frac{\pi}{6} = \frac{1}{x} = \frac{2}{\sqrt{3}} = \frac{2\sqrt{3}}{3}; \quad \cot 30° = \cot\frac{\pi}{6} = \frac{x}{y} = \sqrt{3};$$

and

$$\sin 60° = \sin\frac{\pi}{3} = y = \frac{\sqrt{3}}{2}; \quad \cos 60° = \cos\frac{\pi}{3} = x = \frac{1}{2}; \quad \tan 60° = \tan\frac{\pi}{3} = \frac{y}{x} = \frac{\frac{\sqrt{3}}{2}}{\frac{1}{2}} = \sqrt{3}$$

$$\csc 60° = \csc\frac{\pi}{3} = \frac{1}{y} = \frac{2}{\sqrt{3}} = \frac{2\sqrt{3}}{3}; \quad \sec 60° = \sec\frac{\pi}{3} = \frac{1}{x} = 2; \quad \cot 60° = \cot\frac{\pi}{3} = \frac{x}{y} = \frac{1}{\sqrt{3}} = \frac{\sqrt{3}}{3}$$

Figure 4.31 and Table 4.1 identify the terminal points $P(t) = (x, y)$ on the unit circle for some special values of t in the interval $[0, 2\pi]$. In Table 4.1, we also list the corresponding trigonometric function values of $t = \theta$

TABLE 4.1 Trigonometric function values of common angles

Real Number t	Angle $t = \theta$ Radians	Angle θ in Degrees	$\sin\theta$	$\cos\theta$	$\tan\theta$	$\cot\theta$	$\sec\theta$	$\csc\theta$
0	0	0°	0	1	0	undef	1	undef
$\frac{\pi}{6}$	$\frac{\pi}{6}$	30°	$\frac{1}{2}$	$\frac{\sqrt{3}}{2}$	$\frac{\sqrt{3}}{3}$	$\sqrt{3}$	$\frac{2\sqrt{3}}{3}$	2
$\frac{\pi}{4}$	$\frac{\pi}{4}$	45°	$\frac{\sqrt{2}}{2}$	$\frac{\sqrt{2}}{2}$	1	1	$\sqrt{2}$	$\sqrt{2}$
$\frac{\pi}{3}$	$\frac{\pi}{3}$	60°	$\frac{\sqrt{3}}{2}$	$\frac{1}{2}$	$\sqrt{3}$	$\frac{\sqrt{3}}{3}$	2	$\frac{2\sqrt{3}}{3}$
$\frac{\pi}{2}$	$\frac{\pi}{2}$	90°	1	0	undef	0	undef	1
π	π	180°	0	−1	0	undef	−1	undef
$\frac{3\pi}{2}$	$\frac{3\pi}{2}$	270°	−1	0	undef	0	undef	−1
2π	2π	360°	0	1	0	undef	1	undef

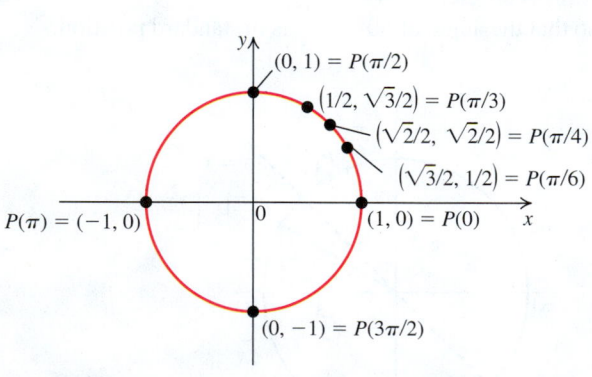

Figure 4.31 Terminal points.

We can use Figure 4.31 and the symmetry of the unit circle (page 352) to find the terminal points for values of t that are integer multiples of $\frac{\pi}{6} = 30°$, $\frac{\pi}{4} = 45°$, and $\frac{\pi}{3} = 60°$. See Figure 4.32.

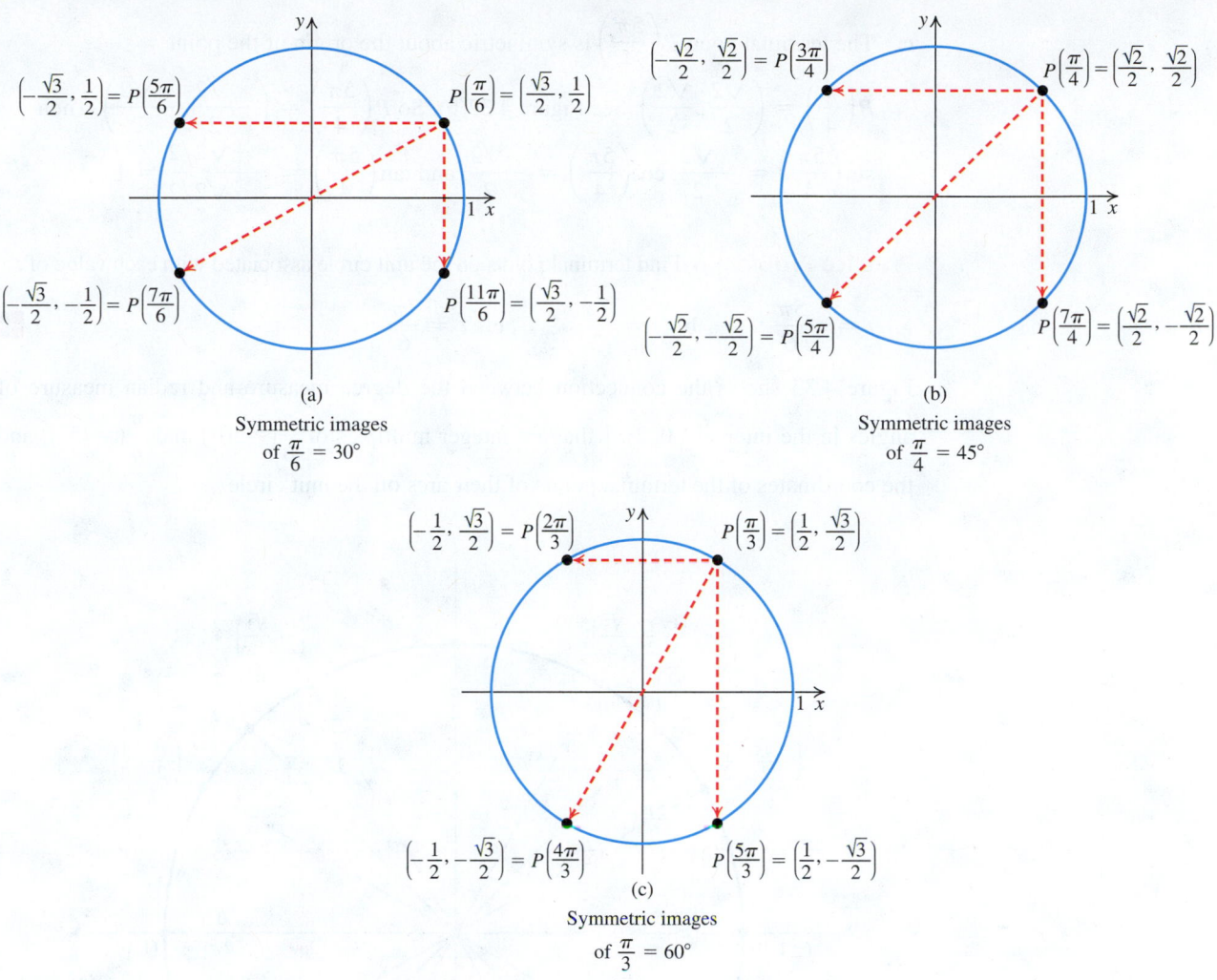

Figure 4.32 If (a, b) is on the unit circle, then so is, $(-a, b)$, $(-a, -b)$, and $(a, -b)$.

EXAMPLE 6 **Finding Terminal Points by Symmetry**

Find terminal points on the unit circle associated with each value of t. Then use symmetries to write the corresponding values of $\sin t$, $\cos t$, and $\tan t$.

a. $t = -\dfrac{\pi}{3}$ **b.** $t = \dfrac{5\pi}{6}$ **c.** $t = \dfrac{5\pi}{4}$

Solution

a. The terminal point $P\left(-\dfrac{\pi}{3}\right) = P\left(\dfrac{5\pi}{3}\right)$ is the symmetric image about the x-axis of

the point $P\left(\dfrac{\pi}{3}\right) = \left(\dfrac{1}{2}, \dfrac{\sqrt{3}}{2}\right)$. See Figure 4.32(c). So, $P\left(-\dfrac{\pi}{3}\right) = \left(\dfrac{1}{2}, -\dfrac{\sqrt{3}}{2}\right)$. Then

$\sin\left(-\dfrac{\pi}{3}\right) = -\dfrac{\sqrt{3}}{2}$; $\cos\left(-\dfrac{\pi}{3}\right) = \dfrac{1}{2}$; and $\tan\left(-\dfrac{\pi}{3}\right) = \dfrac{-\sqrt{3}/2}{1/2} = -\sqrt{3}$.

b. The terminal point $P\left(\dfrac{5\pi}{6}\right)$ is the symmetric image about the y-axis of the point

$P\left(\dfrac{\pi}{6}\right) = \left(\dfrac{\sqrt{3}}{2}, \dfrac{1}{2}\right)$. See Figure 4.32(a). So $P\left(\dfrac{5\pi}{6}\right) = \left(-\dfrac{\sqrt{3}}{2}, \dfrac{1}{2}\right)$. Then

$\sin\left(\dfrac{5\pi}{6}\right) = \dfrac{1}{2}$; $\cos\left(\dfrac{5\pi}{6}\right) = -\dfrac{\sqrt{3}}{2}$; and $\tan\left(\dfrac{5\pi}{6}\right) = \dfrac{1/2}{-\sqrt{3}/2} = -\dfrac{1}{\sqrt{3}} = -\dfrac{\sqrt{3}}{3}$.

c. The terminal point $P\left(\dfrac{5\pi}{4}\right)$ is symmetric about the origin of the point

$P\left(\dfrac{\pi}{4}\right) = \left(\dfrac{\sqrt{2}}{2}, \dfrac{\sqrt{2}}{2}\right)$. See Figure 4.32(b). So $P\left(\dfrac{5\pi}{4}\right) = \left(-\dfrac{\sqrt{2}}{2}, -\dfrac{\sqrt{2}}{2}\right)$. Then

$\sin\left(\dfrac{5\pi}{4}\right) = -\dfrac{\sqrt{2}}{2}$; $\cos\left(\dfrac{5\pi}{4}\right) = -\dfrac{\sqrt{2}}{2}$; and $\tan\left(\dfrac{5\pi}{4}\right) = -\dfrac{-\sqrt{2}/2}{-\sqrt{2}/2} = 1.$

Practice Problem 6 Find terminal points on the unit circle associated with each value of t.

a. $t = -\dfrac{5\pi}{6}$ **b.** $t = \dfrac{2\pi}{3}$ **c.** $t = \dfrac{7\pi}{6}$

Figure 4.33 shows the connection between the degree measure and radian measure of angles in the interval $[0, 2\pi]$ that are integer multiples of $\dfrac{\pi}{6}(=30°)$ and $\dfrac{\pi}{4}(=45°)$, and the coordinates of the terminal points of their arcs on the unit circle.

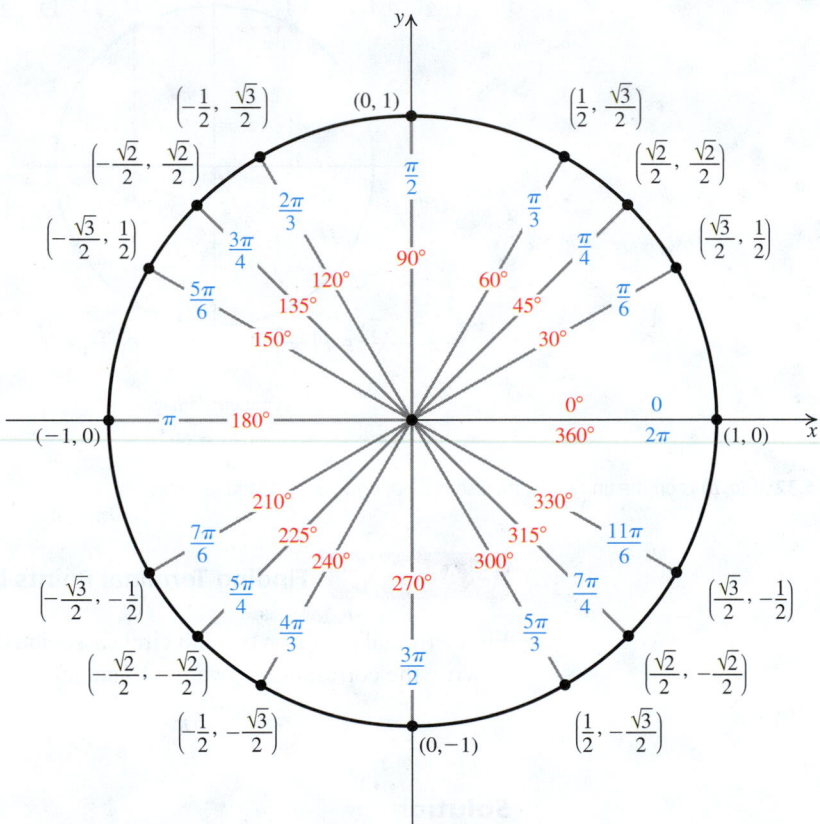

Figure 4.33 Unit circle: degrees, radians, ordered pairs.

EXAMPLE 7 **Finding Terminal Points for $|t| > 2\pi$**

Find the terminal points on the unit circle associated with each value of t. Then write the corresponding values of $\sin t$, $\cos t$, and $\tan t$.

a. $t = \dfrac{33\pi}{4}$ **b.** $t = -\dfrac{19\pi}{6}$

Solution

Recall that for any integer n, $P(t) = P(t + 2n\pi)$. See page 354.

In each case, we first rewrite t as an integer multiple of 2π plus a number in the interval $[0, 2\pi]$.

a. $t = \dfrac{33\pi}{4} = \dfrac{32\pi}{4} + \dfrac{\pi}{4} = 4(2\pi) + \dfrac{\pi}{4}$. So the terminal point associated with

$t = \dfrac{33\pi}{4}$ coincides with the terminal point associated with $t = \dfrac{\pi}{4}$. That is,

$P\left(\dfrac{33\pi}{4}\right) = P\left(\dfrac{\pi}{4}\right) = \left(\dfrac{\sqrt{2}}{2}, \dfrac{\sqrt{2}}{2}\right)$. See Table 4.1. Then,

$\sin\left(\dfrac{33\pi}{4}\right) = \dfrac{\sqrt{2}}{2}; \quad \cos\left(\dfrac{33\pi}{4}\right) = \dfrac{\sqrt{2}}{2}; \quad \tan\left(\dfrac{33\pi}{4}\right) = \dfrac{\sqrt{2}/2}{\sqrt{2}/2} = 1.$

b. Similarly, $t = -\dfrac{19\pi}{6} = -4\pi + \dfrac{5\pi}{6} = 2(-2)\pi + \dfrac{5\pi}{6}$. From Example 6(b),

$P\left(-\dfrac{19\pi}{6}\right) = P\left(\dfrac{5\pi}{6}\right) = \left(-\dfrac{\sqrt{3}}{2}, \dfrac{1}{2}\right)$. Then $\sin\left(-\dfrac{19\pi}{6}\right) = \dfrac{1}{2}$;

$\cos\left(-\dfrac{19\pi}{6}\right) = -\dfrac{\sqrt{3}}{2}; \quad \tan\left(-\dfrac{19\pi}{6}\right) = \dfrac{1/2}{-\sqrt{3}/2} = -\dfrac{1}{\sqrt{3}} = -\dfrac{\sqrt{3}}{3}.$

Practice Problem 7 Repeat Example 7 for **(a)** $t = \dfrac{31\pi}{6}$, **(b)** $t = -\dfrac{28\pi}{3}$.

6 Approximate trigonometric function values using a calculator.

Evaluating Trigonometric Functions Using a Calculator

Suppose we want to find the trigonometric function values of $20° = \dfrac{\pi}{9}$ radian. Unfortu-

nately, the coordinates $(x, y) = \left(\cos\dfrac{\pi}{9}, \sin\dfrac{\pi}{9}\right)$ of the terminal point $P\left(\dfrac{\pi}{9}\right)$ on the unit

circle do not have a nice form. Neither coordinate is a rational number or a radical of a rational number. In such cases, we use a calculator to find their approximate values.

When you are using a calculator to find the values of the trigonometric functions, the first step is to set the *mode* of measurement. If you are working with an angle given in degrees, set the mode to *degrees*; otherwise, set the mode to *radians*.

Your calculator has keys labeled $\boxed{\text{SIN}}$, $\boxed{\text{COS}}$, and $\boxed{\text{TAN}}$ but no keys for directly evaluating the cosecant, secant, or cotangent functions. However, these functions are the reciprocal functions of sine, cosine, and tangent functions, respectively. Therefore, to evaluate these functions, take the reciprocal of the appropriate function value.

TECHNOLOGY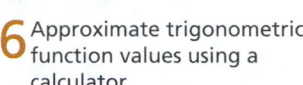
CONNECTION

The first screen shows how sin 71°
is displayed on a graphing calculator in Degree mode.

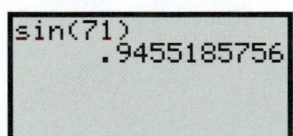

This screen shows how $\tan\dfrac{5\pi}{7}$ and
sec 1.3 are displayed on a graphing calculator in Radian mode.

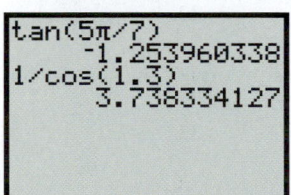

EXAMPLE 8 **Approximating Trigonometric Function Values Using a Calculator**

Use a calculator to find the approximate value of each expression. Round your answers to two decimal places.

a. $\sin 71°$ **b.** $\tan\dfrac{5\pi}{7}$ **c.** $\sec 1.3$

Solution

a. Set the MODE to degrees. $\sin 71° \approx 0.9455185756 \approx 0.95$

b. Set the MODE to radians. $\tan\dfrac{5\pi}{7} \approx -1.253960338 \approx -1.25$

c. Set the MODE to radians. $\sec 1.3 = \dfrac{1}{\cos 1.3} \approx 3.738334127 \approx 3.74$

Practice Problem 8 Repeat Example 8 for each expression.

a. $\cos 114°$ **b.** $\cot 3.6$

WARNING

The calculator keys $\boxed{\text{SIN}^{-1}}$, $\boxed{\text{COS}^{-1}}$, and $\boxed{\text{TAN}^{-1}}$ do not represent the reciprocal functions for the sine, cosine, and tangent functions, respectively. The functions $\sin^{-1}$, $\cos^{-1}$, and $\tan^{-1}$ are inverse functions and are discussed in Section 4.6.

Answers to Practice Problems

1. P is not on the unit circle   2. $\left(-\dfrac{2}{3}, \dfrac{\sqrt{5}}{3}\right)$ and $\left(-\dfrac{2}{3}, -\dfrac{\sqrt{5}}{3}\right)$

3. **a.** $\sin t = \dfrac{3}{5}$, $\cos t = -\dfrac{4}{5}$, $\tan t = -\dfrac{3}{4}$,

   b. $\csc t = \dfrac{3\sqrt{2}}{4}$, $\sec t = 3$, $\cot t = \dfrac{\sqrt{2}}{4}$

4. $\sin \dfrac{5\pi}{2} = 1$, $\cos \dfrac{5\pi}{2} = 0$, $\tan \dfrac{5\pi}{2}$ is undefined

   $\csc \dfrac{5\pi}{2} = 1$, $\sec \dfrac{5\pi}{2}$ is undefined, $\cot \dfrac{5\pi}{2} = 0$

5. $\sin(-270°) = 1$, $\cos(-270°) = 0$, $\tan(-270°)$ is undefined, $\csc(-270°) = 1$, $\sec(-270°)$ is undefined, $\cot(-270°) = 0$

6. **a.** $\left(-\dfrac{\sqrt{3}}{2}, -\dfrac{1}{2}\right)$,   **b.** $\left(-\dfrac{1}{2}, \dfrac{\sqrt{3}}{2}\right)$ **c.** $\left(-\dfrac{\sqrt{3}}{2}, -\dfrac{1}{2}\right)$

7. **a.** $P\left(\dfrac{31\pi}{6}\right) = \left(-\dfrac{\sqrt{3}}{2}, -\dfrac{1}{2}\right)$.

   $\cos \dfrac{31\pi}{6} = -\dfrac{\sqrt{3}}{2}$, $\sin \dfrac{31\pi}{6} = -\dfrac{1}{2}$, $\tan \dfrac{31\pi}{6} = \dfrac{1}{\sqrt{3}} = \dfrac{\sqrt{3}}{3}$

   b. $P\left(-\dfrac{28\pi}{3}\right) = \left(-\dfrac{1}{2}, \dfrac{\sqrt{3}}{2}\right)$.

   $\cos\left(-\dfrac{28\pi}{3}\right) = -\dfrac{1}{2}$, $\sin\left(-\dfrac{28\pi}{3}\right) = \dfrac{\sqrt{3}}{2}$, $\tan\left(-\dfrac{28\pi}{3}\right) = -\sqrt{3}$

8. **a.** -0.41 **b.** 2.03

SECTION 4.2   Exercises

Concepts and Vocabulary

1. A unit circle is a circle of radius _____ centered at the origin. Its equation is _____.

2. Let $P(t) = (x, y)$ be a point on the unit circle. Then

   $\cos t = $ _____, $\sin t = $ _____,

   $\tan t = $ _____; and $\dfrac{1}{x} = $ _____,

   $\dfrac{1}{y} = $ _____, $\dfrac{x}{y} = $ _____.

3. The terminal points $P(t)$ and $P(t + \pi)$ on the unit circle are symmetric about the _____.

4. The terminal points $P(t)$ and $P(\pi - t)$ on the unit circle are symmetric about the _____.

5. **True or False.** The terminal points $P(t)$ and $P(t - 2\pi)$ coincide on the unit circle.

6. **True or False.** The trigonometric values for $t = \pi$ are identical to those for $t = -\pi$.

7. **True or False.** The trigonometric function values for $t = \dfrac{\pi}{2}$ are identical to those for $t = -\dfrac{\pi}{2}$.

8. **True or False.** If $P(t) = (x, y)$, then $P(-t) = (x, -y)$

Building Skills

In Exercises 9–14, determine whether the given point (x, y) is on the unit circle.

9. $\left(\dfrac{3}{5}, \dfrac{4}{5}\right)$ 10. $\left(\dfrac{12}{13}, \dfrac{5}{13}\right)$ 11. $\left(\dfrac{3}{4}, -\dfrac{\sqrt{7}}{4}\right)$

12. $\left(-\dfrac{\sqrt{13}}{7}, \dfrac{6}{7}\right)$ 13. $\left(\dfrac{1}{3}, \dfrac{2}{3}\right)$ 14. $\left(-\dfrac{3}{4}, \dfrac{1}{4}\right)$

In Exercises 15–20, find all numbers u (if any) so that the given point (u, y) or (x, u) is on the unit circle.

15. $\left(u, \dfrac{1}{2}\right)$ 16. $\left(u, -\dfrac{1}{3}\right)$ 17. $\left(-\dfrac{3}{4}, u\right)$

18. $\left(-\dfrac{2}{5}, u\right)$ 19. $\left(u, \dfrac{3}{2}\right)$ 20. $\left(-\dfrac{4}{3}, u\right)$

In Exercises 21–26, $P(t) = (x, y)$ is the terminal point on the unit circle that corresponds to the real number t. Find the values of $\sin t$, $\cos t$, and $\tan t$.

21. $\left(\dfrac{2\sqrt{2}}{3}, \dfrac{1}{3}\right)$ 22. $\left(\dfrac{1}{2}, \dfrac{\sqrt{3}}{2}\right)$ 23. $\left(-\dfrac{1}{3}, \dfrac{2\sqrt{2}}{3}\right)$

24. $\left(-\dfrac{\sqrt{3}}{2}, \dfrac{1}{2}\right)$ 25. $\left(\dfrac{1}{5}, -\dfrac{2\sqrt{6}}{5}\right)$ 26. $\left(\dfrac{1}{2}, -\dfrac{\sqrt{3}}{2}\right)$

In Exercises 27–32, let $P(t) = (x, y)$. Find the values of $\sec t$, $\csc t$, and $\cot t$.

27. $\left(\dfrac{3}{4}, \dfrac{\sqrt{7}}{4}\right)$ 28. $\left(\dfrac{5}{6}, -\dfrac{\sqrt{11}}{6}\right)$ 29. $\left(-\dfrac{\sqrt{3}}{3}, \dfrac{\sqrt{6}}{3}\right)$

30. $\left(-\dfrac{2\sqrt{2}}{3}, -\dfrac{1}{3}\right)$ 31. $\left(-\dfrac{2\sqrt{6}}{5}, -\dfrac{1}{5}\right)$ 32. $\left(-\dfrac{2}{7}, \dfrac{3\sqrt{5}}{7}\right)$

In Exercises 33–38, find the values (if any) of the six trigonometric functions of each value of t.

33. $t = 5\pi$ 34. $t = 3\pi$ 35. $t = -\dfrac{3\pi}{2}$

36. $t = \dfrac{5\pi}{2}$ 37. $t = \dfrac{7\pi}{2}$ 38. $t = -\dfrac{9\pi}{2}$

In Exercises 39–50, find each trigonometric function value.

39. $\tan(4\pi)$ 40. $\sec(7\pi)$ 41. $\sin(-2\pi)$

42. $\cos(-5\pi)$ 43. $\cos\left(-\dfrac{3\pi}{2}\right)$ 44. $\sin\left(-\dfrac{\pi}{2}\right)$

45. $\sin(-180°)$ **46.** $\cos(360°)$ **47.** $\tan(540°)$

48. $\sec(270°)$ **49.** $\csc(-720°)$ **50.** $\cot(-630°)$

In Exercises 51–62, use Table 4.1 to find the exact value of each expression.

51. $\sin 180° - \cos 90°$ **52.** $\cos 180° - \sin 90°$

53. $\sin 30° \sec 60°$ **54.** $\sin 270° \csc 45°$

55. $2 \tan 60° \cos 30° - \cot 30° \csc 60°$

56. $3 \tan 30° \cot 60° + \csc 30° \cos 60°$

57. $\sin \dfrac{\pi}{4} - \cos \pi$ **58.** $\sec \pi + \sin \dfrac{\pi}{6}$

59. $\tan \dfrac{\pi}{4} - \cot \dfrac{\pi}{3}$ **60.** $\csc \dfrac{\pi}{2} + \cos \dfrac{\pi}{3}$

61. $\sin \dfrac{3\pi}{2} \tan \dfrac{\pi}{4}$ **62.** $\cos \dfrac{\pi}{2} \sec \pi$

In Exercises 63–92, use Table 4.1 and symmetry (see Figure 4.32) to find the exact value of each expression.

63. $\sin\left(\dfrac{7\pi}{6}\right)$ **64.** $\cos\left(\dfrac{11\pi}{6}\right)$ **65.** $\tan\left(\dfrac{2\pi}{3}\right)$

66. $\cot\left(\dfrac{3\pi}{4}\right)$ **67.** $\sec\left(\dfrac{4\pi}{3}\right)$ **68.** $\tan\left(\dfrac{7\pi}{6}\right)$

69. $\csc\left(\dfrac{11\pi}{6}\right)$ **70.** $\sin\left(\dfrac{7\pi}{4}\right)$ **71.** $\cos\left(\dfrac{5\pi}{3}\right)$

72. $\sin 120°$ **73.** $\cos 135°$ **74.** $\sec 150°$

75. $\tan 210°$ **76.** $\sin 225°$ **77.** $\cot 240°$

78. $\sin 300°$ **79.** $\cos 315°$ **80.** $\tan 330°$

81. $\sin\left(-\dfrac{5\pi}{6}\right)$ **82.** $\cos\left(-\dfrac{5\pi}{4}\right)$ **83.** $\tan\left(-\dfrac{2\pi}{3}\right)$

84. $\sec\left(-\dfrac{7\pi}{6}\right)$ **85.** $\csc\left(-\dfrac{5\pi}{3}\right)$ **86.** $\tan\left(-\dfrac{7\pi}{4}\right)$

87. $\sin(-30°)$ **88.** $\cos(-45°)$ **89.** $\tan(-120°)$

90. $\sec(-150°)$ **91.** $\csc(-240°)$ **92.** $\cot(-300°)$

In Exercises 93–110, find the exact value of each expression.

93. $\cos\left(\dfrac{13\pi}{6}\right)$ **94.** $\tan\left(\dfrac{9\pi}{4}\right)$ **95.** $\sin\left(\dfrac{10\pi}{3}\right)$

96. $\sin\left(\dfrac{19\pi}{6}\right)$ **97.** $\cos\left(\dfrac{29\pi}{3}\right)$ **98.** $\tan\left(\dfrac{31\pi}{3}\right)$

99. $\tan\left(-\dfrac{9\pi}{4}\right)$ **100.** $\sin\left(-\dfrac{23\pi}{6}\right)$ **101.** $\cos\left(-\dfrac{15\pi}{4}\right)$

102. $\cot\left(-\dfrac{13\pi}{6}\right)$ **103.** $\sin\left(-\dfrac{35\pi}{4}\right)$ **104.** $\cos\left(-\dfrac{35\pi}{6}\right)$

105. $\sin 750°$ **106.** $\cos 1140°$ **107.** $\tan(765°)$

108. $\cos(-675°)$ **109.** $\tan(-1230°)$ **110.** $\sec(-1920°)$

Applying the Concepts

In Exercises 111–126, use a calculator to find the approximate value of each expression. Round your answers to two decimal places.

111. $\cos 14°$ **112.** $\sin 20°$ **113.** $\tan 32°$

114. $\cot 67°$ **115.** $\sec 34°$ **116.** $\csc 72°$

117. $\cos \dfrac{\pi}{5}$ **118.** $\sin \dfrac{\pi}{7}$ **119.** $\tan \dfrac{2\pi}{9}$

120. $\cot \dfrac{3\pi}{7}$ **121.** $\sec \dfrac{\pi}{6}$ **122.** $\csc \dfrac{\pi}{8}$

123. $\sin(-41°)$ **124.** $\cos(-54°)$

125. $\sin(-17.3)$ **126.** $\cos(-38.6)$

127. Blood pressure. Maurice's blood pressure (in millimeters of mercury) while resting is given by the function $R(t) = 25 \sin(2\pi t) + 100$, where t is time in seconds. Find his blood pressure after

   a. 1 second   **b.** 1.75 seconds

128. Deer population. The number of deer in a region is modeled by the equation

$$N(t) = 450 \sin\left(\dfrac{\pi}{6}t\right) + 1550$$

where t is measured in years and $t = 0$ represents 2010. Find the deer population in

   a. 2012

   b. 2018

   c. 2021

129. Tide pattern. The depth of water, d feet, in a channel t hours after midnight is given by

$$d = 3 \cos\left(\dfrac{\pi}{6}t\right) + 10.$$

Find the channel depth at

   a. noon   **b.** 6 P.M.

130. Pollution levels. On a typical day, a particular city's air pollution level (suspended particle levels in mg/m³) is modeled by the equation

$$P(t) = 0.5 + 15t - 0.4t^2 + 10 \sin\left(\dfrac{\pi}{12}t\right)$$

where t is measured in hours and $t = 0$ represents 6 A.M. Find the pollution level at

   a. noon   **b.** 6 P.M.   **c.** midnight

131. Daylight hours in Midland. The monthly average number of daylight hours in Midland is modeled by the equation

$$N(t) = 12.2 + 4.3 \sin\left[\dfrac{\pi}{6}(t - 3)\right]$$

where $t = 1$ represents January. Find the average number of daylight hours in Midland during

   a. March   **b.** June   **c.** December

132. Average temperature in Nashville. The monthly average temperature in degrees Fahrenheit in Nashville is modeled by the equation

$$T(t) = 61.5 + 21.5 \sin\left[\dfrac{\pi}{6}(t - 4)\right]$$

where $t = 1$ represents January. Find the average temperature in Nashville during

   a. April   **b.** August   **c.** January

Beyond the Basics

In Exercises 133–138, find all possible values of t that correspond to the given point $P(x, y)$ on the unit circle.

133. $P(-1, 0)$ **134.** $P(0, -1)$

135. $P\left(\dfrac{\sqrt{2}}{2}, -\dfrac{\sqrt{2}}{2}\right)$ **136.** $P\left(-\dfrac{\sqrt{2}}{2}, -\dfrac{\sqrt{2}}{2}\right)$

137. $P\left(-\dfrac{\sqrt{3}}{2}, \dfrac{1}{2}\right)$ **138.** $P\left(\dfrac{1}{2}, -\dfrac{\sqrt{3}}{2}\right)$

In Exercises 139–144, find the two values of t in the interval $[0, 2\pi)$ that satisfy the given equation.

139. $\sin t = \dfrac{1}{2}$ **140.** $\cos t = -\dfrac{1}{2}$

141. $\tan t = 1$ **142.** $\cot t = \sqrt{3}$

143. $\csc t = -\sqrt{2}$ **144.** $\sec t = 2$

In Exercises 145–148, the point $P(t) = (x, y)$ is the terminal point on the unit circle.

145. If $y = \dfrac{1}{2}$ and $x < 0$, find the value of $\cos t$.

146. If $x = \dfrac{\sqrt{2}}{2}$ and $y < 0$, find the value of $\sin t$.

147. If $x = \dfrac{\sqrt{3}}{2}$ and $y > 0$, find $\cos t$ and $\sin t$.

148. If $xy = 0$ and $x < 0$, find $\cos t$ and $\sin t$.

Critical Thinking / Discussion / Writing

149. Find all values of t in the interval $[3000\pi, 3010\pi]$ for which $\tan t$ is undefined.

150. For how many numbers in the interval $[0, 2000\pi]$ is the function $\sec t$ undefined?

In Exercises 151 and 152, the points (x, y), $(-x, y)$, $(x, -y)$, and $(-x, -y)$ are terminal points on the unit circle of the arcs of length $s, t, u,$ and v, respectively.

151. How are $\cos s$ and $\sin s$ related to $\cos t$ and $\sin t$?

152. How are $\cos(\pi - u)$ and $\sin(\pi - u)$ related to $\cos v$ and $\sin v$?

153. True or False. If $u = v$, then $\sin u = \sin v$.

154. True or False. If $\cos u = \cos v$, then $u = v$.

Getting Ready for the Next Section

In Exercises 155–157, let $r = \sqrt{x^2 + y^2}$, find the missing quantity and then find the desired ratios (rationalize the denominator if necessary).

155. $y = 3, r = 5, x = $?   Find $\dfrac{x}{r}$ and $\dfrac{y}{x}$.

156. $x = -5, r = 15, y = $?

Find $\dfrac{y}{x}$ and $\dfrac{y}{r}$.

157. $x = 3, r = 6, y = $?   Find $\dfrac{x}{y}$ and $\dfrac{y}{x}$.

In Exercises 158–160, draw angle θ in standard position. Then find the acute angle θ' formed by the terminal side of θ and the x-axis.

158. $\theta = 120°$ **159.** $\theta = 250°$ **160.** $\theta = 320°$

In Exercises 161–164, find the angle θ between $0°$ and $360°$ or 0 and 2π radians that is coterminal with the given angle. Then find the acute angle θ' formed by the terminal side of θ and the x-axis.

161. $590°$ **162.** $-835°$

163. $-\dfrac{31\pi}{6}$ **164.** $\dfrac{71\pi}{3}$

In Exercises 165–170, state whether each function is even, odd, or neither.

165. $f(x) = 2x^3 - 4x$

166. $g(x) = \sqrt{1 - x^2}$

167. $h(x) = 3x^4 + 2x^2 - 17$

168. $p(x) = \sqrt[3]{x^5}$

169. $F(x) = x^3 - 2x^2$

170. $G(x) = 5x + 3$

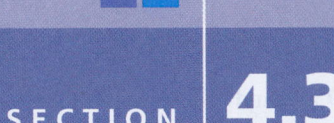

Trigonometric Functions of Angles

BEFORE STARTING THIS SECTION, REVIEW

1   Equation of a circle (Section 1.1, page 12)

2   Angles (Section 4.1, page 337)

3   Trigonometric functions of real numbers. (Section 4.2, page 354)

OBJECTIVES

1   Find trigonometric function values of an angle in standard position.

2   Determine the signs of the trigonometric functions in each quadrant.

3   Find a reference angle.

4   Use fundamental trigonometric identities.

5   Use even–odd properties of trigonometric functions.

◆ Golf and the Sine Function

The sine and cosine functions show up in surprising places.

The angle that a golf ball makes with the ground on takeoff and its initial speed determine the rest of its flight. Moreover, the time it takes for the ball to hit the ground after reaching its maximum height is exactly the same as if the ball had been dropped straight down from that height. The horizontal motion has no effect on the vertical motion. Suppose we ignore air resistance and measure time in seconds and distance in feet. Then the equation $h = v_0 t \sin \theta - 16t^2$ gives the height, h, of the golf ball after t seconds, where θ is the initial angle the ball makes with the ground and v_0 is its initial speed. The horizontal distance, d, the ball travels in t seconds is $d = tv_0 \cos \theta$. In Example 7 we use these equations to investigate the flight of a golf ball.

1 Find trigonometric function values of an angle in standard position.

$$\frac{x_1}{1} = \frac{x}{r}, \frac{y_1}{1} = \frac{y}{r}$$

Figure 4.34

Trigonometric Function Values of an Angle

Suppose the terminal side of an angle θ in standard position intersects the unit circle at the point $P_1(x_1, y_1)$, as illustrated for an acute angle θ in Figure 4.34. Let $P(x, y)$ be any point other than the origin on the terminal side of θ and $r = \sqrt{x^2 + y^2}$, as shown in Figure 4.34.

The triangles POQ and P_1OQ_1 are similar because they have equal angles; so the ratios of their corresponding sides are equal. We have $\sin \theta = y_1 = \dfrac{y_1}{1} = \dfrac{y}{r}$, $\cos \theta = x_1 = \dfrac{x_1}{1} = \dfrac{x}{r}$, $\tan \theta = \dfrac{y_1}{x_1} = \dfrac{y}{x}$, and so on. We can now evaluate the trigonometric functions for any angle θ using any point on its terminal side.

VALUES OF THE TRIGONOMETRIC FUNCTIONS OF AN ANGLE

Let $P(x, y)$ be any point on the terminal ray of an angle θ in standard position (other than the origin) and let $r = \sqrt{x^2 + y^2}$. Then $r > 0$, and

$$\sin\theta = \frac{y}{r} \qquad\qquad \csc\theta = \frac{r}{y} \, (y \neq 0)$$

$$\cos\theta = \frac{x}{r} \qquad\qquad \sec\theta = \frac{r}{x} \, (x \neq 0)$$

$$\tan\theta = \frac{y}{x} \, (x \neq 0) \qquad \cot\theta = \frac{x}{y} \, (y \neq 0)$$

EXAMPLE 1   Finding Exact Trigonometric Function Values

Suppose θ is an angle whose terminal side contains the point $P(-1, 3)$. Find the exact values of the six trigonometric functions of θ.

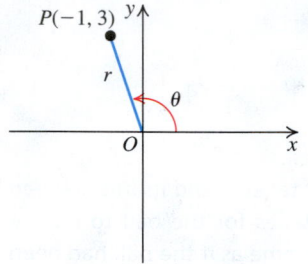

Figure 4.35

Solution

Because $x = -1$ and $y = 3$ (see Figure 4.35), we have

$$r = \sqrt{x^2 + y^2} \qquad\qquad \text{Definition of } r$$
$$= \sqrt{(-1)^2 + 3^2} = \sqrt{10} \qquad \text{Substitute and simplify.}$$

Replacing x with -1, y with 3, and r with $\sqrt{10}$ in the definition of the trigonometric functions, we have

$$\sin\theta = \frac{y}{r} = \frac{3}{\sqrt{10}} = \frac{3\sqrt{10}}{10} \qquad \csc\theta = \frac{r}{y} = \frac{\sqrt{10}}{3}$$

$$\cos\theta = \frac{x}{r} = \frac{-1}{\sqrt{10}} = -\frac{\sqrt{10}}{10} \qquad \sec\theta = \frac{r}{x} = \frac{\sqrt{10}}{-1} = -\sqrt{10}$$

$$\tan\theta = \frac{y}{x} = \frac{3}{-1} = -3 \qquad \cot\theta = \frac{x}{y} = \frac{-1}{3} = -\frac{1}{3}$$

Practice Problem 1 Suppose θ is an angle whose terminal side contains the point $P(2, -5)$. Find the exact values of the six trigonometric functions of θ.

2 Determine the signs of the trigonometric functions in each quadrant.

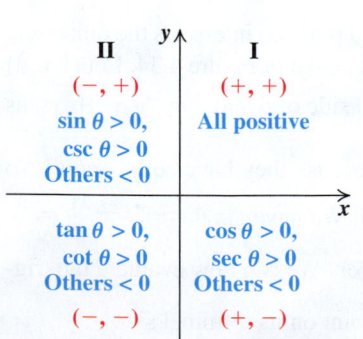

Figure 4.36 Signs of the trigonometric functions.

Signs of the Trigonometric Functions

Suppose an angle θ is not quadrantal and its terminal side contains the point (x, y). The signs of x and y determine the signs of the trigonometric functions. (Notice that $r = \sqrt{x^2 + y^2}$ is always *positive*.)

If θ lies in quadrant I, then both x and y are positive; so all six trigonometric function values are positive. However, if θ lies in quadrant II, then x is negative and y is positive; so only $\sin\theta = \frac{y}{r}$ and $\csc\theta = \frac{r}{y}$ are positive. If θ lies in quadrant III, then x and y are both negative; so only $\tan\theta = \frac{y}{x}$ and $\cot\theta = \frac{x}{y}$ are positive. If θ lies in quadrant IV, then x is positive and y is negative; so only $\cos\theta = \frac{x}{r}$ and $\sec\theta = \frac{r}{x}$ are positive. If θ is not a quadrantal angle, than the sign of a trigonometric function of θ depends on the quadrant in which it lies. Figure 4.36 summarizes the signs of the trigonometric functions.

EXAMPLE 2   **Determining the Quadrant in Which an Angle Lies**

If $\tan \theta > 0$ and $\cos \theta < 0$, in which quadrant does θ lie?

Solution

(i) Because $\tan \theta > 0$, θ lies either in quadrant I or in quadrant III.

(ii) Because $\cos \theta < 0$, θ lies in quadrant II or in quadrant III.

For both conditions (i) and (ii) to be true, θ must lie in quadrant III.

Practice Problem 2 If $\sin \theta > 0$ and $\cos \theta < 0$, in which quadrant does θ lie?

EXAMPLE 3   **Evaluating Trigonometric Functions**

Given that $\tan \theta = \dfrac{3}{2}$ and $\cos \theta < 0$, find the exact values of $\sin \theta$ and $\sec \theta$.

Solution

Because $\tan \theta = \dfrac{3}{2} > 0$ and $\cos \theta < 0$, θ lies in quadrant III. We want to identify a point (x, y) in quadrant III that is on the terminal side of θ. We have

$$\tan \theta = \frac{y}{x} = \frac{3}{2}$$

and because the point (x, y) is in quadrant III, both x and y must be negative. (See Figure 4.37.) If we choose $x = -2$ and $y = -3$, then

$$\tan \theta = \frac{y}{x} = \frac{-3}{-2} = \frac{3}{2}$$

and $r = \sqrt{x^2 + y^2} = \sqrt{(-2)^2 + (-3)^2} = \sqrt{4 + 9} = \sqrt{13}$.

Using $x = -2$, $y = -3$, and $r = \sqrt{13}$, we find $\sin \theta$ and $\sec \theta$:

$$\sin \theta = \frac{y}{r} = \frac{-3}{\sqrt{13}} = -\frac{3\sqrt{13}}{13} \qquad \sec \theta = \frac{r}{x} = \frac{\sqrt{13}}{-2} = -\frac{\sqrt{13}}{2}$$

Practice Problem 3 Given that $\tan \theta = -\dfrac{4}{5}$ and $\cos \theta > 0$, find the exact values of $\sin \theta$ and $\sec \theta$.

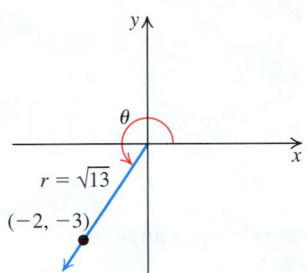

Figure 4.37

3 Find a reference angle.

Reference Angle

For any angle θ, there is a corresponding acute angle called its *reference angle*, whose trigonometric function values are either equal to or opposite in sign to that of θ.

Reference Angle

Let θ be an angle in standard position that is not a quadrantal angle. The **reference angle** for θ is the acute angle θ' ("theta prime") formed by the terminal side of θ and the x-axis.

The reference angle for a positive angle θ ($0° < \theta < 360°$ or $0 < \theta < 2\pi$) in each of the four quadrants is shown in Figure 4.38.

Quadrant I

$\theta = \theta'$
$\theta' = \theta$

Quadrant II

$\theta' = 180° - \theta$ or
$\theta' = \pi - \theta$

Quadrant III

$\theta' = \theta - 180°$ or
$\theta' = \theta - \pi$

Quadrant IV

$\theta' = 360° - \theta$ or
$\theta' = 2\pi - \theta$

Figure 4.38 Reference angles.

▲ **WARNING**

When a reference angle is found, a common error is to use the y-axis as one of the sides. Remember that the two sides of a reference angle are always located using the terminal side of the original angle and the positive or the negative x-axis.

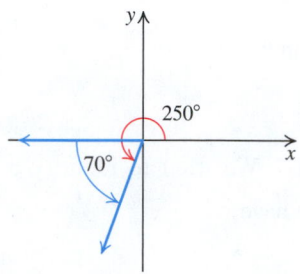

Figure 4.39(a)

EXAMPLE 4   **Identifying Reference Angles**

Find the reference angle θ' for each angle θ.

a. $\theta = 250°$ **b.** $\theta = \dfrac{3\pi}{5}$ **c.** $\theta = -150°$ **d.** $\theta = 5.75$

Solution

a. Because $250°$ lies in quadrant III and $0° < 250° < 360°$, the reference angle is $\theta' = \theta - 180°$. So, $\theta' = 250° - 180° = 70°$. See Figure 4.39(a).

b. Because $\dfrac{3\pi}{5}$ lies in quadrant II and $0 < \theta < 2\pi$, the reference angle is $\theta' = \pi - \theta$.

So, $\theta' = \pi - \dfrac{3\pi}{5} = \dfrac{5\pi}{5} - \dfrac{3\pi}{5} = \dfrac{2\pi}{5}$. See Figure 4.39(b).

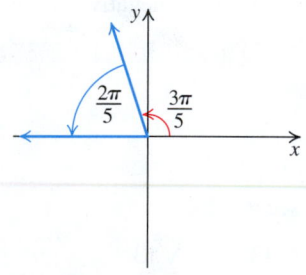

Figure 4.39(b)

c. The angle $\theta = -150°$ is shown in Figure 4.39(c). The figure shows that the acute angle θ' between the terminal side of θ and the x-axis is $30°$.
    Alternatively, note that a negative angle is always coterminal with a positive angle (θ and $\theta + n \cdot 360°$ are coterminal), and they have the same reference angle. So, $-150°$ and $-150° + 360° = 210°$ have the same reference angle. The angle $210°$ lies in quadrant III, and its reference angle $\theta' = 210° - 180° = 30°$.

d. Since no degree symbol appears in $\theta = 5.75$, θ has radian measure. Now $\dfrac{3\pi}{2} \approx 4.71$ and $2\pi \approx 6.28$ shows that θ lies in quadrant IV with $0 < \theta < 2\pi$, so $\theta' = 2\pi - \theta$. Then, $\theta' = 2\pi - 5.75 \approx 6.28 - 5.75 = 0.53$. See Figure 4.39(d).

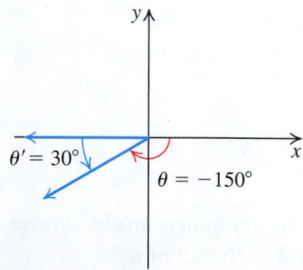

Figure 4.39(c)

Practice Problem 4 Find the reference angle θ' for each angle θ.

a. $\theta = 175°$ **b.** $\theta = \dfrac{5\pi}{3}$ **c.** $\theta = -240°$ **d.** $\theta = 1.94$

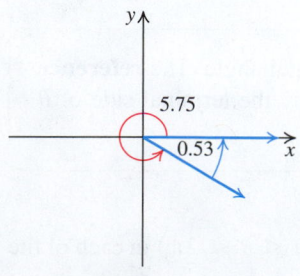

Figure 4.39(d)

Since the value of each trigonometric function of an angle in standard position is completely determined by the position of the terminal side, the following statements are true.

1. Coterminal angles are assigned identical values by the six trigonometric functions.
2. The signs of the values of the trigonometric functions are determined by the quadrant containing the terminal side.

Recall that for any integer n, θ and $\theta + n360°$ (in degree measure) are coterminal angles and θ and $\theta + 2\pi n$ (in radian measure) are coterminal angles.

TRIGONOMETRIC FUNCTION VALUES OF COTERMINAL ANGLES

θ in degrees

$$\sin\theta = \sin(\theta + n360°)$$

$$\cos\theta = \cos(\theta + n360°)$$

θ in radians

$$\sin\theta = \sin(\theta + 2\pi n)$$

$$\cos\theta = \cos(\theta + 2\pi n)$$

These equations hold for any integer n.

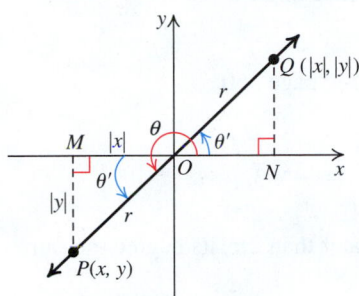

Figure 4.40 Reference angle.

Using Reference Angles A reference angle is used to find the values of trigonometric functions of any angle θ. For example, consider the reference angle θ' of the angle θ in Figure 4.40. Let $P(x, y)$ be a point on the terminal side of θ in quadrant III. Then $Q(|x|, |y|)$ is in quadrant I. From the definition of trigonometric functions, we have

$$\cos\theta = \frac{x}{r} = \frac{-|x|}{r} \qquad \text{is negative, so } x = -|x|.$$

The right triangles POM and QON in Figure 4.40 are congruent by SAS congruence. Since $Q(|x|, |y|)$ is on the terminal side of θ', we have

$$\cos\theta' = \frac{|x|}{r}.$$

So, $\cos\theta = -\dfrac{|x|}{r} = -\cos\theta'$. In general, $\cos\theta = \pm\cos\theta'$. The same is true for the other five trigonometric functions.

We next give a three-step procedure for using reference angles to find trigonometric function values.

PROCEDURE
IN ACTION

EXAMPLE 5 Finding Trigonometric Function Values Using the Reference Angles

OBJECTIVE

Find the value of any trigonometric function of an angle θ.

Step 1 If $\theta \geq 360°$ or $\theta < 0°$, find a coterminal angle between $0°$ and $360°$. Otherwise go to Step 2.

Step 2 Find the reference angle θ' for the angle resulting from Step 1. Write the trigonometric function value of θ'.

Step 3 Choose the correct sign for the trigonometric function based on the quadrant in which θ lies.

EXAMPLE

Find $\sin 1320°$.

1. Because $1320° = 3(360°) + 240°$, so $240°$ is coterminal with $1320°$.

2. Because $240°$ is in quadrant III, its reference angle θ' is

$$\theta' = 240° - 180° = 60°$$

and

$$\sin\theta' = \sin 60° = \frac{\sqrt{3}}{2} \qquad \text{See page 360.}$$

3. Angle θ and its coterminal angle, $240°$, lie in quadrant III, where the sine is *negative*. So,

$$\sin 1320° = \sin 240° = -\sin 60° = -\frac{\sqrt{3}}{2}$$

Practice Problem 5 Find the exact value of $\cos 1035°$.

EXAMPLE 6 Using the Reference Angle to Find Exact Values of Trigonometric Functions

Find the exact value of each expression.

a. $\tan(840°)$ **b.** $\sec \dfrac{59\pi}{6}$

Solution

$$\begin{array}{r} 2 \\ 360\overline{)840} \\ (-)\underline{720} \\ 120 \end{array}$$

$840 = 2(360) + 120$

a. Step 1 We have $840° = 2(360°) + 120°$. So, $840°$ is coterminal with $120°$.

Step 2 Because $120°$ is in quadrant II, its reference angle θ' is:
$$\theta' = 180° - 120° = 60°$$
$$\tan \theta' = \tan 60° = \sqrt{3} \qquad \text{See page 360.}$$

Step 3 In quadrant II, $\tan \theta$ is negative; so
$$\tan(840°) = \tan 120° = -\tan 60° = -\sqrt{3}.$$

$$\begin{array}{r} 4 \\ 2\pi\overline{)59\pi} \approx 9.8\pi \\ \underline{6} \\ (-)8\pi \\ \underline{11\pi} \\ 6 \end{array}$$

$\dfrac{59\pi}{6} = 4(2\pi) + \dfrac{11\pi}{6}$

b. Step 1 Since the radian measure of $\dfrac{59\pi}{6}$ is greater than 2π, its degree measure is greater than $360°$.
$$\frac{59\pi}{6} = \frac{48\pi + 11\pi}{6} = 8\pi + \frac{11\pi}{6};$$
so $\dfrac{11\pi}{6}$ is an angle between 0 and 2π that is coterminal with $\dfrac{59\pi}{6}$.

Step 2 Because $\dfrac{11\pi}{6}$ is in quadrant IV, its reference angle θ' is as follows:
$$\theta' = 2\pi - \frac{11\pi}{6} = \frac{\pi}{6}$$
$$\sec \theta' = \sec \frac{\pi}{6} = \frac{2\sqrt{3}}{3} \qquad \text{See page 360.}$$

Step 3 In quadrant IV, $\sec \theta > 0$; so
$$\sec \frac{59\pi}{6} = \sec \frac{11\pi}{6} = \sec \frac{\pi}{6} = \frac{2\sqrt{3}}{3}.$$

Practice Problem 6 Find the exact value of each expression.

a. $\csc 1035°$ **b.** $\cot \dfrac{17\pi}{6}$

EXAMPLE 7 The Flight of a Golf Ball

A golf ball is hit on a level fairway with an initial velocity of 128 feet per second and an initial angle of flight of $30°$. Find its range (the horizontal distance it traveled before hitting the ground) and its maximum height.

RECALL

The vertex of the graph of a parabola $f(x) = ax^2 + bx + c$ is $\left(-\dfrac{b}{2a}, f\left(-\dfrac{b}{2a}\right)\right)$.

Solution

We use the height equation from the section introduction.

$$h = v_0 t \sin \theta - 16t^2$$
$$h = 128t \sin 30° - 16t^2 \qquad \text{Given } v_0 = 128 \text{ and } \theta = 30°$$
$$= 64t - 16t^2 = -16t(t - 4) \qquad \text{Replace } \sin 30° \text{ with } \frac{1}{2} \text{ and simplify.}$$

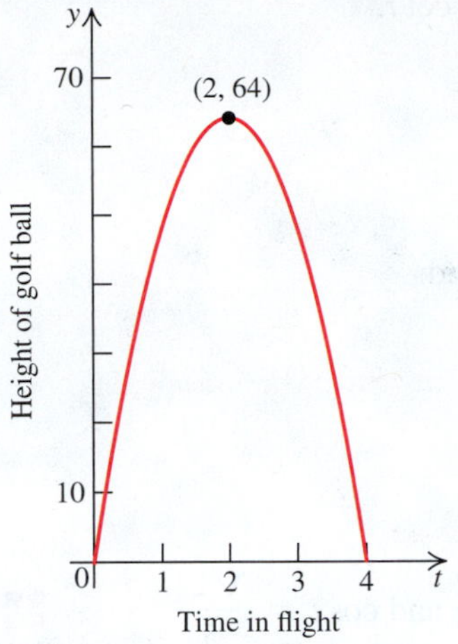

The graph of $h = 64t - 16t^2$ is a parabola; the portion of the graph with $h(t) \geq 0$ represents the flight path of the ball. (See Figure 4.41.) The vertex is $(2, 64)$ because

$$t = \frac{-64}{2(-16)} = 2 \text{ and } h(2) = 64(2) - 16(2)^2 = 64.$$ So the maximum height of the ball

is 64 feet. Because $h(4) = 0$, the ball remains in flight for four seconds. The distance, d, traveled after four seconds is the range, where (again from the section introduction):

$$d = tv_0 \cos \theta \qquad \text{Horizontal distance equation}$$
$$d = 4(128) \cos \theta = 4(128) \cos 30° \qquad \text{Given } v_0 = 128 \text{ and } \theta = 30°$$
$$= 4(128)\frac{\sqrt{3}}{2} \approx 443 \qquad \text{Use a calculator.}$$

The ball reaches a maximum height of 64 feet and has a range of about 443 feet.

Practice Problem 7 Repeat Example 7 for a ball with initial velocity 140 feet per second and initial angle of flight of 45°.

Figure 4.41

4 Use fundamental trigonometric identities.

Fundamental Trigonometric Identities

Based on the definitions of the trigonometric functions (see page 354), we have the following identities.

RECIPROCAL IDENTITIES

$$\csc t = \frac{1}{\sin t} \qquad \sec t = \frac{1}{\cos t} \qquad \cot t = \frac{1}{\tan t}$$
$$\sin t = \frac{1}{\csc t} \qquad \cos t = \frac{1}{\sec t} \qquad \tan t = \frac{1}{\cot t}$$

---

and

QUOTIENT IDENTITIES

$$\tan t = \frac{\sin t}{\cos t} \qquad \cot t = \frac{\cos t}{\sin t}$$

If $\sin t$ and $\cos t$ are given, we can use the reciprocal and quotient identities to find each of the remaining four trigonometric functions.

EXAMPLE 8 **Using Reciprocal and Quotient Identities**

Given $\sin t = \frac{3}{5}$ and $\cos t = -\frac{4}{5}$, find $\csc t$, $\sec t$, $\tan t$, and $\cot t$.

Solution
We first use the quotient identity to find $\tan t$:

$$\tan t = \frac{\sin t}{\cos t} = \frac{\frac{3}{5}}{-\frac{4}{5}} = \frac{3}{5}\left(-\frac{5}{4}\right) = -\frac{3}{4}.$$

We now use the reciprocal identities to find $\csc t$, $\sec t$, and $\cot t$.

$$\csc t = \frac{1}{\sin t} = \frac{1}{\frac{3}{5}} = \frac{5}{3},$$

$$\sec t = \frac{1}{\cos t} = \frac{1}{-\frac{4}{5}} = -\frac{5}{4}, \text{ and}$$

$$\cot t = \frac{1}{\tan t} = \frac{1}{-\frac{3}{4}} = -\frac{4}{3}.$$

Practice Problem 8 Repeat Example 8, given $\sin t = -\frac{2}{5}$ and $\cos t = \frac{\sqrt{21}}{5}$. ■

Because the equation for the unit circle is

$$x^2 + y^2 = 1, \qquad \text{The center is } (0,0); \text{ the radius is } 1.$$

we have the following identity:

$$(\cos t)^2 + (\sin t)^2 = 1. \qquad \text{Replace } x \text{ with } \cos t \text{ and } y \text{ with } \sin t.$$

By agreement, we write $\cos^2 t$ instead of $(\cos t)^2$ and $\sin^2 t$ instead of $(\sin t)^2$. A similar agreement holds for powers of all the trigonometric functions.

The equation $x^2 + y^2 = 1$ yields two other useful identities.

$$\frac{x^2}{x^2} + \frac{y^2}{x^2} = \frac{1}{x^2}, x \neq 0 \qquad \text{Divide both sides by } x^2.$$

$$1 + \left(\frac{y}{x}\right)^2 = \left(\frac{1}{x}\right)^2 \qquad \text{Simplify.}$$

$$1 + \left(\frac{\sin t}{\cos t}\right)^2 = \left(\frac{1}{\cos t}\right)^2 \qquad \text{Replace } x \text{ with } \cos t \text{ and } y \text{ with } \sin t.$$

$$1 + \tan^2 t = \sec^2 t \qquad \frac{\sin t}{\cos t} = \tan t, \frac{1}{\cos t} = \sec t$$

Dividing both sides of the equation $x^2 + y^2 = 1$ by y^2, and then replacing x with $\cos t$ and y with $\sin t$, lead to the identity

$$1 + \cot^2 t = \csc^2 t.$$

The three identities resulting from the equation $x^2 + y^2 = 1$ are called the **Pythagorean identities** because the Pythagorean Theorem is the basis for this equation.

PYTHAGOREAN IDENTITIES

$$\cos^2 t + \sin^2 t = 1 \qquad 1 + \tan^2 t = \sec^2 t \qquad 1 + \cot^2 t = \csc^2 t$$

EXAMPLE 9 Finding the Exact Value of a Trigonometric Function Using a Pythagorean Identity

a. Given $\sin t = \frac{1}{3}$ and $\cos t < 0$, find $\cos t$ and $\tan t$.

b. Given $\sec t = -2$ and $\tan t > 0$, find $\tan t$.

Solution

a. Use the Pythagorean identity involving $\sin t$.

$$\cos^2 t + \sin^2 t = 1 \qquad\qquad \text{Pythagorean identity}$$

$$\cos^2 t + \left(\frac{1}{3}\right)^2 = 1 \qquad\qquad \text{Replace } \sin t \text{ with } \frac{1}{3}.$$

$$\cos^2 t = 1 - \frac{1}{9} = \frac{8}{9} \qquad\qquad \text{Isolate the } \cos^2 t \text{ term and simplify.}$$

$$\cos t = \pm\sqrt{\frac{8}{9}} = \pm\frac{2\sqrt{2}}{3} \qquad\qquad \text{Square root property}$$

$$\cos t = -\frac{2\sqrt{2}}{3} \qquad\qquad \cos t < 0 \text{ is given.}$$

$$\tan t = \frac{\sin t}{\cos t} = \frac{\dfrac{1}{3}}{-\dfrac{2\sqrt{2}}{3}} \qquad\qquad \text{Replace } \sin t \text{ with } \frac{1}{3} \text{ and } \cos t \text{ with } -\frac{2\sqrt{2}}{3}.$$

$$= -\frac{1}{2\sqrt{2}} = -\frac{\sqrt{2}}{4} \qquad\qquad \text{Simplify and rationalize the denominator.}$$

b. Use the Pythagorean identity involving $\sec t$.

$$1 + \tan^2 t = \sec^2 t \qquad \text{Pythagorean identity}$$

$$1 + \tan^2 t = (-2)^2 \qquad \text{Replace } \sec t \text{ with } -2.$$

$$\tan^2 t = 3 \qquad \text{Subtract 1 from both sides and simplify.}$$

$$\tan t = \pm\sqrt{3} \qquad \text{Square root property}$$

$$\tan t = \sqrt{3} \qquad \tan t > 0 \text{ is given}$$

Alternatively, we can do such problems by the method used in Example 3.

Practice Problem 9 Given $\cos t = -\dfrac{2}{3}$ and $\sin t > 0$, find $\sin t$ and $\tan t$.

5 Use even–odd properties of trigonometric functions.

Even–Odd Properties of Trigonometric Functions

Recall (page 10) that for every x in the domain of a function f:

(i) If $f(-x) = f(x)$, then f is an even function.

(ii) If $f(-x) = -f(x)$, then f is an odd function.

Let $P(x, y)$ be the terminal point on the unit circle associated with a real number t. From Figure 4.42 we see that the point $Q(x, -y)$ is the terminal point associated with the number $-t$. Then

$$\cos(-t) = x = \cos t \quad \text{and}$$
$$\sin(-t) = -y = -\sin t.$$

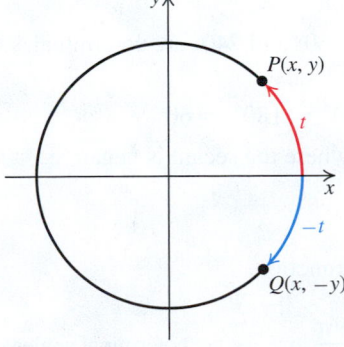

Figure 4.42

Since $\cos(-t) = \cos t$, the cosine function is an *even* function. Similarly, because $\sin(-t) = -\sin t$, the sine function is an *odd* function. These properties along with reciprocal and quotient identities show:

$$\sec(-x) = \frac{1}{\cos(-x)} = \frac{1}{\cos x} = \sec x$$

$$\csc(-x) = \frac{1}{\sin(-x)} = \frac{1}{-\sin x} = -\csc x$$

$$\tan(-x) = \frac{\sin(-x)}{\cos(-x)} = \frac{-\sin x}{\cos x} = -\tan x$$

$$\cot(-x) = \frac{\cos(-x)}{\sin(-x)} = \frac{\cos x}{-\sin x} = -\cot x$$

These results show that $\cos x$ and $\sec x$ are even functions, while $\sin x$, $\csc x$, $\tan x$, and $\cot x$ are odd functions.

We summarize this information in the next box.

EVEN–ODD TRIGONOMETRIC FUNCTIONS

Even Functions	**Odd Functions**
cosine and *secant*	*Sine, cosecant, tangent,* and *cotangent*
$\cos(-x) = \cos x;$	$\sin(-x) = -\sin x; \quad \csc(-x) = -\csc x;$
$\sec(-x) = \sec x.$	$\tan(-x) = -\tan x; \quad \cot(-x) = -\cot x.$

EXAMPLE 10   Using Even–Odd Properties to Find Exact Values

Find the exact value of each expression:

a. $\sin(-30°)$ **b.** $\cos\left(-\dfrac{2\pi}{3}\right)$ **c.** $\sec(-960°)$ **d.** $\cot\left(-\dfrac{19\pi}{4}\right)$

Solution

a. $\sin(-30°) = -\sin 30°$ Sine is an odd function.

$\qquad\qquad\quad = -\dfrac{1}{2}$ See page 360.

b. $\cos\left(-\dfrac{2\pi}{3}\right) = \cos\dfrac{2\pi}{3}$ Cosine is an even function

$\qquad\qquad\quad = \pm\cos\dfrac{\pi}{3}$ For $\theta = \dfrac{2\pi}{3}, \theta' = \pi - \dfrac{2\pi}{3} = \dfrac{\pi}{3}$.

$\qquad\qquad\quad = -\cos\dfrac{\pi}{3}$ $\dfrac{2\pi}{3}$ is in quadrant II where the cosine is negative.

$\qquad\qquad\quad = -\dfrac{1}{2}$ See page 360.

c. $\sec(-960°) = \sec 960°$ Secant is an even function.

$\qquad\qquad\quad = \sec 240°$ $960 = 2(360) + 240$; $960°$ and $240°$ are coterminal angles.

$\qquad\qquad\quad = \pm\sec 60°$ For $\theta = 240°, \theta' = 240° - 180° = 60°$.

$\qquad\qquad\quad = -\sec 60°$ $240°$ is in quadrant III, where the secant is negative.

$\qquad\qquad\quad = -2$ See page 360.

d. $\cot\left(-\dfrac{19\pi}{4}\right) = -\cot\left(\dfrac{19\pi}{4}\right)$ Cotangent is an odd function.

$\qquad\qquad\quad = -\cot\left(\dfrac{3\pi}{4}\right)$ $\dfrac{19\pi}{4} = 4\pi + \dfrac{3\pi}{4}$; $\dfrac{19\pi}{4}$ and $\dfrac{3\pi}{4}$ are coterminal angles.

$\qquad\qquad\quad = -\left(\pm\cot\dfrac{\pi}{4}\right)$ For $\theta = \dfrac{3\pi}{4}, \theta' = \pi - \dfrac{3\pi}{4} = \dfrac{\pi}{4}$

$\qquad\qquad\quad = -\left(-\cot\dfrac{\pi}{4}\right)$ $\dfrac{3\pi}{4}$ is in quadrant II, where the cotangent is negative.

$\qquad\qquad\quad = -(-1)$ See page 360.

$\qquad\qquad\quad = 1$ Simplify.

Practice Problem 10 Find the exact value of each expression.

$$\textbf{a.}\quad \csc\left(-1200°\right) \qquad \textbf{b.}\quad \tan\left(-\frac{19\pi}{3}\right)$$

Answers to Practice Problems

1. $\sin\theta = -\dfrac{5\sqrt{29}}{29}$, $\csc\theta = -\dfrac{\sqrt{29}}{5}$, $\cos\theta = \dfrac{2\sqrt{29}}{29}$,

$\sec\theta = \dfrac{\sqrt{29}}{2}$, $\tan\theta = -\dfrac{5}{2}$, $\cot\theta = -\dfrac{2}{5}$

2. Quadrant II

3. $\sin\theta = -\dfrac{4\sqrt{41}}{41}$; $\sec\theta = \dfrac{\sqrt{41}}{5}$

4. a. 5° **b.** $\dfrac{\pi}{3}$ **c.** 60° **d.** 1.20

5. $\dfrac{\sqrt{2}}{2}$ **6. a.** $-\sqrt{2}$ **b.** $-\sqrt{3}$

7. The ball reaches a maximum height of 153 feet and has a range of 612 feet.

8. $\csc t = -\dfrac{5}{2}$, $\sec t = \dfrac{5\sqrt{21}}{21}$, $\tan t = -\dfrac{2\sqrt{21}}{21}$, $\cot t = -\dfrac{\sqrt{21}}{2}$

9. $\sin t = \dfrac{\sqrt{5}}{3}$; $\tan t = -\dfrac{\sqrt{5}}{2}$

10. a. $-\dfrac{2\sqrt{3}}{3}$ **b.** $-\sqrt{3}$

 SECTION 4.3 **Exercises**

Concepts and Vocabulary

1. Let $P = (x, y)$ be a point on the terminal side of an angle θ in standard position. Let $r = \sqrt{x^2 + y^2}$. Then $\sin\theta =$ _____, $\cos\theta =$ _____, $\tan\theta =$ _____.

2. If $P(x, y)$ is a point on the unit circle corresponding to a quadrantal angle, then either x or y equals _____.

3. Two angles (with radian measure between 0 and 2π) whose sine value is $\frac{1}{2}$ are _____ and _____.

4. The only two trigonometric functions that have positive values in quadrant IV are _____ and _____.

5. If $\theta = 3$, then θ lies in quadrant _____.

6. The reference angle θ' of an angle θ in standard position is the acute angle formed by the _____ side of θ and the _____ axis.

7. **True or False.** The trigonometric function values for $\theta = \pi$ are identical to those for $\theta = -\pi$.

8. **True or False.** The values of a trigonometric function of an angle θ and its reference angle θ' are the same.

Building Skills

In Exercises 9–14, a point on the terminal side of an angle θ in standard position is given. Find the exact values of $\sin\theta$, $\cos\theta$, and $\tan\theta$.

9. $(-4, 3)$ **10.** $(-3, 5)$

11. $(-\sqrt{3}, -1)$ **12.** $(-1, -2)$

13. $(3, 3)$ **14.** $(-2, -2)$

In Exercises 15–20, a point on the terminal side of an angle θ in standard position is given. Find the exact value of $\csc\theta$, $\sec\theta$, and $\cot\theta$.

15. $(12, -5)$ **16.** $(7, -2)$

17. $(7, -24)$ **18.** $(5, -5)$

19. $(-\sqrt{2}, \sqrt{6})$ **20.** $(-\sqrt{5}, -\sqrt{11})$

In Exercises 21–28, use the given information to find the quadrant in which each angle θ lies.

21. $\sin\theta < 0$ and $\cos\theta < 0$ **22.** $\sin\theta < 0$ and $\tan\theta > 0$

23. $\sin\theta > 0$ and $\cos\theta < 0$ **24.** $\tan\theta > 0$ and $\csc\theta < 0$

25. $\cos\theta > 0$ and $\csc\theta < 0$ **26.** $\cos\theta < 0$ and $\cot\theta > 0$

27. $\sec\theta < 0$ and $\csc\theta > 0$ **28.** $\sec\theta < 0$ and $\tan\theta > 0$

In Exercises 29–36, find the exact values of the trigonometric functions of θ from the given information.

29. $\cos\theta = -\dfrac{5}{13}$, θ in quadrant III, find $\tan\theta$.

30. $\tan\theta = -\dfrac{3}{4}$, θ in quadrant IV, find $\sin\theta$.

31. $\cot\theta = -\dfrac{3}{4}$, θ in quadrant II, find $\cos\theta$.

32. $\sec\theta = \dfrac{4}{\sqrt{7}}$, θ in quadrant IV, find $\csc\theta$.

33. $\sin\theta = \dfrac{3}{5}$, $\tan\theta < 0$, find $\sec\theta$.

34. $\cot\theta = \dfrac{3}{2}$, $\sec\theta > 0$, find $\sin\theta$.

35. $\sec\theta = 3$, $\sin\theta < 0$, find $\cot\theta$.

36. $\tan\theta = -2$, $\sin\theta > 0$, find $\cos\theta$.

In Exercises 37–48, find the reference angle for each angle.

37. 120° **38.** 275° **39.** −50°

40. 500° **41.** 420° **42.** −110°

43. $\dfrac{19\pi}{4}$ **44.** $\dfrac{28\pi}{6}$ **45.** $-\dfrac{3\pi}{4}$

46. $\dfrac{31\pi}{6}$ **47.** $\dfrac{5\pi}{6}$ **48.** $-\dfrac{15\pi}{4}$

In Exercises 49–60, use the reference angle to find the exact value of each expression.

49. $\cos 120°$ **50.** $\sin 315°$

51. $\tan 510°$ **52.** $\cot 750°$

53. $\sec 210°$ **54.** $\csc 300°$

55. $\sin \dfrac{7\pi}{6}$ **56.** $\cos \dfrac{4\pi}{3}$

57. $\sec \dfrac{15\pi}{4}$ **58.** $\cot \dfrac{19\pi}{4}$

59. $\sin \dfrac{29\pi}{3}$ **60.** $\tan \dfrac{55\pi}{3}$

In Exercises 61–66, use reciprocal and quotient identities to find the exact value of each expression.

61. Given $\sin t = \dfrac{1}{3}$ and $\cos t = \dfrac{2\sqrt{2}}{3}$, find $\tan t$.

62. Given $\sin t = -\dfrac{1}{4}$ and $\cos t = -\dfrac{\sqrt{15}}{2}$, find $\cot t$.

63. Given $\sin t = \dfrac{2}{3}$ and $\cot t = -\dfrac{\sqrt{5}}{2}$, find $\cos t$.

64. Given $\cos t = -\dfrac{2}{5}$ and $\tan t = -\dfrac{\sqrt{21}}{2}$, find $\sin t$.

65. Given $\csc t = \dfrac{7}{2}$ and $\cos t = -\dfrac{3\sqrt{5}}{7}$, find $\tan t$.

66. Given $\csc t = \dfrac{8}{3}$ and $\sec t = \dfrac{8\sqrt{55}}{55}$, find $\cot t$.

In Exercises 67–78, use fundamental identities to find the exact value of each expression.

67. $\sin 70° \csc 70°$ **68.** $\tan 65° \cot 65°$

69. $\cos 35° \sec 35°$ **70.** $\sin^2 \dfrac{2\pi}{9} + \cos^2 \dfrac{2\pi}{9}$

71. $\cos^2 47° + \sin^2 47°$ **72.** $\sec^2 \dfrac{5\pi}{12} - \tan^2 \dfrac{5\pi}{11}$

73. $\sin^2 \dfrac{\pi}{7} + \cos^2 \dfrac{\pi}{7} + \cot^2 \dfrac{\pi}{6}$ **74.** $\csc^2 32° - \cot^2 32°$

75. $\tan \dfrac{3\pi}{13} - \dfrac{\sin \dfrac{3\pi}{13}}{\cos \dfrac{3\pi}{13}}$ **76.** $\cot 76° - \dfrac{\cos 76°}{\sin 76°}$

77. $\cos 390° \sec 30°$ **78.** $\sin \dfrac{3\pi}{11} \csc \dfrac{47\pi}{11}$

In Exercises 79–86, use the Pythagorean identities to find the exact value of each expression.

79. Given $\sin t = \dfrac{2}{5}$ and $\cos t < 0$, find $\cos t$.

80. Given $\cos t = \dfrac{1}{6}$ and $\sin t < 0$, find $\sin t$.

81. Given $\sec t = -5$ and $\tan t < 0$, find $\tan t$.

82. Given $\cot t = 4$ and $\csc t < 0$, find $\csc t$.

83. Given $\tan t = -3$ and $\sec t > 0$, find $\sec t$.

84. Given $\csc t = -6$ and $\cos t > 0$, find $\cos t$.

85. Given $\sin 15° = \dfrac{\sqrt{2 - \sqrt{3}}}{2}$, find $\cos 15°$ and $\tan 15°$.

86. Given $\cos 22.5° = \dfrac{\sqrt{2 + \sqrt{2}}}{2}$, find $\sin 22.5°$ and $\tan 22.5°$.

In Exercises 87–98, use even–odd identities to find the exact value of each expression.

87. $\cos(-60°)$ **88.** $\sin(-135°)$

89. $\tan(-240°)$ **90.** $\cot(-300°)$

91. $\sin(-1560°)$ **92.** $\cos(-1290°)$

93. $\sin\left(-\dfrac{\pi}{3}\right)$ **94.** $\cos\left(-\dfrac{7\pi}{6}\right)$

95. $\tan\left(-\dfrac{65\pi}{6}\right)$ **96.** $\tan\left(-\dfrac{29\pi}{3}\right)$

97. $\sec\left(-\dfrac{43\pi}{6}\right)$ **98.** $\csc\left(-\dfrac{37\pi}{3}\right)$

Applying the Concepts

99. Polarization. One lens is removed from a pair of polarized sunglasses and put on top of the other lens, both aligned identically. If the top lens is rotated through an angle θ, the fraction of light passing through the aligned lenses is $\cos^2 \theta$. Find θ in degrees so that a rotation through θ allows only $\dfrac{1}{4}$ as much light through the lenses as when they are aligned identically.

100. Cable cost. Cable must be laid from a point on the shore of a lake to an island. The locations and distances are shown in the figure. It costs $700(5 - \cot \theta) + 1200 \csc \theta$ dollars to lay the cable, where θ is the angle the cable to the island makes with the shore. Find the cost if
a. $\theta = 30°$.   **b.** $\theta = 45°$.   **c.** $\theta = 60°$.   **d.** $\theta = 90°$.

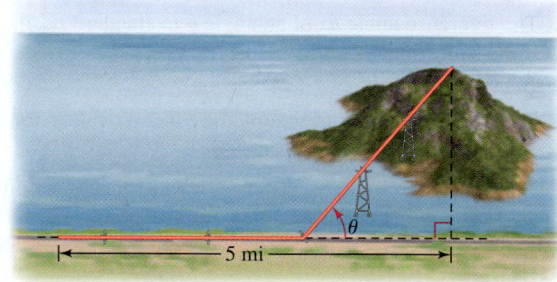

101. Household voltage. The voltage in a common household electrical outlet is given by $V = 166 \cos(120\pi t)$, where V is measured in volts and t in seconds. Find the voltage when t equals
a. $\dfrac{1}{120}$ sec   **b.** $\dfrac{1}{60}$ sec

102. Volume of a cone. The volume, V in feet^3, of the cone in the figure is $9\pi \cos^2 \theta \sin \theta$. Find V if
   a. $\theta = 30°$.   **b.** $\theta = 60°$.

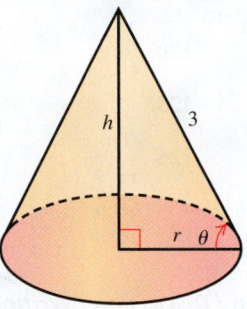

103. Piston action. A rod OA of length r rotates counterclockwise about a fixed point O. A second rod AB with length $L > 2r$ is connected to a piston at point B. If all lengths are given in centimeters, the distance x between the points O and B is $x = r \cos \theta + \sqrt{L^2 - r^2 \sin^2 \theta}$. Find x if
   a. $\theta = 0°$.   **b.** $\theta = 90°$.   **c.** $\theta = 180°$.

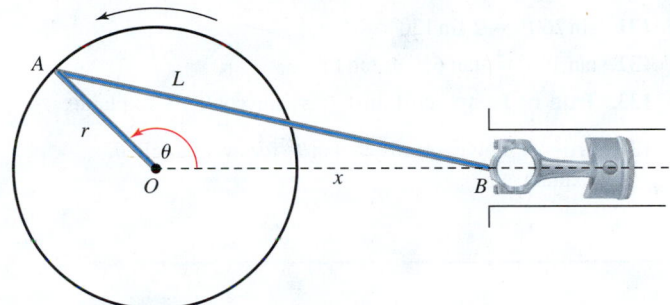

104. Observing London. The Millennium Wheel in London is a large, continuously turning observational wheel with diameter 135 meters (approximately). The height, h, of the passenger capsules above the base is $68 + 67.5 \sin(\theta - 90°)$, where θ is the angle the capsule arm makes with a vertical ray downward from the wheel's center. (See the figure.) How high is the capsule when
   a. $\theta = 0°$?   **b.** $\theta = 90°$?   **c.** $\theta = 180°$?

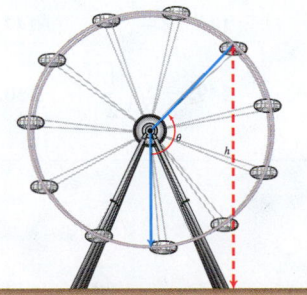

In Exercises 105 and 106, use the following information.

Assuming level ground and neglecting air resistance, the formulas governing the flight of a golf ball in the introduction and Example 7 apply to any object projected upward from ground level at an angle θ. That is, if the initial velocity, v_0, is

given in feet per second and time is measured in seconds, then t seconds into flight gives the following:

$$h = v_0 t \sin \theta - 16t^2 \qquad \text{the object's height}$$
$$d = t v_0 \cos \theta \qquad \text{the horizontal distance traveled (the range)}$$

105. If $\theta = 30°$ and the initial velocity is 80 feet per second, find
   a. the object's maximum height.
   b. the object's range.

106. If $\theta = 30°$ and the initial velocity is 80 feet per second, find the time the object was in flight.

Beyond the Basics

Use the information given for Exercises 105 and 106 for Exercises 107–109.

107. For an angle θ and initial velocity v_0, find an expression for the object's maximum height. [*Hint:* Find the vertex of $h = v_0 t \sin \theta - 16t^2$.]

108. For an angle θ and initial velocity v_0, find an expression for the object's range. [*Hint:* First find the time the object is in flight.]

109. Show that height is a (quadratic) function of distance, that is $h = ad^2 + bd + c$. [*Hint:* First, solve $d = t v_0 \cos \theta$ for t.]

110. In the adjoining figure, by definition
$$P = (\cos\theta, \sin\theta) \text{ and } Q = \left(\cos\left(\theta + \frac{\pi}{2}\right), \sin\left(\theta + \frac{\pi}{2}\right)\right).$$

   a. Show that triangles OMP and ONQ are congruent.
   b. Use the result from part (**a**) to show that $Q = (-\sin\theta, \cos\theta)$.
   c. Use initial definitions and the results from part (b) to show that $\cos\left(\frac{\pi}{2} + \theta\right) = -\sin\theta$ and $\sin\left(\frac{\pi}{2} + \theta\right) = \cos\theta$.
   d. Use quotient and reciprocal identities to find $\tan\left(\frac{\pi}{2} + \theta\right)$, $\cot\left(\frac{\pi}{2} + \theta\right)$, $\sec\left(\frac{\pi}{2} + \theta\right)$, and $\csc\left(\frac{\pi}{2} + \theta\right)$.

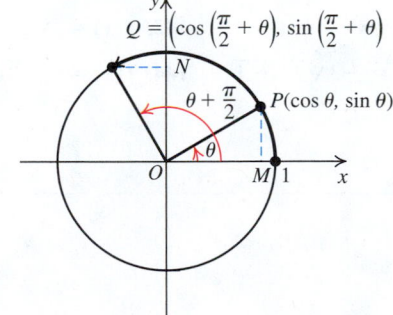

111. Use the adjoining figure to show that $\cos(\pi - \theta) = -\cos\theta$, $\sin(\pi - \theta) = \sin\theta$, and $\tan(\pi - \theta) = -\tan\theta$.

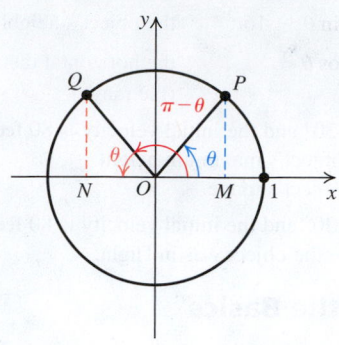

112. Make a figure for angles θ and $\pi + \theta$. Then show that $\cos(\pi + \theta) = -\cos\theta$, $\sin(\pi + \theta) = -\sin\theta$, and $\tan(\pi + \theta) = \tan\theta$.

In Exercises 113–118, find the exact value of each expression.

113. $\sin 150° \cos 240° + \cos 150° \sin 240°$

114. $\cos 135° \cos 225° - \sin 135° \sin 225°$

115. $\sin 150° \tan 120° - \cos 120° \cot 210°$

116. $\cos\dfrac{4\pi}{3} \cos\dfrac{3\pi}{4} + \sin\dfrac{4\pi}{3} \sin\dfrac{3\pi}{4}$

117. $\sin\dfrac{17\pi}{3} \cos\dfrac{13\pi}{3} - \cos\dfrac{17\pi}{3} \sin\dfrac{13\pi}{3}$

118. $\cos\dfrac{5\pi}{3} \tan\dfrac{7\pi}{6} - \sin\dfrac{5\pi}{4} \tan\dfrac{3\pi}{4}$

In Exercises 119–126, determine whether each function is even, odd or neither.

119. $\sin x \cos x$ **120.** $\sin x^2$ **121.** $\dfrac{\sin x}{x}$

122. $x \cos x$ **123.** $\csc x^3$ **124.** $x^2 \sec x$

125. $\dfrac{\cot x}{x^2 + 1}$ **126.** $\dfrac{\tan x}{x^3 + x}$

127. Show that $\ln \sin^2 x - \ln \cos^2 x = 2 \ln \tan x$.

128. Show that $(\sin\theta \sin\alpha)^2 + (\sin\theta \cos\alpha)^2 + \cos^2\theta = 1$.

Critical Thinking / Discussion / Writing

129. Find a positive number x for which
   a. $\cos(\ln x) = 0$.
   b. $\sin(e^x) = 0$.

In Exercises 130–132, without evaluating the functions, explain why each statement is false.

130. $\cos 105° = \cos 60° + \cos 45°$

131. $\sin 260° = 2 \sin 130°$

132. $\tan 123° = \tan 61° + \tan 62°$

133. True or False. $\cos(\sin 0°)° = \sin(\cos 0°)°$. Explain.

134. True or False. $\sec\theta° > \tan\theta°$ for every angle θ. Explain.

Getting Ready for the Next Section

In Exercises 135–138, find the zeros of each function.

135. $f(x) = 3x - 4$ **136.** $g(x) = 2x + 3$

137. $F(x) = x^2 + 3x - 10$ **138.** $G(x) = \dfrac{x^2 - 2x - 3}{x - 4}$

In Exercises 139–144, use transformations on the graph of $f(x) = \sqrt{x}$ to sketch the graph of each function.

139. $f(x) = \sqrt{-x}$ **140.** $f(x) = -\sqrt{x}$

141. $f(x) = \sqrt{x + 1}$ **142.** $f(x) = \sqrt{2x - 1}$

143. $f(x) = -3\sqrt{2x + 1}$ **144.** $f(x) = 3\sqrt{2x - 4} + 5$

In Exercises 145–150, use the given graph of $y = f(x)$ to graph each function.

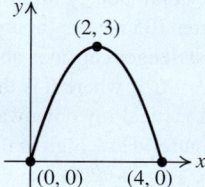

145. $-f(x)$ **146.** $f(-x)$ **147.** $2f(x)$

148. $f(2x)$ **149.** $-2f\left(\dfrac{1}{2}x\right)$ **150.** $-\dfrac{1}{2}f(x) + 3$

Graphs of the Sine and Cosine Functions

BEFORE STARTING THIS SECTION, REVIEW

1 Equation of a circle (Section 1.1, page 12)

2 Transformations of functions (Section 1.5, page 90)

3 Unit circle definitions of the sine and cosine (Section 4.2, page 354)

OBJECTIVES

1 Discuss properties of the sine and cosine functions.

2 Graph the sine and cosine functions.

3 Find the amplitude and period of sinusoidal curves.

4 Find the phase shifts and graph sinusoidal functions of the forms $y = a \sin[b(x - c)]$ and $y = a \cos[b(x - c)]$.

5 Model periodic behavior.

Physical biorhythms
Emotional biorhythms
Intellectual biorhythms

◆ Biorhythms

Biorhythm means "rhythm of life." The theory of biorhythms, a pseudoscience, was developed from the data of several patients in the early twentieth century by Professor Hermann Swoboda (University of Vienna) and Dr. Wilhelm Fliess, a friend of Sigmund Freud. According to the theory, every person has three fundamental biorhythm cycles. All cycles start on a person's day of birth, and each cycle has a fixed period.

1. **Physical cycle** (P) has a period of 23 days. It governs the condition of one's body.
2. **Emotional cycle** (E) has a period of 28 days. It describes one's temperament.
3. **Intellectual cycle** (I) has a period of 33 days. It influences one's thinking capacity.

The periodic functions P, E, and I can each be represented in the form $y = \sin(Bt)$; they are said to determine our biorhythm state on any day, t, since our birth. We experience a "high" or "low" in the corresponding cycle (physical, emotional, or intellectual) on the day if the curve is above or below the midline. We discuss this in Example 11.

---

1 Discuss properties of the sine and cosine functions.

Properties of Sine and Cosine

We first discuss some properties of the sine and cosine functions that will be useful in sketching the graphs of these functions.

Domain and Range of Sine and Cosine

Each real number t determines a terminal point $P(t) = (x, y)$ on the unit circle. Because $x = \cos t$ and $y = \sin t$, both functions are defined for each real number t. So the domain of both the cosine and sine functions in interval notation is the set of real numbers $(-\infty, \infty)$.

For each real number t, the point $P(t) = (\cos t, \sin t)$ lies on the unit circle $x^2 + y^2 = 1$. Substituting $x = \cos t$ and $y = \sin t$ in the equation, we have the **Pythagorean identity:**

$$(\cos t)^2 + (\sin t)^2 = 1 \text{ for every real number } t.$$

So, because both $(\cos t)^2$ and $(\sin t)^2$ are nonnegative, $(\cos t)^2 \leq 1$ and $(\sin t)^2 \leq 1$.

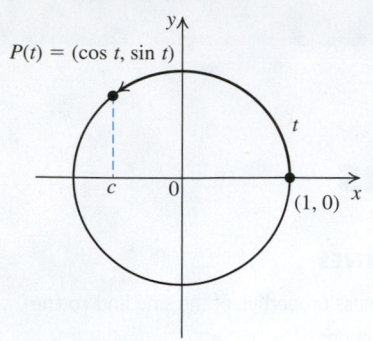

Figure 4.43

These inequalities imply that the values of both cos t and sin t lie between -1 and 1. That is,

$$|\cos t| \le 1 \text{ and } |\sin t| \le 1, \text{ or}$$

$$-1 \le \cos t \le 1 \text{ and } -1 \le \sin t \le 1 \text{ for every number } t.$$

Furthermore, let c be a real number in the interval $[-1, 1]$. Then c is the x-coordinate of some terminal point $P(t) = (\cos t, \sin t)$ on the unit circle. See Figure 4.43. So $c = \cos t$. This shows that the interval $[-1, 1]$ is the range of the cosine function. A similar argument shows that the interval $[-1, 1]$ is also the range of the sine function.

> ### DOMAIN AND RANGE OF SINE AND COSINE
>
> The *domain* of both the sine and cosine functions is $(-\infty, \infty)$.
> The *range* of both the sine and cosine functions is $[-1, 1]$.

Zeros of Sine and Cosine Functions

We know that cos $t = 0$ precisely when the terminal point $P(t) = (\cos t, \sin t)$ lies on the y-axis. This happens when $t = \dfrac{\pi}{2}, -\dfrac{\pi}{2}, \dfrac{3\pi}{2}, -\dfrac{3\pi}{2}, \ldots$; in other words, when t is an odd integer multiple of $\dfrac{\pi}{2}$. That is, the zeros of cos t are given by:

$$t = (2n + 1)\frac{\pi}{2} = \frac{\pi}{2} + n\pi \text{ for any integer } n.$$

Similarly, sin $t = 0$ only if the terminal point $P(t) = (\cos t, \sin t)$ lies on the x-axis. This occurs when $t = 0, \pi, -\pi, 2\pi, -2\pi, \ldots$; that is, when t is an integer multiple of π. So, the zeros of sin t are given by:

$$t = n\pi \text{ for any integer } n.$$

Even–Odd Properties of the Sine and Cosine Functions

From page 376, we have $\sin(-t) = -\sin t$, the sine function is an *odd* function, so its graph is symmetric about the origin. Similarly, $\cos(-t) = \cos t$, so the cosine function is an *even* function and its graph is symmetric about the y-axis.

Periodic Functions

> ### Periodic Functions
>
> A function f is said to be **periodic** if there is a positive number p such that
>
> $$f(x + p) = f(x)$$
>
> for every x in the domain of f.
>
> The smallest value of p (if there is one) for which $f(x + p) = f(x)$ is called the **period** of f. The graph of f over any interval of length p is called one **cycle** of the graph.

We note that the values of a periodic function f of period p repeat themselves every p units. This means that to graph f we may draw only one cycle of the graph of f, since the rest of the graph is generated by repeated copies of it.

Figure 4.44 shows examples of periodic functions.

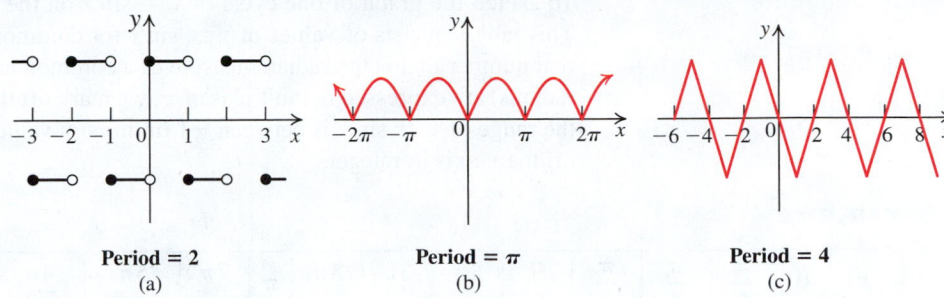

Period = 2	Period = π	Period = 4
(a)	(b)	(c)

Figure 4.44 Periodic functions

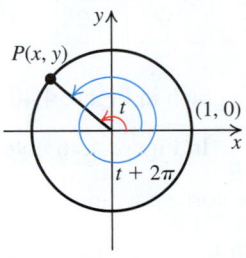

Figure 4.45

Let $P(x, y)$ be a point on the unit circle that corresponds to an angle t, measured in radians. If we add 2π to t, we obtain the same point P on the unit circle. See Figure 4.45. Then $\sin(t + 2\pi) = \sin t = y$ and $\cos(t + 2\pi) = \cos t = x$. Therefore, $\sin t$ and $\cos t$ are periodic functions. The period for each of the sine and cosine functions is 2π. (See Exercise 105.)

PERIOD OF THE SINE AND COSINE FUNCTIONS

For every real number t,

$$\sin(t + 2\pi) = \sin t \quad \text{and} \quad \cos(t + 2\pi) = \cos t.$$

The sine and cosine functions are periodic with **period 2π**.

We summarize the properties of the sine and cosine functions below.

PROPERTIES OF SINE AND COSINE

$y = \sin t$

1. Domain: $(-\infty, \infty)$

2. Range: $[-1, 1]$

3. Zeros at $t = n\pi$, n is any integer

4. (a) Maximum value: $\sin t = 1$

   at $t = \dfrac{\pi}{2} + 2n\pi$

   (b) Minimum value: $\sin t = -1$

   at $t = \dfrac{3\pi}{2} + 2n\pi$

5. Odd function: $\sin(-t) = -\sin t$

6. Period: 2π

$y = \cos t$

1. Domain: $(-\infty, \infty)$

2. Range: $[-1, 1]$

3. Zeros at $t = \dfrac{\pi}{2} + n\pi$, n is any integer

4. (a) Maximum value: $\cos t = 1$ at $t = 2n\pi$

   (b) Minimum value: $\cos t = -1$ at $t = \pi + 2n\pi$

5. Even function: $\cos(-t) = \cos t$

6. Period: 2π

---

2 Graph the sine and cosine functions.

Graphs of Sine and Cosine Functions

To graph the sine and cosine functions on an xy-coordinate plane, we use x as the independent variable and y as the dependent variable. So, we write $y = \sin x$ and $y = \cos x$ rather than $y = \sin t$ and $y = \cos t$, respectively.

In calculus, it can be shown that the trigonometric functions are continuous on their domains. That is, they are continuous wherever they are defined.

Graph of the Sine Function

To sketch the graph of one cycle of $y = \sin x$ on the interval $[0, 2\pi]$, we use Table 4.2. This table consists of values of $y = \sin x$ for common values of x. In other words, x is a real number that is the radian measure of a common angle. Since the common angles x (in radians) are expressed in multiples of π, we mark off the x-axis in multiples of π. However, the range of $y = \sin x$ is between -1 (minimum value) and 1 (maximum value). We mark off the y-axis in integers.

TABLE 4.2

x	0	$\dfrac{\pi}{6}$	$\dfrac{\pi}{4}$	$\dfrac{\pi}{3}$	$\dfrac{\pi}{2}$	$\dfrac{2\pi}{3}$	$\dfrac{3\pi}{4}$	$\dfrac{5\pi}{6}$	π	$\dfrac{7\pi}{6}$	$\dfrac{5\pi}{4}$	$\dfrac{4\pi}{3}$	$\dfrac{3\pi}{2}$	$\dfrac{5\pi}{3}$	$\dfrac{7\pi}{4}$	$\dfrac{11\pi}{6}$	2π
$\sin x$	0	$\dfrac{1}{2}$	$\dfrac{1}{\sqrt{2}}$	$\dfrac{\sqrt{3}}{2}$	1	$\dfrac{\sqrt{3}}{2}$	$\dfrac{1}{\sqrt{2}}$	$\dfrac{1}{2}$	0	$-\dfrac{1}{2}$	$-\dfrac{1}{\sqrt{2}}$	$-\dfrac{\sqrt{3}}{2}$	-1	$-\dfrac{\sqrt{3}}{2}$	$-\dfrac{1}{\sqrt{2}}$	$-\dfrac{1}{2}$	0

Connecting the points $(x, \sin x)$ with a smooth curve gives us the graph in Figure 4.46. We have shown the points $(x, \sin x)$ where x is a multiple of $\dfrac{\pi}{4}$ or $\dfrac{\pi}{6}$. In Figure 4.46 note that for the arc length $\dfrac{\pi}{6}$, shown on the unit circle, the y-value for the corresponding terminal point $P\left(\dfrac{\pi}{6}\right)$ is $\dfrac{1}{2}$, so the point $\left(\dfrac{\pi}{6}, \dfrac{1}{2}\right)$ is on the graph of $y = \sin x$.

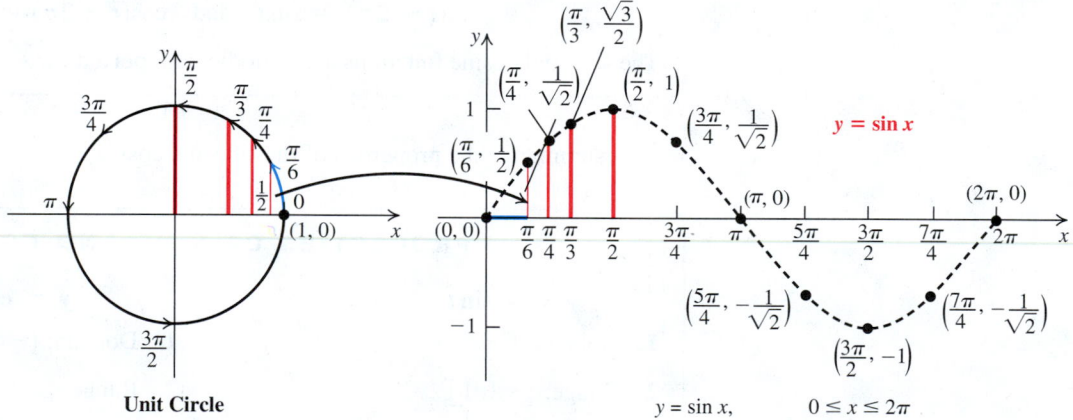

Figure 4.46

Table 4.3 shows the behavior of $y = \sin x$ over the interval $[0, 2\pi]$.

Because the sine function is periodic with period 2π, we can sketch the complete graph of $y = \sin x$ by extending the graph in Figure 4.46 indefinitely to the left and to the right in every successive interval of length 2π. See Figure 4.47.

TABLE 4.3

x	$\sin x$
$0 \to \dfrac{\pi}{2}$	$0 \to 1$
$\dfrac{\pi}{2} \to \pi$	$1 \to 0$
$\pi \to \dfrac{3\pi}{2}$	$0 \to -1$
$\dfrac{3\pi}{2} \to 2\pi$	$-1 \to 0$

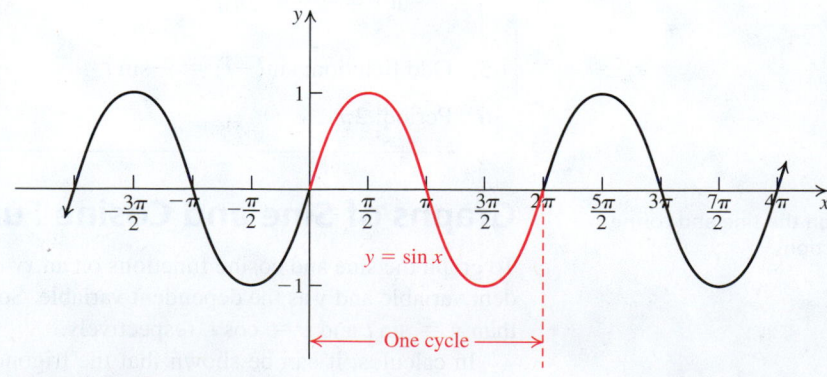

Figure 4.47 The sine curve.

Graph of the Cosine Function

We can use the same method for graphing $y = \cos x$ that we used for $y = \sin x$.

Table 4.4 gives the values for $y = \cos x$, where x has the same values as in Table 4.2.

TABLE 4.4

x	0	$\dfrac{\pi}{6}$	$\dfrac{\pi}{4}$	$\dfrac{\pi}{3}$	$\dfrac{\pi}{2}$	$\dfrac{2\pi}{3}$	$\dfrac{3\pi}{4}$	$\dfrac{5\pi}{6}$	π	$\dfrac{7\pi}{6}$	$\dfrac{5\pi}{4}$	$\dfrac{4\pi}{3}$	$\dfrac{3\pi}{2}$	$\dfrac{5\pi}{3}$	$\dfrac{7\pi}{4}$	$\dfrac{11\pi}{6}$	2π
$\cos x$	1	$\dfrac{\sqrt{3}}{2}$	$\dfrac{1}{\sqrt{2}}$	$\dfrac{1}{2}$	0	$-\dfrac{1}{2}$	$-\dfrac{1}{\sqrt{2}}$	$-\dfrac{\sqrt{3}}{2}$	-1	$-\dfrac{\sqrt{3}}{2}$	$-\dfrac{1}{\sqrt{2}}$	$-\dfrac{1}{2}$	0	$\dfrac{1}{2}$	$\dfrac{1}{\sqrt{2}}$	$\dfrac{\sqrt{3}}{2}$	1

Connecting the points $(x, \cos x)$ with a smooth curve gives the graph of $y = \cos x$, $0 \le x \le 2\pi$, shown in Figure 4.48. In this figure, note that for the arc length $\dfrac{\pi}{6}$, shown on the unit circle, the x-value for the corresponding terminal point $P\left(\dfrac{\pi}{6}\right)$ is $\dfrac{\sqrt{3}}{2}$, so the point $\left(\dfrac{\pi}{6}, \dfrac{\sqrt{3}}{2}\right)$ is on the graph of $y = \cos x$.

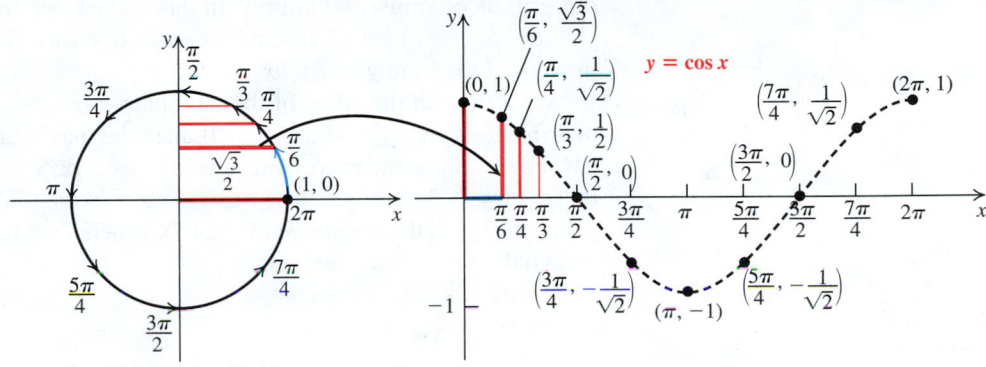

Figure 4.48 $y = \cos x, 0 \le x \le 2\pi$.

TABLE 4.5

x	$\cos x$
$0 \to \dfrac{\pi}{2}$	$1 \to 0$
$\dfrac{\pi}{2} \to \pi$	$0 \to -1$
$\pi \to \dfrac{3\pi}{2}$	$-1 \to 0$
$\dfrac{3\pi}{2} \to 2\pi$	$0 \to 1$

Table 4.5 shows the behavior of $y = \cos x$ over the interval $[0, 2\pi]$.

As with the sine function, the values of x for $y = \cos x$ are not restricted because the domain of the cosine function is $(-\infty, \infty)$. We can draw the complete graph of $y = \cos x$ by extending the graph in Figure 4.48 indefinitely to the left and right in every successive interval of length 2π. See Figure 4.49.

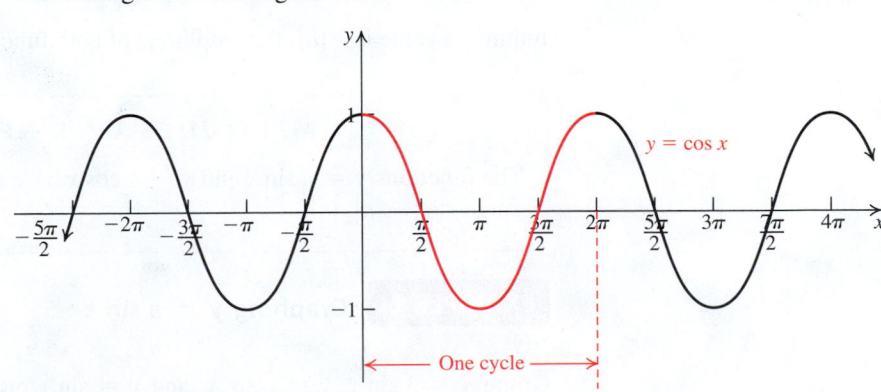

Figure 4.49 The cosine curve.

Five Key Points

There are five *key points* in every cycle of the graphs of the sine and cosine functions. The key points consist of: the **x-intercepts, high points, and low points**. See Figure 4.50. The key points are very helpful for graphing the sine and cosine functions and their

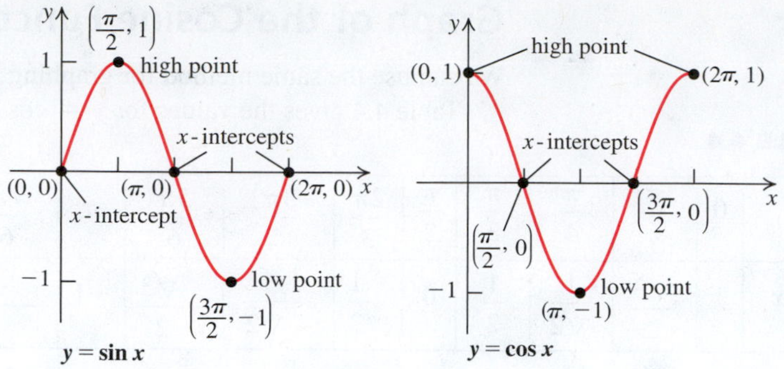

Figure 4.50 Key points of the sine and cosine graphs.

transformations. Note that the x-coordinates of the five key points divide the period into four equal intervals.

Amplitude and Period

3 Find the amplitude and period of sinusoidal curves.

The graphs of the sine and cosine functions and their transformations are called **sinusoidal graphs** or **sinusoidal curves**. In this section, we study sinusoidal curves of the form $y = af[b(x - c)] + d$, where f is a sine or cosine function. We first consider transformations of the form $y = af(bx)$.

We start with the effect of the multiplier a in $y = af(bx)$. If the maximum (or minimum) value of $f(x)$ is M and $a > 0$, then the maximum (or minimum) value of $af(x)$ is aM. For example, the maximum value of $y = 3 \sin x$ is $3(1) = 3$, and its minimum value is $3(-1) = -3$. If the graph of $y = f(x)$ is "centered" about the x-axis, then the maximum value of $f(x)$ is the *amplitude* of $f(x)$. In general, we have the following definition of the amplitude of a periodic function.

SIDE
NOTE

If a function f does not have both a maximum and a minimum value, we say that f has no amplitude or that its amplitude is undefined.

Amplitude

Let f be a periodic function. Suppose M is the maximum value of f and m is the minimum value of f. The amplitude of f is defined by

$$\text{Amplitude} = \frac{1}{2}(M - m).$$

Because the maximum value of both functions $y = a \sin x$ and $y = a \cos x$ is $|a|$ and their minimum value is $-|a|$, the amplitude of both functions is $\frac{1}{2}[|a| - (-|a|)] = |a|$.

AMPLITUDES OF SINE AND COSINE

The functions $y = a \sin x$ and $y = a \cos x$ have **amplitude** $= |a|$ and range $= [-|a|, |a|]$.

EXAMPLE 1 Graphing $y = a \sin x$

Graph $y = 3 \sin x$, $y = \frac{1}{3} \sin x$, and $y = \sin x$ on the same coordinate system over the interval $[-2\pi, 2\pi]$.

Solution

We begin with the graph of $y = \sin x$ and multiply the y-coordinate of each point (including the key points) on this graph by 3 to get the graph of $y = 3 \sin x$. This results in stretching the graph of $y = \sin x$ vertically by a factor of 3. Since $3(0) = 0$, the x-intercepts are unchanged. However, the maximum y value is $3(1) = 3$ and the minimum y value is $3(-1) = -3$. See Figure 4.51. Similarly, we multiply the y-coordinate on the graph by $\frac{1}{3}$ to get the graph of $y = \frac{1}{3} \sin x$. This results in compressing the graph of $y = \sin x$ vertically by a factor of $\frac{1}{3}$. As before, the x-intercepts are unchanged; but here the maximum y value is $\frac{1}{3}$ and the minimum y value is $-\frac{1}{3}$ (Figure 4.51).

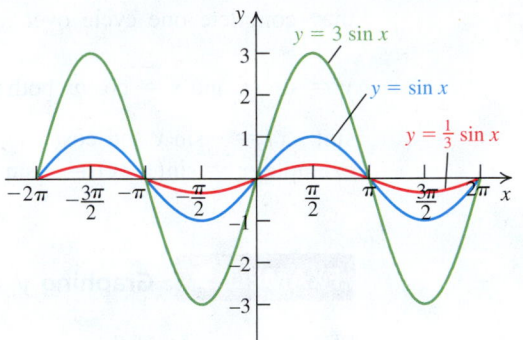

Figure 4.51 Varying a in $y = a \sin x$.

Practice Problem 1 Graph $y = 5 \sin x$, $y = \frac{1}{5} \sin x$, and $y = \sin x$ on the same coordinate system over the interval $[-2\pi, 2\pi]$.

EXAMPLE 2 Graphing $y = a \cos x$

Graph $y = -2 \cos x$ over the interval $[-2\pi, 2\pi]$. Find the amplitude and range of the function.

Solution

Begin with the graph of $y = \cos x$. Multiply the y-coordinate of each point on that graph by 2 to stretch the graph vertically by a factor of 2 to get the graph of $y = 2 \cos x$. Pay particular attention to the key points. Notice that the x-intercepts are unchanged. Next, reflect the graph of $y = 2 \cos x$ about the x-axis to produce the graph of $y = -2 \cos x$. See Figure 4.52. The amplitude is $|-2| = 2$, the largest value attained by $y = -2 \cos x$. The smallest function value attained is -2. Therefore, the range of $y = -2 \cos x$ is $[-2, 2]$.

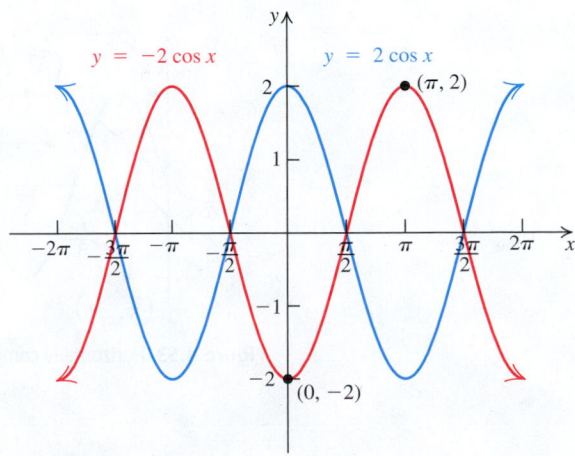

Figure 4.52 Graph of $y = -2 \cos x$.

Practice Problem 2 Graph $y = 5 \cos x$ over the interval $[-2\pi, 2\pi]$. Find the amplitude and range of the function.

Both the sine and cosine functions have period 2π. For $b > 0, 0 \leq bx \leq 2\pi$ is equivalent to $0 \leq x \leq \dfrac{2\pi}{b}$. Therefore, as x takes on the values from 0 to $\dfrac{2\pi}{b}$, the expression bx takes on the values from 0 to 2π.

The graphs of $y = \sin bx$ and $y = \cos bx$ have similar shapes to those of $y = \sin x$ and $y = \cos x$, respectively, but they are compressed or expanded horizontally so that they complete one cycle over any interval of length $\dfrac{2\pi}{b}$. Consequently, the functions $y = \sin bx$ and $y = \cos bx$ both have period $\dfrac{2\pi}{b}$. When $b < 0$, we can use the facts that $\sin(-x) = -\sin x$ and $\cos(-x) = \cos x$ to rewrite the given function with $b > 0$. For example, $y = \sin(-3x) = -\sin 3x$.

EXAMPLE 3   Graphing $y = \sin bx$

Sketch one cycle of the graphs of $y = \sin 3x$, $y = \sin \dfrac{1}{3}x$, and $y = \sin x$ on the same coordinate system.

Solution

We begin with the graph of $y = \sin x$ and adjust its period, 2π, by dividing 2π by 3 to find the period $\dfrac{2\pi}{3}$ of $y = \sin 3x$. We then compress the graph of $y = \sin x$ horizontally so that one cycle of $y = \sin 3x$ is completed on the interval $\left[0, \dfrac{2\pi}{3}\right]$. Divide the interval $0 \leq x \leq \dfrac{2\pi}{3}$ into four equal parts to find the x-coordinates for the key points: $0, \dfrac{\pi}{6}, \dfrac{\pi}{3}, \dfrac{\pi}{2}$ and $\dfrac{2\pi}{3}$. Since the amplitude is 1, the y values of the key points are unchanged. See Figure 4.53.

Similarly, to find the period of $y = \sin \dfrac{1}{3}x$, we divide 2π by $\dfrac{1}{3}$, resulting in $\dfrac{2\pi}{\frac{1}{3}} = 6\pi$.

Therefore, one cycle for $y = \sin \dfrac{1}{3}x$ is completed on the interval $[0, 6\pi]$. Divide the interval $0 \leq x \leq 6\pi$ into four equal parts to get the x-coordinates for the key points: $0, \dfrac{3\pi}{2}, 3\pi, \dfrac{9\pi}{2}$, and 6π. Again, the y values of the key points are unchanged (Figure 4.53).

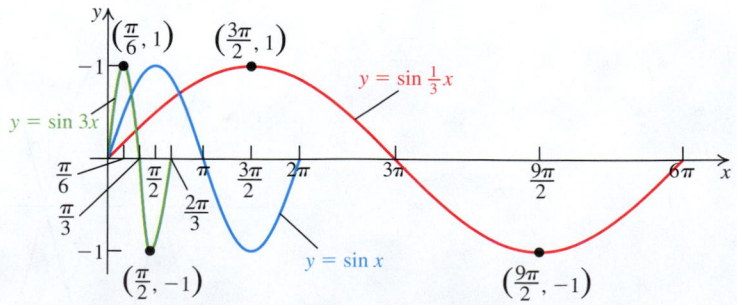

Figure 4.53 Horizontally compressing or expanding the graph of $y = \sin x$.

Practice Problem 3 Sketch one cycle of the graphs of $y = \sin \frac{1}{2}x$ and $y = \sin x$ on the same coordinate system.

The next box summarizes information used to graph $y = a \sin bx$ and $y = a \cos bx$.

CHANGING THE AMPLITUDE AND PERIOD OF THE SINE AND COSINE FUNCTIONS

The functions $y = a \sin bx$ and $y = a \cos bx$ ($b > 0$) have amplitude $|a|$ and period $\frac{2\pi}{b}$. If $a > 0$, the graphs of $y = a \sin bx$ and $y = a \cos bx$ are similar to the graphs of $y = \sin x$ and $y = \cos x$, respectively, with three changes.

1. The range is $[-a, a]$.

2. One cycle is completed over the interval $\left[0, \frac{2\pi}{b}\right]$.

3. The x-coordinates for the key points are $0, \frac{\pi}{2b}, \frac{\pi}{b}, \frac{3\pi}{2b}$, and $\frac{2\pi}{b}$. The y-values at each of these numbers is one of: $-a, 0,$ or a.

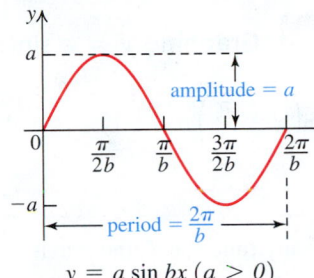
$y = a \sin bx \;(a > 0)$

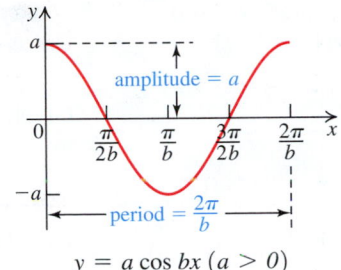
$y = a \cos bx \;(a > 0)$

If $a < 0$, the graphs are the reflections of the graph of $y = |a| \sin bx$ and $y = |a| \cos bx$, respectively, about the x-axis.
If $b < 0$, first use even–odd properties, then graph the function. See page 388.

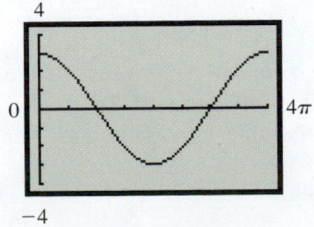
EXAMPLE 4   **Graphing $y = a \cos bx$**

Graph $y = 3 \cos \frac{1}{2}x$ over a one-period interval.

Solution

We begin with the graph of $y = \cos x$ and find the period of $y = \cos \frac{1}{2}x$ by dividing 2π by $\frac{1}{2}$ to get

$$2\pi \div \frac{1}{2} = 4\pi \qquad \text{Replace } b \text{ with } \frac{1}{2} \text{ in } \frac{2\pi}{b}.$$

The period of $y = \cos \frac{1}{2}x$ is 4π. So, the graph of $y = \cos \frac{1}{2}x$ is obtained by horizontally stretching the graph of $y = \cos x$ by a factor of 2.

Next, we multiply the y-coordinate of each point on the graph of $y = \cos \frac{1}{2}x$ by 3 to stretch this graph vertically by a factor of 3. To help sketch the graphs, divide the period, 4π, into four equal parts, starting at 0, to find the x-coordinates for the key points. These

x-coordinates, 0, π, 2π, 3π, and 4π, correspond to the highest points, the lowest point, and the x-intercepts of the graph. The key points on the graph are $(0, 3)$, $(\pi, 0)$, $(2\pi, -3)$, $(3\pi, 0)$ and $(4\pi, 3)$. See Figure 4.54. The graph can then be extended indefinitely to the left and right. The amplitude is 3; so the largest value of y is 3. The range is $[-3, 3]$.

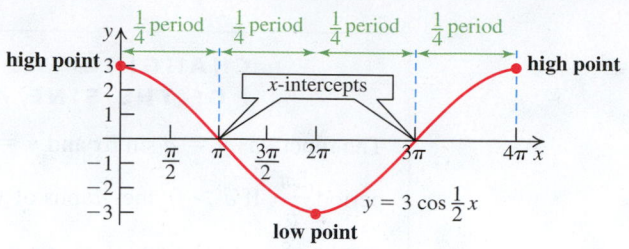

Figure 4.54 One cycle of $y = 3 \cos \dfrac{1}{2}x$.

Practice Problem 4 Graph $y = 2 \cos 2x$ over a one-period interval.

Phase Shift

4 Find the phase shifts and graph sinusoidal functions of the forms $y = a \sin [b(x - c)]$ and $y = a \cos [b(x - c)]$.

We now consider sinusoidal graphs with horizontal shifts.

EXAMPLE 5 Graphing $y = a \sin(x - c)$

Graph $y = \sin\left(x - \dfrac{\pi}{2}\right)$ over a one-period interval.

Solution

Recall that for any function f, the graph of $y = f(x - c)$ is the graph of $y = f(x)$ shifted right c units if $c > 0$ and left $|c|$ units if $c < 0$. Comparing the equation $y = \sin\left(x - \dfrac{\pi}{2}\right)$ with $y = \sin x$, we see that $y = \sin\left(x - \dfrac{\pi}{2}\right)$ has the form $y = \sin(x - c)$, where $c = \dfrac{\pi}{2}$.

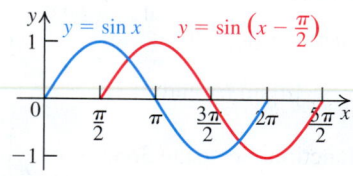

Figure 4.55 Graph of $y = \sin\left(x - \dfrac{\pi}{2}\right)$.

Therefore, the graph of $y = \sin\left(x - \dfrac{\pi}{2}\right)$ is the graph of $y = \sin x$ shifted right $\dfrac{\pi}{2}$ units.

See Figure 4.55.

Practice Problem 5 Graph $y = \cos\left(x - \dfrac{\pi}{4}\right)$ over a one-period interval.

Before describing the general procedure for graphing the sinusoidal functions $y = a \sin [b(x - c)]$ and $y = a \cos [b(x - c)]$, with $b > 0$, we find the value of x for which $b(x - c) = 0$. This number is called the **phase shift**. Let

$$b(x - c) = 0$$
$$x - c = 0 \qquad \text{Divide both sides by } b.$$
$$x = c \qquad \text{Add } c \text{ to both sides.}$$

The phase shift gives the amount by which the graphs of $y = a \sin bx$ or $y = a \cos bx$ are shifted horizontally. These graphs are shifted right if $c > 0$ and left if $c < 0$. To graph a cycle of either function, we start the cycle at $x = c$ and then divide the period into four equal parts. We describe this procedure next.

PROCEDURE
IN ACTION

EXAMPLE 6 Graphing $y = a \sin[b(x - c)]$ and $y = a \cos[b(x - c)]$ with $b > 0$

OBJECTIVE

Graph a function of the form $y = a \sin[b(x - c)]$
or $y = a \cos[b(x - c)]$, *with* $b > 0$, *by finding the
amplitude, period, and phase shift.*

Step 1 Find the amplitude, period, and phase shift.
amplitude $= |a|$

period $= \dfrac{2\pi}{b}$

phase shift $= c$

If $c > 0$, shift to the right.

If $c < 0$, shift to the left.

Step 2 The cycle begins at $x = c$. One complete

cycle occurs over the interval $\left[c, c + \dfrac{2\pi}{b} \right]$.

Step 3 Divide the interval $\left[c, c + \dfrac{2\pi}{b} \right]$ into four

equal parts, each of length $\dfrac{1}{4}(\text{period}) = \dfrac{1}{4}\left(\dfrac{2\pi}{b} \right)$. This

gives the *x*-coordinates for the five key points:

c, $c + \dfrac{1}{4}\left(\dfrac{2\pi}{b} \right)$, $c + \dfrac{1}{2}\left(\dfrac{2\pi}{b} \right)$,

$c + \dfrac{3}{4}\left(\dfrac{2\pi}{b} \right)$, and $c + \dfrac{2\pi}{b}$.

Step 4 If $a > 0$, for $y = a \sin[b(x - c)]$,
sketch one cycle of the sine curve through the

key points $(c, 0)$, $\left(c + \dfrac{\pi}{2b}, a \right)$, $\left(c + \dfrac{\pi}{b}, 0 \right)$,

$\left(c + \dfrac{3\pi}{2b}, -a \right)$, and $\left(c + \dfrac{2\pi}{b}, 0 \right)$.

For $y = a \cos[b(x - c)]$, sketch one cycle of the

cosine curve through the key points (c, a), $\left(c + \dfrac{\pi}{2b}, 0 \right)$,

$\left(c + \dfrac{\pi}{b}, -a \right)$, $\left(c + \dfrac{3\pi}{2b}, 0 \right)$, and $\left(c + \dfrac{2\pi}{b}, a \right)$.

If $a < 0$, reflect the graph of
$y = |a| \sin[b(x - c)]$ or
$y = |a| \cos[b(x - c)]$ about the x-axis.

EXAMPLE

Find the amplitude, period, and phase shift, and sketch the

graph of $y = 3 \sin\left[2\left(x - \dfrac{\pi}{4} \right) \right]$ *over a one-period interval.*

1. $y = 3 \sin\left[2\left(x - \dfrac{\pi}{4} \right) \right]$

$a = 3, b = 2, \quad \dfrac{\pi}{4} = c$

amplitude $= |3| = 3$

period $= \dfrac{2\pi}{2} = \pi$

phase shift $= \dfrac{\pi}{4}$ Shift right because $\dfrac{\pi}{4} > 0$.

2. Begin the cycle at $x = \dfrac{\pi}{4}$ and graph one cycle over

$\left[\dfrac{\pi}{4}, \dfrac{\pi}{4} + \pi \right] = \left[\dfrac{\pi}{4}, \dfrac{5\pi}{4} \right]$

3. Divide the interval $\left[\dfrac{\pi}{4}, \dfrac{5\pi}{4} \right]$ into four equal parts, each

having length $\dfrac{1}{4}(\text{period}) = \dfrac{1}{4}(\pi) = \dfrac{\pi}{4}$.

The *x*-coordinates of the five key points are:

$\dfrac{\pi}{4}, \dfrac{\pi}{4} + \dfrac{\pi}{4} = \dfrac{\pi}{2}, \dfrac{\pi}{4} + \dfrac{\pi}{2} = \dfrac{3\pi}{4},$

$\dfrac{\pi}{4} + \dfrac{3\pi}{4} = \pi, \quad \text{and} \quad \dfrac{\pi}{4} + \pi = \dfrac{5\pi}{4}$

4. Sketch one cycle of the sine curve through the five key

points: $\left(\dfrac{\pi}{4}, 0 \right)$, $\left(\dfrac{\pi}{2}, 3 \right)$, $\left(\dfrac{3\pi}{4}, 0 \right)$, $(\pi, -3)$, and $\left(\dfrac{5\pi}{4}, 0 \right)$.

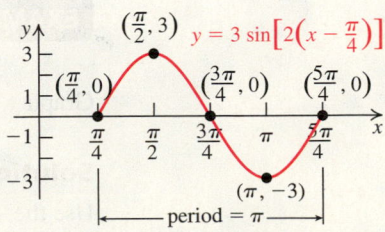

Alternatively, we can use a sequence of transformations (see Section 1.5) on the graph of $y = \sin x$ to obtain the graph of $y = 3 \sin \left[2 \left(x - \dfrac{\pi}{4} \right) \right] = 3 \sin \left(2x - \dfrac{\pi}{2} \right)$. See Figure 4.56(d).

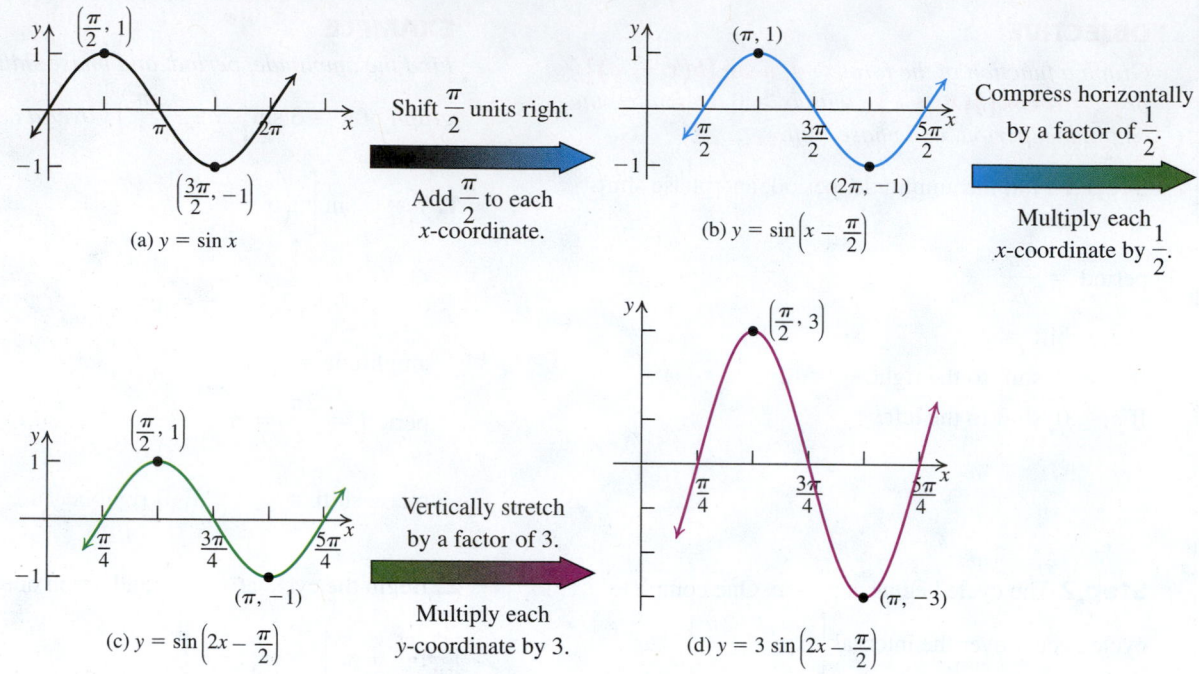

Figure 4.56

Practice Problem 6 Find the amplitude, period, and phase shift and sketch the graph of the function

$$y = \frac{1}{3} \cos \left[\frac{2}{5} \left(x + \frac{\pi}{6} \right) \right].$$

Graphing when $b < 0$ The even–odd properties of the sine and cosine functions allow us to graph $y = a \sin[\, b(x - c)\,]$ and $y = a \cos[\, b(x - c)\,]$ when $b < 0$.

For example, to graph $y = 3 \sin \left[(-2) \left(x - \dfrac{\pi}{2} \right) \right]$, we rewrite and graph the equation $y = -3 \sin \left[2 \left(x - \dfrac{\pi}{2} \right) \right]$, using the fact that $\sin(-x) = -\sin x$.

EXAMPLE 7 **Graphing $y = a \cos[b(x - c)]$**

Graph $y = -\cos \left[2 \left(x + \dfrac{\pi}{2} \right) \right]$ over a one-period interval.

Solution

Use the steps in Example 6 for the procedure for graphing $y = a \cos[\, b(x - c)\,]$.

Step 1 Here $y = -\cos \left[2 \left(x + \dfrac{\pi}{2} \right) \right] = (-1) \cos \left[2 \left(x - \left(-\dfrac{\pi}{2} \right) \right) \right]$

$$a = -1 \quad b = 2 \quad -\dfrac{\pi}{2} = c$$

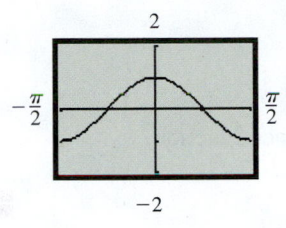
amplitude $= |a| = |-1| = 1$ period $= \dfrac{2\pi}{b} = \dfrac{2\pi}{2} = \pi$

phase shift $= c = -\dfrac{\pi}{2}$ Shift left because $c < 0$.

Step 2   The cycle begins at $x = -\dfrac{\pi}{2}$.

One cycle is graphed over the interval $\left[-\dfrac{\pi}{2}, -\dfrac{\pi}{2} + \pi\right] = \left[-\dfrac{\pi}{2}, \dfrac{\pi}{2}\right]$.

Step 3 $\dfrac{1}{4}(\text{period}) = \dfrac{1}{4}(\pi) = \dfrac{\pi}{4}$

The x-coordinates of the five key points are:

starting point $= -\dfrac{\pi}{2}, \ -\dfrac{\pi}{2} + \dfrac{\pi}{4} = -\dfrac{\pi}{4}, \ -\dfrac{\pi}{2} + \dfrac{\pi}{2} = 0, \ -\dfrac{\pi}{2} + \dfrac{3\pi}{4} = \dfrac{\pi}{4},$

and $-\dfrac{\pi}{2} + \pi = \dfrac{\pi}{2}$

Step 4   Because $a = -1 < 0$, we first graph $y = |-1| \cos\left[2\left(x + \dfrac{\pi}{2}\right)\right]$ by

sketching one cycle, beginning at $\left(-\dfrac{\pi}{2}, 1\right)$ and going through the

points $\left(-\dfrac{\pi}{4}, 0\right)$, $(0, -1)$, $\left(\dfrac{\pi}{4}, 0\right)$, and $\left(\dfrac{\pi}{2}, 1\right)$. We then reflect this graph

about the x-axis to obtain the graph of $y = -\cos\left[2\left(x + \dfrac{\pi}{2}\right)\right]$. See Figure 4.57.

$\left(-\dfrac{\pi}{2}, 1\right)$ $y = \cos\left[2\left(x + \dfrac{\pi}{2}\right)\right]$

$\left(\dfrac{\pi}{2}, 1\right)$

$\left(\dfrac{\pi}{2}, -1\right)$

$\left(-\dfrac{\pi}{2}, -1\right)$ $y = -\cos\left[2\left(x + \dfrac{\pi}{2}\right)\right]$

Figure 4.57

Practice Problem 7 Describe a sequence of transformations on the graph of $y = \cos x$
that result in a sketch of the graph of $y = -3\cos\left(\dfrac{\pi}{2} - \dfrac{x}{2}\right)$.

To graph the function of the form $y = a\sin(bx - k)$ or $y = a\cos(bx - k)$, with
$b > 0$, we first note that

$$bx - k = b\left(x - \dfrac{k}{b}\right) \qquad \text{Factor out } b.$$

Consequently, to graph these functions, we rewrite their equations.

$$y = a\sin(bx - k) = a\sin\left[b\left(x - \dfrac{k}{b}\right)\right]$$

$$y = a\cos(bx - k) = a\cos\left[b\left(x - \dfrac{k}{b}\right)\right]$$

We can then use the procedures for graphing the functions of the form $y = a\sin[b(x - c)]$

and $y = a\cos[b(x - c)]$ by letting $\dfrac{k}{b} = c$.

EXAMPLE 8   **Finding the Period and Phase Shift**

Find the period and the phase shift of each function.

a. $y = 2\cos\left(3x - \dfrac{\pi}{2}\right)$

b. $y = 3\sin(\pi x + 2)$

Solution

a. $y = 2 \cos\left(3x - \dfrac{\pi}{2}\right)$ The given function

$$= 2 \cos\left[3\left(x - \dfrac{\pi}{6}\right)\right]$$ Rewrite, using the distributive property.

       ↑   ↑   ↑
       a b c Compare with $y = a\cos\left[b(x - c)\right]$.

$$\text{Period} = \dfrac{2\pi}{b} = \dfrac{2\pi}{3} \qquad \text{Phase shift} = c = \dfrac{\pi}{6}$$

b. $y = 3 \sin(\pi x + 2)$ The given function

$$y = 3 \sin\left[\pi\left(x - \left(-\dfrac{2}{\pi}\right)\right)\right]$$ Rewrite, using the distributive property.

       ↑   ↑   ↑
       a b c

$$\text{Period} = \dfrac{2\pi}{b} = \dfrac{2\pi}{\pi} = 2 \qquad \text{Phase shift} = c = -\dfrac{2}{\pi}$$

Practice Problem 8 Find the period and the phase shift of each function.

a. $y = 3 \sin\left(2x + \dfrac{\pi}{4}\right)$ **b.** $y = \cos(\pi x - 1)$

EXAMPLE 9   **Graphing $y = a\sin(bx - k)$**

Graph $y = 3 \sin\left(2x - \dfrac{\pi}{2}\right)$ over a one-period interval.

Solution

Because $2x - \dfrac{\pi}{2} = 2\left(x - \dfrac{\pi}{4}\right)$, we can write $y = 3 \sin\left(2x - \dfrac{\pi}{2}\right) = 3 \sin\left[2\left(x - \dfrac{\pi}{4}\right)\right]$.

This is the sinusoidal curve we graphed in Example 6 to show the procedure for graphing the functions $y = a\sin\left[b(x - c)\right]$ and $y = a\cos\left[b(x - c)\right]$. The graph is shown again in Figure 4.58.

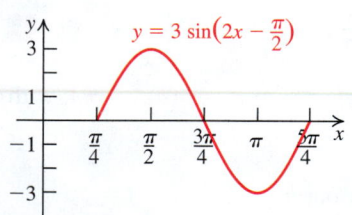

$y = 3\sin\left(2x - \dfrac{\pi}{2}\right)$

Figure 4.58

Practice Problem 9 Graph $y = 2 \cos(3x - \pi)$ over a one-period interval.

Vertical Shifts

Recall that the graph of $y = f(x) + d$ results from shifting the graph of $y = f(x)$ vertically d units up if $d > 0$ and $|d|$ units down if $d < 0$. The next example shows this effect on a sinusoidal graph.

EXAMPLE 10   **Graphing $y = a\cos[b(x - c)] + d$**

Graph $y = \cos\left[2\left(x + \dfrac{\pi}{2}\right)\right] + 3$ over a one-period interval.

Solution

In Example 7, we graphed the function $y = \cos\left[2\left(x + \dfrac{\pi}{2}\right)\right]$ in order to graph $y = -\cos\left[2\left(x + \dfrac{\pi}{2}\right)\right]$. To graph $y = \cos\left[2\left(x + \dfrac{\pi}{2}\right)\right] + 3$, shift the graph of $y = \cos\left[2\left(x + \dfrac{\pi}{2}\right)\right]$, up **three** units. See Figure 4.59.

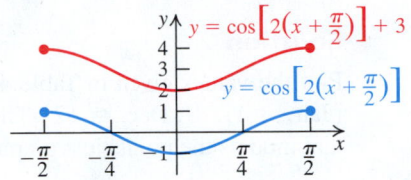

Figure 4.59 Graph of $y = \cos\left[2\left(x + \dfrac{\pi}{2}\right)\right] + 3$.

Practice Problem 10 Graph $y = 4\sin\left(3x - \dfrac{\pi}{3}\right) - 2$ over a one-period interval.

Modeling with Sinusoidal Curves

5 Model periodic behavior.

We begin with the **biorhythm states** discussed in the introduction to this chapter.

The periods for physical (P), emotional (E), and intellectual (I) states are **23**, **28**, and **33** days, respectively. So your biorhythm states on the tth day since birth are given by

$$P = \sin\left(\frac{2\pi}{23}t\right), E = \sin\left(\frac{2\pi}{28}t\right), \text{ and } I = \sin\left(\frac{2\pi}{33}t\right).$$

To calculate t on a given day, use the formula:

$$t = (\text{your age})(365) + (\text{number of leap years since your birth year})$$
$$+ (\text{number of days since your last birthday}).$$

EXAMPLE 11   Modeling Biorhythm States

Find the biorhythm states of Lisa on May 22, 2016, assuming that she was born on March 13, 1997.

Solution

We first calculate t.

Number of years from March 13, 1997 to March 13, 2016 = 2016 − 1997 = 19.

Number of leap years = 5(2000, 2004, 2008, 2012, 2016)

Number of days from
March 13 to May 22 = 18(March) + 30(April) + 22(May) = 70

So $t = (19)(365) + 5 + 70 = 7010$

$$P = \sin\left(\frac{2\pi}{23}t\right) = \sin\left(\frac{2\pi(7010)}{23}\right) \approx -0.98$$

Substitute $t = 7010$ in each equation; use a calculator in *Radian* mode.

$$E = \sin\left(\frac{2\pi}{28}t\right) = \sin\left(\frac{2\pi(7010)}{28}\right) \approx 0.78$$

$$I = \sin\left(\frac{2\pi}{33}t\right) = \sin\left(\frac{2\pi(7010)}{33}\right) \approx 0.46$$

Lisa's physical state is way down, her emotional state is excellent, and her intellectual state is good.

Practice Problem 11 Find your biorhythm states today.

EXAMPLE 12 Modeling the Number of Daylight Hours in Paris

Table 4.6 gives the average number of daylight hours in Paris each month. Let y represent the number of daylight hours in Paris in month x to find a function of the form $y = a \sin[b(x - c)] + d$ that models the hours of daylight throughout the year.

Solution

Plot the values given in Table 4.6, representing the months by the integers 1 through 12 (Jan. = 1, ..., Dec. = 12). Then sketch a function of the form $y = a \sin[b(x - c)] + d$ that models the points just graphed. See Figure 4.60.

TABLE 4.6

Daylight hours in Paris	
January	8.8
February	10.2
March	11.9
April	13.7
May	15.3
June	16.1
July	15.7
August	14.3
September	12.6
October	10.8
November	9.2
December	8.3

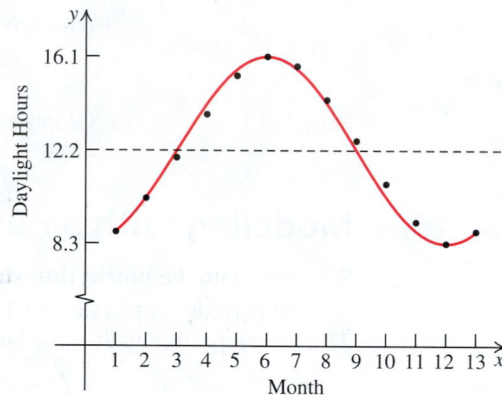

Figure 4.60 Hours of daylight in Paris each month.

We want to find the values for the constants a, b, c, and d that will produce the graph shown in Figure 4.60. From Table 4.6, the range of the function is $[8.3, 16.1]$; so

$$\text{Amplitude} = a = \frac{\text{highest value} - \text{lowest value}}{2} = \frac{(16.1 - 8.3)}{2} = 3.9.$$

The sine graph has been shifted vertically up by $\frac{1}{2}$ (highest value + lowest value), the average of the highest and lowest values. Thus, this average value gives

$$d = \frac{1}{2}(16.1 + 8.3) = 12.2.$$

The weather repeats every 12 months; so the period is 12. Because the period $= \frac{2\pi}{b}$, we write

$$12 = \frac{2\pi}{b} \qquad \text{Replace "period" with 12.}$$

$$b = \frac{2\pi}{12} = \frac{\pi}{6} \qquad \text{Solve for } b.$$

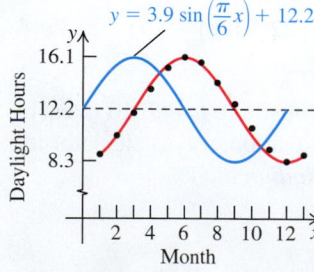

Figure 4.60A

So far, we can write $y = 3.9 \sin\left[\left(\frac{\pi}{6}\right)(x - c)\right] + 12.2$. The horizontally "unshifted" graph of $y = 3.9 \sin\left(\frac{\pi}{6}x\right) + 12.2$ is superimposed on Figure 4.60 in Figure 4.60A. This

TABLE 4.7

Daylight hours in Fargo	
January	9
February	10.3
March	11.9
April	13.6
May	15.1
June	15.8
July	15.4
August	14.2
September	12.5
October	10.9
November	9.4
December	8.6

graph must be shifted to the right to fit the plotted data. To determine the phase shift, find the value of c when y has the greatest value. This value is $y = 16.1$ when $x = 6$ (in June), so we now have

$$16.1 = 3.9 \sin\left[\left(\frac{\pi}{6}\right)(6 - c)\right] + 12.2 \qquad \text{Replace } x \text{ with 6 and } y \text{ with 16.1.}$$

$$1 = \sin\left[\left(\frac{\pi}{6}\right)(6 - c)\right] \qquad \text{Subtract 12.2 from both sides and divide by 3.9.}$$

Now $\sin\left[\left(\frac{\pi}{6}\right)(6 - c)\right]$ will first equal 1 when the argument of sine is $\frac{\pi}{2}$. So

$$\frac{\pi}{6}(6 - c) = \frac{\pi}{2} \qquad \sin t = 1 \text{ when } t = \frac{\pi}{2}.$$

$$c = 3 \qquad \text{Solve for } c.$$

The equation describing the hours of daylight in Paris *in month x* is

$$y = 3.9 \sin\left[\frac{\pi}{6}(x - 3)\right] + 12.2.$$

Practice Problem 12 Rework Example 12 for the city of Fargo, North Dakota, using the values given in Table 4.7.

Simple Harmonic Motion

Trigonometric functions can often describe or model motion caused by vibration, rotation, or oscillation. The motions associated with sound waves, radio waves, alternating electric current, a vibrating guitar string, or the swing of a pendulum are examples of such motion. Another example is the movement of a ball suspended by a spring. If the ball is pulled downward and then released (ignoring the effects of friction and air resistance), the ball will repeatedly move up and down past its rest, or equilibrium, position. See Figure 4.61.

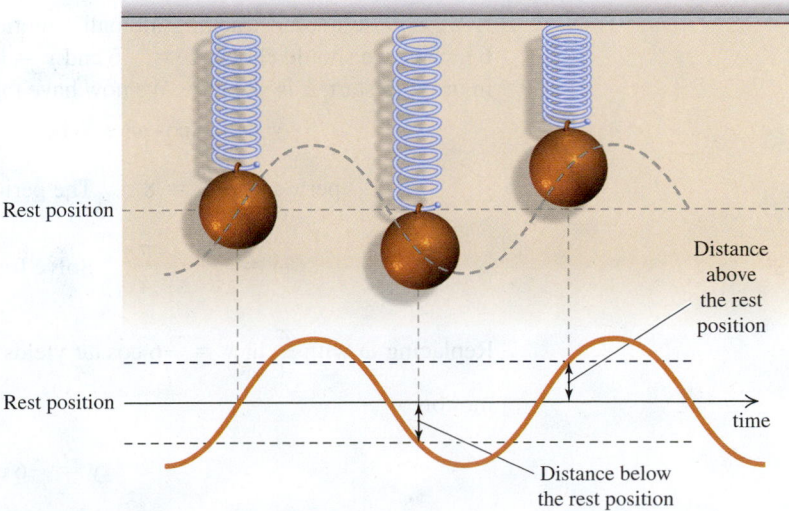

Figure 4.61 Simple harmonic motion.

Simple Harmonic Motion

An object whose position relative to an equilibrium position at time t can be described by either

$$y = a \sin \omega t \quad \text{or} \quad y = a \cos \omega t \quad (\omega > 0)$$

is in simple harmonic motion.

The *amplitude*, $|a|$, is the maximum distance the object travels from its equilibrium position. The *period* of the motion, $\dfrac{2\pi}{\omega}$, is the time it takes the object to complete one full cycle. The **frequency** of the motion is $\dfrac{\omega}{2\pi}$ and gives the number of cycles completed per unit of time.

If the distance above and below the rest position of the ball is graphed as a function of time, a sine or a cosine curve results. The amplitude of the curve is the maximum distance above the rest position the ball travels, and one cycle is completed as the ball travels from its highest position to its lowest position and back to its highest position.

EXAMPLE 13   Modeling Simple Harmonic Motion

Suppose a ball attached to a spring is pulled down 6 inches and released and the resulting simple harmonic motion has a period of eight seconds. Write an equation for the ball's simple harmonic motion.

Solution

We must first choose between an equation of the form $y = a \sin \omega t$ or $y = a \cos \omega t$. We start tracking the motion of the ball when $t = 0$. Note that for $t = 0$,

$$y = a \sin(\omega \cdot 0) = a \cdot \sin 0 = a \cdot 0 = 0 \quad \text{and}$$
$$y = a \cos(\omega \cdot 0) = a \cdot \cos 0 = a \cdot 1 = a.$$

If we choose to start tracking the ball's motion when we release it after pulling it down 6 inches, we should choose $a = -6$ and $y = -6 \cos \omega t$. Because we pulled the ball *down* in order to start, a is negative. We now have the form of the equation of motion.

$$y = -6 \cos \omega t \quad \text{\color{blue}{When } } t = 0, y = -6.$$

$$\text{period} = \frac{2\pi}{\omega} = 8 \quad \text{\color{blue}{The period is given as eight seconds.}}$$

$$\omega = \frac{2\pi}{8} = \frac{\pi}{4} \quad \text{\color{blue}{Solve for } } \omega.$$

Replacing ω with $\dfrac{\pi}{4}$ in $y = -6 \cos \omega t$ yields the equation of the ball's simple harmonic motion:

$$y = -6 \cos \frac{\pi}{4} t$$

Practice Problem 13 Rework Example 13 with a period of three seconds and the ball pulled down 4 inches and released.

Answers to Practice Problems

1.

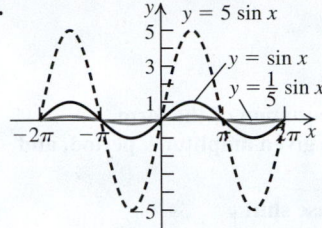

2.

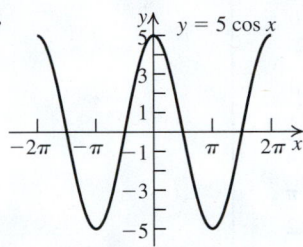

Amplitude = 5; range = $[-5, 5]$

3.

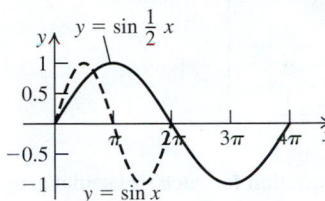

4.

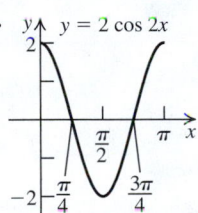

5.

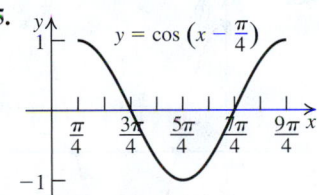

6. Amplitude = $\dfrac{1}{3}$; period = 5π; phase shift = $-\dfrac{\pi}{6}$

$y = \dfrac{1}{3}\cos\left[\dfrac{2}{5}\left(x + \dfrac{\pi}{6}\right)\right]$

$y = \dfrac{1}{3}\cos\left[\dfrac{2}{5}\left(x + \dfrac{\pi}{6}\right)\right]$

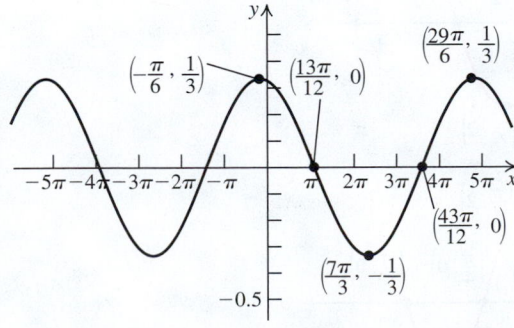

7.
$$\cos x \to \cos\left(\dfrac{\pi}{2} + x\right) \to$$
$$\to \cos\left(\dfrac{\pi}{2} + \dfrac{x}{2}\right) \to \cos\left(\dfrac{\pi}{2} - \dfrac{x}{2}\right) \to$$
$$\to -3\cos\left(\dfrac{\pi}{2} - \dfrac{x}{2}\right)$$

8. a. Period = π; phase shift = $-\dfrac{\pi}{8}$

b. Period = 2; phase shift = $\dfrac{1}{\pi}$

9.

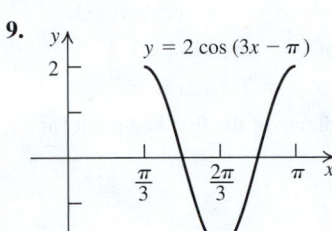

10.

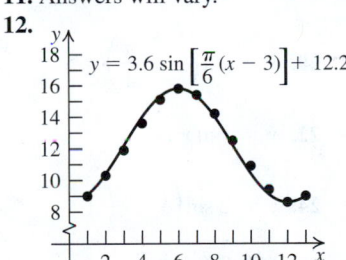

11. Answers will vary.

12.

13. $y = -4\cos\dfrac{2\pi}{3}t$

SECTION 4.4 **Exercises**

Concepts and Vocabulary

1. The lowest point on the graph of $y = \cos x$, $0 \leq x \leq 2\pi$, occurs when $x =$ _____.

2. The highest point on the graph of $y = \sin x$, $0 \leq x \leq 2\pi$, occurs when $x =$ _____.

3. The range of the cosine function is _____.

4. The maximum value of $y = -2 \sin x$ is _____.

5. **True or False.** The amplitude of $y = -3 \cos x$ is -3.

6. **True or False.** The range of $y = \sin 3x$ is $[-3, 3]$.

7. **True or False.** The period of $y = 7 \cos 2x$ is π.

8. **True or False.** The x-coordinate of the first key point for $y = \sin\left(x + \dfrac{\pi}{2}\right)$ is $-\dfrac{\pi}{2}$.

In Exercises 9–28, sketch the graph of each given equation over the interval $[-2\pi, 2\pi]$.

9. $y = 2 \sin x$ 10. $y = 4 \cos x$

11. $y = -\dfrac{1}{2} \sin x$ 12. $y = -2 \sin x$

13. $y = \dfrac{3}{2} \cos x$ 14. $y = \dfrac{5}{4} \sin x$

15. $y = \cos 2x$ 16. $y = \sin 4x$

17. $y = \cos \dfrac{2}{3}x$ 18. $y = \sin \dfrac{4}{3}x$

19. $y = \cos\left(x + \dfrac{\pi}{2}\right)$ 20. $y = \sin\left(x + \dfrac{\pi}{4}\right)$

21. $y = \cos\left(x - \dfrac{\pi}{3}\right)$ 22. $y = \sin(x - \pi)$

23. $y = 2 \cos\left(x - \dfrac{\pi}{2}\right)$ 24. $y = 2 \sin\left(x + \dfrac{\pi}{3}\right)$

25. $y = \sin x + 1$ 26. $y = \cos x - 2$

27. $y = -\cos x + 1$ 28. $y = \sin x - 3$

In Exercises 29–36, find the amplitude, period, and phase shift of each given function.

29. $y = 5 \cos(x - \pi)$ 30. $y = 3 \sin\left(x - \dfrac{\pi}{8}\right)$

31. $y = 7 \cos\left[9\left(x + \dfrac{\pi}{6}\right)\right]$ 32. $y = 11 \sin\left[8\left(x + \dfrac{\pi}{3}\right)\right]$

33. $y = -6 \cos\left[\dfrac{1}{2}(x + 2)\right]$

34. $y = -8 \sin\left[\dfrac{1}{5}(x + 9)\right]$

35. $y = 0.9 \sin\left[0.25\left(x - \dfrac{\pi}{4}\right)\right]$

36. $y = \sqrt{5} \cos[\pi(x + 1)]$

In Exercises 37–44, write an equation of the form $y = a \sin[b(x - c)]$ with the given amplitude, period, and phase shift.

	Amplitude	Period	Phase shift
37.	$\dfrac{1}{4}$	$\dfrac{\pi}{8}$	$\dfrac{\pi}{16}$
38.	2	$\dfrac{\pi}{6}$	$\dfrac{\pi}{10}$
39.	6	3π	$-\dfrac{\pi}{4}$
40.	$\dfrac{1}{2}$	6π	$-\dfrac{\pi}{2}$
41.	2.4	6	3
42.	0.8	10	2
43.	$\dfrac{\pi}{2}$	$\dfrac{5}{8}$	$-\dfrac{1}{4}$
44.	$\dfrac{2}{\pi}$	$\dfrac{2}{3}$	$-\dfrac{1}{6}$

In Exercises 45–52, write an equation for each sinusoidal graph.

45.

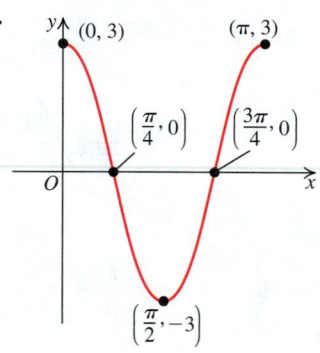

46.

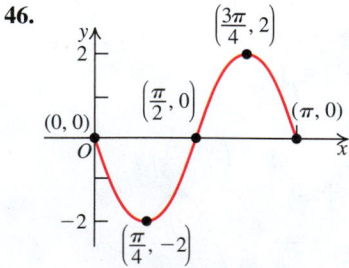

47.

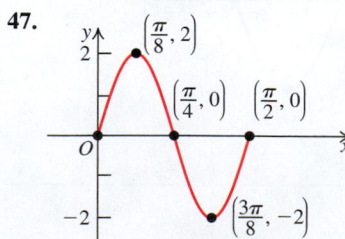

48.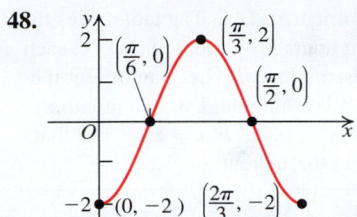

49.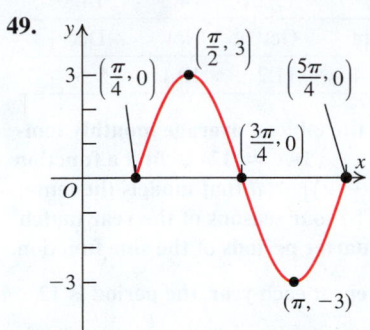

50.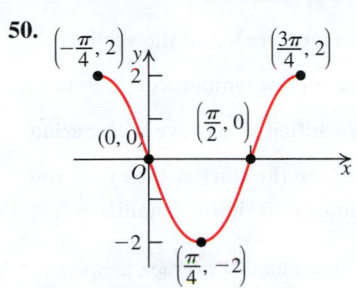

51.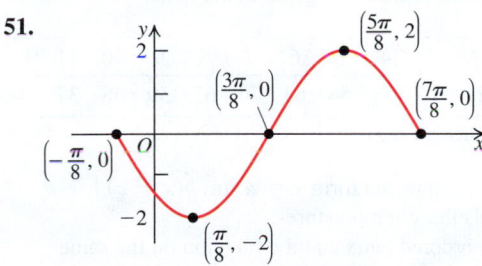

52.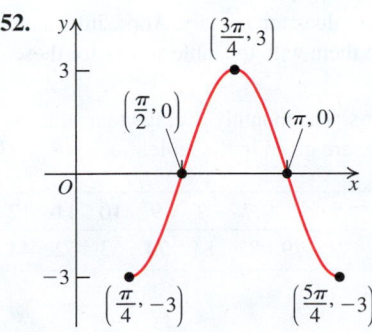

In Exercises 53–60, graph each function over a one-period interval.

53. $y = -4 \cos\left(x + \dfrac{\pi}{6}\right)$

54. $y = -3 \sin\left(x - \dfrac{\pi}{6}\right)$

55. $y = \dfrac{5}{2} \sin\left[2\left(x - \dfrac{\pi}{4}\right)\right]$

56. $y = \dfrac{3}{2} \cos\left[2\left(x + \dfrac{\pi}{3}\right)\right]$

57. $y = -5 \cos\left[4\left(x - \dfrac{\pi}{6}\right)\right]$

58. $y = -3 \sin\left[4\left(x + \dfrac{\pi}{6}\right)\right]$

59. $y = \dfrac{1}{2} \sin\left[4\left(x + \dfrac{\pi}{4}\right)\right] + 2$

60. $y = -\dfrac{1}{2} \cos\left[2\left(x - \dfrac{\pi}{2}\right)\right] - 3$

In Exercises 61–70, write each function in the form $y = a\sin[b(x - c)]$ or $y = a\cos[b(x - c)]$. Find each period and phase shift.

61. $y = 4 \cos\left(2x + \dfrac{\pi}{3}\right)$

62. $y = 5 \sin\left(3x + \dfrac{\pi}{2}\right)$

63. $y = -\dfrac{3}{2} \sin(2x - \pi)$

64. $y = -2 \cos\left(5x - \dfrac{\pi}{4}\right)$

65. $y = 3 \cos \pi x$

66. $y = \sin \dfrac{\pi x}{3}$

67. $y = \dfrac{1}{2} \cos\left(\dfrac{\pi x}{4} + \dfrac{\pi}{4}\right)$

68. $y = -\sin\left(\dfrac{\pi x}{6} + \dfrac{\pi}{6}\right)$

69. $y = 2 \sin(\pi x + 3)$

70. $y = -\cos\left(\pi x - \dfrac{1}{4}\right)$

In Exercises 71–76, each equation describes the simple harmonic motion of an object. In each case, find (a) the period of the motion and (b) its frequency.

71. $y = 3 \sin(4t)$

72. $y = 2 \cos(3t)$

73. $y = 2 \cos\left(\dfrac{\pi}{3} t\right)$

74. $y = 4 \sin\left(\dfrac{2\pi}{5} t\right)$

75. $y = -2 \cos\left(\dfrac{1}{2} t\right)$

76. $y = -3 \sin\left(\dfrac{2}{3} t\right)$

Applying the Concepts

77. Biorhythm. Find the biorhythm states of Tina on April 1, 2015, assuming that she was born on July 22, 1994.

78. Biorhythm. Find the biorhythm states of Dave on August 12, 2016, assuming that he was born on March 31, 1995.

79. Pulse and blood pressure. Blood pressure is given by two numbers written as a fraction: $\dfrac{\text{systolic}}{\text{diastolic}}$. The *systolic* reading is the maximum pressure in an artery, and the *diastolic* reading is the minimum pressure in an artery. As the heart beats, the systolic measurement is taken; when the heart rests, the diastolic measurement is taken. A reading of $\dfrac{120}{80}$ is considered normal. Your pulse is the number of heartbeats per minute. Suppose Desmond's blood pressure after t minutes is given by

$$p(t) = 20 \sin(140\pi t) + 122,$$

where $p(t)$ is the pressure in millimeters of mercury.

a. Find the period and explain how it relates to pulse.
b. Graph the function p over one period.
c. What is Desmond's blood pressure?

80. Electrical circuit voltage. The voltage $V(t)$ in an electrical circuit is given by $V(t) = 4.5 \cos(160\pi t)$, where t is measured in seconds.
   a. Find the amplitude and the period for $V(t)$.
   b. Find the frequency of $V(t)$; that is, the number of cycles completed in one second.
   c. Graph $V(t)$ over the interval $\left[0, \dfrac{1}{40}\right]$.

81. Voltage of an electrical outlet. The voltage across the terminals of a typical electrical outlet is approximated by $V(t) = 156 \sin(120\pi t)$, where t is measured in seconds.
   a. Find the amplitude and the period for $V(t)$.
   b. Find the frequency of $V(t)$; that is, the number of cycles completed in one second.
   c. Graph $V(t)$ over the interval $\left[0, \dfrac{1}{30}\right]$.

82. Simple harmonic motion. Suppose a ball attached to a spring is pulled down 5 inches and released and the resulting simple harmonic motion has a period of ten seconds. Write an equation of the ball's simple harmonic motion.

83. Kangaroo population. The kangaroo population in a certain region is given by the function

$$N(t) = 650 + 150 \sin 2t,$$

where the time t is measured in years.
   a. What is the largest number of kangaroos present in the region at any time?
   b. What is the smallest number of kangaroos present in the region at any time?
   c. How much time elapses between occurrences of the maximum and the minimum kangaroo population?

84. Lion and deer populations. The number of deer in a region is given by $D(t) = 450 \sin\left(\dfrac{3t}{5}\right) + 1200$, where t is in years, and the number of lions in that same region is given by $L(t) = 225 \sin\left[\dfrac{2}{5}(t-2)\right] + 500$, where t is in years.
   a. Graph both functions on the same coordinate plane for $0 \le t \le 10$.
   b. What can you say about the relationship between these populations based on the graphs obtained in (a)?

85. Daylight hours in London. The table gives the average number of daylight hours in London each month, where $x = 1$ represents January. Let y represent the number of daylight hours in London in month x. Find a function of the form $y = a \sin[b(x-c)] + d$ that models the hours of daylight throughout the year.

Jan	Feb	Mar	Apr	May	June
8.4	10	11.9	13.9	15.6	16.6

July	Aug	Sept	Oct	Nov	Dec
16.1	14.6	12.6	10.7	8.9	7.9

86. Daylight hours in Washington, D.C. The table gives the average number of daylight hours in Washington, D.C., each month, where $x = 1$ represents January. Let y represent the number of daylight hours in Washington, D.C., in month x. Find a function of the form $y = a \sin[b(x-c)] + d$ that models the hours of daylight throughout the year.

Jan	Feb	Mar	Apr	May	June
9.8	10.8	12	13.2	14.3	14.8

July	Aug	Sept	Oct	Nov	Dec
14.6	13.6	12.4	11.2	10.1	9.5

In Exercises 87 and 88, use the table of average monthly temperatures (where Jan $= 1, \ldots,$ Dec $= 12$) to find a function of the form $y = a \sin[b(x-c)] + d$ that models the temperature in the given city. The four seasons of the year match remarkably well with the quarter periods of the sine function.

Because the temperatures repeat each year, the period is 12 months; so $b = \dfrac{2\pi}{12} = \dfrac{\pi}{6}$. The amplitude $a = \dfrac{1}{2}$ (high temperature $-$ low temperature), and the vertical shift $d = \dfrac{1}{2}$ (high temperature $+$ low temperature). Because the maximum value a of an unshifted sine curve with period 12 occurs at $\dfrac{1}{4}(12) = 3$ units from the start of the cycle, you can find c by solving the equation $c = $ hottest month -3.

87. Average temperatures. The monthly average temperatures for Augusta, Maine, are given in the table.

Month	1	2	3	4	5	6	7	8	9	10	11	12
°F	19	22	31	43	54	63	70	67	59	48	37	24

Source: www.weatherbase.com

   a. Find a function of the form $y = a \sin[b(x-c)] + d$ that models these temperatures.
   b. Graph the ordered pairs and the function on the same coordinate system.
   c. Compute the function values for January, April, July, and October and compare them with the table values for these months.

88. Average temperatures. The monthly average temperatures for Nashville, Tennessee, are given in the table.

Month	1	2	3	4	5	6	7	8	9	10	11	12
°F	40	45	53	63	71	79	83	81	74	63	52	44

Source: www.weatherbase.com

   a. Find a function of the form $y = a \sin[b(x-c)] + d$ that models these temperatures.
   b. Graph the ordered pairs and the function on the same coordinate system.
   c. Compute the function values for January, April, July, and October and compare them to the table values for these months.

Beyond the Basics

In Exercises 89–98, graph each equation over the interval $[0, 2\pi]$.

89. $y = 3 \sin\left(-x + \dfrac{\pi}{4}\right)$ **90.** $y = -2 \sin\left(-2x + \dfrac{\pi}{3}\right)$

91. $y = 2 \cos\left(-3x + \dfrac{\pi}{2}\right) + 2$

92. $y = -3 \cos\left(-\dfrac{x}{\pi} + \dfrac{1}{2}\right) + 3$

93. $y = |\sin x|$ **94.** $y = |\cos x|$

95. $y = x \sin x$ **96.** $y = x \cos x$

97. $y = \sqrt{x} \sin x$ **98.** $y = |x \sin x|$

99. Graph the equations $y = x^2$, $y = -x^2$, and $y = x^2 \sin x$ on the same coordinate plane.

100. Graph the equations $y = e^{-t}$, $y = -e^{-t}$, and $y = e^{-t} \sin t$ on the same coordinate plane.

101. **Daily temperature.** The average daily temperature T in degrees Fahrenheit for Lithia, Florida, can be modeled by the equation $T = 25 \sin\left[\dfrac{2\pi}{365}(t - 110)\right] + 75$,

where t is the day of the (nonleap) year ($t = 1$ for January 1, . . . , $t = 365$ for December 31).

a. What is the highest average daily temperature? On which day does it occur?

b. What is the lowest average daily temperature? On which day does it occur?

c. What is the average daily temperature on August 15?

102. **Number of daylight hours.** The number of daylight hours D for Saginaw, Michigan, can be modeled by the equation $D = 3 \sin\left[\dfrac{2\pi}{365}(t - 80)\right] + 12$, where t is

the day of the (nonleap) year ($t = 1$ for January 1, . . . , $t = 365$ for December 31).

a. How many hours of daylight are there on the longest day of the year? On which day does it occur?

b. How many hours of daylight are there on the shortest day of the year? On which day does it occur?

c. How many hours of daylight are there on April 10 (the 100th day of the year)?

103. Find the range of the function

$$f(x) = -6 \cos\left(3x - \dfrac{\pi}{4}\right) - 5.$$

104. In Section 5.3, we will discuss the identity $\sin^2 \theta = \dfrac{1}{2}(1 - \cos 2\theta)$. Use this identity to show that the function $f(x) = 1 - \dfrac{1}{4} \sin^2\left(\dfrac{3x}{2} - \dfrac{\pi}{6}\right)$ has period $\dfrac{2\pi}{3}$.

105. We know that $\sin(x + 2\pi) = \sin x$ for all real numbers x. Consequently, if 2π is not the period for the sine function, then $\sin(x + p) = \sin x$ for some real number p, with $0 < p < 2\pi$, and all real numbers x.

a. By evaluating $\sin(x + p) = \sin x$ for $x = 0$, show that if the period is not 2π, then $p = \pi$.

b. Find a value of x for which $\sin(x + \pi) \neq \sin x$ and conclude that the period of the sine function is 2π.

Critical Thinking / Discussion / Writing

106. Suppose a function of the form $y = a \sin[b(x - c)] + d$, with $a > 0$, has period $= 20$.

a. In a one-period interval, how many units are between the value of x at which the maximum value is attained and the value of x at which the minimum value is attained?

b. In a one-period interval, how many units are between the value of x at which y has value d and the value of x at which the maximum value is attained?

107. Suppose the minimum value of a function of the form $y = a \cos[b(x - c)] + d$, with $a > 0$, occurs at a value of x that is five units from the value of x at which the function has the maximum value. What is the period of the function?

Getting Ready for the Next Section

In Exercises 108 and 109, find the equation of each line in slope–intercept form.

108. Passing through the points $(-3, 2)$ and $(1, 4)$

109. Passing through the point $(2, -3)$ with slope 5

In Exercises 110–112, find the x-intercepts and the vertical asymptotes of the graph of each function.

110. $f(x) = \dfrac{(x + 1)(x - 2)}{(x + 2)(x - 3)}$

111. $f(x) = \dfrac{x^2 + 6x + 8}{x^2 + 1}$

112. $f(x) = \dfrac{1}{x^2 - 5x - 14}$

In Exercises 113–120, let $f(x) = x^3 - 2x$ and $g(x) = 3x^2 + 1$. State whether each function is even, odd, or neither.

113. $f(x)$ **114.** $g(x)$

115. $f(x) + g(x)$ **116.** $f(x)g(x)$

117. $\dfrac{f(x)}{g(x)}$ **118.** $\dfrac{g(x)}{f(x)}$

119. $\dfrac{1}{f(x)}$ **120.** $\dfrac{1}{g(x)}$

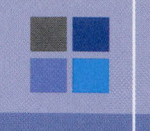

Graphs of the Other Trigonometric Functions

BEFORE STARTING THIS SECTION, REVIEW

1 Slope of a line (Section 1.2, page 23)

2 Vertical asymptotes (Section 2.4, page 189)

3 Transformations of functions (Section 1.5, page 90)

OBJECTIVES

1 Discuss properties of the tangent function.

2 Graph the tangent function.

3 Graph the cosecant, secant, and cotangent functions.

◆ The U.S.S. *Enterprise* and Mach Numbers

Star Trek fans probably know quite a bit about the warp speeds attainable by the Starfleet's U.S.S. *Enterprise*. Near Earth, Mach numbers provide a more relevant measure of speed.

The Mach number (named after the Austrian physicist Ernst Mach) is the ratio of the speed of the aircraft to the speed of sound. For example, an aircraft traveling at Mach 2 is traveling at twice the speed of sound. An airplane flying at less than the speed of sound is traveling at *subsonic* speed; at the speed of sound (Mach 1), the plane is traveling at *transonic* speed; at speeds greater than the speed of sound, it is traveling at *supersonic* speeds; and at speeds greater than 5 times the speed of sound, it is traveling at *hypersonic* speeds. In Example 4, we see how at supersonic and hypersonic speeds Mach numbers are given as values of the cosecant function, and we graph a range of Mach numbers that are attainable by various military aircraft.

1 Discuss properties of the tangent function.

Tangent Function

Recall that if $P(x, y)$ is a point on the terminal side of an angle θ in standard position, then (see Figure 4.62)

$$\tan \theta = \frac{y}{x}$$

Because the point $(0, 0)$ is on the line OP the slope of the line OP is $\dfrac{y - 0}{x - 0} = \dfrac{y}{x} = \tan \theta$.

Since parallel lines have equal slopes (see page 29), we have the following geometric interpretation of the tangent of an angle:

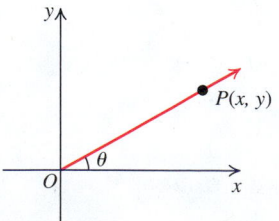

Figure 4.62

TANGENT AS SLOPE

If a nonvertical line makes an angle θ with the positive x-axis, then the slope m of the line is given by $m = \tan \theta$.

EXAMPLE 1   **Equation of a Line**

A line ℓ makes an angle $\theta = 60°$ with positive x-axis and passes through the point $P = (7, 5)$. Find the equation of ℓ in slope–intercept form.

Solution

The slope of the line ℓ is $m = \tan 60° = \sqrt{3}$ (page 360). Then

$$
\begin{aligned}
y - 5 &= \sqrt{3}(x - 7) & &\text{point–slope form: } y - y_1 = m(x - x_1) \\
y - 5 &= \sqrt{3}x - 7\sqrt{3} & &\text{Distributive property} \\
y &= \sqrt{3}x + (5 - 7\sqrt{3}) & &\text{slope–intercept form: } y = mx + b.
\end{aligned}
$$

Practice Problem 1   Repeat Example 1 with $\theta = 30°$ and $P = (2, 3)$.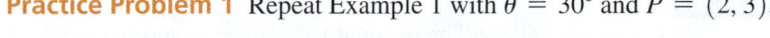

Domain of tan x   From the quotient identity (page 373), we have

$$\tan x = \frac{\sin x}{\cos x}.$$

The function $y = \tan x$ is defined for all real numbers x except where $\cos x = 0$. Since $\cos x = 0$ for $x = \dfrac{\pi}{2} + n\pi$ for any integer n, the domain of $\tan x$ is all real numbers x with $x \neq \dfrac{\pi}{2} + n\pi$. In other words, $\tan x$ is undefined at $x = \pm\dfrac{\pi}{2}, \pm\dfrac{3\pi}{2}, \pm\dfrac{5\pi}{2}, \ldots$.

Range of tan x   We note that the tangent of an angle θ in standard position is the slope of the line containing the corresponding terminal side. Because every real number is the slope of some line through the origin, every number is the tangent of some angle in standard position. This means that the range of the tangent function is $(-\infty, \infty)$.

Zeros of tan x   From $\tan x = \dfrac{\sin x}{\cos x}$, we have $\tan x = 0$ whenever $\sin x = 0$. Since $\sin x = 0$ for $x = n\pi$ for any integer n, the zeros of $y = \tan x$ are $x = n\pi$.

Period of tan x   The tangent function has period π. Figure 4.63 illustrates that $\tan(t + \pi) = \tan t$. Therefore, the values of the tangent function repeat every π units.

Even–Odd Property   The function $y = \tan x$ is an odd function: $\tan(-x) = -\tan x$ (see page 376), and its graph is symmetric about the origin.

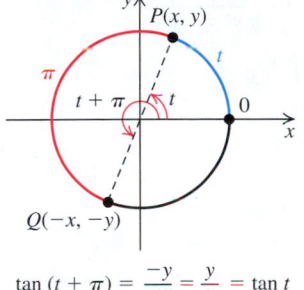

$$\tan(t + \pi) = \frac{-y}{-x} = \frac{y}{x} = \tan t$$

Figure 4.63 $\tan(t + \pi) = \tan t$.

2 Graph the tangent function.

Graph of $y = \tan x$

We first summarize the main facts about $y = \tan x$.

MAIN FACTS ABOUT THE TANGENT FUNCTION

1. *Domain*: All real numbers except odd multiples of $\dfrac{\pi}{2}$.

2. *Range*: $(-\infty, \infty)$

3. *Zeros or x-intercepts*: All integer multiples of π or $x = n\pi$.

4. *Period*: π

5. *Vertical asymptotes*: At $x =$ odd integer multiples of $\dfrac{\pi}{2}$, or $x = (2n + 1)\dfrac{\pi}{2}$.

6. *Symmetry*: Odd function, so the graph is symmetric about the origin.

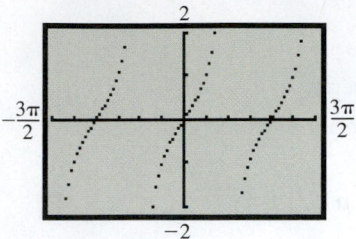
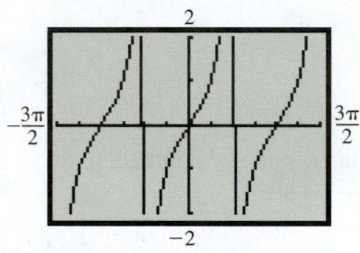
To graph $y = \tan x$, we use the approximate values of $\tan x$ for x in the interval $\left[0, \dfrac{\pi}{2}\right)$. See Table 4.8.

TABLE 4.8

x	0	$\dfrac{\pi}{6}$	$\dfrac{\pi}{4}$	$\dfrac{\pi}{3}$	$\dfrac{7\pi}{18}$	$\dfrac{4\pi}{9}$	$\dfrac{17\pi}{36}$
$y = \tan x$	0	0.6	1	1.7	2.7	5.7	11.4

As x approaches $\dfrac{\pi}{2}$, the values of $\tan x$ continue to increase; in fact, they increase indefinitely. The decimal value of $\dfrac{\pi}{2}$ is approximately 1.5707963. When $x = \dfrac{89\pi}{180} \approx 1.55$, $\tan x > 57$; when $x = 1.57$, $\tan x > 1255$; and when $x = 1.5707$, $\tan x > 10{,}000$. The graph of $y = \tan x$ has a vertical asymptote at $x = \dfrac{\pi}{2}$. See Figure 4.64(a).

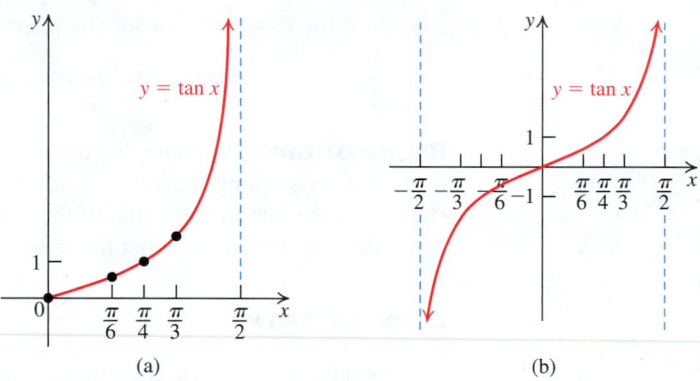

Figure 4.64 Graphing $y = \tan x$.

We can extend the graph of $y = \tan x$ to the interval $\left(-\dfrac{\pi}{2}, \dfrac{\pi}{2}\right)$ by using the fact that the graph is symmetric about the origin. See Figure 4.64(b).

Because the period of the tangent function is π and we have graphed $y = \tan x$ over an interval of length π, we can draw the complete graph of $y = \tan x$ by repeating the graph in Figure 4.64(b) indefinitely to the left and right, over intervals of length π. Figure 4.65 shows three cycles of the graph.

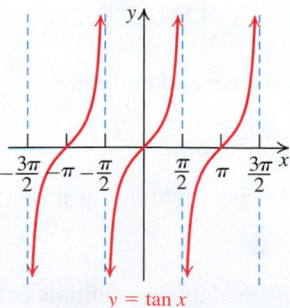

Figure 4.65

Graphing $y = a\tan[b(x - c)]$ The procedure for graphing $y = a\tan[b(x - c)]$ is based on the essential features of the graph of $y = \tan x$.

PROCEDURE
IN ACTION

> **EXAMPLE 2**   Graphing $y = a\tan(bx - k)$

OBJECTIVE

Graph a function of the form $y = a\tan(bx - k)$, *where* $b > 0$, *by finding the period and phase shift.*

Step 1 If necessary, rewrite the equation in the form
$$y = a\tan[b(x - c)], c = \frac{k}{b}.$$
Find the following:
vertical stretch factor $= |a|$

period $= \dfrac{\pi}{b}$

phase shift $= c$

Step 2 Locate two adjacent vertical asymptotes by solving the following equations for x:
$$b(x - c) = -\frac{\pi}{2} \quad \text{and} \quad b(x - c) = \frac{\pi}{2}.$$

Step 3 Divide the interval on the x-axis between the two vertical asymptotes from Step 2 into four equal parts, each of length $\dfrac{1}{4}\left(\dfrac{\pi}{b}\right)$.

Step 4 Evaluate the function at the three x values found in Step 3 that are the division points of the interval.

Step 5 Sketch the vertical asymptotes using the values found in Step 2. Connect the points in Step 4 with a smooth curve in the standard shape of a cycle for the tangent function. Repeat the graph to the left and right over intervals of length $\dfrac{\pi}{b}$.

EXAMPLE

Graph $y = 3\tan\left(2x - \dfrac{\pi}{2}\right)$.

1. $y = 3\tan\left[2\left(x - \dfrac{\pi}{4}\right)\right]$ Rewrite.

$a = 3 \quad b = 2 \quad c = \dfrac{\pi}{4}$

vertical stretch factor $= |a| = |3| = 3$

period $= \dfrac{\pi}{b} = \dfrac{\pi}{2}$ phase shift $= c = \dfrac{\pi}{4}$

2. $2\left(x - \dfrac{\pi}{4}\right) = -\dfrac{\pi}{2}$ $2\left(x - \dfrac{\pi}{4}\right) = \dfrac{\pi}{2}$

$\quad x - \dfrac{\pi}{4} = -\dfrac{\pi}{4}$ $x - \dfrac{\pi}{4} = \dfrac{\pi}{4}$

$\quad x = -\dfrac{\pi}{4} + \dfrac{\pi}{4}$ $x = \dfrac{\pi}{4} + \dfrac{\pi}{4}$

$\quad x = 0$ $x = \dfrac{\pi}{2}$

3. The interval $\left(0, \dfrac{\pi}{2}\right)$ has length $\dfrac{\pi}{2}$, and $\dfrac{1}{4}\left(\dfrac{\pi}{2}\right) = \dfrac{\pi}{8}$.

The division points of the interval $\left(0, \dfrac{\pi}{2}\right)$ are

$0 + \dfrac{\pi}{8} = \dfrac{\pi}{8}, 0 + 2\left(\dfrac{\pi}{8}\right) = \dfrac{\pi}{4},$ and $0 + 3\left(\dfrac{\pi}{8}\right) = \dfrac{3\pi}{8}$.

4.

x	$y = 3\tan\left[2\left(x - \frac{\pi}{4}\right)\right]$
$\dfrac{\pi}{8}$	-3
$\dfrac{\pi}{4}$	0
$\dfrac{3\pi}{8}$	3

5.

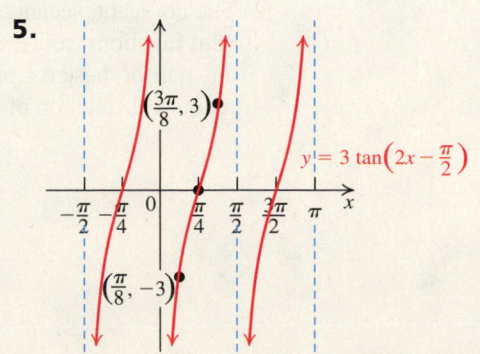

Alternatively, we can use a sequence of transformations on the graph of $y = \tan x$ to obtain the graph of $y = 3 \tan\left(2x - \dfrac{\pi}{2}\right)$. See Figure 4.66.

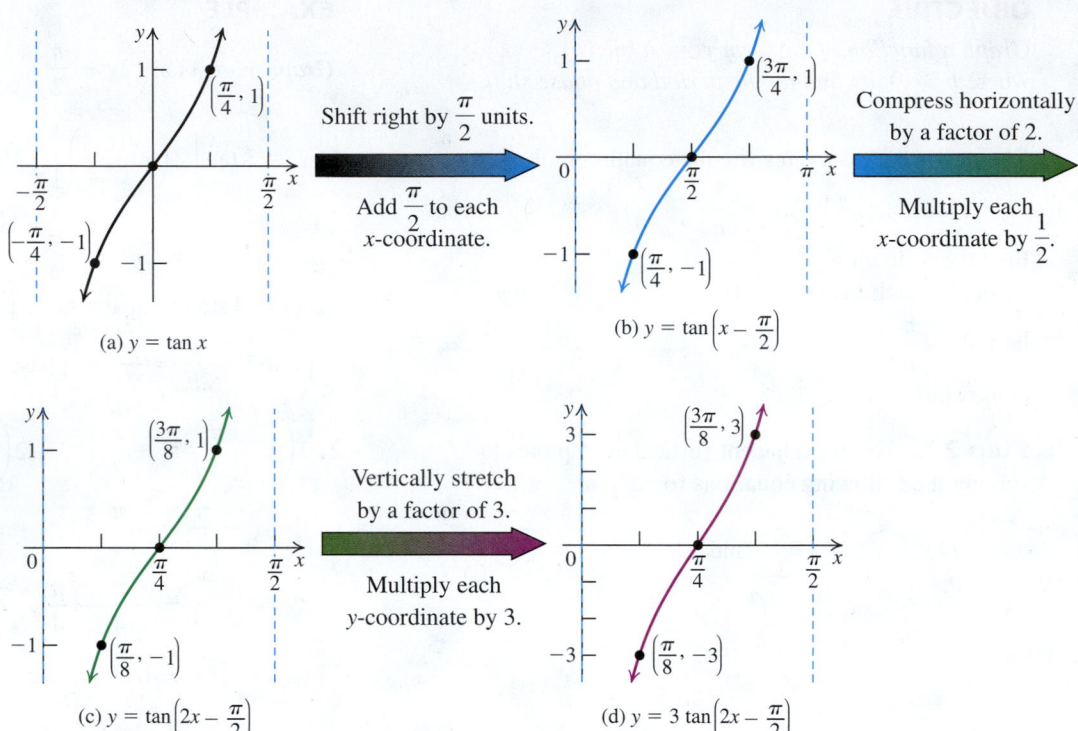

Figure 4.66

Practice Problem 2 Graph $y = -3 \tan\left(x + \dfrac{\pi}{4}\right)$.

The fact that the tangent function is an odd function allows us to graph $y = a \tan[\,b(x - c)\,]$ when $b < 0$. For example, to graph $y = 3 \tan\left[(-2)\left(x - \dfrac{\pi}{2}\right)\right]$, we rewrite the equation as $y = -3 \tan\left[2\left(x - \dfrac{\pi}{2}\right)\right]$. Graph $y = 3 \tan\left[2\left(x - \dfrac{\pi}{2}\right)\right]$ and reflect it about the x-axis.

3 Graph the cosecant, secant, and cotangent functions.

Graphs of the Reciprocal Functions

The cosecant, secant, and cotangent functions are reciprocals of the sine, cosine, and tangent functions, respectively. We first consider some relationships between the properties of any pair of these reciprocal functions. We use these relationships to sketch the graph of the reciprocal function of a trigonometric function $g(x)$.

THE GRAPH OF THE RECIPROCAL OF A TRIGONOMETRIC FUNCTION

Let $f(x)$ be the reciprocal of $g(x)$: $f(x) = \dfrac{1}{g(x)}$, where g is any trigonometric function.

- **Periodicity** If $g(x)$ has period p, then $f(x)$ also has period p.
- **Zeros** If $g(c) = 0$, then $f(c)$ is undefined. So if c is an x-intercept of the graph of g, then the line $x = c$ is a vertical asymptote of the graph of f. Conversely, if $g(d)$ is undefined, then $f(d) = 0$.
- **Even–odd**
  a. If $g(x)$ is odd, then $f(x)$ is odd.
  b. If $g(x)$ is even, then $f(x)$ is even.
- **Special Values**
  a. If $g(x_1) = 1$, then $f(x_1) = 1$. Both graphs pass through the point $(x_1, 1)$.
  b. If $g(x_2) = -1$, then $f(x_2) = -1$. Both graphs pass through the point $(x_2, -1)$.
- **Sign**
  a. If $g(x) > 0$ on an interval (a, b), then $f(x) > 0$ on the interval (a, b). Both graphs are above the x-axis on the interval (a, b).
  b. If $g(x) < 0$ on an interval (c, d), then $f(x) < 0$ on the interval (c, d). Both graphs are below the x-axis on the interval (c, d).
- **Increasing–Decreasing**
  a. If $g(x)$ is increasing on an interval (a, b), then $f(x)$ is decreasing on the interval (a, b).
  b. If $g(x)$ is decreasing on an interval (c, d), then $f(x)$ is increasing on the interval (c, d).
- **Magnitude**
  a. If $|g(x)|$ is small, then $|f(x)|$ is large.
  b. If $|g(x)|$ is large, then $|f(x)|$ is small. If c is in the domain of f and $|g(x)|$ is large as x approaches c, then $f(c) = 0$.

We use the properties of the reciprocal functions to sketch the graphs of the cosecant, secant, and cotangent functions from the graphs of the sine, cosine, and tangent functions. Two or more cycles of these graphs along with their reciprocals are shown in the accompanying box.

MAIN FACTS ABOUT THE COSECANT, SECANT, AND COTANGENT FUNCTIONS

$$y = \csc x = \frac{1}{\sin x} \qquad y = \sec x = \frac{1}{\cos x} \qquad y = \cot x = \frac{1}{\tan x}$$

For any integer n:

	$y = \csc x = \dfrac{1}{\sin x}$	$y = \sec x = \dfrac{1}{\cos x}$	$y = \cot x = \dfrac{1}{\tan x}$
Domain	All real numbers $x \neq n\pi$	All real numbers $x \neq \dfrac{\pi}{2} + n\pi$	All real numbers $x \neq n\pi$
Range	$(-\infty, -1] \cup [1, \infty)$	$(-\infty, -1] \cup [1, \infty)$	$(-\infty, \infty)$
Period	2π	2π	π
x-intercepts	No x-intercepts	No x-intercepts	$x = \dfrac{\pi}{2} + n\pi$
Even–Odd	Odd: $\csc(-x) = -\csc x$	Even: $\sec(-x) = \sec x$	Odd: $\cot(-x) = -\cot x$
Vertical Asymptotes	$x = n\pi$	$x = \dfrac{\pi}{2} + n\pi$	$x = n\pi$
Graph	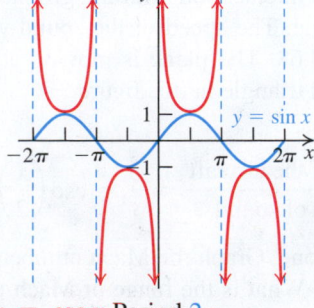	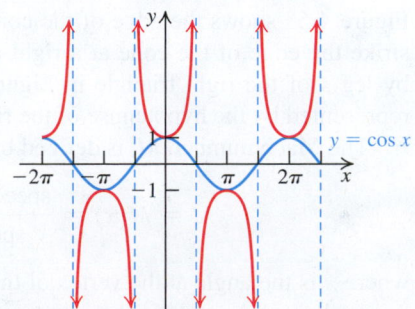	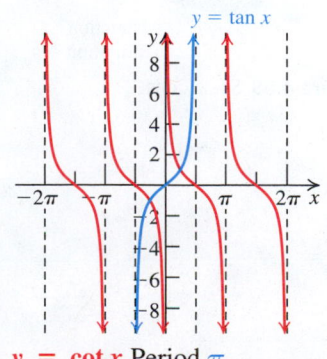

$y = \csc x$ Period 2π $y = \sec x$ Period 2π $y = \cot x$ Period π

We may use a sequence of transformations on the graphs of $y = \csc x$, $y = \sec x$, and $y = \cot x$ to sketch the graphs of $y = a\csc(bx - c) + d$, $y = a\sec(bx - c) + d$, and $y = a\cot(bx - c) + d$, respectively.

EXAMPLE 3 Graphing $y = a\sec(bx - c) + d$

Graph $y = 2\sec(3x - \pi) + 1$

Solution

We obtain the graph of $y = 2\sec(3x - \pi) + 1$ by using a sequence of transformations on the graph of $y = \sec x$. See Figure 4.67.

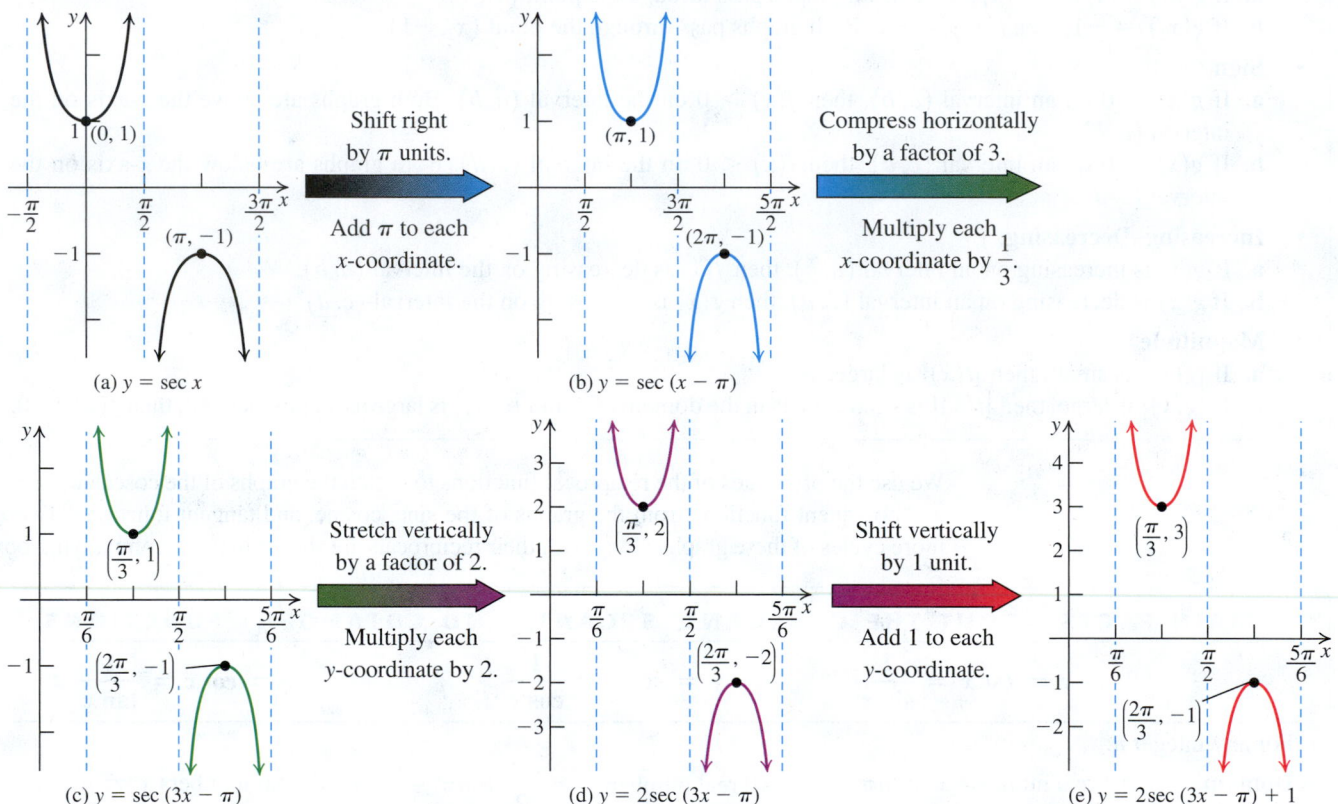

Figure 4.67

Practice Problem 3 Sketch the graph of $y = -3\cot\left(x + \dfrac{\pi}{4}\right)$ over the interval $\left(-\dfrac{3\pi}{4}, \dfrac{5\pi}{4}\right)$.

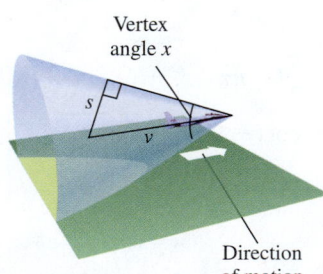

Vertex angle x

Direction of motion

Figure 4.68 Sonic Cone.

◆ **EXAMPLE 4** Graphing a Range of Mach Numbers

When a plane travels at supersonic and hypersonic speeds, small disturbances in the atmosphere are transmitted downstream within a cone. The cone intersects the ground. Figure 4.68 shows the edge of the cone's intersection with the ground. The sound waves strike the edge of the cone at a right angle. The speed of the sound wave is represented by leg s of the right triangle in Figure 4.68. The plane is moving at speed v, which is represented by the hypotenuse of the right triangle in this figure.

The Mach number, M, is defined by

$$M = M(x) = \frac{\text{speed of the aircraft}}{\text{speed of sound}} = \frac{v}{s} = \csc\left(\frac{x}{2}\right),$$

where x is the angle at the vertex of the cone. Graph the Mach number function, $M(x)$, as the angle at the vertex of the cone varies. What is the range of Mach numbers associated with the interval $\left[\dfrac{\pi}{4}, \pi\right)$?

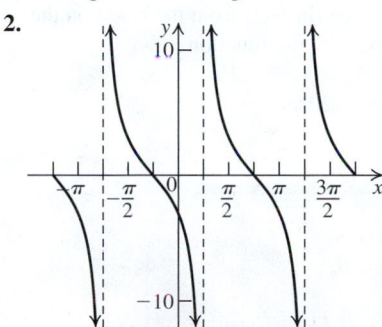

Figure 4.69 Mach numbers.

Solution

Because $\csc \dfrac{x}{2} = \dfrac{1}{\sin \dfrac{x}{2}}$, first graph $y = \sin \dfrac{x}{2}$. For convenience, we have sketched the graph of $y = \sin \dfrac{x}{2}$ over the interval $[0, 2\pi]$ in Figure 4.69. The graph of $y = \csc \dfrac{x}{2}$ is sketched over the interval $(0, 2\pi)$ using the reciprocal relationship between the sine graph and the cosecant graph.

$$\text{For } x = \frac{\pi}{4}, \quad y = \csc \frac{x}{2} = \csc \frac{\pi}{8} \approx 2.6.$$

$$\text{For } x = \pi, \quad y = \csc \frac{x}{2} = \csc \frac{\pi}{2} = 1.$$

The range of Mach numbers associated with the interval $\left[\dfrac{\pi}{4}, \pi\right)$ is $(1, 2.6]$.

Practice Problem 4 In Example 4, what is the range of Mach numbers associated with the interval $\left[\dfrac{\pi}{8}, \dfrac{\pi}{4}\right]$?

Answers to Practice Problems

1. $y = \dfrac{\sqrt{3}}{3}x + 3 - \dfrac{2\sqrt{3}}{3}$

2.

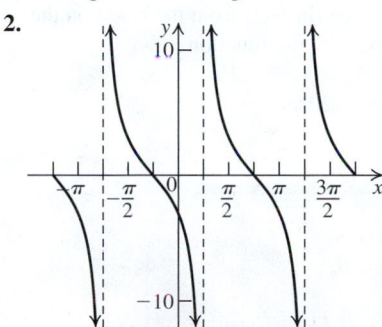

3.

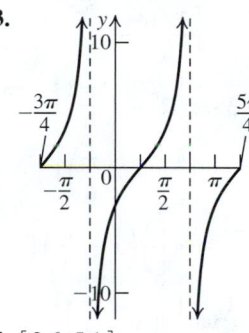

4. $[2.6, 5.1]$

SECTION 4.5 **Exercises**

Concepts and Vocabulary

1. The tangent function has period _____.

2. The value of x with $0 \le x \le \pi$ for which the tangent function is undefined is _____.

3. If $0 \le x \le \pi$ and $\cot x = 0$, then $x =$ _____.

4. The maximum value of $y = \csc x$ for $\pi \le x \le 2\pi$ is _____.

5. **True or False.** The line $y = -x + 5$ makes an angle of $45°$ with the positive x–axis.

6. **True or False.** The graph of $y = 3\sec(2x)$ may be drawn by first sketching the graph of $y = 3\cos(2x)$.

7. **True or False.** The period of $y = \cot(2x)$ is π.

8. **True or False.** Two consecutive asymptotes of $y = 3\cot\left(2x - \dfrac{\pi}{4}\right)$ occur at $x = \dfrac{\pi}{8}$ and $x = \dfrac{5\pi}{8}$.

Building Skills

In Exercises 9–12, find the slope–intercept form of the equation of each line that passes through the point P and makes angle θ with the positive x-axis.

9. $P = (-2, 3), \theta = 45°$

10. $P = (3, -1), \theta = 60°$

11. $P = (-3, -2), \theta = 120°$

12. $P = (2, 5), \theta = 135°$

In Exercises 13–30, graph each function over a one-period interval.

13. $y = \tan\left(x - \dfrac{\pi}{4}\right)$

14. $y = \tan\left(x + \dfrac{\pi}{4}\right)$

15. $y = \cot\left(x + \dfrac{\pi}{4}\right)$

16. $y = \cot\left(x - \dfrac{\pi}{4}\right)$

17. $y = \tan 2x$

18. $y = \tan \dfrac{x}{2}$

19. $y = \cot \dfrac{x}{2}$

20. $y = \cot 2x$

21. $y = -\tan x$

22. $y = -\cot x$

23. $y = 3 \tan x$

24. $y = 3 \cot x$

25. $y = \sec \dfrac{x}{2}$

26. $y = \sec 2x$

27. $y = \csc 3x$

28. $y = \csc \dfrac{x}{3}$

29. $y = \sec(x - \pi)$

30. $y = \csc(x - \pi)$

In Exercises 31–48, graph each function over a two-period interval.

31. $y = \tan\left[2\left(x + \dfrac{\pi}{2}\right)\right]$

32. $y = \tan\left[2\left(x - \dfrac{\pi}{2}\right)\right]$

33. $y = \cot\left[2\left(x - \dfrac{\pi}{2}\right)\right]$

34. $y = \cot\left[2\left(x + \dfrac{\pi}{2}\right)\right]$

35. $y = \tan\left[\dfrac{1}{2}(x + 2\pi)\right]$

36. $y = \tan\left[\dfrac{1}{2}(x - 2\pi)\right]$

37. $y = \cot\left[\dfrac{1}{2}(x - 2\pi)\right]$

38. $y = \cot\left[\dfrac{1}{2}(x + 2\pi)\right]$

39. $y = \sec\left[4\left(x - \dfrac{\pi}{4}\right)\right]$

40. $y = \sec\left[\dfrac{1}{2}\left(x - \dfrac{\pi}{2}\right)\right]$

41. $y = 3 \csc\left(x + \dfrac{\pi}{2}\right)$

42. $y = 3 \sec\left[2\left(x - \dfrac{\pi}{6}\right)\right]$

43. $y = \tan\left[\dfrac{2}{3}\left(x - \dfrac{\pi}{2}\right)\right]$

44. $y = 2 \cot\left[2\left(x - \dfrac{\pi}{6}\right)\right]$

45. $y = -5 \tan\left[2\left(x + \dfrac{\pi}{3}\right)\right]$

46. $y = -3 \cot\left[\dfrac{1}{2}\left(x - \dfrac{\pi}{3}\right)\right]$

47. $y = \dfrac{1}{3} \cot\left[2(x - \pi)\right]$

48. $y = \dfrac{1}{2} \tan\left[4\left(x - \dfrac{\pi}{6}\right)\right]$

In Exercises 49–52, write an equation of the form
$y = a \tan(bx - c) + d$ **for each graph.**

49.

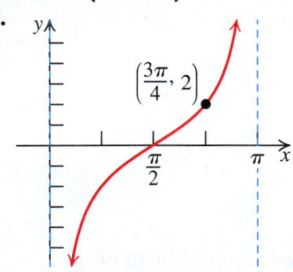

$\left(\dfrac{3\pi}{4}, 2\right)$

50.

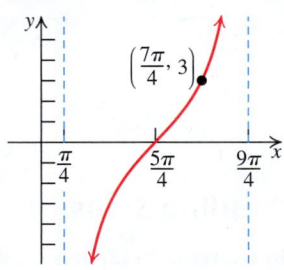

$\left(\dfrac{7\pi}{4}, 3\right)$

51.

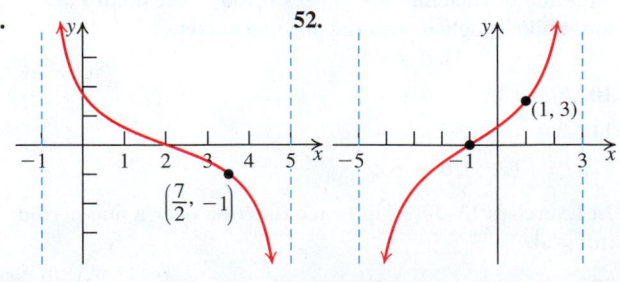

$\left(\dfrac{7}{2}, -1\right)$

52.

$(1, 3)$

In Exercises 53–56, write an equation of the form
$y = a \csc(bx - c) + d$ **for each graph.**

53.

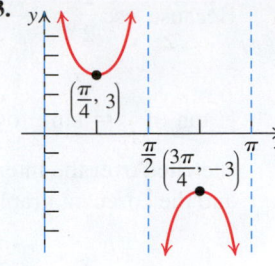

$\left(\dfrac{\pi}{4}, 3\right)$ $\left(\dfrac{3\pi}{4}, -3\right)$

54.

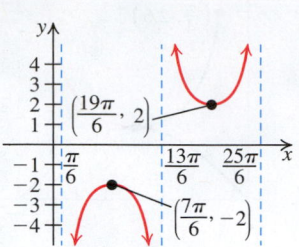

$\left(\dfrac{19\pi}{6}, 2\right)$ $\left(\dfrac{7\pi}{6}, -2\right)$

55.

$(0, 2)$ $(2, -2)$

56.

$\left(\dfrac{5}{4}, 3\right)$ $\left(\dfrac{11}{4}, -1\right)$

Applying the Concepts

57. Prison searchlight. A dual-beam rotating light on a movie set is positioned as a spotlight shining on a prison wall. The light is 20 feet from the wall and rotates clockwise. The light shines on point P on the wall when first turned on ($t = 0$). After t seconds, the distance (in feet) from the beam on the wall to the point P is given by the function

$$d(t) = 20 \tan \dfrac{\pi t}{5}.$$

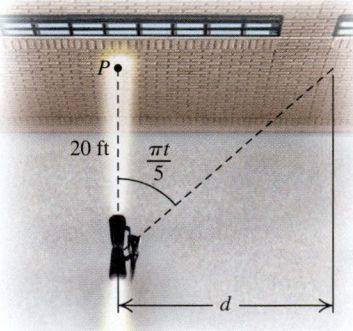

P

20 ft $\dfrac{\pi t}{5}$

d

When the light beam is to the right of P, the value of d is positive, and when the beam is to the left of P, the value of d is negative.

a. Graph d over the interval $0 \leq t \leq 5$.

b. Since $d(t)$ is undefined for $t = 2.5$, where is the rotating light pointing when $t = 2.5$?

58. Prison searchlight. In Exercise 57, find the value for b assuming that the light beam is to sweep the entire wall in ten seconds and $d(t) = 20 \tan bt$.

Circumscribed polygon. A regular **circumscribed *n*-gon** about a circle of radius *r* has *n* vertices and its sides are tangent to *n* equally spaced points on the circumference of the circle. The adjoining figure is a circumscribed hexagon ($n = 6$) about a circle of radius *r*. Use this figure for Exercises 59–62.

59. a. In the figure show that $B = \left(r, r \tan\frac{\pi}{6} \right)$, and conclude

that the length of each side of the hexagon is $2r \tan\frac{\pi}{6}$.

b. Use part **(a)** to find the perimeter of the hexagon.

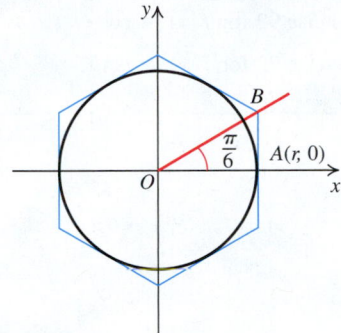

60. a. Use the figure to show that the area of the triangle *AOB* is
$\frac{1}{2}r^2 \tan\frac{\pi}{6}$.

b. Use part **(a)** to find the area of the circumscribed hexagon.

61. a. Show that the perimeter *P* of a regular circumscribed *n*-gon about a circle of radius *r* is:

$$P = 2nr \tan\frac{\pi}{n}.$$

b. What is the circumference of a circle of radius $r = 10$?

c. Calculate *P* in part **(a)** with $r = 10$ and $n = 4, 10, 50,$ and 100. Round your answer to four decimal places. What do you observe?

62. a. Show that the area *A* of a regular *n*-gon circumscribed about a circle of radius *r* is:

$$A = nr^2 \tan\frac{\pi}{n}.$$

b. What is the area of a disc of radius $r = 10$?

c. Calculate *A* by using the formula in part **(a)** with $r = 10$ and $n = 4, 10, 50,$ and 100. Round your answer to four decimal places. What do you observe?

Modeling with Tangent Function. In Exercises 63 and 64, model the given data by using the function $y = a \tan[b(x - c)]$. Use your equation to find *y* when $x = -1.5$ and $x = 1.5$. Round your answers to two decimal places.

63.

x	y
−5	undef. ($-\infty$)
−3	−7
−1	0
1	7
3	undef. ($+\infty$)

64.

x	y
−3	undef. ($+\infty$)
−0.5	6
2	0
4.5	−6
7	undef. ($-\infty$)

Beyond the Basics

65. Show that if *f* is a periodic function with period *p*, then the reciprocal function $\frac{1}{f}$ is also a periodic function with period *p*.

66. Show that if *f* is an odd function, then the reciprocal function $\frac{1}{f}$ is also an odd function.

67. True or False. If *f* is an odd function and *g* is an even function, then

a. Both $\frac{f}{g}$ and $\frac{g}{f}$ are odd functions.

b. Both $f \circ g$ and $g \circ f$ are even functions.

68. True or False. If both *f* and *g* are periodic functions with period *p*, then $\frac{f}{g}$ has period *p*.

In Exercises 69–78, graph each function over a two-period interval.

69. $y = 3 \tan(-2x)$

70. $y = 2 \cot(-3x)$

71. $y = -\frac{1}{2} \csc\left(-\frac{x}{5}\right)$

72. $y = -2 \sec\left(-\frac{x}{2}\right)$

73. $y = -\tan\left(\frac{\pi}{2} - x\right)$

74. $y = \cot(\pi - x)$

75. $y = 2 \sec(\pi - 2x)$

76. $y = -\sec(\pi - 4x)$

77. $y = 4 \cot(\pi - 4x) + 3$

78. $y = 2 \tan\left(\frac{\pi}{2} - 2x\right) + 1$

Critical Thinking / Discussion / Writing

79. For what number *k*, with $-2\pi < k < 0$, is $x = k$ a vertical asymptote for the graph of $y = \cot\left[\frac{1}{2}\left(x - \frac{\pi}{4}\right)\right]$?

80. For what number *b* does $y = \sec bx$ have period $\frac{\pi}{3}$?

81. Write an equation in the form $y = a \csc[b(x - c)]$ that has the same graph as $y = -2 \sec x$.

82. How can you obtain the graph of $y = \cot x$ by using appropriate transformations on the graph of $y = \tan x$?

83. For what number *k*, with $-2\pi < k < 0$, is $x = k$ a vertical asymptote for the graph of $y = 5 \cot\left(\frac{1}{2}x - \frac{\pi}{8}\right)$?

84. For what number *b* do the functions $y = \sec bx$ and $y = \cot 3x$ have the same period?

Getting Ready for the Next Section

85. Show that $f(x) = 3x + 7$ is a one-to-one function. Find its inverse.

86. Show that $f(x) = -2x + 5$ is a one-to-one function. Find its inverse.

87. **True or False.** Every increasing function on an interval (a, b) is a one-to-one function.

88. **True or False.** Every decreasing function on an interval (a, b) is a one-to-one function.

89. **True or False.** If f is a one-to-one function, then it has an inverse function.

90. **True or False.** For a one-to-one function, the graph of f^{-1} is the reflection of the graph of f about the line $y = x$.

91. Show that $f(x) = x^2$ is not one-to-one on its domain $(-\infty, \infty)$. Write a subinterval of the domain of f on which f is one-to-one.

92. Is $f(x) = \sin x$ a one-to-one function on the interval $[-\pi, \pi]$? If your answer is no, find a subinterval on which f is one-to-one.

93. Repeat Exercise 92 for $f(x) = \cos x$.

94. Repeat Exercise 92 for $f(x) = \tan x$.

Inverse Trigonometric Functions

BEFORE STARTING THIS SECTION, REVIEW

1 Inverse functions (Section 1.7, page 118)

2 Composition of functions (Section 1.6, page 102)

3 Exact values of the trigonometric functions (Section 4.2, page 360)

OBJECTIVES

1 Graph the inverse sine function.

2 Graph the inverse cosine function.

3 Graph the inverse tangent function.

4 Evaluate inverse trigonometric functions.

5 Find exact values of composite functions involving the inverse trigonometric functions.

◆ Retail Theft

In 2005, security cameras at Filene's Basement store in Boston filmed a theft coordinated by a thief and an accomplice. The accomplice distracted the salesperson so the thief could steal a $16,000 necklace. Retail theft is a major concern for all retail outlets, from Tiffany & Co. to Walmart.

The National Retail Federation, the industry's largest trade group, and the Retail Industry Leaders Association have instituted password-protected national crime databases online. These databases allow retailers to share information about thefts and determine whether they have been a target of individual shoplifters who steal for themselves or a target of organized crime. In addition to participating in the shared databases, many large retailers have their own organized anticrime squads. One estimate puts loss to organized theft at over $30 billion annually. In Example 10, we see how the methods described in this section can be used to prevent loss from theft.

1 Graph the inverse sine function.

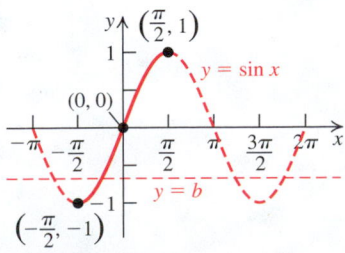

Figure 4.70 $y = \sin x$, $-\dfrac{\pi}{2} \le x \le \dfrac{\pi}{2}$.

The Inverse Sine Function

Recall that a function f has an inverse function if no horizontal line intersects the graph of f in more than one point. Because every horizontal line $y = b$, where $-1 \le b \le 1$, intersects the graph of $y = \sin x$ at more than one point, the sine function fails the horizontal line test, so it is not one-to-one and consequently has no inverse.

The solid portion of the sine graph shown in Figure 4.70 is the graph of $y = \sin x$ for $-\dfrac{\pi}{2} \le x \le \dfrac{\pi}{2}$. If we restrict the domain of $y = \sin x$ to the interval $\left[-\dfrac{\pi}{2}, \dfrac{\pi}{2} \right]$, the resulting function

$$y = \sin x, \quad -\frac{\pi}{2} \le x \le \frac{\pi}{2}$$

is one-to-one (it passes the horizontal line test); so it has an inverse. Notice that the restricted function takes on all values in the range of $y = \sin x$, which is $[-1, 1]$, and that each of these y-values corresponds to exactly one x-value in the restricted domain $\left[-\dfrac{\pi}{2}, \dfrac{\pi}{2}\right]$.

The inverse function for $y = \sin x$, $-\dfrac{\pi}{2} \le x \le \dfrac{\pi}{2}$, is called the **inverse sine**, or **arcsine**, function and is denoted by $\sin^{-1} x$ or by $\arcsin x$. The domain of $y = \sin^{-1} x$ is $[-1, 1]$, and its range is $\left[-\dfrac{\pi}{2}, \dfrac{\pi}{2}\right]$.

Inverse Sine Function

$y = \sin^{-1} x$ means $\sin y = x$, where $-1 \le x \le 1$ and $-\dfrac{\pi}{2} \le y \le \dfrac{\pi}{2}$.

Read $\sin^{-1} x$ as "inverse sine at x."

The equation $\sin \dfrac{\pi}{6} = \dfrac{1}{2}$ can be written as $\sin^{-1} \dfrac{1}{2} = \dfrac{\pi}{6}$ because $-1 \le \frac{1}{2} \le 1$.

Note that $\sin^{-1} \dfrac{1}{2} \ne \dfrac{5\pi}{6}$, even though $\sin \dfrac{5\pi}{6} = \dfrac{1}{2}$, because $\dfrac{5\pi}{6}$ is not in the range:

$\left[-\dfrac{\pi}{2}, \dfrac{\pi}{2}\right]$, of $y = \sin^{-1} x$.

The exact values of $\sin^{-1} x$ are obtained by reading $y = \sin^{-1} x$ as "y is the number in the interval $\left[-\dfrac{\pi}{2}, \dfrac{\pi}{2}\right]$ whose sine is x." See Table 4.9.

TABLE 4.9 Exact values of $\sin y$, $-\dfrac{\pi}{2} \le y \le \dfrac{\pi}{2}$

y	$-\dfrac{\pi}{2}$	$-\dfrac{\pi}{3}$	$-\dfrac{\pi}{4}$	$-\dfrac{\pi}{6}$	0	$\dfrac{\pi}{6}$	$\dfrac{\pi}{4}$	$\dfrac{\pi}{3}$	$\dfrac{\pi}{2}$
$\sin y = x$	-1	$-\dfrac{\sqrt{3}}{2}$	$-\dfrac{\sqrt{2}}{2}$	$-\dfrac{1}{2}$	0	$\dfrac{1}{2}$	$\dfrac{\sqrt{2}}{2}$	$\dfrac{\sqrt{3}}{2}$	1

We can graph $y = \sin^{-1} x$ by plotting points using Table 4.9 or by reflecting the graph of $y = \sin x$, for $-\dfrac{\pi}{2} \le x \le \dfrac{\pi}{2}$, about the line $y = x$. See Figure 4.71.

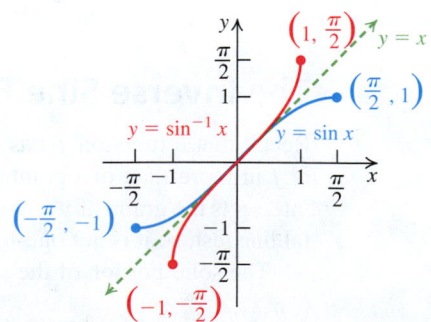

Figure 4.71 Graph of $y = \sin^{-1} x$.

PROCEDURE
IN ACTION

EXAMPLE 1 Finding the Exact Value of $\sin^{-1}x$

OBJECTIVE

Find the exact value of $y = \sin^{-1}x$.

Step 1 Rewrite $y = \sin^{-1}x$ (or $y = \arcsin x$) as $\sin y = x$.

Step 2 From memory or Table 4.9, find the exact value of y in the interval $\left[-\dfrac{\pi}{2}, \dfrac{\pi}{2}\right]$ with $\sin y = x$.

EXAMPLE

Find the exact value of y.

a. $y = \arcsin \dfrac{\sqrt{3}}{2}$ **b.** $y = \sin^{-1}\left(-\dfrac{1}{2}\right)$ **c.** $y = \sin^{-1}(-3)$

1. $\sin y = \dfrac{\sqrt{3}}{2}$ $\sin y = -\dfrac{1}{2}$ $\sin y = -3$

2. $\sin \dfrac{\pi}{3} = \dfrac{\sqrt{3}}{2}$, $\sin\left(-\dfrac{\pi}{6}\right) = -\dfrac{1}{2}$, -3 is not in the domain $[-1, 1]$

so $y = \dfrac{\pi}{3}$ so $y = -\dfrac{\pi}{6}$ of $\sin^{-1}x$, so $\sin^{-1}(-3)$ is undefined.

Practice Problem 1 Find the exact values of y.

a. $y = \sin^{-1}\left(-\dfrac{\sqrt{3}}{2}\right)$ **b.** $y = \sin^{-1}(-1)$ **c.** $y = \sin^{-1}(2)$

2 Graph the inverse cosine function.

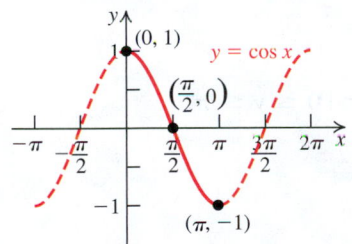

Figure 4.72 $y = \cos x, 0 \le x \le \pi$.

The Inverse Cosine Function

When we restrict the domain of $y = \cos x$ to the interval $[0, \pi]$, the resulting function, $y = \cos x$ (with $0 \le x \le \pi$), is one-to-one. No horizontal line intersects the graph of $y = \cos x$, with $0 \le x \le \pi$, in more than one point. See Figure 4.72. Consequently, the restricted cosine function has an inverse function.

The inverse function for $y = \cos x$, $0 \le x \le \pi$, is called the **inverse cosine**, or **arccosine**, function and is denoted by $\cos^{-1}x$, or by **arccos** x. The domain of $\cos^{-1}x$ is $[-1, 1]$, and its range is $[0, \pi]$.

Inverse Cosine Function

$y = \cos^{-1}x$ means $\cos y = x$ where $-1 \le x \le 1$ and $0 \le y \le \pi$. Read $\cos^{-1}x$ as "inverse cosine at x."

The equation $\cos \dfrac{\pi}{3} = \dfrac{1}{2}$ may be written as $\cos^{-1} \dfrac{1}{2} = \dfrac{\pi}{3}$ because $0 \le \dfrac{\pi}{3} \le \pi$. However, $\cos^{-1} \dfrac{1}{2} \ne -\dfrac{\pi}{3}$, even though $\cos\left(-\dfrac{\pi}{3}\right) = \dfrac{1}{2}$, because $-\dfrac{\pi}{3}$ is not in the range $[0, \pi]$ of $\cos^{-1}x$.

The exact values of $\cos^{-1}x$ are obtained by reading $y = \cos^{-1}x$ as "y is the number in the interval $[0, \pi]$ whose cosine is x." See Table 4.10.

TABLE 4.10 Exact values of $\cos y, 0 \le x \le \pi$

y	0	$\dfrac{\pi}{6}$	$\dfrac{\pi}{4}$	$\dfrac{\pi}{3}$	$\dfrac{\pi}{2}$	$\dfrac{2\pi}{3}$	$\dfrac{3\pi}{4}$	$\dfrac{5\pi}{6}$	π
$\cos y = x$	1	$\dfrac{\sqrt{3}}{2}$	$\dfrac{\sqrt{2}}{2}$	$\dfrac{1}{2}$	0	$-\dfrac{1}{2}$	$-\dfrac{\sqrt{2}}{2}$	$-\dfrac{\sqrt{3}}{2}$	-1

Reflecting the graph of $y = \cos x$, for $0 \le x \le \pi$, about the line $y = x$ produces the graph of $y = \cos^{-1} x$, shown in Figure 4.73.

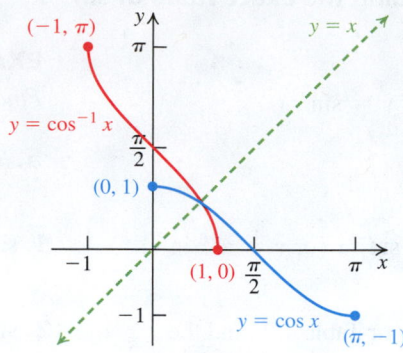

Figure 4.73 Graph of $y = \cos^{-1} x$.

EXAMPLE 2 Finding the Exact Value for $y = \cos^{-1} x$

Find the exact value of y.

a. $y = \cos^{-1} \dfrac{\sqrt{2}}{2}$ **b.** $y = \arccos\left(-\dfrac{1}{2}\right)$

Solution

a. The equation $y = \cos^{-1} \dfrac{\sqrt{2}}{2}$ means $\cos y = \dfrac{\sqrt{2}}{2}$ and $0 \le y \le \pi$.

Since $\cos \dfrac{\pi}{4} = \dfrac{\sqrt{2}}{2}$ and $0 \le \dfrac{\pi}{4} \le \pi$, we have $y = \dfrac{\pi}{4}$.

b. The equation $y = \arccos\left(-\dfrac{1}{2}\right)$ means $\cos y = -\dfrac{1}{2}$ and $0 \le y \le \pi$.

Since $\cos \dfrac{2\pi}{3} = -\dfrac{1}{2}$ and $0 \le \dfrac{2\pi}{3} \le \pi$, we have $y = \dfrac{2\pi}{3}$.

Practice Problem 2 Find the exact value of y.

a. $y = \cos^{-1}\left(-\dfrac{\sqrt{2}}{2}\right)$ **b.** $y = \cos^{-1} \dfrac{1}{2}$ **c.** $\arccos(\pi)$

---

3 Graph the inverse tangent function.

The Inverse Tangent Function

The *inverse tangent function* results from restricting the domain of the tangent function to the interval $\left(-\dfrac{\pi}{2}, \dfrac{\pi}{2}\right)$ to obtain a one-to-one function. The inverse of this restricted tangent function is the **inverse tangent**, or **arctangent**, function and is denoted by **tan⁻¹ x**, or **arctan x**. The domain of $\tan^{-1} x$ is $(-\infty, \infty)$ and its range is $\left(-\dfrac{\pi}{2}, \dfrac{\pi}{2}\right)$.

Inverse Tangent Function

The equation $y = \tan^{-1} x$ means $\tan y = x$, where $-\infty < x < \infty$ and $-\dfrac{\pi}{2} < y < \dfrac{\pi}{2}$.

Read $\tan^{-1} x$ as "the inverse tangent at x."

The exact values of $\tan^{-1} x$ can be obtained by reading $y = \tan^{-1} x$ as "y is the number in the interval $\left(-\dfrac{\pi}{2}, \dfrac{\pi}{2}\right)$ whose tangent is x." See Table 4.11.

TABLE 4.11 Exact values of $\tan y$, $-\dfrac{\pi}{2} < y < \dfrac{\pi}{2}$

y	$-\dfrac{\pi}{3}$	$-\dfrac{\pi}{4}$	$-\dfrac{\pi}{6}$	0	$\dfrac{\pi}{6}$	$\dfrac{\pi}{4}$	$\dfrac{\pi}{3}$
$\tan y = x$	$-\sqrt{3}$	-1	$-\dfrac{\sqrt{3}}{3}$	0	$\dfrac{\sqrt{3}}{3}$	1	$\sqrt{3}$

The graph of $y = \tan^{-1} x$ is obtained by reflecting the graph of $y = \tan x$, with $-\dfrac{\pi}{2} < x < \dfrac{\pi}{2}$, about the line $y = x$. Figure 4.74(a) shows the graph of the restricted tangent function. Figure 4.74(b) shows the graph of $y = \tan^{-1} x$.

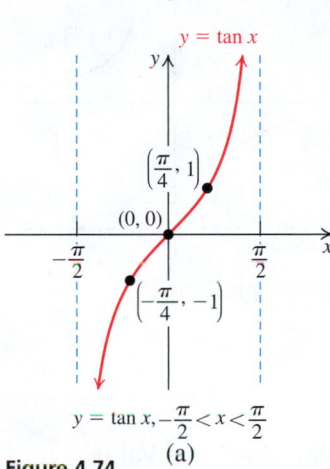

$y = \tan x, -\dfrac{\pi}{2} < x < \dfrac{\pi}{2}$

(a)

Graph of $y = \tan^{-1} x$

(b)

Figure 4.74

EXAMPLE 3 Finding the Exact Value for $y = \tan^{-1} x$

Find the exact value of y.

a. $y = \tan^{-1} 0$ **b.** $y = \arctan(-\sqrt{3})$

Solution

a. Since $\tan 0 = 0$ and $-\dfrac{\pi}{2} < 0 < \dfrac{\pi}{2}$, we have $y = 0$.

b. Since $\tan\left(-\dfrac{\pi}{3}\right) = -\sqrt{3}$ and $-\dfrac{\pi}{2} < -\dfrac{\pi}{3} < \dfrac{\pi}{2}$, we have $y = -\dfrac{\pi}{3}$.

Practice Problem 3 Find the exact value of $y = \tan^{-1} \dfrac{\sqrt{3}}{3}$.

Other Inverse Trigonometric Functions

Sometimes the ranges of the *inverse secant* and *inverse cosecant* functions given in other texts differ from those used in this text. Always check the definitions of the domains of these two functions when they are used outside this course.

Inverse Cotangent, Cosecant, Secant Functions

Inverse cotangent	$y = \cot^{-1} x$ means $\cot y = x$, where $-\infty < x < \infty$ and $0 < y < \pi$.
Inverse cosecant	$y = \csc^{-1} x$ means $\csc y = x$, where $\|x\| \geq 1$ and $-\dfrac{\pi}{2} \leq y \leq \dfrac{\pi}{2}, y \neq 0$.
Inverse secant	$y = \sec^{-1} x$ means $\sec y = x$, where $\|x\| \geq 1$ and $0 \leq y \leq \pi, y \neq \dfrac{\pi}{2}$.

The graphs of $\cot^{-1} x$, $\csc^{-1} x$, and $\sec^{-1} x$ are shown in Figure 4.75.

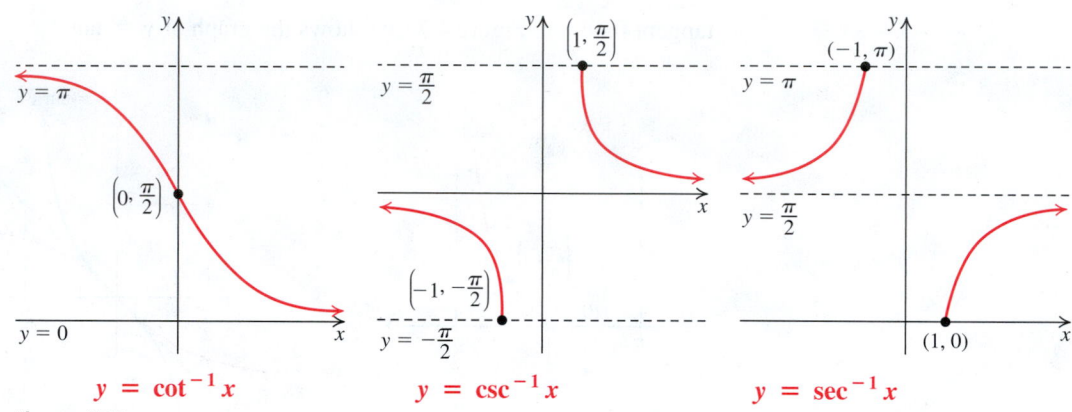

$$y = \cot^{-1} x \qquad y = \csc^{-1} x \qquad y = \sec^{-1} x$$

Figure 4.75

EXAMPLE 4   Finding the Exact Value

Find the exact value of

a. $y = \csc^{-1} 2.$ **b.** $y = \sec^{-1} \dfrac{1}{2}$

Solution

a. We have $\sin \dfrac{\pi}{6} = \dfrac{1}{2}$, so $\csc \dfrac{\pi}{6} = \dfrac{1}{\sin \dfrac{\pi}{6}} = \dfrac{1}{\dfrac{1}{2}} = 2.$

Since $\csc \dfrac{\pi}{6} = 2$ and $-\dfrac{\pi}{2} \leq \dfrac{\pi}{6} \leq \dfrac{\pi}{2}$, we have $y = \csc^{-1} 2 = \dfrac{\pi}{6}.$

b. Since $\dfrac{1}{2}$ is not in the domain $(-\infty, -1] \cup (1, \infty]$ of $\sec^{-1} x$, $\sec^{-1} \dfrac{1}{2}$ is undefined.

Practice Problem 4 Find the exact value of $y = \sec^{-1} 2$.

Inverse Trigonometric Functions			
Inverse Function	**Equivalent to**	**Domain**	**Range**
$y = \sin^{-1} x$	$\sin y = x$	$[-1, 1]$	$\left[-\dfrac{\pi}{2}, \dfrac{\pi}{2}\right]$
$y = \cos^{-1} x$	$\cos y = x$	$[-1, 1]$	$[0, \pi]$
$y = \tan^{-1} x$	$\tan y = x$	$(-\infty, \infty)$	$\left(-\dfrac{\pi}{2}, \dfrac{\pi}{2}\right)$
$y = \cot^{-1} x$	$\cot y = x$	$(-\infty, \infty)$	$(0, \pi)$
$y = \csc^{-1} x$	$\csc y = x$	$(-\infty, -1] \cup [1, \infty)$	$\left[-\dfrac{\pi}{2}, 0\right) \cup \left(0, \dfrac{\pi}{2}\right]$
$y = \sec^{-1} x$	$\sec y = x$	$(-\infty, -1] \cup [1, \infty)$	$\left[0, \dfrac{\pi}{2}\right) \cup \left(\dfrac{\pi}{2}, \pi\right]$

EXAMPLE 5 **Finding the Inverse of a Trigonometric Function**

Use the four-step procedure (page 122) to find the inverse of

$$f(x) = 2\sin(x+1) - 3 \qquad -1-\frac{\pi}{2} \le x \le -1+\frac{\pi}{2}$$

Solution

The function f with domain $\left[-1-\dfrac{\pi}{2}, -1+\dfrac{\pi}{2}\right]$ is one to one, so f has an inverse.

Step 1 $y = 2\sin(x+1) - 3,$

$-1-\dfrac{\pi}{2} \le x \le -1+\dfrac{\pi}{2}$ Replace $f(x)$ with y.

Step 2 $x = 2\sin(y+1) - 3,$

$-1-\dfrac{\pi}{2} \le y \le -1+\dfrac{\pi}{2}$ Interchange x and y.

> **RECALL**
>
> If f is a one-to-one function, then f^{-1} exists and $f^{-1}(f(x)) = x$ for every x in the domain of f. See Section 1.7.

Step 3 Solve $x = 2\sin(y+1) - 3$ for y:

$x + 3 = 2\sin(y+1)$ Add 3 to both sides.

$\dfrac{x+3}{2} = \sin(y+1)$ Divide both sides by 2.

$\sin^{-1}\left(\dfrac{x+3}{2}\right) = y+1$ Take $\sin^{-1}$ of both sides and note   so $\sin^{-1}(\sin(y+1)) = y+1.$

$\sin^{-1}\left(\dfrac{x+3}{2}\right) - 1 = y$ Isolate y.

Step 4 $f^{-1}(x) = \sin^{-1}\left(\dfrac{x+3}{2}\right) - 1$ Replace y with $f^{-1}(x)$.

Practice Problem 5 Repeat Example 5 for $f(x) = 3\cos(x-1) + 2$, $1 \le x \le 1 + \pi$.

4 Evaluate inverse trigonometric functions.

Evaluating Inverse Trigonometric Functions

In Section 4.2, we defined the six trigonometric functions of *real numbers*, and in this section, we have defined the corresponding six inverse trigonometric functions of real numbers. For example,

$$\text{if} \quad \sin \frac{\pi}{4} = \frac{\sqrt{2}}{2}, \quad \text{then} \quad \sin^{-1} \frac{\sqrt{2}}{2} = \frac{\pi}{4}.$$

However, because we also defined the trigonometric functions of *angles* in Section 4.3, it is meaningful when working with angles in degree measure to write a statement such as

$$\sin^{-1} \frac{\sqrt{2}}{2} = 45°.$$

We state some useful identities involving the values of inverse trigonometric functions of $\frac{1}{x}$ and $-x$ in terms of the values of inverse trigonometric functions of x.

INVERSE TRIGONOMETRIC IDENTITIES	
For $\dfrac{1}{x}$	**For $-x$**
$\sin^{-1}\left(\dfrac{1}{x}\right) = \csc^{-1} x, \; \lvert x \rvert \geq 1$	$\sin^{-1}(-x) = -\sin^{-1} x, \; \lvert x \rvert \leq 1$
$\csc^{-1}\left(\dfrac{1}{x}\right) = \sin^{-1} x, \; \lvert x \rvert \leq 1$	$\csc^{-1}(-x) = -\csc^{-1} x, \; \lvert x \rvert \geq 1$
$\cos^{-1}\left(\dfrac{1}{x}\right) = \sec^{-1} x, \; \lvert x \rvert \geq 1$	$\cos^{-1}(-x) = \pi - \cos^{-1} x, \; \lvert x \rvert \leq 1$
$\sec^{-1}\left(\dfrac{1}{x}\right) = \cos^{-1} x, \; \lvert x \rvert \leq 1$	$\sec^{-1}(-x) = \pi - \sec^{-1} x, \; \lvert x \rvert \geq 1$
$\tan^{-1}\left(\dfrac{1}{x}\right) = \begin{cases} \cot^{-1} x, & x > 0 \\ -\pi + \cot^{-1} x, & x < 0 \end{cases}$	$\tan^{-1}(-x) = -\tan^{-1} x, \; -\infty < x < \infty$
$\cot^{-1}\left(\dfrac{1}{x}\right) = \begin{cases} \tan^{-1} x, & x > 0 \\ \pi + \tan^{-1} x, & x < 0 \end{cases}$	$\cot^{-1}(-x) = \pi - \cot^{-1} x, \; -\infty < x < \infty$

Note that the equation $\tan^{-1}\left(\dfrac{1}{x}\right) = \cot^{-1} x$ is true for $x > 0$, since the values of both sides are in the interval $\left(0, \dfrac{\pi}{2}\right)$. But when $x < 0$, $\tan^{-1}\left(\dfrac{1}{x}\right)$ is negative and lies in the interval $\left(-\dfrac{\pi}{2}, 0\right)$ and $\cot^{-1} x$ is positive and lies in the interval $\left(\dfrac{\pi}{2}, \pi\right)$. So for $x < 0$, the correct equation is $\tan^{-1}\left(\dfrac{1}{x}\right) = -\pi + \cot^{-1} x$. In other words, for $x < 0$, we have

$$\cot^{-1} x = \pi + \tan^{-1}\left(\frac{1}{x}\right).$$

EXAMPLE 6   **Verifying an Inverse Trigonometric Identity**

a. Show that $\cos^{-1}(-x) = \pi - \cos^{-1} x$ for $|x| \leq 1$.

b. Use part **a** to evaluate $\sin\left[\cos^{-1}\left(-\dfrac{1}{2}\right)\right]$.

Solution

a. For $|x| \leq 1$, let $\theta = \cos^{-1} x$ with θ in the interval $[0, \pi]$. Then $\cos\theta = x$.
For θ in $[0, \pi]$, $(\pi - \theta)$ is also in $[0, \pi]$ and

$$\cos(\pi - \theta) = -\cos\theta \qquad \text{See page 380.}$$
$$\cos(\pi - \theta) = -x \qquad \text{Replace } \cos\theta \text{ with } x.$$
$$\pi - \theta = \cos^{-1}(-x) \qquad \text{Definition of } \cos^{-1} x$$
$$\pi - \cos^{-1} x = \cos^{-1}(-x). \qquad \text{Replace } \theta \text{ with } \cos^{-1} x.$$

b.
$$\sin\left[\cos^{-1}\left(-\frac{1}{2}\right)\right] = \sin\left[\pi - \cos^{-1}\left(\frac{1}{2}\right)\right] \qquad \text{By part } \mathbf{a}$$
$$= \sin\left[\pi - \frac{\pi}{3}\right] \qquad \cos^{-1}\left(\frac{1}{2}\right) = \frac{\pi}{3}$$
$$= \sin\left(\frac{\pi}{3}\right) \qquad \sin(\pi - \theta) = \sin\theta \text{ (page 380).}$$
$$= \frac{\sqrt{3}}{2}. \qquad \text{See page 360.}$$

Practice Problem 6   a. Show that $\sin^{-1}(-x) = -\sin^{-1} x$ for $|x| \leq 1$.

b. Use part **a** to evaluate $\sin\left[\dfrac{\pi}{3} - \sin^{-1}\left(-\dfrac{1}{2}\right)\right]$.

Using a Calculator with Inverse Functions   When using the inverse trigonometric functions on a calculator to find a real number (or equivalently, an angle measured in radians), make sure you set your calculator to Radian mode.

When using a calculator to find $\csc^{-1} x$ or $\sec^{-1} x$, find $\sin^{-1}\dfrac{1}{x}$ and $\cos^{-1}\dfrac{1}{x}$, respectively. For example, if $\csc^{-1} 5 = \theta$, then $\csc\theta = 5$, or $\dfrac{1}{\sin\theta} = 5$. So $\sin\theta = \dfrac{1}{5}$ and $\theta = \sin^{-1}\left(\dfrac{1}{5}\right)$. However, to find $\cot^{-1} x$, begin by finding $\tan^{-1}\dfrac{1}{x}$; this gives you a value in the interval $\left(-\dfrac{\pi}{2}, \dfrac{\pi}{2}\right)$. If $x > 0$, this is the correct value, but **for $x < 0$, $\cot^{-1}(x) = \pi + \tan^{-1}\dfrac{1}{x}$;** so that $\cot^{-1}(x)$ is in the interval $\left(\dfrac{\pi}{2}, \pi\right)$.

When using a calculator to find an unknown angle measure in degrees, make sure you set your calculator to degree measure.

```
sin-1(√(2)/2)
              45.00
cos-1(0)
              90.00
```

```
tan-1(0.45)
        24.22774532
sin-1(0.2)
        11.53695903
■
```

EXAMPLE 7   **Using a Calculator to Find the Values of Inverse Functions**

Use a calculator to find the value of y in radians rounded to four decimal places.

a. $y = \sin^{-1} 0.75$ **b.** $y = \cot^{-1} 2.8$ **c.** $y = \cot^{-1}(-2.3)$

Solution

Set your calculator to Radian mode.

a. $y = \sin^{-1} 0.75 \approx 0.8481$

b. $y = \cot^{-1} 2.8 = \tan^{-1}\left(\dfrac{1}{2.8}\right) \approx 0.3430$

c. $y = \cot^{-1}(-2.3) = \pi + \tan^{-1}\left(-\dfrac{1}{2.3}\right) \approx 2.7315$

Practice Problem 7   Use a calculator to find the value of y (in degrees) for each expression.

a. $y = \cot^{-1} 0.75$ **b.** $y = \csc^{-1} 13$ **c.** $y = \tan^{-1}(-12)$

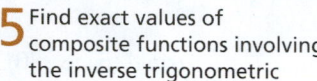

5 Find exact values of composite functions involving the inverse trigonometric functions.

Composition of Trigonometric and Inverse Trigonometric Functions

Recall that if f is a one-to-one function with inverse f^{-1}, then $f^{-1}[f(x)] = x$ for every x in the domain of f and $f[f^{-1}(x)] = x$ for every x in the domain of f^{-1}. This leads to the following formulas for the inverse sine, cosine, and tangent functions.

INVERSE FUNCTION PROPERTIES		
Inverse Sine	**Inverse Cosine**	**Inverse Tangent**
$\sin^{-1}(\sin x) = x,$ $-\dfrac{\pi}{2} \le x \le \dfrac{\pi}{2}$	$\cos^{-1}(\cos x) = x,$ $0 \le x \le \pi$	$\tan^{-1}(\tan x) = x,$ $-\dfrac{\pi}{2} < x < \dfrac{\pi}{2}$
$\sin(\sin^{-1} x) = x,$ $-1 \le x \le 1$	$\cos(\cos^{-1} x) = x,$ $-1 \le x \le 1$	$\tan(\tan^{-1} x) = x,$ $-\infty < x < \infty$

EXAMPLE 8   **Using the properties $f^{-1}(f(x)) = x$ and $f(f^{-1}(x)) = x$**

Find the exact value of

a. $\sin^{-1}\left[\sin\left(-\dfrac{\pi}{8}\right)\right].$ **b.** $\cos^{-1}\left(\cos\dfrac{5\pi}{4}\right).$ **c.** $\cos\left(\cos^{-1}\dfrac{3}{8}\right)$ **d.** $\sin(\sin^{-1} 2)$

Solution

a. Because $-\dfrac{\pi}{2} \le -\dfrac{\pi}{8} \le \dfrac{\pi}{2}$, we have $\sin^{-1}\left[\sin\left(-\dfrac{\pi}{8}\right)\right] = -\dfrac{\pi}{8}$.

b. We cannot use the property $\cos^{-1}(\cos x) = x$ for $x = \dfrac{5\pi}{4}$ because $\dfrac{5\pi}{4}$ is not in the interval $[0, \pi]$. To find the exact value of this expression, we first find θ in the interval $[0, \pi]$ where $\cos\dfrac{5\pi}{4} = \cos\theta$. Figure 4.76 shows that

$\cos \dfrac{5\pi}{4} = x = \cos \dfrac{3\pi}{4}$ and $\dfrac{3\pi}{4}$ is in the interval $[0, \pi]$ where $\cos^{-1}(\cos x) = x$ applies. Therefore,

$$\cos^{-1}\left(\cos \dfrac{5\pi}{4}\right) = \cos^{-1}\left(\cos \dfrac{3\pi}{4}\right) = \dfrac{3\pi}{4}.$$

c. Because $-1 \le \dfrac{3}{8} \le 1$, we have $\cos\left(\cos^{-1}\dfrac{3}{8}\right) = \dfrac{3}{8}$.

d. Here $x = 2$ does not lie in the interval $[-1, 1]$, so $\sin^{-1}(2)$ is undefined. Therefore, the expression $\sin(\sin^{-1} 2)$ is undefined.

Practice Problem 8 Find the exact value of:

a. $\sin^{-1}\left(\sin \dfrac{3\pi}{2}\right)$ **b.** $\cos(\cos^{-1}(2))$

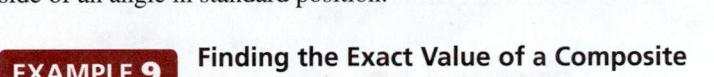

To find the exact values of expressions involving the composition of a trigonometric function and the inverse of a *different* trigonometric function, we use points on the terminal side of an angle in standard position.

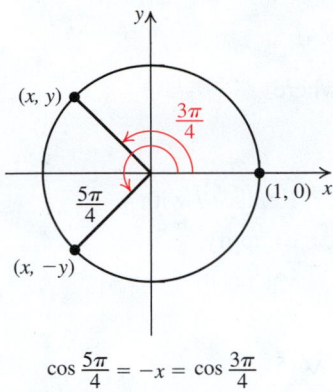

$\cos \dfrac{5\pi}{4} = -x = \cos \dfrac{3\pi}{4}$

Figure 4.76

EXAMPLE 9	**Finding the Exact Value of a Composite Trigonometric Expression**

Find the exact value of

a. $\cos\left(\tan^{-1}\dfrac{2}{3}\right)$.   **b.** $\sin\left[\cos^{-1}\left(-\dfrac{1}{4}\right)\right]$.

Solution

a. Let θ represent the radian measure of the angle in the interval $\left(-\dfrac{\pi}{2}, \dfrac{\pi}{2}\right)$, with $\tan \theta = \dfrac{2}{3}$. Then since $\tan \theta$ is positive, θ must be positive. We have

$$\theta = \tan^{-1}\dfrac{2}{3} \quad \text{and} \quad 0 < \theta < \dfrac{\pi}{2}.$$

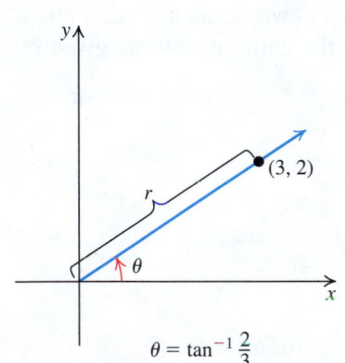

$\theta = \tan^{-1}\dfrac{2}{3}$

Figure 4.77

Figure 4.77 shows θ in standard position. If (x, y) is any point on the terminal side of θ, then $\tan \theta = \dfrac{y}{x}$.

Consequently, we can choose the point with coordinates $(3, 2)$ to determine the terminal side of θ. Then $x = 3$, $y = 2$ and we have

$$\tan \theta = \dfrac{2}{3} \quad \text{and} \quad \cos \theta = \dfrac{x}{r} = \dfrac{3}{r}, \text{ where}$$

$r = \sqrt{x^2 + y^2} = \sqrt{3^2 + 2^2} = \sqrt{9 + 4} = \sqrt{13}$. So

$$\cos\left(\tan^{-1}\dfrac{2}{3}\right) = \cos \theta = \dfrac{3}{r} = \dfrac{3}{\sqrt{13}} = \dfrac{3\sqrt{13}}{13}.$$

b. Let θ represent the radian measure of the angle in $[0, \pi]$, with $\cos \theta = -\dfrac{1}{4}$.

Then since $\cos \theta$ is negative, θ is in quadrant II; so

$$\theta = \cos^{-1}\left(-\dfrac{1}{4}\right) \quad \text{and} \quad \dfrac{\pi}{2} < \theta < \pi.$$

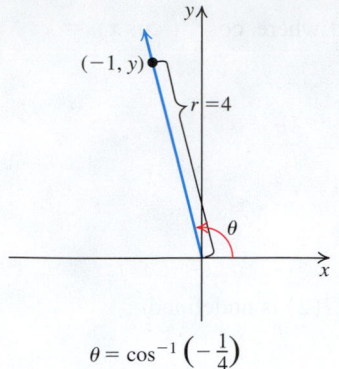

$$\theta = \cos^{-1}\left(-\tfrac{1}{4}\right)$$

Figure 4.78

Figure 4.78 shows θ in standard position. If (x, y) is a point on the terminal side of θ, and r is the distance between (x, y) and the origin, then $\sin \theta = \dfrac{y}{r}$. We choose the point with coordinates $(-1, y)$, a distance of $r = 4$ units from the origin, on the terminal side of θ. Then

$$\cos \theta = -\frac{1}{4} \quad \text{and} \quad \sin \theta = \frac{y}{4}, \text{where}$$

$$r = \sqrt{x^2 + y^2} = \sqrt{(-1)^2 + y^2} \quad \text{or} \quad r^2 = 1 + y^2$$

$$4^2 = 1 + y^2 \qquad \text{Replace } r \text{ with } 4.$$
$$15 = y^2 \qquad \text{Simplify.}$$
$$\sqrt{15} = y \qquad y \text{ is positive.}$$

Thus,

$$\sin\left[\cos^{-1}\left(-\frac{1}{4}\right)\right] = \sin \theta = \frac{y}{r} = \frac{\sqrt{15}}{4}.$$

Practice Problem 9 Find the exact value of $\cos\left[\sin^{-1}\left(-\frac{1}{3}\right)\right]$.

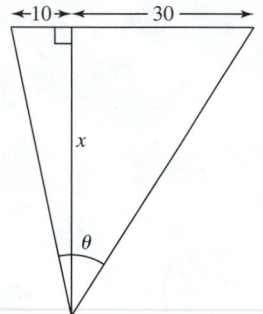

Figure 4.79

◆ **EXAMPLE 10** **Finding the Rotation Angle for a Security Camera**

A security camera (as shown in Figure 4.79) is installed x feet away from a jewelry counter. The angle θ through which the camera rotates to scan the entire counter is given by $\theta = \tan^{-1}\dfrac{30}{x} + \tan^{-1}\dfrac{10}{x}$. Find θ to the nearest degree if

a. $x = 10$ feet **b.** $x = 20$ feet

Solution

a. For $x = 10$ feet,

$$\theta = \tan^{-1}\frac{30}{10} + \tan^{-1}\frac{10}{10} \qquad \text{Substitute } x = 10 \text{ in the equation for } \theta.$$

$$= \tan^{-1}3 + \tan^{-1}1 \qquad \text{Simplify.}$$
$$\approx 71.57° + 45.00° \qquad \text{Use a calculator.}$$
$$= 116.57° \qquad \text{Add.}$$
$$\approx 117° \qquad \text{Round to the nearest degree.}$$

b. For $x = 20$ feet,

$$\theta = \tan^{-1}\frac{30}{20} + \tan^{-1}\frac{10}{20} \qquad \text{Substitute } x = 20.$$

$$= \tan^{-1}\frac{3}{2} + \tan^{-1}\frac{1}{2} \qquad \text{Simplify.}$$
$$\approx 56.31° + 26.56° \qquad \text{Use a calculator.}$$
$$= 82.87° \qquad \text{Add.}$$
$$\approx 83° \qquad \text{Round to the nearest degree.}$$

Practice Problem 10 Rework Example 10 for $x = 25$ feet.

SECTION 4.6 Exercises

Concepts and Vocabulary

1. The domain of $f(x) = \sin^{-1} x$ is _____.

2. The range of $f(x) = \tan^{-1} x$ is _____.

3. The exact value of $y = \cos^{-1}\dfrac{1}{2}$ is _____.

4. $\sin^{-1}(\sin \pi) = $ _____.

5. **True or False.** If $-1 \le x \le 0$, then $\sin^{-1} x \le 0$.

6. **True or False.** If $-1 \le x \le 0$, then $\cos^{-1} x \le 0$.

7. **True or False.** The domain of $f(x) = \cos^{-1} x$ is $0 \le x \le \pi$.

8. **True or False.** The value of $\tan^{-1}\left(\dfrac{1}{\sqrt{3}}\right)$ is $\dfrac{\pi}{3}$.

Building Skills

In Exercises 9–30, find each exact value of y or state that y is undefined.

9. $y = \sin^{-1} 0$

10. $y = \cos^{-1} 0$

11. $y = \sin^{-1}\left(-\dfrac{1}{2}\right)$

12. $y = \cos^{-1}\left(-\dfrac{\sqrt{3}}{2}\right)$

13. $y = \arccos(-1)$

14. $y = \arcsin\dfrac{1}{2}$

15. $y = \arccos\dfrac{\pi}{2}$

16. $y = \arcsin\pi$

17. $y = \tan^{-1}\sqrt{3}$

18. $y = \tan^{-1} 1$

19. $y = \arctan(-1)$

20. $y = \arctan\left(-\dfrac{\sqrt{3}}{3}\right)$

21. $y = \cot^{-1}(-1)$

22. $y = \sin^{-1}\left(-\dfrac{\sqrt{2}}{2}\right)$

23. $y = \cos^{-1}(-2)$

24. $y = \sin^{-1}\sqrt{3}$

25. $y = \sec^{-1}(-2)$

26. $y = \csc^{-1}(-2)$

27. $y = \arcsin 1$

28. $y = \arccos 1$

29. $y = \text{arccot}\left(-\sqrt{3}\right)$

30. $y = \text{arcsec}\left(-\sqrt{2}\right)$

In Exercises 31–36, use the four-step procedure (page 122) to find $f^{-1}(x)$ for the given one-to-one function $f(x)$.

31. $f(x) = 2\sin x + 1, -\dfrac{\pi}{2} \le x \le \dfrac{\pi}{2}$

32. $f(x) = \dfrac{1}{2}\cos x - 1, 0 \le x \le \pi$

33. $f(x) = 3\cos(2x - 1), \dfrac{1}{2} \le x \le \dfrac{\pi}{2} + \dfrac{1}{2}$

34. $f(x) = 4\sin(3x - 2), -\dfrac{\pi}{6} + \dfrac{2}{3} \le x \le \dfrac{\pi}{6} + \dfrac{2}{3}$

35. $f(x) = \tan(x - 1) + 2, 1 - \dfrac{\pi}{2} < x < 1 + \dfrac{\pi}{2}$

36. $f(x) = \cot(x + 1) - 3, -1 \le x \le \pi - 1$

In Exercises 37–52, find each exact value of y or state that y is undefined.

37. $y = \sin\left(\sin^{-1}\dfrac{1}{8}\right)$

38. $y = \cos\left(\cos^{-1}\dfrac{1}{5}\right)$

39. $y = \tan^{-1}\left(\tan\dfrac{\pi}{7}\right)$

40. $y = \tan^{-1}\left(\tan\dfrac{\pi}{4}\right)$

41. $y = \tan(\tan^{-1} 247)$

42. $y = \tan(\tan^{-1} 7)$

43. $y = \sin^{-1}\left(\sin\dfrac{4\pi}{3}\right)$

44. $y = \cos^{-1}\left(\cos\dfrac{5\pi}{3}\right)$

45. $y = \tan^{-1}\left(\tan\dfrac{2\pi}{3}\right)$

46. $y = \tan\left(\tan^{-1}\dfrac{2\pi}{3}\right)$

47. $y = \sin^{-1}\left(\sin\dfrac{3\pi}{4}\right)$

48. $y = \cos^{-1}\left(\cos\dfrac{7\pi}{6}\right)$

49. $y = \sin\left(\sin^{-1}\sqrt{2}\right)$

50. $y = \cos\left(\cos^{-1}(-\sqrt{2})\right)$

51. $y = \cos^{-1}(\cos(-\pi))$

52. $y = \sin^{-1}(\sin 1.2)$

In Exercises 53–68, use the identities on page 422 to find the exact value of each expression.

53. $\cot^{-1}\left(\dfrac{1}{\sqrt{3}}\right)$

54. $\sec^{-1}(\sqrt{2})$

55. $\csc^{-1}(2)$

56. $\csc^{-1}\left(\dfrac{2\sqrt{3}}{3}\right)$

57. $\cot^{-1}\left(-\dfrac{1}{\sqrt{3}}\right)$

58. $\cot^{-1}(-\sqrt{3})$

59. $\sin^{-1}\left(-\dfrac{\sqrt{3}}{2}\right)$

60. $\csc^{-1}\left(-\dfrac{2}{\sqrt{3}}\right)$

61. $\cos^{-1}\left(-\dfrac{1}{2}\right)$

62. $\sec^{-1}\left(-\dfrac{2}{\sqrt{3}}\right)$

63. $\sin\left[\dfrac{\pi}{3} - \sin^{-1}\left(-\dfrac{1}{2}\right)\right]$

64. $\cos\left[\dfrac{\pi}{6} + \cos^{-1}\left(-\dfrac{\sqrt{3}}{2}\right)\right]$

65. $\sin\left[\dfrac{\pi}{2} - \cos^{-1}(-1)\right]$

66. $\tan\left[\dfrac{\pi}{6} + \cot^{-1}\left(-\dfrac{1}{\sqrt{3}}\right)\right]$

67. $\sin\left[\tan^{-1}(-\sqrt{3}) + \cos^{-1}\left(-\dfrac{\sqrt{3}}{2}\right)\right]$

68. $\cos\left[\cot^{-1}(-\sqrt{3}) + \sin^{-1}\left(-\dfrac{1}{2}\right)\right]$

In Exercises 69–78, use a calculator to find each value of y in degrees rounded to two decimal places.

69. $y = \cos^{-1} 0.6$

70. $y = \sin^{-1} 0.23$

71. $y = \sin^{-1}(-0.69)$

72. $y = \cos^{-1}(-0.57)$

73. $y = \sec^{-1}(3.5)$

74. $y = \csc^{-1}(6.8)$

75. $y = \tan^{-1} 14$

76. $y = \tan^{-1} 50$

77. $y = \tan^{-1}(-42.147)$

78. $y = \tan^{-1}(-0.3863)$

In Exercises 79–92, use a sketch to find each exact value of y.

79. $y = \cos\left(\sin^{-1}\dfrac{2}{3}\right)$

80. $y = \sin\left(\cos^{-1}\dfrac{3}{4}\right)$

81. $y = \sin\left[\cos^{-1}\left(-\dfrac{4}{5}\right)\right]$

82. $y = \cos\left(\sin^{-1}\dfrac{3}{5}\right)$

83. $y = \cos\left(\tan^{-1}\dfrac{5}{2}\right)$

84. $y = \sin\left(\tan^{-1}\dfrac{13}{5}\right)$

85. $y = \tan\left(\cos^{-1}\dfrac{4}{5}\right)$

86. $y = \tan\left[\sin^{-1}\left(-\dfrac{3}{4}\right)\right]$

87. $y = \sin(\tan^{-1} 4)$

88. $y = \cos(\tan^{-1} 3)$

89. $y = \tan(\sec^{-1} 2)$

90. $y = \tan\left[\csc^{-1}(-2)\right]$

91. $y = \sin(\cos^{-1} x)$, $|x| < 1$

92. $\tan\left(\sin^{-1}\dfrac{x}{4}\right)$, $|x| < 4$

Applying the Concepts

93. Motorcycle racing. A video camera is set up x feet away from and at a right angle to a straight quarter-mile racetrack, as shown in the figure. The starting line is to the left, and the finish is to the right. The angle through which the camera rotates to film the entire race is given by $\theta = \tan^{-1}\dfrac{880}{x} + \tan^{-1}\dfrac{440}{x}$. Find θ to the nearest degree if $x = 110$ feet.

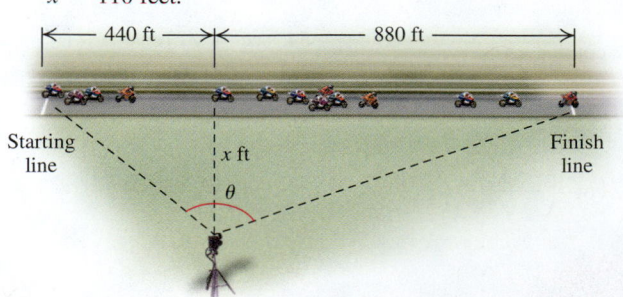

94. Camera's viewing angle. The viewing angle for the 35-millimeter camera is given (in degrees) by $\theta = 2\tan^{-1}\dfrac{18}{x}$, where x is the focal length of the lens.

The focal length on most adjustable cameras is marked in millimeters on the lens mount.

   a. Find the viewing angle, in degrees, if the focal length is 50 millimeters.

   b. Find the viewing angle, in degrees, if the focal length is 200 millimeters.

95. Area under a curve. The area of the region bounded by the curve $y = \dfrac{1}{1 + x^2}$, the x-axis, and the lines $x = a$ and $x = b\,(< b)$ is given by $\tan^{-1} b - \tan^{-1} a$.

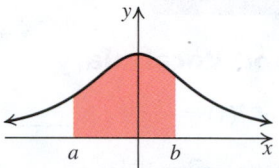

   a. Find the exact area if $a = -1$ and $b = \dfrac{1}{\sqrt{3}}$.

   b. Find the exact area if $a = -\sqrt{3}$ and $b = \sqrt{3}$.

96. Area under a curve. The area of the region bounded by the curve $y = \dfrac{1}{\sqrt{4 - x^2}}\,(-2 < x < 2)$, the x-axis, and the lines $x = a$ and $x = b\,(< b)$ is given by $\sin^{-1}\left(\dfrac{b}{2}\right) - \sin^{-1}\left(\dfrac{a}{2}\right)$.

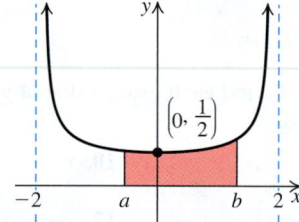

   a. Find the exact area if $a = -1$ and $b = 1$.

   b. Find the exact area if $a = -\sqrt{3}$ and $b = \sqrt{2}$.

97. Area under a curve. The area of the region bounded by the curve $y = \dfrac{x}{\sqrt{4 - x^2}}$, the x-axis, and the lines $x = a$ and $x = b\,(0 \le a < b < 2)$ is given by $2\left[\cos\left(\sin^{-1}\dfrac{a}{2}\right) - \cos\left(\sin^{-1}\dfrac{b}{2}\right)\right]$. Find the exact area if $a = 1$ and $b = \sqrt{3}$.

$$\left[\text{Hint: Show } \cos\left(\sin^{-1}\dfrac{x}{2}\right) = \dfrac{\sqrt{4 - x^2}}{2}.\right]$$

98. Area under a curve. The area of the region bounded by the curve $y = \dfrac{x}{\sqrt{4 + x^2}}$, the x-axis, and the lines $x = a$ and $x = b\,(0 \le a < b)$ is given by

$2\left[\sec\left(\tan^{-1}\dfrac{b}{2}\right) - \sec\left(\tan^{-1}\dfrac{a}{2}\right)\right]$. Find the exact area if $a = 1$ and $b = 3$.

$\left[\textit{Hint}: \text{Show } \sec\left(\tan^{-1}\dfrac{x}{2}\right) = \dfrac{\sqrt{4 + x^2}}{2}\right].$

Shortest distance between two points on the globe. Let A_1 and A_2 be two points on the globe. Let T_1 and N_1 be the latitude, and longitude respectively, of the point A_1. Let T_2 and let N_2 be the latitude, and the longitude respectively, of the point A_2. The shortest distance, d, between the points A_1 and A_2 is given by:

$$d = R\cos^{-1}(x)$$

where $x = \sin T_1 \sin T_2 + \cos T_1 \cos T_2 \cos(N_1 - N_2)$ and $\cos^{-1}x$ is measured in radians;
$R = $ the radius of Earth ≈ 3960 miles.

99. **Shortest distance between two cities.** Find the shortest distance between Mexico City, Mexico, latitude $19°26'$ N, longitude $-99°7'$ (W), and New Delhi, India, latitude $28°35'$ N, longitude $+77°12'$ (E).

100. **Shortest distance between two cities.** Find the shortest distance between London, England, latitude $51°32'$ N, longitude $0°5'$ W, and Havana, Cuba, latitude $28°8'$ N, longitude $82°32'$ W.

Beyond the Basics

In Exercises 101–103, determine whether each function is increasing or decreasing on its domain.

101. $y = \sin^{-1}x$ 102. $y = \cos^{-1}x$

103. $y = \tan^{-1}x$

104. Show that

    a. $\sec^{-1}x \neq \dfrac{1}{\cos^{-1}x}$.   **b.** $\sec^{-1}x = \cos^{-1}\dfrac{1}{x}$.

105. For what values of x is $\cot^{-1}(\cot x) = x$ true?

In Exercises 106–108, find the exact value of each expression.

106. $\sin^{-1}(\sin 10)$ 107. $\cos^{-1}(\cos 10)$

108. $\tan^{-1}(\tan(-6))$

In Exercises 109–114, use transformations to sketch the graph of each function.

109. $y = \sin^{-1}(2x), -\dfrac{1}{2} \leq x \leq \dfrac{1}{2}$

110. $y = \cos^{-1}\left(\dfrac{1}{2}x\right), -2 \leq x \leq 2$

111. $y = \sin^{-1}(2x - 1), 0 \leq x \leq 1$

112. $y = \tan^{-1}(2x - 3), -\infty < x < \infty$

113. $y = \cos^{-1}\left(\dfrac{x}{3} - 1\right), 0 \leq x \leq 6$

114. $y = \sin^{-1}\left(\dfrac{1}{2}x + \dfrac{1}{3}\right), -\dfrac{8}{3} \leq x \leq \dfrac{4}{3}$

Critical Thinking / Discussion / Writing

115. Show that $\sin^{-1}x = \begin{cases} \cos^{-1}(\sqrt{1 - x^2}); & 0 \leq x \leq 1 \\ -\cos^{-1}(\sqrt{1 - x^2}); & -1 \leq x < 0 \end{cases}$

116. Show that: $\sin^{-1}x + \cos^{-1}x = \dfrac{\pi}{2}; |x| \leq 1$.

117. Find all values of x in the interval $[0, 2\pi]$ for which $\sin x \leq \csc x$.

118. Find all values of x for which $\sin^{-1}x = \cos^{-1}x$
     [*Hint*: Take the sine of both sides.]

Preparing for the Next Section

In Exercises 119 and 120, perform the indicated operations and simplify.

119. $\dfrac{1}{1 - x} + \dfrac{1}{1 + x}$ 120. $\dfrac{1}{1 - x} - \dfrac{1}{1 - x^2}$

In Exercises 121 and 122, rationalize the denominator and simplify.

121. $\dfrac{1}{\sqrt{2} + 1} + \dfrac{1}{\sqrt{3} + \sqrt{2}}$

122. $\dfrac{x}{\sqrt{x + 2} - \sqrt{2}}$

123. If $x^2 + y^2 = 1$, simplify $\dfrac{x}{1 + y} - \dfrac{1 - y}{x}$.

124. If $xy = 1$, simplify $\left(x + \dfrac{1}{y}\right)\left(y + \dfrac{1}{x}\right)$.

125. Simplify: $\tan x \cos x \csc x$.

126. Simplify: $(\sin x + \cos x)^2 - 2\sin x \cos x$.

SUMMARY Definitions, Concepts, and Formulas

4.1 Angles and Their Measure

i. An **angle** is formed by rotating a ray around its endpoint.

ii. An angle in a rectangular coordinate system is in **standard position** if its vertex is at the origin and its initial side coincides with the positive x-axis. (See page 338 for positive and negative angles.)

iii. If the terminal side of an angle in standard position lies on the x-axis or the y-axis, the angle is a **quadrantal angle**.

iv. Angles can be measured in **degrees**, where $1° = \dfrac{1}{360}$ of a complete revolution.

v. An **acute angle** is an angle with measure between $0°$ and $90°$, a **right angle** is an angle with measure $90°$, an **obtuse angle** is an angle with measure between $90°$ and $180°$, and a **straight angle** is an angle with measure $180°$.

vi. Angles that have the same initial and terminal sides are **coterminal** angles.

vii. Angles can also be measured in **radians**. The radian measure of a central angle θ that intercepts an arc of length s on a circle of radius r is $\theta = \dfrac{s}{r}$ radians.

viii. To convert from degrees to radians, multiply degrees by $\dfrac{\pi}{180°}$. To convert from radians to degrees, multiply radians by $\dfrac{180°}{\pi}$.

ix. Two positive angles are **complements** if their sum is $90°$. Two positive angles are **supplements** if their sum is $180°$.

x. The formula for the length s of the arc intercepted by a central angle θ in a circle of radius r is $s = r\theta$.

xi. The area of a sector $= \dfrac{1}{2}r^2\theta$, where r is the radius of the circle and θ is in radians.

xii. If an object travels at a constant speed v on a circle of radius r through an angle of θ radians and an arc of length s, in time t, then the (average) **linear speed** of the object is $v = \dfrac{s}{t}$ and the (average) **angular speed** ω of the object is $\omega = \dfrac{\theta}{t}$. Further, $v = r\omega$.

4.2 The Unit Circle; Trigonometric Functions of Real Numbers

i. A number line is wrapped around the unit circle $(x^2 + y^2 = 1)$. See Figure 4.22. For any real number t, let $P(t) = (x, y)$ be the **terminal point** on the unit circle associated with $t = \theta$ radians, We define

$$\sin t = \sin\theta = y, \quad \cos t = \cos\theta = x, \quad \tan t = \tan\theta = \frac{y}{x},$$

$$\csc t = \csc\theta = \frac{1}{y}, \quad \sec t = \sec\theta = \frac{1}{x}, \quad \cot t = \cot\theta = \frac{x}{y}.$$

ii. Trigonometric function values of coterminal angles

θ in degrees	θ in radians
$\sin\theta = \sin(\theta + n360°)$	$\sin\theta = \sin(\theta + 2\pi n)$
$\cos\theta = \cos(\theta + n360°)$	$\cos\theta = \cos(\theta + 2\pi n)$

These equations hold for any integer n.

TABLE 4.12

$\theta°$	θ radians	$\sin\theta$	$\cos\theta$	$\tan\theta$	$\csc\theta$	$\sec\theta$	$\cot\theta$
$30°$	$\dfrac{\pi}{6}$	$\dfrac{1}{2}$	$\dfrac{\sqrt{3}}{2}$	$\dfrac{\sqrt{3}}{3}$	2	$\dfrac{2\sqrt{3}}{3}$	$\sqrt{3}$
$45°$	$\dfrac{\pi}{4}$	$\dfrac{\sqrt{2}}{2}$	$\dfrac{\sqrt{2}}{2}$	1	$\sqrt{2}$	$\sqrt{2}$	1
$60°$	$\dfrac{\pi}{3}$	$\dfrac{\sqrt{3}}{2}$	$\dfrac{1}{2}$	$\sqrt{3}$	$\dfrac{2\sqrt{3}}{3}$	2	$\dfrac{\sqrt{3}}{3}$

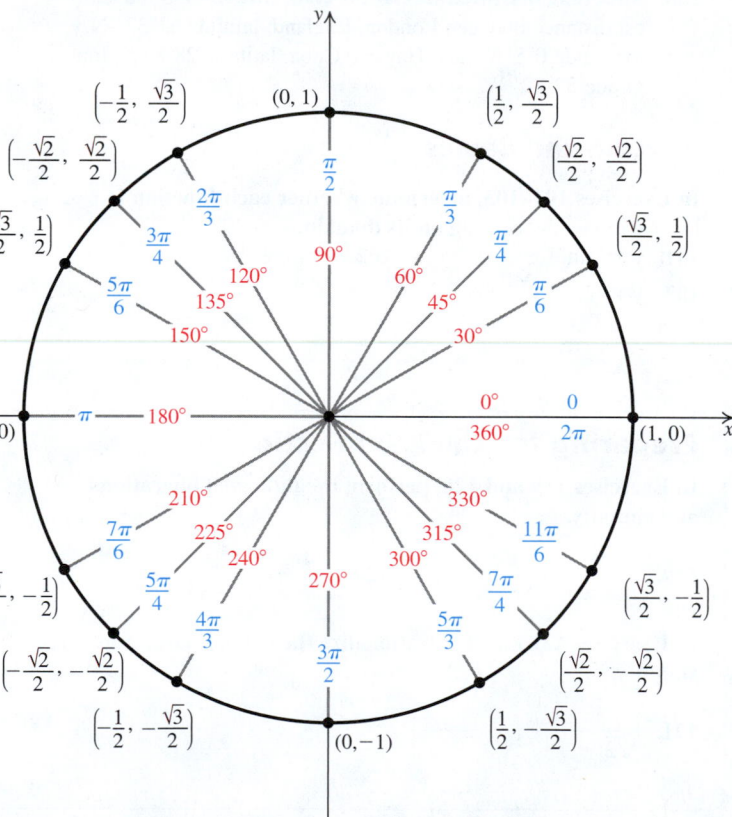

4.3 Trigonometric Functions of Angles

i. Let $P = (x, y)$ be a point on the terminal side of an angle θ in standard position, and let $r = \sqrt{x^2 + y^2}$. Then

$$\sin\theta = \frac{y}{r} \qquad \cos\theta = \frac{x}{r} \qquad \tan\theta = \frac{y}{x}, x \neq 0$$

$$\csc\theta = \frac{r}{y}, y \neq 0 \quad \sec\theta = \frac{r}{x}, x \neq 0 \quad \cot\theta = \frac{x}{y}, y \neq 0$$

ii. *Signs of the trigonometric functions*

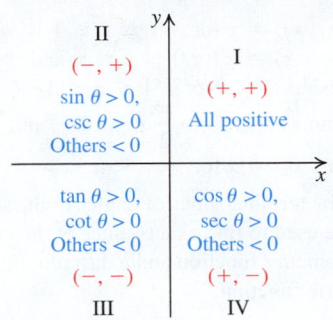

II $(-, +)$ $\sin \theta > 0,$ $\csc \theta > 0$ Others < 0	**I** $(+, +)$ All positive
III $(-, -)$ $\tan \theta > 0,$ $\cot \theta > 0$ Others < 0	**IV** $(+, -)$ $\cos \theta > 0,$ $\sec \theta > 0$ Others < 0

iii. The *reference angle* for θ in standard position is the positive acute angle θ' formed by the terminal side of θ and the x-axis. Reference angles are used to find the values of trigonometric functions for *any* angle θ.

iv. *Fundamental trigonometric identities*
Quotient and reciprocal identities

$$\tan t = \frac{\sin t}{\cos t}, \cot t = \frac{\cos t}{\sin t},$$

$$\csc t = \frac{1}{\sin t}, \sec t = \frac{1}{\cos t}, \cot t = \frac{1}{\tan t}$$

Pythagorean identities

$$\cos^2 t + \sin^2 t = 1, 1 + \tan^2 t = \sec^2 t, 1 + \cot^2 t = \csc^2 t$$

Even–odd identities

$$\sin(-t) = -\sin t \quad \cos(-t) = \cos t \quad \tan(-t) = -\tan t$$

$$\csc(-t) = -\csc t \quad \sec(-t) = \sec t \quad \cot(-t) = -\cot t$$

4.4 Graphs of the Sine and Cosine Functions

i. A function f is periodic if $f(x + p) = f(x)$, for all x, $p > 0$. The smallest such p, if one exists, is called the period of f.

ii. Sine and cosine function graphs and properties

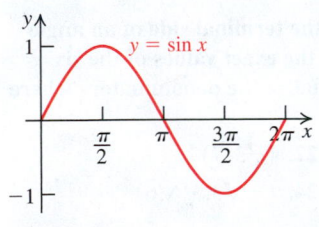

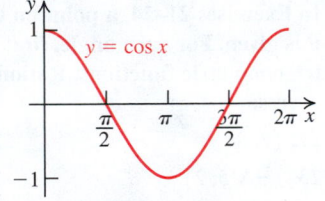

Sine Function

1. Period: 2π
2. Domain: $(-\infty, \infty)$
3. Range: $[-1, 1]$
4. Odd: $\sin(-t) = -\sin t$
5. Zeros at integer multiples of π

Cosine Function

1. Period: 2π
2. Domain: $(-\infty, \infty)$
3. Range: $[-1, 1]$
4. Even: $\cos(-t) = \cos t$
5. Zeros at odd integer multiples of $\frac{\pi}{2}$

iii. Sinusoidal Graphs

The functions $y = a \sin[b(x - c)] + d$ and $y = a \cos[b(x - c)] + d$ ($b > 0$) have amplitude $|a|$, period $\frac{2\pi}{b}$, phase shift c, and vertical shift d.

4.5 Graphs of the Other Trigonometric Functions

i. If a line makes an angle θ with the positive x-axis, its **slope** is $m = \tan \theta$.

ii. Tangent and cotangent function graphs and properties

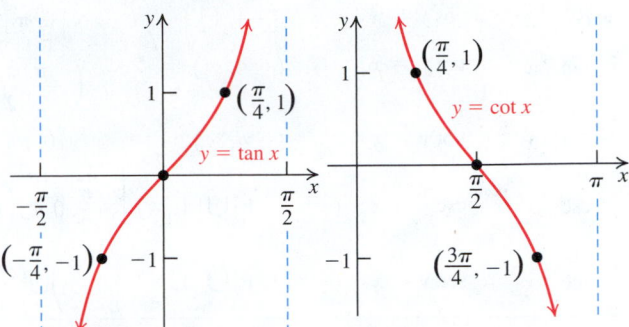

Tangent Function

1. Period: π
2. Domain: All real numbers except for odd multiples of $\frac{\pi}{2}$
3. Range: $(-\infty, \infty)$
4. Odd: $\tan(-x) = -\tan x$
5. Zeros at integer multiples of π
6. Vertical asymptotes at odd integer multiples of $\frac{\pi}{2}$

Cotangent Function

1. Period: π
2. Domain: All real numbers except for integer multiples of π
3. Range: $(-\infty, \infty)$
4. Odd: $\cot(-x) = -\cot x$
5. Zeros at odd integer multiples of $\frac{\pi}{2}$
6. Vertical asymptotes at integer multiples of π

iii. Secant and cosecant function graphs and properties

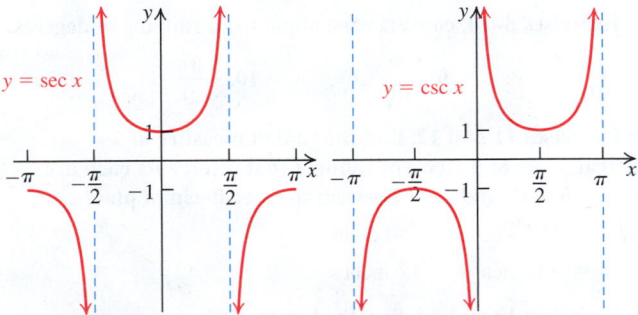

Secant Function

1. Period: 2π
2. Domain: All real numbers except for odd multiples of $\frac{\pi}{2}$
3. Range: $(-\infty, -1] \cup [1, \infty)$
4. Even: $\sec(-x) = \sec x$

Cosecant Function

1. Period: 2π
2. Domain: All real numbers except for integer multiples of π
3. Range: $(-\infty, -1] \cup [1, \infty)$
4. Odd: $\csc(-x) = -\csc x$

4.6 Inverse Trigonometric Functions

i.

Inverse Function	Equivalent to	Domain	Range
$y = \sin^{-1} x$	$\sin y = x$	$[-1, 1]$	$\left[-\dfrac{\pi}{2}, \dfrac{\pi}{2}\right]$
$y = \cos^{-1} x$	$\cos y = x$	$[-1, 1]$	$[0, \pi]$
$y = \tan^{-1} x$	$\tan y = x$	$(-\infty, \infty)$	$\left(-\dfrac{\pi}{2}, \dfrac{\pi}{2}\right)$
$y = \cot^{-1} x$	$\cot y = x$	$(-\infty, \infty)$	$(0, \pi)$
$y = \csc^{-1} x$	$\csc y = x$	$(-\infty, -1] \cup [1, \infty)$	$\left[-\dfrac{\pi}{2}, 0\right) \cup \left(0, \dfrac{\pi}{2}\right]$
$y = \sec^{-1} x$	$\sec y = x$	$(-\infty, -1] \cup [1, \infty)$	$\left[0, \dfrac{\pi}{2}\right) \cup \left(\dfrac{\pi}{2}, \pi\right]$

Inverse Trigonometric Functions

ii. (a) $\sin^{-1}(\sin x) = x$ for $-\dfrac{\pi}{2} \le x \le \dfrac{\pi}{2}$ and
$\sin(\sin^{-1} x) = x$ for $-1 \le x \le 1$
(b) $\cos^{-1}(\cos x) = x$ for $0 \le x \le \pi$ and
$\cos(\cos^{-1} x) = x$ for $-1 \le x \le 1$
(c) $\tan^{-1}(\tan x) = x$ for $-\dfrac{\pi}{2} < x < \dfrac{\pi}{2}$ and
$\tan(\tan^{-1} x) = x$ for $-\infty < x < \infty$

iii. Points on the terminal sides of angles in the standard position are used to find exact values of the composition of a trigonometric function and a different inverse trigonometric function.

REVIEW EXERCISES

Building Skills

In Exercises 1–4, draw each angle in standard position.

1. $240°$
2. $-150°$
3. $-\dfrac{2\pi}{3}$
4. $\dfrac{7\pi}{3}$

In Exercises 5–7, convert each angle from degrees to radians. Express each answer as a multiple of π.

5. $20°$
6. $36°$
7. $-60°$

In Exercises 8–10, convert each angle from radians to degrees.

8. $\dfrac{7\pi}{10}$
9. $\dfrac{5\pi}{18}$
10. $-\dfrac{4\pi}{9}$

In Exercises 11 and 12, find the radian measure of a central angle of a circle of radius r that intercepts each arc of length s. Round your answers to three decimal places.

11. $r = 15$ inches, $s = 40$ inches

12. $r = 9$ inches, $s = 17$ inches

In Exercises 13 and 14, find the length of the arc on each circle of radius r intercepted by a central angle θ. Round your answers to three decimal places.

13. $r = 2$ meters, $\theta = 36°$

14. $r = 0.9$ meter, $\theta = 12°$

In Exercises 15 and 16, find the area of the sector of a circle of radius r formed by the central angle θ. Round your answers to three decimal places.

15. $r = 3$ feet, $\theta = 32°$

16. $r = 4$ meters, $\theta = 47°$

In Exercises 17–20, use each given trigonometric function value of θ to find the five other trigonometric function values of an acute angle θ. Rationalize the denominators where necessary.

17. $\cos \theta = \dfrac{2}{9}$
18. $\sin \theta = \dfrac{1}{5}$
19. $\tan \theta = \dfrac{5}{3}$
20. $\cot \theta = \dfrac{5}{4}$

In Exercises 21–24, a point on the terminal side of an angle θ is given. For each angle, find the exact values of the six trigonometric functions. Rationalize the denominators where necessary.

21. $(2, 8)$
22. $(-3, 7)$
23. $(-\sqrt{5}, 2)$
24. $(-\sqrt{3}, -\sqrt{6})$

In Exercises 25–28, use the given information to find the quadrant in which θ lies.

25. $\sin \theta < 0$ and $\cos \theta > 0$

26. $\sin \theta > 0$ and $\tan \theta > 0$

27. $\sin \theta < 0$ and $\cot \theta > 0$

28. $\cos \theta < 0$ and $\csc \theta > 0$

In Exercises 29–32, find the exact values of the remaining trigonometric functions of θ from the given information.

29. $\cos \theta = -\dfrac{4}{5}$, θ in quadrant III

30. $\tan \theta = -\dfrac{5}{12}$, θ in quadrant IV

31. $\sin \theta = \dfrac{3}{5}$, θ in quadrant II

32. $\csc \theta = -\dfrac{5}{4}$, θ in quadrant III

In Exercises 33–36, find the exact value of each trigonometric function by using reference angles. Do not use a calculator.

33. $\cos 150°$

34. $\sin(-300°)$

35. $\tan 390°$

36. $\cot -405°$

In Exercises 37–40, sketch the graph of each given equation over the interval $[-2\pi, 2\pi]$.

37. $y = -\dfrac{3}{2}\cos x$

38. $y = \dfrac{5}{2}\sin x$

39. $y = 3\sin\left(x - \dfrac{\pi}{3}\right)$

40. $y = 4\cos\left(x + \dfrac{3\pi}{2}\right)$

In Exercises 41–44, find the amplitude or vertical stretch factor, period, and phase shift of each given function.

41. $y = 14\sin\left(2x + \dfrac{\pi}{7}\right)$

42. $y = 21\cos\left(8x + \dfrac{\pi}{9}\right)$

43. $y = 6\tan\left(2x + \dfrac{\pi}{5}\right)$

44. $y = -11\cot\left(6x + \dfrac{\pi}{12}\right)$

In Exercises 45–50, graph each function over a two-period interval.

45. $y = -2\cos\left(x + \dfrac{2\pi}{3}\right)$

46. $y = 3\cos(x - \pi)$

47. $y = 5\tan\left[2\left(x - \dfrac{\pi}{4}\right)\right]$

48. $y = -\cot\left[2\left(x + \dfrac{\pi}{4}\right)\right]$

49. $y = \dfrac{1}{2}\sec\dfrac{x}{2}$

50. $y = -3\csc\dfrac{x}{2}$

In Exercises 51–60, find the exact value of y or state that y is undefined.

51. $y = \cos^{-1}\left(\cos\dfrac{5\pi}{8}\right)$

52. $y = \sin^{-1}\left(\sin\dfrac{7\pi}{6}\right)$

53. $y = \tan^{-1}\left[\tan\left(-\dfrac{2\pi}{3}\right)\right]$

54. $y = \sin\left(\cos^{-1}\dfrac{1}{2}\right)$

55. $y = \cos\left(\sin^{-1}\dfrac{\sqrt{2}}{2}\right)$

56. $y = \cos\left(\tan^{-1}\dfrac{3}{4}\right)$

57. $y = \tan\left[\cos^{-1}\left(-\dfrac{1}{2}\right)\right]$

58. $y = \tan\left[\sin^{-1}\left(-\dfrac{\sqrt{3}}{2}\right)\right]$

59. $y = \sin\left(\sec^{-1}\dfrac{\sqrt{3}}{2}\right)$

60. $y = \sin^{-1}\left(\tan\dfrac{3\pi}{2}\right)$

In Exercises 61–64, write an equation of the form $y = a\sin(bx - c)$ for the given graph.

61.

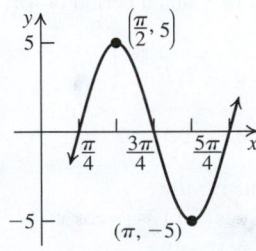

62.

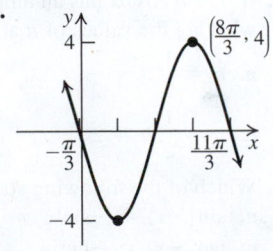

63.

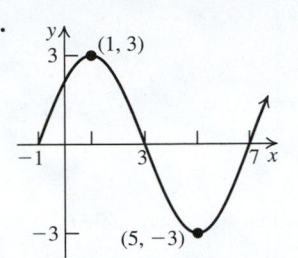

64.

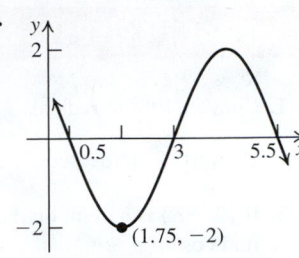

Applying the Concepts

65. Distance. A tractor rolls backward so that its wheels turn a quarter of a revolution. If the tires have a radius of 28 inches, how many inches has the tractor moved?

66. Angle measure. What is the radian measure of the smaller central angle made by the hands of a clock at 5:00? Express your answer as a rational multiple of π.

67. Distance. New Orleans, Louisiana, is due south of Dubuque, Iowa. Find the distance between New Orleans (north latitude 29°59′ N) and Dubuque (north latitude 42°31′ N). Use 3960 miles as the value of the radius of Earth.

68. Angular speed. A horse on the outside row of a merry-go-round is 20 feet from the center, and its linear speed is 314 feet per minute. Find its angular speed.

69. Linear speed. A geostationary satellite has a circular orbit 36,000 kilometers above Earth's surface. It takes 24 hours for the satellite to complete its orbit, appearing to be stationary above a fixed location on Earth. If the radius of Earth is 6400 kilometers, find the linear speed of the satellite.

70. Daylight hours. The table gives the average number of daylight hours in Santa Fe, New Mexico, each month. Let y represent the number of daylight hours in Santa Fe in month x and find a function of the form $y = a\sin[b(x - c)] + d$ that models the hours of daylight throughout the year, where $x = 1$ represents January.

Jan	Feb	Mar	Apr	May	June
10.1	9.9	12.0	12.7	14.1	14.1

July	Aug	Sept	Oct	Nov	Dec
14.3	13.5	12.0	11.3	10.0	9.8

71. Simple harmonic motion. Suppose a ball attached to a spring is pulled down 11 inches and released and the resulting simple harmonic motion has a period of five seconds. Write an equation of the ball's simple harmonic motion.

72. Viewing Angle. The bottom of a 40-foot movie screen is 10 feet above eye level. The viewing angle θ from a distance of x feet from the screen is given by

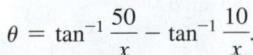

$$\theta = \tan^{-1}\dfrac{50}{x} - \tan^{-1}\dfrac{10}{x}.$$

Find the viewing angle θ (to the nearest degree) from distances of
a. 25 feet,   **b.** 50 feet,   **c.** 75 feet.

1. Convert 195° to radians.

2. Convert $\dfrac{7\pi}{5}$ to degrees.

3. If $(2, -5)$ is a point on the terminal side of an angle θ, find $\cos\theta$.

4. Find the radian measure of a central angle of a circle of radius 10 centimeters that intercepts an arc of length 15 centimeters.

5. Find the area of a sector of a circle of radius 15 centimeters that intersects an arc of length 10 centimeters.

6. If $\cos\theta = \dfrac{2}{7}$ for an acute angle θ, find $\sin\theta$.

7. If $\tan\theta < 0$ and $\csc\theta > 0$, find the quadrant in which θ lies.

8. If $\cot\theta = -\dfrac{5}{12}$ and θ is in quadrant IV, find $\sec\theta$.

9. What is the reference angle for 217°?

10. If $\cos\dfrac{3\pi}{7} = 0.223$, find $\sin\dfrac{\pi}{14}$.

11. From a seat 65 meters high on a Ferris wheel, the angle between the horizontal and the line of sight to a hot dog stand is 30°. How many meters is the hot dog stand from a point directly below the seat?

12. Give the amplitude and range for $y = -7\sin x$.

13. Find the amplitude, period, and phase shift for $y = 19\sin[12(x + 3\pi)]$.

14. Suppose a ball attached to a spring is pulled down 7 inches and released and the resulting simple harmonic motion has a period of four seconds. Write an equation for the ball's simple harmonic motion.

15. Graph $y = -\cos\dfrac{x}{2}$ over a one-period interval.

16. Graph $y = \sin[2(x - \pi)]$ over a one-period interval.

17. Graph $y = \tan\left(x + \dfrac{\pi}{4}\right)$ over a one-period interval.

18. Find the exact value of $y = \sin^{-1}\left(-\dfrac{\sqrt{3}}{2}\right)$.

19. Find the exact value of $y = \cos^{-1}\left[\cos\dfrac{4\pi}{3}\right]$.

20. Find the exact value of $y = \tan\left[\sin^{-1}\left(-\dfrac{1}{3}\right)\right]$.

Multiple choice—select the correct answer.

1. Convert 160° to radians.
   a. $\dfrac{7\pi}{9}$
   b. $\dfrac{8\pi}{9}$
   c. $\dfrac{4\pi}{3}$
   d. $\dfrac{7\pi}{3}$

2. Convert $\dfrac{13\pi}{5}$ to degrees.
   a. 36°
   b. 72°
   c. 108°
   d. 468°

3. If $(-3, 1)$ is a point on the terminal side of an angle θ, find $\cos\theta$.
   a. $-\dfrac{3\sqrt{10}}{10}$
   b. $-\dfrac{\sqrt{10}}{10}$
   c. $\dfrac{\sqrt{10}}{10}$
   d. $\dfrac{3\sqrt{10}}{10}$

4. Find the radian measure of a central angle of a circle of radius 8 that intercepts an arc of length 4.
   a. $\dfrac{1}{2}$
   b. 2
   c. 32
   d. 128

5. Find the area of a sector of a circle of radius 6 centimeters that intercepts an arc of length 3 centimeters.
   a. $\dfrac{9}{2}$ cm²
   b. $\dfrac{27}{2}$ cm²
   c. 6 cm²
   d. 9 cm²

6. If $\cos\theta = \dfrac{3}{4}$ for an acute angle θ, find $\sin\theta$.
   a. $\dfrac{3}{\sqrt{7}}$
   b. $\dfrac{\sqrt{7}}{4}$
   c. $\dfrac{\sqrt{7}}{3}$
   d. $\dfrac{4}{3}$

7. If $\sin\theta < 0$ and $\sec\theta > 0$, find the quadrant in which θ lies.
   a. I
   b. II
   c. III
   d. IV

8. If $\tan\theta = \dfrac{12}{5}$ and θ is in quadrant III, find $\csc\theta$.
   a. $-\dfrac{5}{12}$
   b. $-\dfrac{13}{12}$
   c. $\dfrac{5}{13}$
   d. $\dfrac{5}{12}$

9. What is the reference angle for 640°?
   a. 10°
   b. 60°
   c. 80°
   d. 280°

10. In which quadrant is it true that both $\cos\theta > 0$ and $\csc\theta > 0$?
    a. I
    b. II
    c. III
    d. IV

11. A point on the rim of a wheel with a radius of 50 centimeters has an angular speed of 0.8 radian per second. What is the linear speed of the point?
    a. 4 centimeters per second
    b. 6.25 centimeters per second
    c. 40 centimeters per second
    d. 62.5 centimeters per second

12. If $y = a\cos bx$ has an amplitude of $\dfrac{1}{2}$ and a period of 4π, what are the values of a and b?
    a. $a = \dfrac{1}{2}, b = \dfrac{1}{2}$
    b. $a = 2, b = \dfrac{1}{2}$
    c. $a = \dfrac{1}{2}, b = 2$
    d. $a = 2, b = 2$

13. Which of the following statements is *false*?
    a. $\sin(-x) = -\sin x$
    b. $\cos(-x) = -\cos x$
    c. $\tan(-x) = -\tan x$
    d. $\cot(-x) = -\cot x$

14. The equation $y = -5\cos\dfrac{\pi}{4}t$ describes the harmonic of a ball attached to a spring. What is the period of the motion?

    a. $\dfrac{\pi}{4}$ **b.** -5 **c.** 5 **d.** 8

15. Find the amplitude a, period b, and phase shift c for
$$y = -7\cos\left[\dfrac{1}{6}\left(x - \dfrac{\pi}{12}\right)\right].$$

    a. $a = 7, b = \dfrac{1}{6}\pi, c = \dfrac{\pi}{12}$ **b.** $a = 7, b = \dfrac{1}{6}\pi, c = -\dfrac{\pi}{12}$

    c. $a = 7, b = 12\pi, c = -\dfrac{\pi}{12}$ **d.** $a = 7, b = 12\pi, c = \dfrac{\pi}{12}$

16. This is the graph of which function?

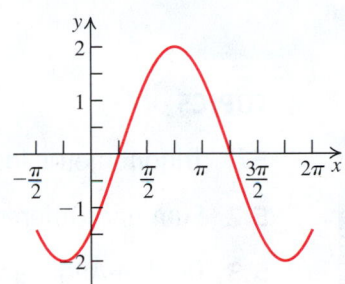

    a. $y = -2\sin\left(x + \dfrac{\pi}{2}\right)$ **b.** $y = 2\sin\left(x + \dfrac{\pi}{2}\right)$

    c. $y = -2\cos\left(x + \dfrac{\pi}{4}\right)$ **d.** $y = 2\cos\left(x + \dfrac{\pi}{4}\right)$

17. Find the exact value of $y = \arccos\left(-\dfrac{1}{2}\right)$.

    a. $\dfrac{\pi}{3}$ **b.** $\dfrac{5\pi}{3}$ **c.** $\dfrac{2\pi}{3}$ **d.** $-\dfrac{\pi}{3}$

18. Find the exact value of $y = \cos\left[\sin^{-1}\left(-\dfrac{4}{5}\right)\right]$.

    a. $\dfrac{3}{5}$ **b.** $-\dfrac{3}{5}$ **c.** $\dfrac{3}{4}$ **d.** $-\dfrac{4}{3}$

19. Find the exact value of $y = \cos^{-1}\left(\cos\dfrac{7\pi}{4}\right)$.

    a. $\dfrac{2\pi}{3}$ **b.** $\dfrac{-2\pi}{3}$ **c.** $\dfrac{7\pi}{4}$ **d.** $\dfrac{\pi}{4}$

20. Find the exact value of $y = \sin\left[\tan^{-1}\left(\dfrac{5}{3}\right)\right]$.

    a. $\dfrac{3}{5}$ **b.** $\dfrac{5}{34}$ **c.** $\dfrac{5\sqrt{34}}{34}$ **d.** $\dfrac{3\sqrt{34}}{34}$

CUMULATIVE REVIEW EXERCISES CHAPTERS 1–4

1. Solve the equation $3x^2 - 30x + 50 = -24$.

2. Find all real number solutions of the equation $(\sqrt{t} + 1)^2 + 2(\sqrt{t} + 1) = 3$.

3. Solve the inequality $\dfrac{4 - x}{2x - 4} > 0$ and write the answer in interval notation.

4. Write the slope–intercept form of the equation of the line containing the points $(3, -1)$ and $(-6, 5)$.

5. Find the domain of the function $f(x) = \ln\left(\dfrac{x}{2 - x}\right)$.

6. Identify the basic function and then use transformations to sketch the graph of the function $f(x) = 2(x - 3)^2 - 5$.

7. Use transformations to sketch the graph of the function $f(x) = 4e^x + 1$.

8. Find and write the domain of $(f \circ g)(x)$ in interval notation if
$f(x) = \ln x$ and $g(x) = -\dfrac{1}{x}$.

9. Determine whether each function has an inverse function. If so, find an equation for $f^{-1}(x)$.
    a. $f(x) = 3x + 7$ **b.** $f(x) = |x|$

10. Use synthetic division to find the quotient and remainder.
$$\dfrac{2x^4 + 4x^3 + x^2 + x - 2}{x + 2}$$

11. Find the horizontal and vertical asymptotes of the graph of
$f(x) = \dfrac{10 - 3x^2}{x^2 + 4x - 5}$.

12. Use rules of logarithms to rewrite each expression in the expanded form.
    a. $\log\sqrt[3]{\dfrac{x^2 y}{z}}$ **b.** $\ln\left(\dfrac{7x^4}{5\sqrt{y}}\right)$

13. Sketch the graph of $f(x) = \begin{cases} -\ln x & \text{if } x > 0 \\ \ln|x| & \text{if } x < 0 \end{cases}$

14. Solve the equation $\log(x^2 + 9x) = 1$.

15. Find all possible rational zeros of the function $f(x) = 2x^4 + 5x^3 + x^2 + 10x - 6$.

16. Use the given trigonometric function value of θ to find the five other trigonometric function values of an acute angle θ. Rationalize the denominators where necessary.
    a. $\cos\theta = \dfrac{3}{7}$ **b.** $\sin\theta = \dfrac{2}{11}$

17. Find the exact values of the remaining trigonometric functions of θ from the given information.
    a. $\cos\theta = -\dfrac{4}{5}, \theta$ in quadrant II
    b. $\cot\theta = \dfrac{5}{12}, \theta$ in quadrant III

18. Sketch the graph of each equation over the interval $[-2\pi, 2\pi]$.
    a. $y = -\dfrac{1}{2}\cos(x + \pi)$
    b. $y = -5\sin\left(x + \dfrac{5\pi}{3}\right)$

19. Find the exact value of y or state that y is undefined.
    a. $y = \cos\left[\sin^{-1}\left(-\dfrac{1}{5}\right)\right]$
    b. $y = \sin\left(\tan^{-1}\dfrac{7}{2}\right)$

20. Find the exact value of $y = \cos^{-1}\left(\cos\dfrac{5\pi}{3}\right)$.

CHAPTER | 5

Analytic Trigonometry

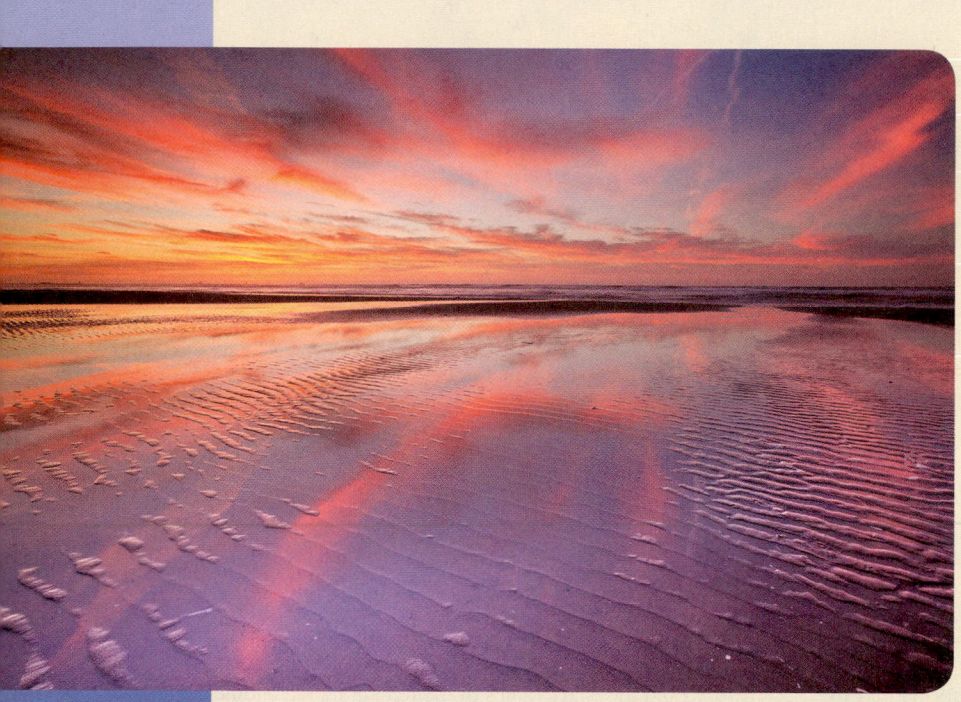

TOPICS

5.1 Trigonometric Identities

5.2 Sum and Difference Formulas

5.3 Double-Angle and Half-Angle Formulas

5.4 Product-to-Sum and Sum-to-Product Formulas

5.5 Trigonometric Equations

Such varied phenomena as sonic booms from aircraft, musical tones, hours of daylight in a given location, current in an electric circuit, tides, and rainfall can be described using trigonometric functions. Solutions of trigonometric equations often provide answers to questions about the conditions under which specific characteristics of these phenomena occur. In this chapter, we study techniques for solving trigonometric equations.

SECTION 5.1

Trigonometric Identities

BEFORE STARTING THIS SECTION, REVIEW

1 Signs of the trigonometric functions (Section 4.3, page 368)

2 LCD and adding fractions (Appendix A.3, page 935)

3 Factoring and special products (Appendix A.2, page 930)

4 Reciprocal, quotient, and Pythagorean identities (Section 4.3, page 373)

OBJECTIVES

1 Use algebra techniques to simplify trigonometric expressions.

2 Prove that a given equation is not an identity.

3 Verify a trigonometric identity.

Hipparchus of Rhodes (190–120 B.C.)
Hipparchus was born in Nicaea (now called Iznik) in Bithynia (northwest Turkey) but spent much of his life in Rhodes (Greece). Many historians consider him the founder of trigonometry. He introduced trigonometric functions in the form of a chord table—the ancestor of the sine table found in ancient Indian astronomical works. With his chord table, Hipparchus could solve the height–distance problems of plane trigonometry. In his time, there was no Greek term for trigonometry because it was not counted as a branch of mathematics. Rather, trigonometry was just an aid to solve problems in astronomy.

◆ Visualizing Trigonometric Functions

Trigonometric functions are commonly defined as the ratios of two sides corresponding to an angle in a right triangle. The Greek mathematician Hipparchus was the first person to relate trigonometric functions to a circle. His work inspired sixteenth-century mathematician Francois Viete (1540–1603) to visualize the trigonometric functions of an acute angle θ as the length of various segments associated with a unit circle. See Figure 5.1. The figure shows several similar triangles. For example, triangles PCT and OCP are similar, so

$$\frac{CT}{PC} = \frac{PC}{OC} \qquad \text{Sides are proportional.}$$

From Figure 5.1, we use the definitions of trigonometric functions to get

$$PC = \sin\theta, \quad OC = \cos\theta, \quad \text{and} \quad CT = OT - OC = \sec\theta - \cos\theta.$$

Substituting these values into the equation $\dfrac{CT}{PC} = \dfrac{PC}{OC}$, we obtain

$$\frac{\sec\theta - \cos\theta}{\sin\theta} = \frac{\sin\theta}{\cos\theta}.$$

In Example 4, we verify that this equation is an identity. Recall that an *identity* is an equation that is true for all numbers for which both sides are defined.

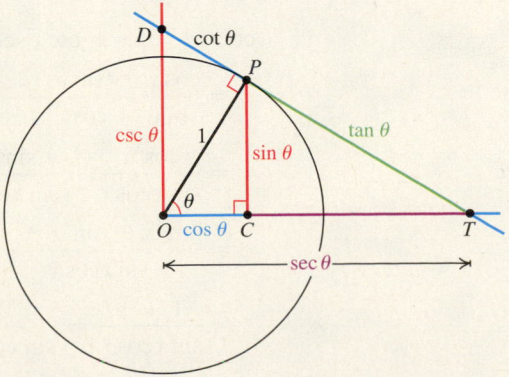

Figure 5.1 The unit circle and trigonometric functions

Fundamental Trigonometric Identities

In Chapter 4, we used the definitions of the trigonometric functions to establish reciprocal, quotient, Pythagorean, and even-odd identities. These are called **fundamental trigonometric identities**.

FUNDAMENTAL TRIGONOMETRIC IDENTITIES

1. Reciprocal Identities

$$\csc x = \frac{1}{\sin x} \qquad \sec x = \frac{1}{\cos x} \qquad \cot x = \frac{1}{\tan x}$$

$$\sin x = \frac{1}{\csc x} \qquad \cos x = \frac{1}{\sec x} \qquad \tan x = \frac{1}{\cot x}$$

2. Quotient Identities

$$\tan x = \frac{\sin x}{\cos x} \qquad \cot x = \frac{\cos x}{\sin x}$$

3. Pythagorean Identities

$$\sin^2 x + \cos^2 x = 1 \qquad 1 + \tan^2 x = \sec^2 x \qquad 1 + \cot^2 x = \csc^2 x$$

4. Even–Odd Identities

$$\sin(-x) = -\sin x \qquad \cos(-x) = \cos x \qquad \tan(-x) = -\tan x$$

$$\csc(-x) = -\csc x \qquad \sec(-x) = \sec x \qquad \cot(-x) = -\cot x$$

---

1 Use algebra techniques to simplify trigonometric expressions.

Simplifying a Trigonometric Expression

Recall that when simplifying algebraic expressions, we can use properties of real numbers, factoring techniques, and special product formulas. We can also rationalize the denominator or find the least common denominator. To simplify trigonometric expressions, we use these same techniques together with the fundamental trigonometric identities.

EXAMPLE 1 **Expressing All Trigonometric Functions in Terms of Sines and Cosines to Simplify an Expression**

Rewrite the given expression in terms of sines and cosines and then simplify the resulting expression.

$$\cot x + \tan x + \csc x \sec x$$

Solution

$$\cot x + \tan x + \csc x \sec x$$

$$= \frac{\cos x}{\sin x} + \frac{\sin x}{\cos x} + \frac{1}{\sin x} \cdot \frac{1}{\cos x} \qquad \text{Quotient and reciprocal identities}$$

$$= \frac{\cos^2 x}{\sin x \cos x} + \frac{\sin^2 x}{\sin x \cos x} + \frac{1}{\sin x \cos x} \qquad \begin{array}{l} \text{Write each fraction with} \\ \text{common denominator } \sin x \cos x. \end{array}$$

$$= \frac{\cos^2 x + \sin^2 x + 1}{\sin x \cos x} \qquad \text{Add numerators.}$$

$$= \frac{1 + 1}{\sin x \cos x} = \frac{2}{\sin x \cos x} \qquad \sin^2 x + \cos^2 x = 1$$

Practice Problem 1 Rewrite the expression in terms of sines and cosines and then simplify the resulting expression.

$$\frac{\tan x}{\sec x + 1} + \frac{\tan x}{\sec x - 1}$$

2 Prove that a given equation is not an identity.

Trigonometric Equations and Identities

To prove or verify that an equation is a trigonometric identity, you must *prove* (*verify*) that both sides of the equation are equal for all values of the variable for which both sides are defined.

Consider the following equation:

$$\cos x = \sqrt{1 - \sin^2 x} \tag{1}$$

Equation (1) is true for all values of x in the interval $\left[-\frac{\pi}{2}, \frac{\pi}{2} \right]$. See Figure 5.2(a).

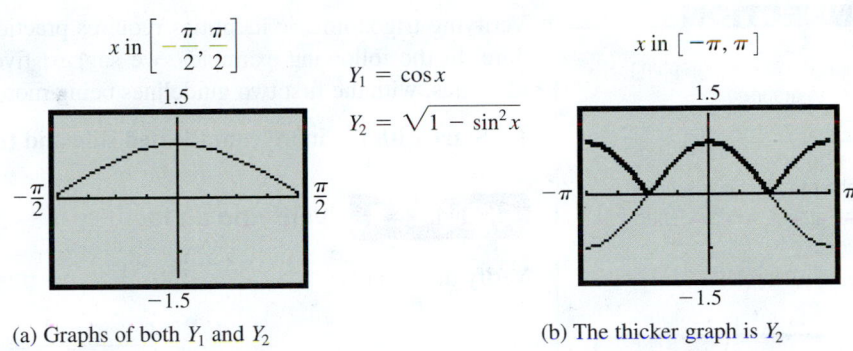

(a) Graphs of both Y_1 and Y_2 (b) The thicker graph is Y_2

Figure 5.2

Since the domain of the sine and cosine functions is the interval $(-\infty, \infty)$, both sides of equation (1) are defined for all real numbers. But for any value of x in the interval $\left(\frac{\pi}{2}, \pi \right]$, the left side of equation (1) has a negative value (because $\cos x$ is negative in quadrant II), while the right side of equation (1) has a positive value. Therefore, equation (1) is *not* an identity. Figure 5.2(b) illustrates that a graphing calculator can help verify that a given equation is *not* an identity. Figure 5.2(a), however, shows that a graphing calculator cannot prove that a given equation *is* an identity because you might happen to use a viewing window where the graphs coincide.

EXAMPLE 2   **Proving That an Equation Is Not an Identity**

Prove that the following equation is not an identity.

$$(\sin x - \cos x)^2 = \sin^2 x - \cos^2 x$$

SIDE
NOTE

A **counter example** is an example that disproves a statement.

Solution

To prove that the equation is not an identity we look for a counter example. That is, we find at least one value of x that results in both sides being defined but not equal.

Let $x = 0$. We know that $\sin 0 = 0$ and $\cos 0 = 1$.

The left side $= (\sin x - \cos x)^2 = (0 - 1)^2 = 1$ Replace x with 0.

The right side $= \sin^2 x - \cos^2 x = (0)^2 - (1)^2 = -1$ Replace x with 0.

For $x = 0$, the equation's two sides are not equal; so it is not an identity.

Practice Problem 2 Prove that the equation $\cos x = 1 - \sin x$ is not an identity.

3 Verify a trigonometric identity.

Process of Verifying Trigonometric Identities

Verifying a trigonometric identity differs from solving an equation. In verifying an identity, we are given an equation and want to show that it is true for *all* values of the variable for which both sides of the equation are defined. We will use the following method.

VERIFYING TRIGONOMETRIC IDENTITIES

To verify that an equation is an identity, transform one side of the equation into the other side by a sequence of steps, each of which produces an identity. The steps involved can be algebraic manipulations or can use known identities. Note that in verifying an identity, we *do not* just perform the same operation on both sides of the equation; see Exercise 113.

TECHNOLOGY
CONNECTION

In Example 3, we use a graphing calculator to graph

$Y_1 = \dfrac{\csc^2 x - 1}{\cot x}$ and $Y_2 = \cot x$.

Although not a definitive method of proof, the graphs appear identical; so the equation in Example 3 *appears* to be an identity.

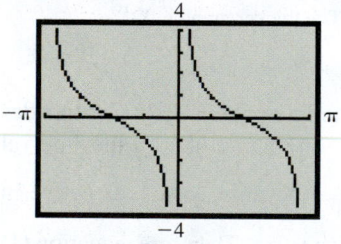

Graph of Y_1

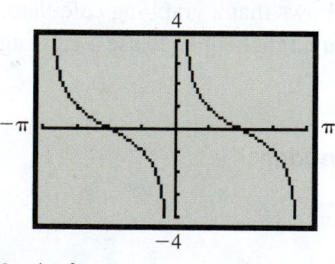

Graph of Y_2

Methods of Verifying Trigonometric Identities

Verifying trigonometric identities requires practice and experience; there is no set procedure. In the following examples, we suggest five guidelines for verifying trigonometric identities, with the first two guidelines being more widely used.

1. Start with the more complicated side and transform it to the simpler side.

EXAMPLE 3   Verifying an Identity

Verify the identity:

$$\frac{\csc^2 x - 1}{\cot x} = \cot x$$

Solution

We start with the left side of the equation because it appears more complicated than the right side.

$$\frac{\csc^2 x - 1}{\cot x} = \frac{\cot^2 x}{\cot x} \qquad \csc^2 x = 1 + \cot^2 x, \text{ so } \csc^2 x - 1 = \cot^2 x$$

$$= \cot x \qquad \text{Remove the common factor, } \cot x.$$

We have shown that the left side of the equation is equal to the right side, which verifies the identity.

Practice Problem 3 Verify the identity:

$$\frac{1 - \sin^2 x}{\cos x} = \cos x$$

2. Stay focused on the final expression.

While working on one side of the equation, stay focused on your goal of converting it to the form on the other side. This often helps you decide what your next step should be.

EXAMPLE 4   Verifying an Identity

Verify the following identity proposed in the introduction to this section:

$$\frac{\sec \theta - \cos \theta}{\sin \theta} = \frac{\sin \theta}{\cos \theta}$$

In Example 4, we use the TABLE feature of a graphing calculator in **Radian** mode to create a table of values for $Y_1 = \dfrac{\sec x - \cos x}{\sin x}$ and $Y_2 = \dfrac{\sin x}{\cos x}$ for different values of x.

X	Y1	Y2
-3	-.288	-.288
-2.5	-1.865	-1.865
-2	-10.5	-10.5
-1.5	-398.7	-398.7
-1	-5.765	-5.765
-.5	-1.245	-1.245
0		ERROR

X=0

X	Y1	Y2
0	0	ERROR
.5	1.245	1.245
1	5.7649	5.7649
1.5	398.7	398.7
2	10.501	10.501
2.5	1.8649	1.8649
3	.28797	.28797

X=0

Note that Y_2 equals 0 when $x = 0$ but that Y_1 is undefined when $x = 0$. While not a definitive method of proof, the table shows that the equation in Example 4 *appears* to be an identity because the table values are identical for values of x for which both Y_1 and Y_2 are defined.

Solution

We start with the more complicated left side; note that the other side involves only $\sin\theta$ and $\cos\theta$.

$$\frac{\sec\theta - \cos\theta}{\sin\theta} = \frac{\frac{1}{\cos\theta} - \cos\theta}{\sin\theta} \qquad \sec\theta = \frac{1}{\cos\theta}$$

$$= \frac{\left(\frac{1}{\cos\theta} - \cos\theta\right)\cos\theta}{\sin\theta\cos\theta} \qquad \text{Multiply numerator and denominator by } \cos\theta.$$

$$= \frac{\frac{1}{\cos\theta}\cdot\cos\theta - \cos^2\theta}{\sin\theta\cos\theta} \qquad \text{Distributive property}$$

$$= \frac{1 - \cos^2\theta}{\sin\theta\cos\theta} \qquad \text{Simplify.}$$

$$= \frac{\sin^2\theta}{\sin\theta\cos\theta} \qquad \sin^2\theta + \cos^2\theta = 1,\ \text{so, } \sin^2\theta = 1 - \cos^2\theta$$

$$= \frac{\sin\theta}{\cos\theta} \qquad \text{Remove the common factor } \sin\theta.$$

The left side is identical to the right side; the given equation is an identity.

Practice Problem 4 In Figure 5.1 on page 439, triangles OPD and TCP are similar. So, $\dfrac{OD}{OP} = \dfrac{PT}{CT}$ leads to

$$\frac{\csc\theta}{1} = \frac{\tan\theta}{\sec\theta - \cos\theta}.$$

Verify that this equation is an identity.

3. Convert to sines and cosines.

It may be helpful to rewrite all trigonometric functions in the equation in terms of sines and cosines and then simplify.

EXAMPLE 5 Verifying by Rewriting with Sines and Cosines

Verify the identity $\cot^4 x + \cot^2 x = \cot^2 x \csc^2 x$.

Solution

We start with the more complicated left side.

$$\cot^4 x + \cot^2 x = \frac{\cos^4 x}{\sin^4 x} + \frac{\cos^2 x}{\sin^2 x} \qquad \cot x = \frac{\cos x}{\sin x}$$

$$= \frac{\cos^4 x}{\sin^4 x} + \frac{\cos^2 x}{\sin^2 x}\cdot\frac{\sin^2 x}{\sin^2 x} \qquad \text{The LCD is } \sin^4 x.$$

$$= \frac{\cos^4 x + \cos^2 x\sin^2 x}{\sin^4 x} \qquad \text{Add rational expressions.}$$

$$= \frac{\cos^2 x(\cos^2 x + \sin^2 x)}{\sin^4 x} \qquad \text{Factor out } \cos^2 x.$$

$$= \frac{\cos^2 x(1)}{\sin^4 x} \qquad \cos^2 x + \sin^2 x = 1$$

$$= \frac{\cos^2 x}{\sin^2 x}\cdot\frac{1}{\sin^2 x} \qquad \text{Factor to obtain the form on the right side.}$$

$$= \cot^2 x \csc^2 x \qquad \cot x = \frac{\cos x}{\sin x}, \csc x = \frac{1}{\sin x}$$

Because the left side is identical to the right side, the given equation is an identity.

Practice Problem 5 Verify the identity $\tan^4 x + \tan^2 x = \tan^2 x \sec^2 x$.

You could also verify the identity in Example 5 as follows:

$$\cot^4 x + \cot^2 x = \cot^2 x(\cot^2 x + 1) \qquad \text{Distributive property}$$
$$= \cot^2 x \csc^2 x \qquad \text{Pythagorean identity}$$

This approach shows that rewriting the expression using only sines and cosines is not always the quickest way to verify an identity. It is a useful approach when you are stuck, however.

4. Work on both sides.

Sometimes it is helpful to work separately on both sides of the equation. To verify the identity $P(x) = Q(x)$, for example, we transform the left side $P(x)$ into $R(x)$ by using algebraic manipulations and known identities. Then $P(x) = R(x)$ is an identity. Next, we transform the right side, $Q(x)$, into $R(x)$ so that the equation $Q(x) = R(x)$ is an identity. It then follows that $P(x) = Q(x)$ is an identity. See the next example.

EXAMPLE 6 **Verifying an Identity by Transforming Both Sides Separately**

Verify the identity:

$$\frac{1}{1 - \sin x} - \frac{1}{1 + \sin x} = \frac{\tan^2 x + \sec^2 x + 1}{\csc x}$$

Solution

We start with the left side of the equation.

$$\frac{1}{1 - \sin x} - \frac{1}{1 + \sin x} \qquad \text{Begin with the left side.}$$

$$= \frac{1}{1 - \sin x} \cdot \frac{1 + \sin x}{1 + \sin x} - \frac{1}{1 + \sin x} \cdot \frac{1 - \sin x}{1 - \sin x} \qquad \begin{array}{l}\text{Rewrite using the LCD,}\\ (1 - \sin x)(1 + \sin x).\end{array}$$

$$= \frac{(1 + \sin x) - (1 - \sin x)}{(1 - \sin x)(1 + \sin x)} \qquad \text{Subtract fractions.}$$

$$= \frac{1 + \sin x - 1 + \sin x}{1 - \sin^2 x} \qquad \begin{array}{l}\text{Rewrite the numerator;}\\ \text{multiply in the denominator.}\end{array}$$

$$= \frac{2 \sin x}{\cos^2 x} \qquad \text{Simplify; } 1 - \sin^2 x = \cos^2 x$$

We now try to convert the right side to the form $\dfrac{2 \sin x}{\cos^2 x}$.

$$\frac{\tan^2 x + \sec^2 x + 1}{\csc x}$$

$$= \frac{(\tan^2 x + 1) + \sec^2 x}{\csc x} \qquad \text{Regroup terms.}$$

$$= \frac{\sec^2 x + \sec^2 x}{\csc x} \qquad \tan^2 x + 1 = \sec^2 x$$

$$= \frac{2 \sec^2 x}{\csc x} \qquad \text{Simplify.}$$

$$= 2\left(\frac{1}{\csc x}\right) \sec^2 x \qquad \text{Rewrite.}$$

$$= 2 \sin x \left(\frac{1}{\cos^2 x}\right) \qquad \text{Reciprocal identities}$$

$$= \frac{2 \sin x}{\cos^2 x} \qquad \text{Multiply.}$$

From the original equation, we see that

Left-Hand Side		Third Expression		Right-Hand Side
$\dfrac{1}{1 - \sin x} - \dfrac{1}{1 + \sin x}$	$=$	$\dfrac{2 \sin x}{\cos^2 x}$	$=$	$\dfrac{\tan^2 x + \sec^2 x + 1}{\csc x}$

Because both sides of the original equation are equal to $\dfrac{2 \sin x}{\cos^2 x}$, the identity is verified.

Practice Problem 6 Verify the identity:

$$\tan \theta + \sec \theta = \frac{\csc \theta + 1}{\cot \theta}$$

5. Use conjugates.

It may be helpful to multiply both the numerator and the denominator of a fraction by the same factor. Recall that the sum $a + b$ and the difference $a - b$ are the *conjugates* of each other and that $(a + b)(a - b) = a^2 - b^2$.

EXAMPLE 7 **Verifying an Identity by Using a Conjugate**

Verify the identity:

$$\frac{\cos x}{1 + \sin x} = \frac{1 - \sin x}{\cos x}$$

Solution

We start with the left side of the equation.

$$\frac{\cos x}{1 + \sin x} = \frac{\cos x(1 - \sin x)}{(1 + \sin x)(1 - \sin x)} \qquad \text{Multiply the numerator and the denominator by } 1 - \sin x, \text{ the conjugate of } 1 + \sin x.$$

$$= \frac{\cos x(1 - \sin x)}{1 - \sin^2 x} \qquad (a + b)(a - b) = a^2 - b^2$$

$$= \frac{\cos x(1 - \sin x)}{\cos^2 x} \qquad 1 - \sin^2 x = \cos^2 x$$

$$= \frac{1 - \sin x}{\cos x} \qquad \text{Remove the common factor } \cos x.$$

Because the left side is identical to the right side, the given equation is an identity.

Practice Problem 7 Verify the identity $\dfrac{\sin \theta}{1 + \cos \theta} = \dfrac{1 - \cos \theta}{\sin \theta}$.

In calculus, it is sometimes helpful to convert algebraic expressions to trigonometric ones. This technique is called *trigonometric substitution* and is illustrated in the next example.

EXAMPLE 8 **Trigonometric Substitution**

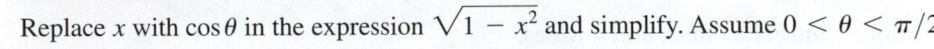 Replace x with $\cos \theta$ in the expression $\sqrt{1 - x^2}$ and simplify. Assume $0 < \theta < \pi/2$.

Solution

$$\sqrt{1 - x^2} = \sqrt{1 - \cos^2 \theta} \qquad \text{Replace } x \text{ with } \cos \theta.$$

$$= \sqrt{\sin^2 \theta} \qquad \text{Pythagorean identity}$$

$$= \sin \theta \qquad \sin \theta > 0 \text{ since } 0 < \theta < \frac{\pi}{2}$$

Practice Problem 8 Replace x with $3 \sin \theta$ in the expression $\sqrt{9 - x^2}$ and simplify. Assume $0 \le \theta \le \dfrac{\pi}{2}$.

In Example 9, we convert trigonometric functions of the forms $\sin^m x \cos^n x$ and $\tan^m x \sec^n x$ to forms that are useful in calculus.

EXAMPLE 9 Useful Identities for Calculus

 Verify each identity:

a. $\sin^2 x \cos^5 x = (\sin^2 x - \sin^4 x + \sin^6 x)\cos x$

b. $\tan^4 x \sec^4 x = (\tan^4 x + \tan^6 x)\sec^2 x$

Solution

a.
$$\sin^2 x \cos^5 x = \sin^2 x \cos^4 x \cos x \qquad \cos^5 x = \cos^4 x \cos x$$
$$= \sin^2 x (\cos^2 x)^2 \cos x \qquad a^4 = (a^2)^2$$
$$= \sin^2 x (1 - \sin^2 x)^2 \cos x \qquad \text{Pythagorean identity}$$
$$= \sin^2 x (1 - 2\sin^2 x + \sin^4 x)\cos x \qquad (a-b)^2 = a^2 - 2ab + b^2$$
$$= (\sin^2 x - 2\sin^4 x + \sin^6 x)\cos x \qquad \text{Distributive property}$$

b.
$$\tan^4 x \sec^4 x = \tan^4 x \sec^2 x \sec^2 x \qquad a^4 = a^2 \cdot a^2$$
$$= \tan^4 x (1 + \tan^2 x)\sec^2 x \qquad \text{Pythagorean identity}$$
$$= (\tan^4 x + \tan^6 x)\sec^2 x \qquad \text{Distributive property}$$

Practice Problem 9 Verify each identity:

a. $\sin^3 x \cos^2 x = (\cos^2 x - \cos^4 x)\sin x$

b. $\tan^4 x = \tan^2 x \sec^2 x - \sec^2 x + 1$

SUMMARY OF **MAIN FACTS**

Guidelines for Verifying Trigonometric Identities

Algebra Operations

Review the procedure for combining fractions by finding the least common denominator.

Fundamental Trigonometric Identities

Review the fundamental trigonometric identities summarized on page 438. Look for an opportunity to apply the fundamental trigonometric identities when working on either side of the identity to be verified. Become thoroughly familiar with alternative forms of fundamental identities. For example, $\sin^2 x = 1 - \cos^2 x$ and $\sec^2 x - \tan^2 x = 1$ are alternative forms of the fundamental identities $\sin^2 x + \cos^2 x = 1$ and $\sec^2 x = 1 + \tan^2 x$, respectively.

1. **Start with the more complicated side.**
   It is generally helpful to start with the more complicated side of an identity and simplify it until it becomes identical to the other side. See Example 3.

2. **Stay focused on the final expression.**
   While working on one side of the identity, stay focused on your goal of converting it to the form on the other side. See Example 4.

   The following three techniques are *sometimes* helpful.

3. **Option: Convert to sines and cosines.**
   Rewrite one side of the identity in terms of sines and cosines. See Example 5.

4. **Option: Work on both sides.**
   Transform each side separately to the same equivalent expression. See Example 6.

5. **Option: Use conjugates.**
   In expressions containing $1 + \sin x$, $1 - \sin x$, $1 + \cos x$, $1 - \cos x$, $\sec x + \tan x$, and so on, multiply the numerator and the denominator by its conjugate and then use the appropriate Pythagorean identity. See Example 7.

Answers to Practice Problems

1. $\dfrac{2}{\sin x}$ **2.** At $x = \dfrac{3\pi}{2}, 0 \ne 2.$ **8.** $3\cos\theta$

SECTION 5.1   Exercises

Concepts and Vocabulary

1. An equation that is true for all values of the variable in its domain is called a(n) _____.

2. To show that an equation is not an identity, we must find at least _____ value(s) of the variable that results in both sides being defined but _____.

3. $\sin^2 x +$ _____ $= 1$;
$1 +$ _____ $= \sec^2 x$; $\csc^2 x - \cot^2 x =$ _____.

4. $\sin x + \sin(-x) =$ _____

5. **True or False.** $\tan^2 x \cot x = \tan x$ is not an identity because $\tan^2 x \cot x$ is not defined for $x = 0$ but $\tan x$ is defined for $x = 0$.

6. **True or False.** The value of $x = \dfrac{3\pi}{2}$ can be used to show that $\sin x = \sqrt{1 - \cos^2 x}$ is not an identity.

7. **True or False.** If $\cos\theta < 0$ and $\pi < \theta < \dfrac{3\pi}{2}$, then $\tan\theta < 0$.

8. **True or False.** $\cos(-x) + \cos x = 0$

Building Skills

In Exercises 9–18, use the fundamental identities and appropriate algebraic operations to simplify each expression.

9. $(1 + \tan x)(1 - \tan x) + \sec^2 x$

10. $(\sec x - 1)(\sec x + 1) - \tan^2 x$

11. $(\sec x + \tan x)(\sec x - \tan x)$

12. $\dfrac{\sec^2 x - 4}{\sec x - 2}$

13. $\csc^4 x - \cot^4 x$

14. $\sin x \cos x (\tan x + \cot x)$

15. $\dfrac{\sec x \csc x (\sin x + \cos x)}{\sec x + \csc x}$

16. $\dfrac{1}{\csc x + 1} - \dfrac{1}{\csc x - 1}$

17. $\dfrac{\tan^2 x - 2\tan x - 3}{\tan x + 1}$

18. $\dfrac{\tan^2 x + \sec x - 1}{\sec x - 1}$

In Exercises 19–24, prove that the given equation is not an identity by finding a value of x for which the two sides have different values. The answers may vary.

19. $\sin x = 1 - \cos x$

20. $\tan x = \sec x - 1$

21. $\cos x = \sqrt{1 - \sin^2 x}$

22. $\sec x = \sqrt{1 + \tan^2 x}$

23. $\sin^2 x = (1 - \cos x)^2$

24. $\cot^2 x = (\csc x + 1)^2$

In Exercises 25–68, verify each identity.

25. $\sin x \tan x + \cos x = \sec x$

26. $\cos x \cot x + \sin x = \csc x$

27. $\dfrac{1 - 4\cos^2 x}{1 - 2\cos x} = 1 + 2\cos x$

28. $\dfrac{9 - 16\sin^2 x}{3 + 4\sin x} = 3 - 4\sin x$

29. $(\cos x - \sin x)(\cos x + \sin x) = 1 - 2\sin^2 x$

30. $(\sin x - \cos x)(\sin x + \cos x) = 1 - 2\cos^2 x$

31. $\sin^2 x \cot^2 x + \sin^2 x = 1$

32. $\cos^2 x \tan^2 x + \cos^2 x = 1$

33. $\tan^2 x - \sin^2 x = \sin^4 x \sec^2 x$

34. $\cot^2 x - \cos^2 x = \cos^4 x \csc^2 x$

35. $\sin^3 x - \cos^3 x = (\sin x - \cos x)(1 + \sin x \cos x)$

36. $\sin^3 x + \cos^3 x = (\sin x + \cos x)(1 - \sin x \cos x)$

37. $\cos^4 x - \sin^4 x = 1 - 2\sin^2 x$

38. $\cos^4 x - \sin^4 x = 2\cos^2 x - 1$

39. $\dfrac{1}{1 - \sin x} + \dfrac{1}{1 + \sin x} = 2\sec^2 x$

40. $\dfrac{1}{1 - \cos x} + \dfrac{1}{1 + \cos x} = 2\csc^2 x$

41. $\dfrac{\sin x}{1 - \sin x} - \dfrac{\sin x}{1 + \sin x} = 2\tan^2 x$

42. $\dfrac{\cos x}{1 - \cos x} - \dfrac{\cos x}{1 + \cos x} = 2\cot^2 x$

43. $\dfrac{1}{\sec x - \tan x} + \dfrac{1}{\sec x + \tan x} = \dfrac{2}{\cos x}$

44. $\dfrac{1}{\csc x + \cot x} + \dfrac{1}{\csc x - \cot x} = \dfrac{2}{\sin x}$

45. $(\sin x + \cos x)^2 = 1 + 2\sin x \cos x$

46. $(\sin x - \cos x)^2 = 1 - 2\sin x \cos x$

47. $(1 + \tan x)^2 = \sec^2 x + 2\tan x$

48. $(1 - \cot x)^2 = \csc^2 x - 2\cot x$

49. $\sec^2 x + \csc^2 x = \sec^2 x \csc^2 x$

50. $\cot^2 x + \tan^2 x = \sec^2 x \csc^2 x - 2$

51. $\dfrac{\sin x}{1 + \cos x} = \csc x - \cot x$

52. $\dfrac{\cos x}{1 + \sin x} = \sec x - \tan x$

53. $\dfrac{\sin x}{1 - \cos x} = \csc x + \cot x$

54. $\dfrac{\cos x}{1 - \sin x} = \sec x + \tan x$

55. $\dfrac{\tan x \sin x}{\tan x + \sin x} = \dfrac{\tan x - \sin x}{\tan x \sin x}$

56. $\dfrac{\cot x \cos x}{\cot x + \cos x} = \dfrac{\cot x - \cos x}{\cot x \cos x}$

57. $(\tan x + \cot x)^2 = \sec^2 x + \csc^2 x$

58. $(\csc x - \sin x)^2 = \cot^2 x - \cos^2 x$

59. $(1 + \cot^2 x)(1 + \tan^2 x) = \dfrac{1}{\sin^2 x \cos^2 x}$

60. $\left(\tan x + \dfrac{1}{\cot x}\right)\left(\cot x + \dfrac{1}{\tan x}\right) = 4$

61. $\dfrac{\sin^2 x - \cos^2 x}{\sec^2 x - \csc^2 x} = \sin^2 x \cos^2 x$

62. $\dfrac{\sin^3 x + \cos^3 x}{\sin^4 x - \cos^4 x} = \dfrac{1 - \sin x \cos x}{\sin x - \cos x}$

63. $\dfrac{\tan x}{1 + \sec x} + \dfrac{1 + \sec x}{\tan x} = 2 \csc x$

64. $\dfrac{\cot x}{1 + \csc x} + \dfrac{1 + \csc x}{\cot x} = 2 \sec x$

65. $\dfrac{\sin x + \tan x}{\cos x + 1} = \tan x$

66. $\dfrac{\sin x}{1 + \tan x} = \dfrac{\cos x}{1 + \cot x}$

67. $\dfrac{\sec x - 1}{\tan x} = \dfrac{\tan x}{\sec x + 1}$

68. $\dfrac{\csc x + 1}{\cot x} = \dfrac{\cot x}{\csc x - 1}$

In Exercises 69–78, make the indicated trigonometric substitution in the given algebraic expression and simplify.

Assume $0 < \theta < \dfrac{\pi}{2}$.

69. $\sqrt{1 + x^2}, x = \tan \theta$

70. $\sqrt{4 - x^2}, x = 2 \cos \theta$

71. $\sqrt{x^2 - 1}, x = \sec \theta$

72. $\sqrt{x^2 - 4}, x = 2 \sec \theta$

73. $\dfrac{x}{\sqrt{1 - x^2}}, x = \cos \theta$

74. $\dfrac{x}{\sqrt{9 - x^2}}, x = 3 \sin \theta$

75. $\dfrac{x}{\sqrt{x^2 - 1}}, x = \sec \theta$

76. $\dfrac{x}{\sqrt{x^2 - 16}}, x = 4 \csc \theta$

77. $\dfrac{x}{\sqrt{x^2 + 3}}, x = \sqrt{3} \tan \theta$

78. $\dfrac{x}{\sqrt{x^2 + 5}}, x = \sqrt{5} \cot \theta$

In Exercises 79–88, verify each identity.

79. $\sin^3 x = (1 - \cos^2 x)\sin x$

80. $\cos^3 x = (1 - \sin^2 x)\cos x$

81. $\cos^5 x = (1 - 2\sin^2 x + \sin^4 x)\cos x$

82. $\sin^5 x = (1 - 2\cos^2 x + \cos^4 x)\sin x$

83. $\cos^2 x \sin^5 x = (\cos^2 x - 2\cos^4 x + \cos^6 x)\sin x$

84. $\sin^5 x \cos^4 x = (\cos^4 x - 2\cos^6 x + \cos^8 x)\sin x$

85. $\cot^4 x = \cot^2 x \csc^2 x - \csc^2 x + 1$

86. $\sec^4 x = \sec^2 x + \tan^2 x \sec^2 x$

87. $\tan^3 x \sec^4 x = \tan^3 x \sec^2 x + \tan^5 x \sec^2 x$

88. $(\sqrt{\tan x} + 1)\sec^4 x = (\tan^{\frac{5}{2}} x + \tan^2 x + \tan^{\frac{1}{2}} x + 1)\sec^2 x$

Applying the Concepts

89. **Length of a ladder.** A ladder x feet long makes an angle θ with the horizontal and reaches a height of 20 feet. Then $x = \dfrac{20}{\sin \theta}$. Use a reciprocal identity to rewrite this formula.

90. **Distance from a building.** From a distance of x feet to a 60-foot-high building, the angle of elevation is θ degrees. Then $x = \dfrac{60}{\tan \theta}$. Use a reciprocal identity to rewrite this formula.

91. **Intersecting lines.** If two nonvertical lines with slopes m_1 and m_2 $(m_1 > m_2)$ intersect, then the acute angle θ between the lines satisfies the equation $m_1 \cos \theta - \sin \theta = m_2 \cos \theta + m_1 m_2 \sin \theta$. Rewrite this equation in terms of a single trigonometric function.

92. **Tower height.** The angles to the top of a tower measured from two locations d units apart are α and β as shown in the figure. The height h of the tower satisfies the equation $h = \dfrac{d \sin \alpha \sin \beta}{\cos \alpha \sin \beta - \sin \alpha \cos \beta}$. Rewrite this equation in terms of cotangents.

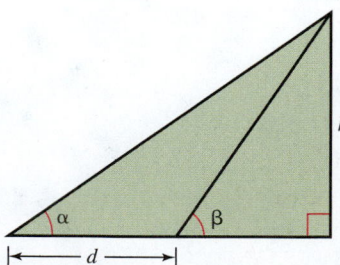

Beyond the Basics

In Exercises 93–98, verify each identity.

93. $\dfrac{1 - \sin x}{1 - \sec x} - \dfrac{1 + \sin x}{1 + \sec x} = 2 \cot x(\cos x - \csc x)$

94. $\dfrac{\sec x + \tan x}{\csc x + \cot x} - \dfrac{\sec x - \tan x}{\csc x - \cot x} = 2(\sec x - \csc x)$

95. $(1 - \tan x)^2 + (1 - \cot x)^2 = (\sec x - \csc x)^2$

96. $\sec^6 x - \tan^6 x = 1 + 3\tan^2 x + 3\tan^4 x$

97. $\dfrac{\cot x + \csc x - 1}{\cot x + \csc x + 1} = \dfrac{1 - \sin x}{\cos x}$

98. $\dfrac{\tan x + \sec x - 1}{\tan x - \sec x + 1} = \dfrac{1 + \sin x}{\cos x}$

In Exercises 99–108, simplify each expression.

99. $(\sin u \cos v + \cos u \sin v)^2 + (\cos u \cos v - \sin u \sin v)^2$

100. $(\sin u \cos v - \cos u \sin v)^2 + (\cos u \cos v + \sin u \sin v)^2$

101. $3(\sin^4 x + \cos^4 x) - 2(\sin^6 x + \cos^6 x)$

102. $\sin^6 x + \cos^6 x + 3\cos^2 x - 3\cos^4 x$

103. $\dfrac{1 - \cos\theta}{1 + \cos\theta} + \dfrac{1 + \cos\theta}{1 - \cos\theta} - 4\cot^2\theta$

104. $\dfrac{\cos^3\theta + \sin^3\theta}{\cos\theta + \sin\theta} + \dfrac{\cos^3\theta - \sin^3\theta}{\cos\theta - \sin\theta}$

105. $\dfrac{\sec^2\theta + 2\tan^2\theta}{1 + 3\tan^2\theta}$

106. $\dfrac{\csc^2\alpha + \sec^2\alpha}{\csc^2\alpha - \sec^2\alpha} - \dfrac{1 + \tan^2\alpha}{1 - \tan^2\alpha}$

107. $\cos^2 x(3 - 4\cos^2 x)^2 + \sin^2 x(3 - 4\sin^2 x)^2$

108. $(\sin\theta + \csc\theta)^2 + (\cos\theta + \sec\theta)^2 - \tan^2\theta - \cot^2\theta$

109. If $x = r\cos u \cos v$, $y = r\cos u \sin v$, and $z = r\sin u$, verify that $x^2 + y^2 + z^2 = r^2$.

110. If $\tan x + \cot x = 2$, show that
a. $\tan^2 x + \cot^2 x = 2$.
b. $\tan^3 x + \cot^3 x = 2$.

111. If $\sec x + \cos x = 2$, show that
a. $\sec^2 x + \cos^2 x = \sec^4 x + \cos^4 x = 2$.
b. $\sec^3 x + \cos^3 x = 2$.

112. If $\sin x + \sin^2 x = 1$, show that $\cos^2 x + \cos^4 x = 1$.

Critical Thinking / Discussion / Writing

113. Consider the equation $x = -x$, which is not an identity. If you square both sides you get $x^2 = (-x)^2$, which is an identity. This shows that performing the same operation on both sides of an equation (a nonidentity) may lead to an identity.

Consider the equation $\sqrt{\sin^2 x} = \sin x$.

a. Find a value of x that shows that this equation is not an identity.

b. Show that squaring both sides of this equation results in an identity.

114. If x and y are real numbers, explain why $\sec\theta$ cannot be equal to $\dfrac{xy}{x^2 + y^2}$.

115. Find values of t for which $\sin\theta = \dfrac{1 + t^2}{1 - t^2}$ is possible.

116. Find values of a and b for which $\cos^2\theta = \dfrac{a^2 + b^2}{2ab}$ is possible.

117. Find values of a and b for which $\csc^2\theta = \dfrac{2ab}{a^2 + b^2}$ is possible.

118. Prove that the following inequalities are true for $0 < \theta < \dfrac{\pi}{2}$.
a. $\sin^2\theta + \csc^2\theta \geq 2$
b. $\cos^2\theta + \sec^2\theta \geq 2$
c. $\sec^2\theta + \csc^2\theta \geq 4$

Getting Ready for the Next Section

In Exercises 119–122, find the distance between P and Q.
119. $P = (3, 4)$, $Q = (-2, 5)$
120. $P = (t, 5)$, $Q = (t - 3, 5 + 4)$
121. $P = (1, 0)$, $Q = (\cos\theta, \sin\theta)$
122. $P = (\cos\alpha, \sin\alpha)$, $Q = (\cos\beta, \sin\beta)$
123. **True or False.** $\cos 30° + \cos 60° = \cos 90°$.
124. **True or False.** $\sin 90° - \sin 30° = \sin(90° - 30°)$.
125. If $\sin\theta = \dfrac{1}{3}$ and $\dfrac{\pi}{2} < \theta < \pi$, find the exact value of $\cos\theta$ and $\tan\theta$.

126. If $\sin\theta = -\dfrac{2}{3}$, find the exact value of $\sin\theta - \sin(-\theta)$.

127. If $\cos\theta = \dfrac{4}{5}$, find the exact value of $\cos\theta - \cos(-\theta)$.

128. If $\sin\alpha = -\dfrac{1}{2}$, $\pi < \alpha < \dfrac{3\pi}{2}$, and $\sin\beta = -\dfrac{2}{3}$, $\dfrac{3\pi}{2} < \beta < 2\pi$, find the exact value of $\sin\alpha\cos\beta + \cos\alpha\sin\beta$.

129. If $u = \sin^{-1}\left(-\dfrac{3}{5}\right)$, find the exact value of $\cos u$.

130. If $u = \cos^{-1}\left(-\dfrac{2}{3}\right)$, find the exact value of $\tan u$.

SECTION | 5.2

Sum and Difference Formulas

BEFORE STARTING THIS SECTION, REVIEW

1. Distance formula (Section 1.1, page 6)

2. Fundamental trigonometric identities (Section 5.1, page 438)

3. Trigonometric functions of common angles (Section 4.2, page 360)

OBJECTIVES

1. Use the sum and difference formulas for cosine.

2. Use cofunction identities.

3. Use the sum and difference formulas for sine.

4. Use the sum and difference formulas for tangent.

5. Use the reduction formula.

◆ Pure Tones in Music

A vibrating part of an instrument transmits its vibrations to the air, and they travel away as sound waves. The number of complete cycles per second traveled by a wave is called its *frequency* and is denoted by *f*. The unit of frequency, cycles per second, is also called hertz (abbreviated Hz) after Gustav Hertz (1887–1975), the discoverer of radio waves. The note A in the octave above middle C has a frequency of 440 Hz, or 440 cycles per second. Therefore, the time between waves is $\frac{1}{440}$ second. We define this time between waves as the period $T: T = \frac{1}{f}$. You hear the sound produced by note A when the vibration from the source travels to your ear and puts pressure on your eardrum.

A *pure tone* is a tone in which the vibration is a simple harmonic of just one frequency. The simple harmonic of a pure tone with frequency *f* can be described by the equation

$$y = a \sin(2\pi f t),$$

where $|a|$ is the amplitude (indicating loudness), *t* is the time in seconds, and *y* is the pressure of a pure tone on an eardrum (in pounds per square foot). In Example 12, we consider the pressure due to two pure tones of the same frequency.

1 Use the sum and difference formulas for cosine.

Sum and Difference Formulas for Cosine

In this section, we develop identities involving the sum and difference of two variables, *u* and *v*. The variables represent any two real numbers, or angles (in radian or degree measure).

> **SUM AND DIFFERENCE FORMULAS FOR COSINE**
>
> $$\cos(u + v) = \cos u \cos v - \sin u \sin v$$
> $$\cos(u - v) = \cos u \cos v + \sin u \sin v$$

To prove the second formula, we assume that $0 < v < u < 2\pi$, although this identity is true for all real numbers u and v. Figure 5.3(a) shows points P and Q on the unit circle on the terminal sides of angles u and v. The definition of the circular functions tells us that $P = (\cos u, \sin u)$ and $Q = (\cos v, \sin v)$. In Figure 5.3(b), we rotated angle $(u - v)$ to standard position and labeled point B on the unit circle and the terminal side of the angle $(u - v)$. So, $B = (\cos(u - v), \sin(u - v))$.

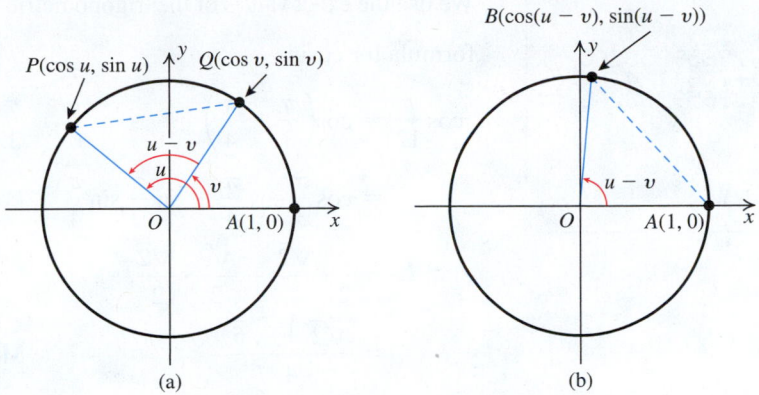

(a) (b)

Figure 5.3

We know that triangle POQ in Figure 5.3(a) is congruent to triangle BOA in Figure 5.3(b) by SAS (side-angle-side) congruence; so

$$d(P, Q) = d(A, B)$$
$$[d(P, Q)]^2 = [d(A, B)]^2 \quad (1) \qquad \text{Square both sides.}$$

Now use the distance formula to simplify the left side of equation (1).

$$
\begin{aligned}
[d(P, Q)]^2 &= (\cos u - \cos v)^2 + (\sin u - \sin v)^2 && \text{Distance formula} \\
&= \cos^2 u - 2\cos u \cos v + \cos^2 v && (a - b)^2 = a^2 - 2ab + b^2 \\
&\quad + \sin^2 u - 2\sin u \sin v + \sin^2 v \\
&= (\cos^2 u + \sin^2 u) + (\cos^2 v + \sin^2 v) && \text{Combine terms.} \\
&\quad - 2(\cos u \cos v + \sin u \sin v) \\
&= 1 + 1 - 2(\cos u \cos v + \sin u \sin v) && \text{Pythagorean identity} \\
[d(P, Q)]^2 &= 2 - 2(\cos u \cos v + \sin u \sin v) && \text{Simplify.}
\end{aligned}
$$

Now simplify the right side of equation (1).

$$
\begin{aligned}
[d(A, B)]^2 &= [\cos(u - v) - 1]^2 + [\sin(u - v) - 0]^2 && \text{Distance formula} \\
&= \cos^2(u - v) - 2\cos(u - v) + 1 + \sin^2(u - v) && \text{Expand binomial.} \\
&= [\cos^2(u - v) + \sin^2(u - v)] + 1 - 2\cos(u - v) && \text{Combine terms.} \\
&= 1 + 1 - 2\cos(u - v) && \text{Pythagorean identity} \\
[d(A, B)]^2 &= 2 - 2\cos(u - v) && \text{Simplify.}
\end{aligned}
$$

Because $[d(A, B)]^2 = [d(P, Q)]^2$,

$$2 - 2\cos(u - v) = 2 - 2(\cos u \cos v + \sin u \sin v) \qquad \text{Substitute in each side.}$$
$$\cos(u - v) = \cos u \cos v + \sin u \sin v \qquad \text{Simplify.}$$

We have proved the formula for the cosine of the difference of two angles or two real numbers.

Although a calculator will give only a decimal approximation for irrational numbers, you can use one to check your work. Remember to use **Radian** mode.

```
cos(π/12)
          .9659258263
(√(2)+√(6))/4
          .9659258263
■
```

EXAMPLE 1 **Using the Difference Formula for Cosine**

Find the exact value of $\cos\dfrac{\pi}{12}$ by using $\dfrac{\pi}{12} = \dfrac{\pi}{3} - \dfrac{\pi}{4}$.

Solution

We use the exact values of the trigonometric functions of $\dfrac{\pi}{3}$ and $\dfrac{\pi}{4}$ as well as the difference formula for cosine.

$$\cos\frac{\pi}{12} = \cos\left(\frac{\pi}{3} - \frac{\pi}{4}\right) \qquad \frac{\pi}{3} - \frac{\pi}{4} = \frac{4\pi}{12} - \frac{3\pi}{12} = \frac{\pi}{12}$$

$$= \cos\frac{\pi}{3}\cos\frac{\pi}{4} + \sin\frac{\pi}{3}\sin\frac{\pi}{4} \qquad \text{Formula for } \cos(u - v),\ u = \frac{\pi}{3} \text{ and } v = \frac{\pi}{4}$$

$$= \frac{1}{2}\cdot\frac{\sqrt{2}}{2} + \frac{\sqrt{3}}{2}\cdot\frac{\sqrt{2}}{2} \qquad \text{Use exact values. (See page 360.)}$$

$$= \frac{\sqrt{2} + \sqrt{6}}{4} \qquad \text{Multiply and add.}$$

Practice Problem 1 Find the exact value of $\cos 15°$ by using $15° = 45° - 30°$. ■

RECALL

The cosine function is even:
$\cos(-x) = \cos(x)$.
The sine function is odd:
$\sin(-x) = -\sin x$.

The difference formula for $\cos(u - v)$ is true for all real numbers and angles u and v. We can use this formula and the even–odd identities to prove the formula for $\cos(u + v)$.

$$\cos(u + v) = \cos[u - (-v)] \qquad u + v = u - (-v)$$

$$= \cos u \cos(-v) + \sin u \sin(-v) \qquad \text{Difference formula for cosine}$$

$$= \cos u \cos v + \sin u(-\sin v) \qquad \cos(-v) = \cos v;\ \sin(-v) = -\sin v$$

$$\mathbf{\cos(u + v) = \cos u \cos v - \sin u \sin v} \qquad \text{Simplify.}$$

We have proved the formula for the cosine of the sum of two angles or two real numbers.

EXAMPLE 2 **Using the Sum Formula for Cosine**

Find the exact value of $\cos 75°$ by using $75° = 45° + 30°$.

Solution

$$\cos 75° = \cos(45° + 30°) \qquad 75° = 45° + 30°$$

$$= \cos 45° \cos 30° - \sin 45° \sin 30° \qquad \text{Sum formula for cosine}$$

$$= \frac{\sqrt{2}}{2}\cdot\frac{\sqrt{3}}{2} - \frac{\sqrt{2}}{2}\cdot\frac{1}{2} \qquad \text{Use exact values.}$$

$$= \frac{\sqrt{6}}{4} - \frac{\sqrt{2}}{4} \qquad \text{Multiply.}$$

$$= \frac{\sqrt{6} - \sqrt{2}}{4} \qquad \text{Simplify.}$$

Practice Problem 2 Find the exact value of $\cos\dfrac{7\pi}{12}$ by using $\dfrac{7\pi}{12} = \dfrac{\pi}{3} + \dfrac{\pi}{4}$. ■

2 Use cofunction identities.

Cofunction Identities

Two trigonometric functions f and g are called **cofunctions** if

$$f\left(\frac{\pi}{2} - x\right) = g(x) \quad \text{and} \quad g\left(\frac{\pi}{2} - x\right) = f(x).$$

We derive two *cofunction identities*.

$$\cos(u - v) = \cos u \cos v + \sin u \sin v \qquad \text{Difference formula for cosine}$$

$$\cos\left(\frac{\pi}{2} - v\right) = \cos\frac{\pi}{2}\cos v + \sin\frac{\pi}{2}\sin v \qquad \text{Replace } u \text{ with } \frac{\pi}{2}.$$

$$= 0 \cdot \cos v + 1 \cdot \sin v = \sin v \qquad \cos\frac{\pi}{2} = 0;\ \sin\frac{\pi}{2} = 1$$

The result is a cofunction identity that holds for any real number v or angle v in radian measure:

$$\cos\left(\frac{\pi}{2} - v\right) = \sin v$$

If we replace v with $\left(\frac{\pi}{2} - v\right)$ in the identity $\cos\left(\frac{\pi}{2} - v\right) = \sin v$, we have

$$\cos\left[\frac{\pi}{2} - \left(\frac{\pi}{2} - v\right)\right] = \sin\left(\frac{\pi}{2} - v\right) \qquad \text{Replace } v \text{ with } \left(\frac{\pi}{2} - v\right).$$

$$\cos v = \sin\left(\frac{\pi}{2} - v\right) \qquad \text{Simplify.}$$

So
$$\sin\left(\frac{\pi}{2} - v\right) = \cos v.$$

EXAMPLE 3 **Using Cofunction Identities**

Verify the identity: $\tan\left(\frac{\pi}{2} - v\right) = \cot v$.

Solution

$$\tan\left(\frac{\pi}{2} - v\right) = \frac{\sin\left(\dfrac{\pi}{2} - v\right)}{\cos\left(\dfrac{\pi}{2} - v\right)} \qquad \text{Quotient identity}$$

$$= \frac{\cos v}{\sin v} \qquad \text{Use cofunction identities.}$$

$$= \cot v \qquad \text{Quotient identity}$$

Practice Problem 3 Verify the identity: $\sec\left(\frac{\pi}{2} - v\right) = \csc v$.

COFUNCTION IDENTITIES

$$\cos\left(\frac{\pi}{2} - v\right) = \sin v \qquad\qquad \sec\left(\frac{\pi}{2} - v\right) = \csc v$$

$$\sin\left(\frac{\pi}{2} - v\right) = \cos v \qquad\qquad \csc\left(\frac{\pi}{2} - v\right) = \sec v$$

$$\tan\left(\frac{\pi}{2} - v\right) = \cot v \qquad\qquad \cot\left(\frac{\pi}{2} - v\right) = \tan v$$

If angle v is measured in degrees, then replace $\dfrac{\pi}{2}$ with $90°$ in these identities.

3 Use the sum and difference formulas for sine.

Sum and Difference Formulas for Sine

To prove the difference formula for the sine function, we start with a cofunction identity.

$$\sin(u - v) = \cos\left[\frac{\pi}{2} - (u - v)\right]$$ Cofunction identity

$$= \cos\left[\left(\frac{\pi}{2} - u\right) + v\right]$$ $\frac{\pi}{2} - (u - v) = \left(\frac{\pi}{2} - u\right) + v$

$$= \cos\left(\frac{\pi}{2} - u\right)\cos v - \sin\left(\frac{\pi}{2} - u\right)\sin v$$ Sum formula for cosine

$$= \sin u \cos v - \cos u \sin v$$ Cofunction identities

We have the difference formula for sine:

$$\sin(u - v) = \sin u \cos v - \cos u \sin v$$

which holds for all real numbers u and v. If we replace v with $-v$, we derive the sum formula for sine.

$$\sin(u + v) = \sin[u - (-v)]$$ $u + v = u - (-v)$

$$= \sin u \cos(-v) - \cos u \sin(-v)$$ Difference formula for sine

$$= \sin u \cos v - \cos u (-\sin v)$$ $\cos(-v) = \cos v;$ $\sin(-v) = -\sin v$

$$= \sin u \cos v + \cos u \sin v$$ Simplify.

This proves the sum formula for sine:

$$\sin(u + v) = \sin u \cos v + \cos u \sin v.$$

SUM AND DIFFERENCE FORMULAS FOR SINE

$$\sin(u + v) = \sin u \cos v + \cos u \sin v$$
$$\sin(u - v) = \sin u \cos v - \cos u \sin v$$

EXAMPLE 4 Using the Sum and Difference Formulas for Sine

Prove the identity: $\sin(\pi - x) = \sin x$.

Solution

$$\sin(\pi - x) = \sin\pi \cos x - \cos\pi \sin x$$ Difference formula for sine

$$= (0)\cos x - (-1)\sin x$$ $\sin\pi = 0; \cos\pi = -1$

$$= \sin x$$ Simplify.

Practice Problem 4 Prove the identity $\sin(\pi + x) = -\sin x$.

EXAMPLE 5 Using the Sum and Difference Formulas for Sine

Find the exact value of $\sin 63° \cos 27° + \cos 63° \sin 27°$ without using a calculator.

Solution

The given expression is the right side of the sum formula for sine:

$$\sin(u + v) = \sin u \cos v + \cos u \sin v$$

$$\sin 63° \cos 27° + \cos 63° \sin 27° = \sin(63° + 27°) \qquad u = 63°, v = 27°$$

$$= \sin 90° = 1$$

Practice Problem 5 Find the exact value of $\sin 43° \cos 13° - \cos 43° \sin 13°$ without using a calculator.

EXAMPLE 6   **Finding the Exact Value of a Sum**

Let $\sin u = -\dfrac{3}{5}$ and $\cos v = \dfrac{12}{13}$, with $\pi < u < \dfrac{3\pi}{2}$ and $\dfrac{3\pi}{2} < v < 2\pi$. Find the exact value of $\sin(u + v)$.

Solution

Given $\sin u = -\dfrac{3}{5}$ with u in quadrant III, we find the exact value of $\cos u$.

SIDE NOTE

Alternatively, you can use the method of Example 3, (Section 4.3, page 369) to find $\cos u$ and $\sin v$.

$$\cos^2 u = 1 - \sin^2 u \qquad \text{Pythagorean identity}$$

$$\cos u = -\sqrt{1 - \sin^2 u} \qquad \text{In quadrant III, } \cos u \text{ is negative.}$$

$$= -\sqrt{1 - \left(-\frac{3}{5}\right)^2} \qquad \text{Replace } \sin u \text{ with } -\frac{3}{5}.$$

$$= -\sqrt{1 - \frac{9}{25}} \qquad \left(-\frac{3}{5}\right)^2 = \frac{9}{25}$$

$$= -\sqrt{\frac{16}{25}} \qquad 1 - \frac{9}{25} = \frac{25}{25} - \frac{9}{25} = \frac{16}{25}$$

$$\cos u = -\frac{4}{5}$$

Similarly, given $\cos v = \dfrac{12}{13}$ with v in quadrant IV, we find the exact value of $\sin v$.

$$\sin v = -\sqrt{1 - \cos^2 v} \qquad \text{In quadrant IV, } \sin v \text{ is negative.}$$

$$= -\sqrt{1 - \left(\frac{12}{13}\right)^2} \qquad \text{Replace } \cos v \text{ with } \frac{12}{13}.$$

$$= -\sqrt{\frac{25}{169}} \qquad 1 - \left(\frac{12}{13}\right)^2 = 1 - \frac{144}{169} = \frac{169 - 144}{169}$$

$$\sin v = -\frac{5}{13}$$

$$\sin(u + v) = \sin u \cos v + \cos u \sin v \qquad \text{Sum formula for sine}$$

$$= \left(-\frac{3}{5}\right)\left(\frac{12}{13}\right) + \left(-\frac{4}{5}\right)\left(-\frac{5}{13}\right) \qquad \begin{array}{l}\text{Replace values of } \sin u, \sin v,\\ \cos u, \text{ and } \cos v.\end{array}$$

$$= -\frac{36}{65} + \frac{20}{65} = -\frac{16}{65} \qquad \text{Multiply and add.}$$

The exact value of $\sin(u + v)$ is $-\dfrac{16}{65}$.

Practice Problem 6 For the angles u and v in Example 6, find the exact value of $\cos(u + v)$.

EXAMPLE 7 **Verifying an Identity by Using a Sum or Difference Formula**

Verify the identity: $\dfrac{\sin(x - y)}{\cos x \cos y} = \tan x - \tan y$.

Solution

We start with the more complicated left side.

$$\dfrac{\sin(x - y)}{\cos x \cos y} = \dfrac{\sin x \cos y - \cos x \sin y}{\cos x \cos y} \qquad \text{Formula for } \sin(u - v)$$

$$= \dfrac{\sin x \cos y}{\cos x \cos y} - \dfrac{\cos x \sin y}{\cos x \cos y} \qquad \dfrac{a - b}{c} = \dfrac{a}{c} - \dfrac{b}{c}$$

$$= \dfrac{\sin x}{\cos x} - \dfrac{\sin y}{\cos y} \qquad \text{Remove common factors.}$$

$$= \tan x - \tan y \qquad \text{Quotient identity}$$

Since the left side is identical to the right side, the given equation is an identity.

Practice Problem 7 Verify the following identity: $\dfrac{\cos(x + y)}{\sin x \sin y} = \cot x \cot y - 1$

4 Use the sum and difference formulas for tangent.

Sum and Difference Formulas for Tangent

We use the quotient identity $\tan x = \dfrac{\sin x}{\cos x}$ and the formulas for $\sin(u - v)$ and $\cos(u - v)$ to derive a difference formula for tangent.

$$\tan(u - v) = \dfrac{\sin(u - v)}{\cos(u - v)} \qquad \text{Quotient identity}$$

$$= \dfrac{\sin u \cos v - \cos u \sin v}{\cos u \cos v + \sin u \sin v} \qquad \text{Formulas for } \sin(u - v) \text{ and } \cos(u - v)$$

$$= \dfrac{\dfrac{\sin u \cos v}{\cos u \cos v} - \dfrac{\cos u \sin v}{\cos u \cos v}}{\dfrac{\cos u \cos v}{\cos u \cos v} + \dfrac{\sin u \sin v}{\cos u \cos v}} \qquad \begin{array}{l}\text{Divide numerator and denominator by}\\ \cos u \cos v.\end{array}$$

$$= \dfrac{\dfrac{\sin u}{\cos u} - \dfrac{\sin v}{\cos v}}{1 + \dfrac{\sin u \sin v}{\cos u \cos v}} \qquad \text{Simplify.}$$

$$= \dfrac{\tan u - \tan v}{1 + \tan u \tan v} \qquad \text{Quotient identity}$$

We have derived the difference formula for the tangent:

$$\tan(u - v) = \dfrac{\tan u - \tan v}{1 + \tan u \tan v}$$

Replacing v with $-v$ in the difference formula, we have

$$\tan(u + v) = \tan[u - (-v)] \qquad u + v = u - (-v)$$

$$= \dfrac{\tan u - \tan(-v)}{1 + \tan u \tan(-v)} \qquad \text{Formula for } \tan(u - v)$$

$$= \dfrac{\tan u - (-\tan v)}{1 + \tan u (-\tan v)} \qquad \tan(-v) = -\tan v$$

$$= \dfrac{\tan u + \tan v}{1 - \tan u \tan v} \qquad \text{Simplify.}$$

We now have the sum formula for tangent:

$$\tan(u + v) = \frac{\tan u + \tan v}{1 - \tan u \tan v}$$

> **SUM AND DIFFERENCE FORMULAS FOR TANGENT**
>
> $$\tan(u + v) = \frac{\tan u + \tan v}{1 - \tan u \tan v}$$
>
> $$\tan(u - v) = \frac{\tan u - \tan v}{1 + \tan u \tan v}$$

EXAMPLE 8 Verifying an Identity Using a Difference Formula

Verify the identity: $\tan(\pi - x) = -\tan x$.

Solution

Apply the difference formula for the tangent to $\tan(\pi - x)$.

$$\tan(\pi - x) = \frac{\tan \pi - \tan x}{1 + \tan \pi \tan x} \qquad \text{Replace } u \text{ with } \pi \text{ and } v \text{ with } x \text{ in the formula for } \tan(u - v).$$

$$= \frac{0 - \tan x}{1 + 0 \cdot \tan x} \qquad \tan \pi = 0$$

$$= -\tan x \qquad \text{Simplify.}$$

Therefore, the given equation is an identity.

Practice Problem 8 Verify the identity: $\tan(\pi + x) = \tan x$.

EXAMPLE 9 Finding the Exact Value Using a Sum Formula

Find the exact value of $\tan\left[\sin^{-1}\left(-\frac{4}{5}\right) + \cos^{-1}\left(-\frac{12}{13}\right)\right]$.

Solution

We find the exact value of $\tan(u + v)$, where $u = \sin^{-1}\left(-\frac{4}{5}\right)$ and $v = \cos^{-1}\left(-\frac{12}{13}\right)$.

Then, because the range of $\sin^{-1}$ is $\left[-\frac{\pi}{2}, \frac{\pi}{2}\right]$ and that of $\cos^{-1}$ is $[0, \pi]$, we have:

$$\sin u = \frac{y}{r} = -\frac{4}{5}, -\frac{\pi}{2} \le u \le \frac{\pi}{2}, P(u) = (x, y) \quad \text{and} \quad \cos v = \frac{x}{r} = -\frac{12}{13}, 0 \le v \le \pi, P(v) = (x, y)$$

$\sin u$ is negative, u lies in quadrant IV and $x > 0$. | $\cos v$ is negative, v lies in quadrant II and $y > 0$.

$$\sin u = -\frac{4}{5} = \frac{-4}{5} = \frac{y}{r}, \qquad\qquad \cos v = -\frac{12}{13} = \frac{-12}{13} = \frac{x}{r}$$

Then

$$x = \sqrt{(5)^2 - (-4)^2} = 3.$$

So,

$$\tan u = \frac{y}{x} = \frac{-4}{3} = -\frac{4}{3}.$$

Then

$$y = \sqrt{(13)^2 - (-12)^2} = 5.$$

So,

$$\tan v = \frac{y}{x} = \frac{5}{-12} = -\frac{5}{12}.$$

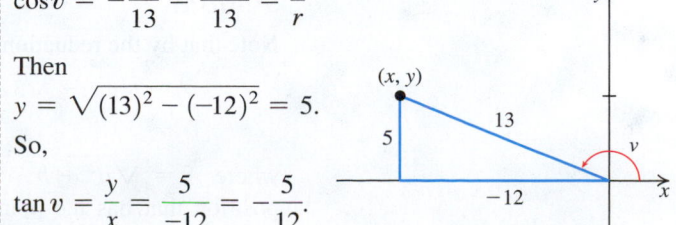

Therefore,

$$\tan\left[\sin^{-1}\left(-\frac{4}{5}\right) + \cos^{-1}\left(-\frac{12}{13}\right)\right] = \tan(u + v)$$

$$= \frac{\tan u + \tan v}{1 - \tan u \tan v} \qquad \text{Formula for } \tan(u + v)$$

$$= \frac{-\dfrac{4}{3} + \left(-\dfrac{5}{12}\right)}{1 - \left(-\dfrac{4}{3}\right)\left(-\dfrac{5}{12}\right)} \qquad \text{Replace } \tan u \text{ with } -\dfrac{4}{3}$$

$$\text{and } \tan v \text{ with } -\dfrac{5}{12}.$$

$$= \frac{-48 - 15}{36 - 20} \qquad \begin{array}{l}\text{Multiply numerator} \\ \text{and denominator by 36.}\end{array}$$

$$= -\frac{63}{16} \qquad \text{Simplify.}$$

Practice Problem 9 Find the exact value of $\tan\left[\cos^{-1}\dfrac{3}{5} - \tan^{-1}\left(-\dfrac{1}{2}\right)\right]$.

5 Use the reduction formula.

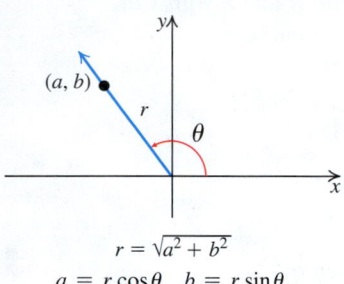

$r = \sqrt{a^2 + b^2}$
$a = r\cos\theta \quad b = r\sin\theta$

Figure 5.4

Reduction Formula

Figure 5.4 suggests that a point (a, b) is on the terminal side of an angle θ in standard position if and only if

$$\cos\theta = \frac{a}{\sqrt{a^2 + b^2}} \quad \text{and} \quad \sin\theta = \frac{b}{\sqrt{a^2 + b^2}}.$$

Using these values for $\cos\theta$ and $\sin\theta$ in the sum formula for the sine, we obtain

$$\sin(x + \theta) = \sin x \cos\theta + \cos x \sin\theta \qquad \text{Sum formula for sine}$$

$$\sin(x + \theta) = \sin x \frac{a}{\sqrt{a^2 + b^2}} + \cos x \frac{b}{\sqrt{a^2 + b^2}}$$

Multiplying both sides of this equation by $\sqrt{a^2 + b^2}$, we have the reduction formula:

$$\sqrt{a^2 + b^2}\,\sin(x + \theta) = a\sin x + b\cos x$$

REDUCTION FORMULA

If (a, b) is any point on the terminal side of an angle θ (radians) in standard position, then

$$a\sin x + b\cos x = \sqrt{a^2 + b^2}\,\sin(x + \theta)$$

for any real number x.

EXAMPLE 10   **Using the Reduction Formula**

Find an angle θ, in radians, and a real number A such that

$$\sin x - \sqrt{3}\cos x = A\sin(x + \theta).$$

Solution

Note that by the reduction formula

$$\sin x - \sqrt{3}\cos x = a\sin x + b\cos x \qquad a = 1 \text{ and } b = -\sqrt{3}$$

$$= A\sin(x + \theta),$$

where $A = \sqrt{a^2 + b^2} = \sqrt{1^2 + (-\sqrt{3})^2} = \sqrt{4} = 2$ and θ is any angle in standard position that has the point $(a, b) = (1, -\sqrt{3})$ on its terminal side. One such angle is

$$\theta = \tan^{-1}\left(\frac{b}{a}\right) = \tan^{-1}(-\sqrt{3}) = -\frac{\pi}{3}.\text{ Then with } A = 2 \text{ and } \theta = -\frac{\pi}{3}, \text{ we have}$$

$$\sin x - \sqrt{3}\cos x = 2\sin\left[x + \left(-\frac{\pi}{3}\right)\right] = 2\sin\left(x - \frac{\pi}{3}\right).$$

Practice Problem 10 Find an angle θ, in radians, and a real number A such that

$$\sin x + \sqrt{3}\cos x = A\sin(x + \theta).$$

EXAMPLE 11 Using the Reduction Formula to Sketch a Graph

Sketch two cycles of the graph of the equation $y = \sin x - \sqrt{3}\cos x$.

Solution

Example 10 shows that $\sin x - \sqrt{3}\cos x = 2\sin\left(x - \dfrac{\pi}{3}\right)$. Therefore, the graphs

of $y = \sin x - \sqrt{3}\cos x$ and $y = 2\sin\left(x - \dfrac{\pi}{3}\right)$ are the same. The graph of

$y = 2\sin\left(x - \dfrac{\pi}{3}\right)$ is a sine wave with amplitude 2, period 2π, and phase shift $\dfrac{\pi}{3}$ units

to the right. Two cycles of the graph of $y = 2\sin\left(x - \dfrac{\pi}{3}\right) = \sin x - \sqrt{3}\cos x$ are

shown in Figure 5.5.

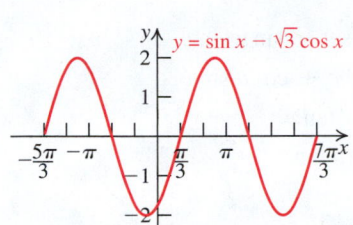

Figure 5.5

Practice Problem 11 Sketch two cycles of the graph of $y = \sin x + \sqrt{3}\cos x$.

EXAMPLE 12 Combining Two Pure Tones

Suppose the pressure exerted by two pure tones (in pounds per square foot) after t seconds is given by

$$y_1 = 0.3\sin(800\pi t) \quad \text{and} \quad y_2 = 0.4\cos(800\pi t).$$

Find the amplitude, period, frequency, and phase shift for the total pressure $y = y_1 + y_2$.

Solution

We have $y = 0.3\sin(800\pi t) + 0.4\cos(800\pi t)$. We use the reduction formula with $a = 0.3$ and $b = 0.4$ to rewrite this equation in the form

$$y = A\sin(800\pi t + \theta) = A\sin\left[800\pi\left(t + \frac{\theta}{800\pi}\right)\right],$$

where

$$A = \sqrt{a^2 + b^2} = \sqrt{(0.3)^2 + (0.4)^2} = 0.5$$

and $(0.3, 0.4)$ is a point on the terminal side of θ. One such angle is $\theta = \tan^{-1}\left(\dfrac{0.4}{0.3}\right)$.

For the total pressure,

$$y = A\sin\left[800\pi\left(t + \frac{\theta}{800\pi}\right)\right]$$

$$= 0.5\sin\left[800\pi\left(t + \frac{\theta}{800\pi}\right)\right] \qquad \textcolor{blue}{A = 0.5}$$

$$\text{amplitude} = 0.5$$

$$\text{period} = \frac{2\pi}{800\pi} = \frac{1}{400}$$

$$\text{frequency} = 400 \qquad \textcolor{blue}{\text{Frequency} = \dfrac{1}{\text{Period}}}$$

$$\text{phase shift} = -\frac{\theta}{800\pi} \approx -0.00037 \qquad \textcolor{blue}{\text{Use a calculator in Radian mode}}$$

$$\textcolor{blue}{\text{with } \theta = \tan^{-1}\left(\dfrac{0.4}{0.3}\right).}$$

Practice Problem 12 Find the amplitude, period, frequency, and phase shift of the total pressure due to the pressure from the two pure tones

$$y_1 = 0.1 \sin(400t) \quad \text{and} \quad y_2 = 0.2 \cos(400t).$$

The important identities involving the sum, difference, and cofunctions of two numbers or angles are summarized next.

SUMMARY OF **MAIN FACTS**

Sum and Difference Formula

$$\cos(u - v) = \cos u \cos v + \sin u \sin v \qquad \cos(u + v) = \cos u \cos v - \sin u \sin v$$

$$\sin(u - v) = \sin u \cos v - \cos u \sin v \qquad \sin(u + v) = \sin u \cos v + \cos u \sin v$$

$$\tan(u - v) = \frac{\tan u - \tan v}{1 + \tan u \tan v} \qquad \tan(u + v) = \frac{\tan u + \tan v}{1 - \tan u \tan v}$$

Cofunction Identities

$$\sin\left(\frac{\pi}{2} - x\right) = \cos x \qquad \cos\left(\frac{\pi}{2} - x\right) = \sin x \qquad \tan\left(\frac{\pi}{2} - x\right) = \cot x$$

$$\csc\left(\frac{\pi}{2} - x\right) = \sec x \qquad \sec\left(\frac{\pi}{2} - x\right) = \csc x \qquad \cot\left(\frac{\pi}{2} - x\right) = \tan x$$

Reduction Formula

If (a, b) is any point on the terminal side of an angle θ (radians) in standard position, then for any real number x,

$$a \sin x + b \cos x = \sqrt{a^2 + b^2} \sin(x + \theta).$$

Answers to Practice Problems

1. $\dfrac{\sqrt{6} + \sqrt{2}}{4}$ **2.** $\dfrac{\sqrt{2} - \sqrt{6}}{4}$ **5.** $\dfrac{1}{2}$ **6.** $-\dfrac{63}{65}$ **9.** $\dfrac{11}{2}$ **10.** $A = 2, \theta = \dfrac{\pi}{3}$

11.

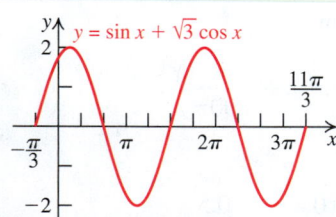

$y = \sin x + \sqrt{3} \cos x$

12. Amplitude $= 0.1\sqrt{5}$; period $= \dfrac{\pi}{200}$; frequency $= \dfrac{200}{\pi}$;

phase shift ≈ -0.0028

SECTION 5.2 **Exercises**

Concepts and Vocabulary

1. $\sin(A + B) = $ _____.

2. $\cos A \cos B - \sin A \sin B = $ _____.

3. $\tan(A + B) = $ _____.

4. $\sin\left(\dfrac{\pi}{2} - x\right) = $ _____, $\cos\left(\dfrac{\pi}{2} - x\right) = $

_____, and $\tan\left(\dfrac{\pi}{2} - x\right) = $ _____.

5. True or False. $\cos\left[\dfrac{\pi}{2} + x\right] = \sin x.$

6. True or False. $\sin\left[\dfrac{\pi}{2} + x\right] = \cos x.$

7. True or False. $\tan(x + y) = \tan x + \tan y.$

8. True or False. $\sin 75° = \sin 45° + \sin 30°.$

Building Skills

In Exercises 9–28, find the exact value of each expression.

9. $\sin(45° + 30°)$ **10.** $\sin(45° - 30°)$

11. $\sin(60° - 45°)$ **12.** $\sin(60° + 45°)$

13. $\sin(-105°)$ **14.** $\cos 285°$

15. $\tan 225°$ **16.** $\tan(-165°)$

17. $\sin\left(\dfrac{\pi}{6} + \dfrac{\pi}{4}\right)$ **18.** $\cos\left(\dfrac{\pi}{3} - \dfrac{\pi}{4}\right)$

19. $\tan\left(\dfrac{\pi}{4} - \dfrac{\pi}{6}\right)$ **20.** $\cot\left(\dfrac{\pi}{3} - \dfrac{\pi}{4}\right)$

21. $\sec\left(\dfrac{\pi}{3} + \dfrac{\pi}{4}\right)$ **22.** $\csc\left(\dfrac{\pi}{4} - \dfrac{\pi}{3}\right)$

23. $\cos \dfrac{-5\pi}{12}$

24. $\sin \dfrac{7\pi}{12}$

25. $\tan \dfrac{19\pi}{12}$

26. $\sec \dfrac{\pi}{12}$

27. $\tan \dfrac{17\pi}{12}$

28. $\csc \dfrac{11\pi}{12}$

In Exercises 29–42, verify each identity.

29. $\sin\left(x + \dfrac{\pi}{2}\right) = \cos x$

30. $\cos\left(x + \dfrac{\pi}{2}\right) = -\sin x$

31. $\sin\left(x - \dfrac{\pi}{2}\right) = -\cos x$

32. $\cos\left(x - \dfrac{\pi}{2}\right) = \sin x$

33. $\tan\left(x + \dfrac{\pi}{2}\right) = -\cot x$

34. $\tan\left(x - \dfrac{\pi}{2}\right) = -\cot x$

35. $\csc(x + \pi) = -\csc x$

36. $\sec(x + \pi) = -\sec x$

37. $\cos\left(x + \dfrac{3\pi}{2}\right) = \sin x$

38. $\cos\left(x - \dfrac{3\pi}{2}\right) = -\sin x$

39. $\tan\left(x - \dfrac{3\pi}{2}\right) = -\cot x$

40. $\tan\left(x + \dfrac{3\pi}{2}\right) = -\cot x$

41. $\cot(3\pi - x) = -\cot x$

42. $\csc\left(\dfrac{5\pi}{2} - x\right) = \sec x$

In Exercises 43–52, find the exact value of each expression without using a calculator.

43. $\sin 56° \cos 34° + \cos 56° \sin 34°$

44. $\cos 57° \cos 33° - \sin 57° \sin 33°$

45. $\cos 331° \cos 61° + \sin 331° \sin 61°$

46. $\cos 110° \sin 70° + \sin 110° \cos 70°$

47. $\dfrac{\tan 129° - \tan 84°}{1 + \tan 129° \tan 84°}$

48. $\dfrac{\tan 28° + \tan 17°}{1 - \tan 28° \tan 17°}$

49. $\sin \dfrac{7\pi}{12} \cos \dfrac{3\pi}{12} - \cos \dfrac{7\pi}{12} \sin \dfrac{3\pi}{12}$

50. $\cos \dfrac{5\pi}{12} \cos \dfrac{\pi}{12} - \sin \dfrac{5\pi}{12} \sin \dfrac{\pi}{12}$

51. $\dfrac{\tan \dfrac{5\pi}{12} - \tan \dfrac{2\pi}{12}}{1 + \tan \dfrac{5\pi}{12} \tan \dfrac{2\pi}{12}}$

52. $\dfrac{\tan \dfrac{5\pi}{12} + \tan \dfrac{7\pi}{12}}{1 - \tan \dfrac{5\pi}{12} \tan \dfrac{7\pi}{12}}$

In Exercises 53–58, find the exact value of each expression, given that $\tan u = \dfrac{3}{4}$, with u in quadrant III, and $\sin v = \dfrac{5}{13}$, with v in quadrant II.

53. $\sin(u - v)$

54. $\sin(u + v)$

55. $\cos(u + v)$

56. $\cos(u - v)$

57. $\tan(u + v)$

58. $\tan(u - v)$

In Exercises 59–64, find the exact value of each expression, given that $\cos \alpha = -\dfrac{2}{5}$, with α in quadrant II, and

$\sin \beta = -\dfrac{3}{7}$, with β in quadrant IV.

59. $\sin(\alpha - \beta)$

60. $\cos(\alpha - \beta)$

61. $\csc(\alpha + \beta)$

62. $\sec(\alpha + \beta)$

63. $\cot(\alpha - \beta)$

64. $\cot(\alpha + \beta)$

In Exercises 65–72, verify each identity.

65. $\dfrac{\sin(x + y)}{\cos x \cos y} = \tan x + \tan y$

66. $\dfrac{\sin(x + y)}{\sin x \sin y} = \cot x + \cot y$

67. $\dfrac{\cos(x + y)}{\cos x \cos y} = 1 - \tan x \tan y$

68. $\dfrac{\cos(x - y)}{\sin x \sin y} = \cot x \cot y + 1$

69. $\dfrac{\cos(x - y)}{\sin(x + y)} = \dfrac{1 + \cot x \cot y}{\cot x + \cot y}$

70. $\dfrac{\cos(x + y)}{\sin(x - y)} = \dfrac{1 - \cot x \cot y}{\cot x - \cot y}$

71. $\dfrac{\sin(x - y)}{\sin(x + y)} = \dfrac{\cot y - \cot x}{\cot y + \cot x}$

72. $\dfrac{\cos(x + y)}{\cos(x - y)} = \dfrac{\cot x \cot y - 1}{\cot x \cot y + 1}$

In Exercises 73–78, find the exact value of each expression.

73. $\sin\left[\tan^{-1}\left(-\dfrac{3}{4}\right) + \cos^{-1}\left(\dfrac{4}{5}\right)\right]$

74. $\cos\left[\sin^{-1}\left(-\dfrac{3}{5}\right) + \cos^{-1}\left(\dfrac{3}{5}\right)\right]$

75. $\sin\left[\sin^{-1}\left(\dfrac{3}{5}\right) - \cos^{-1}\left(\dfrac{4}{5}\right)\right]$

76. $\cos\left[\sin^{-1}\left(\dfrac{3}{5}\right) + \tan^{-1}\left(-\dfrac{4}{3}\right)\right]$

77. $\tan\left[\cos^{-1}\left(\dfrac{4}{5}\right) + \tan^{-1}\left(\dfrac{2}{3}\right)\right]$

78. $\tan\left[\sin^{-1}\left(-\dfrac{3}{5}\right) + \cos^{-1}\left(\dfrac{4}{5}\right)\right]$

In Exercises 79–86, rewrite each equation in the form $y = A \sin(x + \theta)$; graph the resulting equation.

79. $y = \sin x + \cos x$

80. $y = \sin x - \cos x$

81. $y = 3 \sin x + 4 \cos x$

82. $y = 4 \sin x - 3 \cos x$

83. $y = 5 \sin x - 12 \cos x$

84. $y = 12 \sin x + 5 \cos x$

85. $y = \cos x - 3 \sin x$

86. $y = 3 \cos x - 4 \sin x$

Applying the Concepts

87. Analytic geometry. Suppose two nonvertical lines l_1 and l_2 lie in a plane. The slope of l_1 is m_1, and the slope of l_2 is m_2. Note that $m_1 = \tan \theta_1$ and $m_2 = \tan \theta_2$. Show that the tangent of the smallest nonnegative angle α between the lines l_1 and l_2 is given by $\tan \alpha = \left| \dfrac{m_2 - m_1}{1 + m_1 m_2} \right|$.

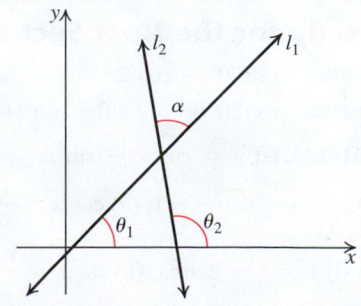

In Exercises 88–90, use Exercise 87 to find the smallest nonnegative angle in radians between the lines l_1 and l_2.

88. Angle between lines. $l_1: y = x + 5, l_2: y = 4x + 2$.

89. Angle between lines. $l_1: y = 2x + 5, l_2: 6x - 3y = 21$.

90. Angle between lines. $l_1: y = 2x - 4, l_2: x + 2y = 7$.

91. Combining two pure tones. The pressure exerted by two pure tones in pounds per square foot is given by the equations $y_1 = 0.4 \sin(400t)$ and $y_2 = 0.3 \cos(400t)$. Find the amplitude and the phase shift for the total pressure $y = y_1 + y_2$.

92. Combining two pure tones. Repeat Exercise 91 for $y_1 = 0.05 \sin(600t)$ and $y_2 = 0.12 \cos(600t)$.

93. Simple harmonic motion. A simple harmonic motion is described by the equation $x = 0.12 \sin 2t + 0.5 \cos 2t$, where x is the distance from the equilibrium position and t is the time in seconds.
 a. Find the amplitude of the motion.
 b. Find the frequency and the phase shift.

94. Simple harmonic motion. Repeat Exercise 93 assuming that the motion is described by the equation
$$x = \frac{1}{2} \sin 3t + \frac{1}{3} \cos 3t.$$

Beyond the Basics

ƒ **95. Difference quotient.** Let $f(x) = \sin x$. Show that
$$\frac{f(x + h) - f(x)}{h} = \sin x\left(\frac{\cos h - 1}{h}\right) + \cos x\left(\frac{\sin h}{h}\right).$$

ƒ **96. Difference quotient.** Let $f(x) = \cos x$. Show that
$$\frac{f(x + h) - f(x)}{h} = \cos x\left(\frac{\cos h - 1}{h}\right) - \sin x\left(\frac{\sin h}{h}\right).$$

In Exercises 97–114, verify each identity.

97. $\cos u \cos(u + v) + \sin u \sin(u + v) = \cos v$

98. $\sin(x + y) \cos y - \cos(x + y) \sin y = \sin x$

99. $\sin 5x \cos 3x - \cos 5x \sin 3x = \sin 2x$

100. $\sin(2x - y) \cos y + \cos(2x - y) \sin y = \sin 2x$

101. $\sin\left(\frac{\pi}{2} + x - y\right) = \cos x \cos y + \sin x \sin y$

102. $\cos\left(\frac{\pi}{2} + x - y\right) = \cos x \sin y - \sin x \cos y$

103. $\sin(x + y) \sin(x - y) = \sin^2 x - \sin^2 y$

104. $\cos(\alpha + \beta) \cos(\alpha - \beta) = \cos^2 \alpha - \sin^2 \beta$

105. $\sin^2\left(\frac{\pi}{4} + \frac{x}{2}\right) - \sin^2\left(\frac{\pi}{4} - \frac{x}{2}\right) = \sin x$

106. $\cos^2\left(\frac{\pi}{4} + \frac{x}{2}\right) - \sin^2\left(\frac{\pi}{4} - \frac{x}{2}\right) = 0$

107. $\dfrac{\sin(\alpha - \beta)}{\sin \alpha \sin \beta} + \dfrac{\sin(\beta - \gamma)}{\sin \beta \sin \gamma} + \dfrac{\sin(\gamma - \alpha)}{\sin \gamma \sin \alpha} = 0$

108. $\dfrac{\sin(\alpha - \beta)}{\cos \alpha \cos \beta} + \dfrac{\sin(\beta - \gamma)}{\cos \beta \cos \gamma} + \dfrac{\sin(\gamma - \alpha)}{\cos \gamma \cos \alpha} = 0$

109. $\sin(x + y) - \sin(x - y) = 2 \cos x \sin y$

110. $\dfrac{\sin(x - y)}{\sin(x + y)} = \dfrac{\tan x - \tan y}{\tan x + \tan y}$

111. $\dfrac{\cos(x - y)}{\cos(x + y)} = \dfrac{1 + \tan x \tan y}{1 - \tan x \tan y}$

112. $\cot(u - v) = \dfrac{\cot u \cot v + 1}{\cot v - \cot u}$

113. $\sec(x + y) = \dfrac{\sec x \sec y}{1 - \tan x \tan y}$

114. $\csc(\alpha - \beta) = \dfrac{\csc \alpha \csc \beta}{\cot \beta - \cot \alpha}$

In Exercises 115–118, find the exact value of each expression.

115. $\cos^2 15° - \cos^2 30° + \cos^2 45° - \cos^2 60° + \cos^2 75°$

116. $\cos^2\left(\dfrac{\pi}{8}\right) + \cos^2\left(\dfrac{3\pi}{8}\right) + \cos^2\left(\dfrac{5\pi}{8}\right) + \cos^2\left(\dfrac{7\pi}{8}\right)$

117. $\cos 60° + \cos 80° + \cos 100°$

118. $\sin 30° - \sin 70° + \sin 110°$

119. Show that $\sin(\theta + n\pi) = (-1)^n \sin\theta$.

120. Show that $\cos(\theta + n\pi) = (-1)^n \cos\theta$.

121. Find the exact value of $\tan 1° \tan 2° \tan 3° \cdots \tan 88° \tan 89°$.

122. Show that $\tan 70° - \tan 20° = 2 \tan 50°$.

Critical Thinking / Discussion / Writing

In Exercises 123–126, assume that x and y are positive numbers in the domain of the inverse function considered. Prove each identity.

123. $\sin^{-1} x \pm \sin^{-1} y = \sin^{-1}(x\sqrt{1 - y^2} \pm y\sqrt{1 - x^2})$ if $x^2 + y^2 < 1$.
 [*Hint:* Let $u = \sin^{-1} x$ and $v = \sin^{-1} y$ and consider $\sin[(u \pm v)]$.]

124. $\cos^{-1} x + \cos^{-1} y = \cos^{-1}(xy - \sqrt{1 - x^2}\sqrt{1 - y^2})$ if $x^2 + y^2 < 1$.

125. $\tan^{-1} x + \tan^{-1} y = \tan^{-1}\left(\dfrac{x + y}{1 - xy}\right)$ if $xy < 1$

126. $\tan^{-1} x + \tan^{-1} y = \pi + \tan^{-1}\left(\dfrac{x + y}{1 - xy}\right)$ if $xy > 1$

127. Explain why we cannot use the formula for $\tan(u - v)$ to verify that $\tan\left(\dfrac{\pi}{2} - x\right) = \cot x$.

128. (a) Verify the identity
$$\tan(A + B + C) = \frac{\tan A + \tan B + \tan C - \tan A \tan B \tan C}{1 - \tan A \tan B - \tan B \tan C - \tan C \tan A}.$$
 (b) Show that $\tan(\tan^{-1} 1 + \tan^{-1} 2 + \tan^{-1} 3) = 0$.
 (c) Use part (b) to conclude that
$$\tan^{-1} 1 + \tan^{-1} 2 + \tan^{-1} 3 = \pi.$$

Getting Ready for the Next Section

129. True or False. $\cos 120° = \cos(2 \cdot 60°) = 2 \cos 60°$

130. True or False. $\cos 120° = \cos^2 60° - \sin^2 60°$

In Exercises 131–133, let $y = \cos^2 x - \sin^2 x$.

131. Show that *(i)* $y = 2\cos^2 x - 1$, *(ii)* $\cos^2 x = \dfrac{1 + y}{2}$.

132. Show that *(i)* $y = 1 - 2\sin^2 x$, *(ii)* $\sin^2 x = \dfrac{1 - y}{2}$.

133. Show that $\tan x = \pm\sqrt{\dfrac{1 - y}{1 + y}}$.

134. *(i)* If θ lies in quadrant III, then $\dfrac{\theta}{2}$ lies in quadrant _____.
 (ii) If θ lies in quadrant IV, then $\dfrac{\theta}{2}$ lies in quadrant _____.

Double-Angle and Half-Angle Formulas

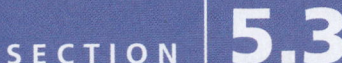

BEFORE STARTING THIS SECTION, REVIEW

1 Sum formulas for sine, cosine, and tangent functions (Section 5.2, page 458)

2 Pythagorean identities (Section 5.1, page 438)

3 Trigonometric functions of common angles (Section 4.2, page 360)

OBJECTIVES

1 Use double-angle formulas.

2 Use power-reducing formulas.

3 Use half-angle formulas.

◆ Cost of Using an Electric Blanket

Three basic properties of electricity—*voltage*, *current*, and *power*—together with the rate charged by your electric company, are used to find the cost of using an electric device.

Voltage (*V*) is measured in *volts*. It is the force that pushes electricity through a wire. *Current* (*I*), measured in *amperes* (amps), measures how much electricity is moving through the device per second. *Power* (*P*), measured in *watts*, gives the energy consumed per second by an electric device and is defined by the equation

$$P = VI \qquad \text{Power} = \text{(Voltage)(Current)}$$

Your electric company bills you by the kilowatt-hour. When you turn on an electric device that consumes 1000 watts for one hour, it consumes 1 kilowatt-hour. Suppose your electric company charges 8¢ per kilowatt-hour. An electric blanket might use 250 watts (depending on the setting). If you turn it on for ten hours, it will consume $250 \times 10 = 2500 = 2.5$ kilowatt-hours. This will cost you $(2.5)(8) = 20$¢.

In Example 5, we use trigonometric identities to compute the *wattage rating* of an electric blanket.

---

1 Use double-angle formulas.

Double-Angle Formulas

Suppose an angle measures x (radians or degrees); then $2x$ is double the measure of x. *Double-angle formulas* express trigonometric functions of $2x$ in terms of functions of x. We begin with the sum formulas for the sine, cosine, and tangent functions.

$$\sin(u + v) = \sin u \cos v + \cos u \sin v$$
$$\cos(u + v) = \cos u \cos v - \sin u \sin v$$
$$\tan(u + v) = \frac{\tan u + \tan v}{1 - \tan u \tan v}$$

To find the double-angle formula for the sine, replace u and v with x in the sum formula for sine.

$\sin(u + v) = \sin u \cos v + \cos u \sin v$	Sum formula for sine
$\sin(x + x) = \sin x \cos x + \cos x \sin x$	Replace both u and v with x.
$\sin 2x = 2 \sin x \cos x$	Simplify.

The identity $$\sin 2x = 2 \sin x \cos x$$

is the double-angle formula for the sine function. We derive double-angle formulas for the cosine and tangent functions by replacing both u and v with x in the sum formulas.

$$\cos 2x = \cos^2 x - \sin^2 x \qquad \tan 2x = \frac{2 \tan x}{1 - \tan^2 x}$$

To derive two other useful forms for $\cos 2x$, first replace $\cos^2 x$ with $1 - \sin^2 x$ and then replace $\sin^2 x$ with $1 - \cos^2 x$ in the formula for $\cos 2x$.

$$\begin{aligned} \cos 2x &= \cos^2 x - \sin^2 x \\ &= (1 - \sin^2 x) - \sin^2 x \\ \cos 2x &= 1 - 2 \sin^2 x \end{aligned} \qquad \begin{aligned} \cos 2x &= \cos^2 x - \sin^2 x \\ &= \cos^2 x - (1 - \cos^2 x) \\ \cos 2x &= 2 \cos^2 x - 1 \end{aligned}$$

DOUBLE-ANGLE FORMULAS

$$\sin 2x = 2 \sin x \cos x \qquad \cos 2x = \cos^2 x - \sin^2 x$$

$$\tan 2x = \frac{2 \tan x}{1 - \tan^2 x} \qquad \cos 2x = 1 - 2 \sin^2 x$$

$$\cos 2x = 2 \cos^2 x - 1$$

EXAMPLE 1 **Using Double-Angle Formulas**

If $\cos \theta = -\dfrac{3}{5}$ and θ is in quadrant II, find the exact value of each expression.

a. $\sin 2\theta$ **b.** $\cos 2\theta$ **c.** $\tan 2\theta$

Solution

We find $\sin \theta$ and $\tan \theta$ from the given information.

Given $\cos \theta = -\dfrac{3}{5} = \dfrac{-3}{5} = \dfrac{x}{y}$, θ in quadrant II, $P(x, y)$ on the terminal side of θ.

See Figure 5.6. Using the Pythagorean Theorem, we have $y = 4$.

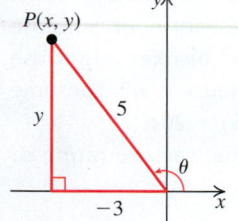

Figure 5.6

So, $$\sin \theta = \frac{y}{r} = \frac{4}{5}$$

$$\tan \theta = \frac{\sin \theta}{\cos \theta} = \frac{4/5}{-3/5} = -\frac{4}{3} \qquad \cos \theta = -\frac{3}{5} \text{ is given}$$

a. $\sin 2\theta = 2 \sin \theta \cos \theta$ Double-angle formula for sine

$$= 2\left(\frac{4}{5}\right)\left(-\frac{3}{5}\right) \qquad \text{Replace } \sin \theta \text{ with } \frac{4}{5} \text{ and } \cos \theta \text{ with } -\frac{3}{5}.$$

$$= -\frac{24}{25} \qquad \text{Simplify.}$$

b. $\cos 2\theta = \cos^2 \theta - \sin^2 \theta$ Double-angle formula for cosines

$$= \left(-\frac{3}{5}\right)^2 - \left(\frac{4}{5}\right)^2 \qquad \text{Replace } \cos \theta \text{ with } -\frac{3}{5} \text{ and } \sin \theta \text{ with } \frac{4}{5}.$$

$$= \frac{9}{25} - \frac{16}{25} = -\frac{7}{25} \qquad \text{Simplify.}$$

c. $\tan 2\theta = \dfrac{2\tan\theta}{1 - \tan^2\theta}$ Double-angle formula for tangent

$= \dfrac{2\left(-\dfrac{4}{3}\right)}{1 - \left(-\dfrac{4}{3}\right)^2}$ Replace $\tan\theta$ with $-\dfrac{4}{3}$.

$= \dfrac{-\dfrac{8}{3}}{1 - \dfrac{16}{9}} = \dfrac{24}{7}$ Simplify.

You can also find $\tan 2\theta$ by using parts **a** and **b**:

$$\tan 2\theta = \frac{\sin 2\theta}{\cos 2\theta} = \frac{-\dfrac{24}{25}}{-\dfrac{7}{25}} = \frac{24}{7}$$

Practice Problem 1 If $\sin x = \dfrac{12}{13}$ and $\dfrac{\pi}{2} < x < \pi$, find the exact value of each expression.

a. $\sin 2x$ **b.** $\cos 2x$ **c.** $\tan 2x$

EXAMPLE 2   Using Double-Angle Formulas

Find the exact value of each expression.

a. $1 - 2\sin^2\left(\dfrac{\pi}{12}\right)$ **b.** $\dfrac{2\tan 22.5°}{1 - \tan^2 22.5°}$

Solution

a. The given expression is the right side of the following formula, where $\theta = \dfrac{\pi}{12}$.

$$\cos 2\theta = 1 - 2\sin^2\theta \qquad \text{A double-angle formula for cosine}$$

$$1 - 2\sin^2\left(\frac{\pi}{12}\right) = \cos\left[2\left(\frac{\pi}{12}\right)\right] \qquad \text{Replace } \theta \text{ with } \frac{\pi}{2}; \text{ interchange sides.}$$

$$= \cos\frac{\pi}{6} = \frac{\sqrt{3}}{2}$$

b. The given expression is the right side of the following formula, where $\theta = 22.5°$.

$$\tan 2\theta = \frac{2\tan\theta}{1 - \tan^2\theta} \qquad \text{Double-angle formula for tangent}$$

$$\frac{2\tan(22.5°)}{1 - \tan^2(22.5°)} = \tan[2(22.5°)] \qquad \text{Replace } \theta \text{ with } 22.5°; \text{ interchange sides.}$$

$$= \tan 45° = 1$$

Practice Problem 2 Find the exact value of each expression.

a. $2\cos^2\left(\dfrac{\pi}{12}\right) - 1$ **b.** $\cos^2 22.5° - \sin^2 22.5°$

EXAMPLE 3 **Finding a Triple-Angle Formula for Sine**

Verify the identity $\sin 3x = 3 \sin x - 4 \sin^3 x$.

Solution

We begin by writing $3x$ as $2x + x$, and we use the sum formula for the sine and the double-angle formulas to verify this identity.

$$
\begin{aligned}
\sin 3x &= \sin (2x + x) \\
&= \sin 2x \cos x + \cos 2x \sin x & \text{Sum formula for sine} \\
&= (2 \sin x \cos x) \cos x + (1 - 2 \sin^2 x) \sin x & \text{Replace } \sin 2x \text{ with} \\
& & 2 \sin x \cos x \text{ and } \cos 2x \text{ with} \\
& & 1 - 2 \sin^2 x. \\
&= 2 \sin x \cos^2 x + \sin x - 2 \sin^3 x & \text{Multiply; distributive property} \\
&= 2 \sin x (1 - \sin^2 x) + \sin x - 2 \sin^3 x & \cos^2 x = 1 - \sin^2 x \\
&= 2 \sin x - 2 \sin^3 x + \sin x - 2 \sin^3 x & \text{Distributive property} \\
&= 3 \sin x - 4 \sin^3 x & \text{Simplify.}
\end{aligned}
$$

We have verified the identity by showing that the left and right sides of the equation are equal.

Practice Problem 3 Verify the identity $\cos 3x = 4 \cos^3 x - 3 \cos x$.

We can use the double-angle formulas to express trigonometric functions of 4θ, 6θ, and 8θ in terms of 2θ, 3θ, and 4θ, respectively. For example, the identities

$$\cos 4\theta = 2 \cos^2 2\theta - 1, \quad \sin 6\theta = 2 \sin 3\theta \cos 3\theta, \quad \text{and} \quad \tan 8\theta = \frac{2 \tan 4\theta}{1 - \tan^2 4\theta}$$

can be directly derived from the appropriate double-angle formulas.

2 Use power-reducing formulas.

Power-Reducing Formulas

The purpose of *power-reducing formulas* is to express $\sin^2 x$, $\cos^2 x$, and $\tan^2 x$ in terms of trigonometric functions with powers not greater than 1. These formulas are useful in calculus.

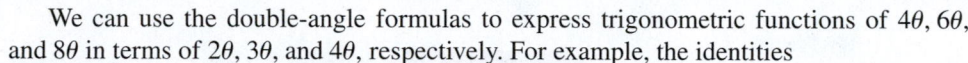

$$
\boxed{
\begin{array}{ccc}
\textbf{POWER-REDUCING FORMULAS} \\
\sin^2 x = \dfrac{1 - \cos 2x}{2} \qquad \cos^2 x = \dfrac{1 + \cos 2x}{2} \qquad \tan^2 x = \dfrac{1 - \cos 2x}{1 + \cos 2x}
\end{array}
}
$$

We can derive the first power-reducing formula by using the appropriate formula for $\cos 2x$.

$$
\begin{aligned}
\cos 2x &= 1 - 2 \sin^2 x & \text{Double-angle formula for } \cos 2x \text{ in terms of sine} \\
2 \sin^2 x &= 1 - \cos 2x & \text{Add } 2 \sin^2 x - \cos 2x \text{ to both sides and simplify.} \\
\sin^2 x &= \frac{1 - \cos 2x}{2} & \text{Divide both sides by 2.}
\end{aligned}
$$

Similarly, we can derive the second formula.

$$
\begin{aligned}
\cos 2x &= 2 \cos^2 x - 1 & \text{Double-angle formula for } \cos 2x \text{ in terms of cosine} \\
1 + \cos 2x &= 2 \cos^2 x & \text{Add 1 to both sides.} \\
\frac{1 + \cos 2x}{2} &= \cos^2 x & \text{Divide both sides by 2.}
\end{aligned}
$$

We begin with the quotient identity to prove the third formula.

$$\tan^2 x = \frac{\sin^2 x}{\cos^2 x} \qquad \text{\color{blue}Quotient identity}$$

$$= \frac{\dfrac{1 - \cos 2x}{2}}{\dfrac{1 + \cos 2x}{2}} \qquad \text{\color{blue}Power-reducing formulas for } \sin^2 x \text{ and } \cos^2 x$$

$$= \frac{1 - \cos 2x}{1 + \cos 2x} \qquad \text{\color{blue}Multiply numerator and denominator by 2. Simplify.}$$

EXAMPLE 4   **Using Power-Reducing Formulas**

Write an equivalent expression for $\cos^4 x$ that contains only first powers of cosines of multiple angles.

Solution

We use the power-reducing formulas repeatedly.

$$\cos^4 x = (\cos^2 x)^2 \qquad \qquad\qquad\qquad\qquad\quad \color{blue}a^4 = (a^2)^2$$

$$= \left(\frac{1 + \cos 2x}{2}\right)^2 \qquad\qquad\qquad\quad \text{\color{blue}Power-reducing formula}$$

$$= \frac{1}{4}(1 + 2\cos 2x + \cos^2 2x) \qquad\quad \text{\color{blue}Expand the binomial.}$$

$$= \frac{1}{4}\left(1 + 2\cos 2x + \frac{1 + \cos 4x}{2}\right) \quad \begin{array}{l}\text{\color{blue}Power-reducing formula for}\\ \text{\color{blue}}\cos^2 x\text{; replace } x \text{ with } 2x.\end{array}$$

$$= \frac{1}{4}\left(1 + 2\cos 2x + \frac{1}{2} + \frac{1}{2}\cos 4x\right) \quad \text{\color{blue}}\frac{a + b}{c} = \frac{a}{c} + \frac{b}{c}$$

$$= \frac{1}{4} + \frac{2}{4}\cos 2x + \frac{1}{8} + \frac{1}{8}\cos 4x \qquad \text{\color{blue}Distributive property}$$

$$= \frac{3}{8} + \frac{1}{2}\cos 2x + \frac{1}{8}\cos 4x \qquad\quad \text{\color{blue}Simplify.}$$

Practice Problem 4 Write an equivalent expression for $\sin^4 x$ that contains only first powers of cosines of multiple angles.

Alternating Current

Alternating current (AC) is electric current that reverses direction, usually many times per second. Most electrical generators produce alternating current. The *wattage rating* of an electric device is $\dfrac{1}{\sqrt{2}} \approx 0.7071$ times the maximum wattage of the device.

◆ **EXAMPLE 5**   **Finding the Wattage Rating of an Electric Blanket**

The voltage V of household current is given by $V = 170 \sin(120\pi t)$ volts, where t is in seconds. Suppose the amount of current passing through an electric blanket is $I = 0.1 \sin(120\pi t)$ amps, where t is in seconds. Find the wattage rating of the blanket.

Solution

We have $P = VI$

$P = [\,170 \sin{(120\pi t)}\,][\,0.1 \sin{(120\pi t)}\,]$ Substitute the given values of V and I.

$= 17 \sin^2{(120\pi t)}$ Multiply.

$= 17 \left[\dfrac{1 - \cos{(240\pi t)}}{2}\right]$ Power-reducing formula

$P = 8.5 - 8.5 \cos{(240\pi t)}$ Simplify.

Power $=$ voltage $\times$ current
watts $=$ volts $\times$ amps

The maximum value of P is 17 watts when $\cos{(240\pi t)} = -1$; so the wattage rating for this blanket is $\dfrac{17}{\sqrt{2}} \approx 12$ watts.

Practice Problem 5 In Example 5, find the wattage rating of a light bulb for which $I = 0.83 \sin{(120\pi t)}$ amps.

3 Use half-angle formulas.

Half-Angle Formulas

If an angle measures θ, then $\dfrac{\theta}{2}$ is half the measure of θ. Half-angle formulas express trigonometric functions of $\dfrac{\theta}{2}$ in terms of functions of θ. To derive the half-angle formulas, we replace x with $\dfrac{\theta}{2}$ in the power-reducing formulas and then take the square root of both sides. For example,

$$\cos^2{x} = \frac{1 + \cos{2x}}{2}$$ Power-reducing formula for cosines

$$\cos^2{\frac{\theta}{2}} = \frac{1 + \cos{\left[2\left(\dfrac{\theta}{2}\right)\right]}}{2}$$ Replace x with $\dfrac{\theta}{2}$.

$$\cos^2{\frac{\theta}{2}} = \frac{1 + \cos{\theta}}{2}$$ Simplify.

$$\cos{\frac{\theta}{2}} = \pm\sqrt{\frac{1 + \cos{\theta}}{2}}$$ Square root property

We call the last equation a *half-angle formula* for cosine. The sign $+$ or $-$ depends on the quadrant in which $\dfrac{\theta}{2}$ lies. We can derive half-angle formulas for sine and tangent in a similar manner.

Exercises 75 and 76 of this section ask you to verify alternate formulas for $\tan{\dfrac{\theta}{2}}$ that do not use $+$ or $-$ signs:

$$\tan{\frac{\theta}{2}} = \frac{\sin{\theta}}{1 + \cos{\theta}} = \frac{1 - \cos{\theta}}{\sin{\theta}}.$$

HALF-ANGLE FORMULAS

$$\sin{\frac{\theta}{2}} = \pm\sqrt{\frac{1 - \cos{\theta}}{2}} \qquad \cos{\frac{\theta}{2}} = \pm\sqrt{\frac{1 + \cos{\theta}}{2}}$$

$$\tan{\frac{\theta}{2}} = \pm\sqrt{\frac{1 - \cos{\theta}}{1 + \cos{\theta}}} = \frac{\sin{\theta}}{1 + \cos{\theta}} = \frac{1 - \cos{\theta}}{\sin{\theta}}$$

The sign $+$ or $-$ depends on the quadrant in which $\dfrac{\theta}{2}$ lies.

EXAMPLE 6   **Using Half-Angle Formulas**

Use a half-angle formula to find the exact value of $\cos 157.5°$.

Solution

Because $157.5° = \dfrac{315°}{2}$, we use the half-angle formula for $\cos \dfrac{\theta}{2}$ with $\theta = 315°$.

Because $\dfrac{\theta}{2} = 157.5°$ lies in quadrant II, $\cos \dfrac{\theta}{2}$ is negative.

$$\cos 157.5° = \cos \frac{315°}{2}$$

$$= -\sqrt{\frac{1 + \cos 315°}{2}} \qquad \text{Half-angle formula for cosine}$$

$$= -\sqrt{\frac{1 + \cos 45°}{2}} \qquad \begin{aligned}&\cos 315° = \cos(360° - 45°) \\ &\quad\ = \cos 45°\end{aligned}$$

$$= -\sqrt{\frac{\left(1 + \dfrac{\sqrt{2}}{2}\right) \cdot 2}{2 \cdot 2}} \qquad \begin{aligned}&\text{Replace } \cos 45° \text{ with } \dfrac{\sqrt{2}}{2}. \text{ Multiply} \\ &\text{numerator and denominator by 2.}\end{aligned}$$

$$= -\sqrt{\frac{2 + \sqrt{2}}{2 \cdot 2}} = -\frac{\sqrt{2 + \sqrt{2}}}{2} \qquad \text{Simplify.}$$

Practice Problem 6 Find the exact value of $\sin 112.5°$.

EXAMPLE 7   **Finding the Exact Value**

Given that $\sin \theta = -\dfrac{5}{13}, \pi < \theta < \dfrac{3\pi}{2}$, find the exact value of each expression.

a. $\sin \dfrac{\theta}{2}$ **b.** $\tan \dfrac{\theta}{2}$

Solution

Because θ lies in quadrant III, $\cos \theta$ is negative; so

$$\cos \theta = -\sqrt{1 - \sin^2 \theta} \qquad \text{Pythagorean identity}$$

$$\cos \theta = -\sqrt{1 - \left(-\frac{5}{13}\right)^2} \qquad \text{Replace } \sin \theta \text{ with } -\frac{5}{13}.$$

$$= -\sqrt{1 - \frac{25}{169}} = -\sqrt{\frac{169 - 25}{169}} = -\frac{12}{13} \qquad \text{Simplify.}$$

a. Because $\pi < \theta < \dfrac{3\pi}{2}, \dfrac{\theta}{2}$ lies in quadrant II $\left(\text{if } \pi < \theta < \dfrac{3\pi}{2}, \text{ then } \dfrac{\pi}{2} < \dfrac{\theta}{2} < \dfrac{3\pi}{4}\right)$.

The half-angle formula gives

$$\sin \frac{\theta}{2} = \sqrt{\frac{1 - \cos \theta}{2}} \qquad \text{In quadrant II, } \sin \frac{\theta}{2} \text{ is positive.}$$

$$= \sqrt{\frac{1 - \left(-\dfrac{12}{13}\right)}{2}} \qquad \text{Replace } \cos \theta \text{ with } -\frac{12}{13}.$$

$$= \sqrt{\frac{\dfrac{25}{13}}{2}} = \frac{5}{\sqrt{26}} = \frac{5\sqrt{26}}{26} \qquad \text{Simplify.}$$

b. From the half-angle formula, we have

$$\tan\frac{\theta}{2} = -\sqrt{\frac{1 - \cos\theta}{1 + \cos\theta}} \qquad \text{In quadrant II, } \tan\frac{\theta}{2} \text{ is negative.}$$

$$= -\sqrt{\frac{1 - \left(-\dfrac{12}{13}\right)}{1 + \left(-\dfrac{12}{13}\right)}} \qquad \cos\theta = -\frac{12}{13}.$$

$$= -\sqrt{\frac{\dfrac{25}{13}}{\dfrac{1}{13}}} = -5 \qquad \text{Simplify.}$$

Practice Problem 7 For θ in Example 7, find $\cos\dfrac{\theta}{2}$.

EXAMPLE 8   Verifying an Identity Containing Half-Angles

Verify the identity $\sin x \cos\dfrac{x}{2} = \sin\dfrac{x}{2}(1 + \cos x)$.

Solution

The left side contains $\sin x$, and the right side contains $\sin\dfrac{x}{2}$. We replace x with $\dfrac{x}{2}$ in the double-angle identity for sine.

$$\sin 2x = 2\sin x \cos x \qquad \text{Double-angle identity for sine}$$

$$\sin x = 2\sin\frac{x}{2}\cos\frac{x}{2} \qquad \text{Replace } x \text{ with } \frac{x}{2}; \ \sin\left[2\left(\frac{x}{2}\right)\right] = \sin x.$$

Begin with the left side of the original equation and use this expression for $\sin x$.

$$\sin x \cos\frac{x}{2} = 2\sin\frac{x}{2}\cos\frac{x}{2}\cos\frac{x}{2} \qquad \text{Replace } \sin x \text{ with } 2\sin\frac{x}{2}\cos\frac{x}{2}.$$

$$= 2\sin\frac{x}{2}\cos^2\frac{x}{2}$$

$$= 2\left(\sin\frac{x}{2}\right)\left(\frac{1 + \cos\left[2\left(\frac{x}{2}\right)\right]}{2}\right) \qquad \text{Power-reducing formula for cosine}$$

$$= \sin\frac{x}{2}(1 + \cos x) \qquad \cos\left[2\left(\frac{x}{2}\right)\right] = \cos x;\ \text{simplify.}$$

Because the left side is identical to the right side of the equation, the identity is verified.

Practice Problem 8 Verify the identity $\sin\dfrac{x}{2}\sin x = \cos\dfrac{x}{2}(1 - \cos x)$.

SUMMARY OF MAIN FACTS

Double-Angle Formulas

$$\sin 2x = 2\sin x \cos x \qquad\qquad \cos 2x = \cos^2 x - \sin^2 x$$

$$\tan 2x = \frac{2\tan x}{1 - \tan^2 x} \qquad\qquad \cos 2x = 1 - 2\sin^2 x$$

$$\cos 2x = 2\cos^2 x - 1$$

Power-Reducing Formulas

$$\sin^2 x = \frac{1 - \cos 2x}{2} \qquad \cos^2 x = \frac{1 + \cos 2x}{2} \qquad \tan^2 x = \frac{1 - \cos 2x}{1 + \cos 2x}$$

Half-Angle Formulas

$$\sin \frac{\theta}{2} = \pm \sqrt{\frac{1 - \cos \theta}{2}} \qquad \cos \frac{\theta}{2} = \pm \sqrt{\frac{1 + \cos \theta}{2}} \qquad \tan \frac{\theta}{2} = \pm \sqrt{\frac{1 - \cos \theta}{1 + \cos \theta}}$$

$$= \frac{\sin \theta}{1 + \cos \theta} = \frac{1 - \cos \theta}{\sin \theta}$$

The sign $+$ or $-$ depends on the quadrant in which $\dfrac{\theta}{2}$ lies.

Answers to Practice Problems

1. a. $-\dfrac{120}{169}$ **b.** $-\dfrac{119}{169}$ **c.** $\dfrac{120}{119}$ **2. a.** $\dfrac{\sqrt{3}}{2}$ **b.** $\dfrac{\sqrt{2}}{2}$ **6.** $\dfrac{\sqrt{2 + \sqrt{2}}}{2}$ **7.** $-\dfrac{\sqrt{26}}{26}$

4. $\dfrac{3}{8} - \dfrac{1}{2}\cos 2x + \dfrac{1}{8}\cos 4x$ **5.** ≈ 99.8 watts

SECTION 5.3 **Exercises**

Concepts and Vocabulary

1. The double-angle formula for $\sin 2x$ is
   $\sin 2x = $ _____.

2. In the double-angle formula $\cos 2x = \cos^2 x - \sin^2 x$, replace $\cos^2 x$ with $1 - \sin^2 x$ to obtain a double-angle formula $\cos 2x = $ _____ in terms of $\sin^2 x$. Solve this formula for $\sin^2 x$ to obtain the power-reducing formula $\sin^2 x = $ _____.

3. The formula for $\cos 2x$ in terms of $\cos^2 x$ is $\cos 2x = $ _____. Solve this formula for $\cos^2 x$ to obtain the power-reducing formula $\cos^2 x = $ _____.

4. The double-angle formula for $\tan 2x$ is $\tan 2x = $ _____.

5. **True or False.** $\tan 2x = \dfrac{1 - \cos 2x}{1 + \cos 2x}$.

6. **True or False.** $\dfrac{1}{2}\tan 2x = \tan x$.

7. **True or False.** $\cos \dfrac{\theta}{2} = -\sqrt{\dfrac{1 + \cos \theta}{2}}$, $\pi < \theta < 2\pi$.

8. **True or False.** $\sin 100° = -\sqrt{\dfrac{1 - \cos 200°}{2}}$.

Building Skills

In Exercises 9–14, use the given information about the angle θ to find the exact value of
a. $\sin 2\theta$ **b. $\cos 2\theta$**   **c. $\tan 2\theta$**

9. $\sin \theta = \dfrac{3}{5}$, θ in quadrant II

10. $\cos \theta = -\dfrac{5}{13}$, θ in quadrant III

11. $\tan \theta = 4$, $\sin \theta < 0$

12. $\sec \theta = -\sqrt{3}$, $\sin \theta > 0$

13. $\tan \theta = -2$, $\dfrac{\pi}{2} < \theta < \pi$

14. $\cot \theta = -7$, $\dfrac{3\pi}{2} < \theta < 2\pi$

In Exercises 15–24, use a double-angle formula to find the exact value of each expression.

15. $1 - 2\sin^2 75°$

16. $\dfrac{2\tan 75°}{1 - \tan^2 75°}$

17. $2\cos^2 105° - 1$

18. $1 - 2\sin^2 165°$

19. $\dfrac{2\tan 165°}{1 - \tan^2 165°}$

20. $2\cos^2 165° - 1$

21. $1 - 2\sin^2 \dfrac{\pi}{8}$

22. $2\cos^2\left(-\dfrac{\pi}{8}\right) - 1$

23. $\dfrac{2\tan\left(-\dfrac{5\pi}{12}\right)}{1 - \tan^2\left(-\dfrac{5\pi}{12}\right)}$

24. $1 - 2\sin^2\left(-\dfrac{7\pi}{12}\right)$

In Exercises 25 and 26, verify each "quadruple-angle" formula.

25. $\sin 4\theta = \cos \theta (4 \sin \theta - 8 \sin^3 \theta)$

26. $\cos 4\theta = 8 \cos^4 \theta - 8 \cos^2 \theta + 1$

In Exercises 27–40, verify each identity.

27. $\cos^4 x - \sin^4 x = \cos 2x$

28. $1 + \cos 2x + 2 \sin^2 x = 2$

29. $(\sin x - \cos x)^2 = 1 - \sin 2x$

30. $(\sin x + \cos x)^2 = 1 + \sin 2x$

31. $\sin 4x = 4 \sin x \cos x \cos 2x$

32. $\sin 4x = 8 \sin x \cos^3 x - 4 \sin x \cos x$

33. $\dfrac{\sin 3x}{\sin x} - \dfrac{\cos 3x}{\cos x} = 2$

34. $\dfrac{\cos 3x}{\sin x} + \dfrac{\sin 3x}{\cos x} = 2 \cot 2x$

35. $\dfrac{1 - \cos 2x}{\sin 2x} = \tan x$ **36.** $\dfrac{1 + \cos 2x}{\sin 2x} = \cot x$

37. $\sin 2x = \dfrac{2 \tan x}{1 + \tan^2 x}$ **38.** $\cos 2x = \dfrac{1 - \tan^2 x}{1 + \tan^2 x}$

39. $\dfrac{1 + \sin 2x}{\cos 2x} = \dfrac{\cos x + \sin x}{\cos x - \sin x}$ **40.** $\cot x - \tan x = 2 \cot 2x$

In Exercises 41–50, use the power-reducing formulas to rewrite each expression that does not contain trigonometric functions of power greater than 1.

41. $4 \sin^2 x \cos^2 x$

42. $\sin^2 x \cos^2 x$

43. $4 \sin x \cos x (1 - 2 \sin^2 x)$

44. $4 \sin x \cos x (2 \cos^2 x - 1)$

45. $2 \sin 3x \cos 3x (2 \cos^2 3x - 1)$

46. $\sin 8x (1 - 2 \sin^2 4x)$

47. $\sin \dfrac{x}{2} \cos \dfrac{x}{2} \left(1 - 2 \sin^2 \dfrac{x}{2}\right)$

48. $\sin x \left(2 \cos^2 \dfrac{x}{2} - 1\right)$

49. $8 \sin^4 \dfrac{x}{2}$

50. $8 \cos^4 \dfrac{x}{2}$

In Exercises 51–62, use half-angle formulas to find the exact value of each expression.

51. $\sin \dfrac{\pi}{12}$ **52.** $\sin \dfrac{\pi}{8}$

53. $\cos \dfrac{\pi}{8}$ **54.** $\tan \dfrac{\pi}{8}$

55. $\sin \left(-\dfrac{3\pi}{8}\right)$ **56.** $\cos \left(-\dfrac{3\pi}{8}\right)$

57. $\tan \left(\dfrac{7\pi}{8}\right)$ **58.** $\sec \left(-\dfrac{7\pi}{8}\right)$

59. $\tan 112.5°$ **60.** $\cos 112.5°$

61. $\sin (-75°)$ **62.** $\tan (-105°)$

In Exercises 63–70, use the information about the angle θ $0 < \theta < 2\pi$ to find the exact value of

a. $\sin \dfrac{\theta}{2}$.   **b.** $\cos \dfrac{\theta}{2}$.   **c.** $\tan \dfrac{\theta}{2}$.

63. $\sin \theta = \dfrac{4}{5}, \dfrac{\pi}{2} < \theta < \pi$

64. $\cos \theta = -\dfrac{12}{13}, \pi < \theta < \dfrac{3\pi}{2}$

65. $\tan \theta = -\dfrac{2}{3}, \dfrac{\pi}{2} < \theta < \pi$

66. $\cot \theta = \dfrac{3}{4}, \pi < \theta < \dfrac{3\pi}{2}$

67. $\sin \theta = \dfrac{1}{5}, \cos \theta < 0$ **68.** $\cos \theta = \dfrac{2}{3}, \sin \theta < 0$

69. $\sec \theta = \sqrt{5}, \sin \theta > 0$ **70.** $\csc \theta = \sqrt{7}, \tan \theta < 0$

In Exercises 71–80, verify each identity.

71. $\left(\sin \dfrac{t}{2} + \cos \dfrac{t}{2}\right)^2 = 1 + \sin t$

72. $\left(\sin \dfrac{t}{2} - \cos \dfrac{t}{2}\right)^2 = 1 - \sin t$

73. $2 \cos^2 \dfrac{x}{2} = \dfrac{\sin^2 x}{1 - \cos x}$

74. $2 \sin^2 \dfrac{x}{2} = \dfrac{\sin^2 x}{1 + \cos x}$

75. $\tan \dfrac{x}{2} = \dfrac{\sin x}{1 + \cos x}$

76. $\tan \dfrac{x}{2} = \dfrac{1 - \cos x}{\sin x}$

77. $\sin^2 \dfrac{x}{2} + \cos x = \cos^2 \dfrac{x}{2}$

78. $\cos^2 x + \cos x = 2 \cos^2 \dfrac{x}{2} - \sin^2 x$

79. $\sin x = \dfrac{2 \tan \dfrac{x}{2}}{1 + \tan^2 \dfrac{x}{2}}$ **80.** $\cos x = \dfrac{1 - \tan^2 \dfrac{x}{2}}{1 + \tan^2 \dfrac{x}{2}}$

Applying the Concepts

In Exercises 81–86, assume that the voltage of the current is given by $V = 170 \sin (120\pi t)$, where t is in seconds. Find the wattage rating $\left(\dfrac{\text{maximum wattage}}{\sqrt{2}}\right)$ of each electric device if the current flowing through the device is I amps.

81. Light bulb. $I = 0.832 \sin (120\pi t)$.

82. Microwave oven. $I = 7.487 \sin (120\pi t)$.

83. Toaster. $I = 9.983 \sin (120\pi t)$.

84. Refrigerator. $I = 6.655 \sin (120\pi t)$.

85. Vacuum cleaner. $I = 4.991 \sin (120\pi t)$.

86. Television. $I = 2.917 \sin (120\pi t)$.

87. Throwing a football. A quarterback throws a ball with an initial velocity of v_0 feet per second at an angle θ with the horizontal. The horizontal distance x in feet that the ball is thrown is modeled by the equation $x = \dfrac{v_0^2}{16} \sin\theta \cos\theta$. For a fixed v_0, find the angle that produces the maximum distance x.

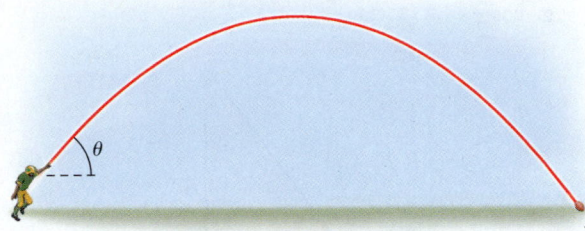

88. **Area.** (a) Show that the area A of a rectangle inscribed in a semicircle of radius R is given by

$$A = R^2 \sin 2\theta.$$

   (b) Find the angle that maximizes this area.

89. **Graphing.** Find the period of $y = \sin^2 x$, and sketch one cycle of its graph.
   [*Hint*: Use power-reducing formula for $\sin^2 x$.]

90. **Graphing.** Find the period of $y = \cos^2 x$, and sketch one cycle of its graph.

Beyond the Basics

In Exercises 91–102, verify each identity.

91. $\tan 3x = \dfrac{3\tan x - \tan^3 x}{1 - 3\tan^2 x}$

92. $\dfrac{\sin 3x + \cos 3x}{\cos x - \sin x} = 1 + 2\sin 2x$

93. $\dfrac{\sin x - \cos x}{\sin x + \cos x} - \dfrac{\sin x + \cos x}{\sin x - \cos x} = 2\tan 2x$

94. $\dfrac{2}{\tan x + \cot x} = \sin 2x$

95. $\dfrac{2\sin x}{\cos 3x} = \tan 3x - \tan x$

96. $\cot x - \tan x = 2\cot 2x$

97. $\dfrac{1 - \tan^2\left(\dfrac{\pi}{4} - x\right)}{1 + \tan^2\left(\dfrac{\pi}{4} - x\right)} = \sin 2x$

98. $\dfrac{1 + \sin 2x - \cos 2x}{1 + \sin 2x + \cos 2x} = \tan x$

99. $\sin^2\dfrac{\pi}{8} + \sin^2\dfrac{3\pi}{8} + \sin^2\dfrac{5\pi}{8} + \sin^2\dfrac{7\pi}{8} = 2$

100. $\sin^2\left(\dfrac{\pi}{8} + \dfrac{x}{2}\right) - \sin^2\left(\dfrac{\pi}{8} - \dfrac{x}{2}\right) = \dfrac{1}{\sqrt{2}}\sin x$

101. $\sqrt{2 + \sqrt{2 + 2\cos 4x}} = 2\cos x, 0 < x < \dfrac{\pi}{4}$

102. $\cos\dfrac{\theta}{2}\cos\dfrac{\theta}{2^2}\cos\dfrac{\theta}{2^3}\cdots\cos\dfrac{\theta}{2^n} = \dfrac{\sin\theta}{2^n \sin\dfrac{\theta}{2^n}}$

In Exercises 103–120, find the exact value of each expression.

103. $\sin\left(2\sin^{-1}\dfrac{1}{2}\right)$

104. $\sin\left(2\sin^{-1}\dfrac{1}{3}\right)$

105. $\cos\left(2\tan^{-1}\dfrac{1}{2}\right)$

106. $\cos\left(2\sin^{-1}\dfrac{2}{3}\right)$

107. $\tan\left(2\sin^{-1}\dfrac{3}{4}\right)$

108. $\tan\left(2\cos^{-1}\dfrac{1}{3}\right)$

109. $\sin\left[2\cos^{-1}\left(-\dfrac{3}{5}\right)\right]$

110. $\sin\left[2\tan^{-1}\left(-\dfrac{5}{12}\right)\right]$

111. $\cos\left[2\sin^{-1}\left(-\dfrac{1}{2}\right)\right]$

112. $\cos\left[2\cos^{-1}\left(-\dfrac{12}{13}\right)\right]$

113. $\tan\left[2\sin^{-1}\left(-\dfrac{2}{3}\right)\right]$

114. $\tan\left[2\cos^{-1}\left(-\dfrac{4}{5}\right)\right]$

115. $\sin\left[\dfrac{1}{2}\sin^{-1}\left(-\dfrac{3}{5}\right)\right]$

116. $\cos\left[\dfrac{1}{2}\sin^{-1}\left(-\dfrac{2}{3}\right)\right]$

117. $\tan\left[\dfrac{1}{2}\cos^{-1}\left(-\dfrac{5}{13}\right)\right]$

118. $\tan\left[\dfrac{1}{2}\tan^{-1}\left(-\dfrac{12}{5}\right)\right]$

119. $\sin^2\left[\dfrac{1}{2}\sin^{-1}\left(-\dfrac{2}{3}\right)\right]$

120. $\cos^2\left[\dfrac{1}{2}\cos^{-1}\left(-\dfrac{3}{5}\right)\right]$

In Exercises 121–123, verify each identity.

121. $2\sin^{-1}x = \sin^{-1}\left(2x\sqrt{1 - x^2}\right), -\dfrac{\sqrt{2}}{2} \le x \le \dfrac{\sqrt{2}}{2}$

122. $2\cos^{-1}x = \cos^{-1}(2x^2 - 1), 0 \le x \le 1$

123. $2\tan^{-1}x = \tan^{-1}\left(\dfrac{2x}{1 - x^2}\right), -1 < x < 1.$

124. Show that (a) $\sin\dfrac{\pi}{10} = \dfrac{\sqrt{5} - 1}{4}$.

   (b) $\cos\dfrac{\pi}{10} = \dfrac{\sqrt{10 + 2\sqrt{5}}}{4}$.

   [*Hint:* Let $x = \dfrac{\pi}{10}$. Then $5x = \dfrac{\pi}{2}$ or $3x = \dfrac{\pi}{2} - 2x$.

   $\cos 3x = \cos\left(\dfrac{\pi}{2} - 2x\right) = \sin 2x$. So,

   $4\cos^3 x - 3\cos x = 2\sin x \cos x$ or $4\cos^2 x - 3 = 2\sin x$ or $4(1 - \sin^2 x) - 3 = 2\sin x$. Solve this quadratic equation for $\sin x$ to obtain (a).]

125. Use Exercise 124 to find the exact value of each expression.

   a. $\sin\dfrac{\pi}{5}$ **b.** $\cos\dfrac{\pi}{5}$ **c.** $\sin\dfrac{3\pi}{10}$ **d.** $\cos\dfrac{7\pi}{10}$

126. Use Exercise 124 to find the exact value of each expression.

   a. $\sin\dfrac{\pi}{20}$ **b.** $\cos\dfrac{\pi}{20}$ **c.** $\sin\dfrac{9\pi}{10}$

127. **a.** Use the identity $\sin 2\theta = 2\sin\theta \cos\theta$ and the cofunction identities to show that

$$\sin\theta = 4\sin\dfrac{\theta}{4}\sin\left(\dfrac{\pi}{2} - \dfrac{\theta}{4}\right)\sin\left(\dfrac{\pi}{2} - \dfrac{\theta}{2}\right)$$

   is an identity.

   b. Solve the equation $\theta = \dfrac{\pi}{2} - \dfrac{\theta}{2}$ for θ; use the result in part (a) to show that $\sin\dfrac{\pi}{12}\sin\dfrac{5\pi}{12} = \dfrac{1}{4}$.

c. Solve the equation $\theta = \dfrac{\pi}{2} - \dfrac{\theta}{4}$ for θ; use the result in

part (a) to show that $\sin\dfrac{\pi}{10}\sin\dfrac{3\pi}{10} = \dfrac{1}{4}$.

Critical Thinking / Discussion / Writing

128. Find the values for x in $[0, 2\pi)$ so that $\sin 2x = 2\sin x$.

129. Explain why Exercise 128 does not show that $\sin 2x = 2\sin x$ is an identity.

130. Without using the identity $\sin^2 x = \dfrac{1 - \cos 2x}{2}$, explain

why the value of $\dfrac{1 - \cos 2x}{2}$ is always positive or zero.

131. Use the figure to find the exact value of each expression.

    a. $\sin\theta$ **b.** $\cos\theta$ **c.** $\sin 2\theta$

    d. $\cos 2\theta$ **e.** $\tan 2\theta$ **f.** $\sin\dfrac{\theta}{2}$

    g. $\cos\dfrac{\theta}{2}$ **h.** $\tan\dfrac{\theta}{2}$

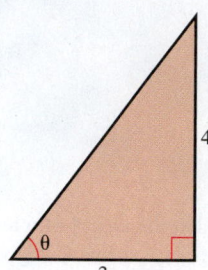

132. Recall that $\cos 60° = \dfrac{1}{2}$. Apply a half-angle identity repeat-

edly to find the exact value of each expression.

    a. $\cos 30°$ **b.** $\cos 15°$ **c.** $\cos 7.5°$

Getting Ready for the Next Section

133. If $f(u + v) + f(u - v) = 2f(u)g(v)$, then

    a. $f(u)g(v) = \dfrac{1}{2}[f(\underline{\quad}) + f(\underline{\quad})]$

    b. $f(x) + f(y) = 2f(\underline{\quad})g(\underline{\quad})$.

134. Use exact values to verify each equation.

    a. $\sin 60° \sin 30° = \dfrac{1}{2}[\cos(60° - 30°) - \cos(60° + 30°)]$

    b. $\sin\dfrac{\pi}{3}\cos\dfrac{\pi}{6} = \dfrac{1}{2}\left[\sin\left(\dfrac{\pi}{3} + \dfrac{\pi}{6}\right) + \sin\left(\dfrac{\pi}{3} - \dfrac{\pi}{6}\right)\right]$

    c. $\cos\dfrac{\pi}{3}\cos\dfrac{\pi}{6} = \dfrac{1}{2}\left[\cos\left(\dfrac{\pi}{3} - \dfrac{\pi}{6}\right) + \cos\left(\dfrac{\pi}{3} + \dfrac{\pi}{6}\right)\right]$

    d. $\cos\dfrac{2\pi}{3} + \cos\dfrac{\pi}{3} = 2\cos\left[\dfrac{1}{2}\left(\dfrac{2\pi}{3} + \dfrac{\pi}{3}\right)\right]\cos\left[\dfrac{1}{2}\left(\dfrac{2\pi}{3} - \dfrac{\pi}{3}\right)\right]$

    e. $\sin 135° - \sin 45° = 2\sin\dfrac{135° - 45°}{2}\cos\dfrac{135° + 45°}{2}$

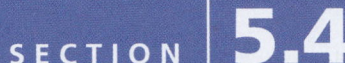

SECTION **5.4**

Product-to-Sum and Sum-to-Product Formulas

BEFORE STARTING THIS SECTION, REVIEW

1 Sum and difference formulas for sines and cosines (Section 5.2, page 458)

2 Fundamental identities (Section 5.1, page 438)

OBJECTIVES

1 Derive product-to-sum formulas.

2 Derive sum-to-product formulas.

3 Verify trigonometric identities involving multiple angles.

◆ Touch-Tone Phones

Dual-tone multi-frequency (DTMF), also known as Touch-Tone, tone dialing, or pushbutton dialing, is a method for instructing a telephone switching network to dial a telephone number. The system was developed by Bell Labs in the late 1950s, and Touch-Tone phones were introduced to the public at the 1964 New York World's Fair. The dual-tone keypad on a Touch-Tone phone has four rows and three columns. Each row represents a *low* frequency, and each column represents a *high* frequency, as shown in the figure. Pressing a button on a Touch-Tone phone produces a unique sound. The sound is produced by the combination of two tones (one of low frequency and the other of high frequency) associated with that button, as shown in the figure. In other words, the sound produced is given by $y = y_1 + y_2$, with $y_1 = \sin(2\pi l t)$ and $y_2 = \sin(2\pi h t)$, where l and h are the button's low and high frequencies (in hertz = Hz), respectively.

For example, if you press a single button, such as "1," it will send a sinusoidal tone of two frequencies 697 Hz and 1209 Hz. The sound produced by pressing "1" is represented by the equation

$$y = \sin[2\pi(697)t] + \sin[2\pi(1209)t]$$

or

$$y = \sin(1394\pi t) + \sin(2418\pi t).$$

The two tones are the reason for calling such phones *dual-tone multi-frequency phones*. In Example 6, we analyze the sound produced by the sum of two sine waves. (*Source:* http://en.wikipedia.org.)

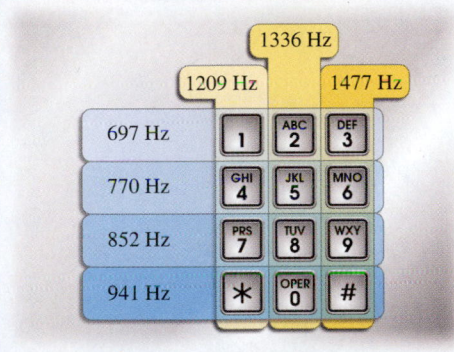

1 Derive product-to-sum formulas.

Product-to-Sum Formulas

The following formulas allow us to rewrite products of sines or cosines as sums or differences.

> ### PRODUCT-TO-SUM FORMULAS
>
> $$\cos x \cos y = \frac{1}{2}[\cos(x - y) + \cos(x + y)]$$
>
> $$\sin x \sin y = \frac{1}{2}[\cos(x - y) - \cos(x + y)]$$
>
> $$\sin x \cos y = \frac{1}{2}[\sin(x + y) + \sin(x - y)]$$
>
> $$\cos x \sin y = \frac{1}{2}[\sin(x + y) - \sin(x - y)]$$

These formulas are fairly easy to prove. For example, to prove the first formula

$$\cos x \cos y = \frac{1}{2}\left[\cos\left(x - y\right) + \cos\left(x + y\right)\right],$$

we write the difference and sum formulas for cosine and add.

$$\cos\left(x - y\right) = \cos x \cos y + \sin x \sin y$$
$$\cos\left(x + y\right) = \cos x \cos y - \sin x \sin y$$

$\cos\left(x - y\right) + \cos\left(x + y\right) = 2\cos x \cos y + 0$	Add.
$2\cos x \cos y = \cos\left(x - y\right) + \cos\left(x + y\right)$	Interchange sides.
$\cos x \cos y = \dfrac{1}{2}\left[\cos\left(x - y\right) + \cos\left(x + y\right)\right]$	Multiply both sides by $\dfrac{1}{2}$.

Similarly, subtracting $\cos\left(x + y\right)$ from $\cos\left(x - y\right)$ gives the second formula:

$$\sin x \sin y = \frac{1}{2}\left[\cos\left(x - y\right) - \cos\left(x + y\right)\right]$$

We can prove the third and fourth formulas in the box on page 473 by using the sum and difference formulas for sine.

EXAMPLE 1 Using a Product-to-Sum Formula

Write $\sin 5\theta \cos 3\theta$ as the sum or difference of two trigonometric functions.

Solution

Use the formula $\sin x \cos y = \dfrac{1}{2}\left[\sin\left(x + y\right) + \sin\left(x - y\right)\right]$.

$\sin 5\theta \cos 3\theta = \dfrac{1}{2}\left[\sin\left(5\theta + 3\theta\right) + \sin\left(5\theta - 3\theta\right)\right]$	Replace x with 5θ and y with 3θ.
$= \dfrac{1}{2}\left[\sin 8\theta + \sin 2\theta\right]$	Simplify.
$= \dfrac{1}{2}\sin 8\theta + \dfrac{1}{2}\sin 2\theta$	Distributive property

Practice Problem 1 Write $\cos 3x \cos x$ as the sum or difference of two trigonometric functions.

EXAMPLE 2 Using a Product-to-Sum Formula

Find the exact value of $\sin 75° \sin 15°$.

Solution

Use the formula $\sin x \sin y = \dfrac{1}{2}\left[\cos\left(x - y\right) - \cos\left(x + y\right)\right]$.

$\sin 75° \sin 15° = \dfrac{1}{2}\left[\cos\left(75° - 15°\right) - \cos\left(75° + 15°\right)\right]$	Replace with 75° and y with 15°.
$= \dfrac{1}{2}\left[\cos 60° - \cos 90°\right]$	Simplify.
$= \dfrac{1}{2}\left[\dfrac{1}{2} - 0\right] = \dfrac{1}{4}$	$\cos 60° = \dfrac{1}{2}; \cos 90° = 0$

Practice Problem 2 Find the exact value of $\sin 15° \cos 75°$.

2 Derive sum-to-product formulas.

Sum-to-Product Formula

The following formulas allow us to do just the opposite—rewrite sums or differences of sines or cosines as products.

SUM-TO-PRODUCT FORMULAS

$$\cos x + \cos y = 2 \cos\left(\frac{x+y}{2}\right) \cos\left(\frac{x-y}{2}\right)$$

$$\cos x - \cos y = -2 \sin\left(\frac{x+y}{2}\right) \sin\left(\frac{x-y}{2}\right)$$

$$\sin x + \sin y = 2 \sin\left(\frac{x+y}{2}\right) \cos\left(\frac{x-y}{2}\right)$$

$$\sin x - \sin y = 2 \sin\left(\frac{x-y}{2}\right) \cos\left(\frac{x+y}{2}\right)$$

We prove these formulas by using the product-to-sum formulas. For example, to prove the first formula, we start with the right side of the equation and use the product-to-sum formulas.

$$2 \cos\left(\frac{x+y}{2}\right) \cos\left(\frac{x-y}{2}\right)$$

$$= 2 \cdot \frac{1}{2}\left[\cos\left(\frac{x+y}{2} - \frac{x-y}{2}\right) + \cos\left(\frac{x+y}{2} + \frac{x-y}{2}\right)\right] \quad \text{Use the formula for the product of cosines.}$$

$$= \cos\left(\frac{x+y-x+y}{2}\right) + \cos\left(\frac{x+y+x-y}{2}\right) \quad \text{Simplify.}$$

$$= \cos\left(\frac{2y}{2}\right) + \cos\left(\frac{2x}{2}\right) \quad \text{Simplify.}$$

$$= \cos y + \cos x \quad \text{Simplify.}$$

You can verify the other three formulas in the box in a similar way. (See Exercise 83.)

EXAMPLE 3   Using Sum-to-Product Formulas

Write each expression as a product of two trigonometric functions and simplify where possible.

a. $\sin 4\theta - \sin 6\theta$ **b.** $\cos 65° + \cos 55°$

Solution

a. Use the formula $\sin x - \sin y = 2 \sin\left(\frac{x-y}{2}\right) \cos\left(\frac{x+y}{2}\right)$.

$$\sin 4\theta - \sin 6\theta = 2 \sin\left(\frac{4\theta - 6\theta}{2}\right) \cos\left(\frac{4\theta + 6\theta}{2}\right) \quad \text{Replace } x \text{ with } 4\theta \text{ and } y \text{ with } 6\theta.$$

$$= 2 \sin(-\theta) \cos(5\theta) \quad \text{Simplify.}$$

$$= -2 \sin\theta \cos 5\theta \quad \sin(-\theta) = -\sin\theta$$

b. Use the formula $\cos x + \cos y = 2 \cos\left(\frac{x+y}{2}\right) \cos\left(\frac{x-y}{2}\right)$.

$$\cos 65° + \cos 55° = 2 \cos\left(\frac{65° + 55°}{2}\right) \cos\left(\frac{65° - 55°}{2}\right) \quad \text{Replace } x \text{ with } 65° \text{ and } y \text{ with } 55°.$$

$$= 2 \cos 60° \cos 5° \quad \text{Simplify.}$$

$$= 2\left(\frac{1}{2}\right) \cos 5° = \cos 5° \quad \cos 60° = \frac{1}{2}$$

Practice Problem 3 Write each expression as a product of two trigonometric functions and simplify where possible.

a. $\cos 2x - \cos 4x$ **b.** $\sin 43° + \sin 17°$

In the next example, we show how to express a sum of a sine and a cosine as a product by first using a cofunction identity.

EXAMPLE 4   **Expressing a Sum of a Sine and a Cosine as a Product**

Write $\sin 5\theta + \cos 3\theta$ as a product of two trigonometric functions.

Solution

Use the formula

$$\cos x + \cos y = 2 \cos \left(\frac{x + y}{2}\right) \cos \left(\frac{x - y}{2}\right).$$

$$\sin 5\theta + \cos 3\theta = \cos \left(\frac{\pi}{2} - 5\theta\right) + \cos 3\theta \qquad \text{Cofunction identity}$$

$$= 2 \cos \left(\frac{\frac{\pi}{2} - 5\theta + 3\theta}{2}\right) \cos \left(\frac{\frac{\pi}{2} - 5\theta - 3\theta}{2}\right) \qquad \begin{array}{l}\text{Sum-to-product} \\ \text{formula for} \\ \text{cosines}\end{array}$$

$$= 2 \cos \left(\frac{\pi}{4} - \theta\right) \cos \left(\frac{\pi}{4} - 4\theta\right) \qquad \text{Simplify.}$$

SIDE NOTE

In Example 4, you could also write $\sin 5\theta + \cos 3\theta =$
$\sin 5\theta + \sin \left(\frac{\pi}{2} - 3\theta\right)$ and
then use the appropriate sum-to-product formula.

Practice Problem 4 Write $\sin 3x - \cos x$ as a product of two trigonometric functions.

3 Verify trigonometric identities involving multiple angles.

Verify Trigonometric Identities

The next example illustrates the use of sum-to-product formulas to verify identities.

EXAMPLE 5   **Verifying an Identity**

Verify the identity $\dfrac{\sin 5\theta + \sin 9\theta}{\cos 5\theta - \cos 9\theta} = \cot 2\theta.$

Solution

$$\frac{\sin 5\theta + \sin 9\theta}{\cos 5\theta - \cos 9\theta} = \frac{2 \sin \dfrac{5\theta + 9\theta}{2} \cos \dfrac{5\theta - 9\theta}{2}}{-2 \sin \dfrac{5\theta + 9\theta}{2} \sin \dfrac{5\theta - 9\theta}{2}} \qquad \begin{array}{l}\text{Use the formulas for} \\ \sin x + \sin y \text{ and } \cos x - \cos y.\end{array}$$

$$= \frac{2 \sin 7\theta \cos (-2\theta)}{-2 \sin 7\theta \sin (-2\theta)} \qquad \text{Simplify.}$$

$$= \frac{\cos (-2\theta)}{-\sin (-2\theta)} \qquad \text{Remove the common factors.}$$

$$= \frac{\cos 2\theta}{\sin 2\theta} \qquad \begin{array}{l}\cos (-x) = \cos x, \text{and} \\ \sin (-x) = -\sin x\end{array}$$

$$= \cot 2\theta \qquad \text{Quotient identity}$$

Practice Problem 5 Verify the identity $\dfrac{\cos 5x - \cos x}{\sin x - \sin 5x} = \tan 3x.$

◆ Analyzing Touch-Tone Phones

When a sound of frequency f_1 is combined with a sound of frequency f_2, the resulting sound has frequency $\dfrac{f_1 + f_2}{2}$. In a Touch-Tone phone, when you press a button with a low frequency l and a high frequency h, the sound produced is modeled by

$$y = \sin(2\pi lt) + \sin(2\pi ht)$$

$$y = 2\sin\left[2\pi\left(\frac{l+h}{2}\right)t\right]\cos\left[2\pi\left(\frac{l-h}{2}\right)t\right] \quad \text{Sum-to-product formula}$$

$$y = 2\cos\left[2\pi\left(\frac{h-l}{2}\right)t\right]\sin\left[2\pi\left(\frac{l+h}{2}\right)t\right] \quad \cos(-\theta) = \cos\theta$$

The last equation suggests that the resulting sound may be thought of as a sine wave with frequency of $\dfrac{l+h}{2}$ and variable amplitude of $2\cos\left[2\pi\left(\dfrac{h-l}{2}\right)t\right]$.

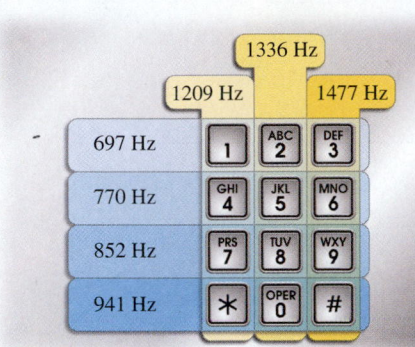

697 Hz   770 Hz   852 Hz   941 Hz

1209 Hz   1336 Hz   1477 Hz

Figure 5.7

EXAMPLE 6 **Touch-Tone Phones**

a. Write an expression that models the tone of the sound produced by pressing the "8" button on your Touch-Tone phone. See Figure 5.7.

b. Rewrite the expression from part **a** as a product of two trigonometric functions.

c. Write the frequency and the variable amplitude of the sound from part **a**.

Solution

a. Pressing the "8" key produces the sound given by

$$y = \sin[2\pi(852)t] + \sin[2\pi(1336)t] \quad l = 852, h = 1336$$

b. $y = 2\sin\left[2\pi\left(\dfrac{852+1336}{2}\right)t\right]\cos\left[2\pi\left(\dfrac{852-1336}{2}\right)t\right]$ Sum-to-product formula

$$= 2\cos[2\pi(242)t]\sin[2\pi(1094)t] \quad \cos(-\theta) = \cos\theta$$

$$= 2\cos(484\pi t)\sin[2\pi(1094)t]$$

c. The frequency is 1094 Hz, and the variable amplitude is $2\cos(484\pi t)$.

Practice Problem 6 Repeat Example 6, assuming that you press the "1" button on your Touch-Tone phone.

continued

TECHNOLOGY
CONNECTION

Graph of
$y = 2\cos(484\pi t)$
$\sin[2\pi(1094)t]$

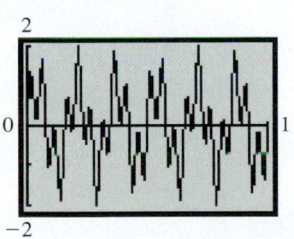

SUMMARY OF **MAIN FACTS**

Product-to-Sum Formulas

$$\cos x \cos y = \frac{1}{2}[\cos(x-y) + \cos(x+y)]$$

$$\sin x \sin y = \frac{1}{2}[\cos(x-y) - \cos(x+y)]$$

$$\sin x \cos y = \frac{1}{2}[\sin(x+y) + \sin(x-y)]$$

$$\cos x \sin y = \frac{1}{2}[\sin(x+y) - \sin(x-y)]$$

Sum-to-Product Formulas

$$\cos x + \cos y = 2\cos\left(\frac{x+y}{2}\right)\cos\left(\frac{x-y}{2}\right)$$

$$\cos x - \cos y = -2\sin\left(\frac{x+y}{2}\right)\sin\left(\frac{x-y}{2}\right)$$

$$\sin x + \sin y = 2\sin\left(\frac{x+y}{2}\right)\cos\left(\frac{x-y}{2}\right)$$

$$\sin x - \sin y = 2\sin\left(\frac{x-y}{2}\right)\cos\left(\frac{x+y}{2}\right)$$

Answers to Practice Problems

1. $\frac{1}{2}\cos 2x + \frac{1}{2}\cos 4x$ 2. $\frac{2-\sqrt{3}}{4}$

3. a. $2\sin 3x \sin x$ b. $\cos 13°$ 4. $2\sin\left(2x - \frac{\pi}{4}\right)\cos\left(x + \frac{\pi}{4}\right)$

6. a. $y = \sin(2\pi \cdot 697t) + \sin(2\pi \cdot 1209t)$
b. $y = 2\cos(512\pi t)\sin(1906\pi t)$
c. $f = 953$ Hz, $A = 2\cos(512\pi t)$

SECTION 5.4 Exercises

Concepts and Vocabulary

1. We can rewrite the product of two sines as a difference of two cosines by using the formula $\sin x \sin y =$ _____.

2. We can rewrite the product of a sine and a cosine as the sum of two sines by using the formula $\sin x \cos y =$ _____.

3. We can rewrite the sum of two cosines as a product of two cosines by using the formula $\cos x + \cos y =$ _____.

4. We can rewrite the sum of two sines by using the formula $\sin x + \sin y =$ _____.

5. **True or False.** $\cos x - \cos y = 2\sin\left(\frac{x+y}{2}\right)\sin\left(\frac{y-x}{2}\right)$.

6. **True or False.** $\cos x \cos y = \frac{1}{2}\left[\cos\left(\frac{x+y}{2}\right) + \cos\left(\frac{x-y}{2}\right)\right]$

7. **True or False.** $\sin x + \sin y = \sin(x+y)$.

8. **True or False.** $\sin(2x+1) - \sin(2x-1)$ $= 2\sin(1)\cos 2x$.

Building Skills

In Exercises 9–24, use the product-to-sum formulas to rewrite each expression as the sum or difference of two functions. Simplify where possible.

9. $\sin x \cos x$
10. $\cos x \cos x$
11. $\sin x \sin x$
12. $\cos x \sin x$

13. $\sin 25° \cos 5°$
14. $\sin 40° \sin 20°$
15. $\cos 140° \cos 20°$
16. $\cos 70° \sin 20°$
17. $\sin\frac{7\pi}{12}\sin\frac{\pi}{12}$
18. $\sin\frac{3\pi}{8}\cos\frac{\pi}{8}$
19. $\cos\frac{5\pi}{8}\sin\frac{\pi}{8}$
20. $\cos\frac{5\pi}{3}\cos\frac{\pi}{3}$
21. $\sin 5\theta \cos\theta$
22. $\cos 3\theta \sin 2\theta$
23. $\cos 4x \cos 3x$
24. $\sin 5x \sin 2x$

In Exercises 25–32, find the exact value of each expression.

25. $\sin 37.5° \sin 7.5°$
26. $\cos 52.5° \cos 7.5°$
27. $\sin 67.5° \cos 22.5°$
28. $\cos 105° \sin 75°$
29. $\sin\frac{5\pi}{24}\cos\frac{\pi}{24}$
30. $\sin\frac{7\pi}{12}\sin\frac{\pi}{12}$
31. $\cos\frac{13\pi}{24}\cos\frac{5\pi}{24}$
32. $\cos\frac{7\pi}{24}\sin\frac{\pi}{24}$

In Exercises 33–52, use sum-to-product formulas to rewrite each expression as a product. Simplify where possible.

33. $\cos 40° - \cos 20°$
34. $\sin 22° + \sin 8°$
35. $\sin 32° - \sin 16°$
36. $\cos 47° + \cos 13°$
37. $\sin\frac{\pi}{5} + \sin\frac{2\pi}{5}$
38. $\cos\frac{\pi}{12} + \cos\frac{\pi}{3}$

39. $\cos\dfrac{1}{2} + \cos\dfrac{1}{3}$

40. $\sin\dfrac{2}{3} - \sin\dfrac{1}{4}$

41. $\cos 3x + \cos 5x$

42. $\sin 5x - \sin 3x$

43. $\sin 7x + \sin(-x)$

44. $\cos 7x - \cos 3x$

45. $\sin x + \cos x$

46. $\cos x - \sin x$

47. $\sin 2x - \cos 2x$

48. $\cos 3x + \sin 3x$

49. $\sin 3x + \cos 5x$

50. $\sin 5x - \cos x$

51. $a(\sin x + \cos x)$

52. $a(\sin bx + \cos bx)$

In Exercises 53–62, verify each identity.

53. $\dfrac{\sin x + \sin 3x}{\cos x + \cos 3x} = \tan 2x$

54. $\dfrac{\sin 2x + \sin 4x}{\cos 2x + \cos 4x} = \tan 3x$

55. $\dfrac{\cos 3x - \cos 7x}{\sin 7x + \sin 3x} = \tan 2x$

56. $\dfrac{\cos 12x - \cos 4x}{\sin 4x - \sin 12x} = \tan 8x$

57. $\dfrac{\cos 2x + \cos 2y}{\cos 2x - \cos 2y} = \cot(y - x)\cot(y + x)$

58. $\dfrac{\sin 2x + \sin 2y}{\sin 2x - \sin 2y} = \dfrac{\tan(x + y)}{\tan(x - y)}$

59. $\sin x + \sin 2x + \sin 3x = \sin 2x(1 + 2\cos x)$

60. $\cos x + \cos 2x + \cos 3x = \cos 2x(1 + 2\cos x)$

61. $\sin 2x + \sin 4x + \sin 6x = 4\cos x\cos 2x\sin 3x$

62. $\cos x + \cos 3x + \cos 5x + \cos 7x = 4\cos x\cos 2x\cos 4x$

Applying the Concepts

63. Touch-Tone phone.

   a. Write an expression that models the tone by pressing the "7" button on your Touch-Tone phone. (See Figure 5.7, page 477.)

   b. Rewrite the expression from part (a) as a product of two functions.

   c. Write the frequency and the variable amplitude of the sound from part (a).

64. Touch-Tone phone. Repeat Exercise 63 assuming that you press the "#" button on your Touch-Tone phone.

65. Beats in music. When two musical tones of slightly different frequencies are played, the sound you hear will fluctuate in volume according to the difference in their frequencies. These are called *beat frequencies*. (Piano tuners use the beat frequency to adjust piano wire until it is at the same frequency as the tuning fork.) Let $y_1 = 0.05\cos(112\pi t)$ and $y_2 = 0.05\cos(120\pi t)$ model two tones.

   a. Write $y = y_1 + y_2$ as the product of two functions.

   b. How many beats are there?

66. Identical waves. Two identical waves (same frequency, velocity, and amplitude) traveling along a string in opposite directions can be represented by the equations $y_1 = a\sin(\omega t + nx)$ and $y_2 = a\sin(\omega t - nx)$. Show that the resultant wave $y = y_1 + y_2$ can be expressed as $y = A\sin\omega t$, where $A = 2a\cos nx$ is the amplitude of the resultant motion.

Beyond the Basics

In Exercises 67 and 68, sketch the graph of f. What are the amplitude and period?

67. $f(x) = \cos(2x + 1) + \cos(2x - 1)$

68. $f(x) = \sin(3x + 1) + \sin(3x - 1)$

In Exercises 69–82, verify each identity.

69. $\cos 40° + \cos 50° + \cos 70° + \cos 80° = \cos 10° + \cos 20°$

70. $\sin 10° + \sin 20° + \sin 40° + \sin 50° = 2\cos 5°\cos 15°$

71. $\cos 4\theta\cos\theta - \cos 6\theta\cos 9\theta = \sin 5\theta\sin 10\theta$

72. $\cos 38°\cos 46° - \sin 14°\sin 22° = \dfrac{1}{2}\cos 24°$

73. $\sin 25°\sin 35° - \sin 25°\sin 85° - \sin 35°\sin 85° = -\dfrac{3}{4}$

74. $\cos 20°\cos 40°\cos 60°\cos 80° = \dfrac{1}{16}$

75. $\dfrac{\sin 3x\cos 5x - \sin x\cos 7x}{\sin x\sin 7x + \cos 3x\cos 5x} = \tan 2x$

76. $\dfrac{\sin 11x\sin x + \sin 7x\sin 3x}{\cos 11x\sin x + \cos 7x\sin 3x} = \tan 8x$

77. $\dfrac{\cos x + \cos 3x + \cos 5x + \cos 7x}{\sin x + \sin 3x + \sin 5x + \sin 7x} = \cot 4x$

78. $\dfrac{\sin 3x + \sin 5x + \sin 7x + \sin 9x}{\cos 3x + \cos 5x + \cos 7x + \cos 9x} = \tan 6x$

79. $\cos 7x = 2\cos 6x\cos x - 2\cos 4x\cos x + 2\cos 2x\cos x - \cos x.$

80. a. $4\sin\theta\sin\left(\dfrac{\pi}{3} + \theta\right)\sin\left(\dfrac{\pi}{3} - \theta\right) = \sin 3\theta.$

     [*Hint:* Recall that $\sin 3\theta = 3\sin\theta - 4\sin^3\theta.$]

   b. Use part (a) to show that

$$\sin 20°\sin 40°\sin 60°\sin 80° = \dfrac{3}{16}.$$

81. $\dfrac{\sin(x + 3y) + \sin(3x + y)}{\sin 2x + \sin 2y} = 2\cos(x + y).$

82. $\sin^3 x\sin 3x + \cos^3 x\cos 3x = \cos^3 2x.$

83. Verify the following sum-to-product formulas.

   a. $\cos x - \cos y = -2\sin\left(\dfrac{x + y}{2}\right)\sin\left(\dfrac{x - y}{2}\right)$

   b. $\sin x + \sin y = 2\sin\left(\dfrac{x + y}{2}\right)\cos\left(\dfrac{x - y}{2}\right)$

   c. $\sin x - \sin y = 2\sin\left(\dfrac{x - y}{2}\right)\cos\left(\dfrac{x + y}{2}\right)$

Critical Thinking / Discussion / Writing

84. Supply the reason for each step in verifying the identity

$$\sin\frac{\pi}{14}\sin\frac{3\pi}{14}\sin\frac{5\pi}{14}=\frac{1}{8}.$$

$$\sin\frac{\pi}{14}\sin\frac{3\pi}{14}\sin\frac{5\pi}{14}=\cos\frac{3\pi}{7}\cos\frac{2\pi}{7}\cos\frac{\pi}{7}$$

$$=\frac{8\sin\frac{\pi}{7}\cos\frac{\pi}{7}\cos\frac{2\pi}{7}\cos\frac{3\pi}{7}}{8\sin\frac{\pi}{7}}$$

$$=\frac{2\sin\frac{4\pi}{7}\cos\frac{3\pi}{7}}{8\sin\frac{\pi}{7}}$$

$$=\frac{\sin\pi+\sin\frac{\pi}{7}}{8\sin\frac{\pi}{7}}$$

$$=\frac{1}{8}$$

In Exercises 85 and 86 assume that $A+B+C=180°$. Verify each identity.

85. $\sin 2A + \sin 2B + \sin 2C = 4\sin A\sin B\sin C$

86. $\cos 2A + \cos 2B + \cos 2C = -1 - 4\cos A\cos B\cos C$

87. Explain how to obtain the identity $\cos^2 x = \dfrac{1+\cos 2x}{2}$ using the product-to-sum formula for $\cos x\cos y$.

88. Use product-to-sum formula for $\sin x\sin y$ to obtain the identity $\sin^2 x = \dfrac{1-\cos 2x}{2}$.

Getting Ready for the Next Section

In Exercises 89–100, identify each equation as an identity, a conditional equation, or an inconsistent equation.

89. $2x - 3 = 9$

90. $x^2 - 6x + 8 = 0$

91. $x^2 + 4 = 0$

92. $2x + 3 = (3x + 1) - (x - 5)$

93. $x^2 - 4 = (x + 2)(x - 2)$

94. $\dfrac{1}{(x-1)(x+2)} = \dfrac{1}{x^2 + x - 2}$

95. $\sin x = \dfrac{1}{2}$

96. $\sin x = \cos x$

97. $\sin x = -2$

98. $\cos x = 3$

99. $\sin 2x = 2\sin x\cos x$

100. $\cos^2 x = 1 - \sin^2 x$

In Exercises 101–106, solve each equation.

101. $(2x - 1) - 3 = 6$

102. $3x - 2 = (x + 1) + (x - 5)$

103. $(x + 2)(x - 3) = 0$

104. $x^2 - x - 2 = 0$

105. $\sqrt{x + 1} = x - 5$

106. $\sqrt{x + 1} + 1 = x$

In Exercises 107–114, find the two values of x in the interval $[0, 2\pi)$ for which the given equation is true.

107. $\sin x = \dfrac{1}{2}$

108. $\sin x = -\dfrac{1}{2}$

109. $\tan x = -1$

110. $\cot x = -\sqrt{3}$

111. $\csc x = -\dfrac{2\sqrt{3}}{3}$

112. $\sec x = 2$

113. $\cos x = -\dfrac{1}{2}$

114. $\cos x = \dfrac{\sqrt{2}}{2}$

Trigonometric Equations

BEFORE STARTING THIS SECTION, REVIEW

1 Reference angle (Section 4.3, page 369)

2 Periods of trigonometric functions (Sections 4.4, page 383)

3 Factoring techniques (Appendix A.2, page 930)

4 Solving linear and quadratic equations (Appendix A.6, page 956)

OBJECTIVES

1 Solve trigonometric equations of the form $a \sin(x - c) = k$, $a \cos(x - c) = k$, and $a \tan(x - c) = k$.

2 Solve trigonometric equations involving multiple angles.

3 Solve trigonometric equations by using the zero-product property.

4 Solve trigonometric equations that contain more than one trigonometric function.

5 Solve trigonometric equations by squaring both sides.

◆ The Phases of the Moon

The moon orbits Earth and completes one revolution in about 29.5 days (a lunar month). During this period, the angle between the sun, Earth, and the moon changes. See Figure 5.8. This causes different amounts of the illuminated moon to face Earth; consequently, the shape of the moon appears to change. The shape varies from a **new moon**—the phase of the moon when the moon is not visible from Earth (this happens when the moon is between the sun and Earth)—to a **half moon**—the phase of the moon when the moon looks like half a circular disk—to a **full moon**—the phase of the moon when the moon looks like a full circular disk in the sky—back to a half moon and then to a new moon. The portion of the moon visible from Earth on any given day is called the **phase of the moon**. In Example 11, we express the phases of the moon by a sinusoidal equation.

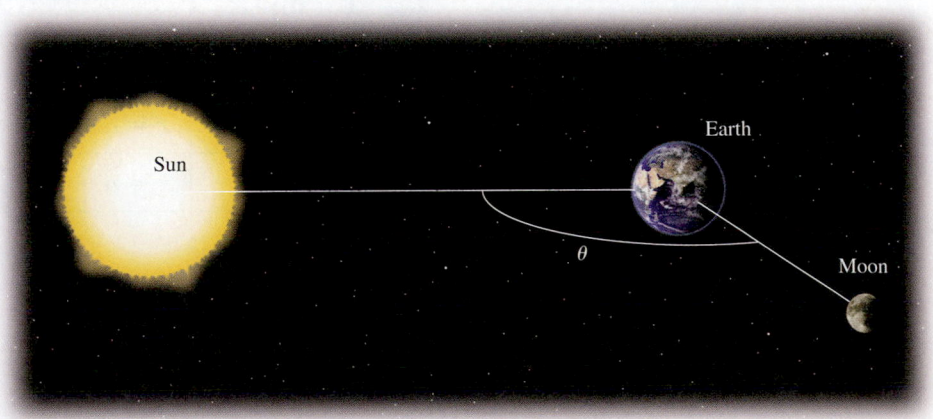

Figure 5.8   Phases of the moon

Trigonometric Equations

Recall that a *trigonometric equation* is an equation that contains a trigonometric function. An identity is an equation that is true for all values in the domain of the variable. We verified trigonometric identities in Section 5.1. Now we will work with trigonometric equations that are satisfied by some but not all values of the variable. For example, $\sin x = \dfrac{1}{2}$ is satisfied by $x = \dfrac{\pi}{6}$ but not by $x = 0$.

Solving a trigonometric equation means finding its *solution set*. Some trigonometric equations do not have a solution. For example, the equation $\sin x = 2$ has no solution because $-1 \le \sin x \le 1$ for all real numbers x. However, if a trigonometric equation such as $\sin x = k$ has a solution, then it has infinitely many solutions because the sine function is periodic. Also note that we cannot find an exact solution for every equation. In such cases, we can approximate the solution by using a calculator.

To solve trigonometric equations, we first recall some facts about the sine, cosine, and tangent functions.

1. For any integer n,

    (i) $\sin \theta = 0$ if any only if $\theta = n\pi$.

    (ii) $\cos \theta = 0$ if any only if $\theta = \dfrac{\pi}{2} + n\pi$.

    (iii) $\tan \theta = 0$ if any only if $\theta = n\pi$.

(These facts follow from the definitions of these functions.)

2. (i) $\sin(\pi - \theta) = \sin \theta$
    (ii) $\cos(2\pi - \theta) = \cos \theta$
    (iii) $\tan(\pi + \theta) = \tan \theta$.

(You can use symmetry or sum and difference formulas to verify these identities.)

We can use these facts and the periodic behavior of the trigonometric functions to find all solutions of the equations in Table 5.1 for any given angle α.

TABLE 5.1 Basic Trigonometric Equations

Equation	All solutions
$\sin x = \sin \alpha$	$x = \alpha + 2n\pi$ or $x = (\pi - \alpha) + 2n\pi$
$\cos x = \cos \alpha$	$x = \alpha + 2n\pi$ or $x = (2\pi - \alpha) + 2n\pi$
$\tan x = \tan \alpha$	$x = \alpha + n\pi$

For α given in degrees, replace π with $180°$ and $2n\pi$ with $n \cdot 360°$.

1 Solve trigonometric equations of the form $a\sin(x - c) = k$, $a\cos(x - c) = k$, and $a\tan(x - c) = k$.

Trigonometric Equations of the Form $a\sin(x - c) = k$, $a\cos(x - c) = k$, and $a\tan(x - c) = k$

We begin with some basic equations.

PROCEDURE
IN ACTION

EXAMPLE **1** **Solving Trigonometric Equations**

OBJECTIVE

Solve trigonometric equations of the form $T(x) = k$ where $T(x)$ is one of the six trigonometric functions.

Step 1 Find $\alpha = T^{-1}(k)$, and write $T(x) = T(\alpha)$.
a. For common angles, find α from memory or page 360 and symmetry; for other angles use a calculator.
b. If $T^{-1}(k)$ does not exist, the equation $T(x) = k$ has no solution.

Step 2 Use Table 5.1 to find all solutions of the equation $T(x) = T(\alpha)$.

Step 3 Try integer values of n to find all solutions of the equation in the given interval.

EXAMPLES

Find all solutions of each equation in the interval $[0, 2\pi)$. Express all solutions in radians.

a. $\sin x = \dfrac{\sqrt{2}}{2}$

b. $\cos x = -\dfrac{\sqrt{3}}{2}$

c. $\tan x = -\sqrt{3}$

1. $\sin^{-1}\left(\dfrac{\sqrt{2}}{2}\right) = \dfrac{\pi}{4}$.
So, $\sin x = \sin\dfrac{\pi}{4}$.

1. $\cos^{-1}\left(-\dfrac{\sqrt{3}}{2}\right) = \dfrac{5\pi}{6}$.
So, $\cos x = \cos\dfrac{5\pi}{6}$.

1. $\tan^{-1}\left(-\sqrt{3}\right) = -\dfrac{\pi}{3}$.
So, $\tan x = \tan\left(-\dfrac{\pi}{3}\right)$.

2. All solutions of
$\sin x = \sin\dfrac{\pi}{4}$ are:

$x = \dfrac{\pi}{4} + 2n\pi$, and

$x = \left(\pi - \dfrac{\pi}{4}\right) + 2n\pi$

$= \dfrac{3\pi}{4} + 2n\pi$.

2. All solutions of
$\cos x = \cos\dfrac{5\pi}{6}$ are:

$x = \dfrac{5\pi}{6} + 2n\pi$, and

$x = \left(2\pi - \dfrac{5\pi}{6}\right) + 2n\pi$

$= \dfrac{7\pi}{6} + 2n\pi$.

2. All solutions of
$\tan x = \tan\left(-\dfrac{\pi}{3}\right)$

are: $x = -\dfrac{\pi}{3} + n\pi$.

3. The solution set in the interval $[0, 2\pi)$, for $n = 0$ is $\left\{\dfrac{\pi}{4}, \dfrac{3\pi}{4}\right\}$.

3. The solution set in the interval $[0, 2\pi)$, for $n = 0$ is $\left\{\dfrac{5\pi}{6}, \dfrac{7\pi}{6}\right\}$.

3. The solution set in the interval $[0, 2\pi)$, for $n = 1$ and $n = 2$ is $\left\{\dfrac{2\pi}{3}, \dfrac{5\pi}{3}\right\}$.

Practice Problem 1 Find all solutions of each equation. Express solutions in radians.

a. $\sin x = 1$ **b.** $\cos x = 1$ **c.** $\tan x = 1$

In Example 1, we were able to find the exact solutions of trigonometric equations because the angles involved were common angles. In the next example, we find approximate solutions of trigonometric equations involving other angles.

EXAMPLE **2** **Finding Approximate Solutions**

a. Find all solutions of $\sin x = 0.3$. Round solutions to the nearest tenth of a degree.
b. Find all solutions of $\cot x = -3.5$ in the interval $[0, 2\pi)$. Round solutions to four decimal places.

Solution

a. **Step 1** We use the $\sin^{-1}$ function to find α.

$$\alpha = \sin^{-1}(0.3) \approx 17.5° \quad \text{Use a calculator in Degree mode.}$$

So, $\sin x = \sin(17.5°)$

Step 2 All solutions are given by
$x \approx 17.5° + n \cdot 360°$ and $x \approx 180° - 17.5° + n \cdot 360° = 162.5° + n \cdot 360°$, for any integer n.

b. Step 1 Because most calculators do not have a key for the cotangent function, we use the equivalent tangent function value.

$$\cot x = -3.5 = -\frac{7}{2}$$

$$\tan x = -\frac{2}{7} \qquad \text{Reciprocal identity}$$

$$\alpha = \tan^{-1}\left(-\frac{2}{7}\right)$$

$$\alpha \approx -0.2783 \qquad \text{Use a calculator in Radian mode.}$$

Step 2 All solutions of $\cot x = -3.5$ or equivalently $\tan x = -\frac{2}{7}$ are given by:

$$x \approx -0.2783 + n\pi.$$

For $n = 1$ and $n = 2$, the two approximate solutions in the interval $[0, 2\pi)$ are:

$$x \approx 2.8633 \quad \text{and} \quad x \approx 6.0049.$$

Practice Problem 2

a. Find all solutions of $\sec x = 1.5$. Round solutions to the nearest tenth of a degree.

b. Find all solutions of $\tan x = -2$ in the interval $[0, 2\pi)$. Round solutions to four decimal places.

EXAMPLE 3 **Solving a Linear Trigonometric Equation**

Find all solutions in the interval $[0, 2\pi)$ of the equation

$$2\sin\left(x - \frac{\pi}{4}\right) + 1 = 2.$$

Solution

Step 1 $2\sin\left(x - \frac{\pi}{4}\right) + 1 = 2$

$$\sin\left(x - \frac{\pi}{4}\right) = \frac{1}{2} \qquad \text{Solve for } \sin\left(x - \frac{\pi}{4}\right).$$

$$\alpha = \sin^{-1}\left(\frac{1}{2}\right) = \frac{\pi}{6}, \text{ so}$$

$$\sin\left(x - \frac{\pi}{4}\right) = \sin\frac{\pi}{6}.$$

Step 2 All solutions of $\sin\left(x - \frac{\pi}{4}\right) = \sin\frac{\pi}{6}$ are:

$$x - \frac{\pi}{4} = \frac{\pi}{6} + 2n\pi \qquad \text{and} \qquad x - \frac{\pi}{4} = \left(\pi - \frac{\pi}{6}\right) + 2n\pi = \frac{5\pi}{6} + 2n\pi$$

$$x = \frac{\pi}{6} + \frac{\pi}{4} + 2n\pi \qquad\qquad x = \frac{5\pi}{6} + \frac{\pi}{4} + 2n\pi \qquad \text{Add } \frac{\pi}{4} \text{ to both sides.}$$

$$x = \frac{5\pi}{12} + 2n\pi \qquad\qquad x = \frac{13\pi}{12} + 2n\pi \qquad \text{Simplify.}$$

Step 3 For $n = 0$, the solution set in the interval $[0, 2\pi)$ is $\left\{\frac{5\pi}{12}, \frac{13\pi}{12}\right\}$.

All other values of n result in solutions outside the interval $[0, 2\pi)$.

Practice Problem 3 Find all solutions in the interval $[0°, 360°)$ of the equation

$$3\sec(x - 30°) - 1 = \sec(x - 30°) + 3.$$

2 Solve trigonometric equations involving multiple angles.

Equations Involving Multiple Angles

If x is the measure of an angle, then for any real number k, the number kx is a **multiple angle** of x. Equations such as

$$\sin 3x = \frac{1}{2}, \quad \cos \frac{1}{2}x = 0.7, \quad \text{and} \quad \sin 2x + \sin 4x = 0$$

involve multiple angles. You must use care when finding all solutions of such equations in the interval $[0, 2\pi)$, as you will see in Example 4.

EXAMPLE 4 Solving a Trigonometric Equation Containing a Multiple Angle

Find all solutions of the equation $\cos 2x = \frac{1}{2}$ in the interval $[0, 2\pi)$.

Solution

Step 1 $\alpha = \cos^{-1}\left(\frac{1}{2}\right) = \frac{\pi}{3}$. So, $\cos 2x = \cos \frac{\pi}{3}$.

Step 2 All solutions of the equation $\cos 2x = \cos \frac{\pi}{3}$ are given by Table 5.1:

$$2x = \frac{\pi}{3} + 2n\pi \quad \text{and} \quad 2x = \left(2\pi - \frac{\pi}{3}\right) + 2n\pi = \frac{5\pi}{3} + 2n\pi \qquad \text{For any integer } n$$

$$x = \frac{\pi}{6} + n\pi \quad \text{and} \quad x = \frac{5\pi}{6} + n\pi \qquad \text{Divide both sides by 2.}$$

Step 3 To find solutions in the interval $[0, 2\pi)$, try the following:

$n = -1$	$x = \frac{\pi}{6} - \pi = -\frac{5\pi}{6}$	$x = \frac{5\pi}{6} - \pi = -\frac{\pi}{6}$
$n = 0$	$x = \frac{\pi}{6}$	$x = \frac{5\pi}{6}$
$n = 1$	$x = \frac{\pi}{6} + \pi = \frac{7\pi}{6}$	$x = \frac{5\pi}{6} + \pi = \frac{11\pi}{6}$
$n = 2$	$x = \frac{\pi}{6} + 2\pi = \frac{13\pi}{6}$	$x = \frac{5\pi}{6} + 2\pi = \frac{17\pi}{6}$

The values resulting from $n = -1$ are too small and those resulting from $n = 2$ are too large to be in the interval $[0, 2\pi)$. The solutions corresponding to $n = 0$ and 1 are:

$$\left\{\frac{\pi}{6}, \frac{5\pi}{6}, \frac{7\pi}{6}, \frac{11\pi}{6}\right\}.$$

Practice Problem 4 Find all solutions of the equation $\sin 2x = \frac{1}{2}$ in the interval $[0, 2\pi)$.

EXAMPLE 5 Solving a Trigonometric Equation Containing a Multiple Angle

Find all solutions of the equation $\sin \frac{x}{2} = -\frac{\sqrt{3}}{2}$ in the interval $[0, 2\pi)$.

Solution

Step 1 $\alpha = \sin^{-1}\left(-\frac{\sqrt{3}}{2}\right) = -\frac{\pi}{3}$. So, $\sin \frac{x}{2} = \sin\left(-\frac{\pi}{3}\right)$.

TECHNOLOGY CONNECTION

On a graphing calculator, you can see that the graphs of

$$Y_1 = \cos 2x \text{ and } Y_2 = \frac{1}{2}$$

intersect in four points, which correspond to the four solutions in the interval $[0, 2\pi)$.

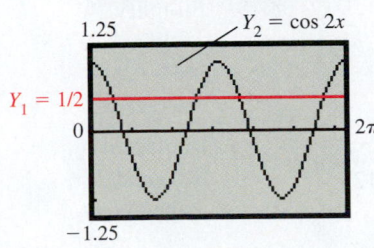

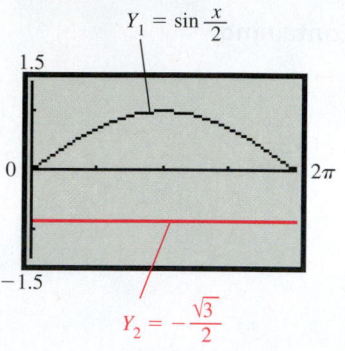
Step 2 All solutions of $\sin \dfrac{x}{2} = \sin\left(-\dfrac{\pi}{3}\right)$ are given by Table 5.1:

$$\dfrac{x}{2} = -\dfrac{\pi}{3} + 2n\pi \quad \text{and} \quad \dfrac{x}{2} = \pi - \left(-\dfrac{\pi}{3}\right) + 2n\pi = \dfrac{4\pi}{3} + 2n\pi \qquad \text{For any integer } n$$

$$x = -\dfrac{2\pi}{3} + 4n\pi \quad \text{and} \quad x = \dfrac{8\pi}{3} + 4n\pi \qquad \begin{array}{l}\text{Multiply both} \\ \text{sides by 2.}\end{array}$$

Step 3 For any integer n, the values of x in both expressions in step 2 lie outside the interval $[0, 2\pi)$. For example, for $n = 0$, $x = -\dfrac{2\pi}{3} < 0$ and for $n = 1$, $x = -\dfrac{2\pi}{3} + 4\pi = \dfrac{10\pi}{3} > 2\pi$. The equation $\sin \dfrac{x}{2} = -\dfrac{\sqrt{3}}{2}$, therefore, has no solution in the interval $[0, 2\pi)$, so the solution set is $\varnothing$.

Practice Problem 5 Find all solutions of the equation $\tan \dfrac{x}{2} = \dfrac{1}{\sqrt{3}}$ in the interval $[0, 2\pi)$.

The next example illustrates that the previous methods can be used to solve equations of the type $\sin u \pm \sin v = 0$, $\sin u \pm \cos v = 0$, and $\cos u \pm \cos v = 0$, where u and v are linear functions of x, that is, $u = ax + b$ and $v = cx + d$.

EXAMPLE 6 **Using Complements to Solve $\sin u \pm \cos v = 0$**

Find all solutions of the equation $\cos 2x + \sin\left(x + \dfrac{\pi}{4}\right) = 0$ in the interval $[0, 2\pi)$.

Solution

$$\cos 2x = -\sin\left(x + \dfrac{\pi}{4}\right) \qquad \begin{array}{l}\text{Add } -\sin\left(x + \dfrac{\pi}{4}\right) \text{ to both sides of the original} \\ \text{equation.}\end{array}$$

$$\cos 2x = \sin\left(-x - \dfrac{\pi}{4}\right) \qquad -\sin\theta = \sin(-\theta)$$

$$\cos 2x = \cos\left(\dfrac{\pi}{2} + x + \dfrac{\pi}{4}\right) \qquad \sin\theta = \cos\left(\dfrac{\pi}{2} - \theta\right); -\left(-x - \dfrac{\pi}{4}\right) = x + \dfrac{\pi}{4}$$

$$\cos 2x = \cos\left(x + \dfrac{3\pi}{4}\right) \qquad \text{Simplify.}$$

$$2x = x + \dfrac{3\pi}{4} + 2n\pi \quad \text{or} \quad 2x = 2\pi - \left(x + \dfrac{3\pi}{4}\right) + 2n\pi \qquad \text{Table 5.1}$$

$$x = \dfrac{3\pi}{4} + 2n\pi \qquad\qquad 3x = \dfrac{5\pi}{4} + 2n\pi \qquad \text{Simplify.}$$

$$x = \dfrac{3\pi}{4} \,(n = 0) \qquad\qquad x = \dfrac{5\pi}{12} + \dfrac{2n\pi}{3} \qquad \begin{array}{l}\text{Divide both} \\ \text{sides by 3.}\end{array}$$

$$x = \dfrac{5\pi}{12} \,(n = 0), x = \dfrac{13\pi}{12} \,(n = 1) \qquad \begin{array}{l}\text{Replace } n \text{ by} \\ 0, 1, \text{ and } 2, \\ \text{respectively.}\end{array}$$

$$x = \dfrac{7\pi}{4} \,(n = 2).$$

All other values of n result in solutions outside the interval $[0, 2\pi)$.

The solution set in the interval $[0, 2\pi)$ is $\left\{\dfrac{5\pi}{12}, \dfrac{3\pi}{4}, \dfrac{13\pi}{12}, \dfrac{7\pi}{4}\right\}$.

Practice Problem 6 Find all solutions of the equation $\sin 2x = \cos\left(x - \dfrac{\pi}{4}\right)$ in the interval $[0, 2\pi)$.

Trigonometric Equations and the Zero-Product Property

3 Solve trigonometric equations by using the zero-product property.

EXAMPLE 7 Solving an Equation by Using the Zero-Product Property

Find all solutions of the equation $(2 \sin x - \sqrt{2})(7 \tan x + 2) = 0$ in the interval $[0, 2\pi)$. Round solutions to multiples of π or four decimal places.

Solution

$$(2 \sin x - \sqrt{2})(7 \tan x + 2) = 0 \qquad \text{Given equation}$$

$$2 \sin x - \sqrt{2} = 0 \quad \text{or} \quad 7 \tan x + 2 = 0 \qquad \text{Zero-product property}$$

$$\sin x = \dfrac{\sqrt{2}}{2} \qquad\qquad \tan x = -\dfrac{2}{7} \qquad \text{Solve for } \sin x \text{ and } \tan x.$$

$$x = \dfrac{\pi}{4} \quad \text{and} \quad x = \dfrac{3\pi}{4} \quad\Big|\quad x \approx 2.8633 \quad \text{and} \quad x \approx 6.0049$$

(See Example **1a**.) $\qquad$ (See Example **2b**.)

So the solution set of the given equation is

$$\left\{ \dfrac{\pi}{4}, \dfrac{3\pi}{4}, 2.8633, 6.0049 \right\}.$$

Practice Problem 7 Find all solutions of the equation

$$(\sin x - 1)(\sqrt{3} \tan x + 1) = 0$$

in the interval $[0, 2\pi)$.

TECHNOLOGY CONNECTION

You can confirm that the equation in Example 8 has the two solutions we found in $[0, 2\pi)$ by graphing $Y_1 = 2 \sin^2 x - 5 \sin x + 2$ and inspecting the x-intercepts using $\boxed{\text{TRACE}}$.

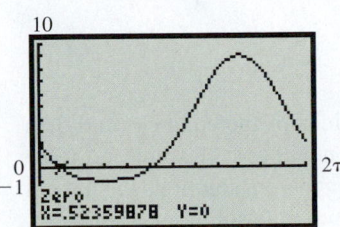

EXAMPLE 8 Solving a Trigonometric Equation of Quadratic Type

Find all solutions of the equation $2 \sin^2 \theta - 5 \sin \theta + 2 = 0$. Express the solutions in radians.

Solution

The equation is quadratic in $\sin \theta$. We will use the zero-product property to solve for $\sin \theta$ and then solve the resulting equation for θ.

$$2 \sin^2 \theta - 5 \sin \theta + 2 = 0 \qquad 2x^2 - 5x + 2 = 0, \text{ if } x = \sin \theta$$

$$(2 \sin \theta - 1)(\sin \theta - 2) = 0 \qquad (2x - 1)(x - 2) = 0; \text{ factor.}$$

$$2 \sin \theta - 1 = 0 \quad \text{or} \quad \sin \theta - 2 = 0 \qquad \text{Zero-product property}$$

$$\sin \theta = \dfrac{1}{2} \qquad\qquad \sin \theta = 2 \qquad \text{Solve for } \sin \theta.$$

$$\alpha = \sin^{-1} 2 \text{ does not exist.}$$

$$\text{No solution.}$$

$\alpha = \sin^{-1}\left(\dfrac{1}{2}\right) = \dfrac{\pi}{6}$. So, $\sin \theta = \sin \dfrac{\pi}{6}$. All solutions of $\sin \theta = \sin \dfrac{\pi}{6}$ are given by:

$$\theta = \dfrac{\pi}{6} + 2n\pi \quad \text{and} \quad \theta = \left(\pi - \dfrac{\pi}{6}\right) + 2n\pi = \dfrac{5\pi}{6} + 2n\pi, \text{ for any integer } n.$$

Practice Problem 8 Find all solutions of the equation $2 \cos^2 \theta - \cos \theta - 1 = 0$. Express the solutions in radians.

4 Solve trigonometric equations that contain more than one trigonometric function.

Equations with More Than One Trigonometric Function

When a trigonometric equation contains more than one trigonometric function, we can sometimes use fundamental trigonometric identities to convert the equation into an equation containing only one trigonometric function.

> **EXAMPLE 9** Solving a Trigonometric Equation Using Identities

Find all solutions of the equation $2\sin^2\theta - \cos\theta - 1 = 0$ in the interval $[0, 2\pi)$.

Solution

Because the equation contains sines and cosines, we will use the Pythagorean identity $\sin^2\theta + \cos^2\theta = 1$ to convert the equation into one containing only cosines.

$$2\sin^2\theta - \cos\theta - 1 = 0 \qquad \text{Given equation}$$
$$2(1 - \cos^2\theta) - \cos\theta - 1 = 0 \qquad \text{Replace } \sin^2\theta \text{ with } 1 - \cos^2\theta.$$
$$2 - 2\cos^2\theta - \cos\theta - 1 = 0 \qquad \text{Distributive property}$$
$$-2\cos^2\theta - \cos\theta + 1 = 0 \qquad \text{Combine terms and rearrange.}$$
$$2\cos^2\theta + \cos\theta - 1 = 0 \qquad \text{Multiply both sides by } -1.$$

So $\quad (2\cos\theta - 1)(\cos\theta + 1) = 0 \qquad \text{Factor}$

SIDE NOTE

From $\cos\theta = -1$, we have $\cos^{-1}(-1) = \pi$. Since $\pi = 2\pi - \pi$, $\pi + 2\pi n$ and $(2\pi - \pi) + 2\pi n$ give identical solutions.

$$2\cos\theta - 1 = 0 \qquad\qquad\qquad \text{or} \quad \cos\theta + 1 = 0 \qquad \text{Zero-product property}$$

$$\cos\theta = \frac{1}{2} \qquad\qquad\qquad\qquad \cos\theta = -1$$

$$\cos\theta = \cos\frac{\pi}{3} \qquad\qquad\qquad\qquad \cos\theta = \cos\pi$$

$$\theta = \frac{\pi}{3} + 2n\pi \quad \text{and} \quad \theta = \left(2\pi - \frac{\pi}{3}\right) + 2n\pi \qquad\qquad \theta = \pi + 2n\pi.$$

$$\theta = \frac{5\pi}{3} + 2n\pi$$

For $n = 0$, $\theta = \dfrac{\pi}{3}$ and $\theta = \dfrac{5\pi}{3}$. $\qquad\qquad\qquad$ For $n = 0$, $\theta = \pi$.

All other values of n result in solutions outside the interval $[0, 2\pi)$.

The solution set for the given equation in the interval $[0, 2\pi)$ is $\left\{\dfrac{\pi}{3}, \pi, \dfrac{5\pi}{3}\right\}$.

Practice Problem 9 Find all solutions of the equation $2\cos^2\theta + 3\sin\theta - 3 = 0$ in the interval $[0, 2\pi)$.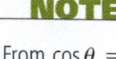

5 Solve trigonometric equations by squaring both sides.

Extraneous Solutions

Next, we solve a trigonometric equation by squaring both sides of the equation and then using an identity. Recall that squaring both sides of an equation will give all solutions of the original equation but may also produce *extraneous* solutions (apparent solutions that do not satisfy the original equation). Therefore, you must check all possible solutions.

> **EXAMPLE 10** Solving a Trigonometric Equation by Squaring

Find all solutions of the equation $\sqrt{3}\cos x = \sin x + 1$ in the interval $[0, 2\pi)$.

TECHNOLOGY CONNECTION

You can confirm that the equation in Example 10 has the two solutions in $[0, 2\pi)$ by graphing $Y_1 = \sqrt{3}\cos x$ and the right side $Y_2 = \sin x + 1$. Find the points of intersection using [INTERSECT].

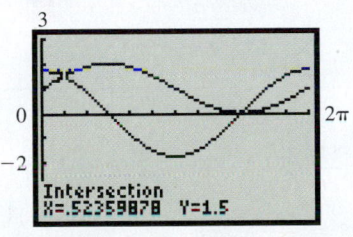

Intersection
X=.52359878   Y=1.5

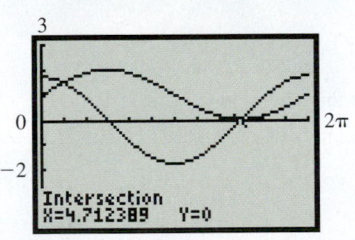

Intersection
X=4.712309   Y=0

Solution

$(\sqrt{3}\cos x)^2 = (\sin x + 1)^2$	Square both sides of the equation.
$3\cos^2 x = \sin^2 x + 2\sin x + 1$	Expand the binomial.
$3(1 - \sin^2 x) = \sin^2 x + 2\sin x + 1$	Pythagorean identity
$3 - 3\sin^2 x = \sin^2 x + 2\sin x + 1$	Distributive property
$-4\sin^2 x - 2\sin x + 2 = 0$	Collect like terms.
$2\sin^2 x + \sin x - 1 = 0$	Divide both sides by -2.
$(2\sin x - 1)(\sin x + 1) = 0$	Factor.
$2\sin x - 1 = 0 \quad$ or $\quad \sin x + 1 = 0$	Zero-product property

$$\sin x = \frac{1}{2} = \sin\frac{\pi}{6} \qquad\qquad \sin x = -1 = \sin\left(-\frac{\pi}{2}\right) \qquad \text{Solve for } \sin x.$$

$$x = \frac{\pi}{6} + 2n\pi \text{ or } x = \pi - \frac{\pi}{6} + 2n\pi \qquad\qquad x = -\frac{\pi}{2} + 2n\pi \text{ or}$$

$$= \frac{5\pi}{6} + 2n\pi \qquad\qquad\qquad x = \pi - \left(-\frac{\pi}{2}\right) + 2n\pi$$

$$= \frac{3\pi}{2} + 2n\pi$$

The possible solutions in the interval $[0, 2\pi)$ are (from $n = 0$): $\dfrac{\pi}{6}, \dfrac{5\pi}{6}$, and $\dfrac{3\pi}{2}$.

All other values of n result in solutions outside the interval $[0, 2\pi)$.

Check

$x = \dfrac{\pi}{6}$	$x = \dfrac{5\pi}{6}$	$x = \dfrac{3\pi}{2}$
$\sqrt{3}\cos\dfrac{\pi}{6} \overset{?}{=} \sin\dfrac{\pi}{6} + 1$	$\sqrt{3}\cos\dfrac{5\pi}{6} \overset{?}{=} \sin\dfrac{5\pi}{6} + 1$	$\sqrt{3}\cos\dfrac{3\pi}{2} \overset{?}{=} \sin\dfrac{3\pi}{2} + 1$
$\sqrt{3}\left(\dfrac{\sqrt{3}}{2}\right) \overset{?}{=} \dfrac{1}{2} + 1$	$\sqrt{3}\left(-\dfrac{\sqrt{3}}{2}\right) \overset{?}{=} \dfrac{1}{2} + 1$	$\sqrt{3}(0) \overset{?}{=} -1 + 1$
$\dfrac{3}{2} \overset{?}{=} \dfrac{3}{2}$ ✓	$-\dfrac{3}{2} \overset{?}{=} \dfrac{3}{2}$	$0 \overset{?}{=} 0$ ✓
Yes	**No**	**Yes**

The solution set of the equation in the interval $[0, 2\pi)$ is $\left\{\dfrac{\pi}{6}, \dfrac{3\pi}{2}\right\}$.

Practice Problem 10 Find all solutions of the equation $\sqrt{3}\cot\theta + 1 = \sqrt{3}\csc\theta$ in the interval $[0, 2\pi)$.

EXAMPLE 11 The Phases of the Moon

The moon orbits Earth in 29.5 days. The portion F of the moon visible from Earth x days after the new moon is given by the equation

$$F = 0.5\sin\left[\frac{\pi}{14.75}\left(x - \frac{14.75}{2}\right)\right] + 0.5.$$

Find x when 75% of the moon is visible from Earth.

Solution

We are asked to find x when $F = 75\% = 0.75$. We solve the equation:

$$0.75 = 0.5 \sin\left[\frac{\pi}{14.75}\left(x - \frac{14.75}{2}\right)\right] + 0.5 \qquad \text{Replace } F \text{ with } 0.75.$$

$$0.25 = 0.5 \sin\left[\frac{\pi}{14.75}\left(x - \frac{14.75}{2}\right)\right] \qquad \text{Subtract 0.5 from both sides.}$$

$$\frac{1}{2} = \sin\left[\frac{\pi}{14.75}\left(x - \frac{14.75}{2}\right)\right] \qquad \text{Divide both sides by 0.5.}$$

$$\sin\left[\frac{\pi}{14.75}\left(x - \frac{14.75}{2}\right)\right] = \sin\frac{\pi}{6} \qquad \sin^{-1}\left(\frac{1}{2}\right) = \frac{\pi}{6}.$$

$$\frac{\pi}{14.75}\left(x - \frac{14.75}{2}\right) = \frac{\pi}{6} + 2n\pi \quad \text{or} \quad \frac{\pi}{14.75}\left(x - \frac{14.75}{2}\right) = \frac{5\pi}{6} + 2n\pi$$

$$\frac{\pi}{14.75}\left(x - \frac{14.75}{2}\right) = \frac{\pi}{6} \quad \text{or} \quad \frac{\pi}{14.75}\left(x - \frac{14.75}{2}\right) = \frac{5\pi}{6} \qquad n = 0$$

$$x - \frac{14.75}{2} = \frac{14.75}{6} \qquad\qquad x - \frac{14.75}{2} = \frac{5(14.75)}{6} \qquad \text{Multiply both sides by } 14.75/\pi.$$

$$x = \frac{14.75}{2} + \frac{14.75}{6} \qquad\qquad x = \frac{14.75}{2} + \frac{5(14.75)}{6} \qquad \text{Solve for } x.$$

$$x \approx 10 \qquad\qquad\qquad\qquad x \approx 20 \qquad \text{Use a calculator and round to the nearest day.}$$

So 75% of the moon is visible 10 days and 20 days after the new moon.

Practice Problem 11 In Example 11, find x when 30% of the moon is visible from Earth.

Answers to Practice Problems

1. a. $x = \dfrac{\pi}{2} + 2n\pi$ **b.** $x = 2n\pi$ **c.** $x = \dfrac{\pi}{4} + n\pi$

2. a. $x \approx 48.2° + n \cdot 360°, x \approx 311.8° + n \cdot 360°$
b. $x \approx 2.0344, x \approx 5.1760$

3. $x = 90°$ or $x = 330°$

4. $\left\{\dfrac{\pi}{12}, \dfrac{5\pi}{12}, \dfrac{13\pi}{12}, \dfrac{17\pi}{12}\right\}$ **5.** $x = \dfrac{\pi}{3}$

6. $\left\{\dfrac{\pi}{4}, \dfrac{11\pi}{12}, \dfrac{19\pi}{12}\right\}$ **7.** $\left\{\dfrac{\pi}{2}, \dfrac{5\pi}{6}, \dfrac{11\pi}{6}\right\}$

8. $\theta = 2\pi n, \theta = \dfrac{2\pi}{3} + 2\pi n$, and $\theta = \dfrac{4\pi}{3} + 2\pi n$

9. $\theta = \dfrac{\pi}{6}, \theta = \dfrac{\pi}{2}, \theta = \dfrac{5\pi}{6}$ **10.** $\theta = \dfrac{\pi}{3}$

11. About 5 days and 24 days after the new moon.

SECTION 5.5 Exercises

Basic Concepts and Vocabulary

1. The equation $\sin x = \dfrac{1}{2}$ has _____ solution(s) in $[0, 2\pi)$.

2. All solutions of $\sin x = \sin \alpha$ are given by _____ and _____, for any integer n.

3. All solutions of $\cos x = \cos \alpha$ are given by _____ and _____.

4. All solutions of $\tan x = \tan \alpha$ are given by _____.

5. **True or False.** All solutions of $\tan x = -1$ are given by $-\dfrac{\pi}{4} + n\pi$, for any integer n.

6. **True or False.** The equation $\sec x = \dfrac{1}{2}$ has two solutions in $[0, 2\pi)$.

7. **True or False.** All solutions of $\sin x = 0$ are given by $x = n\pi$, for any integer n.

8. **True or False.** The equation $\sec x = \dfrac{\sqrt{3}}{2}$ has two solutions in the interval $[0, 2\pi)$.

Building Skills

In Exercises 9–18, find all solutions of each equation. Express the solutions in radians.

9. $\cos x = 0$ **10.** $\sin x = 0$

11. $\tan x = -1$ **12.** $\cot x = -1$

13. $\cos x = \dfrac{\sqrt{2}}{2}$ **14.** $\sin x = \dfrac{\sqrt{3}}{2}$

15. $\cot x = \sqrt{3}$ **16.** $\tan x = -\dfrac{\sqrt{3}}{3}$

17. $\cos x = -\dfrac{1}{2}$ **18.** $\sin x = -\dfrac{\sqrt{3}}{2}$

In Exercises 19–28, find all solutions of each equation. Express the solutions in degrees.

19. $\tan x = \dfrac{\sqrt{3}}{3}$ **20.** $\cot x = 1$

21. $\sin x = -\dfrac{1}{2}$ **22.** $\cos x = \dfrac{1}{2}$

23. $\csc x = 1$ **24.** $\sec x = -1$

25. $\sqrt{3}\csc x - 2 = 0$ **26.** $\sqrt{3}\sec x + 2 = 0$

27. $2\sec x - 4 = 0$ **28.** $2\csc x + 4 = 0$

In Exercises 29–34, find all solutions of each equation in the interval $[0°, 360°)$. Round the answers to the nearest tenth of a degree.

29. $\sin\theta = 0.4$ **30.** $\cos\theta = 0.6$

31. $\sec\theta = 7.2$ **32.** $\csc\theta = -4.5$

33. $\tan(\theta - 30°) = -5$ **34.** $\cot(\theta + 30°) = 6$

In Exercises 35–40, find all solutions of each equation in the interval $[0, 2\pi)$. Round the solutions to four decimal places.

35. $\csc x = -2$ **36.** $3\sin x - 1 = 0$

37. $3\tan x + 4 = 0$ **38.** $2\sec x - 7 = 0$

39. $2\csc x + 5 = 0$ **40.** $\cos x = 0.1106$

In Exercises 41–48, find all solutions of each equation in the interval $[0, 2\pi)$.

41. $\sin\left(x + \dfrac{\pi}{4}\right) = \dfrac{1}{2}$ **42.** $2\cos\left(x - \dfrac{\pi}{4}\right) + 1 = 0$

43. $\sec\left(x - \dfrac{\pi}{8}\right) + 2 = 0$ **44.** $\csc\left(x + \dfrac{\pi}{8}\right) - 2 = 0$

45. $\sqrt{3}\tan\left(x - \dfrac{\pi}{6}\right) - 1 = 0$

46. $\cot\left(x + \dfrac{\pi}{6}\right) + 1 = 0$

47. $2\sin\left(x - \dfrac{\pi}{3}\right) + 1 = 0$

48. $2\cos\left(x + \dfrac{\pi}{3}\right) + \sqrt{2} = 0$

In Exercises 49–68, find all solutions of each equation in the interval $[0, 2\pi)$.

49. $\cos 2x = \dfrac{\sqrt{3}}{2}$

50. $\sin 2x = \dfrac{\sqrt{3}}{2}$

51. $\sin 2x = -\dfrac{1}{2}$

52. $\cos 2x = -\dfrac{1}{2}$

53. $\sec 2x = \dfrac{1}{2}$ **54.** $\csc 2x = \dfrac{\sqrt{3}}{2}$

55. $\tan 2x = \dfrac{\sqrt{3}}{3}$

56. $\cot 2x = \dfrac{\sqrt{3}}{3}$

57. $\sin\left(2x - \dfrac{\pi}{3}\right) = \dfrac{1}{2}$

58. $\cos\left(2x + \dfrac{\pi}{4}\right) = -\dfrac{\sqrt{3}}{2}$

59. $\sin 3x = \dfrac{1}{2}$

60. $\cos 3x = -\dfrac{1}{2}$

61. $\cos 3x = \dfrac{1}{2}$

62. $\sin 3x = -\dfrac{\sqrt{3}}{2}$

63. $\cos\dfrac{x}{2} = \dfrac{1}{2}$ **64.** $\csc\dfrac{x}{2} = 2$

65. $\sin\dfrac{x}{2} = -\dfrac{\sqrt{3}}{2}$ **66.** $\sec\dfrac{x}{2} = -\dfrac{2\sqrt{3}}{3}$

67. $\tan\dfrac{x}{3} = 1$ **68.** $\cot\dfrac{x}{3} = \sqrt{3}$

In Exercises 69–78, find all solutions of each equation in the interval $[0, 2\pi)$.

69. $\sin 2x = \sin x$

70. $\cos 2x = \cos x$

71. $\sin 2x = \cos x$

72. $\cos 2x = \sin x$

73. $\cos\left(2x + \dfrac{\pi}{4}\right) = \cos x$

74. $\sin\left(2x - \dfrac{\pi}{4}\right) = \sin x$

75. $\sin 2x = \cos\left(x - \dfrac{\pi}{4}\right)$

76. $\cos\left(2x - \dfrac{\pi}{4}\right) = \sin x$

77. $\sin\left(2x - \dfrac{\pi}{4}\right) = \cos\left(x + \dfrac{\pi}{4}\right)$

78. $\cos\left(2x + \dfrac{\pi}{4}\right) = \sin\left(x - \dfrac{\pi}{4}\right)$

In Exercises 79–86, find all solutions of each equation in the interval $[0, 2\pi)$.

79. $(\sin x + 1)(\tan x - 1) = 0$

80. $(2\cos x + 1)(\sqrt{3}\tan x - 1) = 0$

81. $(\csc x - 2)(\cot x + 1) = 0$

82. $(\sqrt{3}\sec x - 2)(\sqrt{3}\cot x + 1) = 0$

83. $(\tan x + 1)(2\sin x - 1) = 0$

84. $(2\sin x - \sqrt{3})(2\cos x - 1) = 0$

85. $(\sqrt{2}\sec x - 2)(2\sin x + 1) = 0$

86. $(\cot x - 1)(\sqrt{2}\csc x + 2) = 0$

In Exercises 87–102, find all solutions of each equation in the interval $[0, 2\pi)$.

87. $4\sin^2 x = 1$ **88.** $4\cos^2 x = 1$

89. $\tan^2 x = 1$ **90.** $\sec^2 x = 2$

91. $3\csc^2 x = 4$ **92.** $3\cot^2 x = 1$

93. $3\sin^2 x = \cos^2 x$ **94.** $3\cos^2 x = \sin^2 x$

95. $\cos^2 x - \sin^2 x = 1$

96. $2\sin^2 x + \cos x - 1 = 0$

97. $2\cos^2 x - 3\sin x - 3 = 0$

98. $2\sin^2 x - \cos x - 1 = 0$

99. $2\sin^2 \theta - \sin \theta - 1 = 0$

100. $2\cos^2 \theta - 5\cos \theta + 2 = 0$

101. $\sqrt{3}\sec^2 x - 2\tan x - 2\sqrt{3} = 0$

102. $\csc^2 x - (\sqrt{3} + 1)\cot x + (\sqrt{3} - 1) = 0$

In Exercises 103–108, solve each trigonometric equation in the interval $[0, 2\pi)$ by squaring both sides.

103. $\sqrt{3}\sin x = 1 + \cos x$

104. $1 + \sin x = \sqrt{3}\cos x$

105. $\tan x + 1 = \sec x$

106. $\sqrt{3}\tan \theta + 1 = \sqrt{3}\sec \theta$

107. $\sqrt{3}\cot \theta + 1 = \sqrt{3}\csc \theta$

108. $\sin x + \cos x = \sqrt{1 - \sin 2x}$

Applying the Concepts

109. Electric current. The electric current I (in amperes) produced by an alternator in time t (in seconds) is given by $I = 60\sin(120\pi t)$. Find the smallest possible positive value of t (rounded to four decimal places) such that

    a. $I = 30$ amperes.

    b. $I = -20$ amperes.

110. Simple harmonic motion. A simple harmonic motion is described by the equation $x = 6\sin\left(\dfrac{\pi}{2}t\right)$, where x is the displacement in feet and t is time in seconds. Find positive values of t for which the displacement is 3 feet.

111. Simple harmonic motion. Repeat Exercise 110 assuming that the simple harmonic motion is described by the equation $x = 6\cos\left(\dfrac{\pi}{2}t\right)$.

112. Overcoat sales. The monthly sales of overcoats of the menswear store Suit Yourself in Winterland are approximated by the equation

$$S(x) = 500 + 500\sin\left[\dfrac{\pi}{4}(x - 2)\right],$$

where x is the month, with $x = 1$ for January. Find the months (after appropriate rounding) in which the number of overcoats sold is

    a. 0.

    b. 1000.

    c. 500.

113. Number of tourists. The weekly number of tourists visiting a tropical island is approximated by the equation

$$y = 20 + 10\sin\left[\dfrac{\pi}{26}(x - 14)\right],$$

where y is the number of visitors (in thousands) in the xth week of the year, starting with $x = 1$ for the first week in January. Find the week (after appropriate rounding) in which the number of tourists to the island is

    a. 30,000.

    b. 25,000.

    c. 15,000.

114. Average monthly temperature. The average monthly sea surface temperature T (in degrees Fahrenheit) on Paradise Island is approximated by the equation

$$T = 10.5\sin\left[\dfrac{\pi}{6}(x - 5)\right] + 73.5,$$

where x is the month, with $x = 1$ representing January. Find the months (after appropriate rounding) when the average sea surface temperature is

    a. 78.75° F.

    b. 84° F.

115. Average monthly snowfall. The average monthly snowfall y (in inches) in Rochester, New York, in the xth month can be approximated by

$$y = 12 + 12\sin\left[\dfrac{\pi}{4}(x - 2)\right], 1 \le x \le 8,$$

where October $= 1$, November $= 2, \ldots$, May $= 8$. Find the months (after appropriate rounding) when the average snowfall in Rochester is

    a. 24 inches.

    b. 18 inches.

116. Deer Population. The number of deer in a certain region is given by $D(t) = 300\sin\left[\dfrac{\pi}{12}(t - 2)\right] + 500$, where $t = 0$ respresents 2016. In which years between 2016 and 2050 will the deer population be 650?

Beyond the Basics

In Exercises 117–130, find all solutions of each equation in the interval $[0, 2\pi)$.

117. $\sin^4 2x = 1$
118. $2 \cos 6x = 1$

119. $\tan^2 2\theta = 1$
120. $\sin x \cos x = \sin x$

121. $\sin x \cos x - \dfrac{\sqrt{3}}{2} \sin x + \dfrac{1}{2} \cos x - \dfrac{\sqrt{3}}{4} = 0$
   [*Hint:* Factor by grouping.]

122. $2 \sin x \tan x + 2 \sin x + \tan x + 1 = 0$
   [*Hint:* Factor by grouping.]

123. $\sin^2 \theta - \cos \theta = \dfrac{1}{4}$

124. $3 \cos^2 \theta - 2\sqrt{3} \sin \theta \cos \theta - 3 \sin^2 \theta = 0$

125. $\tan x + \sec x = 2 \cos x$

126. $\sin x + \cos x = 2\sqrt{2} \sin x \cos x$

127. $\sin 2x + \cos 2x = \sqrt{2}$

128. $\sin 2\theta = \cot \theta$

129. $\cos 2\theta = \cot 2\theta$

130. $\sin 2\theta = \tan 2\theta$

In Exercises 131–134, write the left side of each equation in the form $a \sin (bx + c)$, see page 486, then solve for x in the interval $[0, 2\pi)$.

131. $3 \sin 2x + 4 \cos 2x = 0$
132. $3 \cos 2x - 4 \sin 2x = \dfrac{5}{2}$

133. $5 \sin 3x - 12 \cos 3x = \dfrac{13\sqrt{3}}{2}$

134. $12 \sin 3x + 5 \cos 3x = \dfrac{13}{2}$

Critical Thinking / Discussion / Writing

135. Solve the following equation: $\sin x \cos x = \dfrac{1}{2}$.

   [*Hint:* Square both sides and replace $\cos^2 x$ with $1 - \sin^2 x$.]

136. Follow the hint given for Exercise 135 to show that the equation $\sin x \cos x = 1$ has no real-number solutions.

137. Without solving the equation, explain why $\tan^2 x + (\sec x - 1)^2 + 3 = 0$ has no real-number solutions.

138. Write an equation that has exactly one solution in the interval $[0, 2\pi)$.

Getting Ready for the Next Section

In Exercises 139–144, consider triangle ABC shown in the adjoining figure. Use the Pythagorean Theorem to find the third side from the given two sides.

139. $a = 5, b = 12$

140. $a = 12, c = 20$

141. $b = 6, c = 10$

142. $a = 20, b = 21$

143. $b = 3, c = 6$

144. $a = 5, b = 10$

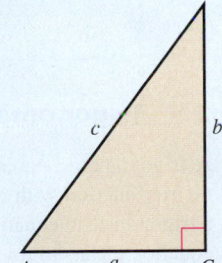

In Exercises 145 and 146, find $\dfrac{a}{b}$ for the given values of a and b. Then rationalize the denominator of the quotient.

145. $a = \dfrac{1}{2}, b = \dfrac{\sqrt{3}}{2}$

146. $a = \dfrac{\sqrt{3}}{2}, b = \sqrt{\dfrac{3}{2}}$

SUMMARY Definitions, Concepts, and Formulas

5.1 Trigonometric Identities

Reciprocal Identities

$$\csc x = \frac{1}{\sin x} \qquad \sec x = \frac{1}{\cos x} \qquad \cot x = \frac{1}{\tan x}$$

Quotient Identities

$$\tan x = \frac{\sin x}{\cos x} \qquad \cot x = \frac{\cos x}{\sin x}$$

Pythagorean Identities

$$\sin^2 x + \cos^2 x = 1$$
$$1 + \tan^2 x = \sec^2 x$$
$$1 + \cot^2 x = \csc^2 x$$

Even–Odd Identities

$$\sin (-x) = -\sin x \quad \cos (-x) = \cos x \quad \tan (-x) = -\tan x$$
$$\csc (-x) = -\csc x \quad \sec (-x) = \sec x \quad \cot (-x) = -\cot x$$

See page 444 for *guidelines for verifying trigonometric identities.*

5.2 Sum and Difference Formulas

Sum and Difference Formulas

$$\cos (u - v) = \cos u \cos v + \sin u \sin v$$
$$\cos (u + v) = \cos u \cos v - \sin u \sin v$$
$$\sin (u - v) = \sin u \cos v - \cos u \sin v$$
$$\sin (u + v) = \sin u \cos v + \cos u \sin v$$

$$\tan(u - v) = \frac{\tan u - \tan v}{1 + \tan u \tan v}$$

$$\tan(u + v) = \frac{\tan u + \tan v}{1 - \tan u \tan v}$$

Cofunction Identities

$$\sin\left(\frac{\pi}{2} - x\right) = \cos x \qquad \cos\left(\frac{\pi}{2} - x\right) = \sin x$$

$$\tan\left(\frac{\pi}{2} - x\right) = \cot x \qquad \csc\left(\frac{\pi}{2} - x\right) = \sec x$$

$$\sec\left(\frac{\pi}{2} - x\right) = \csc x \qquad \cot\left(\frac{\pi}{2} - x\right) = \tan x$$

Reduction Formula

If (a, b) is any point on the terminal side of an angle θ in standard position, then $a \sin x + b \cos x = \sqrt{a^2 + b^2} \sin(x + \theta)$ for any real number x.

5.3 Double-Angle and Half-Angle Formulas

Double-Angle Formulas

$$\sin 2x = 2 \sin x \cos x$$

$$\cos 2x = \cos^2 x - \sin^2 x$$

$$\cos 2x = 1 - 2 \sin^2 x$$

$$\cos 2x = 2 \cos^2 x - 1$$

$$\tan 2x = \frac{2 \tan x}{1 - \tan^2 x}$$

Power-Reducing Formulas

$$\sin^2 x = \frac{1 - \cos 2x}{2}$$

$$\cos^2 x = \frac{1 + \cos 2x}{2}$$

$$\tan^2 x = \frac{1 - \cos 2x}{1 + \cos 2x}$$

Half-Angle Formulas

$$\sin\frac{\theta}{2} = \pm\sqrt{\frac{1 - \cos\theta}{2}}$$

$$\cos\frac{\theta}{2} = \pm\sqrt{\frac{1 + \cos\theta}{2}}$$

$$\tan\frac{\theta}{2} = \pm\sqrt{\frac{1 - \cos\theta}{1 + \cos\theta}} = \frac{\sin\theta}{1 + \cos\theta} = \frac{1 - \cos\theta}{\sin\theta}$$

where the sign $+$ or $-$ depends on the quadrant in which $\frac{\theta}{2}$ lies.

5.4 Product-to-Sum and Sum-to-Product Formulas

Product-to-Sum Formulas

$$\cos x \cos y = \frac{1}{2}\left[\cos(x - y) + \cos(x + y)\right]$$

$$\sin x \sin y = \frac{1}{2}\left[\cos(x - y) - \cos(x + y)\right]$$

$$\sin x \cos y = \frac{1}{2}\left[\sin(x + y) + \sin(x - y)\right]$$

$$\cos x \sin y = \frac{1}{2}\left[\sin(x + y) - \sin(x - y)\right]$$

Sum-to-Product Formulas

$$\cos x + \cos y = 2 \cos\left(\frac{x + y}{2}\right)\cos\left(\frac{x - y}{2}\right)$$

$$\cos x - \cos y = -2 \sin\left(\frac{x + y}{2}\right)\sin\left(\frac{x - y}{2}\right)$$

$$\sin x + \sin y = 2 \sin\left(\frac{x + y}{2}\right)\cos\left(\frac{x - y}{2}\right)$$

$$\sin x - \sin y = 2 \sin\left(\frac{x - y}{2}\right)\cos\left(\frac{x + y}{2}\right)$$

5.5 Trigonometric Equations

A trigonometric equation is an equation that contains a trigonometric function with a variable. Values of the variable that satisfy a trigonometric equation are its *solutions*. Solving a trigonometric equation means to find its *solution set*.

Algebraic techniques are used to solve equations that are in linear, quadratic, or factorable form. Sometimes trigonometric identities are used to solve trigonometric equations.

REVIEW EXERCISES

Building Skills

In Exercises 1–4, use the information given to find the exact value of the remaining trigonometric functions of θ.

1. $\sin\theta = -\dfrac{2}{3}$ and $\cos\theta < 0$

2. $\tan\theta = -\dfrac{1}{2}$ and $\csc\theta > 0$

3. $\sec\theta = 3$ and $\tan\theta < 0$

4. $\csc\theta = 5$ and $\cot\theta < 0$

In Exercises 5–20, verify each identity.

5. $(\sin x + \cos x)^2 + (\sin x - \cos x)^2 = 2$

6. $(1 - \tan x)^2 + (1 + \tan x)^2 = 2 \sec^2 x$

7. $\dfrac{1 - \tan^2\theta}{1 + \tan^2\theta} = \cos^2\theta - \sin^2\theta$

8. $\dfrac{\sin x + \tan x}{\csc x + \cot x} = \sin^2 x \sec x$

9. $\dfrac{\sin\theta}{1+\cos\theta} + \dfrac{\sin\theta}{1-\cos\theta} = 2\csc\theta$

10. $\tan^2 x \sin^2 x = \tan^2 x - \sin^2 x$

11. $\dfrac{\sin\theta}{1-\cot\theta} + \dfrac{\cos\theta}{1-\tan\theta} = \cos\theta + \sin\theta$

12. $\dfrac{\tan\theta - \sin\theta}{\sin^3\theta} = \dfrac{\sec\theta}{1+\cos\theta}$

13. $\dfrac{\tan x}{\sec x - 1} + \dfrac{\tan x}{\sec x + 1} = 2\csc x$

14. $\dfrac{1}{\csc x - \cot x} - \dfrac{1}{\cot x + \csc x} = 2\cot x$

15. $\dfrac{1+\sin\theta}{1-\sin\theta} = (\sec\theta + \tan\theta)^2$

16. $\dfrac{1-\cos\theta}{1+\cos\theta} = (\csc\theta - \cot\theta)^2$

17. $\dfrac{\sec x - \tan x}{\sec x + \tan x} = (\sec x - \tan x)^2$

18. $\dfrac{\csc x + \cot x}{\csc x - \cot x} = (\csc x + \cot x)^2$

19. $\dfrac{\sin x - \cos x + 1}{\sin x + \cos x - 1} = \dfrac{\sin x + 1}{\cos x}$

[*Hint:* Multiply the numerator and the denominator of the left side by $\sin x + (1 - \cos x)$.]

20. $\sin^2 x \cot x = \dfrac{2\cos x}{\sin x + \csc x + \cos^2 x \csc x}$

In Exercises 21–28, find the exact value of each expression.

21. $\cos 15°$

22. $\sin 105°$

23. $\csc 75°$

24. $\tan 75°$

25. $\sin 41° \cos 49° + \cos 41° \sin 49°$

26. $\cos 50° \cos 10° - \sin 50° \sin 10°$

27. $\dfrac{\tan 69° + \tan 66°}{1 - \tan 69° \tan 66°}$

28. $2\cos 75° \cos 15°$

In Exercises 29–32, let $\sin u = \dfrac{4}{5}$, $\cos v = \dfrac{5}{13}$, for

$0 < u \le \dfrac{\pi}{2}, 0 < v \le \dfrac{\pi}{2}$. **Find the value of each expression.**

29. $\sin(u - v)$

30. $\cos(u + v)$

31. $\cos(u - v)$

32. $\tan(u - v)$

In Exercises 33–50, verify each identity.

33. $\sin(x - y)\cos y + \cos(x - y)\sin y = \sin x$

34. $\cos(x - y)\cos y - \sin(x - y)\sin y = \cos x$

35. $\dfrac{\sin(u + v)}{\sin(u - v)} = \dfrac{\tan u + \tan v}{\tan u - \tan v}$

36. $\dfrac{\tan(x + y)}{\cot(x - y)} = \dfrac{\tan^2 x - \tan^2 y}{1 - \tan^2 x \tan^2 y}$

37. $\dfrac{\sin 4x}{\sin 2x} - \dfrac{\cos 4x}{\cos 2x} = \sec 2x$

38. $\dfrac{\sin 3x}{\sin x} - \dfrac{\cos 3x}{\cos x} = 2$

39. $\sin\left(x - \dfrac{\pi}{6}\right) + \cos\left(x + \dfrac{\pi}{3}\right) = 0$

40. $\cos 2x = \cos^4 x - \sin^4 x$

41. $1 + \tan\theta \tan 2\theta = \sec 2\theta$

42. $\dfrac{\sin\theta + \sin 2\theta}{1 + \cos\theta + \cos 2\theta} = \tan\theta$

43. $\dfrac{\sin\theta + \sin 2\theta}{\cos\theta + \cos 2\theta} = \tan\dfrac{3\theta}{2}$

44. $\dfrac{\sin\theta - \sin 2\theta}{\cos\theta - \cos 2\theta} = -\cot\dfrac{3\theta}{2}$

45. $\dfrac{\sin 5x - \sin 3x}{\sin 5x + \sin 3x} = \dfrac{\tan x}{\tan 4x}$

46. $\dfrac{\cos 5x - \cos 3x}{\cos 5x + \cos 3x} = -\tan x \tan 4x$

47. $\dfrac{\tan 3x + \tan x}{\tan 3x - \tan x} = 2\cos 2x$

48. $\dfrac{\sin 4x}{2(1 + \cos 4x)} = \dfrac{\tan x}{1 - \tan^2 x}$

49. $\dfrac{\sin 3x + \sin 5x + \sin 7x + \sin 9x}{\cos 3x + \cos 5x + \cos 7x + \cos 9x} = \tan 6x$

50. $\dfrac{\cos x + \cos 3x + \cos 5x + \cos 7x}{\sin x + \sin 3x + \sin 5x + \sin 7x} = \cot 4x$

In Exercises 51 and 52, sketch the graph of each equation by first converting it to the form $y = A\sin(x + \theta)$.

51. $y = \sqrt{3}\sin x + \cos x$

52. $y = \sin x + \sqrt{3}\cos x$

In Exercises 53 and 54, assume that $A + B + C = 180°$. Verify each identity.

53. $\sin 2A + \sin 2B - \sin 2C = 4\cos A \cos B \sin C$

54. $\tan A + \tan B + \tan C = \tan A \tan B \tan C$

In Exercises 55–60, find all solutions of each equation in the interval $[0, 2\pi)$.

55. $2\cos^2 x - 1 = 0$

56. $3\tan^2 x - 1 = 0$

57. $2\cos^2 x - \cos x - 1 = 0$

58. $2\sin^2 x - 5\sin x - 3 = 0$

59. $2\sin 3x - 1 = 0$

60. $2\cos(2x - 1) - 1 = 0$

In Exercises 61–66, solve each equation for θ in the interval $[0°, 360°)$. If necessary, write the answers to the nearest tenth of a degree.

61. $3\sin\theta - 1 = 0$

62. $4\cos\theta + 1 = 0$

63. $2\sqrt{3}\cos 2\theta - 3 = 0$

64. $(1 - 2\sin 2\theta)(\sqrt{3} + 2\cos 2\theta) = 0$

65. $(\sqrt{3}\tan\theta + 1)(2\cos\theta + 1) = 0$

66. $\sqrt{3}\cos\theta = \sin\theta + 1$

In Exercises 67–70, find all solutions of each equation in the interval $[0, 2\pi)$.

67. $\dfrac{\tan 3x - \tan 2x}{1 + \tan 3x \tan 2x} = 1$

68. $\dfrac{\tan 2\theta + \tan\theta}{1 - \tan 2\theta \tan\theta} = 1$

69. $\sin x = \cos\left(2x + \dfrac{\pi}{6}\right)$

70. $\csc 2\theta = \cot 2\theta$

Applying the Concepts

71. Searchlight. A searchlight beam from a Coast Guard boat shines on a lake at an angle θ to the surface of the water. Viewed by a swimmer under water, the beam appears to make an angle α with the surface of the water, as shown in the figure. Under these circumstances, $\cos\theta = \dfrac{4}{3}\cos\alpha$. Find α if $\cos\theta = \dfrac{2}{3}$.

72. Rabbit population. The rabbit population of an area is described by $N(t) = 2400 + 500 \sin \pi t$, where $N(t)$ gives the number of rabbits in the area after t years. When will the rabbit population first reach 1900?

73. Voltage. The voltage from an alternating current generator is given by $V(t) = 160 \cos 120\pi t$, where t is measured in seconds. Find the time t when the voltage first equals 80.

74. Pendant shape. A jeweler wants to make a gold pendant in the shape of an isosceles triangle. Express the area of the pendant in terms of $\dfrac{\theta}{2}$ and x, where θ is the angle included by two sides of equal length x.

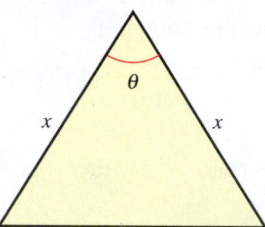

75. Pendant area. If the area of the pendant in Exercise 74 is 1 square inch, find the angle included between the two sides of equal length when each measures 2 inches.

PRACTICE TEST A

1. If $\sin\theta = \dfrac{3}{5}$ and $\cos\theta < 0$, find $\tan\theta$.

2. If $\tan x = \dfrac{2}{3}$ and $\csc x < 0$, find $\cos x$.

In Problems 3–8, verify each identity.

3. $\dfrac{1 - \sin^2 x}{\sin^2 x} = \cot^2 x$

4. $2 \sin x \cos x = (\sin x + \cos x + 1)(\sin x + \cos x - 1)$

5. $\sin x \sin\left(\dfrac{\pi}{2} - x\right) = \dfrac{\sin 2x}{2}$

6. $\dfrac{\sin 2x}{2(\cos x + \sin x)} = \dfrac{\sin x}{1 + \tan x}$

7. $\dfrac{\sin 2x + \sin 4x}{\cos 2x + \cos 4x} = \tan 3x$

8. $\cos(x + y)\cos(x - y) = \cos^2 x + \cos^2 y - 1$

In Problems 9 and 10, find all solutions of each equation.

9. $\tan(-x) = 1$

10. $\sin 4x = \dfrac{1}{2}$

In Problems 11 and 12, find all solutions of each equation in the interval $[0, 2\pi)$.

11. $\cos\dfrac{x}{3} = \dfrac{\sqrt{2}}{2}$

12. $\sin 2x + \cos x = 0$

In Problems 13–15, find the exact value of each expression.

13. $\sin 56° + \cos 146°$

14. $\cos 48° \cos 12° - \sin 48° \sin 12°$

15. $\sin\dfrac{5\pi}{12}$

16. If $\sin\theta = \dfrac{4}{5}$, find the exact value of $\cos 2\theta$.

17. If $\tan\theta = \dfrac{3}{4}$, find the exact value of $\sin 2\theta$.

18. Determine whether the function $y = \cos\left(\dfrac{\pi}{2} - x\right) + \tan x$ is odd, even, or neither.

19. Determine whether $\sin x + \sin y = \sin(x + y)$ is an identity.

20. A golf ball is hit on a level fairway with an initial velocity of $v_0 = 128$ feet per second on an initial angle of flight θ (in degrees). The range (in feet) of the ball is given by the expression $\dfrac{v_0^2 \sin 2\theta}{32}$. Find the value(s) of θ if the range is 512 feet.

PRACTICE TEST B

1. If $\sin\theta = -\dfrac{12}{13}$ and $\pi < \theta < \dfrac{3\pi}{2}$, find $\sec\theta$.

    a. $-\dfrac{13}{5}$ b. $-\dfrac{12}{5}$

    c. $-\dfrac{5}{12}$ d. $\dfrac{13}{5}$

2. Which expression is equal to $(\sin x + \cos x)^2 + (\sin x - \cos x)^2$?

    a. 1 b. 2
    c. $\sin x \cos x$ d. $4\sin x \cos x$

3. Which expression is equal to $\dfrac{\sin x}{1 + \cos x} + \dfrac{\sin x}{1 - \cos x}$?

    a. $2\sin x$ b. $2\cos x$
    c. $2\sec x$ d. $2\csc x$

4. Which expression is equal to $\dfrac{\tan x + \tan y}{\cot x + \cot y}$?

    a. $\cot x \cot y$ b. $\tan x \tan y$
    c. $\sec x \csc y$ d. $\sin x \cos y$

5. Which expression is equal to $\dfrac{\sin x + \sin y}{\cos x + \cos y} + \dfrac{\cos x - \cos y}{\sin x - \sin y}$?

    a. 0 b. 1
    c. $\tan x \tan y$ d. $\tan x + \tan y$

6. Which expression is equal to $\dfrac{\sin 2x}{1 + \cos 2x}$?

    a. $\cot x$ b. $\tan x$
    c. $\tan 2x$ d. $\cot 2x$

7. Which expression is equal to $\dfrac{1 - \cos 2x}{1 + \cos 2x}$?

    a. $\sin^2 x$ b. $\cos^2 x$
    c. $\tan^2 x$ d. $\cot^2 x$

8. Which expression is equal to $\dfrac{\sin 3x - \sin x}{\cos x - \cos 3x}$?

    a. $\tan x$ b. $\tan 2x$
    c. $\cot x$ d. $\cot 2x$

In Problems 9–11, find all solutions of each equation.

9. $\cot(-x) = -1$

    a. $x = \dfrac{\pi}{4} + 2\pi n$, for any integer n

    b. $x = \dfrac{3\pi}{4} + 2\pi n$, for any integer n

    c. $x = \dfrac{\pi}{4} + \pi n$, for any integer n

    d. $x = \dfrac{3\pi}{4} + \pi n$, for any integer n

10. $2\cos 4x = -1$

    a. $x = \dfrac{2\pi}{3} + 2\pi n$ and $x = \dfrac{4\pi}{3} + 2\pi n$, for any integer n

    b. $x = \dfrac{\pi}{3} + \dfrac{n\pi}{2}$ and $x = \dfrac{5\pi}{3} + \dfrac{n\pi}{2}$, for any integer n

    c. $x = \dfrac{\pi}{6} + \dfrac{n\pi}{2}$ and $x = \dfrac{\pi}{3} + \dfrac{n\pi}{2}$, for any integer n

    d. $x = \dfrac{\pi}{12} + 2\pi n$ and $x = \dfrac{5\pi}{12} + 2\pi n$, for any integer n

11. $\sin\dfrac{x}{3} = \dfrac{\sqrt{2}}{2}$

    a. $x = \dfrac{\pi}{4} + 2\pi n$ and $x = \dfrac{3\pi}{4} + 2\pi n$, for any integer n

    b. $x = \dfrac{\pi}{4} + 6n\pi$ and $x = \dfrac{\pi}{12} + 6n\pi$, for any integer n

    c. $x = \dfrac{3\pi}{4} + 6\pi n$ and $x = \dfrac{9\pi}{4} + 6\pi n$, for any integer n

    d. $x = \dfrac{3\pi}{4} + 2\pi n$ and $x = \dfrac{5\pi}{4} + 2\pi n$, for any integer n

12. Find all solutions of $\cos x - \sin 2x = 0$ for the interval $[0, 2\pi)$.

    a. $\left\{ \dfrac{\pi}{6}, \dfrac{\pi}{2}, \dfrac{5\pi}{6}, \dfrac{3\pi}{2} \right\}$

    b. $\left\{ \dfrac{\pi}{6}, \dfrac{\pi}{3}, \dfrac{5\pi}{6}, \dfrac{7\pi}{6} \right\}$

    c. $\left\{ \dfrac{\pi}{2}, \dfrac{5\pi}{6}, \pi, \dfrac{3\pi}{2} \right\}$

    d. $\left\{ \dfrac{\pi}{6}, \dfrac{\pi}{2}, \dfrac{5\pi}{6}, \dfrac{11\pi}{6} \right\}$

In Problems 13 and 14, find the exact value of each expression.

13. $\sin 71° + \cos 161°$

    a. 1.892 b. -1.892
    c. 1 d. 0

14. $\sin 53° \cos 37° + \cos 53° \sin 37°$

    a. $\dfrac{1}{2}$ b. $\dfrac{\sqrt{3}}{2}$

    c. 0 d. 1

15. If $\sin\theta = \dfrac{2}{3}$, find the exact value of $\cos 2\theta$.

    a. $\dfrac{1}{3}$ b. $\dfrac{1}{9}$

    c. $\dfrac{14}{9}$ d. $\dfrac{\sqrt{5}}{3}$

16. If $\tan\theta = \dfrac{4}{3}$, find the exact value of $\sin 2\theta$.

    a. $\dfrac{24}{25}$ b. $\dfrac{4}{5}$

    c. $\dfrac{3}{5}$ d. $\dfrac{8}{5}$

17. Which one of the following best describes the symmetry of the graph of $y = \sin\left(\dfrac{\pi}{2} - x\right) + \cot x$?

    a. Symmetric with respect to the origin
    b. Symmetric with respect to the x-axis
    c. Symmetric with respect to the y-axis
    d. Not symmetric to either axis or the origin

18. Which pair of values results in a false statement for $\tan(x + y) = \tan x + \tan y$?

    a. $x = 0, y = \pi$ b. $x = \dfrac{\pi}{4}, y = -\dfrac{\pi}{4}$

    c. $x = \dfrac{\pi}{4}, y = \dfrac{5\pi}{4}$ d. $x = \dfrac{\pi}{4}, y = \dfrac{3\pi}{4}$

19. Identify all solutions of $2 \cos^2 x = \dfrac{3}{2}$ in the interval

$[0, 2\pi)$.

   a. $\left\{\dfrac{\pi}{6}\right\}$ **b.** $\left\{\dfrac{\pi}{6}, \dfrac{5\pi}{6}\right\}$

   c. $\left\{\dfrac{\pi}{6}, \dfrac{11\pi}{6}\right\}$ **d.** $\left\{\dfrac{\pi}{6}, \dfrac{5\pi}{6}, \dfrac{7\pi}{6}, \dfrac{11\pi}{6}\right\}$

20. A golf ball is hit on a level fairway with an initial velocity of $v_0 = 128$ feet per second and an initial angle of flight θ (in degrees). The range (in feet) of the ball is given by the expression $\dfrac{v_0^2 \sin 2\theta}{32}$. Find the value(s) of θ if the range is $256\sqrt{3}$ feet.

   a. $45°$ **b.** $30°, 60°$

   c. $45°, 135°$ **d.** $150°, 175°$

CUMULATIVE REVIEW EXERCISES CHAPTERS 1–5

In Exercises 1–6, solve each equation or inequality.

1. $x^2 + x - 1 = 0$

2. $\log_2 (x + 1) + \log_2 (x - 1) = 1$

3. $5^{-x} = 9$

4. $\sin 2x - \cos x = 0$ $(0 \le x < 2\pi)$

5. $x^3 - 4x > 0$

6. $\dfrac{2x - 3}{x + 2} - 1 < 0$

7. If $f(x) = 2x^2 - x + 3$, find $\dfrac{f(x + h) - f(x)}{h}$.

8. If $f(x) = \dfrac{x + 2}{2x - 1}$, find $f^{-1}(x)$.

In Exercises 9–16, sketch the graph of each function $f(x)$ by using transformations on the given basic function.

9. $f(x) = x^2 - 4x - 7$; use $y = x^2$.

10. $f(x) = 2\sqrt{x - 3} + 1$; use $y = \sqrt{x}$.

11. $f(x) = 2^{x-1} - 2$; use $y = 2^x$.

12. $f(x) = 2 \ln (x + 1)$; use $y = \ln x$.

13. $f(x) = \dfrac{1}{2} \sin 2x$; use $y = \sin x$.

14. $f(x) = -2 \cos (3x - 1)$; use $y = \cos x$.

15. $f(x) = -3 \csc x$; use $y = 3 \sin x$.

16. $f(x) = 2 \tan 4x$; use $y = \tan x$.

17. Simplify $\dfrac{2 \csc^2 x - 5 \cot x - 5}{2 \cot x + 1}$.

18. Use an identity to find the exact value of $\sec 15°$.

19. Verify the identity $\sin\left(x - \dfrac{3\pi}{2}\right) = \cos x$.

20. Find all solutions of the equation $\sin x = \cos x$ in the interval $[0, 2\pi)$.

CHAPTER | 6

Applications of Trigonometric Functions

TOPICS

6.1 Right-Triangle Trigonometry

6.2 The Law of Sines

6.3 The Law of Cosines

6.4 Vectors

6.5 The Dot Product

6.6 Polar Coordinates

6.7 Polar Form of Complex Numbers; DeMoivre's Theorem

Applications of trigonometry permeate virtually every field, from architecture and bionics to medicine and physics. In this chapter, we introduce vectors and polar coordinates and investigate the many applications of trigonometry to real-world problems.

Right-Triangle Trigonometry

BEFORE STARTING THIS SECTION, REVIEW

1. Rationalizing the denominator (Appendix A.4, page 944)

2. Acute angle definition (Section 4.1, page 338)

3. Pythagorean Theorem (Appendix A.5, page 951)

4. Similar triangles (Appendix A.5, page 950)

5. Functions (Section 1.3, page 40)

OBJECTIVES

1. Express the trigonometric functions using a right triangle.

2. Evaluate trigonometric functions of angles in a right triangle.

3. Solve right triangles.

4. Use right-triangle trigonometry in applications.

Mount Kilimanjaro

◆ Measuring Mount Kilimanjaro

Trigonometry developed as a result of attempts to solve practical problems in astronomy, navigation, and land measurement. The word *trigonometry*, coined by Pitiscus (1561–1613) in 1594, is derived from two Greek words, *trigonon* (triangle) and *metron* (measure), and means "triangle measurement."

Measuring objects that are generally inaccessible is one of the many successes of trigonometry. In Example 7, we use trigonometry to approximate the height of Mount Kilimanjaro, Tanzania. Mount Kilimanjaro is the highest mountain in Africa.

---

1 Express the trigonometric functions using a right triangle.

RECALL

In a right triangle, the side opposite the right angle is called the hypotenuse, and the other two sides are called the legs.

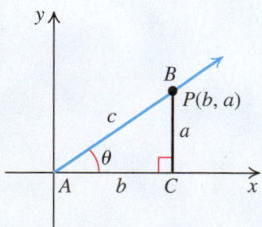

Figure 6.2

Trigonometric Ratios and Functions

A triangle containing a 90° angle is called a **right triangle**. Consider a right triangle with one of its acute angles labeled θ. In Figure 6.1, the capital letters A, B, and C designate the vertices of the triangle, and the lowercase letters a, b, and c, respectively, represent the lengths of the sides opposite these angles.

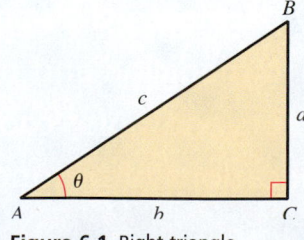

a = length of the side opposite θ
b = length of the side adjacent to θ
c = length of the hypotenuse

Figure 6.1 Right triangle.

Now place the angle θ in standard position, as shown in Figure 6.2. The point corresponding to the vertex B is $P(b, a)$. Because P is on the terminal side of θ, we can use it to express the trigonometric functions of θ as ratios of the sides of a right triangle. We use the words *opposite* for the length of the leg opposite θ, *adjacent* for the length of the leg adjacent to θ, and *hypotenuse* for the length of the hypotenuse.

TRIGONOMETRIC FUNCTIONS OF AN ANGLE θ
IN A RIGHT TRIANGLE

$$\sin\theta = \frac{\text{opposite}}{\text{hypotenuse}} = \frac{a}{c} \qquad \csc\theta = \frac{\text{hypotenuse}}{\text{opposite}} = \frac{c}{a}$$

$$\cos\theta = \frac{\text{adjacent}}{\text{hypotenuse}} = \frac{b}{c} \qquad \sec\theta = \frac{\text{hypotenuse}}{\text{adjacent}} = \frac{c}{b}$$

$$\tan\theta = \frac{\text{opposite}}{\text{adjacent}} = \frac{a}{b} \qquad \cot\theta = \frac{\text{adjacent}}{\text{opposite}} = \frac{b}{a}$$

Because any two right triangles having acute angle θ are similar, the ratios of corresponding sides depend only on the angle θ, not on the size of the triangle. Three right triangles with acute angle θ are shown in Figure 6.3.

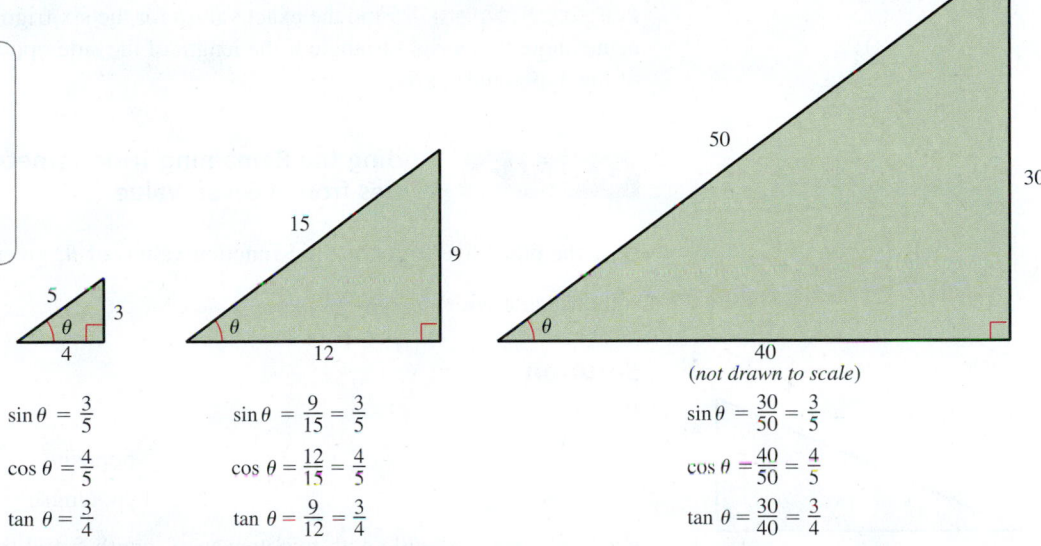

$$\sin\theta = \frac{3}{5} \qquad \sin\theta = \frac{9}{15} = \frac{3}{5} \qquad \sin\theta = \frac{30}{50} = \frac{3}{5}$$

$$\cos\theta = \frac{4}{5} \qquad \cos\theta = \frac{12}{15} = \frac{4}{5} \qquad \cos\theta = \frac{40}{50} = \frac{4}{5}$$

$$\tan\theta = \frac{3}{4} \qquad \tan\theta = \frac{9}{12} = \frac{3}{4} \qquad \tan\theta = \frac{30}{40} = \frac{3}{4}$$

Figure 6.3 Three similar triangles.

Notice that the value of each of the six trigonometric functions is independent of the triangle used to find it. Recall that $\csc\theta$, $\sec\theta$, and $\cot\theta$ are the reciprocals of $\sin\theta$, $\cos\theta$, and $\tan\theta$, respectively. Consequently, for any of the triangles in Figure 6.3, we have

$$\csc\theta = \frac{5}{3}, \sec\theta = \frac{5}{4}, \text{ and } \cot\theta = \frac{4}{3}.$$

2 Evaluate trigonometric functions of angles in a right triangle.

Evaluating Trigonometric Functions

EXAMPLE 1   **Finding the Values of Trigonometric Functions**

Find the exact values for the six trigonometric functions of the angle θ in Figure 6.4.

Solution

To find the values for the six trigonometric functions of θ, we must first find the value of c, the length of the hypotenuse.

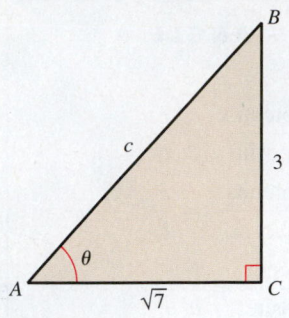

Figure 6.4

$$a^2 + b^2 = c^2 \qquad \text{Pythagorean Theorem}$$
$$(3)^2 + (\sqrt{7})^2 = c^2 \qquad \text{Replace } a \text{ with 3 and } b \text{ with } \sqrt{7}.$$
$$9 + 7 = c^2$$
$$16 = c^2$$
$$4 = c \qquad c \text{ is a positive number.}$$

Now, with $c = 4$, $a = 3$, and $b = \sqrt{7}$, we have:

$$\sin\theta = \frac{\text{opposite}}{\text{hypotenuse}} = \frac{3}{4} \qquad\qquad \csc\theta = \frac{\text{hypotenuse}}{\text{opposite}} = \frac{4}{3}$$

$$\cos\theta = \frac{\text{adjacent}}{\text{hypotenuse}} = \frac{\sqrt{7}}{4} \qquad\qquad \sec\theta = \frac{\text{hypotenuse}}{\text{adjacent}} = \frac{4}{\sqrt{7}} = \frac{4\sqrt{7}}{7}$$

$$\tan\theta = \frac{\text{opposite}}{\text{adjacent}} = \frac{3}{\sqrt{7}} = \frac{3\sqrt{7}}{7} \qquad\qquad \cot\theta = \frac{\text{adjacent}}{\text{opposite}} = \frac{\sqrt{7}}{3}$$

These are exact values; using a calculator would yield approximate values.

Practice Problem 1 Find the exact values for the six trigonometric functions of the acute angle θ in a right triangle if the length of the side opposite θ is 4 and the length of the hypotenuse is 5.

EXAMPLE 2 **Finding the Remaining Trigonometric Function Values from a Given Value**

Find the other five trigonometric function values of θ, given that θ is an acute angle of a right triangle with $\sin\theta = \dfrac{2}{5}$.

Solution

Because

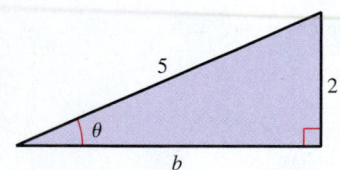

Figure 6.5

$$\sin\theta = \frac{2}{5} = \frac{\text{opposite}}{\text{hypotenuse}}$$

we draw a right triangle with hypotenuse of length 5 and the side opposite θ of length 2. See Figure 6.5. Then

$$5^2 = 2^2 + b^2 \qquad \text{Pythagorean Theorem}$$
$$b^2 = 5^2 - 2^2 \qquad \text{Subtract } 2^2 \text{ from both sides; interchange sides.}$$
$$b^2 = 21 \qquad \text{Simplify.}$$
$$b = \sqrt{21} \qquad b \text{ is a positive number.}$$

For this triangle we have:

$$c = \text{length of the hypotenuse} = 5$$
$$a = \text{length of the opposite side} = 2$$
$$b = \text{length of the adjacent side} = \sqrt{21}$$

So,

$$\sin\theta = \frac{\text{opposite}}{\text{hypotenuse}} = \frac{2}{5} \qquad\qquad \csc\theta = \frac{\text{hypotenuse}}{\text{opposite}} = \frac{5}{2}$$

$$\cos\theta = \frac{\text{adjacent}}{\text{hypotenuse}} = \frac{\sqrt{21}}{5} \qquad\qquad \sec\theta = \frac{\text{hypotenuse}}{\text{adjacent}} = \frac{5}{\sqrt{21}} = \frac{5\sqrt{21}}{21}$$

$$\tan\theta = \frac{\text{opposite}}{\text{adjacent}} = \frac{2}{\sqrt{21}} = \frac{2\sqrt{21}}{21} \qquad\qquad \cot\theta = \frac{\text{adjacent}}{\text{opposite}} = \frac{\sqrt{21}}{2}$$

Practice Problem 2 Find the other five trigonometric function values of θ, given that θ is an acute angle of a right triangle with $\cos\theta = \dfrac{1}{3}$.

Complements

Recall that two acute angles are called *complementary* if their sum is 90°. Because θ and $90° - \theta$ are acute angles in the same right triangle, the same six possible ratios of the lengths of the sides are used to compute the values of the trigonometric functions for both angles. Notice that the leg opposite θ is the leg adjacent to $90° - \theta$; so

$$\sin\theta = \frac{\text{opposite to } \theta}{\text{hypotenuse}} = \frac{\text{adjacent to } (90° - \theta)}{\text{hypotenuse}} = \cos(90° - \theta).$$

Similarly, $\cos\theta = \sin(90° - \theta)$ and $\tan\theta = \cot(90° - \theta)$.

Taking reciprocals of these values extends this relationship to the remaining three trigonometric functions.

The prefix *co-* (in *cosine, cotangent,* and *cosecant*) stands for *complement*. In general, for any acute angle θ in a right triangle, the other acute angle is its complement, $90° - \theta$. See Figure 6.6. The pairs of functions—sine and cosine, tangent and cotangent, secant and cosecant—are cofunctions of each other.

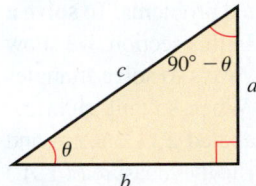

Figure 6.6 *a* is opposite θ and adjacent to $90° - \theta$. *b* is adjacent to θ and opposite $90° - \theta$.

SIDE NOTE

Here is a scheme for remembering the cofunction identities for acute angles A and B with

$$A + B = 90°, \text{ or } A + B = \frac{\pi}{2},$$

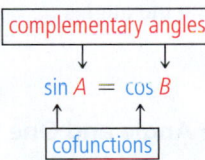

The sine and cosine may be replaced with any other cofunction pairs.

COMPLEMENTARY RELATIONSHIPS

The value of any trigonometric function of an acute angle θ is equal to the cofunction value of the complement of θ. This is true whether θ is measured in degrees or in radians.

θ in degrees

$$\sin\theta = \cos(90° - \theta) \qquad \cos\theta = \sin(90° - \theta)$$
$$\tan\theta = \cot(90° - \theta) \qquad \cot\theta = \tan(90° - \theta)$$
$$\sec\theta = \csc(90° - \theta) \qquad \csc\theta = \sec(90° - \theta)$$

If θ is measured in radians, replace 90° with $\dfrac{\pi}{2}$.

EXAMPLE 3 Finding Trigonometric Function Values of a Complementary Angle

a. Given that $\cot 68° \approx 0.4040$, find $\tan 22°$.

b. Given that $\cos 72° \approx 0.3090$, find $\sin 18°$.

c. Given that $\sec\dfrac{\pi}{12} = \sqrt{6} - \sqrt{2}$, find $\csc\dfrac{5\pi}{12}$.

Solution

a. Note that $22° = 90° - 68°$. So,

$$\tan 22° = \tan(90° - 68°) = \cot 68° \approx 0.4040.$$

b. Here $18° = 90° - 72°$. So,

$$\sin 18° = \sin(90° - 72°) = \cos 72° \approx 0.3090.$$

c. Since $\dfrac{5\pi}{12} = \dfrac{\pi}{2} - \dfrac{\pi}{12}$, we have $\csc\left(\dfrac{5\pi}{12}\right) = \csc\left(\dfrac{\pi}{2} - \dfrac{\pi}{12}\right) = \sec\left(\dfrac{\pi}{12}\right) = \sqrt{6} - \sqrt{2}$.

Practice Problem 3

a. Given that $\csc 21° \approx 2.7904$, find $\sec 69°$.

b. Given that $\tan 75° \approx 3.7321$, find $\cot 15°$.

c. Given that $\sin \dfrac{3\pi}{8} = \dfrac{\sqrt{2 + \sqrt{2}}}{2}$, find $\cos \dfrac{\pi}{8}$.

3 Solve right triangles.

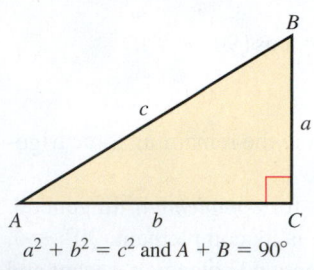

$a^2 + b^2 = c^2$ and $A + B = 90°$

Figure 6.7 Labeling a right triangle.

Solving Right Triangles

We use right-triangle trigonometry to find unknown lengths in triangles and unknown distances in applications. Using inverse trigonometric functions, we also find unknown angle measures in right triangles as well as unknown angle measures in applied problems. To **solve a triangle** means to find all unknown side lengths and angle measures. In this section, we show how to solve right triangles. In the next two sections, we will discuss ways to solve triangles that do not contain right angles, which are called *oblique* triangles. When solving right triangles, we solve each triangle ABC, where the right angle is always labeled $\angle C$ and $\angle A$ and $\angle B$ are the acute angles. (We frequently use A as a shorthand way to write the measure of $\angle A$.) The side lengths are a, b, and c, where a is the length of the side opposite $\angle A$, and so on. Because $\angle C$ is always the right angle, the length of the hypotenuse is always c and the lengths of the legs are a and b. See Figure 6.7. With this notation, $a^2 + b^2 = c^2$ and $A + B = 90°$.

In applications, degree measure is traditionally used for the angles. Here are some useful facts about solving right triangles:

1. The measure of one angle (the right angle) is automatically given: $C = 90°$.

2. Using the Pythagorean Theorem, we can find the length of the third side of any right triangle if two of the side lengths are known.

3. Knowing the measure of one acute angle allows us to find the other because the two acute angles of any right triangle are complementary.

4. We can solve a right triangle if we are given *either* of the following combinations of measurements:

   a. The length of any one side and the measure of either of the acute angles

   b. The lengths of any two sides

EXAMPLE 4 **Solving a Right Triangle, Given One Acute Angle and One Side**

Solve right triangle ABC if $A = 23°$ and $c = 5.8$.

Solution

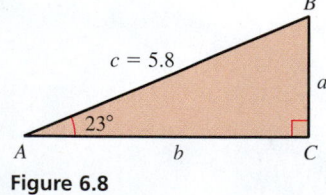

Figure 6.8

Sketch the triangle. See Figure 6.8.

To find a, find a trigonometric function that involves a and the given parts A and c.

$$\sin A = \frac{a}{c} \qquad \text{Definition of sine}$$

$$c \sin A = a \qquad \text{Multiply both sides by } c.$$

$$5.8 \sin 23° = a \qquad \text{Substitute 5.8 for } c \text{ and } 23° \text{ for } A.$$

$$a \approx 2.3 \qquad \text{Use a calculator; round to the nearest tenth.}$$

To find b, find a trigonometric function that involves b and the given parts A and c.

$$\cos A = \frac{b}{c} \qquad \text{Definition of cosine}$$

$$c \cos A = b \qquad \text{Multiply both sides by } c.$$

$$5.8 \cos 23° = b \qquad \text{Substitute 5.8 for } c \text{ and } 23° \text{ for } A.$$

$$b \approx 5.3 \qquad \text{Use a calculator; round to the nearest tenth.}$$

Once a and b have been found, the Pythagorean Theorem can be used as a check.

Finally, because $\angle A$ and $\angle B$ are complementary, $B = 90° - 23° = 67°$.

Practice Problem 4 Solve right triangle ABC if $c = 25.8$ and $A = 56°$.

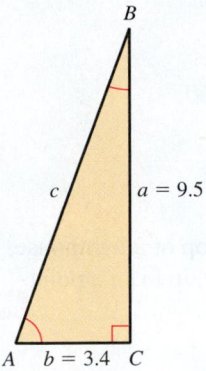

Figure 6.9

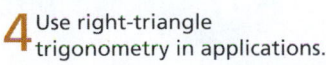

EXAMPLE 5 Solving a Right Triangle, Given Two Sides

Solve right triangle ABC if $a = 9.5$ and $b = 3.4$.

Solution

Sketch the triangle. See Figure 6.9.

To find A, use a trigonometric function that involves A and the given parts a and b.

$$\tan A = \frac{a}{b} \qquad \text{Definition of tangent}$$

$$\tan A = \frac{9.5}{3.4} \qquad \text{Substitute 9.5 for } a \text{ and 3.4 for } b.$$

$$A = \tan^{-1}\left(\frac{9.5}{3.4}\right) \approx 70.3° \qquad \text{Use a calculator; round to one decimal place.}$$

To find c, using the Pythagorean Theorem, we have:

$$c = \sqrt{a^2 + b^2}$$
$$= \sqrt{(9.5)^2 + (3.4)^2} \qquad \text{Substitute 9.5 for } a \text{ and 3.4 for } b.$$
$$c \approx 10.1 \qquad \text{Use a calculator; round to the nearest tenth.}$$

All three sides have been found.

Finally, $B = 90° - A$; so $B \approx 90° - 70.3°$, or $B \approx 19.7°$.

Practice Problem 5 Solve right triangle ABC if $b = 24.5$ and $c = 36.9$.

4 Use right-triangle trigonometry in applications.

Applications

Angles that are measured between a line of sight and a horizontal line occur in many applications. If the line of sight of an observer is *above* the horizontal line, the angle between these two lines is called the **angle of elevation**. If the line of sight of an observer is *below* the horizontal line, the angle between the two lines is called the **angle of depression**.

For example, suppose you are taking a picture of a friend who has climbed to the top of the Arc de Triomphe in Paris. See Figure 6.10. The angle your eyes rotate through as your gaze changes from looking straight ahead to looking at your friend is the angle of elevation.

The angle your friend's eyes rotate through as she changes from looking straight ahead to looking down at you (no pun intended) is the angle of depression.

SIDE
NOTE

Remember that an angle of elevation or an angle of depression is the angle between the line of sight and a *horizontal* line.

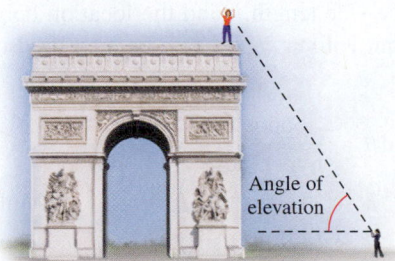
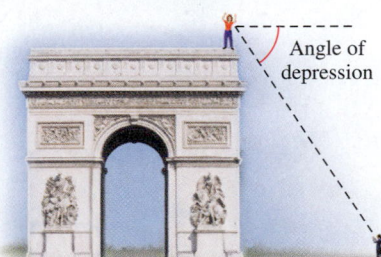

Figure 6.10 Angles of elevation and depression.

EXAMPLE 6 Finding the Height of a Cloud

To determine the height of a cloud at night, a farmer shines a spotlight straight upward to a spot on the cloud. The angle of elevation to this same spot on the cloud from a point located 150 feet horizontally from the spotlight is 68.2°. Find the height of the cloud.

68.2°

h

←— 150 ft —→

Figure 6.11 Height of a cloud.

TECHNOLOGY
CONNECTION

When solving a triangle, to find:

Length, use the SIN, COS, and TAN keys;

Angles, use the SIN⁻¹, COS⁻¹, and TAN⁻¹ keys.

Solution

In Figure 6.11, h represents the height of the cloud, in feet. We have:

$$\tan 68.2° = \frac{h}{150}$$ Definition of tangent

$$h = 150 \tan 68.2°$$ Multiply both sides by 150.

$$h \approx 375$$ Use a calculator.

The height of the cloud is approximately 375 feet.

Practice Problem 6 From an observation deck 425 feet high on the top of a lighthouse, the angle of depression to a ship at sea is 4.2°. How many miles is the ship from a point at sea level directly below the observation deck?

◆ **EXAMPLE 7**   **Measuring the Height of Mount Kilimanjaro**

A surveyor wants to measure the height of Mount Kilimanjaro by using the known height of a nearby mountain. The nearby location is at an altitude of 8720 feet, the distance between that location and Mount Kilimanjaro's peak is 4.9941 miles, and the angle of elevation from the lower location is 23.75°. See Figure 6.12. Use this information to find the approximate height of Mount Kilimanjaro.

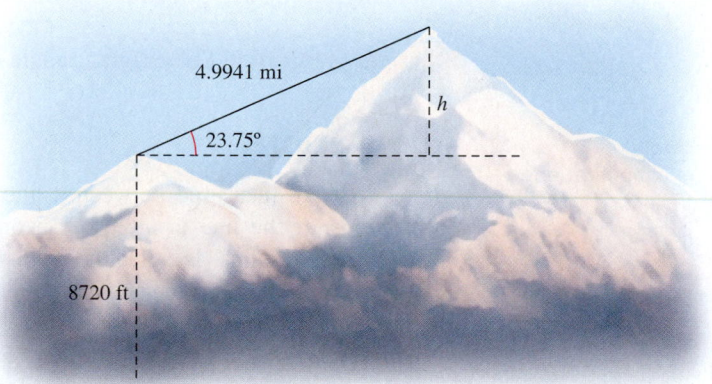

4.9941 mi

23.75°

h

8720 ft

Figure 6.12 Height of Mount Kilimanjaro.

Solution

The sum of the side length h and the location height of 8720 feet gives the approximate height of Mount Kilimanjaro. Let h be measured in miles. Use the definition of $\sin \theta$, for $\theta = 23.75°$.

$$\sin \theta = \frac{\text{opposite}}{\text{hypotenuse}} = \frac{h}{4.9941}$$

$$h = (4.9941) \sin \theta$$ Multiply both sides by 4.9941.

$$= (4.9941) \sin 23.75°$$ Replace θ with 23.75°.

$$h \approx 2.0114$$ Use a calculator.

Because 1 mile = 5280 feet,

$$2.0114 \text{ miles} = (2.0114)(5280)$$

$$\approx 10{,}620 \text{ feet.}$$

Thus, the height of Mount Kilimanjaro

$$\approx 10{,}620 + 8720 = 19{,}340 \text{ feet.}$$

Practice Problem 7 The height of the nearby location to Mount McKinley, Alaska, is 12,870 feet, its distance to Mount McKinley's peak is 3.3387 miles, and the angle of elevation θ is 25°. What is the approximate height of Mount McKinley?

EXAMPLE 8   Finding the Width of a River

To find the width of a river, a surveyor sights straight across the river from a point A on her side to a point B on the opposite side. See Figure 6.13. She then walks 200 feet upstream to a point C. The angle θ that the line of sight from point C to point B makes with the riverbank is 58°. How wide is the river to the nearest foot?

Solution

The points A, B, and C are the vertices of a right triangle with acute angle $\theta = 58°$.
    Let w represent the width of the river. From Figure 6.13, we have:

$$\frac{w}{200} = \tan 58° \qquad \tan \theta = \frac{\text{opposite}}{\text{adjacent}}$$

$$w = 200 \tan 58° \qquad \text{Multiply both sides by 200.}$$

$$w \approx 320.07 \text{ feet} \qquad \text{Use a calculator.}$$

The river is about 320 feet wide at the point A.

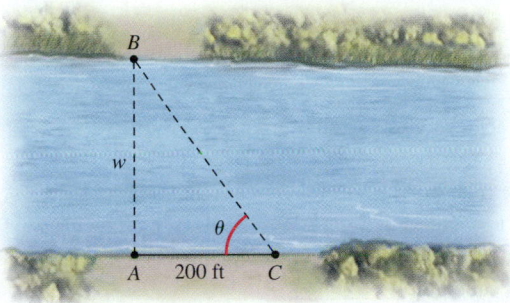

Figure 6.13 Width of a river.

Practice Problem 8 From a point on the edge of one of two parallel rooftop sides, a point directly opposite is identified on the other rooftop. From a second point 25 feet from the first point along the rooftop edge, a second sighting to the point on the opposite rooftop is made, and the angle θ that this line of sight makes with the rooftop is 60°. How far apart are the two buildings? Round your answer to the nearest foot.

EXAMPLE 9   Finding the Rotation Angle for a Security Camera

A security camera is to be installed 40 feet away from the center of a jewelry counter. The counter is 30 feet long. Through what angle, to the nearest degree, should the camera rotate so that it scans the entire counter? See Figure 6.14.

Solution

The counter center B, the camera A, and one end of the counter C form a right triangle. The angle at vertex A is $\frac{\theta}{2}$, where θ is the angle through which the camera rotates. See Figure 6.14. We have:

$$\tan \frac{\theta}{2} = \frac{15}{40} = \frac{3}{8} \qquad \text{Definition of tangent}$$

$$\frac{\theta}{2} = \tan^{-1}\left(\frac{3}{8}\right) \approx 20.556° \qquad \text{Use a calculator in Degree mode.}$$

$$\theta \approx 41.11° \qquad \text{Multiply both sides by 2.}$$

Set the camera to rotate through 41° to scan the entire counter.

Figure 6.14 Security camera.

Practice Problem 9 Rework Example 9 for a counter that is 25 feet long and a camera set 15 feet from the center of the counter.

Answers to Practice Problems

1. $\sin\theta = \dfrac{4}{5}$ $\csc\theta = \dfrac{5}{4}$ **2.** $\sin\theta = \dfrac{2\sqrt{2}}{3}$ $\csc\theta = \dfrac{3\sqrt{2}}{4}$

$\cos\theta = \dfrac{3}{5}$ $\sec\theta = \dfrac{5}{3}$ $\cos\theta = \dfrac{1}{3}$ $\sec\theta = 3$

$\tan\theta = \dfrac{4}{3}$ $\cot\theta = \dfrac{3}{4}$ $\tan\theta = 2\sqrt{2}$ $\cot\theta = \dfrac{\sqrt{2}}{4}$

3. a. $\sec 69° \approx 2.7904$ **b.** $\cot 15° \approx 3.7321$ **c.** $\dfrac{\sqrt{2+\sqrt{2}}}{2}$

4. $B = 34°, a \approx 21.4, b \approx 14.4$ **5.** $A \approx 48.4°, B \approx 41.6°,$

$a \approx 27.6$ **6.** ≈ 1.096 mi   **7.** $\approx 20,320$ ft

8. ≈ 43 feet   **9.** $80°$

 SECTION 6.1 | **Exercises**

Concepts and Vocabulary

1. If θ is an acute angle in a right triangle, then

$\sin\theta = \dfrac{\text{Opposite}}{\text{Hypotenuse}}$, $\cos\theta = $ _____,

$\tan\theta = $ _____.

2. The sum of the two acute angles in a right triangle is

_____.

3. If θ is an acute angle (measured in degrees) in a right triangle and $\cos\theta = 0.7$, then $\sin(90° - \theta) = $ _____.

4. When looking down, the acute angle between the line of sight and the horizontal is called the _____.

5. **True or False.** If θ is an acute angle in a right triangle and $\tan\theta = \dfrac{2}{5}$, then the length of the leg opposite θ is 2.

6. **True or False.** The leg adjacent to angle θ in a right triangle is the leg opposite the angle $90° - \theta$.

7. **True or False.** The length of the hypotenuse of a right triangle having a leg of length 1 must be greater than 1.

8. **True or False.** We can solve a right triangle if the two acute angles are given.

Building Skills

In Exercises 9–14, find the exact values for the six trigonometric functions of the angle θ in each figure. Rationalize the denominator where necessary.

9.

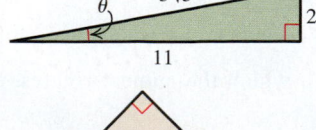

10.

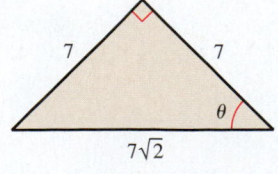

11.

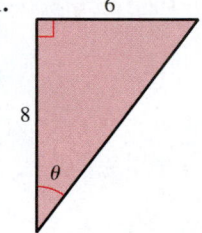

12.

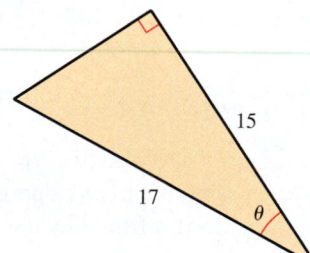

13.

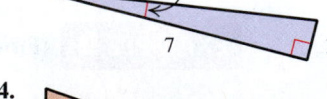

14.

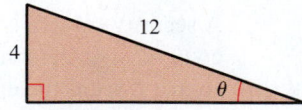

In Exercises 15–26, use each given trigonometric function value of θ to find the requested trigonometric function values. Rationalize the denominators where necessary.

15. $\cos\theta = \dfrac{2}{3}$ **16.** $\sin\theta = \dfrac{3}{4}$

   a. $\sin\theta$ **b.** $\tan\theta$ **a.** $\cos\theta$ **b.** $\tan\theta$

17. $\tan\theta = \dfrac{5}{3}$ **18.** $\cot\theta = \dfrac{6}{11}$

   a. $\sin\theta$ **b.** $\cos\theta$ **a.** $\sin\theta$ **b.** $\cos\theta$

19. $\sec\theta = \dfrac{13}{12}$

    a. $\sin\theta$ **b.** $\tan\theta$

20. $\csc\theta = 4$

    a. $\cos\theta$ **b.** $\tan\theta$

21. $\sin\theta = \dfrac{1}{3}$

    a. $\cos\theta$ **b.** $\cot(90° - \theta)$

22. $\cos\theta = \dfrac{1}{2}$

    a. $\sin\theta$ **b.** $\tan(90° - \theta)$

23. $\sec\theta = \dfrac{5}{4}$

    a. $\sin\theta$ **b.** $\tan(90° - \theta)$

24. $\csc\theta = 2$

    a. $\cos\theta$ **b.** $\cot(90° - \theta)$

25. $\tan\theta = \dfrac{1}{2}$

    a. $\sin\theta$ **b.** $\csc(90° - \theta)$

26. $\cot\theta = \dfrac{3}{4}$

    a. $\cos\theta$ **b.** $\sec(90° - \theta)$

In Exercises 27–32, find the trigonometric function values of each corresponding complementary angle.

27. Given that $\sin 58° \approx 0.8480$, find $\cos 32°$.

28. Given that $\cos 37° \approx 0.7986$, find $\sin 53°$.

29. Given that $\tan 27° \approx 0.5095$, find $\cot 63°$.

30. Given that $\cot 49° \approx 0.8693$, find $\tan 41°$.

31. Given that $\sec 65° \approx 2.3662$, find $\csc 25°$.

32. Given that $\csc 78° \approx 1.0223$, find $\sec 12°$.

In Exercises 33–40, solve each right triangle; angle C is the right angle. Round to the nearest tenth where necessary.

33. $A = 50°, c = 9.2$

34. $B = 28°, c = 10.5$

35. $A = 36°, a = 12.0$

36. $B = 74°, b = 6.3$

37. $a = 12.5, b = 6.2$

38. $a = 4.3, b = 8.1$

39. $b = 9.4, c = 14.5$

40. $a = 6.2, c = 18.7$

In Exercises 41–48, use the figure and the given values to find each specified side length, angle, and trigonometric function value. Round your answer to three decimal places.

41. $a = 8, b = 10$ Find c, $\sin\theta$, and $\tan\theta$.

42. $a = 18, b = 3$ Find c, $\sin\theta$, and $\cos\theta$.

43. $a = 23, b = 7$ Find $\cos\theta$ and $\tan\theta$.

44. $a = 19, b = 27$ Find c, $\cos\theta$, and $\tan\theta$.

45. $\theta = 30°, a = 9$ Find $\sin\theta$, b, and c.

46. $\theta = 30°, a = 5$ Find b and c.

47. $\theta = 60°, b = 7$ Find α, a, and c.

48. $\theta = 60°, b = 10$ Find α, a, and c.

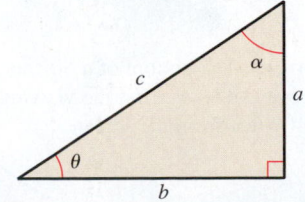

Applying the Concepts

49. **Positioning a ladder.** An 18-foot ladder is resting against a wall of a building in such a way that the top of the ladder is 14 feet above the ground. How far is the foot of the ladder from the base of the building?

50. **Positioning a ladder.** The ladder from Exercise 49 is placed against a wall so that the foot of the ladder is 7 feet from the base of the wall. How high up the wall will the ladder reach?

In Exercises 51 and 52, use the following definitions: The *vertical rise* **of a ski lift is the vertical distance traveled when a lift car goes from the bottom terminal to the top terminal. The** *slope length* **is the linear distance a lift car travels as it moves from the bottom terminal to the top terminal. We treat the slope length as the hypotenuse and the vertical rise as the side opposite the angle formed by the horizontal line at the top terminal and the lift cables.**

51. **Ski lift dimensions.** Find the horizontal distance a lift car passes over if the vertical rise is 295 meters and the slope length is 1070 meters.

52. **Ski lift dimensions.** Find the vertical rise (to the nearest foot) for a ski lift with slope length 2050 feet if the angle formed by a horizontal line at the top terminal and the lift cable is 16°.

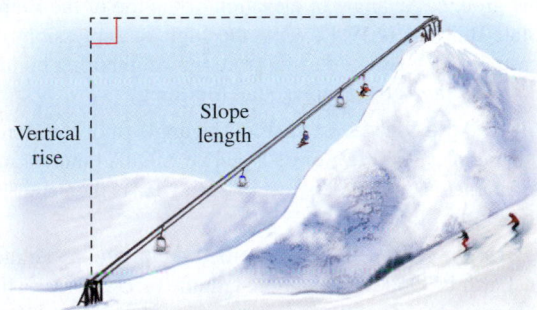

53. **Ski lift height.** If a skier travels 5000 feet up a ski lift whose angle of inclination is 30°, how high above his starting level is he?

54. **Shadow length.** At a time when the sun's rays make a 50° angle with the horizontal, how long is the shadow cast by a 6-foot man standing on flat ground?

55. **Height of a tree.** The angle relative to the horizontal from the top of a tree to a point 110 feet from its base (on flat ground) is 15°. Find the height of the tree to the nearest foot.

56. **Height of a cliff.** The angle of elevation from a rowboat moored 75 feet from a cliff is 73.8°. Find the height of the cliff to the nearest foot.

57. **Height of a flagpole.** A flagpole is supported by a 30-foot tension wire attached to the flagpole and to a stake in the ground. If the ground is flat and the angle the wire makes with the ground is 40°, how high up the flagpole is the wire attached? Round your answer to the nearest foot.

58. Flying a kite. A girl 5 feet tall is flying a kite. How long must the string be in order for her to raise the kite 200 feet above the ground if the string makes a 28° angle with the ground?

59. Width of a river. Sal and his friend are on opposite sides of a river. To find the width of the river, each of them hammered a stake on his bank of the river in such a way that the distance between the two stakes approximates the width of the river. Sal walks 10 feet away from his stake along the river and finds that the line of sight from his new position to his friend's stake across the river and the line of sight to his own stake form a 75° angle. How wide is the river?

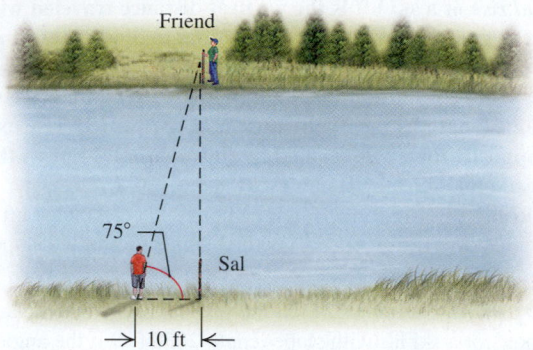

60. Measuring the Empire State Building. From a neighboring building's 13th-floor window that is 128 meters above the ground, the angle of elevation of the top of the Empire State Building is 59.4°, while the angle of depression of the base is 40.2°. How far away from the neighboring building and how tall is the Empire State Building?

61. Sighting a mortar station. The angle of depression from a helicopter 3000 feet directly above an allied mortar station to an enemy supply depot is 13°. How far is the depot from the mortar station?

62. Distance between Earth and the moon. If the radius of Earth is 3960 miles and the latitude of C is 89.05°, find the distance, AB, between the center of Earth and the center of the moon.

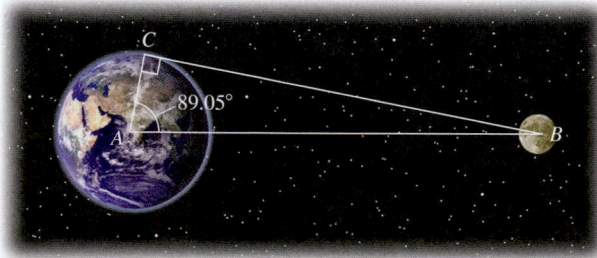

63. Sprinkler rotation. A sprinkler rotates back and forth through an angle θ, as shown in the figure. At a distance of 5 feet from the sprinkler, the rays that form the sides of angle θ are 6 feet apart. Find θ.

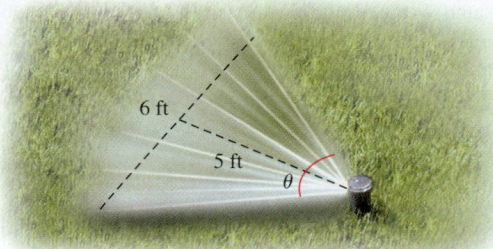

64. Irradiating flowers. A tray of flowers is being irradiated by a beam from a rotating lamp, as shown in the figure. If the tray is 8 feet long and the lamp is 2 feet from the center of the tray, through what angle should the lamp rotate to irradiate the full length of the tray?

In Exercises 65 and 66, use the figure to find the radius r for the given latitudes (values of θ). Use 3960 miles for the radius of Earth.

65. Latitude. Find the radius of the 50th parallel (of latitude).

66. Latitude. Find the radius of the 53rd parallel (of latitude).

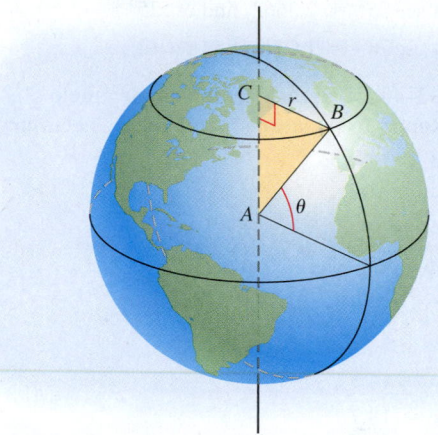

67. Motorcycle racing. A video camera is set up 110 feet at a right angle to a straight quarter-mile racetrack, as shown in the figure. The starting line is to the left, and the finish is to the right. Through what angle must the camera rotate to film the entire race?

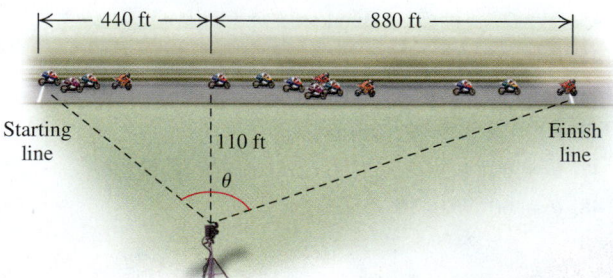

68. Viewing angle. The bottom of a 50-foot movie screen is 20 feet above eye level. Find the viewing angle θ from a distance of 100 feet from the screen.

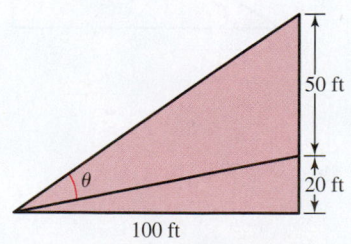

Beyond the Basics

69. Height of an airplane. From a tower 150 feet high with the sun directly overhead, an airplane and its shadow have an angle of elevation of 67.4° and an angle of depression of 8.1°, respectively. Find the height of the airplane.

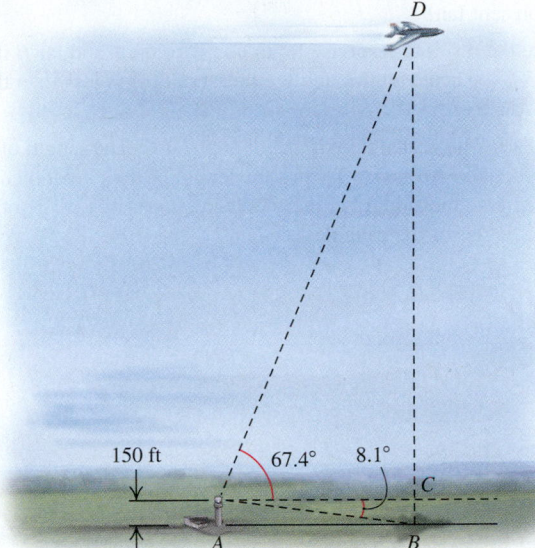

70. Camera angle. Raphael stands 12 feet from a large painting on a wall. He sets his camera 5 feet 6 inches above the floor. If the bottom edge of the painting is 4 feet above the floor and the angle of elevation from the camera to the top of the painting is 50°, find the height of the painting to the nearest tenth of a foot.

71. Height of a building. The WCNC-TV Tower in Dallas, North Carolina, is 600 meters high. Suppose a building is erected such that the base of the building is on the same plane as the base of the tower, the angle of elevation from the top of the building to the top of the tower is 75.24°, and the angle of depression from the top of the building to the foot of the tower is 60.05°. How high would the building have to be? (*Source:* http://en.wikipedia.org)

72. Height of a building. Suppose from the top of the Empire State Building (which is about 1250 feet high) the angle of depression to the top of a second building is 55.8° and the angle of depression to the bottom of the second building is 67.7°. Find the height of the second building.

73. In the figure below, show that $h = \dfrac{d}{\cot \alpha - \cot \beta}$.

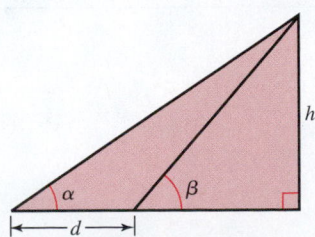

74. Height of a plane. A plane flies directly over two tracking stations that are 500 meters apart on a level plain. The plane flies in the direction from Station *A* to Station *B*. After the plane has passed Station *B*, an observer at Station *B* records the angle of elevation as 38.9°. At the same time, an observer

at station *A* records the angle of elevation as 36.9°. What is the plane's altitude? [*Hint:* Use Exercise 73.]

75. Measuring the Statue of Liberty. While sailing toward the Statue of Liberty, a sailor in a boat observed that at a certain point, the angle of elevation of the tip of the torch was 25°. After sailing another 100 meters toward the statue, the angle of elevation became 41.8°. How tall is the Statue of Liberty?

76. Speed of a plane. A plane flying at an altitude of 12,000 feet finds the angle of depression for a building ahead of the plane to be 10.4°. With the building still straight ahead, two minutes later the angle of depression is 25.6°. Find the speed of the plane relative to the ground (in feet per minute).

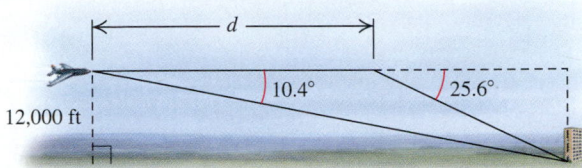

77. Flagpole height. The angle of elevation of the top of a building from a distance of 200 feet from its base is 70°, and the angle of elevation of the top of a flagpole on the top of the building from the same point is 71°. Find the height of the flagpole to the nearest foot.

78. Airplane height. A plane flying at a height of 3000 feet above ground passes vertically above a helicopter at an instant when the angles of the two from the same point on the ground are 60° and 48°, respectively. Find the height of the helicopter to the nearest foot.

79. In the figure below, show that $h = \dfrac{d}{\cot \alpha + \cot \beta}$.

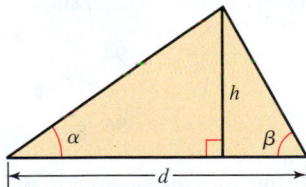

80. Find the area of the triangle in the figure.

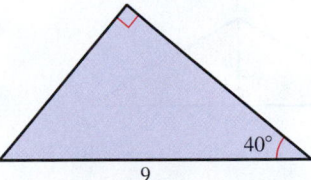

[*Hint:* Use Exercise 79.]

81. Find the area of triangle *ABC* in the figure.

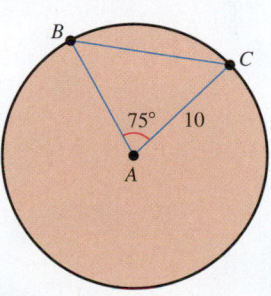

[*Hint:* Show $\angle ACB = 52.5°$.]

82. Find the area of the segment (see the shaded region in the figure) of the circle.

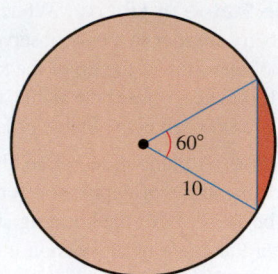

Critical Thinking / Discussion / Writing

83. Draw a right triangle with acute angle θ. Then use the definitions of $\sin\theta$ and $\cos\theta$ together with the Pythagorean Theorem to show that $\sin^2\theta + \cos^2\theta = 1$.

84. In a right triangle with acute angle θ, what value must the adjacent side have for the length of the opposite side to represent $\tan\theta$?

85. Use the result of Exercise 84 to compare $\tan\alpha$ and $\tan\beta$ if α and β are acute angles and the measure of α is less than the measure of β.

86. Let ABC be a triangle with right angle at C. The length of the hypotenuse AB is $2\sqrt{2}$ times the length of the perpendicular from C to the side AB. Find the degree measures of the other two angles of the triangle.

Getting Ready for the Next Section

In Exercises 87–90, let ABC be a triangle. Then
$A + B + C = 180°$.

87. If $A + C = 128°$, find B.

88. If $\dfrac{1}{2}(B + C) = 50°$, find A.

89. If $A = 30°$ and $B = 75°$, find C.

90. If $A = 23.7°$ and $C = 69.8°$, find A.

In Exercises 91–96, solve each equation for x to the nearest tenth. (Note: If $\sin x = r\,(0 < r < 1)$ and $0 \le x \le 180°$, then $x = \sin^{-1}(r)$ or $x = 180° - \sin^{-1}(r)$.)

91. $\dfrac{x}{2} = \dfrac{7}{5}$

92. $\dfrac{x}{\sin 30°} = \dfrac{8}{\sin 75°}$

93. $\dfrac{2.8}{\sin 37.8°} = \dfrac{x}{\sin 63.5°}$

94. $\dfrac{\sin x}{5} = \dfrac{\sin 57°}{8}$

95. $\dfrac{\sin 60°}{6} = \dfrac{\sin x}{5}$

96. $\dfrac{\sin x}{5} = \dfrac{\sin 60°}{3}$

In Exercises 97–100, find the height h to the nearest tenth of a unit in each triangle ABC.

97.

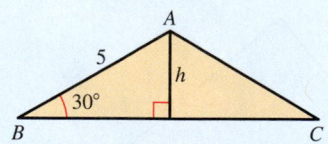

98.

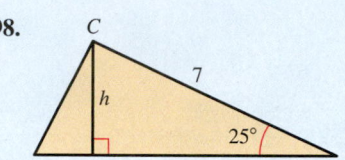

99.

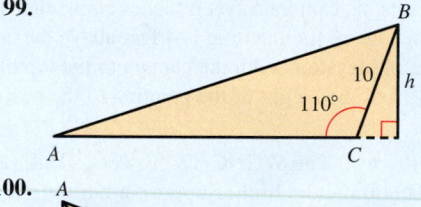

100.

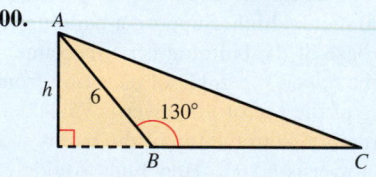

In Exercises 101–104, find the area of each triangle to the nearest tenth of a unit.

Recall: Area of a triangle $= \dfrac{1}{2}(\text{base})(\text{height})$.

101. Triangle of Exercise 97 with $BC = 12$

102. Triangle of Exercise 98 with $AB = 5$

103. Triangle of Exercise 99 with $AC = 6$

104. Triangle of Exercise 100 with $BC = 8$

The Law of Sines

BEFORE STARTING THIS SECTION, REVIEW

1 Trigonometric functions of angles (Section 4.2, page 357)

2 Inverse functions (Section 1.7, page 121)

3 Geometric properties of a triangle (Appendix A.5, page 950)

OBJECTIVES

1 Learn vocabulary and conventions for solving triangles.

2 Derive the Law of Sines.

3 Solve AAS and ASA triangles by using the Law of Sines.

4 Solve for possible triangles in the ambiguous SSA case.

5 Find the area of an SAS triangle.

Mount Everest

In 1852, an unsung hero Radhanath Sickdhar managed to *calculate* the height of Peak XV, an icy peak in the Himalayas. The highest mountain in the world—later named Mount Everest—stood at 29,002 feet. Sickdhar's calculations used triangulations as well as the phenomenon called refraction—the bending of light rays by the density of Earth's atmosphere. Like George Everest, Sickdhar may have never seen Mount Everest.

◆ The Great Trigonometric Survey of India

Triangulation is the process of finding the coordinates and the distance to a point by using the Law of Sines. Triangulation was used in surveying the Indian subcontinent. The survey, which was called "one of the most stupendous works in the whole history of science," was begun by William Lambton (1753–1823), a British army officer, in 1802. It started in the south of India and extended north to Nepal, a distance of approximately 1600 miles. Lasting several decades and employing thousands of workers, the survey was named the Great Trigonometric Survey (GTS). In 1818, George Everest (1790–1866), a Welsh surveyor and geographer, was appointed assistant to Lambton. Everest, who succeeded Lambton in 1823 and was later promoted to surveyor general of India, completed the GTS. Mount Everest was surveyed and named after Everest by his successor Andrew Waugh (1810–1878). In Example 4, we use the Law of Sines to estimate the height of a mountain.

1 Learn vocabulary and conventions for solving triangles.

RECALL

Two triangles are *congruent* if corresponding sides are congruent (equal in length) and corresponding angles are congruent (equal in measure). An *included side* is the side between two angles, and an *included angle* is the angle between two sides.

Solving Oblique Triangles

The process of finding the unknown side lengths and angle measures in a triangle is called *solving a triangle*. In Section 6.1, you learned the techniques of solving right triangles. We now discuss triangles that are not necessarily right triangles. A triangle with no right angle is called an **oblique triangle**. From geometry, we know that two triangles with any of the following sets of equal parts will always be **congruent**. In other words, the following sets of given parts determine a unique triangle:

ASA: Two angles and the included side

AAS: Two angles and a nonincluded side

SAS: Two sides and the included angle

SSS: All three sides

Although **SSA** (two sides and a nonincluded angle) does not guarantee a unique triangle, there can be, at most, two triangles with the given parts. On the other hand, **AAA** (or **AA**) guarantees only *similar* triangles, not congruent ones, so there are infinitely many triangles with the same three angle measures.

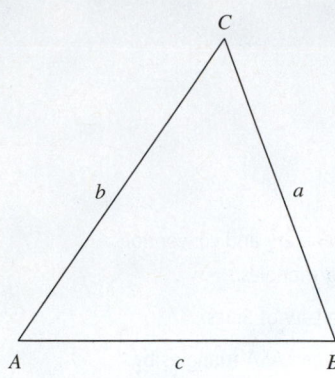

Figure 6.15 Labeling a triangle.

For any triangle ABC, we let each letter, such as A, stand for both the vertex A and the measure of the angle at A. As usual, the angles of a triangle ABC are labeled A, B, and C and the lengths of their opposite sides are labeled, a, b, and c, respectively. See Figure 6.15.

To solve an oblique triangle, given at least one side and two other measures, we consider four possible cases:

Case 1. Two angles and a side are known (**AAS** and **ASA** triangles).

Case 2. Two sides and an angle opposite one of the given sides are known (**SSA** triangles).

Case 3. Two sides and their included angle are known (**SAS** triangles).

Case 4. All three sides are known (**SSS** triangles).

We use the Law of Sines to analyze triangles in Cases 1 and 2. In the next section, we use the *Law of Cosines* to solve triangles in Cases 3 and 4.

2 Derive the Law of Sines.

The Law of Sines

The Law of Sines states that in a triangle ABC with sides of length a, b, and c,

$$\frac{\sin A}{a} = \frac{\sin B}{b} = \frac{\sin C}{c}.$$

We can derive the Law of Sines by using the right-triangle definition of the sine of an acute angle θ:

$$\sin\theta = \frac{\text{opposite}}{\text{hypotenuse}}$$

RECALL

An *altitude* of a triangle is a segment drawn from any vertex of the triangle perpendicular to the opposite side, or to an extension of the opposite side.

We begin with an oblique triangle ABC. See Figures 6.16 and 6.17. Draw altitude CD from the vertex C to the side AB (or its extension, as in Figure 6.17). Let h be the length of segment CD.

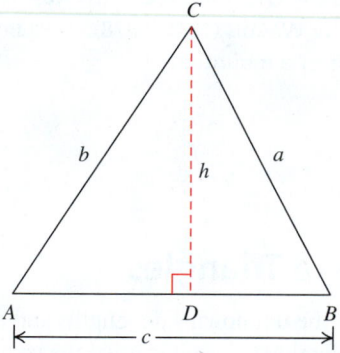

Figure 6.16 Altitude CD with acute $\angle B$.

In right triangle CDA,

$$\frac{h}{b} = \sin A \quad \text{or}$$

$$h = b\sin A.$$

In right triangle CDB,

$$\frac{h}{a} = \sin B \quad \text{or}$$

$$h = a\sin B.$$

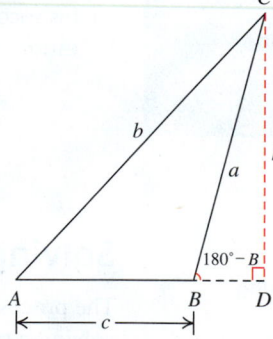

Figure 6.17 Altitude CD with obtuse $\angle B$.

In right triangle CDA,

$$\frac{h}{b} = \sin A \quad \text{or}$$

$$h = b\sin A.$$

In right triangle CDB,

$$\frac{h}{a} = \sin(180° - B) = \sin B \quad \text{or}$$

$$h = a\sin B.$$

Since in each figure $h = b \sin A$ and $h = a \sin B$, we have

$$b \sin A = a \sin B$$

$$\frac{b \sin A}{ab} = \frac{a \sin B}{ab} \qquad \text{Divide both sides by } ab.$$

$$\frac{\sin A}{a} = \frac{\sin B}{b} \qquad \text{Simplify.}$$

Similarly, by drawing altitudes from the other two vertices, we can show that

$$\frac{\sin B}{b} = \frac{\sin C}{c} \quad \text{and} \quad \frac{\sin C}{c} = \frac{\sin A}{a}.$$

We have proved the Law of Sines.

THE LAW OF SINES

In any triangle ABC, with sides of length a, b, and c,

$$\frac{\sin A}{a} = \frac{\sin B}{b}, \quad \frac{\sin B}{b} = \frac{\sin C}{c}, \quad \text{and} \quad \frac{\sin C}{c} = \frac{\sin A}{a}.$$

We can rewrite these relations in compact notation:

$$\frac{\sin A}{a} = \frac{\sin B}{b} = \frac{\sin C}{c}, \quad \text{or equivalently,} \quad \frac{a}{\sin A} = \frac{b}{\sin B} = \frac{c}{\sin C}.$$

In words, the lengths of the sides in a triangle are proportional to the sines of the measures of the angles opposite them.

EXAMPLE 1 Applying the Law of Sines

In a triangle ABC: $b = 4$ centimeters, $B = 40°$, and $C = 75°$. Find c to the nearest tenth of a centimeter.

Solution

$$\frac{c}{\sin C} = \frac{b}{\sin B} \qquad \text{The Law of Sines; unknown side in the numerator}$$

$$\frac{c}{\sin 75°} = \frac{4}{\sin 40°} \qquad \text{Replace } b \text{ with 4, } C \text{ with } 75°, \text{ and } B \text{ with } 40°.$$

$$c = \frac{4 \sin 75°}{\sin 40°} \qquad \text{Solve for } c.$$

$$c \approx 6.0 \text{ centimeters} \qquad \text{Use a calculator.}$$

Practice Problem 1 Find c with $a = 6$, $A = 30°$, and $C = 72°$.

EXAMPLE 2 Applying the Law of Sines

In a triangle ABC: $a = 10$ feet, $c = 20$ feet, and $C = 60°$. Find A to the nearest tenth of a degree.

Solution

We start with the relation

$$\frac{\sin A}{a} = \frac{\sin C}{c} \qquad \text{The Law of Sines; unknown angle in the numerator}$$

$$\frac{\sin A}{10} = \frac{\sin 60°}{20} \qquad \text{Replace } a \text{ with 10, } c \text{ with 20, and } C \text{ with } 60°.$$

$$\sin A = \frac{10 \sin 60°}{20} \quad \text{Solve for } \sin A.$$

$$\sin A = \frac{\sqrt{3}}{4} \quad \sin 60° = \frac{\sqrt{3}}{2}; \text{ simplify.}$$

There are two possible angles A for which $\sin A = \frac{\sqrt{3}}{4}$:

RECALL

$\sin \theta = \sin(180° - \theta)$

$$A_1 = \sin^{-1} \frac{\sqrt{3}}{4} \approx 25.7° \quad \text{and} \quad A_2 = 180° - A_1 \approx 154.3°.$$

However, $A_2 + C = 154.3° + 60° = 214.3° > 180°$, so there is no triangle with vertex A_2. Hence, $A = A_1 = 25.7°$.

Practice Problem 2 Find A with $a = 3$, $b = 5$, and $B = 70°$.

3 Solve AAS and ASA triangles by using the Law of Sines.

Solving AAS and ASA Triangles

We next learn to solve AAS and ASA triangles of Case 1, where two angles and a side are given. Note that if two angles of a triangle are given, then you also know the third angle (since the sum of the angles of a triangle is 180°). So in Case 1, all three angles and one side of a triangle are known.

PROCEDURE
IN ACTION

EXAMPLE 3 Solving the AAS and ASA Triangles

OBJECTIVE

Solve a triangle given two angles and a side.

EXAMPLE

Solve triangle ABC with $A = 62°$, $c = 14$ feet, $B = 74°$. Round side lengths to the nearest tenth.

Step 1 Find the third angle. Find the measure of the third angle by subtracting the measures of the known angles from 180°.

1. $C = 180° - A - B$
$= 180° - 62° - 74°$
$= 44°$

Step 2 Make a chart. Make a chart of the six parts of the triangle, including the known and the unknown parts. Identify two rows in the chart in which three of the four quantities are known.

2.

Angles	Sides
$A = 62°$	$a = ?$
$B = 74°$	$b = ?$
$C = 44°$	$c = 14$

Three quantities are known.
Rows 1 and 3
and
Rows 2 and 3

Step 3 Apply the Law of Sines to each pair of rows selected in Step 2. Use the form of the Law of Sines in which the unknown quantity is in the numerator. Solve for the fourth quantity.

3.
$$\frac{b}{\sin B} = \frac{c}{\sin C} \qquad \frac{a}{\sin A} = \frac{c}{\sin C}$$

$$\frac{b}{\sin 74°} = \frac{14}{\sin 44°} \qquad \frac{a}{\sin 62°} = \frac{14}{\sin 44°}$$

$$b = \frac{14 \sin 74°}{\sin 44°} \qquad a = \frac{14 \sin 62°}{\sin 44°}$$

$$b \approx 19.4 \text{ feet} \qquad a \approx 17.8 \text{ feet}$$

Step 4 Show the solution. Show the solution by completing the chart.

4.

Angles	Sides
$A = 62°$	$a \approx 17.8$ feet
$B = 74°$	$b \approx 19.4$ feet
$C = 44°$	$c = 14$

Practice Problem 3 Solve triangle ABC with $A = 70°$, $B = 65°$, and $a = 16$ inches. Round side lengths to the nearest tenth.

◆ **EXAMPLE 4**   **Height of a Mountain**

From a point on a level plain at the foot of a mountain, a surveyor finds the angle of elevation of the peak of the mountain to be 20°. She walks 3465 meters closer (on a direct line between the first point and the base of the mountain) and finds the angle of elevation to be 23°. Estimate the height of the mountain to the nearest meter.

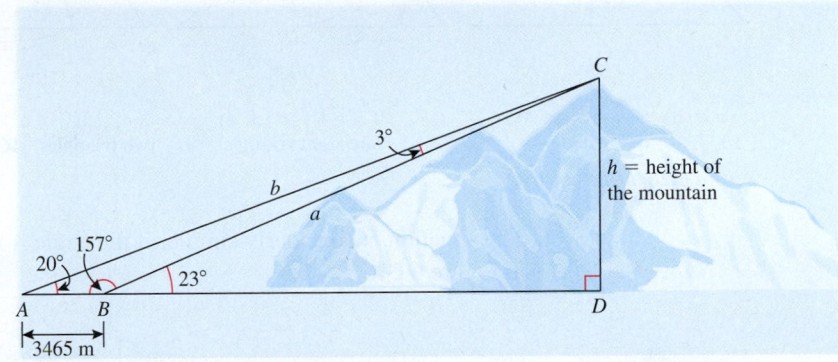

Figure 6.18 Height of a mountain.

Solution

In Figure 6.18, consider triangle *ABC*.

$$\angle ABC = 180° - 23° = 157° \qquad \angle ABC + \angle CBD = 180°$$

Find the third angle in triangle *ABC*.

$$C = 180° - 20° - 157° = 3°$$

Apply the Law of Sines.

$$\frac{a}{\sin 20°} = \frac{3465}{\sin 3°}$$

$$a = \frac{3465 \sin 20°}{\sin 3°} \approx 22,644 \text{ meters}$$

Now in triangle *BCD*, $\sin 23° = \dfrac{h}{a}$. So, $h = a \sin 23° \approx 22,644 \sin 23° \approx 8848$ meters.

The mountain is approximately 8848 meters high.

Practice Problem 4 From a point on a level plain at the foot of a mountain, the angle of elevation of the peak is 40°. If you move 2500 feet farther (on a direct line extending from the first point and the base of the mountain), the angle of elevation of the peak is 35°. How high above the plain is the peak? Round your answer to the nearest foot. ▪▪

4 Solve for possible triangles in the ambiguous SSA case.

Solving SSA Triangles—the Ambiguous Case

In Case 1, where we know two angles and a side, we obtain a unique triangle. However, if we know the lengths of two sides and the measure of the angle opposite one of these sides, then we could have a result that is (1) not a triangle, (2) exactly one triangle, or (3) two different triangles. For this reason, Case 2 is called the **ambiguous case**.

Suppose we want to draw triangle *ABC* with given measures of *A*, *a*, and *b*. We draw angle *A* with a horizontal initial side, and we place the vertex *C* on the terminal side of *A* so that the length of segment *AC* is *b*. We need to locate vertex *B* on the initial side of *A* so that the measure of the segment *CB* is *a*. See Figure 6.19.

The number of possible triangles depends on the length *h* of the altitude from *C* to the initial side of angle *A*. Since $\sin A = \dfrac{h}{b}$, we have $h = b \sin A$. Various possibilities for forming triangle *ACB* are illustrated in Figure 6.20, where *A* is an acute angle.

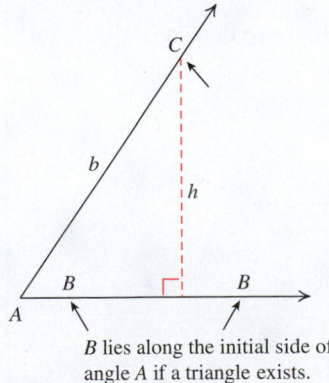

B lies along the initial side of angle *A* if a triangle exists.

Figure 6.19 Need to locate *B*.

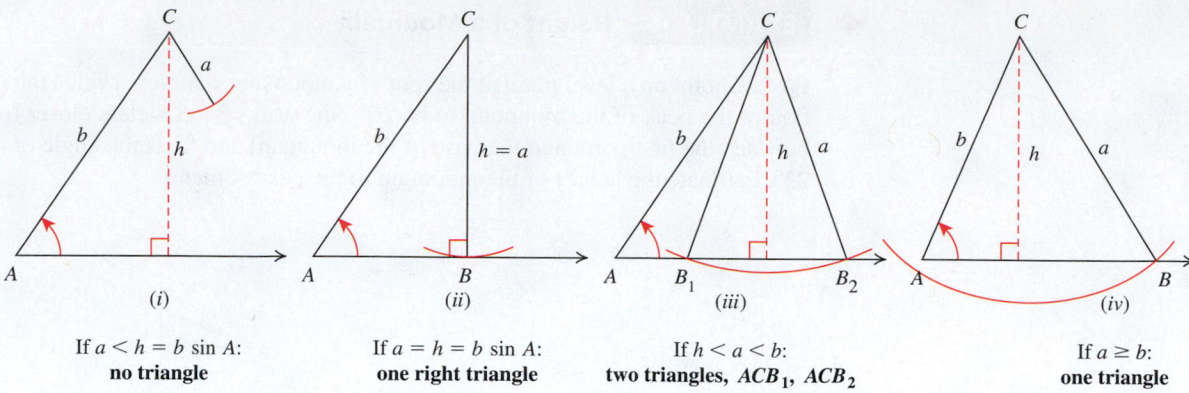

If $a < h = b \sin A$:
no triangle

If $a = h = b \sin A$:
one right triangle

If $h < a < b$:
two triangles, ACB_1, ACB_2

If $a \geq b$:
one triangle

Figure 6.20 Acute angle A.

The situation is considerly simpler if the angle A is obtuse. The possibilities are illustrated in Figure 6.21.

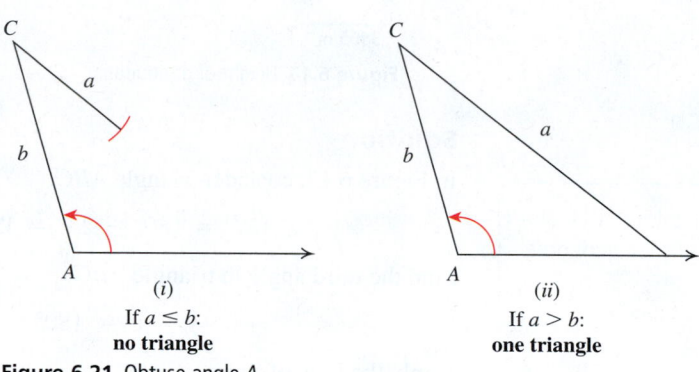

If $a \leq b$:
no triangle

If $a > b$:
one triangle

Figure 6.21 Obtuse angle A.

When solving a triangle, we strongly suggest that you begin with a tentative sketch of the triangle.

PROCEDURE
IN ACTION

EXAMPLE 5 **Solving the SSA Triangles**

OBJECTIVE

Solve a triangle if two sides and an angle opposite one of them are given.

Step 1 Make a chart of the six parts. Identify the known angle θ and its opposite and adjacent sides. Identify the two rows in the chart in which three of the four quantities are known.

EXAMPLE

Solve triangle ABC with B = 32°, b = 100 feet, and c = 150 feet. Round each answer to the nearest tenth.

1. $\theta = B$, opposite side = b, adjacent side = c

Angles	Sides
A = ?	a = ?
B = 32°	b = 100
C = ?	c = 150

Three quantities are known.

Step 2 Apply the Law of Sines to the two rows identified in Step 1. Use the form of the Law of Sines in which the unknown angle is in the numerator. Solve for the sine of the unknown angle.

2. $\dfrac{\sin C}{c} = \dfrac{\sin B}{b}$ The Law of Sines

$\sin C = \dfrac{c \sin B}{b}$ Multiply both sides by c.

$= \dfrac{(150) \sin 32°}{100}$ Substitute values.

$\sin C \approx 0.7949$ Use a calculator.

Step 3 Find two possible angles. For $0 < k < 1$, the equation $\sin x = k$ has two possible solutions:
(i) $\alpha_1 = \sin^{-1}(k)$, and
(ii) $\alpha_2 = 180° - \alpha_1$

Step 4 If both inequalities
(known angle θ) $+ \alpha_1 < 180°$, and
(known angle θ) $+ \alpha_2 < 180°$ are true, then there are two triangles with given measurements. Otherwise there is only one triangle.

Step 5 Find the third angle of the triangle(s). Use the angle sum theorem.

Step 6 Use the Law of Sines to find the remaining side(s). Use the Law of Sines in which the unknown side is in the numerator.

Step 7 Show the solution(s).

3. Two possible values of C are:
$$C_1 \approx \sin^{-1}(0.7949) \approx 52.6°$$
$$C_2 \approx 180° - 52.6° = 127.4°$$

4. $B + C_1 = 32° + 52.6° = 84.6° < 180°$, and
$B + C_2 = 32° + 127.4° = 159.4° < 180°$. We have two triangles with the given measurements.

5. $\angle BAC_1 \approx 180° - 32° - 52.6° = 95.4°$
$\angle BAC_2 \approx 180° - 32° - 127.4° = 20.6°$

6.

$$\frac{a_1}{\sin \angle BAC_1} = \frac{b}{\sin B} \quad \bigg| \quad \frac{a_2}{\sin \angle BAC_2} = \frac{b}{\sin B}$$

$$a_1 = \frac{b \sin \angle BAC_1}{\sin B} \quad \bigg| \quad a_2 = \frac{b \sin \angle BAC_2}{\sin B}$$

$$a_1 = \frac{100 \sin 95.4°}{\sin 32°} \quad \bigg| \quad a_2 = \frac{(100) \sin 20.6°}{\sin 32°}$$

$$a_1 \approx 187.9 \text{ feet} \quad \bigg| \quad a_2 \approx 66.4 \text{ feet}$$

7.

$\triangle BAC_1$

$\angle BAC_1 \approx 95.4°$	$a_1 \approx 187.9$
$B = 32°$	$b = 100$
$C_1 \approx 52.6°$	$c = 150$

$\triangle BAC_2$

$\angle BAC_2 \approx 20.6°$	$a_2 \approx 66.4$
$B = 32°$	$b = 100$
$C_2 \approx 127.4°$	$c = 150$

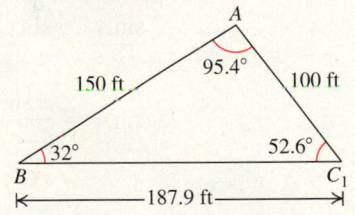

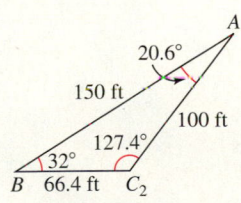

Practice Problem 5 Solve triangle ABC with $C = 35°$, $b = 15$ feet, and $c = 12$ feet. Round each answer to the nearest tenth.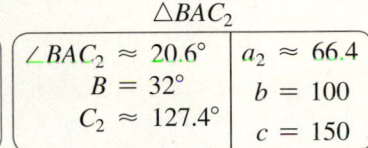

EXAMPLE 6 Solving an SSA Triangle (No Solution)

Solve triangle ABC with $A = 50°$, $a = 8$ inches, and $b = 15$ inches.

Solution

Step 1 Make a chart.

Angles	Sides
$A = 50°$	$a = 8$
$B = ?$	$b = 15$
$C = ?$	$c = ?$

Three quantities are known.

Step 2 Apply the Law of Sines.

$$\frac{\sin B}{b} = \frac{\sin A}{a}$$ The Law of Sines, unknown angle in the numerator

$$\sin B = \frac{b \sin A}{a}$$ Multiply both sides by b.

$$\sin B = \frac{(15) \sin 50°}{8}$$ Substitute values.

$$\sin B \approx 1.44$$ Use a calculator.

Since $\sin B \approx 1.44 > 1$, we conclude that no triangle has the given measures.

Practice Problem 6 Solve triangle ABC with $A = 65°$, $a = 16$ meters, and $b = 30$ meters.

EXAMPLE 7   Solving an SSA Triangle (One Solution)

Solve triangle ABC with $C = 40°$, $c = 20$ meters, and $a = 15$ meters. Round each answer to the nearest tenth.

Solution

Step 1 Make a chart.

Angles	Sides
$A = ?$	$a = 15$
$B = ?$	$b = ?$
$C = 40°$	$c = 20$

Three quantities are known.

Step 2 Apply the Law of Sines.

$$\frac{\sin A}{a} = \frac{\sin C}{c}$$ The Law of Sines, unknown angle in the numerator

$$\sin A = \frac{a \sin C}{c}$$ Multiply both sides by a.

$$= \frac{(15) \sin 40°}{20}$$ Substitute values.

$$\sin A \approx 0.4821$$ Use a calculator.

Step 3 $A = \sin^{-1}(0.4821) \approx 28.8°$. Two possible values of A are $A_1 \approx 28.8°$ and $A_2 \approx 180° - 28.8° = 151.2°$.

Step 4 Since $C + A_2 = 40° + 151.2° = 191.2° > 180°$, there is no triangle with vertex A_2. Thus, only one triangle has measure of angle $A_1 \approx 28.8°$.

Step 5 The third angle B has measure $\approx 180° - 40° - 28.8° = 111.2°$.

Step 6 Find the remaining side length.

$$\frac{b}{\sin B} = \frac{c}{\sin C}$$ The Law of Sines

$$b = \frac{c \sin B}{\sin C}$$ Multiply both sides by $\sin B$.

$$= \frac{(20) \sin (111.2°)}{\sin 40°}$$ Substitute values.

$$b \approx 29.0 \text{ m}$$ Use a calculator.

Step 7 Show the solution. See Figure 6.22.

Angles	Sides
$A_1 \approx 28.8°$	$a = 15$ meters
$B \approx 111.2°$	$b \approx 29.0$ meters
$C \approx 40°$	$c = 20$ meters

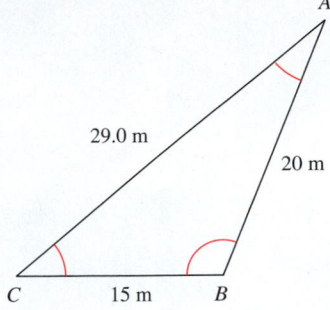

Figure 6.22

Practice Problem 7 Solve triangle ABC with $C = 60°$, $c = 50$ feet, and $a = 30$ feet.

Bearings

In navigation and surveying, directions are usually given by using *bearings*. A **bearing** is the measure of an acute angle from due north or due south. The bearing N 40° E means 40° to the east of due north, whereas S 30° W means 30° to the west of due south. See Figure 6.23.

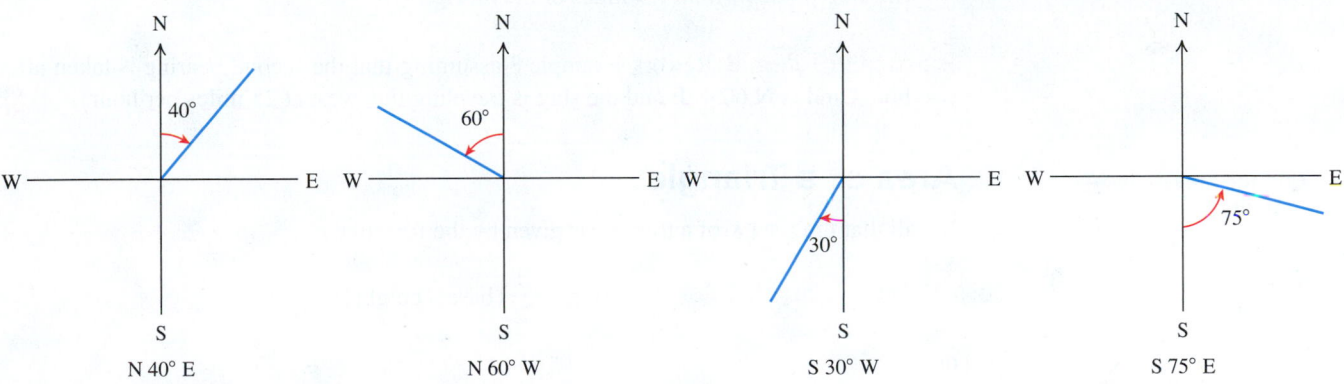

N 40° E   N 60° W   S 30° W   S 75° E

Figure 6.23 Bearings.

EXAMPLE 8   Navigation Using Bearings

A ship sailing due west at 20 miles per hour records the bearing of an oil rig at N 55.4° W. An hour and a half later the bearing of the same rig is N 66.8° E.

a. How far is the ship from the oil rig the second time?

b. How close did the ship pass to the oil rig?

Solution

a. In an hour and a half, the ship travels $(1.5)(20) = 30$ miles due west. The oil rig (C), the starting point for the ship (A), and the position of the ship after an hour and a half (B) form the vertices of the triangle ABC shown in Figure 6.24.

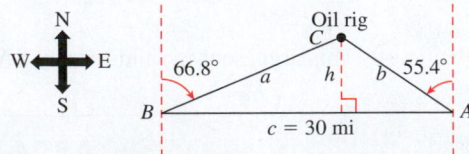

Figure 6.24 Navigation.

Then $A = 90° - 55.4° = 34.6°$ and $B = 90° - 66.8° = 23.2°$. Therefore, $C = 180° - 34.6° - 23.2° = 122.2°$.

$$\frac{a}{\sin A} = \frac{c}{\sin C} \qquad \text{The Law of Sines}$$

$$a = \frac{c \sin A}{\sin C} \qquad \text{Multiply both sides by } \sin A.$$

$$= \frac{30 \sin 34.6°}{\sin 122.2°} \approx 20 \text{ miles} \qquad \text{Substitute values and use a calculator.}$$

The ship is approximately 20 miles from the oil rig when the second bearing is taken.

b. The shortest distance between the ship and the oil rig is the length of the segment h in triangle ABC.

$$\sin B = \frac{h}{a} \qquad \text{Right-triangle definition of the sine}$$

$$h = a \sin B \qquad \text{Multiply both sides by } a.$$

$$h \approx 20 \sin 23.2° \qquad \text{Substitute values.}$$

$$\approx 7.9 \qquad \text{Use a calculator.}$$

The ship passes within 7.9 miles of the oil rig.

Practice Problem 8 Rework Example 8 assuming that the second bearing is taken after two hours and is N 60.4° E and the ship is traveling due west at 25 miles per hour. ▪▪

5 Find the area of an SAS triangle.

Area of a Triangle

Recall that the area K of a triangle is given by the formula

$$\text{Area} = \frac{1}{2}(\text{base})(\text{height}),$$

or

$$K = \frac{1}{2}bh.$$

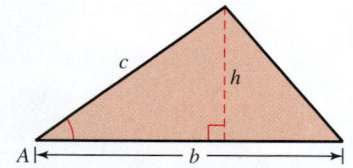

Figure 6.25 Acute angle A.

If A is an acute angle in a triangle (see Figure 6.25), then $\sin A = \dfrac{h}{c}$; so $h = c \sin A$. The area of this triangle is given by

$$K = \frac{1}{2}bh = \frac{1}{2}bc \sin A \qquad \text{Replace } h \text{ with } c \sin A.$$

The angle A shown in Figure 6.26 is an obtuse angle. Here $A' = 180° - A$ is the reference angle for A and $\sin A = \sin A' = \dfrac{h}{c}$. So,

$$h = c \sin A' = c \sin A$$

$$K = \frac{1}{2}bh = \frac{1}{2}bc \sin A \qquad \text{Replace } h \text{ with } c \sin A.$$

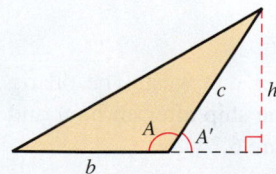

Figure 6.26 Obtuse angle A.

This argument is valid for any SAS triangle.

AREA OF A TRIANGLE (SAS)

In any triangle ABC, the area K of the triangle is given by the formula:

$$K = \frac{1}{2}bc \sin A = \frac{1}{2}ab \sin C = \frac{1}{2}ca \sin B$$

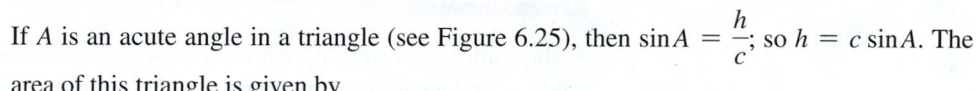

B

29 ft

62°

A   36 ft   C

Figure 6.27

EXAMPLE 9   **Finding the Area of a Triangle**

Find the area of the triangle ABC in Figure 6.27.

Solution

We are given the angle of measure 62° between the sides of lengths 36 feet and 29 feet. The area K of the triangle ABC is given by

$$K = \frac{1}{2} bc \sin A \qquad \text{Area formula}$$

$$= \frac{1}{2} (36)(29) \sin 62° \qquad b = 36, c = 29, A = 62°$$

$$\approx 460.9 \text{ square feet.} \qquad \text{Use a calculator.}$$

Practice Problem 9 Find the area of triangle ABC assuming that the lengths of the sides AC and BC are 27 feet and 38 feet, respectively, and the measure of the angle between these sides is 47°.

Answers to Practice Problem

1. $c = 11.4$ **2.** $A = 34.3°$
3. $C = 45°, b \approx 15.4$ in., $c \approx 12.0$ in.
4. 10,576 ft **5.** Solution 1: $A \approx 99.2°, B \approx 45.8°, a \approx 20.7$ ft
Solution 2: $A \approx 10.8°, B \approx 134.2°, a \approx 3.9$ ft

6. No triangle exists. **7.** $A \approx 31.3°, B \approx 88.7°, b \approx 57.7$ ft
8. a. ≈ 31.5 mi **b.** ≈ 15.6 mi **9.** 375.2 sq ft

SECTION 6.2   **Exercises**

Concepts and Vocabulary

1. If you know two angles of a triangle, then you can determine the third angle because the sum of all three angles is _____ degrees.

2. The Law of Sines states that if a, b, and c are the sides opposite angles A, B, and C, then $\dfrac{a}{\sin A} =$ _____ = _____.

3. If you are given any two angles and one side, then there is exactly _____ triangle(s) possible.

4. For given data a, A, and b, there can be no triangle, exactly _____ triangle, or _____ triangles.

5. True or False. The Law of Sines allows you to solve a triangle if you know two sides and the angle opposite one of them.

6. True or False. A triangle is uniquely determined if any two sides and one angle are given.

7. True or False. In a triangle ABC, $\dfrac{\sin A}{\sin B} = \dfrac{a}{b}$.

8. True or False. The Law of Sines cannot be used to solve SAS triangles.

Building Skills

In Exercises 9–20, the data from a triangle ABC is given. Use the Law of Sines (if applicable) to calculate the requested quantities. Give exact answers where possible, otherwise give answers to the nearest tenth.

9. $A = 45°, B = 60°$, and $a = 10$, find b.

10. $B = 60°, C = 45°$, and $c = 4$, find b.

11. $B = 45°, C = 75°$, and $b = \sqrt{6}$, find a.

12. $A = 105°, C = 45°$, and $b = \sqrt{12}$, find c.

13. $a = 4, b = 6$, and $A = 30°$, find B.

14. $c = 40, a = 70$, and $C = 30°$, find A.

15. $a = 3, b = 5$, and $C = 60°$, find B.

16. $a = 2, b = 3$, and $c = 4$, find A.

17. $a = 2, b = 3$, and $\sin A = \dfrac{2}{3}$, find B.

18. $b = \sqrt{3}, c = \sqrt{2}$, and $B = 60°$, find A.

19. $a = 8, c = 5$, and $C = 60°$, find A.

20. $b = 5, c = 8$, and $B = 135°$, find C.

In Exercises 21–24, solve each triangle. Round each answer to the nearest tenth.

21.

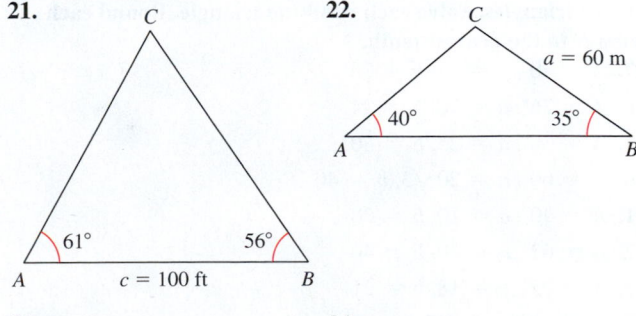

22.

23.

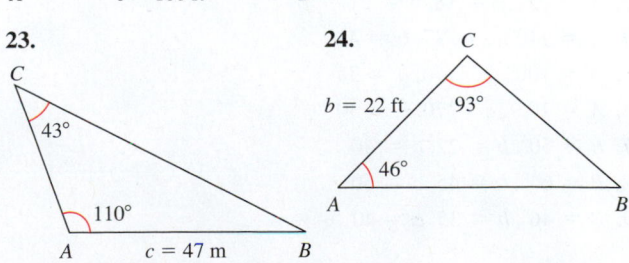

24.

In Exercises 25–36, solve triangle *ABC*. Round each answer to the nearest tenth.

25. $A = 40°, B = 35°, a = 100$ meters

26. $A = 80°, B = 20°, a = 100$ meters

27. $A = 46°, C = 55°, a = 75$ centimeters

28. $A = 35°, C = 98°, a = 75$ centimeters

29. $A = 35°, C = 47°, c = 60$ feet

30. $A = 44°, C = 76°, c = 40$ feet

31. $B = 43°, C = 67°, b = 40$ inches

32. $B = 95°, C = 35°, b = 100$ inches

33. $B = 110°, C = 46°, c = 23.5$ feet

34. $B = 67°, C = 63°, c = 16.8$ feet

35. $A = 35.7°, B = 45.8°, c = 30$ meters

36. $A = 64.5°, B = 54.3°, c = 40$ meters

In Exercises 37–40, solve each triangle. Round each answer to the nearest tenth.

37.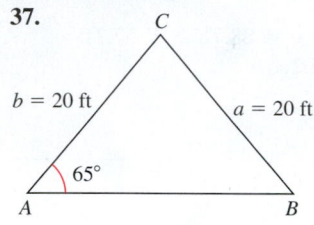

38.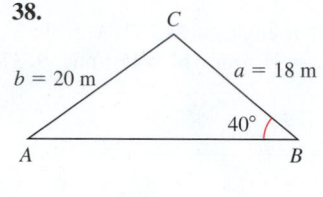

39.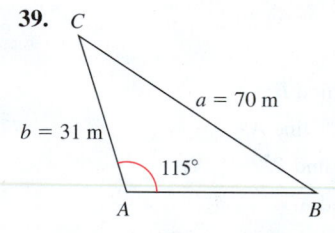

40.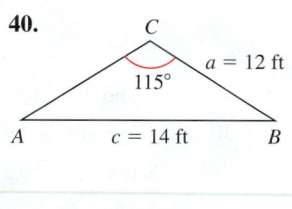

In Exercises 41–46, determine the number of triangles that can be drawn with the given data.

41. $a = 40, b = 70, A = 30°$ 42. $a = 22, b = 32, A = 42°$

43. $b = 15, c = 19, B = 58°$ 44. $b = 75, c = 85, B = 135°$

45. $a = 50, b = 70, B = 120°$ 46. $c = 85, a = 45, C = 150°$

In Exercises 47–66, solve each SSA triangle. Indicate whether the given measurements result in no triangle, one triangle, or two triangles. Solve each resulting triangle. Round each answer to the nearest tenth.

47. $A = 40°, a = 23, b = 20$

48. $A = 36°, a = 30, b = 24$

49. $A = 30°, a = 25, b = 50$

50. $A = 60°, a = 20\sqrt{3}, b = 40$

51. $A = 40°, a = 10, b = 20$

52. $A = 62°, a = 30, b = 40$

53. $A = 95°, a = 18, b = 21$

54. $A = 110°, a = 37, b = 41$

55. $A = 100°, a = 40, b = 34$

56. $A = 105°, a = 70, b = 30$

57. $B = 50°, b = 22, c = 40$

58. $B = 64°, b = 45, c = 60$

59. $B = 46°, b = 35, c = 40$

60. $B = 32°, b = 50, c = 60$

61. $B = 97°, b = 27, c = 30$

62. $B = 110°, b = 19, c = 21$

63. $A = 42°, a = 55, c = 62$

64. $A = 34°, a = 6, c = 8$

65. $C = 40°, a = 3.3, c = 2.1$

66. $C = 62°, a = 50, c = 100$

In Exercises 67–72, find the exact value of the area of each SAS triangle *ABC*.

67. $A = 30°, b = 6, c = 5$

68. $B = 60°, a = 12, c = 7$

69. $C = 120°, a = 8, b = 5$

70. $A = 135°, b = 4, c = 6$

71. $B = 150°, a = 12, c = 9$

72. $C = 45°, a = \sqrt{8}, b = 5$

In Exercises 73–80, find the area of each triangle *ABC*. Round your answers to the nearest tenth.

73. $A = 57°, b = 30$ in., $c = 52$ in.

74. $B = 110°, a = 20$ cm, $c = 27$ cm

75. $C = 46°, a = 15$ km, $b = 22$ km

76. $A = 146.7°, b = 16.7$ ft, $c = 18$ ft

77. $B = 38.6°, a = 12$ mm, $c = 16.7$ mm

78. $B = 112.5°, a = 151.6$ ft, $c = 221.8$ ft

79. $C = 107.3°, a = 271$ ft, $b = 194.3$ ft

80. $A = 131.8°, b = 15.7$ mm, $c = 18.2$ mm

In Exercises 81–84, find an angle θ between the sides a and b of a triangle *ABC* with given area *K*.

81. $K = 12, a = 6, b = 8$ 82. $K = 6, a = 4, b = 2\sqrt{3}$

83. $K = 30, a = 12, b = 5$ 84. $K = 15, a = 5\sqrt{2}, b = 6$

Applying the Concepts

85. **Finding distance.** Angela wants to find the distance from point *A* to her friend Carmen's house at point *C* on the other side of the river. She knows the distance from *A* to Betty's house at *B* is 540 feet. See the figure. The measurement of angles *A* and *B* are 57° and 46°, respectively. Calculate the distance from *A* to *C*. Round to the nearest foot.

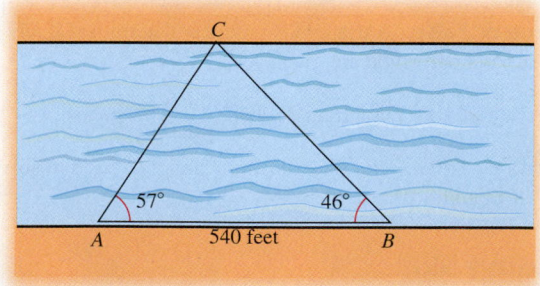

86. **Finding distance.** In Exercise 85, find the width of the river assuming that the houses are on the (very straight) banks of the river.

87. Target. A laser beam with an angle of elevation of 42° is reflected by a target and is received 1200 yards from the point of origin. Assume that the trajectory of the beam forms (approximately) an isosceles triangle.
   a. Find the total distance the beam travels. Round to the nearest yard.
   b. What is the height of the target? Round to the nearest yard.

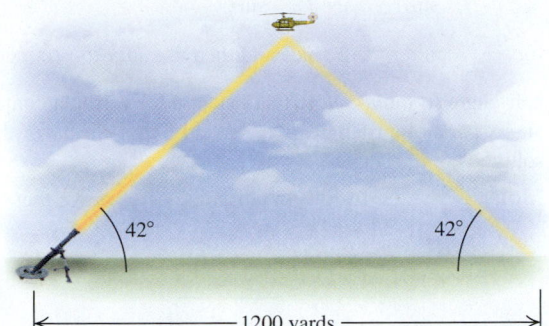

88. Height of a flagpole. Two surveyors stand 200 feet apart with a flagpole between them. Suppose the "transit" at each location is 5 feet high. The transit measures the angles of elevation of the top of the flagpole at the two locations to be 30° and 25°, respectively. Find the height of the flagpole. Round to the nearest foot.

89. Radio beacon. A ship sailing due east at the rate of 16 miles per hour records the bearing of a radio beacon at N 36.5° E. Two hours later, the bearing of the same beacon is N 55.7° W. Round each answer to the nearest tenth of a mile.
   a. How far is the ship from the beacon the second time?
   b. How close to the beacon did the ship pass?

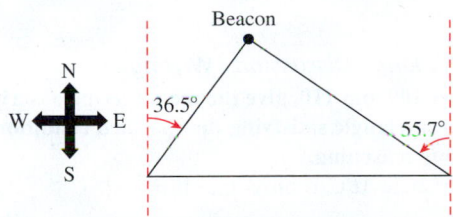

90. Lighthouse. A boat sailing due north at the rate of 14 miles per hour records the bearing of a lighthouse as N 8.4° E. Two hours later the bearing of the same lighthouse is N 30.9° E. Round the answers to the nearest tenth of a mile.
   a. How far is the ship from the lighthouse the second time?
   b. If the boat follows the same course and speed, how close to the lighthouse will it approach?

91. Geostationary satellite. A camel rider was traveling due west at night in the desert. At midnight, his angle of elevation to a satellite was 89°. After traveling for 20 miles, his angle of elevation to the satellite was 89.05°. The satellite is at a constant height above the ground. How high is the satellite? Round the answer to the nearest mile.

92. Height of a tower. A flagpole 20 feet tall is placed on top of a tower. At a point on the ground, a surveyor measures the angle of elevation of the bottom of the flagpole to be 69.6° and that of the top of the flagpole to be 70.9°. What is the height of the tower to the nearest foot?

93. Distance between planets. The angle subtended at Earth by lines joining Venus and the Sun is 31°. If Venus is 67 million miles from the Sun and Earth is 93 million miles from the Sun, what is the distance (to the nearest million) between Earth and Venus? [*Hint:* There are two possible answers.]

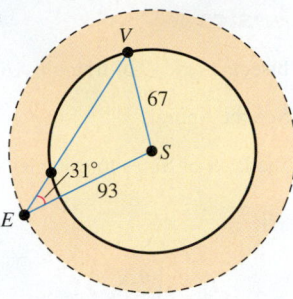

94. Distance between planets. The angle subtended at Mars by the lines joining Mars to Earth and the Sun is 23°. If Mars is 128 million miles from the Sun and Earth is 93 million miles from the Sun, what is the distance (to the nearest million) between Earth and Mars?

95. Navigation. A point is located 30 miles southwest of a dock. A ship leaves the dock traveling due west. Find the distance(s) between the ship and the dock when the ship is 24 miles from the point.

96. Navigation. A lighthouse is 23 miles N 55° E of a dock. A ship leaves the dock at 3 P.M. and sails due east at a speed of 15 mph. Find the time(s) to the nearest minute when the ship will be 18 miles from the lighthouse.

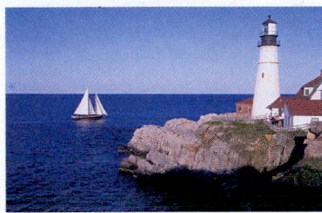

97. Geometry. The sides of a parallelogram are 15 meters and 11 meters, and the longer diagonal makes an angle of 18° with the longer side. Find the length of the longer diagonal.

98. Geometry. In the trapezoid shown, find the length *x* to the nearest tenth.

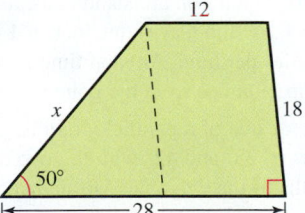

99. Area of the Bermuda Triangle. The angle formed by the lines of sight from Ft. Lauderdale, Florida, to Bermuda and to San Juan, Puerto Rico, is approximately 67°. The distance from Ft. Lauderdale to Bermuda is 1026 miles, and the distance from Ft. Lauderdale to San Juan is 1046 miles. Find the area of the Bermuda Triangle, which is formed with these cities as vertices.

100. Landscaping. A triangular region between the three streets shown in the figure will be landscaped for $30 a square foot. How much will the landscaping cost?

Beyond the Basics

101. Suppose CD bisects angle C of triangle ABC. Show that $\dfrac{AD}{DB} = \dfrac{AC}{CB}$. (See the figure.)

[*Hint:* Apply the Law of Sines to triangles ACD and DCB.]

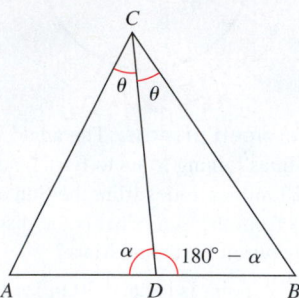

102. Distance. A flagpole 15 feet high stands on a building 75 feet high. To an observer at a height of 90 feet, the building and the flagpole intercept equal angles. Find the distance of the observer from the top of the flagpole. [*Hint:* Use Exercise 101.]

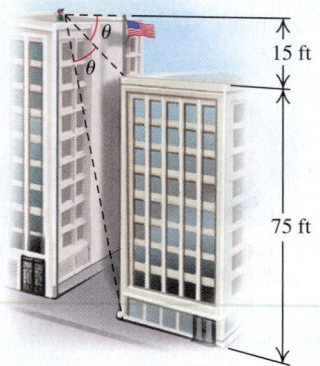

103. Navigation. A point on an island is located 24 miles southwest of a dock. A ship leaves the dock at 1 P.M., traveling west at 12 miles per hour. At what time(s) to the nearest minute is the ship 20 miles from the point?

104. Geometry. A side of a parallelogram is 60 inches long and makes angles of 28° and 42° with the two diagonals. What are the lengths of the diagonals? Round each answer to the nearest tenth of an inch.

105. Mollweide's formula. In a triangle ABC, prove that

$$\frac{b - c}{a} = \frac{\sin\dfrac{B - C}{2}}{\cos\dfrac{A}{2}}.$$

[*Hint:* Let $\dfrac{a}{\sin A} = \dfrac{b}{\sin B} = \dfrac{c}{\sin C} = k$. Write an expression for $\dfrac{b - c}{a}$ in terms of sines and then use sum-to-product and half-angle formulas. Note that because the formula contains all six parts of a triangle, it can be used as a check for the solutions of a triangle.]

106. Mollweide's formula. In a triangle ABC, prove that

$$\frac{b + c}{a} = \frac{\cos\dfrac{B - C}{2}}{\sin\dfrac{A}{2}}.$$

107. Law of Tangents. In a triangle ABC, prove that

$$\frac{b - c}{b + c} = \frac{\tan\left(\dfrac{B - C}{2}\right)}{\tan\left(\dfrac{B + C}{2}\right)}.$$

108. Angle measure. Two sides of a triangle are $\sqrt{3} + 1$ feet and $\sqrt{3} - 1$ feet, and the measure of the angle between these sides is 60°. Use the Law of Tangents (Exercise 107) to find the measure of the difference of the remaining angles.

Critical Thinking / Discussion / Writing

In Exercises 109 and 110, give the most specific description you can for a triangle satisfying the specified conditions. Explain your reasoning.

109. In a triangle ABC, $a \sin A = b \sin B$.

110. In a triangle ABC, $a \cos A = b \cos B$.

111. Use the Law of Sines to show that any isosceles triangle has two equal angles.

112. Use the Law of Sines to show that any triangle with two equal angles is isosceles.

Getting Ready for the Next Section

In Exercises 113–126, write the exact value of each expression without using a calculator.

113. $\cos 135°$

114. $\cos 150°$

115. $\cos^{-1}\left(\dfrac{1}{2}\right)$

116. $\cos^{-1}(2)$

117. $\cos^{-1}\left(-\dfrac{1}{2}\right)$

118. $\cos^{-1}\left(-\dfrac{\sqrt{3}}{2}\right)$

119. If $a = 2$, $b = 3$, and $\cos C = \dfrac{1}{6}$, find $\sqrt{a^2 + b^2 - 2ab \cos C}$.

120. If $b = 4$, $c = 5$, and $\cos A = \dfrac{1}{10}$, find $\sqrt{b^2 + c^2 - 2bc \cos A}$.

121. If $c = 3$, $a = 1$, and $B = 60°$, find $\sqrt{c^2 + a^2 - 2ca \cos B}$.

122. If $a = \sqrt{8}$, $b = 4$, and $C = 45°$, find $\sqrt{a^2 + b^2 - 2ab \cos C}$.

123. If $a = 2$, $b = 3$, and $c = 4$, find $\dfrac{a^2 + b^2 - c^2}{2ab}$.

124. If $a = 5$, $b = 4$, and $c = 3$, find $\dfrac{b^2 + c^2 - a^2}{2bc}$.

125. If $a = 1$, $b = 1$, and $c = \sqrt{3}$, find $\cos^{-1}\left(\dfrac{b^2 + c^2 - a^2}{2bc}\right)$.

126. If $a = 3$, $b = 5$, and $c = 7$, find $\cos^{-1}\left(\dfrac{a^2 + b^2 - c^2}{2ab}\right)$.

SECTION | **6.3**

The Law of Cosines

BEFORE STARTING THIS SECTION, REVIEW

1 Distance formula (Section 1.1, page 6)

2 Inverse functions (Section 1.7, page 121)

3 Area of a triangle (Section 6.2, page 522)

OBJECTIVES

1 Derive the Law of Cosines.

2 Use the Law of Cosines to solve SAS triangles.

3 Use the Law of Cosines to solve SSS triangles.

4 Use Heron's formula to find the area of a triangle.

◆ Generalizing the Pythagorean Theorem

Suppose a Boeing 747 jumbo jet is flying over Orlando, Florida, at 552 miles per hour and is heading due south to Brazil. Twenty minutes later, an F-16 fighter jet heading due east passes over Orlando at a speed of 1250 miles per hour. By using the Pythagorean Theorem, we can easily calculate the distance d between the two planes t hours after the F-16 passes over Orlando. We have:

$$d = \sqrt{(1250t)^2 + \left[\frac{1}{3}(552) + 552t\right]^2} \qquad 20 \text{ minutes} = \frac{1}{3} \text{ hour}$$

Suppose all the other facts in the problem are unchanged except that the F-16 has a bearing of N 37° E. Now we can no longer apply the Pythagorean Theorem. In Example 4, we use the Law of Cosines, described in this section, to solve this problem.

1 Derive the Law of Cosines.

The Law of Cosines

The Pythagorean Theorem states that the relationship $c^2 = a^2 + b^2$ holds in a right triangle ABC, where c represents the length of the hypotenuse. This relationship is not true if the triangle is not a right triangle. The **Law of Cosines** is a generalization of the Pythagorean Theorem that is true in any triangle. This law will be used to solve triangles in which two sides and the included angle are known (SAS triangles), as well as those triangles in which all three sides are known (SSS triangles).

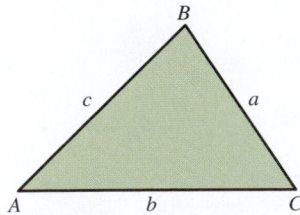

Figure 6.28

> ### THE LAW OF COSINES
>
> In triangle ABC with sides of lengths a, b, and c (as in Figure 6.28),
>
> $$a^2 = b^2 + c^2 - 2bc \cos A$$
> $$b^2 = c^2 + a^2 - 2ca \cos B$$
> $$c^2 = a^2 + b^2 - 2ab \cos C.$$
>
> In words, the square of any side of a triangle is equal to the sum of the squares of the length of the other two sides, less twice the product of the lengths of the other sides and the cosine of their included angle.

EXAMPLE 1 **Using the Law of Cosines to find a Side**

In a triangle ABC, find b, given that $a = 2, c = 3$, and $B = 60°$.

Solution

We use the formula:

$$b^2 = c^2 + a^2 - 2ca \cos B \qquad \text{Law of Cosines}$$
$$= 3^2 + 2^2 - 2(3)(2) \cos 60° \qquad \text{Replace } c \text{ with } 3, a \text{ with } 2, \text{ and } B \text{ with } 60°.$$
$$= 9 + 4 - 12\left(\frac{1}{2}\right) \qquad \cos 60° = \frac{1}{2}$$
$$= 9 + 4 - 6 = 7 \qquad \text{Simplify.}$$

So, $b = \sqrt{7}$.

Practice Problem 1 In a triangle ABC, find c, given that

$$a = 6, b = 3\sqrt{3}, \text{ and } C = 30°.$$

EXAMPLE 2 **Using the Law of Cosines to find an Angle**

In a triangle ABC, find A to the nearest tenth of a degree, $a = 6, b = 3$, and $c = 4$.

Solution

We start with the formula:

$$a^2 = b^2 + c^2 - 2bc \cos A \qquad \text{Law of Cosines}$$
$$\cos A = \frac{b^2 + c^2 - a^2}{2bc} \qquad \text{Solve for } \cos A.$$
$$A = \cos^{-1}\left(\frac{b^2 + c^2 - a^2}{2bc}\right) \qquad \text{Solve for } A.$$
$$= \cos^{-1}\left(\frac{3^2 + 4^2 - 6^2}{2(3)(4)}\right) \qquad \text{Replace } b \text{ with } 3, c \text{ with } 4, \text{ and } a \text{ with } 6.$$
$$= \cos^{-1}\left(-\frac{11}{24}\right) \qquad \text{Simplify.}$$
$$\approx 117.3° \qquad \text{Use a calculator.}$$

Practice Problem 2 In Example 2, find B.

Derivation of the Law of Cosines

To derive the Law of Cosines, place triangle ABC in a rectangular coordinate system with the vertex A at the origin and the side c along the positive x-axis. See Figure 6.29.

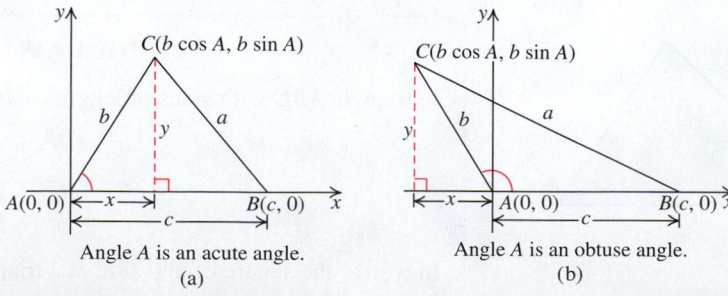

Angle A is an acute angle.
(a)

Angle A is an obtuse angle.
(b)

Figure 6.29 Two cases for an angle A.

Label the coordinates of the vertices as shown in Figure 6.29. In Figures 6.29(a) and 6.29(b), the point $C(x, y)$ on the terminal side of angle A has coordinates $(b \cos A, b \sin A)$, which are found using the cosine and sine definitions.

$$\cos A = \frac{x}{b} \qquad \sin A = \frac{y}{b}$$

$$x = b\cos A \qquad y = b\sin A$$

The point B has coordinates $(c, 0)$. Applying the distance formula to the line segment joining the points $B(c, 0)$ and $C(b\cos A, b\sin A)$, we have

$$a = d(C, B)$$
$$a^2 = [d(C, B)]^2 \qquad\qquad\qquad \text{Square both sides.}$$
$$a^2 = (b\cos A - c)^2 + (b\sin A - 0)^2 \qquad \text{Distance formula}$$
$$a^2 = b^2\cos^2 A - 2bc\cos A + c^2 + b^2\sin^2 A \qquad \text{Expand binomial.}$$
$$a^2 = b^2(\sin^2 A + \cos^2 A) + c^2 - 2bc\cos A \qquad \text{Regroup terms.}$$
$$a^2 = b^2 + c^2 - 2bc\cos A \qquad\qquad \sin^2 A + \cos^2 A = 1$$

The last equation is one of the forms of the Law of Cosines. Similarly, by placing the vertex B and then the vertex C at the origin, we obtain the other two forms. Notice that the Law of Cosines becomes the Pythagorean Theorem if the included angle is $90°$ because $\cos 90° = 0$.

2 Use the Law of Cosines to solve SAS triangles.

Solving SAS Triangles

Let's solve SAS triangles (Case 3 of Section 6.2).

PROCEDURE
IN ACTION

EXAMPLE 3   Solving the SAS Triangles

OBJECTIVE

Solve triangles in which the measures of two sides and the included angle are known.

Step 1 Use the appropriate form of the Law of Cosines to find the side opposite the given angle.

EXAMPLE

Solve triangle ABC with $a = 15$ inches, $b = 10$ inches, and $C = 60°$. Round each answer to the nearest tenth.

1. Find side c opposite angle C.

$$c^2 = a^2 + b^2 - 2ab\cos C \qquad \text{The Law of Cosines}$$
$$c^2 = (15)^2 + (10)^2 - 2(15)(10)\cos 60° \qquad \text{Substitute values.}$$
$$c^2 = 225 + 100 - 2(15)(10)\left(\frac{1}{2}\right) \qquad \cos 60° = \frac{1}{2}$$
$$c^2 = 175 \qquad \text{Simplify.}$$
$$c = \sqrt{175} \approx 13.2 \qquad \text{Use a calculator.}$$

Step 2 Use the **Law of Cosines** to find one of the unknown angles.

2. Find angle B.

$$b^2 = c^2 + a^2 - 2ca\cos B \qquad \text{The Law of Cosines}$$
$$\cos B = \frac{c^2 + a^2 - b^2}{2ca} \qquad \text{Solve for } \cos B.$$
$$\cos B = \frac{175 + 225 - 100}{2(\sqrt{175})(15)} \approx 0.756 \qquad \text{Substitute values and simplify.}$$
$$B = \cos^{-1}(0.756) \approx 40.9° \qquad \text{Use a calculator.}$$

Step 3 Use the angle sum formula to find the third angle.

3. $A \approx 180° - 60° - 40.9° \approx 79.1° \qquad A + B + C = 180°$

Step 4 Write the solution.

4.

Angles	Sides
$A \approx 79.1°$	$a = 15$ inches
$B \approx 40.9°$	$b = 10$ inches
$C = 60°$	$c \approx 13.2$ inches

Practice Problem 3 Solve triangle ABC with $c = 25$, $a = 15$, and $B = 60°$. Round each answer to the nearest tenth.

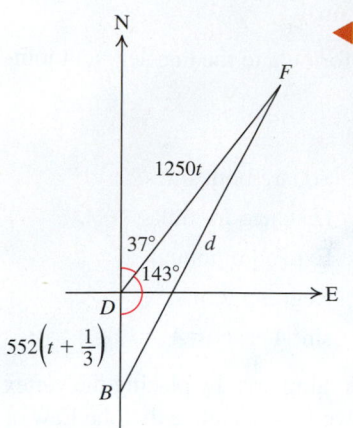

Figure 6.30

◆ EXAMPLE 4 Using the Law of Cosines

Suppose a Boeing 747 is flying over Orlando headed due south at 552 miles per hour. Twenty minutes later an F-16 passes over Orlando with a bearing of N 37° E at a speed of 1250 miles per hour. Find the distance between the two planes three hours after the F-16 passes over Orlando. Round the answer to the nearest tenth.

Solution

Suppose the F-16 has been traveling for t hours after passing over Olando. Then because the Boeing 747 had a head start of 20 minutes $= \dfrac{1}{3}$ hour, the Boeing 747 has been traveling $\left(t + \dfrac{1}{3}\right)$ hours due south. The distance d between the two planes is shown in Figure 6.30. Using the Law of Cosines in triangle FDB, we have

$$d^2 = (1250t)^2 + \left[552\left(t + \frac{1}{3}\right)\right]^2 - 2(1250t) \cdot 552\left(t + \frac{1}{3}\right)\cos 143°$$

$$d^2 \approx 28{,}469{,}270.04 \qquad \text{Substitute } t = 3; \text{ use a calculator.}$$

$$d \approx 5335.7 \text{ miles} \qquad \text{Use a calculator.}$$

Practice Problem 4 Repeat Example 4 assuming that the F-16 is traveling at a speed of 1375 miles per hour due N 75° E and the Boeing 747 is traveling with a bearing of S 12° W at 550 miles per hour.

3 Use the Law of Cosines to solve SSS triangles.

Solving SSS Triangles

Let's solve SSS triangles (Case 4 of Section 6.2).

PROCEDURE
IN ACTION

EXAMPLE 5 Solving SSS Triangles

OBJECTIVE

Solve triangles in which the measures of the three sides are known.

Step 1 Use the Law of Cosines to find any one of the angles.

EXAMPLE

Solve triangle ABC with $a = 3.1$ feet, $b = 5.4$ feet, and $c = 7.2$ feet. Round answers to the nearest tenth.

1. We find angle C.

$$c^2 = a^2 + b^2 - 2ab \cos C \qquad \text{The Law of Cosines}$$

$$2ab \cos C = a^2 + b^2 - c^2 \qquad \begin{array}{l}\text{Add } 2ab \cos C - c^2 \\ \text{to both sides.}\end{array}$$

$$\cos C = \frac{a^2 + b^2 - c^2}{2ab} \qquad \text{Solve for } \cos C.$$

$$\cos C = \frac{(3.1)^2 + (5.4)^2 - (7.2)^2}{2(3.1)(5.4)} \qquad \text{Substitute values.}$$

$$\cos C \approx -0.39 \qquad \text{Use a calculator.}$$

$$C \approx \cos^{-1}(-0.39) \approx 113.0° \qquad 0° < C < 180°$$

(continued)

Step 2 Use the Law of Cosines to find either of the two remaining angles.

2. Find angle B.

$$b^2 = c^2 + a^2 - 2ca \cos B \qquad \text{The Law of Cosines}$$

$$\cos B = \frac{c^2 + a^2 - b^2}{2ca} \qquad \text{Solve for } \cos B.$$

$$B = \cos^{-1}\left(\frac{c^2 + a^2 - b^2}{2ca}\right) \qquad \text{Solve for } B.$$

$$B = \cos^{-1}\left(\frac{7.2^2 + 3.1^2 - 5.4^2}{2(7.3)(3.1)}\right) \qquad \text{Substitute values.}$$

$$B \approx 43.7° \qquad \text{Use a calculator.}$$

Step 3 Use the angle sum formula to find the third angle.

3. $A \approx 180° - 43.7° - 113.0° \qquad A + B + C = 180°$

$A \approx 23.3° \qquad \text{Simplify.}$

Step 4 Write the solution.

4.

Angles	Sides
$A \approx 23.3°$	$a = 3.1$ feet
$B \approx 43.7°$	$b = 5.4$ feet
$C \approx 113.0°$	$c = 7.2$ feet

Practice Problem 5 Solve triangle ABC with $a = 4.5$, $b = 6.7$, and $c = 5.3$. Round each answer to the nearest tenth.

EXAMPLE 6 **Solving an SSS Triangle**

Solve triangle ABC with $a = 2$ meters, $b = 9$ meters, and $c = 5$ meters. Round each answer to the nearest tenth.

Solution

Step 1 We first find B, the angle opposite the longest side.

$$b^2 = c^2 + a^2 - 2ca \cos B \qquad \text{The Law of Cosines}$$

$$\cos B = \frac{c^2 + a^2 - b^2}{2ca} \qquad \text{Solve for } \cos B.$$

$$\cos B = \frac{5^2 + 2^2 - 9^2}{2(5)(2)} \qquad \text{Substitute values.}$$

$$\cos B = -2.6 \qquad \text{Simplify.}$$

RECALL

The Triangle Inequality states that in any triangle, the sum of the lengths of any two sides of a triangle is greater than the length of the third side.

Since the range of the cosine function is $[-1, 1]$, there is no angle B for which $\cos B = -2.6$. This means that a triangle with the given information cannot exist. Since $2 + 5 < 9$, you can also use the Triangle Inequality from geometry to see that there is no such triangle.

Practice Problem 6 Solve triangle ABC with $a = 2$ inches, $b = 3$ inches, and $c = 6$ inches.

4 Use Heron's formula to find the area of a triangle.

Heron's Area Formula

The Law of Cosines can be used to derive a formula for the area of a triangle if the lengths of the *three* sides are known. The formula is called **Heron's formula**.

RECALL

In Section 6.2, you learned that if two sides and their included angle are known, the area K of a triangle ABC is given by

$$K = \frac{1}{2} ab \sin C$$
$$= \frac{1}{2} bc \sin A$$
$$= \frac{1}{2} ca \sin B.$$

HERON'S FORMULA FOR SSS TRIANGLES

The area K of a triangle with sides of lengths, a, b, and c is given by

$$K = \sqrt{s(s-a)(s-b)(s-c)},$$

where $s = \frac{1}{2}(a + b + c)$ is the **semiperimeter**.

In Exercise 72, we ask you to derive Heron's formula.

EXAMPLE 7 **Using Heron's Formula**

Find the area of triangle ABC with $a = 29$ inches, $b = 25$ inches, and $c = 40$ inches. Round the answer to the nearest tenth.

Solution

We first find s:

$$s = \frac{a + b + c}{2}$$
$$= \frac{29 + 25 + 40}{2} = 47.$$

$$\text{Area of the triangle} = \sqrt{s(s-a)(s-b)(s-c)} \qquad \text{Heron's formula}$$
$$= \sqrt{47(47 - 29)(47 - 25)(47 - 40)} \qquad \text{Substitute values.}$$
$$\approx 360.9 \text{ square inches.} \qquad \text{Use a calculator.}$$

Practice Problem 7 Find the area of triangle ABC with $a = 11$ meters, $b = 17$ meters, and $c = 20$ meters.

EXAMPLE 8 **Using Heron's Formula**

A triangular swimming pool has side lengths 23 feet, 17 feet, and 26 feet. How many gallons of water will fill the pool to a depth of 5 feet? Round the answer to the nearest whole number.

Solution

To calculate the volume of water, we first calculate the area of the triangular surface.

We have $a = 23$, $b = 17$, and $c = 26$. So $s = \frac{1}{2}(a + b + c) = 33$.

By Heron's formula, the area K of the triangular surface is

$$K = \sqrt{s(s-a)(s-b)(s-c)}$$
$$= \sqrt{33(33 - 23)(33 - 17)(33 - 26)} \qquad \text{Substitute values for } a, b, c, \text{ and } s.$$
$$\approx 192.2498 \text{ square feet.}$$

The volume of water = surface area × depth

$$\approx 192.2498 \times 5 \approx 961.25 \text{ cubic feet.}$$

One cubic foot contains approximately 7.5 gallons of water. So $961.25 \times 7.5 \approx 7209$ gallons of water will fill the pool.

Practice Problem 8 Repeat Example 8 assuming that the swimming pool has side lengths 25 feet, 30 feet, and 33 feet and the depth of the pool is 5.5 feet.

SECTION 6.3 **Exercises**

Concepts and Vocabulary

1. One form of the Law of Cosines is
$c^2 = a^2 + b^2 - 2ab \cos ($_____$)$.

2. If we take the angle in the Law of Cosines to be 90°, then we get the _____ Theorem.

3. Triangles with SAS given are solved by the Law of Cosines, as are triangles with _____ sides given.

4. When one angle is found by the Law of Cosines, another can be found by using either the Law of _____.

5. **True or False.** The Law of Cosines is used to solve triangles when two angles and a side are given.

6. **True or False.** For a triangle with SSS given, we can find angle A by the formula

$$A = \cos^{-1}\left(\frac{b^2 + c^2 - a^2}{2bc}\right).$$

7. **True or False.** The Pythagorean Theorem is a special case of the Law of Cosines.

8. **True or False.** If $a^2 + b^2 - c^2 \geq 2ab$, then no such triangle ABC exists.

Building Skills

In Exercises 9–18, find the unknown part in triangle ABC to the nearest tenth.

9. $b = 4, c = 5, A = 60°$, find a.

10. $a = 3, b = 4, C = 60°$, find c.

11. $c = 10, a = 5\sqrt{2}, B = 45°$, find b.

12. $b = 2\sqrt{3}, c = 5, A = 30°$, find a.

13. $b = 3\sqrt{2}, a = 4, C = 45°$, find c.

14. $a = 5, c = 4\sqrt{3}, B = 30°$, find b.

15. $a = 2, b = 3, c = 4$, find A.

16. $a = 3, b = 4, c = 6$, find B.

17. $a = 3, b = 5, c = 10$, find C.

18. $a = 2, b = 3, c = 5$, find A.

In Exercises 19–22, solve each triangle. Round each answer to the nearest tenth.

19.

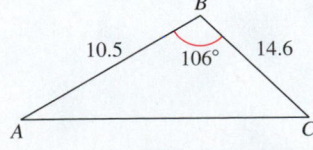

20.

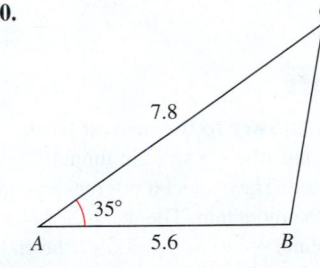

21.

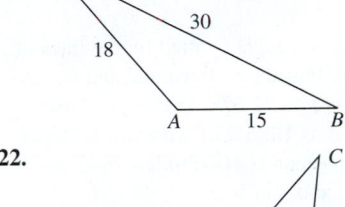

22.

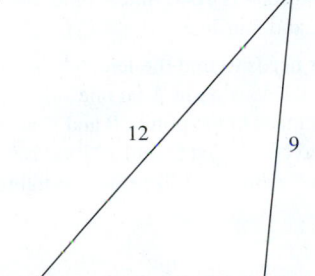

In Exercises 23–38, solve each triangle ABC. Round each answer to the nearest tenth. All sides are measured in feet.

23. $a = 15, b = 9, C = 120°$

24. $a = 14, b = 10, C = 75°$

25. $b = 10, c = 12, A = 62°$

26. $b = 11, c = 16, A = 110°$

27. $c = 12, a = 15, b = 11$

28. $c = 16, a = 11, b = 13$

29. $a = 9, b = 13, c = 18$

30. $a = 14, b = 6, c = 10$

31. $a = 2.5, b = 3.7, c = 5.4$

32. $a = 4.2, b = 2.9, c = 3.6$

33. $b = 3.2, c = 4.3, A = 97.7°$

34. $b = 5.4, c = 3.6, A = 79.2°$

35. $c = 4.9, a = 3.9, B = 68.3°$

36. $c = 7.8, a = 9.8, B = 95.6°$

37. $a = 2.3, b = 2.8, c = 3.7$

38. $a = 5.3, b = 2.9, c = 4.6$

In Exercises 39–46, find the area of each triangle by using Heron's formula. Round each answer to the nearest tenth. All sides are measured in inches.

39. $a = 2, b = 3, c = 4$

40. $a = 50, b = 100, c = 130$

41. $a = 50, b = 50, c = 75$

42. $a = 100, b = 100, c = 125$

43. $a = 7.5, b = 4.5, c = 6.0$

44. $a = 8.5, b = 9.0, c = 4.5$

45. $a = 3.7, b = 5.1, c = 4.2$

46. $a = 9.8, b = 5.7, c = 6.5$

Applying the Concepts

In Exercises 47–60, round each answer to the nearest tenth.

47. **Tunnel length.** Engineers must bore a straight tunnel through the base of a mountain. They select a reference point C on the plain surrounding the mountain. The distance from C to portal A (the starting tunnel point) and portal B (the ending tunnel point) is 2352 yards and 1763 yards, respectively. The measure of $\angle ACB$ is 41°. How long (to the nearest yard) is the tunnel?

48. **The Bermuda Triangle.** The angle formed by the lines of sight from Fort Lauderdale, Florida, to Bermuda and to San Juan, Puerto Rico, is approximately 67°. The distance from Fort Lauderdale to Bermuda is 1026 miles, and the distance from Fort Lauderdale to San Juan is 1046 miles. Find the distance between Bermuda and San Juan.

49. **Pond length.** A surveyor needs to find the length of a pond but does so indirectly. She selects a point A on one side of the pond and measures distances to the points B and C at opposite ends of the pond to be 537 yards and 823 yards, respectively. The measure of $\angle BAC$ is 130°. (See the figure.) How long is the pond?

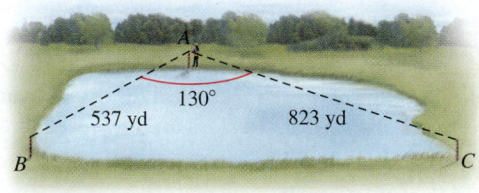

50. **Solar panels.** A roof of a house addition is being built to accept solar energy panels as shown in the figure. Find
   a. the length of the edge AC of the truss.
   b. the measure of $\angle BAC$.

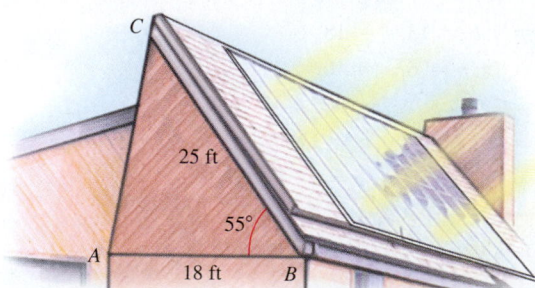

51. **Real estate.** You have inherited a commercial triangular-shaped lot of side lengths 400 feet, 250 feet, and 274 feet. The neighboring property is selling for $1 million per acre. How much is your lot worth? (*Hint:* 1 acre = 43,560 square feet.)

52. **Real estate.** Find the area of the quadrangular lot shown in the figure. All measurements are in feet.

53. **Triangular swimming pool.** Find the number of gallons of water in a triangular swimming pool with sides 11 feet, 16 feet, and 19 feet and a depth of 5 feet. Round your answer to the nearest whole number. [Recall: Approximately 7.5 gallons of water are in 1 cubic foot.]

54. **Quadrantal swimming pool.** Find the number of gallons in a swimming pool whose surface is in the shape of a quadrilateral $ABCD$ with $AB = 10$ feet, $BC = 12$ feet, $CD = 16$ feet, $DA = 14$ feet, $AC = 18$ feet, and a depth of 5 feet. Round your answer to the nearest whole number.

55. **Hikers.** Two hikers, Sonia and Tony, leave the same point at the same time. Sonia walks due east at the rate of 3 miles per hour, and Tony walks 45° northeast at the rate of 4.3 miles per hour. How far apart are the hikers after three hours?

56. **Distance between two vehicles.** A bus and a truck leave the same point O at the same time on two straight roads that intersect at O at an angle of 60°. The bus is traveling at 50 miles per hour, while the truck is on the other road and is traveling at 60 miles per hour. Find the distance (to the nearest mile) between the two vehicles after 2 hours if:
   a. both are traveling in a northeasterly direction.
   b. one is traveling northeast, while the other is traveling southwest.

57. **Distance between ships.** Two ships leave the same port—Ship A at 1.00 P.M. and ship B at 3:30 P.M. Ship A sails on a bearing of S 37° E at 18 miles per hour, and B sails on a bearing of N 28° E at 20 miles per hour. How far apart are the ships at 8.00 P.M.?

58. **Navigation.** A ship is traveling due north. At two different points A and B, the navigator of the ship sites a lighthouse at point C, as shown in the accompanying figure.
   a. Determine the distance from B to C.
   b. How much farther due north must the ship travel to reach the point closest to the lighthouse?

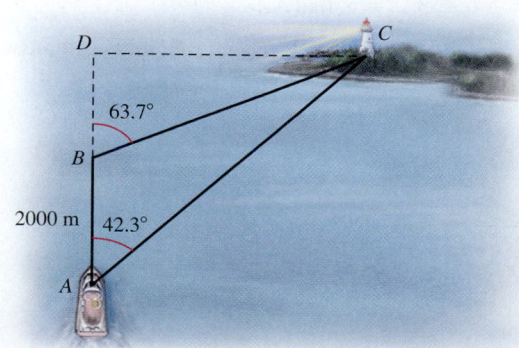

59. Tangential circles. Three circles of radii 1.2 inches, 2.2 inches, and 3.1 inches are tangent to each other externally. Find the angles of the triangle formed by joining the centers of the circles.

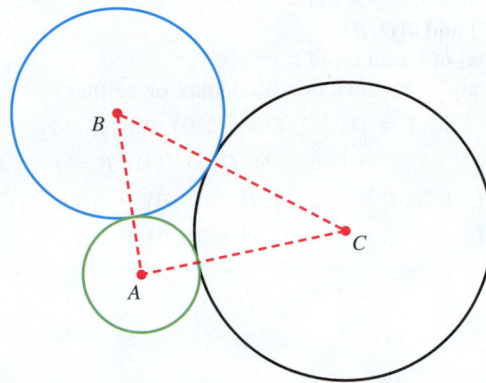

60. Height of a tree. A tree is planted at a point O on horizontal ground. Two points A and B on the ground are 100 feet apart. The angles of elevation of the top of the tree T from the points A and B are 45° and 30°, respectively. The measure of $\angle AOB$ is 60°. Find the height of the tree.

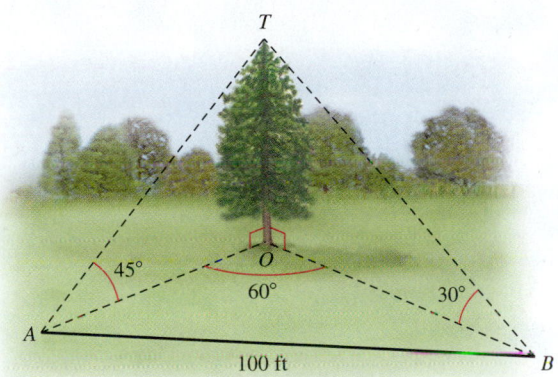

Beyond the Basics

In Exercises 61–64, round each answer to the nearest tenth.

61. A parallelogram has adjacent sides 8 centimeters and 13 centimeters. If the shorter diagonal is 11 centimeters long, find the length of the longer diagonal.

62. Find the area of the parallelogram of Exercise 61.

63. A parallelogram has adjacent sides 10 centimeters and 15 centimeters long, and the angle between them is 40°. Find the lengths of the two diagonals.

64. The length of the two diagonals of a parallelogram are 16 centimeters and 22 centimeters. The acute angle between these diagonals is 63°. Find the lengths of the sides of the parallelogram. [*Hint:* The diagonals of a parallelogram bisect each other.]

In Exercises 65–72, triangle ABC has sides a, b, and c and $s = \dfrac{1}{2}(a + b + c)$. Use the Law of Cosines to prove each identity.

65. $1 - \cos A = \dfrac{(a - b + c)(a + b - c)}{2bc}$

66. $1 - \cos A = \dfrac{2(s - b)(s - c)}{bc}$

67. $1 + \cos A = \dfrac{(b + c + a)(b + c - a)}{2bc}$

68. $1 + \cos A = \dfrac{2s(s - a)}{bc}$

69. Use the half-angle formula and Exercise 66 to prove the following:
$$\sin\frac{A}{2} = \sqrt{\frac{(s - b)(s - c)}{bc}}$$

70. Use the half-angle formula and Exercise 68 to prove the following:
$$\cos\frac{A}{2} = \sqrt{\frac{s(s - a)}{bc}}$$

71. Use Exercises 69 and 70 to prove that
$$\sin A = \frac{2}{bc}\sqrt{s(s - a)(s - b)(s - c)}.$$

72. Use Exercise 71 to prove Heron's formula.

73. Use Heron's formula to show that the area of an equilateral triangle with side length a is $K = \dfrac{\sqrt{3}\,a^2}{4}$.

74. Use the Law of Cosines to show that for a triangle with legs of lengths a, b, and c (with $a \le b$),
$$b - a < c < b + a.$$

Critical Thinking / Discussion / Writing

75. Solve triangle ABC with vertices $A(-2, 1)$, $B(5, 3)$, and $C(3, 6)$.

76. Solve triangle ABC with vertices $A(-3, -5)$, $B(6, 10)$, and $C(3, -2)$.

You can use the Law of Cosines to solve triangle ABC in the ambiguous case of Section 6.2. For example, to solve triangle ABC with $B = 150°$, $b = 10$, and $c = 6$, use the Law of Cosines to write $b^2 = a^2 + c^2 - 2ac \cos B$.

Substitute values of b, c, and B in this equation to obtain a quadratic equation in a. Find the roots of this equation and interpret your results. Again use the Law of Cosines to find A. Then find $C = 180° - A - B$.

In Exercises 77–79, use this technique to solve each triangle ABC.

77. $B = 150°$, $b = 10$, and $c = 6$

78. $A = 30°$, $a = 6$, and $b = 10$

79. $A = 60°$, $a = 12$, and $c = 15$

80. Find values of b, c and A in the form of the Law of Cosines $a^2 = b^2 + c^2 - 2bc \cos A$ so that $a^2 = b^2 + c^2 - 1$.

81. Explain why there are an unlimited number of choices for b and c in Exercise 80.

82. In the equation $a^2 = b^2 + c^2 - 2bc \cos A$, a form of the Law of Cosines, discuss the conditions under which a^2 is less than $b^2 + c^2$.

536 Chapter 6   Applications of Trigonometric Functions

Getting Ready for the Next Section

In Exercises 83 and 84, <u>write each</u> expression in the form $ax + by$ and compute $\sqrt{a^2 + b^2}$.

83. $2(-x + 3y) + 3(x - y)$

84. $3(2x - 3y) - 2(x - 2y)$

85. Find angle θ in quadrant III whose reference angle $\theta' = \tan^{-1}(\sqrt{3})$.

86. Find angle θ in quadrant II whose reference angle $\theta' = \left| \tan^{-1}\left(-\dfrac{1}{\sqrt{3}}\right) \right|$.

In Exercises 87–90, let l_1 be the line passing through the points S and T and l_2 the line passing through the points O and P. In each case, find

a. $d(S, T)$ and $d(O, P)$.

b. slope m_1 of l_1 and m_2 of l_2.

c. Are l_1 and l_2 parallel, perpendicular, or neither?

87. $S = (1, 3), T = (6, 15); O = (0, 0), P = (5, 12)$

88. $S = (2, -3), T = (-1, -7); O = (0, 0), P = (-3, 4)$

89. $S = (-1, 2), T = (-2, 1); O = (0, 0), P = (1, -1)$

90. $S = (-1, 1), T = (3, -2); O = (0, 0), P = (4, -3)$

Vectors

BEFORE STARTING THIS SECTION, REVIEW

1 Trigonometric functions of angles (Section 4.2, page 357)

2 Inverse trigonometric functions (Section 4.6, page 421)

OBJECTIVES

1 Represent vectors geometrically.

2 Represent vectors algebraically.

3 Find a unit vector in the direction of **v**.

4 Write a vector in terms of its magnitude and direction.

5 Use vectors in applications.

◆ The Meaning of Force

A **force** is a push or pull resulting from an object's *interaction* with another object. When the interacting objects are physically in contact with each other, the resulting force is called a *contact force*. Examples of contact forces include friction, tension, air resistance, and applied forces. A noncontact force is called an *action-at-a-distance* force. Examples of such forces include gravitational, electrical, and magnetic forces. The standard units of measurement for the magnitude of a force are pounds (lb) in the English system and newtons (N) in the metric system. The conversion factor between the systems is

$$1 \text{ pound of force} = 4.45 \text{ newtons.}$$

In Examples 8–10, we discuss the *resultant* of two or more forces acting on an object.

SIDE
NOTE

The *magnitude* of a vector is similar to the *absolute value* of a real number. Like absolute value, the magnitude of a vector cannot be negative.

1 Represent vectors geometrically.

Vectors

Many physical quantities such as length, area, volume, mass, and temperature are completely described by their magnitudes in appropriate units. Such quantities are called **scalar quantities**. Other physical quantities such as velocity, acceleration, and force are completely described only if *both* a magnitude (size) and a direction are specified. For example, the movement of wind is usually described by its speed (magnitude) and its direction, say, 15 miles per hour southwest. The wind speed and wind direction together form a **vector quantity** called the *wind velocity*.

Geometric Vectors

We can represent a **vector** geometrically by a directed line segment with an arrowhead. The arrow specifies the direction of the vector, and its length describes the magnitude. The tail of the arrow is the vector's **initial point**, and the tip of the arrow is its **terminal point**. We denote vectors by lowercase boldface type, such as **a**, **b**, **i**, **j**, **u**, **v**, and **w**. When discussing vectors, we refer to real numbers as **scalars**. Scalars will be denoted by lowercase italic type, such as *a*, *b*, *x*, *y*, and *z*.

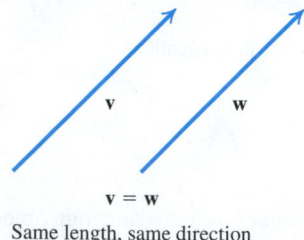

Figure 6.31 A vector.

As in Figure 6.31, if the initial point of a vector **v** is P and the terminal point is Q, we write

$$\mathbf{v} = \overrightarrow{PQ}.$$

The **magnitude** (or **norm**) of a vector $\mathbf{v} = \overrightarrow{PQ}$, denoted by $\|\mathbf{v}\|$ or $\|\overrightarrow{PQ}\|$, is the length of the vector **v** and is a scalar quantity.

Equivalent Vectors

Two vectors having the same length and same direction are called **equivalent vectors**. Since a vector is determined by its length and direction only, equivalent vectors are regarded as **equal** even though they may be located in different positions. If **v** and **w** are equivalent, we write $\mathbf{v} = \mathbf{w}$. See Figure 6.32.

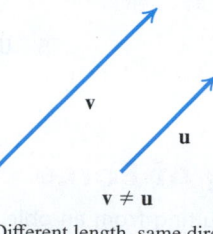
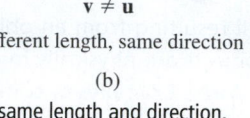
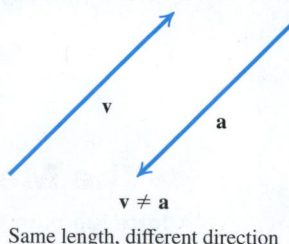
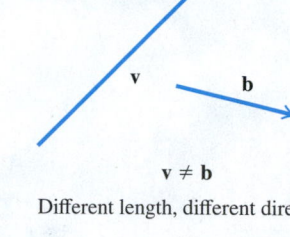

Figure 6.32 Equivalent vectors have the same length and direction.

The vector of length zero is called the **zero vector** and is denoted by **0**. The zero vector has zero magnitude and arbitrary direction. If vectors **v** and **a**, as in Figure 6.32(c), have the same length and opposite direction, then **a** is the **opposite vector** of **v**, and we write $\mathbf{a} = -\mathbf{v}$.

Adding Vectors

We add two vectors **v** and **w**, as shown in Figure 6.33.

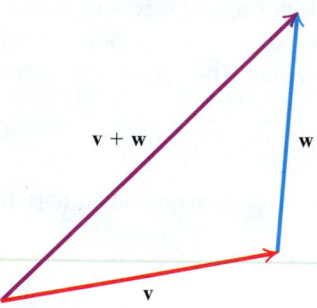

Figure 6.33 Adding **v** and **w**.

Geometric Vector Addition

> Let **v** and **w** be any two vectors. Position the terminal point of **v** so that it coincides with the initial point of **w**. The *sum* $\mathbf{v} + \mathbf{w}$ is the **resultant vector** whose initial point coincides with the initial point of **v**, and its terminal point coincides with the terminal point of **w**.

In Figure 6.34, we have constructed two sums, $\mathbf{v} + \mathbf{w}$ and $\mathbf{w} + \mathbf{v}$. It shows that

$$\mathbf{v} + \mathbf{w} = \mathbf{w} + \mathbf{v}$$

and that the sum coincides with the diagonal of the parallelogram determined by **v** and **w** when **v** and **w** have the same initial point.

Vector subtraction is defined just like the subtraction of real numbers. For any two vectors **v** and **w**, $\mathbf{v} - \mathbf{w} = \mathbf{v} + (-\mathbf{w})$, where $-\mathbf{w}$ is the opposite of **w**. See Figure 6.35.

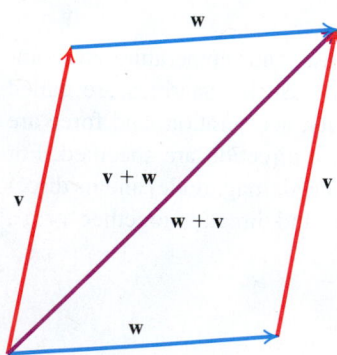

Figure 6.34 $\mathbf{v} + \mathbf{w} = \mathbf{w} + \mathbf{v}$.

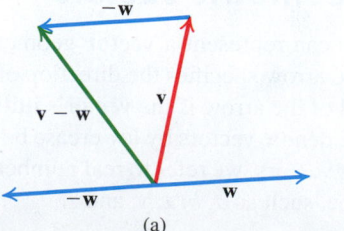
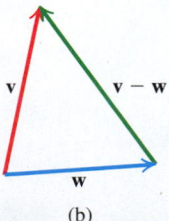

Figure 6.35 Vector $\mathbf{v} - \mathbf{w}$.

Figure 6.35(b) shows that you can construct the difference vector $\mathbf{v} - \mathbf{w}$ without first drawing $-\mathbf{w}$: place the vectors **v** and **w** so that their initial points coincide. Then the vector from the terminal point of **w** to the terminal point of **v** is the vector $\mathbf{v} - \mathbf{w}$.

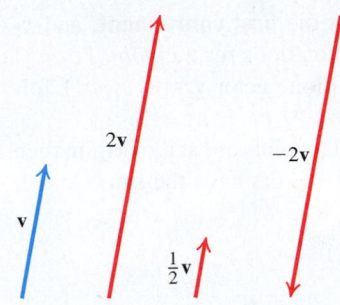

Figure 6.36 Scalar multiples of **v**.

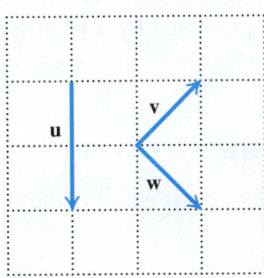

Figure 6.37

A second basic arithmetic operation for vectors is multiplying vectors by real numbers.

Scalar Multiples of Vectors

Let **v** be a vector and c a scalar (a real number). The vector $c\mathbf{v}$ is called the **scalar multiple** of **v**. See Figure 6.36.

If $c > 0$, $c\mathbf{v}$ has the same direction as **v** and magnitude $c\|\mathbf{v}\|$.
If $c < 0$, $c\mathbf{v}$ has the opposite direction of **v** and magnitude $|c|\|\mathbf{v}\|$.
If $c = 0$, $c\mathbf{v} = 0\mathbf{v} = \mathbf{0}$.

EXAMPLE 1   **Geometric Vectors**

Use the vectors **u**, **v**, and **w** in Figure 6.37 to graph each vector.

a. $\mathbf{u} - 2\mathbf{w}$ **b.** $2\mathbf{v} - \mathbf{u} + \mathbf{w}$

Solution

The graphs are shown in Figure 6.38.

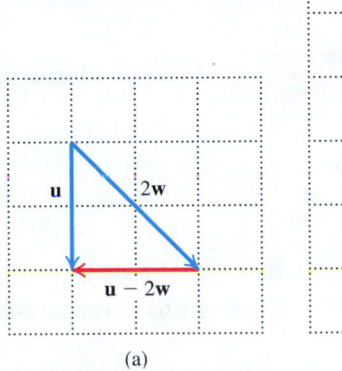
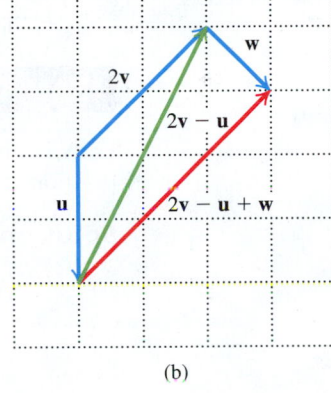

(a)   (b)

Figure 6.38

Practice Problem 1 Use the vectors **u**, **v**, and **w** of Figure 6.37 to graph each vector.

a. $2\mathbf{w} + \mathbf{v}$ **b.** $2\mathbf{w} + \mathbf{v} - \mathbf{u}$

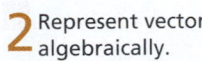 Represent vectors algebraically.

Algebraic Vectors

We now consider vectors in the Cartesian coordinate plane. Since the location of the initial point of a vector is not relevant, we typically draw vectors with their initial point at the origin. Such a vector is called a **position vector**. Note that the terminal point of a position vector will completely determine the vector, so specifying the terminal point will specify the vector. For the position vector **v** (see Figure 6.39) with initial point at the origin O and terminal point at $P(v_1, v_2)$, we denote the vector by

$$\mathbf{v} = \overrightarrow{OP} = \langle v_1, v_2 \rangle.$$

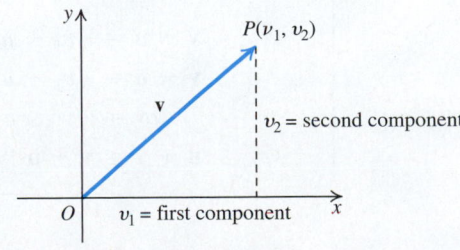

Figure 6.39 A position vector.

SIDE
NOTE

Notice that the point (v_1, v_2) has neither magnitude nor direction, whereas the position vector $\langle v_1, v_2 \rangle$ has both.

We call v_1 and v_2 the **components** of the vector $\mathbf{v}$; v_1 is the **first component**, and v_2 is the **second component**. Note the difference between the notations for the *point* (v_1, v_2) and the *position vector* $\langle v_1, v_2 \rangle$. The magnitude of the position vector $\mathbf{v} = \langle v_1, v_2 \rangle$ follows directly from the Pythagorean Theorem. We have $\|\mathbf{v}\| = \sqrt{v_1^2 + v_2^2}$.

If equivalent vectors $\mathbf{v}$ and $\mathbf{w}$ are located so that their initial points are at the origin, then their terminal points must coincide (because the equivalent vectors have the same length and same direction). Thus, for the vectors

$$\mathbf{v} = \langle v_1, v_2 \rangle \quad \text{and} \quad \mathbf{w} = \langle w_1, w_2 \rangle,$$
$$\mathbf{v} = \mathbf{w} \quad \text{if and only if} \quad v_1 = w_1 \quad \text{and} \quad v_2 = w_2.$$

We can use congruent triangles to show that any vector $\mathbf{v}$ with initial point $P(x_1, y_1)$ and terminal point $Q(x_2, y_2)$ is equivalent to the position vector $\mathbf{w} = \langle x_2 - x_1, y_2 - y_1 \rangle$. See Figure 6.40. Any vector in the Cartesian plane can therefore be represented by a position vector.

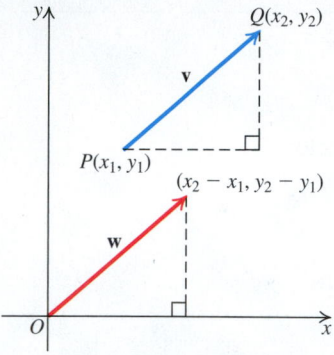

Figure 6.40 Two equivalent vectors.

REPRESENTING A VECTOR AS A POSITION VECTOR

The vector $\overrightarrow{PQ}$ with initial point $P(x_1, y_1)$ and terminal point $Q(x_2, y_2)$ is equal to the position vector

$$\mathbf{w} = \langle x_2 - x_1, y_2 - y_1 \rangle.$$

EXAMPLE 2 **Representing a Vector in the Cartesian Plane**

Let $\mathbf{v}$ be the vector with initial point $P(4, -2)$ and terminal point $Q(-1, 3)$. Write $\mathbf{v}$ as a position vector.

Solution

Because $\mathbf{v}$ has initial point $P(4, -2)$, $x_1 = 4$ and $y_1 = -2$.
Because $\mathbf{v}$ has terminal point $Q(-1, 3)$, $x_2 = -1$ and $y_2 = 3$.

So

$$\mathbf{v} = \langle x_2 - x_1, y_2 - y_1 \rangle \qquad \text{Position vector representation of } \overrightarrow{PQ}$$
$$\mathbf{v} = \langle -1 - 4, 3 - (-2) \rangle \qquad \text{Substitute values.}$$
$$\mathbf{v} = \langle -5, 5 \rangle \qquad \text{Simplify.}$$

Practice Problem 2 Let $\mathbf{w}$ be the vector with initial point $P(-2, 7)$ and terminal point $Q(1, -3)$. Write $\mathbf{w}$ as a position vector.

The zero vector is $\mathbf{0} = \langle 0, 0 \rangle$, and the opposite of the vector $\mathbf{v} = \langle v_1, v_2 \rangle$ is $-\mathbf{v} = \langle -v_1, -v_2 \rangle$. The following properties describe the arithmetic operations on vectors, using components.

ARITHMETIC OPERATIONS AND PROPERTIES OF VECTORS

If $\mathbf{u} = \langle u_1, u_2 \rangle$, $\mathbf{v} = \langle v_1, v_2 \rangle$ and $\mathbf{w} = \langle w_1, w_2 \rangle$ are vectors and c and d are any scalars, then

$$\mathbf{v} + \mathbf{u} = \langle v_1 + u_1, v_2 + u_2 \rangle \qquad (\mathbf{u} + \mathbf{v}) + \mathbf{w} = \mathbf{u} + (\mathbf{v} + \mathbf{w})$$
$$\mathbf{v} - \mathbf{u} = \langle v_1 - u_1, v_2 - u_2 \rangle \qquad (c + d)\mathbf{v} = c\mathbf{v} + d\mathbf{v}$$
$$c\mathbf{v} = \langle cv_1, cv_2 \rangle \qquad c(\mathbf{v} + \mathbf{w}) = c\mathbf{v} + c\mathbf{w}$$
$$\mathbf{u} + \mathbf{v} = \mathbf{v} + \mathbf{u} \qquad c(d\mathbf{v}) = (cd)\mathbf{v}$$

EXAMPLE 3   **Operations on Vectors**

Let $\mathbf{v} = \langle 2, 3 \rangle$ and $\mathbf{w} = \langle -4, 1 \rangle$. Find each expression.

a. $\mathbf{v} + \mathbf{w}$ **b.** $-2\mathbf{v}$ **c.** $2\mathbf{v} - \mathbf{w}$ **d.** $\|2\mathbf{v} - \mathbf{w}\|$

Solution

a. $\mathbf{v} + \mathbf{w} = \langle 2, 3 \rangle + \langle -4, 1 \rangle = \langle 2 - 4, 3 + 1 \rangle = \langle -2, 4 \rangle$

b. $-2\mathbf{v} = -2\langle 2, 3 \rangle = \langle -2 \cdot 2, -2 \cdot 3 \rangle = \langle -4, -6 \rangle$

c. $2\mathbf{v} - \mathbf{w} = 2\langle 2, 3 \rangle - \langle -4, 1 \rangle$

$$= \langle 4, 6 \rangle - \langle -4, 1 \rangle \qquad \text{Scalar multiplication}$$

$$= \langle 4 - (-4), 6 - 1 \rangle \qquad \text{Vector subtraction}$$

$$= \langle 8, 5 \rangle \qquad \text{Simplify.}$$

d. $\|2\mathbf{v} - \mathbf{w}\| = \|\langle 8, 5 \rangle\|$ From part **c**

$$= \sqrt{8^2 + 5^2} \qquad \text{Definition of magnitude}$$

$$= \sqrt{64 + 25} = \sqrt{89} \qquad \text{Simplify.}$$

Practice Problem 3 Let $\mathbf{v} = \langle -1, 2 \rangle$ and $\mathbf{w} = \langle 2, -3 \rangle$. Find each expression.

a. $\mathbf{v} + \mathbf{w}$ **b.** $3\mathbf{w}$ **c.** $3\mathbf{w} - 2\mathbf{v}$ **d.** $\|3\mathbf{w} - 2\mathbf{v}\|$

3 Find a unit vector in the direction of **v**.

Unit Vectors

Recall that if $c > 0$ and $\mathbf{v}$ is any vector, then the vector $c\mathbf{v}$ has the same direction as $\mathbf{v}$ and its length is given by

$$\|c\mathbf{v}\| = c\|\mathbf{v}\|.$$

If $\mathbf{v}$ is a nonzero vector and $c = \dfrac{1}{\|\mathbf{v}\|}$, then $\left\|\dfrac{1}{\|\mathbf{v}\|}\mathbf{v}\right\| = \dfrac{1}{\|\mathbf{v}\|}\|\mathbf{v}\| = 1$. Consequently, $\dfrac{1}{\|\mathbf{v}\|}\mathbf{v}$ is a vector of length 1 in the same direction as $\mathbf{v}$. A vector of length 1 is called a **unit vector**.

A UNIT VECTOR

Let $\mathbf{v}$ be a nonzero vector. The vector

$$\mathbf{u} = \frac{1}{\|\mathbf{v}\|}\mathbf{v}$$

is a unit vector having the same direction as $\mathbf{v}$.

EXAMPLE 4   **Finding a Unit Vector**

Find a unit vector $\mathbf{u}$ in the direction of $\mathbf{v} = \langle 3, -4 \rangle$.

Solution

First, find magnitude of $\mathbf{v} = \langle 3, -4 \rangle$.

$$\|\mathbf{v}\| = \sqrt{(3)^2 + (-4)^2} \qquad \text{Formula for magnitude}$$

$$\|\mathbf{v}\| = \sqrt{25} = 5 \qquad \text{Simplify.}$$

Now let

$$\mathbf{u} = \frac{1}{\|\mathbf{v}\|}\mathbf{v} \qquad \text{Multiply } \mathbf{v} \text{ by the scalar } \frac{1}{\|\mathbf{v}\|}.$$

$$\mathbf{u} = \frac{1}{5}\langle 3, -4 \rangle \qquad \text{Substitute values.}$$

$$\mathbf{u} = \left\langle \frac{3}{5}, -\frac{4}{5} \right\rangle \qquad \text{Scalar multiplication}$$

Check that $\|\mathbf{u}\| = 1$:

$$\|\mathbf{u}\| = \sqrt{\left(\frac{3}{5}\right)^2 + \left(-\frac{4}{5}\right)^2} = \sqrt{\frac{9}{25} + \frac{16}{25}} = \sqrt{\frac{25}{25}} = 1$$

Therefore, $\mathbf{u} = \left\langle \frac{3}{5}, -\frac{4}{5} \right\rangle$ is a unit vector in the same direction as $\mathbf{v}$. (Note that $\mathbf{u}$ is a scalar multiple of $\mathbf{v}$.)

Practice Problem 4 Find a unit vector in the direction of $\mathbf{v} = \langle -12, 5 \rangle$.

Vectors in i, j Form

In a Cartesian coordinate plane, two important unit vectors lie along the positive coordinate axes:

$$\mathbf{i} = \langle 1, 0 \rangle \quad \text{and} \quad \mathbf{j} = \langle 0, 1 \rangle$$

The unit vectors $\mathbf{i}$ and $\mathbf{j}$ are called **standard unit vectors**. See Figure 6.41.

Every vector $\mathbf{v} = \langle v_1, v_2 \rangle$ can be expressed in terms of $\mathbf{i}$ and $\mathbf{j}$ as follows:

$$\begin{aligned} \mathbf{v} = \langle v_1, v_2 \rangle &= \langle v_1, 0 \rangle + \langle 0, v_2 \rangle && \text{Vector addition} \\ &= v_1 \langle 1, 0 \rangle + v_2 \langle 0, 1 \rangle && \text{Scalar multiplication} \\ &= v_1 \mathbf{i} + v_2 \mathbf{j} && \text{Replace } \langle 1, 0 \rangle \text{ with } \mathbf{i} \text{ and } \langle 0, 1 \rangle \text{ with } \mathbf{j}. \end{aligned}$$

The scalars v_1 and v_2 are called the *horizontal* and *vertical components of* $\mathbf{v}$, respectively. A vector $\mathbf{v} = \langle v_1, v_2 \rangle$ from $(0, 0)$ to (v_1, v_2) can therefore be represented in the form

$$\mathbf{v} = v_1 \mathbf{i} + v_2 \mathbf{j}$$

with

$$\|\mathbf{v}\| = \sqrt{v_1^2 + v_2^2}.$$

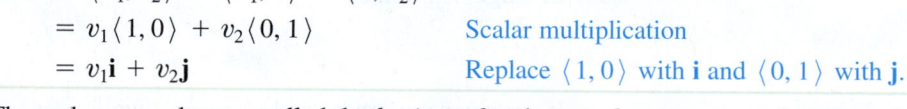

Figure 6.41 Standard unit vectors.

EXAMPLE 5 **Vectors Involving i and j**

Find each expression for $\mathbf{u} = 4\mathbf{i} + 7\mathbf{j}$ and $\mathbf{v} = 2\mathbf{i} + 5\mathbf{j}$.

a. $\mathbf{u} - 3\mathbf{v}$ **b.** $\|\mathbf{u} - 3\mathbf{v}\|$

Solution

a. $\begin{aligned}[t] \mathbf{u} - 3\mathbf{v} &= (4\mathbf{i} + 7\mathbf{j}) - 3(2\mathbf{i} + 5\mathbf{j}) \\ &= 4\mathbf{i} + 7\mathbf{j} - 6\mathbf{i} - 15\mathbf{j} && \text{Scalar multiplication} \\ &= (4 - 6)\mathbf{i} + (7 - 15)\mathbf{j} && \text{Group terms.} \\ &= -2\mathbf{i} - 8\mathbf{j} && \text{Simplify.} \end{aligned}$

b. $\begin{aligned}[t] \|\mathbf{u} - 3\mathbf{v}\| &= \|-2\mathbf{i} - 8\mathbf{j}\| \\ &= \sqrt{(-2)^2 + (-8)^2} && \text{Definition of magnitude} \\ &= \sqrt{68} = 2\sqrt{17} && \text{Simplify.} \end{aligned}$

Practice Problem 5 Find each expression for $\mathbf{u} = -3\mathbf{i} + 2\mathbf{j}$ and $\mathbf{v} = \mathbf{i} + 4\mathbf{j}$.

a. $3\mathbf{u} + 2\mathbf{v}$ **b.** $\|3\mathbf{u} + 2\mathbf{v}\|$

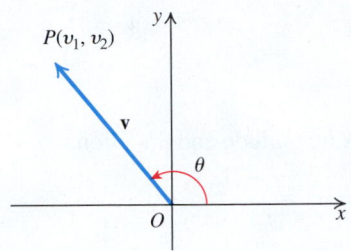

$P(v_1, v_2)$

Figure 6.42 $\mathbf{v} = \|\mathbf{v}\|(\cos\theta\,\mathbf{i} + \sin\theta\,\mathbf{j})$.

4 Write a vector in terms of its magnitude and direction.

Vector in Terms of Magnitude and Direction

Let $\mathbf{v} = \langle v_1, v_2 \rangle$ be a position vector and suppose θ is the smallest nonnegative angle that $\mathbf{v}$ makes with the positive x-axis. The angle θ is called the **direction angle** of $\mathbf{v}$. See Figure 6.42.

The length (or magnitude) of the vector $\overrightarrow{OP}$ is $\|\mathbf{v}\|$; so,

$$\frac{v_1}{\|\mathbf{v}\|} = \cos\theta \quad \text{and} \quad \frac{v_2}{\|\mathbf{v}\|} = \sin\theta \qquad \text{Definitions of cosine and sine}$$

$$v_1 = \|\mathbf{v}\|\cos\theta \qquad v_2 = \|\mathbf{v}\|\sin\theta \qquad \text{Multiply both sides by } \|\mathbf{v}\|.$$

Hence,

$$\mathbf{v} = v_1\mathbf{i} + v_2\mathbf{j} \qquad\qquad \text{Write } \mathbf{v} \text{ in terms of } \mathbf{i} \text{ and } \mathbf{j}.$$

$$\mathbf{v} = \|\mathbf{v}\|\cos\theta\,\mathbf{i} + \|\mathbf{v}\|\sin\theta\,\mathbf{j} \qquad \text{Replace values of } v_1 \text{ and } v_2.$$

$$\mathbf{v} = \|\mathbf{v}\|(\cos\theta\,\mathbf{i} + \sin\theta\,\mathbf{j}) \qquad \text{Factor out } \|\mathbf{v}\|.$$

VECTOR IN TERMS OF MAGNITUDE AND DIRECTION

The magnitude of a vector $\mathbf{v} = \langle v_1,\ v_2 \rangle$ is given by

$$\|\mathbf{v}\| = \sqrt{v_1^2 + v_2^2}.$$

The direction angle θ is the smallest nonnegative angle that satisfies

$$\tan\theta = \frac{v_2}{v_1},\ v_1 \neq 0.$$

The formula

$$\mathbf{v} = \|\mathbf{v}\|(\cos\theta\,\mathbf{i} + \sin\theta\,\mathbf{j})$$

expresses a vector $\mathbf{v}$ in terms of its magnitude $\|\mathbf{v}\|$ and its direction angle θ.

This form is also called the **Trigonometric Form** of the vector $\mathbf{v}$.

Note that any angle coterminal with the direction angle θ of a vector $\mathbf{v}$ can be used in place of θ in the formula $\mathbf{v} = \|\mathbf{v}\|(\cos\theta\,\mathbf{i} + \sin\theta\,\mathbf{j})$.

We can find the direction angle θ of a position vector $\mathbf{v} = \langle v_1, v_2 \rangle$ by using the equation $\tan\theta = \dfrac{v_2}{v_1}$ and the quadrant in which the point (v_1, v_2) lies.

EXAMPLE 6 Finding the Direction Angle of a Vector

Find the magnitude and the direction angle θ of the vector $\mathbf{v} = -4\mathbf{i} + 3\mathbf{j}$. Round θ to the nearest hundredth of a degree.

Solution

$$\mathbf{v} = -4\mathbf{i} + 3\mathbf{j}$$

$$= \langle -4, 3 \rangle \qquad \text{Write } \mathbf{v} \text{ as a position vector.}$$

The magnitude is $\|\mathbf{v}\| = \sqrt{(-4)^2 + 3^2} = \sqrt{25} = 5$. To find the direction angle θ, start with

$$\tan\theta = \frac{3}{-4} = -\frac{3}{4}$$

The reference angle θ' for θ is given by

$$\theta' = \left| \tan^{-1}\left(-\frac{3}{4}\right) \right| \approx |-36.87°| = 36.87° \qquad \text{Use a calculator.}$$

Since the point $(-4, 3)$ lies in quadrant II, we have

$$\theta = 180° - \theta' \approx 180° - 36.87° = 143.13°.$$

The direction angle of $\mathbf{v}$ is $143.13°$. See Figure 6.43.

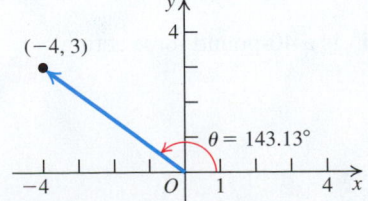

Figure 6.43 The direction angle of $\mathbf{v} = -4\mathbf{i} + 3\mathbf{j}$.

Practice Problem 6 Find the magnitude and the direction angle of $\mathbf{v} = 2\mathbf{i} - 3\mathbf{j}$.

EXAMPLE 7 **Writing a Vector with Given Length and Direction Angle**

Write the vector of magnitude 3 whose direction angle is $\dfrac{\pi}{3}$.

Solution

If **v** is the required vector, then

$$\mathbf{v} = \|\mathbf{v}\|(\cos\theta\,\mathbf{i} + \sin\theta\,\mathbf{j}) \qquad \text{Write } \mathbf{v} \text{ in terms of its magnitude and direction.}$$

$$\mathbf{v} = 3\left(\cos\frac{\pi}{3}\,\mathbf{i} + \sin\frac{\pi}{3}\,\mathbf{j}\right) \qquad \text{Substitute } \|\mathbf{v}\| = 3 \text{ and } \theta = \frac{\pi}{3}.$$

$$\mathbf{v} = 3\left(\frac{1}{2}\,\mathbf{i} + \frac{\sqrt{3}}{2}\,\mathbf{j}\right) \qquad \cos\frac{\pi}{3} = \frac{1}{2}, \sin\frac{\pi}{3} = \frac{\sqrt{3}}{2}$$

$$\mathbf{v} = \frac{3}{2}\,\mathbf{i} + \frac{3\sqrt{3}}{2}\,\mathbf{j} \qquad \text{Distributive property}$$

Practice Problem 7 Write the vector of magnitude 2 that makes an angle of $\dfrac{11\pi}{6}$ with the positive x-axis.

5 Use vectors in applications.

Applications of Vectors

It is an experimental fact that forces behave like vectors. This means that if a system of forces acts on an object, the object will move as though it were acted on by a single force equal to the vector sum of the forces. This single force is called the **resultant** of the system of forces. If the object does not move, the resultant is zero. We say that the particle is in **static equilibrium or equilibrium.** So, for an object to remain in static equilibrium, the sum of all the force vectors must be the zero vector.

EXAMPLE 8 **Finding the Resultant**

Find the magnitude and bearing of the resultant **R** of two forces $\mathbf{F}_1$ and $\mathbf{F}_2$, where $\mathbf{F}_1$ is a 50-pound force acting northward and $\mathbf{F}_2$ is a 40-pound force acting eastward.

Solution

We set up the coordinate system with the y-axis pointing northward. Then $\mathbf{F}_1 = 50\mathbf{j}$ and $\mathbf{F}_2 = 40\mathbf{i}$. See Figure 6.44.

$$\mathbf{R} = \mathbf{F}_1 + \mathbf{F}_2 = \mathbf{F}_2 + \mathbf{F}_1 = 40\mathbf{i} + 50\mathbf{j}$$

$$\|\mathbf{R}\| = \sqrt{(40)^2 + (50)^2} \approx 64.0$$

$$\tan\theta = \frac{50}{40} \qquad\qquad \text{Direction angle}$$

$$\theta = \tan^{-1}\left(\frac{50}{40}\right) \approx 51.3° \qquad \text{Use a calculator.}$$

The angle between **R** and the y-axis (north direction) is $90° - 51.3° = 38.7°$. Therefore, **R** is a force of 64.0 pounds in the direction N 38.7° E.

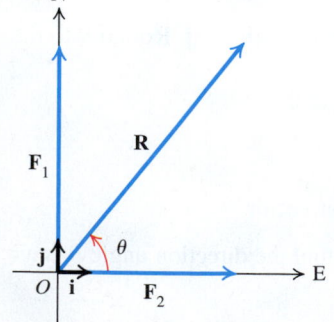

Figure 6.44 The resultant **R**.

Practice Problem 8 Repeat Example 8 assuming that $\mathbf{F}_1$ is a 40-pound force acting northward and $\mathbf{F}_2$ is a 30-pound force acting westward.

EXAMPLE 9 **Using Vectors in Air Navigation**

An F-15 fighter jet is flying over Mount Rushmore at an airspeed (speed in still air) of 800 miles per hour on a bearing of N 30° E. The velocity of wind is 40 miles per hour in the direction of S 45° E. Find the actual speed and direction (relative to the ground) of the plane. Round each answer to the nearest tenth.

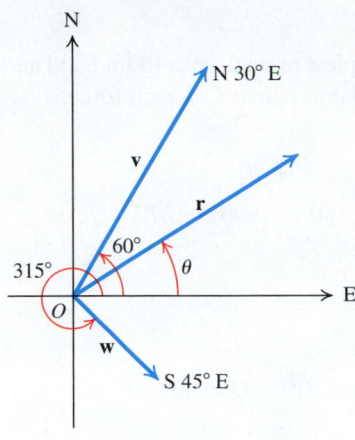

N 30° E

v

r

60°

θ

315°

O

w

S 45° E

E

N

Vectors are not drawn to scale.

Figure 6.45 Velocity vectors.

Solution

Set up a coordinate system with north along the positive *y*-axis. See Figure 6.45.

Let **v** be the air velocity of the plane,

w be the wind velocity, and

r be the resultant ground velocity of the plane.

Writing each vector in terms of its magnitude and the direction angle θ that each vector makes with the positive *x*-axis, we have

$$\mathbf{v} = 800(\cos 60°\mathbf{i} + \sin 60°\mathbf{j}) \qquad 90° - 30° = 60°$$

$$\mathbf{w} = 40[\cos 315°\mathbf{i} + \sin 315°\mathbf{j}] \qquad 360° - 45° = 315°$$

$$= 40[\cos 45°\mathbf{i} - \sin 45°\mathbf{j}] \qquad \cos 315° = \cos 45°; \sin 315° = -\sin 45°$$

The resultant **r** is given by

$$\mathbf{r} = \mathbf{v} + \mathbf{w}$$

$$\mathbf{r} = 800(\cos 60°\mathbf{i} + \sin 60°\mathbf{j}) + 40(\cos 45°\mathbf{i} - \sin 45°\mathbf{j}) \qquad \text{Substitute values of } \mathbf{v} \text{ and } \mathbf{w}.$$

$$\mathbf{r} = (800 \cos 60° + 40 \cos 45°)\mathbf{i} + (800 \sin 60° - 40 \sin 45°)\mathbf{j} \qquad \text{Add vectors.}$$

$$\|\mathbf{r}\| = \sqrt{(800 \cos 60° + 40 \cos 45°)^2 + (800 \sin 60° - 40 \sin 45°)^2} \qquad \|r_1\mathbf{i} + r_2\mathbf{j}\| = \sqrt{r_1^2 + r_2^2}$$

$$\|\mathbf{r}\| \approx 790.6 \qquad \text{Use a calculator.}$$

The ground speed of the F-15 is approximately 790.6 miles per hour. To find the actual direction (bearing) of the plane, we first find the direction angle θ of **r**.

$$\theta = \tan^{-1}\left(\frac{800 \sin 60° - 40 \sin 45°}{800 \cos 60° + 40 \cos 45°}\right) \qquad \text{For } \mathbf{r} = r_1\mathbf{i} + r_2\mathbf{j}, \theta = \tan^{-1}\left(\frac{r_2}{r_1}\right)$$

$$\approx 57.2° \qquad \text{Use a calculator.}$$

The angle between **r** and the *y*-axis (direction north) is

$$90° - 57.2° = 32.8°.$$

The bearing of the F-15 is approximately N 32.8° E.

Practice Problem 9 Repeat Example 9 assuming that the bearing of the plane is N 60° W and the direction of the wind is S 30° W.

EXAMPLE 10 **Obtaining an Equilibrium**

What single force must be added to the system of forces **A**, **B**, **C**, and **D** shown in Figure 6.46 to obtain a system that is in equilibrium?

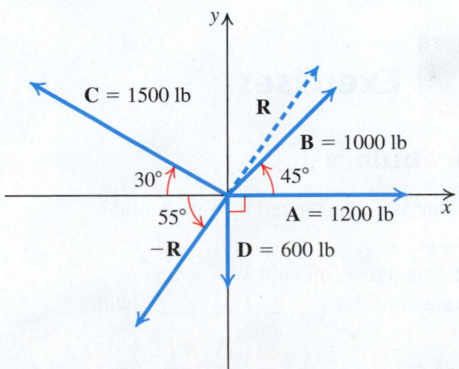

C = 1500 lb

R

B = 1000 lb

30°

45°

55°

A = 1200 lb

−R

D = 600 lb

x

y

Figure 6.46 A system in equilibrium.

Solution

We first find the resultant **R** of the given system. The simplest procedure is to find and add the corresponding horizontal components and the vertical components of each force.

The horizontal component r_1 of **R** is given by

$$r_1 = a_1 \qquad + b_1 \qquad + c_1 \qquad + d_1$$
$$= 1200 \cos 0° + 1000 \cos 45° + 1500 \cos 150° + 600 \cos 270°$$
$$\approx 608.07$$

Similarly, the vertical component r_2 of **R** is given by

$$r_2 = a_2 \qquad + b_2 \qquad + c_2 \qquad + d_2$$
$$= 1200 \sin 0° + 1000 \sin 45° + 1500 \sin 150° + 600 \sin 270°$$
$$\approx 857.11$$
$$\mathbf{R} = r_1\mathbf{i} \qquad + r_2\mathbf{j}$$
$$= 608.07\,\mathbf{i} + 857.11\,\mathbf{j}$$
$$\|\mathbf{R}\| = \sqrt{(608.07)^2 + (857.11)^2} \approx 1051 \text{ lb}$$
$$\theta = \tan^{-1}\left(\frac{r_2}{r_1}\right) = \tan^{-1}\left(\frac{857.11}{608.07}\right)$$
$$\approx 55.0° \qquad \text{Use a calculator.}$$

The force that must be added to obtain equilibrium is the negative of **R**. This is a force of 1051 pounds, and $-\mathbf{R}$ makes an angle of $180° + 55° = 235°$ with the positive x-axis. See Figure 6.46.

Practice Problem 10 Find the resultant of the following system of forces acting simultaneously at a point: $\mathbf{u} = 200$ pounds in the direction N 40° E, $\mathbf{v} = 300$ pounds in the direction N 70° W, and $\mathbf{w} = 400$ pounds in the direction S 20° E.

Answers to Practice Problems

1. a.

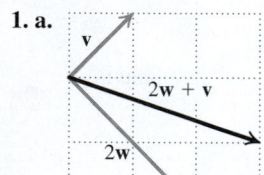

b.

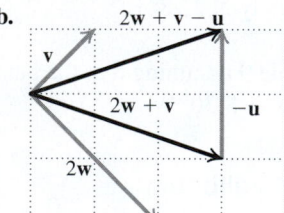

2. $\langle 3, -10 \rangle$ **3. a.** $\langle 1, -1 \rangle$ **b.** $\langle 6, -9 \rangle$ **c.** $\langle 8, -13 \rangle$
d. $\sqrt{233}$ **4.** $\left\langle -\frac{12}{13}, \frac{5}{13} \right\rangle$ **5. a.** $-7\mathbf{i} + 14\mathbf{j}$ **b.** $7\sqrt{5}$
6. $\sqrt{13}, \theta \approx 303.69°$ **7.** $\sqrt{3}\mathbf{i} - \mathbf{j}$ **8.** **R** is a force of 50 lb in the direction N 36.9° W. **9.** The ground speed is approximately 801.0 mph, and the bearing is approximately N 62.9° W.
10. The magnitude of the resultant is approximately 121.2 lb, and the resultant is in the direction S 7.8° W.

SECTION 6.4 **Exercises**

Concepts and Vocabulary

1. A vector is a quantity that is characterized by a magnitude and a(n) _____.

2. The resultant of **v** and **w** is the vector sum $\mathbf{v} + \mathbf{w}$ and is represented by the diagonal of the _____ with adjacent sides **v** and **w**.

3. If $\mathbf{v} = \langle a, b \rangle$, then $\|\mathbf{v}\| =$ _____, and for $a > 0$ its direction angle $\theta = \tan^{-1}($_____$)$.

4. If **v** is a nonzero vector, then the unit vector **u** in the direction of **v** is given by $\mathbf{u} = \dfrac{1}{\|\mathbf{v}\|} ($_____$)$.

5. **True or False.** The zero vector **0** is the only vector with no direction specified.

6. **True or False.** The vector $-\mathbf{v}$ has the same magnitude as **v** but the opposite direction.

7. **True or False.** If the initial point of a vector **v** is (x_1, y_1) and its terminal point is (x_2, y_2), then the position vector equivalent to **v** is $\langle x_1 - x_2, y_1 - y_2 \rangle$.

8. **True or False.** Temperature is an example of a vector.

Building Skills

In Exercises 9–16, use the vectors u, v, and w in the accompanying figure to graph each vector.

9. $\mathbf{u} + \mathbf{v}$

10. $3\mathbf{v}$

11. $2\mathbf{u} - \mathbf{w}$

12. $\mathbf{w} - 2\mathbf{v}$

13. $\mathbf{u} + \mathbf{v} + \mathbf{w}$

14. $2\mathbf{u} - \mathbf{w} + \mathbf{v}$

15. $\mathbf{w} - 2\mathbf{v} + \mathbf{u}$

16. $2\mathbf{v} + 3\mathbf{w} - \mathbf{u}$

In Exercises 17–24, the vector v has initial point P and terminal point Q. Write v as a position vector.

17. $P(3, 6), Q(2, 9)$

18. $P(6, -4), Q(1, 1)$

19. $P(-5, -2), Q(-3, -4)$

20. $P(0, 0), Q(-3, -6)$

21. $P(-1, 4), Q(2, -3)$

22. $P(3.5, 2.7), Q(-1.5, 1.3)$

23. $P\left(\dfrac{1}{2}, \dfrac{3}{4}\right), Q\left(-\dfrac{1}{2}, -\dfrac{7}{4}\right)$

24. $P\left(-\dfrac{2}{3}, \dfrac{4}{9}\right), Q\left(\dfrac{1}{3}, -\dfrac{2}{3}\right)$

In Exercises 25–28, determine whether the vectors $\overrightarrow{AB}$ and $\overrightarrow{CD}$ are equivalent. [*Hint:* Write $\overrightarrow{AB}$ and $\overrightarrow{CD}$ as position vectors.]

25. $A(1, 0), B(3, 4), C(-1, 2), D(1, 6)$

26. $A(-1, 2), B(3, -2), C(2, 5), D(6, 1)$

27. $A(2, -1), B(3, 5), C(-1, 3), D(-2, -3)$

28. $A(5, 7), B(6, 3), C(-2, 1), D(-3, 5)$

In Exercises 29–36, let $\mathbf{v} = \langle -1, 2 \rangle$ and $\mathbf{w} = \langle 3, -2 \rangle$. Find each expression.

29. $\|\mathbf{v}\|$

30. $\|\mathbf{w}\|$

31. $\mathbf{v} - \mathbf{w}$

32. $\mathbf{v} + \mathbf{w}$

33. $2\mathbf{v} - 3\mathbf{w}$

34. $2\mathbf{w} - 3\mathbf{v}$

35. $\|2\mathbf{v} - 3\mathbf{w}\|$

36. $\|2\mathbf{w} - 3\mathbf{v}\|$

In Exercises 37–46, find a unit vector u in the direction of the given vector v.

37. $\mathbf{v} = \langle 1, -1 \rangle$

38. $\mathbf{v} = \langle 1, 3 \rangle$

39. $\mathbf{v} = \langle -4, 3 \rangle$

40. $\mathbf{v} = \langle 5, -12 \rangle$

41. $\mathbf{v} = \langle \sqrt{2}, \sqrt{2} \rangle$

42. $\mathbf{v} = \langle \sqrt{5}, -2 \rangle$

43. $\mathbf{v} = 3\mathbf{i}$

44. $\mathbf{v} = -2\mathbf{j}$

45. $\mathbf{v} = -3\mathbf{i} + 4\mathbf{j}$

46. $\mathbf{v} = -12\mathbf{i} - 5\mathbf{j}$

In Exercises 47–52, find each expression for $\mathbf{u} = 2\mathbf{i} - 5\mathbf{j}$ and $\mathbf{v} = -3\mathbf{i} - 2\mathbf{j}$.

47. $\mathbf{u} + \mathbf{v}$

48. $\mathbf{u} - \mathbf{v}$

49. $2\mathbf{u} - 3\mathbf{v}$

50. $2\mathbf{v} + 3\mathbf{u}$

51. $\|2\mathbf{u} - 3\mathbf{v}\|$

52. $\|2\mathbf{v} + 3\mathbf{u}\|$

In Exercises 53–60, find the magnitude and the direction angle of the vector v.

53. $\mathbf{v} = 10(\cos 60°\mathbf{i} + \sin 60°\mathbf{j})$

54. $\mathbf{v} = -4(\cos 30°\mathbf{i} + \sin 30°\mathbf{j})$

55. $\mathbf{v} = -3(\cos 30°\mathbf{i} - \sin 30°\mathbf{j})$

56. $\mathbf{v} = 2(\cos 300°\mathbf{i} - \sin 300°\mathbf{j})$

57. $\mathbf{v} = 5\mathbf{i} + 12\mathbf{j}$

58. $\mathbf{v} = 12\mathbf{i} - 5\mathbf{j}$

59. $\mathbf{v} = -4\mathbf{i} - 3\mathbf{j}$

60. $\mathbf{v} = -5\mathbf{i} + 12\mathbf{j}$

In Exercises 61–68, write the vector v in the form $v_1\mathbf{i} + v_2\mathbf{j}$, given $\|\mathbf{v}\|$ and the angle θ that v makes with the positive x-axis.

61. $\|\mathbf{v}\| = 2, \theta = 30°$

62. $\|\mathbf{v}\| = 5, \theta = 45°$

63. $\|\mathbf{v}\| = 4, \theta = 120°$

64. $\|\mathbf{v}\| = 3, \theta = 150°$

65. $\|\mathbf{v}\| = 3, \theta = \dfrac{5\pi}{3}$

66. $\|\mathbf{v}\| = 4, \theta = \dfrac{11\pi}{6}$

67. $\|\mathbf{v}\| = 7, \theta = -\dfrac{\pi}{3}$

68. $\|\mathbf{v}\| = 8, \theta = \dfrac{3\pi}{4}$

Applying the Concepts

In Exercises 69–80, write each answer to the nearest tenth of a unit.

69. **Wind and vectors.** A wind is blowing 25 miles per hour in the direction N 67° E. Express the wind velocity **v** in the form $a\mathbf{i} + b\mathbf{j}$.

70. **Wind and vectors.** The velocity **v** of a wind is given by $\mathbf{v} = 5\mathbf{i} - 12\mathbf{j}$, where the vectors **i** and **j** represent 1-mile-per-hour winds blowing east and north, respectively. Find the speed (magnitude of **v**) and the direction of the wind.

71. **Resultant force.** Find the magnitude and bearing of the resultant **R** of two forces $\mathbf{F}_1$ and $\mathbf{F}_2$, where $\mathbf{F}_1$ is a force of 25 pounds acting due south and $\mathbf{F}_2$ is a force of 32 pounds acting due west.

72. **Resultant force.** Repeat Exercise 71 assuming that $\mathbf{F}_1$ is doubled.

73. **Finding components.** A force of 80 pounds acts in the direction N 65° E. Find its east and north components.

74. **Finding components.** Repeat Exercise 73 assuming that a force of 60 pounds acts in the direction S 32° W.

75. **Resultant force.** Two forces of 300 pounds and 400 pounds act at the origin and make angles of 30° and 60°, respectively, with the positive x-axis. Find the resultant.

76. **Resultant force.** Repeat Exercise 75 assuming that the forces have the same magnitude but that the angle each force makes with the positive x-axis is doubled.

77. **Air navigation.** A plane flies at an airspeed (speed in still air) of 500 miles per hour on a bearing of N 35° E. An east wind (wind from east to west) is blowing at 30 miles per hour. Find the plane's ground speed and direction.

78. **Air navigation.** A plane flies at an airspeed of 550 miles per hour on a straight course from an airfield F. A 30-mile-per-hour wind is blowing from the west. What must be the plane's bearing so that after one hour of flying time the plane is due north of F?

79. **River navigation.** A river flowing southward has a current of 4 miles per hour. A motorboat in still water maintains a speed of 15 miles per hour. The motorboat starts at the west shore on a bearing of N 40° E. What is the actual speed and direction of the boat?

80. River crossing. A fisherwoman on the west bank of the river of Exercise 79 sees several people fishing at a spot P directly east of her. If she owns the boat of Exercise 79, what direction should she take to land on the spot P?

In Exercises 81 and 82, find a single force that must be added to the system of forces shown in the figure to obtain a system that is in equilibrium.

81.

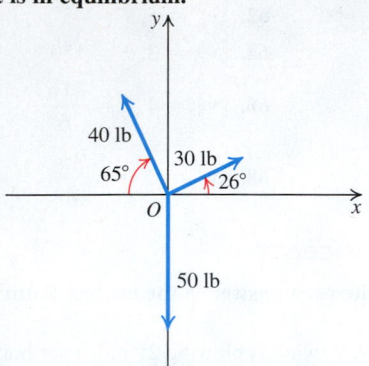

82.

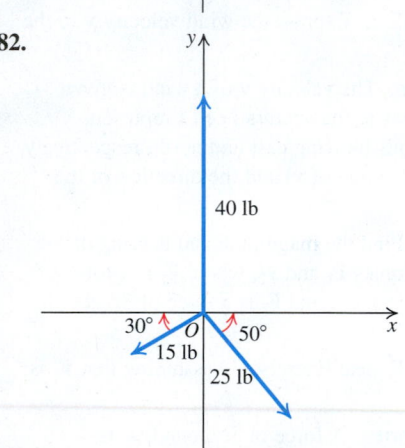

Beyond the Basics

In Exercises 83 and 84, calculate the magnitude of the forces F_1 and F_2 assuming that the system is in equilibrium.

83. **84.**

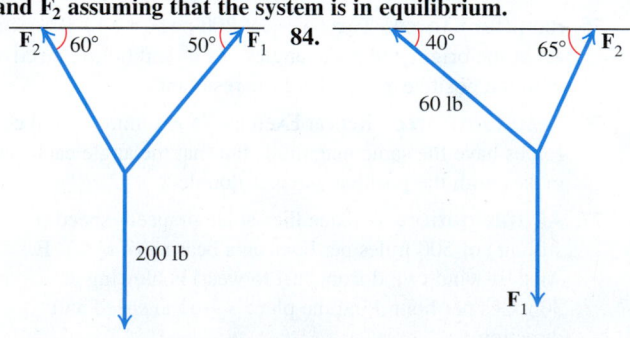

85. Find the terminal point of $v = 4i - 3j$ assuming that the initial point is $(-2, 1)$.

86. Find the initial point of $v = \langle -3, 5 \rangle$ assuming that the terminal point is $(4, 0)$.

87. Let $u = \langle -1, 2 \rangle$ and $v = \langle 3, 5 \rangle$. Find a vector $x = \langle x_1, x_2 \rangle$ that satisfies the equation $2u - x = 2x + 3v$.

88. Let $P(3, 5)$ and $Q(7, -4)$ be two points in a coordinate plane. Use vectors to find the coordinates of the point on the line segment joining P and Q that is $\frac{3}{4}$ of the way from P to Q.

In Exercises 89 and 90, show that $ABCD$ is a parallelogram, given the coordinates of the points A, B, C, and D. [Hint: Show that $\vec{AB} = \pm \vec{CD}$.]

89. $A = (2, 3), B = (1, 4), C = (0, -2)$, and $D = (1, -3)$

90. $A = (-2, -1), B = (3, 0), C = (1, -2)$, and $D = (-4, -3)$

In Exercises 91 and 92, show that the given points P, Q, and R are collinear. [Hint: Show that $\vec{PQ} = c\vec{PR}$ for some scalar c.]

91. $P = (0, -3), Q = (2, 1)$, and $R = (3, 3)$

92. $P = (-2, -6), Q = (2, 0)$, and $R = (4, 3)$

93. Use vectors to find the lengths of the diagonals of the parallelogram determined by $i + 2j$ and $i - j$.

94. Use vectors to find the fourth vertex of a parallelogram, with three vertices at $(0, 0), (2, 3)$, and $(6, 4)$. [Hint: There is more than one answer.]

Critical Thinking / Discussion / Writing

Let $A = (1, 2), B = (4, 3)$, and $C = (6, 1)$ be points in the plane. Find $P = (x, y)$ so that each of the following is true.

95. $\vec{AB} = \vec{CP}$ **96.** $\vec{PA} = \vec{CB}$

97. $\vec{AP} = \vec{CB}$ **98.** $\vec{PA} = \vec{BC}$

99. Explain why two equivalent vectors with the same initial point must have the same terminal point.

100. Give an example of a physical environment in which two equivalent vectors have a different effect in that environment.

Getting Ready for the Next Section

101. True or False. If $\cos \theta > 0$, then θ is an acute angle.

102. True or False. If $\cos \theta = 0$, then $\theta = \frac{\pi}{2}$.

103. Find $\cos^{-1}(-1), \cos^{-1}(0)$, and $\cos^{-1}(1)$.

104. Show that the lines $2x + 3y = 6$ and $3x - 2y = 12$ are perpendicular to each other.

In Exercises 105–108, let $v = \langle 1, -2 \rangle$ and $w = \langle 3, 2 \rangle$.

105. Compute $\|v\|$.

106. Compute $\|w\|$.

107. Compute $\dfrac{(1)(3) + (-2)(2)}{\|v\| \|w\|}$.

108. Verify that $\|v + w\|^2 = \|v\|^2 + \|w\|^2 + 2[(1)(3) + (-2)(2)]$.

The Dot Product

BEFORE STARTING THIS SECTION, REVIEW

1 Pythagorean Theorem (Appendix A.5, page 951)

2 The Law of Sines (Section 6.2, page 515)

3 The Law of Cosines (Section 6.3, page 527)

OBJECTIVES

1 Define the dot product of two vectors.

2 Find the angle between two vectors.

3 Define orthogonal vectors.

4 Find the projection of a vector onto another vector.

5 Decompose a vector into two orthogonal vectors.

6 Use the definition of work.

◆ How to Measure Work

If you move an object a distance of d units by applying a force of magnitude F in the direction of motion, then physicists define the work W done by the force F on the object to be

$$W = Fd$$
$$\text{Work} = (\text{force})(\text{distance}).$$

If the force is measured in pounds and the distance is measured in feet, then the unit of work is *foot-pounds*. Unfortunately, it is not always possible to exert a force in the direction you would like. For example, if you are pulling a child's wagon, then you exert a force in a direction dictated by the position of the hand you use to pull the wagon. In this section, we give a precise definition of work done by a constant force, and in Example 10, we compute the work done in one such instance.

1 Define the dot product of two vectors.

The Dot Product

We know that scalar multiplication and vector addition and subtraction produce vectors. A new operation, the *dot product* of two vectors, produces a *scalar*, not a vector.

The Dot Product ⎯⎯⎯⎯⎯⎯⎯

For two vectors $\mathbf{v} = \langle v_1, v_2 \rangle$ and $\mathbf{w} = \langle w_1, w_2 \rangle$, the **dot product** of $\mathbf{v}$ and $\mathbf{w}$, denoted $\mathbf{v} \cdot \mathbf{w}$, is defined as:

$$\mathbf{v} \cdot \mathbf{w} = \langle v_1, v_2 \rangle \cdot \langle w_1, w_2 \rangle = v_1 w_1 + v_2 w_2$$

Because the dot product produces a scalar, it is sometimes referred to as the **scalar product**, or the **inner product**.

EXAMPLE 1 **Finding the Dot Product**

Find the dot product $\mathbf{v} \cdot \mathbf{w}$.

a. $\mathbf{v} = \langle 2, -3 \rangle$ and $\mathbf{w} = \langle 3, 4 \rangle$

b. $\mathbf{v} = -3\mathbf{i} + 5\mathbf{j}$ and $\mathbf{w} = 2\mathbf{i} + 3\mathbf{j}$

Solution

a. $\mathbf{v} \cdot \mathbf{w} = \langle 2, -3 \rangle \cdot \langle 3, 4 \rangle$

$= (2)(3) + (-3)(4) = -6$ Definition of dot product

b. $\mathbf{v} \cdot \mathbf{w} = (-3\mathbf{i} + 5\mathbf{j}) \cdot (2\mathbf{i} + 3\mathbf{j})$

$= \langle -3, 5 \rangle \cdot \langle 2, 3 \rangle$ Rewrite as position vectors.

$= (-3)(2) + (5)(3) = 9$ Definition of dot product

Practice Problem 1 Find the dot product $\mathbf{v} \cdot \mathbf{w}$.

a. $\mathbf{v} = \langle 1, 2 \rangle$ and $\mathbf{w} = \langle -2, 5 \rangle$ **b.** $\mathbf{v} = 4\mathbf{i} - 3\mathbf{j}$ and $\mathbf{w} = -3\mathbf{i} - 4\mathbf{j}$

2 Find the angle between two vectors.

The Angle Between Two Vectors

We use the Law of Cosines to find another formula for the dot product. The new formula will be used to find the angle between any two vectors.

Let θ be the angle between the two vectors $\mathbf{v} = \langle v_1, v_2 \rangle$ and $\mathbf{w} = \langle w_1, w_2 \rangle$. See Figure 6.47. Apply the Law of Cosines to the triangle OPQ in Figure 6.47.

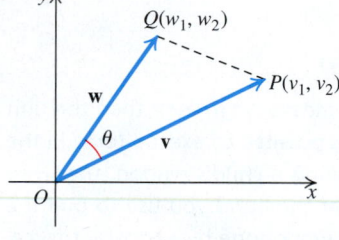

Figure 6.47 Angle between two vectors.

$[d(P, Q)]^2 = \|\mathbf{v}\|^2 + \|\mathbf{w}\|^2 - 2\|\mathbf{v}\| \|\mathbf{w}\| \cos \theta$ The Law of Cosines

$(v_1 - w_1)^2 + (v_2 - w_2)^2$

$\quad = (v_1^2 + v_2^2) + (w_1^2 + w_2^2) - 2\|\mathbf{v}\| \|\mathbf{w}\| \cos \theta$ Use the distance formula and square of the magnitudes of $\mathbf{v}$ and $\mathbf{w}$.

$v_1^2 - 2v_1 w_1 + w_1^2 + v_2^2 - 2v_2 w_2 + w_2^2$

$\quad = v_1^2 + v_2^2 + w_1^2 + w_2^2 - 2\|\mathbf{v}\| \|\mathbf{w}\| \cos \theta$ $(A - B)^2 = A^2 - 2AB + B^2$

$-2v_1 w_1 - 2v_2 w_2 = -2\|\mathbf{v}\| \|\mathbf{w}\| \cos \theta$ Subtract $v_1^2 + w_1^2 + v_2^2 + w_2^2$ from both sides.

$v_1 w_1 + v_2 w_2 = \|\mathbf{v}\| \|\mathbf{w}\| \cos \theta$ Divide both sides by -2.

$\mathbf{v} \cdot \mathbf{w} = \|\mathbf{v}\| \|\mathbf{w}\| \cos \theta$ Definition of $\mathbf{v} \cdot \mathbf{w}$

The last equation provides us with another formula for $\mathbf{v} \cdot \mathbf{w}$. Solving the last equation for $\cos \theta$ gives us a formula for finding the angle between two vectors.

THE DOT PRODUCT AND THE ANGLE BETWEEN TWO VECTORS

If θ $(0° \le \theta \le 180°)$ is the angle between two nonzero vectors $\mathbf{v}$ and $\mathbf{w}$, then

$$\mathbf{v} \cdot \mathbf{w} = \|\mathbf{v}\| \|\mathbf{w}\| \cos \theta \quad \text{or} \quad \cos \theta = \frac{\mathbf{v} \cdot \mathbf{w}}{\|\mathbf{v}\| \|\mathbf{w}\|}.$$

EXAMPLE 2 **Finding the Dot Product**

If $\mathbf{v}$ and $\mathbf{w}$ are two vectors of magnitudes 5 and 7, respectively, and the angle between them is 75°, find $\mathbf{v} \cdot \mathbf{w}$. Round the answer to the nearest tenth.

Solution

$$\mathbf{v} \cdot \mathbf{w} = \|\mathbf{v}\| \|\mathbf{w}\| \cos\theta \qquad \text{Formula for } \mathbf{v} \cdot \mathbf{w}$$
$$= (5)(7)\cos 75° \qquad \text{Substitute given values.}$$
$$\approx 9.1 \qquad \text{Use a calculator.}$$

Practice Problem 2 Repeat Example 2 assuming that the angle between $\mathbf{v}$ and $\mathbf{w}$ is $60°$.

Two vectors $\mathbf{v}$ and $\mathbf{w}$ are **parallel** if there is a nonzero scalar, c, so that $\mathbf{v} = c\mathbf{w}$. The angle θ between parallel vectors is either $0°$ or $180°$. So, $\mathbf{v}$ and $\mathbf{w}$ are parallel if and only if $\mathbf{v} \cdot \mathbf{w} = \pm \|\mathbf{v}\| \|\mathbf{w}\|$.

EXAMPLE 3   Deciding Whether Two Vectors Are Parallel

Determine whether the vectors $\mathbf{v} = 2\mathbf{i} - 3\mathbf{j}$ and $\mathbf{w} = -4\mathbf{i} + 6\mathbf{j}$ are parallel.

Solution

Since $\mathbf{w} = -2(2\mathbf{i} - 3\mathbf{j}) = -2\mathbf{v}$, the vectors $\mathbf{v}$ and $\mathbf{w}$ are parallel. Note that

$$\cos\theta = \frac{\mathbf{v} \cdot \mathbf{w}}{\|\mathbf{v}\| \|\mathbf{w}\|} = \frac{-8 - 18}{\sqrt{13}\sqrt{52}} = \frac{-26}{\sqrt{13}(2\sqrt{13})} = \frac{-26}{26} = -1,$$

so that $\theta = 180°$ is the angle between $\mathbf{v}$ and $\mathbf{w}$. The vectors $\mathbf{v}$ and $\mathbf{w}$ are parallel.

Practice Problem 3 Determine whether $\mathbf{v} = -\mathbf{i} - 2\mathbf{j}$ and $\mathbf{w} = 2\mathbf{i} + \mathbf{j}$ are parallel.

Because the angle θ between two nonzero vectors $\mathbf{v}$ and $\mathbf{w}$ satisfies $\cos\theta = \dfrac{\mathbf{v} \cdot \mathbf{w}}{\|\mathbf{v}\| \|\mathbf{w}\|}$ and $0° \le \theta \le 180°$, we have $\theta = \cos^{-1}\left(\dfrac{\mathbf{v} \cdot \mathbf{w}}{\|\mathbf{v}\| \|\mathbf{w}\|}\right)$.

EXAMPLE 4   Finding the Angle Between Two Vectors

Find the angle θ (in degrees) between the vectors $\mathbf{v} = 2\mathbf{i} + 3\mathbf{j}$ and $\mathbf{w} = -3\mathbf{i} + 4\mathbf{j}$. Round the answer to the nearest tenth of a degree.

Solution

$$\cos\theta = \frac{\mathbf{v} \cdot \mathbf{w}}{\|\mathbf{v}\| \|\mathbf{w}\|} \qquad \text{Formula for the cosine of the angle between } \mathbf{v} \text{ and } \mathbf{w}$$

$$\cos\theta = \frac{(2\mathbf{i} + 3\mathbf{j}) \cdot (-3\mathbf{i} + 4\mathbf{j})}{\|2\mathbf{i} + 3\mathbf{j}\| \|-3\mathbf{i} + 4\mathbf{j}\|} \qquad \text{Substitute expressions for } \mathbf{v} \text{ and } \mathbf{w}.$$

$$\cos\theta = \frac{(2)(-3) + (3)(4)}{\sqrt{2^2 + 3^2}\sqrt{(-3)^2 + 4^2}} \qquad \text{Use formulas for the dot product and the magnitude.}$$

$$\cos\theta = \frac{6}{5\sqrt{13}} \qquad \text{Simplify.}$$

$$\theta = \cos^{-1}\left(\frac{6}{5\sqrt{13}}\right) \qquad 0° \le \theta \le 180°$$

$$\theta \approx 70.6° \qquad \text{Use a calculator.}$$

Practice Problem 4 Find the angle (in degrees) between the vectors $\mathbf{v} = 2\mathbf{i} - 3\mathbf{j}$ and $\mathbf{w} = 3\mathbf{i} + 5\mathbf{j}$. Round the answer to the nearest tenth of a degree.

We can verify the following properties of the dot product by writing the vectors as position vectors and using the definitions of vector addition, scalar multiplication, and the dot product.

Properties of the Dot Product _____

If $\mathbf{u}$, $\mathbf{v}$, and $\mathbf{w}$ are vectors and c is a scalar, then

1. $\mathbf{u} \cdot \mathbf{v} = \mathbf{v} \cdot \mathbf{u}$
2. $\mathbf{u} \cdot (\mathbf{v} + \mathbf{w}) = \mathbf{u} \cdot \mathbf{v} + \mathbf{u} \cdot \mathbf{w}$
3. $\mathbf{0} \cdot \mathbf{v} = 0$
4. $\mathbf{v} \cdot \mathbf{v} = \|\mathbf{v}\|^2$
5. $(c\mathbf{u}) \cdot \mathbf{v} = c(\mathbf{u} \cdot \mathbf{v}) = \mathbf{u} \cdot (c\mathbf{v})$

EXAMPLE 5 **Finding the Magnitude of the Sum of Two Vectors**

Let $\mathbf{v}$ and $\mathbf{w}$ be two vectors in the plane of magnitudes 12 and 7, respectively. The angle between $\mathbf{v}$ and $\mathbf{w}$ is $65°$. Find the magnitude of $\mathbf{v} + 2\mathbf{w}$ to the nearest tenth.

Solution

$$
\begin{aligned}
\|\mathbf{v} + 2\mathbf{w}\|^2 &= (\mathbf{v} + 2\mathbf{w}) \cdot (\mathbf{v} + 2\mathbf{w}) && \text{Property 4}\\
&= (\mathbf{v} + 2\mathbf{w}) \cdot \mathbf{v} + (\mathbf{v} + 2\mathbf{w}) \cdot (2\mathbf{w}) && \text{Property 2}\\
&= \mathbf{v} \cdot \mathbf{v} + 2(\mathbf{w} \cdot \mathbf{v}) + 2(\mathbf{v} \cdot \mathbf{w}) + 4(\mathbf{w} \cdot \mathbf{w}) && \text{Properties 1 and 5}\\
&= \|\mathbf{v}\|^2 + 4(\mathbf{v} \cdot \mathbf{w}) + 4\|\mathbf{w}\|^2 && \text{Properties 1 and 4}\\
&= \|\mathbf{v}\|^2 + 4\|\mathbf{v}\|\|\mathbf{w}\|\cos\theta + 4\|\mathbf{w}\|^2 && \mathbf{v} \cdot \mathbf{w} = \|\mathbf{v}\|\|\mathbf{w}\|\cos\theta\\
&= (12)^2 + 4(12)(7)\cos 65° + 4(7)^2 && \text{Substitute values.}\\
\|\mathbf{v} + 2\mathbf{w}\|^2 &\approx 481.9997 && \text{Use a calculator.}\\
\|\mathbf{v} + 2\mathbf{w}\| &\approx 21.95 \approx 22.0 && \text{Take the positive square root.}
\end{aligned}
$$

Practice Problem 5 Repeat Example 4 assuming that $\|\mathbf{v}\| = 20$ and $\|\mathbf{w}\| = 12$ and the angle between $\mathbf{v}$ and $\mathbf{w}$ is $54°$.

3 Define orthogonal vectors.

Orthogonal Vectors

We know that if $\mathbf{v}$ and $\mathbf{w}$ are two nonzero vectors and θ $(0° \le \theta \le 180°)$ is the measure of the angle between them, then

$$\mathbf{v} \cdot \mathbf{w} = \|\mathbf{v}\|\|\mathbf{w}\|\cos\theta \quad \text{Form of the dot product}$$

From this equation, it follows that

$$\cos\theta = 0 \text{ if and only if } \mathbf{v} \cdot \mathbf{w} = 0.$$

Because $0° \le \theta \le 180°$, we conclude that

$$\theta = 90° \text{ if and only if } \mathbf{v} \cdot \mathbf{w} = 0.$$

Orthogonal Vectors _____

Two vectors $\mathbf{v}$ and $\mathbf{w}$ are **orthogonal** (perpendicular) if and only if $\mathbf{v} \cdot \mathbf{w} = 0$.

Since $\mathbf{0} \cdot \mathbf{w} = 0$ for any vector $\mathbf{w}$, it follows from the definition that the zero vector is orthogonal to every vector.

EXAMPLE 6   **Deciding Whether Vectors Are Orthogonal**

Determine whether $\mathbf{v} = 2\mathbf{i} - 3\mathbf{j}$ and $\mathbf{w} = 3\mathbf{i} + 2\mathbf{j}$ are orthogonal.

Solution

Since $\mathbf{v} \cdot \mathbf{w} = (2\mathbf{i} - 3\mathbf{j}) \cdot (3\mathbf{i} + 2\mathbf{j}) = \langle 2, -3 \rangle \cdot \langle 3, 2 \rangle = 6 - 6 = 0$, $\mathbf{v}$ and $\mathbf{w}$ are orthogonal.

Practice Problem 6 Determine whether $\mathbf{v} = 4\mathbf{i} + 8\mathbf{j}$ and $\mathbf{w} = 2\mathbf{i} - \mathbf{j}$ are orthogonal.

EXAMPLE 7   **Orthogonal Vectors**

Find the scalar c so that the vectors $\mathbf{v} = 3\mathbf{i} + 2\mathbf{j}$ and $\mathbf{w} = 4\mathbf{i} + c\mathbf{j}$ are orthogonal.

Solution

The vectors $\mathbf{v} = 3\mathbf{i} + 2\mathbf{j} = \langle 3, 2 \rangle$ and $\mathbf{w} = 4\mathbf{i} + c\mathbf{j} = \langle 4, c \rangle$ are orthogonal if and only if

$$\begin{aligned}
\mathbf{v} \cdot \mathbf{w} &= 0 && \text{Definition of orthogonality} \\
\langle 3, 2 \rangle \cdot \langle 4, c \rangle &= 0 && \\
(3)(4) + (2)(c) &= 0 && \text{Definition of dot product} \\
12 + 2c &= 0 && \text{Simplify.} \\
c &= -6 && \text{Solve for } c.
\end{aligned}$$

Practice Problem 7 Find the scalar k so that the vectors $\mathbf{v} = \langle k, -2 \rangle$ and $\mathbf{w} = \langle 4, 6 \rangle$ are orthogonal.

4 Find the projection of a vector onto another vector.

Projection of a Vector

A geometric interpretation of the dot product is obtained by considering the *projection* of a vector onto another vector.

Let $\overrightarrow{OP}$ and $\overrightarrow{OQ}$ be representations of the nonzero vectors $\mathbf{v}$ and $\mathbf{w}$, respectively, The projection of $\overrightarrow{OP}$ in the direction of $\overrightarrow{OQ}$ is the directed line segment $\overrightarrow{OR}$, where R is the foot of the perpendicular from P to the line containing $\overrightarrow{OQ}$. See Figure 6.48. Then the vector represented by $\overrightarrow{OR}$ is called the **vector projection of $\mathbf{v}$ onto $\mathbf{w}$** and is denoted by $\text{proj}_{\mathbf{w}}\mathbf{v}$.

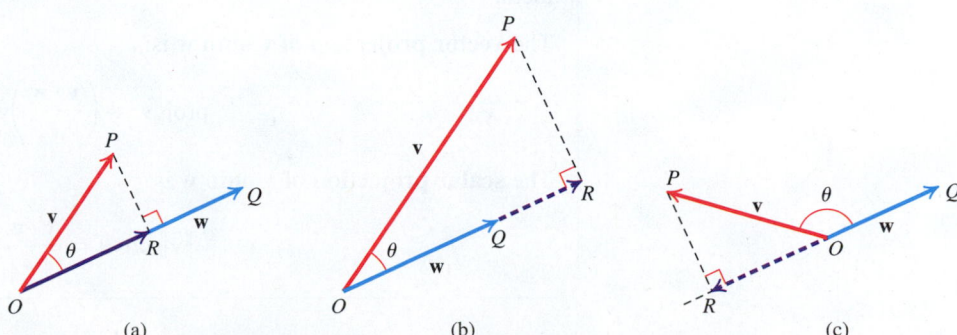

Figure 6.48 Vector projection of $\mathbf{v}$ onto $\mathbf{w}$.

In Figure 6.49, if a force $\mathbf{v}$ is used to pull a box, then the effective force that moves the box in the direction of $\mathbf{w}$ is the vector projection of $\mathbf{v}$ onto $\mathbf{w}$.

Let $\theta\,(0° \leq \theta \leq 180°)$ be the measure of the angle between nonzero vectors $\mathbf{v}$ and $\mathbf{w}$. The number $\|\mathbf{v}\| \cos\theta$ is called the **scalar projection of $\mathbf{v}$ onto $\mathbf{w}$**. We can express this number in terms of the dot product.

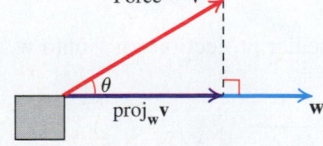

Figure 6.49

$$\text{Scalar projection } \mathbf{v} \text{ onto } \mathbf{w} = \|\mathbf{v}\| \cos\theta$$

$$= \frac{\|\mathbf{v}\|\|\mathbf{w}\|\cos\theta}{\|\mathbf{w}\|} \qquad \text{Multiply numerator and denominator by } \|\mathbf{w}\|.$$

$$= \frac{\mathbf{v}\cdot\mathbf{w}}{\|\mathbf{w}\|} \qquad \mathbf{v}\cdot\mathbf{w} = \|\mathbf{v}\|\|\mathbf{w}\|\cos\theta$$

If θ is acute, then $\text{proj}_{\mathbf{w}}\mathbf{v}$ has magnitude $\|\mathbf{v}\|\cos\theta$ and direction $\dfrac{\mathbf{w}}{\|\mathbf{w}\|}$.

If θ is obtuse, $\cos\theta < 0$ and $\text{proj}_{\mathbf{w}}\mathbf{v}$ has magnitude $-\|\mathbf{v}\|\cos\theta$ and direction $-\dfrac{\mathbf{w}}{\|\mathbf{w}\|}$.
See Figure 6.50.

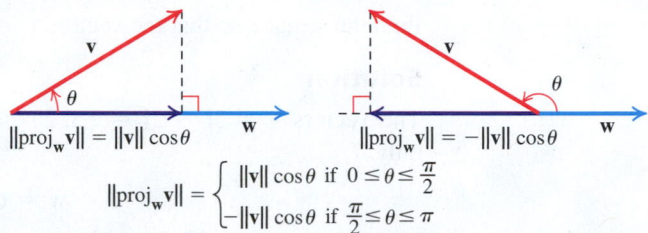

$$\|\text{proj}_{\mathbf{w}}\mathbf{v}\| = \|\mathbf{v}\|\cos\theta \qquad \|\text{proj}_{\mathbf{w}}\mathbf{v}\| = -\|\mathbf{v}\|\cos\theta$$

$$\|\text{proj}_{\mathbf{w}}\mathbf{v}\| = \begin{cases} \|\mathbf{v}\|\cos\theta & \text{if } 0 \le \theta \le \frac{\pi}{2} \\ -\|\mathbf{v}\|\cos\theta & \text{if } \frac{\pi}{2} \le \theta \le \pi \end{cases}$$

Figure 6.50 Magnitude and direction of $\text{proj}_{\mathbf{w}}\mathbf{v}$.

Because $\left(-\|\mathbf{v}\|\cos\theta\right)\left(-\dfrac{\mathbf{w}}{\|\mathbf{w}\|}\right) = \left(\|\mathbf{v}\|\cos\theta\right)\left(\dfrac{\mathbf{w}}{\|\mathbf{w}\|}\right)$, in both cases ($\theta$ acute or obtuse), we have

$$\text{proj}_{\mathbf{w}}\mathbf{v} = \left(\|\mathbf{v}\|\cos\theta\right)\left(\frac{\mathbf{w}}{\|\mathbf{w}\|}\right) \qquad \text{See Side Note.}$$

$$= \left(\frac{\mathbf{v}\cdot\mathbf{w}}{\|\mathbf{w}\|}\right)\frac{\mathbf{w}}{\|\mathbf{w}\|} \qquad \|\mathbf{v}\|\cos\theta = \frac{\mathbf{v}\cdot\mathbf{w}}{\|\mathbf{w}\|}$$

$$= \frac{\mathbf{v}\cdot\mathbf{w}}{\|\mathbf{w}\|^2}\mathbf{w}$$

SIDE NOTE

Any nonzero vector can be written as:
$$\mathbf{R} = \|\mathbf{R}\|\mathbf{u},$$
where $\mathbf{u} = \dfrac{\mathbf{R}}{\|\mathbf{R}\|}$ is the unit vector in the direction of $\mathbf{R}$.

VECTOR AND SCALAR PROJECTIONS OF v ONTO w

Let $\mathbf{v}$ and $\mathbf{w}$ be two nonzero vectors and let θ ($0° \le \theta \le 180°$) be the angle between them.

The **vector projection of v onto w** is:

$$\text{proj}_{\mathbf{w}}\mathbf{v} = \left(\frac{\mathbf{v}\cdot\mathbf{w}}{\|\mathbf{w}\|^2}\right)\mathbf{w}$$

The **scalar projection of v onto w** is:

$$\|\mathbf{v}\|\cos\theta = \frac{\mathbf{v}\cdot\mathbf{w}}{\|\mathbf{w}\|}$$

EXAMPLE 8 Finding the Vector Projection of v onto w

Given $\mathbf{v} = \langle -2, 3\rangle$ and $\mathbf{w} = \langle 3, -4\rangle$, find the vector and scalar projections of $\mathbf{v}$ onto $\mathbf{w}$.

Solution

$$\|\mathbf{v}\| = \sqrt{(-2)^2 + 3^2} = \sqrt{13} \text{ and } \|\mathbf{w}\| = \sqrt{3^2 + (-4)^2} = \sqrt{25} = 5$$

$$\mathbf{v}\cdot\mathbf{w} = \mathbf{w}\cdot\mathbf{v} = \langle -2, 3\rangle \cdot \langle 3, -4\rangle$$

$$= (-2)(3) + 3(-4) = -6 - 12 = -18$$

a. $\text{proj}_{\mathbf{w}} \mathbf{v} = \dfrac{\mathbf{v} \cdot \mathbf{w}}{\|\mathbf{w}\|^2} \mathbf{w}$ Vector projection of **v** onto **w**

$= -\dfrac{18}{25} \langle 3, -4 \rangle$ Substitute values.

$= \left\langle -\dfrac{54}{25}, \dfrac{72}{25} \right\rangle$ Scalar multiplication

b. Scalar projection of **v** onto $\mathbf{w} = \dfrac{\mathbf{v} \cdot \mathbf{w}}{\|\mathbf{w}\|}$

$= -\dfrac{18}{5}$ $\mathbf{v} \cdot \mathbf{w} = -18, \|\mathbf{w}\| = 5$

Practice Problem 8 Given $\mathbf{v} = \langle 1, -2 \rangle$ and $\mathbf{w} = \langle 2, 5 \rangle$, find the vector and scalar projections of **v** onto **w**.

5 Decompose a vector into two orthogonal vectors.

Decomposition of a Vector

Given two vectors **v** and **w**, adding these vectors produces the resultant vector $\mathbf{v} + \mathbf{w}$. In many applications, it is necessary to *decompose* a vector into two orthogonal vectors. We use the vector $\text{proj}_{\mathbf{w}} \mathbf{v}$ to write **v** as the sum of two orthogonal vectors.

Decomposition of a Vector

Let **v** and **w** be two nonzero vectors. The vector **v** can be written in the form

$$\mathbf{v} = \mathbf{v}_1 + \mathbf{v}_2,$$

where $\mathbf{v}_1$ is parallel to **w** and $\mathbf{v}_2$ is orthogonal to **w**. We let

$$\mathbf{v}_1 = \text{proj}_{\mathbf{w}} \mathbf{v} = \left(\dfrac{\mathbf{v} \cdot \mathbf{w}}{\|\mathbf{w}\|^2} \right) \mathbf{w} \text{ and } \mathbf{v}_2 = \mathbf{v} - \mathbf{v}_1.$$

The expression $\mathbf{v} = \mathbf{v}_1 + \mathbf{v}_2$ is called the **decomposition of v with respect to w.** The vectors $\mathbf{v}_1$ and $\mathbf{v}_2$ are called **components of v.**

Because $\mathbf{v}_1$ is a scalar multiple of **w**, $\mathbf{v}_1$ is parallel to **w**. In Exercise 92, we ask you to prove that $\mathbf{v}_2$ is orthogonal to **w**.

PROCEDURE
IN ACTION

EXAMPLE 9   **Decomposing a Vector into Components**

OBJECTIVE

*Express $\mathbf{v} = \mathbf{v}_1 + \mathbf{v}_2$, where $\mathbf{v}_1$ is parallel to **w** and $\mathbf{v}_2$ is orthogonal to **w**.*

Step 1 Compute.

$$\mathbf{v} \cdot \mathbf{w} \text{ and } \|\mathbf{w}\|^2$$

Step 2 Compute.

$$\mathbf{v}_1 = \text{proj}_{\mathbf{w}} \mathbf{v} = \left(\dfrac{\mathbf{v} \cdot \mathbf{w}}{\|\mathbf{w}\|^2} \right) \mathbf{w}$$

EXAMPLE

Find the decomposition $\mathbf{v} = \mathbf{v}_1 + \mathbf{v}_2$ of $\mathbf{v} = \langle -2, 10 \rangle$ with respect to $\mathbf{w} = \langle 3, -2 \rangle$.

1. $\mathbf{v} \cdot \mathbf{w} = \langle -2, 10 \rangle \cdot \langle 3, -2 \rangle$

$= (-2)(3) + (10)(-2) = -26$

$\|\mathbf{w}\|^2 = (3)^2 + (-2)^2 = 9 + 4 = 13$

2. $\mathbf{v}_1 = \left(\dfrac{\mathbf{v} \cdot \mathbf{w}}{\|\mathbf{w}\|^2} \right) \mathbf{w} = \dfrac{-26}{13} \langle 3, -2 \rangle$ From Step 1

$= -2 \langle 3, -2 \rangle = \langle -6, 4 \rangle$ Simplify.

(continued)

Step 3 Compute.

$$\mathbf{v}_2 = \mathbf{v} - \mathbf{v}_1$$

Check that $\mathbf{v}_2$ is orthogonal to $\mathbf{w}$ by showing $\mathbf{v}_2 \cdot \mathbf{w} = 0$.

Step 4 Write the decomposition.

$$\mathbf{v} = \mathbf{v}_1 + \mathbf{v}_2$$

3. $\mathbf{v}_2 = \mathbf{v} - \mathbf{v}_1 = \langle -2, 10 \rangle - \langle -6, 4 \rangle$
$$= \langle -2 - (-6), 10 - 4 \rangle$$
$$= \langle 4, 6 \rangle$$

Check

$$\mathbf{v}_2 \cdot \mathbf{w} = \langle 4, 6 \rangle \cdot \langle 3, -2 \rangle = (4)(3) + (6)(-2) = 0$$

4. $\mathbf{v} = \langle -2, 10 \rangle = \quad \mathbf{v}_1 \quad + \quad \mathbf{v}_2$
$$= \underbrace{\langle -6, 4 \rangle}_{\substack{\text{Parallel} \\ \text{to } \mathbf{w}}} + \underbrace{\langle 4, 6 \rangle}_{\substack{\text{Orthogonal} \\ \text{to } \mathbf{w}}}$$

Practice Problem 9 Repeat Example 9 with $\mathbf{v} = \langle 5, 1 \rangle$ and $\mathbf{w} = \langle 4, 4 \rangle$.

6 Use the definition of work.  ◆ **Work**

Work

> The work W done by a constant force $\mathbf{F}$ in moving an object from a point P to a point Q is defined by
>
> $$W = \mathbf{F} \cdot \overrightarrow{PQ} = \|\mathbf{F}\| \|\overrightarrow{PQ}\| \cos \theta,$$
>
> where θ is the angle between $\mathbf{F}$ and $\overrightarrow{PQ}$.

EXAMPLE 10 **Computing Work**

A child pulls a wagon along level ground, with a force of 40 pounds along the handle on the wagon that makes an angle of $42°$ with the horizontal. How much work has she done by pulling the wagon 150 feet?

Solution

The work done is as follows:

$$
\begin{aligned}
W &= \mathbf{F} \cdot \overrightarrow{PQ} && \text{Definition of work} \\
&= \|\mathbf{F}\| \|\overrightarrow{PQ}\| \cos \theta && \mathbf{u} \cdot \mathbf{v} = \|\mathbf{u}\| \|\mathbf{v}\| \cos \theta \\
&= (40)(150) \cos 42° && \text{Substitute given values.} \\
&\approx 4458.87 \text{ foot-pounds} && \text{Use a calculator.}
\end{aligned}
$$

Practice Problem 10 In Example 10, compute the work done if the handle on the wagon makes an angle of $60°$ with the horizontal.

Answers to Practice Problems

1. a. 8 **b.** 0 **2.** 17.5 **3.** Not parallel
4. $115.3°$ **5.** ≈ 39 **6.** $\mathbf{v}$ and $\mathbf{w}$ are orthogonal
7. $k = 3$ **8.** $\left(-\dfrac{16}{29}, -\dfrac{40}{29} \right); \ -\dfrac{8\sqrt{29}}{29}$

9. $\langle 3, 3 \rangle + \langle 2, -2 \rangle$
10. 3000 foot-pounds

SECTION 6.5   Exercises

Concepts and Vocabulary

1. The dot product of $\mathbf{v} = \langle a_1, a_2 \rangle$ and $\mathbf{w} = \langle b_1, b_2 \rangle$ is defined as $\mathbf{v} \cdot \mathbf{w} = $ _____.

2. If θ is the angle between the vectors $\mathbf{v}$ and $\mathbf{w}$, then $\mathbf{v} \cdot \mathbf{w} = $ _____.

3. If $\mathbf{v}$ and $\mathbf{w}$ are orthogonal, then $\mathbf{v} \cdot \mathbf{w} = $ _____.

4. If $\mathbf{v}$ and $\mathbf{w}$ are nonzero vectors with $\mathbf{v} \cdot \mathbf{w} = 0$, then $\mathbf{v}$ and $\mathbf{w}$ are _____.

5. **True or False.** If $\mathbf{v} \cdot \mathbf{w} < 0$, then the angle between $\mathbf{v}$ and $\mathbf{w}$ is an obtuse angle.

6. **True or False.** To find a vector perpendicular to a given vector, switch components and change the sign of one of them.

7. **True or False.** The dot product of two vectors is a vector.

8. **True or False.** If $\mathbf{u}$ and $\mathbf{v}$ are in the same direction, then $\mathbf{u} \cdot \mathbf{v} = \|\mathbf{u}\|\|\mathbf{v}\|$.

Building Skills

In Exercises 9–16, find $\mathbf{u} \cdot \mathbf{v}$.

9. $\mathbf{u} = \langle 1, -2 \rangle, \mathbf{v} = \langle 3, 5 \rangle$

10. $\mathbf{u} = \langle 1, 3 \rangle, \mathbf{v} = \langle -3, 1 \rangle$

11. $\mathbf{u} = \langle 2, -6 \rangle, \mathbf{v} = \langle 3, 1 \rangle$

12. $\mathbf{u} = \langle -1, -2 \rangle, \mathbf{v} = \langle -3, -4 \rangle$

13. $\mathbf{u} = \mathbf{i} - 3\mathbf{j}, \mathbf{v} = 8\mathbf{i} - 2\mathbf{j}$

14. $\mathbf{u} = 6\mathbf{i} - \mathbf{j}, \mathbf{v} = 2\mathbf{i} + 7\mathbf{j}$

15. $\mathbf{u} = 4\mathbf{i} - 2\mathbf{j}, \mathbf{v} = 3\mathbf{j}$

16. $\mathbf{u} = -2\mathbf{i}, \mathbf{v} = 5\mathbf{j}$

In Exercises 17–24, find $\mathbf{u} \cdot \mathbf{v}$, where θ is the angle between the vectors $\mathbf{u}$ and $\mathbf{v}$. Round answers to the nearest tenth.

17. $\|\mathbf{u}\| = 2, \|\mathbf{v}\| = 5, \theta = \dfrac{\pi}{6}$

18. $\|\mathbf{u}\| = 3, \|\mathbf{v}\| = 4, \theta = \dfrac{\pi}{3}$

19. $\|\mathbf{u}\| = 4, \|\mathbf{v}\| = 5, \theta = \dfrac{2\pi}{3}$

20. $\|\mathbf{u}\| = 2\sqrt{3}, \|\mathbf{v}\| = 7, \theta = \dfrac{5\pi}{6}$

21. $\|\mathbf{u}\| = 5, \|\mathbf{v}\| = 4, \theta = 65°$

22. $\|\mathbf{u}\| = 5, \|\mathbf{v}\| = 3, \theta = 78°$

23. $\|\mathbf{u}\| = 3, \|\mathbf{v}\| = 7, \theta = 120°$

24. $\|\mathbf{u}\| = 6, \|\mathbf{v}\| = 4, \theta = 150°$

In Exercises 25–34, find the angle between the vectors $\mathbf{v}$ and $\mathbf{w}$. Round each answer to the nearest tenth of a degree.

25. $\mathbf{v} = \langle 2, 3 \rangle, \mathbf{w} = \langle 3, 4 \rangle$

26. $\mathbf{v} = \langle -1, 1 \rangle, \mathbf{w} = \langle 5, 12 \rangle$

27. $\mathbf{v} = -2\mathbf{i} - 3\mathbf{j}, \mathbf{w} = \mathbf{i} + \mathbf{j}$

28. $\mathbf{v} = \mathbf{i} - \mathbf{j}, \mathbf{w} = 3\mathbf{i} - 4\mathbf{j}$

29. $\mathbf{v} = 2\mathbf{i} + 5\mathbf{j}, \mathbf{w} = -5\mathbf{i} + 2\mathbf{j}$

30. $\mathbf{v} = 3\mathbf{i} - 7\mathbf{j}, \mathbf{w} = 7\mathbf{i} + 3\mathbf{j}$

31. $\mathbf{v} = \mathbf{i} + \mathbf{j}, \mathbf{w} = 3\mathbf{i} + 3\mathbf{j}$

32. $\mathbf{v} = -2\mathbf{i} + 3\mathbf{j}, \mathbf{w} = -4\mathbf{i} + 6\mathbf{j}$

33. $\mathbf{v} = -2\mathbf{i} - 3\mathbf{j}, \mathbf{w} = -4\mathbf{i} + 6\mathbf{j}$

34. $\mathbf{v} = -3\mathbf{i} + 4\mathbf{j}, \mathbf{w} = 6\mathbf{i} - 8\mathbf{j}$

In Exercises 35–40, determine whether the vectors $\mathbf{v}$ and $\mathbf{w}$ are parallel.

35. $\mathbf{v} = 2\mathbf{i} + 3\mathbf{j}, \mathbf{w} = \dfrac{2}{3}\mathbf{i} + \mathbf{j}$

36. $\mathbf{v} = 5\mathbf{i} - 2\mathbf{j}, \mathbf{w} = -10\mathbf{i} + 4\mathbf{j}$

37. $\mathbf{v} = \langle 3, 5 \rangle, \mathbf{w} = \langle -6, 10 \rangle$

38. $\mathbf{v} = \langle -4, -2 \rangle, \mathbf{w} = \left\langle -1, \dfrac{1}{2} \right\rangle$

39. $\mathbf{v} = \langle 6, 3 \rangle, \mathbf{w} = \langle 2, 1 \rangle$

40. $\mathbf{v} = \langle 2, 5 \rangle, \mathbf{w} = \langle -4, -10 \rangle$

In Exercises 41–48, let v and w be two vectors in the plane of magnitudes 12 and 16, respectively. The angle between v and w is 50°. Find each expression. Round answers to the nearest tenth.

41. $\|\mathbf{v} + \mathbf{w}\|$ 42. $\|\mathbf{v} - \mathbf{w}\|$

43. $\|2\mathbf{v} + \mathbf{w}\|$ 44. $\|2\mathbf{v} - \mathbf{w}\|$

45. Angle between $\mathbf{v} + \mathbf{w}$ and $\mathbf{w}$

46. Angle between $\mathbf{v} - \mathbf{w}$ and $\mathbf{v}$

47. Angle between $2\mathbf{v} + \mathbf{w}$ and $\mathbf{v}$

48. Angle between $2\mathbf{v} - \mathbf{w}$ and $\mathbf{w}$

In Exercises 49–54, determine whether the vectors $\mathbf{v}$ and $\mathbf{w}$ are orthogonal.

49. $\mathbf{v} = 2\mathbf{i} + 3\mathbf{j}, \mathbf{w} = -3\mathbf{i} + 2\mathbf{j}$

50. $\mathbf{v} = -4\mathbf{i} - 5\mathbf{j}, \mathbf{w} = 5\mathbf{i} - 4\mathbf{j}$

51. $\mathbf{v} = \langle 2, 7 \rangle, \mathbf{w} = \langle -7, -2 \rangle$

52. $\mathbf{v} = \langle -3, 4 \rangle, \mathbf{w} = \langle 4, -3 \rangle$

53. $\mathbf{v} = \langle 12, 6 \rangle, \mathbf{w} = \langle 2, -4 \rangle$

54. $\mathbf{v} = \langle -9, 6 \rangle, \mathbf{w} = \langle 4, 6 \rangle$

In Exercises 55–62, determine the scalar c so that the vectors v and w are (a) orthogonal and (b) parallel.

55. $\mathbf{v} = \langle 2, 3 \rangle, \mathbf{w} = \langle 3, c \rangle$

56. $\mathbf{v} = \langle 1, -2 \rangle, \mathbf{w} = \langle c, 4 \rangle$

57. $\mathbf{v} = \langle -4, 3 \rangle, \mathbf{w} = \langle c, -6 \rangle$

58. $\mathbf{v} = \langle -1, -3 \rangle, \mathbf{w} = \langle -2, c \rangle$

59. $\mathbf{v} = \mathbf{i} + \mathbf{j}, \mathbf{w} = 2\mathbf{i} - c\mathbf{j}$

60. $\mathbf{v} = -\mathbf{i} + 2\mathbf{j}, \mathbf{w} = c\mathbf{i} + \mathbf{j}$

61. $\mathbf{v} = c\mathbf{i} + 3\mathbf{j}, \mathbf{w} = 2\mathbf{i} - \mathbf{j}$

62. $\mathbf{v} = 2\mathbf{i} + c\mathbf{j}, \mathbf{w} = 3\mathbf{i} - 5\mathbf{j}$

In Exercises 63–70, find $\text{proj}_{\mathbf{w}}\mathbf{v}$.

63. $\mathbf{v} = \langle 3, 5 \rangle, \mathbf{w} = \langle 4, 1 \rangle$

64. $\mathbf{v} = \langle 3, 5 \rangle, \mathbf{w} = \langle -4, 2 \rangle$

65. $\mathbf{v} = \langle 0, 2 \rangle, \mathbf{w} = \langle 3, 4 \rangle$

66. $\mathbf{v} = \langle 2, 0 \rangle, \mathbf{w} = \langle 4, -3 \rangle$

67. $\mathbf{v} = 5\mathbf{i} - 3\mathbf{j}, \mathbf{w} = \mathbf{i}$

68. $\mathbf{v} = 2\mathbf{i} + 7\mathbf{j}, \mathbf{w} = \mathbf{j}$

69. $\mathbf{v} = \langle a, b \rangle, \mathbf{w} = \mathbf{i} + \mathbf{j}$

70. $\mathbf{v} = \langle a, b \rangle, \mathbf{w} = \mathbf{i} - \mathbf{j}$

In Exercises 71–78, decompose v with respect to w.

71. $\mathbf{v} = \langle 2, 3 \rangle, \mathbf{w} = \mathbf{i}$

72. $\mathbf{v} = \langle 4, -3 \rangle, \mathbf{w} = \mathbf{j}$

73. $\mathbf{v} = \mathbf{i}, \mathbf{w} = \langle 1, 1 \rangle$

74. $\mathbf{v} = \mathbf{j}, \mathbf{w} = \langle 4, -3 \rangle$

75. $\mathbf{v} = \langle 4, 2 \rangle, \mathbf{w} = \langle -2, 4 \rangle$

76. $\mathbf{v} = \langle 5, 2 \rangle, \mathbf{w} = \langle 3, 5 \rangle$

77. $\mathbf{v} = \langle 4, -1 \rangle, \mathbf{w} = \langle 3, 4 \rangle$

78. $\mathbf{v} = \langle -3, 5 \rangle, \mathbf{w} = \langle 3, -4 \rangle$

Applying the Concepts

79. Finding resultant force. Two forces of 6 and 8 pounds act at an angle of 30° to each other. Find the resultant force and the angle of the resultant force to each of the two forces.

80. Finding resultant force. Repeat Exercise 79 assuming that the two forces act at an angle of 60° to each other.

81. Work. A force $\mathbf{F} = 64\mathbf{i} - 20\mathbf{j}$ (in pounds) is applied to an object. The resulting movement of the object is represented by the vector $\mathbf{d} = 10\mathbf{i} + 4\mathbf{j}$ (in feet). Find the work done by the force.

82. Work. A car is pushed 150 feet on a level street by a force of 60 pounds at an angle of 18° with the horizontal. Find the work done.

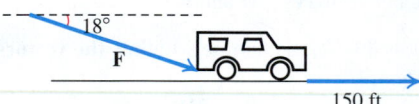

83. Inclined plane. A car is being towed up an inclined plane making an angle of 18° with the horizontal as shown in the figure. Find the force $\mathbf{F}$ needed to make the component of $\mathbf{F}$ parallel to the ground equal to 5250 pounds.

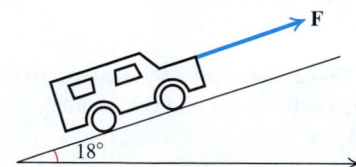

84. Inclined plane. In Exercise 83, find the work done by the force $\mathbf{F}$ to tow the car up 10 feet along the inclined plane.

85. Work. A vector $\mathbf{F}$ represents a force that has a magnitude of 10 pounds, and $\dfrac{2\pi}{3}$ is the angle for its direction. Find the work done by the force in moving an object from the origin to the point $(-6, 3)$. Distance is measured in feet.

86. Work. Two forces $\mathbf{F}_1 = 3\mathbf{i} - 2\mathbf{j}$ and $\mathbf{F}_2 = -4\mathbf{i} + 3\mathbf{j}$ act on an object and move it along a straight line from the point $(8, 4)$ to the point $(3, 5)$. Find the work done by the two forces acting together. The magnitudes of $\mathbf{F}_1$ and $\mathbf{F}_2$ are measured in pounds, and distance is measured in feet.

Beyond the Basics

87. Orthogonality on a circle. In the figure, AB is the diameter of a circle with center $O(0,0)$ and radius a. Show that $\overrightarrow{CA}$ and $\overrightarrow{CB}$ are perpendicular. [*Hint:* $x^2 + y^2 = a^2$.]

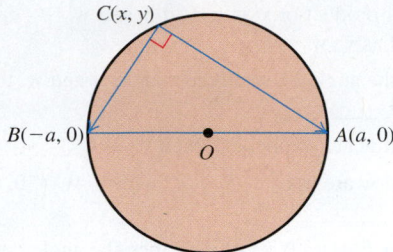

88. Cauchy-Schwarz inequality. If $\mathbf{v}$ and $\mathbf{w}$ are two vectors, then $|\mathbf{v} \cdot \mathbf{w}| \le \|\mathbf{v}\| \|\mathbf{w}\|$. When is $|\mathbf{v} \cdot \mathbf{w}| = \|\mathbf{v}\| \|\mathbf{w}\|$? [*Hint:* $\mathbf{v} \cdot \mathbf{w} = \|\mathbf{v}\| \|\mathbf{w}\| \cos\theta$.]

89. Verify that for any vectors $\mathbf{u}$, $\mathbf{v}$, and $\mathbf{w}$, $\mathbf{u} \cdot (\mathbf{v} + \mathbf{w}) = \mathbf{u} \cdot \mathbf{v} + \mathbf{u} \cdot \mathbf{w}$.

90. Prove that $(\mathbf{v} + \mathbf{w}) \cdot (\mathbf{v} + \mathbf{w}) = \|\mathbf{v}\|^2 + 2(\mathbf{v} \cdot \mathbf{w}) + \|\mathbf{w}\|^2$.

91. Triangle inequality. Use Exercises 88 and 90 to prove that $\|\mathbf{v} + \mathbf{w}\| \le \|\mathbf{v}\| + \|\mathbf{w}\|$.

92. In the decomposition $\mathbf{v} = \mathbf{v}_1 + \mathbf{v}_2$ (see page 555) with respect to $\mathbf{w}$, show that $\mathbf{v}_2 = \mathbf{v} - \text{proj}_\mathbf{w}\mathbf{v}$ is orthogonal to $\mathbf{w}$.

93. Show that the vectors $\mathbf{v} = \langle 2, -5 \rangle$ and $\mathbf{w} = \langle 5, 2 \rangle$ are perpendicular to each other.

94. Let $\mathbf{v} = 3\mathbf{i} + 7\mathbf{j}$. Find a vector perpendicular to $\mathbf{v}$.

95. Orthogonal unit vectors. Find the unit vectors orthogonal to $\mathbf{v} = 3\mathbf{i} - 4\mathbf{j}$.

96. Orthogonal unit vectors. Let $\mathbf{u}_1$ and $\mathbf{u}_2$ be orthogonal unit vectors and $\mathbf{v} = x\mathbf{u}_1 + y\mathbf{u}_2$. Find

 a. $\mathbf{v} \cdot \mathbf{u}_1$ **b.** $\mathbf{v} \cdot \mathbf{u}_2$

97. Right triangle. Use the dot product to show that ABC is a right triangle and identify the right angle assuming that $A = (-2, 1)$, $B = (3, 1)$, and $C = (-1, -1)$.

98. Parallelogram law. Show that for any two vectors $\mathbf{u}$ and $\mathbf{v}$,
$$\|\mathbf{u} + \mathbf{v}\|^2 + \|\mathbf{u} - \mathbf{v}\|^2 = 2\|\mathbf{u}\|^2 + 2\|\mathbf{v}\|^2.$$
[*Hint:* $\|\mathbf{a}\|^2 = \mathbf{a} \cdot \mathbf{a}$.]

99. Polarization identity. Show that for any two vectors $\mathbf{u}$ and $\mathbf{v}$,
$$\|\mathbf{u} + \mathbf{v}\|^2 - \|\mathbf{u} - \mathbf{v}\|^2 = 4(\mathbf{u} \cdot \mathbf{v}).$$

100. Show that $\mathbf{u}$ and $\mathbf{v}$ are orthogonal if and only if $\|\mathbf{u} + \mathbf{v}\| = \|\mathbf{u} - \mathbf{v}\|$.

101. Show that the distance d from a point $P(x_0, y_0)$ to the line $l\colon ax + by + c = 0$ is
$$d = \frac{|ax_0 + by_0 + c|}{\sqrt{a^2 + b^2}}.$$
[*Hint:* **i.** Choose two points $Q(x_1, y_1)$ and $R(x_2, y_2)$ on l. **ii** Show $\mathbf{n} = \langle a, b \rangle$ is perpendicular to l, by showing that $\mathbf{n} \cdot \overrightarrow{QR} = 0$. **iii** Show that $\|\text{proj}_\mathbf{n}\overrightarrow{QP}\| = d$.]

Critical Thinking / Discussion / Writing

102. True or False. If $\mathbf{u} \cdot \mathbf{v} = \mathbf{u} \cdot \mathbf{w}$ and $\mathbf{u} \neq 0$, then $\mathbf{v} = \mathbf{w}$. Give reasons for your answer.

103. Explain why $\dfrac{\mathbf{v} \cdot \mathbf{w}}{\|\mathbf{v}\| \|\mathbf{w}\|}$ is a number between -1 and 1 for any two nonzero vectors $\mathbf{v}$ and $\mathbf{w}$.

104. Is the angle between $-\mathbf{v}$ and $-\mathbf{w}$ the same as the angle between $\mathbf{v}$ and $\mathbf{w}$ for any two nonzero vectors $\mathbf{v}$ and $\mathbf{w}$? Explain your answer.

Getting Ready for the Next Section

In Exercises 105–112, describe the graph of each equation. Find the x- and y-intercepts of each graph.

105. $x = -1$

106. $y = 2$

107. $y = x^2 - x - 6$

108. $x = y^2 + 4y + 3$

109. $x^2 + (y - 1)^2 = 4$

110. $(x + 1)^2 + y^2 = 9$

111. $x^2 + y^2 - 2x + 4y + 2 = 0$

112. $x^2 + y^2 + 6x - 4y - 12 = 0$

In Exercises 113–116, test each equation for symmetry about the x-axis, the y-axis, and the origin.

113. $x^3 + y^2 = 5$

114. $x^3 + y^3 - 3xy^2 = 0$

115. $2x^2 - 3y^3 = 7$

116. $2x^2 + 3y^2 = 6$

Polar Coordinates

BEFORE STARTING THIS SECTION, REVIEW

1 Coordinate plane (Section 1.1, page 2)

2 Degree and radian measure of angles (Section 4.1, pages 338 and 340)

3 Trigonometric functions of angles (Section 4.2, page 357)

OBJECTIVES

1 Plot points using polar coordinates.

2 Convert points between polar and rectangular forms.

3 Convert equations between rectangular and polar forms.

4 Graph polar equations.

◆ Bionics

On large construction sites and maintenance projects, it is common to see mechanical systems that lift, dig, and scoop just like biological systems having arms, legs, and hands. The technique of solving mechanical problems with our knowledge of biological systems is called **bionics**. Bionics is used to make robotic devices for the remote-control handling of radioactive materials and the machines that are sent to the moon or to Mars to scoop soil samples, bring them into the space vehicle, perform analysis on them, and radio the results back to Earth. In Example 4, we use polar coordinates to find the reach of a robotic arm.

Up to now, we have used a rectangular coordinate system to identify points in the plane. In this section, we introduce another coordinate system, called a **polar coordinate system**, that allows us to describe many curves using rather simple equations. The polar coordinate system is believed to have been introduced by Jakob Bernoulli (1654–1705).

---

1 Plot points using polar coordinates.

Figure 6.51 Polar coordinates (r, θ) of a point.

Polar Coordinates

In a rectangular coordinate system, a point in the plane is described in terms of the horizontal and vertical directed distances from the origin. In a polar coordinate system, we draw a horizontal ray, called the **polar axis**, in the plane; its endpoint is called the **pole**. A point P in the plane is described by an ordered pair of numbers (r, θ), the **polar coordinates** of P. Figure 6.51 shows the point $P(r, \theta)$ in the polar coordinate system, where r is the "directed distance" from the pole O to the point P, and θ is a directed angle from the polar axis to the line segment OP. The angle θ may be measured in degrees or radians. As usual, positive angles are measured counterclockwise from the polar axis, and negative angles are measured clockwise from the polar axis. The pole O has polar coordinates $(0, \theta)$ for any θ. For example, $(0, 0°)$, $(0, 50°)$, $(0, -\frac{\pi}{3})$, and $(0, -60°)$ are all coordinates of the pole.

Just as graph paper with rectangular grid lines is available for plotting points in rectangular coordinates, polar graph paper is used to plot points in polar coordinates. Polar graph paper consists of concentric circles centered at the pole and having integer radii. The standard angles are marked off in multiples of $15°$ $\left(=\frac{\pi}{12} \text{ radians}\right)$.

For example, to locate a point $P = (5, 30°)$:

By hand

Draw an angle $\theta = 30°$ counterclockwise from the polar axis and then measure five units along the terminal side of θ. See Figure 6.52(a).

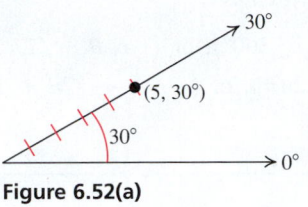

Figure 6.52(a)

By using polar graph paper

Go five units along the polar axis, then move 30° counterclockwise along the circle of radius five units. See Figure 6.52(b).

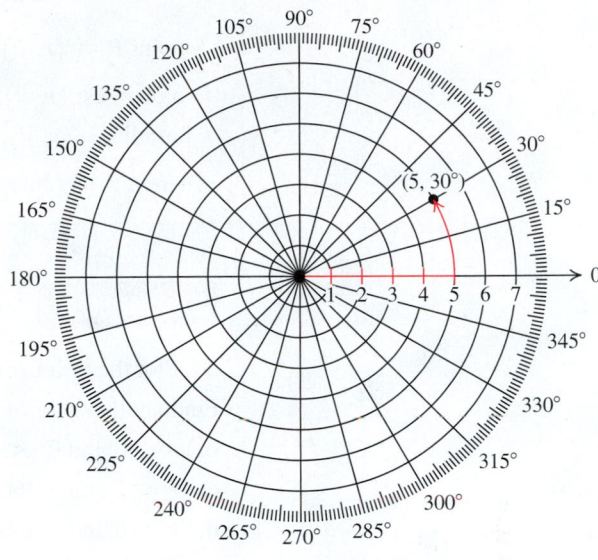

Figure 6.52(b)

Multiple Representations

Unlike rectangular coordinates, the polar coordinates of a point are not unique. For example, the polar coordinates $(3, 60°)$, $(3, 420°)$, and $(3, -300°)$ represent the same point. See Figure 6.53. In general, if a point P has polar coordinates (r, θ), then for any integer n,

$$\underbrace{(r, \theta + n \cdot 360°)}_{\theta \text{ in degrees}} \quad \text{or} \quad \underbrace{(r, \theta + 2n\pi)}_{\theta \text{ in radians}}$$

are also polar coordinates of P.

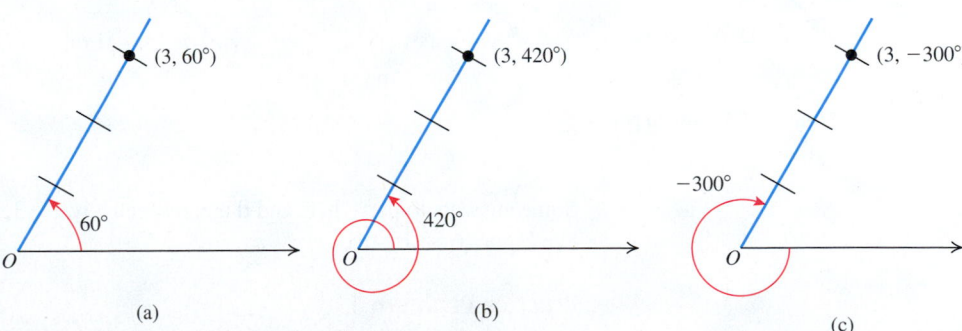

Figure 6.53 Polar coordinates of a point are not unique.

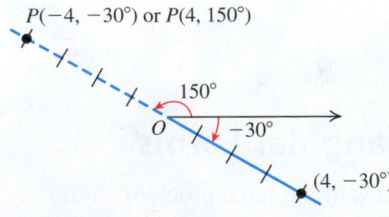

Figure 6.54 Negative r.

Sign of r

When equations are graphed in polar coordinates, it is convenient to allow r to be negative. That's why we defined r as *a directed distance*. If r is negative, then instead of the point being on the terminal side of the angle, it is $|r|$ units on the extension of the terminal side, which is the ray from the pole that extends in the direction opposite the terminal side of θ. So, the polar coordinates (r, θ) and $(-r, \pi + \theta)$ represent the same point. For example, the point $P(-4, -30°)$ shown in Figure 6.54 is the same point as $(4, 150°)$. Some other polar coordinates for the same point P are $(-4, 330°)$, $(4, -210°)$, and $(4, 510°)$.

SUMMARY: **POLAR COORDINATES**

A point P in the plane, with pole at O, can be represented by **polar coordinates** (r, θ), where

$$r = \text{directed distance from } O \text{ to } P$$
$$\theta = \text{directed angle from the polar axis to the segment } OP.$$

A point $P = (r, \theta)$ can be written as

θ in degrees: (r, θ), $(r, \theta + n \cdot 360°)$, or $(-r, \theta + (2n + 1)180°)$;

θ in radians: (r, θ), $(r, \theta + 2n\pi)$, or $(-r, \theta + (2n + 1)\pi)$

where n is any integer.

EXAMPLE 1   **Finding Different Polar Coordinates**

a. Plot the point P with polar coordinates $(3, 225°)$.

Find another pair of polar coordinates of P with the following properties:

b. $r < 0$ and $0° < \theta < 360°$

c. $r < 0$ and $-360° < \theta < 0°$

d. $r > 0$ and $-360° < \theta < 0°$

Solution

a. The point $P(3, 225°)$ is plotted by first drawing $\theta = 225°$ counterclockwise from the polar axis. Since $r = 3 > 0$, the point P is on the terminal side of the angle, three units from the pole. See Figure 6.55(a).

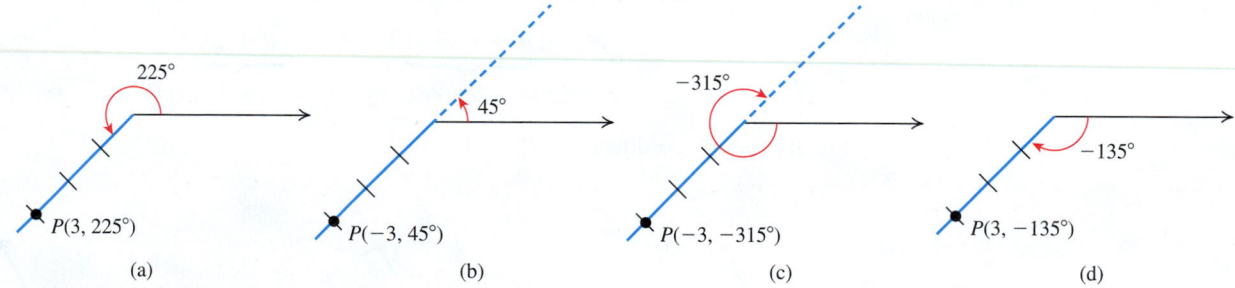

$P(3, 225°)$ $P(-3, 45°)$ $P(-3, -315°)$ $P(3, -135°)$

(a)   (b)   (c)   (d)

Figure 6.55 Plotting points.

Some answers to parts, **b**, **c**, and **d** are, respectively, $(-3, 45°)$, $(-3, -315°)$, and $(3, -135°)$. See Figure 6.55(b)–(d).

Practice Problem 1

a. Plot the point P with polar coordinates $(2, -150°)$.

Find another pair of polar coordinates of P with the following properties:

b. $r < 0$ and $0° < \theta < 360°$

c. $r > 0$ and $0° < \theta < 360°$

d. $r < 0$ and $-360° < \theta < 0°$

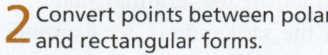

2 Convert points between polar and rectangular forms.

Converting Between Polar and Rectangular Forms

Frequently, we need to use polar and rectangular coordinates in the same problem. To do this, we let the positive x-axis of the rectangular coordinate system serve as the polar axis

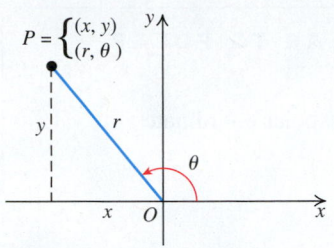

Figure 6.56 $P = (x, y)$ or $P = (r, \theta)$.

and the origin serve as the pole. Each point P then has polar coordinates (r, θ) and rectangular coordinates (x, y), as shown in Figure 6.56.

From Figure 6.56, we see the following relationships: $x^2 + y^2 = r^2$, $\sin \theta = \dfrac{y}{r}$, $\cos \theta = \dfrac{x}{r}$, and $\tan \theta = \dfrac{y}{x}$. These relationships hold whether $r > 0$ or $r < 0$, and they are used to convert a point's coordinates between rectangular and polar forms.

> ### CONVERTING FROM POLAR TO RECTANGULAR COORDINATES
>
> To convert the polar coordinates (r, θ) of a point to rectangular coordinates, use the equations
>
> $$x = r \cos \theta \qquad \text{and} \qquad y = r \sin \theta.$$

EXAMPLE 2 **Converting from Polar to Rectangular Coordinates**

Convert the polar coordinates of each point to rectangular coordinates.

a. $(2, -30°)$ **b.** $\left(-4, \dfrac{\pi}{3}\right)$

Solution

a.

$x = r \cos \theta$	$y = r \sin \theta$	Conversion equations
$x = 2 \cos(-30°)$	$y = 2 \sin(-30°)$	Replace r with 2 and θ with $-30°$.
$x = 2 \cos 30°$	$y = -2 \sin 30°$	$\cos(-\theta) = \cos \theta$, $\sin(-\theta) = -\sin \theta$
$x = 2\left(\dfrac{\sqrt{3}}{2}\right) = \sqrt{3}$	$y = -2\left(\dfrac{1}{2}\right) = -1$	$\cos 30° = \dfrac{\sqrt{3}}{2}$, $\sin 30° = \dfrac{1}{2}$

The rectangular coordinates of $(2, -30°)$ are $(\sqrt{3}, -1)$.

b.

$x = r \cos \theta$	$y = r \sin \theta$	Conversion equations
$x = -4 \cos \dfrac{\pi}{3}$	$y = -4 \sin \dfrac{\pi}{3}$	Replace r with -4 and θ with $\dfrac{\pi}{3}$.
$x = -4\left(\dfrac{1}{2}\right) = -2$	$y = -4\left(\dfrac{\sqrt{3}}{2}\right) = -2\sqrt{3}$	$\cos \dfrac{\pi}{3} = \dfrac{1}{2}$, $\sin \dfrac{\pi}{3} = \dfrac{\sqrt{3}}{2}$

The rectangular coordinates of $\left(-4, \dfrac{\pi}{3}\right)$ are $(-2, -2\sqrt{3})$.

Practice Problem 2 Convert the polar coordinates of each point to rectangular coordinates.

a. $(-3, 60°)$ **b.** $\left(2, -\dfrac{\pi}{4}\right)$

When converting from rectangular coordinates (x, y) of a point to polar coordinates (r, θ), remember that infinitely many representations are possible. We will find the polar coordinates (r, θ) of the point (x, y) that have $r > 0$ and $0° \leq \theta < 360°$ (or $0 \leq \theta < 2\pi$).

TECHNOLOGY CONNECTION

A graphing calculator can be used to check conversions from polar to rectangular coordinates. Choose the appropriate options from the ANGLE menu. The x- and y-coordinates must be calculated separately. The screen shown here is from a calculator set in **Degree** mode.

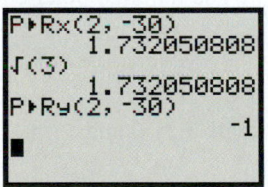

> ### CONVERTING FROM RECTANGULAR TO POLAR COORDINATES
>
> To convert the rectangular coordinates (x, y) of a point to polar coordinates:
>
> 1. Find the quadrant in which the given point (x, y) lies.
> 2. Use $r = \sqrt{x^2 + y^2}$ to find r.
> 3. Find the reference angle θ' by using $\theta' = \tan^{-1}\left|\dfrac{y}{x}\right|$ and then find θ so that it lies in the same quadrant as the point (x, y).

EXAMPLE 3   Converting from Rectangular to Polar Coordinates

Find polar coordinates (r, θ) of the point P with $r > 0$ and $0 \le \theta < 2\pi$, whose rectangular coordinates are $(x, y) = (-2, 2\sqrt{3})$.

Solution

1. The point $P(-2, 2\sqrt{3})$ lies in quadrant II with $x = -2$ and $y = 2\sqrt{3}$.

2. $\begin{aligned} r &= \sqrt{x^2 + y^2} & &\text{Formula for computing } r \\ &= \sqrt{(-2)^2 + (2\sqrt{3}\,)^2} & &\text{Substitute values of } x \text{ and } y. \\ &= \sqrt{4 + 12} = \sqrt{16} = 4 & &\text{Simplify.} \end{aligned}$

3. $\theta' = \tan^{-1}\left|\dfrac{y}{x}\right|$ Formula for computing θ'

$$\theta' = \tan^{-1}\left|\frac{2\sqrt{3}}{-2}\right| = \tan^{-1}\sqrt{3} = \frac{\pi}{3} \qquad \text{Substitute values and simplify.}$$

The point P is in quadrant II, so $\theta = \pi - \theta' = \pi - \dfrac{\pi}{3} = \dfrac{2\pi}{3}$. The required polar coordinates are $\left(4, \dfrac{2\pi}{3}\right)$.

Practice Problem 3 Find the polar coordinates of the point P with $r > 0$ and $0 \le \theta < 2\pi$, whose rectangular coordinates are $(x, y) = (-1, -1)$.

EXAMPLE 4   Positioning a Robotic Hand

For the robotic arm of Figure 6.57, find the polar coordinates of the hand relative to the shoulder. Round all answers to the nearest tenth.

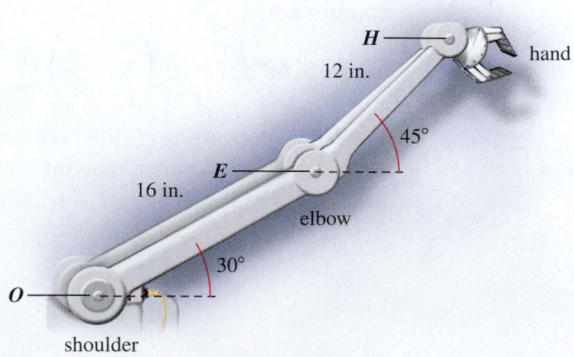

Figure 6.57 Robotic arm.

Solution

We need to find the polar coordinates (r, θ) of the hand H, with the shoulder O as the pole. Since segment $\overline{OE}$ measures 16 inches and makes an angle of 30° with the horizontal, we can write the vector $\overrightarrow{OE}$ in terms of its components:

$$\overrightarrow{OE} = \langle 16 \cos 30°, 16 \sin 30° \rangle$$

Similarly,

$$\overrightarrow{EH} = \langle 12 \cos 45°, 12 \sin 45° \rangle$$

We know that

$$\begin{aligned}\overrightarrow{OH} &= \overrightarrow{OE} + \overrightarrow{EH} &&\text{Resultant vector}\\ &= \langle 16 \cos 30°, 16 \sin 30° \rangle + \langle 12 \cos 45°, 12 \sin 45° \rangle\\ &= \langle 16 \cos 30° + 12 \cos 45°, 16 \sin 30° + 12 \sin 45° \rangle &&\text{Vector addition}\end{aligned}$$

So the rectangular coordinates (x, y) of H are

$$x = 16 \cos 30° + 12 \cos 45°, \quad y = 16 \sin 30° + 12 \sin 45°.$$

We now convert the coordinates (x, y) of H to polar coordinates (r, θ).

$$\begin{aligned}r &= \sqrt{x^2 + y^2} &&\text{Formula for } r\\ &= \sqrt{(16 \cos 30° + 12 \cos 45°)^2 + (16 \sin 30° + 12 \sin 45°)^2} &&\text{Substitute values for } x \text{ and } y.\\ &\approx 27.8 \text{ inches} &&\text{Use a calculator.}\\ \theta = \theta' &= \tan^{-1}\left(\frac{y}{x}\right) &&\text{Definition of inverse tangent}\\ &= \tan^{-1}\left(\frac{16 \sin 30° + 12 \sin 45°}{16 \cos 30° + 12 \cos 45°}\right) \approx 36.4° &&\text{Substitute values and use a calculator.}\end{aligned}$$

The polar coordinates of H relative to the pole O are $(r, \theta) = (27.8, 36.4°)$.

Practice Problem 4 Repeat Example 4 with $\overline{OE} = 15$ inches and $\overline{EH} = 10$ inches.

3 Convert equations between rectangular and polar forms.

Converting Equations Between Rectangular and Polar Forms

An equation that has the rectangular coordinates x and y as variables is called a **rectangular** (or **Cartesian**) **equation**. Similarly, an equation where the polar coordinates r and θ are the variables is called a **polar equation**. Some examples of polar equations are

$$r = \sin \theta, \quad r = 1 + \cos \theta, \quad \text{and} \quad r = \theta.$$

To convert a rectangular equation to a polar equation, we simply replace x with $r \cos \theta$ and y with $r \sin \theta$ and then simplify where possible.

EXAMPLE 5 Converting an Equation from Rectangular to Polar Form

Convert the equation $(x - 1)^2 + (y + 1)^2 = 2$ to polar form.

Solution

$$\begin{aligned}(x - 1)^2 + (y + 1)^2 &= 2 &&\text{Given equation}\\ (r \cos \theta - 1)^2 + (r \sin \theta + 1)^2 &= 2 &&x = r \cos \theta, y = r \sin \theta\\ r^2 \cos^2 \theta - 2r \cos \theta + 1 + r^2 \sin^2 \theta + 2r \sin \theta + 1 &= 2 &&(a \pm b)^2 = a^2 \pm 2ab + b^2\\ r^2(\cos^2 \theta + \sin^2 \theta) - 2r \cos \theta + 2r \sin \theta &= 0 &&\text{Factor out } r^2 \text{ and simplify.}\\ r^2 - 2r \cos \theta + 2r \sin \theta &= 0 &&\cos^2 \theta + \sin^2 \theta = 1\\ r(r - 2 \cos \theta + 2 \sin \theta) &= 0 &&\text{Factor out } r.\\ r = 0 \text{ or } r - 2 \cos \theta + 2 \sin \theta &= 0 &&\text{Set each factor to 0.}\\ r &= 2 \cos \theta - 2 \sin \theta. &&\text{This factor includes } r = 0.\end{aligned}$$

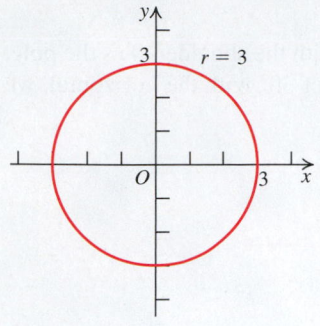

Figure 6.58 $x^2 + y^2 = 9$ or $r = 3$.

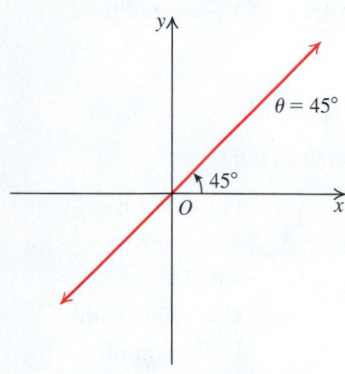

Figure 6.59 $y = x$ or $\theta = 45°$.

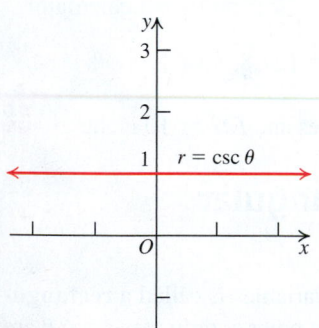

Figure 6.60 $y = 1$ or $r \sin \theta = 1$.

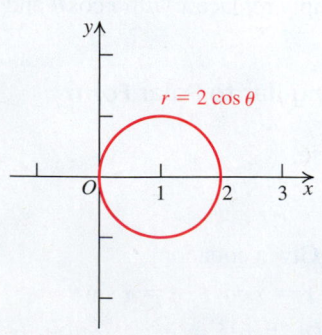

Figure 6.61 $(x - 1)^2 + y^2 = 1$ or $r = 2 \cos \theta$.

The graph of $r = 0$ (pole) is included in the graph of $r = 2 \cos \theta - 2 \sin \theta$ (because $r = 0$ for $\theta = \frac{\pi}{4}$), so it is not necessary to include the equation $r = 0$.

The equation $r = 2 \cos \theta - 2 \sin \theta$ is the polar equation of the given rectangular equation.

Practice Problem 5 Convert the equation $(x + 2)^2 + (y - 2)^2 = 4$ to polar form. ■

Converting an equation from polar to rectangular form will frequently require some ingenuity in order to use the substitutions:

$$r^2 = x^2 + y^2, \quad r \cos \theta = x, \quad r \sin \theta = y, \quad \text{and} \quad \tan \theta = \frac{y}{x}.$$

EXAMPLE 6 Converting an Equation from Polar to Rectangular Form

Convert each polar equation to a rectangular equation and identify its graph.

a. $r = 3$ **b.** $\theta = 45°$ **c.** $r = \csc \theta$ **d.** $r = 2 \cos \theta$

Solution

a.

$r = 3$	Given polar equation
$r^2 = 9$	Square both sides.
$x^2 + y^2 = 9$	Replace r^2 with $x^2 + y^2$

The graph of $r = 3$ (or $x^2 + y^2 = 3^2$) is a circle with center at the pole and radius $= 3$ units. See Figure 6.58.

b.

$\theta = 45°$	Given polar equation
$\tan \theta = \tan 45°$	If $x = y$, then $\tan x = \tan y$.
$\dfrac{y}{x} = 1$	$\tan \theta = \dfrac{y}{x}$ and $\tan 45° = 1$
$y = x$	Multiply both sides by x.

The graph of $\theta = 45°$ (or $y = x$) is a line through the origin, making an angle of $45°$ with the positive x-axis. See Figure 6.59.

c.

$r = \csc \theta$	Given polar equation
$r = \dfrac{1}{\sin \theta}$	Reciprocal identity
$r \sin \theta = 1$	Multiply both sides by $\sin \theta$.
$y = 1$	Replace $r \sin \theta$ with y.

The graph of $r = \csc \theta$ (or $y = 1$) is a horizontal line. See Figure 6.60.

d.

$r = 2 \cos \theta$	Given polar equation
$r^2 = 2r \cos \theta$	Multiply both sides by r.
$x^2 + y^2 = 2x$	Replace r^2 with $x^2 + y^2$ and $r \cos \theta$ with x.
$x^2 - 2x + y^2 = 0$	Subtract $2x$ from both sides and simplify.
$(x^2 - 2x + 1) + y^2 = 1$	Add 1 to both sides to complete the square.
$(x - 1)^2 + y^2 = 1$	Factor perfect square trinomial.

The graph of $r = 2 \cos \theta$ (or $(x - 1)^2 + y^2 = 1$) is a circle with center at $(1, 0)$ and radius 1 unit. See Figure 6.61.

Practice Problem 6 Convert each polar equation to a rectangular equation and identify its graph.

a. $r = 5$ **b.** $\theta = -\dfrac{\pi}{3}$ **c.** $r = \sec\theta$ **d.** $r = 2\sin\theta$

4 Graph polar equations.

The Graph of a Polar Equation

Many important polar equations do not convert "nicely" to rectangular equations. To graph these equations, we plot points in polar coordinates. The graph of a polar equation

$$r = f(\theta)$$

is the set of all points $P(r, \theta)$ that have at least one polar coordinate representation that satisfies the equation. We make a table of several ordered pair solutions (r, θ) of the equation, plot these points, and join them with a smooth curve.

EXAMPLE 7   **Sketching a Graph by Plotting Points**

Sketch the graph of the polar equation $r = \theta \ (\theta \geq 0)$ by plotting points.

Solution

We choose values of θ that are integer multiples of $\dfrac{\pi}{2}$. Since $r = \theta$, the polar coordinates of points on the curve are as follows:

$$(0, 0), \left(\frac{\pi}{2}, \frac{\pi}{2}\right), (\pi, \pi), \left(\frac{3\pi}{2}, \frac{3\pi}{2}\right), (2\pi, 2\pi), \left(\frac{5\pi}{2}, \frac{5\pi}{2}\right), \ldots$$

Plot these points on a polar grid and join them in order of increasing θ to obtain the graph of $r = \theta$ shown in Figure 6.62. This curve is called the **spiral of Archimedes**.

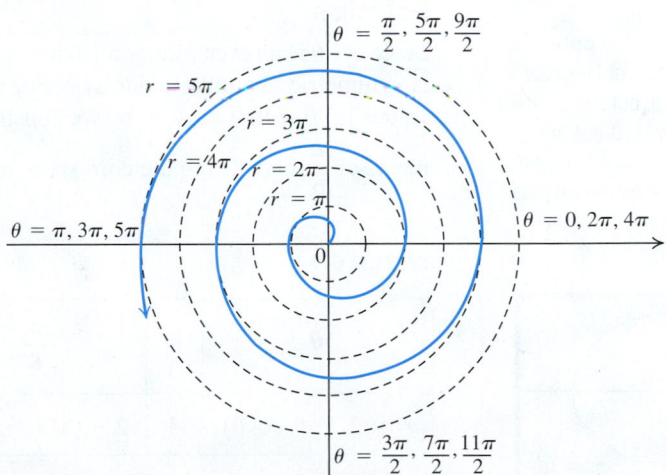

Figure 6.62 Spiral of Archimedes $r = \theta$.

Practice Problem 7 Sketch the graph of the polar equation $r = 2\theta \ (\theta \leq 0)$ by plotting points.

As with Cartesian equations, it is helpful to determine symmetry in the graph of a polar equation. If the graph of a polar equation has symmetry, then we can sketch its graph by plotting fewer points.

TESTS FOR SYMMETRY IN POLAR COORDINATES

The graph of a polar equation has the symmetry described below if an equivalent equation results when making the replacements shown.

Symmetry with respect to the polar axis	**Symmetry with respect to the line $\theta = \dfrac{\pi}{2}$**	**Symmetry with respect to the pole**
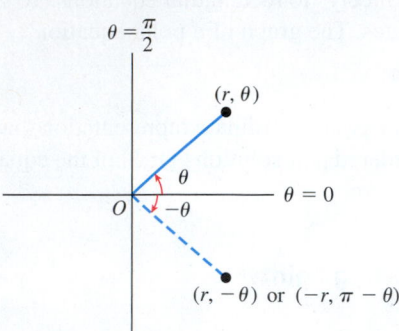	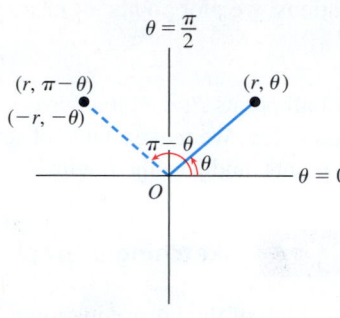	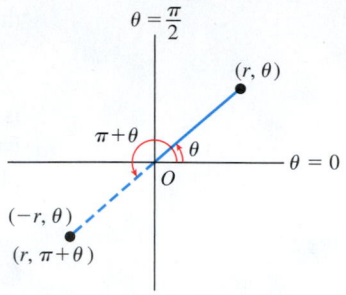
Replace (r, θ) with $(r, -\theta)$ or $(-r, \pi - \theta)$.	Replace (r, θ) with $(r, \pi - \theta)$ or $(-r, -\theta)$.	Replace (r, θ) with $(r, \pi + \theta)$ or $(-r, \theta)$.

Note that if a polar equation has two of these symmetries, then it must have the third symmetry.

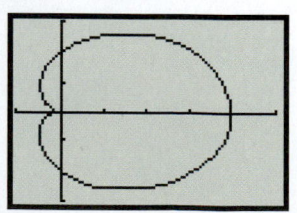
EXAMPLE 8   **Sketching the Graph of a Polar Equation**

Sketch the graph of the polar equation $r = 2(1 + \cos\theta)$.

Solution

Since $\cos\theta$ is an even function (that is, $\cos(-\theta) = \cos\theta$), the graph of $r = 2(1 + \cos\theta)$ is symmetric about the polar axis. So, to graph this equation, it is sufficient to compute values for θ with $0 \leq \theta \leq \pi$. We construct Table 6.1 by substituting values for θ at increments of $\dfrac{\pi}{6}$, calculating the corresponding values of r.

TABLE 6.1

θ	0	$\dfrac{\pi}{6}$	$\dfrac{\pi}{3}$	$\dfrac{\pi}{2}$	$\dfrac{2\pi}{3}$	$\dfrac{5\pi}{6}$	π
$r = 2(1 + \cos\theta)$	4	$2 + \sqrt{3} \approx 3.73$	3	2	1	$2 - \sqrt{3} \approx 0.27$	0

We plot the points (r, θ) and draw a smooth curve through them. Since the graph is symmetric about the polar axis, we also reflect the curve about the polar axis.

The graph of the equation $r = 2(1 + \cos\theta)$ is shown in Figure 6.63. This type of curve is called a *cardioid* because it resembles a heart.

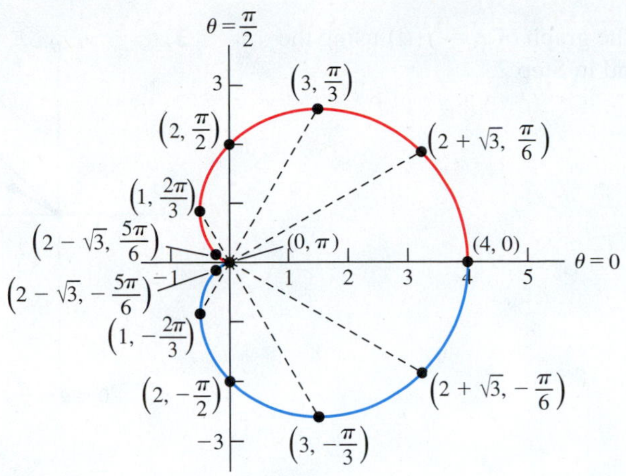

Figure 6.63 $r = 2(1 + \cos\theta)$.

Practice Problem 8 Sketch the graph of the polar equation $r = 2(1 + \sin\theta)$.

PROCEDURE
IN ACTION

EXAMPLE 9 Graphing a Polar Equation

OBJECTIVE

Sketch the graph of a polar equation $r = f(\theta)$, where f is a periodic function.

Step 1 Test for symmetry.

EXAMPLE

Sketch the graph of $r = \cos 2\theta$.

1. Replace θ with $-\theta$,

$$\cos(-2\theta) = \cos 2\theta \tag{1}$$

Replace θ with $\pi - \theta$,

$$\cos[2(\pi - \theta)] = \cos(2\pi - 2\theta) = \cos 2\theta \tag{2}$$

From (1) and (2), we conclude that the graph is symmetric about the polar axis and about the line $\theta = \dfrac{\pi}{2}$. It is also symmetric about the pole.

Step 2 Analyze the behavior of r, symmetries from Step 1, and then find selected points.

2. To get a sense of how r behaves for different values of θ, we make the following table:

As θ goes from:	r varies from:
0 to $\dfrac{\pi}{4}$	1 to 0
$\dfrac{\pi}{4}$ to $\dfrac{\pi}{2}$	0 to -1
$\dfrac{\pi}{2}$ to $\dfrac{3\pi}{4}$	-1 to 0
$\dfrac{3\pi}{4}$ to π	0 to 1

Because the curve is symmetric about $\theta = 0$ and $\theta = \dfrac{\pi}{2}$, we find selected points for $0 \le \theta \le \dfrac{\pi}{2}$:

θ	0	$\dfrac{\pi}{6}$	$\dfrac{\pi}{4}$	$\dfrac{\pi}{3}$	$\dfrac{\pi}{2}$
$r = \cos 2\theta$	1	$\dfrac{1}{2}$	0	$-\dfrac{1}{2}$	-1

(continued)

Step 3 Sketch the graph of $r = f(\theta)$ using the points (r, θ) found in Step 2.

3.

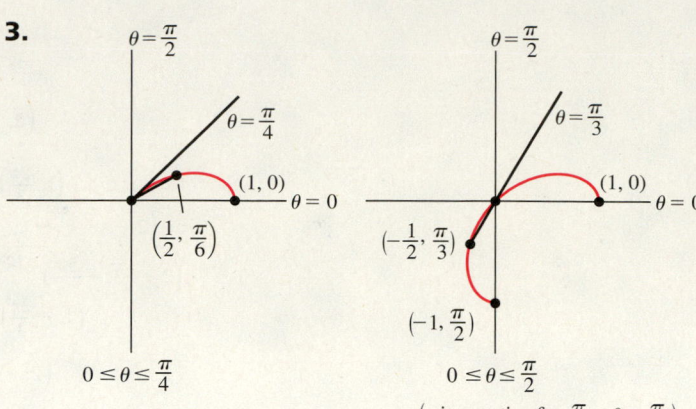

Step 4 Use symmetries to complete the graph.

4.

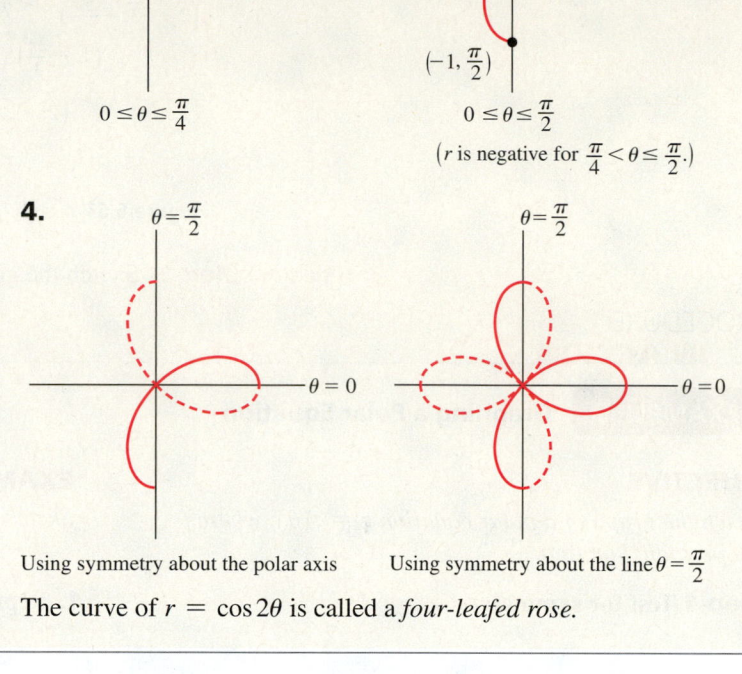

Using symmetry about the polar axis   Using symmetry about the line $\theta = \frac{\pi}{2}$

The curve of $r = \cos 2\theta$ is called a *four-leafed rose*.

Practice Problem 9 Sketch the graph of $r = \sin 2\theta$.

If a curve $r = f(\theta)$ is symmetric about the line $\theta = \frac{\pi}{2}$ (y-axis), you can first sketch the portion of the graph for $-\frac{\pi}{2} \le \theta \le \frac{\pi}{2}$ or for $0 \le \theta \le \frac{\pi}{2}$ and $\pi \le \theta \le \frac{3\pi}{2}$, then complete the graph by symmetry.

EXAMPLE 10   **Graphing a Polar Equation**

Sketch a graph of $r = 1 + 2\sin\theta$.

Solution

Step 1   **Symmetry.** Because $\sin(\pi - \theta) = \sin\theta$, replacing (r, θ) with $(r, \pi - \theta)$, we obtain an equivalent equation. The graph is symmetric with respect to the line $\theta = \frac{\pi}{2}$.

Step 2   **Analyze and make a table of values.** We first analyze the behavior of r:

As θ goes from $-\frac{\pi}{2}$ to $-\frac{\pi}{6}$, r varies from -1 to 0.

As θ goes from $-\frac{\pi}{6}$ to 0, r varies from 0 to 1.

As θ goes from 0 to $\frac{\pi}{2}$, r varies from 1 to 3.

As θ goes from $\frac{\pi}{2}$ to π, r varies from 3 to 1.

So it is sufficient to first sketch the graph for $-\frac{\pi}{2} \le \theta \le \frac{\pi}{2}$. Table 6.2 gives the coordinates of some points on the graph.

TABLE 6.2

θ	$-\dfrac{\pi}{2}$	$-\dfrac{\pi}{3}$	$-\dfrac{\pi}{4}$	$-\dfrac{\pi}{6}$	0	$\dfrac{\pi}{6}$	$\dfrac{\pi}{4}$	$\dfrac{\pi}{3}$	$\dfrac{\pi}{2}$
r	-1	$1-\sqrt{3}$	$1-\sqrt{2}$	0	1	2	$1+\sqrt{2}$	$1+\sqrt{3}$	3

Step 3 Sketch the graph for $-\dfrac{\pi}{2} \le \theta \le \dfrac{\pi}{2}$ using the points (r, θ) found in Step 2. See Figure 6.64.

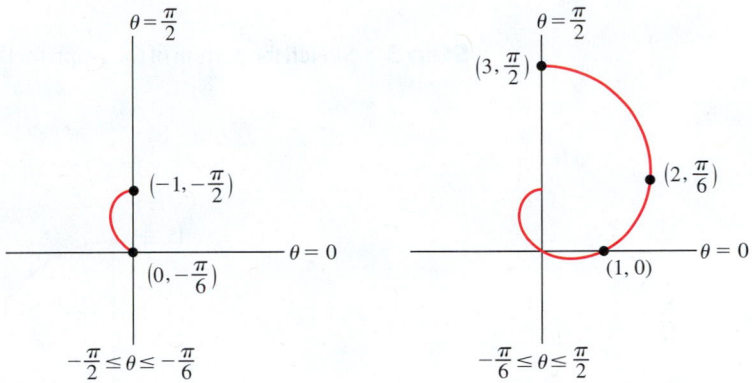

Figure 6.64 Graph of $r = 1 + 2\sin\theta$ for $-\dfrac{\pi}{2} \le \theta \le \dfrac{\pi}{2}$.

Step 4 Complete the graph by symmetry with respect to the line $\theta = \dfrac{\pi}{2}$. The sketch is shown in Figure 6.65. This curve is called a *limaçon* (French for "snail").

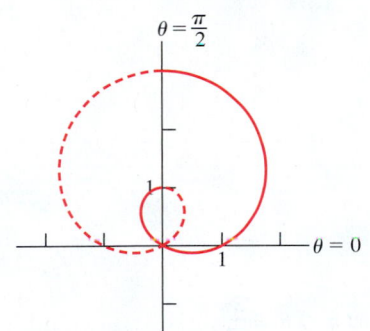

Figure 6.65 Graph of $r = 1 + 2\sin\theta$.

Practice Problem 10 Sketch a graph of $r = 1 + 2\cos\theta$.

The graph of an equation of the form

$$r = a \pm b\cos\theta \qquad \text{or} \qquad r = a \pm b\sin\theta$$

is called a **limaçon of Pascal**. If $b > a$, as in Example 10, the limaçon has a loop. If $b = a$, the limaçon is a **cardioid**, a heart-shaped curve. The curve in Example 8 is a cardioid. If $b < a$, the limaçon has a shape similar to the one shown in Example 11.

EXAMPLE 11 Graphing a Limaçon

Sketch a graph of $r = 3 + 2\sin\theta$.

Solution

Step 1 The graph is symmetric with respect to the line $\theta = \dfrac{\pi}{2}$ because if (r, θ) is replaced with $(r, \pi - \theta)$, an equivalent equation is obtained.

Step 2 Analyze the behavior of $r = 3 + 2\sin\theta$. As θ goes from 0 to $\dfrac{\pi}{2}$, r increases from 3 to 5; as θ goes from $\dfrac{\pi}{2}$ to π, r decreases from 5 to 3; as θ goes from π to $\dfrac{3\pi}{2}$, r decreases from 3 to 1; and as θ goes from $\dfrac{3\pi}{2}$ to 2π, r increases from 1 to 3. Because of symmetry about the line $\theta = \dfrac{\pi}{2}$, we find selected points for

SIDE
NOTE

The symmetry for the curve

$r = 3 + 2\sin\theta$ about $\theta = \dfrac{\pi}{2}$

cannot be decided from the failure of the test: replace (r, θ) with $(-r, -\theta)$.

$0 \leq \theta \leq \dfrac{\pi}{2}$ and $\pi \leq \theta \leq \dfrac{3\pi}{2}$ (see Side Note on page 571). Table 6.3 gives polar coordinates of some points in these intervals on the graph.

TABLE 6.3

θ	0	$\dfrac{\pi}{6}$	$\dfrac{\pi}{3}$	$\dfrac{\pi}{2}$	π	$\dfrac{7\pi}{6}$	$\dfrac{4\pi}{3}$	$\dfrac{3\pi}{2}$
r	3	4	$3 + \sqrt{3}$	5	3	2	$3 - \sqrt{3}$	1

Step 3 Sketch the portion of the graph for $0 \leq \theta \leq \dfrac{\pi}{2}$ and $\pi \leq \theta \leq \dfrac{3\pi}{2}$. See Figure 6.66.

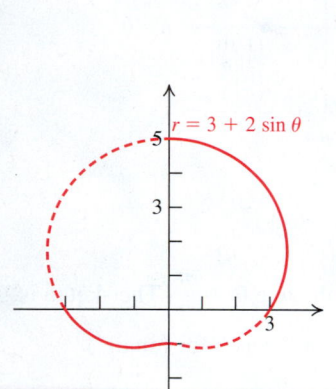

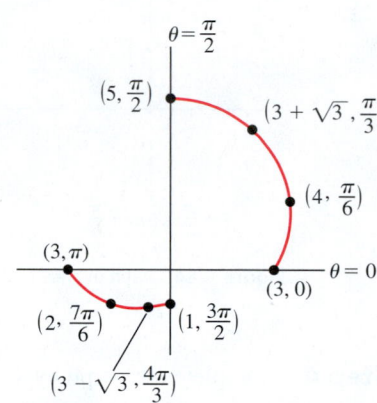

Figure 6.66 $r = 3 + 2\sin\theta$; $0 \leq \theta \leq \dfrac{\pi}{2}$ and $\pi \leq \theta \leq \dfrac{3\pi}{2}$.

Figure 6.67 Graph of $r = a + b\sin\theta$ with $a > b$.

Step 4 Complete the graph by using symmetry about the line $\theta = \dfrac{\pi}{2}$. See Figure 6.67.

Practice Problem 11 Sketch a graph of $r = 2 - \cos\theta$.

Note that even if a polar equation fails a test for a given symmetry, its graph nevertheless may exhibit that symmetry. See the margin Side Note on page 571. In other words, the tests for polar coordinate symmetry are *sufficient* but not *necessary* for showing symmetry. The following graphs have simple polar equations.

LIMAÇONS

The graphs of
$r = a \pm b\cos\theta$,
$r = a \pm b\sin\theta$
$(a > 0, b > 0)$
are called **limaçons**.

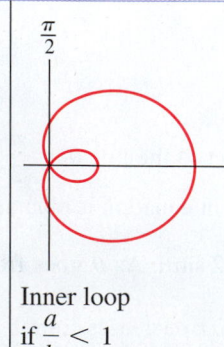

Inner loop
if $\dfrac{a}{b} < 1$

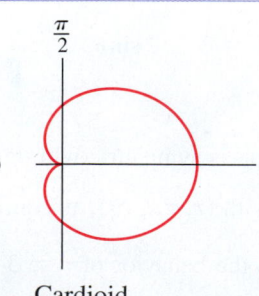

Cardioid
if $\dfrac{a}{b} = 1$

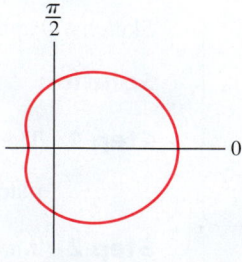

Dimpled
if $1 < \dfrac{a}{b} < 2$

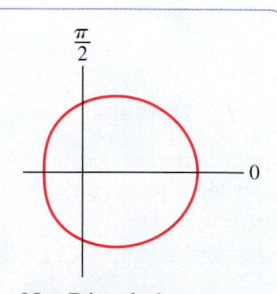

Not Dimpled
if $\dfrac{a}{b} \geq 2$

ROSE CURVES

The graphs of
$r = a \cos n\theta$,
$r = a \sin n\theta$
$(a > 0)$ are called
rose curves.
If n is odd, the rose
has n petals. If n is
even, the rose has
$2n$ petals.

$r = a \cos 3\theta$

$r = a \sin 5\theta$

$r = a \sin 2\theta$

$r = a \cos 4\theta$

CIRCLES AND LEMNISCATES

$(a > 0)$

$r = a \cos \theta$
Circle

$r = a \sin \theta$
Circle

$r^2 = a^2 \sin 2\theta$
Lemniscate

$r^2 = a^2 \cos 2\theta$
Lemniscate

SPIRALS

$(a > 0)$

$r = a\theta, \theta \geq 0$

$r = a^{k\theta}, \theta \geq 0, a \neq 1, k > 0$

Answers to Practice Problems

1. a.

    $P(2, -150°)$

    b. $(-2, 30°)$

c. $(2, 210°)$ **d.** $(-2, -330°)$

2. a. $\left(-\dfrac{3}{2}, -\dfrac{3\sqrt{3}}{2}\right)$ **b.** $(\sqrt{2}, -\sqrt{2})$ **3.** $\left(\sqrt{2}, \dfrac{5\pi}{4}\right)$

4. $(24.8, 36.0°)$ **5.** $r^2 + 4r \cos\theta - 4r \sin\theta + 4 = 0$

6. a. $x^2 + y^2 = 25$; a circle centered at the origin with radius 5

b. $y = -x\sqrt{3}$; a line through the origin with slope $-\sqrt{3}$

c. $x = 1$; a vertical line through $x = 1$

d. $x^2 + (y - 1)^2 = 1$; a circle with center $(0, 1)$ and radius 1

7.

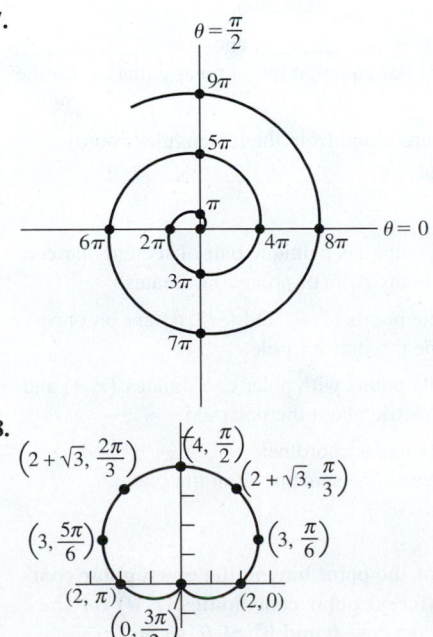

8.

$\left(2 + \sqrt{3}, \dfrac{2\pi}{3}\right)$ $\left(4, \dfrac{\pi}{2}\right)$ $\left(2 + \sqrt{3}, \dfrac{\pi}{3}\right)$

$\left(3, \dfrac{5\pi}{6}\right)$ $\left(3, \dfrac{\pi}{6}\right)$

$(2, \pi)$ $(2, 0)$

$\left(0, \dfrac{3\pi}{2}\right)$

(continued)

9.

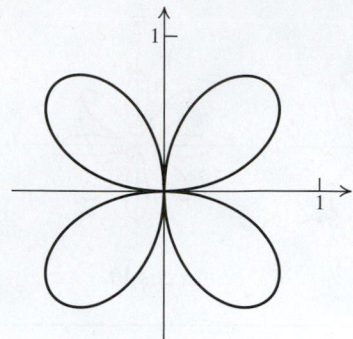

10.

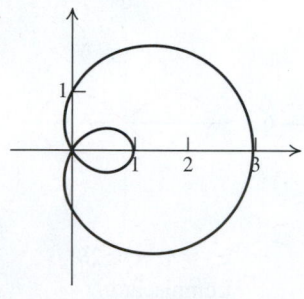

11.

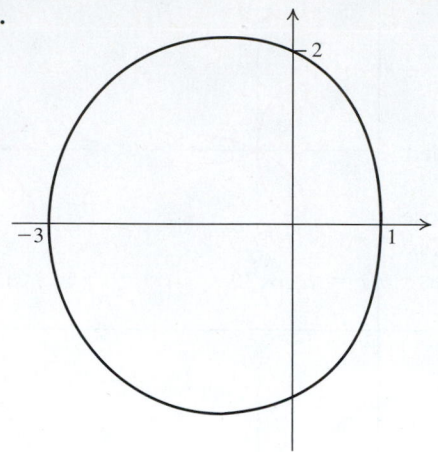

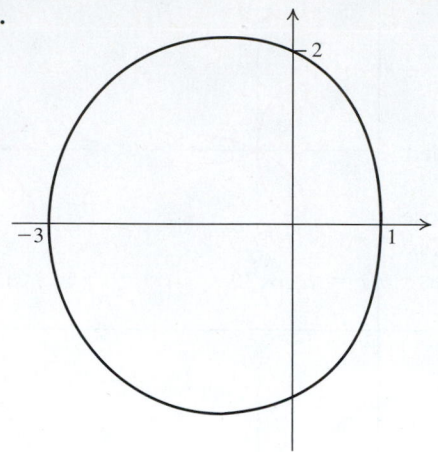

SECTION 6.6   Exercises

Concepts and Vocabulary

1. The polar coordinates (r, θ) of a point are the directed distance r to the _____ and the directed angle θ from the polar _____.

2. The positive value of r corresponds to a distance r on the terminal side of θ, and a negative value corresponds to a distance in the _____ direction.

3. The substitutions $x =$ _____ and $y =$ _____ transform a rectangular equation into a polar equation for the same curve.

4. Polar coordinates are found from the rectangular coordinates by the formulas $r =$ _____ and $\theta =$ _____.

5. **True or False.** A point has a unique pair of rectangular coordinates, but it has many pairs of polar coordinates.

6. **True or False.** The points $(2, \theta)$ and $(-2, \theta)$ are on opposite sides of the line through the pole.

7. **True or False.** The points with polar coordinates (r, θ) and $(-r, -\theta)$ are symmetric about the polar axis.

8. **True or False.** The polar coordinates $(r, \pi - \theta)$ and $(-r, \pi - \theta)$ represent the same point in the plane.

Building Skills

In Exercises 9–16, plot the point having the given polar coordinates. Then find different polar coordinates (r, θ) for the same point for which (a) $r < 0$ and $0° \le \theta < 360°$; (b) $r < 0$ and $-360° < \theta < 0°$; (c) $r > 0$ and $-360° < \theta < 0°$.

9. $(4, 45°)$
10. $(5, 150°)$
11. $(3, 90°)$
12. $(2, 240°)$
13. $(3, 60°)$
14. $(4, 210°)$
15. $(6, 300°)$
16. $(2, 270°)$

In Exercises 17–24, plot the point having the given polar coordinates. Then give two different pairs of polar coordinates of the same point, (a) one with the given value of r and (b) one with r having the opposite sign of the given value of r.

17. $\left(2, -\dfrac{\pi}{3}\right)$
18. $\left(\sqrt{2}, -\dfrac{\pi}{4}\right)$
19. $\left(-3, \dfrac{\pi}{6}\right)$
20. $\left(-2, \dfrac{3\pi}{4}\right)$
21. $\left(-2, -\dfrac{\pi}{6}\right)$
22. $(-2, -\pi)$
23. $\left(4, \dfrac{7\pi}{6}\right)$
24. $(2, 3)$

In Exercises 25–32, convert the given polar coordinates of each point to rectangular coordinates.

25. $(3, 60°)$
26. $(-2, -30°)$
27. $(5, -60°)$
28. $(-3, 90°)$
29. $(3, \pi)$
30. $\left(\sqrt{2}, -\dfrac{\pi}{4}\right)$
31. $\left(-2, -\dfrac{5\pi}{6}\right)$
32. $\left(-1, \dfrac{7\pi}{6}\right)$

In Exercises 33–40, convert the rectangular coordinates of each point to polar coordinates (r, θ) with $r > 0$ and $0 \leq \theta < 2\pi$.

33. $(1, -1)$ **34.** $(-\sqrt{3}, 1)$

35. $(3, 3)$ **36.** $(-4, 0)$

37. $(3, -3)$ **38.** $(2\sqrt{3}, 2)$

39. $(-1, \sqrt{3})$ **40.** $(-2, -2\sqrt{3})$

In Exercises 41–52, convert each rectangular equation to polar form.

41. $x^2 + y^2 = 16$ **42.** $x = 4$

43. $y = -2$ **44.** $x + y = 1$

45. $y^2 = 4x$ **46.** $x^2 = 2y$

47. $x^3 = 3y^2$ **48.** $y^3 = 2x^2$

49. $y^2 = 6y - x^2$

50. $x^2 - y^2 = 1$

51. $x^2 + y^2 - 4x + 6y = 12$

52. $(x + 1)^2 + (y - 2)^2 = 9$

In Exercises 53–62, convert each polar equation to rectangular form. Identify each curve.

53. $r = 2$ **54.** $r = -3$

55. $\theta = \dfrac{3\pi}{4}$ **56.** $\theta = -\dfrac{\pi}{6}$

57. $r \cos \theta = -2$ **58.** $r \sin \theta = 3$

59. $r = 4 \cos \theta$ **60.** $r = 4 \sin \theta$

61. $r = -2 \sin \theta$ **62.** $r = -3 \cos \theta$

In Exercises 63–70, sketch the graph of each polar equation by transforming it to rectangular coordinates.

63. $r = -2$ **64.** $r \cos \theta = -1$

65. $r \sin \theta = 2$ **66.** $r = \sin \theta$

67. $r = 2 \sec \theta$ **68.** $r + 3 \cos \theta = 0$

69. $r = \dfrac{1}{\cos \theta + \sin \theta}$ **70.** $r = \dfrac{6}{2 \cos \theta + 3 \sin \theta}$

In Exercises 71–86, sketch the graph of each polar equation. Identify the curve.

71. $r = \sin \theta$ **72.** $r = \cos \theta$

73. $r = 1 - \cos \theta$ **74.** $r = 1 + \sin \theta$

75. $r = \cos 3\theta$ **76.** $r = \sin 3\theta$

77. $r = \sin 4\theta$ **78.** $r = \cos 4\theta$

79. $r = 1 - 2 \sin \theta$ **80.** $r = 2 - 4 \sin \theta$

81. $r = 2 + 4 \cos \theta$ **82.** $r = 2 - 4 \cos \theta$

83. $r = 3 - 2 \sin \theta$ **84.** $r = 5 + 3 \sin \theta$

85. $r = 4 + 3 \cos \theta$ **86.** $r = 5 + 2 \cos \theta$

Applying the Concepts

In Exercises 87–90, determine the position of the hand relative to the shoulder by considering the robotic arm illustrated in the figure. Round all answers to the nearest tenth.

87. $\alpha = 45°, \beta = 30°$

88. $\alpha = -30°, \beta = 60°$

89. $\alpha = -70°, \beta = 0°$

90. $\alpha = 47°, \beta = 17°$

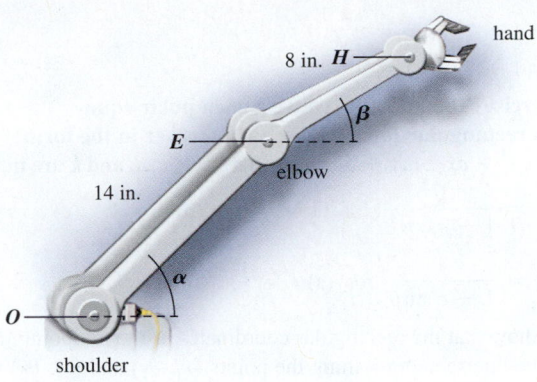

91. Ferris wheel. A ferris wheel has a diameter of 80 feet. If Anne is $\dfrac{1}{3}$ of the way around the wheel, write her polar coordinates.

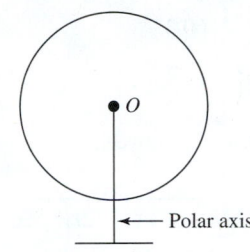

92. Repeat Exercise 91 if Anne is $\dfrac{3}{4}$ of the way around the wheel.

Planetary Orbit

In Exercises 93–96, the polar equation of the orbit of a planet around the sun is modeled by the equation $r = \dfrac{a(1 - e^2)}{1 - e \cos \theta}$. In each case, find (a) the smallest distance and (b) the largest distance of the planet from the sun. The distances are measured in astronomical units (*au*) [one *au* = 93 million miles].

93. Earth. $a = 1.00$ and $e = 0.017$

94. Mercury. $a = 1.39$ and $e = 0.206$

95. Neptune. $a = 30.1$ and $e = 0.009$

96. Mars. $a = 1.52$ and $e = 0.093$

Beyond the Basics

In Exercises 97 and 98, convert each rectangular equation to polar form. Assume that $x \geq 0, y \geq 0$ and that $\sqrt{x^2 + y^2} < \dfrac{\pi}{2}$.

97. $y = x \tan \left(\sqrt{x^2 + y^2} \right)$

98. $y = x \tan \left(\ln \sqrt{x^2 + y^2} \right)$

In Exercises 99–104, convert each polar equation to rectangular form.

99. $r(1 - \sin \theta) = 3$

100. $r(1 + \cos \theta) = 2$

101. $r \left(1 + \dfrac{1}{2} \cos \theta \right) = 1$

102. $r\left(1 + \dfrac{3}{4}\sin\theta\right) = 3$

103. $r(1 - 3\cos\theta) = 5$

104. $r(1 - 2\sin\theta) = 4$

In Exercises 105 and 106, convert each polar equation to rectangular form. Write each answer in the form $ax^2 + by^2 + cx + dy + k = 0$, where $a, b, c, d,$ and k are integers.

105. $r = \dfrac{6}{1 - e\cos\theta}$ for (a) $e = \dfrac{1}{2}$, (b) $e = 1$, (c) $e = 2$.

106. $r = \dfrac{4}{1 - e\sin\theta}$ for (a) $e = \dfrac{1}{2}$, (b) $e = 1$, (c) $e = 2$.

107. Show that the rectangular coordinates of the midpoint M of the line segment joining the points (r_1, θ_1) and (r_2, θ_2) is

$$M = \left(\frac{r_1\cos\theta_1 + r_2\cos\theta_2}{2}, \frac{r_1\sin\theta_1 + r_2\sin\theta_2}{2}\right).$$

108. Use Exercise 107 to find the midpoint of the line segment joining the points:

(a) $(3, 30°)$ and $(5, 60°)$

(b) $\left(5, \dfrac{\pi}{4}\right)$ and $\left(-3, \dfrac{\pi}{3}\right)$.

109. Prove that the distance between the points $P(r_1, \theta_1)$ and $Q(r_2, \theta_2)$ is given by

$$d(P, Q) = \sqrt{r_1^2 + r_2^2 - 2r_1r_2\cos(\theta_1 - \theta_2)}.$$

110. Prove that the equation $r = a\sin\theta + b\cos\theta$ represents a circle. Find its center and radius.

111. Prove that the area K of the triangle whose polar coordinates are $(0, 0)$, (r_1, θ_1), and (r_2, θ_2) is given by

$$K = \frac{1}{2}r_1r_2\sin(\theta_2 - \theta_1).$$ [Assume that $0 \le \theta_1 < \theta_2 \le \pi$

and $r_1 > 0, r_2 > 0.$]

112. Show that the polar equation of a circle with center (a, θ_0) and radius a is given by $r = 2a\cos(\theta - \theta_0)$. Discuss the cases $\theta_0 = 0$ and $\theta_0 = \dfrac{\pi}{2}$.

In Exercises 113–122, use a graphing calculator to graph the polar equation. Identify any horizontal or vertical asymptotes that appear on the screen but are not part of the graph.

113. $r = e^{\theta}$ (logarithmic spiral)

114. $r = e^{\frac{\theta}{3}}$ (logarithmic spiral)

115. $r^2 = 4\sin 2\theta$ (lemniscate)

116. $r^2 = 9\sin 2\theta$ (lemniscate)

117. $r^2 = 16\cos 2\theta$ (lemniscate)

118. $r^2 = 4\cos 2\theta$ (lemniscate)

119. $r = 2\sin\theta\tan\theta$ (cissoid)

120. $(r - 2)^2 = 8\theta$ (parabolic spiral)

121. $r = 2\csc\theta + 3$ (conchoid)

122. $r = 2\sec\theta - 1$ (conchoid)

Critical Thinking / Discussion / Writing

123. Recall that the graph of $r = f(\theta)$ is symmetric about the polar axis if replacing (r, θ) with $(r, -\theta)$ produces an equivalent equation. Show that the graph of $r = \cos 2\theta$ is symmetric about the polar axis.

124. In the equation $r = \sin 2\theta$, replacing (r, θ) with $(r, -\theta)$ does not produce an equivalent equation. Should we conclude that the graph is not symmetric about the polar axis?

125. The graph of $r = f(\theta)$ is symmetric about the y-axis if replacing (r, θ) with $(-r, -\theta)$ produces an equivalent equation. Discuss whether the graph of $r = 2 + 3\sin\theta$ is symmetric about the y-axis.

126. How is the graph of $r = f(\theta)$ related to the graph of $r = f(\theta - \alpha)$? Examine the graphs of $r = 2\cos\theta$ and $r = 2\cos\left(\theta - \dfrac{\pi}{3}\right)$. Find the rectangular forms of both equations.

Getting Ready for the Next Section

In Exercises 127–138, write each complex number in the form $a + bi$.

127. $(2 - 3i) + (-5 + i)$

128. $(5 + 3i) - (4 - 2i)$

129. $i(2 + 5i)$

130. $-i(5 - 3i)$

131. $(2 + i)(3 - i)$

132. $(4 - 3i)(2 - 5i)$

133. $(2 - 3i)^2$

134. $(1 + 2i)^3$

135. $\dfrac{1}{3 - 4i}$

136. $\dfrac{2i}{2 + 3i}$

137. $\dfrac{2 - 3i}{3 + 4i}$

138. $\dfrac{1 + i}{2 - i}$

In Exercises 139–142, write the trigonometric form (see page 543) for each vector.

139. $\mathbf{i} + \mathbf{j}$

140. $2\mathbf{i} - 2\mathbf{j}$

141. $-\sqrt{3}\mathbf{i} + \mathbf{j}$

142. $-\mathbf{i} - \sqrt{3}\mathbf{j}$

SECTION | 6.7

Polar Form of Complex Numbers; DeMoivre's Theorem

BEFORE STARTING THIS SECTION, REVIEW

1 Addition of complex numbers (Appendix A.8, page 976)

2 Conversion between polar and rectangular coordinates (Section 6.6, page 562)

OBJECTIVES

1 Represent complex numbers geometrically.

2 Find the absolute value of a complex number.

3 Write a complex number in polar form.

4 Find products and quotients of complex numbers in polar form.

5 Use DeMoivre's Theorem to find powers of a complex number.

6 Use DeMoivre's Theorem to find the nth roots of a complex number.

◆ Alternating Current Circuits

In the early days of the study of alternate current (AC) circuits, scientists concluded that AC circuits were somehow different from the battery-powered direct current (DC) circuits. However, both types of circuits obey exactly the same physical and mathematical laws. The breakthrough in understanding the AC circuits came in 1893 by Charles Steinmetz (1865–1923). He explained that in AC circuits, the voltage, current, and resistance (called *impedance* in AC circuits) do not act like scalars, but alternate in direction, and possess frequency and phase shift. He advocated the use of the polar form of complex numbers to represent the magnitude, frequency, and phase shift for the AC circuit quantities: voltage, current, and impedance. Steinmetz eventually became known as "the wizard who generated electricity from the square root of minus one." In Example 6, we use complex numbers to compute the total impedance in an AC circuit.

1 Represent complex numbers geometrically.

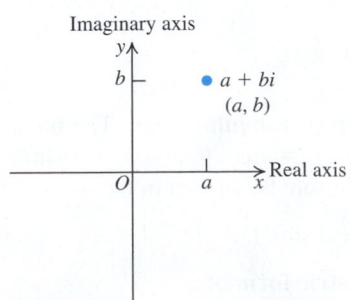

Figure 6.68 Complex plane.

Geometric Representation of Complex Numbers

Because each complex number $a + bi$ determines a unique ordered pair (a, b) of real numbers, we can represent the set of complex numbers geometrically by points in a rectangular coordinate system. Specifically, we can represent the complex number $a + bi$ by the point (a, b) in a rectangular coordinate system. We call the plane in this system the **complex plane**. The x-axis is also called the **real axis** because the real part of a complex number is plotted along the x-axis. Similarly, the y-axis is also called the **imaginary axis**. See Figure 6.68. We can think of a complex number $a + bi$ as a position vector with initial point $(0, 0)$ and terminal point (a, b).

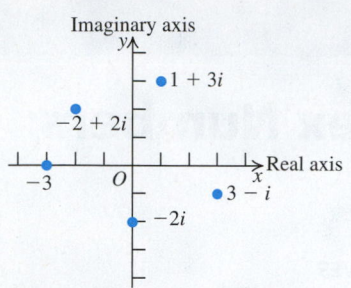

Figure 6.69 Plotting complex numbers.

EXAMPLE 1 **Plotting Complex Numbers**

Plot each complex number in the complex plane.

$$1 + 3i, \quad -2 + 2i, \quad -3, \quad -2i, \quad \text{and} \quad 3 - i$$

Solution

Figure 6.69 shows the points representing the complex numbers $1 + 3i, -2 + 2i, -3, -2i$ and $3 - i$.

Practice Problem 1 Plot each complex number in the complex plane.

$$2 + 3i, \quad -3 + 2i, \quad 4, \quad -2 - 3i, \quad \text{and} \quad 3 - 2i$$

2 Find the absolute value of a complex number.

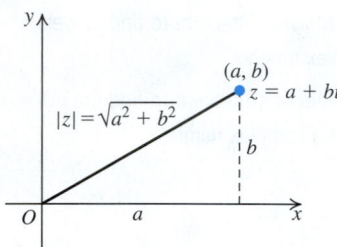

Figure 6.70 Absolute value of z.

The Absolute Value of a Complex Number

Let (a, b) represent the complex number $z = a + bi$ in a rectangular coordinate system. Then the distance from the origin O to the point (a, b) is $\sqrt{a^2 + b^2}$. See Figure 6.70. The number $\sqrt{a^2 + b^2}$ is called the *absolute value* (or **magnitude** or **modulus**) of the complex number $z = a + bi$ and is denoted by $|z|$.

Absolute Value of a Complex Number

The **absolute value** of a complex number $z = a + bi$:

$$|z| = |a + bi| = \sqrt{a^2 + b^2}$$

EXAMPLE 2 **Finding the Absolute Value of a Complex Number**

Find the absolute value of each complex number.

a. $4 + 3i$ **b.** $2 - 3i$ **c.** $-4 + i$ **d.** $-2 - 2i$ **e.** $-3i$

Solution

In each case, we use the formula $|a + bi| = \sqrt{a^2 + b^2}$ and simplify.

a. $|4 + 3i| = \sqrt{4^2 + 3^2} = \sqrt{25} = 5$
b. $|2 - 3i| = \sqrt{2^2 + (-3)^2} = \sqrt{4 + 9} = \sqrt{13}$
c. $|-4 + i| = |-4 + 1 \cdot i| = \sqrt{(-4)^2 + (1)^2} = \sqrt{16 + 1} = \sqrt{17}$
d. $|-2 - 2i| = \sqrt{(-2)^2 + (-2)^2} = \sqrt{4 + 4} = \sqrt{4 \cdot 2} = 2\sqrt{2}$
e. $|-3i| = |0 - 3i| = \sqrt{0^2 + (-3)^2} = \sqrt{0 + 9} = 3$

Practice Problem 2 Find the absolute value of each complex number.

a. $-5 + 12i$ **b.** -7 **c.** i **d.** $a - bi$

3 Write a complex number in polar form.

Polar Form of a Complex Number

A complex number z written as $z = a + bi$ is said to be in **rectangular form**. The point (a, b) has polar coordinates (r, θ), where $r = \sqrt{a^2 + b^2}$, $a = r\cos\theta$, and $b = r\sin\theta$. See Figure 6.71. The complex number $z = a + bi$ can therefore be written in the form

$$z = r\cos\theta + (r\sin\theta)i = r(\cos\theta + i\sin\theta).$$

We call $r(\cos\theta + i\sin\theta)$ the **polar form** or the **trigonometric form** of z.

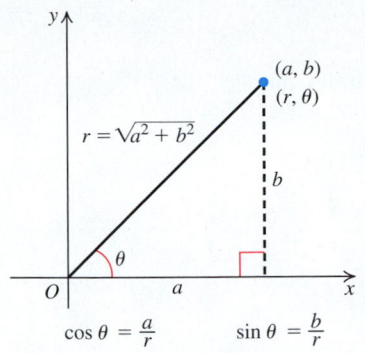

Figure 6.71 Polar form of $z = a + bi$.

POLAR FORM OF A COMPLEX NUMBER

The complex number $z = a + bi$ can be written in **polar form**

$$z = r(\cos\theta + i\sin\theta),$$

where $a = r\cos\theta$, $b = r\sin\theta$, $r = \sqrt{a^2 + b^2}$, and $\tan\theta = \dfrac{b}{a}$.

When a nonzero complex number is written in polar form, the positive number r is the **modulus** or **absolute value** of z; the angle θ is called the **argument** of z (written $\theta = \arg z$).

Note: $\cos\theta + i\sin\theta$ is sometimes abbreviated as $\operatorname{cis}\theta$.

Note that the angle θ in the polar representation of z is not unique because for any integer n,

$$r(\cos\theta + i\sin\theta) = r[\cos(\theta + n \cdot 360°) + i\sin(\theta + n \cdot 360°)].$$

Two complex numbers in polar form are therefore equal if and only if their *moduli* (plural of *modulus*) are equal and their arguments differ by an integer multiple of $360°$ (or of 2π).

PROCEDURE
IN ACTION

EXAMPLE 3 **Writing a Complex Number in Polar Form**

OBJECTIVE

Write a complex number $z = a + bi$ in polar form.

Step 1 Find r. Identify a and b. Use the formula

$$r = \sqrt{a^2 + b^2}.$$

Step 2 Find θ. First find the reference angle θ' by using $\theta' = \tan^{-1}\left|\dfrac{b}{a}\right|$. Then find θ so that it lies in the same quadrant as the point (a, b).

EXAMPLE

Write $z = \sqrt{3} - i$ in polar form. Express the argument θ in degrees, $0° \le \theta < 360°$.

1. $z = \sqrt{3} - i, a = \sqrt{3}, b = -1$

$$r = \sqrt{(\sqrt{3})^2 + (-1)^2} = \sqrt{3 + 1} = 2.$$

2. $\theta' = \tan^{-1}\left|\dfrac{-1}{\sqrt{3}}\right| = \tan^{-1}\dfrac{1}{\sqrt{3}} = 30°.$

Because $(a, b) = (\sqrt{3}, -1)$ lies in quadrant IV, we have $\theta = 360° - \theta' = 360° - 30° = 330°$. See the figure.

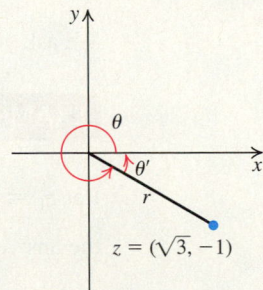

Step 3 Write z in polar form. From Steps 1 and 2, write the polar form $z = r(\cos\theta + i\sin\theta)$.

3. $z = 2(\cos 330° + i\sin 330°)$ $r = 2, \theta = 330°$

Practice Problem 3 Write $z = -1 - i$ in polar form, letting $0 \le \theta < 2\pi$.

EXAMPLE 4 **Writing a Complex Number in Rectangular Form**

Write the complex number $z = 2\left(\cos\dfrac{\pi}{6} + i\sin\dfrac{\pi}{6}\right)$ in rectangular form.

Solution

$$z = 2\left(\cos\frac{\pi}{6} + i\sin\frac{\pi}{6}\right) \qquad \text{Given complex number}$$

$$= 2\cos\frac{\pi}{6} + 2i\sin\frac{\pi}{6} \qquad \text{Distributive property}$$

$$= 2\left(\frac{\sqrt{3}}{2}\right) + 2i\left(\frac{1}{2}\right) \qquad \cos\frac{\pi}{6} = \frac{\sqrt{3}}{2}, \sin\frac{\pi}{6} = \frac{1}{2}$$

$$= \sqrt{3} + i \qquad \text{Simplify.}$$

The rectangular form of $z = 2\left(\cos\dfrac{\pi}{6} + i\sin\dfrac{\pi}{6}\right)$ is $\sqrt{3} + i$.

Practice Problem 4 Write $z = 4\left(\cos\dfrac{5\pi}{3} + i\sin\dfrac{5\pi}{3}\right)$ in rectangular form.

4 Find products and quotients of complex numbers in polar form.

Product and Quotient in Polar Form

The polar representation of complex numbers leads to an interesting interpretation of the product and quotient of two complex numbers.

> **PRODUCT AND QUOTIENT RULES FOR TWO COMPLEX NUMBERS IN POLAR FORM**
>
> Let $z_1 = r_1(\cos\theta_1 + i\sin\theta_1)$ and $z_2 = r_2(\cos\theta_2 + i\sin\theta_2)$ be two complex numbers in polar form. Then
>
> $$z_1 z_2 = r_1 r_2\left[\cos(\theta_1 + \theta_2) + i\sin(\theta_1 + \theta_2)\right] \qquad \text{Product rule}$$
>
> and
>
> $$\frac{z_1}{z_2} = \frac{r_1}{r_2}\left[\cos(\theta_1 - \theta_2) + i\sin(\theta_1 - \theta_2)\right], z_2 \neq 0 \qquad \text{Quotient rule}$$
>
> In words: to multiply two complex numbers in polar form, we multiply their moduli and add their arguments; to divide two complex numbers, we divide their moduli and subtract their arguments.

We ask you to prove these rules in Exercises 103 and 104.

EXAMPLE 5 **Finding the Product and Quotient of Two Complex Numbers**

Let $z_1 = 3(\cos 65° + i\sin 65°)$ and $z_2 = 4(\cos 15° + i\sin 15°)$. Find $z_1 z_2$ and $\dfrac{z_1}{z_2}$. Leave the answers in polar form.

Solution

$$z_1 z_2 = 3(\cos 65° + i\sin 65°) \cdot 4(\cos 15° + i\sin 15°) \qquad \text{Product of given numbers}$$

$$= 3 \cdot 4\left[\cos(65° + 15°) + i\sin(65° + 15°)\right] \qquad \text{Multiply moduli and add arguments.}$$

$$= 12(\cos 80° + i\sin 80°) \qquad \text{Simplify.}$$

$$\frac{z_1}{z_2} = \frac{3(\cos 65° + i\sin 65°)}{4(\cos 15° + i\sin 15°)} \qquad \text{Quotient of given numbers}$$

$$= \frac{3}{4}\left[\cos(65° - 15°) + i\sin(65° - 15°)\right] \qquad \text{Divide moduli and subtract arguments.}$$

$$= \frac{3}{4}(\cos 50° + i\sin 50°) \qquad \text{Simplify.}$$

Practice Problem 5 Let $z_1 = 5(\cos 75° + i \sin 75°)$ and $z_2 = 2(\cos 60° + i \sin 60°)$.
Find $z_1 z_2$ and $\dfrac{z_1}{z_2}$. Leave the answers in polar form.

◆ **EXAMPLE 6** **Using Complex Numbers in AC Circuits**

In a parallel circuit, the total impedance Z_t is given by

$$Z_t = \frac{Z_1 Z_2}{Z_1 + Z_2}.$$

Find Z_t, if

$$Z_1 = 9(\cos 90° + i \sin 90°) \text{ and } Z_2 = 4[\cos(-60°) + i \sin(-60°)].$$

Solution

We first calculate $Z_1 Z_2$ and $Z_1 + Z_2$.

$Z_1 Z_2 = 9(\cos 90° + i \sin 90°) \cdot 4[\cos(-60°) + i \sin(-60°)]$ Substitute values for Z_1 and Z_2.

$\qquad = 9 \cdot 4[\cos(90° - 60°) + i \sin(90° - 60°)]$ Multiply moduli and add arguments.

$\qquad = 36(\cos 30° + i \sin 30°)$ Simplify.

To add complex numbers in polar form, convert to rectangular form, perform addition, and then convert back to polar form.

$Z_1 = 9 \cos 90° + 9i \sin 90° = 9i$ $\cos 90° = 0, \sin 90° = 1$

$Z_2 = 4[\cos(-60°) + i \sin(-60°)]$

$\quad = 4 \cos 60° - 4i \sin 60°$ $\cos(-\theta) = \cos\theta, \sin(-\theta) = -\sin\theta$

$\quad = 4\left(\dfrac{1}{2}\right) - 4i\left(\dfrac{\sqrt{3}}{2}\right)$ $\cos 60° = \dfrac{1}{2}, \sin 60° = \dfrac{\sqrt{3}}{2}$

$\quad = 2 - 2\sqrt{3}i$ Simplify.

$Z_1 + Z_2 = 9i + 2 - 2\sqrt{3}i$

$Z_1 + Z_2 = 2 + (9 - 2\sqrt{3})i$ Regroup.

To write $Z_1 + Z_2$ in polar form, find the values of r and θ.

$r = \sqrt{a^2 + b^2}$

$\quad = \sqrt{2^2 + (9 - 2\sqrt{3})^2}$ Replace a with 2 and b with $9 - 2\sqrt{3}$.

$\quad \approx 5.89$ Use a calculator.

$\tan\theta = \dfrac{b}{a} = \dfrac{9 - 2\sqrt{3}}{2} \approx 2.768$

$\theta \approx \tan^{-1}(2.768) \approx 70°$ Use a calculator.

The polar form of $Z_1 + Z_2$ is approximately $5.89(\cos 70° + i \sin 70°)$. Thus,

$Z_t = \dfrac{Z_1 Z_2}{Z_1 + Z_2} \approx \dfrac{36(\cos 30° + i \sin 30°)}{5.89(\cos 70° + i \sin 70°)}$ Substitute values for $Z_1 Z_2$ and $Z_1 + Z_2$.

$\quad = \dfrac{36}{5.89}[\cos(30° - 70°) + i \sin(30° - 70°)]$ Divide moduli and subtract.

$\quad = \dfrac{36}{5.89}[\cos(-40°) + i \sin(-40°)]$ Simplify.

$\quad = \dfrac{36}{5.89}(\cos 40° - i \sin 40°)$ $\cos(-\theta) = \cos\theta,$ $\sin(-\theta) = -\sin\theta$

$\quad \approx 6.11(\cos 40° - i \sin 40°).$ $\dfrac{36}{5.89} \approx 6.11$

Practice Problem 6 Repeat Example 6 with

$$Z_1 = 4(\cos 45° + i \sin 45°) \quad \text{and} \quad Z_2 = 6(\cos 0° + i \sin 0°).$$

5 Use DeMoivre's Theorem to find powers of a complex number.

Powers of Complex Numbers in Polar Form

Let $z = r(\cos \theta + i \sin \theta)$ be a complex number in polar form. Then

$$
\begin{aligned}
z^2 &= z \cdot z \\
&= r(\cos \theta + i \sin \theta) \cdot r(\cos \theta + i \sin \theta) && \text{Form product } z \cdot z. \\
&= (r \cdot r)[\cos(\theta + \theta) + i \sin(\theta + \theta)] && \text{Multiply moduli and add arguments.} \\
&= r^2(\cos 2\theta + i \sin 2\theta) && \text{Simplify.} \\
z^3 &= z^2 \cdot z \\
&= r^2(\cos 2\theta + i \sin 2\theta) \cdot r(\cos \theta + i \sin \theta) && \text{Polar form of } z^2 \text{ and } z \\
&= (r^2 \cdot r)[\cos(2\theta + \theta) + i \sin(2\theta + \theta)] && \text{Multiply moduli and add arguments.} \\
&= r^3(\cos 3\theta + i \sin 3\theta). && \text{Simplify.}
\end{aligned}
$$

In a similar way, you can show that

$$z^4 = r^4(\cos 4\theta + i \sin 4\theta).$$

DeMoivre's Theorem (whose proof is omitted) follows this pattern.

DEMOIVRE'S THEOREM

Let $z = r(\cos \theta + i \sin \theta)$ be a complex number in polar form. Then for any integer n,

$$z^n = r^n(\cos n\theta + i \sin n\theta).$$

EXAMPLE 7   Finding the Power of a Complex Number

Let $z = 1 + i$. Use DeMoivre's Theorem to find each power of z. Write answers in rectangular form.

a. z^{16} **b.** z^{-10}

Solution

We first convert $z = 1 + i = (1, 1)$ to polar form. We find r and θ.

$$r = \sqrt{a^2 + b^2} = \sqrt{1^2 + 1^2} = \sqrt{2} \quad a = 1, b = 1$$

and

$$\theta' = \tan^{-1}\left|\frac{b}{a}\right| = \tan^{-1}\left|\frac{1}{1}\right| = \tan^{-1}1 = \frac{\pi}{4}.$$

TECHNOLOGY
CONNECTION

A graphing calculator can be used to find powers of complex numbers or to check your work. The calculator must be set in $a + bi$ mode.

```
(1+i)^(-10)
          -.03125i
Ans▶Frac
             -1/32i
■
```

Since $z = (1, 1)$ lies in quadrant I, we have $\theta = \theta'$.

$$\text{So } \theta = \frac{\pi}{4}$$

a.
$$
\begin{aligned}
z &= \sqrt{2}\left(\cos\frac{\pi}{4} + i \sin\frac{\pi}{4}\right) && \text{Polar form of } z = 1 + i \\
z^{16} &= \left[\sqrt{2}\left(\cos\frac{\pi}{4} + i \sin\frac{\pi}{4}\right)\right]^{16} && \text{Raise both sides to the 16th power.} \\
z^{16} &= (\sqrt{2})^{16}\left[\cos\left(16 \cdot \frac{\pi}{4}\right) + i \sin\left(16 \cdot \frac{\pi}{4}\right)\right] && \text{DeMoivre's Theorem} \\
&= 2^8[\cos(4\pi) + i \sin(4\pi)] && (\sqrt{2})^{16} = (2^{\frac{1}{2}})^{16} = 2^{\frac{1}{2} \cdot 16} = 2^8 = 256 \\
&= 256(1 + i \cdot 0) = 256 && \cos 4\pi = 1, \sin 4\pi = 0
\end{aligned}
$$

b.

$$z = \sqrt{2}\left(\cos\frac{\pi}{4} + i\sin\frac{\pi}{4}\right) \qquad \text{Polar form of } z = 1 + i$$

$$z^{-10} = \left[\sqrt{2}\left(\cos\frac{\pi}{4} + i\sin\frac{\pi}{4}\right)\right]^{-10} \qquad \begin{array}{l}\text{Raise both sides to the}\\ -10\text{th power.}\end{array}$$

$$z^{-10} = (\sqrt{2})^{-10}\left[\cos\left(-10\cdot\frac{\pi}{4}\right) + i\sin\left(-10\cdot\frac{\pi}{4}\right)\right] \qquad \text{DeMoivre's Theorem}$$

$$= \frac{1}{32}\left[\cos\left(-\frac{5\pi}{2}\right) + i\sin\left(-\frac{5\pi}{2}\right)\right] \qquad \begin{array}{l}\text{Simplify;}\\ (\sqrt{2})^{-10} = (2^{1/2})^{-10} = 2^{-5} = \dfrac{1}{32}\end{array}$$

$$= \frac{1}{32}[0 + i(-1)] = -\frac{1}{32}i \qquad \cos\left(-\frac{5\pi}{2}\right) = 0, \sin\left(-\frac{5\pi}{2}\right) = -1.$$

Practice Problem 7 Let $z = -1 + i$. Use DeMoivre's Theorem to find each power of z. Write the answers in rectangular form.

a. z^8 **b.** z^{-12}

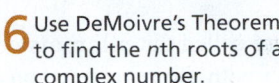

Roots of Complex Numbers

6 Use DeMoivre's Theorem to find the nth roots of a complex number.

Let z and w be two complex numbers and let n be a positive integer. The complex number z is called an **nth root of w** if

$$z^n = w.$$

We can use DeMoivre's Theorem to find the nth roots of a complex number.

DEMOIVRE'S nTH ROOTS THEOREM

The nth roots of a complex number $w = r(\cos\theta + i\sin\theta)$, where $r > 0$ and θ is in degrees, are given by

$$z_k = r^{1/n}\left[\cos\left(\frac{\theta + 360°k}{n}\right) + i\sin\left(\frac{\theta + 360°k}{n}\right)\right], \quad \text{for } k = 0, 1, 2, \ldots, n-1.$$

If θ is in radians, replace $360°$ with 2π in z_k.

The result says that there are exactly n nth roots of a nonzero complex number.

EXAMPLE 8   **Finding the Roots of a Complex Number**

Find the three cube roots of $1 + i$ in polar form, with the argument in degrees.

Solution

In Example 7, we showed that

$$1 + i = \sqrt{2}\left(\cos\frac{\pi}{4} + i\sin\frac{\pi}{4}\right) \qquad \text{Polar form of } 1 + i$$

$$1 + i = \sqrt{2}(\cos 45° + i\sin 45°). \qquad \text{Replace } \frac{\pi}{4} \text{ with } 45°.$$

RECALL

$\sqrt{2} = 2^{1/2}$, so
$(\sqrt{2})^{1/3} = (2^{1/2})^{1/3} = 2^{1/6}$.

From DeMoivre's Theorem for finding complex roots, with $n = 3$, we have

$$z_k = (\sqrt{2})^{1/3}\left[\cos\left(\frac{45° + 360°k}{3}\right) + i\sin\left(\frac{45° + 360°k}{3}\right)\right], \quad k = 0, 1, 2.$$

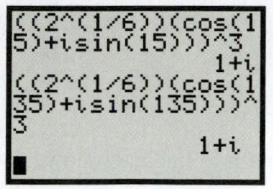
By substituting $k = 0$, 1, and 2 in the expression for z_k and simplifying, we find the three cube roots.

$$z_0 = 2^{1/6}\left[\cos\left(\frac{45° + 360° \cdot 0}{3}\right) + i\sin\left(\frac{45° + 360° \cdot 0}{3}\right)\right] \quad k = 0$$

$$= 2^{1/6}(\cos 15° + i\sin 15°) \qquad \text{Simplify.}$$

$$z_1 = 2^{1/6}\left[\cos\left(\frac{45° + 360° \cdot 1}{3}\right) + i\sin\left(\frac{45° + 360° \cdot 1}{3}\right)\right] \quad k = 1$$

$$= 2^{1/6}(\cos 135° + i\sin 135°) \qquad \text{Simplify.}$$

$$z_2 = 2^{1/6}\left[\cos\left(\frac{45° + 360° \cdot 2}{3}\right) + i\sin\left(\frac{45° + 360° \cdot 2}{3}\right)\right] \quad k = 2$$

$$= 2^{1/6}(\cos 255° + i\sin 255°) \qquad \text{Simplify.}$$

The complex numbers z_0, z_1, and z_2 are the three cube roots of the complex number $1 + i$.

Practice Problem 8 Find the three cube roots of $-1 + i$ in polar form, with the argument in degrees.

The solutions of the so-called *cyclotomic* equation $z^n = 1$ are called the **nth roots of unity**. The name refers to the close association of this equation with the regular n-gons inscribed in the unit circle.

EXAMPLE 9 **Finding nth Roots of Unity**

Find the complex sixth roots of unity. Write the roots in polar form and represent them geometrically.

Solution
The polar form for $1 = 1 + 0i$ is

$$1 = 1(\cos 0° + i\sin 0°).$$

From DeMoivre's Theorem for finding complex roots with $n = 6$, we have

$$z_k = (1)^{1/6}\left[\cos\left(\frac{0° + 360° \cdot k}{6}\right) + i\sin\left(\left(\frac{0° + 360° \cdot k}{6}\right)\right)\right] \quad k = 0, 1, 2, 3, 4, 5$$

$$z_0 = 1(\cos 0° + i\sin 0°) \qquad \text{Let } k = 0 \text{ in } z_k \text{ and simplify.}$$
$$z_1 = 1(\cos 60° + i\sin 60°) \qquad \text{Let } k = 1 \text{ in } z_k \text{ and simplify.}$$
$$z_2 = 1(\cos 120° + i\sin 120°) \qquad \text{Let } k = 2 \text{ in } z_k \text{ and simplify.}$$
$$z_3 = 1(\cos 180° + i\sin 180°) \qquad \text{Let } k = 3 \text{ in } z_k \text{ and simplify.}$$
$$z_4 = 1(\cos 240° + i\sin 240°) \qquad \text{Let } k = 4 \text{ in } z_k \text{ and simplify.}$$
$$z_5 = 1(\cos 300° + i\sin 300°). \qquad \text{Let } k = 5 \text{ in } z_k \text{ and simplify.}$$

Geometrically, the sixth roots of unity form a regular hexagon and are equally spaced at 60° intervals on a unit circle. See Figure 6.72.

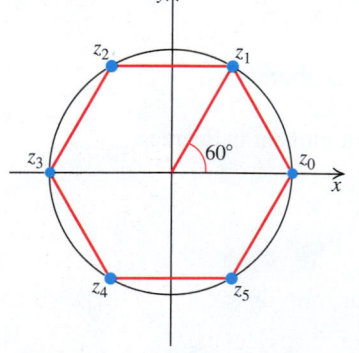

Figure 6.72 Sixth roots of unity.

Practice Problem 9 Find the complex fourth roots of unity. Write the roots in polar form.

Answers to Practice Problems

1.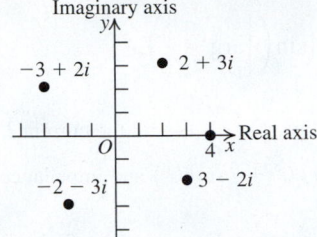

2. a. 13   **b.** 7   **c.** 1   **d.** $\sqrt{a^2 + b^2}$

3. $z = \sqrt{2}\left(\cos\dfrac{5\pi}{4} + i\sin\dfrac{5\pi}{4}\right)$ **4.** $2 - 2\sqrt{3}i$

5. $z_1z_2 = 10(\cos 135° + i\sin 135°); \dfrac{z_1}{z_2} = \dfrac{5}{2}(\cos 15° + i\sin 15°)$

6. $2.6(\cos 27.2° + i\sin 27.2°)$ **7. a.** 16   **b.** $-\dfrac{1}{64}$

8. $2^{\frac{1}{6}}(\cos 45° + i\sin 45°), 2^{\frac{1}{6}}(\cos 165° + i\sin 165°),$
$2^{\frac{1}{6}}(\cos 285° + i\sin 285°)$ **9.** $\cos 0° + i\sin 0°, \cos 90° + i\sin 90°,$
$\cos 180° + i\sin 180°, \cos 270° + i\sin 270°$

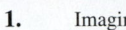 **SECTION 6.7** **Exercises**

Concepts and Vocabulary

1. If $z = a + bi, a > 0$, then $|z| = $ _____ and arg $z = \theta = $ _____.

2. If $|z| = r$ and arg $z = \theta$, then the polar form of z is $z = $ _____.

3. To multiply two complex numbers in polar form, multiply their _____ and _____ their arguments.

4. DeMoivre's Theorem states that $[r(\cos\theta + i\sin\theta)]^n = $ _____.

5. True or False. If $z = r(\cos\theta + i\sin\theta)$, then $\dfrac{1}{z} = z^{-1} = $
$\dfrac{1}{r}(\cos\theta - i\sin\theta)$.

6. True or False. For $n \geq 2$, the n complex nth roots of 1 are equally spaced points on the unit circle.

7. True or False. The complex number $z = a + bi$ is in polar form.

8. True or False. The complex number $z = 0 + 0i$ does not have a polar form.

Building Skills

In Exercises 9–16, plot each complex number and find its absolute value.

9. $z = 2$

10. $z = -3$

11. $z = -4i$

12. $z = 2i$

13. $z = 3 + 4i$

14. $z = -1 + 2i$

15. $z = -2 - 3i$

16. $z = 2 - 5i$

In Exercises 17–28, write each complex number in polar form. Express the argument θ in degrees, with $0 \leq \theta < 360°$.

17. $1 + \sqrt{3}i$

18. $-1 + \sqrt{2}i$

19. $-1 + i$

20. $1 - i$

21. i

22. $-i$

23. 1

24. -1

25. $3 - 3i$

26. $4\sqrt{3} + 4i$

27. $2 - 2\sqrt{3}i$

28. $2 + 3i$

In Exercises 29–40, write each complex number in rectangular form.

29. $2(\cos 60° + i\sin 60°)$

30. $4(\cos 120° + i\sin 120°)$

31. $3(\cos\pi + i\sin\pi)$

32. $5\left(\cos\dfrac{\pi}{2} + i\sin\dfrac{\pi}{2}\right)$

33. $5(\cos 240° + i\sin 240°)$

34. $2(\cos 300° + i\sin 300°)$

35. $8(\cos 0° + i\sin 0°)$

36. $2(\cos(-90°)° + i\sin(-90°))$

37. $6\left(\cos\dfrac{5\pi}{6} + i\sin\dfrac{5\pi}{6}\right)$

38. $4\left(\cos\dfrac{3\pi}{4} + i\sin\dfrac{3\pi}{4}\right)$

39. $3\left(\cos\left(-\dfrac{\pi}{3}\right) + i\sin\left(-\dfrac{\pi}{3}\right)\right)$

40. $5\left(\cos\left(-\dfrac{7\pi}{6}\right) + i\sin\left(-\dfrac{7\pi}{6}\right)\right)$

In Exercises 41–48, find z_1z_2. Write each answer in rectangular form.

41. $z_1 = 4(\cos 75° + i\sin 75°), z_2 = 2(\cos 15° + i\sin 15°)$

42. $z_1 = 6(\cos 90° + i\sin 90°), z_2 = 2(\cos 45° + i\sin 45°)$

43. $z_1 = 5(\cos 240° + i\sin 240°),$
$z_2 = 2(\cos 60° + i\sin 60°)$

44. $z_1 = 10(\cos 135° + i\sin 135°),$
$z_2 = 4(\cos 225° + i\sin 225°)$

45. $z_1 = 3(\cos 40° + i\sin 40°), z_2 = 5(\cos 20° + i\sin 20°)$

46. $z_1 = 5(\cos 65° + i\sin 65°), z_2 = 2(\cos 25° + i\sin 25°)$

47. $z_1 = 3(\cos 140° + i\sin 140°), z_2 = 2(\cos 100° + i\sin 100°)$

48. $z_1 = 4(\cos 190° + i\sin 190°), z_2 = 5(\cos 140° + i\sin 140°)$

In Exercises 49–56, find $\dfrac{z_1}{z_2}$. Write each answer in rectangular form.

49. $z_1 = 8(\cos 72° + i \sin 72°)$, $z_2 = 4(\cos 12° + i \sin 12°)$

50. $z_1 = 9(\cos 56° + i \sin 56°)$, $z_2 = 3(\cos 26° + i \sin 26°)$

51. $z_1 = 5(\cos 102° + i \sin 102°)$, $z_2 = 2.5(\cos 12° + i \sin 12°)$

52. $z_1 = 12(\cos 220° + i \sin 220°)$, $z_2 = 4(\cos 40° + i \sin 40°)$

53. $z_1 = 6(\cos 70° + i \sin 70°)$, $z_2 = 3(\cos 190° + i \sin 190°)$

54. $z_1 = 5(\cos 110° + i \sin 110°)$, $z_2 = 4(\cos 350° + i \sin 350°)$

55. $z_1 = 3\left(\cos \dfrac{\pi}{12} + i \sin \dfrac{\pi}{12}\right)$, $z_2 = 2\left(\cos \dfrac{5\pi}{12} + i \sin \dfrac{5\pi}{12}\right)$

56. $z_1 = 6\left(\cos \dfrac{7\pi}{16} + i \sin \dfrac{7\pi}{16}\right)$, $z_2 = 3\left(\cos \dfrac{27\pi}{16} + i \sin \dfrac{27\pi}{16}\right)$

In Exercises 57–70, use DeMoivre's Theorem to find the indicated power. Write the answers in rectangular form.

57. $\left[2\left(\cos \dfrac{\pi}{3} + i \sin \dfrac{\pi}{3}\right)\right]^{12}$

58. $\left[\sqrt{2}\left(\cos \dfrac{\pi}{4} + i \sin \dfrac{\pi}{4}\right)\right]^{20}$

59. $\left[2\left(\cos\left(-\dfrac{3\pi}{4}\right) + i \sin\left(-\dfrac{3\pi}{4}\right)\right)\right]^{6}$

60. $\left(\cos \dfrac{\pi}{27} + i \sin \dfrac{\pi}{27}\right)^{-9}$

61. $\left[2\left(\cos\left(\dfrac{3\pi}{4}\right) + i \sin\left(\dfrac{3\pi}{4}\right)\right)\right]^{-6}$

62. $\left[3\left(\cos\left(-\dfrac{5\pi}{6}\right) + i \sin\left(-\dfrac{5\pi}{6}\right)\right)\right]^{16}$

63. $\left[2\left(\cos\left(-\dfrac{\pi}{4}\right) + i \sin\left(-\dfrac{\pi}{4}\right)\right)\right]^{-10}$

64. $\left[\dfrac{1}{2}\left(\cos\left(-\dfrac{3\pi}{4}\right) + i \sin\left(-\dfrac{3\pi}{4}\right)\right)\right]^{-6}$

65. $(1 - i)^{12}$

66. $(1 - \sqrt{3}i)^{10}$

67. i^{25}

68. $(3 + 3i)^{8}$

69. $\left(\dfrac{1}{2} - \dfrac{\sqrt{3}}{2}i\right)^{-8}$

70. $\left(\dfrac{\sqrt{3}}{2} + \dfrac{1}{2}i\right)^{10}$

In Exercises 71–78, find all complex roots. Write your answers in polar form, with the argument θ in degrees with $0 \le \theta < 360°$.

71. Cube roots of 64

72. Cube roots of -64

73. Square roots of i

74. Fourth roots of $-i$

75. Sixth roots of -1

76. Eighth roots of unity

77. Square roots of $1 - \sqrt{3}i$

78. Cube roots of $4 + 4\sqrt{3}i$

Applying the Concepts

79. Trigonometry. Use DeMoivre's Theorem to verify each identity.

 a. $\sin 3\theta = 3 \sin \theta - 4 \sin^3 \theta$

 b. $\cos 3\theta = 4 \cos^3 \theta - 3 \cos \theta$

80. Geometry. Show that the area K of a triangle with vertices at the origin O, z_1, and z_2 is given by

$$K = \frac{1}{2}|z_1||z_2|\sin\left(\left|\arg \frac{z_1}{z_2}\right|\right).$$

81. AC circuits. Use Ohm's Law, $I = \dfrac{V}{Z}$, to find the current I, given voltage $V = 120(\cos 60° + i \sin 60°)$ and impedance $Z = 8(\cos 30° + i \sin 30°)$.

82. AC circuits. Use Ohm's Law from Exercise 81 to find the voltage V, given $I = 6(\cos 40° + i \sin 40°)$ and $Z = 16(\cos 110° + i \sin 110°)$.

In Exercises 83 and 84, use the formula $Z_t = \dfrac{Z_1 Z_2}{Z_1 + Z_2}$ to find the total impedance in a parallel circuit for the given values of Z_1 and Z_2.

83. $Z_1 = 16(\cos 180° + i \sin 180°)$ and $Z_2 = 2(\cos 150° + i \sin 150°)$

84. $Z_1 = 12(\cos 270° + i \sin 270°)$ and $Z_2 = 3(\cos 60° + i \sin 60°)$

Beyond the Basics

In Exercises 85–90, use DeMoivre's Theorem to compute each power in polar form, with θ in degrees, with $0 \le \theta < 360°$.

85. $(1 - \sqrt{3}i)^{10}(-2 + 2i)^{-6}$

86. $\left(\dfrac{1}{2} + \dfrac{\sqrt{3}}{2}i\right)^{8}(2 - 2i)^{-6}$

87. $\left(\sin \dfrac{\pi}{6} + i \cos \dfrac{\pi}{6}\right)^{10}$

88. $\left(\sin \dfrac{2\pi}{3} + i \cos \dfrac{2\pi}{3}\right)^{6}$

89. $\left(\sin \dfrac{5\pi}{3} + i \cos \dfrac{5\pi}{3}\right)^{-8}$

90. $\left(\sin \dfrac{7\pi}{6} + i \cos \dfrac{7\pi}{6}\right)^{-10}$

In Exercises 91–96, find all complex solutions of each equation.

91. $z^4 = 1$

92. $z^8 = -1$

93. $z^3 = 1 + i$

94. $z^7 = (1 - i)^2$

95. $z^4 - 2z^2 + 4 = 0$

96. $z^4 + 2z^2 + 4 = 0$

97. Let $z = \cos \theta + i \sin \theta$, $z \ne 0$. Prove that

$$\frac{1}{z} = \cos \theta - i \sin \theta.$$

In Exercises 98–101, let $z = \cos \theta + i \sin \theta$, $z \ne 0$. Use Exercise 97 and DeMoivre's theorem to prove each identity.

98. $z + \dfrac{1}{z} = 2 \cos \theta$

99. $z - \dfrac{1}{z} = 2i \sin \theta$

100. $z^n + \dfrac{1}{z^n} = 2 \cos n\theta$

101. $z^n - \dfrac{1}{z^n} = 2i \sin n\theta$

102. Prove that
$$(1 + \cos \theta + i \sin \theta)^n = \left(2 \cos \dfrac{\theta}{2}\right)^n\left(\cos \dfrac{n\theta}{2} + i \sin \dfrac{n\theta}{2}\right).$$

103. Prove the product rule for two complex numbers in polar form.

104. Prove the quotient rule for two complex numbers in polar form.

105. Find the product $(3 + 2i)(5 + i)$ using FOIL. Then deduce the identity: $\tan^{-1}\left(\dfrac{2}{3}\right) + \tan^{-1}\left(\dfrac{1}{5}\right) = \dfrac{\pi}{4}$. [*Hint:* $\arg(z_1 z_2) = \arg z_1 + \arg z_2$.]

106. Find the product $(p + q + i)(p^2 + pq + 1 + iq)$; then deduce the identity:

$$\tan^{-1}\left(\frac{1}{p + q}\right) + \tan^{-1}\left(\frac{q}{p^2 + pq + 1}\right) = \tan^{-1}\left(\frac{1}{p}\right).$$

107. Expand $(2 + 3i)^4$; then deduce the identity:

$$4\tan^{-1}\left(\frac{3}{2}\right) - \tan^{-1}\left(\frac{120}{119}\right) = \pi.$$

108. Find the product $(5 + i)^4(-239 + i)$; then deduce the identity: $4\tan^{-1}\left(\dfrac{1}{5}\right) - \tan^{-1}\left(\dfrac{1}{239}\right) = \dfrac{\pi}{4}$.

109. Find the three solutions of the equation $x^3 + x^2 + x + 1 = 0$. [*Hint:* Multiply both sides by $(x - 1)$, solve the new equation, and reject the root $x = 1$.]

110. Find the four solutions of the equation $x^4 + x^3 + x^2 + x + 1 = 0$.

Critical Thinking / Discussion / Writing

111. Let n be a positive integer. State whether each of the following is true or false.

a. $\left(\cos\dfrac{\pi}{6} + i\sin\dfrac{\pi}{6}\right)^n = \cos\dfrac{n\pi}{6} + i\sin\dfrac{n\pi}{6}$

b. $\left(\sin\dfrac{\pi}{3} + i\cos\dfrac{\pi}{3}\right)^n = \sin\dfrac{n\pi}{3} + i\cos\dfrac{n\pi}{3}$

c. $\left[\cos\left(-\dfrac{\pi}{3}\right) + i\sin\left(-\dfrac{\pi}{3}\right)\right]^n = \cos\dfrac{n\pi}{3} - i\sin\dfrac{n\pi}{3}$

d. $\left(\cos\dfrac{\pi}{6} + i\sin\dfrac{\pi}{3}\right)^n = \cos\dfrac{n\pi}{6} + i\sin\dfrac{n\pi}{3}$

e. $(\cos\theta + i\sin\theta)^n = \cos n\theta + i\sin n\theta$

f. $(\cos\theta - i\sin\theta)^n = \cos n\theta - i\sin n\theta$

g. $(\cos\theta + i\sin\theta)^{-n} = \cos n\theta - i\sin n\theta$

h. $(\cos\theta - i\sin\theta)^{-n} = \cos n\theta + i\sin n\theta$

i. $(\sin\theta + i\cos\theta)^n = \sin n\theta + i\cos n\theta$

j. $(\sin\theta + i\cos\theta)^n = \cos\left[n\left(\dfrac{\pi}{2} - \theta\right)\right] + i\sin\left[n\left(\dfrac{\pi}{2} - \theta\right)\right]$

Getting Ready for the Next Section

In Exercises 112–120, find the value of a variable that satisfies the given condition.

112. Find x if $3x + 2y = 7$ and $y = \dfrac{1}{2}$.

113. Find x if $-2x + 9y = 5$ and $y = 3$.

114. Find y if $-4x + 7y = 7$ and $x = 0$.

115. Find y if $23x - 14y = -5$ and $x = 1$.

116. Find the slope of the line with equation $10x - 2y = 28$.

117. Find the slope of the line with equation $15x + 5y = 2$.

118. Find an equation of the line through the point $(5, -1)$ parallel to the line with equation $6x - 3y = 7$.

119. Find an equation of the line through the point $(3, 3)$ parallel to the line with equation $x + 2y = 1$.

120. Find values for a and b so that the equation $ax + by = 3$ has the same graph as the equation $2x - 4y = 12$.

SUMMARY Definitions, Concepts, and Formulas

6.1 Right-Triangle Trigonometry

i. The right-triangle definitions for trigonometric functions of an acute angle θ are given by

$$\sin\theta = \frac{\text{opposite side}}{\text{hypotenuse}} = \frac{a}{c} \qquad \csc\theta = \frac{\text{hypotenuse}}{\text{opposite side}} = \frac{c}{a}$$

$$\cos\theta = \frac{\text{adjacent side}}{\text{hypotenuse}} = \frac{b}{c} \qquad \sec\theta = \frac{\text{hypotenuse}}{\text{adjacent side}} = \frac{c}{b}$$

$$\tan\theta = \frac{\text{opposite side}}{\text{adjacent side}} = \frac{a}{b} \qquad \cot\theta = \frac{\text{adjacent side}}{\text{opposite side}} = \frac{b}{a}$$

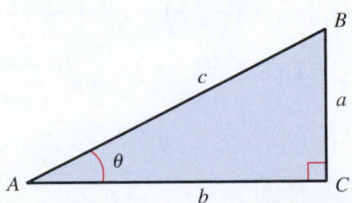

ii. Two acute angles are complements if their sum is $90°$ or $\dfrac{\pi}{2}$.

The value of any trigonometric function of an acute angle θ is equal to the cofunction of the complement of θ.

iii. To solve a right triangle means to find the measurements of missing angles and the sides.

6.2 The Law of Sines

i. In a triangle ABC with sides of length a, b, and c,

$$\frac{\sin A}{a} = \frac{\sin B}{b} = \frac{\sin C}{c}, \text{ or equivalently,}$$

$$\frac{a}{\sin A} = \frac{b}{\sin B} = \frac{c}{\sin C}.$$

ii. The Law of Sines is used to solve AAS, ASA, and SSA triangles. The SSA case is called the *ambiguous case* because the information provided may lead to two triangles, one triangle, or no triangle.

6.3 The Law of Cosines

i. In a triangle ABC with sides of lengths a, b, and c,

$$a^2 = b^2 + c^2 - 2bc\cos A;$$
$$b^2 = c^2 + a^2 - 2ca\cos B;$$
$$c^2 = a^2 + b^2 - 2ab\cos C.$$

ii. The Law of Cosines is used to solve SAS and SSS triangles. This law is also used to prove Heron's formula.

iii. Heron's formula. The area K of a triangle with sides of lengths a, b, and c is given by

$$K = \sqrt{s(s-a)(s-b)(s-c)},$$

where $s = \dfrac{1}{2}(a+b+c)$ is the *semiperimeter*.

6.4 Vectors

i. A vector $\mathbf{v}$ is a quantity that has both magnitude and direction. A vector is geometrically represented by an arrow.

ii. The magnitude of the scalar product $c\mathbf{v}$ is $|c|$ times the magnitude of $\mathbf{v}$. If $c > 0$, $c\mathbf{v}$ has the direction of $\mathbf{v}$. If $c < 0$, $c\mathbf{v}$ has the opposite direction of $\mathbf{v}$. The zero vector $\mathbf{0}$ has magnitude 0 but arbitrary direction.

iii. To add the vectors $\mathbf{v}$ and $\mathbf{w}$ geometrically, place $\mathbf{w}$ so that its initial point coincides with the terminal point of $\mathbf{v}$ and draw a vector from the initial point of $\mathbf{v}$ to the terminal point of $\mathbf{w}$.

iv. Algebraic vectors. A vector $\mathbf{v}$ with initial point at the origin and terminal point (v_1, v_2) is written as $\mathbf{v} = \langle v_1, v_2 \rangle$; v_1 and v_2 are the horizontal and vertical components of $\mathbf{v}$, respectively.

- The magnitude of $\mathbf{v} = \|\mathbf{v}\| = \sqrt{v_1^2 + v_2^2}$. The direction angle of $\mathbf{v}$ is the angle θ that $\mathbf{v}$ makes with the positive x-axis. We have $\cos\theta = \dfrac{v_1}{\|\mathbf{v}\|}$ and $\sin\theta = \dfrac{v_2}{\|\mathbf{v}\|}$, $0° \le \theta < 360°$.

- The horizontal and vertical components of vector $\mathbf{v}$ with magnitude r and direction angle θ are given by

$$v_1 = r\cos\theta, \quad v_2 = r\sin\theta.$$

v. Vector operations. For vectors $\mathbf{v} = \langle v_1, v_2 \rangle = v_1\mathbf{i} + v_2\mathbf{j}$ and $\mathbf{w} = \langle w_1, w_2 \rangle = w_1\mathbf{i} + w_2\mathbf{j}$, the following operations apply:

- **Vector addition:** $\mathbf{v} + \mathbf{w} = \langle v_1 + w_1, v_2 + w_2 \rangle$
- **Scalar multiplication:** $c\mathbf{v} = \langle cv_1, cv_2 \rangle$

6.5 The Dot Product

i. For two vectors $\mathbf{v} = \langle v_1 v_2 \rangle$ and $\mathbf{w} = \langle w_1 w_2 \rangle$, the dot product

$$\mathbf{v} \cdot \mathbf{w} = v_1 w_1 + v_2 w_2 = \|\mathbf{v}\|\|\mathbf{w}\|\cos\theta,$$

where θ $(0° \le \theta \le 180°)$ is the angle between $\mathbf{v}$ and $\mathbf{w}$.

ii. Vectors $\mathbf{v}$ and $\mathbf{w}$ are

1. orthogonal (perpendicular) if $\mathbf{v} \cdot \mathbf{w} = 0$.
2. parallel if $\mathbf{v} \cdot \mathbf{w} = \pm\|\mathbf{v}\|\|\mathbf{w}\|$.

iii. Vector projection of $\mathbf{v}$ onto $\mathbf{w}$ is given by $\text{proj}_{\mathbf{w}}\mathbf{v} = \dfrac{\mathbf{v} \cdot \mathbf{w}}{\|\mathbf{w}\|^2}\mathbf{w}$.

The scalar projection of $\mathbf{v}$ onto $\mathbf{w}$ is $\dfrac{\mathbf{v} \cdot \mathbf{w}}{\|\mathbf{w}\|}$.

iv. A nonzero vector $\mathbf{v}$ can be written as $\mathbf{v} = \mathbf{v}_1 + \mathbf{v}_2$, where $\mathbf{v}_1 = \text{proj}_{\mathbf{w}}\mathbf{v}$ is parallel to $\mathbf{w}$ and $\mathbf{v}_2 = \mathbf{v} - \mathbf{v}_1$ is orthogonal to $\mathbf{w}$.

v. The **work** W done by a constant force $\mathbf{F}$ in moving an object from a point P to a point Q is $W = \mathbf{F} \cdot \overrightarrow{PQ}$.

6.6 Polar Coordinates

i. The ordered pair (r, θ) represents a point P in the plane that is a directed distance of r units from the origin (pole) O, where θ is the directed angle between the polar axis (positive x-axis) and the line segment OP. The ordered pair (r, θ) gives the polar coordinates of P.

ii. The polar coordinates of a point $P(r, \theta)$ are not unique. In fact, if n is any integer, then

$$(r, \theta), (r, \theta + 2n\pi), \text{ and } (-r, \theta + \pi + 2n\pi)$$

are all polar coordinates of the same point.

iii. To convert a point from polar coordinates (r, θ) to rectangular coordinates, use the equations $x = r\cos\theta$, $y = r\sin\theta$.

iv. To convert a point from rectangular coordinates (x, y) to polar coordinates (r, θ), use the procedure on page 564.

v. A polar equation is an equation whose variables are r and θ.

6.7 Polar Form of Complex Numbers; DeMoivre's Theorem

i. Geometric representation. A complex number $z = a + bi$ can be represented as an ordered pair (a, b) in the complex plane.

ii. Modulus. $|z| = |a + bi| = \sqrt{a^2 + b^2}$, $|z|$ is called the **modulus** of z.

iii. Polar form. The polar form of $z = a + bi$ is $z = r(\cos\theta + i\sin\theta)$, where $r = \sqrt{a^2 + b^2}$, $a = r\cos\theta$, $b = r\sin\theta$, and $\tan\theta = \dfrac{b}{a}$. The number r is the modulus of z, and θ is called the argument of z (written arg z).

iv. Product and quotient rules. Let $z_1 = r_1(\cos\theta_1 + i\sin\theta_1)$ and $z_2 = r_2(\cos\theta_2 + i\sin\theta_2)$. Then

$$z_1 z_2 = r_1 r_2 [\cos(\theta_1 + \theta_2) + i\sin(\theta_1 + \theta_2)]$$

and

$$\frac{z_1}{z_2} = \frac{r_1}{r_2}[\cos(\theta_1 - \theta_2) + i\sin(\theta_1 - \theta_2)], \quad z_2 \ne 0.$$

v. DeMoivre's Theorem. Let $z = r(\cos\theta + i\sin\theta)$ be a complex number in polar form. Then for any integer n,

$$[r(\cos\theta + i\sin\theta)]^n = r^n(\cos n\theta + i\sin n\theta).$$

vi. Roots of complex numbers. The n nth roots of a complex number $r(\cos\theta + i\sin\theta)$ are given by

$$z_k = \sqrt[n]{r}\left[\cos\left(\frac{\theta + 2\pi k}{n}\right) + i\sin\left(\frac{\theta + 2\pi k}{n}\right)\right] \text{ or}$$

$$z_k = \sqrt[n]{r}\left[\cos\left(\frac{\theta + 360°k}{n}\right) + i\sin\left(\frac{\theta + 360°k}{n}\right)\right],$$

where $k = 0, 1, 2, \ldots, n - 1$.

REVIEW EXERCISES

Building Skills

In Exercises 1–10, solve right triangle ABC with right angle at C. Round each answer to the nearest tenth.

1. $A = 30°, a = 6$

2. $B = 35°, b = 5$

3. $A = 37°, b = 4$

4. $B = 43°, a = 10$

5. $A = 40°, c = 12$

6. $B = 50°, c = 8$

7. $a = 3, b = 5$

8. $a = 5, b = 10$

9. $a = 4, c = 6$

10. $b = 3, c = 7$

In Exercises 11–18, use the Law of Sines to solve each triangle ABC. Round each answer to the nearest tenth.

11. $A = 40°, B = 35°, c = 100$

12. $B = 30°, C = 80°, a = 100$

13. $A = 45°, a = 25, b = 75$

14. $B = 36°, a = 12.5, b = 8.7$

15. $A = 48.5°, C = 57.3°, b = 47.3$

16. $A = 67°, a = 100, c = 125$

17. $A = 65.2°, a = 21.3, b = 19$

18. $C = 53°, a = 140, c = 115$

19. In triangle ABC, let $c = 20$ and $B = 60°$. Find a value of b such that C has **(i)** two possible values, **(ii)** exactly one value, **(iii)** no value.

20. Repeat Exercise 19 for $c = 20$ and $B = 150°$.

In Exercises 21–28, use the Law of Cosines to solve each triangle ABC. Round each answer to the nearest tenth.

21. $a = 60, b = 90, c = 125$

22. $a = 15, b = 9, C = 120°$

23. $a = 40, c = 38, B = 80°$

24. $a = 10, b = 20, c = 22$

25. $a = 2.6, b = 3.7, c = 4.8$

26. $a = 15, c = 26, B = 115°$

27. $a = 12, b = 7, C = 130°$

28. $b = 75, c = 100, A = 80°$

In Exercises 29–32, find the area of each triangle ABC with the given information. Round each answer to the nearest square unit.

29. $a = 5$ meters, $b = 7$ meters, $c = 10$ meters

30. $a = 2.4$ meters, $b = 3.4$ meters, $c = 4.4$ meters

31. $A = 65°, b = 6$ feet, $c = 4$ feet

32. $A = 115°, b = 20$ inches, $a = 30$ inches

In Exercises 33 and 34, let v be the vector with initial point P and terminal point Q. Write v as a position vector in terms of i and j.

33. $P(3, 5), Q(2, 7)$

34. $P(-1, 3), Q(5, -4)$

In Exercises 35–38, let $\mathbf{v} = \langle -2, 3 \rangle$ and $\mathbf{w} = \langle 5, -6 \rangle$. Find each expression.

35. $5\mathbf{v}$

36. $2\mathbf{v} + \mathbf{w}$

37. $3\mathbf{v} - 2\mathbf{w}$

38. $\|\mathbf{v} + \mathbf{w}\|$

In Exercises 39–42, find the unit vector u in the direction of the vector v.

39. $\mathbf{v} = \mathbf{i} + \mathbf{j}$

40. $\mathbf{v} = 2\mathbf{i} - 7\mathbf{j}$

41. $\mathbf{v} = \langle 3, -5 \rangle$

42. $\mathbf{v} = \langle -5, -2 \rangle$

In Exercises 43–46, write v in terms of i and j for the given magnitude $\|\mathbf{v}\|$ and direction angle θ.

43. $\|\mathbf{v}\| = 6, \theta = 30°$

44. $\|\mathbf{v}\| = 20, \theta = 120°$

45. $\|\mathbf{v}\| = 12, \theta = 225°$

46. $\|\mathbf{v}\| = 10, \theta = -30°$

In Exercises 47–50, find the dot product $\mathbf{v} \cdot \mathbf{w}$ and $\text{proj}_{\mathbf{w}}\, \mathbf{v}$.

47. $\mathbf{v} = \langle 2, -3 \rangle, \mathbf{w} = \langle 3, 4 \rangle$

48. $\mathbf{v} = \langle -1, -2 \rangle, \mathbf{w} = \langle 4, -1 \rangle$

49. $\mathbf{v} = 2\mathbf{i} - 5\mathbf{j}, \mathbf{w} = 5\mathbf{i} + 2\mathbf{j}$

50. $\mathbf{v} = 2\mathbf{i} + \mathbf{j}, \mathbf{w} = 2\mathbf{i} - \mathbf{j}$

In Exercises 51–54, use the dot product to find the angle $\theta\ (0 \le \theta \le \pi)$ between the vectors v and w. Round your answers to the nearest tenth of a degree.

51. $\mathbf{v} = 2\mathbf{i} + 3\mathbf{j}, \mathbf{w} = -\mathbf{i} + 2\mathbf{j}$

52. $\mathbf{v} = \mathbf{i} + 4\mathbf{j}, \mathbf{w} = -4\mathbf{i} - \mathbf{j}$

53. $\mathbf{v} = \langle 1, 1 \rangle, \mathbf{w} = \langle -3, 2 \rangle$

54. $\mathbf{v} = \langle 1, 5 \rangle, \mathbf{w} = \langle 3, -1 \rangle$

In Exercises 55–58, plot each point in polar coordinates and find its rectangular coordinates.

55. $(24, 30°)$

56. $(12, -60°)$

57. $\left(-2, -\dfrac{\pi}{4} \right)$

58. $\left(-3, -\dfrac{3\pi}{4} \right)$

In Exercises 59–62, convert the rectangular coordinates of each point to polar coordinates (r, θ), with $r > 0$ and $0 \le \theta < 2\pi$.

59. $(-2, 2)$

60. $(\sqrt{3}, 1)$

61. $(2\sqrt{3}, -2)$

62. $(-2, -2\sqrt{3})$

In Exercises 63–66, convert each rectangular equation to a polar equation.

63. $3x + 2y = 12$

64. $x^2 + y^2 = 36$

65. $x^2 + y^2 = 8x$

66. $x^2 + y^2 = 6y$

In Exercises 67–72, convert each polar equation to a rectangular equation.

67. $r = -3$

68. $\theta = \dfrac{5\pi}{6}$

69. $r = 3 \csc \theta$

70. $r = 2 \sec \theta$

71. $r = 1 - 2 \sin \theta$

72. $r = 3 \cos \theta$

In Exercises 73–76, write each complex number in polar form. Express the arguments in radians.

73. $-3i$

74. $-1 + i$

75. $5\sqrt{3} - 5i$

76. $-2 - 2\sqrt{3}i$

In Exercises 77–80, write each complex number in rectangular form.

77. $2(\cos 45° + i \sin 45°)$

78. $3(\cos 240° + i \sin 240°)$

79. $6\left(\cos\dfrac{3\pi}{4} + i\sin\dfrac{3\pi}{4}\right)$ **80.** $4\left(\cos\dfrac{7\pi}{6} + i\sin\dfrac{7\pi}{6}\right)$

In Exercises 81–84, find z_1z_2 and $\dfrac{z_1}{z_2}$. Leave your answers in polar form.

81. $z_1 = 3(\cos 25° + \sin 25°)$, $z_2 = 2(\cos 10° + i\sin 10°)$

82. $z_1 = 4(\cos 300° + i\sin 300°)$, $z_2 = 2(\cos 20° + i\sin 20°)$

83. $z_1 = 2\left(\cos\dfrac{5\pi}{6} + i\sin\dfrac{5\pi}{6}\right)$, $z_2 = 3\left(\cos\dfrac{\pi}{3} + i\sin\dfrac{\pi}{3}\right)$

84. $z_1 = 5\left(\cos\dfrac{4\pi}{3} + i\sin\dfrac{4\pi}{3}\right)$, $z_2 = 15\left(\cos\dfrac{\pi}{3} + i\sin\dfrac{\pi}{3}\right)$

In Exercises 85–88, use DeMoivre's Theorem to find the indicated power. Write your answers in polar form.

85. $[3(\cos 40° + i\sin 40°)]^3$ **86.** $\left[4\left(\cos\dfrac{\pi}{6} + i\sin\dfrac{\pi}{6}\right)\right]^6$

87. $(2 - 2\sqrt{3}i)^6$ **88.** $(2 - 2i)^7$

In Exercises 89–92, find all complex roots. Write your answers in polar form, with the arguments in degrees.

89. Cube roots of -125 **90.** Fourth roots of $-16i$

91. Fifth roots of $-1 + \sqrt{3}i$ **92.** Sixth roots of $1 - i$

Applying the Concepts

93. Width of a river. A surveyor wants to measure the width of a river. She stands at point A, with point B opposite her on the other side of the river. From A, she walks 320 feet along the bank to point C. The line CA makes an angle of 40° with the line CB. What is the width of the river?

94. Angle of elevation of a hill. A tower 120 feet tall is located at the top of a hill. At a point 460 feet down the hill, the angle between the surface of the hill and the line of sight to the top of the tower is 12°. Find the angle of elevation of the hill to a horizontal plane.

95. Distance between cars. Two cars start from point A along two straight roads. The angle between the two roads is 72°. The speeds of the two cars are 55 miles per hour and 65 miles per hour. How far apart are the two cars after 80 minutes? Round to the nearest tenth of a mile.

96. Triangular plot. A triangular plot of land has sides of lengths 310 feet, 415 feet, and 175 feet. Find the largest angle between the sides.

97. Area. Find the area of the triangular plot in Exercise 96. Round your answer to the nearest square foot.

98. Resultant force. Two forces of magnitudes 16 and 20 pounds are acting on an object. The bearings of the forces are N 75° E and S 20° E, respectively. Find the magnitude and direction of the resultant force, to the nearest tenth.

99. Work. A force of magnitude 30 pounds at an angle of 48° is used to pull a wagon 60 feet. Find the work done.

100. AC current. The total impedance in an AC circuit is given by $Z = \dfrac{Z_1 Z_2}{Z_1 + Z_2}$. Find Z if $Z_1 = 20\left(\cos\dfrac{\pi}{6} + i\sin\dfrac{\pi}{6}\right)$ and $Z_2 = 10\left(\cos\dfrac{2\pi}{3} + i\sin\dfrac{2\pi}{3}\right)$.

PRACTICE TEST A

1. In Problems 1–4, solve triangle ABC. Round each answer to the nearest tenth.

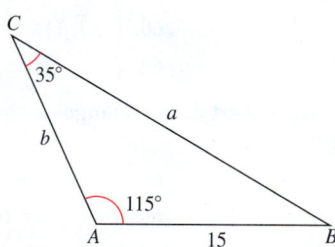

2. $A = 42°$, $B = 37°$, $a = 50$ meters.

3.

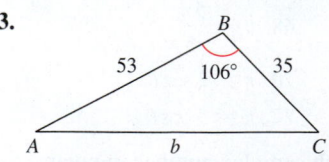

4. $a = 30$ feet, $b = 20$ feet, $c = 25$ feet.

5. Solve right triangle ABC if $a = 5.6$ and $b = 4.1$. Round each answer to the nearest tenth.

6. Find the measure of the central angle of a circle of radius 8 feet that intercepts a chord of length 4.5 feet. Round to the nearest tenth of a degree.

7. A surveyor, starting from point A, walks 580 feet in the direction N 70.0° E. From that point, she walks 725 feet in the direction N 35.0° W. How far, to the nearest foot, is she from her starting point?

8. The vector $\mathbf{v}$ has initial point $P(3, 5)$ and terminal point $Q(2, -7)$. Write $\mathbf{v}$ as a position vector.

9. If $\mathbf{v} = \langle -2, 3\rangle$ and $\mathbf{w} = \langle 1, 5\rangle$, find $\mathbf{v} - \mathbf{w}$.

10. If $\mathbf{v} = -2\mathbf{i} + 3\mathbf{j}$ and $\mathbf{w} = \mathbf{i} + 5\mathbf{j}$, find $2\mathbf{v} - 3\mathbf{w}$.

11. Let $\|\mathbf{v}\| = 3$ and suppose $\mathbf{v}$ makes an angle of $\theta = -30°$ with the positive x-axis. Write the vector $\mathbf{v}$ in the form $v_1\mathbf{i} + v_2\mathbf{j}$.

12. Find the dot product $\mathbf{v} \cdot \mathbf{w}$ if $\mathbf{v} = 4\mathbf{i} + 3\mathbf{j}$ and $\mathbf{w} = -\mathbf{i} + 7\mathbf{j}$.

13. Find the angle $\theta(0 \le \theta \le 180°)$ between the vectors $\mathbf{v} = 3\mathbf{i} - 4\mathbf{j}$ and $\mathbf{w} = -2\mathbf{i} + 5\mathbf{j}$. Round the answer to the nearest tenth of a degree.

14. Convert $(-2, -45°)$ to rectangular coordinates.

15. Convert $(-\sqrt{3}, -1)$ to polar coordinates with $r > 0$ and $0 \le \theta < 2\pi$.

16. Convert the polar equation $r = -3$ to rectangular form. Identify the curve having this equation.

17. Write $3\left(\cos\left(\dfrac{2\pi}{3}\right) + i\sin\left(\dfrac{2\pi}{3}\right)\right)$ in rectangular form.

18. Write $\dfrac{z_1}{z_2}$ in polar form, if $z_1 = 1 - \sqrt{3}i$, $z_2 = -1 + i$, with the argument given in degrees.

19. Use DeMoivre's Theorem to find
$$\left[\sqrt{5}\left(\cos\left(\dfrac{\pi}{4}\right) + i\sin\left(\dfrac{\pi}{4}\right)\right)\right]^{-6}.$$ Write the answer in rectangular form.

20. Find all fourth roots of $1 + i$. Express the argument in degrees.

PRACTICE TEST B

1. In triangle ABC, which is closest to the correct value for b?
   a. $b = 13.4$
   b. $b = 20.8$
   c. $b = 20.2$
   d. $b = 0.05$

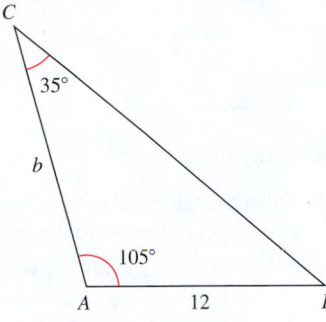

2. In triangle ABC, if $A = 27°$, $B = 50°$, and $a = 25$ meters, which is closest to the correct value for c?
   a. $c = 0.03$ meter
   b. $c = 42.2$ meters
   c. $c = 53.7$ meters
   d. $c = 0.02$ meter

3. In a triangle ABC, which statement is true for the measurements $A = 25°$, $a = 25$ feet, $b = 30$ feet?
   (The angle measurements in choices (c) and (d) are given to the nearest tenth.)
   a. No triangle has these measurements.
   b. One triangle has these measurements.
   c. Two triangles, with $c_1 = 48.7$ and $c_2 = 5.6$, have these measurements (in feet).
   d. Two triangles, with $c_1 = 46$ and $c_2 = 6.4$, have these measurements (in feet).

4. In triangle ABC, which is closest to the correct value for b?
   a. $b = 35.8$
   b. $b = 57.2$
   c. $b = 18.7$
   d. $b = 48.4$

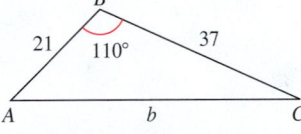

5. A 3-foot shrub is 4.5 feet away from a 12-foot streetlight. What length shadow is cast by the shrub?
   a. 1.5 feet   **b.** 2 feet   **c.** 2.67 feet   **d.** 4 feet

6. In triangle ABC, with $a = 12$ feet, $b = 13$ feet, and $c = 7$ feet, which is closest to the correct value for A?
   a. $24°$ **b.** $66°$ **c.** $81.8°$ **d.** $32.2°$

7. Find the measure of the central angle of a circle of radius 9 feet that intercepts a chord of length 3 feet. Round to the nearest tenth of a degree.
   a. $70.8°$ **b.** $63.6°$ **c.** $26.4°$ **d.** $19.2°$

8. A hunter leaves a straight east–west road at a point A, walking 2 miles at a bearing N 40° W to a point B. From point B, he walks at a bearing of S 20° E and arrives back on the road at a point C. How far, to the nearest tenth of a mile, is point B from point C?
   a. 4.5 miles   **b.** 2.9 miles   **c.** 1.6 miles   **d.** 5.3 miles

9. The vector $\mathbf{v}$ has initial point $P(-2, 5)$ and terminal point $Q(3, -1)$. Write $\mathbf{v}$ as a position vector.
   a. $\langle 5, -6 \rangle$ **b.** $\langle 1, -6 \rangle$ **c.** $\langle 1, -4 \rangle$ **d.** $\langle 5, -4 \rangle$

10. If $\mathbf{v} = \langle 1, -4 \rangle$ and $\mathbf{w} = \langle 3, -2 \rangle$, find $\mathbf{v} - \mathbf{w}$.
    a. $\langle 4, -2 \rangle$ **b.** $\langle 2, 2 \rangle$ **c.** $\langle -2, -2 \rangle$ **d.** $\langle 4, -6 \rangle$

11. Let $\mathbf{v} = 7\mathbf{i} + 3\mathbf{j}$ and $\mathbf{w} = -2\mathbf{i} + \mathbf{j}$. Find $2\mathbf{v} - 3\mathbf{w}$.
    a. $20\mathbf{i} + 3\mathbf{j}$ **b.** $20\mathbf{i} - 3\mathbf{j}$ **c.** $8\mathbf{i} - 6\mathbf{j}$ **d.** $8\mathbf{i} + 6\mathbf{j}$

12. Let $\|\mathbf{v}\| = 6$ and suppose $\mathbf{v}$ makes an angle of $\theta = -60°$ with the positive x-axis. Write the vector $\mathbf{v}$ in the form $v_1\mathbf{i} + v_2\mathbf{j}$.
    a. $3\sqrt{3}\mathbf{i} - 3\mathbf{j}$
    b. $3\sqrt{3}\mathbf{i} + 3\mathbf{j}$
    c. $3\mathbf{i} + 3\sqrt{3}\mathbf{j}$
    d. $3\mathbf{i} - 3\sqrt{3}\mathbf{j}$

13. Find the dot product $\mathbf{v} \cdot \mathbf{w}$ if $\mathbf{v} = -2\mathbf{i} - 5\mathbf{j}$ and $\mathbf{w} = 9\mathbf{i} - 4\mathbf{j}$.
    a. -38 **b.** 38 **c.** $2\mathbf{i} + 20\mathbf{j}$ **d.** 2

14. Find the angle between the vectors $\mathbf{v} = 7\mathbf{i} - \mathbf{j}$ and $\mathbf{w} = \mathbf{i} + \sqrt{2}\mathbf{j}$. Round the answer to the nearest degree.
    a. $84°$ **b.** $63°$ **c.** $32°$ **d.** $17°$

15. Convert $(-4, -30°)$ to rectangular coordinates.
    a. $(-2\sqrt{3}, -2)$
    b. $(2, -2\sqrt{3})$
    c. $(-2\sqrt{3}, 2)$
    d. $(2\sqrt{3}, -2)$

16. Convert $(3, -\sqrt{3})$ to polar coordinates, with $r > 0$ and $0 \le \theta < 360°$.
    a. $(12, 330°)$
    b. $(2\sqrt{3}, 330°)$
    c. $(2\sqrt{3}, 150°)$
    d. $(\sqrt{3}, 150°)$

17. Which polar equation has a circle for its graph?
    a. $r = -4\cos\theta + 1$
    b. $r = -2$
    c. $r\sin\theta = 2$
    d. $r\cos\theta = 1$

18. Write $5\left(\cos\left(\dfrac{7\pi}{3}\right) + i\sin\left(\dfrac{7\pi}{3}\right)\right)$ in rectangular form.
    a. $5 + 5\sqrt{3}i$
    b. $5 - 5\sqrt{3}i$
    c. $\dfrac{5}{2} + \dfrac{5\sqrt{3}}{2}i$
    d. $\dfrac{5}{2} - \dfrac{5\sqrt{3}}{2}i$

19. Write $\dfrac{z_1}{z_2}$ in polar form, if $z_1 = i$ and $z_2 = 1 - i$.
    a. $1(\cos 135° + i\sin 135°)$
    b. $\sqrt{2}(\cos 135° + i\sin 135°)$
    c. $\dfrac{\sqrt{2}}{2}(\cos 135° + i\sin 135°)$
    d. $\dfrac{\sqrt{2}}{2}(\cos 315° + i\sin 315°)$

20. Use DeMoivre's Theorem to find
$$\left[\sqrt{2}\left(\cos\left(\dfrac{\pi}{12}\right) + i\sin\left(\dfrac{\pi}{12}\right)\right)\right]^{4}.$$ Write the answer in rectangular form.
    a. $-2 + 2\sqrt{3}i$
    b. $\dfrac{1}{2} + \dfrac{\sqrt{3}}{2}i$
    c. $2 + 2\sqrt{3}i$
    d. $\dfrac{1}{2} - \dfrac{\sqrt{3}}{2}i$

CUMULATIVE REVIEW EXERCISES CHAPTERS 1–6

1. If $f(x) = \sqrt{x}$ and $g(x) = \dfrac{1}{x-2}$, find the domain of $\dfrac{g}{f}$.

2. If $f(x) = 2x + 7$, find $f^{-1}(x)$.

3. Sketch the graph of $y = -2x^2 + 8x + 10$.

4. Solve the inequality $x^2 - 6x + 8 > 0$.

In Exercises 5 and 6, solve each equation for all solutions in the interval $[0, 2\pi)$. Give your answers in radians.

5. $\sin\theta \cos\theta = -\dfrac{\sqrt{3}}{4}$

6. $4\sin^2\theta - 1 = 0$

In Exercises 7 and 8, find all complex roots. Write your answers in polar form, with the argument in degrees.

7. Fourth roots of $\sqrt{3} + i$

8. Sixth roots of -64

9. Find the equation of the line in slope–intercept form that passes through the point $(-1, 5)$ and is perpendicular to the line $2x + 3y + 6 = 0$.

10. Write the following in condensed form:

$$2 \log x + \frac{1}{2} \log (y + 1) - \log (3x + 1).$$

In Exercises 11–14, use transformations to sketch the graph of each function.

11. $f(x) = 2\sqrt{x - 1} - 3$

12. $f(x) = -|x + 1| + 2$

13. $f(x) = -2 \cos (3x + \pi)$

14. $f(x) = 2 \sec 3x$

15. If $\cos\theta = -\dfrac{5}{13}$ and $\sin\theta > 0$, find $\tan\theta$.

16. Use an identity to find the exact value of $\cos 65° \cos 35° + \sin 65° \sin 35°$.

17. Verify the identity

$$\frac{1}{1 - \cos x} + \frac{1}{1 + \cos x} = 2(1 + \cot^2 x).$$

18. Verify the identity $\dfrac{1 + \cos 2x}{\sin 2x} = \cot x.$

19. Solve triangle ABC with $A = 65°$, $B = 46°$, $c = 60$ meters. Round your answers to the nearest tenth if necessary.

20. A ladder leaning against a vertical wall makes an angle of $47°$ with the ground. The foot of the ladder is 2.3 meters from the wall. Find the length of the ladder. Round to the nearest tenth.

Systems of Equations and Inequalities

TOPICS

7.1 Systems of Equations in Two Variables

7.2 Systems of Linear Equations in Three Variables

7.3 Systems of Inequalities

7.4 Matrices and Systems of Equations

7.5 Determinants and Cramer's Rule

7.6 Partial-Fraction Decomposition

7.7 Matrix Algebra

7.8 The Matrix Inverse

Systems of equations are frequently the tools used to describe relationships among variables representing different interrelated quantities. These same systems of equations are used to predict results associated with changes in the quantities.

The quantities involved can come from the study of medicine, taxes, elections, business, agriculture, city planning, or any discipline in which multiple quantities interact.

Systems of Equations in Two Variables

BEFORE STARTING THIS SECTION, REVIEW

1 Graphs of equations (Section 1.1, page 7)

2 Graphs of linear equations (Section 1.2, page 26)

OBJECTIVES

1 Verify a solution to a system of equations.

2 Solve a system of equations by the graphical method.

3 Solve a system of equations by the substitution method.

4 Solve a system of equations by the elimination method.

5 Solve applied problems by solving systems of equations.

Al-Khwarizmi (780–850)
Al-Khwarizmi came to Baghdad from the town of Khwarizm, which is now called *Khivga* and is part of Uzbekistan. The term *algorithm* is a corruption of the name al-khwarizmi. Originally, the word *algorism* was used for the rules for performing arithmetic by using decimal notation. *Algorism* evolved into the word *algorithm* by the eighteenth century. The word *algorithm* now means a finite set of precise instructions for performing a computation or for solving a problem. (The picture shown is a Soviet postage stamp from 1983 depicting Al-Khwarizmi.)

◆ Algebra–Iraq Connection

The most important contributions of medieval Islamic mathematicians lie in the area of algebra. One of the greatest Islamic scholars was Muhammad ibn Musa al-Khwarizmi (A.D. 780–850). Al-Khwarizmi was one of the first scholars in the House of Wisdom, an academy of scientists established by caliph al-Mamun in the city of Baghdad in what is now Iraq. Jews, Christians, and Muslims worked together in scholarly pursuits in the academy during this period. The eminent scholar al-Khwarizmi wrote several books on astronomy and mathematics. Western Europeans first learned about algebra from his books. The word *algebra* comes from the Arabic *al-jabr*, part of the title of his book *Kitab al-jabr wal-muqabala*.

This book was translated into Latin and was a widely used text. The Arabic word *al-jabr* means "restoration," as in restoring broken parts. (At one time, it was not unusual to see the sign "Algebrista y Sangrador," meaning "bone setter and blood letter," at the entrance of a Spanish barber's shop. The sign informed customers of the barber's side business.) Al-Khwarizmi used the term in the mathematical sense of removing a negative quantity on one side of an equation and restoring it as a positive quantity on the other side.

1 Verify a solution to a system of equations.

System of Equations

A set of equations with common variables is called a **system of equations**. If each equation in a system of equations is linear, then the set of equations is called a **system of linear equations** or a **linear system of equations**. However, if at least one equation in a system of equations is nonlinear, then the set of equations is called a **nonlinear system of equations**. In this section, we will solve systems of two equations in two variables, such as:

$$\begin{cases} 2x - y = 5 \\ x + 2y = 5 \end{cases}$$

Notice that we designate a system of equations by using a left brace. The brace is placed to remind us that the equations must be dealt with simultaneously. A system of equations is sometimes referred to as a set of *simultaneous equations*. A **solution of a system of equations in two variables** x and y is an ordered pair of numbers (a, b), such that when x is replaced with a and y is replaced with b, the resulting equations are true. The **solution set of a system of equations** is the set of all solutions of the system.

EXAMPLE 1 Verifying a Solution

Verify that the ordered pair $(3, 1)$ is a solution of the system of linear equations

$$\begin{cases} 2x - y = 5 & \text{Equation (1)} \\ x + 2y = 5 & \text{Equation (2)} \end{cases}$$

Solution

To verify that $(3, 1)$ is a solution, replace x with 3 and y with 1 in both equations.

Equation (1)	Equation (2)
$2x - y = 5$	$x + 2y = 5$
$2(3) - 1 \overset{?}{=} 5$	$3 + 2(1) \overset{?}{=} 5$
$5 = 5 ✓$	$5 = 5 ✓$

Since the ordered pair $(3, 1)$ satisfies both equations, it is a solution of the system.

Practice Problem 1 Verify that $(1, 3)$ is a solution of the system

$$\begin{cases} x + y = 4 \\ 3x - y = 0 \end{cases}$$

In this section, you will learn three methods of solving a system of two equations in two variables: the *graphical method,* the *substitution method,* and the *elimination method.*

2 Solve a system of equations by the graphical method.

Graphical Method

Recall that the graph of a linear equation

$$ax + by = c \quad (a \text{ and } b \text{ not both zero})$$

is a line. Consider a system of linear equations:

$$\begin{cases} a_1x + b_1y = c_1 \\ a_2x + b_2y = c_2 \end{cases}$$

A solution of this system (if any) is a point in the xy-plane whose coordinates satisfy both equations. To find or estimate the solution set of this system, therefore, we graph both equations on the same coordinate axes and find the coordinates of any points of intersection. This procedure is called the **graphical method**.

EXAMPLE 2 Solving a System by the Graphical Method

Use the graphical method to solve the system of equations.

$$\begin{cases} 2x - y = 4 & \text{Equation (1)} \\ 2x + 3y = 12 & \text{Equation (2)} \end{cases}$$

Solution

Step 1 Graph both equations on the same coordinate axes.
  (i) We first graph equation (1) by finding its intercepts.
    a. To find the y-intercept, set $x = 0$ and solve for y.
      We have $2(0) - y = 4$, or $y = -4$; so the y-intercept is -4.

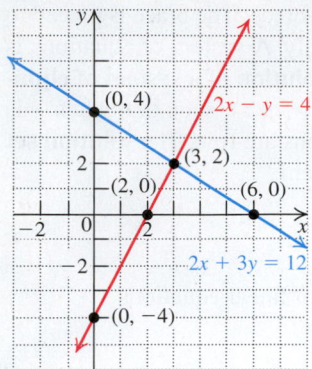

Figure 7.1 Graphical solution.

In Example 2, solve each equation for y. Graph

$$Y_1 = 2x - 4 \quad \text{and}$$

$$Y_2 = -\frac{2}{3}x + 4$$

on the same screen.

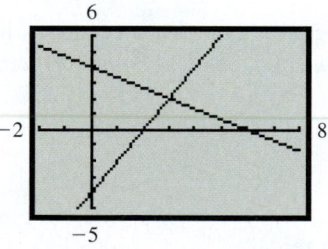

Use the **intersect** feature on your calculator to find the point of intersection.

b. To find the x-intercept, set $y = 0$ and solve for x. We have $2x - 0 = 4$, or $x = 2$; so the x-intercept is 2.

The points $(0, -4)$ and $(2, 0)$ are on the graph of equation (1), which is sketched in red in Figure 7.1.

(ii) We now graph equation (2) using intercepts.

Here the x-intercept is 6 and the y-intercept is 4; so joining the points $(0, 4)$ and $(6, 0)$ produces the line sketched in blue in Figure 7.1.

Step 2 **Find the point(s) of intersection of the two graphs.** We observe in Figure 7.1 that the point of intersection of the two graphs is $(3, 2)$.

Step 3 **Check your solution(s).** Replace x with 3 and y with 2 in equations (1) and (2).

Equation (1)	Equation (2)
$2x - y = 4$	$2x + 3y = 12$
$2(3) - 2 \overset{?}{=} 4$	$2(3) + 3(2) \overset{?}{=} 12$
$4 = 4$ ✓	$12 = 12$ ✓

Step 4 **Write the solution set for the system.**

The solution set is $\{(3, 2)\}$.

Practice Problem 2 Solve the system of equations graphically.

$$\begin{cases} x + y = 2 \\ 4x + y = -1 \end{cases}$$

If a system of equations has at least one solution (as in Example 2), the system is **consistent**. A system of equations with no solution is **inconsistent**, and its solution set is the empty set, $\varnothing$.

A system of two linear equations in two variables must have one of the following types of solution sets:

1. One solution (the lines intersect; see Figure 7.2(a)): the system is consistent, and the equations in the system are said to be **independent**.

2. No solution (the lines are parallel; see Figure 7.2(b)): the system is inconsistent.

3. Infinitely many solutions (the lines coincide; see Figure 7.2(c)): the system is consistent, and the equations in the system are said to be **dependent**.

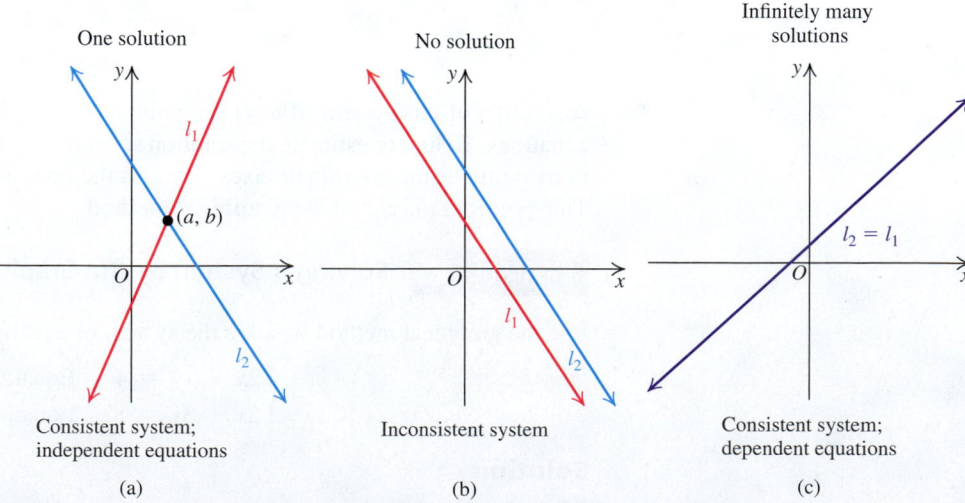

Figure 7.2 Possible solution sets of a system of two linear equations.

3 Solve a system of equations by the substitution method.

Substitution Method

Another way to solve a linear system of equations is called the **substitution method**. Non-graphical methods use a technique that replaces the original system of equations with an *equivalent system*, a system with the same solutions as the original system.

PROCEDURE
IN ACTION

EXAMPLE 3   The Substitution Method

OBJECTIVE

Reduce the solution of the system to the solution of one equation in one variable by substitution.

Step 1 Solve for one variable. Choose one of the equations and express one of its variables in terms of the other variable.

Step 2 Substitute. Substitute the expression found in Step 1 into the other equation to obtain an equation in one variable.

Step 3 Solve the equation obtained in Step 2.

Step 4 Back-substitute. Substitute the value(s) you found in Step 3 back into the expression you found in Step 1. The result is the solution(s).

Step 5 Check. Check your answer(s) in the original equations.

EXAMPLE

Solve the system.

$$\begin{cases} 2x - 5y = 3 & \text{Equation (1)} \\ y - 2x = 9 & \text{Equation (2)} \end{cases}$$

1. We choose equation (2) and express y in terms of x.

   $y = 2x + 9$ (3)   Solve equation (2) for y.

2. $2x - 5y = 3$ Equation (1)

   $2x - 5(2x + 9) = 3$ Substitute $2x + 9$ for y.

   $2x - 10x - 45 = 3$ Distributive property

3. $-8x - 45 = 3$ Combine like terms.

   $-8x = 3 + 45$ Add 45 to both sides.

   $x = -6$ Solve for x.

4. $y = 2x + 9$ Equation (3) from Step 1

   $y = 2(-6) + 9 = -3$ Substitute $x = -6$.

   Because $x = -6$ and $y = -3$, the solution set is $\{(-6, -3)\}$.

5. **Check:** $x = -6$ and $y = -3$.

Equation (1)	Equation (2)
$2(-6) - 5(-3) \overset{?}{=} 3$	$-3 - 2(-6) \overset{?}{=} 9$
$-12 + 15 \quad = 3 \checkmark$	$-3 + 12 \quad = 9 \checkmark$

Practice Problem 3 Solve: $\begin{cases} x - y = 5 \\ 2x + y = 7 \end{cases}$

SIDE
NOTE

It is easy to make arithmetic errors in the many steps involved in the process of solving a system of equations. You should check your solution by substituting it into each equation in the original system.

It is possible that in the process of solving the equation in Step 3, you obtain an equation of the form $0 = k$, where k is a *nonzero* constant. In such cases, the false statement, $0 = k$, indicates that the system is *inconsistent*.

EXAMPLE 4   Attempting to Solve an Inconsistent System of Equations

Solve the system of equations.

$$\begin{cases} x + y = 3 & \text{Equation (1)} \\ 2x + 2y = 9 & \text{Equation (2)} \end{cases}$$

Solution

Step 1 Solve equation (1) for y in terms of x.

$$y = 3 - x \quad \text{Add } -x \text{ to both sides.}$$

Step 2 Substitute this expression into equation (2).

$$2x + 2y = 9 \quad \text{Equation (2)}$$
$$2x + 2(3 - x) = 9 \quad \text{Replace } y \text{ with } 3 - x \text{ (from Step 1).}$$

Step 3 Solve for x.

$$2x + 6 - 2x = 9 \quad \text{Distributive property}$$
$$\underbrace{6 = 9}_{\text{False}} \quad \text{Simplify.}$$

Since the equation $6 = 9$ is false, the system is inconsistent. Figure 7.3 shows that the graphs of the two equations in this system are parallel lines. Because the lines do not intersect, the system has no solution. The solution set is $\varnothing$.

Practice Problem 4 Solve the system of equations.

$$\begin{cases} x - 3y = 1 \\ -2x + 6y = 3 \end{cases}$$

In Step 3, it is also possible to end up with an equation of the form $0 = 0$. In such cases, the equations are *dependent*, and the system has infinitely many solutions.

EXAMPLE 5 **Solving a Dependent System**

Solve the system of equations.

$$\begin{cases} 4x + 2y = 12 & \text{Equation (1)} \\ -2x - y = -6 & \text{Equation (2)} \end{cases}$$

Solution

Step 1 Solve equation (2) for y in terms of x.

$$-2x - y = -6 \quad \text{Equation (2)}$$
$$-y = -6 + 2x \quad \text{Add } 2x \text{ to both sides.}$$
$$y = 6 - 2x \quad \text{Multiply both sides by } -1.$$

Step 2 Substitute $(6 - 2x)$ for y in equation (1).

$$4x + 2y = 12 \quad \text{Equation (1)}$$
$$4x + 2(6 - 2x) = 12 \quad \text{Replace } y \text{ with } 6 - 2x \text{ (from Step 1).}$$

Step 3 Solve for x.

$$4x + 12 - 4x = 12 \quad \text{Distributive property}$$
$$\underbrace{0 = 0}_{\text{True}} \quad \text{Subtract 12 from both sides and simplify.}$$

The equation $0 = 0$ is true for *every* value of x. Thus, *any* value of x can be used in the equation $y = 6 - 2x$ for back-substitution.

The solutions of the system are of the form $(x, 6 - 2x)$, and the solution set is

$$\{(x, 6 - 2x) \mid x \text{ any real number}\}.$$

In other words, the solution set consists of the infinite number of ordered pairs (x, y) lying on the line with equation $4x + 2y = 12$, as shown in Figure 7.4. You can find particular solutions by replacing x with particular real numbers. For example, if we let $x = 0$ in $(x, 6 - 2x)$, we find that $(0, 6 - 2(0)) = (0, 6)$ is a solution of the system. Similarly, letting $x = 1$, we find that $(1, 4)$ is a solution.

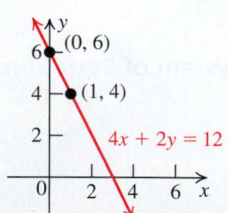

Figure 7.3 Inconsistent system.

Figure 7.4 Dependent equations.

Practice Problem 5 Solve the system of equations.

$$\begin{cases} -2x + y = -3 \\ 4x - 2y = 6 \end{cases}$$

EXAMPLE 6 **Using Substitution to Solve a Nonlinear System**

Solve the system of equations by the substitution method.

$$\begin{cases} 4x + y = -3 & \text{Equation (1)} \\ -x^2 + y = 1 & \text{Equation (2)} \end{cases}$$

Solution

We follow the same steps as those used in the substitution method for solving a system of linear equations.

Step 1 **Solve for one variable.** In equation (2), you can express y in terms of x:

$$y = x^2 + 1 \quad (3) \qquad \text{Solve equation (2) for } y.$$

Step 2 **Substitute.** Substitute $x^2 + 1$ for y in equation (1).

$$4x + y = -3 \qquad \text{Equation (1)}$$
$$4x + (x^2 + 1) = -3 \qquad \text{Replace } y \text{ with } (x^2 + 1).$$
$$4x + x^2 + 4 = 0 \qquad \text{Add 3 to both sides.}$$
$$x^2 + 4x + 4 = 0 \qquad \text{Rewrite in descending powers of } x.$$

Step 3 **Solve** the equation resulting from Step 2.

$$(x + 2)(x + 2) = 0 \qquad \text{Factor.}$$
$$x + 2 = 0 \qquad \text{Zero-product property}$$
$$x = -2 \qquad \text{Solve for } x.$$

Step 4 **Back-substitution.** Substitute $x = -2$ in equation (3) to find the corresponding y-value.

$$y = x^2 + 1 \qquad \text{Equation (3)}$$
$$y = (-2)^2 + 1 \qquad \text{Replace } x \text{ with } -2.$$
$$y = 5 \qquad \text{Simplify.}$$

Since $x = -2$ and $y = 5$, the apparent solution set of the system is $\{(-2, 5)\}$.

Step 5 **Check.** Replace x with -2 and y with 5 in equation (1) and equation (2).

Equation (1) | Equation (2)

$$4x + y = -3 \qquad\qquad -x^2 + y = 1$$
$$4(-2) + 5 \overset{?}{=} -3 \qquad -(-2)^2 + 5 \overset{?}{=} 1$$
$$-8 + 5 = -3 ✓ \qquad\qquad -4 + 5 = 1 ✓$$

The graphs of the line $4x + y = -3$ and the parabola $y = x^2 + 1$ confirm that the solution set is $\{(-2, 5)\}$. See Figure 7.5.

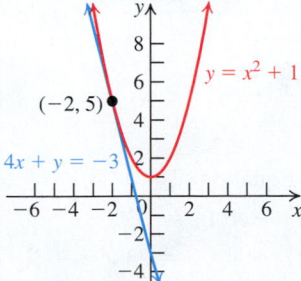

Figure 7.5

Practice Problem 6 Solve the system of equations by the substitution method.

$$\begin{cases} x^2 + y = 2 \\ 2x + y = -1 \end{cases}$$

4 Solve a system of equations by the elimination method.

Elimination Method

The **elimination method** of solving a system of equations is also called the **addition method**.

PROCEDURE
IN ACTION

EXAMPLE 7 The Elimination Method

OBJECTIVE

Solve a system of two linear equations by first eliminating one variable.

EXAMPLE

Solve the system.

$$\begin{cases} 2x + 3y = 21 & \text{Equation (1)} \\ 3x - 4y = 23 & \text{Equation (2)} \end{cases}$$

Step 1 Adjust the coefficients. If necessary, multiply both equations by appropriate numbers to get two new equations in which the coefficients of the variable to be eliminated are opposites.

1. We arbitrarily select y as the variable to be eliminated.

$$\begin{cases} 8x + 12y = 84 & \text{Multiply equation (1) by 4.} \\ 9x - 12y = 69 & \text{Multiply equation (2) by 3.} \end{cases}$$

Coefficients are opposites of each other.

Step 2 Add the equations. Add the resulting equations to get an equation in one variable.

2. $\quad 8x + 12y = 84$ Adding the two equations from
$\quad \underline{9x - 12y = 69}$ Step 1 eliminates the y-terms.
$\quad 17x = 153$

Step 3 Solve the resulting equation.

3. $17x = 153$ Equation from Step 2

$x = \dfrac{153}{17}$ Divide both sides by 17.

$x = 9$ Simplify.

Step 4 Back-substitute the value you found into one of the original equations to solve for the other variable.

4. $\quad 2x + 3y = 21$ Equation (1)
$\quad 2(9) + 3y = 21$ Replace x with 9.
$\quad 18 + 3y = 21$ Simplify.
$\quad 3y = 3$ Subtract 18 from both sides.
$\quad y = 1$ Solve for y.

Step 5 Write the solution set from Steps 3 and 4.

5. The solution set is $\{(9, 1)\}$.

Step 6 Check your solution(s) in the original equations (1) and (2).

6. Check $x = 9$ and $y = 1$.

Equation (1)   Equation (2)
$2(9) + 3(1) \overset{?}{=} 21$ | $3(9) - 4(1) = 23$
$18 + 3 = 21 ✓$ | $27 - 4 = 23 ✓$

Practice Problem 7 Solve the system.

$$\begin{cases} 3x + 2y = 3 \\ 9x - 4y = 4 \end{cases}$$

Just as in the substitution method, if you add the equations in Step 2 and the resulting equation becomes $0 = k$, where $k \neq 0$, then the system is inconsistent: it has no solutions. As an illustration, if you solve the system of equations in Example 4 by the elimination method, you will obtain the equation $0 = 3$. You now conclude that the system is inconsistent. Similarly, in Step 2, if you obtain $0 = 0$, then the equations are dependent. See Example 5.

The next example illustrates how some nonlinear systems can be solved by making a substitution to create a linear system and then solving the new system by the elimination method. We then find the solution(s) to the original equations from the substitution formula.

EXAMPLE 8 **Solving a Nonlinear System by a Linearizing Substitution**

Solve the system.

$$\begin{cases} \dfrac{2}{x} + \dfrac{5}{y} = -5 & \text{Equation (1)} \\ \dfrac{3}{x} - \dfrac{2}{y} = -17 & \text{Equation (2)} \end{cases}$$

Solution

Replace $\dfrac{1}{x}$ with u and $\dfrac{1}{y}$ with v. (Note that $\dfrac{2}{x} = 2\left(\dfrac{1}{x}\right) = 2u$, and so on.) Equations (1) and (2) become

$$\begin{cases} 2u + 5v = -5 & \text{Equation (3)} \\ 3u - 2v = -17 & \text{Equation (4)} \end{cases}$$

Now we solve the new system for u and v using the elimination method.

Step 1 We choose to eliminate the variable u.

$6u + 15v = -15$ (5)	Multiply equation (3) by 3.
$-6u + \ 4v = \ \ 34$ (6)	Multiply equation (4) by -2.

Step 2 $\quad\quad\quad\quad 19v = \ \ 19$ (7) $\quad$ Add equations (5) and (6).

Step 3 $\quad\quad\quad\quad\quad v = 1$ $\quad\quad\quad\quad$ Solve equation (7) for v.

Step 4 $\quad\quad 2u + 5v = -5$ $\quad\quad\quad$ Equation (3)

$\quad\quad\quad\quad 2u + 5(1) = -5$ $\quad\quad\quad$ Back-substitute $v = 1$.

$\quad\quad\quad\quad\quad\quad\quad u = -5$ $\quad\quad\quad$ Solve for u.

Step 5 Now solve for x and y, the variables in the original system.

$$u = \dfrac{1}{x} \quad \text{and} \quad v = \dfrac{1}{y} \quad \text{Original substitution}$$

$$-5 = \dfrac{1}{x} \quad\quad\quad 1 = \dfrac{1}{y} \quad \text{Replace } u \text{ with } -5 \text{ and } v \text{ with } 1.$$

$$x = -\dfrac{1}{5} \quad\quad\quad y = 1 \quad \text{Solve for } x \text{ and } y.$$

The solution set of the original system is $\left\{\left(-\dfrac{1}{5}, 1\right)\right\}$.

Step 6 Verify that the ordered pair $\left(-\dfrac{1}{5}, 1\right)$ is the solution of the original system of equations (1) and (2).

Practice Problem 8 Solve the system.

$$\begin{cases} \dfrac{4}{x} + \dfrac{3}{y} = 1 \\ \dfrac{2}{x} - \dfrac{6}{y} = 3 \end{cases}$$

EXAMPLE 9 **Using Elimination to Solve a Nonlinear System**

Solve the system of equations by the elimination method.

$$\begin{cases} x^2 + y^2 = 25 & \text{Equation (1)} \\ x^2 - y = 5 & \text{Equation (2)} \end{cases}$$

Solution

Step 1 **Adjust the coefficients.** Because x has the same power in both equations, we choose the variable x for elimination. We multiply equation (2) by -1 and obtain the equivalent system.

$$\begin{cases} x^2 + y^2 = 25 & (1) \\ -x^2 + y = -5 & (3) \qquad \text{Multiply equation (2) by } -1. \end{cases}$$

Step 2 $\qquad y^2 + y = 20 \quad (4) \qquad$ Add equations (1) and (3).

Step 3 Solve the equation found in Step 2: $y^2 + y = 20$.

$$y^2 + y - 20 = 0 \qquad \text{Subtract 20 from both sides.}$$
$$(y + 5)(y - 4) = 0 \qquad \text{Factor.}$$
$$y + 5 = 0 \quad \text{or} \quad y - 4 = 0 \qquad \text{Zero-product property}$$
$$y = -5 \quad \text{or} \qquad y = 4 \qquad \text{Solve for } y.$$

Step 4 **Back-substitute** the values in one of the original equations to solve for the other variable.

(i) Substitute $y = -5$ in equation (2) and solve for x.

$$x^2 - y = 5 \qquad \text{Equation (2)}$$
$$x^2 - (-5) = 5 \qquad \text{Replace } y \text{ with } -5.$$
$$x^2 = 0 \qquad \text{Simplify.}$$
$$x = 0 \qquad \text{Solve for } x.$$

So, $(0, -5)$ is a solution of the system.

(ii) Substitute $y = 4$ in equation (2) and solve for x.

$$x^2 - y = 5 \qquad \text{Equation (2)}$$
$$x^2 - 4 = 5 \qquad \text{Replace } y \text{ with 4.}$$
$$x^2 = 9 \qquad \text{Add 4 to both sides.}$$
$$x = \pm 3 \qquad \text{Square root property}$$

So, $(3, 4)$ and $(-3, 4)$ are solutions of the system.
From **(i)** and **(ii)**, the apparent solution set for the system is
$\{(0, -5), (3, 4), (-3, 4)\}$.

Step 5 **Check.**

$(0, -5)$	$(3, 4)$	$(-3, 4)$
$x^2 + y^2 = 25$		
$0^2 + (-5)^2 \stackrel{?}{=} 25$	$3^2 + 4^2 \stackrel{?}{=} 25$	$(-3)^2 + 4^2 \stackrel{?}{=} 25$
$25 = 25$ ✓	$9 + 16 = 25$	$9 + 16 \stackrel{?}{=} 25$
	$25 = 25$ ✓	$25 = 25$ ✓
$x^2 - y = 5$		
$0^2 - (-5) \stackrel{?}{=} 5$	$3^2 - 4 \stackrel{?}{=} 5$	$(-3)^2 - 4 \stackrel{?}{=} 5$
$5 = 5$ ✓	$9 - 4 = 5$ ✓	$9 - 4 = 5$ ✓

The graphs of the circle $x^2 + y^2 = 25$ and the parabola $y = x^2 - 5$ sketched in Figure 7.6 confirm the three solutions.

TECHNOLOGY CONNECTION

To use a graphing calculator to graph the system in Example 9, you must solve $x^2 + y^2 = 25$ for y and enter the two resulting equations.

$$Y_1 = \sqrt{25 - x^2}$$
$$Y_2 = -\sqrt{25 - x^2}$$

Then enter

$$Y_3 = x^2 - 5$$

and use the ZSquare feature.

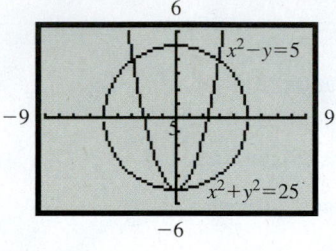

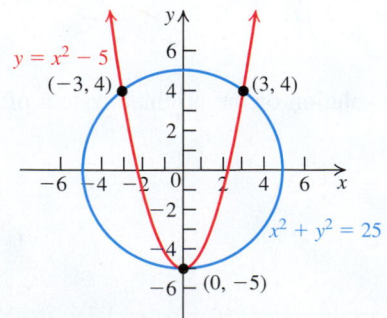

Figure 7.6

Practice Problem 9 Solve the system of equations.

$$\begin{cases} x^2 + 2y^2 = 34 \\ x^2 - y^2 = 7 \end{cases}$$

5 Solve applied problems by solving systems of equations.

SIDE
NOTE

The graph of a system of equations allows you to use geometric intuition to understand algebra. If the geometric and algebraic representations of a solution do not agree, you must go back and find the error(s).

Applications

It is well known that as the price of a product increases, demand for it decreases, and as the price increases, the supply of the product also increases. The **equilibrium point** is the ordered pair (x, p) such that the number of units, x, and the price per unit, p, satisfy *both* the demand and supply equations.

EXAMPLE 10   Finding the Equilibrium Point

Find the equilibrium point if the supply and demand functions for a new brand of digital video recorder (DVR) are given by the system

$$p = 60 + 0.0012x \quad \text{Equation (1)} \quad \text{Supply equation}$$
$$p = 80 - 0.0008x \quad \text{Equation (2)} \quad \text{Demand equation}$$

where p is the price per unit in dollars and x is the number of units.

Solution

We substitute the value of p from equation (1) into equation (2) and solve the resulting equation.

$$p = 80 - 0.0008x \qquad \text{Demand equation}$$
$$60 + 0.0012x = 80 - 0.0008x \qquad \text{Replace } p \text{ with } 60 + 0.0012x.$$
$$0.002x = 20 \qquad \text{Collect like terms and simplify.}$$
$$x = \frac{20}{0.002} = 10{,}000 \qquad \text{Solve for } x.$$

The equilibrium point occurs when the supply and demand for DVRs is 10,000 units. To find the price p, we back-substitute $x = 10{,}000$ into either of the original equations (1) or (2).

$$p = 60 + 0.0012x \qquad \text{Equation (1)}$$
$$= 60 + 0.0012(10{,}000) \qquad \text{Replace } x \text{ with } 10{,}000.$$
$$= 72 \qquad \text{Simplify.}$$

The equilibrium point is $(10{,}000, 72)$. See Figure 7.7.

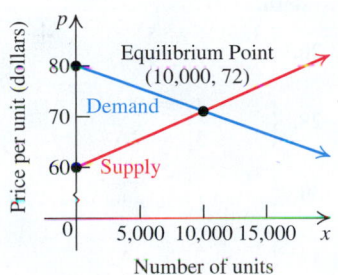

Figure 7.7 Price of DVRs.

Check.   Verify that the ordered pair $(10{,}000, 72)$ satisfies both equations (1) and (2).

Practice Problem 10   Find the equilibrium point.

$$\begin{cases} p = 20 + 0.002x & \text{Supply equation} \\ p = 77 - 0.008x & \text{Demand equation} \end{cases}$$

In general, to solve an applied problem involving two unknowns, we need to set up two equations using two variables to represent the two unknowns. The general rule is that *the number of equations formed must be equal to the number of unknown quantities involved.*

Answers to Practice Problems

1. Equation (1): $(1) + (3) = 4$; equation (2): $3(1) - (3) = 0$
2. $\{(-1, 3)\}$ **3.** $\{(4, -1)\}$ **4.** $\varnothing$
5. $\{(x, 2x - 3) \mid x \text{ any real number}\}$
6. $\{(-1, 1), (3, -7)\}$ **7.** $\left\{\left(\frac{2}{3}, \frac{1}{2}\right)\right\}$ **8.** $\{(2, -3)\}$
9. $\{(4, 3), (4, -3), (-4, 3), (-4, -3)\}$ **10.** $(5700, 31.4)$

SECTION 7.1 **Exercises**

Concepts and Vocabulary

1. The ordered pair (a, b) is a(n) _____ of a system of equations in x and y providing that when x is replaced with a and y is replaced with b, the resulting equations are true.

2. The two nongraphical methods for solving a system of equations are the _____ and _____ methods.

3. If in the process of solving a system of equations you get an equation of the form $0 = k$, where k is not zero, then the system is _____.

4. If in the process of solving a system of equations you get an equation of the form $0 = 0$, then the system has _____.

5. **True or False.** A system consisting of two identical equations has no solution.

6. **True or False.** If in the process of solving a system of two equations in x and y you get the equation $5 = 5$, then the system has exactly one solution.

7. **True or False.** If $x = 4$, $y = -7$ is the solution of a system of two equations in x and y, then the lines determined by the two equations intersect at the point $(4, -7)$.

8. **True or False.** When solving $\begin{cases} 3x - 2y = 14 \\ 9x + 8y = 23 \end{cases}$ by the addition method, we can eliminate x by multiplying the first equation by -3 and adding the equations.

Building Skills

In Exercises 9–14, determine which ordered pairs are solutions of each system of equations.

9. $\begin{cases} 2x + 3y = 3 \\ 3x - 4y = 13 \end{cases}$ $(1, -3), (3, -1), (6, 3), \left(5, \dfrac{1}{2}\right)$

10. $\begin{cases} x + 2y = 6 \\ 3x + 6y = 18 \end{cases}$ $(2, 2), (-2, 4), (0, 3), (1, 2)$

11. $\begin{cases} 5x - 2y = 7 \\ -10x + 4y = 11 \end{cases}$ $\left(\dfrac{5}{4}, 1\right), \left(0, \dfrac{11}{4}\right), (1, -1), (3, 4)$

12. $\begin{cases} x - 2y = -5 \\ 3x - y = 5 \end{cases}$ $(1, 3), (-5, 0), (3, 4), (3, -4)$

13. $\begin{cases} x + y = 1 \\ \dfrac{1}{2}x + \dfrac{1}{3}y = 2 \end{cases}$ $(0, 1), (1, 0), \left(\dfrac{2}{3}, \dfrac{3}{2}\right), (10, -9)$

14. $\begin{cases} \dfrac{2}{x} + \dfrac{3}{y} = 2 \\ \dfrac{6}{x} + \dfrac{18}{y} = 9 \end{cases}$ $(3, 2), (2, 3), (4, 3), (3, 4)$

In Exercises 15–24, estimate the solution(s) (if any) of each system by the graphical method. Check your solution(s). For any dependent equations, write your answer with x being arbitrary.

15. $\begin{cases} x + y = 3 \\ x - y = 1 \end{cases}$

16. $\begin{cases} x + y = 10 \\ x - y = 2 \end{cases}$

17. $\begin{cases} x + 2y = 6 \\ 2x + y = 6 \end{cases}$

18. $\begin{cases} 2x - y = 4 \\ x - y = 3 \end{cases}$

19. $\begin{cases} 3x - y = -9 \\ y = 3x + 6 \end{cases}$

20. $\begin{cases} 5x + 2y = 10 \\ y = -\dfrac{5}{2}x - 5 \end{cases}$

21. $\begin{cases} x + y = 7 \\ y = 2x \end{cases}$

22. $\begin{cases} y - x = 2 \\ y + x = 9 \end{cases}$

23. $\begin{cases} 3x + y = 12 \\ y = -3x + 12 \end{cases}$

24. $\begin{cases} 2x + 3y = 6 \\ 6y = -4x + 12 \end{cases}$

In Exercises 25–38, determine whether each system is *consistent* or *inconsistent*. If the system is consistent, determine whether the equations are dependent or independent. Do not solve the system.

25. $\begin{cases} y = -2x + 3 \\ y = 3x + 5 \end{cases}$

26. $\begin{cases} 3x + y = 5 \\ 2x + y = 4 \end{cases}$

27. $\begin{cases} 2x + 3y = 5 \\ 3x + 2y = 7 \end{cases}$

28. $\begin{cases} 2x - 4y = 5 \\ 3x + 5y = -6 \end{cases}$

29. $\begin{cases} 3x + 5y = 7 \\ 6x + 10y = 14 \end{cases}$

30. $\begin{cases} 3x - y = 2 \\ 9x - 3y = 6 \end{cases}$

31. $\begin{cases} x + 2y = -5 \\ 2x - y = 4 \end{cases}$

32. $\begin{cases} x + 2y = -2 \\ 2x - 3y = 5 \end{cases}$

33. $\begin{cases} 2x - 3y = 5 \\ 6x - 9y = 10 \end{cases}$

34. $\begin{cases} 3x + y = 2 \\ 15x + 5y = 15 \end{cases}$

35. $\begin{cases} -3x + 4y = 5 \\ \dfrac{9}{2}x - 6y = \dfrac{15}{2} \end{cases}$

36. $\begin{cases} 6x + 5y = 11 \\ 9x + \dfrac{15}{2}y = 21 \end{cases}$

37. $\begin{cases} 7x - 2y = 3 \\ 11x - \dfrac{3}{2}y = 8 \end{cases}$

38. $\begin{cases} 4x + 7y = 10 \\ 10x + \dfrac{35}{2}y = 25 \end{cases}$

In Exercises 39–48, solve each system of equations by the substitution method. Check your solutions. For any dependent equations, write your answer in the ordered pair form given in Example 5.

39. $\begin{cases} y = 2x + 1 \\ 5x + 2y = 9 \end{cases}$

40. $\begin{cases} x = 3y - 1 \\ 2x - 3y = 7 \end{cases}$

41. $\begin{cases} 3x - y = 5 \\ x + y = 7 \end{cases}$ **42.** $\begin{cases} 2x + y = 2 \\ 3x - y = -7 \end{cases}$

43. $\begin{cases} 2x - y = 5 \\ -4x + 2y = 7 \end{cases}$ **44.** $\begin{cases} 3x + 2y = 5 \\ -9x - 6y = 15 \end{cases}$

45. $\begin{cases} x - y = 2 \\ x^2 - 4x + y^2 = -2 \end{cases}$ **46.** $\begin{cases} 2x - y = 1 \\ x^2 - 8y + y^2 = -6 \end{cases}$

47. $\begin{cases} x - 2y = 5 \\ -3x + 6y = -15 \end{cases}$ **48.** $\begin{cases} x + y = 3 \\ 2x + 2y = 6 \end{cases}$

In Exercises 49–58, solve each system of equations by the elimination method. Check your solutions. For any dependent equations, write your answer as in Example 5.

49. $\begin{cases} x - y = 1 \\ x + y = 5 \end{cases}$ **50.** $\begin{cases} 2x - 3y = 5 \\ 3x + 2y = 14 \end{cases}$

51. $\begin{cases} x + y = 0 \\ 2x + 3y = 3 \end{cases}$ **52.** $\begin{cases} x + y = 3 \\ 3x + y = 1 \end{cases}$

53. $\begin{cases} x^2 + 2y^2 = 12 \\ -5x^2 + 7y^2 = 8 \end{cases}$ **54.** $\begin{cases} x^2 - 6y^2 = 19 \\ 3x^2 + 2y^2 = 77 \end{cases}$

55. $\begin{cases} x - y = 2 \\ -2x + 2y = 5 \end{cases}$ **56.** $\begin{cases} x + y = 5 \\ 2x + 2y = -10 \end{cases}$

57. $\begin{cases} 4x + 6y = 12 \\ 2x + 3y = 6 \end{cases}$ **58.** $\begin{cases} 4x + 7y = -3 \\ -8x - 14y = 6 \end{cases}$

In Exercises 59–78, use any method to solve each system of equations. For any dependent equations, write your answer as in Example 5.

59. $\begin{cases} 2x + y = 9 \\ 2x - 3y = 5 \end{cases}$ **60.** $\begin{cases} x + 2y = 10 \\ x - 2y = -6 \end{cases}$

61. $\begin{cases} 2x + 5y = 2 \\ x + 3y = 2 \end{cases}$ **62.** $\begin{cases} 4x - y = 6 \\ 3x - 4y = 11 \end{cases}$

63. $\begin{cases} 2x + 3y = 7 \\ 3x + y = 7 \end{cases}$ **64.** $\begin{cases} x = 3y + 4 \\ x = 5y + 10 \end{cases}$

65. $\begin{cases} 2x + 3y = 9 \\ 3x + 2y = 11 \end{cases}$ **66.** $\begin{cases} 3x - 4y = 0 \\ y = \dfrac{2x + 1}{3} \end{cases}$

67. $\begin{cases} \dfrac{x}{4} + \dfrac{y}{6} = 1 \\ x + 2(x - y) = 7 \end{cases}$ **68.** $\begin{cases} \dfrac{x}{3} + \dfrac{y}{5} = 12 \\ x - y = 4 \end{cases}$

69. $\begin{cases} x - y^2 = 2 \\ 2x + 3y = 3 \end{cases}$ **70.** $\begin{cases} x + 2y = 6 \\ y - x^2 = 0 \end{cases}$

71. $\begin{cases} 5x - 2y = 7 \\ x^2 + y^2 = 2 \end{cases}$ **72.** $\begin{cases} x - 2y = -5 \\ x^2 + y^2 = 25 \end{cases}$

73. $\begin{cases} x^2 + y^2 = 20 \\ x^2 - y^2 = 12 \end{cases}$ **74.** $\begin{cases} x^2 - y^2 = 9 \\ 4x^2 + 5y^2 = 180 \end{cases}$

75. $\begin{cases} \dfrac{3}{x} + \dfrac{1}{y} = 4 \\ \dfrac{6}{x} - \dfrac{1}{y} = 2 \end{cases}$ **76.** $\begin{cases} \dfrac{6}{x} + \dfrac{3}{y} = 0 \\ \dfrac{4}{x} + \dfrac{9}{y} = -1 \end{cases}$

77. $\begin{cases} \dfrac{5}{x} + \dfrac{10}{y} = 3 \\ \dfrac{2}{x} - \dfrac{12}{y} = -2 \end{cases}$ **78.** $\begin{cases} \dfrac{3}{x} + \dfrac{4}{y} = 1 \\ \dfrac{6}{x} + \dfrac{4}{y} = 3 \end{cases}$

Applying the Concepts

In Exercises 79–82, the demand and supply functions of a product are given. In each case, p represents the price in dollars per unit and x represents the number of units in hundreds. Find the equilibrium point.

79. $\begin{cases} 2p + x = 140 & \text{Demand equation} \\ 12p - x = 280 & \text{Supply equation} \end{cases}$

80. $\begin{cases} 7p + x = 150 & \text{Demand equation} \\ 10p - x = 20 & \text{Supply equation} \end{cases}$

81. $\begin{cases} 2p + x = 25 & \text{Demand equation} \\ x - p = 13 & \text{Supply equation} \end{cases}$

82. $\begin{cases} p + 2x = 96 & \text{Demand equation} \\ p - x = 39 & \text{Supply equation} \end{cases}$

In Exercises 83–88, use a system of equations to solve each problem.

83. Diameter of a pizza. The sum of the diameters of the largest and the smallest pizza sold at the Monster Pizza Shop is 29 inches. The difference in their diameters is 13 inches. Find the diameters of the largest and the smallest pizza.

84. Calories in hamburgers. The sum of the number of calories in a hamburger from Boston Burger and a hamburger from Carmen's Broiler is 1130. The difference in the number of calories in the hamburgers is 40. If the Boston Burger has the larger number of calories, how many calories are in each restaurant's hamburger?

85. Trash composition. Paper and plastic together account for 48% (by weight) of the total trash collected. If the weight of paper trash collected is five times the weight of plastic trash, what percent of the total trash collected is paper and what percent is plastic?

86. Cost of food and clothing. The average monthly combined cost of food and clothing for the Martínez family is $1000. If they spend four times as much for food as they do for clothes, what is the average monthly cost of each?

87. Mardi Gras parade "throws." Levon paid 40¢ for each string of beads and 30¢ for each doubloon (a special coin) he bought to throw from his Mardi Gras float. He paid a total of $265 for all items. His doubloons and beads combined to a total of 770 items. How many doubloons and how many strings of beads did Levon buy?

88. Halloween candy. Janet bought 135 pieces of candy to give away on Halloween. She bought two kinds of candy, paying 24¢ apiece for one kind and 18¢ apiece for the other. If she spent $26.70 for the candy, how many pieces of each kind did she buy?

In Exercises 89 and 90, use the information in the following table.

Food Description McDonald's	Fat (gms)	Carb (gms)	Protein (gms)
Sausage Burrito	16	21	13
Egg McMuffin	12	27	17

Source: www.shapefit.com/mcdonalds.html

Suppose you ate breakfast at McDonald's during one workweek (Monday–Friday).

89. McDonald's breakfast. If you consumed 123 grams of carbohydrates and 77 grams of protein, how many of each breakfast item did you eat during the workweek?

90. McDonald's breakfast. If you consumed 64 grams of fat and 129 grams of carbohydrates, how many of each breakfast item did you eat during the workweek?

91. Investment. Last year Mrs. García invested $50,000. She put part of the money in a real estate venture that paid 7.5% for the year and the rest in a small-business venture that returned 12% for the year. The combined income from the two investments for the year totaled $5190. How much did she invest at each rate?

92. Investment. Mr. Sharma invested a total of $30,000 in two ventures for a year. The annual return from one of them was 8%, and the other paid 10.5% for the year. He received a total income of $2550 from both investments. How much did he invest at each rate?

93. Tutoring other students. A student earns twice as much per hour for tutoring as she does working at McDougal's. If her average wage is $11.25 per hour, how much does she earn per hour at each job?

94. Plumber's earnings. A plumber earns $15 per hour more than her apprentice. In each 40-hour week, their combined earnings are $2200. What is the hourly rate for each?

95. Mixture problem. An herb that sells for $5.50 per pound is mixed with tea that sells for $3.20 per pound to produce a 100-pound mix that is worth $3.66 per pound. How many pounds of each ingredient does the mix contain?

96. Mixture problem. A chemist had a solution of 60% acid. She added some distilled water, reducing the acid concentration to 40%. She then added 1 more liter of water to further reduce the acid concentration to 30%. How much of the 30% solution did she then have?

97. Airplane speed. With the help of a tail wind, a plane travels 3000 kilometers in 5 hours. The return trip against the wind requires 6 hours. Assume that the direction and the wind speed are constant. Find both the speed of the plane in still air and the wind speed. [*Hint:* Remember that the wind helps the plane in one direction and hinders it in the other.]

98. Motorboat speed. A motorboat travels up a stream a distance of 12 miles in 2 hours. If the current had been twice as strong, the trip would have taken 3 hours. How long should it take for the return trip down the stream?

99. Break-even analysis. A local publishing company publishes *City Magazine*. The production and setup costs are $30,000, and the cost of producing each magazine is $2. Each magazine sells for $3.50. Assume that x magazines are published and sold.
   a. Write the total cost function $y = C(x)$ and the revenue function $y = R(x)$.
   b. Graph both functions from part (a) on the same coordinate plane.
   c. How many magazines must be sold to break even?

100. Break-even analysis. An electronic manufacturer plans to make digital portable music players (MP4s). Fixed costs will be $750,000, and it will cost $50 to manufacture each MP4, which will be sold for $125 to the retailers. Assume that x MP4s are manufactured and sold.
   a. Write the total cost function $y = C(x)$ and the revenue function $y = R(x)$.
   b. Graph both functions from part (a) on the same coordinate plane.
   c. How many MP4s must be sold to break even?

101. Making a job decision. Shanaysha has job offers from department stores A and B to sell major appliances. Store A offers her a fixed salary of $400 per week. Store B offers her $150 per week plus a commission of 4% of her weekly sales. How much should Shanaysha's weekly sales be for the offer from Store B to be better than that from Store A?

102. Making a job decision. Sheena has job offers from companies A and B to sell insurance. Company A offers her $25,000 per year plus 2% of her yearly sales premiums. Company B offers her $30,000 per year plus 1% of her yearly sales premiums. How much must Sheena earn in yearly sales premiums for the offer from Company B to be the better offer?

Beyond the Basics

103. Solve for x and y.

$$\begin{cases} \dfrac{36}{x+y} + \dfrac{12}{x-y} = 3 \\ \dfrac{18}{x+y} + \dfrac{48}{x-y} = 5 \end{cases}$$

[*Hint:* Let $u = \dfrac{1}{x+y}$ and $v = \dfrac{1}{x-y}$, solve for u and v, and then solve for x and y.]

104. Solve for x and y.

$$\begin{cases} 6x + 3y = 7xy \\ 3x + 9y = 11xy \end{cases}$$

[*Hint:* Case (*i*), $xy = 0$; case (*ii*), $xy \neq 0$: divide by xy.]

105. For what value of the constant c is the following system consistent?

$$\begin{cases} x + 2y = 7 \\ 3x + 5y = 11 \\ cx + 3y = 4 \end{cases}$$

[*Hint:* Solve the first two equations for x and y and then substitute the solution into the third equation.]

106. Repeat Exercise 105 for the following system of equations.

$$\begin{cases} 5x + 7y = 11 \\ 13x + 17y = 19 \\ 3x + cy = 5 \end{cases}$$

107. Find the value of c for which the following system of equations represents the same line.

$$\begin{cases} 4x + 7y = 10 \\ 10x + cy = 25 \end{cases}$$

108. Find the value of c for which the following system of equations is inconsistent.

$$\begin{cases} 2x + cy = 11 \\ 5x - 7y = 5 \end{cases}$$

In Exercises 109–111, write your answer in slope–intercept form.

109. Find an equation of the line that passes through the point of intersection of the lines with equations $2x + y = 3$ and $x - 3y = 12$ and that is parallel to the line with equation $3x + 2y = 8$.

110. Find an equation of the line that passes through the point of intersection of the lines with equations $5x + 2y = 7$ and $6x - 5y = 38$ and that passes through the point $(1, 3)$.

111. Find an equation of the line that passes through the point of intersection of the lines with equations $2x + 5y + 7 = 0$ and $13x - 10y + 3 = 0$ and that is perpendicular to the line with equation $7x + 13y = 8$.

Critical Thinking / Discussion / Writing

112. Consider the system of equations

$$\begin{cases} a_1 x + b_1 y = c_1 \\ a_2 x + b_2 y = c_2 \end{cases}$$

where a_1, b_1, c_1, a_2, b_2, and c_2 are constants. Find a relationship (an equation) among these constants such that the system of equations has

a. only one solution. Find the solution in terms of the constants.

b. no solution.

c. infinitely many solutions.

113. Let $3x + 4y - 12 = 0$ be the equation of a line l_1 and let $P(5, 8)$ be a point.

a. Find an equation of a line l_2 passing through the point P and perpendicular to l_1.

b. Find the point Q of the intersection of the two lines l_1 and l_2.

c. Find the distance $d(P, Q)$. This is the distance from the point P to the line l_1.

114. Use the procedure outlined in Exercise 113 to show that the (perpendicular) distance from a point $P(x_1, y_1)$ to the line $ax + by + c = 0$ is given by $\dfrac{|ax_1 + by_1 + c|}{\sqrt{a^2 + b^2}}$.

115. Use the formula from Exercise 114 to find the distance from the given point P to the given line l.

a. $P(2, 3)$, $l: x + y - 7 = 0$

b. $P(-2, 5)$, $l: 2x - y + 3 = 0$

c. $P(3, 4)$, $l: 5x - 2y - 7 = 0$

d. $P(0, 0)$, $l: ax + by + c = 0$

Getting Ready for the Next Section

In Exercises 116–122, find the requested variable.

116. Find x if $3x - (2x - 1) - (4x + 3) = -5$.

117. Find y if $y + 2z = 7$ and $z = 2$.

118. Find x if $x - 3y = 5$ and $y = -3$.

119. Find x if $2x + 3y + 5z = 21$, $y = 1$, and $z = 2$.

120. Find y if $3x - 5y + z = 2$, $x = -2$, and $z = 3$.

121. Find x if $2x + 3y - 2z = 5$, $y - z = 4$, and $z = -3$.

122. Find x if $3x + 2y + z = 2$, $y + 2z = 1$, and $z = 2$.

Systems of Linear Equations in Three Variables

BEFORE STARTING THIS SECTION, REVIEW

1   Solving systems of linear equations (Section 7.1, page 594)

2   Solving a system of equations in two variables (Section 7.1, page 597)

OBJECTIVES

1   Solve a system of linear equations in three variables.

2   Classify systems as consistent and inconsistent.

3   Solve nonsquare systems.

4   Interpret linear systems in three variables geometrically.

5   Use linear systems in applications.

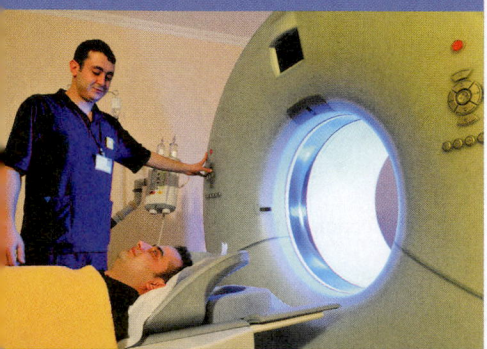

◆ CAT Scans

Computerized axial tomography imaging is more commonly known as a "CAT scan" or "CT scan." Tomography is from the Greek word *tomos*, meaning "slice or section," and *graphia*, meaning "describing." The CAT scan machine was invented in 1972 by British engineer Godfrey Hounsfield and, independently, by physicist Allan Cormack of Tufts University in Massachusetts. Hounsfield never attended a university, but he started experimenting with electrical and mechanical devices as a boy. He shared a Nobel Prize in 1979 with Cormack for inventing the scanner technology and was honored with knighthood in England for his contributions to medicine and science. CAT scanners use X-rays measured outside the patient's body, together with algebraic computations, to construct pictures of the patient's internal body organs, including the brain. The scanner does not "take" any pictures; instead, it *computes* pictures. The idea behind the CAT scanner is simple. When an X-ray passes through an object, it loses intensity. The denser the object, the more intensity the X-ray loses. Since tumors are denser than healthy tissues, an unexpected loss of intensity may indicate the presence of a tumor. To obtain sufficient information, a large number of properly arranged X-rays is needed. Imagine a cross section of body tissue with a superimposed mathematical grid. Each region bounded by the grid lines is called a *grid cell*. In an actual CAT scan, each grid cell is 2 square millimeters and contains 10,000 or more biological cells. The grid shown in Figure 7.8 would require many X-rays. Actual CAT scanners work with grids that have 512×512 grid cells and require hundreds of thousands of X-ray beams. In Example 6, we discuss a simple situation in which X-rays pass through three grid cells.

Reprinted with permission from *Mathematics Teacher*, May 1996. © 1996 by the National Council of Teachers of Mathematics. All rights reserved.

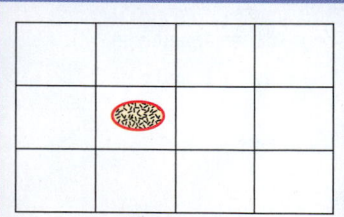

Cross section with tumor

Figure 7.8 Cross section with tumor.

1   Solve a system of linear equations in three variables.

Systems of Linear Equations

A **linear equation in the variables** $x_1, x_2, \ldots, x_n$ is an equation that can be written in the form

$$a_1x_1 + a_2x_2 + \cdots + a_nx_n = b,$$

where b and the *coefficients* $a_1, a_2, \ldots, a_n$ are real numbers. The subscript n can be any positive integer. When n is small—say, between 2 and 5 (as in most textbook examples and exercises)—we normally use $x, y, z, u,$ and v instead of $x_1, x_2, x_3, x_4,$ and x_5. However, in real-life applications, n might be 100 or 1000 or even greater.

A **system of linear equations** (or a **linear system**) in three variables is a collection of two or more linear equations involving the same three variables. For example,

$$\begin{cases} x + 3y + z = 0 \\ 2x - y + z = 5 \\ 3x - 3y + 2z = 10 \end{cases}$$

is a system of three linear equations in the three variables $x, y,$ and z. An ordered triple (a, b, c) is a **solution** of a system of three linear equations in three variables $x, y,$ and z if each equation in the system is a true statement when $a, b,$ and c are substituted for $x, y,$ and z, respectively.

EXAMPLE 1   Verifying a Solution

Determine whether the ordered triple $(2, -1, 3)$ is a solution of the following linear system.

$$\begin{cases} 5x + 3y - 2z = 1 & \text{Equation (1)} \\ x - y + z = 6 & \text{Equation (2)} \\ 2x + 2y - z = -1 & \text{Equation (3)} \end{cases}$$

Solution

We replace x with 2, y with -1, and z with 3 in all three equations.

Equation (1)	Equation (2)	Equation (3)
$5x + 3y - 2z = 1$	$x - y + z = 6$	$2x + 2y - z = -1$
$5(2) + 3(-1) - 2(3) \overset{?}{=} 1$	$2 - (-1) + 3 \overset{?}{=} 6$	$2(2) + 2(-1) - (3) \overset{?}{=} -1$
$10 - 3 - 6 \overset{?}{=} 1$	$2 + 1 + 3 \overset{?}{=} 6$	$4 - 2 - 3 \overset{?}{=} -1$
$1 = 1 ✓$	$6 = 6 ✓$	$-1 = -1 ✓$

Since the ordered triple $(2, -1, 3)$ satisfies all three equations, it is a solution of the system.

Practice Problem 1 Determine whether the ordered triple $(2, -3, 2)$ is a solution of the following linear system.

$$\begin{cases} x + y + z = 1 \\ 3x + 4y + z = -4 \\ 2x + y + 2z = 5 \end{cases}$$

We will now use a procedure called **Gaussian elimination** to convert a system of three equations in the variables $x, y,$ and z into an equivalent system in *triangular form* with leading coefficient 1, as shown below.

$$\begin{cases} x + ay + bz = p & \text{Equation (1)} \\ y + cz = q & \text{Equation (2)} \\ z = z_0 & \text{Equation (3)} \end{cases}$$

We can then solve the system by back-substitution. To get a system in triangular form, we use the following three operations to obtain equivalent systems (systems with the same solution set as the original system).

> ### OPERATIONS THAT PRODUCE EQUIVALENT SYSTEMS
> 1. Interchange the position of any two equations.
> 2. Multiply any equation by a nonzero constant.
> 3. Add a multiple of one equation to another.

PROCEDURE
IN ACTION

EXAMPLE 2 **The Gaussian Elimination Method**

OBJECTIVE

Solve a system of three equations in three variables by first converting the system to the form

$$\begin{cases} x + ay + bz = p \\ \quad\; y + cz = q \\ \qquad\qquad z = z_0 \end{cases}$$

Step 1 Rearrange the equations, if necessary, to obtain an x-term with a nonzero coefficient in the first equation. Then multiply the first equation by the reciprocal of the coefficient of the x-term to get 1 as a leading coefficient.

Step 2 By adding appropriate multiples of the first equation, eliminate any x-terms from the second and third equations. Then multiply the resulting second equation by the reciprocal of the coefficient of the y-term to get 1 as a leading coefficient.

Step 3 Add an appropriate multiple of the second equation from Step 2 to eliminate any y-term from the third equation. Then solve the resulting equation for z to obtain an equivalent system in triangular form.

Step 4 Back-substitute the value of z from Step 3 into one of the equations in Step 3 and solve for y.

EXAMPLE

Solve the system of equations.

$$\begin{cases} \quad\quad 3y - z = 5 & \text{Equation (1)} \\ 2x + \;\; y + z = 9 & \text{Equation (2)} \\ x + 2y + 2z = 3 & \text{Equation (3)} \end{cases}$$

1. Interchange equations (1) and (3).

$$\begin{cases} x + 2y + 2z = 3 & \text{Equation (3)} \\ 2x + \;\; y + z = 9 & \text{Equation (2)} \\ \quad\quad 3y - z = 5 & \text{Equation (1)} \end{cases}$$

The first equation (equation (3)) already has a leading coefficient of 1.

2. To eliminate x from equation (2), add -2 times equation (3) to equation (2).

$$\begin{array}{ll} -2x - 4y - 4z = -6 & -2(x + 2y + 3z = 3) \\ \underline{\;\;2x + \;\;y + \;\;z = \;\;\;9} \quad (2) & \\ \quad\quad -3y - 3z = \;\;\;3 \quad (4) & \text{Add.} \end{array}$$

$$\begin{cases} x + 2y + 2z = 3 & (3) \\ \quad\quad\; y + \;\;z = -1 & (5) \quad \text{Multiply equation (4) by } -\frac{1}{3}. \\ \quad\quad 3y - \;\;z = 5 & (1) \end{cases}$$

3. To eliminate y from equation (1), add -3 times equation (5) to equation (1).

$$\begin{array}{ll} -3y - 3z = \;\;3 & -3(y + z = -1) \\ \underline{\;\;3y - \;\;z = \;\;5} \quad (1) & \\ \quad\quad -4z = \;\;8 & \text{Add.} \\ \quad\quad\quad\; z = -2 & \text{Solve for } z. \end{array}$$

Equivalent system in triangular form is:

$$\begin{cases} x + 2y + 2z = \;\;3 & \text{Equation (3)} \\ \quad\quad\; y + \;\;z = -1 & \text{Equation (5)} \\ \quad\quad\quad\;\; z = -2 \end{cases}$$

4.
$$\begin{array}{ll} y + z = -1 & \text{Equation (5)} \\ y + (-2) = -1 & \text{Substitute } z = -2. \\ y = 1 & \text{Solve for } y. \end{array}$$

Step 5 Back-substitute the values of y and z from Steps 3 and 4 in any equation containing x, y, and z and solve for x.

Step 6 Write the solution set.

Step 7 Check your answer in the original equations.

5.

$$x + 2y + 2z = 3 \quad \text{Equation (3) from Step 1}$$
$$x + 2(1) + 2(-2) = 3 \quad \text{Substitute } y = 1, z = -2.$$
$$x = 5 \quad \text{Solve for } x.$$

6. The solution set is $\{(5, 1, -2)\}$.

7. Check: Verify the solution by substituting $x = 5$, $y = 1$, and $z = -2$ in each of the original equations.

$$3(1) - (-2) = 5 \checkmark \quad \text{Equation (1)}$$
$$2(5) + (1) + (-2) = 9 \checkmark \quad \text{Equation (2)}$$
$$5 + 2(1) + 2(-2) = 3 \checkmark \quad \text{Equation (3)}$$

Practice Problem 2 Solve the system.

$$\begin{cases} 2x + 5y \quad\quad = 1 \\ x - 3y + 2z = 1 \\ -x + 2y + z = 7 \end{cases}$$

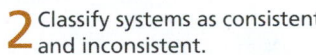

2 Classify systems as consistent and inconsistent.

Number of Solutions of a Linear System

As in the case of two variables, a system of linear equations in three variables may be **consistent** (having at least one solution) or **inconsistent** (having no solution). The equations in a consistent system may be **dependent** (the system has infinitely many solutions) or **independent** (the system has one solution).

> ### INCONSISTENT SYSTEM
>
> If in the process of converting a linear system to triangular form an equation of the form $0 = k$ occurs, where $k \neq 0$, then the system has no solution and is *inconsistent*.

EXAMPLE 3 Attempting to Solve a Linear System with No Solution

Solve the system of equations.

$$\begin{cases} x - y + 2z = 5 \quad \text{Equation (1)} \\ 2x + y + z = 7 \quad \text{Equation (2)} \\ 3x - 2y + 5z = 20 \quad \text{Equation (3)} \end{cases}$$

Solution

Steps 1–2 To eliminate x from equation (2), add -2 times equation (1) to equation (2).

$$\begin{array}{ll} -2x + 2y - 4z = -10 & -2(x - y + 2z = 5) \\ \underline{2x + y + z = \quad 7} \quad (2) & \\ 3y - 3z = \quad -3 \quad (4) & \text{Add.} \end{array}$$

Next, we eliminate x from equation (3) using equations (1) and (3).

$$\begin{array}{ll} -3x + 3y - 6z = -15 & -3(x - y + 2z = 5) \\ \underline{3x - 2y + 5z = \quad 20} \quad (3) & \\ y - z = \quad 5 \quad (5) & \text{Add.} \end{array}$$

We now have the equivalent system:

$$\begin{cases} x - y + 2z = 5 & \text{Equation (1)} \\ 3y - 3z = -3 & \text{Equation (4)} \\ y - z = 5 & \text{Equation (5)} \end{cases}$$

SIDE
NOTE

An alternate method for Step 3 is to solve the 2×2 system of equations (4) and (5), then back-substitute into equation (1) to solve for the third variable.

Step 3 Multiply equation (4) by $\dfrac{1}{3}$ to obtain

$$y - z = -1 \quad \text{Equation (6)}$$

We can eliminate y in equation (5) using equation (6).

$$\begin{array}{ll} -y + z = 1 & \quad -1(y - z = -1) \\ \underline{y - z = 5} \quad (5) & \\ 0 = 6 \quad (7) & \quad \text{Add.} \end{array}$$

We now have the system in triangular form:

$$\begin{cases} x - y + 2z = 5 & \text{Equation (1)} \\ y - z = -1 & \text{Equation (6)} \\ \underline{0 = 6} & \text{Equation (7)} \end{cases}$$

$$\text{False}$$

This system is equivalent to the original system. Because equation (7) is false, we conclude that the solution set of the system is $\varnothing$ and the system is inconsistent.

Practice Problem 3 Solve the system of equations.

$$\begin{cases} 2x + 2y + 2z = 12 \\ -3x + y - 11z = -6 \\ 2x + y + 4z = -8 \end{cases}$$

DEPENDENT EQUATIONS

If in the process of converting a linear system to triangular form,

  (i) an equation of the form $0 = k (k \neq 0)$ *does not* occur but

 (ii) an equation of the form $0 = 0$ *does* occur, then the system of equations has infinitely many solutions and the equations are *dependent*.

EXAMPLE 4   **Solving a System with Infinitely Many Solutions**

Solve the system of equations.

$$\begin{cases} x - y + z = 7 & \text{Equation (1)} \\ 3x + 2y - 12z = 11 & \text{Equation (2)} \\ 4x + y - 11z = 18 & \text{Equation (3)} \end{cases}$$

Solution

Steps 1–2 Eliminate x using equations (1) and (2).

$$\begin{array}{ll} -3x + 3y - 3z = -21 & \quad -3(x - y + z = 7) \\ \underline{3x + 2y - 12z = 11} \quad (2) & \\ 5y - 15z = -10 \quad (4) & \quad \text{Add.} \end{array}$$

Eliminate x using equations (1) and (3).

$$
\begin{array}{rl}
-4x + 4y - 4z = -28 & \qquad -4(x - y + z = 7) \\
\underline{4x + y - 11z = 18} & \quad (3) \\
5y - 15z = -10 & \quad (5) \qquad \text{Add.}
\end{array}
$$

We now have the following equivalent system:

$$
\begin{cases}
x - y + z = 7 & \text{Equation (1)} \\
 5y - 15z = -10 & \text{Equation (4)} \\
 5y - 15z = -10 & \text{Equation (5)}
\end{cases}
$$

Step 3   We eliminate y using equations (4) and (5).

$$
\begin{array}{rl}
-5y + 15z = 10 & \qquad -1(5y - 15z = -10) \\
\underline{5y - 15z = -10} & \quad (5) \\
0 = 0 & \quad (6) \qquad \text{Add.}
\end{array}
$$

We finally have the system in triangular form:

$$
\begin{cases}
x - y + z = 7 & \text{Equation (1)} \\
 5y - 15z = -10 & \text{Equation (4)} \\
\underbrace{0 = 0}_{\text{True}} & \text{Equation (6)}
\end{cases}
$$

The equation $0 = 0$ is equivalent to $0z = 0$, which is true for every value of z. Solving equation (4) for y, we have $y = 3z - 2$. Substituting $y = 3z - 2$ into equation (1) and solving for x, we obtain

$$
\begin{array}{ll}
x = y - z + 7 & \text{Solve equation (1) for } x. \\
x = (3z - 2) - z + 7 & \text{Substitute } y = 3z - 2. \\
x = 2z + 5 & \text{Simplify.}
\end{array}
$$

Thus, for every real number z, the triple $(x, y, z) = (2z + 5, 3z - 2, z)$ is a solution of the system. For example, when $z = 0$, the triple $(5, -2, 0)$ is a solution. When $z = 1$, the triple $(7, 1, 1)$ is a solution. Therefore, the given system has infinitely many solutions, one for each real number z, and the equations are dependent. The solution set for the system is $\{(2z + 5, 3z - 2, z)\}$.

Practice Problem 4 Solve the system of equations.

$$
\begin{cases}
x + y + z = 5 \\
-4x - y - 8z = -29 \\
2x + 5y - 2z = 1
\end{cases}
$$

3 Solve nonsquare systems.

Nonsquare Systems

Sometimes in a linear system, the number of equations is not the same as the number of variables. Such systems are called **nonsquare linear systems**.

EXAMPLE 5   **Solving a Nonsquare Linear System**

Solve the system of linear equations.

$$
\begin{cases}
x + 5y + 3z = 7 & \text{Equation (1)} \\
2x + 11y - 4z = 6 & \text{Equation (2)}
\end{cases}
$$

Solution

Steps 1–2 We eliminate x using equations (1) and (2).

$$
\begin{array}{lll}
-2x - 10y - 6z = -14 & \quad & {\color{blue}-2(x + 5y + 3z = 7)} \\
\underline{2x + 11y - 4z = 6} & \quad (2) & \\
y - 10z = -8 & \quad (3) & {\color{blue}\text{Add.}}
\end{array}
$$

We obtain the equivalent system:

$$
\begin{cases}
x + 5y + 3z = 7 & \text{Equation (1)} \\
y - 10z = -8 & \text{Equation (3)}
\end{cases}
$$

Steps 3–4 Since there is no third equation, Steps 3 and 4 are not needed.

Step 5 Solve equation (3) for y in terms of z to obtain

$$y = 10z - 8.$$

Step 6 Back-substitute $y = 10z - 8$ into equation (1) and solve for x.

$$
\begin{array}{ll}
x + 5y + 3z = 7 & {\color{blue}\text{Equation (1)}} \\
x + 5(10z - 8) + 3z = 7 & {\color{blue}\text{Replace } y \text{ with } 10z - 8.} \\
x = -53z + 47 & {\color{blue}\text{Solve for } x.}
\end{array}
$$

Steps 5 and 6 mean that z determines both the x and y values. The system has infinitely many solutions, one for each real number z. The solution set is $\{(-53z + 47, 10z - 8, z)\}$.

Practice Problem 5 Solve the system of linear equations.

$$
\begin{cases}
x + 3y + 2z = 4 \\
2x + 7y - z = 5
\end{cases}
$$

4 Interpret linear systems in three variables geometrically.

Geometric Interpretation

The graph of a linear equation in three variables, such as $ax + by + cz = d$ (where a, b, and c are not all zero), is a plane in three-dimensional space. Figure 7.9 shows possible situations for a system of three linear equations in three variables.

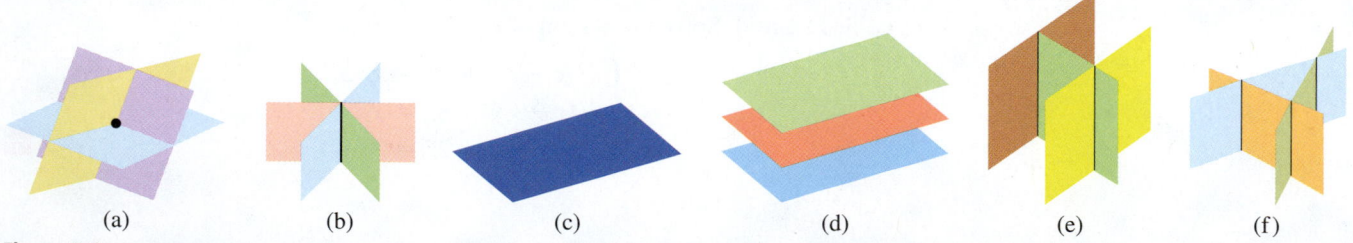

| (a) | (b) | (c) | (d) | (e) | (f) |

Figure 7.9

a. Three planes intersect in a single point. The system has only one solution.

b. Three planes intersect in one line. The system has infinitely many solutions.

c. Three planes coincide with each other. The system has infinitely many solutions.

d. There are three parallel planes. The system has no solution.

e. Two parallel planes are intersected by a third plane. The system has no solution.

f. Three planes have no point in common. The system has no solution.

5 Use linear systems in applications.

◆ An Application to CAT Scans

CAT scanners report X-ray data in units called *linear attenuation units* (LAUs), also called *Hounsfield numbers*. The data are reported in such a way that if an X-ray passes through consecutive grid cells, the total weakening of the beam is the sum of the individual decreases. Table 7.1 gives the range of LAUs that determine whether the grid cell contains healthy tissue, tumorous tissue, a bone, or a metal.

TABLE 7.1 CAT Scanner Ranges

Type of Tissue	LAU Values
Healthy tissue	0.1625–0.2977
Tumorous tissue	0.2679–0.3930
Bone	0.38757–0.5108
Metal	1.54–∞

Reprinted with permission from *Mathematics Teacher*, May 1996. © 1996 by the National Council of Teachers of Mathematics. All rights reserved.

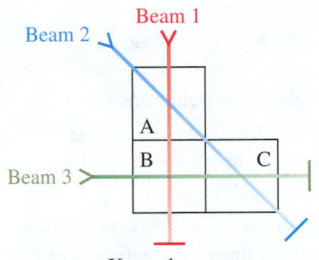

Beam 2 Beam 1

A

B   C

Beam 3

X-ray detector

Three grid cells, three X-rays

Figure 7.10 Reprinted with permission from *Mathematics Teacher*, May 1996. © 1996 by the National Council of Teachers of Mathematics. All rights reserved.

EXAMPLE 6   A CAT Scan with Three Grid Cells

Let A, B, and C be three grid cells as shown in Figure 7.10. A CAT scanner reports the following data for a patient named Monica.

i. Beam 1 is weakened by 0.80 unit as it passes through grid cells A and B.

ii. Beam 2 is weakened by 0.55 unit as it passes through grid cells A and C.

iii. Beam 3 is weakened by 0.65 unit as it passes through grid cells B and C.

Use Table 7.1 to determine which grid cells contain each type of tissue listed.

Solution

Suppose grid cell A weakens the beam by x units, grid cell B weakens it by y units, and grid cell C weakens it by z units. Then we have the system:

$$\begin{cases} x + y \quad\quad = 0.80 & \text{Equation (1)} & \text{(Beam 1)} \\ x \quad\quad + z = 0.55 & \text{Equation (2)} & \text{(Beam 2)} \\ \quad y + z = 0.65 & \text{Equation (3)} & \text{(Beam 3)} \end{cases}$$

To solve this system of equations, we use the elimination procedure. Subtracting equation (1) from equation (2), we obtain the equivalent system:

$$\begin{cases} x + y \quad\quad = \quad 0.80 & (1) \\ \quad -y + z = -0.25 & (4) \quad \text{Subtract equation (1) from equation (2).} \\ \quad\quad y + z = \quad 0.65 & (3) \end{cases}$$

Add equation (4) to equation (3) to get the equivalent system:

$$\begin{cases} x + y \quad\quad = \quad 0.80 & (1) \\ \quad -y + z = -0.25 & (4) \quad \text{Add equation (4) to equation (3).} \\ \quad\quad 2z = \quad 0.40 & (5) \end{cases}$$

Multiply equation (5) by $\frac{1}{2}$ to obtain $z = 0.20$. Then back-substitute $z = 0.20$ in equation (4) to get $-y + 0.20 = -0.25$, or $y = 0.45$. Next, back-substitute $y = 0.45$ in equation (1) to get $x + 0.45 = 0.80$. Solving for x gives $x = 0.35$. Now referring to Table 7.1, we conclude that cell A contains tumorous tissue (because $x = 0.35$), cell B contains a bone (because $y = 0.45$), and cell C contains healthy tissue (because $z = 0.20$).

Practice Problem 6 In Example 6, suppose **(i)** Beam 1 is weakened by 0.65 unit as it passes through grid cells A and B, **(ii)** Beam 2 is weakened by 0.70 unit as it passes through grid cells A and C, and **(iii)** Beam 3 is weakened by 0.55 unit as it passes through grid cells B and C. Use Table 7.1 to determine which grid cells contain each type of tissue listed.

Answers to Practice Problems

1. Yes: $(2) + (-3) + (2) = 1$
$3(2) + 4(-3) + (2) = -4$
$2(2) + (-3) + 2(2) = 5$

2. $\{(-2, 1, 3)\}$ **3.** $\varnothing$ **4.** $\left\{\left(8 - \frac{7}{3}z, \frac{4}{3}z - 3, z\right)\right\}$

5. $\{(13 - 17z, -3 + 5z, z)\}$ **6.** Cell A contains bone (because $x = 0.4$), cell B contains healthy tissue (because $y = 0.25$), and cell C contains tumorous tissue (because $z = 0.3$).

SECTION 7.2 Exercises

Concepts and Vocabulary

1. Systems of equations that have the same solution set are called _____ systems.

2. Adding a(n) _____ multiple of one equation to another equation in the system produces an equivalent system.

3. If any of the equations in a system has no solution, then the system is _____.

4. If the number of equations is different from the number of variables in a linear system, then the system is called a(n) _____ system.

5. **True or False.** The graph of the equation $x - 2y + z = 1$ is a plane in three-dimensional space.

6. **True or False.** If $\{(2z, 1 - z, z + 3)\}$ is the solution set of a linear system in three variables, then $(0, 1, 3)$ is a solution of this system.

7. **True or False.** If in the process of converting a linear system to triangular form an equation of the form $0 = k$ occurs, with $k \neq 0$, then the system has exactly one solution.

8. **True or False.** A dependent system of linear equations has no solutions.

Building Skills

In Exercises 9–12, determine whether the given ordered triple (x, y, z) is a solution of the given linear system.

9. $\begin{cases} 2x - 2y - 3z = 1 \\ 3y + 2z = -1 \quad (1, -1, 1) \\ y + z = 0 \end{cases}$

10. $\begin{cases} 3x - 2y + z = 2 \\ y + z = 5 \quad (2, 3, 2) \\ x + 3z = 8 \end{cases}$

11. $\begin{cases} x + 3y - 2z = 0 \\ 2x - y + 4z = 5 \quad (-10, 8, 7) \\ x - 11y + 14z = 0 \end{cases}$

12. $\begin{cases} x - 4y + 7z = 14 \\ 3x + 8y - 2z = 13 \quad (4, 1, 2) \\ 7x - 8y + 26z = 5 \end{cases}$

In Exercises 13–16, solve each triangular linear system.

13. $\begin{aligned} x + y + z &= 4 \\ y - 2z &= 4 \\ z &= -1 \end{aligned}$

14. $\begin{aligned} x + 3y + z &= 3 \\ y + 2z &= 8 \\ z &= 5 \end{aligned}$

15. $\begin{aligned} x - 5y + 3z &= -1 \\ y - 2z &= -6 \\ z &= 4 \end{aligned}$

16. $\begin{aligned} x + 3y + 5z &= 0 \\ y + 7z &= 2 \\ z &= \frac{1}{2} \end{aligned}$

17. In the system in Exercise 9, interchange equations (2) and (3). Write the new equivalent system. In the new system, eliminate y from the last equation. Convert the system to triangular form.

18. In Exercise 10, interchange equations (1) and (3). In the new equivalent system, eliminate x from the last equation. Convert the system to triangular form.

19. Convert the system of Exercise 11 to triangular form.

20. Convert the system of Exercise 12 to triangular form.

In Exercises 21–24, convert the system to triangular form and then use back-substitution to solve the system.

21. $\begin{cases} x - y + z = 6 \\ 2y + 3z = 5 \\ 2z = 6 \end{cases}$

22. $\begin{cases} 4x + 5y + 2z = -3 \\ 3y - z = 14 \\ -3z = 15 \end{cases}$

23. $\begin{cases} 4x + 4y + 4z = 7 \\ 3x - 8y = 14 \\ 4z = -1 \end{cases}$

24. $\begin{cases} 5x + 10y + 10z = 0 \\ 2y + 3z = -0.6 \\ 4z = 1.6 \end{cases}$

In Exercises 25–46, find the solution set of each linear system. Identify inconsistent systems and dependent equations. For dependent equations, write your answer as in Example 4.

25. $\begin{cases} x + y + z = 6 \\ x - y + z = 2 \\ 2x + y - z = 1 \end{cases}$

26. $\begin{cases} x + y + z = 6 \\ 2x + 3y - z = 5 \\ 3x - 2y + 3z = 8 \end{cases}$

27. $\begin{cases} 2x + 3y + z = 9 \\ x + 2y + 3z = 6 \\ 3x + y + 2z = 8 \end{cases}$

28. $\begin{cases} 4x + 2y + 3z = 6 \\ x + 2y + 2z = 1 \\ 2x - y + z = -1 \end{cases}$

29. $\begin{cases} x - 4y + 7z = 14 \\ 3x + 8y - 2z = 13 \\ 7x - 8y + 26z = 5 \end{cases}$

30. $\begin{cases} x + y + 4z = 6 \\ x + 2y - 2z = 8 \\ 7x + 10y + 10z = 60 \end{cases}$

31. $\begin{cases} 2x + 3y + 2z = 7 \\ x + 3y - z = -2 \\ x - y + 2z = 8 \end{cases}$ **32.** $\begin{cases} x - y + 2z = 3 \\ 2x + 2y + z = 3 \\ x + y + 3z = 4 \end{cases}$

33. $\begin{cases} 4x - 2y + z = 5 \\ 2x + y - 2z = 4 \\ x + 3y - 2z = 6 \end{cases}$ **34.** $\begin{cases} x - 3y + 2z = 9 \\ 2x + 4y - 3z = -9 \\ 3x - 2y + 5z = 12 \end{cases}$

35. $\begin{cases} 2x + y + z = 6 \\ x + y - z = 1 \\ x + y + 2z = 4 \end{cases}$ **36.** $\begin{cases} 2x + y - 3z = 7 \\ x - y - 2z = 4 \\ 3x + 3y + 2z = 4 \end{cases}$

37. $\begin{cases} x - y - z = 3 \\ x + 9y + z = 3 \\ 2x + 3y - z = 6 \end{cases}$ **38.** $\begin{cases} x + y - z = 2 \\ 3x - y - z = 10 \\ 3x + y - 2z = 8 \end{cases}$

39. $\begin{cases} x + y = 0 \\ y + 2z = -4 \\ y + z = 4 - x \end{cases}$ **40.** $\begin{cases} 2x + 4y + 3z = 6 \\ x + 2z = -1 \\ x - 2y + z = -5 \end{cases}$

41. $\begin{cases} 2y - z = -4 \\ x + z = 3 \\ 2x + 3y = -1 \end{cases}$ **42.** $\begin{cases} x + y = 9 \\ 2y + 3z = 7 \\ x - 2z = 4 \end{cases}$

43. $\begin{cases} 3x - 2z = 11 \\ 2x + y = 8 \\ 2y + 3z = 1 \end{cases}$ **44.** $\begin{cases} 2x + y = 4 \\ x + 2z = 3 \\ 3y - z = 5 \end{cases}$

45. $\begin{cases} 2x + 6y + 11 = 0 \\ 6y - 18z + 1 = 0 \end{cases}$

46. $\begin{cases} 3x + 5y - 15 = 0 \\ 6x + 20y - 6z = 11 \end{cases}$

Applying the Concepts

In Exercises 47–54, use a system of equations to solve each problem.

47. Investment. Miguel invested $20,000 in three different funds that paid 4%, 5%, and 6% for the year. The total income for the year from the three funds was $1060. The income from the 6% fund was twice the income from the 5% fund. What amount was invested in each fund?

48. Number problem. The sum of the digits in a three-digit number is 14. The sum of the hundreds digit and the units digit is equal to the tens digit. If the hundreds digit and the units digit are interchanged, the number is increased by 297. What is the original number?

49. Election campaign. Alex, Becky, and Courtney volunteered to stuff 741 envelopes with newsletters for Senator Douglas's reelection campaign. Alex could assemble 124 per hour; Becky, 118 per hour; and Courtney, 132 per hour. They worked a total of 6 hours. The sum of the number of hours that Becky and Courtney spent was twice what Alex spent. How long did each of them work?

50. Age of students. A college algebra class of 38 students at Central State College was made up of people who were 18, 19, and 20 years of age. The average of their ages was 18.5 years. How many of each age were in the class if the number of 18-year-olds was eight more than the combined number of 19- and 20-year-olds?

51. Coins in a machine. A vending machine's coin box contains nickels, dimes, and quarters. The total number of coins in the box is 300. The number of dimes is three times the number of nickels and quarters together. If the box contains $30.05, find the number of nickels, dimes, and quarters that it contains.

52. Sports. The Wildcats scored 46 points in a football game. Twice the number of points resulting from the sum of field goals and extra points equals two more than the number of points from touchdowns. Five times the number of points scored by field goals equals twice the number of points from touchdowns. Find the number of points resulting from touchdowns, field goals, and extra points. [A touchdown = 6 points, a field goal = 3 points, and an extra point = 1 point.]

53. Weekly wage. Amy worked 53 hours one week and was paid at three different rates. She earned $7.40 per hour for her normal daytime work, $9.20 per hour for night work, and $11.75 per hour for holiday work. If her total gross wages for the week were $452.20 and the number of regular daytime hours she worked exceeded the combined hours of night and holiday work by 9 hours, how many hours of each category of work did Amy perform?

54. Components of a product. A manufacturer buys three components—A, B, and C—for use in making a toaster. She used as many units of A as she did of B and C combined. The cost of A, B, and C is $4, $5, and $6 per unit, respectively. If she purchased 100 units of these components for a total of $480, how many units of each did she purchase?

In Exercises 55–58, use Table 7.1 and Figure 7.10 to determine which grid cells of the patients contain healthy tissue, tumorous tissue, bone, or metal.

Beam decrease, in LAUs

Patient	Beam 1	Beam 2	Beam 3
55. Vicky	0.54	0.40	0.52
56. Yolanda	0.65	0.80	0.75
57. Nina	0.51	0.49	0.44
58. Srinivasan	0.44	2.21	2.23

Beyond the Basics

In Exercises 59–62, write a linear equation of the form $x + by + cz = d$ that is satisfied by all three of the given ordered triples.

59. $(1, 0, 0), (0, 1, 0), (0, 0, 1)$

60. $\left(\frac{1}{3}, 0, 0\right), (0, 4, 3), (1, 2, 2)$

61. $(3, -4, 0), \left(0, \frac{1}{4}, \frac{1}{2}\right), (1, 1, -4)$

62. $(0, 1, -10), \left(\frac{1}{8}, 0, \frac{1}{4}\right), \left(1, \frac{1}{3}, -2\right)$

In Exercises 63–66, find an equation of the parabola of the form $y = ax^2 + bx + c$ that passes through the three given points.

63. $(0, 1), (-1, 0), (1, 4)$

64. $(0, 2), (-1, 30), (2, 6)$

65. $(1, 2), (-1, 4), (2, 4)$

66. $(0, 3), (-1, 4), (1, 6)$

In Exercises 67–70, find an equation of the circle of the form $x^2 + y^2 + ax + by + c = 0$ that passes through the three given points.

67. $(0, 4), (2\sqrt{2}, 2\sqrt{2}), (-4, 0)$

68. $(0, 3), (0, -1), (\sqrt{3}, 2)$

69. $(1, 2), (6, -3), (4, 1)$

70. $(5, 6), (-1, 6), (3, 2)$

In Exercises 71 and 72, solve the system of equations.

71. $\begin{cases} \dfrac{1}{x} + \dfrac{3}{y} - \dfrac{1}{z} = 5 \\[2mm] \dfrac{2}{x} + \dfrac{4}{y} + \dfrac{6}{z} = 4 \\[2mm] \dfrac{2}{x} + \dfrac{3}{y} + \dfrac{1}{z} = 3 \end{cases}$

72. $\begin{cases} \dfrac{1}{x} + \dfrac{2}{y} + \dfrac{3}{z} = 8 \\[2mm] \dfrac{2}{x} + \dfrac{5}{y} + \dfrac{9}{z} = 16 \\[2mm] \dfrac{3}{x} - \dfrac{4}{y} - \dfrac{5}{z} = 32 \end{cases}$

$\left[\text{Hint: Let } \dfrac{1}{x} = u, \dfrac{1}{y} = v, \text{ and } \dfrac{1}{z} = w. \text{ Solve for } u, v, \text{ and } w. \right]$

In Exercises 73 and 74, find the value of c for which the given system has a unique solution.

73. $\begin{cases} x + 2y - 5z - 9 = 0 \\ 3x - y + 2z - 14 = 0 \\ 2x + 3y - z - 3 = 0 \\ cx - 5y + z + 3 = 0 \end{cases}$

74. $\begin{cases} x + 5y + z = 42 \\ 3x + y - 3z = c \\ x - y + 3z = 4 \\ x + y - z = 0 \end{cases}$

75. Find a quadratic function of the form $y = ax^2 + bx + c$ whose graph passes through the points $(-1, -1), (0, 5),$ and $(2, 5)$.

76. Use Table 7.1 to investigate four grid cells, arranged in a square, as shown in the given figure. Figures (a) and (b) reprinted with permission from *Mathematics Teacher,* May 1996. © 1996 by National Council of Teachers of Mathematics. All rights reserved.

a. For Figure (a), the following data were observed.

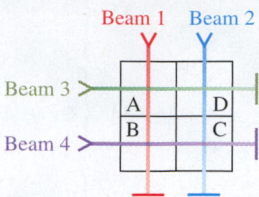

Four grid cells and four X-rays
(a)

1. Beam 1 decreased by 0.60 unit.
2. Beam 2 decreased by 0.75 unit.
3. Beam 3 decreased by 0.65 unit.
4. Beam 4 decreased by 0.70 unit.

Is there sufficient information in Figure (a) to determine which grid cells may contain tumorous tissue?

b. From two additional X-rays, as shown in Figure (b), the following data were observed.

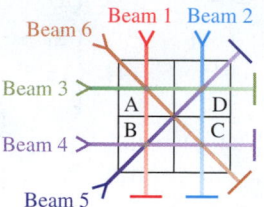

Four grid cells and six X-rays
(b)

5. Beam 5 decreased by 0.85 unit.
6. Beam 6 decreased by 0.50 unit.

Show that the six X-rays represented in the two figures are sufficient to locate tumors in this situation but that, in fact, it is not necessary to use all six rays.

Critical Thinking / Discussion / Writing

77. Write a linear system of three equations in the variables $x, y,$ and z that has $\{(1, -1, 2)\}$ as its solution set.

78. Write a linear system of three equations in the variables $x, y,$ and z such that

a. the system has no solution.

b. the system has infinitely many solutions.

Getting Ready for the Next Section

In Exercises 79–86, sketch the graph of each equation.

79. $2x + y = 0$

80. $2y - x = 0$

81. $3x + 4y = 12$

82. $4x - 3y = 24$

83. $y - 2x^2 = 0$

84. $y - x^2 = 3$

85. $y + x^2 - 2 = 0$

86. $x^2 + y^2 = 16$

In Exercises 87–90, determine which of the given points are on the graph of the equation.

Equation	Points
87. $5x + 3y = 0$	$(0, 0), (-3, 5), (1, -1), (-1, 1)$
88. $3x + 4y = 12$	$(0, 0), (4, 0), (0, 3), (3, 4)$
89. $y = x^2 + 3$	$(0, 0), (1, 4), (-1, 4), (2, 5)$
90. $y + x^2 = 2$	$(0, 0), (0, 2), (2, -2), (1, 1)$

Systems of Inequalities

BEFORE STARTING THIS SECTION, REVIEW

1 Intersection of sets (Appendix A.1, page 919)

2 Graphing a line (Section 1.2, page 26)

3 Solving systems of equations (Section 7.1, page 594)

OBJECTIVES

1 Graph a linear inequality in two variables.

2 Graph systems of linear inequalities in two variables.

3 Apply systems of inequalities to linear programming.

4 Graph a nonlinear inequality in two variables.

5 Graph systems of nonlinear inequalities in two variables.

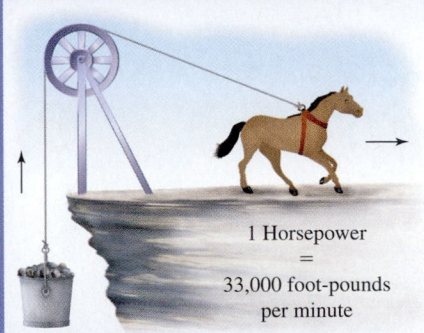

1 Horsepower
=
33,000 foot-pounds per minute

◆ What Is Horsepower?

The definition of horsepower was originated by the inventor James Watt (1736–1819). To demonstrate the power (work capability) of the steam engine that he invented and patented in 1783, he rigged his favorite horse with a rope and some pulleys and showed that his horse could raise a 3300-pound weight 10 feet in the air in one minute. Watt called this amount of work 1 horsepower (hp). So he defined

$$1 \text{ hp} = 33{,}000 \text{ foot-pounds per minute } (3300 \times 10 = 33{,}000)$$

$$= 550 \text{ foot-pounds per second } \left(\frac{33{,}000}{60} = 550\right).$$

Because horses were one of the main energy sources during that time, the public easily understood and accepted the horsepower measure. Watt labeled his steam engines in terms of equivalent horsepower. A 10 hp engine could do the work of 10 horses or could lift 5500 pounds up 1 foot in one second. In the usual definition of power $= \dfrac{(\text{force})(\text{distance})}{\text{time}}$, the horse provides the force. In Exercise 89, we discuss engines with different amounts of horsepower.

---

1 Graph a linear inequality in two variables.

Graph of a Linear Inequality in Two Variables

The statements $x + y > 4$, $2x + 3y < 7$, $y \geq x$, and $x + y \leq 9$ are examples of linear inequalities in the variables x and y. As with equations, a **solution of an inequality in two variables** x and y is an ordered pair (a, b) that results in a true statement when x is replaced with a and y is replaced with b in the inequality. For example, the ordered pair $(3, 2)$ is a solution of the inequality $4x + 5y > 11$ because $4(3) + 5(2) = 22 > 11$ is a true statement. The ordered pair $(1, 1)$ is *not* a solution of the inequality $4x + 5y > 11$ because $4(1) + 5(1) = 9 > 11$ is a false statement. The set of all solutions of an inequality is called the **solution set of the inequality**. The **graph of an inequality in two variables** is the graph of the solution set of the inequality.

EXAMPLE 1 **Graphing a Linear Inequality**

Graph the inequality: $2x + y > 6$

Solution

First, graph the *equation* $2x + y = 6$. See Figure 7.11. Notice that if we solve this equation for y, we have $y = -2x + 6$. Any point $P(a, b)$ whose y-coordinate, b, is *greater* than $6 - 2a$ must be above this line. Therefore, the graph of $y > -2x + 6$, which is equivalent to $2x + y > 6$, consists of all points (x, y) in the plane that lie above the line $2x + y = 6$. The graph of the inequality $2x + y > 6$ is shaded in Figure 7.12.

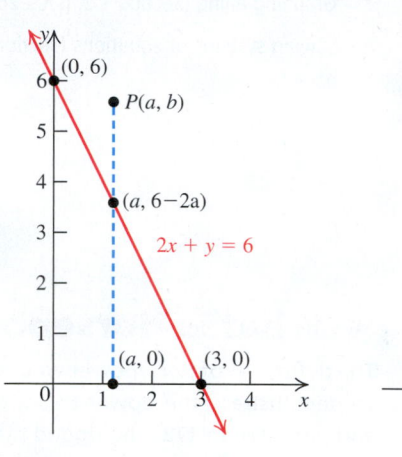

Figure 7.11   **Figure 7.12**

Practice Problem 1 Graph the inequality $3x + y < 6$.

In Figure 7.12, the graph of the equation $2x + y = 6$ is drawn as a dashed line to indicate that points on the line are *not* included in the solution set of the inequality $2x + y > 6$. In Figure 7.13(a), however, the graph of the inequality $2x + y \geq 6$ shows $2x + y = 6$ as a solid line: the solid line indicates that the points on the line *are* included in the solution set. Note that the region below the line $2x + y = 6$ in Figure 7.13(b) is the graph of the inequality $2x + y < 6$.

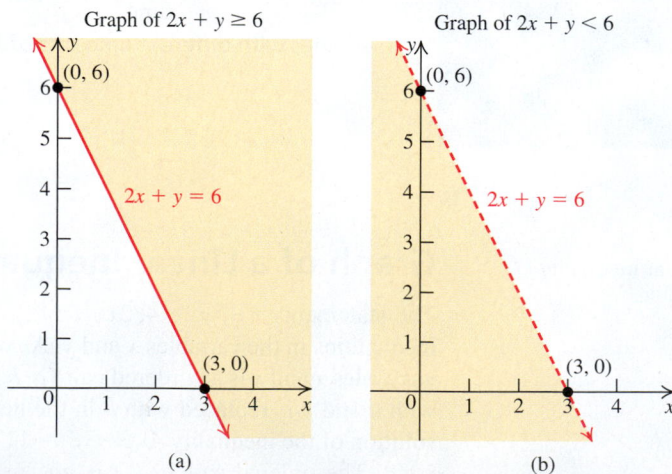

Figure 7.13

In general, the graph of a linear inequality's *corresponding equation* (found by changing the inequality symbol to an equal sign) is a line that separates the plane into two regions, each called a *half plane*. The line itself is called the **boundary** of each region. *All of the* points in one region satisfy the inequality, and *none* of the points in the other region are solutions. You can therefore test one point not on the boundary, called a **test point**, to determine which region represents the solution set.

PROCEDURE
IN ACTION

EXAMPLE 2   **Graphing a Linear Inequality in Two Variables**

OBJECTIVE

Graph an inequality in two variables.

Step 1 Find the corresponding equation. Replace the inequality symbol with the equal $(=)$ sign.

Step 2 Graph the corresponding equation. Sketch the graph of the *corresponding equation* in Step 1. Use a dashed line for the boundary if the given inequality sign is $<$ or $>$ and a solid line if the inequality symbol is $\leq$ or $\geq$.

Step 3 Select a test point. The graph in Step 2 divides the plane into two regions. Select a *test point* in either region but not on the graph of the equation in Step 1.

Step 4 Shade the solution set. (i) If the coordinates of the test point satisfy the inequality, then so do all of the points in that region. Shade that region.

(ii) If the test point's coordinates do not satisfy the inequality, shade the region that does not contain the test point.

The shaded region (including the boundary if it is solid) is the graph of the inequality.

EXAMPLE

Graph the linear inequality $y \geq -2x + 4$.

1. $y = -2x + 4$

2.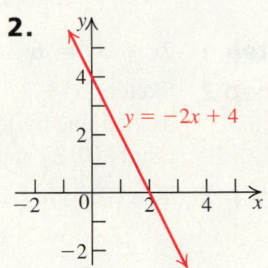

The graph of the equation $y = -2x + 4$ is shown in the figure.

The line drawn is solid because the given inequality sign is $\geq$.

3.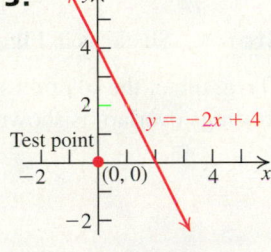

Select $(0, 0)$ as a test point.

4.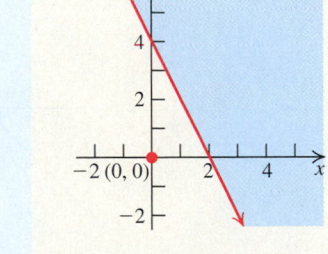

Because $0 \geq -2(0) + 4$ is false, the coordinates of the test point $(0, 0)$ do *not* satisfy the inequality $y \geq -2x + 4$.

The graph of the solution set of $y \geq -2x + 4$ is on the side of the line that does not contain $(0, 0)$. It is shaded in the figure.

Practice Problem 2 Graph the linear inequality $y \leq -2x + 4$.

2 Graph systems of linear inequalities in two variables.

Systems of Linear Inequalities in Two Variables

An ordered pair (a, b) is a **solution of a system of inequalities** involving two variables if it is a solution of each of its inequalities. For example, $(0, 1)$ is a solution of the system

$$\begin{cases} 2x + y \leq 1 \\ x - 3y < 0 \end{cases}$$

because both $2(0) + 1 \leq 1$ and $0 - 3(1) < 0$ are true statements. The **solution set of a system of inequalities** is the *intersection* of the solution sets of all inequalities in the system. To find the graph of this solution set, graph the solution set of each inequality in the system (with different-color shadings) on the same coordinate plane. The region common to all of the solution sets, which is where all of the shaded parts overlap, is the solution set.

EXAMPLE 3 Graphing a System of Two Inequalities

Graph the solution set of the system of inequalities.

$$\begin{cases} 2x + 3y > 6 & (1) \\ y - x \le 0 & (2) \end{cases}$$

Solution

Graph each inequality separately in the same coordinate plane.

Inequality 1	**Inequality 2**
$2x + 3y > 6$	$y - x \le 0$

Step 1 $2x + 3y = 6$ **Step 1** $y - x = 0$

Step 2 Sketch $2x + 3y = 6$ as a dashed line by joining the points $(0, 2)$ and $(3, 0)$.

Step 2 Sketch $y - x = 0$ as a solid line by joining the points $(0, 0)$ and $(1, 1)$.

Step 3 Test $(0, 0)$ in

$$2x + 3y > 6$$
$$2 \cdot 0 + 3 \cdot 0 \overset{?}{>} 6$$
$$0 \overset{?}{>} 6 \quad \text{False}$$

Step 3 Test $(1, 0)$ in

$$y - x \le 0$$
$$0 - 1 \overset{?}{\le} 0$$
$$-1 \overset{?}{\le} 0 \quad \text{True}$$

Step 4 Shade as in Figure 7.14(a).   **Step 4** Shade as in Figure 7.14(b).

The graph of the solution set of the system of inequalities (1) and (2) (the region where the shadings overlap) is shown shaded in purple in Figure 7.14(c).

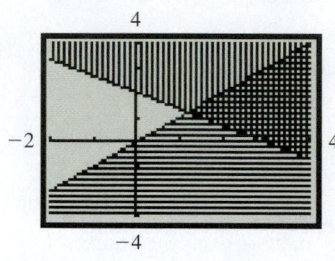
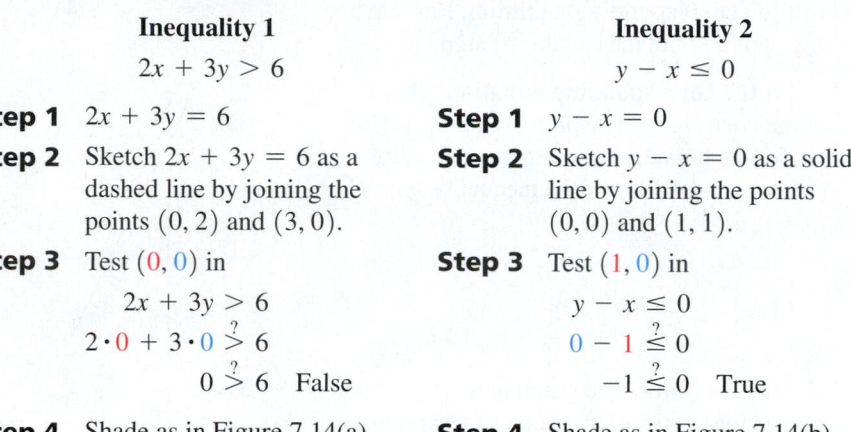

Figure 7.14 Graph of a system of two inequalities.

The point $P\left(\dfrac{6}{5}, \dfrac{6}{5}\right)$ in Figure 7.14(c) is the point of intersection of the two lines $2x + 3y = 6$ and $y - x = 0$. Such a point is called a **corner point**, or **vertex**, of the solution set. It is found by solving the system of equations: $\begin{cases} 2x + 3y = 6 \\ y - x = 0 \end{cases}$

In this case, however, the point P is not in the solution set of the given system because the ordered pair $\left(\dfrac{6}{5}, \dfrac{6}{5}\right)$ satisfies only inequality (2), but not inequality (1), of Example 3. We show P as an open circle in Figure 7.14(c).

Practice Problem 3 Graph the solution set of the system of inequalities.

$$\begin{cases} 3x - y > 3 & (1) \\ x - y \ge 1 & (2) \end{cases}$$

EXAMPLE 4 **Solving a System of Three Linear Inequalities**

Sketch the graph and label the vertices of the solution set of the system of linear inequalities.

$$\begin{cases} 3x + 2y \le 11 & (1) \\ x - y \le 2 & (2) \\ 7x - 2y \ge -1 & (3) \end{cases}$$

Solution

First, on the same coordinate plane, sketch the graphs of the three linear equations that correspond to the three inequalities. Because either $\le$ or $\ge$ occurs in all three inequalities, all of the equations are graphed as *solid* lines. See Figure 7.15.

Notice that $(0, 0)$ satisfies each of the inequalities but none of the equations. You can use $(0, 0)$ as the test point for each inequality.

Make the following conclusions:

(i) Because $(0, 0)$ lies below the line $3x + 2y = 11$, the points on or below the line $3x + 2y = 11$ are in the solution set of inequality (1).

(ii) Because $(0, 0)$ is above the line $x - y = 2$, the points on or above the line $x - y = 2$ are in the solution set of inequality (2).

(iii) Finally, because $(0, 0)$ lies below the line $7x - 2y = -1$, the points on or below the line $7x - 2y = -1$ belong to the solution set of inequality (3).

The region common to the regions in **(i)**, **(ii)**, and **(iii)** (including the boundaries) is the solution set of the given system of inequalities. The solution set is the shaded region in Figure 7.16, including the sides of the triangle.

Figure 7.16 also shows the vertices of the solution set. These vertices are obtained by solving each pair of equations in the system. Because all vertices are solutions of the given system, they are shown as (filled-in) points.

a. To find the vertex $(3, 1)$, solve the system $\begin{cases} 3x + 2y = 11 & (1) \\ x - y = 2 & (2) \end{cases}$

Solve equation (2) for x to obtain $x = 2 + y$. Substitute $x = 2 + y$ in equation (1).

$$3(2 + y) + 2y = 11$$
$$6 + 3y + 2y = 11 \qquad \text{Distributive property}$$
$$y = 1 \qquad \text{Solve for } y.$$

Back-substitute $y = 1$ in equation (2) to find $x = 3$. The vertex is the point $(3, 1)$.

b. Solve the system of equations $\begin{cases} x - y = 2 \\ 7x - 2y = -1 \end{cases}$ to find the vertex $(-1, -3)$.

c. Solve the system of equations $\begin{cases} 3x + 2y = 11 \\ 7x - 2y = -1 \end{cases}$ to find the vertex $(1, 4)$.

Practice Problem 4 Sketch the graph and label the vertices of the solution set of the system of inequalities.

$$\begin{cases} 2x + 3y < 16 \\ 4x + 2y \ge 16 \\ 2x - y \le 8 \end{cases}$$

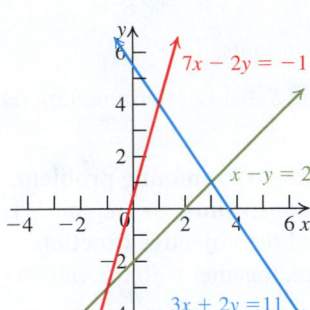

Figure 7.15

SIDE
NOTE

In working with linear inequalities, you must use the test point to graph the solution set. *Do not assume that just because the symbol $\le$ or $<$ appears in the inequality, the graph of the inequality is below the line.*

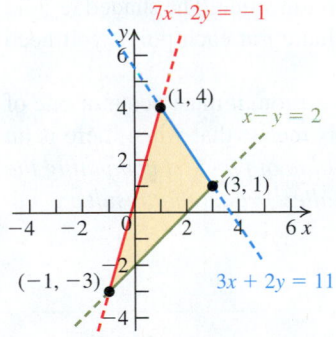

Figure 7.16 Graph of the solution set.

3 Apply systems of inequalities to linear programming.

Applications: Linear Programming

Recall that if a function f with domain $[a, b]$ has a largest and a smallest value, the largest value is called the *maximum value* and the smallest value is called the *minimum value*. The process of finding the maximum or minimum value of a quantity is called **optimization**. For example, for the function

$$f(x) = x^2, \quad -2 \le x \le 3,$$

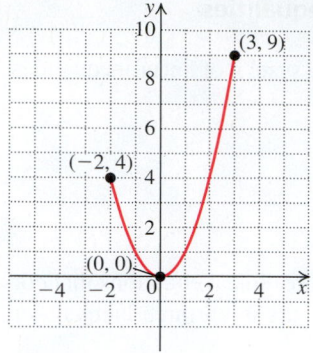

Figure 7.17 Optimization.

the maximum value of f occurs at $x = 3$ and is given by $f(3) = 3^2 = 9$, while the minimum value occurs at $x = 0$ and is given by $f(0) = 0^2 = 0$. See Figure 7.17. This is an example of an optimization problem involving one variable.

Now we will investigate optimization problems involving two variables, say, x and y. Assume the following conditions:

1. The quantity f to be maximized or minimized can be written as a linear expression in x and y; that is,

$$f = ax + by, \text{ where } a \neq 0, b \neq 0 \text{ are constants.}$$

2. The domain of the variables x and y is restricted to a region S that is determined by (is a solution set of) a system of linear inequalities.

A problem that satisfies conditions (1) and (2) is called a **linear programming problem**.

The inequalities that determine the region S are called **constraints**, the region S is called the **set of feasible solutions**, and $f = ax + by$ is called the **objective function**. A point in S at which f reaches its maximum (or minimum) value, together with the value of f at that point, is called an **optimal solution**.

Now consider the following linear programming problem.

Find values of x and y that will make $f = 2x + 3y$ as large as possible (in other words, that will result in a maximum value for f), where x and y are restricted by the following constraints:

$$\begin{cases} 3x + 5y \leq 15 \\ x \geq 0 \\ y \geq 0 \end{cases}$$

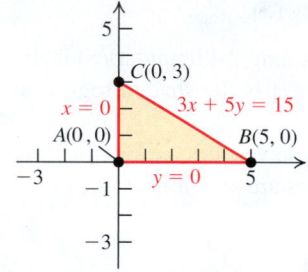

Figure 7.18 Feasible solutions.

To solve this problem, first sketch the graph of the permissible values of x and y; that is, find the set of feasible solutions determined by the constraints of the problem. Using the method explained in Example 4, you find that the set of feasible solutions is the shaded region shown in Figure 7.18.

Now you have to find the values of $f = 2x + 3y$ at each of the points in the shaded region of Figure 7.18 and then see where f takes on the maximum value. The shaded region, however, has infinitely many points. Since you cannot evaluate f at each point, you need another method to solve the problem.

Fortunately, it can be shown that if there is an optimal solution, it must occur at one of the vertices (corner points) of the feasible solution set. This means that when there is an optimal solution, *we can find the maximum (or minimum) value of f by first computing the value of f at each vertex and then picking the largest (or smallest) value that results.*

PROCEDURE
IN ACTION

EXAMPLE 5 **Solving a Linear Programming Problem**

OBJECTIVE

Solve a linear programming problem.

Step 1 Write an expression for the quantity to be maximized or minimized. This expression is the objective function.

Step 2 Write all constraints as linear inequalities.

Step 3 Graph the solution set of the constraint inequalities. This set is the set of feasible solutions.

EXAMPLE

Maximize $f = 2x + 3y$ subject to the constraints given below.

1. Maximize the objective function

$$f = 2x + 3y$$

2. subject to the constraints

$$\begin{cases} 3x + 5y \leq 15 \\ x \geq 0 \\ y \geq 0 \end{cases}$$

3. See Figure 7.18 for the solution set of the constraints.

(continued)

Step 4 Find all vertices of the solution set in Step 3 by solving all pairs of equations corresponding to the constraint inequalities.

4. Find

$A(0, 0)$ by solving $\begin{cases} x = 0 \\ y = 0 \end{cases}$

$B(5, 0)$ by solving $\begin{cases} y = 0 \\ 3x + 5y = 15 \end{cases}$

and

$C(0, 3)$ by solving $\begin{cases} x = 0 \\ 3x + 5y = 15 \end{cases}$

Step 5 Find the values of the objective function at each of the vertices of Step 4.

5. At $A(0, 0)$, $f = 2(0) + 3(0) = 0$.
At $B(5, 0)$, $f = 2(5) + 3(0) = 10$.
At $C(0, 3)$, $f = 2(0) + 3(3) = 9$.

Step 6 The largest of the values (if any) in Step 5 is the maximum value of the objective function, and the smallest of the values (if any) in Step 5 is the minimum value of the objective function.

6. At $B(5, 0)$, the objective function f has a maximum value of 10. Although you are not asked for the minimum value, it is 0 at $A(0, 0)$.

Practice Problem 5 Maximize $f = 4x + 5y$ subject to the constraints

$$\begin{cases} 5x + 7y \le 35 \\ x \ge 0 \\ y \ge 0 \end{cases}$$

EXAMPLE 6 **Nutrition; Minimizing Calories**

Albert wants to go on a diet and needs your help in designing a lunch menu. The menu is to include two items: soup and salad. The vitamin units (milligrams) and calorie counts in each ounce of soup and salad are given in Table 7.2.

TABLE 7.2

Item	Vitamin A	Vitamin C	Calories
Soup	1	3	50
Salad	1	2	40

The menu must provide at least:

10 units of vitamin A.
24 units of vitamin C.

Find the number of ounces of each item in the menu needed to provide the required vitamins with the fewest number of calories.

Solution

a. State the problem mathematically.

Step 1 **Write the objective function.** Let the lunch menu contain x ounces of soup and y ounces of salad. You need to minimize the total number of calories in the menu. Since the soup has 50 calories per ounce (see Table 7.2), x ounces of soup has $50x$ calories. Likewise, the salad has $40y$ calories. The number f of calories in the two items is therefore given by the objective function:

$$f = 50x + 40y$$

Step 2 **Write the constraints.** Since each ounce of soup provides 1 unit of vitamin A, and each ounce of salad provides 1 unit of vitamin A, the two items together provide $1 \cdot x + 1 \cdot y = x + y$ units of vitamin A. The lunch must have at least 10 units of vitamin A; this means that

$$x + y \geq 10.$$

Similarly, the two items provide $3x + 2y$ units of vitamin C. The menu must provide at least 24 units of vitamin C, so

$$3x + 2y \geq 24.$$

The fact that x and y cannot be negative means that $x \geq 0$ and $y \geq 0$. You now have four constraints.

Summarize your information. Find x and y such that the value of

$$f = 50x + 40y \qquad \text{Objective function from Step 1}$$

is a *minimum*, with the restrictions:

$$\begin{cases} x + y \geq 10 \\ 3x + 2y \geq 24 \\ x \geq 0 \\ y \geq 0 \end{cases} \qquad \text{Constraints from Step 2}$$

b. Solve the linear programming problem.

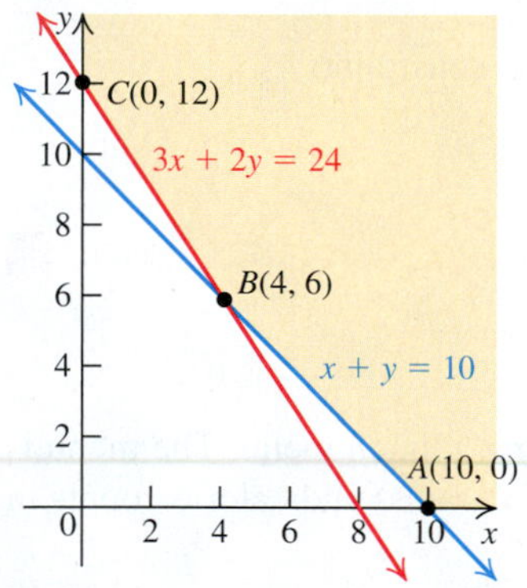

Step 3 **Graph the set of feasible solutions.** The set of feasible solutions is shaded in Figure 7.19. This set is bounded by the lines whose equations are

$$x + y = 10, \quad 3x + 2y = 24, \quad x = 0, \text{ and } \quad y = 0.$$

Step 4 **Find the vertices.** The vertices of the set of feasible solutions are

$$A(10, 0), \text{ found by solving } \begin{cases} y = 0 \\ x + y = 10 \end{cases}$$

$$B(4, 6), \text{ found by solving } \begin{cases} x + y = 10 \\ 3x + 2y = 24 \end{cases}$$

$$C(0, 12), \text{ found by solving } \begin{cases} x = 0 \\ 3x + 2y = 24 \end{cases}$$

Figure 7.19

Step 5 **Find the value of f at the vertices.** The value of f at each of the vertices is given in Table 7.3.

TABLE 7.3

Vertex (x, y)	Value of $f = 50x + 40y$
$(10, 0)$	$50(10) + 40(0) = 500$
$(4, 6)$	$50(4) + 40(6) = 440$
$(0, 12)$	$50(0) + 40(12) = 480$

Step 6 **Find the maximum or minimum value of f.** From Table 7.3, the smallest value of f is 440, which occurs when $x = 4$ and $y = 6$.

c. State the conclusion.

The lunch menu for Albert should contain 4 ounces of soup and 6 ounces of salad. His intake of 440 calories will be as small as possible under the given constraints.

Practice Problem 6 How does the answer to Example 6 change if 1 ounce of soup contains 30 calories and 1 ounce of salad contains 60 calories?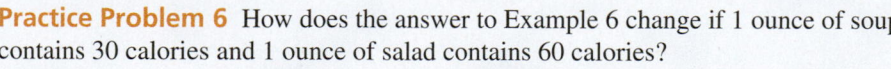

Notice in Example 6 that the value of the objective function $f = 50x + 40y$ can be made as large as you want by choosing x or y (or both) to be very large. So f has a minimum value of 440 but has no maximum value.

4 Graph a nonlinear inequality in two variables.

Nonlinear Inequality

We now discuss a procedure for graphing a nonlinear inequality involving two variables. We follow the same steps that we used in solving a linear inequality.

PROCEDURE
IN ACTION

EXAMPLE 7 Graphing a Nonlinear Inequality in Two Variables

OBJECTIVE	**EXAMPLE**
Graph a nonlinear inequality in two variables.	*Graph* $y > x^2 - 2$.

Step 1 Replace the inequality symbol with the equal $(=)$ sign.

1. $y = x^2 - 2$

Step 2 Sketch the graph of the *corresponding equation* in Step 1. Use a dashed curve if the given inequality sign is $<$ or $>$ and a solid curve if the inequality symbol is $\leq$ or $\geq$.

2. 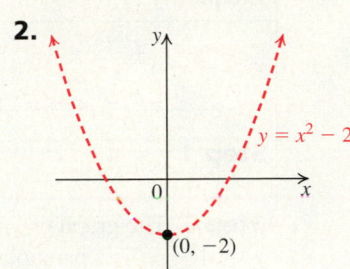 The graph of the parabola with vertex $(0, -2)$ is sketched as a dashed curve in the figure.

Step 3 The graph of the continuous curve in Step 2 divides the plane into several regions. Select a *test point* in each region but not on the graph.

3.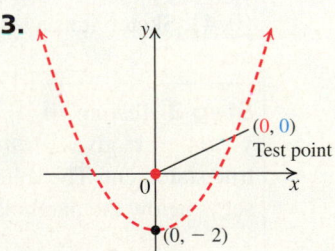

Step 4 **(i)** If the coordinates of the test point satisfy the inequality, then so do all of the points in that region. Shade that region.

(ii) If the test point's coordinates do not satisfy the inequality, shade the region that does not contain the test point.

The shaded region (including the boundary if it is a solid curve) is the graph of the inequality.

4. Because $0 > 0^2 - 2 = -2$, the coordinates of the test point, $(0, 0)$, satisfy the inequality $y > x^2 - 2$.

The graph of the solution set of $y > x^2 - 2$ is shaded.

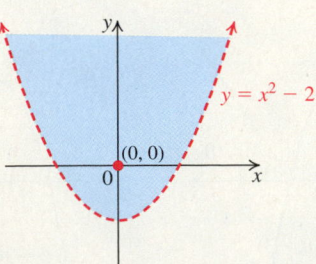

Practice Problem 7 Graph $y \leq x^2 - 2$.

5 Graph systems of nonlinear inequalities in two variables.

Nonlinear Systems

Just as in the definition of a system of nonlinear equations (see page 594), if at least one inequality in a system of inequalities is nonlinear, then the set of inequalities is called a **system of nonlinear inequalities**. The procedure for solving a system of nonlinear inequalities is identical to the procedure for solving a linear system of inequalities.

> **EXAMPLE 8** **Solving a System of Nonlinear Inequalities**

Graph the solution set of the system of inequalities.

$$\begin{cases} y \le 4 - x^2 & \text{Inequality (1)} \\ y \ge \dfrac{3}{2}x - 3 & \text{Inequality (2)} \\ y \ge -6x - 3 & \text{Inequality (3)} \end{cases}$$

Solution

Graph each inequality separately in the same coordinate plane. Because $(0, 0)$ is not a solution of any of the corresponding equations, use $(0, 0)$ as a test point for each inequality.

Inequality (1)	**Inequality (2)**	**Inequality (3)**
$y \le 4 - x^2$	$y \ge \dfrac{3}{2}x - 3$	$y \ge -6x - 3$
Step 1 $y = 4 - x^2$	$y = \dfrac{3}{2}x - 3$	$y = -6x - 3$
Step 2 The graph of $y = 4 - x^2$ is a parabola opening down with vertex $(0, 4)$. Sketch it as a solid curve.	The graph of $y = \dfrac{3}{2}x - 3$ is a line through the points $(0, -3)$ and $(2, 0)$. Sketch a solid line.	The graph of $y = -6x - 3$ is a line through the points $(0, -3)$ and $\left(-\dfrac{1}{2}, 0\right)$. Sketch a solid line.
Step 3 Testing $(0, 0)$ in $y \le 4 - x^2$ gives $0 \le 4$, a true statement. The solution set is below the parabola.	Testing $(0, 0)$ in $y \ge \dfrac{3}{2}x - 3$ gives $0 \ge -3$, a true statement. The solution set is above the line.	Testing $(0, 0)$ in $y \ge -6x - 3$ gives $0 \ge -3$, a true statement. The solution set is above the line.
Step 4 Shade the region as in Figure 7.20(a).	Shade the region as in Figure 7.20(b).	Shade the region as in Figure 7.20(c).

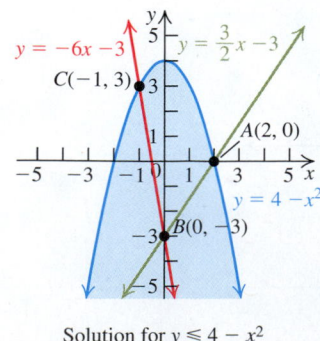

Solution for $y \le 4 - x^2$

(a)

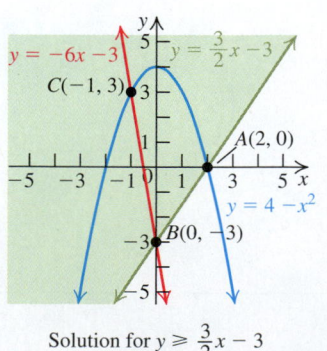

Solution for $y \ge \dfrac{3}{2}x - 3$

(b)

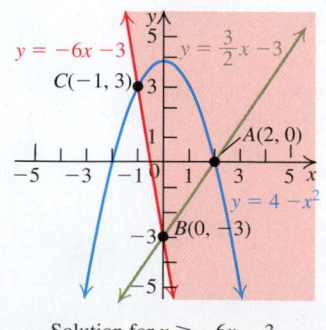

Solution for $y \ge -6x - 3$

(c)

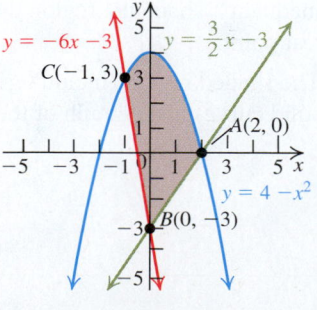

Solution of the system

(d)

Figure 7.20

The region common to all three graphs is the graph of the solution set of the given system of inequalities. See Figure 7.20(d). Use the substitution method to locate the points of intersection A, B, and C of the corresponding equations. See Figure 7.20(d). Solve all pairs of corresponding equations.

The point $A(2, 0)$ is the solution of the system

$$\begin{cases} y = 4 - x^2 & \text{Boundary of inequality (1)} \\ y = \dfrac{3}{2}x - 3 & \text{Boundary of inequality (2)} \end{cases}$$

The point $B(0, -3)$ is the solution of the system

$$\begin{cases} y = \dfrac{3}{2}x - 3 & \text{Boundary of inequality (2)} \\ y = -6x - 3 & \text{Boundary of inequality (3)} \end{cases}$$

Finally, the point $C(-1, 3)$ is the solution of the system

$$\begin{cases} y = -6x - 3 & \text{Boundary of inequality (1)} \\ y = 4 - x^2 & \text{Boundary of inequality (2)} \end{cases}$$

Practice Problem 8 Graph the solution set of the system of inequalities.

$$\begin{cases} y \geq x^2 + 1 \\ y \leq -x + 13 \\ y < 4x + 13 \end{cases}$$

Answers to Practice Problems

1.

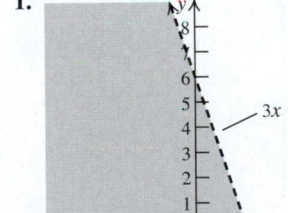

2.

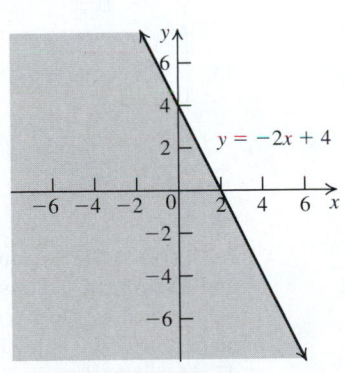

3.

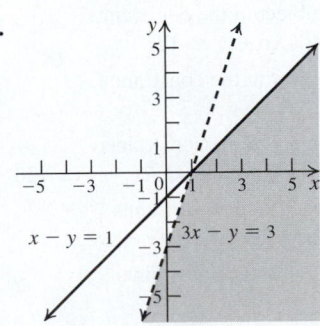

4. $(4, 0)$, $(5, 2)$, and $(2, 4)$

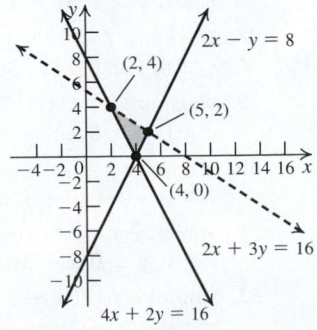

5. The objective function f has a maximum value of 28 at the point $(7, 0)$. **6.** The minimum number of calories is 300. Lunch then consists of 10 oz of soup and no salad.

7.

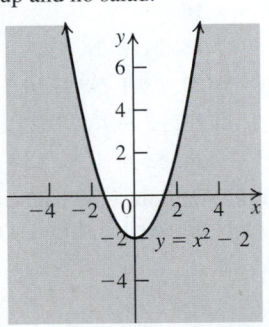

8.

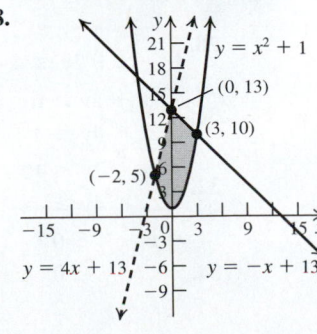

SECTION 7.3 **Exercises**

Concepts and Vocabulary

1. In the graph of $x - 3y > 1$, the corresponding equation $x - 3y = 1$ is graphed as a _____ line.

2. If a test point from one of the two regions determined by an inequality's corresponding equation satisfies the inequality, then _____ in that region satisfy the inequality.

3. In a system of inequalities containing both $2x + y > 5$ and $2x - y \leq 3$, the point of intersection of the lines $2x + y = 5$ and $2x - y = 3$ is _____ solution of the system.

4. In a nonlinear system of equations or inequalities, _____ equation or inequality must be nonlinear.

5. **True or False.** There are always infinitely many solutions of an inequality $ax + by < c$, where a, b, and c are real numbers and $a \neq 0$.

6. **True or False.** If $x^2 + 2y < 5$ is one of the inequalities in a system of inequalities, then the point $(2, 1)$ cannot be used as a test point in graphing the system.

7. **True or False.** None of the points on the graph of the line $3x + 5 = 7$ are solutions of the inequality $3x + 5 > 7$.

8. **True or False.** The intersection point of the graphs of the equations of the system $\begin{cases} 2x + 4y \leq 12 \\ x - 7y < 14 \end{cases}$ is a solution of this system.

Building Skills

In Exercises 9–24, graph each inequality.

9. $x \geq 0$ 10. $y \geq 0$

11. $x > -1$ 12. $y > 2$

13. $x \geq 2$ 14. $y \leq 3$

15. $y - x < 0$ 16. $y + 2x \leq 0$

17. $x + 2y < 6$ 18. $3x + 2y < 12$

19. $2x + 3y \geq 12$ 20. $2x + 5y \geq 10$

21. $x - 2y \leq 4$ 22. $3x - 4y \leq 12$

23. $3x + 5y < 15$ 24. $5x + 7y < 35$

In Exercises 25–40, graph the solution set of each system of linear inequalities and label the vertices (if any) of the solution set.

25. $\begin{cases} x + y \leq 1 \\ x - y \leq -1 \end{cases}$ 26. $\begin{cases} x + y \geq 1 \\ y - x \geq 1 \end{cases}$

27. $\begin{cases} 3x + 5y \leq 15 \\ 2x + 2y \leq 8 \end{cases}$ 28. $\begin{cases} -2x + y \leq 2 \\ 3x + 2y \leq 4 \end{cases}$

29. $\begin{cases} 2x + 3y \leq 6 \\ 4x + 6y \geq 24 \end{cases}$ 30. $\begin{cases} x + 2y \geq 6 \\ 2x + 4y \leq 4 \end{cases}$

31. $\begin{cases} 3x + y \leq 8 \\ 4x + 2y \geq 4 \end{cases}$ 32. $\begin{cases} x + 2y \leq 12 \\ 2x + 4y \geq 8 \end{cases}$

33. $\begin{cases} x \geq 0 \\ y \geq 0 \\ x + y \leq 1 \end{cases}$ 34. $\begin{cases} x \geq 0 \\ y \geq 0 \\ x + y > 2 \end{cases}$

35. $\begin{cases} x \geq 0 \\ y \geq 0 \\ 2x + 3y \geq 6 \end{cases}$ 36. $\begin{cases} x \geq 0 \\ y \geq 0 \\ 3x + 4y \leq 12 \end{cases}$

37. $\begin{cases} x + y \leq 1 \\ x - y \geq 1 \\ 2x + y \leq 1 \end{cases}$ 38. $\begin{cases} x + y \leq 1 \\ x - y \geq 1 \\ 2x + y \geq 1 \end{cases}$

39. $\begin{cases} x + y \leq 1 \\ x - y < 2 \\ -x + y \leq 3 \\ x + y \geq -4 \end{cases}$ 40. $\begin{cases} x + y \leq -1 \\ 2x - y > -2 \\ -2x + y \geq -3 \\ 2x + 2y \geq -4 \end{cases}$

In Exercises 41–46, use the following figure to indicate the regions (A–G) that correspond to the graph of the given system of inequalities.

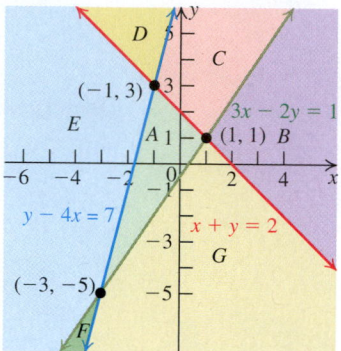

41. $\begin{cases} x + y \leq 2 \\ 3x - 2y \leq 1 \\ y - 4x \leq 7 \end{cases}$ 42. $\begin{cases} x + y \geq 2 \\ 3x - 2y \leq 1 \\ y - 4x \leq 7 \end{cases}$

43. $\begin{cases} x + y \geq 2 \\ 3x - 2y \geq 1 \\ y - 4x \leq 7 \end{cases}$ 44. $\begin{cases} x + y \geq 2 \\ 3x - 2y \leq 1 \\ y - 4x \geq 7 \end{cases}$

45. $\begin{cases} x + y \leq 2 \\ 3x - 2y \leq 1 \\ y - 4x \geq 7 \end{cases}$ 46. $\begin{cases} x + y \leq 2 \\ 3x - 2y \geq 1 \\ y - 4x \geq 7 \end{cases}$

In Exercises 47–54, you are given an objective function f and a set of constraints. Find the set of feasible solutions determined by the given constraints and then maximize or minimize the objective function as directed.

47. Maximize $f = 9x + 13y$, subject to the constraints $x \geq 0, y \geq 0, 5x + 8y \leq 40, 3x + y \leq 12$.

48. Maximize $f = 7x + 6y$, subject to the constraints $x \geq 0, y \geq 0, 2x + 3y \leq 13, x + y \leq 5$.

49. Maximize $f = 5x + 7y$, subject to the constraints $x \geq 0, y \geq 0, x + y \leq 70, x + 2y \leq 100, 2x + y \leq 120$.

50. Maximize $f = 2x + y$, subject to the constraints $x \geq 0, y \geq 0, x + y \geq 5, 2x + 3y \leq 21, 4x + 3y \leq 24$.

51. Minimize $f = x + 4y$, subject to the constraints $x \geq 0, y \geq 0, x + 3y \geq 3, 2x + y \geq 2$.

52. Minimize $f = 5x + 2y$, subject to the constraints $x \geq 0, y \geq 0, 5x + y \geq 10, x + y \geq 6$.

53. Minimize $f = 13x + 15y$, subject to the constraints $x \geq 0, y \geq 0, 3x + 4y \geq 360, 2x + y \geq 100$.

54. Minimize $f = 40x + 37y$, subject to the constraints $x \geq 0, y \geq 0, 10x + 3y \geq 180, 2x + 3y \geq 60$.

In Exercises 55–66, use the procedure for graphing a nonlinear inequality in two variables to graph each inequality.

55. $y > x^2 - 1$

56. $y \geq x^2 + 1$

57. $y \leq (x - 1)^2 + 3$

58. $y > -(x + 1)^2 + 4$

59. $x^2 + y^2 > 4$

60. $x^2 + y^2 \leq 9$

61. $(x - 1)^2 + (y - 2)^2 \leq 9$

62. $(x + 2)^2 + (y - 1)^2 > 4$

63. $y < 2^x$

64. $y \geq e^x$

65. $y < \log x$

66. $y \geq \ln x$

For Exercises 67–72, use the following figure to indicate the region or regions that correspond to the graph of each system of inequalities.

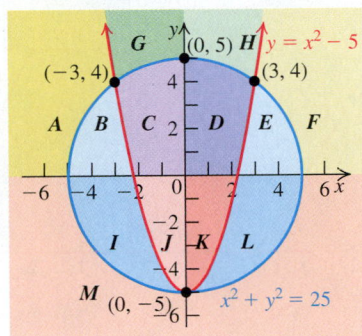

67. $\begin{cases} y \geq x^2 - 5 \\ x^2 + y^2 \leq 25 \\ x \geq 0 \end{cases}$

68. $\begin{cases} y \geq x^2 - 5 \\ x^2 + y^2 \geq 25 \\ x \geq 0 \end{cases}$

69. $\begin{cases} y \leq x^2 - 5 \\ x^2 + y^2 \leq 25 \\ x \geq 0 \end{cases}$

70. $\begin{cases} y \leq x^2 - 5 \\ x^2 + y^2 \geq 25 \\ y \geq 0 \end{cases}$

71. $\begin{cases} y \leq x^2 - 5 \\ x^2 + y^2 \leq 25 \\ y \geq 0 \end{cases}$

72. $\begin{cases} y \geq x^2 - 5 \\ x^2 + y^2 \leq 25 \\ y \geq 0 \end{cases}$

In Exercises 73–80, graph the region determined by each system of inequalities and label all points of intersection.

73. $\begin{cases} x + y \leq 6 \\ y \geq x^2 \end{cases}$

74. $\begin{cases} x + y \geq 6 \\ y \geq x^2 \end{cases}$

75. $\begin{cases} y \geq 2x + 1 \\ y \leq -x^2 + 2 \end{cases}$

76. $\begin{cases} y \leq 2x + 1 \\ y \leq -x^2 + 2 \end{cases}$

77. $\begin{cases} y \geq x \\ x^2 + y^2 \leq 1 \end{cases}$

78. $\begin{cases} y \leq x \\ x^2 + y^2 \leq 1 \end{cases}$

79. $\begin{cases} x^2 + y^2 \leq 25 \\ x^2 + y \leq 5 \end{cases}$

80. $\begin{cases} x^2 + y^2 \leq 25 \\ x^2 + y \geq 5 \end{cases}$

Applying the Concepts

81. **Advertising.** The Fantastic Bread Company allocates a maximum of $24,000 to advertise its product to television and newspaper. Each hour of television costs $4000, and each page of newspaper advertising costs $3000. The exposure from each hour of television time is assumed to be 120,000, and the exposure from each page of newspaper advertising is 80,000. Furthermore, the board of directors requires at least two hours of television time and one page of newspaper advertising. How should the advertising budget be divided to maximize exposure to the advertisements?

82. **Agriculture.** The Florida Juice Company makes two types of fruit punch—Fruity and Tangy—by blending orange juice and apple juice into a mixture. The fruit punch is sold in 5-gallon bottles. A bottle of Fruity earns a profit of $3, and a bottle of Tangy earns a $2 profit. A bottle of Fruity requires 3 gallons of orange juice and 2 gallons of apple juice, while a bottle of Tangy requires 4 gallons of orange juice and 1 gallon of apple juice. If 200 gallons of orange juice and 120 gallons of apple juice are available, find the maximum profit for the company.

83. **Manufacturing.** A company produces a portable global positioning system (GPS) and a portable digital video disc (DVD) player, each of which requires three stages of processing. The length of time for processing each unit is given in the following table.

Stage	GPS (hr/unit)	DVD (hr/unit)	Maximum Process Capacity (hr/day)
I	12	12	840
II	3	6	300
III	8	4	480
Profit per unit	$6	$4	

How many of each product should the company produce per day to maximize profit?

84. **Managing cost.** A production manager can use either of two available crews for up to six hours a day to assemble bikes. Crew A can assemble 50 road bikes and 40 stationary bikes per hour, while crew B only assembles 40 road bikes and 20 stationary bikes per hour. However, crew A costs $110 per hour, and crew B costs $85 per hour. If at least 200 road bikes and 120 stationary bikes must be assembled each day, how many hours per day should each crew work in order to minimize costs?

85. **Return on Investment.** A student in a finance class is tasked with maximizing return on an investment of $40,000 by investing in both Treasury bonds that are paying 2% and in more speculative corporate bonds that are paying 4%. She is required to invest at least $16,000 in Treasury bonds and at most $24,000 in corporate bonds. Additionally, she must invest at least as much in Treasury bonds as in corporate bonds. How much should be invested in each type of bond to maximize the return on her investment?

86. **Hospitality.** Mrs. Adams owns a motel consisting of 300 single rooms and an attached restaurant that serves breakfast. She knows that 30% of the male guests and 50% of the female guests will eat in the restaurant. Suppose she makes a profit of $18.50 per day from every guest who eats in her restaurant and $15 per day from every guest who does not eat in her restaurant. Assuming that Mrs. Adams never has more than 125 female guests and never more than 220 male guests, find the number of male and female guests that would provide the maximum profit.

87. **Transportation.** Major Motors, Inc., must produce at least 5000 luxury cars and 12,000 medium-priced cars. It must

produce, at most, 30,000 units of compact cars. The company owns one factory in Michigan and one in North Carolina. The Michigan factory produces 20, 40, and 60 units of luxury, medium-priced, and compact cars, respectively, per day, while the North Carolina factory produces 10, 30, and 20 luxury, medium-priced, and compact cars, respectively, per day. If the Michigan factory costs $60,000 per day to operate and the North Carolina factory costs $40,000 per day to operate, find the number of days each factory should operate to minimize the cost and meet the requirements.

88. Nutrition. Elisa Epstein is buying two types of frozen meals: a meal of enchiladas with vegetables and a vegetable loaf meal. She wants to guarantee that she gets a minimum of 60 grams of carbohydrates, 40 grams of proteins, and 35 grams of fats. The enchilada meal contains 5, 3, and 5 grams of carbohydrates, proteins, and fats, respectively, per kilogram, while the vegetable loaf contains 2, 2, and 1 gram of carbohydrates, proteins, and fats, respectively, per kilogram. If the enchilada meal costs $3.50 per kilogram and the vegetable loaf costs $2.25 per kilogram, how many kilograms of each should Elisa buy to minimize the cost and still meet the minimum requirement?

89. Engine manufacturing. A company manufactures 3 hp and 5 hp engines. The time (in hours) required to assemble, test, and package each type of engine is shown in the table.

	Process		
Engine	**Assemble**	**Test**	**Package**
3 hp	3	2	0.5
5 hp	4.5	1	0.75

The maximum times (in hours) available for assembling, testing, and packaging are 360 hours, 200 hours, and 60 hours, respectively. Graph the possible processing activities and find the vertices.

90. Pet food. Vege Pet Food manufactures two types of dog food: Vegies and Yummies. Each bag of Vegies contains 3 pounds of vegetables and 2 pounds of cereal; each bag of Yummies contains 1.5 pounds of vegetables and 3.5 pounds of cereal. The total pounds of vegetables and cereal available per week are 20,000 pounds and 27,000 pounds, respectively. Graph the possible number of bags of each type of food and find the vertices.

Beyond the Basics

In Exercises 91–98, write a system of linear inequalities that has the given graph.

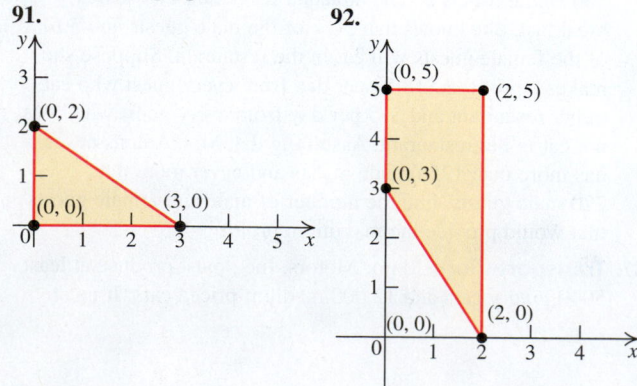

93.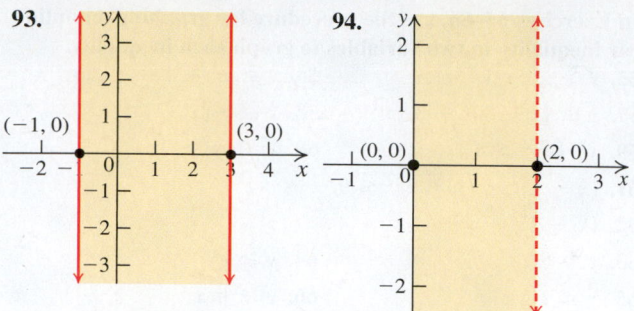

94.

95.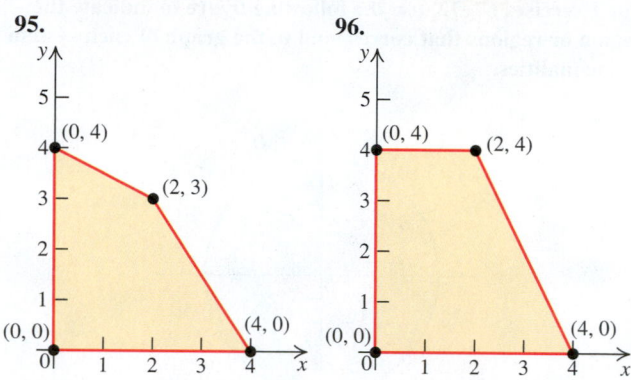

96.

97.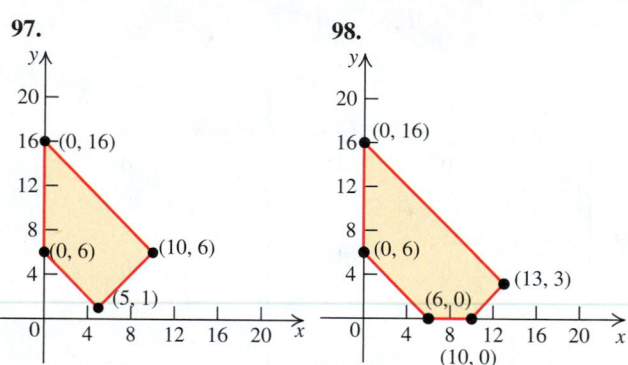

98.

In Exercises 99–108, graph each inequality.

99. $xy \geq 1$ **100.** $xy < 3$

101. $y < 2^x$ **102.** $y \leq 2^{x-1}$

103. $y \geq 3^{x-2}$ **104.** $y > -3^x$

105. $y \geq \ln x$ **106.** $y < -\ln x$

107. $y > \ln(x - 1)$ **108.** $y \leq -\ln(x + 1)$

In Exercises 109 and 110, graph the solution set of each inequality.

109. $1 \leq x^2 + y^2 \leq 9$ **110.** $x^2 \leq y \leq 4x^2$

Critical Thinking / Discussion / Writing

In Exercises 111 and 112, graph the solution set of each inequality.

111. $|x| + |y| < 1$ **112.** $|x + 2y| < 4$

Getting Ready for the Next Section

In Exercises 113–116, solve each system of linear equations by the Gaussian elimination method. (See Page 610.)

113. $\begin{cases} x - 2y = 4 \\ -3x + 5y = -7 \end{cases}$ **114.** $\begin{cases} 3x + 5y = 6 \\ 2x - y = 17 \end{cases}$

115. $\begin{cases} x - 4y - z = 11 \\ 2x - 5y + 2z = 39 \\ -3x + 2y + z = 1 \end{cases}$ **116.** $\begin{cases} x + 3y - 2z = 5 \\ 2x + y + 4z = 8 \\ 6x + y - 3z = 5 \end{cases}$

117. Consider the following rectangular array of numbers:

$$\begin{bmatrix} 2 & 4 & -6 & 10 \\ 3 & -2 & 4 & 12 \end{bmatrix} \begin{matrix} \leftarrow \text{row 1} \\ \leftarrow \text{row 2} \end{matrix}$$

Write the new rectangular array after multiplying each number in row 1 by $\frac{1}{2}$.

118. In the rectangular array obtained in Exercise 117, add -3 times each number in row 1 to the corresponding number in row 2.

119. In the rectangular array obtained in Exercise 118, multiply each number in row 2 by $-\frac{1}{8}$.

120. In the rectangular array obtained in Exercise 119, add -2 times each number in row 2 to the corresponding number in row 1.

Matrices and Systems of Equations

BEFORE STARTING THIS SECTION, REVIEW

1 Systems of equations (Section 7.1, page 594)

2 Equivalent systems (Section 7.1, page 597)

3 Gaussian elimination (Section 7.2, page 610)

OBJECTIVES

1 Define a matrix.

2 Use matrices to solve a system of linear equations.

3 Use Gaussian elimination to solve a system.

4 Use Gauss–Jordan elimination to solve a system.

Wassily Leontief 1906–1999
Leontief studied philosophy, sociology, and economics at the University of Leningrad and earned the degree of Learned Economist in 1925. He continued his studies at the University of Berlin, where he earned his PhD degree. In 1973, the Nobel Prize in Economic Sciences was awarded to Leontief for his work on input–output analysis of the U.S. economy.

◆ Leontief Input–Output Model

In 1949, the Russian-born Harvard Professor Wassily Leontief opened the door to a new era in mathematical modeling in economics. He divided the U.S. economy into 500 "sectors," such as the coal, automotive, communications industries, and agriculture. For each sector, he wrote a linear equation that described how that sector distributed its output to other sectors of the economy. In 1949, the largest computer was the Mark II, and it could not handle the resulting system of 500 equations in 500 variables. Leontief then combined some of the sectors and reduced the system to 42 equations in 42 variables to solve the problem. As computers have become more powerful, scientists and engineers can now work on problems far more complex than they ever dreamed of a few decades ago. In Example 10, we explore Leontief's input–output model applied to a simple economy.

1 Define a matrix.

Definition of a Matrix

Suppose Great Builders Incorporated (GBI) builds ranch, colonial, and modern houses. GBI wants to compare the cost of labor (in millions of dollars) and the cost of material (in millions of dollars) involved in building each type of house during one year. These data could be represented as in Table 7.4.

TABLE 7.4

Type of House	Ranch	Colonial	Modern
Cost of labor (in millions)	12	13	14
Cost of material (in millions)	9	11	10

Or it could be represented more compactly by the following array:

$$\begin{bmatrix} 12 & 13 & 14 \\ 9 & 11 & 10 \end{bmatrix}$$

This rectangular array of numbers is called a *matrix* (plural, *matrices*). This matrix has two rows (the types of costs) and three columns (the types of houses).

Matrix

A **matrix** is a rectangular array of numbers.

$$A = \begin{bmatrix} a_{11} & a_{12} & \cdots & a_{1n} \\ a_{21} & a_{22} & \cdots & a_{2n} \\ \vdots & \vdots & & \vdots \\ a_{m1} & a_{m2} & \cdots & a_{mn} \end{bmatrix} \begin{matrix} \leftarrow \text{Row 1} \\ \leftarrow \text{Row 2} \\ \\ \leftarrow \text{Row } m \end{matrix}$$

Column 1   Column 2   Column n

If a matrix A has m rows and n columns, then we say that A is of **order m by n** (written $m \times n$).

The **entry** or **element** in the ith row and jth column is a real number and is denoted by the *double-subscript* notation a_{ij}. We call a_{ij} the **(i, j)th entry**. For example, a_{24} is the entry in the second row and fourth column of the matrix A. In general, we will use capital letters such as $A, B, \ldots$ to denote matrices, and the corresponding lowercase letter $a_{ij}, b_{ij}, \ldots$ for their entries.

If A has n rows and n columns, then A is called a **square matrix of order n**. The entries $a_{11}, a_{22}, \ldots, a_{nn}$ form the **main diagonal** of A. A $1 \times n$ matrix is called a **row matrix**, and an $n \times 1$ matrix is called a **column matrix**.

EXAMPLE 1   **Determining the Order of Matrices**

Determine the order of each matrix. Identify square, row, and column matrices. Identify entries in the main diagonal of each square matrix.

a. $A = [3]$ **b.** $B = [3 \quad 5 \quad -7]$ **c.** $C = \begin{bmatrix} 0 & 1 \\ -3 & 4 \end{bmatrix}$ **d.** $D = \begin{bmatrix} 1 & 2 & 3 \\ 4 & 5 & 6 \\ 7 & 8 & 9 \end{bmatrix}$

Solution

a. Matrix A with one row and one column is a 1×1 matrix. A is a square matrix of order 1. In A, $a_{11} = 3$ is the main diagonal. A is also a column and a row matrix.

b. Matrix B with one row and three columns is a 1×3 matrix. B is a row matrix.

c. Matrix C, a 2×2 matrix, is a square matrix of order 2. In C, the entries $c_{11} = 0$ and $c_{22} = 4$ form the main diagonal.

d. Matrix D, a 3×3 matrix, is a square matrix of order 3. In D, the entries $d_{11} = 1$, $d_{22} = 5$, and $d_{33} = 9$ form the main diagonal.

Practice Problem 1   Determine the order of each matrix.

a. $\begin{bmatrix} -1 & 3 \\ 7 & 4 \\ 0 & 0 \end{bmatrix}$ **b.** $[3 \quad -8]$

2 Use matrices to solve a system of linear equations.

Using Matrices to Solve Linear Systems

To solve a system of linear equations by the elimination method, the particular symbols used for the variables do not matter; only the coefficients and the constants are important. We can display the constants and coefficients of a system in a matrix called the **augmented**

matrix of the system. The matrix containing only the coefficients of the variables is the **coefficient matrix**. Consider the following system.

$$\begin{cases} x - y - z = 1 \\ 2x - 3y + z = 10 \\ x + y - 2z = 0 \end{cases}$$

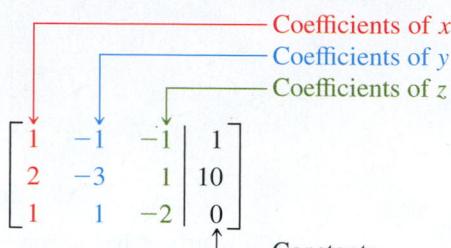

Coefficients of x
Coefficients of y
Coefficients of z

Augmented Matrix:
$$\begin{bmatrix} 1 & -1 & -1 & | & 1 \\ 2 & -3 & 1 & | & 10 \\ 1 & 1 & -2 & | & 0 \end{bmatrix}$$

Constants

Coefficient Matrix:
$$\begin{bmatrix} 1 & -1 & -1 \\ 2 & -3 & 1 \\ 1 & 1 & -2 \end{bmatrix}$$

Notice that the numbers in the first column of the augmented matrix are the coefficients of x, those in the second column are the coefficients of y, and those in the third column are the coefficients of z. The constants on the right side of the equations in the system are found in the fourth column. The vertical line in the augmented matrix is to remind you of the equal signs in the equations. If a variable does not appear in one of the equations, represent it by a zero in the matrix.

EXAMPLE 2 **Writing the Augmented Matrix of a Linear System**

Write the augmented matrix of the linear system.

$$\begin{cases} 2x + 3z = 1 \\ 2z + y = 5 \\ -4x + 5y = 7 \end{cases}$$

Solution

First, write the system with the variables lined up in columns and insert zeros as coefficients of any missing variables.

$$\begin{cases} 2x + 0y + 3z = 1 \\ 0x + 1y + 2z = 5 \\ -4x + 5y + 0z = 7 \end{cases}$$

The augmented matrix of the given system is:

$$\begin{bmatrix} 2 & 0 & 3 & | & 1 \\ 0 & 1 & 2 & | & 5 \\ -4 & 5 & 0 & | & 7 \end{bmatrix}$$

TECHNOLOGY CONNECTION

Many graphing calculators let you first name a matrix, specify the size, and then insert the entries for the matrix. No vertical line separates the column of constants, but this does not affect any of the matrix operations. The augmented matrix from Example 2 has been named A.

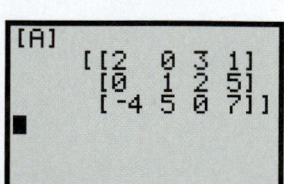

Practice Problem 2 Write the augmented matrix of the linear system.

$$\begin{cases} 3y - z = 8 \\ x + 4y = 14 \\ -2y + 9z = 0 \end{cases}$$

We can reverse this process and write a linear system from a given augmented matrix. For example, the augmented matrix

$$\begin{bmatrix} 9 & 2 & -1 & | & 6 \\ 3 & 4 & 7 & | & -8 \\ 0 & 1 & 5 & | & 11 \end{bmatrix}$$ corresponds to the system of equations $$\begin{cases} 9x + 2y - z = 6 \\ 3x + 4y + 7z = -8. \\ y + 5z = 11 \end{cases}$$

The basic strategy for solving a system of equations is to *replace the given system with an equivalent system* (one with the same solution set) *that is easier to solve.* In the previous section, we used three basic operations to solve a linear system:

1. Interchange two equations.
2. Multiply all of the terms in an equation by a nonzero constant.
3. Add a multiple of one equation to another equation. In other words, replace one equation with the sum of itself and a multiple of another equation.

In matrix terminology, the three corresponding operations are called the **elementary row operations**. Two matrices are **row equivalent** if one can be obtained from the other by a sequence of elementary row operations, given next. The symbol R_i represents the *i*th row of a matrix.

ELEMENTARY ROW OPERATIONS

Row Operation	In Symbols	Description
1. Interchange two rows.	$R_i \leftrightarrow R_j$	Interchange the *i*th and *j*th rows.
2. Multiply a row by a nonzero constant.	cR_j	Multiply the *j*th row by $c, c \neq 0$. constant.
3. Add a multiple of one row to another row.	$cR_i + R_j \rightarrow R_j$	Replace the *j*th row by adding c times the *i*th row to it.

EXAMPLE 3 **Applying Elementary Row Operations**

Perform the indicated row operations (**a**), (**b**), and (**c**) in order on the following matrix:

$$A = \begin{bmatrix} 3 & -2 & -3 \\ 2 & 4 & 14 \end{bmatrix}$$

a. $R_1 \leftrightarrow R_2$, **b.** $\dfrac{1}{2} R_1$, **c.** $-3R_1 + R_2 \rightarrow R_2$

Solution

a. $A = \begin{bmatrix} 3 & -2 & -3 \\ 2 & 4 & 14 \end{bmatrix} \xrightarrow{R_1 \leftrightarrow R_2} \begin{bmatrix} 2 & 4 & 14 \\ 3 & -2 & -3 \end{bmatrix} = B$ Interchange the 1st and 2nd rows.

b. $B = \begin{bmatrix} 2 & 4 & 14 \\ 3 & -2 & -3 \end{bmatrix} \xrightarrow{\frac{1}{2}R_1} \begin{bmatrix} 1 & 2 & 7 \\ 3 & -2 & -3 \end{bmatrix} = C$ Multiply the 1st row by $\frac{1}{2}$.

c. $C = \begin{bmatrix} 1 & 2 & 7 \\ 3 & -2 & -3 \end{bmatrix} \xrightarrow{-3R_1 + R_2 \rightarrow R_2} \begin{bmatrix} 1 & 2 & 7 \\ 0 & -8 & -24 \end{bmatrix}$ Add -3 times the 1st row to the 2nd row.

Practice Problem 3 Perform the row operations of Example 3 in order on the following matrix:

$$A = \begin{bmatrix} 3 & 4 & 5 \\ 2 & 4 & 6 \end{bmatrix}$$

In the next example, we solve a system of equations both with and without matrix notation and place the results side by side for comparison. We solve the system by converting it into triangular form by first using elimination and then back-substitution.

EXAMPLE 4 **Comparing Linear Systems and Matrices**

Solve the system of linear equations.

$$\begin{cases} x - y - z = 1 & \text{Equation (1)} \\ 2x - 3y + z = 10 & \text{Equation (2)} \\ x + y - 2z = 0 & \text{Equation (3)} \end{cases}$$

TECHNOLOGY CONNECTION

Many graphing calculators provide all three row operations. Consider the matrix A.

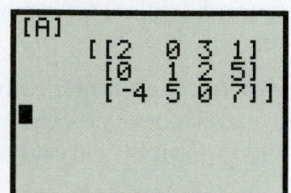

Here are some examples of row operations on the matrix A. Swap rows 1 and 3.

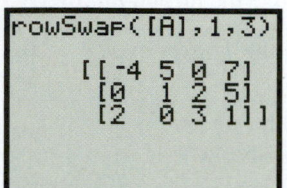

Multiply row 1 by 4.

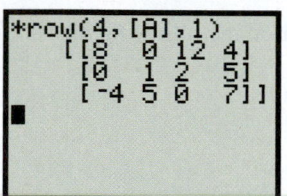

Add 3 times row 1 to row 2.

*row+(3,[A],1,2)

[[2   0   3   1]
 [6   1 11   8]
 [-4 5   0   7]]

Solution

Linear System

$$\begin{cases} x - y - z = 1 & (1) \\ 2x - 3y + z = 10 & (2) \\ x + y - 2z = 0 & (3) \end{cases}$$

Augmented Matrix

$$A = \begin{bmatrix} 1 & -1 & -1 & 1 \\ 2 & -3 & 1 & 10 \\ 1 & 1 & -2 & 0 \end{bmatrix}$$

> Add -2 times equation (1) to equation (2).
> Add -1 times equation (1) to equation (3).

> Use the first row to produce zeros at the (2, 1) and (3, 1) positions.

$$\begin{cases} x - y - z = 1 & (1) \\ -y + 3z = 8 & (4) \\ 2y - z = -1 & (5) \end{cases}$$

$\xrightarrow{-2R_1+R_2\rightarrow R_2}$
$\xrightarrow{(-1)R_1+R_3\rightarrow R_3}$
$$\begin{bmatrix} 1 & -1 & -1 & 1 \\ 0 & -1 & 3 & 8 \\ 0 & 2 & -1 & -1 \end{bmatrix}$$

> Multiply equation (4) by -1.

> Produce a 1 at the (2, 2) position.

$$\begin{cases} x - y - z = 1 & (1) \\ y - 3z = -8 & (6) \\ 2y - z = -1 & (5) \end{cases}$$

$\xrightarrow{(-1)R_2}$
$$\begin{bmatrix} 1 & -1 & -1 & 1 \\ 0 & 1 & -3 & -8 \\ 0 & 2 & -1 & -1 \end{bmatrix}$$

> Add -2 times equation (6) to equation (5).

> Use the second row to produce a zero at the (3, 2) position.

$$\begin{cases} x - y - z = 1 & (1) \\ y - 3z = -8 & (6) \\ 5z = 15 & (7) \end{cases}$$

$\xrightarrow{-2R_2+R_3\rightarrow R_3}$
$$\begin{bmatrix} 1 & -1 & -1 & 1 \\ 0 & 1 & -3 & -8 \\ 0 & 0 & 5 & 15 \end{bmatrix}$$

Equation (7), $5z = 15$, which corresponds to row 3 of the final matrix, gives the value $z = 3$. We find y by back-substitution.

$$y - 3z = -8 \quad \text{Equation (6); final matrix row 2}$$
$$y - 3(3) = -8 \quad \text{Replace } z \text{ with 3.}$$
$$y = 1 \quad \text{Solve for } y.$$

We find x by back-substitution.

$$x - y - z = 1 \quad \text{Equation (1); final matrix row 1}$$
$$x - 1 - 3 = 1 \quad \text{Replace } y \text{ with 1 and } z \text{ with 3.}$$
$$x = 5 \quad \text{Solve for } x.$$

The solution of the system is $x = 5$, $y = 1$, and $z = 3$. You should check this solution by substituting it into the original system of equations. The solution set is $\{(5, 1, 3)\}$.

Practice Problem 4 Solve the system of linear equations.

$$\begin{cases} x - 6y + 3z = -2 \\ 3x + 3y - 2z = -2 \\ 2x - 3y + z = -2 \end{cases}$$

The last matrix in the solution of Example 4 is in *row-echelon form,* which is defined next. In the definition, a **nonzero row** means a row that contains at least one nonzero entry; a **leading entry** of a row is the leftmost nonzero entry in a nonzero row.

> ### ROW-ECHELON FORM AND REDUCED ROW-ECHELON FORM
>
> An $m \times n$ matrix is in **row-echelon form** if it has the following three properties:
>
> 1. The leading entry of each nonzero row is 1.
> 2. The leading entry in any row is to the right of the leading entry in the row above it.
> 3. All nonzero rows are above the rows consisting entirely of zeros.
>
> A matrix in row-echelon form having the following property is in **reduced row-echelon form:** Each leading 1 is the only nonzero entry in its column.

Property 2 says that the leading entries form an echelon (steplike) pattern that moves down and to the right.

The following matrices are in row-echelon form; a starred entry may be any value, including zero.

$$\begin{bmatrix} 1 & * & * \\ 0 & 1 & * \\ 0 & 0 & 1 \end{bmatrix} \quad \begin{bmatrix} 1 & * & * & * \\ 0 & 1 & * & * \\ 0 & 0 & 1 & * \end{bmatrix} \quad \begin{bmatrix} 1 & * & * & * \\ 0 & 0 & 1 & * \\ 0 & 0 & 0 & 1 \\ 0 & 0 & 0 & 0 \end{bmatrix}$$

The following matrices are in *reduced* row-echelon form because the entries above and below each leading 1 are zero. (Again, the starred entries may be any value, including zero.)

$$\begin{bmatrix} 0 & 1 & 0 & 0 \\ 0 & 0 & 1 & 0 \\ 0 & 0 & 0 & 1 \end{bmatrix} \quad \begin{bmatrix} 1 & 0 & 0 & * \\ 0 & 1 & 0 & * \\ 0 & 0 & 1 & * \end{bmatrix} \quad \begin{bmatrix} 1 & * & 0 & 0 \\ 0 & 0 & 1 & 0 \\ 0 & 0 & 0 & 1 \\ 0 & 0 & 0 & 0 \end{bmatrix}$$

An augmented matrix may be transformed into several different row-echelon form matrices by using different sequences of row operations. However, its *reduced* row-echelon form is *unique*. (See Exercise 91.)

3 Use Gaussian elimination to solve a system.

Gaussian Elimination

The method of solving a system of linear equations by transforming the augmented matrix of the system into row-echelon form and then using back-substitution to find the solution set is known as **Gaussian elimination** (see Section 7.2).

PROCEDURE
IN ACTION

EXAMPLE 5 **Solving Linear Systems by Using Gaussian Elimination**

OBJECTIVE

Solve a system of linear equations by Gaussian elimination.

Step 1 Write the augmented matrix of the system.

Step 2 Use elementary row operations to transform the augmented matrix into row-echelon form.

EXAMPLE

Solve the system of equations.

$$\begin{cases} x + 2y = -2 \\ 4x + 3y = 7 \end{cases}$$

1. $A = \begin{bmatrix} 1 & 2 & | & -2 \\ 4 & 3 & | & 7 \end{bmatrix}$ Augmented matrix

2. $\xrightarrow{-4R_1 + R_2 \to R_2} \begin{bmatrix} 1 & 2 & | & -2 \\ 0 & -5 & | & 15 \end{bmatrix}$ This operation produces a 0 in the (2, 1) position.

$\xrightarrow{-\frac{1}{5}R_2} \begin{bmatrix} 1 & 2 & | & -2 \\ 0 & 1 & | & -3 \end{bmatrix}$ This operation produces a 1 in the (2, 2) position.

(continued)

Step 3 Write the system of linear equations that corresponds to the last matrix in Step 2.

Step 4 Use the system of equations from Step 3, together with back-substitution, to find the system's solution set.

3. $\begin{cases} x + 2y = -2 & (1) \\ \qquad y = -3 & (2) \end{cases}$

4.
$$\begin{aligned} x + 2y &= -2 & \text{Equation (1)} \\ x + 2(-3) &= -2 & \text{Replace } y \text{ with } -3. \\ x &= 4 & \text{Solve for } x. \end{aligned}$$

You should check the solution set, $\{(4, -3)\}$, by substituting $x = 4$ and $y = -3$ into the original system of equations.

Practice Problem 5 Solve by Gaussian elimination.

$$\begin{cases} x - 2y = 1 \\ 2x + 3y = 16 \end{cases}$$

EXAMPLE 6 **Solving a System by Using Gaussian Elimination**

Solve by Gaussian elimination.

$$\begin{cases} 2x + y + z = 6 \\ -3x - 4y + 2z = 4 \\ x + y - z = -2 \end{cases}$$

Solution

Step 1 $A = \begin{bmatrix} 2 & 1 & 1 & | & 6 \\ -3 & -4 & 2 & | & 4 \\ 1 & 1 & -1 & | & -2 \end{bmatrix}$ The augmented matrix of the system

Step 2 $\xrightarrow{R_1 \leftrightarrow R_3} \begin{bmatrix} 1 & 1 & -1 & | & -2 \\ -3 & -4 & 2 & | & 4 \\ 2 & 1 & 1 & | & 6 \end{bmatrix}$

$\xrightarrow[\;-2R_1 + R_3 \rightarrow R_3\;]{\;3R_1 + R_2 \rightarrow R_2\;} \begin{bmatrix} 1 & 1 & -1 & | & -2 \\ 0 & -1 & -1 & | & -2 \\ 0 & -1 & 3 & | & 10 \end{bmatrix}$

$\xrightarrow{(-1)R_2} \begin{bmatrix} 1 & 1 & -1 & | & -2 \\ 0 & 1 & 1 & | & 2 \\ 0 & -1 & 3 & | & 10 \end{bmatrix}$

$\xrightarrow{R_2 + R_3 \rightarrow R_3} \begin{bmatrix} 1 & 1 & -1 & | & -2 \\ 0 & 1 & 1 & | & 2 \\ 0 & 0 & 4 & | & 12 \end{bmatrix}$

$\xrightarrow{\frac{1}{4}R_3} \begin{bmatrix} 1 & 1 & -1 & | & -2 \\ 0 & 1 & 1 & | & 2 \\ 0 & 0 & 1 & | & 3 \end{bmatrix}$

The last matrix is in row-echelon form.

Step 3 The system of equations corresponding to the last matrix in Step 2 is

$$\begin{cases} x + y - z = -2 & (1) \\ \qquad y + z = 2 & (2) \\ \qquad\qquad z = 3 & (3) \end{cases}$$

Step 4 Equation (3) in Step 3 gives the value $z = 3$. Back-substitute $z = 3$ in equation (2).

$$y + z = 2 \qquad \text{Equation (2)}$$
$$y + 3 = 2 \qquad \text{Replace } z \text{ with 3.}$$
$$y = -1 \qquad \text{Solve for } y.$$

Now back-substitute $z = 3$ and $y = -1$ in equation(1).

$$x + y - z = -2 \qquad \text{Equation (1)}$$
$$x - 1 - 3 = -2$$
$$x = 2 \qquad \text{Solve for } x.$$

You should check the solution set, $\{(2, -1, 3)\}$, by substituting these values into the original system of equations.

Practice Problem 6 Solve the system of equations.

$$\begin{cases} 2x + y - z = 7 \\ x - 3y - 3z = 4 \\ 4x + y + z = 3 \end{cases}$$

EXAMPLE 7 Attempting to Solve a System with No Solution

Solve the system of equations by Gaussian elimination.

$$\begin{cases} y + 5z = -4 \\ x + 4y + 3z = -2 \\ 2x + 7y + z = 8 \end{cases}$$

Solution

Step 1 $A = \begin{bmatrix} 0 & 1 & 5 & | & -4 \\ 1 & 4 & 3 & | & -2 \\ 2 & 7 & 1 & | & 8 \end{bmatrix}$ The augmented matrix of the system

Step 2 $\xrightarrow{R_1 \leftrightarrow R_2} \begin{bmatrix} 1 & 4 & 3 & | & -2 \\ 0 & 1 & 5 & | & -4 \\ 2 & 7 & 1 & | & 8 \end{bmatrix}$

$\xrightarrow{-2R_1 + R_3 \to R_3} \begin{bmatrix} 1 & 4 & 3 & | & -2 \\ 0 & 1 & 5 & | & -4 \\ 0 & -1 & -5 & | & 12 \end{bmatrix}$ Row 2 already begins with a 0.

$\xrightarrow{R_2 + R_3 \to R_3} \begin{bmatrix} 1 & 4 & 3 & | & -2 \\ 0 & 1 & 5 & | & -4 \\ 0 & 0 & 0 & | & 8 \end{bmatrix}$

$\xrightarrow{\frac{1}{8} R_3} \begin{bmatrix} 1 & 4 & 3 & | & -2 \\ 0 & 1 & 5 & | & -4 \\ 0 & 0 & 0 & | & 1 \end{bmatrix}$ The matrix is now in row-echelon form.

Step 3 The system of equations corresponding to the last matrix in Step 2 is

$$\begin{cases} x + 4y + 3z = -2 \\ y + 5z = -4 \\ 0 = 1 \qquad \text{A false statement} \end{cases}$$

RECALL

A system is called *inconsistent* if it has no solution.

Since the third equation $0 = 1$ is a false statement, we conclude that this system is inconsistent. Because this system is equivalent to the original system, the original system is also inconsistent.

Step 4 The solution set for the system is $\varnothing$.

Practice Problem 7 Solve the following system of equations by first transforming the augmented matrix into row-echelon form.

$$\begin{cases} 6x + 8y - 14z = 3 \\ 3x + 4y - 7z = 12 \\ 6x + 3y + z = 0 \end{cases}$$

4 Use Gauss–Jordan elimination to solve a system.

Gauss–Jordan Elimination

If we continue the Gaussian elimination procedure until we have a *reduced* row-echelon form, the procedure is called **Gauss–Jordan elimination**.

 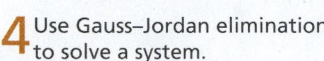
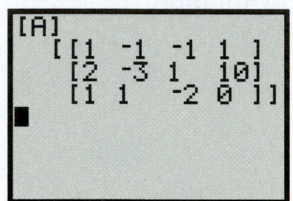
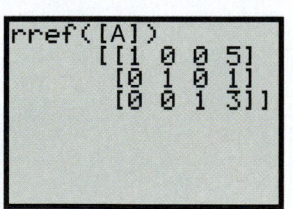
EXAMPLE 8 **Solving a System of Equations by Gauss–Jordan Elimination**

Solve the system given in Example 4 by Gauss–Jordan elimination.

$$\begin{cases} x - y - z = 1 \\ 2x - 3y + z = 10 \\ x + y - 2z = 0 \end{cases} \quad \text{The given system}$$

Solution

$$\begin{bmatrix} 1 & -1 & -1 & | & 1 \\ 0 & 1 & -3 & | & -8 \\ 0 & 0 & 5 & | & 15 \end{bmatrix} \quad \begin{array}{l}\text{The final augmented matrix of the system} \\ \text{in Example 4}\end{array}$$

$$\xrightarrow{\frac{1}{5}R_3} \begin{bmatrix} 1 & -1 & -1 & | & 1 \\ 0 & 1 & -3 & | & -8 \\ 0 & 0 & 1 & | & 3 \end{bmatrix} \quad \text{The matrix is now in row-echelon form.}$$

$$\xrightarrow{R_2 + R_1 \rightarrow R_1} \begin{bmatrix} 1 & 0 & -4 & | & -7 \\ 0 & 1 & -3 & | & -8 \\ 0 & 0 & 1 & | & 3 \end{bmatrix}$$

$$\xrightarrow[3R_3 + R_2 \rightarrow R_2]{4R_3 + R_1 \rightarrow R_1} \begin{bmatrix} 1 & 0 & 0 & | & 5 \\ 0 & 1 & 0 & | & 1 \\ 0 & 0 & 1 & | & 3 \end{bmatrix}$$

We now have an equivalent matrix in reduced row-echelon form. The corresponding system of equations for the last augmented matrix is

$$\begin{cases} x = 5 \\ y = 1 \\ z = 3 \end{cases}$$

The solution set is therefore $\{(5, 1, 3)\}$, as in Example 4.

Practice Problem 8 Solve the system by using Gauss–Jordan elimination.

$$\begin{cases} 2x - 3y - 2z = 0 \\ x + y - 2z = 7 \\ 3x - 5y - 5z = 3 \end{cases}$$

EXAMPLE 9 **Solving a System with Infinitely Many Solutions**

Solve the system of equations.

$$\begin{cases} x + 2y + 5z = 4 \\ y + 4z = 4 \\ 2x + 4y + 10z = 8 \end{cases}$$

Solution

The augmented matrix A of the system is given below. We want to find the equivalent system in reduced row-echelon form.

Since $a_{11} = 1$ and $a_{21} = 0$, we need a zero at the $(3, 1)$ position.

$$A = \begin{bmatrix} 1 & 2 & 5 & | & 4 \\ 0 & 1 & 4 & | & 4 \\ 2 & 4 & 10 & | & 8 \end{bmatrix} \xrightarrow{-2R_1 + R_3 \to R_3} \begin{bmatrix} 1 & 2 & 5 & | & 4 \\ 0 & 1 & 4 & | & 4 \\ 0 & 0 & 0 & | & 0 \end{bmatrix}$$

Next, we need a zero at the $(1, 2)$ position.

$$\xrightarrow{-2R_2 + R_1 \to R_1} \begin{bmatrix} 1 & 0 & -3 & | & -4 \\ 0 & 1 & 4 & | & 4 \\ 0 & 0 & 0 & | & 0 \end{bmatrix}$$

The matrix is now in reduced row-echelon form.

The equivalent system is

$$x - 3z = -4$$
$$y + 4z = 4$$

Solving for x and y in terms of z, we obtain

$$x = 3z - 4$$
$$y = -4z + 4$$

Each real number z results in a solution with $y = -4z + 4$ and $x = 3z - 4$, giving infinitely many solutions of the form $(3z - 4, -4z + 4, z)$. The solution set is $\{(3z - 4, -4z + 4, z)\}$.

Practice Problem 9 Solve the system of equations.

$$\begin{cases} x & + & z = -1 \\ & 3y + 2z = & 5 \\ 3x & - 3y + & z = -8 \end{cases}$$

EXAMPLE 10   Leontief Input–Output Model

Consider an economy that has steel, coal, and transportation industries. There are two types of demands (measured in dollars) on the production of each industry: interindustry demand and external consumer demand. The outputs and requirements of the three industries are shown in Figure 7.21. For example, $1.00 of transportation output requires $0.10 from steel and $0.01 from coal.

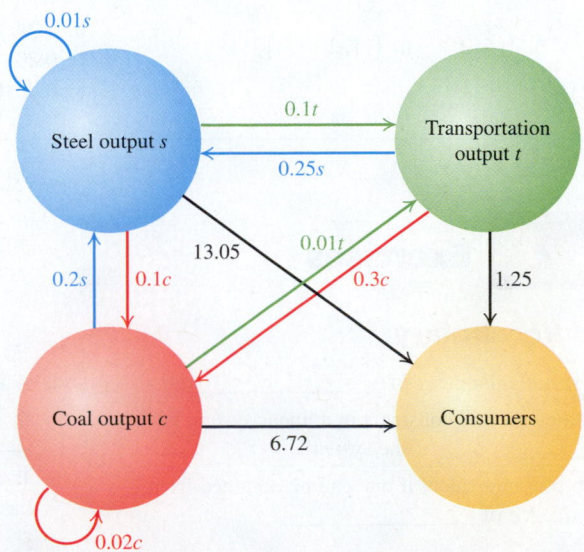

Figure 7.21 Output diagram of an economy.

a. Write a system of equations that expresses the outputs of the three industries. Assume that all quantities are given in millions of dollars.

b. Verify that $s = 15$, $c = 10$, and $t = 8$ will meet both interindustry and consumer demand.

Solution

a. To satisfy both consumer and interindustry demand, we obtain the following system of outputs. (s denotes total steel output; c, total coal output; t, total transportation output.)

$$\begin{cases} s = 0.01s + 0.1t + 0.1c + 13.05 & \text{Distribution of steel output} \\ c = 0.2s + 0.01t + 0.02c + 6.72 & \text{Distribution of coal output} \\ t = 0.25s + 0.3c + 1.25 & \text{Distribution of transportation output} \end{cases}$$

You can rewrite this system as

$$\begin{cases} 0.99s - 0.1c - 0.1t = 13.05 & \text{Equation (1)} \\ -0.2s + 0.98c - 0.01t = 6.72 & \text{Equation (2)} \\ -0.25s - 0.3c + t = 1.25 & \text{Equation (3)} \end{cases}$$

b. To verify that the given numbers (obtained by solving the system above using matrices) satisfy these equations, we substitute $s = 15$, $c = 10$, and $t = 8$ into each equation.

$$(0.99)(15) - (0.1)(10) - (0.1)(8) = 13.05 \checkmark \quad \text{Equation (1)}$$
$$(-0.20)(15) + (0.98)(10) - (0.01)(8) = 6.72 \checkmark \quad \text{Equation (2)}$$
$$(-0.25)(15) - (0.30)(10) + 8 = 1.25 \checkmark \quad \text{Equation (3)}$$

In other words, output levels of $s = 15$, $c = 10$, and $t = 8$ million dollars will meet interindustry and consumer demand.

Practice Problem 10 In Example 10, assume that consumer demand (in millions of dollars) is 12.46 for steel, 3 for coal, and 2.7 for transportation.

a. Write a system of equations to express the outputs for the three industries.

b. Verify that $s = 14$, $c = 6$, $t = 8$ will meet both interindustry and consumer demand.

Answers to Practice Problems

1. a. 3×2 **b.** 1×2 **2.** $\begin{bmatrix} 0 & 3 & -1 & | & 8 \\ 1 & 4 & 0 & | & 14 \\ 0 & -2 & 9 & | & 0 \end{bmatrix}$

3. $B = \begin{bmatrix} 2 & 4 & 6 \\ 3 & 4 & 5 \end{bmatrix}$; $C = \begin{bmatrix} 1 & 2 & 3 \\ 0 & -2 & -4 \end{bmatrix}$

4. $\left\{ \left(-1, -\frac{1}{3}, -1 \right) \right\}$ **5.** $\{(5, 2)\}$ **6.** $\{(1, 2, -3)\}$

7. $\varnothing$ **8.** $\{(1, 2, -2)\}$

9. $\left\{ \left(-z - 1, \frac{5}{3} - \frac{2}{3}z, z \right) \right\}$

10. a. $\begin{cases} s = 0.01s + 0.1c + 0.1t + 12.46 \\ c = 0.2s + 0.02c + 0.01t + 3 \\ t = 0.25s + 0.3c + 2.7 \end{cases}$

b. $(0.99)(14) - (0.1)(6) - (0.1)(8) = 12.46 \checkmark$
$(-0.20)(14) + (0.98)(6) - (0.01)(8) = 3 \checkmark$
$(-0.25)(14) - (0.30)(6) + 8 = 2.7 \checkmark$

SECTION 7.4 Exercises

Concepts and Vocabulary

1. A matrix is any rectangular array of _____.

2. The array of coefficients and constants in a linear system is called the _____ of the system.

3. Two matrices are row-equivalent if one can be obtained from the other by a sequence of _____.

4. If a matrix is in row-echelon form and each leading entry 1 is the only nonzero entry in its column, then the matrix is in _____ form.

5. **True or False.** If A is a 3×4 matrix, then each row of A has three entries.

6. True or False. Every $m \times n$ ($n \geq 2$) matrix is the matrix of some linear system.

7. True or False. The augmented matrix for the system of equations

$$\begin{cases} 2x + 3y & = 5 \\ 3y - 4z = -1 \\ x + & 2z = 3 \end{cases} \quad \text{is} \quad \begin{bmatrix} 2 & 3 & 5 \\ 3 & -4 & -1 \\ 1 & 2 & 3 \end{bmatrix}.$$

8. True or False. The augmented matrix for the system of equations

$$\begin{cases} 2x - y = 1 \\ 3x + y = 9 \\ 5x - 2y = 4 \end{cases} \quad \text{is} \quad \begin{bmatrix} 2x & -y & 1 \\ 3x & y & 9 \\ 5x & -2y & 4 \end{bmatrix}.$$

Building Skills

In Exercises 9–14, determine the order of each matrix.

9. $[7]$

10. $[1 \quad 3 \quad 5 \quad 9]$

11. $\begin{bmatrix} 3 & 4.3 & 7.5 & 8 \\ 2 & -1 & -3 & 0 \end{bmatrix}$

12. $\begin{bmatrix} -1 & 2 & 3 \\ 4 & -5 & 7 \\ 1 & 0 & 0 \end{bmatrix}$

13. $\begin{bmatrix} 4 & 5 & 1 \\ 0 & 0 & 0 \end{bmatrix}$

14. $\begin{bmatrix} -5 & 2 & 3 & 7 \\ 2.5 & e & -\pi & \frac{1}{2} \\ -e & \pi & 0 & 1 \end{bmatrix}$

15. Let $A = \begin{bmatrix} 1 & 2 & 3 & 4 \\ 5 & 6 & 7 & 8 \\ 9 & 10 & 11 & 12 \end{bmatrix}$. Identify the entries $a_{13}, a_{31}, a_{33},$ and a_{34}.

16. In matrix A from Exercise 15, identify the (i, j)th location a_{ij} of each entry.

a. 7   **b.** 10

c. 4   **d.** 12

17. Is the following array a matrix? Why or why not?

$$\begin{bmatrix} 2 & 0 & 4 \\ 1 & -3 & 5 \\ 0 & 0 & \end{bmatrix}$$

18. Write the matrix A with the following entries:

$a_{13} = 5, a_{22} = 2, a_{11} = 6, a_{12} = -5, a_{23} = 4,$ and $a_{21} = 7$

In Exercises 19–24, write the augmented matrix for each system of linear equations.

19. $\begin{cases} 2x + 4y = 2 \\ x - 3y = 1 \end{cases}$

20. $\begin{cases} x + 2y = 7 \\ 3x + 5y = 11 \end{cases}$

21. $\begin{cases} 5x - 11 = 7y \\ 17y - 19 = 13x \end{cases}$

22. $\begin{cases} 2y - 10 = 3x \\ 5x + 7 = y \end{cases}$

23. $\begin{cases} -x + 2y + 3z = 8 \\ 2x - 3y + 9z = 16 \\ 4x - 5y - 6z = 32 \end{cases}$

24. $\begin{cases} x - y & = 2 \\ 2x & + 3z = -5 \\ y - 2z = 7 \end{cases}$

In Exercises 25–28, write the system of linear equations represented by each augmented matrix. Use $x, y,$ and z as the variables.

25. $\begin{bmatrix} 1 & 2 & -3 & 4 \\ -2 & -3 & 1 & 5 \\ 3 & -3 & 2 & 7 \end{bmatrix}$

26. $\begin{bmatrix} -1 & 2 & 3 & 6 \\ 2 & 3 & 1 & 2 \\ 4 & 3 & 2 & 1 \end{bmatrix}$

27. $\begin{bmatrix} 1 & -1 & 1 & 2 \\ 2 & 1 & -3 & 6 \end{bmatrix}$

28. $\begin{bmatrix} 1 & 1 & 1 & 2 \\ 1 & -1 & -1 & 4 \\ 2 & 3 & 1 & 6 \\ -1 & 1 & -1 & 8 \end{bmatrix}$

In Exercises 29–32, perform the indicated elementary row operations in the stated order.

29. $\begin{bmatrix} 2 & 3 & 5 \\ 1 & 2 & 3 \end{bmatrix}$; **(i)** $R_1 \leftrightarrow R_2$, **(ii)** $-2R_1 + R_2 \rightarrow R_2$, **(iii)** $-R_2$

30. $\begin{bmatrix} 2 & 4 & 2 \\ 1 & 5 & 7 \end{bmatrix}$; **(i)** $\frac{1}{2}R_1$, **(ii)** $(-1)R_1 + R_2 \rightarrow R_2$, **(iii)** $\frac{1}{3}R_2$

31. $\begin{bmatrix} 1 & 2 & 3 & 4 \\ 0 & 4 & -3 & 11 \\ 0 & 1 & 5 & -3 \end{bmatrix}$; **(i)** $R_2 \leftrightarrow R_3$, **(ii)** $-4R_2 + R_3 \rightarrow R_3$, **(iii)** $-\frac{1}{23}R_3$

32. $\begin{bmatrix} 1 & \frac{3}{2} & -2 & \frac{1}{2} \\ 0 & 2 & 3 & 4 \\ -2 & -1 & 7 & 3 \end{bmatrix}$; **(i)** $2R_1 + R_3 \rightarrow R_3$, **(ii)** $\frac{1}{2}R_2 + R_3 \rightarrow R_3$

In Exercises 33–36, identify the elementary row operation used and supply the missing entries in each row-equivalent matrix.

33. $\begin{bmatrix} 4 & 5 & -7 \\ 5 & 4 & -2 \end{bmatrix} \rightarrow \begin{bmatrix} 1 & ? & -\frac{7}{4} \\ 5 & 4 & -2 \end{bmatrix} \rightarrow \begin{bmatrix} 1 & ? & -\frac{7}{4} \\ 0 & -\frac{9}{4} & ? \end{bmatrix} \rightarrow$

$\begin{bmatrix} 1 & ? & -\frac{7}{4} \\ 0 & 1 & ? \end{bmatrix}$

34. $\begin{bmatrix} 2 & 6 & -8 \\ 3 & -1 & 2 \end{bmatrix} \rightarrow \begin{bmatrix} 1 & 3 & -? \\ 3 & -1 & 2 \end{bmatrix} \rightarrow \begin{bmatrix} 1 & 3 & ? \\ 0 & -? & 14 \end{bmatrix} \rightarrow$

$\begin{bmatrix} 1 & 3 & -? \\ 0 & 1 & ? \end{bmatrix}$

35. $\begin{bmatrix} 1 & 4 & 3 & 1 \\ 0 & -3 & -2 & 0 \\ 0 & 7 & 5 & -3 \end{bmatrix} \rightarrow \begin{bmatrix} 1 & 4 & 3 & 1 \\ 0 & 1 & ? & 0 \\ 0 & 7 & 5 & -3 \end{bmatrix} \rightarrow$

$\begin{bmatrix} 1 & 4 & 3 & 1 \\ 0 & 1 & ? & 0 \\ 0 & 0 & ? & -3 \end{bmatrix} \rightarrow \begin{bmatrix} 1 & 4 & 3 & 1 \\ 0 & 1 & ? & 0 \\ 0 & 0 & 1 & ? \end{bmatrix}$

36. $\begin{bmatrix} 1 & 1 & 1 & 3 \\ 2 & 3 & 3 & 8 \\ 1 & -3 & -2 & 5 \end{bmatrix} \rightarrow \begin{bmatrix} 1 & 1 & 1 & 3 \\ 0 & 1 & ? & 2 \\ 1 & -3 & -2 & 5 \end{bmatrix} \rightarrow$

$\begin{bmatrix} 1 & 1 & 1 & 3 \\ 0 & 1 & ? & 2 \\ 0 & -4 & ? & 2 \end{bmatrix} \rightarrow \begin{bmatrix} 1 & 1 & 1 & 3 \\ 0 & 1 & ? & 2 \\ 0 & 0 & ? & 10 \end{bmatrix}$

In Exercises 37–44, use the properties on page 639 to determine whether each matrix is in row-echelon form. If your answer is no, explain why. If your answer is yes, is the matrix in reduced row-echelon form?

37. $\begin{bmatrix} 0 & 1 & 2 \\ 1 & 0 & 3 \end{bmatrix}$

38. $\begin{bmatrix} 1 & 2 & 5 \\ 0 & 3 & 6 \end{bmatrix}$

39. $\begin{bmatrix} 1 & 0 & 0 & 2 \\ 0 & 1 & 0 & 3 \\ 0 & 0 & 1 & 4 \end{bmatrix}$

40. $\begin{bmatrix} 0 & 0 & 1 & 2 \\ 0 & 1 & 0 & 1 \\ 1 & 0 & 0 & 3 \end{bmatrix}$

41. $\begin{bmatrix} 1 & 2 & 0 & 2 \\ 0 & 0 & 1 & 5 \end{bmatrix}$

42. $\begin{bmatrix} 0 & 1 & -1 & -2 & 0 & 2 & | & 0 \\ 0 & 0 & 0 & 0 & 1 & 2 & | & 0 \end{bmatrix}$

43. $\begin{bmatrix} 1 & 0 & 0 & | & -2 \\ 0 & 1 & 0 & | & 3 \\ 0 & 0 & 1 & | & 2 \\ 0 & 0 & 0 & | & 0 \end{bmatrix}$

44. $\begin{bmatrix} 1 & 0 & 0 & | & 3 \\ 0 & 1 & 0 & | & 4 \\ 0 & 0 & 0 & | & 5 \\ 0 & 0 & 1 & | & 3 \end{bmatrix}$

In Exercises 45–54, the augmented matrix of a system of equations has been transformed to an equivalent matrix in row-echelon form or reduced row-echelon form. Using $x, y, z,$ and w as variables, write the system of equations corresponding to the matrix. If the system is consistent, solve it.

45. $\begin{bmatrix} 1 & 2 & | & 1 \\ 0 & 1 & | & -2 \end{bmatrix}$

46. $\begin{bmatrix} 1 & 0 & | & 2 \\ 0 & 1 & | & 3 \end{bmatrix}$

47. $\begin{bmatrix} 1 & 4 & 2 & | & 2 \\ 0 & 0 & 1 & | & 3 \end{bmatrix}$

48. $\begin{bmatrix} 1 & 2 & 3 & | & 4 \\ 0 & 0 & 0 & | & 1 \end{bmatrix}$

49. $\begin{bmatrix} 1 & 2 & 3 & | & 2 \\ 0 & 1 & -2 & | & 4 \\ 0 & 0 & 1 & | & -1 \end{bmatrix}$

50. $\begin{bmatrix} 1 & 0 & 2 & | & 12 \\ 0 & 1 & 3 & | & 12 \\ 0 & 0 & 1 & | & 5 \end{bmatrix}$

51. $\begin{bmatrix} 1 & 0 & 0 & 0 & | & 2 \\ 0 & 1 & 0 & 0 & | & -5 \\ 0 & 0 & 1 & 2 & | & 3 \\ 0 & 0 & 0 & 0 & | & 0 \end{bmatrix}$

52. $\begin{bmatrix} 1 & 0 & 0 & 0 & | & 4 \\ 0 & 1 & 0 & 0 & | & 3 \\ 0 & 0 & 0 & 1 & | & 2 \end{bmatrix}$

53. $\begin{bmatrix} 1 & 0 & 0 & 0 & | & -5 \\ 0 & 1 & 0 & 0 & | & 4 \\ 0 & 0 & 1 & 2 & | & 3 \\ 0 & 0 & 0 & 1 & | & 0 \end{bmatrix}$

54. $\begin{bmatrix} 1 & 0 & 0 & 0 & | & 3 \\ 0 & 1 & 0 & 0 & | & 2 \\ 0 & 0 & 1 & 0 & | & 0 \\ 0 & 0 & 0 & 0 & | & 1 \end{bmatrix}$

In Exercises 55–68, solve each system of equations by Gaussian elimination.

55. $\begin{cases} x - 2y = 11 \\ 2x - y = 13 \end{cases}$

56. $\begin{cases} 3x - 2y = 4 \\ 4x - 3y = 5 \end{cases}$

57. $\begin{cases} 2x - 3y = 3 \\ 4x - y = 11 \end{cases}$

58. $\begin{cases} 3x + 2y = 1 \\ 6x + 4y = 3 \end{cases}$

59. $\begin{cases} 3x - 5y = 4 \\ 4x - 15y = 13 \end{cases}$

60. $\begin{cases} -2x + 4y = 1 \\ 3x - 5y = -9 \end{cases}$

61. $\begin{cases} x - y = 1 \\ 2x + y = 5 \\ 3x - 4y = 2 \end{cases}$

62. $\begin{cases} y = 2x + 1 \\ 3x + 2y + 1.5 = 0 \\ 4x - 2y + 2 = 0 \end{cases}$

63. $\begin{cases} x + y + z = 6 \\ x - y + z = 2 \\ 2x + y - z = 1 \end{cases}$

64. $\begin{cases} 2x + 4y + z = 5 \\ x + y + z = 6 \\ 2x + 3y + z = 6 \end{cases}$

65. $\begin{cases} 2x + 3y - z = 9 \\ x + y + z = 9 \\ 3x - y - z = -1 \end{cases}$

66. $\begin{cases} x + y + 2z = 4 \\ 2x - y + 3z = 9 \\ 3x - y - z = 2 \end{cases}$

67. $\begin{cases} 3x + 2y + 4z = 19 \\ 2x - y + z = 3 \\ 6x + 7y - z = 17 \end{cases}$

68. $\begin{cases} 4x + 3y + z = 8 \\ 2x + y + 4z = -4 \\ 3x + z = 1 \end{cases}$

In Exercises 69–74, solve each system of equations by Gauss–Jordan elimination.

69. $\begin{cases} x - y = 1 \\ x - z = -1 \\ 2x + y - z = 3 \end{cases}$

70. $\begin{cases} 4x + 5z = 7 \\ y - 6z = 8 \\ 3x + 4y = 9 \end{cases}$

71. $\begin{cases} x + y - z = 4 \\ x + 3y + 5z = 10 \\ 3x + 5y + 3z = 18 \end{cases}$

72. $\begin{cases} x + y + z = -5 \\ 2x - y - z = -4 \\ y + z = -2 \end{cases}$

73. $\begin{cases} x + 2y - z = 6 \\ 3x + y + 2z = 3 \\ 2x + 5y + 3z = 9 \end{cases}$

74. $\begin{cases} 2x + 4y - z = 9 \\ x + 3y - 3z = 4 \\ 3x + y + 2z = 7 \end{cases}$

In Exercises 75 and 76 the reduced row-echelon forms of the augmented matrices of three systems of equations are given. How many solutions does each system have?

75. a. $\begin{bmatrix} 1 & 0 & | & 2 \\ 0 & 1 & | & 3 \end{bmatrix}$ **b.** $\begin{bmatrix} 1 & 2 & 3 & | & 8 \\ 0 & 0 & 1 & | & 2 \\ 0 & 0 & 0 & | & 0 \end{bmatrix}$

c. $\begin{bmatrix} 1 & 0 & | & 2 \\ 0 & 0 & | & 3 \end{bmatrix}$

76. a. $\begin{bmatrix} 1 & 0 & 0 & | & -1 \\ 0 & 1 & 0 & | & 2 \\ 0 & 0 & 1 & | & 3 \end{bmatrix}$ **b.** $\begin{bmatrix} 1 & 0 & 0 & | & 4 \\ 0 & 1 & 2 & | & 5 \\ 0 & 0 & 0 & | & 6 \end{bmatrix}$

c. $\begin{bmatrix} 1 & 1 & 2 & | & 3 \\ 0 & 0 & 0 & | & 0 \\ 0 & 0 & 0 & | & 0 \end{bmatrix}$

In Exercises 77 and 78, determine whether the statement is true or false and justify your answer.

77. If A is a 5×7 matrix, then each column of A has seven entries.

78. If a matrix A is in reduced row-echelon form, then at least one of the entries in each column must be a 1.

Applying the Concepts

In Exercises 79–82, Leontief's input–output model of a simplified economy is described. (See Example 10.) For each exercise, do the following.

a. Set up (without solving) a linear system whose solution will represent the required production schedule.

b. Write an augmented matrix for the economy.

c. Solve the system of equations by using part (b) to find the production schedule that will meet interindustry and consumer demand.

79. Two-sector economy. Consider an economy with only two industries: A and B. Suppose industry B needs \$0.10 worth of A's product for each \$1.00 of output B produces and that industry A needs \$0.20 worth of B's product for each \$1.00 of output A produces. Consumer demand for A's product is \$1000, and consumer demand for B's product is \$780.

80. A company with two branches. A company has two interacting branches: A and B. Branch A consumes \$0.50 of its own output and \$0.20 of B's output for every \$1.00 it produces. Branch B consumes \$0.60 of A's output and

$0.30 of its own output for $1.00 of output. Consumer demand for A's product is $50,000, and consumer demand for B's product is $40,000.

81. Three-sector economy. An economy has three sectors: labor, transportation, and food industries. Suppose the demand on $1.00 in labor is $0.40 for transportation and $0.20 for food; the demand on $1.00 in transportation is $0.50 for labor and $0.30 for transportation; and the demand on $1.00 in food production is $0.50 for labor, $0.05 for transportation, and $0.35 for food. Outside consumer demand for the current production period is $10,000 for labor, $20,000 for transportation, and $10,000 for food.

82. Three-sector economy. Consider an economy with three industries A, B, and C, with outputs a, b, and c, respectively. Demand on the three industries is shown in the figure.

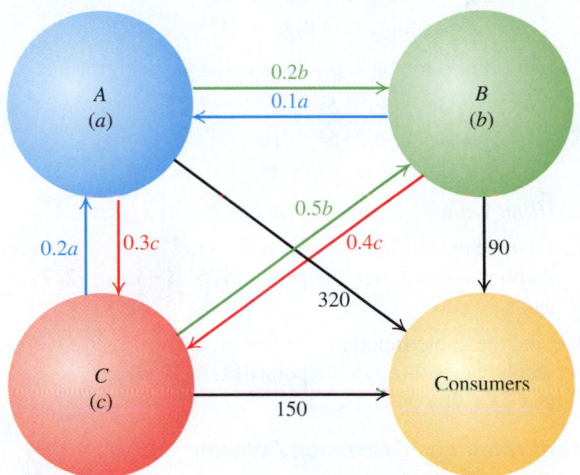

83. Heat transfer. In a study of heat transfer in a grid of wires, the temperature at an exterior node is maintained at a constant value (in °F) as shown in the accompanying figure. When the grid is in thermal equilibrium, the temperature at an interior node is the average of the temperatures at the four adjacent nodes. For instance, $T_1 = \dfrac{0 + 0 + 300 + T_2}{4}$, or $4T_1 - T_2 = 300$. Find the temperatures T_1, T_2, and T_3 when the grid is in thermal equilibrium.

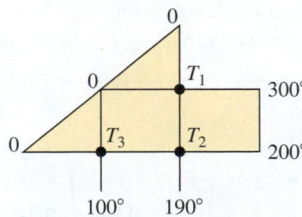

84. Heat transfer. Repeat Exercise 83 for the following grid of wires.

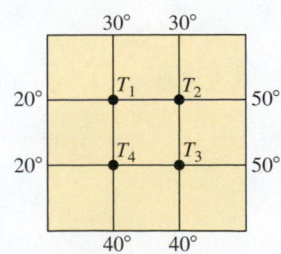

85. Traffic flow. The sketch shows the intersections of certain one-way streets. The traffic volume during one hour was observed. To keep the traffic moving, traffic controllers use the formula that the number of cars per hour entering an intersection equals the number of cars per hour exiting the intersection.
  a. Set up a system of equations that keeps traffic moving.
  b. Solve the system of equations in part (a).

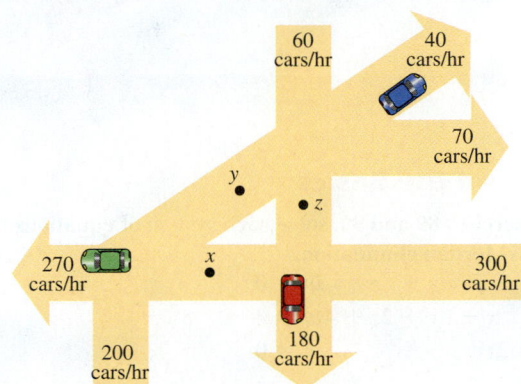

86. Traffic flow. Repeat Exercise 85 for the following pattern of one-way streets.

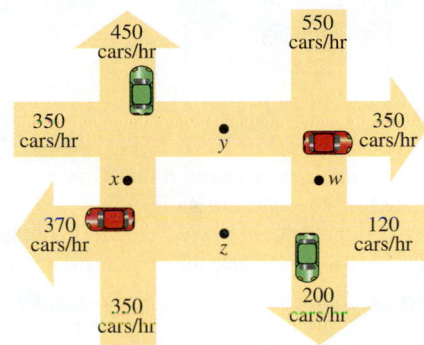

87. Parabola. Find the equation of the parabola $y = ax^2 + bx + c$ that passes through the points $(-1, 9)$, $(1, 3)$, and $(2, 6)$.

88. Modeling. A football was punted. Its height h above the ground at time t is given by the following table:

Time t (seconds)	Height h (feet)
$t = 0.5$	31
$t = 1$	51
$t = 2$	67

  a. Use the system of equations to find the equation of the parabola of the form $h = at^2 + bt + c$.
  b. What is the hang time (the time it takes for the ball to touch the ground) of the punt?
  c. What is the maximum height of the ball?

Beyond the Basics

In Exercises 89 and 90, solve each system of equations by Gauss–Jordan elimination.

89. $\begin{cases} x + y + z + w = 0 \\ x + 3y + 2z + 4w = 0 \\ 2x \qquad + z - w = 0 \end{cases}$

90. $\begin{cases} x - y + z - w = 4 \\ x + 2y + z + w = 2 \\ 2x + 3y + 4z + 5w = 5 \\ 3x + 4y + 2z - w = 8 \end{cases}$

91. Let $A = \begin{bmatrix} 1 & 2 & -3 & 1 \\ 1 & 0 & -3 & 2 \\ 0 & 1 & 1 & 0 \\ 2 & 3 & 0 & -2 \end{bmatrix}$.

a. Find two different matrices B and C in row-echelon form that are row-equivalent to A.

b. Show that the reduced row-echelon form of the matrices B and C of part (a) produce the same matrix.

92. Consider the following system of linear equations:

$$\begin{cases} x + y + z = 6 \\ x - y + z = 2 \\ 2x + y - z = 1 \end{cases}$$

a. Find two different matrices B and C in row-echelon form that yield the same solution set.

b. Show that the reduced row-echelon form of the matrices B and C of part (a) produces the same matrix.

93. a. Use the methods of this section to solve a general 2×2 system:

$$\begin{cases} ax + by = m \\ cx + dy = n \end{cases}$$

b. Under what conditions on the coefficients does the system have **(i)** a unique solution, **(ii)** no solution, and **(iii)** infinitely many solutions?

94. Construct three different augmented matrices for linear systems whose solution set is $x = 1, y = 0, z = -2$. There are many different correct answers.

In Exercises 95 and 96, solve each system of equations by using row operations.

95. $\begin{cases} \log x + \log y + \log z = 6 \\ 3\log x - \log y + 3\log z = 10 \\ 5\log x + 5\log y - 4\log z = 3 \end{cases}$

[*Hint:* Let $u = \log x$, $v = \log y$, and $w = \log z$.]

96. $\begin{cases} 4 \cdot 2^x - 3 \cdot 3^y + 5^z = 1 \\ 2^x + 4 \cdot 3^y - 2 \cdot 5^z = 10 \\ 2 \cdot 2^x - 2 \cdot 3^y + 3 \cdot 5^z = 4 \end{cases}$

[*Hint:* Let $u = 2^x$, $v = 3^y$, and $w = 5^z$.]

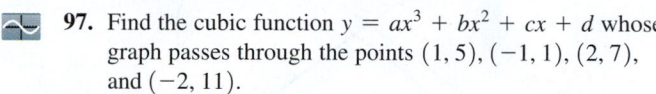 **97.** Find the cubic function $y = ax^3 + bx^2 + cx + d$ whose graph passes through the points $(1, 5), (-1, 1), (2, 7)$, and $(-2, 11)$.

98. Find the cubic function $y = ax^3 + bx^2 + cx + d$ whose graph passes through the points $(1, 8), (-1, 2), (2, 8)$, and $(-2, 20)$.

Critical Thinking / Discussion / Writing

99. Find the form of all 2×1 matrices in reduced row-echelon form.

100. Find the form of all 2×2 matrices in reduced row-echelon form.

101. Suppose a matrix A is transformed to a matrix B by an elementary row operation. Is there an elementary row operation that transforms B to A? Explain.

102. Are the following statements true or false? Justify your answers. Suppose a matrix A is in reduced row-echelon form.
a. If we delete a row of A, then the remaining matrix is in reduced row-echelon form.
b. Repeat part (a), replacing *row* with *column*.

Getting Ready for the Next Section

In Exercises 103–106, state whether each expression is equal to 1 or −1.

103. $(-1)^0$ **104.** $(-1)^{10}$

105. $(-1)^{17}$ **106.** $(-1)^{6+9}$

In Exercises 107–110, solve each system of equations.

107. $\begin{cases} 2x - 3y = 16 \\ x - y = 7 \end{cases}$ **108.** $\begin{cases} 16x - 9y = -5 \\ 10x + 18y = -11 \end{cases}$

109. $\begin{cases} x - y + 5z = -6 \\ 3x + 3y - z = 10 \\ x + 3y + 2z = 5 \end{cases}$ **110.** $\begin{cases} 2x - y + 2z = 3 \\ 2x + 2y - z = 0 \\ -x + 2y + 2z = -12 \end{cases}$

Determinants and Cramer's Rule

BEFORE STARTING THIS SECTION, REVIEW

1 Definition of a matrix (Section 7.4, page 635)

2 Systems of equations (Section 7.1, page 594)

OBJECTIVES

1 Calculate the determinant of a 2 × 2 matrix.

2 Find minors and cofactors.

3 Eval uate the determinant of an $n \times n$ matrix.

4 Apply Cramer's Rule.

Seki Kōwa (1642–1708)

Seki Kōwa has been called "the Japanese Newton." He was born in the same year as Newton. Like Newton, he acquired much of his knowledge by self-study and was a brilliant problem solver. But the greatest similarity between the two men lies in the claim that Seki Kowa was the creator of *yenri* (calculus). Thus, Seki Kōwa probably invented calculus in the East just as Newton and Leibniz invented calculus in the West.

◆ A Little History of Determinants

It appears that determinants were first investigated in 1683 by the outstanding Japanese mathematician Seki Kōwa in connection with systems of linear equations. Determinants were independently investigated in 1693 by the German mathematician Gottfried Leibniz (1646–1716). The French mathematician Alexandre-Théophile Vandermonde (1735–1796) was the first to give a coherent and systematic exposition of the theory of determinants. Determinants were used extensively in the 19th century by eminent mathematicians such as Cauchy, Jacobi, and Kronecker. Today determinants are of little numerical value in the large-scale matrix computations that occur so often. Nevertheless, a knowledge of determinants is useful in some applications of matrices.

---

1 Calculate the determinant of a 2 × 2 matrix.

SIDE NOTE

A determinant can be thought of as a function whose domain is the set of all square matrices and whose range is the set of real numbers.

The Determinant of a 2 × 2 Matrix

Associated with each square matrix A is a number called the *determinant* of A:

Determinant of a 2 × 2 Matrix

The **determinant** of a 2 × 2 matrix

$$A = \begin{bmatrix} a & b \\ c & d \end{bmatrix},$$

denoted by $\det(A)$, $|A|$, or $\begin{vmatrix} a & b \\ c & d \end{vmatrix}$, is called the **2 by 2 determinant** of A and is defined by

$$\det(A) = ad - bc.$$

To remember this definition, note that you form products along diagonal elements as shown by the arrows.

$$\begin{bmatrix} a & b \\ c & d \end{bmatrix}$$
$$-bc \quad +ad$$

You attach a plus sign if you descend from left to right and a minus sign if you descend from right to left. Then you form the sum $ad - bc$.

EXAMPLE 1 **Calculating the Determinant of a 2 × 2 Matrix**

Evaluate each determinant.

a. $\begin{vmatrix} 3 & -4 \\ 1 & 5 \end{vmatrix}$ b. $\begin{vmatrix} -2 & 3 \\ 4 & -5 \end{vmatrix}$

Solution

a. $\begin{vmatrix} 3 & -4 \\ 1 & 5 \end{vmatrix} = (3)(5) - (-4)(1) = 15 - (-4) = 15 + 4 = 19$

b. $\begin{vmatrix} -2 & 3 \\ 4 & -5 \end{vmatrix} = (-2)(-5) - (3)(4) = 10 - 12 = -2$

Practice Problem 1 Evaluate each determinant.

a. $\begin{vmatrix} 5 & 1 \\ 3 & -7 \end{vmatrix}$ b. $\begin{vmatrix} 2 & -9 \\ -4 & 18 \end{vmatrix}$

WARNING Note carefully the notations used for matrices and determinants. Brackets, [ ], denote a matrix, while vertical bars, ||, denote the determinant of a matrix. These bars are *not* the same as the absolute value bars. Remember that *the determinant of a matrix is a number that may be positive, negative, or zero.*

2 Find minors and cofactors.

Minors and Cofactors

To define the determinants of square matrices of order 3 or higher, we need some additional terminology.

Minors and Cofactors in an $n \times n$ Matrix

Let A be an $n \times n$ square matrix. The **minor** M_{ij} of the element a_{ij} is the determinant of the $(n - 1) \times (n - 1)$ matrix obtained by deleting the ith row and the jth column of A. The **cofactor** A_{ij} of the entry a_{ij} is defined by

$$A_{ij} = (-1)^{i+j}M_{ij}.$$

When $i + j$ is an even integer, $(-1)^{i+j} = 1$. When $i + j$ is an odd integer, $(-1)^{i+j} = -1$. Therefore,

$$A_{ij} = \begin{cases} M_{ij} \text{ if } i + j \text{ is an even integer} \\ -M_{ij} \text{ if } i + j \text{ is an odd integer} \end{cases}$$

Notice that to compute a minor in an $n \times n$ matrix, you must know how to compute the determinant of an $(n - 1) \times (n - 1)$ matrix. At this point in the text, because you have learned the definition of the determinant of a 2 × 2 matrix, you can find the minors and cofactors of entries in a 3 × 3 matrix.

> **EXAMPLE 2**   **Finding Minors and Cofactors**

For the matrix $A = \begin{bmatrix} 2 & 3 & 4 \\ 5 & -3 & -6 \\ 0 & 1 & 7 \end{bmatrix}$, find:

a. the minors M_{11}, M_{23}, and M_{32}

b. the cofactors A_{11}, A_{23}, and A_{32}

Solution

a.   **(i)** To find M_{11}, delete the first row and first column of the matrix A.

$$\begin{bmatrix} 2 & 3 & 4 \\ 5 & -3 & -6 \\ 0 & 1 & 7 \end{bmatrix}$$

Now find the determinant of the resulting matrix.

$$M_{11} = \begin{vmatrix} -3 & -6 \\ 1 & 7 \end{vmatrix} = (-3)(7) - (-6)(1) = -21 + 6 = -15$$

(ii) To find the minor M_{23}, delete the second row and third column of the matrix A.

$$\begin{bmatrix} 2 & 3 & 4 \\ 5 & -3 & -6 \\ 0 & 1 & 7 \end{bmatrix}$$

Now find the determinant of the resulting matrix.

$$M_{23} = \begin{vmatrix} 2 & 3 \\ 0 & 1 \end{vmatrix} = (2)(1) - (3)(0) = 2$$

(iii) To find M_{32}, delete the third row and second column of A.

$$\begin{bmatrix} 2 & 3 & 4 \\ 5 & -3 & -6 \\ 0 & 1 & 7 \end{bmatrix}$$

$$M_{32} = \begin{vmatrix} 2 & 4 \\ 5 & -6 \end{vmatrix} = (2)(-6) - (4)(5) = -12 - 20 = -32.$$

b. To find the cofactors, use the formula $A_{ij} = (-1)^{i+j} M_{ij}$.

$A_{11} = (-1)^{1+1} M_{11} = (1)(-15) = -15$ $(-1)^{1+1} = (-1)^2 = 1; M_{11} = -15$

$A_{23} = (-1)^{2+3} M_{23} = (-1)(2) = -2$ $(-1)^{2+3} = (-1)^5 = -1; M_{23} = 2$

$A_{32} = (-1)^{3+2} M_{32} = (-1)(-32) = 32$ $(-1)^{3+2} = (-1)^5 = -1; M_{32} = -32$

Practice Problem 2 For the matrix $A = \begin{bmatrix} 3 & -1 & 2 \\ 4 & 5 & 6 \\ 7 & 1 & 2 \end{bmatrix}$, find

a. the minors M_{11}, M_{23}, and M_{32}

b. the cofactors A_{11}, A_{23}, and A_{32}

3 Evaluate the determinant of an $n \times n$ matrix.

The Determinant of an $n \times n$ Matrix

There are several ways of defining the determinant of a square matrix. Here we give an *inductive definition*. This means that the definition of the determinant of a matrix of order n uses the determinants of the matrices of order $(n - 1)$.

n by *n* Determinant

Let A be a square matrix of order $n \geq 3$. The **determinant** of A is the sum of the entries in any row of A (or column of A) multiplied by their respective cofactors.

Applying the preceding definition to find the determinant of a matrix is called **expanding by cofactors**. Expanding by the cofactors of the *i*th row gives

$$\det(A) = a_{i1}A_{i1} + a_{i2}A_{i2} + \cdots + a_{in}A_{in}$$

Similarly, expanding by the cofactors of the *j*th column yields

$$\det(A) = a_{1j}A_{1j} + a_{2j}A_{2j} + \cdots + a_{nj}A_{nj}$$

We usually say "expanding by the *i*th row (column)" instead of "expanding by the cofactors of the *i*th row (column)."

TECHNOLOGY CONNECTION

Use the Det option on a graphing calculator to find the determinant of a matrix.

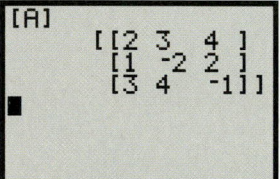

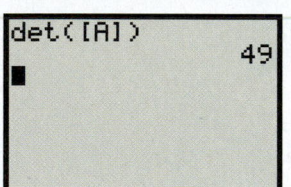

EXAMPLE 3   Evaluating a 3 by 3 Determinant

Find the determinant of $A = \begin{bmatrix} 2 & 3 & 4 \\ 1 & -2 & 2 \\ 3 & 4 & -1 \end{bmatrix}$.

Solution

Expanding by the first row, we have $|A| = a_{11}A_{11} + a_{12}A_{12} + a_{13}A_{13}$ where

$$A_{11} = (-1)^{1+1}M_{11} = (1)\begin{vmatrix} -2 & 2 \\ 4 & -1 \end{vmatrix}$$

$$A_{12} = (-1)^{1+2}M_{12} = (-1)\begin{vmatrix} 1 & 2 \\ 3 & -1 \end{vmatrix}$$

$$A_{13} = (-1)^{1+3}M_{13} = (1)\begin{vmatrix} 1 & -2 \\ 3 & 4 \end{vmatrix}$$

So,

$$\begin{vmatrix} 2 & 3 & 4 \\ 1 & -2 & 2 \\ 3 & 4 & -1 \end{vmatrix} = 2(1)\begin{vmatrix} -2 & 2 \\ 4 & -1 \end{vmatrix} + 3(-1)\begin{vmatrix} 1 & 2 \\ 3 & -1 \end{vmatrix} + 4(1)\begin{vmatrix} 1 & -2 \\ 3 & 4 \end{vmatrix}$$

$$= 2(2 - 8) - 3(-1 - 6) + 4(4 + 6) \qquad \text{Evaluate determinants of order 2.}$$

$$= -12 + 21 + 40 = 49 \qquad \text{Simplify.}$$

Practice Problem 3 Find the determinant of $A = \begin{bmatrix} 2 & -3 & 7 \\ -2 & -1 & 9 \\ 0 & 2 & -9 \end{bmatrix}$.

SIDE NOTE

In the definition, we note that the value det(A) *does not* depend on the choice of a row or a column. The proof of this fact is outside the scope of this book.

The definition says that we can expand the determinant by any row or column. If we expand by the third column of A in Example 3, we have

$$\begin{vmatrix} 2 & 3 & 4 \\ 1 & -2 & 2 \\ 3 & 4 & -1 \end{vmatrix} = 4(1)\begin{vmatrix} 1 & -2 \\ 3 & 4 \end{vmatrix} + 2(-1)\begin{vmatrix} 2 & 3 \\ 3 & 4 \end{vmatrix} + (-1)(1)\begin{vmatrix} 2 & 3 \\ 1 & -2 \end{vmatrix}$$

$$= 4(4 + 6) - 2(8 - 9) - 1(-4 - 3)$$

$$= 40 + 2 + 7 = 49$$

4 Apply Cramer's Rule.

Cramer's Rule

You have already learned several methods for solving a system of linear equations. Another method called **Cramer's Rule** uses determinants. To see a derivation of the rule, consider a system of two linear equations in two variables.

$$\begin{cases} a_1x + b_1y = c_1 & (1) \\ a_2x + b_2y = c_2 & (2) \end{cases}$$

We can solve this system by using the elimination method. Let's first eliminate y and solve for x.

$$\begin{array}{ll} a_1b_2x + b_1b_2y = c_1b_2 & \text{Multiply equation (1) by } b_2 \\ \underline{-a_2b_1x - b_1b_2y = -c_2b_1} & \text{Multiply equation (2) by } -b_1. \\ (a_1b_2 - a_2b_1)x = c_1b_2 - c_2b_1 & \text{Add to eliminate } y. \\ x = \dfrac{c_1b_2 - c_2b_1}{a_1b_2 - a_2b_1} & \begin{array}{l}\text{Divide both sides by}\\ a_1b_2 - a_2b_1 \text{ if } a_1b_2 - a_2b_1 \neq 0.\end{array} \end{array}$$

Similarly, by eliminating x and solving for y, we have

$$y = \dfrac{a_1c_2 - a_2c_1}{a_1b_2 - a_2b_1} \quad \text{if } a_1b_2 - a_2b_1 \neq 0$$

The solution of the system of equations (1) and (2) is therefore given by:

$$x = \dfrac{c_1b_2 - c_2b_1}{a_1b_2 - a_2b_1} \quad y = \dfrac{a_1c_2 - a_2c_1}{a_1b_2 - a_2b_1} \quad \text{provided that } a_1b_2 - a_2b_1 \neq 0.$$

You can write the solution in the form of determinants:

$$x = \dfrac{\begin{vmatrix} c_1 & b_1 \\ c_2 & b_2 \end{vmatrix}}{\begin{vmatrix} a_1 & b_1 \\ a_2 & b_2 \end{vmatrix}}, \quad y = \dfrac{\begin{vmatrix} a_1 & c_1 \\ a_2 & c_2 \end{vmatrix}}{\begin{vmatrix} a_1 & b_1 \\ a_2 & b_2 \end{vmatrix}} \quad \text{provided that } D = \begin{vmatrix} a_1 & b_1 \\ a_2 & b_2 \end{vmatrix} \neq 0.$$

Note that the denominator of both the x- and the y-values in the solution is the determinant D of the coefficient matrix of the linear system.

Suppose we write $D_x = \begin{vmatrix} c_1 & b_1 \\ c_2 & b_2 \end{vmatrix}$ and $D_y = \begin{vmatrix} a_1 & c_1 \\ a_2 & c_2 \end{vmatrix}$, which are obtained by replacing the column containing the coefficients of x in D and the column containing the coefficients of y in D, by the column of constants, respectively. Then the solution just found can be written as:

$$x = \dfrac{D_x}{D}, \quad y = \dfrac{D_y}{D} \quad \text{provided that } D \neq 0$$

This representation of the solution is Cramer's Rule for solving a system of linear equations.

Gabriel Cramer

(1704–1752)
Gabriel Cramer was one of three sons of a medical doctor in Geneva, Switzerland. Gabriel moved rapidly through his education in Geneva, and in 1722, while he was still 18 years old, he was awarded a PhD degree. Two years later he accepted the chair of mathematics at the Académie de Clavin in Geneva.

Cramer taught and traveled extensively and yet found time to produce articles and books covering a wide range of subjects. Although Cramer was not the first person to state the rule, it is universally known as "Cramer's Rule."

CRAMER'S RULE FOR SOLVING TWO EQUATIONS IN TWO VARIABLES

The system

$$\begin{cases} a_1x + b_1y = c_1 \\ a_2x + b_2y = c_2 \end{cases}$$

of two equations in two variables has a unique solution (x, y) given by

$$x = \dfrac{D_x}{D} \quad \text{and} \quad y = \dfrac{D_y}{D}$$

provided that $D \neq 0$, where

$$D = \begin{vmatrix} a_1 & b_1 \\ a_2 & b_2 \end{vmatrix}, \quad D_x = \begin{vmatrix} c_1 & b_1 \\ c_2 & b_2 \end{vmatrix}, \quad \text{and} \quad D_y = \begin{vmatrix} a_1 & c_1 \\ a_2 & c_2 \end{vmatrix}$$

EXAMPLE 4 **Using Cramer's Rule**

Use Cramer's Rule to solve the following system: $\begin{cases} 5x + 4y = 1 \\ 2x + 3y = 6 \end{cases}$

Solution

Step 1 Form the determinant D of the coefficient matrix.

$$D = \begin{vmatrix} 5 & 4 \\ 2 & 3 \end{vmatrix} = 15 - 8 = 7$$

Since $D = 7 \neq 0$, the system has a unique solution.

Step 2 Replace the column of coefficients of x in D by the constant terms to get

$$D_x = \begin{vmatrix} 1 & 4 \\ 6 & 3 \end{vmatrix} = 3 - 24 = -21$$

Step 3 Replace the column of coefficients of y in D by the constant terms to get

$$D_y = \begin{vmatrix} 5 & 1 \\ 2 & 6 \end{vmatrix} = 30 - 2 = 28$$

Step 4 By Cramer's Rule,

$$x = \frac{D_x}{D} = \frac{-21}{7} = -3 \quad \text{and} \quad y = \frac{D_y}{D} = \frac{28}{7} = 4$$

The solution is $x = -3$ and $y = 4$, or $\{(-3, 4)\}$.

Check: You should check the solution in the original system.

Practice Problem 4 Use Cramer's Rule to solve the following system:

$$\begin{cases} 2x + 3y = 7 \\ 5x + 9y = 4 \end{cases}$$

Cramer's Rule, as used in Example 4, can be generalized to a system of equations with more than two variables. We state the rule for three equations in three variables.

CRAMER'S RULE FOR SOLVING THREE EQUATIONS IN THREE VARIABLES

The system

$$\begin{cases} a_1 x + b_1 y + c_1 z = k_1 \\ a_2 x + b_2 y + c_2 z = k_2 \\ a_3 x + b_3 y + c_3 z = k_3 \end{cases}$$

of three equations in three variables has a unique solution (x, y, z) given by

$$x = \frac{D_x}{D}, \quad y = \frac{D_y}{D}, \quad \text{and} \quad z = \frac{D_z}{D}$$

provided that $D \neq 0$, where

$$D = \begin{vmatrix} a_1 & b_1 & c_1 \\ a_2 & b_2 & c_2 \\ a_3 & b_3 & c_3 \end{vmatrix}, \quad D_x = \begin{vmatrix} k_1 & b_1 & c_1 \\ k_2 & b_2 & c_2 \\ k_3 & b_3 & c_3 \end{vmatrix}, \quad D_y = \begin{vmatrix} a_1 & k_1 & c_1 \\ a_2 & k_2 & c_2 \\ a_3 & k_3 & c_3 \end{vmatrix}, \quad \text{and} \quad D_z = \begin{vmatrix} a_1 & b_1 & k_1 \\ a_2 & b_2 & k_2 \\ a_3 & b_3 & k_3 \end{vmatrix}$$

SIDE
NOTE

When you apply Cramer's Rule, make sure that all linear equations are written in the form

$ax + by + cz = k.$

EXAMPLE 5   **Using Cramer's Rule**

Use Cramer's Rule to solve the system of equations.

$$\begin{cases} 7x + \ y + \ z - 1 = 0 \\ \ \ \ \ 5x + 3z - 2y = 4 \\ \ \ \ \ 4x - \ z + 3y = 0 \end{cases}$$

Solution

First, rewrite the system of equations so that the terms on the left side of the equals sign are in the proper order and only the constant terms appear on the right side.

$$\begin{cases} 7x + \ y + \ z = 1 \\ 5x - 2y + 3z = 4 \\ 4x + 3y - \ z = 0 \end{cases}$$

Step 1 $D = \begin{vmatrix} 7 & 1 & 1 \\ 5 & -2 & 3 \\ 4 & 3 & -1 \end{vmatrix}$

D is the determinant of the coefficient matrix.

$= 7\begin{vmatrix} -2 & 3 \\ 3 & -1 \end{vmatrix} - 5\begin{vmatrix} 1 & 1 \\ 3 & -1 \end{vmatrix} + 4\begin{vmatrix} 1 & 1 \\ -2 & 3 \end{vmatrix}$ Expand by column 1.

$= 7(2 - 9) - 5(-1 - 3) + 4(3 + 2)$ Evaluate determinants of order 2.

$= 7(-7) - 5(-4) + 4(5) = -9$ Simplify.

Since $D \neq 0$, the system has a unique solution.

Step 2 $D_x = \begin{vmatrix} 1 & 1 & 1 \\ 4 & -2 & 3 \\ 0 & 3 & -1 \end{vmatrix}$

Replace the x-coefficients in the first column in D with the constants.

$= 1\begin{vmatrix} -2 & 3 \\ 3 & -1 \end{vmatrix} - 4\begin{vmatrix} 1 & 1 \\ 3 & -1 \end{vmatrix} + 0\begin{vmatrix} 1 & 1 \\ -2 & 3 \end{vmatrix}$ Expand by column 1.

$= 1(2 - 9) - 4(-1 - 3)$ Evaluate determinants of order 2.

$= 1(-7) - 4(-4) = 9$ Simplify.

Step 3 $D_y = \begin{vmatrix} 7 & 1 & 1 \\ 5 & 4 & 3 \\ 4 & 0 & -1 \end{vmatrix}$

Replace column 2 in D with the constants.

$= -1\begin{vmatrix} 5 & 3 \\ 4 & -1 \end{vmatrix} + 4\begin{vmatrix} 7 & 1 \\ 4 & -1 \end{vmatrix} - 0\begin{vmatrix} 7 & 1 \\ 5 & 3 \end{vmatrix}$ Expand by column 2 since it contains a zero.

$= -1(-5 - 12) + 4(-7 - 4)$ Evaluate the determinants.

$= 17 - 44 = -27$ Simplify.

Step 4 $D_z = \begin{vmatrix} 7 & 1 & 1 \\ 5 & -2 & 4 \\ 4 & 3 & 0 \end{vmatrix}$

Replace column 3 in D with the constants.

$= 1\begin{vmatrix} 5 & -2 \\ 4 & 3 \end{vmatrix} - 4\begin{vmatrix} 7 & 1 \\ 4 & 3 \end{vmatrix} + 0\begin{vmatrix} 7 & 1 \\ 5 & -2 \end{vmatrix}$ Expand by column 3.

$= (15 + 8) - 4(21 - 4)$ Evaluate the determinants.

$= 23 - 4(17) = -45$ Simplify.

Step 5 Cramer's Rule gives the following values:

$$x = \frac{D_x}{D} = \frac{9}{-9} = -1 \qquad \text{From Steps 1 and 2}$$

$$y = \frac{D_y}{D} = \frac{-27}{-9} = 3 \qquad \text{From Steps 1 and 3}$$

$$z = \frac{D_z}{D} = \frac{-45}{-9} = 5 \qquad \text{From Steps 1 and 4}$$

The solution set is therefore $\{(-1, 3, 5)\}$.

Check: You should check the solution.

Practice Problem 5 Use Cramer's Rule to solve the system.

$$\begin{cases} 3x + 2y + z = 4 \\ 4x + 3y + z = 5 \\ 5x + y + z = 9 \end{cases}$$

▲
WARNING

It is important to note that Cramer's Rule does not apply if $D = 0$. In this case, the system is either consistent and has dependent equations (has infinitely many solutions) or inconsistent (has no solution). When $D = 0$, use a different method for solving systems of equations to find the solution set.

Answers to Practice Problems

1. a. -38 **b.** 0 **2. a.** $M_{11} = 4, M_{23} = 10, M_{32} = 10$ **3.** 8 **4.** $\{(17, -9)\}$ **5.** $\{(2, -1, 0)\}$
b. $A_{11} = 4, A_{23} = -10, A_{32} = -10$

SECTION 7.5 **Exercises**

Concepts and Vocabulary

1. The determinant of $\begin{bmatrix} a & b \\ c & d \end{bmatrix}$ is _____.

2. The minor of an element is the determinant you get by deleting the _____ and the _____ containing that element.

3. To expand an $n \times n$ determinant, you multiply each element of some row (or column) by its _____ and add the result.

4. A system of n linear equations in n variables has a unique solution provided the determinant of the _____ is not zero.

5. **True or False.** A 2 by 2 determinant is always positive.

6. **True or False.** To expand an $n \times n$ determinant, you multiply each element of some row by its minor and add the result.

7. **True or False.** The determinant $\begin{vmatrix} a & 2a \\ a & 2a \end{vmatrix}$ is zero for any real number a.

8. **True or False.** If two rows or two columns in the determinant for a 2 × 2 matrix are identical, the determinant is zero.

Building Skills

In Exercises 9–18, evaluate each determinant.

9. $\begin{vmatrix} 2 & 3 \\ 4 & 5 \end{vmatrix}$

10. $\begin{vmatrix} 3 & -5 \\ 1 & 4 \end{vmatrix}$

11. $\begin{vmatrix} 4 & -2 \\ 3 & -3 \end{vmatrix}$

12. $\begin{vmatrix} -2 & \frac{1}{2} \\ 3 & 1 \end{vmatrix}$

13. $\begin{vmatrix} -1 & -3 \\ -4 & -5 \end{vmatrix}$

14. $\begin{vmatrix} \frac{1}{2} & \frac{1}{3} \\ \frac{1}{4} & \frac{1}{6} \end{vmatrix}$

15. $\begin{vmatrix} \frac{3}{8} & \frac{1}{2} \\ -\frac{1}{9} & 5 \end{vmatrix}$

16. $\begin{vmatrix} -\frac{1}{2} & \frac{1}{3} \\ \frac{1}{4} & -\frac{1}{3} \end{vmatrix}$

17. $\begin{vmatrix} \sqrt{a} & \sqrt{b} \\ \sqrt{b} & \sqrt{a} \end{vmatrix}$

18. $\begin{vmatrix} \sqrt{3} & 1 \\ -1 & \sqrt{3} \end{vmatrix}$

For Exercises 19–22, consider the matrix

$$A = \begin{bmatrix} 2 & -3 & 4 \\ 1 & -1 & 2 \\ 0 & 1 & 2 \end{bmatrix}$$

19. Find the minors.

    a. M_{21} **b.** M_{23} **c.** M_{32}

20. Find the cofactors.

    a. A_{21} **b.** A_{23} **c.** A_{32}

21. Find the minors.

    a. M_{11} **b.** M_{22} **c.** M_{31}

22. Find the cofactors.

    a. A_{11} **b.** A_{22} **c.** A_{31}

In Exercises 23–38, evaluate each determinant.

23. $\begin{vmatrix} 1 & 0 & -1 \\ 0 & 2 & 2 \\ -1 & 0 & 0 \end{vmatrix}$

24. $\begin{vmatrix} 2 & 0 & \frac{1}{2} \\ 1 & 0 & 2 \\ 4 & 0 & -5 \end{vmatrix}$

25. $\begin{vmatrix} 1 & 2 & 3 \\ 0 & 3 & 4 \\ 0 & 0 & 4 \end{vmatrix}$

26. $\begin{vmatrix} 2 & 3 & 4 \\ 0 & -4 & 6 \\ 0 & 0 & -5 \end{vmatrix}$

27. $\begin{vmatrix} 1 & 0 & 0 \\ 2 & 0 & 0 \\ 3 & 4 & 5 \end{vmatrix}$

28. $\begin{vmatrix} -1 & 0 & 0 \\ 3 & 2 & 0 \\ 4 & 5 & -3 \end{vmatrix}$

29. $\begin{vmatrix} 1 & 6 & 0 \\ 2 & 5 & 3 \\ 3 & 4 & 0 \end{vmatrix}$

30. $\begin{vmatrix} 1 & 0 & 2 \\ 0 & 5 & 7 \\ 3 & 1 & 0 \end{vmatrix}$

31. $\begin{vmatrix} 3 & 4 & 1 \\ 1 & 4 & 3 \\ 4 & 3 & 1 \end{vmatrix}$

32. $\begin{vmatrix} 3 & 1 & -2 \\ 4 & 0 & -4 \\ 2 & -1 & -3 \end{vmatrix}$

33. $\begin{vmatrix} 0 & 1 & 6 \\ 1 & 0 & 4 \\ 8 & 3 & 1 \end{vmatrix}$

34. $\begin{vmatrix} 0 & 2 & 3 \\ 1 & 0 & 1 \\ 3 & 2 & 0 \end{vmatrix}$

35. $\begin{vmatrix} a & b & 0 \\ 0 & a & b \\ b & 0 & a \end{vmatrix}$

36. $\begin{vmatrix} 0 & z & y \\ z & 0 & x \\ y & x & 0 \end{vmatrix}$

37. $\begin{vmatrix} a & b & c \\ c & a & b \\ b & c & a \end{vmatrix}$

38. $\begin{vmatrix} u & v & w \\ w & v & u \\ u & v & w \end{vmatrix}$

In Exercises 39–48, use Cramer's Rule (if applicable) to solve each system of equations.

39. $\begin{cases} x + y = 8 \\ x - y = -2 \end{cases}$

40. $\begin{cases} 4x + 3y = -1 \\ 2x - 5y = 9 \end{cases}$

41. $\begin{cases} 5x + 3y = 11 \\ 2x + y = 4 \end{cases}$

42. $\begin{cases} 2x - 7y = 13 \\ 5x + 6y = 9 \end{cases}$

43. $\begin{cases} 2x + 9y = 4 \\ 3x - 2y = 6 \end{cases}$

44. $\begin{cases} 5x + 3y = 1 \\ 2x - 5y = -12 \end{cases}$

45. $\begin{cases} 2x - 3y = 4 \\ 4x - 6y = 8 \end{cases}$

46. $\begin{cases} 3x + y = 2 \\ 6x + 2y = 4 \end{cases}$

47. $\begin{cases} \dfrac{2}{x} + \dfrac{3}{y} = 2 \\ \dfrac{5}{x} + \dfrac{8}{y} = \dfrac{31}{6} \end{cases}$

48. $\begin{cases} \dfrac{3}{x} - \dfrac{6}{y} = 2 \\ \dfrac{4}{x} + \dfrac{7}{y} = -3 \end{cases}$

$$\left[\textit{Hint: } \text{For Exercises 47 and 48, let } u = \frac{1}{x} \text{ and } v = \frac{1}{y}. \right]$$

In Exercises 49–58, solve each system of equations by means of Cramer's Rule. You may use technology to compute the determinants involved.

49. $\begin{cases} x - 2y + z = -1 \\ 3x + y - z = 4 \\ y + z = 1 \end{cases}$

50. $\begin{cases} x + 3y = 4 \\ y + 3z = 7 \\ z + 4x = 6 \end{cases}$

51. $\begin{cases} x + y - z = -3 \\ 2x + 3y + z = 2 \\ 2y + z = 1 \end{cases}$

52. $\begin{cases} 3x + y + z = 10 \\ x + y - z = 0 \\ 5x - 9y = 1 \end{cases}$

53. $\begin{cases} x + y + z = 3 \\ 2x - 3y + 5z = 4 \\ x + 2y - 4z = -1 \end{cases}$

54. $\begin{cases} x + 2y - z = 3 \\ 3x - y + z = 8 \\ x + y + z = 0 \end{cases}$

55. $\begin{cases} 2x - 3y + 5z = 11 \\ 3x + 5y - 2z = 7 \\ x + 2y - 3z = -4 \end{cases}$

56. $\begin{cases} x - 3y - 1 = 0 \\ 2x - y - 4z = 2 \\ y + 2z - 4 = 0 \end{cases}$

57. $\begin{cases} 5x + 2y + z = 12 \\ 2x + y + 3z = 13 \\ 3x + 2y + 4z = 19 \end{cases}$

58. $\begin{cases} 2x + y + z = 7 \\ 3x - y - z = -2 \\ x + 2y - 3z = -4 \end{cases}$

Applying the Concepts

It is known that the *area of a triangle* with vertices (x_1, y_1), (x_2, y_2), and (x_3, y_3) is equal to the absolute value of D, where

$$D = \frac{1}{2} \begin{vmatrix} x_1 & y_1 & 1 \\ x_2 & y_2 & 1 \\ x_3 & y_3 & 1 \end{vmatrix}$$

In Exercises 59–62, use determinants to find the area of each triangle.

59. **Area of a triangle.** The triangle with vertices $(1, 2)$, $(-3, 4)$, and $(4, 6)$.

60. **Area of a triangle.** The triangle with vertices $(3, 1)$, $(4, 2)$, and $(5, 4)$.

61. **Area of a triangle.** The triangle with vertices $(-2, 1)$, $(-3, -5)$, and $(2, 4)$.

62. **Area of a triangle.** The triangle with vertices $(-1, 2)$ and $(2, 4)$ and the origin.

Collinearity. Three points $A(x_1, y_1)$, $B(x_1, y_2)$, and $C(x_3, y_3)$ are *collinear* (lie on the same line) if and only if the area of the triangle ABC is 0. In terms of determinants, three points (x_1, y_1), (x_2, y_2), and (x_3, y_3) are collinear if and only if

$$\begin{vmatrix} x_1 & y_1 & 1 \\ x_2 & y_2 & 1 \\ x_3 & y_3 & 1 \end{vmatrix} = 0$$

In Exercises 63–66, use determinants to find whether the three given points are collinear.

63. $(0, 3)$, $(-1, 1)$, and $(2, 7)$

64. $(2, 0)$, $\left(1, \dfrac{1}{2}\right)$, and $(4, -1)$

65. $(0, -4)$, $(3, -2)$, and $(1, -4)$

66. $\left(0, -\dfrac{1}{4}\right)$, $(1, -1)$, and $(2, -2)$

Equation of a line. Three points (x, y), (x_1, y_1), and (x_2, y_2) all lie on the same line if and only if

$$D = \begin{vmatrix} x & y & 1 \\ x_1 & y_1 & 1 \\ x_2 & y_2 & 1 \end{vmatrix} = 0$$

In other words, a point $P(x, y)$ lies on the line passing through the points (x_1, y_1) and (x_2, y_2) if and only if the determinant $D = 0$. Therefore, the equation

$$\begin{vmatrix} x & y & 1 \\ x_1 & y_1 & 1 \\ x_2 & y_2 & 1 \end{vmatrix} = 0$$

is an equation of the line passing through the points (x_1, y_1) and (x_2, y_2).

In Exercises 67–70, use determinants to write an equation of the line passing through the given points. Then evaluate the determinant and write the equation of the line in slope–intercept form.

67. $(-1, -1)$, $(1, 3)$

68. $(0, 4)$, $(1, 1)$

69. $\left(0, \dfrac{1}{3}\right)$, $(1, 1)$

70. $\left(1, -\dfrac{1}{3}\right)$, $(2, 0)$

Beyond the Basics

In Exercises 71–75, we list some properties of determinants. These properties are true for $n \times n$ matrices. Verify the properties for $n = 2$ or $n = 3$.

71. If each entry in any row or each entry in any column of a matrix A is zero, then $|A| = 0$. Verify each of the following.

a. $\begin{vmatrix} 0 & 0 \\ 2 & 5 \end{vmatrix} = 0$

b. $\begin{vmatrix} 1 & 0 \\ 3 & 0 \end{vmatrix} = 0$

c. $\begin{vmatrix} 1 & -2 & 3 \\ 0 & 0 & 0 \\ 4 & 5 & -7 \end{vmatrix} = 0$

d. $\begin{vmatrix} 4 & 5 & 0 \\ 6 & -7 & 0 \\ 8 & 15 & 0 \end{vmatrix} = 0$

72. If a matrix B is obtained from matrix A by interchanging two rows (or columns), then $|B| = -|A|$. Verify each of the following.

a. $\begin{vmatrix} 2 & 3 \\ 4 & 5 \end{vmatrix} = -\begin{vmatrix} 4 & 5 \\ 2 & 3 \end{vmatrix}$

b. $\begin{vmatrix} -5 & 3 \\ 2 & -4 \end{vmatrix} = -\begin{vmatrix} 3 & -5 \\ -4 & 2 \end{vmatrix}$

c. $\begin{vmatrix} 1 & 3 & 5 \\ 0 & 1 & 2 \\ 3 & -1 & 4 \end{vmatrix} = -\begin{vmatrix} 1 & 3 & 5 \\ 3 & -1 & 4 \\ 0 & 1 & 2 \end{vmatrix}$

73. If two rows (or columns) of a matrix A have corresponding entries that are equal, then $|A| = 0$. Verify the following.

a. $\begin{vmatrix} 2 & 3 & 5 \\ -1 & 4 & 8 \\ 2 & 3 & 5 \end{vmatrix} = 0$

b. $\begin{vmatrix} 3 & 4 & 3 \\ 1 & 2 & 1 \\ -1 & 6 & -1 \end{vmatrix} = 0$

74. If a matrix B is obtained from a matrix A by multiplying every entry of one row (or one column) by a real number c, then $|B| = c|A|$. Verify each of the following.

a. $\begin{vmatrix} -1 & 2 & 3 \\ 1 \cdot 5 & 4 \cdot 5 & 8 \cdot 5 \\ 2 & 3 & 6 \end{vmatrix} = 5 \begin{vmatrix} -1 & 2 & 3 \\ 1 & 4 & 8 \\ 2 & 3 & 6 \end{vmatrix}$

b. $\begin{vmatrix} 1 & 2 & 3 \\ -1 & 5 & 6 \\ 0 & 2 & 9 \end{vmatrix} = 3 \begin{vmatrix} 1 & 2 & 1 \\ -1 & 5 & 2 \\ 0 & 2 & 3 \end{vmatrix}$

75. If a matrix B is obtained from a matrix A by replacing any row (or column) of A by the sum of that row (or column) and c times another row (or column), then $|B| = |A|$. Verify that the following two determinants are equal:

$$\begin{vmatrix} 2 & -3 & 4 \\ -4 & 7 & -8 \\ 5 & -1 & 3 \end{vmatrix} \xrightarrow{2R_1 + R_2 \to R_2} \begin{vmatrix} 2 & -3 & 4 \\ 0 & 1 & 0 \\ 5 & -1 & 3 \end{vmatrix}$$

76. Evaluate the determinant

$$D = \begin{vmatrix} 1 & 2 & 3 & 3 \\ 4 & 5 & 2 & 1 \\ 1 & 2 & 5 & 3 \\ 2 & 4 & 3 & 7 \end{vmatrix}$$

by following these steps:

(i) $(-1)R_1 + R_3 \to R_3$

(ii) Expand by the cofactors of row 3.

(iii) In (ii), you will have a determinant of order 3. Evaluate it to obtain the value of D.

In Exercises 77–80, use the properties of determinants in Exercises 71–75 to obtain three zeros in one row (or column) and then evaluate the given determinant by evaluating the resulting determinant of order 3.

77. $\begin{vmatrix} 2 & 0 & -3 & 4 \\ 0 & 1 & 0 & 5 \\ 5 & 0 & -9 & 8 \\ 1 & 2 & 0 & 7 \end{vmatrix}$

78. $\begin{vmatrix} -3 & 1 & 2 & 3 \\ 0 & -2 & 4 & 7 \\ -9 & 3 & 5 & 2 \\ 2 & -4 & 3 & -12 \end{vmatrix}$

79. $\begin{vmatrix} 5 & 7 & 1 & 2 \\ 6 & 8 & 9 & 3 \\ 24 & 22 & 6 & 10 \\ 21 & 17 & 7 & 10 \end{vmatrix}$

80. $\begin{vmatrix} 1 & 2 & 3 & 0 \\ 2 & -3 & 1 & 0 \\ -1 & -3 & 4 & 5 \\ 1 & 3 & 7 & 4 \end{vmatrix}$

In Exercises 81–88, solve each equation for x.

81. $\begin{vmatrix} 3 & 2 \\ 6 & x \end{vmatrix} = 0$

82. $\begin{vmatrix} 3 & 2 \\ x & 12 \end{vmatrix} = 0$

83. $\begin{vmatrix} x & 2 \\ 1 & x - 1 \end{vmatrix} = 0$

84. $\begin{vmatrix} 2x + 7 & 4 \\ 1 & x \end{vmatrix} = 0$

85. $\begin{vmatrix} 1 & -3 & 1 \\ 4 & 7 & x \\ 0 & 2 & 2 \end{vmatrix} = 0$

86. $\begin{vmatrix} x & x + 1 & x + 2 \\ 2 & 3 & -1 \\ 3 & -2 & 4 \end{vmatrix} = 0$

87. $\begin{vmatrix} x & 0 & 1 \\ 0 & x & 0 \\ 1 & 0 & x \end{vmatrix} = 0$

88. $\begin{vmatrix} x & 2 & 3 \\ x & x & 1 \\ 2 & 0 & 1 \end{vmatrix} = -8$

In Exercises 89 and 90, use the formula on page 657 for the area of a triangle.

89. If the area of the triangle formed by the points $(-4, 2)$, $(0, k)$, and $(-2, k)$ is 28 square units, find k.

90. If the area of the triangle formed by the points $(2, 3)$, $(1, 2)$, and $(-2, k)$ is 3 square units, find k.

Critical Thinking / Discussion / Writing

91. Solve the system by using Cramer's Rule.

$$\begin{cases} \dfrac{1}{x-2} + \dfrac{3}{y+1} = 13 \\ \dfrac{4}{x-2} - \dfrac{5}{y+1} = 1 \end{cases}$$

[*Hint:* Do not try to simplify the equations by clearing the denominators because the resulting equations will not be linear. Instead, let $u = \dfrac{1}{x-2}$ and $v = \dfrac{1}{y+1}$. Solve the system for u and v and then use this solution to solve for x and y.]

In Exercises 92 and 93, use Cramer's Rule to solve each system of equations.

92. $\begin{cases} \dfrac{6}{x+1} + \dfrac{4}{y-1} = 7 \\ \dfrac{8}{x+1} + \dfrac{5}{y-1} = 9 \end{cases}$

93. $\begin{cases} \dfrac{1}{3^x} + \dfrac{4}{4^y} = 25 \\ \dfrac{2}{3^x} - \dfrac{1}{4^y} = 14 \end{cases}$

Getting Ready for the Next Section

In Exercises 94–97, solve each equation for the specified variable.

94. $\dfrac{1}{2 \cdot 3} = \dfrac{1}{2} + \dfrac{B}{3}$, for B

95. $\dfrac{2}{5 \cdot 7} = \dfrac{A}{5} - \dfrac{1}{7}$, for A

96. $\dfrac{1}{n \cdot (n+1)} = \dfrac{A}{n} - \dfrac{1}{n+1}$, for A

97. $\dfrac{4}{n \cdot (n+2)} = \dfrac{2}{n} + \dfrac{B}{n+2}$, for B

98. Find the values of A and B for which the equation

$$2x + 3 = A(x+3) + B(x-1)$$

is an identity (true for all values of x). [*Hint:* Evaluate both sides at $x = 1$ and $x = -3$.]

99. Find the values of A and B for which the equation

$$3x + 2 = A(x-2) + B(x+2)$$

is an identity.

In Exercises 100–105, factor each expression.

100. $x^2 + 5x + 6$

101. $x^2 - 3x - 10$

102. $2x^2 + 5x - 3$

103. $3x^2 + x - 2$

104. $4x^2 - 9$

105. $x^3 - 8$

SECTION | 7.6

Partial-Fraction Decomposition

BEFORE STARTING THIS SECTION, REVIEW

1 Division of polynomials (Section 2.3, page 174)

2 Factoring polynomials (Appendix A.2, page 930)

3 Irreducible polynomials (Appendix A.2, page 931)

4 Rational expressions (Appendix A.3, page 933)

OBJECTIVES

1 Become familiar with partial-fraction decomposition.

2 Decompose $\dfrac{P(x)}{Q(x)}$ when $Q(x)$ has only distinct linear factors.

3 Decompose $\dfrac{P(x)}{Q(x)}$ when $Q(x)$ has repeated linear factors.

4 Decompose $\dfrac{P(x)}{Q(x)}$ when $Q(x)$ has distinct irreducible quadratic factors.

5 Decompose $\dfrac{P(x)}{Q(x)}$ when $Q(x)$ has repeated irreducible quadratic factors.

Georg Simon Ohm (1787–1854)
Georg Ohm was born in Erlangen, Germany. His father, Johann, a mechanic and a self-educated man, was interested in philosophy and mathematics and gave his children an excellent education in physics, chemistry, mathematics, and philosophy through his own teachings. The Ohm's law named in Georg Ohm's honor appeared in his famous pamphlet *Die galvanische Kette, mathematisch bearbeitet* (1827). Ohm's work was not appreciated in Germany, where it was first published. Eventually, it was recognized, first by the British Royal Society, which awarded Ohm the Copley Medal in 1841. In 1849, Ohm became a curator of the Bavarian Academy's science museum and began to lecture at the University of Munich. Finally, in 1852 (two years before his death), Ohm achieved his lifelong ambition of becoming chair of physics at the University of Munich.

◆ Ohm's Law

If a voltage is applied across a resistor (such as a wire), a current will flow. Ohm's law states that in a circuit at constant temperature,

$$I = \frac{V}{R}$$

where V is the voltage (measured in volts), I is the current (measured in amperes), and R is the resistance (measured in ohms). A **parallel circuit** consists of two or more resistors connected as shown in Figure 7.22. In the figure, the current I from the source divides into two currents, I_1 and I_2, with I_1 going through one resistor and I_2 going through the other, so that $I = I_1 + I_2$. Suppose we want to find the total resistance R in the circuit due to the two resistors R_1 and R_2 connected in parallel. Ohm's law can be used to show that in the parallel circuit in Figure 7.22, $\dfrac{1}{R} = \dfrac{1}{R_1} + \dfrac{1}{R_2}$. This result is called the *parallel property of resistors*. In Example 7, we examine the parallel property of resistors by using partial fractions.

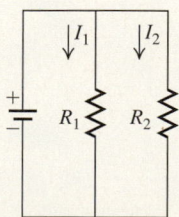

Figure 7.22 Parallel resistors.

1 Become familiar with partial-fraction decomposition.

Partial Fractions

Recall that to add two rational expressions such as $\dfrac{2}{x+3}$ and $\dfrac{3}{x-1}$ required you to find the least common denominator, rewrite expressions with the same denominator, and then add the corresponding numerators. This procedure gives

$$\frac{2}{x+3} + \frac{3}{x-1} = \frac{2(x-1)}{(x+3)(x-1)} + \frac{3(x+3)}{(x-1)(x+3)} = \frac{5x+7}{(x+3)(x-1)}.$$

In some applications of algebra to more advanced mathematics, you need a *reverse* procedure for *splitting* a fraction such as $\dfrac{5x+7}{(x-1)(x+3)}$ into the simpler fractions to obtain

$$\frac{5x+7}{(x+3)(x-1)} = \frac{2}{x+3} + \frac{3}{x-1}.$$

Each of the two fractions on the right is called a **partial fraction**. Their sum is called the **partial-fraction decomposition** of the rational expression on the left.

A rational expression $\dfrac{P(x)}{Q(x)}$ is called **improper** if the degree $P(x) \geq$ degree $Q(x)$ and is called **proper** if the degree $P(x) <$ degree $Q(x)$. Examples of improper rational expressions are

$$\frac{x^3}{x^2-1}, \quad \frac{(x+2)(x-1)}{(x-2)(x+3)}, \quad \text{and} \quad \frac{x^2+x+1}{2x+3}.$$

Using long division, we can express an improper rational fraction $\dfrac{P(x)}{Q(x)}$ as the sum of a polynomial and a proper rational fraction:

$$\underset{\substack{\uparrow \\ \text{Improper} \\ \text{fraction}}}{\frac{P(x)}{Q(x)}} = \underset{\substack{\uparrow \\ \text{Polynomial}}}{S(x)} + \underset{\substack{\uparrow \\ \text{Proper} \\ \text{fraction}}}{\frac{R(x)}{Q(x)}}$$

RECALL

Any polynomial $Q(x)$ with real coefficients can be factored so that each factor is linear or is an irreducible quadratic factor.

We therefore restrict our discussion of the decomposition of $\dfrac{P(x)}{Q(x)}$ into partial fractions to cases involving proper rational fractions. We also assume that $P(x)$ and $Q(x)$ have no common factor.

The problem of decomposing a proper fraction $\dfrac{P(x)}{Q(x)}$ into partial fractions depends on the type of factors in the denominator $Q(x)$. In this section, we consider four cases for $Q(x)$.

2 Decompose $\dfrac{P(x)}{Q(x)}$ when $Q(x)$ has only distinct linear factors.

$Q(x)$ Has Only Distinct Linear Factors

> ### CASE 1: THE DENOMINATOR IS THE PRODUCT OF DISTINCT (NONREPEATED) LINEAR FACTORS
>
> Suppose $Q(x)$ can be factored as
>
> $$Q(x) = c(x - a_1)(x - a_2) \cdots (x - a_n),$$
>
> with no factor repeated. The partial-fraction decomposition of $\dfrac{P(x)}{Q(x)}$ is of the form
>
> $$\frac{P(x)}{Q(x)} = \frac{A_1}{x - a_1} + \frac{A_2}{x - a_2} + \cdots + \frac{A_n}{x - a_n}$$
>
> where $A_1, A_2, \ldots, A_n$ are constants to be determined.

We can find the constants $A_1, A_2, \ldots, A_n$ by using the following procedure. Note that if the number of constants is small, we use the letters $A, B, C, \ldots$, instead of $A_1, A_2, A_3, \ldots$.

PROCEDURE
IN ACTION

EXAMPLE 1 Partial-Fraction Decomposition

OBJECTIVE

Find the partial-fraction decomposition of a rational expression.

Step 1 Write the form of the partial-fraction decomposition with the unknown constants $A, B, C, \ldots$ in the numerators of the decomposition.

Step 2 Multiply both sides of the equation in Step 1 by the original denominator. Use the distributive property and eliminate common factors. Simplify.

Step 3 Write both sides of the equation in Step 2 in descending powers of x and equate the coefficients of like powers of x.

Step 4 Solve the linear system resulting from Step 3 for the constants $A, B, C, \ldots$.

Step 5 Substitute the values you found for $A, B, C, \ldots$ into the equation in Step 1 and write the partial-fraction decomposition.

EXAMPLE

Find the partial-fraction decomposition of $\dfrac{3x + 26}{(x - 3)(x + 4)}$.

1. $\dfrac{3x + 26}{(x - 3)(x + 4)} = \dfrac{A}{x - 3} + \dfrac{B}{x + 4}$

2. Multiply both sides by $(x - 3)(x + 4)$ and simplify to obtain

$$3x + 26 = (x + 4)A + (x - 3)B$$

3. $3x + 26 = (A + B)x + (4A - 3B)$

$$\begin{cases} 3 = A + B & \text{Equate coefficients of } x. \\ 26 = 4A - 3B & \text{Equate constant coefficients.} \end{cases}$$

4. Solving the system of equations in Step 3, we obtain $A = 5$ and $B = -2$.

5. $\dfrac{3x + 26}{(x - 3)(x + 4)} = \dfrac{5}{x - 3} + \dfrac{-2}{x + 4}$ Replace A with 5 and B with -2 in Step 1.

$$\dfrac{3x + 26}{(x - 3)(x + 4)} = \dfrac{5}{x - 3} - \dfrac{2}{x + 4}$$

Practice Problem 1 Find the partial-fraction decomposition of $\dfrac{2x - 7}{(x + 1)(x - 2)}$.

EXAMPLE 2 **Finding the Partial-Fraction Decomposition When the Denominator Has Only Distinct Linear Factors**

Find the partial-fraction decomposition of the expression:

$$\frac{10x - 4}{x^3 - 4x}$$

Solution

First, factor the denominator.

$$x^3 - 4x = x(x^2 - 4) \qquad \text{Distributive property}$$
$$= x(x - 2)(x + 2) \qquad x^2 - 4 = (x - 2)(x + 2)$$

Each of the factors x, $x - 2$, and $x + 2$ becomes a denominator for a partial fraction.

Step 1 The partial-fraction decomposition is given by

$$\frac{10x - 4}{x(x - 2)(x + 2)} = \frac{A}{x} + \frac{B}{x - 2} + \frac{C}{x + 2} \qquad \begin{array}{l} \text{Use } A, B, \text{ and } C \text{ instead of} \\ A_1, A_2, \text{ and } A_3. \end{array}$$

Step 2 Multiply both sides by the common denominator $x(x - 2)(x + 2)$.

Distribute

$$x(x - 2)(x + 2)\left[\frac{10x - 4}{x(x - 2)(x + 2)}\right] = x(x - 2)(x + 2)\left[\frac{A}{x} + \frac{B}{x - 2} + \frac{C}{x + 2}\right]$$

$$10x - 4 = A(x - 2)(x + 2) + Bx(x + 2) + Cx(x - 2) \quad (1) \qquad \begin{array}{l} \text{Distributive property;} \\ \text{simplify.} \end{array}$$

$$= A(x^2 - 4) + B(x^2 + 2x) + C(x^2 - 2x) \qquad \text{Multiply.}$$
$$= Ax^2 - 4A + Bx^2 + 2Bx + Cx^2 - 2Cx \qquad \text{Distributive property}$$
$$= (A + B + C)x^2 + (2B - 2C)x - 4A. \qquad \text{Combine like terms.}$$

Step 3 Now use the fact that *two equal polynomials have equal corresponding coefficients.* Writing $10x - 4 = 0x^2 + 10x - 4$, we have

$$0x^2 + 10x - 4 = (A + B + C)x^2 + (2B - 2C)x - 4A.$$

Equating corresponding coefficients leads to the system of equations.

$$\begin{cases} A + B + C = 0 & \text{Equate coefficients of } x^2. \\ 2B - 2C = 10 & \text{Equate coefficients of } x. \\ -4A = -4 & \text{Equate constant coefficients.} \end{cases}$$

Step 4 Solve the system of equations in Step 3 to obtain $A = 1$, $B = 2$, and $C = -3$. (See Exercise 73.)

Step 5 The partial-fraction decomposition is

$$\frac{10x - 4}{x^3 - 4x} = \frac{1}{x} + \frac{2}{x - 2} + \frac{-3}{x + 2} = \frac{1}{x} + \frac{2}{x - 2} - \frac{3}{x + 2}.$$

Alternative Solution of Example 2

In Example 2, to find the constants A, B, and C, we *equated coefficients* of like powers of x and solved the resulting system. An alternative (and sometimes quicker) method is to *substitute* well-chosen values for x in equation (1) of Step 2:

$$10x - 4 = A(x - 2)(x + 2) + Bx(x + 2) + Cx(x - 2) \qquad (1)$$

You can reinforce your partial-fraction connection graphically:
The graph of

$$Y_1 = \frac{10x - 4}{x^3 - 4x}$$

appears to be identical to the graph of

$$Y_2 = \frac{1}{x} + \frac{2}{x - 2} - \frac{3}{x + 2}.$$

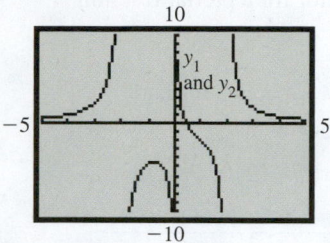

The graphs of Y_1 and Y_2 are the same.

3 Decompose $\frac{P(x)}{Q(x)}$ when $Q(x)$ has repeated linear factors.

Substitute $x = 2$ in equation (1) to cause the terms containing A and C to be 0.

$$10(2) - 4 = A(2 - 2)(2 + 2) + B(2)(2 + 2) + C(2)(2 - 2)$$
$$16 = 8B \qquad \text{Simplify.}$$
$$2 = B \qquad \text{Solve for } B.$$

Now substitute $x = -2$ in equation (1) to get

$$10(-2) - 4 = A(-2 - 2)(-2 + 2) + B(-2)(-2 + 2) + C(-2)(-2 - 2)$$
$$-24 = 8C \qquad \text{Simplify.}$$
$$-3 = C \qquad \text{Solve for } C.$$

Finally, substitute $x = 0$ in equation (1) to get

$$10(0) - 4 = A(0 - 2)(0 + 2) + B(0)(0 + 2) + C(0)(0 - 2)$$
$$-4 = -4A \qquad \text{Simplify.}$$
$$1 = A \qquad \text{Solve for } A.$$

Because we found the same values for A, B, and C, the partial-fraction decomposition will also be the same.

Practice Problem 2 Find the partial-fraction decomposition of:

$$\frac{3x^2 + 4x + 3}{x^3 - x}$$

$Q(x)$ Has Repeated Linear Factors

> ### CASE 2: THE DENOMINATOR HAS A REPEATED LINEAR FACTOR
>
> Let $(x - a)^m$ be the linear factor $(x - a)$ that is repeated m times in $Q(x)$.
>
> Then the portion of the partial-fraction decomposition of $\frac{P(x)}{Q(x)}$ that corresponds to the factor $(x - a)^m$ is
>
> $$\frac{A_1}{x - a} + \frac{A_2}{(x - a)^2} + \cdots + \frac{A_m}{(x - a)^m}$$
>
> where $A_1, A_2, \ldots, A_m$ are constants.

EXAMPLE 3 **Finding the Partial-Fraction Decomposition When the Denominator Has Repeated Linear Factors**

Find the partial-fraction decomposition of: $\dfrac{x + 4}{(x + 3)(x - 1)^2}$

Solution

Step 1 The linear factor $(x - 1)$ is repeated twice, and the factor $(x + 3)$ is nonrepeating. So the partial-fraction decomposition has the form:

$$\frac{x + 4}{(x + 3)(x - 1)^2} = \frac{A}{x + 3} + \frac{B}{x - 1} + \frac{C}{(x - 1)^2}$$

Steps 2–4 Multiply both sides of the equation in Step 1 by the original denominator, $(x + 3)(x - 1)^2$, then use the distributive property on the right side and simplify to get

$$x + 4 = A(x - 1)^2 + B(x + 3)(x - 1) + C(x + 3).$$

Substitute $x = 1$ in the last equation to obtain

$$1 + 4 = A(1 - 1)^2 + B(1 + 3)(1 - 1) + C(1 + 3).$$

This simplifies to $5 = 4C$, or $C = \dfrac{5}{4}$.

To find A, we substitute $x = -3$ to get

$$-3 + 4 = A(-3 - 1)^2 + B(-3 + 3)(-3 - 1) + C(-3 + 3).$$

This simplifies to $1 = 16A$, or $A = \dfrac{1}{16}$.

To find B, we replace x with any convenient number, say, 0. We have

$$0 + 4 = A(0 - 1)^2 + B(0 + 3)(0 - 1) + C(0 + 3)$$
$$4 = A - 3B + 3C \qquad \qquad \text{Simplify.}$$

Now replace A with $\dfrac{1}{16}$ and C with $\dfrac{5}{4}$ in the last equation.

$$4 = \dfrac{1}{16} - 3B + 3\left(\dfrac{5}{4}\right)$$
$$64 = 1 - 48B + 60 \qquad \text{Multiply both sides by 16.}$$
$$B = -\dfrac{1}{16} \qquad \qquad \text{Solve for } B.$$

We have $A = \dfrac{1}{16}, \quad B = -\dfrac{1}{16}, \quad$ and $C = \dfrac{5}{4}$.

Step 5 Substitute $A = \dfrac{1}{16}, B = -\dfrac{1}{16}$, and $C = \dfrac{5}{4}$ in the decomposition in Step 1 to get

$$\frac{x + 4}{(x + 3)(x - 1)^2} = \frac{\frac{1}{16}}{x + 3} + \frac{-\frac{1}{16}}{x - 1} + \frac{\frac{5}{4}}{(x - 1)^2}$$

or

$$\frac{x + 4}{(x + 3)(x - 1)^2} = \frac{1}{16(x + 3)} - \frac{1}{16(x - 1)} + \frac{5}{4(x - 1)^2}$$

Practice Problem 3 Find the partial-fraction decomposition of

$$\frac{x + 5}{x(x - 1)^2}.$$

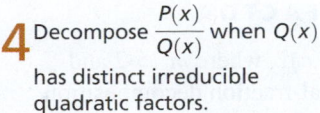
Decompose $\dfrac{P(x)}{Q(x)}$ when $Q(x)$ has distinct irreducible quadratic factors.

$Q(x)$ Has Distinct Irreducible Quadratic Factors

> **CASE 3: THE DENOMINATOR HAS A NONREPEATED IRREDUCIBLE QUADRATIC FACTOR**
>
> Suppose $ax^2 + bx + c$ is an irreducible quadratic factor of $Q(x)$. Then the portion of the partial-fraction decomposition of $\dfrac{P(x)}{Q(x)}$ that corresponds to $ax^2 + bx + c$ has the form:
>
> $$\frac{Ax + B}{ax^2 + bx + c}$$

EXAMPLE 4   **Finding the Partial-Fraction Decomposition When the Denominator Has a Distinct Irreducible Quadratic Factor**

Find the partial-fraction decomposition of: $\dfrac{3x^2 - 8x + 1}{(x - 4)(x^2 + 1)}$

Solution

Step 1   The factor $x - 4$ is linear and $x^2 + 1$ is irreducible; so the partial-fraction decomposition has the form:

$$\frac{3x^2 - 8x + 1}{(x - 4)(x^2 + 1)} = \frac{A}{x - 4} + \frac{Bx + C}{x^2 + 1}$$

Step 2   Multiply both sides of the decomposition in Step 1 by the original denominator, $(x - 4)(x^2 + 1)$, and simplify to get

$$3x^2 - 8x + 1 = A(x^2 + 1) + (Bx + C)(x - 4).$$

Substitute $x = 4$ to find $17 = 17A$, or $A = 1$.

Step 3   Collect like terms and write both sides of the equation in Step 2 in descending powers of x.

$$3x^2 - 8x + 1 = (A + B)x^2 + (-4B + C)x + (A - 4C)$$

Equate corresponding coefficients to get

$$\begin{cases} A + B = 3 & (1) & \text{Equate the coefficients of } x^2. \\ -4B + C = -8 & (2) & \text{Equate the coefficients of } x. \\ A - 4C = 1 & (3) & \text{Equate the constant coefficients.} \end{cases}$$

Step 4   Substitute $A = 1$ from Step 2 in equation (1) to get $1 + B = 3$, or $B = 2$. Substitute $A = 1$ in equation (3) to get $1 - 4C = 1$, or $C = 0$.

Step 5   Substitute $A = 1$, $B = 2$, and $C = 0$ into the decomposition in Step 1 to get:

$$\frac{3x^2 - 8x + 1}{(x - 4)(x^2 + 1)} = \frac{1}{x - 4} + \frac{2x}{x^2 + 1}$$

Practice Problem 4   Find the partial-fraction decomposition of:

$$\frac{3x^2 + 5x - 2}{x(x^2 + 2)}$$

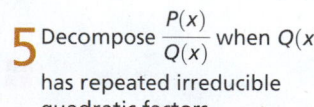 **5** Decompose $\dfrac{P(x)}{Q(x)}$ when $Q(x)$ has repeated irreducible quadratic factors.

$Q(x)$ Has Repeated Irreducible Quadratic Factors

> ### CASE 4: THE DENOMINATOR HAS A REPEATED IRREDUCIBLE QUADRATIC FACTOR
>
> Suppose the denominator $Q(x)$ has a factor $(ax^2 + bx + c)^m$, where $m \geq 2$ and $ax^2 + bx + c$ is irreducible. Then the portion of the partial-fraction decomposition of $\dfrac{P(x)}{Q(x)}$ that corresponds to the factor $(ax^2 + bx + c)^m$ has the form:
>
> $$\frac{A_1 x + B_1}{ax^2 + bx + c} + \frac{A_2 x + B_2}{(ax^2 + bx + c)^2} + \cdots + \frac{A_m x + B_m}{(ax^2 + bx + c)^m}$$

EXAMPLE 5 Finding the Partial-Fraction Decomposition When the Denominator Has a Repeated Irreducible Quadratic Factor

Find the partial-fraction decomposition of: $\dfrac{2x^4 - x^3 + 13x^2 - 2x + 13}{(x-1)(x^2+4)^2}$

Solution

Step 1 The denominator has a nonrepeating linear factor $(x-1)$ and an irreducible quadratic factor (x^2+4) that appears twice. The decomposition therefore has the form:

$$\frac{2x^4 - x^3 + 13x^2 - 2x + 13}{(x-1)(x^2+4)^2} = \frac{A}{x-1} + \frac{Bx+C}{x^2+4} + \frac{Dx+E}{(x^2+4)^2}$$

Step 2 Multiply both sides of the decomposition in Step 1 by the original denominator $(x-1)(x^2+4)^2$, use the distributive property, and eliminate common factors:

$$2x^4 - x^3 + 13x^2 - 2x + 13 =$$
$$A(x^2+4)^2 + (Bx+C)(x-1)(x^2+4) + (Dx+E)(x-1).$$

Substitute $x = 1$ and simplify to obtain $25 = 25A$, or $A = 1$.

Steps 3–4 Multiply the factors on the right side of the equation in Step 2 and collect like terms to obtain

$$2x^4 - x^3 + 13x^2 - 2x + 13 =$$
$$(A+B)x^4 + (-B+C)x^3 + (8A+4B-C+D)x^2 + (-4B+4C-D+E)x + (16A-4C-E).$$

Equate corresponding coefficients to get the system:

$$\begin{cases} A + B & = 2 \quad (1) \quad \text{Coefficient of } x^4 \\ -B+C & = -1 \quad (2) \quad \text{Coefficient of } x^3 \\ 8A+4B-C+D & = 13 \quad (3) \quad \text{Coefficient of } x^2 \\ -4B+4C-D+E & = -2 \quad (4) \quad \text{Coefficient of } x \\ 16A-4C-E & = 13 \quad (5) \quad \text{Constant coefficient} \end{cases}$$

Now back-substitute $A = 1$ from Step 2 in equation (1) to get $B = 1$. Again, back-substitute $B = 1$ in equation (2) to get $C = 0$.

Back-substitute $A = 1$ and $C = 0$ in equation (5) to obtain $E = 3$. Again, back-substitute $A = 1, B = 1$, and $C = 0$ in equation (3) to get $D = 1$. Hence, $A = 1, B = 1, C = 0, D = 1$, and $E = 3$.

Step 5 Substitute $A = 1, B = 1, C = 0, D = 1$, and $E = 3$ in the equation in Step 1 to obtain the partial-fraction decomposition.

$$\frac{2x^4 - x^3 + 13x^2 - 2x + 13}{(x-1)(x^2+4)^2} = \frac{1}{x-1} + \frac{x}{x^2+4} + \frac{x+3}{(x^2+4)^2}$$

Practice Problem 5 Find the partial-fraction decomposition of:

$$\frac{x^2 + 3x + 1}{(x^2+1)^2}$$

∫ **EXAMPLE 6** Learning a Clever Way to Add

Express the sum

as a fraction of whole numbers in lowest terms.

Solution

Each term in the sum is of the form: $\dfrac{1}{k(k+1)}$. From the partial-fraction decomposition, we have:

$$\frac{1}{k(k+1)} = \frac{1}{k} - \frac{1}{k+1}$$

Therefore, by using this decomposition for each term, we get:

$$\frac{1}{2\cdot 3} + \frac{1}{3\cdot 4} + \frac{1}{4\cdot 5} + \cdots + \frac{1}{2019\cdot 2020}$$

$$= \left(\frac{1}{2} - \frac{1}{3}\right) + \left(\frac{1}{3} - \frac{1}{4}\right) + \left(\frac{1}{4} - \frac{1}{5}\right) + \cdots + \left(\frac{1}{2019} - \frac{1}{2020}\right).$$

After removing parentheses, we see that most terms cancel in pairs $\left(\text{such as } -\dfrac{1}{3} \text{ and } \dfrac{1}{3}\right)$, leaving only the first and the last term. The original sum is therefore referred to as a *telescoping sum* that "collapses" to

$$\frac{1}{2} - \frac{1}{2020} = \frac{1010 - 1}{2020} = \frac{1009}{2020}.$$

Practice Problem 6 Express the sum $\dfrac{1}{4\cdot 5} + \dfrac{1}{5\cdot 6} + \dfrac{1}{6\cdot 7} + \cdots + \dfrac{1}{3111\cdot 3112}$ as a fraction of whole numbers in lowest terms.

◆ **EXAMPLE 7** **Calculating the Resistance in a Circuit**

The total resistance R due to two resistors connected in parallel is given by

$$R = \frac{x(x+5)}{3x+10}.$$

Use partial-fraction decomposition to write $\dfrac{1}{R}$ in terms of partial fractions. What does each term represent?

Solution

From $R = \dfrac{x(x+5)}{3x+10}$, we get the following:

$$\frac{1}{R} = \frac{3x+10}{x(x+5)} \qquad \text{Take the reciprocal of both sides.}$$

$$\frac{3x+10}{x(x+5)} = \frac{A}{x} + \frac{B}{x+5} \qquad \text{Form of partial-fraction decomposition}$$

$$3x+10 = A(x+5) + Bx \qquad \text{Multiply both sides by } x(x+5) \text{ and simplify.}$$

RECALL

The total resistance R in a circuit due to two resistors R_1 and R_2 connected in parallel satisfies the equation

$$\frac{1}{R} = \frac{1}{R_1} + \frac{1}{R_2}.$$

Substitute $x = 0$ in the last equation to obtain $5A = 10$, or $A = 2$. Now substitute $x = -5$ in the same equation to get $-5 = -5B$, or $B = 1$. Thus, we have

$$\frac{1}{R} = \frac{3x+10}{x(x+5)} = \frac{2}{x} + \frac{1}{x+5}$$

$$\frac{1}{R} = \frac{1}{\dfrac{x}{2}} + \frac{1}{x+5} \qquad \text{Rewrite the right side.}$$

The last equation says that if two resistors with resistances $R_1 = \dfrac{x}{2}$ and $R_2 = x + 5$ are connected in parallel, they will produce a total resistance R, where $R = \dfrac{x(x + 5)}{3x + 10}$.

Practice Problem 7 Repeat Example 7 assuming that $R = \dfrac{(x + 1)(x + 2)}{4x + 7}$.

Answers to Practice Problems

1. $\dfrac{3}{x + 1} - \dfrac{1}{x - 2}$ **2.** $\dfrac{5}{x - 1} - \dfrac{3}{x} + \dfrac{1}{x + 1}$

3. $\dfrac{5}{x} + \dfrac{6}{(x - 1)^2} - \dfrac{5}{x - 1}$ **4.** $\dfrac{-1}{x} + \dfrac{4x + 5}{x^2 + 2}$

5. $\dfrac{1}{x^2 + 1} + \dfrac{3x}{(x^2 + 1)^2}$ **6.** $\dfrac{777}{3112}$

7. $\dfrac{1}{R} = \dfrac{3}{x + 1} + \dfrac{1}{x + 2} = \dfrac{1}{\dfrac{x + 1}{3}} + \dfrac{1}{x + 2}$

SECTION 7.6   Exercises

Concepts and Vocabulary

1. In a rational expression, if the degree of the numerator, $P(x)$, is less than the degree of the denominator, $Q(x)$, then the expression is _____.

2. The numerators in the partial-fraction decomposition of a rational expression are constants if the denominator, $Q(x)$, can be factored into _____.

3. In a rational expression, if $(x - 8)$ is a linear factor that is repeated three times in the denominator, then the portion of the expression's partial-fraction decomposition that corresponds to $(x - 8)^3$ has _____ terms.

4. True or False. The form of the partial-fraction decomposition of a rational expression depends only on its denominator.

5. True or False. Both $(x - 1)^2$ and $(x + 2)^2$ are irreducible quadratic factors of the polynomial $Q(x) = (x - 1)^2(x + 2)^2$.

6. True or False. If the denominator of a rational expression is $Q(x) = x^3 - 9x$ then x, $x - 3$, and $x + 3$ become denominators in its partial-fraction decomposition.

7. True or False. If $(x - 2)$ is a factor that is repeated exactly 3 times in $Q(x)$, then $(x - 2)$, $(x - 2)^2$, and $(x - 2)^3$ are denominators in the partial-fraction decomposition of $\dfrac{P(x)}{Q(x)}$.

8. True or False. The quadratic expression $x^2 + 2x + 1$ is irreducible.

Building Skills

In Exercises 9–18, write the form of the partial-fraction decomposition of each rational expression. You do not need to solve for the constants.

9. $\dfrac{1}{(x - 1)(x + 2)}$

10. $\dfrac{x}{(x + 1)(x - 3)}$

11. $\dfrac{1}{x^2 + 7x + 6}$

12. $\dfrac{3}{x^2 - 6x + 8}$

13. $\dfrac{2}{x^3 - x^2}$

14. $\dfrac{x - 1}{(x + 2)^2(x - 3)}$

15. $\dfrac{x^2 - 3x + 3}{(x + 1)(x^2 - x + 1)}$

16. $\dfrac{2x + 3}{(x - 1)^2(x^2 + x + 1)}$

17. $\dfrac{3x - 4}{(x^2 + 1)^2}$

18. $\dfrac{x - 2}{(2x + 3)^2(x^2 + 3)^2}$

In Exercises 19–64, find the partial-fraction decomposition of each rational expression.

19. $\dfrac{2x + 1}{(x + 1)(x + 2)}$

20. $\dfrac{7}{(x - 2)(x + 5)}$

21. $\dfrac{1}{x^2 + 4x + 3}$

22. $\dfrac{x}{x^2 + 5x + 6}$

23. $\dfrac{2}{x^2 + 2x}$

24. $\dfrac{x + 9}{x^2 - 9}$

25. $\dfrac{x}{(x + 1)(x + 2)(x + 3)}$

26. $\dfrac{8x - 14}{(x - 1)(x + 2)(x - 3)}$

27. $\dfrac{x^2}{(x - 1)(x^2 + 5x + 4)}$

28. $\dfrac{2x + 14}{(x + 1)(x^2 - 6x + 5)}$

29. $\dfrac{3x + 1}{(x - 5)(x + 3)^2}$

30. $\dfrac{1 - 2x}{(x + 3)(x - 4)^2}$

31. $\dfrac{8x + 1}{(x + 2)^2(x - 3)}$

32. $\dfrac{3x + 2}{x^2(x + 1)}$

33. $\dfrac{x - 1}{(x + 1)^2}$

34. $\dfrac{x}{(x - 2)^2}$

35. $\dfrac{x - 1}{(2x - 3)^2}$

36. $\dfrac{6x + 1}{(3x + 1)^2}$

37. $\dfrac{2x^2 + x}{(x + 1)^3}$

38. $\dfrac{x^2 + 2}{(x - 1)^3}$

39. $\dfrac{-x^2 + 3x + 1}{x^3 + 2x^2 + x}$

40. $\dfrac{5x^2 - 8x + 2}{x^3 - 2x^2 + x}$

41. $\dfrac{3x + 1}{(x + 1)(x^2 - 1)}$

42. $\dfrac{6x - 4}{(x - 2)(x^2 - 4)}$

43. $\dfrac{1}{x^2(x + 1)^2}$

44. $\dfrac{2}{(x - 1)^2(x + 3)^2}$

45. $\dfrac{1}{(x^2 - 1)^2}$

46. $\dfrac{3}{(x^2 - 4)^2}$

47. $\dfrac{x^2 - 5}{(x - 2)(x^2 - 2x + 3)}$

48. $\dfrac{x^2 + 2x + 1}{(x + 2)(x^2 + x + 2)}$

49. $\dfrac{x - 1}{(x + 1)^2(x^2 + 2x + 2)}$

50. $\dfrac{x^2 - 3}{(x - 2)^2(x^2 - x + 1)}$

51. $\dfrac{x^2 + 1}{(x - 2)(x^2 - 3x + 2)}$

52. $\dfrac{x + 1}{x(2x^2 - 5x - 3)}$

53. $\dfrac{6x + 7}{4x^2 + 12x + 9}$

54. $\dfrac{2x + 3}{9x^2 + 30x + 25}$

55. $\dfrac{x - 3}{x^3 + x^2}$

56. $\dfrac{x^2 + 1}{x^4 - x^3}$

57. $\dfrac{x^2 + 2x + 4}{x^3 + x^2}$

58. $\dfrac{x^2 + 2x - 1}{(x - 1)(x^2 + 1)}$

59. $\dfrac{x}{x^4 - 1}$

60. $\dfrac{3x^3 - 5x^2 + 12x + 4}{x^4 - 16}$

61. $\dfrac{1}{x(x^2 + 1)^2}$

62. $\dfrac{x^2}{(x^2 + 2)^2}$

63. $\dfrac{2x^2 + 3x}{(x^2 + 1)(x^2 + 2)}$

64. $\dfrac{x^2 - 2x}{(x^2 + 9)(x^2 + x + 7)}$

Applying the Concepts

In Exercises 65–68, express each sum as a fraction of whole numbers in lowest terms.

65. $\dfrac{1}{1 \cdot 2} + \dfrac{1}{2 \cdot 3} + \dfrac{1}{3 \cdot 4} + \cdots + \dfrac{1}{n(n + 1)}$

66. $\dfrac{2}{1 \cdot 3} + \dfrac{2}{2 \cdot 4} + \dfrac{2}{3 \cdot 5} + \cdots + \dfrac{2}{100 \cdot 102}$

67. $\dfrac{2}{1 \cdot 3} + \dfrac{2}{3 \cdot 5} + \dfrac{2}{5 \cdot 7} + \cdots + \dfrac{2}{(2n - 1)(2n + 1)}$

68. $\dfrac{2}{1 \cdot 2 \cdot 3} + \dfrac{2}{2 \cdot 3 \cdot 4} + \dfrac{2}{3 \cdot 4 \cdot 5} + \cdots + \dfrac{2}{100 \cdot 101 \cdot 102}$

$\Bigg[$ *Hint:* Write

$\dfrac{2}{k(k + 1)(k + 2)} = \dfrac{1}{k} - \dfrac{2}{k + 1} + \dfrac{1}{k + 2}$

$= \dfrac{1}{k} - \dfrac{1}{k + 1} - \dfrac{1}{k + 1} + \dfrac{1}{k + 2}\Bigg]$

In Exercises 69–72, the total resistance R in a circuit is given. Use partial-fraction decomposition to write $\dfrac{1}{R}$ in terms of partial fractions. Interpret each term in the partial fraction.

69. Circuit resistance. $R = \dfrac{(x + 1)(x + 3)}{2x + 4}$

70. Circuit resistance. $R = \dfrac{(x + 2)(x + 4)}{7x + 20}$

71. Circuit resistance. $R = \dfrac{R_1R_2R_3}{R_1R_2 + R_2R_3 + R_3R_1}$

72. Circuit resistance. $R = \dfrac{x(x + 2)(x + 4)}{3x^2 + 12x + 8}$

Beyond the Basics

73. Solve the system of equations in Example 2 to verify the values of the constants A, B, and C.

In Exercises 74–79, find the partial-fraction decomposition of each rational expression.

74. $\dfrac{1}{x^3 + 1}$

75. $\dfrac{4x}{(x^2 - 1)^2}$

76. $\dfrac{2x + 3}{(x^2 + 1)(x + 1)}$

77. $\dfrac{x + 1}{(x^2 + 1)(x - 1)^2}$

78. $\dfrac{x}{(x^2 + 1)^2(x - 1)}$

79. $\dfrac{x^3}{(x + 1)^2(x + 2)^2}$

80. Find the partial-fraction decomposition of:

$$\dfrac{4x}{x^4 + 4}$$

$[$ *Hint:* $x^4 + 4 = x^4 + 4 + 4x^2 - 4x^2 = (x^2 + 2)^2 - (2x)^2 = (x^2 - 2x + 2)(x^2 + 2x + 2).]$

81. Find the partial-fraction decomposition of:

$$\dfrac{x^2 - 8x + 18}{(x - 5)^3}$$

In Exercises 82 and 83, find the partial-fraction decomposition by first using long division.

82. $\dfrac{x^2 + 4x + 5}{x^2 + 3x + 2}$

83. $\dfrac{(x - 1)(x + 2)}{(x + 3)(x - 4)}$

Critical Thinking / Discussion / Writing

84. In the partial-fraction decomposition of an expression, we obtain a system of equations by equating coefficients of the powers of x. Justify the validity of this step.

85. Let $f(x) = \dfrac{x + 8}{x^2 + x - 2} = g(x) + h(x)$, where $g(x)$ and $h(x)$ are the terms in the partial-fraction decomposition of $f(x)$.

a. Sketch the graphs of f, g, and h.
b. Compare the graphs of f and g. What do you observe?
c. Compare the graphs of f and h. What do you observe?

Getting Ready for the Next Section

In Exercises 86–89, solve each equation.

86. $3(x - 1) = 5 - x$

87. $5(x - 2) = 3(x - 3) + 13$

88. $-3(x + 4) + 2 = 8 - x$

89. $6x - 3(5x + 2) = 4 - 5x$

In Exercises 90–93, solve each system.

90. $\begin{cases} 2x - y = 5 \\ x + 2y = 25 \end{cases}$

91. $\begin{cases} x - 3y = 1 \\ 2x + y = -5 \end{cases}$

92. $\begin{cases} x + 3y = 6 \\ 2x + 6y = 8 \end{cases}$

93. $\begin{cases} 2x - y = 3 \\ 6x - 3y = 9 \end{cases}$

In Exercises 94–97, identify each statement as True or False.

94. If $x(3t^2 + \pi\sqrt{t^4 + t^2 + 5}) = 0$ where t is a real number, then $x = 0$.

95. If x, y, and z are any real numbers, then $xy + z = x + yz$.

96. If x, y, and z are real numbers, then $xy(z + 5) = xyz + 5xy$.

97. If $(x^4 + x^2)(y - 3) = 1$ and $y = \dfrac{10}{3}$, then $x^4 + x^2 = 3$.

Matrix Algebra

BEFORE STARTING THIS SECTION, REVIEW

1 Solving systems of equations (Section 7.1, page 594)

OBJECTIVES

1 Define equality of two matrices.

2 Define matrix addition and scalar multiplication.

3 Define matrix multiplication.

4 Apply matrix multiplication to computer graphics.

◆ Computer Graphics

The following fields illustrate the widespread and growing applications of computer graphics:

1. The *automobile industry*. Both computer-aided design (CAD) and computer-aided manufacturing (CAM) have revolutionized the automotive industry. Many months before a new car is built, engineers design and construct a *mathematical car*—a wire-frame model that exists only in computer memory and on graphics display terminals. The mathematical model organizes and influences each step of the design and manufacture of the car.

2. The *entertainment industry*. Computer graphics are used by the entertainment industry with spectacular success. They were used in *Avatar* and other action movies like the latest James Bond thriller for special effects, and they are used in the design and creation of video games.

3. *Military*. From the beginning of the widespread use of computers (in the 1960s), computer graphics have been used by the military to prepare for or defend against war. Modern weaponry depends heavily on computer-generated images.

Engineers use matrix algebra techniques on graphic images to change their orientation of scale, to zoom in on them, or to switch between two- and three-dimensional views. In Example 10 and in the exercises, we examine how basic matrix algebra can be used to manipulate graphical images.

1 Define equality of two matrices.

Equality of Matrices

In Section 7.4, we defined a matrix as a rectangular array of numbers written within brackets. In this and the next sections, you will learn about the operations with and uses of matrices. A matrix may be represented in any of the following three ways:

1. A matrix may be denoted by capital letters, such as A, B, and C.

2. A matrix may be denoted by its representative (i, j)th entry: $[a_{ij}]$, $[b_{ij}]$, $[c_{ij}]$, and so on.

3. A matrix may be written as a rectangular array of numbers:

$$A = [a_{ij}] = \begin{bmatrix} a_{11} & a_{12} & \cdots & a_{1n} \\ a_{21} & a_{22} & \cdots & a_{2n} \\ \vdots & \vdots & \cdots & \vdots \\ a_{m1} & a_{m2} & \cdots & a_{mn} \end{bmatrix}$$

James Joseph Sylvester

(1814–1897)
In 1850, Sylvester coined the term **matrix** to denote a rectangular array of numbers. Sylvester, who was born into a Jewish family in London and studied for several years at Cambridge, was not permitted to take his degree there for religious reasons. Therefore, he received his degree from Trinity College, Dublin. In 1871, he accepted the chair of mathematics at the newly opened Johns Hopkins University in Baltimore. While at Hopkins, Sylvester founded the *American Journal of Mathematics* and helped develop a tradition of graduate education in mathematics in the United States.

We now examine some *algebraic* properties of matrices.

Equality of Matrices

Two matrices $A = [a_{ij}]$ and $B = [b_{ij}]$ are said to be **equal**, written $A = B$, if

1. A and B have the same order $m \times n$ (that is, A and B have the same number m of rows and same number n of columns).

2. $a_{ij} = b_{ij}$ for all i and j; that is, each (i, j)th entry of A is equal to the corresponding (i, j)th entry of B.

EXAMPLE 1 **Determining Equality of Matrices**

Find x and y such that

$$\begin{bmatrix} x + y & 2 \\ 3 & x - y \end{bmatrix} = \begin{bmatrix} 4 & 2 \\ 3 & 1 \end{bmatrix}$$

Solution

Because equal matrices must have identical entries in corresponding positions, we have the following system of equations:

$$\begin{cases} x + y = 4 & (1) \quad \text{Equate entries in the } (1, 1) \text{ position.} \\ x - y = 1 & (2) \quad \text{Equate entries in the } (2, 2) \text{ position.} \end{cases}$$

Solve this system: Add the two equations to obtain $x = \dfrac{5}{2}$. Substitute $x = \dfrac{5}{2}$ in equation $x - y = 1$ and solve for y to obtain $y = \dfrac{3}{2}$. So, $x = \dfrac{5}{2}$ and $y = \dfrac{3}{2}$.

Practice Problem 1 Find x and y such that

$$\begin{bmatrix} 1 & 2x - y \\ x + y & 5 \end{bmatrix} = \begin{bmatrix} 1 & 1 \\ 2 & 5 \end{bmatrix}$$

2 Define matrix addition and scalar multiplication.

Matrix Addition and Scalar Multiplication

Matrix addition differs somewhat from real number addition. You can add any two real numbers, but you can add two matrices only if they have the same order.

Matrix Addition

If $A = [a_{ij}]$ and $B = [b_{ij}]$ are two $m \times n$ matrices, their **sum** $A + B$ is the $m \times n$ matrix defined by

$$A + B = [a_{ij} + b_{ij}],$$

for all i and j.

The definition of matrix addition says that

(i) we can add two matrices of the same order simply by adding their corresponding entries, but

(ii) the sum of two matrices of different orders is not defined.

TECHNOLOGY CONNECTION

To find the sum of two matrices with a graphing calculator, name the matrices and then use the regular addition key with the names of the matrices.

Consider the matrices A and B in Example 2 as follows:

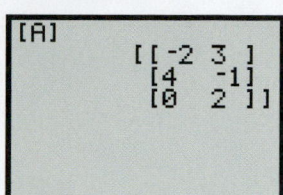

The sum is

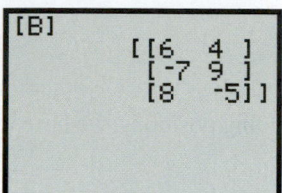

EXAMPLE 2 **Adding Matrices**

Let $A = \begin{bmatrix} -2 & 3 \\ 4 & -1 \\ 0 & 2 \end{bmatrix}$, $B = \begin{bmatrix} 6 & 4 \\ -7 & 9 \\ 8 & -5 \end{bmatrix}$, and $C = \begin{bmatrix} 1 & 2 \\ 3 & 4 \end{bmatrix}$. Find each sum if possible.

a. $A + B$ **b.** $A + C$

Solution

a. Since A and B have the same order, $A + B$ is defined. We have

$$A + B = \begin{bmatrix} -2 & 3 \\ 4 & -1 \\ 0 & 2 \end{bmatrix} + \begin{bmatrix} 6 & 4 \\ -7 & 9 \\ 8 & -5 \end{bmatrix} = \begin{bmatrix} -2+6 & 3+4 \\ 4-7 & -1+9 \\ 0+8 & 2-5 \end{bmatrix} = \begin{bmatrix} 4 & 7 \\ -3 & 8 \\ 8 & -3 \end{bmatrix}$$

b. $A + C$ is not defined because A and C do not have the same order.

Practice Problem 2 Let $A = \begin{bmatrix} 2 & -1 & 4 \\ 5 & 0 & 9 \end{bmatrix}$ and $B = \begin{bmatrix} -8 & 2 & 9 \\ 7 & 3 & 6 \end{bmatrix}$. Find $A + B$.

You can also multiply any matrix A by any real number c. The real number c is called a **scalar**, and the multiplication process is called *scalar multiplication*.

Scalar Multiplication

> Let $A = [a_{ij}]$ be an $m \times n$ matrix and let c be a real number. Then the **scalar product** of A and c is denoted by cA and is defined by
>
> $$cA = [ca_{ij}].$$

In other words, to multiply a matrix A by a real number c, you multiply each entry of A by c. When $c = -1$, we frequently write $-A$ instead of $(-1)A$. The matrix $-A$ is called the **negative** of A. Further, the *difference* of two matrices, $A - B$, is defined to be the sum $A + (-B)$.

Matrix Subtraction

> If A and B are two $m \times n$ matrices, then their **difference** is defined by
>
> $$A - B = A + (-B).$$
>
> Subtraction $A - B$ is performed by subtracting the corresponding entries of B from those of A.

EXAMPLE 3 **Performing Scalar Multiplication and Matrix Subtraction**

Let $A = \begin{bmatrix} 1 & 2 & 0 \\ -1 & 3 & 1 \\ 2 & -1 & 4 \end{bmatrix}$ and $B = \begin{bmatrix} 2 & 1 & 3 \\ 1 & 0 & -2 \\ -3 & 4 & 5 \end{bmatrix}$. Find the following.

a. $3A$ **b.** $2B$ **c.** $3A - 2B$

To find the difference of two matrices with a graphing calculator, name the matrices and then use the regular subtraction key with the names of the matrices. Similarly, to multiply a matrix by a scalar, use the name of the matrix and the regular multiplication key.

Consider the matrices A and B in Example 3 as follows:

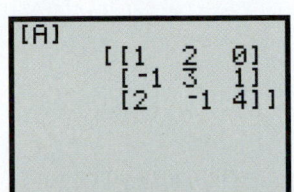

We find $3A - 2B$:

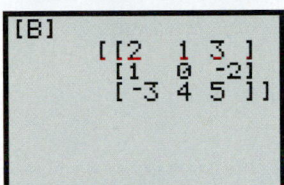

Solution

a. $3A = 3\begin{bmatrix} 1 & 2 & 0 \\ -1 & 3 & 1 \\ 2 & -1 & 4 \end{bmatrix} = \begin{bmatrix} 3(1) & 3(2) & 3(0) \\ 3(-1) & 3(3) & 3(1) \\ 3(2) & 3(-1) & 3(4) \end{bmatrix} = \begin{bmatrix} 3 & 6 & 0 \\ -3 & 9 & 3 \\ 6 & -3 & 12 \end{bmatrix}$

b. $2B = 2\begin{bmatrix} 2 & 1 & 3 \\ 1 & 0 & -2 \\ -3 & 4 & 5 \end{bmatrix} = \begin{bmatrix} 2(2) & 2(1) & 2(3) \\ 2(1) & 2(0) & 2(-2) \\ 2(-3) & 2(4) & 2(5) \end{bmatrix} = \begin{bmatrix} 4 & 2 & 6 \\ 2 & 0 & -4 \\ -6 & 8 & 10 \end{bmatrix}$

c. $3A - 2B = 3A + (-1)2B$

$= \begin{bmatrix} 3 & 6 & 0 \\ -3 & 9 & 3 \\ 6 & -3 & 12 \end{bmatrix} + \begin{bmatrix} -4 & -2 & -6 \\ -2 & 0 & 4 \\ 6 & -8 & -10 \end{bmatrix}$ Substitute from parts **a** and **b**, changing the sign of each entry in **b**.

$= \begin{bmatrix} 3 + (-4) & 6 + (-2) & 0 + (-6) \\ -3 + (-2) & 9 + 0 & 3 + 4 \\ 6 + 6 & -3 + (-8) & 12 + (-10) \end{bmatrix} = \begin{bmatrix} -1 & 4 & -6 \\ -5 & 9 & 7 \\ 12 & -11 & 2 \end{bmatrix}$

Practice Problem 3 Let $A = \begin{bmatrix} 7 & -4 \\ 3 & 6 \\ 0 & -2 \end{bmatrix}$ and $B = \begin{bmatrix} 1 & -3 \\ 2 & 2 \\ 5 & 8 \end{bmatrix}$. Find $2A - 3B$.

An $m \times n$ matrix whose entries are all equal to 0 is called the $m \times n$ **zero matrix** and is denoted by $\mathbf{0}_{mn}$. When m and n are understood from the context, we simply write $\mathbf{0}$. Since the entries of a matrix are real numbers, many of the algebraic properties of matrix addition and scalar multiplication are similar to the familiar properties of real numbers.

MATRIX ADDITION AND SCALAR MULTIPLICATION PROPERTIES

Let A, B, and C be $m \times n$ matrices and c and d be scalars.

1. $A + B = B + A$ Commutative property of addition

2. $A + (B + C) = (A + B) + C$ Associative property of addition

3. $A + \mathbf{0} = \mathbf{0} + A = A$ Additive identity property

4. $A + (-A) = (-A) + A = \mathbf{0}$ Additive inverse property

5. $(cd)A = c(dA)$ Associative property of scalar multiplication

6. $1A = A$ Scalar identity property

7. $c(A + B) = cA + cB$ Distributive property

8. $(c + d)A = cA + dA$ Distributive property

In the next example, we use these properties to solve an equation involving matrices. Such an equation is called a **matrix equation**.

EXAMPLE 4   Solving a Matrix Equation

Solve the matrix equation $3A + 2X = 4B$ for X, where

$$A = \begin{bmatrix} 2 & 0 \\ 4 & 6 \end{bmatrix} \text{ and } B = \begin{bmatrix} 1 & 3 \\ 5 & 2 \end{bmatrix}$$

Solution

$$3A + 2X = 4B \qquad \text{The given equation}$$

$$2X = 4B - 3A \qquad \text{Subtract the matrix } 3A \text{ from both sides.}$$

$$X = \frac{1}{2}(4B - 3A) \qquad \text{Multiply both sides by } \frac{1}{2}.$$

$$= \frac{1}{2}\left(4\begin{bmatrix} 1 & 3 \\ 5 & 2 \end{bmatrix} - 3\begin{bmatrix} 2 & 0 \\ 4 & 6 \end{bmatrix}\right) \qquad \text{Substitute for } A \text{ and } B.$$

$$= \frac{1}{2}\left(\begin{bmatrix} 4 & 12 \\ 20 & 8 \end{bmatrix} - \begin{bmatrix} 6 & 0 \\ 12 & 18 \end{bmatrix}\right) \qquad \text{Scalar multiplication}$$

$$= \frac{1}{2}\begin{bmatrix} -2 & 12 \\ 8 & -10 \end{bmatrix} \qquad \text{Subtract.}$$

$$= \begin{bmatrix} -1 & 6 \\ 4 & -5 \end{bmatrix} \qquad \text{Scalar multiplication}$$

You should check that the matrix $X = \begin{bmatrix} -1 & 6 \\ 4 & -5 \end{bmatrix}$ satisfies the given matrix equation.

Practice Problem 4 Solve the matrix equation $5A + 3X = 2B$ for X, where

$$A = \begin{bmatrix} 1 & -1 \\ 3 & 5 \end{bmatrix} \text{ and } B = \begin{bmatrix} 2 & 7 \\ -3 & -5 \end{bmatrix}$$

3 Define matrix multiplication.

Matrix Multiplication

You know that the sum of two matrices is defined only if both matrices are of the same order. In the case of multiplication, however, the general rule is that if A is an $m \times p$ matrix and B is a $p \times n$ matrix, then the *product* AB is defined; otherwise, it is not defined.

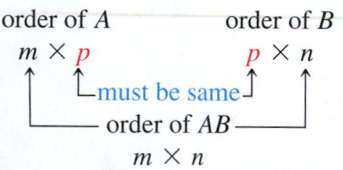

order of A order of B

$m \times p$ $p \times n$

must be same

order of AB

$m \times n$

> ### RULE FOR DEFINING THE PRODUCT *AB*
>
> In order to define the product AB of two matrices A and B, the number of columns of A must be equal to the number of rows of B. If A is an $m \times p$ matrix and B is a $p \times n$ matrix, then the product AB is an $m \times n$ matrix.

EXAMPLE 5 **Determining the Order of the Product**

For each pair of matrices A and B, state whether the product matrix AB is defined. If it is, find the order of AB.

a. $A = \begin{bmatrix} 1 & 3 & 5 \end{bmatrix}, B = \begin{bmatrix} 2 \\ 3 \\ 7 \end{bmatrix}$ **b.** $A = \begin{bmatrix} 2 & 3 \\ -1 & 4 \\ 5 & 6 \end{bmatrix}, B = \begin{bmatrix} 2 & 1 & 3 & 5 \\ 4 & -1 & -2 & 6 \end{bmatrix}$

c. $A = \begin{bmatrix} 1 \\ -1 \\ 2 \end{bmatrix}, B = \begin{bmatrix} 2 & 1 & 4 \\ 0 & 2 & 2 \end{bmatrix}$

Solution

a. Yes, the product AB is defined. A is a 1×3 matrix, and B is a 3×1 matrix; so the number of columns of A equals the number of rows of B. The product AB is a 1×1 matrix, which has a single entry.

b. Yes, AB is defined. A is a 3×2 matrix, and B is a 2×4 matrix; so the number of columns of A equals the number of rows of B. The product AB is a 3×4 matrix.

c. No, AB is not defined. A is a 3×1 matrix, and B is a 2×3 matrix; so the number of columns of A does not equal the number of rows of B.

Practice Problem 5 State whether the product matrix AB is defined. If AB is defined, find the order of the product matrix AB.

$$A = \begin{bmatrix} 1 & 4 & 7 \\ 0 & 2 & -2 \end{bmatrix}, \quad B = \begin{bmatrix} 4 \\ 3 \\ -1 \end{bmatrix}$$

Before you learn how to multiply two matrices A and B, consider the next example.

EXAMPLE 6   **Calculating the Total Revenue from Car Sales**

A car dealership sold 10 sports cars (S), 30 compacts (C), and 45 mini-compacts (M). The average price per car of type S, C, and M was 42, 24, and 17 (in thousands of dollars), respectively. Find the total revenue received by the dealership from the sale of these cars.

Solution

You can represent the number of cars of each type sold by a 1×3 matrix:

$$N = \begin{bmatrix} S & C & M \end{bmatrix} = \begin{bmatrix} 10 & 30 & 45 \end{bmatrix}$$

Using the price of each type of car, you can construct a 3×1 price matrix:

$$P = \begin{bmatrix} 42 \\ 24 \\ 17 \end{bmatrix}$$

Then the total revenue R received by the dealership from the sale of these cars last month is given by $R = 10(42) + 30(24) + 45(17) = 1905$ (or \$1,905,000).

We define the right side of this equation to be the product NP of the two matrices N and P. In matrix notation,

$$NP = \begin{bmatrix} 10 & 30 & 45 \end{bmatrix} \begin{bmatrix} 42 \\ 24 \\ 17 \end{bmatrix} = 10(42) + 30(24) + 45(17) = 1905.$$

Practice Problem 6 In Example 6, suppose the average price per car of types S, C, and M is 41, 26, and 19 (in thousands of dollars), respectively. Find the total revenue received by the dealership from the sale of these cars.

Example 6 leads to the following definition.

PRODUCT OF $1 \times n$ AND $n \times 1$ MATRICES

Suppose A is a $1 \times n$ matrix and B is an $n \times 1$ matrix:

$$A = \begin{bmatrix} a_1 & a_2 & a_3 \cdots a_n \end{bmatrix} \text{ and } B = \begin{bmatrix} b_1 \\ b_2 \\ \vdots \\ b_n \end{bmatrix}$$

We define the product AB by

$$AB = a_1 b_1 + a_2 b_2 + \cdots + a_n b_n.$$

We make the following observations about this definition.

1. Since A is a $1 \times n$ matrix and B is an $n \times 1$ matrix, the product AB is defined and is a 1×1 matrix.

2. To find the product AB, multiply the leftmost entry of A and the top entry of B; then move to the right in A and down in B while multiplying corresponding (first, second, third, and so on) entries together; finally, add all of the resulting products.

EXAMPLE 7 Finding the Product AB

Find AB, where $A = \begin{bmatrix} 2 & 0 & 1 & -2 \end{bmatrix}$ and $B = \begin{bmatrix} 3 \\ 1 \\ -1 \\ 4 \end{bmatrix}$.

Solution

Since A has order 1×4 and B has order 4×1, AB is defined. The product rule gives

$$AB = 2(3) + 0(1) + 1(-1) + (-2)(4) = -3.$$

The product of the 1×4 matrix A and the 4×1 matrix B is therefore the 1×1 matrix $\begin{bmatrix} -3 \end{bmatrix}$.

Practice Problem 7 Find AB:

$$A = \begin{bmatrix} 3 & -1 & 2 & 7 \end{bmatrix} \quad \text{and} \quad B = \begin{bmatrix} -2 \\ 0 \\ 1 \\ 5 \end{bmatrix}$$

In Example 7, you may think of the answer -3 as the 1×1 matrix $\begin{bmatrix} -3 \end{bmatrix}$ or simply as the number -3. We shall have occasion to make use of both points of view.

The rule for multiplying any two matrices can be reduced to the special case of multiplying a $1 \times p$ matrix by a $p \times 1$ matrix repeatedly.

Matrix Multiplication

Let $A = \begin{bmatrix} a_{ij} \end{bmatrix}$ be an $m \times p$ matrix and $B = \begin{bmatrix} b_{ij} \end{bmatrix}$ be a $p \times n$ matrix. Then the **product** AB is the $m \times n$ matrix $C = \begin{bmatrix} c_{ij} \end{bmatrix}$, where the entry c_{ij} of C is obtained by matrix multiplying the ith row of A by the jth column of B.

The definition of the product AB says that

$$c_{ij} = a_{i1}b_{1j} + a_{i2}b_{2j} + \cdots + a_{ip}b_{pj}.$$

Written in full, the matrix product $AB = C$ in the definition is as follows:

$$\begin{bmatrix} a_{11} & a_{12} & \cdots & a_{1p} \\ a_{21} & a_{22} & \cdots & a_{2p} \\ \vdots & \vdots & & \vdots \\ a_{i1} & a_{i2} & \cdots & a_{ip} \\ \vdots & \vdots & & \vdots \\ a_{m1} & a_{m2} & \cdots & a_{mp} \end{bmatrix} \begin{bmatrix} b_{11} & b_{12} & \cdots & b_{1j} & \cdots & b_{1n} \\ b_{21} & b_{22} & \cdots & b_{2j} & \cdots & b_{2n} \\ \vdots & \vdots & \cdots & \vdots & \cdots & \vdots \\ b_{p1} & b_{p2} & \cdots & b_{pj} & \cdots & b_{pn} \end{bmatrix} = \begin{bmatrix} c_{11} & c_{12} & \cdots & c_{1j} & \cdots & c_{1n} \\ \vdots & \vdots & \cdots & \vdots & & \vdots \\ c_{i1} & c_{i2} & \cdots & c_{ij} & \cdots & c_{in} \\ \vdots & \vdots & & \vdots & \cdots & \vdots \\ c_{m1} & c_{m2} & \cdots & c_{mj} & \cdots & c_{mn} \end{bmatrix}$$

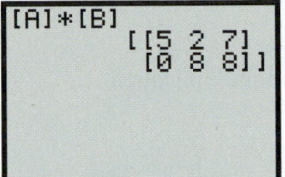
EXAMPLE 8   **Finding the Product of Two Matrices**

Find the products AB and BA, if possible:

$$A = \begin{bmatrix} 1 & 2 \\ -1 & 3 \end{bmatrix} \quad \text{and} \quad B = \begin{bmatrix} 3 & -2 & 1 \\ 1 & 2 & 3 \end{bmatrix}$$

Solution

Since A is of order 2×2 and the order of B is 2×3, the product AB is defined and has order 2×3.

Let $AB = C = [c_{ij}]$. By the definition of the product AB, each entry c_{ij} of C is found by multiplying the ith row of A by the jth column of B. For example, c_{11} is found by multiplying the first row of A by the first column of B:

$$\begin{bmatrix} 1 & 2 \\ * & * \end{bmatrix} \begin{bmatrix} 3 & * & * \\ 1 & * & * \end{bmatrix} = \begin{bmatrix} 1(3) + 2(1) & * & * \\ * & * & * \end{bmatrix}$$

So

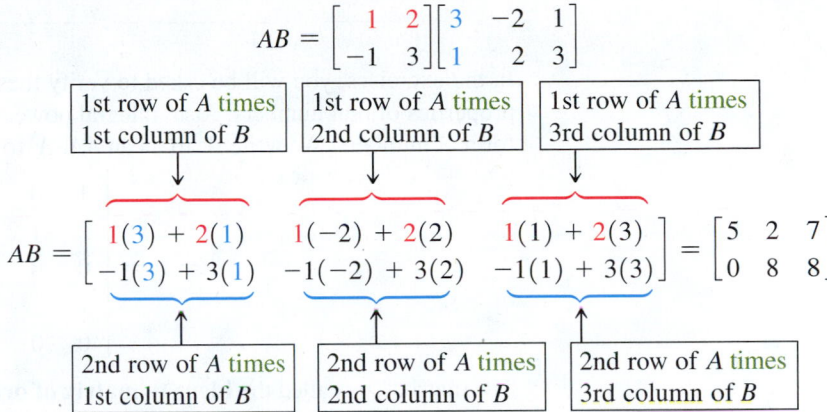

The product BA is not defined because B is of order 2×3 and A is of order 2×2; that is, the number of columns of B is not equal to the number of rows of A.

Practice Problem 8   Find the products AB and BA if possible:

$$A = \begin{bmatrix} 5 & 0 \\ 2 & -1 \end{bmatrix} \quad \text{and} \quad B = \begin{bmatrix} 8 & 1 \\ -2 & 6 \\ 0 & 4 \end{bmatrix}$$

In Example 8, it is obvious that $AB \neq BA$ because BA is undefined. Even if both AB and BA are defined, AB may not equal BA, as the next example illustrates.

EXAMPLE 9   **Finding the Product of Two Matrices**

Find the products AB and BA:

$$A = \begin{bmatrix} 2 & 3 \\ -1 & 0 \end{bmatrix} \quad \text{and} \quad B = \begin{bmatrix} 3 & 4 \\ 5 & 1 \end{bmatrix}$$

Solution

$$AB = \begin{bmatrix} 2 & 3 \\ -1 & 0 \end{bmatrix} \begin{bmatrix} 3 & 4 \\ 5 & 1 \end{bmatrix} = \begin{bmatrix} 2(3) + 3(5) & 2(4) + 3(1) \\ -1(3) + 0(5) & -1(4) + 0(1) \end{bmatrix} = \begin{bmatrix} 21 & 11 \\ -3 & -4 \end{bmatrix}$$

$$BA = \begin{bmatrix} 3 & 4 \\ 5 & 1 \end{bmatrix} \begin{bmatrix} 2 & 3 \\ -1 & 0 \end{bmatrix} = \begin{bmatrix} 3(2) + 4(-1) & 3(3) + 4(0) \\ 5(2) + 1(-1) & 5(3) + 1(0) \end{bmatrix} = \begin{bmatrix} 2 & 9 \\ 9 & 15 \end{bmatrix}$$

Practice Problem 9 Find the products AB and BA:

$$A = \begin{bmatrix} 7 & 1 \\ 0 & 3 \end{bmatrix} \quad \text{and} \quad B = \begin{bmatrix} 2 & -1 \\ 4 & 4 \end{bmatrix}$$

In Example 9, observe that $AB \neq BA$. Thus, in general, *matrix multiplication is not commutative*. Matrix multiplication, however, does share many of the properties of the multiplication of real numbers.

Properties of Matrix Multiplication

Let A, B, and C be matrices and let c be a scalar. Assume that each product and sum is defined. Then

1. $(AB)C = A(BC)$ Associative property of multiplication
2. (i) $A(B + C) = AB + AC$ Distributive property
   (ii) $(A + B)C = AC + BC$ Distributive property
3. $c(AB) = (cA)B = A(cB)$ Associative property of scalar multiplication

In the exercises, you will be asked to verify these properties and compare them to similar properties of real numbers. Also, integral powers of square matrices are defined exactly as for real numbers. We write A^2 to mean AA, A^3 to mean AAA, and so on. We define A^0 by

$$A^0 = I_n = \begin{bmatrix} 1 & 0 & 0 & \cdots & 0 \\ 0 & 1 & 0 & \cdots & 0 \\ 0 & 0 & 1 & \cdots & 0 \\ \vdots & & \vdots & & \vdots \\ 0 & 0 & 0 & \cdots & 1 \end{bmatrix}.$$

The matrix I_n is called the **identity matrix of order n**.

When the order of the matrix is clear from the context, we can drop the subscript n and write I for I_n. For any square matrix A, we have

$$\boxed{IA = A \text{ and } AI = A}$$

So, in matrix multiplication, the matrix I plays a role similar to that of the number 1 in the multiplication of real numbers.

4 Apply matrix multiplication to computer graphics.

◆ Computer Graphics

Letters used for labels on a computer screen are very simple two-dimensional graphics. The next example illustrates how a letter (or a figure) is transformed when the coordinates of the points on the figure are all multiplied by the same matrix.

EXAMPLE 10 **Transforming a Letter**

The capital letter L in Figure 7.23 is determined by six points (or *vertices*) P_1 through P_6. The coordinates of the six points can be stored in a *data matrix D*. We draw line segments connecting these vertices in order.

$$\begin{array}{c} \\ x\text{-coordinate} \\ y\text{-coordinate} \end{array} \begin{array}{cccccc} P_1 & P_2 & P_3 & P_4 & P_5 & P_6 \end{array}$$
$$\begin{array}{c} x\text{-coordinate} \\ y\text{-coordinate} \end{array} \begin{bmatrix} 0 & 4 & 4 & 1 & 1 & 0 \\ 0 & 0 & 1 & 1 & 6 & 6 \end{bmatrix} = D$$

If $A = \begin{bmatrix} 1 & 0.25 \\ 0 & 1 \end{bmatrix}$, compute AD and graph the figure it represents.

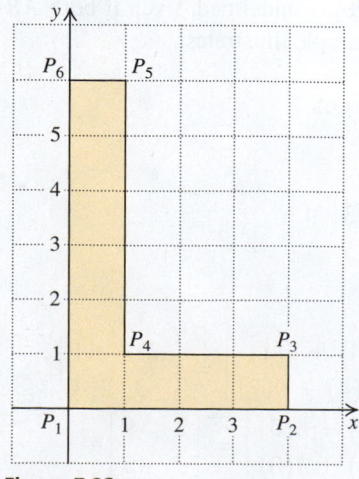

Figure 7.23

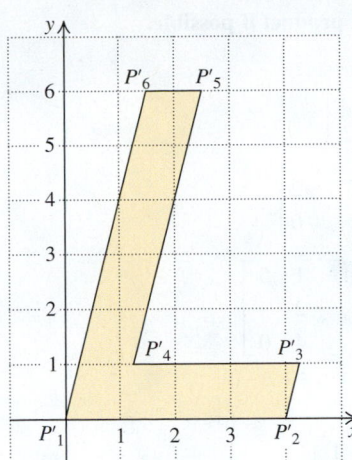

Figure 7.24

Solution

The columns of the product matrix AD represent the transformed vertices of the letter L.

$$AD = \begin{bmatrix} 1 & 0.25 \\ 0 & 1 \end{bmatrix}\begin{bmatrix} 0 & 4 & 4 & 1 & 1 & 0 \\ 0 & 0 & 1 & 1 & 6 & 6 \end{bmatrix}$$

$$= \begin{bmatrix} 1\cdot 0 + 0.25\cdot 0 & 1\cdot 4 + 0.25\cdot 0 & 1\cdot 4 + 0.25\cdot 1 & 1\cdot 1 + 0.25\cdot 1 & 1\cdot 1 + 0.25\cdot 6 & 1\cdot 0 + 0.25\cdot 6 \\ 0\cdot 0 + 1\cdot 0 & 0\cdot 4 + 1\cdot 0 & 0\cdot 4 + 1\cdot 1 & 0\cdot 1 + 1\cdot 1 & 0\cdot 1 + 1\cdot 6 & 0\cdot 0 + 1\cdot 6 \end{bmatrix}$$

$$= \begin{matrix} P'_1 & P'_2 & P'_3 & P'_4 & P'_5 & P'_6 \\ \begin{bmatrix} 0 & 4 & 4.25 & 1.25 & 2.5 & 1.5 \\ 0 & 0 & 1 & 1 & 6 & 6 \end{bmatrix} \end{matrix}$$

Figure 7.24 shows the transformed vertices and the transformed figure "L" formed by these vertices.

Practice Problem 10 In Example 10, use the matrix $A = \begin{bmatrix} 0 & 1 \\ 1 & 0.25 \end{bmatrix}$. Graph the figure represented by the matrix AD.

Answers to Practice Problems

1. $x = 1, y = 1$ **2.** $\begin{bmatrix} -6 & 1 & 13 \\ 12 & 3 & 15 \end{bmatrix}$

3. $\begin{bmatrix} 11 & 1 \\ 0 & 6 \\ -15 & -28 \end{bmatrix}$ **4.** $X = \begin{bmatrix} -\dfrac{1}{3} & \dfrac{19}{3} \\ -7 & -\dfrac{35}{3} \end{bmatrix}$

5. Yes, the product AB is defined with order 2×1.

6. \$2045 thousand   **7.** $[31]$ **8.** The product AB is not defined.

9. $AB = \begin{bmatrix} 18 & -3 \\ 12 & 12 \end{bmatrix}$, $BA = \begin{bmatrix} 14 & -1 \\ 28 & 16 \end{bmatrix}$ $BA = \begin{bmatrix} 42 & -1 \\ 2 & -6 \\ 8 & -4 \end{bmatrix}$

10.

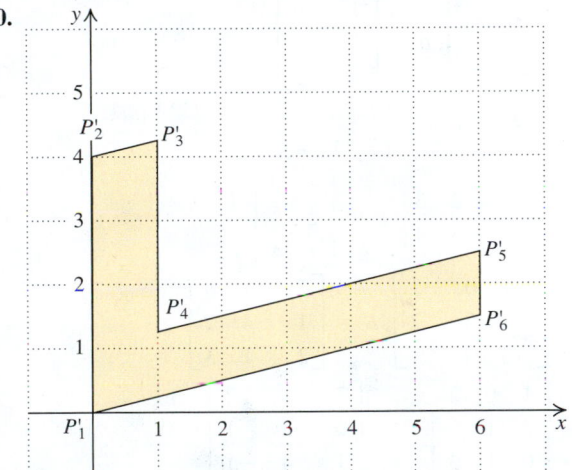

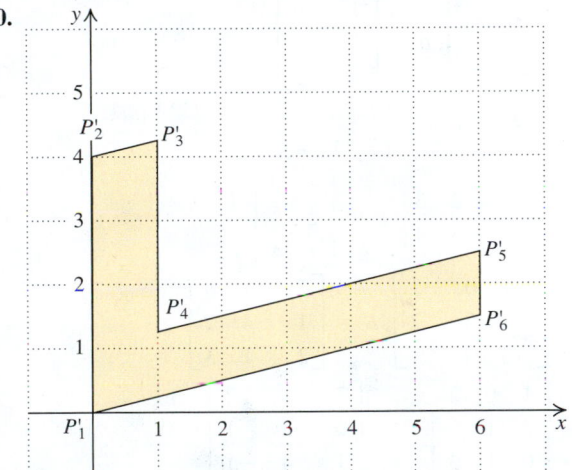   **SECTION 7.7**   **Exercises**

Concepts and Vocabulary

1. Two $m \times n$ matrices $A = [a_{ij}]$ and $B = [b_{ij}]$ are equal if _____ for all i and j.

2. Let $A = [a_{ij}]$, $B = [b_{ij}]$, and $C = [c_{ij}]$. If $A + B = C$, then $c_{ij} =$ _____ for all i and for all j.

3. The product of a $1 \times n$ matrix A and an $n \times 1$ matrix B is a _____ matrix.

4. If A is an $m \times n$ matrix and B is an $n \times p$ matrix, then AB is defined and is a(n) _____ matrix.

5. True or False. If AB and BA are defined, then $AB = BA$.

6. True or False. Any two square matrices can be multiplied.

7. True or False. If A is a 2×2 matrix and $A^2 = 0$, then A is the zero matrix.

8. True or False. If A is the zero matrix and AB is defined, then AB is a zero matrix.

Building Skills

In Exercises 9–16, find the values of all variables.

9. $\begin{bmatrix} 2 \\ -3 \end{bmatrix} = \begin{bmatrix} x \\ u \end{bmatrix}$

10. $\begin{bmatrix} -\dfrac{y}{2} \\ x \end{bmatrix} = \begin{bmatrix} 4 \\ 3 \end{bmatrix}$

11. $\begin{bmatrix} 2 & x \\ y & -3 \end{bmatrix} = \begin{bmatrix} 2 & -1 \\ 3 & -3 \end{bmatrix}$

12. $\begin{bmatrix} 2-x & 1 \\ -2 & 3+y \end{bmatrix} = \begin{bmatrix} 3 & 1 \\ -2 & -3 \end{bmatrix}$

13. $\begin{bmatrix} 2x-3y & -4 \\ 5 & 3x+y \end{bmatrix} = \begin{bmatrix} 1 & -4 \\ 5 & 7 \end{bmatrix}$

14. $\begin{bmatrix} 3x+y & -17 \\ 19 & 2 \end{bmatrix} = \begin{bmatrix} -\frac{1}{2} & -17 \\ 19 & 2x+3y \end{bmatrix}$

15. $\begin{bmatrix} x-y & 1 & 2 \\ 4 & 3x-2y & 3 \\ 5 & 6 & 5x-10y \end{bmatrix} = \begin{bmatrix} -1 & 1 & 2 \\ 4 & -1 & 3 \\ 5 & 6 & 6 \end{bmatrix}$

16. $\begin{bmatrix} x+y & 2 & 3 \\ 2 & x-y & 4 \\ 3 & 4 & 2x+3y \end{bmatrix} = \begin{bmatrix} -1 & 2 & 3 \\ 2 & 5 & 4 \\ 3 & 4 & -5 \end{bmatrix}$

In Exercises 17–24, find each of the following if possible:

a. $A+B$
b. $A-B$
c. $-3A$
d. $3A-2B$
e. $(A+B)^2$
f. A^2-B^2

17. $A = \begin{bmatrix} 1 & 2 \\ 3 & 4 \end{bmatrix}, B = \begin{bmatrix} -1 & 0 \\ 2 & -3 \end{bmatrix}$

18. $A = \begin{bmatrix} \frac{1}{3} & 1 \\ -1 & 2 \end{bmatrix}, B = \begin{bmatrix} 1 & 0 \\ 2 & -\frac{1}{2} \end{bmatrix}$

19. $A = \begin{bmatrix} 2 & 3 \\ -4 & 5 \end{bmatrix}, B = \begin{bmatrix} 1 & 0 & 2 \\ 3 & 1 & 4 \end{bmatrix}$

20. $A = \begin{bmatrix} 1 & 2 & 3 \\ -1 & -3 & 4 \end{bmatrix}, B = \begin{bmatrix} 1 & 0 \\ 2 & 3 \\ 1 & 4 \end{bmatrix}$

21. $A = \begin{bmatrix} 4 & 0 & -1 \\ -2 & 5 & 2 \\ 0 & 0 & 1 \end{bmatrix}, B = \begin{bmatrix} 3 & 1 & 0 \\ 1 & -4 & 2 \\ 2 & 1 & 3 \end{bmatrix}$

22. $A = \begin{bmatrix} 1 & 0 & 2 \\ 2 & 1 & 0 \\ 0 & 1 & 3 \end{bmatrix}, B = \begin{bmatrix} 3 & -1 & 2 \\ 0 & 2 & 1 \\ 1 & 4 & -1 \end{bmatrix}$

23. $A = \begin{bmatrix} 1 & 2 & -3 \\ 3 & 4 & 5 \\ 2 & -1 & 0 \end{bmatrix}, B = \begin{bmatrix} 3 & 1 & 0 \\ 1 & -4 & 2 \\ 2 & 1 & 3 \end{bmatrix}$

24. $A = \begin{bmatrix} 1 & 3 & -2 \\ 0 & 4 & 4 \\ -3 & 5 & 1 \end{bmatrix}, B = \begin{bmatrix} 7 & 0 & -2 \\ -3 & 1 & 1 \\ 2 & 1 & 5 \end{bmatrix}$

In Exercises 25–32, solve each matrix equation for X, where

$$A = \begin{bmatrix} 2 & 3 & -1 \\ 1 & -2 & 4 \end{bmatrix} \text{ and } B = \begin{bmatrix} -2 & 1 & 0 \\ 2 & 3 & 4 \end{bmatrix}.$$

25. $A + X = B$
26. $B + X = A$
27. $2X - A = B$
28. $2X - B = A$
29. $2X + 3A = B$
30. $3X + 2A = B$
31. $2A + 3B + 4X = 0$
32. $3X - 2A + 5B = 0$

In Exercises 33–44, find each product if possible.

a. AB b. BA

33. $A = \begin{bmatrix} 1 & 2 \\ 3 & 4 \end{bmatrix}, B = \begin{bmatrix} -2 & 1 \\ 3 & 5 \end{bmatrix}$

34. $A = \begin{bmatrix} 3 & 2 \\ 1 & 5 \\ 0 & 1 \end{bmatrix}, B = \begin{bmatrix} 1 & 3 & -2 \\ 2 & 5 & 0 \end{bmatrix}$

35. $A = \begin{bmatrix} 2 & -1 & 0 \\ -3 & 1 & 2 \end{bmatrix}, B = \begin{bmatrix} 1 & 5 \\ -2 & 3 \\ 4 & 0 \end{bmatrix}$

36. $A = \begin{bmatrix} 1 & 2 & 3 \\ 4 & 5 & 6 \end{bmatrix}, B = \begin{bmatrix} -1 & -3 & -6 \\ 2 & 5 & 7 \end{bmatrix}$

37. $A = [2 \quad 3 \quad 5], B = \begin{bmatrix} 1 \\ -2 \\ 4 \end{bmatrix}$

38. $A = \begin{bmatrix} 1 \\ 2 \\ -1 \end{bmatrix}, B = [-3 \quad 0 \quad 2]$

39. $A = [1 \quad 2 \quad 3], B = \begin{bmatrix} 1 & 2 & -1 \\ 0 & 3 & 1 \\ 2 & 0 & -3 \end{bmatrix}$

40. $A = \begin{bmatrix} 4 & -6 & 2 \\ 2 & 3 & 0 \\ 1 & 2 & -3 \end{bmatrix}, B = [-2 \quad 0 \quad 1]$

41. $A = \begin{bmatrix} 2 & 0 & 1 \\ 1 & 4 & 2 \\ 3 & -1 & 0 \end{bmatrix}, B = \begin{bmatrix} 3 & 1 & 0 \\ -1 & 2 & 0 \\ 4 & 5 & 2 \end{bmatrix}$

42. $A = \begin{bmatrix} 3 & -1 & 2 \\ 0 & 4 & -3 \\ 1 & -2 & 2 \end{bmatrix}, B = \begin{bmatrix} 2 & -5 & 0 \\ -1 & 2 & -1 \\ 3 & 0 & 2 \end{bmatrix}$

43. Let $A = \begin{bmatrix} 3 & 1 & 2 \\ 0 & 4 & 3 \\ 1 & -2 & 2 \end{bmatrix}$ and $B = \begin{bmatrix} 2 & 5 & 0 \\ 1 & 2 & -1 \\ 3 & 0 & 2 \end{bmatrix}$.

Verify that $AB \neq BA$.

44. Let $A = \begin{bmatrix} 5 & 2 & 4 \\ -6 & -3 & 10 \\ -2 & 6 & 5 \end{bmatrix}$ and $B = \begin{bmatrix} 4 & 1 & 2 \\ -3 & 0 & 5 \\ -1 & 3 & 4 \end{bmatrix}$.

Verify that $AB = BA$.

In Exercises 45–48, let

$$A = \begin{bmatrix} 1 & 2 \\ 3 & 4 \end{bmatrix}, B = \begin{bmatrix} 2 & -3 \\ 3 & 5 \end{bmatrix}, \text{ and } C = \begin{bmatrix} 0 & 1 \\ 2 & 4 \end{bmatrix}.$$

45. Verify the associative property for multiplication:
$(AB)C = A(BC)$.

46. Verify the distributive property: $A(B + C) = AB + AC$.

47. Verify the distributive property: $(A + B)C = AC + BC$.

48. For any real number c, verify that
$c(AB) = (cA)(B) = A(cB)$.

Applying the Concepts

49. Cost matrix. The Build-Rite Co. is building an apartment complex. The cost of purchasing and transporting specific amounts of steel, glass, and wood (in appropriate units) from two different locations is given by the following matrices:

$$A = \begin{bmatrix} 7 & 3 & 18 \\ 4 & 1 & 3 \end{bmatrix} \begin{matrix} \text{Cost of Material} \\ \text{Transportation Cost} \end{matrix}$$

with column headings Steel, Glass, Wood

$$B = \begin{bmatrix} 6 & 2 & 20 \\ 3 & 1 & 4 \end{bmatrix} \begin{matrix} \text{Cost of Material} \\ \text{Transportation Cost} \end{matrix}$$

Find the matrix representing the total cost of material and transportation for steel, glass, and wood from both locations.

50. Cost matrix. Repeat Exercise 49 if the order from location A is tripled and the order from location B is doubled.

51. Stock purchase. Ms. Goodbyer plans to buy 100 shares of computer stock, 300 shares of oil stock, and 400 shares of automobile stock. The computer stock is selling for $60 a share, oil stock is selling for $38 a share, and automobile stock is selling for $17 a share. Use matrix multiplication to calculate the total cost of Ms. Goodbyer's purchases.

52. Product export. The International Export Corporation received an export order for three of its products, say, A, B, and C. The export order is (in thousands) for 50 of product A, 75 of product B, and 150 of product C. The material cost, labor, and profit (each in appropriate units) required to produce each unit of the product are given in the following table:

	Material	Labor	Profit
Product A	20	2	20
Product B	15	3	25
Product C	25	1	15

Use matrix multiplication to compute the total material cost, total labor, and total profit if the entire export order is filled.

53. Executive compensation. The Multinational Oil Corporation pays its top executives a salary, a cash bonus, and shares of its stock annually. In 2016, the chairman of the board received $2.5 million in salary, a $1.5 million bonus, and 50,000 shares of stock; the president of the company received one-half the compensation of the chairman; and each of the four vice presidents was paid $100,000 in salary, a $150,000 bonus, and 5000 shares of stock.
   a. Express payments to these executives in salary, bonus, and stock as a 3×3 matrix.
   b. Express the number of executives of each rank as a column matrix.
   c. Use matrix multiplication to compute the total amount the company paid to each executive in each category in 2015.

54. Calories and protein. The Browns (B) and Newgards (N) are neighboring families. The Brown family has two men, three women, and one child, while the Newgard family has one man, one woman, and two children. Both families are

weight watchers. They have the same dietician, who recommends the following daily allowance of calories and proteins:

Calories
   male adults: 2400 calories
   female adults: 1900 calories
   children: 1800 calories

Protein (in grams)
   male adults: 55
   female adults: 45
   children: 33

Represent the preceding information in matrix form. Use matrix multiplication to compute the total requirements of calories and proteins for each of the two families.

In Exercise 55–58, a matrix A is given. Graph the figure represented by the matrix AD, where D is the matrix representation of the letter L of Example 10.

55. $A = \begin{bmatrix} 1 & 0 \\ 0 & -1 \end{bmatrix}$ **56.** $A = \begin{bmatrix} -1 & 0 \\ 0 & 1 \end{bmatrix}$

57. $A = \begin{bmatrix} 1 & 0 \\ 0.25 & 1 \end{bmatrix}$ **58.** $A = \begin{bmatrix} 1 & -0.25 \\ 0 & 1 \end{bmatrix}$

Beyond the Basics

59. Let $A = \begin{bmatrix} 0 & 3 \\ 0 & 0 \end{bmatrix}$, $B = \begin{bmatrix} 2 & 1 \\ 3 & 0 \end{bmatrix}$, and $C = \begin{bmatrix} 5 & 4 \\ 3 & 0 \end{bmatrix}$.

Verify that $AB = AC$. This example shows that $AB = AC$ does not, in general, imply that $B = C$.

60. Let $A = \begin{bmatrix} 2 & -3 & -5 \\ -1 & 4 & 5 \\ 1 & -3 & -4 \end{bmatrix}$ and $B = \begin{bmatrix} -1 & 3 & 5 \\ 1 & -3 & -5 \\ -1 & 3 & 5 \end{bmatrix}$.

Verify that $AB = \mathbf{0}$. This example shows that $AB = \mathbf{0}$ does not imply that $A = \mathbf{0}$ or $B = \mathbf{0}$.

61. Select appropriate 2×2 matrices A and B to show that, in general, $(A + B)^2 \neq A^2 + 2AB + B^2$.

62. Repeat Exercise 61 to show that, in general, $A^2 - B^2 \neq (A - B)(A + B)$.

63. Let $A = \begin{bmatrix} 1 & -2 & 3 \\ 2 & -4 & 1 \\ 3 & -5 & 2 \end{bmatrix}$. Show that

$$3A^2 - 2A + I = \begin{bmatrix} 17 & -23 & 15 \\ -13 & 30 & 10 \\ -9 & 22 & 21 \end{bmatrix},$$

where $I = \begin{bmatrix} 1 & 0 & 0 \\ 0 & 1 & 0 \\ 0 & 0 & 1 \end{bmatrix}$.

64. Let $A = \begin{bmatrix} 1 & 3 & 2 \\ 2 & 0 & 3 \\ 1 & -1 & 1 \end{bmatrix}$. Show that

$$A^3 - 3A^2 + A - I = \begin{bmatrix} -3 & 14 & -3 \\ 5 & -2 & 8 \\ 5 & -7 & 6 \end{bmatrix},$$

where I is given in Exercise 63.

65. If possible, find a matrix B such that $\begin{bmatrix} 2 & 3 \\ 1 & 2 \end{bmatrix} B = \begin{bmatrix} 1 & 0 \\ 0 & 1 \end{bmatrix}$.

66. If possible find a matrix B such that

$$\begin{bmatrix} 1 & -2 \\ -2 & 4 \end{bmatrix} B = \begin{bmatrix} 1 & 0 \\ 0 & 1 \end{bmatrix}.$$

67. Find x and y if

$$\left(4 \begin{bmatrix} 2 & -1 & 3 \\ 1 & 0 & 2 \end{bmatrix} - \begin{bmatrix} 1 & 2 & -1 \\ 2 & -3 & 4 \end{bmatrix} \right) \begin{bmatrix} 2 \\ -1 \\ 1 \end{bmatrix} = \begin{bmatrix} x \\ y \end{bmatrix}.$$

68. Find $x, y,$ and z if

$$\begin{bmatrix} 3 & 2 & -1 \\ 4 & 9 & 2 \\ 5 & 0 & -2 \end{bmatrix} \begin{bmatrix} x \\ y \\ z \end{bmatrix} = \begin{bmatrix} 0 \\ 7 \\ 2 \end{bmatrix}.$$

Critical Thinking / Discussion / Writing

69. Let A be a $3 \times n$ matrix and B be a $5 \times m$ matrix.
 a. Under what conditions is AB defined? What is the order of AB when this product is defined?
 b. Repeat part (a) for BA.

70. Let $A = \begin{bmatrix} 3 & -1 & 0 \\ -2 & 3 & 1 \end{bmatrix}, B = \begin{bmatrix} 2 \\ 3 \\ 4 \end{bmatrix}$, and $C = \begin{bmatrix} -1 & 2 \end{bmatrix}$.

In which order should all three matrices be multiplied to produce a number?

71. Market share. The Asma Corporation (A), Bronkial Brothers (B), and Coufmore Company (C) simultaneously decide to introduce a new cigarette at a time when each company has one-third of the market. During the year, the following occurs:
 (i) A retains 40% of its customers and loses 30% to B and 30% to C.
 (ii) B retains 30% of its customers and loses 60% to A and 10% to C.
 (iii) C retains 30% of its customers and loses 60% to A and 10% to B.

The foregoing information can be represented by the *transition matrix*

$$P = \begin{matrix} & \begin{matrix} A & B & C \end{matrix} \\ \begin{matrix} A \\ B \\ C \end{matrix} & \begin{bmatrix} 0.4 & 0.3 & 0.3 \\ 0.6 & 0.3 & 0.1 \\ 0.6 & 0.1 & 0.3 \end{bmatrix} \end{matrix}$$

Write the initial market share as $X = \begin{bmatrix} \dfrac{1}{3} & \dfrac{1}{3} & \dfrac{1}{3} \end{bmatrix}$.

 a. Compute XP^2 and interpret your result.
 b. Compute $XP^3, XP^4,$ and XP^5; observe the pattern for the long run. Describe the long-term market share for each company.

Getting Ready for the Next Section

In Exercises 72 and 73, rewrite each expression without exponents.

72. $\left(\dfrac{1}{2} \right)^{-1}$

73. $\left(\dfrac{5}{12} \right)^{-1}$

In Exercises 74–79, solve each equation.

74. $x^{-1} = \dfrac{1}{8}$

75. $\left(\dfrac{x}{35} \right)^{-1} = 7$

76. $\dfrac{1}{4}x - \dfrac{7}{12} = \dfrac{11}{12} - \dfrac{5}{4}x$

77. $\dfrac{2}{3}x - \dfrac{3}{5} = \dfrac{4}{3}x + \dfrac{2}{15}$

78. $\dfrac{2}{4x - 1} = \dfrac{3}{4x + 1}$

79. $\dfrac{2}{x + 1} = \dfrac{1}{x - 1}$

In Exercises 80–83, solve each system.

80. $\begin{cases} 3x + 5y = 2 \\ 6x + 10y = 4 \end{cases}$

81. $\begin{cases} -9x + 2y = 8 \\ 5x + 8y = 32 \end{cases}$

82. $\begin{cases} 3x + 2y = 6 \\ 6x + 4y = -13 \end{cases}$

83. $\begin{cases} 3x - 4y = 13 \\ 2x + 5y = 1 \end{cases}$

The Matrix Inverse

BEFORE STARTING THIS SECTION, REVIEW

1 Equality of matrices (Section 7.7, page 673)

2 Matrix multiplication (Section 7.7, page 674)

3 Gauss–Jordan elimination procedure (Section 7.4, page 642)

4 Leontieff input–output model (Section 7.4, page 643)

OBJECTIVES

1 Verify the multiplicative inverse of a matrix.

2 Find the inverse of a matrix.

3 Find the inverse of a 2×2 matrix.

4 Use matrix inverses to solve systems of linear equations.

5 Use matrix inverses in applied problems.

◆ Coding and Decoding Messages

Imagine you are a secret agent who is looking for spies in different parts of the world. Periodically, you communicate with your headquarters via your laptop or by satellite. You expect that your communications will be intercepted. So before you broadcast any message, you must write it using a secret code. A message written with a secret code is called a **cryptogram**.

First, you might replace each letter in the message with a number and send the message as a sequence of numbers. The problem with this approach is that it is rather easy to crack such a code. For example, if letters such as e are always represented by the same symbol, then a code breaker (hacker) can guess which symbol represents each letter and eventually translate your message. In Example 8, we show you a better way to encode and decode a message using matrix multiplication.

1 Verify the multiplicative inverse of a matrix.

The Multiplicative Inverse of a Matrix

In arithmetic, you know that if a is a nonzero real number, then a has a unique reciprocal, denoted by $\dfrac{1}{a}$, or a^{-1}. The number a^{-1} is called the **multiplicative inverse** of a, and

$$aa^{-1} = a^{-1}a = 1.$$

Similarly, certain square matrices have *multiplicative inverses,* with an identity matrix acting as "1."

The Inverse of a Matrix

Let A be an $n \times n$ matrix and let I be the $n \times n$ identity matrix. If there is an $n \times n$ matrix B such that

$$AB = I \quad \text{and} \quad BA = I,$$

then B is called the **inverse** of A and we write $B = A^{-1}$ (read "A inverse").

Notice that if B is the inverse of A, then A is the inverse of B.

Arthur Cayley

(1821–1895)
English mathematicians of the first third of the nineteenth century attempted to extend the basic ideas of algebra and to determine exactly how much one can generalize the properties of integers to other types of quantities. Although Sylvester (see page 673) coined the term *matrix*, it was Arthur Cayley who developed the algebra of matrices. Cayley studied mathematics at Trinity College, Cambridge, but because there was no suitable teaching job available, he decided to become a lawyer. Although he became skilled in legal work, he regarded the law as just a way to support himself. During his legal career, he was able to write more than 300 mathematical papers. In 1863, Cambridge University esatblished a new post in mathematics and offered it to Cayley. He accepted the job eagerly, even though it paid less money than he made as a lawyer.

Recall that, in general (even for square matrices), $AB \neq BA$. However, if A and B are square matrices with $AB = I$, then it can be shown that $BA = I$. Therefore, to determine whether A and B are inverses, you need only check that either $AB = I$ or $BA = I$.

EXAMPLE 1   Verifying the Inverse of a Matrix

Show that B is the inverse of A:

$$A = \begin{bmatrix} 1 & 3 \\ 2 & 7 \end{bmatrix} \text{ and } B = \begin{bmatrix} 7 & -3 \\ -2 & 1 \end{bmatrix}$$

Solution

You need to verify that $AB = I$ or $BA = I$.

$$AB = \begin{bmatrix} 1 & 3 \\ 2 & 7 \end{bmatrix}\begin{bmatrix} 7 & -3 \\ -2 & 1 \end{bmatrix} = \begin{bmatrix} 1(7) + 3(-2) & 1(-3) + 3(1) \\ 2(7) + 7(-2) & 2(-3) + 7(1) \end{bmatrix} = \begin{bmatrix} 1 & 0 \\ 0 & 1 \end{bmatrix} = I$$

Since $AB = I$, it follows that $B = A^{-1}$.

Practice Problem 1 Show that B is the inverse of A:

$$A = \begin{bmatrix} 3 & 2 \\ 2 & 1 \end{bmatrix} \quad \text{and} \quad B = \begin{bmatrix} -1 & 2 \\ 2 & -3 \end{bmatrix}$$

In working with real numbers, you know that if $a = 0$, then a^{-1} does not exist. So it should not come as a surprise that if A is a square *zero* matrix, then A^{-1} does not exist. Surprisingly, there are also nonzero matrices that do not have inverses.

EXAMPLE 2   Proving That a Particular Nonzero Matrix Has No Inverse

Show that the matrix A does not have an inverse.

$$A = \begin{bmatrix} 2 & 0 \\ 0 & 0 \end{bmatrix}$$

Solution

Suppose A has an inverse B, where

$$B = \begin{bmatrix} x & y \\ z & w \end{bmatrix}$$

Then you have

$$\begin{bmatrix} 2 & 0 \\ 0 & 0 \end{bmatrix}\begin{bmatrix} x & y \\ z & w \end{bmatrix} = \begin{bmatrix} 1 & 0 \\ 0 & 1 \end{bmatrix} \qquad \textcolor{blue}{AB = I}$$

$$\begin{bmatrix} 2x & 2y \\ 0 & 0 \end{bmatrix} = \begin{bmatrix} 1 & 0 \\ 0 & 1 \end{bmatrix} \qquad \textcolor{blue}{\text{Multiply the two matrices on the left side.}}$$

Since these two matrices are equal, you must have

$$0 = 1 \qquad \textcolor{blue}{\text{Equate the entries in the } (2, 2) \text{ position.}}$$

Because $0 = 1$ is a false statement, the matrix A does not have an inverse.

Practice Problem 2 Show that the matrix $A = \begin{bmatrix} 3 & 1 \\ 3 & 1 \end{bmatrix}$ does not have an inverse.

2 Find the inverse of a matrix.

Finding the Inverse of a Matrix

It can be shown that if a square matrix A has an inverse, then the inverse is unique. This means that a square matrix cannot have more than one inverse. If a matrix A has an inverse, then A is called **invertible** or **nonsingular**. A square matrix that does not have an inverse is called a **singular** matrix. The following discussion derives a technique for finding the inverse of an invertible matrix.

Suppose we want to find the inverse of the matrix $A = \begin{bmatrix} 1 & 2 \\ 3 & 5 \end{bmatrix}$. We first let

$B = \begin{bmatrix} x & y \\ z & w \end{bmatrix}$ be the inverse of the matrix A. Then

$$AB = I \qquad \text{Definition of inverse}$$
$$\begin{bmatrix} 1 & 2 \\ 3 & 5 \end{bmatrix}\begin{bmatrix} x & y \\ z & w \end{bmatrix} = \begin{bmatrix} 1 & 0 \\ 0 & 1 \end{bmatrix} \qquad \text{Substitute for } A, B, \text{ and } I.$$
$$\begin{bmatrix} x + 2z & y + 2w \\ 3x + 5z & 3y + 5w \end{bmatrix} = \begin{bmatrix} 1 & 0 \\ 0 & 1 \end{bmatrix} \qquad \text{Multiply matrices } A \text{ and } B.$$

Equating corresponding entries in the last equation yields two systems of linear equations.

$$\begin{cases} x + 2z = 1 \\ 3x + 5z = 0 \end{cases} \quad \begin{array}{l} \text{Equate the entries in the } (1, 1) \\ \text{and } (2, 1) \text{ positions.} \end{array}$$

$$\begin{cases} y + 2w = 0 \\ 3y + 5w = 1 \end{cases} \quad \begin{array}{l} \text{Equate the entries in the } (1, 2) \\ \text{and } (2, 2) \text{ positions.} \end{array}$$

Let's solve the two systems by Gauss–Jordan elimination.

System of Equations	Matrix Form
$\begin{cases} x + 2z = 1 \\ 3x + 5z = 0 \end{cases}$	$\begin{bmatrix} 1 & 2 & \vert & 1 \\ 3 & 5 & \vert & 0 \end{bmatrix}$
$\begin{cases} y + 2w = 0 \\ 3y + 5w = 1 \end{cases}$	$\begin{bmatrix} 1 & 2 & \vert & 0 \\ 3 & 5 & \vert & 1 \end{bmatrix}$

TECHNOLOGY CONNECTION

To use a graphing calculator to find the inverse of an invertible matrix, name the matrix and then use the regular reciprocal key with the name of the matrix.

```
[A]
      [[1 2]
       [3 5]]
■
```

```
[A]⁻¹
      [[-5 2 ]
       [3  -1]]
■
```

Because the coefficient matrix for both systems of equations is the original matrix

$$A = \begin{bmatrix} 1 & 2 \\ 3 & 5 \end{bmatrix},$$

the calculations involved in the Gauss–Jordan elimination on the coefficient matrix A will be the same for both systems. Therefore, you can solve both systems at once by considering the "doubly augmented" matrix

$$C = \begin{bmatrix} 1 & 2 & \vert & 1 & 0 \\ 3 & 5 & \vert & 0 & 1 \end{bmatrix}$$

Note that the matrix C is formed by adjoining the identity matrix I to matrix A:

$$C = [A \,\vert\, I]$$

Now apply Gauss–Jordan elimination to the matrix C.

$$C = \begin{bmatrix} 1 & 2 & \vert & 1 & 0 \\ 3 & 5 & \vert & 0 & 1 \end{bmatrix} \xrightarrow{-3R_1+R_2 \to R_2} \begin{bmatrix} 1 & 2 & \vert & 1 & 0 \\ 0 & -1 & \vert & -3 & 1 \end{bmatrix} \xrightarrow{(-1)R_2} \begin{bmatrix} 1 & 2 & \vert & 1 & 0 \\ 0 & 1 & \vert & 3 & -1 \end{bmatrix}$$
$$\xrightarrow{-2R_2+R_1 \to R_1} \begin{bmatrix} 1 & 0 & \vert & -5 & 2 \\ 0 & 1 & \vert & 3 & -1 \end{bmatrix}$$

Converting the last matrix back into two systems of equations, we have

$$\begin{bmatrix} 1 & 0 & | & -5 \\ 0 & 1 & | & 3 \end{bmatrix} \text{ represents } \begin{cases} 1 \cdot x + 0 \cdot z = -5 \\ 0 \cdot x + 1 \cdot z = 3 \end{cases} \text{ or } \begin{cases} x = -5 \\ z = 3 \end{cases}$$

Similarly,

$$\begin{bmatrix} 1 & 0 & | & 2 \\ 0 & 1 & | & -1 \end{bmatrix} \text{ represents } \begin{cases} 1 \cdot y + 0 \cdot w = 2 \\ 0 \cdot y + 1 \cdot w = -1 \end{cases} \text{ or } \begin{cases} y = 2 \\ w = -1 \end{cases}$$

We conclude that $x = -5$, $y = 2$, $z = 3$, and $w = -1$.

Since $B = \begin{bmatrix} x & y \\ z & w \end{bmatrix}$, it follows that

$$B = \begin{bmatrix} -5 & 2 \\ 3 & -1 \end{bmatrix} = A^{-1}$$

We have shown that to find the inverse of an invertible matrix A, we transform the matrix $[A \mid I]$ by a sequence of elementary row operations into $[I \mid B]$, where $B = A^{-1}$.

PROCEDURE FOR FINDING THE INVERSE OF A MATRIX

Let A be an $n \times n$ matrix.

1. Form the $n \times 2n$ augmented matrix $[A \mid I]$, where I is the $n \times n$ identity matrix.
2. If there is a sequence of elementary row operations that transforms A into I, then this same sequence of row operations will transform $[A \mid I]$ into $[I \mid B]$, where $B = A^{-1}$.
3. Check your work by showing that $AA^{-1} = I$.

If it is not possible to transform A into I by row operations, then A does not have an inverse. (This occurs if, at any step in the process, you obtain a matrix $[C \mid D]$ in which C has a row of zeros.)

EXAMPLE 3   **Finding the Inverse of a Matrix**

Find the inverse (if it exists) of the matrix $A = \begin{bmatrix} 1 & 2 & 3 \\ 2 & 5 & 7 \\ 2 & 4 & 6 \end{bmatrix}$.

Solution

Step 1   Start by adjoining the identity matrix $I = \begin{bmatrix} 1 & 0 & 0 \\ 0 & 1 & 0 \\ 0 & 0 & 1 \end{bmatrix}$ to the matrix A to

form the augmented matrix $[A \mid I]$.

$$[A \mid I] = \begin{bmatrix} 1 & 2 & 3 & | & 1 & 0 & 0 \\ 2 & 5 & 7 & | & 0 & 1 & 0 \\ 2 & 4 & 6 & | & 0 & 0 & 1 \end{bmatrix}$$

Step 2   Perform row operations on $[A \mid I]$ to transform it to a matrix of the form $[I \mid B]$.

$$\begin{bmatrix} 1 & 2 & 3 & | & 1 & 0 & 0 \\ 2 & 5 & 7 & | & 0 & 1 & 0 \\ 2 & 4 & 6 & | & 0 & 0 & 1 \end{bmatrix} \xrightarrow[\substack{-2R_1 + R_3 \to R_3}]{-2R_1 + R_2 \to R_2} \begin{bmatrix} 1 & 2 & 3 & | & 1 & 0 & 0 \\ 0 & 1 & 1 & | & -2 & 1 & 0 \\ 0 & 0 & 0 & | & -2 & 0 & 1 \end{bmatrix} = [C \mid D]$$

Because the third row of matrix C consists of zeros, it is impossible to transform A into I by row operations. Hence, A is *not* invertible.

Practice Problem 3 Find the inverse (if it exists) of the matrix $A = \begin{bmatrix} 1 & 4 & -2 \\ -1 & 1 & 2 \\ 3 & 7 & -6 \end{bmatrix}$. ▦

EXAMPLE 4   **Finding the Inverse of a Matrix**

Find the inverse (if it exists) of the matrix $A = \begin{bmatrix} 1 & 1 & 0 \\ 0 & 3 & 1 \\ 2 & 3 & 3 \end{bmatrix}$.

Solution

Step 1   Start with the matrix $[A \mid I]$:

$$[A \mid I] = \left[\begin{array}{ccc|ccc} 1 & 1 & 0 & 1 & 0 & 0 \\ 0 & 3 & 1 & 0 & 1 & 0 \\ 2 & 3 & 3 & 0 & 0 & 1 \end{array}\right] \qquad I = I_3$$

Step 2   Use row operations on $[A \mid I]$. Then $[A \mid I] \rightarrow$

$$\xrightarrow[\substack{-2R_1+R_3 \rightarrow R_3}]{\frac{1}{3}R_2 \rightarrow R_2} \left[\begin{array}{ccc|ccc} 1 & 1 & 0 & 1 & 0 & 0 \\ 0 & 1 & \frac{1}{3} & 0 & \frac{1}{3} & 0 \\ 0 & 1 & 3 & -2 & 0 & 1 \end{array}\right] \xrightarrow[]{(-1)R_2+R_3 \rightarrow R_3} \left[\begin{array}{ccc|ccc} 1 & 1 & 0 & 1 & 0 & 0 \\ 0 & 1 & \frac{1}{3} & 0 & \frac{1}{3} & 0 \\ 0 & 0 & \frac{8}{3} & -2 & -\frac{1}{3} & 1 \end{array}\right]$$

$$\xrightarrow[]{\frac{3}{8}R_3} \left[\begin{array}{ccc|ccc} 1 & 1 & 0 & 1 & 0 & 0 \\ 0 & 1 & \frac{1}{3} & 0 & \frac{1}{3} & 0 \\ 0 & 0 & 1 & -\frac{6}{8} & -\frac{1}{8} & \frac{3}{8} \end{array}\right] \xrightarrow[]{-\frac{1}{3}R_3+R_2 \rightarrow R_2} \left[\begin{array}{ccc|ccc} 1 & 1 & 0 & 1 & 0 & 0 \\ 0 & 1 & 0 & \frac{2}{8} & \frac{3}{8} & -\frac{1}{8} \\ 0 & 0 & 1 & -\frac{6}{8} & -\frac{1}{8} & \frac{3}{8} \end{array}\right]$$

$$\xrightarrow[]{(-1)R_2+R_1 \rightarrow R_1} \left[\begin{array}{ccc|ccc} 1 & 0 & 0 & \frac{6}{8} & -\frac{3}{8} & \frac{1}{8} \\ 0 & 1 & 0 & \frac{2}{8} & \frac{3}{8} & -\frac{1}{8} \\ 0 & 0 & 1 & -\frac{6}{8} & -\frac{1}{8} & \frac{3}{8} \end{array}\right].$$

Hence, A is invertible and

$$A^{-1} = \begin{bmatrix} \frac{6}{8} & -\frac{3}{8} & \frac{1}{8} \\ \frac{2}{8} & \frac{3}{8} & -\frac{1}{8} \\ -\frac{6}{8} & -\frac{1}{8} & \frac{3}{8} \end{bmatrix} = \frac{1}{8}\begin{bmatrix} 6 & -3 & 1 \\ 2 & 3 & -1 \\ -6 & -1 & 3 \end{bmatrix}.$$

Step 3   You should verify that $AA^{-1} = I$.

Practice Problem 4 Find the inverse (if it exists) of the matrix $A = \begin{bmatrix} 1 & 2 & 3 \\ -2 & 3 & 1 \\ 4 & 5 & -2 \end{bmatrix}$.

▦

3 Find the inverse of a 2×2 matrix.

A Rule for Finding the Inverse of a 2 × 2 Matrix

You can quickly determine whether any 2×2 matrix is invertible and, if so, what its inverse is.

A Rule for Finding the Inverse of a 2 × 2 Matrix

The matrix

$$A = \begin{bmatrix} a & b \\ c & d \end{bmatrix}$$

is invertible if and only if $ad - bc \neq 0$. Moreover, if $ad - bc \neq 0$, then

$$A^{-1} = \frac{1}{ad - bc} \begin{bmatrix} d & -b \\ -c & a \end{bmatrix}.$$

If $ad - bc = 0$, the matrix A does not have an inverse.

EXAMPLE 5 **Finding the Inverse of a 2 × 2 Matrix**

Find the inverse (if it exists) of each matrix.

a. $A = \begin{bmatrix} 5 & 2 \\ 4 & 3 \end{bmatrix}$ **b.** $B = \begin{bmatrix} 4 & 6 \\ 2 & 3 \end{bmatrix}$

Solution

a. For the matrix A, $a = 5, b = 2, c = 4$, and $d = 3$.
Here $ad - bc = (5)(3) - (2)(4) = 15 - 8 = 7 \neq 0$; so A is invertible and

$$A^{-1} = \frac{1}{7} \begin{bmatrix} 3 & -2 \\ -4 & 5 \end{bmatrix} \quad \text{Substitute for } a, b, c, \text{ and } d \text{ in } \frac{1}{ad - bc} \begin{bmatrix} d & -b \\ -c & a \end{bmatrix}.$$

$$= \begin{bmatrix} \dfrac{3}{7} & -\dfrac{2}{7} \\ -\dfrac{4}{7} & \dfrac{5}{7} \end{bmatrix} \quad \text{Scalar multiplication}$$

You should verify that $AA^{-1} = I$.

b. For the matrix B, $a = 4, b = 6, c = 2$, and $d = 3$; so

$$ad - bc = (4)(3) - (6)(2) = 12 - 12 = 0.$$

Therefore, the matrix B does not have an inverse.

Practice Problem 5 Find the inverse (if it exists) of each matrix.

a. $A = \begin{bmatrix} 8 & 2 \\ 4 & 1 \end{bmatrix}$ **b.** $B = \begin{bmatrix} 8 & -2 \\ 3 & 1 \end{bmatrix}$

4 Use matrix inverses to solve systems of linear equations.

Solving Systems of Linear Equations by Using Matrix Inverses

Matrix multiplication can be used to write a system of linear equations in matrix form.

System of Equations	Matrix Form
$\begin{cases} 3x - 2y = 4 \\ 4x - 3y = 5 \end{cases}$	$\begin{bmatrix} 3 & -2 \\ 4 & -3 \end{bmatrix} \begin{bmatrix} x \\ y \end{bmatrix} = \begin{bmatrix} 4 \\ 5 \end{bmatrix}$
$\begin{cases} 2x_1 + 4x_2 - x_3 = 9 \\ 3x_1 + x_2 + 2x_3 = 7 \\ x_1 + 3x_2 - 3x_3 = 4 \end{cases}$	$\begin{bmatrix} 2 & 4 & -1 \\ 3 & 1 & 2 \\ 1 & 3 & -3 \end{bmatrix} \begin{bmatrix} x_1 \\ x_2 \\ x_3 \end{bmatrix} = \begin{bmatrix} 9 \\ 7 \\ 4 \end{bmatrix}$

Solving a system of linear equations amounts to solving a corresponding matrix equation of the form

$$AX = B,$$

where A is the coefficient matrix of the system, X is the column matrix containing the variables, and B is the column matrix of the system's constants.

If A is a square matrix *and* A is invertible, then the matrix equation $AX = B$ has a unique solution, $X = A^{-1}B$, as the following derivation shows.

$AX = B$ Given equation

$A^{-1}(AX) = A^{-1}B$ Be careful to multiply by A^{-1} on the left on both sides of the equation because multiplication of matrices is not commutative.

$(A^{-1}A)X = A^{-1}B$ Associative property

$IX = A^{-1}B$ $A^{-1}A = I$

$X = A^{-1}B$ I is the identity matrix and $IX = X$.

EXAMPLE 6   **Solving a Linear System by Using an Inverse Matrix**

Use a matrix inverse to solve the linear system: $\begin{cases} x + y & = 4 \\ 3y + z = 7 \\ 2x + 3y + 3z = 21 \end{cases}$

Solution

Write the linear system in matrix form:

$$\underbrace{\begin{bmatrix} 1 & 1 & 0 \\ 0 & 3 & 1 \\ 2 & 3 & 3 \end{bmatrix}}_{A} \underbrace{\begin{bmatrix} x \\ y \\ z \end{bmatrix}}_{X} = \underbrace{\begin{bmatrix} 4 \\ 7 \\ 21 \end{bmatrix}}_{B}$$ Use zeros for coefficients of missing variables.

We need to solve the matrix equation $AX = B$ for X. Since the matrix A is invertible (see Example 4), the system has a unique solution $X = A^{-1}B$. Use the following:

$$A^{-1} = \frac{1}{8}\begin{bmatrix} 6 & -3 & 1 \\ 2 & 3 & -1 \\ -6 & -1 & 3 \end{bmatrix}$$ Computed in Example 4

$$X = A^{-1}B$$

$$= \frac{1}{8}\begin{bmatrix} 6 & -3 & 1 \\ 2 & 3 & -1 \\ -6 & -1 & 3 \end{bmatrix}\begin{bmatrix} 4 \\ 7 \\ 21 \end{bmatrix}$$ Substitute for A^{-1} and B.

$$= \frac{1}{8}\begin{bmatrix} 6(4) - 3(7) + 1(21) \\ 2(4) + 3(7) - 1(21) \\ -6(4) - 1(7) + 3(21) \end{bmatrix}$$ Matrix multiplication

$$= \frac{1}{8}\begin{bmatrix} 24 \\ 8 \\ 32 \end{bmatrix} = \begin{bmatrix} 3 \\ 1 \\ 4 \end{bmatrix}$$ Scalar multiplication

The solution set is $\{(3, 1, 4)\}$, which you can check in the original system.

Practice Problem 6 Use a matrix inverse to solve the linear system.

$$\begin{cases} 3x + 2y + 3z = 9 \\ 3x + y & = 12 \\ x & + z = 6 \end{cases}$$

Applications of Matrix Inverses

The Leontief Input–Output Model We can use the inverse of a matrix to analyze input–output models that we studied in Section 7.4.

Suppose a simplified economy depends on two products: energy (E) and food (F). To produce one unit of E requires $\frac{1}{4}$ unit of E and $\frac{1}{2}$ unit of F. To produce one unit of F requires $\frac{1}{3}$ unit of E and $\frac{1}{4}$ unit of F. Then the interindustry consumption is given by the following matrix.

$$
\begin{array}{c}
\textbf{Input} \\
\begin{array}{cc} E & F \end{array}
\end{array}
$$

$$
A = \begin{bmatrix} \dfrac{1}{4} & \dfrac{1}{2} \\[2mm] \dfrac{1}{3} & \dfrac{1}{4} \end{bmatrix} \begin{array}{l} E \\ F \end{array} \quad \textbf{Output}
$$

The matrix A is the **interindustry technology input–output matrix**, or simply the **technology matrix**, of the system. If the system is producing x_1 units of energy and x_2 units of food, then the column matrix $X = \begin{bmatrix} x_1 \\ x_2 \end{bmatrix}$ is called the **gross production matrix**.

Consequently,

$$
AX = \begin{bmatrix} \dfrac{1}{4} & \dfrac{1}{2} \\[2mm] \dfrac{1}{3} & \dfrac{1}{4} \end{bmatrix}\begin{bmatrix} x_1 \\ x_2 \end{bmatrix} = \begin{bmatrix} \dfrac{1}{4}x_1 + \dfrac{1}{2}x_2 \\[2mm] \dfrac{1}{3}x_1 + \dfrac{1}{4}x_2 \end{bmatrix} \begin{array}{l} \leftarrow \text{units consumed by } E \\[4mm] \leftarrow \text{units consumed by } F \end{array}
$$

represents the interindustry consumptions. If the column matrix $D = \begin{bmatrix} d_1 \\ d_2 \end{bmatrix}$ represents consumer demand, then

$$
\begin{aligned}
D &= X - AX && \text{Gross production minus} \\
&&& \text{interindustry consumption} \\
&= IX - AX && I \text{ is the identity matrix.} \\
D &= (I - A)X && \text{Distributive property} \\
(I - A)^{-1}D &= (I - A)^{-1}(I - A)X && \text{Multiply both sides by} \\
&&& (I - A)^{-1} \text{ if } I - A \text{ is invertible.} \\
(I - A)^{-1}D &= IX && (I - A)^{-1}(I - A) = I \\
&= X && I \text{ is the identity matrix.}
\end{aligned}
$$

So if $(I - A)$ is invertible, then the gross production matrix is $X = (I - A)^{-1}D$.

EXAMPLE 7 **Using the Leontief Input–Output Model**

In the preceding discussion, suppose the consumer demand for energy is 1000 units and for food is 3000 units. Find the level of production (X) that will meet interindustry and consumer demand.

Solution

For the matrix A, we have

$$
I - A = \begin{bmatrix} 1 & 0 \\ 0 & 1 \end{bmatrix} - \begin{bmatrix} \dfrac{1}{4} & \dfrac{1}{2} \\[2mm] \dfrac{1}{3} & \dfrac{1}{4} \end{bmatrix} = \begin{bmatrix} \dfrac{3}{4} & -\dfrac{1}{2} \\[2mm] -\dfrac{1}{3} & \dfrac{3}{4} \end{bmatrix}
$$

RECALL

If $ad - bc \neq 0$, then for
$A = \begin{bmatrix} a & b \\ c & d \end{bmatrix}$, we have
$A^{-1} = \dfrac{1}{ad - bc} \begin{bmatrix} d & -b \\ -c & a \end{bmatrix}.$

Use the formula for the inverse of a 2×2 matrix.

$$(I - A)^{-1} = \frac{1}{\dfrac{3}{4} \cdot \dfrac{3}{4} - \left(-\dfrac{1}{2}\right)\left(-\dfrac{1}{3}\right)} \begin{bmatrix} \dfrac{3}{4} & \dfrac{1}{2} \\ \dfrac{1}{3} & \dfrac{3}{4} \end{bmatrix}$$

$$= \frac{48}{19} \begin{bmatrix} \dfrac{3}{4} & \dfrac{1}{2} \\ \dfrac{1}{3} & \dfrac{3}{4} \end{bmatrix} = \frac{1}{19} \begin{bmatrix} 36 & 24 \\ 16 & 36 \end{bmatrix} \qquad \text{Simplify.}$$

The matrix $D = \begin{bmatrix} 1000 \\ 3000 \end{bmatrix}$. Therefore, the gross production matrix X is:

$$X = (I - A)^{-1}D = \frac{1}{19} \begin{bmatrix} 36 & 24 \\ 16 & 36 \end{bmatrix} \begin{bmatrix} 1000 \\ 3000 \end{bmatrix} \qquad \begin{array}{l} \text{Substitute for} \\ (I - A)^{-1} \text{ and } D. \end{array}$$

$$= \frac{1}{19} \begin{bmatrix} 108{,}000 \\ 124{,}000 \end{bmatrix} = \begin{bmatrix} \dfrac{108{,}000}{19} \\ \dfrac{124{,}000}{19} \end{bmatrix}$$

To meet the consumer demand for 1000 units of energy and 3000 units of food, the energy produced must be $\dfrac{108{,}000}{19}$ units and food production must be $\dfrac{124{,}000}{19}$ units.

Practice Problem 7 In Example 7, find the level of production (X) that will meet both the interindustry and consumer demand if consumer demand for energy is 800 units and consumer demand for food is 3400 units.

◆ Cryptography

To encode or decode a message, first associate each letter in the message with a number, in the obvious manner. The blank corresponds to 0.

blank	A	B	C	D	E	F	G	H	I	J	K	L	M
0	1	2	3	4	5	6	7	8	9	10	11	12	13
N	O	P	Q	R	S	T	U	V	W	X	Y	Z	
14	15	16	17	18	19	20	21	22	23	24	25	26	

For example, the message

JOHNSON IS IN DANGER

becomes:

10 15 8 14 19 15 14 0 9 19 0 9 14 0 4 1 14 7 5 18
J   O   H N   S   O   N   I S   I N   D A N   G E   R

This message is easy to decode, so we will make the coding more complicated. First, partition these 20 numbers into groups of three, enclosed in brackets, and insert zeros at the end of the last bracket if necessary.

[10 15 8] [14 19 15] [14 0 9] [19 0 9] [14 0 4] [1 14 7] [5 18 0]
 J   O   H   N   S   O   N   S   I   S   I   N   D A   N   G E   R

Let's write these seven groups of three numbers as the seven columns of a 3×7 matrix:

$$M = \begin{bmatrix} 10 & 14 & 14 & 19 & 14 & 1 & 5 \\ 15 & 19 & 0 & 0 & 0 & 14 & 18 \\ 8 & 15 & 9 & 9 & 4 & 7 & 0 \end{bmatrix}$$

Now choose any invertible 3 × 3 matrix A, say,

$$A = \begin{bmatrix} 1 & 2 & 3 \\ 1 & 3 & 3 \\ 1 & 2 & 4 \end{bmatrix}$$

We can use the matrix A to convert the message into code and then use A^{-1} to decode the message, as in the next example. The matrix A is called the **coding matrix**.

EXAMPLE 8 Encoding and Decoding a Message

Encode and decode the message

JOHNSON IS IN DANGER

by using the matrix A given above.

Solution

Step 1 Express the message numerically and partition the numbers into groups of three:

$$[10 \quad 15 \quad 8][14 \quad 19 \quad 15][14 \quad 0 \quad 9][19 \quad 0 \quad 9][14 \quad 0 \quad 4][1 \quad 14 \quad 7][5 \quad 18 \quad 0]$$

Step 2 Write each group of three numbers in Step 1 as a column of a matrix M.

$$M = \begin{bmatrix} 10 & 14 & 14 & 19 & 14 & 1 & 5 \\ 15 & 19 & 0 & 0 & 0 & 14 & 18 \\ 8 & 15 & 9 & 9 & 4 & 7 & 0 \end{bmatrix}$$

Step 3 Select an invertible 3 × 3 matrix A, such as

$$A = \begin{bmatrix} 1 & 2 & 3 \\ 1 & 3 & 3 \\ 1 & 2 & 4 \end{bmatrix}$$

Step 4 Multiply the matrices in Steps 2 and 3 to form AM.

$$AM = \begin{bmatrix} 1 & 2 & 3 \\ 1 & 3 & 3 \\ 1 & 2 & 4 \end{bmatrix}\begin{bmatrix} 10 & 14 & 14 & 19 & 14 & 1 & 5 \\ 15 & 19 & 0 & 0 & 0 & 14 & 18 \\ 8 & 15 & 9 & 9 & 4 & 7 & 0 \end{bmatrix}$$

$$= \begin{bmatrix} 64 & 97 & 41 & 46 & 26 & 50 & 41 \\ 79 & 116 & 41 & 46 & 26 & 64 & 59 \\ 72 & 112 & 50 & 55 & 30 & 57 & 41 \end{bmatrix}$$ Matrix multiplication

The coded message sent will be the numbers from column 1 of AM, followed by the numbers from column 2, and so on.

64 79 72 97 116 112 41 41 50 46 46 55 26 26 30 50 64 57 41 59 41

Step 5 To decode the message in Step 4, write the numbers received in the message in groups of three and as columns of the matrix AM. Find the **decoding matrix** A^{-1} of the coding matrix A. You can verify that

$$A^{-1} = \begin{bmatrix} 6 & -2 & -3 \\ -1 & 1 & 0 \\ -1 & 0 & 1 \end{bmatrix}$$

Step 6 Compute the message matrix M.

$$M = A^{-1}(AM) = \begin{bmatrix} 6 & -2 & -3 \\ -1 & 1 & 0 \\ -1 & 0 & 1 \end{bmatrix} \begin{bmatrix} 64 & 97 & 41 & 46 & 26 & 50 & 41 \\ 79 & 116 & 41 & 46 & 26 & 64 & 59 \\ 72 & 112 & 50 & 55 & 30 & 57 & 41 \end{bmatrix}$$

$$= \begin{bmatrix} 10 & 14 & 14 & 19 & 14 & 1 & 5 \\ 15 & 19 & 0 & 0 & 0 & 14 & 18 \\ 8 & 15 & 9 & 9 & 4 & 7 & 0 \end{bmatrix} \quad \text{Matrix multiplication}$$

Step 7 Finally, convert the numerical message from the matrix M in Step 6 back into English.

JOHNSON IS IN DANGER.

Practice Problem 8 Rework Example 8, replacing the message JOHNSON IS IN DANGER with the message JACK IS NOW SAFE.

Answers to Practice Problems

1. $\begin{bmatrix} 3 & 2 \\ 2 & 1 \end{bmatrix} \begin{bmatrix} -1 & 2 \\ 2 & -3 \end{bmatrix} = \begin{bmatrix} 1 & 0 \\ 0 & 1 \end{bmatrix}$

2. $\begin{bmatrix} 3x + z & 3y + w \\ 3x + z & 3y + w \end{bmatrix} \neq \begin{bmatrix} 1 & 0 \\ 0 & 1 \end{bmatrix}$ because this implies that $1 = 3x + z = 0$, which is false.

3. The inverse of matrix A does not exist.

4. $A^{-1} = \begin{bmatrix} \dfrac{1}{7} & -\dfrac{19}{77} & \dfrac{1}{11} \\ 0 & \dfrac{2}{11} & \dfrac{1}{11} \\ \dfrac{2}{7} & -\dfrac{3}{77} & -\dfrac{1}{11} \end{bmatrix}$

5. **a.** The inverse of matrix A does not exist

   b. $B^{-1} = \begin{bmatrix} \dfrac{1}{14} & \dfrac{1}{7} \\ \dfrac{3}{14} & \dfrac{4}{7} \end{bmatrix}$.

6. $\left\{ \left(\dfrac{11}{2}, -\dfrac{9}{2}, \dfrac{1}{2} \right) \right\}$

7. $X = \begin{bmatrix} \dfrac{110,400}{19} \\ \dfrac{135,200}{19} \end{bmatrix}$

8. $AM = \begin{bmatrix} 1 & 2 & 3 \\ 1 & 3 & 3 \\ 1 & 2 & 4 \end{bmatrix} \begin{bmatrix} 10 & 11 & 19 & 15 & 19 & 5 \\ 1 & 0 & 0 & 23 & 1 & 0 \\ 3 & 9 & 14 & 0 & 6 & 0 \end{bmatrix}$

   $= \begin{bmatrix} 21 & 38 & 61 & 61 & 39 & 5 \\ 22 & 38 & 61 & 84 & 40 & 5 \\ 24 & 47 & 75 & 61 & 45 & 5 \end{bmatrix}$

■ **SECTION 7.8** **Exercises**

Concepts and Vocabulary

1. For $n \times n$ matrices A and B, if $AB = I = BA$, then B is called the _____ of A.

2. An $n \times n$ matrix A is invertible if there is a matrix B such that _____ $= I$.

3. To find the inverse of an invertible matrix A, we transform $[A \mid I]$ by a sequence of row operations into $[I \mid B]$, where $B =$ _____

4. $A = \begin{bmatrix} a & b \\ c & d \end{bmatrix}$ is invertible if and only if _____.

5. **True or False.** Every square matrix is invertible.

6. **True or False.** $A = \begin{bmatrix} 8 & 16 \\ 4 & 8 \end{bmatrix}$ is invertible.

7. **True or False.** The inverse of $A = \begin{bmatrix} a & b \\ c & d \end{bmatrix}$ is $\dfrac{1}{ad - bc} \begin{bmatrix} d & -b \\ -c & a \end{bmatrix}$ if $ad - bc \neq 0$.

8. **True or False.** If A is a square, invertible matrix and $AX = B$, then $X = A^{-1}B$.

Building Skills

In Exercises 9–18, determine whether B is the inverse of A by computing AB and BA.

9. $A = \begin{bmatrix} 1 & 2 \\ 1 & 3 \end{bmatrix}, B = \begin{bmatrix} 3 & -2 \\ -1 & 1 \end{bmatrix}$

10. $A = \begin{bmatrix} 3 & 2 \\ 4 & 3 \end{bmatrix}, B = \begin{bmatrix} 3 & -2 \\ -4 & 3 \end{bmatrix}$

11. $A = \begin{bmatrix} 3 & 2 \\ 1 & 4 \end{bmatrix}, B = \begin{bmatrix} \dfrac{2}{5} & -\dfrac{1}{5} \\ -\dfrac{1}{10} & \dfrac{3}{10} \end{bmatrix}$

12. $A = \begin{bmatrix} 2 & -3 \\ 4 & -3 \end{bmatrix}, B = \begin{bmatrix} -\dfrac{1}{2} & \dfrac{1}{2} \\ -\dfrac{2}{3} & \dfrac{1}{3} \end{bmatrix}$

13. $A = \begin{bmatrix} 1 & 0 & -1 \\ -1 & 1 & 1 \end{bmatrix}, B = \begin{bmatrix} 2 & 1 \\ 1 & 1 \\ 1 & 1 \end{bmatrix}$

14. $A = \begin{bmatrix} -2 & 1 & 3 \\ 0 & -1 & 1 \\ 1 & 2 & 0 \end{bmatrix}, B = \dfrac{1}{8}\begin{bmatrix} -2 & 6 & 4 \\ 1 & -3 & 2 \\ 1 & 5 & 2 \end{bmatrix}$

15. $A = \begin{bmatrix} 1 & -2 & 1 \\ -8 & 6 & -2 \\ 5 & -3 & 1 \end{bmatrix}, B = \dfrac{1}{2}\begin{bmatrix} 0 & 1 & 2 \\ 1 & 2 & 3 \\ 3 & 1 & 1 \end{bmatrix}$

16. $A = \begin{bmatrix} 2 & 3 & 1 \\ 1 & 2 & 3 \\ 3 & 1 & 2 \end{bmatrix}, B = \dfrac{1}{18}\begin{bmatrix} 1 & -5 & 7 \\ 7 & 1 & -5 \\ -5 & 7 & 1 \end{bmatrix}$

17. $A = \begin{bmatrix} 1 & 1 & 1 \\ 1 & 2 & 3 \\ -1 & 1 & -1 \end{bmatrix}, B = \dfrac{1}{4}\begin{bmatrix} 5 & -2 & -1 \\ 2 & 0 & 2 \\ -3 & 2 & -1 \end{bmatrix}$

18. $A = \begin{bmatrix} 1 & -1 & 0 \\ 0 & 1 & -1 \\ 1 & 0 & 1 \end{bmatrix}, B = \dfrac{1}{2}\begin{bmatrix} 1 & 1 & 1 \\ -1 & 1 & 1 \\ -1 & -1 & 1 \end{bmatrix}$

In Exercises 19–28, find A^{-1} (if it exists) by forming the matrix $[A \mid I]$ and using row operation to obtain $[I \mid B]$, where $B = A^{-1}$.

19. $A = \begin{bmatrix} 2 & 0 \\ 1 & 3 \end{bmatrix}$

20. $A = \begin{bmatrix} 4 & 3 \\ 1 & 0 \end{bmatrix}$

21. $A = \begin{bmatrix} 2 & 4 \\ 3 & 6 \end{bmatrix}$

22. $A = \begin{bmatrix} 9 & 6 \\ 6 & 4 \end{bmatrix}$

23. $A = \begin{bmatrix} 1 & 6 & 4 \\ 0 & 2 & 3 \\ 0 & 1 & 2 \end{bmatrix}$

24. $A = \begin{bmatrix} -2 & 1 & 3 \\ 0 & -1 & 1 \\ 1 & 2 & 0 \end{bmatrix}$

25. $A = \begin{bmatrix} 2 & 4 & 3 \\ 0 & 1 & 1 \\ 2 & 2 & -1 \end{bmatrix}$

26. $A = \begin{bmatrix} 1 & 1 & 5 \\ 4 & 3 & -5 \\ 1 & 1 & 0 \end{bmatrix}$

27. $A = \begin{bmatrix} 1 & 2 & -1 \\ -1 & 1 & 2 \\ 2 & -1 & 1 \end{bmatrix}$

28. $A = \begin{bmatrix} 1 & 2 & 3 \\ 2 & 3 & 0 \\ 0 & 1 & 2 \end{bmatrix}$

In Exercises 29–34, find the inverse (if it exists) of $A = \begin{bmatrix} a & b \\ c & d \end{bmatrix}$ by using the formula $A^{-1} = \dfrac{1}{ad - bc}\begin{bmatrix} d & -b \\ -c & a \end{bmatrix}$. Check that $A^{-1}A = I$.

29. $A = \begin{bmatrix} 1 & 0 \\ 3 & 2 \end{bmatrix}$

30. $A = \begin{bmatrix} 3 & 4 \\ 5 & 6 \end{bmatrix}$

31. $A = \begin{bmatrix} 2 & -3 \\ -3 & 5 \end{bmatrix}$

32. $A = \begin{bmatrix} 3 & -4 \\ 2 & -2 \end{bmatrix}$

33. $A = \begin{bmatrix} a & -b \\ b & -a \end{bmatrix}, a^2 \neq b^2$

34. $A = \begin{bmatrix} 2 & -1 \\ 1 & -1 \end{bmatrix}$

In Exercises 35–38, write the given linear system of equations as a matrix equation $AX = B$.

35. $\begin{cases} 2x + 3y = -9 \\ x - 3y = 13 \end{cases}$

36. $\begin{cases} 5x - 4y = 7 \\ 4x - 3y = 5 \end{cases}$

37. $\begin{cases} 3x + 2y + z = 8 \\ 2x + y + 3z = 7 \\ x + 3y + 2z = 9 \end{cases}$

38. $\begin{cases} x + 3y + z = 4 \\ x - 5y + 2z = 7 \\ 3x + y - 4z = -9 \end{cases}$

In Exercises 39–42, write each matrix equation as a linear system of equations.

39. $\begin{bmatrix} 1 & -2 \\ 2 & 1 \end{bmatrix}\begin{bmatrix} x \\ y \end{bmatrix} = \begin{bmatrix} 0 \\ 5 \end{bmatrix}$

40. $\begin{bmatrix} 2 & 3 \\ 3 & -1 \end{bmatrix}\begin{bmatrix} x_1 \\ x_2 \end{bmatrix} = \begin{bmatrix} 0 \\ 11 \end{bmatrix}$

41. $\begin{bmatrix} 2 & 3 & 1 \\ 5 & 7 & -1 \\ 4 & 3 & 0 \end{bmatrix}\begin{bmatrix} x_1 \\ x_2 \\ x_3 \end{bmatrix} = \begin{bmatrix} -1 \\ 5 \\ 5 \end{bmatrix}$

42. $\begin{bmatrix} 3 & -2 & 3 \\ 5 & 0 & 4 \\ 2 & 7 & 0 \end{bmatrix}\begin{bmatrix} r \\ s \\ t \end{bmatrix} = \begin{bmatrix} 4 \\ 3 \\ -8 \end{bmatrix}$

43. Let $A = \begin{bmatrix} 1 & 2 & 5 \\ 2 & 3 & 8 \\ -1 & 1 & 2 \end{bmatrix}$. Show that $A^{-1} = \begin{bmatrix} 2 & -1 & -1 \\ 12 & -7 & -2 \\ -5 & 3 & 1 \end{bmatrix}$.

In Exercises 44–46, use the result of Exercise 43 to solve the system of equations.

44. $\begin{cases} x + 2y + 5z = 4 \\ 2x + 3y + 8z = 6 \\ -x + y + 2z = 3 \end{cases}$

45. $\begin{cases} x_1 + 2x_2 + 5x_3 = 1 \\ 2x_1 + 3x_2 + 8x_3 = 3 \\ -x_1 + x_2 + 2x_3 = -3 \end{cases}$

46. $\begin{cases} x + 2y + 5z = -4 \\ 2x + 3y + 8z = -6 \\ -x + y + 2z = -\dfrac{5}{2} \end{cases}$

47. a. Find the inverse of the matrix $A = \begin{bmatrix} 1 & 1 & 1 \\ 1 & 2 & 3 \\ 1 & 4 & 9 \end{bmatrix}$.

b. Use the result of part (a) to solve the linear system.

$$\begin{cases} x + y + z = 6 \\ x + 2y + 3z = 14 \\ x + 4y + 9z = 36 \end{cases}$$

48. a. Find the inverse of the matrix $A = \begin{bmatrix} 2 & 4 & -1 \\ 3 & 1 & 2 \\ 1 & 3 & -3 \end{bmatrix}$.

b. Use the result of part (a) to solve the linear system.

$$\begin{cases} 2x + 4y - z = 9 \\ 3x + y + 2z = 7 \\ x + 3y - 3z = 4 \end{cases}$$

Applying the Concepts

In Exercises 49–54, solve each linear system by using an inverse matrix.

49. $\begin{cases} 3x + 7y = 11 \\ -5x + 4y = 13 \end{cases}$

50. $\begin{cases} x - 7y = 3 \\ 2x + 3y = 23 \end{cases}$

51. $\begin{cases} x + y + 2z = 7 \\ x - y - 3z = -6 \\ 2x + 3y + z = 4 \end{cases}$

52. $\begin{cases} x + y + z = 6 \\ 2x - 3y + 3z = 5 \\ 3x - 2y - z = -4 \end{cases}$

53. $\begin{cases} 2x + 2y + 3z = 7 \\ 5x + 3y + 5z = 3 \\ 3x + 5y + z = -5 \end{cases}$

54. $\begin{cases} 9x + 7y + 4z = 12 \\ 6x + 5y + 4z = 5 \\ 4x + 3y + z = 7 \end{cases}$

In Exercises 55–58, use the method presented in this section to solve the problem.

55. Investment. Liz inherited $90,000, and she split it into three investments. Part of the money she invested in a Treasury bill that yields 3% annual interest, part she invested in bonds with an annual yield of 7%, and the rest she invested in a mutual fund. In 2011, when the mutual fund lost 11%, her net income from all three investments was $980. If the mutual fund had gained 8%, her net income from all three investments would have been $5920. How much money did Liz put in each investment?

56. Sandwich sales. The Fresh Sandwich specializes in chicken, fish, and ham sandwiches. In April, May, and June 2012, they sold the following quantities.

	Sales Volume in Hundreds		
	April	**May**	**June**
Chicken	8	7	4
Fish	11	8	6
Ham	6	7	3
Monthly revenue, in hundreds, from the sales of these sandwiches	$164	$141	$86

Find the selling price of each of the three types of sandwiches.

57. Euler's formula. A plane graph connects points called vertices with nonintersecting lines called edges and divides the plane into regions. The accompanying figure has seven vertices, nine edges, and four regions. Euler's theorem states that in a plane graph,

$$V - E + R = 2,$$

where V = the number of vertices,
E = the number of edges, and
R = the number of regions.

In a certain plane graph, twice the number of edges is three times the number of vertices, and twice the number of regions is one less than the number of edges. Find the number of vertices, edges, and regions for this graph. Sketch a plane graph that satisfies the conditions of this exercise.

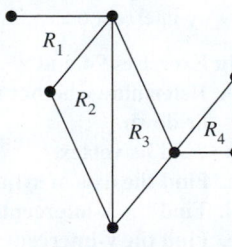

58. Projectile. On a planet the height h of a projectile fired up from an initial height h_0 with initial vertical velocity of v_0 ft/sec after t seconds is given by $h = \dfrac{1}{2}at^2 + v_0 t + h_0$, where a is the gravitational constant. The following measurements were recorded.

Time t in seconds	1	2	3
Height h in feet	100	92	76

a. Find the initial height.
b. Find the height of the projectile after 2.5 seconds.

In Exercises 59–62, technology matrix A, representing inter-industry demand, and matrix D, representing consumer demand for the sectors in a two- or three-sector economy, are given. Use the matrix equation $X = (I - A)^{-1}D$ to find the production level X that will satisfy both demands.

59. Two-sector economy. In a two-sector economy,

$$A = \begin{bmatrix} 0.1 & 0.4 \\ 0.5 & 0.2 \end{bmatrix}, D = \begin{bmatrix} 50 \\ 30 \end{bmatrix}.$$

60. Two-sector economy. In a two-sector economy,

$$A = \begin{bmatrix} 0.2 & 0.5 \\ 0.1 & 0.6 \end{bmatrix}, D = \begin{bmatrix} 11 \\ 18 \end{bmatrix}.$$

61. Three-sector economy. In a three-sector economy,

$$A = \begin{bmatrix} 0.2 & 0.2 & 0 \\ 0.1 & 0.1 & 0.3 \\ 0.1 & 0 & 0.2 \end{bmatrix}, D = \begin{bmatrix} 400 \\ 600 \\ 800 \end{bmatrix}.$$

62. Three-sector economy. In a three-sector economy,

$$A = \begin{bmatrix} 0.5 & 0.4 & 0.2 \\ 0.2 & 0.3 & 0.1 \\ 0.1 & 0.1 & 0.3 \end{bmatrix}, D = \begin{bmatrix} 500 \\ 300 \\ 200 \end{bmatrix}.$$

In Exercises 63 and 64, use a 2 × 2 invertible coding matrix A of your choice. Remember to partition the numerical message into groups of two. (a) Write a cryptogram for each message. (b) Check by decoding.

63. Coding and decoding. CANNOT FIND COLOMBO.

64. Coding and decoding. FOUND HEAD OF TERRORIST NETWORK.

In Exercises 65 and 66, use a 3 × 3 invertible coding matrix A of your choice.

65. Code and then decode the message of Exercise 63.

66. Code and then decode the message of Exercise 64.

Beyond the Basics

67. Suppose A is an invertible square matrix.
a. Is A^{-1} invertible? Why or why not?
b. If your answer to part (a) is yes, then what is $(A^{-1})^{-1}$?

68. Let $A^{-1} = \begin{bmatrix} 2 & 1 \\ 3 & -4 \end{bmatrix}$. Find A.

69. Let A be a square matrix. Show that if A^2 is invertible, then A must be invertible. [Hint: $A^2B = A(AB)$.]

70. Show that if A is invertible, then A^2 is invertible.

71. Show that if A and B are $n \times n$ invertible matrices, then AB is invertible and $(AB)^{-1} = B^{-1}A^{-1}$. (Note the order.)

72. Verify the result of Exercise 71 for

$$A = \begin{bmatrix} 1 & 2 \\ 3 & 4 \end{bmatrix}, B = \begin{bmatrix} -1 & 0 \\ 3 & -2 \end{bmatrix}.$$

73. Let A, B, and C be $n \times n$ matrices such that $AB = AC$. Show that if A is invertible, then $B = C$.

74. Is the result of Exercise 19 true if A is not invertible?

75. Let A and B be $n \times n$ matrices such that $AB = A$. Show that if A is invertible, then $B = I$.

76. Give an example to show that the result of Exercise 75 is not true if A is not invertible.

77. Let $A = \begin{bmatrix} 3 & 4 \\ 2 & 3 \end{bmatrix}$.

  a. Show that the matrix A satisfies the equation

$$A^2 - 6A + I = 0.$$

  b. Use the equation in part (a) to show that $A^{-1} = 6I - A$.

  c. Use the equation in part (b) to find A^{-1}.

 78. Let $A = \begin{bmatrix} 2 & -1 & 1 \\ -1 & 2 & -1 \\ 1 & -1 & 2 \end{bmatrix}$. Assume that A is invertible.

  a. Show that A satisfies the equation
$A^3 - 6A^2 + 9A - 4I = 0.$

  b. Use the equation in part (a) to show that

$$A^{-1} = \frac{1}{4}(A^2 - 6A + 9I).$$

  [*Hint:* Consider the product $\frac{1}{4}(A^2 - 6A + 9I)A$.]

  c. Use part (b) to find A^{-1}.

 In Exercises 79 and 80, use your graphing utility to find the inverse of the coefficient matrix. Then use the inverse to solve each linear system.

79. $\begin{cases} x - y - z - w = -4 \\ x + y + z - w = 2 \\ 2x + y + z - w = 3 \\ x - y + z - w = -2 \end{cases}$

80. $\begin{cases} x + y + z - w = 1 \\ x - y - z + w = 1 \\ x - y + z + 3w = 1 \\ x + y + z + w = 3 \end{cases}$

81. Suppose A and B are 2×2 matrices and $B^{-1} = \begin{bmatrix} 1 & 2 \\ 3 & 4 \end{bmatrix}$.

Find A

  a. if $AB = \begin{bmatrix} -1 & 3 \\ 0 & 2 \end{bmatrix}$.

  b. if $BA = \begin{bmatrix} -1 & 3 \\ 0 & 2 \end{bmatrix}$.

  c. if $BAB^{-1} = \begin{bmatrix} 1 & 1 \\ 1 & 2 \end{bmatrix}$.

  d. if $B^{-1}AB = \begin{bmatrix} 1 & 1 \\ 1 & 2 \end{bmatrix}$.

82. a. Is the matrix $\begin{bmatrix} x & -1 \\ 2 & x + 2 \end{bmatrix}$ invertible for all real numbers x?

  b. Is the matrix $\begin{bmatrix} x - 1 & -1 \\ 3 & x - 5 \end{bmatrix}$ invertible for all real numbers x?

Critical Thinking / Discussion / Writing

83. True or False. Justify your answers.

  a. If A and B are matrices for which $AB = BA$, then the formula $A^2B = BA^2$ must hold.

  b. If A and B are invertible matrices, then $A + B$ must be invertible.

84. Give an example of a 2×2 matrix A such that $A^2 = 0$. Show that the matrix $I + A$ is invertible and that $(I + A)^{-1} = I - A$.

85. Is it true that every nonsingular matrix is a square matrix? Why or why not?

Getting Ready for the Next Section

In Exercises 86–89, solve each equation for y.

86. $x^2 = 8y$

87. $(x + 3)^2 = 4(y - 1)$

88. $(x - 2)^2 = 6(y - 3)$

89. $(x - 5)^2 = -2(y + 7)$

In Exercises 90–93, find the quadratic function, $y = f(x)$, with the given properties.

90. Having form $y = ax^2$, and passing through the point $(-2, -2)$

91. Having form $y = ax^2 + bx + c$, vertex $(0, 0)$, and passing through the point $(2, -20)$

92. Having form $y = ax^2 + bx + c$, vertex $(-3, -8)$, and passing through the point $(0, 10)$

93. Having form $y = ax^2 + bx + c$, x-intercepts 1 and -3, and y-intercept 6

In Exercises 94 and 95,

a. Determine whether the given quadratic function opens up or down.

b. Find its vertex.

c. Find the axis of symmetry.

d. Find the x-intercepts, if any.

e. Find the y-intercept.

94. $f(x) = 2x^2 + 4x - 6$ **95.** $f(x) = -2x^2 + 12x - 10$

SUMMARY Definitions, Concepts, and Formulas

7.1 Systems of Equations in Two Variables

A **system of equations** is a set of equations with common variables.

A **solution** of a system of equations in two variables x and y is an ordered pair of numbers (a, b) that satisfies all equations in the system. The solution set of a system of two linear equations of the form $ax + by = c$ is the point(s) of intersection of the graphs of the equations.

Systems with no solution are called **inconsistent**, and systems with at least one solution are **consistent**. A system of two linear equations in two variables may have one solution (**independent equations**), no solution (**inconsistent system**), or infinitely many solutions (**dependent equations**).

Three methods of solving a system of equations are (1) the graphical method, (2) the substitution method, and (3) the elimination or addition method.

7.2 Systems of Linear Equations in Three Variables

A **linear equation** in n variables $x_1, x_2, \ldots , x_n$ is an equation that can be written in the form

$$a_1x_1 + a_2x_2 + \cdots + a_nx_n = b,$$

where b and the coefficients $a_1, a_2, \ldots , a_n$ are real numbers.

A system of linear equations can be solved by transforming it into an *equivalent system* that is easier to solve.

The following operations produce equivalent systems:

1. Interchange the position of any two equations.
2. Multiply any equation by a nonzero constant.
3. Add a nonzero multiple of one equation to another.

The **Gaussian elimination method** is a procedure for converting a system of linear equations into an equivalent system in *triangular form*.

A linear system may have **(i)** only one solution, **(ii)** infinitely many solutions, or **(iii)** no solution.

7.3 Systems of Inequalities

A statement of the form $ax + by < c$ is a linear inequality in the variables x and y. The symbol $<$ may be replaced by $\leq$, $>$, or $\geq$. A procedure for graphing a linear inequality in two variables is described on page 620.

The graph of the solution set of a system of inequalities is obtained by **(i)** graphing each inequality of the system in the same coordinate plane and then **(ii)** finding the region that is common to every graph in the system.

Linear Programming

In a linear programming model, a quantity f (to be maximized or minimized) that can be expressed in the form $f = ax + by$ is called an **objective function** of the variables x and y. The **constraints** are the restrictions placed on the variables x and y that can be expressed as a system of linear inequalities.

A procedure for solving a linear programming problem is given on page 624.

Systems of Nonlinear Inequalities

A system of equations (inequalities) in which at least one equation (inequality) is nonlinear is called a **system of nonlinear equations (inequalities)**. The methods of substitution, elimination, or graphing are used to solve a nonlinear system.

7.4 Matrices and Systems of Equations

i. **Matrix.** An $m \times n$ matrix is a rectangular array of numbers with m rows and n columns.

ii. **Augmented matrix.** The augmented matrix of a system of linear equations is a matrix whose entries are the coefficients of the variables in the system, together with the constants.

iii. **Elementary row operations.** For an augmented matrix of a system of linear equations, the following elementary row operations will result in the matrix of an equivalent system:

1. Interchange any two rows.
2. Multiply any row by a nonzero constant.
3. Add a multiple of one row to another row.

iv. **Row-echelon form and reduced row-echelon form.** The elementary row operations are used to transform an augmented matrix to row-echelon form or reduced row-echelon form. (See page 639.)

v. **Gaussian elimination** is the method of transforming the augmented matrix to row-echelon form and then using back-substitution to find the solution set of a system of linear equations.

vi. **Gauss–Jordan elimination.** If you continue the Gaussian elimination procedure until a reduced row-echelon form is obtained, the procedure is called Gauss–Jordan elimination.

7.5 Determinants and Cramer's Rule

i. **Determinant.** The determinant of a square matrix A is a real number denoted by $\det(A)$ or $|A|$.

ii. The determinant of a 2×2 matrix $A = \begin{bmatrix} a & b \\ c & d \end{bmatrix}$ is defined by $\det(A) = \begin{vmatrix} a & b \\ c & d \end{vmatrix} = ad - bc$.

iii. The determinant of a matrix of order $n \geq 3$ can be found by expanding by cofactors. (See page 652.)

iv. **Cramer's Rule** gives formulas for solving a square system of linear equations for which the determinant of the coefficient matrix is nonzero. (See page 653 for solving a system of two equations in two variables and page 654 for solving a system of three equations in three variables.)

7.6 Partial-Fraction Decomposition

In partial-fraction decomposition, we reverse the addition of rational expressions. A rational expression $\dfrac{P(x)}{Q(x)}$ is called a proper fraction if $\deg P(x) < \deg Q(x)$.

There are four cases to consider in decomposing $\dfrac{P(x)}{Q(x)}$ into partial fractions:

i. $Q(x)$ has only distinct linear factors. (See page 662.)

ii. $Q(x)$ has repeated linear factors. (See page 664.)

iii. $Q(x)$ has distinct irreducible quadratic factors. (See page 665.)

iv. $Q(x)$ has repeated irreducible quadratic factors. (See page 666.)

7.7 Matrix Algebra

i. **Equality of matrices.** Two matrices $A = [a_{ij}]$ and $B = [b_{ij}]$ are equal if **(i)** A and B have the same order $m \times n$ and **(ii)** all corresponding entries are equal.

ii. **Matrix addition.** The matrices A and B can be added or subtracted if they have the same order. Their sum or difference can be obtained by adding or subtracting the corresponding entries.

iii. **Scalar multiplication.** The product of a matrix by a real number (scalar) is obtained by multiplying every entry of the matrix by the real number.

iv. **Matrix multiplication.** The product AB of two matrices is defined only if the number of columns of A is equal to the number of rows of B. If A is an $m \times p$ matrix and B is a

$p \times n$ matrix, then the product AB is an $m \times n$ matrix. The (i, j)th entry of AB is the sum of the products of the corresponding entries in the ith row of A and the jth column of B.

v. **The identity matrix.** The $n \times n$ identity matrix denoted by I_n or I is a matrix that has 1s on the main diagonal and 0s elsewhere. If A is an $n \times n$ matrix, then $AI = IA = A$.

7.8 The Matrix Inverse

i. **Inverse of a matrix.** Let A be an $n \times n$ matrix and I be the $n \times n$ identity matrix. If there is an $n \times n$ matrix B such that $AB = BA = I$, then A is **invertible** and B is called the **inverse** of A. We write $B = A^{-1}$.

ii. **Finding the inverse.** A procedure for finding A^{-1} (if it exists) is given on page 688.

iii. **Inverse of a 2 × 2 matrix.** The matrix $A = \begin{bmatrix} a & b \\ c & d \end{bmatrix}$ has an inverse if and only if $ad - bc \neq 0$. Moreover, if $ad - bc \neq 0$, then

$$A^{-1} = \frac{1}{ad - bc} \begin{bmatrix} d & -b \\ -c & a \end{bmatrix}.$$

iv. If A is invertible, then the solution of the matrix equation $AX = B$ is $X = A^{-1}B$.

REVIEW EXERCISES

Building Skills

In Exercises 1–18, solve each system of equations by using the method of your choice. Identify systems with no solution and systems with infinitely many solutions.

1. $\begin{cases} 3x - y = -5 \\ x + 2y = 3 \end{cases}$

2. $\begin{cases} x + 3y + 6 = 0 \\ y = 4x - 2 \end{cases}$

3. $\begin{cases} 2x + 4y = 3 \\ 3x + 6y = 10 \end{cases}$

4. $\begin{cases} x - y = 2 \\ 2x - 2y = 9 \end{cases}$

5. $\begin{cases} 3x - y = 3 \\ \dfrac{1}{2}x + \dfrac{1}{3}y = 2 \end{cases}$

6. $\begin{cases} 0.02y - 0.03x = -0.04 \\ 1.5x - y = 3 \end{cases}$

7. $\begin{cases} x + 3y + z = 0 \\ 2x - y + z = 5 \\ 3x - 3y + 2z = 10 \end{cases}$

8. $\begin{cases} 2x + y = 11 \\ 3y - z = 5 \\ x + 2z = 1 \end{cases}$

9. $\begin{cases} x + y = 1 \\ 3y + 2z = 0 \\ 2x - 3z = 7 \end{cases}$

10. $\begin{cases} 2x - 3y + z = 2 \\ x - 3y + 2z = -1 \\ 2x + 3y + 2z = 3 \end{cases}$

11. $\begin{cases} x + y + z = 1 \\ x + 5y + 5z = -1 \\ 3x - y - z = 4 \end{cases}$

12. $\begin{cases} x + 3y - 2z = -4 \\ 2x + 6y - 4z = 3 \\ x + y + z = 1 \end{cases}$

13. $\begin{cases} x + 4y + 3z = 1 \\ 2x + 5y + 4z = 4 \end{cases}$

14. $\begin{cases} 3x - y + 2z = 9 \\ x - 2y + 3z = 2 \end{cases}$

15. $\begin{cases} 3x - y = 2 \\ x + 2y = 9 \\ 3x + y = 10 \end{cases}$

16. $\begin{cases} x - 2y = 1 \\ 3x + 4y = 11 \\ 2x + 2y = 7 \end{cases}$

17. $\begin{cases} x + y + z = 3 \\ 2x - y + z = 4 \\ x + 4y + 2z = 5 \end{cases}$

18. $\begin{cases} x - y + z = 4 \\ x + 2y + 3z = 2 \\ 2x + y + 4z = 6 \end{cases}$

In Exercises 19–22, solve each nonlinear system of equations.

19. $\begin{cases} x + 3y = 1 \\ x^2 - 3x = 7y + 3 \end{cases}$

20. $\begin{cases} 4x - y = 3 \\ y^2 - 2y = x - 2 \end{cases}$

21. $\begin{cases} x - y = 4 \\ 5x^2 + y^2 = 24 \end{cases}$

22. $\begin{cases} x - 2y = 7 \\ 2x^2 + 3y^2 = 29 \end{cases}$

In Exercises 23–34, graph each system of inequalities.

23. $\begin{cases} x + y \leq 1 \\ x - y \leq 1 \\ x \geq 0 \end{cases}$

24. $\begin{cases} 2x + 3y \leq 6 \\ 4x - 3y \leq 12 \\ x \geq 0 \end{cases}$

25. $\begin{cases} 4x + y \leq 7 \\ 2x + 5y \geq -1 \\ x - 2y \geq -5 \end{cases}$

26. $\begin{cases} 7x - 2y + 6 \leq 0 \\ x + y + 14 \geq 0 \\ 2x - 3y + 9 \geq 0 \end{cases}$

27. $\begin{cases} x + y - 3 \leq 0 \\ 3x - y - 9 \leq 0 \\ y + 3 \geq 0 \\ 7x + 4y + 23 \geq 0 \end{cases}$

28. $\begin{cases} x + 5y - 11 \leq 0 \\ 5x + y - 7 \leq 0 \\ x - 5y - 17 \leq 0 \\ 7x + y + 25 \geq 0 \end{cases}$

29. $\begin{cases} x - y \leq 1 \\ x^2 + y^2 \leq 13 \end{cases}$

30. $\begin{cases} x - y \geq 1 \\ x^2 + y^2 \leq 13 \end{cases}$

31. $\begin{cases} x - y \leq 1 \\ x^2 + y^2 \leq 13 \\ x \geq 0 \\ y \geq 0 \end{cases}$

32. $\begin{cases} x - y \geq 1 \\ x^2 + y^2 \geq 13 \\ x \leq 4 \\ y \geq 0 \end{cases}$

33. $\begin{cases} y \leq 3x \\ x^2 + y^2 \leq 25 \\ x \geq 0 \\ y \geq 0 \end{cases}$

34. $\begin{cases} y \geq 3x \\ x^2 + y^2 \leq 25 \\ x \geq 0 \\ y \geq 0 \end{cases}$

In Exercises 35 and 36, convert each matrix to row-echelon form.

35. $\begin{bmatrix} 0 & 1 & 2 & 1 \\ 2 & 0 & 3 & 4 \\ 1 & -2 & 1 & 7 \end{bmatrix}$
36. $\begin{bmatrix} 3 & -1 & 2 & 12 \\ 1 & 1 & -1 & 2 \\ 1 & 2 & 1 & 7 \end{bmatrix}$

In Exercises 37 and 38, convert each matrix to reduced row-echelon form.

37. $\begin{bmatrix} 3 & 1 & 3 & 1 \\ 2 & 1 & 1 & 1 \\ 1 & -1 & -1 & 0 \end{bmatrix}$
38. $\begin{bmatrix} 3 & 4 & -4 & 2 \\ 2 & 1 & -2 & 1 \\ 1 & 1 & 2 & 1 \end{bmatrix}$

39. Let $A = \begin{bmatrix} 1 & 2 \\ -3 & 4 \end{bmatrix}$ and $B = \begin{bmatrix} 2 & -3 \\ -5 & 6 \end{bmatrix}$. Find each of the following.

   a. $A + B$ b. $A - B$ c. $2A$
   d. $-3B$ e. $2A - 3B$

40. Let $A = \begin{bmatrix} 2 & 0 & -1 \\ 1 & 2 & 2 \\ -2 & 0 & 3 \end{bmatrix}$ and $B = \begin{bmatrix} 0 & 1 & -1 \\ 1 & -1 & 0 \\ -1 & 0 & 1 \end{bmatrix}$.

   Find each of the following.

   a. $A + B$ b. $A - B$ c. $2A + 3B$

41. For the matrices A and B of Exercise 39, solve the matrix equation $3A + 2B - 3X = 0$ for X.

42. For the matrices A and B of Exercise 40, solve the matrix equation $A - 2X + 2B = 0$ for X.

In Exercises 43–48, find the partial-fraction decomposition of each rational expression.

43. $\dfrac{x + 4}{x^2 + 5x + 6}$
44. $\dfrac{x + 14}{x^2 + 3x - 4}$

45. $\dfrac{3x^2 + x + 1}{x(x - 1)^2}$
46. $\dfrac{3x^2 + 2x + 3}{(x^2 - 1)(x + 1)}$

47. $\dfrac{x^2 + 2x + 3}{(x^2 + 4)^2}$
48. $\dfrac{2x}{x^4 - 1}$

In Exercises 49–52, find the following products if possible.
a. AB b. BA

49. $A = \begin{bmatrix} 0 & 1 \\ 2 & 3 \end{bmatrix}$, $B = \begin{bmatrix} -1 & -1 \\ -3 & 4 \end{bmatrix}$

50. $A = \begin{bmatrix} 0 & 1 & 2 \\ -1 & 0 & 1 \end{bmatrix}$, $B = \begin{bmatrix} 1 & 2 \\ 3 & -1 \\ 0 & 4 \end{bmatrix}$

51. $A = \begin{bmatrix} 1 & 2 & -1 \end{bmatrix}$, $B = \begin{bmatrix} 2 \\ 3 \\ 1 \end{bmatrix}$

52. $A = \begin{bmatrix} 1 & 0 \\ 2 & -1 \\ 3 & -2 \end{bmatrix}$, $B = \begin{bmatrix} 1 & 2 & 3 & 4 \\ 2 & -1 & 2 & 3 \end{bmatrix}$

53. Find a matrix $A = \begin{bmatrix} x & y \\ z & w \end{bmatrix}$ such that

$$\begin{bmatrix} 1 & 2 \\ 3 & -1 \end{bmatrix} A = \begin{bmatrix} 5 & 6 \\ 1 & -3 \end{bmatrix}$$

54. Find a matrix $B = \begin{bmatrix} x & y \\ z & w \end{bmatrix}$ such that

$$B \begin{bmatrix} 1 & 2 \\ 3 & -1 \end{bmatrix} = \begin{bmatrix} 5 & 6 \\ 1 & -3 \end{bmatrix}$$

   Compare your answers for Exercises 53 and 54.

In Exercises 55–58, find the inverse (if it exists) of each matrix.

55. $\begin{bmatrix} 3 & 1 \\ 2 & 4 \end{bmatrix}$
56. $\begin{bmatrix} 3 & -4 \\ 1 & 2 \end{bmatrix}$

57. $\begin{bmatrix} 1 & 2 & -2 \\ -1 & 3 & 0 \\ 0 & -2 & 1 \end{bmatrix}$
58. $\begin{bmatrix} 1 & 2 & 1 \\ 2 & 2 & 4 \\ 0 & 0 & 3 \end{bmatrix}$

In Exercises 59–62, use an inverse matrix (if possible) to solve each system of linear equations.

59. $\begin{cases} x + 3y + 3z = 3 \\ x + 4y + 3z = 5 \\ x + 3y + 4z = 6 \end{cases}$
60. $\begin{cases} x + 2y + 3z = 6 \\ 2x + 4y + 5z = 8 \\ 3x + 5y + 6z = 10 \end{cases}$

61. $\begin{cases} x - y + z = 3 \\ 4x + 2y = 5 \\ 7x - y - z = 6 \end{cases}$
62. $\begin{cases} x + 2y + 4z = 7 \\ 4x + 3y - 2z = 6 \\ x - 3z = 4 \end{cases}$

In Exercises 63–68, evaluate each determinant and simplify.

63. $\begin{vmatrix} 5 & -4 \\ 2 & 3 \end{vmatrix}$
64. $\begin{vmatrix} \cos\theta & -\sin\theta \\ \sin\theta & \cos\theta \end{vmatrix}$

65. $\begin{vmatrix} x + 1 & x^3 \\ 1 & x^2 - x + 1 \end{vmatrix}$
66. $\begin{vmatrix} \log_b a & 1 \\ 1 & \log_a b \end{vmatrix}$

67. $\begin{vmatrix} 1 & 2 & 1 \\ 0 & 1 & -1 \\ 3 & -2 & 4 \end{vmatrix}$
68. $\begin{vmatrix} 1 & 2 & 3 \\ -1 & 2 & -3 \\ 1 & -2 & 3 \end{vmatrix}$

In Exercises 69–72, use Cramer's Rule to solve the system of equations.

69. $\begin{cases} 2x + y = 17 \\ 3x - 5y = 6 \end{cases}$
70. $\begin{cases} 2x + y = 5 \\ 3x - 5y = 1 \end{cases}$

71. $\begin{cases} x - 4y - z = 11 \\ 2x - 5y + 2z = 39 \\ -3x + 2y + z = 1 \end{cases}$
72. $\begin{cases} x + y = 5 \\ y + z = 3 \\ x + z = 4 \end{cases}$

Applying the Concepts

73. **Investment.** A speculator invested part of $15,000 in a high-risk venture and received a return of 12% at the end of the year. The rest of the $15,000 was invested at 4% annual interest. The combined annual income from the two sources was $1300. How much was invested at each rate?

74. **Agriculture.** A farmer earns a profit of $525 per acre of tomatoes and $475 per acre of soybeans. His soybean acreage is 5 acres more than twice his tomato acreage. If his total profit from the two crops is $24,500, how many acres of tomatoes and how many acres of soybeans does he have?

75. **Geometry.** The area of a rectangle is 63 square feet, and its perimeter is 33 feet. What are the dimensions of the rectangle?

76. Numbers. Twice the sum of the reciprocals of two numbers is 13, and the product of the numbers is $\frac{1}{9}$. Find the numbers.

77. Geometry. The hypotenuse of a right triangle is 17. If one leg of the triangle is increased by 1 and the other leg is increased by 4, the hypotenuse becomes 20. Find the sides of the triangle.

78. Paper route. Chris covers her paper route, which is 21 miles long, by 7:30 A.M. each day. If her average rate of travel were 1 mile faster each hour, she would cover the route by 7 A.M. What time does she start in the morning?

79. Agriculture. A rectangular pasture with an area of 6400 square meters is divided into three smaller pastures by two fences parallel to the shorter sides. The width of two of the smaller pastures is the same, and the width of the third is twice that of the others. Find the dimensions of the original pasture if the perimeter of the larger of the subdivisions is 240 m.

80. Leasing. Budget Rentals leases its compact cars for $23.00 per day plus $0.17 per mile. Dollar Rentals leases the same car for $24.00 per day plus $0.22 per mile.
 a. Find the cost functions describing the daily cost of leasing from each company.
 b. Graph the two functions in part (a) on the same coordinate axes.
 c. From which company should you lease the car if you plan to drive (i) 50 miles per day; (ii) 60 miles per day; (iii) 70 miles per day?

81. Break-even analysis. Auto-Sprinkler Corp. manufactures seven-day, 24-hour variable timers for lawn sprinklers. The corporation has a monthly fixed cost of $60,000 and a production cost of $12 for each timer manufactured. Each timer sells for $20.
 a. Write the cost function and the revenue function for selling x timers per month.
 b. Graph the two functions in part (a) on the same coordinate plane and hence find the break-even point graphically.
 c. Find the break-even point algebraically.
 d. How many timers should be sold in a month to realize a profit (before taxes) of 15% of the cost?

82. Equilibrium quantity and price. The demand equation for a product is $p = \dfrac{4000}{x}$, where p is the price per unit and x is the number of units of the product. The supply equation for the product is $p = \dfrac{x}{20} + 10$.
 a. Find the equilibrium quantity.
 b. Find the equilibrium price.
 c. Graph the supply and demand functions on the same coordinate plane and label the equilibrium point.

83. Apartment lease. Alisha and Sunita signed a lease on an apartment for nine months. At the end of six months, Alisha got married and moved out. She paid the landlady an amount equal to the difference between double-occupancy rental and single-occupancy rental for the remaining three months, and Sunita paid the single-occupancy rate for the same three months. If the nine-month rental cost Alisha $2340 and Sunita $4140, what were the single and double monthly rates?

84. Seating arrangement. The 600 graduating seniors at Central State College are seated in rows, each of which contains the same number of chairs, and every chair is occupied. If five more chairs were in each row, everyone could be seated in four fewer rows. How many chairs are in each row?

85. Passing a final exam. Twenty-six students in a college algebra class took a final exam on which the passing score was 70. The mean score of those who passed was 78, and the mean score of those who failed was 26. The mean of all scores was 72. How many students failed the exam?

86. Finding numbers. The average of a and b is 2.5, the average of b and c is 3.8, and the average of a and c is 3.1. Find the numbers a, b, and c.

87. May–December match. Steve and his wife Janet have the same birthday. When Steve was as old as Janet is now, he was twice as old as Janet was then. When Janet becomes as old as Steve is now, the sum of their ages will be 119. How old are Steve and Janet now?

88. Percentage increase. Let x and y be positive real numbers. Increasing x by y percent gives 46, while increasing y by x percent gives 21. Find x and y.

89. Criminals rob a bank. Three criminals—Butch, Sundance, and Billy—robbed a bank and divided the loot in the following manner: Since Butch planned the job and drove the getaway car, he got 75% as much as Sundance and Billy put together. Since Sundance was an expert safecracker, he got $500 more than 50% of Billy's and Butch's cut put together. Billy got $1000 less than three times the difference of Butch and Sundance's cut. How much did they steal, and what was each criminal's take?

90. Curve fitting. Find all of the curves of the form $y = ax^2 + bx + c$ that contain the points $(0, 1)$, $(1, 0)$, and $(-1, 6)$.

91. Using exponents. Given that $2^x = 8^y$ and $9^y = 3^{x-2}$, find x and y.

92. Area of an octagon. A square of side length 8 has its corners cut off to make it a regular octagon. Find the area of the octagon. (See the figure.)

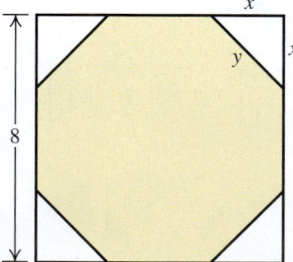

93. Building houses. A builder has 42 units of material and 32 units of labor available during a given period. She builds two-story houses or one-story houses or some of both. Suppose she makes a profit of $10,000 on each two-story house and $4000 on each one-story house. A two-story house requires seven units of material and one unit of labor, whereas a one-story house requires one unit of material and two units of labor. How many houses of each type should she build to make a maximum profit? What is the maximum profit?

94. Minimizing cost. An animal food is to be a mixture of two products X and Y. The content and cost of 1 pound of each product is given in the following table.

Product	Protein grams	Fat grams	Carbohydrates grams	Cost
X	180	2	240	$0.75
Y	36	8	200	$0.56

How much of each product should be used to minimize the cost if each bag must contain at least 612 grams of protein, at least 22 grams of fat, and at most 1880 grams of carbohydrates?

PRACTICE TEST A

In Problems 1–4, solve each system of equations.

1. $\begin{cases} 2x - y = 4 \\ 2x + y = 4 \end{cases}$

2. $\begin{cases} x + 2y = 8 \\ 3x + 6y = 24 \end{cases}$

3. $\begin{cases} -2x + y = 4 \\ 4x - 2y = 4 \end{cases}$

4. $\begin{cases} y = x^2 \\ 3x - y + 4 = 0 \end{cases}$

5. Two gold bars together weigh a total of 485 pounds. One bar weighs 15 pounds more than the other.
   a. Write a system of equations that describes these relationships.
   b. How much does each bar weigh?

6. Use back-substitution to solve the given linear system.
$$\begin{cases} x - 6y + 3z = -2 \\ 9y - 5z = 2 \\ 2z = 10 \end{cases}$$

In Exercises 7–8, solve the system of equations.

7. $\begin{cases} x + y + z = 8 \\ 2x - 2y + 2z = 4 \\ x + y - z = 12 \end{cases}$

8. $\begin{cases} 2x - y + z = 2 \\ x + y - z = -1 \\ x - 5y + 5z = 7 \end{cases}$

9. A vending machine's coin box contains nickels, dimes, and quarters. The total number of coins in the box is 300. The number of dimes is three times the number of nickels and quarters together. If $30.65 is in the box, find the number of nickels, dimes, and quarters that it contains.

For Problems 10 and 11, write the form of the partial-fraction decomposition of the given rational expression. You do not need to solve for the constants.

10. $\dfrac{2x}{(x - 5)(x + 1)}$

11. $\dfrac{-5x^2 + x - 8}{(x - 2)(x^2 + 1)^2}.$

12. Find the partial-fraction decomposition of the rational expression
$$\frac{x + 3}{(x + 4)^2(x - 7)}$$

13. Graph the inequality $3x + y < 6$.

14. Graph the solution set of the system of inequalities
$$\begin{cases} y - x^2 \le 3 \\ y - x > 0 \end{cases}$$

15. Maximize $z = 2x + y$, subject to the constraints $x \ge 0, y \ge 0, x + 3y \le 3$.

16. Find the inverse of the matrix
$$A = \begin{bmatrix} 2 & 1 & 3 \\ 1 & 2 & -1 \\ 3 & 1 & 5 \end{bmatrix}$$

17. Write the matrix equation
$$\begin{bmatrix} 12 & -3 \\ -2 & 7 \end{bmatrix} \begin{bmatrix} x \\ y \end{bmatrix} = \begin{bmatrix} 5 \\ -9 \end{bmatrix}$$

as a system of linear equations without matrices.

18. Solve the matrix equation $A - 5X = 2B$ for X, where
$$A = \begin{bmatrix} 1 & 5 & -2 \\ 4 & -2 & 7 \end{bmatrix} \text{ and } B = \begin{bmatrix} 2 & 5 & -11 \\ 18 & 8 & 11 \end{bmatrix}$$

19. Evaluate the determinant: $\begin{vmatrix} 1 & 3 & 5 \\ 2 & 0 & 10 \\ -3 & 1 & -15 \end{vmatrix}$

20. Use Cramer's Rule to write the solution of the system
$$\begin{cases} 2x - y + z = 3 \\ x + y + z = 6 \\ 4x + 3y - 2z = 4 \end{cases}$$

in the determinant form. (You do not need to find the solution.)

In Problems 1–4, solve each system of equations.

1. $\begin{cases} 6x - 9y = -2 \\ 3x - 5y = -6 \end{cases}$

   a. $\{(-3, 0)\}$
   b. $\left\{\left(0, \dfrac{5}{6}\right)\right\}$
   c. $\left\{\left(\dfrac{44}{3}, 10\right)\right\}$
   d. $\{(1, 1)\}$

2. $\begin{cases} 3x + 5y = 1 \\ -6x - 10y = 2 \end{cases}$

   a. $\left\{\left(0, -\dfrac{1}{5}\right)\right\}$
   b. $\{(2, -1)\}$
   c. $\varnothing$
   d. $\left\{\left(-\dfrac{5}{3}y + \dfrac{1}{3}, y\right)\right\}$

3. $\begin{cases} x - 3y = \dfrac{1}{2} \\ -2x + 6y = -1 \end{cases}$

   a. $\left\{\left(\dfrac{1}{2}, 0\right)\right\}$
   b. $\left\{\left(2, \dfrac{1}{2}\right)\right\}$
   c. $\varnothing$
   d. $\left\{\left(3y + \dfrac{1}{2}, y\right)\right\}$

4. $\begin{cases} 3x + 4y = 12 \\ 3x^2 + 16y^2 = 48 \end{cases}$

   a. $\{(0, 4), (3, 0)\}$
   b. $\{(0, -4)\}$
   c. $\{(0, -4), (3, 0)\}$
   d. $\left\{(4, 0), \left(2, \dfrac{3}{2}\right)\right\}$

5. Student tickets for a dance cost \$2, and nonstudent tickets cost \$5. Three hundred tickets were sold, and the total ticket receipts were \$975. How many of each type of ticket were sold?

   a. 150 student tickets
      150 nonstudent tickets
   b. 200 student tickets
      100 nonstudent tickets
   c. 210 student tickets
      90 nonstudent tickets
   d. 175 student tickets
      125 nonstudent tickets

6. Convert the given system to triangular form.

   $\begin{cases} x + 3y + 3z = 4 \\ 2x + 5y + 4z = 5 \\ x + 2y + 2z = 6 \end{cases}$

   a. $\begin{aligned} x + 3y + 3z &= 4 \\ y + 2z &= 8 \\ z &= 2 \end{aligned}$
   b. $\begin{aligned} x + 3y + 3z &= 4 \\ 2y + z &= 1 \\ z &= 5 \end{aligned}$
   c. $\begin{aligned} x + 3y + 3z &= 4 \\ y + 2z &= 3 \\ z &= 5 \end{aligned}$
   d. $\begin{aligned} x + 3y + 3z &= 4 \\ 2y + z &= -2 \\ z &= 5 \end{aligned}$

7. Solve the given system of equations or state that the system is inconsistent.

   $\begin{cases} 2x + 13y + 6z = 1 \\ 3x + 10y + 11z = 15 \\ 2x + 10y + 8z = 8 \end{cases}$

   a. $\{(3, -1, -4)\}$
   b. $\{(1, -1, 2)\}$
   c. $\varnothing$
   d. $\{(1, 6, 8)\}$

8. Solve the given system of equations or state that the system is inconsistent.

   $\begin{cases} 4x - y + 7z = -2 \\ 2x + y + 11z = 13 \\ 3x - y + 4z = -3 \end{cases}$

   a. $\{(-3z + 1, -5z + 11, z)\}$
   b. $\{(-14, -14, 5)\}$
   c. $\left\{\left(-\dfrac{19}{5}, 3, \dfrac{8}{5}\right)\right\}$
   d. $\varnothing$

9. Forty-six students will go to one of the countries France, Italy, and Spain for six weeks during the summer. The number of students going to France or Italy is four more than the number going to Spain. The number of students going to France is two less than the number going to Spain. How many students are going to each country?

   a. France: 23
      Italy: 21
      Spain: 2
   b. France: 19
      Italy: 6
      Spain: 21
   c. France: 21
      Italy: 6
      Spain: 19
   d. France: 2
      Italy: 21
      Spain: 23

For Problems 10 and 11, write the form of the partial-fraction decomposition of the given rational functions. You do not need to solve for the constants.

10. $\dfrac{x}{(x + 2)(x - 7)}$

    a. $\dfrac{A}{x + 2} + \dfrac{Bx}{x - 7}$
    b. $\dfrac{Ax}{x + 2} + \dfrac{Bx}{x - 7}$
    c. $\dfrac{A}{x + 2} + \dfrac{B}{x - 7}$
    d. $\dfrac{A}{x + 2} + \dfrac{Bx + C}{x - 7}$

11. $\dfrac{7 - x}{(x - 3)(x + 5)^2}$

    a. $\dfrac{A}{x - 3} + \dfrac{B}{x + 5} + \dfrac{C}{(x + 5)^2}$
    b. $\dfrac{A}{x - 3} + \dfrac{B}{x + 5} + \dfrac{Cx + D}{(x + 5)^2}$
    c. $\dfrac{A}{x + 3} + \dfrac{B}{(x + 5)^2}$
    d. $\dfrac{A}{x + 3} + \dfrac{Bx + C}{(x + 5)^2}$

12. Find the partial-fraction decomposition of the rational expression

    $$\dfrac{x^2 + 15x + 18}{x^3 - 9x}.$$

    a. $\dfrac{-2}{x} + \dfrac{26}{x - 9}$
    b. $\dfrac{-2}{x} + \dfrac{4}{x + 3} - \dfrac{1}{x - 3}$
    c. $\dfrac{-2}{x} + \dfrac{4}{x - 3} - \dfrac{1}{x + 3}$
    d. $\dfrac{-2}{x} + \dfrac{3x - 15}{x^2 - 9}$

13. Which of the graphs below is the graph of $x + 3y > 3$?

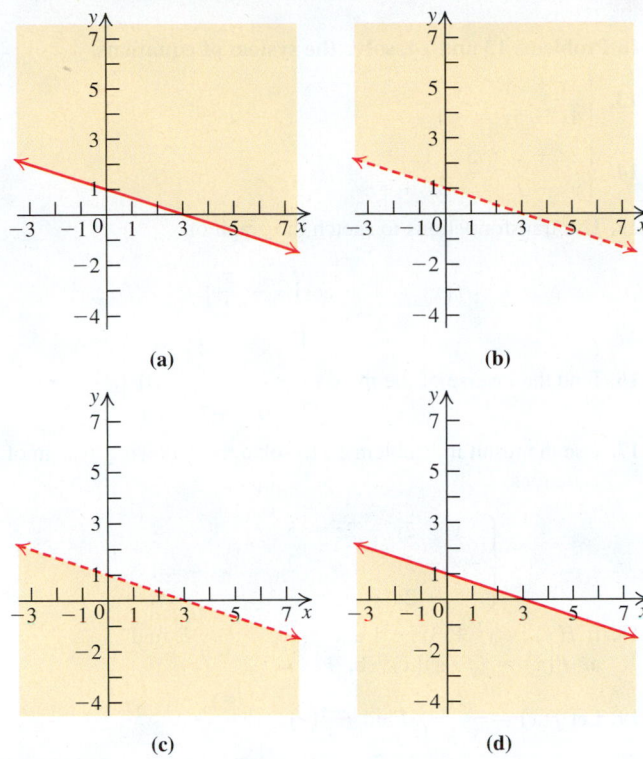

(a)

(b)

(c)

(d)

14. Which of the graphs below is the graph of the solution set of the system of inequalities

$$\begin{cases} y - x^2 + 4 \geq 0 \\ 3x - y \leq 0 \end{cases}?$$

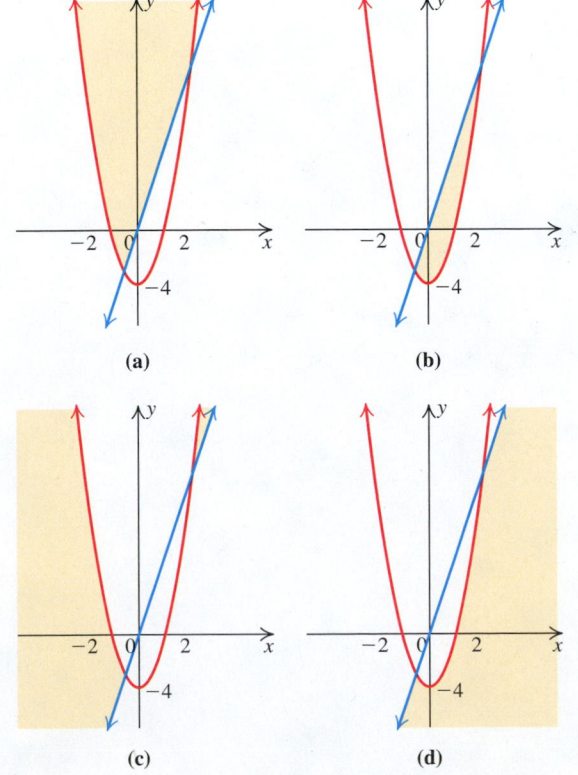

(a)

(b)

(c)

(d)

15. Maximize $z = 3x + 21y$, subject to the constraints

$$x \geq 0, y \geq 0, 2x + y \leq 8, 2x + 3y \leq 16.$$

a. 84

b. 112

c. $\dfrac{16}{3}$

d. 12

16. Find the inverse of the following matrix:

$$A = \begin{bmatrix} 1 & 0 & 0 \\ 2 & 1 & 0 \\ 3 & -4 & 1 \end{bmatrix}$$

a. $\begin{bmatrix} 1 & 0 & 0 \\ -2 & 1 & 0 \\ 3 & 4 & 1 \end{bmatrix}$

b. $\begin{bmatrix} 1 & 0 & 0 \\ -2 & 1 & 0 \\ -8 & 3 & 1 \end{bmatrix}$

c. $\begin{bmatrix} 1 & -2 & -11 \\ 0 & 1 & 4 \\ 0 & 0 & 1 \end{bmatrix}$

d. $\begin{bmatrix} 1 & 0 & 0 \\ -2 & 1 & 0 \\ -11 & 4 & 1 \end{bmatrix}$

17. Write the linear system as a matrix equation

$$\begin{cases} 4x + 2y = 15 \\ -2x + 4y = 9 \end{cases}$$

in the form $AX = B$, where A is the coefficient matrix and B is the constant matrix.

a. $\begin{bmatrix} 4 & 2 \\ -2 & 4 \end{bmatrix}\begin{bmatrix} x \\ y \end{bmatrix} = \begin{bmatrix} 15 \\ 9 \end{bmatrix}$

b. $\begin{bmatrix} 4 & 2 \\ 4 & -2 \end{bmatrix}\begin{bmatrix} x \\ y \end{bmatrix} = \begin{bmatrix} 15 \\ 9 \end{bmatrix}$

c. $\begin{bmatrix} 15 & 2 \\ 9 & -2 \end{bmatrix}\begin{bmatrix} x \\ y \end{bmatrix} = \begin{bmatrix} 4 \\ 4 \end{bmatrix}$

d. $\begin{bmatrix} 4 & -2 \\ 2 & 4 \end{bmatrix}\begin{bmatrix} x \\ y \end{bmatrix} = \begin{bmatrix} 15 \\ 9 \end{bmatrix}$

18. Let $A = \begin{bmatrix} -2 & -3 & 1 \\ -5 & 3 & -2 \end{bmatrix}$ and $B = \begin{bmatrix} -1 & -2 & -1 \\ 1 & 0 & 1 \end{bmatrix}$.

Solve the matrix equation $A - 3X = -5B$ for X.

a. $X = \begin{bmatrix} -\dfrac{5}{2} & -\dfrac{9}{2} & -1 \\ -1 & \dfrac{3}{2} & \dfrac{1}{2} \end{bmatrix}$

b. $X = \begin{bmatrix} \dfrac{7}{3} & -\dfrac{13}{3} & \dfrac{4}{3} \\ 0 & 1 & 1 \end{bmatrix}$

c. $X = \begin{bmatrix} -1 & \dfrac{3}{2} & -1 \\ -\dfrac{5}{2} & \dfrac{9}{2} & -1 \end{bmatrix}$

d. $X = \begin{bmatrix} -\dfrac{1}{3} & \dfrac{1}{3} & \dfrac{8}{3} \\ \dfrac{20}{3} & 3 & -\dfrac{11}{3} \end{bmatrix}$

19. Evaluate the determinant: $\begin{vmatrix} 2 & 3 & -2 \\ 3 & 0 & -3 \\ -3 & 0 & -5 \end{vmatrix}$

a. -72

b. 18

c. 72

d. -18

20. Solve the system of equations

$$\begin{cases} x + y + z = -6 \\ x - y + 3z = -22 \\ 2x + y + z = -10 \end{cases}$$

by using Cramer's Rule.

a. $\varnothing$

b. $\{(-4, 3, -5)\}$

c. $\{(-5, -4, 3)\}$

d. $\{(-5, 3, -4)\}$

CUMULATIVE REVIEW EXERCISES CHAPTERS 1–7

1. If the distance between the points $(2, -3)$ and $(-1, y)$ is five units, find y.

In Problems 2–8, solve each equation or inequality.

2. $\dfrac{4}{x - 1} - \dfrac{3}{x + 2} = \dfrac{18}{(x + 2)(x - 1)}$

3. $|2x - 5| = 3$

4. $4x^2 = 8x - 13$

5. $\left(\dfrac{3x - 1}{x + 5}\right)^2 - 3\left(\dfrac{3x - 1}{x + 5}\right) - 28 = 0$

6. $\log_2|x| + \log_2|x + 6| = 4$

7. $\dfrac{x + 2}{2x - 1} > 0$

8. $\cos x = -\dfrac{1}{2}, 0 \le x \le 2\pi$

9. Write all possible rational zeros of

$$f(x) = 4x^3 + 8x^2 - 11x + 3.$$

10. Show that $\dfrac{1}{2}$ is a zero of multiplicity 2 of the function f in Problem 9.

11. The horsepower required to propel a ship varies as the cube of the ship's speed. If the horsepower required for a speed of 15 miles per hour is 10,125, find the horsepower required for a speed of 20 miles per hour.

12. A city covers an area of 400 square miles. At present, only 2 percent of its area is reserved for parks. In the unincorporated area outside the city limits, 20 percent of the area could be developed into parks. How many square miles of the incorporated area had to be annexed so that the city could develop 12 percent of its area as parks?

In Problems 13 and 14, solve the system of equations.

13. $\begin{cases} 3x + y = 2 \\ 4x + 5y = -1 \end{cases}$

14. $\begin{cases} 2x + y = 5 \\ y^2 - 2y = -3x + 5 \end{cases}$

15. Use transformations to sketch the graph of

$$f(x) = 1 + \cos\left(x - \dfrac{\pi}{4}\right).$$

16. Find the inverse of the matrix $\begin{bmatrix} 1 & 2 & -2 \\ -1 & 3 & 0 \\ 0 & -2 & 1 \end{bmatrix}$.

17. Use the result in Problem 16 to solve the following system of equations:

$$\begin{cases} x + 2y - 2z = 5 \\ -x + 3y = 2 \\ -2y + z = -3 \end{cases}$$

18. If $f(x) = x^2 + 3x - 1$ and $g(x) = x + 2$, find
   a. $F(x) = (f \circ g)(x)$ **b.** $F(4)$.

19. Let $f(x) = \dfrac{x}{x + 4}$. Find $f^{-1}(x)$.

20. Use the power-reducing identity to verify the identity:

$$\sin^4 x = \dfrac{1}{8}(3 - 4\cos 2x + \cos 4x)$$

CHAPTER **8**

Analytic Geometry

TOPICS

8.1 Conic Sections: Overview

8.2 The Parabola

8.3 The Ellipse

8.4 The Hyperbola

8.5 Rotation of Axes

8.6 Polar Equations of Conics

8.7 Parametric Equations

Greek mathematicians were fascinated by certain curves, called conic sections, now used to describe natural phenomena such as planetary orbits. Modern applications also include the construction of lenses, whispering galleries, navigation systems, and even medical devices. In this chapter, we investigate the conic sections and their many uses.

Conic Sections: Overview

Most of the sections in this chapter focus on the plane curves called conics or conic sections. As the name implies, these curves are the sections of a cone (similar to an ice cream cone) formed when a plane intersects the cone. We also discuss parametric equations of conics and other curves in the plane.

Euclid defined a cone as a surface generlated by rotating a right triangle about one of its legs. However, the following description of a cone given by Apollonius is more appropriate.

Right Circular Cone

Draw a circle on a plane (say, horizontal). Draw a line l, called the **axis**, passing through the center of the circle and perpendicular to the plane. Choose a point V above the plane on this line. The surface consisting of all lines that simultaneously pass through the point V and the circle is called a **right circular cone** with **Vertex** V. The vertex V separates the surface into two parts called *nappes of the cone*. See Figure 8.1.

Apollonius of Perga

(c. 262 B.C–190 B.C)

The mathematician Apollonius was born in Perga, in southern Asia Minor (today known as Murtina in Turkey). He went to Alexandria to study with the successors of Euclid. He became famous in ancient times for his work on astronomy and his eight books on conics. He was able to discover and prove hundreds of beautiful and difficult theorems without modern algebraic symbolism. He is credited with introducing the terms *ellipse*, *hyperbola*, and *parabola*.

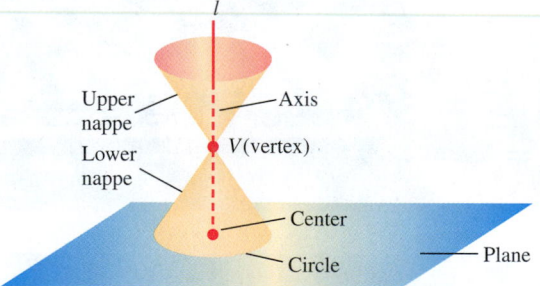

Figure 8.1 Parts of a cone.

If we slice the cone with a plane not passing through the vertex, the intersections of the slicing plane and the cone are **conic sections**. If the slicing plane is horizontal (parallel to the first plane), the intersection is a **circle** (Figure 8.2(a)). If the slicing plane is inclined slightly from the horizontal, intersecting only one nappe, the intersection is an oval-shaped curve called an **ellipse** (Figure 8.2(b)). As the angle of the slicing plane increases, more elongated ellipses are formed (Figure 8.2(b)). If the angle of the slicing plane increases still further so that the slicing plane is parallel to one of the lines generating the cone, the intersection is called a **parabola** (Figure 8.2(c)). If the angle of the slicing plane increases yet further so that the slicing plane intersects both nappes of the cone, the resulting intersection is called a **hyperbola** (Figure 8.2(d)).

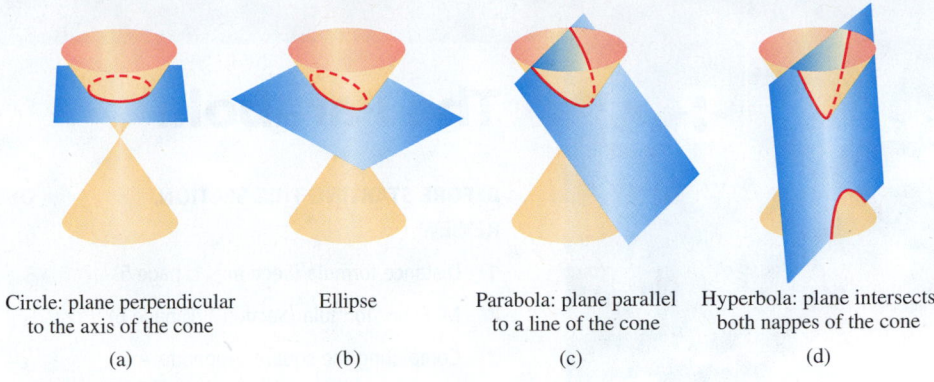

Circle: plane perpendicular to the axis of the cone

(a)

Ellipse

(b)

Parabola: plane parallel to a line of the cone

(c)

Hyperbola: plane intersects both nappes of the cone

(d)

Figure 8.2 Conic sections.

Intersections of the slicing plane through the vertex result in points and lines called **degenerate conic sections**. See Figure 8.3.

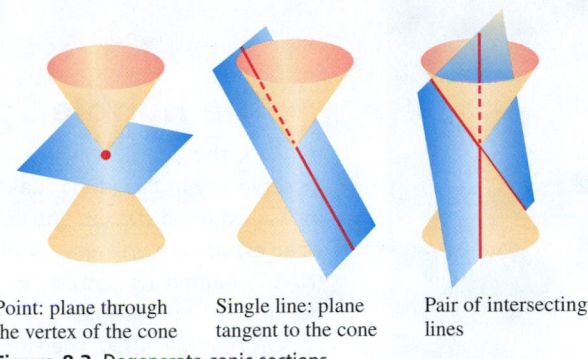

Point: plane through the vertex of the cone

Single line: plane tangent to the cone

Pair of intersecting lines

Figure 8.3 Degenerate conic sections.

Just as a circle (see page 12) is defined as the set of points in the plane at a fixed distance r (the radius) from a fixed point (the center), we can define each of the conic sections just described as a set of points in the plane that satisfy certain geometric conditions. It can be shown that these alternate definitions give the same family of curves described earlier. This enables us to keep our work in a two-dimensional setting. Using these definitions, we will derive the equations of the conic sections and show that an equation of the form $Ax^2 + Cy^2 + Dx + Ey + F = 0$ is (except in degenerate cases represented in Figure 8.3) the equation of a parabola, an ellipse, or a hyperbola. (A circle is a special case of an ellipse.) **Germinal Pierre Dandelin** (1794–1847) was a Belgian/French mathematician who found a relatively straightforward way to establish the equivalence of the algebraic and geometric descriptions of the conics. A web search on **Dandelin spheres** will provide several sources for his proof. In fact, you can even find YouTube videos demonstrating this proof.

We study these conic sections because of their practical applications. For example, a satellite dish, a flashlight lens, and a telescope lens have a parabolic shape; the planets travel in elliptical orbits, and comets travel in orbits that are either elliptical or hyperbolic. A comet with an elliptical orbit can be viewed from Earth more than once, while those with a hyperbolic orbit can be viewed only once!

The Parabola

BEFORE STARTING THIS SECTION, REVIEW

1 Distance formula (Section 1.1, page 5)

2 Midpoint formula (Section 1.1, page 6)

3 Completing the square (Appendix A.6, page 957)

4 Line of symmetry (Section 1.1, page 9)

5 Slopes of perpendicular lines (Section 1.2, page 29)

6 Transformations (Section 1.5, pages 78–87)

OBJECTIVES

1 Define a parabola geometrically.

2 Find an equation of a parabola.

3 Translate a parabola.

4 Use the reflecting property of parabolas.

Hubble Telescope

Edwin Hubble (1889–1953)
Edwin Hubble is renowned for discovering that there are other galaxies in the universe beyond the Milky Way. Hubble then wanted to classify the galaxies according to their content, distance, shape, and brightness patterns. He made the momentous discovery that the galaxies were moving away from each other at a rate proportional to the distance between them (Hubble's Law). This led to the calculation of the point where the expansion began and to confirmation of the Big Bang theory of the origin of the universe. Recent estimates place the age of the universe at about 20 billion years.

◆ The Hubble Space Telescope

In 1918, the 2.5-meter (100-inch) Hooker Telescope was installed at Mount Wilson Observatory in Pasadena, California. With this telescope, astronomer Edwin Hubble measured the distances and velocities of the galaxies. His work led to today's concept of an expanding universe. In 1977, the National Aeronautics and Space Administration (NASA) named its largest, most complex, and most capable orbiting telescope in honor of Edwin Hubble.

On April 24, 1990, the Hubble Space Telescope was deployed into orbit from the space shuttle *Discovery*. However, due to a spherical aberration in the telescope's parabolic mirror, the Hubble had "blurred vision." (See Example 7.) In 1993, astronauts from the space shuttle *Endeavor* installed replacement instruments and supplemental optics to restore the telescope to full optimal performance. In 2015, observations from NASA's Hubble suggested the existence of an underground ocean on Jupiter's largest moon.

Not since Galileo turned his telescope toward the heavens in 1610 has any event so changed our understanding of the universe as did the deployment of the Hubble Space Telescope. The first instrument of any kind that was specifically designed for routine servicing by spacewalking astronauts, the Hubble telescope brought stunning pictures from the ends of the universe to the general public.

1 Define a parabola geometrically.

Geometric Definition of a Parabola

In Section 2.1, we named the graph of a quadratic function $y = ax^2 + bx + c$, $a \neq 0$, a parabola. In this section, we give a *geometric* definition of a parabola. Then we show that this definition leads to the graph of a quadratic function.

RECALL

The distance from a point to a line is defined as the length of the perpendicular line segment from the point to the line.

Parabola

Let l be a line and F a point not on the line l on the plane determined by F and l. Then the set of all points P in the plane that are the same distance from F as they are from the line l is called a **parabola**. See Figure 8.4(b). Thus, a parabola is the set of points P for which $d(F, P) = d(P, l)$, where $d(P, l)$ denotes the perpendicular distance between the point P and the line l.

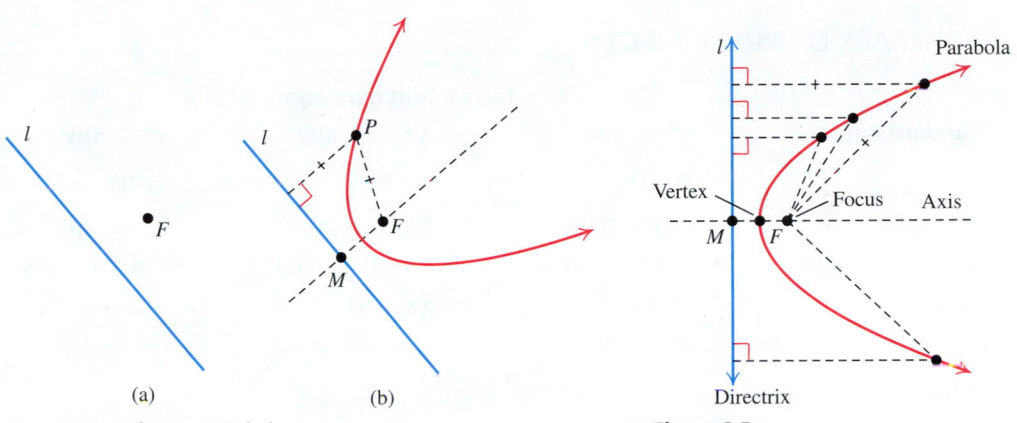

Figure 8.4 Defining a parabola.   **Figure 8.5**

The line l is the **directrix** of the parabola, and the point F is the **focus**. The line through the focus *perpendicular* to the directrix is called the **axis**, or the **axis of symmetry**, of the parabola. The point at which the axis intersects the parabola is called the **vertex**. See Figure 8.5. Note that the vertex of a parabola is the midpoint of the segment $\overline{FM}$.

2 Find an equation of a parabola.

Equation of a Parabola

A coordinate system allows us to transform the geometric description of a parabola into an algebraic *equation of the parabola*. This equation is particularly simple if we place the vertex V at the origin and the focus on one of the coordinate axes. Suppose for a moment that the focus F is on the positive x-axis. Then F has coordinates of the form $(p, 0)$ with $p > 0$, as shown in Figure 8.6.

The vertex V lies on the parabola with focus F and directrix l; so we have the following:

$$d(V, l) = d(V, F) \quad \text{Definition of parabola, where}$$
$$= p \quad \text{V is a point on the parabola}$$

The vertex V at the origin is located midway between the focus and the directrix; so the equation of the directrix is $x = -p$.

Suppose a point $P(x, y)$ lies on the parabola shown in Figure 8.6. Then the distance from the point P to the directrix l is the distance between the points $P(x, y)$ and $M(-p, y)$. So, a point $P(x, y)$ lies on this parabola if and only if

$$d(P, F) = d(P, M).$$

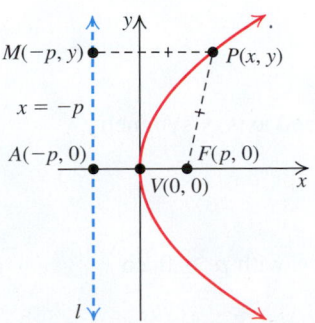

Figure 8.6 Deriving the equation of a parabola.

Using the distance formula on each side gives

$$\sqrt{(x-p)^2 + y^2} = \sqrt{(x+p)^2}$$

$$(x-p)^2 + y^2 = (x+p)^2 \qquad \text{Square both sides.}$$

$$x^2 - 2px + p^2 + y^2 = x^2 + 2px + p^2 \qquad \text{Expand squares.}$$

$$y^2 = 4px \qquad \text{Simplify.}$$

The equation $y^2 = 4px$ is called the **standard equation of a parabola with vertex $(0, 0)$ and focus $(p, 0)$.** Similarly, if the focus of a parabola is placed on the negative x-axis, we obtain the equation $y^2 = -4px$ as the standard equation of a parabola with vertex $(0, 0)$ and focus $(-p, 0)$.

By interchanging the roles of x and y, we find that $x^2 = 4py$ is the standard equation of a parabola with vertex $(0, 0)$ and focus $(0, p)$. Similarly, the equation $x^2 = -4py$ is the standard equation of a parabola with vertex $(0, 0)$ and focus $(0, -p)$.

SUMMARY OF **MAIN FACTS**

Main facts about a parabola with $p > 0$

Standard Equation	$y^2 = 4px$	$y^2 = -4px$	$x^2 = 4py$	$x^2 = -4py$
Vertex	$(0, 0)$	$(0, 0)$	$(0, 0)$	$(0, 0)$
Description	Opens right	Opens left	Opens up	Opens down
Axis of Symmetry	$y = 0$ (x-axis)	$y = 0$ (x-axis)	$x = 0$ (y-axis)	$x = 0$ (y-axis)
Focus	$(p, 0)$	$(-p, 0)$	$(0, p)$	$(0, -p)$
Directrix	$x = -p$	$x = p$	$y = -p$	$y = p$
Graph				

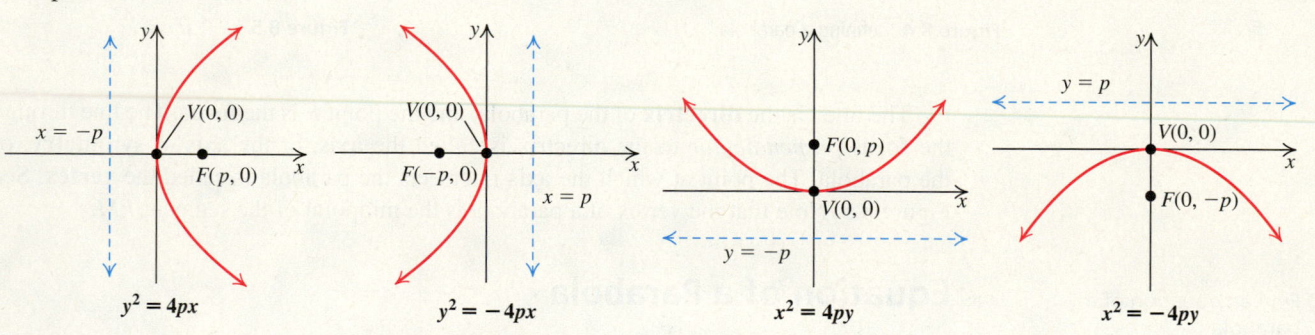

EXAMPLE 1   **Graphing a Parabola**

Graph each parabola and specify the vertex, focus, directrix, and axis of symmetry.

a. $x^2 = -8y$ **b.** $y^2 = 5x$

Solution

a. The equation $x^2 = -8y$ has the standard form $x^2 = -4py$, with $p > 0$, so

$$-4p = -8 \qquad \text{Equate coefficients of } y.$$

$$p = 2 \qquad \text{Divide both sides by } -4.$$

The parabola opens down. The vertex is the origin, and the focus is $(0, -2)$. The directrix is the horizontal line $y = 2$; the axis of the parabola is the y-axis. You should plot some additional points to ensure the accuracy of your sketch. Since the focus is $(0, -2)$, substitute $y = -2$ in the equation $x^2 = -8y$ of the parabola to obtain

$$x^2 = (-8)(-2) \qquad \text{Replace } y \text{ with } -2 \text{ in } x^2 = -8y.$$
$$x^2 = 16 \qquad \text{Simplify.}$$
$$x = \pm 4 \qquad \text{Solve for } x.$$

Thus, the points $(4, -2)$ and $(-4, -2)$ are two symmetric points on the parabola to the right and the left of the focus. The graph of this parabola is sketched in Figure 8.7.

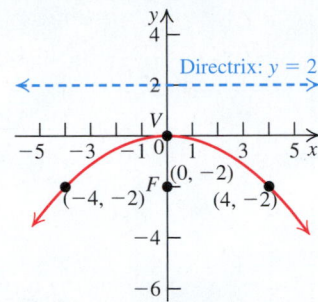

Figure 8.7 Graph of $x^2 = -8y$.

TECHNOLOGY CONNECTION

You can use a graphing calculator to graph parabolas. To graph $x^2 = 4py$, just graph $Y_1 = \dfrac{1}{4p}x^2$.

For example, when $x^2 = -8y$, you have $4p = -8$; so you graph

$$Y_1 = -\frac{1}{8}x^2.$$

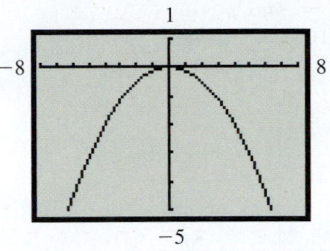

b. The equation $y^2 = 5x$ has the standard form $y^2 = 4px$, with $p > 0$.

$$4p = 5 \qquad \text{Equate coefficients of } x.$$
$$p = \frac{5}{4} \qquad \text{Divide both sides by 4.}$$

The parabola opens to the right. The vertex is the origin, and the focus is $\left(\dfrac{5}{4}, 0\right)$.

The directrix is the vertical line $x = -p = -\dfrac{5}{4}$; the axis of the parabola is the x-axis.

To find two symmetric points on the parabola that are above and below the focus, substitute $x = \dfrac{5}{4}$ in the equation of the parabola.

$$y^2 = 5x \qquad \text{Equation of the parabola}$$
$$y^2 = 5\left(\frac{5}{4}\right) \qquad \text{Replace } x \text{ with } \frac{5}{4}.$$
$$y^2 = \frac{25}{4} \qquad \text{Simplify.}$$
$$y = \pm\frac{5}{2} \qquad \text{Solve for } y.$$

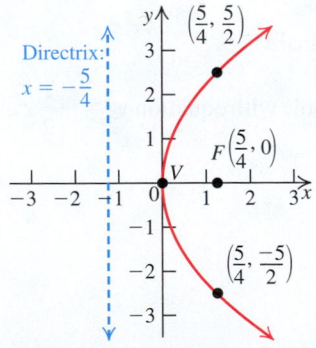

Figure 8.8 Graph of $y^2 = 5x$.

Plot the two additional points $\left(\dfrac{5}{4}, \dfrac{5}{2}\right)$ and $\left(\dfrac{5}{4}, -\dfrac{5}{2}\right)$ on the parabola. The graph of $y^2 = 5x$ is sketched in Figure 8.8.

Practice Problem 1 Graph the parabola and specify the vertex, focus, directrix, and axis.

a. $x^2 = 12y$ **b.** $y^2 = -6x$

The line segment joining the focus and the two points symmetric with respect to the axis on a parabola is called the *latus rectum* of the parabola. Intuitively, the length of the latus rectum determines the "size of the opening" of the parabola.

To graph $y^2 = 4px$, solve the equation for y to obtain

$$y = \pm\sqrt{4px}.$$

Then graph $Y_1 = \sqrt{4px}$ and $Y_2 = -\sqrt{4px}$. For example, to graph $y^2 = 5x$, you graph

$$Y_1 = \sqrt{5x}, Y_2 = -\sqrt{5x}.$$

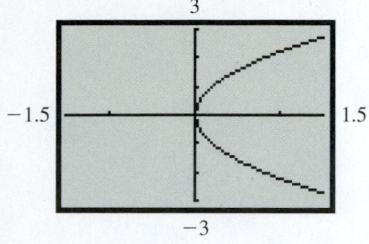

Latus Rectum and Focal Diameter

The line segment passing through the focus of a parabola, perpendicular to the axis, and having endpoints on the parabola is called the **latus rectum of the parabola**. Figure 8.9 shows that the length of the latus rectum, called the **focal diameter**, for each of the graphs $y^2 = \pm 4px$ and $x^2 = \pm 4py$ with $p > 0$ is $4p$.

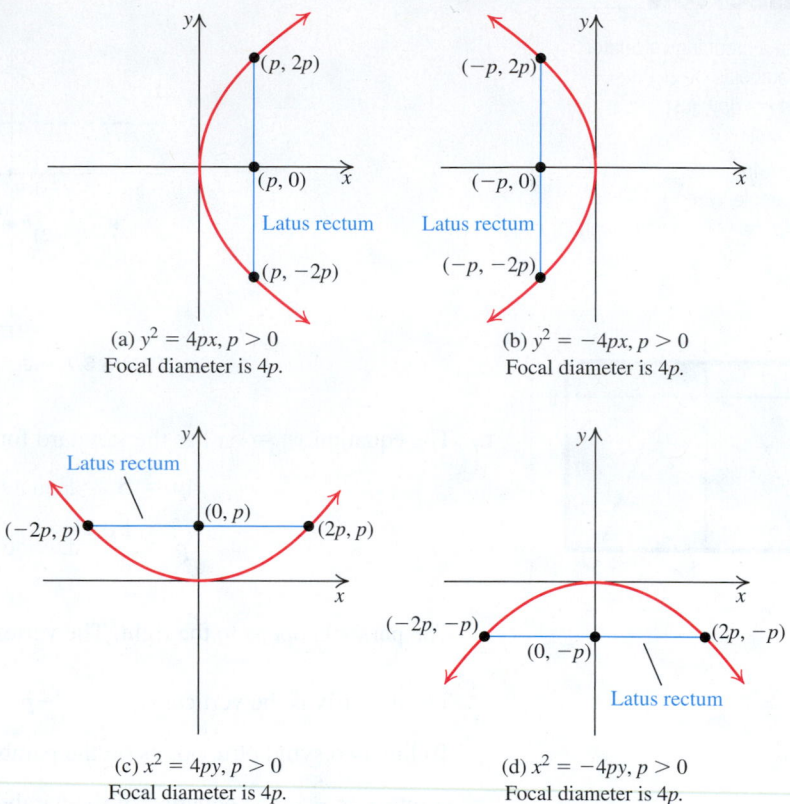

(a) $y^2 = 4px, p > 0$
Focal diameter is $4p$.

(b) $y^2 = -4px, p > 0$
Focal diameter is $4p$.

(c) $x^2 = 4py, p > 0$
Focal diameter is $4p$.

(d) $x^2 = -4py, p > 0$
Focal diameter is $4p$.

Figure 8.9 The latus rectum.

EXAMPLE 2 **Finding the Focal Diameter of a Parabola**

Find the vertex, focus, directrix, and focal diameter of the parabola with equation $y = -\dfrac{1}{2}x^2$, sketch the graph, and show the latus rectum.

Solution

First write the equation in the form $x^2 = -4py$.

$$y = -\frac{1}{2}x^2 \quad \text{Given equation}$$

$$-2y = x^2 \quad \text{Multiply both sides by } -2.$$

The equation $-2y = x^2$ has the standard form $x^2 = -4py$.

So $$-4p = -2 \quad \text{Equate coefficients of } y.$$

$$p = \frac{1}{2} \quad \text{Divide both sides by } -4.$$

The vertex is $(0, 0)$, the focus is $\left(0, -\dfrac{1}{2}\right)$, the directrix is $y = \dfrac{1}{2}$, and the focal diameter is

$4p = 4\left(\dfrac{1}{2}\right) = 2$. Since the focal diameter is 2, the latus rectum extends one unit to the left and one unit to the right of the focus. The graph and the latus rectum (blue) are shown in Figure 8.10.

Figure 8.10 Graph of $y = -\frac{1}{2}x^2$.

Practice Problem 2 Find the vertex, focus, directrix, and focal diameter of the parabola with equation $y^2 = -8x$, sketch the graph, and show the latus rectum.

EXAMPLE 3   Finding the Equation of a Parabola

Find the standard equation of each parabola with vertex $(0, 0)$ and satisfying the given description.

a. The focus is $(-3, 0)$.

b. The axis of the parabola is the y-axis, and the graph passes through the point $(-4, 2)$.

Solution

a. The vertex $(0, 0)$ of the parabola and its focus $(-3, 0)$ are on the x-axis; so the parabola opens to the left. Its equation must be of the form $y^2 = -4px$ with $p = 3$.

$$y^2 = -4px \qquad \text{Form of the equation}$$
$$y^2 = -4(3)x \qquad \text{Replace } p \text{ with 3.}$$
$$y^2 = -12x \qquad \text{Simplify.}$$

The equation of the parabola is $y^2 = -12x$.

b. Since the vertex of the parabola is $(0, 0)$, its axis is the y-axis, and the point $(-4, 2)$ is above the x-axis, so the parabola opens up. The equation of the parabola is of the form

$$x^2 = 4py.$$

This equation is satisfied by $x = -4$ and $y = 2$ because the parabola passes through the point $(-4, 2)$.

$$(-4)^2 = 4p(2) \qquad \text{Replace } x \text{ with } -4 \text{ and } y \text{ with 2.}$$
$$16 = 8p \qquad \text{Simplify.}$$
$$2 = p \qquad \text{Divide both sides by 8.}$$

Therefore, the equation of the parabola is

$$x^2 = 4(2)y \qquad \text{Replace } p \text{ with 2 in } x^2 = 4py.$$
$$x^2 = 8y \qquad \text{Simplify.}$$

The required equation of the parabola is $x^2 = 8y$.

Practice Problem 3 Find the standard equation of each parabola with vertex $(0, 0)$ satisfying the following conditions.

a. The focus is $(0, 2)$.

b. The axis of the parabola is the x-axis, and the graph passes through $(1, 2)$.

3 Translate a parabola.

Translations of Parabolas

The equation of a parabola with vertex $(0, 0)$ and axis on a coordinate axis is of the form $x^2 = \pm 4py$ or $y^2 = \pm 4px$, where p is the distance between the vertex and the focus. You can use translations of the graphs of these equations to find the equation of a parabola with vertex (h, k) and axis of symmetry parallel to a coordinate axis. We obtain these translations by replacing x with $x - h$ (a horizontal shift) and y with $y - k$ (a vertical shift).

SUMMARY OF MAIN FACTS

Main facts about a parabola with vertex (h, k) and $p > 0$

Standard Equation	$(y - k)^2 = 4p(x - h)$	$(y - k)^2 = -4p(x - h)$	$(x - h)^2 = 4p(y - k)$	$(x - h)^2 = -4p(y - k)$
Equation of axis	$y = k$	$y = k$	$x = h$	$x = h$
Description	Opens right	Opens left	Opens up	Opens down
Vertex	(h, k)	(h, k)	(h, k)	(h, k)
Focus	$(h + p, k)$	$(h - p, k)$	$(h, k + p)$	$(h, k - p)$
Directrix	$x = h - p$	$x = h + p$	$y = k - p$	$y = k + p$
Graph				

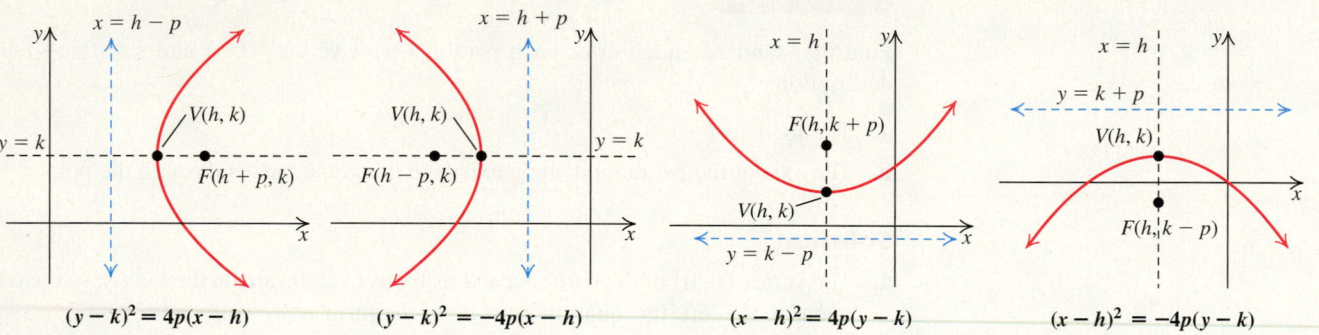

Expanding the standard equations of a parabola and collecting like terms result in an equation of the form

$$Ax^2 + Cy^2 + Dx + Ey + F = 0$$

with $A = 0$ or $C = 0$, but not both. Conversely, except in degenerate cases, any equation of this form can be put into one of the standard forms of a parabola by completing the square on the x- or y- terms.

> ### GENERAL EQUATION OF A PARABOLA
>
> The graph of $Ax^2 + Cy^2 + Dx + Ey + F = 0$ is a parabola if $AC = 0$, $A \neq 0$ or $C \neq 0$. That is, $A = 0$ or $C = 0$, but not both are 0.

EXAMPLE 4 Finding the Equation of a Parabola

Find the standard equation of the parabola that satisfies the given conditions. Also find the focal diameter of each parabola.

a. Focus: $(0, 5)$; directrix: $y = -3$ **b.** Vertex: $(-1, 2)$; directrix: $x = 2$

c. Vertex: $(-5, -2)$; focus: $(1, -2)$

Solution

To find the standard equation of a parabola, we must find the vertex (h, k) and p. Remember that p is both the distance between the vertex and the focus and the distance between the vertex and the directrix. The distance between the focus and the directrix is $2p$.

a. The distance between the focus $(0, 5)$ and directrix $y = -3$ is $2p = 5 - (-3) = 8$ units. Draw a rough sketch to help visualize this. Then the vertex is $\frac{1}{2}(8) = 4$ units below the focus, so $p = 4$, and the vertex is $(0, 5 - 4) = (0, 1)$. We have $(h, k) = (0, 1)$, so $h = 0$ and $k = 1$.

The standard equation when the directrix is below the vertex

$$(x - h)^2 = 4p(y - k) \text{ becomes}$$
$$(x - 0)^2 = 4(4)(y - 1) \qquad \text{Replace } h \text{ with } 0, p \text{ with } 4, \text{ and } k \text{ with } 1.$$
$$\text{or} \quad x^2 = 16(y - 1). \qquad \text{Simplify.}$$

The focal diameter is $4p = 4(4) = 16$.   Replace p with 4.

b. The distance between the vertex $(-1, 2)$ and the directrix $x = 2$ is $p = 2 - (-1) = 3$ units. Draw a rough sketch to help visualize this. The vertex $(-1, 2) = (h, k)$, so $h = -1$ and $k = 2$. The standard equation when the directrix is to the right of the vertex

$$(y - k)^2 = -4p(x - h) \text{ becomes}$$
$$(y - 2)^2 = -4(3)(x - [-1]) \qquad \text{Replace } k \text{ with } 2, p \text{ with } 3, \text{ and } h \text{ with } -1.$$
$$\text{or } (y - 2)^2 = -12(x + 1). \qquad \text{Simplify.}$$

The focal diameter is $4p = 4(3) = 12$.   Replace p with 3.

c. The distance between the vertex $(-5, -2)$ and the focus $(1, -2)$ is $p = 1 - (-5) = 6$ units. Draw a rough sketch to help visualize this. The vertex $(-5, -2) = (h, k)$, so $h = -5$ and $k = -2$. The equation with the vertex to the left of the focus

$$(y - k)^2 = 4p(x - h) \text{ becomes}$$
$$(y - [-2])^2 = 4(6)(x - [-5]) \qquad \text{Replace } k \text{ with } -2, p \text{ with } 6, \text{ and } h \text{ with } -5.$$
$$\text{or } (y + 2)^2 = 24(x + 5). \qquad \text{Simplify.}$$

The focal diameter is $4p = 4(6) = 24$.   Replace p with 6.

Practice Problem 4 Find the standard equation of the parabola that satisfies the given conditions. Also find the focal diameter of each parabola.

a. Focus: $(0, 2)$; directrix: $y = -10$ **b.** Vertex: $(1, 2)$; directrix: $x = 5$

c. Vertex: $(1, 7)$; focus: $(2, 7)$

EXAMPLE 5   Graphing a Parabola

Find the vertex, focus, and directrix of the parabola $2y^2 - 8y - x + 7 = 0$. Sketch the graph of the parabola.

Solution

First, complete the square on y. Then you can compare the resulting equation with one of the standard forms of the equation of a parabola.

$$2y^2 - 8y - x + 7 = 0 \qquad \text{The given equation}$$
$$2y^2 - 8y = x - 7 \qquad \text{Isolate terms containing } y.$$
$$2(y^2 - 4y) = x - 7 \qquad \text{Factor out the common factor, 2.}$$
$$2(y^2 - 4y + 4) = x - 7 + 8 \qquad \text{Add } 2\left(-\frac{4}{2}\right)^2 = 2(-2)^2 = 2 \cdot 4 = 8 \text{ to}$$
$$\text{both sides to complete the square.}$$
$$2(y - 2)^2 = x + 1 \qquad y^2 - 4y + 4 = (y - 2)^2; \text{ simplify.}$$
$$(y - 2)^2 = \frac{1}{2}(x + 1) \qquad \text{Divide both sides by 2.}$$

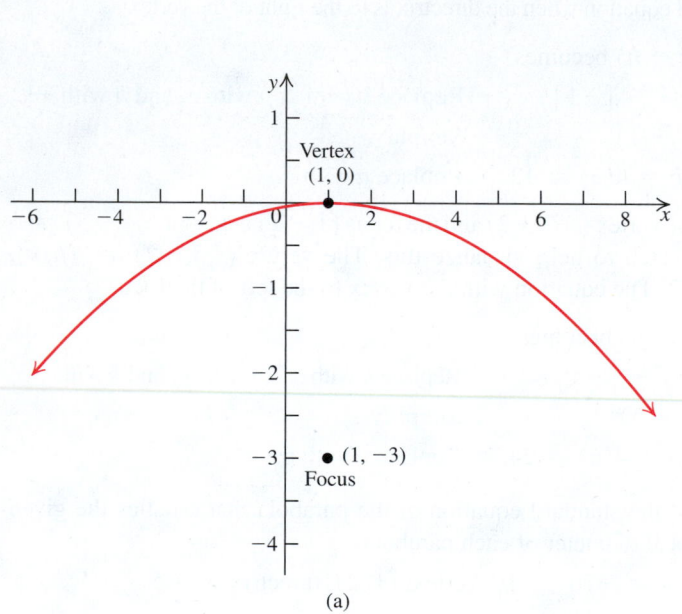

Figure 8.11 Converting $2y^2 - 8y - x + 7 = 0$ to standard form.

The last equation is in one of the standard forms of the parabola. Comparing it with the form $(y - k)^2 = 4p(x - h)$, we have $h = -1$; $k = 2$; and $4p = \dfrac{1}{2}$, or $p = \dfrac{1}{8}$. The parabola opens to the right, the vertex is $(h, k) = (-1, 2)$, the focus is

$$(h + p, k) = \left(-1 + \frac{1}{8}, 2\right) = \left(-\frac{7}{8}, 2\right),$$ and the directrix is the vertical line

$$x = h - p = -1 - \frac{1}{8} = -\frac{9}{8}.$$ The graph is shown in Figure 8.11.

Practice Problem 5 Find the vertex, focus, and directrix of the parabola

$$2x^2 - 8x - y + 7 = 0.$$

Sketch the graph of the parabola.

EXAMPLE 6 Find an Equation in Standard Form of each Parabola whose Graph Is Shown

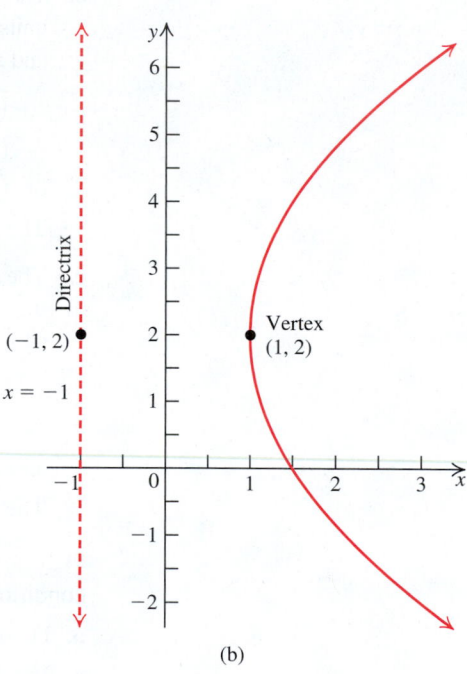

(a)

(b)

Solution

a. The vertex $(h, k) = (1, 0)$, so $h = 1$ and $k = 0$. The distance between the vertex $(1, 0)$ and the focus $(1, -3)$ is 3 units, so $p = 3$.

$$(x - h)^2 = -4p(y - k) \qquad \text{Equation for a parabola that opens down}$$
$$(x - 1)^2 = -4(3)(y - 0) \qquad \text{Replace } h \text{ with 1, } p \text{ with 3, and } k \text{ with 0.}$$
$$(x - 1)^2 = -12y \qquad \text{Simplify.}$$

b. The vertex $(h, k) = (1, 2)$ so $h = 1$ and $k = 2$. The distance, p, between the vertex and the focus is the same as the distance between the vertex and the directrix. That is, p is the distance between the vertex $(1, 2)$ and the point $(-1, 2)$ on the directrix. So $p = 1 - (-1) = 2$ units.

$$(y - k)^2 = 4p(x - h) \qquad \text{Equation for a parabola that opens right}$$
$$(y - 2)^2 = 4(2)(x - 1) \qquad \text{Replace } k \text{ with 2, } p \text{ with 2, and } h \text{ with 1.}$$
$$(y - 2)^2 = 8(x - 1) \qquad \text{Simplify.}$$

Practice Problem 6 Find an equation of the parabola whose graph is given.

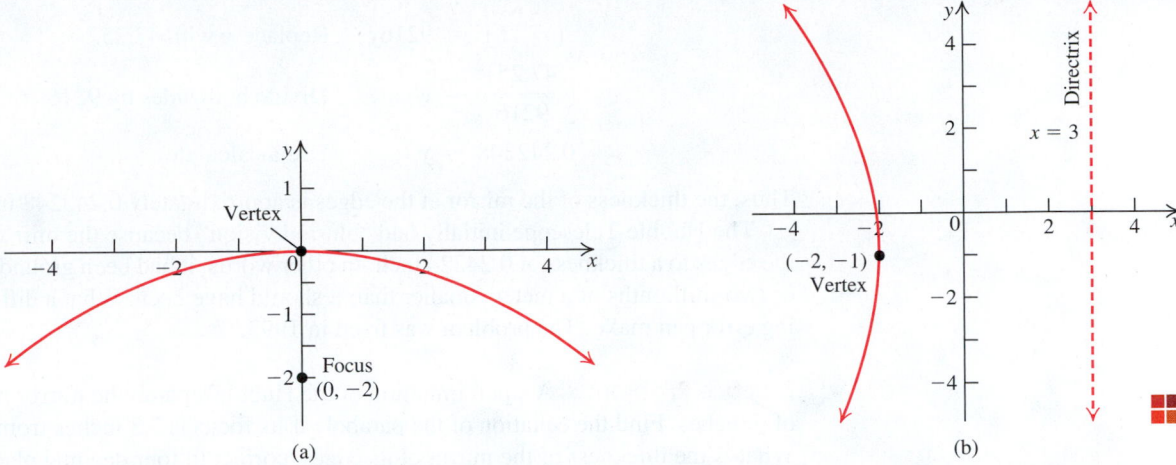

(a)

(b)

4 Use the reflecting property of parabolas.

Reflecting Property of Parabolas

A property of parabolas that is useful in applications is the **reflecting property**: If a reflecting surface has parabolic cross sections with a common focus, then all light rays entering the surface parallel to the axis will be reflected through the focus. See Figure 8.12(a). This property is used in reflecting telescopes and satellite antennas because the light rays or radio waves bouncing off a parabolic surface are reflected to the focus, where they are collected and amplified.

Conversely, if a light source is located at the focus of a parabolic reflector, the reflected rays will form a beam parallel to the axis. See Figure 8.12(b). This principle is used in flashlights, searchlights, and other such devices.

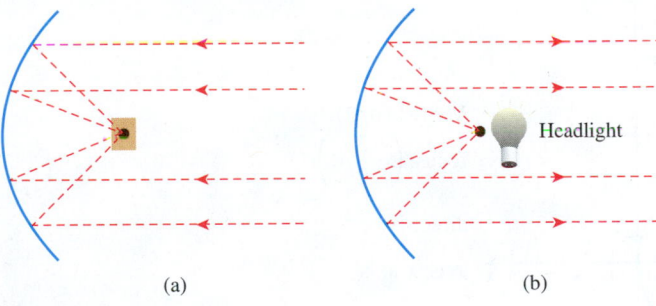

(a)

(b)

Figure 8.12

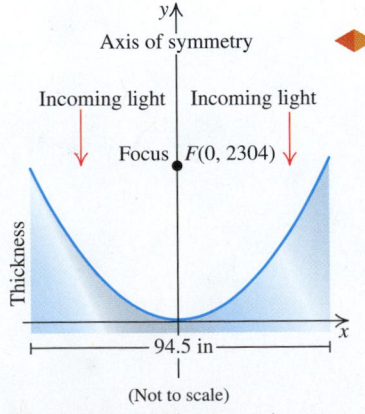

Figure 8.13 Hubble Space Telescope.

◆ **EXAMPLE 7**   **Calculating Some Properties of the Hubble Space Telescope**

The parabolic mirror used in the Hubble Space Telescope has a diameter of 94.5 inches. See Figure 8.13. Find the equation of the parabola if its focus is 2304 inches from the vertex. What is the thickness of the mirror at the edges?

Solution

Position the parabola so that its vertex is at the origin and its focus is on the positive y-axis. The equation of the parabola is of the form

$$x^2 = 4py$$
$$x^2 = 4(2304)y \qquad p = 2304$$
$$x^2 = 9216y \qquad \text{Simplify.}$$

To find the thickness y of the mirror at the edge, substitute $x = 47.25$ (half the diameter) in the equation $x^2 = 9216y$ and solve for y.

$$(47.25)^2 = 9216y \qquad \text{Replace } x \text{ with } 47.25.$$

$$\frac{(47.25)^2}{9216} = y \qquad \text{Divide both sides by 9216.}$$

$$0.242248 \approx y \qquad \text{Use a calculator.}$$

Thus, the thickness of the mirror at the edges is approximately 0.242248 inch.

The Hubble Telescope initially had "blurred vision" because the mirror was ground at the edges to a thickness of 0.24224 inch. In other words, it had been ground 0.000008 inch, or two-millionths of a meter, smaller than it should have been. What a difference a rounding error can make! The problem was fixed in 1993.

Practice Problem 7 A small imitation of the Hubble's parabolic mirror has a diameter of 3 inches. Find the equation of the parabola if its focus is 7.3 inches from the vertex. What is the thickness of the mirror at its edges, correct to four decimal places?

Answers to Practice Problems

1. a.

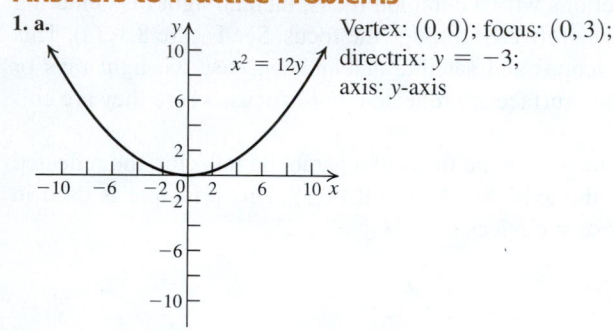

$x^2 = 12y$

Vertex: $(0, 0)$; focus: $(0, 3)$;
directrix: $y = -3$;
axis: y-axis

3. a. $x^2 = 8y$ **b.** $y^2 = 4x$

4. a. $x^2 = 24(y + 4)$; focal diameter = 24
b. $(y - 2)^2 = -16(x - 1)$; focal diameter = 16
c. $(y - 7)^2 = 4(x - 1)$; focal diameter = 4

5. Vertex: $(2, -1)$; focus: $\left(2, -\dfrac{7}{8}\right)$; directrix: $y = -\dfrac{9}{8}$

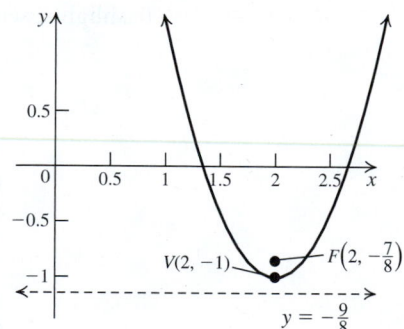

$V(2, -1)$, $F\left(2, -\dfrac{7}{8}\right)$, $y = -\dfrac{9}{8}$

b.

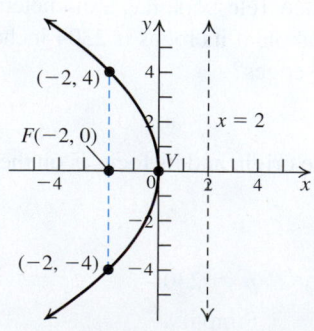

$y^2 = -6x$

Vertex: $(0, 0)$;
focus: $\left(-\dfrac{3}{2}, 0\right)$;
directrix: $x = \dfrac{3}{2}$;
axis: x-axis

6. a. $x^2 = -8y$ **b.** $(y + 1)^2 = -20(x + 2)$
7. Equation: $x^2 = 29.2y$. The thickness of the mirror at the edges is 0.0771 in.

2. Vertex: $(0, 0)$; focus: $(-2, 0)$; directrix: $x = 2$; focal diameter: 8

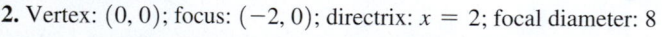

$(-2, 4)$ $F(-2, 0)$ $x = 2$ V $(-2, -4)$

SECTION 8.2   **Exercises**

Concepts and Vocabulary

1. A parabola is the set of all points P in the plane that are equidistant from a fixed line called the _____ and a fixed point not on the line called the _____.

2. The line through the focus perpendicular to the directrix is called the _____ of the parabola.

3. The point at which the axis intersects the parabola is called the _____ of the parabola.

4. The graph of $y + 2 = 3(x - 4)^2$ is obtained by shifting the graph of $y = 3x^2$ to the right _____ units and shifting _____ 2 units.

5. **True or False.** The vertex of a parabola is midway between its focus and its directrix.

6. **True or False.** A parabola is symmetric about its axis.

7. **True or False.** A parabola with focal diameter 3 is narrower than a parabola with focal diameter 2.

8. **True or False.** The latus rectum and the focal diameter are both numbers that indicate the "width" of a parabola.

Building Skills

In Exercises 9–16, find the focus and directrix of the parabola with the given equation. Then match each equation to one of the graphs labeled (a) through (h).

9. $x^2 = 2y$
10. $9x^2 = 4y$
11. $16x^2 = -9y$
12. $x^2 = -2y$
13. $y^2 = 2x$
14. $9y^2 = 16x$
15. $9y^2 = -16x$
16. $y^2 = -2x$

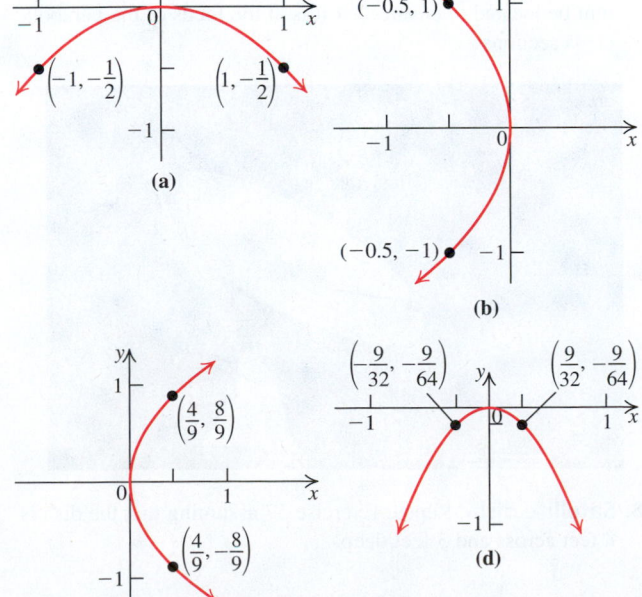

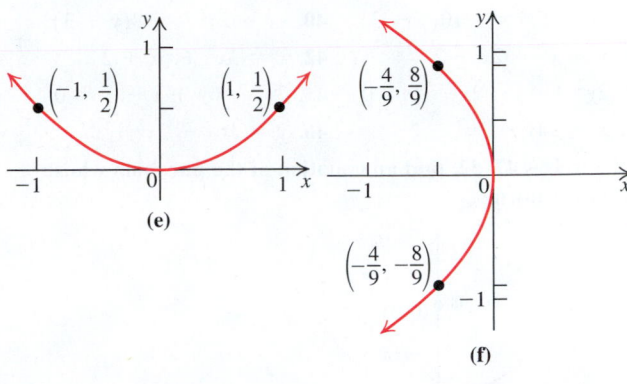

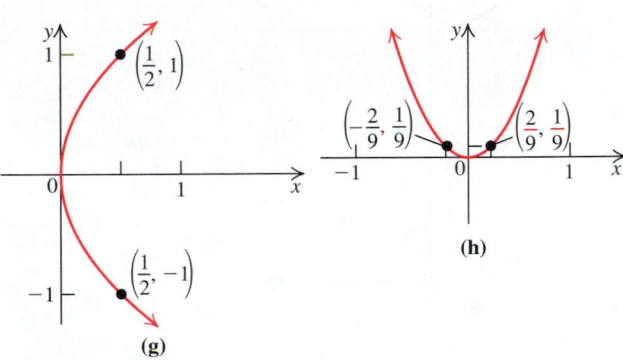

In Exercises 17–32, find the standard equation of the parabola that satisfies the given conditions. Also find the focal diameter of each parabola.

17. Focus: $(0, 2)$; directrix: $y = 4$
18. Focus: $(0, 4)$; directrix: $y = -2$
19. Focus: $(-2, 0)$; directrix: $x = 3$
20. Focus: $(-1, 0)$; directrix: $x = -2$
21. Vertex: $(1, 1)$; directrix: $x = 3$
22. Vertex: $(1, 1)$ directrix: $y = 2$
23. Vertex: $(1, 1)$; directrix: $y = -3$
24. Vertex: $(1, 1)$; directrix: $x = -2$
25. Vertex: $(1, 0)$; focus: $(3, 0)$
26. Vertex: $(0, 1)$; focus: $(0, 2)$
27. Vertex: $(0, 1)$; focus: $(0, -2)$
28. Vertex: $(-1, 0)$; focus: $(-3, 0)$
29. Vertex: $(2, 3)$; directrix: $x = 4$
30. Vertex: $(2, 3)$; directrix: $y = 4$
31. Vertex: $(2, 3)$; directrix: $y = 1$
32. Vertex: $(2, 3)$; directrix: $x = 1$

In Exercises 33–46, find the vertex, focus, and directrix of each parabola. Graph the equation.

33. $(y - 1)^2 = 2(x + 1)$ **34.** $(y + 2)^2 = 8(x - 3)$

35. $(x + 2)^2 = 3(y - 2)$ **36.** $(x - 3)^2 = 4(y + 1)$

37. $(y + 1)^2 = -6(x - 2)$ **38.** $(y - 2)^2 = -12(x + 3)$

39. $(x - 1)^2 = -10(y - 3)$ **40.** $(x + 2)^2 = -8(y + 3)$

41. $y = x^2 + 2x + 2$ **42.** $y = 3x^2 + 6x + 2$

43. $2y^2 + 4y - 2x + 1 = 0$ **44.** $3y^2 - 6y + x - 1 = 0$

45. $x = 24y - 3y^2 - 2$ **46.** $y = 16x - 2x^2 + 3$

In Exercises 47–52, find an equation of the parabola whose graph is shown.

47.

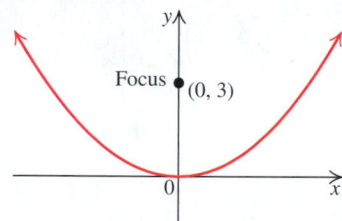

48.

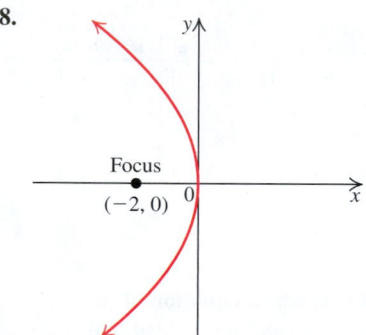

49.

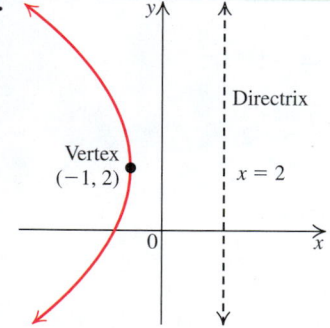

50.

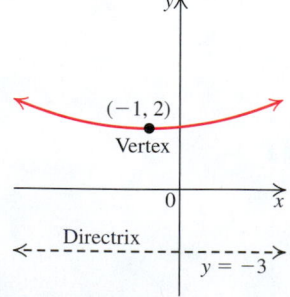

51.

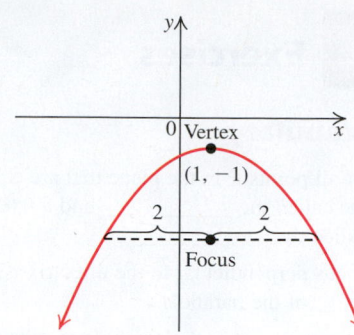

52.

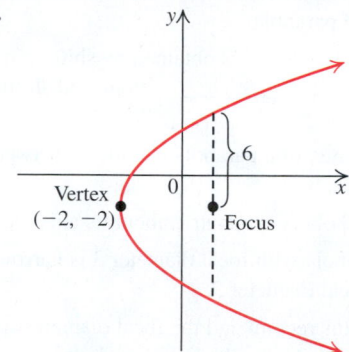

In Exercises 53–56, find two possible equations for the parabola with the given vertex V that passes through the given point P. The axis of symmetry of each parabola is parallel to a coordinate axis.

53. $V(0, 0), P(1, 2)$ **54.** $V(0, 0), P(-3, 2)$

55. $V(0, 1), P(2, 3)$ **56.** $V(1, 2), P(2, 1)$

Applying the Concepts

57. Satellite dish. A parabolic satellite dish is 40 inches across and 20 inches deep (from the vertex to the plane of the rim). How far from the vertex must the signal-receiving (receptor) unit be located to ensure that it is at the focus of the parabolic cross sections?

58. Satellite dish. Repeat Exercise 57 assuming that the dish is 8 feet across and 3 feet deep.

59. Solar heating. A parabolic reflector mirror is to be used for the solar heating of water. The mirror is 18 feet across and 6 feet deep. Where should the heating element be placed to heat the water most rapidly?

60. Flashlight. A parabolic flashlight reflector is to be 4 inches across at its rim and 2 inches deep (from the vertex to the plane of the rim). Where should the light bulb be placed?

61. Flashlight. The shape of a parabolic flashlight reflector can be described by the equation $x = 4y^2$. Where should the light bulb be placed?

62. Flashlight. Repeat Exercise 61 assuming that the parabolic flashlight reflector has the shape described by the equation

$$x = \frac{1}{2}y^2.$$

63. Satellite dish. The shape of a parabolic satellite dish can be described by the equation $y = 4x^2$. Where should the microphone be placed?

64. Satellite dish. Repeat Exercise 63 assuming that the shape of the dish can be described by the equation $y = x^2$.

65. Suspension bridge. The ends of a suspension bridge cable are fastened to two towers, which are 800 feet apart and 120 feet high. Assume that the cable hangs in the shape of a parabola and touches the roadbed midway between the towers. Find the length of the straight wire needed to suspend the roadbed from the cable at a distance of 250 feet from the tower.

66. Suspension bridge. Repeat Exercise 65 assuming that the suspension bridge cable is at a height of 8 feet above the point midway between the towers.

67. Kicking a football. A football is kicked from the ground 70 yards along a parabolic path. The maximum height of the ball is 30 yards. Find the height of the ball 5 yards before it hits the ground.

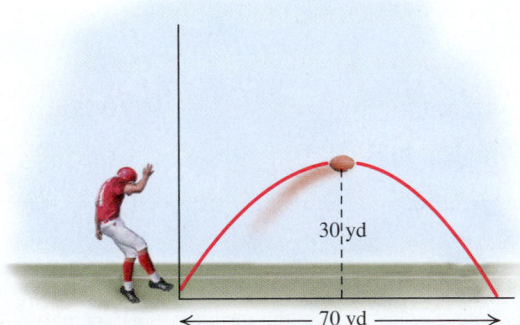

68. Parabolic arch bridge. A parabolic arch of a bridge has a 60-foot base and a height of 24 feet. Find the height of the arch at distances of 5, 10, and 20 feet from the center of the base.

69. Variable cost. The average variable cost y, in dollars, of a monthly output of x tons of a company producing a metal is given by $y = \frac{1}{10}x^2 - 3x + 50$. Find the output and cost at the vertex of the parabola.

70. Variable cost. Repeat Exercise 69 assuming that the variable cost is given by $y = \frac{1}{8}x^2 - 2x + 60$.

Beyond the Basics

71. Find the coordinates of the points at which the line $2x - 3y + 16 = 0$ intersects the parabola $y^2 = 16x$.

72. Show that the line $y = 2x + 3$ intersects the parabola $y^2 = 24x$ at only one point. Find the point of intersection.

73. Find an equation of the parabola with axis parallel to the y-axis, passing through the points $(0, 5)$, $(1, 4)$, and $(2, 7)$.

74. Find an equation of the parabola with axis parallel to the x-axis, passing through the points $(-1, -1)$, $(3, 1)$, and $(4, 0)$.

75. Find the vertex, focus, directrix, and axis of the parabola

$$x^2 - 8x + 2y + 4 = 0.$$

76. Repeat Exercise 75 for the parabola

$$Ax^2 + Dx + Ey + F = 0, \quad A \neq 0, E \neq 0.$$

77. Find the vertex, focus, directrix, and axis of the parabola

$$y^2 + 3x - 6y + 15 = 0.$$

78. Repeat Exercise 77 for the parabola

$$Cy^2 + Dx + Ey + F = 0, \quad C \neq 0, D \neq 0.$$

𝔰 **In Exercises 79–82, use the following definition. A *tangent line* to a parabola is a line that is not parallel to the axis of the parabola and that intersects the parabola at exactly one point.**

79. Find an equation of the tangent line at the point $(1, 3)$ on the parabola $y = 3x^2$ by using the following steps.

Step 1 Let m be the slope of the tangent line. Then the equation of the tangent line is $y - 3 = m(x - 1)$. Solve this equation for y.

Step 2 Substitute the value of y from Step 1 in the equation $y = 3x^2$. You will obtain a quadratic equation in x.

Step 3 For the line in Step 1 to be the tangent line to the parabola, the roots of the quadratic equation in Step 2 must be equal. Set discriminant $= 0$ and solve for m.

Step 4 Use the value of m from Step 3 to write the equation of the tangent line to the parabola $y = 3x^2$.

80. Show that the equation of the tangent line at a point (x_1, y_1) on the parabola $y = 4px^2$ is $y = 8px_1 x - y_1$. [*Hint:* Follow the steps in the previous exercise and note that $y_1 = 4px_1^2$.]

81. Show that the equation of the tangent line at point (x_1, y_1) on the parabola $x = 4py^2$ is $y = \frac{1}{8py_1}x + \frac{y_1}{2}$.

82. Reflection property of a parabola. Let $P(x_1\ y_1)$ be a point on the parabola $y = 4px^2$ with focus $F = (0, p)$ and the directrix $y = -p$. Let AP be the tangent line to the parabola. (See the figure.)

a. Use Exercise 80 to show that $d(F, A) = p + y_1$. Then use the definition of the parabola to show that $d(P, F) = p + y_1$. Conclude that $\angle \beta = \angle \gamma$.

This will suggest the following construction of the tangent line: Let B be a point on the axis of the parabola such that BP is perpendicular to the axis. Let V be the vertex of the parabola. Find the point A on the axis of the parabola such that $d(B, V) = d(A, V)$. Then the line AP is the tangent line to the parabola at the point P.

b. Use geometry to show that the angle between the y-axis and the line AP equals the angle between the lines AP and the line $x = x_1$ (which is parallel to the axis of parabola).

This demonstrates the reflection property of a parabola, which says that the tangent line to a parabola at a point P makes equal angles with

1. the line through P and the focus.
2. the axis of the parabola.

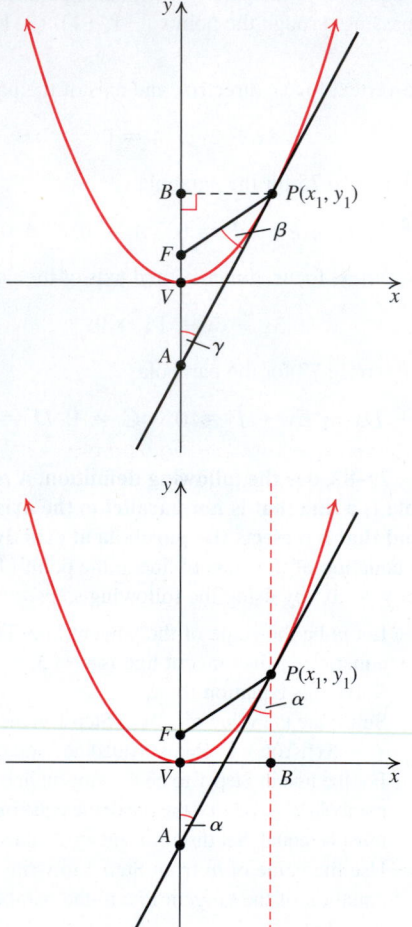

Critical Thinking / Discussion / Writing

83. a. Is a parabola always the graph of a function? Why or why not?

b. Find all possible values of the slope of the directrix of a parabola such that the parabola is always the graph of a function.

84. The equation $y^2 = 4kx$ describes a family of curves for each value of k. Sketch the graphs of the members of this family for $k = \pm 1, \pm 2, \pm 3$. What is the common characteristic of this family?

85. **True or False.** The graph of $x = ay^2 + by + c$ ($a \neq 0$) is a parabola whose axis is parallel to the y-axis.

86. It can be proved (see page 711) that the distance d from a point (x_1, y_1) to a line $ax + by + c = 0$ is given by

$$d = \frac{|ax_1 + by_1 + c|}{\sqrt{a^2 + b^2}}.$$

Use this fact and the definition of a parabola to find an equation of the parabola whose focus is $(4, 5)$ and whose directrix is the line $3x + 4y = 7$. Find an equation for the axis of this parabola.

87. Given that the vertex of a parabola is $(6, -3)$ and the directrix is the line $3x - 5y + 1 = 0$, find the coordinates of the focus.

Getting Ready for the Next Section

88. Compute the distance between $(-1, 4)$ and $(3, -2)$.

89. Find the midpoint of the line segment joining $(-2, 4)$ and $(4, 4)$.

90. Find the midpoint of the line segment joining $(3, 5)$ and $(-7, 5)$.

91. Find the x- and y-intercepts for the equation $x^2 + 2y^2 = 16$.

92. Find the x- and y-intercepts for the equation $\dfrac{x^2}{9} + \dfrac{y^2}{25} = 1$.

93. Solve $c^2 = a^2 - b^2$ for c, when $a = 4$ and $b = 3$.

94. Write $x^2 + 6x + 1 = 0$ in the form $(x + a)^2 = b$ by completing the square.

95. Write $4y^2 - 8y - 7 = 0$ in the form $c(y + a)^2 = b$ by completing the square.

96. Find an equation in point–slope form for the line through $(3, -4)$ having slope -2.

97. Find an equation in point–slope form for the line through $(0, 4)$ perpendicular to the line with equation $3y + x = 5$.

SECTION 8.3

The Ellipse

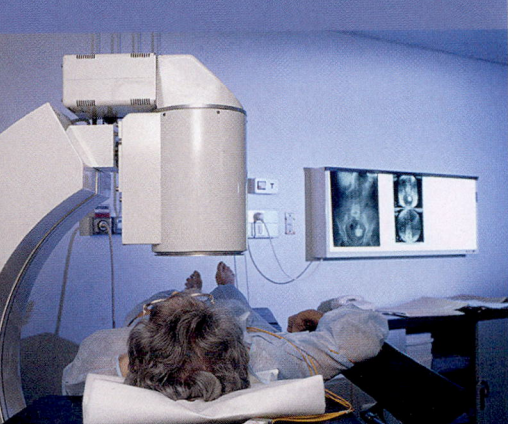

Lithotripter

BEFORE STARTING THIS SECTION, REVIEW

1 Distance formula (Section 1.1, page 5)

2 Completing the square (Appendix A.6, page 957)

3 Midpoint formula (Section 1.1, page 6)

4 Transformations of graphs (Section 1.5, pages 78–87)

5 Symmetry (Section 1.1, page 10)

OBJECTIVES

1 Define an ellipse.

2 Find the equation of an ellipse.

3 Translate ellipses.

4 Use ellipses in applications.

◆ What Is Lithotripsy?

Lithotripsy is a combination of two words: *litho* and *tripsy*. In Greek, the word *lith* denotes a stone and the word *tripsy* means "crushing." The complete medical term for lithotripsy is *extracorporeal shock-wave lithotripsy* (ESWL). Lithotripsy is a medical procedure in which a kidney stone is crushed into small sandlike pieces with the help of ultrasound high-energy shock waves, without breaking the skin.

The lithotripsy machine is called a *lithotripter*. The technology for the lithotripter was developed in Germany. The first successful treatment of a patient by a lithotripter was in February 1980 at Munich University. In the beginning, the patient had to lie in an elliptical tank filled with water and only small kidney stones were crushed. But as the medical research advanced, newer machines with better technology were manufactured. Now there is no need to keep the patient unconscious with an empty stomach, nor is the tub of water required.

In Example 6, you will see how the *reflecting property* of an ellipse is used in lithotripsy.

---

1 Define an ellipse.

Definition of Ellipse

Ellipse

An **ellipse** is the set of all points in the plane, the sum of whose distances from two fixed points is a constant. The fixed points are called the **foci** (the plural of *focus*) of the ellipse.

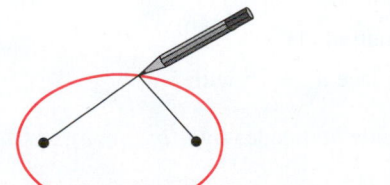

Figure 8.14 Drawing an ellipse.

You can draw an ellipse with the use of thumbtacks, a fixed length of string, and a pencil. Stick a thumbtack at each of the two points (the foci) and tie one end of the string to each tack. Place a pencil inside the loop of the string and pull it taut. Move the pencil, keeping the string taut at all times. The pencil traces an ellipse, as shown in Figure 8.14.

The points of intersection of the ellipse with the line through the foci are called the **vertices** (plural of *vertex*). The line segment connecting the vertices is the **major axis** of the ellipse. The midpoint of the major axis is the **center** of the ellipse. The line segment that is perpendicular to the major axis at the center and with endpoints on the ellipse is called the **minor axis**. See Figure 8.15.

Points F_1 and F_2 are the foci.
Points V_1 and V_2 are the vertices.
Point C is the center.
Segment $\overline{V_1 V_2}$ is the major axis.
Segment $\overline{M_1 M_2}$ is the minor axis.
The sum of the distances
$PF_1 + PF_2$ from any point P on the
ellipse to the foci is constant.

Figure 8.15

Equation of an Ellipse

2 Find the equation of an ellipse.

To find a simple equation of an ellipse, we place the ellipse's major axis along one of the coordinate axes and the center of the ellipse at the origin. We will begin with the case in which the major axis lies along the x-axis. Let $F_1(-c, 0)$ and $F_2(c, 0)$ be the coordinates of the foci. See Figure 8.16. Note that the center at the origin is the midpoint of the line segment having the foci as endpoints. To simplify the equation, it is customary to let $2a$ be the constant distance referred to in the definition. Let $P(x, y)$ be a point on the ellipse. Then

$d(P, F_1) + d(P, F_2) = 2a$	Definition of ellipse
$\sqrt{(x + c)^2 + y^2} + \sqrt{(x - c)^2 + y^2} = 2a$	Distance formula
$\sqrt{(x + c)^2 + y^2} = 2a - \sqrt{(x - c)^2 + y^2}$	Isolate a radical.
$(x + c)^2 + y^2 = (2a - \sqrt{(x - c)^2 + y^2})^2$	Square both sides.
$x^2 + 2cx + c^2 + y^2 = 4a^2 - 4a\sqrt{(x - c)^2 + y^2} + (x - c)^2 + y^2$	Expand squares.
$x^2 + 2cx + c^2 + y^2 = 4a^2 - 4a\sqrt{(x - c)^2 + y^2} + x^2 - 2cx + c^2 + y^2$	Expand squares.
$4cx - 4a^2 = -4a\sqrt{(x - c)^2 + y^2}$	Isolate the radical and simplify.
$cx - a^2 = -a\sqrt{(x - c)^2 + y^2}$	Divide both sides by 4.
$(cx - a^2)^2 = a^2[(x - c)^2 + y^2]$	Square both sides.
$c^2x^2 - 2a^2cx + a^4 = a^2(x^2 - 2cx + c^2 + y^2)$	Expand squares.
$(c^2 - a^2)x^2 - a^2y^2 = a^2c^2 - a^4$	Simplify and rearrange.
$(a^2 - c^2)x^2 + a^2y^2 = a^2(a^2 - c^2). \quad (1)$	Multiply both sides by -1 and factor $a^4 - a^2c^2$.

Now consider triangle F_1PF_2 in Figure 8.16. The *triangle inequality* from geometry states that the sum of the lengths of any two sides of a triangle is greater than the length of the third side; so for this triangle, we have

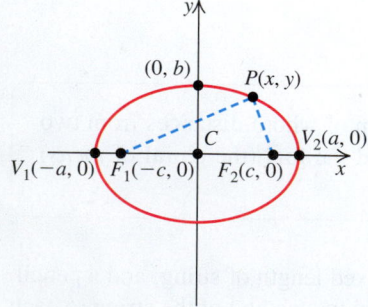

Figure 8.16

$d(P, F_1) + d(P, F_2) > d(F_1, F_2)$	
$2a > 2c$	$2a = d(P, F_1) + d(P, F_2); d(F_1, F_2) = 2c$
$a > c$	Divide both sides by 2.
$a^2 > c^2.$	Square both sides $(c > 0)$.

Therefore, $a^2 - c^2 > 0$. Letting $b^2 = a^2 - c^2$, we obtain

$(a^2 - c^2)x^2 + a^2y^2 = a^2(a^2 - c^2)$	Equation (1)
$b^2x^2 + a^2y^2 = a^2b^2$	Replace $a^2 - c^2$ with b^2.
$\dfrac{x^2}{a^2} + \dfrac{y^2}{b^2} = 1 \quad (2)$	Divide both sides by a^2b^2.

The last equation is the **standard form of the equation of an ellipse** with center $(0, 0)$ and foci $(-c, 0)$ and $(c, 0)$, where $b^2 = a^2 - c^2$.

To find the x-intercepts for the graph of this ellipse, set $y = 0$ in equation (2) to get

$$\frac{x^2}{a^2} = 1.$$

The solutions are $x = \pm a$; so the x-intercepts of the graph are $-a$ and a. The ellipse's vertices are therefore $(-a, 0)$ and $(a, 0)$. The distance between either vertex and the center is a. The distance between the vertices $V_1(-a, 0)$ and $V_2(a, 0)$ is $2a$, so the length of the major axis is $2a$.

The y-intercepts of the ellipse (found by setting $x = 0$ in equation (2)) are $-b$ and b; so the endpoints of the minor axis are $(0, -b)$ and $(0, b)$. Consequently, the length of the minor axis is $2b$. (See the left-hand figure in the table below.) The distance between either endpoint of the minor axis and the center is b.

Similarly, by reversing the roles of x and y, an equation of the ellipse with center $(0, 0)$ and foci $(0, -c)$ and $(0, c)$ on the y-axis is given by

$$\frac{x^2}{b^2} + \frac{y^2}{a^2} = 1, \text{ where } b^2 = a^2 - c^2. \quad (3)$$

In equation (3), the major axis, of length $2a$, is along the y-axis and the minor axis, of length $2b$, is along the x-axis. (See the right-hand figure in the table below.) The distance between either foci and the center is c.

If the major axis of an ellipse is along or parallel to the x-axis, the ellipse is called a **horizontal ellipse**, while an ellipse with major axis along or parallel to the y-axis is called a **vertical ellipse**.

We summarize the facts about horizontal and vertical ellipses with center $(0, 0)$. By the equation of a major or minor axis, we mean the equation of the line on which that axis lies.

SUMMARY OF **MAIN FACTS**

Main facts about an ellipse with center (0, 0)

Standard Equation	$\dfrac{x^2}{a^2} + \dfrac{y^2}{b^2} = 1; a > b > 0$ **(Horizontal ellipse)**	$\dfrac{x^2}{b^2} + \dfrac{y^2}{a^2} = 1; a > b > 0$ **(Vertical ellipse)**
Relationship between a, b, and c	$b^2 = a^2 - c^2$	$b^2 = a^2 - c^2$
Major axis along	x-axis ($y = 0$)	y-axis ($x = 0$)
Length of major axis	$2a$	$2a$
Minor axis along	y-axis ($x = 0$)	x-axis ($y = 0$)
Length of minor axis	$2b$	$2b$
Vertices	$(\pm a, 0)$	$(0, \pm a)$
Foci	$(\pm c, 0)$	$(0, \pm c)$
Endpoints of minor axis	$(0, \pm b)$	$(\pm b, 0)$
Symmetry	The graph is symmetric with respect to the x-axis, y-axis, and origin.	The graph is symmetric with respect to the x-axis, y-axis, and origin.
Graph	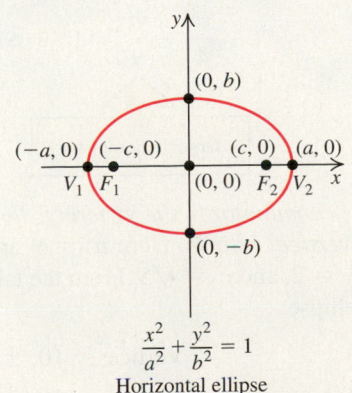 Horizontal ellipse	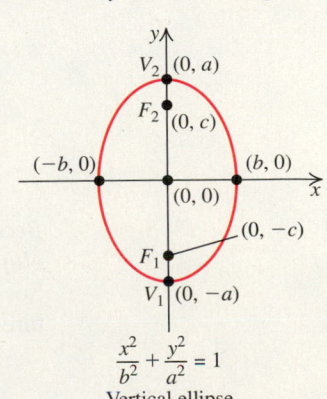 Vertical ellipse

To graph the ellipse
$\frac{x^2}{4} + \frac{y^2}{9} = 1$, first solve the
equation for y

$$\frac{y^2}{9} = 1 - \frac{x^2}{4}$$

or $y^2 = 9\left(1 - \frac{x^2}{4}\right)$

so that $y = \pm 3\sqrt{1 - \frac{x^2}{4}}$.

Then graph both functions

$Y_1 = 3\sqrt{1 - \frac{x^2}{4}}$ and

$Y_2 = -3\sqrt{1 - \frac{x^2}{4}}$.

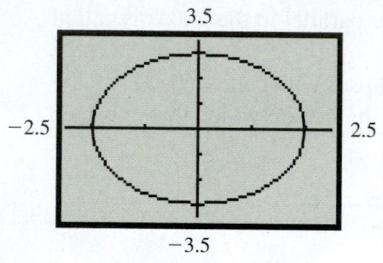

EXAMPLE 1   Finding an Equation of an Ellipse

Find the standard form of the equation of the ellipse satisfying the given conditions.

a. Foci: $(\pm 4, 0)$; vertex: $(5, 0)$

b. Foci: $(0, \pm 3)$; x-intercepts: ± 4

Solution

a. Since the foci are $(-4, 0)$ and $(4, 0)$, the major axis is on the x-axis. You know that $c = 4$ and $a = 5$. Now find b^2:

$$b^2 = a^2 - c^2 \qquad \text{Relationship between } a, b, \text{ and } c$$
$$b^2 = (5)^2 - (4)^2 = 9 \qquad \text{Replace } c \text{ with 4 and } a \text{ with 5.}$$

Substituting 25 for a^2 and 9 for b^2 in the standard form for a horizontal ellipse, we get

$$\frac{x^2}{25} + \frac{y^2}{9} = 1.$$

b. Since the foci are $(0, \pm 3)$, the major axis is on the y-axis. Then the x-intercepts ± 4 are the endpoints of the minor axis, so $b = 4$. Now find a^2:

$$b^2 = a^2 - c^2 \qquad \text{Relationship between } a, b, \text{ and } c.$$
$$(4)^2 = a^2 - (3)^2 \qquad \text{Replace } b \text{ with 4 and } c \text{ with 3.}$$
$$a^2 = 16 + 9 = 25 \qquad \text{Add 9 to both sides; simplify.}$$

Substituting 25 for a^2 and 16 for b^2 in the standard form for a vertical ellipse, we get

$$\frac{x^2}{16} + \frac{y^2}{25} = 1.$$

Practice Problem 1 Find the standard form of the equation of the ellipse satisfying the given conditions.

a. Foci: $(0, \pm 8)$; vertex: $(0, 10)$

b. Foci: $(\pm 2, 0)$; y-intercepts: ± 3

EXAMPLE 2   Graphing an Ellipse

Sketch a graph of the ellipse whose equation is $9x^2 + 4y^2 = 36$. Find the foci of the ellipse.

Solution

First, write the equation in standard form:

$$9x^2 + 4y^2 = 36 \qquad \text{Given equation}$$
$$\frac{9x^2}{36} + \frac{4y^2}{36} = 1 \qquad \text{Divide both sides by 36.}$$
$$\frac{x^2}{4} + \frac{y^2}{9} = 1 \qquad \text{Simplify.}$$

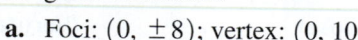

larger denominator

Because the denominator in the y^2-term is larger than the denominator in the x^2-term, the ellipse is a vertical ellipse. Here $a^2 = 9$ and $b^2 = 4$, so $c^2 = a^2 - b^2 = 9 - 4 = 5$. So, $a = 3$, $b = 2$, and $c = \sqrt{5}$. From the table on page 727, we know the following features of the ellipse:

$$\text{Vertices:} \quad (0, \pm a) = (0, \pm 3)$$
$$\text{Foci:} \qquad (0, \pm c) = (0, \pm\sqrt{5})$$

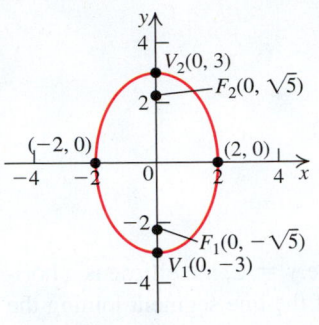

Figure 8.17 Ellipse with equation $9x^2 + 4y^2 = 36$.

3 Translate ellipses.

Length of major axis: $2a = 2(3) = 6$
Length of minor axis: $2b = 2(2) = 4$

The graph of the ellipse is shown in Figure 8.17.

Practice Problem 2 Sketch a graph of the ellipse whose equation is $4x^2 + y^2 = 16$.

Translations of Ellipses

The graph of the equation

$$\frac{x^2}{a^2} + \frac{y^2}{b^2} = 1$$

is an ellipse with center $(0, 0)$ and major axis along a coordinate axis. As in the case of a parabola, horizontal and vertical shifts can be used to obtain the graph of an ellipse whose equation is

$$\frac{(x - h)^2}{a^2} + \frac{(y - k)^2}{b^2} = 1.$$

The center of such an ellipse is (h, k), and its major axis is parallel to a coordinate axis.

SUMMARY OF **MAIN FACTS**

Main facts about horizontal and vertical ellipses with center (h, k)

Standard Equation	$\dfrac{(x - h)^2}{a^2} + \dfrac{(y - k)^2}{b^2} = 1;$ $a > b > 0$ **(Horizontal ellipse)**	$\dfrac{(x - h)^2}{b^2} + \dfrac{(y - k)^2}{a^2} = 1;$ $a > b > 0$ **(Vertical ellipse)**
Center	(h, k)	(h, k)
Major axis along the line	$y = k$	$x = h$
Length of major axis	$2a$	$2a$
Minor axis along the line	$x = h$	$y = k$
Length of minor axis	$2b$	$2b$
Vertices	$(h + a, k), (h - a, k)$	$(h, k + a), (h, k - a)$
Endpoints of minor axis	$(h, k - b), (h, k + b)$	$(h - b, k), (h + b, k)$
Foci	$(h + c, k), (h - c, k)$	$(h, k + c), (h, k - c)$
Equation involving a, b, and c	$c^2 = a^2 - b^2$	$c^2 = a^2 - b^2$
Symmetry	The graph is symmetric about the lines $x = h$ and $y = k$.	The graph is symmetric about the lines $x = h$ and $y = k$.
Graph	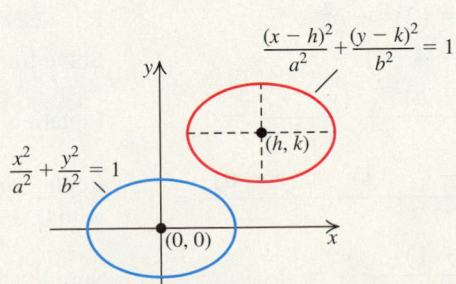 Horizontal ellipse	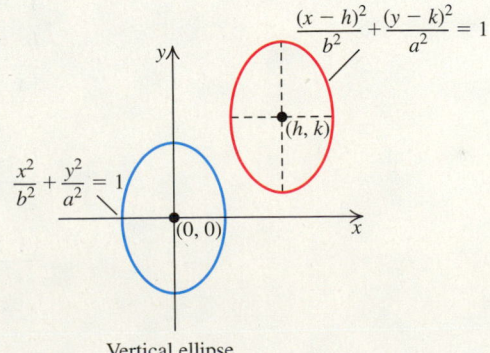 Vertical ellipse

RECALL

The midpoint of the segment joining (x_1, y_1) and (x_2, y_2) is

$$\left(\frac{x_1 + x_2}{2}, \frac{y_1 + y_2}{2} \right).$$

EXAMPLE 3 **Finding the Equation of an Ellipse**

Find an equation of the ellipse that has

a. foci $(-3, 2)$ and $(5, 2)$, and has a major axis of length 10.

b. Vertices $(-1, 2)$ and $(-1, -2)$ and a focus at $(-1, -1)$.

Solution

a. Since the foci $(-3, 2)$ and $(5, 2)$ lie on the horizontal line $y = 2$, the ellipse is a horizontal ellipse. The center of the ellipse is the midpoint of the line segment joining the foci. Using the midpoint formula yields

$$h = \frac{-3 + 5}{2} = 1 \qquad k = \frac{2 + 2}{2} = 2.$$

Center (1, 2)

$F_1(-3, 2)$ $F_2(5, 2)$

Figure 8.18 Finding the center given the foci.

The center of the ellipse is $(1, 2)$. See Figure 8.18. Since the length of the major axis is 10, the vertices must be at a distance $a = 5$ units from the center.

In Figure 8.18, the foci are four units from the center. Thus, $c = 4$. Now find b^2:

$$b^2 = a^2 - c^2 = (5)^2 - (4)^2 = 9$$

Since the major axis is horizontal, the standard form of the equation of the ellipse is

$$\frac{(x - h)^2}{a^2} + \frac{(y - k)^2}{b^2} = 1 \qquad \textcolor{blue}{a > b > 0}$$

$$\frac{(x - 1)^2}{25} + \frac{(y - 2)^2}{9} = 1 \qquad \textcolor{blue}{\text{Replace } h \text{ with 1, } k \text{ with 2,} \\ a^2 \text{ with 25, and } b^2 \text{ with 9.}}$$

Now for this horizontal ellipse with center $(1, 2)$, we have $a = 5, b = 3$, and $c = 4$.

Vertices

$$(h \pm a, k) = (1 \pm 5, 2) = (-4, 2) \text{ and } (6, 2)$$

Endpoints of Minor Axis

$$(h, k \pm b) = (1, 2 \pm 3) = (1, -1) \text{ and } (1, 5)$$

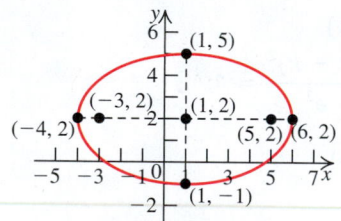

Figure 8.19 Ellipse with foci $(-3, 2)$ and $(5, 2)$.

Use the center and these four points to sketch the graph shown in Figure 8.19.

b. Since the vertices $(-1, 2)$ and $(-1, -2)$ lie on the vertical line $x = -1$, the ellipse is a vertical ellipse. The center (h, k) of the ellipse is the midpoint of the line segment joining the vertices. As in part **a**, use the midpoint formula to find the center (h, k), with $h = -1$ and $k = 0$. The distance, a, from the center $(-1, 0)$ to the vertex $(-1, 2)$ is $\sqrt{(-1 - (-1))^2 + (0 - 2)^2} = \sqrt{0^2 + (-2)^2} = \sqrt{4} = 2$, so $a = 2$. The distance, c, from the center $(-1, 0)$ to the focus $(-1, -1)$ is $\sqrt{(-1 - (-1))^2 + (0 - (-1))^2} = \sqrt{0^2 + 1^2} = \sqrt{1} = 1$, so $c = 1$. Then,

$$c^2 = a^2 - b^2 \qquad \textcolor{blue}{\text{Relationship involving } a, b, \text{ and } c.}$$

$$1^2 = (2)^2 - b^2 \qquad \textcolor{blue}{\text{Replace } c \text{ with 1 and } a \text{ with 2.}}$$

$$b^2 = 4 - 1 = 3 \qquad \textcolor{blue}{\text{Solve for } b^2.}$$

$$\frac{(x - h)^2}{b^2} + \frac{(y - k)^2}{a^2} = 1 \qquad \textcolor{blue}{\text{Equation for a vertical ellipse.}}$$

$$\frac{(x - [-1])^2}{3} + \frac{(y - 0)^2}{4} = 1 \qquad \textcolor{blue}{\text{Replace } h \text{ with } -1, b^2 \text{ with 3, } k \\ \text{with 0, and } a^2 \text{ with 4.}}$$

$$\frac{(x + 1)^2}{3} + \frac{y^2}{4} = 1 \qquad \textcolor{blue}{\text{Simplify.}}$$

Practice Problem 3 Find an equation of the ellipse that has

a. foci at $(2, -3)$ and $(2, 5)$ and has a major axis of length 10
b. vertices at $(0, 0)$ and $(0, 10)$ and a focus at $(0, 8)$.

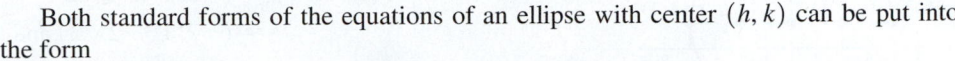

Both standard forms of the equations of an ellipse with center (h, k) can be put into the form

$$Ax^2 + Cy^2 + Dx + Ey + F = 0, \quad (4)$$

where neither A nor C is zero and both have the same sign. For example, consider the ellipse with equation

$$\frac{(x - 1)^2}{4} + \frac{(y + 2)^2}{9} = 1.$$

$9(x - 1)^2 + 4(y + 2)^2 = 36$	Multiply both sides by 36.
$9(x^2 - 2x + 1) + 4(y^2 + 4y + 4) = 36$	Expand squares.
$9x^2 - 18x + 9 + 4y^2 + 16y + 16 = 36$	Distributive property
$9x^2 + 4y^2 - 18x + 16y - 11 = 0$	Simplify.

Comparing this equation with equation (4), we have

$$A = 9, C = 4, D = -18, E = 16, \text{ and } F = -11.$$

Conversely, equation (4) can be put into a standard form for an ellipse by completing the squares in both of the variables x and y.

GENERAL EQUATION OF AN ELLIPSE

The graph of $Ax^2 + Cy^2 + Dx + Ey + F = 0$ is an ellipse if $AC > 0$. That is, neither A nor C is zero, and both have the same sign.

EXAMPLE 4   **Converting to Standard Form**

Find the center, vertices, and foci of the ellipse with equation

$$3x^2 + 4y^2 + 12x - 8y - 32 = 0.$$

Solution

Complete squares on x and y.

$3x^2 + 4y^2 + 12x - 8y - 32 = 0$	Original equation
$(3x^2 + 12x) + (4y^2 - 8y) = 32$	Group like terms.
$3(x^2 + 4x) + 4(y^2 - 2y) = 32$	Factor out 3 and 4.
$3(x^2 + 4x + 4) + 4(y^2 - 2y + 1) = 32 + 12 + 4$	Complete squares on x and y: add $3 \cdot 4$ and $4 \cdot 1$.
$3(x + 2)^2 + 4(y - 1)^2 = 48$	Simplify.
$\dfrac{(x + 2)^2}{16} + \dfrac{(y - 1)^2}{12} = 1$	Divide both sides by 48. (16 is the larger denominator.)

The last equation is the standard form for a horizontal ellipse with center $(-2, 1)$, $a^2 = 16$, $b^2 = 12$, and $c^2 = a^2 - b^2 = 16 - 12 = 4$. Thus, $a = 4, b = \sqrt{12} = 2\sqrt{3}$, and $c = 2$.

From the table on page 729, we have the following information:
The length of the major axis is $2a = 8$.
The length of the minor axis is $2b = 4\sqrt{3}$.

Center: $\quad\quad\quad (h, k) = (-2, 1)$

Foci: $\quad\quad\quad (h \pm c, k) = (-2 \pm 2, 1)$
$\quad\quad\quad\quad\quad\quad\quad = (-4, 1) \text{ and } (0, 1)$

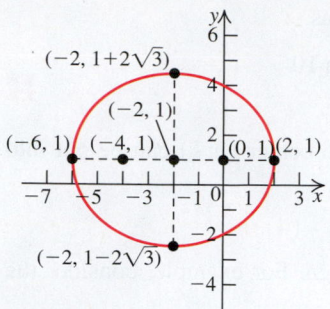

Figure 8.20 Ellipse with equation $3x^2 + 4y^2 + 12x - 8y - 32 = 0$.

Vertices:
$$(h \pm a, k) = (-2 \pm 4, 1)$$
$$= (-6, 1) \text{ and } (2, 1)$$

Endpoints of minor axis:
$$(h, k \pm b) = (-2, 1 \pm 2\sqrt{3})$$
$$= (-2, 1 + 2\sqrt{3}) \text{ and } (-2, 1 - 2\sqrt{3})$$
$$\approx (-2, 4.46) \text{ and } (-2, -2.46)$$

The graph of the ellipse is shown in Figure 8.20.

Practice Problem 4 Find the center, vertices, and foci of the ellipse with equation

$$x^2 + 4y^2 - 6x + 8y - 29 = 0.$$

A *circle* is considered an ellipse in which the two foci coincide at the center. Conversely, if the two foci of an ellipse coincide, then $2c = 0$ (remember, $2c$ is the distance between the two foci of an ellipse); so $c = 0$. Again, from the relationship $c^2 = a^2 - b^2 = 0$, we have $a = b$. Thus, the equation for the ellipse becomes the equation for a circle of radius $r = a$, having the same center as that of the ellipse.

EXAMPLE 5 **Find an Equation for each Ellipse whose Graph Is Shown**

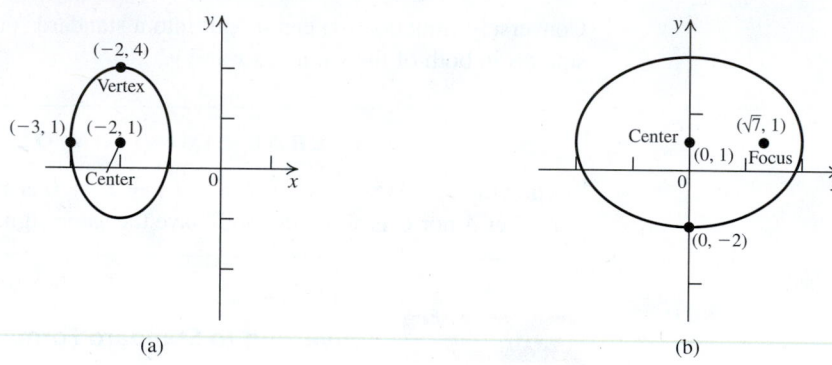

(a) (b)

Solution

a. The center $(h, k) = (-2, 1)$, so $h = -2$ and $k = 1$.

The distance, a, between the center $(-2, 1)$ and the vertex $(-2, 4)$ is
$$\sqrt{(-2 - (-2))^2 + (1 - 4)^2} = \sqrt{0^2 + 3^2} = \sqrt{9} = 3. \text{ So } a = 3.$$
The distance between the center $(-2, 1)$ and $(-3, 1)$ is
$$\sqrt{(-2 - (-3))^2 + (1 - 1)^2} = \sqrt{1^2 + 0^2} = \sqrt{1} = 1. \text{ So } b = 1.$$

$$\frac{(x - h)^2}{b^2} + \frac{(y - k)^2}{a^2} = 1 \qquad \text{Equation for a vertical ellipse}$$

$$\frac{(x - [-2])^2}{1^2} + \frac{(y - 1)^2}{3^2} = 1 \qquad \text{Replace } h \text{ with } -2, b \text{ with } 1, k \text{ with } 1, \text{ and } a \text{ with } 3.$$

$$(x + 2)^2 + \frac{(y - 1)^2}{9} = 1 \qquad \text{Simplify.}$$

b. The center $(h, k) = (0, 1)$, so $h = 0$ and $k = 1$.
The distance, c, between the center $(0, 1)$ and the focus at $\left(\sqrt{7}, 1\right)$ is
$$\sqrt{\left(0 - \sqrt{7}\right)^2 + (1 - 1)^2} = \sqrt{\left(-\sqrt{7}\right)^2 + 0^2} = \sqrt{7}. \text{ So, } c = \sqrt{7}.$$

From the graph we see that the focus at $\left(\sqrt{7}, 1\right)$ and the center at $(0, 1)$ are on the horizontal line $y = 1$, so the ellipse is horizontal. Then $(0, -2)$ is an endpoint of the minor axis. The distance, b, between the center $(0, 1)$ and the endpoint $(0, -2)$ is
$$\sqrt{(0 - 0)^2 + (1 - (-2))^2} = \sqrt{0^2 + 3^2} = \sqrt{9} = 3. \text{ So } b = 3.$$

$$c^2 = a^2 - b^2 \quad \text{Relationship between } a, b, \text{ and } c$$
$$(\sqrt{7})^2 = a^2 - 3^2 \quad \text{Replace } c \text{ with } \sqrt{7} \text{ and } b \text{ with 3.}$$
$$a^2 = 7 + 9 = 16 \quad \text{Solve for } a^2.$$
$$\frac{(x - h)^2}{a^2} + \frac{(y - k)^2}{b^2} = 1 \quad \text{Equation for a horizontal ellipse}$$
$$\frac{(x - 0)^2}{16} + \frac{(y - 1)^2}{3^2} = 1 \quad \text{Replace } h \text{ with 0, } a^2 \text{ with 16, } k \text{ with 1, and } b \text{ with 3.}$$
$$\frac{x^2}{16} + \frac{(y - 1)^2}{9} = 1 \quad \text{Simplify.}$$

Practice Problem 5 Find an equation for the ellipse whose graph is shown.

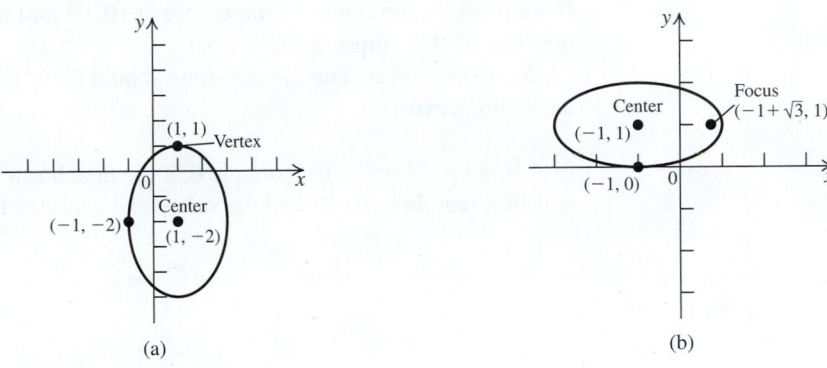

Applications

4 Use ellipses in applications.

Ellipses have many applications. We mention just a few:

1. The orbits of the planets are ellipses with the sun at one focus. This fact alone is sufficient to explain the overwhelming importance of ellipses since the seventeenth century, when Kepler and Newton did their monumental work in astronomy.

2. Newton also reasoned that comets move in elliptical orbits about the sun. His friend Edmund Halley used this information to predict that a certain comet (now called Halley's comet) reappears about every 77 years. Halley saw the comet in 1682 and correctly predicted its return in 1759. The last sighting of the comet was in 1986, and it is due to return in 2061. A recent study shows that Halley's comet has made approximately 2000 cycles so far and has about the same number to go before the sun erodes it away completely. (See Exercise 80.)

3. We can calculate the distance traveled by a planet in one orbit around the sun.

4. **The reflecting property of ellipses.** The *reflecting property* for an ellipse says that a ray of light originating at one focus will be reflected to the other focus. See Figure 8.21. Sound waves also follow such paths. This property is used in the construction of "whispering galleries," such as the gallery at St. Paul's Cathedral in London. Such rooms have ceilings whose cross sections are elliptical with common foci. As a result, sounds emanating from one focus are reflected by the ceiling to the other focus. Thus, a whisper at one focus may not be audible at all at a nearby place but may nevertheless be clearly heard far off at the other focus.

Example 6 illustrates how the *reflecting property* of an ellipse is used in lithotripsy. As the introduction to this section noted, a *lithotripter* is a machine that is designed to crush kidney stones into small sandlike pieces with the help of high-energy shock waves. To focus these shock waves accurately, a lithotripter uses the reflective property of ellipses. A patient is carefully positioned so that the kidney stone is at one focus of the ellipsoid (a three-dimensional ellipse) and the shock-producing source is placed at the other focus. The high-energy shock waves generated at this focus are concentrated on the kidney stone at the other focus, thus pulverizing it without harming the patient.

Edmund Halley

(1656–1742) Edmund Halley was a friend of Isaac Newton. Although he is known primarily for calculating the orbit of Halley's comet, he was also a biologist, a geologist, a sea captain, a spy, and the author of the first actuarial mortality tables.

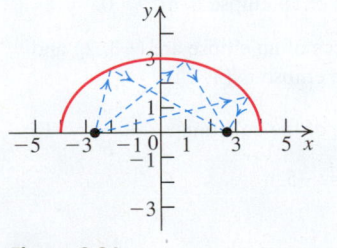

Figure 8.21

EXAMPLE 6 Lithotripsy

An elliptical water tank has a major axis of length 6 feet and a minor axis of length 4 feet. The source of high-energy shock waves from a lithotripter is placed at one focus of the tank. To smash the kidney stone of a patient, how far should the stone be positioned from the source?

Solution

Since the length of the major axis of the ellipse is 6 feet, we have $2a = 6$; so $a = 3$. Similarly, the minor axis of 4 feet gives $2b = 4$, or $b = 2$. To find c, we use the equation $c^2 = a^2 - b^2$. We have

$$c^2 = (3)^2 - (2)^2 = 5. \text{ Therefore, } c = \pm\sqrt{5}.$$

If we position the center of the ellipse at $(0, 0)$ and the major axis along the x-axis, then the foci of the ellipse are $(-\sqrt{5}, 0)$ and $(\sqrt{5}, 0)$. The distance between these foci is $2\sqrt{5} \approx 4.472$ feet. The kidney stone should be positioned 4.472 feet from the source of the shock waves.

Practice Problem 6 In Example 6, if the tank has a major axis of 8 feet and a minor axis of 4 feet, how far should the stone be positioned from the source?

Answers to Practice Problems

1. a. $\dfrac{x^2}{36} + \dfrac{y^2}{100} = 1$ **b.** $\dfrac{x^2}{13} + \dfrac{y^2}{9} = 1$

2.

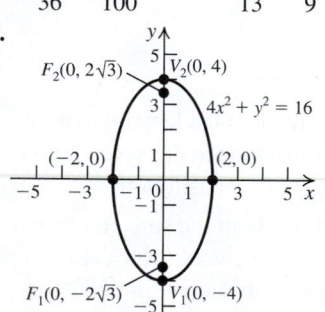

3. a. $\dfrac{(x-2)^2}{9} + \dfrac{(y-1)^2}{25} = 1$ **b.** $\dfrac{x^2}{16} + \dfrac{(y-5)^2}{25} = 1$

4. Center: $(3, -1)$; vertices: $(3 + \sqrt{42}, -1)$ and $(3 - \sqrt{42}, -1)$;

foci: $\left(3 + \dfrac{3\sqrt{14}}{2}, -1\right)$ and $\left(3 - \dfrac{3\sqrt{14}}{2}, -1\right)$

5. a. $\dfrac{(x-1)^2}{4} + \dfrac{(y+2)^2}{9} = 1$ **b.** $\dfrac{(x+1)^2}{4} + (y-1)^2 = 1$

6. $4\sqrt{3} \approx 6.9282$ ft

SECTION 8.3 **Exercises**

Concepts and Vocabulary

1. An ellipse is the set of all points in the plane, the _____ of whose distances from two fixed points is a constant.

2. The points of intersection of the ellipse with the line through the foci are called _____ of the ellipse.

3. The standard equation of an ellipse with center $(0, 0)$, vertices $(\pm a, 0)$, and foci $(\pm c, 0)$ is $\dfrac{x^2}{a^2} + \dfrac{y^2}{b^2} = 1$,

where $b^2 =$ _____.

4. In the equation $\dfrac{x^2}{m^2} + \dfrac{y^2}{n^2} = 1$, if **(i)** $m^2 > n^2$, the graph is a(n) _____ ellipse; **(ii)** if $m^2 < n^2$, the graph is a(n) _____ ellipse; **(iii)** if $m^2 = n^2$, the graph is a(n) _____.

5. True or False. The ellipse $\dfrac{x^2}{a^2} + \dfrac{y^2}{b^2} = 1$ lies inside the box formed by the four lines $x = \pm a, y = \pm b$.

6. True or False. The graph of $Ax^2 + Cy^2 + Dx + Ey + F = 0$ is a degenerate conic or an ellipse if $AC > 0$.

7. True or False. If the vertices of an ellipse are $(-3, 2)$ and $(5, 2)$, then the center of the ellipse is $(1, 2)$.

8. True or False. If the graph of the equation $\dfrac{x^2}{a^2} + \dfrac{y^2}{b^2} = 1$ is a circle and $a = 5$, then $b = -5$.

Building Skills

In Exercises 9–22, find the standard form of the equation for the ellipse satisfying the given conditions. Graph the equation.

9. Foci: $(\pm 1, 0)$; vertex $(3, 0)$

10. Foci: $(\pm 3, 0)$; vertex $(5, 0)$

11. Foci: $(0, \pm 2)$; vertex $(0, 4)$

12. Foci: $(0, \pm 3)$; vertex $(0, -6)$

13. Foci: $(\pm 4, 0)$; y-intercepts: ± 3

14. Foci: $(\pm 3, 0)$; y-intercepts: ± 4

15. Foci: $(0, \pm 2)$; x-intercepts: ± 4

16. Foci: $(0, \pm 3)$; x-intercepts: ± 5

17. Center: $(0, 0)$, vertical major axis of length 10, and minor axis of length 6

18. Center: $(0, 0)$, vertical major axis of length 8, and minor axis of length 4

19. Center: $(0, 0)$; vertices: $(\pm 6, 0)$; $c = 3$

20. Center: $(0, 0)$; vertices: $(0, \pm 5)$; $c = 3$

21. Center: $(0, 0)$; foci: $(0, \pm 2)$; $b = 3$

22. Center: $(0, 0)$; foci: $(\pm 3, 0)$; $b = 2$

In Exercises 23–42, find the vertices and foci for each ellipse. Graph each equation.

23. $\dfrac{x^2}{16} + \dfrac{y^2}{4} = 1$

24. $\dfrac{x^2}{4} + \dfrac{y^2}{16} = 1$

25. $\dfrac{x^2}{9} + y^2 = 1$

26. $x^2 + \dfrac{y^2}{4} = 1$

27. $\dfrac{x^2}{25} + \dfrac{y^2}{16} = 1$

28. $\dfrac{x^2}{25} + \dfrac{y^2}{9} = 1$

29. $\dfrac{x^2}{16} + \dfrac{y^2}{36} = 1$

30. $\dfrac{x^2}{9} + \dfrac{y^2}{16} = 1$

31. $x^2 + y^2 = 4$

32. $x^2 + y^2 = 16$

33. $x^2 + 4y^2 = 4$

34. $9x^2 + y^2 = 9$

35. $9x^2 + 4y^2 = 36$

36. $4x^2 + 25y^2 = 100$

37. $3x^2 + 4y^2 = 12$

38. $4x^2 + 5y^2 = 20$

39. $5x^2 - 10 = -2y^2$

40. $3y^2 = 21 - 7x^2$

41. $2x^2 + 3y^2 = 7$

42. $3x^2 + 4y^2 = 11$

In Exercises 43 and 44, find the vertices and the foci of each ellipse.

43. Vertical ellipse with center at $(-2, 3)$, major axis of length 10 and minor axis of length 6.

44. Horizontal ellipse with center at $(1, -2)$, major axis of length 8 and minor axis of length 4.

In Exercises 45–50, find an equation of the ellipse satisfying the given conditions.

45. Vertices at $(1, 0)$ and $(9, 0)$ and a focus at $(7, 0)$

46. Vertices at $(0, 10)$ and $(0, -2)$ and a focus at $(0, 7)$

47. Vertices at $(2, -5)$ and $(2, 3)$ and a minor axis of length 2

48. Vertices at $(-1, 4)$ and $(7, 4)$ and a minor axis of length 1

49. Foci at $(-2, 3)$ and $(4, 3)$ and a vertex at $(-5, 3)$

50. Foci at $(5, 1)$ and $(13, 1)$ and a vertex at $(-3, 1)$

In Exercises 51–62, find the center, foci, and vertices of each ellipse. Graph each equation.

51. $\dfrac{(x - 1)^2}{4} + \dfrac{(y - 1)^2}{9} = 1$

52. $\dfrac{x^2}{16} + \dfrac{(y + 3)^2}{4} = 1$

53. $4(x + 3)^2 + 5(y - 1)^2 = 20$

54. $25(x - 2)^2 + 4(y + 5)^2 = 100$

55. $5x^2 + 9y^2 + 10x - 36y - 4 = 0$

56. $9x^2 + 4y^2 + 72x - 24y + 144 = 0$

57. $9x^2 + 5y^2 + 36x - 40y + 71 = 0$

58. $3x^2 + 4y^2 + 12x - 16y - 32 = 0$

59. $x^2 + 2y^2 - 2x + 4y + 1 = 0$

60. $2x^2 + 2y^2 - 12x - 8y + 3 = 0$

61. $2x^2 + 9y^2 - 4x + 18y + 12 = 0$

62. $3x^2 + 2y^2 + 12x - 4y + 15 = 0$

In Exercises 63–68, find an equation for the ellipse whose graph is shown.

63.   64.

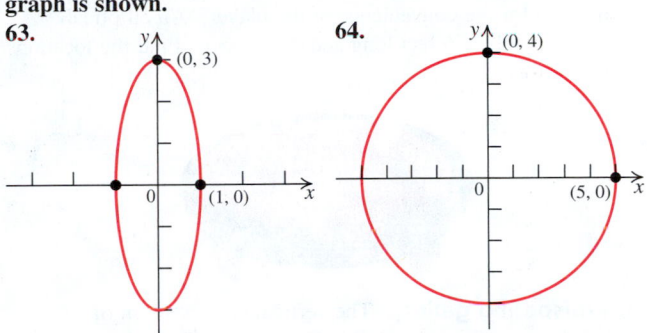

65.   66.

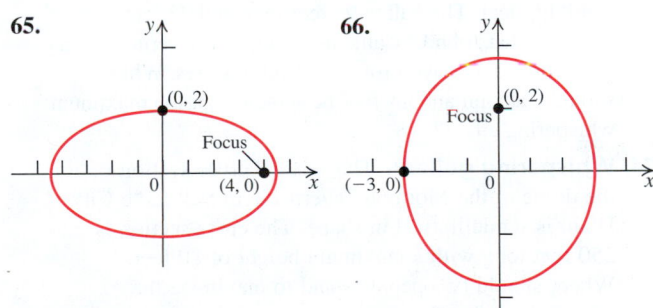

67.   68.

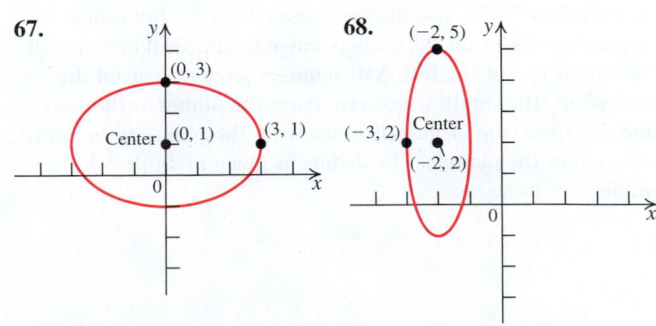

Applying the Concepts

69. Semielliptical arch. A portion of an arch in the shape of a *semiellipse* (half of an ellipse) is 50 feet wide at the base and has a maximum height of 20 feet. Find the height of the arch at a distance of 10 feet from the center. (See the figure.)

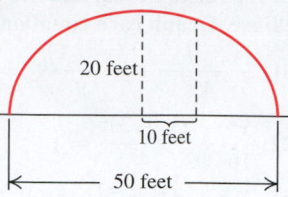

70. Semielliptical arch. In Exercise 69, find the distance between the two points at the base where the height of the arch is 10 feet.

71. Semielliptical arch bridge. A bridge is built in the shape of a semielliptical arch. The span of the bridge is 150 meters, and its maximum height is 45 meters. There are two vertical supports, each at a distance of 25 meters from the central position. Find their heights.

72. Elliptical billiard table. Some billiard tables (similar to pool tables, but without pockets) are manufactured in the shape of an ellipse. The foci of such tables are plainly marked for the convenience of the players. An elliptical billiard table is 6 feet long and 4 feet wide. Find the location of the foci.

73. Whispering gallery. The vertical cross section of the dome of Statuary Hall in Washington, D.C., is semielliptical. The hall is 96 feet long and 23 feet high. It is said that John C. Calhoun used the whispering-gallery phenomenon to eavesdrop on his adversaries. Where should Calhoun and his foes be standing for the maximum whispering effect?

74. Whispering gallery. The vertical cross section of the dome of the Mormon Tabernacle in Salt Lake City, Utah, is semielliptical in shape. The cross section is 250 feet long with a maximum height of 80 feet. Where should two people stand to maximize the whispering effect?

In Exercises 75–78, use the fact (discovered by Johannes Kepler in 1609) that the planets move in elliptical orbits with the sun at one of the foci. Astronomers have measured the *perihelion* (the smallest distance from the planet to the sun) and *aphelion* (the largest distance from the planet to the sun) for each of the planets. The distances given in Table 8.1 are in millions of miles.

TABLE 8.1

Planet	Perihelion	Aphelion
Mercury	28.56	43.88
Venus	66.74	67.68
Earth	91.38	94.54
Mars	128.49	154.83
Jupiter	460.43	506.87
Saturn	837.05	936.37
Uranus	1699.45	1866.59
Neptune	2771.72	2816.42

75. Mars. Find an equation of Mars's orbit about the sun.

> *Hint:* Set up a coordinate system with the sun at one focus and the major axis lying on the *x*-axis. (See the figure.) Calculate a from the equation $2a =$ aphelion + perihelion. Calculate c from the equation $c = a -$ perihelion. Calculate b^2 from the equation $b^2 = a^2 - c^2$. Write the equation $\dfrac{x^2}{a^2} + \dfrac{y^2}{b^2} = 1.$

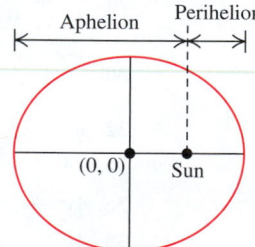

76. Earth. Write an equation for the orbit of Earth about the sun.

77. Mercury. Write an equation for the orbit of the planet Mercury about the sun.

78. Saturn. Write an equation for the orbit of the planet Saturn about the sun.

79. Moon. The moon orbits Earth in an elliptical path with Earth at one focus. The major and minor axes of the orbit have lengths of 768,806 kilometers and 767,746 kilometers, respectively. Find the perihelion and aphelion of the orbit.

80. **Halley's comet.** In 1682, Edmund Halley (1656–1742) identified the orbit of a certain comet (now called Halley's comet) as an elliptical orbit about the sun, with the sun at one focus. He calculated the length of its major axis to be approximately 5.39×10^9 kilometers and its minor axis to be approximately 1.36×10^9 kilometers. Find the perihelion and aphelion of the orbit of Halley's comet.

81. **Orbit of a satellite.** Assume that Earth is a sphere with a radius of 3960 miles. A satellite has an elliptical orbit around Earth with the center of Earth as one of its foci. The maximum and minimum heights of the satellite from the surface of Earth are 164 miles and 110 miles, respectively. Find an equation for the orbit of the satellite.

Beyond the Basics

82. Find an equation of the ellipse in standard form with center $(0, 0)$ and with major axis of length 10, passing through the point $\left(-3, \frac{16}{5}\right)$.

83. Find an equation of the ellipse in standard form with center $(0, 0)$ and with major axis of length 6, passing through the point $\left(1, \frac{3\sqrt{5}}{2}\right)$.

84. Find the lengths of the major and minor axes of the ellipse
$$\frac{x^2}{b^2} + \frac{y^2}{a^2} = 1,$$
given that the ellipse passes through the points $(2, 1)$ and $(1, -3)$.

85. Find an equation of the ellipse in standard form with center $(2, 1)$, passing through the points $(10, 1)$ and $(6, 2)$.

86. The **eccentricity** of an ellipse, denoted by e, is defined as
$$e = \frac{\text{Distance between the foci}}{\text{Distance between the vertices}} = \frac{2c}{2a} = \frac{c}{a}. \text{ What}$$
happens when $e = 0$?

In Exercises 87–91, use the definition of e given in Exercise 86 to determine the eccentricity of each ellipse. (This e has nothing to do with the number e, discussed in Chapter 3, that is the base for natural logarithms.)

87. $\frac{x^2}{16} + \frac{y^2}{9} = 1$

88. $20x^2 + 36y^2 = 720$

89. $x^2 + 4y^2 = 1$

90. $\frac{(x+1)^2}{25} + \frac{(y-2)^2}{9} = 1$

91. $x^2 + 2y^2 - 2x + 4y + 1 = 0$

92. An ellipse has its major axis along the x-axis and its minor axis along the y-axis. Its eccentricity is $\frac{1}{2}$, and the distance between the foci is 4. Find an equation of the ellipse.

93. Find an equation of the ellipse whose major axis has end points $(-3, 0)$ and $(3, 0)$ and that has eccentricity $\frac{1}{3}$.

94. A point $P(x, y)$ moves so that its distance from the point $(4, 0)$ is always one-half its distance from the line $x = 16$. Show that the equation of the path of P is an ellipse of eccentricity $\frac{1}{2}$.

In Exercises 95–98, find all points of intersection of the given curves by solving systems of nonlinear equations. Sketch the graphs of the curves and show the points of intersection.

95. $\begin{cases} x + 3y = -2 \\ 4x^2 + 3y^2 = 7 \end{cases}$

96. $\begin{cases} x^2 = 2y \\ 2x^2 + y^2 = 12 \end{cases}$

97. $\begin{cases} x^2 + y^2 = 20 \\ 9x^2 + y^2 = 36 \end{cases}$

98. $\begin{cases} 9x^2 + 16y^2 = 36 \\ 18x^2 + 5y^2 = 45 \end{cases}$

In Exercises 99 and 100, use the following definition. A *tangent line* to an ellipse is a line that intersects the ellipse at exactly one point.

99. Find the equation of the tangent line to the ellipse with equation $\frac{x^2}{a^2} + \frac{y^2}{b^2} = 1$, with $a = \frac{5}{2}$ and $b = 5$ at the point $(2, 3)$.
[*Hint:* See the steps for Exercise 79 in Section 8.2 and (in Step 3) write the discriminant as a perfect square.]

100. Show that the equation of the tangent line to the ellipse with equation $\frac{x^2}{a^2} + \frac{y^2}{b^2} = 1$, at the point (x_1, y_1), is
$$y = -\frac{b^2 x_1}{a^2 y_1}x + \frac{b^2 x_1^2}{b^2 y_1^2} + y_1.$$
[*Hint:* See the steps for Exercise 80 in Section 8.2 and show that $(a^2 y_1 m + b^2 x_1)^2 = 0$.]

In Exercises 101 and 102, use the fact that the area inside an ellipse is πab, where a and b are the lengths of the semi-major axis and semi-minor axis, respectively. Find the area inside each ellipse.

101. $9(x+1)^2 + 16(y-2)^2 = 144$

102. $5x^2 + 5y^2 + 10x - 20y - 62 = 0$

In Exercises 103 and 104, find the area between the two given ellipses.

103. $16(x-1)^2 + 9(y+2)^2 = 144$
$16(x-1)^2 + 4(y+2)^2 = 64$

104. $x^2 + y^2 + 4x + 2y - 20 = 0$
$16x^2 + 25y^2 + 64x + 50y - 311 = 0$

Critical Thinking / Discussion / Writing

105. Find the number of circles with a radius of three units that touch both of the coordinate axes. Write the equation of each circle.

106. Find the number of ellipses with a major axis and a minor axis of lengths six units and four units, respectively, that touch both axes. Write the equation of each ellipse.

Getting Ready for the Next Section

In Exercises 107 and 108, find the slope–intercept form of the line with the given properties.

107. Slope: -1; passing through $(-3, 5)$.

108. Passing through $(1, -1)$ perpendicular to the line with equation $y = -\dfrac{1}{2}x + 3$.

In Exercises 109 and 110, find the point of intersection, if any.

109. $\begin{cases} y = 3x + 6 \\ y = -3x \end{cases}$

110. $\begin{cases} y = \dfrac{3}{4}x - 2 \\ y = -\dfrac{3}{4}x + 3 \end{cases}$

In Exercises 111 and 112, find the x- and y-intercepts of the graph of the given equation, if any.

111. $4y^2 - x^2 = 1$

112. $16x^2 - 9y^2 = 144$

In Exercises 113 and 114, solve the equation $c^2 = a^2 + b^2$ for the specified variable.

113. Solve for a, when $c = \sqrt{29}$ and $b = 4$.

114. Solve for b, when $a = 3$ and $c = 12$.

In Exercises 115 and 116, find the asymptotes of the graph of the given function.

115. $f(x) = \dfrac{x^2 + 1}{x - 1}$

116. $f(x) = \dfrac{x^2 - x}{x + 1}$

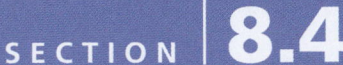

SECTION 8.4

The Hyperbola

BEFORE STARTING THIS SECTION, REVIEW

1 Distance formula (Section 1.1, page 5)

2 Midpoint formula (Section 1.1, page 6)

3 Completing the square (Appendix A.6, page 957)

4 Oblique asymptotes (Section 2.5, page 201)

5 Transformations of graphs (Section 1.5, pages 78–87)

6 Symmetry (Section 1.1, page 10)

OBJECTIVES

1 Define a hyperbola.

2 Find the asymptotes of a hyperbola.

3 Graph a hyperbola.

4 Translate hyperbolas.

5 Use hyperbolas in applications.

Alfred Lee Loomis (1887–1975)
Alfred Loomis was an American lawyer, investment banker, physicist, and patron of scientific research. Besides inventing LORAN, he made significant contributions to the ground-based technology for landing airplanes by instruments. President Roosevelt recognized the value of Loomis's work and described him as second perhaps only to Churchill as the civilian most responsible for the Allied victory in World War II.

◆ What Is LORAN?

LORAN stands for LOng-RAnge Navigation. The federal government runs the LORAN system, which uses land-based radio navigation transmitters to provide users with information on position and timing. During World War II, Alfred Lee Loomis was selected to chair the National Defense Research Committee. Much of his work focused on the problem of creating a light system for plane-carried radar. As a result of this effort, he invented LORAN, the long-range navigation system whose offshoot LORAN-C remains in widespread use. The navigational method provided by LORAN is based on the principle of determining the points of intersection of two hyperbolas to fix the two-dimensional position of the receiver. In Example 9, we illustrate how this is done.

The LORAN-C system is now being supplemented by the global positioning system (GPS), which works on the same principle as LORAN-C. The GPS system consists of 24 satellites that orbit 11,000 miles above Earth. The GPS was created by the U.S. Department of Defense to provide precise navigation information for targeting weapons systems anywhere in the world. The GPS is available to civilian users worldwide in a degraded mode. It is now used in aviation, navigation, and automobiles. The system can provide your exact position on Earth anywhere, anytime.

The GPS and LORAN-C work together, giving a highly accurate, duplicate navigation system for the Defense Department.

1 Define a hyperbola.

Definition of Hyperbola

Hyperbola

A **hyperbola** is the set of all points in the plane, the absolute value of the difference of whose distances from two fixed points is constant. The fixed points are called the **foci** of the hyperbola.

Figure 8.22 shows a hyperbola in *standard position*, with foci $F_1(-c, 0)$ and $F_2(c, 0)$ on the x-axis at equal distances from the origin. The two parts of the hyperbola are called **branches**.

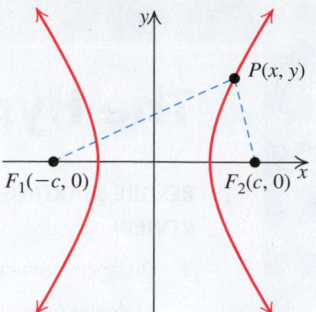

Figure 8.22 $PF_1 - PF_2$ is a constant, for any point P on the hyperbola.

As in the discussion of ellipses in Section 8.3 (see page 726), we take $2a$ as the positive constant referred to in the definition.

A point $P(x, y)$ lies on a hyperbola if and only if

$$|d(P, F_1) - d(P, F_2)| = 2a \qquad \text{Definition of hyperbola}$$
$$\sqrt{(x + c)^2 + y^2} - \sqrt{(x - c)^2 + y^2} = \pm 2a \qquad \text{Distance formula}$$

Just as in the case of an ellipse, we eliminate radicals and simplify (see Exercise 92) to obtain the equation:

$$(c^2 - a^2)x^2 - a^2 y^2 = a^2(c^2 - a^2) \quad (1)$$
$$b^2 x^2 - a^2 y^2 = a^2 b^2 \qquad \text{Replace } (c^2 - a^2) \text{ with } b^2.$$
$$\frac{x^2}{a^2} - \frac{y^2}{b^2} = 1 \quad (2) \qquad \text{Divide both sides by } a^2 b^2.$$

Equation (2) is called the **standard form of the equation of a hyperbola** with center $(0, 0)$. The x-intercepts of the graph of equation (2) are $-a$ and a. The points corresponding to these x-intercepts are the **vertices** of the hyperbola. The distance between the vertices $V_1(-a, 0)$ and $V_2(a, 0)$ is $2a$. The line segment joining the two vertices is called the **transverse axis**. The midpoint of the transverse axis is the **center** of the hyperbola. The *center* is also the midpoint of the line segment joining the foci. The distance between the center and either vertex is a, and the distance between the center and either focus point is c. See Figure 8.23. The line segment joining the points $(0, -b)$ and $(0, b)$, where $b^2 = c^2 - a^2$ and $b > 0$, is called the **conjugate axis**.

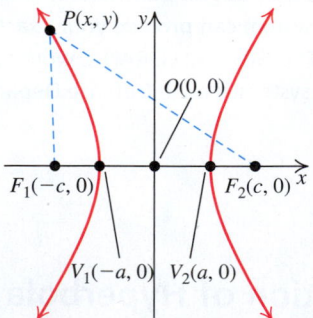

Points F_1 and F_2 are foci.
Points V_1 and V_2 are vertices.
Point O is the center.
Segment $\overline{V_1 V_2}$ is the transverse axis.
The difference of the distances $\left| PF_1 - PF_2 \right|$ from any point P on the hyperbola is constant.

Figure 8.23 Parts of a hyperbola.

Similarly, an equation of a hyperbola with center $(0, 0)$ and foci $(0, -c)$ and $(0, c)$ on the y-axis is given by:

$$\frac{y^2}{a^2} - \frac{x^2}{b^2} = 1, \text{ where } b^2 = c^2 - a^2 \quad (3)$$

Here the vertices are $(0, -a)$ and $(0, a)$. The transverse axis of length $2a$ of the graph of equation (3) lies on the y-axis, and its conjugate axis is the segment joining the points $(-b, 0)$ and $(0, b)$ of length $2b$ that lies on the x-axis. The distance between the center and either endpoint of the conjugate axis is b.

SUMMARY OF **MAIN FACTS**

	Main facts about hyperbolas centered at $(0, 0)$	
Standard Equation	$\dfrac{x^2}{a^2} - \dfrac{y^2}{b^2} = 1; a > 0, b > 0$	$\dfrac{y^2}{a^2} - \dfrac{x^2}{b^2} = 1; a > 0, b > 0$
Transverse axis on	x-axis $(y = 0)$	y-axis $(x = 0)$
Length of transverse axis	$2a$	$2a$
Conjugate axis on	y-axis $(x = 0)$	x-axis $(y = 0)$
Length of conjugate axis	$2b$	$2b$
Vertices	$(\pm a, 0)$	$(0, \pm a)$
Endpoints of conjugate axis	$(0, \pm b)$	$(\pm b, 0)$
Foci	$(\pm c, 0)$, where $c^2 = a^2 + b^2$	$(0, \pm c)$, where $c^2 = a^2 + b^2$
Description	Hyperbola has a *left branch* and a *right branch*. (Hyperbola opens left and right.)	Hyperbola has an *upper branch* and a *lower branch*. (Hyperbola opens up and down.)
Graph	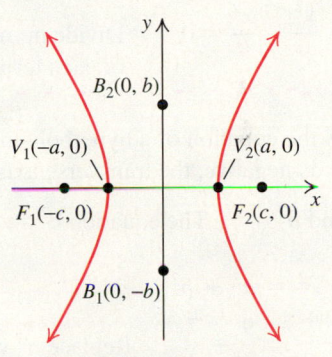	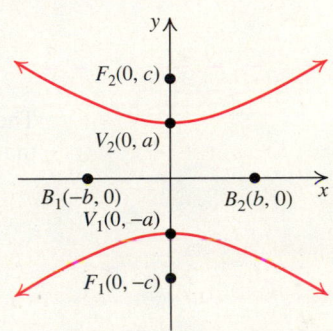
	Horizontal transverse axis	Vertical transverse axis

EXAMPLE 1   Determining the Orientation of a Hyperbola

Does the hyperbola

$$\frac{x^2}{4} - \frac{y^2}{10} = 1$$

have its transverse axis on the x-axis or y-axis?

Solution

The orientation (left–right branches or up–down branches) of a hyperbola is determined *by noting where the minus sign occurs in the standard equation.* From the table above, we see that the transverse axis is on the x-axis when the minus sign precedes the y^2-term and is on the y-axis when the minus sign precedes the x^2-term. In the equation in this example, the minus sign precedes the y^2-term, so the transverse axis is on the x-axis. See Figure 8.24.

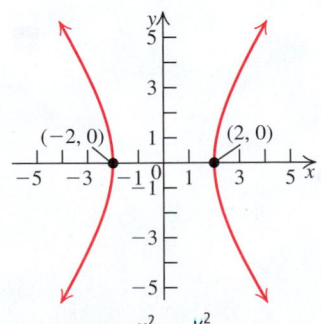

Figure 8.24 $\dfrac{x^2}{4} - \dfrac{y^2}{10} = 1.$

Practice Problem 1 Does the hyperbola $\dfrac{y^2}{8} - \dfrac{x^2}{5} = 1$ have its transverse axis on the x-axis or the y-axis?

EXAMPLE 2   **Finding the Vertices and Foci from the Equation of a Hyperbola**

Find the vertices and foci for the hyperbola

$$28y^2 - 36x^2 = 63.$$

Solution

First, convert the equation to the standard form.

$$28y^2 - 36x^2 = 63 \qquad \text{Given equation}$$

$$\frac{28y^2}{63} - \frac{36x^2}{63} = 1 \qquad \text{Divide both sides by 63.}$$

$$\frac{4y^2}{9} - \frac{4x^2}{7} = 1 \qquad \text{Simplify.}$$

$$\frac{y^2}{\dfrac{9}{4}} - \frac{x^2}{\dfrac{7}{4}} = 1 \qquad \begin{array}{l}\text{Divide numerators and denominators}\\ \text{of each term by 4.}\end{array}$$

The last equation is the equation of a hyperbola in standard form with center $(0, 0)$. Since the coefficient of x^2 is negative, the transverse axis lies on the y-axis.

Here $a^2 = \dfrac{9}{4}$ and $b^2 = \dfrac{7}{4}$. The equation $c^2 = a^2 + b^2$ gives the value of c^2.

$$c^2 = a^2 + b^2$$

$$= \frac{9}{4} + \frac{7}{4} \qquad \text{Replace } a^2 \text{ with } \frac{9}{4} \text{ and } b^2 \text{ with } \frac{7}{4}.$$

$$= \frac{16}{4} = 4 \qquad \text{Simplify.}$$

Now $a^2 = \dfrac{9}{4}$ gives $a = \dfrac{3}{2}$ and $c^2 = 4$ gives $c = 2$ because we require $a > 0$ and $c > 0$.

The vertices of the hyperbola are $\left(0, -\dfrac{3}{2}\right)$ and $\left(0, \dfrac{3}{2}\right)$. The foci of the hyperbola are $(0, -2)$ and $(0, 2)$.

Practice Problem 2 Find the vertices and foci for the hyperbola

$$x^2 - 4y^2 = 8.$$

EXAMPLE 3   **Finding the Equation of a Hyperbola**

Find the standard form of the equation of a hyperbola with

a. vertices $(\pm 4, 0)$ and foci $(\pm 5, 0)$

b. center $(0, 0)$; vertex: $(1, 0)$; focus: $(4, 0)$

c. foci: $(0, \pm 6)$; the length of the transverse axis is 8

Solution

a. Since the foci of the hyperbola, $(-5, 0)$ and $(5, 0)$, are on the x-axis, the transverse axis lies on the x-axis. The center of the hyperbola is midway between the foci, at $(0, 0)$. The standard form of such a hyperbola is

$$\frac{x^2}{a^2} - \frac{y^2}{b^2} = 1.$$

You need to find a^2 and b^2.

The distance a between the center $(0, 0)$ to either vertex, $(-4, 0)$ or $(4, 0)$, is 4; so $a = 4$ and $a^2 = 16$. The distance c between the center $(0, 0)$ to either focus, $(-5, 0)$ or $(5, 0)$, is 5; so $c = 5$ and $c^2 = 25$. Now use the equation $b^2 = c^2 - a^2$ to find $b^2 = c^2 - a^2 = 25 - 16 = 9$. Substitute $a^2 = 16$ and $b^2 = 9$ in $\frac{x^2}{a^2} - \frac{y^2}{b^2} = 1$ to

get the standard form of the equation of the hyperbola

$$\frac{x^2}{16} - \frac{y^2}{9} = 1.$$

b. Since the center $(0, 0)$ and vertex $(1, 0)$ are on the x-axis, the transverse axis is on the x-axis. The distance between the center $(0, 0)$ and the vertex $(1, 0)$ is 1, so $a = 1$. The distance between the center $(0, 0)$ and the focus $(4, 0)$ is 4, so $c = 4$.

$$c^2 = a^2 + b^2 \qquad \text{Relationship between } a, b, \text{ and } c$$
$$4^2 = 1^2 + b^2 \qquad \text{Replace } c \text{ with 4 and } a \text{ with 1.}$$
$$b^2 = 4^2 - 1^2 = 16 - 1 = 15 \qquad \text{Solve for } b^2.$$
$$\frac{x^2}{a^2} - \frac{y^2}{b^2} = 1 \qquad \begin{array}{l}\text{Equation for a hyperbola with}\\\text{horizontal transverse axis}\end{array}$$
$$x^2 - \frac{y^2}{15} = 1 \qquad \begin{array}{l}\text{Replace } a^2 \text{ with 1 and } b^2 \text{ with}\\\text{15 and simplify.}\end{array}$$

c. Since the foci $(0, -6)$ and $(0, 6)$ are on the y-axis, the transverse axis is on the y-axis. The length of the transverse axis is 8, so $2a = 8$ and $a = 4$. The center is midway between the foci $(0, \pm 6)$, so the center is $(0, 0)$. The distance, c, between the center $(0, 0)$ and the focus $(0, 6)$ is 6, so $c = 6$.

$$c^2 = a^2 + b^2 \qquad \text{Relationship between } a, b, \text{ and } c$$
$$6^2 = 4^2 + b^2 \qquad \text{Replace } c \text{ with 6 and } a \text{ with 4.}$$
$$b^2 = 6^2 - 4^2 = 36 - 16 = 20 \qquad \text{Solve for } b^2.$$
$$\frac{y^2}{a^2} - \frac{x^2}{b^2} = 1 \qquad \begin{array}{l}\text{Equation for a hyperbola with}\\\text{vertical transverse axis}\end{array}$$
$$\frac{y^2}{16} - \frac{x^2}{20} = 1 \qquad \begin{array}{l}\text{Replace } a^2 \text{ with 16 and } b^2\\\text{with 20.}\end{array}$$

Practice Problem 3 Find the standard form of the equation of a hyperbola with

a. vertices at $(0, \pm 3)$ and foci at $(0, \pm 6)$

b. center $(0, 0)$; vertex: $(5, 0)$; focus: $(7, 0)$

c. foci: $(0, \pm 6)$; the length of the transverse axis is 10

2 Find the asymptotes of a hyperbola.

The Asymptotes of a Hyperbola

Recall from Section 2.4 that a line $y = mx + b$ is a horizontal asymptote (if $m = 0$) or an oblique asymptote (if $m \neq 0$) of the graph of $y = f(x)$ if the distance from the line to the points on the graph of f approaches 0 as $x \to \infty$ or as $x \to -\infty$. When horizontal or oblique asymptotes exist, they provide us with information about the end behavior of the graph of f. This is especially helpful in graphing a hyperbola.

To find the asymptotes of the hyperbola with equation $\dfrac{x^2}{a^2} - \dfrac{y^2}{b^2} = 1$, we first solve this equation for y.

$$\dfrac{x^2}{a^2} - \dfrac{y^2}{b^2} = 1 \qquad \text{Given equation of hyperbola}$$

$$y^2 = b^2\left(\dfrac{x^2}{a^2} - 1\right) \qquad \text{Isolate } y^2 \text{ term.}$$

$$y^2 = b^2\left(\dfrac{x^2}{a^2} - \dfrac{x^2}{a^2}\cdot\dfrac{a^2}{x^2}\right) \qquad \text{Write } 1 = \dfrac{x^2}{a^2}\cdot\dfrac{a^2}{x^2}.$$

$$y^2 = \dfrac{b^2 x^2}{a^2}\left(1 - \dfrac{a^2}{x^2}\right) \qquad \text{Factor out } \dfrac{x^2}{a^2}.$$

$$y = \pm\dfrac{bx}{a}\sqrt{1 - \dfrac{a^2}{x^2}} \qquad \text{Square root property}$$

As $x \to \infty$ or as $x \to -\infty$, the quantity $\dfrac{a^2}{x^2}$ approaches 0. Thus, $\sqrt{1 - \dfrac{a^2}{x^2}}$ approaches 1 and the value of y approaches $\pm\dfrac{bx}{a}$. Therefore, the lines $y = \dfrac{b}{a}x$ and $y = -\dfrac{b}{a}x$ are the oblique asymptotes for the graph of the hyperbola $\dfrac{x^2}{a^2} - \dfrac{y^2}{b^2} = 1$.

Similarly, you can show that the lines $y = \dfrac{a}{b}x$ and $y = -\dfrac{a}{b}x$ are the oblique asymptotes for the graph of the hyperbola $\dfrac{y^2}{a^2} - \dfrac{x^2}{b^2} = 1$.

SIDE NOTE

There is a convenient device that can be used for obtaining equations of the asymptotes of a hyperbola. Substitute 0 for the 1 on the right side of the equation of the hyperbola in standard form and then solve for y in terms of x. For example, for the hyperbola $\dfrac{x^2}{a^2} - \dfrac{y^2}{b^2} = 1$, replace the 1 on the right side by 0 to obtain $\dfrac{x^2}{a^2} - \dfrac{y^2}{b^2} = 0$. Solve this equation for y:

$$\dfrac{y^2}{b^2} = \dfrac{x^2}{a^2}$$

$$y^2 = \dfrac{b^2}{a^2}x^2$$

The lines $y = \pm\dfrac{b}{a}x$ are the asymptotes of the hyperbola.

THE ASYMPTOTES OF A HYPERBOLA WITH CENTER (0, 0)

1. The graph of the hyperbola $\dfrac{x^2}{a^2} - \dfrac{y^2}{b^2} = 1$ has transverse axis along the x-axis and has the following two asymptotes:

$$y = \dfrac{b}{a}x \qquad \text{and} \qquad y = -\dfrac{b}{a}x$$

2. The graph of the hyperbola $\dfrac{y^2}{a^2} - \dfrac{x^2}{b^2} = 1$ has transverse axis along the y-axis and has the following two asymptotes:

$$y = \dfrac{a}{b}x \qquad \text{and} \qquad y = -\dfrac{a}{b}x$$

EXAMPLE 4 **Finding the Asymptotes of a Hyperbola**

Determine the asymptotes of each hyperbola.

a. $\dfrac{x^2}{4} - \dfrac{y^2}{9} = 1$ **b.** $\dfrac{y^2}{9} - \dfrac{x^2}{16} = 1$

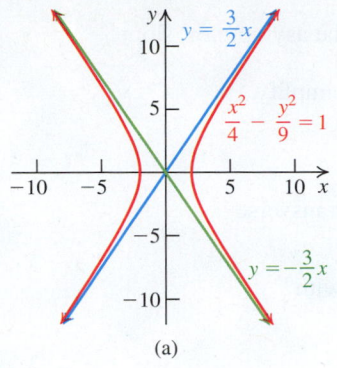

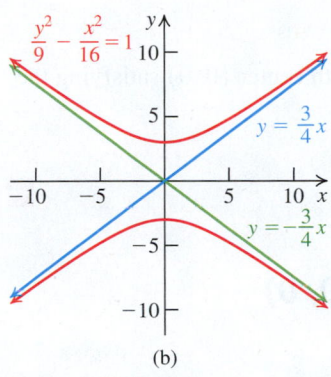

Figure 8.25 Asymptotes.

Solution

a. The hyperbola $\dfrac{x^2}{4} - \dfrac{y^2}{9} = 1$ is of the form $\dfrac{x^2}{a^2} - \dfrac{y^2}{b^2} = 1$; so $a = 2$ and $b = 3$.

Substituting these values into $y = \dfrac{b}{a}x$ and $y = -\dfrac{b}{a}x$, we get the asymptotes

$$y = \frac{3}{2}x \quad \text{and} \quad y = -\frac{3}{2}x.$$

b. The hyperbola $\dfrac{y^2}{9} - \dfrac{x^2}{16} = 1$ has the form $\dfrac{y^2}{a^2} - \dfrac{x^2}{b^2} = 1$; so $a = 3$ and $b = 4$.

Substituting these values into $y = \dfrac{a}{b}x$ and $y = -\dfrac{a}{b}x$, we get the asymptotes

$$y = \frac{3}{4}x \quad \text{and} \quad y = -\frac{3}{4}x.$$

The hyperbolas of **a** and **b** and their asymptotes are shown in Figures 8.25(a) and 8.25(b), respectively.

Practice Problem 4 Determine the asymptotes of the hyperbola $\dfrac{y^2}{4} - \dfrac{x^2}{9} = 1$.

EXAMPLE 5 **Finding an Equation of a Hyperbola**

Find an equation of the hyperbola with center $(0, 0)$ satisfying the given conditions.

a. Foci: $(\pm 5, 0)$; asymptotes $y = \pm 2x$

b. Vertices: $(0, \pm 1)$; asymptotes $y = \pm \dfrac{1}{2}x$

Solution

a. Since the foci are on the x-axis, the transverse axis is on the x-axis and the asymptotes are $y = \pm \dfrac{b}{a}x$. We are given that the asymptotes for this hyperbola are $y = \pm 2x$.

$\dfrac{b}{a} = 2$	Equate the positive values for the asymptotes' slope.
$b = 2a$	Multiply both sides by a.
$c^2 = a^2 + b^2$	Relationship between a, b, and c.
$5^2 = a^2 + (2a)^2$	Foci are $(\pm 5, 0)$, so $c = 5$. Replace c with 5, and b with $2a$.
$25 = a^2 + 4a^2 = 5a^2$	Simplify.
$a^2 = 5$	Divide both sides by 5.
$b^2 = (2a)^2 = 4a^2 = 4(5) = 20$	Square both sides of $b = 2a$ and replace a^2 with 5.
$\dfrac{x^2}{a^2} - \dfrac{y^2}{b^2} = 1$	Equation for a hyperbola with transverse axis on the x-axis
$\dfrac{x^2}{5} - \dfrac{y^2}{20} = 1$	Replace a^2 with 5 and b^2 with 20.

b. Since the vertices are $(0, \pm 1)$, $a = 1$. Because the vertices are on the y-axis, the transverse axis is on the y-axis and the asymptotes are $y = \pm \dfrac{a}{b}x$. We are given that the asymptotes for this hyperbola are $y = \pm \dfrac{1}{2}x$.

$$\frac{a}{b} = \frac{1}{2} \qquad \text{Equate the positive values for the asymptotes' slope.}$$

$$2a = b \qquad \text{Multiply both sides by } 2b \text{ and simplify.}$$

$$2(1) = b \qquad \text{Replace } a \text{ with 1.}$$

$$4 = b^2 \qquad \text{Square both sides.}$$

$$\frac{y^2}{a^2} - \frac{x^2}{b^2} = 1 \qquad \begin{array}{l}\text{Equation for a hyperbola with transverse}\\ \text{axis on the } y\text{-axis.}\end{array}$$

$$\frac{y^2}{1} - \frac{x^2}{4} = 1 \qquad \begin{array}{l} a = 1 \text{ so, } a^2 = 1. \text{ Replace } a^2 \text{ with 1}\\ \text{and } b^2 \text{ with 4.}\end{array}$$

$$y^2 - \frac{x^2}{4} = 1 \qquad \text{Simplify.}$$

Practice Problem 5 Find an equation of the hyperbola with center $(0, 0)$ satisfying the given conditions.

a. Foci: $(\pm 2\sqrt{2}, 0)$; asymptotes $y = \pm x$

b. Vertices: $(0, \pm 1)$; asymptotes $y = \pm\dfrac{1}{3}x$

3 Graph a hyperbola.

Graphing a Hyperbola with Center (0, 0)

Consider the hyperbola with equation

$$\frac{x^2}{a^2} - \frac{y^2}{b^2} = 1.$$

The vertices of this hyperbola are $(-a, 0)$ and $(a, 0)$. The endpoints of the conjugate axis are $(0, -b)$ and $(0, b)$. The rectangle with vertices $(a, b), (-a, b), (-a, -b),$ and $(a, -b)$ is called the **fundamental rectangle** of the hyperbola. See Figure 8.26 on page 747. The diagonals of the fundamental rectangle have slopes $\dfrac{b}{a}$ and $-\dfrac{b}{a}$. Thus, the extensions of these diagonals are the asymptotes of the hyperbola.

PROCEDURE
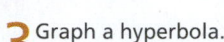

EXAMPLE 6   **Graphing a Hyperbola Centered at (0, 0)**

OBJECTIVE

Sketch the graph of a hyperbola in the form

(i) $\dfrac{x^2}{a^2} - \dfrac{y^2}{b^2} = 1$ *or*   **(ii)** $\dfrac{y^2}{a^2} - \dfrac{x^2}{b^2} = 1$

Step 1 Write the equation in standard form. Determine transverse axis and the orientation of the hyperbola.

EXAMPLE

Sketch the graph of

a. $16x^2 - 9y^2 = 144.$

b. $25y^2 - 4x^2 = 100.$

1.

$$16x^2 - 9y^2 = 144$$

$$\frac{16x^2}{144} - \frac{9y^2}{144} = \frac{144}{144}$$

$$\frac{x^2}{9} - \frac{y^2}{16} = 1$$

Minus sign precedes y^2-term, the transverse axis is on the x-axis; the hyperbola opens left and right.

$$25y^2 - 4x^2 = 100$$

$$\frac{25y^2}{100} - \frac{4x^2}{100} = \frac{100}{100}$$

$$\frac{y^2}{4} - \frac{x^2}{25} = 1$$

Minus sign precedes x^2-term, the transverse axis is on the y-axis; the hyperbola opens up and down.

Step 2 Locate vertices and the endpoints of the conjugate axis.

2. Because $a^2 = 9, a = 3.$ Also, $b^2 = 16,$ so $b = 4.$ The vertices are $(\pm 3, 0),$ and the endpoints of the conjugate axis are $(0, \pm 4).$

Because $a^2 = 4, a = 2.$ Also, $b^2 = 25,$ so $b = 5.$ The vertices are $(0, \pm 2),$ and the endpoints of the conjugate axis are $(\pm 5, 0).$

Step 3 Lightly sketch the fundamental rectangle by drawing dashed lines parallel to the coordinate axes through the points in Step 2.

3. Draw dashed lines $x = 3$, $x = -3$, $y = 4$, and $y = -4$ to form the fundamental rectangle with vertices $(3, 4)$, $(-3, 4)$, $(-3, -4)$, and $(3, -4)$.

Draw dashed lines $y = 2$, $y = -2$, $x = 5$, and $x = -5$ to form the fundamental rectangle with vertices $(5, 2)$, $(-5, 2)$, $(-5, -2)$, and $(5, -2)$.

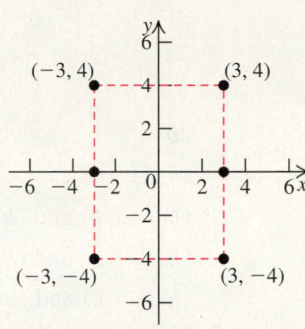

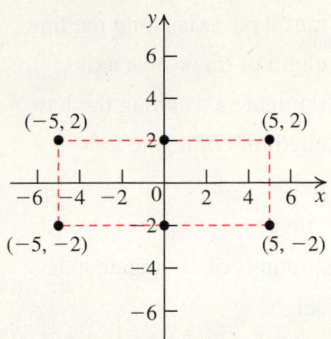

Step 4 Sketch the asymptotes. Extend the diagonals of the fundamental rectangle. These are the asymptotes.

4.

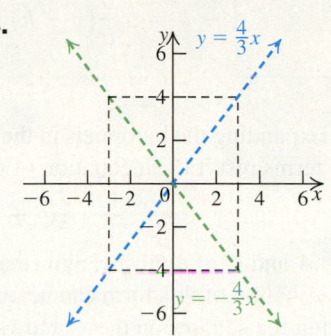

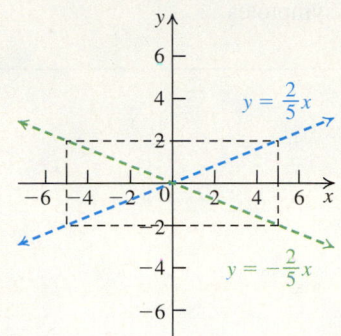

Step 5 Sketch the graph. Draw both branches of the hyperbola through the vertices, approaching the asymptotes. See Figures 8.26 and 8.27. The foci are located on the transverse axis, c units from the center, where $c^2 = a^2 + b^2$.

5.

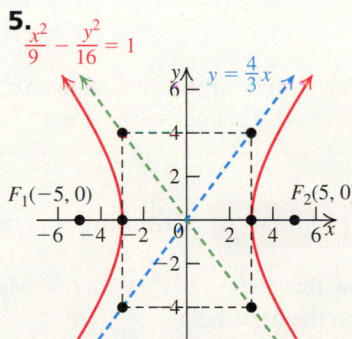

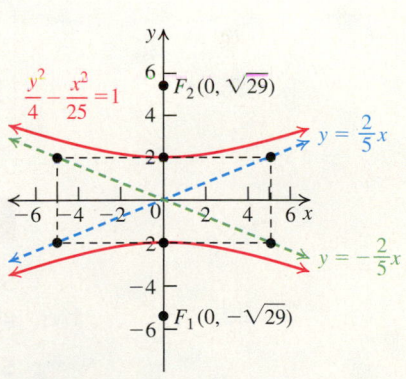

Figure 8.26   **Figure 8.27**

Practice Problem 6 Sketch the graph of each equation.

a. $25x^2 - 4y^2 = 100$ **b.** $9y^2 - x^2 = 1$

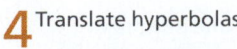 Translate hyperbolas.

Translations of Hyperbolas

We can use horizontal and vertical shifts to find the standard form of the equations of hyperbolas centered at (h, k).

SUMMARY OF **MAIN FACTS**

Main properties of hyperbolas centered at (h, k)

Standard Equation	$\dfrac{(x - h)^2}{a^2} - \dfrac{(y - k)^2}{b^2} = 1;$ $a > 0, b > 0$	$\dfrac{(y - k)^2}{a^2} - \dfrac{(x - h)^2}{b^2} = 1;$ $a > 0, b > 0$
Transverse axis along the line	$y = k$	$x = h$
Length of transverse axis	$2a$	$2a$
Conjugate axis along the line	$x = h$	$y = k$
Length of conjugate axis	$2b$	$2b$
Center	(h, k)	(h, k)
Vertices	$(h - a, k)$ and $(h + a, k)$	$(h, k - a)$ and $(h, k + a)$
Endpoints of conjugate axis	$(h, k - b)$ and $(h, k + b)$	$(h - b, k)$ and $(h + b, k)$
Foci	$(h - c, k)$ and $(h + c, k);$	$(h, k - c)$ and $(h, k + c);$
Equation involving a, b, and c	$c^2 = a^2 + b^2$	$c^2 = a^2 + b^2$
Asymptotes	$y - k = \pm\dfrac{b}{a}(x - h)$	$y - k = \pm\dfrac{a}{b}(x - h)$

Expanding the binomials in the equation of a hyperbola in standard form and collecting like terms result in an equation of the form

$$Ax^2 + Cy^2 + Dx + Ey + F = 0,$$

with A and C of opposite sign (that is, $AC < 0$). Conversely, except in degenerate cases, any equation of this form can be put into one of the standard forms of a hyperbola by completing the squares on the x- and y-terms.

GENERAL EQUATION OF A HYPERBOLA

The graph of the equation $Ax^2 + Cy^2 + Dx + Ey + F = 0$ is a hyperbola if $AC < 0$. That is, neither A nor C is zero and they have opposite signs.

EXAMPLE 7 **Graphing a Hyperbola**

Show that $9x^2 - 16y^2 + 18x + 64y - 199 = 0$ is an equation of a hyperbola and then graph the hyperbola.

Solution

Since we plan on completing squares, first group the x- and y-terms onto the left side and the constants onto the right side.

$9x^2 - 16y^2 + 18x + 64y - 199 = 0$	Given equation
$(9x^2 + 18x) + (-16y^2 + 64y) = 199$	Group terms.
$9(x^2 + 2x) - 16(y^2 - 4y) = 199$	Factor out 9 and -16.
$9(x^2 + 2x + 1) - 16(y^2 - 4y + 4) = 199 + 9 - 64$	Complete the squares: add $9 \cdot 1$ and $-16 \cdot 4$.
$9(x + 1)^2 - 16(y - 2)^2 = 144$	Factor and simplify.
$\dfrac{9(x + 1)^2}{144} - \dfrac{16(y - 2)^2}{144} = 1$	Divide both sides by 144 to obtain 1 on the right side.
$\dfrac{(x + 1)^2}{16} - \dfrac{(y - 2)^2}{9} = 1$	Simplify.

The last equation is the standard form of the equation of a hyperbola with center $(-1, 2)$. Now we sketch the graph of this hyperbola.

Steps 1–2 **Locate the vertices.** For this hyperbola with center $(-1, 2)$, $a^2 = 16$ and $b^2 = 9$. Therefore, $a = 4$ and $b = 3$. From the table on page 747, with $h = -1$ and $k = 2$, the vertices are $(h - a, k) = (-1 - 4, 2) = (-5, 2)$ and $(h + a, k) = (-1 + 4, 2) = (3, 2)$.

Step 3 **Draw the fundamental rectangle.** The vertices of the fundamental rectangle are $(3, -1)$, $(3, 5)$, $(-5, 5)$, and $(-5, -1)$.

Step 4 **Sketch the asymptotes.** Extend the diagonals of the fundamental rectangle to sketch the asymptotes:

$$y - 2 = \frac{3}{4}(x + 1) \quad \text{and} \quad y - 2 = -\frac{3}{4}(x + 1).$$

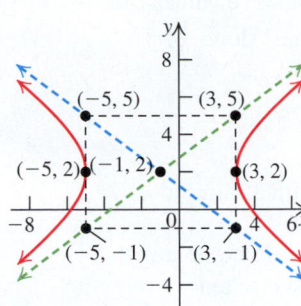

Figure 8.28 The hyperbola $9x^2 - 16y^2 + 18x + 64y - 199 = 0$.

Step 5 **Sketch the graph.** Draw two branches of the hyperbola opening to the left and right, starting from the vertices $(-5, 2)$ and $(3, 2)$ and approaching the asymptotes. See Figure 8.28.

Practice Problem 7 Show that $x^2 - 4y^2 - 2x + 16y - 20 = 0$ is an equation of a hyperbola and graph the hyperbola.

EXAMPLE 8 **Find an Equation for the Hyperbola Whose Graph Is Shown**

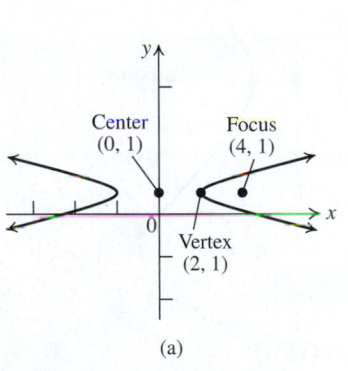

(a)

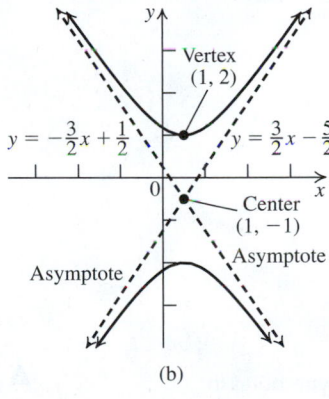

(b)

Solution

a. The center $(h, k) = (0, 1)$, so $h = 0$ and $k = 1$. The distance, a, between the center $(0, 1)$ and the vertex $(2, 1)$ is 2, so $a = 2$. The distance, c, between the center $(0, 1)$ and the focus $(4, 1)$ is 4, so $c = 4$. The transverse axis lies on the horizontal line $y = 1$.

$c^2 = a^2 + b^2$	Relationship between a, b, and c
$4^2 = 2^2 + b^2$	Replace c with 4 and a with 2.
$b^2 = 4^2 - 2^2 = 16 - 4 = 12$	Solve for b^2. Simplify.
$\dfrac{(x - h)^2}{a^2} - \dfrac{(y - k)^2}{b^2} = 1$	Equation for a hyperbola with horizontal transverse axis
$\dfrac{(x - 0)^2}{2^2} - \dfrac{(y - 1)^2}{12} = 1$	Replace h with 0, a with 2, k with 1, and b^2 with 12.
$\dfrac{x^2}{4} - \dfrac{(y - 1)^2}{12} = 1$	Simplify.

b. The center $(h, k) = (1, -1)$, so $h = 1$ and $k = -1$. The distance, a, between the center $(1, -1)$ and the vertex $(1, 2)$ is 3, so $a = 3$. The slope of the asymptote with positive slope is $\dfrac{3}{2}$. The transverse axis is on the vertical line $x = 1$.

$$\frac{a}{b} = \frac{3}{2} \qquad \text{Equate the positive values for the asymptotes' slope, } \frac{a}{b}.$$

$$b = \frac{2}{3}a \qquad \text{Solve for } b.$$

$$b = \frac{2}{3}(3) = 2 \qquad \text{Replace } a \text{ with 3. Simplify.}$$

$$\frac{(y - k)^2}{a^2} - \frac{(x - h)^2}{b^2} = 1 \qquad \text{Equation for a hyperbola with vertical transverse axis}$$

$$\frac{(y - (-1))^2}{3^2} - \frac{(x - h)^2}{2^2} = 1 \qquad \text{Replace } k \text{ with } -1, a \text{ with 3, } h \text{ with 1, and } b \text{ with 2.}$$

$$\frac{(y + 1)^2}{9} - \frac{(x - 1)^2}{4} = 1 \qquad \text{Simplify.}$$

Practice Problem 8 Find an equation for the hyperbola whose graph is shown.

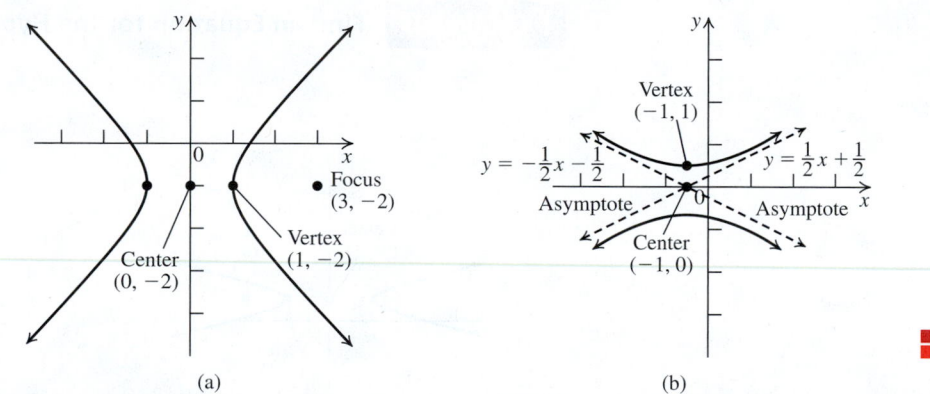

(a)   (b)

5 Use hyperbolas in applications.

Applications

Hyperbolas have many applications. We list a few of them here:

1. Comets that do not move in elliptical orbits around the sun almost always move in hyperbolic orbits. (In theory, they can also move in parabolic orbits.)

2. Boyle's Law states that if a perfect gas is kept at a constant temperature, then its pressure P and volume V are related by the equation $PV = c$, where c is a constant. The graph of this equation is a hyperbola. In this case, the transverse axis is not parallel to a coordinate axis.

3. The hyperbola has the reflecting property that a ray of light from a source at one focus of a hyperbolic mirror (a mirror with hyperbolic cross sections) is reflected along the line through the other focus.

   The reflecting properties of the parabola and hyperbola are combined into one design for a reflecting telescope. See Figure 8.29. The parallel rays from a star are finally focused at the eyepiece at F_2.

4. The definition of a hyperbola forms the basis of several important navigational systems.

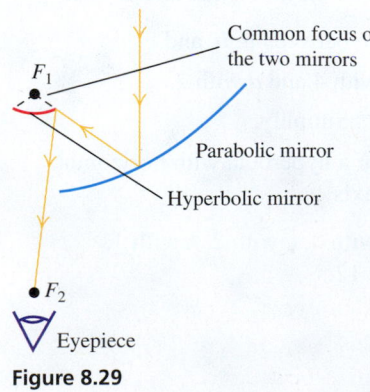

Figure 8.29

5. A three-dimensional solid called a hyperboloid is formed by rotating a hyperbola about the part of its conjugate axis from the center to one endpoint. It is often used by engineers in the construction of nuclear cooling towers and by architects in such structures as the Kobe Port Tower in Kobe, Japan, and the Saint Louis Science Center in St. Louis, Missouri.

◆ How Does LORAN Work?

Suppose two stations A and B (several miles apart) transmit synchronized radio signals. The LORAN receiver in your ship measures the difference in reception times of these synchronized signals. The radio signals travel at a speed of 186,000 miles per second. Using this information, you can determine the difference $2a$ in the distance of your ship's receiver from the two transmitters. By the definition of a hyperbola, this information places your ship somewhere on a hyperbola with foci A and B. With two pairs of transmitters, the position of your ship can be determined at a point of intersection of the two hyperbolas. (See Exercises 86 and 87).

EXAMPLE 9 Using LORAN

LORAN navigational transmitters A and B are located at $(-130, 0)$ and $(130, 0)$, respectively. A receiver P on a fishing boat somewhere in the first quadrant listens to the pair (A, B) of transmissions and computes the difference of the distances from the boat to A and B as 240 miles. Find the equation of the hyperbola on which P is located.

Solution

With reference to the transmitters A and B, P is located on a hyperbola with $2a = 240$, or $a = 120$, and the hyperbola has foci $(-130, 0)$ and $(130, 0)$.

$$\frac{x^2}{a^2} - \frac{y^2}{b^2} = 1 \qquad \text{Standard form of the equation of a hyperbola}$$

$$a^2 = (120)^2 = 14{,}400 \qquad \text{Replace } a \text{ with 120.}$$

$$b^2 = c^2 - a^2 = (130)^2 - (120)^2 = 2500 \qquad \text{Replace } c \text{ with 130 and } a \text{ with 120.}$$

$$\frac{x^2}{14{,}400} - \frac{y^2}{2500} = 1 \qquad \text{Replace } a^2 \text{ with 14,400 and } b^2 \text{ with 2500.}$$

The location of the ship therefore lies on a hyperbola (see Figure 8.30) with equation

$$\frac{x^2}{14{,}400} - \frac{y^2}{2500} = 1.$$

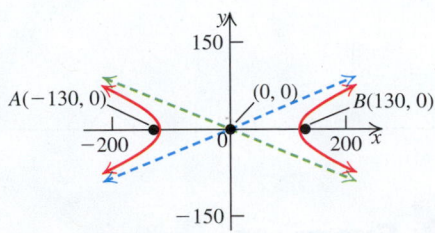

Figure 8.30

Practice Problem 9 In Example 9, the transmitters A and B are located at $(-150, 0)$ and $(150, 0)$, respectively, and the difference of the distances from the boat to A and B is 260 miles. Find the equation of the hyperbola on which P is located.

Answers to Practice Problems

1. y-axis   **2.** Vertices: $(2\sqrt{2}, 0)$ and $(-2\sqrt{2}, 0)$; foci: $(\sqrt{10}, 0)$ and $(-\sqrt{10}, 0)$ **3. a.** $\dfrac{y^2}{9} - \dfrac{x^2}{27} = 1$ **b.** $\dfrac{x^2}{25} - \dfrac{y^2}{24} = 1$

c. $\dfrac{y^2}{25} - \dfrac{x^2}{11} = 1$ **4.** $y = \dfrac{2}{3}x$ and $y = -\dfrac{2}{3}x$

5. a. $\dfrac{x^2}{4} - \dfrac{y^2}{4} = 1$ **b.** $y^2 - \dfrac{x^2}{9} = 1$

6. a. The standard form of the equation is $\dfrac{x^2}{4} - \dfrac{y^2}{25} = 1$. Vertices:

$(2, 0)$ and $(-2, 0)$; endpoints of the conjugate axis: $(0, 5)$ and

$(0, -5)$; asymptotes: $y = \dfrac{5}{2}x$ and $y = -\dfrac{5}{2}x$

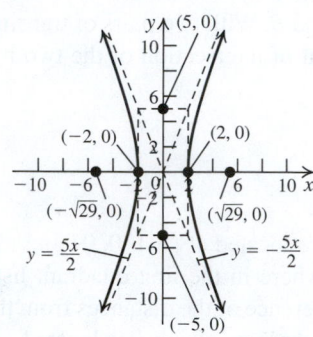

b. The standard form of the equation is $\dfrac{y^2}{\frac{1}{9}} - \dfrac{x^2}{1} = 1$. Vertices:

$\left(0, \dfrac{1}{3}\right)$ and $\left(0, -\dfrac{1}{3}\right)$; endpoints of the conjugate axis: $(1, 0)$ and

$(-1, 0)$; asymptotes: $y = \dfrac{1}{3}x$ and $y = -\dfrac{1}{3}x$

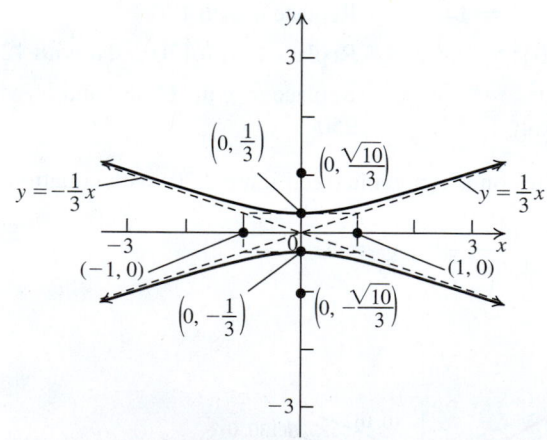

7. $\dfrac{(x-1)^2}{5} - \dfrac{(y-2)^2}{\frac{5}{4}} = 1$. Center: $(1, 2)$; vertices:

$(1 + \sqrt{5}, 2)$ and $(1 - \sqrt{5}, 2)$; endpoints of the conjugate axis:

$\left(1, 2 + \dfrac{\sqrt{5}}{2}\right)$ and $\left(1, 2 - \dfrac{\sqrt{5}}{2}\right)$; asymptotes:

$y - 2 = \pm\dfrac{1}{2}(x - 1)$; foci: $\left(\dfrac{7}{2}, 2\right)$ and $\left(-\dfrac{3}{2}, 2\right)$

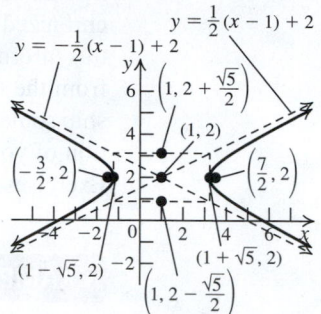

8. a. $x^2 - \dfrac{(y+2)^2}{8} = 1$ **b.** $y^2 - \dfrac{(x+1)^2}{4} = 1$

9. $\dfrac{x^2}{16,900} - \dfrac{y^2}{5,600} = 1$

Concepts and Vocabulary

1. A hyperbola is a set of all points in the plane, the _____ of whose distances from two fixed points is constant.

2. The line segment joining the two vertices of a hyperbola is called the _____ axis.

3. The standard equation of the hyperbola with center $(0, 0)$, vertices $(\pm a, 0)$, and foci $(\pm c, 0)$ is

$$\dfrac{x^2}{a^2} - \dfrac{y^2}{b^2} = 1, \text{ where } b^2 = \underline{\hspace{3cm}}.$$

4. For the hyperbola $\dfrac{y^2}{a^2} - \dfrac{x^2}{b^2} = 1$, the vertices are

_____ and _____ and the

foci are $(0, \pm c)$, where $c^2 =$ _____.

5. **True or False.** The graph of $\dfrac{x^2}{a^2} - \dfrac{y^2}{b^2} = -1$ is a hyperbola.

6. **True or False.** The graph of $Ax^2 + Cy^2 + Dx + Ey + F = 0$ is a degenerate conic or a hyperbola if $AC < 0$.

7. **True or False.** If one asymptote of a hyperbola has equation $y = -2x$, the other must have equation $y = \dfrac{1}{2}x$.

8. **True or False.** If one focus point of a hyperbola is $(0, 5)$, the other must be $(0, -5)$.

Building Skills

In Exercises 9–16, the equation of a hyperbola is given. Match each equation with its graph shown in (a)–(h).

9. $\dfrac{x^2}{9} - \dfrac{y^2}{4} = 1$ 10. $\dfrac{x^2}{4} - \dfrac{y^2}{25} = 1$

11. $\dfrac{y^2}{25} - \dfrac{x^2}{4} = 1$ 12. $\dfrac{y^2}{4} - \dfrac{x^2}{9} = 1$

13. $4x^2 - 25y^2 = 100$ 14. $16x^2 - 49y^2 = 196$

15. $16y^2 - 9x^2 = 100$ 16. $9y^2 - 16x^2 = 144$

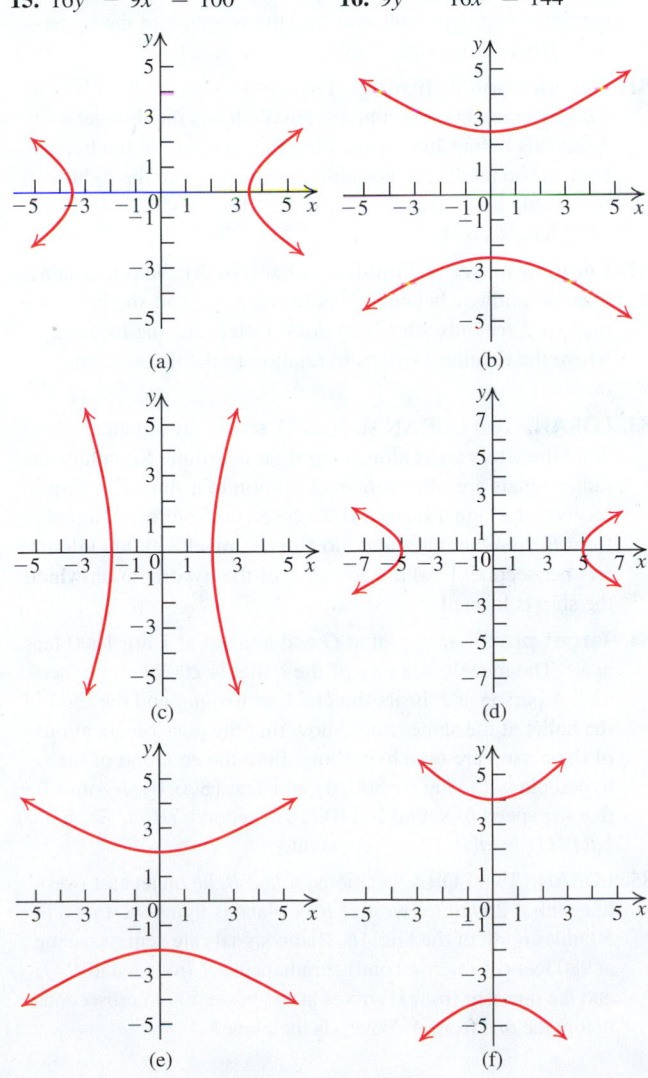

(a) (b) (c) (d) (e) (f)

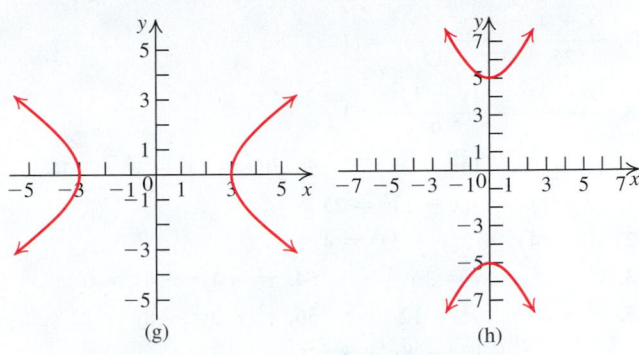

(g) (h)

In Exercises 17–28, the equation of a hyperbola is given.

a. Find and plot the vertices, the foci, and the transverse axis of the hyperbola.

b. State how the hyperbola opens.

c. Find and plot the vertices of the fundamental rectangle.

d. Sketch the asymptotes and write their equations.

e. Graph the hyperbola by using the vertices and the asymptotes.

17. $x^2 - \dfrac{y^2}{4} = 1$ 18. $y^2 - \dfrac{x^2}{4} = 1$

19. $x^2 - y^2 = -1$ 20. $y^2 - x^2 = -1$

21. $9y^2 - x^2 = 36$ 22. $9x^2 - y^2 = 36$

23. $4x^2 - 9y^2 - 36 = 0$ 24. $4x^2 - 9y^2 + 36 = 0$

25. $y = \pm\sqrt{4x^2 + 1}$ 26. $y = \pm 2\sqrt{x^2 + 1}$

27. $y = \pm\sqrt{9x^2 - 1}$ 28. $y = \pm 3\sqrt{x^2 - 4}$

In Exercises 29–42, find an equation of the hyperbola satisfying the given conditions. Graph the hyperbola.

29. Vertices: $(\pm 2, 0)$; foci: $(\pm 3, 0)$

30. Vertices: $(\pm 3, 0)$; foci: $(\pm 5, 0)$

31. Vertices: $(0, \pm 4)$; foci: $(0, \pm 6)$

32. Vertices: $(0, \pm 5)$; foci: $(0, \pm 8)$

33. Center: $(0, 0)$; vertex: $(0, 2)$; focus: $(0, 5)$

34. Center: $(0, 0)$; vertex: $(0, -1)$; focus: $(0, -4)$

35. Center: $(0, 0)$; vertex: $(1, 0)$; focus: $(-5, 0)$

36. Center: $(0, 0)$; vertex: $(-3, 0)$; focus: $(6, 0)$

37. Foci: $(0, \pm 5)$; the length of the transverse axis is 6.

38. Foci: $(\pm 2, 0)$; the length of the transverse axis is 2.

39. Foci: $(\pm\sqrt{5}, 0)$; asymptotes: $y = \pm 2x$

40. Foci: $(0, \pm 10)$; asymptotes: $y = \pm 3x$

41. Vertices: $(0, \pm 4)$; asymptotes: $y = \pm x$

42. Vertices: $(\pm 3, 0)$; asymptotes: $y = \pm 2x$

In Exercises 43–62, the equation of a hyperbola is given.

a. Find and plot the center, vertices, transverse axis, and asymptotes of the hyperbola.

b. Use the vertices and the asymptotes to graph the hyperbola.

43. $\dfrac{(x - 1)^2}{9} - \dfrac{(y + 1)^2}{16} = 1$

44. $\dfrac{(y - 1)^2}{16} - \dfrac{(x + 1)^2}{4} = 1$

45. $\dfrac{(x + 2)^2}{25} - \dfrac{y^2}{49} = 1$

46. $\dfrac{(x + 1)^2}{9} - \dfrac{(y + 2)^2}{36} = 1$

47. $\dfrac{(x+4)^2}{25} - \dfrac{(y+3)^2}{49} = 1$

48. $\dfrac{(x-3)^2}{9} - \dfrac{(y+1)^2}{9} = 1$

49. $4x^2 - (y+1)^2 = 25$ 50. $9(x-1)^2 - y^2 = 144$

51. $(y+1)^2 - 9(x-2)^2 = 25$

52. $6(x-4)^2 - 3(y+3)^2 = 4$

53. $x^2 - y^2 + 6x = 36$ 54. $x^2 - 4y^2 - 4x = 0$

55. $x^2 + 4x - 4y^2 = 12$ 56. $x^2 - 2y^2 - 8y = 12$

57. $2x^2 - y^2 + 12x - 8y + 3 = 0$

58. $y^2 - 9x^2 - 4y - 30x = 33$

59. $3x^2 - 18x - 2y^2 - 8y + 1 = 0$

60. $4y^2 - 9x^2 - 8y - 36x = 68$

61. $y^2 + 2\sqrt{2} - x^2 + 2\sqrt{2}x = 1$

62. $x^2 + x = \dfrac{y^2 - 1}{4}$

In Exercises 63–72, identify the conic section represented by each equation and sketch the graph.

63. $x^2 - 6x + 12y + 33 = 0$

64. $9y^2 = x^2 - 4x$

65. $y^2 - 9x^2 = -1$

66. $4x^2 + 9y^2 - 8x + 36y + 4 = 0$

67. $x^2 + y^2 - 4x + 8y = 16$

68. $2y^2 + 3x - 4y - 7 = 0$

69. $2x^2 - 4x + 3y + 8 = 0$

70. $y^2 - 9x^2 + 18x - 4y = 14$

71. $4x^2 + 9y^2 + 8x - 54y + 49 = 0$

72. $x^2 - 4y^2 + 6x + 12y = 0$

In Exercises 73–78, find an equation (in standard form) for the hyperbola whose graph is shown.

73.

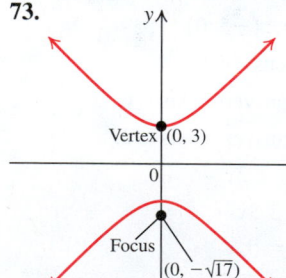

74.

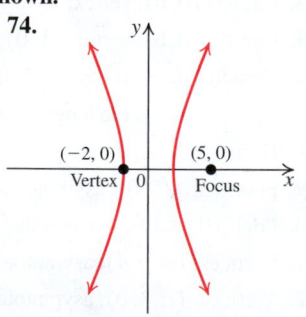

75.   76.

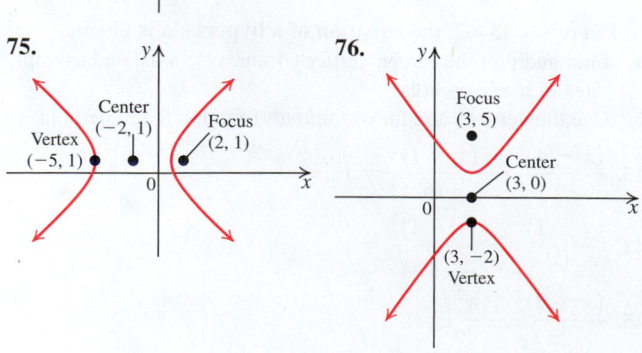

77.   78.

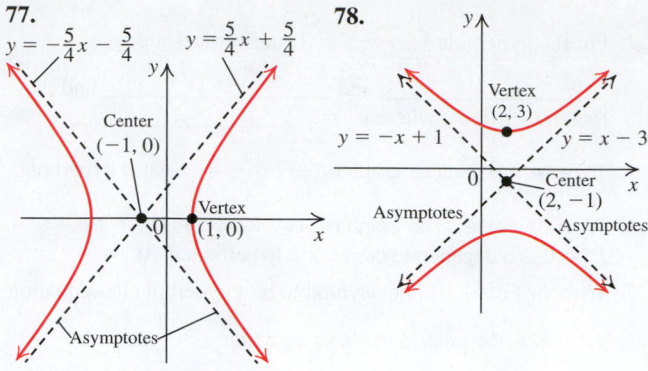

Applying the Concepts

79. **Sound of an explosion.** Points A and B are 1000 meters apart, and it is determined from the sound of an explosion heard at these points at different times that the location of the explosion is 600 meters closer to A than to B. Show that the location of the explosion is restricted to points on a hyperbola and find the equation of the hyperbola. [*Hint:* Let the coordinates of A be $(-500, 0)$ and those of B be $(500, 0)$.]

80. **Sound of an explosion.** Points A and B are 2 miles apart. The sound of an explosion at A was heard 3 seconds before it was heard at B. Assume that the sound travels at 1100 feet/second. Show that the location of the explosion is restricted to a hyperbola and find the equation of the hyperbola. [*Hint:* Recall that 1 mile = 5280 feet.]

81. **Thunder and lightning.** Thunder is heard by Nicole and Juan, who are 8000 feet apart. Nicole hears the thunder 3 seconds before Juan does. Find the equation of the hyperbola whose points are possible locations where the lightning could have struck. Assume that the sound travels at 1100 feet/second.

82. **Locating source of thunder.** In Exercise 81, Valerie is at a location midway between Nicole and Juan, and she hears the thunder 2 seconds after Juan does. Determine the location where the lightning strikes in relation to the three people involved.

83. **LORAN.** Two LORAN stations, A and B, are situated 300 kilometers apart along a straight coastline. Simultaneous radio signals are sent from each station to a ship. The ship receives the signal from A 0.0005 second before the signal from B. Assume that the radio signals travel 300,000 kilometers per second. Find the equation of the hyperbola on which the ship is located.

84. **Target practice.** A gun at G and a target at T are 1600 feet apart. The muzzle velocity of the bullet is 2000 feet per second. A person at P hears the crack of the gun and the thud of the bullet at the same time. Show that the possible locations of the person are on a hyperbola. Find the equation of the hyperbola with G at $(-800, 0)$ and T at $(800, 0)$. Assume that the speed of sound is 1100 feet/second. [*Hint:* Show that $|d(P, G) - d(P, T)|$ is a constant.]

85. **LORAN.** Two LORAN stations, A and B, lie on an east–west line with A 250 miles west of B. A plane is flying west on a line 50 miles north of the line AB. Radio signals are sent (traveling at 980 feet per microsecond) simultaneously from A and B, and the one sent from B arrives at the plane 500 microseconds before the one from A. Where is the plane?

86. Using LORAN. LORAN navigational transmitters A, B, C, and D are located at $(-100, 0)$, $(100, 0)$, $(0, -150)$, and $(0, 150)$, respectively. A navigator P on a ship somewhere in the second quadrant listens to the pair (A, B) of transmitters and computes the difference of the distances from the ship to A and B as 120 miles. Similarly, by listening to the pair (C, D) of transmitters, navigator P computes the difference of the distances from the ship to C and D as 80 miles. Use a calculator to find the coordinates of the ship.

87. Using LORAN. LORAN navigational transmitters A, B, and C are located at $(200, 0)$, $(-200, 0)$, and $(200, 1000)$, respectively. A navigator on a fishing boat listens to the pair (A, B) of transmitters and finds that the difference of the distances from the boat to A and B is 300 miles. She also finds that the difference of the distances from the boat to A and C is 400 miles. Use a calculator to find the possible locations of the boat.

Beyond the Basics

88. Write the standard form of the equation of the hyperbola for which the difference of the distances from any point on the hyperbola to $(-5, 0)$ and $(5, 0)$ is equal to (a) two units; (b) four units; (c) eight units.

89. Sketch the graphs of all three hyperbolas in Exercise 88 on the same coordinate plane.

90. Write the standard form of the equation of the hyperbola for which the difference of the distances from any point on the hyperbola to the points F_1 and F_2 is two units, where

  a. F_1 is $(0, 6)$ and F_2 is $(0, -6)$;
  b. F_1 is $(0, 4)$ and F_2 is $(0, -4)$;
  c. F_1 is $(0, 3)$ and F_2 is $(0, -3)$.

91. Sketch the graphs of all three hyperbolas in Exercise 90 on the same coordinate plane.

92. Rewrite equation

$$\sqrt{(x+c)^2 + y^2} - \sqrt{(x-c)^2 + y^2} = \pm 2a$$

from page 740 as

$$\sqrt{(x+c)^2 + y^2} = \pm 2a + \sqrt{(x-c)^2 + y^2}.$$

Square both sides to obtain

$$(x+c)^2 + y^2 = 4a^2 \pm 4a\sqrt{(x-c)^2 + y^2} + (x-c)^2 + y^2.$$

Simplify and then isolate the radical; then square both sides again to show that the equation can be simplified to

$$(c^2 - a^2)x^2 - a^2 y^2 = a^2(c^2 - a^2).$$

93. A hyperbola for which the lengths of the transverse and conjugate axes are equal is called an **equilateral hyperbola**. Show that the asymptotes of an equilateral hyperbola are perpendicular to one another.

94. The **latus rectum of a hyperbola** is a line segment that passes through a focus, is perpendicular to the transverse axis, and has endpoints on the hyperbola. Show that the length of the latus rectum of the hyperbola

$$\frac{x^2}{a^2} - \frac{y^2}{b^2} = 1 \text{ is } \frac{2b^2}{a}.$$

95. The *eccentricity* **of a hyperbola**, denoted by e, is defined by

$$e = \frac{\text{Distance between the foci}}{\text{Distance between the vertices}} = \frac{2c}{2a} = \frac{c}{a}.$$

Show that for every hyperbola, $e > 1$. What happens when $e = 1$?

In Exercises 96–100, find the eccentricity and the length of the latus rectum of each hyperbola.

96. $\frac{x^2}{16} - \frac{y^2}{9} = 1$ **97.** $36x^2 - 25y^2 = 900$

98. $x^2 - y^2 = 49$ **99.** $8(x-1)^2 - (y+2)^2 = 2$

100. $5x^2 - 4y^2 - 10x - 8y - 19 = 0$

101. For the hyperbola $\frac{x^2}{a^2} - \frac{y^2}{b^2} = 1$ with eccentricity e, show that $b^2 = a^2(e^2 - 1)$.

102. Find an equation of a hyperbola if its foci are $(\pm 6, 0)$ and the length of its latus rectum is 10.

103. A point $P(x, y)$ moves in the plane so that its distance from the point $(3, 0)$ is always twice its distance from the line $x = -1$. Show that the equation of the path of P is a hyperbola of eccentricity 2.

In Exercises 104–107, find all points of intersection of the given curves. Make a sketch of the curves that shows the points of intersection.

104. $y - 2x - 20 = 0$ and $y^2 - 4x^2 = 36$

105. $x - 2y^2 = 0$ and $x^2 = 8y^2 + 5$

106. $x^2 + y^2 = 15$ and $x^2 - y^2 = 1$

107. $x^2 + 9y^2 = 9$ and $4x^2 - 25y^2 = 36$

In Exercises 108 and 109, use the following definition. A *tangent line* **to a hyperbola is a line that is not parallel to either asymptote of the hyperbola and intersects the hyperbola at only one point.**

108. Find the equation of the tangent line to the hyperbola with equation $\frac{x^2}{a^2} - \frac{y^2}{b^2} = 1$, with $a = \sqrt{\frac{5}{3}}$ and $b = \sqrt{\frac{5}{7}}$ at the point $(2, 1)$.
[*Hint:* See the steps for Exercise 79 in Section 8.2 and (in Step 3) write the discriminant as a perfect square.]

109. Show that the equation of the tangent line to the hyperbola with equation $\frac{x^2}{a^2} - \frac{y^2}{b^2} = 1$, at the point (x_1, y_1), is

$$y = \frac{b^2 x_1}{a^2 y_1}x - \frac{b^2 x_1^2}{a^2 y_1} + y_1.$$

[*Hint:* See the steps for Exercise 80 in Section 8.2 and show that $(a^2 y_1 m - b^2 x_1)^2 = 0$.]

Critical Thinking / Discussion / Writing

110. Explain why each of the following represents a degenerate conic section. Where possible, sketch the graph.
  a. $4x^2 - 9y^2 = 0$
  b. $x^2 + 4y^2 + 6 = 0$
  c. $8x^2 + 5y^2 = 0$
  d. $x^2 + y^2 - 4x + 8y = -20$
  e. $y^2 - 2x^2 = 0$

111. Discuss the graph of a quadratic equation in x and y of the form

$$Ax^2 + Cy^2 + Dx + Ey + F = 0, \quad AC \neq 0.$$

Rewrite this equation in the form

$$A\left(x^2 + \frac{D}{A}x + \frac{D^2}{4A^2}\right) + C\left(y^2 + \frac{E}{C}y + \frac{E^2}{4C^2}\right)$$

$$= \frac{D^2}{4A} + \frac{E^2}{4C} - F.$$

Include in your discussion the types of graphs you will obtain in each of the following cases:

(i) $A > 0, C > 0$ (ii) $A > 0, C < 0$
(iii) $A < 0, C > 0$ (iv) $A < 0, C < 0$.

In each case, discuss how the classification is affected by the sign of

$$\frac{D^2}{4A} + \frac{E^2}{4C} - F.$$

Getting Ready for the Next Section

112. Give the exact value for each of the following.
  a. $\cos 45°$ **b.** $\sin 30°$
  c. $\sin 60°$ **d.** $\cos 0°$

113. Find the exact value for $\cos 20° \cos 25° - \sin 20° \sin 25°$

114. Find the exact value for $\sin(60° + 45°)$

115. Fill in the missing trigonometric expression.
  a. $\cos 2\theta = \underline{\quad} - \sin^2\theta$
  b. $2\sin\theta\cos\theta = \underline{\quad}$

116. Find an acute angle θ such that $\cot 2\theta = \frac{\sqrt{3}}{3}$.

117. If $\cot 2\theta = -\frac{7}{24}$ find
  a. $\sin\theta$ **b.** $\cos\theta$

118. Rewrite each equation in standard form for a conic and identify the graph of the equation as a parabola, ellipse, or hyperbola.
  a. $x^2 + 4y^2 + 6x - 8y + 9 = 0$
  b. $y^2 - 2y - 4x - 7 = 0$
  c. $9x^2 - 4y^2 + 54x + 8y + 41 = 0$

119. Each equation has the form
$Ax^2 + Bxy + Cy^2 + Dx + Ey + F = 0$. Compute
$B^2 - 4AC$.
  a. $4x^2 - 4xy + 2y^2 - 8\sqrt{5}x - 16\sqrt{5}y = 0$
  b. $5xy - 2y^2 + 4x + \sqrt{7}y - 14 = 0$

SECTION **8.5**

Rotation of Axes

BEFORE STARTING THIS SECTION, REVIEW

1. Equation of a parabola (Section 8.2, page 716)
2. Equation of an ellipse (Section 8.3, page 729)
3. Equation of a hyperbola (Section 8.4, page 748)
4. Trigonometric functions (Chapters 4–6)

OBJECTIVES

1. Identify a conic.
2. Rotate coordinate axes.
3. Eliminate the xy-term in the equation of a conic.
4. Use the discriminant to identify a conic.

◆ Launching a Space Vehicle

Wind speed and wind direction are usually measured in east–west (EW) and north–south (NS) components. When describing wind characteristics, the standard or natural coordinate system uses the y-axis to represent the NS directions (with north in the direction of increasing positive numbers) and the x-axis to represent the EW directions (with east in the direction of increasing positive numbers). Sometimes information recorded relative to such a standard coordinate system must be interpreted in a coordinate system that is rotated with respect to the system. Consider, for example, space vehicle launches. The launch path may not be exactly due north, east, south, or west. So engineers need to convert the wind speed and direction, measured in EW and NS components, to coordinates in a plane where the axes are parallel and perpendicular to the launch direction. In Example 4, we show how to accomplish this conversion.

1 Identify a conic.

Identifying a Conic

In Sections 8.2–8.4, we learned to identify the graph of an equation of the form

$$Ax^2 + Cy^2 + Dx + Ey + F = 0$$

as a parabola, an ellipse, or a hyperbola by the process of completing the squares. We summarize the results.

RECALL

The graph of equation (1) is a degenerate conic if it results in the following:

A pair of lines, a single line, a single point, or no points at all.

> ### IDENTIFYING CONICS WITH AXES PARALLEL TO THE x- OR y-AXES
>
> If we exclude the degenerate conics, the graph of the equation
>
> $$Ax^2 + Cy^2 + Dx + Ey + F = 0, \quad (1)$$
>
> where A and C are not both zero, is
>
> 1. a parabola if either $A = 0$ or $C = 0$ ($AC = 0$). (See page 716.)
> 2. an ellipse (or a circle if $A = C$) if A and C have the same sign ($AC > 0$). (See page 731.)
> 3. a hyperbola if A and C have opposite signs ($AC < 0$). (See page 748.)

EXAMPLE 1   Identifying a Conic

Determine the type of conic that is the graph of each equation.

a. $5x^2 - 2y + 7 = 0$

b. $7x^2 - 4y^2 - 3x + 8y - 12 = 0$

c. $3x^2 + 2y^2 - 3x + 9 = 0$

Solution

a. We compare each equation with equation (1). We have $A = 5$ and $C = 0$. Since $AC = 0$, we conclude that the graph is a parabola.

b. Here $A = 7$ and $C = -4$. Because A and C have opposite signs, the graph is a hyperbola.

c. Here $A = 3$ and $C = 2$. Because A and C have the same sign, the graph is an ellipse.

Practice Problem 1 Determine the type of conic that is the graph of each equation.

a. $x^2 + 2y^2 - 11x + 3 = 0$ **b.** $8y^2 - x - y + 3 = 0$

2 Rotate coordinate axes.

Rotation of Axes

The graph of the general second-degree equation

$$Ax^2 + Bxy + Cy^2 + Dx + Ey + F = 0 \quad (2)$$

is also a conic (or a degenerate conic). When $B \neq 0$, the axes for the conic are not parallel to the x-axis or the y-axis. We can find the conic's axis (or axes) by rotating the x- and y-axes through a suitable angle θ. We choose θ so that the xy-term in equation (2) is eliminated. We then analyze the graph of the resulting equation.

In **a rotation of axes**, the x- and y-axes are rotated through an angle θ, keeping the origin fixed. The new axes are labeled the x'-axis (read "x-prime axis") and y'-axis (read "y-prime axis"), respectively, as shown in Figure 8.31(a).

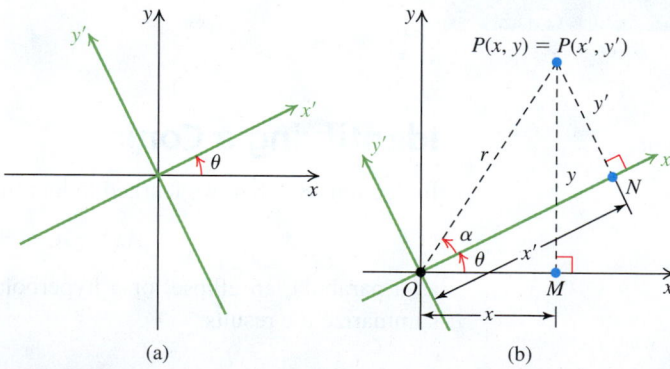

(a)   (b)

Figure 8.31 Rotation of axes.

In Figure 8.31(b), the point P has two sets of coordinates: (x, y) in the old coordinate system and (x', y') in the new coordinate system. We let α denote the angle that the segment OP makes with the new x'-axis and let r denote the distance between P and the origin. In Figure 8.31(b), from the right triangle OPN, we have

$$\cos \alpha = \frac{x'}{r} \qquad \sin \alpha = \frac{y'}{r}.$$

So $x' = r \cos \alpha \qquad y' = r \sin \alpha.$

In the right triangle *OPM* in Figure 8.31(b), we have:

$$\cos(\alpha + \theta) = \frac{x}{r} \qquad\qquad \sin(\alpha + \theta) = \frac{y}{r}$$

So
$$x = r\cos(\alpha + \theta) \qquad\qquad y = r\sin(\alpha + \theta).$$

Now
$$
\begin{aligned}
x &= r\cos(\alpha + \theta) \\
&= r(\cos\alpha\cos\theta - \sin\alpha\sin\theta) && \text{Sum formula for cosine} \\
&= (r\cos\alpha)\cos\theta - (r\sin\alpha)\sin\theta && \text{Distributive and associative properties} \\
&= x'\cos\theta - y'\sin\theta && \text{Replace } r\cos\alpha \text{ with } x' \text{ and } r\sin\alpha \\
&&& \text{with } y'.
\end{aligned}
$$

Similarly,
$$
\begin{aligned}
y &= r\sin(\alpha + \theta) \\
&= r(\sin\alpha\cos\theta + \cos\alpha\sin\theta) && \text{Sum formula for sine} \\
&= x'\sin\theta + y'\cos\theta && \text{Distribute, then replace } r\cos\alpha \\
&&& \text{with } x' \text{ and } r\sin\alpha \text{ with } y'.
\end{aligned}
$$

The equations $x = x'\cos\theta - y'\sin\theta$ and $y = x'\sin\theta - y'\cos\theta$ are two linear equations in x' and y'. This system of equations can be solved for x' and y' (see Exercise 72); the formulas relating the (x, y) and (x', y') coordinates of a point are given next.

ROTATION OF AXES FORMULAS

If the x-axis and y-axis are rotated through an angle θ, then the coordinates (x, y) of a point P determined by the xy-plane and the coordinates (x', y') of the point P determined by the $x'y'$-plane are related by the formulas:

$$x = x'\cos\theta - y'\sin\theta \qquad x' = x\cos\theta + y\sin\theta$$
$$y = x'\sin\theta + y'\cos\theta \qquad y' = -x\sin\theta + y\cos\theta$$

EXAMPLE 2 **Using The Rotation of Axes Formulas**

Find the $x'y'$-coordinates of the point $P(x, y) = (1, 2)$ if the coordinate axes are rotated by 30°

Solution

Since $(x, y) = (1, 2)$, $x = 1$ and $y = 2$. The rotation angle is $\theta = 30°$.

$$x' = x\cos\theta + y\sin\theta; \; y' = -x\sin\theta + y\cos\theta \qquad \text{Rotation of axes formulas}$$

$$x' = 1\cos30° + 2\sin30°; \; y' = -1\sin30° + 2\cos30° \qquad \text{Replace } x \text{ with 1, } y \text{ with 2, and } \theta \text{ with 30°.}$$

$$x' = \frac{\sqrt{3}}{2} + 2\left(\frac{1}{2}\right); \; y' = -1\left(\frac{1}{2}\right) + 2\left(\frac{\sqrt{3}}{2}\right) \qquad \text{Replace } \cos30° \text{ with } \frac{\sqrt{3}}{2} \text{ and } \sin30° \text{ with } \frac{1}{2}.$$

$$x' = 1 + \frac{\sqrt{3}}{2}; \; y' = \sqrt{3} - \frac{1}{2} \qquad \text{Simplify.}$$

Practice Problem 2 Find the $x'y'$-coordinates of the point $P(x, y) = (1, 2)$ if the coordinate axes are rotated by 60°.

EXAMPLE 3 **Rotating Axes to Eliminate *xy*-term**

Rotate the coordinate axes through a 45° angle and express the equation for the conic $xy = 1$ in terms of the new $x'y'$ coordinates. Sketch the graph.

Solution

Substitute $\theta = 45°$ in the rotation equations expressing x and y in terms of x' and y'. We have:

$$x = x' \cos 45° - y' \sin 45° = x'\frac{\sqrt{2}}{2} - y'\frac{\sqrt{2}}{2} = \frac{\sqrt{2}}{2}(x' - y')$$

$$y = x' \sin 45° + y' \cos 45° = x'\frac{\sqrt{2}}{2} + y'\frac{\sqrt{2}}{2} = \frac{\sqrt{2}}{2}(x' + y')$$

Then the equation $xy = 1$ becomes

$$\left[\frac{\sqrt{2}}{2}(x' - y')\right]\left[\frac{\sqrt{2}}{2}(x' + y')\right] = 1 \quad \text{Replace } x \text{ with } \frac{\sqrt{2}}{2}(x' - y')$$

$$\text{and } y \text{ with } \frac{\sqrt{2}}{2}(x' + y'), \text{ in } xy = 1.$$

$$\left(\frac{\sqrt{2}}{2} \cdot \frac{\sqrt{2}}{2}\right)(x' - y')(x' + y') = 1 \quad \text{Rearrange}$$

$$\frac{1}{2}(x'^2 - y'^2) = 1 \quad \frac{\sqrt{2}}{2} \cdot \frac{\sqrt{2}}{2} = \frac{2}{4} = \frac{1}{2}; \text{ difference of squares}$$

$$\frac{x'^2}{2} - \frac{y'^2}{2} = 1 \quad \text{Distributive property}$$

The equation $\dfrac{x'^2}{2} - \dfrac{y'^2}{2} = 1$ is the standard form of the equation of a hyperbola with $a = b = \sqrt{2}$ in the $x'y'$-plane with center $(0,0)$, vertices $(\pm a, 0) = (\pm\sqrt{2}, 0)$, and asymptotes

$$y' = \pm\frac{b}{a}x' = \pm\frac{\sqrt{2}}{\sqrt{2}}x' = \pm x'.$$

Now

$$y' = x' \quad \text{Equation of an asymptote}$$

$$-x \sin 45° + y \cos 45° = x \cos 45° + y \sin 45° \quad \text{Write in } xy\text{-coordinates.}$$

$$-x\frac{\sqrt{2}}{2} + y\frac{\sqrt{2}}{2} = x\frac{\sqrt{2}}{2} + y\frac{\sqrt{2}}{2}$$

$$x = 0 \quad \text{Simplify.}$$

Similarly, $y' = -x'$ leads to $y = 0$.

Therefore, the asymptotes $y' = x'$ and $y' = -x'$ are the y- and x-axes, respectively. The graph is shown in Figure 8.32.

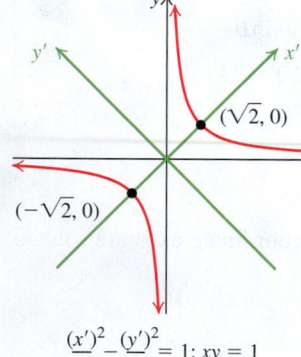

$$\frac{(x')^2}{2} - \frac{(y')^2}{2} = 1; xy = 1$$

Figure 8.32

Practice Problem 3 Rotate the coordinate axes through a 45° angle and express the equation for the conic $xy = 2$ in terms of the new $x'y'$-coordinates. Sketch the graph.

◆ **EXAMPLE 4** **Space Shuttles and Wind Speed**

Suppose that a space shuttle is being launched at a 30° angle to the east of due north as shown in Figure 8.33. Find the wind speed along and perpendicular to the initial launch path caused by a south-to-north wind whose speed is 16 knots.

Solution

We represent the wind vector in the coordinate system in Figure 8.33. We want to find the coordinates of the points $(0, 16)$, which is the wind vector $\mathbf{v} = \langle 0, 16\rangle$, in the $x'y'$ system determined by the shuttle.

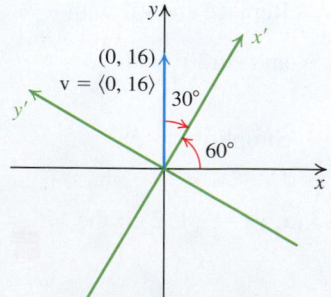

Figure 8.33 Launch direction.

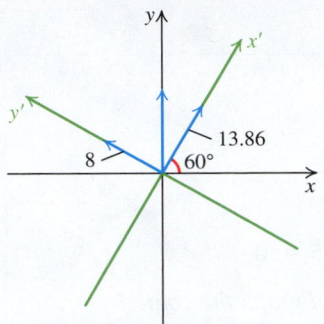

Figure 8.34 Wind speeds.

3 Eliminate the *xy*-term in the equation of a conic.

We use the equations:

$$x' = x\cos\theta + y\sin\theta \qquad\qquad y' = -x\sin\theta + y\cos\theta$$
$$x' = 0\cos 60° + 16\sin 60° \qquad y' = -0\sin 60° + 16\cos 60° \qquad \text{Replace } x \text{ with } 0,$$
$$ y \text{ with } 16, \text{ and } \theta$$
$$ \text{with } 60°.$$
$$x' = 16\left(\frac{\sqrt{3}}{2}\right) \approx 13.86 \qquad y' = 16\left(\frac{1}{2}\right) = 8 \qquad \text{Simplify.}$$

Thus, the wind speed along the launch path is approximately 13.86 knots, and the wind speed perpendicular to the launch path is 8 knots. See Figure 8.34.

Practice Problem 4 Rework Example 4 with a north-to-south wind speed of 12 knots.

Eliminating the *xy*-Term

In Example 3, notice that we eliminated the Bxy term by rotating the axes through 45°.

In general, if we make the substitutions $x = x'\cos\theta - y'\sin\theta$ and $y = x'\sin\theta + y'\cos\theta$ in the equation

$$Ax^2 + Bxy + Cy^2 + Dx + Ey + F = 0$$

in which $B \neq 0$, we obtain an equation of the form

$$A'x'^2 + B'x'y' + C'y'^2 + D'x' + E'y' + F' = 0.$$

The coefficient B' in this equation (see Exercise 73) can be expressed in terms of the original coefficients as follows:

$$B' = B(\cos^2\theta - \sin^2\theta) - 2(A - C)\sin\theta\cos\theta \qquad \text{From Exercise 73}$$
$$ = B\cos 2\theta - (A - C)\sin 2\theta \qquad \text{Replace } \cos^2\theta - \sin^2\theta \text{ with}$$
$$ \cos 2\theta \text{ and replace } 2\sin\theta\cos\theta$$
$$ \text{with } \sin 2\theta.$$

Then setting $B' = 0$, we have

$$B\cos 2\theta - (A - C)\sin 2\theta = 0$$
$$B\cos 2\theta = (A - C)\sin 2\theta \qquad \text{Add } (A - C)\sin 2\theta.$$
$$\frac{\cos 2\theta}{\sin 2\theta} = \frac{A - C}{B}, B \neq 0 \qquad \text{Divide by } B\sin 2\theta.$$
$$\cot 2\theta = \frac{A - C}{B}. \qquad \text{Quotient identity}$$

It is customary to choose an angle θ such that $0° \leq \theta < 90°$.

ELIMINATING AN *xy*-TERM BY ROTATION OF AXES

To transform the second-degree equation

$$Ax^2 + Bxy + Cy^2 + Dx + Ey + F = 0 \quad B \neq 0$$

to an equation of the form $A'x'^2 + C'y'^2 + D'x' + E'y' + F' = 0$ without the $x'y'$ term, rotate the axes through an acute angle θ, where

$$\cot 2\theta = \frac{A - C}{B}.$$

PROCEDURE
IN ACTION

EXAMPLE 5   Rotating Axes to Eliminate the *xy*-term

OBJECTIVE

Eliminate the xy-term in an equation of the form

$$Ax^2 + Bxy + Cy^2 + Dx + Ey + F = 0,$$
$$B \neq 0$$

by rotating the axes. Identify and graph the conic.

Step 1 Find the angle of rotation. Find the acute angle of rotation, θ, using the formula

$$\cot 2\theta = \frac{A - C}{B}.$$

Step 2 Find cos θ and sin θ. Find $\cos \theta$ and $\sin \theta$ by recognizing θ as a standard angle or by finding $\cos 2\theta$ and $\sin 2\theta$ from $\cot 2\theta$ and then using the half-angle formulas for $0 < \theta < 90°$.

$$\cos \theta = \sqrt{\frac{1 + \cos 2\theta}{2}}; \sin \theta = \sqrt{\frac{1 - \cos 2\theta}{2}}$$

Step 3 Convert from *xy*-coordinates to *x'y'*-coordinates. Use Step 2 to replace x with $x' \cos \theta - y' \sin \theta$ and y with $x' \sin \theta + y' \cos \theta$ in the given equation. Collect like terms, and simplify.

Step 4 Identify the conic. Identify the standard form of the conic in the $x'y'$-plane.

Step 5 Sketch the graph. Sketch the graph of the conic in the $x'y'$-plane, using the procedures outlined in Sections 8.2–8.4.

EXAMPLE

Eliminate the xy-term in the equation

$$2x^2 + \sqrt{3}xy + y^2 - 5 = 0$$

by a suitable rotation of axes. Identify and graph the conic.

1. In $2x^2 + \sqrt{3}xy + y^2 - 5 = 0$, we have $A = 2$, $B = \sqrt{3}$, and $C = 1$. So $\cot 2\theta = \dfrac{2 - 1}{\sqrt{3}} = \dfrac{1}{\sqrt{3}} = \dfrac{\sqrt{3}}{3}$. So an appropriate choice for 2θ is $60°$. Since $2\theta = 60°$, we have $\theta = \dfrac{60°}{2} = 30°$.

2. $\cos \theta = \cos 30° = \dfrac{\sqrt{3}}{2}$; $\sin \theta = \sin 30° = \dfrac{1}{2}$.

3. $x = x' \cos 30° - y' \sin 30° = \dfrac{\sqrt{3}}{2}x' - \dfrac{1}{2}y'$ and

$y = x' \sin 30° + y' \cos 30° = \dfrac{1}{2}x' + \dfrac{\sqrt{3}}{2}y'$.

Then $2x^2 + \sqrt{3}xy + y^2 - 5 = 0$ becomes

$$2\left(\frac{\sqrt{3}}{2}x' - \frac{1}{2}y'\right)^2 + \sqrt{3}\left(\frac{\sqrt{3}}{2}x' - \frac{1}{2}y'\right)\left(\frac{1}{2}x' + \frac{\sqrt{3}}{2}y'\right) +$$

$$\left(\frac{1}{2}x' + \frac{\sqrt{3}}{2}y'\right)^2 - 5 = 0, \text{ which (after collecting like}$$

terms) simplifies to $\dfrac{x'^2}{2} + \dfrac{y'^2}{10} = 1$.

4. Using the standard form $\dfrac{(x')^2}{b^2} + \dfrac{(y')^2}{a^2} = 1$, we see that $\dfrac{x'^2}{2} + \dfrac{y'^2}{10} = 1$ is the equation of an ellipse with center $(0, 0)$ in the $x'y'$-coordinate system.
The foci lie on the y'-axis because $10 > 2$.
$a^2 = 10$, $b^2 = 2$, and $c^2 = a^2 - b^2 = 8$

5.

$$\frac{(x')^2}{2} + \frac{(y')^2}{10} = 1$$

Practice Problem 5 Eliminate the xy-term in the equation

$$5x^2 - 8xy + 5y^2 - 9 = 0$$

by a suitable rotation of axes. Identify and graph the conic.

EXAMPLE 6 Identifying a Conic by Rotation of Axes

Eliminate the xy-term in the equation $18x^2 - 48xy + 32y^2 - 20x - 15y - 25 = 0$ by a suitable rotation of axes. Identify and graph the conic.

Solution

Step 1 In the given equation, $18x^2 - 48xy + 32y^2 - 20x - 15y - 25 = 0$, $A = 18, B = -48$, and $C = 32$.
So

$$\cot 2\theta = \frac{A - C}{B} = \frac{18 - 32}{-48} = \frac{-14}{-48} = \frac{7}{24}.$$

Step 2 Since $\cot 2\theta = \dfrac{7}{24} = \dfrac{x}{y}$ is positive, 2θ is in quadrant I.

From the sketch in Figure 8.35 (letting $x = 7$ and $y = 24$), we have

$$r = \sqrt{7^2 + 24^2} = \sqrt{625} = 25.$$

Then $\cos 2\theta = \dfrac{7}{25}$ and

$$\cos\theta = \sqrt{\frac{1 + \cos 2\theta}{2}} \qquad \sin\theta = \sqrt{\frac{1 - \cos 2\theta}{2}}$$

$$= \sqrt{\frac{1 + \dfrac{7}{25}}{2}} \qquad\qquad = \sqrt{\frac{1 - \dfrac{7}{25}}{2}}$$

$$= \sqrt{\frac{16}{25}} = \frac{4}{5} \qquad\qquad = \sqrt{\frac{9}{25}} = \frac{3}{5}$$

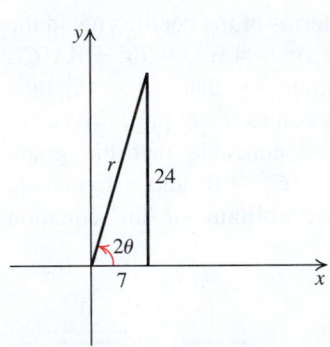

Figure 8.35

Step 3 $x = (\cos\theta)x' - (\sin\theta)y' = \dfrac{4}{5}x' - \dfrac{3}{5}y'$ Replace $\cos\theta$ with $\dfrac{4}{5}$ and $\sin\theta$ with $\dfrac{3}{5}$.

$$y = (\sin\theta)x' + (\cos\theta)y' = \frac{3}{5}x' + \frac{4}{5}y'$$

$18x^2 - 48xy + 32y^2 - 20x - 15y - 25 = 0$ becomes

$$18\left(\frac{4}{5}x' - \frac{3}{5}y'\right)^2 - 48\left(\frac{4}{5}x' - \frac{3}{5}y'\right)\left(\frac{3}{5}x' + \frac{4}{5}y'\right) + 32\left(\frac{3}{5}x' + \frac{4}{5}y'\right)^2$$

$$- 20\left(\frac{4}{5}x' - \frac{3}{5}y'\right) - 15\left(\frac{3}{5}x' + \frac{4}{5}y'\right) - 25 = 0.$$

Step 4 Collecting like terms and simplifying (see Exercise 71) yields

$$50y'^2 - 25x' - 25 = 0 \text{ or } y'^2 = \frac{1}{2}(x' + 1), \text{ which is an equation of a}$$

parabola in the $x'y'$-plane.

Step 5 From $\cos\theta = \dfrac{4}{5}$, we find $\theta = \cos^{-1}\left(\dfrac{4}{5}\right) \approx 37°$.   Use a calculator.

Rotate the axes through approximately $37°$ as in Figure 8.36 and sketch the graph of $y'^2 = \dfrac{1}{2}(x' + 1)$ in $x'y'$-plane. From the standard

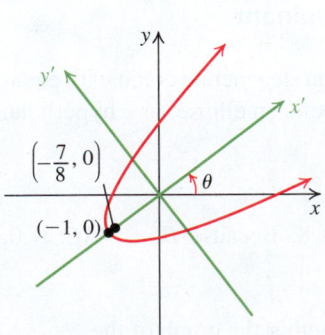

Figure 8.36

form $(y' - k)^2 = 4p(x' - h), p > 0$, we have $k = 0, h = -1,$

and $4p = \dfrac{1}{2}$, or $p = \dfrac{1}{8}$. So the focus in the $x'y'$-plane is

$(h + p, k) = \left(-1 + \dfrac{1}{8}, 0\right) = \left(-\dfrac{7}{8}, 0\right)$, and the vertex is $(-1, 0)$.

Practice Problem 6 Eliminate the xy-term in the equation $32x^2 - 48xy + 18y^2 - 15x - 20y = 0$ by a suitable rotation of axes. Identify and graph the conic.

4 Use the discriminant to identify a conic.

Identifying Conics by Using the Discriminant

By using the rotation of axes formulas, an equation of the form

$$Ax^2 + Bxy + Cy^2 + Dx + Ey + F = 0$$

can be transformed into an equation of the form

$$A'x'^2 + B'x'y' + C'y'^2 + D'x' + E'y' + F' = 0.$$

Each coefficient in the second equation can be expressed in terms of the coefficients in the first equation. (See Exercise 73.) It also can be shown that $B^2 - 4AC = B'^2 - 4A'C'$. (See Exercise 75.) If we can select a suitable rotation so that $B' = 0$, then $B'^2 - 4A'C' = -4A'C'$. From the test for identifying conics (see page 764) for the equation $A'x'^2 + C'y'^2 + D'x' + E'y' + F' = 0$, we conclude that the graph of this equation is a parabola if $A'C' = 0$, an ellipse if $A'C' > 0$, and a hyperbola if $A'C' < 0$. The quantity $B^2 - 4AC$ is called the **discriminant** of the equation $Ax^2 + Bxy + Cy^2 + Dx + Ey + F = 0$.

TECHNOLOGY CONNECTION

To graph a second-degree equation of the form $Ax^2 + Bxy + Cy^2 + Dx + Ey + F = 0$ using a graphing calculator, you must first solve for y. Rewrite the equation as a quadratic equation in y, $Cy^2 + (Bx + E)y + (Ax^2 + Dx + F) = 0$, and use the quadratic formula to find two functions y_1, and y_2. Then graph Y_1 and Y_2 on your calculator in the same viewing window to produce the graph.

For example, to graph $2x^2 + 4xy + y^2 - x - y = 0$, rewrite the equation as $y^2 + (4x - 1)y + (2x^2 - x) = 0$. Solve for y to obtain

$$Y_1 = \frac{1 - 4x + \sqrt{8x^2 - 4x + 1}}{2}$$

and

$$Y_2 = \frac{1 - 4x - \sqrt{8x^2 - 4x + 1}}{2}.$$

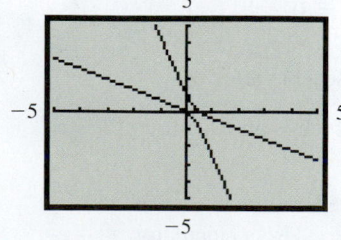

SUMMARY OF **MAIN FACTS**

Except for degenerate cases, the graph of the equation

$$Ax^2 + Bxy + Cy^2 + Dx + Ey + F = 0,$$

where A and C are not both zero, is

1. a parabola if $B^2 - 4AC = 0$.
2. an ellipse (or a circle if $A = C$) if $B^2 - 4AC < 0$.
3. a hyperbola if $B^2 - 4AC > 0$.

EXAMPLE 7 **Identifying a Conic Using the Discriminant**

Use the discriminant to determine whether the graph of the non-degenerate conic with equation $3x^2 - 2\sqrt{5}\,xy + y^2 + 13x - 7y + 11 = 0$ is a parabola, an ellipse, or a hyperbola.

Solution

We have $A = 3, B = -2\sqrt{5}$, and $C = 1$.

So, $B^2 - 4AC = (-2\sqrt{5})^2 - 4(3)(1) = 20 - 12 = 8$. Because $B^2 - 4AC > 0$, the graph of the given equation is a hyperbola.

Practice Problem 7 Use the discriminant to determine whether the graph of the non-degenerate conic with equation $2x^2 + \sqrt{7}\,xy + 5y^2 - 3x + 1 = 0$ is a parabola, an ellipse, or a hyperbola.

Answers to Practice Problems

1. a. ellipse   **b.** parabola   **2.** $x' = \dfrac{1}{2} + \sqrt{3};\ y' = 1 - \dfrac{\sqrt{3}}{2}$

5. $\dfrac{x'^2}{9} + \dfrac{y'^2}{1} = 1$; an ellipse   **6.** $x' = 2y'^2$; a parabola

3. $\dfrac{x'^2}{4} - \dfrac{y'^2}{4} = 1$

4. $x' = -10.39,\ y' = -6$; the wind speed along the launch path is approximately 10.39 knots (opposite the launch direction), and the wind speed perpendicular to the path is 6 knots in the direction 30° south of due east.

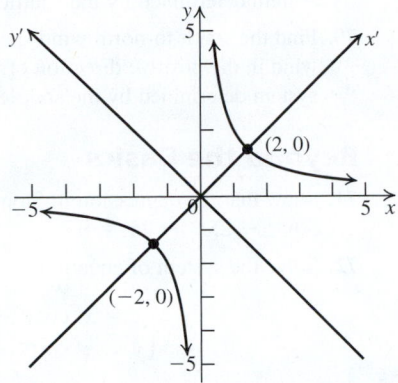

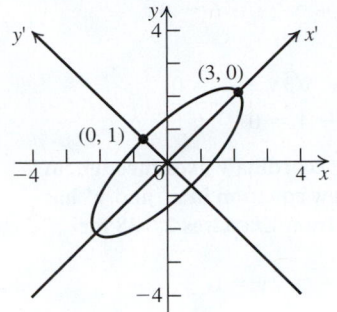

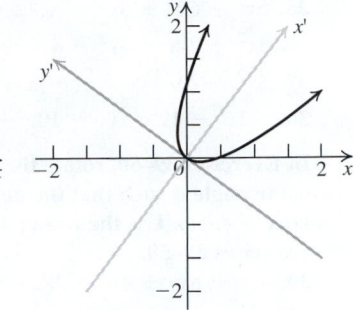

7. an ellipse

Exercises

Concepts and Vocabulary

1. The general form for a second-degree equation is _____.

2. Rotation of axes is performed to eliminate the _____-term in a second-degree equation.

3. If the graph of an equation of the form $Ax^2 + Bxy + Cy^2 + Dx + Ey + F = 0$ is an ellipse, then _____.

4. The graph of the equation $x^2 + 2xy + y^2 - 3x + 4y = 0$ a/an _____.

5. True or False. The graph of a second-degree equation is a degenerate conic if it contains no points.

6. True or False. The graph of $x^2 + xy + y^2 - 3 = 0$ is a hyperbola.

7. True or False. If the coordinate axes are rotated through an acute angle θ and $\cot 2\theta < 0$, then $90° < 2\theta < 180°$.

8. True or False. If $A = 0$ and $B \neq 0$, the graph of $Ax^2 + Bxy + Cy^2 + Dx + Ey + F = 0$ is a hyperbola.

Building Skills

In Exercises 9–14, determine the type of conic that is the graph of the given equation.

9. $7y^2 - 13x + 2 = 0$

10. $-12x^2 + 4x - 8y = 0$

11. $3x^2 - 2y^2 + 4x - 9 = 0$

12. $-5x^2 + 3y^2 + 11x + 6y = 0$

13. $-2x^2 - y^2 + 5x + 2y + 1 = 0$

14. $14x^2 + 3y^2 + 2y - 20 = 0$

In Exercises 15–20, find the $x'y'$-coordinates of the given point if the coordinate axes are rotated through the given angle.

15. $(0, 1);\ \theta = 30°$

16. $(-1, 0);\ \theta = 30°$

17. $(2, 1);\ \theta = 45°$

18. $(1, -1);\ \theta = 45°$

19. $(3, -2);\ \theta = 60°$

20. $(-3, 4);\ \theta = 60°$

In Exercises 21–26, find the equation for the given conic in the $x'y'$-coordinates when the coordinate axes are rotated through an angle θ.

21. $x^2 + 3xy + y^2 + 5 = 0;\ \theta = 45°$

22. $x^2 - 3xy + y^2 + 10 = 0;\ \theta = 45°$

23. $2x^2 - 4\sqrt{3}xy + 6y^2 + 4\sqrt{3}x + 4y = 0;\ \theta = 30°$

24. $2x^2 + 4\sqrt{3}xy + 6y^2 - \sqrt{3}x + y + 2 = 0;\ \theta = 60°$

25. $5x^2 + 4xy + 2y^2 - 6 = 0;\ \cos\theta = \dfrac{2\sqrt{5}}{5},\ \sin\theta = \dfrac{\sqrt{5}}{5}$

26. $10x^2 + 6xy + 2y^2 - 11 = 0;\ \cos\theta = \dfrac{3\sqrt{10}}{10},\ \sin\theta = \dfrac{\sqrt{10}}{10}$

In Exercises 27–38, find $\cos\theta$ and $\sin\theta$ such that rotation through the acute angle θ results in a new equation with no $x'y'$-term.

27. $4x^2 + 8xy + 4y^2 + \sqrt{2}x - \sqrt{2}y = 0$

28. $x^2 - 2xy + y^2 - \sqrt{2}x - \sqrt{2}y + 2 = 0$

29. $2x^2 + \sqrt{3}xy + y^2 - 10 = 0$

30. $2x^2 + 3xy + 2y^2 - 7 = 0$

31. $x^2 + 4xy + y^2 - 3 = 0$

32. $x^2 - 10xy + y^2 - 4 = 0$

33. $16x^2 - 24xy + 9y^2 - 10x + 70y + 25 = 0$

34. $18x^2 - 48xy + 32y^2 - 20x - 15y + 25 = 0$

35. $5x^2 - 6xy + 5y^2 - \sqrt{2}x + \sqrt{2}y - 6 = 0$

36. $3x^2 + 2xy + 3y^2 - 8 = 0$

37. $3x^2 - 2\sqrt{3}xy + y^2 - x - \sqrt{3}y + 2 = 0$

38. $-2x^2 - 12\sqrt{3}xy + 10y^2 - 4 = 0$

In Exercises 39–50, rotate the coordinate axes through an acute angle θ such that the new equation in x' and y' has no $x'y'$-term. Use the results from Exercises 27–38 for Exercises 39–50.

39. $4x^2 + 8xy + 4y^2 + \sqrt{2}x - \sqrt{2}y = 0$

40. $x^2 - 2xy + y^2 - \sqrt{2}x - \sqrt{2}y + 2 = 0$

41. $2x^2 + \sqrt{3}xy + y^2 - 10 = 0$

42. $2x^2 + 3xy + 2y^2 - 7 = 0$

43. $x^2 + 4xy + y^2 - 3 = 0$

44. $x^2 - 10xy + y^2 - 4 = 0$

45. $16x^2 - 24xy + 9y^2 - 10x + 70y + 25 = 0$

46. $18x^2 - 48xy + 32y^2 - 20x - 15y + 25 = 0$

47. $5x^2 - 6xy + 5y^2 - \sqrt{2}x - \sqrt{2}y - 6 = 0$

48. $3x^2 + 2xy + 3y^2 - 8 = 0$

49. $3x^2 - 2\sqrt{3}xy + y^2 - x - \sqrt{3}y + 2 = 0$

50. $-2x^2 - 12\sqrt{3}xy + 10y^2 - 4 = 0$

In Exercises 51–54, sketch the graph of the degenerate conic or state that the graph contains no points.

51. $x^2 - 2xy + y^2 - 2 = 0$

52. $x^2 - 2xy + y^2 + x - y = 0$

53. $7x^2 + 6\sqrt{3}xy + 13y^2 + 4 = 0$

54. $3x^2 - 2\sqrt{3}xy + y^2 - 1 = 0$

In Exercises 55–60, use the discriminant to identify the graph of the given non-degenerate conic as a parabola, an ellipse, or a hyperbola.

55. $3x^2 + 2xy + 3y^2 - 8 = 0$

56. $7x^2 + 2xy + 7y^2 - 16 = 6$

57. $9x^2 - 6xy + y^2 - 2x + 1 = 0$

58. $x^2 + 4xy - 2y^2 - 12 = 0$

59. $4x^2 + 24xy - 3y^2 - 60 = 0$

60. $4x^2 - 4xy + y^2 - 10y - 30 = 0$

In Exercises 61–66, sketch the graph using a graphing utility. These are the same equations given in Exercises 55–60.

61. $3x^2 + 2xy + 3y^2 - 8 = 0$

62. $7x^2 + 2xy + 7y^2 - 16 = 6$

63. $9x^2 - 6xy + y^2 - 2x + 1 = 0$

64. $x^2 + 4xy - 2y^2 - 12 = 0$

65. $4x^2 + 24xy - 3y^2 - 60 = 0$

66. $4x^2 - 4xy + y^2 - 10y - 30 = 0$

Applying the Concepts

In Exercises 67–70, rework Example 4 with the given modifications.

67. Find the wind speed along and perpendicular to the initial launch path caused by an east-to-west wind of 15 knots.

68. Find the wind speed along and perpendicular to the initial launch path caused by a west-to-east wind of 18 knots.

69. Find the south-to-north wind speed caused by an 8-knot wind in the positive direction of the y'-axis of the coordinate system determined by the shuttle.

70. Find the south-to-north wind speed caused by an 8-knot wind in the positive direction of the x'-axis of the coordinate system determined by the shuttle.

Beyond the Basics

71. Show that the expression in Step 3 of Example 6 simplifies to $50y'^2 - 25x' - 25 = 0$.

72. Solve the system of equations

$$\begin{cases} x = x'\cos\theta - y'\sin\theta \\ y = x'\sin\theta + y'\cos\theta \end{cases}$$

for x' and y' in terms of x and y. [*Hint:* Multiply the first equation by $\cos\theta$ and the second equation by $\sin\theta$ and use elimination.]

73. Show that the rotation substitutions $x = x'\cos\theta - y'\sin\theta$ and $y = x'\sin\theta + y'\cos\theta$ in the equation $Ax^2 + Bxy + Cy^2 + Dx + Ey + F = 0$ transform the equation to the equation in the form.

$$A'x'^2 + B'x'y' + C'y'^2 + D'x' + E'y' + F' = 0$$

where

$A' = A\cos^2\theta + B\sin\theta\cos\theta + C\sin^2\theta$

$B' = B(\cos^2\theta - \sin^2\theta) - 2(A - C)\sin\theta\cos\theta$

$C' = A\sin^2\theta - B\sin\theta\cos\theta + C\cos^2\theta$

$D' = D\cos\theta + E\sin\theta$

$E' = -D\sin\theta + E\cos\theta$

$F' = F$

74. Show that the quantity $A + C$ is invariant under rotation of axes; that is, $A + C = A' + C'$.

75. Show that the quantity $B^2 - 4AC$ is invariant under rotation of axes; that is, $B^2 - 4AC = B'^2 - 4A'C'$.. [*Hint:* Replace B', A', and C' with the formulas in Exercise 73 and use $\cos^4\theta + \sin^4\theta = 1 - 2\cos^2\theta\sin^2\theta$.]

76. Use the result of Exercise 75 to show that, excluding degenerate cases, the graph of the equation

$$Ax^2 + Bxy + Cy^2 + Dx + Ey + F = 0 \quad \text{is}$$

a. a parabola if $B^2 - 4AC = 0$.

b. an ellipse (or a circle) if $B^2 - 4AC < 0$.

c. a hyperbola if $B^2 - 4AC > 0$.

77. Explain why the equation $Ax^2 + Bxy + Cy^2 + Dx + Ey + F = 0$ cannot represent a circle if $B \neq 0$.

78. Show that the equation $x^2 + y^2 = r^2$ is transformed into the equation $x'^2 + y'^2 = r^2$ by rotation of axes through any angle θ.

79. It is possible to solve the equation

$$2(C - A) \sin\theta \cos\theta + B(\cos^2\theta - \sin^2\theta) = 0$$

for $\tan\theta$ rather than $\cot 2\theta$. Show that if θ is an acute angle,

then $\tan\theta = \dfrac{(C - A) + \sqrt{(C - A)^2 + B^2}}{B}, B \ne 0.$

[*Hint:* If θ is an acute angle, then $\tan\theta > 0$.]

80. Show that the graph of the equation

$$\sqrt{x} + \sqrt{y} = \sqrt{a}, a > 0$$

is part of the graph of a parabola. [*Hint:* Eliminate radicals and use the discriminant.]

81. Show that the distance between two points $P(x, y)$ and $Q(x_1, y_1)$ in the xy-plane equals the distance $P'(x', y')$ and $Q'(x'_1, y'_1)$ in the $x'y'$-plane.

Critical Thinking / Discussion / Writing

82. Use the result of Exercise 73 to find an angle of rotation such that the equation
$Ax^2 + Bxy + Cy^2 + Dx + Ey + F = 0$ becomes
$Cx^2 - Bxy + Ay^2 + Ex - Dy + F = 0.$

83. Find a value for F so that a rotation through an angle of $30°$ transforms the equation $x^2 + \sqrt{3}xy + x - \sqrt{3}y + F = 0$ into the degenerate conic with equation $3x'^2 = (y' + 2)^2.$

Getting Ready for the Next Section

84. Find the midpoint of the segment with endpoints $(1, 3)$ and $(5, -3)$.

85. Convert the given rectangular coordinates to polar coordinates.

  a. $\left(-2, 2\sqrt{3}\right)$ **b.** $\left(-4\sqrt{3}, -4\right)$

86. Convert the given polar coordinates to rectangular coordinates.

  a. $(6, 120°)$ **b.** $(-3, 315°)$

87. Rewrite $r = \dfrac{6}{3 + \sin\theta}$ by dividing both numerator and denominator by 3.

88. Solve $1 - 2\cos\theta = 0$ for all θ with $-\dfrac{\pi}{2} \le \theta \le \dfrac{\pi}{2}$.

89. Convert the equation $r = 3\sec\theta$ to rectangular form and identify its graph.

90. Convert the equation $r = \dfrac{2}{1 - \sin\theta}$ to rectangular form and identify its graph.

SECTION | **8.6**

Polar Equations of Conics

BEFORE STARTING THIS SECTION, REVIEW

1 Definitions of conics (Sections 8.2–8.4)

2 Polar coordinates (Section 6.6)

OBJECTIVES

1 Define conics using the focus-directrix property.

2 Graph the conic from its polar equation.

3 Apply polar equations to orbits of satellites.

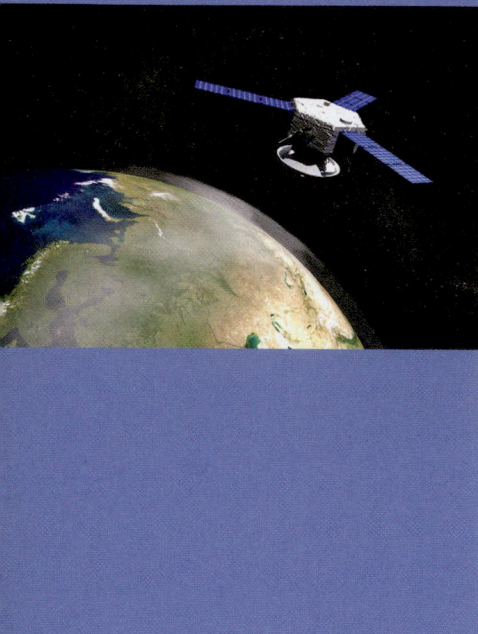

◆ *Sputnik* and the Dawning of the Space Age

On October 4, 1957, the Soviet Union successfully launched the world's first artificial satellite, *Sputnik,* into orbit around Earth. *Sputnik* was about the size of a basketball (diameter 23 inches, weight 184 pounds) and took about 98 minutes to orbit Earth. The *Sputnik* launch occurred during the Cold War, when the Americans and Russians regarded each other as enemies. The United States was shocked at the apparent Soviet superiority in rocket technology. Intense debate about America's standing led to the establishment of the National Aeronautics and Space Administration (NASA). The space race was on! Both the Americans and the Russians raced to develop space-ships to carry humans. On January 31, 1958, the United States successfully launched *Explorer I,* which discovered the radiation belts around Earth. In 1961, President Kennedy challenged the Russians to a race to the moon. In 1969, the United States got there first and the Russians refocused their efforts on a permanent presence in space with a space station.

In Example 6, we will find a polar equation of the elliptical path of *Sputnik.*

---

1 Define conics using the focus-directrix property.

Definition of a Conic

In Sections 8.2–8.4, we defined each conic as the parabola, the ellipse, and the hyperbola individually. We now explore a single property, the *eccentricity*, which allows us to use one definition for all of these curves.

SIDE

NOTE

The symbol e is used both for the eccentricity of a conic and the Euler constant introduced in Chapter 3. The meaning should be clear from the context.

Focus-Directrix Definition of Conics

Let l be a fixed line (called the **directrix**) in the plane and F a fixed point (called the **focus**) not on the line l, and let $e > 0$ be a fixed number. The set of all points P in the plane that satisfy the ratio

$$\frac{d(P, F)}{d(P, l)} = e \quad \text{or} \quad d(P, F) = e \cdot d(P, l)$$

is called a **conic**. The number e in the equation is called the **eccentricity** of the conic. The conic (see Figure 8.37) is

an **ellipse** if $e < 1$.

a **parabola** if $e = 1$.

a **hyperbola** if $e > 1$.

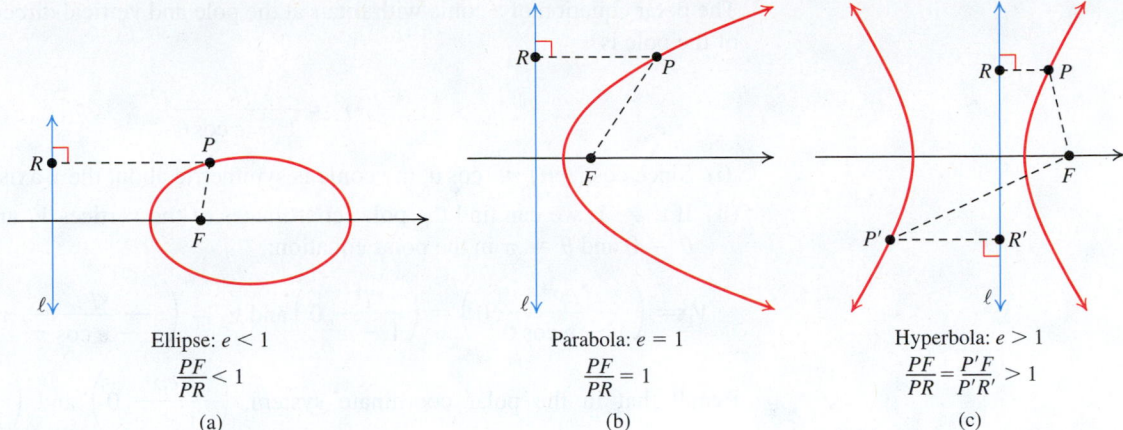

Ellipse: $e < 1$
$$\frac{PF}{PR} < 1$$
(a)

Parabola: $e = 1$
$$\frac{PF}{PR} = 1$$
(b)

Hyperbola: $e > 1$
$$\frac{PF}{PR} = \frac{P'F}{P'R'} > 1$$
(c)

Figure 8.37 Identification of conics by the values of e.

The line through the focus perpendicular to the directrix is called the **principal axis** (or simply the **axis**) of the conic. A point where the conic crosses the axis is called a **vertex**. The parabola has one vertex; the ellipse and hyperbola have two vertices. The point on the axis that is midway between the two vertices of an ellipse or a hyperbola is called the **center**. See Figure 8.38.

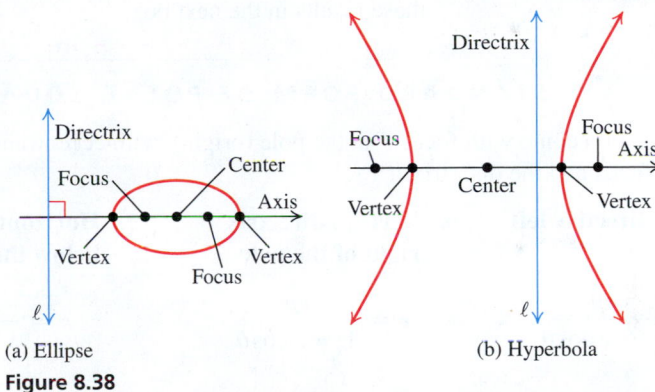

(a) Ellipse

(b) Hyperbola

Figure 8.38

Polar Equation of a Conic

We can find fairly simple polar equations of conics if we place the focus at the pole and let the directrix be vertical or horizontal. For simplicity, we use a mixture of the polar and Cartesian coordinate systems. For example, instead of writing the equation $r \sin \theta = 3$ or $r = 3 \csc \theta$ in polar coordinates, we use the simple form $y = 3$ in Cartesian coordinates.

Consider the case when the directrix is vertical and is p units to the left of the focus F, which is at the pole. The Cartesian equation of the directrix is $x = -p$.

In Figure 8.39, we see that

$$PF = r \text{ and } PR = p + r \cos \theta.$$

A point $P(r, \theta)$ lies on the conic if and only if

$$PF = e \cdot PR \qquad \text{Definition of a conic}$$
$$r = e(p + r \cos \theta) \qquad \text{Substitute for } PF \text{ and } PR.$$
$$r = ep + er \cos \theta \qquad \text{Distributive property}$$
$$r - er \cos \theta = ep \qquad \text{Rearrange}$$
$$r = \frac{ep}{1 - e \cos \theta} \qquad \text{Solve for } r.$$

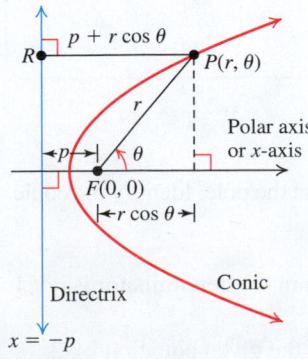

Figure 8.39

The polar equation of a conic with focus at the pole and vertical directrix p units to the left of the pole is

$$r = \frac{ep}{1 - e\cos\theta}.$$

(i) Since $\cos(-\theta) = \cos\theta$, the conic is symmetric about the x-axis.

(ii) If $e \neq 1$, we can find the polar coordinates of the vertices V_1 and V_2 by substituting $\theta = 0$ and $\theta = \pi$ in the polar equation:

$$V_1 = \left(\frac{ep}{1 - e\cos 0}, 0\right) = \left(\frac{ep}{1 - e}, 0\right) \text{ and } V_2 = \left(\frac{ep}{1 - e\cos\pi}, \pi\right) = \left(\frac{ep}{1 + e}, \pi\right)$$

Recall that in the polar coordinate system, $\left(-\frac{ep}{1 + e}, 0\right)$ and $\left(\frac{ep}{1 + e}, \pi\right)$ represent the same point V_2. We can find the polar coordinates of the center by finding the midpoint of $V_1 = \left(\frac{ep}{1 - e}, 0\right)$ and $V_2 = \left(-\frac{ep}{1 + e}, 0\right)$. The center C of the conic is

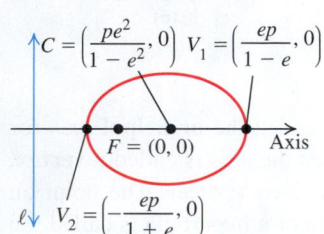

$$C = \left(\cfrac{\cfrac{ep}{1 - e} - \cfrac{ep}{1 + e}}{2}, \frac{0 + 0}{2}\right) = \left(\frac{pe^2}{1 - e^2}, 0\right). \text{ See Figure 8.40.}$$

Figure 8.40 $r = \dfrac{ep}{1 - e\cos\theta}$, $e < 1$.

In a similar manner, we can derive a polar equation of a conic when the vertical directrix is p units to the right of the focus and when the directrix is horizontal. We summarize these results in the next box.

STANDARD FORM OF POLAR EQUATIONS OF CONICS

Polar equation of a conic with focus F at the pole (origin), with eccentricity e, having a distance of p units $(p > 0)$ between the focus and the directrix l:

a. Vertical directrix left of the pole

$$r = \frac{ep}{1 - e\cos\theta}$$

b. Vertical directrix right of the pole

$$r = \frac{ep}{1 + e\cos\theta}$$

c. Horizontal directrix below the pole

$$r = \frac{ep}{1 - e\sin\theta}$$

d. Horizontal directrix above the pole

$$r = \frac{ep}{1 + e\sin\theta}$$

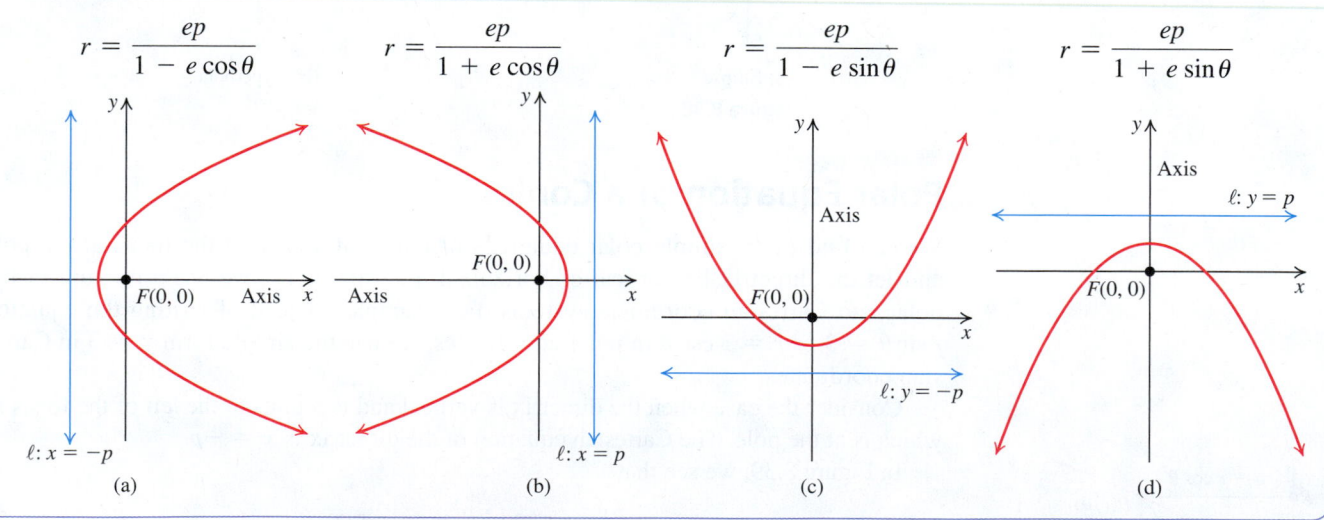

(a)   (b)   (c)   (d)

EXAMPLE 1 Identifying the Conic

The polar equation $r = \dfrac{6}{2 + \cos\theta}$ represents a conic with focus at the pole. Identify the conic.

Solution

The equation is not in standard form because the constant term in the denominator is not 1.

$$r = \frac{3}{1 + \frac{1}{2}\cos\theta} \qquad \text{Divide numerator and denominator of the given equation by 2.}$$

This has the form $r = \dfrac{ep}{1 + e \cos \theta}$ where

$$e = \frac{1}{2}. \qquad \text{\color{blue}Equate the coefficients of } \cos \theta.$$

Because $e = \dfrac{1}{2} < 1$, the conic is an ellipse.

Practice Problem 1 The polar equation $r = \dfrac{3}{3 + 3 \sin \theta}$ represents a conic with focus at the pole. Identify the conic.

EXAMPLE 2 **Writing Polar Equations of Conics**

Write a polar equation of each conic with focus at the pole, eccentricity e, and directrix l. Identify the conic.

a. $e = \dfrac{4}{5}$, l is $x = 3$ **b.** $e = 1$, l is $y = -2$

Solution

a. The vertical directrix $x = 3$ is three units to the right of the focus. We use the following equation:

$$r = \frac{ep}{1 + e \cos \theta} \qquad \text{\color{blue}Standard form}$$

$$= \frac{\left(\dfrac{4}{5}\right)(3)}{1 + \left(\dfrac{4}{5}\right)\cos \theta} \qquad \text{\color{blue}Substitute } e = \dfrac{4}{5} \text{ and } p = 3.$$

$$r = \frac{12}{5 + 4 \cos \theta} \qquad \text{\color{blue}Multiply numerator and denominator by 5 and simplify.}$$

Because $e = \dfrac{4}{5} < 1$, the conic is an ellipse.

b. The horizontal directrix $y = -2$ is two units below the focus. We use the following equation:

$$r = \frac{ep}{1 - e \sin \theta} \qquad \text{\color{blue}Standard form}$$

$$r = \frac{(1)(2)}{1 - (1) \sin \theta} \qquad \text{\color{blue}Substitute } e = 1 \text{ and } p = 2.$$

$$r = \frac{2}{1 - \sin \theta} \qquad \text{\color{blue}Simplify.}$$

Because $e = 1$, the conic is a parabola.

Practice Problem 2 Repeat Example 2 for each conic.

a. $e = \dfrac{3}{4}$, and l is $y = 3$ **b.** $e = \dfrac{3}{2}$, and l is $y = -4$

2 Graph the conic from its polar equation.

Graphing a Conic

The graph of a polar equation of the form

$$r = \frac{ep}{1 \pm e \cos\theta} \qquad \text{Vertical directrix}$$

or

$$r = \frac{ep}{1 \pm e \sin\theta} \qquad \text{Horizontal directrix}$$

is a conic, where e is the eccentricity of the conic and p is the distance between the focus at the pole and the directrix. Remember that the conic is a parabola if $e = 1$, an ellipse if $0 < e < 1$, and a hyperbola if $e > 1$.

PROCEDURE
IN ACTION

EXAMPLE 3   Graphing a Conic from Its Polar Equation

OBJECTIVE

Sketch the graph of a conic given its polar equation.

EXAMPLE

Sketch the graph of the conic

$$r = \frac{5}{3 + 2\cos\theta}.$$

Step 1 Use standard form. Write the equation of the conic in one of the standard forms.

1. The given equation is associated with the standard form

$$r = \frac{ep}{1 + e\cos\theta}$$

$$r = \frac{5}{3 + 2\cos\theta}$$

$$r = \frac{\dfrac{5}{3}}{1 + \dfrac{2}{3}\cos\theta} \qquad \text{Divide numerator and denominator by 3.}$$

Step 2 Find e and p. Use the standard form in Step 1 to find the values of e and p.

2. Comparing $r = \dfrac{\dfrac{5}{3}}{1 + \dfrac{2}{3}\cos\theta}$ with $r = \dfrac{ep}{1 + e\cos\theta}$,

we have $e = \dfrac{2}{3}$ and $ep = \dfrac{5}{3}$.

$$\frac{2}{3}p = \frac{5}{3} \qquad \text{Substitute } e = \frac{2}{3}.$$

$$p = \frac{5}{2} \qquad \text{Solve for } p.$$

Step 3 Identify the conic. Use the value of e from Step 2 to identify the conic. Compare this equation with the standard form to find the directrix.

3. Because $e = \dfrac{2}{3} < 1$, the conic is an ellipse. By comparing our

equation with the standard form $r = \dfrac{ep}{1 + e\cos\theta}$, we see that

the focus of the ellipse is at the pole and the associated directrix

is the vertical line $x = \dfrac{5}{2}\left(\text{or } r\cos\theta = \dfrac{5}{2} \text{ in polar coordinates}\right)$.

Step 4 Find vertices. Locate the vertices by finding the points of intersection of the conic and the principal axis. The principal axis is perpendicular to the directrix. The principal axis is the axis of symmetry for a parabola. It is the major axis for an ellipse and the transverse axis for a hyperbola.

Step 5 Identify symmetry. The conic is symmetric about the principal axis.

Step 6 Sketch the graph. Plot points to sketch half of the conic and use symmetry to sketch the other half.

4. The principal axis is along the x-axis, so $\theta = 0$ and $\theta = \pi$.

When $\theta = 0$, $r = \dfrac{5}{3 + 2\cos 0} = \dfrac{5}{3 + 2} = 1$.

When $\theta = \pi$, $r = \dfrac{5}{3 + 2\cos \pi} = \dfrac{5}{3 - 2} = 5$.

The vertices are at $(1, 0)$ and $(5, \pi)$.

5. The conic is symmetric about the x-axis.

6. Sketch the upper half of the ellipse by plotting points from $\theta = 0$ to $\theta = \pi$. Then use symmetry to sketch the lower half of the ellipse.

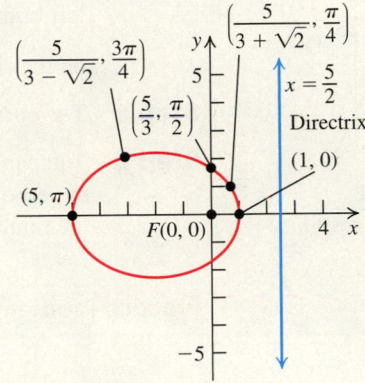

Practice Problem 3 Sketch the graph of the conic $r = \dfrac{5}{3 + 2\sin\theta}$.

EXAMPLE 4   **Graphing a Hyperbola from Its Polar Form**

Sketch the graph of the conic: $r = \dfrac{12}{2 - 4\cos\theta}$

Solution

Step 1 The associated standard form is $r = \dfrac{ep}{1 - e\cos\theta}$, which has a vertical directrix to the left of the focus at the pole.

$r = \dfrac{12}{2 - 4\cos\theta}$ Given equation

$r = \dfrac{6}{1 - 2\cos\theta}$ Divide numerator and denominator by 2.

Step 2 $e = 2$ and $ep = 6$

$2p = 6$ Substitute $e = 2$.

$p = 3$ Solve for p.

Step 3 Because $e = 2 > 1$, the conic is a hyperbola with focus at the pole; because $p = 3$, the directrix is the vertical line $x = -3$.

Step 4 The principal axis is the x-axis. The vertices are as follows:

$$V_1 = \left(\frac{ep}{1 + e}, \pi \right) = (2, \pi) \text{ or } V_1 = (-2, 0)$$

$$V_2 = \left(\frac{ep}{1 - e}, 0 \right) = (-6, 0) \text{ or } V_2 = (6, \pi)$$

The center is at the midpoint of the line segment $\overline{V_2 V_1}$: $(-4, 0)$ or $(4, \pi)$. For the hyperbola $r = \dfrac{6}{1 - 2\cos\theta}$, the denominator $1 - 2\cos\theta$ equals 0 when $\cos\theta = \dfrac{1}{2}$, that is, when $\theta = \pm\dfrac{\pi}{3}$. If $0 \le \theta < \dfrac{\pi}{3}$, then the equation $r = \dfrac{6}{1 - 2\cos\theta}$ gives negative values of r and the points (r, θ) will lie on the left branch of the hyperbola. If $\dfrac{\pi}{3} < \theta \le \pi$, then $r > 0$ and the points will lie on the right branch of the hyperbola.

Step 5 The curve is symmetric about the x-axis.

Step 6 You can plot points for values of θ between 0 and π and sketch a portion of the hyperbola (shown in red in Figure 8.41). Then you can use symmetry to sketch the remaining portion (shown in green in Figure 8.41) of the hyperbola.

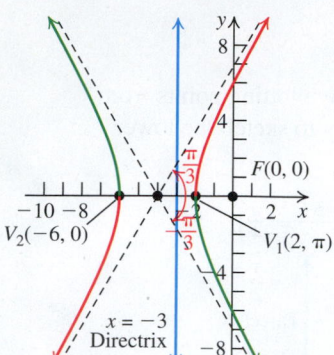

Figure 8.41 $r = \dfrac{12}{2 - 4\cos\theta}$.

Practice Problem 4 Sketch the graph of the conic $r = \dfrac{6}{2 + 2\sin\theta}$.

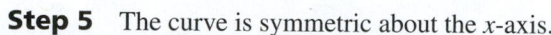

EXAMPLE 5 Finding a Polar Equation of a Conic

Find a polar equation of the hyperbola having a focus at the pole and vertices at $\left(1, \dfrac{\pi}{2} \right)$ and $\left(-3, \dfrac{3\pi}{2} \right)$.

Solution

Because the vertices of the hyperbola lie on the y-axis, the directrix is horizontal. Consider the polar form

$$r = \frac{ep}{1 + e\sin\theta}.$$

We can find its vertices when we substitute $\theta = \dfrac{\pi}{2}$ and $\theta = \dfrac{3\pi}{2}$.

When $\theta = \dfrac{\pi}{2}$, $r = \dfrac{ep}{1 + e}$; so

$$\frac{ep}{1 + e} = 1. \quad (1)$$

When $\theta = \dfrac{3\pi}{2}$, $r = \dfrac{ep}{1 - e}$. Thus,

$$\frac{ep}{1 - e} = -3. \quad (2)$$

Solving equations (1) and (2) for ep and equating them gives

$$1 + e = -3(1 - e) \qquad \text{From (1), } ep = 1 + e; \text{ from (2),}$$
$$ep = -3(1 - e).$$

So

$$e = 2 \qquad \text{Solve for } e.$$

Since $ep = 1 + e$,

$$p = \frac{1 + e}{e} = \frac{1 + 2}{2} = \frac{3}{2} \qquad \text{Solve for } p, \text{ with } e = 2.$$

A polar equation of the hyperbola is therefore

$$r = \frac{ep}{1 + e \sin\theta} = \frac{2\left(\frac{3}{2}\right)}{1 + 2\sin\theta} \qquad \text{Substitute values of } e \text{ and } p.$$

$$r = \frac{3}{1 + 2\sin\theta}$$

Practice Problem 5 Find a polar equation of the hyperbola having a focus at the pole and vertices at $(2, \pi)$ and $(5, \pi)$.

3 Apply polar equations to orbits of satellites.

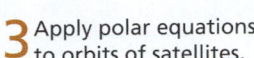

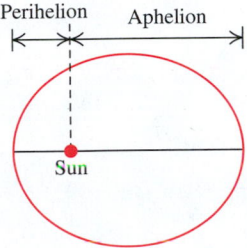

Figure 8.42

Applications

According to Kepler's Law, if an object A moves in the gravitational field of a heavier object B, then the path of the object A must be a conic with a focus at the center of the mass of object B. In elliptical orbits, the closest point to the focus is called **perigee** (or **perihelion**) and the farthest point is called the **apogee** (or **aphelion**). See Figure 8.42.

◆ **EXAMPLE 6** **Polar Equation of *Sputnik I***

Sputnik I had an elliptical orbit around Earth. It reached the maximum and minimum heights of 560 miles and 145 miles above Earth, respectively. Write a polar equation for the path of *Sputnik I* with the pole at the center of Earth. Assume that the radius of Earth is 3960 miles.

Solution

Since the orbit of *Sputnik I* is elliptical, consider the form of its polar equation as

$$r = \frac{ep}{1 - e\cos\theta}, \text{ where } e < 1. \quad (1)$$

The maximum distance of *Sputnik I* from the center of Earth is

$$3960 + 560 = 4520 \text{ miles.}$$

The minimum distance of *Sputnik I* from the center of Earth is

$$3960 + 145 = 4105 \text{ miles.}$$

In equation (1), r has a maximum value when $\theta = 0$; so

$$\frac{ep}{1 - e} = 4520, \quad \text{or} \quad ep = 4520(1 - e)$$

Similarly, r has a minimum value when $\theta = \pi$; so

$$\frac{ep}{1 + e} = 4105, \quad \text{or} \quad ep = 4105(1 + e)$$

Equating the two expressions for ep, we have

$$4520(1 - e) = 4105(1 + e)$$

$$e = \frac{4520 - 4105}{4520 + 4105} \qquad \text{Solve for } e.$$

$$\approx 0.048$$

Because $\dfrac{ep}{1 - e} = 4520$, we have

$$p = \frac{4520(1 - e)}{e} \qquad \text{Solve for } p.$$

$$p = \frac{4520(1 - 0.048)}{0.048} \qquad \text{Replace } e \text{ with } 0.048.$$

$$p \approx 89{,}646.667$$

A polar equation for Sputnik I is

$$r = \frac{ep}{1 - e \cos \theta} = \frac{4303}{1 - 0.048 \cos \theta} \qquad ep \approx 4303$$

Practice Problem 6 The subplanet Pluto has an elliptical orbit around the sun. Its maximum and minimum distances from the center of the sun are 4582.61 million miles and 2749.57 million miles, respectively. Write a polar equation for the path of Pluto with the pole at the center of the sun.

Answers to Practice Problems

1. The conic is a parabola

2. a. $r = \dfrac{9}{4 + 3 \sin \theta}$, ellipse   **b.** $r = \dfrac{12}{2 - 3 \sin \theta}$, hyperbola

3.

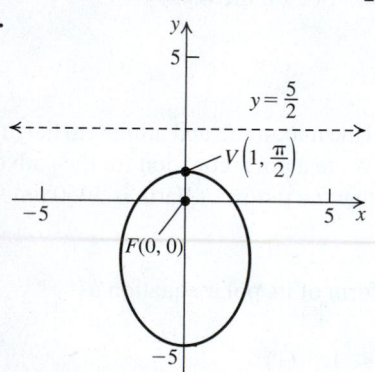

4.

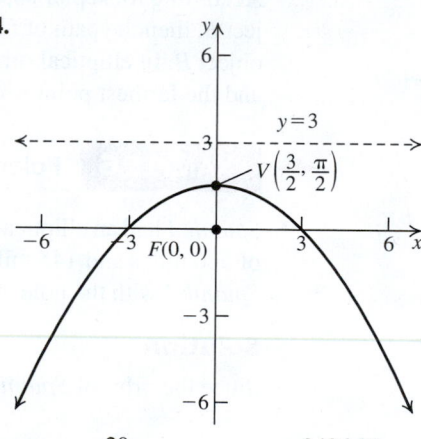

5. $r = \dfrac{20}{3 - 7 \cos \theta}$ **6.** $r = \dfrac{3436.96}{1 - 0.25 \cos \theta}$

SECTION 8.6   Exercises

Concepts and Vocabulary

1. Let p be a point on the graph of a conic with focus F and directrix l. The ratio $\dfrac{d(P, F)}{d(P, l)}$ is called the _____ of the conic. For eccentricity e, if $e < 1$, the conic is a(n) _____; if $e = 1$, the conic is a(n) _____; and if $e > 1$, the conic is a(n) _____.

In Exercises 2 and 3, write the polar form of a conic with focus at the pole, eccentricity e, and directrix p units from the focus.

2. Vertical directrix left of the pole

3. Horizontal directrix above the pole

4. The graph of a conic in the form $r = \dfrac{ep}{1 \pm e \cos \theta}$ is symmetric about the _____-axis. The graph of a conic in the form $r = \dfrac{ep}{1 \pm e \sin \theta}$ is symmetric about the _____-axis.

5. **True or False.** In all the standard forms of the polar equations the focus is at the pole.

6. **True or False.** The graph of the equation $r = \dfrac{3}{1 - \sin \theta}$ is an ellipse.

7. **True or False.** The equation $r = \dfrac{6}{5 + 2\cos\theta}$ is not in standard form because the constant in the denominator is not 1.

8. **True or False.** The directrix for the conic whose polar equation is $r = \dfrac{3}{1 - \sin\theta}$ is left of the pole.

Building Skills

In Exercises 9–16, each polar equation represents a conic with a focus at the pole. Identify the conic.

9. $r = \dfrac{4}{1 - \cos\theta}$ 10. $r = \dfrac{5}{4 + 6\cos\theta}$

11. $r = \dfrac{5}{4 - \sin\theta}$ 12. $r = \dfrac{1}{1 + \sin\theta}$

13. $r = \dfrac{3}{2 + 3\sin\theta}$ 14. $r = \dfrac{2}{3 + \sin\theta}$

15. $r = \dfrac{4}{1 + \sin\theta}$ 16. $r = \dfrac{4}{1 - \cos\theta}$

In Exercises 17–28, identify the conic and sketch its graph.

17. $r = \dfrac{2}{1 - \cos\theta}$ 18. $r = \dfrac{4}{1 + \cos\theta}$

19. $r = \dfrac{7}{3 + 4\cos\theta}$ 20. $r = \dfrac{9}{4 + 5\cos\theta}$

21. $r = \dfrac{5}{3 + 2\cos\theta}$ 22. $r = \dfrac{5}{3 - 2\cos\theta}$

23. $r = \dfrac{6}{2 + \sin\theta}$ 24. $r = \dfrac{6}{2 - \sin\theta}$

25. $r = \dfrac{6}{1 - 2\sin\theta}$ 26. $r = \dfrac{6}{1 + 2\sin\theta}$

27. $r = \dfrac{9}{7 + 2\sin\theta}$ 28. $r = \dfrac{8}{5 - 3\sin\theta}$

In Exercises 29–40, find a polar equation of the conic that has a focus at the pole and that satisfies the given conditions.

29. Parabola; vertex $\left(6, \dfrac{3\pi}{2}\right)$

30. Parabola; vertex $\left(4, \dfrac{\pi}{2}\right)$

31. Parabola; vertex $(3, 0)$

32. Parabola; vertex $(4, \pi)$

33. Ellipse; vertices $(4, 0)$ and $(6, \pi)$

34. Ellipse; vertices $(2, 0)$ and $(10, \pi)$

35. Ellipse; vertices $\left(3, \dfrac{\pi}{2}\right)$ and $\left(5, \dfrac{3\pi}{2}\right)$

36. Ellipse; vertices $\left(2, \dfrac{\pi}{2}\right)$ and $\left(6, \dfrac{3\pi}{2}\right)$

37. Hyperbola; vertices $(2, 0)$ and $(4, 0)$

38. Hyperbola; vertices $(3, \pi)$ and $(7, \pi)$

39. Hyperbola; vertices $\left(2, \dfrac{\pi}{2}\right)$ and $\left(6, \dfrac{\pi}{2}\right)$

40. Hyperbola; vertices $\left(1, \dfrac{3\pi}{2}\right)$ and $\left(5, \dfrac{3\pi}{2}\right)$

Applying the Concepts

Planetary motion. The planets travel in elliptical orbits with the sun at one focus. The table below lists the *perihelion* (the smallest distance from the planet to the sun) and the *aphelion* (the largest distance from the planet to the sun) for each of the planets. The distances given are in millions of miles. In Exercises 41–44, let the sun be the pole and the major axis be along the polar axis.

Planet	Perihelion	Aphelion
Earth	91.38	94.54
Mars	128.49	154.83
Mercury	28.56	43.88
Saturn	837.05	936.37

41. **Earth.** Find the polar equation for the orbit of Earth.

42. **Mars.** Find the polar equation for the orbit of Mars.

43. **Mercury.** Find the polar equation for the orbit of Mercury.

44. **Saturn.** Find the polar equation for the orbit of Saturn.

45. **Comet motion.** A comet travels in a parabolic orbit around the sun at the focus. When the comet is 60 million kilometers from the sun, the line segment from the sun to the comet makes an angle of $\dfrac{\pi}{3}$ radians with the axis of the orbit.

(a) Find a polar equation of the comet's orbit. (b) How close does the comet come to the sun?

46. **Comet motion.** A comet travels in a hyperbolic orbit around the sun at the focus. When the comet is 30 and 150 million miles from the sun, the line segment from the sun to the comet makes angles of $\dfrac{\pi}{3}$ and $\dfrac{4\pi}{3}$ radians, respectively, with the axis of the orbit.

a. Find a polar equation of the comet's orbit.
b. How close does the comet come to the sun?

Beyond the Basics

47. Graph and interpret the conic with equation
$$r = \dfrac{6}{2 - \sin\left(\theta - \dfrac{\pi}{3}\right)}.$$

48. Show that the equation $r = 2\sec^2\dfrac{\theta}{2}$ is a polar equation of a parabola. Find an equation of the directrix.

49. Show that the equation $r = 2\csc^2\dfrac{\theta}{2}$ is a polar equation of a parabola. Find an equation of the directrix.

50. A **focal chord** of a conic is a chord that passes through one focus of the conic. (See the figure.) Let the focus F divide the chord into two segments of lengths r_1 and r_2. Show that for a fixed ellipse or a parabola, $\dfrac{1}{r_1} + \dfrac{1}{r_2}$ is a constant.

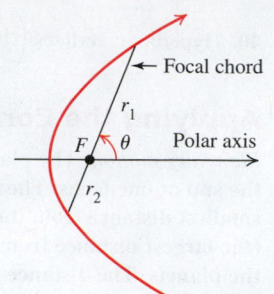

Focal chord

r_1

F θ Polar axis

r_2

51. Show that for a hyperbola, the result in Exercise 50 is false unless we stay on one branch. $\Big[$ *Hint:* Use the hyperbola $r = \dfrac{15}{1 - 4\cos\theta}$ and show what happens when $\theta = 0$ and when $\theta = \dfrac{\pi}{3}.\Big]$

52. Consider the polar equation $r = \dfrac{ep}{1 - e\cos\theta}$ of an ellipse ($e < 1$). Show that in the figure, $c = ae$.

$\Big[$ *Hint:* Write V_1, V_2, and C in the form $(r_1, 0)$, $(r_2, 0)$, and $(r_3, 0)$, respectively; then $r_3 = \dfrac{r_1 + r_2}{2}$ and $a = r_1 - r_3 = \dfrac{r_1 - r_2}{2}.\Big]$

y

Directrix l

Center

V_2 F C V_1 x

$\leftarrow p \rightarrow \leftarrow a \rightarrow$

c

$x = -p$

53. Show that the polar equation of the ellipse in Exercise 52 can be expressed in the form $r = \dfrac{a(1 - e^2)}{1 - e\cos\theta}$.

54. Use Exercise 52 to write a Cartesian equation of the ellipse $r = \dfrac{3.2}{1 - 0.6\cos\theta}$.

$\Big[$ *Hint:* Show $a = 5$ and $c = 3$. Find $b = \sqrt{a^2 - c^2}$. Then write the equation in the form $\dfrac{(x - c)^2}{a^2} + \dfrac{y^2}{b^2} = 1.\Big]$

Critical Thinking / Discussion / Writing

55. True or False. The graph of $r = \dfrac{ep}{1 - e\sin\theta}$ can be obtained by rotating the graph of $r = \dfrac{ep}{1 - e\cos\theta}$ counterclockwise through an angle of $\dfrac{\pi}{2}$ radians.

56. True or False. The graph of $r = -\dfrac{4}{2 - \cos\theta}$ is the same as the graph of $r = \dfrac{4}{2 + \cos\theta}$.

Getting Ready for the Next Section

57. Find $\sin\theta$ and $\cos\theta$ from the triangle

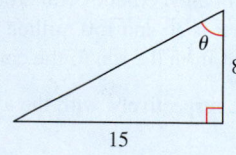

θ

8

15

58. Graph $y = 2x^2 + 1$, $-2 \le x \le 1$.

59. Find the maximum value of $f(x) = -3x^2 + 6x$.

60. Find the minimum value of $f(x) = x^2 - 6x + 7$.

61. Find the amplitude and period of $y = 2\cos 4x$.

62. Find the amplitude and period of $y = -3\sin 2x$.

63. Graph the equation $\dfrac{x^2}{4} + \dfrac{y^2}{9} = 1$.

64. Graph the equation $\dfrac{y^2}{16} - \dfrac{x^2}{4} = 1$.

Parametric Equations

BEFORE STARTING THIS SECTION, REVIEW

1. Equations of conics (Sections 8.2–8.4)

2. Fundamental trigonometric identities (Section 5.1, page 438)

3. Radian measure and arc length (Section 4.1, page 341)

4. Conversion of polar equations to Cartesian equations (Section 6.6, page 563)

OBJECTIVES

1. Graph plane curves described by parametric equations.

2. Eliminate the parameter.

3. Find parametric equations of a curve.

◆ The Brachistochrone Problem

Brachistochrone is a combination of two Greek words, *brachistos* meaning "shortest" and *chronos* meaning "time." In 1692, Johann Bernoulli posed the *brachistochrone problem* as a challenge to other mathematicians: Find the shape of a wire (or slide) along which a bead might slide in the least amount of time if it starts from a point *A* with zero speed and ends at a point *B* (not directly below *A*) under the influence of gravity and ignoring friction. The problem was solved by Isaac Newton, Jakob Bernoulli (Johann's brother), Gottfried Leibniz, Guillaume de L' Hôpital, and Johann Bernoulli. Its solution was instrumental in the development of a branch of mathematics known as the *calculus of variations*.

You might think that the wire would form a straight line because a straight line is the shortest distance between the two points *A* and *B*. However, the correct shape is half of one arch of an inverted *cycloid*, shown in Figure 8.43. The brachistochrone curve (or the curve of fastest descent) is the cycloid investigated in Example 5.

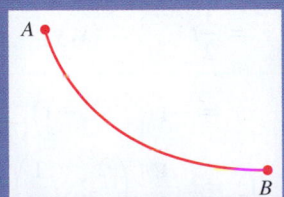

Figure 8.43

1 Graph plane curves described by parametric equations.

Parametric Equations

So far, we have used **Cartesian** or **rectangular equations** such as $2x^2 + 3y^2 = 6$ and $y^2 = 8x$ to represent curves in the plane. We have also represented curves using **polar equations** such as $r = 2\sin\theta$ and $r = 1 - \cos\theta$. When describing the path of a particle moving in the plane, it is sometimes more helpful and informative to express each of the x- and y-coordinates of the path as a function of a third variable, called a **parameter**.

Parametric Equations of a Plane Curve

Suppose $f(t)$ and $g(t)$ are functions defined on an interval I. A **plane curve** C is the set of points (x, y) where

$$x = f(t) \text{ and } y = g(t).$$

The equations $x = f(t)$ and $y = g(t)$ are called the **parametric equations** for the curve C, and t is called the **parameter** for C.

If I is the closed interval $[a, b]$, then $a \le t \le b$. The point $(f(a), g(a))$ is the **initial point** of the curve C, and the point $(f(b), g(b))$ is the **terminal point** of C. In applications involving the motion of a particle, the parameter t often represents time, but we also use other variables, such as θ (for an angle) and s (for distance), as parameters.

Graphing a Plane Curve

To sketch the graph of a plane curve C represented by the parametric equations $x = f(t)$ and $y = g(t)$, you plot points in the xy-plane corresponding to successive values of t. You then connect these points with a smooth curve. The path of a particle moving along the curve C with increasing values of t describes the **direction of increasing parameter** or the **orientation** of the curve C.

PROCEDURE IN ACTION

EXAMPLE 1 Graphing a Plane Curve

OBJECTIVE

Sketch the graph of a plane curve represented by parametric equations.

EXAMPLE

Sketch the graph represented by the parametric equations

$$x = t^2 - 3, \quad y = \frac{1}{2}t, \text{ where } -2 \le t \le 3.$$

Step 1 Select values of t. Select some values of t (in increasing order) in the given interval.

1. Select integer values $t = -2, -1, 0, 1, 2,$ and 3 for t in the interval $[-2, 3]$.

Step 2 Make a table. For each value of t in Step 1, use the given parametric equations to compute the x and y coordinates of the points (x, y) in the plane.

2.

t	$x = t^2 - 3$	$y = \dfrac{1}{2}t$	(x, y)
-2	$(-2)^2 - 3 = 1$	$\dfrac{1}{2}(-2) = -1$	$(1, -1)$
-1	$(-1)^2 - 3 = -2$	$\dfrac{1}{2}(-1) = -\dfrac{1}{2}$	$\left(-2, -\dfrac{1}{2}\right)$
0	$0^2 - 3 = -3$	$\dfrac{1}{2}(0) = 0$	$(-3, 0)$
1	$1^2 - 3 = -2$	$\dfrac{1}{2}(1) = \dfrac{1}{2}$	$\left(-2, \dfrac{1}{2}\right)$
2	$2^2 - 3 = 1$	$\dfrac{1}{2}(2) = 1$	$(1, 1)$
3	$3^2 - 3 = 6$	$\dfrac{1}{2}(3) = \dfrac{3}{2}$	$\left(6, \dfrac{3}{2}\right)$

Step 3 Sketch the curve. Plot the points (x, y) from Step 2 in an xy-plane. Connect the points by the order of increasing values of t with a smooth curve. Show the orientation by placing arrowheads on the curve.

3.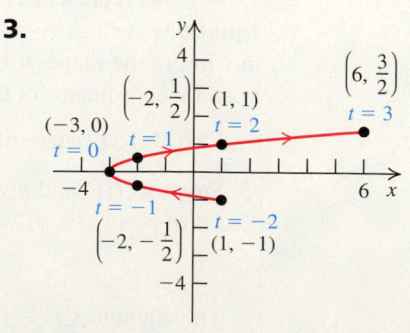

Practice Problem 1 Sketch the graph of $x = 2t$, $y = t^2 - 3$, $-2 \le t \le 2$.

2 Eliminate the parameter.

Eliminating the Parameter

In Example 1, we sketched the graph of a curve represented by parametric equations by plotting points. Sometimes we can identify or sketch the curve by first finding its Cartesian representation. We do this by eliminating the parameter in the given parametric equations. If the parametric equations involve trigonometric functions, we often can use trigonometric identities (see Example 3) to eliminate the parameter.

PROCEDURE
IN ACTION

EXAMPLE 2 Eliminating the Parameter

OBJECTIVE

Convert parametric equations with parameter t to a Cartesian equation by eliminating the parameter. Sketch the graph and indicate its orientation.

Step 1 Solve for t. Solve one of the parametric equations for the parameter.

Step 2 Substitute. Substitute the expression for the parameter obtained in Step 1 into the other parametric equation.

Step 3 Simplify. Simplify the resulting Cartesian equation.

Step 4 Graph. Sketch the graph of the Cartesian equation in Step 3. Sketch the portion of the graph for the given values of the parameter and indicate its orientation.

EXAMPLE

Identify parametric curve

$$x = \sqrt{t+1} \quad y = \sqrt{t}, \text{ where } t \geq 0,$$

by converting it to a Cartesian equation. Sketch its graph and show the orientation.

1. $y = \sqrt{t}$ A parametric equation
 $y^2 = t$ Square both sides.

2. $x = \sqrt{t+1}$ The other parametric equation
 $x = \sqrt{y^2 + 1}$ Substitute y^2 for t from Step 1.

3. $x^2 = y^2 + 1$ Square both sides.
 $x^2 - y^2 = 1$ Subtract y^2 from both sides.

4. The graph of $x^2 - y^2 = 1$ is a horizontal hyperbola. Because $t \geq 0$, we have $x = \sqrt{t+1} \geq 1, y = \sqrt{t} \geq 0$. The curve starts at the initial point $(1, 0)$ when $t = 0$ and rises along the hyperbola into the first quadrant as t increases. The parameter interval is $[0, \infty)$, so there is no terminal point. The graph, along with its orientation, is shown by a solid curve in the figure.

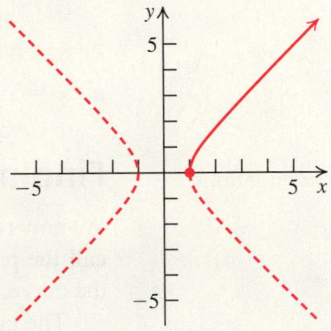

Practice Problem 2 Identify parametric curve $x = t + 1, y = 3 + t^2, t \leq 0$ by eliminating the parameter and indicate the orientation of the graph.

EXAMPLE 3 **Eliminating the Parameter**

Identify the parametric curve and indicate its orientation.

a. $x = \sin t, y = \cos^2 t, 0 \le t \le \dfrac{\pi}{2}$ **b.** $x = \sin t, y = \cos^2 t, t \ge 0$

SIDE
NOTE

Be careful that the process of elimination does not include more (or fewer) points than those given by the parametric equations.

Solution

$$y = \cos^2 t$$
$$= 1 - \sin^2 t \qquad \text{Pythagorean identity}$$
$$= 1 - x^2 \qquad \sin t = x$$

The graph of the Cartesian equation $y = 1 - x^2$ is shown in Figure 8.44(a).

a. When $t = 0$, $x = \sin 0 = 0$ and $y = \cos^2 0 = 1$. When $t = \dfrac{\pi}{2}$, $x = \sin \dfrac{\pi}{2} = 1$ and $y = \cos^2 \dfrac{\pi}{2} = 0$. As t ranges from 0 to $\dfrac{\pi}{2}$, x values increase from 0 to 1, while y values decrease from 1 to 0. Thus, the parametric equations describe the parabolic arc $y = 1 - x^2$ with initial point $(0, 1)$ and terminal point $(1, 0)$. See Figure 8.44(b) for the oriented graph.

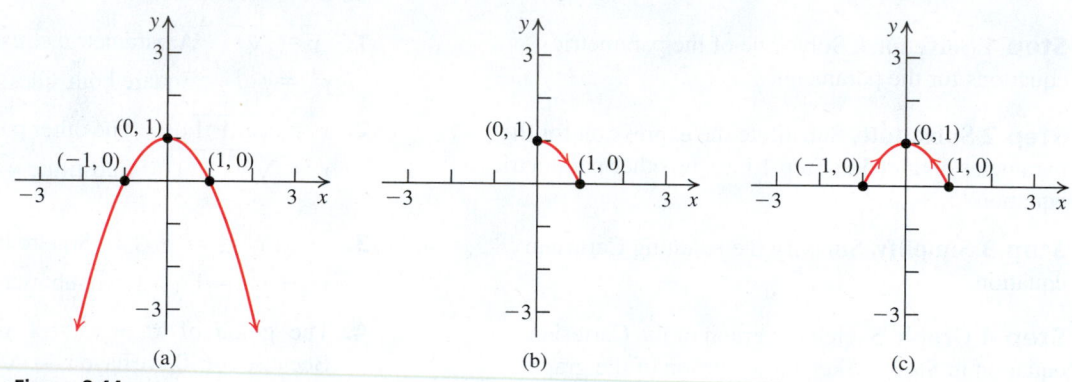

(a)   (b)   (c)

Figure 8.44

b. Note that for any value of t, we have $-1 \le x \le 1$ and $0 \le y \le 1$. Thus, as t ranges over the interval $[0, \infty)$, the parabolic arc $y = 1 - x^2$ shown in Figure 8.44(c) is traced back and forth an infinite number of times. The initial point is $(0, 1)$, and there is no terminal point.

Practice Problem 3 Identify the parametric curve and indicate its orientation.

a. $x = \cos t, y = \sin t, 0 \le t \le \pi$ **b.** $x = \cos t, y = \sin t, t \ge 0$

3 Find parametric equations of a curve.

Finding Parametric Equations

We now consider the reverse problem of finding parametric equations for a given curve. We call the parametric equations together with a parameter interval I a **parameterization** of the curve. A given curve can be parameterized in infinitely many ways.

The easiest way to write the graph of any Cartesian equation $y = f(x)$ in parametric form is $x = t, y = f(t)$, with t in the domain of f. For example, a parameterization of the parabola $y = x^2 + 1$ is

$$x = t, \qquad y = t^2 + 1$$

with $I = (-\infty, \infty)$.

You can convert any polar equation $r = g(\theta)$ to a parametric form as follows:

First, convert the polar coordinates to Cartesian coordinates (see page 563):

$$x = r\cos\theta, \qquad y = r\sin\theta$$
$$x = g(\theta)\cos\theta \qquad y = g(\theta)\sin\theta \qquad \color{blue}{\text{Replace } r \text{ with } g(\theta).}$$

The equations $x = g(\theta)\cos\theta, y = g(\theta)\sin\theta$ give the parametric equations of $r = g(\theta)$ with parameter θ. For example, a parameterization of the polar equation $r = 1 + \cos\theta$ of a cardioid is

$$x = (1 + \cos\theta)\cos\theta$$
$$y = (1 + \cos\theta)\sin\theta$$

with $I = [0, 2\pi]$.

The next example illustrates that the parametric equations of a curve are *not unique*.

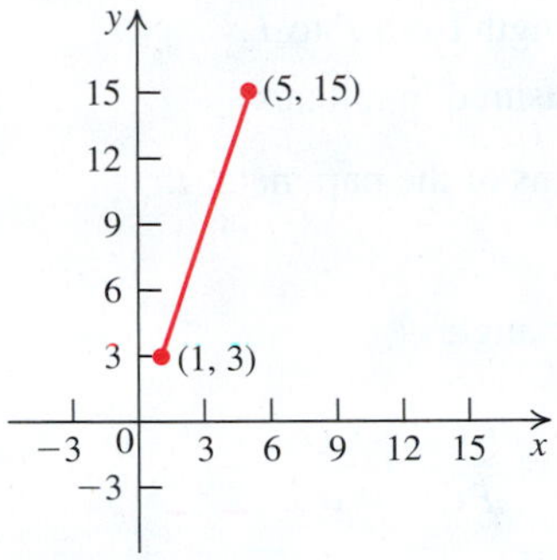

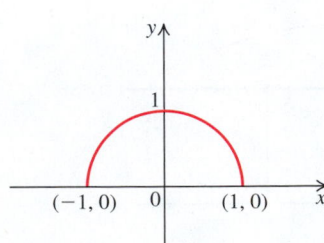

Figure 8.45

EXAMPLE 4   Parameterizing a Line Segment

The equation $y = 3x, 1 \le x \le 5$ represents the line segment joining the points $(1, 3)$ and $(5, 15)$. See Figure 8.45. Find a set of parametric equations and interpret the motion of a particle moving along this segment by using the following parameters.

a. $x = t$ **b.** $x = 1 + 4t$ **c.** $x = 5 - t$

Solution

a. With $x = t$, we obtain the following parameterization of the line segment $y = 3x$, $1 \le x \le 5$:

$$x = t, y = 3t, 1 \le t \le 5$$

At time $t = 1$, the particle is at $(1, 3)$. It travels up the line segment and reaches the point $(5, 15)$ in $t = 5$ time units.

b. If $x = 1 + 4t$, then $y = 3x, 1 \le x \le 5$ becomes

$$y = 3(1 + 4t), 1 \le 1 + 4t \le 5 \qquad \color{blue}{\text{Replace } x \text{ with } 1 + 4t.}$$
$$y = 3 + 12t, 0 \le t \le 1 \qquad \color{blue}{\text{Solve } 1 \le 1 + 4t \le 5 \text{ for } t.}$$

The parametric equations are: $x = 1 + 4t, y = 3 + 12t; 0 \le t \le 1$. At time $t = 0$, the particle is at $(1, 3)$. It travels the same line segment upward and reaches the point $(5, 15)$ in $t = 1$ time unit.

c. If $x = 5 - t$, then $y = 3x, 1 \le x \le 5$ becomes

$$y = 3(5 - t), 1 \le 5 - t \le 5 \qquad \color{blue}{\text{Replace } x \text{ with } 5 - t.}$$
$$= 15 - 3t, -4 \le -t \le 0 \qquad \color{blue}{\text{Subtract 5 from all parts of } 1 \le 5 - t \le 5.}$$
$$= 15 - 3t, 4 \ge t \ge 0 \qquad \color{blue}{\text{Multiply the inequality } -4 \le -t \le 0 \text{ by } -1 \text{ and reverse the inequality signs.}}$$
$$y = 15 - 3t, 0 \le t \le 4 \qquad \color{blue}{\text{Rewrite.}}$$

The parametric equations are: $x = 5 - t, y = 15 - 3t; 0 \le t \le 4$. At time $t = 0$, the particle is at $(5, 15)$. It travels the same line segment (in the opposite direction) and reaches the point $(1, 3)$ in $t = 4$ time units.

Practice Problem 4 The graph of the equation $y = \sqrt{1 - x^2}, -1 \le x \le 1$ is the upper semicircle of radius 1 shown in Figure 8.46. Find a set of parametric equations and interpret the motion of a particle moving along this arc using the following parameters.

a. $x = \cos t$ **b.** $x = \cos 2t$ **c.** $x = -\cos t$

Figure 8.46

Remark. Example 4 shows the advantages of parametric equations over Cartesian equations in describing the motion of a particle. A Cartesian equation only indicates the path on which a particle is traveling. However, parametric equations of a curve give not only the path, but also information about the location of the particle at any specific time as well as the direction in which the particle is traveling.

◆ **EXAMPLE 5** **Parametric Equations of a Cycloid**

If a wheel (circle) of radius a rolls along a straight line, then a point P on the rim of the wheel traces a curve called a **cycloid**. Find parametric equations for a cycloid.

Solution

Suppose the wheel rolls on the x-axis and the point P touches the x-axis at the origin. We let $t = 0$ when P is at the origin. Let $P(x, y)$ denote the location of the point P after the wheel rolls through an angle of t radians. In Figure 8.47, the measure of $\angle PCT$ is t radians.

When the wheel in Figure 8.47 rolls the distance OT, the radial line from the center C to the point P rotates through angle t. We treat t as a positive angle. So

$$OT = \overset{\frown}{PT} \qquad \text{The distance } OT \text{ equals the arc length from } P \text{ to } T.$$
$$= at \qquad \text{arc length} = (\text{radius})(\text{angle measured in radians})$$

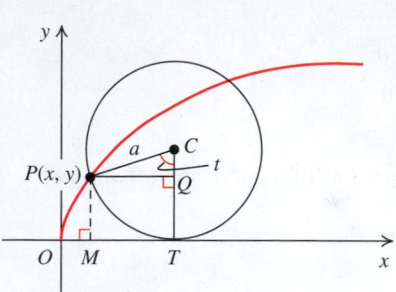

Figure 8.47 Graph of a cycloid.

We now compute the coordinates of the point $P(x, y)$ in terms of the parameter t.

$$
\begin{aligned}
y &= PM \\
&= QT && \text{Opposite sides of a rectangle} \\
&= CT - CQ && \text{Figure 8.47} \\
&= a - a\cos t && \text{In triangle } PCQ, \cos t = \frac{CQ}{PC} = \frac{CQ}{a}. \\
&= a(1 - \cos t) && \text{Factor out } a. \\
x &= OM \\
&= OT - MT && \text{Figure 8.47} \\
&= at - PQ && OT = at, MT = PQ \\
&= at - a\sin t && \text{In triangle } PCQ, \sin t = \frac{PQ}{PC} = \frac{PQ}{a}. \\
&= a(t - \sin t)
\end{aligned}
$$

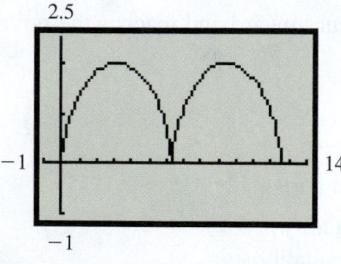
Thus, the parametric equations of a cycloid generated by a circle of radius a are given by

$$x = a(t - \sin t), y = a(1 - \cos t), -\infty < t < \infty$$

Practice Problem 5 A wheel of radius a rolls along a horizontal straight line; then a point P that is b units from the center with $b < a$ traces a curve called a **curtate cycloid**. See Figure 8.48. Find parametric equations for a curtate cycloid.

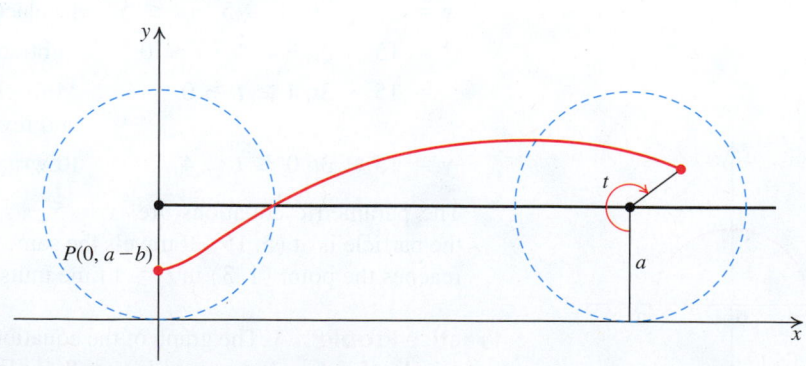

Figure 8.48 Graph of a curtate cycloid.

Answers to Practice Problems

1.

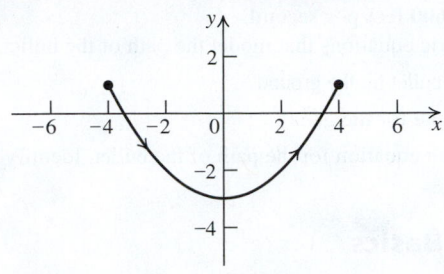

2. The graph is $y = (x - 1)^2 + 3$ starting at the vertex $(1, 3)$ and continuing to the left for $x < 1$.

3. a. A semicircle centered at the origin, starting at $(1, 0)$ and continuing counterclockwise to $(-1, 0)$
b. Circles of radius 1 around the origin, starting at $(1, 0)$ and continuing counterclockwise for infinitely many rotations
4. a. $x = \cos t, y = \sin t$; the particle moves counterclockwise starting at $t = 0$ and finishing at $t = \pi$.
b. $x = \cos 2t, y = \sin 2t$; the particle moves counterclockwise starting at $t = 0$ and finishing at $t = \frac{\pi}{2}$.
c. $x = -\cos t, y = \sin t$; the particle moves clockwise starting at $t = 0$ and finishing at $t = \pi$.
5. $x = at - b \sin t, y = a - b \cos t$

SECTION 8.7 Exercises

Concepts and Vocabulary

In Exercises 1–8, let f and g be functions of t with $a \leq t \leq b$.

1. The set of points in the plane defined by the ordered pairs (x, y), where $x = f(t)$ and $y = g(t)$, is called a(n) _____.

2. The equations $x = f(t), y = g(t), a \leq t \leq b$ are called the _____ for the curve.

3. The point $(f(a), g(a))$ is called the _____.

4. The point $(f(b), g(b))$ is called the _____.

5. True or False. The terminal point for the plane curve represented by the parametric equations: $x = 3 \cos t$, $y = 2 \sin t, 0 \leq t \leq \pi$ is $(3,0)$.

6. True or False. The path along a parameterized curve C with increasing values of t is called the orientation of the curve.

7. True or False. Every plane curve has a unique parameterization.

8. True or False. When you eliminate the parameter in a set of parametric equations, you obtain a Cartesian equation.

Building Skills

In Exercises 9–12, sketch the graph of each curve by plotting points. Indicate the points on the curve that correspond to the integer values of t in the given interval.

9. $x = t - 1, y = 2t - 1, -2 \leq t \leq 2$

10. $x = t + 1, y = 3t + 2, -2 \leq t \leq 3$

11. $x = \frac{1}{2}t, y = t^2 + 1, -2 \leq t \leq 3$

12. $x = 2t^2 - 1, y = 2t, -2 \leq t \leq 2$

In Exercises 13–16, sketch the graph of each curve by plotting points. Indicate the points on the curve that correspond to

$\theta = 0, \frac{\pi}{4}, \frac{\pi}{2}, \frac{3\pi}{4}, \pi.$

13. $x = \cos \theta, y = \sin \theta, 0 \leq \theta \leq \pi$

14. $x = \sin \theta, y = \cos \theta, 0 \leq \theta \leq \pi$

15. $x = 2 \cos \theta, y = \sin \theta, 0 \leq \theta \leq \pi$

16. $x = 2 \sin \theta, y = 3 \cos \theta, 0 \leq \theta \leq \pi$

In Exercises 17–28, eliminate the parameter to obtain a Cartesian equation of the curve. Graph the curve and indicate its orientation.

17. $x = t, y = 2t, -\infty < t < \infty$

18. $x = 3t, y = t, -\infty < t < \infty$

19. $x = 2t, y = 6t, 0 \leq t \leq 4$

20. $x = 2t + 1, y = -3t + \frac{3}{2}, -1 \leq t \leq \frac{1}{2}$

21. $x = \sqrt{t}, y = 3 - t, 0 \leq t \leq 4$

22. $x = t - 2, y = \sqrt{t}, 0 \leq t \leq 6$

23. $x = \sqrt{t}, y = \sqrt{4 - t}, 0 \leq t \leq 4$

24. $x = \sqrt{9 - t}, y = -\sqrt{t}, 0 \leq t \leq 9$

25. $x = 2\sqrt{2 - t}, y = 3\sqrt{t - 1}, 1 \leq t \leq 2$

26. $x = 3\sqrt{t - 2}, y = 4\sqrt{3 - t}, 2 \leq t \leq 3$

27. $x = t + 2, y = \sqrt{t^2 + 4t}, t \geq 0$

28. $x = 3\sqrt{t - 1}, y = 2\sqrt{t}, t \geq 1$

In Exercises 29–44, eliminate the parameter to obtain a Cartesian equation of the curve. Graph the curve and indicate its orientation.

29. $x = \sin t, y = \cos t, 0 \leq t \leq 2\pi$

30. $x = \cos t, y = \sin t, 0 \leq t \leq 2\pi$

31. $x = \cos^2 t, y = \sin^2 t, 0 \leq t \leq \frac{\pi}{2}$

32. $x = \sin^2 t, y = \cos^2 t, 0 \leq t \leq \pi$

33. $x = 3 \cos t, y = 2 \sin t, 0 \leq t \leq 2\pi$

34. $x = 3 \sin t, y = 2 \cos t, 0 \leq t \leq 4\pi$

35. $x = \cos \theta, y = \cos^2 2\theta, -\infty < \theta < \infty$

36. $x = \sin \theta, y = \cos^2 2\theta, -\infty < \theta < \infty$

37. $x = 1 - \cos t, y = 2 - \sin t, 0 \leq t \leq 2\pi$

38. $x = 3 - 2 \cos t, y = -1 + 5 \sin t, 0 \le t \le 2\pi$

39. $x = 2 \tan t - 1, y = 3 \sec t + 2, 0 \le t < \dfrac{\pi}{2}$

40. $x = 3 \csc t + 2, y = 5 \cot t - 1, 0 < t < \pi$

41. $x = e^t, y = e^{2t}, t \ge 0$ **42.** $x = 4e^{2t}, y = -e^t, t \ge 0$

43. $x = t^2, y = 2 \ln t, t \ge 1$ **44.** $x = 5 \ln t, y = t^5, t \ge 1$

In Exercises 45–48, consider the Cartesian equation
$y = 3x + 2$ **of a line. Find a set of parametric equations and interpret the motion of a particle moving along this line by using the parameter** t **with** $-\infty < t < \infty$.

45. $x = t$ **46.** $x = \sin t$

47. $x = e^t$ **48.** $x = e^{-t}$

In Exercises 49–52, consider the Cartesian equation
$y = x^2 - 4$ **of a parabola. Find a set of parametric equations and interpret the motion of a particle moving along this parabola by using the parameter** t **with** $-\infty < t < \infty$.

49. $x = -t$ **50.** $x = t^2$

51. $x = \cos t$ **52.** $x = e^t$

In Exercises 53–56, consider the Cartesian equation $x = \sqrt{1 - y^2}, -1 \le y \le 1$ **of a right semicircle of radius 1. Find a set of parametric equations and interpret the motion of a particle moving along this semicircle by using the following parameters.**

53. $x = \sin t$ **54.** $x = \sin 2t$

55. $x = -\sin t$ **56.** $x = -\sin 4t$

Applying the Concepts

Projectile motion. Suppose an object is launched upward with an initial speed v_0 **at an angle of** θ **with the horizontal from a height** h **above the ground. The resulting motion is called** *projectile motion*. **It can be shown that the parametric equations of the path of a projectile are**

$$x = (v_0 \cos \theta)t, \quad y = (v_0 \sin \theta)t - \frac{1}{2}gt^2 + h,$$

where t **is the time and** g **is the constant acceleration due to gravity (**$g \approx 32 \text{ ft/sec}^2$ **or** 9.8 m/sec^2**).**

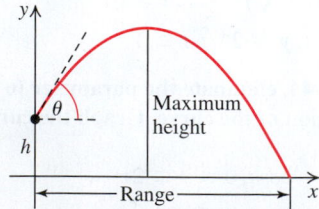

In Exercises 57–60, a golf ball is hit at an angle of 30° with an initial speed of 64 feet per second (Here $h = 0$**.)**

57. Write parametric equations that model the path of the golf ball.

58. What is the location of the ball 2 seconds later?

59. When does the ball land on the ground?

60. What is the range of the ball?

In Exercises 61–64, a gun is fired from a tower 96 feet above the ground. The angle of elevation of the gun is 60°, and its muzzle speed is 1600 feet per second.

61. Write parametric equations that model the path of the bullet.

62. When does the bullet hit the ground?

63. What is the range for the gun?

64. Find a Cartesian equation for the path of the bullet. Identify the curve.

Beyond the Basics

65. a. Verify that the parametric equations

$$x = x_1 + (x_2 - x_1)t, y = y_1 + (y_2 - y_1)t, -\infty < t < \infty$$

represent the line passing through the points (x_1, y_1) and (x_2, y_2).

    b. Find parametric equations of the line segment from $(2, -3)$ to $(-1, 4)$.

66. Find parametric equations of the line segment from $(4, 1)$ to $(-2, -3)$.

67. Verify that the parametric equations

$$x = h + a \cos t, y = k + a \sin t, 0 \le t \le 2\pi$$

represent the curve that starts at the point $(h + a, k)$ and goes counterclockwise once around the circle with center (h, k) and radius a.

68. Find a parametric equation of the curve that starts at the point $(3, 0)$ and goes counterclockwise once around the circle with center at $(1, 0)$ and radius 2.

69. Find parametric equations of the curve that starts at $(-1, 5)$ and goes clockwise once around the circle with center $(-1, 2)$ and radius 3.

70. Verify that the parametric equations of the curve that starts at the point $(h + a, k)$ and goes counterclockwise once around the ellipse $\dfrac{(x - h)^2}{a^2} + \dfrac{(y - k)^2}{b^2} = 1$ are $x = h + a \cos t, y = k + b \sin t, 0 \le t \le 2\pi$.

71. Verify that the parametric equations of the hyperbola $\dfrac{(x - h)^2}{a^2} - \dfrac{(y - k)^2}{b^2} = 1$ are $x = h + a \sec t$, $y = k + b \tan t$. One branch is traced for $-\dfrac{\pi}{2} < t < \dfrac{\pi}{2}$; the other branch is traced for $\dfrac{\pi}{2} < t < \dfrac{2\pi}{2}$.

72. Verify that the parametric equations of the hyperbola $\dfrac{(y - k)^2}{a^2} - \dfrac{(x - h)^2}{b^2} = 1$ are $x = h + b \tan t$, $y = k + a \sec t$. One branch is traced for $-\dfrac{\pi}{2} < t < \dfrac{\pi}{2}$; the other branch is traced for $\dfrac{\pi}{2} < t < \dfrac{3\pi}{2}$.

In Exercises 73–77, describe the curve represented by the parametric equations.

73. $x = 2 + t^2, y = 2t + 1, t \ge 0$

74. $x = \dfrac{e^t + e^{-t}}{2}, y = \dfrac{e^t - e^{-t}}{2}, -\infty < t < \infty$

75. $x = \dfrac{1 - t^2}{1 + t^2}, y = \dfrac{2t}{1 + t^2}, t \geq 0$

76. $x = \dfrac{t^2 + 1}{t^2 - 1}, y = \dfrac{2t}{t^2 - 1}, t \geq 0; t \neq 1$

77. $x = 8(\cos t + \sin t), y = 15(\cos t - \sin t),$
$-\infty < t < \infty$

78. **a.** Verify that the equations $x = at^2, y = 2at, -\infty < t < \infty$ represent parametric equations of the parabola $y^2 = 4ax$.
   b. Suppose the points $t = t_1$ and $t = t_2$ of the parabola $y^2 = 4ax$ are the endpoints of a focal chord (a chord through the focus). Show that $t_1 t_2 = -1$.

Critical Thinking / Discussion / Writing

79. Which of the following parametric equations represent the line $y = x + 1$? Explain.
   a. $x = 2t - 1, y = 2t$
   b. $x = t^2 - 1, y = t^2$
   c. $x = t^3 + 1, y = t^3 + 3$
   d. $x = e^t - 1, y = e^t$

80. **Projectile motion.** Eliminate the parameter t from the parametric equations $x = (v_0 \cos \theta)t, y = (v_0 \sin \theta)t - \dfrac{1}{2}gt^2$ to obtain a Cartesian equation.
   a. Show that the trajectory of a projectile is a parabola.
   b. Find the vertex of the parabola.
   c. Find the time of flight of the projectile.
   d. Find the range.
   e. Show that for a fixed initial speed v_0, the maximum range is obtained when $\theta = 45°$. Find a formula for this maximum range.
   f. Show that doubling the initial speed of a projectile has the effect of multiplying both the range and the maximum height by a factor of four.

Getting Ready for the Next Section

In Exercises 81–84, find the value of each function for the given number.

81. $f(x) = \dfrac{1}{1 + x}; x = 3$

82. $g(x) = 2x + 1; x = 10$

83. $h(x) = (-1)^3 3^{x-1}; x = 5$

84. $f(x) = \dfrac{16x}{x^2 + 12}; x = 6$

In Exercises 85–88, find the values of the given expression for the specified integer, n.

85. $(-1)^n; n = 17$

86. $(-1)^n \dfrac{1}{n + 1}; n = 8$

87. $(-1)^{n+1} 2^n; n = 5$

88. $(-1)^{n-2}(2n + 1) + (-1)^n; n = 7$

In Exercises 89 and 90, specify the integer that is the result of the computation, where a represents a nonzero real number.

89. $\dfrac{3 \cdot \pi \cdot e^2 \cdot 7 \cdot 11 \cdot 19 \cdot a}{a \cdot 3 \cdot \pi \cdot e^2 \cdot 19}$

90. $\dfrac{a \cdot 2 \cdot 3 \cdot 4 \cdot 5 \cdot 6 \cdot 7 \cdot 8 \cdot 9 \cdot 10}{3 \cdot 5 \cdot 7 \cdot 8 \cdot 9 \cdot 10 \cdot a}$

In Exercises 91 and 92, specify whether the given statement is true or false.

91. $c(1 - 5a + 3b) = c - 5ac + 3bc$

92. $3a - 2b + 6c - 5d = (3a + 6c) - (2b + 5d)$

SUMMARY Definitions, Concepts, and Formulas

8.1 Conic Sections: Overview

Conic sections are the curves formed when a plane intersects the surface of a right circular cone. The curves formed are the circle, the ellipse, the parabola, and the hyperbola, except in some degenerate cases.

8.2 The Parabola

i. Definition. A **parabola** is the set of all points in the plane that are the same distance from a fixed line and a fixed point not on the line. The fixed line is called the **directrix**, and the fixed point is called the **focus**.

ii. Axis or axis of symmetry. This is the line that passes through the focus of the parabola, perpendicular to the directrix.

iii. Vertex. This is the point at which the parabola intersects its axis. The vertex is halfway between the focus and the directrix.

iv. Standard forms of equations of parabolas with vertex at $(0, 0)$ and $p > 0$ are

$$y^2 = \pm 4px \text{ and } x^2 = \pm 4py.$$

(See the table on page 712.)

v. The **latus rectum** of a parabola is the line segment that passes through the focus, is perpendicular to the axis of the parabola, and has endpoints on the parabola.

vi. Standard forms of equations of parabolas with vertex (h, k) and $p > 0$ are $(y - k)^2 = \pm 4p(x - h)$ and $(x - h)^2 = \pm 4p(y - k)$. (See the table on page 716.)

8.3 The Ellipse

i. Definition. An **ellipse** is the set of all points in the plane, the sum of whose distances from two fixed points is a constant. The fixed points are called the **foci** of the ellipse.

ii. Vertices. These are the two points where the line through the foci intersects the ellipse.

iii. Major axis. This is the line segment joining the two vertices of the ellipse.

iv. Center. This is the midpoint of the line segment joining the foci of the ellipse. It is also the midpoint of the major axis.

v. Minor axis. This is the line segment that passes through the center of the ellipse, is perpendicular to the major axis, and has endpoints on the ellipse.

vi. Standard forms of equations of ellipses with center $(0, 0)$, $a > b > 0$, $c < a$, and $b^2 = a^2 - c^2$ are

$$\frac{x^2}{a^2} + \frac{y^2}{b^2} = 1, \text{ with foci } (\pm c, 0) \text{ and vertices } (\pm a, 0),$$

and

$$\frac{x^2}{b^2} + \frac{y^2}{a^2} = 1, \text{ with foci } (0, \pm c) \text{ and vertices } (0, \pm a).$$

(See the table on page 727.)

vii. Standard forms of equations of ellipses with center (h, k), $a > b > 0$, and $b^2 = a^2 - c^2$ are

$$\frac{(x - h)^2}{a^2} + \frac{(y - k)^2}{b^2} = 1 \text{ and } \frac{(x - h)^2}{b^2} + \frac{(y - k)^2}{a^2} = 1.$$

(See the table on page 729.)

8.4 The Hyperbola

i. Definition. A hyperbola is the set of all points in the plane, the difference of whose distances from two fixed points is a constant. The fixed points are called the **foci** of the hyperbola.

ii. Vertices. These are the two points where the line through the foci intersects the hyperbola.

iii. Transverse axis. This is the line segment joining the two vertices of the hyperbola.

iv. Center. This is the midpoint of the line segment joining the foci of the hyperbola. It is also the midpoint of the transverse axis.

v. Standard forms of equations of hyperbolas with center $(0, 0)$, $c > a$, and $b^2 = c^2 - a^2$ are

$$\frac{x^2}{a^2} - \frac{y^2}{b^2} = 1, \text{ with foci } (\pm c, 0) \text{ and vertices } (\pm a, 0),$$

and

$$\frac{y^2}{a^2} - \frac{x^2}{b^2} = 1, \text{ with foci } (0, \pm c) \text{ and vertices } (0, \pm a).$$

(See the table on page 741.)

vi. Conjugate axis. This is a line segment of length $2b$ that passes through the center of the hyperbola and is perpendicularly bisected by the transverse axis.

vii. Asymptotes. The asymptotes of the hyperbola $\frac{x^2}{a^2} - \frac{y^2}{b^2} = 1$ are the two lines $y = \pm \frac{b}{a}x$; the asymptotes of the hyperbola $\frac{y^2}{a^2} - \frac{x^2}{b^2} = 1$ are the lines $y = \pm \frac{a}{b}x$.

viii. Graphing. A procedure for graphing a hyperbola centered at $(0, 0)$ is given on page 746.

ix. Standard forms of equations of hyperbolas centered at (h, k) are

$$\frac{(x - h)^2}{a^2} - \frac{(y - k)^2}{b^2} = 1 \text{ and } \frac{(y - k)^2}{a^2} - \frac{(x - h)^2}{b^2} = 1.$$

(See the table on page 748. Also see page 748 for a procedure for graphing such hyperbolas.)

8.5 Rotation of Axes

i. If we exclude the degenerate conics, the graph of the equation $Ax^2 + Cy^2 + Dx + Ey + F = 0$, where A and C are not both zero, is
(1) a parabola if $AC = 0$, (2) an ellipse if $AC > 0$ (or a circle if $A = C$), and (3) a hyperbola if $AC < 0$.

ii. Rotation of axes formulas. Suppose the x-axis is rotated through an angle θ and the new axes are labeled the x'-axis and y'-axis, respectively. If $P = (x, y)$ and $P = (x', y')$, then $x = x' \cos\theta - y' \sin\theta$ and $y = x' \sin\theta + y' \cos\theta$.

iii. Eliminating an xy-term by rotation of axes. To transform a second-degree equation $Ax^2 + Bxy + Cy^2 + Dx + Ey + F = 0$ to the form $A'x'^2 + C'y'^2 + D'x' + E'y' + F' = 0$ without the $x'y'$ term, rotate the axes through an acute angle θ, where

$$\cot 2\theta = \frac{A - C}{B}.$$

iv. Identify conics using the discriminant $B^2 - 4AC$. Except for degenerate cases, the graph of an equation $Ax^2 + Bxy + Cy^2 + Dx + Ey + F = 0$, with $AC \neq 0$, is (1) a parabola if $B^2 - 4AC = 0$, (2) an ellipse if $B^2 - 4AC < 0$, and (3) a hyperbola if $B^2 - 4AC > 0$.

8.6 Polar Equations of Conics

i. Focus-Directrix definition of conics. The set of all points P in the plane for which the ratio of the distance from a fixed point (focus) and the fixed line (directrix) is a constant. The constant is denoted by e and is called the **eccentricity**. This conic is
(1) an ellipse if $e < 1$, (2) a parabola if $e = 1$, and (3) a hyperbola if $e > 1$.

ii. Standard forms of polar equations of conics. Polar equations of conics with focus at the pole, eccentricity e, and a distance of p units between the focus and directrix are as follows:

$$r = \frac{ep}{1 \pm e\cos\theta} \text{ and } r = \frac{ep}{1 \pm e\sin\theta}$$

(See page 770.)

iii. A procedure for graphing a conic given its polar equation is found on page 772.

8.7 Parametric Equations

i. Suppose $f(t)$ and $g(t)$ are functions defined on interval I. A **plane curve** C is a set of points (x, y), where $x = f(t)$ and $y = g(t)$. The equations $x = f(t)$ and $y = g(t)$ are the **parametric equations** for the curve C, and t is the **parameter**. The path of a particle moving along the curve C with increasing values of t describes the **orientation** (direction) of C.

ii. We can graph a plane curve represented by parametric equations by plotting points. (A procedure is described on page 780.)

iii. We can eliminate the parameter t in the parametric equations to convert to a Cartesian equation of the curve. It may be necessary to change the domain of the Cartesian equation to be consistent with the domain of the parameter t. (See page 782.)

iv. The easiest way to convert any Cartesian equation $y = f(x)$ in parametric form is $x = t$ and $y = f(t)$, with t in the domain of f. A polar equation $r = f(\theta)$ in parametric form is $x = f(\theta) \cos\theta$ and $y = f(\theta) \sin\theta$ with parameter θ.

REVIEW EXERCISES

Building Skills

In Exercises 1–8, sketch the graph of each parabola. Determine the vertex, focus, axis, and directrix of each parabola.

1. $y^2 = -6x$

2. $y^2 = 12x$

3. $x^2 = 7y$

4. $x^2 = -3y$

5. $(x - 2)^2 = -(y + 3)$

6. $(y + 1)^2 = 5(x + 2)$

7. $y^2 = -4y + 2x + 1$

8. $-x^2 + 2x + y = 0$

In Exercises 9–12, find an equation of the parabola satisfying the given conditions.

9. Vertex: $(0, 0)$; focus: $(-3, 0)$

10. Vertex: $(0, 0)$; focus: $(0, 4)$

11. Focus: $(0, 4)$; directrix: $y = -4$

12. Focus: $(-3, 0)$; directrix: $x = 3$

In Exercises 13–20, sketch the graph of each ellipse. Determine the foci, vertices, and endpoints of the minor axis of the ellipse.

13. $\dfrac{x^2}{25} + \dfrac{y^2}{4} = 1$

14. $\dfrac{x^2}{9} + \dfrac{y^2}{36} = 1$

15. $4x^2 + y^2 = 4$

16. $16x^2 + y^2 = 64$

17. $16(x + 1)^2 + 9(y + 4)^2 = 144$

18. $4(x - 1)^2 + 3(y + 2)^2 = 12$

19. $x^2 + 9y^2 + 2x - 18y + 1 = 0$

20. $4x^2 + y^2 + 8x - 10y + 13 = 0$

In Exercises 21–24, find an equation of the ellipse satisfying the given conditions.

21. Vertices: $(\pm 4, 0)$; endpoints of minor axis: $(0, \pm 2)$

22. Vertices: $(0, \pm 6)$; endpoints of minor axis: $(\pm 2, 0)$

23. Length of major axis 20; foci: $(\pm 5, 0)$

24. Length of minor axis 16; foci: $(0, \pm 6)$

In Exercises 25–32, sketch the graph of each hyperbola. Determine the vertices, the foci, and the asymptotes of each hyperbola.

25. $\dfrac{y^2}{16} - \dfrac{x^2}{4} = 1$

26. $\dfrac{x^2}{16} - \dfrac{y^2}{9} = 1$

27. $8x^2 - y^2 = 8$

28. $4y^2 - 4x^2 = 1$

29. $\dfrac{(x + 2)^2}{9} - \dfrac{(y - 3)^2}{4} = 1$

30. $\dfrac{(y + 1)^2}{6} - \dfrac{(x - 2)^2}{8} = 1$

31. $4y^2 - x^2 + 40y - 4x + 60 = 0$

32. $4x^2 - 9y^2 + 16x - 54y - 29 = 0$

In Exercises 33–36, find an equation of the hyperbola satisfying the given condition.

33. Vertices: $(\pm 1, 0)$; foci: $(\pm 2, 0)$

34. Vertices: $(0, \pm 2)$; foci: $(0, \pm 4)$

35. Vertices: $(\pm 2, 0)$; asymptotes: $y = \pm 3x$

36. Vertices: $(0, \pm 3)$; asymptotes: $y = \pm x$

In Exercises 37–48, identify each equation as representing a circle, a parabola, an ellipse, or a hyperbola. Sketch the graph of the conic.

37. $5x^2 - 4y^2 = 20$

38. $x^2 - x + y = 1$

39. $3x^2 + 4y^2 + 8y - 12x - 6 = 0$

40. $2y - x^2 = 0$

41. $x^2 + y^2 + 2x - 3 = 0$

42. $2x + y^2 = 0$

43. $y^2 = x^2 + 3$

44. $x^2 = 10 - 3y^2$

45. $3x^2 + 3y^2 - 6x + 12y + 5 = 0$

46. $y + 2x - x^2 + 2y^2 = 0$

47. $9x^2 + 8y^2 = 36$

48. $2y^2 + 4y - 3x^2 - 6x + 9$

49. Find an equation of the hyperbola whose foci are the vertices of the ellipse $4x^2 + 9y^2 = 36$ and whose vertices are the foci of this ellipse.

50. Find an equation of the ellipse whose foci are the vertices of the hyperbola $9x^2 - 16y^2 = 144$ and whose vertices are the foci of this hyperbola.

In Exercises 51–54, find all points of intersection of the given curves and make a sketch.

51. $x^2 - 4y^2 = 36$ and $x - 2y - 20 = 0$

52. $y^2 - 8x^2 = 5$ and $y - 2x^2 = 0$

53. $3x^2 - 7y^2 = 5$ and $9y^2 - 2x^2 = 1$

54. $x^2 - y^2 = 1$ and $x^2 + y^2 = 7$

In Exercises 55–58, eliminate the xy-term in the equation by a suitable rotation. Identify and graph the conic.

55. $xy = 8$

56. $x^2 + xy + y^2 = 8$

57. $x^2 + 2xy + y^2 + x - y - 4 = 0$

58. $24xy + 7y^2 + 36 = 0$

In Exercises 59–66, the polar equation of a conic is given. Sketch the graph of the conic.

59. $r = \dfrac{3}{3 - \sin\theta}$

60. $r = \dfrac{6}{3 + 3\sin\theta}$

61. $r = \dfrac{6}{3 + 2\sin\theta}$

62. $r = \dfrac{6}{2 + 3\sin\theta}$

63. $r = \dfrac{4}{2 - 3\cos\theta}$

64. $r = \dfrac{4}{3 - 2\cos\theta}$

65. $r = \dfrac{6}{2 + 3\cos\theta}$

66. $r = \dfrac{6}{3 + 2\cos\theta}$

In Exercises 67–74, eliminate the parameter to obtain a Cartesian equation of the curve. Graph the curve and indicate its orientation.

67. $x = t, y = 1 - t, 0 \le t \le 1$

68. $x = 3\cos t, y = 3\sin t, 0 \le t \le 2\pi$

69. $x = 3\sin^2 t, y = 3\cos^2 t, 0 \le t \le 2\pi$

70. $x = 3\cos\theta, y = -3\sin\theta, 0 \le \theta \le \pi$

71. $x = \sin t, y = \cos 2t, 0 \le t \le 2\pi$

72. $x = 3\sec\theta, y = 3\tan\theta, -\dfrac{\pi}{2} < \theta < \dfrac{\pi}{2}$

73. $x = t^3, y = t^2, -\infty < t < \infty$

74. $x = t^2 - 2t, y = t, -\infty < t < \infty$

75. **Parabolic curve.** Water flowing from the end of a horizontal pipe 20 feet above the ground describes a parabolic curve whose vertex is at the end of the pipe. If at a point 6 feet below the line of the pipe the flow of water has curved outward 8 feet beyond a vertical line through the end of the pipe, how far beyond this vertical line will the water strike the ground?

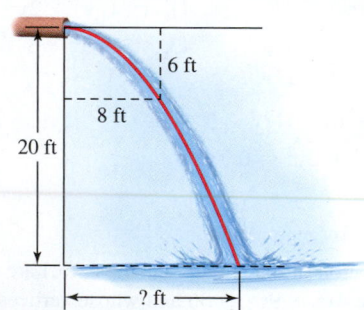

6 ft

8 ft

20 ft

? ft

76. **Parabolic arch.** A parabolic arch has a height of 20 meters and a width of 36 meters at the base. If the vertex of the parabola is at the top of the arch, find the height of the arch at a distance of 9 meters from the center of the base.

77. **Football.** A football is 12 inches long, and a cross section containing a seam is an ellipse with a minor axis of length 7 inches. Suppose every cross section of the ball formed by a plane perpendicular to the major axis of the ellipse is a circle. Find the circumference of such a circular cross section located 2 inches from an end of the ball.

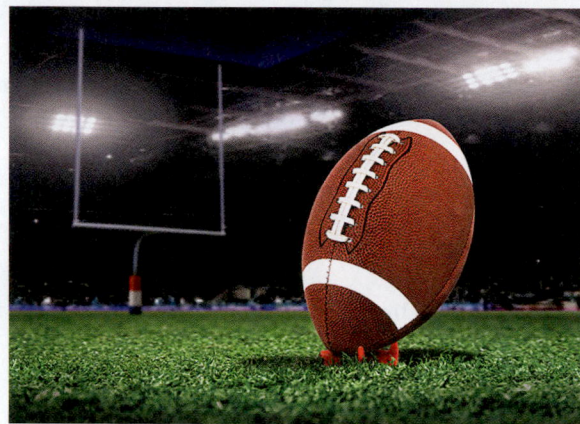

78. **Production cost.** A company assembles computers at two locations A and B that are 1000 miles apart. The per unit cost of production at location A is \$20 less than at location B. Assume that the route of delivery of the computers is along a straight line and that the delivery cost is 25¢ per unit per mile. Find the equation of the curve at any point of which the computers can be supplied from either location at the same cost. [*Hint*: Take A at $(-500, 0)$ and B at $(500, 0)$.]

PRACTICE TEST A

1. Find the standard form of the parabola with focus $(0, 12)$ and directrix with equation $x = -12$.

2. Convert the equation $y^2 - 2y + 8x + 25 = 0$ to the standard form for a parabola by completing the square.

3. Find the vertex, focus, and directrix of the parabola with equation $(x + 2)^2 = -8(y - 1)$.

4. The base of a water slide is parabolic in shape and is 6 feet wide and 2 feet deep. Find the height of the slide 1 foot from its center.

5. Find an equation of the parabola with vertex $(3, -1)$ and directrix $x = -3$.

In Problems 6 and 7, find the standard form of the equation of the ellipse satisfying the given conditions.

6. Foci: $(0, -2), (0, 2)$; vertices: $(0, -4), (0, 4)$

7. Major axis horizontal with length 18; minor axis of length 4; center $(0, 0)$

8. Graph the ellipse with equation $9x^2 + y^2 = 9$.

9. Find the standard form of the equation of the hyperbola with foci $(0, -\sqrt{45}), (0, \sqrt{45})$ and vertices $(0, -6), (0, 6)$.

In Problems 10 and 11, convert the equation to the standard form for a hyperbola by completing the squares on x and y.

10. $y^2 - x^2 + 2x = 2$

11. $x^2 - 2x - 4y^2 - 16y = 19$

In Problems 12–20, identify the conic section and sketch its graph.

12. $x^2 - 9y^2 = -16$

13. $x^2 + 2y + 1 = 0$

14. $x^2 + y^2 + 2x - 6y = 3$

15. $25x^2 + 4y^2 = 100$

16. $xy = 1$

17. $2x^2 + 3xy + 2y^2 = 7$

18. $r = \dfrac{6}{4 - 3\cos\theta}$

19. $r = \dfrac{6}{3 + 4\cos\theta}$

20. $x = 4 - t, y = t^2$

1. Find the standard form of the equation of the parabola with focus $(-10, 0)$ and directrix $x = 10$.
   a. $y^2 = -10x$
   b. $y^2 = -40x$
   c. $x^2 = -40y$
   d. $y^2 = 40x$

2. Convert the equation $y^2 - 4y - 5x + 24 = 0$ to the standard form for a parabola by completing the square.
   a. $(y + 2)^2 = 5(x - 4)$
   b. $(y - 2)^2 = 5(x - 4)$
   c. $(y + 2)^2 = -5(x - 4)$
   d. $(y - 2)^2 = 5(x + 4)$

3. Find the vertex, focus, and directrix of the parabola with equation $(x - 3)^2 = 12(y - 1)$.
   a. vertex: $(3, 1)$ focus: $(3, 4)$ directrix: $y = -2$
   b. vertex: $(-3, -1)$ focus: $(-3, 2)$ directrix: $y = -4$
   c. vertex: $(3, 1)$ focus: $(3, -2)$ directrix: $x = 4$
   d. vertex: $(1, 3)$ focus: $(1, 6)$ directrix: $y = 0$

4. The parabolic arch of a bridge has a 180-foot base and a height of 25 feet. Find the height of the arch 45 feet from the center of the base.
   a. 12.5 feet
   b. 6.25 feet
   c. 16.7 feet
   d. 18.75 feet

5. Find an equation of the parabola with vertex $(-2, 1)$ and directrix $x = 2$.
   a. $(y - 1)^2 = -16(x + 2)$
   b. $(y - 1)^2 = 16(x + 2)$
   c. $(y + 2)^2 = -16(x - 1)$
   d. $(y + 2)^2 = 16(x - 1)$

In Problems 6 and 7, find the standard form of the equation of the ellipse satisfying the given conditions.

6. Foci $(-3, 0), (3, 0)$; vertices $(-5, 0), (5, 0)$
   a. $\dfrac{x^2}{9} + \dfrac{y^2}{16} = 1$
   b. $\dfrac{x^2}{25} + \dfrac{y^2}{16} = 1$
   c. $\dfrac{x^2}{9} + \dfrac{y^2}{25} = 1$
   d. $\dfrac{x^2}{16} + \dfrac{y^2}{25} = 1$

7. Major axis vertical with length 16; length of minor axis = 8; center $(0, 0)$
   a. $\dfrac{x^2}{16} + \dfrac{y^2}{64} = 1$
   b. $\dfrac{x^2}{64} + \dfrac{y^2}{256} = 1$
   c. $\dfrac{x^2}{8} + \dfrac{y^2}{64} = 1$
   d. $\dfrac{x^2}{64} + \dfrac{y^2}{16} = 1$

8. Find the vertices and foci of the hyperbola with equation $\dfrac{x^2}{121} - \dfrac{y^2}{4} = 1$.
   a. vertices: $(-11, 0), (11, 0)$ foci: $(-2, 0), (2, 0)$
   b. vertices: $(0, -1), (0, 11)$ foci: $(-5\sqrt{5}, 0), (5\sqrt{5}, 0)$
   c. vertices: $(-11, 0), (11, 0)$ foci: $(-5\sqrt{5}, 0), (5\sqrt{5}, 0)$
   d. vertices: $(-2, 0), (2, 0)$ foci: $(-5\sqrt{5}, 0), (5\sqrt{5}, 0)$

9. Find the standard form of the equation of the hyperbola with foci $(0, -10), (0, 10)$ and vertices $(0, -5), (0, 5)$.
   a. $\dfrac{y^2}{25} - \dfrac{x^2}{100} = 1$
   b. $\dfrac{y^2}{25} - \dfrac{x^2}{75} = 1$
   c. $\dfrac{x^2}{25} - \dfrac{y^2}{100} = 1$
   d. $\dfrac{x^2}{25} - \dfrac{y^2}{75} = 1$

In Problems 10 and 11, convert the equation to the standard form for a hyperbola by completing the squares on x and y.

10. $y^2 - 4x^2 - 2y - 16x - 19 = 0$
    a. $\dfrac{(y - 2)^2}{4} - (x + 4)^2 = 1$
    b. $\dfrac{(x - 1)^2}{4} - (y + 2)^2 = 1$
    c. $\dfrac{(y - 1)^2}{4} - (x + 2)^2 = 1$
    d. $(x + 2)^2 - \dfrac{(y - 1)^2}{4} = 1$

11. $4y^2 - 9x^2 - 16y - 36x - 56 = 0$
    a. $\dfrac{(y - 2)^2}{4} - \dfrac{(x + 2)^2}{9} = 1$
    b. $\dfrac{(x - 2)^2}{4} - \dfrac{(y + 2)^2}{9} = 1$
    c. $\dfrac{(y - 2)^2}{9} - \dfrac{(x + 2)^2}{4} = 1$
    d. $\dfrac{(y + 2)^2}{9} - \dfrac{(x - 2)^2}{4} = 1$

In Problems 12–20, identify the conic section.

12. $3x^2 + 2y^2 - 6x + 4y - 1 = 0$
    a. parabola
    b. circle
    c. ellipse
    d. hyperbola

13. $x^2 + 2x + 4y + 5 = 0$
    a. parabola
    b. circle
    c. ellipse
    d. hyperbola

14. $4x^2 - y^2 + 16x + 2y + 11 = 0$
    a. parabola
    b. circle
    c. ellipse
    d. hyperbola

15. $5y^2 - 3x^2 + 20y - 6x + 2 = 0$
    a. parabola
    b. circle
    c. ellipse
    d. hyperbola

16. $2x^2 - 5xy + y^2 = 7$
    a. parabola
    b. circle
    c. ellipse
    d. hyperbola

17. $x^2 + xy + y^2 = 10$
    a. parabola
    b. circle
    c. ellipse
    d. hyperbola

18. $r = \dfrac{2}{2 + \cos\theta}$
    a. parabola
    b. circle
    c. ellipse
    d. hyperbola

19. $r = \dfrac{8}{2 + 3\sin\theta}$
    a. parabola
    b. circle
    c. ellipse
    d. hyperbola

20. $x = 3\sec t, \quad y = 2\tan t$
    a. parabola
    b. circle
    c. ellipse
    d. hyperbola

CUMULATIVE REVIEW EXERCISES CHAPTERS 1–8

1. Let $f(x) = x^2 - 3x + 2$. Find $\dfrac{f(x+h) - f(x)}{h}$.

2. Sketch the graph of $f(x) = \sqrt{x+1} - 2$.

3. Let $f(x) = 2x - 3$. Find the inverse function f^{-1}. Verify that $f(f^{-1}(x)) = x$.

4. Sketch the graph of $f(x) = \left(\dfrac{1}{2}\right)^{x+2}$.

5. Solve the equation
$\log_5(x-1) + \log_5(x-2) = 3\log_5 \sqrt[3]{6}$.

6. Express
$$\log_a \sqrt[3]{x\sqrt{yz}}$$
in terms of logarithms of x, y, and z.

7. Solve the inequality
$$\frac{x}{x-2} \geq 1.$$

In Problems 8–10, solve each system of equations.

8. $\begin{cases} 1.4x - 0.5y = 1.3 \\ 0.4x + 1.1y = 4.1 \end{cases}$

9. $\begin{cases} 2x + y - 4z = 3 \\ x - 2y + 3z = 4 \\ -3x + 4y - z = -2 \end{cases}$

10. $\begin{cases} y = x^2 - 1 \\ 3x^2 + 8y^2 = 8 \end{cases}$

11. If $\cot\theta = \dfrac{5}{2}$ and $\sec\theta < 0$, find $\cos\theta$.

12. If $\cos\theta = -\dfrac{1}{3}$ and θ is in quadrant III, find $\sin 2\theta$.

13. Convert the equation $x^2 + y^2 - 6x + 3 = 0$ to polar form.

14. Find the determinant of the matrix $\begin{bmatrix} 1 & 4 & 7 \\ 2 & 5 & 8 \\ 3 & 6 & 9 \end{bmatrix}$.

15. Use Cramer's rule to solve the system of equations:
$$\begin{cases} 2x - 3y = -4 \\ 5x + 7y = 1 \end{cases}$$

16. Find the inverse of the matrix $A = \begin{bmatrix} 3 & -2 \\ -5 & 4 \end{bmatrix}$.

17. Solve the equation $2x^4 - 5x^2 + 3 = 0$.

In Problems 18–19, an equation of a conic section is given. Identify the conic and sketch its graph.

18. $x^2 - y^2 = -4$

19. $9x^2 + 9y^2 = 144$

20. Sketch the graph of the rational function $f(x) = \dfrac{x}{x^2 - 16}$.

CHAPTER | 9

Further Topics in Algebra

TOPICS

9.1 Sequences and Series

9.2 Arithmetic Sequences; Partial Sums

9.3 Geometric Sequences and Series

9.4 Mathematical Induction

9.5 The Binomial Theorem

9.6 Counting Principles

9.7 Probability

Probability was studied in ancient Eastern civilizations, including those in Babylonia and Egypt. Probability is most famously used in gambling, and Egyptians even used objects made from the heel bones of animals, called *tali*, as a type of dice. Today, probability is used not only in gambling, but also in polling, insurance, business, the life sciences, and many other areas. In this chapter, we learn to understand and compute the probability of an event.

Sequences and Series

BEFORE STARTING THIS SECTION, REVIEW

1 Simplifying algebraic expressions (Appendix A.1, page 923)

2 Finding functional values (Section 1.3, page 43)

OBJECTIVES

1 Use sequence notation and find specific and general terms in a sequence.

2 Use factorial notation.

3 Use summation notation to write partial sums of a series.

◆ The Family Tree of the Honeybee

The bee that most of us know best is the honeybee, an insect that lives in a colony and has an unusual family tree. Perhaps the most surprising fact about honeybees is that not all of them have two parents. This phenomenon begins with a special female in the colony called the *queen*. Many other female bees, called *worker bees*, live in the colony, but unlike the queen bee, they produce no eggs. Male bees do no work and are produced from the queen's unfertilized eggs. The female bees are produced as a result of the queen bee mating with a male bee. Consequently, all female bees have two parents—a male and a female—whereas male bees have just one parent—a female. In this section, we study sequences, and in Example 4, we see how a famous sequence accurately counts a honeybee's ancestors.

---

1 Use sequence notation and find specific and general terms in a sequence.

Sequences

The word *sequence* is used in mathematics in much the same way it is used in ordinary English. If someone saw a *sequence* of bad movies, you know that the person could list the first bad movie he or she saw, the second bad movie, and so on.

Sequence _____

An **infinite sequence** is a function whose domain is the set of positive integers. The function values, written as

$$a_1, a_2, a_3, a_4, \ldots, a_n, \ldots,$$

are called the **terms** of the sequence. The **nth term**, a_n, is called the **general term** of the sequence.

If the domain of a function consists of only the first n positive integers, the sequence is called a **finite sequence**.

In computer science, finite sequences of the form $a_1, a_2, \ldots, a_n$ are called *strings*. Sometimes for convenience, we include 0 in the domain of the function that defines a sequence, and we write the sequence terms as $a_0, a_1, a_2, a_3, \ldots$. The nth term of such a sequence is a_{n-1}.

We can use subscripts on variables other than a to represent the terms of a sequence. In Example 1(c), we use the variable b.

EXAMPLE 1 **Writing the Terms of a Sequence from the General Term**

Write the first four terms of each sequence.

a. $a_n = 5n - 1$ **b.** $a_n = \dfrac{1}{n + 1}$ **c.** $b_n = (-1)^{n+1}\left(\dfrac{1}{n}\right)$

Solution

The first four terms of each sequence are found by replacing n with the integers 1, 2, 3, and 4 in the equation defining a_n.

a. $a_n = 5n - 1$ General term of the sequence
$a_1 = 5(1) - 1 = 4$ 1st term: Replace n with 1.
$a_2 = 5(2) - 1 = 9$ 2nd term: Replace n with 2.
$a_3 = 5(3) - 1 = 14$ 3rd term: Replace n with 3.
$a_4 = 5(4) - 1 = 19$ 4th term: Replace n with 4.

The first four terms of the sequence are: 4, 9, 14, and 19.

b. $a_n = \dfrac{1}{n + 1}$

$a_1 = \dfrac{1}{1 + 1} = \dfrac{1}{2}$

$a_2 = \dfrac{1}{2 + 1} = \dfrac{1}{3}$

$a_3 = \dfrac{1}{3 + 1} = \dfrac{1}{4}$

$a_4 = \dfrac{1}{4 + 1} = \dfrac{1}{5}$

The first four terms are: $\dfrac{1}{2}, \dfrac{1}{3}, \dfrac{1}{4},$ and $\dfrac{1}{5}$.

c. $b_n = (-1)^{n+1}\left(\dfrac{1}{n}\right)$

$b_1 = (-1)^{1+1}\left(\dfrac{1}{1}\right) = (-1)^2(1) = 1$

$b_2 = (-1)^{2+1}\left(\dfrac{1}{2}\right) = (-1)^3\left(\dfrac{1}{2}\right) = -\dfrac{1}{2}$

$b_3 = (-1)^{3+1}\left(\dfrac{1}{3}\right) = (-1)^4\left(\dfrac{1}{3}\right) = \dfrac{1}{3}$

$b_4 = (-1)^{4+1}\left(\dfrac{1}{4}\right) = (-1)^5\left(\dfrac{1}{4}\right) = -\dfrac{1}{4}$

The first four terms are: $1, -\dfrac{1}{2}, \dfrac{1}{3},$ and $-\dfrac{1}{4}$.

Practice Problem 1 Write the first four terms of the sequence with general term $a_n = -2^n$.

Frequently, the first few terms of a sequence exhibit a pattern that *suggests* a natural choice for the general term of the sequence.

EXAMPLE 2 **Finding a General Term of a Sequence from a Pattern**

Write the general term a_n for a sequence whose first five terms are given.

a. $1, 4, 9, 16, 25, \ldots$ **b.** $0, \dfrac{1}{2}, -\dfrac{2}{3}, \dfrac{3}{4}, -\dfrac{4}{5}, \ldots$

Solution

Write the position number of the term above each term of the sequence and look for a pattern that connects the term to the position number of the term.

a.
n: 1, 2, 3, 4, 5, $\ldots$, n
term: 1, 4, 9, 16, 25, $\ldots$, a_n

Apparent pattern: Here $1 = 1^2, 4 = 2^2, 9 = 3^2, 16 = 4^2$, and $25 = 5^2$. Each term is the square of the position number of that term. This suggests $a_n = n^2$.

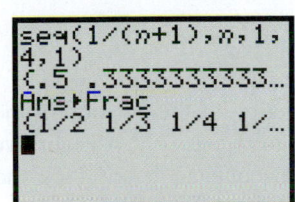
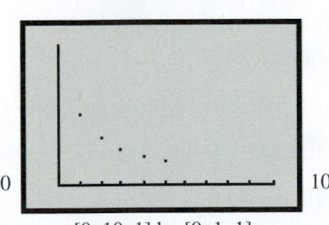

b.

$$n:\qquad 1,\quad 2,\quad 3,\quad 4,\quad 5,\quad \ldots,\quad n$$

$$\text{term:}\quad 0,\quad \frac{1}{2},\quad -\frac{2}{3},\quad \frac{3}{4},\quad -\frac{4}{5},\quad \ldots,\quad a_n$$

Apparent pattern: When the terms alternate in sign and $n = 1$, we use factors such as $(-1)^n$ if we want to begin with the factor -1 or we use factors such as $(-1)^{n+1}$ if we want to begin with the factor 1. Notice that each term can be written as a quotient with denominator equal to the position number and numerator equal to one less than the position number, suggesting the general term $a_n = (-1)^n\left(\dfrac{n-1}{n}\right)$.

Practice Problem 2 Write a general term for a sequence whose first five terms are

$$0, -\frac{1}{2}, \frac{2}{3}, -\frac{3}{4}, \frac{4}{5}, \ldots.$$

WARNING

In Example 2, we found the general term of a sequence by finding a pattern in the first few terms of the sequence. However, when only a finite number of successive terms are given without a rule that defines the general term, a *unique* general term cannot be found.

To indicate why this is so, consider the two sequences $a_n = n^2$ and $b_n = n^2 + (n-1)(n-2)(n-3)(n-4)$. The first four terms of these two sequences are identical: 1, 4, 9, 16. However, the fifth term is different: $a_5 = 25$, but $b_5 = 49$. Thus, either a_n or b_n could correctly describe a sequence whose first four terms are 1, 4, 9, and 16. You can *never* find a unique general term from a finite number of terms in a sequence.

Leonardo of Pisa (Fibonacci)

(1175–1250)
Leonardo, often known today by the name Fibonacci (son of Bonaccio), was born around 1175. He traveled extensively through North Africa and the Mediterranean, probably on business with his father. At each location, he met with Islamic scholars and absorbed the mathematical knowledge of the Islamic world. After returning to Pisa around 1200, he started writing books on mathematics. He was one of the earliest European writers on algebra. In his most famous book *Liber abaci*, or *Book of Calculations*, he introduced the Western world to the rules of computing with the new Hindu–Arabic numerals.

Recursive Formulas

Up to this point, we have expressed the general term as a function of the position of the term in the sequence. A second approach is to define a sequence *recursively*. A **recursive formula** requires that one or more of the first few terms of the sequence be specified and all other terms be defined in relation to previously defined terms.

For example, the **Fibonacci sequence** is a famous sequence that is defined recursively and shows up often in nature. In this sequence, we specify the first two terms as $a_0 = 1$ and $a_1 = 1$; each subsequent term is the sum of the two terms immediately preceding it. So we have

$a_0 = 1, a_1 = 1, a_2 = a_0 + a_1 = 1 + 1 = 2$	Add the two given terms.
$a_1 = 1, a_2 = 2, a_3 = a_1 + a_2 = 1 + 2 = 3$	Add the two previous terms.
$a_2 = 2, a_3 = 3, a_4 = a_2 + a_3 = 2 + 3 = 5$	Add the two previous terms.
$a_3 = 3, a_4 = 5, a_5 = a_3 + a_4 = 3 + 5 = 8$	Add the two previous terms.

The first six terms of the sequence are then

$$1, 1, 2, 3, 5, 8.$$

This sequence can also be defined in subscript notation:

$$a_0 = 1, a_1 = 1, a_k = a_{k-2} + a_{k-1}, \text{ for all } k \geq 2$$

Replacing k with 2, 3, 4, and 5, respectively, in the equation $a_k = a_{k-2} + a_{k-1}$ will produce the values $a_2 = 2, a_3 = 3, a_4 = 5,$ and $a_5 = 8$.

EXAMPLE 3   Finding Terms of a Recursively Defined Sequence

Write the first five terms of the recursively defined sequence

$$a_1 = 4, a_{n+1} = 2a_n - 9.$$

Solution

We are given the first term of the sequence: $a_1 = 4$.

$a_2 = 2a_1 - 9$	Replace n with 1 in $a_{n+1} = 2a_n - 9$.
$a_2 = 2(4) - 9 = -1$	Replace a_1 with 4.
$a_3 = 2a_2 - 9$	Replace n with 2 in $a_{n+1} = 2a_n - 9$.
$a_3 = 2(-1) - 9 = -11$	Replace a_2 with -1.
$a_4 = 2a_3 - 9$	Replace n with 3 in $a_{n+1} = 2a_n - 9$.
$a_4 = 2(-11) - 9 = -31$	Replace a_3 with -11.
$a_5 = 2a_4 - 9$	Replace n with 4 in $a_{n+1} = 2a_n - 9$.
$a_5 = 2(-31) - 9 = -71$	Replace a_4 with -31.

Thus, the first five terms of the sequence are

$$4, -1, -11, -31, -71.$$

Practice Problem 3 Write the first five terms of the recursively defined sequence
$a_1 = -3, a_{n+1} = 2a_n + 5$.

EXAMPLE 4   Diagramming the Family Tree of a Honeybee

Find a sequence that accurately counts the ancestors of a male honeybee. The first term of the sequence should give the number of parents of a male honeybee, the second term the number of grandparents, the third term the number of great-grandparents, and so on.

Solution

Recall from the introduction to this section that female honeybees have two parents—a male and a female—but that male honeybees have just one parent—a female.

First, we consider the family tree of a male honeybee. A male bee has one parent: a female. Since his mother had two parents, he has two grandparents. Because his grandmother had two parents and his grandfather had only one parent, he has three great-grandparents.

Now, to count the bee's ancestors, we count backward using Figure 9.1. You should recognize the first six terms of the Fibonacci sequence: 1, 1, 2, 3, 5, 8. This sequence is defined by $a_0 = 1$, $a_1 = 1$, and $a_k = a_{k-2} + a_{k-1}$, for all $k \geq 2$.

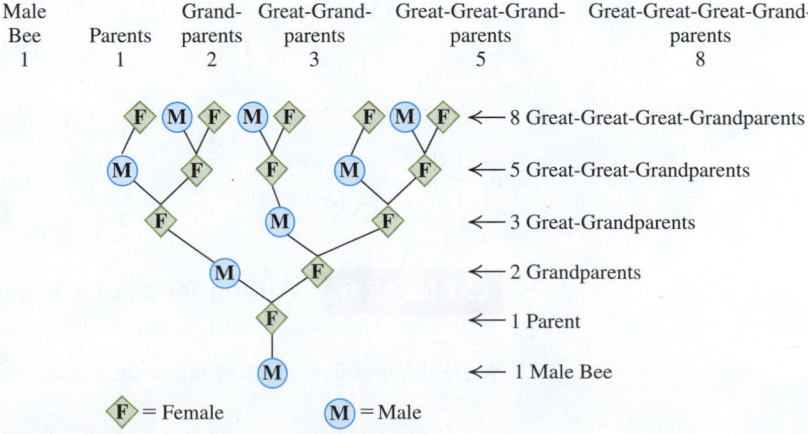

Figure 9.1

Practice Problem 4 Find a sequence that accurately counts the ancestors of a female honeybee.

2 Use factorial notation.

Factorial Notation

Special types of products, called factorials, appear in some very important sequences.

Factorial

For any positive integer n, **n factorial** (written as $n!$) is defined as

$$n! = n \cdot (n-1) \cdots 4 \cdot 3 \cdot 2 \cdot 1.$$

As a special case, **zero factorial** $(0!)$ is defined by

$$0! = 1.$$

Here are the first eight values of $n!$

$$0! = 1 \qquad\qquad 4! = 4 \cdot 3 \cdot 2 \cdot 1 = 24$$
$$1! = 1 \qquad\qquad 5! = 5 \cdot 4 \cdot 3 \cdot 2 \cdot 1 = 120$$
$$2! = 2 \cdot 1 = 2 \qquad 6! = 6 \cdot 5 \cdot 4 \cdot 3 \cdot 2 \cdot 1 = 720$$
$$3! = 3 \cdot 2 \cdot 1 = 6 \qquad 7! = 7 \cdot 6 \cdot 5 \cdot 4 \cdot 3 \cdot 2 \cdot 1 = 5040$$

The values of $n!$ get large very quickly. For example, $10! = 3{,}628{,}800$ and $12! = 479{,}001{,}600$.

In simplifying some expressions involving factorials, it is helpful to note that

$$\boxed{n! = n \cdot (n-1)!}$$

TECHNOLOGY CONNECTION

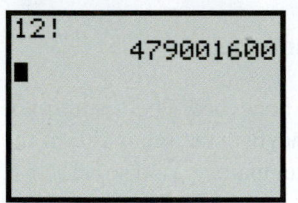

EXAMPLE 5 **Simplifying a Factorial Expression**

Simplify.

a. $\dfrac{16!}{14!}$ **b.** $\dfrac{(n+1)!}{(n-1)!}$

Solution

a. $\dfrac{16!}{14!} = \dfrac{16 \cdot 15 \cdot 14!}{14!}$ $16! = 16 \cdot 15! = 16 \cdot 15 \cdot 14!$

     $= 16 \cdot 15 = 240$ Divide out the factor $14!$

b. $\dfrac{(n+1)!}{(n-1)!} = \dfrac{(n+1) \cdot n \cdot (n-1)!}{(n-1)!}$ $(n+1)! = (n+1) \cdot n! = (n+1) \cdot n \cdot (n-1)!$

     $= (n+1)n$ Divide out the common factor $(n-1)!$

Practice Problem 5 Simplify.

a. $\dfrac{13!}{12!}$ **b.** $\dfrac{n!}{(n-3)!}$

EXAMPLE 6 **Writing Terms of a Sequence Involving Factorials**

Write the first five terms of the sequence whose general term is $a_n = \dfrac{(-1)^{n+1}}{n!}$.

Solution

Replace n in the formula for the general term with each positive integer from 1 through 5.

$$a_1 = \frac{(-1)^{1+1}}{1!} = \frac{(-1)^2}{1} = 1 \qquad a_4 = \frac{(-1)^{4+1}}{4!} = \frac{(-1)^5}{24} = -\frac{1}{24}$$

$$a_2 = \frac{(-1)^{2+1}}{2!} = \frac{(-1)^3}{2} = -\frac{1}{2} \qquad a_5 = \frac{(-1)^{5+1}}{5!} = \frac{(-1)^6}{120} = \frac{1}{120}$$

$$a_3 = \frac{(-1)^{3+1}}{3!} = \frac{(-1)^4}{6} = \frac{1}{6}$$

Using these five terms, we could write this sequence as

$$1, -\frac{1}{2}, \frac{1}{6}, -\frac{1}{24}, \frac{1}{120}, \ldots .$$

Practice Problem 6 Write the first five terms of the sequence whose general term is

$$a_n = \frac{(-1)^n 2^n}{n!}.$$

3 Use summation notation to write partial sums of a series.

Summation Notation

If we know the general term of a sequence, we can represent a *sum* of terms of the sequence by using *summation* (or *sigma*) *notation*. In this notation, the Greek letter Σ (capital sigma) indicates that we are to *add* the given terms. The letter Σ corresponds to the English letter S, the first letter of the word *sum*.

SUMMATION NOTATION

The sum of the first $n\,(\geq 1)$ terms of a sequence $a_1, a_2, a_3, \ldots, a_n, \ldots$ is denoted by

$$\sum_{i=1}^{n} a_i = a_1 + a_2 + a_3 + \cdots + a_n.$$

The letter i in the summation notation is called the **index of summation**, n is called the **upper limit**, and 1 is called the **lower limit** of the summation.

In summation notation, the index need not start at $i = 1$. Also, any letter may be used in place of the index i. In general, we have

$$\sum_{j=k}^{n} a_j = a_k + a_{k+1} + \cdots + a_n, \quad n \geq k \geq 0.$$

 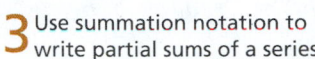
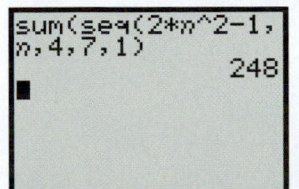
EXAMPLE 7 **Evaluating Sums Given in Summation Notation**

Find each sum.

a. $\displaystyle\sum_{i=1}^{9} i$ **b.** $\displaystyle\sum_{j=4}^{7} (2j^2 - 1)$ **c.** $\displaystyle\sum_{k=0}^{4} \frac{2^k}{k!}$

Solution

a. Replace i with successive integers from 1 to 9 inclusive; then add.

$$\sum_{i=1}^{9} i = 1 + 2 + 3 + 4 + 5 + 6 + 7 + 8 + 9 = 45$$

b. The index of summation is j. Evaluate $(2j^2 - 1)$ for all consecutive integers from 4 through 7 inclusive; then add.

$$\sum_{j=4}^{7} (2j^2 - 1) = [2(4)^2 - 1] + [2(5)^2 - 1] + [2(6)^2 - 1] + [2(7)^2 - 1]$$

$$= 31 + 49 + 71 + 97 = 248$$

c. The index of summation is k. Evaluate $\dfrac{2^k}{k!}$ for consecutive integers 0 through 4, inclusive, then add.

$$\sum_{k=0}^{4}\frac{2^k}{k!}=\frac{2^0}{0!}+\frac{2^1}{1!}+\frac{2^2}{2!}+\frac{2^3}{3!}+\frac{2^4}{4!}$$

$$=\frac{1}{1}+\frac{2}{1}+\frac{4}{2}+\frac{8}{6}+\frac{16}{24}=1+2+2+\frac{4}{3}+\frac{2}{3}=7$$

Practice Problem 7 Find the following sum: $\displaystyle\sum_{k=0}^{3}(-1)^k k!$

The familiar properties of real numbers, such as the distributive, commutative, and associative properties, can be used to prove the summation properties listed next.

Summation Properties

Let a_k and b_k represent the general terms of two sequences and let c represent any real number. Then

1. $\displaystyle\sum_{k=1}^{n}c=c\cdot n$ **2.** $\displaystyle\sum_{k=1}^{n}ca_k=c\sum_{k=1}^{n}a_k$

3. $\displaystyle\sum_{k=1}^{n}(a_k+b_k)=\sum_{k=1}^{n}a_k+\sum_{k=1}^{n}b_k$ **4.** $\displaystyle\sum_{k=1}^{n}(a_k-b_k)=\sum_{k=1}^{n}a_k-\sum_{k=1}^{n}b_k$

5. $\displaystyle\sum_{k=1}^{n}a_k=\sum_{k=1}^{j}a_k+\sum_{k=j+1}^{n}a_k,\text{ for }1\le j<n$

Series

We all know how to add numbers, but do you think it is possible to add infinitely many numbers? It is sometimes possible to add infinitely many numbers, although a complete explanation of this kind of addition must be put off until a calculus course. For now, we need some definitions.

Series

Let $a_1,a_2,a_3,\ldots,a_k,\ldots$ be an infinite sequence. Then

1. The sum of all terms of the sequence is called a **series** and is denoted by

$$a_1+a_2+a_3+\cdots=\sum_{i=1}^{\infty}a_i.$$

2. The sum of the first n terms of the sequence

$$a_1+a_2+a_3+\cdots+a_n=\sum_{i=1}^{n}a_i$$

is called the **nth partial sum** of the series.

EXAMPLE 8 **Writing a Sum in Summation Notation**

Write each sum in summation notation.

a. $3+5+7+\cdots+21$ **b.** $\dfrac{1}{4}+\dfrac{1}{9}+\cdots+\dfrac{1}{49}$

Solution

a. The finite series $3+5+7+\cdots+21$ is a sum of consecutive odd numbers from 3 to 21. Each of these numbers can be expressed in the form $2k+1$, starting with $k=1$ and ending with $k=10$. With k as the index of summation, 1 as the lower limit, and 10 as the upper limit, we write

$$3+5+7+\cdots+21=\sum_{k=1}^{10}(2k+1).$$

b. The finite series $\dfrac{1}{4} + \dfrac{1}{9} + \cdots + \dfrac{1}{49}$ is a sum of fractions, each of which has numerator 1 and denominator k^2, starting with $k = 2$ and ending with $k = 7$. We write

$$\frac{1}{4} + \frac{1}{9} + \cdots + \frac{1}{49} = \sum_{k=2}^{7} \frac{1}{k^2}.$$

Practice Problem 8 Write the following in summation notation: $2 - 4 + 6 - 8 + 10 - 12 + 14$

Answers to Practice Problems

1. $a_1 = -2, a_2 = -4, a_3 = -8, a_4 = -16$

2. $a_n = (-1)^{n+1}\left(1 - \dfrac{1}{n}\right)$ **3.** $a_1 = -3, a_2 = -1, a_3 = 3,$
$a_4 = 11, a_5 = 27$ **4.** $a_0 = 2, a_1 = 3,$ and $a_k = a_{k-2} + a_{k-1},$
for all $k \geq 2$ **5. a.** 13 **b.** $n(n - 1)(n - 2)$

6. $a_1 = -2, a_2 = 2, a_3 = -\dfrac{4}{3}, a_4 = \dfrac{2}{3}, a_5 = -\dfrac{4}{15}$ **7.** -4

8. $\displaystyle\sum_{k=1}^{7} (-1)^{k+1}(2k)$

 SECTION 9.1 **Exercises**

Concepts and Vocabulary

1. An infinite sequence is a function whose domain is the set of _____.

2. If a sequence is defined by $a_n = 2n - \dfrac{10}{n}$, then $a_5 = $

_____.

3. By definition, $0! = $ _____.

4. The notation $\displaystyle\sum_{k=1}^{5} k^2 = 1^2 + 2^2 + 3^2 + 4^2 + 5^2$ is an example of _____ notation.

5. True or False. If $a_1 = 3$ and $a_{n+1} = a_n^2$, then $a_3 = 27$.

6. True or False. $\dfrac{(2n)!}{n!} = 2$.

7. True or False. $\displaystyle\sum_{k=0}^{5} \frac{2k + 1}{2} = 18$

8. True or False. $(2n + 1)! = (2n + 1) \cdot (2n)!$

Building Skills

In Exercises 9–28, write the first four terms of each sequence.

9. $a_n = 3n - 2$ **10.** $a_n = 2n + 1$

11. $a_n = 1 - \dfrac{1}{n}$ **12.** $a_n = 1 + \dfrac{1}{n}$

13. $a_n = -n^2$ **14.** $a_n = n^3$

15. $a_n = \dfrac{2n}{n + 1}$ **16.** $a_n = \dfrac{3n}{n^2 + 1}$

17. $a_n = (-1)^{n+1}$ **18.** $a_n = (-3)^{n-1}$

19. $a_n = 3 - \dfrac{1}{2^n}$ **20.** $a_n = \left(\dfrac{3}{2}\right)^n$

21. $a_n = 0.6$ **22.** $a_n = -0.4$

23. $a_n = \dfrac{(-1)^n}{n!}$ **24.** $a_n = \dfrac{n}{n!}$

25. $a_n = (-1)^n 3^{-n}$ **26.** $a_n = (-1)^n 3^{n-1}$

27. $a_n = \dfrac{e^n}{2n}$ **28.** $a_n = \dfrac{2^n}{e^n}$

In Exercises 29–42, write a general term a_n for each sequence. Assume that n begins with 1.

29. $1, 4, 7, 10, \ldots$ **30.** $7, 9, 11, 13, \ldots$

31. $\dfrac{1}{2}, \dfrac{1}{3}, \dfrac{1}{4}, \dfrac{1}{5}, \ldots$ **32.** $\dfrac{2}{3}, \dfrac{3}{4}, \dfrac{4}{5}, \dfrac{5}{6}, \ldots$

33. $2, -2, 2, -2, \ldots$ **34.** $-3, 6, -9, 12, \ldots$

35. $\dfrac{1}{2}, \dfrac{3}{4}, \dfrac{9}{8}, \dfrac{27}{16}, \ldots$ **36.** $-\dfrac{1}{2}, \dfrac{1}{4}, -\dfrac{1}{8}, \dfrac{1}{16}, \ldots$

37. $1 \cdot 2, 2 \cdot 3, 3 \cdot 4, 4 \cdot 5, \ldots$

38. $-\dfrac{1}{1 \cdot 2}, \dfrac{1}{2 \cdot 3}, -\dfrac{1}{3 \cdot 4}, \dfrac{1}{4 \cdot 5}, \ldots$

39. $2 + \dfrac{1}{2}, 2 - \dfrac{1}{3}, 2 + \dfrac{1}{4}, 2 - \dfrac{1}{5}, \ldots$

40. $1 + \dfrac{1}{2}, 1 + \dfrac{1}{3}, 1 + \dfrac{1}{4}, 1 + \dfrac{1}{5}, \ldots$

41. $\dfrac{3^2}{2}, \dfrac{3^3}{3}, \dfrac{3^4}{4}, \dfrac{3^5}{5}, \ldots$ **42.** $\dfrac{e}{2}, \dfrac{e^2}{4}, \dfrac{e^3}{8}, \dfrac{e^4}{16}, \ldots$

In Exercises 43–52, write the first five terms of each recursively defined sequence.

43. $a_1 = 2, a_{n+1} = a_n + 3$ **44.** $a_1 = 5, a_{n+1} = a_n - 1$

45. $a_1 = 3, a_{n+1} = 2a_n$ **46.** $a_1 = 1, a_{n+1} = \dfrac{1}{2}a_n$

47. $a_1 = 7, a_{n+1} = -2a_n + 3$

48. $a_1 = -4, a_{n+1} = -3a_n - 5$

49. $a_1 = 2, a_{n+1} = \dfrac{1}{a_n}$ **50.** $a_1 = -1, a_{n+1} = \dfrac{-1}{a_n}$

51. $a_1 = 25, a_{n+1} = \dfrac{(-1)^n}{5a_n}$ **52.** $a_1 = 12, a_{n+1} = \dfrac{(-1)^n}{3a_n}$

In Exercises 53–58, use a graphing calculator to (a) find the first ten terms of the sequence and (b) graph the first ten terms of the sequence.

53. $a_n = 3n^2 - 1$ **54.** $a_n = 4 - \dfrac{3}{n}$

55. $a_n = n\left(1 - \dfrac{1}{n}\right)$ **56.** $a_n = n^3 - n^2$

57. $a_n = 1 - \dfrac{1}{a_{n-1}}, a_1 = \dfrac{1}{2}$ **58.** $a_n = a_{n-1}^2, a_1 = 1$

In Exercises 59–66, simplify the factorial expression.

59. $\dfrac{3!}{5!}$ **60.** $\dfrac{8!}{10!}$

61. $\dfrac{12!}{11!}$ **62.** $\dfrac{20!}{18!}$

63. $\dfrac{n!}{(n+1)!}$ **64.** $\dfrac{(n-1)!}{(n-2)!}$

65. $\dfrac{(2n+1)!}{(2n)!}$ **66.** $\dfrac{(2n+1)!}{(2n-1)!}$

In Exercises 67–78, find each sum.

67. $\displaystyle\sum_{k=1}^{7} 5$ **68.** $\displaystyle\sum_{j=1}^{4} 12$

69. $\displaystyle\sum_{j=0}^{5} j^2$ **70.** $\displaystyle\sum_{k=0}^{4} k^3$

71. $\displaystyle\sum_{i=1}^{5} (2i - 1)$ **72.** $\displaystyle\sum_{k=0}^{6} (1 - 3k)$

73. $\displaystyle\sum_{j=3}^{7} \dfrac{j + 1}{j}$ **74.** $\displaystyle\sum_{k=3}^{8} \dfrac{1}{k + 1}$

75. $\displaystyle\sum_{i=2}^{6} (-1)^i 3^{i-1}$ **76.** $\displaystyle\sum_{k=2}^{4} (-1)^{k+1} k$

77. $\displaystyle\sum_{k=4}^{7} (2 - k^2)$ **78.** $\displaystyle\sum_{j=4}^{9} (j^3 + 1)$

In Exercises 79–86, write each sum in summation notation.

79. $1 + 3 + 5 + 7 + \cdots + 101$

80. $2 + 4 + 6 + 8 + \cdots + 102$

81. $\dfrac{1}{5(1)} + \dfrac{1}{5(2)} + \dfrac{1}{5(3)} + \dfrac{1}{5(4)} + \cdots + \dfrac{1}{5(11)}$

82. $1 + \dfrac{2}{2 \cdot 3} + \dfrac{2}{3 \cdot 4} + \dfrac{2}{4 \cdot 5} + \cdots + \dfrac{2}{9 \cdot 10}$

83. $1 - \dfrac{1}{2} + \dfrac{1}{3} - \dfrac{1}{4} + \cdots - \dfrac{1}{50}$

84. $1 - 2 + 3 - 4 + \cdots + (-50)$

85. $\dfrac{1}{2} + \dfrac{2}{3} + \dfrac{3}{4} + \dfrac{4}{5} + \cdots + \dfrac{10}{11}$

86. $2 + \dfrac{2^2}{2} + \dfrac{2^3}{3} + \dfrac{2^4}{4} + \cdots + \dfrac{2^{10}}{10}$

In Exercises 87–92, use a graphing calculator to find each sum.

87. $\displaystyle\sum_{i=1}^{10} 12i^2$ **88.** $\displaystyle\sum_{i=1}^{50} (2i + 7)$

89. $\displaystyle\sum_{k=5}^{30} \dfrac{7}{1 - k^2}$ **90.** $\displaystyle\sum_{k=10}^{25} \dfrac{(k + 1)^2}{k}$

91. $\displaystyle\sum_{j=1}^{100} (-1)^j j$ **92.** $\displaystyle\sum_{j=8}^{50} \left(7 + \dfrac{(-1)^j}{j}\right)$

Applying the Concepts

93. Free fall. In the absence of friction, a freely falling body will fall about 16 feet the first second, 48 feet the next second, 80 feet the third second, 112 feet the fourth second, and so on. How far has it fallen during
  a. the seventh second?
  b. the nth second?

94. Bacterial growth. A colony of 1000 bacteria doubles in size every hour. How many bacteria will there be in
  a. 2 hours?
  b. 5 hours?
  c. n hours?

95. Workplace fines. A contractor who quits work on a house was told she would be fined $50 if she did not resume work on Monday, $75 if she failed to resume work on Tuesday, $100 if she did not resume work on Wednesday, and so on (including weekends). How much will her fine be on the ninth day she fails to show up for work?

96. Appreciating value. A painting valued at $30,000 is expected to appreciate $1280 the first year, $1240 the second year, $1200 the third year, and so on. How much will the painting appreciate in
  a. the seventh year?
  b. the tenth year?

97. Cell phone use. At the end of the first six months after a company began providing cell phones to its sales force, it averaged 600 cell minutes per month. For the next four years, its monthly cell phone use doubled every six months. How many cell minutes per month were being used at the end of three years?

98. Motorcycle acceleration. A motorcycle travels 10 yards the first second and then increases its speed by 20 yards per second in each succeeding second. At this rate, how far will the motorcycle travel during
  a. the eighth second?
  b. the nth second?

99. Compound interest. Suppose $10,000 is deposited into an account that earns 6% interest compounded semiannually. The balance in the account after n compounding periods is given by the sequence

$$A_n = 10,000\left(1 + \dfrac{0.06}{2}\right)^n, n = 1, 2, 3, \ldots.$$

  a. Find the first six terms of this sequence.
  b. Find the balance in the account after 8 years.

100. Compound interest. Suppose $100 is deposited at the beginning of a year into an account that earns 8% interest compounded quarterly. The balance in the account after n compounding periods is

$$A_n = 100\left(1 + \frac{0.08}{4}\right)^n, n = 1, 2, 3, \ldots.$$

    a. Find the first six terms of this sequence.
    b. Find the balance in the account after ten years.

101. Real estate value. Analysts estimate that a $100,000 condominium will increase 5% in value each year for the next seven years. Find the value of the condominium in each of those years. Write a formula for a sequence whose first seven terms give these values.

102. Diminishing savings. Laura withdraws 10% of her $50,000 savings each year. Find the amount remaining in her account for each of the next ten years. Write a formula for a sequence whose first ten terms give these values.

Beyond the Basics

103. Find a formula for the nth term of the sequence defined recursively by $a_1 = \sqrt{2}, a_{n+1} = \sqrt{2a_n}$. [*Hint:* Write each of the first five terms as a power of 2.]

104. The triangular tiles used in the figures shown have white interiors and gold edges. A sequence of figures is obtained by adding one triangle to the previous figure.

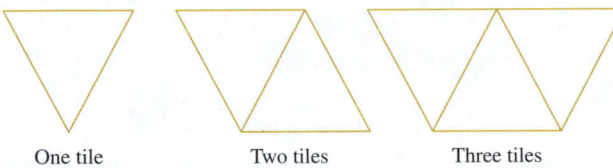

    One tile   Two tiles   Three tiles

    a. Write a recursive sequence whose nth term, a_n, gives the number of gold edges in the nth figure.
    b. Write a recursive sequence whose nth term, b_n, gives the number of gold edges that lie on the perimeter of the nth figure.

105. The figure shows a sequence of successively smaller squares formed by connecting the midpoints of the sides of the preceding larger square. The area of the largest square is 1.

    a. Write the first five terms of the sequence whose general term, a_n, gives the area of the nth square.
    b. Write the general term, a_n, of the sequence.

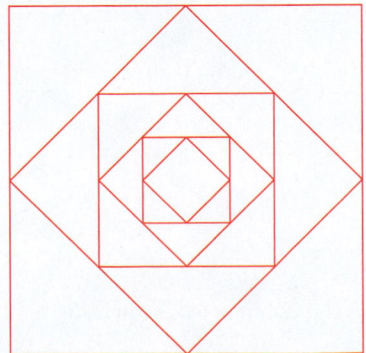

106. A sequence of concentric circles is designed so that the radius of the first circle is 1 and each successive circle has a radius that is twice the length of the radius of the preceding circle. Write a recursive sequence whose nth term, a_n, gives the area of the nth circle.

107. Find the first ten terms of the sequence defined recursively by $a_1 = a_2 = 1, a_n = a_{a_{n-1}} + a_{n-a_{n-1}}$. In 1989, a prize of $1000 was offered to whoever first discovered (for this sequence) a value of n for which $\left|\dfrac{a_i}{i}\right| - \left|\dfrac{1}{2}\right| < \dfrac{1}{20}$ for all $i > n$. A mathematician named Colin L. Mallows of AT&T/Bell Laboratories found that $n = 1489$ and claimed the prize.

    (*Source:* http://el.media.mit.edu/logo-foundation/pubs/papers/easy_as_11223.html)

108. A sequence of numbers $a_0, a_1, a_2, a_3, \ldots$ satisfies the equation

$$a_n^2 = (-1)^n a_{n-1} + a_{n+1}.$$

    If $a_0 = 1$ and $a_1 = 3$, find a_3.

109. Find a sequence with general term a_m such that

$$\sum_{n=1}^{20} n^2 = \sum_{m=0}^{19} a_m.$$

110. Find a real number c such that

$$\sum_{n=1}^{50} (n - 3)^2 = \sum_{n=1}^{50} (n^2 - 6n) + \sum_{n=1}^{50} c.$$

111. Find a lower limit p and an upper limit q such that

$$\sum_{n=0}^{10} n^3 = \sum_{m=p}^{q} (m + 2)^3.$$

112. Give an example of two sequences a_k and b_k such that

$$\sum_{k=1}^{20} a_k b_k \neq \left(\sum_{k=1}^{20} a_k\right)\left(\sum_{k=1}^{20} b_k\right).$$

Critical Thinking / Discussion / Writing

113. Find the largest integer value for the upper limit k if

$$\sum_{n=1}^{k} (n^2 + n)$$

$$= \sum_{n=1}^{k} \left[(n - 1)(n - 2)(n - 3)(n - 4)(n - 5) + n^2 + n \right].$$

114. The Ulam conjecture. Define the sequence a_n recursively by

$$a_n = \begin{cases} \dfrac{a_{n-1}}{2} & \text{if } a_{n-1} \text{ is even.} \\ 3a_{n-1} + 1 & \text{if } a_{n-1} \text{ is odd} \end{cases}$$

The Ulam conjecture says that if your first term a_0 is any positive integer, the sequence a_n will eventually reach the integer 1. Verify the conjecture for $a_0 = 13$.

Getting Ready for the Next Section

For each sequence in Exercises 115–118, find $a_2 - a_1, a_3 - a_2,$ $a_4 - a_3,$ and $a_5 - a_4.$

115. 1, 5, 9, 13, 17,

116. 2, 5.2, 8.4, 11.6, 14.8.

117. 6, 1, −4, −9, −14

118. 5, 1.5, −2, −5.5, −9

In Exercises 119–122, the first term a_1 and a number d are given. Write the next four terms of the sequence defined by $a_2 = a_1 + d, a_3 = a_2 + d, a_4 = a_3 + d,$ and $a_5 = a_4 + d.$

119. $a_1 = 3, d = 4$

120. $a_1 = 4, d = 2.5$

121. $a_1 = 7, d = -3$

122. $a_1 = 8, d = -4.5$

In Exercises 123–128, the nth term a_n is given. Find $a_2 - a_1, a_3 - a_2, a_4 - a_3,$ and $a_5 - a_4.$

123. $a_n = 3n + 5$

124. $a_n = 5n - 7$

125. $a_n = -2n - 6$

126. $a_n = -3n + 8$

127. Let $a_n = 2n - 5.$ Find a_{n+1} and $a_{n-1}.$

128. Let $a_n = -3n + 4.$ Find a_{n+1} and $a_{n-1}.$

9.2

Arithmetic Sequences; Partial Sums

BEFORE STARTING THIS SECTION, REVIEW

1 The general term of a sequence (Section 9.1, page 794)

2 Partial sums (Section 9.1, page 800)

OBJECTIVES

1 Identify an arithmetic sequence and find its common difference.

2 Find the sum of the first n terms of an arithmetic sequence.

◆ Falling Space Junk

One April, more than 300 kilograms (700 pounds) of "space junk" crashed to the ground in South Africa. The material was eventually identified as part of a Delta 2 rocket used to launch a global positioning satellite. Once an object begins a free fall toward Earth, it falls (in the absence of friction) 16 feet in the first second, 48 feet in the next second, 80 feet in the third second, and so on. The number of feet traveled in succeeding seconds is the sequence

$$16, 48, 80, 112, \ldots .$$

This is an example of an *arithmetic sequence*. We investigate such a sequence in Example 5.

---

1 Identify an arithmetic sequence and find its common difference.

Arithmetic Sequence

When the difference $(a_{n+1} - a_n)$ between any two consecutive terms of a sequence is always the same number, the sequence is called an *arithmetic sequence*. In other words, the sequence $a_1, a_2, a_3, a_4, \ldots$ is an arithmetic sequence if $a_2 - a_1 = a_3 - a_2 = a_4 - a_3 = \cdots$.

Arithmetic Sequence

The sequence

$$a_1, a_2, a_3, a_4, \ldots, a_n, \ldots$$

is an **arithmetic sequence**, or an **arithmetic progression**, if there is a number d such that each term in the sequence except the first is obtained from the preceding term by adding d to it. The number d is called the **common difference** of the arithmetic sequence. We have

$$d = a_{n+1} - a_n, \quad n \geq 1.$$

EXAMPLE 1 **Finding the Common Difference**

Find the common difference of each arithmetic sequence.

a. $3, 23, 43, 63, 83, \ldots$

b. $29, 19, 9, -1, -11, \ldots$

Solution

a. The common difference is $23 - 3 = 20$ (or $43 - 23, 63 - 43$, or $83 - 63$).

b. The common difference is $19 - 29 = -10$ (or $9 - 19, -1 - 9$, or $-11 - (-1)$).

Practice Problem 1 Find the common difference of the arithmetic sequence

$$3, -2, -7, -12, -17, \ldots .$$

An arithmetic sequence can be completely specified by giving the first term a_1 and the common difference d. Let's see why. Suppose we are told that the sequence $a_1, a_2, a_3, a_4, \ldots$ is an arithmetic sequence and that $a_1 = 5$ and $d = 4$. Then

$$a_2 - a_1 = 4, \quad \text{so} \quad a_2 = a_1 + 4 = 5 + 4 = 9;$$
$$a_3 - a_2 = 4, \quad \text{so} \quad a_3 = a_2 + 4 = 9 + 4 = 13;$$
$$a_4 - a_3 = 4, \quad \text{so} \quad a_4 = a_3 + 4 = 13 + 4 = 17; \text{ and so on.}$$

Rewriting $d = a_{n+1} - a_n$ as $a_{n+1} = a_n + d$, we have a recursive definition of an arithmetic sequence.

Recursive Definition of an Arithmetic Sequence

An arithmetic sequence $a_1, a_2, a_3, a_4, \ldots, a_n, \ldots$ can be defined recursively. The recursive formula

$$a_{n+1} = a_n + d \text{ for } n \geq 1$$

defines an arithmetic sequence with first term a_1 and common difference d.

Consider an arithmetic sequence with first term a_1 and common difference d. Using the recursive definition $a_{n+1} = a_n + d$, for $n \geq 1$, we write

$a_1 = a_1$ — The first term is a_1.

$a_2 = a_1 + d$ — Replace n with 1 in $a_{n+1} = a_n + d$.

$a_3 = a_2 + d = (a_1 + d) + d = a_1 + 2d$ — Replace a_2 with $a_1 + d$; simplify.

$a_4 = a_3 + d = (a_1 + 2d) + d = a_1 + 3d$ — Replace a_3 with $a_1 + 2d$; simplify.

$a_5 = a_4 + d = (a_1 + 3d) + d = a_1 + 4d$ — Replace a_4 with $a_1 + 3d$; simplify.

$$\vdots$$

$a_n = a_{n-1} + d$

$\quad = [a_1 + (n - 2)d] + d$ — Replace a_{n-1} with $a_1 + (n - 2)d$.

$\quad = a_1 + (n - 1)d$. — Simplify.

Thus, we can find the general term a_n from the values a_1 and d.

nTH TERM OF AN ARITHMETIC SEQUENCE

If a sequence $a_1, a_2, a_3, \ldots$ is an arithmetic sequence, then its nth term, a_n, is given by

$$a_n = a_1 + (n - 1)d,$$

where a_1 is the first term and d is the common difference.

DO YOU KNOW?

Notice that converting $a_n = a_1 + (n - 1)d$ to standard function notation gives $f(n) = d(n - 1) + a_1$, a linear function with slope d.

EXAMPLE 2 Finding the nth Term of an Arithmetic Sequence

Find an expression for the nth term of the arithmetic sequence

$$3, 7, 11, 15, \ldots .$$

Solution

Since $a_1 = 3$ and $d = 7 - 3 = 4$, we have

$$a_n = a_1 + (n - 1)d \qquad \text{Formula for the } n\text{th term}$$
$$= 3 + (n - 1)4 \qquad \text{Replace } a_1 \text{ with 3 and } d \text{ with 4.}$$
$$= 3 + 4n - 4 = 4n - 1 \qquad \text{Simplify.}$$

Note that in the expression $a_n = 4n - 1$, we get the sequence $3, 7, 11, 15, \ldots$ by substituting $n = 1, 2, 3, 4, \ldots$.

Practice Problem 2 Find an expression for the nth term of the arithmetic sequence

$$-3, 1, 5, 9, 13, 17, \ldots.$$

EXAMPLE 3 **Finding the Common Difference of an Arithmetic Sequence**

Find the common difference d and the nth term a_n of the arithmetic sequence whose 5th term is 15 and whose 20th term is 45.

Solution

$$a_n = a_1 + (n - 1)d \qquad \text{Formula for the } n\text{th term}$$
$$45 = a_1 + (20 - 1)d \qquad \text{Replace } n \text{ with 20; } a_n = a_{20} = 45.$$
$$(1) \quad 45 = a_1 + 19d. \qquad \text{Simplify.}$$

Also,

$$15 = a_1 + (5 - 1)d \qquad \text{Replace } n \text{ with 5; } a_n = a_5 = 15.$$
$$(2) \quad 15 = a_1 + 4d. \qquad \text{Simplify.}$$

Solving the system of equations

$$\begin{cases} a_1 + 19d = 45 & (1) \\ a_1 + 4d = 15 & (2) \end{cases}$$

gives $d = 2$ and $a_1 = 7$. We substitute $a_1 = 7$ and $d = 2$ in the formula for the nth term.

$$a_n = a_1 + (n - 1)d \qquad \text{Formula for the } n\text{th term}$$
$$a_n = 7 + (n - 1)2 \qquad \text{Replace } a_1 \text{ with 7 and } d \text{ with 2.}$$
$$= 7 + 2n - 2 = 2n + 5 \qquad \text{Simplify.}$$

The nth term of this sequence is given by

$$a_n = 2n + 5, \quad n \geq 1.$$

Practice Problem 3 Find the common difference d and the nth term a_n of the arithmetic sequence whose 4th term is 41 and whose 15th term is 8.

2 Find the sum of the first n terms of an arithmetic sequence.

Sum of a Finite Arithmetic Sequence

We can use the following formula in many applications involving the sum of a finite arithmetic sequence.

$$1 + 2 + 3 \cdots + n = \frac{n(n + 1)}{2}$$

Karl Friedrich Gauss

(1777–1855)

Karl Friedrich Gauss discovered this formula in his arithmetic class at the age of 8. One day, the story goes, the teacher became so incensed with the class that he assigned the students the task of adding up all of the numbers from 1 to 100. As Gauss's classmates dutifully began to add, Gauss walked up to the teacher and presented the answer, 5050. The story goes that the teacher was neither impressed nor amused. Here are Gauss's calculations:

$S = 1 + 2 + 3 + \cdots + 100$
$S = 100 + 99 + 98 + \cdots + 1$
$2S = \underbrace{101 + 101 + 101 + \cdots + 101}_{100 \text{ terms}}$
$2S = 100 \times 101$
$S = \dfrac{100 \times 101}{2} = 5050$

It is said that Karl Friedrich Gauss discovered this formula when he realized that any sum of numbers added in reverse order produces the same sum. Therefore, if S denotes the sum of the first n natural numbers, then

$$S = \quad 1 \quad + \quad 2 \quad + \quad 3 \quad + \cdots + \quad n$$
$$S = \quad n \quad + (n-1) + (n-2) + \cdots + \quad 1$$
$$2S = \underbrace{(n+1) + (n+1) + (n+1) + \cdots + (n+1)}_{n \text{ terms}} \qquad \text{Add } S + S = 2S.$$

$$2S = n(n+1)$$
$$S = \frac{n(n+1)}{2} \qquad \text{Divide both sides by 2.}$$

For example, if $n = 100$, we have

$$S = 1 + 2 + 3 + \cdots + 100 = \frac{100(100 + 1)}{2}$$
$$= \frac{100(101)}{2} = 5050.$$

We can use this method to calculate the sum S_n of the first n terms of any arithmetic sequence.

First, we write the sum S_n of the first n terms:

$$S_n = a_1 + (a_1 + d) + (a_1 + 2d) + (a_1 + 3d) + \cdots + a_n$$

We can also write S_n (with the terms in reverse order) by starting with a_n and *subtracting* the common difference d:

$$S_n = a_n + (a_n - d) + (a_n - 2d) + (a_n - 3d) + \cdots + a_1$$

Adding the two equations for S_n, we find that the d's in the sums "drop out" (for example, $(a_1 + d) + (a_n - d) = a_1 + a_n$) and we have

$$2S_n = \underbrace{(a_1 + a_n) + (a_1 + a_n) + \cdots + (a_1 + a_n)}_{n \text{ terms}}$$

$$2S_n = n(a_1 + a_n)$$

$$S_n = n\left(\frac{a_1 + a_n}{2}\right) \qquad \text{Divide both sides by 2.}$$

SUM OF THE FIRST n TERMS OF AN ARITHMETIC SEQUENCE

Let $a_1, a_2, a_3, \ldots a_n$ be the first n terms of an arithmetic sequence with common difference d. The sum S_n of these n terms is given by

$$S_n = n\left(\frac{a_1 + a_n}{2}\right)$$

where $a_n = a_1 + (n-1)d$.

EXAMPLE 4 **Finding the Sum of the Terms of a Finite Arithmetic Sequence**

Find the sum of the following arithmetic sequence of numbers:

$$1 + 4 + 7 + \cdots + 25$$

Solution

In this arithmetic sequence, $a_1 = 1$ and $d = 3$. Let's first find the number of terms.

$$a_n = a_1 + (n-1)d \qquad \text{Formula for } n\text{th term}$$

$$25 = 1 + (n-1)3 \qquad \text{Replace } a_n \text{ with 25, } a_1 \text{ with 1, and } d \text{ with 3.}$$

$$24 = (n-1)3 \qquad \text{Subtract 1 from both sides.}$$

$$8 = n - 1 \qquad \text{Divide both sides by 3.}$$

$$n = 9 \qquad \text{Solve for } n.$$

$$S_n = n\left(\frac{a_1 + a_n}{2}\right) \qquad \text{Formula for } S_n$$

$$S_9 = 9\left(\frac{1 + 25}{2}\right) \qquad \text{Replace } n \text{ with 9, } a_1 \text{ with 1, and } a_n(=a_9) \text{ with 25.}$$

$$= 9(13) = 117 \qquad \text{Simplify.}$$

So $1 + 4 + 7 + \cdots + 25 = 117$.

Practice Problem 4 Find the sum $\dfrac{2}{3} + \dfrac{5}{6} + 1 + \dfrac{7}{6} + \dfrac{4}{3} + \dfrac{3}{2} + \dfrac{5}{3} + \dfrac{11}{6} + 2 + \dfrac{13}{6}$.

◆ **EXAMPLE 5** **Calculating the Distance Traveled by a Freely Falling Object**

In the introduction to this section, we described the arithmetic sequence 16, 48, 80, 112, ... whose terms gave the number of feet that freely falling space junk falls in successive seconds. For this sequence, find the following:

a. The common difference d

b. The nth term a_n

c. The distance a piece of space junk travels in 10 seconds

Solution

a. The common difference $d = a_2 - a_1 = 48 - 16 = 32$.

b. $a_n = a_1 + (n-1)d \qquad \text{Formula for the } n\text{th term}$

$$= 16 + (n-1)32 \qquad \text{Replace } a_1 \text{ with 16 and } d \text{ with 32.}$$

$$= 32n - 16 \qquad \text{Simplify.}$$

c. The sum of the first ten terms of this sequence gives the distance the piece travels in ten seconds. We find a_{10} from the formula for a_n.

$$a_n = 32n - 16 \qquad \text{From part b}$$

$$a_{10} = 32(10) - 16 \qquad \text{Replace } n \text{ with 10.}$$

$$= 304$$

$$S_n = n\left(\frac{a_1 + a_n}{2}\right) \qquad \text{Formula for } S_n$$

$$S_{10} = 10\left(\frac{16 + 304}{2}\right) \qquad \text{Replace } n \text{ with 10, } a_1 \text{ with 16, and } a_n(=a_{10}) \text{ with 304.}$$

$$= 1600 \qquad \text{Simplify.}$$

The piece falls 1600 feet in ten seconds.

Practice Problem 5 In Example 5, find the distance the piece travels during the eleventh through fifteenth seconds.

SECTION 9.2   Exercises

Concepts and Vocabulary

1. If 5 is the common difference of an arithmetic sequence with general term a_n, then $a_{17} - a_{16} =$ _____.

2. If 5 is the common difference of an arithmetic sequence with general terms a_n and $a_{21} = 7$, then $a_{22} =$ _____.

3. If 14 is the term immediately following the sequence term 17 in an arithmetic sequence, then the common difference is _____.

4. If $a_1 = 2$ and $a_{11} = 22$ for an arithmetic sequence, then the common difference, d, is _____.

5. **True or False.** The common difference of an arithmetic sequence is always positive.

6. **True or False.** The sum of the first n terms of an arithmetic sequence always equals n times the average of its first and nth terms.

7. **True or False.** If a_1 is the first term of an arithmetic sequence with common difference d, the nth term is given by $a_n = a_1 + nd$.

8. **True or False.** The points of the graph of an arithmetic sequence with common difference d lie on a line with slope d.

Building Skills

In Exercises 9–22, determine whether each sequence is arithmetic. For those that are, find the first term a_1 and the common difference d.

9. $1, 2, 3, 4, 5, \ldots$.

10. $1, 3, 5, 7, 9, \ldots$

11. $2, 5, 8, 11, 14, \ldots$

12. $10, 7, 4, 1, -2, \ldots$

13. $1, \dfrac{1}{2}, 0, -\dfrac{1}{2}, -\dfrac{1}{4}, \ldots$

14. $2, 4, 8, 16, 32, \ldots$

15. $1, -1, 2, -2, 3, \ldots$

16. $-\dfrac{1}{4}, \dfrac{1}{4}, \dfrac{3}{4}, \dfrac{5}{4}, \dfrac{7}{4}, \ldots$

17. $0.6, 0.2, -0.2, -0.6, -1, \ldots$

18. $2.3, 2.7, 3.1, 3.5, 3.9, \ldots$

19. $a_n = 2n + 6$

20. $a_n = 1 - 5n$

21. $a_n = 1 - n^2$

22. $a_n = 2n^2 - 3$

In Exercises 23–32, find an expression for the nth term of the arithmetic sequence.

23. $5, 8, 11, 14, 17, \ldots$

24. $4, 7, 10, 13, 16, \ldots$

25. $11, 6, 1, -4, -9, \ldots$

26. $9, 5, 1, -3, -7, \ldots$

27. $\dfrac{1}{2}, \dfrac{1}{4}, 0, -\dfrac{1}{4}, -\dfrac{1}{2}, \ldots$

28. $\dfrac{2}{3}, \dfrac{5}{6}, 1, \dfrac{7}{6}, \dfrac{4}{3}, \ldots$

29. $-\dfrac{3}{5}, -1, -\dfrac{7}{5}, -\dfrac{9}{5}, -\dfrac{11}{5}, \ldots$

30. $\dfrac{1}{2}, 2, \dfrac{7}{2}, 5, \dfrac{13}{2}, \ldots$

31. $e, 3 + e, 6 + e, 9 + e, 12 + e, \ldots$

32. $2\pi, 2(\pi + 2), 2(\pi + 4), 2(\pi + 6), 2(\pi + 8) \ldots ,$

In Exercises 33–38, find the common difference d and the nth term a_n of the arithmetic sequence with the specified terms.

33. 4th term 21; 10th term 60

34. 3rd term 15; 21st term 87

35. 7th term 8; 15th term -8

36. 5th term 12; 18th term -1

37. 3rd term 7; 23rd term 17

38. 11th term -1; 31st term 5

In Exercises 39–48, find the sum of each arithmetic sequence.

39. $1 + 2 + 3 + \cdots + 50$

40. $2 + 4 + 6 + \cdots + 102$

41. $1 + 3 + 5 + \cdots + 99$

42. $5 + 10 + 15 + \cdots + 200$

43. $3 + 6 + 9 + \cdots + 300$

44. $4 + 7 + 10 + \cdots + 301$

45. $2 - 1 - 4 - \cdots - 34$

46. $-3 - 8 - 13 - \cdots - 48$

47. $\dfrac{1}{3} + 1 + \dfrac{5}{3} + \cdots + 7$

48. $\dfrac{3}{5} + 2 + \dfrac{17}{5} + \cdots + \dfrac{101}{5}$

In Exercises 49–54, find the sum of the first n terms of the given arithmetic sequence.

49. $2, 7, 12, \ldots ; n = 50$

50. $8, 10, 12, \ldots ; n = 40$

51. $-15, -11, -7, \ldots ; n = 20$

52. $-20, -13, -6, \ldots ; n = 25$

53. $3.5, 3.7, 3.9, \ldots ; n = 100$

54. $-7, -6.5, -6, \ldots ; n = 80$

In Exercises 55–60, find n for the given value of a_n in each arithmetic sequence.

55. $a_n = 75; 1, 3, 5, \ldots$

56. $a_n = 120; 2, 4, 6, \ldots$

57. $a_n = 95; -1, 3, 7, \ldots$

58. $a_n = 83; -5, -3, -1, \ldots$

59. $a_n = 50\sqrt{3}; 2\sqrt{3}, 4\sqrt{3}, 6\sqrt{3}, \ldots$

60. $a_n = 73\pi; 3\pi, 5\pi, 7\pi, \ldots$

Applying the Concepts

In Exercises 61–67, assume that the indicated sequence is arithmetic.

61. Orange picking. When Eric started work as an orange picker, he picked 10 oranges in the first minute, 12 in the second minute, 14 in the third minute, and so on. How many oranges did Eric pick in the first half hour?

62. Exercise. Walking up a steep hill, Jan walks 60 feet in the first minute, 57 feet in the second minute, 54 feet in the third minute, and so on.
   a. How far will Jan walk in the nth minute?
   b. How far will Jan walk in the first 15 minutes?

63. Contest winner. A contest winner will receive money each month for three years. The winner receives $50 the first month, $75 the second month, $100 the third month, and so on. How much money will the winner have collected after 30 months?

64. Salary. Darren took a 12-month temporary job with a monthly salary that increased a fixed amount each month. He can't recall the starting salary, but does remember that he was paid $820 at the end of the third month and $910 for his last month's work. How much was Darren's total pay for the entire 12 months?

65. Hourly wage. Antonio's new weekend job started at an hourly wage of $12.75. He is guaranteed a raise of 25¢ an hour every three months for the next four years. What will Antonio's hourly wage be at the end of four years?

66. Competing job offers. Denzel is considering offers from two companies. A marketing company pays $32,500 the first year and guarantees a raise of $1300 each year; an exporting company pays $36,000 the first year, with a guaranteed raise of $400 each year. Over a five-year period, which company will pay more? How much more?

67. Theater seating. A theater has 25 rows of seats. The first row has 20 seats, the second row has 22 seats, the third row has 24 seats, and so on. How many seats are in the theater?

68. Bricks in a driveway. A brick driveway has 50 rows of bricks. The first row has 16 bricks, and the fiftieth row has 65 bricks. Assuming that the sequence of numbers giving the number of bricks in rows 1 through 50 is arithmetic, how many bricks does the driveway contain?

69. Stacked boxes. A stack of boxes has 17 rows. The bottom row has 40 boxes, and the top row has 8 boxes. Assuming the sequence of numbers that gives the number of boxes in succeeding rows is arithmetic, how many boxes does the stack contain?

Beyond the Basics

70. Elena wanted to teach her young daughter Sophia about saving. She gave Sophia 10¢ the first day and an additional 5¢ per day on each subsequent day. After a while, Sophia counted her money and found that she had saved a total of $3.25. How many days had she been saving?

71. Only 3 customers showed up for the opening day of a flea market. However, 9 came the second day, and an additional 6 customers came on each subsequent day. After several days, there was a total of 192 customers. How many days had the flea market been open?

72. Find the sum of all natural numbers between 45 and 100 that are divisible by 3.

73. Find the sum of all natural numbers between 26 and 120 that are divisible by 7.

74. A sequence is **harmonic** if the reciprocals of the terms of the sequence form an arithmetic sequence. Is the sequence $\dfrac{1}{2}, \dfrac{3}{5}, \dfrac{3}{4}, 1, \dfrac{3}{2}, 3 \ldots$ harmonic? Explain your reasoning.

75. Consider the harmonic sequence $\dfrac{2}{5}, \dfrac{2}{7}, \dfrac{2}{9}, \ldots$
   a. Find the fourth term.
   b. Find the nth term.

76. Show that if the numbers a, m, b are consecutive terms in an arithmetic sequence, then $m = \dfrac{a + b}{2}$. We call m the **arithmetic mean** of a and b.

77. If the numbers $a, m_1, m_2, \ldots, m_k, b$ form an arithmetic sequence, we say that the numbers $m_1, m_2, \ldots, m_k$ are k **arithmetic means** between a and b. Insert k arithmetic means between 1 and 30 so that the sum of the resulting series is 465.

Critical Thinking / Discussion / Writing

78. Find the nth term of an arithmetic sequence whose first term is 10 and whose 21st term is 0.

79. If a_1 and d are the first term and common difference, respectively, of an arithmetic sequence, find the first term and difference of an arithmetic sequence whose terms are the negatives of the terms in the given sequence.

80. How many terms are in the series $\displaystyle\sum_{i=22}^{100} 17i^3$?

81. The sum of the first n counting numbers is 12,403. Find the value of n.

Getting Ready for the Next Section

For each sequence in Exercises 82–85, find $\dfrac{a_2}{a_1}, \dfrac{a_3}{a_2}, \dfrac{a_4}{a_3}$, and $\dfrac{a_5}{a_4}$.

82. 3, 6, 12, 24, 48

83. $2, \dfrac{4}{3}, \dfrac{8}{9}, \dfrac{16}{27}, \dfrac{32}{81}$

84. $4, -12, 36, -108, 324$

85. $3, -\dfrac{3}{2}, \dfrac{3}{4}, -\dfrac{3}{8}, \dfrac{3}{16}$

In Exercises 86–89, the first term a_1 and a number r are given. Write the next four terms of the sequence defined by $a_2 = a_1 r, a_3 = a_2 r, a_4 = a_3 r,$ and $a_5 = a_4 r$.

86. $a_1 = 2, r = 3$

87. $a_1 = 3, r = \dfrac{2}{5}$

88. $a_1 = 1, r = -2$

89. $a_1 = 1, r = -\dfrac{3}{2}$

In Exercises 90–95, the nth term a_n of a sequence is given. Find $\dfrac{a_2}{a_1}, \dfrac{a_3}{a_2}, \dfrac{a_4}{a_3},$ and $\dfrac{a_5}{a_4}$.

90. $a_n = 5 \cdot 2^n$

91. $a_n = 4 \cdot 3^n$

92. $a_n = 2(-3)^n$

93. $a_n = 5(-2)^{-n}$

94. Let $a_n = -3 \cdot 2^n$. Find a_{n+1} and a_{n-1}.

95. Let $a_n = 5(-3)^{-n}$. Find a_{n+1} and a_{n-1}.

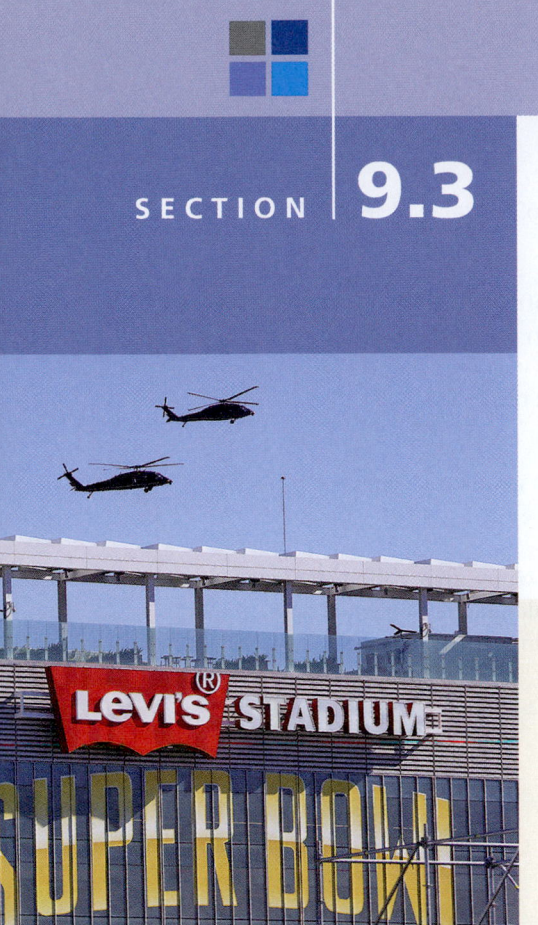

Geometric Sequences and Series

BEFORE STARTING THIS SECTION, REVIEW

1   The general term of a sequence (Section 9.1, page 794)

2   Partial sums (Section 9.1, page 800)

3   Compound interest (Section 3.1, page 260)

OBJECTIVES

1   Identify a geometric sequence and find its common ratio.

2   Find the sum of a finite geometric sequence.

3   Solve annuity problems.

4   Find the sum of an infinite geometric sequence.

◆ Spreading the Wealth

Cities across the nation compete to host a Super Bowl. Cities across the world compete to host the Olympic Games. What is the incentive? The *multiplier effect* is often listed among the benefits for hosting one of these events. The effect refers to the phenomenon that occurs when money spent directly on an event has a ripple effect on the economy that is many times the amount spent directly on the event.

Suppose, for example, that $10 million is spent directly by tourists for hotels, meals, tickets, cabs, and so on. Some portion of that amount will be respent by its recipients on groceries, clothes, gas, and so on.

This process is repeated over and over, affecting the economy much more than the initial $10 million spent. Under certain assumptions, the spending just described results in an *infinite geometric series*, whose sum is called the *multiplier*. Example 8 illustrates this effect.

---

1 Identify a geometric sequence and find its common ratio.

Geometric Sequence

When the *ratio* $\left(\dfrac{a_{n+1}}{a_n}\right)$ of any two consecutive terms of a sequence is always the same number, the sequence is called a *geometric sequence*. In other words, the sequence $a_1, a_2, a_3, a_4, \ldots$ is a geometric sequence if

$$\frac{a_2}{a_1} = \frac{a_3}{a_2} = \frac{a_4}{a_3} = \cdots .$$

Geometric Sequence

The sequence

$$a_1, a_2, a_3, a_4, \ldots, a_n, \ldots$$

is a **geometric sequence**, or a **geometric progression**, if there is a number r such that each term except the first in the sequence is obtained by multiplying the previous term by r. The number r is called the **common ratio** of the geometric sequence. We have

$$\frac{a_{n+1}}{a_n} = r, \quad n \geq 1$$

> **EXAMPLE 1** **Finding the Common Ratio**

Find the common ratio for each geometric sequence.

a. 2, 10, 50, 250, 1250, ... **b.** $-162, -54, -18, -6, -2, \ldots$

Solution

a. The common ratio is $\dfrac{10}{2} = 5 \left(\text{or } \dfrac{50}{10}, \dfrac{250}{50}, \text{ or } \dfrac{1250}{250} \right).$

b. The common ratio is $\dfrac{-54}{-162} = \dfrac{1}{3} \left(\text{or } \dfrac{-18}{-54}, \dfrac{-6}{-18}, \text{ or } \dfrac{-2}{-6} \right).$

Practice Problem 1 Find the common ratio for the geometric sequence 6, 18, 54, 162, 486, ...

A geometric sequence can be completely specified by giving the first term a_1 and the common ratio r. Suppose, for example, that we are told that the sequence $a_1, a_2, a_3, a_4, \ldots$ is geometric and that $a_1 = 3$ and $r = 4$. Then we can find a_2:

$$\frac{a_2}{a_1} = 4, \text{ so } a_2 = 4a_1 = 4 \cdot 3 = 12$$

Now we continue:

$$\frac{a_3}{a_2} = 4, \text{ so } a_3 = 4a_2 = 4 \cdot 12 = 48;$$

$$\frac{a_4}{a_3} = 4, \text{ so } a_4 = 4a_3 = 4 \cdot 48 = 192; \text{ and so on.}$$

By rewriting $\dfrac{a_{n+1}}{a_n} = r$ as $a_{n+1} = ra_n$, we have a recursive definition of a geometric sequence.

Recursive Definition of a Geometric Sequence

A geometric sequence $a_1, a_2, a_3, a_4, \ldots, a_n, \ldots$ can be defined recursively. The recursive formula

$$a_{n+1} = ra_n, \quad n \geq 1$$

defines a geometric sequence with the first term a_1 and the common ratio r.

> **EXAMPLE 2** **Determining Whether a Sequence Is Geometric**

Determine whether each sequence is geometric. If it is, find the first term and the common ratio.

a. $a_n = 4^n$ **b.** $b_n = \dfrac{3}{2^n}$ **c.** $c_n = 1 - 5^n$

Solution

a. The sequence $4, 16, 64, 256, \ldots, 4^n, \ldots$ appears to be geometric, with $a_1 = 4$ and $r = 4$. To verify this, we check that $\dfrac{a_{n+1}}{a_n} = 4$ for all $n \geq 1$.

$$a_{n+1} = 4^{n+1} \qquad \text{Replace } n \text{ with } n + 1 \text{ in the general term } a_n = 4^n.$$

$$\frac{a_{n+1}}{a_n} = \frac{4^{n+1}}{4^n} = 4 \qquad \frac{4^{n+1}}{4^n} = \frac{\cancel{4^n} \cdot 4}{\cancel{4^n}} = 4$$

Consequently, $\dfrac{a_{n+1}}{a_n} = 4$ for all $n \geq 1$, and the sequence is geometric.

b. The sequence $\dfrac{3}{2}, \dfrac{3}{4}, \dfrac{3}{8}, \dfrac{3}{16}, \ldots$ appears to be geometric, with $b_1 = \dfrac{3}{2}$ and $r = \dfrac{1}{2}$.

To verify this, we check that $\dfrac{b_{n+1}}{b_n} = \dfrac{1}{2}$ for all $n \geq 1$.

$$b_{n+1} = \frac{3}{2^{n+1}} \qquad \text{\color{blue}Replace } n \text{ with } n+1 \text{ in the general term } b_n = \frac{3}{2^n}.$$

$$\frac{b_{n+1}}{b_n} = \frac{3}{2^{n+1}} \div \frac{3}{2^n} \qquad {\color{blue}\frac{b_{n+1}}{b_n} = b_{n+1} \div b_n}$$

$$= \frac{3}{2^{n+1}} \cdot \frac{2^n}{3} = \frac{1}{2}$$

Consequently, $\dfrac{b_{n+1}}{b_n} = \dfrac{1}{2}$ for all $n \geq 1$, and the sequence is geometric.

c. The sequence $-4, -24, -124, -624, \ldots, 1 - 5^n, \ldots$ is *not* a geometric sequence because not all pairs of consecutive terms have identical quotients, $\dfrac{a_{n+1}}{a_n}$. The quotient of the first two terms is $\dfrac{-24}{-4} = 6$, but the quotient of the third and second terms is $\dfrac{-124}{-24} = \dfrac{31}{6}$.

Practice Problem 2 Determine whether the sequence $a_n = \left(\dfrac{3}{2}\right)^n$ is geometric. If so, find the first term and the common ratio.

From the equation $a_{n+1} = a_n r$ used in the recursive definition of a geometric sequence, we see that each term after the first is obtained by multiplying the preceding term by the common ratio r.

THE GENERAL TERM OF A GEOMETRIC SEQUENCE

Every geometric sequence can be written in the form

$$a_1, a_1 r, a_1 r^2, a_1 r^3, \ldots, a_1 r^{n-1}, \ldots,$$

where r is the common ratio. Since $a_1 = a_1(1) = a_1 r^0$, the **nth term of the geometric sequence** is

$$a_n = a_1 r^{n-1}, \text{ for } n \geq 1.$$

EXAMPLE 3 **Finding Terms in a Geometric Sequence**

For the geometric sequence $1, 3, 9, 27, \ldots$, find each of the following:

a. a_1 **b.** r **c.** a_n

Solution

a. The first term of the sequence is given: $a_1 = 1$.

b. The sequence is geometric, so we can take the ratio of any two consecutive terms, $\dfrac{a_{n+1}}{a_n}$. The ratio of the first two terms gives us

$$r = \frac{3}{1} = 3.$$

c. $a_n = a_1 r^{n-1}$ Formula for the nth term

     $= (1)(3^{n-1})$ Replace a_1 with 1 and r with 3.

     $= 3^{n-1}$

Practice Problem 3 For the geometric sequence $2, \dfrac{6}{5}, \dfrac{18}{25}, \dfrac{54}{125}, \ldots$, find the following:

a. a_1 **b.** r **c.** a_n

Notice in Example 3 that converting $a_n = 3^{n-1}$ to standard function notation gives us $f(n) = 3^{n-1}$, an exponential function. **A geometric sequence with first term a_1 and common ratio r, other than 1, can be viewed as an exponential function $f(n) = a_1 r^{n-1}$ with the set of natural numbers as its domain.**

EXAMPLE 4   **Finding a Particular Term in a Geometric Sequence**

Find the 23rd term of a geometric sequence whose first term is 10 and whose common ratio is 1.2.

Solution

$a_n = a_1 r^{n-1}$ Formula for the nth term

$a_{23} = 10(1.2)^{23-1}$ Replace n with 23, a_1 with 10, and r with 1.2.

     $= 10(1.2)^{22}$

     ≈ 552.06 Use a calculator.

Practice Problem 4 Find the 18th term of a geometric sequence whose first term is 7 and whose common ratio is 1.5.

2 Find the sum of a finite geometric sequence.

Finding the Sum of a Finite Geometric Sequence

We can find a formula for the sum S_n of the first n terms in a geometric sequence. (We have already found a similar formula for arithmetic sequences.)

$S_n = a_1 + a_1 r + a_1 r^2 + a_1 r^3 + \cdots + a_1 r^{n-1}$ Definition of S_n

$rS_n = a_1 r + a_1 r^2 + a_1 r^3 + a_1 r^4 + \cdots + a_1 r^n$ Multiply both sides by r.

$S_n - rS_n = a_1 - a_1 r^n$ Subtract the second equation from the first.

$(1 - r)S_n = a_1(1 - r^n)$ Factor both sides.

$S_n = \dfrac{a_1(1 - r^n)}{1 - r}, r \neq 1$ Divide both sides by $1 - r$.

SUM OF THE TERMS OF A FINITE GEOMETRIC SEQUENCE

Let $a_1, a_2, a_3, \ldots, a_n$ be the first n terms of a geometric sequence with first term a_1 and common ratio r. The sum S_n of these terms is

$$S_n = \sum_{i=1}^{n} a_1 r^{i-1} = \frac{a_1(1 - r^n)}{1 - r}, r \neq 1.$$

A geometric sequence with $r = 1$ is a sequence in which all terms are identical; that is, $a_n = a_1(1)^{n-1} = a_1$. Consequently, the sum of the first n terms, S_n, is

$$\underbrace{a_1 + a_1 + \cdots + a_1}_{n \text{ terms}} = na_1.$$

EXAMPLE 5 Finding the Sum of the Terms of a Finite Geometric Sequence

Find each sum.

a. $\displaystyle\sum_{i=1}^{15} 5(0.7)^{i-1}$ **b.** $\displaystyle\sum_{i=1}^{15} 5(0.7)^{i}$

Solution

a. We can evaluate $\displaystyle\sum_{i=1}^{15} 5(0.7)^{i-1}$ by using the formula for $S_n = \displaystyle\sum_{i=1}^{n} a_1 r^{i-1} = \dfrac{a_1(1-r^n)}{1-r}$,

where $a_1 = 5$, $r = 0.7$, and $n = 15$.

$$S_{15} = \sum_{i=1}^{15} a_1 r^{i-1} = 5\left[\frac{1-(0.7)^{15}}{1-0.7}\right] \qquad \text{Replace } a_1 \text{ with } 5, r = 0.7, \text{ and } n = 15.$$

$$\approx 16.5875 \qquad\qquad\qquad \text{Use a calculator.}$$

b. $\displaystyle\sum_{i=1}^{15} 5(0.7)^{i} = (0.7)\sum_{i=1}^{15} 5(0.7)^{i-1}$ Factor out 0.7 from each term.

$$\approx (0.7)(16.5875) \qquad \text{From part } \mathbf{a}$$

$$= 11.6113$$

Practice Problem 5 Find the sum $\displaystyle\sum_{i=1}^{17} 3(0.4)^{i}$.

3 Solve annuity problems.

Annuities

One of the most important applications of finite geometric series is the computation of the value of an *annuity*. An **annuity** is a sequence of equal periodic payments. When a fixed rate of compound interest applies to all payments, the sum of all the payments made plus all interest can be found by using the formula for the sum of a finite geometric sequence. The **value of an annuity** (also called the **future value of an annuity**) is the sum of all payments and interest. To develop a formula for the value of an annuity, we suppose P is deposited at the end of each year at an annual interest rate i compounded annually.

The compound interest formula $A = P(1 + i)^t$ gives the total value after t years when P earns an annual interest rate i (in decimal form) compounded once a year. At the end of the first year, the initial payment of P is made and the annuity's value is P. At the end of the second year, P is again deposited. At this time, the first deposit has earned interest during the second year. The value of the annuity after two years is

$$P \qquad + \qquad P(1 + i).$$

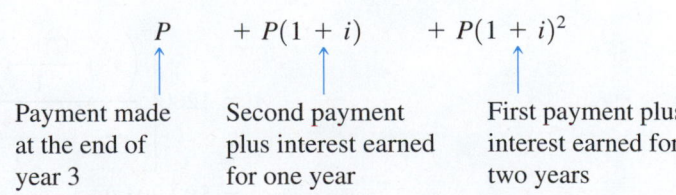

Payment made at the   First payment plus interest
end of year 2   earned for one year

The value of the annuity after three years is

$$P \qquad + P(1 + i) \qquad + P(1 + i)^2$$

Payment made   Second payment   First payment plus
at the end of   plus interest earned   interest earned for
year 3   for one year   two years

The value of the annuity after t years is

$$P + P(1 + i) + P(1 + i)^2 + P(1 + i)^3 + \cdots + P(1 + i)^{t-1}.$$

Payment made at
the end of year t

First payment plus interest
earned for $t - 1$ year

The value of the annuity after t years is the sum of a geometric sequence with first term $a_1 = P$ and common ratio $r = 1 + i$. We compute this sum A as follows:

$$A = S_n = \frac{a_1(1 - r^n)}{1 - r} \qquad \text{Formula for } S_n$$

$$A = \frac{P[1 - (1 + i)^t]}{1 - (1 + i)} \qquad \text{Replace } n \text{ with } t, a_1 \text{ with } P, \text{ and } r \text{ with } 1 + i.$$

$$= \frac{P[1 - (1 + i)^t]}{-i} \qquad \text{Simplify.}$$

$$= P\frac{[(1 + i)^t - 1]}{i} \qquad \text{Multiply numerator and denominator by } -1.$$

If interest is compounded n times per year and equal payments are made at the end of each compounding period, we adjust the formula as we did in Section 3.1 to account for the more frequent compounding.

VALUE OF AN ANNUITY

Let P represent the payment in dollars made at the end of each of n compounding periods per year and let i be the annual interest rate. Then the value A of the annuity after t years is:

$$A = P\left[\frac{\left(1 + \dfrac{i}{n}\right)^{nt} - 1}{\dfrac{i}{n}}\right]$$

EXAMPLE 6 **Finding the Value of an Annuity**

An individual retirement account (IRA) is a common way to save money to provide funds after retirement. Suppose you make payments of $1200 into an IRA at the end of each year at an annual interest rate of 4.5% per year, compounded annually. What is the value of this annuity after 35 years?

Solution

Each annuity payment is $P = \$1200$. The annual interest rate is 4.5% (so $i = 0.045$), and the number of years is $t = 35$. Because interest is compounded annually, $n = 1$. The value of the annuity is

$$A = 1200\left[\frac{\left(1 + \dfrac{0.045}{1}\right)^{(1)35} - 1}{\dfrac{0.045}{1}}\right]$$

$$= \$97,795.94 \qquad \text{Use a calculator.}$$

The value of the IRA after 35 years is $97,795.94.

Practice Problem 6 If in Example 6 the end-of-year payments are $1500, the annual interest rate remains 4.5%, compounded annually, and the payments are made for 30 years, what is the value of the annuity?

4 Find the sum of an infinite geometric sequence.

Infinite Geometric Series

The sum S_n of the first n terms of a geometric series is given by the formula $S_n = \dfrac{a_1(1 - r^n)}{1 - r}$.

If this finite sum S_n approaches a number S as $n \to \infty$ (that is, as n gets larger and larger), we say that S is the **sum of the infinite geometric series**, and we write

$$S = \sum_{i=1}^{\infty} a_1 r^{i-1}.$$

If r is any real number with $-1 < r < 1$ (equivalently, $|r| < 1$), then the value of r^n, like the value of $\left(\dfrac{1}{2}\right)^n$, gets closer and closer to 0 as n gets larger and larger. We indicate that the values of r^n approach 0 by writing

$$\lim_{n \to \infty} r^n = 0 \text{ if } |r| < 1.$$

The expression $\lim\limits_{n \to \infty} r^n$ is read "the limit, as n approaches infinity, of r to the n" or "the limit, as n approaches infinity, of r to the nth power."

When $|r| < 1$,

$$S_n = \frac{a_1(1 - r^n)}{1 - r} \to \frac{a_1(1 - 0)}{1 - r} = \frac{a_1}{1 - r} \quad \text{as } n \to \infty.$$

SUM OF THE TERMS OF AN INFINITE GEOMETRIC SEQUENCE

If $|r| < 1$, the infinite sum

$$a_1 + a_1 r + a_1 r^2 + a_1 r^3 + \cdots + a_1 r^{n-1} + \cdots$$

is given by

$$S = \sum_{i=1}^{\infty} a_1 r^{i-1} = \frac{a_1}{1 - r}.$$

When $|r| \geq 1$, the infinite geometric series does not have a sum. This is because the value of r^n does not approach 0 as $n \to \infty$. For example, if $r = 2$, we have

$$2^1 = 2, 2^2 = 4, 2^3 = 8, 2^4 = 16, 2^5 = 32, 2^6 = 64, \ldots.$$

EXAMPLE 7   **Finding the Sum of an Infinite Geometric Series**

Find the sum $2 + \dfrac{3}{2} + \dfrac{9}{8} + \dfrac{27}{32} + \cdots$.

Solution

The first term is $a_1 = 2$, and the common ratio is

$$r = \frac{3/2}{2} = \frac{3}{4}.$$

Because $|r| = \dfrac{3}{4} < 1$, we can use the formula for the sum of an infinite geometric series.

$$S = \frac{a_1}{1 - r} \qquad \text{Formula for } S$$

$$= \frac{2}{1 - \dfrac{3}{4}} \qquad \text{Replace } a_1 \text{ with 2 and } r \text{ with } \dfrac{3}{4}.$$

$$= 8 \qquad \text{Simplify.}$$

Practice Problem 7   Find the sum $3 + \dfrac{6}{3} + \dfrac{12}{9} + \dfrac{24}{27} + \cdots$.

EXAMPLE 8   Calculating the Multiplier Effect

The host city for the Super Bowl expects that tourists will spend $10,000,000. Assume that 80% of this money is spent again in the city, then 80% of this second round of spending is spent again, and so on. In the introduction to this section, we said that such a spending pattern results in a geometric series whose sum is called the *multiplier*. Find this series and its sum.

Solution

We start our series with the $10,000,000 brought into the city and add the subsequent amounts spent.

$$10,000,000 + 10,000,000\,(0.80) + 10,000,000\,(0.80)^2 + 10,000,000\,(0.80)^3 + \cdots$$

original 10,000,000   80% of 10,000,000   80% of previous amount   80% of previous amount

Using the formula $\displaystyle\sum_{i=1}^{\infty} ar^{i-1} = \frac{a}{1-r}$ for the sum of an infinite geometric series, we have

$$\sum_{i=1}^{\infty} (10,000,000)(0.80)^{i-1} = \frac{\$10,000,000}{1 - 0.80} = \$50,000,000.$$

This amount should shed some light on why there is so much competition to serve as the host city for a major sports event.

Practice Problem 8   Find the series and the sum that results in Example 8 assuming that 85%, rather than 80%, occurs in the repeated spending.

Answers to Practice Problems

1. 3   **2.** It is geometric; $a_1 = \dfrac{3}{2}, r = \dfrac{3}{2}$.

3. a. $a_1 = 2$ **b.** $r = \dfrac{3}{5}$ **c.** $a_n = 2\left(\dfrac{3}{5}\right)^{n-1}$

4. $7(1.5)^{17} \approx 6896.8288$ **5.** ≈ 2 **6.** $91,510.60$

7. 9   **8.** $\displaystyle\sum_{i=1}^{\infty} (10,000,000)(0.85)^i \approx \$66,666,666.67$

SECTION 9.3 **Exercises**

Concepts and Vocabulary

1. If 5 is the common ratio of a geometric sequence with general term a_n, then $\dfrac{a_{63}}{a_{62}} =$ _____.

2. If 5 is the common ratio of a geometric sequence with general term a_n and $a_{15} = -2$, then $a_{16} =$ _____.

3. If 24 is the term immediately following the sequence term 8 in a geometric sequence, then the common ratio is

   _____.

4. An infinite geometric series $a_1 + a_1 r + a_1 r^2 + \cdots$ does not have a sum if _____ ≥ 1.

5. **True or False.** If the sum of an infinite geometric series having the common ratio $r = \dfrac{1}{3}$ is $S = 15$, then $a_1 = 10$.

6. **True or False.** The sum of an infinite geometric series can never be negative.

7. **True or False.** The general term for a geometric sequence with first term a_1 and common ratio r is $a_n = a_1 r^n$.

8. **True or False.** $\displaystyle\sum_{k=5}^{n} a_n = \sum_{k=1}^{n} a_n - \sum_{k=1}^{4} a_n$.

Building Skills

In Exercises 9–28, determine whether each sequence is geometric. If it is, find the first term and the common ratio.

9. $3, 6, 12, 24, \ldots$

10. $2, 4, 8, 16, \ldots$

11. $1, 5, 10, 20, \ldots$

12. $1, 1, 3, 3, 9, 9, \ldots$

13. $1, -3, 9, -27, \ldots$

14. $-1, 2, -4, 8, \ldots$

15. $7, -7, 7, -7, \ldots$

16. $1, -2, 4, -8, \ldots$

17. $9, 3, 1, \dfrac{1}{3}, \ldots$

18. $5, 2, \dfrac{4}{5}, \dfrac{8}{25}, \ldots$

19. $a_n = \left(-\dfrac{1}{2}\right)^n$

20. $a_n = 5\left(\dfrac{2}{3}\right)^n$

21. $a_n = 2^{n-1}$

22. $a_n = -(1.06)^{n-1}$

23. $a_n = 7n^2 + 1$

24. $a_n = 1 - (2n)^2$

25. $a_n = 3^{-n}$

26. $a_n = 50(0.1)^{-n}$

27. $a_n = 5^{n/2}$

28. $a_n = 7^{\sqrt{n}}$

In Exercises 29–36, find the first term a_1, the common ratio r, and the nth term a_n for each geometric sequence.

29. $2, 10, 50, 250, \ldots$

30. $-3, -6, -12, -24, \ldots$

31. $5, \dfrac{10}{3}, \dfrac{20}{9}, \dfrac{40}{27}, \ldots$

32. $1, \sqrt{3}, 3, 3\sqrt{3}, \ldots$

33. $0.2, -0.6, 1.8, -5.4, \ldots$

34. $1.3, -0.26, 0.052, -0.0104, \ldots$

35. $\pi^4, \pi^6, \pi^8, \pi^{10}, \ldots$

36. $e^2, 1, e^{-2}, e^{-4}, \ldots$

In Exercises 37–46, find the indicated term of each geometric sequence.

37. a_7 when $a_1 = 5$ and $r = 2$

38. a_7 when $a_1 = 8$ and $r = 3$

39. a_{10} when $a_1 = 3$ and $r = -2$

40. a_{10} when $a_1 = 7$ and $r = -2$

41. a_6 when $a_1 = \dfrac{1}{16}$ and $r = 3$

42. a_6 when $a_1 = \dfrac{1}{81}$ and $r = 3$

43. a_9 when $a_1 = -1$ and $r = \dfrac{5}{2}$

44. a_9 when $a_1 = -4$ and $r = \dfrac{3}{4}$

45. a_{20} when $a_1 = 500$ and $r = -\dfrac{1}{2}$

46. a_{20} when $a_1 = 1000$ and $r = -\dfrac{1}{10}$

In Exercises 47–52, find the sum S_n of the first n terms of each geometric sequence.

47. $\dfrac{1}{10}, \dfrac{1}{2}, \dfrac{5}{2}, \dfrac{25}{2}, \ldots; n = 10$

48. $6, 2, \dfrac{2}{3}, \dfrac{2}{9}, \ldots; n = 10$

49. $\dfrac{1}{25}, -\dfrac{1}{5}, 1, -5, \ldots; n = 12$

50. $-10, \dfrac{1}{10}, \dfrac{-1}{1000}, \dfrac{1}{100{,}000}, \ldots; n = 12$

51. $5, \dfrac{5}{4}, \dfrac{5}{4^2}, \dfrac{5}{4^3}, \ldots; n = 8$

52. $2, \dfrac{2}{5}, \dfrac{2}{5^2}, \dfrac{2}{5^3}, \ldots; n = 8$

In Exercises 53–60, find each sum.

53. $\displaystyle\sum_{i=1}^{5} \left(\dfrac{1}{2}\right)^{i-1}$

54. $\displaystyle\sum_{i=1}^{5} \left(\dfrac{1}{5}\right)^{i-1}$

55. $\displaystyle\sum_{i=1}^{8} 3\left(\dfrac{2}{3}\right)^{i-1}$

56. $\displaystyle\sum_{i=1}^{8} 2\left(\dfrac{3}{5}\right)^{i-1}$

57. $\displaystyle\sum_{i=3}^{10} \dfrac{2^{i-1}}{4}$

58. $\displaystyle\sum_{i=3}^{10} \dfrac{5^{i-1}}{2}$

59. $\displaystyle\sum_{i=1}^{20} \left(-\dfrac{3}{5}\right)\left(-\dfrac{5}{2}\right)^{i-1}$

60. $\displaystyle\sum_{i=1}^{20} \left(-\dfrac{1}{4}\right)(3^{2-i})$

In Exercises 61–66, use a calculator as needed to find each sum.

61. $\dfrac{1}{5} + \dfrac{1}{10} + \dfrac{1}{20} + \cdots + \dfrac{1}{320}$

62. $2 + 6 + 18 + \cdots + 1458$

63. $\displaystyle\sum_{n=1}^{10} 3^{n-2}$

64. $\displaystyle\sum_{n=1}^{15} \left(\dfrac{1}{7}\right)^{n-1}$

65. $\displaystyle\sum_{n=1}^{10} (-1)^{n-1} 3^{2-n}$

66. $\displaystyle\sum_{n=1}^{20} (-4)^{3-n}$

In Exercises 67–76, find each sum.

67. $\dfrac{1}{3} + \dfrac{1}{9} + \dfrac{1}{27} + \dfrac{1}{81} + \cdots$

68. $\dfrac{5}{2} + \dfrac{5}{4} + \dfrac{5}{8} + \dfrac{5}{16} + \cdots$

69. $-\dfrac{1}{2} + \dfrac{1}{4} - \dfrac{1}{8} + \dfrac{1}{16} - \cdots$

70. $-\dfrac{3}{2} + \dfrac{3}{4} - \dfrac{3}{8} + \dfrac{3}{16} + \cdots$

71. $8 - 2 + \dfrac{1}{2} - \dfrac{1}{8} + \cdots$

72. $1 - \dfrac{3}{5} + \dfrac{9}{25} - \dfrac{27}{125} + \cdots$

73. $\displaystyle\sum_{n=0}^{\infty} 5\left(\dfrac{1}{3}\right)^{n}$

74. $\displaystyle\sum_{n=0}^{\infty} 3\left(\dfrac{1}{4}\right)^{n}$

75. $\displaystyle\sum_{n=0}^{\infty} \left(-\dfrac{1}{4}\right)^{n}$

76. $\displaystyle\sum_{n=0}^{\infty} \left(-\dfrac{1}{3}\right)^{n}$

In Exercises 77–82, find n for the given value of a_n in each geometric sequence.

77. $a_n = 512;\ 2, 4, 8, \ldots$

78. $a_n = \dfrac{3}{1024};\ 3, \dfrac{3}{2}, \dfrac{3}{4}, \ldots$

79. $a_n = -\dfrac{1}{4096};\ 1, -\dfrac{1}{2}, \dfrac{1}{4}, \ldots$

80. $a_n = \dfrac{1}{2187};\ 3, -1, \dfrac{1}{3}, \ldots$

81. $a_n = \dfrac{5}{5,764,801};\ 5, -\dfrac{5}{7}, \dfrac{5}{49}, \ldots$

82. $a_n = \dfrac{1}{1,000,000};\ 1000, 100, 10, \ldots$

Applying the Concepts

83. **Population growth.** The population in a small town is increasing at the rate of 3% per year. If the present population is 20,000, what will the population be at the end of five years?

84. **Growth of a zoo collection.** The butterfly collection at a local zoo started with only 25 butterflies. If the number of butterflies doubled each month, how many butterflies would be in the collection after 12 months?

85. **Savings growth.** Ramón deposits $100 on the last day of each month into a savings account that pays 6% annually, compounded monthly. What is the balance in the account after 36 compounding periods?

86. **Savings growth.** Kat deposits $300 semiannually into a savings account for her son. If the account pays 8% annually, compounded semiannually, what is the balance in the account after 20 compounding periods?

87. **Ancestors.** Every person has two parents, four grandparents, eight great-grandparents, and so on. Find the number of ancestors a person has in the tenth generation back.

88. **Bacterial growth.** A colony of bacteria doubles in number every day. If there are 1000 bacteria now, how many will there be
    a. on the 7th day?
    b. on the nth day?

89. **Investment.** An actuarial firm budgets a new position at $36,000 for the first year and a 5% raise each year thereafter. Find the total compensation a successful applicant for the job would receive during the first 20 years of employment.

90. **Investment.** A realtor offers two choices of payment for a small condominium:
    a. Pay $4000 per month for the next 25 months.
    b. Pay 1¢ the first month, 2¢ the second month, 4¢ the third month, and so on for 25 months.
    Which option, a or b, is the better choice for the buyer? Explain your reasoning.

91. **Pendulum motion.** The distance traveled by any point on a certain pendulum is 20% less than in the preceding swing. If the length traveled by the pendulum bob during the first swing is 56.25 centimeters, find the total distance the bob has traveled at the end of the fifth swing.

92. **Depreciating a tractor.** Every year a tractor loses 20% of the value it had at the beginning of the year. If the tractor now has a value of $120,000, what will its value be in five years?

93. **Rebounding ball.** A particular ball always rebounds $\dfrac{3}{5}$ the distance it falls. If the ball is dropped from a height of 5 meters, how far will it travel before coming to a stop?

94. **Rebounding ball.** Repeat Exercise 93 assuming that the ball is dropped from a height of 9 meters.

Beyond the Basics

95. The accompanying figure shows the first six squares in an infinite sequence of successively smaller squares formed by connecting the midpoints of the sides of the preceding larger square. The area of the largest square is 1. Find the total area of the indicated sections that are shaded as the number of squares increases to infinity.

96. The accompanying figure shows the first three equilateral triangles in an infinite sequence of successively smaller equilateral triangles formed by connecting the midpoints of the sides of the preceding larger triangle. The largest triangle has sides of length 4. Find the total perimeter of all the triangles.

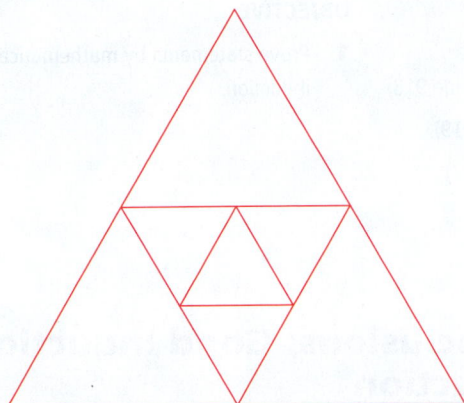

97. Prove that if $a_1, a_2, a_3, \ldots$ is a geometric sequence, then the sequence $\ln a_1, \ln a_2, \ln a_3, \ldots$ is an arithmetic sequence.

98. Prove that if $a_1, a_2, a_3, \ldots$ is a geometric sequence, then $a_1^2, a_2^2, a_3^2, \ldots$ is also a geometric sequence.

99. Show that if $a_1, a_2, a_3, \ldots$ is a geometric sequence with common ratio r, then $\dfrac{1}{a_1}, \dfrac{1}{a_2}, \dfrac{1}{a_3}, \ldots$ is also a geometric sequence. Also find its common ratio.

100. Show that if c is any nonzero real number, then
$$c^2, c, 1, \frac{1}{c}, \ldots \text{ is a geometric sequence.}$$

101. Show that if x is any nonzero real number, then
$$x, 2, \frac{4}{x}, \frac{8}{x^2}, \ldots \text{ is a geometric sequence.}$$

102. Show that if a and x are any two nonzero real numbers, then $\dfrac{x}{a}, -1, \dfrac{a}{x}, -\dfrac{a^2}{x^2}, \ldots$ is a geometric sequence.

103. Show that if $a, x,$ and y are any three nonzero real numbers, then $\dfrac{a}{x}, -\dfrac{a}{xy}, \dfrac{a}{xy^2}, \dfrac{-a}{xy^3}, \ldots$ is a geometric sequence.

104. Show that if $a_1, a_2, a_3, \ldots$ is an arithmetic sequence with common difference d, then $2^{a_1}, 2^{a_2}, 2^{a_3}, \ldots$ is a geometric sequence. Also find its common ratio.

Critical Thinking / Discussion / Writing

105. If a_n denotes the nth term of an arithmetic sequence with $a_1 = 10$ and $d = 2.7$ and if b_n denotes the nth term of a geometric sequence with $b_1 = 10$ and $r = 2$, which number is larger, a_{1001} or b_{1001}?

106. If the sum $\dfrac{a(1 - r^n)}{1 - r}$ of the first n terms of a geometric sequence is 1023, find a when $n = 10$ and $r = \dfrac{1}{2}$.

107. Find n and k so that the sum
$5 + 5 \cdot 2 + 5 \cdot 2^2 + \cdots + 5 \cdot 2^{15}$ can be written as
$$\sum_{i=0}^{n} 5 \cdot 2^i \text{ or as } \sum_{i=1}^{k} 5 \cdot 2^{i-1}.$$

108. The sum of three consecutive terms of a geometric sequence is 35, and their product is 1000. Find the numbers.

[*Hint:* Let $\dfrac{a}{r}, a,$ and ar be the three terms of a geometric sequence.]

109. The sum of three consecutive terms of an arithmetic sequence is 15. If 1, 4, and 19 are respectively added to the three terms, the resulting numbers form three consecutive terms of a geometric sequence. Find the numbers.

Getting Ready for the Next Section

In Exercises 110–113, a statement P_n involving positive integers n is given. Write the statements P_3 and P_4. State whether P_3 and P_4 are *true* or *false*.

110. P_n: $n(n + 1)$ is even.

111. P_n: $n^3 + n$ is divisible by 3.

112. P_n: $1 + 2 + 3 + \cdots + n = \dfrac{n(n + 1)}{2}$

113. P_n: $2^n > 3n$

In Exercises 114–117, a statement P_n involving a positive integer n is given. Write the statement P_{n+1}.

114. P_n: $n^2 + n$ is divisible by 2.

115. P_n: $2^{3n} - 1$ is divisible by 7.

116. P_n: $3^n > 5n$

117. P_n: $1 + 4 + 7 + \cdots + (3n - 2) = \dfrac{1}{2}n(3n - 1)$

118. Show that $\dfrac{k(k + 1)}{2} + (k + 1) = \dfrac{(k + 1)(k + 2)}{2}$.

119. Show that $\dfrac{1}{6}k(k + 1)(2k + 1) + (k + 1)^2 = \dfrac{1}{6}(k + 1)(k + 2)(2k + 3)$.

SECTION | 9.4

Mathematical Induction

BEFORE STARTING THIS SECTION, REVIEW

1 Natural numbers (Appendix A.1, page 918)

2 Inequalities (Appendix A.1, page 919)

3 Exponents (Appendix A.1, page 923)

4 Series (Section 9.1, page 800)

OBJECTIVE

1 Prove statements by mathematical induction.

◆ Jumping to Conclusions; Good Induction Versus Bad Induction

A scientist had two large jars before him on the laboratory table. The jar on his left contained a hundred fleas; the jar on his right was empty. The scientist carefully lifted a flea from the jar on the left; placed the flea on the table between the two jars; stepped back; and said in a loud voice, "Jump!" The flea jumped and was put into the jar on the right. A second flea was carefully lifted from the jar on the left and placed on the table between the two jars. Again, the scientist stepped back and said in a loud voice, "Jump!" The flea jumped and was put into the jar on the right. In the same manner, the scientist treated each of the hundred fleas in the jar on the left, and each flea jumped as ordered. The two jars were then interchanged, and the experiment was continued with a slight difference. This time the scientist carefully lifted a flea from the jar on the left; *yanked off its hind legs;* placed the flea on the table between the jars; stepped back; and said in a loud voice, "Jump!" The flea did not jump and was put into the jar on the right. A second flea was carefully lifted from the jar on the left, its hind legs yanked off, and then placed on the table between the two jars. Again the scientist stepped back and said in a loud voice, "Jump!" The flea did not jump and was put into the jar on the right. In this manner, the scientist treated each of the hundred fleas in the jar on the left, and in no case did a flea jump when ordered. So the scientist recorded the following inductive statement in his notebook: "A flea, if its hind legs are yanked off, cannot hear." (*Source:* Modified from Howard Eves, *Mathematical Circles.*)

This, of course, is an example of truly bad reasoning. The principle of mathematical induction, demonstrated in Example 2, is the gold standard for reasoning in proving that statements about natural numbers are true.

1 Prove statements by mathematical induction.

Mathematical Induction

Suppose your boss wants you to verify that each of a thousand bags contains two specific gold coins. There is only one way to be absolutely sure: You must check every bag. That would take a very long time, but you can eventually finish the job. However, consider the problem of verifying that a statement about the natural numbers such as

$$1^2 + 2^2 + 3^2 + \cdots + n^2 = \frac{n(n + 1)(2n + 1)}{6}$$

is true for all natural numbers n. Let's first test the statement for several values of n by comparing the value of $1^2 + 2^2 + 3^2 + \cdots + n^2$ in the second column of Table 9.1 with the value of $\frac{n(n + 1)(2n + 1)}{6}$ in the third column.

824 Chapter 9 Further Topics in Algebra

TABLE 9.1

n	$1^2 + 2^2 + 3^2 + \cdots + n^2$	$\dfrac{n(n+1)(2n+1)}{6}$	Equal?
1	$1^2 = 1$	$\dfrac{1(1+1)[2(1)+1]}{6} = 1$	Yes
2	$1^2 + 2^2 = 5$	$\dfrac{2(2+1)[2(2)+1]}{6} = 5$	Yes
3	$1^2 + 2^2 + 3^2 = 14$	$\dfrac{3(3+1)[2(3)+1]}{6} = 14$	Yes
4	$1^2 + 2^2 + 3^2 + 4^2 = 30$	$\dfrac{4(4+1)[2(4)+1]}{6} = 30$	Yes
5	$1^2 + 2^2 + 3^2 + 4^2 + 5^2 = 55$	$\dfrac{5(5+1)[2(5)+1]}{6} = 55$	Yes

We see that the statement is correct for the natural numbers 1, 2, 3, 4, and 5; furthermore, you can continue to verify that the statement is correct for any particular natural number n you try. But is it *always* correct? Clearly, we can't check every natural number!

The principle of *mathematical induction* provides a method for proving that statements about natural numbers are true for *all* natural numbers. The principle is based on the fact that after any natural number n, there is a next-larger natural number $n + 1$ and that any specified natural number can be reached by a finite number of such steps, starting from the natural number 1.

> ### THE PRINCIPLE OF MATHEMATICAL INDUCTION
>
> Let P_n be a statement that involves the natural number n with the following two properties:
>
> 1. P_1 is true (the statement is true for the natural number 1).
> 2. If P_k is a true statement, then P_{k+1} is a true statement.
>
> Then the statement P_n is true for every natural number n.

A common physical interpretation is helpful in understanding why this principle works. Imagine an unending line of dominoes, as shown in Figure 9.2. Suppose the dominoes are arranged so that if *any* domino is knocked down, it will knock down the next domino behind it as well. What will happen if the first domino is knocked down? *All* of the dominoes will fall because

1. knocking down the first domino causes the second domino to be knocked down,
2. when the second domino is knocked down, it knocks down the third domino, and so on.

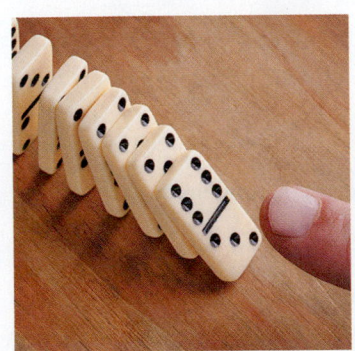

Figure 9.2

Determining the Statement P_{k+1} from the Statement P_k

To successfully use the principle of mathematical induction, you must be able to determine the statement P_{k+1} from a given statement P_k. Suppose the given statement is

$$P_k: k \geq 1;$$

then we have

$$P_{k+1}: k + 1 \geq 1. \qquad \text{Replace } k \text{ with } k + 1 \text{ in } P_k.$$

It is important to recognize that since P_k says "k is greater than or equal to 1," the statement P_{k+1} says, "$k + 1$ is greater than or equal to 1." That is, P_{k+1} asserts the same property for $k + 1$ that P_k asserts for k.

EXAMPLE 1   Determining P_{k+1} from P_k

Find the statement P_{k+1} from the given statement P_k.

a. $P_k\colon k < 2^k$

b. $P_k\colon S_k = 3k^2 + 9$

c. $P_k\colon 1 + 2 + 2^2 + 2^3 + \cdots + 2^{k-1} = 2^k - 1$

d. $P_k\colon 3 + 6 + 9 + 12 + \cdots + 3k = \dfrac{3k(k+1)}{2}$

Solution

a. $P_k\colon k < 2^k$, $P_{k+1}\colon k + 1 < 2^{k+1}$ Replace k with $k + 1$.

b. $P_k\colon S_k = 3k^2 + 9$, $P_{k+1}\colon S_{k+1} = 3(k+1)^2 + 9$ Replace k with $k + 1$.

c. $P_k\colon 1 + 2 + 2^2 + 2^3 + \cdots + 2^{k-1} = 2^k - 1$

    $P_{k+1}\colon 1 + 2 + 2^2 + 2^3 + \cdots + 2^{(k+1)-1} = 2^{k+1} - 1$ Replace k with $k + 1$.

d. $P_k\colon 3 + 6 + 9 + 12 + \cdots + 3k = \dfrac{3k(k+1)}{2}$

    $P_{k+1}\colon 3 + 6 + 9 + 12 + \cdots + 3(k+1) = \dfrac{3(k+1)[(k+1)+1]}{2}$ Replace k with $k + 1$.

Practice Problem 1 Find P_{k+1} for $P_k\colon (k + 3)^2 > k^2 + 9$.

EXAMPLE 2   Using Mathematical Induction

Use mathematical induction to prove that, for all natural numbers n,

$$2 + 4 + 6 + \cdots + 2n = n(n + 1).$$

Solution

First, we verify that this statement is true for $n = 1$.

Check: $2(1) = 1(1 + 1)$ Replace n with 1 in the original statement.

          $2 = 2.$ ✓

The given statement is true for $n = 1$, so the first condition for the principle of mathematical induction holds.

The second condition for the principle of mathematical induction requires two steps.

Step 1   Assume that the formula is true for some unspecified natural number k:

$$P_k\colon 2 + 4 + 6 + \cdots + 2k = k(k + 1) \quad \text{\color{blue}Assumed true}$$

Step 2   On the basis of the assumption that P_k is true, show that P_{k+1} is true, which is the same as showing that

$$P_{k+1}\colon 2 + 4 + 6 + \cdots + 2(k+1) = (k+1)[(k+1)+1]. \quad \text{\color{blue}Replace k with $k + 1$ in P_k.}$$

Begin by using P_k, the statement assumed to be true. Add $2(k + 1)$ to both sides of P_k in Step 1, which again results in a true statement.

$$2 + 4 + 6 + \cdots + 2k = k(k + 1) \qquad \text{Assumed true}$$

$$2 + 4 + 6 + \cdots + 2k + 2(k + 1) = k(k + 1) + 2(k + 1) \qquad \begin{array}{l}\text{Add } 2(k+1) \text{ to}\\ \text{both sides.}\end{array}$$

$$2 + 4 + 6 + \cdots + 2k + 2(k + 1) = (k + 1)(k + 2) \qquad \begin{array}{l}\text{Factor out the}\\ \text{common factor}\\ (k+1).\end{array}$$

$$2 + 4 + 6 + \cdots + 2(k + 1) = (k + 1)[(k + 1) + 1] \qquad \begin{array}{l}\text{Rewrite } (k+2) \text{ as}\\ (k+1)+1.\end{array}$$

> It is not *necessary* to write $2k$ here because it is understood that $2k$ is the even integer that precedes $2(k + 1)$.

This last equation says that P_{k+1} is true if P_k is assumed to be true. Therefore, by the principle of mathematical induction, the statement

$$2 + 4 + 6 + \cdots + 2n = n(n + 1)$$

is true for every natural number n.

Practice Problem 2 Use mathematical induction to prove that for all natural numbers n,

$$1 + 2 + 3 + \cdots + n = \frac{n(n + 1)}{2}.$$

EXAMPLE 3 Using Mathematical Induction

Use mathematical induction to prove that

$$2^n > n$$

for all natural numbers n.

Solution

First, we show that the given statement is true for $n = 1$.

$$2^1 > 1 \qquad \text{Replace } n \text{ with 1 in the original statement.}$$

Thus, the inequality is true for $n = 1$.

Next, we assume that for some unspecified natural number k,

$$P_k: 2^k > k \quad \text{is true.}$$

Then we must use P_k to prove that P_{k+1} is true; that is,

$$P_{k+1}: 2^{k+1} > k + 1.$$

Now, by the product rule of exponents, $2^{k+1} = 2^k \cdot 2^1 = 2^k \cdot 2$; so we can get some information about 2^{k+1} by multiplying both sides of $P_k: 2^k > k$ by 2.

$$2^k > k \qquad P_k \text{ is assumed true.}$$

$$2^k \cdot 2 > 2 \cdot k \qquad \text{Multiply both sides by 2.}$$

$$2^{k+1} = 2^k \cdot 2 > 2k = k + k \geq k + 1 \qquad \text{Since } k \geq 1, k + k \geq k + 1.$$

Thus, $2^{k+1} > k + 1$ is true.

By the principle of mathematical induction, the statement

$$2^n > n$$

is true for every natural number n.

Practice Problem 3 Use mathematical induction to prove that $3^n > n$ for all natural numbers n.

SECTION 9.4 **Exercises**

Concepts and Vocabulary

1. Mathematical induction can only be used to prove statements about the _____.

2. The first step in a mathematical induction proof that a statement P_n is true for all natural numbers n is that _____.

3. The second step in a mathematical induction proof is to assume that P_k is true for a natural number k and then show that _____.

4. The first step when using mathematical induction to prove that $\sum_{k=1}^{n} 6 \cdot 7^k = 7(7^k - 1)$ is to prove _____.

5. True or False. If P_n is a statement about natural numbers and there is exactly one natural number, r for which P_r is false, then P_n is false.

6. True or False. The statement S about all natural numbers n, $n \geq 4$, can be proved true by showing that the statement $P_n = S_{n+3}$ is true for all $n \geq 1$.

7. True or False. The statement $e^n \geq n$ cannot be proved by mathematical induction because e is not a natural number.

8. True or False. The first step in proving that $2n \neq n$ for all natural numbers n is to note that $2 \neq 1$.

Building Skills

In Exercises 9–12, find P_{k+1} from the given statement P_k.
9. $P_k: (k + 1)^2 - 2k = k^2 + 1$

10. $P_k: (1 + k)(1 - k) = 1 - k^2$

11. $P_k: 2^k > 5k$

12. $P_k: 1 + 3 + 5 + \cdots + (2k - 1) = k^2$

In Exercises 13–42, use mathematical induction to prove that each statement is true for all natural numbers n.
13. $2 + 4 + 6 + \cdots + 2n = n^2 + n$

14. $1 + 3 + 5 + \cdots + (2n - 1) = n^2$

15. $4 + 8 + 12 + \cdots + 4n = 2n(n + 1)$

16. $3 + 6 + 9 + \cdots + 3n = \dfrac{3n(n + 1)}{2}$

17. $1 + 5 + 9 + \cdots + (4n - 3) = n(2n - 1)$

18. $3 + 8 + 13 + \cdots + (5n - 2) = \dfrac{n(5n + 1)}{2}$

19. $3 + 9 + 27 + \cdots + 3^n = \dfrac{3(3^n - 1)}{2}$

20. $5 + 25 + 125 + \cdots + 5^n = \dfrac{5(5^n - 1)}{4}$

21. $1 \cdot 2 + 3 \cdot 4 + 5 \cdot 6 + \cdots + (2n - 1)(2n) = \dfrac{1}{3}n(n + 1)(4n - 1)$

22. $1 \cdot 3 + 2 \cdot 4 + 3 \cdot 5 + \cdots + (n)(n + 2) = \dfrac{1}{6}n(n + 1)(2n + 7)$

23. $\dfrac{1}{1 \cdot 2} + \dfrac{1}{2 \cdot 3} + \dfrac{1}{3 \cdot 4} + \cdots + \dfrac{1}{n(n + 1)} = \dfrac{n}{n + 1}$

24. $\dfrac{1}{2 \cdot 4} + \dfrac{1}{4 \cdot 6} + \dfrac{1}{6 \cdot 8} + \cdots + \dfrac{1}{2n(2n + 2)} = \dfrac{n}{4(n + 1)}$

25. $2 \leq 2^n$ **26.** $n(n + 2) < (n + 1)^2$

27. $\dfrac{n!}{n} = (n - 1)!$ **28.** $\dfrac{n!}{(n + 1)!} = \dfrac{1}{n + 1}$

29. $1^2 + 2^2 + 3^2 + \cdots + n^2 = \dfrac{n(n + 1)(2n + 1)}{6}$

30. $1^3 + 2^3 + 3^3 + \cdots + n^3 = \dfrac{n^2(n + 1)^2}{4}$

31. 2 is a factor of $n^2 + n$.

32. 2 is a factor of $n^3 + 5n$.

33. 6 is a factor of $n(n + 1)(n + 2)$.

34. 3 is a factor of $n(n + 1)(n - 1)$.

35. $(1 + a)^n \geq na, a > 1$

36. $(1 + a)^n \geq 1 + na, a > 0$

37. $\sum_{i=1}^{n} 3 \cdot 4^i = 4(4^n - 1)$

38. $\sum_{i=1}^{n} 8 \cdot 9^i = 9(9^n - 1)$

39. $\left(1 + \dfrac{1}{1}\right)\left(1 + \dfrac{1}{2}\right)\left(1 + \dfrac{1}{3}\right)\left(1 + \dfrac{1}{n}\right) = n + 1$

40. $\dfrac{1}{\sqrt{1}} + \dfrac{1}{\sqrt{2}} + \dfrac{1}{\sqrt{3}} + \cdots + \dfrac{1}{\sqrt{n}} \geq \sqrt{n}$

41. $(ab)^n = a^n b^n$ **42.** $\left(\dfrac{a}{b}\right)^n = \dfrac{a^n}{b^n}$

Applying the Concepts

43. Counting hugs. At a family reunion of n people ($n \geq 2$), each person hugs everyone else. Use mathematical induction to show that the number of hugs is $\dfrac{n^2 - n}{2}$.

44. Winning prizes. In a contest in which n prizes are possible, a contestant can win any number of the prizes (from 0 to n). Use mathematical induction to show that there are 2^n possible outcomes for the sets of prizes the contestant might win. [*Hint:* If $n = 1$, there are 2 ($= 2^1$) outcomes: winning 0 prizes or winning the one prize.]

45. Koch's snowflake. A geometric shape called the Koch snowflake can be formed by starting with an equilateral triangle, as in the figure. Assume that each side of the triangle has length 1. On the middle part of each side, construct an equilateral triangle with sides of length $\frac{1}{3}$ and erase the side of the smaller triangle that is on the side of the original triangle. Repeat this procedure for each of the smaller triangles. This process produces a figure that resembles a snowflake.

a. Find a formula for the number of sides of the nth figure. Use mathematical induction to prove that this formula is correct.

b. Find a formula for the perimeter of the nth figure. Use mathematical induction to prove that this formula is correct.

46. Sierpinski's triangle. A geometric figure known as Sierpinski's triangle is constructed by starting with an equilateral triangle as in the figure. Assume that each side of the triangle has length 1. Inside the original triangle, draw a second triangle by connecting the midpoints of the sides of the original triangle. This divides the original triangle into four identical equilateral triangles. Repeat the process by constructing equilateral triangles inside each of the three smaller blue triangles, again using the midpoints of the sides of the triangle as vertices. Continuing this process with each group of smaller triangles produces Sierpinski's triangle. Coloring each set of smaller triangles helps in following the construction. (Be sure to count blue as well as white triangles.)

a. When a process is repeated over and over, each repetition is called an **iteration**. How many triangles will the fourth iteration have?

b. How many triangles will the fifth iteration have?

c. Find a formula for the number of triangles in the nth iteration. Use mathematical induction to prove that this formula is correct.

47. Towers of Hanoi. Three pegs are attached vertically to a horizontal board, as shown in the figure. One peg has n rings stacked on it, each ring smaller than the one below it. In a game known as the Tower of Hanoi puzzle, all of the rings must be moved to a different peg, with only one ring moved at a time, and no ring can be moved on top of a smaller ring. Determine the least number of moves that will accomplish this transfer. Use mathematical induction to prove that your answer is correct.

48. Exam answer sheets. An exam in which each question is answered with either "True" or "False" has n questions.

a. If every question is answered, how many answer sheets are possible
  (i) if $n = 1$?
  (ii) if $n = 2$?

b. Find a formula for the number of possible answer sheets for an exam with n questions (for any natural number n) and use mathematical induction to prove that the formula is correct.

Beyond the Basics

In Exercises 49–57, use mathematical induction to prove each statement for all natural numbers n.

49. 5 is a factor of $8^n - 3^n$.

50. 24 is a factor of $5^{2n} - 1$.

51. 64 is a factor of $3^{2n+2} - 8n - 9$.

52. 64 is a factor of $9^n - 8n - 1$.

53. 3 is a factor of $2^{2n+1} + 1$.

54. 5 is a factor of $2^{4n} - 1$.

55. $a - b$ is a factor of $a^n - b^n$.

[*Hint:* $a^{k+1} - b^{k+1} = a(a^k - b^k) + b^k(a - b)$.]

56. If $a \neq 1$, then $1 + a + a^2 + \cdots + a^{n-1} = \dfrac{a^n - 1}{a - 1}$.

57. $\displaystyle\sum_{k=1}^{n+1} \frac{1}{n + k} \leq \frac{5}{6}$

Critical Thinking / Discussion / Writing

58. Consider the statement "2 is a factor of $4n - 1$ for all natural numbers n." Then P_k: 2 is a factor of $4k - 1$.

Assume that P_k is true so that $4k - 1 = 2m$ for some integer m. Then

$$
\begin{aligned}
P_{k+1} &= 4(k + 1) - 1 \\
&= 4k + 4 - 1 \\
&= (4k - 1) + 4 \\
&= 2m + 4 \\
&= 2(m + 2)
\end{aligned}
$$

Thus, if P_k is true, then P_{k+1} is also true. However, 2 is never a factor of $4n - 1$ because $4n - 1$ is always an odd integer. Explain why this "proof" is not valid.

Extended principle of mathematical induction. This principle states that if a statement P_n about natural numbers satisfies the two conditions,

(1) P_m is true for some natural number m.

(2) P_k is true implies that P_{k+1} is also true.

Then P_n is true for all natural numbers n, with $n \geq m$.

In Exercises 59 and 60, use the extended principle of mathematical induction to prove the statement.

59. Prove that $2^n > n^2$ for all natural numbers n, $n > 4$.

60. Prove that $n! > n^3$ for all natural numbers n, $n \geq 6$.

Getting Ready for the Next Section

In Exercises 61–64, verify each product by using the product rules.

61. $(x + y)^2 = (x + y)(x + y) = x^2 + 2xy + y^2$

62. $(x + y)^3 = (x + y)(x + y)^2 = x^3 + 3x^2y + 3xy^2 + y^3$

63. $(x + y)^4 = (x + y)(x + y)^3$
$= x^4 + 4x^3y + 6x^2y^2 + 4xy^3 + y^4$

64. $(x + y)^5 = (x + y)(x + y)^4$
$= x^5 + 5x^4y + 10x^3y^2 + 10x^2y^3 + 5xy^4 + y^5$

65. In Exercises 61–64, what is the pattern on the exponents of x?

66. In Exercises 61–64, what is the pattern on the exponents of y?

From the pattern of expansions in Exercises 61–64, if we expand $(x + y)^n$ for a positive integer n, what is your answer to each of the following?

67. The sum of the exponents on x and y in each term

68. The number of terms in the expansion

SECTION 9.5

The Binomial Theorem

BEFORE STARTING THIS SECTION, REVIEW

1 Special products (Appendix A.2, page 929)

2 Factorials (Section 9.1, page 798)

3 Summation notation (Section 9.1, page 799)

OBJECTIVES

1 Use Pascal's Triangle to compute binomial coefficients.

2 Use Pascal's Triangle to expand a binomial power.

3 Use the Binomial Theorem to expand a binomial power.

4 Find the coefficient of a term in a binomial expansion.

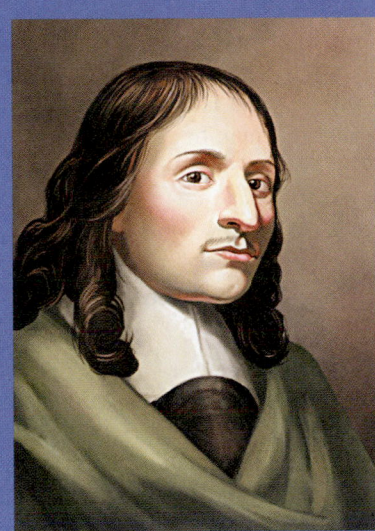

Blaise Pascal (1623–1662)

Pascaline

◆ Blaise Pascal

Blaise Pascal was a French mathematician. An evident genius, he was 14 when he began to accompany his father to gatherings of mathematicians that were arranged by Father Marin Mersenne. At the age of 16, Pascal presented his own results at one of Mersenne's meetings.

Pascal's desire to help his father with his work collecting taxes led him to invent the first digital calculator, called the Pascaline. Pascal was also interested in atmospheric pressure, and in 1648, he observed that the pressure of the atmosphere decreased with height and that a vacuum existed above the atmosphere.

Pascal's intense interest in mathematics led him to produce important results related to conic sections and to engage in correspondence with Fermat in which he formulated the foundations for the *theory of probability*. Pascal died a painful death from cancer at the age of 39.

In this section, we study *Pascal's Triangle*. This triangle of numbers contains the important *binomial coefficients*. See page 835. Pascal's work was influential in Newton's discovery of the general *Binomial Theorem* for fractional and negative powers.

Binomial Expansions Recall that a polynomial that has exactly two terms is called a *binomial*. In Appendix A.2, the formulas for expanding the binomial powers $(x + y)^2$ and $(x + y)^3$ are given. In the current section, we study a method for expanding $(x + y)^n$ for any positive integer n.

First consider these binomial expansions:

$$(x + y)^1 = x + y$$
$$(x + y)^2 = x^2 + 2xy + y^2$$
$$(x + y)^3 = x^3 + 3x^2y + 3xy^2 + y^3$$
$$(x + y)^4 = x^4 + 4x^3y + 6x^2y^2 + 4xy^3 + y^4$$
$$(x + y)^5 = x^5 + 5x^4y + 10x^3y^2 + 10x^2y^3 + 5xy^4 + y^5$$

The expansions of $(x + y)^2$ and $(x + y)^3$ should be familiar to you; the last two are left for you to verify. Each is a product of the previous binomial and $(x + y)$. For example, $(x + y)^4 = (x + y)^3(x + y)$.

The patterns for expansions of $(x + y)^n$ (with $n = 1, 2, 3, 4, 5$) suggest the following:

1. The expansion of $(x + y)^n$ has $n + 1$ terms.

2. The sum of the exponents on x and y in each term equals n.

3. The exponent on x starts at $n (x^n = x^n \cdot y^0)$ in the first term and decreases by 1 for each term until it is 0 in the last term $(x^0 \cdot y^n = y^n)$.

4. The exponent on y starts at 0 $(x^n = x^n \cdot y^0)$ in the first term and increases by 1 for each term until it is n in the last term $(x^0 \cdot y^n = y^n)$.

5. The variables x and y have symmetrical roles. That is, replacing x with y and y with x in the expansion of $(x + y)^n$ yields the same terms, just in reversed order.

You may also have noticed that the coefficients of the first and last terms are both 1 and the coefficients of the second and the next-to-last terms are equal. In general, the coefficients of

$$x^{n-j}y^j \quad \text{and} \quad x^j y^{n-j}$$

are equal for $j = 0, 1, 2, \ldots, n$.

The coefficients in the expansion of $(x + y)^n$ are called the **binomial coefficients**.

---

1 Use Pascal's Triangle to compute binomial coefficients.

Pascal's Triangle

As early as A.D. 1100, the Chinese scholar Chia Hsien had discovered the secret of the binomial coefficients that was later rediscovered by the French philosopher and mathematician Blaise Pascal (1623–1662). To understand Chia Hsien's and Pascal's construction of binomial coefficients, let's first look at the coefficients in these binomial expansions:

$$(x + y)^0 = 1$$
$$(x + y)^1 = 1x + 1y$$
$$(x + y)^2 = 1x^2 + 2xy + 1y^2$$
$$(x + y)^3 = 1x^3 + 3x^2y + 3xy^2 + 1y^3$$
$$(x + y)^4 = 1x^4 + 4x^3y + 6x^2y^2 + 4xy^3 + 1y^4$$
$$(x + y)^5 = 1x^5 + 5x^4y + 10x^3y^2 + 10x^2y^3 + 5xy^4 + 1y^5$$

If we remove the variables and the plus signs and list only the coefficients, we get a triangle of numbers composed of the binomial coefficients. This triangle of numbers is known as *Pascal's Triangle.*

$(x + y)^0$:						1							Row 0
$(x + y)^1$:					1		1						Row 1
$(x + y)^2$:				1		2		1					Row 2
$(x + y)^3$:			1		3		3		1				Row 3
$(x + y)^4$:		1		4		6		4		1			Row 4
$(x + y)^5$:	1		5		10		10		5		1		Row 5
	1	6		15		20		15		6	1		Row 6

$$1 + 5 \quad 5 + 10 \quad 10 + 10 \quad 10 + 5 \quad 5 + 1$$

Note the symmetry in Pascal's Triangle. If the triangle were folded vertically down the middle, the numbers on each side of the crease would match. To create a new bottom row in the triangle, put the number 1 in the first and last places of the new row and add two neighboring entries in the previous row.

The top row is called the *zeroth row* because it corresponds to the binomial expansion of $(x + y)^0$. The next row is called the *first row* because it corresponds to the binomial expansion of $(x + y)^1$. All of the rows are named so that the nth *row* corresponds to the coefficients of $(x + y)^n$.

2 Use Pascal's Triangle to expand a binomial power.

EXAMPLE 1 **Using Pascal's Triangle to Expand a Binomial Power**

Expand $(4y - 2x)^5$.

Solution

From the fifth row of Pascal's Triangle, we see that the binomial coefficients are

$$1, 5, 10, 10, 5, 1.$$

We must make some changes in the expansion

$$(x + y)^5 = x^5 + 5x^4y + 10x^3y^2 + 10x^2y^3 + 5xy^4 + y^5$$

to get the expansion for $(4y - 2x)^5$:

1. Replace x with $4y$.
2. Replace y with $-2x$.

$$
\begin{aligned}
(4y - 2x)^5 &= [4y + (-2x)]^5 = (4y)^5 + 5(4y)^4(-2x) \\
&\quad + 10(4y)^3(-2x)^2 + 10(4y)^2(-2x)^3 + 5(4y)(-2x)^4 + (-2x)^5 \\
&= 1024y^5 - 2560y^4x + 2560y^3x^2 - 1280y^2x^3 + 320yx^4 - 32x^5
\end{aligned}
$$

We note that expanding a *difference* results in alternating signs between terms.

Practice Problem 1 Expand $(3y - x)^6$.

3 Use the Binomial Theorem to expand a binomial power.

The Binomial Theorem

The coefficients in a binomial expansion can be computed by using ratios of certain factorials. We first introduce the symbol $\binom{n}{r}$.

The Symbol $\binom{n}{r}$

If r and n are integers with $0 \leq r \leq n$, then we define

$$\binom{n}{r} = \frac{n!}{r!(n - r)!}, \quad \binom{n}{0} = 1 \text{ and } \binom{n}{n} = 1$$

The symbol $\binom{n}{r}$ is read "n choose r." It can be shown that $\binom{n}{r}$ is the number of ways of choosing a subset containing exactly r elements from a set with n elements.

EXAMPLE 2 **Evaluating $\binom{n}{r}$**

Evaluate each expression.

a. $\binom{4}{1}$ b. $\binom{5}{3}$ c. $\binom{9}{0}$ d. $\binom{35}{35}$

Solution

a. $\binom{4}{1} = \dfrac{4!}{1!(4-1)!} = \dfrac{4!}{1!\,3!} = \dfrac{4 \cdot 3 \cdot 2 \cdot 1}{1(3 \cdot 2 \cdot 1)} = \dfrac{4}{1} = 4$

b. $\binom{5}{3} = \dfrac{5!}{3!(5-3)!} = \dfrac{5!}{3!\,2!} = \dfrac{5 \cdot 4 \cdot 3!}{3!\,2!} = \dfrac{5 \cdot 4}{2} = 5 \cdot 2 = 10$

 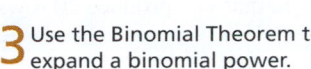
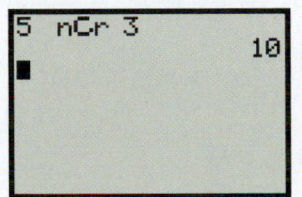

c. $\dbinom{9}{0} = \dfrac{9!}{0!(9-0)!} = \dfrac{9!}{0!\,9!} = \dfrac{1}{1} = 1$ Recall that 0! = 1.

d. $\dbinom{35}{35} = \dfrac{35!}{35!(35-35)!} = \dfrac{35!}{35!\,0!} = \dfrac{1}{1} = 1$

Practice Problem 2 Evaluate each expression.

a. $\dbinom{6}{2}$ **b.** $\dbinom{12}{9}$

The numbers $\dbinom{n}{r}$ show up as the coefficients in the expansion of a binomial power. For example, $(x+y)^4$ can be written as either

$$(x+y)^4 = x^4 + 4x^3y + 6x^2y^2 + 4xy^3 + y^4$$

or

$$(x+y)^4 = \binom{4}{0}x^4 + \binom{4}{1}x^3y + \binom{4}{2}x^2y^2 + \binom{4}{3}xy^3 + \binom{4}{4}y^4.$$

This result is the *Binomial Theorem* (for $n = 4$), which can be proved by mathematical induction. It provides an efficient method for expanding a binomial power. (See Exercise 86.) The Binomial Theorem can be used to expand a binomial power directly, without reference to Pascal's Triangle. This technique is particularly useful in expanding large powers of a binomial. For example, the expansion of $(x+y)^{20}$ would require that you produce 20 rows of the Pascal Triangle.

SIDE NOTE

The kth term in the expansion of $(x+y)^n$ has coefficient $\dbinom{n}{k-1}$

THE BINOMIAL THEOREM

If n is a natural number, then the binomial expansion of $(x+y)^n$ is given by

$$(x+y)^n = \binom{n}{0}x^n + \binom{n}{1}x^{n-1}y + \binom{n}{2}x^{n-2}y^2 + \cdots + \binom{n}{r}x^{n-r}y^r + \cdots + \binom{n}{n}y^n$$

$$= \sum_{r=0}^{n}\binom{n}{r}x^{n-r}y^r.$$

The coefficient of $x^{n-r}y^r$ is $\dbinom{n}{r} = \dfrac{n!}{r!(n-r)!}$.

EXAMPLE 3 **Expanding a Binomial Power by Using the Binomial Theorem**

Find the binomial expansion of $(x-3y)^4$.

Solution

Replace y with $-3y$ in the expansion of $(x+y)^4$.

$$(x-3y)^4 = [x+(-3y)]^4$$

$$= \binom{4}{0}x^4 + \binom{4}{1}x^3(-3y) + \binom{4}{2}x^2(-3y)^2 + \binom{4}{3}x(-3y)^3 + \binom{4}{4}(-3y)^3$$

$$= \frac{4!}{0!\,4!}x^4 + \frac{4!}{1!\,3!}x^3(-3y) + \frac{4!}{2!\,2!}x^2(-3y)^2 + \frac{4!}{3!\,1!}x(-3y)^3 + \frac{4!}{4!\,0!}(-3y)^4$$

$$= x^4 + \frac{4\cdot 3!}{1!3!}x^3(-3y) + \frac{4\cdot 3\cdot 2!}{2!2!}x^2(-3y)^2 + \frac{4\cdot 3!}{3!1!}x(-3y)^3 + (-3y)^4$$

$$= x^4 + 4x^3(-3y) + \frac{4\cdot 3}{2!}x^2(9y^2) + 4x(-27y^3) + (81y^4)$$

$$= x^4 - 12x^3y + 54x^2y^2 - 108xy^3 + 81y^4$$

Practice Problem 3 Find the binomial expansion of $(3x-y)^4$.

4 Find the coefficient of a term in a binomial expansion.

Binomial Coefficients

The binomial coefficients $\binom{n}{r}$ are useful in finding a particular coefficient or term in a binomial expansion.

EXAMPLE 4 Finding a Particular Coefficient in a Binomial Expansion

Find the coefficient of x^9y^3 in the expansion of $(x + y)^{12}$.

Solution

The coefficient of $x^{n-r}y^r$ is $\binom{n}{r}$. Here $n = 12$, $r = 3$, and $n - r = 9$, so the coefficient of x^9y^3 is

$$\binom{n}{r} = \binom{12}{3} = \frac{12!}{3!(12 - 3!)} = \frac{12!}{3!\,9!} = \frac{12 \cdot 11 \cdot 10 \cdot 9!}{3!\,9!} = 220.$$

Practice Problem 4 Find the coefficient of x^3y^9 in the expansion of $(x + y)^{12}$.

The method illustrated in Example 4 allows us to find any particular term in a binomial expansion without writing out the complete expansion.

PARTICULAR TERM IN A BINOMIAL EXPANSION

The term containing the factor x^r in the expansion of $(x + y)^n$ is

$$\binom{n}{n - r}x^ry^{n-r}.$$

This term also contains the factor y^{n-r}.

EXAMPLE 5 Finding a Particular Term in a Binomial Expansion

Find the term containing x^{10} in the expansion of $(x + 2a)^{15}$.

Solution

We begin with the formula for the term containing the factor x^r.

$$\binom{n}{n - r}x^ry^{n-r} = \binom{15}{15 - 10}x^{10}(2a)^{15-10}$$ Replace n with 15, r with 10, and y with $2a$.

$$= \binom{15}{5}x^{10}(2a)^5$$ Simplify.

$$= \frac{15!}{5!(15 - 5)!}x^{10}2^5a^5$$ $\binom{n}{r} = \frac{n!}{r!(n - r)!}; (2a)^5 = 2^5a^5$

$$= \frac{15!}{5!\,10!} \cdot 32x^{10}a^5$$ $2^5 = 32$

$$= \frac{15 \cdot 14 \cdot 13 \cdot 12 \cdot 11 \cdot 10!}{5! \cdot 10!} \cdot 32x^{10}a^5$$ Use a calculator.

$$= 96,096x^{10}a^5$$

Practice Problem 5 Find the term containing x^3 in the expansion of $(x + 2a)^{15}$.

Recall that, assuming decreasing powers of x, the kth term in the expansion of $(x + y)^n$ is $\binom{n}{k-1}x^{n-k+1}y^{k-1}$.

EXAMPLE 6 Finding a Specific Term in the Expansion of $(x + y)^n$

Find the fifteenth term in the expansion of $(2x - 1)^{18}$. Assume decreasing powers of x.

Solution

The k_{th} term in the expansion of $(x + y)^n$ is $\binom{n}{k-1}x^{n-k-1}y^{k-1}$. Note that $(2x - 1)^{18} = (2x + (-1))^{18}$. If we replace x with $2x$ and y with -1, then n with 18, and k with 15, we get:

$\binom{18}{14}x^{18-15+1}y^{15-1}$ Coefficient of the 15th term of $(x + y)^{18}$

$\binom{18}{14}(2x)^{18-14}(-1)^{14}$ Replace x with $2x$ and y with -1 in $\binom{18}{14}x^{18-14}y^{14}$.

$\binom{18}{14}(2x)^{18-14}(-1)^{14} = \frac{18!}{14!4!}2^4x^4$ $(2x)^{18-14} = (2x)^4 = 2^4x^4;\ (-1)^{14} = 1.$

The fifteenth term in the expansion of $(2x - 1)^{18}$ is $\frac{18!}{14!4!}2^4x^4 = 48960x^4$.   Simplify

Practice Problem 6 Find the fourth term in the expansion of $(x - 3)^{12}$. Assume decreasing powers of x.

Answers to Practice Problems

1. $(3y - x)^6 = 729y^6 - 1458y^5x + 1215y^4x^2 - 540y^3x^3 + 135y^2x^4 - 18yx^5 + x^6$

2. a. $\binom{6}{2} = 15$ **b.** $\binom{12}{9} = 220$

3. $(3x - y)^4 = 81x^4 - 108x^3y + 54x^2y^2 - 12xy^3 + y^4$
4. 220   **5.** $1,863,680x^3a^{12}$ **6.** $-5940x^9$

SECTION 9.5   **Exercises**

Concepts and Vocabulary

1. The expansion of $(x + y)^n$ has _____ terms.

2. In the expansion of $(x + y)^5$, the coefficient of x^2y^3 is $\binom{n}{r}$ when $n =$ _____ and $r =$ _____.

3. Expanding a difference such as $(2x - y)^{10}$ results in _____ between terms.

4. For any positive integer n, $\binom{n}{n} =$ _____.

5. **True or False.** Every coefficient in the expansion of $(x + y)^n$, for $n > 1$, appears exactly twice.

6. **True or False.** If $n > 3$, then $2\binom{n}{2} = n$.

7. **True or False.** If $n \geq 1$, then $(x - y)^n$ has $n + 1$ terms.

8. **True or False.** The coefficient of the sixth term in the expansion of $(x + y)^{55}$ is $\binom{55}{5}$, assuming descending powers of x.

Building Skills

In Exercises 9–24, evaluate each expression.

9. $\dfrac{6!}{3!}$ **10.** $\dfrac{11!}{9!}$

11. $\dfrac{12!}{11!}$ **12.** $\dfrac{3!}{0!}$

13. $\binom{6}{4}$ **14.** $\binom{6}{3}$

15. $\dbinom{9}{0}$

16. $\dbinom{12}{0}$

17. $\dbinom{7}{1}$

18. $\dbinom{7}{3}$

19. $\dbinom{45}{45}$

20. $\dbinom{45}{0}$

21. $\dbinom{100}{98}$

22. $\dbinom{100}{2}$

23. $\dbinom{21}{19}$

24. $\dbinom{15}{2}$

In Exercises 25–32, use Pascal's Triangle to expand each binomial.

25. $(x + 2)^4$

26. $(x + 3)^4$

27. $(x - 2)^5$

28. $(3 - x)^5$

29. $(2 - 3x)^3$

30. $(3 - 2x)^3$

31. $(2x + 3y)^4$

32. $(2x + 5y)^4$

In Exercises 33–54, use either the Binomial Theorem or Pascal's Triangle to expand each binomial.

33. $(x + 1)^4$

34. $(x + 2)^4$

35. $(x - 1)^5$

36. $(1 - x)^5$

37. $(y - 3)^3$

38. $(2 - y)^5$

39. $(x + y)^6$

40. $(x - y)^6$

41. $(1 + 3y)^5$

42. $(2x + 1)^5$

43. $(2x + 1)^4$

44. $(3x - 1)^4$

45. $(x - 2y)^3$

46. $(2x - y)^3$

47. $(2x + y)^4$

48. $(3x - 2y)^4$

49. $\left(\dfrac{x}{2} + 2\right)^7$

50. $\left(2 - \dfrac{x}{2}\right)^7$

51. $\left(a^2 - \dfrac{1}{3}\right)^4$

52. $\left(\dfrac{1}{2} - a^2\right)^4$

53. $\left(\dfrac{1}{x} + y\right)^3$

54. $\left(x + \dfrac{2}{y}\right)^3$

In Exercises 55–62, find the specified coefficient.

55. for x^2y^7 in $(x + y)^9$

56. for $x^{10}y^5$ in $(x - y)^{15}$

57. for x^5 in $(x + 3)^8$

58. for x^7 in $(x - 2)^{10}$

59. for x^2y^7 in $(2x + 3y)^9$

60. for x^3y^4 in $(x - 2y)^7$

61. for $x^{10}y^{12}$ in $(x^2 + y^3)^9$

62. for $x^{18}y^6$ in $(x^3 + 2y^2)^9$

In Exercises 63–76, find the specified term. In Exercises 71–76 assume descending powers of x.

63. $(x + y)^{10}$; term containing x^7

64. $(x + y)^{10}$; term containing y^7

65. $(x - 2)^{12}$; term containing x^3

66. $(2 - x)^{12}$; term containing x^3

67. $(2x + 3y)^8$; term containing x^6

68. $(2x + 3y)^8$; term containing y^6

69. $(5x - 2y)^{11}$; term containing y^9

70. $(7x - y)^{15}$; term containing x^9

71. $(x - 1)^9$; seventh term.

72. $(x - 1)^9$; fourth term.

73. $(2x - y)^7$; fifth term.

74. $(-2x + y)^7$; seventh term.

75. $(x + 3y)^{10}$; third term.

76. $(x + 0.5y)^8$; sixth term.

77. Find the value of $(1.2)^5$ by writing it in the form $(1 + 0.2)^5$ and applying the Binomial Theorem.

78. Use the method in Exercise 77 to evaluate each expression.
  a. $(2.9)^4$
  b. $(10.4)^3$

Beyond the Basics

79. Find the middle term in the expansion of $\left(\sqrt{x} - \dfrac{2}{x^2}\right)^{10}$.

80. Find the middle term in the expansion of $\left(\sqrt{x} + \dfrac{1}{x^2}\right)^{10}$.

81. Find the middle term in the expansion of $(1 - x^2y^{-3})^{12}$.

82. Show that the middle term in the expansion of $(1 + x)^{2n}$ is

$$\frac{1 \cdot 3 \cdot 5 \cdot \cdots \cdot (2n - 1)}{n!} 2^n x^n.$$

83. Prove that $\dbinom{n}{0} + \dbinom{n}{1} + \dbinom{n}{2} + \cdots + \dbinom{n}{n} = 2^n$.

84. Prove that $\dbinom{n}{0} - \dbinom{n}{1} + \dbinom{n}{2} - \dbinom{n}{3} + \cdots + (-1)^n \dbinom{n}{n} = 0$.

85. Prove that $\dbinom{k}{j} + \dbinom{k}{j-1} = \dbinom{k+1}{j}$.

86. Prove the Binomial Theorem by using the principle of mathematical induction.

Hint:

$$(x + y)^{k+1} = (x + y)(x + y)^k = (x + y) \sum_{j=0}^{k} \binom{k}{j} x^{k-j} y^j$$

$$= \sum_{j=0}^{k} \binom{k}{j} x^{k+1-j} y^j + \sum_{j=0}^{k} \binom{k}{j} x^{k-j} y^{j+1}.$$

Use Exercise 85 to show that

$$(x + y)^{k+1} = \sum_{j=0}^{k+1} \binom{k+1}{j} x^{k+1-j} y^j.$$

In Exercises 87–89, find the value of the expression without expanding any term.

87. $(2x - 1)^4 + 4(2x - 1)^3(3 - 2x) + 6(2x - 1)^2(3 - 2x)^2 + 4(2x - 1)(3 - 2x)^3 + (3 - 2x)^4$

88. $(x + 1)^4 - 4(x + 1)^3(x - 1) + 6(x + 1)^2(x - 1)^2 - 4(x + 1)(x - 1)^3 + (x - 1)^4$

89. $(3x - 1)^5 + 5(3x - 1)^4(1 - 2x) + 10(3x - 1)^3(1 - 2x)^2 + 10(3x - 1)^2(1 - 2x)^3 + 5(3x - 1)(1 - 2x)^4 + (1 - 2x)^5$

90. Find the constant term in each expansion.
  a. $\left(x^2 - \dfrac{1}{x}\right)^9$
  b. $\left(\sqrt{x} - \dfrac{2}{x^2}\right)^{10}$

91. If the constant term in the expansion of $\left(kx - \dfrac{1}{x^2}\right)^6$ is 240, find k.

92. If the constant term in the expansion of $\left(x^3 + \dfrac{k}{x^8}\right)^{11}$ is 1320, find k.

93. Show that there is no constant term in the expansion of $\left(2x^2 - \dfrac{1}{4x}\right)^{11}$.

Critical Thinking / Discussion / Writing

94. Use the binomial expansion of $(x + y)^6$, with $x = 1$ and $y = 1$, to show that

$$2^6 = \binom{6}{0} + \binom{6}{1} + \binom{6}{2} + \binom{6}{3} + \binom{6}{4} + \binom{6}{5} + \binom{6}{6}.$$

95. Use the binomial expansion of $(x + y)^{10}$ to show that

$$\binom{10}{0} - \binom{10}{1} + \binom{10}{2} - \binom{10}{3} + \binom{10}{4} - \binom{10}{5}$$
$$+ \binom{10}{6} - \binom{10}{7} + \binom{10}{8} - \binom{10}{9} + \binom{10}{10} = 0.$$

96. Use the binomial expansion of $(x + y)^2$ to show that if $x > 0$ and $y > 0$, then $(x + y)^2 > x^2 + y^2$.

97. Use the binomial expansion of $(x + y)^n$ to show that if $x > 0$ and $y > 0$, then $(x + y)^n > x^n + y^n$.

98. Use the binomial expansion of $(1 + x)^n$ to show that if $x > 0$, then $(1 + x)^n > 1 + nx$.

Getting Ready for the Next Section

In Exercises 99–106, write each expression using factorial notation.

99. $5 \cdot 6 \cdot 7 \cdot 8$

100. $10 \cdot 9 \cdot 8 \cdot 7 \cdot 6$

101. $2 \cdot 4 \cdot 6 \cdot 8 \cdot 10 \cdot 12$

102. $3 \cdot 6 \cdot 9 \cdot 12 \cdot 15$

103. Compute $\dfrac{12!}{10!}$.

104. Compute $\dfrac{n!}{(n - r)!}$. for $n = 100$ and $r = 2$.

105. Compute $\dfrac{8!}{5!3!}$.

106. Compute $\dfrac{n!}{(n - r)!r!}$ for $n = 10$ and $r = 3$.

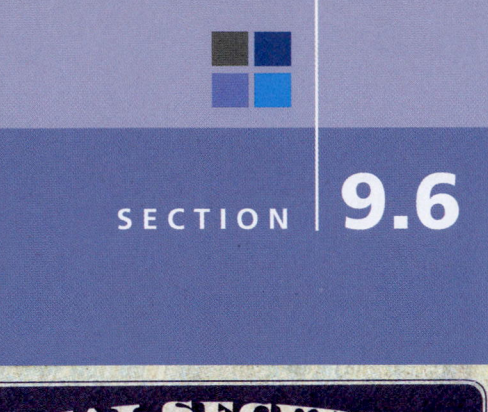

Counting Principles

BEFORE STARTING THIS SECTION, REVIEW

1 Exponents (Appendix A.1, page 923)

2 Factorials (Section 9.1, page 798)

OBJECTIVES

1 Use the Fundamental Counting Principle.

2 Use the formula for permutations.

3 Use the formula for combinations.

4 Use the formula for distinguishable permutations.

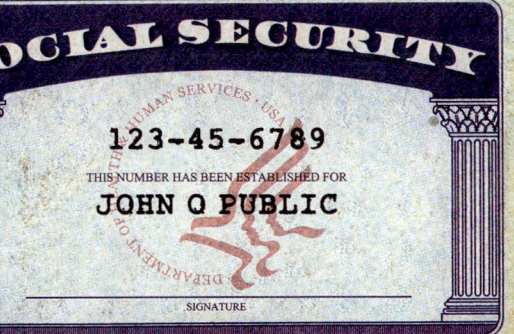

◆ The Mystery of Social Security Numbers

Social Security numbers may seem to be assigned randomly, but they are not. Although the details are complicated, here is a brief introduction: First, the format for all Social Security numbers is NNN-NN-NNNN, where N must be one of the numbers 0, 1, 2, 3, 4, 5, 6, 7, 8, 9. U.S. citizens, permanent residents, and certain temporary residents are issued these numbers. For working people, the numbers help track tax contributions and Social Security benefits.

The first three digits in a Social Security number make up an *area number* assigned to a geographic location. The area number indicates the state in which the card was issued or the ZIP Code in the mailing address provided in the original application.

The middle two digits constitute a *group number* and are designed simply to break the Social Security number into convenient blocks for issuance. The exact procedure is not very enlightening.

The last four digits are *serial numbers* and are assigned in each group from 0001 through 9999.

Various restrictions limit the numbers that can be used. For example, numbers with all zeros in any of the digit blocks (000-xx-xxxx, xxx-00-xxxx, or xxx-xx-0000) are never issued. Even if there were no restrictions, at some point, there might be *more people than available numbers.* How many numbers can be issued in the Social Security format with no restrictions? The answer is 1,000,000,000 (1 billion). (The U.S. population in October 2006 was about 300,000,000 [300 million].)

In Example 3, you learn how to calculate the total number of possible Social Security numbers.

1 Use the Fundamental Counting Principle.

Fundamental Counting Principle

You may think that you already know all there is to know about counting. However, mathematicians have developed a variety of techniques that allow you to determine the number of items that are of interest to you without making the effort of listing and counting all of them. We start with an example that easily lists the items to be counted but that also demonstrates a general counting procedure.

EXAMPLE 1 **Counting Possible Car Selections**

Suppose you are trying to decide between buying a sport-utility vehicle (SUV) and a four-door sedan. The SUV is available in black, red, and silver. The sedan is available in black, blue, and green. In how many ways can you choose a type of car and its color?

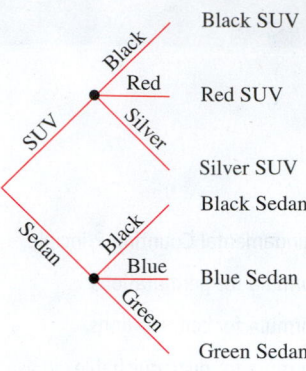

Figure 9.3 Tree Diagram.

Solution

We can represent the possible solutions by using a *tree diagram*, as shown in Figure 9.3 There are two initial choices: an SUV and a sedan. For each of these two choices, there are three additional choices of color; so a total of $2 \cdot 3 = 6$ car selections are possible.

Practice Problem 1 Construct a tree diagram and determine the number of ways you can choose to take one of three courses—English, French, or Math—in the morning, afternoon, or evening.

The car selection process in Example 1 involves making two choices. First, choose a type of car; second, choose a color. It is important in this process that the same *number* of choices be available for the second choice regardless of the result of the first choice. The exact rules for counting by this method are given next.

FUNDAMENTAL COUNTING PRINCIPLE

If a first choice can be made in p different ways, a second choice can be made in q different ways, a third choice can be made in r different ways, and so on, then the sequence of choices can be made in $p \cdot q \cdot r \cdots$ different ways.

EXAMPLE 2 Counting Music Programs

A disc jockey wants to open a program by picking a song from among 40 songs by one artist and close the program by picking a song from among 32 songs by a second artist. How many different choices are possible?

Solution

Each choice of an opening and ending song sequence can be obtained as follows:

First, choose an opening song in 40 ways.
Second, choose the song to end the program in 32 ways.

Since no matter which song is chosen to open the program, all 32 songs are available to end the program, the counting principle applies, and there are a total of

$$40 \cdot 32 = 1280$$

possible choices for the opening and ending sequences.

Practice Problem 2 Reba gets to choose both dinner and a movie for a Saturday night date. She can choose from any of seven restaurants and five movies. How many different choices are possible?

WARNING

Be careful when using the Fundamental Counting Principle. Suppose you are trying to decide between buying an SUV available only in black, red, or silver, and a convertible available only in black or red. In how many ways can you choose a type of car and its color? The car and color can be selected by making two choices: type (SUV or convertible) followed by color. But in this situation, the number of ways you can make the second choice depends on the first choice. You can choose from *three* colors for an SUV but only *two* colors for a convertible. The Fundamental Counting Principle *does not apply*. Fortunately, there are so few different choices here that you can easily list all five: black SUV, red SUV, silver SUV, black convertible, and red convertible.

◆ **EXAMPLE 3**   **Counting Possible Social Security Numbers**

Social Security numbers have the format NNN-NN-NNNN, where each N must be one of the integers 0, 1, 2, 3, 4, 5, 6, 7, 8, and 9. Assuming that there are no other restrictions, how many such numbers are possible?

Solution

There are nine positions in the format

$$NNN\text{-}NN\text{-}NNNN.$$

Since the number of choices available at each position is not affected by previous choices, the Fundamental Counting Principle can be used. Replace each of the letters N with a blank to be filled in with one of the ten digits 0, 1, 2, 3, 4, 5, 6, 7, 8, and 9.

We have

$$\underline{10} \cdot \underline{10} \cdot \underline{10} \cdot \underline{10} \cdot \underline{10} \cdot \underline{10} \cdot \underline{10} \cdot \underline{10} \cdot \underline{10} = 10^9.$$

Thus, there are $10^9 = 1{,}000{,}000{,}000$ possible Social Security numbers.

Practice Problem 3 In Example 3, suppose we fix the area number for the Social Security number to be 433. How many Social Security numbers can be assigned to this area—that is, numbers having the form 433-NN-NNNN?

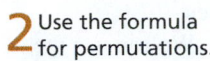

Permutations

2 Use the formula for permutations.

Permutation

A **permutation** is an arrangement of n distinct objects in a fixed order in which no object is used more than once. The specific order is important: Each different ordering of the same objects is a different permutation.

EXAMPLE 4   **Counting Possible Arrangements for a Group Photograph**

How many different ways can five people be arranged in a row for a group photograph?

Solution

To count the number of ways we can arrange five people in a row for a group photograph, we can assign numbers to each of the five photograph positions.

| 1 | 2 | 3 | 4 | 5 |

Then we proceed as follows:

First, choosing any of the five people for Position 1 gives 5 ways.
Second, choosing one of the four remaining people for Position 2 gives 4 ways.
Third, choosing one of the three remaining people for Position 3 gives 3 ways.
Fourth, choosing one of the remaining two people for Position 4 gives 2 ways.
Fifth, choosing the last person remaining for Position 5 gives 1 way.
Using the Fundamental Counting Principle, we see that there are

$$5 \cdot 4 \cdot 3 \cdot 2 \cdot 1 = 5!, \text{ or } 120,$$

possible arrangements.

Practice Problem 4 How many different ways can seven books be arranged on a shelf?

This technique gives a general formula for counting permutations of n distinct objects.

> ### NUMBER OF PERMUTATIONS OF *n* OBJECTS
>
> The number of permutations of *n* distinct objects is
>
> $$n! = n(n-1) \cdots \cdot 4 \cdot 3 \cdot 2 \cdot 1.$$
>
> That is, *n* distinct objects can be arranged in *n*! different ways.

Sometimes only some, but not all, of the available objects are to be arranged in a specific order, again with no object being used more than once. You may want to choose *r* objects from among *n* available objects. This type of ordering is called a **permutation of *n* objects taken *r* at a time**.

EXAMPLE 5 Counting Prizewinners

Three prizes (for first, second, and third place) are to be given out in a dog show having 27 contestants. In how many different ways can dogs come in first, second, and third?

Solution

Consider how we might list the prize winners:

Any dog might win first prize; there are 27 possibilities.
Any remaining dog might win second prize; there are 26 possibilities.
Any remaining dog might win third prize; there are 25 possibilities. We can use the Fundamental Counting Principle: There are

$$27 \cdot 26 \cdot 25 = 17,550$$

1st   2nd   3rd
prize   prize   prize

distinct ways that the dogs can come in first, second, and third. This is an example of permutations of 27 objects taken 3 at a time.

Practice Problem 5 There are nine different rides at a state fair. A group has to decide which four rides they will go on and in what order. How many possibilities are there?

> ### PERMUTATIONS OF *n* OBJECTS TAKEN *r* AT A TIME
>
> The number of permutations of *n* distinct objects taken *r* at a time is denoted by $P(n, r)$, where
>
> $$P(n, r) = \frac{n!}{(n-r)!}$$
> $$= n(n-1)(n-2) \cdots (n-r+1).$$

EXAMPLE 6 Using the Permutations Formula

Use the formula for $P(n, r)$ to evaluate each expression.

a. $P(7, 3)$ **b.** $P(6, 0)$

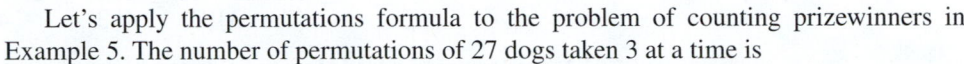
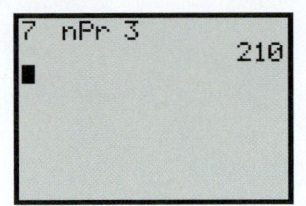
Solution

a. $P(7, 3) = \dfrac{7!}{(7 - 3)!}$ Replace *n* with 7 and *r* with 3 in $P(n, r)$.

$\qquad = \dfrac{7!}{4!} = \dfrac{7 \cdot 6 \cdot 5 \cdot 4!}{4!} = 7 \cdot 6 \cdot 5 = 210$

b. $P(6, 0) = \dfrac{6!}{(6 - 0)!}$ Replace *n* with 6 and *r* with 0 in $P(n, r)$.

$\qquad = \dfrac{6!}{6!} = 1$

Practice Problem 6 Evaluate.

a. $P(9, 2)$ **b.** $P(n, 0)$

Let's apply the permutations formula to the problem of counting prizewinners in Example 5. The number of permutations of 27 dogs taken 3 at a time is

$$P(27, 3) = \dfrac{27!}{(27 - 3)!}$$

$$= \dfrac{27!}{24!} = \dfrac{27 \cdot 26 \cdot 25 \cdot 24!}{24!} = 27 \cdot 26 \cdot 25 = 17{,}550.$$

This is the same answer we found in Example 5.

3 Use the formula for combinations.

Combinations

Order is not always important in selecting objects. For example, if someone is bringing six movies on a vacation, the order in which the movies were chosen is unimportant.

A Combination of *n* Distinct Objects Taken *r* at a Time

When *r* objects are chosen from *n* distinct objects without regard to order, we call the set of *r* objects a **combination of *n* objects taken *r* at a time**. The symbol $C(n, r)$ denotes the total number of combinations of *n* objects taken *r* at a time.

EXAMPLE 7 **Listing Combinations**

List all of the combinations of the four mascot names "bears," "bulls," "lions," and "tigers," taken two at a time. Find $C(4, 2)$.

Solution

If "bears" is one of the mascots chosen, the possibilities are {bears, bulls}, {bears, lions}, and {bears, tigers}.

If "bulls" is one of the mascots chosen, the remaining possibilities (excluding {bears, bulls}) are {bulls, lions} and {bulls, tigers}.

The final choice of two mascot names is {lions, tigers}.

The combinations of the four mascot names taken two at a time are {bears, bulls}, {bears, lions}, {bears, tigers}, {bulls, lions}, {bulls, tigers}, and {lions, tigers}.

Since there are six such combinations, we have $C(4, 2) = 6$.

Practice Problem 7 List all of the combinations of the four mascot names "bears," "bulls," "lions," and "tigers," taken three at a time. Find $C(4, 3)$.

To find a formula for $C(n, r)$, we begin by comparing $C(n, r)$ with $P(n, r)$, the number of permutations of *n* objects taken *r* at a time. Suppose, for example, we have seven distinct objects—*A, B, C, D, E, F,* and *G*—and we pick three of them—*A, D,* and *E* (a combination of seven objects taken three at a time). This one combination *ADE* gives $3! = 6$ permutations of seven objects taken three at a time. Here are all of the possible arrangements of the letters *A, D,* and *E: ADE, AED, DAE, DEA, EAD,* and *EDA*. Since one combination generates six permutations, there are six (3!) times as many permutations of seven objects taken three at a time as there are combinations of seven objects taken three at a time. Thus, $3!(\text{number of combinations}) = (\text{number of permutations})$.

$$3!\, C(7, 3) = P(7, 3)$$

$$C(7, 3) = \frac{P(7, 3)}{3!}$$

$$= \frac{\dfrac{7!}{(7 - 3)!}}{3!} \qquad \text{Replace } P(7, 3) \text{ with } \frac{7!}{(7 - 3)!}.$$

$$= \frac{7!}{(7 - 3)!\,3!} = \frac{7!}{4!\,3!} = \frac{7 \cdot 6 \cdot 5 \cdot 4!}{4!\,3!} = \frac{7 \cdot 6 \cdot 5}{3 \cdot 2 \cdot 1} = 35$$

A similar process leads to a more general formula.

> ### THE NUMBER OF COMBINATIONS OF *n* DISTINCT OBJECTS TAKEN *r* AT A TIME
>
> The number of combinations of *n* distinct objects taken *r* at a time is
>
> $$C(n, r) = \frac{n!}{(n - r)!\, r!}.$$

EXAMPLE 8   **Using the Combinations Formula**

Use the formula $C(n, r)$ to evaluate each expression.

a. $C(7, 5)$ **b.** $C(8, 0)$ **c.** $C(4, 4)$

Solution

a. $C(7, 5) = \dfrac{7!}{(7 - 5)!\,5!}$ Replace *n* with 7 and *r* with 5 in $C(n, r)$.

$$= \frac{7!}{2!\,5!} = \frac{7 \cdot \overset{3}{6} \cdot 5!}{2 \cdot 1 \cdot 5!} = 21.$$

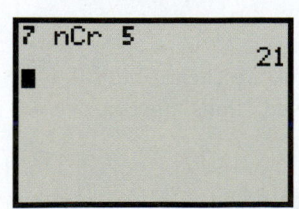
b. $C(8, 0) = \dfrac{8!}{(8 - 0)!\,0!}$ Substitute $n = 8$ and $r = 0$ in $C(n, r)$.

$$= \frac{8!}{8!\,0!} = 1.$$

c. $C(4, 4) = \dfrac{4!}{(4 - 4)!\,4!}$ Substitute $n = 4$ and $r = 4$ in $C(n, r)$.

$$= \frac{4!}{0!\,4!} = 1.$$

Practice Problem 8 Evaluate.

a. $C(9, 2)$ **b.** $C(n, 0)$ **c.** $C(n, n)$

EXAMPLE 9 **Choosing Pizza Toppings**

How many different ways can five pizza toppings be chosen from the following choices: pepperoni, onions, mushrooms, green peppers, olives, tomatoes, mozzarella, and anchovies?

Solution

Eight toppings are listed, so the question is, how many ways can we pick five things from a collection of eight things? That is, we need to find the number of combinations of eight objects taken five at a time, or $C(8, 5)$.

Use the formula $C(n, r) = \dfrac{n!}{(n - r)!\, r!}$.

$$C(8, 5) = \frac{8!}{(8 - 5)!\, 5!}$$ Replace n with 8 and r with 5.

$$= \frac{8!}{3!\, 5!} = \frac{8 \cdot 7 \cdot 6}{3 \cdot 2 \cdot 1} = 56.$$

There are 56 ways that five of the eight pizza toppings can be chosen.

Practice Problem 9 A box of assorted chocolates contains 12 different kinds of chocolates. In how many ways can three chocolates be chosen?

4 Use the formula for distinguishable permutations.

Distinguishable Permutations

Suppose you were to stand in a line with identical twins. Although the three of you can arrange yourselves in $3! = 6$ different ways, someone who could not tell the twins apart could distinguish only three different arrangements. These three arrangements are called *distinguishable permutations*. To better understand the idea of distinguishable permutations, suppose we have three tiles with the numbers 1, 2, and 3 printed on the fronts and the letters *M O M* printed on the backs, respectively.

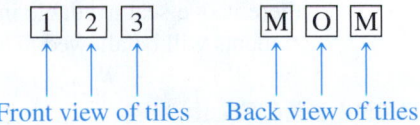

Front view of tiles   Back view of tiles

Front View of Tiles	Back View of Tiles
1 2 3	M O M
1 3 2	M M O
2 1 3	O M M
2 3 1	O M M
3 1 2	M M O
3 2 1	M O M

There are $3! = 6$ permutations of the numbers 1, 2, 3.

Now consider the corresponding arrangements of the letters *M, O,* and *M* on the back sides of the tiles as the fronts of the tiles are arranged to show each of the six permutations of 1, 2, 3. (See margin.)

Only three distinguishable arrangements appear when the tiles are viewed from the back: MOM, MMO, and OMM.

The front view shows the permutations of three distinct objects, but the back view shows the permutations of three objects of which one is of one kind (the letter *O*) and two are of a second kind (the letter *M*).

DISTINGUISHABLE PERMUTATIONS

The number of distinguishable permutations of n objects of which n_1 are of one kind, n_2 are of a second kind, . . . , and n_k are of a kth kind is

$$\frac{n!}{n_1!\cdot n_2!\cdot\,\cdots\,\cdot n_k!},$$

where $n_1 + n_2 + \cdots + n_k = n$.

EXAMPLE 10 **Distributing Gifts**

In how many ways can nine gifts be distributed among three children if each child receives three gifts?

Solution

If we number the gifts 1 through 9 and use *A, B,* and *C* to represent the children, we can visualize any distribution of gifts as an arrangement of the letters *A, B,* and *C* (each letter being used three times) above the numbers 1 through 9. Here is one possible arrangement:

$$
\begin{array}{ccccccccc}
B & A & A & C & C & B & A & C & B \\
1 & 2 & 3 & 4 & 5 & 6 & 7 & 8 & 9
\end{array}
$$

In this arrangement, child *A* gets gifts 2, 3, and 7; child *B* gets gifts 1, 6, and 9; and child *C* gets gifts 4, 5, and 8.

To count all of the possible distributions of gifts, we count the number of permutations of nine objects, of which three are of one kind, three are of a second kind, and three are of a third kind:

$$
\frac{9!}{3!\,3!\,3!} = 1680
$$

There are 1680 ways the nine gifts can be distributed among the three children.

Practice Problem 10 In how many ways can six counselors be sent in pairs to three different locations?

Deciding Whether to Use Permutations, Combinations, or the Fundamental Counting Principle

Suppose 23 students show up late for a private screening of a new movie. Suppose also that there is one seat available in each of the first, second, eighth, and tenth rows and that five students will be allowed to stand in the back of the theater.

SUMMARY OF **MAIN FACTS**

Permutations

If order is important, use permutations.

In how many ways can 4 of the 23 students be assigned the available seats in Rows 1, 2, 8, and 10?

Answer:

$$
\underset{\text{Row 1}}{23} \cdot \underset{\text{Row 2}}{22} \cdot \underset{\text{Row 8}}{21} \cdot \underset{\text{Row 10}}{20} \cdot = 212{,}520 \text{ ways}
$$

Combinations

If order is not important, use combinations.

In how many ways can 5 of the 19 students *not given seats* be chosen to stand in the back?

Answer: $C(19, 5) = 11{,}628$ ways

Fundamental Counting Principle

Whenever you are making consecutive choices and the number of choices at each stage is not affected by the way earlier choices were made, you can use the Fundamental Counting Principle.

In how many ways can students be assigned to the available seats and five students be chosen to stand in the back?

Answer: $(212{,}520 \times 11{,}628)$ ways $= 2{,}471{,}182{,}560$ ways.

Assign the four seats. Pick five students.

Notice that

- Since the Fundamental Counting Principle is used to count permutations, you can always use that principle instead of the formula for permutations.

• Frequently, there are equally good, though different, ways to use counting techniques to solve a problem. For example, we could first choose 5 students to stand in the back of the theater in $C(23, 5)$ ways and then assign the seats in Rows 1, 2, 8, and 10 to the 18 remaining students in $18 \cdot 17 \cdot 16 \cdot 15$ ways. The total number of ways students can be assigned to stand in the back and assigned to the available seats is then $C(23, 5) \cdot 18 \cdot 17 \cdot 16 \cdot 15$. We verify that

$$C(23, 5) \cdot 18 \cdot 17 \cdot 16 \cdot 15 = 33,649 \times 73,440 = 2,471,182,560 \text{ ways.}$$

Answers to Practice Problems

1.

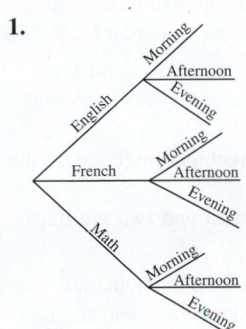

There are $(3)(3) = 9$ ways to choose a course.

2. 35   **3.** 1,000,000   **4.** 5040   **5.** 3024   **6. a.** 72   **b.** 1
7. { bears, bulls, lions }, { bears, bulls, tigers },
{ bears, lions, tigers }, { bulls, lions, tigers }; $C(4, 3) = 4$
8. a. 36   **b.** 1   **c.** 1   **9.** 220   **10.** 90

SECTION 9.6   **Exercises**

Concepts and Vocabulary

1. Any arrangement of n distinct objects in a fixed order in which no object is used more than once is called a(n) _____.

2. The number of permutations of five distinct objects is _____.

3. When r objects are chosen from n distinct objects, the set of r objects is called a(n) _____ of n objects taken r at a time.

4. The number of distinguishable permutations of 10 objects of which 3 are of one kind and 7 are of a second kind is _____.

5. True or False. When order is important in counting objects, we use permutations.

6. True or False. There are more permutations of n objects taken r at a time than there are combinations of n objects taken r at a time.

7. True or False. The number of distinguishable ways that 3 Coke, 2 Sprite, and 3 Pepsi cans can be arranged in a row is 56.

8. True or False. The number of ways a 5-person committee can be selected from 12 people is 5!.

Building Skills

In Exercises 9–16, use the formula for $P(n, r)$ to evaluate each expression.

9. $P(6, 1)$

10. $P(7, 3)$

11. $P(8, 2)$

12. $P(10, 6)$

13. $P(9, 9)$

14. $P(5, 5)$

15. $P(7, 0)$

16. $P(4, 0)$

In Exercises 17–24, use the formula for $C(n, r)$ to evaluate each expression.

17. $C(8, 3)$

18. $C(7, 2)$

19. $C(9, 4)$

20. $C(10, 5)$

21. $C(5, 5)$

22. $C(6, 6)$

23. $C(3, 0)$

24. $C(5, 0)$

In Exercises 25–28, use the Fundamental Counting Principle to solve each problem.

25. How many different 2-letter codes can be made up from the 26 capital letters of the alphabet if (a) repeated letters are allowed; (b) repeated letters are not allowed.

26. How many possible answer sheets are there for a ten-question true–false exam if no answer is left blank?

27. In how many ways can a president and vice president be chosen from an organization with 50 members?

28. How many four-digit numbers can be formed with the digits 0, 1, 2, 3, 4, 5, 6, 7, 8, and 9 if the first digit cannot be 0?

In Exercises 29–32, use the formula for counting permutations to solve each problem.

29. In how many different ways can the members of a family of four be seated in a row of four chairs?

30. Ten teams are entered in a bowling tournament. In how many ways can first, second, and third prizes be awarded?

31. Frederico has five shirts, three pairs of pants, and four ties that are appropriate for an interview. If he wears one of each type of apparel, how many different outfits can he wear?

32. A choreographer has to arrange eight pieces for a dance program. In how many ways can this be done?

In Exercises 33–36, use the formula for counting combinations to solve each problem.

33. A pizza menu offers pepperoni, peppers, mushrooms, sausage, and meatballs as extra toppings that can be added to your pizza. How many different pizzas can you order with two additional toppings?

34. In how many ways can 2 student representatives be chosen from a class of 15 students?

35. Students are allowed to choose 9 out of 11 problems to work for credit on an exam. How many ways can this be done?

36. You have to choose 3 out of 18 potential teammates to help you with a group assignment. How many ways can this be done?

In Exercises 37–42, solve the problem by any appropriate counting method.

37. How many ways can Ashley choose 4 out of 11 different canned goods to donate to charity?

38. How many ways can a computer store hire a salesclerk and a technician from seven applicants for the clerk position and four applicants for the technician position?

39. How many ways can six people be seated in a row with eight seats, leaving two seats vacant?

40. Seven students arrive to rent two canoes and three kayaks. How many ways can two students be selected to rent the canoes and three to rent the kayaks?

41. Dakota is trying to decide in which order to visit five prospective colleges. How many ways can this be done?

42. How many ways can the first 4 numbers be called from the 75 possible bingo numbers?

Applying the Concepts

43. **Area codes.** How many three-digit area codes for telephones are possible if
    a. the first digit cannot be a 0 or 1 and the second digit must be a 0 or 1?
    b. there are no restrictions on the second digit? (Previous restrictions were abandoned in 1995.)

44. **Radio station call letters.** Radio stations in the United States have call letters that begin with either K or W (for example, WXYZ in Detroit). Some have a total of three letters, and others have four. How many different call-letter selections are possible?

45. **Fraternity letters.** The Greek alphabet contains 24 letters. How many 3-letter fraternity names can be made with Greek letters
    a. if no repetitions are allowed?
    b. if repetitions are allowed?

46. **Codes.** How many three-letter codes can be made from the first ten letters of the alphabet if
    a. no letters are repeated?
    b. letters may be repeated?

47. **Supreme Court decisions.** In how many ways can the nine justices of the U.S. Supreme Court reach a majority decision? (The U.S. Supreme Court has nine justices who are appointed for life.)

48. **Committee selection.** A club consists of ten members. How many ways can a committee of four be selected?

49. **Wardrobe choices.** Al's wardrobe consists of eight pairs of slacks, eight shirts, and eight pairs of shoes. He does not want to wear the same outfit on any two days of the year. Will his current wardrobe allow him to do this?

50. **Menu selection.** A French restaurant offers two choices of soup, three of salad, eight of an entree, and five of a dessert. How many different complete dinners can you order?

51. **Housing development.** A developer wants to build seven houses, each of a different design. How many ways can she arrange these homes on a street if
    a. four lots are on one side of the street and three are on the opposite side?
    b. five lots are on one side of the street and two are on the opposite side?

52. **Designing exercise sets.** The author of a mathematics textbook has seven problems for an exercise set, and she is trying to arrange them in increasing order of difficulty. How many ways can she select the problems if no two of them are equally difficult? How many ways can the author fail?

53. **Letter selection.** From the letters of the word *CHARITY*, groups of four letters are formed (without regard to order). How many of them will contain
    a. both A and R?
    b. A but not R?
    c. neither A nor R?

54. **Committee selection.** How many ways can a committee of 4 students and 2 professors be formed from 20 students and 8 professors?

55. **Inventory orders.** A manufacturer makes shirts in five different colors, seven different neck sizes, and three different sleeve lengths. How many shirts should a department store order to have one of each type?

56. **International book codes.** All recent books are identified by their International Standard Book Number (ISBN), a ten-digit code assigned by the publisher. A typical ISBN is 0-321-75526-2. The first digit, 0, represents the language of the book (English); the second block of digits, 321, represents the publishing company (Pearson); the third block of digits, 75526, is the number assigned by the publishing company to that book; and the final digit, 2, is the check digit, used to detect the most commonly made errors when ISBNs are copied. How many ISBNs are possible? (Remember that the previous nine digits determine the last digit.)

57. **Political polling.** A senator sends out questionnaires asking his constituents to rank 10 issues of concern to them, such as crime, education, and taxes, in order of importance. Replies are filed in folders, and two replies are put in the same folder if and only if they give the same ranking to all 10 issues. Find the maximum number of folders that might be needed.

58. **Parcel delivery.** A UPS driver has to deliver 12 parcels to 12 different addresses. In how many orders can the driver make these deliveries?

59. Circus visit. A father with seven children takes four of them at a time to a circus as often as he can, without taking the same four children more than once. What is the maximum number of times each child will go to the circus, and how many times will the father go?

60. Arranging books. From six history books, four biology books, and five economics books, how many ways can a person select three history books, two biology books, and four economics books and arrange them on a shelf?

61. Bingo. Each of four columns on a bingo card contains 5 different numbers chosen in any order from 15. Column "B" chooses from 1 to 15, column "I" from 16 to 30, . . . , and column "O" from 61 to 75. Only four numbers are under column "N" because of the free space. How many different cards are possible?

62. Football conferences. The Grid-Iron Football League in Sardonia is divided into two conferences—East and West— each having six members. Each team in the league must play each team in its conference twice and each team in the other conference once during a season. What is the total number of games played in the league?

63. Married couples prohibited. How many ways can a committee of four people be selected from five married couples if no committee is to include husband-and-wife pairs?

64. Social club committees. A social club contains seven women and four men. The club wants to select a committee of three to represent it at a state convention. How many of the possible committees contain at least one man?

65. Coin gifts. Among Corey's collection of early American coins, he has a half-dollar, a quarter, a nickel, and a penny. He wants to give his younger sister some or all of these four coins. How many different sets of coins can Cory give?

66. Meal possibilities. The university cafeteria offers two choices of soups, three of salad, five of main dishes, and six of desserts. How many different four-course meals can you choose?

In Exercises 67–70, how many distinguishable ways can the letters of each word be arranged?
67. TAIWAN
68. AMERICA
69. SUCCESS
70. ARRANGE

Beyond the Basics

71. A company that supplies temporary help has a contract for 20 weeks to provide three workers per week to an insurance firm. The agreement specifies that in no two weeks can the same three workers be sent to work at the firm. How many different workers are necessary?

72. An FM radio station wants to start its evening programming with five songs every day for 21 days without using the same five songs on any two days. What is the smallest number of songs that can be used to accomplish this result?

73. In geometry, any 2 points determine a line. How many lines are determined by 30 given points, no 3 of which lie on the same line?

74. In geometry, any 3 noncollinear points determine a triangle. How many triangles can be determined by 21 given points, no 3 of which are collinear?

In Exercises 75–82, solve each equation.

75. $\binom{m+1}{m-1} = 3!$

76. $6\binom{n-1}{2} = \binom{n+1}{4}$

77. $2\binom{n-1}{2} = \binom{n}{3}$

78. $\frac{4}{3}\binom{k}{2} = \binom{k+1}{3}$

79. $\binom{n}{0} + \binom{n}{1} + \binom{n}{2} + \cdots + \binom{n}{n} = 64$

80. $\binom{k}{1} + \binom{k}{2} + \binom{k}{3} + \cdots + \binom{k}{k-1} = 126$

81. $\binom{m}{4} = \binom{m}{5}$

82. $\binom{n}{7} = \binom{n}{3}$

Critical Thinking / Discussion / Writing

83. How many terms of the form $x^n y^m$ are possible if n and m can be any integer such that $1 \le n \le 5$ and $1 \le m \le 5$?

84. Use the Fundamental Counting Principle to determine the number of terms in the product

$$(a + b)(x^2 + 2xy + y^2).$$

Explain your reasoning.

[*Hint:* Each term in the product can be obtained by multiplying one term chosen from each factor.]

Getting Ready for the Next Section

In Exercises 85–88, the notation $n(E)$ means the number of elements in the set E.
85. If $E = \{2, 5, 7\}$, find $n(E)$.
86. If $E = \{-3, -1\}$, find $n(E)$.
87. If $E = \{0\}$, find $n(E)$.
88. If $E = \emptyset$, find $n(E)$.

In Exercises 89–92 let $A = \{1, 2, 3, 5\}$ and $B = \{3, 4, 5, 6, 7\}$.
89. Find $n(A)$ and $n(B)$.
90. Find $A \cup B$ and $n(A \cup B)$.
91. Find $A \cap B$ and $n(A \cap B)$.
92. Verify that $n(A \cup B) = n(A) + n(B) - n(A \cap B)$.

In Exercises 93–96, let $A = \{a, b, c\}$ and $B = \{2, 4, 6, 8\}$.
93. Find $n(A)$ and $n(B)$.
94. Find $A \cup B$ and $n(A \cup B)$.
95. Find $A \cap B$ and $n(A \cap B)$.
96. Verify that $n(A \cup B) = n(A) + n(B) - n(A \cap B)$.

Probability

BEFORE STARTING THIS SECTION, REVIEW

1 Set notation (Appendix A.1, page 919)

2 Fundamental Counting Principle (Section 9.6, page 839)

3 Permutations (Section 9.6, page 841)

4 Combinations (Section 9.6, page 843)

OBJECTIVES

1 Find the probability of an event.

2 Use the Additive Rule for finding probabilities.

3 Find the probability of mutually exclusive events.

4 Find the probability of the complement of an event.

5 Find experimental probabilities.

◆ Double Lottery Winner

Many people dream of winning a lottery, but only the most optimistic dream of winning it twice. When Evelyn Marie Adams won the New Jersey state lottery for the second time, *The New York Times* (February 14, 1986) claimed that the chances of one person winning the lottery twice were about 1 in 17 trillion. Two weeks later a letter from two statisticians appeared in the *Times* challenging this claim. Although they agreed that any particular person had very little chance of winning a lottery twice, they said that it was almost certain that someone in the United States would win twice. Further, they said that the odds were even for another double lottery win to occur within seven years. In less than two years, Robert Humphries won his second Pennsylvania lottery prize.

Evaluating how likely an event is to occur is the realm of probability, which we study in this section. In Example 4, we determine the probability of winning a lottery in which 6 numbers are randomly selected from the numbers 1 through 53.

1 Find the probability of an event.

The Probability of an Event

Any process that terminates in one or more outcomes (results) is an **experiment**. For example, if we ask people what time of the day they typically eat their first meal or if we pull balls from a bag of colored balls, we have performed an experiment. The outcomes of the first experiment are the times of day given as responses to our question. The outcomes of the second experiment are the colors of the balls drawn from the bag.

The set of all possible outcomes of a given experiment is called the **sample space** of the experiment. Suppose the experiment is tossing a coin. Then we could denote the outcome "heads" by H; denote the outcome "tails" by T; and write the sample space, S, in set notation as

$$S = \{H, T\}.$$

A single die is a cube, each face of which contains one, two, three, four, five, or six dots, and no two faces contain the same number of dots. A roll of the die is an experiment, the outcome of which is the number of dots showing on the upper face, and the sample space S can be written in set notation as

$$S = \{1, 2, 3, 4, 5, 6\}.$$

An event is any set of possible outcomes; that is, an **event** is any subset of a sample space. For the die-rolling experiment, one possible event is "The number showing is even." In set notation, we write this event as $E = \{2, 4, 6\}$. The event "The number showing is a 6" is written in set notation as $A = \{6\}$.

Outcomes of an experiment are said to be **equally likely** if no outcome should result more often than any other outcome when the experiment is run repeatedly. In both the coin-tossing and die-rolling experiments, all of the outcomes are equally likely.

Probability of an Event

If all of the outcomes in a sample space S are equally likely, then the **probability** of an event E, denoted $P(E)$, is the ratio of the number of outcomes in E, denoted $n(E)$, to the total number of outcomes in S, denoted $n(S)$. That is:

$$P(E) = \frac{n(E)}{n(S)}$$

Because E is a subset of S, $0 \le n(E) \le n(S)$. Dividing by $n(S)$ yields $0 \le \dfrac{n(E)}{n(S)} \le 1$.

So the probability $P(E)$ of an event E is a number between 0 and 1 inclusive.

EXAMPLE 1 Finding the Probability of an Event

A single die is rolled. Write each event in set notation and find the probability of the event.

a. E_1: The number 2 is showing.

b. E_2: The number showing is odd.

c. E_3: The number showing is less than 5.

d. E_4: The number showing is greater than or equal to 1.

e. E_5: The number showing is 0.

Solution

The sample space for the experiment of rolling a die is $S = \{1, 2, 3, 4, 5, 6\}$, so $n(S) = 6$. For any event E, $P(E) = \dfrac{n(E)}{n(S)}$.

a. $E_1 = \{2\}$, so $n(E_1) = 1$. $P(E_1) = \dfrac{n(E_1)}{n(S)} = \dfrac{1}{6}$

b. $E_2 = \{1, 3, 5\}$, so $n(E_2) = 3$. $P(E_2) = \dfrac{n(E_2)}{n(S)} = \dfrac{3}{6} = \dfrac{1}{2}$

c. $E_3 = \{1, 2, 3, 4\}$, so $n(E_3) = 4$. $P(E_3) = \dfrac{n(E_3)}{n(S)} = \dfrac{4}{6} = \dfrac{2}{3}$

d. $E_4 = \{1, 2, 3, 4, 5, 6\}$, so $n(E_4) = 6$. $P(E_4) = \dfrac{n(E_4)}{n(S)} = \dfrac{6}{6} = 1$

e. $E_5 = \varnothing$, so $n(E_5) = 0$. $P(E_5) = \dfrac{n(E_5)}{n(S)} = \dfrac{0}{6} = 0$

Practice Problem 1 A single die is rolled. Write each event in set notation and give the probability of each event.

a. E_1: The number showing is odd.

b. E_2: The number showing is greater than 4.

Notice in Example 1d that $E_4 = S$; so the event E_4 is certain to occur on every roll of the die. Every **certain event** has probability 1. On the other hand, in Example 1(e), the event $E_5 = \varnothing$, so E_5 never occurs, and $P(E_5) = 0$. An **impossible event** always has probability 0.

EXAMPLE 2 **Tossing a Coin**

What is the probability of getting at least one tail when a coin is tossed twice?

Solution

Tossing a coin once results in two equally likely outcomes: heads (H) and tails (T). Tossing the coin twice results in two equally likely outcomes for the first toss and, regardless of the first outcome, two equally likely outcomes for the second toss. By the Fundamental Counting Principle, there are $2 \cdot 2$, or 4, outcomes when we toss the coin twice. We can represent these outcomes as (first toss, second toss) pairs:

$$S = \{(H, H), (H, T), (T, H), (T, T)\}.$$

Because three of these outcomes, $E = \{(H, T), (T, H), (T, T)\}$, result in at least one tail, we have

$$P(E) = \frac{n(E)}{n(S)} = \frac{3}{4}.$$

Practice Problem 2 What is the probability of getting exactly one head when a coin is tossed twice?

Equally likely outcomes occur in most games of chance that involve tossing coins, rolling dice, or drawing cards. Lotteries and games involving spinning wheels also generate equally likely outcomes.

EXAMPLE 3 **Rolling a Pair of Dice**

What is the probability of getting a sum of 8 when a pair of dice is rolled?

Solution

There are six equally likely outcomes for each die; so by the Fundamental Counting Principle, there are $6 \cdot 6$, or 36, equally likely outcomes when a pair of dice is rolled. It is useful to think of having two dice of different colors—say, red and green—in distinguishing the 36 outcomes. In this way, 2 on the red die and 6 on the green die is easily seen to be a different physical outcome from 6 on the red die and 2 on the green die, even though the same sum results.

We can represent the 36 outcomes as (red die, green die) pairs:

$$S = \{(1, 1), (1, 2), (1, 3), (1, 4), (1, 5), (1, 6),$$
$$(2, 1), (2, 2), (2, 3), (2, 4), (2, 5), (2, 6),$$
$$(3, 1), (3, 2), (3, 3), (3, 4), (3, 5), (3, 6),$$
$$(4, 1), (4, 2), (4, 3), (4, 4), (4, 5), (4, 6),$$
$$(5, 1), (5, 2), (5, 3), (5, 4), (5, 5), (5, 6),$$
$$(6, 1), (6, 2), (6, 3), (6, 4), (6, 5), (6, 6)\}$$

The highlighted ordered pairs in the sample space are the outcomes that have a sum of 8. In set notation, the event "The sum of the numbers on the dice is 8" is

$$E = \{(6, 2), (5, 3), (4, 4), (3, 5), (2, 6)\}.$$

Since $n(E) = 5$ and $n(S) = 36$,

$$P(E) = \frac{n(E)}{n(S)} = \frac{5}{36}.$$

So the probability of getting a sum of 8 is $\frac{5}{36}$.

Practice Problem 3 What is the probability of getting a sum of 7 when a pair of dice is rolled?

◆ **EXAMPLE 4**   **Winning the Florida Lottery**

Table tennis balls numbered 1 through 53 are mixed together in a turning drum or basket. Six of the balls are chosen in a random way. To win the lottery, a player must have selected all six of the numbers chosen. The order in which the numbers are chosen does not matter. What is the probability of matching all six numbers? (*Source:* www.fllottery.com)

Solution

Because the order in which the numbers are selected is not important, the sample space S consists of all sets of 6 numbers that can be selected from 53 numbers. Therefore, $n(S) = C(53, 6) = 22,957,480$. Because only 1 set of 6 numbers matches the 6 numbers drawn, the probability of the event "guessed all 6 numbers" is

$$P(\text{winning the lottery}) = P(\text{guessed all 6 numbers})$$

$$= \frac{1}{22,957,480} \approx 0.00000004.$$

Whether you say your chances of winning are 1 in 22,957,480 or about 0.00000004, you definitely know they are not very good!

Practice Problem 4 A state lottery requires a winner to correctly select 6 out of 50 numbers. The order of the numbers does not matter. What is the probability of matching all six numbers?

2 Use the Additive Rule for finding probabilities.

The Additive Rule

Consider the experiment of rolling one die. The sample space is $S = \{1, 2, 3, 4, 5, 6\}$, and $n(S) = 6$.

Let E be the event "The number rolled is greater than 4," and let F be the event "The number rolled is even." Then

$$E = \{5, 6\} \quad \text{and} \quad F = \{2, 4, 6\}.$$

The event "The number rolled is greater than 4 *and* even" is $E \cap F$.

$$E \cap F = \{6\}$$

Because $n(E) = 2, n(F) = 3, n(E \cap F) = 1$, and $n(S) = 6$, we have

$$P(E) = \frac{2}{6}, P(F) = \frac{3}{6}, P(E \cap F) = \frac{1}{6}.$$

The event "The number rolled is either greater than 4 *or* even" is $E \cup F$.

$$E \cup F = \{2, 4, 5, 6\}$$

Note that 6 is in *both* E and F, so it is counted once when $n(E)$ is counted and a second time when $n(F)$ is counted. However, 6 is counted only once when $n(E \cup F)$ is counted. We have $n(E \cup F) = n(E) + n(F) - n(E \cap F)$. That is, since 6 was counted twice when adding $n(E) + n(F)$, we must subtract $n(E \cap F)$ to get an accurate count for $E \cup F$.

$$n(E \cup F) = n(E) + n(F) - n(E \cap F)$$

$$\frac{n(E \cup F)}{n(S)} = \frac{n(E)}{n(S)} + \frac{n(F)}{n(S)} - \frac{n(E \cap F)}{n(S)} \quad \text{Divide both sides by } n(S).$$

$$P(E \cup F) = P(E) + P(F) - P(E \cap F) \quad \text{Definition of probability}$$

This example illustrates the *Additive Rule*.

> ### THE ADDITIVE RULE
>
> If E and F are events in a sample space, then
>
> $$P(E \cup F) = P(E) + P(F) - P(E \cap F)$$

3 Find the probability of mutually exclusive events.

Mutually Exclusive Events

Two events are *mutually exclusive* if it is impossible for both to occur simultaneously. That is, E and F are **mutually exclusive** events if $E \cap F$ contains no outcomes. In this case, $P(E \cap F) = 0$ and the Additive Rule is somewhat simplified.

> ### MUTUALLY EXCLUSIVE EVENTS
>
> If E and F are any events in a sample space and $E \cap F = \varnothing$, then
>
> $$P(E \cup F) = P(E) + P(F).$$

A standard deck of 52 playing cards uses four different designs called suits:

clubs, 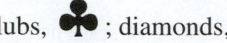 ; diamonds, 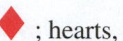 ; hearts, 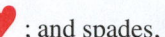 ; and spades,

Each suit has 13 cards: an ace, king, queen, and jack, and cards numbered 2 through 10. The diamond and heart symbols are red, and the club and spade symbols are black. For the purposes of this section, we assume that all decks of cards are standard and well shuffled, so that all cards drawn or dealt from a deck are equally likely.

EXAMPLE 5 **Drawing a Card from a Deck**

A card is drawn from a deck. What is the probability that the card is

a. a king?

b. a spade?

c. a king or a spade?

d. a heart or a spade?

Solution

The sample space S is the 52-card deck; so $n(S) = 52$.

a. There are four kings, one in each suit; so

$$P(\text{king}) = \frac{4}{52} = \frac{1}{13}.$$

b. There are 13 spades in the deck; so

$$P(\text{spade}) = \frac{13}{52} = \frac{1}{4}.$$

c. By the Additive Rule, we have

$$P(\text{king or spade}) = P(\text{king}) + P(\text{spade}) - P(\text{king and spade}).$$

Because there is only one card that is both a king and a spade, $P(\text{king and spade}) = \frac{1}{52}$. From parts **a** and **b** in this example,

$$P(\text{king or spade}) = \frac{4}{52} + \frac{13}{52} - \frac{1}{52}$$

$$= \frac{16}{52} = \frac{4}{13}.$$

d. Because no card is both a heart and a spade, the events "Draw a heart" and "Draw a spade" are mutually exclusive. Thus,

$$P(\text{heart or spade}) = P(\text{heart}) + P(\text{spade})$$
$$= \frac{13}{52} + \frac{13}{52} = \frac{1}{2}.$$

Practice Problem 5 What is the probability that a card pulled from a deck is a jack or a king?

The Complement of an Event

4 Find the probability of the complement of an event.

The **complement of an event** E is the set of all outcomes in the sample space that are *not* in E. The complement of E is denoted by E'. Because every outcome in the sample space is in E or in E', the event $E \cup E'$ is a certain event and $P(E \cup E') = 1$. Also, E and E' are mutually exclusive events, so $P(E \cup E') = P(E) + P(E')$. Thus, $P(E) + P(E') = 1$ and $P(E') = 1 - P(E)$.

PROBABILITY OF THE COMPLEMENT

If E is any event and E' is its complement, then
$$P(E') = 1 - P(E).$$

EXAMPLE 6 Finding the Probability of the Complement

Brandy has applied to receive funding from her company to attend a conference. She knows that her name, along with the names of nine other deserving candidates, was put in a box from which two names will be drawn. What is the probability that Brandy will be one of the two selected?

Solution

It is easier to determine the probability that Brandy will *not* be selected. Since there are nine candidates other than Brandy, there are $C(9, 2)$ ways of selecting two employees not including Brandy. There are $C(10, 2)$ ways of selecting two employees from all ten candidates. Hence, the probability that Brandy is not selected is

$$P(\text{Brandy is not selected}) = \frac{C(9, 2)}{C(10, 2)}$$
$$= \frac{9!}{2! \, 7!} \cdot \frac{2! \, 8!}{10!} = \frac{8}{10} = \frac{4}{5}.$$

The probability that Brandy is selected is

$$P(\text{Brandy is selected}) = 1 - P(\text{Brandy is not selected})$$
$$= 1 - \frac{4}{5} = \frac{1}{5}.$$

Practice Problem 6 In Example 6, what is the probability of Brandy being selected if three names are drawn from the box?

Experimental Probabilities

5 Find experimental probabilities.

The probabilities we have seen up to this point have been determined by some form of reasoning, not data. Data support the decision to assign the value $\frac{1}{2}$ as the probability of a head resulting from one toss of a coin, but the assignment is not *based* on data. This is also true for the probabilities associated with drawing a card, rolling a pair of dice, drawing a ball from an urn, and so on. Probability assigned this way is called **theoretical probability**.

In contrast, we can assign probability values based on observation and data. If an experiment is performed n times (where n is large) and a particular outcome occurs k times, then the number $\dfrac{k}{n}$ is the value for the probability of that outcome. Probability determined this way is called **experimental probability**. A statement such as "The probability that a 10-year-old child will live to be 95 years old is $\dfrac{3}{100,000}$" refers to experimental probability.

RELATIVE FREQUENCY PRINCIPLE

If an experiment is performed n times and an event E occurs k times, then we say that the experimental (or statistical) probability of the event, $P(E)$, is

$$P(E) = \frac{k}{n}.$$

EXAMPLE 7 Using Data to Estimate Probabilities

Table 9.2 shows the gender distribution of students at four California City colleges.

TABLE 9.2

College	Male	Female
Los Angeles	6,995	8,179
San Diego	12,251	14,914
San Francisco	16,857	2,000
Sacramento	9,040	11,838

Source: www.areaconnect.com.

Find the probability that a student selected *at random* is:

a. a female.

b. attending Los Angeles City College.

c. a female attending Los Angeles City College.

Round your answers to the nearest hundredth.

Solution

The term *randomly selected* means that every possible selection is equally likely. In our example, this means that each student is equally likely to be selected. With random selection, the percentage of students in a given category gives the probability that a student in that category will be chosen.

a. The total number of students is 82,074. The total number of female students is 36,931. Thus, $P(\text{female}) = \dfrac{36,931}{82,074} \approx 0.45.$

b. The total number of students is 82,074. The total number of students attending Los Angeles City College is 15,174. Hence,

$$P(\text{attends Los Angeles City College}) = \frac{15,174}{82,074} \approx 0.18.$$

c. There are 8179 female students attending Los Angeles City College. Consequently,

$$P(\text{female and attending Los Angeles City College}) = \frac{8179}{82,074} \approx 0.10.$$

Practice Problem 7 In Example 7, find the probability that a student selected at random is a male or attends Sacramento College.

 **SECTION 9.7** **Exercises**

Concepts and Vocabulary

1. If no outcome of an experiment results more often than any other outcome, the outcomes are said to be _____.

2. A *certain* event has probability _____, and an *impossible* event has probability _____.

3. If it is impossible for two events to occur simultaneously, then the events are said to be _____.

4. If the probability that an event occurs is 0.7, then the probability that the event does not occur is _____.

5. **True or False.** The probability of any event is a number between 0 and 1.

6. **True or False.** If an experiment is performed 100 times and an event E occurs 65 times, then the *experimental probability* of the event is 0.65.

7. **True or False.** If an experiment has 10 possible outcomes, then each outcome must have probability $\dfrac{1}{10}$ of occurring.

8. **True or False.** If two events are mutually exclusive, then the probability that one or the other occurs is 1.

Building Skills

In Exercises 9–14, write the sample space for each experiment.

9. Students are asked to rank Wendy's, McDonald's, and Burger King in order of preference.

10. Students are asked to choose either a hamburger or a cheeseburger and then pick Wendy's, McDonald's, or Burger King as the restaurant from which they would prefer to order it.

11. Two albums are selected from among *Thriller* (Michael Jackson), *The Wall* (Pink Floyd), *Eagles: Their Greatest Hits* (Eagles), and *Led Zeppelin IV* (Led Zeppelin).

12. Three types of potato chips are selected from among regular salted, regular unsalted, barbecue, cheddar cheese, and sour cream and onion.

13. Two blanks in a questionnaire are filled in, the first identifying race as white, African-American, Native American, Asian, other, or multiracial, and the second indicating gender as male or female.

14. A six-sided die is chosen from a box containing a white, a red, and a green die. Then the die is tossed.

In Exercises 15–18, find the probability of the given event if a year is selected at random.

15. Thanksgiving falls on a Sunday.

16. Thanksgiving falls on a Thursday.

17. Christmas in the United States falls on December 25.

18. Christmas falls on January 1.

In Exercises 19–28, classify each statement as an example of *theoretical probability* or *experimental probability*.

19. The probability that Ken will go to church next Sunday is 0.85.

20. The probability of exactly two heads when a fair coin is tossed twice is $\dfrac{1}{4}$.

21. The probability of drawing a heart from a deck of cards is 0.25.

22. The probability that an American adult will watch a major network's evening newscast is 0.17.

23. The probability that a person driving during the Memorial Day weekend will die in an automobile accident is $\dfrac{1}{58,000}$.

24. The probability that a student will get a D or an F in a college statistics course is 0.21.

25. The probability of drawing an ace from a deck of cards is $\dfrac{1}{13}$.

26. The probability that a person will be struck by lightning this year is $\dfrac{1}{727,000}$.

27. The probability that a plane will take off on time from JFK International airport is 0.78.

28. The probability that a college student cheats on an exam is 0.32.

29. Match each given sentence with an appropriate probability.

An event that	has probability
is very likely to happen	0
will surely happen	0.5
is a rare event	0.001
equally likely to happen or not to happen	1
will never happen	0.999

30. A coin is tossed three times. Rearrange the order of the following events from the most probable event (first) to the least probable event (last).
{exactly two heads}, {the sure event}, {at least one head}, {at least two heads}.

In Exercises 31–36, a die is rolled. Write each event in set notation and give the probability of each event.

31. The number showing is a 1 or a 6.

32. The number 5 is showing.

33. The number showing is greater than 4.

34. The number showing is a multiple of 3.

35. The number showing is even.

36. The number showing is less than 1.

In Exercises 37–42, a card is drawn randomly from a standard 52-card deck. Write each event in set notation and find its probability.

37. A queen is drawn.   38. An ace is drawn.

39. A heart is drawn.

40. An ace of hearts or a king of hearts is drawn.

41. A face card (jack, queen, or king) is drawn.

42. A heart that is not a face card is drawn.

In Exercises 43–48, a digit is chosen randomly from the digits 0 through 9, and then 3 is added to the digit and the sum recorded. Write each event in set notation and find the probability of the event.

43. The sum is 11.   44. The sum is 0.

45. The sum is less than 4.   46. The sum is greater than 8.

47. The sum is even.   48. The sum is odd.

Applying the Concepts

49. **A blind date.** The probability that Tony will like his blind date is 0.3. What is the probability that he will not like his blind date?

50. **Drawing cards.** The probability of drawing a spade from a standard 52-card deck is 0.25. What is the probability of drawing a card that is not a spade?

51. **Gender and the Air Force.** In 2006, about 20% of members of the U.S. Air Force were women. What is the probability that an Air Force member chosen at random is a woman? (*Source:* www.af.mil/news)

52. **Movie attendance.** If 85% of people between the ages of 18 and 24 go to a movie at least once a month, what is the probability that a randomly chosen person in this age group will go to a movie this month?

53. **Gender and the Marines.** In 2006, about 6% of members of the U.S. Marines were women. What is the probability that a Marine chosen at random is a man? (*Source:* DMDC-June 2005, TFDW-June 2005)

54. **Movie attendance.** If 20% of people over 65 go to a movie at least once a month, what is the probability that a randomly chosen person over 65 will *not* go to a movie this month?

55. **Raffles.** One hundred twenty-five tickets were sold at a raffle. If you have ten tickets, what is the probability of your winning the (only) prize? If there are two prizes, what is the probability of your winning both?

56. **Committee selection.** To select a committee with two members, two people will be selected at random from a group of four men and four women. What is the probability that a man and a woman will be selected?

57. **Gender of children.** A couple has two children. Assuming that each child is equally likely to be a boy or a girl, find the probability of having
    a. two boys.   b. two girls.
    c. one boy and one girl.

58. **Gender of children.** A couple has three children. Assuming that each child is equally likely to be a boy or a girl, find the probability that the couple will have
    a. all girls.   b. at least two girls.
    c. at most two girls.   d. exactly two girls.

59. **Car ownership** The following table shows the number of cars owned by each of the 4440 families in a small town.

No. of families	37	1256	2370	526	131	120
No. of cars owned	0	1	2	3	4	5

A family is selected at random. Find the probability that this family owns
    a. exactly two cars.   b. at most two cars.
    c. exactly six cars.   d. at most six cars.
    e. at least one car.

60. **Mammograms.** According to the American Cancer Society, 199 of 200 mammograms (for breast cancer screening) turn out to be normal. Find the probability that a mammogram of a woman chosen at random is not normal.

61. **Lactose intolerance.** Lactose intolerance (the inability to digest sugars found in dairy products) affects about 20% of non-Hispanic white Americans and about 75% of African, Asian, and Native Americans. What is the probability that a non-Hispanic white American has no lactose intolerance? Answer the same question for a person in the second group.

62. **Guessing on an exam.** A student has just taken a ten-question true–false test. If he must guess to answer each of the ten questions, what is the probability that
    a. he scores 100% on the test.
    b. he scores 90% on the test.
    c. he scores at least 90% on the test.

63. **Committee selection.** A five-person committee is choosing a subcommittee of two of its members. Each member of the committee is a different age. The subcommittee members are chosen at random.
    a. What is the probability that the subcommittee will consist of the two oldest members?
    b. What is the probability that the subcommittee will consist of the oldest and the youngest members?

64. **How many children do mothers have?** The following data are based on statistics published by the U.S. Census Bureau on the eve of Mother's Day in 2000: Among the 35 million mothers in the United States in the age group from 15 to 44 in 1998, 10.7 million had one child, 14.3 million had two children, 6.7 million had three, and 3.3 million had four or more. If a mother from this age group is chosen at random, what is the probability that she has more than one child?

65. A test for depression. Suppose that 10% of the U.S. population suffers from depression. A pharmaceutical company is believed to have developed a test with the following results: The test works for 90% of the people who are tested and are actually depressed (as diagnosed through other acceptable methods). In other words, if a patient is actually depressed, there is a probability of 0.9 that this patient will be diagnosed as a depressed patient by the pharmaceutical company. The test also works for 85% of people who are tested and not depressed. Suppose the company tests 1000 people for depression. Complete the following table.

Number of people who are	Depressed according to the test	Normal according to the test
actually depressed		
actually normal		

66. Playing marbles. Natasha and Deshawn are playing a game with marbles. Natasha has one marble (call it N), and Deshawn has two (call them M_1 and M_2). They roll their marbles toward a stake in the ground. The person whose marble is closest to the stake wins. Assume that measurements are accurate enough to eliminate ties. List all of the sample points. Use the notation M_1NM_2 to denote the sample point when M_1 is closest to the stake and M_2 is farthest from the stake. If the children are equally skillful (when the sample points are all equally likely), find the probability that Deshawn wins.

67. Gender among siblings. If a family with four children is chosen at random, are the children more likely to be three of one gender and one of the other gender than two boys and two girls? (Assume that each child in a family is equally likely to be a boy or a girl.)

68. Hospital treatments. Hospital records show that 11% of all patients admitted are admitted for surgical treatment, 15% are admitted for obstetrics, and 3% are admitted for both obstetrics and surgery. What is the probability that a new patient admitted is an obstetrics patient or a surgery patient or both?

69. Pain relief medicine. A pharmaceutical company testing a new over-the-counter pain relief medicine gave the new medicine, their old medicine, and placebos to different groups of patients. (A placebo is a sugar pill with no medicinal ingredients. However, it sometimes works, supposedly for psychological reasons.) The following table shows the results of the tests.

Medicine	Amount of Pain Relief		
	Some	Complete	None
Placebo	12	4	34
Old	25	25	10
New	18	40	12
Total	55	69	56

Find the probability that a randomly chosen person from the group of 180 people receiving one of the three types of medicines

a. received the placebo or had no pain relief.

b. received the new medicine or had complete pain relief.

Beyond the Basics

70. A famous problem called the *birthday problem* asks for the probability that at least two people in a group of people have the same birthday (the same day and the same month but not necessarily the same year). The best way to approach this problem is to consider the complementary event that no two people have the same birthday. Assume that all birthdays are equally likely.

a. Find the probability that in a group of two people, both have the same birthday.

b. Find the probability that in a group of three people, at least two have the same birthday.

c. Find the probability that in a group of 50 people, at least two have the same birthday.

71. What is the probability that a number between 1 and 1,000,000 contains the digit 3? [*Hint:* Consider the complementary event.]

72. A business executive types four letters to her clients and then types four envelopes addressed to the clients. What is the probability that if the four letters are randomly placed in envelopes, they are placed in correct envelopes?

73. Juan gave his telephone number to Maggie. She remembers that the first three digits are 407 and the remaining four digits consist of two 3s, one 6, and one 8. She is not certain about the order of the digits, however. Maggie dials 407-3638. What is the probability that Maggie dialed the correct number?

Critical Thinking / Discussion / Writing

74. The March 2005 population survey by the U.S. Census Bureau found that among the 187 million people aged 25 or older, 59,840,000 reported high school graduation as the highest level of education attained. Find the probability that a person chosen at random from the 187 million in the survey would report his or her highest level of education as high school graduation.

75. Find the probability of randomly choosing a dark caramel from a box that contains only light and dark caramels and has twice as many light as dark caramels.

76. If a single die is rolled two times, what is the probability that the second roll will result in a number larger than the first roll?

SUMMARY Definitions, Concepts, and Formulas

9.1 Sequences and Series

i. An infinite sequence is a function whose domain is the set of positive integers. The function values $a_1, a_2, a_3, \ldots, a_n, \ldots$ are called the **terms** of the sequence. The term a_n is the **nth term**, or **general term**, of the sequence.

ii. *Recursive formula*
A sequence may be defined recursively, with the nth term of the sequence defined in relation to previous terms.

iii. *Factorial notation*

$$n! = n(n-1)\cdots 3\cdot 2\cdot 1 \text{ and } 0! = 1$$

iv. *Summation notation*

$$\sum_{i=1}^{n} a_i = a_1 + a_2 + a_3 + \cdots + a_n$$

v. *Series*
Given an infinite sequence $a_1, a_2, a_3, \ldots, a_n, \ldots$, the sum $\sum_{i=1}^{\infty} a_i = a_1 + a_2 + a_3 + \cdots$ is called a **series** and the sum $\sum_{i=1}^{n} a_i$ is called the **nth partial sum** of the series.

9.2 Arithmetic Sequences; Partial Sums

i. A sequence is an **arithmetic sequence** if each term after the first differs from the preceding term by a constant. The constant difference d between the consecutive terms is called the **common difference**. We have $d = a_n - a_{n-1}$ for all $n \geq 2$.

ii. The **nth term** a_n of the arithmetic sequence $a_1, a_2, a_3 \ldots, a_n, \ldots$ is given by $a_n = a_1 + (n-1)d, n \geq 1$, where d is the common difference.

iii. The **sum** S_n of the first n terms of the arithmetic sequence $a_1, a_2, \ldots, a_n, \ldots$ is given by the formula

$$S_n = n\left(\frac{a_1 + a_n}{2}\right).$$

9.3 Geometric Sequences and Series

i. A sequence is a **geometric sequence** if each term after the first is a constant multiple of the preceding term. The constant ratio r between the consecutive terms is called the **common ratio**. We have $\frac{a_{n+1}}{a_n} = r, n \geq 1$.

ii. The **nth term** a_n **of the geometric sequence** $a_1, a_2, a_3, \ldots, a_n, \ldots$ is given by $a_n = a_1 r^{n-1}, n \geq 1$, where r is the common ratio.

iii. The **sum** S_n of the first n terms of the geometric sequence $a_1, a_2, \ldots, a_n, \ldots$ with common ratio $r \neq 1$ is given by the formula

$$S_n = \frac{a_1(1 - r^n)}{1 - r}.$$

The **sum** S **of an infinite geometric sequence** is given by

$$S = \sum_{i=1}^{\infty} a_1 r^{i-1} = \frac{a_1}{1 - r} \text{ if } |r| < 1.$$

iv. An **annuity** is a sequence of equal payments made at equal time intervals. Suppose \$$P$ is the payment made at the end of each of n compounding periods per year and i is the annual interest rate. Then the value A of the annuity after t years is

$$A = P\frac{\left(1 + \frac{i}{n}\right)^{nt} - 1}{\frac{i}{n}}.$$

v. If $|r| < 1$, the infinite geometric series $a_1 + a_1 r + a_1 r^2 + \cdots + a_1 r^{n-1} + \cdots$ has the sum $S = \frac{a_1}{1 - r}$. When $|r| \geq 1$, the series does not have a finite sum.

9.4 Mathematical Induction

The statement P_n is true for all positive integers n if the following properties hold:

i. P_1 is true.

ii. If P_k is a true statement, then P_{k+1} is a true statement.

9.5 The Binomial Theorem

i. Binomial coefficient: $\binom{n}{j} = \frac{n!}{j!(n-j)!}, 0 \leq j \leq n$

$$\binom{n}{0} = 1 = \binom{n}{n}$$

ii. Binomial theorem:

$$(x + y)^n = \binom{n}{0}x^n + \binom{n}{1}x^{n-1}y + \binom{n}{2}x^{n-2}y^2 + \cdots$$
$$+ \binom{n}{j}x^{n-j}y^j + \cdots + \binom{n}{n}y^n$$
$$= \sum_{j=0}^{n} \binom{n}{j}x^{n-j}y^j$$

iii. The term containing the factor x^j in the expansion of $(x + y)^n$ is

$$\binom{n}{n-j}x^j y^{n-j}.$$

9.6 Counting Principles

i. **Fundamental Counting Principle:** If a first choice can be made in p different ways, a second choice in q different ways, a third choice in r different ways, and so on, then the sequence of choices can be made in $p \cdot q \cdot r \ldots$ different ways.

ii. A **permutation** is an arrangement of n distinct objects in a fixed order in which no object is used more than once. The order in an arrangement is important.

iii. **Permutation formula:** The number of permutations of n distinct objects taken r at a time is

$$P(n, r) = \frac{n!}{(n-r)!} = n(n-1)(n-2)\ldots(n-r+1).$$

iv. Combination formula: When r objects are chosen from n distinct objects without regard to order, we call the set of r objects a **combination** of n objects taken r at a time. The number of combinations of n distinct objects taken r at a time is denoted by $C(n, r)$, where

$$C(n, r) = \frac{n!}{(n - r)! \, r!}.$$

v. Distinguishable permutations: The number of permutations of n objects of which n_1 are of one kind, n_2 are of a second kind, . . . , and n_k are of a kth kind is

$$\frac{n!}{n_1! \, n_2! \ldots n_k!},$$

where $n_1 + n_2 + \cdots + n_k = n$.

9.7 Probability

i. Experiment: Any process that terminates in one or more **outcomes** (results) is an experiment.

ii. Sample space: The set of all possible outcomes of an experiment is called the sample space of the experiment.

iii. Event: An event is any subset of a sample space.

iv. Equally likely events: Outcomes of an experiment are said to be equally likely if no outcome should result more often than any other outcome when the experiment is performed repeatedly.

v. Probability: If all of the outcomes in a sample space S are equally likely, the probability of an event E, denoted $P(E)$, is defined by

$$P(E) = \frac{n(E)}{n(S)},$$

where $n(E)$ is the number of outcomes in E and $n(S)$ is the total number of outcomes in S.

a. The Additive Rule: If E and F are events in a sample space S, then

$$P(E \text{ or } F) = P(E \cup F) = P(E) + P(F) - P(E \cap F).$$

b. If $E \cap F = \varnothing$, E and F are called **mutually exclusive** events. For mutually exclusive events E and F, $P(E \cup F) = P(E) + P(F)$.

c. $P(E') = P(\text{not } E) = 1 - P(E)$; the event E' is called the **complement** of the event E.

REVIEW EXERCISES

In Exercises 1–4, write the first five terms of each sequence.

1. $a_n = 2n - 3$

2. $a_n = \dfrac{n(n - 2)}{2}$

3. $a_n = \dfrac{n}{2n + 1}$

4. $a_n = (-2)^{n-1}$

In Exercises 5 and 6, write a general term a_n for each sequence.

5. $30, 28, 26, 24, \ldots$

6. $-1, 2, -4, 8, \ldots$

In Exercises 7–10, simplify each expression.

7. $\dfrac{9!}{8!}$

8. $\dfrac{10!}{7!}$

9. $\dfrac{(n + 1)!}{n!}$

10. $\dfrac{n!}{(n - 2)!}$

In Exercises 11–14, find each sum.

11. $\displaystyle\sum_{k=1}^{4} k^3$

12. $\displaystyle\sum_{j=1}^{5} \frac{1}{2j}$

13. $\displaystyle\sum_{k=1}^{7} \frac{k + 1}{k}$

14. $\displaystyle\sum_{k=1}^{5} (-1)^k 3^{k+1}$

In Exercises 15 and 16, write each sum in summation notation.

15. $1 + \dfrac{1}{2} + \dfrac{1}{3} + \dfrac{1}{4} + \cdots + \dfrac{1}{50}$

16. $2 - 4 + 8 - 16 + 32$

In Exercises 17–20, determine whether each sequence is arithmetic. If a sequence is arithmetic, find the first term a_1 and the common difference d.

17. $11, 6, 1, -4, \ldots$

18. $\dfrac{2}{3}, \dfrac{5}{6}, 1, \dfrac{7}{6}, \ldots$

19. $a_n = n^2 + 4$

20. $a_n = 3n - 5$

In Exercises 21–24, find an expression for the nth term of each arithmetic sequence.

21. $3, 6, 9, 12, 15, \ldots$

22. $5, 9, 13, 17, 21, \ldots$

23. $x, x + 1, x + 2, x + 3, x + 4, \ldots$

24. $3x, 5x, 7x, 9x, 11x, \ldots$

In Exercises 25 and 26, find the common difference d and the nth term a_n for each arithmetic sequence.

25. 3rd term: 7; 8th term: 17

26. 5th term: -16; 20th term: -46

In Exercises 27 and 28, find each sum.

27. $7 + 9 + 11 + 13 + \cdots + 37$

28. $\dfrac{1}{4} + \dfrac{1}{2} + \dfrac{3}{4} + 1 + \cdots + 15$

In Exercises 29 and 30, find the sum of the first n terms of each arithmetic sequence.

29. $3, 8, 13, \ldots; n = 40$

30. $-6, -5.5, -5, \ldots; n = 60$

In Exercises 31–34, determine whether each sequence is geometric. If a sequence is geometric, find the first term a_1 and the common ratio r.

31. $4, -8, 16, -32, \ldots$

32. $\dfrac{1}{5}, \dfrac{1}{10}, \dfrac{1}{20}, \dfrac{1}{40}, \ldots$

33. $\dfrac{1}{3}, 1, \dfrac{5}{3}, 9, \ldots$

34. $a_n = 2^{-n}$

In Exercises 35 and 36, for each geometric sequence, find the first term a_1, the common ratio r, and the nth term a_n.

35. $16, -4, 1, -\dfrac{1}{4}, \ldots$

36. $-\dfrac{5}{6}, -\dfrac{1}{3}, -\dfrac{2}{15}, -\dfrac{4}{75}, \ldots$

In Exercises 37 and 38, find the indicated term of each geometric sequence.

37. a_{10} when $a_1 = 2, r = 3$

38. a_{12} when $a_1 = -2, r = \dfrac{3}{2}$

In Exercises 39 and 40, find the sum S_n of the first n terms of each geometric sequence.

39. $\dfrac{1}{10}, \dfrac{1}{5}, \dfrac{2}{5}, \dfrac{4}{5}, \ldots; n = 12$

40. $2, -1, \dfrac{1}{2}, -\dfrac{1}{4}, \ldots; n = 10$

In Exercises 41–44, find each sum.

41. $\dfrac{1}{2} + \dfrac{1}{6} + \dfrac{1}{18} + \dfrac{1}{54} + \cdots$

42. $-5 - 2 - \dfrac{4}{5} - \dfrac{8}{25} - \cdots$

43. $\displaystyle\sum_{i=1}^{\infty} \left(\dfrac{3}{5}\right)^i$

44. $\displaystyle\sum_{i=1}^{\infty} 7\left(-\dfrac{1}{4}\right)^i$

In Exercises 45 and 46, use mathematical induction to prove that each statement is true for all natural numbers n.

45. $\displaystyle\sum_{k=1}^{n} 2^k = 2^{n+1} - 2$

46. $\displaystyle\sum_{k=1}^{n} k(k + 1) = \dfrac{n(n + 1)(n + 2)}{3}$

In Exercises 47 and 48, evaluate each binomial coefficient.

47. $\dbinom{12}{7}$

48. $\dbinom{11}{0}$

In Exercises 49 and 50, find each binomial expansion.

49. $(x - 3)^4$

50. $\left(\dfrac{x}{2} + 2\right)^6$

In Exercises 51 and 52, find the specified term.

51. $(x + 2)^{12}$; term containing x^5

52. $(x + 2y)^{12}$; term containing x^7

In Exercises 53–64, use any method to solve the problem.

53. How many different numbers can be written with the digits 1, 2, 3, and 4 if no digit may be repeated?

54. How many ways can horses finish in first, second, and third place in a ten-horse race?

55. How many ways can seven people line up at a ticket window?

56. In a building with seven entrances, how many ways can you enter the building and leave by a different entrance?

57. How many ways can five movies be listed on a display (vertically, one movie per line)?

58. How many different amounts of money can you make with a nickel, a dime, a quarter, and a half-dollar?

59. How many ways can three candies be taken from a box of ten candies?

60. How many ways can 2 face cards be drawn from a standard 52-card deck?

61. How many ways can 2 shirts and 3 pairs of pants be chosen from 12 shirts and 8 pairs of pants?

62. How many doubles teams can be formed from eight tennis players?

63. In how many distinguishable ways can the letters of the word R E I T E R A T E be arranged?

64. A bookshelf has room for four books. How many different arrangements can be made on the shelf from six available books?

In Exercises 65–74, find the probability requested.

65. Four silver dollars, all with different dates, are randomly arranged on a horizontal display. What is the probability that the two with the most recent dates will be next to each other?

66. A jar contains five white, three black, and two green stones. What is the probability of drawing a black stone on a single draw?

67. The word R A N D O M has been spelled in a game of Scrabble. If two letters are chosen from this word at random, what is the probability that they are both consonants?

68. The names of all 30 party guests are placed in a box, and one name is drawn to award a door prize. If you and your date are among the guests, what is the probability that one of you will win the prize?

69. If two people are chosen at random from a group consisting of five men and three women, what is the probability that one man and one woman are chosen?

70. A company determines that 32% of the e-mail it receives is junk mail. What is the probability that a randomly chosen e-mail at this company is not junk mail?

71. If 2 cards are drawn from a standard 52-card deck, what is the probability that both are clubs?

72. A battery manufacturer inspects 500 batteries and finds that 30 are defective. What is the probability that a randomly selected battery is not defective?

73. A ball is taken at random from a pool table containing balls numbered 1 through 9. What is the probability of obtaining (a) ball number 7? (b) an even-numbered ball? (c) an odd-numbered ball?

74. A clothes dryer load contains two shirts, four pairs of socks, and three nightgowns. You pull one item randomly from the dryer.
  a. What is the probability that it is a sock?
  b. What is the probability that it is a nightgown?

In Problems 1 and 2, write the first five terms of each sequence. State whether the sequence is arithmetic or geometric.

1. $a_n = 3(5 - 4n)$ 2. $a_n = -3(2^n)$

3. Write the first five terms of the sequence defined by $a_1 = -2$ and $a_n = 3a_{n-1} + 5$ for $n \geq 2$.

4. Evaluate $\dfrac{(n-1)!}{n!}$.

5. Find a_7 for the arithmetic sequence with $a_1 = -3$ and $d = 4$.

6. Find a_8 for the geometric sequence with $a_1 = 13$ and $r = -\dfrac{1}{2}$.

In Problems 7–9, find each sum.

7. $\displaystyle\sum_{k=1}^{20} (3k - 2)$ 8. $\displaystyle\sum_{k=1}^{5} \left(\dfrac{3}{4}\right)(2^{-k})$

9. $\displaystyle\sum_{k=1}^{\infty} 18\left(\dfrac{1}{100}\right)^k$; write the answer as a fraction in lowest terms.

10. Evaluate $\dbinom{13}{0}$.   11. Expand $(1 - 2x)^4$.

12. Find the term containing x^1 in the expansion of $(2x + 1)^4$.

13. In how many ways can you answer every question on a true–false test containing ten questions?

In Problems 14 and 15, evaluate each expression.

14. $P(9, 2)$

15. $C(7, 5)$

16. A firm needs to fill two positions—one in accounting and one in human resources. How many ways can these positions be filled if there are six applicants for the accounting position and ten applicants for the human resources position?

17. A state is considering using license plates consisting of two letters followed by a four-digit number. How many distinct license plate numbers can be formed?

18. A box contains five red balls and seven black balls. What is the probability that a ball drawn at random will be red?

19. What is the probability that a 4 does *not* come up on one roll of a die?

20. A box contains seven quarters, six dimes, and five nickels. If two coins are drawn at random, what is the probability that both are quarters?

In Problems 1 and 2, write the first five terms of each sequence. State whether the sequence is arithmetic or geometric.

1. $a_n = 4(2n - 3)$
   a. $-1, 1, 3, 5, 7$; arithmetic
   b. $-1, 1, 3, 5, 7$; geometric
   c. $-4, 4, 12, 20, 28$; arithmetic
   d. $-4, 4, 12, 20, 28$; geometric

2. $a_n = 2(4^n)$
   a. $2, 8, 32, 128, 512$; arithmetic
   b. $2, 8, 32, 128, 512$; geometric
   c. $8, 32, 128, 512, 2048$; arithmetic
   d. $8, 32, 128, 512, 2048$; geometric

3. Write the first five terms of the sequence defined by $a_1 = 5$ and $a_n = 2a_{n-1} + 4$, for $n \geq 2$.
   a. $5, 14, 32, 68, 140$ b. $5, 14, 19, 24, 29$
   c. $5, 9, 13, 17, 21$ d. $5, 8, 10, 12, 14$

4. Evaluate $\dfrac{(n+2)!}{n+2}$.
   a. $2!$ b. 1
   c. $(n+1)!$ d. $n+1$

5. Find a_8 for the arithmetic sequence with $a_1 = -6$ and $d = 3$.
   a. 15 b. -271
   c. 30 d. 18

6. Find a_{10} for the geometric sequence with $a_1 = 247$ and $r = \dfrac{1}{3}$.
   a. 250 b. $\dfrac{247}{19{,}683}$
   c. $\dfrac{247}{59{,}049}$ d. $\dfrac{247}{177{,}147}$

In Problems 7 and 8, find each sum.

7. $\displaystyle\sum_{k=1}^{45} (4k - 7)$
   a. 4185 b. 3735
   c. 4027.5 d. 3825

8. $\displaystyle\sum_{k=1}^{5} \left(\dfrac{4}{3}\right)(2^k)$
   a. $\dfrac{248}{3}$ b. $\dfrac{251}{3}$
   c. $\dfrac{242}{3}$ d. $\dfrac{287}{3}$

9. Evaluate $\displaystyle\sum_{k=1}^{\infty} 8(-0.3)^{k-1}$.
   a. 6 b. $\dfrac{80}{13}$
   c. $-\dfrac{80}{13}$ d. $-\dfrac{83}{20}$

10. Evaluate $\begin{pmatrix} 12 \\ 11 \end{pmatrix}$.
    a. $1.\overline{09}$ b. 12
    c. 1   d. 11!

11. Expand $(3x - 1)^4$.
    a. $81x^4 - 108x^3 + 54x^2 - 12x + 1$
    b. $-81x^4 + 108x^3 - 54x^2 + 12x - 1$
    c. $-81x^4 + 108x^3 + 54x^2 + 12x + 1$
    d. $81x^4 - 108x^2 + 54x - 12$

12. Find the coefficient of x in the expansion of $(2x + 3)^3$.
    a. 108   b. 9
    c. 36   d. 54

13. Karen has four necklaces, ten pairs of earrings, and three bracelets. How many ways can she choose a necklace, a pair of earrings, and a bracelet?
    a. 40   b. 17
    c. 120   d. 240

In Problems 14 and 15, evaluate each expression.

14. $P(7, 3)$
    a. 210   b. 840
    c. 1680   d. 5040

15. $C(10, 7)$
    a. 3   b. 720
    c. 620,400   d. 120

16. How many two-digit numbers can be formed from the digits 0, 1, 2, 3, 4, 5, 6, 7, 8, and 9? No digit can be used more than once.
    a. 45   b. 81
    c. 3,628,880   d. 100

17. You have selected eight CDs for your dad, but have only enough money to buy five of them. In how many ways can you select the five if you decide that a particular CD is a "must buy"?
    a. 56   b. 70
    c. 1680   d. 35

18. What is the probability of rolling a 5 on one roll of a die?
    a. $\dfrac{1}{5}$ b. 4
    c. $\dfrac{1}{6}$ d. 0.05

19. What is the probability of getting a sum greater than 9 when a pair of dice is rolled?
    a. $\dfrac{1}{6}$ b. $\dfrac{1}{4}$
    c. $\dfrac{1}{12}$ d. $\dfrac{1}{18}$

20. What is the probability that a card drawn at random from a standard 52-card deck is *not* a spade?
    a. $\dfrac{2}{5}$ b. $\dfrac{1}{4}$
    c. $\dfrac{4}{13}$ d. $\dfrac{3}{4}$

CUMULATIVE REVIEW EXERCISES CHAPTERS 1–9

In Problems 1–9, solve each equation or inequality.

1. $|3x - 8| = |x|$
2. $x^2(x^2 - 5) = -4$
3. $\log_2 (3x - 5) + \log_2 x = 1$
4. $e^{2x} - e^x - 2 = 0$
5. $\dfrac{6}{x + 2} = \dfrac{4}{x}$
6. $2(x - 8)^{-1} = (x - 2)^{-1}$
7. $\sqrt{x - 3} = \sqrt{x} - 1$
8. $x^2 + 4x \geq 0$
9. $18 \leq x^2 + 6x$

In Problems 10–11, solve each system.

10. $\begin{cases} 5x + 4y = 6 \\ 4x - 3y = 11 \end{cases}$

11. $\begin{cases} x - 3y + 6z = -8 \\ 5x - 6y - 2z = 7 \\ 3x - 2y - 10z = 11 \end{cases}$

12. Graph the function $y = \cos \dfrac{2}{3} x$ over one period.

13. Use identities to find the exact value of $\sin \left(-\dfrac{7\pi}{12} \right)$.

14. Write the complex number $2 + 2\sqrt{3}i$ in polar form.

15. Use transformations to sketch the graph of $f(x) = \sqrt{x - 3}$.

16. Find the domain of $f(x) = \dfrac{x + 3}{x(x + 1)}$.

17. Write $7 \ln x - \ln (x - 5)$ in condensed form.

18. If $A = \begin{bmatrix} 1 & 0 & -2 \\ 4 & 1 & 0 \\ 1 & 1 & 7 \end{bmatrix}$, find A^{-1}.

19. Four students are pairing up to ride two tandem bicycles, each of which accommodates two riders, one in back and one in front. How many ways can this be done if we count arrangements as different if either the front or back rider is different?

20. A room has 6 people who are 25 years old or older and 12 people whose age is between 17 and 24. Half the members of each group have gotten speeding tickets at some time. What is the probability that a person chosen at random from the room will be a person who is 25 years old or older or has gotten a speeding ticket at some time?

CHAPTER 10

An Introduction to Calculus

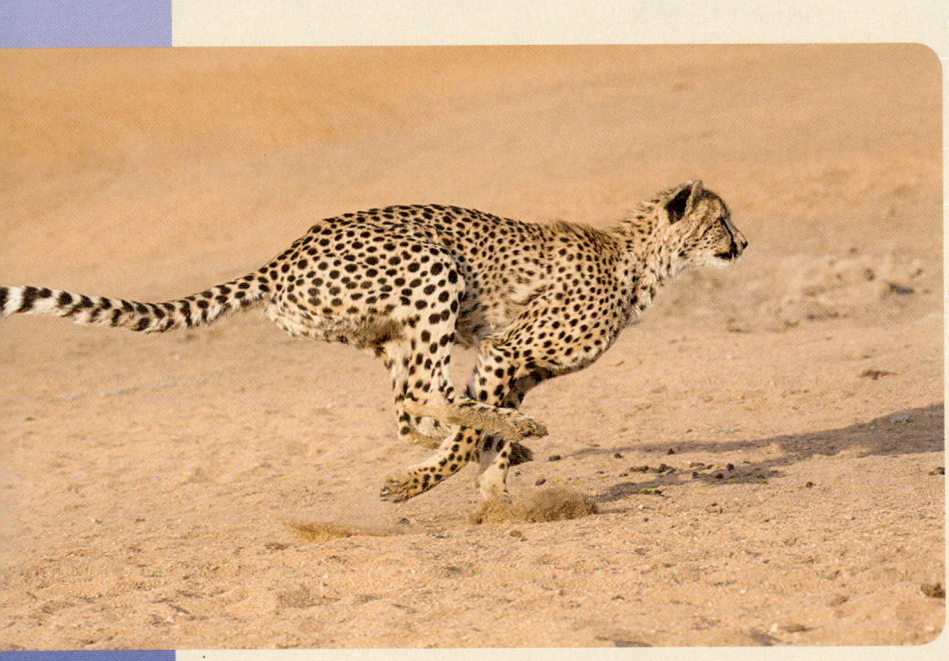

TOPICS

10.1 Finding Limits Using Tables and Graphs

10.2 Finding Limits Algebraically

10.3 Infinite Limits and Limits at Infinity

10.4 Introduction to Derivatives

10.5 Area and the Integral

Calculus is the mathematics of change and motion. It was developed in the seventeenth century by Gottfried Wilhelm Leibniz and Isaac Newton, two of the greatest minds of that era. A major question of that time was brilliantly solved by Isaac Newton, who used calculus to show that the orbit of the Earth around the sun formed an ellipse. Calculus actually provides solutions to a great many challenging problems, both mathematical and physical. The first and most basic tool in calculus is the "limit," which is used to formally define the notion of instantaneous rate of change. That is, if you "freeze" a moving object, how fast was it moving at the moment it was "frozen"?

Finding Limits Using Tables and Graphs

BEFORE STARTING THIS SECTION, REVIEW

1 Evaluating a function (Section 1.3, page 42).

2 Symbols $x \rightarrow a^+$ and $x \rightarrow a^-$ (Section 2.4, page 191).

3 Library of functions (Section 1.4, page 70).

OBJECTIVES

1 Define the limit of a function.

2 Find a limit using a table.

3 Find a limit using a graph.

◆ **Karl W.T. Weierstrass**

Both Newton and Leibniz considered the concept of limit, but their definition lacked mathematical precision. It was the great mathematician Karl Weierstrass who finally gave the notion of limit a rigorous mathematical foundation. Weierstrass was born in Ennigerloh, Germany, in 1815. He was the eldest in a family of two brothers and two sisters. At 19 he entered the University of Bonn, where he indulged in fencing and beer drinking. He returned home after four years without a degree. At age 26, he earned his teachers certificate and taught in secondary schools until 1853.

Weierstrass is called the father of modern analysis. His early publications in a high school paper went unnoticed. In 1854, he published a memoir on Abelian functions that created a sensation and won him an honorary doctorate from the University of Königsberg. He was elected to the Berlin Academy and was given a professorship at the University of Berlin, where he stayed until his death in 1897.

Karl Weierstrass (1815–1897)

1 Define limit of a function.

The Limit Concept

The concept of the limit of a function is fundamental in the development of calculus. The rigorous definition of limits is rather complicated, so we will discuss limits intuitively.

Recall (Section 2.5) that x approaches c is written: $x \rightarrow c$. It means that:

i. x is getting arbitrarily close to c with $x < c$, written $x \rightarrow c^-$; and

ii. x is getting arbitrarily close to c with $x > c$, written $x \rightarrow c^+$.

DEFINITION OF THE LIMIT OF A FUNCTION

The notation

$$\lim_{x \to c} f(x) = L$$

is read "the limit of $f(x)$ as x approaches c, equals L." It means that as x approaches c, the corresponding values of $f(x)$ approach the number L.

We add a few details to the definition of $\lim_{x \to c} f(x) = L$.

1. The function f is defined on an open interval containing c.
2. The function f may or may not be defined at c. That is, we consider $x \to c^+$ or $x \to c^-$, but $x \neq c$.
3. $f(x)$ approaches L as $x \to c^+$ (Written: $\lim\limits_{x \to c^+} f(x) = L$.).
4. $f(x)$ approaches L as $x \to c^-$ (Written: $\lim\limits_{x \to c^-} f(x) = L$.).
5. $\lim\limits_{x \to c} f(x)$ **does not exist if**

   a. $\lim\limits_{x \to c^+} f(x)$ does not approach a fixed value, or

   b. $\lim\limits_{x \to c^-} f(x)$ does not approach a fixed value, or

   c. $\lim\limits_{x \to c^+} f(x) \neq \lim\limits_{x \to c^-} f(x)$.

2 Find a limit using a table.

Limits Using Tables

We can use a calculator to create a table of values to find $\lim\limits_{x \to c} f(x)$.

> **EXAMPLE 1** Finding a Limit Using a Table

Find: $\lim\limits_{x \to 2} (3x^2 - 1)$.

Solution

Let us see what happens to the values of $f(x) = 3x^2 - 1$ as x gets closer and closer to 2 but remains unequal to 2.

We select values of x both as x approaches 2 from the right $(x \to 2^+)$ and as x approaches 2 from the left $(x \to 2^-)$. See Table 10.1.

TABLE 10.1

	x approaches 2 from the left: $x \to 2^-$				$\to$ $\leftarrow$	x approaches 2 from the right: $x \leftarrow 2^+$			
x	1.9	1.99	1.999	1.9999	$\to$ $\leftarrow$	2.0001	2.001	2.01	2.1
$f(x) = 3x^2 - 1$	9.8300	10.8803	10.9880	10.9988	$\to$ $\leftarrow$	11.0012	11.0120	11.1203	12.2300

$f(x)$ approaches 11 $f(x)$ approaches 11

From Table 10.1, we see that as x gets closer to 2 from the left $(x \to 2^-)$, the values of $f(x)$ approach 11, and we write

$$\lim_{x \to 2^-} f(x) = 11 \quad \text{or} \quad \lim_{x \to 2^-} (3x^2 - 1) = 11.$$

Similarly, as x approaches 2 from the right $(x \to 2^+)$, the values of $f(x)$ approach 11, and we write

$$\lim_{x \to 2^+} f(x) = 11 \quad \text{or} \quad \lim_{x \to 2^+} (3x^2 - 1) = 11.$$

Since the values of $f(x) = 3x^2 - 1$ approach the same number, 11, as x approaches 2 from the left and from the right, we say that the limit of $f(x)$ as x approaches the number 2 is the number 11, and we write

$$\lim_{x \to 2} f(x) = 11 \quad \text{or} \quad \lim_{x \to 2} (3x^2 - 1) = 11.$$

Practice Problem 1 Find $\lim\limits_{x \to 1} (2x + 3)$

In Example 1, you may have noticed that instead of making tables of values for $f(x)$ as x gets closer to 2, we could have computed the limit by substituting 2 for x in $f(x)$.

In other words,

$$\lim_{x \to 2} (3x^2 - 1) = 3(2)^2 - 1 = 3(4) - 1 = 12 - 1 = 11.$$

This technique is quite valid and will work for this type of problem. Unfortunately, this is not always the case. Let's consider another example.

EXAMPLE 2 Finding the Limit Using a Table

Find: $\lim\limits_{x \to 1} \dfrac{x^2 - 1}{x - 1}$

Solution

The function

$$f(x) = \frac{x^2 - 1}{x - 1}$$

is not defined at $x = 1$ because substituting $x = 1$ causes division by zero. So, we cannot take the shortcut of substituting 1 for x in $f(x)$.

We again construct a table of values as $x \to 1^-$ and as $x \to 1^+$. See Table 10.2.

TABLE 10.2

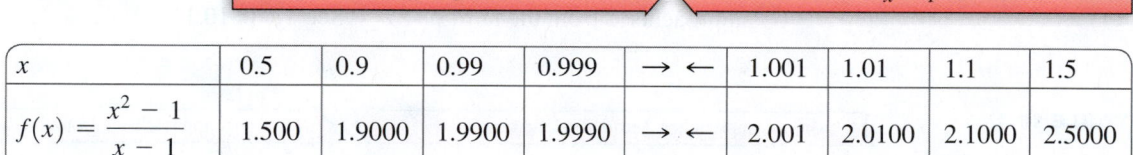

x	0.5	0.9	0.99	0.999	→ ←	1.001	1.01	1.1	1.5
$f(x) = \dfrac{x^2 - 1}{x - 1}$	1.500	1.9000	1.9900	1.9990	→ ←	2.001	2.0100	2.1000	2.5000

The values of $f(x)$ in Table 10.2 suggest that

$$\lim_{x \to 1^-} f(x) = 2 \quad \text{and} \quad \lim_{x \to 1^+} f(x) = 2.$$

We conclude that

$$\lim_{x \to 1} f(x) = 2 \quad \text{or} \quad \lim_{x \to 1} \frac{x^2 - 1}{x - 1} = 2.$$

Practice Problem 2 Find: $\lim\limits_{x \to -1} \dfrac{x^2 - 1}{x + 1}$.

EXAMPLE 3 One-Sided Limits

For the function $f(x) = \begin{cases} x - 1, & x > 2 \\ x^2, & x \le 2 \end{cases}$

find

a. $\lim\limits_{x \to 2^-} f(x)$ **b.** $\lim\limits_{x \to 2^+} f(x)$ **c.** $\lim\limits_{x \to 2} f(x)$ **d.** $f(2)$

Solution

a. The function f is a piecewise-defined function. To find $\lim\limits_{x \to 2^-} f(x)$, we look for values of x that are less than 2. The equation $f(x) = x^2$ is the valid formula for such values. We construct Table 10.3.

TABLE 10.3　　　　　　　　　　$x \to 2^-$

x	1.5	1.9	1.99	1.999	1.9999	$\to$
$f(x) = x^2$	2.2500	3.6100	3.9601	3.9960	3.9996	$\to$

We note that $\lim\limits_{x \to 2^-} f(x) = 4$.

b. To find $\lim\limits_{x \to 2^+} f(x)$, we are interested in the values of x that are greater than 2. The equation $f(x) = x - 1$ is the valid formula for such values. We construct Table 10.4.

TABLE 10.4　　　　　　　　　　$x \to 2^+$

x	2.1	2.01	2.001	2.0001	$\to$
$f(x) = x - 1$	1.1	1.01	1.001	1.0001	$\to$

We conclude that $\lim\limits_{x \to 2^+} f(x) = 1$.

c. Since $\lim\limits_{x \to 2^-} f(x) \neq \lim\limits_{x \to 2^+} f(x)$, $\lim\limits_{x \to 2} f(x)$ does not exist.

d. Because $f(x) = x^2$ for $x \le 2$, and $2 \le 2$, we have $f(2) = 2^2 = 4$.

Practice Problem 3 For the function $f(x) = \begin{cases} 2x - 1, & x \le 1 \\ 2 - x^2, & x > 1 \end{cases}$

Find　**a.** $\lim\limits_{x \to 1^-} f(x)$　　**b.** $\lim\limits_{x \to 1^+} f(x)$　　**c.** $\lim\limits_{x \to 1} f(x)$　　**d.** $f(1)$

EXAMPLE 4　**Finding a Limit Using a Table of Values**

Find:

$$\lim_{x \to 0} \frac{\sin x}{x}.$$

Solution

Using a calculator we construct a table of values, as $x \to 0^+$ and as $x \to 0^-$. See Table 10.5. The values of x in the table are measured in radians.

TABLE 10.5

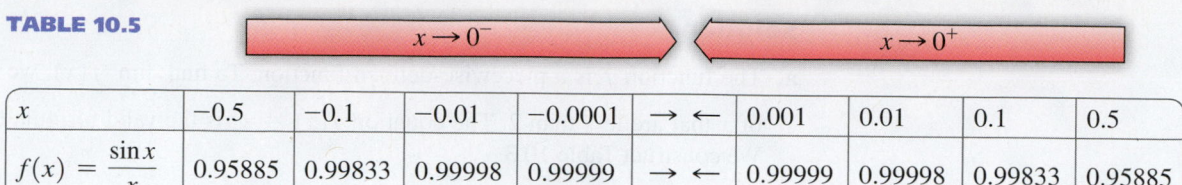

x	-0.5	-0.1	-0.01	-0.0001	$\rightarrow$ $\leftarrow$	0.001	0.01	0.1	0.5
$f(x) = \dfrac{\sin x}{x}$	0.95885	0.99833	0.99998	0.99999	$\rightarrow$ $\leftarrow$	0.99999	0.99998	0.99833	0.95885

The values of $f(x)$ in Table 10.5 suggest that

$$\lim_{x \to 0^-} \frac{\sin x}{x} = 1 \quad \text{and} \quad \lim_{x \to 0^+} \frac{\sin x}{x} = 1.$$

We conclude that

$$\lim_{x \to 0} \frac{\sin x}{x} = 1.$$

Practice Problem 4 Find: $\lim\limits_{x \to 0} \dfrac{1 - \cos x}{x^2}$.

3 Find a limit using a graph.

Finding Limits Graphically

We next find the limit of a function by using its graph. Recall that if (a, b) is a point on the graph of a function $y = f(x)$, then $b = f(a)$. This means that for an input a, we get the output b. The input a is represented on the x-axis, and the output b, the height or depth of the graph, is represented on the y-axis. See Figure 10.1(a).

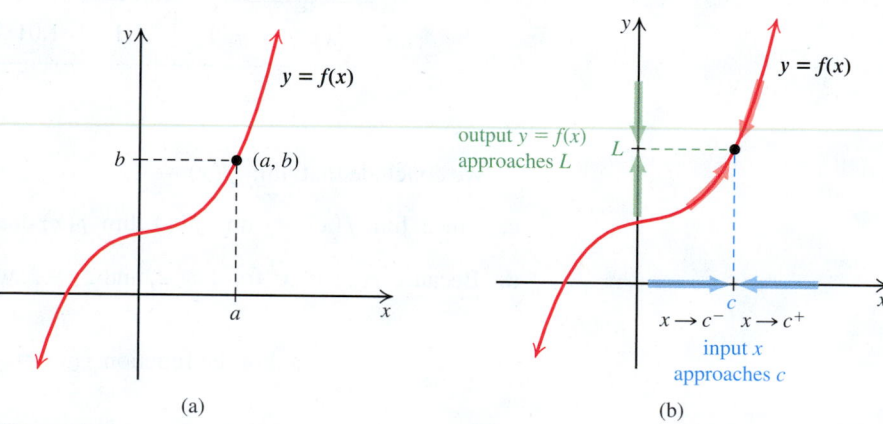

(a)   (b)

Figure 10.1

So, the statement

$$\lim_{x \to c} f(x) = L$$

means that as x approaches c (both as $x \to c^-$ and $x \to c^+$) on the x-axis, the height of the graph near $x = c$ approaches the number L. See Figure 10.1(b).

Figure 10.2 shows that in each case $\lim\limits_{x \to c} f(x) = L$.

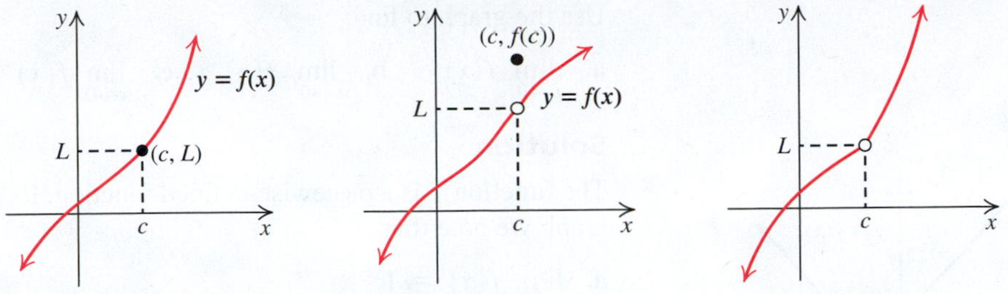

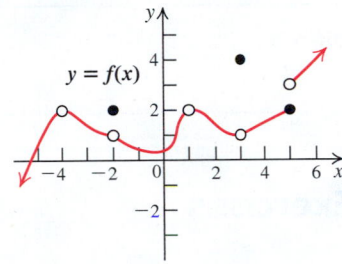

$$\lim_{x \to c} f(x) = L,$$
and $f(c) = L$

(a)

$$\lim_{x \to c} f(x) = L,$$
but $f(c) \neq L$

(b)

$$\lim_{x \to c} f(x) = L,$$
but $f(c)$ is undefined

(c)

Figure 10.2

Figure 10.2 illustrates that in determining the limit of a function at $x = c$, the most important point is to see how the function behaves near $x = c$, rather than at $x = c$.

EXAMPLE 5 **Finding the Limit Graphically**

Find the requested limits for the graph of f shown in Figure 10.3.

Figure 10.3

a. $\displaystyle \lim_{x \to -4} f(x)$ b. $\displaystyle \lim_{x \to 3} f(x)$ c. $\displaystyle \lim_{x \to 5} f(x)$

Solution

a. As we trace the graph of f near $x = -4$ from both sides, we see that the height of the graph approaches 2. That is, $\displaystyle \lim_{x \to -4^-} f(x) = 2$ and $\displaystyle \lim_{x \to -4^+} f(x) = 2$. So, $\displaystyle \lim_{x \to -4} f(x) = 2$.

b. Even though the height of f at $x = 3$ is 4 ($f(3) = 4$), the height of the graph approaches 1 from both directions near $x = 3$.
   So, $\displaystyle \lim_{x \to 3} f(x) = 1$.

c. i. The height of the graph approaches 2 as we approach $x = 5$ from the left. So,
    $\displaystyle \lim_{x \to 5^-} f(x) = 2$.

   ii. The height of the graph approaches 3 as we approach $x = 5$ from the right. So,
    $\displaystyle \lim_{x \to 5^+} f(x) = 3$.

Since the heights do not approach the single number as $x \to 5$, we conclude that $\displaystyle \lim_{x \to 5} f(x)$
does not exist.

Practice Problem 5 In Figure 10.3, find

a. $\displaystyle \lim_{x \to -2} f(x)$ b. $\displaystyle \lim_{x \to 1} f(x)$

EXAMPLE 6 **Finding a Limit by Graphing**

Sketch the graph of $f(x) = \begin{cases} x & \text{if } x \geq 0 \\ 1 - x & \text{if } x < 0. \end{cases}$

Use the graph to find:

a. $\lim\limits_{x \to 0^-} f(x)$ **b.** $\lim\limits_{x \to 0^+} f(x)$ **c.** $\lim\limits_{x \to 0} f(x)$ **d.** $f(0)$.

Solution

The function f is a piecewise-defined function. Its graph is shown in Figure 10.4. From the graph we note that

a. $\lim\limits_{x \to 0^-} f(x) = 1$

b. $\lim\limits_{x \to 0^+} f(x) = 0$

c. From parts **a** and **b**, we have $\lim\limits_{x \to 0^-} f(x) \neq \lim\limits_{x \to 0^+} f(x)$. We conclude that $\lim\limits_{x \to 0} f(x)$ does not exist.

d. $f(0) = 0$.

Practice Problem 6 Sketch the graph of $g(x) = \begin{cases} x + 1 & \text{if } x > 1 \\ 3x - 1 & \text{if } x < 1. \end{cases}$

Use the graph to find:

a. $\lim\limits_{x \to 1^-} g(x)$ **b.** $\lim\limits_{x \to 1^+} g(x)$ **c.** $\lim\limits_{x \to 1} g(x)$ **d.** $g(1)$.

y↑ graph with y = f(x)

Figure 10.4

Answers to Practice Problems

1. 5   **2.** -2 **3. a.** 1 **b.** 1 **c.** 1 **d.** 1   **4.** $\frac{1}{2}$

5. a. 1 **b.** 2   **6. a.** 2 **b.** 2 **c.** 2 **d.** undefined

---

SECTION 10.1 **Exercises**

Concepts and Vocabulary

1. The limit of $f(x)$ as x approaches c equals the number L is written as _____.

2. The notation $\lim\limits_{x \to c^+} f(x) = L$ means that as _____ approaches c from the right, but remains greater than c and unequal to c, the corresponding values of _____ get closer to _____.

3. If $\lim\limits_{x \to c^-} f(x) = L$ and $\lim\limits_{x \to c^+} f(x) = L$, then $\lim\limits_{x \to c} f(x)$ _____.

4. If $\lim\limits_{x \to c^-} f(x) = L$ and $\lim\limits_{x \to c^+} f(x) = M$, but $L \neq M$, then $\lim\limits_{x \to c} f(x)$ _____.

5. **True or False.** If $f(c) = L$, then $\lim\limits_{x \to c} f(x) = L$.

6. **True or False.** $\lim\limits_{x \to c^-} f(x) \leq \lim\limits_{x \to c^+} f(x)$.

7. **True or False.** If $\lim\limits_{x \to c} f(x) = L$, then $\lim\limits_{x \to c^-} f(x) = \lim\limits_{x \to c^+} f(x)$.

8. **True or False.** If $\lim\limits_{x \to c} f(x) = L$ and $f(c) = M$, then $L = M$.

Building Skills

In Exercises 9–14, complete each table and use the table to estimate each limit.

9. $\lim\limits_{x \to 2} (2x + 3)$

x			1.9	1.99	1.999	→ ←	2.001	2.01	2.1
$f(x) = 2x + 3$									

10. $\lim\limits_{x \to 1} (3x - 4)$

x	0.9	0.99	0.999	→ ←	1.001	1.01	1.1
$f(x) = 3x - 4$							

11. $\lim\limits_{x \to -3} (2x^2 - 1)$

x	-3.1	-3.01	-3.001	→ ←	-2.999	-2.99	-2.9
$f(x) = 2x^2 - 1$							

12. $\lim\limits_{x \to -1} (x^2 + 2)$

x	-1.1	-1.01	-1.001	→ ←	-0.999	-0.99	-0.9
$f(x) = x^2 + 2$							

13. $\lim\limits_{x \to 0} \dfrac{\sin 2x}{x}$

x	-0.1	-0.01	-0.001	→ ←	0.001	0.01	0.1
$f(x) = \dfrac{\sin 2x}{x}$							

14. $\lim\limits_{x \to 0} \dfrac{\tan x}{x}$

x	-0.1	-0.01	-0.001	→ ←	0.001	0.01	0.1
$f(x) = \dfrac{\tan x}{x}$							

In Exercises 15–30, make a table to find each limit.

15. $\lim\limits_{x \to 2} (3x + 1)$

16. $\lim\limits_{x \to 3} (2x - 1)$

17. $\lim\limits_{x \to -1} (3x + 2)$

18. $\lim\limits_{x \to -2} (3x - 4)$

19. $\lim\limits_{x \to 1} \dfrac{2x}{x + 1}$

20. $\lim\limits_{x \to -1} \dfrac{x - 1}{3x}$

21. $\lim\limits_{x \to 2} \dfrac{x^2 - 4}{x - 2}$

22. $\lim\limits_{x \to -2} \dfrac{x + 2}{x^2 - 4}$

23. $\lim\limits_{x \to 2} \dfrac{x^3 - 8}{x - 2}$

24. $\lim\limits_{x \to -2} \dfrac{x^3 + 8}{x + 2}$

25. $\lim\limits_{x \to 0} \dfrac{\sin x^2}{x}$

26. $\lim\limits_{x \to 0} \dfrac{\sin 2x}{\sin x}$

27. $\lim\limits_{x \to 0} \dfrac{\sin x}{\tan x}$

28. $\lim\limits_{x \to 0} \dfrac{2x}{\sin x + 3x}$

29. $\lim\limits_{x \to -1} f(x)$, where $f(x) = \begin{cases} 2x + 1 & x < 1 \\ -x^2, & x \geq 1 \end{cases}$

30. $\lim\limits_{x \to 2^+} f(x)$, where $f(x) = \begin{cases} x - 1, & x < 2 \\ x^2 - 3, & x > 2 \end{cases}$

In Exercises 31–58, use the following graph of $y = f(x)$ to state the value of each quality if it exists.

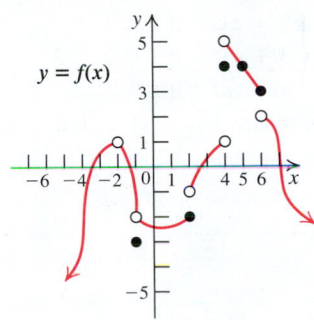

31. $\lim\limits_{x \to -4^+} f(x)$

32. $\lim\limits_{x \to -4} f(x)$

33. $\lim\limits_{x \to -4} f(x)$

34. $f(-4)$

35. $\lim\limits_{x \to -2^-} f(x)$

36. $\lim\limits_{x \to -2^+} f(x)$

37. $\lim\limits_{x \to -2} f(x)$

38. $f(-2)$

39. $\lim\limits_{x \to -1^+} f(x)$

40. $\lim\limits_{x \to -1^-} f(x)$

41. $\lim\limits_{x \to -1} f(x)$

42. $f(-1)$

43. $\lim\limits_{x \to 2^-} f(x)$

44. $\lim\limits_{x \to 2^+} f(x)$

45. $\lim\limits_{x \to 2} f(x)$

46. $f(2)$

47. $\lim\limits_{x \to 4^+} f(x)$

48. $\lim\limits_{x \to 4^-} f(x)$

49. $\lim\limits_{x \to 4} f(x)$

50. $f(4)$

51. $\lim\limits_{x \to 5^-} f(x)$

52. $\lim\limits_{x \to 5^+} f(x)$

53. $\lim\limits_{x \to 5} f(x)$

54. $f(5)$

55. $\lim\limits_{x \to 6^+} f(x)$

56. $\lim\limits_{x \to 6^-} f(x)$

57. $\lim\limits_{x \to 6} f(x)$

58. $f(6)$

In Exercises 59–76, sketch the graph of each function. Use the graph to find the value of each quantity.

59. $f(x) = 2x + 4; \lim\limits_{x \to 1} f(x)$

60. $f(x) = -2x + 1; \lim\limits_{x \to -1} f(x)$

61. $f(x) = -x^2 + 1; \lim\limits_{x \to 2} f(x)$

62. $f(x) = x^2 + 1; \lim\limits_{x \to -2} f(x)$

63. $f(x) = |x - 1|; \lim\limits_{x \to -2} f(x)$

64. $f(x) = |x + 1|; \lim\limits_{x \to -2} f(x)$

65. $f(x) = \dfrac{x^2 - 4}{x - 2}; \lim\limits_{x \to 2} f(x)$

66. $f(x) = \dfrac{x^2 - 4}{x + 2}; \lim\limits_{x \to -2} f(x)$

67. $g(x) = e^x; \lim\limits_{x \to 0} g(x)$

68. $g(x) = \ln x; \lim\limits_{x \to e} g(x)$

69. $h(x) = \sin x; \lim\limits_{x \to 0} h(x)$

70. $h(x) = \cos x; \lim\limits_{x \to 0} h(x)$

71. $f(x) = \begin{cases} x + 1, & \text{if } x \geq 1 \\ 2, & \text{if } x < 1 \end{cases}; \lim\limits_{x \to 1} f(x)$

72. $f(x) = \begin{cases} 2x + 4, & \text{if } x > 2 \\ 2x^2, & \text{if } x \leq 2 \end{cases}; \lim\limits_{x \to 2} f(x)$

73. $g(x) = \begin{cases} \cos x & \text{if } x > 0 \\ e^x, & \text{if } x \leq 0 \end{cases}; \lim\limits_{x \to 0} g(x)$

74. $g(x) = \begin{cases} x & \text{if } x < 0 \\ \sin x, & \text{if } x \geq 0 \end{cases}; \lim\limits_{x \to 0} g(x)$

75. $h(x) = \begin{cases} \ln x & \text{if } x > 1 \\ 1 - x, & \text{if } x \leq 1 \end{cases}; \lim\limits_{x \to 1} h(x)$

76. $h(x) = \begin{cases} 1 + x & \text{if } x > 0 \\ \cos x, & \text{if } x \leq 0 \end{cases}; \lim\limits_{x \to 0} h(x)$

Applying the Concepts

77. Electric Rates. An electric company charges $0.12 per kilowatt hour (kwh) of electricity used for the first 50 kwh and $0.18 per kwh above the 50 kwh used.

(a) Write a piecewise-defined function $c(x)$ for the cost of x kwh of electricity used.

(b) Find $\lim\limits_{x \to 50^-} c(x)$, $\lim\limits_{x \to 50^+} c(x)$, and $\lim\limits_{x \to 50} c(x)$.

(c) Find $c(15)$.

78. Volume Discount. A dairy farm sells milk to retailers at the rate of $2.50 per gallon for the first 100 gallons, $2.25 per gallon for the number of gallons purchased between 100 (not included) and 500 (included) gallons, and $2.00 per gallon for the number of gallons purchased in excess of 500.

a. Write a piecewise-defined function $P(x)$ for the purchase price of x gallons purchased.

b. Find $\lim\limits_{x \to 100^-} P(x)$, $\lim\limits_{x \to 100^+} P(x)$, and $\lim\limits_{x \to 100} P(x)$.

c. Find $P(700)$.

79. Hourly Employees. Annette is allowed to work at CBD Pizza for a maximum of 60 hours per week. For the first 40 hours she is paid $10 per hour, and for the next 20 hours she is paid overtime wages at the rate of $15 per hour.

a. Write the function $W(x)$ that represents her wages if she works x hours per week.

b. Find $\lim\limits_{x \to 40^-} W(x)$, $\lim\limits_{x \to 40^+} W(x)$, and $\lim\limits_{x \to 40} W(x)$.

c. Find $W(50)$.

80. Parking cost. The cost $c(x)$ of parking a car at the metropolitan airport is \$8 for the first hour and \$4 for each additional hour or fraction thereof.

a. Write the function $c(x)$ if x is the number of parking hours.

b. Find $\lim\limits_{x \to 1^-} c(x)$, $\lim\limits_{x \to 1^+} c(x)$, and $\lim\limits_{x \to 1} c(x)$

c. Find $c(5.5)$.

Beyond the Basics

In Exercises 81–84, find each indicated quantity if it exists.

81. Let $f(x) = \sqrt{x}$. Find

a. $\lim\limits_{x \to 0^+} f(x)$ **b.** $\lim\limits_{x \to 0^-} f(x)$

c. $\lim\limits_{x \to 0} f(x)$ **d.** $f(0)$.

82. Let $f(x) = \dfrac{|x|}{x}$. Find

a. $\lim\limits_{x \to 0^+} f(x)$ **b.** $\lim\limits_{x \to 0^-} f(x)$

c. $\lim\limits_{x \to 0} f(x)$ **d.** $f(0)$.

83. Let $f(x) = \dfrac{x^2 + 4x + 3}{x + 1}$. Find

a. $\lim\limits_{x \to -1^+} f(x)$ **b.** $\lim\limits_{x \to -1^-} f(x)$

c. $\lim\limits_{x \to -1} f(x)$ **d.** $f(-1)$.

84. Let $f(x) = x - [x]$. Find

a. $\lim\limits_{x \to 3^+} f(x)$ **b.** $\lim\limits_{x \to 3^-} f(x)$

c. $\lim\limits_{x \to 3} f(x)$ **d.** $f(3)$.

In Exercises 85–88, write a possible function $f(x)$ that satisfies the given conditions. Answers may vary.

85. $f(1) = \dfrac{1}{2}$, $\lim\limits_{x \to 1} f(x) = 2$.

86. $f(2) = 2$, $\lim\limits_{x \to 2^-} f(x) = 2$, $\lim\limits_{x \to 2^+} f(x) = 3$.

87. $f(4) = 2$, $\lim\limits_{x \to 4^-} f(x) = -1$, $\lim\limits_{x \to 4^+} f(x) = 1$.

88. $f(1) = 0$, $\lim\limits_{x \to 1^-} f(x) = 0$, $\lim\limits_{x \to 1^+} f(x)$ does not exist.

Critical Thinking / Discussion Writing

89. Let $f(x) = \begin{cases} 3x + ax, & \text{if } x < 1 \\ 7x - ax, & \text{if } x > 1. \end{cases}$

Define $f(1)$, so that $f(1) = \lim\limits_{x \to 1^-} f(x) = \lim\limits_{x \to 1^+} f(x)$.

90. Let $f(x) = \dfrac{x^2 + x - 2}{x + 2}$, $x \neq -2$

Define $f(-2)$, so that $f(-2) = \lim\limits_{x \to -2} f(x)$.

91. Let $f(x) = \begin{cases} 2 + ax, & \text{if } x < 0 \\ 3 - ax, & \text{if } x > 0. \end{cases}$

Define $f(0)$, so that $f(0) = \lim\limits_{x \to 0^-} f(x) = \lim\limits_{x \to 0^+} f(x)$.

92. Let $f(x) = \begin{cases} 2 + ax, & \text{if } x < 1 \\ 3 + bx, & \text{if } 1 < x < 3 \\ 4x - a, & \text{if } x > 3. \end{cases}$

Find a and b, so that $\lim\limits_{x \to 1^-} f(x) = \lim\limits_{x \to 1^+} f(x)$ and $\lim\limits_{x \to 3^-} f(x) = \lim\limits_{x \to 3^+} f(x)$.

Getting Ready for the Next Section

In Exercises 93–96, factor each expression.

93. $x^2 - 5x - 24$

94. $x^2 - 8x + 12$

95. $x^2 - 36$

96. $x^3 - 8$

In Exercises 97–100, rationalize each numerator.

97. $\dfrac{\sqrt{3} - 1}{2}$

98. $\dfrac{\sqrt{5} - \sqrt{3}}{4}$

99. $\dfrac{\sqrt{x + 1} - 1}{x}$

100. $\dfrac{\sqrt{x^2 + x + 1} - \sqrt{x^2 + 1}}{x}$

In Exercises 101–104, form and simplify the difference quotient $\dfrac{f(x) - f(a)}{x - a}$ for given function f and the number a.

101. $f(x) = x$, $a = 2$

102. $f(x) = 2x - 1$, $a = 3$

103. $f(x) = x^2$, $a = 1$

104. $f(x) = -x^2$, $a = 2$

SECTION **10.2**

Finding Limits Algebraically

BEFORE STARTING THIS SECTION, REVIEW

1   Factoring polynomials (Appendix A.4, page 930).

2   Rationalizing a numerator (Appendix A.4, page 944).

3   Library of functions (Section 1.4, page 70).

OBJECTIVES

1   State properties of limits.

2   Use algebraic manipulations.

3   Find limits of difference quotients.

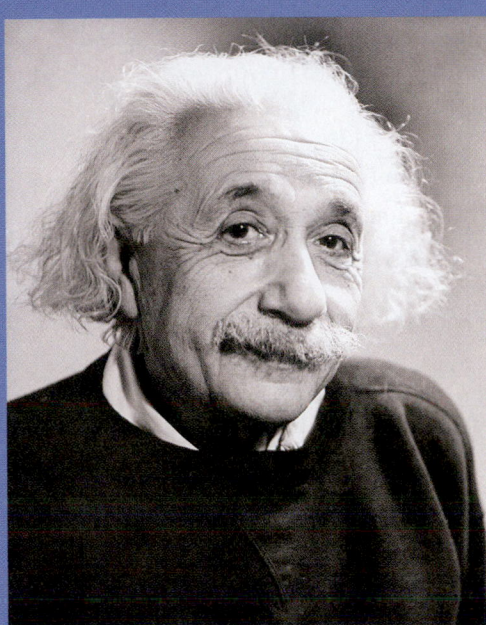

Albert Einstein 1879–1955

◆ **Time Dilation Equation**

In his Special Theory of Relativity, Einstein asserted that time itself is not absolute. He meant that time passes at different rates in different reference frames. If a spaceship leaves Earth and is traveling at v m/sec, he derived the following time dilation equation:

$$\Delta t_0 = \Delta t \sqrt{1 - \frac{v^2}{c^2}} \qquad \text{Time dilation equation}$$

where

$c =$ the speed of light ($\approx 3.000 \times 10^8$ m/s),

$\Delta t_0 =$ the "proper time" measured on the spaceship, and

$\Delta t =$ the time observed on Earth.

In Exercises 67 and 68, we state some of the consequences of the time dilation equation.

---

1 State properties of limits.

Properties of Limits

In Section 10.1, we learned to find the limit of a function by making a table of values or sketching the graph of the function. However, many limits can be found quickly by algebraic manipulations and by the basic properties of limits listed below.

Properties of Limits

Here c and k are real numbers, and we assume that both $\lim\limits_{x \to c} f(x)$ and $\lim\limits_{x \to c} g(x)$ exist.

Property	In Words	Examples
1. $\lim\limits_{x \to c} k = k$	The limit of a constant is that constant.	$\lim\limits_{x \to 1} 6 = 6;\ \lim\limits_{x \to -2} 8 = 8;\ \lim\limits_{x \to -3} (-12) = -12.$
2. $\lim\limits_{x \to c} x = c$	The limit of $f(x) = x$ as $x \to c$ is obtained by substituting c for x.	$\lim\limits_{x \to 3} x = 3;\ \lim\limits_{x \to -7} x = -7;\ \lim\limits_{x \to 0} x = 0.$
3. $\lim\limits_{x \to c} (kf(x)) = k \lim\limits_{x \to c} f(x)$	The limit of a constant times $f(x)$ is the constant times the limit of $f(x)$.	$\lim\limits_{x \to 3} (2x) = 2(\lim\limits_{x \to 3} x) = 2(3) = 6;$ $\lim\limits_{x \to -4} (-3x) = -3(\lim\limits_{x \to -4} x) = -3(-4) = 12.$
4. $\lim\limits_{x \to c} (f(x) + g(x)) =$ $\lim\limits_{x \to c} f(x) + \lim\limits_{x \to c} g(x)$	The limit of the sum of two functions is the sum of their limits.	$\lim\limits_{x \to 3} (2x + 5) = \lim\limits_{x \to 3} (2x) + \lim\limits_{x \to 3} 5$ $= 2 \lim\limits_{x \to 3} x + 5$ $= 2(3) + 5 = 6 + 5 = 11$
5. $\lim\limits_{x \to c} (f(x) - g(x)) =$ $\lim\limits_{x \to c} f(x) - \lim\limits_{x \to c} g(x)$	The limits of the difference of two functions is the difference of their limits.	$\lim\limits_{x \to -2} (x - 7) = \lim\limits_{x \to -2} x - \lim\limits_{x \to -2} 7$ $= -2 - 7 = -9$
6. $\lim\limits_{x \to c} (f(x) \cdot g(x)) =$ $\lim\limits_{x \to c} f(x) \cdot \lim\limits_{x \to c} g(x)$	The limit of the product of two functions is the product of their limits.	$\lim\limits_{x \to 3} (x \cdot x) = \lim\limits_{x \to 3} x \cdot \lim\limits_{x \to 3} x$ $= (3) \cdot (3) = 9$
7. $\lim\limits_{x \to c} \dfrac{f(x)}{g(x)} = \dfrac{\lim\limits_{x \to c} f(x)}{\lim\limits_{x \to c} g(x)},$ if $\lim\limits_{x \to c} g(x) \neq 0$	The limit of the quotient of two functions is the quotient of their limits, provided that the limit of the denominator is not zero.	$\lim\limits_{x \to 3} \dfrac{x}{2x + 5} = \dfrac{\lim\limits_{x \to 3} x}{\lim\limits_{x \to 3} (2x + 5)} = \dfrac{3}{11};$ $\lim\limits_{x \to -2} \dfrac{x - 7}{x^2} = \dfrac{\lim\limits_{x \to -2} (x - 7)}{\lim\limits_{x \to -2} x^2} = \dfrac{-9}{(-2)^2} = -\dfrac{9}{4}$
8. $\lim\limits_{x \to c} \sqrt[n]{f(x)} = \sqrt[n]{\lim\limits_{x \to c} f(x)}$ n is a positive integer, and $\lim\limits_{x \to c} f(x) > 0$ if n is even.	The limit of the nth root of a function is the nth root of its limit.	$\lim\limits_{x \to -2} \sqrt[3]{x} = \sqrt[3]{\lim\limits_{x \to -2} x} = \sqrt[3]{-2};$ $\lim\limits_{x \to 2} \sqrt[4]{2x - 7} = \sqrt[4]{\lim\limits_{x \to 2} (2x - 7)}$ $= \sqrt[4]{2(2) - 7} = \sqrt[4]{-3}$ Does not exist because $n = 4$ is even and $\lim\limits_{x \to 2} f(x) = -3$ is negative.

Notice that by applying Property 6 with $g(x) = f(x)$ we get

$$\lim_{x \to c} [f(x)]^2 = \lim_{x \to c} [f(x) \cdot f(x)]$$

$$= \left[\lim_{x \to c} f(x)\right]\left[\lim_{x \to c} f(x)\right] = \left[\lim_{x \to c} f(x)\right]^2.$$

Similarly,

For any positive integer n, we have

9. $\lim\limits_{x \to c} [f(x)]^n = \left[\lim\limits_{x \to c} f(x)\right]^n.$

If we let $f(x) = x$ in Property 9, we get

$$\lim_{x \to c} x^n = \left[\lim_{x \to c} x\right]^n = c^n.$$

$\uparrow$ Property 2 $\uparrow$

The last equation says that to compute the limit of any positive power of x as $x \to c$, you simply substitute c for x.

> **10.** $\lim_{x \to c} x^n = c^n$

EXAMPLE 1 Finding the Limit of a Polynomial

Find: $\lim_{x \to 2} (5x^2 - 3x + 4)$.

Solution

$$\lim_{x \to 2} (5x^2 - 3x + 4) = \lim_{x \to 2} (5x^2) - \lim_{x \to 2} (3x) + \lim_{x \to 2} 4 \quad \text{Properties 4 and 5}$$

$$= 5 \lim_{x \to 2} x^2 - 3 \lim_{x \to 2} x + 4 \quad \text{Property 3}$$

$$= 5(2)^2 - 3(2) + 4 \quad \text{Property 10}$$

$$= 5(4) - 6 + 4 = 18 \quad \text{Simplify.}$$

Practice Problem 1 Find $\lim_{x \to -2} (2x^3 + 3x^2 + 5)$.

In Example 1, we note that $\lim_{x \to 2} (5x^2 - 3x + 4) = 5(2)^2 - 3(2) + 4 = 18$ is obtained by simply substituting 2 for x. Similarly, imitating the steps in Example 1, we can show that the limit of a polynomial $P(x)$ as $x \to c$ is simply the value of the polynomial at $x = c$. That is,

> For any polynomial $P(x)$ of degree $n \geq 0$,
>
> **11.** $\lim_{x \to c} P(x) = P(c)$.

EXAMPLE 2 Finding the Limit of a Rational Function

Find: $\lim_{x \to -2} \dfrac{x^3 - 7x + 3}{x^2 - 1}$.

Solution

$$\lim_{x \to -2} \frac{x^3 - 7x + 3}{x^2 - 1} = \frac{\lim_{x \to -2} (x^3 - 7x + 3)}{\lim_{x \to -2} (x^2 - 1)} \quad \text{Property 7}$$

$$= \frac{(-2)^3 - 7(-2) + 3}{(-2)^2 - 1} \quad \text{Property 11}$$

$$= \frac{-8 + 14 + 3}{4 - 1} = \frac{9}{3} = 3 \quad \text{Simplify.}$$

Practice Problem 2 Find: $\lim_{x \to 2} \dfrac{2x^2 - 3x + 4}{x^3 + 1}$.

Example 2 illustrates that the limit of a rational function is obtained by substituting c for x. We have

> For any rational function $R(x) = \dfrac{P(x)}{Q(x)}$,
>
> **12.** $\lim_{x \to c} \dfrac{P(x)}{Q(x)} = \dfrac{P(c)}{Q(c)}$, if $Q(c) \neq 0$

Each of the limit properties **1 – 12** is valid if $x \to c$ is replaced everywhere by $x \to c^+$ or by $x \to c^-$.

EXAMPLE 3 Finding Limits of a Piecewise-Defined Function

Let $f(x) = \begin{cases} 3x^2 + 4x - 5 & \text{if } x < 2 \\ 2x - 1, & \text{if } x \geq 2. \end{cases}$

find

a. $\lim\limits_{x \to 2^-} f(x)$ **b.** $\lim\limits_{x \to 2^+} f(x)$ **c.** $\lim\limits_{x \to 2} f(x)$ **d.** $f(2)$

Solution

a. $\lim\limits_{x \to 2^-} f(x) = \lim\limits_{x \to 2^-} (3x^2 + 4x - 5)$ $f(x) = 3x^2 + 4 - 5$ if $x < 2$

$= 3(2)^2 + 4(2) - 5$ Property 11

$= 15$ Simplify.

b. $\lim\limits_{x \to 2^+} f(x) = \lim\limits_{x \to 2^+} (2x - 1)$ $f(x) = 2x - 1$ for $x \geq 2$

$= 2(2) - 1$ Property 11

$= 3$ Simplify.

c. Since $\lim\limits_{x \to 2^-} f(x) = 15 \neq 3 = \lim\limits_{x \to 2^+} f(x)$, we conclude that $\lim\limits_{x \to 2} f(x)$ does not exist.

d. $f(2) = 2(2) - 1$ $f(x) = 2x - 1$ for $x \geq 2$

$= 3.$

Practice Problem 3 Let $f(x) = \begin{cases} x^2 + 1, & \text{if } x > 3 \\ 3x + 1, & \text{if } x < 3. \end{cases}$

Find

a. $\lim\limits_{x \to 3^-} f(x)$ **b.** $\lim\limits_{x \to 3^+} f(x)$ **c.** $\lim\limits_{x \to 3} f(x)$ **d.** $f(3)$

2 Use algebraic manipulations.

Algebraic Methods

Property 7 of the limits says that:

$$\text{If } \lim\limits_{x \to c} f(x) = L \text{ and } \lim\limits_{x \to c} g(x) = M, \text{ then}$$

$$\lim\limits_{x \to c} \frac{f(x)}{g(x)} = \frac{L}{M} \text{ if } M \neq 0.$$

Let us now consider the situations when $M = 0$. We consider the two cases:

i. $L \neq 0, M = 0$

ii. $L = 0, M = 0.$

Case (i) $L \neq 0, M = 0$
In this case, the limit of the quotient does not exist.

---

13. If $\lim\limits_{x \to c} f(x) = L \neq 0$ and $\lim\limits_{x \to c} g(x) = 0$, then

$$\lim\limits_{x \to c} \frac{f(x)}{g(x)} \qquad \text{does not exist.}$$

We shall learn more about Property 7 in the next section.

EXAMPLE 4   **A Limit That Does Not Exist**

Find: $\lim\limits_{x \to 2} \dfrac{x^2 + 3}{x - 2}$.

Solution

We have $\lim\limits_{x \to 2} (x^2 + 3) = 2^2 + 3 = 7 \neq 0$ and $\lim\limits_{x \to 2} (x - 2) = 2 - 2 = 0$.

So, by Property 13, $\lim\limits_{x \to 2} \dfrac{x^2 + 3}{x - 2}$ does not exist.

Practice Problem 4 Find: $\lim\limits_{x \to -2} \dfrac{x^2 + 5}{x^2 + 3x + 2}$.

Case (ii) $L = 0$ and $M = 0$
This case occurs so frequently in calculus that it has earned a special name.

INDETERMINATE FORM

If $\lim\limits_{x \to c} f(x) = 0$ and $\lim\limits_{x \to c} g(x) = 0$, then $\lim\limits_{x \to c} \dfrac{f(x)}{g(x)}$ is called a $\dfrac{0}{0}$ **indeterminate form**.

Indeterminate form means that the limit may or may not exist. We consider some examples.

EXAMPLE 5   **Finding a Limit by Factoring**

Find: $\lim\limits_{x \to 2} \dfrac{x^2 - 5x + 6}{x - 2}$.

Solution

If x is replaced by 2, we have

$$\lim\limits_{x \to 2} \dfrac{x^2 - 5x + 6}{x - 2} = \dfrac{\lim\limits_{x \to 2} (x^2 - 5x + 6)}{\lim\limits_{x \to 2} (x - 2)} = \dfrac{(2)^2 - 5(2) + 6}{2 - 2} = \dfrac{0}{0}.$$

Before we jump to the conclusion that the limit does not exist, we try factoring:

$$\dfrac{x^2 - 5x + 6}{x - 2} = \dfrac{(x - 2)(x - 3)}{x - 2}.$$

Since $x \to 2$ means x is close to 2 but $x \neq 2$ (or $x - 2 \neq 0$), we can remove the common factor $(x - 2)$.

So, $\lim\limits_{x \to 2} \dfrac{x^2 - 5x + 6}{x - 2} = \lim\limits_{x \to 2} \dfrac{(x - 2)(x - 3)}{(x - 2)} = \lim\limits_{x \to 2} (x - 3)$

$$= 2 - 3 = -1.$$

Practice Problem 5 Find: $\lim\limits_{x \to -4} \dfrac{x^2 - 16}{x + 4}$.

EXAMPLE 6   **Finding a Limit by Rationalizing the Numerator**

Find: $\lim\limits_{h \to 0} \dfrac{\sqrt{3 + h} - \sqrt{3}}{h}$.

Solution

If we substitute 0 for h, we get

$$\frac{\sqrt{3+h}-\sqrt{3}}{h} = \frac{\sqrt{3}-\sqrt{3}}{0} = \frac{0}{0}.$$

Again we have the $\frac{0}{0}$ indeterminate form. Unlike Example 5, we cannot factor the numerator. We try rationalizing the numerator:

$$\frac{\sqrt{3+h}-\sqrt{3}}{h} = \frac{\left(\sqrt{3+h}-\sqrt{3}\right)\left(\sqrt{3+h}+\sqrt{3}\right)}{h\left(\sqrt{3+h}+\sqrt{3}\right)}$$

Multiply numerator and denominator by $\left(\sqrt{3+h}+\sqrt{3}\right)$.

$$= \frac{3+h-3}{h\left(\sqrt{3+h}+\sqrt{3}\right)}$$

$(a-b)(a+b) = a^2 - b^2$

$$= \frac{h}{h\left(\sqrt{3+h}+\sqrt{3}\right)}$$

Simplify.

$$= \frac{1}{\sqrt{3+h}+\sqrt{3}}$$

Remove common factor $h \neq 0$.

So,

$$\lim_{h\to 0}\frac{\sqrt{3+h}-\sqrt{3}}{h} = \lim_{h\to 0}\frac{1}{\sqrt{3+h}+\sqrt{3}}$$

$$= \frac{1}{\sqrt{3+0}+\sqrt{3}}$$

$$= \frac{1}{\sqrt{3}+\sqrt{3}} = \frac{1}{2\sqrt{3}}.$$

Practice Problem 6 Find $\lim_{h\to 0}\dfrac{\sqrt{h+2}-\sqrt{2}}{h}$

3 Find limits of difference quotients.

Limits of Difference Quotients

One of the most important limits is the limit of the **difference quotient**:

$$\lim_{h\to 0}\frac{f(a+h)-f(a)}{h}$$

We consider some examples.

EXAMPLE 7 Finding the Limit of a Difference Quotient

Let $f(x) = 3x - 4$. Find: $\lim_{h\to 0}\dfrac{f(2+h)-f(2)}{h}$

Solution

$$\lim_{h\to 0}\frac{f(2+h)-f(2)}{h} = \lim_{h\to 0}\frac{[3(2+h)-4]-[3(2)-4]}{h}$$

$f(x) = 3x + 4$
$f(2+h) = 3(2+h) - 4$
$f(2) = 3(2) - 4$

$$= \lim_{h\to 0}\frac{6+3h-4-6+4}{h}$$

$$= \lim_{h\to 0}\frac{3h}{h}$$

Simplify $(h \neq 0)$.

$$= \lim_{h\to 0} 3 = 3.$$

Property 1

Practice Problem 7 Let $f(x) = 5 - 4x$. Find: $\lim\limits_{h \to 0} \dfrac{f(3 + h) - f(3)}{h}$. ▪▪

EXAMPLE 8 Finding the Limit of a Difference Quotient

Let $f(x) = 1 - 2x^2$. Find: $\lim\limits_{h \to 0} \dfrac{f(3 + h) - f(3)}{h}$.

Solution

$$
\begin{aligned}
\lim_{h \to 0} \frac{f(3 + h) - f(3)}{h} &= \lim_{h \to 0} \frac{[1 - 2(3 + h)^2] - [1 - 2 \cdot 3^2]}{h} \\
&= \lim_{h \to 0} \frac{[1 - 2(9 + 6h + h^2)] - [1 - 18]}{h} && (a + b)^2 = \\
&&& a^2 + 2ab + b^2 \\
&= \lim_{h \to 0} \frac{1 - 18 - 12h - 2h^2 - 1 + 18}{h} && \text{Simplify.} \\
&= \lim_{h \to 0} \frac{-12h - 2h^2}{h} && \text{Simplify.} \\
&= \lim_{h \to 0} \frac{\cancel{h}(-12 - 2h)}{\cancel{h}} && \text{Factor numerator.} \\
&= \lim_{h \to 0} (-12 - 2h) \\
&= -12 - 2(0) && \text{Substitute } h = 0. \\
&= -12 && \text{Simplify.}
\end{aligned}
$$

Practice Problem 8 Let $f(x) = 3x^2 - 7$. Find: $\lim\limits_{h \to 0} \dfrac{f(1 + h) - f(1)}{h}$. ▪▪

EXAMPLE 9 Limit of a Difference Quotient Does not Exist

Let $f(x) = |x - 2|$. Find: $\lim\limits_{h \to 0} \dfrac{f(2 + h) - f(2)}{h}$.

Solution

$$
\begin{aligned}
\lim_{h \to 0} \frac{f(2 + h) - f(2)}{h} &= \lim_{h \to 0} \frac{|(2 + h) - 2| - |2 - 2|}{h} \\
&= \lim_{h \to 0} \frac{|h|}{h} && \text{Simplify.}
\end{aligned}
$$

We consider one-sided limits:

$$
\begin{aligned}
\lim_{h \to 0^+} \frac{|h|}{h} &= \lim_{h \to 0^+} \frac{h}{h} && |h| = h \text{ if } h > 0 \\
&= \lim_{h \to 0^+} 1 = 1. \\
\lim_{h \to 0^-} \frac{|h|}{h} &= \lim_{h \to 0^-} \frac{-h}{h} && |h| = -h \text{ if } h < 0 \\
&= \lim_{h \to 0^-} -1 = -1.
\end{aligned}
$$

Since the limits from the right and the left of $\dfrac{|h|}{h}$ are not the same, we conclude that $\lim\limits_{h \to 0} \dfrac{f(2 + h) - f(2)}{h} = \lim\limits_{h \to 0} \dfrac{|h|}{h}$ does not exist.

Practice Problem 9 Let $f(x) = |2x - 3|$. Find: $\displaystyle\lim_{h \to 0} \frac{f(1.5 + h) - f(1.5)}{h}$.

Answers to Practice Problems

1. 1 **2.** $\frac{2}{3}$ **3. a.** 10 **b.** 10 **c.** 10 **d.** Does not exist **4.** Does not exist **5.** -8 **6.** $\frac{1}{2\sqrt{2}}$ **7.** -4 **8.** 6 **9.** Does not exist

SECTION 10.2 Exercises

Concepts and Vocabulary

1. $\displaystyle\lim_{x \to 3} 5 =$ _____.

2. $\displaystyle\lim_{x \to c} [f(x)]^3 = [$ _____ $]^3$.

3. $\displaystyle\lim_{x \to c} \frac{f(x)}{g(x)} = \frac{\displaystyle\lim_{x \to c} f(x)}{\displaystyle\lim_{x \to c} g(x)}$, if _____.

4. $\displaystyle\lim_{x \to c} \sqrt{f(x)} = \sqrt{\displaystyle\lim_{x \to c} f(x)}$, if _____.

5. True or False. The limit of any polynomial function $P(x)$ as $x \to c$ is the value of the polynomial at $x = c$.

6. True or False. The limit of any rational function $R(x)$ as $x \to 2$ is $R(2)$.

7. True or False. If $\displaystyle\lim_{x \to c} f(x) = 0$ and $\displaystyle\lim_{x \to c} g(x) = 0$, then $\displaystyle\lim_{x \to c} \frac{f(x)}{g(x)} = \frac{0}{0} = 1$.

8. True or False. If $\displaystyle\lim_{x \to c} f(x) = L \neq 0$, and $\displaystyle\lim_{x \to c} g(x) = 0$, then $\displaystyle\lim_{x \to c} \frac{f(x)}{g(x)}$ does not exist.

Building Skills

In Exercises 9–52, find each indicated limit if it exists.

9. $\displaystyle\lim_{x \to 2} (3 + 5x)$

10. $\displaystyle\lim_{x \to 3} (-7x + 1)$

11. $\displaystyle\lim_{x \to -2} (5x^2 - 2x)$

12. $\displaystyle\lim_{x \to -1} (-3x^2 + 5x + 1)$

13. $\displaystyle\lim_{x \to 2} (3x^4)$

14. $\displaystyle\lim_{x \to -3} (4x^3)$

15. $\displaystyle\lim_{x \to 1} (7x^4 - 3x^3 + 4x + 3)$

16. $\displaystyle\lim_{x \to -1} (3x^5 + 5x^4 - 3x^3 + 2x^2 + 1)$

17. $\displaystyle\lim_{x \to 2} (x^3 - 5)^2$

18. $\displaystyle\lim_{x \to -2} (3x^2 - 5x - 20)^4$

19. $\displaystyle\lim_{x \to 2} \frac{x^2 + 1}{x + 1}$

20. $\displaystyle\lim_{x \to -2} \frac{3x + 1}{x^2 + 2}$

21. $\displaystyle\lim_{x \to -1} \frac{3x^2 + 5}{x^2 - x}$

22. $\displaystyle\lim_{x \to 0} \frac{-x^4 + 4x - 8}{x^3 + 3x + 2}$

23. $\displaystyle\lim_{x \to 1} f(x); f(x) = \begin{cases} 2x & \text{if } x < 1 \\ 1 + x & \text{if } x > 1 \end{cases}$

24. $\displaystyle\lim_{x \to 0} f(x); f(x) = \begin{cases} 1 + x^2, & x < 0 \\ 1 - x^2, & x > 0 \end{cases}$

25. $\displaystyle\lim_{x \to 0} f(x); f(x) = \begin{cases} 1 & \text{if } x > 0 \\ -1 & \text{if } x < 0 \end{cases}$

26. $\displaystyle\lim_{x \to 1} f(x); f(x) = \begin{cases} 2x & \text{if } x \geq 1 \\ 1 - x & \text{if } x < 1 \end{cases}$

27. $\displaystyle\lim_{x \to 2} f(x); f(x) = \begin{cases} 3 & \text{if } x > 1 \\ -1 & \text{if } x \leq 1 \end{cases}$

28. $\displaystyle\lim_{x \to -1} f(x); f(x) = \begin{cases} 2 & \text{if } x > 0 \\ -3 & \text{if } x < 0 \end{cases}$

29. $\displaystyle\lim_{x \to 2} \sqrt{3x + 3}$

30. $\displaystyle\lim_{x \to 1} \sqrt[3]{5x + 3}$

31. $\displaystyle\lim_{x \to 1} (3x^2 + 4x + 1)^{2/3}$

32. $\displaystyle\lim_{x \to 2} (5x - 6)^{3/2}$

33. $\displaystyle\lim_{x \to 0} \frac{x}{x^2 + x}$

34. $\displaystyle\lim_{x \to 2} \frac{x^2 - 4}{x^2 + x - 6}$

35. $\displaystyle\lim_{x \to -3} \frac{x^2 + 2x - 3}{x^2 - 9}$

36. $\displaystyle\lim_{x \to -2} \frac{x^2 + 5x + 6}{x^2 - 2x - 8}$

37. $\displaystyle\lim_{x \to 1} \frac{x^3 - 1}{x - 1}$

38. $\displaystyle\lim_{x \to -2} \frac{x^3 + 8}{x + 2}$

39. $\displaystyle\lim_{x \to 1} \frac{x^4 - 1}{x - 1}$

40. $\displaystyle\lim_{x \to 2} \frac{x^2 - 4}{x^3 - 8}$

41. $\displaystyle\lim_{x \to 2} \frac{2x}{x - 2}$

42. $\displaystyle\lim_{x \to -2} \frac{5x + 1}{x^2 + x - 2}$

43. $\displaystyle\lim_{x \to 1^+} \frac{2x + 1}{\sqrt{x - 1}}$

44. $\displaystyle\lim_{x \to 1^-} \frac{3x + 2}{\sqrt{1 - x}}$

45. $\displaystyle\lim_{x \to 0} \frac{\sqrt{x + 9} - 3}{x}$

46. $\displaystyle\lim_{x \to 0} \frac{\sqrt{x + 16} - 4}{x}$

47. $\displaystyle\lim_{x \to 0} \frac{1 - \sqrt{1 - x^2}}{x^2}$

48. $\displaystyle\lim_{x \to 0} \frac{\sqrt{2 + 3x} - \sqrt{2 - 5x}}{4x}$

49. $\displaystyle\lim_{x \to 1} \frac{x^2 - 1}{\sqrt{5x - 1} - \sqrt{3x + 1}}$

50. $\displaystyle\lim_{x \to -1} \frac{x + 1}{\sqrt{8x^2 + 1} + 3x}$

51. $\displaystyle\lim_{x \to 0} \frac{\sqrt{1 + x} - \sqrt{1 - x}}{x}$

52. $\displaystyle\lim_{x \to 3} \frac{x - 3}{\sqrt{4 - x} - \sqrt{x - 2}}$

In Exercises 53–64, for each function f, form the difference quotient:

$$F(h) = \frac{f(c + h) - f(c)}{h},$$

and then find $\displaystyle\lim_{h \to 0} F(h)$ for the specified value of c.

53. $f(x) = x, c = 2$

54. $f(x) = -2x, c = 1$

55. $f(x) = 2x + 3, c = 1$

56. $f(x) = 3 - 2x, c = 0$

57. $f(x) = x^2, c = 2$

58. $f(x) = -x^2, c = -1$

59. $f(x) = \sqrt{x}, c = 4$

60. $f(x) = \sqrt{x + 1}, c = 3$

61. $f(x) = 2, c = 1$

62. $f(x) = -2, c = -1$

63. $f(x) = |x|, c = 0$

64. $f(x) = |x - 1|, c = 1$

Applying the Concepts

65. State Income Tax. A state's income tax directive specifies that the tax liability $T(x)$ on x dollars of taxable income is given by the piecewise-defined function:

$$T(x) = \begin{cases} 0.05x & \text{if } 0 \le x < 30{,}000 \\ 1500 + .08(x - 30{,}000) & \text{if } x \ge 30{,}000 \end{cases}$$

Find

a. $\lim\limits_{x \to 30{,}000^-} T(x)$

b. $\lim\limits_{x \to 30{,}000^+} T(x)$

c. $\lim\limits_{x \to 30{,}000} T(x)$

d. $T(30{,}000)$.

66. Federal Income Tax. The partial tax table for the 2014 U.S. income tax for a single taxpayer is as follows:

If Taxable income is over	But not over	The tax is
$0	$9225	10%
$9225	$37,450	15%
$37,450	$90,750	25%
$90,750	$189,300	28%

a. Write a piecewise-defined function $T(x)$ for your taxable income of x dollars, $0 \le x \le 189{,}300$.

b. Find $\lim\limits_{x \to 37450^-} T(x)$, $\lim\limits_{x \to 37450^+} T(x)$, $\lim\limits_{x \to 37450} T(x)$.

c. Is it necessary for fairness that for $x = c$
$$\lim\limits_{x \to c^-} T(x) = \lim\limits_{x \to c^+} T(x)?$$

67. Length Contraction. One consequence of the time dilation equation of the introduction is the formula

$$L = L_0 \sqrt{1 - \frac{v^2}{c^2}}$$

where L_0 is the length of the object in a spaceship moving at velocity v and c is the velocity of light. *The length of an object is shorter when it is moving than when it is at rest.*

a. A spaceship is moving away from Earth at velocity $v = 0.80 \, c$. What is the length of a meter stick in the spaceship as it appears to an observer on Earth?

b. Find $\lim\limits_{v \to c^-} L$.

c. In part (b), should we consider $\lim\limits_{v \to c^+} L$? Explain.

68. Length Expansion. The flight crew of a spaceship traveling at a speed of $0.90 \, c$ measures its length to be 100 m. What is its length at rest?
[*Hint:* Use the formula from Exercise 67.]

Beyond the Basics

In Exercises 69–72, find each indicated limit, if it exists.

69. Let $f(x) = \begin{cases} 5 + 3x & \text{if } x < 0 \\ 5 - 3x & \text{if } 0 \le x < \dfrac{2}{3} \\ -5 - 3x & \text{if } x \ge \dfrac{2}{3} \end{cases}$.

Find

a. $\lim\limits_{x \to 0} f(x)$

b. $\lim\limits_{x \to \frac{2}{3}} f(x)$.

70. Find $\lim\limits_{x \to 0} \dfrac{5x + |x|}{2x + |x|}$.

71. Find $\lim\limits_{x \to 1} \dfrac{x^2 - 1}{\sqrt{3x + 1} - \sqrt{5x - 1}}$.

72. Find $\lim\limits_{x \to -1} \dfrac{x + 1}{\sqrt{6x^2 + 3} + 3x}$.

A function f is said to be **continuous** at c if
i. f is defined at c; that is, $f(c)$ exists.
ii. $\lim\limits_{x \to c} f(x)$ exists, and
iii. $\lim\limits_{x \to c} f(x) = f(c)$.

In Exercises 73–80, state whether the given function f is continuous at c.

73. $f(x) = \begin{cases} 1 & \text{if } x \le 2 \\ -1 & \text{if } x > 2 \end{cases}$ at $c = 2$

74. $f(x) = \begin{cases} |x - 3| & \text{if } x \ne 3 \\ 1 & \text{if } x = 3 \end{cases}$ at $c = 3$

75. $f(x) = \begin{cases} \dfrac{x^2 + x - 2}{x - 1} & \text{if } x \ne 1 \\ 3 & \text{if } x = 1 \end{cases}$ at $c = 1$

76. $f(x) = \begin{cases} \dfrac{x^2 - 9}{x + 3} & \text{if } x \ne -3 \\ -6 & \text{if } x = -3 \end{cases}$ at $c = -3$

77. $f(x) = \sqrt{x - 1}$ at $c = 1$

78. $f(x) = \sqrt[3]{x + 1}$ at $c = 7$

79. $f(x) = \begin{cases} x - 1, & \text{if } x \le 1 \\ \ln x, & \text{if } x > 1 \end{cases}$ at $c = 1$

80. $f(x) = \begin{cases} e^x, & \text{if } x < 0 \\ \ln(e + x) & \text{if } x \ge 0 \end{cases}$ at $c = 0$

In Exercises 81–88, use the properties of limits and the following facts: $\lim\limits_{\theta \to 0} \dfrac{\sin \theta}{\theta} = 1$, $\lim\limits_{\theta \to 0} \cos \theta = 1$, $\lim\limits_{\theta \to \frac{\pi}{2}} \cos \theta = 0$

to find the following limits

81. $\lim\limits_{x \to \frac{\pi}{2}} \tan x$

82. $\lim\limits_{x \to 0} \cot x$

83. $\lim\limits_{x \to 0} \dfrac{\tan x}{x}$

84. $\lim\limits_{x \to 0} x \cot x$

85. $\lim\limits_{x \to 0} \dfrac{\sin 2x}{x}$

86. $\lim\limits_{x \to 0} \dfrac{\tan 2x}{x}$

87. $\lim\limits_{x \to 0} \dfrac{\sin 2x}{\sin 4x}$

88. $\lim\limits_{x \to 0} \dfrac{\sin 2x}{\tan 4x}$

Critical Thinking / Discussion / Writing

89. Distinguish between the statements $\lim\limits_{x \to c} f(x)$ and $f(c)$.

90. Give an example of a function in each case.
  i. $\lim\limits_{x \to c} f(x)$ exists but $f(c)$ does not exist.
  ii. $\lim\limits_{x \to c} f(x)$ does not exist, but $f(c)$ exists.

  iii. $\lim\limits_{x \to c} f(x)$ exists and $\lim\limits_{x \to c} g(x)$ does not exist, but
  $\lim\limits_{x \to c} [f(x)g(x)]$ exists.
  iv. $\lim\limits_{x \to c} f(x)$ does not exist and $\lim\limits_{x \to c} g(x)$ does not exist but
  $\lim\limits_{x \to c} [f(x)g(x)]$ exists.

Getting Ready for the Next Section

In Exercises 91–94, find the end behavior of each polynomial. (See Section 2.2, page 156.)

91. $P(x) = 2x^2 - 3x + 1$

92. $P(x) = -3x^2 + 5x + 19$

93. $P(x) = -x^3 + 5x^2 + 7$

94. $P(x) = 2x^3 - 7x^2 + 50$

In Exercises 95–98, find the vertical and horizontal asymptotes for the graphs of each rational function $R(x)$. (See Section 2.4, page 189.)

95. $R(x) = \dfrac{x}{x - 2}$

96. $R(x) = \dfrac{x + 1}{x - 1}$

97. $R(x)\dfrac{(x - 1)(x + 1)}{(x - 2)(x + 2)}$

98. $R(x) = \dfrac{x^2 + 6x + 8}{x^2 - 4}$

Infinite Limits and Limits at Infinity

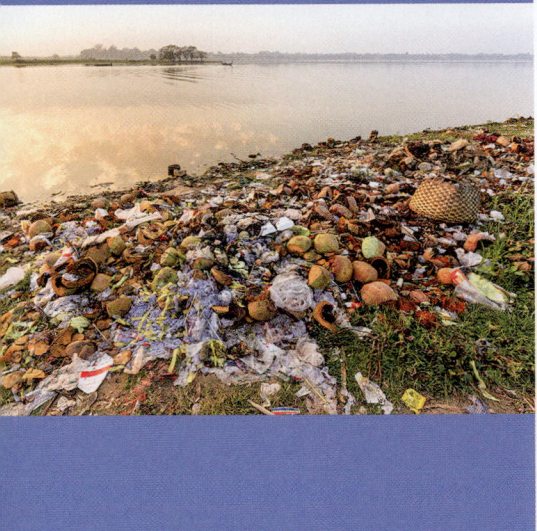

BEFORE STARTING THIS SECTION, REVIEW

1 Factoring polynomials (Appendix A.2, page 930)

2 Rationalizing a numerator (Appendix A.4, page 944)

3 Library of functions (Section 1.4, page 70)

OBJECTIVES

1 Find infinite limits.

2 Find limits at infinity.

◆ Environmental Pollution

Pollution is the process of making air, land, water, or other parts of the environment dangerous and unsafe for life on a planet. Pollution is usually caused by the introduction of impurities into a natural environment.

A common type of air pollution occurs when we release dangerous gases such as sulfur dioxide and carbon monoxide from burning fuel. Land is polluted by household garbage and industrial waste. Water pollution happens when chemicals, sewage, pesticides, and metals such as lead or mercury are introduced to water.

In Exercise 67, we discuss impurities in a polluted river.

1 Find infinite limits.

Infinite Limit

Consider the graphs of the functions $f(x) = \dfrac{1}{x}$ and $g(x) = \dfrac{1}{x^2}$. In Section 10.2, we indicated that both $\lim\limits_{x \to 0} f(x)$ and $\lim\limits_{x \to 0} g(x)$ do not exist because the numerator in both cases is 1 and the denominator approaches 0 as $x \to 0$. From the table on page 189, we observed that as $x \to 0^+$, the values of $f(x) = \dfrac{1}{x}$ are positive and become larger and larger ($f(x)$ increases without bound). Symbolically, we write

$$f(x) = \frac{1}{x} \to \infty \text{ as } x \to 0^+ \text{ or } \lim_{x \to 0^+} f(x) = \infty \, (\text{or} + \infty).$$

Similarly, as $x \to 0^-$, the values of $f(x)$ are negative and become larger and larger in absolute value. We express this by writing

$$f(x) = \frac{1}{x} \to -\infty \text{ as } x \to 0^- \text{ or } \lim_{x \to 0^-} f(x) = -\infty.$$

You can construct a table of values to show that

$$g(x) = \frac{1}{x^2} \to \infty \text{ as } x \to 0^+ \text{ and } g(x) = \frac{1}{x^2} \to \infty \text{ as } x \to 0^-.$$

So, $\lim_{x \to 0^+} g(x) = \infty$ and $\lim_{x \to 0^-} g(x) = \infty$.

We use the symbols ∞ (or $+\infty$) and $-\infty$ to describe the manner in which a function $f(x)$ behaves near $x = c$. We refer to situations such as $\lim_{x \to c^+} f(x) = \pm \infty$ or $\lim_{x \to c^-} f(x) = \pm \infty$ as **infinite limits**.

EXAMPLE 1 **Finding Limits When the Denominator Approaches Zero**

Let

$$f(x) = \frac{x^2 - x - 6}{x^2 - 4}.$$

Describe the behavior of f near the zeros of the denominator. Use the symbols ∞ and $-\infty$ where appropriate.

Solution

Let $N(x) = x^2 - x - 6$ and $D(x) = x^2 - 4$. Factoring the denominator $D(x)$, we have

$$D(x) = x^2 - 4 = (x + 2)(x - 2).$$

The denominator has two zeros: $x = -2$ and $x = 2$.

i. Consider the zero: $x = -2$. We have

$$
\begin{aligned}
N(-2) &= (-2)^2 - (-2) - 6 \quad \text{Replace } x \text{ by } -2 \text{ in } N(x) = x^2 - x - 6 \\
&= 4 + 2 - 6 \\
&= 0.
\end{aligned}
$$

Since $N(-2) = 0$ and $D(-2) = 0$ $\left(\dfrac{0}{0} \text{ indeterminate form}\right)$, the limit of $f(x)$ as $x \to -2$ may exist.

We use the algebraic manipulations to find the behavior of f near $x = -2$. We have

$$
\begin{aligned}
\lim_{x \to -2} f(x) &= \lim_{x \to -2} \frac{x^2 - x - 6}{x^2 - 4} \\
&= \lim_{x \to -2} \frac{(x + 2)(x - 3)}{(x + 2)(x - 2)} \quad \text{Factor.} \\
&= \lim_{x \to -2} \frac{x - 3}{x - 2} \quad \text{Remove the common factor } (x + 2). \\
&= \frac{-2 - 3}{-2 - 2} = \frac{-5}{-4} = \frac{5}{4}. \quad \text{Simplify.}
\end{aligned}
$$

Since $f(-2)$ is undefined, but $\lim_{x \to -2} f(x)$ exists, the graph of $y = f(x)$ does not have a vertical asymptote at $x = -2$. The graph has a hole at $\left(-2, \dfrac{5}{4}\right)$.

ii. Now we consider the zero: $x = 2$.
Since $D(2) = 0$ and $N(2) = 2^2 - 2 - 6$

$$= 4 - 8 = -4 \neq 0.$$

We conclude that $\lim_{x \to 2} f(x)$ does not exist. However, to analyze the behavior of f near $x = 2$, we make a table of values:

TABLE 10.6

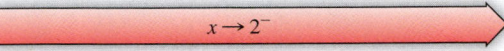
$x \to 2^-$

x	1.9	1.99	1.999	1.9999	$\to 2^-$
$f(x) = \dfrac{x^2 - x - 6}{x^2 - 4}$	11	101	1001	10001	$\to \infty$

TABLE 10.7

$x \to 2^+$

x	2.1	2.01	2.001	2.0001	$\to 2^+$
$f(x) = \dfrac{x^2 - x - 6}{x^2 - 4}$	-9	-99	-999	-9999	$\to -\infty$

From Tables 10.6 and 10.7, we see that

$$\lim_{x \to 2^-} \frac{x^2 - x - 6}{x^2 - 4} = \infty \text{ and } \lim_{x \to 2^+} \frac{x^2 - x - 6}{x^2 - 4} = -\infty.$$

The vertical line $x = 2$ is a vertical asymptote for the graph of $y = \dfrac{x^2 - x - 6}{x^2 - 4}$. See Figure 10.5.

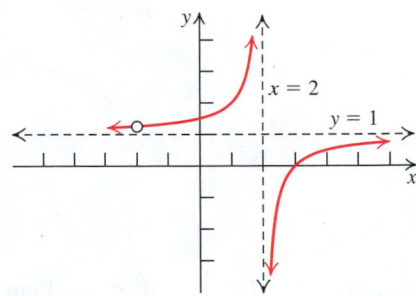

Figure 10.5

Practice Problem 1 Repeat Example 1 for $f(x) = \dfrac{x^2 + 8x + 15}{x^2 - 9}$.

2 Find limits at infinity.

Limits at Infinity

Limits at infinity describe the end behavior of $f(x)$ as the independent variable x increases without bound. Recall that the notation $x \to \infty$ means that x increases without bound through positive values, and the notation $x \to -\infty$ means that x decreases without bound through negative values.

The notation

$$\lim_{x \to c} f(x) = L$$

can be used with $c = \pm \infty$ and $L = \pm \infty$.

The following box lists the properties of limits at infinity that we discussed in Chapters 2 and 3.

PROPERTIES OF LIMITS AT INFINITY

For any positive integer number n, and any real number k,

1. $\lim\limits_{x \to \infty} x^n = \infty$

2. $\lim\limits_{x \to -\infty} x^n = \begin{cases} \infty & \text{if } n \text{ is even} \\ -\infty & \text{if } n \text{ is odd} \end{cases}$

3. $\lim\limits_{x \to \infty} kx^n = \begin{cases} \infty & \text{if } k > 0 \\ -\infty & \text{if } k < 0 \end{cases}$

4. $\lim\limits_{x \to -\infty} kx^n = \begin{cases} \infty & \text{if } n \text{ is even and } k > 0 \\ -\infty & \text{if } n \text{ is even and } k < 0 \\ -\infty & \text{if } n \text{ is odd and } k > 0 \\ \infty & \text{if } n \text{ is odd and } k < 0 \end{cases}$

5. $\lim\limits_{x \to \infty} \dfrac{k}{x^n} = 0$

6. $\lim\limits_{x \to -\infty} \dfrac{k}{x^n} = 0$

7. $\lim\limits_{x \to \infty} a^x = \begin{cases} 0 & \text{if } 0 < a < 1 \\ \infty & \text{if } a > 1 \end{cases}$

8. $\lim\limits_{x \to \infty} \log_a x = \begin{cases} \infty & \text{if } a > 1 \\ 0 & \text{if } 0 < a < 1 \end{cases}$

EXAMPLE 2 **Finding the Limits at Infinity for Polynomials**

Find the limit: $\lim\limits_{x \to \infty} (x^3 + 3x^2 + 7x + 5)$.

Solution

$$\lim_{x \to \infty} (x^3 + 3x^2 + 7x + 5) = \lim_{x \to \infty} \left[x^3 \left(1 - \frac{3}{x} + \frac{7}{x^2} + \frac{5}{x^2} \right) \right]$$

$$= \left[\left(\lim_{x \to \infty} x^3 \right) \right] \left[\lim_{x \to \infty} \left(1 - \frac{3}{x} + \frac{7}{x^2} + \frac{5}{x^2} \right) \right]$$

$$= \left(\lim_{x \to \infty} x^3 \right) \left(\lim_{x \to \infty} 1 - \lim_{x \to \infty} \frac{3}{x} + \lim_{x \to \infty} \frac{7}{x^2} + \lim_{x \to \infty} \frac{5}{x^2} \right)$$

$$= \left(\lim_{x \to \infty} x^3 \right) (1 - 0 + 0 + 0) \qquad \text{Property 5}$$

$$= \lim_{x \to \infty} x^3$$

$$= \infty \qquad \text{Property 1}$$

Practice Problem 2 Find the limit: $\lim\limits_{x \to -\infty} (7x^3 - 5x + 11)$.

Example 1 illustrates that as $|x|$ gets very large, the function values of a polynomial $P(x)$ are determined by the term with the highest power.

---

9. Limits at Infinity for Polynomials

For any polynomial $P(x) = a_n x^n + a_{n-1} x^{n-1} + \ldots + a_0$,

$$\lim_{x \to \pm \infty} P(x) = \lim_{x \to \pm \infty} a_n x^n.$$

EXAMPLE 3 **Finding Limits at Infinity for Rational Functions**

Find the limits

a. $\lim\limits_{x \to \infty} \dfrac{3x^2 + 4x - 5}{2 - 3x + 6x^2}$

b. $\lim\limits_{x \to -\infty} \dfrac{2x^3 - 3x + 17}{4x^2 - 5x + 1}$

c. $\lim\limits_{x \to -\infty} \dfrac{2x + 3}{x^2 - 5x + 7}$

Solution

a. $\displaystyle\lim_{x\to\infty}\frac{3x^2+4x-5}{2-3x+6x^2}=\lim_{x\to\infty}\left[\frac{x^2\left(3+\dfrac{4}{x}-\dfrac{5}{x^2}\right)}{x^2\left(\dfrac{2}{x^2}-\dfrac{3}{x}+6\right)}\right]$ Factor out the highest power of x.

$$=\lim_{x\to\infty}\left[\frac{3+\dfrac{4}{x}-\dfrac{5}{x^2}}{\dfrac{2}{x^2}-\dfrac{3}{x}+6}\right]$$ Remove the common factor x^2.

$$=\frac{\displaystyle\lim_{x\to\infty}3+\lim_{x\to\infty}\frac{4}{x}-\lim_{x\to\infty}\frac{5}{x^2}}{\displaystyle\lim_{x\to\infty}\frac{2}{x^2}-\lim_{x\to\infty}\frac{3}{x}+\lim_{x\to\infty}6}$$ Properties of Limits

$$=\frac{3+0-0}{0-0+6}=\frac{1}{2}$$ Simplify.

b. $\displaystyle\lim_{x\to-\infty}\frac{2x^3-3x+17}{4x^2-5x+1}=\lim_{x\to-\infty}\left[\frac{x^3\left(2-\dfrac{3}{x^2}+\dfrac{17}{x^3}\right)}{x^2\left(4-\dfrac{5}{x}+\dfrac{1}{x^2}\right)}\right]$ Factor out the highest power of x.

$$=\lim_{x\to-\infty}\left[\frac{2x^3}{4x^2}\right]=\lim_{x\to-\infty}\left[\frac{1}{2}x\right]$$

$$=-\infty$$

c. $\displaystyle\lim_{x\to\infty}\frac{2x+3}{x^2-5x+7}=\lim_{x\to\infty}\left[\frac{x\left(2+\dfrac{3}{x}\right)}{x^2\left(1-\dfrac{5}{x}+\dfrac{7}{x^2}\right)}\right]$ Property 4

$$=\lim_{x\to\infty}\left[\frac{2x}{x^2}\right]$$ $\displaystyle\lim_{x\to\infty}\left[\frac{k}{x^n}\right]=0$

$$=\lim_{x\to\infty}\frac{2}{x}$$ Simplify

$$=0$$ $\displaystyle\lim_{x\to+\infty}\frac{k}{x^n}=0$

Practice Problem 3 Find the limits:

a. $\displaystyle\lim_{x\to\infty}\frac{5-3x-8x^2}{4x^2+7x-11}$ **b.** $\displaystyle\lim_{x\to-\infty}\frac{-x^3+2x^2+5}{3x^2+17}$ **c.** $\displaystyle\lim_{x\to\infty}\frac{5x^2-2x+7}{3x^3+16x}$ ◼◼

Recall that if for a rational function $R(x)$, if $\displaystyle\lim_{x\to+\infty}R(x)=L$, then the line $y=L$ is a horizontal asymptote for the graph of $y=R(x)$.

EXAMPLE 4   **Finding Limits at Infinity by Rationalizing**

Find: $\displaystyle\lim_{x\to-\infty}\left(\sqrt{x^2+3x+2}-\sqrt{x^2+2}\right).$

Solution

$$\lim_{x \to -\infty} \left(\sqrt{x^2 + 3x + 2} - \sqrt{x^2 + 2} \right)$$

$$= \lim_{x \to -\infty} \frac{\left(\sqrt{x^2 + 3x + 2} - \sqrt{x^2 + 2} \right)\left[\sqrt{x^2 + 3x + 2} + \sqrt{x^2 + 2} \right]}{\sqrt{x^2 + 3x + 2} + \sqrt{x^2 + 2}}$$ Rationalize the numerator.

$$= \lim_{x \to -\infty} \frac{(x^2 + 3x + 2) - (x^2 + 2)}{\sqrt{x^2 + 3x + 2} + \sqrt{x^2 + 2}}$$

$$= \lim_{x \to -\infty} \frac{3x}{\sqrt{x^2\left(1 + \dfrac{3}{x} + \dfrac{2}{x^2}\right)} + \sqrt{x^2\left(1 + \dfrac{2}{x^2}\right)}}$$

$$= \lim_{x \to -\infty} \frac{3x}{|x|\sqrt{1 + \dfrac{3}{x} + \dfrac{2}{x^2}} + |x|\sqrt{1 + \dfrac{2}{x^2}}}$$ $\sqrt{x^2} = |x|$

$$= \lim_{x \to -\infty} \frac{3x}{|x|\left(\sqrt{1 + \dfrac{3}{x} + \dfrac{2}{x^2}} + \sqrt{1 + \dfrac{2}{x^2}} \right)}$$

$$= \lim_{x \to -\infty} \frac{3x}{-x\left(\sqrt{1 + \dfrac{3}{x} + \dfrac{2}{x^2}} + \sqrt{1 + \dfrac{2}{x^2}} \right)}$$ Since $x \to -\infty$, x is negative, so $|x| = -x$.

$$= \lim_{x \to -\infty} \frac{3}{-1\left(\sqrt{1 + \dfrac{3}{x} + \dfrac{2}{x^2}} + \sqrt{1 + \dfrac{2}{x^2}} \right)}$$ Divide numerator and denominator by x.

$$= \frac{3}{-1\left(\sqrt{1} + \sqrt{1} \right)}$$ As $x \to -\infty$, $\dfrac{3}{x}$ and $\dfrac{2}{x^2} \to 0$

$$= \frac{3}{-1(2)} = -\frac{3}{2}.$$

Practice Problem 4 Find $\displaystyle\lim_{x \to \infty} \left(\sqrt{x^2 + 2x} - x \right)$

Answers to Practice Problems

1. $\displaystyle\lim_{x \to -3} f(x) = -\frac{1}{3}$, hole at $\left(-3, -\frac{1}{3}\right)$ and vertical asymptote

at $x = 3$; $\displaystyle\lim_{x \to 3^+} f(x) = \infty$, $\displaystyle\lim_{x \to 3^-} f(x) = -\infty$

2. $-\infty$ **3. a.** -2 **b.** ∞ **c.** 0 **4.** 1

Concepts and Vocabulary

1. $\displaystyle\lim_{x \to c} f(x) = \infty$ means that as x approaches c the values of $f(x)$ increase without _____.

2. $\displaystyle\lim_{x \to c} g(x) = -\infty$ means that as x approaches c the values of _____ decrease without bound.

3. If $\displaystyle\lim_{x \to c^+} f(x) = \infty$, then the line $x = c$ is a _____ to the graph of $y = f(x)$.

4. $\displaystyle\lim_{x \to \infty} f(x)$ describes the _____ for the graph of $y = f(x)$.

5. **True or False.** For any polynomial function $P(x)$, $\displaystyle\lim_{x \to \infty} P(x) = \pm\infty$.

6. **True or False.** For a rational function $R(x)$, $\displaystyle\lim_{x \to \infty} R(x)$ can be any real number or $\pm\infty$.

7. **True or False.** If $\displaystyle\lim_{x \to \infty} f(x) = L$, then $\displaystyle\lim_{x \to -\infty} f(x)$ is also L.

8. **True or False.** $\displaystyle\lim_{x \to \infty} \sin(x)$ does not exist.

Building Skills

For Exercises 9–16, use the following graph of $y = f(x)$ to find each limit.

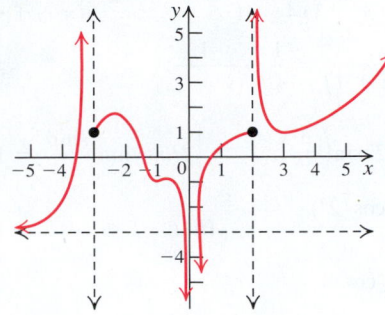

9. $\lim\limits_{x \to -3^-} f(x)$

10. $\lim\limits_{x \to -3^+} f(x)$

11. $\lim\limits_{x \to 0^-} f(x)$

12. $\lim\limits_{x \to 0^+} f(x)$

13. $\lim\limits_{x \to 2^-} f(x)$

14. $\lim\limits_{x \to 2^+} f(x)$

15. $\lim\limits_{x \to -\infty} f(x)$

16. $\lim\limits_{x \to \infty} f(x)$

In Exercises 17–28, find each limit. Use $-\infty$ and ∞ where appropriate.

17. $\lim\limits_{x \to 2^-} \dfrac{x}{x - 2}$

18. $\lim\limits_{x \to 3^-} \dfrac{2x}{x - 3}$

19. $\lim\limits_{x \to 5^+} \dfrac{x - 1}{x - 5}$

20. $\lim\limits_{x \to 4^+} \dfrac{3x + 1}{x - 4}$

21. $\lim\limits_{x \to 7^-} \dfrac{2}{7 - x}$

22. $\lim\limits_{x \to 8^+} \dfrac{x}{8 - x}$

23. $\lim\limits_{x \to -3^-} \dfrac{3x + 3}{(x + 3)^2}$

24. $\lim\limits_{x \to 2^-} \dfrac{2x - 2}{(x - 2)^2}$

25. $\lim\limits_{x \to 3^+} \dfrac{x}{(x - 3)^2}$

26. $\lim\limits_{x \to 1^+} \dfrac{2x + 1}{(x - 1)^2}$

27. $\lim\limits_{x \to 1^-} \dfrac{x^2 + x + 1}{1 - x}$

28. $\lim\limits_{x \to 2^+} \dfrac{2x^2 - x - 3}{2 - x}$

In Exercises 29–36, describe the behavior of each function near the zeros of its denominator. Use $-\infty$ and ∞ where apopropriate.

29. $f(x) = \dfrac{x - 1}{x^2 - 1}$

30. $f(x) = \dfrac{x + 3}{x^2 - 9}$

31. $g(t) = \dfrac{t + 1}{t^2 - t - 2}$

32. $g(u) = \dfrac{u - 1}{u^2 + u - 2}$

33. $f(x) = \dfrac{x + 2}{4 - x^2}$

34. $f(u) = \dfrac{u - 6}{36 - u^2}$

35. $f(t) = \dfrac{2 - t}{6 - t - t^2}$

36. $f(x) = \dfrac{x + 1}{5 + 4x - x^2}$

In Exercises 37–40, find the limit of each polynomial $P(x)$.
a. as $x \to -\infty$ **b.** as $x \to \infty$

37. $P(x) = 3x^5 - 5x^2 + 17$

38. $P(x) = -2x^3 + 12x^2 + 81$

39. $P(x) = -2x^4 + 7x^3 + 11$

40. $P(x) = -2x^6 + 5x^4 + x^2$

In Exercises 41–64, evaluate each limit.

41. $\lim\limits_{x \to \infty} \dfrac{x^2 + x + 1}{2x^2 + 3x - 17}$

42. $\lim\limits_{x \to \infty} \dfrac{x^2 - 3x + 7}{3x^2 + x + 15}$

43. $\lim\limits_{x \to -\infty} \dfrac{2x^3 + x^2 + 1}{4x^3 - x - 1}$

44. $\lim\limits_{x \to -\infty} \dfrac{2x^2 + x + 1}{5 + 3x - 2x^2}$

45. $\lim\limits_{x \to \infty} \dfrac{3x + 1}{x^2 + 7x - 8}$

46. $\lim\limits_{x \to \infty} \dfrac{3x^2 + 11}{x^3 + 2x^2 - 13}$

47. $\lim\limits_{x \to -\infty} \dfrac{2x^2 - 11x + 17}{x^3 - 12x + 15}$

48. $\lim\limits_{x \to -\infty} \dfrac{5x^2 - 19}{x^4 + x^3 + 18}$

49. $\lim\limits_{x \to \infty} \dfrac{2x^2 + 3x}{3x + 15}$

50. $\lim\limits_{x \to \infty} \dfrac{5x^3 + 2x^2 + 1}{x^2 - x + 1}$

51. $\lim\limits_{x \to -\infty} \dfrac{3x^3 + 15x + 11}{x^2 - 4}$

52. $\lim\limits_{x \to -\infty} \dfrac{5x^4 + 12x^2 + 1}{2x^2 + 3}$

53. $\lim\limits_{x \to \infty} \dfrac{2x^2 + 3}{\sqrt{5 + 4x^4}}$

54. $\lim\limits_{x \to \infty} \dfrac{\sqrt{5 + 4x^6}}{3x^3 - 11}$

55. $\lim\limits_{x \to -\infty} \dfrac{\sqrt{9x^2 + 12}}{x + 4}$

56. $\lim\limits_{x \to -\infty} \dfrac{x}{\sqrt{9x^2 + 1}}$

57. $\lim\limits_{x \to \infty} \left(\sqrt{x + 1} - \sqrt{x} \right)$

58. $\lim\limits_{x \to -\infty} \left(\sqrt{2 - x} - \sqrt{-x} \right)$

59. $\lim\limits_{x \to \infty} \left(\sqrt{x^2 + x} - x \right)$

60. $\lim\limits_{x \to \infty} \left(\sqrt{x^2 - 3x} - x \right)$

61. $\lim\limits_{x \to \infty} \left(\sqrt{x^2 + x + 1} - \sqrt{x^2 + 1} \right)$

62. $\lim\limits_{x \to \infty} \left(\sqrt{x^2 - 5x + 3} - \sqrt{x^2 + 2} \right)$

63. $\lim\limits_{x \to -\infty} \left(\sqrt{x^2 - 7x} - \sqrt{x^2 + 1} \right)$

64. $\lim\limits_{x \to -\infty} \left(\sqrt{4x^2 + 3x} - \sqrt{4x^2 + 1} \right)$

Applying the Concepts

65. Geometry. Find the horizontal and vertical asymptotes of the graph of

$$f(x) = \dfrac{2x}{\sqrt{x^2 + 1}}.$$

[*Hint*: Find $\lim\limits_{x \to \infty} f(x)$ and $\lim\limits_{x \to -\infty} f(x)$.]

66. Package Delivery. An online seller estimates that the number $N(x)$ of packages a new employee can prepare for delivery per day after x days of training is given by:

$$N(x) = 20 + \dfrac{30x}{x + 1}.$$

How many packages does the seller expect a new employee to prepare for delivery per day in the long run?

67. Environment. The estimated cost of removing $x\%$ of the impurities from a polluted lake is given by

$$c(x) = \dfrac{2x^2 + 25}{x(100 - x)} \text{ million dollars.}$$

Find $\lim\limits_{x \to 100^-} c(x)$. Interpret your answer.

68. Drug Concentration. A drug is injected in the bloodstream of a patient. The concentration $c(t)$ of the drug (in milligrams per liter) after t hours of injection is given by:

$$c(t) = \frac{7t}{2t^2 + 3}.$$

Find $\lim\limits_{t \to \infty} c(t)$. Interpret your result.

Beyond the Basics

In Exercises 69–72, evaluate each limit. Use $-\infty$ and ∞ where appropriate.

69. $\lim\limits_{x \to \infty} \dfrac{\sqrt{4x^2 - 1} - \sqrt{x^2 + 1}}{4 + 3x}$

70. $\lim\limits_{x \to \infty} \dfrac{\sqrt{x^2 + 4} - \sqrt{x^2 + 1}}{\sqrt{x^2 + 36} - \sqrt{x^2 + 9}}$

71. $\lim\limits_{x \to -\infty} \left(\sqrt{4x^2 + 7x} + 2x \right)$

72. $\lim\limits_{x \to -\infty} \left(\sqrt{x^2 + 3x} + x \right)$

73. Let $f(x) = \dfrac{ax + b}{x + 1}$.

If $\lim\limits_{x \to 0} f(x) = 3$ and $\lim\limits_{x \to \infty} f(x) = 5$, find $f(3)$.

74. Let $f(x) = \dfrac{ax^2 + b}{x^2 + x + 1}$.

If $\lim\limits_{x \to 0} f(x) = 2$ and $\lim\limits_{x \to \infty} f(x) = 3$, find $f(2)$.

75. Find $\lim\limits_{n \to \infty} a_n$, if $a_n = \dfrac{1 + 2 + 3 + \dots + n}{n^2}$.

$$\left[Hint: 1 + 2 + \dots + n = \frac{n(n + 1)}{2}. \right]$$

76. Consider an infinite series: $a_1 + a_2 + \dots + a_n + \dots$ We say that this series has sum s if $\lim\limits_{n \to \infty} s_n = s$, where $s_n = a_1 + a_2 + \dots + a_n$. Find the sum of the infinite series

$$\frac{1}{1 \cdot 2} + \frac{1}{2 \cdot 3} + \frac{1}{3 \cdot 4} + \frac{1}{4 \cdot 5} + \dots + \frac{1}{n(n + 1)} + \dots.$$

$$\left[Hint: \frac{1}{k(k + 1)} = \frac{1}{k} - \frac{1}{k + 1}. \right]$$

77. Find $\lim\limits_{n \to \infty} 2^n \sin\left(\dfrac{3}{2^n} \right)$. $\quad \left[Hint: \lim\limits_{\theta \to 0} \dfrac{\sin \theta}{\theta} = 1. \right]$

78. Find $\lim\limits_{n \to \infty} \dfrac{\cos (2^n)}{n}$. $\quad [\, Hint: -1 \le \cos \theta \le 1. \,]$

79. Find $\lim\limits_{n \to \infty} (\cos x)$

80. Find $\lim\limits_{x \to \infty} \left(\cos \dfrac{1}{x} \right)$

Critical Thinking / Discussion / Writing

81. Which types of rational functions do not have horizontal asymptotes? Give an example of such functions.

82. Which types of rational functions do not have vertical asymptotes? Give an example of such functions.

83. To find $\lim\limits_{x \to \infty} f(x)$, we can replace x by $\dfrac{1}{h}$ and rewrite the limit as $\lim\limits_{h \to 0^+} f\left(\dfrac{1}{h} \right)$.

Use this technique to find: $\lim\limits_{x \to \infty} \dfrac{x^2 - 3x + 7}{2x^2 + 5x + 11}$.

84. Find $\lim\limits_{x \to -\infty} \left(\sqrt{4x^2 + 3x} + 2x \right)$ by replacing x with $\dfrac{1}{h}$ and letting $h \to 0^-$.

Getting Ready for the Next Section

In Exercises 85–88, find the average rate of change $\dfrac{f(b) - f(a)}{b - a}$ for given f, a, and b.

85. $f(x) = 5; a = 2, b = 4$

86. $f(x) = 3x + 1; a = 2, b = 5$

87. $f(x) = x^2; a = 1, b = 3$

88. $f(x) = -x^2 + 3; a = 2, b = 4$

In Exercises 89–92, find the equation of the line with slope m, in slope-intercept form $y = mx + b$, that passes through the point $P = (x_1, y_1)$.

89. $m = 2, P = (1, 3)$

90. $m = -2, P = (-3, 4)$

91. $m = -\dfrac{1}{2}, P = (-2, -4)$

92. $m = \dfrac{2}{3}, P = (3, -6)$

Introduction to Derivatives

BEFORE STARTING THIS SECTION, REVIEW

1 Average rate of change of a function (Section 1.3, page 49)

2 Point-slope form of the equation of a line (Section 1.2, page 25)

OBJECTIVES

1 Find instantaneous rates of change.

2 Find an equation of the tangent line to the graph of a function.

3 Find the derivative of a function.

4 Find instantaneous velocity and speed.

◆ Average and Instantaneous Speed

The Cheetah Hunt roller coaster operating at Busch Gardens in Tampa, Florida, was built to replicate the movements of a cheetah pursuing its prey. The attraction features three launches (30 mph, 60 mph, and 40 mph), inversion, brake run, and numerous short banked corkscrew turns to simulate "near misses" experienced by cheetahs in nature. The ride is 90 seconds long, and the length of the ride is estimated to be 4429 feet. That gives the average speed of the ride at about 33 mph. As you can imagine, this average speed does not detail all the thrills and rapid changes in speed you are experiencing when riding this roller coaster. The safety of the riders is paramount, and to ensure safety the design team needs to compute and account for the *instantaneous speed* at every moment of the ride. In Example 5, we compute the instantaneous speed of an object.

To describe changes in natural processes, we must understand and prescribe how these changes occur. One of the main goals and success stories of calculus is to provide a mathematical study of change. In this section, we use a limit process to introduce the notion of a derivative. Derivatives measure the rate of change at a given moment. We also develop the geometric meaning of derivatives in relation to the tangent lines to curves.

1 Find instantaneous rates of change.

Instantaneous Rates of Change

Assume that the relation between two variables x and y is given by a function f; that is, $y = f(x)$. In Section 1.3, we defined the average rate of change (ARC) of f as x changes from a to b as

$$ARC \text{ of } f \text{ from } a \text{ to } b = \frac{\text{change in } y}{\text{change in } x} = \frac{\Delta y}{\Delta x} = \frac{f(b) - f(a)}{b - a}, \quad b \neq a.$$

Average rate of change measures how fast, on average, the function changes on a given interval. Geometrically, the average rate of change of f from $x = a$ to $x = b$ is the slope of the **secant line** joining the points $P = (a, f(a))$ and $Q = (b, f(b))$ on the graph of f. See Figure 10.6.

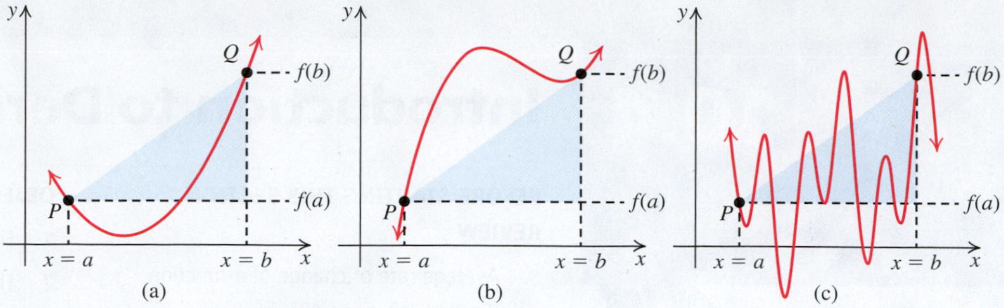

Figure 10.6 Average rates of change.

The average rate of change does not provide us with enough information to precisely determine how a function f changes between two given points. Three functions shown in (a), (b), and (c) in Figure 10.6 have the same average rate of change as x changes from a to b, but their behavior is completely different.

Our goal is to measure how a function f changes at a given point. Note that the value of a function at the given point does not determine its behavior near the point and that it does not provide us with enough information to determine whether the function is increasing or decreasing and how steep the function is near that point. Three functions shown in (a), (b), and (c) in Figure 10.6 have the same value at $x = a$, but their behavior near the point $P = (a, f(a))$ is completely different.

To get a good approximation of the behavior of a function f near $x = a$, we need to study the ARC of f on very small intervals around $x = a$, that is, when $\Delta x = b - a$ is close to zero. The limit process allows us to define the instantaneous rate of change of the function f at $x = a$ as a limiting value of the average rates of change of the function f from $x = a$ to $x = b$ when $\Delta x = b - a \to 0$.

INSTANTANEOUS RATE OF CHANGE OF A FUNCTION f

Instantaneous rate of change of the function f with respect to x at $x = a$ is given by

$$IRC \text{ of } f \text{ at } a = \lim_{\Delta x \to 0} \frac{\Delta y}{\Delta x} = \lim_{\Delta x \to 0} \frac{f(a + \Delta x) - f(a)}{\Delta x} = \lim_{h \to 0} \frac{f(a + h) - f(a)}{h}$$

provided that this limit exists.

EXAMPLE 1 **Finding the Average and Instantaneous Rates of Change**

Let $f(x) = \frac{1}{3}x^2 + 1$. Find

a. the average rate of change of f as x changes from 0.5 to 4.

b. the average rate of change of f as x changes from 0.5 to 2.5.

c. the average rate of change of f as x changes from 0.5 to 1.

d. the instantaneous rate of change of f with respect to x at 0.5.

Solution

a. As x changes from $a = 0.5$ to $b = 4$, we compute $f(a) = f(0.5) = \frac{13}{12}$ and $f(b) = f(4) = \frac{19}{3}$, so

$$ARC \text{ of } f \text{ from } 0.5 \text{ to } 4 = \frac{f(4) - f\left(\frac{1}{2}\right)}{4 - \frac{1}{2}} = \frac{\frac{19}{3} - \frac{13}{12}}{4 - \frac{1}{2}} = \frac{\frac{21}{4}}{\frac{7}{2}} = \frac{3}{2} = 1.5$$

b. As x changes from $a = 0.5$ to $b = 2.5$, we compute $f(a) = f(0.5) = \frac{13}{12}$ and $f(b) = f(2.5) = f\left(\frac{5}{2}\right) = \frac{37}{12}$, so

$$ARC \text{ of } f \text{ from } 0.5 \text{ to } 2.5 = \frac{f\left(\frac{5}{2}\right) - f\left(\frac{1}{2}\right)}{\frac{5}{2} - \frac{1}{2}} = \frac{\frac{37}{12} - \frac{13}{12}}{\frac{5}{2} - \frac{1}{2}} = \frac{2}{2} = 1$$

c. As x changes from $a = 0.5$ to $b = 1$, we compute $f(a) = f(0.5) = \frac{13}{12}$ and $f(b) = f(1) = \frac{4}{3}$ and determine the average rate of change:

$$ARC \text{ of } f \text{ from } 0.5 \text{ to } 1 = \frac{f(1) - f\left(\frac{1}{2}\right)}{1 - \frac{1}{2}} = \frac{\frac{4}{3} - \frac{13}{12}}{1 - \frac{1}{2}} = \frac{\frac{1}{4}}{\frac{1}{2}} = \frac{1}{2} = 0.5$$

d. At $a = 0.5$, the instantaneous rate of change is

$$IRC \text{ of } f \text{ at } 0.5 = \lim_{h \to 0} \frac{f(a + h) - f(a)}{h} \qquad \text{General formula}$$

$$= \lim_{h \to 0} \frac{f\left(\frac{1}{2} + h\right) - f\left(\frac{1}{2}\right)}{h} \qquad \text{Substitute } a = 0.5 = \frac{1}{2}.$$

$$= \lim_{h \to 0} \frac{\left[\frac{1}{3}\left(\frac{1}{2} + h\right)^2 + 1\right] - \left[\frac{1}{3}\left(\frac{1}{2}\right)^2 + 1\right]}{h} \qquad \text{Evaluate } f\left(\frac{1}{2} + h\right) \text{ and } f(1).$$

$$= \lim_{h \to 0} \frac{\left[\frac{1}{12} + \frac{1}{3}h + \frac{1}{3}h^2 + 1\right] - \left[\frac{13}{12}\right]}{h} \qquad \text{Expand.}$$

$$= \lim_{h \to 0} \frac{\frac{1}{3}h + \frac{1}{3}h^2}{h} \qquad \text{Simplify.}$$

$$= \lim_{h \to 0} \frac{\frac{1}{3}h(h + 1)}{h} \qquad \text{Factor.}$$

$$= \lim_{h \to 0} \frac{1}{3}h + \frac{1}{3} \qquad \text{Remove common factor } h.$$

$$= 0 + \frac{1}{3} = \frac{1}{3} \approx 0.3333\ldots \qquad \text{Use limit properties and simplify.}$$

In this example, the closer the values of b are to the number $a = 0.5$, the closer the values of the corresponding average rates of change of f (computed in **a**, **b**, and **c**) are to the value of the instantaneous rate of change of f at $a = 0.5$. See Figure 10.7 for a geometric interpretation. In every case, when b is sufficiently close to a, the average rate of change is a good estimate of the instantaneous rate of change at $x = a$.

Practice Problem 1 Repeat Example 1 for the function $f(x) = \frac{1}{2}x^2 - 1$. ■

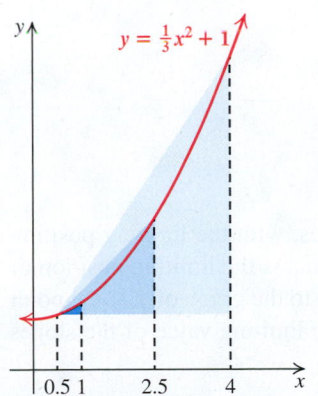

$y = \frac{1}{3}x^2 + 1$

0.5 1   2.5   4 x

Figure 10.7

2 Find an equation of the tangent line to the graph of a function.

Tangent Line to the Graph of a Function

In this section, we develop the geometric meaning of the instantaneous rate of change of f, which will lead us to construction of the tangent line to the graph of a function f at a given point.

Geometrically, the average rate of change of f from $x = a$ to $x = b$ is the slope of the **secant line** joining the points $P = (a, f(a))$ and $Q = (b, f(b))$ on the graph of f. For simplicity we set $h = \Delta x = b - a$, so $b = a + h$. See Figure 10.8.

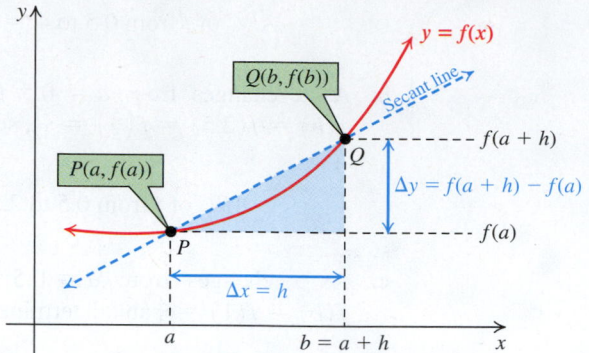

Figure 10.8 Secant line and average rate of change.

Suppose we fix the point $x = a$ and examine what happens if we move the point $x = b = a + h$ toward $x = a$. What happens to the secant line if the distance Δx between a and b approaches 0? As h approaches 0, $a + h$ approaches a. Geometrically, the point $Q(a + h, f(a + h))$ moves along the graph of f and approaches the point $P(a, f(a))$. See Figure 10.9.

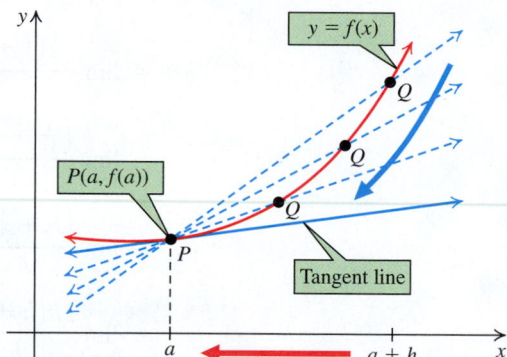

Figure 10.9 Tangent line.

As Q moves toward P, we obtain a succession of secant lines, with the limiting position being a particular line passing through the point P. The line that is the limiting position of these secant lines passing through P is called the **tangent line** to the graph of f at the point $P = (a, f(a))$. Therefore, the slope of the tangent line is the limiting value of the slopes of these secant lines. Thus,

---

THE SLOPE OF A TANGENT LINE

The slope of the tangent line to the graph of the function $y = f(x)$ at the point $P = (a, f(a))$ is

$$m_{tan} = \lim_{h \to 0} m_{sec} = \lim_{\Delta x \to 0} \frac{\Delta y}{\Delta x} = \lim_{h \to 0} \frac{f(a + h) - f(a)}{h},$$

provided that the limit exists.

The slope–intercept equation of a tangent line is provided next.

EQUATION OF A TANGENT LINE

If m_{tan} exists, then the equation of the tangent line to the graph of the function $y = f(x)$ at the point $P = (a, f(a))$ is

$$y = m_{tan}(x - a) + f(a).$$

We solved the problem of determining a tangent line to the graph of a function f at a given point. The tangent line is a very special line, as it touches the graph of f at P and is also the best linear approximation of the graph at the point P. In simple terms, around the point P, the graph of f looks very much like its tangent line. This is a profound observation that allows us to study curves using their tangent lines. By way of example, if the slope of a tangent line is positive at P, the function is increasing, and if the slope of the tangent line is negative, the function is decreasing.

EXAMPLE 2 **Finding an Equation of the Tangent line**

Find the slope–intercept form of the equation of the tangent line to the graph of $f(x) = 3x^2 - 8x$ at the point $(1, -5)$.

Solution

At $(1, -5)$ we have $a = 1$. The slope of the tangent line at $a = 1$ is:

$$m_{tan} = \lim_{h \to 0} \frac{f(a + h) - f(a)}{h} \qquad \text{General formula}$$

$$\lim_{h \to 0} \frac{f(1 + h) - f(1)}{h} \qquad \text{Replace } a \text{ by } 1.$$

$$= \lim_{h \to 0} \frac{[3(1 + h)^2 - 8(1 + h)] - [-5]}{h} \qquad \text{Evaluate } f(1 + h) \text{ and } f(1).$$

$$= \lim_{h \to 0} \frac{3 + 6h + 3h^2 - 8 - 8h + 5}{h} \qquad \text{Expand.}$$

$$= \lim_{h \to 0} \frac{3h^2 - 2h}{h} \qquad \text{Simplify.}$$

$$= \lim_{h \to 0} \frac{h(3h - 2)}{h} \qquad \text{Factor out } h.$$

$$= \lim_{h \to 0} (3h - 2) \qquad \text{Remove the common factor } h.$$

$$= 0 - 2 = -2 \qquad \text{Use limit properties and simplify.}$$

Now we can write the slope–intercept form of the equation for the tangent line to the graph of f at $P = (1, -5)$. Observe that for $a = 1$, $f(a) = f(1) = -5$.

$$y = m_{tan}(x - a) + f(a) \qquad \text{General formula}$$

$$y = -2(x - 1) - 5 \qquad a = 1, f(a) = f(1) = -5 \text{ and } m_{tan} = -2$$

$$y = -2x - 3 \qquad \text{Simplify.}$$

The slope–intercept form of the equation of the tangent line to the graph of $f(x) = 3x^2 - 8x$ at the point $(1, -5)$ is: $y = -2x - 3$. (See Figure 10.10.)

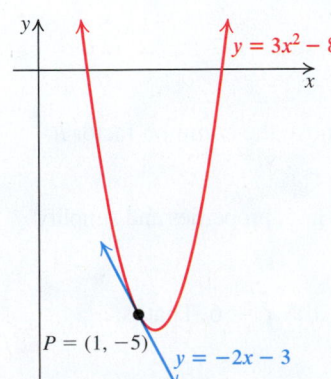

Figure 10.10

Practice Problem 2 Find the slope–intercept form of the equation of the tangent line to the graph of $f(x) = -2x^2 + x$ at the point $(-1, -3)$.

3 Find the derivative of a function.

Derivative of a Function

The instantaneous rate of change and the slope of the tangent line are both found using the same limiting process. We unify the underlying idea in the concept of the **derivative**.

DERIVATIVE OF A FUNCTION $y = f(x)$ **AT THE POINT** $x = a$

If $x = a$ is in the domain of f, then the **derivative of f at the point** $x = a$, denoted by $f'(a)$ and read "f prime of a," is the number defined by

$$f'(a) = \lim_{h \to 0} \frac{f(a + h) - f(a)}{h}$$

provided that this limit exists. The number $f'(a)$ also describes

- the **slope of the tangent line** to the graph of f at $(a, f(a))$.
- the **instantanenous rate of change** of f with respect to x at a.

EXAMPLE 3   **Finding the Derivative of a Function at the Given Point**

Find the derivative of $f(x) = \sqrt{x + 1}$ at $x = 2$. That is, find $f'(2)$.

Solution

At $x = 2$, the derivative of a function f is:

$$f'(2) = \lim_{h \to 0} \frac{f(a + h) - f(a)}{h} \qquad \text{General formula}$$

$$= \lim_{h \to 0} \frac{f(2 + h) - f(2)}{h} \qquad \text{Replace } a \text{ with 2.}$$

$$= \lim_{h \to 0} \frac{\sqrt{3 + h} - \sqrt{3}}{h} \qquad \text{Evaluate } f(2 + h) \text{ and } f(2).$$

$$= \lim_{h \to 0} \frac{\left(\sqrt{3 + h} - \sqrt{3}\right)\left(\sqrt{3 + h} + \sqrt{3}\right)}{h\left(\sqrt{3 + h} + \sqrt{3}\right)} \qquad \begin{array}{l}\text{Multiply numerator and denomi-} \\ \text{nator by } \left(\sqrt{3 + h} + \sqrt{3}\right).\end{array}$$

$$= \lim_{h \to 0} \frac{3 + h - 3}{h\left(\sqrt{3 + h} + \sqrt{3}\right)} \qquad (a + b)(a - b) = a^2 - b^2$$

$$= \lim_{h \to 0} \frac{h}{h\left(\sqrt{3 + h} + \sqrt{3}\right)} \qquad \text{Simplify.}$$

$$= \lim_{h \to 0} \frac{1}{\sqrt{3 + h} + \sqrt{3}} \qquad \text{Remove the common factor } h.$$

$$= \frac{1}{\sqrt{3 + 0} + \sqrt{3}} = \frac{1}{2\sqrt{3}} \approx 0.2886\ldots \qquad \text{Use limit properties and simplify.}$$

Practice Problem 3 Find the derivative of $f(x) = \sqrt{x - 1}$ at $x = 6$. That is, find $f'(6)$.

To study how the function f changes, we need to know its derivative (slope of the tangent line) at every point in its domain.

> ### DERIVATIVE FUNCTION OF *f*
>
> The derivative function of f, denoted by f', is a function that assigns the value $f'(a)$ to any point a in the domain of f. For any point x in the domain of f:
>
> $$f'(x) = \lim_{h \to 0} \frac{f(x + h) - f(x)}{h}$$
>
> provided that this limit exists.

The domain of the function f' is the set of real numbers, a, in the domain of f for which $f'(a)$ exists.

EXAMPLE 4   **Finding the Derivative Function of a Given Function**

Find the derivative function $f'(x)$ of $f(x) = \dfrac{1}{x}$.

Solution

We have

$$f'(x) = \lim_{h \to 0} \frac{f(x + h) - f(x)}{h} \qquad \text{General formula}$$

$$= \lim_{h \to 0} \frac{\dfrac{1}{x + h} - \dfrac{1}{x}}{h} \qquad \text{Evaluate } f(x + h) \text{ and } f(x).$$

$$= \lim_{h \to 0} \frac{\dfrac{x}{(x + h)x} - \dfrac{(x + h)}{x(x + h)}}{h} \qquad \text{Use } (x + h)x \text{ as a common denominator.}$$

$$= \lim_{h \to 0} \frac{\dfrac{x - (x + h)}{x(x + h)}}{h} \qquad \text{Combine rational expressions.}$$

$$= \lim_{h \to 0} \frac{\dfrac{-h}{x(x + h)}}{h} \qquad \text{Simplify the numerator.}$$

$$= \lim_{h \to 0} \frac{-h}{x(x + h)} \cdot \frac{1}{h} \qquad \text{Simplify.}$$

$$= \lim_{h \to 0} -\frac{1}{x(x + h)} \qquad \text{Remove the common factor } h.$$

$$= -\frac{1}{x(x + 0)} = -\frac{1}{x^2} \qquad \text{Use limit properties and simplify.}$$

The domain of $f'(x) = -\dfrac{1}{x^2}$ is the set $(-\infty, 0) \cup (0, \infty)$.

Practice Problem 4 Repeat Example 4 for the function $f(x) = \sqrt{x}$.

4 Find instantaneous velocity and speed.

Applications

Many applications concern the motion of various objects. The velocity $v(t)$ of an object at time t is the instantaneous rate of change of its position function $s(t)$.

In other words,

$$v(t) = \lim_{\Delta t \to 0} \frac{\Delta s}{\Delta t} = s'(t).$$

Velocity provides the direction and magnitude of the motion. In our next example, we see that velocity can be a positive or negative number. *Instantaneous speed* is the magnitude of velocity and is not related to the direction of the movement.

EXAMPLE 5 Finding Instantaneous Velocity and Speed

A stone is thrown vertically upward from the edge of a bridge 96 feet high. The height $s(t)$ of the stone (in feet) after t seconds is given by the function

$$s(t) = -16t^2 + 40t + 96.$$

a. Find the instantaneous velocity and speed of the stone 1 second after it is thrown.

b. Find the instantaneous velocity and speed of the stone 3 seconds after it is thrown.

c. Find the instantaneous velocity and speed of the stone when it hits the ground.

Solution

The instantaneous velocity of the stone at time t is the derivative $s'(t)$:

$$s'(t) = \lim_{h \to 0} \frac{s(t+h) - s(t)}{h} \qquad \text{General formula}$$

$$= \lim_{h \to 0} \frac{[-16(t+h)^2 + 40(t+h) + 96] - [-16t^2 + 40t + 96]}{h} \qquad \begin{array}{l}\text{Evaluate } s(t+h) \\ \text{and } s(t).\end{array}$$

$$= \lim_{h \to 0} \frac{-32th - 16h^2 + 40h}{h} \qquad \text{Simplify.}$$

$$= \lim_{h \to 0} \frac{h(-32t - 16h + 40)}{h} \qquad \text{Factor out } h.$$

$$= \lim_{h \to 0} (-32t - 16h + 40) \qquad \begin{array}{l}\text{Remove the} \\ \text{common factor } h.\end{array}$$

$$= -32t - 0 + 40 \qquad \begin{array}{l}\text{Use limit} \\ \text{properties.}\end{array}$$

$$= -32t + 40. \qquad \text{Simplify.}$$

The instantaneous velocity of the stone at time t is

$$s'(t) = -32t + 40.$$

a. At $t = 1$, the instantaneous velocity is

$$s'(1) = -32 \cdot 1 + 40 = 8 \text{ ft/s}.$$

The stone is moving at a speed of 8 feet per second in the upward direction when $t = 1$ second.

b. At $t = 3$, the instantaneous velocity is

$$s'(3) = -32 \cdot 3 + 40 = -56 \text{ ft/s}.$$

In this situation, the negative sign of the velocity means that the object is traveling downward. The stone is moving at a speed of 56 feet per second in the downward direction when $t = 3$ seconds.

   c. First, we find the exact time the rock will hit the ground. We set $s(t)$ equal to 0.

$$-16t^2 + 40t + 96 = 0 \quad \text{Set } s(t) = 0.$$
$$-8(2t + 3)(t - 4) = 0 \quad \text{Factor quadratic equation.}$$
$$2t + 3 = 0 \quad t - 4 = 0 \quad \text{Zero-product property}$$
$$t = -1.5 \quad\quad t = 4 \quad \text{Solve for } t.$$

Since the time $t \geq 0$, we discard the solution $t = -1.5$. The stone hits the ground after 4 seconds. At $t = 4$, the instantaneous velocity is

$$s'(4) = -32 \cdot 4 + 40 = -88 \text{ ft/s.}$$

In this situation, the negative sign of the velocity means that the object is traveling downward. The stone is moving at a speed of 88 feet per second in the downward direction when it hits the ground.

Practice Problem 5 Repeat Example 5 for the bridge 125 feet high; the height of the stone is given by the function $s(t) = -16t^2 + 55t + 125$.

Answers to Practice Problems

1. a. $\frac{9}{4}$ **b.** $\frac{3}{2}$ **c.** $\frac{3}{4}$ **d.** $\frac{1}{2}$

2. $y = 5x + 2$ **3.** $f'(6) = \frac{1}{2\sqrt{5}}$

4. $f'(x) = \frac{1}{2\sqrt{x}}$

5. a. velocity: 23 ft/s; speed: 23 ft/s
  b. velocity: −41 ft/s; speed: 41 ft/s
  c. velocity: −105 ft/s; speed: 105 ft/s

SECTION 10.4 **Exercises**

Concepts and Vocabulary

1. The average rate of change (ARC) of f as x changes from a to b ($a \neq b$) is given by _____.

2. The derivative of the function f at the point $x = a$ is $\lim\limits_{h \to 0}$ _____, provided that this limit exists.

3. Provided it exists, $f'(a)$ describes the _____ of the tangent line to the graph of f at $(a, f(a))$.

4. Provided it exists, $f'(a)$ describes the _____ rate of change of the function f at $x = a$.

5. **True or False.** The instantaneous rate of change is found as a limiting value of the corresponding average rates of change.

6. **True or False.** The slope of the tangent line to the graph of a function is found as a limiting value of the slopes of the corresponding secant lines.

7. **True or False.** The derivative is used to find the average velocity.

8. **True or False.** The derivative of a function f at a given point is always positive.

Building Skills

In Exercises 9–10, use the graph to find the average rate of change of the function between points P and Q.

9.

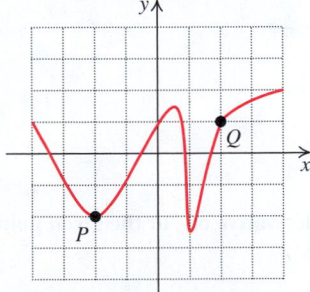

10.

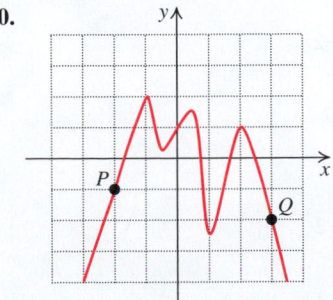

11. The area S of a square with edge length a inches is
$S = S(a) = a^2$.
    a. Find the average rate of change of the area S as the edge length a changes from $a = 1$ to $a = 3$.
    b. Find the average rate of change of the area S as the edge length a changes from $a = 1$ to $a = 2$.
    c. Find the average rate of change of the area S as the edge length a changes from $a = 1$ to $a = 1.5$.
    d. Find the instantaneous rate of change of the area S with respect to the edge length a at $a = 1$.

12. The edge length a of a square with the area S (in in^2) is
$a = a(S) = \sqrt{S}$.
    a. Find the average rate of change of the edge a as the area S changes from $S = 1$ to $S = 3$.
    b. Find the average rate of change of the edge a as the area S changes from $S = 1$ to $S = 2$.
    c. Find the average rate of change of the edge a as the area S changes from $S = 1$ to $S = 1.5$.
    d. Find the instantaneous rate of change of the edge a with respect to the area S at $S = 1$.

In Exercises 13 and 14, use the graph to find the derivative of the function at the point P.

13.

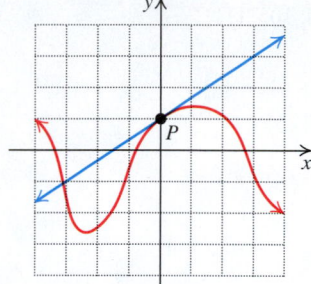

14.

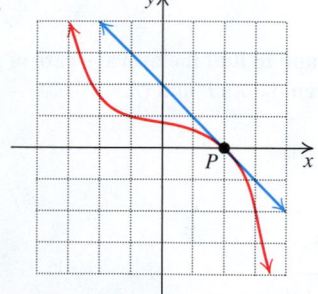

In Exercises 15–28, find the derivative of f at the given point.

15. $f(x) = 2x - 1$ at $x = -2$

16. $f(x) = -3x + 4$ at $x = 3$

17. $f(x) = 2x^2 + 1$ at $x = 2$

18. $f(x) = 3x^2 - 2$ at $x = -2$

19. $f(x) = 4x^2 - 3x$ at $x = 1$

20. $f(x) = 3x^2 - 5x$ at $x = 2$

21. $f(x) = 2x - x^2$ at $x = 2$

22. $f(x) = x - 2x^2$ at $x = 1$

23. $f(x) = 2x^3 - 1$ at $x = -2$

24. $f(x) = 3x^3 + 4$ at $x = 2$

25. $f(x) = \sqrt{2x - 3}$ at $x = 3$

26. $f(x) = \sqrt{3x + 1}$ at $x = 2$

27. $f(x) = \dfrac{1}{2x - 1}$ at $x = 2$

28. $f(x) = \dfrac{1}{4x + 3}$ at $x = -2$

In Exercises 29 and 30, use the graph to find the slope–intercept form of the equation for the tangent to the graph of the function at the point P.

29.

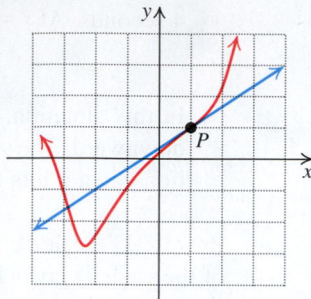

30.

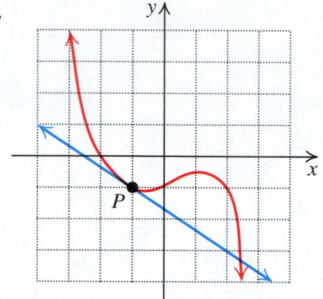

In Exercises 31–42, find the slope–intercept form of the equation for the tangent line to the graph of f at the given point.

31. $f(x) = x^2 + 1$ at $(3, 10)$

32. $f(x) = x^2 - 1$ at $(-2, 3)$

33. $f(x) = 3 - 2x^2$ at $(2, -5)$

34. $f(x) = 1 - 5x^2$ at $(-1, -4)$

35. $f(x) = x^2 - 3x + 4$ at $(-1, 8)$

36. $f(x) = x^2 - x - 5$ at $(2, -3)$

37. $f(x) = x^3 - 4$ at $(2, 4)$

38. $f(x) = x^3 + 3$ at $(-2, -5)$

39. $f(x) = \sqrt{x}$ at $(9, 3)$

40. $f(x) = \sqrt{x}$ at $(1/4, 1/2)$

41. $f(x) = \frac{1}{x}$ at $(3, 1/3)$

42. $f(x) = \frac{1}{x}$ at $(1/3, 3)$

In Exercises 43–54, find the derivative function of the given function.

43. $f(x) = 3x - 2$

44. $f(x) = 1 - 4x$

45. $f(x) = \dfrac{1}{3}x^2 - 1$

46. $f(x) = 1 - \dfrac{2}{3}x^2$

47. $f(x) = 2x^2 - 3x + 1$

48. $f(x) = 3x^2 - 4x - 7$

49. $f(x) = x^3 - 4$

50. $f(x) = x^3 + 3$

51. $f(x) = \sqrt{3x - 2}$

52. $f(x) = \sqrt{2x + 3}$

53. $f(x) = \dfrac{1}{2x + 1}$ **54.** $f(x) = \dfrac{1}{2x - 1}$

Applying the Concepts

55. Instantaneous Speed. A ball is dropped from the edge of a building 64 feet above the ground. The height of the ball (in feet) after t seconds is given by the function

$$s(t) = 64 - 16t^2.$$

a. How long does it take the ball to hit the ground?
b. Find the instantaneous speed of the ball when it hits the ground.

56. Instantaneous Speed. A ball is thrown vertically upward from the height of 4 feet, with an initial speed of 48 feet per second. The height of the ball (in feet) after t seconds is given by the function

$$s(t) = -16t^2 + 48t + 4.$$

a. Find the instantaneous speed of the ball 1 second after it is thrown and determine the direction the ball is traveling.
b. Find the instantaneous speed of the ball 3 seconds after it is thrown and determine the direction the ball is traveling.

57. Instantaneous Rate of Change. Air is being pumped into a spherical balloon so that its radius increases at the constant rate of 3 centimeters per second (so $r = 3t$). Find the instantaneous rate of change of the balloon's volume V with respect to time t at $t = 2$ seconds after the process started. (The volume V of a sphere of radius r is $\dfrac{4\pi}{3}r^3$.)

58. Instantaneous Rate of Change. Air is being pumped into a spherical balloon so that its radius increases at the constant rate of 3 centimeters per second (so $r = 3t$). Find the instantaneous rate of change of the balloon's surface area S with respect to time t at $t = 2$ seconds after the process started. (The surface area S of a sphere of radius r is $4\pi r^2$.)

59. Action Camera Sales. The number of sales S of a particular model of sports action camera is modeled by the function

$$S(d) = -0.2d^2 + 40d,$$

where d represents the number of days since the sale began.
a. Find the instantaneous rate of change of sales 70 days after the sale began and determine whether the sales were increasing or decreasing.
b. Find the instantaneous rate of change of sales 110 days after the sale began and determine whether the sales were increasing or decreasing.

60. Solution Concentration. A tank with a total capacity of 80 gallons contains 15 gallons of 20% salt solution. As water is pumped into the tank, the concentration of salt (in percent points) is modeled by the function

$$C(x) = \frac{300}{15 + x},$$

where x represents the amount of water added.

a. Find the instantaneous rate of change of the concentration of salt after 5 gallons of water has been pumped and determine whether the concentration of salt is increasing or decreasing.
b. Find the instantaneous rate of change of the concentration of salt after 10 gallons of water has been pumped and whether the concentration is increasing or decreasing.
c. In which situation, described in a) and b), is the concentration of salt in the solution changing most rapidly.

61. As an object moves, the distance s traveled from the origin in time t is given by the following graph.

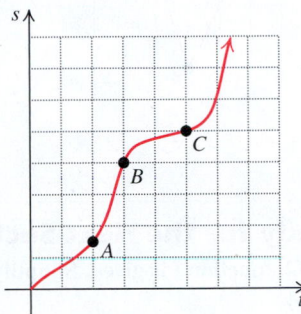

Arrange the points A, B, C in order from the highest velocity to the lowest.

62. The total population p of a particular species as a function of time t is given by the following graph.

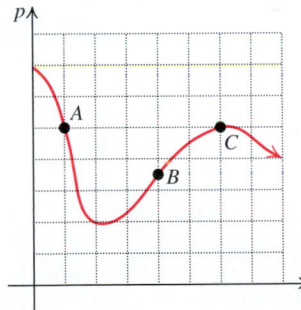

Arrange the points A, B, C in order from the highest growth rate to the lowest.

Beyond the Basics

63. Find the derivative of $f(x) = \frac{1}{x^2}$ at $x = a$.

64. Find the derivative of $f(x) = \frac{1}{\sqrt{x}}$ at $x = a$.

65. Use the facts $\lim\limits_{h \to 0} \dfrac{\sin h}{h} = 1$ and $\lim\limits_{h \to 0} \dfrac{\cos h - 1}{h} = 0$ to find the derivative of f at the given point.
a. $f(x) = \sin x$ at $x = a$
b. $f(x) = \cos x$ at $x = a$

66. Find the derivative function of a quadratic function $f(x) = ax^2 + bx + c$ and use it to determine all points P on the graph of f such that the tangent line at P is horizontal.

Critical Thinking / Discussion / Writing

67. Show that the function $f(x) = |x|$ does not have a tangent line at the point $(0, 0)$.

68. Show that the function $f(x) = \sqrt[3]{x}$ does not have a tangent line at the point $(0, 0)$.

69. The graph of f is as follows.

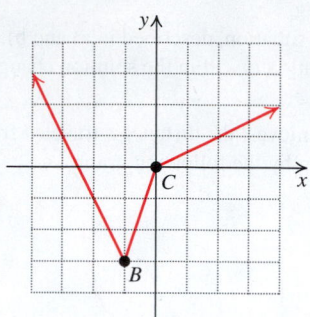

Explain why the graph of f does not have a tangent line at points B and C. Draw the graph of the derivative function $f'(x)$.

70. Let the domain of a function f be $(-\infty, \infty)$ and let $f'(1) = 3$. Find $f'(-1)$ if

a. we assume additionally that f is even.

b. we assume additionally that f is odd.

Getting Ready for the Next Section

In Exercises 71–74, function f is given. Simplify $f\left(1 + \frac{k}{2}\right)$.

71. $f(x) = 2$

72. $f(x) = 2x + 1$

73. $f(x) = 1 - x$

74. $f(x) = x^2 - 1$

In Exercises 75–78, find each sum.

75. $\displaystyle\sum_{i=1}^{4} (3i - 1)$

76. $\displaystyle\sum_{i=1}^{3} (1 - 2i)$

77. $\displaystyle\sum_{i=1}^{4} \left(1 + \frac{i-1}{4}\right)$

78. $\displaystyle\sum_{i=1}^{3} \left(1 + \frac{i-1}{3}\right)^2$

Area and the Integral

BEFORE STARTING THIS SECTION, REVIEW

1 Geometry Formulas (Appendix A.5, page 952)

2 Arithmetic Sequences (Section 9.2, page 805)

3 Summation Notation (Section 9.1, page 799)

OBJECTIVES

1 Approximate the area under the graph of a function.

2 Define the integral of a function.

3 Approximate an integral using a graphing utility.

◆ Path of a Hummingbird

If you drive on a straight road at 60 mph for 30 minutes, you have traveled a distance of 30 miles. However, when a hummingbird flies over an erratic path at a constant rate for a specific amount of time, determining the distance it has traveled is considerably more difficult. When velocity is constant on an interval of time, the distance traveled during this time is the definite integral of this (constant) velocity function on the interval. We introduce this important integral in this section, and in Exercises 21 and 22 we ask you to compute the distance traveled over a path by evaluating an integral.

---

1 Approximate the area under the graph of a function.

Area

We have formulas for finding the area of a rectangle or a triangle, and the area of a polygonal region can be found by dividing it into triangles. (See Figure 10.11).

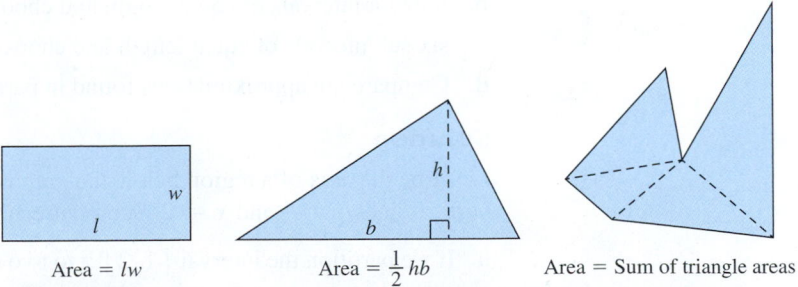

Area = lw Area = $\frac{1}{2} hb$ Area = Sum of triangle areas

Figure 10.11

However, determining the area of a region with one or more curved boundaries presents quite a challenge. The concept of limit, used to solve the geometric problem of defining tangent lines, comes to our aid again in defining the area of certain regions of the plane.

We investigate the problem of finding the area of a region below the graph of $y = f(x)$, above the x-axis, and between the vertical lines $x = a$ and $x = b$, where f is continuous and $f(x) \geq 0$ for all x in $[a, b]$. We first approximate the area of this region by summing the areas of rectangles that are formed by first subdividing, or partitioning, the interval $[a, b]$ into subintervals of equal length. Each subinterval is the base for a rectangle of height $f(t)$ for some number t in the subinterval. In Figure 10.12(a), the interval $[a, b]$

is divided into two subintervals, each of length $\dfrac{b-a}{2}$. In Figure 10.12(b), the interval

$[a, b]$ is divided into four subintervals, each of length $\dfrac{b-a}{4}$.

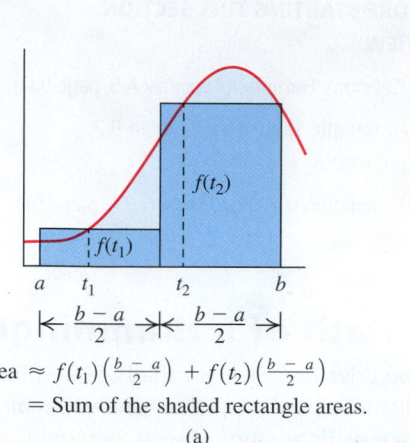

Area $\approx f(t_1)\left(\dfrac{b-a}{2}\right) + f(t_2)\left(\dfrac{b-a}{2}\right)$

= Sum of the shaded rectangle areas.

(a)

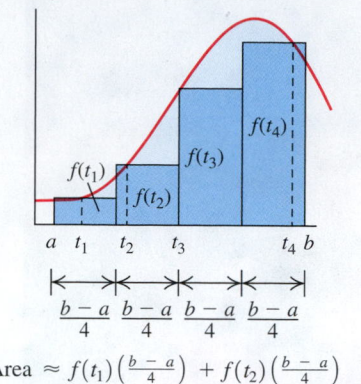

Area $\approx f(t_1)\left(\dfrac{b-a}{4}\right) + f(t_2)\left(\dfrac{b-a}{4}\right)$

$\qquad + f(t_3)\left(\dfrac{b-a}{4}\right) + f(t_4)\left(\dfrac{b-a}{4}\right)$

= Sum of the shaded rectangle areas.

(b)

Figure 10.12

Any number of equal-length subintervals may be used to approximate the area under the graph of a function from a to b. The choice of the number t in each subinterval is arbitrary; however, we will always choose t as either the left endpoint of each subinterval or the right endpoint of each subinterval. This choice will not affect the result of our area calculations.

EXAMPLE 1   **Approximating the Area under the Graph of $f(x)$**

Approximate the area under the graph of $f(x) = x$ from $x = 1$ to $x = 2$ by partitioning the interval $[1, 2]$ into:

a. two subintervals of equal length and choosing t to be the left endpoint.

b. four subintervals of equal length and choosing t to be the left endpoint.

c. six subintervals of equal length and choosing t to be the left endpoint.

d. Compare the approximations found in parts a–c with the actual area.

Solution

Let A be the area of a region below the graph of $y = x$, above the x-axis, and between the vertical lines $x = 1$ and $x = 2$. See Figure 10.13(d).

a. If we partition the interval $[1, 2]$ into two subintervals of equal length, then each inter-

val will have length $\dfrac{2-1}{2} = \dfrac{1}{2}$. The subintervals will be $\left[1, 1 + \dfrac{1}{2}\right]$ and $\left[\dfrac{3}{2}, 2\right]$, with

left endpoints 1 and $\dfrac{3}{2}$, respectively. See Figure 10.13(a).

$A \approx f(1) \cdot \dfrac{1}{2} + f\left(\dfrac{3}{2}\right) \cdot \dfrac{1}{2}$ Sum of the areas of the two rectangles

$= 1 \cdot \dfrac{1}{2} + \dfrac{3}{2} \cdot \dfrac{1}{2}$ Replace $f(1)$ with 1 and $f\left(\dfrac{3}{2}\right)$ with $\dfrac{3}{2}$.

$= \dfrac{5}{4} = 1.25$ Simplify.

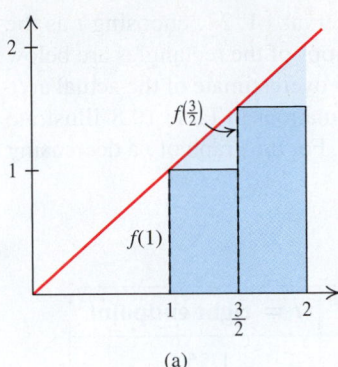

(a)

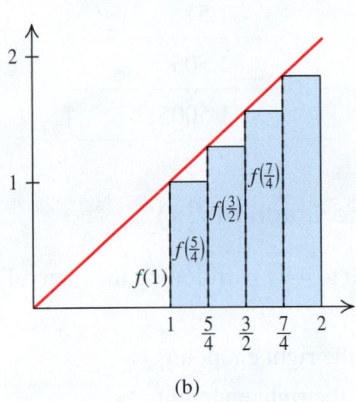

(b)

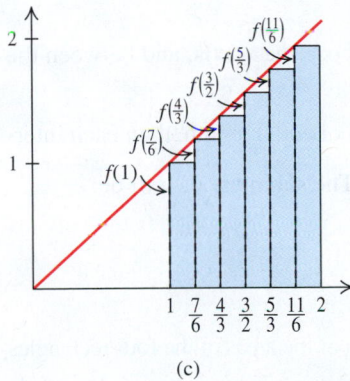

(c)

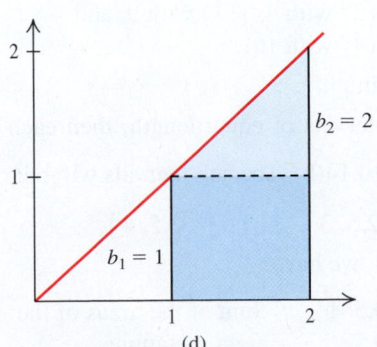

(d)

Figure 10.13

b. If we partition the interval $[1, 2]$ into four subintervals of equal length, then each interval will have length $\dfrac{2-1}{4} = \dfrac{1}{4}$. The subintervals will be $\left[1, 1 + \dfrac{1}{4}\right]$, $\left[\dfrac{5}{4}, \dfrac{5}{4} + \dfrac{1}{4}\right]$, $\left[\dfrac{3}{2}, \dfrac{3}{2} + \dfrac{1}{4}\right]$, and $\left[\dfrac{7}{4}, \dfrac{7}{4} + \dfrac{1}{4}\right]$, with left endpoints 1, $\dfrac{5}{4}$, $\dfrac{3}{2}$, and $\dfrac{7}{4}$, respectively. See Figure 10.13(b). Then

$$A \approx f(1) \cdot \dfrac{1}{4} + f\left(\dfrac{5}{4}\right) \cdot \dfrac{1}{4} + f\left(\dfrac{3}{2}\right) \cdot \dfrac{1}{4} + f\left(\dfrac{7}{4}\right) \cdot \dfrac{1}{4}$$
Sum of the areas of the four rectangles

$$= 1 \cdot \dfrac{1}{4} + \dfrac{5}{4} \cdot \dfrac{1}{4} + \dfrac{3}{2} \cdot \dfrac{1}{4} + \dfrac{7}{4} \cdot \dfrac{1}{4}$$
Replace $f(1)$ with 1, $f\left(\dfrac{5}{4}\right)$ with $\dfrac{5}{4}$, $f\left(\dfrac{3}{2}\right)$ with $\dfrac{3}{2}$, and $f\left(\dfrac{7}{4}\right)$ with $\dfrac{7}{4}$.

$$= \dfrac{22}{16} = 1.375$$
Simplify.

c. If we partition the interval $[1, 2]$ into six subintervals of equal length, then each interval will have length $\dfrac{2-1}{6} = \dfrac{1}{6}$. The subintervals will be

$$\left[1, 1 + \dfrac{1}{6}\right], \left[\dfrac{7}{6}, \dfrac{7}{6} + \dfrac{1}{6}\right], \left[\dfrac{4}{3}, \dfrac{4}{3} + \dfrac{1}{6}\right], \left[\dfrac{3}{2}, \dfrac{3}{2} + \dfrac{1}{6}\right], \left[\dfrac{5}{3}, \dfrac{5}{3} + \dfrac{1}{6}\right], \text{ and } \left[\dfrac{11}{6}, \dfrac{11}{6} + \dfrac{1}{6}\right]$$

with left endpoints 1, $\dfrac{7}{6}$, $\dfrac{4}{3}$, $\dfrac{3}{2}$, $\dfrac{5}{3}$, and $\dfrac{11}{6}$, respectively. See Figure 10.13(c). Then

$$A \approx f(1) \cdot \dfrac{1}{6} + f\left(\dfrac{7}{6}\right) \cdot \dfrac{1}{6} + f\left(\dfrac{4}{3}\right) \cdot \dfrac{1}{6} + f\left(\dfrac{3}{2}\right) \cdot \dfrac{1}{6} + f\left(\dfrac{5}{3}\right) \cdot \dfrac{1}{6} + f\left(\dfrac{11}{6}\right) \cdot \dfrac{1}{6}$$
Sum of the areas of the six rectangles

$$= 1 \cdot \dfrac{1}{6} + \dfrac{7}{6} \cdot \dfrac{1}{6} + \dfrac{4}{3} \cdot \dfrac{1}{6} + \dfrac{3}{2} \cdot \dfrac{1}{6} + \dfrac{5}{3} \cdot \dfrac{1}{6} + \dfrac{11}{6} \cdot \dfrac{1}{6}$$
Replace $f(1)$ with 1, $f\left(\dfrac{7}{6}\right)$ with $\dfrac{7}{6}$, $f\left(\dfrac{4}{3}\right)$ with $\dfrac{4}{3}$, $f\left(\dfrac{3}{2}\right)$ with $\dfrac{3}{2}$, $f\left(\dfrac{5}{3}\right)$ with $\dfrac{5}{3}$, and $f\left(\dfrac{11}{6}\right)$ with $\dfrac{11}{6}$.

$$= \dfrac{17}{12} \approx 1.417 \quad \text{Simplify.}$$

d. The actual area A under the graph of $f(x) = x$ from 1 to 2 can be computed as either:

i. the sum of the areas of a square of side 1 and a right triangle having two legs of length 1 $\left(\text{that is, } 1 + \dfrac{1}{2} = \dfrac{3}{2} = 1.5\right)$ or

ii. the area of a trapezoid with height $h = 1$ and bases $b_1 = 1$ and $b_2 = 2$ $\left(\text{that is, } \dfrac{1}{2} \cdot 1 \cdot (1 + 2) = \dfrac{3}{2} = 1.5\right)$. See Figure 10.13(d). The three approximations are increasingly better underestimates of the area A (which is exactly 1.5).

Practice Problem 1 Approximate the area under the graph of $f(x) = x$ from 1 to 2 by partitioning the interval $[1, 2]$ into:

a. two subintervals of equal length and choosing t to be the right endpoint.

b. four subintervals of equal length and choosing t to be the right endpoint.

c. six subintervals of equal length and choosing t to be the right endpoint.

d. Compare the approximations found in parts a–c with the actual area.

Because the graph of $f(x) = x$ is increasing on the interval $[1, 2]$, choosing t as the left endpoint gives an underestimate of the actual area (the tops of the rectangles are below the graph), whereas choosing t as the right endpoint gives an overestimate of the actual area (the tops of the rectangles are above the graph). The approximations in Table 10.8 illustrate this for several choices of the number n of subintervals used. For the graph of f a decreasing function the reverse would be true.

TABLE 10.8

n = number of subintervals	t = left endpoint	t = right endpoint
10	1.45	1.55
50	1.49	1.51
100	1.495	1.505
1000	1.4995	1.5005

EXAMPLE 2 **Approximating the Area under the Graph of $f(x)$**

Approximate the area under the graph of $f(x) = x^2$ from 0 to 4 by partitioning the interval $[0, 4]$ into:

a. four subintervals of equal length and choosing t to be the right endpoint.

b. eight subintervals of equal length and choosing t to be the right endpoint.

Solution

Let A be the area of a region below the graph of $y = x^2$, above the x-axis, and between the vertical lines $x = 0$ and $x = 4$.

a. If we partition the interval $[0, 4]$ into four subintervals of equal length, then each interval will have length $\dfrac{4 - 0}{4} = 1$. See Figure 10.14(a). The subintervals will be:

$[0, 1]$ $[1, 2]$ $[2, 3]$ $[3, 4]$.

Choosing right endpoints 1, 2, 3, 4, we have:

$A \approx f(1) \cdot 1 + f(2) \cdot 1 + f(3) \cdot 1 + f(4) \cdot 1$ Sum of the areas of the four rectangles

$= 1 \cdot 1 + 4 \cdot 1 + 9 \cdot 1 + 16 \cdot 1$ Since $f(x) = x^2$, replace $f(1)$ with 1, $f(2)$ with 4, $f(3)$ with 9, and $f(4)$ with 16.

$= 30$ Simplify.

b. If we partition the interval $[0, 4]$ into eight subintervals of equal length, then each interval will have length $\dfrac{4 - 0}{8} = 0.5$. See Figure 10.14(b). The subintervals will be:

$[0, 0.5]$ $[0.5, 1]$ $[1, 1.5]$ $[1.5, 2]$ $[2, 2.5]$ $[2.5, 3]$ $[3, 3.5]$ $[3.5, 4]$.

Choosing right endpoints 0.5, 1, 1.5, 2, 2.5, 3, 3.5, 4, we have:

$A \approx f(0.5) \cdot 0.5 + f(1) \cdot 0.5 + f(1.5) \cdot 0.5 + f(2) \cdot 0.5 +$ Sum of the areas of the
$f(2.5) \cdot 0.5 + f(3) \cdot 0.5 + f(3.5) \cdot 0.5 + f(4) \cdot 0.5$ eight rectangles

$= 0.25 \cdot 0.5 + 1 \cdot 0.5 + 2.25 \cdot 0.5 + 4 \cdot 0.5 +$ Since $f(x) = x^2$, replace $f(0.5)$
$6.25 \cdot 0.5 + 9 \cdot 0.5 + 12.25 \cdot 0.5 + 16 \cdot 0.5$ with 0.25, $f(1)$ with 1, $f(1.5)$ with 2.25, $f(2)$ with 4, $f(2.5)$ with 6.25, $f(3)$ with 9, $f(3.5)$ with 12.25, and $f(4)$ with 16.

$= 25.5$ Simplify.

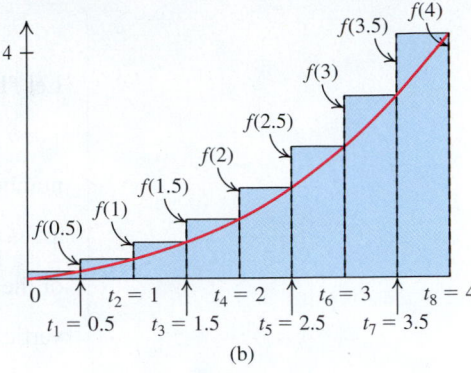

Figure 10.14

Practice Problem 2 Approximate the area under the graph of $f(x) = x^2$ from 0 to 4 by partitioning the interval $[0, 4]$ into:

a. four subintervals of equal length and choosing t to be the left endpoint.

b. eight subintervals of equal length and choosing t to be the left endpoint.

As noted earlier, when the graph is increasing on an interval, choosing t as the right endpoint gives an overestimate of the actual area (the tops of the rectangles are above the graph), whereas choosing t as the left endpoint gives an underestimate of the actual area (the tops of the rectangles are below the graph). The approximations in Table 10.9 illustrate this for $f(x) = x^2$ from 0 to 4, with several choices of the number n of subintervals used.

TABLE 10.9

n = number of subintervals	t = left endpoint	t = right endpoint
10	18.24	24.64
50	20.6976	21.9776
100	21.0144	21.6544
1000	21.3013	21.3653

There is no simple way to calculate the actual area under the graph of $f(x) = x^2$ from 0 to 4, but it turns out that this area is $21\frac{1}{3}$. Notice that the average of the left endpoint and right endpoint approximations in Table 10.7 with 1000 subintervals is 21.3333.

2 Define the integral of a function.

Definition of Integral

The preceding examples suggest that as we increase the number n of subintervals of equal length that we use, we get increasingly better approximations of the area under the graph. In fact, if we let the number n increase without bound, we obtain the exact area under the graph of f from a to b.

AREA UNDER THE GRAPH OF A FUNCTION

Let f be a continuous function defined on an interval $[a, b]$, with $f(x) \geq 0$ on $[a, b]$. Partition $[a, b]$ into n subintervals, each of length $\Delta x = \dfrac{b-a}{n}$. Choose a number t_i for $i = 1, 2, 3 \dots n$ in each subinterval. The numbers $f(t_i) \cdot \Delta x$ are the areas of approximating rectangles, and the sum $\sum\limits_{i=1}^{n} f(t_i) \cdot \Delta x$ approximates the area of the region R below the graph of $y = f(x)$, above the x-axis, and between the vertical lines $x = a$ and $x = b$. If the limit $\lim\limits_{n\to\infty} \sum\limits_{i=1}^{n} f(t_i) \cdot \Delta x$ exists, it is defined to be the area of the region R. This limit is denoted by the symbol $\displaystyle\int_a^b f(x)\,dx$, read "the integral from a to b of $f(x)$." That is, $\lim\limits_{n\to\infty} \sum\limits_{i=1}^{n} f(t_i) \cdot \Delta x = \displaystyle\int_a^b f(x)\,dx.$

To see how this limiting process works, consider the known area under the graph of $f(x) = x$ from 1 to 2 examined in Example 1. First, we partition the interval $[1, 2]$ into n subintervals, each of length $\Delta x = \dfrac{2-1}{n} = \dfrac{1}{n}$. We then pick t_i as the left endpoint of each subinterval. Adding the length $\dfrac{1}{n}$ repeatedly to the first left endpoint 1, we have endpoints:

$$t_1 = 1, \quad t_2 = 1 + \frac{1}{n}, \quad t_3 = 1 + \frac{2}{n}, \quad t_4 = 1 + \frac{3}{n} \dots, t_n = 1 + \frac{n-1}{n}.$$

The inscribed rectangles have areas:

$f(t_1) \cdot \Delta x = t_1 \cdot \Delta x = 1 \cdot \dfrac{1}{n}$ Replace $f(t_1)$ by t_1, then t_1 by 1.

$f(t_2) \cdot \Delta x = t_2 \cdot \Delta x = \left(1 + \dfrac{1}{n}\right) \cdot \dfrac{1}{n}$ Replace $f(t_2)$ by t_2, then t_2 by $1 + \dfrac{1}{n}$

$f(t_3) \cdot \Delta x = t_3 \cdot \Delta x = \left(1 + \dfrac{2}{n}\right) \cdot \dfrac{1}{n}$ Replace $f(t_3)$ by t_3, then t_3 by $1 + \dfrac{2}{n}$

$\vdots$

$f(t_n) \cdot \Delta x = t_n \cdot \Delta x = \left(1 + \dfrac{n-1}{n}\right) \cdot \dfrac{1}{n}$ Replace $f(t_n)$ by t_n, then t_n by $1 + \dfrac{n-1}{n}$

Their sum is

$$\sum_{i=1}^{n} f(t_i) \cdot \Delta x = \sum_{i=1}^{n}\left(\left(1 + \frac{i-1}{n}\right)\cdot\frac{1}{n}\right) = \frac{1}{n}\cdot\sum_{i=1}^{n}\left(1 + \frac{i-1}{n}\right)$$

Replace $f(t_i) \cdot \Delta x$ by $\left(1 + \dfrac{i-1}{n}\right)\cdot\dfrac{1}{n}$, factor out $\dfrac{1}{n}$.

$$= \frac{1}{n}\cdot\left(1 + \left(1+\frac{1}{n}\right) + \left(1+\frac{2}{n}\right) + \left(1+\frac{3}{n}\right) + \dots + \left(1+\frac{n-1}{n}\right)\right)$$

Expand the sum.

$$= \frac{1}{n}\cdot\left(n + \frac{1}{n} + \frac{2}{n} + \frac{3}{n} + \dots + \frac{(n-1)}{n}\right)$$

Rearrange terms.

$$= 1 + \frac{1}{n^2}\cdot(1 + 2 + 3 + \dots + (n-1))$$

$\dfrac{1}{n}\cdot n = 1$, Factor out $\dfrac{1}{n}$.

$$= 1 + \frac{1}{n^2} \cdot \frac{(n-1)\cdot n}{2}$$ Replace $1 + 2 + 3 + \ldots + (n-1)$
with $\dfrac{(n-1)\cdot n}{2}$ (see page 827)

$$= \frac{3}{2} - \frac{1}{2n}$$ Simplify.

The area under the graph of $f(x) = x$ from 1 to 2 is:

$$\int_1^2 x \, dx = \lim_{n\to\infty} \sum_{i=1}^{n} f(t_i)\cdot \Delta x$$

$$= \lim_{n\to\infty}\left(\frac{3}{2} - \frac{1}{2n}\right) = \frac{3}{2} - \lim_{n\to\infty}\frac{1}{2n} = \frac{3}{2}.$$ Recall $\lim_{n\to\infty}\dfrac{1}{2n} = 0$.

This agrees with our result in Example 1. In integral notation, $\displaystyle\int_1^2 x\,dx = \frac{3}{2}$.

3 Approximate an integral using a graphing utility.

Evaluating Integrals Using a Graphing Utility

EXAMPLE 3   **Approximating an Integral Using a Graphing Utility**

Use a graphing utility to approximate $\displaystyle\int_0^4 x^2 \, dx$, that is, the area under the graph of $f(x) = x^2$ from 0 to 4. This is the integral approximated in Example 2.

Solution

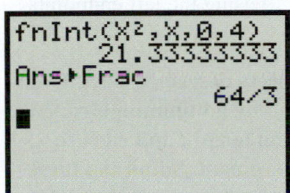

Figure 10.15

We show the result of using the integration option on a TI 84 Plus in Figure 10.15. Since the keystrokes will vary for different graphing utilities, you will need to consult the owner's manual. This is the integral that was approximated in Example 2. Here the graphing utility gave the exact value of this integral $\dfrac{64}{3}$, which is unusual. The graphing utility will not give an exact value in Practice Problem 3 where value of the integral is not a rational number.

Practice Problem 3 Use a graphing utility to approximate $\displaystyle\int_0^2 \sqrt{x}\,dx$, that is, the area under the graph of $f(x) = \sqrt{x}$ from 0 to 2. Round the result to four decimal places.

Answers to Practice Problems

1. a. 1.75 **b.** 1.625 **c.** 1.583 **d.** All three approximations are increasingly better overestimates of the area (which is exactly 1.5).

2. a. 14 **b.** 17.5 **3.** 1.8856

Exercises

Concepts and Vocabulary

1. Approximations using the left endpoints of the subintervals for a decreasing function result in a/an _____ of the area under the graph.

2. Approximations using the _____ endpoints of the subintervals for an increasing function result in an underestimate of the area under the graph.

3. In the approximation $\displaystyle\sum_{i=1}^{n} f(t_i)\cdot \Delta x$, the length of each subinterval is denoted by _____.

4. $\displaystyle\int_0^3 x^3 \, dx$ calculates the area under the graph of $f(x) = $ _____ over the interval _____.

5. **True or False.** Increasing the number of equal-length subintervals in a left endpoint approximation results in an equal or better approximation of the area under the graph.

6. **True or False.** If the interval $[a, b]$ is divided into 25 subintervals of equal length, the 12th left endpoint is $a + \dfrac{12}{25}$.

Building Skills

In Exercises 7 and 8, approximate the shaded area under the graph of a function f by using the given rectangles. In each case, the values of f for the subinterval endpoints are given.

7.

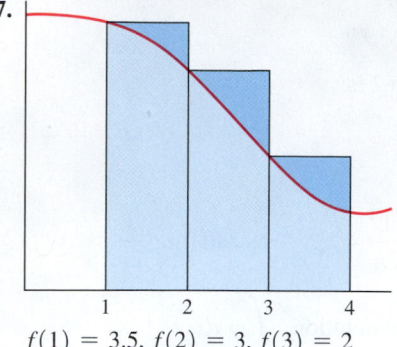

$f(1) = 3.5, f(2) = 3, f(3) = 2$

8.

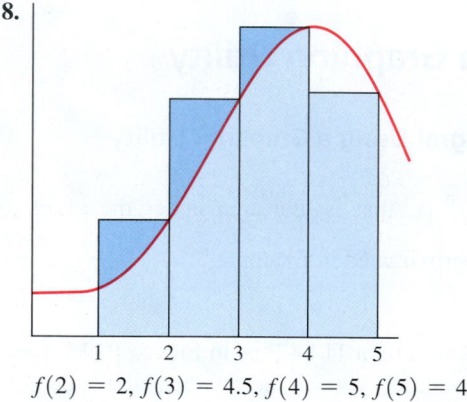

$f(2) = 2, f(3) = 4.5, f(4) = 5, f(5) = 4.5$

In Exercises 9–20, round your noninteger answers to three decimal places.

9. Approximate the area under the graph of $f(x) = 2x + 1$ from 0 to 2 by partitioning the interval $[0, 2]$ into:
   a. two subintervals of equal length and choosing t to be the right endpoint.
   b. four subintervals of equal length and choosing t to be the right endpoint.
   c. six subintervals of equal length and choosing t to be the right endpoint.
   d. Use geometry to find the actual area.

10. Repeat parts a–c of Exercise 9 using the left endpoints of the subintervals.

11. Approximate the area under the graph of $f(x) = 3 - x$ from 0 to 3 by partitioning the interval into:
   a. two subintervals of equal length and choosing t to be the left endpoint.
   b. four subintervals of equal length and choosing t to be the left endpoint.
   c. six subintervals of equal length and choosing t to be the left endpoint.
   d. Use geometry to find the actual area.

12. Repeat parts a–c of Exercise 11 using the right endpoints of the subintervals.

13. Approximate the area under the graph of $f(x) = 2x^2 + 1$ from 0 to 6 by partitioning the interval into:
   a. three subintervals of equal length and choosing t to be the left endpoint.
   b. six subintervals of equal length and choosing t to be the left endpoint.

14. Repeat parts a and b of Exercise 13 using the right endpoints of the subintervals

15. Approximate the area under the graph of $f(x) = 10 - x^2$ from 0 to 3 by partitioning the interval into:
   a. three subintervals of equal length and choosing t to be the right endpoint.
   b. six subintervals of equal length and choosing t to be the right endpoint.

16. Repeat parts a and b of Exercise 15 using the left endpoints of the subintervals.

17. Approximate the area under the graph of $f(x) = \dfrac{1}{x}$ from 1 to 5 by partitioning the interval into:
   a. four subintervals of equal length and choosing t to be the left endpoint.
   b. six subintervals of equal length and choosing t to be the left endpoint.

18. Repeat parts a and b of Exercise 17 using the right endpoints of the subintervals.

19. Approximate the area under the graph of $f(x) = \sqrt{x}$ from 0 to 4 by partitioning the interval into:
   a. four subintervals of equal length and choosing t to be the right endpoint.
   b. six subintervals of equal length and choosing t to be the right endpoint.

20. Repeat parts a and b of Exercise 19 using the left endpoints of the subintervals

Exercises 21 and 22, rely on the process described next. Even an erratic path, such as that traveled by a hummingbird, can be divided into short sections of equal length, and each section can be approximated by a line segment. Summing these line segments approximates the length of the path, and taking the limit of these approximations as the number of approximating line segments increases without bound, yields the length of the path, or the distance traveled. We use v, instead of f, for the velocity function and t, instead of x, to represent time. Because "velocity × time = distance" on each approximating line segment, the integral of v over the interval of time traveled gives the distance traveled.

21. Suppose the velocity of a moving object at time t is $v(t) = 3t$, where t is in seconds and velocity is in meters/second. The integral of v over the interval of time traveled gives the distance traveled.
   a. Write the integral that gives the distance traveled by the object during the time interval $I = [0, 1]$.
   b. Use geometry to evaluate the integral identified in part a.

22. Suppose the velocity of a moving object at time t is $v(t) = \sin(t)$, where t is in seconds and velocity is in meters/second. The integral of v over the interval of time traveled gives the distance traveled.
   a. Write the integral that gives the distance traveled by the object during the time interval $I = [0, \pi]$.
   b. Approximate the integral that gives the distance traveled by the object during the time interval $I = [0, \pi]$ by partitioning the interval into 12 subintervals of equal length and choosing t to be the right endpoint (round to three decimal places).

In Exercises 23–28, a function f is defined over an interval $[a, b]$.

a. Graph f, shading the area A under the graph of f from a to b.

b. Approximate the area A by partitioning $[a, b]$ into four subintervals of equal length and choosing t to be the left endpoint.

c. Approximate the area A by partitioning $[a, b]$ into four subintervals of equal length and choosing t to be the right endpoint.

d. Express the area A as an integral.

 e. Use a graphing utility to approximate the area A (round your answer to three decimal places).

23. $f(x) = 2x^2$, $[0, 2]$ 24. $f(x) = x^2 + 3$, $[1, 5]$

25. $f(x) = x^3 + 1$, $[0, 4]$ 26. $f(x) = \dfrac{x^3}{125}$, $[1, 5]$

27. $f(x) = \ln(x)$, $[1, 5]$ 28. $f(x) = e^x$, $[-2, 2]$

In Exercises 29–32, an integral is given.

a. Specify the area, A, that this integral represents by identifying the function and interval.

b. Graph the function identified in part a, shading the area A.

 c. Use a graphing utility to approximate the area A (round your noninteger answers to three decimal places).

29. $\displaystyle\int_{0.5}^{2} \dfrac{1}{x}\,dx$ 30. $\displaystyle\int_{1}^{3} (x^2 + x)\,dx$

31. $\displaystyle\int_{0}^{\pi} \sin(x)\,dx$ 32. $\displaystyle\int_{0}^{\frac{\pi}{2}} \cos(x)\,dx$

Beyond the Basics

Review the definition of area and the discussion following it before beginning Exercises 33 and 34. Use the limiting process illustrated there and the formula $\displaystyle\sum_{i=1}^{n} i = \dfrac{n \cdot (n+1)}{2}$.

33. a. Identify the area below the graph of $f(x) = 2x + 1$ over $[0, 3]$ and compute this area using geometry.

b. Show that the definition $\displaystyle\int_{a}^{b} f(x)\,dx = \lim_{n \to \infty} \sum_{i=1}^{n} f(t_i) \cdot \Delta x$ gives the same value for the area found in part a.

34. a. Identify the area below the graph of $f(x) = 4 - x$ over $[0, 4]$ and compute this area using geometry.

b. Show that the definition $\displaystyle\int_{a}^{b} f(x)\,dx = \lim_{n \to \infty} \sum_{i=1}^{n} f(t_i) \cdot \Delta x$ gives the same value for the area found in part a.

Critical Thinking / Discussion / Writing

35. The area below the graph of $f(x) = \sqrt{1 - x^2}$, domain $[-1, 1]$, is the upper half of the unit circle, so

$$\int_{-1}^{1} \sqrt{1 - x^2}\,dx = \frac{1}{2}(\text{area of the unit cirle}) = \frac{\pi}{2} \approx 1.57$$

(when rounded to two decimal places). Into how many subintervals must $[-1, 1]$ be partitioned to get an approximation of 1.52 (when rounded to two decimal places) for the area under the graph of $f(x) = \sqrt{1 - x^2}$ from $x = -1$ to $x = 1$ using the right endpoints of the subintervals?

SUMMARY Definitions, Concepts, and Formulas

10.1 Finding Limits Using Tables and Graphs

i. **Definition.** The notation $\lim_{x \to c} f(x) = L$ is read "the limit of $f(x)$ as x approaches c, equals L." This means that as x approaches c, the corresponding values of $f(x)$ approach the number L.

ii. To find the limit L:

(a) We use tables of values of x close to c and examine the corresponding values of $f(x)$. See Example 1 (page 867).

(b) We graph the function $y = f(x)$ and examine the height of the graph for values of x close to c.

iii. If $\lim_{x \to c^+} f(x) \neq \lim_{x \to c^-} f(x)$, then $\lim_{x \to c} f(x)$ does not exist.

10.2 Finding Limits Algebraically

i. Thirteen properties of limits are listed on pages 876–878.

ii. The properties of limits are used to find limits.

iii. The properties of limits are also valid for one-sided limits.

iv. The quotient property: $\lim_{x \to c} \dfrac{f(x)}{g(x)} = \dfrac{\lim_{x \to c} f(x)}{\lim_{x \to c} g(x)}$ cannot be used if $\lim_{x \to c} g(x) = 0$.

To find the limit (if it exists) in such cases, it may be helpful to rewrite the expression by (a) factoring or (b) rationalizing the numerator or the denominator.

10.3 Infinite Limits and Limits at Infinity

i. **Infinite Limits.** The symbols ∞ (or $+\infty$) and $-\infty$ are used to describe the behavior of $f(x)$ near $x = c$. Situations such as $\lim_{x \to c^-} f(x) = \pm\infty$ or $\lim_{x \to c^+} f(x) = \pm\infty$ are called infinite limits. We obtain infinite limits when the vertical line $x = c$ is a vertical asymptote to the graph of $y = f(x)$.

ii. **Limits at Infinity.** Limits at infinity describe the behavior of $f(x)$ as x approaches $\pm\infty$.

(a) Properties of limits at infinity are given on page 888.

(b) Algebraic manipulations, including the distributive property, factoring, and rationalizing the numerator or the denominator, can be helpful in evaluating limits at infinity.

10.4 Introduction to Derivatives

i. **Instantaneous Rate of Change** of a function f at a (denoted by $IRC[f]_a$) is given by

$$IRC[f]_a = \lim_{h \to 0} \frac{f(a + h) - f(a)}{h}$$

provided this limit exists.

ii. The Slope of the Tangent Line to the graph of the function $y = f(x)$ at the point $P = (a, f(a))$ is given by

$$m_{\tan} = \lim_{\Delta x \to 0} \frac{\Delta y}{\Delta x} = \lim_{h \to 0} \frac{f(a + h) - f(a)}{h}$$

provided this limit exists.

The equation of the tangent line to the graph of the function $y = f(x)$ at the point $P = (a, f(a))$ is:

$$y = m_{\tan}(x - a) + f(a)$$

iii. The derivative of f at $x = a$, denoted by $f'(a)$, is defined by:

$$f'(a) = \lim_{h \to 0} \frac{f(a + h) - f(a)}{h}$$

provided this limit exists.

The number $f'(a)$ describes:

a. the slope of the tangent line to the graph of f at $(a, f(a))$.

b. the instantaneous rate of change of f at $x = a$.

iv. The derivative function of $f(x)$, denoted by $f'(x)$, is defined by:

$$f'(x) = \lim_{h \to 0} \frac{f(x + h) - f(x)}{h}$$

provided this limit exists. The derivative function $f'(x)$ assigns the value $f'(a)$ to any point a in the domain of f, provided this limit exists.

v. Velocity. If a function $s(t)$ gives the position of an object at time t, then the derivative $s'(t)$ gives the instantaneous velocity of the object at time t.

10.5 Area and the Integral

Integral. Let f be a continuous function defined on the interval $[a, b]$, with $f(x) \geq 0$ on $[a, b]$. Partition $[a, b]$ into n subintervals, each of length $\Delta x = \dfrac{b - a}{n}$. Choose a number t_i for $i = 1, 2, 3 \ldots n$, in each subinterval. The numbers $f(t_i) \cdot \Delta x$ are the areas of approximating rectangles, and the sum

$\displaystyle\sum_{i=1}^{n} f(t_i) \cdot \Delta x$ approximates the area of the region R between the graph of f, the x-axis, and the lines $x = a$ and $x = b$.

If the limit $\displaystyle\lim_{n \to \infty} \sum_{i=1}^{n} f(t_i) \cdot \Delta x$ exists, it is defined to be the area of the region R and is denoted by the symbol $\displaystyle\int_a^b f(x)\, dx$ (read "the integral from a to b of $f(x)$").

That is, $\displaystyle\lim_{n \to \infty} \sum_{i=1}^{n} f(t_i) \cdot \Delta x = \int_a^b f(x)\, dx.$

REVIEW EXERCISES

In Exercises 1–4, construct a table to find the limit if it exists.

1. $\displaystyle\lim_{x \to 1} (2x^2 + 5)$

2. $\displaystyle\lim_{x \to 1} \frac{x^2 + x - 6}{x - 2}$

3. $\displaystyle\lim_{t \to 0} \frac{\sqrt{t + 1} - 1}{t}$

4. $f(t) = \begin{cases} t + 1, & \text{if } x < 2 \\ t^2 - 1, & \text{if } x > 2 \end{cases}; \displaystyle\lim_{x \to 2} f(t)$

In Exercises 5–20, use the graph of the function f to state the value of each limit if it exists.

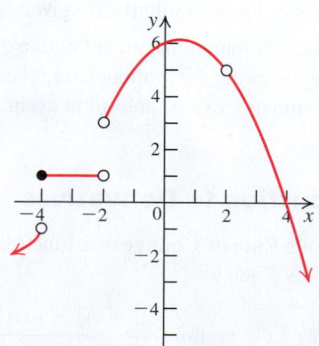

5. $\displaystyle\lim_{x \to -4^-} f(x)$

6. $\displaystyle\lim_{x \to -4^+} f(x)$

7. $\displaystyle\lim_{x \to -4} f(x)$

8. $f(-4)$

9. $\displaystyle\lim_{x \to -2^-} f(x)$

10. $\displaystyle\lim_{x \to -2^+} f(x)$

11. $\displaystyle\lim_{x \to -2} f(x)$

12. $f(-2)$

13. $\displaystyle\lim_{x \to 2^-} f(x)$

14. $\displaystyle\lim_{x \to 2^+} f(x)$

15. $\displaystyle\lim_{x \to 2} f(x)$

16. $f(2)$

17. $\displaystyle\lim_{x \to 4^-} f(x)$

18. $\displaystyle\lim_{x \to 4^+} f(x)$

19. $\displaystyle\lim_{x \to 4} f(x)$

20. $f(4)$

In Exercises 21–24, find each limit, or state that it doesn't exist.

21. $f(x) = \dfrac{x^2 - 25}{x - 5}; \displaystyle\lim_{x \to 5} f(x)$

22. $f(x) = |x - 2|; \displaystyle\lim_{x \to 2} f(x)$

23. $f(x) = [\![x]\!]; \displaystyle\lim_{x \to 1} f(x)$

24. $f(x) = \begin{cases} \sin(x), & \text{if } x > 0 \\ x, & \text{if } x < 0 \end{cases}; \displaystyle\lim_{x \to 0} f(x)$

In Exercises 25–42, use the properties of limits to find each limit, or state that it doesn't exist.

25. $\displaystyle\lim_{x \to 3} (3x^2 - 4x + 7)$

26. $\displaystyle\lim_{x \to -2} (-2x^3 + 5x + 18)$

27. $\displaystyle\lim_{x \to 1} (2x^2 + 1)^3$

28. $\displaystyle\lim_{x \to -1} (3x^2 - 1)^2$

29. $\displaystyle\lim_{x \to 3} \sqrt{25 - x^2}$

30. $\displaystyle\lim_{x \to 4} \sqrt[3]{2x^2 - 5}$

31. $\displaystyle\lim_{x \to 0} \frac{4x}{2x + 3}$

32. $\displaystyle\lim_{x \to 3} \frac{x^2 - 4}{x + 2}$

33. $\lim\limits_{x \to 2} \dfrac{x^2 - 2x}{x - 2}$

34. $\lim\limits_{x \to -2} \dfrac{x^3 + 8}{x + 2}$

35. $\lim\limits_{x \to 4} \dfrac{\sqrt{x} - 2}{x - 4}$

36. $\lim\limits_{x \to 9} \dfrac{x - 9}{\sqrt{x} - 3}$

37. $\lim\limits_{x \to 0} \dfrac{\frac{1}{x + 1} - 1}{x}$

38. $\lim\limits_{x \to 0} \dfrac{\frac{1}{\sqrt{1 + x}} - 1}{x}$

39. $\lim\limits_{x \to 1} f(x); f(x) = \begin{cases} (x^2 + 1)^3, & x < 1 \\ 5x + 3, & x > 1 \end{cases}$

40. $\lim\limits_{x \to 2} f(x); f(x) = \begin{cases} (2x + 1)^2, & x < 2 \\ 5x + 3, & x \geq 2 \end{cases}$

41. $\lim\limits_{x \to 1} \dfrac{x - 1}{|x - 1|}$

42. $\lim\limits_{x \to 2} \dfrac{[\![x]\!]}{x}$

In Exercises 43–58, find each limit and write $-\infty$ or ∞ where appropriate.

43. $\lim\limits_{x \to 3^-} \dfrac{x}{x - 3}$

44. $\lim\limits_{x \to 4^-} \dfrac{x + 1}{4 - x}$

45. $\lim\limits_{x \to 2^+} \dfrac{x + 3}{x^2 + x - 6}$

46. $\lim\limits_{x \to 3^+} \dfrac{x + 3}{9 - x^2}$

47. $\lim\limits_{x \to \infty} (2x^3 - 5x + 7)$

48. $\lim\limits_{x \to \infty} (-3x^3 + 5x + 19)$

49. $\lim\limits_{x \to -\infty} (5x^2 + 81x + 785)$

50. $\lim\limits_{x \to -\infty} (2x^3 + 11x^2 + 83)$

51. $\lim\limits_{x \to \infty} \dfrac{x^2 - 2x + 5}{2x^2 + 17x + 11}$

52. $\lim\limits_{x \to \infty} \dfrac{x^2 - 1}{x^2 - 12}$

53. $\lim\limits_{x \to -\infty} \dfrac{2x + 3}{x^2 - 5x + 7}$

54. $\lim\limits_{x \to \infty} \dfrac{2x^2 + 3x + 11}{2x + 7}$

55. $\lim\limits_{x \to \infty} (\sqrt{x + 1} - \sqrt{x})$

56. $\lim\limits_{x \to \infty} (\sqrt{x^2 + x + 1} - \sqrt{x^2 + 1})$

57. $\lim\limits_{x \to -\infty} (\sqrt{x^2 + 2x} - \sqrt{x^2 + 1})$

58. $\lim\limits_{x \to -\infty} (\sqrt{x^2 + x} + x)$

In Exercises 59–66, find: (a) the slope of the tangent line to the graph of f at the given point. (b) the slope–intercept form of the equation of the tangent line to the graph of f at the given point.

59. $f(x) = 2x$ at $(3, 6)$

60. $f(x) = -3x$ at $(2, -6)$

61. $f(x) = x^2 + 5x$ at $(1, 6)$

62. $f(x) = -2x^2 + 4x$ at $(1, 2)$

63. $f(x) = 2x^2 - 4x + 1$ at $(-1, 7)$

64. $f(x) = -3x^2 - x + 1$ at $(-2, -9)$

65. $f(x) = \sqrt{x}$ at $(4, 2)$

66. $f(x) = \dfrac{1}{x}$ at $\left(2, \dfrac{1}{2}\right)$

In Exercises 67–70, approximate $\int_a^b f(x)\, dx$ by partitioning the interval $[a, b]$ into six subintervals of equal length and choosing t to be the right endpoint.

67. $\int_0^1 (1 + x)\, dx$

68. $\int_0^3 \dfrac{1}{2} x\, dx$

69. $\int_0^3 x^2\, dx$

70. $\int_1^9 \dfrac{1}{x}\, dx$

71. The distance in meters traveled by an object traveling in a straight line in t seconds is given by $s(t) = t^2 + 5t + 4$. Find the velocity of the object at the end of 2 seconds.

72. The position of a particle moving on the x-axis is given by $s(t) = 2t^2 - 4t + 5$, where t is in seconds and s is in feet. Find:
   a. the initial position of the particle.
   b. the average velocity of the particle between $t = 2$ and $t = 4$.
   c. the instantaneous velocity at $t = 2$ and $t = 4$.

73. The height of a ball thrown upward from the ground is given by $s(t) = -16t^2 + 128t$, where s is measured in feet and t is measured in seconds. Find:
   a. the average velocity of the ball over the first 2 seconds.
   b. the instantaneous velocity at $t = 0$ and $t = 2$.

74. The height of a rock thrown upward from the roof of a building is given by $s(t) = -16t^2 + 144t + 160$. Find:
   a. the initial velocity of the rock. [*Hint*: Find $s'(0)$.]
   b. the speed at which the rock hits the ground.
   [First find t_0 when $s(t_0) = 0$, then find $s'(t_0)$.]

PRACTICE TEST A

In Problems 1–5, find each indicated limit or state that it does not exist.

1. $f(x) = |2 - x|;\quad \lim\limits_{x \to -3} f(x)$

2. $f(x) = \dfrac{x^2 - 9}{x - 3};\quad \lim\limits_{x \to 3} f(x)$

3. $h(x) = \begin{cases} 1 - \ln x & \text{if } x < 1 \\ 2 - x, & \text{if } x \geq 1 \end{cases};\quad \lim\limits_{x \to 1} h(x)$

4. $\lim\limits_{x \to -2} \dfrac{1 - 4x}{x^3 - x + 6}$

5. $\lim\limits_{x \to 2} \dfrac{x^2 - 4}{\sqrt{3x - 2} - \sqrt{x + 2}}$

6. Form the difference quotient $f(h) = \dfrac{f(c + h) - f(c)}{h}$ and then find $\lim\limits_{h \to 0} f(h)$ if $f(x) = 3x^2, c = -1$.

In Problems 7–12, find each indicated limit or state that it does not exist. Use $-\infty$ and ∞ where appropriate.

7. $\lim\limits_{x \to 4^-} \dfrac{5}{12 - 3x}$

8. $\lim\limits_{x \to 3^+} \dfrac{2x}{3 - x}$

9. $\lim\limits_{x \to \infty} \dfrac{6x^2 - 5x + 1}{8 - 3x - 6x^2}$

10. $\lim\limits_{x \to -\infty} (\sqrt{4 - x} - \sqrt{-x})$

In Problems 11 and 12, find the derivative of f at the given point.

11. $f(x) = 1 - 2x^2$ at $x = 2$

12. $\dfrac{1}{1 - 3x}$ at $x = 1$

In Problems 13 and 14, find the slope–intercept equation for the tangent line to the graph of f at the given point.

13. $f(x) = 2x^2 + 3$ at $(-1, 5)$

14. $f(x) = -\sqrt{x}$ at $(4, -2)$

15. Find the derivative function of $f(x) = 6x^2 - 4x + 9$.

16. A hammer is dropped from the roof of the building 32 feet above the ground. The height of the hammer (in feet) after t seconds is given by the function $s(t) = 32 - 16t^2$:
   a. How long does it take the ball to hit the ground?
   b. Find the instantaneous speed of the ball when it hits the ground.

In Problems 17–18, a function f is defined over an interval $[a, b]$. Approximate the area A by partitioning $[a, b]$ into four subintervals of equal length and choosing t to be the given endpoint.

17. $f(x) = x^3$, $[0, 4]$, left endpoint
18. $f(x) = 25 - x^2$, $[1, 5]$, right endpoint

In Problems 19–20, an integral is given. Specify the area, A, that this integral represents by identifying the function and interval.

19. $\int_{-1}^{5}(x^2 + x)\, dx$

20. $\int_{-6}^{0}e^x\, dx$

PRACTICE TEST B

In Problems 1–5, find each indicated limit or state that it does not exist.

1. $f(x) = -|x - 3|$; $\lim_{x \to -3} f(x)$
   a. 6 **b.** -6
   c. 0 **d.** Does not exist

2. $f(x) = \dfrac{25 - x^2}{x - 5}$; $\lim_{x \to 5} f(x)$
   a. -10 **b.** 10
   c. 5 **d.** -5

3. $\lim_{x \to \infty} f(x); f(x) = \begin{cases} e^x & \text{if } x < 2 \\ 8 - x & \text{if } x \geq 2 \end{cases}$
   a. e^2 **b.** 6
   c. Does not exist **d.** 8

4. $\lim_{x \to -2} \dfrac{1 - 3x}{x^3 + 1}$
   a. -1 **b.** 1
   c. $\dfrac{5}{7}$ **d.** $-\dfrac{5}{7}$

5. $\lim_{x \to 10} \sqrt[3]{6x + 4}$
   a. Does not exist **b.** 8
   c. 6 **d.** 4

6. **Form the difference quotient** $f(h) = \dfrac{f(c + h) - f(c)}{h}$
   and then find $\lim_{h \to 0} f(h)$ **for c if** $f(x) = -x^2, c = 2$
   a. 4 **b.** -4
   c. 2 **d.** -2

In Problems 7–10, find each indicated limit or state that it does not exist. Use $-\infty$ and ∞ where appropriate.

7. $\lim_{x \to 9^+} \dfrac{x - 2}{x - 9}$
   a. ∞ **b.** $-\infty$
   c. 1 **d.** $\dfrac{2}{9}$

8. $\lim_{x \to 3^+} \dfrac{3x^2 - 2x - 5}{3 - x}$
   a. ∞ **b.** 3
   c. 1 **d.** $-\infty$

9. $\lim_{x \to -\infty} \dfrac{4x^2 - 2x + 1}{3x^2 + 5x + 12}$
   a. ∞ **b.** $-\infty$
   c. $\dfrac{4}{3}$ **d.** $\dfrac{3}{4}$

10. $\lim_{x \to \infty} \left(\sqrt{x - 5} - \sqrt{x} \right)$
   a. ∞ **b.** $-\infty$
   c. 0 **d.** Does not exist

In Problems 11 and 12, find the derivative of f at the given point.

11. $f(x) = 1 - x^3$ at $x = 3$
   a. -27 **b.** -26
   c. 9 **d.** -9

12. $f(x) = -\sqrt{x}$ at $x = 4$
   a. 2 **b.** -2
   c. $\dfrac{1}{4}$ **d.** $-\dfrac{1}{4}$

In Problems 13 and 14, find the slope–intercept form of the equation of the tangent line to the graph of f at the given point.

13. $f(x) = x^2 + 2x - 15$ at $(-2, -15)$
   a. $y = 6x - 19$ **b.** $y = -2x - 19$
   c. $y = 2x - 19$ **d.** $y = -2x + 11$

14. $f(x) = 3 - x^3$ at $(1, 2)$
   a. $y = -3x + 5$ **b.** $y = 3x + 5$
   c. $y = -3x + 1$ **d.** $y = -2x + 4$

15. Find the derivative function of the function $f(x) = x^2 - 4x + 9$.
   a. $f'(x) = 2x$ **b.** $f'(x) = 2x - 4$
   c. $f'(x) = x - 4$ **d.** $f'(x) = 9$

16. The distance in meters traveled by an object traveling in a straight line in t seconds is given by $s(t) = t^2 + 3t + 8$. Find the velocity of the object at the end of 3 seconds.
   a. 6 m/s **b.** 15 m/s **c.** 9 m/s **d.** 8 m/s

In Problems 17–18, a function f is defined over an interval $[a, b]$. Approximate the area A by partitioning $[a, b]$ into four subintervals of equal length and choosing t to be the given endpoint.

17. $f(x) = 1 + x^3$, $[0, 4]$, left endpoint
   a. 104 **b.** 40 **c.** 100 **d.** 36

18. $f(x) = 2x^2$, $[1, 5]$, right endpoint
   a. 108 **b.** 60 **c.** 100 **d.** 56

In Problems 19–20, an integral is given. Specify the area, A, that this integral represents by identifying the function and interval.

19. $\int_{1}^{4} 2x^2 + 3\, dx$
   a. $f(x) = 2x$, $[1, 4]$ **b.** $f(x) = 2x^2 + 3$, $[1, 4]$
   c. $f(x) = 2x^2 + 3$, $[0, 4]$ **d.** $f(x) = 2x^2$, $[1, 4]$

20. $\int_{.05}^{3.6} \ln(x)\, dx$
   a. $f(x) = \ln(x)$, $[0.05, 4]$ **b.** $f(x) = \ln(x)$, $[1, 4]$
   c. $f(x) = \ln(x)$, $[1, 3.6]$ **d.** $f(x) = \ln(x)$, $[0.05, 3.6]$

Review

TOPICS

A.1 The Real Numbers; Integer Exponents

A.2 Polynomials

A.3 Rational Expressions

A.4 Radicals and Rational Exponents

A.5 Topics in Geometry

A.6 Equations

A.7 Inequalities

A.8 Complex Numbers

The Real Numbers; Integer Exponents

OBJECTIVES

1 Classify sets of real numbers.

2 Specify sets of numbers in roster or set-builder notation.

3 Use interval notation.

4 Relate absolute value and distance on the real number line.

5 Evaluate algebraic expressions.

6 Define and use integer exponents.

Classifying Sets of Numbers

The idea of a set is familiar to us all. We regularly refer to "a set of baseball cards," a "set of CDs," and "a set of dishes." In mathematics, as in everyday life, a **set** is a collection of objects. The objects in the set are called the **elements**, or **members**, of the set. In the study of algebra, we are interested primarily in sets of numbers. In listing the elements of a set, it is customary to enclose the listed elements in braces, { }, and to separate them by commas.

We distinguish among various sets of numbers. The numbers we use to count with constitute the set of **natural numbers**: $\{1, 2, 3, 4, \ldots\}$. The three dots $\ldots$ can be read as "and so on" and indicate that the pattern continues indefinitely. The **whole numbers** are formed by adjoining the number 0 to the set of natural numbers to obtain $\{0, 1, 2, 3, 4, \ldots\}$. The **integers** consist of the set of natural numbers together with their opposites and 0: $\{\ldots, -4, -3, -2, -1, 0, 1, 2, 3, 4, \ldots\}$. The **rational numbers** consist of all numbers that *can* be expressed as the quotient, $\frac{a}{b}$, of two integers, where $b \neq 0$.

Examples of rational numbers are $\frac{1}{2}, \frac{5}{3}, -\frac{4}{17}$, and $0.07 = \frac{7}{100}$. Any integer a can be expressed as the quotient of two integers by writing $a = \frac{a}{1}$. Consequently, every integer is also a rational number. In particular, $0 = \frac{0}{1}$ is a rational number. The rational number $\frac{a}{b}$ can be written as a decimal by using long division. When you take two integers a and b and divide a by b, the result is either a **terminating decimal**, such as $\frac{1}{2} = 0.5$, or a **nonterminating repeating decimal**, such as $\frac{2}{3} = 0.666\ldots$.

There are decimals that neither terminate nor repeat. Decimals that neither terminate nor repeat represent **irrational numbers**. The rational numbers together with the irrational numbers form the **real numbers**.

Real Number Line

We associate the real numbers with points on a geometric line (imagined to be extended indefinitely in both directions) in such a way that each real number corresponds to exactly one point and each point corresponds to exactly one real number. The point is called the **graph** of the corresponding real number and labeled accordingly, and the real number is called the **coordinate** of the point. By agreement, the graphs of **positive numbers** lie to the right of the point corresponding to 0, and the graphs of **negative numbers** lie to the left of 0. See Figure A.1.

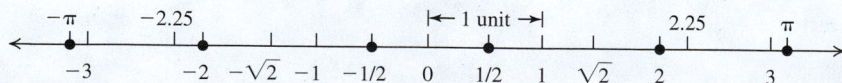

Figure A.1

Notice that $\frac{1}{2}$ and $-\frac{1}{2}$, 2 and -2, and π and $-\pi$ correspond to pairs of points exactly the same distance from 0 but on opposite sides of 0.

When coordinates have been assigned to points on a line in the manner just described, the line is called a **real number line**, a **coordinate line**, a **real line**, or simply a **number line**. The point corresponding to 0 is called the **origin**. We will treat real numbers either as numbers or their graphs on the number line depending on whether algebraic or geometric properties are under discussion.

Inequalities

The real numbers are **ordered** by their size. We say that **a is less than b**, and we write $a < b$, provided that $b = a + c$ for some *positive* number c. We also write $b > a$, meaning the same thing as $a < b$, and we say that **b is greater than a**. On the real line, the numbers increase from left to right. Consequently, **a is to the left of b on the number line when $a < b$.** Similarly, a is to the right of b on the number line when $a > b$. We sometimes want to indicate that at least one of two conditions is correct: either $a < b$ or $a = b$. In this case, we write $a \le b$ or $b \ge a$. The four symbols $<, >, \le$, and $\ge$ are called **inequality symbols**.

The following order properties of inequalities are used throughout this text:

The *trichotomy property* says that if two real numbers are not equal, then one is larger than the other. The *transitive property* says that "less than" works like "smaller than" or "lighter than."

Frequently, we read $a > 0$ as "a is positive" instead of "a is greater than 0." We can also read $a < 0$ as "a is negative." If $a \ge 0$, then either $a > 0$ or $a = 0$ and we may say that "a is nonnegative."

Sets

2 Specify sets of numbers in roster or set-builder notation.

To specify a set, we do one of the following:

1. List the elements of the set (**roster method**)

2. Describe the elements of the set (often using **set-builder notation**)

Variables are helpful in describing sets when we use set-builder notation. The notation $\{x \mid x \text{ is a natural number less than } 6\}$ is in set-builder notation and describes the set $\{1, 2, 3, 4, 5\}$, using the roster method. We read $\{x \mid x \text{ is a natural number less than } 6\}$ as "the set of all x such that x is a natural number less than six." Generally, $\{x \mid x \text{ has property } P\}$ designates the set of all x such that (the vertical bar is read "such that") x has property P. In this notation, we may describe the set of rational numbers as:

$$\left\{ x \,\middle|\, x = \frac{a}{b}, a \text{ and } b \text{ integers with } b \ne 0 \right\}.$$

It may happen that a description fails to describe any number. For example, consider $\{x \mid x < 2 \text{ and } x > 7\}$. Of course, no number can be simultaneously less than 2 and greater than 7; so this set has no members. We refer to a set with no elements as the **empty set**, or **null set**, and use the special symbol $\varnothing$ to denote it. In set notation, the empty set is written as: $\{\ \}$.

The **union** of two sets A and B, denoted $A \cup B$, is the set consisting of all elements that are in A *or* B (or both). The **intersection** of A and B, denoted $A \cap B$, is the set consisting of all elements that are in both A *and* B. In other words, $A \cap B$ consists of the elements common to both A and B.

EXAMPLE 1 **Forming Set Unions and Intersection**

Find $A \cap B$ and $A \cup B$ assuming that $A = \{-2, -1, 0, 1, 2\}$ and $B = \{-4, -2, 0, 2, 4\}$.

Solution

$A \cap B = \{-2, 0, 2\}$, the set of elements common to both A and B.
$A \cup B = \{-4, -2, -1, 0, 1, 2, 4\}$, the set of elements that are in A or in B (or in both).

Practice Problem 1 Find $A \cap B$ and $A \cup B$ if $A = \{-3, -1, 0, 1, 3\}$ and $B = \{-4, -2, 0, 2, 4\}$.

3 Use interval notation.

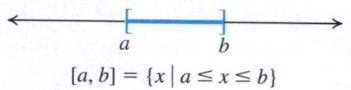

$(a, b) = \{x \mid a < x < b\}$

Figure A.2

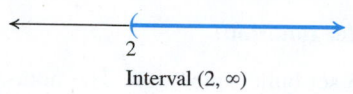

$[a, b] = \{x \mid a \le x \le b\}$

Figure A.3

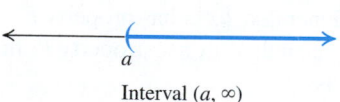

2

Interval $(2, \infty)$

Figure A.4

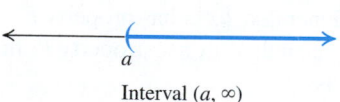

a

Interval (a, ∞)

Figure A.5

Intervals

We now turn our attention to graphing certain sets of numbers. That is, we graph each number in a given set. We are particularly interested in sets of real numbers, called **intervals**, whose graphs correspond to special sections of the number line.

If $a < b$, then the set of real numbers between a and b, but not including either a or b, is called the **open interval** from a to b and is denoted by (a, b). See Figure A.2. Using set-builder notation, we can write

$$(a, b) = \{x \mid a < x < b\}.$$

We indicate graphically that the endpoints a and b are excluded from the open interval by drawing a left parenthesis at a and a right parenthesis at b. These parentheses enclose the numbers between a and b.

The **closed interval** from a to b is the set

$$[a, b] = \{x \mid a \le x \le b\}.$$

The closed interval includes both endpoints a and b. We replace the parentheses by square brackets both in the interval notation and on the graph. See Figure A.3. Sometimes we want to include only one endpoint of an interval and exclude the other. Table A.1 shows how this is done.

We also want to graph all numbers to the left or right of a given number, such as

$$\{x \mid x > 2\}.$$

The number 2 is not included in this set, and we again indicate this fact graphically by drawing a left parenthesis at 2. We then graph all points to the right of 2, as shown in Figure A.4.

For any number a, we call $\{x \mid x > a\}$ an **unbounded** (or infinite) **interval** and denote this interval by (a, ∞). The symbol ∞ ("infinity") does not represent a number. The notation (a, ∞) is used to indicate the set of all real numbers that are greater than a, and the symbol ∞ is used to indicate that the interval extends indefinitely to the right of a. See Figure A.5.

The symbol $-\infty$ is another symbol that does not represent a number. The notation $(-\infty, a)$ is used to indicate the set of all real numbers that are less than a. The notation $(-\infty, \infty)$ represents the set of all real numbers.

Table A.1 lists various types of intervals that we use in this book. In the table, when two points a and b are involved, we assume that $a < b$.

SIDE NOTE

The symbols ∞ and $-\infty$ are always used with parentheses, not square brackets. Also note that $<$ and $>$ correspond to parentheses and that $\le$ and $\ge$ correspond to square brackets.

TABLE A.1

Interval Notation	Set-Builder Notation	Graph
(a, b)	$\{x \mid a < x < b\}$	
$[a, b]$	$\{x \mid a \le x \le b\}$	
$(a, b]$	$\{x \mid a < x \le b\}$	
$[a, b)$	$\{x \mid a \le x < b\}$	
(a, ∞)	$\{x \mid x > a\}$	
$[a, \infty)$	$\{x \mid x \ge a\}$	
$(-\infty, b)$	$\{x \mid x < b\}$	
$(-\infty, b]$	$\{x \mid x \le b\}$	
$(-\infty, \infty)$	$\{x \mid x \text{ is a real number}\}$	

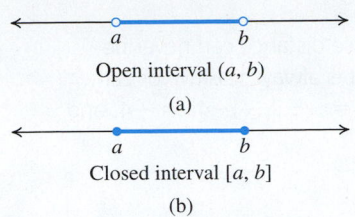

Open interval (a, b)

(a)

Closed interval $[a, b]$

(b)

Figure A.6

An alternative notation for indicating whether endpoints are included uses closed circles to show inclusion and open circles to show exclusion. See Figure A.6.

We will sometimes find it useful to work with the union and intersection of intervals.

EXAMPLE 2 Union and Intersection of Intervals

Consider the two intervals: $I_1 = (-3, 4)$ and $I_2 = [2, 6]$.

Find:

a. $I_1 \cup I_2$ **b.** $I_1 \cap I_2$

Solution

a. From the figure, we see that $I_1 \cup I_2 = (-3, 6]$. We note that every number in the interval $(-3, 6]$ is either in I_1 or in I_2 or in both I_1 and I_2.

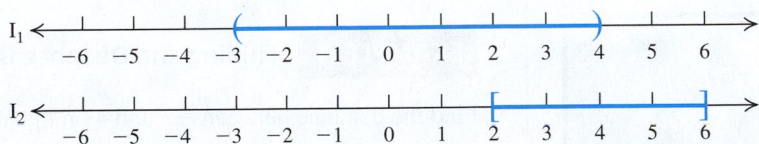

b. We see in the figure that $I_1 \cap I_2 = [2, 4)$ because a number is in the interval $[2, 4)$ if and only if it is both in I_1 and in I_2.

Practice Problem 2 Let $I_1 = (-\infty, 5)$ and $I_2 = [2, \infty)$. Find

a. $I_1 \cup I_2$ **b.** $I_1 \cap I_2$

4 Relate absolute value and distance on the real number line.

Absolute Value

The **absolute value** of a number a, denoted by $|a|$, is the distance between the origin and the point on the number line with coordinate a. The point with coordinate -3 is three units from the origin, so we write $|-3| = 3$ and say that the absolute value of -3 is 3. See Figure A.7.

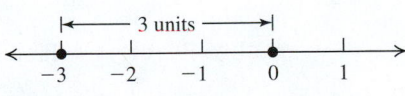

Figure A.7

Absolute Value

For any real number a, the **absolute value** of a, denoted $|a|$, is defined by

$$|a| = a \quad \text{if } a \geq 0 \qquad \text{and} \qquad |a| = -a \quad \text{if } a < 0.$$

EXAMPLE 3 Determining Absolute Value

Find the value of each of the following expressions.

a. $|4|$ **b.** $|-4|$ **c.** $|0|$ **d.** $|(-3) + 1|$

Solution

a. $|4| = 4$ **b.** $|-4| = -(-4) = 4$

c. $|0| = 0$ **d.** $|(-3) + 1| = |-2| = -(-2) = 2$

Practice Problem 3 Find the value of each of the following.

a. $|-10|$ **b.** $|3 - 4|$ **c.** $|2(-3) + 7|$

 **WARNING**

The absolute value of a number represents a distance. Because distance can never be negative, the absolute value of a number is never negative; it is always positive or zero. However, if a is not 0, $-|a|$ is always negative. Thus, $-|5.3| = -5.3$, $-|-4| = -4$, and $-|1.18| = -1.18$.

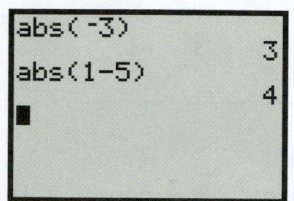

TECHNOLOGY CONNECTION

The absolute value function on your graphing calculator will find the value of the expression entered and then compute its absolute value.

Distance Between Two Points on a Real Number Line

> **DISTANCE FORMULA ON A NUMBER LINE**
>
> If a and b are the coordinates of two points on a number line, then the distance between a and b, denoted by $d(a, b)$, is $|a - b|$.

EXAMPLE 4 Finding the Distance Between Two Points

Find the distance between -3 and 4 on the number line.

Solution

Figure A.8 shows that the distance between -3 and 4 is seven units. The distance formula gives the same answer.

$$d(-3, 4) = |-3 - 4| = |-7| = -(-7) = 7.$$

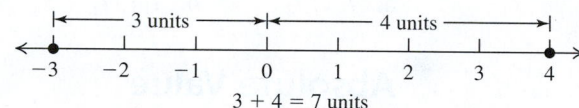

Figure A.8

Notice that reversing the order of -3 and 4 in this computation gives the same answer. That is, the distance between 4 and -3 is $|4 - (-3)| = |4 + 3| = |7| = 7$. It is always true that $|a - b| = |b - a|$. See Figure A.9.

Figure A.9

Practice Problem 4 Find the distance between -7 and 2 on the number line.

5 Evaluate algebraic expressions.

Algebraic Expressions

In algebra, we use letters such as a, b, x, and y to represent numbers. A letter that is used to represent one or more numbers is called a **variable**. A **constant** is either a specific number, such as 3 or $\frac{1}{2}$, or a letter that represents a fixed (but not necessarily specified) number.

Any number, constant, variable, or parenthetical group (or any product of them) is called a **term**. A term or combination of terms using the ordinary operations of addition, subtraction, multiplication, division, and absolute value (as well as exponentiation and roots, which are discussed later in this appendix) is called an **algebraic expression**, or simply an **expression**. Here are some examples of algebraic expressions:

$$6; \quad y; \quad x - 2; \quad \frac{1}{x} + \sqrt{7}; \quad \frac{10}{y + 3}; \quad \sqrt{x} + |y| \div 5$$

If we replace each variable in an expression with a specific number and get a real number, the resulting number is called the **value of the algebraic expression**.

The following two properties of real numbers are used frequently in algebra. The first property is used in simplifying expressions, and the second is used in solving equations.

DISTRIBUTIVE AND ZERO-PRODUCT PROPERTIES

Distributive property $\qquad a \cdot (b + c) = a \cdot b + a \cdot c$

Zero-product property $\qquad$ If $a \cdot b = 0$, then $a = 0$ or $b = 0$, or both.

EXAMPLE 5 **Evaluating an Algebraic Expression**

Evaluate the following expressions for the given values of the variables.

a. $[(9 + x) \div 7] \cdot 3 - x$ for $x = 5$

b. $|x| - \dfrac{2}{y}$ for $x = 2$ and $y = -2$

Solution

a.
$$
\begin{aligned}
[(9 + x) \div 7] \cdot 3 - x &= [(9 + 5) \div 7] \cdot 3 - 5 \qquad \text{Replace } x \text{ with 5.} \\
&= [14 \div 7] \cdot 3 - 5 \\
&= 2 \cdot 3 - 5 \\
&= 6 - 5 \\
&= 1
\end{aligned}
$$

b.
$$
\begin{aligned}
|x| - \frac{2}{y} &= |2| - \frac{2}{-2} \qquad \text{Replace } x \text{ with 2 and } y \text{ with } -2. \\
&= 2 - \frac{2}{-2} \\
&= 2 - (-1) \\
&= 3
\end{aligned}
$$

Practice Problem 5 Evaluate.

a. $(x - 2) \div 3 + x$ for $x = 3$ $\qquad$ **b.** $7 - \dfrac{x}{|y|}$ for $x = -1, y = 3$

Integer Exponents

6 Define and use integer exponents.

The area of a square with side 5 feet is $5 \cdot 5 = 25$ square feet. The volume of a cube whose sides are each 5 feet is $5 \cdot 5 \cdot 5 = 125$ cubic feet. A shorter notation for $5 \cdot 5$ is 5^2, and for $5 \cdot 5 \cdot 5$ it is 5^3. The number 5 is called the *base* for both 5^2 and 5^3. The number 2 is called the *exponent* in the expression 5^2 and indicates that the base 5 appears as a factor twice.

Positive Integer Exponent

If a is a real number and n is a positive integer, then

$$a^n = \underbrace{a \cdot a \cdot \cdots \cdot a}_{n \text{ factors}},$$

where a is the base and n is the exponent. We adopt the convention that $a^1 = a$.

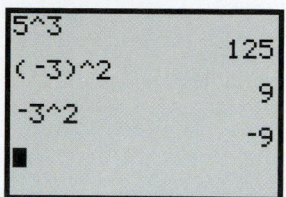
We can evaluate exponents as follows:

$$5^3 = 5 \cdot 5 \cdot 5 = 125$$
$$(-3)^2 = (-3)(-3) = 9$$
$$-3^2 = -(3 \cdot 3) = -9$$
$$(-2)^3 = (-2)(-2)(-2) = -8$$

Pay careful attention to the fact that $(-3)^2 \neq -3^2$.

Negative exponents indicate the ***reciprocal*** of the expression with the corresponding positive exponent.

Zero and Negative Integer Exponents

For any nonzero number a and any positive integer n,

$$a^0 = 1 \text{ and } a^{-n} = \frac{1}{a^n}.$$

EXAMPLE 6 **Evaluating Zero or Negative Exponents**

Evaluate.

a. $(-5)^{-2}$ **b.** -5^{-2} **c.** 8^0 **d.** $\left(\frac{2}{3}\right)^{-3}$

Solution

a. $(-5)^{-2} = \dfrac{1}{(-5)^2} = \dfrac{1}{25}$ **b.** $-5^{-2} = -\dfrac{1}{5^2} = -\dfrac{1}{25}$

c. $8^0 = 1$ **d.** $\left(\dfrac{2}{3}\right)^{-3} = \dfrac{1}{\left(\dfrac{2}{3}\right)^3} = \dfrac{1}{\dfrac{8}{27}} = \dfrac{27}{8}$

Practice Problem 6 Evaluate.

a. 2^{-1} **b.** $\left(\dfrac{4}{5}\right)^0$ **c.** $\left(\dfrac{3}{2}\right)^{-2}$

Rules of Exponents

The following rules of exponents can be proved using the previous definitions. All expressions are assumed to be defined.

RULES OF EXPONENTS

$$a^m \cdot a^n = a^{m+n} \qquad \frac{a^m}{a^n} = a^{m-n} \qquad (a^m)^n = a^{mn}$$

$$(a \cdot b)^n = a^n \cdot b^n \qquad \left(\frac{a}{b}\right)^n = \frac{a^n}{b^n}$$

EXAMPLE 7   Using the Rules of Exponents

Use the rules of exponents to write each expression without negative exponents.

a. $(-4x^2y^3)(7x^3y)$ **b.** $\left(\dfrac{x^5}{2y^{-3}}\right)^{-3}$

Solution

a. $(-4x^2y^3)(7x^3y) = (-4)(7)x^2x^3y^3y$
$= -28x^{2+3}y^{3+1}$
$= -28x^5y^4$

b. $\left(\dfrac{x^5}{2y^{-3}}\right)^{-3} = \dfrac{(x^5)^{-3}}{(2y^{-3})^{-3}}$ $\left(\dfrac{a}{b}\right)^n = \dfrac{a^n}{b^n}$

$= \dfrac{(x^5)^{-3}}{2^{-3}(y^{-3})^{-3}}$ $(ab)^n = a^nb^n$

$= \dfrac{x^{-15}}{2^{-3}y^{(-3)(-3)}}$ $(a^m)^n = a^{mn}$

$= \dfrac{x^{-15}}{2^{-3}y^9}$ Simplify.

$= \dfrac{x^{15}x^{-15}2^3}{x^{15}2^32^{-3}y^9}$ $\dfrac{a}{b} = \dfrac{ca}{cb}, c \neq 0$

$= \dfrac{2^3}{x^{15}y^9}$ $a^0 = 1$

$= \dfrac{8}{x^{15}y^9}$

Practice Problem 7   Simplify each expression.

a. $(2x^4)^{-2}$ **b.** $\dfrac{x^2(-y)^3}{(xy^2)^3}$

Answers to Practice Problems
1. $A \cap B = \{0\}$
$A \cup B = \{-4, -3, -2, -1, 0, 1, 2, 3, 4\}$
2. a. $I_1 \cup I_2 = (-\infty, \infty)$ **b.** $I_1 \cap I_2 = [2, 5)$
3. a. 10   **b.** 1   **c.** 1   **4.** 9   **5. a.** $\dfrac{10}{3}$ **b.** $\dfrac{22}{3}$ **6. a.** $\dfrac{1}{2}$ **b.** 1
c. $\dfrac{4}{9}$ **7. a.** $\dfrac{1}{4x^8}$ **b.** $-\dfrac{1}{xy^3}$

SECTION A.1   **Exercises**

Basic Concepts and Skills

In Exercises 1–4, write each of the following rational numbers as a decimal and state whether the decimal is repeating or terminating.

1. $-\dfrac{4}{5}$ **2.** $-\dfrac{3}{12}$

3. $\dfrac{3}{11}$ **4.** $\dfrac{41}{15}$

In Exercises 5–10, classify each of the following numbers as rational or irrational.

5. -207 **6.** $-\sqrt{25}$
7. $\sqrt{32}$ **8.** $5 + \sqrt{18}$
9. 0.321 **10.** $5.8\overline{2}$

In Exercises 11–14, find each set, given
$A = \{-4, -2, 0, 2, 4\}, B = \{-3, 0, 1, 2, 3, 4\}$, and
$C = \{-4, -3, -2, -1, 0, 2\}$.

11. $A \cup B$

12. $A \cap B$

13. $(A \cup B) \cap C$

14. $(A \cup B) \cup C$

In Exercises 15 and 16, convert each decimal to a quotient of two integers in lowest terms.

15. 3.75

16. -2.35

In Exercises 17–24, find the union and the intersection of the given intervals.

17. $I_1 = (-2, 3]; I_2 = [1, 5)$

18. $I_1 = [1, 7]; I_2 = (3, 5)$

19. $I_1 = (-6, 2); I_2 = [2, 10)$

20. $I_1 = (-\infty, -3]; I_2 = (-3, \infty)$

21. $I_1 = (-\infty, 7); I_2 = (-\infty, 3)$

22. $I_1 = (-2, \infty); I_2 = (0, \infty)$

23. $I_1 = (-\infty, 10); I_2 = (10, \infty)$

24. $I_1 = (-5, 6); I_2 = (-\infty, \infty)$

In Exercises 25–38, rewrite each expression without absolute value bars.

25. $|20|$

26. $|12|$

27. $-|-4|$

28. $-|-17|$

29. $\left|\dfrac{5}{7}\right|$

30. $\left|\dfrac{-3}{5}\right|$

31. $|5 - \sqrt{2}|$

32. $|\sqrt{2} - 5|$

33. $\dfrac{8}{|-8|}$

34. $\dfrac{-8}{|8|}$

35. $|5 + |-7||$

36. $|5 - |-7||$

37. $||7| - |4||$

38. $||4| - |7||$

In Exercises 39–46, use the absolute value to express the distance between the points with coordinates a and b on the number line. Then determine this distance by evaluating the absolute value expression.

39. $a = 3$ and $b = 8$

40. $a = 2$ and $b = 14$

41. $a = -6$ and $b = 9$

42. $a = -12$ and $b = 3$

43. $a = -20$ and $b = -6$

44. $a = -14$ and $b = -1$

45. $a = \dfrac{22}{7}$ and $b = -\dfrac{4}{7}$

46. $a = \dfrac{16}{5}$ and $b = -\dfrac{3}{5}$

In Exercises 47–58, graph each of the given intervals on a separate number line and write the inequality notation for each.

47. $[1, 4]$

48. $[-2, 2]$

49. $(14, 28)$

50. $\left(\dfrac{1}{2}, \dfrac{9}{2}\right)$

51. $(-3, 1]$

52. $[-6, -2)$

53. $[-3, \infty)$

54. $[0, \infty)$

55. $(-\infty, 5]$

56. $(-\infty, -1]$

57. $\left(-\dfrac{3}{4}, \dfrac{9}{4}\right)$

58. $\left(-3, -\dfrac{1}{2}\right)$

In Exercises 59–68, evaluate each expression for $x = 3$ and $y = -5$.

59. $2(x + y) - 3y$

60. $-2(x + y) + 5y$

61. $3|x| - 2|y|$

62. $7|x - y|$

63. $\dfrac{x - 3y}{2} + xy$

64. $\dfrac{y + 3}{x} - xy$

65. $\dfrac{2(1 - 2x)}{y} - (-x)y$

66. $\dfrac{3(2 - x)}{y} - (1 - xy)$

67. $\dfrac{\dfrac{14}{x} + \dfrac{1}{2}}{\dfrac{-y}{4}}$

68. $\dfrac{\dfrac{4}{-y} + \dfrac{8}{x}}{\dfrac{y}{2}}$

In Exercises 69–78, name the exponent and the base.

69. 17^3

70. 10^2

71. 9^0

72. $(-2)^0$

73. $(-5)^5$

74. $(-99)^2$

75. -10^3

76. $-(-3)^7$

77. a^2

78. $(-b)^3$

In Exercises 79–100, evaluate each expression.

79. 6^1

80. 3^4

81. 7^0

82. $(-8)^0$

83. $(2^3)^2$

84. $(3^2)^3$

85. $(3^2)^{-2}$

86. $(7^2)^{-1}$

87. $(5^{-2})^3$

88. $(5^{-1})^3$

89. $(4^{-3}) \cdot (4^5)$

90. $(7^{-2}) \cdot (7^3)$

91. $3^{-2} + \left(\dfrac{1}{3}\right)^2$

92. $5^{-2} + \left(\dfrac{1}{5}\right)^2$

93. $\dfrac{2^{11}}{2^{10}}$

94. $\dfrac{3^6}{3^8}$

95. $\dfrac{(5^3)^4}{5^{12}}$

96. $\dfrac{(9^5)^2}{9^8}$

97. $\dfrac{2^5 \cdot 3^{-2}}{2^4 \cdot 3^{-3}}$

98. $\dfrac{4^{-2} \cdot 5^3}{4^{-3} \cdot 5}$

99. $\left(\dfrac{11}{7}\right)^{-2}$

100. $\left(\dfrac{13}{5}\right)^{-2}$

In Exercises 101–134, simplify each expression. Write your answers without negative exponents. Whenever an exponent is negative or zero, assume that the base is not zero.

101. $x^4 y^0$

102. $x^{-1} y^0$

103. $x^{-1} y$

104. $x^2 y^{-2}$

105. $-8x^{-1}$

106. $(-8x)^{-1}$

107. $x^{-1}(3y^0)$

108. $x^{-3}(3y^2)$

109. $x^{-1}y^{-2}$

110. $x^{-3}y^{-2}$

111. $(x^{-3})^4$

112. $(x^{-5})^2$

113. $(x^{-11})^{-3}$

114. $(x^{-4})^{-12}$

115. $-3(xy)^5$

116. $-8(xy)^6$

117. $4(xy^{-1})^2$

118. $6(x^{-1}y)^3$

119. $3(x^{-1}y)^{-5}$

120. $-5(xy^{-1})^{-6}$

121. $\dfrac{(x^3)^2}{(x^2)^5}$

122. $\dfrac{x^2}{(x^3)^4}$

123. $\left(\dfrac{2xy}{x}\right)^3$

124. $\left(\dfrac{5xy}{x^3}\right)^4$

125. $\left(\dfrac{-3x^2y}{x}\right)^5$

126. $\left(\dfrac{-2xy^2}{y}\right)^3$

127. $\left(\dfrac{-3x}{5}\right)^{-2}$

128. $\left(\dfrac{-5y}{3}\right)^{-4}$

129. $\left(\dfrac{4x^{-2}}{xy^5}\right)^3$

130. $\left(\dfrac{3x^2y}{y^3}\right)^5$

131. $\dfrac{x^3y^{-3}}{x^{-2}y}$

132. $\dfrac{x^2y^{-2}}{x^{-1}y^2}$

133. $\dfrac{27x^{-3}y^5}{9x^{-4}y^7}$

134. $\dfrac{15x^5y^{-2}}{3x^7y^{-3}}$

Applying the Concepts

135. Media players. Let A = the set of people who own MP3 players and B = the set of people who own DVD players.
  a. Describe the set $A \cup B$.
  b. Describe the set $A \cap B$.

136. Standard car features. The table indicates whether certain features are "standard" for each of three types of car.

	Navigation System	Automatic Transmission	Leather Seats
2016 Lexus GS 350	yes	yes	yes
2016 Lincoln MKZ	no	yes	yes
2016 Cadillac ATS-V	no	no	yes

Use the roster method to describe each of the following sets.

  a. A = cars in which a navigation system is standard.
  b. B = cars in which an automatic transmission is standard.
  c. C = cars in which leather seats are standard.
  d. $A \cap B$
  e. $B \cap C$
  f. $A \cup B$
  g. $A \cup C$

137. Blood pressure. A group of college students had systolic blood pressure readings that ranged from 119.5 to 134.5 inclusive. Let x represent the value of the systolic blood pressure readings. Use inequalities to describe this range of values and graph the corresponding interval on a number line.

138. Population projections. Population projections suggest that by the year 2050, the number of people 60 years old and older in the United States will be about 107 million. In 1950, the number of people 60 years old and older in the United States was 30 million. Let x represent the number (in millions) of people in the United States who are 60 years old and older. Use inequalities to describe this population range from 1950 to 2050 and graph the corresponding interval on a number line. *Source:* U.S. Census Bureau

139. Heart rate. For exercise to be most beneficial, the optimum heart rate for a 20-year-old person is 120 beats per minute. Use absolute value notation to write an expression that describes the difference between the heart rate achieved by each of the following 20-year-old people and the ideal exercise heart rate. Then evaluate that expression.
  a. Latasha: 124 beats per minute
  b. Frances: 137 beats per minute
  c. Ignacio: 114 beats per minute

140. Downloading music. To download a 4 MB song with a 56 Kbs modem takes an average of 15 minutes. Use absolute value notation to write an expression that describes the difference between this average time and the actual time it took to download the following songs. Then evaluate that expression.
  a. *Believe* (Cher): 14 minutes
  b. *Caged Bird* (Alicia Keys): 17.5 minutes
  c. *Somewhere* (Barbra Streisand): 15 minutes

141. Square area. The area A of a square with side of length x is given by $A = x^2$. Use this relationship to
  a. verify that doubling the length of the side of a square floor increases the area of the floor by a factor of 2^2.
  b. verify that tripling the length of the side of a square floor increases the area of the floor by a factor of 3^2.

142. Circle area. The area A of a circle with diameter d is given by $A = \pi\left(\dfrac{d}{2}\right)^2$. Use this relationship to
  a. verify that doubling the length of the diameter of a circular skating rink increases the area of the rink by a factor of 2^2.
  b. verify that tripling the length of the diameter of a circular skating rink increases the area of the rink by a factor of 3^2.

Polynomials

Polynomial Vocabulary

1 Use polynomial vocabulary.

We begin by reviewing the basic vocabulary of polynomials. A **monomial** is the simplest polynomial in one variable, say, x; it contains one term and has the form ax^k, where a is a constant and k is either a positive integer or zero. The constant a is called the **coefficient** of the monomial. For $a \neq 0$, the integer k is called the **degree** of the monomial.

Consider the following monomials:

$2x^5$	The coefficient is 2, and the degree is 5.
$-3x^2$	The coefficient is -3, and the degree is 2.
$-3x^2$	The coefficient is -3, and the degree is 2.
-7	$-7 = -7(1) = -7x^0$. The coefficient is -7, and the degree is 0.
$8x$	The coefficient is 8, and the degree is $1(x = x^1)$.
x^{12}	The coefficient is $1(x^{12} = 1 \cdot x^{12})$, and the degree is 12.
$-x^4$	The coefficient is $-1(-x^4 = (-1) \cdot x^4)$, and the degree is 4.

The expression $5x^{-3}$ is not a monomial because the exponent of x is negative. Two monomials in the same variable with the same degree can be added or subtracted using the distributive property. For example, $-3x^5 - 9x^5 = (-3 - 9)x^5 = -12x^5$. Two monomials in the same variable with the same degree are called **like terms**.

Polynomials in One Variable

A **polynomial** in x is any sum of monomials in x. By combining like terms, we can write any polynomial in the form

$$a_n x^n + a_{n-1} x^{n-1} + \cdots + a_2 x^2 + a_1 x + a_0,$$

where n is either a positive integer or zero and $a_n, a_{n-1}, \ldots a_1, a_0$ are constants, called the **coefficients** of the polynomial. If $a_n \neq 0$, then n, the largest exponent on x, is called the **degree** and a_n is called the **leading coefficient** of the polynomial. The monomials $a_n x^n, a_{n-1} x^{n-1} \ldots, a_2 x^2, a_1 x$, and a_0 are the **terms** of the polynomial. The monomial $a_n x^n$ is the **leading term** of the polynomial, and a_0 is the **constant term**.

Polynomials can be classified according to the number of terms they have. Polynomials with one term are called **monomials**, polynomials with two *unlike* terms are called **binomials**, and polynomials with three *unlike* terms are called **trinomials**. Polynomials with more than three unlike terms do not have special names.

By agreement, the only polynomial that has *no* degree is the **zero polynomial**, which results when all of the coefficients are 0. It is easiest to find the degree of a polynomial when it is written in **descending order**—that is, when the exponents decrease from left to right. In this case, the degree is the exponent of the leading term. A polynomial written in descending order is said to be in **standard form**.

The symbols $a_0, a_1, a_2, \ldots a_n$ in the general notation for a polynomial are just constants; the numbers to the lower right of a are called subscripts. We read the notation a_2 as "a sub 2." This type of notation is used when a large or indefinite number of constants are required.

2 Use special-product formulas.

Special Products

The basic multiplication rule for polynomials can be stated as follows:

MULTIPLYING POLYNOMIALS

To multiply two polynomials, multiply each term of one polynomial by every term of the other polynomial and combine like terms.

Particular polynomial products called *special products* occur frequently enough to deserve special attention. We introduce a very useful method, called F O I L (First, Outer, Inner, Last), for multiplying two binomials.

FOIL METHOD FOR $(A + B)(C + D)$

If A, B, C, and D represent any algebraic expressions, then

$$(A + B)(C + D) = \underset{F}{A \cdot C} + \underset{O}{A \cdot D} + \underset{I}{B \cdot C} + \underset{L}{B \cdot D}$$

Outer, Inner, First, Last

EXAMPLE 1 Using the FOIL Method

Use the FOIL method to find the following products.

a. $(2x + 3)(x - 1) = \underset{F}{(2x)(x)} + \underset{O}{(2x)(-1)} + \underset{I}{(3)(x)} + \underset{L}{(3)(-1)}$

$$= 2x^2 - 2x + 3x - 3$$
$$= 2x^2 + x - 3$$

b. $(3x - 5)(4x - 6) = \underset{F}{(3x)(4x)} + \underset{O}{(3x)(-6)} + \underset{I}{(-5)(4x)} + \underset{L}{(-5)(-6)}$

$$= 12x^2 - 18x - 20x + 30$$
$$= 12x^2 - 38x + 30$$

c. $(x + a)(x + b) = \underset{F}{x \cdot x} + \underset{O}{x \cdot b} + \underset{I}{a \cdot x} + \underset{L}{a \cdot b}$

$$= x^2 + bx + ax + a \cdot b$$
$$= x^2 + (b + a)x + ab$$

Practice Problem 1 Use FOIL to find each product.

a. $(4x - 1)(x + 7)$ **b.** $(3x - 2)(2x - 5)$

The following special-product formulas, such as the formulas for squaring a binomial sum or difference, are used often and should be memorized. However, you should be able to derive them using FOIL if you forget them. We list several of these formulas next.

SPECIAL-PRODUCT FORMULAS

A and B represent any algebraic expression.

Formula	Example

Product giving a difference of squares

$(A + B)(A - B) = A^2 - B^2$

$$(5x + 2)(5x - 2) = (5x)^2 - 2^2$$
$$= 25x^2 - 4$$

Squaring a binomial sum or difference

$(A + B)^2 = A^2 + 2AB + B^2$

$$(3x + 2)^2 = (3x)^2 + 2(3x)(2) + 2^2$$
$$= 9x^2 + 12x + 4$$

$(A - B)^2 = A^2 - 2AB + B^2$

$$(2x - 5)^2 = (2x)^2 - 2(2x)(5) + 5^2$$
$$= 4x^2 - 20x + 25$$

Cubing a binomial sum or difference

$(A + B)^3 = A^3 + 3A^2B + 3AB^2 + B^3$

$$(x + 5)^3 = x^3 + 3x^2(5) + 3x(5)^2 + 5^3$$
$$= x^3 + 15x^2 + 75x + 125$$

$(A - B)^3 = A^3 - 3A^2B + 3AB^2 - B^3$

$$(x - 4)^3 = x^3 - 3x^2(4) + 3x(4)^2 - 4^3$$
$$= x^3 - 12x^2 + 48x - 64$$

Products giving a sum or difference of cubes

$(A + B)(A^2 - AB + B^2) = A^3 + B^3$ $(x + 2)(x^2 - 2x + 4) = x^3 + 2^3 = x^3 + 8$

$(A - B)(A^2 + AB + B^2) = A^3 - B^3$ $(x - 3)(x^2 + 3x + 9) = x^3 - 3^3 = x^3 - 27$

---

3 Factor polynomials.

Factoring Polynomials

In this section, we discuss only polynomials having integer coefficients. This restriction is called **factoring over the integers**.

To factor a polynomial means to write it as a *product of factors*. Consider the product

$$(x + 3)(x - 3) = x^2 - 9.$$

The polynomials $(x + 3)$ and $(x - 3)$ are called **factors** of the polynomial $x^2 - 9$.

When factoring a polynomial, the first thing to look for is a factor that is common to every term. This **common factor** can be "factored out" by the distributive property, $a(b + c) = ab + ac$.

Polynomial	Common Factor	Factored Form
$9 + 18y$	9	$9(1 + 2y)$
$2y^2 + 6y$	$2y$	$2y(y + 3)$
$7x^4 + 3x^3 + x^2$	x^2	$x^2(7x^2 + 3x + 1)$
$5x + 3$	No common factor	$5x + 3$

You can verify that any factorization is correct by multiplying the factors. We factored these polynomials by finding a *common factor* of their terms.

We frequently factor by reversing a product, such as $(x + a)(x + b) = x^2 + (a + b)x + ab$, to get the factoring form

$$x^2 + (a + b)x + ab = (x + a)(x + b).$$

For example, to factor $x^2 - 6x - 16$, we must find two integers a and b with $ab = -16$ and $a + b = -6$. Because 2 and -8 are factors of -16 and give a sum of -6,

$$x^2 - 6x - 16 = (x + 2)[x + (-8)] = (x + 2)(x - 8).$$

Polynomials that cannot be factored as a product of two polynomials (excluding the constant polynomials 1 and −1) are said to be **irreducible**. A polynomial is said to be **factored completely** when it is written as a product consisting of only irreducible factors.

EXAMPLE 2 Factoring Polynomials

Factor.

a. $16x^2 - 8x + 1$ **b.** $25x^2 - 49$ **c.** $x^4 - 16$ **d.** $x^3 - 64$ **e.** $x^3 + 2x^2 + 3x + 6$

Solution

a. $16x^2 - 8x + 1 = (4x)^2 - 2(4x)(1) + 1^2 = (4x - 1)^2$ (perfect square)

b. $25x^2 - 49 = (5x)^2 - 7^2 = (5x + 7)(5x - 7)$ (difference of squares)

c. $x^4 - 16 = (x^2 + 4)(x^2 - 4) = (x^2 + 4)(x + 2)(x - 2)$ (difference of squares)

d. $x^3 - 64 = x^3 - 4^3 = (x - 4)(x^2 + 4x + 16)$ (difference of cubes)

e. $x^3 + 2x^2 + 3x + 6 = (x^3 + 2x^2) + (3x + 6)$ Group terms with a common factor.

$= x^2(x + 2) + 3(x + 2)$ Factor out the common factor.

$= (x^2 + 3)(x + 2)$ Distributive property

You should use FOIL to check that $x^2 + 4$ is irreducible in part **c** and that $x^2 + 4x + 16$ is irreducible in part **d**.

SIDE NOTE

You should always check your answer to a factoring problem by multiplying the factors to verify that the result is the original expression.

Practice Problem 2 Factor.

a. $x^2 + 4x + 4$ **b.** $4x^2 - 25$ **c.** $x^4 - 81$ **d.** $x^3 - 125$ **e.** $x^3 + 3x^2 + x + 3$

To factor the trinomial $Ax^2 + Bx + C$ as $(ax + b)(cx + d)$, we use FOIL and combine like terms to get $Ax^2 + Bx + C = acx^2 + (ad + bc)x + bd$. We can factor $Ax^2 + Bx + C$ if we can find integer factors of the product $A \cdot C$ whose sum is B. We illustrate the procedure in the next example.

EXAMPLE 3 Factoring Using FOIL and Grouping

Factor.

a. $6x^2 + 17x + 7$ **b.** $20x^2 - 7x - 6$

Solution

a. $Ax^2 + Bx + C = 6x^2 + 17x + 7$. We have $A \cdot C = 6 \cdot 7 = 42$. We are looking for two factors of 42 whose sum is 17. We have $42 = 3 \cdot 14$ and $3 + 14 = 17$. So,

$6x^2 + 17x + 7 = 6x^2 + 3x + 14x + 7$ $17x = 3x + 14x$

$= (6x^2 + 3x) + (14x + 7)$ Group terms.

$= 3x(2x + 1) + 7(2x + 1)$ Factor out the common factor.

$= (3x + 7)(2x + 1)$ Distributive property

b. $20x^2 - 7x - 6$. We have $(20)(-6) = -120$. The two factors of -120 whose sum is -7 are -15 and 8. So,

$20x^2 - 7x - 6 = 20x^2 - 15x + 8x - 6$ $-7x = -15x + 8x$

$= (20x^2 - 15x) + (8x - 6)$ Group terms.

$= 5x(4x - 3) + 2(4x - 3)$ Factor out the common factor.

$= (5x + 2)(4x - 3)$ Distributive property

Practice Problem 3 Factor.

a. $5x^2 + 11x + 2$ **b.** $4x^2 + x - 3$

SECTION A.2 Exercises

Basic Concepts and Skills

In Exercises 1–4, determine whether the given expression is a polynomial. If it is, write it in standard form.

1. $1 + x^2 + 2x$

2. $x - \dfrac{1}{x}$

3. $x^{-2} + 3x + 5$

4. $3x^4 + x^7 + 3x^5 - 2x + 1$

In Exercises 5–8, find the degree and list the terms of the polynomial.

5. $7x + 3$

6. $-3x^2 + 7$

7. $x^2 - x^4 + 2x - 9$

8. $x + 2x^3 + 9x^7 - 21$

In Exercises 9–16, perform the indicated operations. Write the resulting polynomial in standard form.

9. $(x^3 + 2x^2 - 5x + 3) + (-x^3 + 2x - 4)$

10. $(x^3 - 3x + 1) + (x^3 - x^2 + x - 3)$

11. $(2x^3 - x^2 + x - 5) - (x^3 - 4x + 3)$

12. $(-x^3 + 2x - 4) - (x^3 + 3x^2 - 7x + 2)$

13. $-2(3x^2 + x + 1) + 6(-3x^2 - 2x - 2)$

14. $2(5x^2 - x + 3) - 4(3x^2 + 7x + 1)$

15. $(3y^3 - 4y + 2) + (2y + 1) - (y^3 - y^2 + 4)$

16. $(5y^2 + 3y - 1) - (y^2 - 2y + 3) + (2y^2 + y + 5)$

In Exercises 17–50, perform the indicated operations.

17. $6x(2x + 3)$

18. $7x(3x - 4)$

19. $(x + 1)(x^2 + 2x + 2)$

20. $(x - 5)(2x^2 - 3x + 1)$

21. $(3x - 2)(x^2 - x + 1)$

22. $(2x + 1)(x^2 - 3x + 4)$

23. $(x + 1)(x + 2)$

24. $(x + 2)(x + 3)$

25. $(3x + 2)(3x + 1)$

26. $(x + 3)(2x + 5)$

27. $(-4x + 5)(x + 3)$

28. $(-2x + 1)(x - 5)$

29. $(3x - 2)(2x - 1)$

30. $(x - 1)(5x - 3)$

31. $(2x - 3a)(2x + 5a)$

32. $(5x - 2a)(x + 5a)$

33. $(x + 2)^2 - x^2$

34. $(x - 3)^2 - x^2$

35. $(x + 3)^3 - x^3$

36. $(x - 2)^3 - x^3$

37. $(4x + 1)^2$

38. $(3x + 2)^2$

39. $(3x + 1)^3$

40. $(2x + 3)^3$

41. $(5 - 2x)(5 + 2x)$

42. $(3 - 4x)(3 + 4x)$

43. $\left(x + \dfrac{3}{4}\right)^2$

44. $\left(x + \dfrac{2}{5}\right)^2$

45. $(2x - 3)(x^2 - 3x + 5)$

46. $(x - 2)(x^2 - 4x - 3)$

47. $(1 + y)(1 - y + y^2)$

48. $(y + 4)(y^2 - 4y + 16)$

49. $(x - 6)(x^2 + 6x + 36)$

50. $(x - 1)(x^2 + x + 1)$

In Exercises 51–60, perform the indicated operations.

51. $(x + 2y)(3x + 5y)$

52. $(2x + y)(7x + 2y)$

53. $(2x - y)(3x + 7y)$

54. $(x - 3y)(2x + 5y)$

55. $(x - y)^2(x + y)^2$

56. $(2x + y)^2(2x - y)^2$

57. $(x + y)(x - 2y)^2$

58. $(x - y)(x + 2y)^2$

59. $(x - 2y)^3(x + 2y)$

60. $(2x + y)^3(2x - y)$

In Exercises 61–108, factor each polynomial completely. If a polynomial cannot be factored, state that it is irreducible.

61. $3x^3 - x^2$

62. $2x^3 + 2x^2$

63. $x^2 + 7x + 12$

64. $x^2 + 8x + 15$

65. $x^2 - 6x + 8$

66. $x^2 - 9x + 14$

67. $6x^2 + 17x + 12$

68. $8x^2 - 10x - 3$

69. $x^2 + 6x + 9$

70. $x^2 + 8x + 16$

71. $9x^2 + 6x + 1$

72. $36x^2 + 12x + 1$

73. $x^2 - 64$

74. $x^2 - 121$

75. $16x^2 - 9$

76. $25x^2 - 49$

77. $x^3 - 27$

78. $x^3 - 216$

79. $8 - x^3$

80. $27 - x^3$

81. $x^3 - 3x^2 + x - 3$

82. $x^3 + 5x^2 + x + 5$

83. $x^3 - 5x^2 + x - 5$

84. $x^3 - 7x^2 + x - 7$

85. $x^4 - 1$

86. $x^4 - 81$

87. $20x^4 - 5$

88. $12x^4 - 75$

89. $1 - 16x^2$

90. $4 - 25x^2$

91. $x^2 - 6x + 9$

92. $x^2 - 8x + 16$

93. $4x^2 + 4x + 1$

94. $16x^2 + 8x + 1$

95. $2x^2 - 8x - 10$

96. $5x^2 - 10x - 40$

97. $2x^2 + 3x - 20$

98. $2x^2 - 7x - 30$

99. $x^2 - 12x + 36$

100. $x^2 - 20x + 25$

101. $3x^5 + 12x^4 + 12x^3$

102. $2x^5 + 16x^4 + 32x^3$

103. $9x^2 - 1$

104. $16x^2 - 25$

105. $16x^2 + 24x + 9$

106. $4x^2 + 20x + 25$

107. $x^2 + 15$

108. $x^2 + 24$

Rational Expressions

Rational Expressions

Recall that the quotient of two integers, $\frac{a}{b}$ ($b \neq 0$), is a rational number. When we form the quotient of two polynomials, the result is called a **rational expression**. Following are some examples of rational expressions.

$$\frac{10}{17} \qquad \frac{x+2}{3} \qquad \frac{x+6}{(x-3)(x+4)} \qquad \frac{7}{x^2+1} \qquad \frac{x^2+2x-3}{x^2+5x}$$

We use the same language to describe rational expressions that we use to describe rational numbers. For $\frac{x^2+2x-3}{x^2+5x+6}$, we call x^2+2x-3 the **numerator** and x^2+5x+6 the **denominator**. As with rational numbers, we do not allow the denominator to be 0. We write $\frac{x+6}{(x-3)(x+4)}, x \neq 3, x \neq -4$ to indicate that 3 and -4 are not in the domain of this expression; the domain in interval notation is $(-\infty, -4) \cup (-4, 3) \cup (3, \infty)$.

1 Reduce a rational expression to lowest terms.

Lowest Terms for a Rational Expression

EXAMPLE 1 **Reduce a Rational Expression to Lowest Terms**

Simplify each expression.

a. $\dfrac{x^4+2x^3}{x+2}$ **b.** $\dfrac{x^2-x-6}{x^3-3x^2}$

Solution

Factor each numerator and denominator and remove the common factors.

a. $\dfrac{x^4+2x^3}{x+2} = \dfrac{x^3(x+2)}{x+2} = \dfrac{x^3\cancel{(x+2)}}{\cancel{(x+2)}} = x^3$

b. $\dfrac{x^2-x-6}{x^3-3x^2} = \dfrac{(x+2)(x-3)}{x^2(x-3)} = \dfrac{(x+2)\cancel{(x-3)}}{x^2\cancel{(x-3)}} = \dfrac{x+2}{x^2}$

Practice Problem 1 Simplify each expression.

a. $\dfrac{2x^3+8x^2}{3x^2+12x}$ **b.** $\dfrac{x^2-4}{x^2+4x+4}$

WARNING

Because x^2 is a *term*, not a *factor*, in both the numerator and denominator of $\dfrac{x^2-1}{x^2+2x+1}$, you cannot remove x^2. You can remove only *factors* common to both numerator and denominator.

----

2 Multiply and divide rational expressions.

Multiplication and Division of Rational Expressions

We use the same rules for multiplying and dividing rational expressions as we do for rational numbers.

> **MULTIPLICATION AND DIVISION**
>
> For rational expressions $\dfrac{A}{B}$ and $\dfrac{C}{D}$
>
> $$\frac{A}{B} \cdot \frac{C}{D} = \frac{AC}{BD} \quad \text{and} \quad \frac{\dfrac{A}{B}}{\dfrac{C}{D}} = \frac{A}{B} \div \frac{C}{D} = \frac{A}{B} \cdot \frac{D}{C} = \frac{AD}{BC}$$
>
> if $B \neq 0, D \neq 0$.   if $B \neq 0, C \neq 0, D \neq 0$.

SIDE NOTE

The best strategy to use when multiplying and dividing rational expressions is to factor each numerator and denominator completely and then remove the common factors.

EXAMPLE 2   **Multiplying and Dividing Rational Expressions**

Multiply or divide as indicated. Simplify your answer and leave it in factored form.

a. $\dfrac{x^2 + 3x + 2}{x^3 + 3x} \cdot \dfrac{2x^3 + 6x}{x^2 + x - 2}$

b. $\dfrac{\dfrac{3x^2 + 11x - 4}{8x^3 - 40x^2}}{\dfrac{3x + 12}{4x^4 - 20x^3}}$

Solution

Factor each numerator and denominator and remove the common factors.

a.
$$\frac{x^2 + 3x + 2}{x^3 + 3x} \cdot \frac{2x^3 + 6x}{x^2 + x - 2} = \frac{(x + 2)(x + 1)}{x(x^2 + 3)} \cdot \frac{2x(x^2 + 3)}{(x - 1)(x + 2))}$$
$$= \frac{\cancel{(x + 2)}(x + 1)(2\cancel{x})\cancel{(x^2 + 3)}}{x\cancel{(x^2 + 3)}(x - 1)\cancel{(x + 2)}} = \frac{2(x + 1)}{x - 1}$$

b.
$$\frac{\dfrac{3x^2 + 11x - 4}{8x^3 - 40x^2}}{\dfrac{3x + 12}{4x^4 - 20x^3}} = \frac{3x^2 + 11x - 4}{8x^3 - 40x^2} \cdot \frac{4x^4 - 20x^3}{3x + 12} = \frac{(x + 4)(3x - 1)}{8x^2(x - 5)} \cdot \frac{4x^3(x - 5)}{3(x + 4)}$$

$$= \frac{\cancel{(x + 4)}(3x - 1)(\overset{x}{\cancel{4x^3}})\cancel{(x - 5)}}{\underset{2}{\cancel{8}x^2}\cancel{(x - 5)}(3)\cancel{(x + 4)}} = \frac{(3x - 1)x}{6} = \frac{x(3x - 1)}{6}$$

SIDE NOTE

In the quotient of two rational expressions $\dfrac{\dfrac{A}{B}}{\dfrac{C}{D}} = \dfrac{A}{B} \cdot \dfrac{D}{C}$, note that three polynomials appear as denominators: B and D from $\dfrac{A}{B}$ and $\dfrac{C}{D}$, and C from $\dfrac{A}{B} \cdot \dfrac{D}{C}$.

Practice Problem 2   Divide as indicated. Simplify your answer.

$$\frac{\dfrac{x^2 - 2x - 3}{7x^3 + 28x^2}}{\dfrac{4x + 4}{2x^4 + 8x^3}}$$

3 Add and subtract rational expressions.

Addition and Subtraction of Rational Expressions

To add and subtract rational expressions, we use the same rules as for adding and subtracting rational numbers.

> **ADDITION AND SUBTRACTION OF RATIONAL EXPRESSIONS**
>
> For rational expressions $\dfrac{A}{D}$ and $\dfrac{C}{D}$ (same denominator $D \neq 0$),
>
> $$\frac{A}{D} + \frac{C}{D} = \frac{A+C}{D} \quad \text{and} \quad \frac{A}{D} - \frac{C}{D} = \frac{A-C}{D}.$$
>
> For rational expressions $\dfrac{A}{B}$ and $\dfrac{C}{D}$ $(B \neq 0, D \neq 0)$,
>
> $$\frac{A}{B} + \frac{C}{D} = \frac{AD+BC}{BD} \quad \text{and} \quad \frac{A}{B} - \frac{C}{D} = \frac{AD-BC}{BD}.$$

EXAMPLE 3 **Adding and Subtracting Rational Expressions with the Same Denominator**

Add or subtract as indicated. Simplify your answer and leave both numerator and denominator in factored form.

a. $\dfrac{x-6}{(x+1)^2} + \dfrac{x+8}{(x+1)^2}$

b. $\dfrac{3x-2}{x^2-5x+6} - \dfrac{2x+1}{x^2-5x+6}$

Solution

a. $\dfrac{x-6}{(x+1)^2} + \dfrac{x+8}{(x+1)^2} = \dfrac{x-6+x+8}{(x+1)^2} = \dfrac{2x+2}{(x+1)^2} = \dfrac{2(x+1)}{(x+1)(x+1)} = \dfrac{2}{x+1}$

b. $\dfrac{3x-2}{x^2-5x+6} - \dfrac{2x+1}{x^2-5x+6} = \dfrac{(3x-2)-(2x+1)}{x^2-5x+6} = \dfrac{3x-2-2x-1}{x^2-5x+6}$

$$= \dfrac{x-3}{(x-2)(x-3)} = \dfrac{x-3}{(x-2)(x-3)} = \dfrac{1}{x-2}$$

Practice Problem 3 Add or subtract as indicated. Simplify your answers.

a. $\dfrac{5x+22}{x^2-36} + \dfrac{2(x+10)}{x^2-36}$ **b.** $\dfrac{4x+1}{x^2+x-12} - \dfrac{3x+4}{x^2+x-12}$

When adding or subtracting rational expressions with different denominators, we proceed (as with fractions) by finding a common denominator. The one most convenient to use is called the **least common denominator (LCD)**. It is the polynomial of least degree that contains each denominator as a factor. In the simplest case, the LCD is just the product of the denominators.

EXAMPLE 4 Adding and Subtracting When Denominators Have No Common Factor

Add or subtract as indicated. Simplify your answer and leave it in factored form.

a. $\dfrac{x}{x+1} + \dfrac{2x-1}{x+3}$ **b.** $\dfrac{2x}{x+1} - \dfrac{x}{x+2}$

Solution

a. $\dfrac{x}{x+1} + \dfrac{2x-1}{x+3} = \dfrac{x(x+3)}{(x+1)(x+3)} + \dfrac{(2x-1)(x+1)}{(x+1)(x+3)}$

$= \dfrac{x(x+3) + (2x-1)(x+1)}{(x+1)(x+3)}$ $\quad$ LCD $= (x+1)(x+3)$

$= \dfrac{x^2 + 3x + 2x^2 + 2x - x - 1}{(x+1)(x+3)}$

$= \dfrac{3x^2 + 4x - 1}{(x+1)(x+3)}$

b. $\dfrac{2x}{x+1} - \dfrac{x}{x+2} = \dfrac{2x(x+2)}{(x+1)(x+2)} - \dfrac{x(x+1)}{(x+1)(x+2)}$

$= \dfrac{2x(x+2) - x(x+1)}{(x+1)(x+2)}$ $\quad$ LCD $= (x+1)(x+2)$

$= \dfrac{2x^2 + 4x - x^2 - x}{(x+1)(x+2)}$

$= \dfrac{x^2 + 3x}{(x+1)(x+2)} = \dfrac{x(x+3)}{(x+1)(x+2)}$

Practice Problem 4 Add or subtract as indicated. Simplify your answers.

a. $\dfrac{2x}{x+2} + \dfrac{3x}{x-5}$ **b.** $\dfrac{5x}{x-4} - \dfrac{2x}{x+3}$

TO FIND THE LCD FOR RATIONAL EXPRESSIONS

1. Factor each denominator polynomial completely.
2. Form a product of the different irreducible factors of each polynomial. (Each distinct factor is used exactly once.)
3. Attach to each factor in this product the largest exponent that appears on this factor in *any* of the factored denominators.

EXAMPLE 5 Finding the LCD for Rational Expressions

Find the LCD for each pair of rational expressions.

a. $\dfrac{x+2}{x(x-1)^2(x+2)}, \quad \dfrac{3x+7}{4x^2(x+2)^3}$

b. $\dfrac{x+1}{x^2-x-6}, \quad \dfrac{2x-13}{x^2-9}$

Solution

a. The denominators are already completely factored.

$4x(x - 1)(x + 2)$ Product of the different factors

$\text{LCD} = 4x^2(x - 1)^2(x + 2)^3$ Use the largest exponent on each factor.

b. $x^2 - x - 6 = (x + 2)(x - 3)$

      $x^2 - 9 = (x + 3)(x - 3)$

   $(x + 2)(x - 3)(x + 3)$ Product of the different factors

   $\text{LCD} = (x + 2)(x - 3)(x + 3)$ The largest exponent on each factor is 1.

Practice Problem 5 Find the LCD for each pair of rational expressions.

a. $\dfrac{x^2 + 3x}{x^2(x + 2)^2(x - 2)}, \quad \dfrac{4x^2 + 1}{3x(x - 2)^2}$ **b.** $\dfrac{2x - 1}{x^2 - 25}, \quad \dfrac{3 - 7x^2}{(x^2 + 4x - 5)}$

In adding or subtracting rational expressions with different denominators, the first step is to find the LCD. Here is the general procedure.

PROCEDURE FOR ADDING OR SUBTRACTING RATIONAL EXPRESSIONS WITH DIFFERENT DENOMINATORS

Step 1 Find the LCD.

Step 2 Using $\dfrac{A}{B} = \dfrac{AC}{BC}$ write each rational expression as a rational expression with the LCD from Step 1 as the denominator.

Step 3 Following the order of operations, add or subtract numerators, keeping the LCD as the denominator.

Step 4 Simplify.

EXAMPLE 6   **Using the LCD to Add and Subtract Rational Expressions**

Add or subtract as indicated. Simplify your answer and leave it in factored form.

a. $\dfrac{3}{x^2 - 1} + \dfrac{x}{x^2 + 2x + 1}$ **b.** $\dfrac{x + 2}{x^2 - x} - \dfrac{3x}{4(x - 1)^2}$

Solution

a. Note that $x^2 - 1 = (x - 1)(x + 1)$ and $x^2 + 2x + 1 = (x + 1)^2$; the LCD is $(x + 1)^2(x - 1)$.

$\dfrac{3}{x^2 - 1} = \dfrac{3}{(x - 1)(x + 1)} = \dfrac{3(x + 1)}{(x - 1)(x + 1)^2}$ Multiply numerator and denominator by $x + 1$.

$\dfrac{x}{x^2 + 2x + 1} = \dfrac{x}{(x + 1)^2} = \dfrac{x(x - 1)}{(x + 1)^2(x - 1)}$ Multiply numerator and denominator by $x - 1$.

Now add.

$$\dfrac{3}{x^2 - 1} + \dfrac{x}{x^2 + 2x + 1} = \dfrac{3(x + 1)}{(x - 1)(x + 1)^2} + \dfrac{x(x - 1)}{(x - 1)(x + 1)^2}$$

$$= \dfrac{3(x + 1) + x(x - 1)}{(x - 1)(x + 1)^2} = \dfrac{3x + 3 + x^2 - x}{(x - 1)(x + 1)^2}$$

$$= \dfrac{x^2 + 2x + 3}{(x - 1)(x + 1)^2}$$

b. The LCD is $4x(x - 1)^2$. $\qquad\qquad\qquad\qquad\qquad\qquad x^2 - x = x(x - 1)$

$$\frac{x + 2}{x^2 - x} = \frac{x + 2}{x(x - 1)} = \frac{(x + 2) \cdot 4(x - 1)}{x(x - 1) \cdot 4(x - 1)} = \frac{4(x + 2)(x - 1)}{4x(x - 1)^2}$$ Multiply numerator and denominator by $4(x - 1)$.

$$\frac{3x}{4(x - 1)^2} = \frac{(3x)(x)}{4(x - 1)^2(x)} = \frac{3x^2}{4x(x - 1)^2}$$ Multiply numerator and denominator by x.

Now subtract.

$$\frac{x + 2}{x^2 - x} - \frac{3x}{4x(x - 1)^2} = \frac{4(x + 2)(x - 1)}{4x(x - 1)^2} - \frac{3x^2}{4x(x - 1)^2} = \frac{4(x + 2)(x - 1) - 3x^2}{4x(x - 1)^2}$$

$$= \frac{4x^2 + 4x - 8 - 3x^2}{4x(x - 1)^2} = \frac{x^2 + 4x - 8}{4x(x - 1)^2}$$

Practice Problem 6 Add or subtract as indicated. Simplify your answers.

a. $\dfrac{4}{x^2 - 4x + 4} + \dfrac{x}{x^2 - 4}$ $\qquad$ **b.** $\dfrac{2x}{3(x - 5)^2} - \dfrac{6x}{2(x^2 - 5x)}$

4 Identify and simplify complex fractions.

Complex Fractions

A rational expression that contains another rational expression in its numerator or denominator (or both) is called a **complex rational expression** or **complex fraction**. To simplify a complex fraction, we write it as a rational expression in lowest terms.

There are two effective methods of simplifying complex fractions.

PROCEDURES FOR SIMPLIFYING COMPLEX FRACTIONS

Method 1 If necessary, simplify the numerator and/or denominator of the complex fraction, so that each contains one rational expression. Then multiply the resulting numerator by the reciprocal of the denominator.

Method 2 Multiply the numerator and the denominator of the complex fraction by the LCD of all rational expressions that appear in either the numerator or the denominator. Simplify the result.

EXAMPLE 7 **Simplifying a Complex Fraction**

Simplify $\dfrac{\dfrac{1}{2} + \dfrac{1}{x}}{\dfrac{x^2 - 4}{2x}}$ using each of the two methods.

Solution

Method 1 $\dfrac{\dfrac{1}{2} + \dfrac{1}{x}}{\dfrac{x^2 - 4}{2x}} = \dfrac{\dfrac{x + 2}{2x}}{\dfrac{x^2 - 4}{2x}} = \dfrac{x + 2}{2x} \cdot \dfrac{2x}{x^2 - 4} = \dfrac{x + 2}{2x} \cdot \dfrac{2x}{(x + 2)(x - 2)}$

$$= \frac{(x + 2)(2x)}{(2x)(x + 2)(x - 2)} = \frac{1}{x - 2}$$

Method 2 The LCD of $\dfrac{1}{2}$, $\dfrac{1}{x}$, and $\dfrac{x^2-4}{2x}$ is $2x$.

$$\frac{\dfrac{1}{2}+\dfrac{1}{x}}{\dfrac{x^2-4}{2x}}=\frac{\left(\dfrac{1}{2}+\dfrac{1}{x}\right)(2x)}{\left(\dfrac{x^2-4}{2x}\right)(2x)}\qquad\text{Multiply numerator and denominator by the LCD.}$$

$$=\frac{\dfrac{1}{2}(2x)+\dfrac{1}{x}(2x)}{\left[\dfrac{(x^2-4)}{2x}\right](2x)}=\frac{x+2}{x^2-4}=\frac{x+2}{(x-2)(x+2)}=\frac{1}{x-2}$$

Practice Problem 7 Simplify $\dfrac{\dfrac{5}{3x}+\dfrac{1}{3}}{\dfrac{x^2-25}{3x}}$.

EXAMPLE 8 **Simplifying a Complex Fraction**

Simplify $\dfrac{x}{2x-\dfrac{7}{3-\dfrac{1}{2}}}$.

Solution

We will use Method 1.

Begin with $\dfrac{7}{3-\dfrac{1}{2}}=\dfrac{7}{\dfrac{5}{2}}=7\cdot\dfrac{2}{5}=\dfrac{14}{5}$.   Replace $3-\dfrac{1}{2}$ with $\dfrac{5}{2}$.

Then $\dfrac{x}{2x-\dfrac{7}{3-\dfrac{1}{2}}}=\dfrac{x}{2x-\dfrac{14}{5}}$ Replace $\dfrac{7}{3-\dfrac{1}{2}}$ with $\dfrac{14}{5}$.

$$=\frac{x}{\dfrac{10x-14}{5}}=x\cdot\frac{5}{10x-14}=\frac{5x}{10x-14}=\frac{5x}{2(5x-7)}$$

Practice Problem 8 Simplify $\dfrac{5x}{3x-\dfrac{4}{2-\dfrac{1}{3}}}$.

Answers to Practice Problems

1. a. $\dfrac{2x}{3}$ **b.** $\dfrac{x-2}{x+2}$ **2.** $\dfrac{x(x-3)}{14}$ **3. a.** $\dfrac{7}{x-6}$

b. $\dfrac{1}{x+4}$ **4. a.** $\dfrac{x(5x-4)}{(x+2)(x-5)}$ **b.** $\dfrac{x(3x+23)}{(x+3)(x-4)}$

5. a. $3x^2(x+2)^2(x-2)^2$ **b.** $(x-5)(x+5)(x-1)$

6. a. $\dfrac{x^2+2x+8}{(x+2)(x-2)^2}$ **b.** $\dfrac{45-7x}{3(x-5)^2}$ **7.** $\dfrac{1}{x-5}$ **8.** $\dfrac{25x}{15x-12}$

SECTION A.3 Exercises

Basic Concepts and Skills

In Exercises 1–10, reduce each rational expression to lowest terms. Identify all numbers that must be excluded from the domain of the given rational expression.

1. $\dfrac{2x + 2}{x^2 + 2x + 1}$

2. $\dfrac{3x - 6}{x^2 - 4x + 4}$

3. $\dfrac{3x + 3}{x^2 - 1}$

4. $\dfrac{15 + 3x}{x^2 - 25}$

5. $\dfrac{x^2 - 6x + 9}{4x - 12}$

6. $\dfrac{x^2 - 10x + 25}{3x - 15}$

7. $\dfrac{7x^2 + 7x}{x^2 + 2x + 1}$

8. $\dfrac{x^2 + 2x - 15}{x^2 - 7x + 12}$

9. $\dfrac{6x^4 + 14x^3 + 4x^2}{6x^4 - 10x^3 - 4x^2}$

10. $\dfrac{3x^3 + x^2}{3x^4 - 11x^3 - 4x^2}$

In Exercises 11–20, multiply or divide as indicated. Simplify and leave the numerator and denominator in your answer in factored form.

11. $\dfrac{x - 3}{2x + 4} \cdot \dfrac{10x + 20}{5x - 15}$

12. $\dfrac{25x^2 - 9}{4 - 2x} \cdot \dfrac{4 - x^2}{10x - 6}$

13. $\dfrac{x^2 - x - 6}{x^2 + 3x + 2} \cdot \dfrac{x^2 - 1}{x^2 - 9}$

14. $\dfrac{2 - x}{x + 1} \cdot \dfrac{x^2 + 3x + 2}{x^2 - 4}$

15. $\dfrac{x^2 - 9}{x} \div \dfrac{2x + 6}{5x^2}$

16. $\dfrac{x^2 - 1}{3x} \div \dfrac{7x - 7}{x^2 + x}$

17. $\dfrac{x^2 + 2x - 3}{x^2 + 8x + 16} \div \dfrac{x - 1}{3x + 12}$

18. $\dfrac{x^2 + 5x + 6}{x^2 + 6x + 9} \div \dfrac{x^2 + 3x + 2}{x^2 + 7x + 12}$

19. $\left(\dfrac{x^2 - 9}{x^3 + 8} \div \dfrac{x + 3}{x^3 + 2x^2 - x - 2}\right) \dfrac{1}{x^2 - 1}$

20. $\left(\dfrac{x^2 - 25}{x^2 - 3x - 4} \div \dfrac{x^2 + 3x - 10}{x^2 - 1}\right) \dfrac{x - 2}{x - 5}$

In Exercises 21–30, add and subtract as indicated. Simplify and leave the numerator and denominator in your answer in factored form.

21. $\dfrac{x}{2x + 1} + \dfrac{4}{2x + 1}$

22. $\dfrac{x^2}{x + 1} - \dfrac{x^2 - 1}{x + 1}$

23. $\dfrac{4}{3 - x} + \dfrac{2x}{x - 3}$

24. $\dfrac{5x}{x^2 + 1} + \dfrac{2x}{x^2 + 1}$

25. $\dfrac{x}{2(x - 1)^2} + \dfrac{3x}{2(x - 1)^2}$

26. $\dfrac{x}{x^2 - 4} - \dfrac{2}{x^2 - 4}$

27. $\dfrac{5x}{x^2 - 1} - \dfrac{5}{x^2 - 1}$

28. $\dfrac{x - 2}{2x + 1} - \dfrac{x}{2x - 1}$

29. $\dfrac{-x}{x + 2} + \dfrac{x - 2}{x} - \dfrac{x}{x - 2}$

30. $\dfrac{3x}{x - 1} + \dfrac{x + 1}{x} - \dfrac{2x}{x + 1}$

In Exercises 31–36, find the LCD for each pair of rational fractions.

31. $\dfrac{5}{3x - 6}, \dfrac{2x}{4x - 8}$

32. $\dfrac{3 - x}{4x^2 - 1}, \dfrac{7x}{(2x + 1)^2}$

33. $\dfrac{1 - x}{x^2 + 3x + 2}, \dfrac{3x + 12}{x^2 - 1}$

34. $\dfrac{5x + 9}{x^2 - x - 6}, \dfrac{x + 5}{x^2 - 9}$

35. $\dfrac{7 - 4x}{x^2 - 5x + 4}, \dfrac{x^2 - x}{x^2 + x - 2}$

36. $\dfrac{13x}{x^2 - 2x - 3}, \dfrac{2x^2 - 4}{x^2 + 3x + 2}$

In Exercises 37–52, perform the indicated operations and simplify the result. Leave the numerator and denominator in your answer in factored form.

37. $\dfrac{5}{x - 3} + \dfrac{2x}{x^2 - 9}$

38. $\dfrac{3x}{x - 1} + \dfrac{x}{x^2 - 1}$

39. $\dfrac{2x}{x^2 - 4} - \dfrac{x}{x + 2}$

40. $\dfrac{3x - 1}{x^2 - 16} - \dfrac{2x + 1}{x - 4}$

41. $\dfrac{x - 2}{x^2 + 3x - 10} + \dfrac{x + 3}{x^2 + x - 6}$

42. $\dfrac{x + 3}{x^2 - x - 2} + \dfrac{x - 1}{x^2 + 2x + 1}$

43. $\dfrac{2x - 3}{9x^2 - 1} + \dfrac{4x - 1}{(3x - 1)^2}$

44. $\dfrac{3x + 1}{(2x + 1)^2} + \dfrac{x + 3}{4x^2 - 1}$

45. $\dfrac{x - 3}{x^2 - 25} - \dfrac{x - 3}{x^2 + 9x + 20}$

46. $\dfrac{2x}{x^2 - 16} - \dfrac{2x - 7}{x^2 - 7x + 12}$

47. $\dfrac{3}{x^2 - 4} + \dfrac{1}{2 - x} - \dfrac{1}{2 + x}$

48. $\dfrac{2}{5 + x} + \dfrac{5}{x^2 - 25} + \dfrac{7}{5 - x}$

49. $\dfrac{x + 3a}{x - 5a} - \dfrac{x + 5a}{x - 3a}$

50. $\dfrac{3x - a}{2x - a} - \dfrac{2x + a}{3x + a}$

51. $\dfrac{1}{x + h} - \dfrac{1}{x}$

52. $\dfrac{1}{(x + h)^2} - \dfrac{1}{x^2}$

In Exercises 53–66, perform the indicated operations and simplify the result. Leave the numerator and denominator in your answer in factored form.

53. $\dfrac{\dfrac{2}{x}}{\dfrac{3}{x^2}}$

54. $\dfrac{\dfrac{-6}{x}}{\dfrac{2}{x^3}}$

55. $\dfrac{\dfrac{1}{x}}{1 - \dfrac{1}{x}}$

56. $\dfrac{\dfrac{1}{x^2}}{\dfrac{1}{x^2} - 1}$

57. $\dfrac{\dfrac{1}{x} - 1}{\dfrac{1}{x} + 1}$

58. $\dfrac{2 - \dfrac{2}{x}}{1 + \dfrac{2}{x}}$

59. $\dfrac{\dfrac{1}{x} - x}{1 - \dfrac{1}{x^2}}$

60. $\dfrac{x - \dfrac{1}{x}}{\dfrac{1}{x^2} - 1}$

61. $x - \dfrac{x}{x + \dfrac{1}{2}}$

62. $x + \dfrac{x}{x + \dfrac{1}{2}}$

63. $\dfrac{\dfrac{1}{x + h} - \dfrac{1}{x}}{h}$

64. $\dfrac{\dfrac{1}{(x + h)^2} - \dfrac{1}{x^2}}{h}$

65. $\dfrac{\dfrac{1}{x - a} + \dfrac{1}{x + a}}{\dfrac{1}{x - a} - \dfrac{1}{x + a}}$

66. $\dfrac{\dfrac{1}{x - a} + \dfrac{1}{x + a}}{\dfrac{x}{x - a} - \dfrac{a}{x + a}}$

Applying the Concepts

67. Bearing length. The length (in centimeters) of a bearing in the shape of a cylinder is given by $\dfrac{125.6}{\pi r^2}$, where r is the radius of the base of the cylinder. Find the length of a bearing with a diameter of 4 centimeters, using 3.14 as an approximation of π.

68. Toy box height. The height (in feet) of an open toy box that is twice as long as it is wide is given by $\dfrac{13.5 - 2x^2}{6x}$, where x is the width of the box. Find the height of a toy box that is 1.5 feet wide.

69. Diluting a mixture. A 100-gallon mixture of citrus extract and water is 3% citrus extract.
  a. Write a rational expression in x whose values give the percentage (in decimal form) of the mixture that is citrus extract when x gallons of water are added to the mixture.
  b. Find the percentage of citrus extract in the mixture if 50 gallons of water are added to it.

70. Acidity in a reservoir. A half-full 400,000-gallon reservoir is found to be 0.75% acid.
  a. Write a rational expression in x whose values give the percentage (in decimal form) of acid in the reservoir when x gallons of water are added to it.
  b. Find the percentage of acid in the reservoir if 100,000 gallons of water are added to it.

71. Diet lemonade. A company packages its powdered diet lemonade mix in containers in the shape of a cylinder. The top and bottom are made of a tin product that costs 5 cents per square inch. The side of the container is made of a cheaper material that costs 1 cent per square inch. The height of any cylinder can be found by dividing its volume by the area of its base.
  a. If x is the radius of the base (in inches), write a rational expression in x whose values give the cost of each container, assuming that the capacity of the container is 120 cubic inches.
  b. Find the cost of a container whose base is 4 inches in diameter, using 3.14 as an approximation of π.

72. Storage containers. A company makes storage containers in the shape of an open box with a square base. The sides and base of the box are made of a plastic that costs 40 cents per square foot. The height of any box can be found by dividing its volume by the area of its base.
  a. If x is the length of the base (in feet), write a rational expression in x whose values give the cost of each container, assuming that the capacity of the container is 2.25 cubic feet.
  b. Find the cost of a container whose base is 1.5 feet long.

Radicals and Rational Exponents

Square Roots

When we raise the number 5 to the power 2, we write $5^2 = 25$. The reverse of this squaring process is called finding a **square root**. We know that 5 is a square root of 25, and we write $\sqrt{25} = 5$. It is also true that $(-5)^2 = 25$; so -5 is *another* square root of 25.

The symbol $\sqrt{}$, called a **radical sign**, is used to distinguish between 5 and -5, the two square roots of 25. We write $5 = \sqrt{25}$ and $-5 = -\sqrt{25}$ and call $\sqrt{25}$ the *principal square root* of 25.

Principal Square Root

$$\sqrt{a} = b \quad \text{means} \quad (1) \ b^2 = a \quad \text{and} \quad (2) \ b \geq 0$$

1 Define and evaluate *n*th roots.

If $\sqrt{a} = b$, then $a = b^2 \geq 0$. Consequently, the symbol $\sqrt{a}$ denotes a real number only when $a \geq 0$. The domain of the expression $\sqrt{x}$ is then $x \geq 0$, or in interval notation, $[0, \infty)$.

SQUARE ROOT PROPERTIES

1. For any real number x, $\sqrt{x^2} = |x|$ and $(\sqrt{x})^2 = x$ if $x \geq 0$.
2. For any real number x, if $x^2 = a$ then $x = \pm\sqrt{a}$.
3. If $a \geq 0$ and $b > 0$ are real numbers, then $\sqrt{ab} = \sqrt{a}\sqrt{b}$, $\sqrt{\dfrac{a}{b}} = \dfrac{\sqrt{a}}{\sqrt{b}}$.

For example, $\sqrt{121} = \sqrt{11^2} = 11$, $\sqrt{\dfrac{25}{81}} = \sqrt{\dfrac{5^2}{9^2}} = \sqrt{\left(\dfrac{5}{9}\right)^2} = \dfrac{5}{9}$, $(\sqrt{3})^2 = 3$.

Other Roots

If n is a positive integer, we say that b is an *n*th root of a if $b^n = a$. We define the **principal *n*th root** of a real number a, denoted by the symbol $\sqrt[n]{a}$, as follows.

THE PRINCIPAL *n*TH ROOT OF A REAL NUMBER

1. If a is positive ($a > 0$), then $\sqrt[n]{a} = b$ provided that $b^n = a$ and $b > 0$.
2. If a is negative ($a < 0$) and n is odd, then $\sqrt[n]{a} = b$ provided that $b^n = a$.
3. If a is negative ($a < 0$) and n is even, then $\sqrt[n]{a}$ is not a real number.
4. If $a = 0$, then $\sqrt[n]{a} = 0$.

We use the same vocabulary for *n*th roots that we used for square roots. That is, $\sqrt[n]{a}$ is called a **radical expression**, $\sqrt{}$ is the **radical sign**, and a is the **radicand**. The positive integer n is called the **index**. The index 2 is not written; we write $\sqrt{a}$ rather than $\sqrt[2]{a}$ for the principal square root of a. It is also common practice to call $\sqrt[3]{a}$ the **cube root** of a.

EXAMPLE 1 **Finding Principal *n*th Roots**

Find each root.

a. $\sqrt[3]{27}$ **b.** $\sqrt[3]{-64}$ **c.** $\sqrt[4]{16}$ **d.** $\sqrt[4]{(-3)^4}$ **e.** $\sqrt[8]{-46}$

Solution

a. The radicand, 27, is *positive*; so $\sqrt[3]{27} = 3$ because $3^3 = 27$ and $3 > 0$.

b. The radicand, -64, is *negative*; so $\sqrt[3]{-64} = -4$ because $(-4)^3 = -64$.

c. The radicand, 16, is *positive*; so $\sqrt[4]{16} = 2$ because $2^4 = 16$ and $2 > 0$.

d. The radicand, $(-3)^4$, is *positive*; so $\sqrt[4]{(-3)^4} = 3$ because $3^4 = (-3)^4$ and $3 > 0$.

e. The radicand, -46, is *negative*; and the index, 8, is even; so $\sqrt[8]{-46}$ is not a real number.

Practice Problem 1 Find each root.

a. $\sqrt[3]{-8}$ **b.** $\sqrt[5]{32}$ **c.** $\sqrt[4]{81}$ **d.** $\sqrt[6]{-4}$

 **2** Simplify expressions involving radicals.

Simplifying Radical Expressions

The following properties are often used to simplify expressions involving radicals.

> ### PROPERTIES OF RADICALS
>
> Let m and n be natural numbers with $n > 1$, and a and b be real numbers. Assume all expressions are defined.
>
> **1.** $(\sqrt[n]{a})^n = a$
>
> **2.** $\sqrt[n]{a^n} = \begin{cases} |a|, & \text{If } n \text{ is even} \\ a, & \text{if } n \text{ is odd} \end{cases}$
>
> **3.** $\sqrt[n]{ab} = \sqrt[n]{a} \cdot \sqrt[n]{b}$
>
> **4.** $\sqrt[n]{\dfrac{a}{b}} = \dfrac{\sqrt[n]{a}}{\sqrt[n]{b}}$
>
> **5.** $\sqrt[n]{a^m} = (\sqrt[n]{a})^m$
>
> **6.** $\sqrt[m]{\sqrt[n]{a}} = \sqrt[mn]{a}$

EXAMPLE 2 Simplifying nth Roots

Simplify.

a. $\sqrt[3]{135}$ **b.** $\sqrt[4]{162a^4}$ **c.** $\sqrt[5]{(32)^3}$

Solution

a. Because $135 = 27 \cdot 5$ and 27 is a perfect cube ($3^3 = 27$), by property 3 we have

$$\sqrt[3]{135} = \sqrt[3]{27 \cdot 5} = \sqrt[3]{27} \cdot \sqrt[3]{5} = 3\sqrt[3]{5}.$$

b. Because $162 = 81 \cdot 2$ and 81 is a perfect fourth power ($3^4 = 81$), by properties 2 and 3 we have

$$\sqrt[4]{162a^4} = \sqrt[4]{162}\sqrt[4]{a^4} = \sqrt[4]{81 \cdot 2}|a| = \sqrt[4]{81}\sqrt[4]{2}|a| = 3\sqrt[4]{2}|a|.$$

c. Because $2^5 = 32$, $\sqrt[5]{32} = 2$. Using property 5, we have

$$\sqrt[5]{(32)^3} = (\sqrt[5]{32})^3 = (2)^3 = 8.$$

Practice Problem 2 Simplify.

a. $\sqrt[3]{72}$ **b.** $\sqrt[4]{48a^2}$ **c.** $\sqrt[4]{a^3}\sqrt[4]{a^5}, a \geq 0$

Like Radicals

Radicals that have the same index and the same radicand are called **like radicals**. Like radicals can be combined with the use of the distributive property.

EXAMPLE 3 Adding and Subtracting Radical Expressions

Simplify.

a. $\sqrt{45} + 7\sqrt{20}$ **b.** $5\sqrt[3]{80x} - 3\sqrt[3]{270x}$

Solution

a. We find perfect-square factors for both 45 and 20.

$$\sqrt{45} + 7\sqrt{20} = \sqrt{9 \cdot 5} + 7\sqrt{4 \cdot 5}$$
$$= 3\sqrt{5} + 14\sqrt{5} = (3 + 14)\sqrt{5} = 17\sqrt{5}$$

b. This time we find perfect-cube factors for both 80 and 270.

$$5\sqrt[3]{80x} - 3\sqrt[3]{270x} = 5\sqrt[3]{8 \cdot 10x} - 3\sqrt[3]{27 \cdot 10x}$$
$$= 10\sqrt[3]{10x} - 9\sqrt[3]{10x} = (10 - 9)\sqrt[3]{10x} = \sqrt[3]{10x}$$

Practice Problem 3 Simplify.

a. $3\sqrt{12} + 7\sqrt{3}$ **b.** $2\sqrt[3]{135x} - 3\sqrt[3]{40x}$

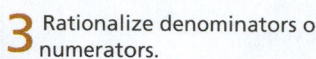

3 Rationalize denominators or numerators.

Rationalizing

Removing radicals in the denominator or the numerator of a fraction is called **rationalizing the denominator** or **rationalizing the numerator**. The procedure for rationalizing involves multiplying the fraction by 1 in a special way so as to obtain a perfect *n*th power.

EXAMPLE 4 Rationalizing the Denominator

Rationalize each denominator

a. $\sqrt{\dfrac{5}{2}}$ **b.** $\sqrt[3]{\dfrac{2x^4}{9y^5}}$

Solution

a. We multiply the numerator and the denominator of the radicand so that the denominator becomes a perfect square. So

$$\sqrt{\frac{5}{2}} = \sqrt{\frac{5 \cdot 2}{2 \cdot 2}} = \sqrt{\frac{10}{2^2}} = \frac{\sqrt{10}}{\sqrt{2^2}} = \frac{\sqrt{10}}{2}.$$

b. We multiply the numerator and the denominator of the radicand so that the denominator becomes a perfect cube.

$$\sqrt[3]{\frac{2x^4}{9y^5}} = \sqrt[3]{\frac{2x^4 \cdot 3y}{9y^5 \cdot 3y}} \qquad 9 \cdot 3 = 27 = (3)^3 \text{ and } y^5 \cdot y = y^6 = (y^2)^3$$

$$= \frac{\sqrt[3]{6x^4 y}}{\sqrt[3]{27y^6}} \qquad \sqrt[n]{\frac{a}{b}} = \frac{\sqrt[n]{a}}{\sqrt[n]{b}}$$

$$= \frac{\sqrt[3]{x^3(6xy)}}{\sqrt[3]{(3y^2)^3}} \qquad x^4 = x^3 \cdot x; \ 27y^6 = (3y^2)^3$$

$$= \frac{\sqrt[3]{x^3}\sqrt[3]{6xy}}{\sqrt[3]{(3y^2)^3}} \qquad \sqrt[3]{ab} = \sqrt[3]{a}\sqrt[3]{b}$$

$$= \frac{x\sqrt[3]{6xy}}{3y^2} \qquad \sqrt[n]{a^n} = a, \text{ if } n \text{ is odd}$$

Practice Problem 4 Rationalize each denominator.

a. $\dfrac{7}{\sqrt{8}}$ b. $\sqrt[3]{\dfrac{a}{4b^4}}$

Use Conjugates

Pairs of expressions of the form $\sqrt{x} + \sqrt{y}$ and $\sqrt{x} - \sqrt{y}$ are called **conjugates**. We form the product

$$(\sqrt{x} + \sqrt{y})(\sqrt{x} - \sqrt{y}) = (\sqrt{x})^2 - (\sqrt{y})^2 \qquad (a+b)(a-b) = a^2 - b^2$$
$$= x - y.$$

We see that the product contains no radicals. So, if the denominator or the numerator of a fraction is an expression of the form $\sqrt{x} + \sqrt{y}$ or $\sqrt{x} - \sqrt{y}$, conjugates can be used to rationalize the denominator or the numerator.

EXAMPLE 5 **Rationalizing the Numerator**

Rationalize the numerator and simplify:

$$\frac{\sqrt{x+h+2} - \sqrt{x+2}}{h}.$$

Solution

$$\frac{\sqrt{x+h+2} - \sqrt{x+2}}{h}$$

$$= \frac{(\sqrt{x+h+2} - \sqrt{x+2})(\sqrt{x+h+2} + \sqrt{x+2})}{h(\sqrt{x+h+2} + \sqrt{x+2})} \qquad \text{Multiply and divide by conjugate.}$$

$$= \frac{(x+h+2) - (x+2)}{h(\sqrt{x+h+2} + \sqrt{x+2})} \qquad \begin{aligned}&(\sqrt{a} - \sqrt{b})(\sqrt{a} + \sqrt{b})\\ &= a - b\end{aligned}$$

$$= \frac{x+h+2-x-2}{h(\sqrt{x+h+2} + \sqrt{x+2})} \qquad \text{Distribute}$$

$$= \frac{h}{h(\sqrt{x+h+2} + \sqrt{x+2})} \qquad \text{Simplify}$$

$$= \frac{1}{\sqrt{x+h+2} + \sqrt{x+2}}. \qquad \text{Remove } h$$

Practice Problem 5 Rationalize the numerator and simplify:

$$\frac{\sqrt{x} - \sqrt{3}}{x^2 - 9}.$$

4 Define and evaluate rational exponents.

Rational Exponents

We know what 2^3 means, but what about $2^{1/3}$?

THE EXPONENT $\dfrac{1}{n}$

For any real number a and any integer $n > 1$,

$$a^{1/n} = \sqrt[n]{a}.$$

When n is even and $a < 0$, $\sqrt[n]{a}$ and $a^{1/n}$ are not real numbers.

We use this definition to evaluate some expressions with rational exponents.

$$16^{1/2} = \sqrt{16} = 4$$

$$(-27)^{1/3} = \sqrt[3]{-27} = \sqrt[3]{(-3)^3} = -3$$

$$\left(\frac{1}{16}\right)^{1/4} = \sqrt[4]{\frac{1}{16}} = \sqrt[4]{\left(\frac{1}{2}\right)^4} = \frac{1}{2}$$

$$32^{1/5} = \sqrt[5]{32} = \sqrt[5]{2^5} = 2$$

$(-5)^{1/4}$ is not a real number because the base is negative and 4 is even.

We next define $a^{m/n}$, where m and n are integers, when $n > 1$ and $\sqrt[n]{a}$ is a real number.

> ### RATIONAL EXPONENTS
>
> $$a^{m/n} = (\sqrt[n]{a})^m = \sqrt[n]{a^m}$$
>
> provided that m and n are integers with no common factors, $n > 1$, and $\sqrt[n]{a}$ is a real number.

Note that the numerator m of the exponent $\dfrac{m}{n}$ is the *exponent of the radicand* and the denominator n is the *index of the radical expression*. When n is even and $a < 0$, the symbol $a^{m/n}$ is not a real number.

We change rational exponents with negative denominators such as $\dfrac{3}{-5}$ and $\dfrac{-2}{-7}$ to $-\dfrac{3}{5}$ and $\dfrac{2}{7}$, respectively, before applying the definition. Note that $(-8)^{2/6} \neq [(-8)^{1/6}]^2$ because $(-8)^{1/6} = \sqrt[6]{-8}$ is not a real number but $(-8)^{2/6}$ is a real number. Writing $\dfrac{2}{6} = \dfrac{1}{3}$, we have

$$(-8)^{2/6} = (-8)^{1/3} = \sqrt[3]{-8} = -2.$$

We can now evaluate expressions having rational exponents.

$$8^{2/3} = (\sqrt[3]{8})^2 = 2^2 = 4$$

$$-16^{5/2} = -(\sqrt{16})^5 = -4^5 = -1024$$

$$100^{-3/2} = ((100)^{1/2})^{-3} = (\sqrt{100})^{-3} = \frac{1}{(\sqrt{100})^3} = \frac{1}{10^3} = \frac{1}{1000}$$

$$(-25)^{7/2} = [(-25)^{1/2}]^7 = (\sqrt{-25})^7, \text{ which is not a real number.}$$

EXAMPLE 6 Simplifying Expressions Having Rational Exponents

Simplify. Express your answer with only positive exponents. Assume that x represents a positive real number.

a. $2x^{1/3} \cdot 5x^{1/4}$ 　　**b.** $\dfrac{21x^{-2/3}}{7x^{1/5}}$ 　　**c.** $(x^{3/5})^{-1/6}$

Solution

a. $2x^{1/3} \cdot 5x^{1/4} = 2 \cdot 5x^{1/3} \cdot x^{1/4} = 10x^{1/3+1/4} = 10x^{4/12+3/12} = 10x^{7/12}$

b. $\dfrac{21x^{-2/3}}{7x^{1/5}} = \left(\dfrac{21}{7}\right)\left(\dfrac{x^{-2/3}}{x^{1/5}}\right) = 3x^{-2/3-1/5} = 3x^{-10/15-3/15} = 3x^{-13/15} = 3 \cdot \left(\dfrac{1}{x^{13/15}}\right)$

$$= \dfrac{3}{x^{13/15}}$$

c. $(x^{3/5})^{-1/6} = x^{(3/5)(-1/6)} = x^{-1/10} = \dfrac{1}{x^{1/10}}$

Practice Problem 6 Simplify. Express the answer with only positive exponents.

a. $4x^{1/2} \cdot 3x^{1/5}$ **b.** $\dfrac{25x^{-1/4}}{5x^{1/3}}, x > 0$ **c.** $(x^{2/3})^{-1/5}, x \neq 0$

EXAMPLE 7 **Simplifying an Expression Involving Negative Rational Exponents**

Simplify $x(x + 1)^{-1/2} + 2(x + 1)^{1/2}$. Express your answer with only positive exponents and then rationalize the denominator. Assume that $(x + 1)^{-1/2}$ is defined.

Solution

$$x(x + 1)^{-1/2} + 2(x + 1)^{1/2} = \frac{x}{(x + 1)^{1/2}} + \frac{2(x + 1)^{1/2}}{1} \qquad (x + 1)^{-1/2} = \frac{1}{(x + 1)^{1/2}}$$

$$= \frac{x}{(x + 1)^{1/2}} + \frac{2(x + 1)^{1/2}(x + 1)^{1/2}}{(x + 1)^{1/2}}$$

$$= \frac{x + 2(x + 1)}{(x + 1)^{1/2}} = \frac{3x + 2}{(x + 1)^{1/2}} \qquad \begin{array}{c}(x + 1)^{1/2}(x + 1)^{1/2} \\ = x + 1\end{array}$$

To rationalize the denominator $(x + 1)^{1/2} = \sqrt{x + 1}$, multiply both numerator and denominator by $(x + 1)^{1/2}$.

$$\frac{3x + 2}{(x + 1)^{1/2}} = \frac{(3x + 2)(x + 1)^{1/2}}{(x + 1)^{1/2}(x + 1)^{1/2}} = \frac{(3x + 2)(x + 1)^{1/2}}{x + 1}$$

Practice Problem 7 Simplify $x(x + 3)^{-1/2} + (x + 3)^{1/2}$.

EXAMPLE 8 **Factoring an Expression Involving Rational Exponents**

Factor $\dfrac{4}{3}x^{1/3}(3x - 1) + 3x^{4/3}$.

Solution

Begin by recognizing that $x^{4/3} = x \cdot x^{1/3}$; then factor out $\dfrac{1}{3}x^{1/3}$.

$$\frac{4}{3}x^{1/3}(3x - 1) + 3x^{4/3} = \frac{4}{3}x^{1/3}(3x - 1) + \frac{9}{3}x \cdot x^{1/3}$$

$$= \frac{1}{3}x^{1/3}[4(3x - 1) + 9x] = \frac{1}{3}x^{1/3}(21x - 4)$$

Practice Problem 8 Factor $\dfrac{4}{3}x^{1/3}(x + 5) + x^{4/3}$.

Answers to Practice Problems

1. a. -2 **b.** 2 **c.** 3 **d.** Not a real number
2. a. $2\sqrt[3]{9}$ **b.** $2\sqrt[4]{3a^2}$ **c.** a^2 **3. a.** $13\sqrt{3}$ **b.** 0
4. a. $\dfrac{7\sqrt{2}}{4}$ **b.** $\dfrac{\sqrt[3]{2ab^2}}{2b^2}$ **5.** $\dfrac{1}{(x + 3)(\sqrt{x} + \sqrt{3})}$

6. a. $12x^{7/10}$ **b.** $\dfrac{5}{x^{7/12}}$ **c.** $\dfrac{1}{x^{2/15}}$ **7.** $\dfrac{(2x + 3)(x + 3)^{1/2}}{x + 3}$
8. $\dfrac{1}{3}x^{1/3}(7x + 20)$

SECTION A.4 | Exercises

Basic Concepts and Skills

In Exercises 1–16, evaluate each root or state that the root is not a real number.

1. $\sqrt{64}$

2. $\sqrt{100}$

3. $\sqrt[3]{64}$

4. $\sqrt[3]{125}$

5. $\sqrt[3]{-27}$

6. $\sqrt[3]{-216}$

7. $\sqrt[3]{-\dfrac{1}{8}}$

8. $\sqrt[3]{\dfrac{27}{64}}$

9. $\sqrt{(-3)^2}$

10. $-\sqrt{4(-3)^2}$

11. $\sqrt[4]{-16}$

12. $\sqrt[6]{-64}$

13. $-\sqrt[5]{-1}$

14. $\sqrt[3]{-1}$

15. $\sqrt[5]{(-7)^5}$

16. $-\sqrt[5]{(-4)^5}$

In Exercises 17–30, simplify each expression. Assume that all variables represent nonnegative real numbers.

17. $2\sqrt{3} + 5\sqrt{3}$

18. $7\sqrt{2} - \sqrt{2}$

19. $6\sqrt{5} - \sqrt{5} + 4\sqrt{5}$

20. $5\sqrt{7} + 3\sqrt{7} - 2\sqrt{7}$

21. $\sqrt{98x} - \sqrt{32x}$

22. $\sqrt{45x} + \sqrt{20x}$

23. $\sqrt[3]{24} - \sqrt[3]{81}$

24. $\sqrt[3]{54} + \sqrt[3]{16}$

25. $\sqrt[3]{3x} - 2\sqrt[3]{24x} + \sqrt[3]{375x}$

26. $\sqrt[3]{16x} + 3\sqrt[3]{54x} - \sqrt[3]{2x}$

27. $\sqrt{2x^5} - 5\sqrt{32x} + \sqrt{18x^3}$

28. $2\sqrt{50x^5} + 7\sqrt{2x^3} - 3\sqrt{72x}$

29. $\sqrt{48x^5y} - 4y\sqrt{3x^3y} + y\sqrt{3xy^3}$

30. $x\sqrt{8xy} + 4\sqrt{2xy^3} - \sqrt{18x^5y}$

In Exercises 31–40, rationalize the denominator of each expression.

31. $\dfrac{2}{\sqrt{3}}$

32. $\dfrac{10}{\sqrt{5}}$

33. $\dfrac{1}{\sqrt{2} + x}$

34. $\dfrac{1}{\sqrt{5} + 2x}$

35. $\dfrac{1}{\sqrt{3} + \sqrt{2}}$

36. $\dfrac{6}{\sqrt{5} - \sqrt{3}}$

37. $\dfrac{\sqrt{5} - \sqrt{2}}{\sqrt{5} + \sqrt{2}}$

38. $\dfrac{\sqrt{7} + \sqrt{3}}{\sqrt{7} - \sqrt{3}}$

39. $\dfrac{\sqrt{x + h} - \sqrt{x}}{\sqrt{x + h} + \sqrt{x}}$

40. $\dfrac{a\sqrt{x} + 3}{a\sqrt{x} - 3}$

In Exercises 41–48, rationalize the numerator of each expression.

41. $\dfrac{\sqrt{4 + h} - 2}{h}$

42. $\dfrac{\sqrt{y + 9} - 3}{y}$

43. $\dfrac{2 - \sqrt{4 - x}}{x}$

44. $\dfrac{\sqrt{x + 2} - \sqrt{2}}{x}$

45. $\dfrac{\sqrt{x} - 2}{x - 4}$

46. $\dfrac{\sqrt{x} - \sqrt{5}}{x^2 - 25}$

47. $\dfrac{\sqrt{x^2 + 4x} - x}{2}$

48. $\dfrac{\sqrt{x^2 + 2x + 3} - \sqrt{x^2 + 3}}{2}$

In Exercises 49–58, evaluate each expression without using a calculator.

49. $25^{1/2}$

50. $144^{1/2}$

51. $(-8)^{1/3}$

52. $(-27)^{1/3}$

53. $8^{2/3}$

54. $16^{3/2}$

55. $-25^{-3/2}$

56. $-9^{-3/2}$

57. $\left(\dfrac{9}{25}\right)^{-3/2}$

58. $\left(\dfrac{1}{27}\right)^{-2/3}$

In Exercises 59–70, simplify each expression, leaving your answer with only positive exponents. Assume that all variables represent positive numbers.

59. $x^{1/2} \cdot x^{2/5}$

60. $x^{3/5} \cdot 5x^{2/3}$

61. $x^{3/5} \cdot x^{-1/2}$

62. $x^{5/3} \cdot x^{-3/4}$

63. $(8x^6)^{2/3}$

64. $(16x^3)^{1/2}$

65. $(27x^6y^3)^{-2/3}$

66. $(16x^4y^6)^{-3/2}$

67. $\dfrac{15x^{3/2}}{3x^{1/4}}$

68. $\dfrac{20x^{5/2}}{4x^{2/3}}$

69. $\left(\dfrac{x^{-1/4}}{y^{-2/3}}\right)^{-12}$

70. $\left(\dfrac{27x^{-5/2}}{y^{-3}}\right)^{-1/3}$

In Exercises 71–78, convert each radical expression to its rational exponent form and then simplify. Assume that all variables represent positive numbers.

71. $\sqrt[4]{3^2}$

72. $\sqrt[4]{5^2}$

73. $\sqrt[3]{x^9}$

74. $\sqrt[3]{x^{12}}$

75. $\sqrt[3]{x^6 y^9}$

76. $\sqrt[3]{8x^3 y^{12}}$

77. $\sqrt[4]{9\sqrt{3}}$

78. $\sqrt[4]{49\sqrt{7}}$

In Exercises 79–84, factor each expression. Use only positive exponents in your answer.

79. $\dfrac{4}{3}x^{1/3}(2x - 3) + 2x^{4/3}$

80. $\dfrac{4}{3}x^{1/3}(7x + 1) + 7x^{4/3}$

81. $4(3x + 1)^{1/3}(2x - 1) + 2(3x + 1)^{4/3}$

82. $4(3x - 1)^{1/3}(x + 2) + (3x - 1)^{4/3}$

83. $3x(x^2 + 1)^{1/2}(2x^2 - x) + 2(x^2 + 1)^{3/2}(4x - 1)$

84. $3x(x^2 + 2)^{1/2}(x^2 - 2x) + (x^2 + 2)^{3/2}(2x - 2)$

Applying the Concepts

85. **Sign dimensions.** A sign in the shape of an equilateral triangle of area A has sides of length $\sqrt{\dfrac{4A}{\sqrt{3}}}$ centimeters. Find the length of a side if the sign has an area of 692 square centimeters. (Use the approximation $\sqrt{3} \approx 1.73$.)

86. **Return on investment.** For an initial investment of
P dollars to mature to S dollars after two years when the
interest is compounded annually, an annual interest rate of
$r = \sqrt{\dfrac{S}{P}} - 1$ is required. Find r if $P = \$1,210,000$ and
$S = \$1,411,344$.

87. **Terminal speed.** The terminal speed V of a steel ball that
weighs w grams and is falling in a cylinder of oil is given by
the equation $V = \sqrt{\dfrac{w}{1.5}}$ centimeters per second. Find the terminal speed of a steel ball weighing 6 grams.

88. **Electronic game current.** The current I (in amperes) in
a circuit for an electronic game using W watts and having a
resistance of R ohms is given by $I = \sqrt{\dfrac{W}{R}}$. Find the current
in a game circuit using 1058 watts and having a resistance of
2 ohms.

Topics in Geometry

Triangles

The intuitive idea that two triangles have exactly the same shape (but not necessarily the same size) is captured in the concept of *similar triangles:*

Similar Triangles   Two triangles are **similar** if and only if they have equal corresponding angles and their corresponding sides are proportional. See Figure A.10.

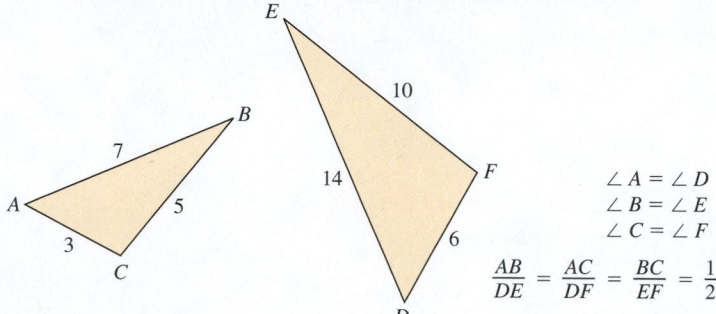

$\angle A = \angle D$
$\angle B = \angle E$
$\angle C = \angle F$

$$\frac{AB}{DE} = \frac{AC}{DF} = \frac{BC}{EF} = \frac{1}{2}$$

Figure A.10

Congruent Triangles   Two triangles are *congruent* if and only if they have equal corresponding angles and sides.

Included Side   A side of a triangle that lies between two angles is called the **included side** of the angles.

Included Angle   The angle formed by two sides is the **included angle** of the sides.

Fortunately, not every part of the definition for congruent and similar triangles must be verified to decide whether two triangles are congruent, or similar.

CONGRUENT-TRIANGLE THEOREMS

SAS (Side–Angle–Side). If two sides and the included angle of one triangle are equal to the corresponding sides and the included angle of a second triangle, the two triangles are congruent.

ASA (Angle–Side–Angle). If two angles and the included side of one triangle are equal to the corresponding two angles and the included side of a second triangle, the two triangles are congruent.

SSS (Side–Side–Side). If the three sides of one triangle are equal to the corresponding sides of a second triangle, the two triangles are congruent.

SIMILAR-TRIANGLE THEOREMS

SAS (Side–Angle–Side). If two included sides are proportional and the angles are equal, the two triangles are similar.

AA (Angle–Angle). If two angles are equal, the two triangles are similar.

SSS (Side–Side–Side). If the lengths of all three sides of two triangles are proportional, the triangles are similar.

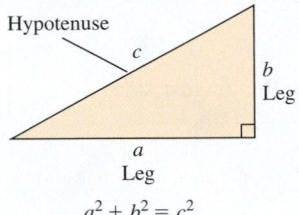

1 Use the Pythagorean Theorem and its converse.

Hypotenuse

c

b
Leg

a
Leg

$a^2 + b^2 = c^2$

Figure A.11

Pythagorean Theorem and Its Converse

A **right triangle** is a triangle one of whose angles is a right angle—that is, an angle of 90°. The side of the triangle opposite the 90° angle is called the **hypotenuse**; the other two sides are the **legs** of the triangle. Figure A.11 shows a right triangle with legs of length a and b and hypotenuse of length c. The symbol ⌐ is used to identify the right angle.

The *Pythagorean Theorem* says that the square of the length of a right triangle's hypotenuse is equal to the sum of the squares of the lengths of its legs.

> **PYTHAGOREAN THEOREM**
>
> If a, b, and c represent the lengths of the two legs and the hypotenuse, respectively, of a right triangle, then
>
> $$a^2 + b^2 = c^2.$$

EXAMPLE 1 **The Pythagorean Theorem**

A right triangle has one leg of length 15 and a hypotenuse of length 39. Find the length of the remaining leg.

Solution

Let c represent the length of the hypotenuse and a and b represent the lengths of the legs of the right triangle. Then $c = 39$, $a = 15$, and $a^2 + b^2 = c^2$ can be written as follows:

$$15^2 + b^2 = 39^2$$
$$b^2 = 39^2 - 15^2 = 1296$$
$$b = 36$$

The length of the remaining leg is 36.

Practice Problem 1 A right triangle has legs of lengths 8 and 6. Find the length of the hypotenuse.

EXAMPLE 2 **The Pythagorean Theorem**

You want to change some prices advertised on a letter display fastened 15 feet high on the front wall of your business building. A garden along the front wall forces you to place a ladder 8 feet from the wall. What length ladder is required?

Solution

If a ladder is placed 8 feet from the wall and rests 15 feet high on the wall, the ladder can be viewed as the hypotenuse of a right triangle with legs having lengths 8 feet and 15 feet. See Figure A.12.

Let c represent the length of the ladder.

$$c^2 = a^2 + b^2 \qquad \text{Pythagorean Theorem}$$
$$c^2 = 8^2 + 15^2 \qquad \text{Replace } a \text{ with 8 and } b \text{ with 15.}$$
$$c^2 = 289$$
$$c = 17$$

A 17-foot ladder will allow you to change the display.

Practice Problem 2 In Example 2, suppose the display is fastened 8 feet high and the ladder must be placed 6 feet from the wall. What length ladder is required?

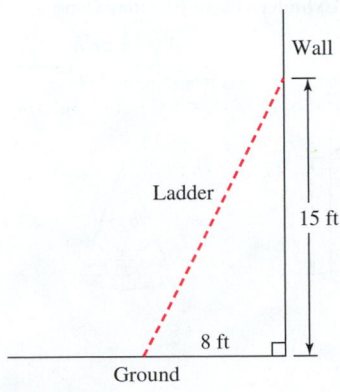

Wall

Ladder

15 ft

8 ft

Ground

Figure A.12

> ## CONVERSE OF THE PYTHAGOREAN THEOREM
>
> Suppose a triangle has sides of length a, b, and c. If
>
> $$c^2 = a^2 + b^2,$$
>
> then the triangle is a right triangle, with a right angle opposite the longest side.

EXAMPLE 3 Converse of the Pythagorean Theorem

A triangle has sides of lengths 20, 21, and 29. Is it a right triangle?

Solution

To determine whether the triangle is a right triangle, we first square the length of each side.

$$20^2 = 400, \quad 21^2 = 441, \quad \text{and} \quad 29^2 = 841$$
$$841 = 400 + 441$$

Because the square of the length of the largest side equals the sum of the squares of the lengths of the other two sides, the triangle is a right triangle.

Practice Problem 3 A triangle has sides of lengths 2, 5, and 6. Is it a right triangle?

2 Use geometry formulas.

Geometry Formulas

Figure A.13 lists some useful formulas from geometry.

Formulas for area (A), circumference (C), perimeter (P), volume (V), and surface area (S)

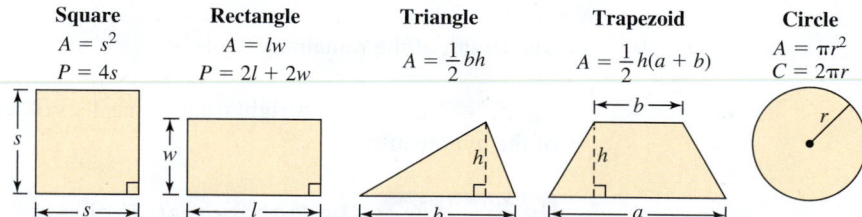

Square	Rectangle	Triangle	Trapezoid	Circle
$A = s^2$	$A = lw$	$A = \frac{1}{2}bh$	$A = \frac{1}{2}h(a + b)$	$A = \pi r^2$
$P = 4s$	$P = 2l + 2w$			$C = 2\pi r$

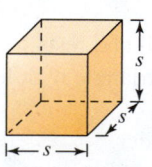 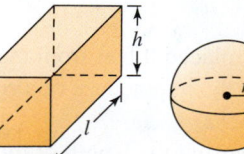 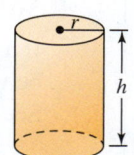 

Cube	Rectangular Solid	Sphere	Right Circular Cylinder	Right Circular Cone
$V = s^3$	$V = lwh$	$V = \frac{4}{3}\pi r^3$	$V = \pi r^2 h$	$V = \frac{1}{3}\pi r^2 h$
$S = 6s^2$	$S = 2(lw + wh + lh)$	$S = 4\pi r^2$	$S = 2\pi rh$	$S = \pi rl = \pi r \sqrt{h^2 + r^2}$
			$S = 2\pi rh + 2\pi r^2$ (Solid)	$S = \pi rl + \pi r^2$ (Solid)

Figure A.13

Answers to Practice Problems
1. 10 **2.** 10 ft **3.** No

SECTION A.5 Exercises

Basic Concepts and Skills

In Exercises 1–6, the lengths of the legs of a right triangle are given. Find the length of the hypotenuse.

1. $a = 7, b = 24$ 2. $a = 5, b = 12$

3. $a = 9, b = 12$ 4. $a = 10, b = 24$

5. $a = 3, b = 3$ 6. $a = 2, b = 4$

In Exercises 7–10, the lengths of one leg, a, and the hypotenuse, c, of a right triangle are given. Find the length of the remaining leg.

7. $a = 15, c = 17$ 8. $a = 35, c = 37$

9. $a = 14, c = 50$ 10. $a = 10, c = 26$

In Exercises 11–14, find the area A and the perimeter P of a rectangle with length l and width w.

11. $l = 3, w = 5$

12. $l = 4, w = 3$

13. $l = 10, w = \dfrac{1}{2}$

14. $l = 16, w = \dfrac{1}{4}$

In Exercises 15–18, find the area A of a triangle with base b and altitude h.

15. $b = 3, h = 4$ 16. $b = 6, h = 1$

17. $b = \dfrac{1}{2}, h = 12$ 18. $b = \dfrac{1}{4}, h = \dfrac{2}{3}$

In Exercises 19–22, find the area A and the circumference C of a circle with radius r. (Use $\pi \approx 3.14$.)

19. $r = 1$

20. $r = 3$

21. $r = \dfrac{1}{2}$

22. $r = \dfrac{3}{4}$

In Exercises 23–26, find the volume V of a box with length l, width w, and height h.

23. $l = 1, w = 1, h = 5$ 24. $l = 2, w = 7, h = 4$

25. $l = \dfrac{1}{2}, w = 10, h = 3$ 26. $l = \dfrac{1}{3}, w = \dfrac{3}{7}, h = 14$

In Exercises 27 and 28, find the area A of a trapezoid having parallel sides a and b, and height h with

27. $a = 3, b = 7, h = 5$ 28. $a = 4, b = 5, h = 6$

In Exercises 29 and 30, consider a closed can in the form of a right circular cylinder with radius $r = 4$ in. and height $h = 6$ in.

29. Find the volume of the can.

30. Find the surface area of the can.

Applying the Concepts

31. **Garden area.** Find the area of a rectangular garden if the width is 12 feet and the diagonal measures 20 feet.

32. **Perimeter of a mural.** A mural is 6 feet wide, and its diagonal measures 10 feet. Find the perimeter of the mural.

33. **Dimensions of a computer monitor.** The monitor on a laptop computer has a rectangular shape with a 15-inch diagonal. If the width of the monitor is 12 inches, what is its height?

34. **Dimensions of a baseball diamond.** A baseball diamond is actually a square with 90-foot sides. How far does the catcher have to throw the ball to reach second base from home plate?

35. **Repair problem.** A repairperson has to work on an air conditioner's overflow vent located 25 feet above the ground on the side of a house. Shrubs along the house require that a ladder be placed 10 feet from the house. How long must the ladder be for the repairperson to reach the overflow vent?

36. **Building a ramp.** You must build a ramp that allows you to roll a cart from a storeroom into the back of a truck. The rear of the truck must remain 8 feet from the storeroom because of a curb. If the back of the truck is 4 feet above the storeroom floor, how long (in feet) must the ramp be?

37. **Football field.** The playing surface of a football field (including the end zones) is 120 yards long and 53 yards wide. How far will a player jogging around the perimeter travel?

38. **Reflecting pool.** The reflecting pool of the Lincoln Memorial has an approximate perimeter of 4400 feet. The length is approximately 1860 feet more than the width. Find the area of the reflecting pool.

39. **Crop circle.** A crop circle with a diameter of 787 feet occurred at Milk Hill in Wiltshire, UK, August 2001. If you walk around the entire edge of the circle, how far will you have walked? (Use $\pi \approx 3.14$.)

40. **Bicycle tires.** The diameter of a bicycle tire is 26 inches. How far will the bicycle travel when the wheel makes one complete turn? (Use $\pi \approx 3.14$.)

41. **Fishpond border.** A rectangular border a foot and a half wide is built around a rectangular fishpond. If the pond is 6 feet long and 4 feet wide, what is the area of the border?

42. **Garden fence.** Casper wants to put a decorative fence around his rectangular garden. The garden is 5 feet wide, and its diagonal measures 13 feet. How many feet of fencing is required?

43. **Oven dimensions.** An oven in the shape of a rectangular box is 2 feet wide, 2.5 feet high, and 3 feet deep. What is the volume of the oven?

Equations

OBJECTIVES

1 Solve equations in one variable.

2 Solve linear equations.

3 Solve quadratic equations.

4 Solve equations by factoring.

5 Solve rational equations.

6 Solve equations involving radicals.

7 Solve equations that are quadratic in form.

Definitions

An **equation in one variable** is a statement that two expressions, with at least one containing the variable, are equal. For example, $2x - 3 = 7$ is an equation in the variable x. The expressions $2x - 3$ and 7 are the *sides* of the equation. The **domain** of the variable in an equation is the set of all real numbers for which both sides of the equation are defined.

When the variable in an equation is replaced by a specific value from its domain, the resulting statement may be true or false. For example, in the equation $2x - 3 = 7$, if we let $x = 1$, the equation is false. However, if we replace x with 5, the equation is true. Those values (if any) of the variable that result in a true statement are called **solutions** or **roots** of the equation. So, 5 is a solution (or root) of the equation $2x - 3 = 7$. We also say that 5 **satisfies** the equation $2x - 3 = 7$. To **solve** an equation means to find all solutions of the equation; the set of all solutions of an equation is called its **solution set**.

An equation that is satisfied by every real number in the domain of the variable is called an **identity**. The equations

$$2(x + 3) = 2x + 6 \qquad \text{Distributive property}$$
$$x^2 - 9 = (x + 3)(x - 3) \qquad \text{Difference of squares}$$
$$\text{and} \quad \frac{1}{x^2 - 3x + 2} = \frac{1}{(x - 1)(x - 2)} \qquad \text{Factor}$$

are examples of identities. Some equations, such as $x = x + 5$, have no solution. Such equations are called **inconsistent equations**, and their solution set is designated by the empty set symbol, $\varnothing$.

1 Solve equations in one variable.

RECALL

When finding the domain of a variable, remember that

1. Division by 0 is undefined.

2. The square root (or any even root) of a negative number is not a real number.

Equivalent Equations

Equations that have the same solution set are called **equivalent equations**. For example, equations $x = 4$ and $3x = 12$ are equivalent equations with solution set $\{4\}$.

In general, to solve an equation in one variable, we replace the given equation with a sequence of equivalent equations until we obtain an equation whose solution is obvious, such as $x = 4$. The following operations yield equivalent equations.

GENERATING EQUIVALENT EQUATIONS

Operations	Given Equation	Equivalent Equation
1. Simplify expressions on either side by eliminating parentheses, combining like terms, etc.	$(2x - 1) - (x + 1) = 4$	$2x - 1 - x - 1 = 4$ or $x - 2 = 4$
2. Add (or subtract) the same expression on *both* sides of the equation.	$x - 2 = 4$	$x - 2 + 2 = 4 + 2$ or $x = 6$
3. Multiply (or divide) *both* sides of the equation by the same *nonzero* expression.	$2x = 8$	$\frac{1}{2} \cdot 2x = \frac{1}{2} \cdot 8$ or $x = 4$
4. Interchange the two sides of the equation.	$-3 = x$	$x = -3$

2 Solve linear equations.

Solving Linear Equations in One Variable

Linear Equations

A **linear equation in one variable**, such as x, is an equation that can be written in the *standard form*

$$ax + b = 0,$$

where a and b are real numbers with $a \neq 0$.

EXAMPLE 1 Solving a Linear Equation

Solve: $6x - [3x - 2(x - 2)] = 11$

Solution

$6x - [3x - 2(x - 2)] = 11$	Original equation
$6x - [3x - 2x + 4] = 11$	Remove innermost parentheses by distributing $-2(x - 2) = -2x + 4$.
$6x - 3x + 2x - 4 = 11$	Remove square brackets and change the sign of each enclosed term.
$5x - 4 = 11$	Combine like terms.
$5x - 4 + 4 = 11 + 4$	Add 4 to both sides.
$5x = 15$	Simplify both sides.
$\dfrac{5x}{5} = \dfrac{15}{5}$	Divide both sides by 5.
$x = 3$	Simplify.

The apparent solution is 3.

Check: Substitute $x = 3$ into the original equation. You should obtain $11 = 11$, Thus, 3 is the only solution of the original equation; so $\{3\}$ is its solution set.

Practice Problem 1 Solve $3x - [2x - 6(x + 1)] = -1$.

EXAMPLE 2 Solving an Equation

Solve:

a. $1 - 5y + 2(y + 7) = 2y + 5(3 - y)$

b. $2(x + 1) = 4x - 2(x + 2)$

Solution

(a) $1 - 5y + 2(y + 7) = 2y + 5(3 - y)$	Original equation
$1 - 5y + 2y + 14 = 2y + 15 - 5y$	Distributive property
$15 - 3y = 15 - 3y$	Collect like terms and combine constants on each side.
$15 - 3y + 3y = 15 - 3y + 3y$	Add 3y to both sides.
$15 = 15$	Simplify.
$0 = 0$	Subtract 15 from both sides.

We have shown that $0 = 0$ is equivalent to the original equation. The equation $0 = 0$ is always true; its solution set is the set of all real numbers. Therefore, the solution set of the original equation is also the set of all real numbers. Thus, the original equation is an identity.

(b)

$$2(x + 1) = 4x - 2(x + 2) \qquad \text{Original equation}$$
$$2x + 2 = 4x - 2x - 4 \qquad \text{Distributive property}$$
$$2x + 2 = 2x - 4 \qquad \text{Collect like terms.}$$
$$2 = -4 \qquad \text{Subtract } 2x \text{ from both sides.}$$

The false statement $2 = -4$ shows that the equivalent original equation has no solution. So, the original equation is inconsistent, and its solution set is $\varnothing$.

Practice Problem 2 Solve the equation:

a. $2(3x - 6) + 5 = 6x - 7$ **b.** $3(2x - 1) = 2(3x + 1)$

TYPES OF EQUATIONS

There are three types of linear equations.

1. An equation that is satisfied by all values in the domain of the variable is an *identity*. For example, $2(x - 1) = 2x - 2$. When you try to solve an identity, you get a true statement such as $3 = 3$ or $0 = 0$.
2. An equation that is not an identity but is satisfied by at least one number in the domain of the variable is a *conditional* equation. For example, $2x = 6$ is a linear conditional equation.
3. An equation that is not satisfied by any value of the variable is an *inconsistent* equation. For example, $x = x + 2$ is an inconsistent equation. Because no number is 2 more than itself, the solution set of the equation $x = x + 2$ is $\varnothing$. When you try to solve an inconsistent equation, you obtain a false statement such as $0 = 2$.

---

3 Solve quadratic equations.

DO YOU
KNOW?

The word *quadratic* comes from the Latin *quadratus*, meaning square.

Solving Quadratic Equations

Quadratic Equation

A **quadratic equation in the variable** x is an equation equivalent to the equation

$$ax^2 + bx + c = 0,$$

where a, b, and c are real numbers and $a \neq 0$.

A quadratic equation written in the form $ax^2 + bx + c = 0$ is said to be in **standard form**. Here are some examples of quadratic equations that are not in standard form.

$$2x^2 - 5x + 7 = x^2 + 3x - 1$$
$$3y^2 + 4y + 1 = 6y - 5$$
$$x^2 - 2x = 3$$

We discuss two methods for solving quadratic equations: (1) by factoring and (2) by using the quadratic formula.

Factoring Method

Some quadratic equations in the standard form $ax^2 + bx + c = 0$ can be solved by factoring and using the **zero-product property**.

ZERO–PRODUCT PROPERTY

Let A and B be two algebraic expressions. Then $AB = 0$ if and only if $A = 0$ or $B = 0$ or both.

EXAMPLE 3   **Solving a Quadratic Equation by Factoring**

Solve by factoring: $2x^2 + 5x = 3$

Solution

$$2x^2 + 5x = 3 \qquad \text{Original equation}$$
$$2x^2 + 5x - 3 = 0 \qquad \text{Subtract 3 from both sides.}$$
$$(2x - 1)(x + 3) = 0 \qquad \text{Factor the left side.}$$

We now set each factor equal to 0 and solve the resulting linear equations. The vertical line separates the computations leading to the two solutions.

$$
\begin{array}{c|c l}
2x - 1 = 0 & x + 3 = 0 & \text{Zero-product property} \\
2x = 1 & x = -3 & \text{Isolate the } x \text{ term on one side.} \\
x = \dfrac{1}{2} & x = -3 & \text{Solve for } x.
\end{array}
$$

Check: You should check the solutions $x = \dfrac{1}{2}$ and $x = -3$ in the original equation.

The solution set is $\left\{ \dfrac{1}{2}, -3 \right\}$.

Practice Problem 3 Solve by factoring: $x^2 + 25x = -84$

The Quadratic Formula

We solve the standard form of the quadratic equation by *completing the square*.

$$ax^2 + bx + c = 0$$

$$ax^2 + bx = -c \qquad \text{Isolate the constant on the right side.}$$

$$x^2 + \frac{b}{a}x = -\frac{c}{a} \qquad \text{Divide both sides by } a.$$

$$x^2 + \frac{b}{a}x + \left(\frac{b}{2a}\right)^2 = \left(\frac{b}{2a}\right)^2 - \frac{c}{a} \qquad \begin{array}{l}\text{Add the square of one-half} \\ \text{the coefficient of } x \text{ to both sides to} \\ \text{complete the square.}\end{array}$$

$$\left(x + \frac{b}{2a}\right)^2 = \frac{b^2}{4a^2} - \frac{c}{a} \qquad \text{The left side is a perfect square.}$$

$$\left(x + \frac{b}{2a}\right)^2 = \frac{b^2 - 4ac}{4a^2} \qquad \text{Combine fractions on the right side.}$$

$$x + \frac{b}{2a} = \pm\sqrt{\frac{b^2 - 4ac}{4a^2}} \qquad \text{Take the square roots of both sides.}$$

$$x = -\frac{b}{2a} \pm \frac{\sqrt{b^2 - 4ac}}{2a} \qquad \begin{array}{l}\text{Add } -\dfrac{b}{2a} \text{ to both sides;} \\ \pm\sqrt{4a^2} = \pm 2|a| = \pm 2a.\end{array}$$

$$x = \frac{-b \pm \sqrt{b^2 - 4ac}}{2a} \qquad \text{Combine fractions.}$$

This formula is called the **quadratic formula**.

THE QUADRATIC FORMULA

The solutions of the quadratic equation in the standard form $ax^2 + bx + c = 0$ with $a \neq 0$ are given by the formula:

$$x = \frac{-b \pm \sqrt{b^2 - 4ac}}{2a}$$

WARNING   To use the quadratic formula, write the given quadratic equation in standard form. Then determine the values of a (coefficient of x^2), b (coefficient of x), and c (constant term).

EXAMPLE 4   **Solving a Quadratic Equation by Using the Quadratic Formula**

Solve $3x^2 = 5x + 2$ by using the quadratic formula.

Solution

We first rewrite the equation in standard form.

$$3x^2 - 5x - 2 = 0 \qquad \text{Subtract } 5x + 2 \text{ from both sides.}$$

$$\underset{\underset{a}{\uparrow}}{3x^2} + \underset{\underset{b}{\uparrow}}{(-5)x} + \underset{\underset{c}{\uparrow}}{(-2)} = 0 \qquad \text{Identify values of } a, b, \text{ and } c \text{ to be used in the quadratic formula.}$$

$$x = \frac{-b \pm \sqrt{b^2 - 4ac}}{2a} \qquad \text{Quadratic formula}$$

$$x = \frac{-(-5) \pm \sqrt{(-5)^2 - 4(3)(-2)}}{2(3)} \qquad \text{Substitute 3 for } a, -5 \text{ for } b, \text{ and } -2 \text{ for } c.$$

$$= \frac{5 \pm \sqrt{25 + 24}}{6} \qquad \text{Simplify.}$$

$$= \frac{5 \pm \sqrt{49}}{6} = \frac{5 \pm 7}{6} \qquad \text{Simplify.}$$

Then,

$$x = \frac{5 + 7}{6} = \frac{12}{6} = 2 \text{ or } x = \frac{5 - 7}{6} = \frac{-2}{6} = -\frac{1}{3}; \text{ the solution set is } \left\{-\frac{1}{3}, 2\right\}.$$

Practice Problem 4   Solve by using the quadratic formula: $6x^2 - x - 2 = 0$

---

4 Solve equations by factoring.

Solving Polynomial Equations by Factoring

The method of factoring used in solving quadratic equations can also be used to solve certain nonquadratic equations. Specifically, we use the zero-product property for solving those equations that can be expressed in factored form with 0 *on one side*.

EXAMPLE 5   **Solving an Equation by Factoring**

Solve by factoring:

a. $x^4 = 9x^2$ **b.** $x^3 - 2x^2 = x - 2$

Solution

a. We move all terms to the left side, factor, and set each factor equal to zero.

$$x^4 = 9x^2 \qquad \text{Original equation}$$
$$x^4 - 9x^2 = 0 \qquad \text{Subtract } 9x^2 \text{ from both sides.}$$
$$x^2(x^2 - 9) = 0 \qquad x^4 - 9x^2 = x^2(x^2 - 9)$$
$$x^2(x + 3)(x - 3) = 0 \qquad x^2 - 9 = (x + 3)(x - 3)$$

$x^2 = 0$ or $x + 3 = 0$ or $x - 3 = 0$ Zero-product property

$x = 0$ or $x = -3$ or $x = 3$ Solve each equation for x.

Check each possible solution.

Let $x = 0$	Let $x = -3$	Let $x = 3$
$(0)^4 \overset{?}{=} 9(0)^2$	$(-3)^4 \overset{?}{=} 9(-3)^2$	$(3)^4 \overset{?}{=} 9(3)^2$
$0 = 0 \checkmark$	$81 = 81 \checkmark$	$81 = 81 \checkmark$

The solution set is $\{0, 3, -3\}$.

b. Move all terms to the left side and factor by grouping.

$$x^3 - 2x^2 = x - 2 \qquad \text{Original equation}$$
$$x^3 - 2x^2 - x + 2 = 0 \qquad \text{Subtract } x - 2 \text{ from both sides and simplify.}$$
$$(x^3 - 2x^2) - (x - 2) = 0 \qquad \text{Group terms.}$$
$$x^2(x - 2) - 1 \cdot (x - 2) = 0 \qquad \text{Factor } x^2 \text{ from the first two terms and rewrite } x - 2 \text{ as } 1 \cdot (x - 2).$$
$$(x - 2)(x^2 - 1) = 0 \qquad \text{Factor out the common factor } (x - 2).$$
$$(x - 2)(x + 1)(x - 1) = 0 \qquad \text{Factor } x^2 - 1.$$

$x - 2 = 0$	$x + 1 = 0$	$x - 1 = 0$	
$x = 2$	$x = -1$	$x = 1$	Set each factor equal to zero.

We ask you to check each possible solution.
The solution set of the equation is $\{2, 1, -1\}$.

Practice Problem 5 Solve by factoring:

a. $x^4 = 4x^2$ **b.** $x^3 - 5x^2 = 4x - 20$

RECALL

$A^2 - B^2 = (A + B)(A - B)$

Difference of two squares

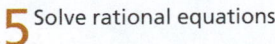

Rational Equations

If at least one algebraic expression with the variable in the denominator appears in an equation, the equation is a **rational equation**. When we multiply a rational equation by the least common denominator (LCD) of the fractions involved, we obtain a polynomial equation. However, some of the solutions to the new equation may not satisfy the original equation. Any such solution is called an **extraneous solution** or **extraneous root**. See Exercises 29 and 30.

EXAMPLE 6 Solving a Rational Equation

Solve: $\dfrac{1}{6} + \dfrac{1}{x+1} = \dfrac{1}{x}$

Solution

The LCD from the denominators 6, $x+1$, and x is $6x(x+1)$. Multiply both sides of the equation by the LCD and make the right side 0.

$$6x(x+1)\left[\dfrac{1}{6} + \dfrac{1}{1+x}\right] = 6x(x+1)\left[\dfrac{1}{x}\right] \qquad \text{Multiply both sides by the LCD.}$$

$$\dfrac{6x(x+1)}{6} + \dfrac{6x(x+1)}{x+1} = \dfrac{6x(x+1)}{x} \qquad \text{Distributive property}$$

$$x(x+1) + 6x = 6(x+1) \qquad \text{Simplify.}$$

$$x^2 + x + 6x = 6x + 6 \qquad \text{Distributive property}$$

$$x^2 + x - 6 = 0 \qquad \text{Subtract } 6x + 6 \text{ from both sides.}$$

$$(x+3)(x-2) = 0 \qquad \text{Factor.}$$

$$x + 3 = 0 \quad \text{or} \quad x - 2 = 0 \qquad \text{Set each factor equal to zero.}$$

$$x = -3 \quad \text{or} \quad x = 2 \qquad \text{Solve for } x.$$

Check the solutions, -3 and 2, in the original equation. The solution set is $\{-3, 2\}$.

Practice Problem 6 Solve: $\dfrac{1}{x} - \dfrac{12}{5x+10} = \dfrac{1}{5}$

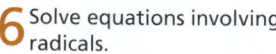 **6** Solve equations involving radicals.

Equations Involving Radicals

If an equation involves radicals or rational exponents, the method of raising both sides to a positive integer power is often used to obtain a polynomial equation. When this is done, the set of solutions of the new equation always contains the solutions of the original equation. However, in some cases, the new equation has *more* solutions than the original equation. For instance, consider the equation $x = 3$. If we square both sides, we get $x^2 = 9$. Notice that the given equation, $x = 3$, has only one solution, namely, 3, whereas the new equation, $x^2 = 9$, has two solutions: 3 and -3. Extraneous solutions may be introduced whenever we raise both sides of an equation to an *even* power. So *we must check all solutions* after doing so.

EXAMPLE 7 Solving an Equation Involving a Radical

Solve $\sqrt{2x+1} + 1 = x$.

Solution

First, we isolate the radical on one side of the equation.

$$\sqrt{2x+1} = x - 1. \qquad \text{Subtract 1 from both sides.}$$

Next, we square both sides and simplify.

$$(\sqrt{2x+1})^2 = (x-1)^2$$

$$2x + 1 = x^2 - 2x + 1 \qquad (\sqrt{a})^2 = a$$

$$2x + 1 - 2x - 1 = x^2 - 2x + 1 - 2x - 1 \qquad \text{Subtract } 2x + 1 \text{ from both sides.}$$

$$0 = x^2 - 4x \qquad \text{Simplify.}$$

$$0 = x(x-4) \qquad \text{Factor.}$$

$$x = 0 \quad \text{or} \quad x - 4 = 0 \qquad \text{Set each factor equal to zero.}$$

$$x = 0 \quad \text{or} \quad x = 4 \qquad \text{Solve for } x.$$

Check:　$x = 0;$　$\sqrt{2(0) + 1} + 1 \overset{?}{=} 0$　　　$x = 4;$　$\sqrt{2(4) + 1} + 1 \overset{?}{=} 4$

$$\sqrt{1} + 1 \overset{?}{=} 0 \qquad\qquad\qquad \sqrt{9} + 1 \overset{?}{=} 4$$

$$2 \neq 0 \qquad\qquad\qquad\qquad\qquad 4 = 4\ \checkmark$$

Because 0 is an extraneous root, the only solution is 4; the solution set is $\{4\}$.

Practice Problem 7 Solve $\sqrt{6x + 4} + 2 = x$.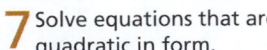

EXAMPLE 8　Solving an Equation Involving a Radical

Solve: $\sqrt[3]{2x + 1} + 4 = 3$

Solution

$$\sqrt[3]{2x + 1} = -1 \qquad \text{Subtract 4 from both sides.}$$
$$(\sqrt[3]{2x + 1})^3 = (-1)^3 \qquad \text{Cube both sides.}$$
$$2x + 1 = -1 \qquad (\sqrt[3]{a})^3 = a$$
$$2x = -2 \qquad \text{Isolate the } x \text{ term.}$$
$$x = -1 \qquad \text{Solve for } x.$$

You should always check your answers to avoid errors. The solution set is $\{-1\}$.

Practice Problem 8 Solve $\sqrt[5]{3x - 7} = 2$.

---

7 Solve equations that are quadratic in form.

Equations That Are Quadratic in Form

An equation in a variable x is **quadratic in form** if it can be written as

$$au^2 + bu + c = 0 \qquad (a \neq 0),$$

where u is an expression in the variable x. We solve the equation $au^2 + bu + c = 0$ for u. Then the solutions of the original equation can be obtained by replacing u with the expression in x that u represents and solving for x.

EXAMPLE 9　Solving an Equation That Is Quadratic in Form

Solve $x^{2/3} - 7x^{1/3} + 6 = 0$.

Solution

The equation becomes quadratic if we replace $x^{1/3}$ with u.

$$x^{2/3} - 7x^{1/3} + 6 = 0 \qquad \text{Original equation}$$
$$u^2 - 7u + 6 = 0 \qquad \text{Replace } x^{1/3} \text{ with } u.$$
$$(u - 1)(u - 6) = 0 \qquad \text{Factor}$$
$$u - 1 = 0 \quad \text{or} \quad u - 6 = 0 \qquad \text{Zero-product property}$$
$$u = 1 \qquad\qquad u = 6 \qquad \text{Solve for } u.$$
$$x^{1/3} = 1 \qquad\qquad x^{1/3} = 6 \qquad \text{Replace } u \text{ with } x^{1/3}.$$
$$x = 1^3 = 1 \qquad x = 6^3 = 216 \qquad \text{Raise both sides to power 3;}$$
$$\qquad\qquad\qquad\qquad\qquad (x^{1/3})^3 = x.$$

You should check that both $x = 1$ and $x = 216$ are solutions of the original equation. The solution set is $\{1, 216\}$.

Practice Problem 9 Solve $x - 6\sqrt{x} + 8 = 0$.

SECTION A.6   **Exercises**

Basic Concepts and Skills

In Exercises 1–46, find the solution set of each equation. If the equation has no solution write $\varnothing$, and if the equation is an identity so state.

1. $5 - (6y + 9) + 2y = 2(y + 1)$

2. $3(4y - 3) = 4[y - (4y - 3)]$

3. $2(3x + 4) = 6(x + 2) - 4$

4. $2(x - 2) + 3x = 6x + 7 - (x - 3)$

5. $\dfrac{2x + 1}{9} - \dfrac{x + 4}{6} = 1$

6. $\dfrac{2 - 3x}{7} + \dfrac{x - 1}{3} = \dfrac{3x}{7}$

7. $\dfrac{1}{3x} + \dfrac{1}{2x} = \dfrac{1}{6} - \dfrac{1}{x}$

8. $\dfrac{2}{x - 1} = \dfrac{3}{x + 1}$

9. $\dfrac{t}{t - 2} = -\dfrac{2}{3}$

10. $\dfrac{2}{x + 1} + 3 = \dfrac{8}{x + 1}$

11. $\dfrac{5x}{x - 1} = \dfrac{5}{x - 1} + 3$

12. $\dfrac{3x}{x + 1} - 3 + \dfrac{3}{2(x + 1)} = 0$

13. $\dfrac{1 + x}{x} = \dfrac{1}{x} + 1$

14. $\dfrac{x - 1}{x - 2} - 1 = \dfrac{1}{x - 2}$

15. $x^2 - 5x = 0$

16. $x^2 - 5x + 4 = 0$

17. $x^2 = 5x + 6$

18. $x = x^2 - 12$

19. $x^2 + 2x - 5 = 0$

20. $x^2 - x - 3 = 0$

21. $3(t^2 - 1) = 2t^2 + 4t + 1$

22. $8(y^2 - y) = y^2 + 3$

23. $x^3 = 2x^2$

24. $3x^4 - 27x^2 = 0$

25. $x^4 - x^3 = x^2 - x$

26. $x^3 - 36x = 16(x - 6)$

27. $\dfrac{x + 3}{x - 1} + \dfrac{x + 4}{x + 1} = \dfrac{8x + 5}{x^2 - 1}$

28. $\dfrac{6}{2x - 2} - \dfrac{1}{x + 1} = \dfrac{2}{2x + 2} + 1$

29. $\dfrac{x}{x - 4} - \dfrac{4}{x + 4} = \dfrac{2x}{x^2 - 16}$

30. $\dfrac{x}{x - 3} + \dfrac{3}{x + 3} = \dfrac{6x}{x^2 - 9}$

31. $\dfrac{1}{x - 1} + \dfrac{x}{x + 3} = \dfrac{4}{x^2 + 2x - 3}$

32. $\dfrac{2x}{x + 3} - \dfrac{x}{x - 1} = \dfrac{14}{x^2 + 2x - 3}$

33. $t - \sqrt{3t + 6} = -2$

34. $\sqrt{5y^2 - 10y + 9} = 2y - 1$

35. $\sqrt{6t - 11} = 2t - 7$

36. $\sqrt{3x + 1} = x - 1$

37. $\sqrt{x - 3} = \sqrt{2x - 5} - 1$

38. $\sqrt{7x + 1} - \sqrt{5x + 4} = 1$

39. $x^{2/3} - 6x^{1/3} + 8 = 0$

40. $x^{2/5} + x^{1/5} - 2 = 0$

41. $2x^{1/2} - x^{1/4} - 1 = 0$

42. $81y^4 + 1 = 18y^2$

43. $(x^2 - 4)^2 - 3(x^2 - 4) - 4 = 0$

44. $(x^2 - 4x)^2 + 7(x^2 - 4x) + 12 = 0$

45. $\left(\dfrac{x}{x + 1}\right)^2 - \dfrac{2x}{x + 1} - 8 = 0$

46. $8\sqrt{\dfrac{x}{x + 3}} - \sqrt{\dfrac{x + 3}{x}} = 2$

47. Solve for x: $\dfrac{x + b}{x - b} = \dfrac{x - 5b}{2x - 5b}$

48. If $a > 0$ solve for x: $3a - \sqrt{ax} = \sqrt{a(3a + x)}$

49. Solve for x: $x = \sqrt{1 + \sqrt{1 + \sqrt{1 + \cdots}}}$

    (Hint: $x = \sqrt{1 + x}$)

50. Solve for x: $x = 2 + \dfrac{1}{2 + \dfrac{1}{2 + \cdots}}$

    $\left(\text{Hint: } x = 2 + \dfrac{1}{x}\right)$

In Exercises 51 and 52, find the values of k for which the given equation has equal roots.

51. $2x^2 + kx + k = 0$

52. $x^2 + k^2 = 2(k + 1)x$

53. If r and s are the roots of the quadratic equation $ax^2 + bx + c = 0$, show that

$$r + s = -\dfrac{b}{a} \quad \text{and} \quad r \cdot s = \dfrac{c}{a}.$$

54. Use Exercise 53 to find the sum and the product of the roots of the following equations without solving the equations.
    a. $3x^2 + 5x - 5 = 0$
    b. $3x^2 - 7x = 1$
    c. $\sqrt{3}x^2 = 3x + 4$
    d. $(1 + \sqrt{2})x^2 - \sqrt{2}x + 5 = 0$

In Exercises 55 and 56, determine k so that the sum and the product of the roots are equal.

55. $2x^2 + (k - 3)x + 3k - 5 = 0$

56. $5x^2 + (2k - 3)x - 2k + 3 = 0$

57. Suppose r and s are the roots of the quadratic equation $ax^2 + bx + c = 0$. Show that you can write $ax^2 + bx + c = a(x - r)(x - s)$. Thus, the quadratic trinomial can be written in factored form.
[*Hint:* From Exercise 53, $b = -a(r + s)$ and $c = ars$. Substitute in $ax^2 + bx + c$ and factor.]

58. Use Exercise 57 to factor each trinomial.
    a. $4x^2 + 4x - 5$
    b. $25x^2 + 40x + 11$
    c. $72x^2 + 95x - 1000$

Applying the Concepts

59. Manufacturing. The surface area A of a cylinder with height h and radius r is given by the equation $A = 2\pi rh + 2\pi r^2$. A company makes soup cans by using 32π square inches of aluminum sheet for each can. If the height of the can is 6 inches, find the radius of the can.

60. Making a box. A 2-inch square is cut from each corner of a rectangular piece of cardboard whose length exceeds the width by 4 inches. The sides are then turned up to form an open box. If the volume of the box is 64 cubic inches, find the dimensions of the box.

61. Travel time. Aya rode her bicycle from her house to a friend's house, averaging 16 kilometers per hour. It became dark before Aya was ready to return home, so her friend drove her home at a rate of 80 kilometers per hour. If Aya's total traveling time was 3 hours, how far away did her friend live?

62. Interest rates. Mr. Kaplan invests one sum of money at a certain rate of interest and half that sum at twice the first rate of interest. This arrangement turns out to yield 8% on the total investment. What are the two rates of interest?

63. Mixture. A mixture contains alcohol and water in a 5:1 ratio. Upon adding 5 liters of water, the ratio of alcohol to water becomes 5:2. Find the quantity of alcohol in the original mixture.

64. Age problem. A man and a woman have the same birthday. When he was as old as she is now, the man was twice as old as the woman. When she becomes as old as he is now, the sum of their ages will be 119. How old is the man now?

Inequalities

OBJECTIVES

1 Learn the vocabulary for discussing inequalities.

2 Solve and graph linear inequalities.

3 Solve and graph a compound inequality.

4 Solve a nonlinear inequality.

5 Solve inequalities involving absolute value.

Inequalities

An **inequality** is a statement that one algebraic expression is less than (denoted by $<$) or less than or equal (denoted by $\leq$) to another algebraic expression. Some examples of inequalities are

$$4 < 5, \quad 3x + 2 \leq 14, \quad 5x + 7 > 3x + 23, \quad \text{and} \quad x^2 \geq 0.$$

The real numbers that result in a true statement when those numbers are substituted for a variable in an inequality are called **solutions** of the inequality. Because replacing x with 1 in the inequality $3x + 2 \leq 14$ results in the true statement $3(1) + 2 \leq 14$, the number 1 is a solution of the inequality $3x + 2 \leq 14$. We also say that 1 *satisfies* the inequality $3x + 2 \leq 14$. To **solve** an inequality means to find all solutions of the inequality—that is, its solution set.

The following properties of inequalities will be helpful in solving inequalities.

1 Learn the vocabulary for discussing inequalities.

INEQUALITY PROPERTIES

For real numbers, $a, b, c,$ and d

1. **Trichotomy Property:** Exactly one of the following is true:

$$a < b, a = b, \text{ or } a > b.$$

2. **Transitive Property:** If $a < b$ and $b < c$, then $a < c$.

3. **Addition Property:** **(i)** If $a < b$, then $a + c < b + c$.
   (ii) If $a < b$, and $c < d$, then $a + c < b + d$.

4. **Multiplication Property:** Suppose $a < b$.
   a. If $c > 0$, then $ac < bc$.
   b. If $c < 0$, then $ac > bc$.

Similar results hold when $<$ is replaced throughout properties 2–4 by any of the symbols $\leq, >,$ or $\geq$.

The inequality properties can be verified by considering the relative positions of the graphs of the real numbers $a, b, c,$ and d on a number line.

WARNING

Note that in Property 4, the inequality symbol must be reversed ($<$ to $>$ or $>$ to $<$) when we multiply both sides of the inequality by a negative number. For example, $2 < 5$ so $(-3)(2) > (-3)(5)$ or $-6 > -15$. The statement $-6 > -15$ can be seen to be true geometrically because -15 is to the left of -6 on a number line.

The inequality properties can be used to establish inequalities involving opposites and inverses of a and b.

INEQUALITY	EXAMPLE
1. If $a < b$, then $-a > -b$.	$2 < 5$, so $-2 > -5$
2. If $a > b$, then $-a < -b$.	$6 > 3$, so $-6 < -3$
3. Suppose a and b are both positive or both negative. If $a < b$, then $\dfrac{1}{a} > \dfrac{1}{b}$.	$2 < 5$, so $\dfrac{1}{2} > \dfrac{1}{5}$ $\quad -3 < -2$, so $-\dfrac{1}{3} > -\dfrac{1}{2}$
4. Suppose a is negative and b is positive (that is, $a < 0 < b$), then $\dfrac{1}{a} < \dfrac{1}{b}$.	$-2 < 3$, so $-\dfrac{1}{2} < \dfrac{1}{3}$

Two inequalities that have exactly the same solution set are called **equivalent inequalities**. The basic method of solving inequalities is similar to the method for solving equations: we replace a given inequality by a series of equivalent inequalities until we arrive at an equivalent inequality, such as $x < 5$, whose solution set we already know.

> The following operations produce equivalent inequalities.
>
> 1. Simplifying one or both sides of an inequality by combining like terms and eliminating parentheses.
> 2. Using the addition or multiplication properties on both sides of the inequality.

2 Solve and graph linear inequalities.

Linear Inequalities

A **linear inequality in one variable** is an inequality that is equivalent to one of the forms

$$ax + b < 0 \quad \text{or} \quad ax + b \le 0,$$

where a and b represent real numbers and $a \ne 0$.

Inequalities such as $2x - 1 > 0$ and $2x - 1 \ge 0$ are linear inequalities because they are equivalent to $-2x + 1 < 0$ and $-2x + 1 \le 0$, respectively.

EXAMPLE 1 **Solving and Graphing Linear Inequalities**

Solve each inequality and graph its solution set.

a. $7x - 11 < 2(x - 3)$ **b.** $8 - 3x \le 2$

Solution

a. $7x - 11 < 2(x - 3)$

$$
\begin{aligned}
7x - 11 &< 2x - 6 && \text{Distributive property} \\
7x - 11 + 11 &< 2x - 6 + 11 && \text{Add 11 to both sides.} \\
7x &< 2x + 5 && \text{Simplify.} \\
7x - 2x &< 2x + 5 - 2x && \text{Subtract } 2x \text{ from both sides.} \\
5x &< 5 && \text{Simplify.} \\
\frac{5x}{5} &< \frac{5}{5} && \text{Divide both sides by 5.} \\
x &< 1 && \text{Simplify.}
\end{aligned}
$$

The solution set is $\{x \mid x < 1\}$, or in interval notation, $(-\infty, 1)$. The graph is shown in Figure A.14.

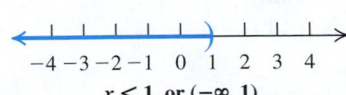

$x < 1$, or $(-\infty, 1)$

Figure A.14

b. $8 - 3x \le 2$

$$
\begin{aligned}
8 - 3x - 8 &\le 2 - 8 && \text{Subtract 8 from both sides.} \\
-3x &\le -6 && \text{Simplify.} \\
\frac{-3x}{-3} &\ge \frac{-6}{-3} && \text{Divide both sides by } -3. \text{ (Reverse the direction of the inequality symbol.)} \\
x &\ge 2 && \text{Simplify.}
\end{aligned}
$$

The solution set is $\{x \mid x \ge 2\}$, or in interval notation, $[2, \infty)$. The graph is shown in Figure A.15.

$x \ge 2$, or $[2, \infty)$

Figure A.15

Practice Problem 1 Solve each inequality and graph the solution set.

a. $4x + 9 > 2(x + 6) + 1$ **b.** $7 - 2x \ge -3$

If an inequality is true for all real numbers, its solution set is $(-\infty, \infty)$. If an inequality has no solution, its solution set is $\varnothing$.

EXAMPLE 2 **Solving an Inequality**

Write the solution set of each inequality.

a. $7(x + 2) - 20 - 4x < 3(x - 1)$
b. $2(x + 5) + 3x < 5(x - 1) + 3$

Solution

a.

$7(x + 2) - 20 - 4x < 3(x - 1)$	Original inequality
$7x + 14 - 20 - 4x < 3x - 3$	Distributive property
$3x - 6 < 3x - 3$	Combine like terms.
$-6 < -3$	Add $-3x$ to both sides and simplify.

The last inequality is equivalent to the original inequality and is always true. So the solution set of the original inequality is $(-\infty, \infty)$.

b.

$2(x + 5) + 3x < 5(x - 1) + 3$	Original inequality
$2x + 10 + 3x < 5x - 5 + 3$	Distributive property
$5x + 10 < 5x - 2$	Combine like terms.
$10 < -2$	Add $-5x$ and simplify.

The resulting inequality is equivalent to the original inequality and is always false. So the solution set of the original inequality is $\varnothing$.

Practice Problem 2 Write the solution set of each inequality.

a. $2(4 - x) + 6x < 4(x + 1) + 7$
b. $3(x - 2) + 5 \geq 7(x - 1) - 4(x - 2)$

3 Solve and graph a compound inequality.

Combining Two Inequalities

Sometimes we are interested in the solution set of two or more inequalities. The combination of two or more inequalities is called a **compound inequality**.

Suppose E_1 is an inequality with solution set (interval) I_1 and E_2 is another inequality with solution set I_2. Then the solution set of the compound inequality "E_1 and E_2" is $I_1 \cap I_2$, and the solution set of the compound inequality "E_1 or E_2" is $I_1 \cup I_2$.

EXAMPLE 3 **Solving and Graphing a Compound Inequality**

Graph and write the solution set of the compound inequality.

$$2x + 7 \leq 1 \quad \text{or} \quad 3x - 2 < 4(x - 1)$$

Solution

$2x + 7 \leq 1$ or $3x - 2 < 4(x - 1)$	Original inequalities
$2x + 7 \leq 1$ or $3x - 2 < 4x - 4$	Distributive property
$2x \leq -6$ or $-x < -2$	Isolate variable term.
$x \leq -3$ or $x > 2$	Solve for x: reverse second inequality symbol.

Graph each inequality and select the *union* of the two intervals.

$x \le -3$

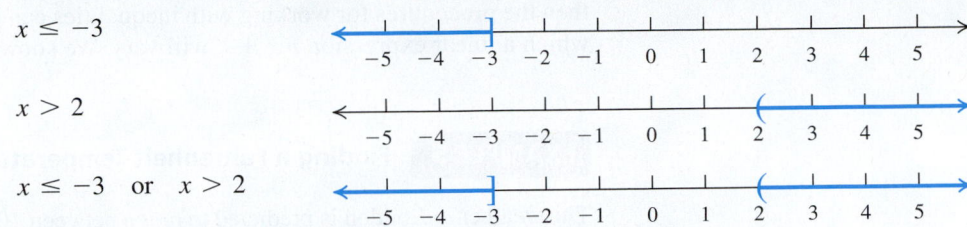

$x > 2$

$x \le -3$ or $x > 2$

The solution set of the compound inequality is $(-\infty, -3\,] \cup (2, \infty)$.

Practice Problem 3 Write the solution set of the compound inequality.

$$3x - 5 \ge 7 \quad \text{or} \quad 5 - 2x \ge 1$$

Sometimes we are interested in a joint inequality such as $-5 < 2x + 3 \le 9$. This use of two inequality symbols in a single expression is shorthand for $-5 < 2x + 3$ *and* $2x + 3 \le 9$. Fortunately, solving such inequalities requires no new principles, as we see in the next example.

EXAMPLE 4 **Solving and Graphing a Compound Inequality**

Solve the inequality $-5 < 2x + 3 \le 9$ and graph its solution set.

Solution

We must find all real numbers that are solutions of *both* inequalities

$$-5 < 2x + 3 \quad \text{and} \quad 2x + 3 \le 9.$$

Let's see what happens when we solve these inequalities separately.

$-5 < 2x + 3$	$2x + 3 \le 9$	Original inequalities
$-5 - 3 < 2x + 3 - 3$	$2x + 3 - 3 \le 9 - 3$	Subtract 3 from both sides.
$-8 < 2x$	$2x \le 6$	Simplify.
$\dfrac{-8}{2} < \dfrac{2x}{2}$	$\dfrac{2x}{2} \le \dfrac{6}{2}$	Divide both sides by 2.
$-4 < x$	$x \le 3$	Simplify.

The solution of the original pair of inequalities consists of all real numbers x such that $-4 < x$ *and* $x \le 3$. This solution may be written more compactly as $\{x \mid -4 < x \le 3\}$. In interval notation, we write $(-4, 3\,]$. The graph is shown in Figure A.16.

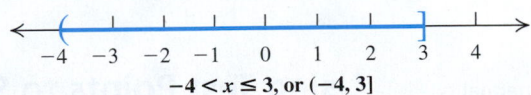

$-4 < x \le 3$, or $(-4, 3]$

Figure A.16

SIDE NOTE

If you multiply (or divide) a compound inequality by a negative number, then you must reverse *both* inequality symbols (and rewrite the resulting compound inequality in the correct format).

Notice that we did the same thing to both inequalities in each step of the solution process. We can accomplish this simultaneous solution more efficiently by working on both inequalities at the same time as follows.

$-5 < 2x + 3 \le 9$	Original inequality
$-5 - 3 < 2x + 3 - 3 \le 9 - 3$	Subtract 3 from each part.
$-8 < 2x \le 6$	Simplify each part.
$\dfrac{-8}{2} < \dfrac{2x}{2} \le \dfrac{6}{2}$	Divide each part by 2.
$-4 < x \le 3$	Simplify each part.

The solution set is $(-4, 3\,]$, the same solution set we found previously.

Practice Problem 4 Solve and graph $-6 \le 4x - 2 < 4$.

If you know that a particular quantity, represented by *x*, can vary between two values, then the procedures for working with inequalities can be used to find the values between which a linear expression *mx* + *k* will vary. We show how to do this in Example 5.

EXAMPLE 5   **Finding a Fahrenheit Temperature from a Celsius Range**

The weather in London is predicted to range between 10° and 20° Celsius during the three-week period you will be working there. To decide what kind of clothes to take, you want to convert the temperature range to Fahrenheit temperatures. The formula for converting Celsius temperature *C* to Fahrenheit temperature *F* is $F = \frac{9}{5}C + 32$. What range of Fahrenheit temperatures might you find in London during your stay there?

Solution

First, we express the Celsius temperature range as an inequality. Let *C* be the temperature in Celsius degrees.

For the three weeks under consideration, $10 \leq C \leq 20$.

We want to know the range of $F = \frac{9}{5}C + 32$ when $10 \leq C \leq 20$.

$$10 \leq C \leq 20$$

$$\left(\frac{9}{5}\right)(10) \leq \frac{9}{5}C \leq \left(\frac{9}{5}\right)(20) \qquad \text{Multiply each part by } \frac{9}{5}.$$

$$18 \leq \frac{9}{5}C \leq 36 \qquad \text{Simplify.}$$

$$18 + 32 \leq \frac{9}{5}C + 32 \leq 36 + 32 \qquad \text{Add 32 to each part.}$$

$$50 \leq \frac{9}{5}C + 32 \leq 68 \qquad \text{Simplify.}$$

$$50 \leq F \leq 68 \qquad F = \frac{9}{5}C + 32$$

So the temperature range from 10° to 20° Celsius corresponds to a range from 50° to 68° Fahrenheit.

Practice Problem 5 If $-2 < x < 5$, find real numbers *a* and *b* so that $a < 3x - 5 < b$.

4 Solve a nonlinear inequality.

Using Test Points to Solve Inequalities

The **test-point** method, also known as the **sign-chart** method, involves writing an inequality (by rearranging if necessary) so that the expression on the left side of the inequality symbol is in factored form and the right side is 0.

PROCEDURE
IN ACTION

EXAMPLE 6   **Using the Test-Point Method to Solve an Inequality**

OBJECTIVE

Solve an inequality by using the test-point method.

Step 1 Rewrite (if necessary) the inequality so that the right side of the inequality is 0. Simplify the left side.

Step 2 Factor the left side.

Step 3 Draw a number line. Find and plot the points where each factor of the left side is 0.

The n points so obtained divide the number line into $(n + 1)$ intervals.

Step 4 The final expression on the left side of the inequality in Step 1 is always positive or always negative on each of the $(n + 1)$ intervals of Step 3. Select convenient "test points" in each interval to determine the sign of the inequality.

Step 5 Draw a sign chart. Show the information from Steps 3 and 4 on a number line.

Step 6 From the sign chart in Step 5, write the solution set of the inequality. Graph the solution set.

EXAMPLE

Solve $x^2 + 4x \geq 5x + 6$

1. $x^2 + 4x - 5x - 6 \geq 0$ Subtract $5x + 6$ from both sides.
   $x^2 - x - 6 \geq 0$ Simplify.

2. $(x + 2)(x - 3) \geq 0$ Factor.

3. $x + 2 = 0$ for $x = -2$, and $x - 3 = 0$ for $x = 3$.

   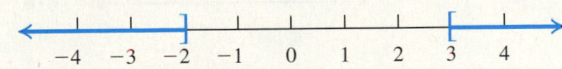

   The three intervals determined by the two points -2 and 3 are: $(-\infty, -2)$, $(-2, 3)$, and $(3, \infty)$.

4.

Test Interval	Test-Point	Value of $x^2 - x - 6$	Result
$(-\infty, -2)$	-3	$(-3)^2 - (-3) - 6 = 6$	Positive
$(-2, 3)$	0	$0^2 - 0 - 6 = -6$	Negative
$(3, \infty)$	4	$4^2 - 4 - 6 = 6$	Positive

5. $(x^2 - x - 6)$

6. To find where $x^2 - x - 6 \geq 0$ means to find where $x^2 - x - 6 > 0$ (indicated by "+" signs) or $x^2 - x - 6 = 0$ (indicated by "0") on the sign chart. So the solution set of the inequality is $(-\infty, -2] \cup [3, \infty)$. This set is shown below.

Practice Problem 6   Solve: $x^2 + 2 < 3x + 6$

EXAMPLE 7   **Solving a Rational Inequality**

Solve: $\dfrac{x^2 + 2x - 15}{x - 1} \geq 3$

Step 1

$\dfrac{x^2 + 2x - 15}{x - 1} - 3 \geq 0$ Rearrange so the right side is 0.

$\dfrac{x^2 + 2x - 15}{x - 1} - \dfrac{3(x - 1)}{x - 1} \geq 0$ Common denominator

$\dfrac{(x^2 + 2x - 15) - 3(x - 1)}{x - 1} \geq 0$ $\dfrac{a}{c} \pm \dfrac{b}{c} = \dfrac{a \pm b}{c}$

$\dfrac{x^2 + 2x - 15 - 3x + 3}{x - 1} \geq 0$ Distribute.

$\dfrac{x^2 - x - 12}{x - 1} \geq 0$ Simplify.

Step 2 $\dfrac{(x+3)(x-4)}{x-1} \geq 0$ Factor the numerator.

Step 3 $x + 3 = 0$ for $x = -3$

$x - 4 = 0$ for $x = 4$

$x - 1 = 0$ for $x = 1$ The rational expression is undefined at $x = 1$.

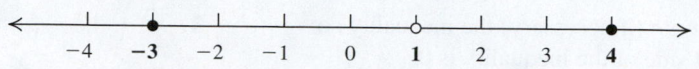

The four intervals determined by the three points -3, 1, and 4 are:

$$(-\infty, -3), (-3, 1), (1, 4) \text{ and } (4, \infty).$$

Step 4

Test Interval	Test-Point	Value of $\dfrac{(x+3)(x-4)}{(x-1)}$	Result
$(-\infty, -3)$	-4	$\dfrac{(-)(-)}{(-)}$	Negative
$(-3, 1)$	0	$\dfrac{(+)(-)}{(-)}$	Positive
$(1, 4)$	2	$\dfrac{(+)(-)}{(+)}$	Negative
$(4, \infty)$	5	$\dfrac{(+)(+)}{(+)}$	Positive

Step 5

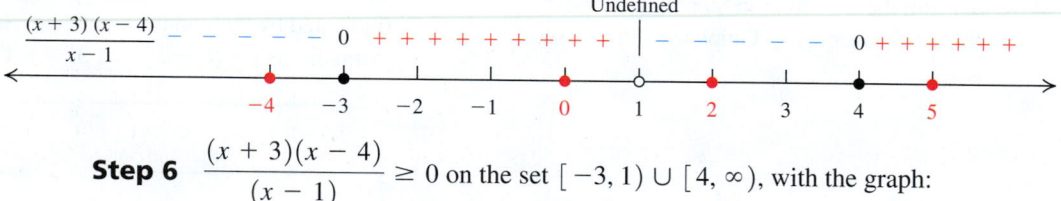

Step 6 $\dfrac{(x+3)(x-4)}{(x-1)} \geq 0$ on the set $[-3, 1) \cup [4, \infty)$, with the graph:

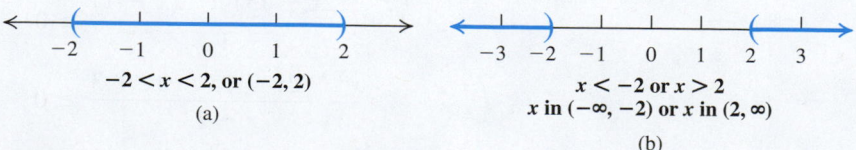

Practice Problem 7 Solve: $\dfrac{2x+5}{x-1} \leq 1$

5 Solve inequalities involving absolute value.

Inequalities Involving Absolute Value

If $|x| < 2$, then x is closer than two units to the origin. See Figure A.17(a). If $|x| > 2$, then x is farther than two units from the origin and is in either the interval $(-\infty, -2)$ or the interval $(2, \infty)$, that is, either $x < -2$ or $x > 2$. See Figure A.17(b).

$-2 < x < 2$, or $(-2, 2)$

(a)

$x < -2$ or $x > 2$
x in $(-\infty, -2)$ or x in $(2, \infty)$

(b)

Figure A.17

This discussion suggests the following rules.

RULES FOR SOLVING ABSOLUTE VALUE INEQUALITIES

If $a > 0$ and u is an algebraic expression, then

1. $|u| < a$ is equivalent to $-a < u < a$.
2. $|u| \leq a$ is equivalent to $-a \leq u \leq a$.
3. $|u| > a$ is equivalent to $u < -a$ or $u > a$.
4. $|u| \geq a$ is equivalent to $u \leq -a$ or $u \geq a$.

EXAMPLE 8 Solving an Inequality Involving Absolute Value

Solve the inequality $|4x - 1| \leq 9$ and graph the solution set.

Solution

Rule 2 applies here, with $u = 4x - 1$ and $a = 9$.

$$|4x - 1| \leq 9 \text{ is equivalent to}$$

$-9 \leq 4x - 1 \leq 9$	Rule 2, $-a \leq u \leq a$
$1 - 9 \leq 4x - 1 + 1 \leq 9 + 1$	Add 1 to each part.
$-8 \leq 4x \leq 10$	Simplify.
$-\dfrac{8}{4} \leq \dfrac{4x}{4} \leq \dfrac{10}{4}$	Divide by 4: the sense of the inequality is unchanged.
$-2 \leq x \leq \dfrac{5}{2}$	Simplify.

The solution set is $\left\{ x \mid -2 \leq x \leq \dfrac{5}{2} \right\}$; that is, the solution set is the closed interval $\left[-2, \dfrac{5}{2} \right]$. See Figure A.18.

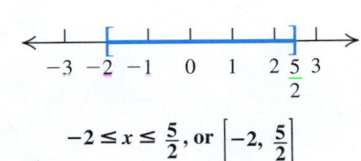

$-2 \leq x \leq \dfrac{5}{2}$, or $\left[-2, \dfrac{5}{2} \right]$

Figure A.18

Practice Problem 8 Solve $|3x + 3| \leq 6$ and graph the solution set.

EXAMPLE 9 Solving an Inequality Involving Absolute Value

Solve the inequality $|2x - 8| \geq 4$ and graph the solution set.

Solution

$$|2x - 8| \geq 4 \text{ is equivalent to}$$

$2x - 8 \leq -4$	or	$2x - 8 \geq 4$.	Rule 4, with $u = 2x - 8$ and $a = 4$.
$2x - 8 + 8 \leq -4 + 8$		$2x - 8 + 8 \geq 4 + 8$	Add 8 to each part.
$2x \leq 4$		$2x \geq 12$	Simplify.
$\dfrac{2x}{2} \leq \dfrac{4}{2}$		$\dfrac{2x}{2} \geq \dfrac{12}{2}$	Divide both sides of each part by 2.
$x \leq 2$		$x \geq 6$	Simplify.

The solution set is $\{ x \mid x \leq 2 \text{ or } x \geq 6 \}$; that is, the solution set is the set of all real numbers x in either of the intervals $(-\infty, 2]$ or $[6, \infty)$. This set can also be written as $(-\infty, 2] \cup [6, \infty)$. See Figure A.19.

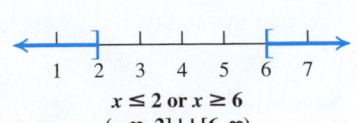

$x \leq 2$ or $x \geq 6$
$(-\infty, 2] \cup [6, \infty)$

Figure A.19

Practice Problem 9 Solve $|2x + 3| \geq 6$ and graph the solution set.

EXAMPLE 10 **Solving Special Cases of Absolute Value Inequalities**

Solve each inequality.

a. $|3x - 2| > -5$ **b.** $|5x + 3| \leq -2$

Solution

a. Because the absolute value is always nonnegative, $|3x - 2| > -5$ is true for all real numbers x. The solution set is the set of all real numbers; in interval notation, we write $(-\infty, \infty)$.

b. There is no real number with absolute value ≤ -2 because the absolute value of any number is nonnegative. The solution set for $|5x + 3| \leq -2$ is the empty set, $\varnothing$.

Practice Problem 10 Solve.

a. $|5 - 9x| > -3$ **b.** $|7x - 4| \leq -1$

More complex inequalities, containing two or more separate absolute value terms, have to be solved by considering cases.

PROCEDURE
IN ACTION

EXAMPLE 11 **Solving an Inequality with Two or More Absolute Value Terms**

OBJECTIVE

Solve an inequality containing more than one absolute value term.

Step 1 Obtain boundary points
Set each expression within the absolute value symbols equal to zero and solve it. The solutions are boundary points.

Step 2 Locate the boundary points on the number line. These boundary points divide the number line into intervals in which the expressions within the absolute value symbols have constant sign.

Step 3 Make a diagram to indicate the value of each absolute value term on every interval obtained in Step 2. These intervals will be used to solve the original inequality.

Step 4 Rewrite the original inequality for each interval in Step 2 without the absolute value symbols, and solve it in each case.

EXAMPLE

Solve: $|x + 1| + |x - 2| \geq 6$

$$x + 1 = 0 \quad | \quad x - 2 = 0$$
$$x = -1 \quad | \quad x = 2 \qquad \text{Solve for } x.$$

The boundary points are -1 and 2.

The boundary points -1 and 2 divide the number line into three intervals: $I_1 = (-\infty, -1], I_2 = [-1, 2]$, and $I_3 = [2, \infty)$.

	$(-\infty, -1]$	$[-1, 2]$	$[2, \infty)$
$|x - 2|$	$-(x - 2)$	$-(x - 2)$	$(x - 2)$
$|x + 1|$	$-(x + 1)$	$(x + 1)$	$(x + 1)$

Case (i); for x in the interval $I_1 = (-\infty, -1]$:
In this case, $|x + 1| = -(x + 1)$ and $|x - 2| = -(x - 2)$.

$$|x + 1| + |x - 2| \geq 6 \qquad \text{Original inequality}$$
$$-(x + 1) + (-(x - 2)) \geq 6 \qquad \text{Case (i)}$$
$$-2x + 1 \geq 6 \qquad \text{Simplify.}$$

$$x \leq -\frac{5}{2} \qquad \text{Solve for } x.$$

$$\text{or} \left(-\infty, -\frac{5}{2}\right] \qquad \text{Write the solution in interval notation.}$$

The solution set in this case is:

$$J_1 = (-\infty, -1] \cap \left(-\infty, -\tfrac{5}{2}\right] = \left(-\infty, -\tfrac{5}{2}\right].$$

Case (ii); for x in the interval $I_2 = [-1, 2]$:
In this case, $|x + 1| = (x + 1)$ and $|x - 2| = -(x - 2)$.

$|x + 1| + |x - 2| \geq 6$	Original inequality
$(x + 1) + (-(x - 2)) \geq 6$	Case (ii)
$3 \geq 6$	Simplify.

The false statement $3 \geq 6$ means that there is no solution. The solution set in this case is:

$$J_2 = \phi.$$

Case (iii); for x in the interval $I_3 = [2, \infty)$:
In this case, $|x + 1| = (x + 1)$ and $|x - 2| = (x - 2)$.

$|x + 1| + |x - 2| \geq 6$	Original inequality
$(x + 1) + (x - 2) \geq 6$	Case (iii)
$2x - 1 \geq 6$	Simplify.
$x \geq \dfrac{7}{2}$	Solve for x.
or $\left[\dfrac{7}{2}, \infty\right)$	Write the solution in interval notation.

The solution set in this case is:

$$J_3 = [2, \infty) \cap \left[\tfrac{7}{2}, \infty\right) = \left[\tfrac{7}{2}, \infty\right).$$

Step 5 Combine the solution sets of each case in step 4 to obtain the solution of the given inequality.

The solution set of the original inequality is:

$$J_1 \cup J_2 \cup J_3 = \left(-\infty, -\tfrac{5}{2}\right] \cup \phi \cup \left[\tfrac{7}{2}, \infty\right) = \left(-\infty, -\tfrac{5}{2}\right] \cup \left[\tfrac{7}{2}, \infty\right).$$

Practice Problem 11 Solve $\left|\frac{x-2}{x+4}\right| < 4$. [*Hint*: $\left|\frac{x-2}{x+4}\right| < 4$ can be rewritten as $|x - 2| - 4|x + 4| < 0, x \neq -4$.]

Answers to Practice Problems

1. a. $(2, \infty)$
b. $(-\infty, 5]$
2. a. $(-\infty, \infty)$ **b.** $\varnothing$ **3.** $(-\infty, 2] \cup [4, \infty)$
4. $\left(-1, \dfrac{3}{2}\right)$
5. $a = -11, b = 10$ **6.** $(-1, 4)$ **7.** $[-6, 1)$
8. $[-3, 1]$
9. $\left(-\infty, -\dfrac{9}{2}\right] \cup \left[\dfrac{3}{2}, \infty\right)$
10. a. $(-\infty, \infty)$ **b.** $\varnothing$ **11.** $(-\infty, -6) \cup \left(-\dfrac{14}{5}, \infty\right)$

SECTION A.7 Exercises

Basic Concepts and Skills

In Exercises 1–6, solve each inequality. Write the solution in interval notation and graph the solution set.

1. $3(x + 2) < 2x + 5$

2. $4(x - 4) > 3(x - 5)$

3. $3(x + 2) \geq 5x + 18$

4. $5(x + 2) \leq 3(x + 1) + 10$

5. $3(x - 1) + (7 - x) < 2(x + 11) - 10$

6. $2(3x + 5) - 2(x + 1) > 4(x + 5) - 2$

In Exercises 7–12, solve each compound *or* inequality.

7. $2x + 5 < 1$ or $2 + x > 4$

8. $3x - 2 > 7$ or $2(1 - x) > 1$

9. $\dfrac{2x - 3}{4} \leq 2$ or $\dfrac{4 - 3x}{2} \geq 2$

10. $\dfrac{5 - 3x}{5} \geq \dfrac{1}{6}$ or $\dfrac{x - 1}{3} \leq 1$

11. $\dfrac{2x + 1}{3} \geq x + 1$ or $\dfrac{x}{2} - 1 > \dfrac{x}{3}$

12. $\dfrac{x - 1}{2} < \dfrac{x}{3} - 1$ or $\dfrac{2x + 5}{3} \leq \dfrac{x + 1}{6}$

In Exercises 13–18, solve each compound *and* inequality.

13. $3x - 5 < 10$ and $3 - 2x \leq 1$

14. $6 - x \leq 3x + 10$ and $7x - 14 \leq 3x + 14$

15. $2(x + 1) - 3 > 7$ and $3(2x + 1) + 1 < 10$

16. $5(x + 2) + 7 < 2$ and $2(5 - 3x) + 3 < 17$

17. $5 + 3(x - 1) < 3 + 3(x + 1)$ and $3x - 7 \leq 8$

18. $3(x + 3) + 7 > 2(x - 1) + 5$ and
    $2x + 1 \geq 3(5 - x) + 11$

In Exercises 19–24, solve each combined inequality.

19. $3 < x + 5 < 4$

20. $-4 \leq x - 2 < 2$

21. $0 \leq 1 - \dfrac{x}{3} < 2$

22. $0 < 5 - \dfrac{x}{2} \leq 3$

23. $5x \leq 3x + 1 < 4x + 2$

24. $3x + 2 < 2x + 3 < 4x - 1$

In Exercises 25–30, find a and b.

25. If $-2 < x < -1$, then $a < x + 7 < b$.

26. If $1 < x < 5$, then $a < 2x + 3 < b$.

27. If $-1 < x < 1$, then $a < 2 - x < b$.

28. If $3 < x < 7$, then $a < 1 - 3x < b$.

29. If $0 < x < 4$, then $a < 5x - 1 < b$.

30. If $-4 < x < 0$, then $a < 3x + 4 < b$.

In Exercises 31–42, solve each inequality. Write the solution set in interval notation.

31. $x^2 + 4x - 12 \leq 0$

32. $x^2 - 8x + 7 \geq 0$

33. $x^3 - x^2 - 4x + 4 > 0$

34. $x^3 + x^2 - 9x - 9 < 0$

35. $x^4 < 2x^2$

36. $x^3 < -8$

37. $\dfrac{(x - 3)(x + 1)}{(x + 2)(x - 4)} \leq 0$

38. $\dfrac{x(x - 4)}{(x - 2)(x + 3)} \leq 0$

39. $\dfrac{2x - 3}{x + 3} \leq 1$

40. $\dfrac{2x + 6}{2x + 1} \geq 3$

41. $\dfrac{x - 1}{x + 1} \leq \dfrac{x + 2}{x - 3}$

42. $\dfrac{x + 3}{x - 1} \geq \dfrac{x - 1}{x - 2}$

In Exercises 43–52, solve each inequality.

43. $|x + 1| < 3$

44. $|x - 4| < 1$

45. $|2x - 3| < 4$

46. $|4x - 6| \leq 6$

47. $|2x - 5| > 3$

48. $|3x - 3| \geq 15$

49. $|2x - 15| < 0$

50. $|x + 5| \leq -3$

51. $|2x - 3| > -4$

52. $|2x - 6| \leq 0$

In Exercises 53–66, solve each inequality.

53. $|x - 1| + |x - 2| \leq 4$

54. $|x - 7| + |2 - x| \leq 5$

55. $|x - 2| + |x - 4| \geq 8$

56. $|x + 3| + |x - 1| \geq 7$

57. $\left|\dfrac{x - 2}{x + 3}\right| < 1$

58. $\left|\dfrac{x + 3}{x - 1}\right| < 2$

59. $\left|\dfrac{2x - 3}{x + 1}\right| \leq 3$

60. $\left|\dfrac{2x - 1}{3x + 2}\right| \leq 1$

61. $\left|\dfrac{x - 1}{x + 2}\right| \geq 2$

62. $\left|\dfrac{x + 3}{x - 2}\right| \geq 3$

63. $\left|\dfrac{2x + 1}{x - 1}\right| > 4$

64. $\left|\dfrac{2x - 1}{3x + 2}\right| > 5$

65. $|x - 1| + |x - 2| + |x - 3| \geq 6$

66. $|x + 2| + |x - 1| + |x - 4| \geq 12$

Applying the Concepts

67. **Appliance markup.** The markup over the dealer's cost on a new refrigerator ranges from 15% to 20%. If the dealer's cost is $1750, over what range will the selling price vary?

68. **Return on investment.** An investor has $5000 to invest for a period of 1 year. Find the range of per annum simple interest rates required to generate interest between $200 and $275 inclusive.

69. Average grade. Sean has taken three exams and earned scores of 85, 72, and 77 out of a possible 100 points. He needs an average of at least 80 to earn a B in the course. What range of scores on the fourth (and last) 100-point test will guarantee a B in the course?

70. Temperature conversion. The formula for converting Fahrenheit temperature F to Celsius temperature C is $C = \dfrac{5}{9}(F - 32)$. What range in Celsius degrees corresponds to a range of 68° to 86° Fahrenheit?

71. Parking expense. The parking cost at the local airport (in dollars) is $C = 2 + 1.75(h - 1)$, where h is the number of hours the car is parked. For what range of hours is the parking cost between $37 and $51?

72. Plumbing charges. A plumber charges $42 per hour. A repair estimate includes a fixed cost of $147 for parts and a labor cost that is determined by how long it takes to complete the job. If the total estimate is for at least $210 and at most $294, what interval was estimated for the job?

SECTION A.8

Complex Numbers

OBJECTIVES

1 Define a complex number.

2 Add and subtract complex numbers.

3 Multiply and divide complex numbers.

Complex Numbers

In Section A-6, you learned methods for solving a quadratic equation $ax^2 + bx + c = 0$. Because the squares of real numbers are nonnegative (that is, $x^2 \geq 0$ for any real number x), the quadratic equation $x^2 = -1$ has no solution in the set of real numbers. To solve this equation, we extend the real number system to a larger system called the complex number system. We define a new number, i, with the following properties:

$$i = \sqrt{-1}, \quad i^2 = -1$$

1 Define a complex number.

Complex Numbers

A **complex number** is a number of the form

$$a + bi,$$

where a and b are real numbers and $i^2 = -1$.

The **real part** of the complex number $a + bi$ is a.

The **imaginary part** of the complex number $a + bi$ is b.

Two complex numbers are **equal** if and only if their real parts are equal and their imaginary parts are equal.

TECHNOLOGY
CONNECTION

Most graphing calculators allow you to work with complex numbers by changing from "real" to **a + bi** mode. Once the **a + bi** mode is set, the $\sqrt{-1}$ is recognized as i. The i key is used to enter complex numbers such as $2 + 3i$.

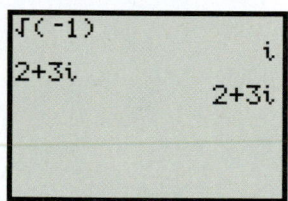

A complex number written in the form $a + bi$, where a and b are real numbers, is said to be in **standard form**. A complex number with real part 0, written as just bi, is called a **pure imaginary number**. Real numbers form a subset of complex numbers with imaginary part 0. For example, $-3 = -3 + 0i$.

We can express the **principal square root** of any negative number as the product of a real number and i.

If $a > 0$ is any real number, then $\sqrt{-a} = (\sqrt{a})i$.

The real and imaginary parts of a complex number can then be found.

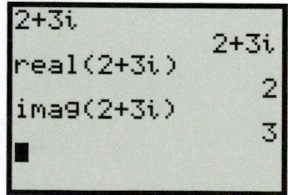

We identify the real and the imaginary parts of each complex number.

	Real Part	Imaginary Part
$2 + 5i$	2	5
$7 - 8i$	7	-8
$3i$	0	3
-9	-9	0
$3 + \sqrt{-25}$	3	5 (since $\sqrt{-25} = i\sqrt{25} = 5i$)

Addition and Subtraction

2 Add and subtract complex numbers.

ADDITION AND SUBTRACTION OF COMPLEX NUMBERS

For all real numbers a, b, c, and d,

$$(a + bi) + (c + di) = (a + c) + (b + d)i \quad \text{and}$$
$$(a + bi) - (c + di) = (a - c) + (b - d)i.$$

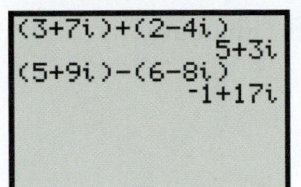

EXAMPLE 1 Adding and Subtracting Complex Numbers

Write the sum or difference of two complex numbers in standard form.

a. $(3 + 7i) + (2 - 4i)$ **b.** $(5 + 9i) - (6 - 8i)$

Solution

a. $(3 + 7i) + (2 - 4i) = (3 + 2) + [7 + (-4)]i = 5 + 3i$

b. $(5 + 9i) - (6 - 8i) = (5 - 6) + [9 - (-8)]i = -1 + 17i$

Practice Problem 1 Write the following complex numbers in standard form.

a. $(1 - 4i) + (3 + 2i)$ **b.** $(4 + 3i) - (5 - i)$

Recall that if *a* and *b* are nonnegative real numbers, then

$$\sqrt{a}\sqrt{b} = \sqrt{ab}.$$

However, this property is not true for nonreal numbers. For example,

$$\sqrt{-9}\sqrt{-9} = (3i)(3i) = 9i^2 = 9(-1) = -9,$$

but

$$\sqrt{(-9)(-9)} = \sqrt{81} = 9.$$

Thus,

$$\sqrt{-9}\sqrt{-9} \neq \sqrt{(-9)(-9)}.$$

---

3 Multiply and divide complex numbers.

Multiplying and Dividing Complex Numbers

We multiply complex numbers by first using FOIL (as we did with binomials) and then replacing i^2 with -1. The result, given next, need not be memorized.

> **MULTIPLYING COMPLEX NUMBERS**
>
> For all real numbers *a*, *b*, *c*, and *d*,
>
> $$(a + bi)(c + di) = (ac - bd) + (ad + bc)i.$$

EXAMPLE 2 Multiplying Complex Numbers

Write the following products in standard form.

a. $(3 - 5i)(2 + 7i)$ **b.** $-2i(5 - 9i)$

Solution

$$
\begin{array}{ll}
\textbf{a.} \; (3 - 5i)(2 + 7i) = \overset{F}{6} + \overset{O}{21i} - \overset{I}{10i} - \overset{L}{35i^2} & \\
\qquad\qquad\qquad\quad = 6 + 11i + 35 & \text{Because } i^2 = -1, -35i^2 = 35. \\
\qquad\qquad\qquad\quad = 41 + 11i & \text{Combine terms.}
\end{array}
$$

$$
\begin{array}{ll}
\textbf{b.} \; -2i(5 - 9i) = -10i + 18i^2 & \text{Distributive property} \\
\qquad\qquad\quad = -10i - 18 & \text{Because } i^2 = -1, 18i^2 = -18. \\
\qquad\qquad\quad = -18 - 10i & \text{Standard form}
\end{array}
$$

Practice Problem 2 Write the following products in standard form.

a. $(2 - 6i)(1 + 4i)$ **b.** $-3i(7 - 5i)$

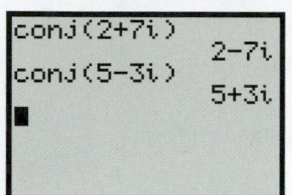
The Conjugate of a Complex Number

> If $z = a + bi$, then the conjugate (or complex conjugate) of z is denoted by $\bar{z}$ and defined by $\bar{z} = \overline{a + bi} = a - bi$.

For example, $\overline{2 + 7i} = 2 - 7i$ and $\overline{5 - 3i} = \overline{5 + (-3)i} = 5 - (-3)i = 5 + 3i$.

EXAMPLE 3 Multiplying a Complex Number by Its Conjugate

Find the product $z\bar{z}$ for the complex number $z = 2 + 5i$.

Solution

If $z = 2 + 5i$, then $\bar{z} = 2 - 5i$.

$$z\bar{z} = (2 + 5i)(2 - 5i) = 2^2 - (5i)^2 \qquad \text{Difference of squares}$$
$$= 4 - 25i^2 \qquad (5i)^2 = 5^2 i^2 = 25i^2$$
$$= 4 - (-25) \qquad \text{Because } i^2 = -1, 25i^2 = -25.$$
$$= 29 \qquad \text{Simplify.}$$

Practice Problem 3 Find the product $z\bar{z}$ for the complex number $z = 1 + 6i$.

Notice that the product $z\bar{z}$ in Example 3 is a real number. To write the reciprocal of a nonzero complex number, or the quotient of two complex numbers in standard form, multiply the numerator and denominator by the conjugate of the denominator. The resulting denominator is a real number, as the next theorem states.

> ### COMPLEX CONJUGATE PRODUCT THEOREM
>
> If $z = a + bi$, then
>
> $$z\bar{z} = a^2 + b^2.$$

The complex conjugate product theorem provides a procedure for writing the quotient of two complex numbers in standard form. You do not need to memorize the final result; just perform the indicated operations.

> ### DIVIDING COMPLEX NUMBERS
>
> Let $z = a + bi \ (z \neq 0)$ and $w = c + di$ be two complex numbers in standard form. Then
>
> $$\frac{w}{z} = \frac{w\bar{z}}{z\bar{z}} \qquad \text{Multiply numerator and denominator by } \bar{z}.$$
> $$= \frac{(c + di)(a - bi)}{a^2 + b^2} = \frac{ac + bd + (ad - bc)i}{a^2 + b^2}$$
> $$= \left(\frac{ac + bd}{a^2 + b^2}\right) + \left(\frac{ad - bc}{a^2 + b^2}\right)i$$

EXAMPLE 4   **Dividing Complex Numbers**

Write the following quotients in standard form.

a. $\dfrac{1}{2 + i}$ **b.** $\dfrac{4 + \sqrt{-25}}{2 - \sqrt{-9}}$

Solution

a. The denominator is $2 + i$, so its conjugate is $2 - i$.

$$\frac{1}{2 + i} = \frac{1(2 - i)}{(2 + i)(2 - i)} \qquad \text{Multiply numerator and denominator by } 2 - i.$$

$$= \frac{2 - i}{2^2 + 1^2} = \frac{2 - i}{5} = \frac{2}{5} - \frac{1}{5}i$$

b. Write $\dfrac{4 + \sqrt{-25}}{2 - \sqrt{-9}}$ as $\dfrac{4 + 5i}{2 - 3i}$.

$$\frac{4 + 5i}{2 - 3i} = \frac{(4 + 5i)(2 + 3i)}{(2 - 3i)(2 + 3i)} \qquad \text{Multiply numerator and denominator by } 2 + 3i.$$

$$= \frac{8 + 12i + 10i + 15i^2}{2^2 + 3^2} \qquad \text{Use FOIL in the numerator; } (2 - 3i)(2 + 3i) = 2^2 + 3^2.$$

$$= \frac{-7 + 22i}{13} = -\frac{7}{13} + \frac{22}{13}i \qquad i^2 = -1, 8 + 15i^2 = 8 - 15 = -7$$

Practice Problem 4 Write the following quotients in standard form.

a. $\dfrac{2}{1 - i}$ **b.** $\dfrac{-3i}{4 + \sqrt{-25}}$

Powers of *i*

We already know the first two powers of i: $i^1 = i$ and $i^2 = -1$. For the following powers of i, notice the pattern.

$$i^1 = i \qquad\qquad i^5 = i^4 \cdot i = (1)i = i$$
$$i^2 = -1 \qquad\qquad i^6 = i^4 \cdot i^2 = (1) \cdot i^2 = i^2 = -1$$
$$i^3 = i^2 \cdot i = (-1)i = -i \qquad\qquad i^7 = i^4 \cdot i^3 = (1) \cdot i^3 = i^3 = -i$$
$$i^4 = i^2 \cdot i^2 = (-1)(-1) = 1 \qquad\qquad i^8 = i^4 \cdot i^4 = (1)(1) = 1$$

After i^4, the powers of i cycle indefinitely through the values $i, -1, -i,$ and 1. If n is a positive integer, a quick way to evaluate i^n is to divide n by 4; then $i^n = i^r$, where r is the remainder: 0, 1, 2, or 3.

To see how this works, we evaluate i^{1003}. Dividing 1003 by 4, we get $1003 = 4(250) + 3$. Then $i^{1003} = i^{4(250)+3} = i^{4(250)} \cdot i^3 = (i^4)^{250} \cdot i^3 = (1)^{250} \cdot i^3 = 1 \cdot i^3 = i^3 = -i$. Similarly, dividing 26 by 4 gives the remainder 2; so $i^{26} = i^2 = -1$.

Answers to Practice Problems

1. a. $4 - 2i$ **b.** $-1 + 4i$ **2. a.** $26 + 2i$ **b.** $-15 - 21i$ **3.** 37   **4. a.** $1 + i$ **b.** $-\dfrac{15}{41} - \dfrac{12}{41}i$

SECTION A.8   **Exercises**

Basic Concepts and Skills

In Exercises 1–4, use the definition of equality of complex numbers to find the real numbers x and y such that the equation is true.

1. $2 + xi = y + 3i$

2. $x - 2i = 7 + yi$

3. $x - \sqrt{-16} = 2 + yi$

4. $3 + yi = x - \sqrt{-25}$

In Exercises 5–22, perform each operation and write the result in the standard form $a + bi$.

5. $(5 + 2i) + (3 + i)$

6. $(4 - 3i) - (5 + 3i)$

7. $(3 - 5i) - (3 + 2i)$

8. $(-2 - 3i) + (-3 - 2i)$

9. $3(5 + 2i)$

10. $-4(2 - 3i)$

11. $3i(5 + i)$

12. $-3i(5 - 2i)$

13. $(3 + i)(2 + 3i)$

14. $(4 + 3i)(2 + 5i)$

15. $(2 - 3i)(2 + 3i)$

16. $(4 - 3i)(4 + 3i)$

17. $(3 + 4i)(4 - 3i)$

18. $(-2 + 3i)(-3 + 10i)$

19. $(\sqrt{3} - 12i)^2$

20. $(-\sqrt{5} - 13i)^2$

21. $(1 + 3i)^3$

22. $(1 - 2i)^3$

In Exercises 23–28, write the conjugate $\bar{z}$ of each complex number z. Then find $z\bar{z}$.

23. $z = 2 - 3i$

24. $z = \dfrac{1}{2} - 2i$

25. $z = 4 + 5i$

26. $z = \dfrac{2}{3} + \dfrac{1}{2}i$

27. $z = \sqrt{2} - 3i$

28. $z = \sqrt{5} + \sqrt{3}i$

In Exercises 29–34, write each quotient in the standard form $a + bi$.

29. $\dfrac{5}{-i}$

30. $\dfrac{-1}{1 + i}$

31. $\dfrac{5i}{2 + i}$

32. $\dfrac{2 + 3i}{1 + i}$

33. $\dfrac{2 - 5i}{4 - 7i}$

34. $\dfrac{2 + \sqrt{-4}}{1 + i}$

In Exercises 35–38, find each power of i and simplify the expression.

35. i^{17}

36. i^{125}

37. i^{-7}

38. i^{-24}

In Exercises 39–42, let $z_1 = a + bi$ and $z_2 = c + di$.

39. Prove that $\overline{\overline{z_1}} = z_1$.

40. Prove that $\overline{z_1 + z_2} = \overline{z_1} + \overline{z_2}$.

41. Prove that $\overline{z_1 z_2} = \overline{z_1}\,\overline{z_2}$. Use this fact to prove that $\overline{z^2} = (\bar{z})^2$.

42. Prove that $z_1 + \overline{z_1} = 2a$ and that $z_1 - \overline{z_1} = 2bi$.

In Exercises 43–46, let $z = 2 + 4i$ and $w = 3 - 2i$.

43. $\overline{wz}$

44. $\overline{w}z$

45. $\bar{z}w$

46. $\dfrac{w}{z}$

Applying the Concepts

47. Series circuits. If the impedance of a resistor in a circuit is $Z_1 = 4 + 3i$ ohms and the impedance of a second resistor is $Z_2 = 5 - 2i$ ohms, find the total impedance of the two resistors when they are placed in series (the sum of the two impedances).

Series circuit

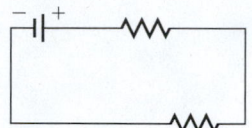

48. Parallel circuits. If the two resistors in Exercise 47 are connected in parallel, the total impedance is given by

$$\frac{Z_1 Z_2}{Z_1 + Z_2}.$$

Find the total impedance if the resistors in Exercise 47 are connected in parallel.

Parallel circuit

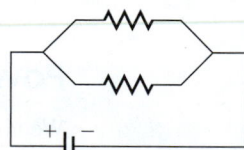

As with impedance, the current I and voltage V in a circuit can be represented by complex numbers. The three quantities (voltage V, impedance Z, and current I) are related by the equation $Z = \dfrac{V}{I}$. Thus, if two of these values are given, the value of the third can be found from this equation. In Exercises 49–54, find the value that is not specified.

49. Finding impedance. $I = 7 + 5i$ $V = 35 + 70i$

50. Finding impedance. $I = 7 + 4i$ $V = 45 + 88i$

51. Finding voltage. $Z = 5 - 7i$ $I = 2 + 5i$

52. Finding voltage. $Z = 7 - 8i$ $I = \dfrac{1}{3} + \dfrac{1}{6}i$

53. Finding current. $V = 12 + 10i$ $Z = 12 + 6i$

54. Finding current. $V = 29 + 18i$ $Z = 25 + 6i$

Answers

CHAPTER 1

Section 1.1

Concepts and Vocabulary:

1. that satisfy the equation   **3.** $\left(\dfrac{x_1 + x_2}{2}, \dfrac{y_1 + y_2}{2}\right)$ **5.** False   **7.** False

Building Skills:

9.

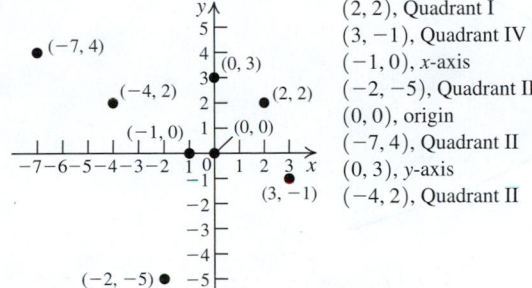

(2, 2), Quadrant I
(3, −1), Quadrant IV
(−1, 0), x-axis
(−2, −5), Quadrant III
(0, 0), origin
(−7, 4), Quadrant II
(0, 3), y-axis
(−4, 2), Quadrant II

11. a. $(1, 0,) (2, 0), (3, 0), (4, 0), (5, 0)$ The y-coordinate is 0.
b. Horizontal line intersecting the y-axis at 1

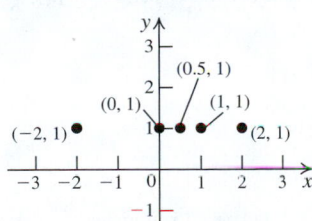

13. a. $\sqrt{17}$ **b.** $(2.5, 3)$ **15. a.** $\sqrt{58}$ **b.** $(2.5, 1.5)$ **17. a.** $\sqrt{13}$
b. $(0.5, -4)$ **19. a.** $\sqrt{(x + 2)^2 + (y - 3)^2}$ **b.** $\left(\dfrac{x - 2}{2}, \dfrac{y + 3}{2}\right)$

21. Yes   **23.** No: $9\sqrt{2} + 5 + \sqrt{61}$ **25.** Isosceles   **27.** Scalene
29. Equilateral   **31.** On the graph: $(-3, -4), (1, 0), (4, 3)$ Not on the
graph: $(2, 3)$ **33.** On the graph: $(3, 2), (0, 1), (8, 3)$; not on the graph:
$(8, -3)$ **35.** On the graph: $(1, 0), (2, \sqrt{3}), (2, -\sqrt{3})$; not on
the graph $(0, -1)$

37.

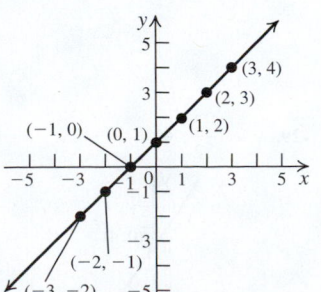

39.

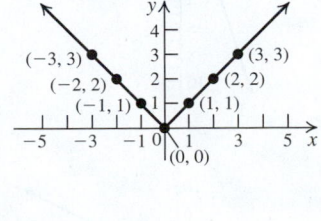

41.

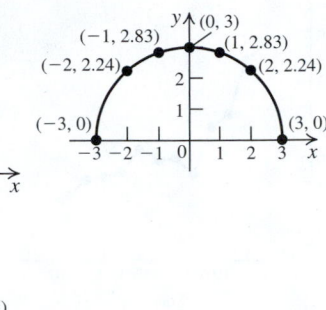

43.

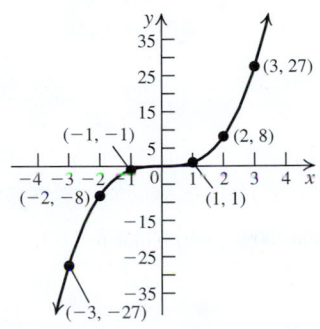

45.

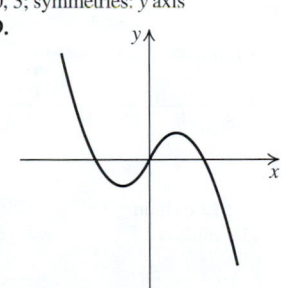

47. x-intercepts: $-1, 1$; y-intercepts: none; symmetries: x axis, y axis, origin
49. x-intercepts: $-\pi, 0, \pi$; y-intercept: 0; symmetries: origin
51. x-intercepts: $-3, 3$; y-intercepts: $-2, 2$; symmetries: x axis, y axis, origin
53. x-intercepts: $-2, 0, 2$; y-intercept: 0; symmetries: origin
55. x-intercepts: $-2, 0, 2$; y-intercepts: 0, 3; symmetries: y axis

57.   **59.**

61. x-intercept: 4; y-intercept: 3   **63.** x-intercept: $\dfrac{5}{2}$; y-intercept: $\dfrac{5}{3}$
65. x-intercept: 2, y intercept: -2 **67.** x-intercepts: 2, 4; y-intercept: 8
69. x-intercepts: $-2, 2$; y-intercepts: $-2, 2$ **71.** x-intercepts: $-1, 1$; no
y-intercept   **73.** Not symmetric with respect to the x-axis, symmetric with
respect to the y-axis, not symmetric with respect to the origin   **75.** Not
symmetric with respect to the x-axis, not symmetric with respect to the
y-axis, symmetric with respect to the origin   **77.** Not symmetric with
respect to the x-axis, symmetric with respect to the y-axis, symmetric with
respect to the origin   **79.** Not symmetric with respect to x-axis, not symmetric with respect to y-axis, symmetric with respect to origin.
81. Center: $(-1, 3)$; radius: 4   **83.** center: $(1, 0)$, radius: 2
85. $(x - 3)^2 + (y - 1)^2 = 9$ **87.** $(x - 3)^2 + (y + 4)^2 = 97$
89. $(x + 3)^2 + (y + 2)^2 = 9$ **91.** $(x - 5)^2 + (y - 1)^2 = 25$

93. a. Circle; center: $(-2, 3)$, radius: 2; **b.** x-intercept: none, y-intercept: 3 **95. a.** Circle; center: $(1, 1)$; radius: $\sqrt{6}$ **b.** x-intercepts: $1 \pm \sqrt{5}$; y-intercepts: $1 \pm \sqrt{5}$ **97. a.** Circle; center: $(0, -1)$; radius: 1 **b.** x-intercept: 0; y-intercepts: $-2, 0$ **99.** Point $(1, -2)$

Applying the Concepts:
101. 51.5% **103.** June 2014
105.

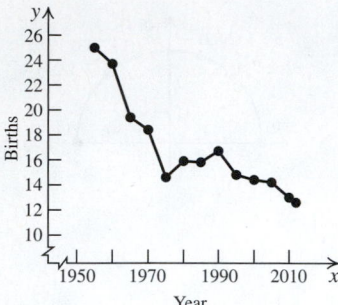

107.

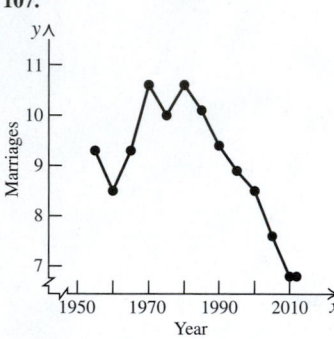

109. \$251 billion in 2008; \$274 billion in 2009; \$297 billion in 2010

111. $|x| = |y|$ **113.** $x = \dfrac{y^2}{4} + 1$
115. a.

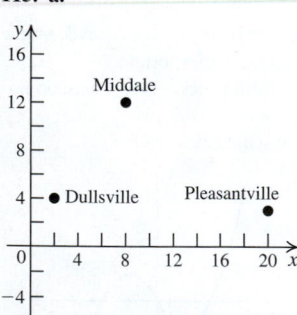

b. 2500 mi
c. $5\sqrt{13} \approx 1802.78$ mi

117. a. \$12 million
b. $-$\$5.5 million

c.

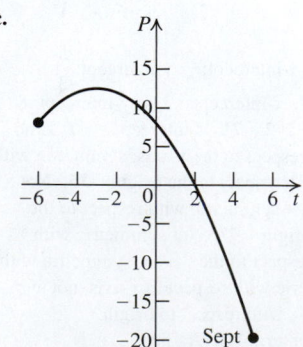
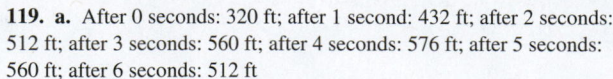

d. 2. It represents the month (Sept. 2012) when there is neither profit nor loss. **e.** 8. It represents the profit in July 2012.

119. a. After 0 seconds: 320 ft; after 1 second: 432 ft; after 2 seconds: 512 ft; after 3 seconds: 560 ft; after 4 seconds: 576 ft; after 5 seconds: 560 ft; after 6 seconds: 512 ft

b.

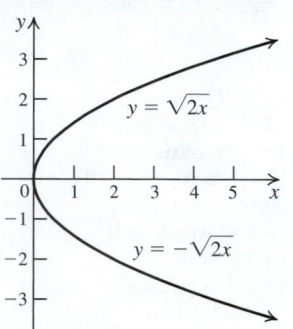

c. $0 \le t \le 10$
d. t-intercept 10. It represents the time when the object hits the ground. y-intercept; 320. It shows the height of the buillding

Beyond the Basics:
125. c. $(2/3, 5/3), (7/3, 4/3)$ **127.** y-axis **131.** Quadrants I and III

Critical Thinking / Discussion / Writing
133. The graph is the union of the graphs of $y = \sqrt{2x}$ and $y = -\sqrt{2x}$.

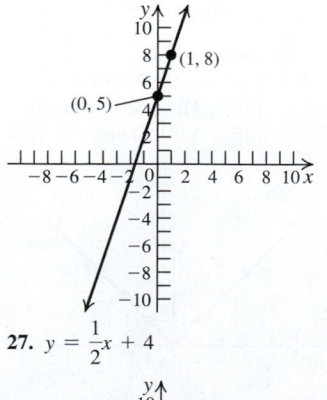

137. a. $(r, r), (3r, r)$ **b.** $\left(2 - \dfrac{\pi}{2}\right)r^2$

Getting Ready for the Next Section
139. $\dfrac{1}{2}$ **141.** $-\dfrac{1}{2}$ **143.** $2 - \dfrac{2}{3}x$ **145.** $y = \dfrac{2}{3}x + \dfrac{8}{3}$ **147.** $-\dfrac{1}{2}$ **149.** $\dfrac{3}{2}$

Section 1.2

Concepts and Vocabulary:
1. 0; undefined **3.** 3 **5.** False **7.** True

Building Skills:
9. $\dfrac{4}{3}$, rising **11.** 0, horizontal **13.** -2.2, falling **15.** 4, rising

17. ℓ_3 **19.** ℓ_4 **21.** 1 **23.** 2
25. $y = 3x + 5$

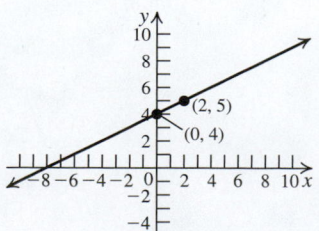

27. $y = \dfrac{1}{2}x + 4$

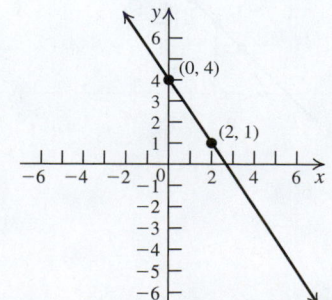

29. $y = -\dfrac{3}{2}x + 4$

31. $y = -4$

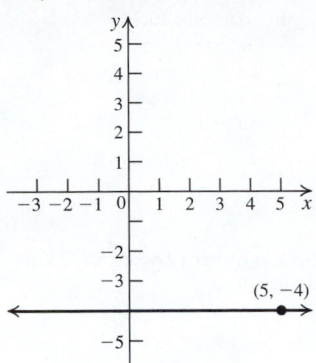

33. $y = -x + 1$ **35.** $y = 3$ **37.** $y = \frac{2}{3}x + \frac{1}{3}$ **39.** $y = -\frac{7}{2}x + 2$

41. $x = 5$ **43.** $y = 0$ **45.** $y = 14$ **47.** $y = -\frac{2}{3}x - 4$

49. $y = \frac{4}{3}x + 4$ **51.** $y = 7$ **53.** $y = -5$

55. $m = 3$ x intercept: $2/3$,
y intercept -2

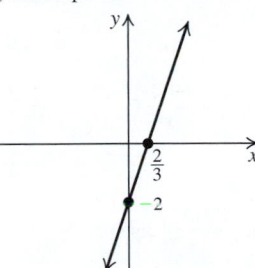

57. $m = -\frac{1}{2}$; y-intercept $= 2$;
x-intercept $= 4$

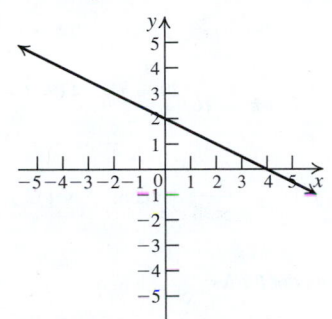

59. $m = \frac{3}{2}$; y-intercept $= 3$;
x-intercept $= -2$

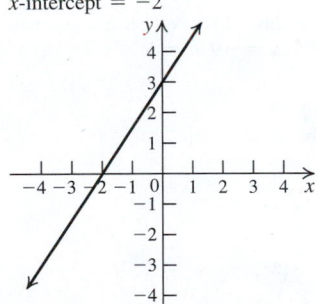

61. m is undefined; no y-intercept;
x-intercept $= 5$

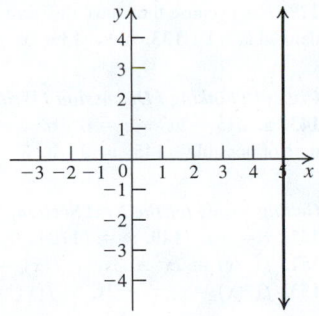

63. m is undefined;
y-intercept $= y$-axis;
x-intercept $= 0$

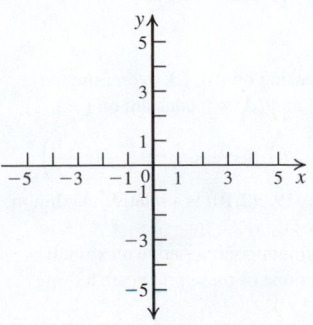

67. $\frac{x}{3} + \frac{y}{2} = 1$; x-intercept $= 3$; y-intercept $= 2$

69. x-intercept: 6, y-intercept -4 **71.** x-intercept: -2, y-intercept 5

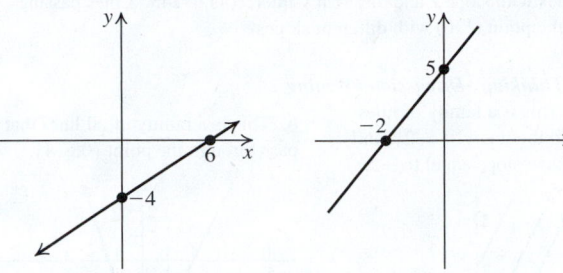

73. $y = x + 2$; because the coordinates of the point $(-1, 1)$ satisfy this equation, $(-1, 1)$ also lies on this line.   **75.** $y = -\frac{3}{4}x$

77. $y = (3/2)x - 4$ **79.** parallel   **81.** perpendicular   **83.** neither
85. perpendicular   **87.** parallel   **89.** $y = 3x - 9$ **91.** $y = 2x + 4$
93. $y = -x - 7$ **95.** $y = -(1/3)x - 5/3$ **97.** $y = 6x + 4$

99. $y = -\frac{1}{6}x + 4$ **101.** $y = -x + 2$ **103.** $y = -3x - 2$

Applying the Concepts:
105. a. initial expenses   **b.** breaking even   **c.** \$640

d. $P = (695/8)n - 750$ **107.** $\frac{1}{10}$ **109.** He should trim the hedge

every 10 days.   **111. a.** x: time in weeks; y: amount of money after x
week in the account; $y = 7x + 130$ **b.** Slope: weekly deposit in the
account; y-intercept: initial deposit   **113. a.** x: number of months since
purchase of the refrigerator; y: amount owed after x months;
$y = -15x + 600$ **b.** Slope: monthly payment; y-intercept: initial charge
115 a. x: number of years after 2010; y: life expectancy of a female born
in the United States in the year $x + 2010$; $y = 0.17x + 80.8$ **b.** Slope:
rate of increase in the life expectancy of a female born in the United
States; y-intercept: current life expectancy of a female born in the United
States in 2010   **117.** No extra buying needed. Estimated 20 MB left.

119. $y = 5x + 40,000$ **121. a.** $F = \frac{9}{5}C + 32$ **b.** 1 degree Celsius

change in the temperature is equal to $\frac{9}{5}$ degrees change in Fahrenheit.

c. 104°F, 77°F, 23°F, 14°F   **d.** 37.78°C, 32.22°C, 23.89°C, −23.33°C,
−28.89°C   **e.** 36.44°C to 37.56°C   **f.** At −40°
123. a. $y = 266.8t + 2425$
b.   **c.** 3225   **d.** 5627

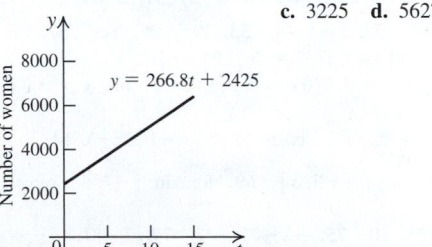

125. 13.7%   **127. a.** $y = -2x + 12.4$
b.   **c.** 0 papers

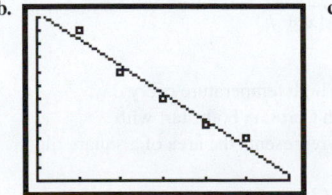

Beyond the Basics:

129. $c = 12$ **139.** $(9.5, 12)$ or $(-5.5, -8)$ **141.** $y = \dfrac{5}{12}x + \dfrac{169}{12}$

143. Lines with slope 2 and different y-intercepts.   **145.** Lines passing through the point $(1, 0)$ with different slopes.

Critical Thinking / Discussion / Writing

147. a. This is a family of lines parallel to the line $y = -2x$, and they all have slope equal to -2.

b. This is a family of all lines that pass through the point $(0, -4)$.

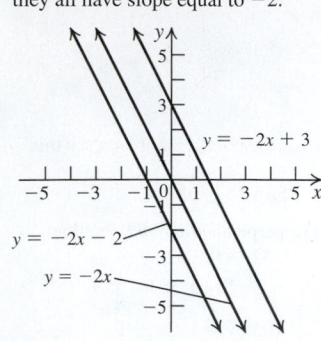

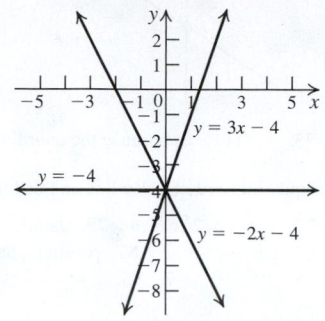

Getting Ready for the Next Section:

149. $-2, 2$ **151.** $-1, 2$ **153.** $3h$ **155.** $-2a - h$

Section 1.3

Concepts and Vocabulary:

1. independent   **3.** -14 **5.** False   **7.** True

Building Skills:

9. Domain: $\{a, b, c\}$; range: $\{d, e\}$; function   **11.** Domain; $\{a, b, c\}$; range: $\{1, 2\}$; function   **13.** Domain: $\{-3, -1.0, 1, 2, 3\}$: range: $\{-8, -3, 0, 1\}$; function   **15.** Yes   **17.** Yes   **19.** No   **21.** Yes   **23.** Yes   **25.** No   **27.** Yes   **29.** $f(0) = 1, g(0)$ is not defined, $h(0) = \sqrt{2}, f(a) = a^2 - 3a + 1, f(-x) = x^2 + 3x + 1$
31. $f(-1) = 5, g(-1)$ is not defined, $h(-1) = \sqrt{3}, h(c) = \sqrt{2 - c}, h(-x) = \sqrt{2 + x}$ **33. a.** 0 **b.** $\dfrac{2\sqrt{3}}{3}$ **c.** Not defined
d. Not defined   **e.** $\dfrac{-2x}{\sqrt{4 - x^2}}$ **35.** 5 **37.** $(-\infty, \infty)$

39. $(-\infty, 9) \cup (9, \infty)$ **41.** $(-\infty, -1) \cup (-1, 1) \cup (1, \infty)$
43. $(-\infty, 4)$ **45.** $[1, 2) \cup (2, \infty)$ **47.** $(-\infty, -2) \cup (-2, -1) \cup (-1, \infty)$
49. $(-\infty, 0) \cup (0, \infty)$ **51.** $[-1, 1]$ **53.** Yes   **55.** Yes   **57.** No
59. $f(-4) = -2, f(-1) = 1, f(3) = 5, f(5) = 7$
61. $h(-2) = -5, h(-1) = 4, h(0) = 3, h(1) = 4$ **63.** $x = -2$ or 3

65. a. No   **b.** $x = -1 \pm \sqrt{3}$ **c.** 5 **d.** $x = -1 \pm \dfrac{1}{2}\sqrt{14}$

67. domain: $[-3, 2]$, range: $[-3, 3]$ **69.** domain: $[-4, 4]$, range:

$[-2, 3]$ **71.** -2 **73.** 10 **75.** -2 **77.** -2 **79.** 7 **81.** $-\dfrac{1}{12}$

83. 2 **85.** $-x - 1$ **87.** $3x + 7$ **89.** $-\dfrac{4}{x}$ **91.** -2 **93.** $2x + h$

95. $6x + 3h - 2$ **97.** 0 **99.** $-\dfrac{1}{x(x + h)}$

Applying the Concepts:

101. Yes, because there is only one high temperature every day.
103. No, because Nevada and North Carolina both start with N.
105. $A(x) = x^2; A(4) = 16; A(4)$ represents the area of a square tile with side of length 4.
107. It is a function. $S(x) = 6x^2; S(3) = 54$
109. a. $[0, 8]$ **b.** $h(2) = 192, h(4) = 256, h(6) = 192$ **c.** 8 sec

d.

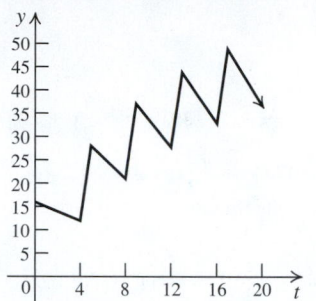

111. After 4 hours: 12 ml; after 8 hours: 21 ml; after 12 hours: 27. 75 ml; after 16 hours: 32. 81 ml; after 20 hours: 36. 61 ml

113. $P = 28x - x^2$ **115.** $S = 2x^2 + \dfrac{256}{x}$

117. $A = \dfrac{x^2}{4\pi} + \dfrac{1}{16}(20 - x)^2$ **119.** $C = 2x^2 + \dfrac{6912}{x}$

121. $d = [(x - 2)^2 + (x^3 - 3x + 5)^2]^{1/2}$

123. $f(x) = \dfrac{x}{x - 1}$: domain: $(-\infty, 1) \cup (1, \infty); f(4) = \dfrac{4}{3}$

Beyond the Basics:

125. $f(x) = \dfrac{4x + 2}{x}$; domain: $(-\infty, 0) \cup (0, \infty); f(4) = \dfrac{9}{2}$

127. $f(x) = \dfrac{2 - x}{x^2 + 1}$; domain: $(-\infty, \infty); f(4) = -\dfrac{2}{17}$
129. No, because they have different domains   **131.** No, because g is not defined at -1 **133.** Yes   **135.** 3 **137.** $a = -9, b = 6$

Critical Thinking / Discussion / Writing

143. a. $ax^2 + bx + c = 0$ **b.** $y = c$ **c.** $b^2 - 4ac < 0$
d. Not possible   **145. a.** 9 **b.** 8

Getting Ready for the Next Section:

147. $y = -x$ **149.** $y = (1/2)x + 7/2$
151. $f(-x) = 2x^2 + 3x, -f(x) = -2x^2 + 3x$.
153. $f(-x) = -x^3 + 2x, -f(x) = -x^3 + 2x$

Section 1.4

Concepts and Vocabulary:

1. $f(x_1) > f(x_2)$ **3.** $f(-x) = f(x)$ **5.** True   **7.** False

Building Skills:

9. Increasing on $(-\infty, \infty)$ **11.** Increasing on $(\infty, 2)$, decreasing on $(2, \infty)$ **13.** Increasing on $(-\infty, -2)$ and $(2, \infty)$, constant on $(-2, 2)$

15. Increasing on $(-\infty, -3)$ and $\left(-\dfrac{1}{2}, 2\right)$, decreasing on $\left(-3, -\dfrac{1}{2}\right)$
and $(2, \infty)$ **17.** No relative extrema   **19.** $(2, 10)$ is a relative maximum point and a turning point.   **21.** Any point $(x, 2)$, for $-2 < x < 2$, is a relative maximum and a relative minimum point; relative maximum at $(-2, 2)$, relative minimum at $(2, 2)$; none of these points are turning points.

23. $(-3, 4)$ and $(2, 5)$ are relative maximum points and turning points; $\left(-\dfrac{1}{2}, -2\right)$ is a relative minimum and a turning point.   **25.** Odd

27. Neither   **29.** Odd   **31.** Even   **33.** Odd   **35.** Even   **37.** Odd
39. Neither   **41.** Even   **43.** Odd   **45.** Even   **47.** Neither
49. a. $f(1) = 2, f(2) = 2, f(3) = 3$ **51. a.** $f(-15) = -1, f(12) = 1$
b.

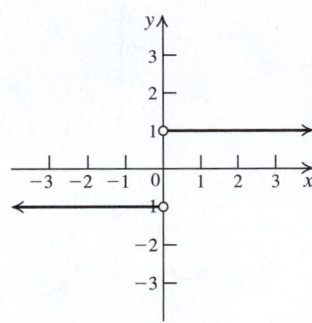

b.

c. Domain: $(-\infty, 0) \cup (0, \infty)$; range: $\{-1, 1\}$

53. Range: $(-\infty, \infty)$
55.

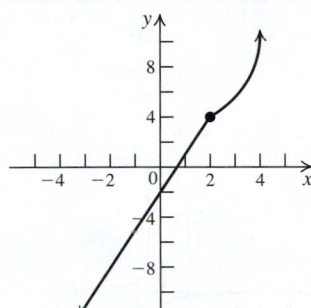

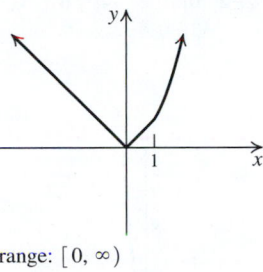

range: $[0, \infty)$

57.

59.

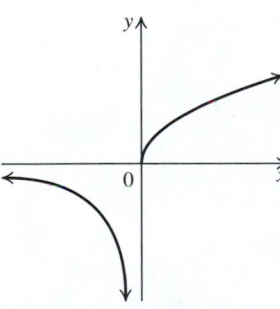

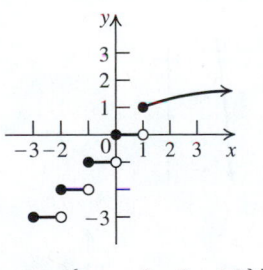

range: $\{\ldots, -3, -2, -1.0\} \cup [1, \infty)$

range: $(-\infty, \infty)$

61.

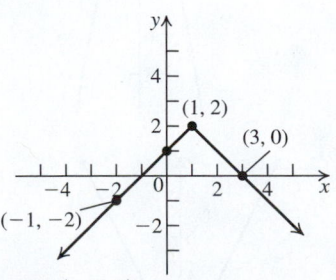

range: $(-\infty, 2)$

63. $g(x) = \begin{cases} 3x - 1 & \text{if } 1 \le x \le 2 \\ -(4/3)x + 23/3 & \text{if } 2 < x \le 5 \end{cases}$

65. $g(x) = \begin{cases} (2/3)x + 1/3 & \text{if } 1 \le x \le 4 \\ x - 1 & \text{if } 4 < x \le 6 \end{cases}$

67. $g(x) = \begin{cases} 4x - 3 & \text{if } 1 \le x \le 2 \\ (1/4)x + 9/2 & \text{if } 2 < x \le 6 \end{cases}$

69. $g(x) = \begin{cases} -2x + 8 & \text{if } 1 \le x \le 3 \\ -(1/3)x + 3 & \text{if } 3 < x \le 6 \end{cases}$

71. a. Increasing: (2006, 2009) (2011, 2012), (2013, 2014), Decreasing: (2009, 2011), (2012, 2013)   **b.** Relative maximum of 251.1 in 2009, Relative maximum of 293.2 in 2012, Relative minimum of 21.5 in 2011, Relative minimum of 187.0 in 2013
73.

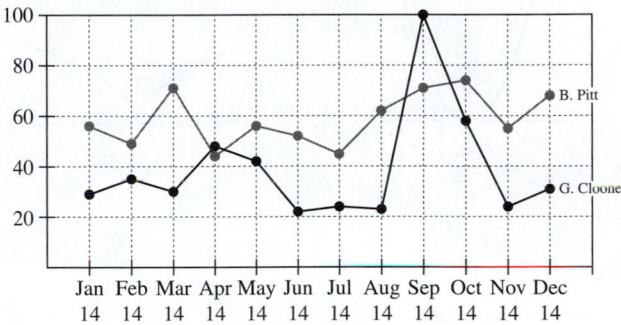

Applying the Concepts:
75. a. $f(x) = \dfrac{1}{33.81}x$; domain: $[0, \infty)$; range: $[0, \infty)$ **b.** 0.0887. It means that 3 oz = 0.0887 liters.   **c.** 0.3549
77. a. No d-intercept since $y \ge 1$ y-intercept: 1. It means that the pressure at sea level ($d = 0$) is 1 atm.
b. $P(0) = 1, P(10) \approx 1.3, P(33) = 2, P(100) \approx 4.03$ **c.** 132 ft
79. a. $C = 50x + 6000$
b. The y-intercept is the fixed overhead cost.
c. 110

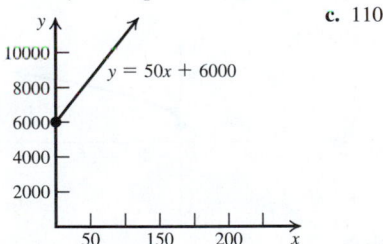

81. a. $R = 900 - 30x$ **b.** \$720   **c.** Ten days after the first of the month   **83.** 70

85. a. $y = \dfrac{2}{27}(x - 150) + 30$ **b.** 45   **c.** 352.5 mg/m³
87. a.

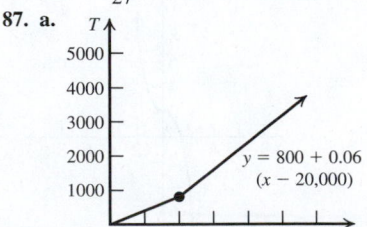

b. (i) \$480   **(ii)** \$800   **(iii)** \$2,600   **c. (i)** \$15,000   **(ii)** \$30,000   **(iii)** \$45,000

89. $a = 14$

Beyond the Basics:

91. a. Domain: $(-\infty, \infty)$; range $[0, 1)$ **b.** Increasing on $(n, n + 1)$ for every integer n **c.** Neither   **93. a. (i)** 40   **(ii)** 21   **(iii)** 9
b. (i) $-4°$ F   **(ii)** $28°$ F   **95.** $C(x) = 2[[x]] + 4$

Critical Thinking / Discussion / Writing:

97. d   **99. a.** f is increasing on $(-\infty, 0)$ and decreasing on $(0, 1)$
b. f is decreasing on $(-\infty, 0)$ and increasing on $(0, 1)$

101. $f(x) = \dfrac{x + |x|}{2}$ **103.** $f(x) = [[x + 0.5]]$ **105.** $f(x) + 3$

107. $f(-x)$

109.

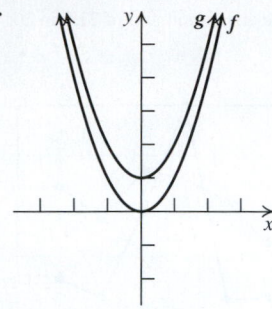

111.

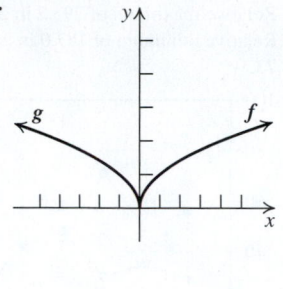

Section 1.5

Concepts and Vocabulary:

1. down   **3.** y-axis   **5.** False   **7.** False

Building Skills:

9. a. Shift two units up   **b.** Shift one unit down   **11. a.** Shift one unit
left   **b.** Shift two units right   **13. a.** Shift one unit left and two units
down   **b.** Shift one unit right and three units up   **15. a.** Reflect about
x-axis   **b.** Reflect about y-axis   **17. a.** Stretch vertically by a factor of 2
b. Compress horizontally by a factor of 2   **19. a.** Reflect about x-axis,
shift one unit up   **b.** Reflect about y-axis, shift one unit up   **21. a.** Shift
one unit up   **b.** Shift one unit left   **23.** e   **25.** g   **27.** i   **29.** b   **31.** l
33. d

35.

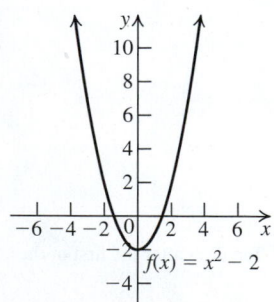

$f(x) = x^2 - 2$

37.

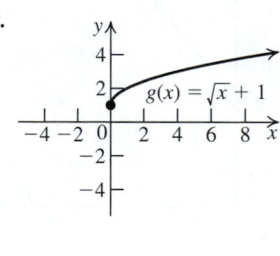

$g(x) = \sqrt{x} + 1$

39.

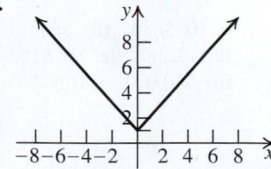

41.

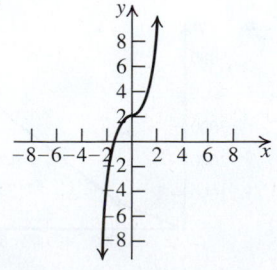

43.

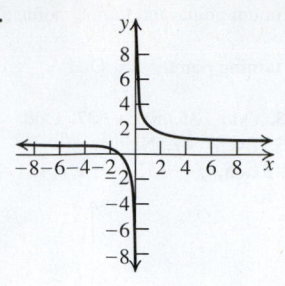

45.

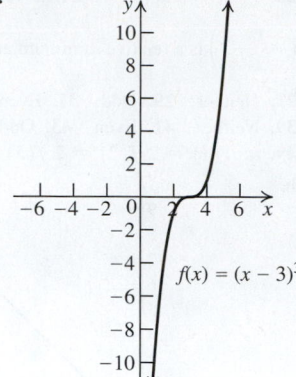

$f(x) = (x - 3)^3$

47.

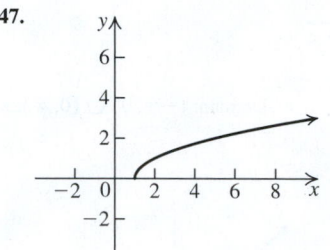

49.

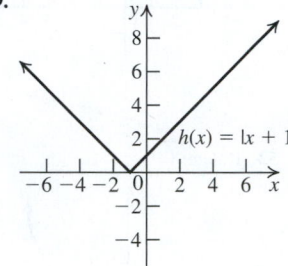

$h(x) = |x + 1|$

51.

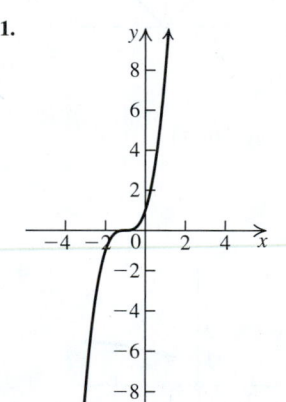

53.

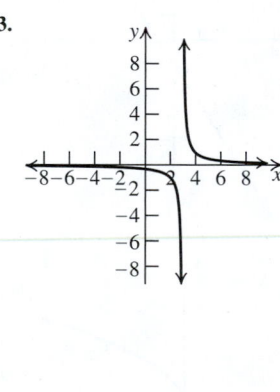

55.

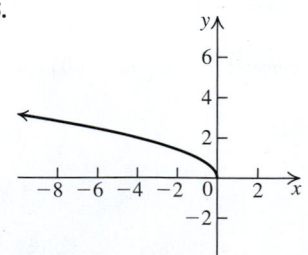

57.

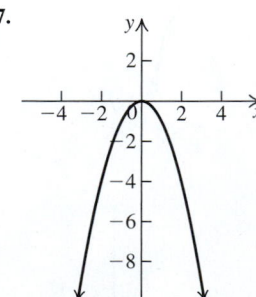

59.

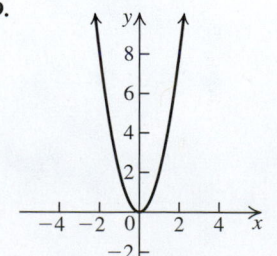

61.

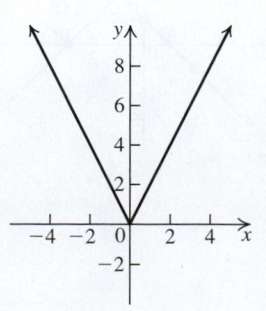

63.
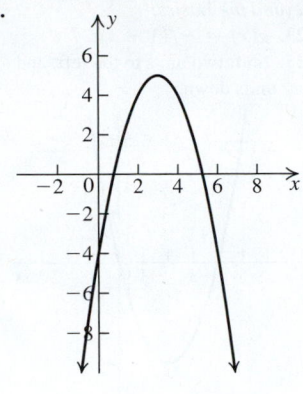

65.
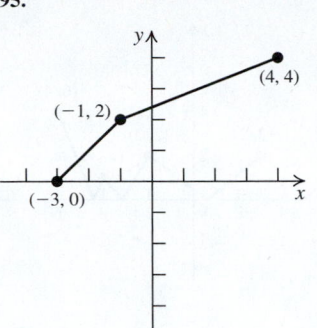

95.

97.
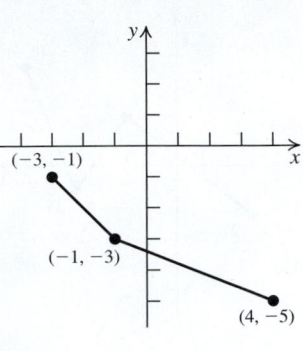

67.

69.
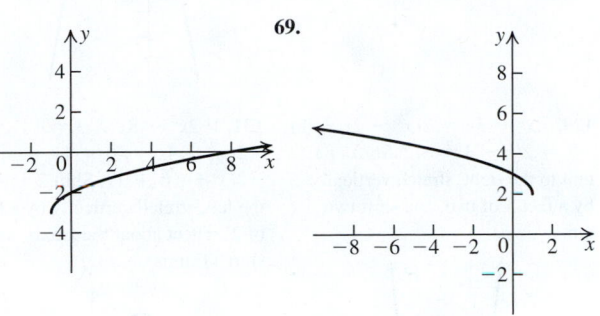

99.

101.
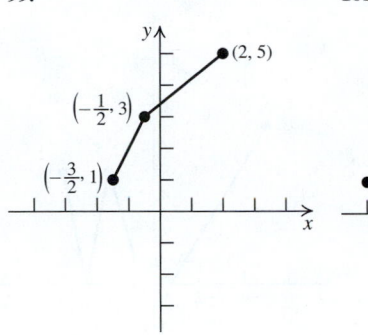
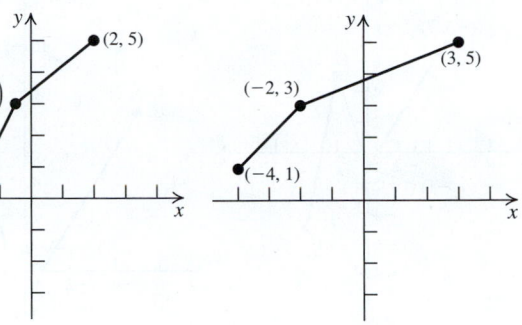

71.

73.
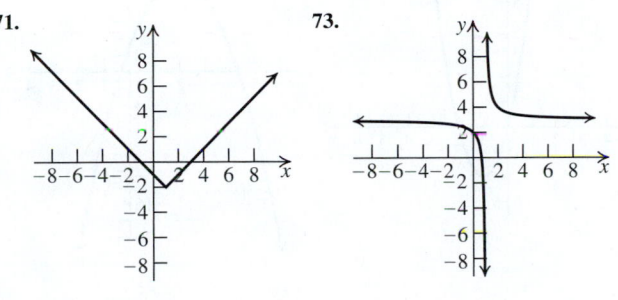

103.
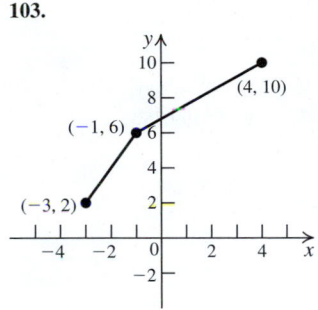

75.

77.
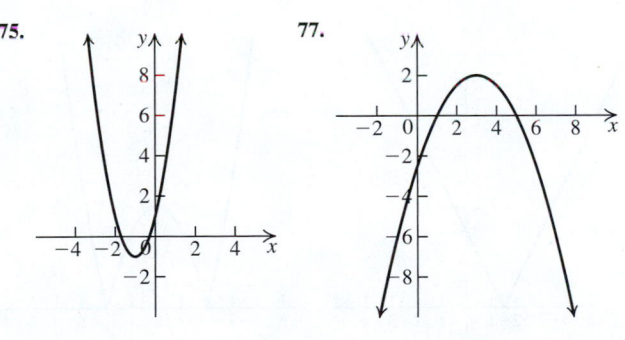

105. a.

b.
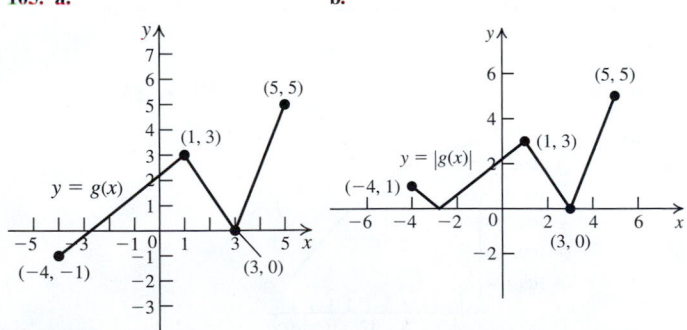

79.

81.
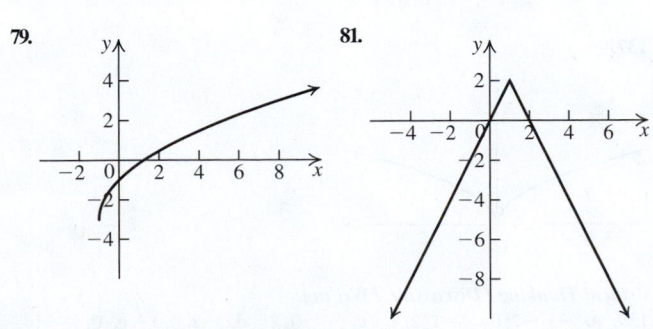

107. a.

b.
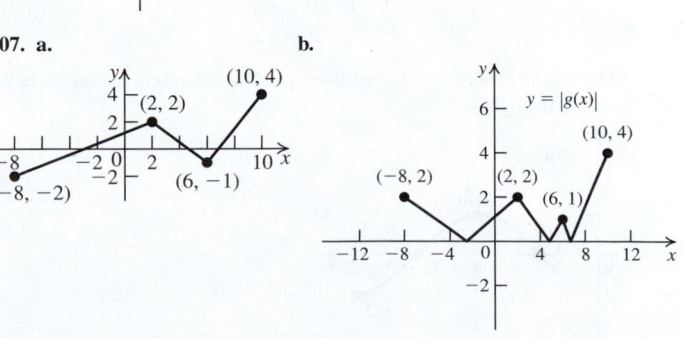

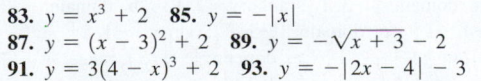

83. $y = x^3 + 2$ **85.** $y = -|x|$
87. $y = (x - 3)^2 + 2$ **89.** $y = -\sqrt{x + 3} - 2$
91. $y = 3(4 - x)^3 + 2$ **93.** $y = -|2x - 4| - 3$

109. a.

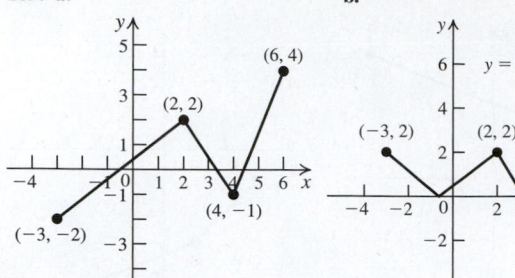

b.

Beyond the Basics:

123. $g(x) = -f(1 - x)$

125. Shift two units to the left, and four units down.

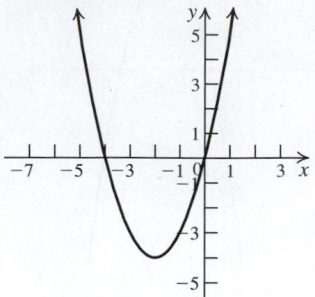

127. $-x^2 + 2x = -(x^2 - 2x + 1) + 1 = -(x - 1)^2 + 1$. Shift one unit to the right, reflect about the x-axis, and shift one unit up.

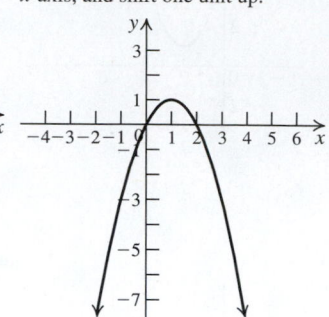

111. a.

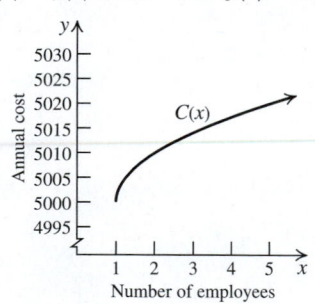

b.

129. $2x^2 - 4x = 2(x^2 - 2x + 1) - 2 = 2(x - 1)^2 - 2$. Shift one unit to the right, stretch vertically by a factor of two, and shift two units down.

131. $-2x^2 - 8x + 3 = -2(x^2 + 4x + 4) + 11 = -2(x + 2)^2 + 11$. Shift 2 units to the left, stretch vertically by a factor of 2, reflect about the x-axis, and shift 11 units up.

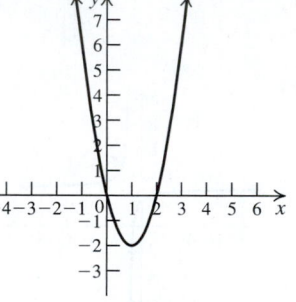

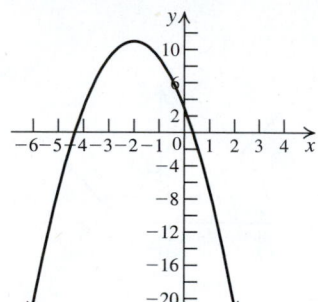

Applying the Concepts:

113. $g(x) = f(x) + 800$ **115.** $p(x) = 1.02(f(x) + 500)$

117. a.

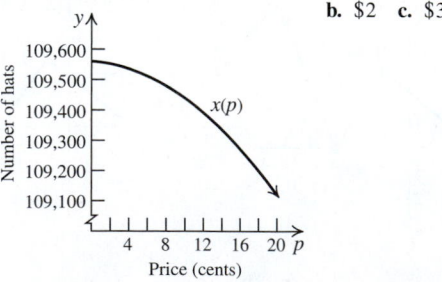

b. $5199.75

133.

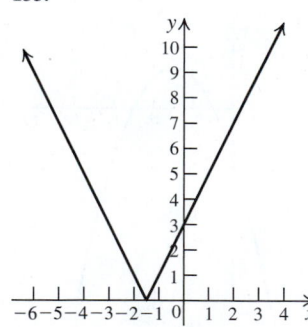

135.

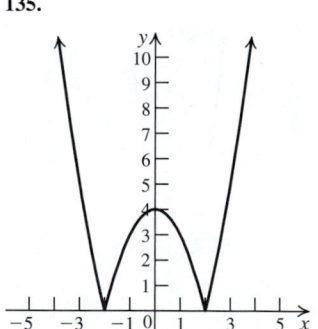

119. a.

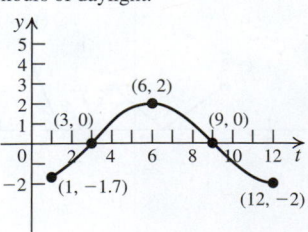

b. $2 **c.** $3.30

137.

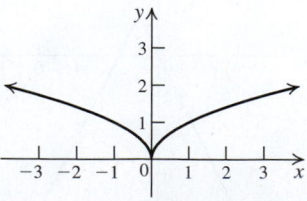

121. The first coordinate gives the month. The second coordinate gives the hours of daylight.

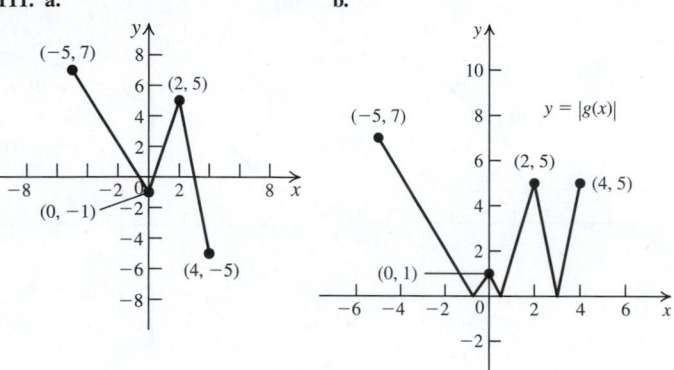

Critical Thinking / Discussion / Writing:

139. a. $-3, -2, 0$ **b.** $1, 2, 4$ **c.** $-1, 0, 2$ **d.** $-2, 0, 1$ **e.** $0, 1, -1/2$ **f.** $-2, 0, 4$ **141. a.** domain: $[-3, 1]$, range: $[-2, 1]$ **b.** domain: $[-1, 3]$, range: $[-4, -1]$ **c.** domain: $[-1, 3]$, range: $[-1, 2]$ **d.** domain: $[-3, 1]$, range: $[-2, 1]$ **e.** domain: $[-1, 3]$, range: $[-4, 2]$ **f.** domain: $[-1, 3]$, range: $[-1, 1/2]$

Getting Ready for the Next Section:
143. $6x^3 + x^2 + 5$ **145.** $x^3 - 8$ **147.** $(-\infty, 2) \cup (2, 3) \cup (3, \infty)$

149. $\left[\dfrac{3}{2}, \infty\right)$ **151.** $f(2x) = 4x^2 - 3, f(x+1) = x^2 + 2x - 2, f(2x - 1)$
$= 4x^2 - 4x - 2$ **153.** $f(2x) = \sqrt{4 - 2x}, f(x+1) =$
$\sqrt{3 - x}, f(2x - 1) = \sqrt{5 - 2x}$

Section 1.6

Concepts and Vocabulary:
1. $f(x) \cdot g(x)$ **3.** $f(g(x))$ **5.** False **7.** False

Building Skills:

9. 3 **11.** -3 **13.** -4 **15.** $\dfrac{1}{2}$ **17.** 1 **19.** 0

21. a. $(f + g)(-1) = -1$ **b.** $(f - g)(0) = 0$ **c.** $(f \cdot g)(2) = -8$

d. $\left(\dfrac{f}{g}\right)(1) = -2$ **23. a.** $(f + g)(-1) = 0$

b. $(f - g)(0) = \dfrac{1}{2}(\sqrt{2} - 2)$ **c.** $(f \cdot g)(2) = \dfrac{5}{2}$ **d.** $\left(\dfrac{f}{g}\right)(1) = \dfrac{1}{3\sqrt{3}}$

25. a. $(f + g)(x) = x^2 + x - 3$; domain: $(-\infty, \infty)$
b. $(f - g)(x) = -x^2 + x - 3$; domain: $(-\infty, \infty)$
c. $(f \cdot g)(x) = x^3 - 3x^2$; domain: $(-\infty, \infty)$
d. $\left(\dfrac{f}{g}\right)(x) = \dfrac{x - 3}{x^2}$; domain: $(-\infty, 0) \cup (0, \infty)$
e. $\left(\dfrac{g}{f}\right)(x) = \dfrac{x^2}{x - 3}$; domain: $(-\infty, 3) \cup (3, \infty)$

27. a. $(f + g)(x) = x^3 + 2x^2 + 4$; domain: $(-\infty, \infty)$
b. $(f - g)(x) = x^3 - 2x^2 - 6$; domain: $(-\infty, \infty)$
c. $(f \cdot g)(x) = 2x^5 + 5x^3 - 2x^2 - 5$; domain: $(-\infty, \infty)$
d. $\left(\dfrac{f}{g}\right)(x) = \dfrac{x^3 - 1}{2x^2 + 5}$; domain: $(-\infty, \infty)$
e. $\left(\dfrac{g}{f}\right)(x) = \dfrac{2x^2 + 5}{x^3 - 1}$; domain: $(-\infty, 1) \cup (1, \infty)$

29. a. $(f + g)(x) = 2x + \sqrt{x} - 1$; domain: $[0, \infty)$
b. $(f - g)(x) = 2x - \sqrt{x} - 1$; domain: $[0, \infty)$
c. $(f \cdot g)(x) = 2x\sqrt{x} - \sqrt{x}$; domain: $[0, \infty)$
d. $\left(\dfrac{f}{g}\right)(x) = \dfrac{2x - 1}{\sqrt{x}}$; domain: $(0, \infty)$
e. $\left(\dfrac{g}{f}\right)(x) = \dfrac{\sqrt{x}}{2x - 1}$; domain: $\left[0, \dfrac{1}{2}\right) \cup \left(\dfrac{1}{2}, \infty\right)$

31. $(f + g)(x) = x - 6 + \sqrt{x - 3}$; Domain $[3, \infty)$
$(f - g)(x) = x - 6 - \sqrt{x - 3}$; Domain $[3, \infty)$
$(f \cdot g)(x) = (x - 6)\sqrt{x - 3}$; Domain $[3, \infty)$
$\left(\dfrac{f}{g}\right)(x) = (x - 6)/\sqrt{x - 3}$; Domain $(3, \infty)$
$\left(\dfrac{g}{f}\right)(x) = \sqrt{x - 3}/(x - 6)$; Domain $[3, 6) \cup (6, \infty)$

33. a. $(f + g)(x) = 1 - \dfrac{2}{x + 1} + \dfrac{1}{x} = \dfrac{x^2 + 1}{(x + 1)x}$;
  Domain $(\infty, -1) \cup (-1, 0) \cup (0, \infty)$
b. $(f + g)(x) = 1 - \dfrac{2}{x + 1} - \dfrac{1}{x} = \dfrac{x^2 - 2x - 1}{(x + 1)x}$;
  Domain $(-\infty, -1) \cup (-1, 0) \cup (0, \infty)$
c. $(f \cdot g)(x) = \left(1 - \dfrac{2}{x + 1}\right) \cdot \left(\dfrac{1}{x}\right) = \dfrac{x - 1}{(x + 1)x}$;
  Domain $(-\infty, -1) \cup (-1, 0) \cup (0, \infty)$

d. $\left(\dfrac{f}{g}\right)(x) = \dfrac{1 - \frac{2}{x + 1}}{\frac{1}{x}} = \dfrac{(x - 1)x}{x + 1}$;
  Domain $(-\infty, -1) \cup (-1, 0) \cup (0, \infty)$
e. $\left(\dfrac{g}{f}\right)(x) = \dfrac{\frac{1}{x}}{1 - \frac{2}{x + 1}} = \dfrac{x + 1}{(x - 1)x}$;
  Domain $(-\infty, -1) \cup (-1, 0) \cup (0, 1) \cup (1, \infty)$

35. a. $(f + g)(x) = \dfrac{2 + x}{x + 1}$; domain: $(-\infty, -1) \cup (-1, \infty)$
b. $(f - g)(x) = \dfrac{2 - x}{x + 1}$; domain: $(-\infty, -1) \cup (-1, \infty)$
c. $(f \cdot g)(x) = \dfrac{2x}{(x + 1)^2}$; domain: $(-\infty, -1) \cup (-1, \infty)$
d. $\left(\dfrac{f}{g}\right)(x) = \dfrac{2}{x}$; domain: $(-\infty, -1) \cup (-1, 0) \cup (0, \infty)$
e. $\left(\dfrac{g}{f}\right)(x) = \dfrac{x}{2}$; domain: $(-\infty, -1) \cup (-1, \infty)$

37. a. $(f + g)(x) = \dfrac{x^3 - x^2 + 2x}{x^2 - 1}$; domain:
  $(-\infty, -1) \cup (-1, 1) \cup (1, \infty)$
b. $(f - g)(x) = \dfrac{x^2 - 2x}{x - 1}$; domain: $(-\infty, -1) \cup (-1, 1) \cup (1, \infty)$
c. $(f \cdot g)(x) = \dfrac{2x^3}{x^3 + x^2 - x - 1}$; domain:
  $(-\infty, -1) \cup (-1, 1) \cup (1, \infty)$
d. $\left(\dfrac{f}{g}\right)(x) = \dfrac{x^2 - x}{2}$; domain:
  $(-\infty, -1) \cup (-1, 0) \cup (0, 1) \cup (1, \infty)$
e. $\left(\dfrac{g}{f}\right)(x) = \dfrac{2}{x^2 - x}$; domain:
  $(-\infty, -1) \cup (-1, 0) \cup (0, 1) \cup (1, \infty)$

39. a. $[1, 5]$ **b.** $[1, 5)$ **41 a.** $[-2, 3]$ **b.** $[-2, 3)$
43. $g(f(x)) = 2x^2 + 1; g(f(2)) = 9; g(f(-3)) = 19$ **45.** 11
47. 31 **49.** -5 **51.** $4c^2 - 5$ **53.** $8a^2 + 8a - 1$ **55.** 7
57. $\dfrac{2x}{x + 1}; (-\infty, -1) \cup (-1, 0) \cup (0, \infty)$
59. $\sqrt{-3x - 1}; \left(-\infty, -\dfrac{1}{3}\right]$ **61.** $|x^2 - 1|; (-\infty, \infty)$

63. a. $(f \circ g)(x) = 2x + 5$; domain: $(-\infty, \infty)$
b. $(g \circ f)(x) = 2x + 1$; domain: $(-\infty, \infty)$
c. $(f \circ f)(x) = 4x - 9$; domain: $(-\infty, \infty)$
d. $(g \circ g)(x) = x + 8$; domain: $(-\infty, \infty)$
65. a. $(f \circ g)(x) = -2x^2 - 1$; domain: $(-\infty, \infty)$
b. $(g \circ f)(x) = 4x^2 - 4x + 2$; domain: $(-\infty, \infty)$
c. $(f \circ f)(x) = 4x - 1$; domain: $(-\infty, \infty)$
d. $(g \circ g)(x) = x^4 + 2x^2 + 2$; domain: $(-\infty, \infty)$
67. a. $(f \circ g)(x) = 8x^2 - 2x - 1$; domain: $(-\infty, \infty)$
b. $(g \circ f)(x) = 4x^2 + 6x - 1$; domain: $(-\infty, \infty)$
c. $(f \circ f)(x) = 8x^4 + 24x^3 + 24x^2 + 9x$; domain: $(-\infty, \infty)$
d. $(g \circ g)(x) = 4x - 3$; domain: $(-\infty, \infty)$
69. a. $(f \circ g)(x) = x$; domain: $[0, \infty)$
b. $(g \circ f)(x) = |x|$; domain: $(-\infty, \infty)$
c. $(f \circ f)(x) = x^4$; domain: $(-\infty, \infty)$
d. $(g \circ g)(x) = \sqrt[4]{x}$; domain: $[0, \infty)$
71. a. $(f \circ g)(x) = -\dfrac{x^2}{x^2 - 2}$;
  domain: $(-\infty, -\sqrt{2}) \cup (-\sqrt{2}, 0) \cup (0, \sqrt{2}) \cup (\sqrt{2}, \infty)$
b. $(g \circ f)(x) = (2x - 1)^2$; domain: $\left(-\infty, \dfrac{1}{2}\right) \cup \left(\dfrac{1}{2}, \infty\right)$

c. $(f \circ f)(x) = \dfrac{2x - 1}{2x - 3}$; domain: $\left(-\infty, \dfrac{1}{2}\right) \cup \left(\dfrac{1}{2}, \dfrac{3}{2}\right) \cup \left(\dfrac{3}{2}, \infty\right)$

d. $(g \circ g)(x) = x^4$; domain: $(-\infty, 0) \cup (0, \infty)$

73. a. $(f \circ g)(x) = \sqrt{\sqrt{4 - x} - 1}$; domain: $(-\infty, 3]$
  b. $(g \circ f)(x) = \sqrt{4 - \sqrt{x - 1}}$; domain: $[1, 17]$
  c. $(f \circ f)(x) = \sqrt{\sqrt{x - 1} - 1}$; domain: $[2, \infty)$
  d. $(g \circ g)(x) = \sqrt{4 - \sqrt{4 - x}}$; domain: $[-12, 4]$

75. a. $(f \circ g)(x) = -7/(3x - 5)$
  Domain $(-\infty, 5/3) \cup (5/3, 4) \cup (4, \infty)$
  b. $(g \circ f)(x) = -(2x + 7)/(5x + 7)$
  Domain $(-\infty, -2) \cup (-2, -7/5) \cup (-7/5, \infty)$
  c. $(f \circ f)(x) = (2x + 1)/(x + 5)$
  Domain $(-\infty, -5) \cup (-5, -2) \cup (-2, \infty)$
  d. $(g \circ g)(x) = -(4x - 9)/(3x - 19)$
  Domain $(-\infty, 4) \cup (4, 19/3) \cup (19/3, \infty)$

77. a. $(f \circ g)(x) = \dfrac{2}{1 + x}$; domain: $(-\infty, -1) \cup (-1, 1) \cup (1, \infty)$
  b. $(g \circ f)(x) = -2x - 1$; domain: $(-\infty, 0) \cup (0, \infty)$
  c. $(f \circ f)(x) = \dfrac{2x + 1}{x + 1}$; domain: $(-\infty, -1) \cup (-1, 0) \cup (0, \infty)$
  d. $(g \circ g)(x) = -\dfrac{1}{x}$; domain: $(-\infty, 0) \cup (0, 1) \cup (1, \infty)$

79. domain: $[0, 1] \cup (3, 6]$ **81.** domain: $[0, 4]$

83. $g(t) = \begin{cases} 2 & \text{if } 0 \le x \le 1 \\ 14 & \text{if } 1 < x \le 3 \\ 7 & \text{if } 3 < x \le 6 \end{cases}$ **85.** $g(t) = \begin{cases} 1/4 & \text{if } 0 \le x \le 1 \\ 1/6 & \text{if } 1 < x \le 3 \\ 1/5 & \text{if } 3 < x \le 6 \end{cases}$

87. $f(x) = \sqrt{x}; g(x) = x + 2$ **89.** $f(x) = x^{10}; g(x) = x^2 - 3$

91. $f(x) = \dfrac{1}{x}; g(x) = 3x - 5$ **93.** $f(x) = \sqrt[3]{x}; g(x) = x^2 - 7$

95. $f(x) = \dfrac{1}{|x|}; g(x) = x^3 - 1$ **97.** 8

99. -27 **101.** $-\dfrac{3}{28}$

Applying the Concepts:
103. a. Cost function   **b.** Revenue function   **c.** Selling price of x shirts after sale tax   **d.** Profit function
105. a. $P(x) = 20x - 350$ **b.** $P(20) = 50$; profit made after 20 radios were sold   **c.** 43   **d.** $(R \circ x)(C) = 5C - 1{,}750$. The function represents the revenue in terms of the cost C.
107. a. $f(x) = 0.7x$ **b.** $g(x) = x - 5$ **c.** $(g \circ f)(x) = 0.7x - 5$ **d.** $(f \circ g)(x) = 0.7(x - 5)$ **e.** \$1.50
109. a. $f(x) = 1.1x; g(x) = x + 8$ **b.** $(f \circ g)(x) = 1.1x + 8.8$; a final test score, which originally was x, after adding 8 points and then increasing the score by 10%   **c.** $(g \circ f)(x) = 1.1x + 8$; a final test score, which originally was x, after increasing the score by 10% and then adding 8 points   **d.** 85.8 and 85.0   **e.** No   **f. (i)** 73.82 **(ii)** 74.55
111. a. $f(x) = \pi x^2$ **b.** $g(x) = \pi(x + 30)^2$ **c.** The area between the fountain and the fence   **d.** \$16,052   **113. a.** $A(t) = 4\pi t^2$
  b. 452 square miles

Beyond the Basics:
115. a. $f + g = \{(-2, 3), (1, 0), (3, 2)\}$
  b. $fg = \{(-2, 0), (1, -4), (3, 0)\}$ **c.** $\dfrac{f}{g} = \{(1, -1), (3, 0)\}$
  d. $f \circ g = \{(0, 1), (1, 3), (3, 1)\}$ **117.** 2

119. c. $h(x) = \dfrac{h(x) + h(-x)}{2} + \dfrac{h(x) - h(-x)}{2}$

Critical Thinking / Discussion / Writing:
121. a. Domain: $(-\infty, 0) \cup [1, \infty)$ **b.** Domain: $[0, 2]$ **c.** Domain: $[1, 2]$ **d.** Domain: $[1, 2)$ **123. a.** Even function   **b.** Odd function   **c.** Neither   **d.** Even function   **e.** Even function   **f.** Odd function

Getting Ready for the Next Section:
125. $y = \dfrac{x - 3}{2}$ **127.** $x = -\sqrt{4 - y^2}$ **129.** $\dfrac{x + 2}{2x + 1}$

Section 1.7

Concepts and Vocabulary:
1. $f(x_1) \ne f(x_2)$ **3.** x **5.** True   **7.** False

Building Skills:
9. One-to-one   **11.** Not one-to-one   **13.** Not one-to-one   **15.** One-to-one   **17.** 2   **19.** -1 **21. a.** 3   **b.** 3   **c.** 19   **d.** 5   **23. a.** 2   **b.** 1   **c.** 269   **25. a.** not one-to-one   **b.** not one-to-one   **c.** one-to-one
27. First: Put the milk back to the fridge. Second: Close the fridge door.
29. First: Remove makeup. Second: Go to sleep.
31. First: Subtract 3. Second: Divide by 2. $f(x) = 2x + 3, f^{-1}(x) = \dfrac{x - 3}{2}$
33. First: Subtract 2. Second: Take a cube root of it. $f(x) = x^3 + 2, f^{-1}(x) = \sqrt[3]{x - 2}$

43.
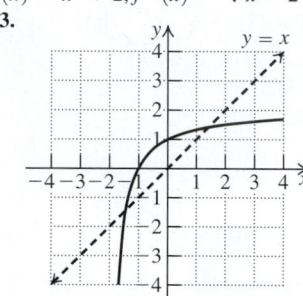

45.
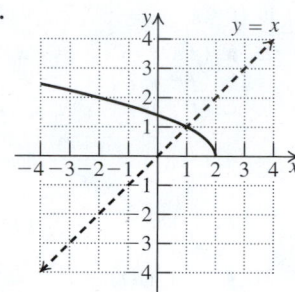

47.
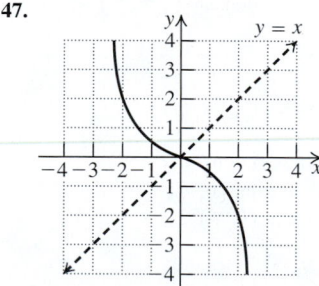

49.
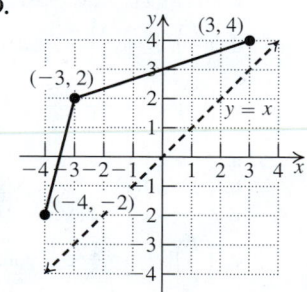

51. Domain: $(-\infty, \infty)$, Range: $(-\infty, \infty)$, $f^{-1} = 1/3 + (1/3)x$
53. Domain: $(-\infty, \infty)$, Range: $(-\infty, \infty)$, $f^{-1} = 3(x - 2)^3 - 1$
55. Domain: $(-\infty, \infty)$, Range: $= (-\infty, \infty)$, $f^{-1} = \dfrac{\sqrt[3]{x - 2} + 1}{3}$
57. Domain: $(-\infty, -1) \cup (-1, \infty)$, Range:
  $= (-\infty, 0) \cup (0, \infty), f^{-1} = \dfrac{2}{x} - 1$
59. $(-\infty, 2) \cup (2, \infty)$, Range: $= (-\infty, 1) \cup (1, \infty)$, $f^{-1} = \dfrac{1 + 2x}{x - 1}$
61. Domain $f: [0, \infty)$, Range $f: = [1, \infty), f^{-1} = \sqrt{x - 1}$ Domain $f^{-1}: [1, \infty)$, Range $f^{-1}: = [0, \infty)$
63. Domain $f: (-\infty, 0]$, Range $f: = (-\infty, 2], f^{-1} = -\sqrt{2 - x}$ Domain $f^{-1}: (-\infty, 2]$, Range $f^{-1}: = (-\infty, 0]$

65.

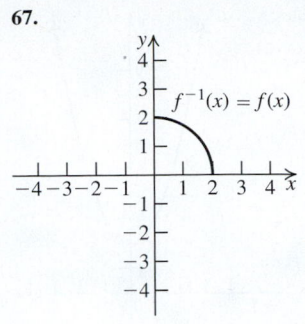

67.

69.

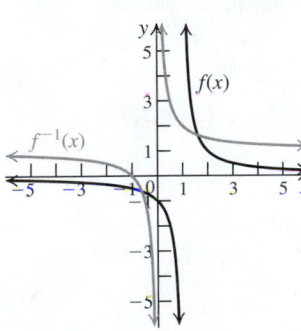

71.

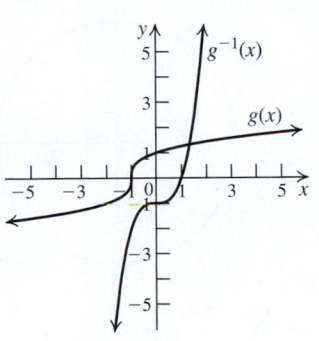

73.

75.

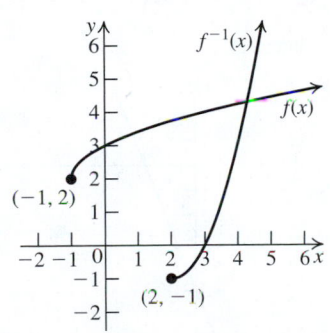

77. $f^{-1}(x) = \dfrac{2x + 1}{x - 1}$ Domain: $(-\infty, 2)\cup(2, \infty)$;
range: $(-\infty, 1)\cup(1, \infty)$

79. $f^{-1}(x) = \dfrac{1 - x}{x + 2}$ Domain: $(-\infty, -1)\cup(-1, \infty)$;
range: $(-\infty, -2)\cup(-2, \infty)$ **81.** 1/9
83. −8 **85.** −3/7

Applying the Concepts:
87. a. $C(K) = K - 273$. It represents the Celsius temperature corresponding to a given Kelvin temperature. **b.** 27°C **c.** 295 k

89. a. $F = \dfrac{9}{5}C + 32$ **b.** $C = \dfrac{5}{9}F - \dfrac{160}{9}$

91. a. Dollars to euros: $f(x) = 0.75x$ (x is the number of dollars; $f(x)$ is the number of euros). Euros to dollars: $g(x) = 1.25x$ (x is the number of euros; $g(x)$ is the number of dollars) **b.** $g(f(x)) = 0.9375x \neq x$; therefore, g and f are not inverse functions **c.** She loses money.

93. a. $w = \begin{cases} 4 + 0.05x, & \text{if } x > 60 \\ 7, & \text{if } x \le 60 \end{cases}$

b. It does not have an inverse because it is constant on $(0, 60)$ and hence is not one-to-one. **c.** Restriction of the domain: $[60, \infty)$

95. a. $x = \dfrac{1}{64}V^2$; it gives the height of the water in terms of the velocity.

b. (i) 14.0625 ft **(ii)** 6.25 ft **97. a.** The balance due after x months

b. $f^{-1}(x) = 60 - \dfrac{1}{600}x$; it shows the number of months that have passed from the first payment until the balance due is $x.
c. $23.333 \approx 24$ months

99. a. $f(x) = GMm\dfrac{1}{x^2}$ is decreasing on $(0, \infty)$

b. $r = \sqrt{\dfrac{GMm}{F}}$ **c.** $F^{-1}(x) = \sqrt{\dfrac{GMm}{x}}$

Beyond the Basics:
103. a.

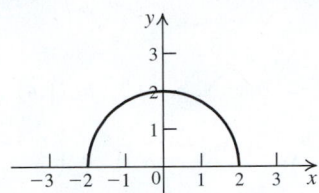

b. No **c.** Domain: $[-2, 2]$; range: $[0, 2]$
105. a.

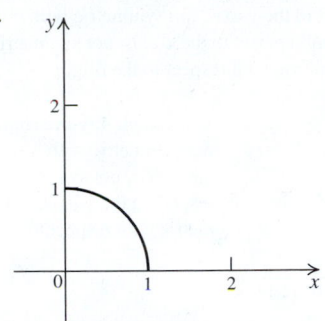

g satisfies the horizontal-line test
b. $g^{-1}(x) = g(x) = \sqrt{1 - x^2}$
c. Domain = range = $[0, 1]$
107. Each of the functions $1, 2x^3, 3x^5$ and $4x^7$ are increasing therefore f is increasing as a sum of increasing functions. Since f is increasing f is one-to-one. To find f^{-1} algebraically we would have to solve the equation $x = 1 + 2y^3 + 3y^5 + 4y^7$ for y. This is not possible.
109. a. The midpoint is $M(5, 5)$. Because its coordinates satisfy the equation $y = x$, it lies on the line $y = x$. **b.** The slope of the line segment $\overline{PQ}$ is −1, and the slope of the line $y = x$ is 1. Because $(-1)(1) = -1$, the two lines are perpendicular.

111. a. (i) $f^{-1}(x) = \dfrac{1}{2}x + \dfrac{1}{2}$ **(ii)** $g^{-1}(x) = \dfrac{1}{3}x - \dfrac{4}{3}$

(iii) $(f \circ g)(x) = 6x + 7$ **(iv)** $(g \circ f)(x) = 6x + 1$

(v) $(f \circ g)^{-1}(x) = \dfrac{1}{6}x - \dfrac{7}{6}$ **(vi)** $(g \circ f)^{-1}(x) = \dfrac{1}{6}x - \dfrac{1}{6}$

(vii) $(f^{-1} \circ g^{-1})(x) = \dfrac{1}{6}x - \dfrac{1}{6}$

(viii) $(g^{-1} \circ f^{-1})(x) = \dfrac{1}{6}x - \dfrac{7}{6}$

b. (i) $(f \circ g)^{-1}(x) = \dfrac{1}{6}x - \dfrac{7}{6} = (g^{-1} \circ f^{-1})(x)$

(ii) $(g \circ f)^{-1}(x) = \dfrac{1}{6}x - \dfrac{1}{6} = (f^{-1} \circ g^{-1})(x)$

Critical Thinking / Discussion / Writing:
113. No. $f(x) = x^3 - x$ is odd, but it does not have an inverse because $f(0) = f(1)$. So it is not one-to-one. **115.** Yes, because increasing and decreasing functions are one-to-one

Getting Ready for the Next Section:
117. $(x + 3)(x - 4)$ **119.** $(x + 4)(x - 2)$
121. $3, 4$ **123.** $-2, -1/3$
125.
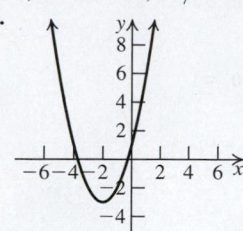
127.
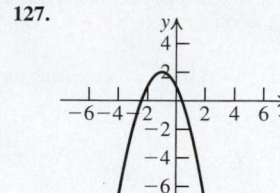

Review Exercises

1. False **3.** True **5.** False **7.** True **9. a.** $2\sqrt{5}$ **b.** $(1, 4)$ **c.** $\dfrac{1}{2}$

11. a. $5\sqrt{2}$ **b.** $\left(\dfrac{13}{2}, -\dfrac{11}{2}\right)$ **c.** -1 **13. a.** $\sqrt{34}$ **b.** $\left(\dfrac{7}{2}, -\dfrac{9}{2}\right)$

c. $\dfrac{5}{3}$ **17.** $(4, 5)$ **19.** $\left(\dfrac{31}{18}, 0\right)$ **21.** Not symmetric with respect to the

x-axis, symmetric with respect to the y-axis, not symmetric with respect to the origin **23.** Symmetric with respect to the x-axis, not symmetric with respect to the y-axis, not symmetric with respect to the origin

25.
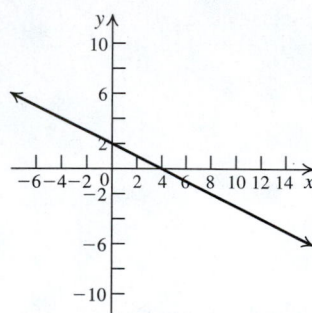
x-intercept: 4; y-intercept: 2; not symmetric with respect to the x-axis, not symmetric with respect to the y-axis, not symmetric with respect to the origin

27.
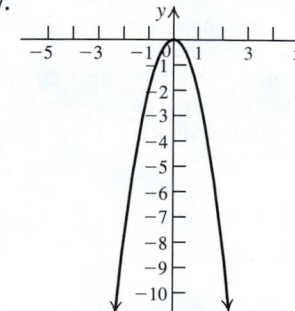
x-intercept: 0; y-intercept: 0; not symmetric with respect to the x-axis, symmetric with respect to the y-axis, not symmetric with respect to the origin

29.
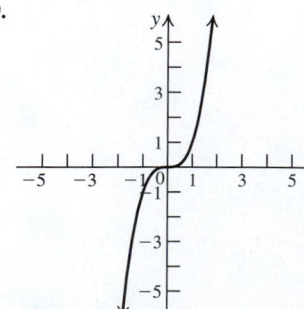
x-intercept: 0; y-intercept: 0; not symmetric with respect to the x-axis, not symmetric with respect to the y-axis, symmetric with respect to the origin

31.
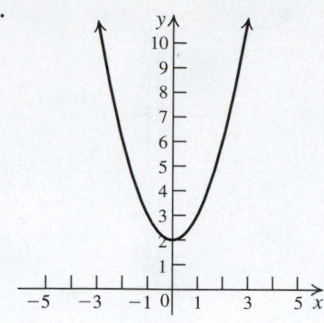
No x-intercept; y-intercept: 2; not symmetric with respect to the x-axis, symmetric with respect to the y-axis, not symmetric with respect to the origin

33.
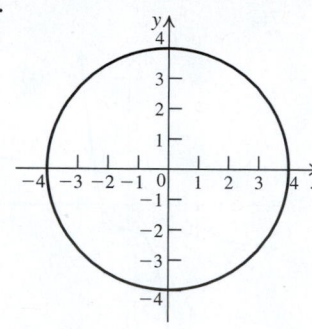
x-intercepts: -4 and 4; y-intercepts: -4 and 4; symmetric with respect to the x-axis, symmetric with respect to the y-axis, symmetric with respect to the origin

35. $(x - 2)^2 + (y + 3)^2 = 25$ **37.** $(x + 2)^2 + (y + 5)^2 = 4$

39. $\dfrac{x}{2} - \dfrac{y}{5} = 1 \Rightarrow y = \dfrac{5}{2}x - 5$; line with slope $\dfrac{5}{2}$; y-intercept: -5, x-intercept: 2

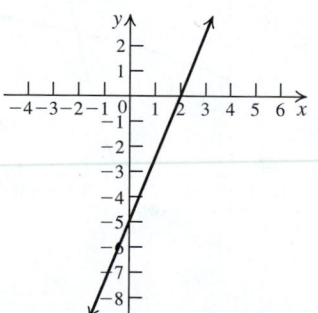

41. $x^2 + y^2 - 2x + 4y - 4 = 0 \Rightarrow (x - 1)^2 + (y + 2)^2 = 9$; circle centered at $(1, -2)$ and with radius 3

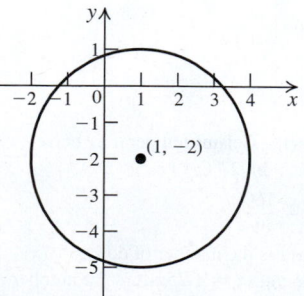

x-intercepts: $1 \pm \sqrt{5}$
y-intercepts: $-2 \pm 2\sqrt{2}$

43. $y = -2x + 4$ **45.** $y = -2x + 5$ **47.** $y = -\dfrac{4}{3}x + \dfrac{13}{3}$

49. Not a function

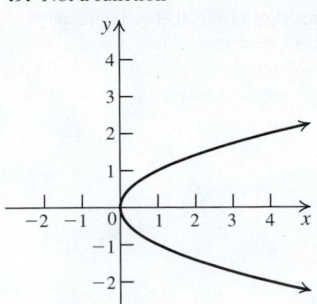

51. A function

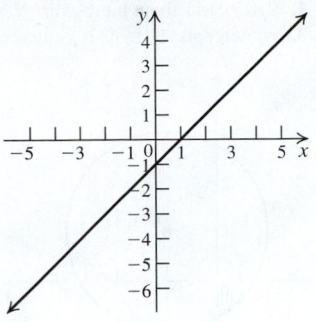

53. Not a function

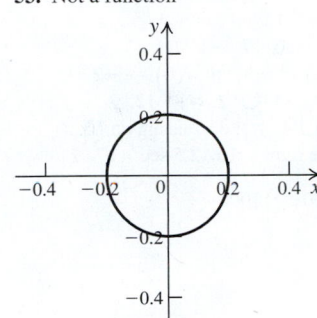

55. Not a function

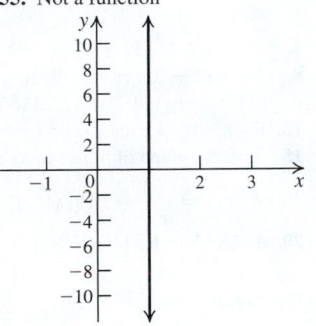

57. A function

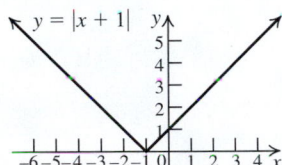

$y = |x + 1|$

59. -5 **61.** $x = 1$ **63.** 3 **65.** -10 **67.** 22 **69.** $3x^2 - 5$
71. $3a + 3h + 1$ **73.** 3 **75.** $f \circ g = \{(-1, 1), (1, 5)\}$
77.

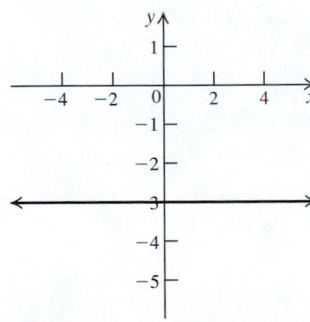

Domain: $(-\infty, \infty)$; range: $\{-3\}$;
constant on $(-\infty, \infty)$

79.

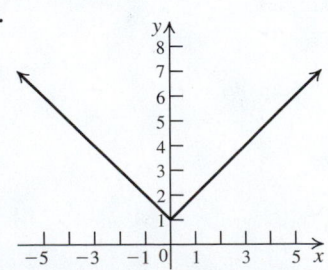

Domain: $\left[\dfrac{2}{3}, \infty\right)$; range: $[0, \infty)$;

increasing on $\left(\dfrac{2}{3}, \infty\right)$

81.

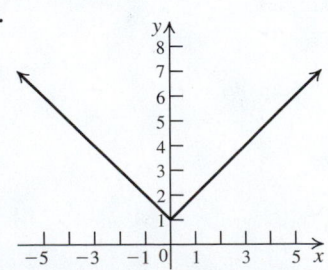

Domain: $(-\infty, \infty)$; range: $[1, \infty)$; decreasing on $(-\infty, 0)$, increasing on $(0, \infty)$

83. The graph of g is the graph of f shifted one unit to the left.

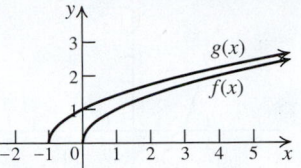

85. The graph of g is the graph of f shifted two units to the right and reflected about the x-axis.

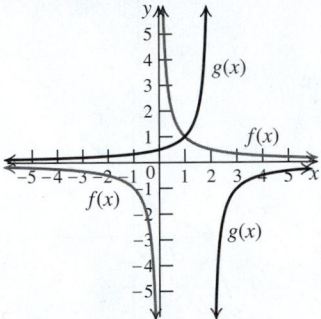

87. Even; not symmetric with respect to the x-axis, symmetric with respect to the y-axis, not symmetric with respect to the origin
89. Even; not symmetric with respect to the x-axis, symmetric with respect to the y-axis, not symmetric with respect to the origin
91. Neither even nor odd; not symmetric with respect to the x-axis, not symmetric with respect to the y-axis, not symmetric with respect to the origin
93. $f(x) = (g \circ h)(x)$, where $g(x) = \sqrt{x}$ and $h(x) = x^2 - 4$
95. $h(x) = (f \circ g)(x)$, where $f(x) = \sqrt{x}$ and $g(x) = \dfrac{x - 3}{2x + 5}$
97. One-to-one; $f^{-1}(x) = x - 2$

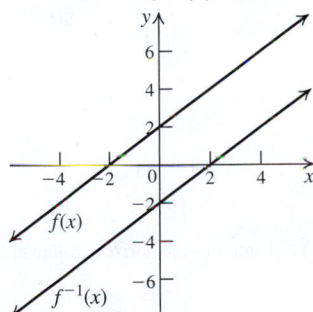

99. One-to-one; $f^{-1}(x) = x^3 + 2$

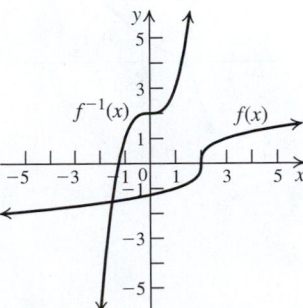

101. $f^{-1}(x) = x^2 - 8x + 17$; $x \geq 4$; Domain f: $[1, \infty)$; Range f: $[4, \infty)$

103. a. $f(x) = \begin{cases} 3x + 6, & \text{if } -3 \leq x \leq -2 \\ \dfrac{1}{2}x + 1, & \text{if } -2 < x < 0 \\ x + 1, & \text{if } 0 \leq x \leq 3 \end{cases}$

b. Domain: $[-3, 3]$; range: $[-3, 4]$
c. x-intercept: -2; y-intercept: 1

d.

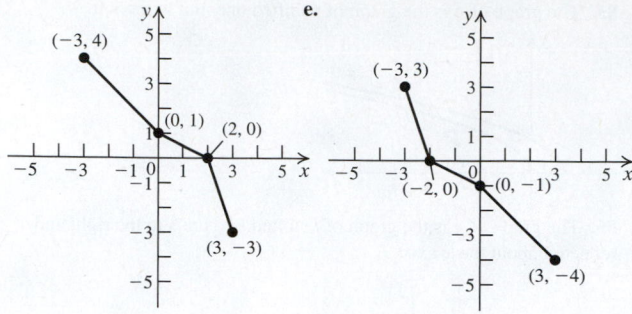

e.

f.

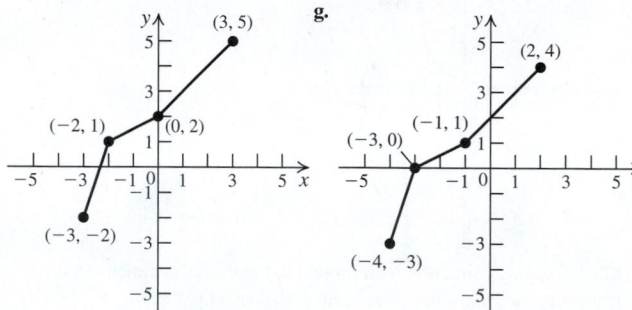

g.

h.

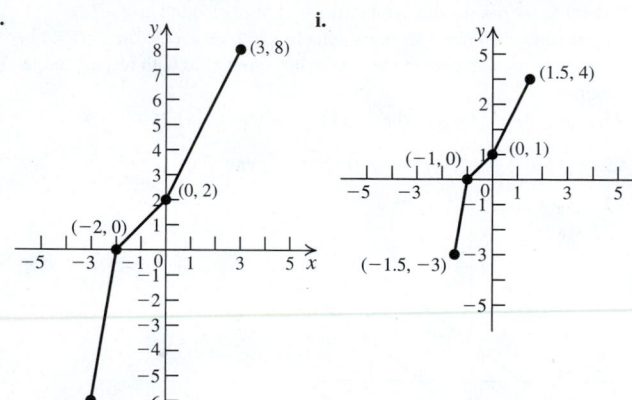

i.

j.

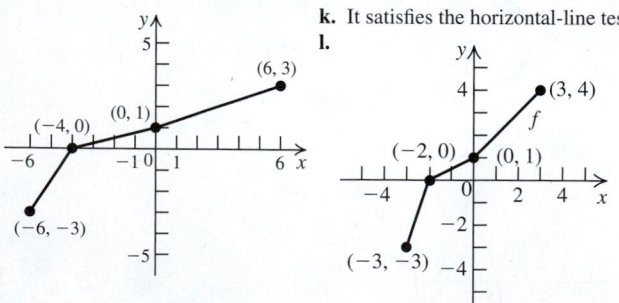

k. It satisfies the horizontal-line test

l.

Applying the Concepts:
105. a. $C = 0.875w - 22{,}125$
b. Slope: the cost of disposing of 1 pound of waste; x-intercept: the amount of waste that can be disposed of with no cost; y-intercept: the fixed cost   **c.** \$510,750   **d.** $\approx 1{,}168{,}000$ pounds   **107. a.** She won \$98   **b.** She was winning at a rate of \$49/h.   **c.** Yes. After 20 hours   **d.** \$5/hr   **109. a.** $0.5\sqrt{0.000004t^4 + 0.004t^2 + 5}$ **b.** 1.13

Practice Test A
1. Symmetric about the x-axis   **2.** x-intercepts: $-1, 0, 3$; y-intercept: 0
3. x-intercepts: $1 \pm \sqrt{3}$; y-intercepts: $1 \pm \sqrt{3}$

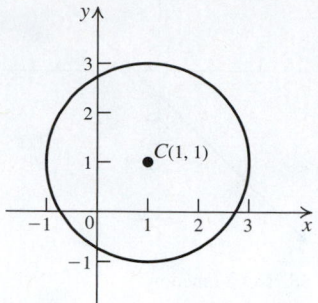

4. $y = -x + 9$ **5.** $y = 4x - 9$ **6.** -36 **7.** -1
8. $x^4 - 4x^3 + 2x^2 + 4x$ **9. a.** $f(-1) = -3$ **b.** $f(0) = -2$
c. $f(1) = -1$ **10.** $[0, 1)$ **11.** $(-\infty, -3]\cup[2, \infty)$ **12.** 2
13. Even   **14.** Increasing on $(-\infty, 0)\cup(2, \infty)$; decreasing on $(0, 2)$
15. Shift the graph of f three units to the right.   **16.** 2.5 sec   **17.** 2

18. $f^{-1}(x) = \dfrac{1}{x - 1}$ **19.** $A(x) = 100x + 1000$

20. a. \$87.50   **b.** 110 mi

Practice Test B
1. d   **2.** b   **3.** d   **4.** d   **5.** c   **6.** b   **7.** d   **8.** b   **9.** a   **10.** c
11. a   **12.** b   **13.** a   **14.** a   **15.** b   **16.** d   **17.** c   **18.** c
19. b   **20.** a

CHAPTER 2

Section 2.1

Concepts and Vocabulary:
1. Parabola   **3.** $x = h$ **5.** True   **7.** False

Basic Skills:
9. f   **11.** a   **13.** h   **15.** g   **17.** $y = -2x^2$ **19.** $y = 5x^2$
21. $y = 2x^2$ **23.** $y = -(x + 3)^2$ **25.** $y = 2(x - 2)^2 + 5$

27. $y = \dfrac{11}{49}(x - 2)^2 - 3$ **29.** $y = -12\left(x - \dfrac{1}{2}\right)^2 + \dfrac{1}{2}$

31. $y = \dfrac{3}{4}(x + 2)^2$ **33.** $y = \dfrac{3}{4}(x - 3)^2 - 1$

35.

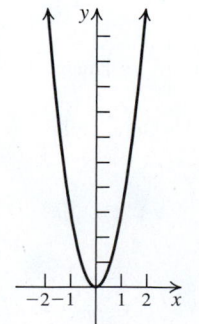

The graph is the graph of $y = x^2$ stretched vertically by a factor of 3.

37.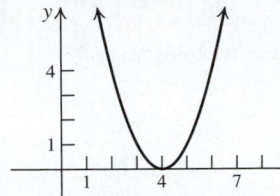

The graph is the graph of $y = x^2$ shifted four units to the right.

39.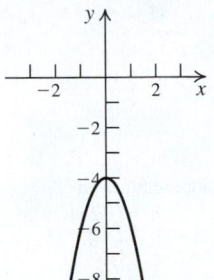

The graph is the graph of $y = x^2$ stretched vertically by a factor of 2, reflected about the x-axis, and shifted down 4 units.

41.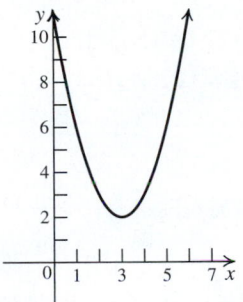

The graph is the graph of $y = x^2$ shifted three units right and two units up.

43. The graph is the graph of $y = x^2$ shifted two units right, stretched vertically by a factor of 3, reflected about the x-axis, and shifted four units up.

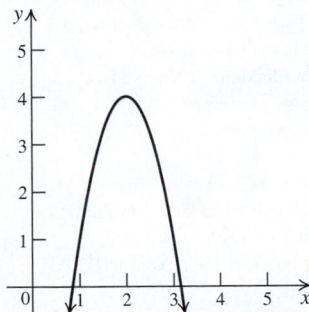

45. $y = (x + 2)^2 - 4$. The graph is the graph of $y = x^2$ shifted two units to the left and four units down. Vertex: $(-2, -4)$; axis: $x = -2$; x-intercepts: -4 and 0; y-intercept: 0

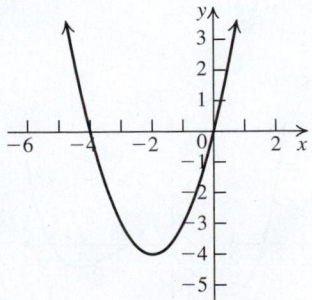

47. $y = -(x - 3)^2 - 1$. The graph is the graph of $y = x^2$ shifted three units to the right, reflected about the x-axis, and shifted one unit down. Vertex: $(3, -1)$; axis: $x = 3$; no x-intercept; y-intercept: -10

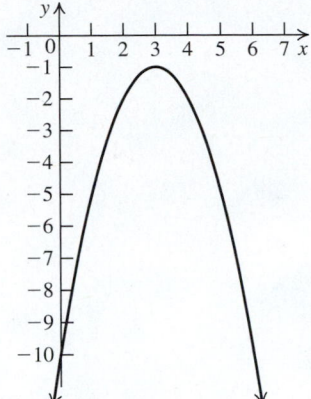

49. $y = 2(x - 2)^2 + 1$. The graph is the graph of $y = x^2$ shifted two units to the right, stretched vertically by a factor of 2, and shifted one unit up.

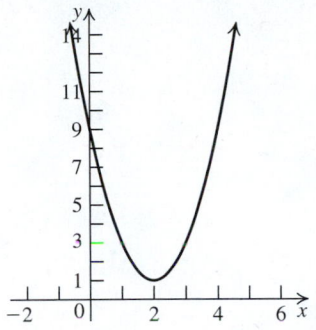

Vertex: $(2, 1)$; axis: $x = 2$; no x-intercept; y-intercept: 9

51. $y = -3(x - 3)^2 + 16$. The graph is the graph of $y = x^2$ shifted three units to the right, stretched vertically by a factor of 3, reflected about the x-axis, and shifted 16 units up.

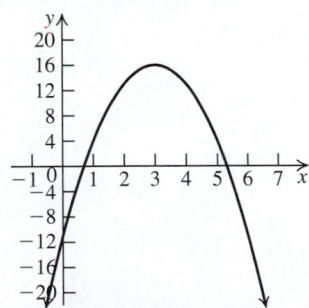

Vertex: $(3, 16)$; axis: $x = 3$; x-intercepts: $3 \pm \dfrac{4}{3}\sqrt{3}$; y-intercept: -11

53. a. Opens up **b.** $(4, -1)$ **c.** $x = 4$
d. x-intercepts: 3 and 5; y-intercept: 15
e.

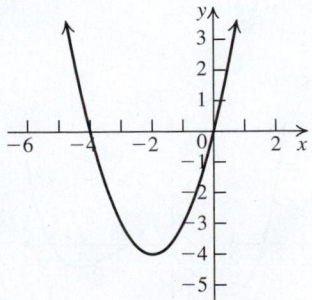

55. a. Opens up **b.** $\left(\dfrac{1}{2}, \dfrac{-25}{4}\right)$ **c.** $x = \dfrac{1}{2}$

d. x-intercepts: -2 and 3; y-intercept: -6

e.

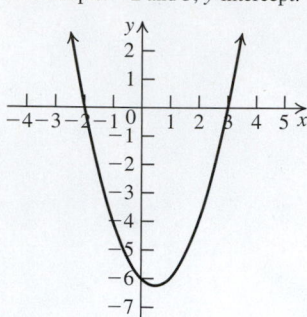

57. a. Opens up **b.** $(1, 3)$ **c.** $x = 1$

d. x-intercepts: none; y-intercept: 4

e.

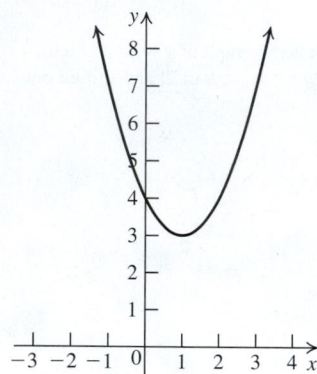

59. a. Opens down **b.** $(-1, 7)$ **c.** $x = -1$

d. x-intercepts: $-1 \pm \sqrt{7}$; y-intercept: 6

e.

61. a. Minimum: -1 **b.** $[-1, \infty)$ **63. a.** Maximum: 0 **b.** $(-\infty, 0]$
65. a. Minimum: -5 **b.** $[-5, \infty)$ **67. a.** Maximum: 16 **b.** $(-\infty, 16]$
69. $[-2, 2]$ **71.** $(1, 3)$
73. $(-\infty, 3/2 - \sqrt{37}/2] \cup [3/2 + \sqrt{37}/2, \infty)$
75. $(-\infty, \infty)$ **77.** $\varnothing$ **79.** $[-5, 2]$
81. $(-\infty, 1) \cup (7/6, \infty)$

Applying the Concepts:
83. a. ≈ 550 ft **b.** ≈ 140 ft **c.** 547 ft: 139 ft **85.** 19 **87.** 25
89. Square with 20-unit sides; maximum area: 400 square units
91. Dimensions: 75 m × 100 m; area: 7500 m² **93.** 38
95. 28 students; maximum revenue: $1568
97. a. $h(t) = -0.8t^2 + 8t + 5$ (meters) **b.** 5 seconds
99. a. 64 ft **b.** After 4 sec

101. Dimensions of the rectangle height $= \dfrac{36}{\pi + 4}$ ft; width $= \dfrac{18}{\pi + 4}$ ft

103. a. $(59, 36.81)$ **b.** 120 ft **c.** 37 ft **d.** 9 ft **e.** 113.6 ft
105. a. $0.029(x - 3.79)^2 + 1.09$
b. At $x = 3.79 \approx 4$ (year 2010), which fits the original data
c. $f(5) = 1.14$ millions

Beyond the Basics:
107. $f(x) = 3x^2 + 12x + 15$ **109.** $f(x) = 6x^2 - 12x + 4$
111. $f(x) = -2x^2 + 12x + 14$
113. Opening up: $x^2 - 4x - 12$; opening down: $-x^2 + 4x + 12$
115. Opening up: $x^2 + 8x + 7$; opening down: $-x^2 - 8x - 7$
117. minimum is 4

Critical Thinking/Discussion/Writing:
121. a. $\left(\dfrac{h - b}{m}, k\right)$ **b.** $(h, mk + b)$
123. $x^2 + (a - x)^2 = 2x^2 - 2ax + a^2$. Quadratic function

$f(x) = 2x^2 - 2ax + a^2$ has minimum value at $x = \dfrac{-(-2a)}{2 \cdot 2} = \dfrac{a}{2}$.

This minimum is $f\left(\dfrac{a}{2}\right) = \dfrac{a^2}{2}$. Therefore

$f(x) = x^2 + (a - x)^2 \geq f\left(\dfrac{a}{2}\right) = \dfrac{a^2}{2}$.

Getting Ready for the Next Section:
125. -1 **127.** $-6x^6$ **129.** $\dfrac{9x^2}{4}$ **131.** $(2x + 3)(2x - 3)$

133. $(3x + 4)(5x - 3)$ **135.** $(x - 2)(x - 1)(x + 2)$

Section 2.2

Concepts and Vocabulary:
1. 5, $2x^5$, 2, -6 **3.** touches **5.** False **7.** False

Building Skills:
9. Polynomial function; degree: 5; leading term: $2x^5$; leading coefficient: 2

11. Polynomial function; degree: 3; leading term: $\dfrac{2}{3}x^3$; leading coefficient: $\dfrac{2}{3}$

13. Polynomial function; degree: 4; leading term: πx^4; leading coefficient: π
15. Presence of $|x|$ **17.** Domain not $(-\infty, \infty)$ **19.** Presence of $\sqrt{x}$
21. Domain not $(-\infty, \infty)$ **23.** Graph is not continuous
25. Graph is not continuous **27.** Not a function **29.** c **31.** a **33.** d
35. $f(x) \to \infty$, as $x \to -\infty$; $f(x) \to -\infty$, as $x \to \infty$
37. $f(x) \to \infty$, as $x \to -\infty$; $f(x) \to \infty$, as $x \to \infty$
39. $f(x) \to -\infty$, as $x \to -\infty$; $f(x) \to \infty$, as $x \to \infty$
41. $f(x) \to \infty$, as $x \to -\infty$; $f(x) \to -\infty$, as $x \to \infty$
43. zero $x = -2$ multiplicity 1 crosses x-axis; zero $x = 1$ multiplicity
1 x-axis; zero $x = 3$ multiplicity 1 crosses x-axis;
45. zero $x = -2$ multiplicity 2 touches but does not cross x-axis;
zero $x = 1/2$ multiplicity 1 crosses x-axis;
47. zero $x = -3$ multiplicity 1 crosses x-axis; zero $x = -2/3$ multiplicity 3 crosses x-axis; zero $x = 0$ multiplicity 2 touches but does not cross
x-axis; zero $x = 3$ multiplicity 1 crosses x-axis;
49. zero $x = 2/3$ multiplicity 2 touches but does not cross x-axis
51. 2.09 **53.** 2.28 **55.** $(x - 1)(x - 3)(x + 2)$
57. $(x - 2)^2(x + 3)$ **59.** $(x - 2)^2(x + 1)^2$
61. $(x - 2)(x - 3)(x + 2)^2$
63.   **65.**

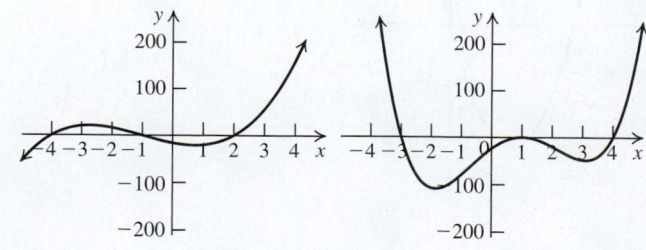

67.

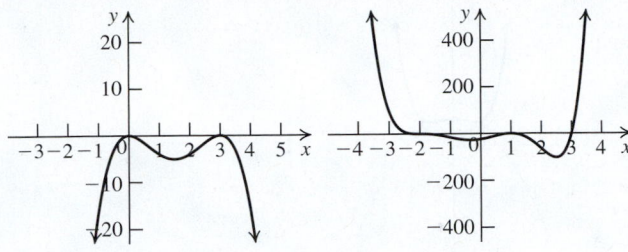

69.

71.

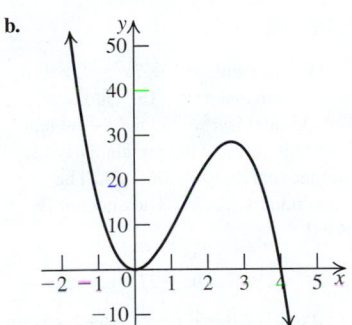

73.

75. a. 4.14 hp   **b.** 17.76 hp   **c.** 33.12 hp   **d.** $\dfrac{p(2v)}{p(v)} = 8$

Applying the Concepts:

77. a. Zeros: $x = 0$, multiplicity 2; $x = 4$, multiplicity 1

b.

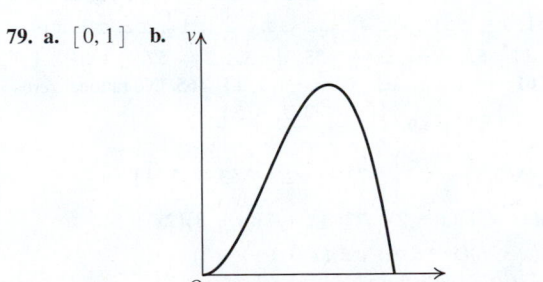

c. 2   **d.** Domain: $[0, 4]$. The portion between the x-intercepts constitutes the graph of $R(x)$.

79. a. $[0, 1]$ **b.**

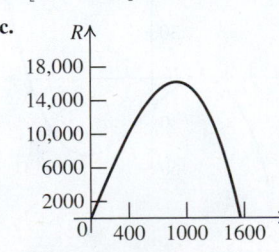

81. a. $R(x) = 27x - \dfrac{1}{90{,}000}x^3$

b. $[0, 900\sqrt{3}]$

c.

83. a. $N(x) = (x + 12)(400 - 2x^2)$

b. $[0, 10\sqrt{2}]$

c.

85. a. $V(x) = x(8 - 2x)(15 - 2x)$

b.

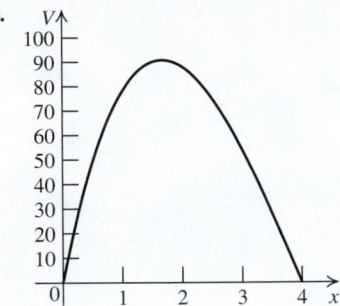

87. a. $V(x) = x^2(108 - 4x)$

b.

89. a. $V(x) = x^2(62 - 2x)$

b.

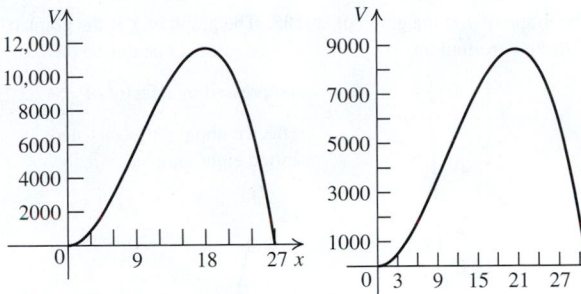

91. a. $V(x) = 2x^2(45 - 3x)$

b.

93.

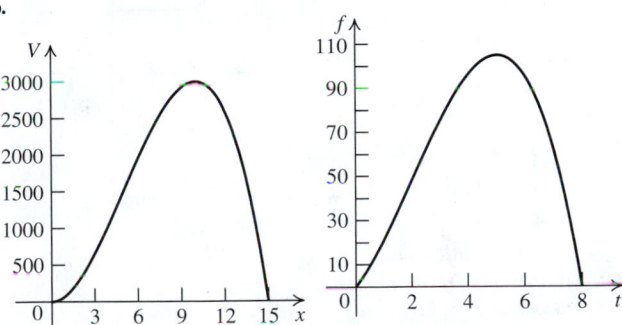

95. a. $f(7) = 1.73$ trillions   **b.** two turning points, one zero
c. At $x = 3.27 \approx 3$ (year 2011), which fits the original data

Beyond the Basics:

97. The graph of f is the graph of $y = x^4$ shifted one unit to the right.

99. The graph of f is the graph of $y = x^4$ shifted two units up.

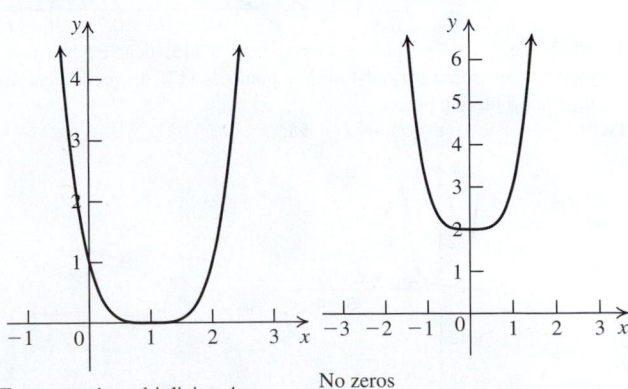

Zero: $x = 1$, multiplicity: 4

No zeros

101. The graph of f is the graph of $y = x^4$ shifted one unit to the right and reflected about the x-axis.

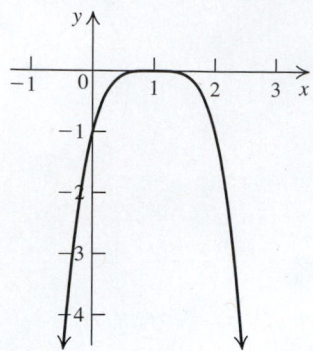

Zero: $x = 1$, multiplicity: 4

103. The graph of f is the graph of $y = x^5$ shifted one unit up.

105. The graph of f is the graph of $y = x^5$ shifted one unit to the left, compressed by a factor of $\dfrac{1}{4}$, reflected about the x-axis, and shifted eight units up.

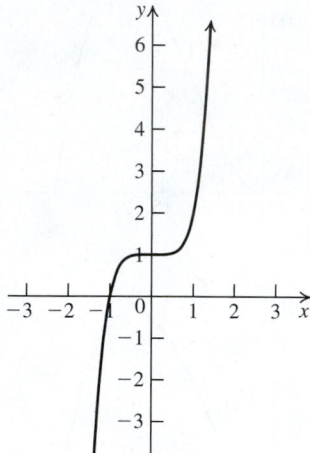

Zero: $x = -1$, multiplicity: 1

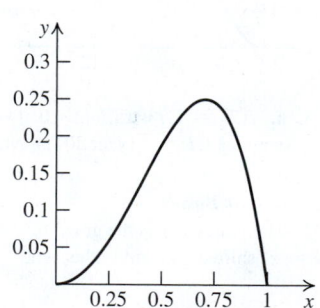

Zero: $x = 1$, multiplicity: 1

109. Maximum vertical distance: 0.25. It occurs at $x \approx 0.71$.

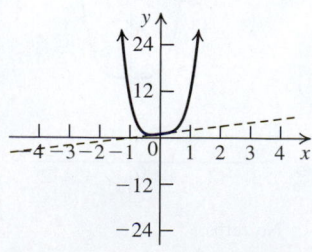

111. $f(x) = x^2(x + 1)$ **113.** $f(x) = 1 - x^4$ **115.** degree 5 because the graph has 5 x-intercepts and 4 turning points.   **117.** degree 6 because the graph has 5 turning points
119. $f(x) \approx 10x^4$ and $f(x) \approx x + 1$

121. $f(x) \approx x^4$ and $f(x) \approx 1 - x^2$

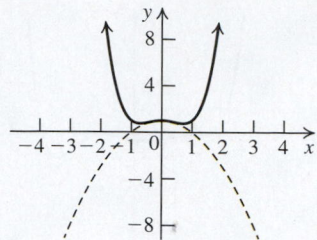

Critical Thinking/Discussion/Writing:
123. The total weight will increase $20 \times 20 \times 20 = 8000$ times, but the bone strength will only increase $20 \times 20 = 400$ times, so its bones would break under its weight.   **125.** No because the domain of any polynomial function is $(-\infty, \infty)$, which includes the point $x = 0$.   **127.** Not possible; this is inconsistent with its end behavior.   **129.** $P \circ Q$ is a polynomial of degree $m \cdot n$

Getting Ready for the Next Section:
131. $f(-2) = -33$ **133.** $g(13) = 0$ **135.** $(x + 2)(x - 5)$
137. $(2x + 3)(x - 1)(x + 7)$

Section 2.3

Concepts and Vocabulary:
1. $x^3 - 3x + 5$, $x^2 + 1$, x, $-4x + 5$ **3.** $F(a)$ **5.** True   **7.** True

Building Skills:
9. Quotient: $3x - 2$; remainder: 0   **11.** Quotient: $3x^3 - 3x^2 - 3x + 6$; remainder: -13 **13.** Quotient: $2x + 1$; remainder: 0   **15.** Quotient: $z^2 + 2z + 1$; remainder: 0   **17.** The quotient is $x^2 - 7$, and the remainder is -5.   **19.** The quotient is $x^2 + 2x - 11$, and the remainder is 12.
21. The quotient is $x^3 - x^2 + 4$, and the remainder is 13.   **23.** The quotient is $2x^2 + 5x - \frac{1}{2}$, and the remainder is $\frac{3}{4}$.   **25.** The quotient is $2x^2 - 6x + 6$, and the remainder is -1.

27. a. 5   **b.** 3   **c.** $\dfrac{15}{8}$ **d.** 1301   **29. a.** -17 **b.** -27 **c.** -56
d. 24   **39.** $k = -1$ **41.** $k = -7$ **43.** Remainder: 1   **45.** Remainder: 6
47. $\left\{ \pm\frac{1}{3}, \pm 1, \pm\frac{5}{3}, \pm 5 \right\}$ **49.** $\left\{ \pm\frac{1}{4}, \pm\frac{1}{2}, \pm\frac{3}{4}, \pm 1, \pm\frac{3}{2}, \pm 2, \pm 3, \pm 6 \right\}$
51. $\{-2, 1, 2\}$ **53.** $\{-1, 2, 3\}$ **55.** $\left\{ -3, \frac{1}{2}, 2 \right\}$ **57.** $\left\{ -2, -\frac{1}{2}, \frac{1}{3} \right\}$
59. $\left\{ -\frac{1}{3} \right\}$ **61.** $\{-1, 2\}$ **63.** $\{-3, -1, 1, 4\}$ **65.** No rational zeros
67. $\{1, -3 \pm \sqrt{11}\}$ **69.** $\left\{ 1, -1, \dfrac{3 \pm \sqrt{5}}{2} \right\}$
71. $\left\{ \dfrac{1}{2}, 2 - \sqrt{3}, 2 + \sqrt{3} \right\}$ **73.** $\{-2, -\sqrt{3}, 1, \sqrt{3}\}$
75. $(x - 1)(2x + 1)(x - 2)$ **77.** $(x + 2)(x - 3)(2x + 5)$
79. $(x - 2)(x + 3)(x - 1 + \sqrt{2})(x - 1 - \sqrt{2})$
81.

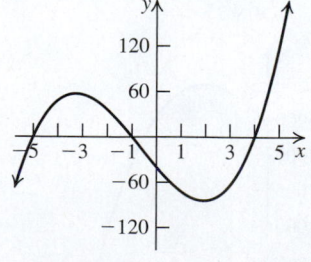

83.

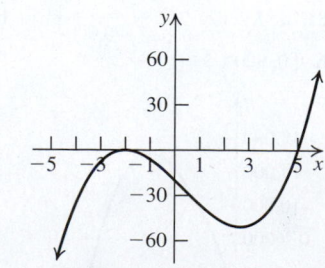

85. **87.**

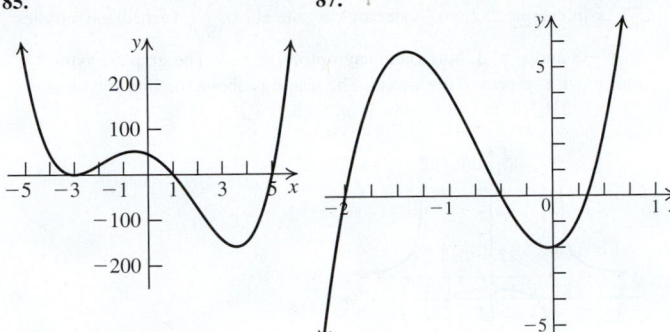

Applying the Concepts:

89. $2x^2 + 1$ cm **91. a.** $R(x) = -2x^2 + 200x - 4800$
b. $R(x) = -2x^2 + 200x - 1800$ **c.** The maximum weekly revenue is
$3200 if the phone is priced at $50. **93.** 2008 **95.** 2006 **97.** $x = 4$
99. 1500 **101.** The edge of the taller box is 11 inches, and shorter box
is 8 inches **103.** $x = 3$ in **111.** -10 **113.** $\left\{ -\dfrac{3}{2}, \dfrac{1}{2} \right\}$

Critical Thinking/Discussion/Writing:

115. $c = -1, 3, 5$ **121. a.** n is an odd integer. **b.** n is an even
integer. **c.** There is no such n. **d.** n is a positive integer.

Getting Ready for the Next Section:

123. **125.**

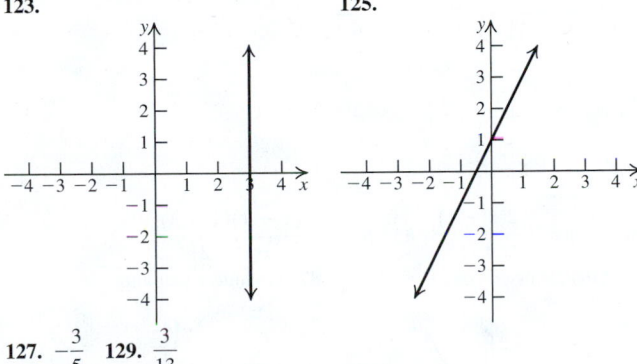

127. $-\dfrac{3}{5}$ **129.** $\dfrac{3}{13}$

Section 2.4

Concepts and Vocabulary:

1. polynomials **3.** $k, \infty, -\infty$ **5.** True **7.** True

Building Skills:

9. $(-\infty, -4)\cup(-4, \infty)$ **11.** $(-\infty, \infty)$
13. $(-\infty, -2)\cup(-2, 3)\cup(3, \infty)$ **15.** $(-\infty, 2)\cup(2, 4)\cup(4, \infty)$
17. ∞ **19.** ∞ **21.** 1 **23.** $(-\infty, -2)\cup(-2, 1)\cup(1, \infty)$
25. $x = -2, x = 1$
27. Domain: $(-\infty, 4)\cup(4, \infty)$
Range: $(-\infty, 0)\cup(0, \infty)$
Vertical Asymptote: $x = 4$
Horizontal Asymptote: $y = 0$

29. Domain: $\left(-\infty, -\dfrac{1}{3}\right)\cup\left(-\dfrac{1}{3}, \infty\right)$
Range: $\left(-\infty, -\dfrac{1}{3}\right)\cup\left(-\dfrac{1}{3}, \infty\right)$
Vertical Asymptote: $x = -\dfrac{1}{3}$
Horizontal Asymptote: $y = -\dfrac{1}{3}$

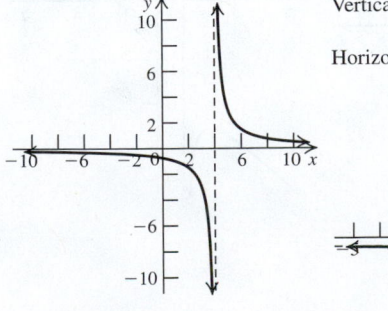
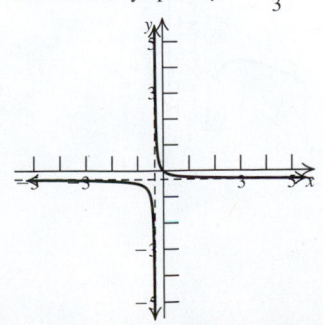

31. Domain: $(-\infty, -2)\cup(-2, \infty)$
Range: $(-\infty, -3)\cup(-3, \infty)$
Vertical Asymptote: $x = -2$
Horizontal Asymptote: $y = -3$

33. Domain: $(-\infty, 4)\cup(4, \infty)$
Range: $(-\infty, 5)\cup(5, \infty)$
Vertical Asymptote: $x = 4$
Horizontal Asymptote: $y = 5$

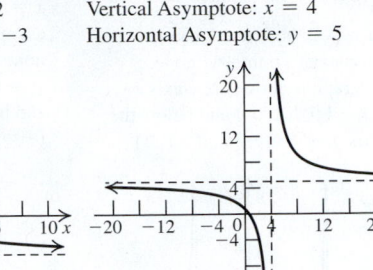

35. $x = 1$ **37.** $x = -4$ and $x = 3$ **39.** $x = -3$ and $x = 2$
41. $x = -3$ **43.** No vertical asymptote **45.** $y = 0$ **47.** $y = \dfrac{2}{3}$
49. No horizontal asymptote **51.** $y = 0$ **53.** d **55.** e **57.** a
59. x-intercept: 0; y-intercept: 0;
vertical asymptote: $x = 3$;
horizontal asymptote: $y = 2$;
symmetry: none. The graph is
above the line $y = 2$ on $(3, \infty)$
and below it on $(-\infty, 3)$.

61. x-intercept: 0; y-intercept: 0;
vertical asymptotes: $x = -2$ and
$x = 2$; horizontal asymptote: x-axis.
The graph is symmetric with respect
to the origin. The graph is above the
x-axis on $(-2, 0)\cup(2, \infty)$ and below the x-axis on $(-\infty, -2)\cup(0, 2)$.

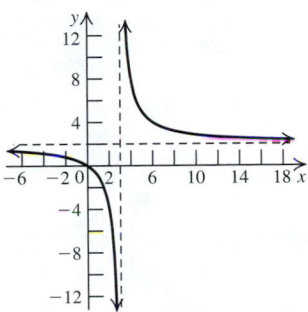
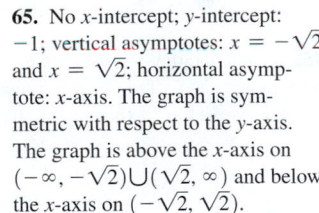

63. x-intercept: 0; y-intercept:
0; vertical asymptotes: $x = -3$
and $x = 3$; horizontal asymptote: $y = -2$. The graph is
symmetric about the y-axis. The
graph is above the line $y = -2$
on $(-3, 3)$ and below it on
$(-\infty, -3)\cup(3, \infty)$.

65. No x-intercept; y-intercept:
-1; vertical asymptotes: $x = -\sqrt{2}$
and $x = \sqrt{2}$; horizontal asymptote: x-axis. The graph is symmetric with respect to the y-axis.
The graph is above the x-axis on
$(-\infty, -\sqrt{2})\cup(\sqrt{2}, \infty)$ and below
the x-axis on $(-\sqrt{2}, \sqrt{2})$.

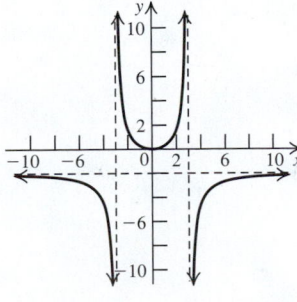
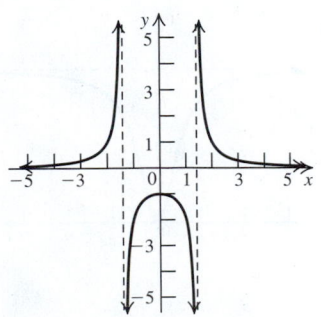

67. x-intercept: -1; y-intercept: $-\frac{1}{6}$; vertical asymptotes: $x = -3$ and $x = 2$; horizontal asymptote: x-axis; symmetry: none. The graph is above the x-axis on $(-3, -1) \cup (2, \infty)$ and below the x-axis on $(-\infty, -3) \cup (-1, 2)$.

69. x-intercept: 0; y-intercept: 0; vertical asymptote: $x = -2$, $x = -1$; horizontal asymptote: $y = 0$; symmetry: none; the graph is above the line $y = 0$ on $(-2, -1) \cup (0, \infty)$ and below it on $(-\infty, -2) \cup (-1, 0)$;

77. x-intercept: ± 2; no y-intercept: a hole at $\left(0, \frac{4}{9}\right)$; vertical asymptotes: $x = -3$ and $x = 3$; horizontal asymptote: $y = 1$. The graph is symmetric with respect to the y-axis. The graph is above the line $y = 1$ on $(-\infty, -3) \cup (3, \infty)$ and below it on $(-3, 0) \cup (0, 3)$.

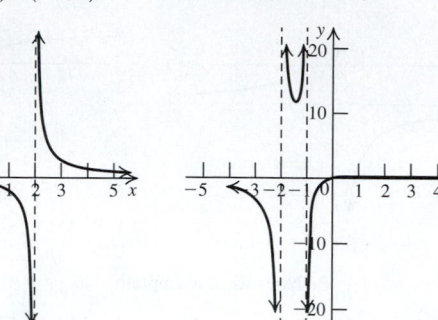

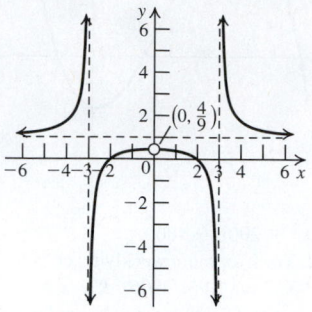

71. x-intercept: 1; y-intercept: none; vertical asymptote: $x = -1$, $x = 0$; horizontal asymptote: $y = 0$; symmetry: none; the graph is above the line $y = 0$ on $(-\infty, -1) \cup (1, \infty)$ and below it on $(-1, 0) \cup (0, 1)$;

73. x-intercept: none; y-intercept: $-\frac{1}{4}$; vertical asymptote: $x = 4$; horizontal asymptote: $y = 0$; a hole at $(3, -1)$; symmetry: none; the graph is above the line $y = 0$ on $(4, \infty)$ and below it on $(-\infty, 3) \cup (3, 4)$;

79. No x-intercept; y-intercept: -2; hole at $(2, 0)$ no vertical asymptote; no horizontal asymptote; symmetry: none. The graph is above the x-axis on $(2, \infty)$ and below the x-axis on $(-\infty, 2)$.

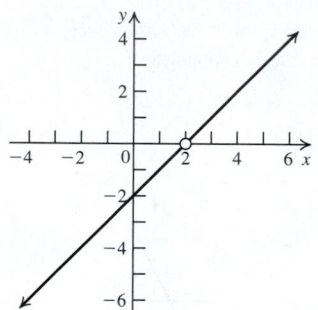

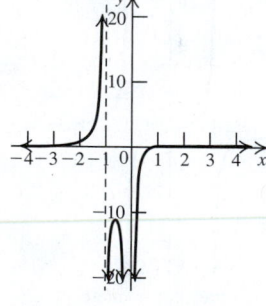

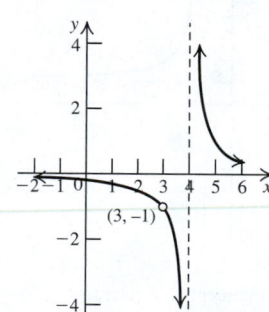

81. $f(x) = \dfrac{-2(x - 1)}{x - 2}$ **83.** $f(x) = \dfrac{(x - 1)(x - 3)}{x(x - 2)}$

85. Oblique asymptote: $y = 2x$ **87.** Oblique asymptote: $y = x$

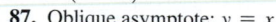

75. x-intercept: 0; y-intercept: 0; no vertical asymptote; horizontal asymptote: $y = 1$. The graph is symmetric with respect to the y-axis. The graph is above the x-axis on $(-\infty, 0) \cup (0, \infty)$, and below the line $y = 1$ on $(-\infty, \infty)$.

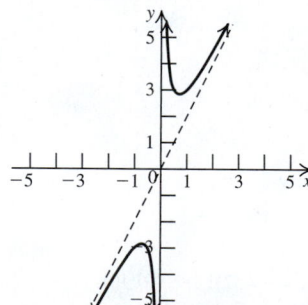

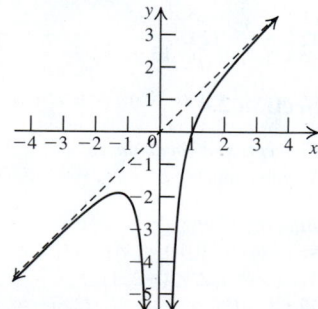

89. Oblique asymptote: $y = x - 2$ **91.** Oblique asymptote: $y = x - 2$

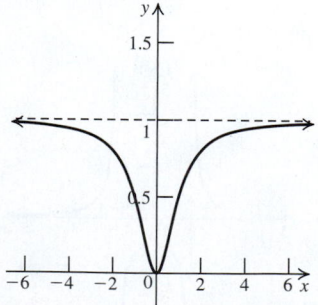

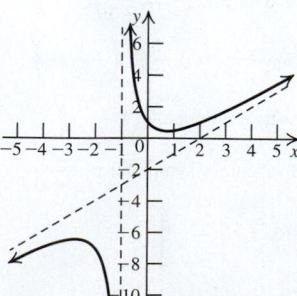

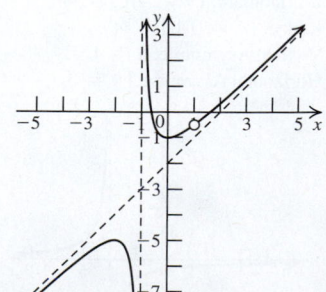

Applying the Concepts:

93. a. $\overline{C}(x) = 0.5x + 2000$ **b.** $\overline{C}(x) = 0.5 + \dfrac{2000}{x}$

c. $\overline{C}(100) = 20.5, \overline{C}(500) = 4.5, \overline{C}(1000) = 2.5$. These show the average cost of producing 100, 500, and 1000 trinkets, respectively.
d. Horizontal asymptote: $y = 0.5$. It means that the average cost (the fixed daily cost of producing each trinket) if the number of trinkets produced approaches ∞.

95. a.

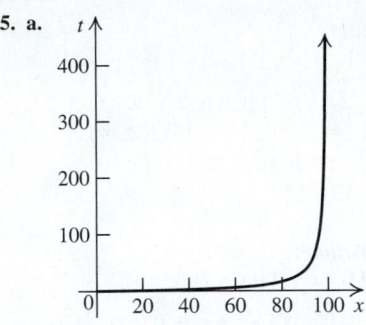

b. i) about 4 min; **ii)** about 12 min; **iii)** about 76 min; **iv)** 397 min; **c. i)** ∞; **ii)** Not applicable; the domain is $x < 100$.

97. a. $3.02 billion

b.

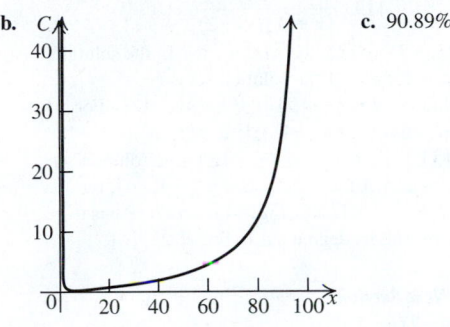

c. 90.89%

99. a. 16,000 **b.** The population will stabilize at 4000.
101. a. $f(x) = (10x + 200,000)/(x - 2500)$ **b.** $40 **c.** More than 25,000 **d.** $x = 2500$ (free books); $y = 10$ (cost per book)

Beyond the Basics:

103. Stretch the graph of $y = \dfrac{1}{x}$ vertically by a factor of 2 and reflect the graph about the x-axis.

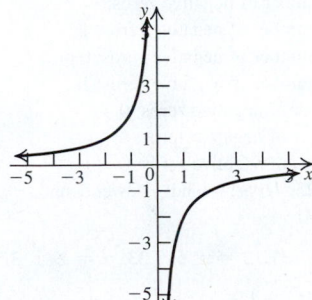

105. Shift the graph of $y = \dfrac{1}{x^2}$ two units to the right.

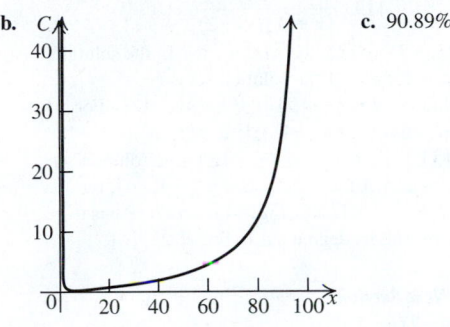

107. Shift the graph of $y = \dfrac{1}{x^2}$ one unit to the right and two units down.

109. Shift the graph of $y = \dfrac{1}{x^2}$ six units to the left.

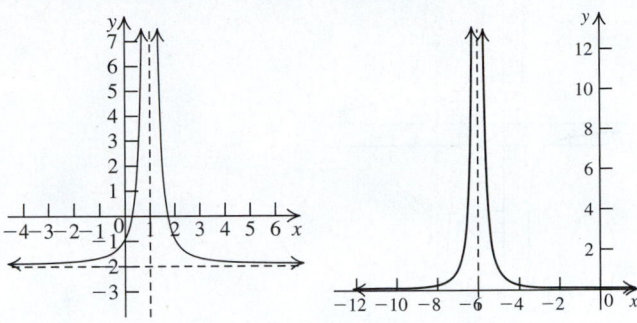

111. Shift the graph of $y = \dfrac{1}{x^2}$ one unit to the right and one unit up.

113. a. If c is a zero of $f(x)$, then because c is not a factor of the numerator, $x = c$ is a vertical asymptote. **b.** Because the numerator is positive, the signs of $f(x)$ and $g(x)$ are the same. **c.** If the graphs intersect for some value x, then $f(x) = g(x)$. It implies that $f(x) = \dfrac{1}{f(x)}$, from which it follows that $f(x) = \pm 1$. **d.** If $f(x)$ increases (decreases, remains constant), the denominator of $g(x)$ increases (decreases, remains constant). Therefore, $g(x)$ decreases (increases, remains constant).

115. $[f(x)]^{-1} = \dfrac{1}{2x + 3}$ and $f^{-1}(x) = \dfrac{1}{2}x - \dfrac{3}{2}$. The two functions are different.

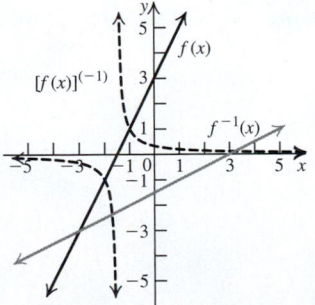

117. $g(x)$ has the oblique asymptote $y = 2x + 3$; $y \to -\infty$ as $x \to -\infty$, and $y \to \infty$ as $x \to \infty$.

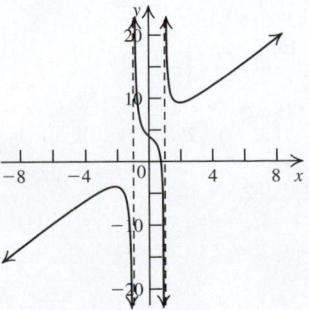

119. Point of intersection: $\left(-\dfrac{1}{3}, 1\right)$; horizontal asymptote $y = 1$

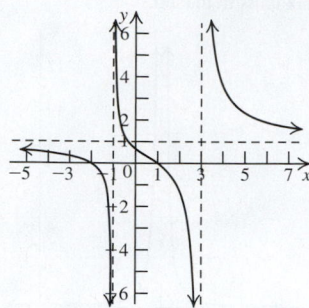

121. $f(x) = \dfrac{2 - x}{x - 3}$ **123.** $f(x) = \dfrac{-x^2 + 1}{x}$

125. $f(x) = (x + 4)/(x - 2)$ **127.** $f(x) = (4x^2 - 1)/(x - 1)^2$

129. $f(x) = (3x^2 - x + 1)/(x - 1)$

Critical Thinking / Discussion / Writing:

131. a. $R_T = \dfrac{5R_1}{R_1 + 5}$

b.

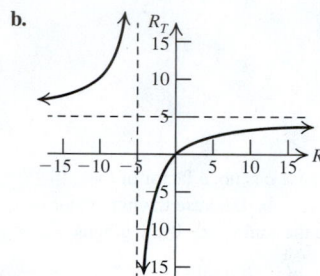

c. $R_1 > 0$ so $0 < R_T < 5$. Horizontal asymptote $R_T = 5$ and as $R_T \to \infty$, $R_T \to 5$.

133. a. $f(x) = \dfrac{1}{x}$ **b.** $f(x) = \dfrac{x^2 + 2}{x^2 + x - 2}$

c. $f(x) = \dfrac{x^3 + 2x^2 - 7x - 1}{x^3 + x^2 - 6x + 5}$

135. $R(x) = \dfrac{(ax + b)D(x) + K(x - c_1)(x - c_2)\cdots(x - c_n)}{D(x)}$ where $K \neq 0$ and $D(x)$ is a polynomial of degree $n + 1$ and none of its zeros are at $c_1, c_2, \ldots, c_n$.

Getting Ready for the Next Section:

137. $-\dfrac{2(x + 5)}{x + 3}$ **139.** $\dfrac{x + 13}{(x - 2)(x + 1)}$

141. $(-\infty, -2)\cup(1, \infty)$ **143.** $(-\infty, -5]\cup[2, \infty)$

Section 2.5

Concepts and Vocabulary:

1. included **3.** zeros, vertical **5.** True **7.** False

Building Skills:

9. $[-2, 1]\cup[3, \infty)$ **11.** $(-3, 2)\cup(2, \infty)$

13. $(-\infty, -2)\cup(-2, 2)\cup(3, \infty)$ **15.** $[-3, 2]$

17. $(-\infty, 3)\cup(4, \infty)$ **19.** $(-\infty, -1]\cup[8, \infty)$

21. $(-\infty, -5)\cup(1, 3)$ **23.** $[-7, 1]\cup[4, \infty)$

25. $(-2, 0)\cup(0, 1)$ **27.** $(-\infty, 1]\cup[3, \infty)$ **29.** $(-\infty, 0)\cup(1, \infty)$

31. $(-\infty, -1)\cup(2, \infty)$ **33.** $(-\infty, \infty)$ **35.** $(-3, -2)\cup(1, \infty)$

37. $\{-2\}\cup[4, \infty)$ **39.** $(-\infty, -1]\cup[2, 4]$

41. $(-\infty, -\sqrt{3}]\cup\{0\}\cup[\sqrt{3}, \infty)$ **43.** $(-\infty, 0]\cup(1, \infty)$

45. $(-1, 0)\cup(1, \infty)$ **47.** $(-1, 0]\cup[2, \infty)$

49. $(-\infty, -1)\cup(-1, 0)\cup(1, \infty)$ **51.** $(-\infty, -2)\cup(1, \infty)$

53. $(-\infty, -2]\cup(8, \infty)$ **55.** $[2/3, 4/3)$ **57.** $(-2, -1)\cup(1, \infty)$

59. $[-5, -3)\cup[2, \infty)$ **61.** $(-4, 1]\cup(3, \infty)$

63. $(-1, 2]\cup[3, \infty)$ **65.** $(-\infty, -7)\cup(4, \infty)$ **67.** $(-3, 2]$

69. $(-7, -2)\cup(3, \infty)$ **71.** $(-\infty, -2)\cup[-1/4, 5)$

73. $(-\infty, 0)\cup[2, 4]$ **75.** $(-2, 0]\cup(2, \infty)$ **77.** Yes

79. Q between 3000 and 17000 **81.** Between 0 and 1000 and between 4000 and 6000 **83.** $v \geq 18$ **85.** $0 < I < 4$ **87.** Between 2007 and 2013

Beyond the Basics:

89. $(-\infty, -2]\cup[6, \infty)$ **91.** $(-\infty, -2]\cup[0, 2]$ **93.** $(-5, 2]$

95. $(-\infty, 2]\cup(3, \infty)$ **97.** $(-\infty, -2)\cup[-1, 1]\cup(2, \infty)$

99. $[-\sqrt{11}, -2]\cup[2 - \sqrt{7}, 2]\cup[\sqrt{11}, 2 + \sqrt{7}]$

101. $[4, 8]$ **103.** $(-\infty, -3]\cup[2, 5]$

Critical Thinking/Discussion/Writing:

109. $(x + 4)(x - 5) < 0$ **111.** $(x + 1)(x - 2)(x - 3) > 0$

113. $x(x + 1)(x - 4)(x - 5) < 0$ **115.** $\dfrac{x + 2}{x - 6} \leq 0$

117. $\dfrac{(x + 1)(x - 3)}{x - 2} \geq 0$ **119.** $\dfrac{x - 3}{(x + 1)(x - 2)} \geq 0$

121. $(x + 2)(x - 1)(x - 3)^2$ **123.** For $a \in (-\infty, 1)$ the solution set is $(-\infty, a]\cup[1, \infty)$; for $a = 1$ the solution set is $(-\infty, \infty)$; For $a \in (1, \infty)$ the solution set is $(-\infty, 1]\cup[a, \infty)$ **125.** For $a \in (-\infty, -1)$ the solution set is $[a, -1]\cup[1, \infty)$; for $a = -1$ the solution set is $\{-1\}\cup[1, \infty)$; for $a \in (-1, 1)$ the solution set is $[-1, a]\cup[1, \infty)$; for $a = 1$ the solution set is $[-1, \infty)$; for $a \in (1, \infty)$ the solution set is $[-1, 1]\cup[a, \infty)$; **127.** P has to be polynomial of odd degree, and its degree has to be $\geq 2k + 1$.

Getting Ready for the Next Section:

129. $x^2 + 1$ **131.** $x^2 - 2x + 5$

133. $x^4 + x^3 + 38x - 40 = (x^2 - 2x + 10)(x^2 + 3x - 4)$

135. $x^4 - 7x^3 + 9x^2 + 13x - 4 = (x^4 - 4x + 1)(x^2 - 3x - 4)$

Section 2.6

Concept and Vocabulary:

1. consecutive **3.** complex **5.** True **7.** True

Building Skills:

9. Number of positive zeros: 0 or 2; number of negative zeros: 1

11. Number of positive zeros: 0 or 2; number of negative zeros: 1

13. Number of positive zeros: 1 or 3; number of negative zeros: 0 or 2

15. Number of positive zeros: 1 or 3; number of negative zeros: 1

17. Number of positive zeros: 0; number of negative zeros: 0

19. Number of positive zeros: 0; number of negative zeros: 0

21. Number of positive zeros: 1; number of negative zeros: 0

23. Upper bound: $\frac{1}{3}$; lower bound $-\frac{1}{3}$ **25.** Upper bound: $\frac{1}{3}$; lower bound: $-\frac{7}{3}$

27. Upper bound: 31; lower bound: -31

29. Upper bound: $\dfrac{7}{2}$; lower bound: $-\dfrac{7}{2}$ **31.** $x = \pm 5i$ **33.** $x = 2 \pm 3i$

35. $x = -2 \pm 3i$ **37.** $\{2, -1 + i\sqrt{3}, -1 - i\sqrt{3}\}$

39. $\{2, 3i, -3i\}$ **41.** $3 - i$ **43.** $2 + i$ **45.** $-i, -3i$

47. $P(x) = 2x^4 - 20x^3 + 70x^2 - 180x + 468$

49. $P(x) = 7x^5 - 119x^4 + 777x^3 - 2415x^2 + 3500x - 1750$

51. $-1, 0, 3i, -3i$ **53.** $-2, 0, 1, 3 + i, 3 - i$ **55.** $1, 4 + i, 4 - i$

57. $-\dfrac{4}{3}, 1 + 3i, 1 - 3i$ **59.** $1, \dfrac{3}{2} + \dfrac{3}{2}i, \dfrac{3}{2} - \dfrac{3}{2}i$ **61.** $-3, 1, 3 + i, 3 - i$

63. $\dfrac{1}{2}, 2, 3, i, -i$ **65.** $f(x) = \dfrac{1}{3}(x^2 + 9)$

67. $f(x) = \dfrac{1}{2}(2 - x)(x^2 + 1)(x^2 + 4)$

Beyond the Basics:

69. There are three cube roots: $1, -\dfrac{1}{2} + \dfrac{1}{2}\sqrt{3}i$, and $-\dfrac{1}{2} - \dfrac{1}{2}\sqrt{3}i$.

75. $f(x) = -4(x-2)(x^2 - 2x + 5); y \to \infty$ as $x \to -\infty$,
$y \to -\infty$ as $x \to \infty$.

77. $f(x) = -2(x^2 - 1)(x^2 - 6x + 10); y \to -\infty$ as $x \to -\infty$,
$y \to -\infty$ as $x \to \infty$.

79. $P(x) = (x-2)(x+i)$ **81.** $P(x) = (x-4)(x+1)(x-i)$

Critical Thinking/Discussion/Writing:

85. a. According to Descartes's Rule of Signs, the polynomial
$x^3 + 6x - 20$ has one positive zero and no negative zero.
Therefore, there is exactly one real solution.
  b. Substituting $v - u$ for x, we obtain
$$(v-u)^3 + 6(v-u) = v^3 - 3v^2u + 3vu^2 - u^3 + 6(v-u)$$
$$= (v^3 - u^3) + (6 - 3uv)(v-u)$$
$$= 20 - (6 - 3(2))(v-u) = 20.$$
Therefore, $x = v - u$ is the solution.
  c. $u = \sqrt[3]{-10 \pm 6\sqrt{3}}, v = \dfrac{2}{\sqrt[3]{-10 \pm 6\sqrt{3}}}, v = \sqrt[3]{10 \pm 6\sqrt{3}}$

Getting Ready for the Next Section:

87. a. $y = -\dfrac{2}{3}x - \dfrac{1}{3}$ **b.** $y = -1$ **89.** $k = \dfrac{3}{2}, y = 7$

Section 2.7

Concepts and Vocabulary:

1. $y = kx$ **3.** $y = kx^n$ **5.** True **7.** False

Building Skills:

9. $k = \dfrac{1}{2}, x = 14$ **11.** $k = 16, s = 400$ **13.** $k = 33, r = 99$

15. $k = 8, B = \dfrac{1}{8}$ **17.** $k = 7, y = 4$ **19.** $k = 2, z = 50$

21. $k = \dfrac{1}{17}, P = \dfrac{324}{17}$ **23.** $k = 54, x = 4$ **25.** $y = 6$ **27.** $y = 200$

Applying the Concepts:

29. $y = Hx$, where y is the speed of the galaxies and x is the distance between them. **31. a.** $y = 30.5x$, where y is the length measured in centimeters and x is the length measured in feet. **b. (i)** 244 cm **(ii)** 162.67 cm
c. (i) 1.87 ft **(ii)** 4.07 ft **33.** 35 g **35.** $\frac{3}{4}$ sec **37.** 60 lb/in.^{2}
39. a. 18.97 lb **b.** 201.01 lb **41.** 1.63 m/s^2 **43 a.** 1280 candela
b. 8.94 ft from the source **45.** The length is multiplied by 4.
47. a. $H = kR^2N$, where k is a constant. **b.** The horsepower is multiplied by 4. **c.** The horsepower is doubled. **d.** The horsepower is cut in half.

Beyond the Basics:

49. a. $E = kl^2v^3$ **b.** $k = 0.0375$ **c.** 37,500 W **d.** E is multiplied by 8. **e.** E is multiplied by 4. **f.** E is multiplied by 32. **51. a.** $k = 2.94$
b. 287.25 W **c.** The metabolic rate is multiplied by 2.83. **d.** 373.93 kg

53. a. $T^2 = \left(\dfrac{4\pi^2}{G}\right)\left(\dfrac{r^3}{M_1 + M_2}\right)$ **b.** 2.01×10^{30}kg

55. a. $R = kN(P - N)$, where k is the constant of variation. **b.** $k = \dfrac{5}{10^6}$

c. 125 people per day **d.** 2764 or 7236

Critical Thinking / Discussion / Writing:

57. a. \$480 **b.** \$15,920 **c.** The value of the original diamond is \$112,500. The value of a diamond whose weight is twice that of the original is \$450,000. **59.** 1:2

Getting Ready for the Next Section:

61. 8 **63.** $\dfrac{1}{8}$ **65.** 4 **67.** 2^{2x+3}

Review Exercises

1. (i) Opens up
(ii) Vertex: (1, 2)
(iii) Axis: $x = 1$
(iv) No x-intercept
(v) y-intercept: 3
(vi) Decreasing on $(-\infty, 1)$, increasing on $(1, \infty)$

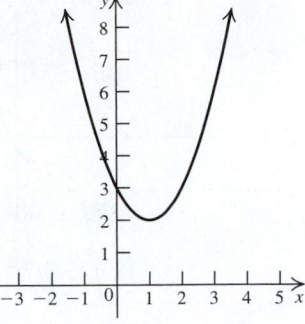

3. (i) Opens down
(ii) Vertex: (3, 4)
(iii) Axis: $x = 3$
(iv) x-intercepts: $3 \pm \sqrt{2}$
(v) y-intercept: -14
(vi) Increasing on $(-\infty, 3)$, decreasing on $(3, \infty)$

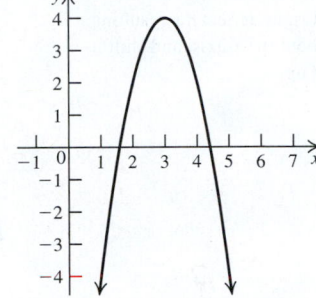

5. (i) Opens down
(ii) Vertex: (0, 3)
(iii) Axis: y-axis
(iv) x-intercepts: $\pm\dfrac{\sqrt{6}}{2}$
(v) y-intercept: 3
(vi) Increasing on $(-\infty, 0)$, decreasing on $(0, \infty)$

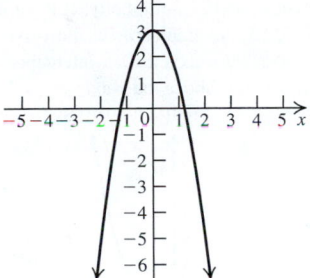

7. (i) Opens up
(ii) Vertex: (1, 1)
(iii) Axis: $x = 1$
(iv) No x-intercept
(v) y-intercept: 3
(vi) Decreasing on $(-\infty, 1)$, increasing on $(1, \infty)$

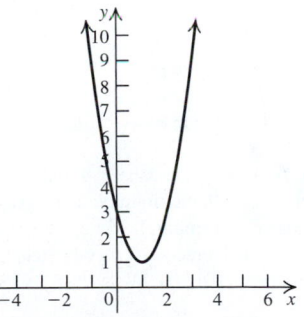

9. (i) Opens up
(ii) Vertex: $\left(\dfrac{1}{3}, \dfrac{2}{3}\right)$
(iii) Axis: $x = \dfrac{1}{3}$
(iv) No x-intercept
(v) y-intercept: 1
(vi) Decreasing on $\left(-\infty, \dfrac{1}{3}\right)$, increasing on $\left(\dfrac{1}{3}, \infty\right)$

11. Minimum: (2, −1) **13.** Maximum: $\left(-\dfrac{3}{4}, \dfrac{25}{8}\right)$

15. Shift the graph of $y = x^3$ one unit to the left and two units down.

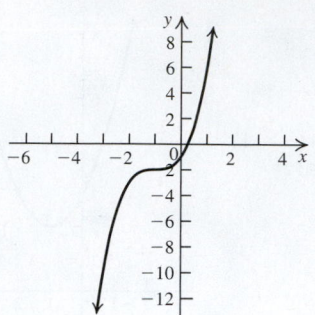

17. Shift the graph of $y = x^3$ one unit right, reflect the resulting graph about the x-axis, and shift it one unit up.

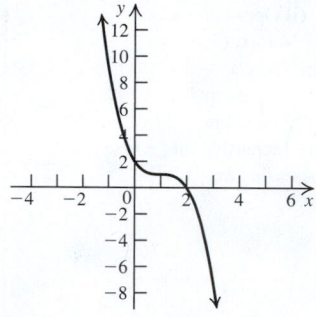

19. (i) $f(x) \to -\infty$ as $x \to -\infty$ and $f(x) \to \infty$ as $x \to \infty$.
(ii) Zeros: $x = -2$, multiplicity: 1, the graph crosses the x-axis; $x = 0$, multiplicity: 1, the graph crosses the x-axis; $x = 1$, multiplicity: 1, the graph crosses the x-axis. **(iii)** x-intercepts: $-2, 0, 1$; y-intercept: 0.
(iv) The graph is above the x-axis on $(-2, 0) \cup (1, \infty)$ and below the x-axis on $(-\infty, -2) \cup (0, 1)$. **(v)** Symmetry: none.
(vi)

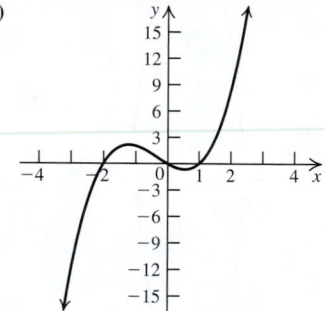

21. (i) $f(x) \to -\infty$ as $x \to -\infty$ and $f(x) \to -\infty$ as $x \to \infty$.
(ii) Zeros: $x = 0$, multiplicity: 2, the graph touches but does not cross the x-axis; $x = 1$, multiplicity: 2, the graph touches but does not cross the x-axis. **(iii)** x-intercepts: $0, 1$; y-intercept: 0. **(iv)** The graph is below the x-axis on $(-\infty, 0) \cup (0, 1) \cup (1, \infty)$. **(v)** Symmetry: none.
(vi)

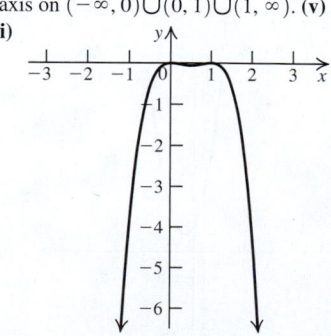

23. (i) $f(x) \to -\infty$ as $x \to -\infty$ and $f(x) \to -\infty$ as $x \to \infty$.
(ii) Zeros: $x = -1$, multiplicity: 1, the graph crosses the x-axis; $x = 0$; multiplicity: 2, the graph touches but does not cross the x-axis; $x = 1$ multiplicity: 1, the graph crosses the x-axis. **(iii)** x-intercepts: $-1, 0, 1$; y-intercept: 0. **(iv)** The graph is above the x-axis on $(-1, 0) \cup (0, 1)$ and below the x-axis on $(-\infty, -1) \cup (1, \infty)$. **(vi)** The graph is symmetric with respect to the y-axis.
(vi)

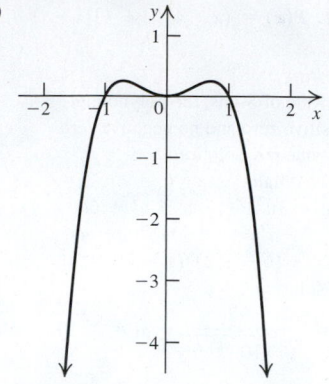

25. Quotient: $2x + 3$; remainder: -7
27. Quotient: $8x^3 - 12x^2 + 14x - 21$; remainder: 186
29. Quotient: $x^2 + 3x - 3$; remainder: -6
31. Quotient: $2x^3 - 5x^2 + 10x - 17$; remainder: 182 **33.** -11 **35.** 88

37. $1, 2, 4$ **39.** $-3, -2, \dfrac{1}{3}$ **41.** $-6, -3, -2, -1, 1, 2, 3, 6$

43. $-2, -\dfrac{1}{5}, 0$ **45.** $-3, -2, 2$ **47.** $-3, 1, 2$ **49.** $-1, 3, 3i, -3i$

51. $2i, -2i, -1 + 2i, -1 - 2i$ **53.** $\{-2, 1, 2\}$ **55.** $\left\{-1, -\dfrac{1}{2}, \dfrac{3}{2}\right\}$

57. $\{-1, 2, i, -i\}$ **59.** The only possible rational roots are $-2, -1, 1$, and 2. None of them satisfies the equation. **61.** 1.88

63. x-intercept: -1; no y-intercept; vertical asymptote: y-axis; horizontal asymptote: $y = 1$; symmetry: none. The graph is above the line $y = 1$ on $(0, \infty)$ and below it on $(-\infty, 0)$.

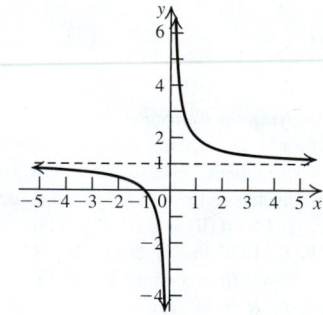

65. x-intercept: 0; y-intercept: 0; vertical asymptotes: $x = -1$ and $x = 1$; horizontal asymptote: x-axis; symmetric about the origin. The graph is above the x-axis on $(-1, 0) \cup (1, \infty)$ and below the x-axis on $(-\infty, -1) \cup (0, 1)$.

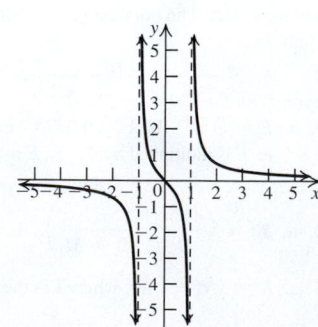

67. x-intercept: 0; y-intercept: 0; vertical asymptotes: $x = -3$ and $x = 3$; oblique asymptote: $y = x$; symmetric about origin. The graph is above the line $y = x$ on $(-3, 0) \cup (3, \infty)$ and below it on $(-\infty, -3) \cup (0, 3)$.

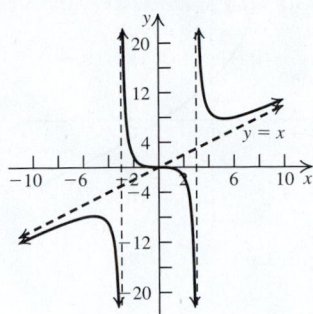

69. x-intercept: 0; y-intercept: 0; vertical asymptotes: $x = -2$ and $x = 2$; no horizontal asymptote. The graph is symmetric with respect to the y-axis. The graph is above the x-axis on $(-\infty, -2) \cup (2, \infty)$ and below the x-axis on $(-2, 0) \cup (0, 2)$.

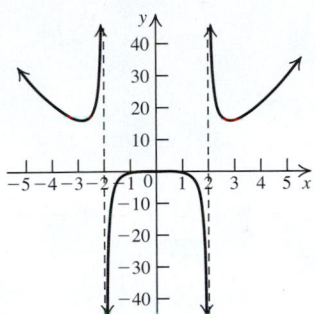

71. $(-\infty, -2] \cup [2, 3]$ **73.** $(3, \infty)$ **75.** $[-3, -2) \cup [1, \infty)$
77. $(-\infty, -15) \cup (-4, \infty)$

Applying the Concepts:
79. $y = 15$ **81.** $s = 45$
83. Maximum height: 1000. The missile hits the ground at $x = 200$.

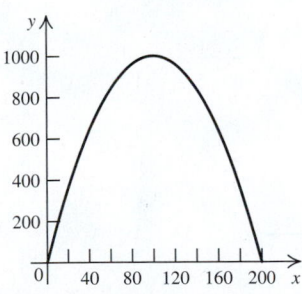

85. $x \approx 24.5$ ft, $y \approx 16.3$ ft

87. a.

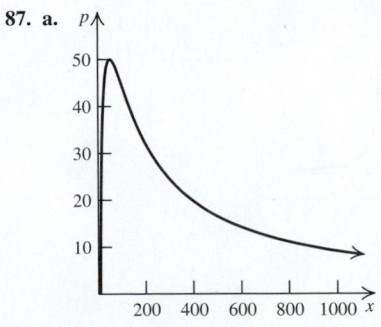

b. $\dfrac{150}{7} < x < \dfrac{350}{3}$
c. $x = 50$

89. a. $P(x) = -\dfrac{3}{1000}x^2 + 20.1x - 150$ **b.** 3350

c. $\overline{C}(x) = \dfrac{3}{1000}x + 3.9 + \dfrac{150}{x}$

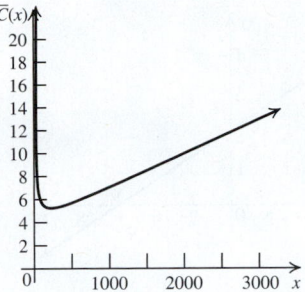

91. \$245   **93.** $6\sqrt{2}$ in.   **95. a.** 36 amp   **b.** 150 Ω
97. a. $k = \dfrac{1}{200,000}$ **b.** 500 people per day   **c.** 1000 and 19,000

Practice Test A:
1. x-intercepts: $3 \pm \sqrt{7}$
2.

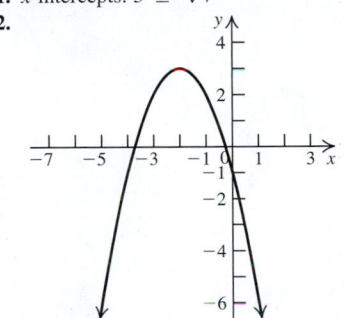

3. $(1, 10)$ **4.** $(-\infty, -4) \cup (-4, 1) \cup (1, \infty)$.
5. Quotient: $x^2 - 4x + 3$; remainder: 0
6.

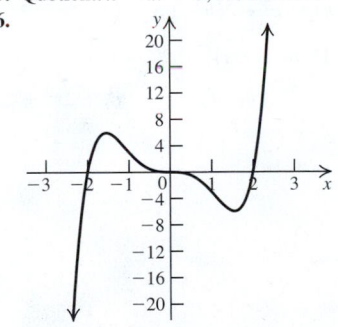

7. $-2, 1, 2$ **8.** Quotient: $-3x^2 + 5x + 1$ **9.** $P(-2) = -53$
10. $-2, 2, 5$ **11.** $0, -\dfrac{1}{2} + \dfrac{\sqrt{61}}{2}, -\dfrac{1}{2} - \dfrac{\sqrt{61}}{2}$
12. $-9, -\dfrac{9}{2}, -3, -\dfrac{3}{2}, -1, -\dfrac{1}{2}, \dfrac{1}{2}, 1, \dfrac{3}{2}, 3, \dfrac{9}{2}, 9$
13. $f(x) \to -\infty$ as $x \to -\infty$ and $f(x) \to \infty$ as $x \to \infty$.
14. $x = -2$, multiplicity: 3; $x = 2$, multiplicity: 1
15. Number of positive zeros: 0 or 2; number of negative zeros: 1
16. Horizontal asymptote: $y = 2$; vertical asymptotes:
$x = -4$ and $x = 5$
17. $[-7, -4) \cup (2, \infty)$ **18.** $y = 4$ **19.** $x = 15$
20. $V(x) = x(8 - 2x)(17 - 2x)$

Practice Test B:
1. b   **2.** d   **3.** a   **4.** b   **5.** d   **6.** c   **7.** c   **8.** b   **9.** c   **10.** a   **11.** a
12. d   **13.** c   **14.** b   **15.** c   **16.** c   **17.** c   **18.** b   **19.** b   **20.** d

Cumulative Review Exercises (Chapters 1 and 2)

1. $\sqrt{13}$ **2.** $M(-3, -4)$

3. x-intercepts: 4, -2 y-intercept: -8

4. Slope is $-\dfrac{1}{3}$; x-intercept: 6; y-intercept: 2

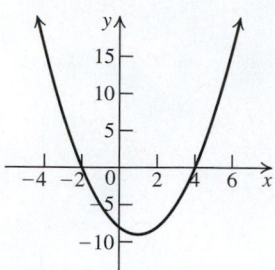

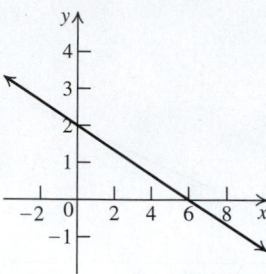

5. Slope is $\dfrac{1}{2}$; x-intercept: -6; y-intercept: 3

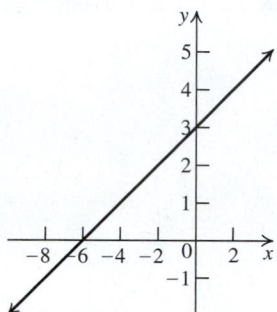

6. $(x-2)^2 + (y + 3)^2 = 16$ **7.** Center: $(-1, 2)$; radius: 3

8. $y = 3x - 5$ **9.** $y = -\dfrac{2}{3}x + \dfrac{11}{3}$

10. Domain: $\left(-\infty, -\dfrac{3}{2}\right) \cup \left(-\dfrac{3}{2}, \infty\right)$ **11.** Domain $(-\infty, 2)$.

12. $f(-2) = 11, f(3) = 6, f(x + h) = x^2 + 2(h-1)x + h^2 - 2h + 3,$ $\dfrac{f(x + h) - f(x)}{h} = 2x + h - 2$ **13 a.** $f(g(x)) = \sqrt{x^2 + 1}$

b. $g(f(x)) = x + 1$ **c.** $f(f(x)) = \sqrt[4]{x}$ **d.** $g(g(x)) = x^4 + 2x^2 + 2$

14. a. $f(1) = 5, f(3) = 11, f(4) = 6$

b.

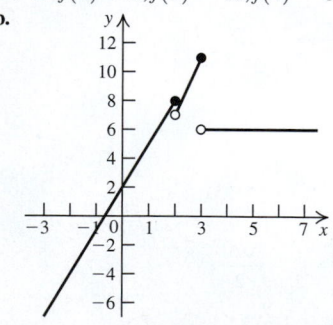

15. $f^{-1}(x) = \dfrac{1}{2}x + \dfrac{3}{2}$

16. a. Shift the graph of $y = \sqrt{x}$ two units to the left.

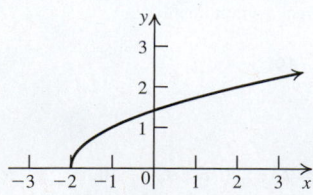

b. Shift the graph of $y = \sqrt{x}$ one unit to the left, stretch the resulting graph horizontally by a factor of 2, reflect it in the x-axis and shift it three units up.

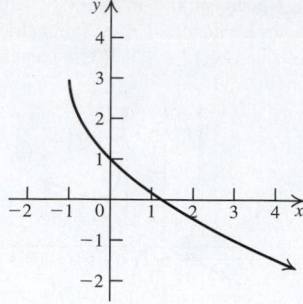

17. $-6, -3, -2, -\dfrac{3}{2}, -1, -\dfrac{1}{2}, \dfrac{1}{2}, 1, \dfrac{3}{2}, 2, 3, 6$

18. a.

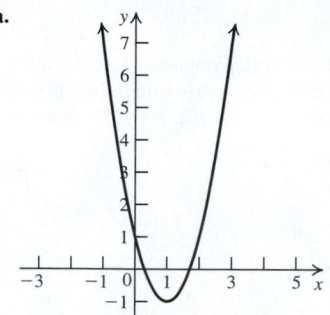

b. $\left(-\infty, 1 - \sqrt{\tfrac{1}{2}}\right] \cup \left[1 + \sqrt{\tfrac{1}{2}}, \infty\right)$

19. a.

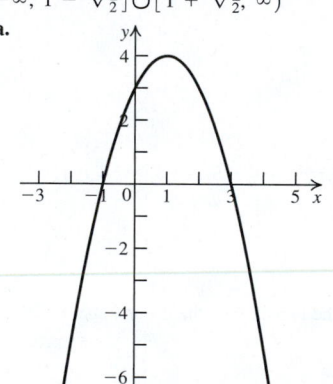

b. $[-1, 3]$

20. a.

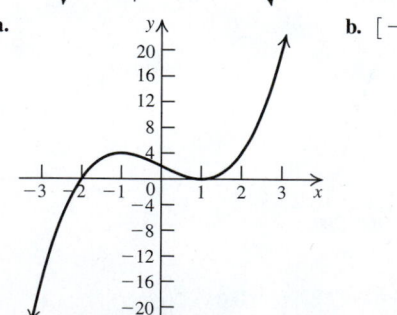

b. $[-2, \infty)$

21. a.

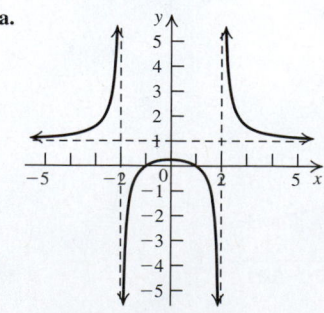

b. $(-\infty, -2) \cup [-1, 1] \cup (2, \infty)$

22. $-1, 2, 1 + i, 1 - i$ **23.** 9   **24.** 1250, \$28,250   **25.** $x = 50$

CHAPTER 3

Section 3.1

Concepts and Vocabulary:
1. $(-\infty, \infty), (0, \infty)$ **3.** *x*-axis   **5.** False   **7.** True

Building Skills:

9. No, the base is not a constant.   **11.** Yes, the base is $\frac{1}{2}$.

13. No, the base is not a constant.   **15.** No, the base is not a positive constant.

17. $\frac{1}{5}$ **19.** 0.0892, 11.2116   **21.** $\frac{4}{9}, \frac{729}{64}$

23.
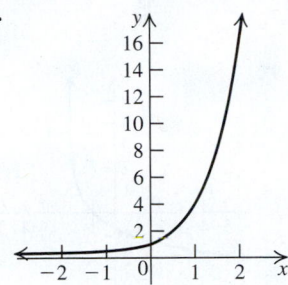

25.
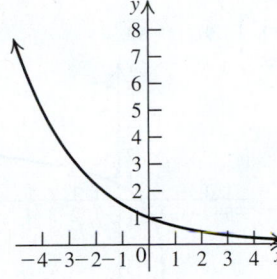

27.
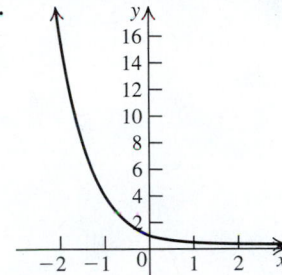

29.
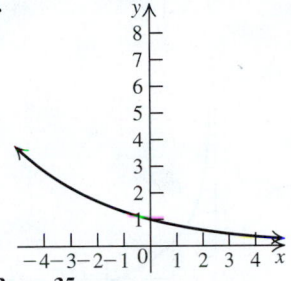

31. Yes, reflection about the *y*-axis
33. c **35.** a
37. Domain: $(-\infty, \infty)$; Range: $(0, \infty)$; Horizontal asymptote: $y = 0$
39. Domain: $(-\infty, \infty)$; Range: $(0, \infty)$; Horizontal asymptote: $y = 0$

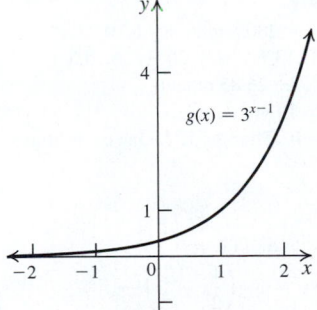

41. Domain: $(-\infty, \infty)$; Range: $(-\infty, 4)$; Horizontal asymptote: $y = 4$
43. Domain: $(-\infty, \infty)$; Range: $(-\infty, 3)$; Horizontal asymptote: $y = 3$

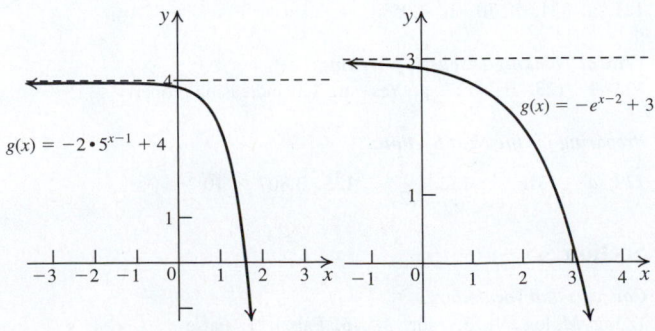

45. Domain: $(-\infty, \infty)$; Range: $[1, \infty)$. No horizontal asymptote.

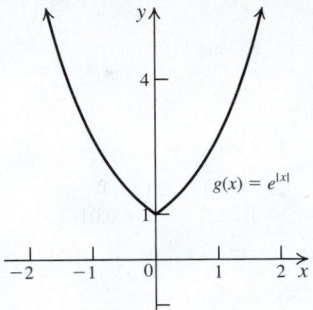

47. 4^x **49.** $3 \cdot 2^x$ **51.** $\frac{1}{5} \cdot 5^x$ **53.** $\frac{1}{5} \cdot 25^x$ **55.** $f(x) = (1.5)^x + 2$, $f(2) = 4.25$ **57.** $f(x) = \left(\frac{1}{3}\right)^x - 2, f(2) = -\frac{17}{9}$ **59.** $y = 2^{x+2} + 5$

61. $y = 2\left(\frac{1}{2}\right)^x - 5$ **63.** \$2500   **65.** \$5764.69   **67. a.** \$7936.21

b. \$4436.21   **69. a.** \$12,365.41   **b.** \$4865.41   **71.** \$4631.93
73. \$4493.68
75. $f(x) = e^{-x}; y = 0$ **77.** $f(x) = e^{x-2}; y = 0$

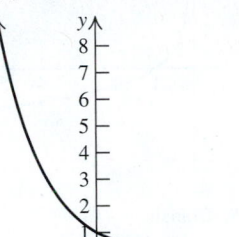

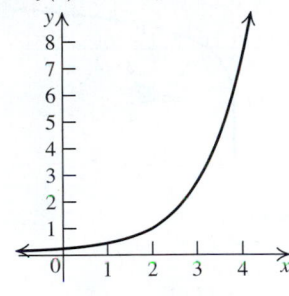

79. $f(x) = 1 + e^x; y = 1$ **81.** $f(x) = -e^{x-2} + 3; y = 3$

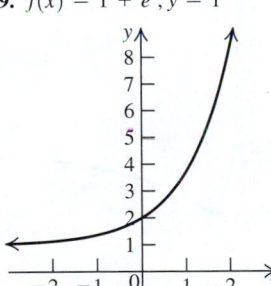

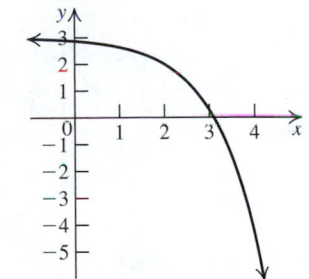

Applying the Concepts:
83. a. (i) 176.6°C; **(ii)** 148.1°C **b.** 5 hours **c.** 25°C **85.** \$220,262
87. \$35,496.43 **89.** 16.81% **91.** 0.1357 mm² **93.** 3.9%
95. a. $0.015 \times 2^{30} \approx 16,106,127$ cm
b. $0.015 \times 2^{40} \approx 16,492,674,420$ cm
c. $0.015 \times 2^{50} \approx 1.689 \times 10^{13}$ cm
99. $S_5 = 2.716667, S_{10} = 2.718282, S_{15} = 2.718282$
103. $y = e^x \rightarrow y = e^{x-1} \rightarrow y = e^{2x-1} \rightarrow y = 3e^{2x-1}$
105. $y = e^x \rightarrow y = e^{2+x} \rightarrow y = e^{2+3x} \rightarrow y = 5e^{2+3x} \rightarrow y = 5e^{2-3x} \rightarrow y = 5e^{2-3x} + 4$

Critical Thinking/Discussion/Writing

Preparing for the Next Section:
113. 1 **115.** -2 **117.** 49 **119.** $3\sqrt{2}$

121. $f^{-1}(x) = \dfrac{x-4}{3}$; Domain and range of f^{-1} is $(-\infty, \infty)$.

123. $f^{-1}(x) = x^2, x \geq 0$; Domain and range of f^{-1} is $[0, \infty)$.

125. $f^{-1}(x) = 1 + \dfrac{1}{x}$; Domain of $f^{-1} = (-\infty, 0) \cup (0, \infty)$; range of $f^{-1} = (-\infty, 1) \cup (1, \infty)$.

Section 3.2

Concept and Vocabulary:
1. $(0, \infty), (-\infty, \infty)$ **3.** common, natural **5.** False **7.** False

Building Skills:

9. $\log_5 25 = 2$ **11.** $\log_{1/16} 4 = -\dfrac{1}{2}$ **13.** $\log_{10} 1 = 0$

15. $\log_{10} 0.1 = -1$ **17.** $\log_a 5 = 2$ **19.** $\log_a (13/2) = 3$
21. $2^5 = 32$ **23.** $10^2 = 100$ **25.** $10^0 = 1$ **27.** $10^{-2} = 0.01$

29. $8^{1/3} = 2$ **31.** $e^x = 2$ **33.** 3 **35.** 4 **37.** -3 **39.** $\dfrac{3}{2}$ **41.** $\dfrac{1}{4}$

43. 0 **45.** 1 **47.** 7 **49.** 5 **51.** 4 **53.** 8 **55.** $(-1, \infty)$
57. $(1, \infty)$ **59.** $(2, \infty)$ **61.** $(1, 2)$ **63.** $(-\infty, 0) \cup (0, \infty)$
65. $(12, \infty)$ **67.** $(-\infty, -1) \cup (0, \infty)$ **69.** $(-\infty, -1) \cup (2, \infty)$
71. a. f **b.** a **c.** d **d.** b **e.** e **f.** c
73. Domain $= (-3, \infty)$
Range $= (-\infty, \infty)$
Asymptote: $x = -3$

75. Domain $= (0, \infty)$
Range $= (-\infty, \infty)$
Asymptote: $x = 0$

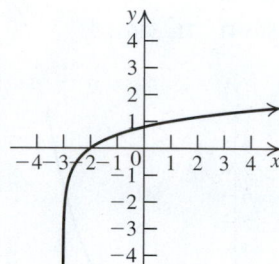
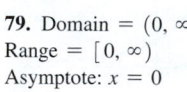

77. Domain $= (-\infty, 0)$
Range $= (-\infty, \infty)$
Asymptote: $x = 0$

79. Domain $= (0, \infty)$
Range $= [0, \infty)$
Asymptote: $x = 0$

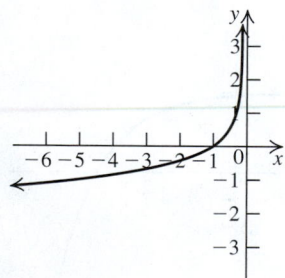
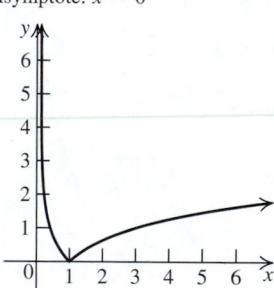

81. Domain $= (1, \infty)$
Range $= (-\infty, \infty)$
Asymptote: $x = 1$

83. Domain $= (-\infty, 3)$
Range $= (-\infty, \infty)$
Asymptote: $x = 3$

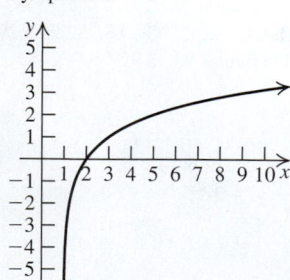
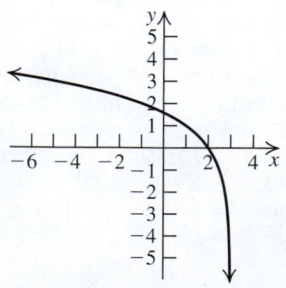

85. Domain $= (-\infty, 3)$
Range $= (-\infty, \infty)$
Asymptote: $x = 3$

87. Domain $= (-\infty, 0) \cup (0, \infty)$
Range $= (-\infty, \infty)$
Asymptote: $x = 0$

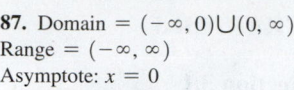
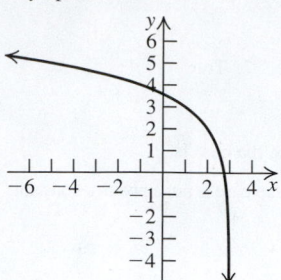
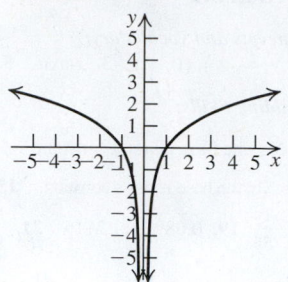

89. 1 **91.** 2 **93.** 4
95.   **97.**

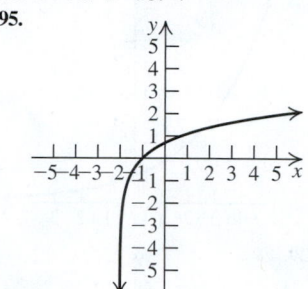
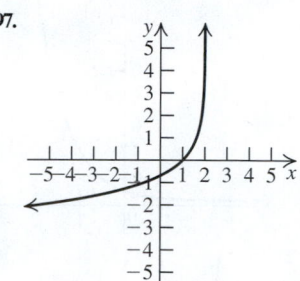

99.

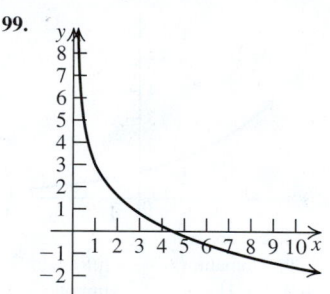

Applying the Concepts:
101. 8.66 yr **103.** 11.55% **105. a.** ≈ 280.8 million **b.** 0.93%
c. 2039 **107. a. (i)** 22.66°F. **(ii)** 20.13°F. **(iii)** 20°F. **b.** 5.5
min **109.** 17.6 min **111. a.** 0.27% **b.** 26.85 months
113. a. $P \approx 1500e^{0.231t}$ **b.** 7557 **c.** 10 yr
115. a. $\log(4/3) \approx 0.125 = 12.5\%$ **b.** The sum is 1. One of the digits
$1 \ldots 9$ will appear at the first place.

Beyond the Basics:
117. a. $(1, \infty)$ **b.** $(2, \infty)$ **c.** $(2, \infty)$ **d.** $(11, \infty)$

119. $y = \log x \rightarrow y = \log(x - 2) \rightarrow y = \log\left(\dfrac{1}{2}x - 2\right) \rightarrow$

$y = 3\log\left(\dfrac{1}{2}x - 2\right) \rightarrow y = -3\log\left(\dfrac{1}{2}x - 2\right) \rightarrow$

$y = -3\log\left(\dfrac{1}{2}x - 2\right) + 4$

121. a. \$24,659.70 **b.** 4.05%

Critical Thinking/Discussion/Writing:
123. 1 **125.** 16 **127. a.** Yes **b.** The increasing property

Preparing for the Next Section:
129. a^9 **131.** a^4 **133.** $\left(\dfrac{3}{2}\right)^4$ **135.** 3.807×10^{13}

Section 3.3

Concepts and Vocabulary:
1. $\log_a M, \log_a N$ **3.** $r \log_a M$ **5.** False **7.** False

Building Skills:
9. 0.78 **11.** 0.7

13. −1.7 **15.** 7.3 **17.** $\dfrac{16}{3}$ **19.** 0.56 **21.** $\ln(x) + \ln(x-1)$

23. $\ln(x+1) - 2\ln(x-2)$ **25.** $\dfrac{1}{2}\log(x^2+1) - \log(x+3)$

27. $\log_3(x-1) + \log_3(x+1) - \log_3(x-2) - \log_3(x+2)$

29. $2\log_b x + 3\log_b y + \log_b z$ **31.** $\ln x + \dfrac{1}{2}\ln(x-1) - \ln(x^2+2)$

33. $2\ln(x+1) - \ln(x-3) - \dfrac{1}{2}\ln(x+4)$

35. $\ln(x+1) + \dfrac{1}{2}\ln(x^2+2) - \dfrac{1}{2}\ln(x^2+5)$

37. $3\ln x + 4\ln(3x+1) - \dfrac{1}{2}\ln(x^2+1) + 5\ln(x+2) - 2\ln(x-3)$

39. $\log_2(7x)$ **41.** $\log\dfrac{3x+2}{x}$ **43.** $\ln y^2\sqrt{x}$ **45.** $\log\left(\dfrac{z\sqrt{x}}{y}\right)$

47. $\log_2(zy^2)^{1/5}$ **49.** $\ln(xy^2z^3)$ **51.** $\log\dfrac{x^3\sqrt{x^2+4}}{y}$ **53.** $\ln\left(\dfrac{x^2}{\sqrt{x^2+1}}\right)$

55. $1.4036 \cdot 10^{217}$ **57.** $9.3762 \cdot 10^{1897}$ **59.** $8.3803 \cdot 10^{2097}$ **61.** 234^{567}

63. 362 **65.** 2.322 **67.** −1.585 **69.** 1.760 **71.** 3.6 **73.** $\dfrac{1}{2}$ **75.** 1

77. 18 **79.** 2 **81.** $y = 2 - \log x$ **83.** $y = 2 - \ln x$
85. $y = 1 + 3\log_5 x$ **87.** $y = -1 + 5\log_2 x$ **89.** ≈ 10.7 yrs
91. ≈ 10.4 hrs

Applying the Concepts:
93. $k \approx 0.012775$ **95.** 0.010498 **97.** 12.53 yr **99.** 12.969 g
101. 12.60 yr **103.** $k = 0.1204$ **105.** 13.7 g **107.** $P = \$859.72$;
Interest $\approx \$86,332,14$ **109.** He can afford a mortgage of $110,545.60.

Beyond the Basics:
115.

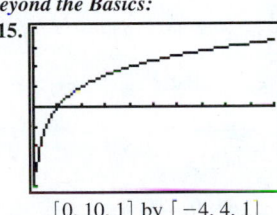

$[0, 10, 1]$ by $[-4, 4, 1]$

117. a. 0 **b.** 0 **c.** 2 **d.** 1 **119.** (8, 10) **121. a.** False **b.** True
c. False **d.** False **e.** True **f.** True **g.** False **h.** True **i.** False
j. True **125.** Because $\log\dfrac{1}{2}$ is a negative number, $3 < 4$ implies that
$3\log\dfrac{1}{2} > 4\log\dfrac{1}{2}$ **127.** 12,978,189 **129.** 11 **131.** 2
133. $t^2 - t + 1 = 0$ **135.** $t^2 + 4 = 0$ **137.** −2 **139.** −4, 1
141. $(-\infty, 7)$ **143.** $(2, \infty)$

Section 3.4

Concepts and Vocabulary:
1. exponential **3.** logistic **5.** True **7.** True

Building Skills:
9. $x = 4$ **11.** $x = 5/3$ **13.** $x = \pm\dfrac{7}{2}$ **15.** $\varnothing$ **17.** $x = 1$ **19.** $x = \dfrac{1}{2}$

21. $x = \pm9$ **23.** $x = 10,000$ **25.** $x = \dfrac{\ln 3}{\ln 2} \approx 1.585$

27. $x = \dfrac{(\ln 15 - 3\ln 2)}{2\ln 2} \approx 0.453$ **29.** $x = -1 + \ln 3 \approx 0.099$

31. $x = \dfrac{\ln 17 - \ln 5}{\ln 2} \approx 1.766$

33. $x = \dfrac{\ln 10 - \ln 3 + \ln 4}{2\ln 4} \approx 0.934$ **35.** $x = 2 + \ln 2 \approx 2.693$

37. $x = \dfrac{\ln 5}{\ln 5 + \ln 2} \approx 0.699$ **39.** $x = \dfrac{\ln 2 - 6\ln 3}{4\ln 3 + \ln 2} \approx -1.159$

41. $x = \dfrac{\ln 2 - \ln 3 - \ln 5}{\ln 5 - \ln 3} \approx -3.944$ **43.** $t = \dfrac{\ln 2}{\ln(1.065)} \approx 11.007$

45. $x = \dfrac{\ln 7}{\ln 2} \approx 2.807$ **47.** $x = \dfrac{\ln 2}{\ln 3}, \dfrac{\ln 4}{\ln 3} \approx 0.631, 1.262$

49. $x = \dfrac{\ln 4}{\ln 3} \approx 1.262$ **51.** $x = \ln 3 \approx 1.099$

53. $x = \dfrac{\ln 5 - \ln 3}{2\ln 3} \approx 0.232$ **55.** $x = \dfrac{\ln 2}{\ln 3} \approx 0.631$

57. $x = \dfrac{\ln 18 - \ln 7}{\ln 3} \approx 0.860$ **59.** $\varnothing$ **61.** $x = -\dfrac{49}{20}$ **63.** $x = 3, -2$

65. $x = 5, 2$ **67.** $x = 8$ **69.** $x = 1$ **71.** $x = 3$ **73.** $x = 4$

75. $\varnothing$ **77.** $x = \dfrac{2}{3}, \dfrac{5}{2}$ **79.** $a = 30, k = 2; 500$ **81.** $a = 5, k = \dfrac{1}{2}; 31$

83. $a = 2, k = \ln\left(\dfrac{17}{2}\right); 0.068$ **85.** $a = -2, k = \ln\left(\dfrac{11}{18}\right); -7.902$

87. $x \geq \dfrac{\ln 2}{\ln 0.3}$ **89.** $x \leq 0$ **91.** $(-3, 17)$ **93.** $x \geq e + 5$ **95.** $\left(\dfrac{7}{3}, 5\right)$

97. $x > \dfrac{1}{2}\ln 2$ **99.** $x \geq -\ln 2$

Applying the Concepts:
101. a. 10.087 yr **b.** 9.870 yr **c.** 9.821 yr **d.** 9.797 yr **e.** 9.796 yr
103. $r \approx 8.66\%$; $30.837.52 **105. a.** 2.134 lm **b.** 19.08 ft
107. a. 18,542 **b.** 5 wk **109. a.** 20 **b.** 4490
c.

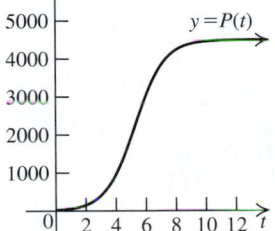

111. $h \geq 2.9$ miles

Beyond the Basics:
113. $t = \dfrac{1}{k}\ln\left(\dfrac{P}{M - P}\right)$ **117.** $x = 1, 10$ **119.** $\dfrac{1}{9}, 3$ **121.** $x = 2$

123. $\varnothing$ **125.** $f^{-1}(x) = \log_3(x - 5)$ **127.** $f^{-1}(x) = \log_4\dfrac{x - 7}{3}$

129. $f^{-1}(x) = 2^{x-1} + 1$ **131.** $f^{-1}(x) = \dfrac{1 + e^{2x}}{1 - e^{2x}}$ **133.** 130

135. 209 **137.** 380 **139.** 1

Critical Thinking/Discussion/Writing:
141. $t = \dfrac{\ln a}{k}$

Preparing for the Next Section:
143. $\log_9 81 = 2$ **145.** $\log 3 = x$ **147.** $64 = 2^6$ **149.** $A = 2 \cdot 10^3$

151. $x = \dfrac{1}{125}$ **153.** $x = 10$ **155.** $x = 4$ **157.** $x = \dfrac{\ln 3 + 3\ln 5}{2\ln 5 - \ln 3}$

Section 3.5

Concepts and Vocabulary:
1. $\log\left(\dfrac{I}{I_0}\right)$, $4.4 + 1.5 M$ **3.** $1200\log_2\left(\dfrac{f}{f_0}\right)$ **5.** False **7.** True

Building Skills:
9. pH $= 8$, base **11.** pH ≈ 4.64, acid
13. $[\text{H}^+] = 10^{-6}, [\text{OH}^-] = 10^{-8}$

15. $[H^+] = 3.16 \cdot 10^{-10}$ $[OH^-] = 3.16 \cdot 10^{-5}$ **17. a.** $10^5 I_0$;
b. $7.94 \cdot 10^{11}$ **19. a.** $6.3 \cdot 10^7 I_0$; **b.** $1.26 \cdot 10^{16}$ **21. a.** 6; **b.** $10^6 I_0$
23. a. 5.1; **b.** $1.26 \cdot 10^5 I_0$ **25.** 40 **27.** 55.4 **29.** 10^{-4}
31. $2.95 \cdot 10^{-6}$ **33.** A#466 Hz, C523 Hz **35.** ≈ 18 cents; noticeable
37. A is $\approx 2.5 \times 10^6$ times brighter than B.
39. B is ≈ 6.31 times brighter than A.

Applying the Concepts:
41. 7.4; basic **43. a.** $7.1 \cdot 10^{-4}$ **b.** $6.3 \cdot 10^{-8}$ **c.** $1.7 \cdot 10^{-8}$ **d.** 10^{-3}
45. $1.6 \cdot 10^{-4}$; about 160 times more **47.** $pH_A = pH_B - 2$
49. pH decreased by $\log 50 \approx 1.7$; more acidic **51. a.** $6.3 \cdot 10^7 I_0$
b. $1.259 \cdot 10^{16}$ **53. a.** $I_A = 10 I_B$ **b.** $E_A = 31.6 E_B$
55. $\log 150 \approx 2.18$ **57.** ≈ 77 dB **59.** $\approx 3.16 \cdot 10^6$ times as intense
61. 160 dB **63.** $I_2 \approx 1.26 I_1$ **65.** ≈ 4 cents more **67.** $\approx 3.98 \cdot 10^5$
69. $\approx 1.58 \cdot 10^9$ **71.** ≈ -5.4 **73.** $y = 25.72 e^{0.511x}$
75. $y = 66.654 - 10.367 \ln x$

Beyond the Basics:
77. a. 26 dB; 18 dB; 12 dB; 6 dB **b.** 200 ft

c. $a = 10(4 + \log 4) \approx 46$; $b = -20$ **d.** $r = \dfrac{200}{10^{L/20}}$

79. 5×10^{15} vs 2.5×10^{16}; earthquake is 5 times more powerful
81. $[875.885]$ Hz **83.** -2.2

Preparing for the Next Section:

85. $d = 45$ **87.** $x = \dfrac{\pi}{6}$ **89.** Quadrant II **91.** Quadrant I

Review Exercises

Basic Concepts and Skills:
1. False **3.** False **5.** True **7.** False **9.** True **11.** h
13. f **15.** d **17.** a
19. Domain $= (-\infty, \infty)$ **21.** Domain $= (-\infty, \infty)$
Range $= (0, \infty)$ Range $= (3, \infty)$
Asymptote: $y = 0$ Asymptote: $y = 3$

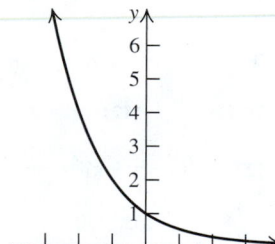

23. Domain $= (-\infty, \infty)$ **25.** Domain $= (-\infty, 0)$
Range $= (0, 1]$ Range $= (-\infty, \infty)$
Asymptote: $y = 0$ Asymptote: $x = 0$

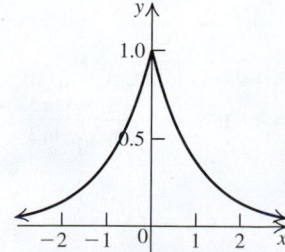

27. Domain $= (1, \infty)$ **29.** Domain $= (-\infty, 0)$
Range $= (-\infty, \infty)$ Range $= (-\infty, \infty)$
Asymptote: $x = 1$ Asymptote: $x = 0$

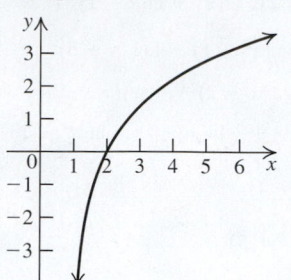

31. a. x-intercept $= \ln(2/3)$,   **33. a.** x-intercept: none,
y-intercept $= 1$ y-intercept $= 1$
b. $\lim_{x \to \infty} f(x) = 3$ $\lim_{x \to -\infty} f(x) = -\infty$ **b.** as $x \to \infty, y \to 0$
  as $x \to -\infty, y \to 0$

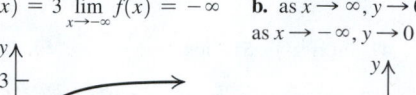
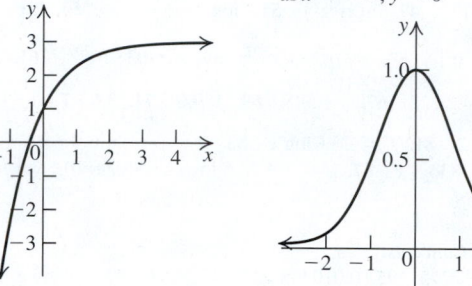

35. $a = 10, k = 2, f(2) = 160$
37. $a = 3, k = -0.2824, f(4) = 0.38898$ **39.** $y = 3 \cdot 2^x$
41. $y = \log_5 (x - 1)$ **43. a.** $y = -2^{(x-1)} - 3$ **b.** $y = -2^{(x-1)} + 3$
45. $\ln x + 2 \ln y + 3 \ln z$ **47.** $\ln x + \dfrac{1}{2} \ln(x^2 + 1) - 2 \ln(x^2 + 3)$
49. $y = 3x$ **51.** $y = -1 + \sqrt{x - 2}$ **53.** $y = \dfrac{\sqrt{x^2 - 1}}{x^2 + 1}$ **55.** 4
57. $-1 \pm \sqrt{5}$ **59.** $\dfrac{\ln 23}{\ln 3} \approx 2.854$ **61.** $\dfrac{\ln 19}{\ln 273} \approx 0.525$
63. 0 **65.** $\dfrac{-\ln 3}{\ln\left(\dfrac{(1.7)^3}{9}\right)} \approx 1.815$ **67.** $\dfrac{5}{2}$ **69.** 0 **71.** 8
73. 5 **75.** 9 **77.** 3 **79.** $x \ge -1$ **81.** $\left(-\dfrac{7}{2}, \dfrac{93}{2}\right)$

Applying the Concepts:
83. 11.4 yr **85.** 40 hr **87.** At 5% compounded yearly, $A = \$9849.70$.
At 4.75% compounded monthly, $A = \$9754.74$. The 5% investment will
provide the greater return. **89. a.** 34 million **b.** In 2206
(after 199.3 yr)
91. a.   **b.** ≈ 5 hr

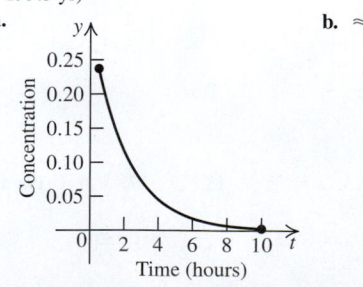

93. $t = 6.57$ min **95. a.** 5000 people/square mile
b. 7459 people/square mile **c.** ≈ 13.73 miles
97. $I = 0.2322 I_0$ **99. a.** 11.27 ft/sec
b. 5.29 ft/sec **c.** 201 **101. a.** 1 g **b.** Remains stationary at 6 g
c. 4.6 days **103.** 16 cents; Yes, it is noticeable.
105. $4.0 \cdot 10^{-3}$; 1585 times more **107.** 100 times more intense
109. ≈ 9.7

Practice Test A

1. $x = -3$ **2.** $x = 32$

3. Range $= (-\infty, 1)$; horizontal asymptote: $y = 1$ **4.** -3 **5.** $x = 3$

6. -3 **7.** $\ln 3x^5$ **8.** $\dfrac{\ln\left(\dfrac{5}{2}\right)}{\ln 2}$ **9.** $\ln 2$ **10.** $\ln 2 + 3\ln x - 5\ln(x+1)$

11. -5 **12.** $y = \ln(x-1) + 3$

13.

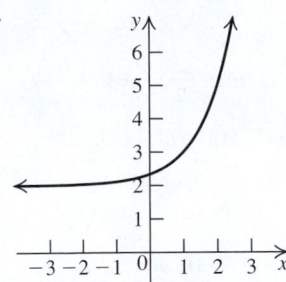

14. $(-\infty, 0)$ **15.** $\ln\dfrac{x^3(x^3 + 2)}{\sqrt{3x^2 + 2}}$ **16.** $x = 3$ **17.** $x = 3$

18. $A(t) = 15{,}000(1.0175)^{4t}$ **19.** In 1983 (after 23.03 yr)

20. The Nicaragua earthquake was about 2.5 times the intensity of the Morocco earthquake.

Practice Test B

1. b **2.** b **3.** b **4.** d **5.** d **6.** b **7.** b **8.** c **9.** d **10.** d

11. b **12.** d **13.** b **14.** a **15.** a **16.** d **17.** b **18.** d

19. b **20.** b

Cumulative Review Exercises (Chapters 1–3)

1. x-intercept: 0; y-intercepts: 0, 2; symmetric about y-axis

2.

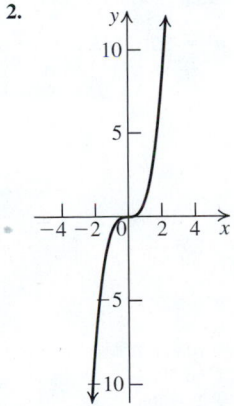

3. Center: $(-1, 2)$; radius: 5

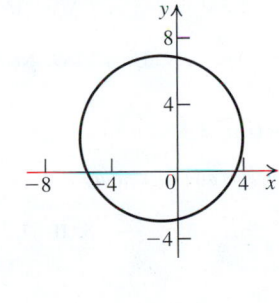

4. $y = \dfrac{3}{2}x + 7$ **5.** $(-\infty, -3)\cup(2, \infty)$ **6. a.** -7 **b.** -3 **c.** 19

7. Start with $y = \sqrt{x}$, shift one unit left, stretch vertically by a factor of 3, and shift two units down.

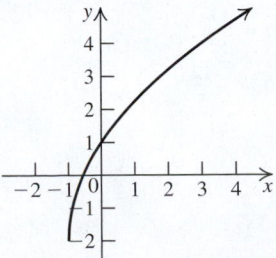

8. $(f \circ g)(x) = 3\sqrt{-\dfrac{1}{x}}$. Domain of $f \circ g$ is $(-\infty, 0)$.

9. a. Yes, $f^{-1}(x) = \left(\dfrac{x-1}{2}\right)^{1/3}$ **b.** No **c.** Yes, $f^{-1}(x) = e^x$

10. a. $Q = 2x, R = x + 1, D = x^2 + 1$; $2x + \dfrac{x+1}{x^2+1}$

b. $Q = 2x^3 + x^2 - 3x + 7, R = -18, D = x + 3$;

$2x^3 + x^2 - 3x + 7 - \dfrac{18}{x+3}$

11. a. $(x-1)(x-2)(x+3) = x^3 - 7x + 6$ **b.** $x^4 - 1$

12. a. $\log_2 5 + 3\log_2 x$ **b.** $\frac{1}{3}(\log_a x + 2\log_a y - \log_a z)$

c. $\ln 3 + \frac{1}{2}\ln x - \ln 5 - \ln y$ **13. a.** $\log\sqrt{xy}$ **b.** $\ln\dfrac{x^3}{y^2}$

14. a. 1.289 **b.** 2.930 **c.** 1.737 **15. a.** $\{3\}$ **b.** $\{1, 3\}$ **c.** 3.77

16. Possible rational zeros $\{\pm 1, \pm 2, \pm 4\}$. Rational zeros: $\{1, 2\}$; upper bound: 3; lower bound: -1 **17. a.** 1, 1, 1, -2, -2, 3

b. At $x = 1$ and $x = 3$, the graph crosses the x-axis. At $x = -2$, the graph touches but does not cross the x-axis.

c. $f(x) \to \infty$ as $x \to \infty$; $f(x) \to \infty$ as $x \to -\infty$

d.

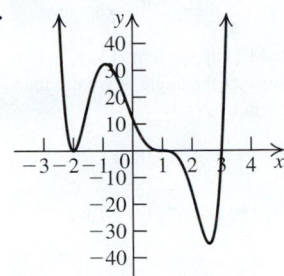

18. a. Vertical asymptotes: $x = 3, x = -4$ **b.** Horizontal asymptotes: $y = 1$ **c.** $f(x)$ above the line $y = 1$ on $(-\infty, -4)\cup(3, \infty)$ and below it on $(-4, 3)$.

d.

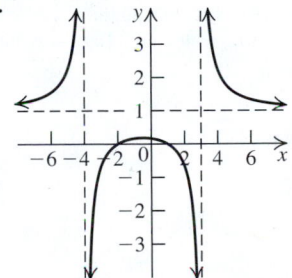

CHAPTER 4

Section 4.1

Concepts and Vocabulary

1. clockwise **3.** coterminal **5.** False **7.** True

Building Skills

9.　　　　　　　　　　**11.**

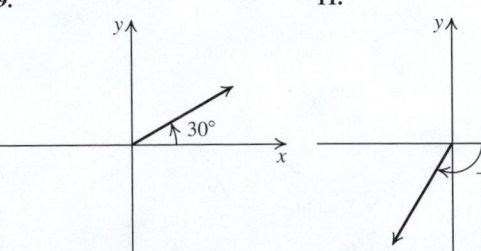

13. **15.**

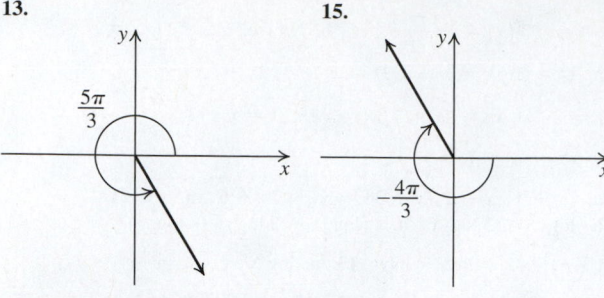

17. 70.75° **19.** 23.71° **21.** −15.72° **23.** 27° 19′ 12″ **25.** 13° 20′ 49″

27. 19° 3′ 4″ **29.** $\dfrac{\pi}{9}$ **31.** $-\pi$ **33.** $\dfrac{7\pi}{4}$ **35.** $-\dfrac{17\pi}{6}$ **37.** 15°

39. −100° **41.** 300° **43.** −495° **45.** 0.21 radian **47.** −1.47 radians

49. 53.86° **51.** −470.40° **53.** $\dfrac{7\pi}{4}$ **55.** $\dfrac{\pi}{4}$ **57.** $\dfrac{4\pi}{3}$ **59.** 295°

61. 160° **63.** 340° **65.** Complement: 43°; supplement: 133°
67. Complement: none because the measure of the angle is greater than
90°; Supplement: 60° **71.** 7/25 ≈ 0.28 radian

73. $\dfrac{44}{21}$ ≈ 2.095 radians **75.** $\dfrac{5\pi}{12}$ ≈ 1.309 m **77.** 78 m **79.** 6 cm

81. 9 m **83.** 200 in.² **85.** $\sqrt{\dfrac{120}{\pi}}$ ≈ 6.180 ft **87.** 60 m/min

89. 2 rad/sec

Applying the Concepts

91. 70.69 in. **93.** $\dfrac{2\pi}{3}$ radians **95.** 18 in. **97.** 86°

99. a. 120 rad/min **b.** 300 rad/min **101.** 132,000 rad/hr
103. 84,823 in./min **105.** ≈ 508 mi **107.** ≈ 462 mi **109.** ≈ 6.366°

Beyond the Basics

111. ≈ 7 m **113.** 250 **115.** $\dfrac{99}{112}$ ≈ 0.884 ft²

Critical Thinking/Discussion/Writing

117. $3, \pi, 4, \dfrac{3\pi}{2}$

Getting Ready for the Next Section

121. $c = 13$ **123.** $a = 8$ **125.** $a = 3\sqrt{3}$ **127.** $\dfrac{\sqrt{3}}{3}$

Section 4.2

Basic Concepts and Vocabulary:
1. 1, $x^2 + y^2 = 1$ **3.** origin **5.** True **7.** False **9.** Yes **11.** Yes
13. No **15.** $\pm\dfrac{\sqrt{3}}{2}$ **17.** $\pm\dfrac{\sqrt{7}}{4}$ **19.** None

21. $\sin t = \dfrac{1}{3}$, $\cos t = \dfrac{2\sqrt{2}}{3}$, $\tan t = \dfrac{\sqrt{2}}{4}$

23. $\sin t = \dfrac{2\sqrt{2}}{3}$, $\cos t = -\dfrac{1}{3}$, $\tan t = -2\sqrt{2}$

25. $\sin t = -\dfrac{2\sqrt{6}}{5}$, $\cos t = \dfrac{1}{5}$, $\tan t = -2\sqrt{6}$,

27. $\csc t = \dfrac{4\sqrt{7}}{7}$, $\sec t = \dfrac{4}{3}$, $\cot t = \dfrac{3\sqrt{7}}{7}$

29. $\csc t = \dfrac{\sqrt{6}}{2}$, $\sec t = -\sqrt{3}$, $\cot t = -\dfrac{\sqrt{2}}{2}$

31. $\csc t = -5$, $\sec t = -\dfrac{5\sqrt{6}}{12}$, $\cot t = 2\sqrt{6}$

33. $\sin t = \tan t = 0$; $\cos t = \sec t = -1$; $\cot t, \csc t$ *undef.*

35. $\sin t = \csc t = 1$; $\cos t = \cot t = 0$; $\tan t, \sec t$ *undef.*
37. $\sin t = \csc t = -1$; $\cos t = \cot t = 0$; $\tan t, \sec t$ *undef.*
39. 0 **41.** 0 **43.** 0 **45.** 0 **47.** 0 **49.** undef. **51.** 0 **53.** 1 **55.** 1

57. $1 + \dfrac{\sqrt{2}}{2}$ **59.** $1 - \dfrac{\sqrt{3}}{3}$ **61.** −1 **63.** $-\dfrac{1}{2}$ **65.** $-\sqrt{3}$ **67.** −2

69. −2 **71.** $\dfrac{1}{2}$ **73.** $-\dfrac{\sqrt{2}}{2}$ **75.** $\dfrac{\sqrt{3}}{3}$ **77.** $\dfrac{\sqrt{3}}{3}$ **79.** $\dfrac{\sqrt{2}}{2}$

81. $-\dfrac{1}{2}$ **83.** $\sqrt{3}$ **85.** $\dfrac{2\sqrt{3}}{3}$ **87.** $-\dfrac{1}{2}$ **89.** $\sqrt{3}$ **91.** $\dfrac{2\sqrt{3}}{3}$

93. $\dfrac{\sqrt{3}}{2}$ **95.** $-\dfrac{\sqrt{3}}{2}$ **97.** $\dfrac{1}{2}$ **99.** −1 **101.** $\dfrac{\sqrt{2}}{2}$ **103.** $-\dfrac{\sqrt{2}}{2}$

105. $\dfrac{1}{2}$ **107.** 1 **109.** $\dfrac{\sqrt{3}}{3}$ **111.** 0.97 **113.** 0.62 **115.** 1.21

117. 0.81 **119.** 0.84 **121.** 1.15 **123.** −0.66 **125.** 1.00

Applying the Concepts:
127. a. 100 **b.** 75 **129. a.** 13ft **b.** 7ft **131. a.** 12.2
b. 16.5 **c.** 7.9

Beyond the Basics:

133. $\pi + 2\pi k, k = 0, \pm 1, \dots$ **135.** $-\dfrac{\pi}{4} + 2\pi k, k = 0, \pm 1, \dots$

137. $\dfrac{5\pi}{6} + 2\pi k, k = 0, \pm 1, \dots$ **139.** $\dfrac{\pi}{6}, \dfrac{5\pi}{6}$ **141.** $\dfrac{\pi}{4}, \dfrac{5\pi}{4}$

143. $\dfrac{5\pi}{4}, \dfrac{7\pi}{4}$ **145.** $-\dfrac{\sqrt{3}}{2}$ **147.** $\dfrac{\sqrt{3}}{2}, \dfrac{1}{2}$

149. $3000\pi + \dfrac{\pi}{2}(2n + 1), n = 0, 1, \dots 9$

Critical Thinking / Discussion / Writing:

151. $\cos t = -\cos s$, $\sin t = \sin s$ **153.** True **155.** $\pm 4; \pm\dfrac{4}{5}, \pm\dfrac{3}{4}$

157. $\pm 3\sqrt{3}; \pm\dfrac{\sqrt{3}}{3}; \pm\sqrt{3}$ **159.** 70° **161.** $\theta' = 50°$ **163.** $\theta' = \dfrac{\pi}{6}$

165. Odd **167.** Even **169.** Neither

Section 4.3

Concepts and Vocabulary:

1. $\dfrac{y}{r}; \dfrac{x}{r}; \dfrac{y}{x}$ **3.** $\dfrac{\pi}{6}; \dfrac{5\pi}{6}$ **5.** II **7.** True

9. $\sin\theta = \dfrac{3}{5}$ **11.** $\sin\theta = -\dfrac{1}{2}$

$\cos\theta = -\dfrac{4}{5}$ $\cos\theta = -\dfrac{\sqrt{3}}{2}$

$\tan\theta = -\dfrac{3}{4}$ $\tan\theta = \dfrac{\sqrt{3}}{3}$

13. $\sin\theta = \dfrac{\sqrt{2}}{2}$ **15.** $\csc\theta = -\dfrac{13}{5}$

$\cos\theta = \dfrac{\sqrt{2}}{2}$ $\sec\theta = \dfrac{13}{12}$

$\tan\theta = 1$ $\cot\theta = -\dfrac{12}{5}$

17. $\csc\theta = \dfrac{25}{24}$, $\sec\theta = \dfrac{25}{7}$, $\cot\theta = \dfrac{24}{7}$

19. $\csc\theta = \dfrac{2\sqrt{3}}{3}$, $\sec\theta = -2$, $\cot\theta = -\dfrac{7}{24}$

21. Quadrant III **23.** Quadrant II **25.** Quadrant IV **27.** Quadrant II

29. $\tan\theta = \dfrac{12}{5}$ **31.** $\cos\theta = -\dfrac{3}{5}$ **33.** $\sec\theta = -\dfrac{5}{4}$ **35.** $\cot\theta = -\dfrac{\sqrt{2}}{4}$

37. 60° **39.** 50° **41.** 60° **43.** $\dfrac{\pi}{4}$ **45.** $\pi/4$ **47.** $\dfrac{\pi}{6}$ **49.** $-\dfrac{1}{2}$

51. $-\dfrac{\sqrt{3}}{3}$ **53.** $-\dfrac{2\sqrt{3}}{3}$ **55.** $-\dfrac{1}{2}$ **57.** $\sqrt{2}$ **59.** $-\dfrac{\sqrt{3}}{2}$ **61.** $\dfrac{\sqrt{2}}{4}$

63. $-\dfrac{\sqrt{5}}{3}$ **65.** $-\dfrac{2\sqrt{5}}{15}$ **67.** 1 **69.** 1 **71.** 1 **73.** 4 **75.** 0

77. 1 **79.** $-\dfrac{\sqrt{21}}{5}$ **81.** $-2\sqrt{6}$ **83.** $\sqrt{10}$

85. $\cos 15° = \dfrac{\sqrt{2+\sqrt{3}}}{2}$, $\tan 15° = 2 - \sqrt{3}$ **87.** $\dfrac{1}{2}$ **89.** $-\sqrt{3}$

91. $-\dfrac{\sqrt{3}}{2}$ **93.** $-\dfrac{\sqrt{3}}{2}$ **95.** $\dfrac{\sqrt{3}}{3}$ **97.** $-\dfrac{2\sqrt{3}}{3}$

99. $\theta = 60°$ or $\theta = 300°$ **101. a.** -166 volts **b.** 166 volts

103. a. $L + r$ cm **b.** $\sqrt{L^2 - r^2}$ cm **c.** $L - r$ cm

105. a. 25 ft **b.** 173 ft

Beyond the Basics:

107. $\dfrac{(v_0 \sin \theta)^2}{64}$ **109.** $h = (\tan \theta)d - (16 \sec^2 \theta)\dfrac{d^2}{v_0^2}$ **113.** $\dfrac{1}{2}$

115. 0 **117.** $-\dfrac{\sqrt{3}}{2}$ **119.** Odd **121.** Even **123.** Odd **125.** Even

Critical Thinking / Discussion / Writing:

129. a. $e^{\frac{\pi}{2}}$ **b.** $\ln \pi$ **133.** False; $1 \neq \sin 1°$ **135.** $\dfrac{4}{3}$ **137.** $-5, 2$

139. 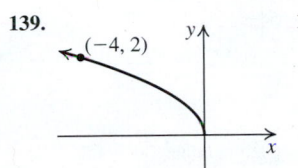 **141.**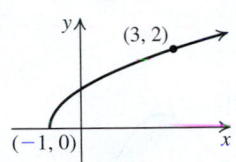

143. 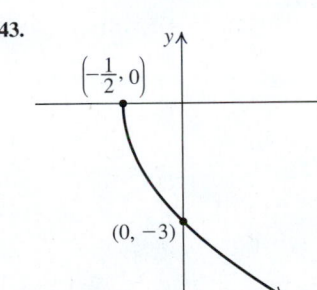 **145.**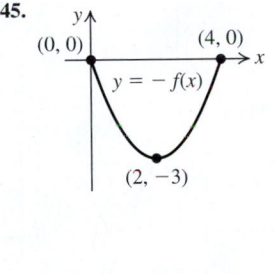

147. 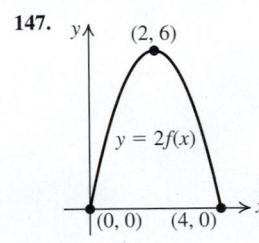 **149.**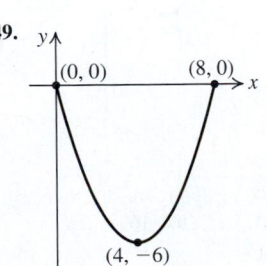

Section 4.4

Concepts and Vocabulary:

1. π **3.** $[-1, 1]$ **5.** False **7.** True

9. 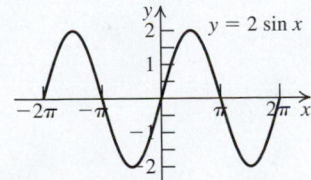 **11.**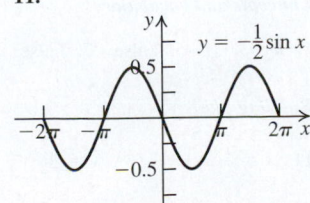

13. 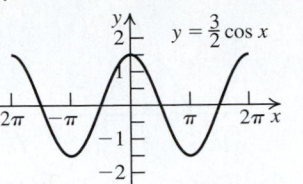 **15.**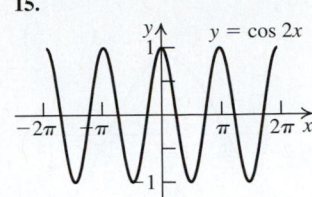

17.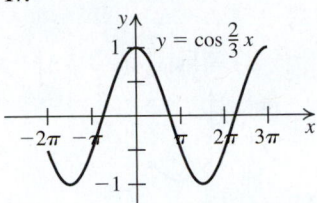

19.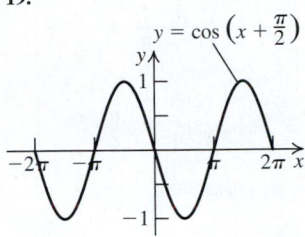

21. 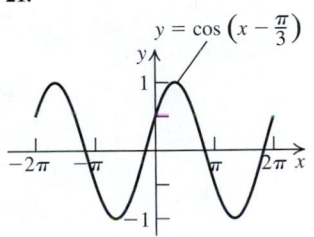 **23.**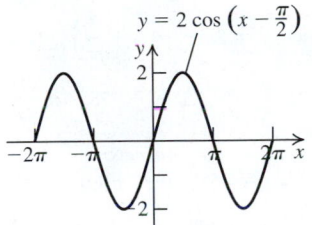

25. 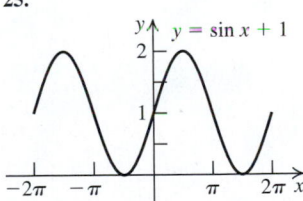 **27.**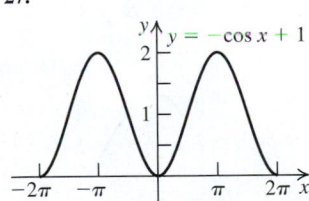

29. Amplitude $= 5$; period $= 2\pi$; phase shift $= \pi$

31. Amplitude $= 7$; period $= \dfrac{2\pi}{9}$; phase shift $= -\dfrac{\pi}{6}$

33. Amplitude $= 6$; period $= 4\pi$; phase shift $= -2$

35. Amplitude $= 0.9$; period $= 8\pi$; phase shift $= \dfrac{\pi}{4}$

37. $y = \dfrac{1}{4} \sin \left[16\left(x - \dfrac{\pi}{16} \right) \right]$ **39.** $y = 6 \sin \left[\dfrac{2}{3}\left(x + \dfrac{\pi}{4} \right) \right]$

41. $y = 2.4 \sin \left[\dfrac{\pi}{3}(x - 3) \right]$ **43.** $y = \dfrac{\pi}{2} \sin \left[\dfrac{16\pi}{5}\left(x + \dfrac{1}{4} \right) \right]$

45. $y = 3 \cos 2x$ **47.** $y = 2\sin 4x$

49. $y = 3 \sin \left[2\left(x - \dfrac{\pi}{4} \right) \right]$ **51.** $y = -2 \sin \left[2\left(x + \dfrac{\pi}{8} \right) \right]$

53.

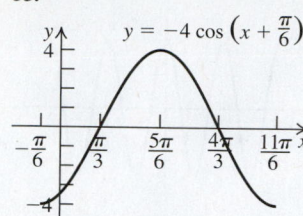

$y = -4\cos\left(x + \frac{\pi}{6}\right)$

55.

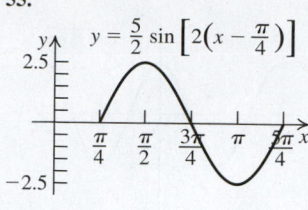

$y = \frac{5}{2}\sin\left[2\left(x - \frac{\pi}{4}\right)\right]$

b.

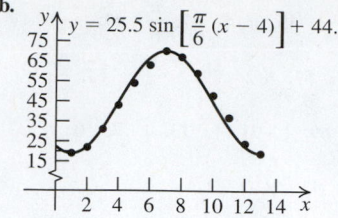

$y = 25.5\sin\left[\frac{\pi}{6}(x - 4)\right] + 44.5$

c. January $= f(1) = 19$;
April $= f(4) = 44.5$;
July $= f(7) = 70$;
October $= f(10) = 44.5$.
The computed values are
very close to the measured
values.

57.

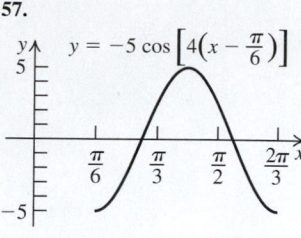

$y = -5\cos\left[4\left(x - \frac{\pi}{6}\right)\right]$

59.

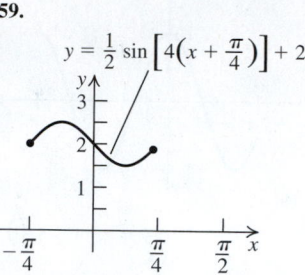

$y = \frac{1}{2}\sin\left[4\left(x + \frac{\pi}{4}\right)\right] + 2$

Beyond the Basics:

89.

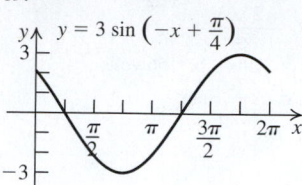

$y = 3\sin\left(-x + \frac{\pi}{4}\right)$

91.

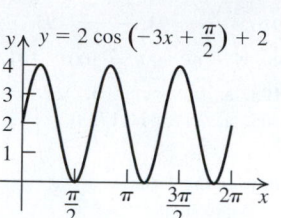

$y = 2\cos\left(-3x + \frac{\pi}{2}\right) + 2$

93.

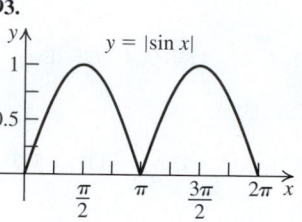

$y = |\sin x|$

95.

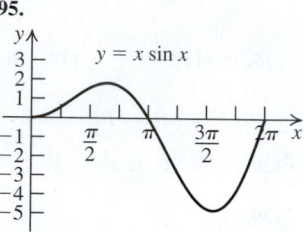

$y = x\sin x$

61. $y = 4\cos\left[2\left(x + \frac{\pi}{6}\right)\right]$; period $= \pi$; phase shift $= -\frac{\pi}{6}$

63. $y = -\frac{3}{2}\sin\left[2\left(x - \frac{\pi}{2}\right)\right]$; period $= \pi$; phase shift $= \frac{\pi}{2}$

65. $y = 3\cos\pi x$; period $= 2$; phase shift $= 0$

67. $y = \frac{1}{2}\cos\left[\frac{\pi}{4}(x + 1)\right]$; period $= 8$; phase shift $= -1$

69. $y = 2\sin\left[\pi\left(x + \frac{3}{\pi}\right)\right]$; period $= 2$; phase shift $= -\frac{3}{\pi}$

71. $\frac{\pi}{2}, \frac{2}{\pi}$ **73.** $6, \frac{1}{6}$ **75.** $4\pi, \frac{1}{4\pi}$

97.

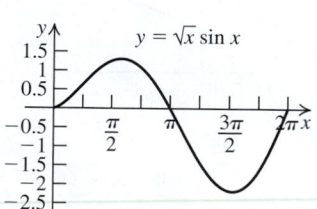

$y = \sqrt{x}\sin x$

Applying the Concepts:

77. $t = 7558$; $P \approx -0.63$; $E \approx -0.43$; $I \approx 0.19$

79. a. Period $= \frac{1}{70}$. The pulse is the frequency of the function; it says
how many times the heart beats in one minute.

b.

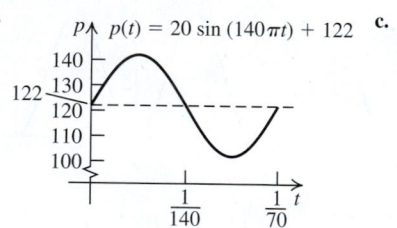

$p(t) = 20\sin(140\pi t) + 122$ **c.** $\frac{142}{102}$

81. a. Amplitude $= 156$; period $= \frac{1}{60}$ **b.** Frequency $= 60$

c.

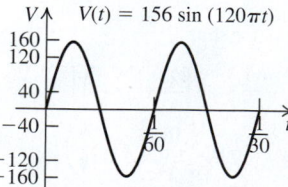

$V(t) = 156\sin(120\pi t)$

99.

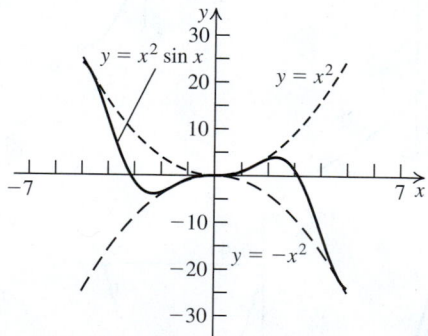

$y = x^2\sin x$, $y = x^2$, $y = -x^2$

101. a. $100°$, July 21 **b.** $50°$, January 19 **c.** $97.6°$
103. $[-11, 1]$ **107.** 10

Getting Ready for the Next Section:
109. $y = 5x - 3$ **111.** $-2, -4$, no vertical asymptotes **113.** Odd
115. Neither **117.** Odd **119.** Odd

83. a. 800 kangaroos **b.** 500 kangaroos **c.** $\frac{\pi}{2}$ yrs ≈ 1.57 yr

85. $y = 4.35\sin\left[\frac{\pi}{6}(x - 3)\right] + 12.25$

87. a. $y = 25.5\sin\left[\frac{\pi}{6}(x - 4)\right] + 44.5$

Section 4.5

Concepts and Vocabulary:

1. π **3.** $\frac{\pi}{2}$ **5.** False **7.** False

Building Skills:
9. $y = x + 5$
11. $y = -\sqrt{3}x - (3\sqrt{3} + 2)$

13.

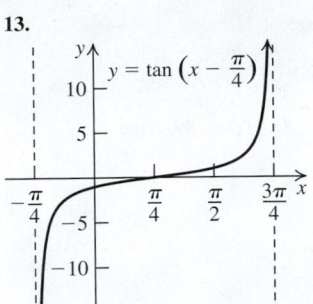

$y = \tan\left(x - \frac{\pi}{4}\right)$

15.

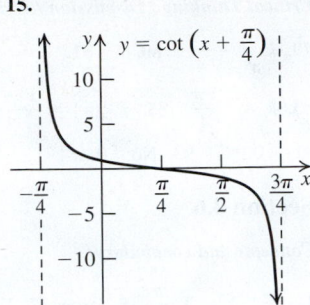

$y = \cot\left(x + \frac{\pi}{4}\right)$

33.

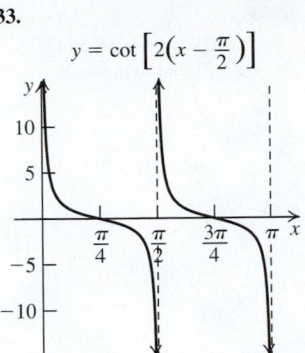

$y = \cot\left[2\left(x - \frac{\pi}{2}\right)\right]$

35.

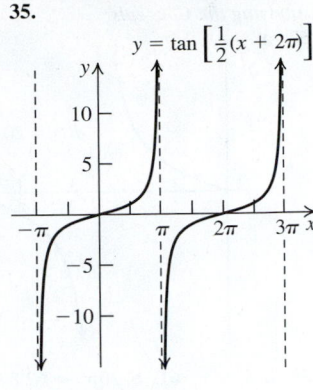

$y = \tan\left[\frac{1}{2}(x + 2\pi)\right]$

17.

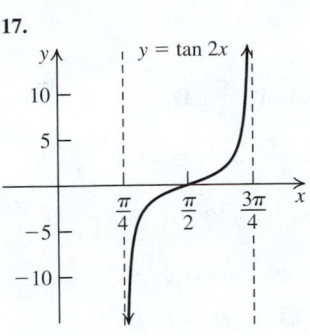

$y = \tan 2x$

19.

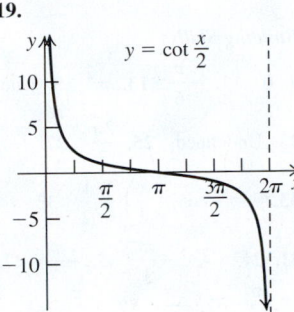

$y = \cot\frac{x}{2}$

37.

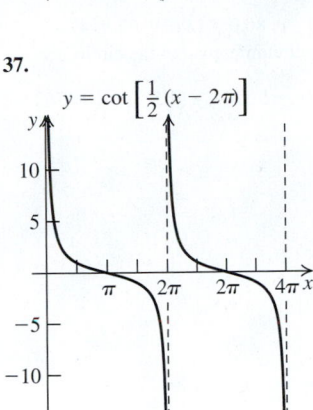

$y = \cot\left[\frac{1}{2}(x - 2\pi)\right]$

39.

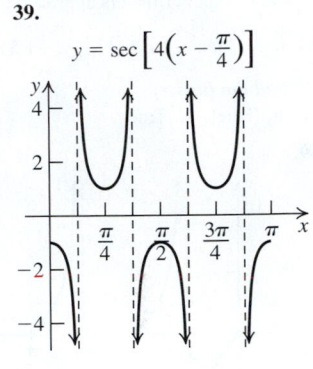

$y = \sec\left[4\left(x - \frac{\pi}{4}\right)\right]$

21.

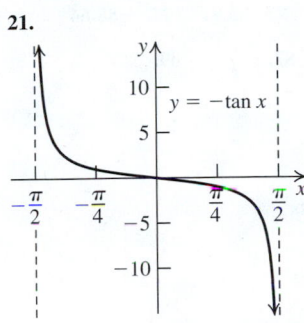

$y = -\tan x$

23.

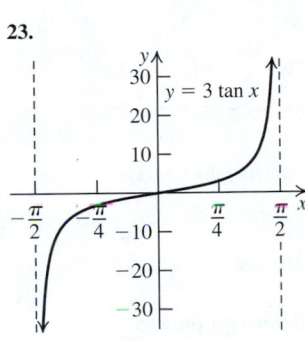

$y = 3\tan x$

41.

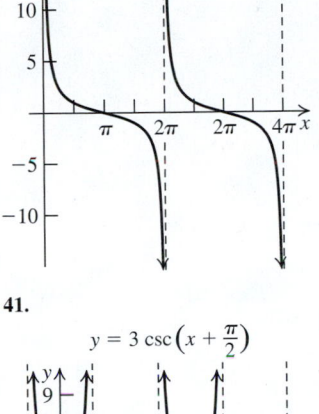

$y = 3\csc\left(x + \frac{\pi}{2}\right)$

43.

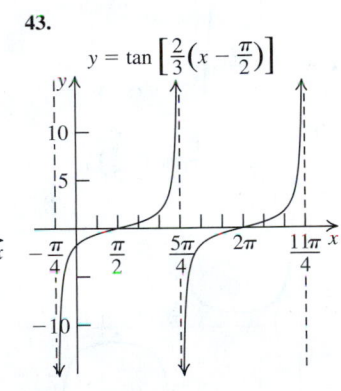

$y = \tan\left[\frac{2}{3}\left(x - \frac{\pi}{2}\right)\right]$

25.

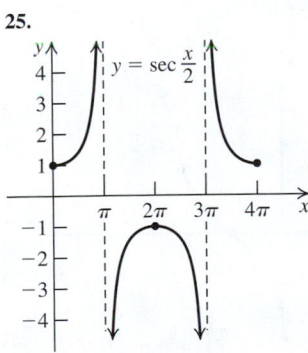

$y = \sec\frac{x}{2}$

27.

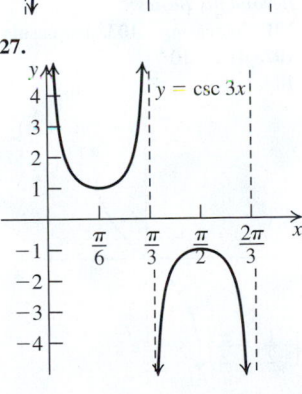

$y = \csc 3x$

45.

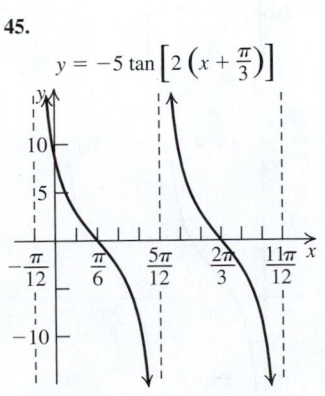

$y = -5\tan\left[2\left(x + \frac{\pi}{3}\right)\right]$

47.

$y = \frac{1}{3}\cot[2(x - \pi)]$

29.

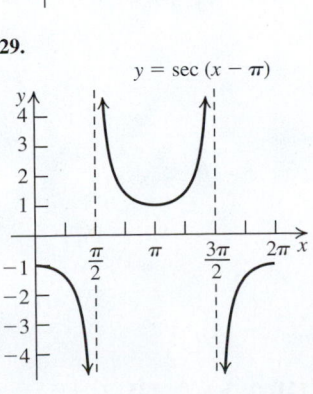

$y = \sec(x - \pi)$

31.

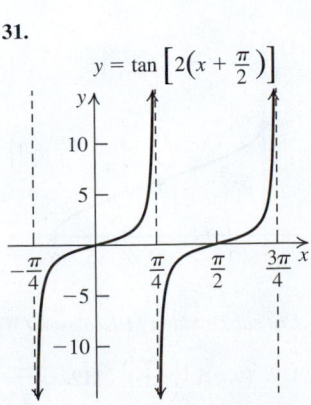

$y = \tan\left[2\left(x + \frac{\pi}{2}\right)\right]$

49. $y = 2\tan\left(x - \frac{\pi}{2}\right)$ **51.** $y = -\tan\left(\frac{\pi}{6}x - \frac{\pi}{3}\right)$

53. $y = 3\csc(2x)$ **55.** $y = 2\csc\left(\frac{\pi}{2}x + \frac{\pi}{2}\right)$

Applying the Concepts:

57. a.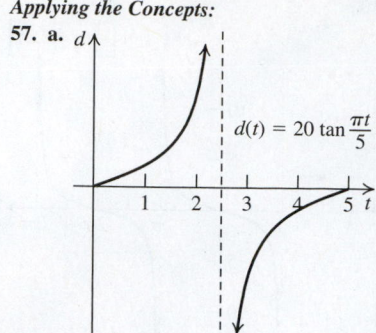

b. The light is pointing parallel to the wall.

$d(t) = 20 \tan \dfrac{\pi t}{5}$

59. b. $4r\sqrt{3}$ **61. b.** $20\pi \approx 62.8319$ **c.** $80.0, 64.9839, 62.9147,$ 62.8525, resp. Perimeters approach the circumference of the circle.

63. $y = 7 \tan\left[\dfrac{\pi}{8}(x+1)\right]; y(-1.5) \approx -1.3924, y(1.5) \approx 10.4762$

Beyond the Basics:

67. a. True **b.** True

69.

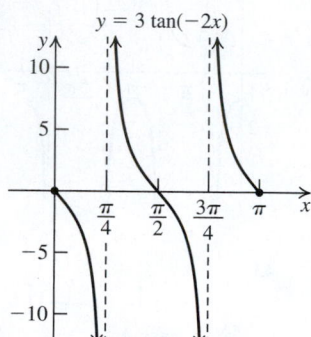

$y = 3 \tan(-2x)$

71.

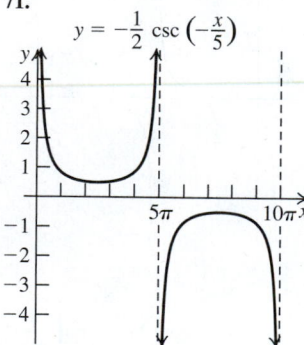

$y = -\dfrac{1}{2} \csc\left(-\dfrac{x}{5}\right)$

73.

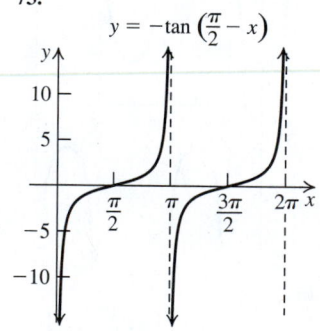

$y = -\tan\left(\dfrac{\pi}{2} - x\right)$

75.

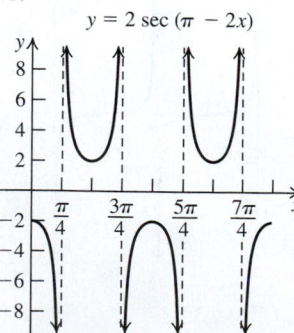

$y = 2 \sec(\pi - 2x)$

77.

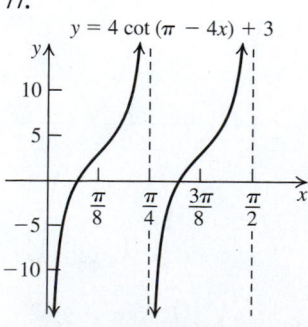

$y = 4 \cot(\pi - 4x) + 3$

79. $k = -\dfrac{7\pi}{4}$ **80.** 6 **81.** $y = -2 \csc\left(x + \dfrac{\pi}{2}\right)$

83. $k = -\dfrac{7\pi}{4}$ **85.** $f^{-1}(x) = \dfrac{x-7}{3}$ **87.** True **89.** True

91. $[0, \infty)$ **93.** No; $[0, \pi]$

Section 4.6

Concepts and Vocabulary:

1. $[-1, 1]$ **3.** $\dfrac{\pi}{3}$ **5.** True **7.** False

Building Skills:

9. 0 **11.** $-\dfrac{\pi}{6}$ **13.** π **15.** Undefined **17.** $\dfrac{\pi}{3}$ **19.** $-\dfrac{\pi}{4}$ **21.** $\dfrac{3\pi}{4}$

23. Undefined **25.** $\dfrac{2\pi}{3}$ **27.** $\dfrac{\pi}{2}$ **29.** $\dfrac{5\pi}{6}$ **31.** $y = \sin^{-1}\left(\dfrac{x-1}{2}\right)$

33. $y = \dfrac{1}{2}\cos^{-1}\left(\dfrac{x}{3}\right) + \dfrac{1}{2}$ **35.** $y = \tan^{-1}(x-2) + 1$ **37.** $\dfrac{1}{8}$ **39.** $\dfrac{\pi}{7}$

41. 247 **43.** $-\dfrac{\pi}{3}$ **45.** $-\dfrac{\pi}{3}$ **47.** $\dfrac{\pi}{4}$ **49.** Undefined **51.** π **53.** $\dfrac{\pi}{3}$

55. $\dfrac{\pi}{6}$ **57.** $\dfrac{2\pi}{3}$ **59.** $-\dfrac{\pi}{3}$ **61.** $\dfrac{2\pi}{3}$ **63.** 1 **65.** -1 **67.** 1

69. $53.13°$ **71.** $-43.63°$ **73.** $73.40°$ **75.** $85.91°$ **77.** $-88.64°$

79. $\dfrac{\sqrt{5}}{3}$ **81.** $\dfrac{3}{5}$ **83.** $\dfrac{2\sqrt{29}}{29}$ **85.** $\dfrac{3}{4}$ **87.** $\dfrac{4\sqrt{17}}{17}$ **89.** $\sqrt{3}$

91. $\sqrt{1 - x^2}$

Applying the Concepts:

93. $159°$ **95. a.** $\dfrac{5\pi}{12}$ units² **b.** $\dfrac{2\pi}{3}$ units²

97. $\left(\sqrt{3} - 1\right)$ units² **99.** ≈ 9113 miles

Beyond the Basics:

101. Increasing **103.** Increasing **105.** $(0, \pi)$

107. $4\pi - 10$

109.

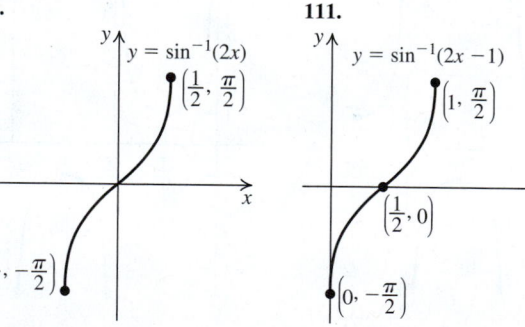

$y = \sin^{-1}(2x)$

$\left(\dfrac{1}{2}, \dfrac{\pi}{2}\right)$

$\left(-\dfrac{1}{2}, -\dfrac{\pi}{2}\right)$

111.

$y = \sin^{-1}(2x - 1)$

$\left(1, \dfrac{\pi}{2}\right)$

$\left(\dfrac{1}{2}, 0\right)$

$\left(0, -\dfrac{\pi}{2}\right)$

113.

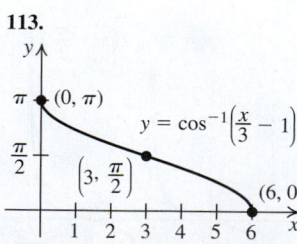

$(0, \pi)$

$y = \cos^{-1}\left(\dfrac{x}{3} - 1\right)$

$\left(3, \dfrac{\pi}{2}\right)$

$(6, 0)$

Critical Thinking / Discussion / Writing:

117. $(0, \pi) \cup \left\{\dfrac{3\pi}{2}\right\}$ **119.** $\dfrac{2}{1 - x^2}$ **121.** $\sqrt{3} - 1$ **123.** 0 **125.** 1

Review Exercises

Building Skills:

1.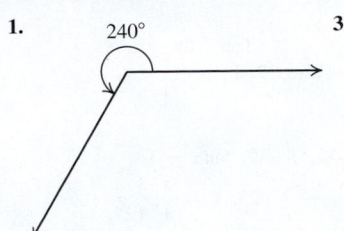

3.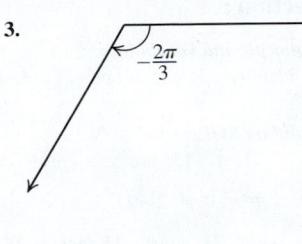

5. $\dfrac{\pi}{9}$ **7.** $-\dfrac{\pi}{3}$ **9.** $50°$ **11.** 2.667 **13.** 1.257 m   **15.** 2.513 ft^2

17. $\sin\theta = \dfrac{\sqrt{77}}{9}$, $\tan\theta = \dfrac{\sqrt{77}}{2}$, $\cot\theta = \dfrac{2\sqrt{77}}{77}$, $\sec\theta = \dfrac{9}{2}$,

$\csc\theta = \dfrac{9\sqrt{77}}{77}$ **19.** $\sin\theta = \dfrac{5\sqrt{34}}{34}$, $\cos\theta = \dfrac{3\sqrt{34}}{34}$, $\cot\theta = \dfrac{3}{5}$,

$\sec\theta = \dfrac{\sqrt{34}}{3}$, $\csc\theta = \dfrac{\sqrt{34}}{5}$ **21.** $\sin\theta = \dfrac{4\sqrt{17}}{17}$, $\cos\theta = \dfrac{\sqrt{17}}{17}$,

$\tan\theta = 4$, $\cot\theta = \dfrac{1}{4}$, $\sec\theta = \sqrt{17}$, $\csc\theta = \dfrac{\sqrt{17}}{4}$ **23.** $\sin\theta = \dfrac{2}{3}$,

$\cos\theta = -\dfrac{\sqrt{5}}{3}$, $\tan\theta = -\dfrac{2\sqrt{5}}{5}$, $\cot\theta = -\dfrac{\sqrt{5}}{2}$, $\sec\theta = -\dfrac{3\sqrt{5}}{5}$,

$\csc\theta = \dfrac{3}{2}$ **25.** IV   **27.** III   **29.** $\sin\theta = -\dfrac{3}{5}$, $\tan\theta = \dfrac{3}{4}$,

$\cot\theta = \dfrac{4}{3}$, $\sec\theta = -\dfrac{5}{4}$, $\csc\theta = -\dfrac{5}{3}$ **31.** $\cos\theta = -\dfrac{4}{5}$, $\tan\theta = -\dfrac{3}{4}$,

$\cot\theta = -\dfrac{4}{3}$, $\sec\theta = -\dfrac{5}{4}$, $\csc\theta = \dfrac{5}{3}$ **33.** $-\dfrac{\sqrt{3}}{2}$ **35.** $\dfrac{\sqrt{3}}{3}$

37.   **39.**

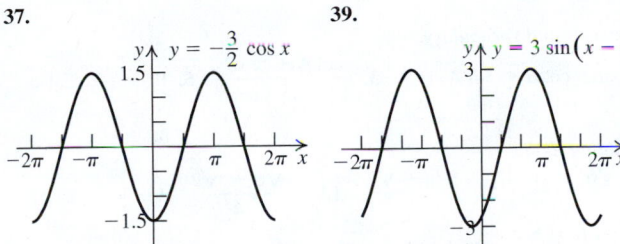

41. Amplitude $= 14$; period $= \pi$; phase shift $= -\dfrac{\pi}{14}$

43. Vertical stretch $= 6$; period $= \dfrac{\pi}{2}$; phase shift $= -\dfrac{\pi}{10}$

45.   **47.**

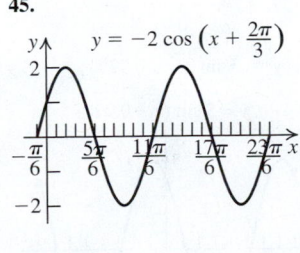   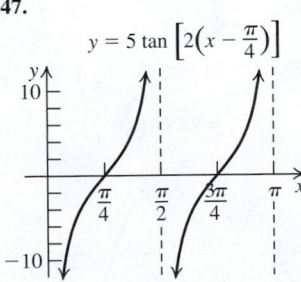

49.

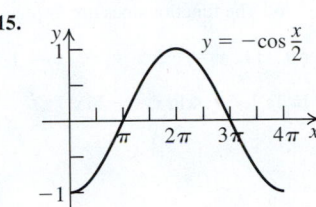

51. $\dfrac{5\pi}{8}$ **53.** $\dfrac{\pi}{3}$ **55.** $\dfrac{\sqrt{2}}{2}$ **57.** $-\sqrt{3}$ **59.** Undefined

61. $y = 5\sin\left(2x - \dfrac{\pi}{2}\right)$ **63.** $y = 3\sin\left(\dfrac{\pi}{4}x + \dfrac{\pi}{4}\right)$

Applying the Concepts:

65. ≈ 44 inches   **67.** ≈ 866 miles   **69.** $11,100$ km/hr

71. $y = -11\cos\dfrac{2\pi}{5}t$

Practice Test A:

1. $\dfrac{13\pi}{12}$ **2.** $252°$ **3.** $\dfrac{2\sqrt{29}}{29}$ **4.** 1.5 **5.** 75 cm^2 **6.** $\dfrac{3\sqrt{5}}{7}$

7. Quadrant II   **8.** $\dfrac{13}{5}$ **9.** $37°$ **10.** 0.223 **11.** $65\sqrt{3} \approx 112.58$ ft

12. Amplitude $= 7$; range $= [-7, 7]$ **13.** Amplitude $= 19$; period $= \dfrac{\pi}{6}$;

phase shift $= -3\pi$ **14.** $y = -7\cos\left(\dfrac{\pi}{2}t\right)$

15.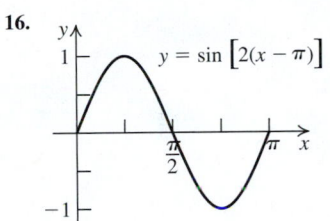

16.

17. 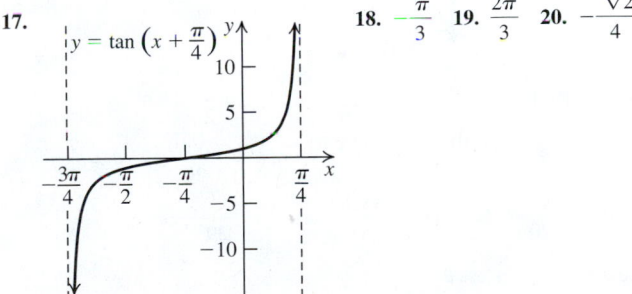   **18.** $-\dfrac{\pi}{3}$ **19.** $\dfrac{2\pi}{3}$ **20.** $-\dfrac{\sqrt{2}}{4}$

Practice Test B:

1. b   **2.** d   **3.** a   **4.** a   **5.** d   **6.** b   **7.** d   **8.** b   **9.** c   **10.** a   **11.** c
12. a   **13.** b   **14.** d   **15.** d   **16.** c   **17.** c   **18.** a   **19.** d   **20.** c

Cumulative Review Exercises (Chapters 1–4):

1. $5 \pm \dfrac{\sqrt{3}}{3}$ **2.** 0 **3.** $(2, 4)$ **4.** $y = -\dfrac{2}{3}x + 1$ **5.** $(0, 2)$

6. $f(x) = x^2$

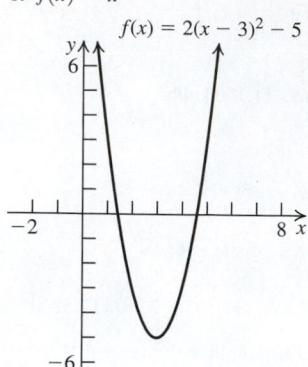

$f(x) = 2(x - 3)^2 - 5$

7.

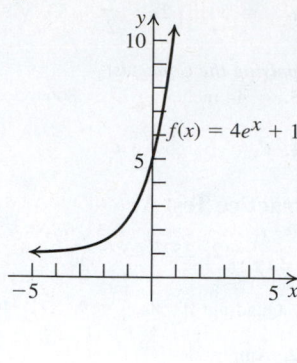

$f(x) = 4e^x + 1$

8. $(-\infty, 0)$ **9. a.** $f^{-1}(x) = \dfrac{x}{3} - \dfrac{7}{3}$ **b.** The function does not have an

inverse.   **10.** $2x^3 + x - 1$, remainder 0   **11.** $y = -3, x = -5, x = 1$

12. a. $\dfrac{2}{3}\log x + \dfrac{1}{3}\log y - \dfrac{1}{3}\log z$ **b.** $\ln 1.4 + 4\ln x - \dfrac{1}{2}\ln y$

13.

$$f(x) = \begin{cases} -\ln x & \text{if } x > 0 \\ \ln |x| & \text{if } x < 0 \end{cases}$$

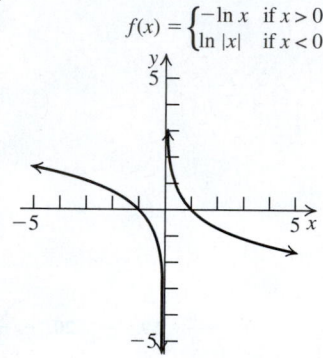

14. $\{-10, 1\}$ **15.** $\pm 1, \pm 2, \pm 3, \pm 6, \pm \dfrac{1}{2}, \pm \dfrac{3}{2}$

16. a. $\sin \theta = \dfrac{2\sqrt{10}}{7}$, $\tan \theta = \dfrac{2\sqrt{10}}{3}$, $\cot \theta = \dfrac{3\sqrt{10}}{20}$, $\sec \theta = \dfrac{7}{3}$,

$\csc \theta = \dfrac{7\sqrt{10}}{20}$ **b.** $\cos \theta = \dfrac{3\sqrt{13}}{11}$, $\tan \theta = \dfrac{2\sqrt{13}}{39}$, $\cot \theta = \dfrac{3\sqrt{13}}{2}$,

$\sec \theta = \dfrac{11\sqrt{13}}{39}$, $\csc \theta = \dfrac{11}{2}$

17. a. $\sin \theta = \dfrac{3}{5}$, $\tan \theta = -\dfrac{3}{4}$, $\cot \theta = -\dfrac{4}{3}$, $\sec \theta = -\dfrac{5}{4}$, $\csc \theta = \dfrac{5}{3}$

b. $\sin \theta = -\dfrac{12}{13}$, $\cos \theta = -\dfrac{5}{13}$, $\tan \theta = \dfrac{12}{5}$, $\sec \theta = -\dfrac{13}{5}$, $\csc \theta = -\dfrac{13}{12}$

18.

a.

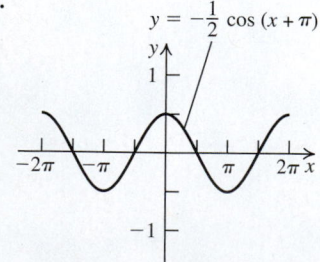

$y = -\dfrac{1}{2}\cos(x + \pi)$

b.

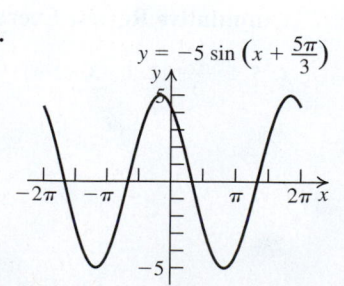

$y = -5\sin\left(x + \dfrac{5\pi}{3}\right)$

19. a. $\dfrac{2\sqrt{6}}{5}$ **b.** $\dfrac{7\sqrt{53}}{53}$ **20.** $\dfrac{\pi}{3}$

CHAPTER 5

Section 5.1

Concepts and Vocabulary:

1. identity   **3.** $\cos^2 x$; $\tan^2 x$; 1   **5.** False   **7.** False

Building Skills:

9. 2   **11.** 1   **13.** $\csc^2 x + \cot^2 x$ **15.** 1   **17.** $\tan x - 3$

19. $x = \pi$; $0 \neq 2$ **21.** $x = \pi$; $-1 \neq 1$ **23.** $x = \dfrac{\pi}{3}$; $\dfrac{3}{4} \neq \dfrac{1}{4}$

69. $\sec \theta$ **71.** $\tan \theta$ **73.** $\cot \theta$ **75.** $\csc \theta$ **77.** $\sin \theta$

Applying the Concepts:

89. $20 \csc \theta$ **91.** $m_1 - \tan \theta = m_2 + m_1 m_2 \tan \theta$

Beyond the Basics:

99. 1   **101.** 1   **103.** 2   **105.** 1   **107.** 1

Critical Thinking / Discussion / Writing:

113. a. $x = -\dfrac{\pi}{2}$ **115.** $t = 0$ **117.** $a = b$, and $a, b \neq 0$

Getting Ready for the Next Section:

119. $\sqrt{26}$ **121.** $\sqrt{2(1 - (\cos \alpha \cos \beta + \sin \alpha \sin \beta))}$ **123.** False

125. $-\dfrac{2\sqrt{2}}{3}$, $-\dfrac{\sqrt{2}}{4}$ **127.** 0   **129.** $\dfrac{4}{5}$

Section 5.2

Concepts and Vocabulary:

1. $\sin A \cos B + \cos A \sin B$ **3.** $\dfrac{\tan A + \tan B}{1 - \tan A \tan B}$ **5.** False   **7.** False

Building Skills:

9. $\dfrac{\sqrt{6} + \sqrt{2}}{4}$ **11.** $\dfrac{\sqrt{6} - \sqrt{2}}{4}$ **13.** $-\dfrac{\sqrt{6} + \sqrt{2}}{4}$ **15.** 1

17. $\dfrac{\sqrt{6} + \sqrt{2}}{4}$ **19.** $2 - \sqrt{3}$ **21.** $-\sqrt{6} - \sqrt{2}$ **23.** $\dfrac{\sqrt{6} - \sqrt{2}}{4}$

25. $-2 - \sqrt{3}$ **27.** $2 + \sqrt{3}$ **43.** 1   **45.** 0   **47.** 1   **49.** $\dfrac{\sqrt{3}}{2}$ **51.** 1

53. $\dfrac{56}{65}$ **55.** $\dfrac{63}{65}$ **57.** $\dfrac{16}{63}$ **59.** $\dfrac{2\sqrt{210} - 6}{35}$ **61.** $\dfrac{35\sqrt{210} - 105}{402}$

63. $\dfrac{-4\sqrt{10} - 3\sqrt{21}}{2\sqrt{210} - 6}$ **73.** 0   **75.** 0   **77.** 17/6

79. $y = \sqrt{2}\sin\left(x + \dfrac{\pi}{4}\right)$ **81.** $y = 5\sin(x + 0.927)$

$y = \sqrt{2}\sin\left(x + \dfrac{\pi}{4}\right)$

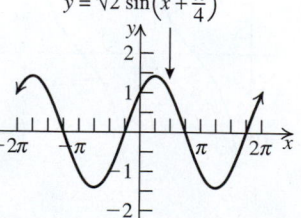

$y = 5\sin(x + 0.927)$

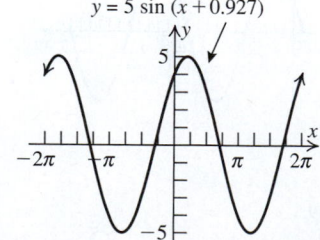

83. $y = 13 \sin(x - 1.176)$

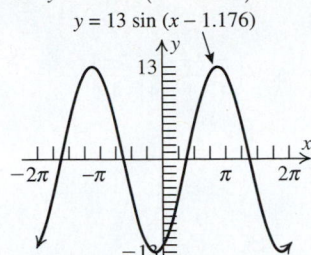

85. $y = \sqrt{10} \sin(x + 2.820)$

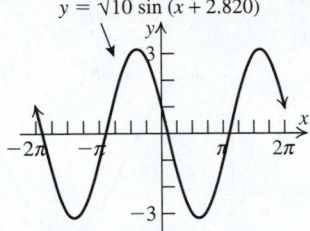

Applying the Concepts:

89. 0   **91.** $A = 0.5; \theta \approx -0.00161$ **93. a.** $A \approx 0.5142$

b. Frequency $= \dfrac{1}{\pi}$; phase shift ≈ -0.6676

Beyond the Basics:

115. $\dfrac{1}{2}$ **117.** $\dfrac{1}{2}$ **121.** 1

Getting Ready for the Next Section:

129. False

Section 5.3

Concepts and Vocabulary:

1. $2 \sin x \cos x$ **3.** $2 \cos^2 x - 1; \dfrac{1 + \cos 2x}{2}$ **5.** False   **7.** True

Building Skills:

9. a. $-\dfrac{24}{25}$ **b.** $\dfrac{7}{25}$ **c.** $-\dfrac{24}{7}$ **11. a.** $\dfrac{8}{17}$ **b.** $-\dfrac{15}{17}$ **c.** $-\dfrac{8}{15}$

13. a. $-\dfrac{4}{5}$ **b.** $-\dfrac{3}{5}$ **c.** $\dfrac{4}{3}$ **15.** $-\dfrac{\sqrt{3}}{2}$ **17.** $-\dfrac{\sqrt{3}}{2}$ **19.** $-\dfrac{\sqrt{3}}{3}$

21. $\dfrac{\sqrt{2}}{2}$ **23.** $\dfrac{\sqrt{3}}{3}$ **41.** $\dfrac{1 - \cos 4x}{2}$ **43.** $\sin 4x$ **45.** $\dfrac{\sin 12x}{2}$

47. $\dfrac{\sin 2x}{4}$ **49.** $\cos 2x - 4 \cos x + 3$ **51.** $\dfrac{\sqrt{2 - \sqrt{3}}}{2}$

53. $\dfrac{\sqrt{2 + \sqrt{2}}}{2}$ **55.** $-\dfrac{\sqrt{2 + \sqrt{2}}}{2}$ **57.** $-\dfrac{\sqrt{2 - \sqrt{2}}}{\sqrt{2 + \sqrt{2}}}$ **59.** $-\sqrt{2} - 1$

61. $-\dfrac{\sqrt{2 + \sqrt{3}}}{2}$ **63. a.** $\dfrac{2\sqrt{5}}{5}$ **b.** $\dfrac{\sqrt{5}}{5}$ **c.** 2

65. a. $\sqrt{\dfrac{13 + 3\sqrt{13}}{26}}$ **b.** $\sqrt{\dfrac{13 - 3\sqrt{13}}{26}}$ **c.** $\sqrt{\dfrac{13 + 3\sqrt{13}}{13 - 3\sqrt{13}}}$

67. a. $\sqrt{\dfrac{5 + 2\sqrt{6}}{10}}$ **b.** $\sqrt{\dfrac{5 - 2\sqrt{6}}{10}}$ **c.** $\sqrt{\dfrac{5 + 2\sqrt{6}}{5 - 2\sqrt{6}}}$

69. a. $\sqrt{\dfrac{5 - \sqrt{5}}{10}}$ **b.** $\sqrt{\dfrac{5 + \sqrt{5}}{10}}$ **c.** $\sqrt{\dfrac{5 - \sqrt{5}}{5 + \sqrt{5}}}$

Applying the Concepts:

81. 100 watts   **83.** 1200 watts   **85.** 600 watts   **87.** 45°

89.

Period $= \pi$

$\left(\dfrac{\pi}{2}, 1\right)$

$\left(\dfrac{3\pi}{4}, \dfrac{1}{2}\right)$

$\left(\dfrac{\pi}{4}, \dfrac{1}{2}\right)$

$(0, 0)$ $(\pi, 0)$

103. $\dfrac{\sqrt{3}}{2}$ **105.** $\dfrac{3}{5}$ **107.** $-3\sqrt{7}$ **109.** $-\dfrac{24}{25}$ **111.** $\dfrac{1}{2}$ **113.** $-4\sqrt{5}$

115. $\dfrac{\sqrt{10}}{10}$ **117.** $\dfrac{3}{2}$ **119.** $\dfrac{3 - \sqrt{5}}{6}$ **125. a.** $\dfrac{(\sqrt{5} - 1)(\sqrt{10 + 2\sqrt{5}})}{8}$

b. $\dfrac{\sqrt{5} + 1}{4}$ **c.** $\dfrac{\sqrt{5} + 1}{4}$ **d.** $\dfrac{(\sqrt{5} - 1)(\sqrt{10 + 2\sqrt{5}})}{8}$

Critical Thinking / Discussion / Writing:

131. a. $\dfrac{4}{5}$ **b.** $\dfrac{3}{5}$ **c.** $\dfrac{24}{25}$ **d.** $-\dfrac{7}{25}$ **e.** $-\dfrac{24}{7}$ **f.** $\dfrac{\sqrt{5}}{5}$ **g.** $\dfrac{2\sqrt{5}}{5}$ **h.** $\dfrac{1}{2}$

Getting Ready for the Next Section:

133. a. $u + v, u - v$ **b.** $\dfrac{x + y}{2}, \dfrac{x - y}{2}$

Section 5.4

Concepts and Vocabulary:

1. $\dfrac{1}{2}[\cos(x - y) - \cos(x + y)]$ **3.** $2 \cos\left(\dfrac{x + y}{2}\right) \cos\left(\dfrac{x - y}{2}\right)$

5. True   **7.** False

Building Skills:

9. $\dfrac{1}{2} \sin 2x$ **11.** $\dfrac{1}{2} - \dfrac{1}{2} \cos 2x$ **13.** $\dfrac{1}{4} + \dfrac{1}{2} \sin 20°$

15. $-\dfrac{1}{4} - \dfrac{1}{2} \cos 160°$ **17.** $\dfrac{1}{4}$ **19.** $\dfrac{\sqrt{2}}{4} - \dfrac{1}{2}$ **21.** $\dfrac{1}{2} \sin 6\theta + \dfrac{1}{2} \sin 4\theta$

23. $\dfrac{1}{2} \cos x + \dfrac{1}{2} \cos 7x$ **25.** $\dfrac{\sqrt{3}}{4} - \dfrac{\sqrt{2}}{4}$ **27.** $\dfrac{1}{2} + \dfrac{\sqrt{2}}{4}$ **29.** $\dfrac{1}{4} + \dfrac{\sqrt{2}}{4}$

31. $\dfrac{1}{4} - \dfrac{\sqrt{2}}{4}$ **33.** $-\sin 10°$ **35.** $2 \sin 8° \cos 24°$ **37.** $2 \sin \dfrac{3\pi}{10} \cos \dfrac{\pi}{10}$

39. $2 \cos \dfrac{5}{12} \cos \dfrac{1}{12}$ **41.** $2 \cos 4x \cos x$ **43.** $2 \sin 3x \cos 4x$

45. $\sqrt{2} \cos\left(x - \dfrac{\pi}{4}\right)$ **47.** $\sqrt{2} \sin\left(2x - \dfrac{\pi}{4}\right)$

49. $2 \sin\left(\dfrac{\pi}{4} - x\right) \cos\left(4x - \dfrac{\pi}{4}\right)$ **51.** $a\sqrt{2} \cos\left(x - \dfrac{\pi}{4}\right)$

Applying the Concepts:

63. a. $y = \sin[2\pi(852)t] + \sin[2\pi(1209)t]$

b. $y = 2 \cos(357\pi t) \sin[2\pi(1030.5)t]$

c. $f = 1030.5$ Hz, $A = 2 \cos(357\pi t)$

65. a. $y = 0.1 \cos(116\pi t) \cos(4\pi t)$ **b.** $f = 4$ beats per unit time.

Beyond the Basics:

67. The amplitude is $2 \cos 1 \approx 1.081$. The period is π.

$y = \cos(2x + 1) + \cos(2x - 1)$

Getting Ready for the Next Section:

89. Conditional   **91.** Inconsistent   **93.** Identity   **95.** Conditional

97. Inconsistent   **99.** Identity   **101.** $\{5\}$ **103.** $\{-2, 3\}$ **105.** $\{8\}$

107. $\dfrac{\pi}{6}, \dfrac{5\pi}{6}$ **109.** $\dfrac{3\pi}{4}, \dfrac{7\pi}{4}$ **111.** $\dfrac{4\pi}{3}, \dfrac{5\pi}{3}$ **113.** $\dfrac{2\pi}{3}, \dfrac{4\pi}{3}$

Section 5.5

Basic Concepts and Vocabulary:

1. two   **3.** $\alpha + 2n\pi; (2\pi - \alpha) + 2n\pi$ **5.** True   **7.** True

Building Skills:

9. $\dfrac{\pi}{2} + n\pi$ **11.** $-\dfrac{\pi}{4} + n\pi$ **13.** $\dfrac{\pi}{4} + 2n\pi, \dfrac{7\pi}{4} + 2n\pi$ **15.** $\dfrac{\pi}{6} + n\pi$

17. $\dfrac{2\pi}{3} + 2n\pi, \dfrac{4\pi}{3} + 2n\pi$ **19.** $30° + 180°n$

21. $210° + 360°n, 330° + 360°n$ **23.** $90° + 360°n$

25. $60° + 360°n, 120° + 360°n$ **27.** $60° + 360°n, 300° + 360°n$

29. $\{23.6°, 156.4°\}$ **31.** $\{82.0°, 278.0°\}$ **33.** $\{131.3°, 311.3°\}$

35. $\{3.6652, 5.7596\}$ **37.** $\{2.2143, 5.3559\}$ **39.** $\{3.5531, 5.8717\}$

41. $\left\{\dfrac{7\pi}{12}, \dfrac{23\pi}{12}\right\}$ **43.** $\left\{\dfrac{19\pi}{24}, \dfrac{35\pi}{24}\right\}$ **45.** $\left\{\dfrac{\pi}{3}, \dfrac{4\pi}{3}\right\}$ **47.** $\left\{\dfrac{\pi}{6}, \dfrac{3\pi}{2}\right\}$

49. $\left\{\dfrac{\pi}{12}, \dfrac{11\pi}{12}, \dfrac{13\pi}{12}, \dfrac{23\pi}{12}\right\}$ **51.** $\left\{\dfrac{7\pi}{12}, \dfrac{11\pi}{12}, \dfrac{19\pi}{12}, \dfrac{23\pi}{12}\right\}$ **53.** $\varnothing$

55. $\left\{\dfrac{\pi}{12}, \dfrac{7\pi}{12}, \dfrac{13\pi}{12}, \dfrac{19\pi}{12}\right\}$ **57.** $\left\{\dfrac{\pi}{4}, \dfrac{7\pi}{12}, \dfrac{5\pi}{4}, \dfrac{19\pi}{12}\right\}$

59. $\left\{\dfrac{\pi}{18}, \dfrac{5\pi}{18}, \dfrac{13\pi}{18}, \dfrac{17\pi}{18}, \dfrac{25\pi}{18}, \dfrac{29\pi}{18}\right\}$ **61.** $\left\{\dfrac{\pi}{9}, \dfrac{5\pi}{9}, \dfrac{7\pi}{9}, \dfrac{11\pi}{9}, \dfrac{13\pi}{9}, \dfrac{17\pi}{9}\right\}$

63. $\left\{\dfrac{2\pi}{3}\right\}$ **65.** $\varnothing$ **67.** $\left\{\dfrac{3\pi}{4}\right\}$ **69.** $\left\{0, \dfrac{\pi}{3}, \pi, \dfrac{5\pi}{3}\right\}$

71. $\left\{\dfrac{\pi}{6}, \dfrac{\pi}{2}, \dfrac{5\pi}{6}, \dfrac{3\pi}{2}\right\}$ **73.** $\left\{\dfrac{7\pi}{12}, \dfrac{5\pi}{4}, \dfrac{7\pi}{4}, \dfrac{23\pi}{12}\right\}$ **75.** $\left\{\dfrac{\pi}{4}, \dfrac{11\pi}{12}, \dfrac{19\pi}{12}\right\}$

77. $\left\{\dfrac{\pi}{6}, \dfrac{5\pi}{6}, \pi, \dfrac{3\pi}{2}\right\}$ **79.** $\left\{\dfrac{\pi}{4}, \dfrac{5\pi}{4}\right\}$ **81.** $\left\{\dfrac{\pi}{6}, \dfrac{3\pi}{4}, \dfrac{5\pi}{6}, \dfrac{7\pi}{4}\right\}$

83. $\left\{\dfrac{\pi}{6}, \dfrac{3\pi}{4}, \dfrac{5\pi}{6}, \dfrac{7\pi}{4}\right\}$ **85.** $\left\{\dfrac{\pi}{4}, \dfrac{7\pi}{6}, \dfrac{7\pi}{4}, \dfrac{11\pi}{6}\right\}$

87. $\left\{\dfrac{\pi}{6}, \dfrac{5\pi}{6}, \dfrac{7\pi}{6}, \dfrac{11\pi}{6}\right\}$ **89.** $\left\{\dfrac{\pi}{4}, \dfrac{3\pi}{4}, \dfrac{5\pi}{4}, \dfrac{7\pi}{4}\right\}$

91. $\left\{\dfrac{\pi}{3}, \dfrac{2\pi}{3}, \dfrac{4\pi}{3}, \dfrac{5\pi}{3}\right\}$ **93.** $\left\{\dfrac{\pi}{6}, \dfrac{5\pi}{6}, \dfrac{7\pi}{6}, \dfrac{11\pi}{6}\right\}$ **95.** $\{0, \pi\}$

97. $\left\{\dfrac{7\pi}{6}, \dfrac{3\pi}{2}, \dfrac{11\pi}{6}\right\}$ **99.** $\left\{\dfrac{\pi}{2}, \dfrac{7\pi}{6}, \dfrac{11\pi}{6}\right\}$

101. $\left\{\dfrac{\pi}{3}, \dfrac{5\pi}{6}, \dfrac{4\pi}{3}, \dfrac{11\pi}{6}\right\}$ **103.** $\left\{\dfrac{\pi}{3}, \pi\right\}$ **105.** $\{0\}$ **107.** $\left\{\dfrac{\pi}{3}\right\}$

Applying the Concepts:

109. a. ≈ 0.0014 sec;   **b.** ≈ 0.0092 sec   **111.** $2/3 + 4n$ sec $10/3 +$ $4n$ sec, n an integer   **113. a.** 27   **b.** 18, 36   **c.** 44, 10

115. a. January   **b.** February, December

Beyond the Basics:

117. $\left\{\dfrac{\pi}{4}, \dfrac{3\pi}{4}, \dfrac{5\pi}{4}, \dfrac{7\pi}{4}\right\}$ **119.** $\left\{\dfrac{\pi}{8}, \dfrac{3\pi}{8}, \dfrac{5\pi}{8}, \dfrac{7\pi}{8}, \dfrac{9\pi}{8}, \dfrac{11\pi}{8}, \dfrac{13\pi}{8}, \dfrac{15\pi}{8}\right\}$

121. $\left\{\dfrac{\pi}{6}, \dfrac{7\pi}{6}, \dfrac{11\pi}{6}\right\}$ **123.** $\left\{\dfrac{\pi}{3}, \dfrac{5\pi}{3}\right\}$ **125.** $\left\{\dfrac{\pi}{6}, \dfrac{5\pi}{6}\right\}$

127. $\left\{\dfrac{\pi}{8}, \dfrac{9\pi}{8}\right\}$ **129.** $\left\{\dfrac{\pi}{4}, \dfrac{3\pi}{4}, \dfrac{5\pi}{4}, \dfrac{7\pi}{4}\right\}$

131. $5\sin\left[2x + \tan^{-1}\left(\dfrac{4}{3}\right)\right]$; $\{1.107, 2.678, 4.249, 5.820\}$

133. $13\sin\left[3x + \tan^{-1}\left(-\dfrac{12}{5}\right)\right]$; $\{0.741, 1.090, 3.185, 4.930, 5.279\}$

135. $\left\{\dfrac{\pi}{4}, \dfrac{5\pi}{4}\right\}$ **137.** A square of a real number is nonnegative.

Getting Ready for the Next Section:

139. $c = 13$ **141.** $a = 8$ **143.** $a = 3\sqrt{3}$ **145.** $\dfrac{\sqrt{3}}{3}$

Review Exercises

Building Skills:

1. $\cos\theta = -\dfrac{\sqrt{5}}{3}, \tan\theta = \dfrac{2\sqrt{5}}{5}, \cot\theta = \dfrac{\sqrt{5}}{2},$

$\sec\theta = -\dfrac{3\sqrt{5}}{5}, \csc\theta = -\dfrac{3}{2}$ **3.** $\sin\theta = -\dfrac{2\sqrt{2}}{3}, \cos\theta = \dfrac{1}{3},$

$\tan\theta = -2\sqrt{2}, \cot\theta = -\dfrac{\sqrt{2}}{4}, \csc\theta = -\dfrac{3\sqrt{2}}{4}$

21. $\dfrac{\sqrt{2+\sqrt{3}}}{2}$ **23.** $\sqrt{6} - \sqrt{2}$ **25.** 1   **27.** -1 **29.** $-\dfrac{16}{65}$ **31.** $\dfrac{63}{65}$

51. 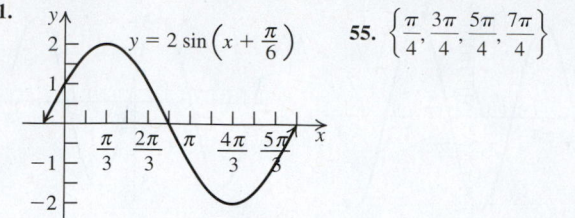   **55.** $\left\{\dfrac{\pi}{4}, \dfrac{3\pi}{4}, \dfrac{5\pi}{4}, \dfrac{7\pi}{4}\right\}$

57. $\left\{0, \dfrac{2\pi}{3}, \dfrac{4\pi}{3}\right\}$ **59.** $\left\{\dfrac{\pi}{18}, \dfrac{5\pi}{18}, \dfrac{13\pi}{18}, \dfrac{17\pi}{18}, \dfrac{25\pi}{18}, \dfrac{29\pi}{18}\right\}$

61. $\{19.5°, 160.5°\}$ **63.** $\{15°, 165°, 195°, 345°\}$

65. $\{120°, 150°, 240°, 330°\}$ **67.** $\varnothing$ **69.** $\left\{\dfrac{\pi}{9}, \dfrac{7\pi}{9}, \dfrac{4\pi}{3}, \dfrac{13\pi}{9}\right\}$ **71.** $60°$

73. $\dfrac{1}{360}$ sec   **75.** $30°$

Practice Test A:

1. $-\dfrac{3}{4}$ **2.** $-\dfrac{3\sqrt{13}}{13}$ **9.** $-\dfrac{\pi}{4} + n\pi$ **10.** $\dfrac{\pi}{24} + n\dfrac{\pi}{2}, \dfrac{5\pi}{24} + n\dfrac{\pi}{2}$

11. $\left\{\dfrac{3\pi}{4}\right\}$ **12.** $\left\{\dfrac{\pi}{2}, \dfrac{7\pi}{6}, \dfrac{3\pi}{2}, \dfrac{11\pi}{6}\right\}$ **13.** 0   **14.** $\dfrac{1}{2}$ **15.** $\dfrac{\sqrt{6}+\sqrt{2}}{4}$

16. $-\dfrac{7}{25}$ **17.** $\dfrac{24}{25}$ **18.** Odd   **19.** No; let $x = y = \dfrac{\pi}{2}$.   **20.** $\theta = 45°$

Practice Test B:

1. a   **2.** b   **3.** d   **4.** b   **5.** a   **6.** b   **7.** c   **8.** d   **9.** c   **10.** c   **11.** c

12. a   **13.** d   **14.** d   **15.** b   **16.** a   **17.** d   **18.** c   **19.** d   **20.** b

Cumulative Review Exercises (Chapters 1–5):

1. $x = -\dfrac{1}{2} \pm \dfrac{\sqrt{5}}{2}$ **2.** $x = \sqrt{3}$ **3.** $x = -\dfrac{\log 9}{\log 5}$ **4.** $\left\{\dfrac{\pi}{6}, \dfrac{\pi}{2}, \dfrac{5\pi}{6}, \dfrac{3\pi}{2}\right\}$

5. $(-2, 0) \cup (2, \infty)$ **6.** $(-2, 5)$ **7.** $4x - 1 + 2h$ **8.** $f^{-1}(x) = \dfrac{x+2}{2x-1}$

9. 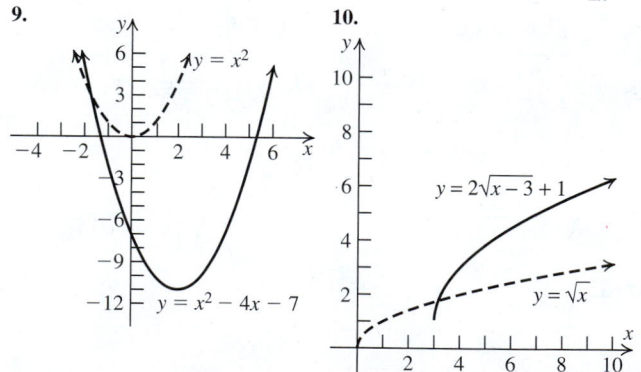   **10.**

11. 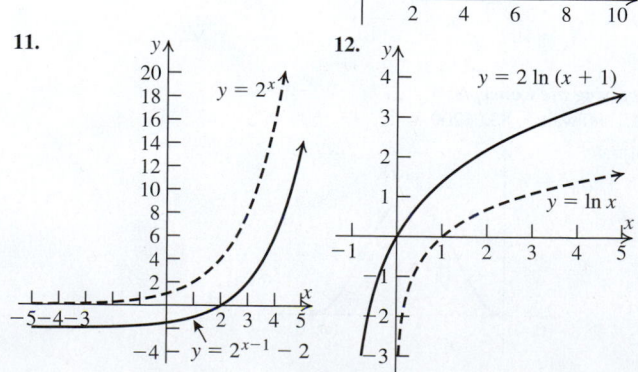   **12.**

13.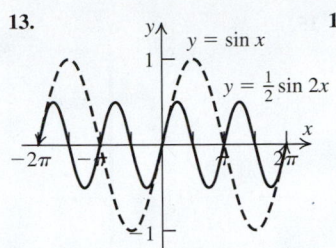

14.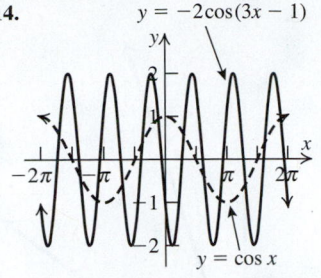

15.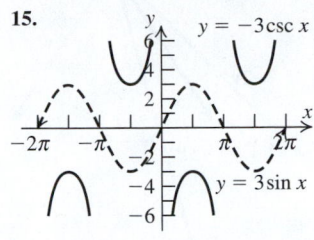

16.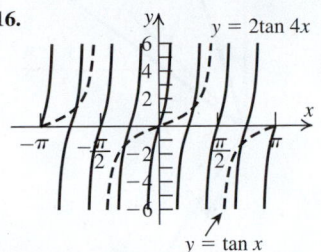

17. $\cot x - 3$ **18.** $\sqrt{6} - \sqrt{2}$ **20.** $\{\pi/4,\ 5\pi/4\}$

CHAPTER 6

Section 6.1

Concepts and Vocabulary:

1. $\dfrac{\text{Adjacent}}{\text{Hypotenuse}}, \dfrac{\text{Opposite}}{\text{Adjacent}}$ **3.** 0.7   **5.** False   **7.** True

Building Skills:

9. $\sin\theta = \dfrac{2\sqrt{5}}{25}, \csc\theta = \dfrac{5\sqrt{5}}{2}$ **11.** $\sin\theta = \dfrac{3}{5}, \csc\theta = \dfrac{5}{3}$

$\cos\theta = \dfrac{11\sqrt{5}}{25}, \sec\theta = \dfrac{5\sqrt{5}}{11},$ $\cos\theta = \dfrac{4}{5}, \sec\theta = \dfrac{5}{4}$

$\tan\theta = \dfrac{2}{11}, \cot\theta = \dfrac{11}{2}$ $\tan\theta = \dfrac{3}{4}, \cot\theta = \dfrac{4}{3}$

13. $\sin\theta = \dfrac{\sqrt{2}}{10}, \csc\theta = 5\sqrt{2};$ $\cos\theta = \dfrac{7\sqrt{2}}{10}, \sec\theta = \dfrac{5\sqrt{2}}{7}$

$\tan\theta = \dfrac{1}{7}$ $\cot\theta = 7$ **15. a.** $\sin\theta = \dfrac{\sqrt{5}}{3}$ **b.** $\tan\theta = \dfrac{\sqrt{5}}{2}$

17. a. $\sin\theta = \dfrac{5\sqrt{34}}{34}$ **b.** $\cos\theta = \dfrac{3\sqrt{34}}{34}$ **19. a.** $\sin\theta = \dfrac{5}{13}$

b. $\tan\theta = \dfrac{5}{12}$ **21.** $\cos\theta = \dfrac{2\sqrt{2}}{3}, \cot(90° - \theta) = \dfrac{\sqrt{2}}{4}$

23. $\sin\theta = \dfrac{3}{5}, \tan(90° - \theta) = \dfrac{4}{3}$

25. $\sin\theta = \dfrac{\sqrt{5}}{5}, \csc(90° - \theta) = \dfrac{\sqrt{5}}{2}$ **27.** 0.8480   **29.** 0.5095

31. 2.3662   **33.** $B = 40°, a \approx 7.0, b \approx 5.9$

35. $B = 54°, c \approx 20.4, b \approx 16.5$ **37.** $c \approx 14.0, A \approx 63.6°, B \approx 26.4°$

39. $a \approx 11, A \approx 49.6°, B \approx 40.4°$

41. $c \approx 12.806, \sin\theta \approx 0.625, \tan\theta = 0.800$

43. $\cos\theta \approx 0.291, \tan\theta \approx 3.286$

45. $\sin\theta = 0.5, b \approx 15.588, c = 18$

47. $\alpha = 30°, a = 12.124, c = 14$

Applying the Concepts:

49. ≈ 11 feet   **51.** ≈ 1029 meters   **53.** 2500 feet   **55.** ≈ 29 feet

57. ≈ 19 feet   **59.** ≈ 37 feet   **61.** $\approx 12{,}994$ feet   **63.** 62°

65. ≈ 2545 miles   **67.** 159°

Beyond the Basics:

69. ≈ 2682 feet   **71.** ≈ 188.27 meters   **75.** ≈ 97.5 meters   **77.** 31 feet

81. ≈ 48 square units

Critical Thinking / Discussion / Writing:

85. $\tan\beta > \tan\alpha$

Getting Ready for the Next Section:

87. 52°   **89.** 75°   **91.** 2.8   **93.** 4.1

95. 46.2°, 133.8°   **97.** 2.5   **99.** 9.4   **101.** 15 units   **103.** 28.2 units

Section 6.2

Concepts and Vocabulary:

1. 180   **3.** one   **5.** True   **7.** True

Building Skills:

9. $5\sqrt{6}$ **11.** 3   **13.** 48.6°, 131.4°   **15.** Not applicable   **17.** 90°

19. No such triangle   **21.** $C = 63°, a \approx 98.2$ ft, $b \approx 93.0$ ft

23. $B = 27°, a \approx 64.8$ m, $b \approx 31.3$ m

25. $C = 105°, b \approx 89.2$ m, $c \approx 150.3$ m

27. $B = 79°, b \approx 102.3$ cm, $c \approx 85.4$ cm

29. $B = 98°, a \approx 47.1$ ft, $b \approx 81.2$ ft

31. $A = 70°, a \approx 55.1$ in., $c \approx 54.0$ in.

33. $A = 24°, a \approx 13.3$ ft, $b \approx 30.7$ ft

35. $C = 98.5°, a \approx 17.7$ m, $b \approx 21.7$ m

37. $B = 65°, C = 50°, c \approx 16.9$ ft

39. $B \approx 23.7°, C \approx 41.3°, c \approx 51.0$ m

41. two   **43.** none   **45.** one

47. $B \approx 34°, C \approx 106°, c \approx 34.4$

49. $C = 60°, B = 90°, c = 25\sqrt{3}$ **51.** No triangle exists.

53. No triangle exists.   **55.** $B \approx 56.8°, C \approx 23.2°, c \approx 16.0$

57. No triangle exists.

59. Solution 1: $C \approx 55.3°, A \approx 78.7°, a \approx 47.7$

Solution 2: $C \approx 124.7°, A \approx 9.3°, a \approx 7.9$

61. No triangle exists

63. Solution 1: $C \approx 49.0°, B \approx 89.0°, b \approx 82.2$

Solution 2: $C \approx 131.0°, B \approx 7.0°, b \approx 10.0$

65. No triangle exists   **67.** 7.5   **69.** $10\sqrt{3}$ **71.** 27   **73.** 654.2 sq in

75. 118.7 sq km   **77.** 62.5 sq mm   **79.** 25,136.6 sq ft   **81.** 30° or 150°

83. 90°

Applying the Concepts:

85. ≈ 399 ft   **87. a.** 1615 yd   **b.** 540 yd   **89. a.** 25.7 mi

b. 14.5 mi   **91.** 22,912 mi   **93.** 33 or 127 million miles

95. 9.98 miles, 32.44 miles   **97.** 24.2 m   **99.** 493,941 square miles

Beyond the Basics:

103. 1:32 P.M., 3:18 P.M.

Critical Thinking / Discussion / Writing:

109. An isosceles triangle with $\angle A = \angle B$

Getting Ready for the Next Section:

113. $-\dfrac{\sqrt{2}}{2}$ **115.** 60°   **117.** 120°   **119.** $\sqrt{11}$ **121.** $\sqrt{7}$

123. $-\dfrac{1}{4}$ **125.** 30°

Section 6.3

Concepts and Vocabulary:

1. C   **3.** three   **5.** False   **7.** True

Building Skills:

9. $\sqrt{21} \approx 4.6$ **11.** $\sqrt{50} \approx 7.1$ **13.** $\sqrt{10} \approx 3.2$

15. $\cos^{-1}\left(\dfrac{7}{8}\right) \approx 29.0°$ **17.** No triangle

19. $b \approx 20.2, A \approx 44°, C \approx 30°$

21. $A \approx 130.5°, B \approx 27.1°, C \approx 22.4°$

23. $c \approx 21, A \approx 38.2°, B \approx 21.8°$

25. $a \approx 11.5, B \approx 50.4°, C \approx 67.6°$
27. $A \approx 81.3°, B \approx 46.4°, C \approx 52.3°$
29. $A \approx 28.3°, B \approx 43.3°, C \approx 108.4°$
31. $A \approx 23.7°, B \approx 36.4°, C \approx 119.9°$
33. $a \approx 5.7, B \approx 33.8°, C \approx 48.5°$
35. $b \approx 5.0, A \approx 46.3°, C \approx 65.4°$
37. $A \approx 38.4°, B \approx 49.1°, C \approx 92.5°$ **39.** 2.9 square inches
41. 1240.2 square inches **43.** 13.5 square inches **45.** 7.7 square inches

Applying the Concepts:
47. 1543.1 yards **49.** 1238.5 yards **51.** \$780,000 **53.** 3297 gallons
55. 9.1 miles **57.** 183.2 miles **59.** $A \approx 86.2°, B \approx 54.0°, C \approx 39.8°$

Beyond the Basics:
61. 18.6 cm **63.** 9.8 cm, 23.6 cm

Critical Thinking / Discussion / Writing:
75. $a = \sqrt{13}, b = 5\sqrt{2}, c = \sqrt{53}, A \approx 29.1°, B \approx 72.2°, C \approx 78.7°$
77. $a \approx 4.3, A \approx 12.5°, C \approx 17.5°$ **79.** No triangle can be formed.

Getting Ready for the Next Section:
83. $x + 3y, \sqrt{10}$ **85.** $\theta = \dfrac{4\pi}{3}$ **87. a.** $d(S, T) = d(O, P) = 13$

b. $m_1 = m_2 = \dfrac{12}{5}$ **c.** Parallel **89. a.** $d(S, T) = d(O, P) = \sqrt{2}$

b. $m_1 = 1, m_2 = 1$ **c.** Perpendicular

Section 6.4

Concepts and Vocabulary:
1. direction **3.** $\sqrt{a^2 + b^2}, \dfrac{b}{a}$ **5.** True **7.** False

Building Skills:
9.
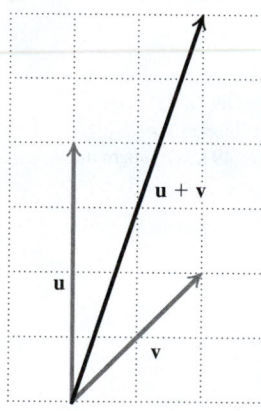
11.
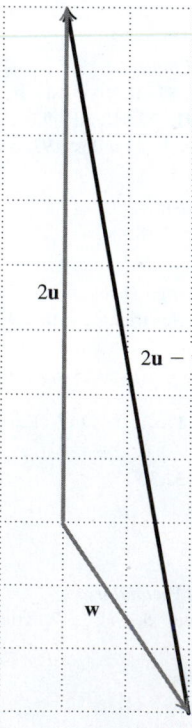

13.
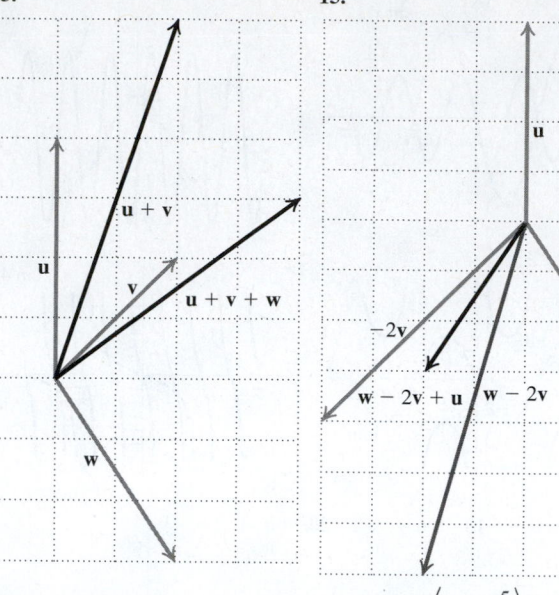
15.

17. $\langle -1, 3 \rangle$ **19.** $\langle 2, -2 \rangle$ **21.** $\langle 3, -7 \rangle$ **23.** $\left\langle -1, -\dfrac{5}{2} \right\rangle$
25. Equivalent **27.** Not equivalent **29.** $\sqrt{5}$ **31.** $\langle -4, 4 \rangle$
33. $\langle -11, 10 \rangle$ **35.** $\sqrt{221}$ **37.** $\left\langle \dfrac{\sqrt{2}}{2}, -\dfrac{\sqrt{2}}{2} \right\rangle$ **39.** $\left\langle -\dfrac{4}{5}, \dfrac{3}{5} \right\rangle$
41. $\left\langle \dfrac{\sqrt{2}}{2}, \dfrac{\sqrt{2}}{2} \right\rangle$ **43.** $\mathbf{i}$ **45.** $-\dfrac{3}{5}\mathbf{i} + \dfrac{4}{5}\mathbf{j}$ **47.** $-\mathbf{i} - 7\mathbf{j}$ **49.** $13\mathbf{i} - 4\mathbf{j}$
51. $\sqrt{185}$ **53.** $10, 60°$ **55.** $3, 150°$ **57.** $13, 67.38°$ **59.** $5, 216.87°$
61. $\sqrt{3}\mathbf{i} + \mathbf{j}$ **63.** $-2\mathbf{i} + 2\sqrt{3}\mathbf{j}$ **65.** $\dfrac{3}{2}\mathbf{i} - \dfrac{3\sqrt{3}}{2}\mathbf{j}$ **67.** $\dfrac{7}{2}\mathbf{i} - \dfrac{7\sqrt{3}}{2}\mathbf{j}$

Applying the Concepts:
69. $23.0\mathbf{i} + 9.8\mathbf{j}$ **71.** $40.6, 218.0°$ **73.** $72.5, 33.8$
75. The resultant is a force of 676.64 pounds making an angle of 47.19° with the positive x-axis. **77.** 483.4 mi/hr, N 32.1° E **79.** 12.2 mi/hr, N 52.2° E
81. A force of magnitude 10.08 pounds making an angle of 176.61° with the positive x-axis

Beyond the Basics:
83. $\|\mathbf{F}_1\| = 106.42$ lb, $\|\mathbf{F}_2\| = 136.81$ lb **85.** $(2, -2)$
87. $\left\langle -\dfrac{11}{3}, -\dfrac{11}{3} \right\rangle$ **89.** $\overrightarrow{AB} = -\overrightarrow{CD}$ **91.** $\overrightarrow{PQ} = \dfrac{2}{3}\overrightarrow{PR}$ **93.** $3, \sqrt{5}$

Critical Thinking / Discussion / Writing:
95. $(9, 2)$ **97.** $(-1, 4)$

Getting Ready for the Next Section:
101. false **103.** $\pi, \dfrac{\pi}{2}, 0$ **105.** $\sqrt{5}$ **107.** $-\dfrac{1}{\sqrt{65}}$

Section 6.5

Concepts and Vocabulary:
1. $a_1 b_1 + a_2 b_2$ **3.** 0 **5.** True **7.** False

Building Skills:
9. -7 **11.** 0 **13.** 14 **15.** -6 **17.** $5\sqrt{3} \approx 8.7$ **19.** -10.0 **21.** 8.5
23. -10.5 **25.** $3.2°$ **27.** $168.7°$ **29.** $90°$ **31.** $0°$ **33.** $112.6°$
35. Parallel **37.** Not parallel **39.** Parallel **41.** 25.4 **43.** 36.4
45. $21.2°$ **47.** $19.7°$ **49.** Orthogonal **51.** Not orthogonal
53. Orthogonal **55.** $c = -2, c = \dfrac{9}{2}$ **57.** $c = -\dfrac{9}{2}, c = 8$
59. $c = 2, c = -2$ **61.** $c = \dfrac{3}{2}, c = -6$ **63.** $\langle 4, 1 \rangle$ **65.** $\left\langle \dfrac{24}{25}, \dfrac{32}{25} \right\rangle$
67. $\langle 5, 0 \rangle$ **69.** $\left\langle \dfrac{a}{2} + \dfrac{b}{2}, \dfrac{a}{2} + \dfrac{b}{2} \right\rangle$ **71.** $\langle 2, 0 \rangle + \langle 0, 3 \rangle$

73. $\left\langle \dfrac{1}{2}, \dfrac{1}{2} \right\rangle + \left\langle \dfrac{1}{2}, -\dfrac{1}{2} \right\rangle$ **75.** $\langle 0, 0 \rangle + \langle 4, 2 \rangle$

77. $\left\langle \dfrac{24}{25}, \dfrac{32}{25} \right\rangle + \left\langle \dfrac{76}{25}, -\dfrac{57}{25} \right\rangle$

Applying the Concepts:
79. 13.5 lb, 12.8°, 17.2° **81.** 560 foot-pounds **83.** 5520.2 pounds.
85. 56 foot-pounds

Beyond the Basics:
95. $\pm \left\langle \dfrac{4}{5}, \dfrac{3}{5} \right\rangle$

Getting Ready for the Next Section:
105. Vertical line; x-intercept: -1, no y-intercept
107. Parabola; x-intercepts: $-2, 3$, y-intercept: -6
109. Circle; x-intercepts: $\pm\sqrt{3}$, y-intercepts: $-1, 3$
111. Circle; no x-intercepts, y-intercept: $-2 \pm \sqrt{2}$
113. Only the x-axis **115.** Only the y-axis

Section 6.6

Concepts and Vocabulary:
1. origin, axis **3.** $r\cos\theta, r\sin\theta$ **5.** True **7.** False

Building Skills:
9.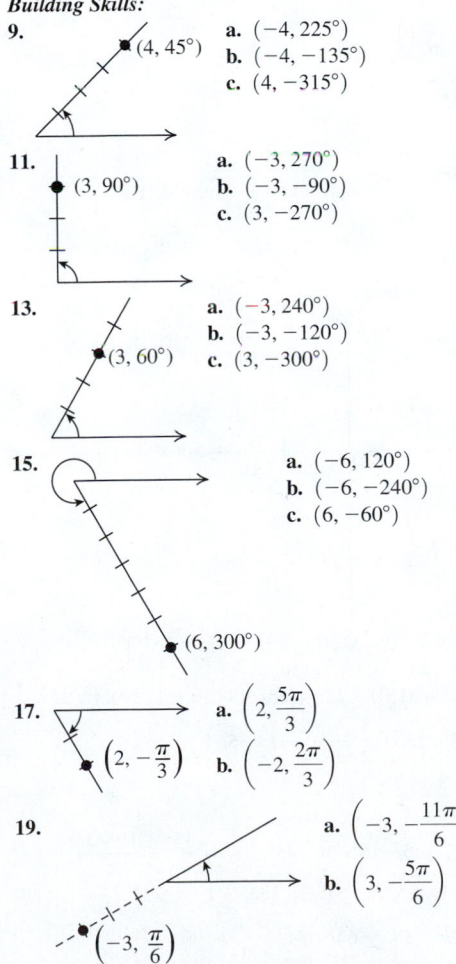

a. $(-4, 225°)$
b. $(-4, -135°)$
c. $(4, -315°)$

11. $(3, 90°)$
a. $(-3, 270°)$
b. $(-3, -90°)$
c. $(3, -270°)$

13. $(3, 60°)$
a. $(-3, 240°)$
b. $(-3, -120°)$
c. $(3, -300°)$

15.
a. $(-6, 120°)$
b. $(-6, -240°)$
c. $(6, -60°)$
$(6, 300°)$

17. $\left(2, -\dfrac{\pi}{3}\right)$
a. $\left(2, \dfrac{5\pi}{3}\right)$
b. $\left(-2, \dfrac{2\pi}{3}\right)$

19. $\left(-3, \dfrac{\pi}{6}\right)$
a. $\left(-3, -\dfrac{11\pi}{6}\right)$
b. $\left(3, -\dfrac{5\pi}{6}\right)$

21. $\left(-2, -\dfrac{\pi}{6}\right)$
a. $\left(-2, \dfrac{11\pi}{6}\right)$
b. $\left(2, \dfrac{5\pi}{6}\right)$

23.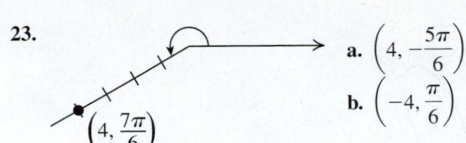
a. $\left(4, -\dfrac{5\pi}{6}\right)$
b. $\left(-4, \dfrac{\pi}{6}\right)$
$\left(4, \dfrac{7\pi}{6}\right)$

25. $\left(\dfrac{3}{2}, \dfrac{3\sqrt{3}}{2}\right)$ **27.** $\left(\dfrac{5}{2}, -\dfrac{5\sqrt{3}}{2}\right)$ **29.** $(-3, 0)$ **31.** $(\sqrt{3}, 1)$

33. $\left(\sqrt{2}, \dfrac{7\pi}{4}\right)$ **35.** $\left(3\sqrt{2}, \dfrac{\pi}{4}\right)$ **37.** $\left(3\sqrt{2}, \dfrac{7\pi}{4}\right)$ **39.** $\left(2, \dfrac{2\pi}{3}\right)$

41. $r = 4$ **43.** $r = -2\csc\theta$ **45.** $r = 4\cot\theta\csc\theta$
47. $r = 3\tan^2\theta\sec$ **49.** $r = 6\sin\theta$
51. $r^2 - 4r\cos\theta + 6r\sin\theta = 12$
53. $x^2 + y^2 = 4$; circle centered at $(0, 0)$ with radius 2
55. $y = -x$; a line with slope -1 and y-intercept 0 **57.** $x = -2$
59. $(x - 2)^2 + y^2 = 4$; a circle centered at $(2, 0)$ with radius 2
61. $x^2 + (y + 1)^2 = 1$; a circle centered at $(0, -1)$ with radius 1

63.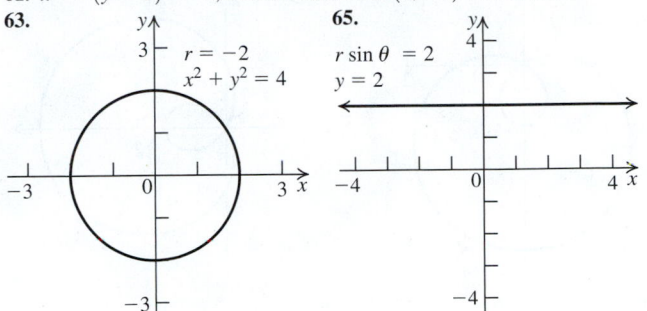

65.

67.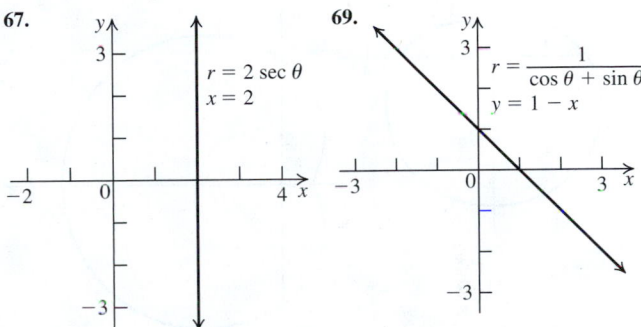

69.

71.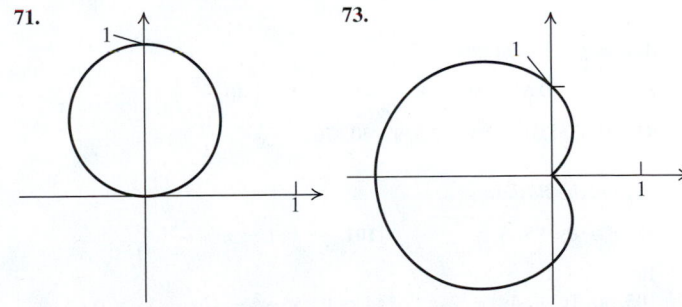

73.

75.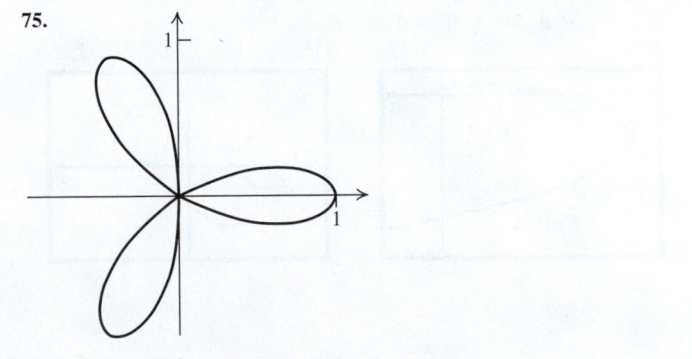

77.

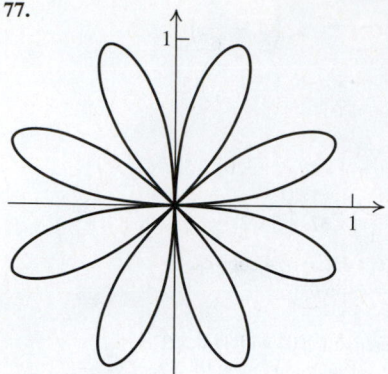

79.

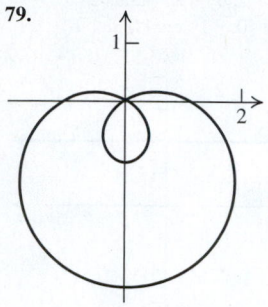

81.

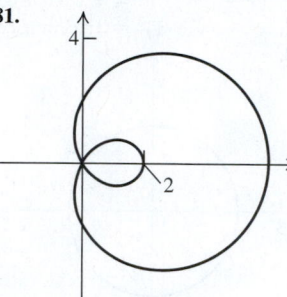

83. **85.**

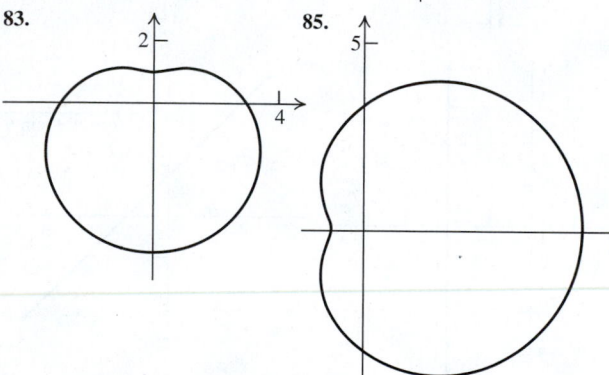

Applying the Concepts:

87. $(21.8, 39.6°)$ **89.** $(18.3, -45.8°)$ **91.** $\left(40, \dfrac{2\pi}{3}\right)$

93. $0.983, 1.017$ **95.** $29.8291, 30.3709$

Beyond the Basics:

97. $r = \theta$ **99.** $y = \dfrac{x^2}{6} - \dfrac{3}{2}$ **101.** $y^2 = 1 - x - \dfrac{3x^2}{4}$

103. $y^2 - 8x^2 - 30x - 25 = 0$

105. a. $3x^2 + 4y^2 - 24x - 144 = 0$ **b.** $y^2 - 12x - 36 = 0$
c. $3x^2 - y^2 + 24x + 36 = 0$

113. **115.**

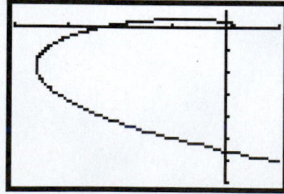 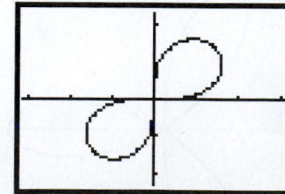

117. **119.**

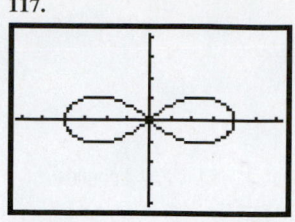 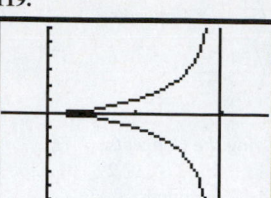

121.

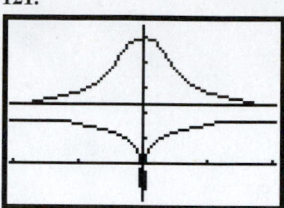

Critical Thinking / Discussion / Writing:

125. Replacing (r, θ) with $(-r, -\theta)$ yields no conclusion. Replacing (r, θ) with $(r, \pi - \theta)$ shows that the graph is symmetric about the y-axis.

Getting Ready for the Next Section:

127. $-3 - 2i$ **129.** $-5 + 2i$ **131.** $7 + i$ **133.** $-5 - 12i$

135. $\dfrac{3}{25} + \dfrac{4}{25}i$ **137.** $-\dfrac{6}{25} - \dfrac{17}{25}i$ **139.** $\sqrt{2}\left(\cos \dfrac{\pi}{4}\mathbf{i} + \sin \dfrac{\pi}{4}\mathbf{j}\right)$

141. $2\left(\cos \dfrac{5\pi}{6}\mathbf{i} + \sin \dfrac{5\pi}{6}\mathbf{j}\right)$

Section 6.7

Concepts and Vocabulary:

1. $\sqrt{a^2 + b^2}, \tan^{-1}\left(\dfrac{b}{a}\right)$ **3.** moduli, add   **5.** True   **7.** False

Building Skills:
9–16.

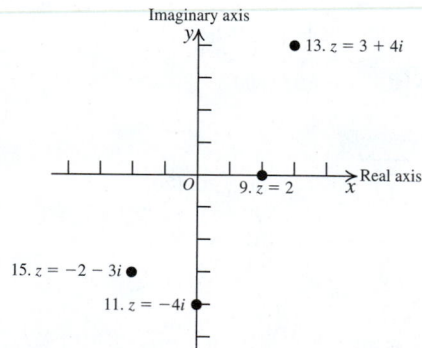

17. $2(\cos 60° + i \sin 60°)$ **19.** $\sqrt{2}(\cos 135° + i \sin 135°)$
21. $\cos 90° + i \sin 90°$ **23.** $\cos 0° + i \sin 0°$
25. $3\sqrt{2}(\cos 315° + i \sin 315°)$ **27.** $4(\cos 300° + i \sin 300°)$
29. $1 + \sqrt{3}i$ **31.** -3 **33.** $-\dfrac{5}{2} - \dfrac{5\sqrt{3}}{2}i$ **35.** 8
37. $-3\sqrt{3} + 3i$ **39.** $\dfrac{3}{2} - \dfrac{3\sqrt{3}}{2}i$
41. $z_1 z_2 = 8i$ **43.** $z_1 z_2 = 5 - 5\sqrt{3}i$ **45.** $z_1 z_2 = \dfrac{15}{2} + \dfrac{15\sqrt{3}}{2}i$
47. $-3 - 3\sqrt{3}i$ **49.** $1 + \sqrt{3}i$ **51.** $2i$ **53.** $-1 - \sqrt{3}i$
55. $\dfrac{3}{4} - \dfrac{3\sqrt{3}}{4}i$ **57.** 2^{12} **59.** $-64i$ **61.** $-\dfrac{1}{64}i$
63. $\dfrac{1}{2^{10}}i$ **65.** -64 **67.** i **69.** $-\dfrac{1}{2} + \dfrac{\sqrt{3}}{2}i$
71. $4(\cos 0° + i \sin 0°), 4(\cos 120° + i \sin 120°),$
$4(\cos 240° + i \sin 240°)$ **73.** $\cos 45° + i \sin 45°, \cos 225° + i \sin 225°$
75. $\cos 30° + i \sin 30°, \cos 90° + i \sin 90°, \cos 150° + i \sin 150°,$
$\cos 210° + i \sin 210°, \cos 270° + i \sin 270°, \cos 330° + i \sin 330°$
77. $\sqrt{2}(\cos 150° + i \sin 150°), \sqrt{2}(\cos 330° + i \sin 330°)$

Applying the Concepts:
81. $15(\cos 30° + i\sin 30°)$ **83.** $1.8(\cos 153.2° + i\sin 153.2°)$

Beyond the Basics:
85. $2(\cos 30° + i\sin 30°)$ **87.** $\cos 240° + i\sin 240°$
89. $\cos 240° + i\sin 240°$ **91.** $\cos 0° + i\sin 0°$, $\cos 90° + i\sin 90°$,
$\cos 180° + i\sin 180°$, $\cos 270° + i\sin 270°$ **93.** $2^{\frac{1}{6}}(\cos 15° + i\sin 15°)$,
$2^{\frac{1}{6}}(\cos 135° + i\sin 135°)$, $2^{\frac{1}{6}}(\cos 255° + i\sin 255°)$
95. $\pm\left(\dfrac{\sqrt{6}}{2} + \dfrac{\sqrt{2}}{2}i\right)$, $\pm\left(\dfrac{\sqrt{6}}{2} - \dfrac{\sqrt{2}}{2}i\right)$ **109.** $\{-1, i, -i\}$

Critical Thinking / Discussion / Writing:
111. a. True **b.** False **c.** True **d.** False **e.** True **f.** True
g. True **h.** True **i.** False **j.** True

Getting Ready for the Next Section:
113. 11 **115.** 2 **117.** -3 **119.** $x + 2y = 9$

Review Exercises

Building Skills:
1. $B = 60°, b \approx 10.4, c = 12$ **3.** $B = 53°, a \approx 3.0, c \approx 5.0$
5. $B = 50°, a \approx 7.7, b \approx 9.2$ **7.** $A \approx 31.0°, B \approx 59.0°, c \approx 5.8$
9. $A \approx 41.8°, B \approx 48.2°, b \approx 4.5$ **11.** $C = 105°, a \approx 66.5, c \approx 59.4$
13. No triangle exists. **15.** $B \approx 74.2°, a \approx 36.8, c \approx 41.4$
17. $B \approx 54.1°, C \approx 60.7°, c \approx 20.5$ **19.** Answers will vary. Sample
answers: **i.** 18, **ii.** $10\sqrt{3}$, **iii.** 10 **21.** $A \approx 26.6°, B \approx 42.1°$,
$C \approx 111.3°$ **23.** $A \approx 51.7°, C \approx 48.3°, b \approx 50.2$
25. $A \approx 32.5°, B \approx 49.8°, C \approx 97.7°$ **27.** $A \approx 32.0°, B \approx 18.0°$,
$c \approx 17.3$ **29.** 16 square meters **31.** 11 square feet **33.** $-i + 2j$
35. $\langle -10, 15 \rangle$ **37.** $\langle -16, 21 \rangle$ **39.** $\dfrac{\sqrt{2}}{2}i + \dfrac{\sqrt{2}}{2}j$
41. $\left\langle \dfrac{3\sqrt{34}}{34}, -\dfrac{5\sqrt{34}}{34} \right\rangle$ **43.** $3\sqrt{3}i + 3j$ **45.** $-6\sqrt{2}i - 6\sqrt{2}j$
47. -6; $\text{proj}_w \mathbf{v} = \left\langle -\dfrac{18}{25}, -\dfrac{24}{25} \right\rangle$ **49.** 0; $\text{proj}_w \mathbf{v} = \langle 0, 0 \rangle$
51. $60.3°$ **53.** $101.3°$
55.

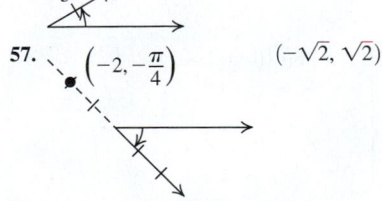

57.

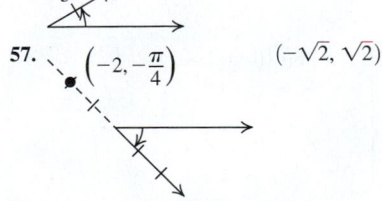

59. $\left(2\sqrt{2}, \dfrac{3\pi}{4}\right)$ **61.** $\left(4, \dfrac{11\pi}{6}\right)$ **63.** $r = \dfrac{12}{3\cos\theta + 2\sin\theta}$
65. $r = 8\cos\theta$ **67.** $x^2 + y^2 = 9$ **69.** $y = 3$
71. $(x^2 + y^2 + 2y)^2 = x^2 + y^2$ **73.** $3\left(\cos\dfrac{3\pi}{2} + i\sin\dfrac{3\pi}{2}\right)$
75. $10\left(\cos\dfrac{11\pi}{6} + i\sin\dfrac{11\pi}{6}\right)$ **77.** $\sqrt{2} + \sqrt{2}i$ **79.** $-3\sqrt{2} + 3\sqrt{2}i$
81. $z_1 z_2 = 6(\cos 35° + i\sin 35°)$; $\dfrac{z_1}{z_2} = \dfrac{3}{2}(\cos 15° + i\sin 15°)$
83. $z_1 z_2 = 6\left(\cos\dfrac{7\pi}{6} + i\sin\dfrac{7\pi}{6}\right)$; $\dfrac{z_1}{z_2} = \dfrac{2}{3}\left(\cos\dfrac{\pi}{2} + i\sin\dfrac{\pi}{2}\right)$
85. $27(\cos 120° + i\sin 120°)$ **87.** $4096(\cos 0 + i\sin 0)$
89. $5(\cos 60° + i\sin 60°)$, $5(\cos 180° + i\sin 180°)$,
$5(\cos 300° + i\sin 300°)$ **91.** $2^{\frac{1}{5}}(\cos 24° + i\sin 24°)$,
$2^{\frac{1}{5}}(\cos 96° + i\sin 96°)$, $2^{\frac{1}{5}}(\cos 168° + i\sin 168°)$,
$2^{\frac{1}{5}}(\cos 240° + i\sin 240°)$, $2^{\frac{1}{5}}(\cos 312° + i\sin 312°)$

Applying the Concepts:
93. 268.5 feet **95.** 94.7 miles **97.** 24,625 square feet
99. 1204.4 foot-pounds

Practice Test A:
1. $B = 30°, a \approx 23.7, b \approx 13.1$
2. $C = 101°, b \approx 45.0\text{ m}, c \approx 73.4\text{ m}$
3. $A \approx 28.2°, C \approx 45.8°, b \approx 71.1$
4. $A \approx 82.8°, B \approx 41.4°, C \approx 55.8°$
5. $c \approx 6.9, A \approx 53.8°, B \approx 36.2°$ **6.** $32.7°$ **7.** 803 feet
8. $\langle -1, -12 \rangle$ **9.** $\langle -3, -2 \rangle$ **10.** $-7i - 9j$ **11.** $\dfrac{3\sqrt{3}}{2}i - \dfrac{3}{2}j$
12. 17 **13.** $164.9°$ **14.** $(-\sqrt{2}, \sqrt{2})$ **15.** $\left(2, \dfrac{7\pi}{6}\right)$
16. $x^2 + y^2 = 9$; circle with center $(0,0)$ and radius 3.
17. $-\dfrac{3}{2} + \dfrac{3\sqrt{3}}{2}i$ **18.** $\sqrt{2}(\cos 165° + i\sin 165°)$ **19.** $\dfrac{1}{125}i$
20. $2^{\frac{1}{8}}(\cos 11.25° + i\sin 11.25°)$, $2^{\frac{1}{8}}(\cos 101.25° + i\sin 101.25°)$,
$2^{\frac{1}{8}}(\cos 191.25° + i\sin 191.25°)$, $2^{\frac{1}{8}}(\cos 281.25° + i\sin 281.25°)$

Practice Test B:
1. a **2.** c **3.** c **4.** d **5.** a **6.** b **7.** d **8.** c **9.** a **10.** c **11.** a
12. d **13.** d **14.** b **15.** c **16.** b **17.** b **18.** c **19.** c **20.** c

Cumulative Review Exercises (Chapters 1–6):
1. $(0, 2)\cup(2, \infty)$ **2.** $f^{-1}(x) = \dfrac{x - 7}{2}$
3. **4.** $(-\infty, 2)\cup(4, \infty)$

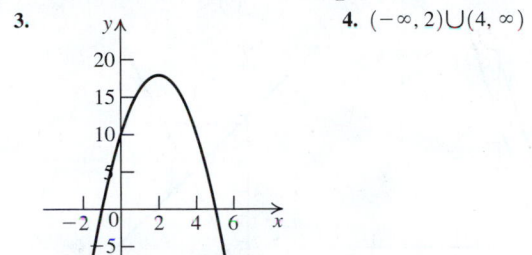

5. $\left\{\dfrac{2\pi}{3}, \dfrac{5\pi}{6}, \dfrac{5\pi}{3}, \dfrac{11\pi}{6}\right\}$ **6.** $\left\{\dfrac{\pi}{6}, \dfrac{5\pi}{6}, \dfrac{7\pi}{6}, \dfrac{11\pi}{6}\right\}$
7. $2^{\frac{1}{4}}(\cos 7.5° + i\sin 7.5°)$, $2^{\frac{1}{4}}(\cos 97.5° + i\sin 97.5°)$,
$2^{\frac{1}{4}}(\cos 187.5° + i\sin 187.5°)$, $2^{\frac{1}{4}}(\cos 277.5° + i\sin 277.5°)$
8. $2(\cos 30° + i\sin 30°)$, $2(\cos 90° + i\sin 90°)$,
$2(\cos 150° + i\sin 150°)$, $2(\cos 210° + i\sin 210°)$,
$2(\cos 270° + i\sin 270°)$, $2(\cos 330° + i\sin 330°)$
9. $y = \dfrac{3}{2}x + \dfrac{13}{2}$ **10.** $\log\dfrac{x^2\sqrt{y + 1}}{3x + 1}$
11. **12.**

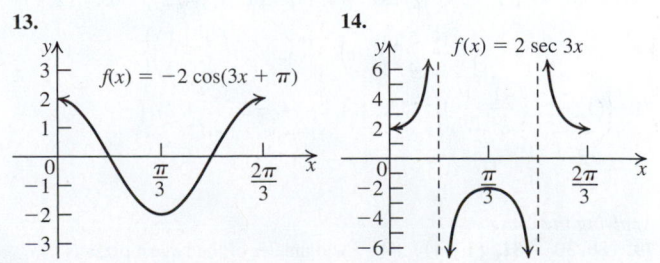

13. **14.**

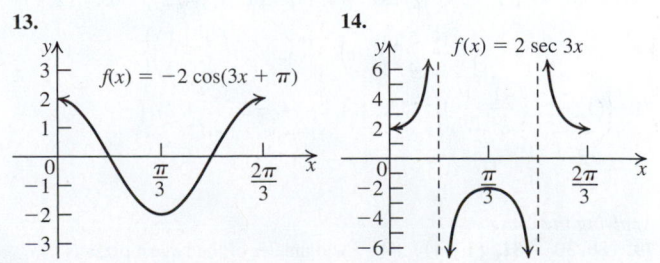

15. $-\dfrac{12}{5}$ **16.** $\dfrac{\sqrt{3}}{2}$ **19.** $C = 69°, a \approx 58.2\text{ m}, b \approx 46.2\text{ m}$ **20.** 3.4 m

CHAPTER 7

Section 7.1

Concepts and Vocabulary:

1. solution **3.** inconsistent **5.** False **7.** True

Building Skills:

9. $(3, -1)$ **11.** None of them **13.** $(10, -9)$
15. $(2, 1)$ **17.** $(2, 2)$

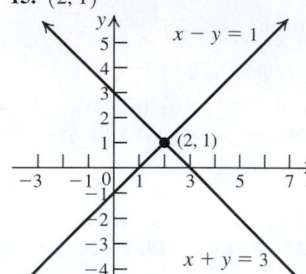
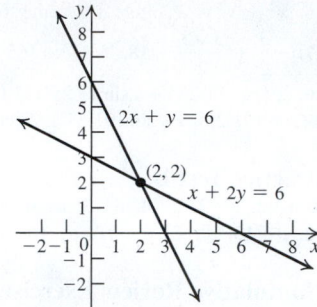

19. Inconsistent **21.** $\left(\dfrac{7}{3}, \dfrac{14}{3}\right)$

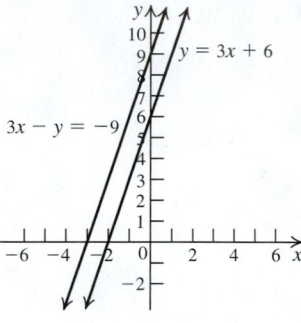
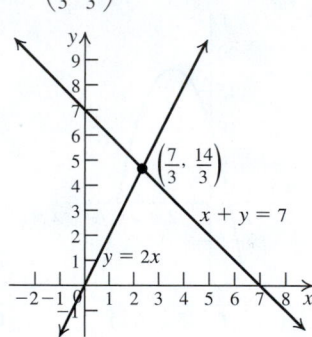

23. $\{(x, 12 - 3x)\}$

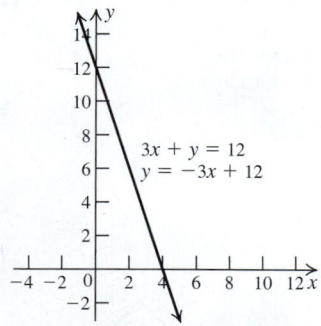

25. Independent **27.** Independent **29.** Dependent **31.** Independent

33. Inconsistent **35.** Inconsistent **37.** Independent **39.** $\left\{\left(\dfrac{7}{9}, \dfrac{23}{9}\right)\right\}$

41. $\{(3, 4)\}$ **43.** $\varnothing$ **45.** $\{(1, -1)(3, 1)\}$ **47.** $\left\{\left(x, \tfrac{1}{2}(x - 5)\right)\right\}$

49. $\{(3, 2)\}$ **51.** $\{(-3, 3)\}$ **53.** $\{(2, 2), (2, -2), (-2, 2), (-2, -2)\}$

55. $\varnothing$ **57.** $\left\{\left(x, 2 - \dfrac{2}{3}x\right)\right\}$ **59.** $\{(4, 1)\}$ **61.** $\{(-4, 2)\}$ **63.** $\{(2, 1)\}$

65. $\{(3, 1)\}$ **67.** $\left\{\left(\dfrac{19}{6}, \dfrac{5}{4}\right)\right\}$ **69.** $\left\{(3, -1), \left(\dfrac{9}{4}, -\dfrac{1}{2}\right)\right\}$

71. $\left\{(1, -1), \left(\dfrac{41}{29}, \dfrac{1}{29}\right)\right\}$ **73.** $\{(4, 2), (4, -2), (-4, 2), (-4, -2)\}$

75. $\left\{\left(\dfrac{3}{2}, \dfrac{1}{2}\right)\right\}$ **77.** $\{(5, 5)\}$

Applying the Concepts:

79. $(80, 30)$ **81.** $(17, 4)$ **83.** The diameter of the largest pizza is 21 inches, and the diameter of the smallest pizza is 8 inches.

85. 8% of the total trash collected is plastic, and 40% is paper.
87. 340 beads and 430 doubloons **89.** 3 Egg McMuffins and 2 Sausage Burritos **91.** \$18,000 at 7.5% and \$32,000 at 12% **93.** \$7.50 per hour at McDougal's and \$15 per hour for tutoring **95.** 20 pounds of herb and 80 pounds of tea **97.** The speed of the plane is 550 km/hr, and the wind speed is 50 km/hr.
99. a. $C(x) = 30,000 + 2x; R(x) = 3.5x$
b.

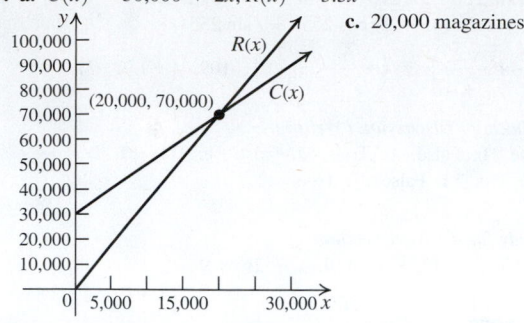

c. 20,000 magazines

101. Her weekly sales should be more than \$6250.

Beyond the Basics:

103. $\{(15, 3)\}$ **105.** $c = 2$ **107.** $c = \dfrac{35}{2}$

109. $y = -\dfrac{3}{2}x + \dfrac{3}{2}$ **111.** $y = \dfrac{13}{7}x + \dfrac{6}{7}$

Critical Thinking / Discussion / Writing:

113. a. $3y - 4x = 4$ **b.** $\left(\dfrac{4}{5}, \dfrac{12}{5}\right)$ **c.** 7
115. a. $\sqrt{2}$ **b.** $\dfrac{6\sqrt{5}}{5}$ **c.** 0 **d.** $\dfrac{|c|}{\sqrt{a^2 + b^2}}$ if $a \neq 0$ or $b \neq 0$

Getting Ready for the Next Section:

117. 3 **119.** 4 **121.** -2

Section 7.2

Concepts and Vocabulary:

1. equivalent **3.** inconsistent **5.** True **7.** False

Building Skills:

9. Yes **11.** No **13.** $\{(3, 2, -1)\}$ **15.** $\{(-3, 2, 4)\}$

17. $\begin{cases} x - y - \dfrac{3}{2}z = \dfrac{1}{2} \\ y + z = 0 \\ z = 1 \end{cases}$ **19.** $\begin{cases} x + 3y - 2z = 0 \\ y - \dfrac{8}{7}z = -\dfrac{5}{7} \\ 0 = k, k \neq 0 \end{cases}$

21. $\{(1, -2, 3)\}$ **23.** $\left\{\left(\dfrac{30}{11}, -\dfrac{8}{11}, -\dfrac{1}{4}\right)\right\}$ **25.** $\{(1, 2, 3)\}$

27. $\left\{\left(\dfrac{35}{18}, \dfrac{29}{18}, \dfrac{5}{18}\right)\right\}$ **29.** Inconsistent **31.** $\{(3, -1, 2)\}$

33. $\{(2, 2, 1)\}$ **35.** $\{(3, -1, 1)\}$ **37.** $\left\{\left(3 + \dfrac{4}{5}z, -\dfrac{1}{5}z, z\right)\right\}$

39. $\{(12, -12, 4)\}$ **41.** $\{(1, -1, 2)\}$ **43.** $\{(3, 2, -1)\}$

45. $\{(9z - 5, -3z - \tfrac{1}{6}, z)\}$

Applying the Concepts:

47. \$4000 invested at 4%, \$6000 invested at 5%, and \$10,000 invested at 6% **49.** Alex worked 2 hours, Becky worked 2.5 hours, and Courtney worked 1.5 hours. **51.** 56 nickels, 225 dimes, and 19 quarters
53. 31 hours for normal daytime work, 14 hours at night, and 8 hours on a holiday **55.** Cell A contains healthy tissue (because $x = 0.21$), cell B contains tumorous tissue (because $y = 0.33$), and cell C contains healthy tissue (because $z = 0.19$). **57.** Cells A, B, and C contain healthy tissue (because $x = 0.28$, $y = 0.23$, and $z = 0.21$).

Beyond the Basics:

59. $x + y + z = 1$　**61.** $x + \dfrac{2}{3}y + \dfrac{1}{3}z = \dfrac{1}{3}$

63. $y = x^2 + 2x + 1$　**65.** $y = x^2 - x + 2$　**67.** $x^2 + y^2 - 16 = 0$

69. $x^2 + y^2 - 2x + 6y - 15 = 0$　**71.** $\left\{\left(-\dfrac{9}{14}, \dfrac{9}{19}, -\dfrac{9}{2}\right)\right\}$

73. $c = -\dfrac{16}{5}$　**75.** $y = -2x^2 + 4x + 5$

Critical Thinking / Discussion / Writing:

77. $\begin{cases} x + y - z = -2 \\ 2x - y + 3z = 9 \\ x + y + z = 2 \end{cases}$　Answers may vary.

Getting Ready for the Next Section:

79.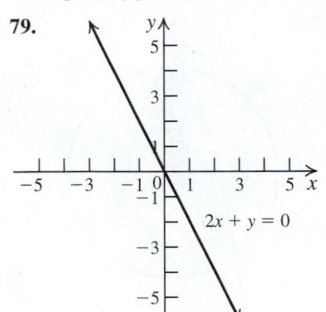

81.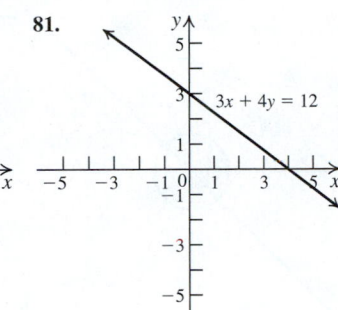

83. 85.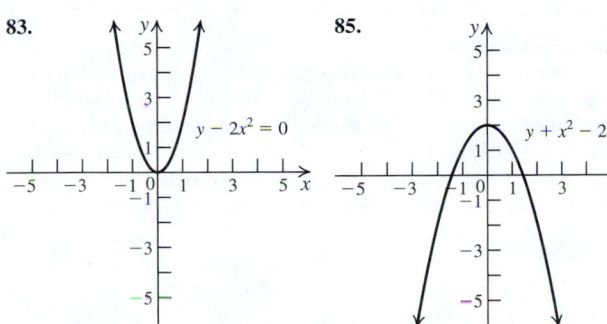

87. $(0, 0), (-3, 5)$　**89.** $(1, 4), (-1, 4)$

Section 7.3

Concepts and Vocabulary:

1. dashed　**3.** not a solution of the system　**5.** True　**7.** True

Building Skills:

9. 11.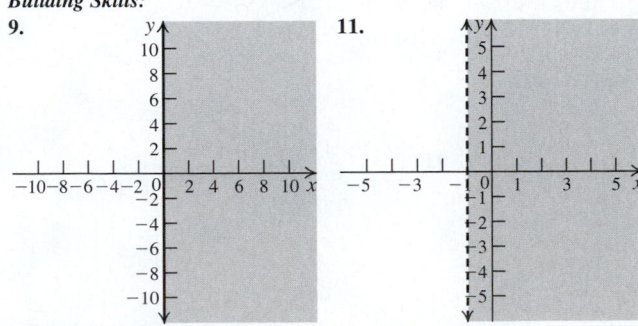

13.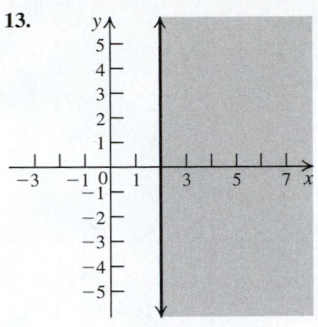

15.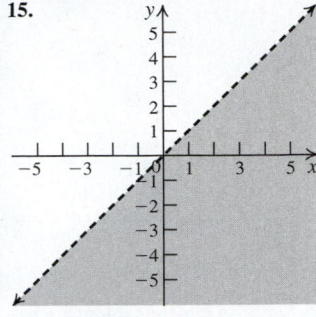

17.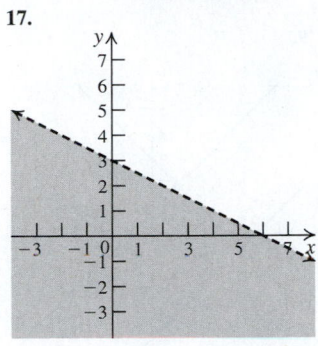

19.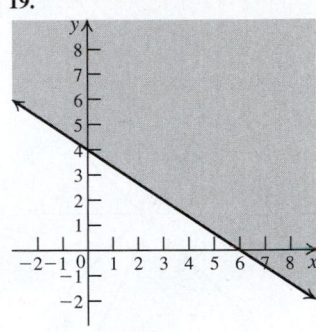

21.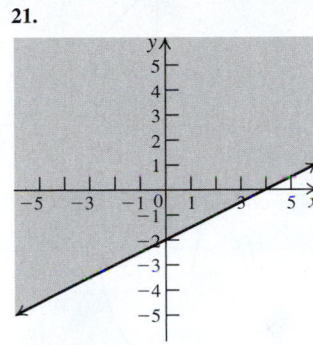

23.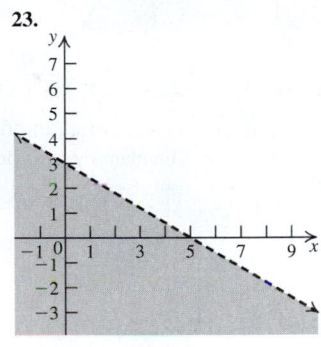

25.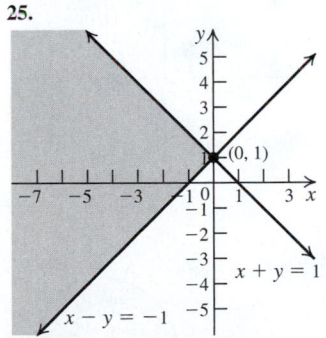

27.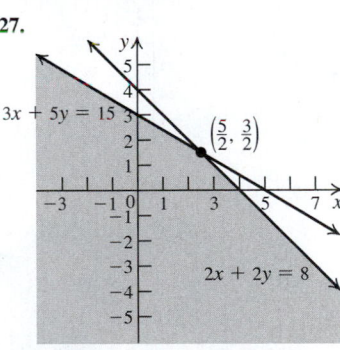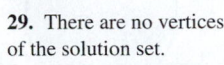

29. There are no vertices of the solution set.

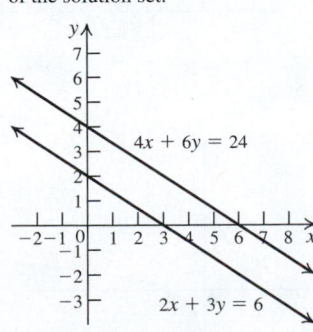

31.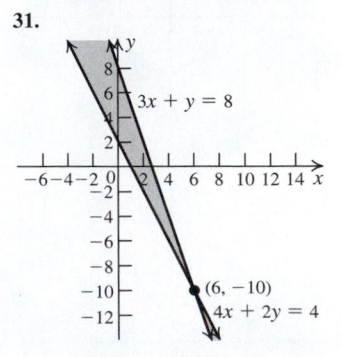

33.

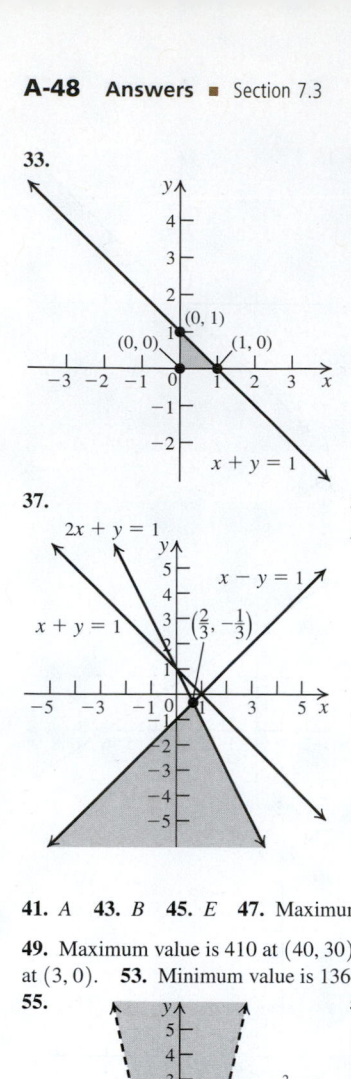

35.

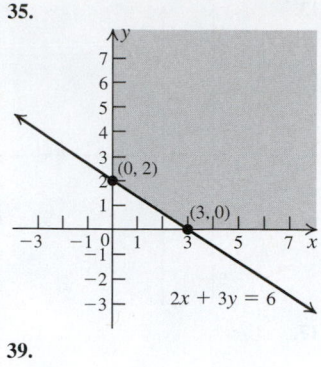

67. D, K **69.** E, L **71.** E, B
73.

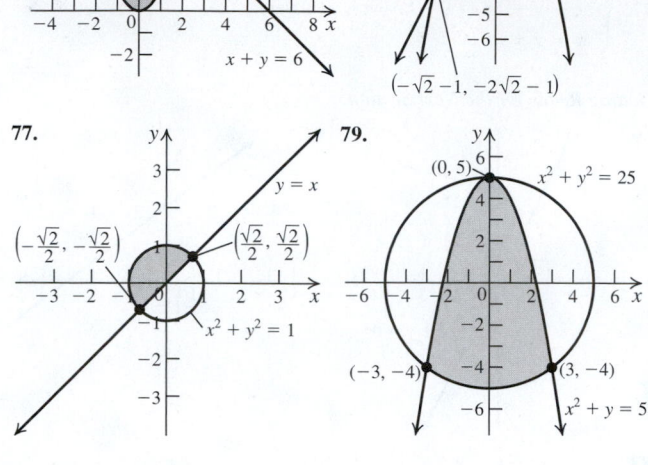

75.

37.

39.

77.

79.

41. A **43.** B **45.** E **47.** Maximum value is $\dfrac{1284}{19}$ at $\left(\dfrac{56}{19}, \dfrac{60}{19}\right)$.

49. Maximum value is 410 at $(40, 30)$.   **51.** Minimum value is 3 at $(3, 0)$.   **53.** Minimum value is 1364 at $(8, 84)$.

55.

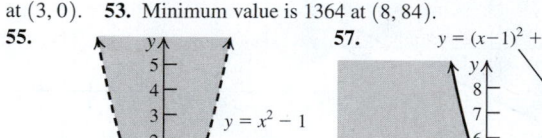

57.

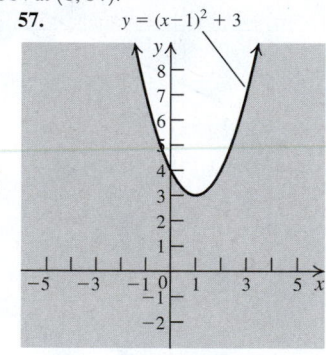

Applying the Concepts:
81. 5.25 minutes of television and one page of newspaper advertisement to obtain a maximum of 710,000 viewers.   **83.** 50 GPSs and 20 DVD players should be made to obtain a maximum profit of $380.   **85.** She should invest $20,000 in Treasury bonds and $20,000 in corporate bonds.
87. The Michigan factory should operate 150 days and the North Carolina factory should operate 200 days for a minimum cost of 17,000,000.
89.

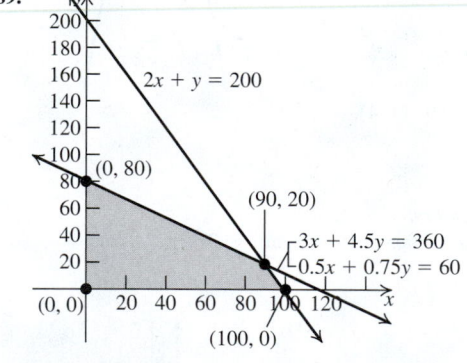

59.

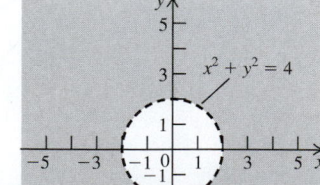

61.

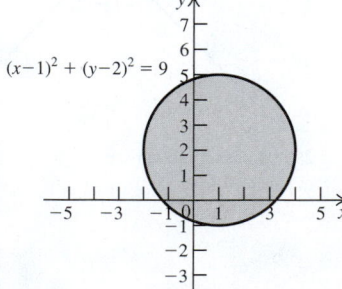

Beyond the Basics:
91. $\begin{cases} x \geq 0 \\ y \geq 0 \\ y \leq \dfrac{-2}{3}x + 2 \end{cases}$ **93.** $\begin{cases} x \geq -1 \\ x \leq 3 \end{cases}$ **95.** $\begin{cases} x \geq 0 \\ y \geq 0 \\ y \leq \dfrac{-1}{2}x + 4 \\ y \leq \dfrac{-3}{2}x + 6 \end{cases}$

63.

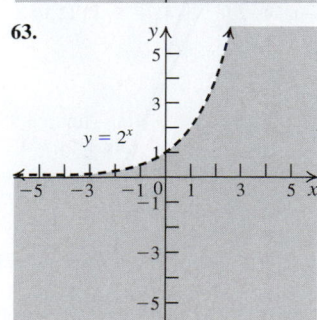

65.

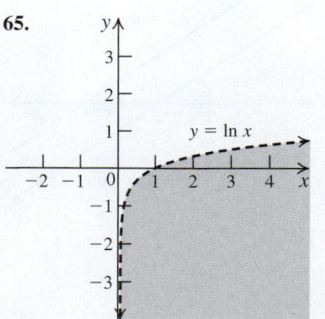

97. $\begin{cases} x \geq 0 \\ y \geq x - 4 \\ y \geq -x + 6 \\ y \leq -x + 16 \end{cases}$ **99.**

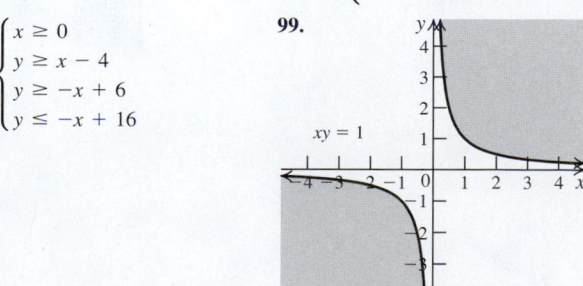

101.

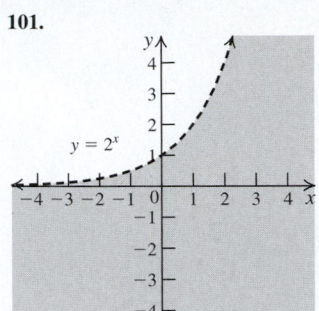

103.

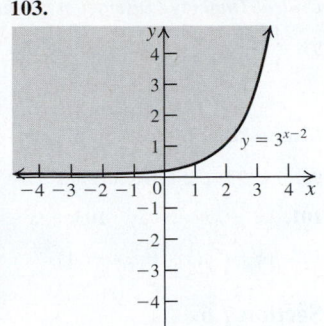

105.

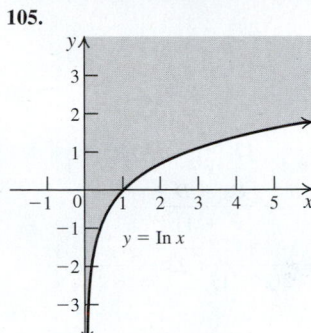

107.

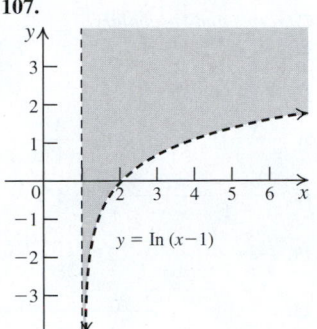

109.

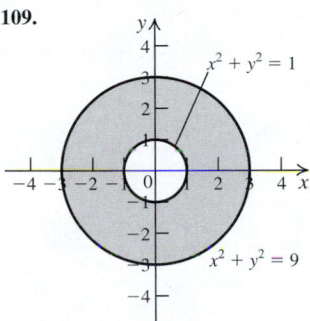

Critical Thinking / Discussion / Writing:

111.

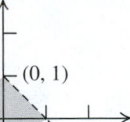
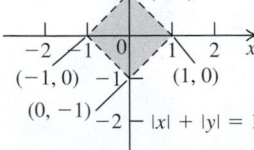

112.

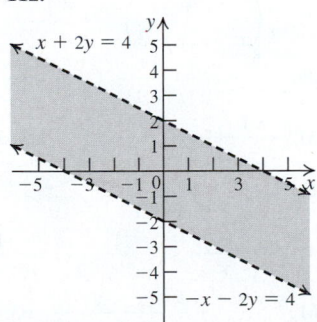

Getting Ready for the Next Section:

113. $(-6, -5)$ **115.** $(-1, -5, 8)$ **117.** $\begin{bmatrix} 1 & 2 & -3 & 5 \\ 3 & -2 & 4 & 12 \end{bmatrix}$

119. $\begin{bmatrix} 1 & 2 & -3 & 5 \\ 0 & 1 & -\frac{13}{8} & \frac{3}{8} \end{bmatrix}$

Section 7.4

Concepts and Vocabulary:
1. numbers   **3.** row operations   **5.** False   **7.** False

Building Skills:
9. 1×1 **11.** 2×4 **13.** 2×3
15. $a_{13} = 3, a_{31} = 9, a_{33} = 11, a_{34} = 12$ **17.** No, it is not a rectangular array of numbers.

19. $\begin{bmatrix} 2 & 4 & 2 \\ 1 & -3 & 1 \end{bmatrix}$ **21.** $\begin{bmatrix} 5 & -7 & 11 \\ -13 & 17 & 19 \end{bmatrix}$ **23.** $\begin{bmatrix} -1 & 2 & 3 & 8 \\ 2 & -3 & 9 & 16 \\ 4 & -5 & -6 & 32 \end{bmatrix}$

25. $\begin{cases} x + 2y - 3z = 4 \\ -2x - 3y + z = 5 \\ 3x - 3y + 2z = 7 \end{cases}$ **27.** $\begin{cases} x - y + z = 2 \\ 2x + y - 3z = 6 \end{cases}$

29. $\begin{bmatrix} 1 & 2 & 3 \\ 0 & 1 & 1 \end{bmatrix}$ **31.** $\begin{bmatrix} 1 & 2 & 3 & 4 \\ 0 & 1 & 5 & -3 \\ 0 & 0 & 1 & -1 \end{bmatrix}$

33. $\begin{bmatrix} 4 & 5 & -7 \\ 5 & 4 & -2 \end{bmatrix} \xrightarrow{\frac{1}{4}R_1} \begin{bmatrix} 1 & \frac{5}{4} & -\frac{7}{4} \\ 5 & 4 & -2 \end{bmatrix} \xrightarrow{-5R_1 + R_2 \rightarrow R_2}$

$\begin{bmatrix} 1 & \frac{5}{4} & -\frac{7}{4} \\ 0 & -\frac{9}{4} & \frac{27}{4} \end{bmatrix} \xrightarrow{-\frac{4}{9}R_2} \begin{bmatrix} 1 & \frac{5}{4} & -\frac{7}{4} \\ 0 & 1 & -3 \end{bmatrix}$

35. $\begin{bmatrix} 1 & 4 & 3 & 1 \\ 0 & -3 & -2 & 0 \\ 0 & 7 & 5 & -3 \end{bmatrix} \xrightarrow{-\frac{1}{3}R_2} \begin{bmatrix} 1 & 4 & 3 & 1 \\ 0 & 1 & \frac{2}{3} & 0 \\ 0 & 7 & 5 & -3 \end{bmatrix}$

$\xrightarrow{-7R_2 + R_3 \rightarrow R_3} \begin{bmatrix} 1 & 4 & 3 & 1 \\ 0 & 1 & \frac{2}{3} & 0 \\ 0 & 0 & \frac{1}{3} & -3 \end{bmatrix} \xrightarrow{3R_3} \begin{bmatrix} 1 & 4 & 3 & 1 \\ 0 & 1 & \frac{2}{3} & 0 \\ 0 & 0 & 1 & -9 \end{bmatrix}$

37. No, because Property 2 is not satisfied.   **39.** The matrix is in reduced row-echelon form because Properties 1–4 are satisfied.   **41.** The matrix is in reduced row-echelon form because Properties 1–4 are satisfied.
43. The matrix is in reduced row-echelon form because Properties 1–4 are satisfied.

45. $(5, -2)$ $\begin{cases} x + 2y = 1 \\ y = -2 \end{cases}$

47. $(-4 - 4y, y, 3)$ $\begin{cases} x + 4y + 2z = 2 \\ z = 3 \end{cases}$

49. $(1, 2, -1);$ $\begin{cases} x + 2y + 3z = 2 \\ y - 2z = 4 \\ z = -1 \end{cases}$

51. $(2, -5, -2w + 3, w);$ $\begin{cases} x = 2 \\ y = -5 \\ z + 2w = 3 \end{cases}$

53. $(-5, 4, 3, 0)$ $\begin{cases} x = -5 \\ y = 4 \\ z + 2w = 3 \\ w = 0 \end{cases}$ **55.** $\{(5, -3)\}$

57. $\{(3, 1)\}$ **59.** $\left\{\left(-\frac{1}{5}, -\frac{23}{25}\right)\right\}$ **61.** $\{(2, 1)\}$ **63.** $\{(1, 2, 3)\}$

65. $\{(2, 3, 4)\}$ **67.** $\{(1, 2, 3)\}$ **69.** $\left\{\left(\frac{5}{2}, \frac{3}{2}, \frac{7}{2}\right)\right\}$

71. $\{(1 + 4z, 3 - 3z, z)\}$ **73.** $\{(1, 2, -1)\}$
75. a. One solution **b.** Infinitely many solutions **c.** No solution
77. False. Each row of A has seven entries.

Applying the Concepts:
79. a. $\begin{cases} a = 0.1b + 1000 \\ b = 0.2a + 780 \end{cases}$ **b.** $\begin{bmatrix} 1 & -0.1 & 1000 \\ -0.2 & 1 & 780 \end{bmatrix}$
c. $a = 1100, b = 1000$

81. a. $l = 0.4t + 0.2f + 10{,}000$
$t = 0.5l + 0.3t + 20{,}000$
$f = 0.5l + 0.05t + 0.35f + 10{,}000$

b. $\begin{bmatrix} 1 & -0.4 & -0.2 & 10{,}000 \\ -0.5 & 0.7 & 0 & 20{,}000 \\ -0.5 & -0.05 & 0.65 & 10{,}000 \end{bmatrix}$

c. food, \$55,000; transportation, \$61,000; labor, \$45,400
83. $T_1 = 110, T_2 = 140, T_3 = 60$

85. a. $\begin{cases} x - y & = 70 \\ -x + & z = -120 \\ y - z = & 50 \end{cases}$

b. $\{(120 + z, 50 + z, z)\} : 0 \le z \le 130$
87. $y = 2x^2 - 3x + 4$

Beyond the Basics:

89. $\{(y + 2w, y, -2y - 3w, w)\}$

91. a. $B = \begin{bmatrix} 1 & \frac{3}{2} & 0 & -1 \\ 0 & 1 & 0 & \frac{1}{2} \\ 0 & 0 & 1 & \frac{1}{2} \\ 0 & 0 & 0 & 1 \end{bmatrix}, C = \begin{bmatrix} 1 & 2 & -3 & 1 \\ 0 & 1 & 0 & -\frac{1}{2} \\ 0 & 0 & 1 & \frac{1}{2} \\ 0 & 0 & 0 & 1 \end{bmatrix}$ **b.** The

reduced row-echelon form of the matrices B and C is $\begin{bmatrix} 1 & 0 & 0 & 0 \\ 0 & 1 & 0 & 0 \\ 0 & 0 & 1 & 0 \\ 0 & 0 & 0 & 1 \end{bmatrix}$.

93. a. $\left\{ \left(\dfrac{dm - nb}{ad - bc}, \dfrac{an - cm}{ad - bc} \right) \right\}$ **b. (i)** $cb \ne ad$ **(ii)** $cb = ad$ and

$\dfrac{m}{b} \ne \dfrac{n}{d}$ **(iii)** $cb = ad$ and $\dfrac{m}{b} = \dfrac{n}{d}$ **95.** $\{(10, 100, 1000)\}$
97. $y = -x^3 + 2x^2 + 3x + 1$

Critical Thinking / Discussion / Writing:

99. $\begin{bmatrix} 0 \\ 0 \end{bmatrix}, \begin{bmatrix} 1 \\ 0 \end{bmatrix}$ **101.** Yes, the inverse of the operation used to transform A to B
103. 1 **105.** −1

Getting Ready for the Next Section:

107. $\{(5, -2)\}$ **109.** $\{(1, 2, -1)\}$

Section 7.5

Concepts and Vocabulary:
1. $ad - bc$ **3.** cofactor **5.** False **7.** True

Building Skills:

9. −2 **11.** −6 **13.** −7 **15.** $\dfrac{139}{72}$ **17.** $a - b$ **19. a.** −10 **b.** 2
c. 0 **21. a.** −4 **b.** 4 **c.** −2 **23.** −2 **25.** 12 **27.** 0 **29.** 42
31. 16 **33.** 49 **35.** $a^3 + b^3$ **37.** $a^3 + b^3 + c^3 - 3abc$

39. $\{(3, 5)\}$ **41.** $\{(1, 2)\}$ **43.** $\{(2, 0)\}$ **45.** $\left\{\left(\dfrac{3}{2}y + 2, y\right)\right\}$

47. $\{(2, 3)\}$ **49.** $\{(1, 1, 0)\}$ **51.** $\{(1, -1, 3)\}$ **53.** $\{(1, 1, 1)\}$
55. $\{(1, 2, 3)\}$ **57.** $\{(1, 2, 3)\}$

Applying the Concepts:
59. 11 **61.** 10.5 **63.** Yes **65.** No **67.** $y = 2x + 1$

69. $y = \dfrac{2}{3}x + \dfrac{1}{3}$

Beyond the Basics:

77. 21 **79.** 476 **81.** 4 **83.** −1, 2 **85.** 23 **87.** −1, 0, 1 **89.** 30

Critical Thinking / Discussion / Writing:

91. $\left\{\left(\dfrac{9}{4}, -\dfrac{2}{3}\right)\right\}$ **93.** $\{(-2, -1)\}$

Getting Ready for the Next Section:

95. 1. **97.** −2 **99.** $A = 1\, B = 2$

101. $(x + 2)(x - 5)$ **103.** $(3x - 2)(x + 1)$

105. $(x - 2)(x^2 + 2x + 4)$

Section 7.6

Concepts and Vocabulary:
1. proper **3.** three **5.** False **7.** True

Building Skills:

9. $\dfrac{A}{x - 1} + \dfrac{B}{x + 2}$ **11.** $\dfrac{A}{x + 6} + \dfrac{B}{x + 1}$ **13.** $\dfrac{A}{x^2} + \dfrac{B}{x} + \dfrac{C}{x - 1}$

15. $\dfrac{A}{x + 1} + \dfrac{Bx + C}{x^2 - x + 1}$ **17.** $\dfrac{Ax + B}{x^2 + 1} + \dfrac{Cx + D}{(x^2 + 1)^2}$

19. $\dfrac{3}{x + 2} - \dfrac{1}{x + 1}$ **21.** $\dfrac{-1}{2(x + 3)} + \dfrac{1}{2(x + 1)}$ **23.** $\dfrac{-1}{x + 2} + \dfrac{1}{x}$

25. $\dfrac{2}{x + 2} - \dfrac{3}{2(x + 3)} - \dfrac{1}{2(x + 1)}$

27. $\dfrac{16}{15(x + 4)} + \dfrac{1}{10(x - 1)} - \dfrac{1}{6(x + 1)}$

29. $\dfrac{1}{(x + 3)^2} - \dfrac{1}{4(x + 3)} + \dfrac{1}{4(x - 5)}$

31. $\dfrac{-1}{x + 2} + \dfrac{3}{(x + 2)^2} + \dfrac{1}{x - 3}$ **33.** $\dfrac{-2}{(x + 1)^2} + \dfrac{1}{x + 1}$

35. $\dfrac{1}{2(2x - 3)} + \dfrac{1}{2(2x - 3)^2}$

37. $\dfrac{2}{x + 1} - \dfrac{3}{(x + 1)^2} + \dfrac{1}{(x + 1)^3}$ **39.** $\dfrac{1}{x} + \dfrac{3}{(x + 1)^2} - \dfrac{2}{x + 1}$

41. $\dfrac{1}{x - 1} - \dfrac{1}{x + 1} + \dfrac{1}{(x + 1)^2}$ **43.** $\dfrac{-2}{x} + \dfrac{1}{x^2} + \dfrac{1}{(x + 1)^2} + \dfrac{2}{x + 1}$

45. $\dfrac{1}{4(x - 1)^2} + \dfrac{1}{4(x + 1)^2} + \dfrac{1}{4(x + 1)} - \dfrac{1}{4(x - 1)}$

47. $\dfrac{2}{3}\left(\dfrac{2x + 3}{x^2 - 2x + 3}\right) - \dfrac{1}{3(x - 2)}$

49. $\dfrac{1}{x + 1} - \dfrac{2}{(x + 1)^2} + \dfrac{1 - x}{x^2 + 2x + 2}$

51. $\dfrac{-1}{x - 2} + \dfrac{5}{(x - 2)^2} + \dfrac{2}{x - 1}$ **53.** $\dfrac{-2}{(2x + 3)^2} + \dfrac{3}{2x + 3}$

55. $\dfrac{4}{x} - \dfrac{3}{x^2} - \dfrac{4}{x + 1}$ **57.** $\dfrac{-2}{x} + \dfrac{3}{x + 1} + \dfrac{4}{x^2}$

59. $\dfrac{-x}{2(x^2 + 1)} + \dfrac{1}{4(x + 1)} + \dfrac{1}{4(x - 1)}$

61. $\dfrac{-x}{x^2 + 1} + \dfrac{1}{x} - \dfrac{x}{(x^2 + 1)^2}$ **63.** $\dfrac{4 - 3x}{x^2 + 2} + \dfrac{-2 + 3x}{x^2 + 1}$

Applying the Concepts:

65. $\dfrac{n}{n + 1}$ **67.** $\dfrac{2n}{2n + 1}$ **69.** $\dfrac{1}{R} = \dfrac{1}{x + 1} + \dfrac{1}{x + 3}$

71. $\dfrac{1}{R} = \dfrac{1}{R_1} + \dfrac{1}{R_2} + \dfrac{1}{R_3}$

Beyond the Basics:

75. $\dfrac{1}{(x - 1)^2} - \dfrac{1}{(x + 1)^2}$ **77.** $\dfrac{-1}{2(x - 1)} + \dfrac{x - 1}{2(x^2 + 1)} + \dfrac{1}{(x - 1)^2}$

79. $\dfrac{-4}{x + 2} - \dfrac{8}{(x + 2)^2} - \dfrac{1}{(x + 1)^2} + \dfrac{5}{x + 1}$

81. $\dfrac{3}{(x - 5)^3} + \dfrac{1}{x - 5} + \dfrac{2}{(x - 5)^2}$ **83.** $1 - \dfrac{4}{7(x + 3)} + \dfrac{18}{7(x - 4)}$

Getting Ready for the Next Section:
87. $\{7\}$ **89.** $\left\{-\dfrac{5}{2}\right\}$ **91.** $\{(-2,-1)\}$ **93.** $\{(x, 2x-3)\}$
95. false   **97.** true

Section 7.7

Concepts and Vocabulary:
1. $a_{ij} = b_{ij}$ **3.** 1×1 **5.** False   **7.** False

Building Skills:
9. $x = 2, u = -3$ **11.** $x = -1, y = 3$ **13.** $x = 2, y = 1$ **15.** $\varnothing$
17. a. $\begin{bmatrix} 0 & 2 \\ 5 & 1 \end{bmatrix}$ **b.** $\begin{bmatrix} 2 & 2 \\ 1 & 7 \end{bmatrix}$ **c.** $\begin{bmatrix} -3 & -6 \\ -9 & -12 \end{bmatrix}$ **d.** $\begin{bmatrix} 5 & 6 \\ 5 & 18 \end{bmatrix}$
e. $\begin{bmatrix} 10 & 2 \\ 5 & 11 \end{bmatrix}$ **f.** $\begin{bmatrix} 6 & 10 \\ 23 & 13 \end{bmatrix}$

19. a, b, d, e, and f are not defined.   **c.** $\begin{bmatrix} -6 & -9 \\ 12 & -15 \end{bmatrix}$
21. a. $\begin{bmatrix} 7 & 1 & -1 \\ -1 & 1 & 4 \\ 2 & 1 & 4 \end{bmatrix}$ **b.** $\begin{bmatrix} 1 & -1 & -1 \\ -3 & 9 & 0 \\ -2 & -1 & -2 \end{bmatrix}$ **c.** $\begin{bmatrix} -12 & 0 & 3 \\ 6 & -15 & -6 \\ 0 & 0 & -3 \end{bmatrix}$
d. $\begin{bmatrix} 6 & -2 & -3 \\ -8 & 23 & 2 \\ -4 & -2 & -3 \end{bmatrix}$ **e.** $\begin{bmatrix} 46 & 7 & -7 \\ 0 & 4 & 21 \\ 21 & 7 & 18 \end{bmatrix}$ **f.** $\begin{bmatrix} 6 & 1 & -7 \\ -21 & 6 & 16 \\ -13 & -1 & -10 \end{bmatrix}$
23. a. $\begin{bmatrix} 4 & 3 & -3 \\ 4 & 0 & 7 \\ 4 & 0 & 3 \end{bmatrix}$ **b.** $\begin{bmatrix} -2 & 1 & -3 \\ 2 & 8 & 3 \\ 0 & -2 & -3 \end{bmatrix}$ **c.** $\begin{bmatrix} -3 & -6 & 9 \\ -9 & -12 & -15 \\ -6 & 3 & 0 \end{bmatrix}$
d. $\begin{bmatrix} -3 & 4 & -9 \\ 7 & 20 & 11 \\ 2 & -5 & -6 \end{bmatrix}$ **e.** $\begin{bmatrix} 16 & 12 & 0 \\ 44 & 12 & 9 \\ 28 & 12 & -3 \end{bmatrix}$ **f.** $\begin{bmatrix} -9 & 14 & 5 \\ 22 & -2 & 13 \\ -14 & -1 & -22 \end{bmatrix}$

25. $X = \begin{bmatrix} -4 & -2 & 1 \\ 1 & 5 & 0 \end{bmatrix}$ **27.** $X = \begin{bmatrix} 0 & 2 & -\dfrac{1}{2} \\ \dfrac{3}{2} & \dfrac{1}{2} & 4 \end{bmatrix}$
29. $X = \begin{bmatrix} -4 & -4 & \dfrac{3}{2} \\ -\dfrac{1}{2} & \dfrac{9}{2} & -4 \end{bmatrix}$ **31.** $X = \begin{bmatrix} \dfrac{1}{2} & -\dfrac{9}{4} & \dfrac{1}{2} \\ -2 & -\dfrac{5}{4} & -5 \end{bmatrix}$

33. a. $AB = \begin{bmatrix} 4 & 11 \\ 6 & 23 \end{bmatrix}$ **b.** $BA = \begin{bmatrix} 1 & 0 \\ 18 & 26 \end{bmatrix}$
35. a. $AB = \begin{bmatrix} 4 & 7 \\ 3 & -12 \end{bmatrix}$ **b.** $BA = \begin{bmatrix} -13 & 4 & 10 \\ -13 & 5 & 6 \\ 8 & -4 & 0 \end{bmatrix}$
37. a. $AB = \begin{bmatrix} 16 \end{bmatrix}$ **b.** $BA = \begin{bmatrix} 2 & 3 & 5 \\ -4 & -6 & -10 \\ 8 & 12 & 20 \end{bmatrix}$
39. a. $AB = \begin{bmatrix} 7 & 8 & -8 \end{bmatrix}$ **b.** The product BA is not defined.
41. a. $AB = \begin{bmatrix} 10 & 7 & 2 \\ 7 & 19 & 4 \\ 10 & 1 & 0 \end{bmatrix}$ **b.** $BA = \begin{bmatrix} 7 & 4 & 5 \\ 0 & 8 & 3 \\ 19 & 18 & 14 \end{bmatrix}$
43. $AB = \begin{bmatrix} 13 & 17 & 3 \\ 13 & 8 & 2 \\ 6 & 1 & 6 \end{bmatrix} \neq \begin{bmatrix} 6 & 22 & 19 \\ 2 & 11 & 6 \\ 11 & -1 & 10 \end{bmatrix} = BA$

Applying the Concepts:
49.
	Steel	Glass	Wood	
$C = \Big[$	13	5	38	$\Big]$ Cost of material
	7	2	7	Transportation cost

51. $24,200
53. a.
	Chairman	President	Vice president
Salary	2,500,000	1,250,000	100,000
Bonus	1,500,000	750,000	150,000
Stock	50,000	25,000	5000

b. $\begin{bmatrix} 1 \\ 1 \\ 4 \end{bmatrix}$ Chairman President Vice president   **c.** $\begin{bmatrix} 4{,}150{,}000 \\ 2{,}850{,}000 \\ 95{,}000 \end{bmatrix}$ Total salary Total bonuses Total stocks

55. $AD = \begin{bmatrix} 0 & 4 & 4 & 1 & 1 & 0 \\ 0 & 0 & -1 & -1 & -6 & -6 \end{bmatrix}$

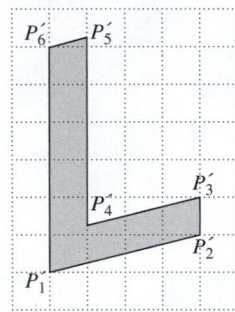

57. $AD = \begin{bmatrix} 0 & 4 & 4 & 1 & 1 & 0 \\ 0 & 1 & 2 & 1.25 & 6.25 & 6 \end{bmatrix}$

Beyond the Basics:
59. $AB = \begin{bmatrix} 9 & 0 \\ 0 & 0 \end{bmatrix} = AC$
61. Let $A = \begin{bmatrix} 1 & 2 \\ 3 & 4 \end{bmatrix}$ and $B = \begin{bmatrix} -1 & 0 \\ 2 & -3 \end{bmatrix}$. Then $(A+B)^2 = \begin{bmatrix} 10 & 2 \\ 5 & 11 \end{bmatrix} \neq \begin{bmatrix} 14 & -2 \\ 17 & 7 \end{bmatrix} = A^2 + 2AB + B^2.$
65. $B = \begin{bmatrix} 2 & -3 \\ -1 & 2 \end{bmatrix}$ **67.** $x = 33, y = 5$

Critical Thinking / Discussion / Writing:
69. a. AB is defined when $n = 5$, and the order of AB when this product is defined is $3 \times m$.   **b.** BA is defined when $m = 3$, and the order of BA when this product is defined is $5 \times n$.   **68.** $(CA)B$
71. a. $XP^2 = \begin{bmatrix} \dfrac{37}{75} & \dfrac{19}{75} & \dfrac{19}{75} \end{bmatrix}$
b. $XP^3 = \begin{bmatrix} \dfrac{188}{375} & \dfrac{187}{750} & \dfrac{187}{750} \end{bmatrix}$
$XP^4 = \begin{bmatrix} \dfrac{937}{1875} & \dfrac{469}{1875} & \dfrac{469}{1875} \end{bmatrix}$
$XP^5 = \begin{bmatrix} \dfrac{4688}{9375} & \dfrac{4687}{18{,}750} & \dfrac{4687}{18{,}750} \end{bmatrix}$
$XP^n = \begin{bmatrix} 0.5 & 0.25 & 0.25 \end{bmatrix}$

Getting Ready for the Next Section:
73. $\dfrac{12}{5}$ **75.** $\{5\}$ **77.** $\left\{\dfrac{-11}{10}\right\}$ **79.** $\{3\}$ **81.** $\{(0,4)\}$
83. $\{(3,-1)\}$

Section 7.8

Concepts and Vocabulary:
1. inverse   **3.** A^{-1} **5.** False   **7.** True

Building Skills:

9. Yes   **11.** Yes   **13.** No   **15.** No   **17.** Yes

19. $A^{-1} = \begin{bmatrix} \dfrac{1}{2} & 0 \\ -\dfrac{1}{6} & \dfrac{1}{3} \end{bmatrix}$ **21.** Does not exist

23. $A^{-1} = \begin{bmatrix} 1 & -8 & 10 \\ 0 & 2 & -3 \\ 0 & -1 & 2 \end{bmatrix}$ **25.** $A^{-1} = \begin{bmatrix} \dfrac{3}{4} & -\dfrac{5}{2} & -\dfrac{1}{4} \\ -\dfrac{1}{2} & 2 & \dfrac{1}{2} \\ \dfrac{1}{2} & -1 & -\dfrac{1}{2} \end{bmatrix}$

27. $A^{-1} = \begin{bmatrix} \dfrac{3}{14} & -\dfrac{1}{14} & \dfrac{5}{14} \\ \dfrac{5}{14} & \dfrac{3}{14} & -\dfrac{1}{14} \\ -\dfrac{1}{14} & \dfrac{5}{14} & \dfrac{3}{14} \end{bmatrix}$ **29.** $A^{-1} = \begin{bmatrix} 1 & 0 \\ -\dfrac{3}{2} & \dfrac{1}{2} \end{bmatrix}$

31. $A^{-1} = \begin{bmatrix} 5 & 3 \\ 3 & 2 \end{bmatrix}$ **33.** $A^{-1} = \dfrac{1}{-a^2 + b^2}\begin{bmatrix} -a & b \\ -b & a \end{bmatrix}$, where $a^2 \neq b^2$

35. $\begin{bmatrix} 2 & 3 \\ 1 & -3 \end{bmatrix}\begin{bmatrix} x \\ y \end{bmatrix} = \begin{bmatrix} -9 \\ 13 \end{bmatrix}$ **37.** $\begin{bmatrix} 3 & 2 & 1 \\ 2 & 1 & 3 \\ 1 & 3 & 2 \end{bmatrix}\begin{bmatrix} x \\ y \\ z \end{bmatrix} = \begin{bmatrix} 8 \\ 7 \\ 9 \end{bmatrix}$

39. $\begin{cases} x - 2y = 0 \\ 2x + y = 5 \end{cases}$ **41.** $\begin{cases} 2x_1 + 3x_2 + x_3 = -1 \\ 5x_1 + 7x_2 - x_3 = 5 \\ 4x_1 + 3x_2 = 5 \end{cases}$

43. $\begin{bmatrix} 1 & 2 & 5 \\ 2 & 3 & 8 \\ -1 & 1 & 2 \end{bmatrix}\begin{bmatrix} 2 & -1 & -1 \\ 12 & -7 & -2 \\ -5 & 3 & 1 \end{bmatrix} = \begin{bmatrix} 1 & 0 & 0 \\ 0 & 1 & 0 \\ 0 & 0 & 1 \end{bmatrix}$

45. $\{(2, -3, 1)\}$ **47. a.** $A^{-1} = \begin{bmatrix} 3 & -\dfrac{5}{2} & \dfrac{1}{2} \\ -3 & 4 & -1 \\ 1 & -\dfrac{3}{2} & \dfrac{1}{2} \end{bmatrix}$ **b.** $\{(1, 2, 3)\}$

Applying the Concepts:

49. $\{(-1, 2)\}$ **51.** $\{(2, -1, 3)\}$ **53.** $\{(-5, 1, 5)\}$
55. Treasury: \$16,000; bonds: \$48,000; fund: \$26,000
57. $V = 10, E = 15, R = 7$

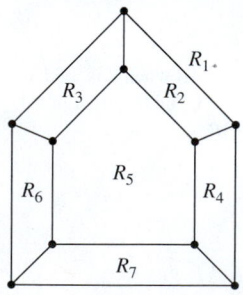

59. $X = \begin{bmatrix} 100 \\ 100 \end{bmatrix}$ **61.** $X = \begin{bmatrix} \dfrac{216,000}{277} \\ \dfrac{310,000}{277} \\ \dfrac{304,000}{277} \end{bmatrix}$

Beyond the Basics:

67. a. Yes. By the definition of inverse, if $AB = I$, then $BA = I$ and A and B are inverses.   **b.** $(A^{-1})^{-1} = A$ **73.** $\dfrac{12}{5}$

75. If A is invertible, there exists some matrix C such that $AC = I$ and $CA = I$. Then $AB = A \Leftrightarrow CAB = CA \Leftrightarrow IB = I \Leftrightarrow B = I$.

77. c. $A^{-1} = \begin{bmatrix} 3 & -4 \\ -2 & 3 \end{bmatrix}$ **79.** $\{(1, 2, 1, 2)\}$

81. a. $A = \begin{bmatrix} 8 & 10 \\ 6 & 8 \end{bmatrix}$ **b.** $A = \begin{bmatrix} -1 & 7 \\ -3 & 17 \end{bmatrix}$ **c.** $A = \begin{bmatrix} -1 & -2 \\ \dfrac{5}{2} & 4 \end{bmatrix}$

d. $A = \begin{bmatrix} \dfrac{3}{2} & \dfrac{1}{2} \\ \dfrac{5}{2} & \dfrac{3}{2} \end{bmatrix}$

Critical Thinking / Discussion / Writing:

83. a. True. $A^2B = AAB = ABA = BAA = BA^2$
b. False. Let $A = I$ and $B = -I$.   **85.** True

Getting Ready for the Next Section:

87. $y = \dfrac{1}{4}x^2 + \dfrac{3}{2}x + \dfrac{13}{4}$ **89.** $y = -\dfrac{1}{2}x^2 + 5x - \dfrac{39}{2}$

91. $y = -5x^2$ **93.** $y = -2x^2 - 4x + 6$ **95. a.** Opens down
b. $(3, 8)$ **c.** $x = 3$ **d.** 1 and 5   **e.** -10

Review Exercises

Building Skills:

1. $\{(-1, 2)\}$ **3.** $\varnothing$ **5.** $\{(2, 3)\}$ **7.** $\{(1, -1, 2)\}$
9. $\{(-1, 2, -3)\}$ **11.** $\varnothing$ **13.** $\{(x, 2x - 8, 11 - 3x)\}$ **15.** $\varnothing$

17. $\left\{\left(\dfrac{7}{3} - \dfrac{2}{3}z, \dfrac{2}{3} - \dfrac{1}{3}z, z\right)\right\}$ **19.** $\left\{(-2, 1), \left(\dfrac{8}{3}, \dfrac{-5}{9}\right)\right\}$

21. $\left\{(2, -2), \left(\dfrac{-2}{3}, \dfrac{-14}{3}\right)\right\}$

23.

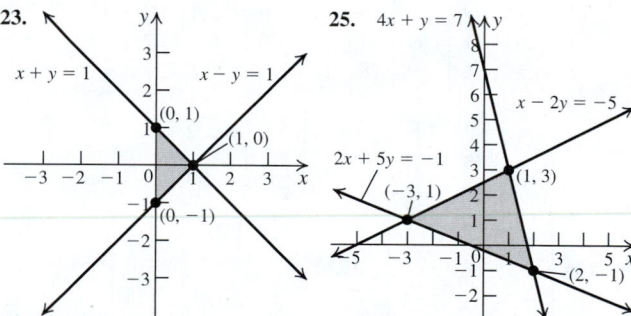

25. $4x + y = 7$

27.

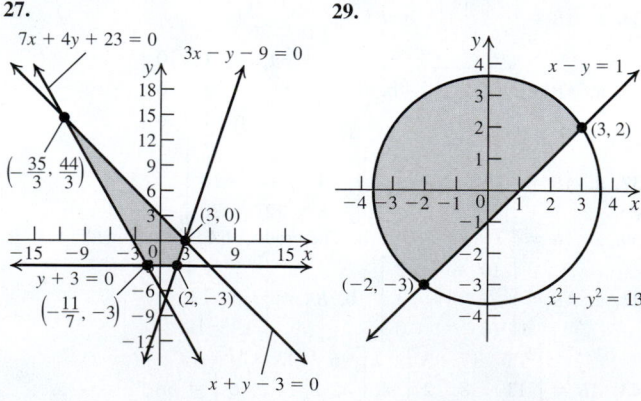

29.

31.

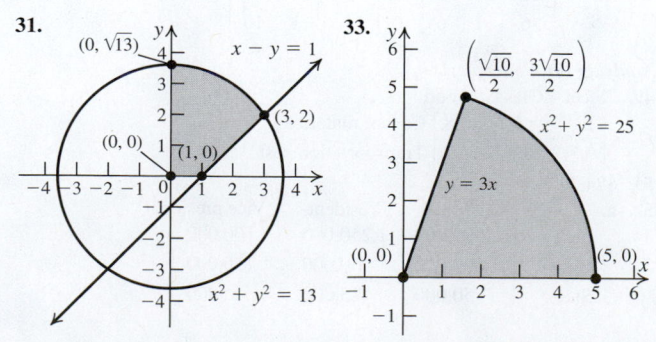

33.

35. $\begin{bmatrix} 1 & -2 & 1 & 7 \\ 0 & 1 & 2 & 1 \\ 0 & 0 & 1 & 2 \end{bmatrix}$ **37.** $\begin{bmatrix} 1 & 0 & 0 & \frac{1}{3} \\ 0 & 1 & 0 & \frac{1}{2} \\ 0 & 0 & 1 & -\frac{1}{6} \end{bmatrix}$ **39. a.** $\begin{bmatrix} 3 & -1 \\ -8 & 10 \end{bmatrix}$

b. $\begin{bmatrix} -1 & 5 \\ 2 & -2 \end{bmatrix}$ **c.** $\begin{bmatrix} 2 & 4 \\ -6 & 8 \end{bmatrix}$ **d.** $\begin{bmatrix} -6 & 9 \\ 15 & -18 \end{bmatrix}$ **e.** $\begin{bmatrix} -4 & 13 \\ 9 & -10 \end{bmatrix}$

41. $X = \begin{bmatrix} \frac{7}{3} & 0 \\ -\frac{19}{3} & 8 \end{bmatrix}$ **43.** $\dfrac{2}{x+2} - \dfrac{1}{x+3}$

45. $\dfrac{2}{x-1} + \dfrac{1}{x} + \dfrac{5}{(x-1)^2}$ **47.** $\dfrac{-1+2x}{(x^2+4)^2} + \dfrac{1}{x^2+4}$

49. $AB = \begin{bmatrix} -3 & 4 \\ -11 & 10 \end{bmatrix}, BA = \begin{bmatrix} -2 & -4 \\ 8 & 9 \end{bmatrix}$

51. $AB = [7], BA = \begin{bmatrix} 2 & 4 & -2 \\ 3 & 6 & -3 \\ 1 & 2 & -1 \end{bmatrix}$ **53.** $\begin{bmatrix} 1 & 0 \\ 2 & 3 \end{bmatrix}$ **55.** $\begin{bmatrix} \frac{2}{5} & -\frac{1}{10} \\ -\frac{1}{5} & \frac{3}{10} \end{bmatrix}$

57. $\begin{bmatrix} 3 & 2 & 6 \\ 1 & 1 & 2 \\ 2 & 2 & 5 \end{bmatrix}$ **59.** $\{(-12, 2, 3)\}$ **61.** $\{(\frac{7}{6}, \frac{1}{6}, 2)\}$ **63.** 23 **65.** 1

67. -7 **69.** $(7, 3)$ **71.** $(-1, -5, 8)$

Applying the Concepts:
73. $8750 was invested at 12%, and $6250 was invested at 4%.
75. 10.5 ft by 6 ft **77.** 15 and 8 **79.** The original width is 40 m, and the original length is 160 m.
81. a. $C(x) = 60{,}000 + 12x, R(x) = 20x$
b.

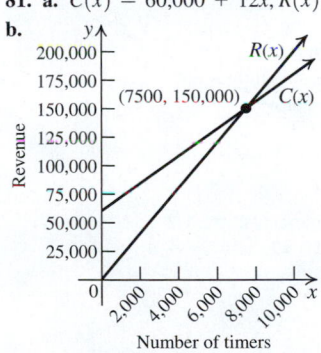

Number of timers

c. (7500, 150,000) **d.** 11,129 timers

83. $660 for the single and $720 for the double **85.** 3
87. Janet is 34, and Steve is 51. **89.** They stole $35,000, and $15,000 went to Butch, $12,000 went to Sundance, and $8000 went to Billy.
91. $x = 6, y = 2$ **93.** She should build 4 two-story houses and 14 one-story houses to obtain a maximum profit of $96,000.

Practice Test A:
1. $\{(2, 0)\}$ **2.** $\{(8 - 2y, y)\}$ **3.** $\varnothing$ **4.** $\{(-1, 1), (4, 16)\}$
5. a. $\begin{cases} x + y = 485 \\ x - y = 15 \end{cases}$ **b.** $x = 250, y = 235$ **6.** $\{(1, 3, 5)\}$
7. $(7, 3, -2)$ **8.** $\left(\dfrac{1}{3}, \dfrac{-4}{3} + z, z\right)$ **9.** 53 nickels, 225 dimes, and 22 quarters **10.** $\dfrac{A}{(x-5)} + \dfrac{B}{(x+1)}$
11. $\dfrac{A}{x-2} + \dfrac{Bx + C}{x^2 + 1} + \dfrac{Dx + E}{(x^2 + 1)^2}$

12. $-\dfrac{10}{121(x+4)} + \dfrac{1}{11(x+4)^2} + \dfrac{10}{121(x-7)}$
13.

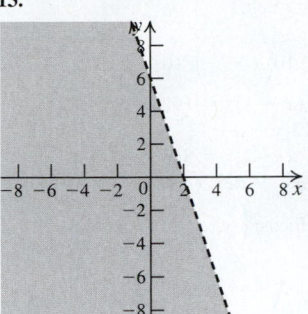

14.

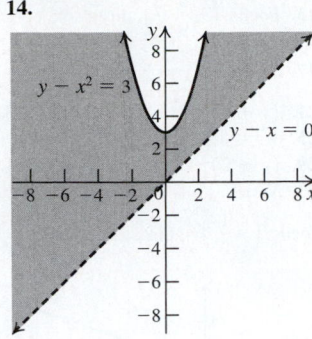

15. Maximum value is 6 at $(3, 0)$.

17. $\begin{cases} 12x - 3y = 5 \\ -2x + 7y = -9 \end{cases}$ **19.** 0

Practice Test B:
1. c **2.** c **3.** d **4.** d **5.** d **6.** c **7.** b **8.** d **9.** b **10.** c **11.** a
12. c **13.** b **14.** a **15.** b **16.** d **17.** a **18.** b **19.** c **20.** b

Cumulative Review Exercises (Chapters 1–7):
1. $-7, 1$ **3.** $4, 1$ **5.** $-9, \dfrac{-19}{7}$
7. $x < -2$ or $x > \dfrac{1}{2}$
9. $\pm 1, \pm \dfrac{1}{2}, \pm \dfrac{1}{4}, \pm 3, \pm \dfrac{3}{2}, \pm \dfrac{3}{4}$ **11.** 24,000 horsepower
13. $\{(1, -1)\}$
15.

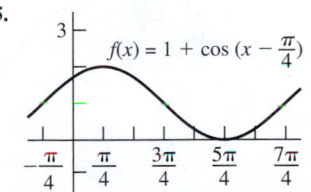

$f(x) = 1 + \cos\left(x - \dfrac{\pi}{4}\right)$

17. $\{(1, 1, -1)\}$ **19.** $f^{-1}(x) = \dfrac{-4x}{x-1}$

CHAPTER 8

Section 8.2

Concepts and Vocabulary:
1. directrix, focus **3.** vertex **5.** True **7.** False

Building Skills:
9. Focus: $\left(0, \dfrac{1}{2}\right)$; directrix: $y = -\dfrac{1}{2}$; graph: e
11. Focus: $\left(0, -\dfrac{9}{64}\right)$; directrix: $y = \dfrac{9}{64}$; graph: d

13. Focus: $\left(\frac{1}{2}, 0\right)$; directrix: $x = -\frac{1}{2}$; graph: g

15. Focus: $\left(-\frac{4}{9}, 0\right)$; directrix: $x = \frac{4}{9}$; graph: f

17. $x^2 = -4(y - 3)$; 4   **19.** $y^2 = -10\left(x - \frac{1}{2}\right)$; 10

21. $(y - 1)^2 = -8(x - 1)$; 8   **23.** $(x - 1)^2 = 16(y - 1)$; 16

25. $y^2 = 8(x - 1)$; 8   **27.** $x^2 = -12(y - 1)$; 12

29. $(y - 3)^2 = -8(x - 2)$; 8   **31.** $(x - 2)^2 = 8(y - 3)$; 8

33. Vertex: $(-1, 1)$;   **35.** Vertex: $(-2, 2)$;

focus: $\left(-\frac{1}{2}, 1\right)$; directrix: $x = -\frac{3}{2}$ focus: $\left(-2, \frac{11}{4}\right)$; directrix: $y = \frac{5}{4}$

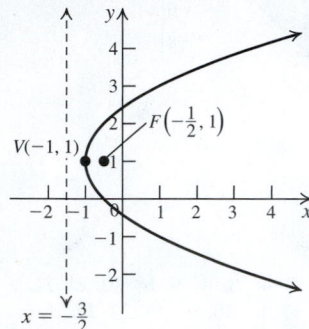   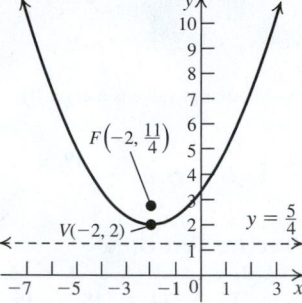

37. Vertex: $(2, -1)$;   **39.** Vertex: $(1, 3)$; focus: $\left(1, \frac{1}{2}\right)$;

focus: $\left(\frac{1}{2}, -1\right)$; directrix: $x = \frac{7}{2}$ directrix: $y = \frac{11}{2}$

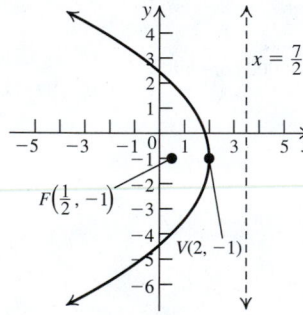   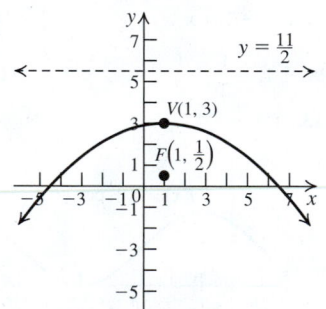

41. Vertex: $(-1, 1)$;   **43.** Vertex: $\left(-\frac{1}{2}, -1\right)$; focus:

focus: $\left(-1, \frac{5}{4}\right)$; directrix: $y = \frac{3}{4}$ $\left(-\frac{1}{4}, -1\right)$; directrix: $x = -\frac{3}{4}$

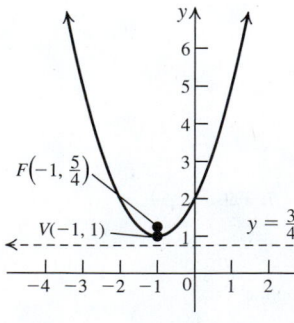   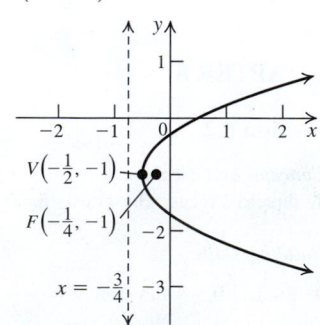

45. Vertex: $(46, 4)$; focus: $\left(\frac{551}{12}, 4\right)$; directrix: $x = \frac{553}{12}$

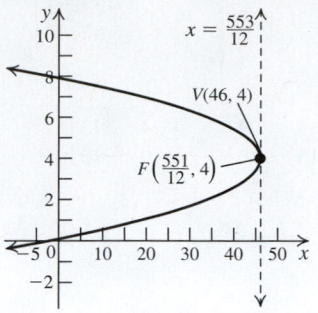

47. $x^2 = 12y$ **49.** $(y - 2)^2 = -12(x + 1)$

51. $(x - 1)^2 = -4(y + 1)$ **53.** $y^2 = 4x$ and $x^2 = \frac{1}{2}y$

55. $(y - 1)^2 = 2x$ and $x^2 = 2(y - 1)$

Applying the Concepts:

57. 5 in.   **59.** $3\frac{3}{8}$ ft from the vertex   **61.** $\left(\frac{1}{16}, 0\right)$ **63.** $\left(0, \frac{1}{16}\right)$

65. 16.875 ft   **67.** 7.96 yd   **69.** Output: 15 tons; cost: $27.50

Beyond the Basics:

71. $(4, 8)$ and $(16, 16)$ **73.** $\left(x - \frac{3}{4}\right)^2 = \frac{1}{2}\left(y - \frac{31}{8}\right)$

75. Vertex: $(4, 6)$; focus: $\left(4, \frac{11}{2}\right)$; directrix: $y = \frac{13}{2}$; axis: $x = 4$

77. Vertex: $(-2, 3)$; focus: $\left(-\frac{11}{4}, 3\right)$; directrix: $x = -\frac{5}{4}$; axis: $y = 3$

79. $y = 6x - 3$

Critical Thinking / Discussion / Writing:

83. a. No, because if the axis is horizontal, it does not pass the vertical line test.   **b.** 0   **85.** False

87. $(9, -8)$

Getting Ready for the Next Section:

89. $(1, 4)$ **91.** x-intercepts; ± 4; y-intercepts: $\pm 2\sqrt{2}$

93. $c = \pm\sqrt{7}$ **95.** $4(y - 1)^2 = 11$ **97.** $y = 3x + 4$

Section 8.3

Concepts and Vocabulary:

1. sum   **3.** $a^2 - c^2$ **5.** True   **7.** True

Building Skills:

9. $\dfrac{x^2}{9} + \dfrac{y^2}{8} = 1$ **11.** $\dfrac{x^2}{12} + \dfrac{y^2}{16} = 1$

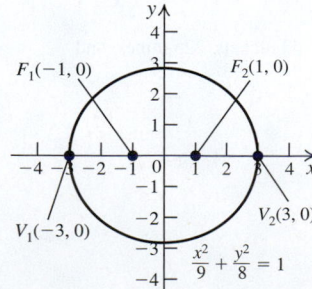   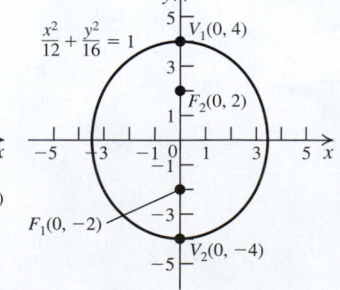

13. $\dfrac{x^2}{25} + \dfrac{y^2}{9} = 1$

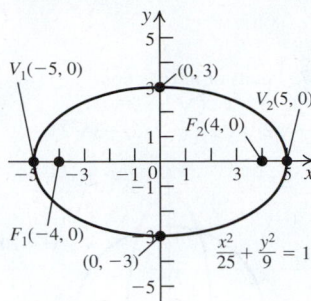

15. $\dfrac{x^2}{16} + \dfrac{y^2}{20} = 1$

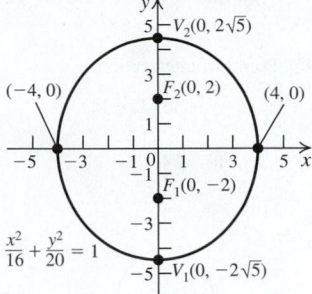

29. Vertices: $(0, 6)$ and $(0, -6)$ foci: $(0, 2\sqrt{5})$ and $(0, -2\sqrt{5})$

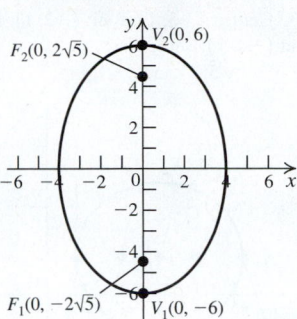

31. Circle centered at the origin with radius 2

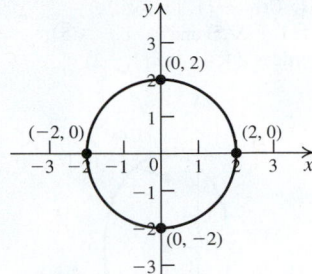

17. $\dfrac{x^2}{9} + \dfrac{y^2}{25} = 1$

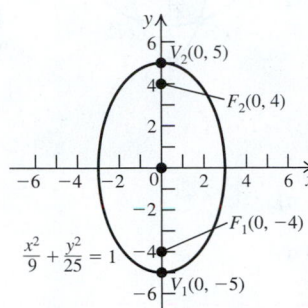

19. $\dfrac{x^2}{36} + \dfrac{y^2}{27} = 1$

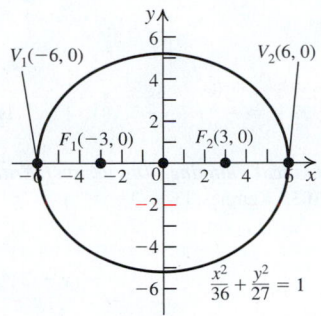
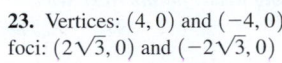

33. Vertices: $(2, 0)$ and $(-2, 0)$ foci: $(\sqrt{3}, 0)$ and $(-\sqrt{3}, 0)$

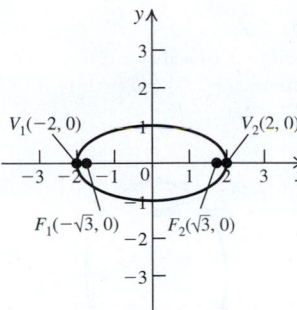

35. Vertices: $(0, 3)$ and $(0, -3)$ foci: $(0, \sqrt{5})$ and $(0, -\sqrt{5})$

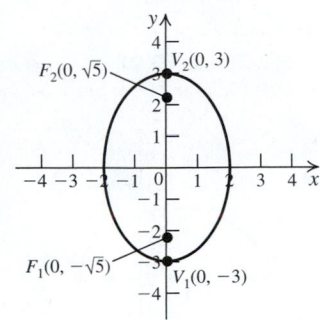
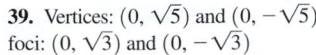

21. $\dfrac{x^2}{9} + \dfrac{y^2}{13} = 1$

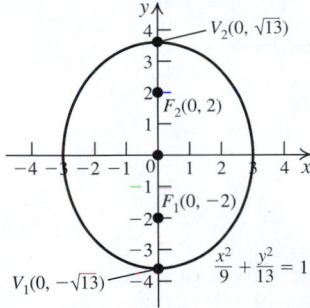

23. Vertices: $(4, 0)$ and $(-4, 0)$ foci: $(2\sqrt{3}, 0)$ and $(-2\sqrt{3}, 0)$

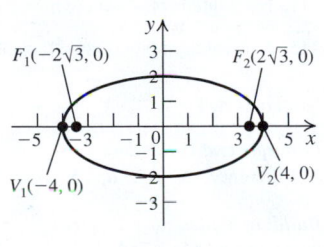

37. Vertices: $(2, 0)$ and $(-2, 0)$ foci: $(1, 0)$ and $(-1, 0)$

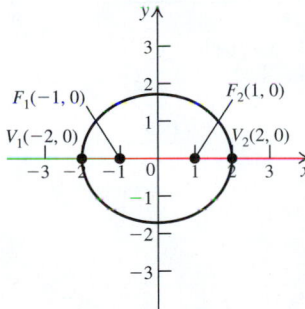

39. Vertices: $(0, \sqrt{5})$ and $(0, -\sqrt{5})$ foci: $(0, \sqrt{3})$ and $(0, -\sqrt{3})$

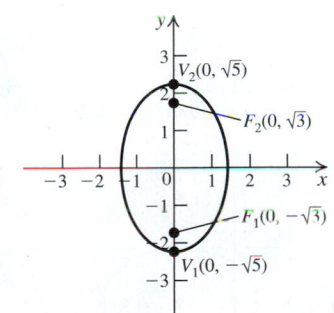

25. Vertices: $(3, 0)$ and $(-3, 0)$ foci: $(2\sqrt{2}, 0)$ and $(-2\sqrt{2}, 0)$

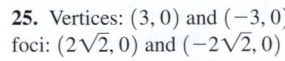
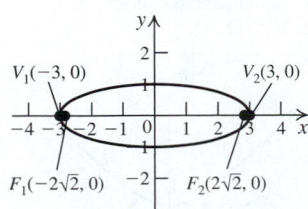

27. Vertices: $(5, 0)$ and $(-5, 0)$ foci: $(3, 0)$ and $(-3, 0)$

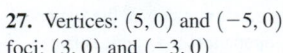
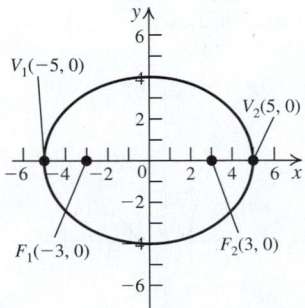

41. Vertices: $\left(\dfrac{\sqrt{14}}{2}, 0\right)$ and $\left(-\dfrac{\sqrt{14}}{2}, 0\right)$; foci: $\left(\dfrac{\sqrt{42}}{6}, 0\right)$ and $\left(-\dfrac{\sqrt{42}}{6}, 0\right)$

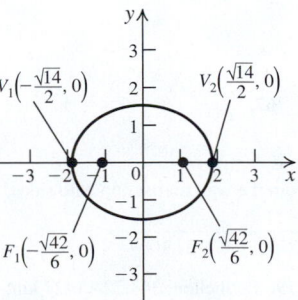

43. $(-2, -2), (-2, 8); (-2, -1), (-2, 7)$

45. $\dfrac{(x - 5)^2}{16} + \dfrac{y^2}{12} = 1$ **47.** $(x - 2)^2 + \dfrac{(y + 1)^2}{16} = 1$

49. $\dfrac{(x-1)^2}{36} + \dfrac{(y-3)^2}{27} = 1$

51. Center: $(1, 1)$; foci: $(1, 1 + \sqrt{5})$ and $(1, 1 - \sqrt{5})$; vertices: $(1, 4)$ and $(1, -2)$

53. Center: $(-3, 1)$; foci: $(-2, 1)$ and $(-4, 1)$; vertices: $(-3 + \sqrt{5}, 1)$ and $(-3 - \sqrt{5}, 1)$

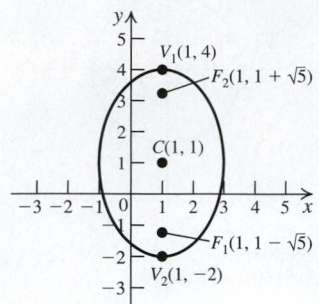

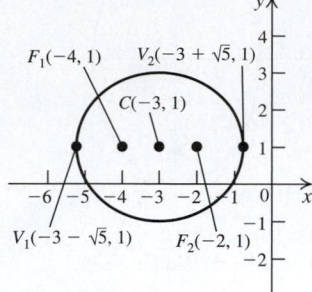

55. Center: $(-1, 2)$; foci: $(1, 2)$ and $(-3, 2)$; vertices: $(2, 2)$ and $(-4, 2)$

57. Center: $(-2, 4)$; foci: $(-2, 6)$ and $(-2, 2)$; vertices: $(-2, 7)$ and $(-2, 1)$

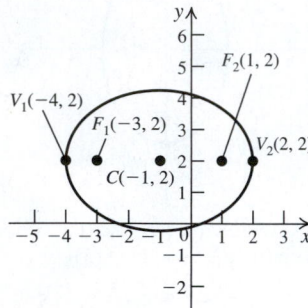

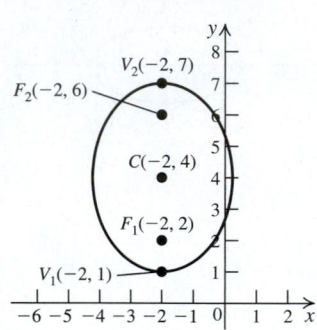

59. Center: $(1, -1)$; foci: $(2, -1)$ and $(0, -1)$; vertices: $(1 + \sqrt{2}, -1)$ and $(1 - \sqrt{2}, -1)$

61. No graph

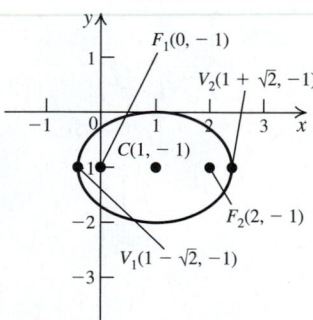

63. $x^2 + \dfrac{y^2}{9} = 1$ **65.** $\dfrac{x^2}{20} + \dfrac{y^2}{4} = 1$ **67.** $\dfrac{x^2}{9} + \dfrac{(y-1)^2}{4} = 1$

Applying the Concepts:

69. 18.3 ft   **71.** 42.4 m   **73.** 5.9 ft from the wall on the opposite sides, along the major access   **75.** $\dfrac{x^2}{20{,}067.5556} + \dfrac{y^2}{19{,}894.1067} = 1$

77. $\dfrac{x^2}{1{,}311.8884} + \dfrac{y^2}{1{,}253.2128} = 1$ **79.** Perihelion: 364,224.1427 km;

aphelion: 404,581.8573 km   **81.** $\dfrac{x^2}{16{,}785{,}409} + \dfrac{y^2}{16{,}784{,}680} = 1$

Beyond the Basics:

83. No solution.   **85.** $(x-2)^2 + 48(y-1)^2 = 64$ **87.** $\dfrac{\sqrt{7}}{4}$

89. $\dfrac{\sqrt{3}}{2}$ **91.** $\dfrac{\sqrt{2}}{2}$ **93.** $\dfrac{x^2}{9} + \dfrac{y^2}{8} = 1$

95. Points of intersection: $\left(-\dfrac{17}{13}, -\dfrac{3}{13}\right)$ and $(1, -1)$

97. Points of intersection: $(\sqrt{2}, 3\sqrt{2}), (\sqrt{2}, -3\sqrt{2})$, $(-\sqrt{2}, 3\sqrt{2}), (-\sqrt{2}, -3\sqrt{2})$

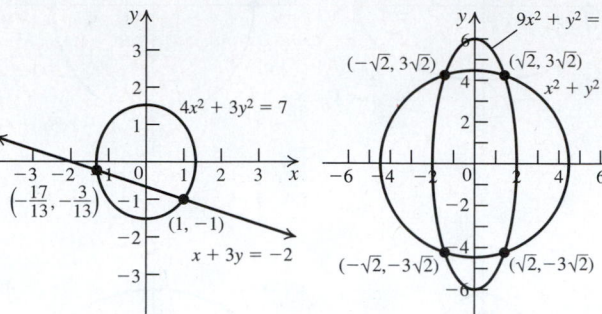

99. $y = -\dfrac{8}{3}x + \dfrac{25}{3}$ **101.** 12π **103.** 4π

Critical Thinking / Discussion / Writing:

105. 4 circles: $(x-3)^2 + (y-3)^2 = 9$
$(x+3)^2 + (y-3)^2 = 9$
$(x+3)^2 + (y+3)^2 = 9$
$(x-3)^2 + (y+3)^2 = 9$

Getting Ready for the Next Section:

107. $y = -x + 2$ **109.** $(-1, 3)$

111. No x-intercepts; y-intercepts: $\pm\dfrac{1}{2}$ **113.** $a = \pm\sqrt{13}$

115. Vertical asymptote: $x = 1$; slant asymptote: $y = x + 1$

Section 8.4

Concepts and Vocabulary:

1. difference   **3.** $c^2 - a^2$ **5.** True   **7.** False

Building Skills:

9. g   **11.** h   **13.** d   **15.** b

17. Vertices: $(1, 0)$ and $(-1, 0)$; foci: $(\sqrt{5}, 0)$ and $1(-\sqrt{5}, 0)$; transverse axis: x-axis; the hyperbola opens left and right; vertices of the fundamental rectangle: $(1, 2), (-1, 2), (-1, -2), (1, -2)$; asymptotes: $y = \pm 2x$

19. Vertices: $(0, 1)$ and $(0, -1)$; foci: $(0, \sqrt{2})$ and $(0, -\sqrt{2})$; transverse axis: y-axis; the hyperbola opens up and down; vertices of the fundamental rectangle: $(1, 1), (-1, 1), (-1, -1), (1, -1)$; asymptotes: $y = \pm x$

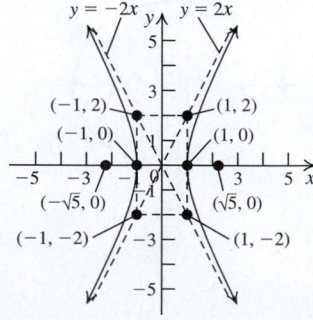

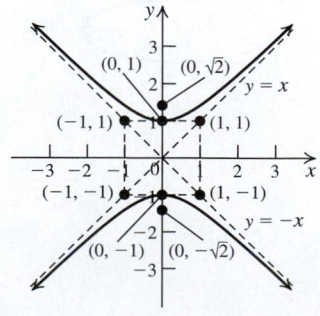

21. Vertices: $(0, 2)$ and $(0, -2)$; foci: $(0, 2\sqrt{10})$ and $(0, -2\sqrt{10})$; transverse axis: y-axis; the hyperbola opens up and down; vertices of the fundamental rectangle: $(6, 2)$, $(-6, 2)$, $(-6, -2)$, $(6, -2)$; asymptotes: $y = \pm\frac{1}{3}x$

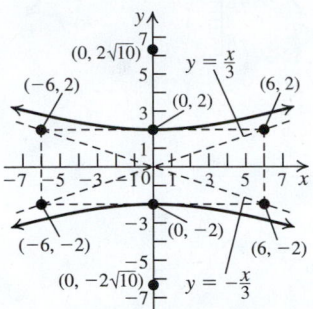

23. Vertices: $(3, 0)$ and $(-3, 0)$; foci: $(\sqrt{13}, 0)$ and $(-\sqrt{13}, 0)$; transverse axis: x-axis; the hyperbola opens left and right; vertices of the fundamental rectangle: $(3, 2)$, $(-3, 2)$, $(-3, -2)$, $(3, -2)$; asymptotes: $y = \pm\frac{2}{3}x$

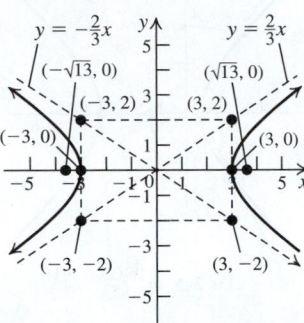

25. Vertices: $(0, 1)$ and $(0, -1)$; foci: $\left(0, \frac{\sqrt{5}}{2}\right)$ and $\left(0, -\frac{\sqrt{5}}{2}\right)$; transverse axis: y-axis; the hyperbola opens up and down; vertices of the fundamental rectangle: $\left(\frac{1}{2}, 1\right)$, $\left(-\frac{1}{2}, 1\right)$, $\left(-\frac{1}{2}, -1\right)$, $\left(\frac{1}{2}, -1\right)$; asymptotes: $y = \pm 2x$

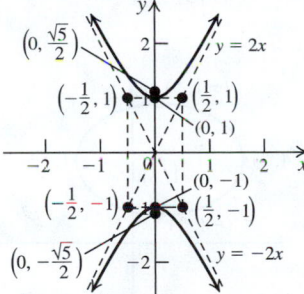

27. Vertices: $\left(\frac{1}{3}, 0\right)$ and $\left(-\frac{1}{3}, 0\right)$; foci: $\left(\frac{\sqrt{10}}{3}, 0\right)$ and $\left(-\frac{\sqrt{10}}{3}, 0\right)$; transverse axis: x-axis; the hyperbola opens left and right; vertices of the fundamental rectangle: $\left(\frac{1}{3}, 1\right)$, $\left(-\frac{1}{3}, 1\right)$, $\left(-\frac{1}{3}, -1\right)$, $\left(\frac{1}{3}, -1\right)$; asymptotes: $y = \pm 3x$

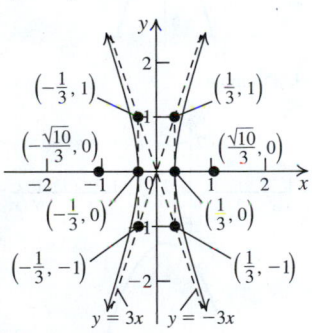

29. $\frac{x^2}{4} - \frac{y^2}{5} = 1$

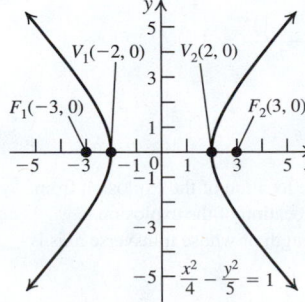

31. $\frac{y^2}{16} - \frac{x^2}{20} = 1$

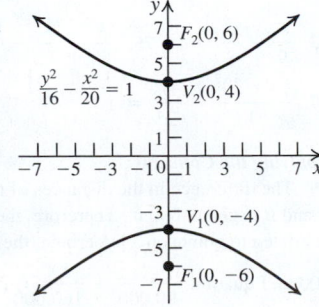

33. $\frac{y^2}{4} - \frac{x^2}{21} = 1$

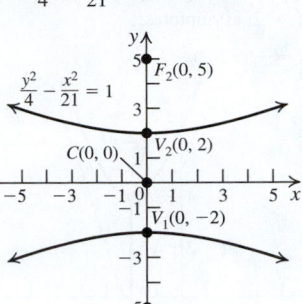

35. $x^2 - \frac{y^2}{24} = 1$

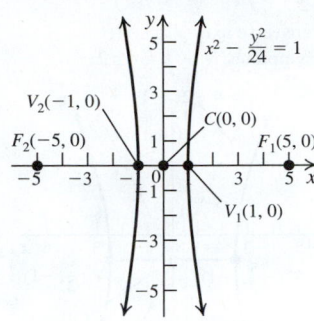

37. $\frac{y^2}{9} - \frac{x^2}{16} = 1$

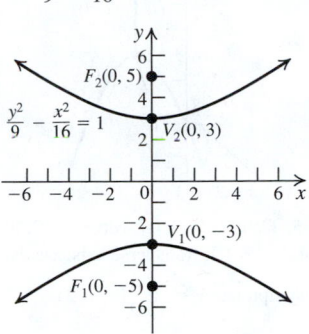

39. $x^2 - \frac{y^2}{4} = 1$

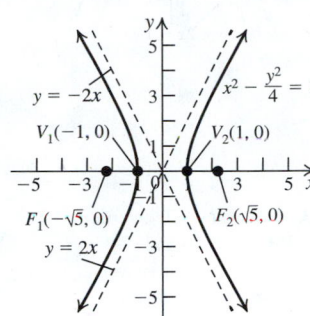

41. $\frac{y^2}{16} - \frac{x^2}{16} = 1$

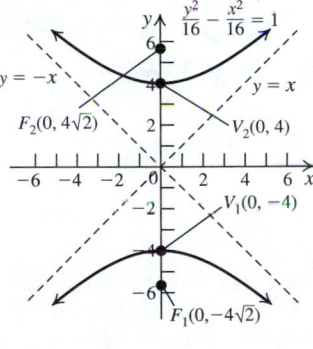

43. Center: $(1, -1)$; vertices: $(4, -1)$ and $(-2, -1)$; transverse axis: $y = -1$; asymptotes: $y + 1 = \pm\frac{4}{3}(x - 1)$

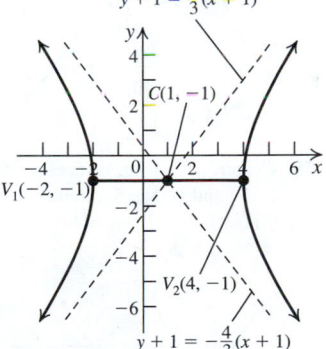

45. Center: $(-2, 0)$; vertices: $(3, 0)$ and $(-7, 0)$; transverse axis: x-axis; asymptotes: $y = \pm\frac{7}{5}(x + 2)$

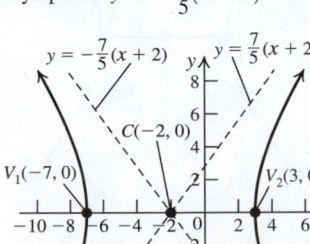

47. Center: $(-4, -3)$; vertices: $(1, -3)$ and $(-9, -3)$; transverse axis: $y = -3$; asymptotes: $y + 3 = \pm\frac{7}{5}(x + 4)$

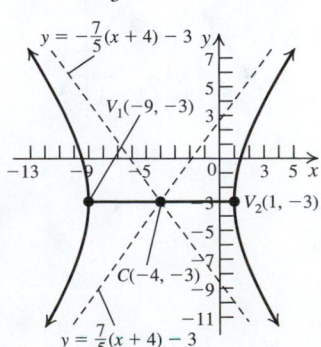

49. Center: $(0, -1)$; vertices: $\left(\frac{5}{2}, -1\right)$ and $\left(-\frac{5}{2}, -1\right)$; transverse axis: $y = -1$; asymptotes: $y + 1 = \pm 2x$

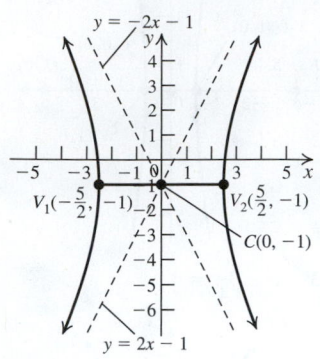

51. Center: $(2, -1)$; vertices: $(2, 4)$ and $(2, -6)$; transverse axis: $x = 2$; asymptotes: $y = -3(x - 2) - 1$

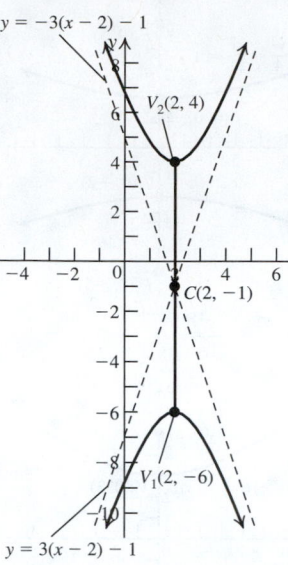
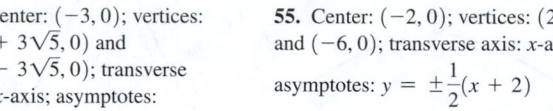

53. Center: $(-3, 0)$; vertices: $(-3 + 3\sqrt{5}, 0)$ and $(-3 - 3\sqrt{5}, 0)$; transverse axis: x-axis; asymptotes: $y = \pm(x + 3)$

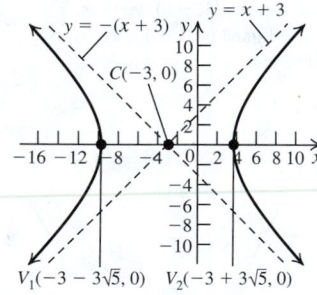

55. Center: $(-2, 0)$; vertices: $(2, 0)$ and $(-6, 0)$; transverse axis: x-axis; asymptotes: $y = \pm\frac{1}{2}(x + 2)$

57. Center: $(-3, -4)$; vertices: $(-3, -3)$ and $(-3, -5)$; transverse axis: $x = -3$; asymptotes: $y + 4 = \pm\sqrt{2}(x + 3)$

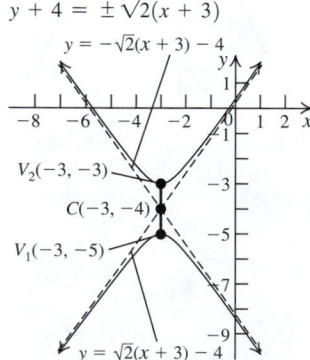

59. Center: $(3, -2)$; vertices: $(3 + \sqrt{6}, -2)$ and $(3 - \sqrt{6}, -2)$; transverse axis: $y = -2$; asymptotes: $y + 2 = \pm\frac{\sqrt{6}}{2}(x - 3)$

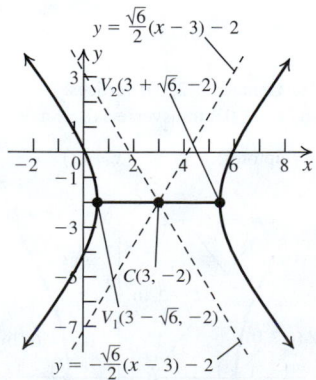

61. Center: $(\sqrt{2}, 0)$; vertices: $(\sqrt{2} + \sqrt{2\sqrt{2} + 1}, 0)$ and $(\sqrt{2} - \sqrt{2\sqrt{2} + 1}, 0)$; transverse axis: x-axis; asymptotes: $y = \pm(x - \sqrt{2})$

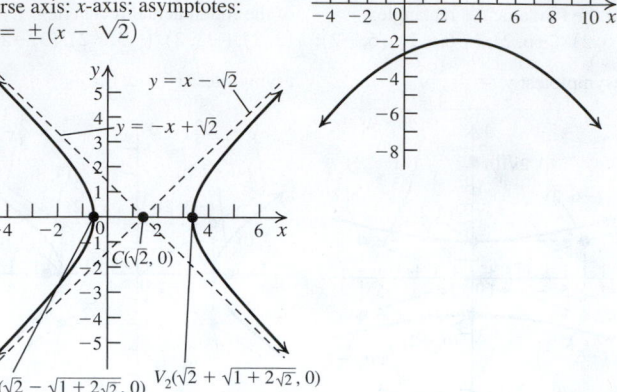

63. Parabola

65. Hyperbola

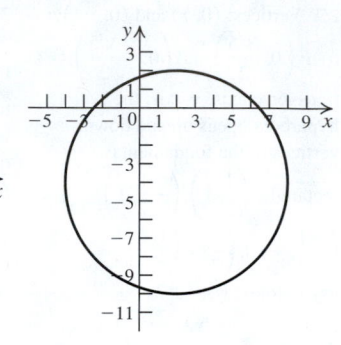

67. Circle

69. Parabola

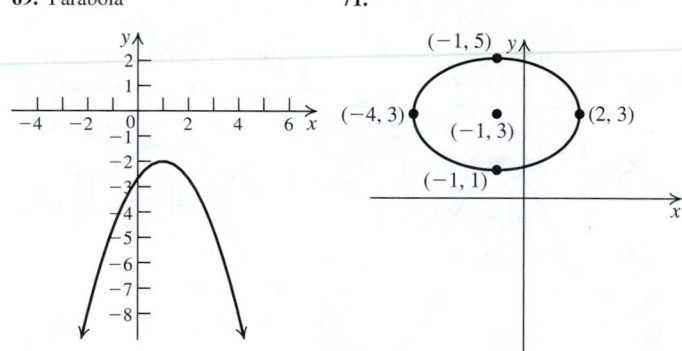

71.

73. $\dfrac{y^2}{9} - \dfrac{x^2}{8} = 1$ **75.** $\dfrac{(x + 2)^2}{9} - \dfrac{(y - 1)^2}{7} = 1$

77. $\dfrac{(x + 1)^2}{4} - \dfrac{4y^2}{25} = 1$

Applying the Concepts:
79. The difference in the distances of the location of the explosion from A and B is given: 600 m. Therefore, the location of the explosion is restricted to points on a hyperbola, the length of whose transverse axis is 600 m. Equation: $\dfrac{x^2}{90,000} - \dfrac{y^2}{160,000} = 1$

81. Let the coordinates of Nicole be $(-4000, 0)$ and those of Juan be $(4000, 0)$.

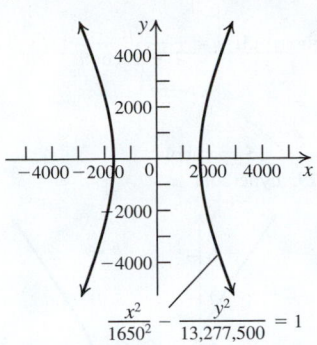

$$\frac{x^2}{1650^2} - \frac{y^2}{13,277,500} = 1$$

83. Let the coordinates of A be $(-150, 0)$ and the coordinates of B be $(150, 0)$. Equation: $\frac{x^2}{5625} - \frac{y^2}{16,875} = 1$

85. Let the coordinates of A be $(-125, 0)$ and the coordinates of B be $(125, 0)$. The plane is flying on the line $y = 50$. At the time of receiving the signals, the plane is at $(50.5238, 50)$.

87. Possible locations of the boat: $(-1397.427, 1225.2945)$, $(-281.3655, 209.9383)$, $(359.5894, 288.219)$, $(1059.7444, 925.19691)$

Beyond the Basics:

89.

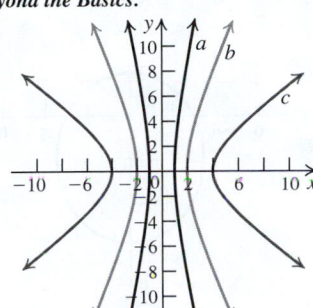

91.

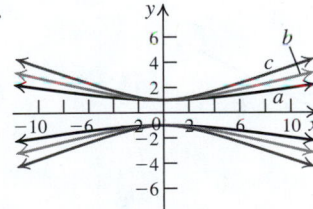

97. $e = \frac{\sqrt{61}}{5}$; length of the latus rectum: $\frac{72}{5}$

99. $e = 3$; length of the latus rectum: 8

105. Points of intersection: $\left(5, \frac{1}{2}\sqrt{10}\right), \left(5, -\frac{1}{2}\sqrt{10}\right)$

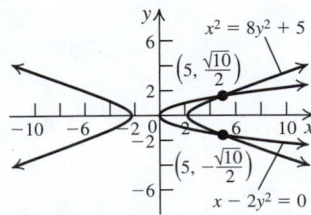

107. Points of intersection: $(-3, 0)$, $(3, 0)$

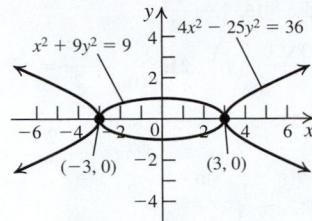

Critical Thinking / Discussion / Writing:

111. The rewritten form of the equation is equivalent to

$$A\left(x + \frac{D}{2A}\right)^2 + C\left(y + \frac{E}{2C}\right)^2 = \frac{D^2}{4A} + \frac{E^2}{4C} - F.$$ If the term on the right-hand side is not zero, then the equation can be written as follows:

$$\frac{\left(x + \dfrac{D}{2A}\right)^2}{\dfrac{D^2}{4A^2} + \dfrac{E^2}{4AC} - AF} + \frac{\left(y + \dfrac{E}{2C}\right)^2}{\dfrac{D^2}{4AC} + \dfrac{E^2}{4C^2} - CF} = 1$$

i. If $\frac{D^2}{4A} + \frac{E^2}{4C} - F = 0$, then the graph consists of one point:

$\left(-\frac{D}{2A}, -\frac{E}{2C}\right)$. If $\frac{D^2}{4A} + \frac{E^2}{4C} - F > 0$, then the graph is an ellipse.

If $\frac{D^2}{4A} + \frac{E^2}{4C} - F < 0$, then there is no solution and the graph is empty.

ii. If $\frac{D^2}{4A} + \frac{E^2}{4C} - F = 0$, then the graph consists of two lines:

$\sqrt{A}\left(x + \frac{D}{2A}\right) = \pm\sqrt{-C}\left(y + \frac{E}{2C}\right)$. If $\frac{D^2}{4A} + \frac{E^2}{4C} - F > 0$, then the graph is a hyperbola with a horizontal transverse axis. If $\frac{D^2}{4A} + \frac{E^2}{4C} - F < 0$, then the graph is a hyperbola with a vertical

transverse axis. **iii.** If $\frac{D^2}{4A} + \frac{E^2}{4C} - F = 0$, then the graph consists of

two lines: $\sqrt{-A}\left(x + \frac{D}{2A}\right) = \pm\sqrt{C}\left(y + \frac{E}{2C}\right)$. If $\frac{D^2}{4A} + \frac{E^2}{4C} - F > 0$,

then the graph is a hyperbola with a vertical transverse axis. If $\frac{D^2}{4A} + \frac{E^2}{4C} - F < 0$, then the graph is a hyperbola with a horizontal trans-

verse axis. **iv.** If $\frac{D^2}{4A} + \frac{E^2}{4C} - F = 0$, then the graph consists of one

point: $\left(-\frac{D}{2A}, -\frac{E}{2C}\right)$. If $\frac{D^2}{4A} + \frac{E^2}{4C} - F > 0$, then there is no solution and

the graph is empty. If $\frac{D^2}{4A} + \frac{E^2}{4C} - F < 0$, then the graph is an ellipse.

Getting Ready for the Next Section:

113. $\frac{\sqrt{2}}{2}$ **115. a.** $\cos^2\theta$ **b.** $\sin 2\theta$

117. a. $\frac{4}{5}$ **b.** $\frac{3}{5}$ **119. a.** -16 **b.** 25

Section 8.5

Concepts and Vocabulary:

1. $Ax^2 + Bxy + Cy^2 + Dx + Ey + F = 0$ **3.** $B^2 - 4AC < 0$

5. True   **7.** True

Building Skills:

9. parabola **11.** hyperbola **13.** ellipse **15.** $\left(\frac{1}{2}, \frac{\sqrt{3}}{2}\right)$

17. $\left(\frac{3\sqrt{2}}{2}, \frac{\sqrt{2}}{-2}\right)$ **19.** $\left(\frac{3}{2} - \sqrt{3}, -\frac{3\sqrt{3}}{2} - 1\right)$ **21.** $\frac{y'^2}{10} - \frac{x'^2}{2} = 1$

23. $-x' = y'^2$ **25.** $\frac{x'^2}{1} + \frac{y'^2}{6} = 1$ **27.** $\cos\theta = \frac{\sqrt{2}}{2}$, $\sin\theta = \frac{\sqrt{2}}{2}$

29. $\cos\theta = \frac{\sqrt{3}}{2}$, $\sin\theta = \frac{1}{2}$ **31.** $\cos\theta = \frac{\sqrt{2}}{2}$, $\sin\theta = \frac{\sqrt{2}}{2}$

33. $\cos\theta = \frac{3}{5}$, $\sin\theta = \frac{4}{5}$ **35.** $\cos\theta = \frac{\sqrt{2}}{2}$, $\sin\theta = \frac{\sqrt{2}}{2}$

37. $\cos\theta = \frac{1}{2}$, $\sin\theta = \frac{\sqrt{3}}{2}$ **39.** $y' = 4x'^2$ **41.** $\frac{x'^2}{4} + \frac{y'^2}{20} = 1$

43. $\frac{x'^2}{1} - \frac{y'^2}{3} = 1$ **45.** $x' = -\frac{1}{2}(y' + 1)^2$ **47.** $x'^2 + y' + 4y'^2 = 3$

49. $x' = 2y'^2 + 1$

51.

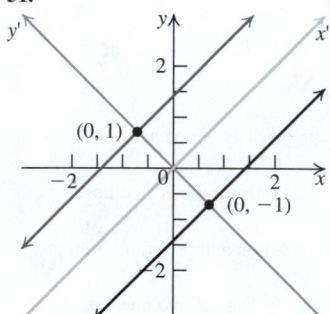

53. no points **55.** ellipse

57. parabola **59.** hyperbola

61.

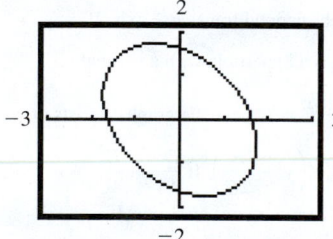

63.

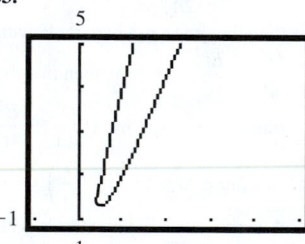

65.

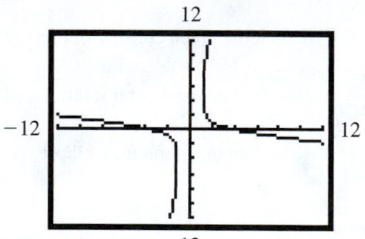

Applying the Concepts:
67. 7.5 knots opposite the shuttle's direction and 12.99 knots at 30° north of west **69.** 4 knots

Critical Thinking / Discussion / Writing
83. $F = -2$

Getting Ready for the Next Section:
85. **a.** $(4, 120°)$ **b.** $(8, 210°)$

87. $r = \dfrac{2}{1 + \dfrac{1}{3}\sin\theta}$ **89.** $x = 3$; a vertical line

Section 8.6

Concepts and Vocabulary:

1. eccentricity, ellipse, parabola, hyperbola **3.** $r = \dfrac{ep}{1 + e\cos\theta}$
5. True **7.** True

Building Skills:

9. parabola **11.** ellipse **13.** hyperbola **15.** parabola
17. parabola **19.** hyperbola

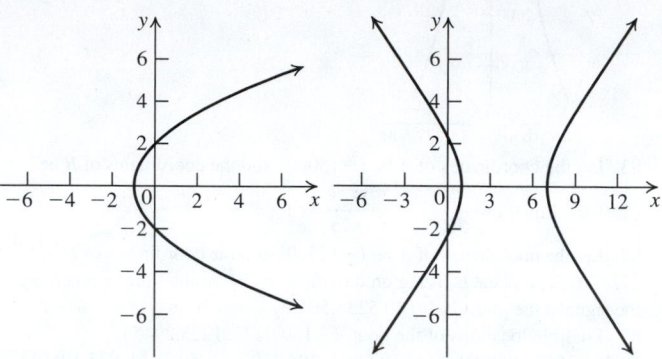

21. ellipse **23.** ellipse

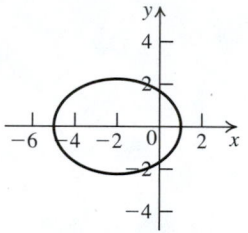

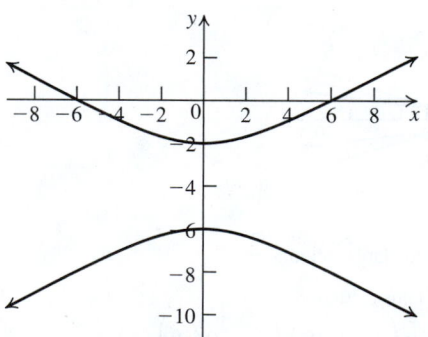

25. hyperbola

27. ellipse

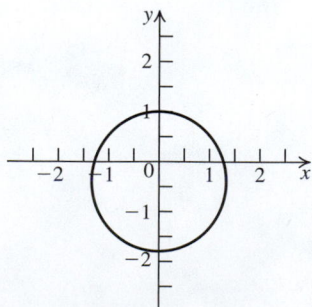

29. $r = \dfrac{12}{1 - \sin\theta}$ **31.** $r = \dfrac{6}{1 + \cos\theta}$ **33.** $r = \dfrac{24}{5 + \cos\theta}$

35. $r = \dfrac{15}{4 + \sin\theta}$ **37.** $r = \dfrac{8}{1 + 3\cos\theta}$ **39.** $r = \dfrac{6}{1 + 2\sin\theta}$

Applying the Concepts:

41. $r = \dfrac{92.933}{1 - 0.016997\cos\theta}$ **43.** $r = \dfrac{34.600}{1 - 0.211485\cos\theta}$

45. a. $r = \dfrac{30}{1 - \cos\theta}$ with r in millions of miles; **b.** the closest

approach to the sun is 15 million miles.

Beyond the Basics:
47. Rotated ellipse **49.** $x = 4$

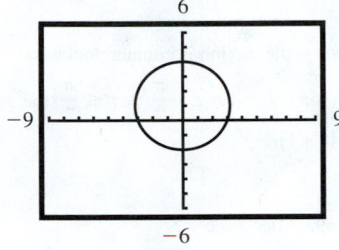

Critical Thinking / Discussion / Writing:
55. True

Getting Ready for the Next Section:

57. $\sin\theta = \dfrac{15}{17}$; $\cos\theta = \dfrac{8}{17}$ **59.** 3 **61.** amplitude: 2; period $\dfrac{\pi}{2}$

63.

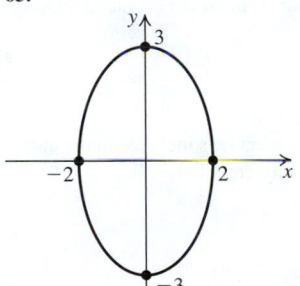

Section 8.7

Concepts and Vocabulary:
1. plane curve **3.** initial point **5.** False **7.** False

Building Skills:
9.   **11.**

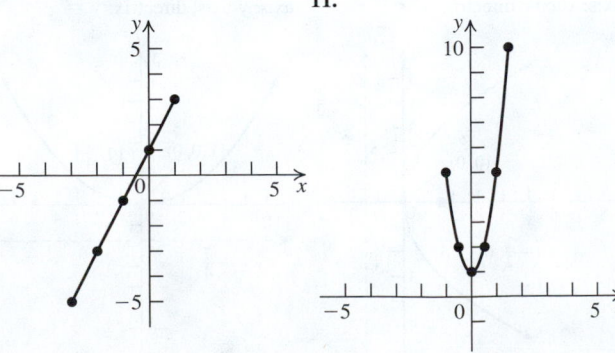

13.   **15.**

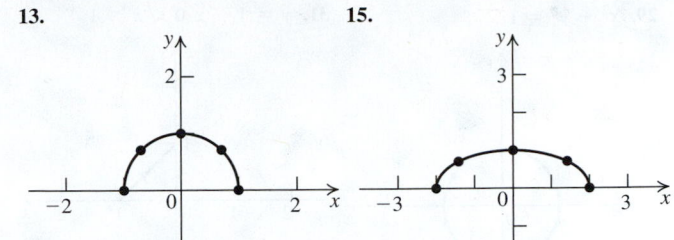

17. $y = 2x$ **19.** $y = 3x, 0 \le x \le 8$

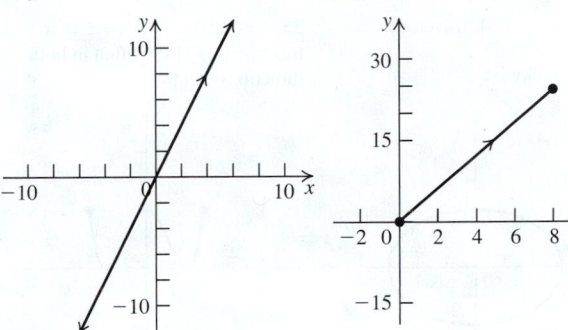

21. $y = 3 - x^2, 0 \le x \le 2$ **23.** $y = \sqrt{4 - x^2}, 0 \le x \le 2$

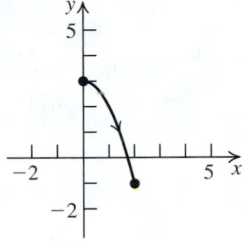
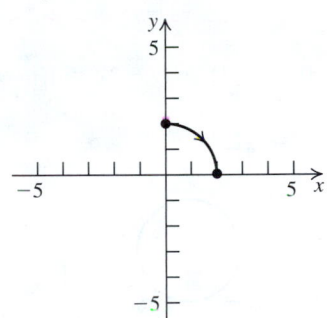

25. $y = 3\sqrt{1 - \dfrac{x^2}{4}}, 0 \le x \le 2$

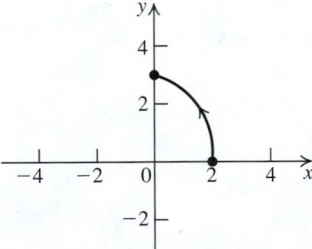

27. $y = \sqrt{x^2 - 4}, x \ge 2$

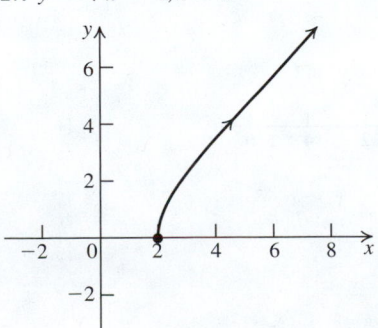

29. $x^2 + y^2 = 1$

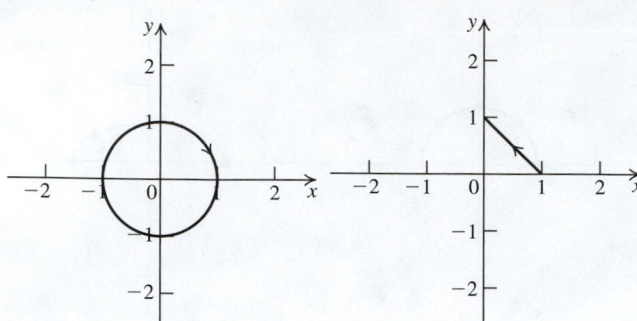

31. $y = 1 - x, 0 \le x \le 1$

33. $\dfrac{x^2}{9} + \dfrac{y^2}{4} = 1$, traversed counterclockwise

35. $y = (2x^2 - 1)^2, -1 \le x \le 1$, traversed infinitely often in both directions

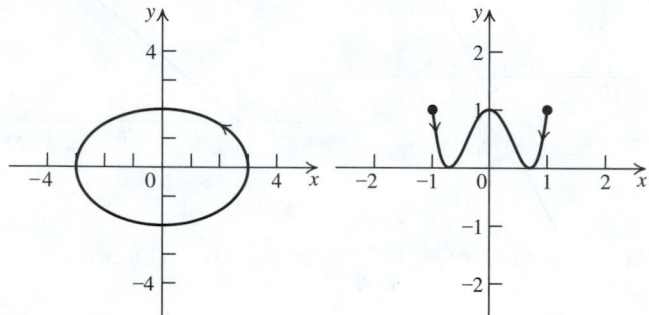

37. $(x - 1)^2 + (y - 2)^2 = 1$, traversed counterclockwise

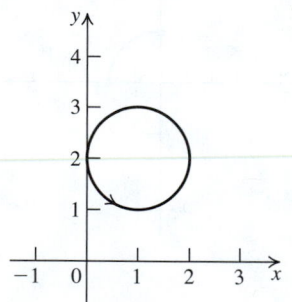

39. $\dfrac{(y - 2)^2}{9} - \dfrac{(x + 1)^2}{4} = 1, x \ge -1, y \ge 5$

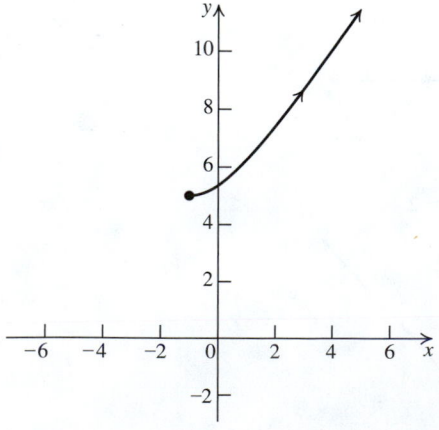

41. $y = x^2, x \ge 1$

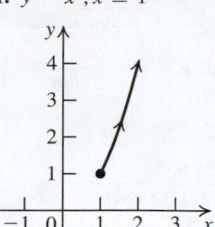

43. $y = \ln x, x \ge 1$

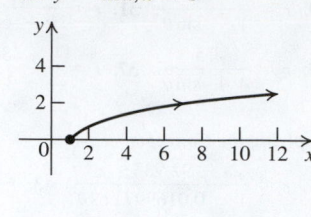

45. $x = t, y = 3t + 2$; the particle moves along the entire line from left to right.   **47.** $x = e^t, y = 3e^t + 2$; the particle moves to the right on the x-interval $0 < x < \infty$.   **49.** $x = -t, y = t^2 - 4$; the particle moves along the entire parabola from right to left.   **51.** $x = \cos t$, $y = \cos^2 t - 4$; the particle oscillates on the parabola between the points $(-1, -3)$ and $(1, -3)$.

53. $x = \sin t, y = \cos t, -\dfrac{\pi}{2} \le t \le \dfrac{\pi}{2}$; the motion is counterclockwise from $(0, -1)$ to $(0, 1)$.   **55.** $x = -\sin t, y = \cos t, -\dfrac{\pi}{2} \le t \le \dfrac{\pi}{2}$; the motion is clockwise from $(0, 1)$ to $(0, -1)$.

Applying the Concepts:
57. $x = 32\sqrt{3}t, y = 32t - 16t^2$ **59.** After 2 sec
61. $x = 800t, y = 800\sqrt{3}t - 16t^2 + 96$
63. Approximately 13.13 mi

Beyond the Basics:
65. b. $x = 2 - 3t, y = -3 + 7t, 0 \le t \le 1$
69. $x = -1 + 3\sin t, y = 2 + 3\cos t, 0 \le t \le 2\pi$

73. The upper half of the parabola $x = \dfrac{1}{4}(y - 1)^2 + 2$

75. The upper half of the circle $x^2 + y^2 = 1$, traversed counterclockwise starting at $(1, 0)$ **77.** The ellipse $\dfrac{x^2}{128} + \dfrac{y^2}{450} = 1$

Critical Thinking / Discussion / Writing:
79. (a) represents the line; (b) and (d) represent rays included in the line but not the entire line, and (c) represents a different line.

Getting Ready for the Next Section:
81. $\dfrac{1}{4}$ **83.** -81 **85.** -1 **87.** 32   **89.** 77   **91.** True

Review Exercises

Building Skills:

1. Vertex: $(0, 0)$; focus: $\left(-\dfrac{3}{2}, 0\right)$; axis: x-axis; directrix: $x = \dfrac{3}{2}$

3. Vertex: $(0, 0)$; focus: $\left(0, \dfrac{7}{4}\right)$; axis: y-axis; directrix: $y = -\dfrac{7}{4}$

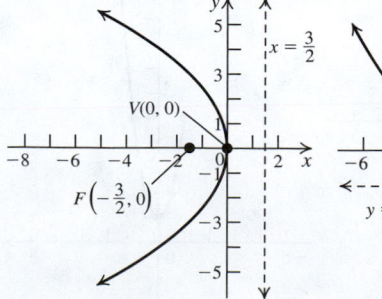

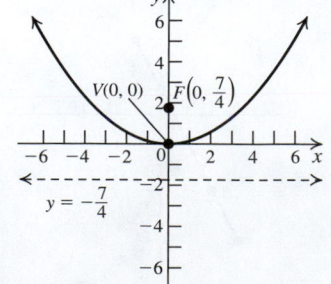

5. Vertex: $(2, -3)$ Focus: $\left(2, -\dfrac{13}{4}\right)$

Axis: $x = 2$ Directrix: $y = -\dfrac{11}{4}$

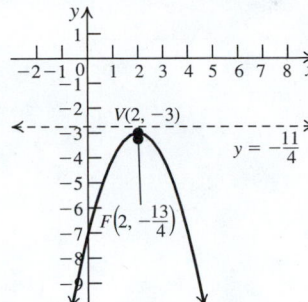

7. Vertex: $\left(-\dfrac{5}{2}, -2\right)$ Focus:

$(-2, -2)$ Axis: $y = -2$ Directrix:
$x = -3$

9. $y^2 = -12x$ **11.** $x^2 = 16y$

13. Foci: $(\sqrt{21}, 0)$ and $(-\sqrt{21}, 0)$
Vertices: $(5, 0)$ and $(-5, 0)$
Endpoints of the minor axis: $(0, 2)$
and $(0, -2)$

15. Foci: $(0, \sqrt{3})$ and $(0, -\sqrt{3})$
Vertices: $(0, 2)$ and $(0, -2)$
Endpoints of the minor axis: $(1, 0)$
and $(-1, 0)$

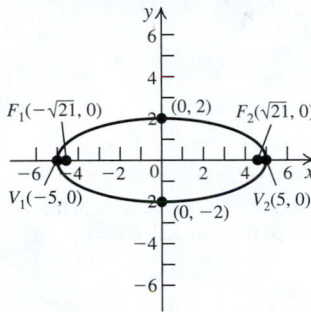

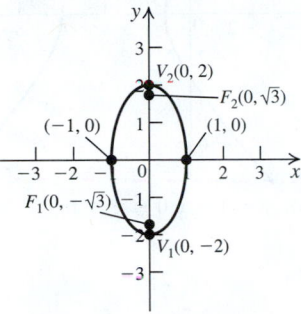

17. Foci: $(-1, -4 + \sqrt{7})$ and
$(-1, -4 - \sqrt{7})$; vertices: $(-1, 0)$
and $(-1, -8)$; endpoints of the
minor axis: $(2, -4)$ and $(-4, -4)$

19. Foci: $(-1 + 2\sqrt{2}, 1)$ and
$(-1 - 2\sqrt{2}, 1)$; vertices: $(2, 1)$ and
$(-4, 1)$; endpoints of the minor
axis: $(-1, 2)$ and $(-1, 0)$

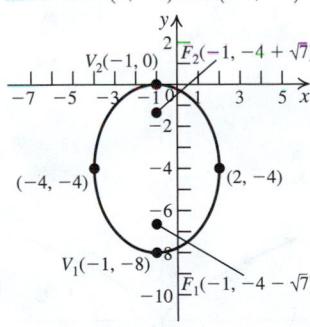

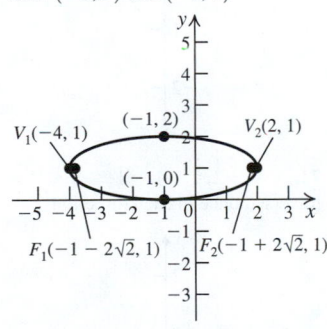

21. $\dfrac{x^2}{16} + \dfrac{y^2}{4} = 1$ **23.** $\dfrac{x^2}{100} + \dfrac{y^2}{75} = 1$

25. Vertices: $(0, 4)$ and $(0, -4)$
Foci: $(0, 2\sqrt{5})$ and $(0, -2\sqrt{5})$
Asymptotes: $y = \pm 2x$

27. Vertices: $(1, 0)$ and $(-1, 0)$
Foci: $(3, 0)$ and $(-3, 0)$
Asymptotes: $y = \pm 2\sqrt{2}x$

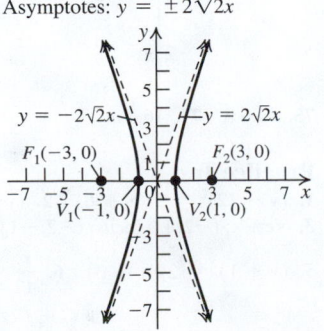

29. Vertices: $(1, 3)$ and $(-5, 3)$; foci: $(-2 + \sqrt{13}, 3)$ and
$(-2 - \sqrt{13}, 3)$; asymptotes: $y - 3 = \pm\dfrac{2}{3}(x + 2)$

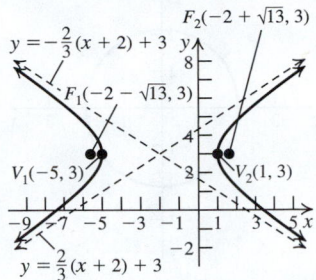

31. Vertices: $(-2, -2)$ and $(-2, -8)$; foci: $(-2, -5 + 3\sqrt{5})$
and $(-2, -5 - 3\sqrt{5})$; asymptotes: $y + 5 = \pm\dfrac{1}{2}(x + 2)$

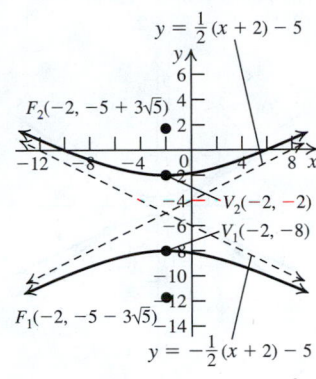

33. $x^2 - \dfrac{y^2}{3} = 1$ **35.** $\dfrac{x^2}{4} - \dfrac{y^2}{36} = 1$

37. Hyperbola

39. Ellipse

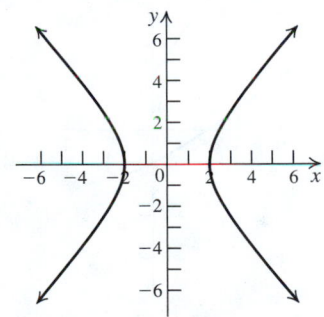

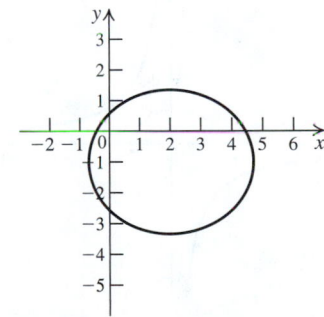

41. Circle

43. Hyperbola

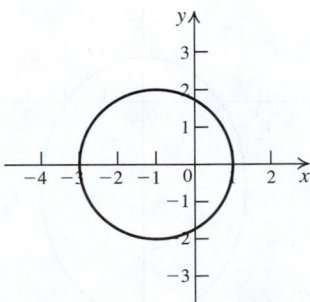

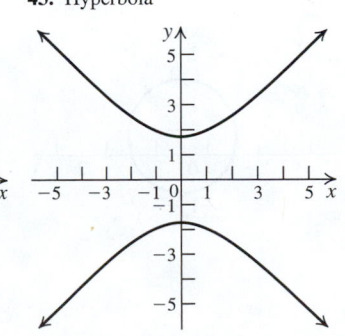

45. Circle

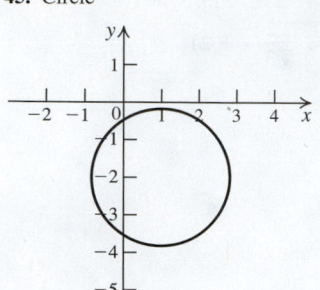

47. Ellipse

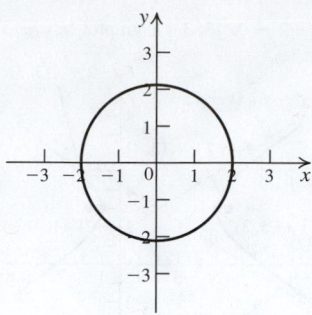

49. $\dfrac{x^2}{5} - \dfrac{y^2}{4} = 1$

51. Point of intersection: $\left(\dfrac{109}{10}, -\dfrac{91}{20}\right)$

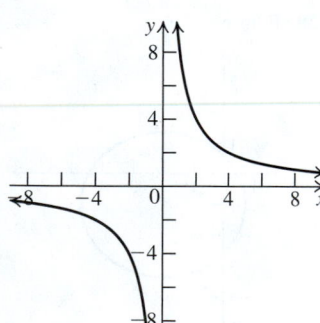

53. Points of intersection:
$(2, 1), (2, -1), (-2, 1), (-2, -1),$

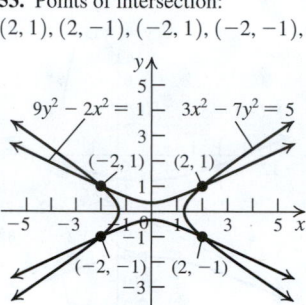

55. $\dfrac{x'^2}{16} - \dfrac{y'^2}{16} = 1$; hyperbola

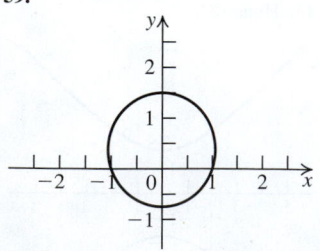

57. $y' = \sqrt{2}x'^2 - 2\sqrt{2}$; parabola

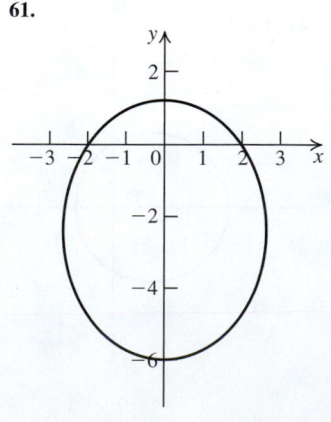

59.

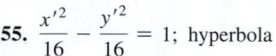

61.

63.

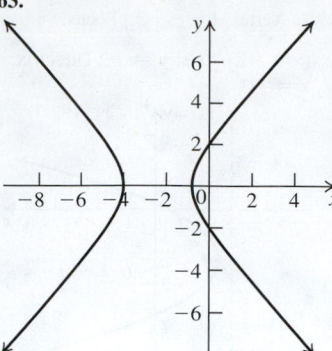

65.

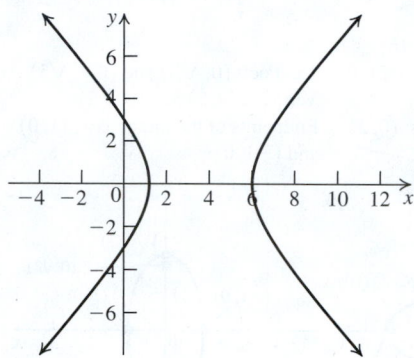

67. $y = 1 - x, 0 \le x \le 1$

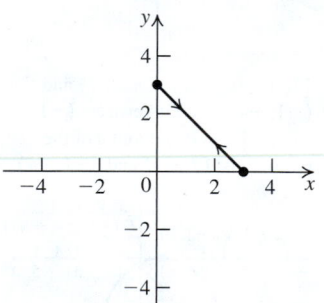

69. $y = 3 - x, 0 \le x \le 3$, traversed twice in each direction

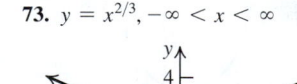

71. $y = -2x^2 + 1, -1 \le x \le 1$, traversed once in each direction

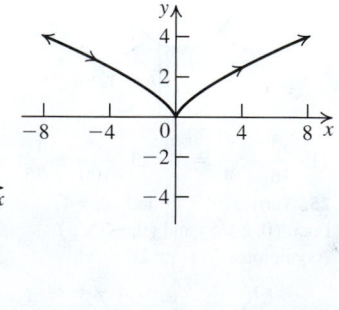

73. $y = x^{2/3}, -\infty < x < \infty$

75. 14.6 ft   **77.** 16.4 in

Practice Test A

1. $(y - 12)^2 = 24(x + 6)$ **2.** $(y - 1)^2 = -8(x + 3)$

3. Vertex: $(-2, 1)$; focus: $(-2, -1)$; directrix: $y = 3$ **4.** 2/9 ft

5. $(y + 1)^2 = 24(x - 3)$ **6.** $\dfrac{x^2}{12} + \dfrac{y^2}{16} = 1$ **7.** $\dfrac{x^2}{81} + \dfrac{y^2}{4} = 1$

8.
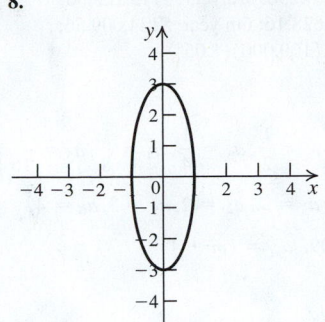

9. $\dfrac{y^2}{36} - \dfrac{x^2}{9} = 1$

10. $y^2 - (x-1)^2 = 1$ **11.** $\dfrac{(x-1)^2}{4} - (y+2)^2 = 1$

12. Hyperbola
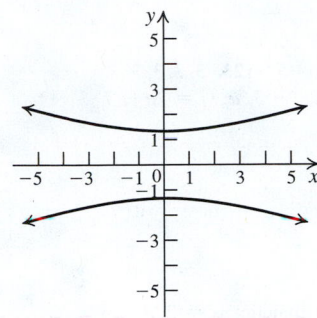

13. Parabola
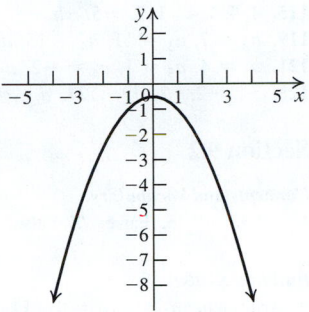

14. Circle
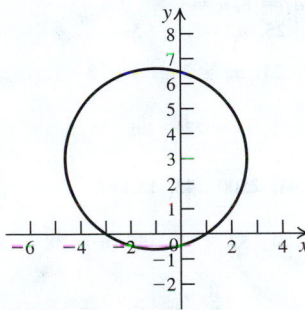

15. Ellipse
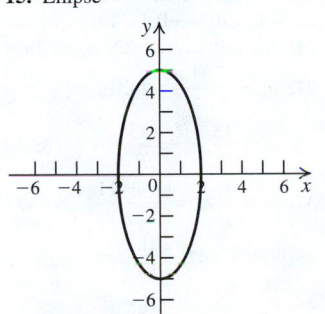

16. Hyperbola
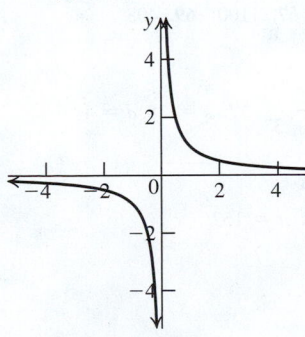

17. Ellipse
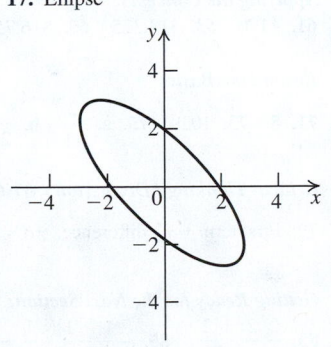

18. Ellipse
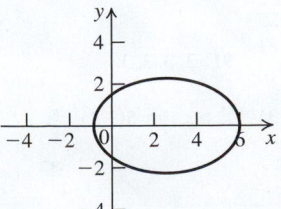

19. Hyperbola
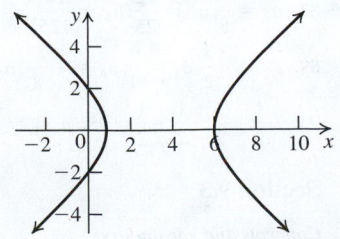

20. Parabola
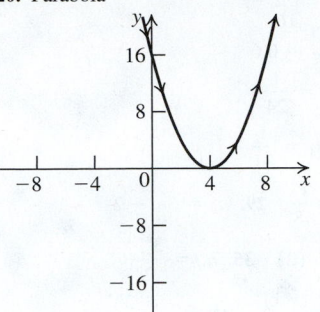

Practice Test B

1. b **2.** b **3.** a **4.** d **5.** a **6.** b **7.** a **8.** c **9.** b **10.** c
11. c **12.** c **13.** a **14.** d **15.** d **16.** d **17.** c **18.** c
19. d **20.** d

Cumulative Review Exercises (Chapters 1–8)

1. $\dfrac{f(x+h)-f(x)}{h} = 2x + h - 3$

3. $f^{-1}(x) = \dfrac{1}{2}x + \dfrac{3}{2}$

$\quad f(f^{-1}(x)) = f\left(\dfrac{1}{2}x + \dfrac{3}{2}\right)$

$\quad\quad = 2\left(\dfrac{1}{2}x + \dfrac{3}{2}\right) - 3$

$\quad\quad = x + 3 - 3 = x$

5. Solution: $x = 4$ **7.** $(2, \infty)$ **9.** $\{(4,3,2)\}$ **11.** $-\dfrac{5\sqrt{29}}{29}$

13. $r^2 - 6r\cos\theta + 3 = 0$ **15.** $\left\{\left(-\dfrac{25}{29}, \dfrac{22}{29}\right)\right\}$

17. $\left\{1, -1, \dfrac{1}{2}\sqrt{6}, -\dfrac{1}{2}\sqrt{6}\right\}$ **19.** Circle
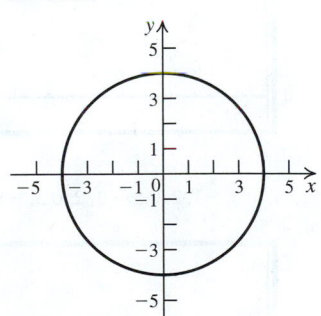

CHAPTER 9

Section 9.1

Concepts and Vocabulary:
1. positive integers **3.** 1 **5.** False **7.** True

Building Skills:
9. $a_1 = 1, a_2 = 4, a_3 = 7, a_4 = 10$
11. $a_1 = 0, a_2 = \dfrac{1}{2}, a_3 = \dfrac{2}{3}, a_4 = \dfrac{3}{4}$
13. $a_1 = -1, a_2 = -4, a_3 = -9, a_4 = -16$
15. $a_1 = 1, a_2 = \dfrac{4}{3}, a_3 = \dfrac{3}{2}, a_4 = \dfrac{8}{5}$
17. $a_1 = 1, a_2 = -1, a_3 = 1, a_4 = -1$

19. $a_1 = \dfrac{5}{2}, a_2 = \dfrac{11}{4}, a_3 = \dfrac{23}{8}, a_4 = \dfrac{47}{16}$

21. $a_1 = 0.6, a_2 = 0.6, a_3 = 0.6, a_4 = 0.6$

23. $a_1 = -1, a_2 = \dfrac{1}{2}, a_3 = -\dfrac{1}{6}, a_4 = \dfrac{1}{24}$

25. $a_1 = -\dfrac{1}{3}, a_2 = \dfrac{1}{9}, a_3 = -\dfrac{1}{27}, a_4 = \dfrac{1}{81}$

27. $a_1 = \dfrac{e}{2}, a_2 = \dfrac{e^2}{4}, a_3 = \dfrac{e^3}{6}, a_4 = \dfrac{e^4}{8}$ **29.** $a_n = 3n - 2$

31. $a_n = \dfrac{1}{n+1}$ **33.** $a_n = (-1)^{n+1}(2)$ **35.** $a_n = \dfrac{3^{n-1}}{2^n}$

37. $a_n = n(n+1)$ **39.** $a_n = 2 - \dfrac{(-1)^n}{n+1}$ **41.** $a_n = \dfrac{3^{n+1}}{n+1}$

43. $a_1 = 2, a_2 = 5, a_3 = 8, a_4 = 11, a_5 = 14$

45. $a_1 = 3, a_2 = 6, a_3 = 12, a_4 = 24, a_5 = 48$

47. $a_1 = 7, a_2 = -11, a_3 = 25, a_4 = -47, a_5 = 97$

49. $a_1 = 2, a_2 = \dfrac{1}{2}, a_3 = 2, a_4 = \dfrac{1}{2}, a_5 = 2$

51. $a_1 = 25, a_2 = -\dfrac{1}{125}, a_3 = -25, a_4 = \dfrac{1}{125}, a_5 = 25$

53. a. 2, 11, 26, 47, 74, 107, 146, 191, 242, 299

b. 400

55. a. 0, 1, 2, 3, 4, 5, 6, 7, 8, 9

b. 12

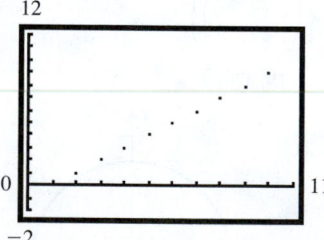

57. a. 0.5, -1, 2, 0.5, -1, 2, 0.5, -1, 2, 0.5

b. 5

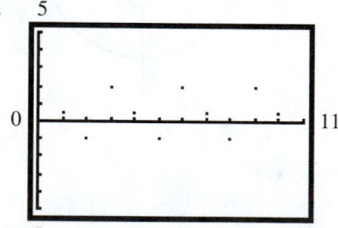

59. $\dfrac{1}{20}$ **61.** 12 **63.** $\dfrac{1}{n+1}$ **65.** $2n + 1$ **67.** 35 **69.** 55

71. 25 **73.** $\dfrac{853}{140}$ **75.** 183 **77.** -118 **79.** $\displaystyle\sum_{k=1}^{51} (2k - 1)$

81. $\displaystyle\sum_{k=1}^{11} \dfrac{1}{5k}$ **83.** $\displaystyle\sum_{k=1}^{50} \dfrac{(-1)^{k+1}}{k}$ **85.** $\displaystyle\sum_{k=1}^{10} \left(\dfrac{k}{k+1}\right)$ **87.** 4620

89. -1.345 (rounded to three decimal places) **91.** 50

Applying the Concepts:

93. a. 208 ft **b.** $(16 + (n-1)32)$ ft **95.** $250 **97.** 19,200 min

99. a. $A_1 = 10,300, A_2 = 10,609, A_3 = 10,927.27, A_4 = 11,255.09,$
$A_5 = 11,592.74, A_6 = 11,940.52$ **b.** $16,047.10

101. 1st year: $105,000; 2nd year: $110,250; 3rd year: $115,762.50; 4th year: $121,550.63; 5th year: $127,628.16; 6th year: $134,009.56; 7th year: $140,710.04; formula: $a_n = (100,000)(1.05^n)$

Beyond the Basics:

103. $a_n = 2^{1-(1/2^n)}$ **105. a.** $a_1 = 1, a_2 = \dfrac{1}{2}, a_3 = \dfrac{1}{4}, a_4 = \dfrac{1}{8}, a_5 = \dfrac{1}{16}$

b. $a_n = \dfrac{1}{2^{n-1}}$ **107.** $a_1 = 1, a_2 = 1, a_3 = 2, a_4 = 2, a_5 = 3, a_6 = 4,$
$a_7 = 4, a_8 = 4, a_9 = 5, a_{10} = 6$ **109.** $a_m = (m+1)^2$
111. $p = -2, q = 8$

Critical Thinking / Discussion / Writing:

113. $k = 5$

Getting Ready for the Next Section:

115. 4, 4, 4, 4 **117.** $-5, -5, -5, -5$
119. $a_2 = 7, a_3 = 11, a_4 = 15, a_5 = 19$
121. $a_2 = 4, a_3 = 1, a_4 = -2, a_5 = -5$ **123.** 3, 3, 3, 3
125. $-2, -2, -2, -2$ **127.** $a_{n+1} = 2n - 3, a_{n-1} = 2n - 7$

Section 9.2

Concepts and Vocabulary:

1. 5 **3.** -3 **5.** False **7.** False

Building Skills:

9. Arithmetic; $a_1 = 1, d = 1$ **11.** Arithmetic; $a_1 = 2, d = 3$
13. Not Arithmetic **15.** Not Arithmetic **17.** Arithmetic;
$a_1 = 0.6, d = -0.4$ **19.** Arithmetic; $a_1 = 8, d = 2$
21. Not arithmetic **23.** $a_n = 3n + 2$ **25.** $a_n = 16 - 5n$

27. $a_n = \dfrac{3-n}{4}$ **29.** $a_n = -\dfrac{2n+1}{5}$ **31.** $a_n = 3n + e - 3$

33. $d = \dfrac{13}{2}, a_n = \dfrac{13}{2}n - 5$ **35.** $d = -2, a_n = 22 - 2n$

37. $d = \dfrac{1}{2}, a_n = \dfrac{n+11}{2}$ **39.** 1275 **41.** 2500 **43.** 15,150

45. -208 **47.** $\dfrac{121}{3}$ **49.** 6225 **51.** 460 **53.** 1340 **55.** $n = 38$

57. $n = 25$ **59.** $n = 25$

Applying the Concepts:

61. 1170 **63.** $12,375 **65.** $16.75 **67.** 1100 **69.** 408

Beyond the Basics:

71. 8 **73.** 1029 **75. a.** $\dfrac{2}{11}$ **b.** $\dfrac{2}{2n+3}$ **77.** $k = 28, d = 1$

Critical Thinking / Discussion / Writing:

79. First term $-a_1$; difference: $-d$ **81.** $n = 157$

Getting Ready for the Next Section:

83. $\dfrac{2}{3}, \dfrac{2}{3}, \dfrac{2}{3}, \dfrac{2}{3}$ **85.** $-\dfrac{1}{2}, -\dfrac{1}{2}, -\dfrac{1}{2}, -\dfrac{1}{2}$

87. $a_2 = \dfrac{6}{5}, a_3 = \dfrac{12}{25}, a_4 = \dfrac{24}{125}, a_5 = \dfrac{48}{625}$

89. $a_2 = -\dfrac{3}{2}, a_3 = \dfrac{9}{4}, a_4 = -\dfrac{27}{8}, a_5 = \dfrac{81}{16}$ **91.** 3, 3, 3, 3

93. $-\dfrac{1}{2}, -\dfrac{1}{2}, -\dfrac{1}{2}, -\dfrac{1}{2}$ **95.** $a_{n+1} = 5(-3)^{-n-1}, a_{n-1} = 5(-3)^{1-n}$

Section 9.3

Concepts and Vocabulary:

1. 5 **3.** 3 **5.** True **7.** False

Building Skills:
9. Geometric; $a_1 = 3, r = 2$ **11.** Not geometric
13. Geometric; $a_1 = 1, r = -3$ **15.** Geometric; $a_1 = 7, r = -1$
17. Geometric; $a_1 = 9, r = \frac{1}{3}$ **19.** Geometric; $a_1 = -\frac{1}{2}, r = -\frac{1}{2}$
21. Geometric; $a_1 = 1, r = 2$ **23.** Not geometric
25. Geometric; $a_1 = \frac{1}{3}, r = \frac{1}{3}$ **27.** Geometric; $a_1 = \sqrt{5}, r = \sqrt{5}$
29. $a_1 = 2, r = 5, a_n = (2)(5)^{n-1}$
31. $a_1 = 5, r = \frac{2}{3}, a_n = (5)\left(\frac{2}{3}\right)^{n-1}$
33. $a_1 = 0.2, r = -3, a_n = 0.2(-3)^{n-1}$
35. $a_1 = \pi^4, r = \pi^2, a_n = \pi^{2n+2}$
37. $a_7 = 320$ **39.** $a_{10} = -1536$ **41.** $a_6 = \frac{243}{16}$ **43.** $a_9 = -\frac{390,625}{256}$
45. $a_{20} = -\frac{125}{131,072}$ **47.** $S_{10} = \frac{1,220,703}{5}$ **49.** $S_{12} = -\frac{40,690,104}{25}$
51. $S_8 = \frac{109,225}{16,384}$ **53.** $\frac{31}{16}$ **55.** $\frac{6305}{729}$ **57.** 255 **59.** $\frac{3(5^{20} - 2^{20})}{35(2)^{19}}$
61. $\frac{127}{320}$ **63.** 9841.333 (rounded to three decimal places)
65. 2.250 (rounded to three decimal places) **67.** $\frac{1}{2}$ **69.** $-\frac{1}{3}$ **71.** $\frac{32}{5}$
73. $\frac{15}{2}$ **75.** $\frac{4}{5}$ **77.** $n = 9$ **79.** $n = 13$ **81.** $n = 9$

Applying the Concepts:
83. 23,185 **85.** $3933.61 **87.** 1024 **89.** $1,190,374.35
91. 189.09 cm **93.** 20 m

Beyond the Basics:
95. $\frac{2}{3}$

Critical Thinking / Discussion / Writing:
105. b_{1001} is larger. **106.** $a = 512$ **107.** $n = 15, k = 16$
108. 5, 10, and 20 **109.** 2, 5, and 8; or 26, 5, and -16

Getting Ready for the Next Section:
111. P_3: $3^3 + 3$ is divisible by 3; true
P_4: $4^3 + 4$ is divisible by 3; false
113. P_3: $2^3 > 3 \cdot 3$; false
P_4: $2^4 > 3 \cdot 4$; true
115. $P_n + 1$: $2^{3(n+1)} - 1$ is divisible by 7.
117. $P_n + 1$: $1 + 4 + 7 + \ldots + (3n - 2) + (3n + 1) = \frac{1}{2}(n + 1)(3n + 2)$

Section 9.4

Concepts and Vocabulary:
1. natural numbers **3.** P_{k+1} is true **5.** True **7.** False

Building Skills:
9. P_{k+1}: $(k + 2)^2 - 2(k + 1) = (k + 1)^2 + 1$
11. P_{k+1}: $2^{k+1} > 5(k + 1)$

Applying the Concepts:
45. a. The number of sides of the nth figure is $3(4^{n-1})$. **b.** The perimeter of the nth figure is $3\left(\frac{4}{3}\right)^{n-1}$.
47. The smallest number of moves to accomplish the transfer is $2^n - 1$.

Getting Ready for the Next Section:
65. The exponents of x decrease from n to 1 when expanding $(x + y)^n$.
67. n

Section 9.5

Concepts and Vocabulary:
1. $n + 1$ **3.** alternating signs **5.** False **7.** True

Building Skills:
9. 120 **11.** 12 **13.** 15 **15.** 1 **17.** 7 **19.** 1 **21.** 4950 **23.** 210
25. $(x + 2)^4 = x^4 + 8x^3 + 24x^2 + 32x + 16$
27. $(x - 2)^5 = x^5 - 10x^4 + 40x^3 - 80x^2 + 80x - 32$
29. $(2 - 3x)^3 = -27x^3 + 54x^2 - 36x + 8$
31. $(2x + 3y)^4 = 16x^4 + 96x^3y + 216x^2y^2 + 216xy^3 + 81y^4$
33. $(x + 1)^4 = x^4 + 4x^3 + 6x^2 + 4x + 1$
35. $(x - 1)^5 = x^5 - 5x^4 + 10x^3 - 10x^2 + 5x - 1$
37. $(y - 3)^3 = y^3 - 9y^2 + 27y - 27$
39. $(x + y)^6 = x^6 + 6x^5y + 15x^4y^2 + 20x^3y^3 + 15x^2y^4 + 6xy^5 + y^6$
41. $(1 + 3y)^5 = 243y^5 + 405y^4 + 270y^3 + 90y^2 + 15y + 1$
43. $(2x + 1)^4 = 16x^4 + 32x^3 + 24x^2 + 8x + 1$
45. $(x - 2y)^3 = x^3 - 6x^2y + 12xy^2 - 8y^3$
47. $(2x + y)^4 = 16x^4 + 32x^3y + 24x^2y^2 + 8xy^3 + y^4$
49. $\left(\frac{x}{2} + 2\right)^7 = \frac{1}{128}x^7 + \frac{7}{32}x^6 + \frac{21}{8}x^5 + \frac{35}{2}x^4 + 70x^3 + 168x^2 + 224x + 128$ **51.** $\left(a^2 - \frac{1}{3}\right)^4 = a^8 - \frac{4}{3}a^6 + \frac{2}{3}a^4 - \frac{4}{27}a^2 + \frac{1}{81}$
53. $\left(\frac{1}{x} + y\right)^3 = y^3 + 3\frac{y^2}{x} + 3\frac{y}{x^2} + \frac{1}{x^3}$ **55.** 36 **57.** 1512
59. 314,928 **61.** 126 **63.** $120x^7y^3$ **65.** $-112,640x^3$
67. $16,128x^6y^2$ **69.** $-704,000x^2y^9$ **71.** $84x^3$ **73.** $280x^3y^4$
75. $405x^8y^5$ **77.** 2.48832

Beyond the Basics:
79. $-\frac{8064\sqrt{x}}{x^8}$ **81.** $924\frac{x^{12}}{y^{18}}$ **87.** 16 **89.** x^5 **91.** $k = \pm 2$

Getting Ready for the Next Section:
99. $\frac{8!}{4!}$ **101.** $2^6 \cdot 6!$ **103.** 132 **105.** 56

Section 9.6

Concepts and Vocabulary:
1. permutation **3.** combination **5.** True **7.** False

Building Skills:
9. 6 **11.** 56 **13.** 362,880 **15.** 1 **17.** 56 **19.** 126 **21.** 1 **23.** 1
25. a. 676 **b.** 650 **27.** 2450 **29.** 24 **31.** 60 **33.** 10 **35.** 55
37. 330 **39.** 20,160 **41.** 120

Applying the Concepts:
43. a. 160 **b.** 800 **45. a.** 12,144 **b.** 13,824 **47.** 256 **49.** Yes
51. a. 5040 **b.** 5040 **53. a.** 10 **b.** 10 **c.** 5 **55.** 105
57. 3,628,800 **59.** The maximum number of times each child will go to the circus is 20. The father goes 35 times. **61.** 111,007,923,832,370,565
63. 80 **65.** 15 **67.** 360 **69.** 420

Beyond the Basics:
71. 6 **73.** 435 **75.** $m = 3$ **77.** $n = 6$ **79.** $n = 6$ **81.** $m = 9$

Critical Thinking / Discussion / Writing:
83. 25

Getting Ready for the Next Section:
85. 3 **87.** 1 **89.** 4, 5 **91.** $\{3, 5\}, 2$ **93.** 3, 4 **95.** $\varnothing, 0$

Section 9.7

Concepts and Vocabulary:
1. equally likely **3.** mutually exclusive **5.** True (if "between 0 and 1" includes 0 and 1) **7.** False

Building Skills:

9. $S = \{$ (Wendy's, McDonald's, Burger King), (Wendy's, Burger King, McDonald's), (McDonald's, Wendy's, Burger King), (McDonald's, Burger King, Wendy's), (Burger King, Wendy's, McDonald's), (Burger King, McDonald's, Wendy's) $\}$

11. $S = \{\{$ Thriller, The Wall $\}$, $\{$ Thriller, Eagles: Their Greatest Hits $\}$, $\{$ Thriller, Led Zeppelin IV $\}$, $\{$ The Wall, Eagles: Their Greatest Hits $\}$, $\{$ The Wall, Led Zeppelin IV $\}$, $\{$ Eagles: Their Greatest Hits, Led Zeppelin IV $\}\}$

13. $S = \{$ (white, male), (white, female), (African American, male), (African American, female), (Native American, male), (Native American, female), (Asian, male), (Asian, female), (other, male), (other, female), (multiracial, male), (multiracial, female) $\}$

15. 0 **17.** 1 **19.** experimental **21.** theoretical **23.** experimental **25.** theoretical **27.** experimental

29. An event that is very likely to happen has probability 0.999
An event that will surely happen has probability 1.
An event that is a rare event has probability 0.001.
An event that may or may not happen has probability 0.5.
An event that will never happen has probability 0.

31. $\dfrac{1}{3}$ **33.** $\dfrac{1}{3}$ **35.** $\dfrac{1}{2}$ **37.** $\dfrac{1}{13}$ **39.** $\dfrac{1}{4}$ **41.** $\dfrac{3}{13}$ **43.** $\dfrac{1}{10}$

45. $\dfrac{1}{10}$ **47.** $\dfrac{1}{2}$

Applying the Concepts:

49. 0.7 **51.** $\dfrac{1}{5}$ **53.** $\dfrac{47}{50}$ **55.** The probability of winning one prize is $\dfrac{2}{25}$. The probability of winning both is $\dfrac{9}{1550}$. **57. a.** $\dfrac{1}{4}$ **b.** $\dfrac{1}{4}$ **c.** $\dfrac{1}{2}$

59. a. $\dfrac{79}{148}$ **b.** $\dfrac{33}{40}$ **c.** 0 **d.** 1 **e.** $\dfrac{119}{120}$ **61.** The probability that a non-Hispanic white American has no lactose intolerance is 0.8. The probability for the second group is $\dfrac{1}{4}$. **63. a.** $\dfrac{1}{10}$ **b.** $\dfrac{1}{10}$

65. 90, 10, 135, 765 **67.** Three of one gender and one of the other is more likely. **69. a.** $\dfrac{2}{5}$ **b.** $\dfrac{89}{180}$

Beyond the Basics:

71. $\dfrac{468,559}{1,000,000}$ **73.** $\dfrac{1}{12}$

Critical Thinking / Discussion / Writing:

75. $\dfrac{1}{3}$

Review Exercises

1. $a_1 = -1, a_2 = 1, a_3 = 3, a_4 = 5, a_5 = 7$

3. $a_1 = \dfrac{1}{3}, a_2 = \dfrac{2}{5}, a_3 = \dfrac{3}{7}, a_4 = \dfrac{4}{9}, a_5 = \dfrac{5}{11}$

5. $a_n = 32 - 2n$ **7.** 9 **9.** $n + 1$ **11.** 100 **13.** $\dfrac{1343}{140}$ **15.** $\displaystyle\sum_{k=1}^{50} \dfrac{1}{k}$

17. Arithmetic; $a_1 = 11, d = -5$ **19.** Not arithmetic **21.** $a_n = 3n$ **23.** $a_n = x + n - 1$ **25.** $d = 2, a_n = 2n + 1$ **27.** 352 **29.** 4020 **31.** Geometric: $a_1 = 4, r = -2$ **33.** Not geometric

35. $a_1 = 16, r = -\dfrac{1}{4}, a_n = \dfrac{(-1)^{n-1}}{4^{n-3}}$ **37.** $a_{10} = 39,366$

39. $S_{12} = \dfrac{819}{2}$ **41.** $\dfrac{3}{4}$ **43.** $\dfrac{3}{2}$ **47.** $\dbinom{12}{7} = 792$

49. $(x - 3)^4 = x^4 - 12x^3 + 54x^2 - 108x + 81$ **51.** $101,376x^5$

53. 24 **55.** 5040 **57.** 120 **59.** 120 **61.** 3696 **63.** 15,120

65. $\dfrac{1}{2}$ **67.** $\dfrac{2}{5}$ **69.** $\dfrac{15}{28}$ **71.** $\dfrac{1}{17}$ **73. a.** $\dfrac{1}{9}$ **b.** $\dfrac{4}{9}$ **c.** $\dfrac{5}{9}$

Practice Test A:

1. $a_1 = 3, a_2 = -9, a_3 = -21, a_4 = -33, a_5 = -45$; arithmetic sequence **2.** $a_1 = -6, a_2 = -12, a_3 = -24, a_4 = -48, a_5 = -96$; geometric sequence **3.** $a_1 = -2, a_2 = -1, a_3 = 2, a_4 = 11, a_5 = 38$

4. $\dfrac{1}{n}$ **5.** $a_7 = 21$ **6.** $a_8 = -\dfrac{13}{128}$ **7.** 590 **8.** $\dfrac{93}{128}$ **9.** $\dfrac{2}{11}$ **10.** 1

11. $16x^4 - 32x^3 + 24x^2 - 8x + 1$ **12.** $8x$ **13.** 1024 **14.** 72

15. 21 **16.** 60 **17.** 6,760,000 **18.** $\dfrac{5}{12}$ **19.** $\dfrac{5}{6}$ **20.** $\dfrac{7}{51}$

Practice Test B:

1. c **2.** d **3.** a **4.** c **5.** a **6.** b **7.** d **8.** a **9.** b **10.** b **11.** a **12.** d **13.** c **14.** a **15.** d **16.** b **17.** d **18.** c **19.** a **20.** d

Cumulative Review Exercises (Chapters 1–9)

1. $\{2, 4\}$ **3.** $x = 2$ **5.** $x = 4$ **7.** $x = 4$

9. $(-\infty, -3 - 3\sqrt{3}\,] \cup [-3 + 3\sqrt{3}, \infty)$

11. $x = 10, y = 7, z = \dfrac{1}{2}$ **13.** $\dfrac{-\sqrt{6} - \sqrt{2}}{4}$

15. Shift the graph of $g(x) = \sqrt{x}$ three units to the right.

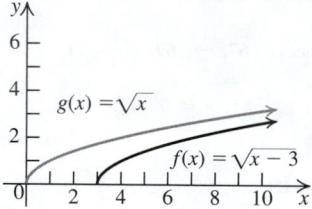

16. $(-\infty, -1) \cup (-1, 0) \cup (0, \infty)$ **17.** $\ln\left(\dfrac{x^7}{x - 5}\right)$

18. $A^{-1} = \begin{bmatrix} 7 & -2 & 2 \\ -28 & 9 & -8 \\ 3 & -1 & 1 \end{bmatrix}$ **19.** 12 **20.** $\dfrac{2}{3}$

CHAPTER 10

Section 10.1

Concepts and Vocabulary:

1. $\lim\limits_{x \to c} f(x) = L$ **3.** L **5.** False **7.** True

Building Skills:

9. 7 **11.** 17 **13.** 2 **15.** 7 **17.** -1 **19.** 1 **21.** 4 **23.** 12 **25.** 0 **27.** 1 **29.** -1 **31.** -3 **33.** -3 **35.** 1 **37.** 1 **39.** -2 **41.** -2 **43.** -2 **45.** Does not exist **47.** 5 **49.** Does not exist **51.** 4 **53.** 4 **55.** 2 **57.** Does not exist **59.** 6 **61.** -3 **63.** 3 **65.** 4 **67.** 1 **69.** 0 **71.** 2 **73.** 1 **75.** 0

Applying the Concepts:

77. a. $c(x) = \begin{cases} 0.12x & \text{if } 0 \le x \le 50 \\ 6 + 0.18(x - 50) & \text{if } x > 50 \end{cases}$ **b.** 6, 6, 6

c. \$1.80 **79. a.** $W(x) = \begin{cases} 10x & \text{if } 0 \le x \le 40 \\ 400 + 15(x - 40) & \text{if } 40 < x \le 60 \end{cases}$

b. 400, 400, 400 **c.** \$550

Beyond the Basics:

81. a. 0 **b.** Does not exist **c.** Does not exist **d.** 0
83. a. 2 **b.** 2 **c.** 2 **d.** Does not exist

85. $f(x) = \begin{cases} x + 1, & \text{if } x \ne 1 \\ \dfrac{1}{2}, & \text{if } x = 1 \end{cases}$ **87.** $f(x) = \begin{cases} -1, & \text{if } x < 4 \\ 2, & \text{if } x = 4 \\ 1, & \text{if } x > 4 \end{cases}$

Critical Thinking / Discussion / Writing:
89. $f(1) = 5$ **91.** Not possible

Getting Ready for the Next Section:
93. $(x - 8)(x + 3)$ **95.** $(x - 6)(x + 6)$ **97.** $\dfrac{1}{\sqrt{3} + 1}$

99. $\dfrac{1}{\sqrt{x + 1} + 1}$ **101.** 1   **103.** $x + 1$

Section 10.2

Concepts and Vocabulary:
1. 5   **3.** $\lim\limits_{x \to c} g(x) \neq 0$ **5.** True   **7.** False

Building Skills:
9. 13   **11.** 24   **13.** 48   **15.** 11   **17.** 9   **19.** $\dfrac{5}{3}$ **21.** 4   **23.** 2

25. Does not exist   **27.** 3   **29.** 3   **31.** 4   **33.** 1   **35.** $\dfrac{2}{3}$ **37.** 3

39. 4   **41.** Does not exist   **43.** Does not exist   **45.** $\dfrac{1}{6}$ **47.** $\dfrac{1}{2}$ **49.** 4

51. 1   **53.** 1   **55.** 2   **57.** 4   **59.** $\dfrac{1}{4}$ **61.** 0   **63.** Does not exist

Applying the Concepts:
65. a. 1500   **b.** 1500   **c.** 1500   **d.** 1500
67. a. 0.6 m   **b.** 0   **c.** No, then L will not be a real number.
69. a. 5   **b.** Does not exist   **71.** -4 **73.** No   **75.** Yes   **77.** No
79. Yes   **81.** Does not exist   **83.** 1   **85.** 2   **87.** $\dfrac{1}{2}$

Critical Thinking / Discussion / Writing:
89. The output value of f for input c is $f(c)$, whereas $\lim\limits_{x \to c} f(x)$ is the value that $f(x)$ is near when x is near c, if this value exists.

Getting Ready for the Next Section:
91. $\lim\limits_{x \to \infty} P(x) = \infty$, $\lim\limits_{x \to -\infty} P(x) = \infty$
93. $\lim\limits_{x \to \infty} P(x) = -\infty$, $\lim\limits_{x \to -\infty} P(x) = \infty$
95. vert.: $x = 2$
   hor. $y = 1$
97. vert.: $x = -2, x = 2$
   hor. $y = 1$

Section 10.3

Concepts and Vocabulary:
1. bound   **3.** vertical asymptote   **5.** True   **7.** False

Building Skills:
9. ∞ **11.** $-\infty$ **13.** 1   **15.** -3 **17.** $-\infty$ **19.** ∞ **21.** ∞ **23.** $-\infty$
25. ∞ **27.** ∞

29. $\lim\limits_{x \to 1} f(x) = \dfrac{1}{2}$, $\lim\limits_{x \to -1^{-}} f(x) = -\infty$, $\lim\limits_{x \to -1^{+}} f(x) = \infty$

31. $\lim\limits_{t \to -1} g(t) = -\dfrac{1}{3}$, $\lim\limits_{t \to 2^{-}} g(t) = -\infty$, $\lim\limits_{t \to 2^{+}} g(t) = \infty$

33. $\lim\limits_{x \to -2} f(x) = \dfrac{1}{4}$, $\lim\limits_{x \to 2^{-}} f(x) = \infty$, $\lim\limits_{x \to 2^{+}} f(x) = -\infty$

35. $\lim\limits_{t \to 2} f(t) = \dfrac{1}{5}$, $\lim\limits_{t \to -3^{-}} f(t) = -\infty$, $\lim\limits_{t \to -3^{+}} f(t) = \infty$

37. $\lim\limits_{x \to -\infty} P(x) = -\infty$, $\lim\limits_{x \to \infty} P(x) = \infty$

39. $\lim\limits_{x \to -\infty} P(x) = -\infty$, $\lim\limits_{x \to \infty} P(x) = -\infty$ **41.** $\dfrac{1}{2}$ **43.** $\dfrac{1}{2}$ **45.** 0

47. 0   **49.** ∞ **51.** $-\infty$ **53.** 1   **55.** -3 **57.** 0   **59.** $\dfrac{1}{2}$

61. $\dfrac{1}{2}$ **63.** $\dfrac{7}{2}$

Applying the Concepts:
65. $y = -2$ and $y = 2$; no vertical asymptotes   **67.** $-\infty$, The lake will not be 100% pollution free.

Beyond the Basics:
69. $\dfrac{1}{3}$ **71.** $-\dfrac{7}{4}$ **73.** $\dfrac{9}{2}$ **75.** $\dfrac{1}{2}$ **77.** 3   **79.** Does not exist

Critical Thinking / Discussion / Writing:
81. When the degree of the numerator is greater than the degree of the denominator, such as $f(x) = \dfrac{x^2 + 1}{2x - 1}$ **83.** $\dfrac{1}{2}$

Getting Ready for the Next Section:
85. 0   **87.** 4   **89.** $y = 2x + 1$ **91.** $y = -\dfrac{1}{2}x - 5$

Section 10.4

Concepts and Vocabulary:
1. $\dfrac{f(b) - f(a)}{b - a}$ **3.** slope   **5.** True   **7.** False

Building Skills:
9. $\dfrac{3}{4}$ **11. a.** 4   **b.** 3   **c.** 2.5   **d.** 2   **13.** 2/3   **15.** 2   **17.** 8

19. 5   **21.** -2 **23.** 24   **25.** $\dfrac{\sqrt{3}}{3}$ **27.** $-\dfrac{2}{9}$ **29.** $y = (2/3)x + 1/3$

31. $y = 6x - 8$ **33.** $y = -8x + 11$ **35.** $y = -5x + 3$
37. $y = 12x - 20$ **39.** $y = (1/6)x + 3/2$ **41.** $y = -(1/9)x + 2/3$

43. $f'(x) = 3$ **45.** $f'(x) = \dfrac{2}{3}x$ **47.** $f'(x) = 4x - 3$

49. $f'(x) = 3x^2$ **51.** $f'(x) = \dfrac{3}{2\sqrt{3x - 2}}$ **53.** $f'(x) = -\dfrac{2}{(2x + 1)^2}$

55. a. 2 sec   **b.** 64 ft/sec   **57.** $432\pi \text{ cm}^3/\text{sec}$ **59. a.** 12 units per day, increasing   **b.** -4 units per day, decreasing   **61.** B, A, C

63. $f'(a) = -\dfrac{2}{a^3}$ **65. a.** $f'(a) = \cos(a)$ **b.** $f'(a) = -\sin(a)$

Critical Thinking / Discussion / Writing:
69.

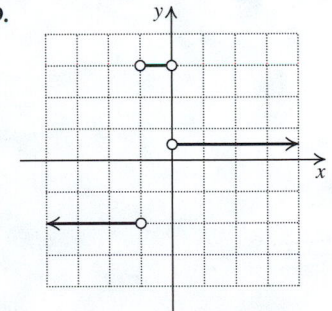

Getting Ready for the Next Section:
71. 2   **73.** $-\dfrac{k}{2}$ **75.** 26   **77.** $\dfrac{11}{2}$

Section 10.5

Concepts and Vocabulary:
1. overestimate   **3.** Δx **5.** True

Building Skills:
7. 8.5   **9. a.** 8   **b.** 7   **c.** 6.667   **d.** 6   **11. a.** 6.750   **b.** 5.625
c. 5.250   **d.** 4.500   **13. a.** 86   **b.** 116   **15. a.** 16   **b.** 18.625

17. a. 2.083   **b.** 1.910   **19. a.** 6.146   **b.** 5.896   **21. a.** $\displaystyle\int_0^1 3t \, dt$

b. 1.5m

23. a.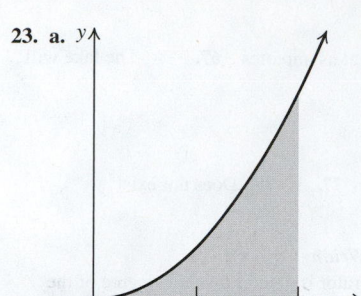

b. 3.5   **c.** 7.5

d. $\displaystyle\int_0^2 2x^2\,dx$ **e.** 5.333

25. a.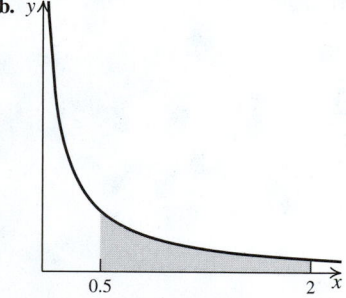

b. 40   **c.** 104

d. $\displaystyle\int_0^4 (x^3 + 1)\,dx$ **e.** 68

27. a.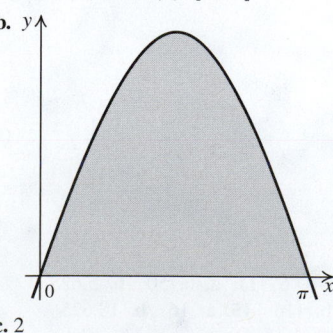

b. 3.178   **c.** 4.787

d. $\displaystyle\int_1^5 \ln(x)\,dx$ **e.** 4.047

29. a. $f(x) = \dfrac{1}{x}$, $[0.5, 2]$

b. y

c. 1.386

31. a. $f(x) = \sin(x)$, $[0, \pi]$

b. y

c. 2

Beyond the Basics:
33. 12

Critical Thinking / Discussion / Writing:
35. 10

Review Exercises

1. 7   **3.** $\dfrac{1}{2}$ **5.** -1 **7.** Does not exist   **9.** 1   **11.** Does not exist

13. 5   **15.** 5   **17.** 0   **19.** 0   **21.** 10   **23.** Does not exist   **25.** 22

27. 27   **29.** 4   **31.** 0   **33.** 2   **35.** $\dfrac{1}{4}$ **37.** -1 **39.** 8

41. Does not exist   **43.** $-\infty$ **45.** ∞ **47.** ∞ **49.** ∞ **51.** $\dfrac{1}{2}$

53. 0   **55.** 0   **57.** -1 **59. a.** 2   **b.** $y = 2x$ **61. a.** 7

b. $y = 7x - 1$ **63. a.** -8 **b.** $y = -8x - 1$ **65. a.** $\dfrac{1}{4}$

b. $y = \dfrac{1}{4}x + 1$ **67.** $\dfrac{19}{12}$ **69.** $\dfrac{91}{8}$ **71.** 9 m/s   **73. a.** 96 ft/s

b. 128 ft/s, 64 ft/s

Practice Test A:

1. 5   **2.** 6   **3.** 1   **4.** Does not exist   **5.** 8   **6.** -6 **7.** ∞ **8.** $-\infty$

9. -1 **10.** 0   **11.** -8 **12.** $\dfrac{3}{4}$ **13.** $y = -4x + 1$ **14.** $-\dfrac{1}{4}x - 1$

15. $f'(x) = 12x - 4$ **16. a.** $\sqrt{2}$ sec   **b.** $32\sqrt{2} \approx 45.25$ ft/sec
17. 36   **18.** 46   **19.** $f(x) = x^2 + x$, $[-1, 5]$
20. $f(x) = e^x$, $[-6, 0]$

Practice Test B:

1. b   **2.** a   **3.** c   **4.** a   **5.** d   **6.** b   **7.** a   **8.** d   **9.** c   **10.** c   **11.** a
12. d   **13.** b   **14.** a   **15.** b   **16.** c   **17.** b   **18.** a   **19.** b   **20.** d

APPENDIX A

Section A.1

Basic Concepts and Skills:
1. -0.8 terminating   **3.** $0.\overline{27}$ repeating   **5.** Rational   **7.** Irrational
9. Rational   **11.** $\{-4, -3, -2, 0, 1, 2, 3, 4\}$ **13.** $\{-4, -3, -2, 0, 2\}$

15. $\dfrac{15}{4}$ **17.** $I_1 \cup I_2 = (-2, 5), I_1 \cap I_2 = [1, 3]$

19. $I_1 \cup I_2 = (-6, 10), I_1 \cap I_2 = \varnothing$
21. $I_1 \cup I_2 = (-\infty, 7), I_1 \cap I_2 = (-\infty, 3)$
23. $I_1 \cup I_2 = (-\infty, 10) \cup (10, \infty), I_1 \cap I_2 = \varnothing$ **25.** 20   **27.** -4 **29.** $\dfrac{5}{7}$
31. $5 - \sqrt{2}$ **33.** 1   **35.** 12   **37.** 3   **39.** $|3 - 8| = 5$

41. $|-6 - 9| = 15$ **43.** $|-20 - (-6)| = 14$ **45.** $\left|\dfrac{22}{7} - \left(-\dfrac{4}{7}\right)\right| = \dfrac{26}{7}$

47. $1 \le x \le 4$

49. $14 < x < 28$

51. $-3 < x \le 1$

53. $x \ge -3$

55. $x \le 5$

57. $-\dfrac{3}{4} < x < \dfrac{9}{4}$

59. 11   **61.** -1 **63.** -6 **65.** -13 **67.** $\dfrac{62}{15}$

69. Exponent: 3; base: 17   **71.** Exponent: 0; base: 9
73. Exponent: 5; base: -5 **75.** Exponent: 3; base: 10

77. Exponent: 2; base: a **79.** 6 **81.** 1 **83.** 64 **85.** $\dfrac{1}{81}$ **87.** $\dfrac{1}{15{,}625}$

89. 16 **91.** $\dfrac{2}{9}$ **93.** 2 **95.** 1 **97.** 6 **99.** $\dfrac{49}{121}$ **101.** x^4 **103.** $\dfrac{y}{x}$

105. $-\dfrac{8}{x}$ **107.** $\dfrac{3}{x}$ **109.** $\dfrac{1}{xy^2}$ **111.** $\dfrac{1}{x^{12}}$ **113.** x^{33} **115.** $-3x^5y^5$

117. $\dfrac{4x^2}{y^2}$ **119.** $\dfrac{3x^5}{y^5}$ **121.** $\dfrac{1}{x^4}$ **123.** $8y^3$ **125.** $-243x^5y^5$ **127.** $\dfrac{25}{9x^2}$

129. $\dfrac{64}{x^9y^{15}}$ **131.** $\dfrac{x^5}{y^4}$ **133.** $\dfrac{3x}{y^2}$

Applying the Concepts:
135. a. people who own either MP3 or DVD players or both
b. people who own both MP3 and DVD players
137. $119.5 \le x \le 134.5$

139. a. $|124 - 120| = 4$
b. $|137 - 120| = 17$
d. $|114 - 120| = 6$

141. a. $(2x)(2x) = 4x^2 = 2^2A$ **b.** $(3x)(3x) = 3x^2 = 3^2A$

Section A.2
Basic Concepts and Skills

1. Yes; $x^2 + 2x + 1$ **3.** No **5.** Degree 1: terms: $7x$, 3
7. Degree 4; terms: $-x^4, x^2, 2x, -9$ **9.** $2x^2 - 3x - 1$
11. $x^3 - x^2 + 5x - 8$ **13.** $-24x^2 - 14x - 14$
15. $2y^3 + y^2 - 2y - 1$ **17.** $12x^2 + 18x$ **19.** $x^3 + 3x^2 + 4x + 2$
21. $3x^3 - 5x^2 + 5x - 2$ **23.** $x^2 + 3x + 2$ **25.** $9x^2 + 9x + 2$
27. $-4x^2 - 7x + 15$ **29.** $6x^2 - 7x + 2$ **31.** $4x^2 + 4ax - 15a^2$
33. $4x + 4$ **35.** $9x^2 + 27x + 27$ **37.** $16x^2 + 8x + 1$
39. $27x^3 + 27x^2 + 9x + 1$ **41.** $-4x^2 + 25$ **43.** $x^2 + \dfrac{3}{2}x + \dfrac{9}{16}$
45. $2x^3 - 9x^2 + 19x - 15$ **47.** $y^3 + 1$ **49.** $x^3 - 216$
51. $3x^2 + 11xy + 10y^2$ **53.** $6x^2 + 11xy - 7y^2$ **55.** $x^4 - 2x^2y^2 + y^4$
57. $x^3 - 3x^2y + 4y^3$ **59.** $x^4 - 4x^3y + 16xy^3 - 16y^4$ **61.** $x^2(3x - 1)$
63. $(x + 3)(x + 4)$ **65.** $(x - 2)(x - 4)$ **67.** $(2x + 3)(3x + 4)$
69. $(x + 3)^2$ **71.** $(3x + 1)^2$ **73.** $(x - 8)(x + 8)$
75. $(4x - 3)(4x + 3)$ **77.** $(x - 3)(x^2 + 3x + 9)$
79. $(2 - x)(4 + 2x + x^2)$ **81.** $(x - 3)(x^2 + 1)$ **83.** $(x - 5)(x^2 + 1)$
85. $(x - 1)(x + 1)(x^2 + 1)$ **87.** $5(2x^2 - 1)(2x^2 + 1)$
89. $(1 - 4x)(1 + 4x)$ **91.** $(x - 3)^2$ **93.** $(2x + 1)^2$
95. $2(x + 1)(x - 5)$ **97.** $(2x - 5)(x + 4)$ **99.** $(x - 6)^2$
101. $3x^3(x + 2)^2$ **103.** $(3x + 1)(3x - 1)$ **105.** $(4x + 3)^2$
107. Irreducible

Section A.3
Basic Concepts and Skills:

1. $\dfrac{2}{x + 1}, x \neq -1$ **3.** $\dfrac{3}{x - 1}, x \neq -1, x \neq 1$ **5.** $\dfrac{x - 3}{4}, x \neq 3$

7. $\dfrac{7x}{x + 1}, x \neq -1$ **9.** $\dfrac{x + 2}{x - 2}, \ x \neq -\dfrac{1}{3}, x \neq 0, x \neq 2$ **11.** 1

13. $\dfrac{x - 1}{x + 3}$ **15.** $\dfrac{5x(x - 3)}{2}$ **17.** $\dfrac{3(x + 3)}{x + 4}$ **19.** $\dfrac{x - 3}{x^2 - 2x + 4}$

21. $\dfrac{x + 4}{2x + 1}$ **23.** $\dfrac{2(x - 2)}{x - 3}$ **25.** $\dfrac{2x}{(x - 1)^2}$ **27.** $\dfrac{5}{x + 1}$

29. $\dfrac{x^3 + 2x^2 + 4x - 8}{x(x - 2)(x + 2)}$ **31.** $12(x - 2)$ **33.** $(x - 1)(x + 1)(x + 2)$

35. $(x - 1)(x - 4)(x + 2)$ **37.** $\dfrac{7x + 15}{(x - 3)(x + 3)}$

39. $\dfrac{x(4 - x)}{(x - 2)(x + 2)}$ **41.** $\dfrac{2x + 3}{(x - 2)(x + 5)}$ **43.** $\dfrac{2(9x^2 - 5x + 1)}{(3x + 1)(3x - 1)^2}$

45. $\dfrac{9(x - 3)}{(x + 4)(x + 5)(x - 5)}$ **47.** $\dfrac{2x - 3}{(2 + x)(2 - x)}$

49. $\dfrac{16a^2}{(x - 5a)(x - 3a)}$ **51.** $-\dfrac{h}{x(x + h)}$ **53.** $\dfrac{2x}{3}$ **55.** $\dfrac{1}{x - 1}$

57. $\dfrac{1 - x}{1 + x}$ **59.** $-x$ **61.** $\dfrac{x(2x - 1)}{2x + 1}$ **63.** $-\dfrac{1}{x(x + h)}$ **65.** $\dfrac{x}{a}$

Applying the Concepts
67. 10 cm **69. a.** $\dfrac{3}{100 + x}$ **b.** 2% **71. a.** $\dfrac{10\pi x^3 + 240}{x}$ **b.** \$2.46

Section A.4
Basic Concepts and Skills:

1. 8 **3.** 4 **5.** -3 **7.** $-\dfrac{1}{2}$ **9.** 3 **11.** Not a real number **13.** 1

15. -7 **17.** $7\sqrt{3}$ **19.** $9\sqrt{5}$ **21.** $3\sqrt{2x}$ **23.** $-\sqrt[3]{3}$ **25.** $2\sqrt[3]{3x}$

27. $\sqrt{2x}(x^2 + 3x - 20)$ **29.** $(2x - y)^2\sqrt{3xy}$ **31.** $\dfrac{2\sqrt{3}}{3}$

33. $\dfrac{\sqrt{2} - x}{2 - x^2}$ **35.** $\sqrt{3} - \sqrt{2}$ **37.** $\dfrac{7 - 2\sqrt{10}}{3}$

39. $\dfrac{2x + h - 2\sqrt{x(x + h)}}{h}$ **41.** $\dfrac{1}{\sqrt{4 + h} + 2}$ **43.** $\dfrac{1}{2 + \sqrt{4 - x}}$

45. $\dfrac{1}{\sqrt{x} + 2}$ **47.** $\dfrac{2x}{\sqrt{x^2 + 4x} + x}$ **49.** 5 **51.** -2 **53.** 4 **55.** $-\dfrac{1}{125}$

57. $\dfrac{125}{27}$ **59.** $x^{9/10}$ **61.** $x^{1/10}$ **63.** $4x^4$ **65.** $\dfrac{1}{9x^4y^2}$ **67.** $5x^{5/4}$ **69.** $\dfrac{x^3}{y^8}$

71. $3^{1/2}$ **73.** x^3 **75.** x^2y^3 **77.** 3 **79.** $\dfrac{2}{3}x^{1/3}(7x - 6)$

81. $2(3x + 1)^{1/3}(7x - 1)$ **83.** $(x^2 + 1)^{1/2}(14x^3 - 5x^2 + 8x - 2)$

Applying the Concepts:
85. 40 cm **87.** 2 cm/sec

Section A.5
Basic Concepts and Skills:

1. 25 **3.** 15 **5.** $3\sqrt{2}$ **7.** 8 **9.** 48 **11.** $A = 15, P = 16$
13. $A = 5, P = 21$ **15.** 6 **17.** 3 **19.** $A = 3.14; C = 6.28$
21. $A = 0.785; C = 3.14$ **23.** 5 **25.** 15 **27.** 25 **29.** 96π in.^{3}

Applying the Concepts:
31. 192 ft^2 **33.** 9 in. **35.** $5\sqrt{29} \approx 27$ ft **37.** 346 yd **39.** 2471 ft
41. 39 ft^2 **43.** 15 ft^3

Section A.6
Basic Concepts and Skills:

1. $\{-1\}$ **3.** Identity **5.** $\{28\}$ **7.** $\{11\}$ **9.** $\left\{\dfrac{4}{5}\right\}$ **11.** $\varnothing$

13. Identity **15.** $\{0, 5\}$ **17.** $\{-1, 6\}$ **19.** $\{-1 \pm \sqrt{6}\}$

21. $\{2 \pm 2\sqrt{2}\}$ **23.** $\{0, 2\}$ **25.** $\{-1, 0, 1\}$ **27.** $\left\{-\dfrac{3}{2}, 2\right\}$ **29.** $\varnothing$

31. $\{-1\}$ **33.** $\{-2, 1\}$ **35.** $\{6\}$ **37.** $\{3, 7\}$ **39.** $\{8, 64\}$

41. $\{1\}$ **43.** $\{\pm 2\sqrt{2}, \pm \sqrt{3}\}$ **45.** $\left\{-\dfrac{4}{3}, -\dfrac{2}{3}\right\}$ **47.** $\{-5b, 2b\}$

49. $\left\{\dfrac{1 + \sqrt{5}}{2}\right\}$ **51.** 0, 8 **55.** 2

Applying the Concepts:
59. 2 in. **61.** 40 km **63.** 25 liters

Section A.7
Basic Concepts and Skills:

1. $(-\infty, -1)$

3. $(-\infty, -6)$

5. $(-\infty, \infty)$

7. $(-\infty, -2) \cup (2, \infty)$ **9.** $\left(-\infty, \dfrac{11}{2}\right]$ **11.** $(-\infty, -2] \cup (6, \infty)$

13. $[1, 5)$ **15.** $\varnothing$ **17.** $(-\infty, 5]$ **19.** $(-2, -1)$ **21.** $(-3, 3]$

23. $\left(-1, \dfrac{1}{2}\right]$ **25.** $a = 5, b = 6$ **27.** $a = 1, b = 3$

29. $a = -1, b = 19$ **31.** $[-6, 2]$ **33.** $(-2, 1) \cup (2, \infty)$

35. $(-\sqrt{2}, 0) \cup (0, \sqrt{2})$ **37.** $(-2, -1] \cup [3, 4)$ **39.** $(-3, 6]$

41. $\left(-1, \dfrac{1}{7}\right] \cup (3, \infty)$ **43.** $(-4, 2)$ **45.** $\left(-\dfrac{1}{2}, \dfrac{7}{2}\right)$

47. $(-\infty, 1) \cup (4, \infty)$ **49.** $\varnothing$ **51.** $(-\infty, \infty)$ **53.** $\left[-\dfrac{1}{2}, \dfrac{7}{2}\right]$

55. $(-\infty, -1] \cup [7, \infty)$ **57.** $\left(-\dfrac{1}{2}, \infty\right)$ **59.** $(-\infty, -6] \cup [0, \infty)$

61. $[-5, -2) \cup (-2, -1]$ **63.** $\left(\dfrac{1}{2}, 1\right) \cup \left(1, \dfrac{5}{2}\right)$

65. $(-\infty, 0] \cup [4, \infty)$

Applying the Concepts:
67. \$2012.50 to \$2100.00 **69.** 86 to 100 **71.** 21 to 29 hr

Section A.8

Basic Concepts and Skills:

1. $x = 3, y = 2$ **3.** $x = 2, y = -4$ **5.** $8 + 3i$ **7.** $-7i$ **9.** $15 + 6i$

11. $-3 + 15i$ **13.** $3 + 11i$ **15.** 13 **17.** $24 + 7i$ **19.** $-141 - 24\sqrt{3}i$

21. $-26 - 18i$ **23.** $\bar{z} = 2 + 3i; z\bar{z} = 13$ **25.** $\bar{z} = 4 - 5i; z\bar{z} = 41$

27. $\bar{z} = \sqrt{2} + 3i; z\bar{z} = 11$ **29.** $5i$ **31.** $1 + 2i$ **33.** $\dfrac{43}{65} - \dfrac{6}{65}i$

35. i **37.** i **43.** $14 - 8i$ **45.** $-2 - 16i$

Applying the Concepts:

47. $9 + i$ **49.** $Z = \dfrac{595}{74} + \dfrac{315}{74}i$ **51.** $V = 45 + 11i$

53. $I = \dfrac{17}{15} + \dfrac{4}{15}i$

Credits

INDEX

A

AAA triangles, 513
AAS triangles, 513, 514
  solving, 516–517
AA triangles, 950
Absolute value, 921–922
  of complex numbers, 578
  inequalities involving, 970–973
Absolute value function, 70
Acid(s), 315
Acid rain, 314
Action-at-a-distance forces, 537
Acute angles, 338
  common, trigonometric function values of, 358–363
Adams, Evelyn Marie, 850
Addition. *See also* Sum(s)
  of complex numbers, 976–977
  of matrices, 673–674, 675
  of rational expressions, 935–938, 944
  of vectors, 538–539
Addition method, for solving systems of equations, 600–602
Addition property, 964
Additive Rule, 853–854
Air navigation, vectors in, 544–545
Algebraic expressions, 922–923
Algebraic numbers, 265
Algebraic vectors, 539–541
Algorithms
  division, 173–176
  origin of word, 594
Al-Khwarizmi, Muhammad ibn Musa, 594
All Students Take Calculus mnemonic, 369
Al-Mamum, caliph, 594
Alternating current, 465–466, 577, 581
Altitude, of triangles, 514
Ambiguous case, 517–521
Amenhotep III, 287
American Journal of Mathematics, 673
American Museum of Natural History, 60
Amperes, 461
Amplitude
  in simple harmonic motion, 398
  of sinusoidal functions, 386–390
Angles, 337–347
  acute, 338
    common, trigonometric function values of, 358–363
  arc length formula and, 344–345
  area of sectors and, 345–346
  central, 340–341
  complementary, 344, 503–504
  coterminal, 338, 343
  of depression, 505
  direction, of vectors, 543–544
  of elevation, 505–507
  linear and angular speed and, 346–347
  measuring, 338–344

  multiple, trigonometric equations involving, 485–487
  obtuse, 338
  positive and negative, 337–338
  quadrantal, 338, 357
  reference, 369–373
  right, 338
  in standard position, 338, 339
  straight, 338
  supplementary, 344
  trigonometric functions of, 356–363
  between two vectors, 550–552
Angular speed, 346–347
Annuities, 817–819
Apollonius of Perga, 708
Applied problems
  composite functions in, 108–110
  derivatives in, 899–901
  dot product in, 556
  ellipses in, 733–734
  exponential equations in, 304–305
  exponential functions in
    growth and decay and, 265–266
    interest and, 259–265
  geometric sequences in, 817–819
  hyperbolas in, 750–751
  inverse functions in, 124–125
  linear inequalities in two variables in, 623–627
  linear programming in, 623–627
  logarithms in, 294–296
  matrix inverses in, 692–693
  matrix multiplication in, 677
  polar equations of conic sections in, 775–776
  polynomials in, 179
  quadratic functions in, 145–147
  right-triangle trigonometry in, 505–508
  systems of equations in two variables in, 603
  systems of linear equations in three variables in, 614–615
  trigonometric equations in, 489
  trigonometric formulas in, 477
  variation in, 236–237
  vectors in, 544–546
Arccosine function, 417–418
Archimedes, spiral of, 567
Arc length formula, 344–345
Arcsine function, 415–417, 421
Arctangent function, 418–419, 421
Area
  under graphs, 905–909
  maximum, finding, 146–147
  of rectangles, 44
  of sectors, 345–346
  of triangles, 522–523
Arguments, of complex numbers, 579
Arithmeticorum Libri Duo (Maurolico), 826

Arithmetic sequences, 805–810
  finite, sum of, 807–809
  recursive definition of, 806
Ars Magna (Cardano), 225
ASA triangles, 513, 514, 950
  solving, 516–517
Asymptotes
  horizontal, 190–191, 194–195
  of hyperbolas, 743–746
  oblique, 200–202
  vertical, 189–190, 192, 196, 197
Augmented matrices, 635–636
Average rate of change, of functions, 49–51
  composite, 107–108
Axes
  coordinate, 2
  of ellipses, major and minor, 725, 727, 729
  of hyperbolas, conjugate and transverse, 740, 741, 748
  imaginary, 577
  polar, 560
  principal, of conics, 769
  real, 577
  rotation of, 757–765
    eliminating the xy-term by, 761–764
    identifying conic sections and, 757–758, 764
  of symmetry, 9
  of parabolas, 139, 711, 712, 716
  x and y, 2
    reflections about, 83

B

Basal metabolic rate, 324
Bases
  different, exponential equations with, 303
  of exponential functions, 254
  of a logarithmic function, 272
  of logarithms, change of, 292–294
Bearings, 521–522
Bees, 794, 797
Bendel, Hannskarl, 138
Benford, Frank, 279
Benford's Law, 279
Bernoulli, Jakob, 560, 779
Bernoulli, Johann, 264, 779
Binomial(s), 928
Binomial coefficients, 832, 835–836
Binomial expansions, 831–832
Binomial Theorem, 831, 833–836
Bionics, 560
Biorhythms, 381, 395–396
Blood pressure measurement, 78, 92–93, 351
Blue moon, 481
Body weight, metabolic rate versus, for small animals, 324–325
Bonaccio, 796
Bounds on real zeros, 226–227

Brachistochrone problem, 779
Brahe, Tycho, 153
Branches, of hyperbolas, 739
British Petroleum (BP) oil spill, 99

C

Calculators. *See also* Graphing calculators
  digital, first, 831
  evaluating trigonometric functions using, 363–364
  inverse trigonometric functions using, 423–424
Calculus
  identities useful for, 444
  invention of, 649
  of variations, 779
Calories
  burn rate of, 108–109
  minimizing, 625–627
Carcharodon megalodon, 60, 61
Cardano, Gerolamo, 225
Cardioids, 571
Carnarvon, Lord, 287
Carrying capacity, 301
Car sales, 677
  applying composition to, 109–110
Car selections, 839–840
Carter, Howard, 287
Cartesian equations, 779
  converting between polar and rectangular forms and, 565–567
Cartesian plane, 2–5
  distance formula in, 6
  representing vectors in, 540
CAT scans, 608, 614–615
Cavendish, Henry, 240
Cayley, Arthur, 686
Celsius temperature, finding a Fahrenheit temperature from, 968
Centers, of conics, 769
  circles, 12
  ellipses, 725, 727, 729
  hyperbolas, 740
Central angles, 340–341
Change-of-base formula, 292–294
Chapman, Helen, 22
Chia Hsien, 832
Circles, 12–13, 573, 709, 732
  arc length formula and, 344–345
  area of sectors of, 345–346
  unit, 13, 351–355
Circular functions, 354. *See also* Trigonometric functions
Climate change, 173
Closed intervals, 920
Coefficient(s)
  binomial, 832, 835–836
  leading
    of polynomial functions, 153
    of polynomials, 928

of monomials, 928
of polynomials, 928
Coefficient matrices, 636
Cofactors
expanding by, 652
in matrices, 650–651
Cofunction identities, 450–451
Cogito ergo sum, 3
Coin tosses, 852
Column matrices, 635
Combinations, 843–845, 846–847
Combinations formula, 844
Combined variation, 239–241
Commentari Academia Petropolitanae (Euler), 40
Common difference, of arithmetic sequences,
    805–806
Common factors, factoring out, 930
Common logarithms, 278–279
    number of digits and, 292
Common ratio, of geometric sequences, 813–814
Complement(s), of events, 855
Complementary angles, 344, 503–504
Completing the square, 957–958
Complex Conjugate Product Theorem, 978
Complex numbers, 577–585, 976–979
    absolute value of, 578
    adding, 976–979
    conjugates of, 978
    defined, 976
    dividing, 978–979
    equal, 976
    geometric representation of, 577–578
    multiplying, 977–978
    polar form of, 578–580
        powers in, 582–583
        product and quotient in, 580–582
    in rectangular (trigonometric) form, 578, 579–580
    roots of, 583–584
    subtracting, 976–979
Complex plane, 577
Complex polynomials, 228
Complex rational expressions (fractions), 938–939
Complex zeros of polynomials, 228–229
Components of vectors, 540
Composite functions, 102–110
    applications of, 108–110
    composition of functions and, 102–104
    decomposition of, 108–110
    domain of, 104–106
    trigonometric, 424–426
Compound inequalities, 966–968
Compound interest, 260–263
Compound interest formula, 262
Compressing graphs, 85–88
Computer graphics, 672, 680–681
Computerized axial tomography scans, 608, 614–615
Congruent triangles, 513, 950

Conic sections, 708–709. *See also* Circles; Ellipses;
    Hyperbolas; Parabolas
    defined, 768–769
    identifying, 757–758, 764
    polar equations of, 769–776
        applications involving, 775–776
        graphing, 772–775
Conjugate(s)
    of complex numbers, 978
    rationalizing the numerator or denominator using, 945
    verifying trigonometric identities using, 443
Conjugate axis, of hyperbolas, 740, 741, 748
Conjugate Pairs Theorem, 229–232
Consistent systems of equations, 596, 611
Constant(s), 922
    of variation (proportionality), 236
Constant functions, 60, 62, 70
Constant terms, of polynomial functions, 153
Constraints, 624
Contact forces, 537
Continuous compounding, 264–265, 280–281
Continuous compound interest formula, 263–265
Converse, of Pythagorean Theorem, 951
Conway, John, 26
Coordinate(s), of points on real number line, 918
Coordinate axes, 2
Coordinate line, 918
Coordinate plane, 2–5
    distance formula in, 6
    representing vectors in, 540
Cormack, Allan, 608
Corner points, of solution sets of systems of inequalities, 622
Correspondence diagram, 40
Cosecant function, 354
    properties of, 409
Cosine(s), Law of. *See* Law of Cosines
Cosine function, 354. *See also* Sinusoidal functions
    graphs of, 385–386
    inverse, 321, 417–418
    properties of, 381–383
    sum and difference formulas for, 448–450
Cotangent function, 354
    properties of, 409
Coterminal angles, 338, 343
Counting principles, 839–847
    combinations and, 843–845, 846–847
    deciding which method to use, 846–847
    fundamental, 839–841, 846–847
    permutations and, 841–843, 846–847
        distinguishable, 845–846
Cramer, Gabriel, 653
Cramer's Rule, 653–656
Crime databases, 415
Crude oil consumption, 179–180
Cryptography, 685, 693–695
Cube root, 942
Cube root function, 71

Cubic polynomials, 154
Cubing function, 70
Current (electric), 461, 465–466
Cycloids, 779
    parametric equations of, 784

D

Dandelin, Germinal Pierre, 709
Dandelin spheres, 709
Daylight hours in Paris, 396–397
Decay factor, 257
    exponential, 265–266
Decay formula, 294
Decibel scale, 318–322
Decimals, 918
Decoding matrices, 694–695
Decomposition
    of functions, 108–110
    partial-fraction. *See* Partial-fraction decomposition
    of vectors, 555–556
Decreasing functions, 61–63
Deepwater Horizon, 99
Degenerate conic sections, 709
Degrees
    measuring angles using, 338–340
    of monomials, 928
    radians related to, 341–344
DeMoivre's Theorem, 582–585
De Motu Cordis (Harvey), 351
Denominators, least common, 935, 936–938
Dependent systems of equations, 596, 611, 612
Dependent variables, 7, 39
Depressed equations, 178
De Propria Vita (Cardano), 225
Derivatives, 898–901
    applications involving, 899–901
    tangent lines to graphs of functions and, 895–897
De Rossi, Michele Stefano, 315
Descartes, René
    biographical information on, 3
    coordinate plane discovered by, 2
    Rule of Signs and, 225–226
Descending order, 928
Determinants, 649–656
    Cramer's rule and, 653–656
    history of, 649
    minors and cofactors and, 650–651
    of $n \times n$ matrices, 651–652
    of 2×2 matrices, 649–650
Dice rolls, 852
Difference(s). *See also* Subtraction
    common, of arithmetic sequences, 805–806
Difference formulas. *See* Sum and difference formulas
Difference function, 100
Difference quotients, 51
    limits of, 880–882

Direction angle, of vectors, 543–544
Directrix, of parabolas, 711, 712, 716
Direct variation, 236–238
Discriminants, identifying conics by using, 764
Distance formula, 6
    on a number line, 922
Distinguishable permutations, 845–846
Distributive property, 923
Dividend, 174
Division. *See also* Quotient(s)
    of complex numbers, 978–979
    of polynomials, 173–176
        long, 174–175
        synthetic, 174, 175–176
    of rational expressions, 934
Division algorithm, 173–176
Divisor, 174
Domains
    agreement on, 44
    of combined functions, 100
    of composite functions, 104–106
    of cosine function, 381–382
    of exponential functions, 254
    of functions, 40, 44–45, 47, 100, 101–102
    of logarithmic functions, 274–275
    of one-to-one functions, 123–124
    of rational functions, 188
    of relations, 41
    of sine function, 381–382
    of tangent function, 405
    of variables, 954
Dot product, 549–556
    angle between two vectors and, 550–552
    decomposition of a vector and, 555–556
    defined, 549
    orthogonal vectors and, 552–553
    projection of a vector and, 553–555
    properties of, 552
    work and, 556
Double-angle formulas, 461–464
Drawing cards, 854–855
Drug bioavailability, 39, 52–53
Dual-tone multi-frequency (DTMF), 473, 477

E

Earthquake intensity, 315–318
Economics, Leontief Input-Output Model in, 634, 643–644, 692–693
Einstein, Albert, 875
Electrical circuits, direct variation in, 236–237
Electric blankets, 461, 465–466
Electricity
    alternating current, 465–466, 577, 581
    basic properties of, 461

Element(s)
  in matrices, 635
  of sets, 918
Elementary row operations, 637
Elimination
  Gaussian, 639–642
  Gauss-Jordan, 642–644
Elimination method, for solving systems of equations, 600–602
Ellipses, 708, 709, 725–734
  applications of, 733–734
  defined, 725–726
  equation of, 726–729
  graphing, 728–729
  reflecting property of, 725, 733–734
  translations of, 733–734
Empty set, 919
End behavior
  of polynomial functions 156-158, 155–156
  of rational functions, 200–202
Endpoints, of angles, 337
Entries, in matrices, 635
Environmental pollution, 885
Equality
  of complex numbers, 976
  of matrices, 672–673
Equally likely outcomes, 851
Equal ordered pairs, 2
Equal-tempered scale, 321
Equations, 954–962
  Cartesian (rectangular), 565–567, 779
  of circles, 12–13, 14–15
  converting between rectangular and polar forms, 565–567
  depressed, 178
  of ellipses, 726–729
  equivalent, 954
  exponential. *See* Exponential equations
  functions defined by, 42–44
  general. *See* General equations
  graphs of, 7–8
  of hyperbolas, 740, 742–743, 745
  inconsistent, 954
  involving radicals, 960–961
  linear. *See* Linear equations
  in one variable, 954
    solving, 955–956
  of parabolas, 711–715
  parametric. *See* Parametric equations
  polar. *See* Polar equations
  polynomial, factoring method for solving, 958–959
  quadratic, 956–957
  of quadratic functions, 140
  quadratic in form, 961
  rational, 959–960
  rectangular (Cartesian), 565–567, 779
  satisfying, 7
  solutions of, 7

  standard. *See* Standard equations
  systems of. *See* Systems of equations; Systems of linear equations
  Time Dilation, 875
  trigonometric. *See* Trigonometric equations
Equilibrium, static, 544
Equilibrium point, 603
Equivalent equations, 954
Equivalent vectors, 538
Euclid, 708
Euler, Leonhard, 40
  biographical information, 264
Euler number, 264
Even degree, power functions of, 155
Even functions, 65
Even-odd identities, 438
Even trigonometric functions, 375–376, 382
Events, probability of. *See* Probability(ies)
Everest, George, 513
Eves, Howard, 824
Expanding by cofactors, 652
Experiment(s), 850
Experimental probabilities, 855–857
Exponent(s). *See also* Power(s)
  integer, 923–924
  rational, 945–947
  rules of, 255, 924–925
Exponential equations
  applications of, 304–305
  in quadratic form, 303–304
  solving, 302–304
Exponential functions, 253–267
  defined, 254
  evaluating, 254–255
  graphing, 255–257
  growth and decay, 265–266
  interest and
    compound, 260–263
    continuous compounding and, 263–265
    simple, 259–260
  natural, 265–266
  properties of, 257, 276
  transformations on, 257–259
Exponential inequalities, 308–309
Expressions
  algebraic, 922–923
  improper, 174, 661
  proper, 174, 661
  radical, 942
    simplifying, 943
  rational. *See* Rational expressions
  trigonometric, simplifying, 438–439
Extracorporeal shock-wave lithotripsy (ESWL), 725, 733–734
Extraneous roots, 959
Extraneous solutions, 959

F

Factor(s), of polynomials, 174
Factorial notation, 798–799
Factoring
    finding limits by, 879
    over the integers, 930
    of polynomials, 930–931
    solving polynomial equations using, 958–959
    solving quadratic equations using, 956–957
Factorization Theorem for a Polynomial with Real
        Coefficients, 231
Factorization Theorem for Polynomials, 228–229
Factor Theorem, 178–179
Fahrenheit temperature, finding from a Celsius range, 968
Feasible solutions, set of, 624
Federal taxes and revenues, 187
Femur, inferring height from, 22, 29
Fiber optics cable cost, 52
Fibonacci, 796
Fibonacci sequences, 796
Filene's Basement store, 415
Finance. See Interest
Finite sequences, 794
    arithmetic, sum of, 807–809
    geometric, sum of, 816–817
Fiore, Antonio Maria, 225
First component
    of ordered pairs, 2, 3
    of vectors, 540
Fliess, Wilhelm, 381
Focal diameter, of parabolas, 714–715
Focus(i)
    of ellipses, 725
    of hyperbolas, 741, 742, 748
    of parabolas, 711, 712, 716
FOIL method, 929, 931
Foot-pounds, 549
Force, 537
    resultant, 544
Formulas
    arc length, 344–345
    change-of-base, 292–294
    combinations, 844
    compound interest, 262
    continuous compound interest, 263–265
    decay, 294
    difference. See Sum and difference formulas
    distance, 6, 922
    double-angle, 461–464
    geometry, 952
    growth, 294
    half-angle, 466–468
    half-life, 294
    Heron's, 531–532
    midpoint, 6
    permutations, 842–843

power-reducing, 464–465
product-to-sum, 473–474
quadratic, 957–958
recursive, 796–797
reduction, 456–458
rotation of axes, 759
simple interest, 260
sum and difference. See Sum and difference formulas
sum-to-product, 475–476
Four-leafed rose, 570
Fractions
    complex, 938–939
    partial. See Partial-fraction decomposition
Fraud detection, 271
Frequency
    of musical tones, 448
    simple harmonic motion and, 398
Full moon, 481
Functions, 39–53
    absolute value, 70
    average rate of change of, 49–51
    building, 52–53
    circular, 354. See also Trigonometric functions
    combining, 99–102
    composite. See Composite functions
    composition of, 102–104
    constant, 60, 62, 70
    cube root, 71
    cubing, 70
    decomposition of, 108–110
    decreasing, 61–63
    defined, 40
    difference, 100
    domain of, 40, 44–45, 47, 100, 101–102
    even, 65
    exponential. See Exponential functions
    graphs of, 46–49
    greatest integer, 69, 71
    identity, 60, 70
    increasing, 61–63
    inverse. See Inverse function(s)
    limits of. See Limits, of a function
    linear, 60–61
    logarithmic. See Logarithmic functions
    objective, 624
    odd, 65
    one-to-one, 116–119
        domains of, 123–124
        range of, 123–124
    periodic, 382–383
    piecewise, 65–70
    piecewise-defined, limits of, 875
    polynomial. See Polynomial functions
    power, 155–156
        rational, 71
    product, 100
    quadratic. See Quadratic functions

quotient, 100
range of, 40, 45, 47
rational. *See* Rational functions
reciprocal, 71
  of trigonometric functions, 408–411
reciprocal square, 71
relative maximum/minimum values of, 63–64
representations of, 41–44
square root, 70
squaring, 70
step, 69–70
sum, 100
transformations of. *See* Transformations
trigonometric. *See* Trigonometric functions
Fundamental counting principles, 839–841, 846–847
Fundamental rectangle, of hyperbolas, 746
Fundamental Theorem of Algebra, 228
Fundamental trigonometric identities, 373–375, 438
Future value, 262
  of an annuity, 817–819

G

Die galvanische Kette, mathematisch bearbeitet (Ohm), 660
Gateway Arch, 138
Gauss, Karl Friedrich, 228, 808
Gaussian elimination
  with matrices, 639–642
  with systems of equations in three variables, 609–611
Gauss-Jordan elimination, 642–644
General equations
  of ellipses, 731
  of hyperbolas, 748
  of parabolas, 716
General form
  of equation of a circle, 14–15
  of equation of a line, 27–29
General term. *See* *n*th term
Geometric sequences, 813–817
  finite, sum of, 816–817
  infinite, sum of, 819–820
  recursive definition of, 814
Geometric vectors, 537–539
Geometry formulas, 952
Gift distribution, 846
Global warming, 173
Golf ball flight, 367, 372–373
Graphical method, for solving systems of equations, 595–596
Graphing calculators
  absolute value on, 922
  angle measurement in degrees using, 423
  binomial coefficients on, 833
  circles on, 13
  combinations using, 844
  complex numbers using
    adding, 977
    conjugates of, 978

dividing, 978
entering, 976
powers of, 582
real and imaginary parts of, 976
reciprocals of, 978
roots of, 584
subtracting, 977
conversion between polar and rectangular coordinates
  using, 563, 564
conversions between angle measurements
  using, 340
conversions between degrees and radians using, 342
difference formula using, 450
exponential functions on
  compound interest and, 264
  Euler number and, 265
  evaluating, 254
exponents on, 924
factorials using, 798
graphing ellipses using, 728
graphing parabolas using, 713, 714
graphing polar equations using, 568
graphing second-degree equations using, 764
greatest integer function using, 69
horizontal shifts using, 80
integrals using, 911
linear equations using, 27, 28
linear regression using, 32
logarithmic functions on, 293
logarithms on
  common, 278
  natural, 279
matrices on, 642
  adding, 674
  creating, 636
  find determinants of, 652
  invertible, inverses of, 687
  multiplying, 679
  row operations and, 638
  subtracting, 675
one-to-one functions using, 123
partial fractions on, 664
permutations using, 843
polynomial functions on, 157
radian and parametric modes of, 784
rational functions on
  connected vs. dot mode and, 194
  with hole, 192
  revenue curve and, 202
scales on, 5
sequences using, 795, 817
series using, 799
solving triangles to find length and angles
  using, 506
square roots on, 942
sum-to-product formula using, 477
symmetry on, 11, 123

Graphing calculators (*continued*)
 systems of equations in two variables on
  linear, 596
  nonlinear, 602
 systems of linear inequalities in two variables on, 622
 trigonometric equations using
  with multiple angles, 485, 486
  of quadratic type, 487
  solving by squaring, 489
 trigonometric functions on
  cosine, 389
  in degree and radian mode, 363
  inverse, 420
  phase shifts and, 393
  tangent, 406
 trigonometric identities on, 439, 440, 441
 turning points on, 163
 vertical shifts using, 79
 viewing rectangle of, 5
 viewing window of, 784
 zeros of a polynomial function on, 159
Graphs and graphing
 of absolute value function, 70
 area under, 905–910
 of circles, 13
 of compound inequalities, 966–968
 of constant function, 70
 of cosecant, secant, and cotangent functions, 409
 of cosine function, 385–386
 of cube root function, 71
 of cubing function, 70
 of ellipses, 728–729
 of equations, 7–8
 of exponential functions, 255–257
 finding limits using, 870–872
 of functions, 46–49
  tangent lines to, 895–897
 functions defined by, 42
 of greatest integer function, 71
 of hyperbolas, 746–747
 of identity function, 70
 intercepts of, 8–9
 of intervals, 920–921
 of inverse functions, 118, 121
 of linear inequalities in one variable, 965
 of linear inequalities in two variables, 619–621
 of logarithmic functions, 275–278
 of nonlinear inequalities, 627
 of ordered pairs, 3, 4–5
 of parabolas, 712–715, 717–719
 of piecewise functions, 68–70
 of plane curves, 780
 by plotting points, 7–8, 567
 of points, 3–4
 of polar equations, 567–573
  of conics, 772–775
 of polynomial functions, 163–165

 of power functions, 71
 of quadratic functions, 140–145
  in standard form, 140–141
 of rational functions, 191–192, 193, 195–200, 201–203
  with holes, 193
 of rational power functions, 71
 of real numbers, 918
 of reciprocal functions, 71
  square, 71
  of trigonometric functions, 409–411
 of sine function, 384, 385–386
 sinusoidal, 386
 solving rational inequalities using, 217–218
 of square root function, 70
 of squaring function, 70
 stretching/compressing of, 85–88
 symmetry of, 9–12
 of systems of nonlinear inequalities, 628–629
 of tangent function, 405–408
 vertical and horizontal shifts of, 78–80, 82, 394–395
Greatest integer function, 69, 71
Great Trigonometric Survey (GTS), 513
Greenhouse effect, 173
Gross production matrix, 692
Group photograph arrangements, 841
Growth
 exponential, 265–266
 linear, 260
Growth factor, 253, 257
Growth formula, 294
Growth model, logistic, 301–302

H

Hales, Stephen, 78, 351
Half-angle formulas, 466–468
Half-life formula, 294
Half-lives, 294–295
Half moon, 481
Halley, Edmund, 733
Harmonic motion, simple, 397–398
Harmony, in music, 321–322
Harvey, William, 351
Hawass, Zahi, 287
Hearing, threshold of, 318
Height
 of a cloud, 505–506
 inferring from femur, 22, 29
 of a mountain, 517
Heron's formula, 531–532
Hertz, Gustav, 448
Hertz (Hz), 448
Hipparchus, 322
Hipparchus of Rhodes, 437
Honeybees, 794, 797
Hooker Telescope, 710
Horizontal asymptotes, 190–191, 194–195

Horizontal lines, 27
Horizontal-line test, 117–118
Horizontal shifts, 78, 80–82
Horizontal stretching/compressing, 87–88
Horsepower, 619
Hounsfield, Godfrey, 608
Hounsfield numbers, 614–615
Hubble, Edwin, 710
Hubble Space Telescope, 710, 719–720
Hummingbird's path, 905
Hyperbolas, 708, 709, 739–752
    applications involving, 750–751
    asymptotes of, 743–746
    defined, 739
    equation of, 740, 742–743, 745
    graphing, 746–747
    orientation of, 741–742
    translations of, 747–750

I

Identities, 437
    cofunction, 450–451
    containing half-angles, verifying, 468
    Pythagorean, 374–375, 381–382
    trigonometric. *See* Trigonometric identities
    useful for calculus, 444
Identity functions, 60, 70
Identity matrices, 680
Imaginary axis, 577
Imaginary numbers, pure, 976
Imaginary part of a complex number, 976
Impedance, 577
Impossible events, 851
Improper expressions, 174
Improper rational expressions, 661
Included sides/angles, 513, 950
Inconsistent equations, 954
Inconsistent systems of equations, 596, 611
Increasing functions, 61–63
Independent systems of equations, 596, 611
Independent variable, 7, 39
Indeterminate form, 879
Index of summation, 799
Inequalities, 919, 964–973
    combining, 966–968
    compound, 966–968
    exponential, 308–309
    involving absolute value, 970–973
    linear. *See* Linear inequalities
    logarithmic, 308–309
    nonlinear
        graphing, 627
        systems of, 628–629
    polynomial, 210–214
    properties of, 964
    quadratic, 144–145
    rational, 214–220

    systems of, 619–629
    test points for solving, 968–970
Inequality symbols, 919
Infinite geometric series, 813
    sum of, 819–820
Infinite limits, 885–887
Infinite sequences, 794
Infinity, limits at, 887–890
Initial point
    of plane curves, 780
    of vectors, 537
Initial side, of angles, 337
Inner product. *See* Dot product
Input-output diagram, 118
Instantaneous rates of change, 893–895
Integer(s), 918
Integer exponents, 923–924
Integrals
    defined, 909–911
    evaluating using a graphing calculator, 911
Intercepts, 8–9
Interest
    compound, 260–263
    compounded annually, 261
    continuous compounding of, 264–265, 280–281
    defined, 259
    simple, 259–260
Interest rate, 259
    nominal, 261
Interindustry input-output matrix, 692
Intermediate Value Theorem, 159–160
Intersections
    of intervals, 921
    of sets, 919
Interval(s), 920–921
Interval notation, 920
Inverse(s), of matrices. *See* Matrix inverses
Inverse function(s), 115–125
    applications of, 124–125
    finding, 121–122
    one-to-one, 116–119, 123–124
    trigonometric, 415–427
        composition of, 424–426
        evaluating, 422–424
        verifying, 120
Inverse function property, 118
Inverse trigonometric identities, 422–423
Inverse variation, 236, 238–239
Invertible matrices, 687
Investments, continuous compounding and, 264–265, 280–281
Irrational numbers, 918
Irreducible polynomials, 931

J

Joint variation, 239–241

K

Kennedy, John F., 768
Kepler, Johannes
  biographical information on, 153
  Third Law of Planetary Motion, 235
  volume of wine barrels and, 153, 166
King Tut, 287, 295–296
Kitab al-jabr wal-muqabala (Al-Khwarizmi), 594
Köwa, Seki, 649
Kramp, Christian, 798

L

Lambton, William, 513
Latitude, 337, 345
Latus rectum, 714
Law of Cooling, of Newton, 281–282
Law of Cosines, 527–533
  derivation of, 528–529
  Heron's formula and, 531–532
  solving SAS triangles and, 529–531
Law of Sines, 514–523
  area of a triangle and, 522–523
  bearings and, 521–522
  solving AAS and ASA triangles and, 516–517
  solving SSA triangles and, 517–521
Law of Universal Gravitation, 235, 240
LCD (Least common denominator), 935, 936–938
Leading coefficients
  of polynomial functions, 153
  of polynomials, 928
Leading entries, of rows in matrices, 638
Leading term(s), of polynomial functions, 153
Leading-term test, 158
Leaning ladder, 181–182
Least common denominator (LCD), 935, 936–938
Least-squares line, 32
Leibniz, Gottfried Wilhelm, 649, 779, 865, 866
Lemniscates, 573
Leonardo of Pisa, 796
Leontief, Wassily, 634
Leontief Input-Output Model, 634, 643–644, 692–693
L'Hôpital, Guillaume de, 779
Libby, Willard Frank, 295
Liber abaci (Fibonacci), 796
Light years, 239
Like radicals, 943–944
Like terms, 928
Limaçon(s), 571–572
Limaçon of Pascal, 571
Limits
  of difference quotients, 878–880
  of functions, 866–872
    algebraic methods for finding, 878–880
    defined, 866
    finding graphically, 870–872
    finding using tables, 867–870

  piecewise-defined, 875
  properties of, 875–878
  rational, 875
infinite, 885–887
at infinity, 887–890
of polynomials, 875
of summations, 799
Line(s), 22–33
  least-squares, 32
  parallel and perpendicular, 29–31
  slope of, 22–24
  tangent, 353
Linear attenuation units (LAUs), 614–615
Linear equations
  general form of, 27–29
  of horizontal and vertical lines, 27
  in one variable, solving, 955–956
  of parallel and perpendicular lines, 30
  point-slope form of, 24–25
  slope-intercept form, 25–26
  systems of. *See* Systems of linear equations
  types of, 956
Linear functions, 60–61
Linear growth, 260
Linear inequalities
  in one variable, 965–966
    graphing, 965
  in two variables
    applications of, 623–627
    graph of, 619–621
    systems of, 621–623
Linear programming, 623–627
Linear regression, 31–32
Linear speed, 346–347
Line of symmetry, 9
Line segments
  finding midpoint of, 6
  parameterizing, 783–784
Lithotripsy, 725, 733–734
Local maxima/minima of polynomial functions, 162
Logarithm(s)
  basic properties of, 273–274
  change of base and, 292–294
  common, 278–279
    number of digits and, 292
  evaluating, 273
  natural, 279–280
  number of digits and, 291–292
  one-to-one property of, 306
  rules of, 287–296
Logarithmic equations, 305–308
Logarithmic functions, 271–282
  converting between logarithmic and exponential form,
    272–273
  domains of, 274–275
  graphs of, 275–278
  investments and, 280–281

Newton's Law of Cooling and, 281–282
properties of, 276
Logarithmic inequalities, 308–309
Logarithmic scales, 313–323
  apparent brightness, 322–323
  decibel, 318–322
  pH, 313–315
  Richter, 315–318
Logistic growth model, 301–302, 307–308
Long division, of polynomials, 174–175
Longitude, 337, 345
Longley, Wild Bill, 22, 29
Loomis, Alfred Lee, 739
LORAN (long-range navigation), 739, 751
Lottery wins, 850, 853
Loudness of sound, 318–322
Lower bound on real zeros, 226–227
Lower limit of summations, 799

M

Mach, Ernst, 404
Mach numbers, 404, 410–411
Magnitude
  of complex numbers, 578
  of vectors, 537, 538, 543–544
Main diagonal, of a matrix, 635
Maisey, John, 60
Major axis, of ellipses, 725, 727, 729
Mastlin, Michael, 153
Mathematical induction, 824–828
Matrices, 634–639
  addition and subtraction of, 673–675
  decoding, 694–695
  defined, 634–635
  determinants of. *See* Determinants
  equality of, 672–673
  Gaussian elimination with, 639–642
  identity, 680
  invertible (nonsingular), 687
  multiplication of, 676–680
  scalar multiplication of, 674–675
  singular, 687
  solving linear systems using, 635–639
  zero, 675
Matrix inverses, 685–695
  applications of, 692–693
  finding, 687–690
  solving linear systems using, 690–691
  verifying, 686
Maurolico, Francesco, 826
Maximum area, finding, 146–147
Maximum/minimum values of functions, relative, 63–64
Maximum value of quadratic functions, 139
McDonald's coffee, 282
"Megatooth" shark, 60, 61
Members of sets, 918

Meridians, 337
Metabolic rate, body weight versus, for small animals, 324–325
Midpoint formula, 6
Minimum value of quadratic functions, 139
Minor(s), in matrices, 650–651
Minor axis, of ellipses, 725, 727, 729
Minutes (angular), 339–340
Modulus, of complex numbers, 578, 579
Monomials, 928
Monter (to climb), 26
Mount Everest, 513
Mount Kilimanjaro, 500, 506
Multiple angles, trigonometric equations involving, 485–487
Multiplication. *See also* Product(s)
  of complex numbers, 977–978
  of matrices, 676–680
  of polynomials, 929–930
  of rational expressions, 934
  scalar, 674–675
Multiplication property, 964
Multiplicative inverses, 685. *See also* Matrix inverses
Multiplier, 820
Multiplier effect, 813, 820
Music, pure tones in, 448
Musical pitch, 320–321
Music programs, 840
Mutual funds, 99
Mutually exclusive events, 854–855

N

National Aeronautics and Space Administration (NASA), 768
Natural exponential functions, 265–266
Natural logarithm, 279–280
Natural numbers, 918
Negative angles, 338
Negative integer exponents, 924
Negative numbers, 918
New moon, 481
Newton, Isaac, 733, 779, 831, 865, 866
  biographical information on, 235
  Law of Cooling of, 281–282
  Law of Universal Gravitation of, 235, 240
Nominal interest rate, 261
Nonlinear inequalities
  graphing, 627
  systems of, 628–629
Nonrigid transformations, 85–88
Nonsingular matrices, 687
Nonsquare linear systems, 613–614
Nonterminating repeating decimals, 918
Nonzero rows, 638
Notation
  factorial, 798–799
  interval, 920
  set-builder, 919
  summation, 799–801

Nova Astronomia (Kepler), 153
Nova Stereometria Doliorum Vinariorum (Kepler), 153
*n*th roots of unity, 585
*n*th term
    of sequences, 794, 795–796
        arithmetic, 806–807
        geometric, 815
Null set, 919
Number(s)
    algebraic, 265
    complex. *See* Complex numbers
    imaginary, pure, 976
    irrational, 918
    large, estimating, 292
    natural, 918
    negative, 918
    positive, 918
    rational, 918
    real. *See* Real numbers
    transcendental, 265
    whole, 918
Number line, 918
    distance between two points on, 918
Number of Zeros Theorem, 229
Numerator, rationalizing, finding limits by, 879–880, 889–890
Nutrition, 625–627

O

Objective function, 624
Oblique triangles, solving, 513–514
Obtuse angles, 338
Octaves, 320–321
Odd degree, power functions of, 155–156
Odd functions, 65
    trigonometric, 375–376, 382
Ohm's Law, 660
One-to-one functions, 116–119
    domains of, 123–124
    range of, 123–124
One-to-one property of logarithms, 306
Open intervals, 920
Opposite vectors, 538
Optiks (Newton), 235
Optimal solutions, 624
Optimization, 623–624
Ordered pairs, 2, 3
    functions defined by, 41
    graph of, 3, 4–5
Origin, 2
    of real number line, 918
Orthogonal vectors, 552–553

P

Parabolas, 708, 709, 710–720
    equation of, 711–715
    geometric definition of, 711

graphing, 712–715, 717–719
    reflecting property of, 719–720
    translations of, 716–720
Parallel circuits, 660
Parallel lines, 29–31
Parallel property of resistors, 660
Parallel vectors, 551
Parameter(s), 779
Parameterization of curves, 782–784
Parametric equations, 779–785
    of cycloids, 784
    eliminating the parameter, 781–782
    finding, 782–784
    graphing, 780
    of plane curves, 779–780
Partial-fraction decomposition, 660–669
    with irreducible quadratic factors, 665–666
    with only distinct linear factors, 662–664
    with repeated irreducible quadratic factors, 666–669
    with repeated linear factors, 664–665
Pascal, Blaise, 831, 832
Pascaline, 831
Pascal's Triangle, 832–833
Period(s)
    of sine and cosine functions, 383
    of sinusoidal functions, 389, 393–394
Periodic functions, 382–383
Permutations, 841–843, 846–847
    distinguishable, 845–846
Permutations formula, 842–843
Perpendicular lines, 29–31
Petroleum consumption, 179–180
Pharmacokinetics, 39, 52–53
Phase shift, of sinusoidal functions, 390, 392–394
Phases of the moon, 481
Photosynthesis, 210
pH scale, 313–315
Piecewise-defined functions, limits of, 875
Piecewise functions, 65–70
    graphing, 68–70
Pitiscus, 500
Pizza topping choices, 845
Plane curves
    graphing, 780
    parametric equations of, 779–780
Points
    graphing, 3–4
        by plotting, 7–8, 567
    symmetry about, 9
    transition, 69
    turning, 64
    zeros of polynomial functions and, 162–163
Point-slope form, 24–25
Poiseuille, Jean Louis Marie, 78
Poiseuille's law, 92–93
Polar axis, 560

Polar coordinates, 560–574
    converting between polar and rectangular forms and, 562–567
    graphs of polar equations and, 567–573
    multiple representations and, 561
    sign of r and, 561–562
Polar coordinate system, 560
Polar equations, 779
    of conic sections, 769–776
       applications involving, 775–776
       graphing, 772–775
    converting between polar and rectangular forms and, 565–567
    graphs of, 567–573
Pole, 560
Polynomial(s), 928–932
    complex, 228
    complex zeros of, 228–229
    Conjugate Pairs Theorem and, 229–232
    cubic, 154
    dividing, 173–176
    factored completely, 931
    factoring, 930–931
    Factorization Theorem for, 228–229
    Factorization Theorem for a Polynomial with Real Coefficients and, 231
    Factor Theorem and, 178–179
    finding limits at infinity for, 888
    limits of, 875
    multiplying, 929–930
    odd-degree, with real zeros, 230
    in one variable, 928
    quartic, 154
    quintic, 154
    rational zeros test and, 180–182
    Remainder Theorem and, 177–178, 179–180
    special products and, 929–930
Polynomial equations, factoring method for solving, 958–959
Polynomial factors, 174
Polynomial functions, 153–167
    common properties of, 154
    end behavior of, 156–158
    graphing, 163–165
    power functions, 155–156
    sign of, 211
    zeros of, 158–160
       complex, 228–229
       real, 159–161, 226–227
       turning points and, 160–163
Polynomial inequalities, 210–214
Population growth, 304–305, 307–308
Position vectors, 539–540
Positive angles, 337–338
Positive integer exponents, 923–924
Positive numbers, 918
Power(s). *See also* Exponent(s); Exponential equations; Exponential functions; Exponential inequalities
    of complex numbers in polar form, 582

of i, 979
    inverse variation with, 239
Power functions, 155–156, 461
    rational, 71
Power-reducing formulas, 464–465
Power rule, of logarithms, 288
Principal, 259
Principal axis, of conics, 769
Principal nth root, 942–943
Principal square root, of a negative number, 976
Prizewinners, counting, 842
Probability(ies), 850–857
    Additive Rule for finding, 853–854
    of an event, 850–853
    of complement of an event, 855
    experimental, 855–857
    of mutually exclusive events, 854–855
    theory of, 831
Product(s). *See also* Multiplication
    of complex numbers, 580–582
    scalar, 674
    special, 929–930
Product function, 100
Product rule, for logarithms, 288, 307
Product-to-sum formulas, 473–474
Proper expressions, 174
Proper rational expressions, 661
Proportionality, constant of, 236
Ptolemy, 322
Pure imaginary numbers, 976
Pure tones in music, 448
Pythagorean identities, 374–375, 381–382, 438
Pythagorean Theorem, 5, 527, 950
    converse of, 6, 951

Q

Quadrant(s), 2, 3
Quadrantal angles, 338, 357
Quadratic equations, 956–957
    factoring method for solving, 956–957
    solving using the quadratic formula, 958
Quadratic form, exponential equations in, 303–304
Quadratic formula, 957–958
Quadratic functions, 138–147
    applications of, 145–147
    defined, 138
    graphing, 140–145
    standard form of, 139–141, 142
Quadratic inequalities, 144–145
Quadratic in form equations, 961
Quadratic type, trigonometric equations of, 487
Quartic polynomials, 154
Queen bees, 794
Quetelet, Lambert, 308
Quintic polynomials, 154

Quotient(s), 174. *See also* Division
  of complex numbers, 580–582
  difference, 51
  sign of, 215
Quotient function, 100
Quotient identities, 373–375, 438
Quotient rule
  for logarithms, 307
  of logarithms, 288

R

Radians
  degrees related to, 341–344
  measuring angles using, 340–341
Radical(s)
  equations involving, 960–961
  like, 943–944
  properties of, 943
Radical expressions, 942
  simplifying, 943
Radical sign, 942
Radicand, 942
Radioactive half-lives, 294–295
Radiocarbon dating, 295–296
Radius, of a circle, 12
Ranges
  of cosine function, 381–382
  of functions, 40, 45, 47
  of logarithmic functions, 274
  of one-to-one functions, 123–124
  of relations, 41
  of sine function, 381–382
  of tangent function, 405
Rates of change, instantaneous, 893–895
Ratio, common, of geometric sequences, 813–814
Rational equations, 959–960
Rational exponents, 945–947
Rational expressions, 933–939
  adding, 935–938, 944
  complex, 938–939
  dividing, 934
  finding the least common denominator
    for, 936
  lowest terms for, 933
  multiplying, 934
  proper and improper, 661
  subtracting, 935–938, 944
Rational functions, 187–204
  defined, 188
  domain of, 188–189
  finding limits at infinity for, 888–889
  graphing, 191–192, 193, 195–200, 201–203
    with holes, 193
  limits of, 875
  with oblique asymptotes, 200–202
  sign of, 215

  translations of, 191–195
  with vertical and horizontal asymptotes, 189–191, 192,
    194–196
Rational inequalities, 214–220
Rationalizing the denominator, 944–945
Rationalizing the numerator
  finding limits by, 879–880, 889–890
  in radical expressions, 944–945
Rational numbers, 918
Rational zeros test, 180–182
Rays, of angles, 337
Real axis, 577
Real number line (real line), 918
Real numbers, 911–924
  absolute value and, 921–922
  algebraic expressions and, 922–923
  inequalities and, 919
  integer exponents and, 923–924
  intervals and, 920–921
  sets and, 919
  trigonometric functions of, 353–355
Real part of a complex number, 976
Real zeros of polynomial functions, 159–160
Reciprocal(s), of expressions, 924
Reciprocal functions, 71
  of trigonometric functions, 408–411
    graphs of, 409–411
    properties of, 409
Reciprocal identities, 373–375, 438
Reciprocal square function, 71
Rectangles
  area of, 44
  fundamental, of hyperbolas, 746
Rectangular coordinates, converting between polar and rect-
    angular forms and, 562–567
Rectangular equations, 779
  converting between polar and rectangular forms and,
    565–567
Rectangular form, complex numbers in, 578, 579–580
Recursive definitions, 52
  of arithmetic sequences, 806
  of geometric sequences, 814
Recursive formulas, 796–797
Reduced row-echelon form, 639
Reference angles, 369–373
Reflecting property
  of ellipses, 725, 733–734
  of parabolas, 719–720
Reflections, 82–85
Regression line, 32
Relations, 41
Relative maxima/minima of polynomial functions, 162
Relative maximum/minimum values of functions, 63–64
Remainder Theorem, 174, 177–178, 179–180
Resistors, parallel property of, 660
Resultant (force), 544
Resultant vectors, 538

Retail theft, 415
Revenue curve, 202–203
Richter, Charles Francis, 315
Richter scale, 315–318
Right angles, 338
Right circular cones, 708
Right transformations, 78
Right triangles, 5, 500, 950. *See also* Pythagorean
    Theorem
  solving, 504–505
  trigonometric functions of, 501
Rise, 22
Robotic hand, positioning, 564–565
Roots
  of complex numbers, 583–584
  cube, 942
  of equations in one variable, 954
  extraneous, 959
  square, 942
Rose curves, 570, 573
Roster method, 919
Rotation of axes, 757–765
  eliminating the xy-term by, 761–764
  identifying conic sections and, 757–758, 764
Rotation of axes formulas, 759
Row-echelon form, 638, 639
Row equivalent matrices, 637
Row matrices, 635
Run, 22

S

Saarinen, Eero, 138
Sample space, 850
SAS triangles, 513, 514, 950
  solving, 529–530
Scalar(s), 537
Scalar multiples, of vectors, 539
Scalar multiplication, 674–675
Scalar product. *See* Dot product
Scalar projection of vectors, 554
Scalar quantities, 537
Scatterplots, 42
Secant function, 354
  inverse, 419–421
  properties of, 409
Second(s) (angular), 339–340
Second component
  of ordered pairs, 2, 3
  of vectors, 540
Sectors, area of, 345–346
Security camera rotation angle, 426, 507
Semicircles, 14–15
Sequences, 794–800
  arithmetic, 805–810
    recursive definition of, 806
  factorial notation and, 798

  geometric, 813–817
    recursive definition of, 814
  recursively defined, 796–797
  summation notation and, 799–800
Series, 800–801
Set(s), 918, 919
Set-builder notation, 919, 920
Sickdhar, Radhanath, 513
Sign(s)
  of cofactors, 651
  Descartes Rule of Signs, 225–226
  of polynomial functions, 211
  of quotients, 215
  of rational functions, 215, 561–562
  of trigonometric functions, 368–369
  variation of, 225–226
Sign-chart method, for solving inequalities, 968–970
Similar triangles, 501, 950
Simple harmonic motion, 397–398
Simple interest, 259–260
Simple interest formula, 260
Simplification, of trigonometric expressions, 438–439
Sine(s), Law of. *See* Law of Sines
Sine function, 354. *See also* Sinusoidal functions
  graph of, 384, 385–386
  inverse, 415–417, 421
  properties of, 381–383
  sum and difference formulas for, 452–454
Singular matrices, 687
Sinusoidal functions
  amplitude of, 386–390
  modeling with, 395–397
  period of, 389, 393–394
  phase shift of, 390, 392–394
  vertical shifts of, 394–395
Sinusoidal graphs (curves), 386
Slope, 22–24
  tangent as, 404
Slope-intercept form, 25–26
Social Security numbers, 839, 841
SOHCAHTOA mnemomic, 501
Solutions
  of equations, 7
    in one variable, 954
  extraneous, 959
    of trigonometric equations, 488–490
  feasible, set of, 624
  optimal, 624
  of systems of equations, 595
  of systems of inequalities, 621
Solution sets
  of equations in one variable, 954
  of systems of equations, 595
  of systems of inequalities, 621–622
Sørensen, Søren, 314
Sound, loudness of, 318–322
Space junk, 805

Space shuttles and wind speed, 760–761
Special products, 929–930
Special Theory of Relativity, 875
Speed
    instantaneous, 900–901
    linear and angular, 346–347
Spirals, 573
    of Archimedes, 567
Sputnik, 768, 775–776
Square matrices, 635
Square root(s), 942
    principal, of a negative number, 976
Square root function, 70
Squaring function, 70
SSA triangles, solving, 517–521
SSS triangles, 513, 514, 950
    Heron's formula for, 532
    solving, 530–531
Standard equations
    of hyperbolas, 740
    of parabolas, 712
    of polar equations of conics, 770
Standard form
    complex numbers in, 976
    of ellipses, 731–732
    of equation of a circle, 12, 15
    polynomials in, 928
    of quadratic equations, 956
    of quadratic functions, 139–141, 142
Standard position, angles in, 338, 339
Standard unit vectors, 542
Star brightness, 322–323
Static equilibrium, 544
Statins, 52–53
Steinmetz, Charles, 577
Step functions, 69–70
Straight angles, 338
Stretching graphs, 85–88
Substitution, trigonometric, 443
Substitution method, for solving systems of equations, 597–599
Subtraction. *See also* Difference(s)
    of complex numbers, 976–977
    of matrices, 674–675
    of rational expressions, 935–938, 944
Sum(s). *See also* Addition
    of arithmetic sequences, 807–809
    of finite geometric sequences, 816–817
    of infinite geometric sequences, 819–820
Sum and difference formulas, 448–458
    cofunction identities and, 450–451
    for cosine function, 448–450
    reduction formula and, 456–458
    for sine function, 452–454
    for tangent function, 454–456
Sum function, 100
Summation notation, 799–801

Sum-to-product formulas, 475–476
Sylvester, James Joseph, 673
Symmetry, 9–12
    even-odd functions and, 65
    finding terminal points by, 361–362
    inverse functions and, 121
    tests for, 10–11
        in polar coordinates, 568
Synthetic division, of polynomials, 174, 175–176
Systems of equations, 594–595
    linear. *See* Systems of linear equations
    nonlinear, 594
    in two variables, 594–603
        applications of, 603
        elimination method for solving, 600–602
        graphical method for solving, 595–596
        solutions of, 595
        substitution method for solving, 597–599
Systems of inequalities, 619–629
Systems of linear equations, 594, 608–609
    Cramer's Rule for solving, 653–656
    matrices compared with, 637–638
    solving using Gaussian elimination, 639–642
    solving using Gauss-Jordan elimination, 642–644
    solving using matrices, 635–639
    solving using matrix inverses, 690–691
    in three variables, 608–616
        application of, 614–615
        geometric interpretation of, 614
        nonsquare, 613–614
        number of solutions of, 611–613
Systems of nonlinear inequalities, 628–629

T

Tables, functions defined by, 42
Tangent function, 354, 404–408
    graph of, 405–408
    inverse, 418–419, 421
    properties of, 405
    sum and difference formulas for, 454–456
Tangent lines, to graphs of functions, 895–897
Tartaglia, Nicolo, 225
Technology matrix, 692
Temperature
    conversion between Fahrenheit and Celsius, 968
    inside a parked car, 323–324
Term(s), 922
    lowest, for rational expressions, 933
    of polynomials, 928
    of sequences, 794, 795–796
        arithmetic, 806–807
        geometric, 815
Terminal points
    of plane curves, 780
    on unit circles, 353
    of vectors, 537

Terminal sides, of angles, 337
Terminating decimals, 918
Test-point method, for solving inequalities, 968–970
Theoretical probabilities, 855
Third Law of Planetary Motion, 235
30°-60°-90° triangles, 359
Threshold of hearing, 318
Time Dilation Equation, 875
Touch-tone phones, 473, 477
Tracking moving objects, 146
Transcendental numbers, 265
Transformations, 78–94
    combining, 84, 88–90
    on exponential functions, 257–259
    of logarithmic functions, 277–278
    multiple, in sequence, 88–90
    nonrigid, 85–88
    reflections, 82–85
    stretching and compressing, 85–88
    vertical and horizontal shifts, 78–82
Transition points, 69
Transitive property, 919, 964
Translations
    of ellipses, 729–733
    of hyperbolas, 747–750
    of parabolas, 716–720
    of rational functions, 191–195
Transverse axis, of hyperbolas, 740, 741, 748
Triangle Inequality, 531
Triangles, 950
    AA, 950
    AAA, 513
    AAS, 513, 514
      solving, 516–517
    altitude of, 514
    area of, 522–523
    ASA, 513, 514, 950
      solving, 516–517
    congruent, 513, 950
    right. *See* Right triangles
    SAS, 513, 514, 950
      solving, 529–530
    similar, 501, 950
    solving, 504
    SSA, solving, 517–521
    SSS, 950
      Heron's formula for, 532
      solving, 530–531
    30°-60°-90°, 359
Triangulation, 513
Trichotomy property, 919, 964
Trigonometric equations, 439, 481–490
    extraneous solutions to, 488–490
    of form $a \sin(x - c) = k$, $a \cos(x - c) = k$,
        $a \tan(x - c) = k$, 482–484
    involving multiple angles, 485–487
    with more than one trigonometric function, 488

of quadratic type, 487
    solving, 483–484
    zero-product property and, 487
Trigonometric expressions, simplifying, 438–439
Trigonometric form, complex numbers in, 578, 579–580
Trigonometric functions, 336–435. *See also specific functions, e.g.* Sine function
    of angles, 356–363, 367–377
      even-odd properties of, 375–376
      finding values of, 358–363, 367–369, 371–372
      fundamental trigonometric identities and, 373–375
      reference angle and, 369–373
      signs of, 368–369
    composition of, 424–426
    evaluating, 501–503
    evaluating using a calculator, 363–364
    finding exact values of, 355–356
    finding values of complementary angles, 503
    inverse, 415–427
      composition of, 424–426
      evaluating, 422–424
    multiple, trigonometric equations with, 488
    of real numbers, 353–355
      unit circle definitions of, 354–355
    reciprocal functions of, 408–411
      graphs of, 409–411
      properties of, 409
Trigonometric identities, 373–375, 437–445
    fundamental, 373–375, 438
    inverse, 422–423
    simplifying trigonometric expressions and, 438–439
    trigonometric equations and, 439
    verifying, 440–444, 476
Trigonometric ratios, 501
Trinomials, 928
Turning points, 64
    zeros of polynomial functions and, 162–163
Tutankhamun, 287, 295–296
Two-intercept form, 28–29

U

Unbounded intervals, 920
Union
    of intervals, 921
    of sets, 919
Unit circles, 13, 351–355
Unit vectors, 541–542
    standard, 542
Universal gravitational constant, 240
Upper bound on real zeros, 226–227
Upper limit of summations, 799

V

Value of an algebraic expression, 922
Value of an annuity, 817–819
Vandermonde, Alexandre-Théophile, 649

Variables, 922
  dependent and independent, 7, 39
Variation
  constant of, 236
  direct, 236–238
  inverse, 236, 238–239
  joint and combined, 239–241
Variation of signs, 225–226, 235–241
Vector(s), 537–546
  adding, 538–539
  algebraic, 539–541
  angle between, 550–552
  applications of, 544–546
  components of, 540
  decomposition of, 555–556
  dot product of. *See* Dot product
  equivalent, 538
  geometric, 537–539
  in i, j form, 542
  magnitude of, 537, 538, 543–544
  orthogonal, 552–553
  parallel, 551
  position, 539–540
  projection of, 553–555
  resultant, 538
  in terms of magnitude and direction, 543–544
  unit, 541–542
Vector projection of vectors, 554–555
Vector quantities, 537
Velocity, instantaneous, 900–901
Verhulst, Pierre François, 301, 307, 308
Verhulst model, 301–302
Vertical asymptotes, 189–190, 192, 196, 197
Vertical lines, 27
Vertical-line test, 46
Vertical shifts, 78–80, 82
  of sinusoidal functions, 394–395
Vertical stretching/compressing, 85–88
Vertices
  of angles, 337
  of conics, 769
  of ellipses, 725, 727, 729
  of hyperbolas, 740, 741, 742, 748
  of parabolas, 139, 711, 712, 716

of right circular cones, 708
of solution sets of systems of inequalities, 622
Viete, Francois, 437
Viewing rectangle, 5
Voltage, 461
Volume, of wine barrels, 153, 166

W

Water pressure on underwater devices, 115, 125
Watt, James, 619
Watts, 461, 465–466
Waugh, Andrew, 513
Weierstrass, Karl W. T., 866
Whole numbers, 918
Width of a river, 507
Work, measuring, 549
Worker bees, 794

X

x-axis, 2
  reflections about, 83
x-intercept, 8, 9
xy-plane, 2

Y

y-axis, 2
  reflections about, 83
y-intercept, 8, 9

Z

Zero(s)
  of cosine function, 382
  of multiplicity, 161–162
  of polynomial functions, 158–160
    complex, 228–229
    real, 159–160, 161, 226–227
    turning points and, 160–163
  of sine function, 382
Zero exponents, 924
Zero polynomials, 928
Zero-product property, 923, 956–957
Zero vector, 538

Standard Form of a Parabola

$(y - k)^2 = \pm 4p(x - h)$ vertex: (h, k)
                                        and $(p > 0)$

$(x - h)^2 = \pm 4p(y - k)$

Standard Form of an Ellipse

$\dfrac{(x - h)^2}{a^2} + \dfrac{(y - k)^2}{b^2} = 1;$

$a > b > 0$ (Horizontal ellipse)

$\dfrac{(x - h)^2}{b^2} + \dfrac{(y - k)^2}{a^2} = 1;$

$a > b > 0$ (Vertical ellipse)

Standard Form of a Hyperbola

$\dfrac{(x - h)^2}{a^2} - \dfrac{(y - k)^2}{b^2} = 1;$

$a > 0, b > 0$ (Horizontal)

$\dfrac{(y - k)^2}{a^2} - \dfrac{(x - h)^2}{b^2} = 1;$

$a > 0, b > 0$ (Vertical)

Functions/Graphs

Constant Function $f(x) = c$	Identity Function $f(x) = x$	Squaring Function $f(x) = x^2$	Cubing Function $f(x) = x^3$
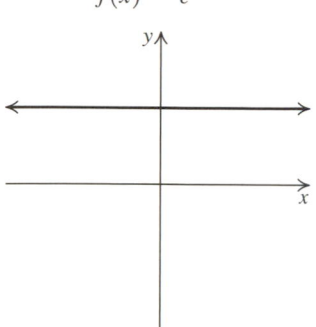	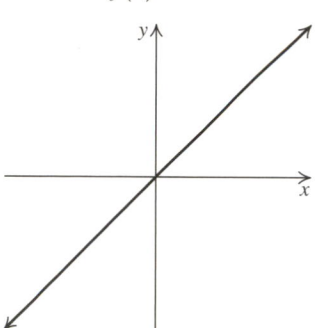	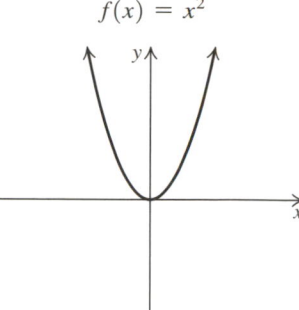	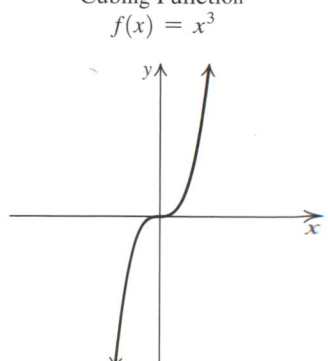

Absolute Value Function $f(x) = |x|$	Square Root Function $f(x) = \sqrt{x}$	Cube Root Function $f(x) = \sqrt[3]{x}$	Reciprocal Function $f(x) = \dfrac{1}{x}$
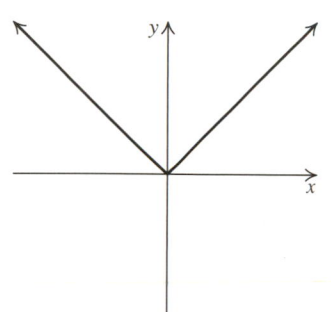	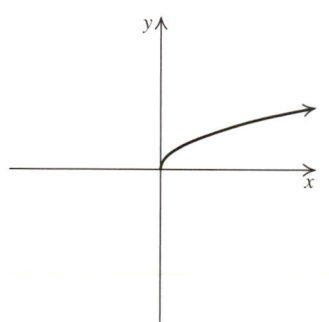	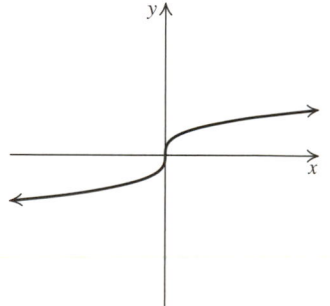	

Greatest Integer Function
$f(x) = [\![x]\!]$

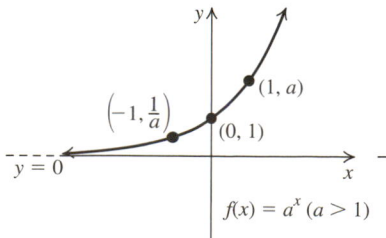

Reciprocal Square Function
$f(x) = \dfrac{1}{x^2}$

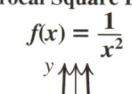
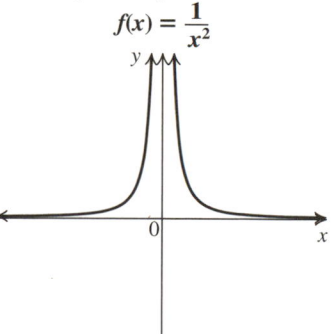

Exponential Functions

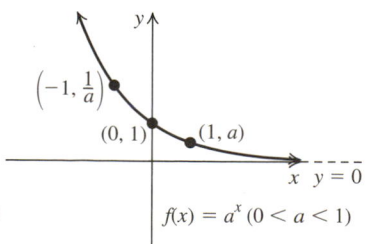

$f(x) = a^x \ (a > 1)$

$f(x) = a^x \ (0 < a < 1)$

Logarithmic Functions

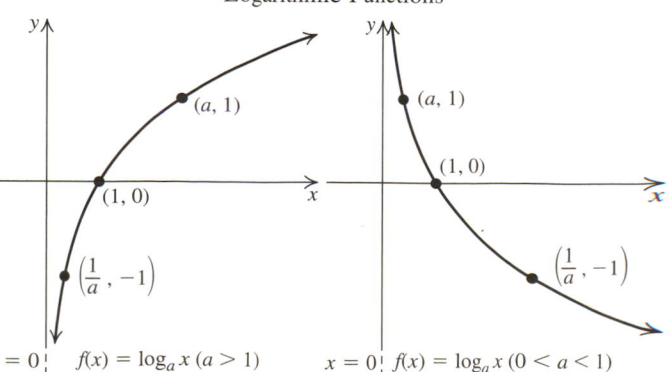

$x = 0 \quad f(x) = \log_a x \ (a > 1)$

$x = 0 \quad f(x) = \log_a x \ (0 < a < 1)$

(a)

(b)